CLINICAL PARASITOLOGY

CLINICAL PARASITOLOGY

9TH EDITION

PAUL CHESTER BEAVER, Ph.D., Sc.D. (HON.)

Emeritus Professor of Parasitology, School of Public Health and Tropical Medicine, Tulane University, New Orleans, Louisiana

RODNEY CLIFTON JUNG, M.D., Ph.D., F.A.C.P.

Clinical Professor of Medicine, School of Medicine, and Clinical Professor of Tropical Medicine, School of Public Health and Tropical Medicine, Tulane University, New Orleans; Physician in Internal Medicine, Touro Infirmary, New Orleans; Senior Visiting Physician, The Charity Hospital of Louisiana at New Orleans; Consultant to U.S. Quarantine Division of Centers for Disease Control, Atlanta, Georgia; Former Director of Health, City of New Orleans, Louisiana

EDDIE WAYNE CUPP, Ph.D.

Associate Professor of Medical Entomology, Cornell University, Ithaca, New York

Lea & Febiger *Philadelphia 1984*

Lea & Febiger
600 Washington Square
Philadelphia, PA 19106
U.S.A.

Library of Congress Cataloging in Publication Data

Beaver, Paul C. (Paul Chester), 1905–
Clinical parasitology.

Rev. ed. of: Craig and Faust's Clinical parasitology/
Charles Franklin Craig. 8th ed. 1970.
Includes bibliographies and index.
1. Medical parasitology. I. Jung, Rodney Clifton.
II. Cupp, E. W. III. Craig, Charles Franklin, 1872–1950.
Craig and Faust's Clinical parasitology. IV. Title.
[DNLM: 1. Parasites. 2. Parasitic diseases. WC 695
B386c]
RC119.B35 1984 616.9′6 83-11338
ISBN 0-8121-0876-0

First Edition, 1937
Second Edition, 1940
Third Edition, 1943
Reprinted, 1943, 1944
Fourth Edition, 1945
Reprinted, 1946, 1947, 1948, 1949
Fifth Edition, 1951
Reprinted, 1951, 1953
Sixth Edition, 1957
Reprinted, 1958, 1961
Seventh Edition, 1964
Reprinted, 1965, 1968
Eighth Edition, 1970
Reprinted, 1974, 1976, 1977
Ninth Edition, 1984

Printed in the United States of America

Print No. 4 3 2 1

Preface

Originally, and through its several editions, this volume was intended to be a handbook source of information for physicians and for graduate and postgraduate students in human parasitology. In the preparation of the new edition, our aim was to retain the format and general purpose of Craig and Faust's *Clinical Parasitology* as stated in the preface of the first edition: "to set down . . . the essential facts concerning the animal parasites of man and the diseases they produce, together with approved methods of diagnosis, treatment and control." In the preface to the eighth edition, attention was called to the expansion of scientific knowledge since the appearance of the first edition 33 years earlier, and to the difficulty of preparing a new edition of acceptable size, yet encompassing the greatly expanded subject. The past decade has not only produced scientific advances at an accelerated pace, but it has witnessed extraordinary changes in the way of life around the globe.

Rapid and increasingly frequent travel is usually identified as a major factor in bringing patients with exotic parasitic infections to the offices of European and North American physicians. Perhaps more important is the easy availability of intercontinental travel, which permits large numbers of young people with casual or cultist attitudes toward disease prevention to wander about the world. Drug abuse by the intravenous route is now commonplace in all segments of society and may be expected to be involved increasingly in the transmission of parasitic infections of the blood. A new permissiveness in the sexual mores has resulted in remarkable changes in the epidemiology of certain parasitic infections. Giardiasis and amebiasis are now in some cities more commonly transmitted sexually than by contaminated water or food. The mysterious disorder of the immune mechanisms known as the acquired immune deficiency syndrome (AIDS), first noted in male homosexuals, but now recognized in ever more groups, is responsible for outbreaks of pneumocystosis, formerly a relatively rare condition. One of the most striking changes in the United States, Canada, and some European countries has been the influx of large numbers of refugees and immigrants from less developed countries. Many of these people have parasitic infections that should be detected and eliminated if possible. Many continue the food and sanitation practices that are traditional in their native cultures and that may result in new zoonotic infections in the new environment.

There has been a tremendous multiplication of chemotherapeutic agents that give promise of providing safe, effective treatment for many of the infections that only a decade ago had no effective therapy or were treated with relatively toxic drugs. At the same time, legitimate governmental concern over the safety and efficacy of drugs has led to regulations that limit the availability of useful drugs. The risk of lawsuit for damages as much as concern for the patient's safety has forced the practicing physician to revise his office routine and to modify his attitudes toward drug selection. Carcinogenicity, teratogenicity, and mutagenicity of drugs were hardly considered ten years ago. Now they must be explained to the patient in order that he may give "informed consent" for treatment. We have endeavored to provide the user of this book with up-to-date information regarding these changes in a fashion sufficiently concise that the book is not appreciably increased in size.

Recently, numerous species known only as parasites of animals have become recognized as occasional parasites of man, and thus it became necessary to expand the number of zoonotic species to be described and discussed. Freeliving animals that colonize and cause serious disease in man, the amphizoic forms, have assumed new importance. The amphizoic amebae, *Naegleria* and *Acanthamoeba* species, only briefly mentioned in the previous edition, now constitute a separate chapter. To include the additional zoonotic and amphizoic agents of human disease, and to add essential new information about familiar forms, without expanding the volume, required rewriting parts of all chapters and the greater part of some. Many of the old illustrations were discarded or replaced and new ones were added. Numerous references are cited, both to sources of additional information and to reports of findings that are subject to different interpretations.

Studies on parasites and parasitic diseases have moved more actively into the fields of biochemistry, immunology, and electron microscopy. Although basic information derived from these newer studies has been incorporated in some of the descriptions, our chief aim has been to draw from the massive and widely scattered literature the essential facts about the individual parasites needed by workers in those fields who would apply their specialized skills, knowledge, and instruments in research efforts toward a better understanding of parasitism as a biological phenomenon and its relation to human disease.

The two senior authors have collaborated on all subjects considered in this volume. The junior author assumed primary resonsibility for the section on arthropods and collaborated in preparing the material on arthropod vectors of malaria, filariasis, and some other infections. Work on the revision began in January 1979 and manuscript copy was completed in May 1982. Additions and further minor revisions were made in galley proof up to April 1983.

For advice and assistance we are indebted to departmental colleagues and others. For permission to use published or original illustrations we are indebted to numerous publishers, authors, and others, all of whom are acknowledged individually in the legends. For assistance in obtaining essential literature and in manuscript preparation and editing, especial thanks are due Miss Helen Day and Mrs. Ena Castillo. We acknowledge with special thanks also the guidance and cooperation of the publisher, Lea & Febiger.

New Orleans, Louisiana Paul Chester Beaver
New Orleans, Louisiana Rodney Clifton Jung
Ithaca, New York Eddie Wayne Cupp

Contents

Section IV
ARTHROPODS AND HUMAN DISEASE

Section V
TECHNICAL APPENDIX

Section I

General Information

Chapter 1

The Parasite and Its Environment

All animal and plant forms originated and developed as free-living organisms that were obliged to compete with others for their existence. Only those that developed satisfactory adjustments and adaptations were able to survive. Among this group were many species in different phyla of the Animal and Plant Kingdoms that came to depend on their associates for shelter and food. In some instances, the remarkable adaptations suggest that this interrelationship has existed for a long time, probably thousands of years. Other groups of parasites appear to have more recently acquired a parasitic mode of life, and a few of these have not yet become irreversibly committed to parasitism. Still others are only now developing the earliest adaptations to parasitism.

DEFINITION OF TERMS

Parasitology is the area of biology concerned with the phenomenon of dependence of one living organism on another. The *parasite* lives on or in its *host,* which is usually a larger organism that provides physical protection and nourishment. If the parasite derives benefit without reciprocating and without injury to the host, the relationship is referred to as *commensalism;* if the relationship is beneficial to both associates, it is *mutualism*. Close association of the two is *symbiosis*. Hosts that transmit parasites to man are *vectors*. Those that are essential in the life cycle are *biologic vectors;* those that are not essential are *phoretic* or *mechanical vectors*. *Predators,* which attack and kill animals in order to obtain food, usually feed on smaller or weaker animals who are their *prey.*

Many animals that are related to parasitic species are free-living. Among the parasitic forms, some depend on plants as their hosts and are frequently of great economic importance. Others live on or within invertebrate animals, while still others parasitize vertebrate hosts. *Clinical parasitology* is concerned primarily with the animal parasites of man and their medical significance, as well as their importance in human communities.

HOST-PARASITE INTERRELATIONSHIPS

In a majority of instances, only one host is required for a parasite to carry on its life cycle. There may be relatively "fast" host specificity for a particular species of parasite, as occurs with the pinworm of man *(Enterobius vermicularis)* and the human body louse *(Pediculus humanus humanus),* for example. Certain anopheline mosquitoes have a preference for human blood when it is available—i.e., they are *anthropophilic*. Other parasites are less discriminating, and man constitutes only one of several satisfactory hosts. The Chinese liver fluke (*Clonorchis sinensis*), the Oriental blood fluke (*Schistosoma japonicum*), and the trichina worm (*Trichinella spiralis*) are representative of this group. In many instances, man is only incidentally involved as a host, while domestic or wild animals serve as *reservoirs* of the parasite.

Susceptibility to a parasite by several host species introduces the concept of *zoonosis,* which literally means "disease of animals," but today the term is employed for those diseases of animals that are transmissible to man. Hoare (1962) credits Virchow with the invention of the term. Several classifications of zoonoses have been proposed. For example, if the parasitosis is common to man and reservoir hosts, it may be referred to as a *euzoonosis;* if man is an infrequent (hence, incidental) host, it is a *parazoonosis*. Koegel (1951) proposed the term "anthropozoonosis" for infections acquired by man from other vertebrates; e.g., those infections acquired directly, as in the case of trichinosis or cysticercosis, or the large number of infections transmitted by arthropod and molluscan intermediate hosts. Diseases primarily of human origin that may be acquired by other vertebrates are *zooanthroponoses* (Wagener, 1957), and diseases common to man and other vertebrates are *amphixenoses* (Nelson, 1960).

Hoare (1962) adds that zoonoses constitute "a chapter in the natural history of parasitic infections which concerns the ecological and evolutionary aspects of the host-parasite relations . . . has led to intensive studies of the ecological conditions under which infections are maintained among wild animal hosts and vectors." Garnham (1959) stated that the zoonotic process is dynamic and that the best zoonotic species are the least differentiated ones. Modern civilization tends to disrupt the natural course of many euzoonoses so that man becomes involved in their cycles, at times with disastrous consequences (van der Hoeden, 1964; Hubbert et al., 1975).

Organisms that cannot exist without a host are *obligate parasites*. Other organisms that—under favorable circumstances—may live either a parasitic or free-living existence are called *facultative* or "opportunist" parasites. For example, in some circumstances, *Micronema* and certain free-living amebae (*Naegleria* and *Acanthamoeba*) may invade and colonize in the human brain or other sites; Page (1974) coined the term "amphizoic" for such organisms. In addition, some free-living animals or organisms that parasitize other hosts, having merely passed through the intestine, are recovered in a living or dead state from human feces; they are referred to as *spurious* parasites of man.

In microbiology, the relationship of the parasite to its host is designated as an *infection*. This is also the appropriate designation for all animal species that are endoparasites, whereas those that are *ectoparasites* (those that are attached to the skin or that temporarily invade the superficial tissues of the host's body) produce an *infestation*. This distinction is employed irrespective of the size of the invader, but infestation of the host is most frequently applied to species of arthropods. In the more limited sense, *parasitosis* is the state of infection or infestation with an animal parasite.

Some animal parasites of man are normally harmless commensals (Baer, 1951; Henry, 1966). Notable among these are several intestinal amebae and flagellates. On the other hand, a great majority of the animal parasites are harmful, frequently causing local and systemic damage of one type or another; these species are called *pathogens*.

When an individual harboring a parasite is reinfected with the same species of parasite, this is *superinfection*. In case the infected person is his own direct source of the re-exposure, it is *autoinfection*, which may be external (namely, perianal or anus-to-mouth via fingers) or internal in the mode of infection.

SOURCES OF EXPOSURE TO INFECTION OR INFESTATION

Exposure may occur from one or more of the following sources: (1) contaminated soil or water; (2) food containing the immature infective stage of the parasite; (3) a blood-sucking insect; (4) a domestic or wild animal harboring the parasite; (5) another person, his clothing, bedding, or the immediate environment that he has contaminated; or (6) one's self. Each of these sources will be briefly illustrated.

Soil polluted with human excreta is commonly responsible for exposure to infection with *Ascaris lumbricoides*, *Trichuris trichiura*, human hookworms, and *Strongyloides stercoralis*. Water may contain viable cysts of parasitic amebae, intestinal flagellates, *Taenia solium* eggs, and the infective cercarial stage of the human blood flukes. Freshwater fishes constitute the source for fish tapeworm (*Diphyllobothrium latum*), as well as for several types of intestinal and liver flukes; crabs and crayfishes are sources for the Oriental lung fluke; raw pork is the source for *Trichinella spiralis* and *Taenia solium;* beef is the source for *Taenia saginata;* buffalo nuts (*Trapa* spp.) are the source for the giant intestinal fluke (*Fasciolopsis buski*); and watercress (*Nasturtium officinale*) is the source for the sheep liver fluke (*Fasciola hepatica*). Blood-sucking arthropods transmit malaria parasites, leishmanias, trypanosomes, filariae, viruses, rickettsiae, bacteria, and spirochetes. Dogs are the direct source for human infection with the hydatid cyst of *Echinococcus granulosus*, as well as for visceral larva migrans due to *Toxocara canis*, while herbivorous animals commonly constitute the source for human infection with *Trichostrongylus* spp. Other human beings are directly responsible for all or a considerable amount of infection with the pathogenic ameba (*Entamoeba histolytica*), the pinworm (*Enterobius vermicularis*), and the dwarf tapeworm (*Hymenolepis nana*). Autoinfection accounts for some of these parasitoses and for some reinfections with *Strongyloides stercoralis*.

PORTAL OF ENTRY INTO THE BODY

In the case of internal parasites, the most common portal of invasion is the mouth. This is the entrance for the intestinal protozoa (for most species in the encysted stage); the common roundworms *Ascaris lumbricoides, Trichuris trichiura* and *Enterobius vermicularis* and the dwarf tapeworm *Hymenolepis nana* (all in the fully embryonated egg stage); *Trichostrongylus* spp. as a free-living infective larva; and *Trichinella spiralis, Taenia solium, T. saginata, Diphyllobothrium latum,* intestinal flukes and the lung fluke, from eating food containing the mature larval stages of these worms.

A few important roundworms—namely, species of hookworms and *Strongyloides*—actively enter the body from the soil via the skin, as do the blood flukes (*Schistosoma* spp.) from fresh water. A large number of parasites that have a required developmental stage in blood-sucking arthropods are introduced percutaneously when the arthropod punctures the skin to feed. Among these parasites are the agents of malaria, the leishmanias, trypanosomes, filariae, many viruses and rickettsiae, and a few species of bacteria and spirochetes.

Other methods of transmission include the following: (1) inhalation of air-borne eggs of *Enterobius vermicularis* into the posterior pharynx; (2) transplacental (congenital) infection with *Toxoplasma gondii,* and occasionally with malaria parasites, blood flukes and others (Loke, 1982); (3) transmammary (milk) infection with species of *Strongyloides, Ancylostoma,* and certain trematodes (Miller, 1981); and (4) by sexual intercourse, in the case of *Trichomonas vaginalis.*

BIOLOGIC INCUBATION PERIOD

Exposure vs. Infection. Exposure to infection is the act or process of inoculation, whereas *infection* connotes a "take," whereby the infective agent becomes established in the host. Often the term "infection" is employed carelessly when the concept "inoculation" is intended. For example, a person may be exposed by ingesting cysts of *Entamoeba histolytica* without colonization of the ameba, so that no infection is produced.

Development of the Infection. Once the successful parasite has entered the body of the host, characteristically it is carried or actively migrates to a location where it matures and produces progeny. This may be a relatively simple procedure, or it may be lengthy and complicated. For example, for *Entamoeba histolytica,* it consists of excystation of the protoplast in the lumen of the ileum and division into as many small metacystic trophozoites as there were nuclei in the encysted stage, then passive carriage in the fecal stream through the ileocecal sphincter into the large intestine, lodgement in the glandular crypt, growth to normal size, and multiplication by asexual binary division. Similarly, for fully embryonated eggs of *Trichuris trichiura* and *Enterobius vermicularis,* it consists of hatching in the duodenum, migration of the larva down to the cecal level, attachment to the mucosa, and development to the adult stage, which is followed by the production of eggs of the new generation.

In the case of *Ascaris lumbricoides,* the ingested egg hatches in the duodenum. The emerging larva enters the mucosa and migrates via blood vessels to the lungs; then, after intrinsic development in the tissues, it breaks out into the air sacs and passes up the respiratory tree to the epiglottis, passes over into the digestive tract, and is carried to the small intestine, where it develops into the adult form. Hookworms and *Strongyloides stercoralis,* which actively invade the skin, are carried in the blood stream to the lungs, after which their course of migration parallels that of *Ascaris.*

Organisms that are introduced in the encysted larval stage in food characteristically become freed of their cyst wall or capsule in the duodenum. Intestinal helminths soon become attached to the mucosa, usually in the upper or middle levels of the small intestine, where they mature; inhabitants of the bile ducts enter via the ampulla of Vater (e.g., *Clonorchis sinensis*) or utilize a more indirect route (e.g., *Fasciola hepatica*); and the lung fluke *Paragonimus westermani* takes a devious route through tissues and body cavities before arriving at its destination near a bronchiole.

The blood flukes (*Schistosoma* spp.), after actively invading the skin, burrow to the cutaneous lymphatics and blood vessels. Once they have entered the blood stream, they are carried to the lungs, where they develop for several days before migrating to the liver. They grow in the liver and, on reaching sexual maturity, mate and migrate in copula to the smaller mesenteric venules (*S. japonicum* or *S. mansoni*), or via the mesenteric and rectal venules into the vesical venules (*S. haematobium*).

Malaria parasites, when introduced into human

skin by an anopheline mosquito, are rapidly carried in the blood stream, from which they disappear in about 30 minutes. Later, after 2 or 3 days, they are found in the liver, where they undergo asexual multiplication in parenchyma cells before being discharged into the blood to initiate infection in the red blood cells. Trypanosomes may rapidly multiply in circulating blood (*Trypanosoma gambiense*) or may develop as intracellular parasites in macrophages near the site of inoculation or in the viscera (*Trypanosoma cruzi*). Species of *Leishmania* invariably colonize intracellularly in the reticuloendothelial system.

The filaria worms, once introduced as filariform larvae into the skin by the infected insect, enter lymphatic vessels in which the immature worms develop for several weeks or months, finally reaching maturity in various sites depending on the species of filaria; for example, *Wuchereria bancrofti* and *Brugia malayi* inhabit the lymphatic system, and *Loa loa* and *Onchocerca volvulus* inhabit the subcutaneous tissues.

End of the Biologic Incubation Period. Biologic incubation is terminated as soon as the parasites or their products can be demonstrated in the feces or other excreta or in the circulating blood (*parasitemia*), by aspiration, biopsy, or other diagnostic procedure. The biologic incubation period, also referred to as the *prepatent period,* varies from 1 or more days to weeks or months, depending on the particular species of parasite and its ability to develop in the particular host. *Biologic incubation* is related to the development of the parasite; *clinical incubation* refers to the interval between exposure and the earliest evidence of symptoms produced as a result of the infection (or infestation).

ENVIRONMENT AND METABOLISM OF PARASITES

The habitats of parasites are extremely varied. Among the parasites of man, there are examples of protozoa, helminths, and arthropods that normally inhabit the intestine, liver, lungs, muscles, brain, blood, and lymphatic tissues. Many species of parasites have complicated life cycles involving developmental stages that live in soil or water, or use various kinds of intermediate hosts, including invertebrates and vertebrates and cold and warm-blooded animals. In such varied environments, parasites have become adapted to using or tolerating widely differing oxygen, carbon dioxide, and hydrogen ion concentrations and temperatures, and their nutritional requirements are varied, as are their means of obtaining and utilizing the nutrients required for growth, motility, and reproduction. Current knowledge of the biochemistry and physiology of parasitic protozoa and helminths is reviewed in a recent treatise by von Brand (1979), who was the recognized founder and leader in the field of parasite biochemistry for 40 years. In their evolutionary adaptations, parasites have found ways to deal with the immune reactions of their hosts and to induce physical and physiologic host changes that are advantageous to themselves and tolerated by their hosts. The immunology of parasitic infections now constitutes a major subdiscipline of parasitology (Cohen and Sadun, 1976; Miller et al., 1977; Warren et al., 1977).

REFERENCES

Baer, J.G. 1951. *Ecology of Animal Parasites*. Urbana, University of Illinois Press.

Cohen, S., and Sadun, E.H. 1976. *Immunology of Parasitic Infections*. Oxford, Blackwell Scientific Publications.

Garnham, P.C.C. 1959. The evolution of the zoonoses. Med. Press, *242:* 251–256.

Henry, S.M., editor. 1966. *Symbiosis*. Vol. I; Vol. II. New York, Academic Press.

Hoare, C.A. 1962. Reservoir hosts and natural foci of human protozoal infections. Acta Trop., *19:* 281–317.

Hubbert, W.T., McCulloch, W.F., and Schnurrenberger, P.R., editors. 1975. *Diseases Transmitted from Animals to Man,* 6th ed. Springfield, Ill., Charles C Thomas.

Joint FAO/WHO Expert Committee on Zoonoses. 1967. Third Report. W.H.O. Tech. Rep. Ser., no. 378.

Koegel, A. 1951. *Zoonosen (Anthropozoonosen)*. Basel, E. Reinhardt.

Loke, Y.W. 1982. Transmission of parasites across the placenta, pp. 155–228. In *Advances in Parasitology* vol. 21. Edited by J.R. Baker and R. Muller. New York, Academic Press.

Miller, G.C. 1981. Helminths and the transmammary route of infection. Parasitology, *82*:335–342.

Miller, L.H., Pino, J.A., and McKelvey, J.J., Jr. 1977. *Immunity to Blood Parasites of Animals and Man*. New York, Plenum Press.

Nelson, G.S. 1960. Schistosome infections as zoonoses in Africa. Trans. R. Soc. Trop. Med. Hyg., *54:* 301–316.

Page, F.C. 1974. *Rosculus ithacus* Hawes, 1963 (Amoebida, Flabellulidae) and the amphizoic tendency in amoebae. Acta Protozool., *13:* 143–154.

van der Hoeden, J. 1964. *Zoonoses*. Amsterdam, Elsevier Publishing Co.

von Brand, T. 1979. *Biochemistry and Physiology of Endoparasites*. New York, Elsevier North-Holland.

Wagener, K. 1957. Zoonosen-Anthropozoonosen-Zooanthroponosen. Munch. Tierarztl. Wschr., *70:* 12.

Warren, K.S., David, J., Neva, F.A., and Phifer, K.O. 1977. *Immunology of Parasitic Infections: Report of a Workshop*. Am. J. Trop. Med. Hyg., *26* (suppl.).

Chapter 2

Epidemiology and Distribution of Diseases Caused by Animal Parasites

Epidemiology is the body of knowledge that concerns disease in human populations or communities rather than in individuals. Jungblut (1935) states, "In its widest definition the term epidemiology may be understood to include the study of the manifestations of any disease which happens to attack groups of individuals at any time." Maxcy (1956) refers to the subject as "that field of medical science which is concerned with the relationships of the various factors and conditions which determine frequencies and distributions of an infectious process, a disease, or a physiological state in a human community." Thus, epidemiology as a branch of medicine not only incorporates data in the area of infectious diseases in population groups, but also those resulting from anatomical deformities, genetic constitution, metabolic dysfunction, malnutrition, neoplasms, occupational pursuits, and the process of aging.

In spite of this broad concept, until recent decades the mass of epidemiologic information has accumulated mostly from a study of *communicable diseases*. Yet it is frequently difficult, except in epidemics, to isolate the effects of an infectious agent in a community from other disease entities that are almost invariably operating simultaneously.

When disease in the human population maintains a relatively steady, moderate level, it is said to be *endemic;* if the prevalence is high, it is *hyperendemic*; if there is a sharp rise in the incidence or an outbreak of considerable intensity occurs, it is considered to be *epidemic*; and if it appears only occasionally in one or at most a few members of a community, it is reported as *sporadic*. Endemicity is commonly associated with a certain degree of tolerance to the pathogen; epidemicity typically results from introduction of an agent or a new strain into a community that is essentially nonimmune, the sudden development of enhanced pathogenicity by the infectious agent, or the markedly increased prevalence of a vector. For example, certain communicable diseases, such as smallpox and measles, were endemic in Europe at the time Columbus discovered America, but when the American Indian population was exposed, these diseases became epidemic in the Americas, with decimating effect. At times communicable diseases have been disseminated over extensive areas of the world, producing *pandemics*. Notable among these have been influenza, plague, and cholera.

The term referring to the knowledge of the frequency of disease in animal communities is "epizootiology." Moderate prevalence of a disease in animals is designated as *enzootic,* a sudden outbreak as *epizootic,* and wide dissemination as *panzootic*.

EPIDEMIOLOGIC METHODS

Methods of epidemiologic study are applicable irrespective of the type of disease-producing agent. Useful information may be obtained by two principal approaches. First of all, observation and study of an individual patient by a competent investigator may provide accurate clues to the way a particular disease has been acquired. This can lead to an epidemiologic study to determine the extent and prevalence of the disease, the nature of vectors or other factors involved in transmission in the community in which the patient contracted it, and the chances of its dissemination. Second, one may use the statistical approach, which utilizes vital statistics accumulated by local, state, and national government health agencies that incorporate morbidity data on infectious diseases.

In 1923 the Health Organization of the League of Nations came into existence; it was assigned the responsibility of notifying of epidemic diseases, providing consultation service, standardizing biologicals, drugs, health education, and training and research in the field of infectious and epidemic diseases. In 1948, 3 years after the United Nations came into existence, the World Health Organization (WHO) was established as the special agency of the United Nations. WHO took over the functions of the Health Organization of the League of Nations.

Biostatistical Methods. Mortality rates are usually expressed in deaths per 100,000 per year. Morbidity rates are commonly reported as cases of notifiable diseases per 1000 per year. The crude rates require adjustment to account for variables in the community, such as racial composition, age, sex, occupation, hospitalized or nonhospitalized persons, and any other factors that may reasonably affect the data. Moreover, in order to be significant, the statistics must have been obtained from a sufficient sampling of the population that it is intended to represent. The information derived from the adjusted data of a representative sample requires treatment to determine the mean, standard deviation and standard error of the mean before it is possible to judge the reliability of the data in making generalizations about the population under investigation. Even then it represents a group and is not applicable to a particular individual of the group sampled or to another group in the same community or elsewhere. Nevertheless, vital statistics, when they are properly accumulated and analyzed, may be compared with similar data from other areas to determine whether or not there is a significant correlation; likewise, when accumulated year by year, such data provide valuable evidence as to trends, including epidemic cycles, and increases and declines in the rates.

EPIDEMIOLOGIC INVESTIGATION OF PARASITOSES

Basic epidemiologic information is available on practically all important parasitic infections that affect man. This information includes the stage of the parasite that is discharged from the human body (or reservoir host); the stage of the parasite that is available for an arthropod to pick up while feeding on the skin; extrinsic development of the parasite, whether on the soil, in water, or in the body of required alternate host (or hosts); and the way in which each parasite gains entry into the human body. Moreover, there is relatively good information concerning the geographic distribution of these infections, their frequency in different population groups, their tendency to endemicity, hyperendemicity, epidemicity, or sporadic occurrence in different areas and, in some infections, trends or cycles of increase or decrease. Relatively abundant data have been accumulated for amebiasis, malaria, leishmaniasis, African trypanosomiasis, Chagas' disease, trichinosis, hookworm disease, ascariasis, enterobiasis, Bancroft's filariasis, onchocerciasis, and schistosomiasis.

In undertaking an investigation on the prevalence of malaria or intestinal parasites in a community, a sufficient number of individuals must be examined to constitute a representative sample of the population. Frequently, school children are surveyed, and while the results obtained may be dependable insofar as this age group is concerned, they are not necessarily referable to older age groups. However, children found infected in a school-age survey may be utilized as leads for follow-up studies on parasite frequency in each child's family. In this way, it may be possible to obtain significant information distinguishing between the environmental sanitation and personal hygiene responsible for exposure to the infections.

Essential epidemiologic data have been contributed by medical entomologists and biologists for the parasitoses of man; arthropod-transmitted viral diseases, such as yellow fever, dengue, and several encephalitides; the rickettsioses, plague, tularemia, and relapsing fevers.

The epidemiologic aspects of each important disease considered in this text will be presented in chapters on the etiologic agents or vectors.

GEOGRAPHIC DISTRIBUTION OF HUMAN PARASITOSES

Geographical Areas of the Earth. Maps of the world usually divide the land and sea into a central band called the "Tropics," two broad bands, one on either side of the Tropics, called the "Temperate Zones," and a cap around each of the poles, called the "Arctic" and "Antarctic Regions." For general purposes, these geographic delimitations are relatively satisfactory, but they are not accurate insofar as climate and the existence of plant and animal life are concerned. Astronomically and geographically, the Tropics lie between two charted circles engirdling the globe, one 23°28′ North and one 23°28′ South of the Equator. The former is the Tropic of Cancer, and the latter is the Tropic of Capricorn. Similarly, the Arctic and Antarctic Circles lie 23°28′ from the North and South Poles, respectively.

The Tropics may be defined as the warm band engirdling the globe in which the mean temperature *for the coldest month of the year* is never less than 20° C (68° F), while the subtropical zones are those on either side of the Tropics in which the mean temperature *for the entire year* is never lower than 20° C (68° F). The plotted lines (frequently referred to as "Supan's lines") delimiting these areas, and their divergence from the Tropic of Cancer and Tropic of Capricorn, respectively, are shown on the accompanying map (Fig. 2–1). The differences between the geographic and the isothermal boundaries of the several regions of the globe are due to various factors, of which winds, water currents, the relative size of land and water mass, and mountain ranges are probably the most important.

Types of Tropical and Subtropical Habitats. As a result of prevailing winds, ocean currents, relation of land and sea masses, elevation of land, and numerous other contributing factors, the climates of the globe may be classed as "dry," "dry-wet," and "wet." "Dry" designates an essentially desert area, with minimal rainfall. "Dry-wet" indicates seasonal alternation between periods of relative dryness and precipitation, as in India, for example, where rainfall is limited to the estivo-autumnal monsoons. "Wet" constitutes an area in which rainfall occurs rather abundantly throughout the year. Moisture, temperature, and elevation determine the type of *flora,* and indirectly, the *fauna.* In the tropical belt, there may be a luxuriant tropical rain-forest (essentially impenetrable jungle), either coastal, as in the Guianas, or far inland, as in the Amazon and Congo basins. In the higher valleys of warm climates, there may be temperate-zone deciduous and evergreen forests, and on the higher elevations, alpine forests and tundra. Moist lowlands may be essentially free of trees but covered with luxuriant swampy savannas. In the more temperate regions, there may be extensive grassy plains (for example, the midwestern plains of the United States and the pampas of Argentina). Areas of minimal rainfall may be covered with xerophytic scrub growth or may be entirely barren save for occasional oases. The climatic features of essentially all major localities of the world have been described in graphic form in an atlas by Walter and Lieth (1960).

Faunistic Areas of the World. There are six main faunistic areas of the Earth, determined primarily by the essentially different types of mammals in the respective areas. These areas are as follows (Fig. 2–1):

1. *Palearctic,* which includes Europe, Africa north of the Sahara, northern Arabia, and northern and central Asia
2. *Nearctic,* which includes North America as far south as central Mexico. In many respects, this region is the counterpart in the Western Hemisphere of the Palearctic in the Eastern Hemisphere
3. *Neotropical,* which includes southern Mexico, Central America, the West Indies, and South America
4. *Ethiopian,* which includes all of Africa but the Mediterranean coast; also, the adjacent islands and the southwestern tip of Arabia
5. *Oriental,* which includes all of India except the northern mountainous regions; Burma; Thailand; Malaysia; the Indonesian Islands exclusive of New Guinea; the Philippines; Vietnam; Taiwan; and southern mainland China
6. *Australian,* which includes Australia, New Guinea, and other Melanesian islands as well as New Zealand.

In addition to the above areas, there is the Micronesian region of the Central and South Pacific (Oceania), which has a relatively distinct fauna.

CLIMATE AND ITS EFFECTS ON MAN

Various physical factors influence the climate in different parts of the globe, and climate, in turn, is responsible for different flora and fauna. Thus, consistently dry conditions produce a desert, and consistently heavy precipitation in warm climates or during warm seasons in cooler climates is responsible for luxuriant growth of vegetation. Moreover, in much of the temperate and some of the subtropical areas, as in the United States, Europe, and the U.S.S.R., there are frequent periodic cyclonic disturbances that have a general direction from west to east and produce an alternation of high and low atmospheric pressures. At times these are modified by storm paths from the outside; in

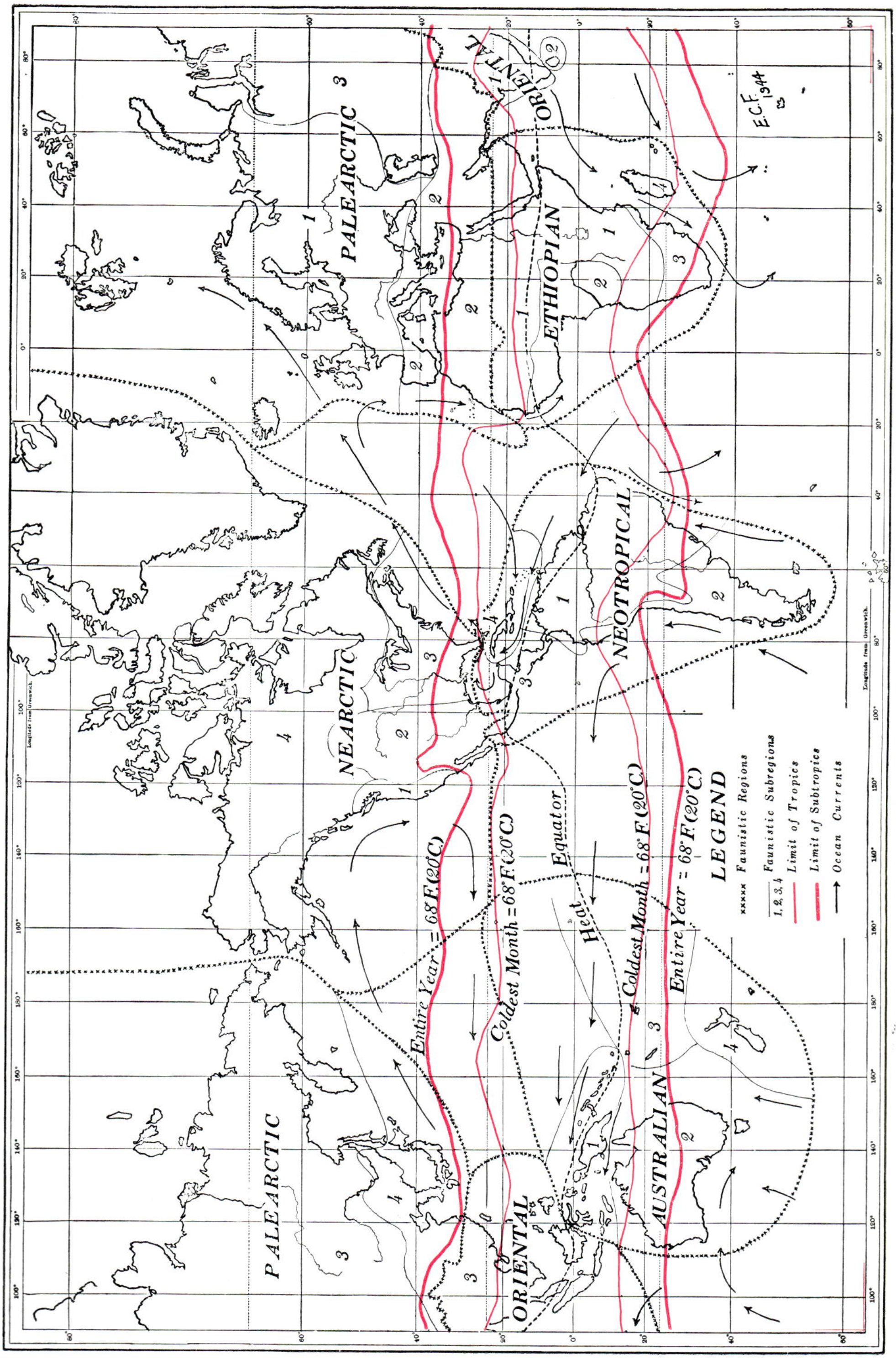

Fig. 2–1. World map, Mercator projection, showing the equatorial isothermic boundaries of the Tropics and Subtropics (Supan's lines in red: thin line shows limits of the Tropics, thick line limits of the Subtropics), principal ocean currents, and the six faunistic regions (see text) and subregions (1–4) of the earth. (By E.C. Faust.)

the United States, for example, atmospheric pressure might be affected by storm paths from Alaska and Northwest Canada or from the Gulf of Mexico. In addition, there is another type of climate, in which a considerable part of the year is very dry and rainfall is seasonal (i.e., *monsoon*). In this dry-wet climate, vegetation and animal life and the existence of man himself are dependent on the regularity and amount of the annual rainfall, together with the fertility of the soil and its ability to hold moisture for crops and pasturage. If the run-off of the water, its sinkage into subsurface streams, or its evaporation are in excess of the rainfall, the soil is arid and will not sustain life. These latter conditions are well illustrated by much of the land in Australia.

Temperature is the most important single extrinsic factor that influences man's existence. This is particularly true in hot climates, in which sustained environmental heat is the most detrimental of all climatic conditions.

When the temperature of the environment reaches 95° F (35° C), radiation and convection cease to function and vaporization must bear practically the entire burden of heat loss. With a high humidity and no breeze, vaporization is greatly hampered and the body temperature rises.

Although *sunlight* is essential for human existence, the ill effects of too much direct sunlight are well known. The shorter length rays of the sun cause hyperemia in fair skins, and later may produce pigmentation and hyperkeratosis as a protective mechanism. The rays of medium length (luminous rays) irritate the nervous system and, if continued, may cause tropical neurasthenia. These rays are responsible for "light stroke," with dizziness and headache, acute or chronic conjunctivitis, retinitis, and photophobia. The sun's long rays are the heat rays that are directly or indirectly responsible for unfavorable heat effects.

In addition to climatic factors, *dietary deficiencies* play a very significant role in human welfare. Both occur in hot and cold climates, but they are peculiarly significant in the Tropics because they constitute a common background for infectious diseases by providing a lowered threshold of resistance. Except in over-populated areas like Java, India, and China, there is usually a quantitative abundance of food; yet most of the foods of warm climates are low in vitamins, essential minerals, and proteins, and high in carbohydrates, thus providing a heat-producing, poorly fortified dietary.

Parasitic Diseases with Reference to Climate. While many infectious diseases result from direct person-to-person contact (e.g., viral respiratory infections, smallpox, pneumonic plague, and Q fever), this mode of transmission is uncommon for parasitic infections. Pinworm infection (enterobiasis) is an example of a parasitic infection transmitted by contact. Other parasitic infections reach man directly from human excreta (e.g., amebiasis and giardiasis) or from food or water contaminated with feces, while others require alternate hosts or vectors, or a period of development in the soil or water. The incidence and mass of human infection are roughly commensurate with the degree of personal and group sanitation, as well as the individual's resistance to pathogens to which he is exposed. It is not surprising, therefore, that most infectious diseases are particularly prevalent in areas with warm climates in which man exerts himself least in developing sanitary protection and typically has a low threshold of resistance to invading organisms.

Introduction of Parasitic Diseases by Human Migration. The migrations of populations have contributed largely to the development of parasites in new localities. Evidence favors the view that yellow fever, dengue, falciparum malaria, broad fish tapeworm infection, the hookworm infection produced by *Necator americanus,* Manson's blood-fluke infection, Bancroft's and other types of filariasis, and dracunculosis were brought to the Western Hemisphere by the white colonists and their slaves imported from Africa. Typhus fever, leprosy, smallpox, measles, mumps, syphilis, yaws, and probably influenza were brought to the Western Hemisphere in the same manner (Scott, 1943; Marr, 1982). Wherever climate, necessary intermediate hosts, and customs of the population were favorable, these diseases became established in the new soil. On the other hand, the chigoe (*Tunga penetrans*) was probably originally indigenous to the warmer areas of the Western Hemisphere and was later carried to Africa, where today it is a much more serious parasite of the skin than in its old home. Carrion's disease, which is due to infection with *Bartonella bacilliformis,* and infestation with the tropical warble-fly, *Dermatobia hominis,* were most likely indigenous in the Amerind populations before the first voyage of Columbus. While the hookworm *Ancylostoma duodenale* was originally confined to the North Temperate zone and *Necator americanus* to the tropical belt

Table 2–1. Basic Epidemiologic Data on the Important Protozoan and Helminthic Parasites of Man.

Parasite	*Geographic Distribution*	*Common Infective Stage and Portal of Entry*	*Reservoir Hosts (R) or Vectors (V)*
Protozoa			
Entamoeba histolytica	Cosmopolitan, especially prevalent in warm climates	Cyst via mouth	Monkey, dog (R)
Balantidium coli	Warm climates	Cyst via mouth	Pig, monkey (R)
Giardia lamblia	Cosmopolitan, especially prevalent in children in warm climates	Cyst via mouth	None
Trichomonas vaginalis	Cosmopolitan	Trophozoite via vulva or urethra	None
Leishmania tropica	Mediterranean area to W. India	Leptomonas via skin	Dog, rodents (R); *Phlebotomus* (V)
Leishmania braziliensis	Mexico to N. Argentina	Leptomonas via skin	Dog (R); *Lutzomyia* (V)
Leishmania donovani	China, India, Africa, Mediterranean area, Continental Latin America	Leptomonas via skin	Dog, fox, rodents (R); *Phlebotomus* and *Lutzomyia* (V)
Trypanosoma gambiense	West and Central Africa	Trypanosome via skin	Cattle? (R); *Glossina* (V)
Trypanosoma rhodesiense	Central and East Africa	Trypanosome via skin	Wild game mammals (R); *Glossina* (V)
Trypanosoma cruzi	Continental Latin America	Trypanosome via skin	Domestic and wild mammals (R); triatomid bugs (V)
Plasmodium vivax	Warm and cooler climates	Sporozoite via skin	*Anopheles* (V)
Plasmodium malariae	Warm climates	Sporozoite via skin	*Anopheles* (V)
Plasmodium falciparum	Warm climates	Sporozoite via skin	*Anopheles* (V)
Isospora belli	Cosmopolitan	Sporocyst via mouth	None known
Sarcocystis suihominis	Cosmopolitan	Merozoite in pork via mouth	Pig (R)
Sarcocystis hominis	Cosmopolitan	Merozoite in beef via mouth	Cow (R)
Sarcocystis spp.	Varied	Sporocyst via mouth	Unknown final hosts (R)
Toxoplasma gondii	Cosmopolitan	Oocyst or merozoite via mouth	Cat (oocysts), meat animals (R)
Nematodes (Roundworms)			
Trichinella spiralis	Cosmopolitan; common in the United States	Encysted larva in pork via mouth	Pig, bear, walrus, rat (R)
Trichuris trichiura	Warm, moist climates	Embryonated egg via mouth	None
Strongyloides stercoralis	Warm, moist climates	Filariform larva via skin	Chimpanzee?, dog (R)
Necator americanus	Common in warm climates	Filariform larva via skin	None
Ancylostoma duodenale	Palearctic Region, W. South America	Filariform larva via skin or mouth	None
Ancylostoma ceylanicum	Limited distribution in warm climates	Filariform larva via skin or mouth	Dog, cat (R)
Enterobius vermicularis	Cosmopolitan; common in children	Embryonated egg via mouth	None
Ascaris lumbricoides	Cosmopolitan, especially in warm climates	Embryonated egg via mouth	Pig occasionally (R)

Wuchereria bancrofti	Prevalent in warm climates	Filariform larva via skin	*Aedes* spp., *Culex* spp., *Mansonia* spp., *Anopheles* spp. (V)
Brugia malayi	Oriental	Filariform larva via skin	*Mansonia* spp., *Anopheles* spp. (V)
Brugia timori	Timor and adjacent islands	Filariform larva via skin	*Anopheles* spp. (V)
Onchocerca volvulus	Tropical Africa, Mexico, Guatemala, Venezuela, Surinam, Colombia, Ecuador	Filariform larva via skin	*Simulium* spp. (V)
Loa loa	Tropical Africa	Filariform larva via skin	*Chrysops* spp. (V)
Dracunculus medinensis	Tropical Eastern Hemisphere	Larva in arthropod host via mouth	*Cyclops* spp. (V)
Cestodes (Tapeworms)			
Taenia saginata	Cosmopolitan	Cysticercus in beef via mouth	Cow (V)
Taenia solium			
1. Adult worm	Cosmopolitan	1. Cysticercus in pork via mouth	Pig (V)
2. Cysticercus stage		2. Eggs from human infections via mouth	None
Echinococcus granulosus	Cosmopolitan	Eggs from canines via mouth	Sheep, cow, pig, other (R); dog and other canines (V)
Echinococcus multilocularis	Central Europe, Asia, Alaska	Eggs from foxes via mouth	Small rodents of forest and tundra (R); foxes (V)
Hymenolepis nana	Prevalent in warm climates	Eggs from human infections via mouth	None (heterologous strains in rodents)
Hymenolepis diminuta	Prevalent in warm climates	Larva in arthropod host via mouth	Rodents (R)
Dipylidium caninum	Prevalent in warm climates	Larva in insect host via mouth	Dog, cat (R)
Diphyllobothrium latum	North Temperate Zone, Argentina, Chile, Australia	Sparganum larva in fish flesh via mouth	Dog, bear (R)
Trematodes (Flukes)			
Fasciola hepatica	Sheep-raising countries	Larva encysted on vegetation via mouth	Sheep, et al. (R)
Fasciolopsis buski	Oriental	Larva encysted on water nuts and bulbs via mouth	Pig (R)
Clonorchis sinensis	Oriental	Larva encysted in fresh-water fish via mouth	Many fish-eating mammals (R)
Opisthorchis felineus	Europe, Asia	Larva encysted in fresh-water fish via mouth	Dog, cat, et al. (R); fresh-water fish (V)
Opisthorchis viverrini	Thailand	Larva encysted in fresh-water fish via mouth	Dog, cat (R); fresh-water fish (V)
Paragonimus westermani	Primarily Oriental	Larva encysted in crabs or crayfishes via mouth	Felines and other crab-eating mammals (R)
Schistosoma japonicum	Oriental	Cercarial larva in fresh water via skin	Many mammals (R)
Schistosoma mansoni	Africa, Latin America	Cercarial larva in fresh water via skin	None important
Schistosoma haematobium	Africa to India, S. Portugal	Cercarial larva in fresh water via skin	None important

of the Eastern Hemisphere, the north and south migrations of people, particularly from China and northern India to the Malay Archipelago, have caused an intermingling of these hookworm species in both the indigenous and immigrant populations. Distinguished pediculologists state that the races or strains of human lice originally had color differences as an adaptation to the skin colors of the several races of man. Migration, intermingling, and interbreeding of human beings have long since caused these color differences to disappear.

Even today, isolated aboriginal groups of the human population have preserved unmixed species of certain parasites such as hookworms, whereas the surrounding populations of more recent advent into the country harbor a different species. Furthermore, contact with certain infections such as yellow fever, malaria, and amebiasis has produced marked immunity or resistance to these diseases among native populations, while newcomers to the area are much more susceptible to these infections or manifest more serious symptoms.

In addition to peaceful migrations of people, wars and conquests have done much to modify the picture of disease distribution and epidemiology in the human race. Smallpox, typhus, plague, and syphilis have had a marked effect on social and economic development. As a result of exposure in highly endemic areas, military personnel during 1941 to 1945 contracted many parasitic diseases not indigenous, or at least not prevalent, in their home countries. Among these diseases were malaria, amebiasis, kala-azar, cutaneous leishmaniasis, hookworm infection, strongyloidiasis, Bancroft's filariasis, and schistosomiasis. Nevertheless, the public health services of most countries, with the cooperation of WHO, have set up safeguards so that these diseases have had little opportunity to become established in new foci.

Epidemiologic data on the important animal parasites of man are summarized in Table 2-1.

REFERENCES

Hare, R. 1955. *Pomp and Pestilence. Infectious Disease, Its Origins and Conquest.* New York, Philosophical Library.

Jungblut, C.W. 1935. General principles of epidemiology. In *Agents of Disease and Host Resistance.* Edited by F.P.Gay. Springfield, Ill., Charles C Thomas.

Last, J.M., editor. 1980. *Maxcy-Rosenau Public Health and Preventive Medicine,* 11th ed. New York, Appleton-Century-Crofts.

Lilienfeld, A.M., and Lilienfeld, D.E. 1980. *Foundations of Epidemiology,* 2nd ed. New York, Oxford University Press.

Marr, J.J. 1982. Merchants of death: The role of the slave trade in the transmission of disease from Africa to the Americas. Pharos, *45:* 31–35.

Maxcy, K.E. 1956. *Rosenau's Preventive Medicine,* 8th ed. New York, Appleton-Century-Crofts.

McKay, D.M. 1981. *Preservation of Personal Health in Warm Climates,* 8th ed. London, Ross Institute of Tropical Hygiene.

Scott, H.H. 1943. The influence of the slave trade in the spread of tropical disease. Trans. R. Soc. Trop. Med. Hyg., *37:* 169–183.

Simmons, J.S., Whayne, T.F., Anderson, G.W., and Horack, H.M. *Global Epidemiology. India and the Far East; The Pacific Area,* vol. 1, 1944. *Africa and the Adjacent Islands,* vol. 2, 1951. *The Near and Middle East,* vol. 3, 1954. Philadelphia, J.B. Lippincott.

Walter, H., and Lieth, H. 1960. *Klimadiagramm-Weltatlas.* Jena, Gustav Fischer Verlag.

Chapter 3

Nomenclature of Animal Species Responsible for Human Disease

All animals and plants must have names by which they can be distinguished. Although common names are frequently used for this purpose, these are not universally understood, partly because of language barriers and partly because a common name is not necessarily applied to the same organism in different countries or even in different parts of the same country. For example, in the United States, "threadworm" usually refers to *Strongyloides stercoralis,* while in England it has been employed in connection with *Enterobius vermicularis*. To overcome this difficulty, a binomial scientific name is used, consisting of a generic and a specific designation. By agreement, these names are of Greek or Latin derivation or conform to classical endings if they are of nonclassical origin; thus the names *Strongyloides stercoralis* and *Enterobius vermicularis* are universally understood to refer to two different species of nematodes.

The first name in the binomial is that of the genus to which the organism belongs, and the second is that of the species. This combination in designating an animal or plant species is termed "binomial nomenclature." For animals this dates from the tenth edition of Linnaeus' *Systema Naturae,* which was published in 1758. In earlier editions of this compendium, Linnaeus (born 1717, died 1778) used relatively long descriptions to characterize each species. Because of the rapidly increasing number of known species and the increasing cost of publication, for the tenth edition Linnaeus abbreviated the descriptions to two classical words, a genus and a species, which incorporated the essential concept of the particular species.

CLASSIFICATION

Although conforming to the same fundamentals of binomial nomenclature, species of animals are classified in the Animal Kingdom and species of plants in the Plant Kingdom. Intermediate kingdoms are Monera, Protista, and Fungi. On the basis of morphology and life histories wherever known, systems of classification have been developed within each of these kingdoms to express the intimate or more distant relationship of the different species. Those that are closely akin are placed in the same genus. Related genera are grouped in the same family, related families in the same order, related orders in the same class, and related classes in the same phylum, the rank that ordinarily constitutes one of the major divisions of the kingdom. The following, for example, is a classification of one of the common parasites that cause malaria:

Subkingdom Protozoa
- Phylum Apicomplexa
  - Class Sporozoea
    - Subclass Coccidia
      - Order Eucoccidida
        - Suborder Haemosporina
          - Family Plasmodiidae
            - Genus *Plasmodium*
              - Species *vivax*

ZOOLOGICAL NOMENCLATURE

Although the concept of binomial nomenclature is relatively simple, many complications have developed in its application (Mahr, 1969; Jeffrey, 1977; Taylor and Muller, 1979). Real difficulties have arisen that have confused the student of parasitology and the physician.

In the first place, if all of the earliest descriptions of species following the 1758 edition of *Systema Naturae* had been accurate and sufficiently complete so that subsequent workers could recognize the species, serious complications would have been avoided. But many of the early descriptions were inaccurate and inadequate, so that later it became impossible to differentiate related species, and some descriptions were equally applicable to two or more species. Linnaeus himself (1758) had one name, *Taenia solium*, for the pork tapeworm *(T. solium)*, the beef tapeworm *(T. saginata)*, and a taenia of the dog *(T. hydatigena)*.

In the second place, subsequent investigation of organisms originally assigned to a particular genus has indicated that, in many instances, the original generic concept needed to be expanded into several related genera grouped within a family or even a superfamily. Thus the genus name *Distoma* Retzius, 1790, is not sufficiently comprehensive for such different trematodes as *Fasciola hepatica, Dicrocoelium dendriticum, Clonorchis sinensis, Heterophyes heterophyes*, and *Paragonimus westermani*, to mention only a few species that parasitize man. Each of these is now allocated to a separate superfamily group.

Another difficulty arises when two different names are proposed by different workers for the same species or for the same genus. It is then necessary to discover which name with a recognizable description was published first—that is, which has *priority*. For example, *Fasciolopsis buski* (Lankester, 1857) has priority over *F. crassum* (Cobbold, 1860), and *Echinococcus granulosus* (Goeze, 1786) Rudolphi, 1805 over *Echinococcifer echinococcus* (Zeder, 1803). However, in applying the law of priority, a real hardship develops when a specific or generic name commonly employed for many decades is found to be "illegal" when a search in the older literature, at times in an obscure publication, reveals a designation antedating one that has been in use for a long time and is familiar to many persons who refer to this species in professional publications.

Although generic names can be duplicated when applied to different groups of organisms, one of which is in the Animal Kingdom and one in the Plant Kingdom, *no two genera in the Animal Kingdom may have the same name*. When it was discovered that the name *Trichina*, in combination with the species *spiralis* for the nematode that parasitizes pigs and man, had been previously employed to designate a group of flies, and hence, had priority, the nematode genus was renamed *Trichinella* (Railliet, 1895).

Certain specific or generic names are today recognized as synonyms of others that have priority. Examples of these are found when species described under different names from different hosts or different localities are found to be identical. Consequently, provided the parasitologist recognizes them as designations for the same species, *Paragonimus westermani* (Kerbert, 1878), the lung fluke of the tiger, takes precedence over *P. ringeri* (Cobbold, 1880) described for the human host; *Hymenolepis nana* (von Siebold, 1852), the dwarf tapeworm of man, over *H. fraterna* Stiles, 1906 of the rat, and *Schistosoma* Weinland, 1858 over *Bilharzia* Cobbold, 1859 for the genus of blood flukes parasitizing man and domestic mammals.

Nevertheless, recent studies on the biology of several parasites indicate that physiologic dissimilarities have developed in species that are morphologically indistinguishable. These strains are at times distinguished with respect to the host, as illustrated by *Hymenolepis nana* of man, the mouse, and the rat, in which the strain adapted to any one of these hosts will not readily produce infection in either of the other hosts. Another example is that of *Ascaris* of man and pig. Moreover, geographic strains of Manson's blood fluke, *Schistosoma mansoni*, have been developed within recent historical times. In this species, differences in intermediate (molluscan) host susceptibility have apparently resulted from geographic isolation, so that distinct strains have been discovered in Egypt, West Africa, Puerto Rico, and several localities in South America.

As biologic investigation and life cycle studies are pursued on an increasingly intensive scale, and the genetics of parasites become more satisfactorily understood, the new information will more accurately portray the species concept. Originally, species descriptions were based on apparent morphologic similarities and differences; hence, the species concept was *phenotypic*. The newer approach provides a more natural concept—namely, a *genotypic* one. Many, possibly most, so-called species are polytypic, consisting of several to many races, varieties, or subspecies that may differ morphologically, physiologically, or immunologically as a result of exposure to differences in environmental conditions over thousands, or even millions, of years.

THE INTERNATIONAL CODE OF ZOOLOGICAL NOMENCLATURE

For more than a century following the publication of the tenth edition of Linnaeus' *Systema Naturae* (1758), zoologists tried unsuccessfully to modify and supplement the original system of nomenclature. Without mutual agreements, the difficulties became so serious that in 1889 R. Blanchard presented to the First International Zoological Congress in Paris a comprehensive code that was adopted by the Congress but failed to receive universal sanction. Growing out of the recognized need for an acceptable code, plans were developed that were adopted by the Fifth Congress (1901) and were consummated in 1904 for establishment of a permanent Commission to serve as a quasijuridical body on zoological names. In this manner, *The International Commission* and *International Code of Zoological Nomenclature* were developed.

Amendments, additions, and opinions were rendered by the Commission and approved by successive Congresses. In 1961 a new, complete official transcript of the *International Code of Zoological Nomenclature,* adopted by the XV International Congress of Zoology in July, 1958, was published in London, with alternate pages in French and English. It is described as "the set of criteria to be met in giving an animal, or to a taxonomic group of animals, a scientific name, with its proper reference to author and date, and to regulate *inter se* names that have been given in the past . . . . Thus the meaning of each name available for a species . . . is defined by the characters of an individual specimen—its type, that of a genus by those of its type-species, that of a family by those of its type-genus."

REFERENCES

Hertig, M., Taliaferro, W.H., and Schwartz, B. 1937. Report of the Committee on Terminology, American Society of Parasitologists. J. Parasitol., *23:* 325, 329.

International Commission on Zoological Nomenclature. 1961. *International Code of Zoological Nomenclature,* adopted by the XV International Congress of Zoology. London.

Jeffrey, C. 1977. *Biological Nomenclature,* 2nd ed. New York, Crane Russak.

Linnaeus, Carolus. 1758. *Systema Naturae per Regna Tria Naturae,* 10th ed. Holmiae.

Mahr, E. 1969. *Principles of Systematic Zoology.* New York, McGraw-Hill Book Co.

Railliet, A. 1895. *Traite de Zoologie Medicale et Agricole.,* 2nd ed. Paris, Asselin & Houzeau.

Taylor, A.E.R., and Muller, R., editors. 1979. *Problems in the Identification of Parasites and Their Vectors.* Symposia of the British Society of Parasitology, Vol. 17. Oxford, Blackwell Scientific Publications.

Chapter 4

Disease Processes Due to Animal Agents and Their Clinical Manifestations

HOST-PARASITE RELATIONSHIPS

As outlined in Chapter 1, a parasite may live in or on the tissues of its host without causing evident harm, and in some instances, the relationship may be mutually advantageous. In a majority of cases, however, the parasite has the capacity to produce damage. If infection exists with the production of little or no damage to the host and without symptoms, the host is referred to as a *carrier*. Under other conditions, and invariably in some parasitoses, the damage is considerable, resulting in varying degrees of disease. The parasite is then designated as a *pathogen*.

Pathogenesis

The dynamics of any disease process are its *pathogenesis*. The lesion may be localized at the site where the parasite has become established, or it may extend to distant parts of the host's body. The ways in which damage may be produced include the following: trauma or physical damage; lytic necrosis; stimulation of host-tissue reactions; toxic and allergic phenomena; and the opening of pathways for entry of other pathogens into the tissues (Spencer et al., 1973; Binford and Connor, 1976).

Inapparent Infection. Parasitic infections, like those caused by bacteria, rickettsiae, and viruses, may be inapparent, causing no symptoms and producing no detectable harm. Among the protozoan infections, there are some (malaria, for example) that can be inapparent continuously, for long periods, or between short periods of relapse (Garnham, 1977). In the helminthic infections, it usually is only when the worms are present in large numbers that characteristic signs and symptoms are produced (Beaver, 1975).

Traumatic Damage. This occurs when the parasite invades the skin, as in the case of the scabies mite (*Sarcoptes scabiei*) and of several species of fly maggots, or even as a result of the puncture wound made by a tick. Relatively slight physical damage is produced by entry of hookworm or blood-fluke larvae into the skin, or the usually minute lesion made by a mosquito. Internally, during the migration of larval stages of several helminths through the lungs, there is rupture of the pulmonary capillaries, which may cause appreciable extravasation of blood if many organisms are in migration simultaneously. Similar damage in the cerebral, retinal, or renal capillaries may precipitate serious, even fatal injury. Eggs of *Schistosoma* spp. cause extensive trauma with hemorrhage as they escape from mesenteric or vesical venules, respectively, into the intestinal canal or the lumen of the urinary bladder. Moreover, the attachment of hookworms to the intestinal wall invariably results in traumatic destruction of the villi. Large worms, such as *Ascaris* or *Taenia saginata,* may produce acute intestinal obstruction, while *Ascaris* may occlude the lumen of the appendix or the common bile duct, may cause perforation of the intestinal wall, or may penetrate into the parenchyma of the liver and even of the lungs.

Lytic Necrosis. Enzymes elaborated by many parasites make it possible for them to digest food available in the immediate environment and to transform this nutriment into their own protoplasm or store it for the production of energy. One of the most striking examples of tissue digestion and necrosis is that of *Entamoeba histolytica,* which not only lyses tissues for nutritional needs but also as a means of penetration into tissues of the colon and extraintestinal viscera. Several species of protozoa that are obligate intracellular parasites (i.e., plasmodia, leishmaniae, *Trypanosoma cruzi, Toxoplasma gondii*) cause necrosis of the parasitized host cells during the parasite's growth and multiplication.

Stimulation of Host Tissue Reaction. With a few exceptions, all animal parasites provoke host tissue reaction. This may consist of cellular proliferation and infiltration at the site of the parasite, and it may involve systemic increase in certain types of cells, especially those circulating in the blood. Characteristically, in many parasitoses the host reaction walls off the parasite by fibrous encapsulation.

In metazoan parasitoses, but infrequently in protozoan infections, there is a moderate-to-notable eosinophilia, which is most often observed in infections with helminths that have an intimate host-tissue association, as occurs in, for example, trichinosis, strongyloidiasis, trichuriasis, and larval toxocariasis. However, arthropod infestation may also be responsible for marked eosinophilia.

To a certain degree, increased erythropoiesis may be stimulated in infections that cause mechanical loss or destruction of the erythrocytes. This may be observed following deep and extensive amebic invasion of the colon or attachment of hookworms to the intestinal wall, although it will not compensate for prolonged, extensive hemorrhage. In malaria, with the destruction of the red blood cells, there is evidence of increased red blood cell production, but this is frequently not sufficient to replace the cells that are being destroyed. When an infection such as hookworm disease causes chronic blood loss, iron deficiency results that interferes with erythropoiesis. Likewise, in chronic malaria, although iron is not lost from the body, it is deposited in the tissues in a relatively unusable form; thus anemia is commonly associated with chronic hookworm disease, malaria, and a number of other parasitoses.

The only infection in which almost no host-cell stimulation is produced is uncomplicated amebic colitis. Nevertheless, host tissue reaction to *Entamoeba histolytica* may result in amebic granulomas. This relatively rare type of lesion may develop at any level of the large intestine that has been invaded by *Entamoeba histolytica*. The tumor (the so-called ameboma) may be polypoid with an undamaged surface or ulcerative.

In visceral leishmaniasis (kala-azar), the etiological agent (*Leishmania donovani*) parasitizes fixed macrophages in the spleen, liver, bone marrow, and visceral lymph nodes. In these organs, it stimulates a remarkable increment of the reticuloendothelium that is perhaps comparable in degree to leukemia, but completely different in its prognosis since specific drugs administered to kill the protozoan agent will provide a gradual return to the normal cytologic picture.

In the medical literature, attention frequently has been focused on parasitic worms as causative agents of neoplastic growth in the digestive tract, liver, and urinary bladder. The spiruroid nematode of rodents, *Gongylonema neoplasticum*, provokes malignant gastric tumors in rats that have been experimentally inoculated with the larval stage of this worm. *Cysticercus fasciolaris*, the larval stage of the cat tapeworm *Taenia taeniaeformis*, stimulates the development of sarcoma in the liver of the rodent intermediate host. Although no such oncogenic association of nematodes or tapeworms has been demonstrated for man, several species of trematodes have reportedly been involved in the development of carcinoma.

Flukes in the bile ducts provoke hyperplasia of the biliary epithelium, formation of new ducts, and periductal fibrosis with progressive hepatic cirrhosis in heavy infections. Occasionally, adenocarcinoma and solid primary carcinoma of the liver have been found at postmortem of such cases.

In blood fluke infections involving primarily the mesenteric-portal system, extensive perivascular infiltration of eggs in the wall of the large intestine and the liver has been found to be associated at times with colonic, rectal, and hepatic carcinoma. Moreover, in adult male farmers of Lower Egypt, the statistical relationship of *Schistosoma haematobium* eggs trapped in the bladder wall to vesical carcinoma is impressive, while several cases of carcinoma involving the male and female genitalia have been diagnosed in persons with blood-fluke involvement of these organs (Cheever et al., 1978).

Toxic and Allergic Phenomena. Several of the arthropods harmful to man introduce toxins into the skin, inducing host reaction. The volume of the toxin may be minute and the potency hardly measurable (as in the case of some of the smaller insects), or the amount may be appreciable with varying degrees of potency. Spiders and ticks introduce venom when they insert their mouthparts into the skin; scorpions, bees, wasps, and ants have their venom apparatus at the caudal end of the body.

Many species of caterpillars elaborate toxins in glands at the base of delicate hollow hairs that become filled with the secretions; in contact with human skin or mucous membranes, the hairs break off, introducing the toxin. Jellyfishes (Coelenterata) are provided with multiple stinging cells, each of which contains a nematocyst. When these cells touch the skin, the nematocysts explode, which causes a painful nettling dermatitis of the contact area and occasionally systemic reactions. Blister beetles have a vesicating fluid that produces painful blisters if the fluid is discharged on body surfaces.

When protein or, at times, other metabolites of the parasite are introduced into the skin in minute drops of an arthropod's saliva or from active penetration of a living helminth larva, sensitization to the foreign substance is initiated. Similarly, when an active helminth larva makes intimate contact with the host's viscera, it stimulates production of antibody. Wherever migration through the tissues normally occurs following active entry of a parasite through the skin or after its peroral introduction, the relatively prolonged association of the parasite with the host increases the probability of specific *antibody* response to introduction of the foreign *antigen*, and in some instances, *anaphylactic shock*, as in bee sting. Likewise, the sudden release of the foreign protein in the host's body may precipitate anaphylaxis, as when hydatid fluid is released by rupture of an *Echinococcus* cyst in a body cavity.

Immunity

Immunity in animal parasitoses is rarely solid, as it is in some of the viral diseases (e.g., measles and yellow fever). A notable exception is found in cutaneous leishmaniasis, in which a natural infection followed by spontaneous cure appears to ensure immunity for life against subsequent attack. In trichinosis, light exposure in experimental animals prevents infection of clinical grade when a large inoculum is administered later. There is similar suggestive information with respect to hookworm infection, blood-fluke infection, and infestation with the fly maggot, *Cordylobia anthropophaga*.

In malaria there is substantial evidence concerning human immunity or susceptibility to different strains of the same species of plasmodium and to different species of plasmodia. Individuals who have had malaria are highly resistant to the homologous strain of the parasite, less so to heterologous strains of the same species, and demonstrate little or no resistance to other species of plasmodia that are infective for man.

With regard to parasitic infections in general, the immune mechanisms of the host may or may not be protective to a significant degree. In some instances, the immune mechanism is indeed harmful to the host, as in the allergic phenomena mentioned above or in the case of the pneumonitis produced by migrating *Ascaris*. In other instances, the host's immune mechanisms may be helpful to the parasite, as in the proliferation of leishmania within phagocytic cells.

The physiologic role of the human immune system is to differentiate intrinsic (self) from foreign (non-self) molecules and to react with the latter to eliminate them. The system consists of two specific arms (the antibody and cellular mediation arms) and two nonspecific arms (complement and inflammatory cells). The specific arms are associated with lymphocytes, those derived from the thymus (T cells), which provide specificity to the cellular immune system, and those derived from elsewhere in the body (B cells), which provide specific humoral substances or antibodies. Complement and inflammatory cells are specifically activated by the lymphocytes.

The antigens on the surface of infecting parasites or excreted by them stimulate the production of specific antibodies by cells of the B-cell line. In the case of allergic reactivity, IgE is usually thought to be involved, but antibody of the IgG class has been shown to be produced as a consequence of *Ascaris* infection (Parish, 1969). In visceral leishmaniasis, serum gammaglobulin levels are characteristically greatly elevated; these elevations are attributable to increases in both IgG and IgM. In African trypanosomiasis, IgM is so characteristically elevated as to make the finding diagnostic.

The characteristic cellular reaction to a tissue parasite is a granuloma. This consists of macrophages or epithelioid cells that have arranged a connective layer about the parasite object. Multinuclear giant cells and eosinophils frequently are seen in these lesions.

Secondary Invaders

In assessing the over-all damage produced in parasitic diseases, one must keep in mind that entry of an animal parasite may open pathways in the skin or the intestinal tract for invasion by other pathogenic microorganisms. "Ground itch" in the skin at sites where hookworm larvae have penetrated is characteristically complicated by pyogenic bacteria. Similarly, in chronic amebic colitis and in balantidiasis, the architecture of the enlarged ulcer and its histologic appearance indicate that enteric bacteria have entered the lesion and have altered the picture, with dense infiltration of neutrophilic leukocytes and fibroblasts around the margins of the lesion. Regardless of whether the parasite entered the tissues through the skin or the intestinal mucosa, it is probably the exception rather than the rule that bacteria reach the deeper visceral organs via pathways produced by the parasites. Recent investigations suggest that viruses may be introduced into the viscera by invading and migrating animal parasites.

Finally, in communities in which conditions are favorable for high prevalence of parasites in the population, several species of protozoa and helminths are usually found together in a majority of individuals. Under these circumstances, the pathologic effect of the multiple infections (*polyparasitism*) is the sum of the individual effects, to which an underlying state of malnutrition must frequently be added as an important contributor to the total disease picture.

CLINICAL EVIDENCE OF PARASITISM

The physician obtains information concerning the effect of parasitic diseases in his patient on the basis of the subjective and objective symptoms evoked, the patient's history, and physical examination. A majority of the symptoms attributable to the parasitoses are not diagnostically specific—i.e., they are not pathognomonic. On the basis of his scientific knowledge and clinical experience, the physician will consider several possible agents that characteristically produce the pattern of disease observed in the patient. The presenting symptoms may be acute or of a chronic nature. In parasitic diseases, as in infection with microbial agents, there may be initial toxic manifestations, high fever, evidence of inflammation, generalized or localized pain and tenderness, and physical signs of disease. The condition may be less acute, with gradual loss of weight, anemia, lethargy, and other vague symptoms. Moreover, evidence is accumulating that parasitic organisms, even though they may not produce clinically demonstrable symptoms, may alter the metabolic balance of the host, causing abnormal responses.

In contrast to many bacterial and viral diseases that typically have a sudden onset, the clinical effects of most animal parasitoses often develop slowly; there are exceptions, however. Within a few hours after exposure to encysted larvae of *Trichinella spiralis,* the irritation due to their excystation and invasion of the duodenum may produce symptoms that are suggestive of acute food poisoning. Pneumonitis may develop within a few days after a heavy inoculum of *Ascaris* eggs has been ingested (Phills et al., 1972). Chills and fever will characteristically occur within 2 weeks in nonimmune subjects who are exposed to vivax malaria.

Intoxication and sensitization constitute important segments of the total clinical picture in most helminthiases. While these evidences of disease occur in hookworm infection, the mechanical loss of blood, particularly in poorly nourished patients with hypoproteinemia, is usually the precipitating cause of the disease syndrome. In schistosomiasis, the cellular reaction around eggs increasingly infiltrating the perivascular tissues contributes substantially to the seriousness of the symptoms. In filaria infections involving the lymphatics (*Wuchereria bancrofti, Brugia malayi*), sensitization to the presence of the worms and their metabolites may provide the basis for early and recurrent bouts of lymphangitis and fever and eventually fibrotic blockage of lymphatic vessels, resulting in varicose lymphatics and elephantiasis.

The salivary secretions and venoms of arthropods, when they are introduced into the skin, set up local reactions and, at times, systemic evidence of allergy or neurologic symptoms. These manifestations vary within wide limits, depending on the species of arthropod, the amount and potency of the inoculum, and the reactivity of the inoculated individual (Harwood and James, 1979).

The Blood Picture In Parasitic Infections. This depends both on the parasite and the patient. If there is no underlying blood dyscrasia, the picture in early amebic colitis is essentially normal unless there is extensive hemorrhage; with bacterial invasion of the colonic lesions, and typically in hepatic or other extraintestinal amebiasis, there is a moderate neutrophilic leukocytosis. In visceral leishmaniasis, a significant monocytosis with neutropenia and, subsequently, an erythropenia, develops. Moderate to severe anemia occurs with malaria. A demonstrable peripheral eosinophilia is not typically present in any of the protozoan diseases.

In the helminthiases, an early polymorphonuclear leukocytosis may occur, followed by development of a relative monocytosis and leukopenia as the infection becomes chronic. Eosinophilia is characteristic of many helminthic diseases, although the absolute eosinophil count may be low, moderate, or high, depending on the species of helminth and the individual host's reaction. In schistosomiasis, fascioliasis, fasciolopsiasis, trichinosis, strongyloidiasis, and visceral larva migrans, hypereosinophilia (i.e., 25 to 90% of total white count) is relatively characteristic. Heavy hookworm infection produces a microcytic hypochromic anemia. Fish tapeworm (*Diphyllobothrium latum*) at times precipitates a macrocytic anemia (Von Bonsdorff, 1948).

Leukocytosis and eosinophilia may occur in arthropod infestations, but they are not always present.

DELUSIONAL PARASITOSIS

People generally are eager to be rid of their parasites even when they are essentially harmless—or nonexistent (Beaver, 1981). Among the most difficult patients encountered by the clinical parasitologist are those with delusional parasitosis. This is a mental disorder that exhibits features of obsessive-compulsive neurosis and is characterized by delusions of infection or infestation with bizarre parasites. The patients frequently describe complicated life-cycles of parasites, their emission and entry through the skin or by way of body orifices and their migration through the organs and cavities of the body. The patient usually experiences feelings of guilt and depression and frequently has paranoid ideas.

Many of these patients may continue to have relatively well integrated personalities, lead productive lives in gainful employment, and function satisfactorily as parents. Others become so preoccupied with their delusions and their efforts to secure scientific confirmation and treatment that they are unable to function acceptably as employees, parents, or spouses.

Nearly all of the patients are adults, although the onset may be as early as the second decade of life. Often it begins with an experience involving exposure or fear of exposure to a parasitic infection. Rarely, the delusion may be passed from one person to another (McAndrews et al., 1956).

Diagnosis is based primarily on the patient's description of his illness, which consists of a narrative of the activities of his parasites, rather than of his symptoms. Objective evidence of the abnormal personality should be looked for. Obsessive-compulsive behavior such as preparation and submission to the physician of lists of parasite appearances, drawings of parasite forms, and a collection of numerous specimens support the diagnosis of delusional parasitosis. Those with delusions involving dermal parasitosis are apt to probe into their skin with needles and tweezers, producing factitious lesions. These lesions are recognizable because their distribution is limited to those parts of the body and extremities that can be reached by the patient's hands.

Efforts must be made to rule out *bona-fide* infection. Conditions that might be mistaken for delusional parasitosis are infestation with rat or bird mites (*vide infra,* p. 595) and migratory parasitoses, such as loiasis, sparganosis, and gnathostomiasis.

Management of these patients is extremely difficult; thus they are likely to take up much of the sympathetic physician's time and make unreasonable impositions upon him. It is not surprising that these patients are generally rejected by the physicians and parasitologists from whom they seek assistance.

There is little that can be done for the patient with delusional parasitosis, since such patients firmly resist all efforts to convince them that they do not have a parasitic infection. In fact, it is probable that the removal of the delusion would result in further disintegration of the patient's personality, or suicide.

The physician should endeavor to keep the patient from poisoning himself with toxic drugs or chemicals, wasting his financial resources by going from one laboratory or clinic to another, and repeating diagnostic studies and treatments; the physician must also be certain that organic disease actually present is not neglected. To do this the physician must not deny the existence of the imaginary parasite, since this merely causes the patient to move on to another physician or laboratory. The best procedure is to report that no parasite was found in the specimens that were submitted. The patient should not be given any false information.

The prognosis for recovery is poor.

REFERENCES

Beaver, P.C. 1975. Biology of soil-transmitted helminths: The massive infection. Health Lab. Sci., *12*: 116–125.

———. 1981. What is parasitology? pp. 1–6. In *The Current Status and Future of Parasitology.* Edited by K.S. Warren and E.F. Purcell. New York, Josiah Macy Jr. Foundation.

Binford, C.H., and Connor, D.H. 1976. *Pathology of Tropical and Extraordinary Diseases: An Atlas,* Vols. 1 and 2. Washington, D.C., Armed Forces Institute of Pathology.

Cheever, A.W., Kamel, I.A., Elwi, A.M., Mosimann, J.E., Danner, R., and Sippel, J.E. 1978. *Schistosoma mansoni* and *S. haematobium* infections in Egypt. III. Extrahepatic pathology, Am. J. Trop. Med. Hyg., *27*:55–75.

Garnham, P.C.C. 1977. Charles Nicolle and inapparent infections. Am. J. Trop. Med. Hyg., *26*:1101–1104.

Harwood, R.F., and James, M.T. 1979. *Entomology in Human and Animal Health,* 7th ed. New York, Macmillan.

McAndrews, J., Jung, R.C., and Derbes, V. 1956. Delusions of dermal parasitosis (acarophobia) manifested by folie à deux. J. La. State Med. Soc., *108*:279–286.

Parish, W.E. 1969. Detection of reagins, IgG, IgA and IgM antibodies in human sera. Int. Arch. Allergy Appl. Immunol., *36*:245–265.

Phills, J.A., Harrold, J., Whiteman, G.V., and Perelmutter, L. 1972. Pulmonary infiltrates, asthma and eosinophilia due to *Ascaris suum* infestation in man. N. Engl. J. Med., *286*:965–970.

Spencer, H., et al., editors. 1973. *Tropical Pathology.* New York, Springer-Verlag.

Von Bonsdorff, V. 1948. Pernicious anemia caused by *Diphyllobothrium latum* in the light of recent investigations. Blood, *3*:91–102.

Chapter 5
Diagnosis and Treatment of Diseases Due to Animal Parasites

THE PROBLEM OF DIAGNOSIS

Diagnosis has two methods of approach—clinical and laboratory. In the case of the individual patient, either type of diagnosis may precede the other. Assuming that the physician lives in an endemic area of parasitic diseases and by training and experience is familiar with their cardinal manifestations, he may have developed considerable skill in making a relatively accurate presumptive diagnosis of the infection provided that the symptoms are typical. For the physician who practices in an area in which parasitic diseases are relatively uncommon, the problem is more difficult. The patient may have acquired infection in a different part of the country or while traveling or resident in a foreign land several weeks, months, or years earlier, yet seeks medical attention because symptoms of disease have recently developed. It is imperative that the physician question the patient about where he has been and about the environmental conditions under which he has lived.

It is important to know what type of specimen should be obtained from the patient for laboratory analysis—stool, blood, urine, sputum, aspirate, or biopsy. It is equally important to know how and when the specimen is to be obtained and what precautions are to be taken so that it will be satisfactory for examination.

It is essential that the parasitologic specimen be processed and examined by a technician who has been specially trained and has developed experience and skill in this field of diagnosis. With rare exceptions, medical technologists require several months of intensive training in this field before they can be considered competent.

Detailed information on the most useful techniques employed in parasitological diagnosis is presented in Section V of the Technical Appendix (pp. 733–780).

In epidemiologic studies of parasitic infections, the problem of diagnosis is carried from the individual patient to an entire community. Advantages of such a field study are obvious: (1) It will provide information on the prevalence of parasites in a population; (2) it will indicate whether the infections are endemic, hyperendemic, epidemic, or sporadic; (3) it will provide clues as to the sources of exposure; and (4) as an important corollary, it will direct attention to the amount of illness in the community apparently resulting from these infections.

Diagnosis in connection with an epidemiologic survey near a good clinical laboratory offers no particular problem, but in rural and other areas difficult to reach provision must be made for the collection and transportation of specimens so that they reach the laboratory in a condition suitable for satisfactory examination.

TREATMENT

When an accurate, specific diagnosis has been made, before proceeding with treatment one must consider the severity, duration, and intensity of infection and the probability of reinfection, and the efficacy, availability, toxicity, and acceptability of the treatment. In many instances, treatment may not be indicated, either because the toxicity of the only available effective drug may present a greater hazard to the patient than the infection, or because the probability of reinfection makes treatment of a relatively harmless infection useless.

Chemotherapy. Today many synthetic chemical compounds are awaiting the tests of pharmacologists and physicians to determine their potential value in the treatment of parasitic and many other varieties of human and animal diseases. Burkholder (1959) has pointed out that in past eras "cinchona bark and ipecacuanha rested on their own merits as simple remedies, produced in nature and discovered through early empiricism." Not so in modern chemotherapeusis. The potential drug is first tested for its effects on laboratory animals and for its ability to interfere with the metabolism of various etiologic agents of disease at dosage levels well within the tolerance of the test animal. If both of these criteria prove to be satisfactory and the chemical appears to have specific action against a particular microorganism, clinical trials are then undertaken under carefully controlled conditions. In this way, quinacrine was first discovered to be active against *Giardia* in the mouse; it then became accepted as a standard for treatment of human giardiasis. Modern malaricidal drugs were winnowed out of many thousand synthetic compounds that were considered to have possible value in malaria therapy (Coatney, 1963).

A few years ago there were many parasitic diseases for which there was no specific treatment or only partially satisfactory medication, and others for which only a single drug was available; currently, there is an increasing number of parasitic infections for which there is now a choice of two or more efficient chemotherapeutics. This makes it possible to select the drug that is relatively free of side effects and that has maximum efficacy for treatment of individual patients as well as for mass treatment of a population or community. While some of the older chemotherapeutics have retained their place in the pharmacopeia—e.g., emetine and diiodohydroxyquin in amebiasis—others such as gentian violet, at one time used for *Strongyloides* infection, have been replaced by more effective parasiticides. Likewise, two recent products, bithionol and praziquantel, have been found to be effective against paragonimiasis.

As in the case of resistance of flies and other arthropods to insecticides, so resistance to chemotherapeutics has developed in some protozoan infections. Earliest evidence of this tolerance of a parasite for a drug was first discovered in Gambian trypanosomiasis and more recently in malaria. This difficulty is likely to increase in the future as a greater number of parasiticidal chemicals will be employed.

Because of laws and regulations currently in force, drugs known to be effective against certain parasitic infections are not permitted to be commercially distributed in the United States. In some instances, a drug not available commercially for the treatment of a parasitic infection may be obtained from the Parasitic Disease Drug Service, Parasitic Disease Section, Centers for Disease Control, Atlanta, Georgia 30333.

Surgical Intervention. For some parasitoses in which chemotherapy is unavailing or has proved inadequate, surgical removal of the parasite is indicated. The best known examples are found in lesions of the brain or eye produced by *Cysticercus cellulosae* and the hydatid cyst. Surgical removal of the parasite occurs less frequently in patients suffering from paragonimiasis and deep myiasis resulting from infestation with fly maggots. Today surgeons are much more conservative than in earlier decades in the treatment of amebiasis of the appendix and liver, since satisfactory results can often be obtained with the newer chemotherapeutics; there has been a corresponding decrease in prolonged convalescence and fatalities. Nevertheless, exploratory surgery or biopsy may be necessary to obtain evidence of the cause of a tumor mass or a suspicious lesion, and excision of the lesion may provide the only therapeutic measure available.

REFERENCES

Anonymous. 1980. Drugs for parasitic infections, pp. 61–72. In *Handbook of Antimicrobial Therapy.* New Rochelle, N.Y., The Medical Letter.

Ash, L.R., and Orihel, T.C. 1980. *Atlas of Human Parasitology.* Chicago, American Society of Clinical Pathologists.

Burkholder, P.R. 1959. Antibiotics. The exploitation of microbial antagonism is having a challenging impact on medicine and society. Science, *129*:1457–1465.

Coatney, G.R. 1963. Pitfalls in a discovery: The chronicle of chloroquine. Am. J. Trop. Med. Hyg., *12*:121–128.

Faust, E.C., Beaver, P.C., and Jung, R.C. 1975. *Animal Agents and Vectors of Human Disease,* 4th ed. Philadelphia, Lea & Febiger.

Garcia, L.S., and Ash, L.R. 1979. Diagnostic Parasitology, 2nd ed. St. Louis, C.V. Mosby.

Markell, E.K., and Voge, M. 1976. *Medical Parasitology,* 4th ed. Philadelphia, W.B. Saunders.

Rollo, I.M. 1980. Chemotherapy of parasitic diseases, pp. 1013–1037. In *The Pharmacological Basis of Therapeutics,* 6th ed. Edited by L.S. Goodman and A.G. Gilman. New York, Macmillan.

Warren, K.S., and Mahmoud, A.A.F. 1978. *Geographic Medicine for the Practitioner.* Chicago, University of Chicago Press.

World Health Organization. 1979. The selection of essential drugs. Geneva, W.H.O. Tech. Rep. Ser. no. 641.

Chapter 6

Control and Prevention of Diseases Caused by Animal Parasites

Animal agents and vectors are responsible for a large proportion of the communicable diseases. Except for the contagious diseases of viral and bacterial origin, no other group of pathogenic agents has caused so large an amount of morbidity, mortality, economic loss, human suffering, and degradation as animal parasites and animal transmitters of disease-producing agents.

Diseases resulting from animal parasites or transmitted by arthropods as biological vectors vary considerably in their public-health importance in different areas, but malaria and hookworm infection are among the ten top causes of morbidity in most tropical areas, while amebiasis, schistosomiasis, filariasis, and arthropod-borne viral diseases are rated as moderately important in one or more regions.

In general, the prevalence of parasitic diseases is an indication of environmental conditions. A high index of intestinal parasitoses shows deficiencies in sanitation, the standard of living, and personal habits of cleanliness. Taeniasis, trichinosis, and hydatid disease are indicative of poor conditions and lack of proper control in slaughterhouses, personal and social behavior patterns that promote the spreading of these diseases, and inadequate sanitation. Chagas' disease and the presence of triatomids inside houses are determined by poor housing conditions, lack of hygiene, and low social and economic status of the population (Neghme, 1961). The control and prevention of animal parasitoses are inseparable from the causative agent, man as the ignorant sufferer, and his unwholesome environment.

Control of a disease refers to its restraint. In the patient, it may constitute symptomatic relief or repression of the activities of the causative agent but not necessarily its elimination. In public-health practice, the term implies a check that is sufficient to remove the likelihood of further dissemination of an epidemic or to reduce and maintain at a low level the prevalence of disease in a population. In contrast, *prevention* implies interception, so that the individuals in a community are no longer exposed to the hazards of the disease.

A prerequisite to control and prevention of the parasitoses is a clear understanding of the epidemiology of each causative organism. It is essential to know in advance its distinctive characteristics, so that it can be recognized, the conditions under which it survives and propagates, and how man is exposed to infection. With this information, as well as reliable data on the incidence and significance of the disease in each area in which control and prevention are to be undertaken, the next steps to be considered are the most practical and economic ways by which the goal can be achieved.

Experience in the field of preventive medicine has demonstrated the need for trained personnel. These individuals must have not only theoretical knowledge but a practical understanding of public-health principles and techniques and experience in their application. They must have a working knowledge of the topography and climatology of the region and of the particular locality, familiarity with the social and religious customs of the people, their intelligence and educational levels, their economic standards, their willingness to cooperate, and their ability to help finance the project on a continuing basis and to maintain a suitable public health organization.

The practicing physician can and should contribute to the control of disease in the following ways: (1) detection, accurate diagnosis, and evaluation of the clinical importance of the disease in the patient; (2) adequate treatment of the patient; (3) search for and treatment of other cases in the patient's family; (4) determination, if possible, of the source of the infection, reporting it promptly to health officials; (5) advising patients and their families how to avoid further exposure; (6) supporting and cooperating in community preventive measures; and (7) educating patients in ways of utilizing and strengthening local health departments.

THE FIELD PROGRAM OF CONTROL AND PREVENTION

In practically all countries today there are public health agencies at different political echelons that are responsible for the health of their respective communities. A full complement of personnel consists of administrators, epidemiologists, engineers, biologists, statisticians, sanitarians, and public health nurses, all trained in their particular specialties. As a result of previous experience and the rapid transmission of public health intelligence, they are usually prepared to deal with epidemics, but they frequently are less effective in reducing the incidence of endemic parasitoses.

METHODS OF CONTROL AND PREVENTION

In order to set up effective control of a disease that is due to an animal agent or one that is arthropod-transmitted, it is important to have a clear picture of the entire natural history of the disease and then to determine at what place or places barriers can be most effectively erected. All of these infections are preventable, but if the control program is to succeed it must be applicable to the area in which it is to be developed. At times, interception at a single location in the cycle is adequate; at other times, two or more loci should be attacked simultaneously. Examples can be cited to illustrate the need for different approaches to the problem. Not a single human being need be infected with any of these parasites but, as Morison (1963) has stated, eradication or control of any of them requires an appreciation of the balance of nature and often a reorganization of social and political institutions.

Amebiasis results from swallowing viable cysts of *Entamoeba histolytica* in water or food, or on fingertips. Human stools are the usual sources of contamination. Poor personal and group hygiene and dirty food handlers result in a high prevalence of amebiasis in a community, providing opportunity for epidemic outbreaks in temperate as well as warm climates. Sanitary sewage disposal and unpolluted water supplies frequently will control or forestall epidemics; the fundamental problem is improved personal and environmental hygiene coupled with antiamebic therapy for food handlers and other carriers of the disease.

Kala-azar (visceral leishmaniasis) concerns clinicians and public health workers in extensive areas of the world. Chemotherapy alone is not adequate for control, as has been shown by epidemic outbreaks of the disease within control areas. While in some endemic regions there are no active human cases, dogs, foxes, and rodents provide a reservoir of the infection. Fortunately the demonstrated effectiveness of residual insecticide spraying in and around human habitations to kill the adult sand fly vector has pointed the way to prevention by a simple, inexpensive, practical technique.

Malaria is the most widespread scourge of mankind throughout warm climates. Until recent years there were no satisfactory methods of control, but with the development and use of new, more effective malaricidal drugs and the spraying of residual *Anopheles* toxicants around dwellings, success has been achieved in the previously heavily malarious areas of the southeastern United States and many other countries. The newer methods have resulted in reducing the incidence of infection in the human population, the breeding of mosquito vectors, and the number of infective adult mosquitoes. Many countries claim to have completed malaria eradication and others have programs in progress, but in the less developed countries, particularly in Africa, the eradication of malaria remains a formidable task.

Trichinosis results from ingesting the flesh of another host harboring the encysted larval stage of the causative organism, *Trichinella spiralis*. Man acquires infection principally from pigs, occasionally from bears, and rarely from other mammals. Measures directed against human disease consist of thorough cooking or deep-freezing of pork and feeding only well-heated garbage to hogs. Although human trichinosis is relatively uncommon in the United States today, no substantial success has attended the attempts to control the disease.

Hookworm disease is widely distributed

throughout regions of warm, moist climate. Shortly after the life cycle of the etiologic agents became well known, studies were initiated to determine the prevalence and intensity of the infection and to discover methods of control. In many countries, extensive control projects were undertaken that employed anthelmintic treatment, construction of sanitary facilities in an attempt to discourage promiscuous defecation, and education concerning methods of exposure and means of prevention. In some regions, as in the southern United States, intensive hookworm infection of clinical grade has been eliminated from most communities, although light infection persists over wide rural areas. In many tropical regions, however, little progress has been made in educating the population to use sanitary latrines, and in these countries, hookworm infection remains a major public health challenge.

Ascariasis is hyperendemic in the tropics, and in certain areas of the southeastern United States, it is prevalent. Young children who defecate in and around the home are most frequently parasitized; they and their playmates serve both as the sources of the eggs and later as the victims of the infection when the fully embryonated eggs are picked up from soil and swallowed. Moreover, in the Orient and European countries in which human excreta are used as fertilizer, persons who cultivate the fields and others who eat raw vegetables contaminated with eggs are commonly exposed to infection. Attempts to reduce ascariasis in endemic areas by providing sanitary conveniences for general use have had little success.

Bancroft's filariasis is a disease of warm climates. Domestic and semi-domestic mosquitoes serve as essential intermediate hosts and vectors, and recent control procedures have been twofold, involving (1) treatment of human carriers with diethylcarbamazine (Hetrazan) to kill microfilariae circulating in the blood and to reduce the chances of the infection being acquired by mosquitoes, and (2) use of insecticides and modifications of the physical environment to eliminate the vector mosquitoes (Curtis and Feachem, 1981). Encouraging results are being reported in some areas of high endemicity.

Hydatid disease is produced by *Echinococcus granulosus, E. multilocularis,* and *E. vogeli.* Dogs and their wild relatives are the definitive hosts, while sheep, hogs, cattle, and man are parasitized by the larval stage (hydatid cyst). In countries of high endemicity, young boys with infected pet dogs are most frequently exposed; in several of these countries, attempts to educate the population to the dangers of exposure have met with indifferent results; in Iceland, however, where human infection was common 50 years ago, the last human case was seen in 1960, the last in sheep in 1953, and postmortem examinations of dogs have revealed no infection since 1950.

Schistosomiasis in man has three principal etiologic agents: *Schistosoma japonicum, S. mansoni,* and *S. haematobium.* When the eggs of these worms are evacuated in human stools and urine, the emerging larvae hatch in water, then invade certain snails in which they undergo two generations of development, with multiplication in each. The fork-tailed larvae (cercariae) that escape from the snails swim about in the water and invade human skin on contact. Possible methods of control include: (1) sanitary disposal of human excreta; (2) use of toxicants in the water to kill the snail hosts or the free-living cercariae; (3) educating the population of endemic areas to avoid contact with infested water; and (4) anthelmintic treatment. Except in Japan, none of these measures or any combination of measures has thus far provided effective control; therefore, in many endemic regions, the disease is spreading to contiguous areas as new land is developed and irrigated.

CONTROL OF DISEASES HAVING RESERVOIR HOSTS

Reservoir hosts constitute serious obstacles to prevention in several of the diseases that have been cited, as well as in many others. In these parasitoses, control in the human population does not break the basic chain and thus does not eliminate human exposure. In kala-azar, trichinosis, Oriental schistosomiasis, and sylvatic plague, safeguards must be erected to reduce the possibility of human contact with the stage of the microorganism infective for man. While much can be achieved along this line by public health authorities, active cooperation and participation by members of the community are necessary for lasting results.

SUMMARY

Successful efforts to control and eventually to prevent human disease caused by animal parasites or transmitted by an arthropod depend basically on an adequate understanding of the epidemiology of each infection and the conditions that favor or hinder the propagation of the agent. In the development and utilization of such information, there is need for close cooperation between the practicing physician and the public health workers in any given area. Control programs must be planned to meet the practical aspects in each particular region in order to be successful. Methods applicable in one area or to a particular parasitosis are frequently not applicable to other parasitoses. The examples cited indicate how varied are the natural histories of the parasites and, likewise, how varied are the results of attempts to control the diseases.

REFERENCES

Benenson, A.S., editor. 1980. *The Control of Communicable Diseases in Man,* 13th ed. New York, American Public Health Association.

Bruce-Chwatt, L.J. 1973. L'eradication du paludisme en Afrique. Ann. Parasitol. Hum. Comp., *48*:221–229.

Curtis, C.F., and Feachem, R.G. 1981. Sanitation and *Culex pipiens* mosquitoes: A brief review. J. Trop. Med. Hyg., *84*:17–25.

Gabaldon, A. 1969. Global eradication of malaria: Changes of strategy and future outlook. Am. J. Trop. Med. Hyg., *18*:641–656.

Gemmell, M.A. 1968. The Styx field-trial. A study on the application of control measures against hydatid disease caused by *Echinococcus granulosus*. Bull. W.H.O., *39*:73–100.

Lloyd, S. 1981. Progress in immunization against parasitic helminths. Parasitology, *83*:225–242.

Molineaux, L., and Gramiccia, G. 1980. *The Garki Project. Research on the Epidemiology and Control of Malaria in the Sudan Savanna of West Africa*. Geneva, World Health Organization.

Morishita, K. 1972. Studies on epidemiological aspects of ascariasis in Japan and basic knowledge concerning its control, pp. 1–153. In *Progress of Medical Parasitology in Japan,* vol. 3. Edited by K. Morishita, Y. Kamiya, and H. Matsubayashi. Tokyo, Meguro Parasitological Museum.

Morison, R.S. 1963. The concept of eradication and public education. Pharos, *26*:3–6, 27.

Neghme, A. 1961. Ecological problems in parasitic diseases. Bol. Chil. Parasitol., *16*:53.

World Health Organization. 1970. Insecticide resistance and vector control: Seventeenth report of the WHO Expert Committee on Insecticides. Geneva, W.H.O. Tech. Rep. Ser., no. 443.

———. 1973. Schistosomiasis control. Report of a WHO Expert Committee. Geneva, W.H.O. Tech. Rep. Ser., no. 515.

Chapter 7

Landmarks and Literature in Parasitology

Parasitology in Ancient Times. According to Chinese folklore, the god of creation brushed the lice off his own body and from them created human beings. Ancient Chinese physicians differentiated tertian, quartan, and estivo-autumnal malaria by their febrile manifestations, and they associated "ague cake" (large, fibrotic spleen) with malaria just as the Greek and Roman physicians did many centuries later. The Ebers papyrus of Egypt (ca. 1600 B.C.) described a worm, probably the beef tapeworm (*Taenia saginata*), as pathogenic in man and prescribed infusion of the bark of the pomegranate tree for its evacuation. Moses, who received medical instruction from Egyptian priests, prescribed sanitary laws to guard against plagues from insect pests as well as from infected animal flesh. The Greeks of Aristotle's day recognized tapeworms. Hippocrates diagnosed hydatid disease and described a technique for removal of the cyst from the human body. The Persian physician Avicenna (980–1037 A.D.) described worms probably referable to *Ascaris lumbricoides, Taenia saginata, Enterobius vermicularis,* and possibly *Ancylostoma duodenale,* listed the symptoms produced by them, and prescribed remedies that even today are recognized as satisfactory anthelmintics. Hoeppli (1959) has given an informative general review of the early knowledge of parasites and parasitic infections up to the seventeenth century.

Medieval and Modern Times. The earliest book devoted to parasitology was published in 1684 by Redi, who included descriptions of reproductive organs and eggs of *Ascaris*. In 1817 Lancisi recorded studies of mosquitoes and vague surmises about their role in the cause of intermittent fevers. Goldfuss, in 1817, first used the word "*protozoa,*" which was given its modern meaning in 1845 by Siebold. Leeuwenhoek (1632–1723) had devised and used simple microscopes, but achromatic objectives in a compound microscope were not used in England until 1824. Gross, in 1849, was the first to describe an ameba parasitic in man—*Entamoeba gingivalis*—and Lösch identified *E. histolytica* in 1875. Then came the discovery of mosquito hosts for filariae by Manson (1877–1878) and plasmodia by Laveran in 1880, transmission of babesiosis by ticks by Smith and Kilbourne in 1894, trypanosomes and their transmission by tsetse flies by Bruce (1895–1896), and mosquito transmission of plasmodia by Ross (1897–1898)(Scott, 1939; Foster, 1965).

After these discoveries, the science of parasitology expanded rapidly. With the aid of microscopes, morphological characters of various parasites were first studied and species and group characteristics were determined. Then the several stages of the organisms were related to one another in life-cycle sequence. This provided important information on extrinsic as well as intrinsic development and paved the way for epidemiologic studies. Moreover, the relation of parasite to host provided a background for studying the pathogenesis of infection in man and reservoir hosts and, indirectly, for understanding clinical aspects. Recent investigations have been largely concerned with the ecology of parasitic infections, anatomy and physiology as revealed by electron microscopy, metabolism of the parasite in its host, immunologic phenomena, and the rationale of chemotherapy. Meanwhile, practical methods are being developed to control these infections and to reduce human exposure to them.

A landmark of great significance was the publication in 1978 of a two-volume work by B.H. Kean, K.E. Mott, and A.J. Russell, which brought together in the original English or English translation the articles on the major discoveries and earliest descriptions by workers in tropical medicine and parasitology.

GENERAL REFERENCES

Binford, C.H., and Connor, D.H. 1976. *Pathology of Tropical and Extraordinary Diseases. An Atlas,* vols. 1 and 2. Washington, D.C., Armed Forces Institute of Pathology.

Brown, H.W., and Neva, F.A. 1983. *Basic Clinical Parasitology,* 5th ed. Norwalk, Appleton-Century-Crofts.

Chernin, E., editor. 1977. A Bicentennial Sampler: Milestones in the History of Tropical Medicine and Hygiene. Am. J. Trop. Med. Hyg., *26:*1053–1104.

Faust, E.C. 1949. *Human Helminthology,* 3rd ed. Philadelphia, Lea & Febiger.

———, Beaver, P.C., and Jung, R.C. 1975. *Animal Agents and Vectors of Human Disease,* 4th ed. Philadelphia, Lea & Febiger.

Foster, W.D. 1965. *A History of Parasitology.* London, E. & S. Livingstone.

Harwood, R.F., and James, M.T. 1979. *Entomology in Human and Animal Health.* New York, Macmillan.

Hoeppli, R. 1959. *Parasites and Parasitic Infections in Early Medicine and Science.* Singapore, University of Malaya Press.

Horsfall, W.R. 1962. *Medical Entomology. Arthropods and Human Disease.* New York, The Ronald Press Co.

Hubbert, W.T., editor. 1975. *Diseases Transmitted from Animals to Man,* 6th ed. Springfield, Ill., Charles C Thomas.

Hunter, G.W., Frye, W.W., and Swartzwelder, J.C. 1975. *Tropical Medicine,* 5th ed. Philadelphia, W.B. Saunders.

Kean, B.H., Mott, K.E., and Russell, A.J., editors. 1978. *Tropical Medicine and Parasitology. Classic Investigations,* 2 vols. Ithaca, Cornell University Press.

Last, J.M., editor. 1980. *Maxcy-Rosenau Public Health and Preventive Medicine,* 11th ed. New York, Appleton-Century-Crofts.

Manson-Bahr, P.E.C., and Apted, F.I.C. 1982. *Manson's Tropical Diseases,* 18th ed. London, Baillière Tindall.

Markell, E.K., and Voge, M. 1981. *Medical Parasitology,* 5th ed. Philadelphia, W.B. Saunders.

Morishita, I., Komiya, Y., and Matsubayashi, H., editors. 1964–1978. *Progress of Medical Parasitology in Japan,* 6 vols. Tokyo, Meguro Parasitological Museum.

Noble, E.R., and Noble, G.A. 1982. *Parasitology. The Biology of Animal Parasites,* 5th ed. Philadelphia, Lea & Febiger.

Schmidt, G.D., and Roberts, L.S. 1977. *Foundations of Parasitology.* St. Louis, C.V. Mosby.

Scott, H.H. 1939. *A History of Tropical Medicine.* 2 vols. London, Edward Arnold & Co.

Smith, K.C.V. 1973. *Insects and Other Arthropods of Medical Importance.* New York, John Wiley & Sons.

Soulsby, E.J.L. 1982. *Helminths, Arthropods and Protozoa of Domesticated Animals,* 7th ed. Philadelphia, Lea & Febiger.

Spencer, F.M., and Monroe, L.S. 1982. *The Color Atlas of Intestinal Parasites,* 2nd ed. Springfield, Ill., Charles C Thomas.

Spencer, H., et al., editors. 1973. *Tropical Pathology.* New York, Springer-Verlag.

Taylor, A.E.R., and Baker, J.R. editors. 1978. *Methods of Cultivating Parasites in Vitro.* New York, Academic Press.

Von Brand, T. 1979. *Biochemistry and Physiology of Endoparasites.* New York, Elsevier/North Holland.

JOURNALS AND PERIODICALS

Acarologia, 1959 to date.

Acta Tropica, 1944 to date.

Advances in Parasitology, 1963 to date.

American Journal of Tropical Medicine, 1921–1951.

American Journal of Tropical Medicine and Hygiene, 1952 to date.

Anales del Instituto de Biologia (Mexico), 1930 to date.

Annales de la Société Belge de Medecine Tropicale, 1920 to date.

Annales de Parasitologie humaine et comparée, 1922 to date.

Annals of the Entomological Society of America, 1908 to date.

Annals of Tropical Medicine and Parasitology, 1907 to date.

Annual Review of Entomology, 1956 to date.

Archiv für Schiffs- und Tropen-hygiene, 1897–1941. Continued as Deutsche Tropenzeitschrift.

Archives de Parasitologie, 1898–1919.

Boletin Chileno de Parasitologia, 1945 to date.

Bulletin de la Société de Pathologie Exotique, 1908 to date.

Bulletin of Entomological Research, 1910 to date.

Cahiers O.R.S.T.O.M., serie Entomologie Medicale et Parasitologie, 1963 to date.

China Medical Journal, 1887–1931. Continued as Chinese Medical Journal, 1932 to date.

Experimental Parasitology, 1951 to date.

Indian Journal of Malariology, 1946–1963, 1983 to date.

International Journal for Parasitology, 1971 to date.

Japanese Journal of Parasitology, 1951 to date.

Journal of Economic Entomology, 1908 to date.

Journal of Helminthology, 1923 to date.

Journal of Medical Entomology, 1964 to date.

Journal of Tropical Medicine and Hygiene, 1898 to date.

Korean Journal of Parasitology, 1963 to date.

Memorias do Instituto Oswaldo Cruz, 1909 to date.

Molecular and Biochemical Parasitology, 1980 to date.

Mosquito News, 1941 to date.

Mosquito Systematics, 1969 to date.

Parasite Immunology, 1979 to date.

Parasitology, 1908 to date.

Parassitologia, 1956 to date.

Proceedings of the Helminthological Society of Washington, 1934 to date.

Puerto Rico Journal of Public Health and Tropical Medicine, 1925–1950.

Revista Brasileira de Malariologia e Doenças Tropicais, 1949 to date.

Revista de Biologia Tropical (Costa Rica), 1932 to date.

Revista do Instituto de Medicina Tropical de São Paulo, 1959 to date.

Revista Iberica de Parasitologia, 1940 to date.

Rivista di Parassitologia, 1937 to date.

Southeast Asian Journal of Tropical Medicine and Public Health, 1970 to date.

Systematic Parasitology, 1979 to date.

Transactions of the American Microscopical Society, 1892 to date.

Transactions of the Royal Society of Tropical Medicine and Hygiene, 1907 to date.

Tropenmedizin und Parasitologie, 1974 to date.

Tropical and Geographical Medicine, 1949 to date. Continuation of Documenta de Medicina Geographica et Tropica.

Wiadomosci Parazytologiczne, 1955 to date.

World Health Organization: Monographs, Technical Reports, Reports of Expert Committees, Bulletin, and Chronicle, 1948 to date.

Zeitschrift fur Parasitenkunde, 1929 to date.

Zeitschrift fur Tropenmedizin und Parasitologie, 1950–1973. Continued as Tropenmedizin und Parasitologie.

ABSTRACTS AND BIBLIOGRAPHIC SOURCES

Biological Abstracts, 1929 to date.

Bulletin Analytique d'Entomologie Medicale et Veterinaire, 1954 to date.
Excerpta Medica, Section 4. Microbiology, 1948 to date.
Helminthological Abstracts, 1932 to date.
Index Catalogue of Medical and Veterinary Zoology (Author Catalogue), 1932–1952 with current supplements.
Index Medicus, 1878–1889, 1903–1927, 1960 to date.
Quarterly Cumulative Index, 1916–1926, 1927–1956.
Quarterly Bibliography of Major Tropical Diseases, 1978 to date.
Protozoological Abstracts, 1977 to date.
Review of Applied Entomology, Section B (Medical and Veterinary), 1913 to date.
Tropical Diseases Bulletin, 1913 to date.

Section II

Protozoa and Protozoan Infections

Chapter 8

Morphology, Biology, Life Cycles, and Classification of Protozoa

Protozoa consist of a vast assemblage of single cell-like microorganisms that zoologists have placed in the subkingdom, or phylum, Protozoa. For convenience these organisms are referred to as "unicellular," although many protozoologists prefer to consider them as "acellular" because, unlike the Metazoa (many-celled animals), the single unit that the protozoa typically consists of performs all of the functions of life. Some of the free-living flagellated protozoa contain green plastids that enable them to synthesize carbohydrate from inorganic substances in their environment, and for this reason they are claimed by botanists as belonging to the Plant Kingdom. Although the medical parasitologist and the physician may regard this as a purely academic question, it does have considerable merit since it provides rather convincing evidence that animal and plant life had a common origin, that the flagellate protozoa probably represent the most primitive extant type of animal life, and that animal forms gradually lost their ability to photosynthesize carbohydrates and hence, became dependent on previously synthesized foods.

Manwell (1961) stated that "the Protozoa rival anything in the Animal Kingdom for uniqueness and novelty. Could they be magnified to the size of vertebrates we are most familiar with, some might even be terrifying. Their habits would be equally frightening; some Protozoa manage to capture and swallow animals larger than themselves; others are cannibals. Many are unconscionably greedy, gorging themselves twenty-four hours a day."

MORPHOLOGY

The single unit that constitutes the average protozoan organism has a remarkable range in size and form, depending principally on the group to which it belongs. Some species are visible to the naked eye; others require magnification up to a thousandfold before they can be seen. While some are spherical or ovoidal, others have bizarre contours. Some have radial symmetry, some bilateral symmetry, and some have longitudinal torsion to their body. These characteristics are relatively fixed in certain groups while in others there is remarkable change in form during the active (trophic) stage of the organism. Moreover, a number of the free-living flagellates have developed into colonial aggregates, although it need not be assumed that this constitutes an intermediate step to the Metazoa, in which groups of the component cells have become morphologically and functionally differentiated.

Although protozoa exhibit marked differences in size and form, there are certain fundamental structures (organelles) common to the entire group. The most important of these is the relatively viscous nucleus containing the chromosomes that are essential for life, reproduction, and transmission of the genetic constitution of the organism. In groups having a vesicular nucleus (Fig. 8–1), a single mass or aggregate of granules referred to as the *karyosome* (endosome, nucleolus) frequently may be near the center of the cell or at times surrounded by chromatin particles that appear to be arranged on an achromatic network, or chromatin granules may line the nuclear membrane. All of these structures lie in a relatively dense nucleoplasm. In other groups there is a "compact" nucleus that contains a large amount of granular chromatin material and much less nucleoplasm. Limiting the nucleus is a relatively tough *nuclear membrane,* which has

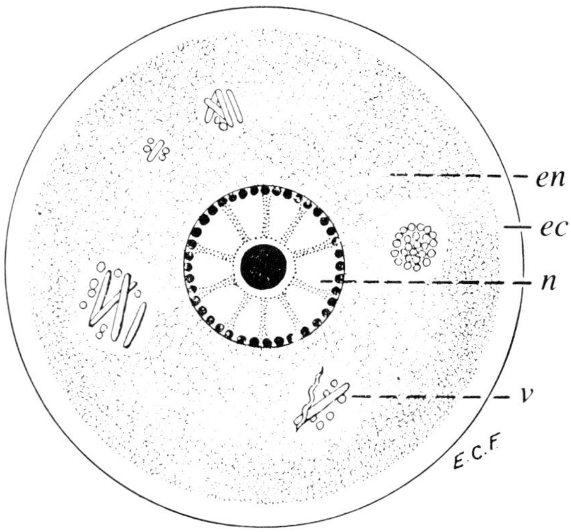

Fig. 8–1. Schematic diagram of a protozoan cell that has a vesicular type of nucleus *(n)* with central karyosome (endosome, nucleolus), chromatin beads within nuclear membrane, and *achromatic fibrils* extending from karyosome to periphery; endoplasm *(en)*, made up of a denser, granular meshwork and intermediate lighter cytoplasm, with numerous food vacuoles *(v)*; and ectoplasm *(ec)*, consisting of agranular or very finely granular cytoplasm. The entire organism is bounded by a cell membrane. (By E.C. Faust.)

minute channels connecting nucleoplasm with endoplasm. Some protozoa during their trophic stage have two nuclei; these may be similar in structure and function (e.g., *Giardia, Dientamoeba*) or dissimilar (e.g., *Balantidium*). Other species have many nuclei, occasionally numbering a hundred or more.

The *nucleus* lies within the cytoplasm of the cell. The inner portion of the cell is the *endoplasm,* which immediately surrounds the nucleus and consists of moderately dense granular protoplasm in which undigested food is contained in food vacuoles. Food synthesis also takes place in the endoplasm, and in it food may be stored in the form of glycogen or protein (*chromatoidal bodies*). *Mitochondria,* as well as *Golgi apparatus, microsomes,* and *endoplasmic reticulum*—organelles that are common to protozoan cells and single metazoan cells—can usually be demonstrated in the endoplasm.

Enveloping the endoplasm is the less granular, more homogeneous *ectoplasm,* which in many species exhibits form and organelles as remarkably varied as the structures that differentiate the phyla of the Metazoa. The ectoplasm functions as the locomotor apparatus for the procurement and ingestion of food, in respiration, in discharge of metabolic wastes, and in the protection of the organism. One or more contractile (pulsating) vacuoles, believed to be osmo-regulators, are generally present in the protozoa but are usually not found in marine and parasitic species of Sarcodina and Mastigophora and are not developed in any Sporozoa. In the trophic stage, there is a limiting cell boundary, the semipermeable *plasma membrane,* which controls the intake and output of foods, secretions, and excretions, and maintains normal concentration of the plasma substance by being permeable to some substances and impermeable to others. In some groups (like the amebae), it has no constant shape but changes form by means of the extension and retraction of temporary *pseudopodia.* In other groups, the contour is more or less constant. In some free-living protozoa (Heliozoa, Radiolaria, etc.), there is a nonliving calcareous or siliceous shell secreted by the ectoplasm that has numerous symmetrical perforations through which pseudopodia may be projected and retracted. In the encysted stage, which is characteristic of several groups of protozoa, there is a relatively tough membrane called the "cyst wall," which is secreted by the ectoplasm to protect the organism.

Locomotion in the amebae is accomplished by a flowing movement of pseudopodia. In the flagellates, one or more delicate, hair-like projections of the cytoplasm called *"flagella"* arise from a kinetoplast within the cytoplasm and serve as locomotor organelles. In the ciliates, numerous short threads called "*cilia,*" which arise from basal granules within the ectoplasm and are distributed over the surface of the body, perform the function of locomotion.

Some protozoa have no special site for the intake of food; they absorb liquid nutriment from the medium in which they live or ingest bacteria and microscopic food particles at any site on the body surface. Other protozoa have a specialized "cell mouth," the *cytostome,* which is usually located laterally near the anterior end of the body. Ciliate protozoa have a "cell anus," the *cytopyge,* through which particulate food wastes are discharged. *Excretory vacuoles* at times serve for the collection of fluid wastes that are then ejected from the cell, usually at the aboral end. *Contractile vacuoles* in certain protozoa that live in a hypotonic medium maintain normal osmotic pressure by collecting excess water within their cytoplasm and expelling it to the outside.

Most protozoan trophozoites lack specialized

protective organelles; they depend on the irritability of their protoplasm to avoid anisotonic media or other injurious environmental conditions. Many protozoa are able to encyst and thus survive under unfavorable conditions. Some with calcareous or siliceous shells have a protective armament. Several of the ciliates are provided with multiple ectoplasmic trichocysts, which on mechanical or chemical stimulation are discharged into entangling or nettling filaments; this closely resembles the mechanism developed by certain classes of coelenterates.

Cytologic study with special stains has demonstrated that flagellates and ciliates have specialized primitive nervous systems. The kinetoplast of flagellates, which typically consists of a parabasal body and blepharoplast joined by a delicate fibril, constitutes the energizing part of the neuromotor apparatus; this is connected with the axoneme, which is the intracytoplasmic portion of the flagellum. In some ciliates, the basal granules at the insertion ends of the cilia, which are immediately under the cell membrane, are joined by, or are adjacent to, fibrils that connect with a complex meshwork of neuronemes ("silver-line system"). The neuronemes may be exclusively ectoplasmic or may penetrate into the endoplasm, in which case they are referred to as the neuromotor apparatus.

BIOLOGY

As the oldest forms of animal life, protozoa have become adapted to practically all types of environment except those that are extremely dry. Free-living forms are found in the sea in both pelagic and littoral zones, in deep and shallow fresh water, and in moist soil and subsoil. Many species are parasites of plants, insects, other invertebrates, and of all groups of vertebrate hosts. Protozoa have certain optima for their development but are at times able to adjust to wide ranges of temperature, pH, salinity, light, and nutriment. Some species live in thermal hot springs, while others live in water at 0° C or thrive in snow at high altitudes. Hydrogen ion concentrations of habitats of forms that have been investigated vary from pH 3.0 to 9.5. Investigation of tolerated salinity and osmotic pressure indicates equally remarkable adjustments. Moreover, species that secrete a thick, relatively impermeable cyst wall are able to survive long periods of desiccation. While parasitic species are more limited than free-living ones in their tolerance to environmental conditions, the former also exhibit considerable adaptability. For example, species of *Leishmania* in the lumen of the midgut of the sandfly live in a poikilothermic environment, multiplying in the flagellated leptomonas (promastigote) stage, with a sodium concentration that is much lower and a potassium concentration that is much higher than in mammals, and reduced chlorides and increased phosphates. On introduction into the susceptible mammalian host, however, these protozoa live and multiply in the more condensed leishmania (amastigote) form in an intracellular medium that has greater osmotic tension and a considerably higher, but relatively stable, temperature. The sewage ameba, *Entamoeba moshkovskii*, normally lives at a temperature range of 20 to 28° C but thrives in the test tube at 37° C; in this latter environment, however, it metabolizes much more rapidly than at the lower temperatures.

Availability of nourishment is the most important factor governing the existence and prevalence of protozoan life in different microclimates. Relatively few present-day protozoa are able to synthesize their food from inorganic materials; hence, they must depend on available organic substances. Such food may be disintegrating plants or animals or living miroorganisms, such as bacteria, that are present in the immediate environment. The parasitic species obtain food from their host by using the predigested substances present in the host's digestive tract, blood stream, or lymphatic vessels; or the parasite may actively digest the tissue cells of its host, as do *Entamoeba histolytica*, malaria parasites, *Leishmania* spp., *Trypanosoma cruzi* and *Toxoplasma*. As protozoa gradually lost their ability to synthesize metabolites from inorganic substances they developed complex enzyme systems capable of transforming preformed organic materials into amino acids and then synthesizing them into their own protoplasm or into substances needed for energy production. Most free-living protozoa have retained aerobic pathways of metabolism, but many parasitic species have become facultative or obligative anaerobes.

LIFE CYCLES AMONG PROTOZOA

Among the protozoa, the process of reproduction may be relatively simple. It is accomplished by repeated asexual multiplication by binary fission, whereby the nucleus of the parent organism divides mitotically so that each of the two daughter cells receives qualitatively the exact proportion of the genetic material, one consisting of the original and the other of the newly formed replica. This process is followed by cytoplasmic separation into two essentially equal parts, one for each daughter nucleus. In the amebae, which have no axial gradient, there is no characteristic plane of division; in the flagellates, division is along a longitudinal axis; in the ciliates, it is in the transverse plane; and in the malaria parasites, the production of merozoites is accomplished by several to many binary nuclear divisions within the schizont before the daughter organisms are fully formed. In these groups, this asexual method of propagation may be continued indefinitely as long as the environmental conditions are satisfactory. For all but a few species of the parasitic protozoa, however, there are special safeguards for survival. These consist of direct host-to-host transfer of the trophic stage, cyst formation, and sexual phases in the life cycle.

Person-to-Person Transfer. Two of the parasitic protozoa of man—*Entamoeba gingivalis,* which lives in the mouth, and *Dientamoeba fragilis,* which lives in the large intestine—have only a trophozoite stage. The former is commonly transmitted from person to person by droplet spray and, in this way, is able to perpetuate itself. There is no definite information about the method of host-to-host transmission employed by *D. fragilis*. All of the three trichomonads of man lack a cystic stage. *Trichomonas tenax* of the mouth is transmitted in the same way as *E. gingivalis,* and *T. vaginalis* is transmitted primarily by sexual intercourse, but the mechanism of person-to-person transfer for the intestinal species, *T. hominis,* has not been satisfactorily explained.

Encystation. Many protozoa, both free-living and parasitic, have a cystic stage that is developed when conditions in the environment are unfavorable for continued trophic multiplication. Factors that have been predicated as responsible for encystation include: (1) deficiency or overabundance of food supply; (2) excess of catabolic products of the organism or of associated bacteria; (3) marked change in the pH; (4) desiccation of the medium; (5) depletion or excess supply of oxygen; and (6) overpopulation. Possibly two or more of these combine to cause encystation. In general, the cyst wall secreted by encysting free-living forms is thicker and can provide protection for a longer time than that of the parasitic species. Even among the parasitic amebae of man, the cyst wall of *Entamoeba coli* is more resistant to desiccation than that of *E. histolytica*.

Two types of encystment are recognized—protective and reproductive. Among the parasitic species, the former is represented by *Balantidium coli,* in which the cyst wall protects the organism outside the body of its host and for transfer to a new host, but no profound morphologic change takes place within the cyst. The reproductive type is found in several of the free-living and parasitic amebae and in many of the flagellates. In these species, the nuclei characteristically divide one or more times during the cystic stage, with a corresponding increase in the number of trophozoites following excystation. Reproductive cysts typically contain reserves of glycogen, are condensed, have resorbed their locomotor organelles, and in other respects, exhibit a dedifferentiation of their protoplasm.

The following factors are possibly involved in excystation: (1) osmotic changes in the medium; (2) enzymatic action of the enclosed organism on the inner surface of the cyst wall; and (3) among the parasitic protozoa, favorable pH and enzymatic action of the host tissues.

Intermediate Hosts. Host-to-host transfer of the blood and tissue flagellates, the hemogregarines, and hemosporidia is typically accomplished through the intermediary of a hematophagous invertebrate that probably constitutes the primary (original) host. Representative invertebrate hosts are leeches (*Trypanosoma rotatorium* of frogs), mites (*Hepatozoon muris* of rodents), bugs (*T. cruzi*), tsetse flies (*T. gambiense*), mosquitoes (malaria parasites), and sandflies (*Leishmania* spp.). The parasite multiplies in the intestine of the invertebrate host, so that there is an increased chance for introduction of the parasite into the susceptible vertebrate host on which the vector next feeds. In one species of trypanosome (*T. equiperdum*), the arthropod host-vector has been suppressed; transmission is by coitus.

Sexual Stages in the Life Cycle. No convincing evidence has been produced to support the older view that sexual stages take place in the life cycle of parasitic amebae, although syngamy occurs in a number of free-living flagellates. In the ciliates, sexual reproduction is accomplished by conjugation, in which two similar organisms come together on their oral surfaces and reciprocally exchange a mature microgamete (division product of the micronucleus), so that a micronucleus with diploid number of chromosomes (zygote) is produced in each of the pairing individuals. All sporozoa have an alternation of asexual and sexual generations in their life cycle. In the coccidia, all essential stages (schizogony, gametocyte-gamete production, zygote formation) occur in one host, with extrinsic ripening of the encysted sexual product (oocyst). In certain groups, two hosts are required: a vertebrate, in which repeated schizogony takes place and gametogony is begun, and a blood-sucking invertebrate (usually an arthropod) in which gametogony is completed and sporogony occurs.

CLASSIFICATION OF PROTOZOA

The classification and definitions given here are taken from the revised classification of the Protozoa prepared for the Society of Protozoologists by a *Committee on Systematics and Evolution* (Levine, et al., 1980). The definitions, which are somewhat abbreviated, apply mainly to species that are parasites of man.

The Protozoa are single-celled eukaryotic organisms—i.e., they have the nuclear structures enclosed in a membrane, as opposed to the prokaryotic bacteria in which such structures are not separated from the cytoplasm. Formerly, all Protozoa were grouped under one phylum. A more modern classification assigns living things to five kingdoms: Monera, Protista, Plantas, Fungi, and Animalia, with the Protozoa a subkingdom of the Animalia. The major groups of Protozoa—SARCOMASTIGOPHORA, APICOMPLEXA, MICROSPORIDA, CILIOPHORA, and three others—become phyla, whereas in earlier classifications each was a subphylum, superclass, or class. While their ranks have changed with advances in knowledge, the stem-names and endings of taxa generally have remained the same as they were when a list of phyla, classes, and orders was prepared for the Zoology Section of the American Association for the Advancement of Science and reported by Pearse in 1936. The endings for orders and suborders are *-ida* and *-ina,* respectively; for class and subclass, *-ea* and *-ia,* and for higher taxa, *-a*.

Phylum **SARCOMASTIGOPHORA** Honigberg and Balamuth, 1963. Protozoa with single-type nucleus; sexual reproduction, when present, essentially syngamous; flagella, pseudopodia, or both types of locomotor organelles present.

Subphylum MASTIGOPHORA Diesing, 1866. Trophozoites typically with one or more flagella; asexual reproduction basically by longitudinal binary fission; sexual reproduction in some groups.

Class **ZOOMASTIGOPHOREA** Calkins, 1909. Chloroplasts absent; one to many flagella; ameboid forms with or without flagella in some groups.

Order **KINETOPLASTIDA** Honigberg, 1963 emend. Vickerman, 1976. With one or two flagella arising from depression, typically with paraxial rod in addition to axoneme; with single mitochondrion extending length of body, usually containing conspicuous Feulgen-positive kinetoplast (nucleoid) located near base of flagellum; Golgi apparatus typically in region of flagellar depression not connected to flagellum.

Suborder TRYPANOSOMATINA Kent, 1880. Flagellum single, free, or attached to body by undulating membrane; kinetoplast small and compact.
Leishmania, Trypanosoma

Order **RETORTAMONADIDA** Grassé, 1952. With two or four flagella, one turned posteriorly and associated with ventrally located cytostomal area bordered by fibril; mitochondria and Golgi apparatus absent; cysts present.
Retortamonas, Chilomastix

Order **DIPLOMONADIDA** Wenyon, 1926 emend. Brugerolle, 1975. With one or two karyomastigonts (sets of flagella arising near the nucleus); genera with two karyomastigonts with two-fold rotational symmetry or, in one genus, primarily mirror symmetry; individual mastigonts with one to four flagella, typically one recurrent associated with cytostome or, in more advanced genera, with organelles forming cell axis; mitochondria and Golgi apparatus absent; cysts present.

Suborder ENTEROMONADINA Brugerolle, 1975 emend. Kulda and Nohynkova, 1978. With single karyomastigont bearing one to four flagella; in genera having more than single flagellum, one recurrent; cysts in at least one genus.
Enteromonas

Suborder DIPLOMONADINA Wenyon, 1926 emend. Brugerolle, 1975. With two karyomastigonts; body with two-fold rotational symmetry or bilateral symmetry; each mastigont with four flagella, one of them recurved; cysts present.
Giardia

Superorder **PARABASALIDEA** Honigberg, 1973. Typically with at least some kinetosomes, bearing flagella or barren; flagella arranged in pattern and associated with rootlet filaments characteristic of trichomonads; mitochondria absent; division spindle extranuclear.

Order **TRICHOMONADIDA** Kirby, 1947 emend. Honigberg, 1974. Typically with four to six flagella (one flagellum in one genus and none in another); in typical genera, one flagellum recurved, free, or with proximal or entire length adherent to body surface; undulating membrane, if present, associated with adherent segment of recurved flagellum; true cysts known in very few species.
Trichomonas, Dientamoeba

Subphylum SARCODINA Schmarda, 1871. Pseudopodia or locomotive protoplasmic flow without discrete pseudopodia; flagella, when present, usually restricted to developmental or other temporary stages; body naked or with test or skeleton; asexual reproduction by fission; sexual reproduction, if present, associated with flagellate or, more rarely, ameboid gametes.

Superclass **RHIZOPODA** von Siebold, 1845. Locomotion by lobopodia, filopodia, or reticulopodia, or by protoplasmic flow without production of discrete pseudopodia.

Class **LOBOSEA** Carpenter, 1861. Pseudopodia lobose or more or less filiform but extended from a broader hyaline lobe; usually uninucleate; multinucleate forms not flattened or with much-branched plasmodia.

Subclass GYMNAMOEBIA Haeckel, 1862. Without test.

Order **AMOEBIDA** Ehrenberg, 1830. Typically uninucleate; mitochondria typically present; no flagellate stage.

Suborder TUBULINA Bovee and Jahn, 1966. Body a branched or unbranched cylinder; no bidirectional flow of cytoplasm; nuclear division mesomitotic.
Entamoeba, Endolimax, Iodamoeba

Suborder ACANTHOPODINA Page, 1976. Subpseudopodia more or less finely tipped, sometimes filiform, often furcate hyaline, produced from a broad hyaline lobe; not regularly discoid; cysts usually formed; nuclear division mesomitotic or metamitotic.
Acanthamoeba

Order **SCHIZOPYRENIDA** Singh, 1952. Body a monopodial cylinder, usually moving with more or less eruptive, hyaline, hemispherical bulges; typically uninucleate; nuclear division promitotic; temporary flagellate stage in most.
Naegleria

Phylum **APICOMPLEXA** Levine, 1970. Apical complex (visible with electron microscope) generally consisting of polar ring(s), rhoptries, micronemes, conoid and subpellicular microtubules present at some stage; micropore(s) generally present at some stage; sexual reproduction by syngamy.

Class **SPOROZOEA** Leuckart, 1879. Conoid, if present, a complete cone; reproduction generally both sexual and aexual; oocysts generally contain infective sporozoites produced by sporogony; locomotion by body flexion, gliding, or undulation of longitudinal ridges; flagella present only in microgametes of some groups; pseudopods, if present, used for feeding, not locomotion; homoxenous or heteroxenous.

Subclass COCCIDIA Leuckart, 1879. Gamonts ordinarily present, small when mature, typically intracellular; syzygy, if present, involves gametes; anisogamy marked; life cycle characteristically consists of merogony, gametogony, and sporogony; most species in vertebrates.

Order **EUCOCCIDIIDA** Léger and Duboseq, 1910. Merogony present; in vertebrates or invertebrates, or both.

Suborder EIMERIINA Léger, 1911. Macrogamete and microgamont develop independently; microgamont typically produces many microgametes; zygote not motile; sporozoites typically enclosed in sporocyst within oocyst; homoxenous or heteroxenous.
Cryptosporidium, Isospora, Sarcocystis, Toxoplasma

Suborder HAEMOSPORINA Danilewsky, 1885. Macrogamete and microgamont develop independently; conoid ordinarily absent; microgamont produces eight flagellated microgametes; zygote motile (ookinete); sporozoites naked, with three-membraned wall; merogony in vertebrate, sporogony in invertebrate host; transmitted by blood-sucking insects.
Plasmodium

Subclass PIROPLASMIA Levine, 1961. Merozoite pyriform, round, rod-shaped or amoeboid; conoid absent; oocysts, spores, and pseudocysts absent; usually without subpellicular microtubules, with polar ring and rhoptries; locomotion by body flexion, gliding, or in sexual stages (in Babesiidae and Theileriidae, at least) by large axopodium-like "Strahlen"; asexual and probably sexual reproduction present; parasitic in erythrocytes and sometimes in other circulating and fixed cells; heteroxenous, with merogony in vertebrate and sporogony in invertebrate; sporozoites with one-membraned wall; transmitted by ticks.

Order **PIROPLASMIDA** Wenyon, 1926. With the characters of the subclass.
Babesia, Entopolypoides

Phylum **MICROSPORA** Sprague, 1977. Minute intracellular parasites with spores of unicellular origin, in undivided membranous capsule, containing a single ameboid germ discharged via tubular polar filament; a polaroplast fills apical part of spore, and a posterosome the basal part.

Class **MICROSPOREA** Delphy, 1963. Microsporidia with long, polar filament, polar opening apical, spores spherical, oval, or tubular; sporoplasm uninuclear or binuclear; pansporoblast thinwalled.

Order **MICROSPORIDA** Balbiani, 1882. Sporogony and spores diplokaryotic.
Nosema

Phylum **CILIOPHORA** Doflein, 1901. Simple cilia or compound ciliary organelles typical in at least one stage of life cycle; with subpellicular infraciliature present even when cilia absent; usually with two types of nuclei; binary fission transverse; contractile vacuole typically present.

Class **KINETOFRAGMINOPHOREA** de Puytorac et al., 1974. Oral cilia only slightly distinct from somatic cilia; cytostome often apical (or subapical) or mid-ventral, on surface of body or at bottom of atrium or vestibulum; cytopharyngeal apparatus commonly prominent.

Subclass VESTIBULIFERIA de Puytorac et al., 1974. Vestibulum, if present, apical or near-apical, with cilia derived from anterior parts of somatic kineties (rows of cilia) leading to cytostome.

Order **TRICHOSTOMATIDA** Bütschli, 1889. Somatic kineties not reorganized at level of vestibulum other than more packed alignment or addition of supernumerary segments of kineties; many species commensal in vertebrate hosts.

Suborder TRICHOSTOMATINA Bütschli, 1889. Ciliature not reduced.
Balantidium

REFERENCES

Kudo, R.R. 1966. *Protozoology,* 5th ed. Springfield, Ill., Charles C Thomas.

Levine, N.D. 1973. *Protozoan Parasites of Domestic Animals and of Man,* 2nd ed. Minneapolis, Burgess Publishing Co.

———, et al. 1980. A newly revised classification of the Protozoa. J. Protozool., *27:* 37–58.

Manwell, R.D. 1961. *Introduction to Protozoology.* New York, St. Martin's Press.

Pearse, A.S., editor. 1936. *Zoological Names—A List of Phyla, Classes, and Orders.* Durham, Duke University Press.

Wenyon, C.M. 1926. *Protozoology: A Manual for Medical Men, Veterinarians and Zoologists,* 2 vols. London, Bailliere, Tindall and Cox.

Chapter 9

Flagellate Protozoa of the Digestive and Urogenital Tracts

The parasitic flagellate Protozoa in earlier classifications were placed in the class or superclass Mastigophora together with all species of one-celled animals that bear one to several long, delicate, thread-like extensions of the cytoplasm, which are termed *flagella* (singular, *flagellum*). Species that have chromatophores or plastids, which are used for the synthesis of carbohydrates (i.e., free-living species that are autotrophic) were placed in the class Phytomastigophora, and the parasitic species and others that do not have plastids were classified as Zoomastigophora. Protozoologists now group the parasitic flagellates together with the amebae and certain ameba-like species in the phylum Sarcomastigophora, subphylum Mastigophora, class Zoomastigophorea (see p. 39).

CLASS ZOOMASTIGOPHOREA

Calkins, 1909

Members of this class lack chromatophores and thus depend on previously manufactured plant and animal foods. Their nutrition is holozoic or parasitic, and their form may be simple or quite complex, but as a rule they have a single nucleus and a neuromotor apparatus. The neuromotor apparatus consists of a blepharoplast (and at times a parabasal body), which constitutes the kinetoplast (energizing component), and an axoneme, with or without a free flagellum (motor component, possibly a modified centriole). Although the finer structure of the flagellum is not completely known, studies by electron microscopy have confirmed and supplemented earlier histologic studies. The axial structure of the flagellum is a continuation of the axoneme, which is composed of one or more fibrils. Occasionally the flagellum may terminate in a small knob (e.g., *Trypanosoma rhodesiense*). Surrounding the flagellum is a sheath, which may be circular, oval, or compressed in cross section. Usually, but not in all species, the sheath extends to the inner end of the axoneme and may be provided with a spirally coiled filament or have one or more rows of short lateral filaments (mastigonemes) parallel to one another. These delicate structures are believed to provide contractility for the flagellum. In case there are two or more flagella, there will be a comparable number of axonemes and typically a similar number of blepharoplasts.

Some species of flagellate protozoa have a rudimentary mouth, the *cytostome*. Reproduction is by longitudinal binary fission, in which mitotic division of the nucleus and binary division of the kinetoplast precede that of the cytoplasm. In a majority of the family groups, the life cycle is simple, but in the trypanosomes, the organisms are dimorphic or polymorphic.

FAMILY BODONIDAE

Bütschli, 1878

Members of this family have two flagella, one of which is directed anteriorly, the other posteriorly and trailing; both flagella arise from a blepharoplast immediately in front of the anteriorly situated nucleus. A cytostome has been described for most species. Contractile vacuoles are present in free-living species. Division is by binary fission. Species parasitic in man belong to the genus *Retortamonas*.

Genus Retortamonas

Grassi, 1879

Members of this genus have a relatively plastic body that is pyriform, ovoidal, or fusiform and is attenuated posteriorly. They are found in the intestine of the cricket, cockroach, guinea pig, and man.

Retortamonas intestinalis (Wenyon and O'Connor, 1917) Wenrich, 1932

Synonyms. *Waskia intestinalis* Wenyon and O'Connor, 1917; *Embadomonas intestinalis* Wenyon and O'Connor, 1917.

Historical and Geographic Notes. *Retortamonas intestinalis* was first observed in the feces of two persons in Egypt by Wenyon and O'Connor and was described by these investigators under the name *Waskia intestinalis*. In the same year, the species was transferred by them to the genus *Embadomonas* Mackinnon, 1911. Wenrich (1932) recommended that the species *intestinalis* from man be placed in the genus *Retortamonas*, which Grassi had erected for closely related flagellates. This protozoan has been reported from mainland China, Malaysia, the Philippines, Egypt, Brazil, and the United States and is probably cosmopolitan in warm climates and elsewhere in certain groups such as inmates of mental hospitals. It has never been found to occur commonly in any locality.

Morphology, Biology, and Life Cycle. *Retortamonas intestinalis* is small, measuring 4 to 9 μm in length by 3 to 4 μm in breadth in its trophozoite stage and 4 to 7 μm by 3 to 4 μm in the encysted stage. The trophozoites (Fig. 9–1,*A*) are elongate-pyriform in the active condition and more or less ovoidal when inactive. The cytoplasm is finely granular and vacuolated. Near the anterior end there is a cleft-like cytostome and a vesicular spherical nucleus with a central karyosome. Near the nucleus are two minute blepharoplasts, each of which gives rise to a flagellum; the longer flagellum is directed anteriorly and the slightly shorter one passes backwards through the cytostome before it becomes free. The pyriform cysts (Fig. 9–1,*B*) appear to have a doubly contoured wall due to separation of the cytoplasm from the cyst wall. Both trophozoite and cyst have a single nucleus. Multiplication of the trophozoite is accomplished by longitudinal binary fission. No nuclear multiplication has been described for the cyst.

Epidemiology. Cysts passed in the stools of infected individuals no doubt constitute the source for infection of other persons when taken into the mouth in food or drink or on unclean objects. The infection rate is so low in all populations that the organism may not be discovered unless thousands of fecal samples are examined. This protozoan has not been reported from other hosts although it could be mistaken for one of the species that have been described from monkeys, rabbits, and guinea pigs.

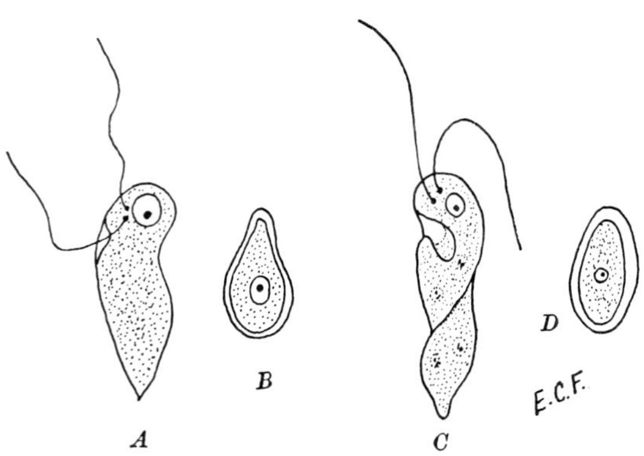

Fig. 9–1. *Retortamonas intestinalis* and *R. sinensis*. *A*, Trophozoite of *R. intestinalis*. *B*, Cyst of *R. intestinalis*. *C*, Trophozoite of *R. sinensis*. *D*, Cyst of *R. sinensis*. (× 2000.) (Adapted from Faust, E.C. 1932. Examination and identification of protozoa. J. Lab. Clin. Med., *17*. Courtesy of C.V. Mosby Company.)

Pathogenesis, Pathology, and Symptomatology. There is no evidence that *R. intestinalis* is pathogenic, although it is commonly discovered only in diarrheic stools.

Diagnosis. A diagnosis is made by examination of fresh fecal films and hematoxylin-stained preparations and following inoculation into appropriate culture media (see Technical Appendix, p. 731).

Treatment. No data are available on treatment and there is no indication for chemotherapy.

Prevention. *R. intestinalis* has been diagnosed only in groups of individuals who have a low level of personal and group hygiene. Since it barely survives under these conditions, any substantial improvement in hygienic conditions should cause its disappearance.

Retortamonas sinensis (Faust and Wassell, 1921) was discovered in diarrheic stools at Wuchang, China, and was found again in diarrheic stools of patients in Peking, China (Watt, 1933). This species (Fig. 9–1,*C, D*) differs from *R. intestinalis* in the greater length and breadth of the trophozoite (13 by 5 μm), a more cavernous cytostome, no evidence that the shorter flagellum passes through the cytostome, and a definite torsion to the body. The cyst is broadly ellipsoidal rather than pyriform.

FAMILY CHILOMASTIGIDAE Wenyon, 1926

Species belonging to this family have trophozoites with three anteriorly directed free flagella, a delicate fourth flagellum lying within a cytostomal cleft, and pear-shaped cysts in which the cytostome is clearly visible. All described species belong to the genus *Chilomastix*.

Genus Chilomastix Alexeieff, 1910

Trophozoites of this genus are rounded anteriorly, attenuated posteriorly, have three anteriorly directed free flagella and a more delicate one within the prominent cytostome. The cyst is pear- or lemon-shaped with a rather thick wall and shows clearly the large cytostome. Species have been described from man, monkeys, rabbits, guinea pigs, rats and mice, goats, chickens and turkeys, tadpoles, lizards and fishes.

Chilomastix mesnili (Wenyon, 1910) Alexeieff, 1912

Synonyms. *Cercomonas intestinalis* Marchand, 1875; *Monocercomonas hominis* Grassi, 1881; *Macrostoma mesnili* Wenyon, 1910; *Tetramitus mesnili* Alexeieff, 1910; *Chilomastix hominis* von Prowazek and Werner, 1914; *Chilomastix davainei* (Moquin-Tandon, 1860) Kofoid, 1920.

Historical and Geographic Notes. This organism was probably first seen by Davaine (1854), who in 1860 called it *Cercomonas hominis* var. A, but the first accurate description was provided by Wenyon (1910). Alexeieff (1912) created the genus *Chilomastix* for Wenyon's species *mesnili* after discovering that it did not conform to the generic description of *Macrostoma* or *Tetramitus*. *C. mesnili* has a cosmopolitan distribution but is more prevalent in warm than in cool climates.

Morphology, Biology, and Life Cycle. *Chilomastix mesnili* has well-defined trophic and cystic stages. The living trophozoite (Fig. 9–2,*A,B,C*) is asymmetrically pear-shaped as a result of a spiral groove extending through the middle half of the body. When in active progressive movement, the posterior end is drawn out into a long cone; when quiescent, the organism is more nearly pyriform or obovoidal. Depending on the state of nourishment and activity, *C. mesnili* trophozoites measure 6 to 20 μm in length by 3 to 10 μm in breadth. The spherical nucleus, measuring 3 to 4 μm, is situated medially near the

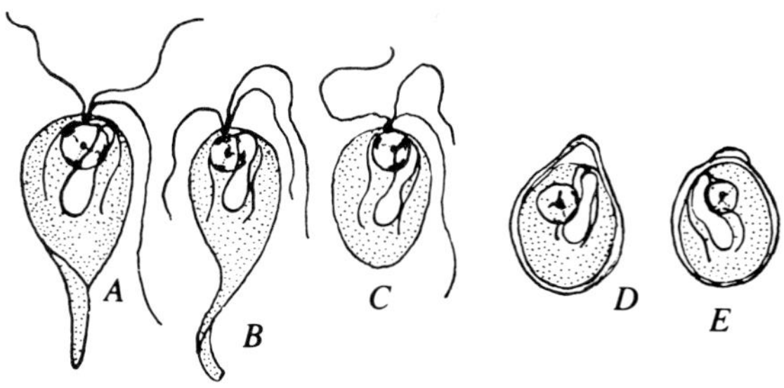

Fig. 9–2. *Chilomastix mesnili. A, B, C,* Trophozoites. *D, E,* Cysts. (× 1200.) (By E.C. Faust.)

anterior pole; it has a small, distinct, central karyosome, a few achromatic fibrils extending to the nuclear membrane, and chromatin plaques lining the membrane. A well-defined cytostome, which is seen on one side of the nucleus, is rounded anteriorly and posteriorly and has a median constriction. It is about half the width and twice the length of the nuclear diameter. Just anterior to the nucleus there is a complex of six minute blepharoplasts; from three of these, the three free anterior flagella (two short, one long) originate; from another, a delicate flagellum lying within the cytostome; and from the other, two axonemic (?) fibrils that encompass the lateral margins of the cytostome. These detailed structures can be observed only after careful fixation and iron-hematoxylin staining of fecal films containing the living organisms. The cytoplasm is delicately granular and contains numerous food vacuoles.

The cysts are characteristically pear- or lemon-shaped, rather broadly rounded at one end, and somewhat bluntly conical at the other end (Fig. 9–2,*D,E*). They are colorless, measure 7 to 10 μm in length by 4 to 6 μm in breadth, and have a thick, tough wall. Usually the densely granular cytoplasm is separated some distance from the wall at the narrower end of the cyst. Internally, hematoxylin-stained films clearly demonstrate the single large vesicular nucleus (occasionally two nuclei) and the cytostome, which is almost a long as the encysted organism. Good preparations reveal a fibril on either side of the cytostome.

Chilomastix mesnili is a normal inhabitant of the cecal region of the large intestine, where the trophozoites live on enteric bacteria in the lumen of the glands and multiply by binary fission. In freshly passed liquid stools, only trophozoites are observed; in semi-formed stools, both trophozoites and cysts may be seen; and in well-formed stools, only cysts are recovered. Moreover, encystation may occur in unformed fecal specimens and is relatively common during centrifugation. Although cysts are typically uninucleate, two-nucleate forms have been demonstrated by several workers. The trophozoite may be cultivated on a variety of media.

Epidemiology. Transmission from person to person undoubtedly occurs when cysts of *C. mesnili* in the stools of an infected individual get into the mouth of another and are swallowed. Although monkeys harbor a species of *Chilomastix* morphologically indistinguishable from *C. mesnili,* this source of exposure does not seem to be a common one for man. Depending on the particular population group and the age of the individuals surveyed, the prevalence of *C. mesnili* infection ranges from possibly less than 1 to 10% or more.

Pathogenesis, Pathology, and Symptomatology. *C. mesnili* is a harmless commensal and is not responsible for symptoms.

Treatment. No treatment is indicated.

Prevention. Better sanitation and personal hygiene will reduce the prevalence of this infection.

FAMILY TETRAMITIDAE
Bütschli, 1878

Members of this family have three anterior flagella and a fourth trailing flagellum, and lack an axostyle or other axial organelles. The only genus of this group reported from man is *Enteromonas*.

Genus Enteromonas
da Fonseca, 1915

Species of this genus have been described from man and guinea pig, and possibly from the pig (*Tricercomonas suis* Knowles and Das Gupta, 1929).

Enteromonas hominis
da Fonseca, 1915

Synonym. *Tricercomonas intestinalis* Wenyon and O'Connor, 1917.

Historical and Geographic Notes. This species was established by da Fonseca (1915) for a flagellate discovered in diarrheic stools in Brazil. It was next found by Wenyon and O'Connor (1917) in Egypt. Since that time it has been reported from Spain, France, Holland, Morocco, Egypt, southwest Sudan, Turkey, India, Malaysia, Java, Sumatra, Brazil, Guyana, Peru, and United States. It appears to be widely distributed throughout tropical areas and has been identified in temperate regions.

Morphology, Biology, and Life Cycle. The trophozoite of *E. hominis* (Fig. 9–3,*A*) is a hyaline, pear-shaped, rounded or ovoidal organism measuring 4 to 10 μm by 3 to 6 μm. By means of its three anterior flagella, the active organism has a jerky forward movement. A fourth flagellum may be seen along the flattened surface of one side, then extending posterior for an appreciable distance as a free organelle. All four flagella originate in a group of blepharoplasts just in front of the ovoidal nucleus, which is near the anterior end. There is no cytostome. The entire cytoplasm is finely vacuolated and contains numerous bacteria. The trophozoite divides by longitudinal binary fission. The cyst (Fig. 9–3,*C*) is elongate-ovoidal, measures 6 to 8 μm by 4 to 6 μm and, as it matures, first contains two nuclei, with one at each end (Fig. 9–3,*C*), then four nuclei, with two at each end. There is a well-defined cyst wall.

Enteromonas hominis is a commensal that lives in the lumen of the human digestive tract (for the most part probably in the cecal region). It is readily cultured on Boeck and Drbohlav's LES medium (see Technical Appendix, p. 760.

Epidemiology. Although no proof has been provided, it seems probable that infection with *E. hominis* results from ingesting the cysts in contaminated food or drink.

Pathogenesis, Pathology, and Symptomatology. *E. hominis* is considered to be a harmless commensal. No symptoms

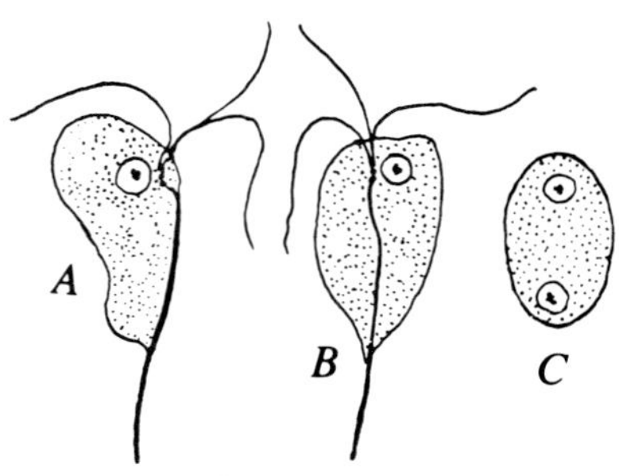

Fig. 9–3. *Enteromonas hominis. A, B,* Trophozoites. *C,* Cyst. (× 1600.) (By E.C. Faust.)

have been attributed to it, even though it is detected most often in diarrheic stools.

Diagnosis. This is based on demonstration of the characteristic living trophozoite or cyst, confirmed by hematoxylin-stained fecal films.

Treatment. No therapy is indicated.

Prevention. This requires better personal and group hygiene.

FAMILY HEXAMITIDAE
Kent, 1880

Members of this family differ from all other flagellates in having two nuclei lying side-by-side in the same transverse plane, six or eight flagella (three or four pairs), in some genera, paired axonemes, and in general, bilateral symmetry. There are five recognized genera with a few species that are free-living, and a majority that are parasitic in invertebrate and vertebrate hosts. The species parasitic in man belongs to the genus *Giardia*.

Genus Giardia
Kunstler, 1882

The species of *Giardia* are morphologically rather similar flagellates that, with one exception (described from a nematode), inhabit the intestinal tract of vertebrates. They have been described from man, monkey, many species of rodents, dog, cat, ox, horse, goat, many species of birds, tadpole, monitor lizard, and fish. The trophozoites are rounded anteriorly, tapered posteriorly, have a convex dorsal surface and a flattened ventral side with a shallow sucking disc in the anterior portion. They have eight (four pairs) flagella arising from a complicated system of axonemes. The cysts are ovoidal to ellipsoidal and have a thin, tough wall from which the cytoplasm is characteristically separated.

Giardia lamblia
Stiles, 1915

Synonyms. *Cercomonas intestinalis* Lambl, 1859; *Megastoma enterica* Grassi, 1881; *Lamblia intestinalis* (Lambl, 1858) Blanchard, 1888; *Giardia enterica* (Grassi, 1881) Kofoid, 1920.

Historical and Geographic Notes. This flagellate was discovered by Leeuwenhoek (1681), who found it in his own stools, but was first recognizably described by Lambl (1859), who gave it the name *intestinalis*. Because of confusion with reference to the availability of the names *intestinalis* and *enterica* for this species, Stiles (1915) created a new binomial, *Giardia lamblia*, in honor of Professor A. Giard of Paris and Doctor F. Lambl of Prague. *Giardia lamblia* is a cosmopolitan parasite, characteristically more prevalent in children than in adults, and more common in warm than in cool climates. It is the most commonly diagnosed flagellate of the human intestinal tract. The subject was reviewed recently by Meyer and Jarroll (1980).

Morphology, Biology, and Life Cycle. Like other species of this genus, *G. lamblia* has both a trophic and a cystic stage. The living trophozoite (Fig. 9–4) is rounded anteriorly and pointed posteriorly. Dorsally it is convex, and ventrally it is provided with a shallow posteriorly notched concavity (sucking disc) that embraces practically all of the anterior half of the organism. In profile it is much more slender than when observed *en face*. The size of the trophozoite varies within wide limits (9 to 21 μm in length by 5 to 15 μm in breadth and 2 to 4 μm in thickness). The pair of nuclei, one on each side of the midline about one-fourth the body length from the anterior end, are ovoidal and contain a central karyosome consisting of a single dense chromatin mass or a large number of relatively discrete granules (chromosomes?) that may be dispersed throughout the nucleoplasm. No chromatin lines the delicate nuclear membrane. The four pairs of flagella arise from superficial organelles on the ventral side of the body. According to Wenyon (1936), there are probably four pairs of blepharoplasts, the motivating organelles. Two pairs of blepharoplasts (one lateral and one more median) are situated on each side of the midline, between and slightly anterior to the two nuclei. From the two lateral blepharoplasts there originate two axonemes that proceed forward in a curved course towards the midline, cross one another, then describe a wide arc and at their greatest distance from the midline, give rise to the *lateral pair of crossed flagella*. Somewhat thicker axonemes (sometimes incorrectly referred to as "axostyles") originate from the median pair of blepharoplasts; these axonemes pass directly to the posterior end of the body and nearly converge at the tip, where they give rise to the *posterior pair of flagella*. A third pair of blepharoplasts near the center of the sucking disc is associated with a pair of very short

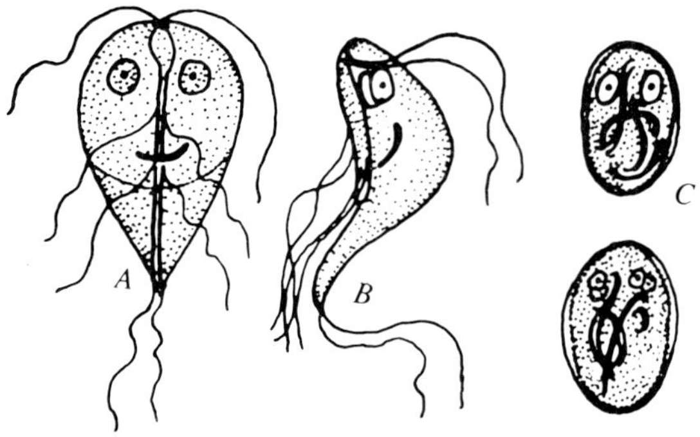

Fig. 9–4. *Giardia lamblia*. Trophozoite, *A*, ventral view. *B*, profile view. *C*, Cyst, immature (above) and mature with four nuclei (below). (× 1360.) (Adapted from original of E.C. Faust.)

axonemes that give rise to the *central pair of flagella*. The fourth pair of blepharoplasts has not been definitely located, although the pair of axonemes that originate from them have been traced forward along the margins of the notch of the sucking disc, and each appears to have fibrillar connections with the anterior lateral blepharoplast. This fourth pair of axonemes gives rise to the *lateral pair of uncrossed flagella*. In addition to the complex blepharoplast-axoneme-flagellar organelles, two closely associated, slightly curved, sausage-shaped bodies are often observed lying transversely or obliquely just posterior to the sucking disc and dorsal to the posterior axonemes. These may be parabasal bodies, but this has not been definitely demonstrated.

By the rapid movement of the flagella, the trophozoites move from place to place, and by applying their sucking discs to epithelial surfaces, they become firmly attached.

The trophozoites divide by a complicated process of longitudinal binary fission involving division of the nuclei, then the neuromotor apparatus and the sucking disc, which is followed by separation of the cytoplasm so that two daughter trophozoites are formed. The characteristic location of *G. lamblia* is in the crypts at the duodenal level of the small intestine.

In frankly diarrheic stools, it is usual to find only the trophozoites. Encystation occurs commonly in transit down the colon. In preparation for encystment, the trophozoites retract their flagella into the axonemes, which assume the appearance of four pairs of curved bristles; the cytoplasm becomes condensed, and a thin, tough hyaline membrane (cyst wall) is secreted. The cysts (Fig. 9–4,*C*) are ovoidal and measure 8 to 12 μm in length and 7 to 10 μm in breadth. They have a finely granular cytoplasm that is clearly separated from the thin cyst wall. When stained with iodine, the cysts normally are yellow or pale brown; occasionally they are pale blue or green and are small and relatively slim. Both types may be seen in the same specimen. Degenerate cysts with the cell small and dense in a normal cyst wall may be misdiagnosed or escape notice (Collins et al., 1978). Recently formed cysts have two nuclei; mature cysts have four nuclei (Fig. 9–4,*C*). These are situated at one end of the organism until it is ready to divide to form two individuals. As the cyst ripens, the internal structures and the sucking disc are doubled, so that when excystation occurs in the duodenum or in an appropriate culture medium all that is required for the production of identical twins is division of the cytoplasm and growth of the flagella from the axonemes. Karapetyan (1962) reported cultivation of *G. lamblia* in a medium of human, horse, or beef serum containing a seeding of *Saccharomyces*. Excystation in vitro was described by Bingham and Meyer (1979), who observed that cell division was essentially a part of the excystation process and normally required 5 to 30 minutes. Meyer (1976) reported the successful isolation and axenic cultivation of *G. lamblia* from duodenal aspirates.

Epidemiology. Transmission of *Giardia lamblia* is by viable cysts that are swallowed. Food or drink may be the source of exposure, but in some circumstances it is more likely that intimate contact of an infected with an uninfected individual is the usual mechanism, as in pinworm (*Enterobius vermicularis*) infection (p. 303). *G. lamblia* is typically more prevalent in children than in adults, especially in children of large families and those in orphan asylums and elementary schools (Ormiston et al., 1942). Dancescu and Tintareanu (1963) reported 14% prevalence in the 3- to 6-month age group in a children's home in Romania; the prevalence increased with age to 100% in the 2-year-old group. In 45% of the children, there was some evidence of interference with growth. Thus, infection typically reaches its maximum frequency during childhood or at puberty, then declines in prevalence. In warm climates, the prevalence peak is higher and is reached earlier than in cool climates. In Norway, *Giardia* is found in fewer than 5% of the people examined (Petersen, 1972).

In the United States, the epidemiologic pattern has changed from the typical picture, possibly because of a decrease of immunity acquired in childhood within the population as a consequence of improved sanitation and standard of living. Today epidemics of giardiasis are being observed among adults, presumably due to fecal contamination of food or drink, as occurred in Aspen and Boulder, Colorado, in 1969 and 1971 (Moore et al., 1969; Schultz, 1975). More commonly, giardiasis is seen as a cause of travelers' diarrhea in adults, especially among visitors to certain centers, notably Leningrad. One aspect of the so-called sexual revolution in the United States is the increasingly high prevalence of giardiasis ascribable to homosexual anal/oral sexual practices among male adults (Schmerin et al., 1978; Phillips et al., 1981). In summary, while outbreaks of giardiasis among adults and high rates of infection among travelers in certain countries can usually be traced to contaminated drinking water, among children of preschool age endemicity depends mainly on direct person-to-person transmission (Keystone et al., 1978).

Although giardiasis is highly contagious and may be acquired in early infancy, in a considerable proportion of infected persons of all ages the infection eventually is lost spontaneously.

An exception to the usual pattern of endemic *Giardia* infection occurred in the years 1954 to 1955 in Portland, Oregon, where an epidemic of several hundred cases developed in young and middle-aged adults. The source of the epidemic was not located. Similar outbreaks, attributed to contamination of resort or municipal water supplies have occurred in Colorado, Utah, Washington, Oregon, New Hampshire, and New York (Kirner; 1978; Shaw et al., 1977; Wright et al., 1977; Dykes et al., 1980). High frequencies of infection have been reported among hikers and campers who drank from mountain streams (Schultz, 1975; Meyer and Jarroll, 1980). In some instances, infections were acquired in remote areas where contamination by people was unlikely; therefore, infected wild animals such as beavers have been suspected as the source (Dykes et al., 1980). Cats, dogs, and other animals can carry *Giardia* infection (Hewlett et al., 1982), and in cats, it may cause disease (Brightman and Slonka, 1976). Groups of travelers in various parts of the world have experienced high frequencies of diarrhea while abroad or soon after return to their home countries that was apparently caused by *Giardia* (Kettis and Magnius, 1973; Thompson et al., 1974; Lopez et al., 1978). The usual prepatent period in such cases was 2 to 3 weeks. Due in part to a greater awareness of the infection, as well as to more frequent foreign travel, the incidence of symptomatic giardiasis in the United States and Europe has increased sharply in recent years, especially among adults (Weiss et al., 1977).

Numerous protozoologic surveys in representative regions of the world show 2 to 25% or more prevalence, depending on the age group examined, environmental hygiene, and climatic conditions.

Pathogenesis, Pathology, and Symptomatology. The presence of *G. lamblia* in the glandular crypts of the duodenal-jejunal mucosa, even when myriads of the organisms are involved, ordinarily causes no apparent irritation. These flagellates do not invade the tissues but feed on the mucous secretions, and in the average infection are strictly commensal in their relationship to the host. However, although it has not been shown that *G. lamblia* can invade the intestinal mucosa, *G. muris* in mice have been observed in the epithelium overlying Peyer's patches, with remnants of trophozoites contained in macrophages in the deeper tissues (Owen et al., 1981). In an appreciable number of both juvenile and adult cases, there is duodenal irritation with excess secretion of mucus and dehydration, accompanied by dull epigastric pain, flatulence, and chronic diarrhea with a gruelly or steatorrheic type of stool containing a large amount of mucus and fat but no blood, and a celiac syndrome in small children (Chaudhuri, 1943; Welch, 1944; Peterson, 1957; Craft, 1982). These individuals lose weight as a result of constant dehydration and poor appetite (Hartman and Kyser, 1941). Veghelyi (1940) suggested and Erlandsen (1974) has demonstrated by scanning electron microscopy that the coating of large surface areas of the upper small bowel by *Giardia* may act as a barrier to fat absorption and thus cause the steatorrhea; this does not explain the failure of the host intestine to absorb other substances such as vitamin B12 that are absorbed at levels of the intestine below those inhabited by *Giardia*. It has been observed that subjects with giardiasis and steatorrhea have bacterial colonization of the small intestine and significantly increased free bile acids in the intestine. It has been suggested that either or both of these phenomena may be responsible for the steatorrhea (Tandon et al., 1977). The bacteria present in giardiasis are the same enterobacteria previously found to be present in patients with tropical sprue, suggesting a possible relationship and offering a clue to the etiology of tropical sprue (Tomkins et al., 1978).

In addition to duodenal involvement, occasionally the gallbladder may become parasitized by *G. lamblia*. In these patients, there may be associated gallbladder colic and jaundice, due to obstruction of the bile passages or irritation and edema of the ampulla of Vater.

Diagnosis. Fluoroscopy may demonstrate hypermotility at the duodenal and jejunal levels, and x-ray films may reveal mucosal defects. In this infection, diarrheic stools will usually reveal numerous active trophozoites, and semi-formed or formed stools will contain cysts of this organism. If there are few cysts, the zinc sulfate centrifugal flotation technique is satisfactory for concentration (see Technical Appendix, p. 742). Occasionally the stools will be negative when duodenal aspiration will provide material teeming with active trophozoites. It is usual for infections to have alternating periods of high and low excretion of cysts (Danciger and Lopez, 1975). Nevertheless, demonstration of the flagellate is not evidence as such of its causal relationship to a syndrome of duodenitis. Peptic ulcer, carcinoma, celiac disease of other etiology, and strongyloidiasis (p. 253) must be excluded. The blood picture in uncomplicated giardiasis shows no abnormality. Craft and Nelson (1982) recommended a technique of counterimmunoelectrophoresis of feces for the diagnosis of giardiasis.

Treatment. In about 90% of cases, *G. lamblia* can be eradicated by the administration of *quinacrine* (Atabrine) in the dosage for adults of 0.1 g three times daily for 5 days. Children receive a reduced dosage. Rapid amelioration of symptoms associated with giardiasis following quinacrine therapy suggests the flagellate as the causal agent. *Metronidazole* too is highly effective against giardiasis but is regarded as second choice because it is carcinogenic in rats and mice and mutagenic in bacteria. Although these adverse characteristics have never been demonstrated in human beings, metronidazole should not be given to pregnant women. When taken with alcoholic beverages it may produce a disulfiram-like effect (abdominal distress, nausea, vomiting, headache, and modification of the taste of the alcoholic beverage). Alcohol should not be permitted for 24 hours after administration of the drug. Since it is a nitroimidazole, it should be used with caution in patients with a history of blood dyscrasia. The dosage for adults is 250 mg three times daily, and for children 5 mg/kg body weight three times daily, for 10 days. Furazolidone, which is available as a suspension, can be used for treatment of small children (Wolfe, 1978).

Prevention. Since *Giardia* infection is readily contracted and, like pinworm (p. 303) and dwarf tapeworm (p. 509) infections, is especially prevalent in children, prevention is primarily a question of group, family, and personal hygiene. Travelers can be warned to avoid local water supplies when possible, but they need not forego the pleasures of travel even in high-risk areas (Knaus, 1974) or resorts where outbreaks are known to have occurred (Lopez et al., 1978). Iodine is an effective disinfectant of unsafe water for drinking. A solution obtained from crystalline iodine has been recommended for water purification, but its use for this purpose is not without risk (Zemlyn et al., 1981).

FAMILY TRICHOMONADIDAE
Wenyon, 1926

Members of this family have a cytostome, three to five free flagella, an additional flagellum on the margin of an undulating membrane, and an axostyle that usually protrudes through the posterior end of the body.

Genus Trichomonas
Donné, 1837

The trophozoites of *Trichomonas* typically have four free flagella and a fifth along the outer margin of the undulating membrane, a costa at the base of the undulating membrane, and a conspicuous axostyle. All described species of the genus (*sensu stricto*) are parasitic, and without exception none produces cysts. Species of *Trichomonas* are found in man, monkeys, rodents, fowls, pigeons, doves, termites, and slugs. Three species occur in man: *T. tenax, T. hominis,* and *T. vaginalis*.

Trichomonas tenax
(O.F. Müller, 1773) Dobell, 1939

Synonyms. *Trichomonas elongata* Steinberg, 1862; *Tetratrichomonas buccalis* Goodey, 1917; *Tetratrichomonas hominis* Ohira and Noguchi, 1917; *Trichomonas buccalis* (Goodey, 1917) Kofoid, 1920.

Historical and Geographic Notes. Müller (1773) observed a flagellate in a 4-day aqueous culture of tartar from teeth and named the organism *Cercaria tenax*. Höffle (1850) apparently first observed trichomonads removed directly from the human mouth, and Steinberg (1862) employed species names (*elongata, caudata,* and *flagellata*) for varieties of the oral flagellate that he studied. Although protozoologists have generally accepted Dobell's opinion that *tenax* is the correct species name for this trichomonad there is more than a reasonable doubt that Müller actually observed the buccal flagellate, since his culture may have been seeded with a free-living infusorian. *T. tenax* is worldwide in its distribution.

Morphology, Biology, and Life Cycle. *Trichomonas tenax* (Fig. 9–5) is a pyriform flagellate known only in the trophozoite stage. It has a length measurement of 5 to 12 μm (average 6.5 to 7.5 μm), is smaller and more slender than *T. vaginalis* (Fig. 9–7), has four free flagella of approximately equal length and a fifth one on the margin of an undulating membrane (which does not reach the posterior end of the body and lacks a free posterior extension). It also has the following characteristics: a costa (i.e., chromatic basal rod) of the same length as the undulating membrane; a single blepharoplast or closely associated group of granules from which the motor organelles arise; a parabasal apparatus consisting of a fibril and a shorter, thicker body; a relatively thick axostyle originating near the blepharoplast and extending as a sharp object a considerable distance behind the body; an ovoidal vesicular nucleus with relatively few chromatin granules; and a cytostome near the anterior end of the side opposite the undulating membrane. The cytoplasm

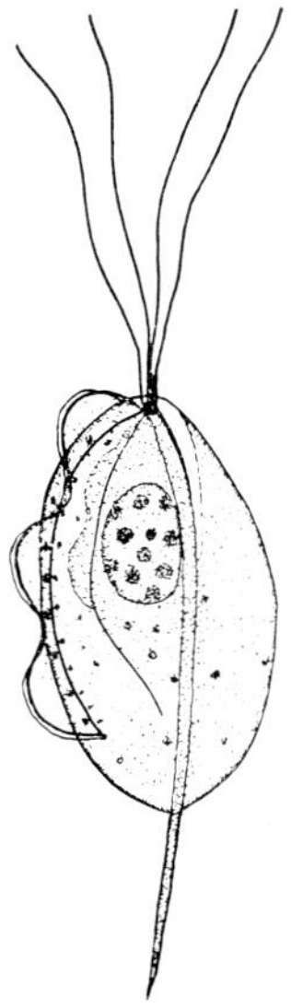

Fig. 9–5. *Trichomonas tenax.* (× ca. 1600.) (Adapted from Wenrich, D.H. 1944. Comparative morphology of the trichomonad flagellates of man. Am. J. Trop. Med., *23*:125–127.)

is delicately granular. Cytochemical tests show varying amounts of glycogen, depending on the activity of the organism, lipids in the marginal areas of the parabasal body and axostyle, cholesterol in minute granules throughout the cytoplasm, and heavy concentration of DNA in the nuclear wall chromatin network and the endosome. There is considerable plasticity to the body and moderate pseudopodial activity of the protoplasm. *T. tenax* feeds on microorganisms in its environment, grows in cultures (see Technical Appendix, p. 759) at room and body temperatures, does not survive through the intestinal tract, and cannot be permanently established in the vagina (Stabler and Feo, 1942). It multiplies by longitudinal binary fission, which is initiated by division of the nucleus, followed by division of the neuromotor apparatus, and finally separation of the cytoplasm into two daughter organisms.

Epidemiology. Although the exact method of transmission of *T. tenax* is unknown, exposure undoubtedly results from droplet spray from the mouth, kissing, or common use of contaminated dishes and drinking glasses. The prevalence of infection varies from 0 to more than 25%, depending on the opportunities for exposure and state of oral hygiene of the individual examined. De Carneri (1957) compiled incidence figures from Europe that are based on his own and previous reports: Milan, 46%, Hamburg, 53%, Bonn, 4%, Prague, 20% and Zlin, 13.5%. *T. tenax* does not occur in the mouths of persons who observe good oral hygiene.

Pathogenesis, Pathology, and Symptomatology. *Trichomonas tenax* is a harmless commensal of the human mouth, living in the tartar around the teeth, in cavities of carious teeth, in necrotic mucosal cells in the gingival margins of gums, in pus pockets in tonsillar follicles, or associated with oral spirochetes in Vincent's angina. It is quite resistant to changes in temperature and will survive for several hours in drinking water.

Walton and Bacharach (1963) reported three cases of respiratory tract infection with *Trichomonas (T. tenax?)* in one United States hospital in 6 months; the infection was identified first in mouth washings, then in sputum smears. Thirty previous records, mostly from Western Europe, were associated with lung or thoracic abscesses of pleural exudates.

Diagnosis. This is made on recovery of a trichomonad from the tartar between the teeth, the gingival margin of the gums, tonsillar crypts, or following culture of suspected material in appropriate media.

Treatment. No specific treatment is required, but better oral hygiene is indicated.

Prevention. Presence of this trichomonad in the mouth indicates poor oral hygiene. When better care is observed, the flagellate will disappear.

Trichomonas hominis (Davaine, 1860) Leuckart, 1879

Synonyms. *Cercomonas* Davaine, 1854; *Cercomonas hominis* Davaine, 1860; *Monocercomonas hominis* Grassi, 1879; *Trichomonas confusa Stiles, 1902; Tritrichomonas hominis* Kofoid, 1920; *Pentatrichomonas ardin delteili* (Derrieu and Raynaud, 1914) Kofoid and Swezy, 1923.

Historical and Geographic Notes. It is generally believed that this flagellate was first observed by Davaine (1854) and described by him (1860) under the species name *hominis*. Although some protozoologists have doubted that Davaine saw and described this intestinal trichomonad as such, his species name has been accepted by most workers. Next to *Giardia lamblia* (p. 44), *T. hominis* is probably the most common and most cosmopolitan of the intestinal flagellates of man, although it is usually identified only in diarrheic stools. The following prevalence figures were recorded by E. C. Faust: U.S.A., college students, 0.16%; Georgia (rural), 0.3%; New Orleans clinic population, 1935, 1.2%; Peking, China, clinic patients, 3.5%; Cali, Colombia survey (1956–1961), 2.0 to 13.5%.

Morphology, Biology, and Life Cycle. Only a trophozoite stage has been observed. The motile organism (Fig. 9–6) is pyriform, measuring 5 to 14 μm (average 7 to 8 μm) in length by 7 to 10 μm in breadth. It has three to five (usually four) anterior free flagella and an additional one on the margin of the undulating membrane with a free-trailing posterior end. When there are five anterior flagella, there is a twinned blepharoplast. There is a conspicuous semi-rigid axostyle with a spiked posterior end protruding through the posterior portion of the cytoplasm, a relatively thick costa that extends the full length of the undulating membrane, a conspicuous cytostomal cleft opposite the undulating membrane, an ovoidal nucleus with scattered fine chromatin granules lining the nuclear membrane, and a relatively conspicuous central karyosome (Wenrich, 1944). No parabasal body has been observed.

The common habitat of *T. hominis* is the cecal area of the large intestine of man and several other primate species, where the organism characteristically feeds on enteric bacteria taken in through its cytostome, but if opportunity permits it will ingest

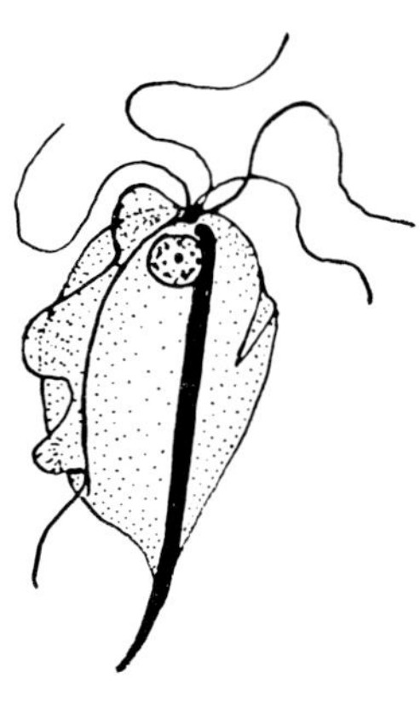

Fig. 9–6. *Trichomonas hominis* from a diarrheic stool. (× 1600.) (By E.C. Faust.)

red blood corpuscles. In the living host, *T. hominis* does not invade the intestinal mucosa, although it has occasionally been identified in tissues at necropsy. In freshly passed diarrheic stools, it manifests not only rapid forward movement but remarkable pseudopodial extension of the cytoplasm, especially on the side of the undulating membrane. Like the other species of *Trichomonas*, *T. hominis* multiplies by longitudinal binary fission.

When it becomes relatively quiescent in a thin fecal film, one can observe the movement of the individual anterior flagella and the successive undulations of the membrane, as well as the protruding posterior end of the axostyle. In formed stools, the organism is difficult to identify because it is rounded up (but not encysted) and exhibits none of the characteristic movements of the active stage.

Kott and Adler (1961) demonstrated distinct serotype differences between *T. hominis* and *T. vaginalis* by means of agglutination and hemagglutination techiques. The method employed also served to differentiate different strains of the same species.

Epidemiology. Since there is no cystic stage of *T. hominis*, transmission necessarily occurs in the trophic condition, most likely in a quiescent form. Wenyon and O'Connor (1917) found that filth flies may serve as mechanical vectors of this organism. Upon reaching the mouth and being swallowed in contaminated food or drink, it survives passage through the stomach and small intestine, provided it is ingested in a medium such as milk or gruel, or in cases of achlorhydria. *T. hominis* cannot be transplanted into the vagina. The incidence of *T. hominis* in different population groups is relatively low compared with that of *T. tenax*, ranging from a small fraction of 1% to somewhat less than 12%. The infection is diagnosed more frequently in warm than in cool climates, in children under 10 years than in older age groups. It is more readily identified when individuals eat large amounts of fresh vegetables and fruits, which produce unformed stools and hence, allow the trichomonads to be active in the fecal specimen.

Pathogenesis, Pathology, and Symptomatology. There is no convincing evidence that *T. hominis as such* is pathogenic or causes intestinal disturbance. Its presence, however, may indicate an unnatural condition that requires medical attention.

Diagnosis. This is readily made on recognition of the cardinal diagnostic features of the active or semi-active trophozoite in fecal films made from unformed stools, or following saline purgation or enema—namely, the successive wave-like movements of the undulating membrane and the spike-like posterior projection of the axostyle.

Treatment. There is no indication for specific treatment.

Prevention. This involves improvement in community sanitation and personal cleanliness.

Trichomonas vaginalis
Donné, 1837

Historical and Geographic Notes. Donné (1836) first observed "animalcules dans les matière purulentes et le produit des sécrétions des organes genitaux de l'homme et de la femme," and the next year named the object *vaginalis*, for which he created the genus *Trichomonas*. This species, therefore, is the type of the genus by original designation. Infection produced by this organism has been found wherever search has been made.

Morphology, Biology, and Life Cycle. *T. vaginalis* exists only in the trophozoite stage, and it is morphologically like *T. tenax* with the following exceptions: *T. vaginalis* (Fig. 9–7) is considerably larger (length 7 to 23 μm, average 13 μm vs. 5 to 12 μm, average about 7 μm); the undulating membrane of *T. vaginalis* is relatively shorter; in large specimens of *T. vaginalis*, the anterior portion of the axostyle may be split into several fibrils; nuclear chromatin in *T. vaginalis* is uniformly distributed and the cytoplasm contains a remarkably large amount of siderophil granules, which are numerous around the delicate costa and the axostyle; and finally, the cytostome in *T. vaginalis* is much less conspicuous (see *T. tenax*, p. 47).

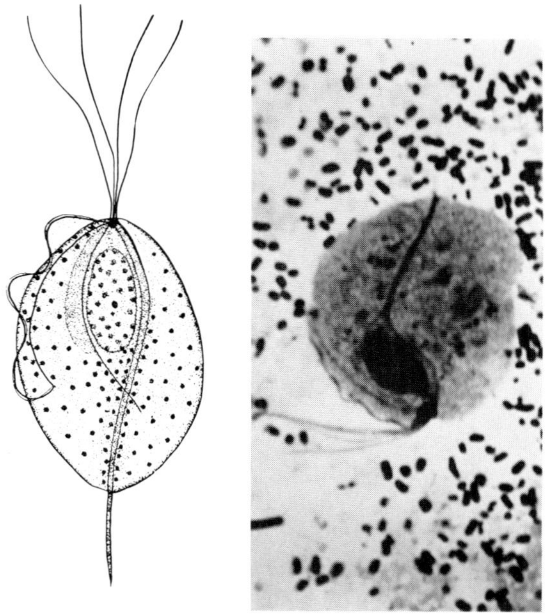

Fig. 9–7. *Trichomonas vaginalis*. *A* (× ca. 1200). (Adapted from Wenrich, D.H. 1944. Comparative morphology of the trichomonad flagellates of man. Am. J. Trop. Med., *23*:125–127.) *B* (× 1600). (Adapted from Jírovec, O. 1948. *Parasitologie pro Zuěrolékaře*. Prague, Melantrich.)

In bacteria-free cultures, Kott and Adler (1961) demonstrated distinct serotypes in 19 strains of *T. vaginalis*, which did not change in two strains tested after 68 passages.

The normal habitats of *T. vaginalis* are the human vagina and prostate gland. In the female host, the organism typically feeds on the mucosal surface of the vagina, ingesting bacteria and leukocytes and, at times, being phagocytosed by macrophages. It prefers a medium slightly alkaline or somewhat more acid than that of the healthy vagina. Similar information is not available for this trichomonad in the infected male, from whom it may be demonstrated in the urine, urethral discharge, or following massage of the prostate with emission of prostatic fluid. Not infrequently the organism is of sedimented or centrifuged urine of male and female hosts, in which it may remain active for hours with progressive undulating and pseudopodial movements. *T. vaginalis* divides by longitudinal binary fission in a manner similar to that described for *T. tenax*. Degenerate forms of *T. vaginalis* are commonly seen, but no rounded-up viable stage like that of *T. hominis* in formed feces has been observed.

Epidemiology. Lacking a cystic stage, *T. vaginalis* is necessarily transmitted as a trophozoite, typically during sexual intercourse. This view is subscribed to by obstetricians, gynecologists, and microbiologists who have studied the problem (Feo, 1944; Young, 1949; Jírovec, 1965). The peak incidence of vaginal trichomoniasis occurs between the ages of 16 and 35, the period of greatest sexual activity. It is found in highest incidence in populations at high risk for other venereal diseases. Half of the women in Great Britain presenting with gonorrhea also had trichomoniasis. Similarly, in Prague 50% of the women in venereal disease clinics, but only 30 to 40% of those in gynecological clinics, had trichomoniasis. One hundred per cent of the female sexual partners of men with trichomoniasis have the infection. Appropriate treatment of women with metronidazole cures 60 to 80% with a single course, but if the male partner is treated simultaneously the cure rate in females is 95%.

Outside the body, this trichomonad rapidly succumbs to temperatures above 40° C (104° F), drying, or direct sunlight, and dies in not more than 35 to 40 minutes in water. Nevertheless, trichomonads can survive on wet sponges for several hours and in urine for more than 24 hours. Therefore, while infection during bathing is highly unlikely, nonvenereal transmission between adults could occur through communal bathing or sharing of douche equipment. Spread from infected mothers to female infants has been noted.

Trichomoniasis is possibly the most commonly acquired sexually transmitted infection with an estimated 2.5 million cases annually in the United States. In asymptomatic women seen in private practice, the prevalence of positive cultures is 3 to 15%, while among prostitutes it is 50 to 75%.

Pathogenesis, Pathology, and Symptomatology. *Trichomonas vaginalis* cannot live naturally without close association with the vaginal, urethral, or prostatic tissues. A few days after introduction of viable *T. vaginalis* into the vagina, the proliferating colonies of the flagellate cause degeneration and desquamation of the vaginal epithelium, soon to be followed by leukocytic inflammation of the tissue layer. Very large numbers of trichomonads and leukocytes are now present in the vaginal secretions, which are liquid, greenish or yellow, and cover the mucosa down to the urethral orifice, vestibular glands, and clitoris. Gram-negative rods and gram-positive cocci may be seen in the microscopic field of vaginal smears. As the acute condition changes to the chronic stage, the secretion loses its purulent appearance due to decrease in the number of trichomonads and leukocytes, increase in epithelial cells, and establishment of a mixed bacterial flora. Investigators in Polish penal institutions found that, in both men and women in isolation, *T. vaginalis* was diagnosed for at least 2 years following commitment to prison.

Symptomatology. Based on experimental evidence, the incubation period is 4 to 28 days. The onset of vulvar pruritus and vaginal pruritus and vaginal discharge is usually acute and occurs with or immediately following menstruation, probably due to the increased vaginal acidity at that time.

Aside from the common symptoms of vaginal discharge, vulvitis, and dysuria, trichomoniasis appears to be associated with an increased incidence of postpartum endometritis, the rate in women with trichomoniasis being double that in women free of the infection. Furthermore, 90% of women with trichomoniasis have cervical erosions, and this has been interpreted as indicating a predisposition to cervical carcinoma. In men it can cause prostatitis and, rarely, reversible sterility. About 20% of women with vaginal trichomoniasis have dysuria, and this may be the presenting symptom.

Trichomonas infection in the male may be latent and hence, essentially symptomless, or it may be responsible for an irritating persistent or recurring urethritis. In nonspecific urethritis, the flagellate has been detected in 10 to 20% of subjects and in 20 to 30% of those whose consorts had trichomonal vaginitis (Brown, 1972).

Frost and Honigberg (1962) found that when axenically cultured *T. vaginalis* were introduced aseptically into the subcutaneous tissues of mice there was repetitive infiltration of polymorphonuclear leukocytes, multiplication of the flagellates, then death of the leukocytes and spread of the trichomonads, which resulted in pure mantles of leukocytes or parasites against the inner wall of the injection pocket. The degree of abscess formation was related to the pathogenic index of the flagellate.

Trichomonas vaginalis was found in a perinephric abscess that developed following trauma in a 58-year-old Thai woman who had a history of vulvo-vaginitis and obstructive uropathy associated with renal calculi (Suriyanon et al., 1975).

Diagnosis. In the female, *T. vaginalis* may be found in sedimented urine, vaginal secretions, or vaginal scrapings; in the male, it may be found in urine, prostatic secretions, or following massage of the prostate gland. Care should be taken to prevent contamination of the specimen with feces, since *T. hominis* may be seen and misdiagnosed as *T. vaginalis*. Hoffman (1966) used the indirect hemagglutination test, which employed glycoprotein obtained by the Fuller method with a mixture of various serologic types of *T. vaginalis*, as an antigen for adsorption by sheep blood corpuscles, and considered this test to be highly specific.

Treatment. Urogenital trichomoniasis is best treated with metronidazole. The most effective dosage is 250 mg administered orally three times daily; in the case of trichomonas vaginitis, 250 mg should also be administered vaginally daily. A course of treatment lasts 5 to 7 days. Resistance of *T. vaginalis* to metronidazole and other 5-nitroimidazoles has been reported (Meingassner and Thurner, 1979).

To avoid prompt reinfection, sexual partners should be treated simultaneously.

Prevention. Since infection is contracted almost exclusively through sexual intercourse, the infections in symptomless males must be diagnosed and cured. While trichomoniasis of the genitalia does not have the stigma of syphilis or Neisserian disease, its control and prevention require both patience and strict discipline.

Dientamoeba fragilis
Jepps and Dobell, 1918

Historical and Geographic Notes. *Dientamoeba fragilis* was discovered by Wenyon (1907) and described by Jepps and Dobell (1918). It has been identified in practically all regions of the world in which satisfactory iron hematoxylin-stained films have been carefully examined. Dobell (1940) recognized the close morphologic similarities between *D. fragilis* and *Histomonas meleagridis*, the ameboflagellate parasite of turkeys. Based on light and electron microscopic observations on culture forms, Camp et al. (1974) redescribed the species and Honigberg (1974) revised the definition and taxonomic position of the genus. A major scientific advance was made when Tyzzer in 1934 reported that *Histomonas meleagridis* is transmitted in the eggs of *Heterakis*, the cecal worm of chickens and turkeys.

Morphology, Biology, and Life Cycle. Only the trophozoite stage—an active, ameba-like organism—is known. It is relatively small, varying from 3 to 22 μm in diameter (Wenrich, 1944) but usually with a narrower size range of 5 to 12 μm. In fresh preparations, the organisms may be actively motile, exhibiting distinct differentiation between cytoplasm and endoplasm, with hyaline pseudopodia that may be lobose or angular. At times there is progressive movement. Food vacuoles containing bacteria may be present (Fig. 9–8).

In well-stained permanent preparations, one, two, or rarely three or four nuclei can be seen, two being the usual number (arrested telophase); and occasionally, portions of a spindle can be seen extending between the nuclei. The nuclear membrane is without peripheral chromatin; and the karyosome typically is relatively large and central and consists of several, frequently four (tetrad-like), discrete granules. Electron micrographs show an extranuclear spindle and other structures characteristic of trichomonad flagellates (Camp et al., 1974).

Dientamoeba fragilis lives in the lumen of the cecum and upper colon. Examination of 15 appendices in which *D. fragilis* was identified in histopathologic sections revealed no evidence of tissue penetration (Burrows et al., 1954). In fresh fecal specimens, the organisms most frequently are found in clumps of mucus and may be present in relatively large numbers, even in semi-formed stools.

Epidemiology. The mode of transmission of *D. fragilis* is unknown. Dobell (1940) failed to infect himself and two monkeys (*Macaca mulatta*) with cultured organisms taken by mouth. He suggested that *D. fragilis*, like the ameboflagellate *Histomonas meleagridis*, which causes a disease (blackhead) in young turkeys, may be carried inside the egg of some common nematode parasite. Burrows and Swerdlow (1956) reported the presence of small structures that they thought could be *D. fragilis* in the eggs of the common pinworm, *Enterobius vermicularis*. Ockert (1972a) noted that *D. fragilis* and *E. vermicularis* were highly prevalent com-

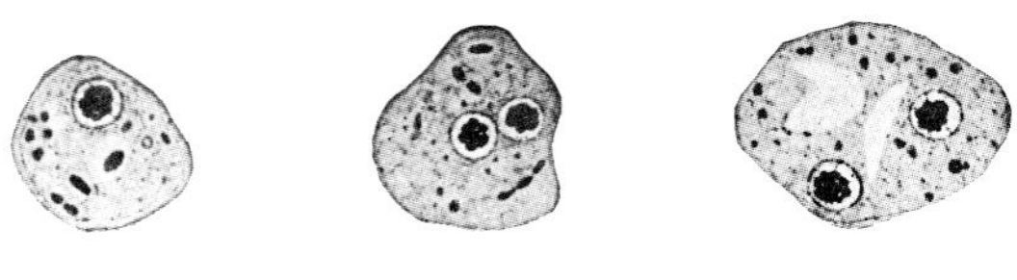

Fig. 9–8. *Dientamoeba fragilis*. Uninucleate and binucleate trophozoites (× ca. 1200). (Adapted from Dobell, C.C., and O'Connor, F.W. 1921. *The Intestinal Protozoa of Man*. London, John Bale, Sons and Danielsson.)

panion parasites of preschool children in Germany. Desser and Yang (1976), in a report of *D. fragilis* infection in two children who had pinworm infections, recounted the experience of Ockert (1972b), who collected eggs of *E. vermicularis,* washed them in water and suspended them in pepsin and hydrochloric acid, then ingested them and 25 days later developed patent infections with both *E. vermicularis* and *D. fragilis*. *D. fragilis* and *E. vermicularis* infections occur together in the same individual more frequently than would be expected by random occurrence (Yang and Scholten, 1977). If it is eventually demonstrated that *D. fragilis,* like *Histomonas meleagridis,* is transferred from host to host in the egg of a common nematode, perhaps attention should then turn to the taxonomically related *Trichomonas hominis,* which also has no cyst stage in its life cycle (p. 48).

The prevalence of *D. fragilis* in most localities is poorly documented, as the methods commonly used in intestinal parasite surveys are unsuitable for detecting trophozoites. Reported rates vary from less than 1% to more than 5% (Yang and Scholten, 1977), except in mental hospitals, in which it is much higher; for example, 36% in Holland, 51% in Sweden, and 42% in Panama. In a semicommunal group in California, *D. fragilis* was the most frequently diagnosed intestinal protozoan—76% of 138 children and 56% of 82 adults (Millet et al., 1983).

A parasite morphologically similar to *D. fragilis* has been reported in monkeys in the Philippines (Hegner and Chu, 1930).

Pathogenesis, Pathology, and Symptomatology. *Dientamoeba fragilis* does not invade tissues but there is some evidence that its presence in the intestine occasionally produces irritation of the mucosa, with secretion of excess mucus and hypermotility of the bowel. The manifestations usually attributed to this infection consist mostly of a mucous diarrhea, with vague to moderate regional abdominal pain and tenderness. In a series of cases in children and adults, Knoll and Howell (1945) reported nausea, vomiting, low-grade fever, diarrhea, and abdominal discomfort, with symptomatic cure following appropriate antiamebic therapy. Kean and Malloch (1966) reported on 100 diagnosed cases among 10,000 patients in a New York hospital over a 6-year period. Several had physical signs consistent with intestinal disturbance. Yang and Scholten (1977) reported that more than half of 255 patients in whom *D. fragilis* was the only parasite found and 187 recorded in 19 published reports had diarrhea, abdominal pain, or a combination of these. The next most common symptom, which was reported in 11% of the patients, was anal pruritus.

Diagnosis. Unless the laboratory examiner is aware of the possibility that *D. fragilis* may be present in fresh fecal films that are prepared either from unformed stools with mucus or from formed specimens, this ameba is apt to be overlooked. Saline catharsis will provide more suitable material for examination than the average formed stool. Even when found, *D. fragilis* many be misdiagnosed as small *Entamoeba histolytica, E. hartmanni,* or *Endolimax nana*. Systematic search of a well-fixed and stained permanent fecal film mounted with coverglass and examined with oil immersion objective will often reveal the organisms when they were not observed in fresh preparations. Prompt fixation of the fresh fecal specimen is essential. The trophozoites grow readily in ameba culture media, and they can survive for 2 hours in distilled water (Matsubayasi, 1942; Camp et al., 1974).

Treatment. Standard antiamebic therapy appears to be satisfactory for *D. fragilis*. If it is responsible for diarrhea and abdominal discomfort, such treatment should relieve the symptoms. Drugs that apparently are effective are oxytetracycline (Yoeli, 1955), diphetarsone (Desser and Yang, 1976), diiodohydroxyquin, and metronidazole (Spencer et al., 1979).

Prevention. Definite recommendations for control cannot be made until there is more specific information concerning the method of transmission.

REFERENCES

Bingham, A.K., and Meyer, E.A. 1979. *Giarda* encystation can be induced *in vitro* in acidic solutions. Nature, *277*:301–302.

Brightman, A.H., II, and Slonka, G.F. 1976. A review of five clinical cases of giardiasis in cats. J. Am. Anim. Hosp. Assoc., *12*:492–497.

Brown, M.T. 1972. Trichomoniasis. Practitioner, *209*:639–644.

Burrows, R.B., and Swerdlow, M.A. 1956. *Enterobius vermicularis* as a probable vector of *Dientamoeba fragilis*. Am. J. Trop. Med. Hyg., *5*:258–265.

———, Frost, J.K., and Leeper, C.K. 1954. Pathology of *Dientamoeba fragilis* infections of the appendix. Am.J. Trop. Med. Hyg., *3*:1033–1039.

Camp, R.R., Mattern, C.F.T., and Honigberg, B.M. 1974. Study of *Dientamoeba fragilis* Jepps and Dobell. I. Electronmicroscopic observations of the binucleate stages. J. Protozool., *21*:69–79.

Chaudhuri, R.N. 1943. A note on giardiasis with steatorrhea. Indian Med. Gaz., *78*:284–285.

Collins, J.P., Keller, K.F., and Brown, L. 1978. "Ghost"

forms of *Giardia lamblia* cysts initially misdiagnosed as *Isospora*. Am. J. Trop. Med. Hyg., *27*:334–335.

Craft, J.C. 1982. Giardia and giardiasis in children. Pediat. Infect. Dis., *1*:196–211.

———, and Nelson, J.D. 1982. Diagnosis of giardiasis by counterimmunoelectrophoresis of feces. J. Infect. Dis., *145*:499–504.

Dancescu, P., and Tintareanu, J. 1963. Investigation concerning the spread of giardiasis in a children's community. (English summary.) Microbiol. Parazitol. Epidemiol. Bucharest, *9*:343–350.

Danciger, M., and Lopez, M. 1975. Numbers of *Giardia* in the feces of infected children. Am. J. Trop. Med. Hyg., *24*:237–242.

De Carneri, I. 1957. Frequenza delle infezioni da *Entamoeba gingivalis* e *Trichomonas tenax* in un campione della popolazione attiva di Milano. Arch. Ital. Sci. Med. Trop. e Parasitol., *8*:420–424.

Desser, S.S., and Yang, Y.J. 1976. *Dientamoeba fragilis in* idiopathic gastrointestinal disorders. Can. Med. Assoc. J., *114*:290–293.

Dobell, C. 1940. Researches on the intestinal protozoa of monkeys and man. X. The life history of *Dientamoeba fragilis*. Parasitology, *32:* 417–461.

Durel, P., Roiron, V., Siboulet, A., and Borel, L.I. 1959. Clinical trial of a trichomonacidal derivative of imidazole. C.R. Soc. Franc. Gynecol., *29*:36.

Dykes, A.C., Juranek, D.D., Lorenz, R.A., Sinclair, S., Jakubowski, W., and Davies, R. 1980. Municipal waterborne giardiasis. An epidemiologic investigation. Ann. Intern. Med., *93*:165–170.

Erlandsen, S.L. 1974. Scanning electron microscopy of intestinal giardiasis: Lesions of the microvillous border of villus epithelial cells produced by trophozoites of *Giardia*. *Scanning Electron Microscopy 1974* (Part III). Proceedings of the Workshop on Advances in Biomedical Applications of the SEM. Chicago, IIT Research Institute.

Faust, E.C., and Wassell, C.M. 1921. Preliminary survey of the intestinal parasites of man in the central Yangtse Valley. China Med. J. , *35*:532–561.

Feo, L.G. 1944. The incidence and significance of *Trichomonas vaginalis* infestation in the male. Am. J. Trop. Med., *24*:195–198.

Frost, J.K., and Honigberg, B.W. 1962. Comparative pathogenicity of *Trichomonas vaginalis* and *Trichomonas gallinae* to mice. II. Histopathology of subcutaneous lesions. J. Parasitol., *48*:898–918.

Hartman, H.R., and Kyser, F.A. 1941. Giardiasis and its treatment. J.A.M.A., *116*:2835–2839.

Hegner, R., and Chu, H.J. 1930. A comparative study of the intestinal protozoa of wild monkeys and man. Am. J. Hyg., *12*:62–108.

Hewlett, E.L., Andrews, J.S., Ruffier, J., and Shaeffer, F.W. 1982. Experimental infection in mongrel dogs with *Giardia lamblia* cysts and cultured trophozoites. J. Infect. Dis., *145*:89–93.

Hoffman, B. 1966. An evaluation of the use of the indirect hemagglutination method in the serodiagnostic of trichomonadosis. (English summary.) Wiadomosci Parazytol., *12*:392–397.

Honigberg, B.M. 1974. Study of *Dientamoeba fragilis* Jepps and Dobell. II. Taxonomic position and revision of the genus. J. Protozool., *21*:79–82.

Jírovec, O. 1965. Neuere Forschungen über *Trichomonas vaginalis* und vaginale Trichomonosis. Angew. Parasitol., *6*:202–210.

Karapetyan, A. 1962. *In vitro* cultivation of *Giardia duodenalis*. J. Parasitol., *48*:337–340.

Kean, B.H., and Malloch, C.L. 1966. The neglected ameba: *Dientamoeba fragilis*. A report of 100 "pure" infections. Am. J. Dig. Dis., *11*:735–745.

Kettis, A.A., and Magnius, L. 1973. *Giardia lamblia* infection in a group of students after a visit to Leningrad in March 1970. Scand. J. Infect. Dis., *5*:289–292.

Keystone, J.S., Krajden, S., and Warren, M.R. 1978. Person-to-person transmission of *Giardia lamblia* in day-care nurseries. Can. Med. Assoc. J., *119*:242–244.

Kirner, J.C. 1978. A waterborne outbreak of giardiasis in Camas, Washington. J. Am. Waterworks Assoc., January 1978: 35–40.

Knaus, W.A. 1974. Reassurance about Russian giardiasis. N. Engl. J. Med., *291*:156.

Knoll, E.W., and Howell, K.M. 1945. Studies on *Dientamoeba fragilis:* Its incidence and possible pathology. Am. J. Clin. Med., *15*:178–183.

Kott, W., and Adler, S. 1961. A serological study of *Trichomonas* sp. parasitic in man. Trans. R. Soc. Trop. Med. Hyg., *55*:333–344.

Lopez, C.E., Juranek, D.D., Sinclair, S.P., and Schultz, M.A. 1978. Giardiasis in American travelers to Madeira Island, Portugal. Am. J. Trop. Med. Hyg., *27*:1128–1132.

Matsubayasi, H. 1942. Studies on *Dientamoeba fragilis* with special reference to viability outside the host. Jpn. J. Med. Sci. Biol., *2*:67–83.

Meingassner, J.G., and Thurner, J. 1979. Strain of *Trichomonas vaginalis* resistant to metronidazole and other 5-nitroimidazoles. Antimicrob. Agents Chemother., *15*:254–257.

Meyer, E.A. 1976. *Giardia lamblia:* Isolation and axenic cultivation. Exp. Parasitol., *39*:101–105.

———, and Jarroll, E.L. 1980. Giardiasis. Am. J. Epidemiol., *111*:1–12.

Millet, V., Spencer, M., Chapin, M., Garcia, R., Yatabe, J., and Stewart, M. 1983. Intestinal protozoa infection in a semicommunal group. Am. J. Trop. Med. Hyg., *32*:54–60.

Moore, G.T., Gross, W.W., McGuire, D., Mollohan, C.S., Gleason, N.N., Healy, G.R., and Newton, L.H. 1969. Epidemic giardiasis at a ski resort. N. Engl. J. Med., *281*:402–407.

Ockert, G. 1972a. Zur Epidemiologie von *Dientamoeba fragilis* Jepps et Dobell 1918. 1. Mitteilung. Die Verbreitung der Art in Kinderkollektiven. J. Hyg., Epidemiol., Microbiol., Immunol., *16*:213–221.

———. 1972b. Zur Epidemiologie von *Dientamoeba fragilis* Jepps et Dobell, 1918. 2. Mitteilung: Versuch der Ubertragung der Art mit *Enterobius*-Eiern. J. Hyg., Epidemiol., Microbiol., Immunol., *16*:222–225.

Ormiston, G., Taylor, J., and Wilson, G.S. 1942. Enteritis in a nursery home associated with *Giardia lamblia*. Br. Med. J., *2*:151–154.

Owen, R.L., Allen, C.L., and Stevens, D.P. 1981. Phagocytosis of *Giardia muris* by macrophages in Peyer's patch epithelium in mice. Infect. Immun., *33*:591–601.

Pereyra, A.J., and Lansing, J.D. 1964. Urogenital trichomoniasis: Treatment with metronidazole in 2002 incarcerated women. Obstet. Gynecol., *24*:499–508.

Petersen, H. 1972. Giardiasis (lambliasis). Scand. J. Gastroenterol., *7* (suppl. 14).

Peterson, J.M. 1957. Intestinal changes in *Giardia lamblia* infestation. Am. J. Roentgenol., *77*:670–677.

Phillips, S.C., Mildvan, D., William, D.C., Gelb, A.M., and White, M.C. 1981. Sexual transmission of enteric protozoa and helminths in a venereal-disease-clinic population. N. Engl. J. Med., *305*:603–606.

Schmerin, M.J., Jones, T.C., and Klein, H. 1978. Giardiasis: Association with homosexuality. Ann. Intern. Med., *88*:801–803.

Schultz, M.G. 1975. Giardiasis. J.A.M.A., *233*:1383–1384.

Shaw, P.K., Brodsky, R.E., Lyman, D.D., Wood, B.T., Hibler, C.P., Healy, G.R., MacLeod, K.I.E., Stahl, W., and Shultz, M.G. 1977. A communitywide outbreak of giardiasis with evidence of transmission by a municipal water supply. Ann. Intern. Med., *87*:426-432.

Spencer, M.J., Garcia, L.S., and Chapin, M.R. 1979. *Dientamoeba fragilis*. An intestinal pathogen in children? Am. J. Dis. Child., *133*:390–393.

Stabler, R.M., and Feo, L.G. 1942. Inoculation of the oral trichomonad (*T. tenax*) into the human vagina. Am. J. Trop. Med., *22*:639–642.

Suriyanon, V., Nelson, K.E., and Ayudhya, V.C.N. 1975. *Trichomonas vaginalis* in a perinephric abscess. A case report. Am. J. Trop. Med. Hyg., *24*:776–780.

Tandon, B.N., Tandon, R.K., Satipathy, B.K., and Shriniwas. 1977. Mechanism of malabsorption in giardiasis: A study of bacterial flora and bile salt deconjugation in upper jejunum. Gut, *18*:176–181.

Thompson, R.G., Karandikar, D.S., and Leek, J. 1974. Giardiasis, an unusual cause of epidemic diarrhoea. Lancet, *1*:615–616.

Tomkins, A.M., Wright, S.G., Drasar, B.S., and James, W.P.T. 1978. Bacterial colonization of jejunal mucosa in giardiasis. Trans. R. Soc. Trop. Med. Hyg., *72*:33–36.

Tyzzer, E.E. 1934. Studies on histomoniasis or "blackhead" infection, in the chicken and the turkey. Proc. Am. Acad. Arts Sci., *69*:190–264.

Veghelyi, P. 1940. Giardiasis. Am. J. Dis. Child., *59*:793–804.

Walton, B.C., and Bacharach, T. 1963. Occurrence of trichomonads in the respiratory tract. Report of three cases. J. Parasitol., *49*:35–38.

Weiss, H.B., Winegar, M.S., Levy, B.S., and Washburn, J.W. 1977. Giardiasis in Minnesota, 1971–1975. Minnesota Med., *60*:815–820.

Welch, P.B. 1944. Giardiasis with unusual findings. Gastroenterology, *3*:98–102.

Wenrich, D.H. 1932. The relation of the protozoan flagellate, *Retortomonas gryllotalpae* (Grassi, 1879) Stiles, 1902 to the species of the genus *Embadomonas* Mackinnon, 1911. Trans. Am. Microsc. Soc., *51*:225–238.

———. 1944. Studies on *Dientamoeba fragilis* (Protozoa). IV. Further observations, with an outline of present-day knowledge of this species. J. Parasitol., *30*:322–338.

Wenyon, C.M. 1936. *Protozoology*, vol. 1. London, Bailliere, Tindall and Cox.

Wenyon, C.M., and O'Connor, F.W. 1917. *Human Intestinal Protozoa in the Near East*. London, John Bole, Sons & Danielsson, Ltd.

Wolfe, M.S. 1978. Current concepts in parasitology: Giardiasis. N. Engl. J. Med., *298*:319–321.

World Health Organization. 1981. Intestinal protozoan and helminthic infections. Geneva, W.H.O. Tech. Rep. Ser. no. 666.

Wright, R.A., Spencer, H.C., Brodsy, R.E., and Vernon, T.M. 1977. Giardiasis in Colorado: An epidemiologic study. Am. J. Epidemiol., *105*:330–336.

Yang, J., and Scholten, T. 1977. *Dientamoeba fragilis:* A review with notes on its epidemiology, pathogenicity, mode of transmission, and diagnosis. Am. J. Trop. Med. Hyg., *26*:16–22.

Yoeli, M. 1955. A report on intestinal disorders accompanied by large numbers of *Dientamoeba fragilis*. J. Trop. Med. Hyg., *58*:38–41.

Young, R.V. 1949. Trichomoniasis in the male. Rocky Mt. Med. J., *46*:928–931.

Zemlyn, S., Wilson, W.W., and Hillweg, P.A. 1981. A caution on iodine water purification. West. J. Med., *135*:166–167.

Chapter 10

The Blood and Tissue Flagellates

FAMILY TRYPANOSOMATIDAE
Doflein, 1901

The flagellate protozoa that live in the blood and tissues of the human host all belong to the order Kinetoplastida, family Trypanosomatidae.

This family includes species that have a single flagellum, a nucleus, and a kinetoplast from which the flagellum arises. In stained preparations, the *kinetoplast* (sometimes incorrectly called kinetonucleus or micronucleus) consists of a deeply stained *parabasal body* and an adjacent dot-like *blepharoplast*. The blepharoplast and parabasal body are connected by one or possibly more delicate fibrils. The inner portion of the flagellum extending from the blepharoplast to the surface of the body is the *axoneme*, or axial filament.

Species of this family may exist in two or more of four forms or states, which are named after genera exemplifying these forms: leishmania, leptomonas, crithidia, and trypanosoma. During the past decade, many authorities have preferred to utilize synonyms for the parasite stages to avoid confusion between the above designations and the genera with the same names. These terms are ''amastigote,'' ''promastigote,'' epimastigote,'' and ''trypomastigote,'' respectively.

The family Trypanosomatidae is divided into six genera:

1. Genus *Leptomonas*. These species have an *amastigote (leishmania)* stage and a *promastigote (leptomonad)* stage in the life cycle and occur only in invertebrates. Transmission is through rounded-up, small amastigote forms that are resistant to desiccation and are passed in the dejecta. Example: *L. ctenocephali*, in intestine of the dog flea, *Ctenocephalides canis*.
2. Genus *Crithidia*. These species have an amastigote, a promastigote, and an epimastigote *(crithidia)* stage in the life cycle, occur only in invertebrate hosts, and are transmitted by means of rounded-up, small, resistant amastigote forms, as in *Leptomonas*. Example: *C, gerridis*, in the intestine of water bugs.
3. Genus *Herpetomonas*. These species have an amastigote, a promastigote, an epimastigote, and a *trypomastigote (trypanosoma)* stage in the life cycle, occur only in invertebrate hosts, and are transmitted in the same manner as *Leptomonas* and *Crithidia*. Example: *H. muscarum*, in the intestine of *Musca* and other filth flies.
4. Genus *Leishmania*. Like the genus *Leptomonas*, these species have only an amastigote and a promastigote stage in the life cycle, but *Leishmania* spp. have both a vertebrate and an invertebrate host. They are transmitted by certain species of sand flies *(Phlebotomus* and *Lutzomyia)*.
5. Genus *Phytomonas*. These species have an amastigote stage and a promastigote stage in the life cycle but have both an invertebrate and a plant host. Example: *P. elmassiani*, in latex of milkweeds, transmitted by heteropterous bugs.
6. Genus *Trypanosoma*. These species may have an amastigote, a promastigote, an epimastigote, and a trypomastigote stage in the life cycle (e.g., *Trypanosoma cruzi*) and typically have both a vertebrate and invertebrate host. So far as is known, amastigote and promastigote stages of development do not occur in the human trypanosomes *T. gambiense* and *T. rhodesiense* and their relatives. Transmission of most African species is by tsetse flies. *T. cruzi* utilizes sucking bugs of the

family group Triatomidae as biologic transmitters.

The relationship of genera and the morphology of amastigote, promastigote, epimastigote, and trypomastigote forms of species parasitizing man is shown in Table 10–1.

During the amastigote stage of their life cycle in the mammalian host, species of *Leishmania* and *Trypanosoma cruzi* are more or less ovoidal in shape as an adaptation to intracellular existence in the vertebrate host. The free flagellum is retracted into the axoneme. Species of *Leptomonas*, and related organisms in their promastigote stage, are elongate

Table 10–1. Stages of Leishmania and Trypanosoma in Man and Insect Host*

Stage of the Parasite	Amastigote *(Leishmania)*	Promastigote *(Leptomonas)*	Epimastigote *(Crithidia)*	Trypomastigote *(Trypanosoma)*
Name of Parasite	*A* 	*B* 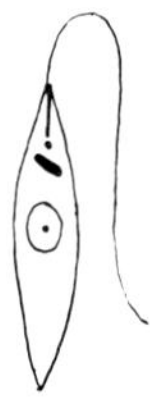	*C* 	*D* 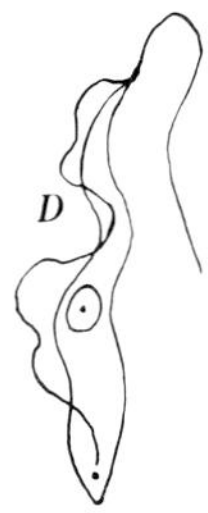
Leishmania tropica	*Intracellular* in macrophages of skin and subcutaneous tissue	In midgut, later in proboscis of sand fly *(Phlebotomus); transfer stage* to man	Lacking	Lacking
Leishmania braziliensis	*Intracellular* in macrophages of skin; may be carried to mucocutaneous junctions	In midgut and later in proboscis of sand fly *(Lutzomyia); transfer stage* to man	Lacking	Lacking
Leishmania donovani	*Intracellular* in macrophages; predominantly in liver, spleen, bone marrow and lymph nodes	In midgut and later in proboscis of sand fly *(Phlebotumus); transfer stage* to man	Lacking	Lacking
Trypanosoma rhodesiense	Lacking	Lacking	In salivary glands of tsetse fly *(Glossina)*	In proboscis of tsetse fly; *transfer stage* to man; first in blood stream, then in lymph nodes
Trypanosoma gambiense	Lacking	Lacking	In salivary glands of tsetse fly *(Glossina)*	In proboscis of tsetse fly; *transfer stage* to man; first in blood stream, then in lymph nodes, later in central nervous system
Trypanosoma cruzi	*Intracellular* in macrophages, especially in skin, lymph nodes, liver and spleen; also in myocardium, brain and endocrine glands	Transitional stage only	In midgut of triatomid bug	In feces of triatomid bug; *transfer stage* to man; present in blood stream only during acute attacks
Trypanosoma rangeli	No intracellular stage in macrophages demonstrated.	Transitional stage only	In midgut of triatomid bug	In hemolymph, salivary glands and proboscis of triatomid bug for transfer to definitive host

*From Faust, E.C., Beaver, P.C., and Jung, R.C. 1975. *Animal Agents and Vectors of Human Disease*, 4th ed. Philadelphia, Lea & Febiger.

spindle-shaped, have pointed ends and a free flagellum that arises from the kinetoplast at the anterior end of the organism. Species of *Crithidia,* and trypanosomes in their epimastigote stage, are more or less morphologically like *Leptomonas* except that the free flagellum is continued from the anterior end backwards along the margin of an undulating membrane and ends internally at the kinetoplast, which is situated anterior to the nucleus. Species of *Trypanosoma,* and the trypomastigote stage of *Herpetomonas,* are morphologically like *Crithidia* except that the flagellum on the margin of the undulating membrane continues to a post-nuclear position, where it ends internally at the kinetoplast. All stages with a free flagellum are adapted for living in a liquid medium of relatively low viscosity. All members of the family Trypanosomatidae multiply by longitudinal binary fission.

Genus Leishmania

Ross, 1903.
(Named after Sir William Leishman)

This genus was established by Ross in 1903 to include *Leishmania donovani,* the flagellate causing kala-azar. Species belonging to this genus have an amastigote and a promastigote stage in their life cycle and both a mammalian and an insect (i.e., sand fly) host. They are morphologically similar to those belonging to the genus *Leptomonas.*

Species of *Leishmania* have a nucleus, a kinetoplast, and an axoneme; in the promastigote stage, they have a single flagellum that arises from the axoneme but they do not have an undulating membrane. The morphology in man or other mammalian host is that of a typical *Leishmania,* consisting of an ovoidal body containing a nucleus and kinetoplast; in the invertebrate host, the body becomes spindle-shaped and has a single flagellum arising from the axoneme at the anterior end. Reproduction occurs by binary fission; the blepharoplast and parabasal body (kinetoplast) divide first, then the nucleus and cytoplasm, resulting in two daughter individuals.

The *Leishmania* that infect man have long been grouped into three species: *L. tropica,* which causes Old World cutaneous leishmaniasis or Oriental sore; *L. braziliensis,* which causes New World mucocutaneous leishmaniasis; and *L. donovani,* which causes visceral leishmaniasis or kala-azar. Recently it has become apparent that this simple classification is inadequate. There are at least four clinical types of cutaneous leishmaniasis in the New World, each probably caused by a distinct species or variety of *Leishmania,* and two distinct types of cutaneous leishmaniasis of the Old World that differ clinically, epidemiologically, and immunologically. Visceral leishmaniasis, too, is known to vary in its clinical picture in different geographical areas, and the respective agents can be distinguished by a number of laboratory procedures.

Morphologically, the species are identical by light microscopy. Although morphologic differentiation may be possible by electron microscopy (Gardener et al., 1977), this mode of examination is not ordinarily available for the clinical laboratory. Therefore, in addition to the relatively distinct clinical picture produced in man by each, serologic reactions are employed to differentiate them. Nevertheless, there is considerable overlapping in the various clinical aspects of the types of leishmaniasis and immunologic tests may at times be equivocal. Adler et al. (1966) studied Sudanese strains of *Leishmania* from human cases of kala-azar and cutaneous leishmaniasis from naturally infected sand flies *(P. orientalis)* and from five species of naturally infected rodents, serologically by growth on medium containing immune serum. With this method, all strains were indistinguishable from each other and were considered to belong to the same species. Three cases of cutaneous leishmaniasis studied were considered to have been either abortive kala-azar or primary leishmaniomas. The fact that three types of human leishmaniasis occur in southern Iran is of great significance, particularly since the hot, dry climate of the region is not considered favorable for *L. braziliensis* (Zial et al., 1968). Mucocutaneous leishmaniasis has been reported from northeast China (Tientsin) (Wang and Wu, 1959). Several factors may play a role in the pathogenicity of these organisms and their transformation from one form to another.

Another species, *L. enriettii* Muniz and Medina, 1948, has been found as a natural infection of the Brazilian guinea pig. Its host-tissue relationship is, in general, comparable to that of *L. braziliensis* (see p. 64), but its origin is unknown. *L. hertigi* causes cutaneous infection in a Panamanian porcupine, *Coendu rothschildi.* Other species parasitize lizards but none have been reported from birds.

The life cycle of species of *Leishmania* is illustrated in Figure 10–1.

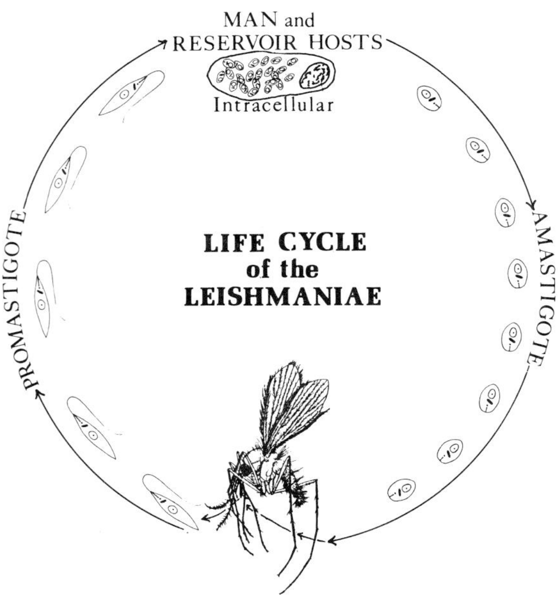

Fig. 10–1. *Leishmania.* Diagrammatic representation of the life cycle of species that parasitize man. *Above,* intracellular amastigotes in macrophage of infected mammalian host. *Below,* sand fly *(Phlebotomus* or *Lutzomyia),* intermediate host and vector. *Right,* amastigotes taken up with a blood meal from infected mammalian host. *Left,* promastigotes in salivary secretions of sand fly, to be introduced into mammalian skin at time of subsequent blood meal. (Adapted by R.C. Jung from original of E.C. Faust.)

OLD WORLD CUTANEOUS LEISHMANIASIS

Leishmania tropica Complex

Leishmania tropica minor Kozevnikov, 1963

(Producing dry or urban cutaneous leishmaniasis, Oriental sore, Aleppo button, Jericho boil, Delhi boil, bouton de Biskra.)

Synonyms. *Helicosoma tropica* Wright, 1903; *Herpetomonas tropica* Patton, 1912; *Herpetomonas furunculosa* (Wright) Patton, 1922.

History. It is probable that *Leishmania tropica,* the cause of Oriental sore, was first seen by Cunningham (1885) and by Firth (1891), who gave it the name *Sporozoa furunculosa;* however, Wright (1903) of Boston is usually credited with the first complete description of the organism, although it was accurately described in 1898 by Borovsky, a Russian military surgeon (Hoare, 1938). Wright observed the parasite in an Armenian patient under treatment at the Massachusetts General Hospital and gave it the name *Helicosoma tropica.* Lühe (1906) removed the species to the genus *Leishmania.* The role of the sand fly *Phlebotomus papatasi* in the transmission of the infection was suggested by Wenyon (1911) and demonstrated by the Sergent brothers with Parrot, Donatien, and Begue (1921), by Adler and Theodor (1925–1929), and by Adler and Ber (1941). Elucidation of the nature of Oriental sore was provided by Soviet workers who distinguished wet or rural cutaneous leishmaniasis caused by *L. tropica major* and dry or urban cutaneous leishmaniasis caused by *L. tropica minor* (Kozevnikov, 1963). Some authorities believe that these two agents may merit the rank of species rather than subspecies, because *L. tropica* is the agent of urban Old World cutaneous leishmaniasis and *L. major* is the agent of rural cutaneous leishmaniasis. In addition, the name *L. aethiopica* has been suggested for the agent causing cutaneous and diffuse cutaneous leishmaniasis of Ethiopia (Bray, 1974).

Geographical Distribution. Urban cutaneous leishmaniasis caused by *L. tropica* is endemic in the countries of Portugal, Spain, southern France, Italy, Yugoslavia, Sardinia, Sicily, Greece, Bulgaria, Crete, Turkey, Saudi Arabia, southern Yemen; the cities of Aleppo, Jericho, Damascus, Tripoli, Mosul, Baghdad, Tehran, Meshed, Erevan, Karovabad, Tabriz, Shiraz, Kabul, Delhi, Cambay, Ashkabad, Merv, Samarkand, Bokhara, Balkh in Asia; and the oasis towns of Morocco, Algeria, and Tunisia (Fig. 10–2).

Morphology. The morphology of *L. tropica minor* is identical with that of *L. donovani* (p. 70) and the agents of New World human cutaneous and mucocutaneous leishmaniasis. It is identical in morphology to *L. tropica major,* the agent of rural cutaneous leishmaniasis of the Old World, except that, in mouse tissues, the amastigotes of *L. tropica major* are consistently larger than those of *L. tropica minor* (Fig. 10–3).

Habitat. In man *L. tropica minor* is a parasite of the skin. It is found in the endothelial cells of the capillaries of the infected area, in nearby lymph nodes, within large mononuclear cells, in neutrophilic leukocytes, and free in the serum exuding from the ulcerated areas. This species is not found in the peripheral blood unless it is obtained from the immediate vicinity of the ulcer, and it rarely metastasizes to other sites or invades the endothelium of the viscera.

Dogs of Iraq, Iran, Turkestan, Bombay, Northern Africa, and elsewhere suffer from Oriental sore, produced by a leishmania apparently identical with *L. tropica.* Adler and Theodor (1930) produced an Oriental sore in man by the subcutaneous injection of cultures of *L. tropica* from a naturally infected dog in Baghdad and concluded that this flagellate is a parasite of both man and the dog. In South Central Asia, gerbils are natural reservoirs. Reported *L. tropica* infection in cats, bears, and bullocks requires confirmation.

Cultivation. Nicolle (1908) first cultivated *L. tropica,* using the N.N.N. medium; since his observation, this organism has been cultivated by numerous observers upon several culture media (see Technical Appendix). When introduced into a suitable culture medium, the amastigote transforms into the promastigote, just as it does in the midgut of the appropriate sand-fly host, and proceeds to colonize (Fig. 10–4). The thermal death point of *L.*

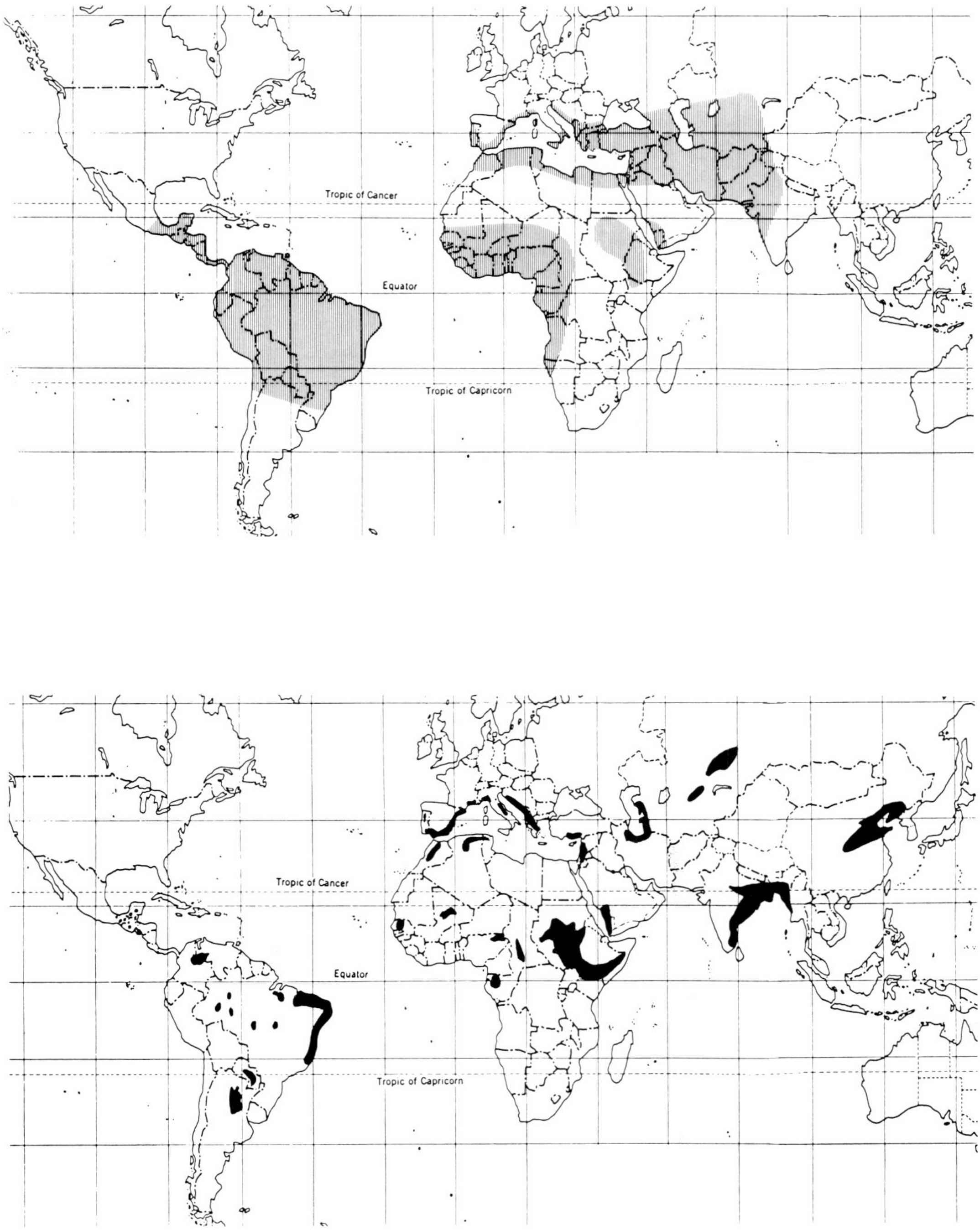

Fig. 10–2. *Leishmaniasis*. Approximate world distribution of cutaneous and mucocutaneous infection above, and visceral infection, below. The endemic area in China probably is less extensive than shown. (Adapted by R.C. Jung from original of E.C. Faust.)

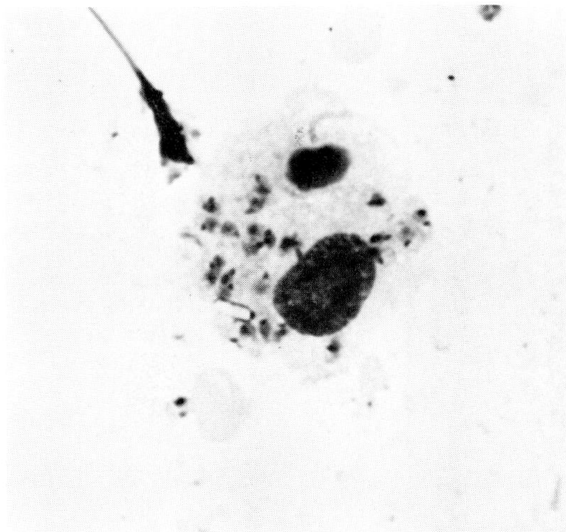

Fig. 10–3. *Leishmania tropica.* Amastigotes in macrophage and free in smear from Oriental sore on child from Saudi Arabia. Giemsa stain. (× 1000.) (By P. Beaver; preparation by Dr. R.G. Yaeger.)

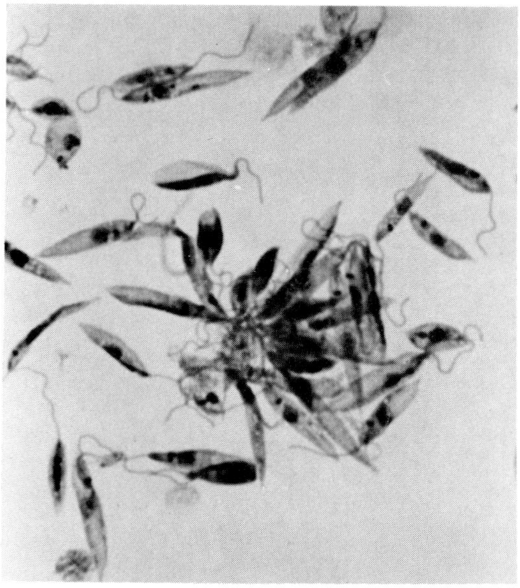

Fig. 10–4. *Leishmania tropica.* Promastigotes (leptomonas) in smear from culture of forms inoculated directly from scrapings of skin lesion. Giemsa stain. (× 1000.) (By P. Beaver; preparation by Dr. R.G. Yaeger.)

tropica, like that of other leishmaniae, is 40° C within 15 to 30 minutes and almost instantaneously at 45° C. Infectivity for man may be retained in cultures for very long periods of time. Adler and Zuckerman (1948) described a strain of this leishmania infective for man after a period of 22 years of cultivation upon Locke-serum and Locke-blood agar media, including 800 transfers.

Life Cycle. The life cycle of *L. tropica* is similar to that of *L. donovani* in man and in sand flies (p. 70), except that in man the amastigote forms live only in the reticuloendothelial cells and lymphoid tissue of the skin and do not invade the viscera, as does *L. donovani*. The method of reproduction in both the amastigote and promastigote stages is by binary longitudinal division.

Epidemiology. Urban cutaneous leishmaniasis is an anthroponosis. It is common in man and endemic in cities, being practically universal in some areas. It is commonly transmitted from man to man by the vector fly but *contact infection* is possible. The infection is inoculable and autoinoculable. A wild reservoir host is lacking, but domesticated dogs are usually involved. Man—not dog—is the probable reservoir host however. The urban anthroponotic cycle probably evolved from a rural zoonotic cycle (see *L. tropica major,* p. 62), with subsequent adaptation of the parasite to the human host resulting in decreased pathogenicity.

Sand flies are the natural vectors of the disease. The most common transmitters are *Phlebotomus papatasi* and *P. sergenti*. The former is involved in western India, West Pakistan, Afghanistan, Transcaucasus, Turkmenistan, the countries around the Mediterranean, Ethiopia, and West Africa. *P. sergenti* is involved in Iran, Iraq, Crete, Portugal, Algeria, Israel, Lebanon, Syria, Turkey, Greece, and Yugoslavia. In Italy and Sicily, the vector is *P. perfiliewi (macedonicus);* in Central Asia, *P. caucasicus;* and in Spain, *P. perniciosus*. Distinct biological races of *L. tropica* differ in their infectability for different species of *Phlebotomus*. The stable fly or biting dog fly *(Stomoxys calcitrans)* may transmit the organisms mechanically from an open ulcer through unbroken skin.

In endemic areas, mothers sometimes inoculate their children with serum from a sore, thereby producing an immunizing infection. Multiple sores may be produced by scratching the serum that is exuding from a primary sore into normal skin. The amastigotes do not penetrate unbroken skin, however. One attack probably confers life-long immunity, at least to the homologous strain.

Pathogenesis and Pathology. When the promastigote stage of *Leishmania tropica* is introduced into the skin of a person while an infected sand fly is taking a blood meal, the parasites are engulfed by local white cells. In neutrophilic leukocytes, phagocytosis is usually successful, but in macrophages, the parasites round up to the amastigote form, multiply, and, on escape from rupturing macrophages, are taken up by others in the immediate vicinity. In the early stages, the lesion is characterized by the proliferation of macrophages that contain numerous amastigotes. There is a variable infiltration of lymphocytes and plasma cells. The overlying epithelium shows acanthosis and hyperkeratosis, which are later followed by necrosis and ulceration. As the lesion heals, the macrophages and organisms gradually diminish in number and the lymphocytes and plasma cells increase. Ulceration follows, and healing occurs in the presence of a nonspecific granulation tissue (Hutt et al., 1973). Grossly, a raised, crateriform lesion with a depressed ulcerated center that is covered with granulating tissue and has an indurated periphery is produced. Rarely, metastasis to secondary foci takes place but visceral involvement is unknown. Secondary infections with bacteria are common. There is no anemia or marked increase in the mononuclear leukocytes.

Clinical Picture. The incubation period of Oriental sore has been reported as varying from a few days to as long as 6 months. The briefer incubation periods probably have been in cases of rural cutaneous leishmaniasis due to *L. tropica major.* The symptoms are variable, and the sores, which usually occur on the exposed surface of the body, may be single or multiple (Fig. 10–5). The primary lesion develops at the site of the bite of the vector sand fly. This anatomic location has a relationship to customs of dress.

Papules and Nodules. The smallest lesions are the papules, which are nonulcerating, superficial, purplish, and approximately 2 to 10 mm in diameter (Fig. 10–6). They last for several months, after which they heal without scarring. Nodules are larger and deeper than papules, tend not to ulcerate even though they are present for many months, and are not painful. These lesions, which are characteristic of the dry, or urban, Oriental sore, also occur in other endemic areas. Some lesions that begin as papules or nodules ulcerate and leave scars upon healing, while others may extend broadly without ulcerating, producing ugly, large, dry, crusty lesions.

Ulcers. A macule appears, then develops a slightly raised center that ulcerates. The exudate is usually scanty. Granulation appears at the base of the growing ulcer, and the margin becomes raised and hard. Typically, a shallow ulcer is produced with a granulated base and bluish-red margins, unless secondary bacterial infection alters its appearance. Such secondarily infected lesions are frequently painful. If not complicated by significant secondary infection or trauma, lesions usually heal

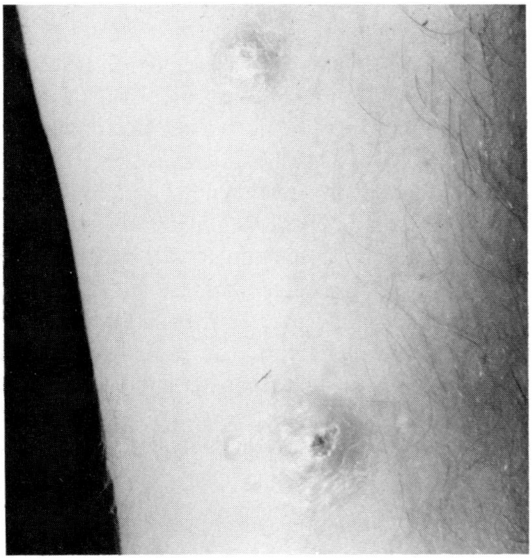

Fig. 10–5. Dermal leishmaniasis on the forearm of man in Mali, West Africa. Note typical satellite lesions around the primary lesion below. (Courtesy of Dr. P.E.C. Manson-Bahr.)

Fig. 10–6. *Leishmania tropica*. Early lesion, on face. North Africa. (Adapted from Sergent, E., et al. 1925. Le clou de Mila. Arch. Inst. Pasteur Alger., *3*.)

in 9 months, leaving scars. A case of mutilating mucocutaneous leishmaniasis was reported from southern Iran (Zial et al., 1968).

Relapsing Lesions (leishmaniasis recidiva, lupoid leishmaniasis). In cases of relapses or recurrences, there usually are low papules resembling lupus vulgaris. They are yellowish in color, covered by whitish scales, and are located at the margins of scars of healed cutaneous leishmaniasis. They may represent a new or persistent residual lesion in the scar and tend to persist for years. Amastigotes are sparse in these lesions, and the histological picture resembles that of tuberculosis.

Diagnosis. The diagnosis of infection with *L. tropica minor* is usually made by the microscopic examination of material obtained by puncture of the indurated edge of the sore and stained with Giemsa's or Wright's stain. (see p. 752). The parasites are found in cells of the granulating tissue and in endothelial cells and large mononuclear leukocytes, but not in peripheral blood. Smears made by scraping the floor of the ulcer are often negative due to bacteria with which *L. tropica minor* cannot live. If smears are negative, biopsy of lymphoid material from under the ulcer's edge at times provides specific proof of infection and in all cases, cultures should be made upon N.N.N.-type medium to which 200 to 500 units of penicillin per ml of overlay are added to control bacteria; if these, too, are negative, treatment with antimony may be instituted, for if the lesion is due to infection with *L. tropica minor,* it will respond to such treatment. The leishmanin reaction (Montenegro immunologic test) can be invoked by intradermal inoculation of dead and washed promastigotes of any species of mammalian *Leishmania* or *Trypanosoma cruzi*. This delayed hypersensitivity reaction is in itself no index of immunity; it may be elicited in early stages of the disease or years after spontaneous cure.

Treatment. Antimony sodium gluconate (pentavalent)—Pentostam and Solustibosan—is the drug of choice. Ten daily 600-mg intramuscular injections are given. Injection of 600 mg of the drug in 5 ml of water around the edge of the lesion may give additional benefit and may be repeated up to four times on alternate days; injection of 200 mg quinacrine hydrochloride (Atabrine) with 5 ml of water into the base of the ulcer also has been used. Cryosurgery may be used to remove lesions with negligible scarring (Barriouny et al., 1982).

In leishmaniasis recidiva, the same course of treatment is used, but several courses (usually three) are required. Parasites may persist at the site of the lesion after multiple courses of treatment, including treatment with amphotericin B (Neva et al., 1979). In a trial reported from Saudi Arabia, levamisole therapy of chronic Oriental sore (150 mg daily for 2 days per week over a period of 37 weeks) appeared to have accelerated healing (Butler, 1978). Metronidazole and rifampicin have been reported to be effective in the treatment of cutaneous leishmaniasis, but their efficacy has not been established (Neva, 1982).

Prevention. If possible, the lesions should be protected from insects by a gauze bandage or other covering to prevent transmission to other persons. The patient should be warned regarding the possibility of autoinfection.

Residual spraying of inside walls of human dwellings and adjacent buildings with residual insecticides is an effective control measure. Spraying around the windows and doors is especially important. Insect repellents such as dimethylphthalate, when applied to exposed skin, will protect for a few hours against the "bites" of *Phlebotomus*.

To be effective, netting must be fine mesh, with 45 holes to the square inch. This interferes considerably with moving air but may be useful in endemic areas. Inasmuch as *Phlebotomus* is nocturnal and does not fly high above ground level, some interference with transmission of the disease may be achieved by sleeping at night on the roof or second floor of unsprayed houses.

In southern U.S.S.R. and Lebanon, prophylactic inoculation with cultures of *Leishmania tropica* in the promastigote stage has been used. But such vaccinations, as well as mass prophylactic treatment, are much less practicable and useful than the residual spraying described previously.

Wherever dogs, rodents, or other mammals serve as natural reservoirs of *L. tropica minor* infection, the destruction of these animals will materially reduce the incidence of the disease in the human population; thus Pavlovsky (1937–1939) found that the destruction of rodents in their burrows in the endemic foci of the disease in Turkmenistan by the use of chloropicrin was followed by a reduction of the incidence in nearby human habitations from 70 to 0.4%.

Leishmania tropica major
Yakimov and Schockov, 1915
(Producing rural or wet cutaneous leishmaniasis)

Historical Notes. Rural, or wet, cutaneous leishmaniasis was formerly believed to be merely a different clinical variety of Oriental sore. In 1963, Soviet workers described the cultural, immunologic, epidemiologic, and clinical differences between the agent of urban, or dry, Oriental sore, which they called *L. tropica minor*, and that of wet, or rural, disease, which was called *L. tropica major.* Bray et al. (1973) presented reasons for regarding *L. tropica major* as a distinct species, *L. major.*

Geographical Distribution. Old World wet cutaneous leishmaniasis due to *L. tropica major* is believed to occur as a sporadic infection in humans, in rural areas of Israel, Jordan, U.S.S.R., Iran, Senegal, Mali, Sudan, Niger, northern Nigeria, Angola, and Namibia. The distribution is based on the presumption that clinically similar infections occurring as zoonoses in rural areas are due to the same agent, *L. tropica major.* Future investigation may reveal that there are more than one such *Leishmania*.

Morphology. The morphology of *L. tropica major* is identical to that of *L. tropica minor* (see p. 58), except that, in mouse tissues, the amastigotes of *L. tropica major* are consistently larger than those of *L. tropica minor.*

Life Cycle. The life cycle is the same as for *L. tropica minor,* except for the insect and mammalian hosts, which are discussed in the following section.

Epidemiology. In contrast with the disease caused by *L. tropica minor,* rural Oriental sore is a zoonosis with human infections occurring only sporadically. The principal reservoir host in the U.S.S.R. and Iran is the gerbil *Rhombomys opimus,* a ground-dwelling rodent. Secondary reservoirs are *Meriones lybicus erythrourus, M. meridianus, Spermophilopsis leptodactylus, Mus musculus, Allactaga servertzovi, Hemiechinus albulus,* and *Erinaceus auritus*. In Israel and Jordan, the principal reservoir is the sand rat *Psammomys obesus,* and the secondary reservoir is a gerbil, *Meriones* sp. In Africa, the reservoir has been identified as *Arvicanthis niloticus*.

The principal vector in the U.S.S.R. and Iran is *Phlebotomus papatasi*. *P. caucasicus* is important in maintaining the infection in the reservoir hosts but does not have a predilection for man. Other flies transmitting the disease are *P. mongolensis, P. ansarii, P. andrejevi,* and perhaps *P. alexandri*. In the Jordan Valley, the vector is *P. papatasi*. *P. grovei* is the vector in Namibia, and the vector in West Africa is unknown.

The infection is maintained in the mammalian reservoir population by transmission by the sand fly vectors that frequent the rodents' burrows. In *Rhombomys opimus,* infection rates approach 100%. Skin lesions of the reservoir hosts may be insignificant and involve only the nose, ears, or eyelids.

Man becomes infected upon invading the enzootic area, particularly if infected anthropophilic sand flies, such as *P. papatasi,* are in the vicinity.

Human beings recovering from wet Oriental sore have a high degree of immunity that protects against both the wet and dry types of infection.

Pathogenesis, Pathology, and Clinical Picture. These are essentially the same as for the disease caused by *L. tropica minor,* except that the lesion caused by *L. tropica major* develops more rapidly and with a more severely ulcerating lesion. The incubation period is relatively brief, sometimes as short as 6 days. The course is rapid, with spontaneous healing in 6 months. The lesion exudes serous or serosanguineous fluid; hence the name "wet" Oriental sore.

Diagnosis. This is the same as for *L. tropica minor* infection (see p. 62).

Treatment. Rural cutaneous leishmaniasis may not require specific treatment since the condition is relatively brief and self-limiting. In the case of multiple or cosmetically undesirable lesions or metastases, treatment is necessary. This is the same as for urban Oriental sore (see p. 62).

Prognosis. Healing takes place spontaneously, and immunity is protective against reinfection.

Prevention. There is no practical way of preventing rural cutaneous leishmaniasis since it is a zoonosis. Persons inoculated with urban leishmaniasis are not protected from the rural disease. Those persons who usually acquire this infection would find efforts at personal protection such as insect repellents and netting too inconvenient to use.

Leishmania aethiopica
Bray, Ashford and Bray, 1973

(Producing cutaneous and diffuse or disseminated cutaneous leishmaniasis of Ethiopia, anergic cutaneous leishmaniasis)

Synonym. *L. tropica* (Wright, 1903) Lühe, 1906.

Historical Notes. Ethiopian cutaneous leishmaniasis was regarded as a variety of the disease produced by *L. tropica* until Bray, Ashford, and Bray (1973) separated *L. aethiopica* on nosologic, serologic, and behavioral grounds. While the taxonomic status of this agent as a species is not yet broadly accepted, it facilitates discussion of the clinical aspects to assign a name to it.

Geographical Distribution. Disease caused by *L. aethiopicus* occurs in the highlands of Ethiopia, Mount Elgon, Kenya, and possibly southern Yemen.

Morphology. This is indistinguishable from that of *L. tropica minor* (see p. 58).

Life Cycle and Epidemiology. These are similar to those of *L. tropica major.* Ethiopian cutaneous leishmaniasis is probably a zoonosis with the animal reservoirs being rock hyraxes *Procavia habessinica* and *Heterohyrax brucei* and the sand fly vectors *Phlebotomus longipes* and *P. pedifer.*

Pathology and Symptomatology. The relationship of Ethiopian diffuse cutaneous leishmaniasis to leishmaniasis tegumentaria diffusa of the New World (see p. 68) has not been established. Clinically, Price and Fitzherbert (1965) classify the cases into three types: lepromatoid, intermediate, and tuberculoid. The lepromatoid type is identical clinically with disseminated cutaneous leishmaniasis of the New World (Fig. 10–7). The intermediate and tuberculoid forms bear superficial resemblance to the respective variety of leprosy.

The histologic picture of diffuse cutaneous leishmaniasis resembles that of leprosy. The lepromatoid variety is identical with disseminated cutaneous leishmaniasis of the New World (see p. 68). The intermediate cases are characterized histologically by tuberculoid follicles and scanty amastigotes adjacent to histiocytic clusters rich in parasites. In the tuberculoid type, lymphangiitis and elephantiasis sometimes occur (Edington and Gilles, 1969).

Diagnosis. The disease is principally confused with leprosy. Diagnosis must be established by demonstration of the etiological agent in scrapings or biopsy.

Treatment. Disseminated cutaneous leishmaniasis of the lepromatoid type is resistant to treatment with antimonials. The drug of choice is amphotericin B (Fungizone), if it is tolerated. It is administered by *slow* intravenous infusion in 5% dextrose solution in a concentration not exceeding 0.01% (1 mg/10 ml); each dose is given over a period of approximately 6 hours. The dosage must be adjusted to the specific requirements of each patient, since tolerance to amphotericin B varies with each individual. Therapy is preferably begun with a test dose of 1 mg. If this is tolerated, the dose is increased by stepwise increments such as to 2 mg, 5 mg, 10 mg, then by 5-mg increments until a daily dosage of 1.0 mg/kg body weight, or the alternate day dosage of 1.5 mg/kg is attained. Treatment may then be continued at this level for up to 8 weeks. Adverse reactions including fever, nausea, anorexia, vomiting, chest or abdominal pain, headache, diarrhea, local thrombophlebitis, and renal damage are common. Since the drug is highly nephrotoxic, daily or alternate-day monitoring of urea nitrogen and creatinine of the blood and of urine is required. The appearance of signs of renal impairment is an indication for cessation of treatment or decrease in dosage. Giving the drug on alternate days rather than daily tends to decrease anorexia and phlebitis. Administration of small doses of adrenal corticosteroids intravenously just before or during the infusion of amphotericin tends to decrease the febrile reactions.

If amphotericin B cannot be used, pentamidine isethionate is given intravenously in daily doses of 200 mg for ten doses. Several courses usually have to be given over a prolonged period, and relapses are frequent.

NEW WORLD CUTANEOUS AND MUCOCUTANEOUS LEISHMANIASIS

Leishmania mexicana Complex and *Leishmania braziliensis* Complex

(Producing American leishmaniasis, espundia, uta, ulcera de los chicleros pro parte, bubas brazilianas, or cutaneous-mucocutaneous leishmaniasis)

Synonym. *Leishmania tropica* var. *americana* Laveran and Nattan-Larrier, 1912.

Historical Notes. Within a few decades following the Spanish conquest of Peru, Spaniards who entered the high Andean valleys, where small blood-sucking gnats were common, developed mutilating lesions of the nares, while native tribes in the forested areas rarely showed evidence of the disease (Weiss, 1943).

Lindenberg and also Carini and Paranhos in 1909 found amastigotes in ulcers occurring in man in Brazil. Their work was followed by that of Carini in 1911, who found amastigotes in the nasopharyngeal mucous membrane as well as in ulcerative lesions in patients in that country. Vianna studied this type of leishmania, considered it a new species, and in 1911 named it *Leishmania braziliensis,* while later in the same year Laveran studied it and named it *Leishmania tropica* var. *americana.* Morphologically, this parasite is identical with *L. donovani* and *L. tropica minor.* The serological investigations of Kligler (1926) and Noguchi (1926) demonstrated that this is a true species (or at least a true variety). Kligler found that *L. donovani, L. tropica,* and *L. braziliensis* agglutinated only with homologous immune serum, and Noguchi showed that the three parasites have distinct agglutination and complement-fixation reactions.

During the past two decades, extensive life cycle and epidemiologic studies in the New World tropics have provided evidence that the various types of clinical and epidemiological pictures are produced by several distinct strains of *Leishmania.* These are now generally accorded species or subspecies status. Two or three species complexes are recognized as producing human disease: *L. braziliensis, L. mexicana,* and *L. peruviana* (Lainson and Shaw, 1978).

Geographical Distribution. New World cutaneous and mucocutaneous leishmaniasis are found in all of the Americas from southern Mexico (states of Vera Cruz, Tabasco, Oaxaca, Chiapas, and nearby Yucatan) to Paraguay and the northern part of Argentina. Canine dermal leishmaniasis is known to be endemic as far north as Oklahoma (Anderson et al., 1980). Rare cases of autochthonous human infection have been reported from Texas (Shaw et al., 1976). It is not indigenous in Canada, Uruguay, or Chile. The type of leishmaniasis common in Yu-

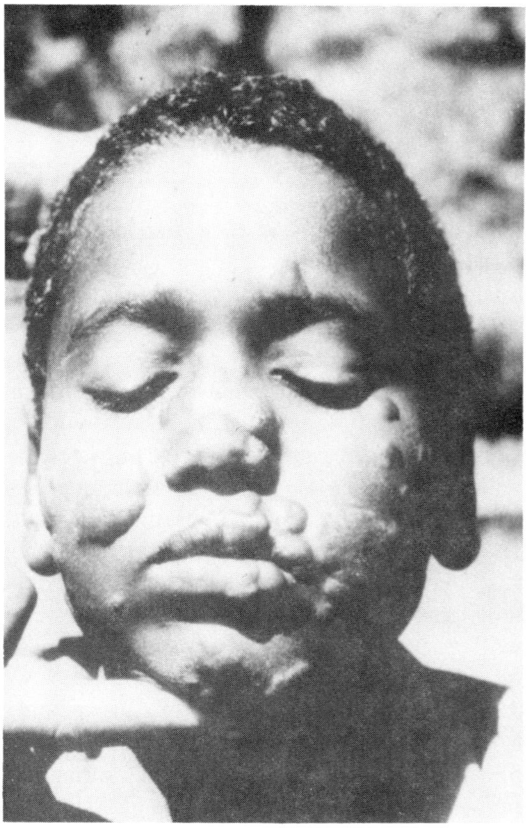

Fig. 10–7. Dermal leishmaniasis. Disseminated anergic cutaneous leishmaniasis in Ethiopia, showing confluent nodules. (Courtesy of Dr. Donald Minter, London School of Hygiene and Tropical Medicine.)

catan, Tabasco State, and southern Vera Cruz State, Mexico, and many of the cases in Central America and northern South America do not have secondary mucocutaneous manifestations and clinically correspond to Oriental sore. The disease is usually restricted to hot, humid forest areas and commonly to altitudes under 2500 feet, although the variety in Peru occurs at elevations of 3600 to 9000 feet.

Taxonomy and Morphology. The classification of the parasites that cause the various clinical types of American cutaneous leishmaniasis is not yet firmly established. Garnham (1962) has recommended that the organisms associated with each clinical disease type be given species status (at least temporarily) so that data collected may relate to accurately identified entities whatever their eventual taxonomic rank may be.

Since human infections (except for *uta*) are sporadic and derived from zoonotic diseases, all of the *Leishmania* infecting forest animals such as rodents, sloths, and porcupines must be considered as potential agents of human cutaneous leishmaniasis. Several types of *Leishmania* distinguishable by serologic methods have been isolated from human infections in Panama alone. *L. enriettii,* a species commonly used in laboratory research, was isolated from naturally infected laboratory guinea pigs and will not infect the wild Brazilian guinea pig. Its natural host in the wild state is not known.

The following classification has been proposed by Lainson and Shaw (1978) of the species of *Leishmania* that cause cutaneous or mucocutaneous leishmaniasis in the New World.

Leishmania mexicana Complex

1. *L. mexicana mexicana* (Mexico, Guatemala, British Honduras; cause of chiclero ulcer)
2. *L. mexicana amazonensis* (Amazon basin; rarely infects man; no nasopharyngeal involvement)
3. *L. mexicana pifanoi* (cause of disseminated [anergic] cutaneous leishmaniasis)
4. *L. enriettii* (known only in the laboratory guinea pig)

Leishmania braziliensis Complex

1. *L. braziliensis braziliensis* (Brazil and adjacent forested areas east of the Andes; cause of espundia)
2. *L. braziliensis guyanensis* (the Guianas and neighboring Brazil and Venezuela; cause of pian bois)
3. *L. braziliensis panamensis* (Panama and neighboring Central America and Colombia; nasopharyngeal involvement rare; does not metastasize)
4. *L. peruviana* (Peru, western slope of Andes; cause of uta; no nasopharyngeal involvement)

As determined by light microscopy, these parasites are morphologically identical to each other and to the other species of *Leishmania* in general. The two major complexes, *mexicana* and *braziliensis,* correspond to the fast-growing and slow-growing strains of *Leishmania,* respectively. The fast-growing strain, when inoculated into the skin of hamsters or mice, causes large histocytoma tumors containing abundant amastigotes without causing significant host cellular reaction. These parasites grow prolifically in blood agar media such as N.N.N. and are easily maintained. Most of the strains isolated from wild animals are fast growing; with the exception of *L. mexicana pifanoi,* they produce benign or negligible disease in man. The validity of *L. mexicana pifanoi* as a separate subspecies for the agent of disseminated cutaneous leishmaniasis is doubtful, since organisms taken from a patient with disseminated disease produced ordinary nondisseminated cutaneous leishmaniasis when inoculated into a volunteer (Convit et al., 1972). Furthermore, there is evidence that both *L. mexicana mexicana* and *L. mexicana amazonensis* may cause the disseminated disease.

Cultivation. Pedrosa and da Silva first cultivated *L. braziliensis* in 1911, using the N.N.N. medium of Novy, MacNeal, and Nicolle; since then, many others have cultivated the parasite.

Life Cycle. The life cycle and methods of reproduction in the *L. braziliensis* group are essentially the same as in the other *Leishmaniae (L. donovani* and *L. tropica)*. Since about 1940, studies of the life cycle of American *Leishmania* by workers in several different areas have led to the description of many new species of sand fly and the recognition of a variety of enzootic cycles relating to various ecologic conditions. The course of the infection in man has been found to be related to the enzootic system and ecologic type, possibly because of the involvement of different species or strains of the etiologic agent.

There are four genera of American sand flies (Phlebotominae): *Hertigia, Warileya, Brumptomyia,* and *Lutzomyia,* which collectively contain about 310 species. *Lutzomyia,* which contains 90% of the species and almost all of those biting man, is the most important. The great majority of the species of *Lutzomyia* are forest forms that feed on wild mammals. Although they must be vectors of enzootic leishmaniasis of wild animals, they rarely transmit human infections. There are species that have not yet been described. Some can survive in situations more arid and less protected than in the forest. In each habitat there exist species that bite man under some circumstances.

The usual vector of chiclero ulcer is *Lu. olmeca*. Its range is from Southern Mexico (especially the Yucatan) to Guatemala and Belize and possibly farther southward. Espundia is probably transmitted in Northern Brazil by *Lu. paraensis* and other species, and in Venezuela and Southern Brazil by *Lu. migonei, Lu. whitmani,* and *Lu. anduzei*. *Lu. anduzei* is also the vector of pian bois. Panamanian cutaneous leishmaniasis is carried by *Lu. trapidoi, Lu. ylephiletor, Lu. gomezi,* and *Lu. panamensis*. The vectors of uta are *Lu. verrucarum* and probably *Lu. peruensis*. (For definitions of the clinical types see pp. 67–68.)

Epidemiology. Most of the cutaneous and mucocutaneous leishmaniasis of the New World exists in enzootic cycles of infection involving wild animals, especially forest rodents, the infection being transmitted from rodent to rodent by forest sand flies. Human infections in such enzootic zones are sporadic zoonoses due to rodent-to-sand fly-to-man transmission when man invades the habitat of the forest rodents.

Although espundia is known to be acquired in the forest and is probably zoonotic, its animal reservoirs are poorly known. There have been only a few isolations from naturally infected forest rodents, including the paca, the agouti, and other species.

The exceptional American leishmaniasis is uta, the cutaneous leishmaniasis of the Peruvian Andes. This infection is not associated with the forest and involves no known wild-animal reservoir. It is found in villages in arid areas. Its reservoir is the domestic dog. The agent of uta, *Leishmania peruviana,* was probably once transmitted by its present vectors in an enzootic cycle involving some wild mammal until increasing aridity and/or clearing for agriculture so modified the habitat that the original mammalian host disappeared. The sand flies survived by utilizing the damp crevices in stone fences (*pircas*) for the production of larval stages and man and his dog as sources of blood. Since the latter were susceptible to infection with *L. peruviana,* the anthropozoonotic domestic cycle was established.

The rhesus monkey is at times susceptible to infection but Syrian hamsters have been refractory. Natural infection occurs in dogs, occasionally in cats, and in several species of wild rodents. Investigators in the Gorgas Memorial Laboratory, Panama, found about 10% natural infection in the spiny rat *(Proechimys semispinosus panamensis)* in numerous endemic sites in that country (Hertig et al., 1958). In Panama the kinkajou, *Potos flavus,* has been found naturally infected (Thatcher et al., 1965).

Adler and Theodor (1927) found that, although the promastigote of *L. braziliensis* develops in the stomach of *P. papatasi,* it never migrates to the buccal cavity as does the flagellate form of *L. tropica;* this suggests that *P. papatasi* is not an appropriate intermediate host of *L. braziliensis*.

Pathogenesis. The pathogenesis of American cutaneous leishmaniasis is the same as for Oriental sore (see p. 58). Some of the American strains of *Leishmania* tend to invade the mucous membranes of the mouth, nose, pharynx, and larynx either initially by direct extension or by metastasis. The latter is usually via lymphatic channels but occasionally may be by the bloodstream. The lesions so produced may be minimal and readily overlooked or may consist of progressive mutilation of the face. There may be many years between the appearance of the primary and the metastatic lesions. Secondary lesions occur on the skin as well as on mucous membranes. Nasal, oral, and pharyngeal lesions may be polypoid initially, then erode to form ulcers that expand to destroy the soft tissue and cartilage about the face and larynx. Regional lymphadenopathy is common.

In a recent study of espundia in the eastern Andes of Bolivia, it was observed that mutilation of the face was found mainly in Negroes, although many more Amerinds were infected. Skin testing with two dilutions of antigen produced greater reactions among the Negroes, suggesting that mutilating lesions are due to necrosis resulting from exaggerated inflammatory immune reactions (Walton and Valverde, 1979).

Pathology. Simple papular lesions and ulcers have the same pathology as Oriental sore.

In espundia, there are conspicuous lesions of the nose, mouth, and larynx. Pendulous, sessile, fibrous, or fleshy polyps form in the nasal cavity and pharynx. Fungating and indurated ulcers destroy the superficial tissues and cartilage of the nose and larynx. The tongue, gums, and buccal mucosa also may be destroyed. Masses of histiocytes containing innumerable amastigotes are seen microscopically (Fig. 10–8) (Edington and Gilles, 1969).

Clinical Manifestations. The types of lesions are more varied than those of Oriental sore and include at least chiclero ulcer, pian bois, Panamanian leishmaniasis, uta, espundia, and disseminated cutaneous leishmaniasis. Some individual lesions, namely papules, nodules, cutaneous ulcers, and relapsing lesions, appear the same as in Oriental sore.

Chiclero ulcer is a condition typically seen in chicle collectors who are exposed as they harvest the gum from sapodilla trees in the forest. It is caused by *L. mexicana mexicana*. Lesions occur about the face and ears, which are exposed to the bites of the vector *Lutzomyia olmeca* (Fig. 10–9,*A*). This disease is relatively mild, usually consisting of a single self-limiting cutaneous papule, nodule,

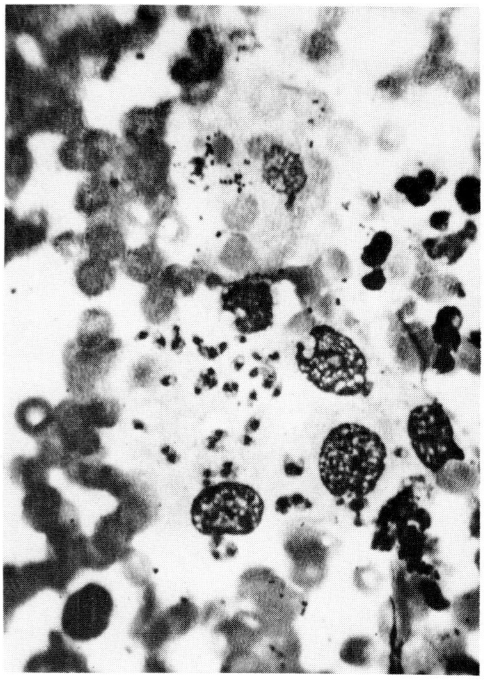

Fig. 10–8. *Leishmania braziliensis.* Amastigotes in an impression smear from a skin lesion. A cluster of the parasites is near the center of the field, having escaped from a disrupted macrophage. Within a leucocyte (above) are several cocci. (× ca. 800.) (Preparation by Dr. S. Zuñiga; photomicrograph by Dr. G.M. Carrera.)

or ulcer. A significant percentage of ulcerative lesions of the ear extensively destroy the pinna. Nasopharyngeal involvement does not occur.

Pian bois or forest yaws is caused by *Leishmania braziliensis guyanensis* and transmitted by *Lu. anduzei*. Although it may produce a single ulcerative lesion, it often metastasizes, probably via the lymphatic channels, producing multiple ulcers over the torso and extremities (Fig. 10–9,*B*). Metastases along lymphatic channels may simulate sporotrichosis. Nasopharyngeal lesions do not occur unless the sand fly's bite is on the nose.

Panamanian leishmaniasis is caused by *L. braziliensis panamensis* and transmitted by several species of *Lutzomyia,* including *Lu. trapidoi, Lu. ylephiletor, Lu. gomezi,* and *Lu. panamensis*. It causes one or a few shallow ulcers that may metastasize along lymphatic channels, forming secondary nodules. Metastatic involvement of the nasopharynx is not believed to occur.

Uta is caused by *L. peruviana* and is carried by *Lu. verrucarum* and probably *Lu. peruensis*. It causes one or a few lesions that are usually self-limiting and do not metastasize. The number of

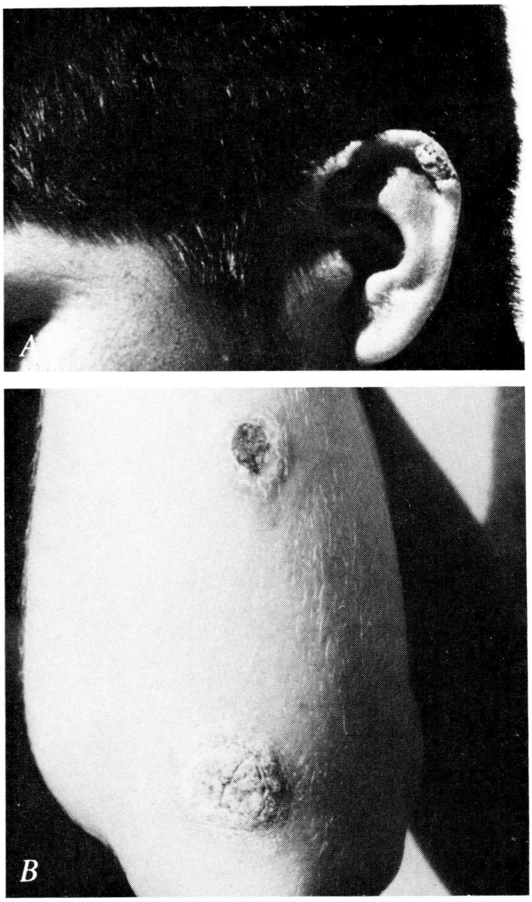

Fig. 10–9. American cutaneous leishmaniasis. A, Chiclero ulcer, showing characteristic lesion of the pinna. B, Pian bois, dermal lesions on forearm, Belize. (*A* by R.C. Jung. *B* courtesy of Dr. P.E.C. Manson-Bahr.)

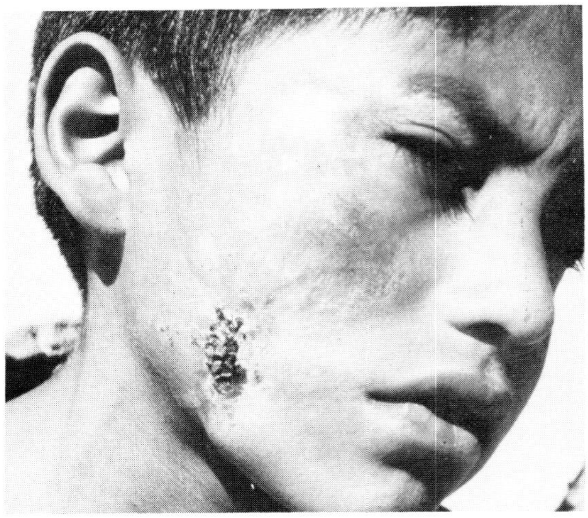

Fig. 10–10. American cutaneous leishmaniasis. Uta in Surco, Peru, showing leishmaniasis recidiva. Ulceration has recurred at the margin of a large scar marking the site of a previously healed facial lesion extending downward from the outer canthus of the eye. (By R.C. Jung. From *Dermatology in General Practice* by T.B. Fitzpatrick et al. (editors). Copyright © 1971 by McGraw-Hill, Inc. Used with permission of McGraw-Hill Book Company.)

skin lesions is therefore an index of the prevalence of infected sand flies. Lesions of the nasopharynx do not occur. Relapsing lesions (leishmaniasis recidiva) (Fig. 10–10) are not uncommon both clinically and epidemiologically.

Espundia (mucocutaneous leishmaniasis) is caused by *L. braziliensis braziliensis* and carried by *Lu. paraensis, Lu. migonei, Lu. whitmani, Lu. anduzei* and other species. It is the most notorious of the American leishmaniases because of its mutilation, resistance to treatment, and persistence (Fig. 10–11). Patients suffer pain and interference with feeding, which leads to starvation. It is characterized by metastases to the nasopharynx and larynx. Mucosal lesions, which begin as swellings, appear from 6 months to years after the primary lesions. Cartilage of the face and larynx may be destroyed but bone is not. Two intersecting furrows sometimes may be formed on the palate, forming a cross. The nose, lips, pharynx, larynx, trachea, bronchi, and esophagus may all be involved. Although mostly confined to South America, mucocutaneous leishmaniasis was reported once from Texas (Barlow et al., 1977).

Disseminated cutaneous leishmaniasis (leishmaniasis tegumentaria diffusa, anergic cutaneous leishmaniasis) appears the same as lepromatoid Ethiopian leishmaniasis (see p. 63). It is believed to be caused by *L. mexicana mexicana* and *L. mexicana amazonensis* and occurs in Venezuela and Brazil. Immunosuppression caused by the infection may in turn cause dissemination of the parasite (Nussenzweig, 1982). The disease begins with discrete papules that slowly evolve into plaques and nodules. These are at first smooth, firm, and bluish, then scaly and rough. Ulceration does not occur. Infiltrative lesions about the eyebrows, nose, and ears produce a leonine facies indistinguishable from lepromatous leprosy by inspection alone.

Diagnosis. Clinically this infection must be distinguished from furuncles, tropical ulcer, eczema, impetigo, pemphigus, fusospirochetal lesions of the mucosa, sporotrichosis, South American blastomycosis, syphilis, leprosy, yaws, and tuberculosis of the skin. The diagnosis is established by demonstration of *Leishmania* in material obtained by puncture of the edge of the initial ulcer, or from nodules or ulcerations in the mucous membrane,

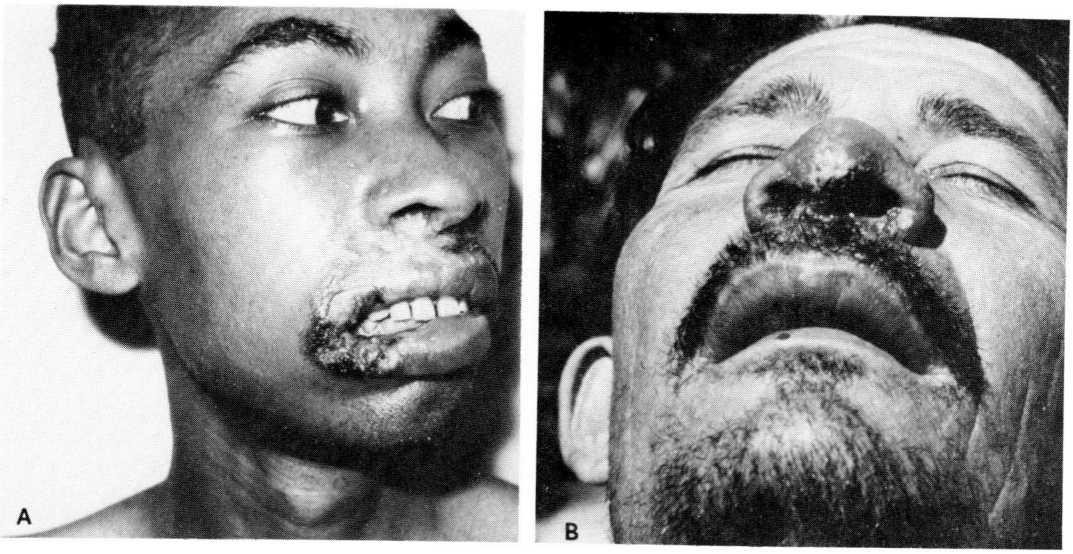

Fig. 10–11. *Leishmania braziliensis.* Espundia causing mucocutaneous lesions. *A,* Ulceration involving the nose, palate, and pharynx as well as the lips of a man in Colombia. *B,* Characteristic edema of the nose and destruction of the nasal septum of a man in Brazil. (*A* courtesy of Dr. A D'Alessandro. *B* by R.C. Jung.)

that are stained with Giemsa's or Wright's stain or cultured on N.N.N. medium. Serologic tests are sometimes used. For example, Pessôa and Pestana (1940) utilized a modified Montenegro intradermal reaction. The organisms were cultured on N.N.N. medium for 8 to 12 days, then suspended in physiologic salt solution to which 0.4% phenol was added. The suspension (0.1 ml) was injected intracutaneously. In positive cases, it produced an erythematous wheal in 48 hours, which lasted at times for 4 or 5 days. In the center of the wheal a small papule developed that became vesicular or pustular. The Montenegro skin test is the method of choice for diagnosis of mucocutaneous leishmaniasis in clinics and health centers. Pellegrino (1951) employed a polysaccharide fraction of cultured *L. braziliensis* for comparison with 1:5000 saline suspension of cultured organisms; he obtained strongly positive reactions in infected persons and negative reactions in uninfected controls. (For immunologic methods, see Technical Appendix, p. 765.)

Treatment. Responses to treatments may be evaluated by means of an indirect fluorescent antibody test (Walton, 1980).

Chiclero ulcer ordinarily needs no treatment, since it is self-limiting and benign (except on the ear). Lesions of the pinna are more persistent and destructive and therefore should be treated. Cycloguanil pamoate (Camolar) in a dose of 5 mg (base)/kg of body weight intramuscularly in a suspension of 40% benzyl benzoate and 60% castor oil is the treatment of choice. Usually one injection suffices, but if the infection relapses the dose may be repeated.

Pian bois and Panamanian leishmaniasis also respond to cycloguanil pamoate initially but are more likely to relapse than is chiclero ulcer. If treatment with cycloguanil pamoate fails, antimony sodium gluconate (pentavalent) should be used as for Oriental sore (see p. 62).

Treatment of uta, including relapsing lesions, is the same as for Oriental sore.

Espundia and disseminated cutaneous leishmaniasis are the most resistant to treatment of the clinical types. For espundia, pentavalent antimony sodium gluconate (Pentostam) should be given as it is for Oriental sore (see p. 62). If the treatment fails or the infection relapses, the course should be repeated twice before other drugs are tried.

If pentavalent antimony fails, pyrimethamine (Daraprim) should be given orally in a dosage of 25 mg daily for 2 weeks followed by a 1-week interval without medication. The course may be repeated twice. One-fourth of the adult dosage is given to children 8 to 14 years of age. Folic acid administered as 10 mg three times daily helps to prevent side effects due to the antifolic acid action of the drug.

If both pentavalent antimony and pyrimethamine

fail, amphotericin B should be given as described for Ethiopian leishmaniasis (see p. 64).

Disseminated cutaneous leishmaniasis is treated as the Ethiopian variety (Manson-Bahr, 1971).

Prognosis. Except for disseminated cutaneous leishmaniasis and espundia, all forms have a good prognosis including leishmaniasis recidiva, which may need several courses of treatment. The disseminated disease responds to treatment with amphotericin B but is likely to relapse. If espundia is cured, reconstructive surgery may be needed.

Prevention. Since American leishmaniasis is almost exclusively confined to persons who frequent or are associated with forests or other nonurban areas in which the sand flies *(Lutzomyia)* breed, the problem of prevention concerns almost exclusively these individuals. Protection can be provided for a few hours by repellents such as dimethylphthalate applied to the exposed skin. Direct contact with infected persons can be avoided. Pessôa and Pestana (1940) had considerable success in vaccinating susceptible individuals, and Pessôa (1941) recommended persistent treatment of infected individuals together with vaccination.

Since American leishmaniasis is generally contracted outdoors, usually away from homes in forest areas, insecticides have limited usefulness. Power spraying from trucks or planes may be practicable in forest camps.

Community control programs are effective only in the case of uta, since this disease is primarily anthroponotic and domestic. Residual insecticide should be applied on the surfaces of homes and stone fences to eliminate the vectors. Treatment of all human and canine infections or destruction of infected dogs eliminates the sources of infection.

Leishmania donovani

(Laveran and Mesnil, 1903) Ross, 1903
(Producing visceral leishmaniasis or kala-azar; "death fever"; Dum-Dum fever; tropical splenomegaly; ponos—Greece)

Synonyms. *Piroplasma donovani* Laveran and Mesnil, 1903; *Herpetomonas donovani* Ross 1904; *Leishmania chagasi* E. Chagas et al., 1937. Opinions differ concerning *Leishmania infantum* Nicolle 1908.

Historical Notes. In 1900, William Leishman discovered *Leishmania donovani* in spleen smears of a soldier who died from a fever contracted at Dum-Dum, India, known as Dum-Dum fever or kala-azar. His observations were published in 1903, in which year Charles Donovan found the same parasite in smears made from splenic puncture biopsy. Leonard Rogers (1904) first succeeded in culturing the parasite and showed that flagellated forms developed in cultures, while Patton (1907) proved that the amastigote stage could be found in wandering histiocytes in the peripheral blood and that promastigote forms occurred in the intestine of insects fed upon kala-azar patients. Cathoire (1904), Pianese (1905), and Nicolle (1908) discovered an identical parasite in children in the Mediterranean region who were suffering from splenic anemia of infants, and Nicolle and Compte (1908) found the parasite in dogs in Tunis. The transmission of *L. donovani* by the sand fly *Phlebotomus argentipes* was demonstrated by the Indian Kala-azar Commission (1931–1934). More recently it has been proposed that the agent of the Mediterranean visceral leishmaniasis of children be assigned a distinct species, *L. infantum* (Adler, 1964; Hommel, 1978).

Geographical Distribution. Visceral leishmaniasis is widely distributed, but local endemicity is usually rather sharply delimited. In the *Americas,* the disease occurs in northern Argentina, Paraguay, Bolivia, northeastern Brazil (heavy human and canine infection), scattered foci in central Brazil, and to a lesser extent in Guyana, Venezuela, Colombia, El Salvador, Guatemala, and southern Mexico. In *Africa,* it is endemic in the Sudan, Ethiopia, Somalia, northern Kenya, Equatorial and West Africa, Morocco, Algeria, and Tunisia. In *Asia,* it has extensive distribution in north, northwest and west China, Manchuria, Turkestan, and eastern India, and is present in Iran, Iraq, Turkey, Jordan, Syria, Lebanon, Israel, Yemen, and Saudi Arabia. In *Europe,* it is endemic in Portugal, Spain, Corsica, Italy, Albania, Greece, Bulgaria, and the U.S.S.R. (Tashkent, Turkmenistan, Uzbeck, and in the Caspian Sea littoral).

Morphology. Two stages in the life history of *L. donovani* are known: the aflagellar amastigote stage in man and reservoir mammals, and the promastigote stage in the sand fly and in cultures (Fig. 10–4).

In man, this organism is always in the amastigote stage, appearing as an ovoidal or rounded body, measuring about 2 to 3 μm in length, and living intracellularly in monocytes, polymorphonuclear leukocytes, or endothelial cells. In preparations stained with Giemsa's or Wright's stain, the cytoplasm is pale blue in color and there is a limiting membrane. Lying within the cytoplasm is a relatively large nucleus that stains red, while at right angles to it is a deep red or violet rod-like body called the "parabasal body." In well-stained specimens, a dot-like blepharoplast may be seen near the parabasal body, as well as a delicate thread connecting the two organelles and an axoneme that arises from the blepharoplast and extends to the anterior tip. Elongated slender forms, the so-called "torpedo forms," may also be observed in man; they have pointed or rounded ends, and the nucleus and kinetoplast are closely associated. When hematoxylin stains are used after wet fixation, the nucleus is observed to have a delicate membrane and a large central karyosome. Dividing forms may be seen in spleen smears.

In liver or spleen of kala-azar patients, *L. donovani* lives and multiplies within the fixed macrophages, which it destroys. It also appears in smaller numbers in wandering histiocytes in the blood or tissue juices. The parasite appears to be phagocytosed and destroyed by neutrophilic leukocytes.

In cultures on N.N.N. medium (Technical Appendix, p. 762), the morphology of *L. donovani* differs from that seen in man and other mammalian hosts; here promastigote forms develop that have a single flagellum and marked motility. The average length of fully developed specimens is 15 to 25 μm and the breadth 1.5 to 3.5 μm, while the flagellum measures 15 to 28 μm in length. Short, broad, flagellated forms also occur, while rounded promastigote forms, from 4 to 5 μm in diameter but with a long flagellum, may also be seen in cultures. In old cultures, there are many degenerative organisms in which the morphology is atypical.

In the digestive tract of appropriate insects, the promastigote forms are identical in morphology with those grown in cultures.

Habitat. The natural habitat of *L. donovani* in man is the reticuloendothelial system of the viscera, especially of spleen, liver, bone marrow, intestinal mucosa, and mesenteric lymph nodes. This parasite may be found in endothelial cells of the kidneys, suprarenal capsules, lungs, meninges, and in cerebrospinal fluid. Not infrequently the organisms are present in considerable numbers in macrophages in the intestinal wall, and they have also been found in the feces, urine, blood, and nasal secretions.

Life Cycle. The life cycle of *L. donovani* in the mammalian host is a simple one. The organism is engulfed by reticuloendothelial cells, in which it multiplies as an amastigote by simple binary fission until the host cell is destroyed, whereupon new macrophage cells are parasitized.

Division in *L. donovani* is initiated in the kinetoplast, which elongates, and both the blepharoplast and parabasal body divide into two portions; the nucleus then elongates and divides mitotically. After division of the blepharoplast, a new axoneme, or flagellar root, is formed by the daughter blepharoplast that is not attached to the old axoneme. Eventually the cytoplasm divides and two individuals are produced.

In the intestine of insects, the cycle of development is also simple. The amastigotes become promastigote forms and divide by longitudinal fission into two individuals. The process of division is similar to that in the amastigote forms in man; the flagellum does not split, but a second flagellum is produced by an axoneme developing from one daughter blepharoplast and grows out into a new flagellum.

The amastigote of *L.donovani* is taken up by the sand fly in a blood meal that clots in the midgut ("stomach"). Here the parasites transform into promastigotes and multiply, producing enormous numbers. They migrate to the pharynx, which they block on about the fifth day. On the seventh day, they reach the hypostome, where blocking also occurs. The sand fly now has difficulty in getting a blood meal; nevertheless, it pricks the skin of the victim and loosens the plug of adherent parasites, some of which remain in the puncture wound and thus initiate the infection.

In cultures reproduction occurs in the same manner as in the intestine of insects. Rapid multiplication in the insect host and in cultures gives rise to the so-called "rosette forms," in which several organisms are arranged with their flagella entangled so that they are seen with the flagellar end innermost.

Epidemiology. The vectors of visceral leishmaniasis are known or suspected to be the following:

Asia (India, East Pakistan, China north of Yangtse River, Tashkent, Turkestan, Transcaucasus): *Phlebotomus argentipes, P. chinensis chinensis, P. chinensis longiductus, P. major syriacus,* and *P. saheli.*

Middle East (Turkey, Israel, Lebanon, Syria, Jordan, Saudi Arabia, Iran, U.S.S.R.): *P. perfiliewi, P. papatasi, P. alexandri, P. ansarii, P. caucasicus, P. mongolensis, P. perniciosus, P. sergenti, P. simii,* and *P. tobbi.*

Africa (Sudan, Ethiopia, North Kenya, Somalia, Tripoli, Morocco, Algeria, Tunisia): *P. sergenti, P. orientalis, P. bergeroti, P. chabaudi, P. duboscqi, P. longeroni orientalis, P. longicuspis, P.longipes,* and *P. martini.*

Europe (Greece, Yugoslavia, Cyprus, Italy, Corsica, France, Spain, Malta, Portugal): *P. perniciosus, P. major, P. papatasi, P. tobbi, P. perfiliewi,* and *P. ariasi.*

South America (Brazil, Venezuela, northern Argentina, Paraguay, Bolivia): *Lutzomyia intermedia, Lu. longipalpis,* and *Lu. paraensis.*

Central America (Panama, Guatemala, British Honduras): *Lutzomyia cruciatus*

Mexico: *Lutzomyia cruciatus*

Kala-azar was first recognized as a specific disease in India, yet the Indian variety of visceral leishmaniasis is the most atypical epidemiologically. As in cutaneous leishmaniasis, there are several different life cycles, each related to a different epidemiologic and clinical variety of visceral leishmaniasis. These cycles may be arranged in a series that suggests that visceral leishmaniasis evolved from an enzootic disease through intermediate steps to a wholly anthroponotic disease in which the only hosts are vector and human. These steps are as follows:

1. *Enzootic foci in wild animals*. The cycle is jackal to sand fly to jackal in sparsely inhabited areas of Southeast Asia. Sporadic human cases are due to man's invasions of the enzootic areas.
2. *Rural endemic foci*. Here the domestic dog acquires infection from the sylvatic cycle and becomes a local reservoir. The pattern is fox- or jackal to sand fly to dog to sand fly to man. Human cases are more numerous than in cycle 1. This cycle occurs in the Middle East, Transcaucasia, and parts of the Mediterranean littoral.
3. *Urban endemic foci*. In this situation, the wild mammalian reservoir is absent. Transmission involving both man and dog via the sand-fly vector has existed continously over a long period of time in a dense population of both dog and man, producing extremely numerous infections in humans as well as in dogs. Foci with this cycle occur in Middle Asia, Transcaucasia, and Southeastern China.
4. *Indian kala-azar*. This is the final stage of conversion of the cycle to anthroponosis. The dog having been eliminated from the cycle, man and the sand fly *Phlebotomus argentipes* are the only known hosts (Hoogstraal and Heyneman, 1969). There is no known reservoir host. Epidemics occur presumably because of lowered resistance due to famine: an increased number of susceptibles, due to children born since the last epidemic or immigrants; or increased virulence of the parasite.

American visceral leishmaniasis is zoonotic, with the common wild mammalian reservoir host being the fox, *Lycalopex vetulus*, and the vector being *Lutzomyia longipalpis*. Sporadic human infection may occur within patches of enzootic disease, or the disease may be introduced into domestic dogs resulting in a dog-to-sand fly-to-man cycle that produces more human cases. American visceral leishmaniasis may have originated when the Mediterranean form was introduced into America by infected settlers and their dogs and subsequently passed into the wild canine population. This would be a reversal of the sequence of steps described previously. Although it is possible that the disease could have existed in South American foxes in preColumbian times, the poor tolerance of the foxes to the infection suggests that their association with visceral leishmaniasis is of relatively recent origin.

Recent studies in the Sudan Republic suggest that human visceral leishmaniasis was not introduced from India through human migration to East Africa, but rather that the outbreaks in man are derived from enzootic foci involving local rodents such as *Arvicanthis niloticus*. The only likely vector to man is *Phlebotomus orientalis*.

In Kenya the reservoirs include the mongoose, dog, and genet cat (Mutinga et al., 1982). The vectors are probably *Phlebotomus martini* and *P. vansomerenae*. The human disease occurs in rural areas in the proximity of eroded termite hills in which the flies breed.

There is a seasonal variation in the incidence of new cases of kala-azar in endemic areas. Peak incidence tends to be approximately 3 months after the onset of the rains, which hasten the emergence of the sand flies.

The parasite is not very host-specific. A wide range of mammals may be experimentally infected, although the nature of the disease produced in the various species differs as to the relative degree of cutaneous and visceral involvement. Many species of sand fly may become infected with *L. donovani*, but there is wide variation in abundance and location of promastigotes that form and their location in the insects' gastrointestinal tracts. The most efficient vectors are those that bite both man and reservoir hosts and produce many parasites that migrate to the hypostome.

The age distribution of human visceral leishmaniasis varies greatly from one geographic region to another. In the Mediterranean region and northwestern China (Kansu), the disease occurs in children from 1 to 4 years of age; in eastern China (Kiangsu) it is older children and in India, young adults who are most commonly infected. In general, males are more commonly infected than females, due probably to greater occupational exposure. Lowered resistance due to malnutrition or other disease, such as intestinal parasitism, probably predisposes to clinical disease.

Recovery from visceral leishmaniasis is followed by lasting immunity.

The geographical varieties of visceral leishmaniasis are so distinct that the respective etiologic agents have been given subspecies or even species rank by some authorities (Hoogstraal and Heyneman, 1969; Lainson and Shaw, 1978). They are as follows:

1. Indian, caused by *L. donovani donovani*, producing only anthroponotic disease; no mammalian reservoir; vector *Phlebotomus argentipes*.
2. Mediterranean-Middle Asian, caused by *L. donovani infantum* (or *L. infantum*); zoonotic, involving canine reservoirs; vectors of the subgenera *Larroussius* and *Adlerius*, especially *Phlebotomus major, P. perniciosus* and *P. longicuspis*.
3. East African, caused by *L. donovani archibaldi;* zoonotic, involving wild mammalian reservoirs; vectors *Phlebotomus orientalis* and *P. martini*.
4. South American, caused by *L. donovani chagasi*, which may be synonymous with *L. donovani infantum*; zoonotic, involving wild and domestic canines; vector *Lutzomyia longipalpis*. Clinical features in man are similar to those of the Mediterranean disease (see p. 75).

In China, the disease resembles the Mediterranean type in the northwest (Kansu) and is like the Indian type in the east (Kiangsu) (Wang and Wu, 1959).

The validity of these subspecies and species is still under debate, since the organisms are morphologically indistinguishable and immunologic differences are not consistent. The variations in clinical picture may be due to differences in host-parasite reactions caused by differences in age at exposure and passage through various reservoir hosts.

Pathogenesis. When the promastigote stage of the parasite is inoculated into the victim's skin by an infected sand fly, nearby macrophages engulf the organisms that metamorphose into the amastigote stage within the cytoplasm of these host cells. Here the amastigotes multiply slowly and may remain more or less quiescent for weeks or months. In time, some of the parasitized macrophages are set free into the blood stream and are carried from the skin to the viscera, where they lodge and the parasites develop rapidly. In spleen, liver, bone marrow, and other centers of reticuloendothelial activity, the amastigotes are now taken up by fixed macrophages, such as Kupffer's cells in the liver, multiply in these cells, and destroy them.

Host cellular and humoral defense mechanisms are stimulated. The former stimulus results in marked proliferation of macrophages. These cellular elements make up a large part of the bone marrow, compromising both the erythropoietic and the granulocytic series. The effect of this in the peripheral blood is leukopenia with granulocytopenia and relative monocytosis, anemia (usually normocytic), and thrombocytopenia.

The spleen and liver become markedly enlarged, and hypersplenism contributes to the production of anemia (Fig. 10–12). Observations by Agu et al. (1982) suggested that the erythrocytes adsorb immune complexes and become subject to enhanced phagocytosis by the macrophages of the liver and spleen. Lymphadenopathy is also produced. There is a greatly increased production of globulin resulting in marked hyperglobulinemia and reversal of the albumin-globulin ratio.

Pathologic and Microscopic Anatomy. Patients dying of kala-azar are anemic and much emaciated, and the principal lesions are observed in spleen, liver, bone marrow, heart, kidneys, and lymph nodes. The disease is reticuloendotheliosis, due to invasion of the reticular syncytium by leishmaniae. Free histiocytes are gradually formed and also become parasitized. Late in the infection, the lymphatic reticulum of the spleen becomes involved. Reticuloendothelial blockage is regarded

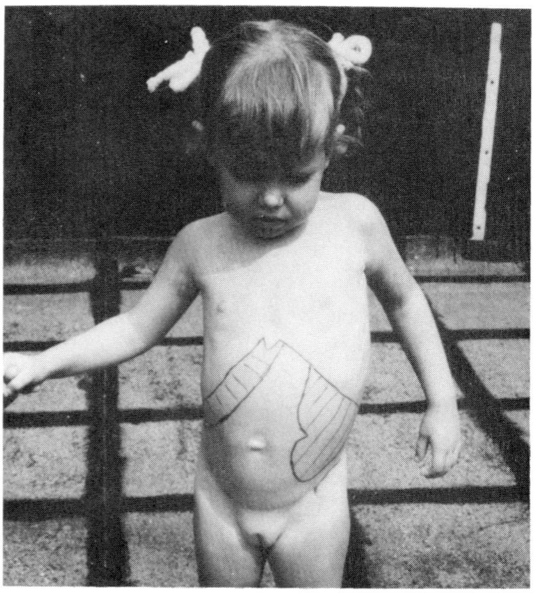

Fig. 10–12. Kala-azar of the Mediterranean (childhood) type. The disconsolate expression and enlarged liver and spleen are typical. (Courtesy of Dr. P.E.C. Manson-Bahr.)

as the probable cause of the anemia and of the tendency towards inflammatory complications.

Microscopically, one sees a marked increase in parasitized reticuloendothelial cells. The *spleen* is enormously enlarged, congested, purple or brown in color, with prominent Malpighian corpuscles. In the enlarged *liver,* there is fatty infiltration of Kupffer's cells and the endothelial cells of the blood vessels are invaded by amastigotes. The *red bone marrow* contains numerous macrophages filled with amastigotes together with myelocytes and neutrophilic leukocytes that may also be crowded with the parasites. The *heart* is pale and microscopically may show evidence of myocardial degeneration, while the *kidneys* show cloudy swelling and macrophages containing amastigotes are present. The *lymph nodes* are usually enlarged, and parasitized cells may be found in the lymph spaces. The *intestines* are often ulcerated, usually due to some other cause but the submucosa is infiltrated with numerous macrophages that are filled with amastigotes that are especially numerous around Peyer's patches and the solitary lymph follicles.

The *peripheral blood* shows a progressive leukopenia with first a relative and, later, an absolute monocytosis. The total leukocyte count may vary within wide limits (2200 to 5500 per mm^3) with 10 to 50% neutrophils. The total red cell count also ranges from 1 to 5 million per mm^3. Commenting on the anemia in kala-azar, Martins et al. (1965) stated that in 190 cases in Brazil the mean value for the red cell count was 2.6 million per mm^3. Microcytic anemia was present in 58% of the cases; bone marrow exhibited hyperplasia of the erythroid series. The anemia was more severe in females, in persons less than 10 years old, in those with greatly enlarged spleens, and when associated with hookworm infection; it was less severe in patients with combined kala-azar and schistosomiasis.

Clinical Picture. The symptomatology of infection with *L. donovani* is characteristic. In some forms of the disease (e.g., Sudanese), there is a primary cutaneous lesion resembling that of cutaneous leishmaniasis (leishmanoma), but usually this is lacking. The *incubation period* may be as short as 10 days or longer than a year, but it is usually limited to 2 to 4 months. The infection is typically insidious, but at times it may have a sudden onset. The first symptoms consist of malaise, headache, and fever, which occur at irregular intervals, with progressive enlargement of the spleen, and occasionally acute abdominal pain. In the beginning, the temperature may spike two or three times during the first 24 hours and quotidian exacerbations or great irregularity of temperature may be noted. After infection is fully established, the most important symptoms are irregular undulant fever, at times with two peaks daily, possibly reaching 104 to 105° F (40 to 40.6° C); anemia with leukopenia; edema of skin and periosteum; emaciation, especially of the chest; dysentery or diarrhea; bleeding of mucous membranes of the gums and nares; and cachexia, which is accompanied by a progressive and marked enlargement of the spleen and a lesser enlargement of the liver (Fig. 10–12). Unlike intestinal schistosomiasis (see pp. 416 and 427) or Banti's disease, there is no ascites in kala-azar. The tongue is almost invariably clean—contrasted with its condition in typhoid fever—and there are relatively few physical signs characteristic of the disease except pallor of mucous membranes, splenomegaly, and possibly enlargement of lymph nodes.

In the most acute infections, death may occur within a few weeks; in subacute cases, within a year; and in chronic cases, from 2 to 3 years. Yet there are probably many asymptomatic individuals who recover spontaneously. Death frequently occurs from complications that, in the advanced form of the infection, include cancrum oris, hemorrhages, bacillary and amebic dysentery, pernicious malaria, lobar or bronchopneumonia, nephritis, septicemia, cerebral hemorrhage, myocardial degeneration, anemia, or extreme toxemia. In certain geographic areas, mucosal lesions occur. These are ordinarily mere erosions of the palate or nasal mucosa and do not resemble the destructive lesions of mucocutaneous leishmaniasis.

A clinical condition known as *post-kala-azar dermal leishmaniasis* at times develops after incomplete treatment with antimony; it is caused by the reversal of the agent from viscerotropic to dermatotropic. The patient may be considered cured of visceral leishmaniasis during post-kala-azar dermal leishmaniasis. It has been observed most often in India but has been noted in other endemic areas.

The rash consists of hypopigmented macules that may occur anywhere on the body, especially the upper trunk, arms, thighs, forearms, legs, abdomen, and neck, in that order (Fig. 10–13). The face, hands, and feet are less commonly involved. The lesions at first appear as punctate hypopigmented macules that grow to about 1 cm in diameter. They may coalesce to form irregular

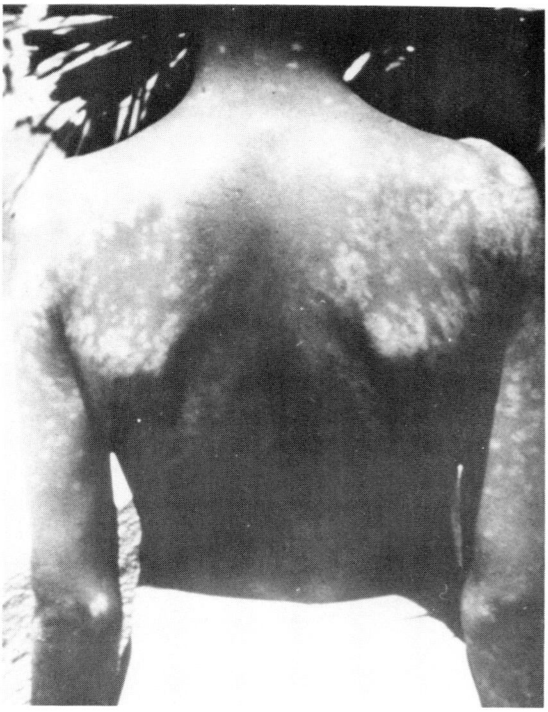

Fig. 10–13. Depigmented lesions of Indian post-kala-azar dermal leishmaniasis. (By R.C. Jung. From Spittell, J.A. (editor). 1980. *Clinical Medicine*. Philadelphia, J.B. Lippincott.)

patches. Usually only one crop of lesions appears but sometimes depigmented areas and nodules may occur for months.

Early in the course of dermal leishmaniasis, an erythematous facial rash that resembles discoid lupus may appear. Later nodules appear mostly on the face—especially on the nose, chin, cheek, lips, forehead, and ears. When the nodules are abundant, the clinical apearance may resemble that of lepromatous leprosy or disseminated cutaneous leishmaniasis (Fig. 10–14). Epidemic waves of kala-azar are followed by a marked increase in post-treatment dermal leishmaniasis.

The phenomenon of dermal leishmaniasis probably represents a modification of the host's immune status rather than the inherent dermatotropism of the etiological agent.

Several geographical types of visceral leishmaniasis have been described. The relationships of these types to the epidemiological varieties described above is obvious.

1. *Indian type*. This is usually a disease of young adults. Cutaneous lesions, as well as a primary papule, are very rare in the untreated patient. The leishmanin test is often negative after complete cure, but it is positive in about one-half of the cases of post-treatment dermal leishmaniasis, which commonly occurs a year or more after completion of treatment. Parasites are common in the peripheral blood. The disease usually responds to pentavalent antimony therapy.
2. *Mediterranean type*. This disease occurs only in infants and juveniles (usually under 3 years of age),. Primary and cutaneous lesions are not common but have been reported from the U.S.S.R. Post-treatment dermal leishmaniasis does not occur.
3. *Chinese type*. This disease is more common in juveniles but does occur in adults. Dermal and mucocutaneous lesions have been reported. Post-treatment dermal leishmaniasis is rare.
4. *Sudanese type*. This disease usually occurs in adolescents and young adults. Cutaneous, oral, and sometimes nasal lesions containing parasites commonly occur. Parasites in peripheral blood are less common. This disease is resistant to treatment with pentavalent antimonials. Post-treatment dermal leishmaniasis is common and appears promptly after completion of treatment.
5. *South American type*. This disease occurs sporadically in humans of all ages who are exposed. Post-treatment dermal leishmaniasis does not occur.

Just as the epidemiologic cycle of visceral leishmaniasis varies, the clinical picture varies as well depending largely on the dermatotropism of the agent. In the case of a highly dermatotropic strain, such as the Sudanese, a dermal primary lesion, as well as secondary cutaneous and mucocutaneous lesions, occur. In Indian kala-azar, cutaneous lesions do not occur in the untreated patient.

Diagnosis. Animal inoculation as a diagnostic test is not usually employed, since it requires several months. The nonspecific aldehyde test of Napier, antimony test of Chopra, finger-prick blood test of Chopra, and precipitation test of Sia are all based on increase in serum gamma globulin and decrease in serum albumin, and they should not be relied upon to the exclusion of demonstration of the parasite. (See Technical Appendix, p. 762.)

Complement fixation tests using the human tubercle bacillus or leprosy bacillus of Kedrowsky

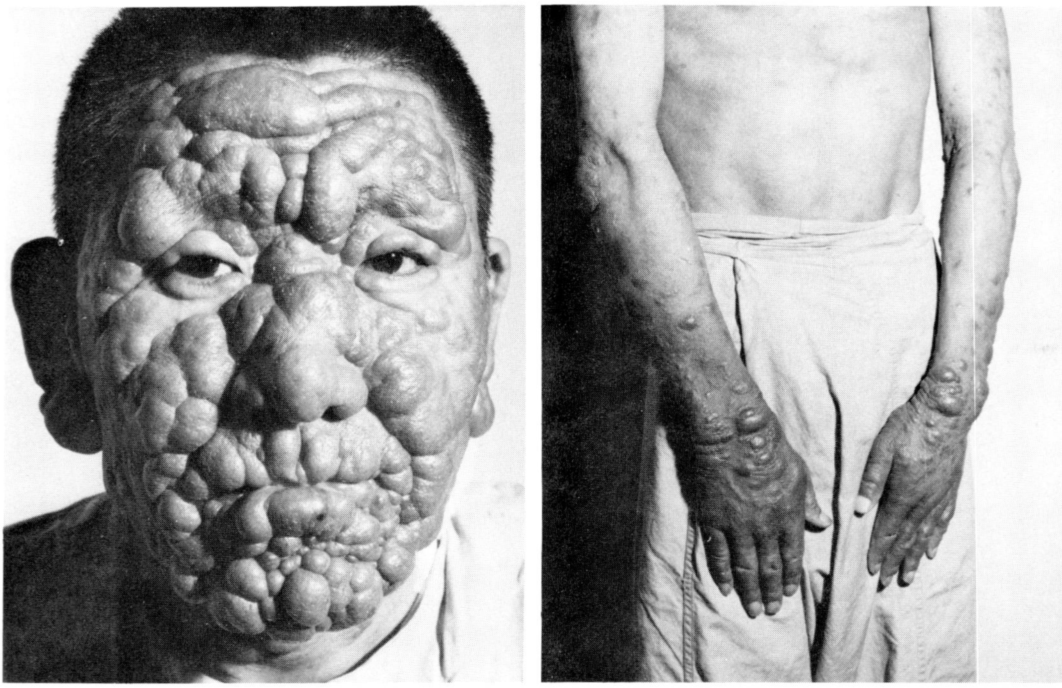

Fig. 10–14. Post-kala-azar dermal leishmaniasis. Confluent nodular lesions of face and hands in a Taiwanese patient. (Courtesy of Dr. R.E. Kuntz.)

as antigen have been employed successfully for diagnosis of kala-azar in India, especially in suspected early cases when the aldehyde test is negative.

The *Montenegro (leishmanin) test* for immunity was employed by P.E.C. Manson-Bahr (1961) to aid in kala-azar studies in endemic foci of Kenya. In this test, 0.1 to 0.2 ml of antigen (containing 6 to 10 million living promastigotes of *Leishmania braziliensis, L. donovani* isolated from the ground squirrel, or East African, Sudanese, North African, or Indian strains of human *L. donovani* or *L. tropica)* is injected intradermally and the results are read after 72 hours. A delayed reaction develops in cured individuals; the test is negative in early cases, but it becomes positive 6 to 8 weeks after completion of treatment. Manson-Bahr stated that when more than 5% of the people in a community show a positive leishmanin reaction, kala-azar exists in the area. Subjects immunized with the ground squirrel strain of *L. donovani* (which does not produce visceral disease in man) resist challenge from East African, Sudanese, Mediterranean, and Indian strains of this parasite, but there appears to be no cross-immunity between strains of *L. donovani* and *L. tropica*.

For *recovery of amastigotes from tissues,* biopsy techniques are used. Of these techniques, splenic puncture has been employed for many decades, and although attended by considerable risk, it is still employed in many outpatient clinics and on hospital wards. Taking bone marrow specimens obtained by sternal puncture, or preferably from the iliac crest, is advised as a safer procedure, while some physicians depend principally on lymph-node aspiration. Presence of the organisms frequently can be demonstrated in cultures of biopsy specimens when direct examination fails (Hockmeyer et al., 1981).

Kala-azar must be differentiated from malaria, typhoid and paratyphoid fevers, relapsing fevers, undulant fevers, dysentery of various etiologies, visceral syphilis, influenza, septic conditions of unknown etiology, and tuberculosis, but the clinical pictures of these infections, together with the laboratory methods for their diagnosis, should allow one to make a differential diagnosis without much difficulty.

Treatment. If possible, treatment should be administered in a hospital. The drug of choice, *pentavalent antimony sodium gluconate* (Pentostam or Solustibosan), is given intravenously or intramuscularly in a dosage of 600 mg daily for 6 days for treatment of Indian kala-azar and 30 days for treatment of all other forms of the infection. Children under 14 years of age are given 400 mg daily and those under 2 years are given 200 mg daily.

Pentamidine isethionate (Lomidine), a diamidine, is given intramuscularly in a dosage of 4 mg of base/kg of body weight daily for 10 days. The injections should be given deeply in the upper outer quadrant of the gluteus maximus muscle. The course may be repeated after a 10-day interval and, if necessary, a third course may be given. An immediate hypotensive reaction and local pain are major side effects. Hypoglycemia, a less common effect, may require intravenous administration of glucose and is an indication to discontinue the drug. The diamidine, hydroxystilbamidine, is preferred by some workers but is not available in the United States.

Amphotericin B (Fungizone) is used when the disease does not respond to pentavalent antimonial or diamidine therapy. It is given as described for disseminated cutaneous leishmaniasis (see p. 64).

If the disease resists all drug therapy, splenectomy may result in a favorable response. After splenectomy, another course of treatment is given, preferably with pentavalent antimony (antimony sodium gluconate). Since splenectomy markedly impairs resistance to malaria, splenectomies should be followed by antimalarial therapy and, if the patient resides in an area endemic for malaria, by continuous malaria suppression.

Prognosis. Response to therapy is usually good. The best prognosis is for Indian kala-azar, in which there is less than a 20% failure to respond to one course of pentavalent antimony. Less favorable results are obtained in Mediterranean and American visceral leishmaniasis. Resistance to treatment is greatest in the African disease, in which 10 to 15% of treated cases tend to relapse.

Prevention. Until recently, the best preventive measures known were the following:

1. Treat adequately with antimony all human patients harboring *L. donovani,* including inapparent or postvisceral types.
2. Avoid and destroy infected dogs, and control the movements of all dogs. Even with an abundant sand-fly population, destruction of infected dogs and general reduction in the dog population was found to produce a marked decline of kala-azar in Canea, Crete (Hertig, 1949).
3. Remove from the grounds surrounding houses all rank and decaying vegetation, which may harbor sand flies or in which the flies may breed.
4. Use the finest mesh mosquito screens or netting.
5. Avoid houses that are surrounded by dense vegetation.
6. Have houses well lit and ventilated.

In addition, it is desirable to have sleeping quarters above the ground floor whenever possible, since sand flies do not commonly fly many feet above the ground; to use effective repellents, such as dimethyl phthalate, when unavoidably exposed to sand flies; and to consider the possible use of diamidines for infections refractory to antimony.

Controlling sand flies by spraying residual insecticides in houses and in the outdoor resting places of the adult sand flies has been found to be both practical and effective. It provides virtually complete indoors protection for several weeks to many months by gradually reducing the sand-fly population to the vanishing point and is accompanied by a rapid reduction in leishmaniasis. The cessation of antimalarial spraying has resulted in a return of kala-azar in India (Chance, 1981).

In Azerbaijan, U.S.S.R., where the disease appears to be transmitted from man to man without the intervention of another mammalian host, control measures taking this fact into account have achieved noticeable success.

In rural areas in which the reservoirs are wild canines or rodents, attempts at control are less successful. These control methods are based on detecting those areas in which sand flies breed and spraying them with residual insecticide. Efforts to separate the human residences from sites where sand flies breed may be helpful when these sites are easily identified; for example, in Kenya termite hills are the common breeding sites for *Phlebotomus martini*.

Genus Trypanosoma
Gruby, 1843

The genus *Trypanosoma* was established for a flagellate protozoon recovered from the blood of frogs, but many other species of this generic group have since been found to parasitize the blood and tissues of vertebrates, including fishes, amphibia, reptiles, birds, and mammals. All flagellates of this genus at some time in their life history have the typical trypanosome structure—namely, an elongate spindle-shaped body that more or less tapers at both ends, a centrally situated nucleus, a kinetoplast posterior to the nucleus, an undulating membrane arising from the kinetoplast and proceeding forward along the margin of the cell membrane, and a single free flagellum at the anterior end (Fig. 10–15).

Volutin granules are probably present in all trypanosomes, but their number varies. Usually they are not observed during the acute stage of infection, but they do appear in chronic and relapsed cases. Differences are also seen in strains of the same species from different areas. Their number and appearance may possibly be correlated with the antigenic response to the host's reaction to the trypanosome (Ormerod, 1961a).

A blood-sucking invertebrate (a leech in the case of a few aquatic vertebrates; otherwise, an arthropod) constitutes the intermediate host and vector of the parasite in the natural life cycle. *Trypanosoma equiperdum* of horses and mules has dispensed with its intermediate host, and transmission takes place only by coitus.

Immunity is produced by some of the trypanosome infections, and infection with many species may be differentiated by serologic reactions. Development of some species in the invertebrate host includes migration of the parasite to the mouth parts of this host (*anterior station*), so that transmission can occur by the "bite" of the insect, while in other species the trypanosomes are evacuated only in the feces, in which case the organisms are said to utilize the *posterior station*. The trypanosomes that cause African trypanosomiasis in man (namely, *T. gambiense* and *T. rhodesiense*) are transmitted by the "bites" of *Glossina* flies; *T. cruzi*, which causes Chagas' disease, is transmitted by introduction of the infected feces of a bug into the wound made by the "bite." Trypanosomes are only pathogenic for hosts that have not developed a relative immunity to them. In Africa, for example, the big game animals are practically immune to the deleterious effects of *T. brucei*, the wild relative of *T. rhodesiense* that is highly pathogenic for man.

Studies on the ultrastructure of members of the *T. brucei* group have shown marked differences in the mitochondria of the stages in the intestine of the tsetse fly and in the mammalian host. The former have elaborate mitochondrial tubes (both anterior to and posterior to the kinetoplast), while the latter have a simple anterior tubular mitochondrion and no posterior mitochondrion.

With few exceptions, trypanosomes in the trypomastigote stage live in the blood stream and interstitially in the tissues of the vertebrate host, where they multiply by longitudinal binary fission. A notable exception is *T. cruzi* of mammalian hosts, which is adapted to intracellular life in the amastigote stage and does not multiply in the trypomastigote stage.

Stages of division fall into the following sequence: (1) basal body of flagellum replicates and a new flagellum grows out alongside the old one; (2) kinetoplast divides into two, reflecting division of the single mitochondrion; (3) nucleus divides promitotically, the chromosomes at anaphase migrating along the elongating karyosome and the nuclear envelope remaining intact; (4) longitudinal cleavage commences at the anterior end, resulting in two daughter flagellates (Luckins et al., 1967).

Electron-microscopic studies of blood stream forms of *T. rhodesiense* show tubular fibrils paralleling the karyosome at the elongating nucleus; these appear to arise from the persisting nuclear envelope and not at the centrioles. Nuclear division in trypanosomes differs from that of animal cells in that the chromatin material (*chromatin*) is not attached to the spindle fibrils but appears to adhere to the nuclear envelope (Luckins et al., 1967).

Hoare (1948) classified the mammalian trypanosomes into two major sections: those presently (or

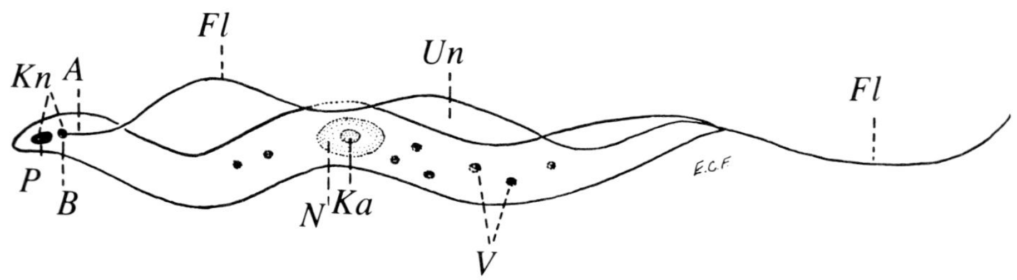

Fig. 10–15. Morphology of a typical trypanosome. *Kn*, Kinetoplast. *P*, parabasal body. *B*, Blepharoplast. *A*, Axoneme. *Fl*, Flagellum. *Un*, Undulating membrane. *N*, Nucleus. *Ka*, Karyosome. *V*, Volutin granules. (By E.C. Faust.)

originally) with migration of the metacyclic trypanosomes to the anterior station of the biological vector and transmitted through the "bite" of a blood-sucking fly; and those transmitted through fecal contamination by the arthropod vector. All of the species pathogenic for man and domestic animals except *T. cruzi* belong to the first group. Later (1967) he suggested that the species in the "*brucei* group" (*T. brucei, T. gambiense,* and *T. rhodesiense*) should be considered as subspecies of *T. brucei*. Ormerod (1967) considered that *T. gambiense* and *T. rhodesiense* should be made synonymous with *T. brucei* and classified at a subtaxonomic level as "strains" or "demes."

Hoare (1967) emphasized "the multiformity of variation and the heterogeneity among mammalian trypanosomes, which have furnished abundant material for deductions regarding the evolution of these hemoflagellates . . . indications of incipient speciation are found among infrasubspecific strains or demes exhibiting minor biological divergence. The gradual transition from one morphological and/or biological type of trypanosome to another, as well as the progressive complication of the life cycles and mode of transmission of these parasites, supply clues to their phylogenetic relations."

According to Nash (1960), human trypanosomiasis in Africa is of considerable antiquity. Epidemics were described by Arab writers of the fourteenth century, long before European exploration of the endemic regions. Cattle trypanosomiasis (*nagana*) was probably more widespread than it is today, due to extensive wild-game reservoirs. This prevented the use of ox-drawn vehicles, which necessitated the use of the primitive hoe and head loads. It also limited the invasion of Islam from the north, and Portuguese and Boer immigration from the south.

Trypanosomiasis in man and domestic animals in Africa is limited to the tsetse-fly belt of Central Africa, an area of 4½ million square miles in which there are 22 species of tsetse flies breeding in a wide variety of habitats, such as swamps, woodlands, savannas, near-desert conditions, sea level and high plateaus. *Glossina* flies and the trypanosomes they host have effectively excluded man and his cattle from one-fourth of the African continent (Foulkes, 1981). Only relatively few species of *Glossina* attack man and transmit human trypanosomiasis; a majority prefer to take the blood of wild game animals.

The two species of trypanosomes parasitizing

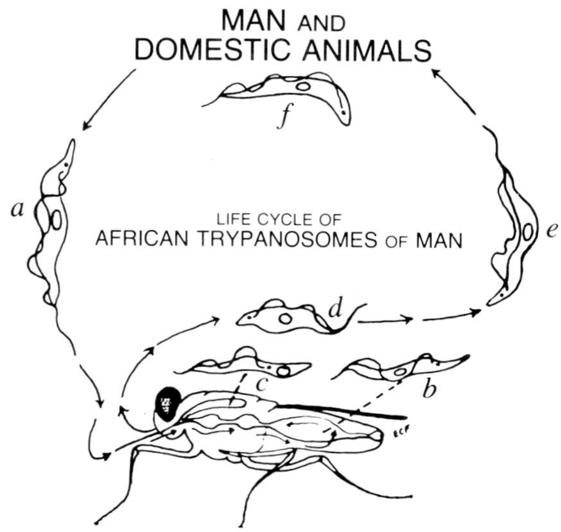

Fig. 10–16. *Trypanosoma brucei-rhodesiense-gambiense* complex. Diagram illustrating the life cycle in the tsetse fly, in mammals, and man. *a,* Trypanosome stage from mammalian host sucked up by fly. *b,* Multiplication of trypanosome stage in fly's midgut. *c,* Multiplication of crithidial stage in salivary glands, followed by transformation into metacyclic trypanosome stage. *d.* Metacyclic trypanosomes introduced by fly into mammal. *e.* Intercellular trypanosome stage in mammal. *f.* Trypanosome stage in blood stream of mammal. (From Faust, E.C., Beaver, P.C., and Jung, R.C. 1968. *Animal Agents and Vectors of Human Disease,* 3rd ed. Philadelphia, Lea & Febiger.)

man in Africa, *T. gambiense* and *T. rhodesiense,* belong to the group complex *brucei-rhodesiense-gambiense* of which *T. brucei,* the ancestral type, is the form in wild game animals, *T. rhodesiense* the form that has fairly recently become adapted to the human host, and *T. gambiense* the form with better human-host adaptation. Members of this group are morphologically indistinguishable. The only distinguishing characteristic of *T. brucei* is its inability to infect man under natural conditions; yet there is evidence suggesting that through lowered resistance of the exposed human host or stepped-up virulence of a particular strain of *T. brucei,* transient human infection may result. The life cycle of this complex is illustrated in Figure 10–16.

Trypanosoma gambiense
Dutton, 1902
(Producing Gambian trypanosomiasis, or Mid- and West African sleeping sickness)

Synonyms. *Trypanosoma ugandense* Castellani 1903; *Trypanosoma hominis* Manson 1903; *Trypanosoma nigeriense* Macfie 1913; *Castellanella gambiense* (Dutton 1902) Chalmers 1918.

Historical Notes. Forde first saw *T. gambiense* in the blood of a European in Gambia, West Africa, in 1901, and Dutton, in 1902, proposed a name for it—*Trypanosoma gambiense*. In 1903, Castellani discovered a trypanosome in the spinal fluid of a patient suffering from sleeping sickness and named it *Trypanosoma ugandense*. It was soon evident that *T. gambiense* and *T. ugandense* were identical, although they were found in different geographical parts of Africa, different anatomic sites of the victims, and different stages of the disease.

Bruce and Nabarro (1903) and Kleine (1909) proved that this trypanosome is transmitted from man to man by the tsetse fly, *Glossina palpalis*, after a period of development in the insect; the exact method of this transmission was elucidated by Kleine.

Geographical Distribution. The endemic home of *T. gambiense* is the tropical heart of Africa, where it is limited on the West Coast to districts between 15° N and 18° S of the equator; Lakes Victoria and Tanganyika are the central foci of the eastern limit (Fig. 10–17). Infection with this trypanosome occurs along the West Coast of Africa from Senegal to northern Angola; toward the east in the regions around Lakes Victoria, Albert, Bangweulu, and Tanganyika; in Equatorial and Central Africa; in Angola; and in the former Belgian and French Congos. In contrast, the regions of *T. rhodesiense* prevalence extend from the southernmost limits of the Sudan through Uganda, Kenya, Tanzania, Zambia, Zimbabwe, Malawi to Botswana, and Northern Mozambique (World Health Organization, 1979). In some of these regions, the incidence of disease is high. Differences in prevalence depend principally upon the number of *Glossina* transmitters present and the opportunities afforded these flies for biting persons harboring trypanosomes in the peripheral blood.

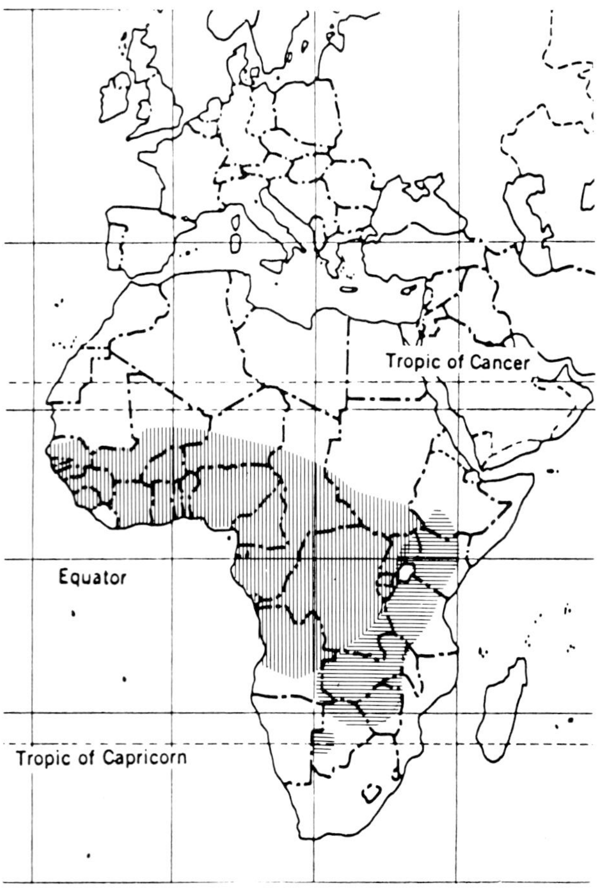

Fig. 10–17. African trypanosomiases, distribution as of 1979. Vertical shading indicates area in which Gambian trypanosomiasis (due to *Trypanosoma gambiense)* occurs. Horizontal shading shows area in which Rhodesian trypanosomiasis (due to *Trypanosoma rhodesiense)* occurs. The areas overlap at the northwest corner of Lake Victoria in Uganda. Recent information from Mozambique and Equatorial Guinea was not available. (Adapted from original of E.C. Faust by R.C. Jung.)

Morphology. In the blood, *T. gambiense* occurs in three forms: a long, slender trypanosome form having a flagellum (trypomastigote); a short, broad one without a flagellum; and an intermediate one. Because of these morphologic differences in the blood of the vertebrate host, this trypanosome is said to be polymorphic.

In fresh blood the trypanosomes may be seen as colorless, spindle-shaped bodies that move rapidly among the red blood corpuscles to which they often give a spinning motion. As motility is reduced, the undulating membrane and flagellum may be differentiated, but not the nucleus or kinetoplast. In fresh wet films, this organism disintegrates rapidly and for study should be fixed immediately as a dry preparation, then suitably stained.

In Giemsa's or Wright's stained preparations, *T. gambiense* measures 14 to 33 μm in length and 1.5 to 3.5 μm in breadth. In the blood of man, the long typical trypomastigote is predominant, as it is in *T. rhodesiense* (Fig. 10–20). The cytoplasm is granular and may be vacuolated; it stains a pale blue and contains dark-blue volutin granules. Centrally located is the nucleus, which stains reddish or reddish-purple; in specimens that are wet-fixed and stained with hematoxylin, a definite nuclear membrane and large central karyosome may be distinguished. The undulating membrane, which stains a pale blue, originates near the blepharoplast and terminates by merging with the body of the trypanosome at the anterior end. It is rather broad and thrown into folds, but it is not striated and does not appear to contain myonemes. At the posterior end of the body, there is a kinetoplast consisting of a blepharoplast and parabasal body. In most preparations, the kinetoplast appears as a dark-red dot, and it is impossible to distinguish between the blepharoplast and parabasal body. Arising from the blepharoplast is the red-staining flagellum, which runs along the edge of the undulating membrane and becomes a free flagellum at the anterior end of the body.

The morphological changes occurring in *T. gambiense* within tsetse flies *(Glossina)* are discussed in the section "Life Cycle."

Habitat. During the early, or febrile, stage of sleeping sickness, *T. gambiense* occurs in the blood and lymph nodes; after the development of cerebral symptoms, however, it is primarily found in cerebrospinal fluid, It does not invade or live within tissue cells but inhabits connective tissue spaces of the various organs, the reticular tissue of the lymph nodes and spleen, and the intercellular spaces in the brain, and is present in large numbers in the lymph channels throughout the body.

Cultivation. *Trypanosoma gambiense* cannot be maintained in N.N.N. medium but will grow in Weinman's medium (see Technical Appendix, p. 762). It can also be maintained by serial passage in laboratory animals, but this method, like culture techniques, causes the organism to lose its infectivity for the tsetse-fly vector and its ability to infect the natural vertebrate hosts.

Trypanosoma gambiense and other salivarian trypanosomes have recently been grown in the presence of bovine fibroblast-like cells in a defined medium in which tney retain the characteristics of blood stream forms (Hirumi et al., 1977; Evans, 1978). The organisms can be produced in the large volumes essential for biochemical and immunologic research, but not for diagnosis.

Life Cycle. In its vertebrate hosts, man and domestic mammals, *T. gambiense* lives in the blood, reticular interstices of the lymph nodes and spleen, and cerebrospinal fluid, in which it multiplies by longitudinal binary division. Reproduction is initiated by the division of the blepharoplast and parabasal body, followed by that of the nucleus, the undulating membrane, and finally, the body of the organism. The axoneme and flagellum do not divide, but a new daughter axoneme develops from one of the daughter blepharoplasts, runs along the edge of the newly formed undulating membrane, and becomes a flagellum at the anterior end of the body of one daughter individual.

In the invertebrate host *(Glossina),* the trypanosomes that reach the intestine of the fly reproduce in the lumen of the mid- and hind-gut both inside and outside the peritrophic membrane, but never intracellularly or attached to the gut wall. After 10 days large numbers of trypomastigote and sometimes epimastigote forms are present. Characteristic slender organisms, produced from the broad forms and called the proventricular forms, appear about the fifteenth day and multiply by binary fission. After migrating back through the esophagus, pharynx, and labrum-epipharynx to the prestomum (i.e., labial cavity), they gain access to the salivary glands via the hypopharynx and salivary ducts. In the salivary glands they become attached to the epithelium and metamorphose into broad epimastigotes that multiply and fill the cavity of the gland. Metacyclic trypanosomes are developed from the epimastigote forms in 2 to 5 days after the slender forms reach the salivary glands, whereupon the fly becomes infective. Once infective, the fly introduces the trypanosomes into the next victim in saliva injected into the puncture wound during the act of feeding. The entire cycle of development in the fly takes about 3 weeks.

Epidemiology. Transmission of *T. gambiense* from man to man typically occurs through the bites of *Glossina* (tsetse) flies after a cycle of development within the insects. Both male and female flies bite man and may serve as vectors. *Glossina palpalis* is the usual species inolved but in certain regions *G. tachinoides* is more frequently naturally infected than is *G. palpalis,* or is the only species of tsetse fly present in these endemic areas.

The fly may transmit the infection mechanically. *Direct mechanical transmission* is accomplished through the soiled proboscis of a fly that, after an interrupted blood meal on an infected person, "bites" a healthy one within 2 to 3 hours. This method of transmission may be of great importance during epidemics when infected individuals, noninfected individuals, and the flies are all numerous. Mechanical transmission by other blood-sucking arthropods may occur, but the extent of such transmission to man is unknown.

The usual method of transmission is through the *"bite" of the fly* after the trypanosomes have gone through the metacyclic phase of the cycle. Under various conditions of temperature and moisture, the development period in the fly may be as short as 12 or as long as 30 days, but is usually about 20 days. Less than 10% of tsetse flies become infective after biting sleeping sickness patients. The infection is not congenital in the fly.

Congenital transmission of the trypanosome in man is possible. In 1929, a case was reported by Mühlens of a child, born in Germany of an infected mother, who died less than 3 months after birth as a result of proven trypanosomiasis. Reports of seven more recent cases of congenital African trypanosomiasis were summarized by Burke et al. (1974).

While *T. gambiense* may be transmitted to most mammals, including game animals, domestic species, and laboratory animals, the virulence of the infection depends upon the particular strain of trypanosome employed, the animal inoculated, and upon successive passages through certain host species. In monkeys, the infection is fatal, but the higher apes are more resistant. Infection in the dog may last from 30 days to 8 or 9 months, according to the virulence of the strain inoculated. In rats, mice, guinea pigs, and rabbits, the infection runs a chronic course, and in horses, cattle, and sheep, it is chronic and does not result in fatalities.

More important from an economic viewpoint is the fact that the same tsetse flies that transmit Gambian trypanosomiasis from man to man also transmit other trypanosome infections that are not infective for man but are pathogenic and frequently fatal for domestic animals. In tsetse-fly zones, this makes it practically impossible for natives to raise livestock, so that they have little or no animal protein in their food. Consequently, their level of resistance is lowered and they are more seriously affected by *T. gambiense* when exposure occurs.

Reservoir Hosts. There is no proof that any of the game animals of Africa act as reservoirs of infection for man, even though various species of antelopes may be experimentally infected. However, domestic animals, particularly cattle, pigs, and goats, carry this infection for long periods of time without apparent symptoms.

Aside from the influence exerted by exposure to infection by flies, *age, sex, race,* or *occupation* have no relation to susceptibility to Gambian trypanosomiasis. In regions where the disease is endemic, the natives appear to develop a greater resistance to infection than is observed in newcomers to the district, but beyond this there is no evidence of natural immunity, although spontaneous recovery may occur in some cases. The breeding, flight, and biting habits of *Glossina palpalis* and other natural vectors are intimately associated with the spread of infection. (For a discussion of these subjects, the reader is referred to Section IV, pp. 667–670.)

Pathogenesis, Pathology, and Symptomatology. *T. gambiense* typically produces a chronic disease in which there are three progressive states following an incubation period of 2 to 23 days (average 6–14 days): trypomastigotes exclusively or predominantly in the blood stream; trypomastigotes predominantly in the lymph nodes; and invasion of the central nervous system by the trypomastigote.

When the metacyclic trypomastigote stage of *T. gambiense* is introduced into the skin of a person from the proboscis of an infected tsetse fly, the parasites first lodge in the tissues at the site of injection. Here they set up a local inflammatory reaction that subsides in 1 or 2 weeks. Gradually they gain entrance to the blood stream and, as they multiply, produce a parasitemia. They never typically invade tissue cells, but produce injurious effects on every tissue and organ of the body. This is first noted in the lymph nodes, where there is marked proliferation of endothelial cells lining the sinuses and infiltration of leukocytes around blood vessels. Later, when the central nervous system is involved, trypomastigotes enter the arachnoid spaces and then the brain substance, with infiltration of plasma cells and lymphocytes, and perivascular proliferation of endothelial and neuroglia cells between the blood vessel and perivascular sheath.

There are two forms of human trypanosomiasis in Africa (Fig. 10–17): the one caused by *Trypanosoma rhodesiense* has a short course and ends fatally within a year; and the other, which is caused by *T. gambiense,* has a much longer, milder chronic course and ends fatally with central nervous system involvement after several years' duration. The pathologic anatomy and general symptomatology of the two forms are practically the same except that the more rapidly fatal form does not usually allow time for the marked changes in the nervous system characteristic of *T. gambiense* infection.

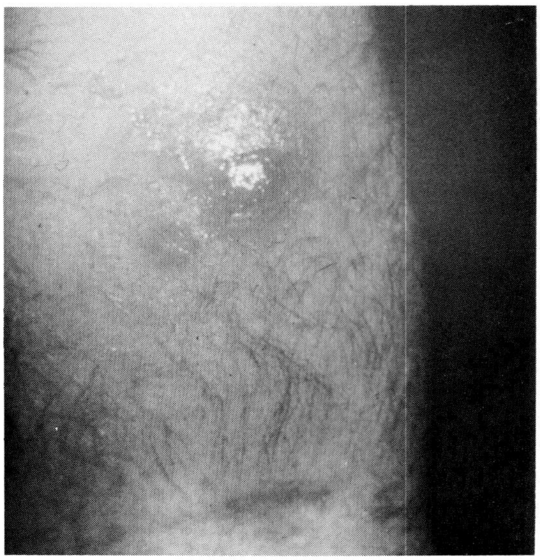

Fig. 10–18. The primary lesion of African trypanosomiasis on the forearm. (Courtesy of Dr. P.E.C. Manson-Bahr.)

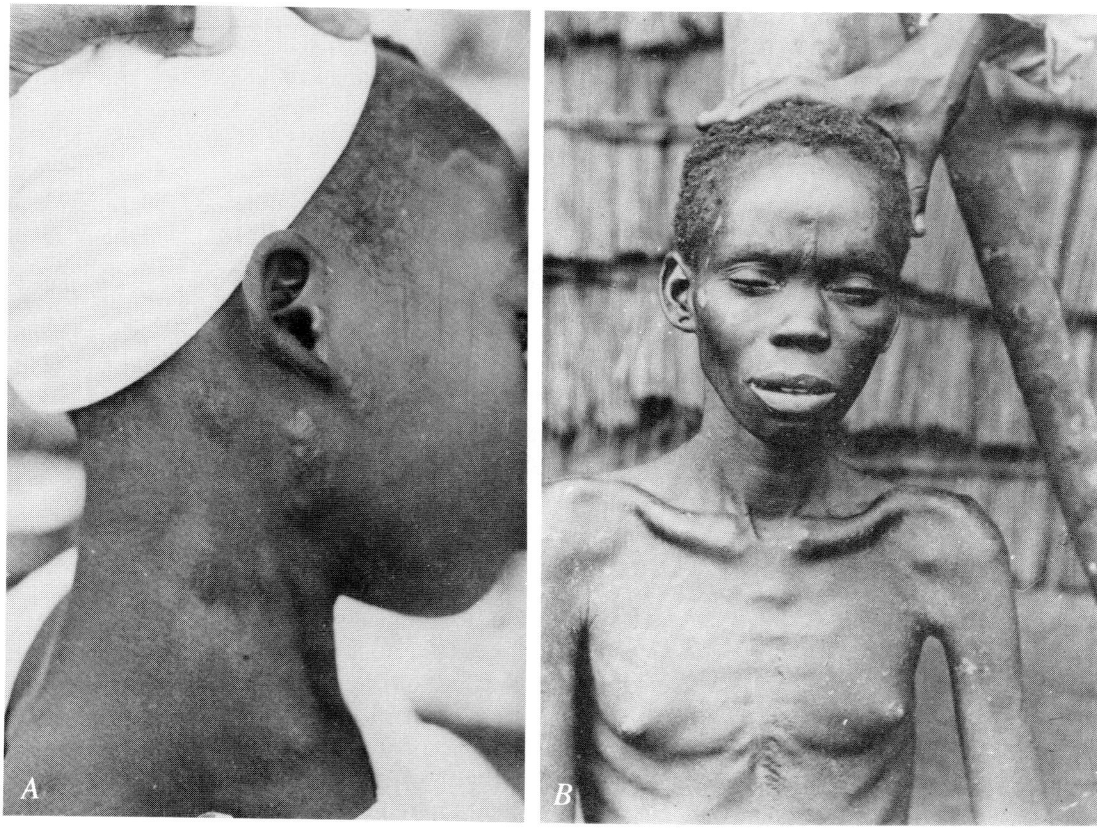

Fig. 10–19. Gambian trypanosomiasis. *A,* Posterior cervical lymphadenopathy (Winterbottom's sign). Swellings due to enlarged lymph nodes are visible back of the neck. On the side of the face and neck are scars that were caused by the treatment of a tribal healer. Scars of this type in West Africa were recognized by slave traders as an indication to look more closely for enlarged lymph nodes. *B,* Typical somnolence and emaciation. (*A* courtesy of Wellcome Museum of Medical Science, London. *B* courtesy of Dr. Cuthbert Christie.)

The *pathologic and microscopic anatomy* of Gambian trypanosomiasis is characteristic and most evident in the lymphatic and nervous systems. In Europeans (but seldom in African natives), an initial trypanosome chancre, which is somewhat elevated, indurated, and painful to the touch (Fig. 10–18), develops at the site of the *G. palpalis* "bite." During the first stage, there is little gross or microscopic change in the tissues except for the presence of trypomastigotes in the circulating blood. With the invasion of lymph nodes, a generalized enlargement occurs, especially in postcervical (Fig. 10–19*A*), submaxillary, inguinal, and femoral nodes. In the febrile stage, the nodes are swollen, soft in consistency, and enlarged; in later stages of infection, they may be only slightly enlarged and fibrous. The spleen is enlarged, and the liver may be enlarged and congested. Macroscopically, the brain shows the usual pathologic changes of a chronic leptomeningitis, and the cerebrospinal fluid is turbid, increased in amount, and contains trypomastigotes.

The pathology of sleeping sickness is essentially that of a chronic inflammation of the lymphatic system and perivascular infiltration around blood vessels. This consists of endothelial cells, the plasma cells of Marschalko (cells having a wheel-like arrangement of the nuclear chromatin), proliferated neuroglia cells, lymphocytes, and macrophages that may contain red blood cells. Where pressure is present due to infiltration or hemorrhage, the dendrons in ganglion cells may be atrophied. Trypanosomes may leave the blood vessels and lymphatics and penetrate surrounding tissues; they are found within connective and nervous tissues.

The *symptomatology* of infections with *T. gambiense* involves an incubation period, an essentially symptomless patent period, an acute febrile period, toxic exacerbations following a mild chronic period, and the sleeping sickness stage with central nervous system involvement. The *period of incubation* has been estimated to be from 2 days to several weeks; the period was determined to be about 6 to 14 days when it was possible to calculate accurately the time elapsing between the infective fly "bite" and appearance of the first symptoms. The "bite" of an infected fly is followed by local irritation of varying severity and in Europeans is at times accompanied by headache, fever, chills, and loss of appetite; in Africans, this is essentially a symptomless period.

The *second stage* in Africans is also without symptoms or any awareness of infection unless routine blood examination reveals the trypomastigotes. The individual is still able to do hard work. From this stage, the disease may have repeated relapses, become abortive, change into the lymphatic stage, or directly transform into the central nervous system type.

The first distinct symptoms in Africans parallel the *invasion* of lymph nodes by trypomastigotes and are characterized by an attack of fever that usually lasts about a week. This is followed, after an afebrile period of variable duration, by another febrile period. The trypomastigotes are released into the blood stream only during the febrile state; even then they are rapidly killed by circulating trypanocidal antibodies. Fever may become continuous but is usually irregular and reaches a maximum in the evening. With fever the following usually occur: enlargement of the spleen, liver, and lymph nodes, especially those of the postcervical triangle *(Winterbottom's sign)* (Fig. 10–19,*A*), but later also of the axilla and groin; headache; neuralgic pains, especially in the joints; weakness in the legs; cramps; and erythematous eruptions of the skin. Delayed sensation to pain *(Kerandel's sign)* is frequently present, and localized edema of hips, hands, legs, and especially the eyes may occur. Febrile attacks alternate with quiescent periods in which most of the symptoms may disappear. There is typically a developing dyspnea and an increasing sense of cardiac pain, disturbed vision, anemia, and loss of strength. At times there is nephritis, and in women, dysmenorrhea is characteristic. The pulse ranges from 100 to 140. In some cases, spontaneous recovery may occur during this acute febrile period. At this stage, the enlarged lymph nodes are soft, nonsuppurative, and may be very tender; trypomastigotes are typically found in the juice.

With *invasion of the nervous system* by the trypomastigotes, the chronic or sleeping-sickness stage of infection is initiated. In this stage, headache is severe and mental dullness and apathy gradually develop. The patient becomes morose or excitable, is disinclined to work, and has a weary or shuffling gait. Typically there are muscle spasms, trembling of hands and trunk, disturbance of coordination and of the reflexes, pain and stiffness in the neck, and spastic or flaccid paralysis—all of which increase from day to day. Sleepiness develops and becomes so pronounced that the patient will fall asleep while eating or wherever he may be standing or sitting. Mania or melancholia are often noted, and delusions and symptoms resembling those of general paresis may occur. The knee jerks are increased at first but lost before death. There are no characteristic eye changes, with normal fundus oculi. Tremor of tongue and limbs and choreic movements of the latter are common. As the disease progresses, the patient becomes extremely emaciated, sleeps continuously, cannot be aroused to take food, and has convulsions. Death follows from asthenia, which is preceded by a profound coma (Fig. 10–19,*B*). Complications, such as malaria, ancylostomiasis, schistosomiasis, amebic and bacillary dysentery, or lobar or bronchopneumonia are common and often the direct cause of death. Death may also result from meningitis, heart failure, or a severe fall.

Infection with *T. gambiense* is not always chronic in type, for acute and rapidly fatal infections in man have been reported, especially during epidemic outbreaks.

Diagnosis. The diagnosis of infection with *T. gambiense* is based on the patient's characteristic symptoms and his history of living or having lived in an endemic area. The diagnosis is established by demonstration of the parasite in blood, lymph node juice, sternal bone marrow, or cerebrospinal fluid. Methods available to assist in diagnosis are direct microscopic examination of stained or unstained preparations of blood, lymph node aspirate, or cerebrospinal fluid, cultivation upon suitable media, and animal inoculation (Lanham, 1977). Early diagnosis is essential since early treatment is most successful.

Not only are lymph-node puncture and spinal-fluid examination essential in diagnosis as the disease progresses, they also serve as indices of this advance even in relatively quiescent cases. Many individuals in the large relatively asymptomatic group would not be found positive and would die if reliance were placed solely on blood-film examination.

Serologic techniques can be applied to the large-scale screening of populations for African trypanosomiasis (Lanham, 1977). An enzyme-linked immunosorbent assay (ELISA) technique appears to be a good alternative to the immunofluorescence (IF) technique that is now widely used (Ruitenberg and Buys, 1977).

The symptoms of diagnostic importance in early stages of sleeping sickness are attacks of irregular fever, enlargement of lymph nodes, especially those of the posterior triangle of the neck *(Winterbottom's sign),* erythematous skin eruptions, and delayed sensation to pain *(Kerandel's sign).* The differential diagnosis between Gambian trypanosomiasis and malaria, relapsing fever, enteric fever, undulant fever, ancylostomiasis, and syphilis may be made with the various laboratory methods available for each infection.

Treatment. Frequently, it is not known whether the patient has Gambian or Rhodesian (see p. 86) trypanosomiasis when treatment is instituted; therefore, the treatments of both types of infection will be described. Early treatment is essential, since results are poor once central nervous system involvment has occurred. General supportive measures such as proper hygiene, good nutrition, and adequate fluid intake should be instituted when possible.

All drugs that are useful in the management of African trypanosomiasis are toxic. To begin therapy, the least toxic drug must be selected unless there is central nervous system invasion. Effective therapy for such invasion—i.e., when there is increased protein and more than five lymphocytic cells/cu mm of cerebrospinal fluid—requires the use of the more toxic drugs.

Available in the United States, from the Parasitic Disease Drug Service of the Centers for Disease Control in Atlanta, Georgia are the following: pentamidine, suramin, and melarsoprol.

Pentamidine isethionate usually cures Gambian trypanosomiasis in its first stage and gives fair results in the early phases of the Rhodesian type. It is administered in a dosage of 4 mg/kg intramuscularly every day for 10 days, as for visceral leishmaniasis (see p. 77). Common side effects are an immediate hypotensive reaction and local pain. Less commonly, hypoglycemia may occur.

Suramin sodium, which is more toxic than pentamidine, is also valuable if used before the central nervous system is severely damaged. It is preferable to pentamidine in the treatment of early Rhodesian infection or if there is only slight cerebrospinal fluid abnormality. The initial dosage is 100 to 200 mg as a test dose, then 1.0 g intravenously on days 1, 3, 7, 14, and 21.

Melarsoprol, which is more toxic than suramin, offers the greatest hope of improvement in the later stages of the infection. It is administered intravenously in a dosage of 18 mg initially with a progressive daily increase until 2 to 3.6 mg/kg is given daily. This dose is administered for 3 successive days. After 1 week, 3.6 mg/kg is given daily for three doses. The last 3-day course may be repeated after 10 days.

Tryparsamide is administered intravenously as a 20% solution every 5 days. The initial dose is 1.0 g; this is gradually increased to 35 mg/kg (maximum 3.0 g), which is repeated for 10 to 12 injections. It has some value in late *T. gambiense* infection but is unacceptably toxic and can precipitate optic neuritis. The development of the latter complication is preceded by optic pain, photophobia, and lacrimation, symptoms that should be carefully watched for in the course of treatment. Tryparsamide is of little value in Rhodesian trypanosomiasis. It is not available in the United States.

Disodium-p-melaminylphenylarsonate (Melarsen, Melarsen oxide), *trimelarsan* (Mel W), and *berenil* are other drugs that have been used with some success in febrile infections in the early stages of trypanosomiasis.

Prognosis. In cases of Gambian trypanosomiasis that are recognized early and treated with specific trypanocidal drugs, the outlook is usually good, but the prognosis is poor after central nervous system involvement has developed.

Prevention. Earliest control measures consisted in the removal of entire villages to a safe area. This was later integrated with rural development in tsetse-free areas and mass survey and treatment, especially with pentamidine. Clearing of riverine vegetation has been supplemented by game destruction in various places; in other places, killing potential wild mammalian hosts on which the tsetses feed has, by itself, given relief from trypanosomiasis. Aircraft spraying of two blocks of woodland in Tanzania with dieldrin and isobenzan, respectively, resulted in elimination of *G. morsitans* and *G. pallidipes* from both areas (Burnett et al., 1964). In Northern Nigeria, where rivers flowing northeastwards into Lake Chad pass through areas infested by *G. morsitans submorsitans* and *G. tachinoides,* campaigns were undertaken from 1955 to 1964 to eradicate both species by selective spraying of the vegetation on which the flies rest. This method of control has been discontinued or decreased in some areas, because the widespread use of insecticides is costly and contaminates the environment, affecting not only the tsetse population but the other insects, and directly or indirectly the birds, fish, and wild and domestic animals. However, effective alternative methods of control are not yet available (Jordan, 1978).

With the ever-changing political situation in Africa, the application of mass treatment and prophylaxis will become increasingly difficult and rigorous control of population movement will no longer be possible. Complete control can never be attained by chemotherapy alone; regular and continuous occupation of sleeping sickness-endemic regions by control teams will be necessary until the developing countries have acquired the trained personnel necessary to organize their health and other services.

Trypanosoma rhodesiense Stephens and Fantham, 1910 (Producing Rhodesian trypanosomiasis or East African sleeping sickness)

Historical Notes. This trypanosome is the cause of a disease that typically runs a rapidly fatal course and often ends in death before the development of the nervous symptoms so typical of *T. gambiense* infection. The trypanosome was discovered by Stephens and Fantham in 1910 in the blood of a patient in Rhodesia who was suffering from symptoms suggestive of "sleeping sickness." These workers named it *Trypanosoma rhodesiense*. In 1912, Kinghorn and Yorke proved that it is transmitted to man by the tsetse fly *Glossina morsitans* and that it undergoes a cycle of development in the transmitter similar to that of *T. gambiense* in *G. palpalis*.

Numerous investigators regard *T. rhodesiense* as the human strain of *T. brucei,* a common trypanosome that is found in the antelope and produces a fatal disease in cattle known as "nagana." Ormerod (1961b) suggested that the original strains of *T. rhodesiense,* which were observed in epidemic form in Malawi, Tanzania, and Uganda, probably arose in the Zambesi basins as mutants from *T. brucei*. Both species—the former in man, the latter in domestic stock—are found sporadically today in Zimbabwe. In support of its specific nature, Taute inoculated himself, his colleague Huber, and 129 volunteers with *T. brucei* derived from naturally infected horses and mules without a single human infection resulting (Taute and Huber, 1919). Heisch et al. (1958) and Ashcroft (1959) demonstrated that *T. rhodesiense* may be maintained alternately in domestic mammals or wild game animals and *G. morsitans,* and Van Hoeve et al. (1967) achieved successful cyclical transmission of a strain isolated from a cow through *G. morsitans* to a man, a cow, and sheep. It seems more practical at this time, however, to consider that there are two species, *T. rhodesiense* and *T. brucei,* which are morphologically indistinguishable, and that together with *T. gambiense* they constitute a closely related biologic group, the *brucei* complex.

Geographical Distribution. The distribution of *T. rhodesiense* (Fig. 10–17) is much more limited than that of *T. gambiense,* as it has been found only in the countries of Zimbabwe, Malawi, Tanzania, Rwanda, Burundi, eastern Uganda, southernmost Sudan, around the shores of Lake Victoria in Kenya, and in the northeastern portions of Mozambique and Botswana.

Morphology. The morphology of *T. rhodesiense,* both in man and in the transmitting flies, is identical to that of *T. gambiense* (see p. 80; although, when it is inoculated into rats, mice, or guinea pigs, posterior nucleate forms are apparently more common in *T. rhodesiense* than they are in *T. gambiense* (Fig. 10–20).

Habitat. The habitat of *T. rhodesiense* is in the blood, lymph nodes, cerebrospinal fluid, and tissues of man, and in the blood of cattle and various species of antelope. It also lives and multiplies within the intestinal tract of flies belonging to the genus *Glossina* and within their salivary duct and glands.

Cultivation. This species of trypanosome may

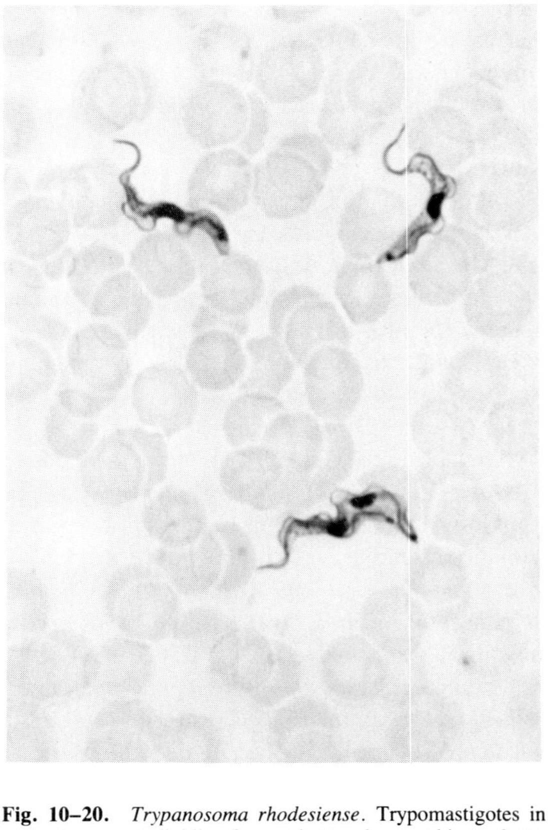

Fig. 10–20. *Trypanosoma rhodesiense*. Trypomastigotes in a blood smear. A dividing form at bottom has two kinetoplasts, two nuclei, and two flagella. Giemsa stain. (× 1200.) (Courtesy of Dr. R.G. Yaeger.)

be cultivated on Weinman's medium or that of Tobie et al. (see Technical Appendix, p. 762).

Life Cycle. The life cycle and methods of reproduction of *T. rhodesiense* are similar to those of *T. gambiense* in man, susceptible mammals, and the transmitting flies (see p. 81).

Epidemiology. Under ordinary conditions, *T. rhodesiense* is transmitted from man to man by the bite of infected male or female *Glossina morsitans* and, in some regions, by *G. swynnertoni* or *G. pallidipes,* and at times, by *G. palpalis*. The Rhodesian disease has lower overall prevalence in its endemic territory, is less often epidemic than is the Gambian infection, and is more prevalent in males than females. Development within the fly is controlled largely by temperature. Kinghorn and Yorke (1912) found that the ratio of infective flies to non-infective ones in hot valleys was 1 to 534, while in the cool plateau regions it was 1 to 1312. The cycle of development in the fly is identical with that of *T. gambiense* (see p. 79); the fly becomes infective in 10 to 31 days, usually 18 to 20 days. Other blood-sucking insects have been suspected of transmitting the trypanosome to man mechanically, but this has not been proven. *Mechanical transmission* by *G. morsitans* may occur, but the fly must bite within a few hours after feeding on an infected individual. *Sex, age, race, and social condition* have no influence in the epidemiology of the infection beyond their possible relationship to exposure to the "bite" of the transmitting fly.

The *incidence of infection* with *T. rhodesiense* is considerably less than that with *T. gambiense*. Characteristically, the distribution of Rhodesian sleeping sickness is in small foci, and the infection usually occurs in sporadic form with few epidemics, which are of small extent. In Tanzania, fly belts and the prevalence of cattle trypanosomiasis have been influenced by the rinderpest epizootic of 1890 to 1896 (Ford, 1965). *T. rhodesiense* in man has followed the expanded or shrunken tsetse belts and is of zoonotic *(T. brucei),* not *T. gambiense* derivation. In the past, epidemics have been self-terminating, but medical services have encouraged persistence of zoonotically derived sleeping sickness by preventing population extermination and encouraging social behavior favoring continuation of transmissions involving the human host (Ford, 1965).

The *reservoirs of infection* for man are believed to be game animals. Heisch et al. (1958) demonstrated that a strain of *T. brucei* complex obtained from the bush buck *(Tragelaphus scriptus)* produced a *T. rhodesiense* type of disease when introduced into a human volunteer.

Domestic and laboratory animals are easily infected experimentally with *T. rhodesiense,* and the infection is usually rapidly fatal in laboratory rats.

Pathogenesis, Pathology, and Symptomatology. *Trypanosoma rhodesiense* produces acute infection. The earlier stages in the development of the pathologic state parallel those of the Gambian infection (see p. 82). However, in the Rhodesian type, the toxic process is characteristically more rapid and fulminating and the disease reaches its climax before any considerable involvement of the nervous system occurs. In untreated cases, the disease is typically fatal. Nevertheless, some patients may remain symptomless ("carriers") for many months before developing clinical evidence of the disease.

In man, *T. rhodesiense* is the cause of a type of trypanosomiasis that is characterized by its rapid course, severe symptoms, and resistance to treatment in its more advanced stage. The incubation period in experimental infections of man has been from 1 to 2 weeks.

The *pathologic and microscopic anatomy* of infections with *T. rhodesiense* in man is similar to that of *T. gambiense* infections (see p. 83), but the changes noted in the nervous system are usually lacking.

The general *symptomatology* of infection with *T. rhodesiense* is similar to that of *T. gambiense* infections (see p. 83), but the symptoms develop more rapidly and death occurs after months rather than years, as in the latter infection. Febrile paroxysms are more frequent, but lymph-node enlargement is less pronounced, and edema, weakness, rapid loss of weight, myocarditis (Manson-Bahr and Charters, 1963), and fever are the striking symptoms. Mania and delusions are often noted, but the profound somnolence and other marked nervous symptoms of the "sleeping sickness" stage are lacking or not so evident in this infection as in *T. gambiense* infection.

Diagnosis. The same laboratory methods are available for the diagnosis of *T. rhodesiense* infection as for *T. gambiense* infection. The trypomastigotes are found most frequently in peripheral blood during the febrile periods, but puncture of the enlarged lymph nodes and examination of the material so obtained for trypomastigotes is perhaps the best diagnostic method (see p. 84). Typically, there is fulminating development of the infection when *T. rhodesiense* is introduced into laboratory rats.

Treatment. This infection is more resistant to treatment than the infection due to *T. gambiense* (see p. 85, under Gambian trypanosomiasis). Tryparsamide is of little value, while Atoxyl and antimony preparations are of no value. *Suramin* U.S.P. is effective *if treatment is begun early.* In advanced stages of the disease, Mel B cures otherwise hopeless cases.

Pentamidine is employed to control epidemics.

Prognosis. This is fair to excellent in cases treated with suramin early in the infection, but poor if treatment is delayed. So far as is known, untreated cases always end in death.

Prevention. Even though pentamidine is employed in epidemic outbreaks of Rhodesian trypanosomiasis, it does not seem likely that it alone will eliminate human infection, particularly when animal reservoirs are near a hamlet and enter into the vicious cycle in addition to the man-tsetse-man cycle. The man-tsetse contact must be broken by concentration of the population in large villages in open country with a circumference free of tsetse flies and game animals. Vector control involves the clearing of vegetation around human settlements, employment of insecticidal measures (especially spraying insecticides in attempts to exterminate *G. morsitans, G. swynnertoni,* and *G. pallidipes),* and prophylactic treatment of cattle and other domestic mammals to prevent trypanosomiasis in these animals from endangering the human food supply. In Tanzania, *G. swynnertoni* was eliminated from an area 35 miles square by selective application of 3% dieldrin in emulsified solution to the resting sites of the flies (Chadwick et al., 1964).

Trypanosoma cruzi
Chagas, 1909
(Producing Chagas' disease or South American trypanosomiasis)

Synonyms. *Schizotrypanum cruzi* Chagas, 1909; *Trypanosoma triatomae* Kofoid and McCulloch, 1916.

Historical Notes. In 1909, while investigating malaria in Brazil, Carlos Chagas discovered *Trypanosoma cruzi* in the intestine of a triatomid bug, *Panstrongylus megistus.* Infected bugs were allowed to bite a monkey, and the trypomastigote form was afterwards found in the blood of this animal. Chagas later found the same parasite in the blood of a child suffering from fever, anemia, and enlargement of the lymph nodes and proved it to be the cause of a common disease endemic in certain parts of Brazil. This is the only instance known in medicine in which an animal parasite causing a disease and the insect transmitting it were discovered before the disease itself was differentiated.

Chagas erroneously believed that, during a certain stage in its life cycle in man, *T. cruzi* multiplied by schizogony; for this reason, he established for it the genus *Schizotrypanum,* renaming it *Schizotrypanum cruzi,* a name that is still used by some workers. As this name was based upon a misconception, it was withdrawn by Chagas, who reassigned the species to the genus *Trypanosoma.* Nevertheless, the marked differences between *T. cruzi* and the African parasites *T. gambiense* and *T. rhodesiense* could eventually result in the assignment of *T. cruzi* to a separate genus for which the original name, *Schizotrypanum,* would have priority, despite the erroneous reason for so naming the parasite.

Geographical Distribution. Insofar as is known, Chagas' disease is confined to the Western Hemisphere (Fig. 10–21). Originally it was believed to be limited to Brazil, but extensive investigations have shown that it has a much wider distribution. In addition to Brazil, it has been found in human beings in Chile, Argentina (northern provinces), Uruguay (inland), Paraguay (primarily the Chaco area), Bolivia, Peru, Ecuador, Colombia, French Guiana, Venezuela, Panama, Costa Rica, El Salvador, Guatemala, in Oaxaca, Nayarit, and Yucatan states of Mexico, and at Corpus Christi, Texas, United States. The most highly endemic center in Brazil is in the State of Minas Gerais. It has been estimated that 35 million people in Latin America are exposed to this infection. Naturally infected triatomids and mammals have been found in several countries of the Americas outside areas in which human cases have been recognized, including the southern United States from Maryland to California.

Trypanosoma cruzi may occur in the Eastern Hemisphere, but more intensive study is needed to determine the autochthonicity of reported cases.

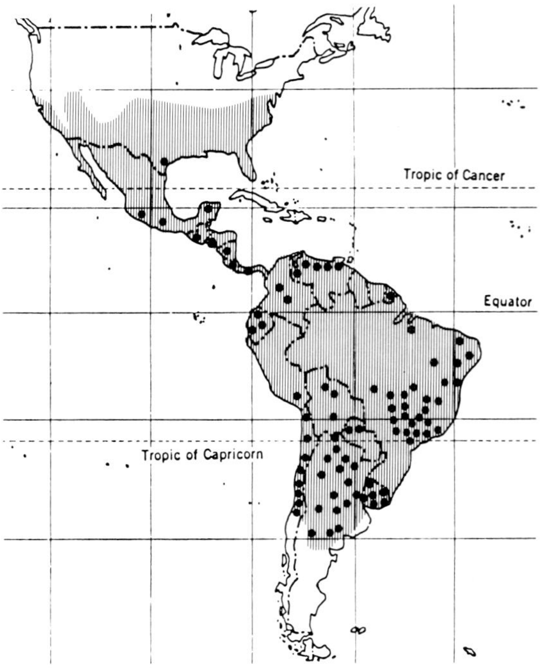

Fig. 10–21. *Trypanosoma cruzi,* distribution of infection (Chagas' disease). Shading indicates estimated extent of infection in reservoir host. Dots indicate distribution of human infections. (Adapted by R.C. Jung from original of E.C. Faust.)

Morphology. *Trypanosoma cruzi* is a pleomorphic trypanosome that has two phases in its life cycle—one in man or reservoir host, and one in the transmitting insects. The morphology is best studied in preparations stained with Giemsa's stain. In the mammalian host, the organism occurs in the blood as a typical trypomastigote and in the reticuloendothelial and other tissue cells as a typical amastigote; during transformation the organism occurs in promastigote, epimastigote, and trypomastigote forms, while in the transmitting insects, amastigote, epimastigote, and trypomastigote forms occur. (See Table 10–1, p. 56, and Fig. 10–22.)

In the blood of man, *T. cruzi* is a spindle-shaped trypomastigote measuring about 20 μm in length. The cytoplasm stains blue and the nucleus, kinetoplast, and flagellum stain red or violet with Giemsa's stain. Two forms occur in the blood—a long, slender one and a short, broad one—but they probably have no sexual significance. Both forms have a nucleus situated at the center of the body and a kinetoplast, which in the shorter forms is large and appears as a deep red or violet mass at the posterior extremity. In well-stained preparations, the kinetoplast is seen to consist of a dot-like blepharoplast and a larger, oval parabasal body. The root of the flagellum, the axoneme, arises from the blepharoplast and extends along the edge of a narrow undulating membrane having few convolutions, and at the anterior end of the body projects as a free flagellum. In stained preparations, the trypanosomes assume a characteristic C-shape (Fig. 10–23,*D*). This trypomastigote does not divide in the blood stream; it multiplies only in reticuloendothelial and other tissue cells as amastigote forms or as a promastigote or epimastigote form when it is about to escape from its intracellular location. The undulating membrane and flagellum disappear after entry into the cell, and the organism divides by binary fission, producing amastigotes that, by repeated division, eventually fill and destroy the invaded cell. The amastigote forms (Fig. 10–23,*A*) are round or ovoidal in shape and measure from 1.5 to 4 μm in diameter. Each has a large nucleus that is ruby-red in color and a rod-like or spherical kinetoplast that is stained deep violet or almost black with Giemsa stain. In fixed and wandering histiocytes, especially in spleen, liver, lymph nodes, and bone marrow, the amastigote stage behaves like *Leishmania donovani,* from which it is almost indistinguishable as an intracellular parasite. In neuroglia cells and myocardium, the amastigote forms are collected within a cyst-like cavity in the invaded cells and, after further multiplication and development, are liberated (with destruction of the cells) as amastigotes, promastigotes, epimastigotes, or trypomastigotes; only the trypomastigotes are found free in peripheral blood.

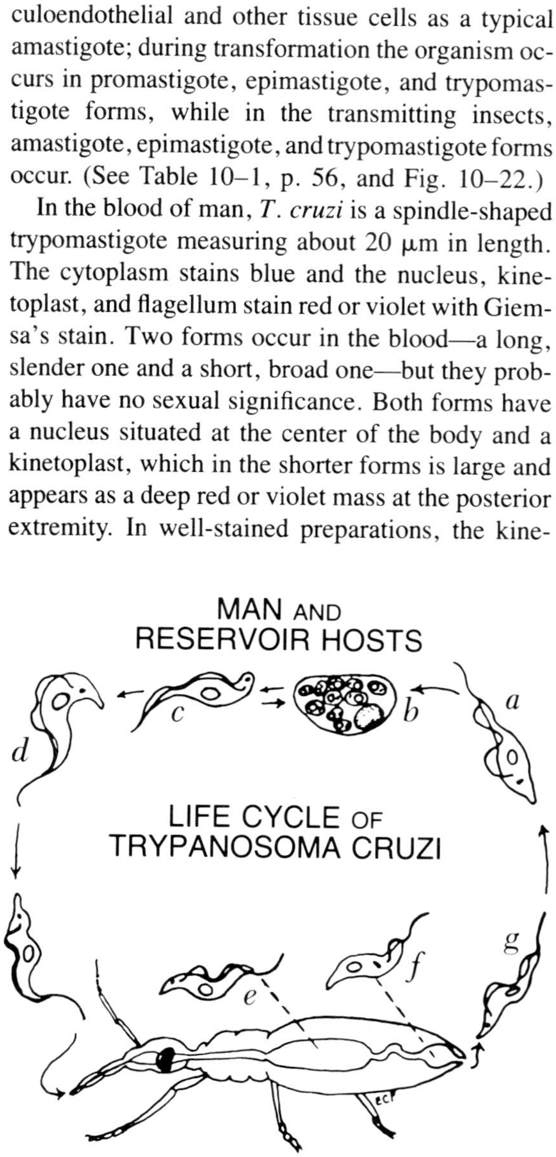

Fig. 10–22. *Trypanosoma cruzi*. Diagram illustrating the life-cycle stages in the vector (triatomid bug), in wild and domestic mammals, and in man. *a,* Trypanosome in mammalian blood. *b,* Amastigote in histiocytes. *c,* Trypanosome in blood stream derived from intracellular amastigotes. *d,* Trypanosome sucked up by bug. *e,* Trypanosome in mid-gut, transforming into epimastigote *(f)* in hindgut. *g,* Metacyclic trypanosome derived from epimastigote, excreted in bug's feces, infective for mammalian host. (Adapted from original of E.C. Faust. In Faust, E.C., Beaver, P.C., and Jung, R.C. 1968. *Animal Agents and Vectors of Human Disease,* 3rd ed. Philadelphia, Lea & Febiger.)

In the transmitting insects, *T. cruzi* has the general morphology common to the epimastigotes and metacyclic trypomastigote forms of other trypanosomes. Short and long epimastigote forms are seen, having the kinetoplast at the anterior end with the nucleus central in position; intermediate forms having the kinetoplast at various levels of the body; and metacyclic trypomastigotes having the kinetoplast at the posterior end and a well-developed undulating membrane and flagellum.

Habitat. This species of trypanosome lives as a morphologically typical trypomastigote in the blood; as an amastigote in reticuloendothelial and other tissue cells of man and many mammals; and typically as an epimastigote and metacyclic trypomastigote in the intestinal tract of certain insects. In man the most frequent locations of the parasite are in reticuloendothelial cells of spleen, liver,

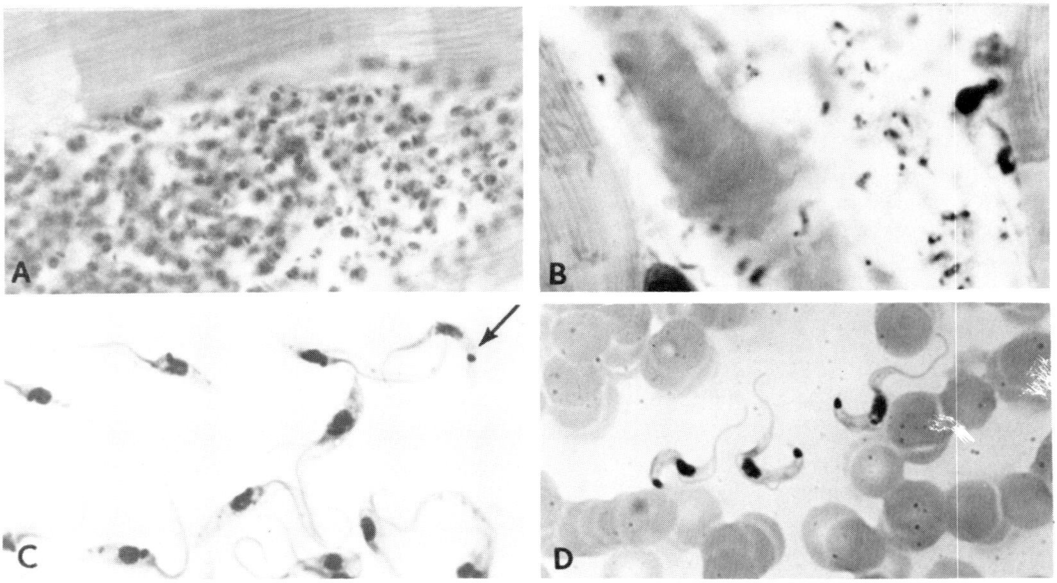

Fig. 10–23. *Trypanosoma cruzi. A,* Intracellular amastigotes in skeletal muscle. H & E stain. *B,* Slender trypomastigotes breaking out of a muscle fiber. H & E stain. *C,* Smear preparation of culture forms, one of which is a metacyclic trypomastigote (arrow) with the kinetoplast posterior to the nucleus. The others are epimastigotes in which the kinetoplast is immediately in front of the nucleus. These and transitional forms are found in the intestine of the triatomid bug as well as in cultures. Giemsa stain. *D,* Trypomastigotes in blood. Giemsa stain. (× 1200.) (Courtesy of Dr. R.G. Yaeger.)

lymph nodes, and lymphatic tissues, and in myocardium. In addition, cells of the striated muscles, bone marrow, suprarenal glands, testes, ovaries, and nervous system may be invaded. The parasites also occur in histiocytes of cutaneous tissue and in cells of the epidermis, as well as in the intestinal mucous membrane.

Cultivation. *T. cruzi* is easily cultivated on N.N.N. medium (see Technical Appendix, p. 762). The forms occurring in culture, morphologically similar to those found in the transmitting insects, are infective to experimental animals.

Romaña and Meyer (1942), using chick-embryo cell cultures (Carrel technique), observed luxuriant growth of *T. cruzi* in fibroblasts, histiocytes, macrophages, epithelial cells and cells of the nervous system. They followed the flagellate's transformation into intracellular amastigotes, binary fission, and the epimastigote and trypomastigote stages. In some cultures, there was evidence suggestive of direct change from the amastigote to the trypomastigote stage without an intermediate epimastigote stage. A medium consisting of autoclaved trypticase, liver fraction L, inorganic salts, and blood was developed by Nakamura (1967). It is suitable for routine cultivation of stock cultures of trypanosomes.

Life Cycle. The life history of *T. cruzi* comprises two cycles of development—one in man or a reservoir mammalian host, and one within certain insects (Fig. 10–22). In man, the trypomastigotes do not usually multiply while they are in blood, but after entry into tissue cells, these forms lose their flagellum and undulating membrane to become amastigote forms and divide by binary fission, which results in the production of numerous forms; these eventually may pass through promastigote and epimastigote forms to the trypomastigote forms, which then appear in peripheral blood, thus completing the cycle.

In the most common vectors (*Panstrongylus megistus, Triatoma infestans,* and *Rhodnius prolixus*), the cycle of development may occur in nymphs of various instars or the adult and always takes place in the bug's intestinal tract. Typical trypomastigote forms ingested by the insect become short epimastigote forms that multiply and produce long epimastigote forms in the posterior portion of the midgut. In from 8 to 10 days, small trypomastigote forms that developed from the epimastigote forms appear in the rectum. These metacyclic trypomastigotes are passed in the feces (posterior station), and are infective to man and reservoir animals when rubbed into the puncture wound made by the insect or into an abrasion of the skin. Infection can be acquired by the oral route when infective bugs are eaten by animals (Yaeger 1971).

The life cycle of *T. cruzi* is diagrammatically shown in Figure 10–22.

Epidemiology. Chagas' disease is a typical zoonosis that occurs in the moist, warm regions of the Western Hemisphere and basically involves reduviid bugs living in close association with various wild mammals, such as armadillos, opossums, wood rats, and raccoons; it also occurs in inhabited areas where the bugs have become adapted to living in the walls of mud and thatch dwellings and emerging to feed on human blood at night. In the feral reservoir hosts, the trypomastigotes are always present in the peripheral blood, whereas they are only periodically seen in the peripheral blood of people.

The usual method of transmission of *T. cruzi* from man to man in southern Brazil is by the triatomid bug *Panstrongylus megistus;* in northern Brazil, by *Triatoma sordida* and *T. brasiliensis;* in northern Argentina, Chile, and Uruguay, by *Triatoma infestans;* in northern South America and Central America, by *Rhodnius prolixus;* in Mexico, principally by species of the *T. protracta-phylosoma-pallidipennis* group, but numerous other triatomids have been found infected throughout endemic or enzootic territory and are capable of transmitting *T. cruzi*. In the United States, *T. sanguisuga* and other species have been found naturally infected.

Many other arthropods have been experimentally infected, including bedbugs and ticks.

One natural infection has been diagnosed in the United States in a 10-month-old infant at Corpus Christi, Texas, in 1955 (Woody and Woody, 1955), and other cases have probably been unrecognized.

The incidence of infection among insect vectors of *T. cruzi* may be very high. In Brazil, for instance, 41% of 2412 *P. megistus* dissected were infected; in Chile, 50% of *T. infestans* and *T. spinolai;* and in California, 37.5% of *T. protracta*.

Monkeys have been infected with strains of *T. cruzi* isolated from triatomid bugs in the United States. Packchanian (1943), using a strain obtained from *T. heidemanni* in Texas, produced the disease in a volunteer by inoculating the uninjured conjunctiva. The patient developed fever, severe edema and hyperemia of the eyelid and conjunctiva, and enlarged axillary lymph nodes. *T. cruzi* was found in the blood and in cultures, and the infection was transmitted from the patient to susceptible animals by direct blood inoculation and through the insect vector.

Various mammals are *reservoirs of T. cruzi infection*. In South America, the following have been found naturally infected: dogs, cats, armadillos, bats, ferrets, foxes, opossums, anteaters, squirrels, and monkeys. In Panama, bats appear to be the common natural host but armadillos, opossums, and the spiny rat *(Proechymys semispinosus panamensis)* have also been found positive. In the southwestern United States, wood rats are most frequently infected, but opossums and armadillos have also been incriminated as reservoirs. In the deep south and southeastern United States, the raccoon is a frequent natural mammalian host.

Dogs and cats may be common reservoirs of infection, as in South America. In Chile, 302 of 3321 dogs (9%) and 216 of 1805 cats (12%) that were examined over a 20-year period were found to be naturally infected.

The usual method of transmission of *T. cruzi* is contaminative, from feces of the arthropod containing the metacyclic trypomastigotes rubbed into the wound. This occurs easily, because the arthropod defecates while biting and the contaminated fecal material is rubbed or scratched into the puncture wound. The triatomid bug becomes infective in about 8 to 10 days after biting an infected host and remains infective for as long as 2 years. The adult triatomids are usually the transmitting agents. Insofar as is known, the infection is not congenital in arthropods.

Transmission of *T. cruzi* from vertebrate to vertebrate may occur by the placental route (Eberhard and D'Alessandro, 1982), or by accidental ingestion of parasitized arthropods. In Santiago, Chile, Howard et al. (1957) and Rubio et al. (1962) reported premature births of *T. cruzi*-infected infants from asymptomatic mothers. Moreover, there is an obvious hazard of inapparent *T. cruzi*-positives serving as donors for blood transfusions. In areas such as Guanabara State, Brazil, where natural human infection is uncommon, blood transfusions may be the most important source of Chagas' disease (Coura et al., 1966).

Social conditions have much to do with infections, since Chagas' disease chiefly affects the poorest classes in the endemic-enzootic regions. The transmitting insects live in cracks, holes, and between the thatch of primitive adobe, grass, and thatched houses principally in suburban and rural localities. Age is important in epidemiology; the infection is most frequently acquired by children between a few months and 2 years of age. The triatomid bites preferably at the junction of a mucous and cutaneous surface, as at the outer canthus of the eye or angle of the lip (the bite is usually painless); hence the common name "kissing-bug." Children are frequently observed sleeping peacefully while these bugs are feeding upon them; thus, exposure usually occurs without the knowledge of the individual. The bugs "bite" at night and retire to their hiding places before daylight.

Pathogenesis and Pathology. While it is possible that, under favorable conditions, entry of *Trypanosoma cruzi* takes place through intact skin, a majority of human infections result from deposition of semiliquid feces of the infected triatomid bug on exposed mucous membranes, particularly at the outer canthus of the eye or around the nares or lips. Deposition of the metacyclic stage of the parasite characteristically occurs while the bug is taking a blood meal from the victim. The invading organisms actively enter or are engulfed by histiocytes in the corium and invade adipose cells of the subcutaneous tissues and nearby muscle cells immediately below the site of inoculation. For about 3 days they multiply as amastigotes; then on the fourth or fifth day, with the rupture of parasitized cells, the first infiltration of polymorphonuclear leukocytes, monocytes, and lymphocytes takes place, with a mobilization and proliferation of regional histiocytes, particularly in nearby lymph nodes. The local inflammatory process continues, with histiocytes assuming the major central role but with a periphery of neutrophilic leukocytes and eventual fibrotic encapsulation. This is the characteristic primary lesion *(chagoma)*, which blocks lymphatic capillaries and produces edema of the area.

From the primary site of development, amastigote forms metastasize to satellite lymph nodes but sooner or later are distributed through blood or lymph channels to axillary and inguinal lymph nodes, lungs, spleen, liver, bone marrow, and mesenteric lymph nodes; in these areas they multiply in fixed histiocytes, muscle fibers (especially in the myocardium), neuroglia, microglia, and pyramidal cells of the brain cortex, adrenal cortex, thyroid gland, and intestinal mucosa. In these visceral tissues, the parasites multiply as amastigotes. Only when large numbers escape simultaneously from rupturing host tissue cells into the blood stream are they found temporarily circulating as trypomastigotes, typically accompanied by fever. Since the parasites do not multiply in the trypomastigote stage, the fundamental development of the parasite—and the fundamental pathogenesis of the disease—must be sought in the tissue cells, which harbor the parasites. This accounts for the protean manifestations in Chagas' disease.

The *pathologic and microscopic anatomy* of *T. cruzi* infection consists essentially of the destruction of reticuloendothelial and other tissue cells of the body by the growth and development of the parasite within these cells. Practically every organ of the body may be invaded, but the organisms have a predilection for cells of mesenchymal origin, especially adipose, myocardial, reticuloendothelial, and neuroglial cells.

Mazza (1939) described nodular formations in the liver, usually located in the center of the lobules, that are composed of lymphocytes lying in a network of reticular histiocytes. These microscopic nodules are considered pathognomonic of Chagas' disease. Fat cells are invaded and destroyed by the multiplying parasites. The released fat droplets become incorporated into firm nodular lipogranulomas or lipochagomas.

Leukocytosis and lymphocytosis often occur early in acute Chagas' disease; however, leukopenia soon supervenes. Anemia characteristically occurs.

Destruction of cells of various organs and blockage of the reticuloendothelial system by *T. cruzi* lead to functional disability and may finally cause death. Amastigote forms of the parasite may be seen lying within invaded cells that are swollen and degenerating, and eventually are destroyed.

It appears that geographic strains of *T. cruzi* may vary with regard to the organ system that they attack and thus may be described as, for example, neurotropic, viscerotropic, or cardiotropic.

Neuroglia are especially invaded by neutrotropic strains. Upon rupture of these cells, a substance is released that, in a manner yet to be clarified, kills nearby nerve cells. Depending on the principal localization of this phenomenon, encephalitis, encephalomyelitis, or meningoencephalitis may occur. These manifestations of the disease are severe and commonly fatal, especially in children. In some geographic areas, nerve cells situated in autonomic ganglia in the walls of hollow viscera are commonly destroyed, resulting in disruption of the neural coordination necessary for the initiation or continuation of peristaltic waves in the involved viscus. The segment of viscus that is above that which is denervated later becomes greatly dilated, producing the ''mega'' disease that is characteristic of chronic Chagas' disease in some areas. This includes megaesophagus, megacolon, and megaureter. The megacolon of Chagas' disease is thus physiologically equivalent to that of Hirschsprung's disease; both are aganglionic, but the former is acquired rather than congenital.

The heart is always invaded. Myocardial cells are parasitized and destroyed, producing characteristically an acute myocarditis in acute Chagas' disease. The conduction system of the heart tends to be specifically damaged, producing conduction defects of various kinds. Most typical is right bundle branch block, but left bundle branch and bifascicular blocks, as well as various degrees of atrioventricular block, may occur. Damage to the sinoauricular node may cause bradycardia.

As the acute phase of Chagas' disease subsides, the chronic phase is entered. The organisms tend to disappear from the peripheral blood, reappearing from time to time in scanty numbers associated with febrile relapses.

In chronic Chagas' heart disease, there is marked cardiomegaly, with dilatation and thickening of the ventricular muscle towards the base but with marked thinning and fibrosis at the apex. The latter frequently progresses to aneurysm formation and accounts for the fact that one of the important causes of sudden death in Chagas' disease is rupture of a ventricular aneurysm. When the ventricular wall is cut through, fibrous streaks are found in the myocardium.

As mentioned previously, megaesophagus, megacolon, and megaureter are commonly found in patients with chronic Chagas' disease in certain endemic areas.

Symptomatology. The symptomatology of infection with *T. cruzi* is as varied as the lesions produced by the etiologic agent. The *incubation period* in man is probably 7 to 14 days, although some authorities place it at 12 to 30 days. In a series of 26 experimentally inoculated patients suffering from incurable cancer, Galliard et al. (1950) found that parasitemia developed 14 to 21 days following introduction of *T. cruzi* subcutaneously or onto scarified skin.

Acute and chronic forms of Chagas' disease have been described. The *acute form,* which is usually found in children but may occur in adults, lasts for 20 to 30 days and is characterized by a high fever that is frequently followed in about 2 weeks by a

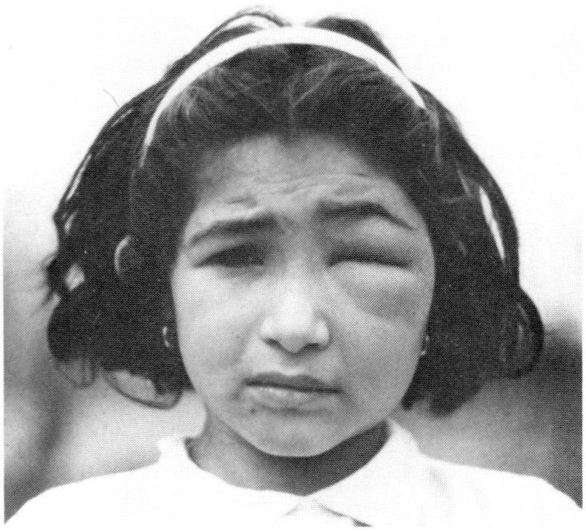

Fig. 10–24. *Trypanosoma cruzi.* Romaña's sign in a Paraguayan girl with acute Chagas' disease. The edema of both left eyelids and the edema of the cheek were accompanied by preauricular lymph node enlargement and conjunctivitis. (Courtesy of Dr. A. D'Alessandro.)

deposit of mucoid material in the tissues that causes swelling—often of the entire body. The face is swollen, there is marked edema of the eyelids (Romaña's sign) of one or both eyes, and the patient appears to be suffering from myxedema (Fig. 10–24). The edema does not pit on pressure and the skin is dry. Unilateral palpebral edema is the most common early evidence of this disease and is almost pathognomonic, although Dias (1955), in an analysis of 300 patients with known initial stage, found only 48.7% involving the eye (57% on the head), 23.3% with no detectable primary site, and the balance mostly on the extremities. The thyroid, preauricular, parotid and submaxillary glands, lymph nodes, spleen, and liver are usually enlarged. The fever is intermittent, remitting, or continuous. Keratitis is frequently observed. In other cases, there is a typhoid-like syndrome with fever and enlargement of the spleen and liver, but without edema. Central nervous system involvement causes anxiety, irritability, mental dullness, and sometimes coma. There may be signs of meningeal irritation.

Myocarditis is manifested by tachycardia, weakness, chest pain, and electrocardiographic changes. Congestive heart failure that responds poorly to digitalis may occur. Death may be precipitated by myocardial insufficiency or cardiac arrest.

Due to the invasion of adipose cells, numerous tender subcutaneous nodules appear that are called ''lipochagomas.'' In infants, the fat pads of the cheeks commonly are involved. The resulting painful lesions interfere with feeding and, thus, cause malnutrition. Anemia, a marked loss of strength, and frequently tachycardia occur. During the acute stage and in subsequent febrile exacerbations, the trypomastigotes may be found in the peripheral blood.

The acute stage lasts weeks or months in survivors and may be followed by apparent complete recovery or by the chronic stage of the disease. Patients infected with strains of low virulence may harbor the infection for years with few or no apparent ill effects. In others, a fulminating course ends fatally in 3 to 4 weeks. Probably the majority develop some evidence of chronic disease.

A *subacute stage* is recognized by some authorities and is characterized by tachycardia, enlargement of liver and spleen and general glandular enlargement.

Chronic Chagas' disease is by far more commonly diagnosed than the acute disease and is seen predominantly in adults. Chronic Chagas' disease has been observed to follow acute disease in children, and in both children and adults, recovery from acute Chagas' disease may be followed by a prolonged latent and asymptomatic period; therefore, one must conclude that many cases of acute Chagas' disease in children are overlooked or misdiagnosed. More than 1300 of the patients studied by Laranja et al. (1956) had chronic Chagas' disease. Approximately 50% of these persons were asymptomatic, and the others had myocardial damage leading to congestive heart failure.

The symptoms of the chronic form are probably related to damage sustained during the acute phase. Thus, the character of the symptoms depends upon the principal localization of the intracellular parasites at that time. Cardiac, meningoencephalitic, myxedematous, pseudomyxedematous, and suprarenal types have been described. It is now apparent that myxedema, once thought to be a common manifestation of chronic Chagas' disease, was primarily due to the coexistence of endemic goiter in the first recognized areas of endemic Chagas' disease. Nevertheless, Chagas' disease may occasionally precipitate thyroiditis. Evidence has been obtained that shows that Chagas' disease causes dilatation of the esophagus *(megaesophagus,* cardiac achalasia or esophageal aperistalsis) and of the colon *(megacolon)* (Koeberle, 1963). These manifestations of chronic Chagas' disease are due to deranged peristalsis as a result of the destruction of autonomic ganglia within the wall of the viscus.

Dysphagia is the principal symptom in megaesophagus, and there may be regurgitation of undigested food long after eating. Use of the muscles of deglutition to force food downward in the absence of esophageal peristalsis results in a characteristic prominence of the cheeks and submandibular area due to hypertrophy. Constipation is the primary symptom of megacolon; this may be marked and can result in massive accumulation of feces in the colon.

It is noteworthy that the principal organs or organ systems involved in chronic Chagas' disease vary from one geographic region to another. Thus, in Argentina, the cardiac form is extremely common but megacolon is rare; in Chile, neurological disorders are noted; in the states of São Paulo and Minas Gerais, Brazil, megacolon is common.

In the *cardiac form,* which characterizes about 50% of the victims with chronic symptoms, there is evidence of myocarditis, with A-V block and Adams-Stokes syndrome. The x-ray picture at times shows a greatly enlarged heart without evidence of aortic abnormality. Chronic Chagas' cardiac disease may cause not only congestive heart failure but also sudden death due to cardiac arrest or ventricular rupture. A survey in two rural areas of Venezuela, where Chagas' disease is highly endemic, showed a form of chronic myocardial disease to be prevalent in patients with positive specific serology and good functional capacity. Electrocardiographic findings included ventricular hypertrophy with incomplete bundle branch block, apical thinning of left ventricle and dilatation of the ventricles and auricles, inflammatory lesions of myocardium, and varying degrees of fibrosis. In some cases there were pericardial lesions (Puigbó et al., 1966).

The *neurotropic forms* produce aphasia, paraplegia, diplegia, and spastic paralysis—all due to destruction of motor centers; at times psychotic states are noted. During the chronic stage, acute febrile paroxysms may occur, at which time the trypomastigotes will be found in the blood, but usually blood examination is negative unless fever is present.

The presence or absence of clinical symptoms may be due to differences in virulence of different strains of *T. cruzi.* Mazzotti (1940) found that, in mice and guinea pigs, different results followed inoculation with strains obtained from different species of triatomids, and that strains varied in virulence even when obtained from the same species of bug in different localities. The number of trypanosomes inoculated did not seem to affect the virulence. Likewise, differences in age and individual susceptibility must be taken into consideration. Young children are highly susceptible.

Diagnosis. Diagnosis of this infection rests upon demonstration of *T. cruzi* in blood or tissues or the presence of a positive serologic reaction. The methods of examination of blood for demonstration of *T. gambiense* and *T. rhodesiense* are equally applicable in this infection (see Technical Appendix, p. 751), but it should be remembered that it is only during the acute febrile stage or during febrile exacerbations of the chronic stage that trypomastigotes are ordinarily found in peripheral blood films. If the blood is negative for *T. cruzi,* inoculation of guinea pigs may be used as a diagnostic measure; this is often successful in chronic cases. The examination of juice from enlarged lymph nodes, just as in African trypanosomiasis, may be effective.

If nervous symptoms are present, inoculation of cerebrospinal fluid into susceptible animals is sometimes successful in demonstrating the presence of trypomastigotes. If chagomas are present, one should be enucleated and examined for the amastigote stage of the parasites by making smears, or sections, and then staining.

Xenodiagnosis has been employed frequently, allowing trypanosome-free laboratory-bred triatomids to "bite" the individual suspected of having the disease. If infection is present in the blood meal, the parasites multiply rapidly in the intestine of the bug and an examination of the intestinal contents in 10 to 30 days will reveal flagellated stages. This method may be successful even when the trypomastigotes in the blood are so few that they cannot be demonstrated microscopically. Pifano (1941), in Venezuela, employing xenodiagnosis and using *Rhodnius prolixus* as the diagnostic bug, found that 14 of 40 persons examined were infected with *Trypanosoma cruzi,* although no positive individual manifested symptoms at the time of examination. For routine xenodiagnosis in Brazil, third-instar *Triatoma infestans* is ordinarily used, but Marsden et al. (1979) have found that, for detecting *T. cruzi* in patients with chronic infections, first-instar *Dipetalogaster maximus* was equally effective and more practical.

The *complement-fixation test* with antigen prepared from cultures of *T. cruzi* is both specific and practical. Of 1626 persons tested in Panama, 60 (3.7%) gave a positive reaction. Modified techniques employing purified antigen have been demonstrated to provide up to 90% diagnostic accuracy in use of the complement-fixation test (Kagan and Norman, 1976). A complement-fixing antigen prepared from amastigotes and trypomastigotes of *T. cruzi* grown in beef embryo cell culture (Neva and Gam, 1977) was found to be more sensitive but somewhat less specific than epimastigote antigens (Gam and Neva, 1977). Three serologic tests are commonly in use for the diagnosis of Chagas' disease—complement-fixation, indirect hemagglutination, and indirect immunofluorescence. An advantage of the second is that eluates from filter-paper blood samples are adequate and it is therefore suitable for surveys (Kagan and Norman, 1976). In a small intensive survey for *T. cruzi* infection in a rural population in Brazil where parasitemia was found in 20% of blood cultures from 116 persons, xenodiagnosis, complement-fixation, and indirect immunofluorescence tests were used; each test contributed specific information (Hoff et al., 1979).

In view of the endemicity of kala-azar in Brazil and elsewhere in Latin America, it is necessary that the utmost care be taken to differentiate this infection from Chagas' disease by culture, immunological tests, or xenodiagnosis.

Differential Diagnosis. Acute Chagas' disease should be suspected in any individual, especially a child, who has been in an endemic area and who develops an acute febrile illness with lymphadenopathy and myocarditis. The presence of a primary lesion, especially Romaña's sign, strengthens the suspicion of *T. cruzi* infection, but allergic reactions to, or secondary infections of, insect bites may produce lesions similar to the chagoma. The febrile illness must be distinguished from malaria, typhoid fever, rheumatic fever, and acute leukemia.

Chronic Chagas' disease should be suspected in the individual who has been in an endemic area, has defects of cardiac conduction, marked cardiomegaly, or fluoroscopic evidence of an apical ventricular aneurysm, or dysphagia, severe constipation, or spastic paralysis. The disease must be distinguished from primary cardiomyopathy, arteriosclerotic heart disease, syphilis, carcinoma of the esophagus or colon, and hypothyroidism.

Treatment. There is no satisfactory treatment for Chagas' disease although many drugs have been tried, including most of those used with some success in African trypanosomiasis. The *8-aminoquinolines,* such as primaquine and related drugs, apparently reduce the parasitemia in acute Chagas' disease, and thus perhaps decrease the degree of dissemination. There is no evidence, however, that they have any effect on the ultimate course, and they do not eliminate the infection.

Bayer 2502 (nifurtimox), an experimental nitrofurane, has been more promising in amelioration of acute Chagas' disease, but it is also more toxic and must be administered over a prolonged course. It has not been shown to cure the chronic disease.

Acute Chagas' myocarditis does not respond well to digitalis, which may in fact accentuate conduction defects. Diuretics are preferable for treatment of congestive heart failure due to either acute or chronic Chagas' disease.

Installation of an electric pacemaker may be necessary for management of sinus bradycardia or atrioventricular block.

The pains of megaesophagus are relieved to some extent by anticholinergic drugs such as belladonna, but such drugs tend to increase the dilatation and achalasia. Nitrites are effective early in the course of the disease, but in advanced cases surgery is required. Similarly, treatment of megacolon is dependent on removal, if possible, of the aganglionic undilated segment and the redundant portion of the dilated segment. Administration of corticosteroids results in intensification of the infection and should be avoided.

Prognosis. A fatal outcome is not uncommon in acute Chagas' disease in children. Adults appear to be somewhat more resistant. Laranja et al. (1956) reported 21 fatalities in 180 acute cases (11.7%). Deaths usually occur during the first 30 days of the acute disease. Those surviving acute Chagas' disease are believed to develop chronic infections that are either relatively asymptomatic or symptomatic. It seems probable that some of those who are asymptomatic may later develop symptomatic or clinical chronic disease. It is also probable that many cases of acute infection are not diagnosed. It seems reasonable to believe that, of individuals contracting acute Chagas' disease, 40 to 50% may escape death during the acute phase and escape significant symptomatic illness during the chronic phase.

Prevention. Elimination of the feral reservoirs of Chagas' disease is not practical. The problem, therefore, concerns the semidomestic triatomids living in human habitations. Neghme and Schenone (1961) considered that man himself is basically responsible for the present endemic state of *Trypanosoma cruzi* infection in the Americas. Habitations of precarious or inadequate construction, together with the ignorance of the rural or suburban population, are the major factors in the development of triatomid vector adaptation to human domiciles. This is greatly aggravated by the poverty of the people. The immediate control method is to destroy the bugs by application of effective insecticides to the walls and thatched roofs of the dwellings. Usually, BHC (Gammexane) has been the insecticide of choice. Since a single spray application is not sufficient to rid a home or a community of the triatomids, one or more subsequent sprayings must be carried out between 30 and 180 days later. This prophylactic activity against the bugs should be accompanied by education of the individuals, families, and communities in the endemic zones concerning the role of the triatomids in the maintenance and spread of Chagas' disease. The more fundamental requirement consists in the repair of the habitations at minimum cost, so that the walls and roofs will not provide breeding places for the vectors. Additionally, care must be taken to prevent infection due to a blood transfusion from an infected to a noninfected person.

Trypanosoma (Herpetosoma) rangeli Tejera, 1920*

Synonyms. *Trypanosoma escomeli* Yorke 1920; *Trypanosoma guatamalense* De Léon 1946; *Trypanosoma ariarii* Groot, Renjifo, and Uribe 1951.

Distribution and Biology. *Trypanosoma rangeli,* a flagellate harmless to man and wild and domestic animals but damaging to the triatomine bug, coincides in many areas with the pathogenic *T. cruzi*. Consequently, the two parasites can be confused and must be differentiated since simultaneous infections are common. So far, the parasite has been identified in Guatemala, Honduras, El Salvador, Costa Rica, Panama, Colombia, Venezuela, Peru, and possibly French Guiana, Brazil, Paraguay, Uruguay, Chile, and Ecuador.

Transmission, first observed by Groot (1952), is by inoculation of metacyclic trypomastigotes. The flagellates from the bug's intestine invade the hemolymph and then reach and develop in the salivary glands rather than in the rectum, as occurs with *T. cruzi*. Transmission by fecal contamination is controversial. The few positive trials are probably spu-

*Section written by Dr. Antonio D'Alessandro.

rious due to manipulation of the material used in the inoculation of animals (D'Alessandro, 1976).

The morphology of the blood trypomastigote in the vertebrate host has the general characteristics of *T. lewisi* and other members of the subgenus *Herpetosoma*. It is a slender trypanosome, averaging 26 to 34 μm in length including the free flagellum. The undulating membrane is well developed and the nucleus is located near the middle, in the anterior half of the body; the kinetoplast is small, round, and subterminal (Fig. 10–25). Although human infections are known to last for more than a year, the reproductive phase of *T. rangeli* in the vertebrate host is not known; some dividing blood trypanosomes have been described.

In contrast with *T. cruzi, T. rangeli* has a limited number of proven natural biological vectors: *Rhodnius prolixus* (Guatemala, Honduras, El Salvador, Costa Rica, Colombia, Venezuela); *R. pallescens* (Panama); *R. ecuadoriensis* (Peru); *R. pictipes* and *R. robustus* (Venezuela); *R. dalessandroi* (previously considered to be *R. brethesi)* and *Triatoma dimidiata* (Colombia). In addition, *R. neglectus, T. patagonica, T. protracta,* and *T. infestans* have been shown to be experimental vectors of the parasite (D'Alessandro, 1972). Several other species of triatomines have been reported naturally or experimentally infected with *T. rangeli*-like flagellates in the intestine or the hemolymph but not in the salivary glands. Their vector capacity is uncertain, however, because transmission trials to vertebrates, either by the anterior (inoculative) or posterior (contaminative) stations, are lacking.

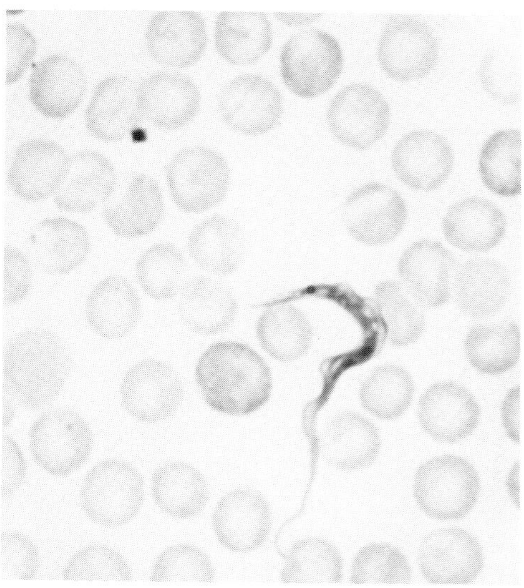

Fig. 10–25. *Trypanosoma rangeli* in blood smear. Giemsa stain. (× 1200.) (By P. Beaver; preparation courtesy of Dr. A. D'Alessandro.)

The parasite has been demonstrated frequently in man, especially in Venezuela and Guatemala; several hundred cases have been reported. In two Panama states, *T. rangeli* was found to be more than five times as frequent as *T. cruzi* among hemocultures from 10,570 persons (Sousa and Johnson, 1971). Laboratory animals susceptible to the infection range from newborn mice to a horse, and the wild animals naturally or experimentally infected include marsupials, edentates, rodents, carnivores, and nonhuman primates (Hoare, 1972; D'Alessandro, 1976).

Diagnosis. The diagnosis of *T. rangeli* infection in man and animals is based on the morphologic characteristics of the parasite in blood smears, hemoculture, xenodiagnosis, behavior of the flagellates in susceptible triatomines (invasion of hemolymph and salivary glands and bite transmission), and the vector host in different geographical areas. For instance, the type triatomine host, *R. prolixus,* is not a vector for Panamanian and Peruvian strains of *T. rangeli*. There is no specific serologic test for this parasite, but apparently those used to recognize *T. cruzi* do not cross-react with *T. rangeli* infections if pure *T. cruzi* antigen is employed.

The most characteristic forms of *T. rangeli* in hemoculture and triatomines are elongate epimastigotes and trypomastigotes (20 to 100 μm) with small kinetoplasts in the latter, located far from the posterior end; very small metatrypomastigotes (about 10 μm) are observed in the salivary glands and seldom seen in the hemolymph. Since diagnosis based on morphologic criteria requires experience and is often not feasible, the examination of hemolymph and salivary glands is recommended together with the usual study of extracted feces. In laboratories in which it is common practice to identify any flagellate coming from man or from triatomines as *T. cruzi,* errors may be made.

Blastocrithidia triatominae, a monoxenous flagellate of triatomids, has been described in the intestine of laboratory-bred *T. infestans* and wild *T. maculata* (Cerisola et al., 1971; Humbsch et al., 1977). Infections due to this parasite and to *T. rangeli* should be differentiated from those due to *T. cruzi* in individual cases and in a given population to determine the true prevalence of Chagas' disease (D'Alessandro and del Prado, 1977). Whether *T. rangeli* and *T. rangeli*-like flagellates, due to their known behavioral variations, should be regarded as a complex of species or a complex of variant strains is uncertain (D'Alessandro, 1976).

REFERENCES

Adler, S. 1964. *Leishmania*. Adv. Parasitol., *2*:35–96.

———, Foner, A., and Montiglio, B. 1966. The relationship between human and animal strains of *Leishmania* from the Sudan. Trans. R. Soc. Trop. Med. Hyg., *60*:380–386.

———, and Theodor, O. 1927. The transmission of *Leishmania tropica* from artificially infected sandflies to man. Ann. Trop. Med. Parasitol., *21*: 89–110.

———. 1930. The inoculation of canine cutaneous leishmaniasis into man and the behaviour of various strains of *Leishmania* in mice. Ann. Trop. Med. Parasitol., *24*:197–210.

———, and Zuckerman, A. 1948. Observations on a strain of *Leishmania tropica* after prolonged cultivation. Ann. Trop. Med. Parasitol., *42*:173–183.

Agu, W.E., Farrell, J.P., and Soulsby, E.J.L. 1982. Pathogenesis of anemia in hamsters infected with *Leishmania donovani*. Z. Parasitenkd., *68*:27–32.

Anderson, D.C., Buckner, R.G., Glenn, B.L., and MacVean, D.W. 1980. Endemic canine leishmaniasis. Vet. Pathol., *17*:94–96.

Ashcroft, M.T. 1959. The Tinde experiment: A further study of the long-term cyclical transmission of *Trypanosoma rhodesiense*. Ann. Trop. Med. Parasitol., *53*:137–146.

Barlow, D., Lonergan, M., Tio, F., and Meyerhoff, W. 1977. American mucocutaneous leishmaniasis. South. Med. J., *70*:246–248.

Bassiouny, A., El Meshad, M., Talaat, M., Kutty, K., and Metawaa, B. 1982. Cryosurgery in cutaneous leishmaniasis. Br. J. Dermatol., *107*:467–474.

Bray, R.S. 1974. *Leishmania*. Annu. Rev. Microbiol., *28*:189–218.

———, Ashford, R.W., and Bray, M.A. 1973. The parasite causing cutaneous leishmaniasis in Ethiopia. Trans. R. Soc. Trop. Med. Hyg., *67*:345–348.

Burke, J., Bengosi, K., and Diantete, N.L. 1974. Un cas de trypanosomiase africaine *(T. gambiense)* congenitale. Ann. Soc. Belg. Med. Trop., *54*:1–4.

Burnett, G.F., Chadwick, P.R., Miller, A.W.D., and Beesley, J.S.S. 1964. Aircraft applications of insecticides in East Africa. XIV. Very-low-volume aerosols of dieldrin and isobenzan for the control of *Glossina morsitans* Westw. Bull. Entomol. Res., *55*:527–539.

Butler, P.G. 1978. Levamisole therapy of chronic *Leishmania tropica*. J. Trop. Med. Hyg., *81*:221–224.

Cerisola, J.A., del Prado, C.E., Rohwedder, R., and Bozzini, J.P. 1971. *Blastocrithidia triatomae* n. sp. found in *Triatoma infestans* from Argentina. J. Protozool., *18*:503–506.

Chadwick, P.R., Beesley, J.S.S., White, P.J., and Matechi, H.T. 1964. An experiment on the eradication of *Glossina swynnertoni* Aust. by insecticidal treatment of its resting sites. Bull. Entomol. Res., *33*:411–419.

Chance, M.L. 1981. The six diseases of WHO: Leishmaniasis. Br. Med. J., *2*:1245–1247.

Convit, J., Pinardi, M.E., and Rondon, A.J. 1972. Diffuse cutaneous leishmaniasis: A disease due to an immunological defect of the host. Trans. R. Soc. Trop. Med. Hyg., *66*:603–610.

Coura, J.R., Nogueira, E.S., and Rodrigues da Silva, J. 1966. Indices de transmissão da doença de Chagas por transfusão de sangue de doadores na fase crônica da doença. O Hospital, *69*:991–998.

D'Alessandro, A. 1972. New experimental vectors of Colombian *Trypanosoma rangeli*. J. Med. Entomol., *9*:187–195.

———. 1976. Biology of *Trypanosoma (Herpetosoma) rangeli* Tejera, 1920, pp. 327–403. In *Biology of the Kinetoplastida*, vol. 1. Edited by W.H.R. Lumsden and D.A. Evans. London, E. Arnold.

———, and del Prado, C.E. 1977. Search for *Trypanosoma rangeli* in endemic areas of *Trypanosoma cruzi* in Argentina and Brazil. Am. J. Trop. Med. Hyg. *26*:623–627.

Dias, E. 1955. Informaçoes acérca de 300 casos de doença de Chagas con periodo inicial conhecido, fichado no Centro de Estudos de Bambui. O Hospital, *47*:9–17.

Eberhard, M.L., and D'Alessandro, A. 1982. Congenital *Trypanosoma cruzi* infection in a laboratory-born squirrel monkey, *Saimiri sciureus*. Am. J. Trop. Med. Hyg., *31*:931–933.

Edington, G.M., and Gilles, H.M. 1969. *Pathology in the Tropics*. Baltimore, Williams & Wilkins Co.

Evans, D.A. 1978. Kinetoplastida, pp. 55–88. In *Methods of Cultivating Parasites in Vitro*. Edited by A.E.R. Taylor and J.R. Baker. New York, Academic Press.

Ford, J. 1965. Distribution of *Glossina* and epidemiological patterns in the African trypanosomiases. J. Trop. Med. Hyg., *68*:211–225.

Foulkes, J.R. 1981. The six diseases of WHO: Human trypanosomiasis in Africa. Br. Med. J., *2*:1172–1174.

Galliard, H., Brumpt, L.C., and Martinez, R. 1950. Infections expérimentales a *Trypanosoma cruzi* Chagas chez l'homme à propos de la biothérapie du cancer. Bull. Soc. Pathol. Exot., *43*:204–216.

Gam, A.A., and Neva, F.A. 1977. Comparison of cell culture with epimastigote antigens of *Trypanosoma cruzi*. Am. J. Trop. Med. Hyg., *26*:47–57.

Gardener, P.J., Shchory, L., and Chance, M.L. 1977. Species differentiation in the genus *Leishmania* by morphometric studies with the electron microscope. Ann. Trop. Med. Parasitol., *71*:147–155.

Garnham, P.C.C. 1962. Cutaneous leishmaniasis in the New World with special reference to *Leishmania mexicana*. Sci. Rep. Ist. Sup. Sanita, *2*:76–82.

Groot, H. 1952. Further observations on *Trypanosoma ariarii* of Colombia, South America. Am. J. Trop. Med. Hyg., *1*:585–592.

Heisch, R.B., McMahon, M.P., and Manson-Bahr, P.E.C. 1958. The isolation of *Trypanosoma rhodesiense* from a bush-buck. Br. Med. J., *2*:1203–1204.

Hertig, M. 1949. Phlebotomus and residual DDT in Greece and Italy. Am. J. Trop. Med., *29*:773–800.

———, Fairchild, G.B., and Johnson, C.M. 1958. Washington, D.C., Annual Report of the Gorgas Memorial Laboratory, 1957.

Hirumi, H., Doyle, J.J., and Hirumi, K. 1977. African trypanosomes: Cultivation of animal-infective *Trypanosoma brucei in vitro*. Science, *196*:992–994.

Hoare, C.A. 1948. The relationship of the hemoflagellates. Proceedings of the 3rd International Congress on Tropical Medicine. Malaria, Sect. 2, *7*:1110–1118.

———. 1967. Evolutionary trends in mammalian trypanosomes. Adv. Parasitol., *5*:47–91.

———. 1972. *The Trypanosomes of Mammals. A Zoological Monograph*. Oxford, Blackwell Scientific Publications.

Hockmeyer, W.T., Kager, P.A., Rees, P.H., and Hendricks, L.D. 1981. The culture of *Leishmania donovani* in Schneider's insect medium: Its value in the diagnosis and management of patients with visceral leishmaniasis. Trans. R. Soc. Trop. Med. Hyg., *75*:861–863.

Hoff, R., Mott, K.E., Silva, J.F., Meneges, V., Hoff, J.N., Barrett, T.V., and Sherlock, I. 1979. Prevalence of parasitemia and seroreactivity to *Trypanosoma cruzi* in a rural population of northeast Brazil. Am. J. Trop. Med. Hyg., *28*:461–466.

Hommel, M. 1978. The genus *Leishmania*: Biology of the parasites and clinical aspects. Bull. Inst. Pasteur, *76*:5–102.

Hoogstraal, H., and Heyneman, D. 1969. Leishmaniasis in the Sudan Republic. 30. Final epidemiologic report. Am. J. Trop. Med. Hyg., *18*:1087–1210.

Howard, J.E., Rios, C., Ebensberger, I., and Olivos, P. 1957. Enfermedad de Chagas congénita. Bol. Chil. Parasitol., *12*:42–45.

Humbsch, R. de, Nuñez, V., Mora, E., and Carrasquero, B. 1977. *Blastocrithidia triatomae torrealbai* n. subsp. encontrado en *Triatoma maculata* (Hemiptera, Reduviidae) de Venezuela. Bol. Dir. Malariol. San. Amb. (Venezuela), *17*:14–19.

Hutt, M.S.R., Köberle, F., and Salfelder, K. 1973. Leishmaniasis and trypanosomiasis, pp. 351–398. In *Tropical Pathology*. Edited by H. Spencer et al. New York, Springer-Verlag.

Jordan, A.M. 1978. Principles of the eradication or control of tsetse flies, Nature, *273*:607–609.

Kagan, I.G., and Norman, L. 1976. Serodiagnosis of parasitic diseases, pp. 382–409. In *Manual of Clinical Immunology*. Edited by N.R. Rose and H. Friedman. Washington, D.C., American Society for Microbiology.

Kinghorn, A., and Yorke, W. 1912. On the transmission of human trypanosomes by *Glossina morsitans*, etc. Ann. Trop. Med. Parasitol., *6*:1–23.

Kligler, I.J. 1926. The cultural and serological relationship of *Leishmania*. Trans. R. Soc. Trop. Med. Hyg., *19*:330–335.

Koeberle, F. 1963. Patogenia do megaesôfago brasileiro e europeu. Rev. Goiana Med., *9*:79–116.

Kozevnikov, P.V. 1963. Two nosological forms of cutaneous leishmaniasis. Am. J. Trop. Med. Hyg., *12*:719-724.

Lainson, R., and Shaw, J.J. 1978. Epidemiology and ecology of leishmaniasis in Latin-America. Nature, *273*:595–600.

Lanham, S.M. 1977. Field diagnosis of trypanosomiasis: Protozoological methods in the field diagnosis of trypanosomiasis. Trans. R. Soc. Trop. Med. Hyg., *71*:8–10.

Laranja, F.S., Dias, E., Nobrega, G., and Miranda, A. 1956. Chagas' disease: A clinical, epidemiological and pathological study. Circulation, *14*:1035–1060.

Luckins, A.G., Preston, T.M., and Vickerman, K. 1967. Recent work on cell division in trypanosomes. Trans. R. Soc. Trop. Med. Hyg., *61*:449–450.

Manson-Bahr, P.E.C. 1961. The leishmanin test and immunity in kala-azar. East Afr. Med. J., *38*:165–167.

———. 1971. Leishmaniasis. Int. Rev. Trop. Med., *4*:123–140.

———, and Charters, A.D. 1963. Myocarditis in African trypanosomiasis. Trans. R. Soc. Trop. Med. Hyg., *57*:119–121.

Marsden, P.D., Barreto, A.C., Cuba, C.C., Gama, M.B., and Ackers, J. 1979. Improvements in routine xenodiagnosis with first instar *Dipetalogaster maximus* (Uhler 1894) (Triatominae). Am. J. Trop. Med. Hyg., *28*:649–652.

Martins, J.M., de Alencar, J.E., and Magalhaes, V.B. 1965. The anemia of kala-azar. Rev. Inst. Med. Trop. S. Paulo, *7*:47–64.

Mazza, S. 1939. Inexistencia de un sintoma patognomónico en formas agudas de enfermedad de Chagas. Prensa Med. Argent., *26*:1569–1580.

Mazzotti, L. 1940. Variations in virulence for mice and guinea-pigs in strains of *Trypanosoma cruzi* Chagas from different species of bugs (Triatomidae) from different localities in Mexico. Am. J. Hyg., *31*:67–85.

Mühlens, P., 1929. Trypanosomiasis bei Mutter und Säugling. Arch. Schiffs- u. Tropen. Hyg., *33*:181–187.

Mutinga, M.J., Kaddu, J.B., Sand, D., and Arapsiongok, T.K. 1982. Studies of animal reservoirs of visceral leishmaniasis in Kenya. East Afr. Med. J., *59*:446–450.

Nakamura, M. 1967. An autoclaved medium for routine cultivation of *Trypanosoma cruzi*. Trans. R. Soc. Trop. Med. Hyg., *61*:792–794.

Nash, T.A.M. 1960. A review of the African trypanosomiasis problem. Trop. Dis. Bull., *57*:974–1003.

Neghme, A., and Schenone, F.H. 1961. Conceptos basicos en la profilaxis de la enfermedad de Chagas. Bol. Ofic. Sanit. Panam., *51*:465–469.

Neva, F. 1982. Diagnosis and treatment of cutaneous leishmaniasis, pp. 364–380. In *Current Clinical Topics in Infectious Diseases*, vol. 3. Edited by J.S. Remington and M.N. Swartz. New York, McGraw-Hill.

———, and Gam, A.A. 1977. A complement-fixing antigen from *Trypanosoma cruzi* grown in cell cultures. Am. J. Trop. Med. Hyg., *26*:37–46.

———, Wyler, D., and Nash, T. 1979. Cutaneous leishmaniasis—a case with persistent organisms after treatment in presence of normal immune response. Am. J. Trop. Med. Hyg., *28*:467–471.

Nicolle, C. 1908. Culture du parasite du bouton d'Orient. C.R. Acad. Sci. (Paris), *146*:842–843.

———. 1908. Isolement et culture des corps de *Leishmania*. Arch. Inst. Pasteur Tunis, *4*:55–56.

Noguchi, H. 1926. Comparative studies on herpetomonads and leishmanias. II. Differentiation of the organisms by serological reactions and fermentation tests. J. Exp. Med., *44*:305–314.

Nussenzweig, R. 1982. Parasitic disease as a cause of immunosuppression. N. Engl. J. Med., *306*:423–424.

Ormerod, W.E. 1961a. A study of volutin granules in trypanosomes. Trans. R. Soc. Trop. Med. Hyg., *55*:313–327.

———. 1961b. The epidemic spread of Rhodesian sleeping sickness, 1908–1960. Trans. R. Soc. Trop. Med. Hyg., *55*:525–538.

———. 1967. Taxonomy of the sleeping sickness trypanosomes. J. Parasitol., *53*:824–830.

Packchanian, A. 1943. Infectivity of the Texas strain of *Trypanosoma cruzi* to man. Am. J. Trop. Med., *23*:309–314.

Pavlovsky, E. 1937–1939. Reservoir hosts of cutaneous leishmaniasis, pp. 108–109. In Report of the Scientific Research Works of All-union Institute of Experimental Medicine (Viem) for 1938–1939. (in Russian). (Abstract in Trop. Dis. Bull., *39*:538, 1942.)

Pedrosa, A., and da Silva, P.D. 1911. Botão do Oriente (leishmoniase ulcerosa). Cultura da *Leishmania tropica (Leishmania wrightii)*. Arch. Soc. Med. Cirurg. S. Paulo, *1*:305–314.

Pellegrino, J. 1951. Nota preliminar sobre a reação intradérmica feita com a fração polissacaridea isolada de formas de cultura da *Leishmania brasiliensis* em casos de leishmaniose tegumentar americana. O Hospital, *39*:859-863.

Pessôa, S.B. 1941. Dados sobre la leishmaniose tegumentar en S. Paulo. O Hospital, *19*:389–409.

———, and Pestana, B.R. 1940. A introdermoreacção de Montenegro nas campanhas sanitarias contra a leishmaniose. Sao Paulo Med., *13*:133–151.

Pifano, F. 1941. La enfermedad de Chagas en el Estado Yaracuy, Venezuela. Caracas Med., *8*:1103–1166.

Price, E.W., and Fitzherbert, M. 1965. Cutaneous leishmaniasis in Ethiopia. A clinical study and review of literature. Ethiop. Med. J., *3*:57–83.

Puigbó, J.J., Nava Rhode, J.R., Garcia Barrios, H., Suárez, J.A., and Gil Yépez, C. 1966. Clinical and epidemiological study of chronic heart involvement in Chagas' disease. Bull. WHO, *34*:655–659.

Romaña, C., and Meyer, H. 1942. Estudo do ciclo evolutivo de "*Schizotrypanum cruzi*" em cultura de tecidos de embrias de galinha. Mem. Inst. Osw. Cruz, *37*:19–27.

Rubio, M., Ebensperger, I., Howard, J., Knierim, F., and Naquira, F. 1962. Búsqueda de enfermedad de Chagas en

100 madres de prematuros, con hallazgo de un caso de enfermedad de Chagas congénita. Bol. Chil. Parasitol., *17*:13–16.

Ruitenberg, E.J., and Buys, J. 1977. Application of the enzyme-linked immunosorbent assay (ELISA) for the serodiagnosis of human African trypanosomiasis (sleeping sickness). Am. J. Trop. Med. Hyg., *26*:31–36.

Shaw, P.K., Quigg, L.T., Allain, D.S., Juranek, D., and Healy, G. 1976. Autochthonous dermal leishmaniasis in Texas. Am. J. Trop. Med. Hyg., *25*:788–796.

Sousa, O.E., and Johnson, C.M. 1971. Frequency and distribution of *Trypanosoma cruzi* and *Trypanosoma rangeli* in the Republic of Panama. Am. J. Trop. Med. Hyg., *20*:405–410.

Stephens, J.W.W., and Fantham, H.B. 1910. On the peculiar morphology of a trypanosome from a case of sleeping sickness and the possibility of its being a new species *(T. rhodesiense)*. Proc. R. Soc. Lond., Ser. B., *83*:28–33.

Taute, M., and Huber, F. 1919. Die Unterscheidung des *Trypanosoma rhodesiense* vom *Trypanosoma brucei*. Beobachtungen und Experimente aus dem Kriege in Ostafrika. Arch. Schiffs- u. Tropenhyg., *23*:211–226.

Thatcher, V.E., Eisenmann, C., and Hertig, M. 1965. A natural infection of *Leishmania* in the kinkajou, *Potos flavus*, in Panama. J. Parasitol., *51*:1022–1023.

Van Hoeve, K., Onyango, R.J., Harley, J.M.B., and De Raadt, P., 1967. The epidemiology of *Trypanosoma rhodesiense* sleeping sickness in Alego Location, Central Nyanza, Kenya. II. The cyclical transmission of *Trypanosoma rhodesiense* isolated from cattle to a man, a cow and to sheep. Trans. R. Soc. Trop. Med. Hyg., *61*:684–687.

Walton, B.C. 1980. Evaluation of chemotherapy of American leishmaniasis by the indirect fluorescent antibody test. Am. J. Trop. Med. Hyg., *29*:747–752.

———, and Valverde, L. 1979. Racial differences in espundia. Ann. Trop. Med. Parasitol., *73*:23–29.

Wang, C.T., and Wu, C.C. 1959. Studies on kala-azar in new China. Chinese Med. J., *78*:55–71.

Weiss, P. 1943. Epidemiología y clínica de la leishmaniasis tegumentarias en el Peru. Rev. Med. Exp. Lima, *2*:209–248.

Woody, N.C., and Woody, H.B. 1955. American trypanosomiasis (Chagas' disease). First indigenous case in the United States. J.A.M.A., *159*:476–477.

World Health Organization. 1979. The African trypanosomiases. Report of a Joint WHO Expert Committee and FAO Expert Consultation. Geneva, W.H.O. Tech. Rep. Ser., no. 635.

Yaeger, R.G. 1971. Transmission of *Trypanosoma cruzi* infection to opossums via the oral route. J. Parasitol., *57*:1375–1376.

Zial, M., Bowman, J.E., McMillan, C.W., and Tabatabai, M. 1968. Leishmaniasis in southern Iran: The occurrence of all three varieties in the same area . Trans. R. Soc. Trop. Med. Hyg., *62*:668–671.

Chapter 11

Amebae Inhabiting the Alimentary Canal

CLASSIFICATION AND NOMENCLATURE

Amebae belong to the subphylum Sarcodina, superclass Rhizopoda, order Amoebida, which includes species that have a naked protoplast during their trophic stage and characteristically form lobose pseudopodia that constitute the locomotor organelles. Most free-living species belong to the family Amoebidae, while all strictly parasitic forms are classified in the family Endamoebidae Calkins, 1926. With few exceptions, members of the latter family are exclusively parasitic in the alimentary canal of their hosts. They are comparatively small organisms, typically lack contractile vacuoles, multiply by binary fission, and undergo encystation.

A common characteristic of free-living and parasitic amebae is so-called "ameboid motion," which Villee et al. (1958) defined as "the irregular flowing of protoplasm. Not all protoplasm streams simultaneously. There is a non-moving layer at the surface of the cell which surrounds a core of liquid flowing protoplasm. At the rear of the moving ameba, protoplasmic gel is converted to sol to flow forward. At the front end, the streaming protoplasm bulges out in a projection known as a pseudopodium and changes from a sol to a gel. The tip of the pseudopodium is covered with a thinner gel layer than that elsewhere in the cell and hence is the part to bulge when contraction occurs. . . . Ameboid motion is a crawling motion, not a swimming one; the cell must be attached to some physical substrate in order to move."

The genera of Endamoebidae are differentiated primarily by their nuclear structure (Fig. 11–1). In species of *Endamoeba,* parasites in the intestine of invertebrate hosts (e.g., *E. blattae* of the cockroach), the large central area is devoid of chromatin material and is surrounded by a wide zone of chromatin granules, while the nuclear membrane is thick. In species of *Entamoeba,* a comparatively small karyosome is situated near the center of the nucleus and numerous chromatin granules line the nuclear membrane. *Endolimax* has a relatively large karyosome of irregular shape and several achromatic threads connecting it with a delicate nuclear membrane. *Iodamoeba* has a large karyosome (endosome) rich in chromatin that is typically surrounded by a single layer of periendosomal granules and attached to the karyosome and the nuclear membrane by radiating achromatic fibrils.

The genus *Endamoeba* Leidy, 1879, was established for the species *blattae* reported by Bütschli in 1878 from the intestine of the cockroach *Periplaneta (Blatta) orientalis*. The generic name *Entamoeba* Casagrandi and Barbagallo, 1895, was proposed for the two species of ameba then known to be parasitic in the human intestine (i.e., *histolytica* and *coli*). In spite of the distinct difference in nuclear structure of these two kinds of amebae, in 1928 The International Commission on Zoological Nomenclature ruled (Opinion 99) that *Entamoeba* is a synonym (homonym) of *Endamoeba*. On December 28, 1954, however, the Commission reversed its ruling, validating *Entamoeba* as a generic name, with *E. coli* as the type species. *Entamoeba* should be used in place of *Endamoeba* when reference is made to species of amebae having the generic characteristics of *Entamoeba,* including *histolytica, hartmanni, coli,* and *gingivalis,* which commonly parasitize man.

Genus *Entamoeba*

Casagrandi and Barbagallo, 1895

This genus includes all endoparasitic amebae that have a more or less spherical nucleus, a distinct nuclear membrane lined with chromatin granules, and a relatively small karyosome situated at or near

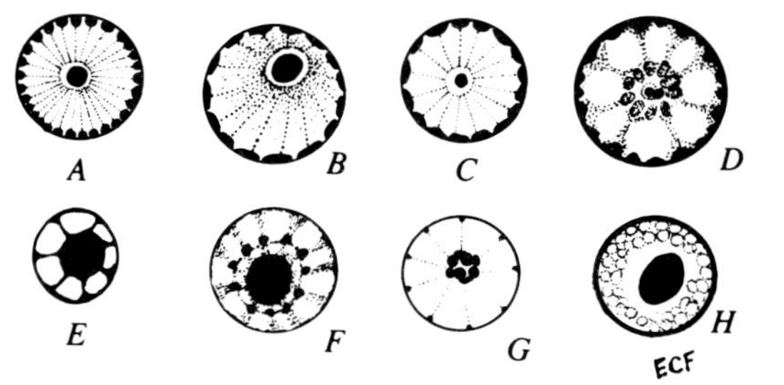

Fig. 11–1. Nuclei of several genera and species of amebae encountered in human stools and the mouth. *A, Entamoeba histolytica. B, E. coli. C, E. gingivalis. D, E. polecki. E, Endolimax nana. F, Iodamoeba buetschlii. G, Dientamoeba fragilis. H, Hartmanella hyalina.* Not drawn to same magnification. (By E.C. Faust.)

the center of the nucleus. Generally, the trophozoite has a single nucleus.

Morphologically, most species of *Entamoeba* fall into three groups according to the number of nuclei in the mature cysts (namely, 8, 4, and 1). A fourth category would include *Entamoeba gingivalis*, which has no encysted stage. Each species of *Entamoeba* probably consists of a series of strains or races adapted to a particular host or hosts. It is expected that biochemical, physiological, and immunological methods may soon provide means of distinguishing species and strains of *Entamoeba* (Sargeaunt and Williams, 1979).

Species of *Entamoeba* have been described as natural parasites of the alimentary tract of man, monkeys, and many other species of vertebrate and invertebrate hosts. In addition, *Entamoeba moshkovskii* has been found only in sewage. With two known exceptions (*E. gingivalis* of man and monkeys and *E. bovis* of cattle), the various species of *Entamoeba* in vertebrate hosts inhabit the large intestine and are exclusively lumen dwellers (with the two known exceptions of *E. histolytica* and *E. invadens* of reptiles).

In the trophozoite stage, the members of this genus divide by binary fission. All except *E. gingivalis* have a cystic stage during which two to several nuclear divisions occur in most species, with comparable division of the cytoplasm following excystation.

Entamoeba histolytica
Schaudinn, 1903
(Producing amebiasis)

Synonyms. *Amoeba coli* Lösch, 1875; *Amoeba dysenteriae* Councilman and Lafleur, 1891; *Entamoeba dysenteriae* (Councilman and Lafleur, 1891) Craig, 1905; *Entamoeba tetragena* Hartmann, 1908; *Endamoeba histolytica* (Schaudinn, 1903) Hickson, 1909; *Endamoeba dysenteria* Kofoid, 1920; *Entamoeba dispar* Brumpt, 1925.

Historical and Geographical Notes. *Entamoeba histolytica* was first described by Lösch (1875) from dysenteric stools of a patient in St. Petersburg (now Leningrad), Russia. Although Lösch found the trophozoites of this ameba in ulcers of the colon at autopsy and induced dysentery in a dog that he inoculated *per rectum* with the patient's blood-mucous stools containing the organism, he failed to recognize the causal relationship between the ameba and the acute colitis. Subsequent investigations of Kartulis (1886) in Cairo, Hlava (1887) in Prague, and Councilman and Lafleur (1891) in Baltimore provided clinical and pathological evidence that this ameba is responsible for a special type of dysentery and liver abscess. In 1893, Quincke and Roos discovered the cysts, while Schaudinn (1903) named the species *Entamoeba histolytica* and differentiated it from the common colon ameba, *E. coli*. Ten years later Walker and Sellards in the Philippines obtained experimental evidence with human volunteers that *E. histolytica* is the cause of amebic colitis and *E. coli* is a harmless commensal of the large intestine. These and other historical events were reviewed more fully by Imperato (1981).

The nomenclatural difficulty concerning *Entamoeba* vs. *Endamoeba* when referring to *E. histolytica*, which was discussed earlier (see p. 101), was reviewed in detail by Kirby (1951).

Entamoeba histolytica has been found in all populations throughout the world where searches have been conducted. It is more prevalent in the tropics and subtropics than in cooler climates, but in insanitary communities of temperate and subarctic areas, the incidence is at times as high as in the tropics. *E. histolytica* has been found in the Western Hemisphere, from Anchorage, Alaska (61°N), to the Strait of Magellan (52°S), and in the Eastern Hemisphere, from Finland (60°N) to Australia and South Africa (30°S).

Morphology, Biology, and Life Cycle. *Entamoeba histolytica* has the following successive stages in its life cycle: trophozoite, precyst, cyst, metacyst, and metacystic trophozoite.

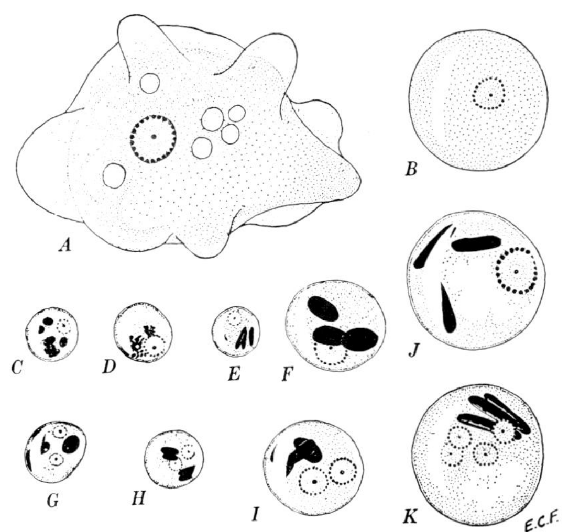

Fig. 11–2. *Entamoeba histolytica* and *Entamoeba hartmanni*. *A, E. histolytica* trophozoite. *B, E. histolytica* precyst. *F, I, J, K,* 1-, 2-, and 4-nucleate cysts with various types of chromatoidals. *C, D, E, G, H, E. hartmanni*, formerly referred to by some authors as "small race *E. histolytica*." (By E.C. Faust.)

Trophozoite. The living trophozoites (Fig. 11–2) vary in size from about 10 to 60 μm in diameter, depending on their degree of activity and various other conditions. Specimens in a freshly evacuated dysenteric stool generally are relatively large (Plate I,*1*), and those from an unformed stool of an asymptomatic carrier may be no larger than a small-sized *Endolimax nana* trophozoite.

The active trophozoite exhibits rather remarkable locomotion, which can be observed in organisms obtained from freshly passed dysenteric or diarrheic stools or in cultured amebae. In clear culture medium, the amebae may be seen crawling up the sides of the culture tube. Movement results from long finger-like or broadly rounded pseudopodial extensions of the ectoplasm into which the endoplasm flows (Fig. 11–2,*A*). However, the direction of movement changes rapidly in response to conditions in the micro-climate. At one moment, a pseudopodium may be actively progressing in one direction, then almost instantaneously it may be retracted as other pseudopodia are formed and alter the course of locomotion. The movement may be continuous or intermittent, may appear to be erratic, but is rarely consistently in a straight line.

The protoplasm of *E. histolytica* is differentiated into a clear outer ectoplasm and a more central, finely granular endoplasm in which food vacuoles may be observed, and the nucleus in advance of the center of the protoplasmic mass can occasionally be seen in the living trophozoite. In specimens fixed by the Schaudinn technique and stained with iron-hematoxylin, the structure of the trophozoite (Fig. 11–2,*A*) can be more satisfactorily studied. The nucleus is spherical, with a diameter about one-fifth or one-sixth that of the entire ameba. It contains a small, distinct central karyosome surrounded by an unstained "halo" and anchored by numerous delicate, radiating, achromatic fibrils to the inner surface of the nuclear membrane. Lining the membrane is a rim of chromatin that is frequently distributed in small round granules but may be aggregated into larger plaque-like masses. The membrane itself is thin and the nucleoplasm is considerably more viscous than the endoplasm. In dysenteric stools, the food vacuoles often contain red blood corpuscles in the process of digestion, but these do not constitute essential food for the ameba. Amebae in tissues frequently contain no particulate matter, and good growth of colonies is obtained in cell-free, particle-free (axenic) culture media (Diamond, 1968).

Electron microscopic study of *E. histolytica* from the colon and liver of man and from experimental animals has revealed that the trophozoites have no mitochondria or Golgi complex, and the endoplasmic reticulum is poorly developed (El-Hashimi and Pittman, 1970). The cytoplasm contains surface lysosomes, glycogen particles, small cylindrical electron-dense bodies of an undetermined nature, and sometimes crystal-like ribosomes of helically arranged particles on stacks or rods (Sepulveda, 1982). The crystalline structures were seen earlier in cysts, aggregated in the chromatoidal bodies that, both by light and electron microscopy, were noted to be more abundant during periods of nuclear division (Griffin, 1972).

Encystation. Under natural conditions, encystation does not occur in the tissues. In the lumen of the colon, under conditions not yet known, the amebic trophozoite discharges undigested food and condenses into a spherical mass; this is the *precyst* (Fig. 11–2,*B*). Then a relatively thin, tough wall is secreted, so that the unripe cyst is formed. At this stage (Figs. 11–2,*F* and *J*), a single nucleus like that of the trophozoite and precyst is formed. Usually there are two types of inclusions in the unripe and ripening cyst: (1) a mass of *glycogen* with hazy margins and (2) highly refractile objects, the *chromatoidals,* which are long or short rods usually with rounded ends. Although the chromatoidals take on a deep blue-black color with iron-hematoxylin stain, they give a negative Feulgen reaction and are not considered to contain ribonucleic acid. The cysts range in diameter from about 10 μm to 20 μm, with the average diameter around 12 or 13 μm. Cysts under 10 μm in diameter may be seen but usually these are *E. hartmanni,* formerly called small race *E. histolytica* (Fig. 11–2,*C, D, E, G, H*). Cysts of *E. histolytica* are usually spherical, but they may be subspherical or ovoidal; they are seldom as varied in contour as cysts of *E. coli,* however.

Ripening of the cysts of *E. histolytica* consists of two consecutive mitotic divisions of the nucleus to produce four nuclei, each of which is a small replica of the nucleus at the time of encystation. During this maturing process, the glycogen is expended and the chromatoidals become less conspicuous or completely disappear. Rarely, as many as eight nuclei may be found in the ripe cysts.

At times, the nuclei, and more often the chromatoidals, may be observed in unstained living cysts. In iodine-stained fresh material, the glycogen stains a golden brown; when glycogen does not obstruct the view, the nuclei can be easily demonstrated, with the karyosome and peripheral chromatin appearing as brilliant points of yellow light against the somewhat darker yellowish-brown of the nucleoplasm. In iron-hematoxylin preparations, the nuclear chromatin and the chromatoidals take on a deep blue-black color, while the glycogen is unstained. In cysts containing a large amount of glycogen and chromatoidal material, the nuclei may be obscured, one or more nuclei may be concealed, or the karyosome may appear to be eccentric in position.

Trophozoites do not become encysted after evacuation from the bowel. In semiformed stools, it is sometimes possible to find precysts, uninucleate, binucleate, and occasionally 4-nucleate cysts. In well-formed stools, it is customary to find the ripe (4-nucleate) cysts. The 1-, 2-, and 4-nucleate cysts are infective; these constitute the transfer stage to the next host.

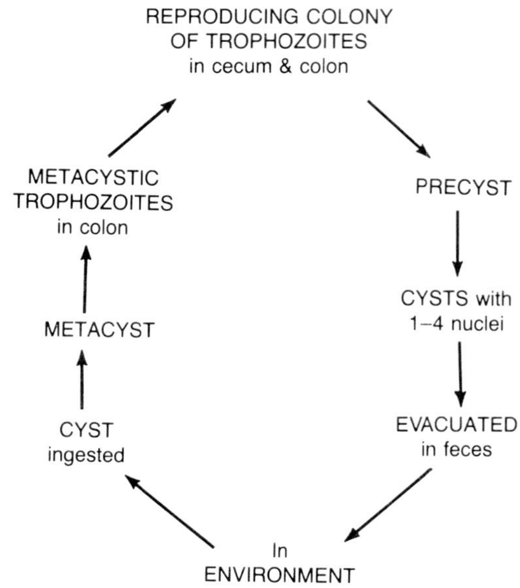

Fig. 11–3. *Entamoeba histolytica*. Diagram of the life cycle. (Adapted by P.C. Beaver from original of E.C. Faust.)

Excystation. Excystation has been observed in vitro under conditions approximating those within the digestive tract of an appropriate host. Once a viable cyst is taken into the mouth and swallowed, it passes through the stomach into the small intestine. No apparent change occurs at levels at which an acid reaction takes place, but as soon as the medium becomes neutral or slightly alkaline, the encysted ameba becomes active. This—possibly combined with the effect of the digestive juices—stimulates the multinucleate ameba *(metacyst)* to emerge from the cyst. Almost immediately the cytoplasm becomes separated into as many parts as there are nuclei, so that each nucleus now becomes the center of a small *metacystic trophozoite*. Thus, four small trophozoites typically result from the process of excystation.

Colonization. The metacystic trophozoites of *E. histolytica* do not colonize in the small intestine. They are carried in the fecal stream into the cecum, where they may first become established, provided they are sufficiently numerous that one or more make contact with the mucosa or become lodged in the glandular crypts. Once the amebae begin to feed and grow, they develop into normal trophozoites and the cycle of development is completed. The life cycle is shown diagrammatically in Figure 11–3.

Cultivation. Credit for the first reproducible culture technique for *E. histolytica* belongs to Boeck and Drbohlav (1925). The medium they employed consisted of an egg slant base with an isotonic overlay; many modifications of this culture medium have been developed. All-liquid media were described by Balamuth (1946) and Nelson (1947). In all of these media, it has been found necessary to include certain associates (enteric bacteria, organism *t*, or the flagellate *Trypanosoma cruzi*), as well as starch or rice flour, in order for the amebae to grow and multiply. A more nearly axenic medium was developed by Shaffer et al. (1949) in sealed tubes with Thioglycollate and living, but not multiplying, streptobacillus. Axenic cultivation of *E. histolytica* was achieved in 1961, and soon thereafter methods of mass cultivation of different strains of *E. histolytica* and *E. histolytica*-like amebae were developed for physiological, pathogenicity, and immunological studies (Diamond, 1968; Diamond et al., 1978).

While a few clinical laboratories employ amebic cultures in routine diagnosis, the greatest usefulness of in vitro cultivation of *E. histolytica* has been in the study of its metabolism and pathogenicity, and in the production of antigens for serodiagnosis.

Epidemiology. As noted earlier, *E. histolytica* infections are worldwide, more numerous in the tropics and subtropics than elsewhere. The prevalence of infection varies greatly, and local findings will depend largely on methods employed and the number of examinations made in individual cases.

For instance if, as commonly in the past, there is but one microscopic examination of feces by an unconcentrated film technique, at least two-thirds of the infections may be missed; indeed, 10 or 12 such examinations may not discover all of the infections. Three to four examinations by a combination of iodine-stained and hematoxylin-stained unconcentrated fecal films together with the zinc sulfate centrifugal flotation technique will detect a high percentage of the total *E. histolytica* positives. (See Technical Appendix, pp. 735–743).

Prevalence. Previous editions of this volume included tabular data on the prevalence of *Entamoeba histolytica* in a variety of human subjects throughout the United States, covering the years 1930–1955. Although the technique employed and the number and types of population groups examined were remarkably varied, the rate of infection averaged about 10% with a range from 1 to 40, depending on prevailing levels of personal and environmental hygiene. Today, in the United States, not only do practicing physicians have an awareness of the potential clinical importance of *E. histolytica* infection, but medical technologists are increasingly able to recognize this ameba. Coupled with these developments has been the gradual improvement in providing potable water supplies and sanitary sewage disposal. All of these measures have resulted in a reduction in the prevalence of infectious diseases, including amebiasis, except in isolated communities such as some institutionalized groups. One may conclude that, since the purposes of these survey data have been fulfilled, their inclusion is no longer justified. Moreover, while it is true that technologists are better trained and diagnostic methods have been improved, published reports still have a degree of unreliability. This is especially true of survey data when they are based on fecal examination by concentration techniques, and more particularly, when they are done by workers who do not attempt to distinguish between *E. histolytica* and *E. hartmanni*. Survey data in general cannot be used to challenge old concepts or to generate new ideas and often cannot be accepted as reflecting high or low standards of sanitation. The stained permanent fecal film is indispensable for the accurate routine identification of *E. histolytica* and other intestinal protozoa.

In the Western Hemisphere, infection with *Entamoeba histolytica* has been diagnosed from Alaska to the southern tip of Argentina. In the United States, *E. histolytica* prevalence in the healthy population in 1961 was estimated to be 3 to 7% (Burrows, 1961). A recent tabulation of findings from fecal examination at public health laboratories in various regions of the United States showed *E. histolytica* in less than 2% of specimens examined in all but six states; 2.0 to 3.9% in California, Texas, Illinois, and Pennsylvania; 4.0 to 5.9% in Oklahoma and New York City, and 8% in Arizona (Centers for Disease Control, 1979). In a small community in Arkansas with a population of about 500, which may be somewhat typical of rural communities, *E. histolytica* was found in 22 of 223 blacks (roughly 10%) and in 1 of 162 white persons (Spencer et al., 1976). Family clustering of positive cases suggested person-to-person rather than food or water transmission. A survey in New Orleans in 1974 revealed six infections among 887 kindergarten children. In New York City, 1000 to 1600 cases of amebiasis are reported annually (Imperato et al., 1977). In Alaska, only indigenous populations have relatively high rates of infection. In Canada, except in institutionalized groups, the prevalence figure is less than 1%.

In Mexico, Central and South America, and the West Indies, the following percentage findings in diverse population groups provide some indication of the infection prevalences: Argentina, 13 to 22%; Brazil, 10 to 47%; Chile, 23 to 26%; Colombia, 4 to 54%; Cuba, 1 to 31%; Ecuador, 2 to 35%; El Salvador, 20%; Haiti, 16 to 50%; Honduras, 20 to 46%; Jamaica, 15 to 48%; Nicaragua, 7 to 28%; Paraguay, 17%; Peru, 2 to 37%; Puerto Rico, 12 to 14%; Dominican Republic, 14 to 34%; Uruguay, 10 to 15%; and Venezuela, 7 to 43%. Recent reports give *E. histolytica* percentage rates of 16% for Mexico (Tay et al., 1978), 11% for Costa Rica (Morales and Lizano, 1978), and 16% for Panama (Cutting, 1975). With general improvement in public health facilities and public health education during the next decade or two, the average prevalence may be expected to decline in these countries.

In the Eastern Hemisphere, various surveys provide evidence of wide variation in the prevalence of *E. histolytica* infection. In Northern Europe, 10 to 15 years ago, it ranged from 5 to 20%, and in southern Europe, from 20 to 51%. In Holland, the number of cyst passers varied from 5 to 19%. In Yugoslavia before 1970, direct examination supplemented by culture of 7266 stools of school children revealed the following percentage rates: In Banat, 22 to 67%; in Bosnia and Herzegovina, 22 to 51%; in Dalmatia, 22 to 63%; in Macedonia, 3 to 72%; in Montenegro, 2 to 39%; in Serbia, 1 to 54%; and in Slovenia, 6 to 29%. For the U.S.S.R. the following percentage figures were reported: Moscow, 3%; Leningrad, 14 to 25%; Baku, 34%; Armenia, 34%; Samarkand, 10%; Tashkent, 25%; and Turkmenistan, 26%.

Country-wide surveys by Bray and Harris (1977) in West Africa (The Gambia) revealed prevalence rates of *E. histolytica* ranging from 14% in the interior to 52% near the coast, and a longitudinal survey showed nearly 100% infected over a 1-year period; however, in those surveys *Entamoeba hartmanni* was not recognized and presumably was included in the tabulations of *E. histolytica*. In Ibadan, Nigeria, from 1967 to 1976, *E. histolytica* was found in 13.0% of 360,000 stool specimens from hospital patients and in 19.6% of stools from 3400 school children and 621 adults from the same population (Ogunba, 1977). Findings for the Central African Republic were 37 and 35.6% (Brumpt et al., 1972; Pampiglione and Ricciardi, 1974) and for Rhodesia (Zimbabwe), 3.9%. In eastern Turkey, fecal examinations revealed *E. histolytica* in 5% of 600 school children (Saygi and Ogutman, 1973).

In Singapore, the reported rates are around 1% (Kan and Cheah, 1970), while in Indonesia, rates reach up to 18% (Carney et al., 1974). In Malaysia rates are up to 16% for aborigines and 6% for the populations in neighboring areas (Chiu et al., 1971). In the Mekong River Delta of Vietnam, reported rates were approximately 2% for both rural and urban populations (Colwell et al., 1971), while in Korea, infection rates ranged from 3 to 11% (Lee, 1973). In Tahiti, French Oceania, 22% of 560 persons examined in 1953 were infected. During World War II, at least 20% of American troops in the Southwest Pacific acquired *E. histolytica* infection, but 4 years after return to civilian life the rate had dropped to 5% (Lincicome et al., 1950).

Age influences the prevalence of infection. Children below the age of 5 have a much lower rate of infection than do older children and adults. In the southeastern United States, the highest prevalence has been found in people between the ages of 26 and 30 with a marked decline after 35 years. Regarding the influence of *sex* upon the prevalence of infection, males, for reasons that are not primarily occupational, usually are somewhat more frequently infected than females; in Korea, however, Soh (1967) found that females showed a higher rate of infection than males. *Climatic influences* have already been mentioned. The greater prevalence of infection in hot regions generally results from poorer sanitation in the tropics and subtropics, but the greater severity of manifestations in hotter areas seems to be due to lessened resistance to infection in individuals from temperate localities who are residing in areas of higher temperature and humidity. *Race* has no influence on the rate of infection; all races are susceptible, but those living under conditions that subject them to repeated and long continued exposure to particular strains of *E. histolytica* may acquire considerable tolerance for the infection. *Social conditions* are important, and disease rates are higher among the poor, who are more frequently exposed to infection and less able to resist it. Disease incidence, particularly the prevalence of amebic infection, in mental hospitals, prisons, and institutions for the care of children is significantly higher than in the general population in the same locality. In recent years, intestinal amebiasis has been noted to be prevalent among male homosexuals. This mode of transmission probably has been prevalent for a long time (Imperato, 1980).

Transmission. Food or drink contaminated with feces containing the cysts of *E. histolytica* is a common source of infection. Most cases arise from human carriers, or "cyst passers," who pass cysts in formed or semiformed stools, with only slight diarrhea or often no detectable symptoms. Fluid or semifluid dysenteric stools usually contain trophozoites rather than cysts; as a rule, only cysts are infective, since the trophozoites are destroyed by gastric secretions.

Cysts of *E. histolytica* may remain viable in damp soil for at least 8 days (Beaver and Deschamps, 1949a); in other moist, cool situations, for over 12 days; in water, from 9 to 30 days; and in water at 4°C, for about 3 months, varying with the amount of bacterial contamination. The greater the numbers of bacteria and the higher the temperature, the shorter the period of survival (Neal, 1974). Cysts are killed quickly by drying and have a thermal death point of 50°C (122°F), but trophozoites stored in liquid nitrogen remain viable for at least 7 years (Neal et al., 1974). Cysts have been found to resist a 1:2500 solution of bichloride of mercury and a 5% formalin solution, in each case for 30 minutes. Cysts in suspension are killed within 15 minutes by a 1:20 watery solution of cresol, within 30 minutes by a 1% solution of carbolic acid, and within 15 minutes at 30°C in 5% acetic acid (Beaver and Deschamps, 1949b). Cysts have considerable resistance to chlorine but can be destroyed in drinking water by hyperchlorination or with iodine. A method of iodine disinfection of water suitable for travelers and hikers was described by Kahn and Visscher (1975). A glass bottle of about 30 ml capacity with a nonmetallic screw-cap stopper is one-quarter filled with iodine crystals and water is added to fill. When saturated, 12 to 15 ml of the iodine solution will disinfect 1 L of water in about 15 minutes, or less. The bottle can be refilled with water and the procedure repeated as long as crystals remain. This method and others were also found to be effective in killing *Giardia* cysts (Jarroll et al., 1980). However, handling crystals of iodine is hazardous.

Viable cysts and trophozoites of *E. histolytica* may be recovered from the vomitus and feces of *flies*. The cysts may remain viable in the fly's excreta for as long as 48 hours after the insect has fed on contaminated feces. Trophozoites remain cultivatable from fly vomitus for between 28 minutes and 1 hour. Filth flies can retain cysts for periods up to 4 hours and then deposit them in a viable state (Pipkin, 1949). Cysts have been found in the intestine of flies caught in houses where human cases resided. They have also been found to survive in the intestine of the *cockroach*. (Frye and Meleney, 1936).

Food and drink may become contaminated with cysts of *E. histolytica* due to the following: (1) a polluted water supply; (2) unclean handling by infected individuals; (3) droppings of flies and other insects; (4) use of human excrement in vegetable gardens; and (5) gross carelessness in personal hygiene in children's asylums, mental hospitals, prisons, and the general population. In well-sanitated areas, the infected food handler is a possible source of danger, and the incidence of such carriers is high in some localities. Familial infections are often traceable to a food handler in the home.

The transmission of *E. histolytica* by water is common when inhabitants or transients depend on untreated wells, springs, streams, and storage tanks. *E. histolytica* has been recovered from irrigation canals in Chile and grown in culture (Horwitz et al., 1954). When a community obtains a filtered water supply, the incidence of amebiasis, if the disease is present, invariably shows a great decline.

The use of human excreta for the fertilization of vegetable gardens, which is widely practiced in the Orient and in many communities of Europe and Latin America, is an important source of infection. In some areas where such fertilizer is prohibited, the vegetables may be freshened in ditch or stream water that is almost equally dangerous. Cysts may remain alive in water-diluted feces for several days provided the temperature does not exceed 25°C.

Intestinal amebiasis found in male homosexuals presumably is acquired by ingestion of the amebic cysts or by their introduction into the rectum. Amebiasis of the cervix, vulva, and the penis possibly is acquired through sexual, including anal, intercourse (Phillips et al., 1981).

Outbreaks. *E. histolytica* infections sometimes occur in several to many people more or less at the same time and from a common source, usually as a result of fecal contamination of piped water. In the United States, no major outbreak or epidemic of amebiasis has been reported since the South Bend outbreak of 1956 (LeMaistre et al., 1956). The most notable outbreak ever recorded occurred in Chicago in 1933; over 1000 cases were diagnosed among persons exposed at two hotels (Bundesen et al., 1936). In this instance, and in the South Bend outbreak, the cause was gross pollution of the piped water supply.

Strain Differences. Strains of *E. histolytica* vary greatly in virulence. It was once thought that all strains are potentially pathogenic, because those from symptomless carriers were reported to be pathogenic to experimental animals (Krupp and Faust, 1959). The question of why some strains are highly pathogenic while others are essentially avirulent remains open, its solution awaiting further knowledge of pathogenicity factors and the character of so-called *E. histolytica*-like amebae. Numerous strains and isolates of *E. histolytica* and other species of amebae from man and from nonhuman hosts have been characterized by their isoenzyme patterns and classified as pathogenic or nonpathogenic zymodemes (Sargeant et al., 1982a, b). Whether such characterization can assist in the identification of pathogenicity factors is not yet clear (Jones, 1981).

An *E. histolytica*-like strain of ameba known as the Laredo strain was isolated from a patient with recurrent diarrhea from Laredo, Texas, by F.H. Connell in 1956. It was first described by Dreyer in 1961. It was morphologically indistinguishable from *E. histolytica* but differed in its temperature requirements for growth in culture; it was able to grow and colonize at room temperature, as well as at body temperatures, and in hypotonic solutions.

Numerous similar strains have been isolated (Nelson and Jones, 1964; Entner and Most, 1965; Vinayak and Chitkara, 1976). Five strains of the low-temperature (25–27°C) amebae, designated Laredo, Huff, 403, JA, and AG, were studied by Richards et al. (1966) and compared with seven high-temperature (35–37°C) strains that had been isolated from patients—four of whom had dysentery and one a brain abscess. Differences in temperature and hypotonicity tolerance were sharp and constant. Excepting the original Laredo strain, all low-temperature strains have come from persons with few or no gastrointestinal symptoms. Since there are antigenic differences between virulent and Laredo types of amebae (Goldman et al., 1962), it is probable that they represent distinct though morphologically similar species. A report that *E. histolytica* can adapt to growth at 29°C (Cabrera, 1958) has not been confirmed; however, it was found that although 202 of 206 isolates in a tropical community (Cali, Colombia) grew in cultures at both 30°C and 37°C, only 5 isolates grew at 27°C (Bong, 1980).

Immunity. Even in communities in which amebic infection is hyperendemic, certain individuals escape or resist infection. Immunity to *E. histolytica* may exist, but it has not been clearly demonstrated. Undoubtedly, symptomless carriers may develop considerable tolerance to the parasite, and it is possible that indigenous residents of highly endemic areas may become more or less immune to local strains. Recovery from amebiasis does not confer an immunity against reinfection, and persons who lose the infection spontaneously are susceptible after 1 year or less (Beaver et al., 1956).

Animal Reservoirs. Natural infections with *E. histolytica* occur in monkeys, dogs, and possibly pigs, but these animals constitute, at the most, a minor source of human exposure compared with man himself.

Pathogenesis and Pathology. Among the amebae parasitizing man, *E. histolytica* is unique in its ability to invade tissues; hence, the appropriateness of its species name, *histolytica*. Once the small metacystic trophozoites have developed from the metacyst after its escape from the cyst, growth and reproduction probably begin in the cecum, where clinical, pathologic, and experimental data indicate that the amebae have the first opportunity to colonize.

Pathogenicity Factors. Colonization is dependent on several factors, including the number of active trophozoites initially present. If there are many, the chance that some of them will make contact with the mucosa or lodge in the glandular crypts long enough to feed, to grow into normal-sized trophozoites, and to multiply is relatively good.

Pathologists occasionally have found amebic ulcers in the colon of persons who, during life, manifested no distinct symptoms of colitis. An appreciable number of cases of amebic hepatitis and liver abscess develop in individuals with no previous history of amebic colitis. Finally, considerable information indicates that the pathogenicity of *E. histolytica* may be enhanced by rapid passage from person to person during epidemics. Pathogenicity that has been lost in cultured amebae can be restored by passage through the liver of hamsters.

Some strains of this ameba that have a low pathogenic index, particularly those long indigenous to a community (i.e., homologous strains), live superficially in the glandular crypts of the large intestine, eroding the mucosal surface. Under favorable conditions, an equilibrium is established, and no distinctive manifestation of disease is exhibited. At times, as noted above, the amebae are spontaneously eliminated. Among 23 infected human volunteers who were under observation for 9 to 14 months, 7 lost the infections after having passed cysts in the feces regularly for 3 weeks to 2 months (Beaver et al., 1956a). When a low threshold of resistance develops due to a variety of conditions, actual invasion of the tissues may occur, with consequent clinical evidence of disease. Strains of *E. histolytica* with a high pathogenic index, or those with lesser pathogenicity in hosts with low resistance, may soon invade the tissues and produce serious, even fatal, destruction of the intestinal wall.

It has been shown that *E. histolytica* has a high requirement for iron and is able to utilize iron in hemoglobin and other complex proteins as well as simpler organic and inorganic compounds. These findings suggest a relationship between the pathogenicity of the parasite and the amounts and sources of iron available to it (Diamond et al., 1978). In experimental infections of axenically cultured *E. histolytica* in newborn mice and hamsters, virulence does not depend on the presence of bacteria; some strains are highly virulent whereas others are essentially nonpathogenic. A highly virulent strain can colonize and produce lesions in the mouse brain or hamster liver when the inoculum contains as few as 20 amebae (Mattern and Keister, 1977a, b). It still is frequently stated that the pathogenesis of amebic ulceration of the intestine probably requires the presence of certain bacteria (Krogstad et al., 1978a); convincing evidence in support of this hypothesis has not been found. Experimentally, the cytolethal action of *E. histolytica* requires direct contact with a target cell. The effect, therefore, is extracellular and precedes phagocytosis (Ravdin et al., 1980). The essential pathogenic factors in amebiasis are still largely unknown (Jones, 1981).

When it was discovered that certain pathogenic strains of *E. histolytica* were infected with transmissible viruses (Diamond et al., 1972; Mattern et al., 1972), it was expected that their presence might enhance or otherwise modify the ameba's virulence. Subsequent studies have failed to show such a relationship (Mattern et al., 1979); however, viruses that are pathogenic for the amebae may prove to be a factor in the spontaneous loss of amebic infection. The same might be said for other organisms that live in *E. histolytica* and other intestinal amebae. Not infrequently, organisms known as *Sphaerita* (Fig. 11–4) are seen in cysts or trophozoites of *E. histolytica, E. coli,* or *Endolimax nana* of humans, or *Entamoeba muris* of rats and mice (Wenyon, 1926). Another such parasite is *Nucleophaga,* which lives in the nucleus of parasitic and free-living protozoa. These coccus-like fungus spores are highly refringent in the unstained amebae and black when stained with hematoxylin. These and other parasites of amebae in certain areas could be responsible for the spontaneous loss of infection, and thus, may account for the exceptionally low prevalence of infection in communities in which sanitation is poor and the frequency of transmission of enteric infection is expected to be exceptionally high.

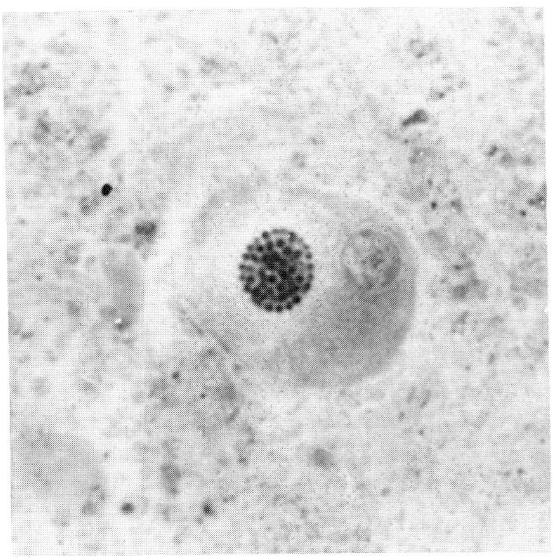

Fig. 11–4. *Sphaerita,* a parasitic fungus, in a trophozoite of *Entamoeba coli*. Celestine Blue B. (× 1000.) (By P.C. Beaver; preparation courtesy of Dr. T. Scholten.)

The Intestinal Lesion. The typical lesion is most satisfactorily observed in human or experimental infections in which the amebae rapidly enter the intestinal mucosa. The mode of entry of the mucosa by amebae is unknown. Findings in light and electronmicroscopic studies of colonic mucosal biopsies in cases of amebic dysentery suggest that the mucosal epithelium is lysed by substances liberated from lysosomal-like vesicles at the surface of the amebae, or from ruptured amebae (Pittman et al., 1973b), or possibly by cytotoxic substances released from lysed leukocytes directly or excreted by amebae after digestion of ingested leukocytes (Griffin, 1972). Observations leading to similar conclusions were made on experimentally infected guinea pigs, and it was further observed that lysis of the mucosa is due in part to tissue ischemia following endothelial damage, and thrombosis of capillaries and venules caused by degenerating leukocytes (Takeuchi and Phillips, 1976; Takeuchi et al., 1977). The cytolytic action of *E. histolytica* on cells in tissue culture, and the digestion of collagen, are dependent on contact (Ravdin et al., 1980; Munoz et al., 1982). Some workers report that sonicated amebae cause no damage to cells (Knight, 1977; Lushbaugh et al., 1979); others have found that extracts of virulent, but not avirulent, strains of *E. histolytica* have a marked cytotoxic effect on cells in tissue cultures (McGowan et al., 1982).

The *primary ulcer* most often develops in the cecum, appendix, or adjacent portion of the ascending colon, but such lesions may occur at lower levels, particularly in the sigmoidorectal area (Fig. 11–5).

There may be only one or several points of penetration. In the latter case, the ulcers are typically separate on the mucosal surface, somewhat elevated, and have a pin-head center through which blood-tinged mucus, amebae, and necrotic cells will be extruded, with undamaged surface between the lesions.

Frequently, the lesions are confined to the mucosa and tend to spread out laterally. Repair may

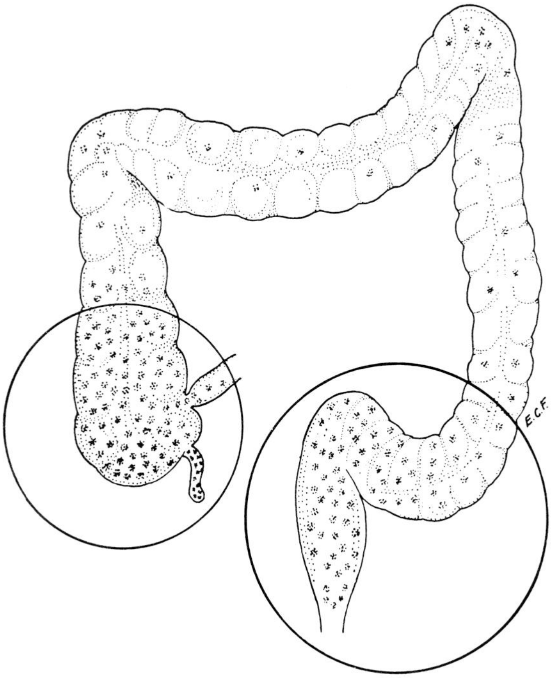

Fig. 11–5. Diagram illustrating the relative numbers of amebic lesions in the large intestine. The most frequently invaded level is the cecal area and the second most common site is the sigmoido-rectal area. (Adapted from Faust, E.C. 1943. Some modern conceptions of amebiasis. Transactions & Studies, College of Physicians of Philadelphia, Series 4, *2*:101–113.)

keep pace with erosion of the epithelium or it may exceed the destruction at a particular level or throughout the invaded area, resulting in spontaneous healing of the lesion. Probably more often than not, the amebae in the depth of the glands or in the stroma break through the muscularis mucosae interstitially or via the lymphatics and blood vessels and reach the submucosa. Here they characteristically spread out radially through relatively nonresistant tissue, so that nests of amebae may accumulate or may honeycomb this layer. This step completes the typical early amebic lesion, which has a minute site of entry into the mucosa, a neck of small diameter, and a bottle-like base. In this early lesion, there is frequently no bacterial invasion and almost no cellular reaction; it is a lesion almost exclusively of lytic necrosis. As the ulcer becomes older, the amebae may penetrate the muscular coats and occasionally the serosa, with perforation into the peritoneal cavity, or the subsurface portions of adjacent colonies may coalesce, undermining the mucosa. By this time, the neck of the ulcer becomes enlarged with overhanging edges and the ulcer is invaded by bacteria, provoking infiltration of neutrophilic leukocytes so that the gross and microscopic picture is remarkably altered (Fig. 11–6).

From these primary sites of invasion, *secondary lesions* may be produced at other levels of the intestine and in extraintestinal organs and tissues. Amebae carried down the bowel invade and produce additional ulceration at lower levels of the colon and in the rectum; by regurgitation, they may become established in the mucosa of the ileocecal area. Thus, as the infection changes from the early to the chronic state, increased numbers of lesions are developed in lower segments of the large intestine, particularly in the sigmoid colon and rectum. Observations on experimental infections in germ-free guinea pigs have suggested that lysis of colonic mucosa in amebic infection is due in part to tissue ischemia caused by vascular thrombosis, and in part to the lytic action of degenerating neutrophils (Takeuchi et al., 1977).

Lesions in substantial autopsy series were first graphically presented by Clark (1925) in Panama, and were confirmed by Faust (1943) in New Orleans (Fig. 11–5). Duque's (1965) figures for fatal cases of amebiasis in Medellin, Colombia, were as follows: total cases, 176; 114 male, 62 female; 55 under 10 years of age. The clinical diagnosis was correct in 69 cases (33%) only. The large intestine was generally involved from the cecum (141 cases) to and including the rectum (129 cases). There was ameboma of the colon in 12 cases, segmental involvement of the ileum (terminal) in 18, perforation of amebic ulcers in 52, dysentery in 33, acute dehydration in 22, acute abdominal pain in 22, and acute ileus in 9. There were single perforations in 25 cases, multiple in 27; 21 of these were in the transverse colon, 20 in the sigmoid, 15 in the cecum, and 1 in the appendix. There was associated liver abscess in 66 cases (47 of them males), one in an infant of 2 months. The abscesses were single in 41 cases, multiple in 25. There were 32 single in the right lobe, 7 in the left lobe, and 2 in the median area. The hepatic abscess ruptured into the peritoneum in 12 cases, into the pleura in 12, and into the pericardium in 5. There were pulmonary abscesses in 17 cases, 15 of these associated with hepatic abscesses, and in 2 cases, there were cerebral abscesses. Rowland (1967) reported that of 1500 consecutive patients referred to a single medical firm in Freetown, Sierra Leone, 54 (3.6%) suffered from amebic infection causing disease. Of those with amebiasis, 31 (57%) had a

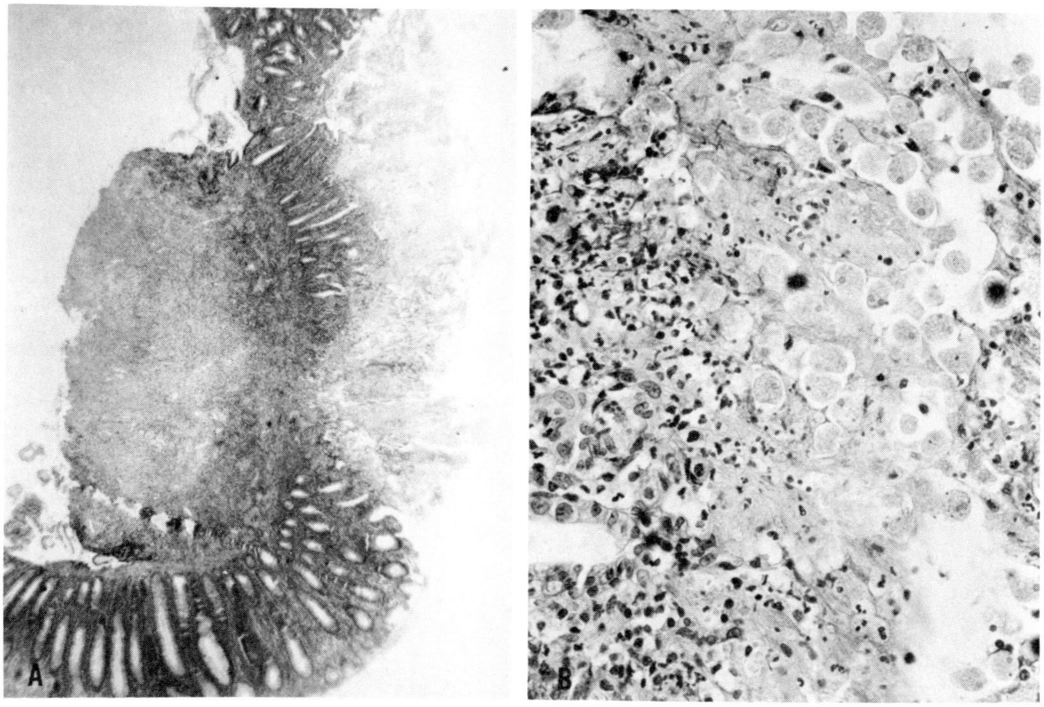

Fig. 11–6. *Entamoeba histolytica*. Trophozoites in typical lesion of mucosa in rectal biopsy specimen. *A*, Lesion at low magnification. (× 20.) *B*, Higher magnification (× 400) showing the number and distribution of amebae. (By P.C. Beaver.)

liver abscess and 21 (39%) died. Severe bowel ulceration was found to be much more dangerous than liver abscess, and both intestinal perforation and large abscesses of the liver often were remarkably silent.

Amebic Granuloma. Occasionally, a granulomatous tumor mass *(ameboma)* develops at some locus on the wall of the large intestine as a sequel to an amebic ulcer. In a series of 3013 cases of amebic dysentery reported from Durban, South Africa, ameboma was a complication in 15 (0.5%) (Adams and MacLeod, 1977a). Amebomas are relatively firm and nodular, with a fibrous outer wall under an edematous mucosa, an intermediate granulating zone with eosinophils, lymphocytes and few fibroblasts, and one or more internal abscesses containing necrotic tissue and amebic trophozoites. Amebomas, when demonstrated roentgenographically or by endoscopic visualization, bear a strong gross resemblance to carcinoma. The patient with colonic amebiasis may be spared a hazardous operation if this is appreciated.

Extraintestinal Lesions. Although *E. histolytica* has been demonstrated in practically every soft organ and tissue of the body, extraintestinal amebiasis, with certain exceptions mentioned below, is secondary to earlier colonization in the wall of the large intestine, even though this latter condition may not be diagnosed clinically. Corresponding to the early predominance of lesions in the cecal area, a high incidence of extraintestinal infection occurs in the liver, and of this number, a majority develop in the right lobe. Of 48,780 patients treated for amebiasis in the University College Hospital, Ibadan, Nigeria, during the years 1958 to 1962, 120 (approximately 1 in 400) were suffering from hepatic amebiasis (Adi, 1965). Age and sex distributions were almost identical with those distributions reported by DeBakey and Ochsner (1951) for cases in the United States. Extraintestinal amebiasis, now rare in the United States and Europe except in immigrants and travelers (Abuabara et al., 1982), is frequently encountered in the tropics (Adams and MacLeod, 1977b).

Liver. Colonies of *E. histolytica* that reach the muscularis mucosae frequently digest the walls of the mesenteric venules that they enter and are carried into the intrahepatic portal vessels. Many of these amebae fail to lodge, but if thrombi composed of fibrinous filaments and leukocytes occur in the small branches of the portal vein or in the interlobular veins, the amebae caught in the occlusion

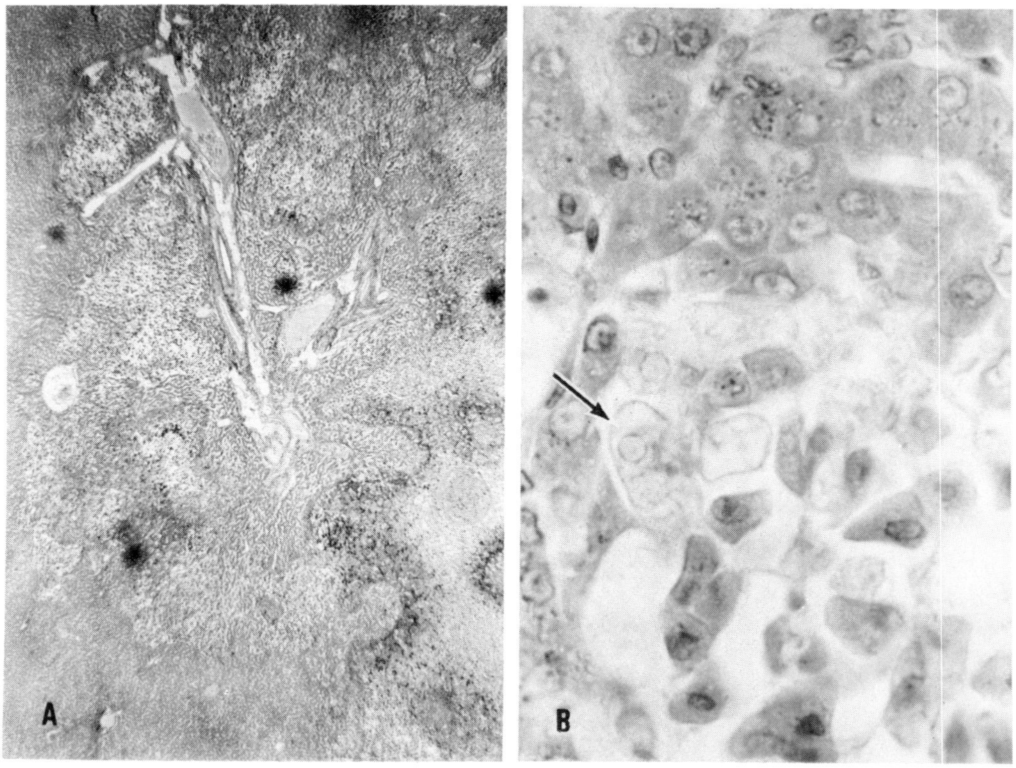

Fig. 11–7. *Entamoeba histolytica*. Early amebic invasion of liver of kitten 2 weeks after rectal inoculation. *A*, Area of invasion extending from branch of portal vein. (× 25.) *B*, Margin of the lesion where amebae (at center) have caused destruction of the stroma and necrosis of free hepatic cells (at lower center). (× 1000.) (By P.C. Beaver. Preparation courtesy of Dr. G.M. Carrera.)

produce lytic necrosis of the wall of the vessels, then enter periportal sinusoids and digest pathways into the lobules (Fig. 11–7). By this mechanism, one to many small secondary colonies are formed in the liver (Meleney, 1934). At this early stage, essentially no inflammatory reaction occurs, but as the colonies increase in size, with corresponding lytic necrosis of the liver tissues, considerable infiltration of leukocytes takes place. This has been termed "amebic hepatitis." Frequently, one or occasionally more of these initial lesions will enlarge to develop into an amebic liver abscess. The smallest of these may be only a few millimeters in diameter, solid and whitish, but it soon increases in diameter, with a yellowish-gelatinous content and later a cavity filled with liver-colored fluid and cellular debris. If the abscess has developed rapidly, there is no limiting capsule, but in the more chronic lesions, there is a fibrous wall. The following three zones may be recognized grossly and under low power of the microscope: (1) a necrotic center filled with thick fluid; (2) a median zone consisting only of coarse stroma; and (3) an outer zone of nearly normal tissue just being invaded by the amebae (Fig. 11–8). The solitary abscesses may be as large as a grapefruit, with trabeculae extending to the center. Most of these develop in the right lobe of the liver, and about half eventually involve the diaphragm and usually rupture into the right lung or pleural cavity. Some, however, discharge into the colon, peritoneal cavity, other abdominal organs, or through the body wall.

Lungs. Amebiasis in the lungs usually develops as an extension of a hepatic abscess by rupture through the diaphragm, but it may originate concomitantly yet independently of the liver involvement by metastasis from the intestine without colonization in the liver. Perforation of a hepatic abscess through the diaphragm and hence, into a bronchus, results in a hepatobronchial fistula. The characteristic sign of this condition is expectoration of typical liver-colored sputum. Among 2074 cases of amebic liver abscess seen in Durban, South Africa, extension into the chest was a complication in 146 (7%), and in 27 of these cases, the pericardium was invaded (Adams and MacLeod,

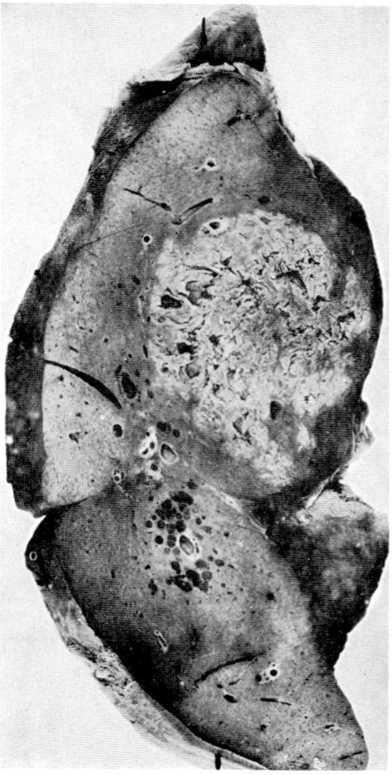

Fig. 11–8. Large, solitary amebic abscess of the liver. Note that the cytolysis of the liver tissue is not complete and that there is no well-defined abscess wall. (From Craig, C.F. 1934. *Amebiasis and Amebic Dysentery.* Springfield, Ill., Charles C Thomas.)

1977b). Pericardial involvement is more prone to occur with abscess in the left lobe of the liver.

Skin. Amebiasis cutis appears as an edematous elevation of the involved skin with induration of the irregular margins of the spreading ulcer, gangrenous epidermis, and a necrotic base in which the amebae are found advancing peripherally. It develops as a perianal extension of acute amebic colitis, or on the abdominal wall through rupture or open drainage of a colonic, appendiceal, or hepatic lesion. It also occurs as a venereal infection of the penis as a consequence of anal intercourse. Vulvar amebiasis is less common.

Brain and Other Uncommon Sites. Uncommon or rare is amebic abscess of the brain, which is hematogenous in origin but usually arises from, or is concomitant with, amebiasis of the liver or lungs. In the Durban series of 2074 cases of amebic liver abscess referred to previously, brain abscess was a complication in only one case. Other rare ectopic sites of amebiasis include the spleen, adrenals, kidneys, ureters, urinary bladder, urethra, clitoris, pericardium, and in a nasal polyp. In a case involving the eye, the amebae colonized the broken skin of the eyelid of a 4-month-old child and the dermal lesion expanded into the tissues of the orbit (Fig. 11–9). The source of the infection was not determined in this case, nor in a similar case involving the nose (Beaver et al., 1978).

In all of the extraintestinal types of amebiasis involving the viscera, the abscess is usually bacteriologically sterile. Characteristically, however, there is moderate infiltration of neutrophilic leukocytes in response to necrotic destruction of the invaded organ. In about 15% of abscesses, there is secondary bacterial infection. In these, the character of the contents is more purulent and the acute inflammatory infiltration tends to be more intense.

The relationship of extraintestinal types of amebiasis to one another and to the primary infection in the large intestine is diagrammatically shown in Figure 11–10.

Pathologic Anatomy. In amebiasis of the intestinal tract, commonly referred to as amebic colitis, erosive ulceration may occur at any or all levels of the large intestine and in the terminal segment of the ileum. The earliest—hence, the oldest and most advanced—lesions are commonly found in the cecal region (cecum, ileocecal valve, appendix, and ascending colon); second in frequency and intensity of involvement are the sigmoid colon and rectum; and third, the sigmoid and hepatic flexures. As stated earlier in this chapter (p. 109), the initial ulcer consists of minute entry sites into the mucosa, exhibiting hyperemia or nodular elevations, and separated from one another by intact epithelium. These lesions are commonly overlooked at autopsy. When detected and incised, they are found to be confined to the mucosa (Faust, 1941) or to have a small neck through the mucosa leading to an enlarged base in the submucosa. They contain a brownish-yellow gelatinous material that consists of mucus, necrotic tissue cells, and amebic trophozoites. The more advanced amebic lesions are characterized by total destruction of the area immediately around the sites of colonization, with thick overhanging ragged edges and a floor at the base of the mucosa or in the submucosa covered with necrotic gelatinous material. These ulcers vary in size; they may be the size of a pinhead or may coalesce and encircle the entire bowel wall. When still discrete, they are usually oval, with the longer diameter lying at right angles to the long axis of the intestine. Gangrenous sloughing may be evident due to subsurface undermining of the mucosa cutting off arterial blood supply, or the wall may

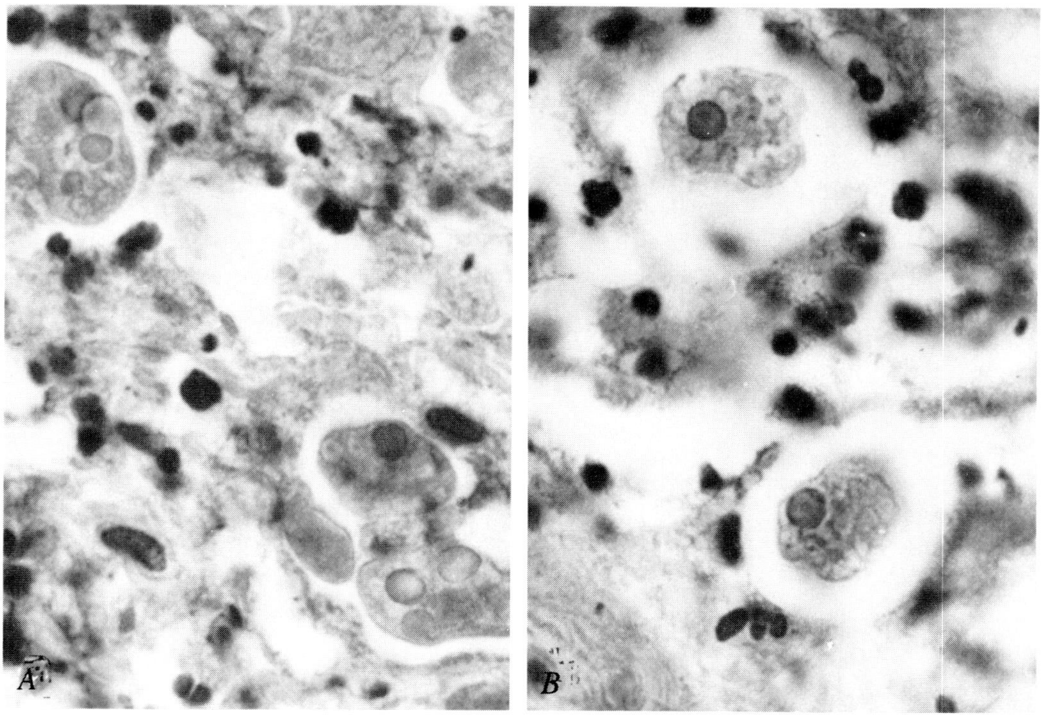

Fig. 11–9. *Entamoeba histolytica*. Trophozoites with ingested red blood cells *(A)* and showing characteristic nuclei *(B)* in tissues removed with the enucleated eye of a child in Colombia. The primary lesion was in the skin of the eyelid. H & E. (× 1200.) (From Beaver, P.C., Lopez-Villegas, A., Cuello, C., and D'Alessandro, A. 1978. Cutaneous amebiasis of the eyelid with extension into the orbit. Am. J. Trop. Med. Hyg., *27*:1133–1136.)

be paper-thin in spots due to deep invasion by the amebae. Occasionally, perforation through the serosa may be demonstrated. Intestinal damage of major proportions is often found *postmortem* in cases with no specific clinical diagnosis.

The more advanced amebic ulcers typically show signs of inflammatory reaction as a result of bacterial invasion. Amebic and *Shigella* infections rarely occur together (Powell et al., 1961; Lara et al., 1974).

The gross pathology of *amebiasis in the extraintestinal viscera* depends on the anatomical location and the stage of development of the lesion. In early *amebic hepatitis,* there may be no gross evidence of damage superficially or on the cut surface (Fig. 11–7). At a somewhat later stage (amebic liver abscess), the section will show one or more rounded or ovoid abscesses varying from 1 mm to 10 cm in diameter. The smaller ones contain grayish-brown necrotic material, while the larger ones (Fig. 11–8) have a cavernous center filled with cytolyzed yellowish-brown or anchovy sauce-colored coagulative necrotic debris, at times with undigested stromal fibers more marginally, with or without a circumscribing wall of infiltrated fibroblasts separating the abscess from the essentially undamaged tissue.

Development of an *amebic abscess in the pleural or pulmonary region* is commonly an extension of an abscess in the upper level of the right lobe of the liver (Fig. 11–10). In case the pulmonary abscess arises directly from an intestinal lesion via the blood stream, development is similar to that described for the liver. In *amebic involvement of the brain,* the abscess is perivascular, usually single, of relatively small size, and with rare exceptions, is located in the cerebrum. At autopsy, there may be evidence of rupture into a lateral ventricle.

In *amebiasis cutis,* the lesion may be an extension of an intestinal infection into the perianal or perineal skin; it may occur at the site of rupture of an abdominal amebic abscess or of a colostomy drainage; or it may be of several origins. It shows signs of necrosis and gangrenous sloughing, or at times, granulomatous ulceration.

Symptomatology. The manifestations of amebiasis are as varied in type and intensity as the lesions produced by the etiologic agent. Symptoms differ depending on the site of tissue damage and on the extent of such damage.

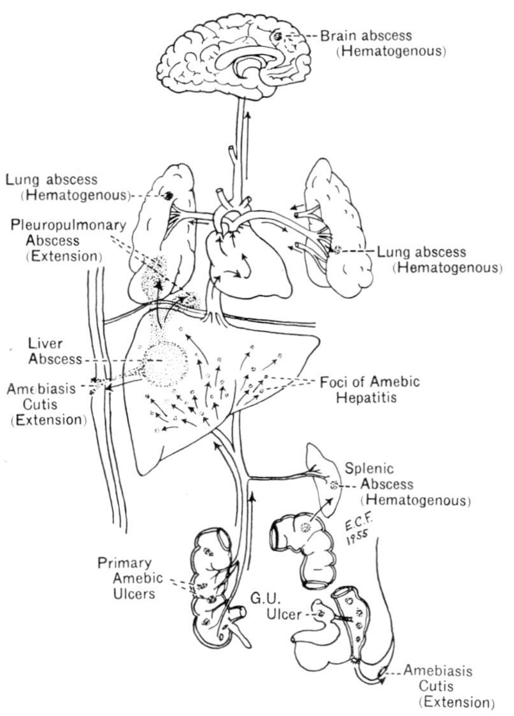

Fig. 11–10. Routes of origin and locations of extraintestinal amebiasis. (Adapted by E.C. Faust from Chatterjee, K.C. 1952. *Human Parasites and Parasitic Diseases*. Calcutta. Published by author.)

Incubation Period. Although the *biological incubation period* varies from 2 to 5 days or longer, the clinical incubation period, by comparison, may be as short as 4 days or as long as 1 year; in many instances, in areas with opportunity for repeated infection, it is impossible to determine the interval between exposure and onset of symptoms, or the loss of an earlier infection and the acquisition of a subsequent one. Furthermore, many infected persons remain apparently symptomless for years. Nevertheless, the expected incubation time is 1 to 4 months.

Onset. In an area of heavy endemicity of a highly pathogenic strain of *E. histolytica,* the onset may occur suddenly only a few days after exposure. This situation is most frequent in water-borne epidemics. More often the symptoms develop gradually, with one or more prodromal episodes of diarrhea, accompanied or followed by abdominal cramps. A much larger proportion of infected persons may experience several weeks of vague abdominal discomfort or generalized malaise, loss of appetite and weight, and apathy with no definite suggestion of localization in the colon. Again, the patient may harbor the parasite for years in apparent good health, then have a sudden attack of acute colitis or develop a liver abscess.

Amebic Colitis. Symptoms may be mild or severe, or the individual may be a "symptomless carrier," depending on the extent and intensity of lesions. If the entire length of the large intestine is involved, there may be exquisite tenderness of the entire abdomen, with fulminating dysentery, rapid loss of weight, dehydration, and serious incapacitation. These symptoms may frequently be accompanied or succeeded by hepatic involvement, or at times, by perforation of the bowel. If the cecal area is exclusively or primarily involved, the symptoms often mimic acute or subacute appendicitis, with pain and tenderness at McBurney's point, nausea, and vomiting. There may be moderate or profuse diarrhea, or the stools may be formed or semiformed, but frank dysentery does not occur unless there is sigmoidorectal disease. In cases with involvement of the sigmoidorectal region, the most conspicuous symptom is dysentery, in which relatively small amounts of blood and mucus are evacuated with formed or semiformed feces in frequent stools, and there is some degree of tenesmus.

Cases in which symptoms have been present for less than a month are classified as *acute*. If symptoms persist for more than a month or if there are relatively long intervals of mild symptoms that are usually accompanied by formed stools or by constipation with periodic reactivation of the acute state, the condition is classified as *chronic*. In other individuals, there are no distinctive manifestations except generalized indisposition, with presence of *E. histolytica* in the stools. Except during epidemics, the incidence of acute and subacute amebic colitis is low compared with the chronic variety.

Bacterial Complications. Reference has been made earlier (p. 110) to the alteration of the pathologic picture following entry of bacteria into amebic ulcers of the large intestine. Clinically, this is significant, since the bacteria alter the structure and hence, the functioning of the involved areas. This is particularly important in chronic amebic infections of long duration. Where therapy has eliminated both the amebae and bacteria and healing is complete, normal peristalsis may not be possible because of extensive scarring. Even when scarring is not demonstrable, bowel function may remain abnormal in the *postdysenteric syndrome*.

Hepatic Involvement. When colonies of *E. histolytica* become established in the submucosa and

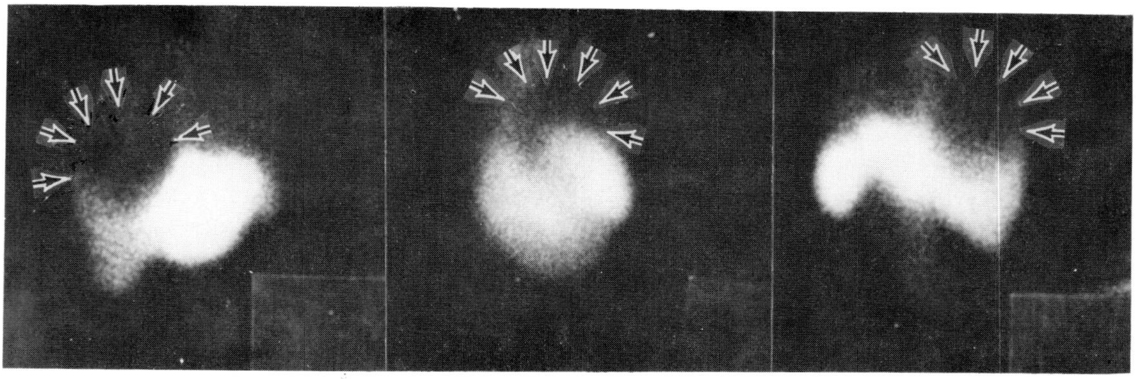

Fig. 11–11. Amebic liver abscess. Man from India, age 33, with a large abscess deforming the dome of the right lobe of the liver. Arrows define upper margin of the abnormality viewed from anterior (left), right lateral (middle), and posterior aspects (right). The scintigrams were made with a gamma camera following intravenous injection of 3 millicuries of Technetium^{99m} sulfur colloid. (Courtesy of Dr. Roy Staub, Touro Infirmary, New Orleans, La.)

erode mesenteric venules, particularly those draining the cecal area, numerous pathways are opened for continued invasion of the liver. The diagnosed cases of amebic hepatitis and amebic liver abscess are relatively few compared with the occasions when amebic trophozoites enter the portal vessel. In a considerable number of patients with amebic colitis, there is an associated enlargement and tenderness of the liver, without fever, subphrenic bulging, or diaphragmatic impairment, and with negative or equivocal liver function tests (Wilmot, 1962). This possibly is due to constant invasion of the amebae from an active intestinal process, with their rapid destruction; however, it is just as likely that the transport of bacteria or toxic substances to the liver from intestinal ulcers is responsible. In foci where amebic colonization continues, the acute hepatic process becomes transformed into an amebic liver abscess. This results from infection of the portal venules, with lytic necrosis of adjacent hepatic parenchyma (Fig. 11–7). The enlarged abscess can usually be visualized radiologically, sonically, or by radionuclear scanning (Fig. 11–11). In a preponderance of cases, the upper right lobe of the liver is involved, and the infected person experiences upper right abdominal pain. These patients may be weak and underweight as a result of absorption of toxic metabolites from necrotic tissue within the abscess pocket, but in the great majority of cases, the lesions are bacteriologically sterile. These symptoms and signs in amebic liver abscess are relatively pathognomonic. Abnormalities of liver function tests are usually slight and nonspecific (Carneiro Chaves et al., 1977). Statistically, most patients with hepatic amebiasis are males in the fourth and fifth decades of life. Except in certain localities, *E. histolytica* is a relatively uncommon cause of liver abscess in the United States (de la Maza et al., 1974).

Pleuropulmonary Infection. The most common type of pleuropulmonary infection consists of an extension of a hepatic abscess through the diaphragm. The cardinal symptom is sustained intense pain in the lower right region of the chest due to inflammation of the diaphragmatic pleura that is accompanied by a persistent, unproductive cough. If the abscess breaks through into a bronchus, the reddish and gray-brown contents of the abscess are typically coughed up. Marked dyspnea occurs with considerable involvement of the right lung. In addition, there is the clinical syndrome of the hepatic abscess. Occasionally, a primary pulmonary amebic abscess resulting from hematogenous extension directly from the colon provides a picture lacking hepatic or pleural complications (Wilmot, 1962). Another common type of presentation for pleuropulmonary infection is effusion in the right pleural space, which is due to diaphragmatic pleuritis overlying the abscess without rupture through the diaphragm. The pleural fluid is serosanguineous. If rupture occurs thereafter, the character of the fluid changes to that of the abscess contents.

Amebic Abscess of Brain. This is commonly an extension of preexisting amebiasis in extraintestinal viscera, but it may be a direct hematogenous extension from the colon (Swartzwelder and McGill, 1949). It is invariably fulminating, with clinical evidence of a destructive brain lesion that is usually in the cerebral hemispheres, and is not specifically diagnosed until autopsy. Fatal termination usually occurs in 7 to 10 days.

Amebiasis of Other Viscera and Skin. The symptoms vary depending on the organs and tissues involved. Localized pain with accompanying fever, leukocytosis, and other manifestations of an internal abscess, but usually lacking the intensity of bacterial abscesses, will occur in the affected viscera. In *amebiasis cutis,* there is a tender, swollen, ulcerated area with elevated indurated margin that is dark red in color, and sometimes more centrally an abscess pocket with semiliquid content with fetid odor.

Diagnosis. With the exception of hepatic amebiasis, in which the symptoms and signs are relatively pathognomonic, accurate diagnosis of amebiasis requires demonstration of *Entamoeba histolytica* in its trophic or cystic stage, and its specific differentiation from other objects with which it may be confused. A positive laboratory finding may support the clinical view that *E. histolytica* is primarily responsible for the patient's illness, or that the amebic infection may be contributory to the condition. On the other hand, the demonstration of *E. histolytica* may be an incidental finding.

The x-ray picture may furnish useful information concerning the anatomic location of lesions, and the hemogram may indicate systemic reaction to a pathologic process suspected of being amebiasis, but actual proof depends on identification of *E. histolytica*. At times, one may interpret the filling defects in the cecum and ascending colon seen with the x-ray picture following a barium meal as being due to amebic ulceration, but similar appearances are seen in other forms of dysentery and ulcerative colitis. In amebiasis of the right upper lobe of the liver, the doming up of the diaphragm and its restricted movement provide considerable suspicion of amebic liver abscess in this portion of the liver. Similarly, shadows on the x-ray film of the pleural cavity or lung may suggest amebic abscess. The demonstration of a triangular shadow with the base toward the liver and the apex extending toward the hilum may suggest hepato-bronchial fistula.

Radionuclide scan has proved of great value in diagnosing and locating an amebic liver abscess. It is also useful in detecting unsuspected multiple abscesses and in following the course of the lesion under therapy. More recent techniques that have proven of great value in detecting intrahepatic lesions are sonography (Sukov et al., 1980) and computerized tomography (Sepulveda, 1982).

Ellman et al. (1965) employed pneumoperitoneal insufflation with carbon dioxide for the differential diagnosis of subphrenic amebic liver abscesses and reported that this seems to be a safe, well tolerated, and valuable technique.

The *blood picture* in uncomplicated amebic infection of the large intestine is little altered from normal, except in instances of profuse dysentery when anemia may develop. In dehydrated patients, however, this may be masked by hemoconcentration. Moderate neutrophilic leukocytosis occurs with chronic amebic colitis or in amebic appendicitis, with invasion of the amebic ulcers by enteric bacteria. There is a rise in the white cell count in uncomplicated amebic abscess of the extraintestinal viscera, but if the count rapidly reaches an unusual elevation, pyogenic infection should be considered.

Laboratory Diagnosis. The detection and identification of *E. histolytica* in fecal specimens are among the most difficult tasks assigned to the diagnostic laboratory. In the United States, and presumably elsewhere, laboratories tend to overdiagnose amebiasis because leukocytes and other host cells are frequently mistaken for amebae, and *E. coli* and *E. hartmanni* also are mistaken for *E. histolytica* (Krogstad et al., 1978b). The prevalence of amebiasis in any community is always open to doubt because the detection of amebae depends largely on the selection of appropriate methods of obtaining, transporting, and processing specimens, and the identification of amebae in feces, exudates, or tissues depends on the skill and diligence of the medical technologist (Foley et al., 1978).

The specimen submitted for diagnosis is commonly feces, but it may be obtained from a saline-purged or enema specimen, aspirate, surgical biopsy, or necropsy material. Falure to find cysts on a particular day should not be taken as conclusive evidence of their absence, since subsequent examination may disclose an abundance of cysts or trophozoites.

Fecal Specimen. The stool may be formed, and the specimen may consist almost exclusively of feces. The specimen may be mushy, or it may be unformed (diarrheic, or with considerable mucus so that it is semigelatinous, or it may have blood, mucus, and tissue shreds sloughed from the bowel wall). In the formed stool, one may expect to find cysts of *E. histolytica* and other intestinal amebae (Plate I); occasionally, a formed or semiformed specimen will contain trophozoites and no cysts. In trials of a method of preserving fecal specimens, 27% of the protozoa identified in formed stools were present as trophozoites only (Goldman and Brooke, 1953).

Formed stools may be left overnight at room temperature, or for several days if kept cool, with the expectation that identifiable cysts will be found if they are present in the sample. Unformed stools that are positive for *E. histolytica* commonly contain trophozoites. These stools should be examined as soon as possible after they have been passed, preferably within 30 minutes, or they should be preserved for later examination by appropriate techniques such as PVA or MIF. The portion of the specimen chosen for examination should be specially selected; for formed stools, this may be any part of the feces but should include a portion of any adherent mucus; for unformed stools, the mucus or fecal material is more likely to contain *E. histolytica* than is the liquid part. Stool specimens to be submitted for examination should be passed in a clean container, be uncontaminated with urine, and be kept in a closed container, free from air contaminants until examination is made.

Direct examination is recommended for all stool specimens. First of all, the gross appearance, consistency, and odor of the specimen should be noted. Then a coverglass mount of the specimen should be made on a clean microscope slide. A small portion of the stool (about 2 mg) is placed in a drop of physiologic salt solution on the slide, thoroughly comminuted, and mounted with a 22 mm-square coverglass. This fecal film should be thin enough so that all particles will be suspended and easily seen under the microscope. If protozoa are found or suspected, a similar preparation is made and a drop of aqueous iodine stain is added before the coverglass is mounted, or the stain can be placed at the edge of the covered saline smear and allowed to run under the coverglass. The stain is a saturated solution of iodine crystals in 1% potassium iodide; Lugol's solution is too strong and damages the amebae.

Motility of active trophozoites in the saline preparation can be studied in the unstained film; at times, cysts of *E. histolytica* can be identified by the characteristics of the nuclei and the highly refractive chromatoidals. In iodine-stained cysts, the nuclei can be studied more easily and the species of ameba identified (Plate I). In some specimens, however, it is not possible to make a specific diagnosis of *E. histolytica* on the basis of the temporary wet film. In these cases, Schaudinn-fixed, iron-hematoxylin- or trichrome-stained fecal films should be prepared (See Technical Appendix, p. 737). For surveys requiring the collection of stools in distant areas, the fecal specimens should be preserved in special fixatives for examination later. Hospitals and clinics may also require examination of fixed specimens collected from freshly passed stools as a part of the laboratory routine (Markell and Quinn, 1977), although in most laboratories, this would be unacceptable. Special fixatives are required, because formalin is unsatisfactory for specimens to be examined for protozoa.

If cysts of *E. histolytica* are in the stool specimen but too few to be present in direct fecal films, the cysts should be concentrated. The most satisfactory method for clinical laboratory diagnosis of *E. histolytica* cysts, as well as eggs and larvae of many intestinal helminths, is zinc sulfate centrifugal flotation (p. 742). In lightly endemic areas for *E. histolytica,* a single direct fecal film is not likely to discover more than 20 to 25% of the infections; with the addition of zinc sulfate concentration, the number of positive diagnoses will usually be doubled. Examination of three or four stool specimens passed every second or third day, employing direct wet or hematoxylin-stained film and zinc sulfate concentration, should detect 85 to 95% of all infections.

Saline-Purged Specimens. At times, the saline-purged specimen will provide material for positive diagnosis when routine fecal examination has been unrewarding. Sodium sulfate (Glauber salts) or phosphosoda is preferred to magnesium sulfate (Epsom salts). Mineral oil, powdered magnesia or other particulate cathartics, or barium used in x-ray examination produce globules or granules and mask the amebae. Such specimens are unsatisfactory for diagnostic purposes. Following saline purgation, the earlier fecal evacuations from the bowel are discarded; sedimented elements of mucus and tissue cells from the second or third bowel movement are pipetted onto a slide, mounted with a coverglass without staining, and examined for trophozoites. If saline purgation is contraindicated, the same type of material for examination may be obtained by a high retention enema of tepid physiologic salt solution. The chief advantage of the purged or enema specimen is that it can be obtained on schedule for immediate examination.

Aspirates. During proctoscopy, *scrapings or aspirates from suspected sites* of amebic ulceration may be obtained, or shallow *biopsies* secured. The aspirated material should be microscopically examined for motile trophozoites immediately, under a coverglass, but great care must be exercised not to mistake macrophages or other tissue cells for amebae. Under no circumstance is it safe to diagnose a nonmotile rounded cell in the aspirate as a "precyst" or "cyst" of *E. histolytica,* since encystation of intestinal amebae does not normally occur in the intestinal wall. The biopsy specimen may be pressed between slide and coverglass in an attempt to observe motile trophozoites, but it is more satisfactory to fix, section, and stain the specimen before attempting to examine it.

Aspirates from an amebic abscess of the liver or other extraintestinal viscera should be secured from the margin of the abscess pocket, where the amebic trophozoites are relatively abundant compared with the necrotic material in the center of the abscess. Enzymatic liquefaction of viscous aspirates will facilitate the search for amebae.

Biopsy and Necropsy. Surgical biopsy and postmortem material should first be examined with the unaided eye or hand lens, then fixed in 10% formalin for sectioning and staining. For satisfactory study, autopsy material should be obtained within 4 hours after death. If older postmortem tissues are secured, the amebae usually fail to show satisfactory nuclear details and cannot be identified with routine hematoxylin-eosin staining. Counterstaining with Best's carmine is selective for the amebic trophozoites because it stains the glycogen in the cytoplasm.

(For detailed recommendations on preservation and staining of fecal films, bulk feces, and tissues, see the Technical Appendix, pages 738 to 754.)

Seroimmunologic Diagnosis. The usefulness of a serological method in the diagnosis of amebiasis was first shown by Craig (1928–1931), who extracted an amebic antigen with absolute ethanol and demonstrated that *E. histolytica* produces complement-fixing antibodies that are specific, most readily demonstrated when the amebae have invaded the tissues, and rapidly disappear when the infection has been eradicated. Craig's complement-fixation was not widely used, however, largely because a satisfactory standard antigen was not available. Numerous other methods have been developed, and with the advent of axenic cultivation of *E. histolytica* (Diamond, 1961, 1968), the preparation of antigens for these tests was greatly facilitated.

A fluorescent antibody technique was used by Goldman (1962, 1964) in a study of antigenic differences between *E. histolytica* and *E. histolytica*-like amebae—*E. hartmanni, E. moshkovskii,* the Laredo ameba, and others. Although its application to those studies was successful, it did not come into general use in the diagnostic laboratory. Use of the indirect fluorescent antibody (IFA) test in the identification of amebae in stools (Stamm et al., 1976) and in tissues (Hoffman and Miller, 1975) is satisfactory as a screening test, but in most instances, it would not be as reliable or convenient as the conventional methods of preparation and direct microscopical examination. The most practical application of serodiagnostic methods is for the detection or confirmation of amebic liver abscess and amebic infection of the intestine when present and not detected by stool examination in cases of suspected carcinoma or ulcerative colitis (Ridley and Warhurst, 1978). The tests that have come into general use are indirect hemagglutination (IHA), gel diffusion precipitin (GDP), and latex agglutination (LA). The IHA and LA tests are about equally reliable but they are too nonspecific, and the antibodies on which they depend are too persistent. The GDP test, on the other hand, generally is positive only in the presence of active amebic disease (Stamm et al., 1976). The gel diffusion precipitin test, now commercially available, easy to use, and relatively inexpensive, is slightly less sensitive than the IHA test, but whereas the latter remains positive for many years after successful treatment, the former becomes negative in a high percentage of cases within 6 months (Patterson et al., 1979).

Counter-current immunoelectrophoresis and particularly enzyme-linked immunosorbent tests, which are now commercially available, are promising. Thus, several dependable serological tests for the presence of occult *E. histolytica* infection or for epidemiological surveys are available. None is suitable for detecting carriers (Knight and Wright, 1978). The choice may depend on various considerations. Tests that combine the features of adequate sensitivity, speed of performance, availability at reasonable cost, and suitability for routine testing of the occasional serum specimen, are cellulose acetate diffusion and agar gel diffusion; the latter may be favored because it provides results the same day, whereas the former requires 2 to 3 days (Sodeman and Dowda, 1973; Kagan and Norman, 1976). Tests that detect *E. histolytica* antigen in an aqueous suspension of feces rather than antibodies in the blood have been introduced (Grundy, 1982; Pillai and Mohimen, 1982).

Isolation in Culture. In vitro cultivation of *E. histolytica* for maintenance of strains and for various research purposes can be done by a number of different methods with a wide range of media and is a relatively simple procedure. Because it is generally easy to obtain positive cultures by inoculation with feces containing numerous cysts or trophozoites, it is tempting to believe that reproduction in culture media would frequently make it possible to detect the presence of amebae when there are too few to be found by direct examination or by concentration methods. Experience soon demonstrates, however, that cultures rarely are positive unless the amebae are so abundant in the inoculum that only a brief examination of a simple direct smear under the microscope is needed to detect their presence. On the other hand, the amebae in feces may be more readily or more accurately diagnosed when grown in a culture than they are in the fecal specimen (Nelson, 1947). Therefore, cultivation of *E. histolytica* is seldom used as a primary diagnostic procedure, but is instead reserved for special cases by most workers. Routine cultures for diagnosis may be successful in some laboratories, however (Robinson, 1968).

Differential Diagnosis. Clinically, amebiasis of the cecal area must be differentiated from appendicitis of other etiologies. Sigmoido-rectal amebiasis and amebomas require differentiation principally from carcinoma and tuberculosis. Shigellosis, brucellosis of the colon, and idiopathic ulcerative colitis must also be excluded. Amebic colitis may be indistinguishable clinically and sigmoidoscopically from idiopathic ulcerative colitis (Pittman et al., 1973a). Also, the exudates and other elements of the dysenteric stool in amebiasis and ulcerative colitis are notably similar (Beaver, 1958). Ulcerative colitis is foremost among the conditions that are most closely mimicked. Not more than 5% of persons with amebic colitis have demonstrable liver involvement. Therefore, liver disease is more likely to be due to other causes, including those of viral origin, relapsing fever, Weil's disease, syphilis, tuberculosis, hydatid cyst, schistosomiasis, carcinoma, acute yellow atrophy, fatty degeneration, and obstructive jaundice. Amebiasis is not likely to be suspected as a cause of brain disease and has been specifically diagnosed only postmortem. Amebiasis in the perianal area will probably not be considered until chemotherapeutic procedures directed against more common skin diseases fail to produce healing of the lesion.

In *laboratory diagnosis,* trophozites and cysts of *E. histolytica* must be differentiated from those of other intestinal amebae and *Blastocystis hominis.* Summary information on the differential characters of the intestinal amebae of man is provided in Table 11–1.

The most common type of fulminating colitis with which acute amebic colitis may be confused is acute shigellosis (''bacillary dysentery''). The essential characteristics of the two diseases are compared in Table 11–2.

Treatment. Different therapeutic procedures are employed for intestinal and extraintestinal amebiasis. Management depends upon the disease picture, since the form of the parasite—cyst or trophozoite—varies with regard to susceptibility to chemical agents. Access of drugs to the parasite varies with regard to absorbability, peristaltic rate, and presence or absence of intestinal ulceration.

In fulminating amebic dysentery, symptomatic treatment may be indicated. If dehydration or fluid and electrolyte deficits are present, fluid and electrolytes should be replaced by oral or parenteral administration. Cramps may be controlled by administration of atropine (0.6 mg) intramuscularly, powdered opium (30 mg) or tincture of opium (10–15 drops) three times daily until relief is obtained. The currently popular drugs loperamide hydrochloride, diphenoxylate hydrochloride, and thephenamil hydrochloride, which are somewhat more effective in relieving diarrhea and cramps, are contraindicated in severe, acute ulcerative colitis because they may cause toxic megacolon.

The drug of choice in amebic dysentery is metronidazole, which is readily absorbed and able to reach, *via* the bloodstream, trophozoites invading the tissues. It is given orally in a dosage of 750 mg three times daily for 5 to 10 days. Metronidazole is available for intravenous injection for use in patients who are comatose or otherwise unable to take oral medication. The dosage is the same intravenously as orally. Metronidazole causes relatively minor side effects, mostly nausea, headache, and metallic aftertaste, but ingestion of alcohol by patients taking metronidazole may cause

Table 11–1. Differential Characters of *Entamoeba histolytica* and Other Intestinal Amebae.

Character	*Entamoeba histolytica*	*Entamoeba hartmanni*	*Entamoeba coli*	*Endolimax nana*	*Iodamoeba buetschlii*	*Dientamoeba fragilis*
			"Vegetative" or Trophozoite Stage, Unstained			
Size	8–30 μm	4–12 μm	15–50 μm	6–15 μm	8–20 μm	5–12 μm
Motility	Active, progressive, directional	Active, directional	Sluggish, rarely progressive and directional	Sluggishly progressive	Sluggishly progressive	Active, progressive
Pseudopodia	Finger-shaped, hyaline and glass-like; rapidly extruded	Finger-shaped; rapidly extruded	Short and blunt; granular; slowly extruded	Blunt and hyaline; rapidly extruded	Blunt and hyaline; slowly extruded	Blunt and leaf-like; hyaline
Inclusions	Red blood corpuscles; no bacteria in fresh specimens	Bacteria and other material; no blood corpuscles	Bacteria and other material; no blood corpuscles	Bacteria; no blood corpuscles	Bacteria; no blood corpuscles	Bacteria; no blood corpuscles
Nucleus	Usually invisible	Invisible	Rarely visible	Rarely visible	Invisible	Invisible
			"Vegetative" or Trophozoite Stage, Iron Hematoxylin Stain			
Nuclear membrane	Delicate; lined with single layer of chromatin dots	Relatively thick; lined with coarse chromatin dots and bars	Thick; lined with coarse chromatin dots and bars	Thin; seldom lined with chromatin	Thick; often lined with chromatin dots	Delicate; not lined with chromatin dots
Karyosome	Minute; location is central	Usually large; location variable	Large; location is eccentric	Large; location may be central or eccentric	Large and granular; location is central or somewhat eccentric	Large; composed of chromatin granules lying in dimly stained matrix
			Cyst Stage, Iodine Smear Preparations			
Size	10–20 μm	5–10 μm	10–33 μm	5–14 μm	5–20 μm	Cysts not described
Shape	Usually spherical	Usually spherical	Usually spherical	Spherical or ovoid	Irregular	—
Cytoplasm	Bright greenish-yellow	Yellowish-brown	Yellowish-brown	Pale green with refractile vacuoles	Yellowish-green	—
Glycogen mass	Diffuse; reddish-brown	Usually diffuse; brown	Indefinite or definite; brown	Usually diffuse; brownish	Usually present; dark brown; definite	—
Nuclei	1 to 4; minute central karyosome refractile; nuclear membrane beaded	1 to 4; chromatin *E. coli*-like	1 to 8; nuclear membrane refractile; karyosome with eccentric location	1 to 4; indistinct	1; indistinct but usually evident	
			Cyst Stage, Iron Hematoxylin Stain			
Size	10–20 μm	5–10 μm	10–33 μm	5–14 μm	5–20 μm	Cysts not described
Shape	Usually spherical	Usually spherical	Usually spherical	Ovoid or spherical	Irregular	—
Cytoplasm	Alveolar; often vacuolated; finely granular	Granular; vacuolated	Granular; vacuolated	Vacuolated; finely granular	Granular, fine and coarse; vacuolated	—
Chromatoidal bodies	Bar, oval, or thick rod-like masses with rounded ends	Often numerous; shape variable	Splinter-like with angular ends	Small, spherical, or bacilliform	Usually absent; when present, small, round or granular	—
Nuclei	1 to 4; membrane delicate; lined with minute granules; minute karyosome with central location	1 to 4; membrane thick, lined with coarse chromatin; karyosome with central or eccentric location	1 to 8; membrane thick, lined with large dots or bars; large karyosome with eccentric location	1 to 4; membrane indistinct; karyosome irregular, near nuclear membrane	1, with nuclear membrane often indistinct; karyosome surrounded by granules	—

Table 11–2. Differential Diagnosis of Dysenteric Stage of Amebiasis and of Shigellosis.

	Amebiasis	*Shigellosis*
Epidemiology	Typically endemic; may be epidemic	Typically epidemic, but with endemic foci
	Onset insidious, preceded by diarrheal attacks	*Onset* acute
	Incubation period usually long and often indefinite; it may be short in rare instances	*Incubation period* short, usually 7 days or less
	Acute infections rare in *children*	Common *in children*
Pathology	Superficial diffuse inflammation of mucous membrane absent; bottleneck or crateriform ulcers characteristic and involving all coats of the intestine	Superficial inflammation with necrosis of mucous membrane, diffuse in type, with characteristic ulcers
	Stools consisting of blood, mucus, necrotic tissue cells, and fecal material, but with few leukocytes; erythrocytes may be agglutinated; *E. histolytica* trophozoites may have ingested erythrocytes; Charcot-Leyden crystals usually present	Stools filled with cellular exudate; pus cells numerous and erythrocytes not agglutinated; Charcot-Leyden crystals absent
	Blood shows no *leukocytosis* or one of mild degree; leukocytosis marked if amebic abscess of the liver also present	*Marked leukocytosis* present at onset but disappears later
Symptomatology	Local abdominal tenderness over sigmoid, cecum, or colon; no fever in most cases; tenesmus moderate	Generalized abdominal tenderness; fever usually present; tenesmus severe
	Prostration not evident, except in very acute cases	Prostration marked in most cases
Complications	Severe hemorrhage, peritonitis abscess of organs, especially of the liver	Polyarthritis may occur
Diagnostic tests	Blood serum negative for agglutination with dysentery bacilli	Blood serum positive for agglutination with one of the dysentry bacilli
	Dysentery bacilli not demonstrable on culture	Dysentery bacilli demonstrable on culture
	E. histolytica trophozoites in dysenteric discharge	*E. histolytica* trophozoites not present in uncomplicated shigellosis

Note: One should not forget that the dysenteric stage of amebiasis and shigellosis may, in some instances, coexist. The presence of one of these, however, may tend to exclude the other. In a series of 100 consecutive patients seen with amebic dysentery and 50 with bacillary dysentery, the two infections occurred together in none, while in a control group, *E. histolytica* infection was demonstrated in 11% (Powell et al., 1961). The value of examining colonic exudates was emphasized recently by Korzeniowski et al. (1979).

abdominal pain, nausea, vomiting, and headache. These symptoms may be severe and have an abrupt onset that may be precipitated by a very small amount of alcohol. Experimental evidence indicates that metronidazole is mutagenic in bacteria and tumorigenic in rodents; no such effects have been reported in humans, even though the drug has been widely used for about 15 years.

Emetine hydrochloride (6% solution) given subcutaneously or intramuscularly in a dose of 1 mg/kg of body weight daily (maximum daily dose, 60 mg) for 5 days may produce dramatic cessation of symptoms, although it is not highly effective in eradicating the infection. Dehydroemetine (experimental drug) is perhaps somewhat less toxic as compared to its efficacy. Emetine and dehydroemetine are potent toxins that affect both the myocardium and peripheral nerves. Although in a course of less than 7 days serious toxic effects are not likely to occur when the recommended dosage is not exceeded, the patient should be monitored closely for signs of cardiac toxic reaction and preferably be kept at bed rest until 2 days after completion of the course.

Several antibiotics may quickly stop dysentery due to amebiasis, although these, too, do not regularly eradicate the parasitic infection. These antibiotics include the tetracyclines and erythromycin, which are given in their usual dosages. Once dysentery has ceased, or in cases without diarrhea

PLATE I

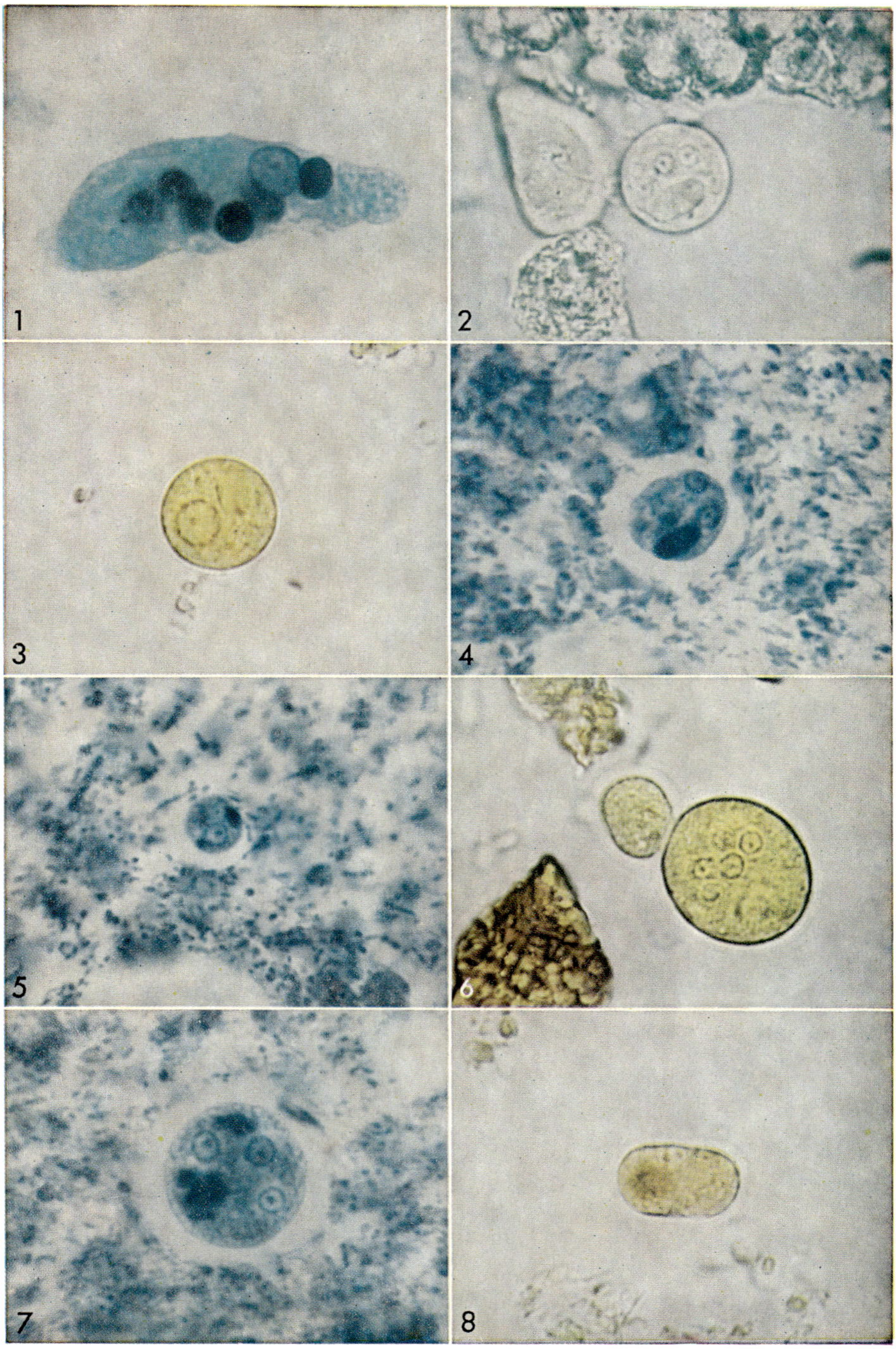

Intestinal amebae of man. Photomicrographs illustrating diagnostic characteristics as seen under oil-immersion magnification. (× ca. 1300.) 1, *Entamoeba histolytica,* trophozoite containing six red blood cells. 2, *E. histolytica,* 4-nucleate cyst in which two of the nuclei and a chromatoidal bar are evident. 3, *E. histolytica,* uninucleate cyst (iodine stain). 4, *E. histolytica,* 4-nucleate cyst with deeply stained, large chromatoidal (hematoxylin stain). 5, *Entamoeba hartmanni* (hematoxylin stain). 6. *Entamoeba coli* and *Endolimax nana* cysts (iodine stain). 7, *E. coli* cyst with 3 of 8 nuclei and several chromatoidal bars evident (hematoxylin stain). 8, *Iodamoeba buetschlii* cyst (iodine stain). (Originals provided by Dr. F.M. Spencer, selected from a larger series in Spencer, F.M., and Monroe, L.S. 1961. *The Color Atlas of Intestinal Parasites*. Springfield, Ill., Charles C Thomas.)

or dysentery, a so-called luminal agent may be used. These are drugs that are poorly absorbed by the gastrointestinal tract and thus act within the intestinal lumen. The most effective of these drugs is diiodohydroxyquin, which is given in a dosage of 650 mg three times daily for 2 to 3 weeks. The halogenated oxyquinoline drugs, including diiodohydroxyquin, may cause serious damage to the optic nerves. Therefore, the recommended dose should not be exceeded nor the course prolonged. The patient should be cautioned to note and report disturbances in vision in the course of therapy.

If diiodohydroxyquin is not available, metronidazole may be used as it is for amebic dysentery, but it is less effective in the nondysenteric patient. An alternative drug that is effective is diloxanide furoate, which is given in a dosage of 500 mg three times daily for 10 days. This drug is not commercially available in the United States, but it can be obtained from the Parasitic Diseases Division, Centers for Disease Control, Atlanta, Ga.

The first choice of drugs for the treatment of *amebic liver abscess* is, by reason of efficacy and safety, metronidazole. It is given in a dosage of 750 mg three times daily for 10 days. Unfortunately, metronidazole is not always effective, nor are any of the other amebicidal drugs.

Emetine hydrochloride U.S.P. was, until 1948, the only effective drug. It is given by subcutaneous or intramuscular injection in a dosage of 1 mg/kg of body weight daily (maximum daily dose is 60 mg) for a period preferably under 5 days but not to exceed 10 days in any case. The patient should be hospitalized at bed rest if feasible and cardiotoxicity monitored by physical examination and electrocardiography.

Dehydroemetine (experimental drug) is perhaps somewhat less toxic as compared to its efficacy. It is obtainable from the Parasitic Diseases Division, Centers for Disease Control, Atlanta, Ga.

Chloroquine phosphate U.S.P., although it is not effective in the treatment of intestinal amebiasis, is useful in treatment of amebic liver abscess because it is concentrated in the liver. The recommended course of treatment consists of a loading dose of 1 g chloroquine diphosphate (0.6 g base) administered orally for 2 successive days followed by a dose of 0.5 g (0.3 g base)/day for 2 to 3 weeks (Conan, 1949; Sodeman et al., 1951).

It appears that best results are obtained by using metronidazole plus emetine or dehydroemetine, or that combination plus chloroquine.

In the case of large abscesses, the content should be aspirated. This not only provides confirmation of the diagnosis by demonstration of the parasite or characteristic pus but also improves the probability of cure. Open surgical drainage should not be done if it can be avoided. If there is evidence of bacterial infection in the abscess an appropriate antibiotic should be given dependent on the sensitivity of the involved organism.

Surgical Intervention in Amebiasis. Surgical intervention, with the exception of aspiration of abscesses by needle, should be avoided if possible in the presence of intestinal or extraintestinal amebiasis. If surgery is necessary and amebiasis is present, treatment by appropriate drugs should be initiated before the operation is carried out.

Prognosis. Fatalities are in most cases due to a failure to diagnose the infection promptly and accurately (Thuse, 1979). In at least 90% of all cases of amebiasis, specific and adjuvantal drugs will be curative provided the infection is accurately diagnosed and appropriate treatment is instituted early. In case several courses of treatment with modern antiamebic drugs fail to effect clinical cure, some other etiology should be considered. In chronic amebic colitis of many years' duration, however, even with eradication of the amebae and healing of the ulcers, permanent fibrous scarring and thickening of the wall may have developed so that normal intestinal function cannot be obtained.

Control and Prevention of Amebiasis. Control of amebiasis and its eventual prevention depend on acceptance of the infection as a public health problem. First of all, it is necessary to discover the basic cause of endemicity or epidemic outbreak of amebiasis for each community. Adequate, safe supplies of water for drinking, household purposes, and bathing may be obtained by filtration, sedimentation, or by sterilization employing iodinization (Kahn and Visscher, 1975). Food supplies must be protected from contamination due to use of human excreta as fertilizer and from infected food handlers, filth flies, or cockroaches. No practical way has been found to prevent direct person-to-person transmission in institutionalized groups such as children's asylums, prisons, and mental hospitals and in backward populations (Thacker et al., 1981).

In an amebiasis control campaign, the *physician* can effectively carry out his part by adequately treating infected persons under his care, not only those obviously suffering from the disease, but also food handlers and other relatively asymptomatic *E. histolytica* cyst passers who are responsible for maintaining the infection in the community. In the United States, amebiasis is a notifiable infection, yet most cases are unreported. Much more can be accomplished if physicians report their cases promptly. The services of the *medical parasitologist* are twofold—to train laboratory diagnosticians in the accurate diagnosis of *E. histolytica* and to conduct surveys to determine the prevalence of amebic infection in the community and in particular segments of the population when this seems advisable. The *epidemiologist* will trace the cause for high endemicity or epidemic outbreaks of amebiasis and will advise on practical methods of control. The *sanitarian* will inspect food services in restaurants, note the presence of filth flies and cockroaches on the premises, and carelessness of food handlers and their toilet habits—all of these items should be reported to the public health department. The *sanitary engineer* will see that water supplies are uncontaminated and that there is sanitary sewage disposal for the community and for individual installations. During a survey of four areas in and near Little Rock, Arkansas, by Brooke et al. (1963), the beneficial effects of improvements in sanitary conditions were reflected in the lower amebic prevalence rate in the under 5-year age group in one of the areas in which a piped, indoor city-water supply and sewage facilities had been installed 6 years previously. The *public health nurse* and *health educator* will provide the community with accurate, easily comprehended information concerning the ways in which amebic infection is contracted. The *public health administrator* will coordinate the program. He will set up the project on a practical basis to justify the expenditure of funds and services. Too often no real effort is made until an epidemic develops that could have been prevented if water supplies, sewage disposal, food contamination, and filth flies had been given proper attention.

Group Treatment. The control measures just outlined refer to diagnosis and treatment of the individual who is infected with *E. histolytica,* together with measures to provide better environmental and personal hygiene. Prevention in a public health sense can be achieved by mass treatment in institutionalized or other relatively isolated communities with high incidence of amebic disease in which attainment of adequate personal hygiene is not feasible. In mental hospitals, prisons, and in small isolated communities with high prevalence of amebiasis, the administration of antiamebic drugs to all infected individuals greatly reduces clinical manifestations and the overall frequency of amebiasis.

Chemoprophylaxis. The administration of an amebicidal drug such as diiodohydroxyquine can prevent infection with *E. histolytica*. Because of the risks of drug intoxication, this practice is not generally recommended; under special circumstances, however, it may be indicated. The dose required for prophylaxis is only one-sixth the recommended therapeutic dose (Beaver et al., 1956b).

Entamoeba hartmanni von Prowazek, 1912

Synonyms. *Entamoeba minuta* Woodcock and Penfold, 1916; *Entamoeba tenuis* Kuenen and Swellengrebel, 1917; *Entamoeba minutissima* Brug, 1917; "small race *Entamoeba histolytica*" of Wenyon and O'Connor, 1917, and other authors.

Historical and Geographical Notes. *Entamoeba hartmanni* is cosmopolitan in distribution, having been distinguished from other minute amebae and reported wherever it has been among the species looked for. Until 1957, when it was redescribed by Burrows (1959), *E. hartmanni* generally was regarded as a small race of *E. histolytica,* and in survey reports, frequently was not distinguished from the so-called "large race" *E. histolytica*. Thus, the prevalence data for both *E. histolytica* and *E. hartmanni* were inaccurate. Faust (1958) reported in detail the morphological differences observed between the large and small "races" of *E. histolytica* and noted that the two kinds of amebae were clearly distinct and easily recognized as such on morphological features that were constant. He gave them subspecies rather than species status, however, and referred to them as *E. histolytica histolytica* and *E. histolytica hartmanni,* respectively.

Morphology, Biology, and Life Cycle. In all of its life-cycle stages, *E. hartmanni* resembles *E. histolytica* except in size, and even in size, there is a slight overlap in range. In size, it most closely resembles *Endolimax nana,* and when seen in fresh, unstained or iodine-stained preparations, its cysts are often indistinguishable from those of the spherical forms of *E. nana*. In diameter, the trophozoites range from around 4 to 12 μm, the cysts from 5 to 10 μm.

The trophozoites do not ingest red blood cells, and their motility is generally less vigorous than that of *E. histolytica*. The nucleus, *E. coli*-like in the character of its chromatin and karyosome, may not be evident in iodine-stained preparations but is clearly seen in permanently stained fecal smears (Fig. 11–3 and Plate I). The cysts have a rather coarse, *E. coli*-like cytoplasm. Immature cysts with one or two nuclei frequently have a distinct glycogen mass and several to numerous chromatoidals that are short with tapered ends (rice-grain shaped)

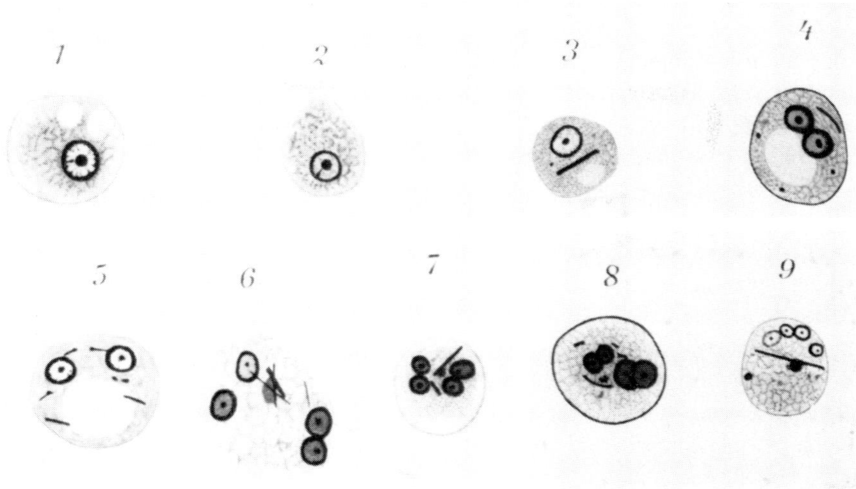

Fig. 11–12. *Entamoeba hartmanni.* Legends indicate that diameters of the amebae, all drawn to the same scale, range from 7 to 13 μm. (Photocopies of original drawings by S.J.M. von Prowazek. 1912. Arch. Protistenk., *22*; 241. From Faust, E.C., Beaver, P.C., and Jung, R.C. 1975. *Animal Agents and Vectors of Human Disease,* 4th ed. Philadelphia, Lea & Febiger.)

(Faust, 1958, 1961) or thin and bar-like. The presence of one to several minute spherical chromatoidals even in the mature cyst with four nuclei is characteristic (Fig. 11–12).

Epidemiology. As in other amebae, infection is acquired by ingestion of food or water contaminated with cyst-bearing feces. In localities where it has been accurately identified, its prevalence is roughly the same as that of *E. histolytica,* or less. In a recent study in Cali, Colombia, for example, in which the diagnoses were based on iron hematoxylin-stained smears, among 115 infected persons 12 to 73 years of age, *E. histolytica* was found in 65, *E. hartmanni* in 20, and both species in 30 (Spillmann et al., 1976).

Pathogenesis, Pathology, and Symptomatology. Early studies on pathogenesis in experimental animals indicated that *E. hartmanni* (synonym, "small race" *E. histolytica*) caused amebic colitis in dogs (Tobie, 1940) and cats (Frye and Meleney, 1938), but the experiments were inadequately controlled and the findings have not been confirmed. The reported transformation of the "small races" of *E. histolytica* to "large races" (Meleney and Zuckerman, 1948; Faust, 1954) also awaits confirmation.

Diagnosis. Identifications based on examination of small amebae in unstained or iodine-stained temporary mounts of concentrates or saline fecal suspensions are untrustworthy. *E. hartmanni* and *E. histolytica* can best be distinguished by the size of their cysts, if accurately measured, but *E. hartmanni* is often misdiagnosed as *E. nana* and *vice versa.*

Treatment. Usually no treatment is indicated. Among 40 asymptomatic carriers of *E. hartmanni,* treatment with metronidazole failed in 11 of 24 (46%) cases, and it failed with tinidazole in 11 of 16 cases (69%) (Spillmann et al., 1976).

Prevention. Measures generally effective against fecal-borne infections will control this amebic infection.

Entamoeba coli (Grassi, 1879) Casagrandi and Barbagallo, 1895

Synonyms. *Amoeba coli* Grassi, 1879; *Endamoeba hominis* Pestana, 1917; *Loschia coli* Chatton and Lalung-Bonnaire, 1912; *Councilmania lafleuri* Kofoid and Swezy, 1921.

Historical and Geographical Notes. Although *Entamoeba coli* was probably recognized by Lewis (1870) and described by Cunningham (1871) in Calcutta, Grassi (1878, 1888) first provided a recognizable description and reported the presence of this species in healthy as well as sick individuals. Additional information was obtained by Quincke and Roos (1893), Casagrandi and Barbagallo (1895), Schaudinn (1903), and by Walker and Sellards (1913) who proved that, in human volunteers, *E. coli,* unlike *E. histolytica,* is a harmless inhabitant of the colon.

Entamoeba coli is a worldwide parasite of man. Although it is most prevalent in populations native to warm climates, it is likewise commonly found in some groups even in cold climates. For example, Hitchcock (1950) reported 51% incidence among Eskimos in Anchorage, Alaska.

Morphology, Biology, and Life Cycle. The following are stages in the life cycle of *E. coli:* trophozoite, precyst, cyst, metacyst, and metacystic trophozoite. The living *trophozoite* of *E. coli,* as observed in diarrheic or dysenteric stools or following saline purgation and enema evacuation of the feces, is colorless, varies in size from 15 to 50 μm, has a viscous cytoplasm in which the ectoplasm is usually difficult to distinguish from the endoplasm and a nucleus that is not easily visualized. Typically its movements are sluggish, with broad short pseudopodia and little locomotion, but at times the living specimens cannot be distinguished from the active trophozoites of *E. histolytica*. In iron-hematoxylin preparations (Fig. 11–13,*A*) the organism is observed as a somewhat condensed, thickened, rounded mass, with a relatively nongranular ectoplasmic rim and a densely granular endoplasm. Within the endoplasm, there are few to numerous food vacuoles that usually contain bacteria, and the spherical nucleus has a relatively thick membrane that is lined internally with plaques or coarse granules of chromatin and a moderately large karyosome that is eccentrically placed. *E. coli* occasionally ingests red blood cells or *Giardia* cysts, and, at times, spores of *Sphaerita* may be observed in the endoplasm (Fig. 11–4).

In preparation for encystation, the trophozoite extrudes undigested food and becomes more definitely spherical or subspherical in contour. This is the *precyst* (Fig. 11–13,*B*). Almost immediately, a tough cyst membrane is secreted and encystation is accomplished, providing some protection from an unfavorable environment. Within the cytoplasm of the *unripe cyst* (Fig. 11–13,*C*), there is commonly a rather dense mass of glycogen lying within a vacuole with diffuse borders that is easily demonstrated by iodine stain. Splinter-like irregular masses of hematoxylin-staining material also sometimes appear in the cytoplasm; these are the *chromatoidals*. In the newly formed cyst, just as in the trophozoite and the precyst, there is a single nucleus; as the cyst matures, however, the nucleus divides so that there are two, then four, and finally eight nuclei in the *ripe cyst* (Fig. 11–13,*D, E, F*). Occasionally, cysts of large size may contain 16 or more nuclei. As the cyst ripens, the glycogen and the chromatoidal material appear less dense or entirely disappear. It is believed that these cytoplasmic inclusions are used as a source of energy during this stage. The definitive nuclei are much smaller than the single nucleus of the trophozoite but in all other respects they are identical. The cysts are remarkably variable in size, even from the same host, measuring 10 to 31 μm in diameter.

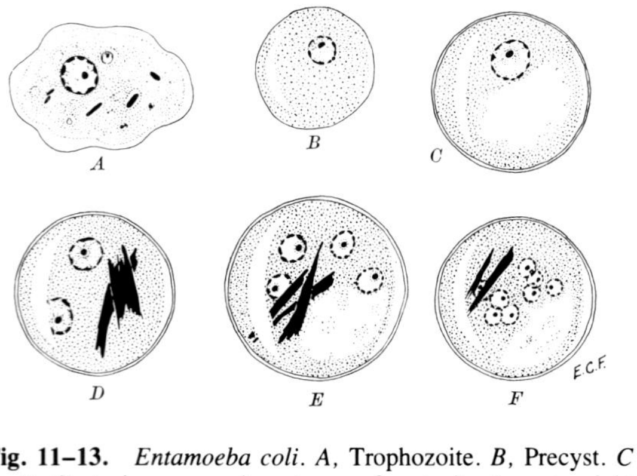

Fig. 11–13. *Entamoeba coli. A,* Trophozoite. *B,* Precyst. *C* to *F,* Cysts in progressive stages of maturity (one to eight nuclei). Note chromatoidal bodies in *D, E,* and *F,* and glycogen vacuoles in *C, E,* and *F*. Drawn from hematoxylin preparations. (× 1000.) (By E.C. Faust.)

When evacuated in the stool of an infected person, the mature cyst can withstand moderate putrefaction or desiccation. When taken into the mouth as a contamination of food or drink, on feces-soiled fingers, or on other objects, the cyst is swallowed; on reaching the intestine, the enclosed 8-nucleated ameba escapes. It is now the *metacyst* that undergoes a maximum number of cytoplasmic divisions corresponding to the number of nuclei, although Dobell's study of *E. coli* in culture (1938) suggests that the usual number of metacystic trophozoites produced is less than the total number of nuclei in the ripe cyst. The little amebae become established in the cecum where, as lumen dwellers, they grow to normal trophozoite size by feeding on enteric bacteria and other elements and proceed to multiply by binary division.

Epidemiology. *Entamoeba coli* is transmitted in its viable cystic stage through fecal contamination that reaches the mouth and is swallowed. The infection is readily acquired (Walker and Sellards, 1913), accounting in part for the relatively high prevalence. In tropical countries, as well as in communities with primitive hygiene and sanitation in cooler climates, infection rates at times are near saturation. Although monkeys and occasionally the dog have been found naturally infected with an *Entamoeba* morphologically similar to *E. coli,* human infection results almost exclusively from human sources.

Pathogenesis, Pathology, and Symptomatology. *Entamoeba coli* as a lumen parasite is nonpathogenic and produces no symptoms.

Diagnosis. *Entamoeba coli* requires differentiation from the pathogenic ameba, *E. histolytica*.

Although this is usually not difficult in the mature cystic stage (with more than four nuclei), at times it is almost impossible to differentiate the trophozoites passed in a liquid stool.

In fecal material containing a very large number of *E. coli* cysts, scanty *E. histolytica* cysts may not be discovered.

Treatment. Specific treatment is not indicated since this ameba is nonpathogenic. However, it is less susceptible to antiamebic drugs than is *E. histolytica*. In a group of about 300 institutionalized patients, Sodeman and Beaver (1952) found standard antiamebic treatments equally effective against *E. histolytica* and *Iodamoeba buetschlii* but relatively ineffective against *E. coli* and *Endolimax nana*.

Prevention. The presence of *Entamoeba coli* in a stool specimen is evidence that fecal contamination has been introduced into the mouth. Frequently, but not consistently, *E. coli* and *E. histolytica* are found in the same individual. Reduction in the prevalence of these and other intestinal protozoa in an individual or a community depends on better personal hygiene and sanitary disposal of human excreta.

Entamoeba polecki von Prowazek, 1912

This relatively cosmopolitan parasite of the large intestine of the hog and monkey is rarely diagnosed in man. The trophozite (Fig. 11–14,*A*) resembles that of *E. coli* in its viscosity and sluggish movement, and the typically uninucleate cyst (Fig. 11–14,*B*) contains ovoid chromatoidal bodies (trophozoite 10 to 12 μm, cyst 5 to 11 μm in diameter) (Burrows, 1959). There is no evidence of pathogenicity in the reservoir hosts or in man.

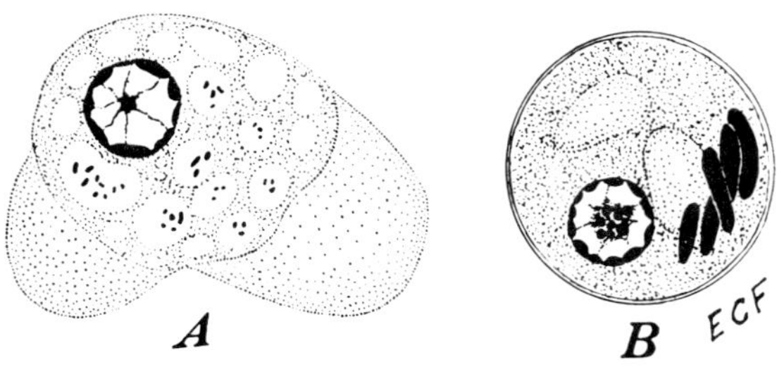

Fig. 11–14. *Entamoeba polecki. A,* Trophozoite. *B,* Cyst. (× 1800.) (Adapted by E.C. Faust from Kessel, J.F. 1928. Intestinal protozoa of the domestic pig. Am. J. Trop. Med., *8* and from Kessell, J.F., and Johnstone, H.G. 1949. Occurrence of *Endamoeba polecki,* Prowazek 1912, in *Macaca mulatta* and in man. Am. J. Trop. Med., *29.*)

Entamoeba gingivalis (Gros, 1849) Brumpt, 1913

Synonyms. *Amoeba gingivalis* Gros, 1849; *Amoeba buccalis* Steinberg, 1862; *Entamoeba buccalis* von Prowazek, 1904.

Historical and Geographical Notes. This was the first parasitic ameba of man to be described. Gros (1849) recovered and described the organism from the soft tartar between the teeth, and von Prowazek (1904) provided the first detailed description. *E. gingivalis* is prevalent in all populations in which careful search has been made.

Morphology, Biology, and Life Cycle. Only the trophozoite stage has been found, and encystation probably does not occur. The living trophozoite (Fig. 11–15) measures 5 to 15 μm in diameter (average 10 to 20 μm). Like *E. histolytica,* during its characteristic active movement this species exhibits a definite zone of demarcation between the clear ectoplasm and the more granular endoplasm. The multiple pseudopodia are at one time long and lobose, at another time, short, and blunt. The endoplasm contains food vacuoles in which phagocytosed and partly digested host leukocytes and epithelial cells are seen. At times, bacteria and spirilla, but rarely red blood cells, are also seen. The nucleus is nearly spherical, measures 2 to 4 μm in diameter, has a distinct membrane lined with closely packed chromatin granules, and near its center, a moderately small, well-defined karyosome with delicate achromatic strands extending to the nuclear membrane.

Entamoeba gingivalis lives in the gingival tissues around the teeth, particularly if there is suppuration, as in pyorrhea alveolaris, but it also thrives in apparently hygienic mouths or on dental plates if they are not kept clean. Occasionally this ameba has been identified from the crypts and in histologic sections of diseased tonsils. Recently, amebae identified as *E. gingivalis* were found in vaginal and cervical smears from women using intrauterine devices (De Moraes-Ruehsen et al., 1980). Apparently the same species has been recovered from the gums of dogs, cats, and monkeys (Hegner and Chu, 1930).

Only the trophozoite stage of *E. gingivalis* has been observed either in material removed from its natural habitat or in culture. It divides by binary fission.

Epidemiology. In the absence of cyst formation and on the basis of experimental infections, it is concluded that transmission in man is via droplet spray from the mouth of an infected individual to another during close contact, or from contaminated drinking utensils. In unhygienic mouths, the incidence of in-

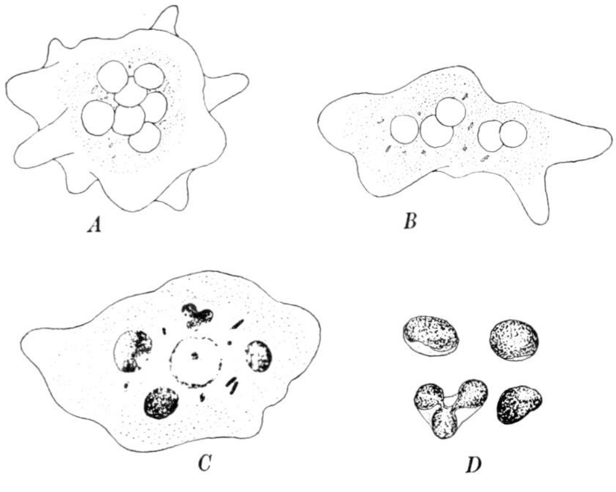

Fig. 11–15. *Entamoeba gingivalis. A, B,* Unstained living trophozoites. *C,* Hematoxylin-stained ameba with ingested bodies. *D,* Characteristic nuclear debris in food-vacuoles. (× 750.) (By E.C. Faust.)

fection varies from 71 to 96% (Fisher, 1927; Westphal, 1941); in "healthy" mouths, it has been reported in 10 to 50% of individuals examined.

Pathogenesis, Pathology, and Symptomatology. *Entamoeba gingivalis* is a commensal, living primarily on exuviae from the margins of the gums, and thrives best on unhealthy gums. Infections observed in the genital tract apparently were nonpathogenic and disappeared spontaneously when the IUDs were removed (De Moraes-Ruehsen et al., 1980).

Diagnosis. Demonstration of the trophozoite stage of an ameba in material removed from the gingival margin of the gums, from between the teeth, or from dentures indicates the presence of *E. gingivalis*. It may be associated with *Trichomonas tenax*, oral spirochetes, fusiform bacilli, or monilias. In the female genital tract, it frequently was associated with *Actinomyces*.

Treatment. No specific therapy is indicated. However, the presence of *E. gingivalis* suggests a need for better oral hygiene.

Prevention. The infection can be prevented by proper care of the teeth and gums.

Entamoeba moshkovskii Tshalaia, 1941

This species closely resembles *E. histolytica* in its trophozoite and cystic stages (Fig. 11–16). It has been recovered from sewage in several countries. It exhibits anaerobic metabolism with optima between 20° and 28°C but grows well in cultures at 37°C. Although it seems highly improbable that *E. moshkovskii* parasitizes man, its presence in sewage may cause diagnostic difficulties if treated sewage effluents are examined for *E. histolytica* cysts.

Genus *Endolimax* Kuenen and Swellengrebel, 1917

This genus includes parasitic amebae of small size that have a vesicular nucleus with a comparatively large, irregularly-shaped karyosome anchored by several achromatic threads to the delicate nuclear membrane. Species of *Endolimax* have been described from man, guinea pig, fowl, turtle, frog, and cockroach.

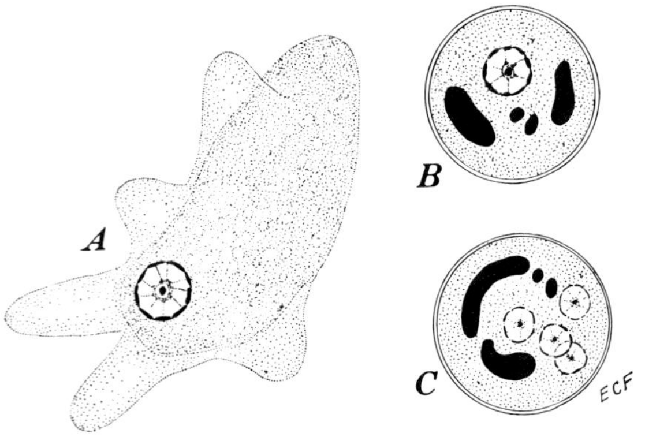

Fig. 11–16. *Entamoeba moshkovskii. A*, Trophozoite. *B*, Uninuclear cyst. *C*, Ripe cyst. (× 1400.) (By E.C. Faust.)

Endolimax nana (Wenyon and O'Connor, 1917) Brug, 1918

Synonyms. *Entamoeba nana* Wenyon and O'Connor, 1917; *Endolimax intestinalis* Kuenen and Swellengrebel, 1917.

Historical and Geographical Notes. *E. nana*, as its species name suggests, is a small ameba. It was distinguished as a separate ameba about 1908 but was first given a species designation by Wenyon and O'Connor (1917). It is cosmopolitan in distribution, and is as prevalent in most populations as *Entamoeba coli*.

Morphology, Biology, and Life Cycle. *Endolimax nana* (Fig. 11–17) has all of the stages described for *Entamoeba coli* and *E. histolytica*—trophozoite, precyst, cyst, and metacystic forms. The trophozoite is small, measuring 6 to 15 μm in diameter (average under 10 μm). The cytoplasm is delicately granular and vacuolated, with a narrow rim of clean ectoplasm. When exhibiting movement, the ameba projects short, blunt hyaline pseudopodia. Motility is typically sluggish (hence, the generic designation, "slug-like"), but in freshly passed diarrheic stools, the movement may be more active, with evidence of locomotion. In hematoxylin-stained fecal films (Fig. 11–17,*A*), the vacuoles are seen to contain bacteria, small vegetable cells, and crystals. Near the active center of the ameba is the minute spherical or subspherical nucleus, with its conspicuous karyosome, which may be centrally or eccentrically placed and from which several achromatic strands extend to the nuclear membrane.

Endolimax nana is a lumen dweller in the large intestine, primarily at the cecal level, where it feeds mostly on bacteria. Iodine-stained cysts often demonstrate glycogen with diffuse margins, and in good hematoxylin preparations, the nuclear structure can be studied. Occasionally there will be one to a few small, slightly curved chromatoidal rods in the cy-

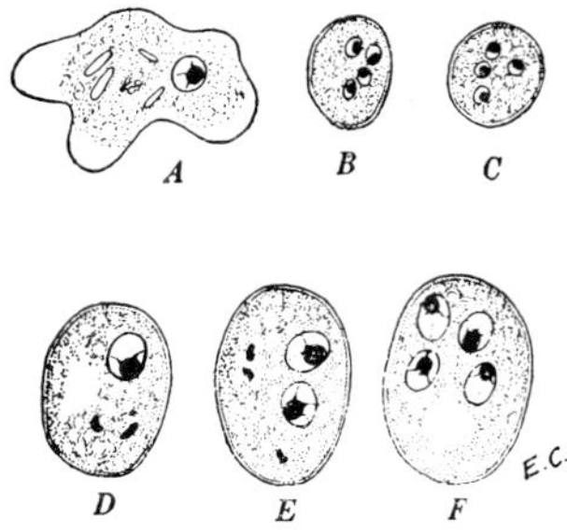

Fig. 11–17. *Endolimax nana. A*, Trophozoite. *B* to *F*, Cysts with one, two and four nuclei. *D* and *F* show glycogen vacuoles; *D* and *E* have characteristic chromatoidal bodies. (× 1000.) (By E.C. Faust.)

toplasm. By two nuclear divisions, the cysts mature to the 4-nucleate *ripe cyst* (Fig. 11–17,*F*). These cysts have a remarkable range in size, from 5 to 14 μm in greater diameter, depending on the size of the trophozoite.

Human infection results from ingestion of viable cysts. Dobell (1933) demonstrated this by swallowing washed cysts of *E. nana* obtained from the feces of a naturally infected monkey. He also found all stages of development in culture with a favorable bacterial associate: "trophic amoebae in abundance, multiplying by binary fission; precystic amoebae, and cysts at all stages of development; excystation, and all of the metacyclic stages."

Epidemiology. The method of infection with *E. nana* is similar to that of the other intestinal amebae that form cysts—by ingestion of the viable cysts in polluted water or food or from contaminated objects. *E. nana* cysts are more susceptible to desiccation and other unfavorable extrinsic conditions than are those of *E. coli*. The prevalence of infection roughly parallels that of *E. coli:* it is high in warm, moist climates and in other areas in which there is a low standard of personal hygiene; it is lower in most temperate climates and in regions in which hygienic and sanitary conditions are relatively good. Many species of monkeys harbor an *Endolimax* that appears to be the same as *E. nana* of man (Dobell, 1933).

Pathogenesis, Pathology, and Symptomatology. *Endolimax nana* is nonpathogenic and produces no symptoms as such, although at times there may be associated symptoms that are erroneously attributed to this ameba.

Diagnosis. Typical ovoid cysts of *E. nana* in hematoxylin-stained fecal films are easily diagnosed. Rounded cysts and living trophozoites may be difficult to differentiate from *E. hartmanni* and small *E. histolytica*.

Treatment. No treatment is indicated.

Prevention. As in the case of other intestinal protozoa, infection with *E. nana* indicates contaminated food or drink or poor personal hygiene. Control can be effected through personal cleanliness and community sanitation.

Genus *Iodamoeba*
Dobell, 1919

This genus was created by Dobell (1919) for the intestinal ameba of man characterized by having a large karyosome rich in chromatin that is surrounded by a layer of achromatic globules and anchored to the nuclear membrane by achromatic fibrils. In the endoplasm, there are one or, at times, two dense, well-circumscribed glycogen vacuoles that are invariably present in the cyst and only occasionally seen in the trophozoite.

Iodamoeba buetschlii
(von Prowazek, 1911) Dobell, 1919

Synonyms. *Entamoeba williamsi* von Prowazek, 1911; *Entamoeba bütschlii* von Prowazek, 1912; "iodine cysts" of Wenyon, 1916; *Endolimax williamsi* (von Prowazek, 1911) Taliaferro and Becker, 1922.

Historical and Geographical Notes. The species name *williamsi* would have priority over *buetschlii* provided the earlier designation refers to the same species of ameba. Many protozoologists, however, are convinced that von Prowazek's "*williamsi*" actually referred to atypical cysts of *E. coli* and that his "*bütschlii*" fairly well described the species that Dobell placed in the new genus *Iodamoeba*.

Iodamoeba buetschlii is worldwide in its distribution, but it is ordinarily much less common than *E. coli* or *E. nana*.

Morphology, Biology, and Life Cycle. This ameba has trophozoite, precyst, cyst, and metacystic stages. The *trophozoite* (Fig. 11–18,*A*) is small to medium sized (diameter, 6–25 μm) and fairly active, with progressive movement in freshly evacuated unformed stools and sluggish movement in older stools. The clear ectoplasm is not usually well differentiated from the denser endoplasm that contains coarse and fine granules and has bacteria and yeast cells in food vacuoles. The nucleus can be observed in both living and stained specimens; it is provided with a densely chromatic karyosome that has a diameter about half that of the nucleus. In hematoxylin-stained fecal films, the karyosome is observed to be either centrally placed or some-

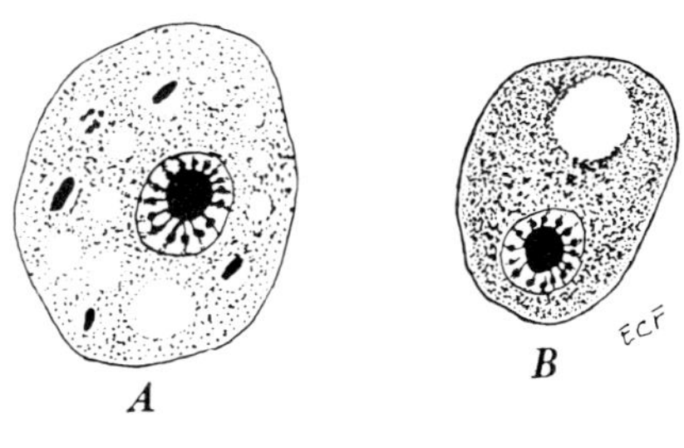

Fig. 11–18. *Iodamoeba buetschlii*. *A*, Trophozoite. *B*, Cyst. In the nucleus, the distinctive pattern of the chromatin granules radiating out from the karyosome is characteristic in both stages of this ameba. The dense glycogen vacuole figured in the cyst may sometimes also be observed in the trophozoite. (× 1500.) Iron-hematoxylin stain of Schaudinn-fixed fecal film. (Adapted by E.C. Faust from Wenrich, D.H. 1937. Studies on *Iodamoeba bütschlii* (Protozoa) with special reference to nuclear structure. Proceedings of the American Philosophical Society, *77* and from Pan, C.-T. 1959. Nuclear division in the trophic stages of *Iodamoeba bütschlii* (Prowazek, 1912) Dobell, 1919. Parasitology, *49*.)

what eccentric in position. It may appear to have a thin, unstained "halo" around it and to lie in a mass of achromatic globules, all of which are suspended in a rather intricate meshwork of achromatic strands that extend to the nuclear membrane. Occasionally, a discrete glycogen vacuole can be demonstrated in the trophozoite by iodine staining.

In preparation for encystation, *I. buetschlii* discards undigested food and becomes the somewhat more compact precyst, but it rarely assumes a spherical form. It then secretes a wall that may appear to be somewhat separated from the enclosed cytoplasm. The *cyst* (Fig. 11–18,*B*) is much more likely to be irregularly pyriform or ovoid than regularly spherical and measures 6 to 15 μm in longest diameter. Typically, there is no increase in number of nuclei within the cyst, although rarely the ripened cyst may have two nuclei. The most conspicuous feature in the cyst is the large vacuole that is densely packed with glycogen and distinctly outlined as an ovoid, polygonal, or broadly reniform mass that stains golden brown with iodine but is devoid of content in fixed hematoxylin preparations. Occasionally, there may be two glycogen vacuoles.

The natural habitat of *I. buetschlii* is the lumen of the large intestine, the principal site probably being the cecum. Here the trophozoite feeds on enteric bacteria. It is a natural parasite of man and lower primates. An ameba described as *Iodamoeba suis* from the hog is believed to be the same species.

Epidemiology. *Iodamoeba buetschlii* is transmitted from man to man when the viable cysts are ingested in fecal contamination of food and drink or on soiled objects. This ameba is usually much less common than *Entamoeba coli* and *Endolimax nana*. In temperate climates, its prevalence in representative population samplings ranges from 0.25 to 5.0%; in tropical regions, it is appreciably higher. It is possible that some infections are contracted from hog feces.

Pathogenesis, Pathology, and Symptomatology. *Iodamoeba buetschlii* is generally regarded as a nonpathogenic lumen parasite of the large intestine. Only the report of a case by Derrick (1948) suggests that the trophozoites of this species may rarely invade the intestinal wall and gain entry to extraintestinal viscera such as the lungs and brain, where these amebae were found in multiple lesions. However, the amebae in this case were later identified as *Naegleria fowleri* (see p. 136).

Diagnosis. In fresh iodine-stained preparations, one can usually make a specific diagnosis on the type of glycogen vacuole, but at times the glycogen vacuole in *E. coli* may resemble that of *I. buetschlii*. In hematoxylin-stained preparations the nuclear structure and the clean-cut empty glycogen vacuole are species-diagnostic.

Treatment. No treatment is ordinarily indicated for *I. buetschlii* infection.

Prevention. This is based on better personal hygiene and sanitation in the community.

Genus *Dientamoeba*

Jepps and Dobell, 1918

The genus *Dientamoeba* is characterized by its minute size, the frequent preponderance of binucleate over uninucleate forms, the presence of several distinct granules in the karyosome, and the lack of a cystic stage. Formerly regarded as an ameba, it is now classified as an ameba-like flagellate that does not bear a flagellum, related to the *Trichomonas* group in the class Zoomastigophorea, order Trichomonadida, family Monocercomonadidae, subfamily Dientamoebinae (Camp et al., 1974). (See Chap. 9, "Flagellate Protozoa of the Digestive and Urogenital Tracts," p. 51.)

REFERENCES

Abuabara, S.F., Barrett, J.A., Hau, T., and Jonasson, O. 1982. Amebic liver abscess. Arch. Surg., *117*:239–244.

Adams, E.B., and MacLeod, I.N. 1977a. Invasive amebiasis. I. Amebic dysentery and its complications. Medicine, *56*:315–323.

———. 1977b. Invasive amebiasis. II. Amebic liver abscess and its complications. Medicine, *56*:325–334.

Adi, F.C. 1965. Clinical features of hepatic amoebiasis. West Afr. Med. J., *14*:181–197.

Balamuth, W. 1946. Improved egg yolk infusion for cultivation of *Entamoeba histolytica* and other intestinal protozoa. Am. J. Clin. Pathol., *16*:380–384.

Beaver, P.C. 1959. The exudates in amebic colitis. Proceedings of the 6th International Congresses on Tropical Medicine and Malaria (Lisbon, 1958), *3*:419–434.

———, and Deschamps, G. 1949a. The viability of *E. histolytica* cysts in soil. Am. J. Trop. Med., *29*:189–191.

———, and Deschamps, G. 1949b. The effect of acetic acid on the viability of *Endamoeba histolytica* cysts. Am. J. Trop. Med., *29*:193–197.

———, Jung, R., Sherman, H., Read, T., and Robinson, T. 1956a. Experimental *Entamoeba histolytica* infections in man. Am. J. Trop. Med. Hyg., *5*:1000–1009.

———, Jung, R., Sherman, H., Read, T., and Robinson, T. 1956b. Experimental chemoprophylaxis of amebiasis. Am. J. Trop. Med. Hyg., *5*:1015–1021.

———, Lopez-Villegas, A., Cuello, C., and D'Alessandro, A. 1978. Cutaneous amebiasis of the eyelid with extension into the orbit. Am. J. Trop. Med. Hyg., *27*:1133–1136.

Boeck, W.C., and Drbohlav, J. 1925. The cultivation of *Endamoeba histolytica*. Am. J. Hyg., *5*:371–407.

Bong, M.Q. 1980. Viability of *Entamoeba histolytica* cysts and morphology of five low-temperature *Entamoeba histoly-*

tica-like amebae. Sc.D. diss. New Orleans, Tulane University.

Bray, R.S., and Harris, W.G. 1977. The epidemiology of infections with *Entamoeba histolytica* in the Gambia, West Africa. Trans. R. Soc. Trop. Med. Hyg., *71*:401–407.

Brooke, M.M., Healy, G.R., Levy, P., Kaiser, R.L., and Bunch, W.L. 1963. A sample survey of selected areas in and near Little Rock, Arkansas, to assess the prevalence of *Entamoeba histolytica*. Bull. W.H.O., *29*:813–822.

Brumpt, L.C., Sang, H.T., and Jaeger, G. 1972. Quelques réflexions a propos du parasitisme sanguin et intestinal dans deux villages d'Afrique centrale. Bull. Soc. Pathol. Exot., *65*:263–270.

Bundesen, H.N., et al. 1936. Epidemic amebic dysentery. The Chicago outbreak of 1933. Natl. Inst. Health Bull., no 166.

Burrows, R.B. 1957. *Endamoeba hartmanni*. Am. J. Hyg., *65*:172–188.

———. 1959. Morphological differentiation of *Entamoeba hartmanni* and *E. polecki* from *E. histolytica*. Am. J. Trop. Med. Hyg., *8*:583–589.

———. 1961. Prevalence of amebiasis in the United States and Canada. Am. J. Trop. Med. Hyg., *10*:172–184.

Cabrera, H.A. 1958. Temperature adaptation of *Entamoeba histolytica* and its effect on virulence. Exp. Parasitol., *7*:276–284.

Camp, R.R., Mattern, C.F.T., and Honigberg, B.H. 1974. Study of *Dientamoeba fragilis* Jepps & Dobell. I. Electronmicroscopic observations of the binucleate stages. II. Taxonomic position and revision of the genus. J. Protozool., *21*:69–82.

Carneiro Chaves, F.J.Z., Cruz, I., Gomes, C., Domingues, W., Marques da Silva, E., and Tavarela Veloso, E. 1977. Hepatic amebiasis—analysis of 56 cases. Am. J. Gastroenterol., *68*:273–277.

Carney, W.P., Putrali, J., and Caleb, J.M. 1974. Intestinal parasites and malaria in the Poso Valley, Central Sulawesi, Indonesia. Southeast Asian J. Trop. Med. Public Health, *5*:368–373.

Casagrandi, O., and Barbagallo, P. 1895. Ricerche biologiche e cliniche sull'*Amoeba coli* (Lösch). Boll. Acad. Gioënia Sci. Nat. Catania, *39*:4–14.

Centers for Disease Control. 1979. *Intestinal Parasite Surveillance. Annual Summary, 1978*. Atlanta, Ga.

Chiu, J.-K., Su, K.-E., Kao, C.-T., and Chiu, P.-C. 1971. Schistosome skin testing and intestinal parasite survey among inhabitants of Yuan-Shan, Ilan County, Taiwan. Chin. J. Microbiol., *4*:106–115.

Clark, H.C. 1925. The distribution and complications of amebic lesions found in 186 post-mortem examinations. Am. J. Trop. Med., *5*:157–171.

Colwell, E.J., Welsh, J.D., Boone, S.C., and Legters, L.J. 1971. Intestinal parasitism in residents of the Mekong Delta of Vietnam. Southeast Asian J. Trop. Med. Public Health, *2*:25–28.

Conan, N.J. 1949. The treatment of hepatic amebiasis with chloroquine. Bull. N.Y. Acad. Med., *24*:545–546.

Craig, C.F. 1928. Observations upon complement fixation in infections with *Endamoeba histolytica*. Proc. Natl. Acad. Sci., *14*:520–526.

———. 1931. Complement fixation in the diagnosis of amebiasis. Ann. Intern. Med., *5*:170–177.

Cutting, J.W. 1975. A survey of intestinal parasitism in a community on the Pan American Highway route in eastern Panama. Pan. Am. Health Org. Bull., *9*:13–18.

DeBakey, M.E., and Ochsner, A. 1951. Hepatic amebiasis. A 20-year experience and analysis of 263 cases. Int. Abstr. Surg., *92*:210–231.

de la Maza, L.M., Naeim, F., and Berman, L.D. 1974. The changing etiology of liver abscess. J.A.M.A., *227*:161–163.

De Moraes-Ruehsen, M., McNeill, R.E., Frost, J.K., Gupta, P.K., Diamond, L.S., and Honigberg, B.M., 1980. Amebae resembling *Entamoeba gingivalis* in the genital tracts of IUD users. Acta Cytol., *24*:413–420.

Derrick, E.H. 1948. A fatal case of generalized amoebiasis due to a protozoon resembling, if not identical with, *Iodamoeba bütschlii*. Trans. R. Soc. Trop. Med. Hyg., *42*:191–202.

Diamond, L.S. 1961. Axenic cultivation of *Entamoeba histolytica*. Science, *134*:336–337.

———. 1968. Techniques of axenic cultivation of *Entamoeba histolytica* Schaudinn, 1903 and *E. histolytica*-like amebae. J. Parasitol., *54*:1047–1056.

———, Harlow, D.R., and Cunnick, C.C. 1978. A new medium for the axenic cultivation of *Entamoeba histolytica* and other *Entamoeba*. Trans. R. Soc. Trop. Med. Hyg., *72*:431–432.

———, Harlow, D.R., Phillips, B.P., and Keister, D.B. 1978. *Entamoeba histolytica:* Iron and nutritional immunity. Arch. Invest. Med. (Mexico), *9* (suppl. 1):329–338.

———, Mattern, C.F.T., and Bartgis, I.L. 1972. Viruses of *Entamoeba histolytica*. I. Identification of transmissible virus-like agents. J. Virol., *9*:326–341.

Dobell, C. 1919. *The Amoebas Living in Man. A Zoological Monograph*. London, Bale, Sons and Danielsson, Ltd.

———. 1933. Researches on the intestinal protozoa of monkeys and man. V. The *Endolimax* of macaques. Parasitology, *25*:436–467.

———. 1938. Researches on the intestinal protozoa of monkeys and man. IX. The life history of *Entamoeba coli* with special reference to metacystic development. Parasitology, *30*:195–238.

Dreyer, D.A. 1961. Growth of a strain of *Entamoeba histolytica* at room temperature. Texas Rep. Biol. Med., *19*:393–396.

Duque, H. 1965. Amibiasis fatal en Colombia. Antioquia Med., *15*:300–301.

El-Hashimi, W., and Pittman, F. 1970. Ultrastructure of *Entamoeba histolytica* trophozoites obtained from the colon and from in vitro cultures. Am. J. Trop. Med. Hyg., *19*:215–226.

Ellman, B., McLeod, L.N., and Powell, S.J. 1965. Diagnostic pneumoperitoneum in amoebic liver abscess. Br. Med. J., *2*:1406–1407.

Entner, N., and Most, H.E. 1965. Genetics of *Entamoeba:* Characterization of two new parasitic strains which grow at room temperature (and at 37°C). J. Protozool., *12*: 10–13.

Faust, E.C. 1932. Experimental amebiasis in dogs. Am. J. Trop. Med., *12*:37–47.

———. 1941. Amebiasis in the New Orleans population as revealed by autopsy examination of accident cases. Am. J. Trop. Med., *21*:35–48.

———. 1943. Some modern conceptions of amebiasis. College of Physicians of Philadelphia, Transactions and Studies, Ser. 4, *2*:101–113.

———. 1954. *Amebiasis*. Springfield, Charles C Thomas.

———. 1958. Parasitologic surveys in Cali, Departamento del Valle, Colombia. I. Incidence and morphologic characteristics of strains of *Entamoeba histolytica*. Am. J. Trop. Med. Hyg., *7*:4–15.

———. 1961. The multiple facets of *Entamoeba histolytica* infection. Int. Rev. Trop. Med., *1*:43–75.

Fisher, J.H. 1927. Pyorrhea alveolaris: The role of certain microorganisms found in the lesions. Am. J. Pathol., *3*:169–178.

Foley, T.H., Price, J.L., and Foley, F. 1978. Misdiagnosis of amebiasis. Br. Med. J., *2*:428–429.

Frye, W.W., and Meleney, H.E. 1936. The viability of *Endamoeba histolytica* after passage through the cockroach. J. Parasitol., *22*:221–222.

———, and Meleney, H.E. 1938. The pathogenicity of a strain of small race *Endamoeba histolytica*. Am. J. Hyg., *27*:580–589.

Garcia, L.S., and Ash, L.R. 1980. *Diagnostic Parasitology. Clinical Laboratory Manual*. St. Louis, C.V. Mosby Co.

Goldman, M. 1964. Identification and diagnosis of *Entamoeba histolytica*. Am. J. Gastroenterol., *41*:362–365.

———, and Brooke, M.M. 1953. Protozoans in stools unpreserved and preserved in PVA-fixative. Public Health Rep., *68*:703–706.

———, and Carver, R.K. 1962. Antigenic analysis of *Entamoeba histolytica* by means of fluorescent antibody. III. Reactions of the Laredo strain with five anti-*histolytica* sera. Am. J. Trop. Med. Hyg., *11*:341–346.

———, and Gleason, N.N. 1962. Antigenic analysis of *Entamoeba histolytica* by means of fluorescent antibody. IV. Relationship of two strains of *E. histolytica* and one of *E. hartmanni* demonstrated by cross-absorption techniques. J. Parasitol., *48*:778–783.

Griffin, J.L. 1972. Human amebic dysentery: Electron microscopy of *Entamoeba histolytica* contacting, ingesting, and digesting inflammatory cells. Am. J. Trop. Med. Hyg., *21*:895–906.

Grundy, M.S. 1982. Preliminary observations using a multilayer ELISA method for the detection of *Entamoeba histolytica* trophozoite antigens in stool samples. Trans. R. Soc. Trop. Med. Hyg., *76*:396–400.

Hegner, R., and Chu, H.J. 1930. A comparative study of the intestinal protozoa of wild monkeys and man. Am. J. Hyg., *12*:62–108.

Hitchcock, D.J. 1950. Parasitological study on the Eskimos in the Bethel area of Alaska. J. Parasitol., *36*:232–234.

Hoffmann, E.O., and Miller, M.J. 1975. Immunofluorescent staining of amebae in routine paraffin-embedded tissues. J. Parasitol., *61*:1104–1105.

Horwitz, E., Artigas, J., and Silva, R. 1954. Algunas observaciones sobre entero-parasitos en agua de riego. Bol. Chil. Parasitol., *9*:99–105.

Imperato, P.J. 1980. Conclusions about amebiasis. Bull. N.Y. Acad. Med., *57*:240–242.

———. 1981. A historical overview of amebiasis. Bull. N.Y. Acad. Med., *57*:175–187.

———, Shookhoff, H.B., Marr, J.S., Friedman, S., and Hwa, C.L. 1977. Parasitic infections in New York City. N.Y. State J. Med., *77*:50–56.

Jarroll, E.L., Jr., Bingham, A.K., and Meyer, E.A. 1980. *Giardia* cyst destruction: Effectiveness of six small-quantity water disinfection methods. Am. J. Trop. Med. Hyg., *29*:8–11.

Jones, T.C. 1981. Current research in amebiasis. Bull. N.Y. Acad. Med., *57*:227–233.

Kagan, I.G., and Norman, L. 1976. Serodiagnosis of parasitic diseases, pp. 382–409. In *Manual of Clinical Immunology*. Edited by N.R. Rose and H. Friedman. Washington, D.C., American Society for Microbiology.

Kahn, F.H., and Visscher, B.R. 1975. Water disinfection in the wilderness—a simple, effective method of iodination. West. J. Med., *122*:450–453.

Kan, S.P., and Cheah, J.S. 1970. Prevalence and intensity of helminthic infections in adults among the ethnic groups in Singapore. Singapore Med. J., *11*:283–286.

Kirby, H. 1951. "*Entamoeba coli*" versus "*Endamoeba coli*." Bull. Zool. Nomencl., *2*:9–10, 243–281.

Knight, R. 1977. An in vitro model for measuring the cytopathic effect of *Entamoeba histolytica*. J. Parasitol., *63*:388–389.

———, and Wright, S.G. 1978. Intestinal protozoa. Gut, *10*:940–953.

Korzeniowski, O., Barada, F.A., Rouse, J.D., and Guerrant, R.L. 1979. Value of examination for fecal leukocytes in the early diagnosis of shigellosis. Am. J. Trop. Med. Hyg., *28*:1031–1035.

Krogstad, D.J., Spencer, H.C., and Healy, G.R. 1978a. Current concepts in parasitology. Amebiasis. N. Engl. J. Med., *298*:262–265.

———, Gleason, N.N., Sexton, D.J., and Herron, C.A. 1978b. Amebiasis: Epidemiologic studies in the United States, 1971–1974. Ann. Intern. Med., *88*:89–97.

Krupp, I., and Faust, E.C. 1959. Parasitologic surveys in Cali, Departamento del Valle, Colombia. IV. Experimental study in guinea pigs of the pathogenic capacity of *Entamoeba histolytica* from infected persons in Ward Siloe. J. Parasitol., *45*:449–455.

Lara, R., Galindo, E., Olarte, J., and Hernandez, G. 1974. Infecciones mixtas por *E. histolytica, Shigella* y otras enteropatogenis encontradas en niños con diarrea. Arch. Invest. Med. (Mexico), *5*(suppl. 2):515–518.

Lee, S.H. 1973. Study on the status of intestinal protozoan infections in Koreans. J. Health Fellowship Foundation, *3*:1–6.

LeMaistre, C.A., Sappenfield, R., Culbertson, C., Carter, F.R.N., Offutt, A., Black, H., and Brooke, M.M. 1956. Studies of a water-borne outbreak of amebiasis, South Bend, Indiana. I. Epidemiological aspects. Am. J. Hyg., *64*:30–45.

Lincicome, D.R., Thiede, W.H., and Carpenter, E. 1950. An evaluation of the influence of World War II on the incidence of amebiasis. Am. J. Trop. Med., *30*:171–179.

Lushbaugh, W.B., Bairalla, A.B., Cantey, J.R., Hofbauer, A.F., and Pittman, F.E., 1979. Isolation of a cytotoxin-enterotoxin from *Entamoeba histolytica*. J. Infect. Dis., *139*:9–17.

Markell, E.K., and Quinn, P.M. 1977. Comparison of immediate polyvinyl alcohol (PVA) fixation with delayed Schaudinn's fixation for the demonstration of protozoa in stool specimens. Am. J. Trop. Med. Hyg., *26*:1139–1142.

Mattern, C.F.T., Diamond, L.S., and Daniel, W.A. 1972. Viruses of *Entamoeba histolytica*. II. Morphogenesis of the polyhedral particle ($BARM_2 \rightarrow$ HK:9) $\rightarrow$ HB·301 and the filamentous agent $(ABRM)_2 \rightarrow$ HK·9. J. Virol., *9*:342–358.

———, and Keister, D.B. 1977a. Experimental amebiasis. I. Pathogenicity of axenically cultured *Entamoeba histolytica* in the brain of the newborn mouse. Am. J. Trop. Med. Hyg., *26*:393–401.

———, and Keister, D.B. 1977b. Experimental amebiasis. II. Hepatic amebiasis in the newborn hamster. Am. J. Trop. Med. Hyg., *26*:402–411.

———, Keister, D.B., and Caspar, P.C. 1977. Experimental amebiasis. III. A rapid in vitro assay for virulence of *Entamoeba histolytica*. Am. J. Trop. Med. Hyg., *27*:882–887.

———, Keister, D.B., and Diamond, L.S. 1979. Experimental amebiasis. IV. Amebal viruses and the virulence of *Entamoeba histolytica*. Am. J. Trop. Med. Hyg., *28*:653–657.

McGowan, K., Deneke, C.F., Thorne, G.M., and Gorbach, S.L. 1982. *Entamoeba histolytica* cytotoxin: Purification, characterization, strain virulence, and protease activity. J. Infect. Dis., *146*:616–625.

Meleney, H.E., and Zuckerman, L.K. 1948. Note on a strain of a small race of *E. histolytica* which became large in cultures. Am. J. Hyg., *47*:187–188.

Morales, M.T., and Lizano, C. 1978. Cambios observados en la problematica del parasitismo intestinal en las ultimas dos decadad. Rev. Med. Hosp. Natl. Ninos, *13*:71–78.

Munoz, Ma. de L., Calderon, J., and Rojkind, M. 1982. The

collagenase of *Entamoeba histolytica*. J. Exp. Med., *155*:42–51.

Neal, R.A. 1974. Survival of *Entamoeba* and related amoebae at low temperature. I. Viability of *Endamoeba* cysts at 4°C. Int. J. Parasitol., *4*:227–229.

———, Latter, V.S., and Richards, W.H.G. 1974. Survival of *Entamoeba* and related amoebae at low temperature. II. Viability of amoebae and cysts stored in liquid nitrogen. Int. J. Parasitol., *4*:353–360.

Nelson, E.C. 1947. Alcoholic extract medium for the diagnosis and cultivation of *Endamoeba histolytica*. Am. J. Trop. Med., *27*:545–552.

———, and Jones, M.M. 1964. Cultivation of *Entamoeba histolytica* in carbon dioxide-bicarbonate buffer system media. Am. J. Trop. Med. Hyg., *13*:667–673.

Ogunba, E.O. 1977. The prevalence of human intestinal protozoa in Ibadan, Nigeria. J. Trop. Med. Hyg., *80*:187–191.

Pampiglione, S., and Ricciardi, M.L. 1974. Parasitological survey on pygmies in Central Africa. I. Babinga group (Central African Republic). Riv. Parassitol., *35*:161–188.

Patterson, M., Healy, G.R., and Shabot, J.M. 1979. Serologic testing for amoebiasis. Gastroenterology, *78*:136–141.

Phillips, S.C., Mildvan, D., William, D.C., Gelb, A.M., and White, M.C. 1981. Sexual transmission of enteric protozoa and helminths in a venereal-disease-clinic population. N. Engl. J. Med., *305*:603–606.

Pillai, S., and Mohimen, A. 1982. A solid-phase sandwich radioimmunoassay for *Entamoeba histolytica* proteins and the detection of circulating antigens in amoebiasis. Gastroenterology, *83*:1210–1216.

Pipkin, A.C. 1949. Experimental studies on the role of filth flies in the transmission of *Endamoeba histolytica*. Am. J. Hyg., *49*:255–275.

Pittman, F.E., El-Hashimi, W.K., and Pittman, J.C. 1973a. Studies of human amebiasis. I. Clinical and laboratory findings in eight cases of acute amebic colitis. Gastroenterology, *65*:581–587.

———. 1973b. Studies of human amebiasis. II. Light and electron-microscopic observations of colonic mucosa and exudate in acute amebic colitis. Gastroenterology, *65*:588–603.

Powell, S.J., Hennessy, E., Wilmot, A.J., and Elsdon-Dew, R. 1961. The incidence of intestinal parasites in amebic and bacillary dysentery. Am. J. Trop. Med. Hyg., *10*:22–24.

Ravdin, J.I., Croft, B.Y., and Guerrant, R.L. 1980. Cytopathogenic mechanisms of *Entamoeba histolytica*. J. Exp. Med., *152*:377–390.

Rees, C.W. 1955. *Problems in Amoebiasis*. Springfield, Charles C Thomas.

Richards, C.S., Goldman, M., and Cannon, L.T. 1966. Cultivation of *Entamoeba histolytica* and *Entamoeba histolytica*-like strains at reduced temperature and the behavior of the amebae in diluted media. Am. J. Trop. Med. Hyg., *15*:648–656.

Ridley, D.S., and Warhurst, D.C. 1978. Misdiagnosis of amoebiasis. Br. Med. J., *2*:632–633.

Robinson, G.L. 1968. The laboratory diagnosis of human parasitic amoebae. Trans. R. Soc. Trop. Med. Hyg., *62*:285–294.

Rowland, K.A.K. 1967. Amoebiasis in Freetown, Sierra Leone. Trans. R. Soc. Trop. Med. Hyg., *61*:706–709.

Sargeaunt, P.G., and Williams, J.E. 1979. Electrophoretic isoenzyme patterns of the pathogenic and non-pathogenic intestinal amoebae of man. Trans. R. Soc. Trop. Med. Hyg., *73*:225–227.

———, Jackson, T., and Simjee, A. 1982a. Biochemical homogeneity of *Entamoeba histolytica* isolates, especially those from liver abscess. Lancet, *June 19*:1386–1388.

———, Williams, J.E., Jackson, T., and Simjee, A.E. 1982b. A zymodeme study of *Entamoeba histolytica* in a group of South African schoolchildren. Trans. R. Soc. Trop. Med. Hyg., *76*:401–402.

Saygi, G., and Ogutman, R. 1973. A copro-parasitologic survey in a primary school of Erzurum. Ataturk Universitesi Tip Fakultesi Tip Dergisi Cilt: 5, Sayl: 20, 1973 (Ayr Baskl), pp. 331–337.

Sepulveda, B. 1982. Progress in amebiasis. Scand. J. Gastroenterol., Suppl. *77*:153–164.

Shaffer, J.G., Ryden, F.W., and Frye, W.W. 1949. Studies on the growth requirements of *Endamoeba histolytica* and other intestinal protozoa in a clear medium without demonstrable bacterial multiplication. Some modifications and simplifications of the medium. Am. J. Hyg., *49*:127–133.

Sodeman, W.A., and Beaver, P.C. 1952. A study of the therapeutic effects of some amebicidal drugs. Am. J. Med., *12*:440–446.

———, Doerner, A.A., Gordon, E.M., and Gillikin, C.M. 1951. Chloroquine in hepatic amebiasis. Ann. Intern. Med., *35*:331–341.

Sodeman, W.A., Jr., and Dowda, M.C. 1973. Rapid serological methods for the demonstration of *Entamoeba histolytica* activity. Gastroenterology, *65*:604–607.

Spencer, H.C., Jr., Hermos, J.A., Healy, G.R., Melvin, D.M., and Shmunes, E. 1976. Endemic amebiasis in an Arkansas community. Am. J. Epidemiol., *104*:93–99.

Spillmann, R., Ayala, S.C., and Sanchez, C.E. de. 1976. Double-blind test of metronidazole and tinidazole in the treatment of asymptomatic *Entamoeba histolytica* and *Entamoeba hartmanni* carriers. Am. J. Trop. Med. Hyg., *25*:549–551.

Stamm, W.P., Ashley, M.J., and Bell, K. 1976. The value of amoebic serology in an area of low endemicity. Trans. R. Soc. Trop. Med. Hyg., *70*:49–53.

Sukov, R.J., Cohen, L.J., and Sample, W.F. 1980. Sonography of hepatic amebic abscesses. Am. J. Roentgenol., *134*:911–915.

Swartzwelder, J.C., and McGill, H.S., Jr. 1949. Intestinal amebiasis complicated by abscess of the brain. Report of a case without hepatic or pulmonary involvement. Rev. Kuba Trop. Med. Parasitol., *5*:27–29.

Takeuchi, A., Jervis, H.R., and Phillips, B.P. 1977. Electron microscope studies of experimental *Entamoeba histolytica* infection in the guinea pig. III. Histolysis of the cecum. Virchows Arch. B Cell Pathol., *24*:263–272.

———, and Phillips, B.P. 1976. Electron microscope studies of experimental *Entamoeba histolytica* infection in the guinea pig. II. Early cellular and vascular changes accompanying invasion of the lamina propria. Virchows Arch. B Cell Pathol., *20*:1–13.

Tay, J., Salazar Schettino, P.M., de Haro Arteaga, I., and Ruiz Hernandez, A.L. 1978. Frecuencia de las protozoosis intestinales en Mexico. Salud Publica Mex., *20*:297–337.

Thacker, S.B., Kimball, A.M., Wolfe, M., Choi, K., and Gilmore, L. 1981. Parasitic disease control in a residential facility for the mentally retarded: Failure of selected isolation procedures. Am. J. Pub. Health, *71*:303–305.

Thuse, M.G. 1979. Fatal intestinal amoebiasis. Postgrad. Med. J., *55*:548–552.

Tobie, J.E. 1940. Pathogenicity of "carrier" strains of *Endamoeba histolytica* in the experimental dog. Proc. Soc. Exp. Biol. Med., *45*:691–693.

Villee, C.A., Walker, W.F., Jr., and Smith, F.E. 1958. *General Zoology*. Philadelphia, W.B. Saunders.

Vinayak, V.K., and Chitkara, N.L. 1976. Search for Laredo type amoebae among *Entamoeba histolytica*. Indian J. Med. Res., *64*:1446–1450.

Walker, E.L., and Sellards, A.W. 1913. Experimental entamoebic dysentery. Philip. J. Sci., *8*:253–331.

Wenyon, C.M. 1926. *Protozoology,* vol. II. London, Bailliere, Tindal and Cox.

Westphal, A. 1941. Ein Kulturverfahren für *Entamoeba gingivalis* und dessen Anwendung für die differentialdiagnose von *E. gingivalis* und *E. histolytica*. Deutsche Tropenmed. Zeitschr., *45*:685–690.

Wilmot, A.J. 1962. *Clinical Amoebiasis*. Oxford, Blackwell Scientific Publications.

Chapter 12

Pathogenic Free-Living Amebae: *Naegleria* and *Acanthamoeba*

Among the numerous free-living amebae of soil and water habitats, certain species belonging to two genera—*Acanthamoeba* and *Naegleria*—are facultative parasites of man. Human infections with amebae of this type were first recognized by Fowler and Carter, who, in 1965, reported four fatal cases in Australia. The following year, four cases were reported in the United States—one in Texas by Patras and Andujar (1966) and three in Florida by Butt (1966). Butt named the disease *primary amebic meningoencephalitis*. Distinctive features of *Naegleria* and the related genus *Acanthamoeba* were established by the studies of several workers in the late 1960s (Culbertson, 1971); within a few years, the reported cases of primary amebic meningoencephalitis reached a total of more than 130 (Stevens et al., 1981), the majority of which were known or thought to be caused by *Naegleria*, although the identification of *Acanthamoeba* as the pathogen in at least 15 cases was unquestionable (Martinez, 1980). The term "amphizoic" was proposed by Page (1974) for these amebae and other such organisms that feed and multiply well in both free-living *(exozoic)* and endozoic conditions.

The species most frequently identified in human tissues are *Naegleria fowleri* and *Acanthamoeba culbertsoni*. *N. fowleri*, first isolated in Australia from two cases of primary amebic meningoencephalitis, was described and named by Carter in 1970. *A. culbertsoni*, first recognized in tissue sections from experimental mice and monkeys that had died following inoculation with tissue culture fluid that was thought to contain a virus, was first isolated by Culbertson and associates in 1958. Now known as the Lilly A-1 strain, it was described as *Hartmannella culbertsoni* by Singh and Das (1970) and redescribed by Visvesvara and Balamuth (1975), who placed it in the genus *Acanthamoeba*.

Classification. Various classifications of the free-living amebae have been proposed (Page, 1976; Levine et al., 1980). *Naegleria* and *Acanthamoeba*, along with the parasitic amebae, are placed in the subphylum Sarcodina Hertwig and Lesser, 1874, superclass Rhizopoda von Siebold, 1845, class Lobosea Carpenter, 1861, subclass Gymnamoebia Haeckel, 1862. The *Entamoeba* species and others that are natural intestinal parasites constitute the family *Endamoebidae* Calkins, 1926, in the order Amoebida Kent, 1860. *Acanthamoeba* is given a separate family, Acanthamoebidae Sawyer and Griffin, 1975, and *Naegleria* is a member of the family Vahlkampfiidae (Jollos) Zulueta, 1917, in the order Schizopyrenida Singh, 1952, as outlined below.

- Order Amoebida
  - Family Endamoebidae
    - *Entamoeba*
  - Family Acanthamoebidae
    - *Acanthamoeba*
- Order Schizopyrenida
  - Family Vahlkampfiidae
    - *Naegleria*

In the Vahlkampfiidae, the active trophozoites commonly assume the limax (monopodial) form, the pseudopodia are formed eruptively, nuclear division is by promitosis (mitosis inside the nuclear membrane) (Fig. 12–1), and in some *(Naegleria)*, there are transient flagellate forms. The cysts generally are round with smooth walls. In the Acanthamoebidae, the type and only genus is *Acanthamoeba* and the type species is *A. castellanii*. *Acanthamoeba* trophozoites have several to many

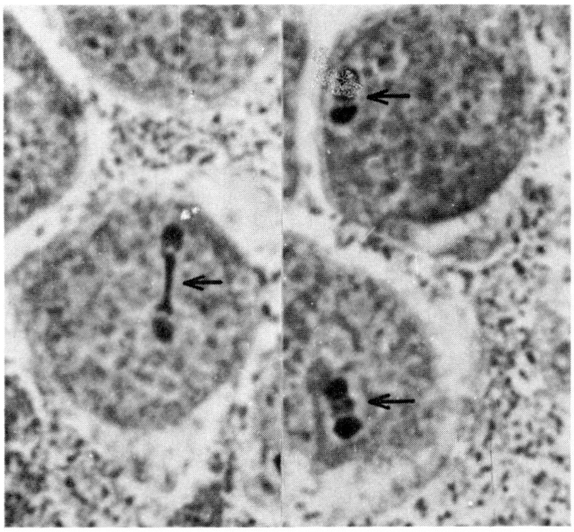

Fig. 12–1. *Naegleria fowleri.* Culture forms in mitosis. Arrows in left figure point to polar masses within the nuclear membrane; arrow in right figure points to interzonal body. The nucleus occupies a large part of the ameba. Iron-alam hematoxylin. (× 1200.) (From Lawande, R.V., Macfarlane, J.T., Weir, W.R.C., and Awunor-Renner, C. 1980. A case of primary amebic meningoencephalitis in a Nigerian farmer. Am. J. Trop. Med. Hyg., *29*:21–25.)

subpseudopodia (acanthopodia), which are usually sharply pointed. The active amebae are more or less triangular, elongate, or irregular; the cysts are polyhedral or thickly biconvex, with the wall consisting of a polygonal or stellate endocyst and a more or less rippled ectocyst; the nuclear membrane and nucleolus (endosome) disappear during mitosis, and no flagellate stage is known.

Naegleria fowleri Carter, 1970 (Producing primary amebic meningoencephalitis)

Synonyms. *Naegleria aerobia* Singh and Das, 1970; *Naegleria invadens* Chang, 1971.

Historical Notes. Pathogenic species referred to in the early literature as *Naegleria gruberi* or *Naegleria* sp. by numerous workers (Callicott et al., 1968; Jadin et al., 1971; Carter, 1972) and as "slime mould" amebae (Myxomycetale) by Mandal et al. (1970) were apparently, in all instances, *N. fowleri*. The ameba reported as *Hartmannella (Acanthamoeba)* by Patras and Andujar (1966), subsequently was identified by Stamm (1974) and Culbertson (1975) as apparently *N. fowleri*. In a case reported by Derrick (1948), the ameba described as resembling *Iodamoeba buetschlii* probably was *N. fowleri* instead (Culbertson et al., 1972; Stamm, 1974); it was apparently the first published description of a *Naegleria* infection in man.

Morphology, Biology, and Life Cycle. The life cycle stages of *Naegleria* are motile trophozoites and nonmotile resistant cysts. Reproduction is by simple binary fission, and the trophozoites can colonize in water or moist soil and in tissue-culture or other artificial media. The trophozoites of *N. fowleri,* which occur in two forms—ameboid and flagellate—may be observed in distilled water hanging-drop preparations from cultures or cerebrospinal fluid. When in motion, the ameboid form is elongate, broad anteriorly, distinctly tapered posteriorly, and forms a single pseudopod of clear ectoplasm at the forward end and a small tapered or knob-like process, often with attached debris (uroid process), at the posterior end (Fig. 12–2, *A*); the dimensions vary around 7 by 20 μm. The diameter of rounded forms ranges from 10 μm to 35 μm, but generally is 15 μm or less. In the living, motile ameba, a nucleus may be seen as a small clear area about 3 μm in diameter with a large central nucleolus (endosome) in the anterior part of the endoplasm. The endoplasm is filled with granules and vacuoles, and one or more contractile vacuoles may be seen in water mounts.

Flagellate forms with two flagella begin to appear among the ameboid trophozoites a few hours after being transferred to water from a culture or from teased tissue, and maintained at temperatures between 27° and 37°C. The changes may occur as late as 20 hours, and usually, some of the flagellate forms persist for 2 days or longer (Carter, 1970). The transformed ameba is usually pear-shaped, with the flagella at the broader end; it may move rapidly forward or spin slowly in circles, at which time the individual flagella in proper light and sharp focus become evident. Flagellate forms do not divide, but they readily shed the flagella, then resume ameboid motility and reproduction.

Cysts are formed on agar cultures, the same as in nature. They are uninucleate, circular in outline, and 7 to 10 μm in diameter, with a smooth, heavy wall that is approximately 1 μm thick. In the unstained cyst, only a few fine granules can be seen; the nucleus is not evident and cysts often appear empty (Fig. 12–2,*B*). When stained, the nucleus has the same appearance as in the trophozoite, but it is smaller (around 1.5 μm). Structures evident on electron micrographs include pores in the cyst wall and mitochondria, endoplasmic reticulum, vesicles, and secretory granules (Jadin et al., 1974; Schuster, 1975).

In stained tissue sections, *Naegleria* trophozoites are poorly differentiated, having much the same staining qualities as phagocytes and tissue cells. Distinctive, however, is the round clear nucleus with a large central endosome which usually is stained purple or slightly red in formalin-fixed, H

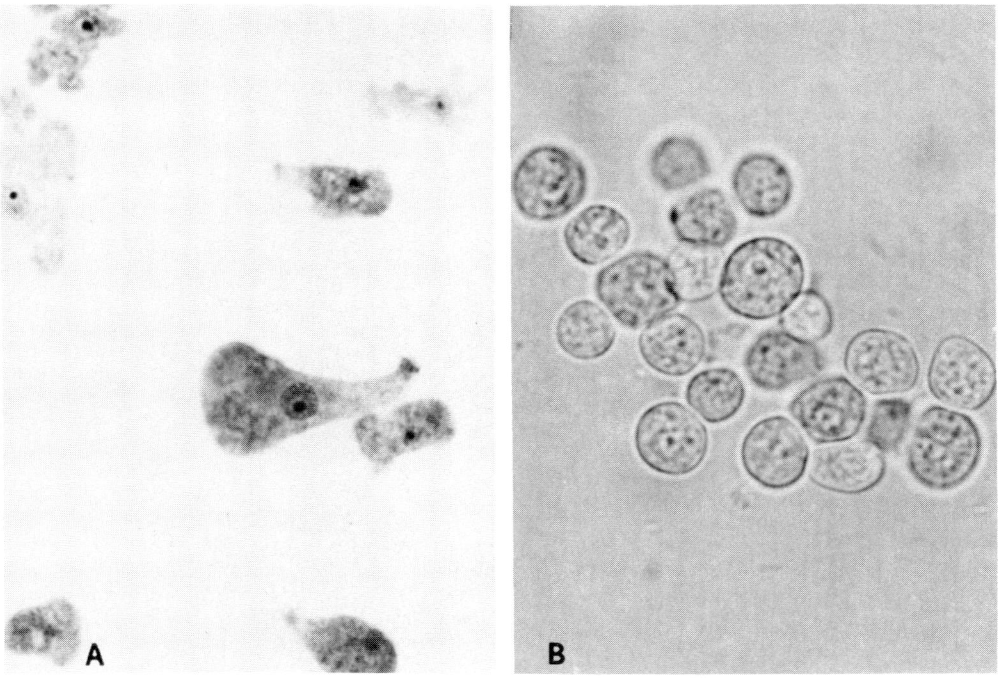

Fig. 12–2. *Naegleria fowleri. A,* Trophozoites from culture, fixed while actively motile on a glass surface, illustrating typical shapes and size variation. Hematoxylin. (× 330.) (By P.C. Beaver. Preparation courtesy of Dr. E.C. Nelson.) *B,* Cysts in wet preparation from culture. Unstained. (× 800.) (From Lawande, R.V., Macfarlane, J.T., Weir, W.R.C., and Awunor-Renner, C. 1980. A case of primary amebic meningoencephalitis in a Nigerian farmer. Am. J. Trop. Med. Hyg., *29:*21–25.)

& E-stained tissue. Also, the ameba usually is set-off from the surrounding tissue by a clear zone formed by shrinkage (Fig. 12–3).

Epidemiology. Through 1981, more than 100 cases of primary amebic meningoencephalitis had been reported. Identification was not always possible, but more often than not the ameba identified was *Naegleria fowleri*. Numerous cases have been reported from the United States (mainly from Virginia, Florida, and California, but also from Georgia, New York, Puerto Rico, Arizona, South Carolina, and Texas). Case reports have been most numerous from Belgium, Czechoslovakia, Australia, New Zealand, India, and Nigeria, and one or more have been reported from England, Ireland, Venezuela, Panama, and New Guinea. Several cases that were reported as *Acanthamoeba* were corrected to *Naegleria* when the amebae could be more accurately identified by immunoperoxidase (Culbertson, 1975) or indirect immunofluorescence staining (Stamm, 1974).

Usually, the infected persons had a history of swimming in a freshwater lake or pond a few days before onset of symptoms. In the cases reported from Czechoslovakia, the infected persons had come from ten different localities but all had been swimming in the same indoor pool, which had been filled with river water that had been filtered, heated, and chlorinated (Cerva et al., 1968). In cases reported from Virginia (United States), the infections were acquired in small ponds or dirt-lined pools filled with chlorinated water (Duma et al., 1971; Duma, 1972). The chlorination in such cases not only failed as a protection but may have been a predisposing factor (Griffin, 1978). In one or more of four cases reported from northern Nigeria, the infection was acquired from snuffing pond water into the nose as a prayer ritual (Lawande et al., 1980).

A study in South Australia showed that heavy chlorination to 10 ppm failed to eradicate *Naegleria* from a pool, but salination to 0.7% resulted in negative cultures of the water for a period of 5 months (Anderson and Jamieson, 1972). In the same area, infections apparently were contracted from tap water that was taken unfiltered from a river and chlorinated. Evidence indicated that, in some instances, the amebae *(N. fowleri)* proliferated in the water system where pipes were exposed to high air temperatures (Anderson and Jamieson, 1972). In Belgium, where five cases of primary amebic meningoencephalitis had been seen in one locality, it was found that *N. fowleri* was widely distributed in streams and canals and was associated with thermal pollution by the effluent from several factories, especially the older ones (De Jonckheere and Van de Voorde, 1977). Growth of *N. fowleri* is enhanced by high temperatures; it tolerates temperatures up to about 45° C. In two cases reported from California (United States), the victims had bathed in a natural hot spring 5 to 7 days before admission to the hospital (Hecht et al., 1972; Centers for Disease Control, 1978). In the first three cases reported from the United States, the infected persons had been water-skiing, swimming, or diving in late summer in a small freshwater lake of central Florida (Butt, 1966). Two fatal infections in New Zealand were acquired in a natural thermal stream (Cursons et al., 1976) and in an indoor swimming pool (Cursons et al., 1979).

From experimental and epidemiologic observations, it is apparent that, in the usual case, infection results from organisms being taken into the nose in water. However, pathogenic *Naegleria* have been isolated from the nasal cavity of persons who had not been swimming (Chang et al., 1975), suggesting that viable cysts may be air borne. Sheep and probably other domestic and wild animals are susceptible to infection with *N. fowleri,* but such infections have no known epidemiologic significance (Young et al., 1980).

As the flagellate stage of *Naegleria* is a distinctly aquatic and very motile stage, it probably is the form that most frequently enters the nasal cavity, where it reverts to the ameboid form before invading the olfactory tissues and the brain. Cysts instilled intranasally are not infective in experimental animals (Carter, 1970).

Pathogenesis, Pathology, and Symptomatology. Necropsy findings in cases of *Naegleria* infection have been remarkably uniform, essentially limited to the brain, and not markedly different grossly from those of bacterial meningitis (Culbertson, 1971; Duma et al., 1971; Carter, 1972; John, 1982). The maxillary and sphenoid sinuses may be filled with amebae and exudates (Darby et al., 1979). Affected areas of the brain are soft, and the meninges are hyperemic and moderately purulent. The olfactory bulbs are congested or hemorrhagic and necrotic. Histopathologic sections show amebae throughout the affected gray matter, particularly near the subarachnoid space. Usually, amebae can be found in the exudate, although they are difficult to recognize among inflammatory cells;

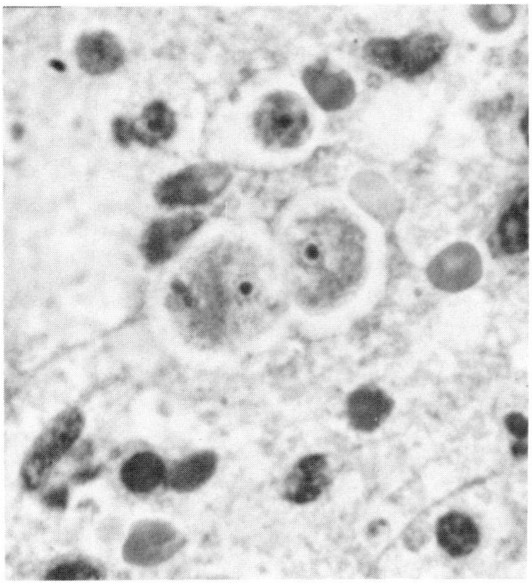

Fig. 12–3. *Naegleria fowleri.* Two typical trophozoites in brain impression smear from a 15-year-old girl who died 30 hours after admission to hospital, 8 days after swimming in a freshwater lake. (× 1300.) (By P.C. Beaver. Preparation courtesy of Dr. E.C. Nelson.)

they are most evident in the perivascular tissues, where inflammatory cells are scarce or absent. The main recognizable feature of the ameba is its characteristic open nucleus with its large, dense, central endosome. Although the amebae resemble degenerating inflammatory cells, they are generally more regularly rounded and uniform in size (10 to 12 μm). Usually, there is a halo around the nucleus that is produced by aggregates of small vacuoles, and a conspicuous space around the ameba that is produced by fixation shrinkage (Fig. 12–3). Amebae may also be found in ulcerated areas of the olfactory bulbs, where invasion may be massive, and in the spinal cord (Rothrock and Buchsbaum, 1980). Extension of invasion to other parts of the body has not been reported, though changes in the lungs, spleen, and occasionally the heart, secondary to severe cerebral infection, have been noted (Carter, 1972). The symptomatology is essentially the same as that of fulminant bacterial meningitis.

Diagnosis. A patient's history of having been swimming in thermal or stagnant water 3 to 6 days prior to onset of symptoms of severe meningitis or meningoencephalitis should suggest a possible diagnosis of *Naegleria* infection. An immediate attempt should be made to detect and to isolate the amebae from cerebrospinal fluid. The amebae usually are not recognizable after gram stain but are readily distinguished from inflammatory cells in

Wright-stained preparations (John, 1982). Although the amebae may be difficult to recognize in stained smears, they can be identified by their motility in unstained wet preparations. The amebae can be detected and identified in histopathologic sections by indirect immunofluorescence and immunoperoxidase methods (Culbertson and Harper, 1980).

Cultivation. A medium described by Culbertson et al. (1968) is recommended as satisfactory for the isolation of either *Naegleria* or *Acanthamoeba* from cerebrospinal fluid or from brain or other tissues. It is a plate of 1.5% non-nutrient agar seeded with living *Escherichia coli* that is grown separately and added. This approximates natural habitats where the amebae feed on bacteria in aerobic environments. As colonies grow and expand, resistant cysts that survive moderate desiccation are formed; thus, strains can be maintained by transfer of either trophic or cyst forms.

Nonpathogenic *Naegleria* (and *Acanthamoeba*) in cultures tolerate temperatures up to 42° C and saline up to 0.85%, but pathogenic *Naegleria* will tolerate even higher temperatures (up to 45° to 46° C) and grow poorly in saline concentrations above 0.4% (De Jonckheere and Van de Voorde, 1977). *Naegleria* requires living bacteria or tissue cells for good growth from small inocula. For maintenance, sustained tissue culture cells grown at 37° C are satisfactory (Culbertson, 1971). Because meningoencephalitis from *Naegleria* infection develops rapidly, in suspected cases with an elevated cell count, it is worthwhile to culture cerebrospinal fluid even if no amebae are recognized in the direct or wet mount examination (Griffin, 1978).

Treatment. No satisfactory treatment for primary amebic meningoencephalitis is known (John, 1982). The drug most commonly referred to as effective against *Naegleria* infection is amphotericin B. A 2-year-old boy in Great Britain who was treated for *Naegleria* meningoencephalitis with amphotericin B and sulfadiazine survived longer than any previously reported case. Two companion children who had been exposed to the same possible source of infection and had developed mild meningitis were similarly treated and survived. Identification of the amebae in these cases has been questioned (Griffin, 1978). In two cases reported by Duma et al. (1971) and two by Carter (1972), the drug's effect on the amebae was evident even though the patients died. In a successful treatment of a 14-year-old boy in Australia, amphotericin B was given intravenously, 1 mg/kg daily, beginning on the fourth day of illness; penicillin, ampicillin, and sulfadiazine, which had been started 3 days previously, were continued. There was marked improvement within 2 days, but when it was noted that even after 5 days amebae persisted in the cerebrospinal fluid, amphotericin B was given intrathecally and later intraventricularly in doses of 0.1 mg on alternate days. The cerebrospinal fluid gradually cleared and the patient recovered. Although it seemed clear that survival in that case was due to the amphotericin B, not to the other drugs given, it was recommended that sulfadiazine, as well as amphotericin B, be given in all such cases. Sulfadiazine is apparently effective against *Acanthamoeba,* and it is not always possible to distinguish between *Naegleria* and *Acanthamoeba* in cerebrosinal fluid. Also, during treatment with amphotericin B, the use of corticosteroids is not recommended (Carter, 1972). A marked synergism between amphotericin B and tetracycline has been observed when given together to *Naegleria fowleri*-infected mice (Thong et al., 1978, 1979). A recent report giving few details described the case of a 9-year-old girl with severe meningoencephalitis due to *Naegleria fowleri* infection, which was acquired while she was swimming in a hot spring in California. She was given intravenous and intrathecal amphotericin B, miconazole (an experimental drug), and oral rifampin. Her condition improved after 48 hours, and she recovered completely within a month (Centers for Disease Control, 1978).

Despite early and intensive treatment with amphotericin B, the infection is usually fatal. Nevertheless, amphotericin B remains the only known effective agent in the treatment of primary amebic meningoencephalitis (Stevens et al., 1981).

Prognosis. *Naegleria* meningoencephalitis is usually fatal within a week of onset. Survival depends on early suspicion of infection, which is based largely on the patient's history of contact with water; early diagnosis based on demonstration of amebae in cerebrospinal fluid; and immediate treatment.

Prevention. The only known preventable risk of infection is contact with stagnant or thermal waters. Hyperchlorination of river water for indoor or outdoor swimming pools is not protective. Chlorinated water was implicated in a high proportion of cases reported from the United States and Europe. However, no *Naegleria* infection is known to have been acquired in a standard swimming pool in the United States. When there is doubt about the safety of water used in swimming pools, salination to 0.7% (w/v) may be a useful measure (Anderson and Jamieson, 1972).

Acanthamoeba culbertsoni (Singh and Das, 1970) Sawyer and Griffin, 1975, and Other Pathogenic *Acanthamoeba* Species (Producing amebic meningoencephalitis, uveitis, and ulceration of the cornea)

Historical Notes. Formerly, the genus *Acanthamoeba* was placed along with *Hartmannella* Alexeieff, 1912, in the family Hartmannellidae. In 1975, Sawyer and Griffin proposed a new family, Acanthamoebidae, for the spiny or filose amebae, with *Acanthamoeba* as the type genus and *Acanthamoeba castellanii* (Douglas, 1930) Volkonsky, 1931, as the type species. The most distinctive feature of this family is the presence of acanthopodia. Other family characteristics include the length of trophozoites in locomotion (30 to 60 μm) and the shape variable, with acanthopodia arising from all surfaces of the body and continuously being formed and resorbed. The advancing hyaline pseudopodium is wide and tongue-shaped, with irregular margins and filopodia. A posterior uroid with short filaments is present in some species. The nucleolus (endosome) and nuclear membrane disappear during mitosis. The cysts have a double wall, with the ectocyst generally wrinkled; the endocyst is variable among species. All known species can be grown on bacterized agar media and axenically in media containing proteose-peptone and glucose.

Hartmannella culbertsoni Singh and Das, 1970, earlier known as *Hartmannella-Acanthamoeba* Strain A (also A-1 H-A, and Lily A1), became *Acanthamoeba culbertsoni* (Singh and Das) Sawyer and Griffin 1975 when it was assigned to the genus *Acanthamoeba*. References to *Hartmanella* species causing disease generally refer to amebae that are now known to be *Acanthamoeba;* some probably were *Naegleria fowleri*. Until recently, no species of *Hartmannella*, a very common genus of free-living amebae, was known to be pathogenic in mammals, although several species have been isolated from the throat (Wang and Feldman, 1967) and from the nasal passages of children (Lawande et al., 1979). However, in 1982, Cleland et al. reported a case of chronic amebic meningoencephalitis in a man in Northern Nigeria in which the ameba was identified as *Hartmannella rhysodes*, seen in and cultured from cerebrospinal fluid three times over a period of several days.

Of special interest among the species of *Acanthamoeba* are *A. culbertsoni*, the species most frequently identified in cases of *Acanthamoeba* infection in man, and *A. castellanii*, type species of the genus and one that occasionally is pathogenic; *A. castellanii* also is frequently a contaminant in bacterial and mycologic cultures. A third species of interest as a human pathogen causing keratitis and uveitis is *A. polyphaga* (Puschkarew, 1913) Page, 1967. *A. astronyxis* (Ray and Hayes, 1954) Page, 1967, has been identified as the cause of fatal encephalitis in one case (Gullett et al., 1979).

Morphology, Biology, and Life Cycle. The life-cycle stages are the active trophic forms and resistant cysts. There is no flagellate form. When active, all *Acanthamoeba* species have an irregular appearance with spine-like pseudopodia, and acanthopodia, arising from lobopodia and from other areas of the body (Fig. 12–4). There is perceptible, slow movement of the acanthopodia as they are formed and resorbed continuously, but progressive place-to-place movement of the trophozoite generally is not readily evident.

A. culbertsoni trophozoites are extremely variable in size and shape, the average diameter of rounded forms being approximately 30 μm (10–45 μm). The nucleus and its large nucleolus are easily observed in the unstained ameba but less so than the contractile vacuole when present. Cysts of *A. culbertsoni* are spherical, slightly irregular in outline, about 20 μm or more in diameter, with a double wall forming a smooth or slightly wrinkled ectocyst and a roughly polyhedral endocyst (Fig. 12–4,*B*). A nucleus is often visible, centrally located with a large central endosome, as in the trophozoite. Precysts, at first smooth and spherical with a single wall, later show gradations that are hard to identify (Culbertson, 1971).

Acanthamoeba polyphaga was redescribed by Visvesvara et al. (1975) from cultures and material obtained in corneal scrapings from a 59-year-old farmer in southern Texas. The trophozoites usually produce a broad hyaline lobopodium in the direction of movement and bear numerous delicate acanthopodia. A uroid is often present on trophozoites associated with bacteria, but not on those axenically grown. A single contractile vacuole is prominent. Trophozoites in tissue cultures may assume the limax shape (Nagington et al., 1974). The mean length of trophozoites is 23 μm; the range, 16 to 48 μm. The nucleus is vesicular, 3 to 5 μm in diameter with a large, centrally located endosome. Cysts formed on agar plates are double-walled with a wrinkled outer wall (ectocyst), a polyhedral or stellate inner wall (endocyst), and a mean diameter of 10 μm (range 7 to 18 μm).

Acanthamoeba castellanii trophozoites, when motile, are less than three times as long as they are wide. Cysts are somewhat mammillated, either thickly biconvex or polyhedral. The endocyst is polyhedral or, if stellate, usually has six or more truncate rays of the star in one plane. The ectocyst is loosely applied to the endocyst or, occasionally, nearly circular, rippled, or wrinkled. Each point of the endocyst is situated at the bottom of a depression surrounded by a circular ridge formed by the ectocyst, not projected above the general level of the outer wall. The average cyst diameter is about 16 μm or more, but smaller ones are not rare (Page, 1967).

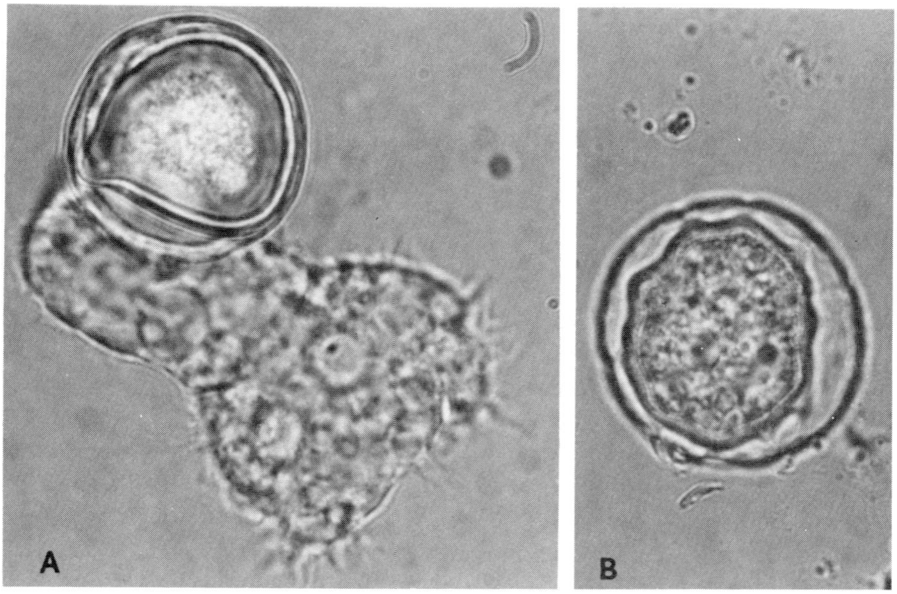

Fig. 12–4. *Acanthamoeba culbertsoni*. Culture forms in wet preparation. *A*, Trophozoite exhibiting numerous acanthopodia, a nucleus and a contractile vacuole, and attached to a cyst with somewhat atypical endocyst wall. *B*, Cyst with typical polyhedral endocyst wall and spherical ectocyst. Unstained. (× 1200.) (By P.C. Beaver.)

Acanthamoeba astronyxis trophozoites are large; their length in motion (25–60 μm) is approximately three times their breadth; food vacuoles are usually conspicuous and often have many small yellowish refractile bodies in the endoplasm. The cysts are biconcave with an ectocyst that is folded inconspicuously and a stellate endocyst with 4 to 10 conically rounded rays that are mostly in one plane. Diameter of the cyst is 14 to 25 μm; diameter of the nucleus is 4 to 5 μm. The cyst form closely resembles that of *A. castellanii,* except that its endocyst rays protrude slightly above the general cyst wall contour.

Species differentiation in the genus *Acanthamoeba* is difficult. Except with the aid of immunoperoxidase and immunofluorescent staining techniques, few of the *Acanthamoeba* diagnosed in man would have been identified specifically.

Epidemiology. About 30 reports of disease in man have implicated free-living amebae that were known or believed to be species of *Acanthamoeba*. In a number of instances, the identification was retrospective, based on immunoperoxidase or immunofluorescent staining of amebae in tissue sections or cultures. Cases in which the amebae invaded the central nervous system have been reported from England (Apley et al., 1970), India (Pan and Ghosh, 1971), Zambia (Bhagwandeen et al., 1975), Korea (Ringsted et al., 1976), Venezuela and Peru (Martinez, 1977). Cases have also been reported in the United States, including Arizona (Kernohan et al., 1960), New York (Kenney, 1971; Jager and Stamm, 1972; Centers for Disease Control, 1978), Pennsylvania (Robert and Rorke, 1973), Louisiana (Hoffmann et al., 1978), Virginia (Duma et al., 1978), and California (Gullett et al., 1979). Findings in 15 cases were summarized by Martinez (1980). Invasion of the cornea or interior of the eye, or both, has been reported in 10 cases in England, Germany, United States, and Netherlands (Key et al., 1980; Bos et al., 1981). Four cases in which the amebae were identified in exudates from the vagina have been reported, all in the United States (Griffin, 1978; McNeill and Moraes-Ruehsen, 1978). Invasion of the ear in Israel (Lengy et al., 1971) and of the skin in Africa (Spencer et al., 1973) have been reported. In some cases, colonization in the skin or eye was followed by invasion of the brain (Kernohan et al., 1960; Ringsted et al., 1976; Sotelo-Avila et al., 1974). In reporting a California case of disseminated *Acanthamoeba* infection that presented as an unusual skin lesion, Gullett et al. (1979) listed 10 other such cases, not including one reported from Louisiana (Hoffmann et al., 1978).

In a few cases, the infected persons had a history

of contact with ditch water or puddles, but usually they did not. Frequently, the probable source of the infection was not identified but was presumed to be dust or water. Cysts of *Acanthamoeba* species are resistant to chlorine and can withstand drying; thus, transport by water and air is possible. Amebae have been isolated in nasal swabs and in large numbers of throat-swab cultures taken in surveys for respiratory virus infections (Wang and Feldman, 1967). Whether the organisms isolated were cysts or trophozoites, or both, was not determined. Pathogenic amebae *(A. culbertsoni* and *A. hatchetti)* have been isolated from brackish and marine sediments (Sawyer et al., 1977), from a humidifier (Edwards et al., 1976), and from a dialysis apparatus (Casemore, 1977).

Pathogenesis, Pathology, and Symptomatology. *Acanthamoeba* infections in man differ from *Naegleria* infections in a number of respects; they also are variable as a group. Although only one species of *Naegleria (N. fowleri)* is known to cause disease in man, four *Acanthamoeba* species—*A. culbertsoni, A. polyphaga, A. castellanii, A. astronyxis*—and possibly others may do so. Also, although the only known portal of entry for *Naegleria* is the olfactory neuroepithelium, the *Acanthamoeba* species may enter through the broken or ulcerated skin or eye, and possibly through the lungs and genitourinary tract.

Tissue invasion by *Acanthamoeba* species is relatively slow and tends to stimulate granuloma formation. The disease, therefore, is usually of gradual onset and takes a prolonged chronic course described in terms of months or weeks rather than days (Martinez et al., 1980). This is in sharp contrast to *Naegleria* infections.

Meningoencephalitis caused by *Acanthamoeba* frequently has occurred in chronically ill and immunosuppressed patients. Death occurred in one case after 11 years of treatment with radiation and immunosuppressive drugs for Hodgkin's disease; postmortem examination revealed a brain abscess and mild meningitis caused by an *Acanthamoeba,* probably *A. culbertsoni.* When the primary site of invasion was the skin, the period between exposure and onset of central nervous system symptoms was several months to 1 year or longer (Kernohan et al., 1960; Ringsted et al., 1976), or was unknown (Bhagwandeen et al., 1975). In a number of cases, the symptoms at onset were sore throat and fever, followed by headache and other indications of meningeal irritation. The period from onset of symptoms to death in several cases was about 3 weeks (Apley et al., 1970; Martinez et al., 1977; Hoffmann et al., 1978). The shortest recorded period thus far was in an 11-year-old girl in New York who, after 2 days of headache, vomiting, fever, and stiff neck, was admitted to a hospital where she died 6 days after admission. *Acanthamoeba* were found in sections of the brain (Centers for Disease Control, 1978).

Gross findings at autopsy in deaths caused by *Acanthamoeba* were purulent leptomeningitis, brain edema, and foci of necrosis in various locations both at the surface and in the deeper parts (the thalamus and cerebellum, in one case; on the frontotemporal occipital lobes, thalamus, cerebellar folia, and pons, in another; and on the frontoparietal lobe, cerebral peduncle and medulla of a third). The olfactory nerves and lobes were not affected in any of the cases (Martinez et al., 1977) (Fig. 12–5). In these cases and others (Hoffmann et al., 1978; Martinez et al., 1980), the histopathologic

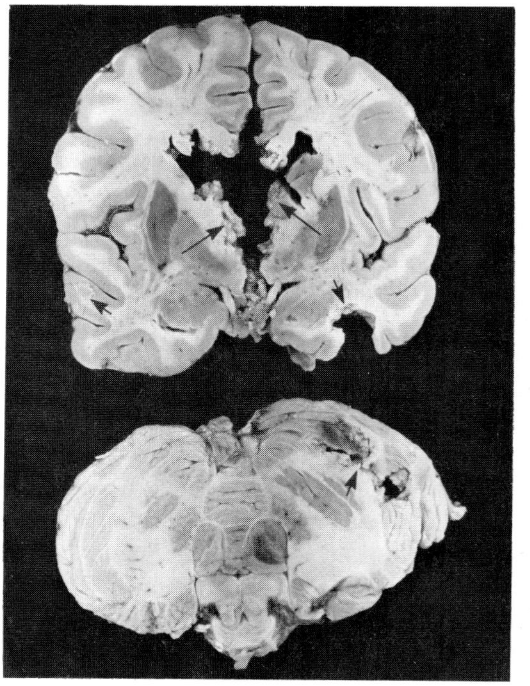

Fig. 12–5. *Acanthamoeba.* Gross appearance of sectioned brain from a man who died 3 weeks after onset of meningoencephalitis. Amebae were found in areas of necrosis in the midline and adjacent structures (above, arrows) and in the temporal lobes, left parietal cortex, and cerebellum (below, arrow). In other sections of the brain, cysts as well as trophozoites were present. (From Hoffman, E.O., et al. 1978. A case of primary amebic meningoencephalitis. Light and electron microscopy and immunohistologic studies. Am. J. Trop. Med. Hyg., *27*:29–38.)

findings were necrotizing chronic, granulomatous encephalitis with focal hemorrhages, multinucleated giant cells, and large amebic trophozoites (20–30 μm) and smaller cysts. Both trophozoites and cysts were scattered throughout the lesions, but the trophozoites were seen most abundantly around and near blood vessels at the margins of areas of acute inflammation and necrosis (Fig. 12–6).

Although the amebae are initially difficult to differentiate among the inflammatory and necrotic tissue cells, once they are recognized, they are readily distinguished by their large, dense endosome, which is surrounded by a clear nuclear halo, and abundant spongy cytoplasm, which is often set off from the surrounding tissue by a clear space that is caused by contraction of the surrounding tissues. The cysts, most numerous in the granulomatous areas, generally can be recognized by their double-contoured, thick, often somewhat wrinkled walls and dense granular cytoplasm. The diameter of cysts is around 10 μm or more.

Chronic granulomatous lesions caused by *Acanthamoeba* elsewhere than in the brain have been

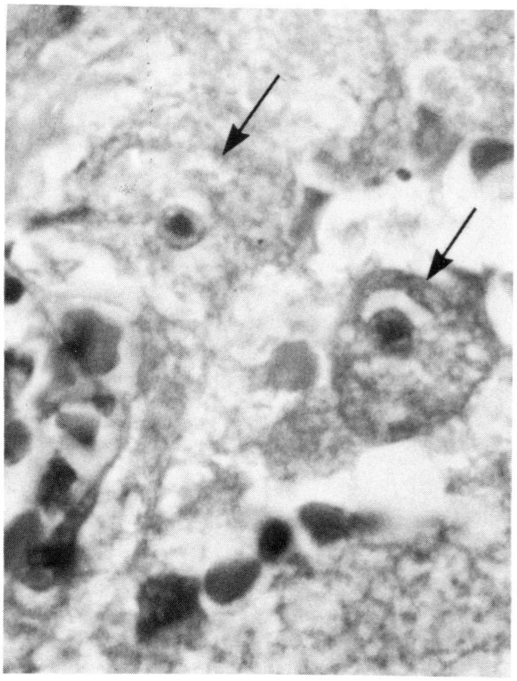

Fig. 12–6. *Acanthamoeba*. Two trophozoites (arrows) near a small blood vessel in tissue from the brain shown in Figure 12–5. Hematoxylin and eosin. (× 1200.) (From Hoffman, E.O., et al. 1978. A case of primary amebic meningoencephalitis. Light and electron microscopy and immunohistologic studies. Am. J. Trop. Med. Hyg., *27*:29–38.)

reported in skin, kidneys, liver, spleen, uterus, and prostate (Martinez et al., 1977). Microabscesses up to 1 cm in diameter in the pancreas with areas of necrotizing vasculitis containing amebae were seen in one case (Hoffmann et al., 1978). In others, the lungs, or pancreas and lungs, contained fibrous lesions in which amebae were not detected (Jager and Stamm, 1977; Duma, 1978).

Skin lesions caused by *Acanthamoeba* have been described in six cases in which there was secondary invasion of the central nervous system (Kernohan et al., 1960; Bhagwandeen et al., 1975; Ringsted et al., 1976; Gullett et al, 1979; Martinez, 1982) and one in which the primary skin lesion apparently remained focal (Spencer, 1973). In two of the four cases, a persistent granulomatous reaction involving the subcutaneous tissues had developed at the site of a head wound in a 5-year-old girl in Arizona and a 6-year-old boy in Korea. Amebae were not seen in the skin at necropsy. In an adult who died from amebic meningoencephalitis in Zambia, an early lesion had developed around the umbilicus and amebae were seen in sections of both biopsy and necropsy specimens from a large, crusting superficial ulcer with a granulomatous, tuberculoid reaction extending to the lower dermis. In Malawi, a chronic *Acanthamoeba* granuloma was removed from the forehead of a woman, and it was suggested that the infection may have been acquired from the water she carried daily from the village pond in a container on her head (Spencer, 1973).

Eye lesions caused by *Acanthamoeba* infection were first reported in 1974 in two cases in England and one in Texas. Since then, seven others have been described (Key et al., 1980; Bos et al., 1981). In one case, a 7-year-old boy who had waded in muddy water had a sore throat for 3 weeks, followed by conjunctivitis and iritis of the left eye, which, after 1 week, was followed by headache and indications of meningoencephalitis. The child died 3 weeks later and abundant amebae tentatively identified as *Naegleria* were found in sections of the brain (Sotelo-Avila et al., 1974). The eye also was sectioned revealing inflammation of the iris, ciliary body, and vitreous, with necrosis of the ciliary body and the presence of amebic trophozoites. Later review of the sections established that the amebae seen in the brain and eye were *Acanthamoeba castellanii,* and this was confirmed by different techniques (Jones et al., 1975; Willaert and Stevens, 1976). It was not determined whether the primary site of infection was the eye or nasopharynx.

In all other cases of *Acanthamoeba* invasion of the eye, the lesions were primarily refractory corneal ulcers. Two healthy adults in England had keratoconjunctivitis and uveitis that did not respond to treatment (Nagington et al., 1974). In one adult, a school teacher, corneal ulceration, pain, and loss of vision led to a corneal graft that was rejected. Histological sections of the cornea showed *Acanthamoeba* cysts throughout the stroma. In the other adult, a farmer, essentially identical clinical conditions resulted in enucleation of the affected eye. *Acanthamoeba* was repeatedly isolated in cultures from the first case and from the post-enucleation fragments of the cornea in the second case. Whether amebae had invaded the inner tissues was not determined in either case. The amebae were identified as *A. polyphaga* in the first case, as *A. castellanii* in the second (Warhurst and Thomas, 1975). In a similar case reported from England, the ameba was identified as *A. polyphaga* based on trophozoite and cyst morphology (Warhurst et al., 1976).

In two reports from the United States, the amebae were identified as *A. polyphaga* (Jones et al., 1975). In one case, trophozoites and cysts were repeatedly seen in scrapings and later in sections from the ulcerated cornea of a Texas rancher who had been struck in the eye by a straw. In the other case, a nurse in Texas developed herpes simplex epithelial keratitis. The lesions failed to heal under conventional treatment. In a few weeks, there was dendritic ulceration with multifocal stromal keratitis and uveitis, and for more than a year, there was inflammation and ulceration of the cornea and recurrent uveitis, during which the presence of *A. polyphaga* was repeatedly demonstrated. Other cases of amebic keratitis have been reported from Germany, New York, and Texas (Key et al., 1980). With few exceptions, the infections have caused enucleation or serious permanent damage to the eye. An exceptional case in which a corneal transplant was successful was reported from the Netherlands (Bos et al., 1981).

Diagnosis. A firm diagnosis of *Acanthamoeba* infection depends on the detection and identification of the amebae in cerebrospinal fluid in cases of central nervous system infection, in scrapings from lesions in cases of corneal or cutaneous infections, or in cultures of material from those sources. Amebae may be recognized in stained vaginal smears or, as in one instance, in purulent discharge from an infected ear (Lengy et al., 1971). An *Acanthamoeba* isolation has not been made from a fatal case of meningoencephalitis (Gullett et al., 1979). Isolation was successful in a nonfatal case (Callicot et al., 1968), and repeated isolations were made in cases of corneal infections (Jones et al., 1975; Visvesvara et al., 1975; Nagington et al., 1974). Serological tests possibly can be useful in detecting the presence of infection. A micro-Kolmer complement fixation test detected two infected persons among 1000 whose sera were tested, and in one case, the diagnosis was confirmed at autopsy a few months later; in the other case, serum titers decreased after treatment with dehydroemetine and chloroquine (Kenney, 1971). Successful use of indirect immunofluorescent staining for the demonstration and generic identification of free-living amebae was described by Stamm (1974) and by Willaert and Stevens (1976). Examination of sections of formalin-fixed tissues from three cases of meningoencephalitis showed that indirect immunofluoroescent staining may be useful in the localization and identification of amebae (Martinez et al. 1977). This test can be used for retrospective species identification of amebae in tissue sections (Willaert et al., 1978) or from cultures (Visvesvara et al., 1975). With plasma-membrane antisera, agglutination and fluorescence of homologous amebae can be obtained at very high titers (Stevens et al., 1977). The indirect immunoenzymatic (peroxidase) staining of formalin-fixed sections also can be used for the localization of amebae and their specific identification (Culbertson, 1975). Testing for salinity- and temperature-tolerance, may assist in distinguishing *Acanthamoeba* from *Naegleria*.

Treatment. Treatment of *Acanthamoeba* meningoencephalitis has not been successful in any well-documented case. A child who died from amebic meningoencephalitis and two companions who survived had been treated with amphotericin B and sulfadiazine (Apley et al., 1970). The treatments were credited with prolonging life in the fatal case and possibly saving life in another. Amebae cultured from the cerebrospinal fluid of the child who died and from one of the survivors were identified as *Naegleria*, but it now seems possible that the ameba in both cases was a species of *Acanthamoeba*, and that the effective drug was the sulfadiazine, not the amphotericin B (Griffin, 1978). Sulfa drugs are known to be effective against *Acanthamoeba* infections in experimental animals (Culbertson, 1971). Therapy of amebic keratitis and uveitis likewise remains problematical (Key et al., 1980). Cysts of *A. polyphaga* are extremely resistant and have been seen repeatedly and abundantly in corneal scrapings and sections from patients under treatment; it seems probable, therefore, that effective treatment will require eradication of trophozoites and at the same time, require prolonged inhibition of excystation. Numerous drugs, including sulfadiazine, sulfacetamide, and sulfamethoxazole, were given with little or no benefit in two cases of amebic keratitis. Observations on cases of amebic keratitis and laboratory studies have suggested the possible effectiveness of sulfonamides, hydroxystilbamidine isethionate, fluocytosine, paromomycin, and other antimicrobial agents. Clinical observations suggest beneficial effects of natamycin (primaricin) and rifampin, and in a few cases, keratoplasty has been successful (Key et al., 1980; Bos et al., 1981).

Prognosis. In view of the demonstrated susceptibility of *Acanthamoeba* species to various drugs *in vitro* and in experimental animals, the outlook in cases diagnosed early would seem to be favorable. This is problematical, however.

Prevention. As the involvement of the central nervous system is secondary to infection elsewhere in the body, preventive measures might be effectively directed toward prompt treatment of lesions in the skin, eye, or the genitourinary or respiratory tract. Since fatal *Acanthamoeba* meningoencephalitis has occurred in chronically ill or debilitated patients or in those whose cell-mediated responses have been impaired as a result of either a chronic disease or its treatment by immunosuppressive methods, caution in the use of immunosuppressive therapy may be indicated, as in diseases caused by other parasites such as *Strongyloides, Pneumocystis,* and *Cryptosporidium.*

REFERENCES

Anderson, K., and Jamieson, A. 1972. Primary amoebic meningoencephalitis. Lancet, *1*:902–903.

———, Jadin, J.B., and Willaert, E. 1973. Primary amoebic meningoencephalitis. Lancet, *1*:672.

Apley, J., Clarke, S.K.R., Roome, A.P.C.H., Sandry, S.A., Saygi, G., Silk, B., and Warhurst, D.C. 1970. Primary amoebic meningoencephalitis in Britain. Br. Med. J., *1*:596–599.

Bhagwandeen, S.B., Path, M.R.C., Carter, R.F., Naik, K.G., and Levitt, D. 1975. A case of hartmanellid amebic meningoencephalitis in Zambia. Am. J. Clin. Pathol., *63*:483–492.

Bohle, A.L., Friedman, T.A., Braunstein, H., and Tomasulo, M. 1978. Rapid diagnosis of primary meningoencephalitis due to *Naegleria:* Detection of organisms with bacterial stains. J. Clin. Pathol., *32*:36–37.

Bos, H.J., Völker-Dieben, H.J.M., and Alphen, C.C. 1981. A case of *Acanthamoeba* keratitis in the Netherlands. Trans. R. Soc. Trop. Med. Hyg., *75*:86–91.

Brass, K. 1973. Meningo-encefalitis amebiasica primaria (por *Naegleria*). Arch. Venezol. Med. Trop. Parasitol. Med., *5*:291–301.

Butt, C.G. 1966. Primary amebic meningoencephalitis. N. Engl. J. Med., *274*:1473–1476.

———, Baro, C., and Knorr, R.W. 1968. *Naegleria* (sp.) identified in amebic encephalitis. Am. J. Clin. Pathol., *50*:568–574.

Callicott, J.H. 1968. Amebic meningoencephalitis due to free-living amebas of the *Hartmannella (Acanthamoeba)-Naegleria* group. Am. J. Clin. Pathol., *49*:84–91.

———, Nelson, E.C., Jones, M.M., Dos Santos, J.G., Utz, J.P., Duma, R.J., and Morison, J.V. 1968. Meningoencephalitis due to pathogenic free-living amoebae: Report of two cases. J.A.M.A., *206*:579–582.

Carter, R.F. 1968. Primary amoebic meningo-encephalitis: Clinical, pathological and epidemiological features of six fatal cases. J. Pathol. Bacteriol., *96*:1–25.

———. 1970. Description of a *Naegleria* sp. isolated from two cases of primary amoebic meningoencephalitis, and of the experimental pathological changes induced by it. J. Pathol., *100*:217–244.

———. 1972. Primary amoebic meningo-encephalitis—an appraisal of present knowledge. Trans. R. Soc. Trop. Med. Hyg., *66*:193–213.

Casemore, D.P. 1977. Free-living amoebae in home dialysis unit. Lancet, *2*:1078.

Centers for Disease Control. 1978. Primary amebic meningoencephalitis—California, Florida, New York. Morbidity & Mortality Weekly Rep., *27*:343–344.

Cerva, L., Novak, K., and Culbertson, C.G. 1968. An outbreak of acute, fatal amebic meningoencephalitis. Am. J. Epidemiol., *88*:436–444.

Cerva, L., Serbus, C., and Skocil, V. 1973. Isolation of limax amoebae from the nasal mucosa of man. Folia Parasitol., *20*:97–103.

Chang, S.-L. 1971. Small free-living amebas: Cultivation, quantitation, identification, classification, pathogenesis, and resistance. Curr. Topics Comp. Pathobiol., *1*:201–254.

———, Healy, G.R., McCabe, L., Shumaker, J.B., and Schultz, M.G. 1975. A strain of pathogenic *Naegleria* isolated from a human nasal swab. Health Lab. Sci., *12*:1–7.

Cleland, P.G., Lawande, R.V., Onyemelukwe, G., and Whittle, H.C. 1982. Chronic amebic meningoencephalitis. Arch. Neurol., *39*:56–57.

Culbertson, C.G. 1971. The pathogenicity of soil amebas. Annu. Rev. Microbiol., *25*:231–254.

———. 1975. Soil ameba infections—specific indirect immunoenzymatic (peroxidase) staining of formalin-fixed paraffin sections. Am. J. Clin. Pathol., *63*:475–482.

———, Ensminger, P.W., and Overton, W.M. 1968. Pathogenic *Naegleria* sp.—study of a strain isolated from human cerebrospinal fluid. J. Protozool., *15*:353–363.

———. 1972. Amebic cellulocutaneous invasion by *Naegleria aerobia* with generalized visceral lesions after subcutaneous inoculation: An experimental study in guinea pigs. Am. J. Clin. Pathol., *57*:375–386.

Culbertson, C.G., and Harper, K. 1977. Immunoperoxidase staining of *E. histolytica* and soil amebas in formalin-fixed tissue. Am. J. Clin. Pathol., *68*:529.

———. 1980. Surface coagglutination with formalinized, stained protein A staphylococci in the immunologic study of three pathogenic amebae. Am. J. Trop. Med. Hyg., *29*:785–794.

Culbertson, C.G., Smith, J.W., Cohen, H.K., and Minner, J.R.

1959. Experimental infection of mice and monkeys by *Acanthamoeba*. Am. J. Pathol., *35*:185–197.

Culbertson, C.G., Smith, J.W., and Minner, J.R. 1958. *Acanthamoeba:* Observations on animal pathogenicity. Science, *127*:1506.

Cursons, R.T.M., Brown, T.J., Bruns, B.J., and Taylor, D.E.M. 1976. Primary amoebic meningoencephalitis contracted in a thermal tributary of the Waikato River—Taupo: A case report. N.Z. Med. J., *84*:479–481.

Cursons, R.T.M., Keys, E.A., Gordon, E.H., Leng, R.H., Havill, J.H., and Hyne, B.E.B. 1979. Primary amoebic meningoencephalitis in an indoor heat-exchange swimming pool. N.Z. Med. J., *89*:330–331.

Darby, C.P., Conradi, S.E., Holbrook, T.W., and Chatellier, C. 1979. Primary amebic meningoencephalitis. Am. J. Dis. Child., *133*:1025–1027.

Das, S.R. 1971. Chemotherapy of experimental amoebic meningoencephalitis in mice infected with *Naegleria aerobia*. Trans. R. Soc. Trop. Med. Hyg., *65*:106–107.

———. 1974. Pathogenicity of flagellate stage of *Naegleria aerobia* and its bearing on the epidemiology of exogenous amoebiasis. Ann. Soc. Belge Med. Trop., *54*:327–332.

De Jonckheere, J. 1977. Use of an axenic medium for differentiation between pathogenic and nonpathogenic *Naegleria fowleri* isolates. Appl. Environ. Microbiol., *33*:751–757.

———, and Van de Voorde, H. 1977. The distribution of *Naegleria fowleri* in man-made thermal waters. Am. J. Trop. Med. Hyg., *26*:10–15.

Derrick, E.H. 1948. A fatal case of generalized amoebiasis due to a protozoon closely resemblng, if not identical with, *Iodamoeba bütschlii*. Trans. R. Soc. Trop. Med. Hyg., *42*:191–198.

Deschiens, R. 1974. Discrimination des meningo-encephalites provoquées par les amibes "limax" et des abces amibiens encéphaliques à *Entamoeba histolytica*. Bull. Soc. Pathol. Exot., *67*:194–199.

Dos Santos, J.G. 1970. Fatal primary amebic meningoencephalitis: A retrospective study in Richmond, Virginia. Am. J. Clin. Pathol., *54*:737–743.

Duma, R.J. 1972. Primary amebic meningoencephalitis. CRC Rev. Clin. Lab. Sci., *3*:163–192.

———, Ferrell, H.W., Nelson, E.C., and Jones, M.M. 1969. Primary amebic meningoencephalitis. N. Engl. J. Med., *281*:1315–1323.

———, Helwig, W.B., and Martinez, A.J. 1978. Meningoencephalitis and brain abscess due to a free-living amoeba. Ann. Intern. Med., *88*:468–473.

———, Rosenblum, W.I., McGehee, R.F., Jones, M.M., and Nelson, E.C., 1971. Primary amoebic meningoencephalitis caused by *Naegleria*—two new cases, response to amphotericin B, and a review. Ann. Intern. Med., *74*:923–931.

Dunnebacke, T.H., and Schuster, F.L. 1974. An infectious agent associated with amebas of the genus *Naegleria*. J. Protozool., *21*:327–329.

———. 1977. The nature of a cytopathogenic material present in amebae of the genus *Naegleria*. Am. J. Trop. Med. Hyg., *26*:412–421.

Edwards, J.H., Griffiths, A.J., and Mullins, J. 1976. Protozoa as sources of antigen in 'humidifier fever.' Nature, *264*:438–439.

Fowler, M., and Carter, R.F. 1965. Acute pyogenic meningitis probably due to *Acanthamoeba* sp.: A preliminary report. Br. Med. J., *2*:740–742.

Griffin, J.L. 1976. *Acanthamoeba* meningoencephalitis in Britain. Br. Med. J., *1*:153.

———. 1978. Pathogenic free-living amoebae, pp. 507–549. In *Parasitic Protozoa,* vol. 2. Edited by J.P. Kreier. New York, Academic Press.

Grundy, R., and Blowers, R. 1970. A case of primary amoebic meningoencephalitis treated with chloroquine. E. Afr. Med. J., *47*:153–158.

Gullett, M.J., Hadley, K., Podemski, B., Pitts, L., and Gelber, R. 1979. Disseminated granulomatous *Acanthamoeba* infection presenting as unusual skin lesion. Am. J. Med., *67*:891–895.

Hecht, R.H., Cohen, A.H., Stoner, J., and Irwin, C. 1972. Primary amebic meningoencephalitis. Calif. Med., *117*:69–73.

Hoffmann, E.O., Garcia, C., Lunseth, J., and Coover, J. 1978. A case of primary amebic meningoencephalitis. Light and electron microscopy, and immunohistologic studies. Am. J. Trop. Med. Hyg., *27*:29–38.

Jadin, J.M., Eschbach, H.L., Verheyen, F., and Willaert, E. 1974. Etudes comparatives des kystes de *Naegleria* et d'*Acanthamoeba*. Ann. Soc. Belg. Med. Trop., *54*:259–264.

———, Hermanne, J., Robyn, G., and Willaert, E. 1971. Trois cas de meningo-encephalite amibienne primitive en Europe occidentale, à Anvers. Bull. Acad. Natl. Med., *155*:232–239.

Jager, B.V., and Stamm, W.P. 1972. Brain abscesses caused by free-living amoeba probably of the genus *Hartmannella* in a patient with Hodgkin's disease. Lancet, *2*:1343–1345.

Jahnes, W.G., Fullmer, M., and Li, C.P. 1957. Free-living amoebae as contaminants in monkey kidney tissue culture. Proc. Soc. Exp. Biol. Med., *96*:484–488.

Jamieson, A., and Anderson, K. 1974. Primary amoebic meningoencephalitis. Lancet, *1*:261.

John, D.T. 1982. Primary amebic meningoencephalitis and the biology of *Naegleria fowleri*. Ann. Rev. Microbiol., *36*:101–123.

Johnson, C.M. 1977. Un caso de meningoencefalitis primaria amibiana. Rev. Med. Panam., *2*:141–144.

Jones, D.B., Visvesvara, G.S., and Robinson, N.M. 1975. *Acanthamoeba polyphaga* keratitis and *Acanthamoeba* uveitis associated with fatal meningoencephalitis. Trans. Ophthalmol. Soc. U.K., *95*:221–232.

Josephson, S.L., Weik, R.R., and John, D.T. 1977. Concanavalin A-induced agglutination of *Naegleria*. Am. J. Trop. Med. Hyg., *26*:856–858.

Kenney, M. 1971. The micro-Kolmer complement fixation test in routine screening for soil ameba infection. Health Lab. Sci., *8*:5–10.

Kernohan, J.W., Magath, T.B., and Schloss, G.T. 1960. Granuloma of brain due to *Endolimax williamsi (Iodamoeba butschlii)*. Arch. Pathol., *70*:576–580.

Key, S.N., Green, W.R., Willaert, E., Stevens, A.R., and Key, S.N. Jr. 1980. Keratitis due to *Acanthamoeba castellani*. A clinico-pathologic case report. Arch. Ophthalmol., *98*:475–479.

Lawande, R.V., Abraham, S.N., John, I., and Egler, L.J. 1979. Recovery of soil amebas from the nasal passages of children during the dusty harmattan period in Zaria. Am. J. Clin. Pathol., *71*:201–203.

———, Macfarlane, J.T., Weir, W.R.C., and Awunor-Renner, C. 1980. A case of primary amebic meningoencephalitis in a Nigerian farmer. Am. J. Trop. Med. Hyg., *29*:21–25.

Lengy, J., Jakovljevich, R., and Talis, B. 1971. Recovery of a hartmanelloid amoeba from a purulent ear discharge. Harefuah, *80*:23–24. (In Hebrew. English summary in Trop. Dis. Bull., *68*:818–819, 1971.)

Levine, N.D., Corliss, J.O., Cox, E.E., Deroux, G., Grain, J., Honigberg, B.M., Leedale, G.C., Loeblich, A.R., Lom, J., Merinfield, L.T., Page, F.C., Poljansky, L., Sprague, V., Vavra, J., and Wallace, F.C. 1980. A newly revised classification of the Protozoa. J. Protozool., *27*:37–58.

Lockey, M.W. 1978. Primary amoebic meningoencephalitis. Laryngoscope, *88*:484–503.

Malhotra, K.K., Kakar, P.N., Pillay, P., Pathak, L.R., and Chuttani, H.K. 1978. Unusual presentation of primary amoebic meningoencephalitis—a serious diagnostic and therapeutic problem. J. Trop. Med. Hyg., *81*:113–115.

Mandal, B.N., Gudex, D.J., Fitchett, M.R., Pullon, D.H.H., Malloch, J.A., David, C.M., and Apthorp, J. 1970. Acute meningoencephalitis due to amoebae of the order *Myxomycetale* (slime mould). N.Z. Med. J., *71*:16–23.

Markowitz, S.M., Martinez, A.J., Duma, R.J., and Shiel, F.O.M. 1974. Myocarditis associated with primary amebic *(Naegleria)* meningoencephalitis. Am. J. Clin. Pathol., *62*:619–628.

———, Sobieski, T., Martinez, A.J., and Duma, R.J. 1978. Experimental *Acanthamoeba* infections in mice pretreated with methylprednisolone or tetracycline. Am. J. Pathol., *93*:733–744.

Martin, J.J. 1974. Neuropathological study of three cases of primary amoebic meningo-encephalitis. Ann. Soc. Belg. Med. Trop., *54*:305–319.

Martinez, A.J. 1977. Amebic meningoencephalitides due to *Naegleria* and *Acanthamoeba*. Proceedings of the International Conference on Amebiasis. Mexico, Instituto Mexicano del Seguro Social, pp. 64–81.

———. 1980. Is *Acanthamoeba* encephalitis an opportunistic infection? Neurology, *30*:567–574.

———. 1982. Acanthamoebiasis and immunosuppression. A case report. J. Neuropathol. Exp. Neurol., *41*:548–557.

———, Duma, R.J., Nelson, E.C., and Moretta, F.L. 1973. Experimental *Naegleria* meningoencephalitis in mice. Penetration of the olfactory mucosal epithelium by *Naegleria* and pathologic changes produced: A light and electron microscope study. Lab. Invest., *29*:121–133.

———, Fultz, D.G., and Armstrong, C. 1976. The value of electron microscopy in the differential diagnosis of amebic meningoencephalitides, pp. 226–227. In *34th Annual Proceedings of the Electron Microscopy Society of America*. Edited by G.W. Bailey. Miami Beach, Florida.

———, Garcia, C.A., Halks-Miller, M., and Arce-Vela, R. 1980. Granulomatous amebic encephalitis presenting as a cerebral mass lesion. Acta Neuropathol., *51*:85–91.

———, Nelson, E.C., Fultz, D.G., and Geeraets, R. 1975. Trophozoites of pathogenic *Naegleria:* Scanning electron microscope study, pp. 652–653. In *33rd Annual Proceedings of the Electron Microscopy Society of America*. Edited by G.W. Bailey. Las Vegas, Nevada.

———, Nelson, E.C., Jones, M.M., Duma, R.J., and Rosenblum, W.I. 1971. Experimental *Naegleria* meningoencephalitis in mice. An electron microscope study. Lab. Invest., *25*:465–475.

McNeill, R.E., and de Moraes-Ruehsen, M. 1978. Ameba trophozoites in cervico-vaginal smear of a patient using an intrauterine device. A case report. Acta Cytol., *22*:91–92.

Nagington, J., Watson, P.G., Playfair, T.J., McGill, J., Jones, B.R., and Steel, A.D. 1974. Amoebic infection of the eye. Lancet, *2*:1537–1540.

Napolitano, J.J., Grillo, R.S., and Gamble, H.R. 1978. Primary amebic meningoencephalitis. Occurrence in New York State. N.Y. State J. Med., *78*:1883–1885.

Neva, F.A. 1970. Amebic meningoencephalitis—a new disease? N. Engl. J. Med., *282*:450–452.

Nicoll, A.M. 1973. Fatal primary amoebic meningoencephalitis. N.Z. Med. J., *78*:108–112.

Page, F.C. 1967. Redefinition of the genus *Acanthamoeba* with description of three species. J. Protozool., 14:709–714.

———. 1974. *Rosculus ithacus* Hawes, 1963 (Amoebida, Flabelluidae) and the amphizoic tendency in amoebae. Acta Protozool., *13*:143–154.

———. 1976. A revised classification of the Gymnamoebia (Protozoa: Sarcodina). Zool. J. Linnean Soc., *58*:61–77.

Pan, N.R., and Ghosh, T.N. 1971. Primary amoebic meningoencephalitis in two Indian children. J. Indian Med. Assoc., *56*:134–137.

Patras, D., and Andujar, J.J. 1966. Meningoencephalitis due to *Hartmanella (Acanthamoeba)*. Am. J. Clin. Pathol., *46*:226–233.

Phillips, B.P. 1974. *Naegleria:* Another pathogenic ameba. Studies in germfree guinea pigs. Am. J. Trop. Med. Hyg., *23*:850–855.

Preston, T.M., and O'Dell, D.S. 1974. Studies on the amoeboid and flagellate stages of *Naegleria*. Ann. Soc. Belg. Med. Trop., *54*:279–286.

Richards, C.S. 1968. Two new species of *Hartmannella* amebae infecting freshwater molluscs. J. Protozool., *15*:651–656.

Ringsted, J., Jager, B.V., Suk, D., and Visvesvara, G.S. 1976. Probable *Acanthamoeba* meningoencephalitis in a Korean child. Am. J. Clin. Pathol., *66*:723–730.

Robert, V.B., and Rorke, L.B. 1973. Primary amebic encephalitis, probably from *Acanthamoeba*. Ann. Intern. Med., *79*:174–179.

Rothrock, J.F., and Buchsbaum, H.W. 1980. Primary amebic meningoencephalitis. J.A.M.A., *243*:2329–2330.

Sawyer, T.K., and Griffin, J.L. 1975. A proposed new family, Acanthamoebidae n. fam. (Order Amoebida), for certain cyst-forming filose amoebae. Trans. Am. Microsc. Soc., *94*:93–98.

———, Visvesvara, G.S., and Harke, B.A. 1977. Pathogenic amoebas from brackish and ocean sediments, with a description of *Acanthamoeba hatchetti*, n. sp. Science, 196:1324–1325.

Schuster, F.L. 1975. Ultrastructure of cysts of *Naegleria* spp.: A comparative study. J. Protozool., *22*:352–359.

Singh, B.N., and Das, S.R. 1970. Studies on pathogenic and nonpathogenic small free-living amoebae and the bearing of nuclear division on the classification of the order Amoebida. Philos. Trans. R. Soc. Lond., *259B*:435–476.

Sotelo-Avila, C., Taylor, F.M., and Ewing, C.W. 1974. Primary amebic meningoencephalitis in a healthy 7-year-old boy. J. Pediatr., *85*:131–136.

Spencer, H., et al. (editors). 1973. *Tropical Pathology*. New York, Springer-Verlag.

Stamm, W.P. 1974. The staining of free-living amoebae by indirect immunofluorescence. Ann. Soc. Belg. Med. Trop., *54*:321–326.

Stevens, A.R., Kilpatrick, T., Willaert, E., and Capron, A. 1977. Serologic analyses of cell-surface antigens of *Acanthamoeba* spp. with plasma membrane antisera. J. Protozool., *24*:316–324.

———, Shulman, S.T., Lansen, T.A., Cichon, M.J., and Willaert, E. 1981. Primary amoebic meningoencephalitis. A report of two cases and antibiotic and immunologic studies. J. Infect. Dis., *143*:193–199.

Thong, Y.H., Rowan-Kelly, B., and Ferrante, A. 1979. Treatment of experimental *Naegleria* meningoencephalitis with a combination of amphotericin B and rifamycin. Scand. J. Infect. Dis., *11*:151–153.

———, and Shepherd, C. 1978. Synergism between tetracycline and amphotericin B in experimental amoebic meningoencephalitis. Med. J. Aust., *1*:663–664.

Visvesvara, G.S., and Balamuth, W. 1975. Comparative studies on related free-living and pathogenic amebae with special reference to *Acanthamoeba*. J. Protozool., *22*:245–256.

———, Jones, D.B., and Robinson, N.M. 1975. Isolation, identification, and biological characterization of *Acanthamoeba polyphaga* from a human eye. Am. J. Trop. Med. Hyg., *24*:784–790.

Wang, S.S., and Feldman, H.A. 1967. Isolation of *Hartmannella* species from human throats. N. Engl. J. Med., *277*:1174–1179.

Warhurst, D.C., Stamm, W.P., and Phillips, E.A. 1976. *Acanthamoeba* from a new case of corneal ulcer. Trans. R. Soc. Trop. Med. Hyg., *70*:279.

———, and Thomas, S.C. 1975. *Acanthamoeba* spp. from corneal ulcers. J. Protozool., *22*:57A.

Willaert, E. 1971. Isolement et culture *in vitro* des amibes du genre *Naegleria*. Ann. Soc. Belg. Med. Trop., *51*:701–708.

———, and Stamm, W.P. 1977. Meningoencephalitis due to *Acanthamoeba* sp. Pathogenesis and clinico-pathological study. Acta Neuropathol., *37*:183–191.

———, and Stevens, A.K. 1976. Isolation of pathogenic amoeba from thermal-discharge water. Lancet, *2*:741.

———, Stevens, A.K., and Healy, G.R. 1978. Retrospective identification of *Acanthamoeba culbertsoni* in a case of amoebic meningoencephalitis. J. Clin. Pathol., *31*:717–720.

Young, M.D., Willaert, E., Neal, F.C., Simpson, C.F., and Stevens, A.R. 1980. Experimental infection of sheep with *Naegleria fowleri* of human origin. Am. J. Trop. Med. Hyg., *29*:476–477.

Chapter 13

Coccidia, Microsporidia, and *Pneumocystis*

The Coccidia are classified as members of the class Sporozoea, phylum Apicomplexa; the Microsporidia are classified separately, as phylum Microspora. The taxonomic position of *Pneumocystis* is undetermined.

The phylum Apicomplexa includes parasitic protozoa that live in the body fluids or tissues of the host. It takes its name from the apical complex that is generally present at some stage and consists of elements that are visible with the electron microscope—polar ring, rhoptries, micronemes, conoid, and subpellicular microtubules (Fig. 14–3). Micropores generally are present at some stage; cilia are absent.

Most of the Apicomplexa that are parasites of man belong to the class Sporozoea, in which the life cycle is characterized by an alternation of generations, one sexual and one asexual, occurring in the same host, or requiring an alternation of hosts. In the asexual reproductive cycle of development, multiplication is by *schizogony (vide infra),* while in the asexual cycle, it is by *sporogony,* which involves fertilization or syngamy and produces oocysts containing sporozoites, the infective forms. Locomotion of the mature organisms is by body flexion, gliding, or undulation of longitudinal ridges; flagella are present only on the microgametes of some groups; pseudopodia, if present, are used for feeding, not for locomotion. The term "sporozoa," which is now obsolete as a taxon, was formerly used to include many of those forms now classified as Apicomplexa, which are characterized by the production of spores. It is still used loosely in that sense.

Classification. Formerly classified with the Sporozoa is a group of intracellular parasites, the Microsporidia (Microspora), which, as the name implies, characteristically produces minute spores. Microsporidia are common parasites of arthropods and fishes, and one genus, *Nosema,* rarely occurs in man.

Species of the above groups that parasitize man may be classified as follows:

- Phylum Apicomplexa Levine, 1970
  - Class Sporozoea Leuckart, 1879
    - Subclass Coccidia Leuckart, 1879
      - Order Eucoccidiida Léger and Duboscq, 1910
        - Suborder Eimeriina Léger, 1911
          - Family Cryptosporiidae Léger, 1911
            - Genus *Cryptosporidium* Tyzzer, 1907
          - Family Eimeriidae Minchin, 1903
            - Genus *Isospora* Schneider, 1881
          - Family Sarcocystidae Poche, 1913
            - Genus *Sarcocystis* Lankester, 1882
            - Genus *Toxoplasma* Nicolle and Manceaux, 1909
        - Suborder Haemosporina Danilewsky, 1885
          - Family Plasmodiidae Mesnil, 1908
            - Genus *Plasmodium* Marchiafava and Celli, 1885
    - Subclass Piroplasmia Levine, 1961
      - Order Piroplasmida Wenyon, 1926
        - Family Babesiidae Poche, 1913
          - Genus *Babesia* Starcovici, 1893
          - Genus *Entopolypoides* Mayer, 1933
- Phylum Microspora Sprague, 1977
  - Class Microsporea Delphy, 1963
    - Order Microsporida Balbiani, 1882
      - Family Nosematidae Labbé, 1899
        - Genus *Nosema* Nageli, 1857
- *Incertae sedis*
  - Genus *Pneumocystis* Delanoë and Delanoë, 1912.

Considered here are the Eimeriina of the subclass Coccidia (Cryptosporiidae, Eimeriidae, and Sarcocystidae), the microsporidan *Nosema,* and *Pneumocystis.* The malaria parasites (Haemosporina) and piroplasms (Piroplasmia) will be discussed in the following chapter.

THE COCCIDIA

The term "coccidia," which was formerly more restricted in meaning, is now somewhat synonymous with Eimeriina. However, it is still used to refer primarily to the species of the genera *Eimeria* and *Isospora,* family Eimeriidae. By the present grouping, the coccidia that cause disease in man are species of the genera *Cryptosporidium, Isospora, Sarcocystis,* and *Toxoplasma*. Electron microscopy and life-cycle studies on these forms have revealed new morphological features, and this information has led to new insights into taxonomic relationships. Around 1970, it became known that *Toxoplasma* and *Sarcocystis* species, earlier recognized only in the asexual tissue cyst stage, have a coccidia-like sexual stage of development in the intestine of cats and other predatory animals, thus demonstrating that these and certain related forms should be joined with *Eimeria* and *Isospora* to constitute the suborder Eimeriina. *Cryptosporidium* is also placed in the Eimeriina, but it is given a separate family designation because the sporocyst stage of development is omitted in its life cycle.

In the asexual reproductive cycle (Fig. 13–1) of the coccidia, the infective stage, which is either a merozoite or sporozoite, enters a host cell, where, as a trophozoite, it grows to a certain size that is characteristic for each species. It then reproduces by multiple division, in which the nucleus first divides into several small nuclei, followed by division of the cytoplasm into a portion for each individual nucleus. The reproducing cell is called a segmenter or *schizont,* and the reproductive process is known as *schizogony.* The fusiform, motile daughter cells produced by schizogony, called "merozoites," are set free by disintegration of the schizont. The merozoites then invade other host cells, and in this way, the schizogonic cycle may be repeated through two or more generations. Eventually some of the merozoites become differentiated into male cells (microgametocytes) and female cells (macrogametocytes), and the sexual cycle of development, called "sporogony," is initiated. The nucleus of the microgametocyte divides repeatedly to produce several to many daughter nuclei, each of which becomes a motile microgamete. At the same time, the macrogametocyte becomes a mature macrogamete that is ready for fertilization. When the microgamete enters the macrogamete and initiates fertilization, the resulting zygote secretes a cyst wall around itself and is then referred to as an *oocyst*. Developing in the oocyst are one or more *sporoblasts,* which produce *sporocysts,* and these in turn produce *sporozoites*. For each subgroup of the Eimeriina, there is a characteristic number of sporoblasts, sporocysts, and sporozoites. The transfer stage is either the oocyst or sporocyst, being first shed from the intestinal wall into the fecal stream, then from the infected host to the environment in the feces. Sporozoites in some families are infective when passed in the feces; in others, essential development occurs outside the body and requires special environmental conditions.

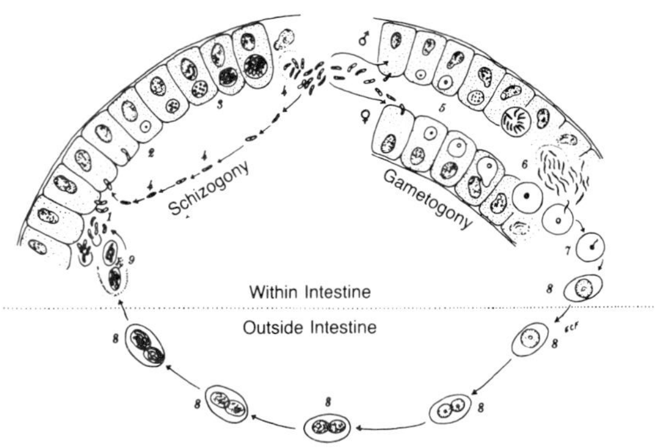

Fig. 13–1. *Isospora belli.* Diagram of life cycle. *1,* Sporozoites entering intestinal mucosa. *2,* Young trophozoite in mucosal epithelial cell. *3,* Development of a schizont. *4,* Merozoites shed from disrupted cell. *5,* Gametocytes, male (above) and female (below), developed from merozoites. *6,* Gametes. *7,* Union of gametes to form zygote. *8,* Maturing stages of sporoblasts and sporocysts in oocyst. *9,* Sporocysts shed from oocyst. (By E.C. Faust.)

FAMILY CRYPTOSPORIIDAE

Léger, 1911

In the coccidian family Cryptosporiidae, the life cycle stages in both asexual and sexual cycles develop just under the surface membrane or within the brush-border of the host cell. The sporocyst stage is omitted, and the sporozoites (four) develop free in the oocyst. Development is complete in one host.

Genus *Cryptosporidium*

Tyzzer, 1907

Historical and Geographical Notes. The first described species of *Cryptosporidium, C. muris,* was discovered by Tyzzer, in 1907, in the gastric crypts of a laboratory mouse. Because all stages of the parasite are minute (about 5 μm or less), it is probable that this and other species of *Cryptosporidium* frequently have been overlooked both before and since 1907. The parasite has been reported from various parts of the world in chickens, turkeys, mice, guinea pigs, horses, pigs, calves, sheep, and rhesus monkeys (Vetterling et al., 1971; Barker and Carbonell, 1974). By early 1982, eight cases of human infection had been reported, three from the United States (Meisel et al., 1976; Nime et al., 1976; Reese et al., 1982), and five from England or unspecified locations (Bird et al., 1980; Bird, 1981). There is sharp disagreement regarding host specificity in the *Cryptosporidium* species. Although transmission from calf to man and to rats and mice was described in one recent report, (Reese et al., 1982), and the experimental transmission from calf to pigs and to specific pathogen-free rats, mice, and lambs was described in another (Tzipori et al., 1981), a third report described the parasite observed in man as a new species *(Cryptosporidium garnhami)* "in view of the fact that species of *Cryptosporidium* are known to be host specific . . ." (Bird, 1981).

Morphology, Biology, and Life Cycle. In the life cycle as described in experimental animals, all known developmental stages of the parasite are found in the brush border of the mucosal epithelium of the stomach or intestine. Trophozoites and schizonts, 2 to 5 μm in diameter, are attached to the host-cell membrane (Fig. 13–2,*A*, *B*, *C*). Schizonts produce eight falciform merozoites that are released to form a new schizogonic cycle (Fig. 13–2,*D*) or to initiate a sporozonic cycle by forming macro- and microgametocytes that form mature gametes. The oocyte develops into an oocyst, which on reaching full maturity is 4 to 5 μm long. In

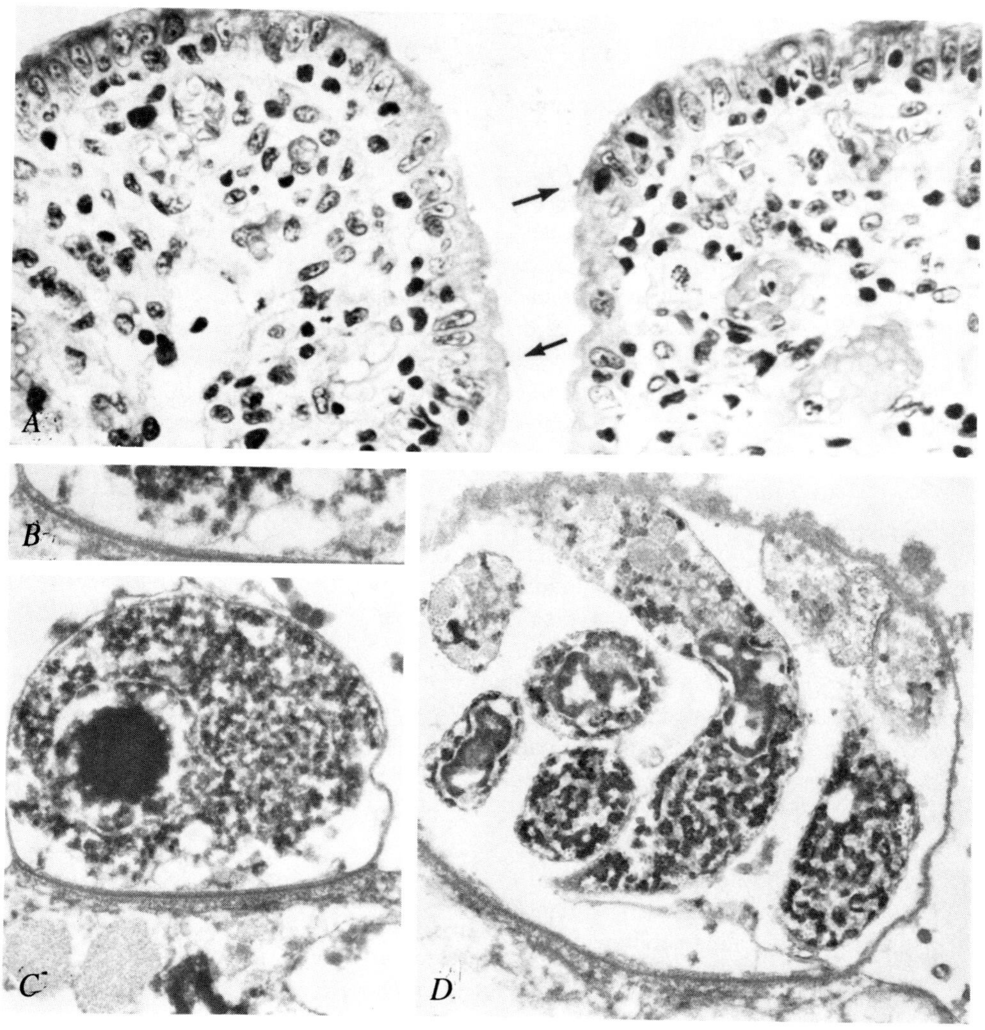

Fig. 13–2. *Cryptosporidium*. Stages in a jejunal biopsy from an immunosuppressed patient with watery diarrhea. *A*, Minute 2- to 4-μm trophozoites attached to surface of abnormal epithelium (arrows) and presence of inflammatory cells in the lamina propria. (× 736.) *B* to *D*, Electron micrographs. *B*, Detail of the attachment zone of a trophozoite on the host cell microvillous membrane. (× 48,000.) *C*, Trophozoite with large nucleus and condensed cytoplasm. (× 22,500.) *D*, Schizont with merozoites. (× 22,500.) (From Meisel, J.L., Perera, D.R., Meligro, C., and Rubin, C.E. 1976. Overwhelming watery diarrhea associated with a *Cryptosporidium* in an immunosuppressed patient. Gastroenterology, *70*:1156–1170. Courtesy of Dr. C.E. Rubin and American Gastroenterological Society.)

some studies, oocysts could not be demonstrated (Vetterling et al., 1971), but in others, well-formed oocysts containing four sporozoites without a sporocyst wall have been described and confirmed by electron microscopy (Barker and Carbonell, 1974). Electron micrographs also have demonstrated that all developmental stages are actually intracellular, because the outer membrane that covers the organism is of host-cell origin (Vetterling et al., 1971b). Until recently, the infective stage of *Cryptosporidium* had not been recognized in the feces or otherwise demonstrated in any host. In experimental infections, host-to-host transmission was accomplished with mucosal scrapings introduced by gavage (Vetterling et al., 1971a). Presumably infective sporozoites in oocysts (not sporocysts) were present in the scrapings.

Pathology, Pathogenesis, and Symptomatology. The first two cases of *Cryptosporidium* infection were reported in 1976—one in a 3-year-old girl in Tennessee (Nime et al., 1976), the other in an immunosuppressed 39-year-old man in Washington State (Meisel et al., 1976). In each case, the infection was accompanied by acute diarrhea. In both instances, the organisms were observed in mucosal biopsies (rectal from the child; jejunal and ileal from the adult). First observed in these cases under the light microscope, the trophozoites, schizonts, and gametocytes were later described in detail with the aid of electron micrographs. Under the light microscope, the organisms were seen as minute round or ovoid bodies, 2 to 4 μm in diameter, lying in the striated border attached to the luminal surface of mucosal epithelial cells, in the crypts of the colon in one case, on the villi of the small intestine in the other (Fig. 13–2,*A*). Electron micrographs showed well-differentiated trophozoites, schizonts, and macro- and microgametocytes, all enveloped in a membrane of host origin and thus, intracellular. Oocysts were not seen in either case. Both of the infections were self-limited. Among the other *Cryptosporidium* infections, one was in an immunodeficient patient and another was in a child with congenital hypogammaglobulinemia; no details regarding the latter were reported (Bird et al., 1980).

In a discussion of the former, four other cases (eight total) were referred to, and it was stated that diarrhea was a prominent feature with all of them (Bird, 1981). This was also true in subsequent cases. Cryptosporidiosis is now recognized as a significant cause of acute diarrhea in man and may be acquired as an occupational disease from domestic animals (Reese et al., 1982). In persons with sound immune functions the illness is self-limited, lasting 1 to 2 weeks (Babb et al., 1982; Fletcher et al., 1982). In immunodeficient persons, including those with congenital deficiency (Sloper et al., 1982), those with an acquired immunodeficiency syndrome (AIDS), and those receiving immunosuppressant drugs, the diarrhea may become chronic.

Treatment. In one case, treatment consisted of discontinuing immunosuppressive therapy (Meisel et al., 1976), while in another the infection disappeared under supportive therapy. In a third case, the enterocolitis disappeared with pyrimethamine and sulphadiazine treatment, but it relapsed when the treatment was withdrawn. No safe and effective chemotherapeutic agent has been identified, but in the patient with normal immune function, since cure is spontaneous, only supportive treatment is necessary.

Prognosis. Limited experience indicates that, with appropriate supportive therapy, cryptosporidiosis can be expected to terminate spontaneously.

Prevention. Until the question of transmission factors has been clarified, preventive measures cannot be identified. However, the infection presumably is fecal borne, acquired from food or water contaminated with the feces of an animal reservoir.

FAMILY EIMERIIDAE

Minchin, 1903

The family Eimeriidae includes coccidia in which all of the asexual stages, and sexual stages through the unsporulated oocyst, develop within the host cell. Oocysts contain sporocysts in numbers characteristic of individual genera. Microgametes have two or three flagella. Sporogony is completed outside the host. The only member of the family that is known to develop in man is *Isospora belli*.

Genus *Isospora*

Schneider, 1881

In species of the genus *Isospora*, the oocysts produce two sporocysts, each with four sporozoites. In addition to *Isospora belli*, which is parasitic in man, there are numerous species of *Isospora* that are parasitic in wild and domestic animals.

Isospora belli

Wenyon, 1923

(Producing human coccidiosis)

Historical and Geographical Notes. Before it was discovered that *Isospora hominis* (Rivolta, 1878), Dobell, 1919, has a life cycle that differs markedly from that of *Isospora belli* (and must now be classified with the Sarcocystidae), the two species were regarded as similar in all respects except the morphology of their oocysts. Numerous proposals have been offered for revising the classification of Eimeriidae and renaming certain species of *Isospora* and related genera (Frenkel et al., 1979). Such proposals are based largely on recent and rapidly expanding information about life cycles, ultrastructure, and host specificity. Also, because the earlier reports frequently failed to indicate clearly whether the species reported was *I. belli* or *I. hominis,* or both, some of the records published before 1975 are confusing.

Isospora belli was named by Wenyon in 1923. He credited Virchow as having first described it, in 1860, in villi of intestinal mucosa obtained at autopsy. Virchow believed he had seen the same species of ''Psorosperms'' in the intestinal villi of a dog. *I. belli* possibly was seen in material obtained at autopsy in two cases reported from Berlin by Eimer in 1870. It was not until the period of 1914 to 1921, however, when it was frequently detected in the feces of military personnel, that *I. belli* became a well-established species. Wenyon and others had seen it often in troops stationed in Egypt, Palestine, and especially in Turkey, as well as other eastern Mediterranean countries. It was this association with the war that suggested to Wenyon the name *I. belli* for the coccidia that he and others had seen often during the war years (Wenyon, 1923). Faust et al. (1961) reported 33 cases of *I. belli* infection in Colombia and tabulated numerous reported cases from all countries of the Western Hemisphere except Canada. It was later reported from Canada (Eaton, 1966). In 1970, notable observations were reported on mucosal biopsies of the small intestine from patients with symptomatic *I. belli* infection.

Morphology, Biology, and Life Cycle. Schizogonic and sporogonic stages in the life cycle of *I. belli* have been described from specimens of intestinal mucosa obtained by peroral biopsy from six patients in one study (Brandborg et al., 1970) and from one in another (Trier et al., 1974).

Development in the intestine is intracellular and occurs in the epithelium of the distal duodenum and proximal jejunum, and presumably at other levels as well (Fig. 13–1). Round trophozoites grow and, by schizogony, produce numerous slender falciform merozoites that, when liberated from the disintegrating schizonts, invade other epithelial cells and undergo further schizogony; by repeated schizogonic cycles, the population of trophozoites, schizonts, and merozoites is expanded. After a minimum period of about 1 week, the sexual cycle (sporogony) starts producing macro- and microgametocytes, the macrogametes are fertilized by microgametes, and the zygote becomes an oocyst that, when shed into the fecal stream, is an oval or flask-shaped cyst (capsule) containing a single slightly elongate sporoblast. Oocysts observed in the fresh fecal specimen are elongate-ovoidal in shape and measure 20 to 33 μm by 10 to 19 μm (Fig. 13–3,*A*). Some are long and slender, while others are somewhat shorter. One end of the oocyst may be narrower, presenting a neck-like appearance. The cyst wall consists of two layers and is smooth, thin, and colorless, the inner layer being membranous, while the outer is hard and relatively impermeable to fluids. Usually, the oocyst in fresh feces contains only one sporoblast, but two may be present. Further development of the sporoblast stage occurs in the feces after passage, forming two sporoblasts (Fig. 13–3,*B*); each sporoblast secretes a cyst wall, thus becoming a sporocyst measuring 12 to 14 μm by 7 to 9 μm. The sporoblast then divides twice to form four sporozoites within

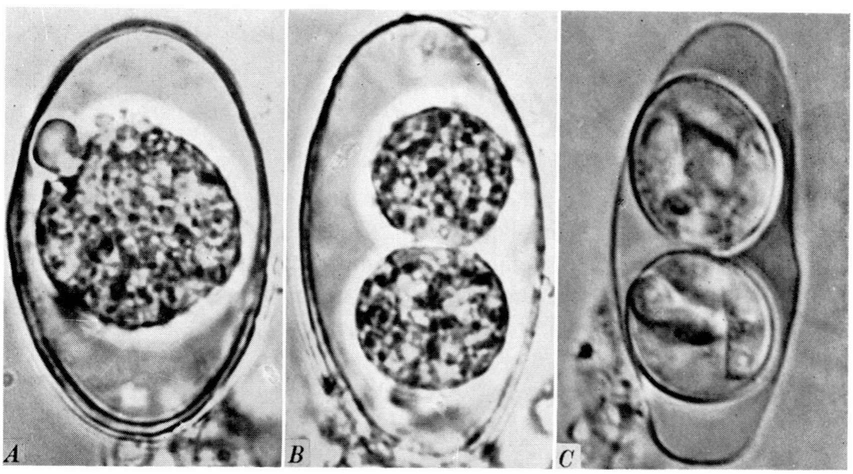

Fig. 13–3. *Isospora belli. A,* Oocyst from freshly passed feces. *B,* Maturing oocyst with two sporoblasts. *C,* Mature oocyst with two sporocysts, each containing four sporozoites. (× ca. 2000.) (Photomicrographs by J.E. Gullberg, from material furnished by Dr. E.K. Markell.)

each sporocyst (Fig. 13–3,*B*). Sporozoites are long, slender, somewhat crescent-shaped bodies, and the four lie clumped together along with a granular residual mass within the sporocyst wall (Fig. 13–3,*C*). The sporocyst is the infective stage that, on being swallowed, excysts in the small intestine releasing the sporozoites, which then penetrate epithelial cells and initiate schizogonic development.

Morphological observations on *I. belli* by light microscopy (Brandborg et al., 1970) and by electron microscopy (Trier et al., 1974) have shown that the schizonts, merozoites, gametocytes, gametes, and oocysts have the same general arrangement of nucleus and other structures, including the apical complex in the merozoite, as that seen in other coccidia. *I. belli* has not been cultivated *in vitro,* although its oocysts mature after evacuation in the feces.

Epidemiology. Human infection with coccidia is less common in temperate climates than in the tropics. Infection is acquired by ingestion of food or drink contaminated with feces containing ripe oocysts. A laboratory worker acquired the infection from an aqueous suspension of feces splashed from a broken flask (Connal, 1922), and two others were infected while attempting to inoculate a rabbit with a suspension of sporocysts (Cook, 1961). Dogs and other mammals have been reported as harboring *I. belli,* but no convincing evidence has been obtained that any animal is a true reservoir of human *Isospora*. On the other hand, the infection appears to be too uncommon in most localities to maintain endemicity. This may only be apparent, because diagnostic methods ordinarily employed in routine fecal examinations frequently fail to reveal the infection when it is present.

Pathogenesis, Pathology, and Symptomatology. A history of diarrhea over periods of several months to 15 years was present in all of the six cases of coccidiosis reported by Brandborg et al. (1970). Four of the patients had lost weight; three of these and one other patient had abdominal colic and fever. Intestinal biopsy showed a flat mucosa in two patients, and tall villi with expanded tips and excessive collagen in the lamina propria or broad villi with focal flattening in others. Various endogenous developmental stages of the parasite were seen in the mucosal epithelium. Plasma cells and lymphocytes were present in increased numbers in the lamina propria, and in some, the presence of numerous eosinophils was a prominent feature. Charcot-Leyden crystals derived from eosinophils are frequently observed in the feces of patients with *I. belli* infection. Two patients with a flat intestinal mucosa died, and coccidia were present at all levels of the small intestine. In a patient with a 7-year history of intermittent diarrhea, documented hypereosinophilia, and probable malabsorption, a series of biopsies showed endogenous stages of the parasite in a markedly abnormal intestinal mucosa with short villi, hypertrophied crypts, and infiltration of the lamina propria with eosinophils, neutrophils, and round cells (Trier et al., 1974). In essentially all symptomatic cases, the main complaint was diarrhea of short or long duration. An infection thought to have been acquired in the South Pacific was believed to have persisted for 26 years (Ravenel et al., 1976). Diarrhea attributed to *I. belli* has been severe in some cases, with up to 20 stools per day for 3 to 4 days intermittently over periods of several months (Syrkis et al., 1975). Nausea and vomiting have accompanied diarrhea in some cases (Miller et al., 1971; McCracken, 1972).

Experimental and laboratory-acquired infections caused abdominal discomfort, low-grade fever, and diarrhea about 1 week after exposure (Matsubayashi and Nozawa, 1948; McCracken, 1972). The illness lasted only 5 to 10 days, but oocysts of *I. belli* were present in the stools for an additional 10 to 20 days. Chronic diarrhea with recurrent episodes of severe diarrhea with dehydration was reported in a patient immunosuppressed for leukemia. Oocysts of *I. belli* continued to be passed in the stools, and symptoms failed to resolve when immunosuppressive therapy was discontinued (Westerman and Christensen, 1979). Apparently, both schizogonic and sporogonic developmental cycles may continue simultaneously for indefinite periods.

The prevalence of *I. belli* infection in asymptomatic carriers is not well known, largely because, in general, the oocysts are either not looked for or not recognized when seen, the technique of fecal examination is unsuitable, or the sampling of the population is inadequate (Faust et al., 1961).

Diagnosis. The demonstration of *I. belli* oocysts in the feces usually is accomplished by examination of unstained or iodine-stained direct smear preparations, or concentrates, of fresh fecal specimens. No permanent staining method is available for this parasite. Oocysts usually occur in small numbers and may be found with difficulty or not at all when infection is readily demonstrated in intestinal biopsy specimens (Brandborg et al., 1970). Oocysts are so transparent that they are apt to be overlooked in direct fecal films, but they are easily concentrated and readily detected following concentration by zinc sulfate centrifugal flotation. Such preparations should be examined with the microscope light reduced to a level at which other protozoa and bacteria stand out with sharp borders. Satisfactory concentration with formalin-ether sedimentation has been reported (Miller et al., 1971). Oocysts can be seen in zinc sulfate concentration preparations when they are examined under low-power magnification.

Treatment. The parasite was eradicated and symptoms relieved in a patient treated with co-trimoxazole (trimethoprim, 80 mg; sulfamethoxazole, 400 mg) in dosages of two tablets every 6 hours for 10 days, then two tablets twice a day for 3 weeks (Westerman and Christensen, 1979). In another patient, 400 mg trimethoprim and 2 g sulfamethoxazole daily for 3 weeks gave prompt relief and eradicated parasites within 1 week (Syrkis et al., 1975). Dramatic results were obtained in a patient treated with pyrimethamine and sulfadiazine at full dosage (75 mg plus 4 g every day) for 3 weeks and at half dosage for an additional 4 weeks (Trier et al., 1974). Primaquine phosphate plus chloroquine phosphate, in one case, and nitrofurantoin, in another, may have been beneficial (Brandborg et al., 1970; Trier et al., 1974). Trials with dithiazanine, tetracycline, metronidazole, phanquone, and quinacrine hydrochloride have failed.

Prognosis. Successful chemotherapeutic trials have been too few to establish any drug as curative. In a high percentage of cases, the infection is self-limited or persists without causing symptoms. Deaths have been reported (Brandborg et al., 1970), but with early diagnosis and treatment, the prospect of cure in most cases is probably favorable.

Prevention. Because the infection is acquired through ingestion of food or drink that is contaminated with feces containing viable oocysts, the preventive measures that are successful against other fecal-borne infections should effectively prevent *I. belli* infection.

FAMILY SARCOCYSTIDAE
Poche, 1913

The family Sarcocystidae includes parasites of known medical importance belonging to two genera—*Sarcocystis* and *Toxoplasma*. Information based on fine structure and life cycles has become available only within the last decade, making it possible to recognize the taxonomic relationships between these forms. This has resulted in the discovery that two developmental forms of the same species of parasite have, in several instances, been placed in two different genera (e.g., *Isospora* and *Sarcocystis,* and *Isospora* and *Toxoplasma*).

In the Sarcocystidae, there is an alteration of a sexual generation of the parasite in the intestinal mucosa of a flesh-eating (carnivorous, omnivorous predator or scavenger) final host, with an asexual generation in the tissues of a prey animal. The transfer stage from final host to intermediate host is a sporocyst containing four sporozoites (Fig. 13–4). In *Toxoplasma,* the sporocysts develop in oocysts after passage in the feces, while in *Sarcocystis,* the sporocysts contain four sporozoites when passed. The transfer stage that is infective to the final host is the merozoite, which is contained (often in massive numbers) in cysts in the muscles or other tissues of the intermediate host.

Genus *Sarcocystis*
Lankester, 1882

Synonyms. *Isospora* Schneider, 1881, in part; *Endorimospora* Tadros and Laarman, 1976; ? *Octoplasma* Dissanaike and Poopalachelvan, 1975.

Host Relationships. Levine and Tadros (1980) listed 93 species of *Sarcocystis;* the intermediate and final hosts of 29 of these species were known. For at least two species—*Sarcocystis hominis* and *Sarcocystis suihominis*—man is the *final* host, and sporogonic development occurs in the lamina propria of the intestinal mucosa. For an undetermined number of unidentified species referred to collectively as *Sarcocystis ''lindemanni,''* man serves as accidental *intermediate* host, and asexual reproduction occurs first in the vascular endothelial cells (one or two generations), then in the skeletal or cardiac muscle fibers. For *S. hominis* and *S. suihominis,* the intermediate hosts are cattle and swine, respectively. For none of the several species that develop in the muscles of man is the final host or the natural intermediate host known (Beaver et al., 1979).

Sarcocystis hominis
(Railliet and Lucet, 1891)
Levine, 1977

Synonyms. *Miescheria cruzi* Hasselmann, 1926, *pro parte; Sarcocystis fusiformis* Railliet, 1897 of Babudieri (1932) and auctores, *pro parte,* (non) *S. fusiformis* Railliet, 1897, Bernard

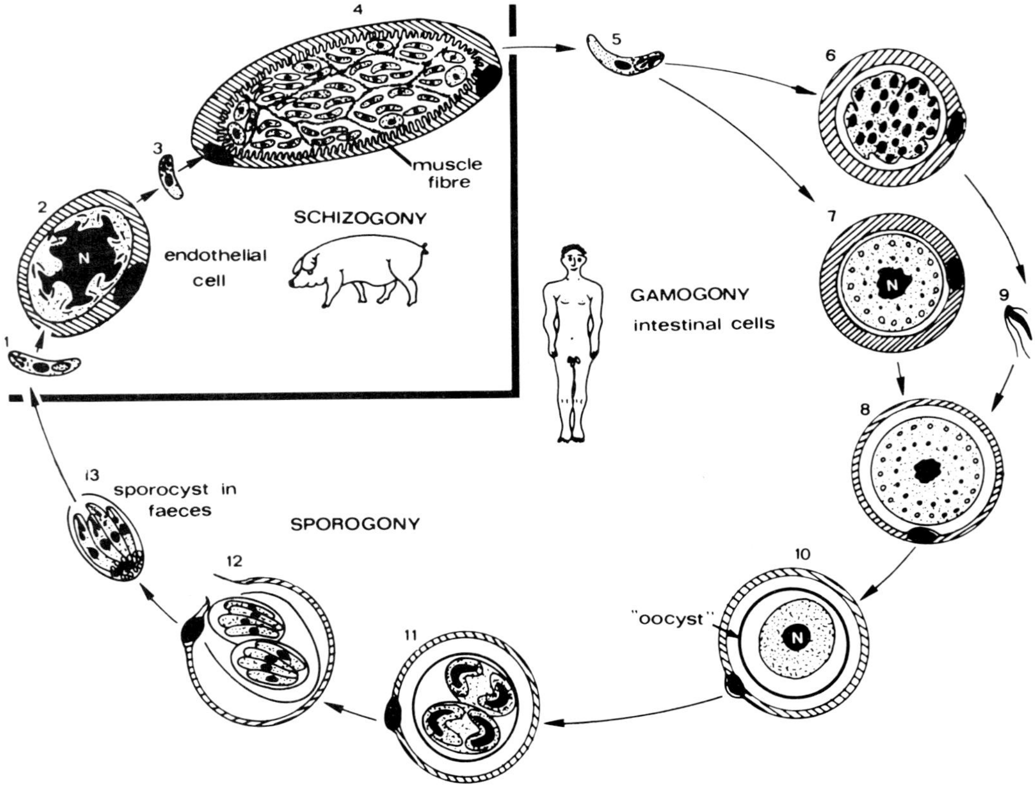

Fig. 13–4. *Sarcocystis suihominis*. Diagram of the life cycle. 1, Sporozoite. 2, Schizont formed from sporozoite within endothelial cell of the pig, giving rise to one or more generations of merozoites by multiple simultaneous division of the nucleus. 3, Merozoite, which enters a muscle fiber. 4, Cyst formed by merozoite within the muscle fiber with production of large round metrocytes and elongate merozoites (also called bradyzoites, or zoites). 5, When man eats raw pork containing cysts, merozoites are set free in the intestine. 6,7, Micro- and macrogamonts formed from merozoites develop in a parasitophorous vacuole within cells of the lamina propria. 8,9, The nonmotile macrogamete (8) is fertilized by a motile microgamete (9) and becomes a zygote. 10, The zygote is surrounded by a wall and becomes an oocyst. 11, Two sporocysts are formed, each developing four sporozoites while still in the oocyst in the host cell. 12, The oocyst wall is broken and the two sporocysts, each containing four sporozoites, are set free and discharged in the feces. 13, When ingested by a pig, the sporocyst liberates its four sporozoites (1), which then find their way to cells of the vascular endothelium (2). (Adapted by P.C. Beaver from Frenkel, J.K., et al. 1979. Sarcocystinae: *Nomina dubia* and available names. Z. Parasitenk., *58*: 115–139.)

and Bauche, 1912; *Coccidium bigeminum* var. *hominis* Railliet and Lucet, 1891 *pro parte; Isospora hominis* (Railliet and Lucet, 1891) Wenyon 1923 *pro parte; Sarcocystis bovihominis* Heydorn, Gestrich, Mehlhorn, and Rommel, 1975.

Historical and Geographical Notes. The small mature sporocysts in human feces, which were referred to as *Isospora* in early reports, are now presumed to have been either *S. hominis* or *S. suihominis*. Virchow, in 1860, may have been the first to record such an observation (Wenyon, 1926). Sporocysts that Reichenow (1925) saw in the stool of a man with amebic dysentery were 10.5 by 16.0 μm. They were probably *S. hominis*, because the maximum length of sporocysts of that species obtained in experimental infections is 17 μm (that of *S. suihominis* is 12.6 μm). Studies reported in 1972 by Rommel and Heydorn and confirmed by others demonstrated that sarcocysts ingested in rare beef produce infections of *S. hominis* in human volunteers, rhesus monkeys *(Macaca rhesus)*, and baboons *(Papio cynocephalus)* (Heydorn et al., 1976; Tadros and Laarman, 1976; Aryeetey and Piekarski, 1976).

Morphology, Biology, and Life Cycle. The sporocysts of *S. hominis* are mature and infective when passed in the feces. Occasionally, they are seen in pairs contained in a delicate, transparent oocyst wall. The sporocysts are broadly ovoid, with a firm, colorless wall, about 15 by 9 μm, and usually single, having been released by rupture of the oocyst wall. The range in size of sporocysts passed in the feces was 13.6 to 16.4 μm by 8.3 to 10.6 μm (Mehlhorn et al., 1975). Mean dimensions, as reported by Rommel and Heydorn (1972), were 14.7 by 9.3 μm, and, as reported by Heydorn et al. (1976), 14.9 by 11.0 μm in monkeys and 14.2 by 9.6 μm in baboons. Four elongate sporozoites with a large nucleus, as well as a coarsely

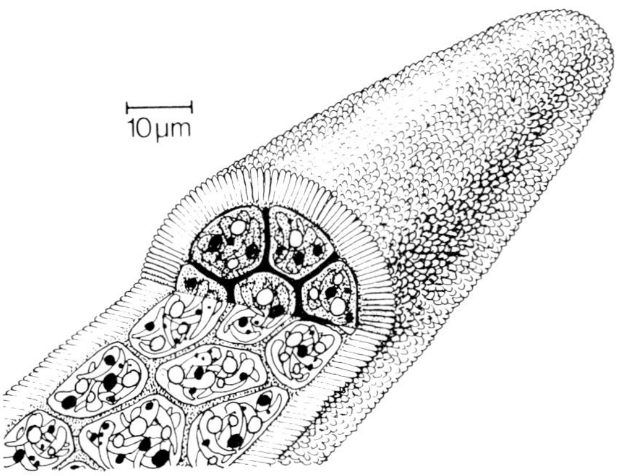

Fig. 13–5. *Sarcocystis hominis*. Diagrammatic representation of the cyst as seen with the light microscope in muscle of a cow. The wall is a thick layer of closely set villi. Inside are compartments formed by septa and filled with zoites (bradyzoites). (From Boch, J., Laupheimer, K.E., and Eber, M. 1978. Drei Sarkosporidienarten bei Schlachtrindern in Süddeutschland. Berl. Munch. Tierarztl. Wschr., *91*:426–431.)

granular residual body, are contained in each sporocyst. In the skeletal muscles of calves that were infected with sporocysts excreted by volunteers, sarcocysts containing large round cells called "metrocytes" were observed at 8 weeks; both metrocytes and merozoites were present at 2 months; and only merozoites were evident at 3 months. Electron micrographs showed that both metrocytes and merozoites multiply by endodyogeny. Countless numbers of merozoites are contained in mature cysts, which, in this species, are spindle-shaped when small, narrowly cylindrical with tapering ends when fully grown (Mehlhorn et al., 1975; Boch et al., 1978). The cyst wall is covered with closely set, thin villi, giving the appearance of being striated and about 7 μm thick (Fig. 13–5). First of all, when ingested by man, the merozoites, which are essentially gamonts, invade the lamina propria of the intestinal mucosa and initiate the formation of gametes. Second, the oocysts contain one and then two sporoblasts, each of which produces a sporocyst containing four sporozoites. This completes, in the human final host, a cycle that began in a bovine intermediate host with the ingestion of viable sporozoites in sporocysts.

Epidemiology. In southern Germany, 99.7% of 1007 slaughtered cattle of various age-groups and of both sexes were found to harbor sarcocysts (Boch et al., 1978). Approximately two-thirds of the infections were *S. hominis* acquired from human fecal contamination of food or water; an equal percentage had acquired infection from dog feces *(S. bovicanis)* and about one-third from cat feces *(S. bovifelis)*. The prevalence, in the human population, of infection that is acquired by eating uncooked beef is unknown. Susceptibility is not universal among people, because only 5 of 12 who ingested raw beef in an experiment passed sporocysts in the feces 18 to 39 days later. Sporocysts were passed in the stools for 9 to 179 days without reinfection, but when the volunteers ate infective raw beef again, the number of sporocysts in the feces was markedly increased (Aryeetey and Piekarski, 1976). Sarcocysts in cattle contain infective merozoites as early as 2 months after ingestion of sporozoites; at 3 months of infection and later, the merozoites are present in countless numbers. In heavily infected animals, the number of individual sarcocysts that may be present are countless also.

Pathogenesis, Pathology, and Symptomatology. The proliferative stages in calves are apparently only slightly pathogenic, and the cyst stage is essentially nonpathogenic (Mehlhorn et al., 1975). Apparently, the sporogonic stages in the human intestine usually are only slightly pathogenic. In one experiment, neither of the two volunteers reported symptoms (Rommel and Heydorn, 1972), and in a later experiment, five infected volunteers reported no symptoms (Aryeetey and Piekarski, 1976). Although 100 g of infected raw beef was ingested daily for 3 to 5 days, the inocula of infective stages (zoites) may have been relatively light. This seems probable, because in the second experiment, 7 of 12 persons who ate the raw beef failed to pass sporocysts; among the infected group, there was wide variation in the prepatent period (18 to 39 days). Six patients in Bangkok who were hospitalized with acute abdominal symptoms diagnosed as segmental eosinophilic enteritis, or segmental necrotizing enteritis, had massive numbers of sporogonic stages of *Sarcocystis* in the lamina propria of resected segments of ileum or jejunum (Bunyaratvej et al., 1982). The species in these cases was believed to be *S. hominis*.

Diagnosis. Infection is detected by finding the characteristic sporocysts in the feces. Concentration methods of fecal examination aid in the search.

Prognosis. Observed infections have been both self-limited and nonsymptomatic.

Prevention. Cattle can be protected from *S. hominis* infection by proper disposal of human feces. People can protect themselves by not eating uncooked beef.

Sarcocystis suihominis Heydorn, 1977

Synonyms. *Psorospermium* sp. Virchow, 1860, *pro parte; Synchytrium mieschcrianum* Kuhn, 1865, *pro parte; Sarcocystis miescheriana* (Kuhn, 1865) Lankester, 1882, *pro parte; Coccidium bigeminum* var. *hominis* Railliet and Lucet, 1891, *pro parte; Isospora hominis* (Railliet and Lucet, 1891) Wenyon, 1923 *pro parte; Lucetina hominis* (Railliet and Lucet, 1891) Henry and Leblois, 1923, *pro parte.*

Historical and Geographical Notes. Virchow reported observations on sporocysts of *Sarcocystis* in human feces in 1861, but it is not known whether the species was *S. suihominis* or *S. hominis* (Wenyon, 1926). In 1865, Kuhn described sarcocysts in the muscles of domestic and wild pigs, and until recently, these and the sarcocysts observed by other workers in pigs were thought to be one species that is generally referred to as *S. miescheriana.* Transmission experiments by Rommel and Heydorn (1972) and Rommel et al. (1974) showed that pigs are infected with more than one species of sarcocysts; for one of them—*S. suihominis* Heydorn, 1977—man is the final host. Further observations on experimental infections in human and porcine hosts (Mehlhorn and Heydorn, 1977; Heydorn and Mehlhorn, 1978; Piekarski et al., 1978), in swine and nonhuman primates (Fayer et al., 1979), and in human tissue cultures (Mehlhorn and Heydorn, 1979) have provided detailed descriptions of the life-cycle stages and their pathogenicity. The geographic range is known to include central Europe and the Netherlands. An unknown proportion of the infections reported as *Isospora hominis* in other regions were *S. suihominis.* In two cases reported from the United States (Georgia), the infections can be identified as *S. suihominis* by the size of the sporocysts (Routh et al., 1955).

Morphology, Biology, and Life Cycle. The life cycle resembles that of *S. hominis,* except that the intermediate host is the pig (Fig. 13–4). The stage of the parasite found in human feces is the sporocyst, which is ovoid with a firm, transparent wall enclosing four banana-shaped sporozoites; it is usually single, but it may be paired (Fig. 13–6). When paired, the sporozoites are snugly enveloped in a thin, colorless, transparent oocyst wall that is constricted between them. The oocysts and sporocysts of *S. suihominis* resemble those of *S. hominis* except in size. The oocysts and sporocysts of *S. suihominis* are smaller, the mean dimensions of the oocysts being approximately 19 by 13 μm, and the dimensions of the sporocysts, 13 by 10 μm (Heydorn, 1977). Sporocysts ingested by pigs liberate sporozoites in the small intestine. Without multiplication in the intestinal mucosa, sporozoites migrate to the bloodstream and produce two generations of schizonts in the vascular endothelium (Fayer, 1980). Merozoites liberated from schizonts (Fig. 13–7) enter the skeletal and cardiac muscle cells at approximately 1 month of infection and produce cysts that contain zoites at 2 months and are mature at 3 to 4 months. At 142 days, mature cysts are spindle-shaped and up to 1.5 mm long, with an interior divided into compartments and a wall that is up to 8 μm thick and striated. Infective zoites (merozoites or bradyzoites) in the cysts are banana-shaped and relatively large (4.5 × 15 μm). When ingested by human volunteers, these forms initiate the sporogonic cycle directly; sporocysts appear in the feces in about 10 days (Rommel and Heydorn, 1972) or 11 to 13 days (Piekarski et al., 1978) and continue for at least 2 months. The prepatent period in the chimpanzee and rhesus and

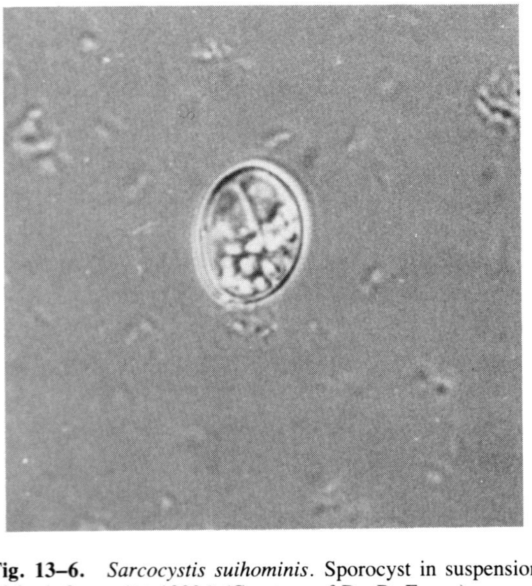

Fig. 13–6. *Sarcocystis suihominis.* Sporocyst in suspension of fresh feces. (× 1200.) (Courtesy of Dr. R. Fayer.)

Fig. 13–7. *Sarcocystis suihominis.* Merozoite in blood smear from pig 14 days after oral inoculation with sporocysts. (× ca. 1600.) (Courtesy of Dr. R. Fayer.)

cynomolgus monkeys was found to be 13 to 15 days, and sporocysts were shed for a minimum of 30 days (Fayer et al., 1979).

Pathology, Pathogenesis, and Symptomatology. One of three infected volunteers experienced mild fever and diarrhea (Rommel and Heydorn, 1972). In another group of eight medical students, however, all suffered from acute diarrhea and vomiting with chills and perspiration 6 to 24 hours after eating raw pork from an experimentally infected pig; the symptoms decreased within 12 to 24 hours. No effects of the infection were observed in the chimpanzee or monkeys. The proliferative stages (schizonts) in the vascular endothelial cells of heavily infected pigs were highly pathogenic, causing death in at least 5 of 7 young pigs on the fifteenth to seventeenth day after swallowing 1 to 5 million sporocysts (Heydorn and Mehlhorn, 1978).

Epidemiology. Natural infections of *S. suihominis* cysts in the muscles of pigs have been reported from Germany (Rommel and Heydorn, 1972), and *Sarcocystis* sporocysts have been observed in fecal samples from persons presumed to have acquired the infection locally in the Netherlands, Mediterranean region, Egypt, Sudan, West Africa, Indonesia, and South America. Some of these records doubtless include *S. hominis* as well as *S. suihominis* (reported together as *Isospora hominis*); the existence of the two species was unknown when the records were reported (Laarman, 1968). Sporocysts may be excreted in the feces for several weeks or months. Sarcocysts persist in the muscles of pigs for at least 7 months (Mehlhorn and Heydorn, 1977), probably for 1 or 2 years.

Diagnosis. Infection in man is detected by demonstrating the presence of sporocysts, or occasionally oocysts, in the feces. The method most frequently reported to be effective in fecal examination for sporocysts is zinc sulfate centrifugal flotation (Laarman, 1968). Examination of several specimens may be required, since the sporocysts may be scanty in the feces. Serologic tests on infected volunteers by an indirect immunofluorescence technique showed high titers at the fourteenth and forty-third days of infection, with a decrease in titer on the seventy-first day (Piekarski et al., 1978). The presence of sarcocysts in cardiac and skeletal muscles of pigs can be detected by microscopic examination of small samples pressed between glass or by digestion, as for *Trichinella*, but species differentiation would be difficult.

Treatment. No specific treatment is known for either sporogonic or schizogonic infection.

Prognosis. Infection is self-limited.

Prevention. Cooking of pork will prevent intestinal infection in man. Protection against human fecal contamination of food and water will prevent infection in pigs.

Sarcocystis "lindemanni"
(Producing human sarcocystosis or sarcosporidiosis)

Historical and Geographical Notes. Before the relationship of *Sarcocystis* to the Coccidia was known, the sarcocysts observed in muscle fibers of vertebrates were classified as Sarcosporidia, a subgroup of the Sporozoa, although some authorities regarded them as fungi. All of the recognized species were placed in the genus *Sarcocystis* as defined by Lankester in 1882, and all *Sarcocystis* species described and reported from man, though widely different, were identified as *Sarcocystis lindemanni* Rivolta, 1878. However, the organisms that Lindemann described as "psorospermien" in 1868 and that Rivolta named *Gregarina lindemanni* in 1878 were nonprotozoan organisms having no relationship to Sarcosporidia (Beaver et al., 1979). This was convincingly pointed out in 1894 by Baraban and St. Remy, but it was generally ignored. Similarly, other nonprotozoan or nonparasite objects were erroneously interpreted and reported as sarcosporidia. Four such cases were reported by Rosenberg in 1892, Cone in 1922, Felsenfeld in 1951, and Kremmydas et al. in 1964. In a report from Romania, an encapsulated larval nematode was described as a sarcocyst (Bonciu et al., 1981). Other reports have described organisms that unquestionably were protozoan parasites in or among muscle fibers, but all were proliferative stages presumed to be *Sarcocystis* because they were found in muscle (Lorenz, 1904; Darling, 1909; Price, 1933; Hertig, 1934). With a recent report of five new cases, the number of authentic records of sarcocystosis in humans was brought to 40, 8 of which were reported from Europe, 8 from Southeast Asia, 7 from India, 5 from the United States, 5 from Central and South America, 4 from Africa, 2 from Indonesia, and 1 from China (Beaver et al., 1979). Three of the cases reported from Europe and all of those from the United States probably were acquired in Southeast Asia or India.

Morphology, Biology, and Life History. In the 40 recorded cases of sarcocystosis, the sarcocysts were found in either skeletal or cardiac muscle, not in both. In some instances, a species found in cardiac muscle may also occur in the skeletal muscles, and vice versa. The number of species represented by the 40 case records is unknown. Four distinct types can be recognized among those found in skeletal muscle.

In the part of the life cycle that occurs in man as intermediate host, infection is acquired by the ingestion of sporozoites passed in the feces of a final host. The final host from which humans acquire infection is as yet unknown, but in most cases, it probably is a carnivorous mammal, such as a dog or cat, although a raptorial bird or a reptile would be a possibility. Sporozoites excysting from sporocysts in the intestine find their way to the vascular endothelium where they develop into schizonts and produce one or more generations of merozoites (also called tachyzoites). The merozoites then invade striated muscle cells and produce sarcocysts with the characteristic size and structure of individual species. Young cysts contain relatively large, round cells called "metrocytes," which produce and maintain the cyst wall and generate fusiform merozoites (also called bradyzoites, cystozoites, or zoites) within the cyst. Cysts in skeletal muscle may be spindle-shaped or cylindrical, measure up to several millimeters in length with tapered ends, have diameters ranging from medium (100–200 μm) to large (over 200 μm)—as compared with those of small size (diameter less than 70 μm)—and have rounded ends in cardiac muscle. In some species, the cyst wall, as viewed with the light microscope, is thick and striated, the striation being due to a covering of thickly set villi (also called cytophaneres); in others, the cyst wall is thin and apparently smooth, although electron micrographs may show protrusions of various shapes and sizes. Also, in some species, metrocytes persist in the mature cyst, and septa separate the zoites into compartments; in others, the metrocytes and septa, if present, are inapparent. Zoites are banana-shaped and fall into large, medium, and small sizes, with diameters of 2 to 3 μm, approximately 1.5 μm, and 1 μm, respectively.

Types of Sarcocysts in Skeletal Muscle. Four morphological types of sarcocysts have been observed in human autopsy or biopsy specimens of *skeletal muscle* (Beaver et al., 1979).

Type 1 sarcocysts are large with villi, which give the appearance of a thick striated wall, septa that are usually evident, or compartments indicated by the grouping of zoites. These sarcocysts have metrocytes that are usually evident, especially at the tips, and large zoites that are often sparse in the central area of old cysts (Fig. 13–8).

Cysts of this type were found in a biopsy of skeletal muscle from a man in China, and at autopsy in two cases reported from France and in one from East Africa. The cysts in the East African case closely resemble those of an undescribed species seen in an African monkey, *Cercopithecus talapoin*.

Type 2 sarcocysts are large to medium-sized with a thin, smooth wall, medium-sized zoites that are separated into compartments by evident septa, and central compartments that are often empty. Cysts of this type in skeletal muscle were seen in biopsy specimens from five patients who were then living or had earlier lived in India, and in autopsy or biopsy specimens from two in Africa. A species with the above features *(S. nesbitti)* is found in the rhesus monkey, *Macaca mulatta,* in India.

Type 3 sarcocysts are small to medium sized, have a thin, smooth wall, medium-sized zoites, and septa that are inapparent except by grouping of zoites. Biopsy specimens of skeletal muscle from three residents of India and one each of Southeast Asia and the Sudan contained sarcocysts of this type. These may represent several species, but at least two of the three from India (Fig. 13–9,*A*) apparently are the same as a well-described but unnamed species commonly seen in *Macaca mulatta* from India (Dubin and Wilcox, 1947).

Type 4 sarcocysts are medium sized, and have a thin wall, small zoites that generally are not distinctly grouped, and septa that are often evident in the central area, where zoites are sparse. Of the 13 infections of this type, all were probably, or possibly, acquired in Malaysia, Indonesia, or elsewhere in Southeast Asia. Unidentified sarcocysts indistinguishable from those in this group were described by Kan et al. (1979) from a naturally infected long-tailed macaque, *Macaca fascicularis*. Closely similar forms also occur in two other species of monkeys from Southeast Asia (Beaver et al., 1979). An additional case of type 4 infection was reported in a man in Malaysia (Pathmanathan and Kan, 1981).

Types in Myocardium. Sarcocysts in cardiac muscle, except when located in Purkinje fibers, are small (70 μm in diameter or less). Among 11 reported cardiac infections, three distinct types of sarcocysts were described—three with large, relatively basophilic zoites, 1.5 to 3 μm in diameter (Fig. 13–9,*B*); six with zoites about 1 μm in diameter; and two with zoites that were poorly described but apparently of intermediate size. Two of those with large zoites, which were reported from Brazil and Costa Rica, resemble *Sarcocystis cruzi (S. bovicanis),* a species commonly seen in cardiac muscle of cattle and whose final hosts are the dog, coyote, wolf, red fox, raccoon, and hyena. This species is cosmopolitan in distribution. The other types from human heart apparently represent several different species, each morphologically indistinguishable from a well-known species in animals.

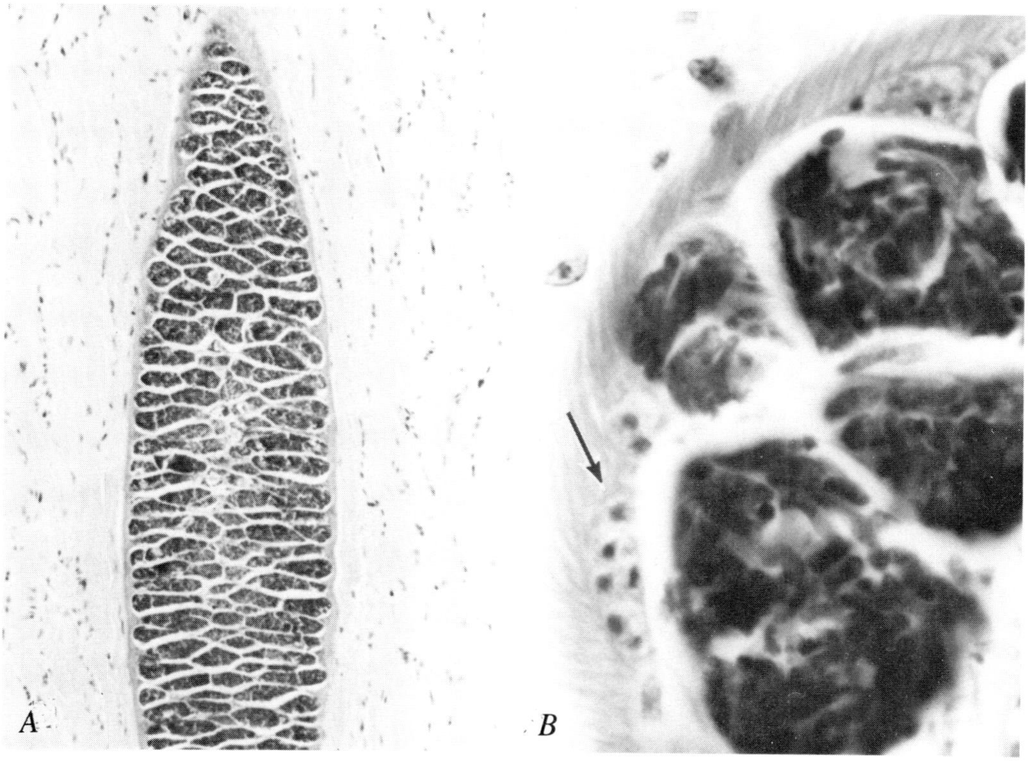

Fig. 13–8. *Sarcocystis.* Cyst in leg muscle of man in Uganda, East Africa. *A,* Longitudinal, slightly oblique section showing a thick cyst wall and many compartments filled with zoites. Hematoxylin and eosin. (× 100.) *B,* Portion of transverse section of cyst showing villi on the outer surface, metrocytes (arrow) adjacent to the wall and large zoites in compartments formed by septa. (× 1000.) (From Beaver, P.C., Gadgil, R.K., and Morera, P. 1979. *Sarcocystis* in man: A review and report of five cases. Am. J. Trop. Med. Hyg., *28:*819–844.)

Epidemiology. The sources of *Sarcocystis* infection in human heart muscle are unknown except that the natural hosts, both final and intermediate, are nonhuman. Whether any of the sarcocysts found in the human heart were or were not also present in skeletal muscle is likewise unknown. In two cases, the skeletal muscles were examined extensively with negative results. It is well established that some species of *Sarcocystis* occur in skeletal muscle only, and some are reported to be found in both cardiac and skeletal muscle. Whether any occur exclusively in cardiac muscle is unknown. Zoites of sarcocysts developing in the two locations probably do not differ in size. Sarcocysts with medium-sized zoites in cardiac muscle were reported from England and Italy, and those with small zoites were seen in New York in immigrants from Puerto Rico, West Indies, and Germany, and in residents of Panama and Brazil.

Pathogenesis, Pathology, and Symptomatology. No inflammation or other evidence of pathogenicity of mature sarcocysts have been reported. Pathological conditions associated with the presence of cysts in some cases apparently were coincidental. However, observations on experimental infections in animals indicate that, in the early, proliferative stage, sarcocystosis is often highly pathogenic, causing massive destruction of the vascular endothelium.

Diagnosis. Sarcocysts in biopsy or autopsy specimens can be identified by their characteristic morphology, which differs from that of *Toxoplasma* by having a more distinct cyst wall and larger zoites. In *Toxoplasma,* almost all of the individual zoite, except its nucleus, is PAS-positive; in *Sarcocystis,* however, areas in the cytoplasm are PAS-negative.

Treatment. There is no known specific treatment for the muscle stage of sarcocystosis.

Prognosis. Prognosis is excellent for infection in the muscles.

Prevention. Protection of uncooked food from contamination with feces of flesh-eating animals is indicated.

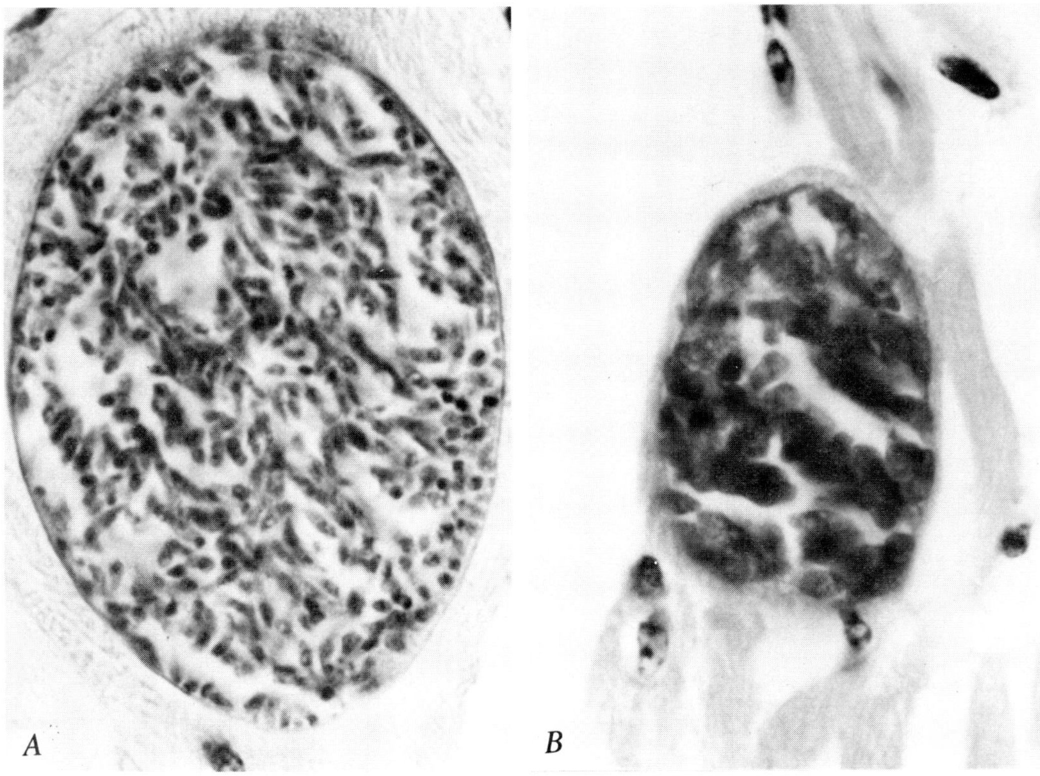

Fig. 13–9. *Sarcocystis* in muscle of man. *A*, Oblique-transverse section of a cyst in a biopsy of muscle from the arm of a man in India. The cyst wall is moderately thin and apparently smooth; the zoites show a geometric arrangement indicating the presence of septa. (× 1000.) *B*, Small cyst with thin, apparently smooth wall and large zoites, in cardiac muscle of a child in Costa Rica. (× 1000.) (From Beaver, P.C., Gadgil, R.K., and Morera, P. 1979. *Sarcocystis* in man: A review and report of five cases. Am. J. Trop. Med. Hyg., *28*:819–844.)

Genus *Toxoplasma*
Nicolle and Manceaux, 1909

The genus *Toxoplasma* was established by Nicolle and Manceaux (1909) for an organism they had discovered in the gundi *(Ctenodactylus gundi)*, a small North African rodent. They named it *Toxoplasma gondii*, the type species of the genus. More than 20 other species of *Toxoplasma* have been described, but according to Levine (1977a, b), only seven are valid, the remainder being synonyms of *T. gondii*. Of the six species other than *T. gondii*, five are found in reptiles and amphibians; the sixth, *T. hammondi*, (= *Hammondia hammondi* Frenkel and Dubey, 1975) is a parasite of the house mouse and cat and was recognized as distinct from *T. gondii* only recently (Frenkel and Dubey, 1975).

In the *Toxoplasma* life cycle, oocysts produce two sporocysts, each with four sporozoites. Sporogony is completed outside the host. The cycle can be either heteroxenous or monoxenous, with the definitive host apparently always a felid and the intermediate host, if any, a bird or mammal. Merogony occurs in both the intermediate and definitive host, and both the merozoites and oocysts can cause infection in either intermediate or definitive hosts. Thin-walled cysts (meronts) without metrocytes are formed in cells of the brain, cardiac muscle, skeletal muscle, and other tissues. Schizogony and sporogony occur in the intestinal cells of the definitive host (Levine, 1977).

Toxoplasma gondii
Nicolle and Manceaux, 1909
(Producing toxoplasmosis)

Historical and Geographical Notes. *Toxoplasma gondii* was discovered by Nicolle and Manceaux, in 1908, in smears of the spleen and liver of a small rodent *(Ctenodactylus gundi)* that had been kept for some time at the Institut Pasteur of Tunis. In retrospect, it is not unlikely that the infection was acquired from cats also housed at the Institut (Wenyon, 1926). At first, the organism was thought to be a species of *Leishmania*, but a year later, after further study, it was recognized as distinct, and the new genus *Toxoplasma* was founded for it. The species was soon found to have a cosmopolitan distribution, with records of infection in over 200 species of birds and mammals, including man. Human infections have been reported from numerous locations in the United States and from all major regions of the world. It was observed by Hutchison in 1965—and confirmed by others—that when cats were fed *Toxoplasma*-infected mice, the infection could be transmitted back to mice through the cat's feces even after storage in water up to a year or more (Work, 1971). Initially, it was presumed that the infectious *Toxoplasma* organism was carried in the egg of *Toxocara cati*, an ascarid commonly found in cats. It was soon shown, however, that *T. gondii* is a coccidian parasite of the domestic cat, that *Toxocara* had no role in the *Toxoplasma* life cycle, and that transmission from cat to mouse was accomplished by sporocysts passed in the cat's feces (Frenkel et al., 1970; Sheffield and Melton, 1970). It became evident that natural infections in cattle, sheep, and other herbivores were also due to sporocysts from cats (Beverley, 1976).

Morphology, Biology, and Life Cycle. There are five main stages in the life cycle of *T. gondii*. Although all five occur in cats, only two stages are found in man, other mammals, and in birds. These two stages are (1) the intracellular trophozoite or proliferative form (tachyzoite) usually seen during the early acute stage of infection (Fig. 13–10), and (2) an encysted form, which is found during chronic or latent infection (Fig. 13–11,*A*). Reproduction is by endodyogeny, a process of division in which two daughter trophozoites are formed within the parent cell, which disintegrates when the young trophozoites are released. The trophozoites in stained films are crescentic in shape, 4 to 8 μm in length, 2 to 3 μm in breadth, and have one end that is more rounded than the other. Evident in Giemsa-stained preparations of various types of smears are a delicate azure cytoplasm and a reddish, spherical, or ovoid nucleus that is usually nearer the blunt end of the parasite (Fig. 13–10). Electron micrographs reveal a complex system of organelles that clearly demonstrate the taxonomic relationship of this organism to the Apicomplexa. One of these structures is a rather short, truncate, hollow conoid that is located at the more pointed (anterior) end. In histological sections, the trophozoites often appear to be ovoid. The cysts that occur in chronic infection are formed when the parasites multiply and produce a wall within a host cell. The cyst wall is eosinophilic, argyrophilic, and weakly PAS-positive; the organisms within the cyst are strongly PAS-positive. During acute infection, groups of proliferative stages may be seen in a wide range of host-cell types. These have been termed "pseudocysts" and "terminal colonies" and can be differentiated from the true cysts in that the

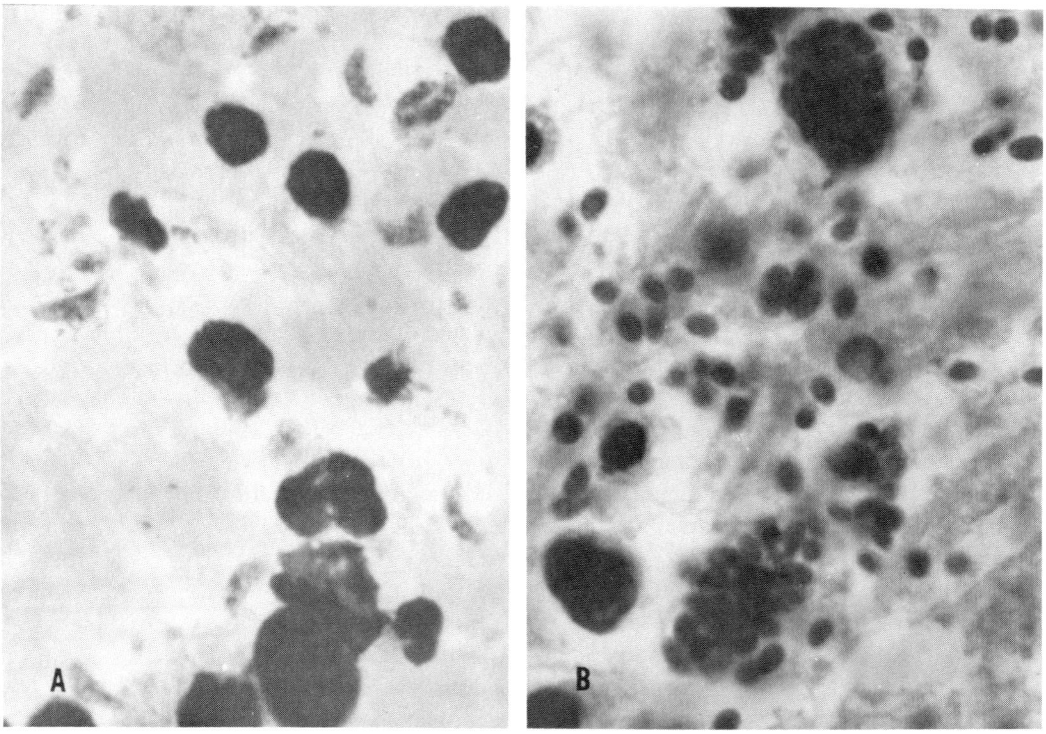

Fig. 13–10. *Toxoplasma gondii*, proliferative forms (tachyzoites). *A*, Crescent-shaped organisms in impression smear of spleen. Giemsa stain. (× 1200.) *B*, Rounded forms, free and in a pseudocyst, in a section of brain of congenitally infected infant. Hematoxylin and eosin stain. (× 1200.) (Courtesy of Dr. R.G. Yaeger.)

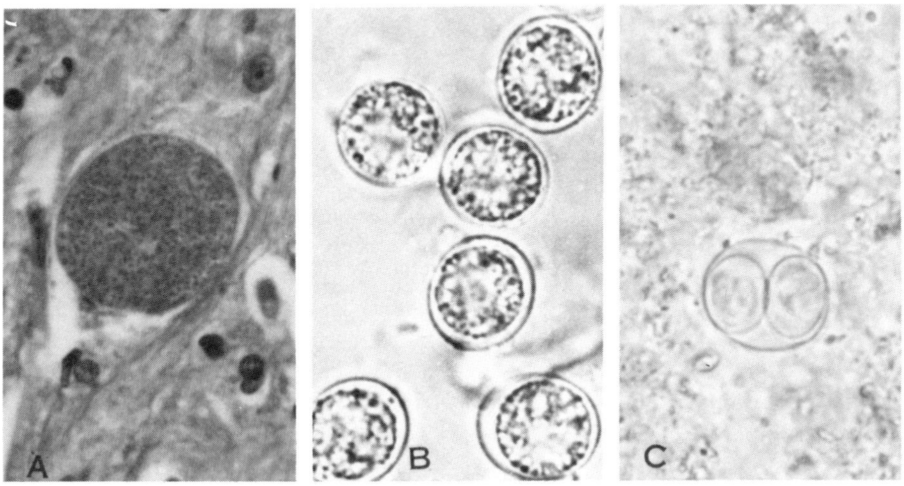

Fig. 13–11. *Toxoplasma gondii*. *A*, Cyst in the brain of a chronically infected mouse. (× 480.) *B*, Oocysts recovered from fresh feces of a cat during the fourth day after eating a chronically infected mouse containing cyst forms in the brain and other tissues. Unstained. (× 1200.) *C*, Oocyst containing two sporocysts, each containing four sporozoites, after 7 days of development in a culture of washed feces from an experimentaly infected cat. Unstained. (× 1200.) (Courtesy of Dr. R.G. Yaeger.)

organisms are at most only slightly PAS-positive and the cyst membrane is neither argyrophilic nor PAS-positive. Frenkel (1973) proposed the term "tachyzoite" for the rapidly dividing, proliferative forms of trophozoite seen in acute infection, and the term "bradyzoite" for the slowly multiplying forms contained in true cysts.

During primary infection for a period of about 2 weeks the cat sheds unsporulated oocysts that measure approximately 10 by 12 μm, contain a sporoblast, and are not infective (Fig. 13–11,*B*). Sporulation at room temperature (20 to 22°C) requires 3 to 4 days; during this time, the primary sporoblast divides into two sporoblasts, and four sporozoites are formed within each of these. The ripe, infective oocyst thus contains two sporocysts, each with four sporozoites (Fig. 13–11,*C*). These sporocysts are relatively tolerant of environmental conditions and may remain infective in the soil for at least 1 year. Ingestion of the infective oocyst by a susceptible bird or mammal initiates an acute infection that usually subsides to a chronic infection, or the initial infection may be relatively mild and unrecognized. During the acute stage, proliferative forms occur in various tissues (Fig. 13–10), but in the chronic phase of the disease, it is the cyst form that is found in the muscles and various other tissues, and is most readily demonstrated in the central nervous system (Fig. 13–11,*A*).

The cycle that occurs in cats includes both asexual multiplication (schizogony) and sexual reproduction (gametogony) within the mucosal epithelial cells of the small intestine (Fig. 13–12). The final product of the sexual phase of the cycle is the fertilized macrogamete or zygote, which then protects itself with a thin but remarkably resistant wall before elimination in the feces as the unsporulated oocyst referred to above. If a susceptible cat ingests sporulated oocysts and develops intestinal infection, the animal will pass oocysts in 21 to 24 days. However, if a cat is fed an acutely ill mouse with proliferative forms of *T. gondii* in its tissues, oocysts will appear in the cat's feces in 9 to 11 days. Finally, when a chronically infected mouse with cysts in its tissues is fed to a cat, oocysts are shed in the cat's feces after only 3 to 5 days. These observations suggest that reproduction in the proliferative stage and in the cyst stage in the tissues of nonfelines that ingest oocysts fulfills an essential part of the life cycle of the parasite; hence, when a cat ingests tissues infected with one or the other of these, the cycle is completed in a much shorter time. It has been reported that the cat may also have an extraintestinal infection—proliferative or cyst forms in various tissues that are infective to animals eating the cat (Dubey, 1977).

Epidemiology. Evidence accumulated during recent years helps to explain the high prevalence of *Toxoplasma* infection. It is estimated that toxoplasmosis exists in chronic asymptomatic form in approximately half of the population in the United States (Krick and Remington, 1978). Prevalence also increases with age, a reflection of continued exposure to the risk of infection either by eating

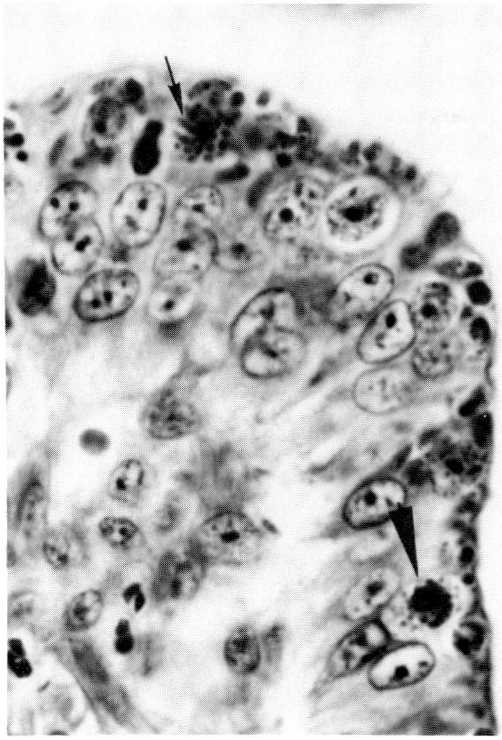

Fig. 13–12. *Toxoplasma gondii*. Intracellular enteric stages in the mucosal epithelium of an experimentally infected cat showing merozoites in mature schizont (small arrow), microgametocyte (large arrow), and macrogametocytes in various stages of development near luminal surface of the host cell. (× 1000.) (By P.C. Beaver. Preparation courtesy of Dr. R.G. Yaeger.)

raw or undercooked meat or through ingestion of oocysts introduced into the environment by cats. Congenital transmission of toxoplasmosis accounts for relatively few of the many infections seen in man, even though such infections possibly account for the majority of acute, fatal infections. Toxoplasmosis acquired by the mother during pregnancy may be fatal to the fetus (2 to 6 cases per 1000 pregnancies in the United States), but a woman does not transmit the infection to her unborn child unless her primary infection occurs during that pregnancy, and it has not been proved that toxoplasmosis is a cause of spontaneous abortion (Krick and Remington, 1978). Blood transfusion, granulocyte transfusion, and organ transplants from donors who are asymptomatic but have latent forms of the parasite in their tissues may transmit infection, especially when the recipients are on immunosuppressive drugs. It is just as likely, however, that the recipient would already have a latent infection, and immunosuppressive therapy would thus produce a toxoplasmosis relapse.

Some of the *Toxoplasma* infections in humans and cats are acquired by eating meat containing viable trophozoites in cysts. The principal mode of infection, however, the kind of transmission necessary to maintain *Toxoplasma* in a country, community or neighborhood, is that resulting from the ingestion of oocysts shed into the environment in the feces of cats. In urban and rural communities surveyed in Costa Rica where the overall antibody prevalence in people was 60% (Frenkel and Ruiz, 1980), over 20% of 237 cats were shedding oocysts when examined, and 60% overall were infected as shown by oocyst shedding or by antibody (Ruiz and Frenkel, 1980).

Oocysts shed into the environment find their way into intermediate hosts, such as mice, rats, and birds, that, when eaten as prey, carry the infection back to cats. In rural and suburban communities, a variable but generally high percentage of cats are infected at least once in their lives, each shedding millions of oocysts into the environment. As the oocysts can survive in moist, shaded soil for more than a year, the soil represents an important source of infection (Ruiz and Frenkel, 1980).

The house mouse and other small rodents that are eaten by cats are important as reservoirs of *Toxoplasma* infection. Among 100 mice *(Mus musculus)* examined in San Jose, Costa Rica (Chinchilla, 1978), *Toxoplasma* antibodies were found in 5 and the parasite was isolated from 2 of these. Antibodies were present in 7 of 23 cats (30.4%). It has become evident that toxoplasmosis is, to a large extent, a soil-transmitted infection that is acquired in the same manner and often at the same time as toxocariasis (visceral larva migrans) (Stagno et al., 1980). Also, as cats in general tend to defecate into water, toxoplasmosis may be a waterborne infection. An outbreak of acute toxoplasmosis among American soldiers in Panama provided epidemiologic evidence that the infections were acquired through drinking water from a jungle stream (Benenson et al., 1982).

Pathogenesis, Pathology, and Symptomatology. The large number of healthy people and animals with serum antibodies to *Toxoplasma* suggests that most infections are relatively asymptomatic (Krick and Remington, 1978; Ruiz and Frenkel, 1980). The most seriously affected among humans is the newborn with infection acquired transplacentally, often during the second or third trimester of fetal development, from a primary infection acquired by a symptomless mother. At birth or shortly thereafter, these infants commonly have evidence of retinochoroiditis, cerebral calcification, and occasionally hydrocephalus or microcephaly; psychomotor disturbances may be detected at this time. Occasionally, there may be no signs of congenitally acquired infection during the neonatal period, but the complications of central nervous system involvement appear several years later.

The most common form of toxoplasmosis acquired postnatally is manifested by lymphadenitis, fever, headache, and myalgia; during the first week or two, splenomegaly and a fleeting erythematous rash may also be present (Jones et al., 1969). A typhus-like exanthematous form of the disease may produce myocarditis, meningoencephalitis, and an atypical pneumonia; death often occurs in such cases. A third, rare form primarily involves the central nervous system, usually with a fatal outcome. A fourth type of noncongenital infection is retinochoroiditis, in which ocular lesions originate in the retina and spread to the choroid. Retinochoroiditis may be severe, requiring enucleation (Rao and Font, 1977).

The most common clinical manifestation in adults is lymphadenopathy, local or generalized, superficial or deep, and the nodes most commonly involved are those of the neck. Only rarely can the parasite be demonstrated in a lymph gland biopsy, and while the histologic appearance is considered diagnostic by some authorities (Dorfman and Remington, 1973), others report that histologic changes are not diagnostic, being primarily a follicular hyperplasia with infiltration of the subcapsular sinus and outer parts of the medullary sinuses with histiocytic cells. At deeper levels, the medullary sinuses contain collections of pale, enlarged histiocytic cells suggestive of sarcoid lesions (Spencer, 1973).

Diagnosis. Occasionally *Toxoplasma* trophozoites can be detected and identified in smears of lymph nodes, bone marrow, spleen, brain, or other material. The organisms are readily identified in impression smears, but in histologic preparations of fixed tissues, they do not exhibit the typical morphology. Body fluids or ground tissues may be inoculated intraperitoneally into young laboratory mice from a *Toxoplasma*-free colony, and after 7 to 10 days, the peritoneal fluid and smears of lung, spleen, and liver examined for the proliferative forms. Serum from the inoculated animals may also be tested for the presence of antibodies. Serologic tests should be done in all cases of suspected toxoplasmosis, and a variety of good tests are now available (e.g., the Sabin-Feldman dye test, the indirect fluorescent antibody test, the indirect hemagglutination test, and the complement-fixation test). Serum titers for the dye, indirect fluorescent antibody, and indirect hemagglutination tests appear earlier, reach much higher levels, and drop much later than do those obtained with the complement-fixation test. A high IgM indirect fluorescent antibody titer or a serial two-tube rise in titer to a high level in any standard test establishes the diagnosis of acute infection. Positive results with other serologic tests may be obtained in chronic asymptomatic infections, and high titers may persist for more than a year. Tests of serum taken at intervals of several weeks can be useful in determining the course of infection.

Treatment. There is no entirely satisfactory treatment for toxoplasmosis. Although clinical improvement or cure has been obtained by the combined use of pyrimethamine and sulfonamides, which act synergistically, there is evidence that the parasite may not be eliminated. It is probable that the resistant cyst forms persist, and these may later initiate an active infection. Specific treatment is not required for otherwise normal patients with acute toxoplasmosis, but it is indicated for those with severe symptoms or active retinochoroiditis and for immunologically compromised patients. In the treatment of active retinochoroiditis, administration of a corticosteroid drug such as prednisone is helpful in reducing the inflammatory process and consequent scarring of the retina. Corticosteroids are otherwise contraindicated in the treatment of toxoplasmosis. Pyrimethamine is not recommended in cases of infection acquired during pregnancy. A combination of clindamycin and sulfadiazine is reported to be an effective regimen for the treatment of ocular toxoplasmosis (Tabbara and O'Connor, 1980), but the potential side effects of clindamycin (such as pseudomembranous colitis) must be weighed against those of pyrimethamine in considering such a regimen.

Prognosis. Acute toxoplasmosis in children is often fatal, although mothers of infected newborns are characteristically without symptoms. Other children born to mothers with inapparent infection subsequent to the birth of a diseased infant will not be at risk of infection. In older children and adults, the prognosis depends on the sites and degree of tissue damage. In many cases, the infection is asymptomatic, but in some nonfatal cases with infection contracted prenatally, there is permanent invalidism.

Prevention. Proper cooking of meat is essential in view of various reports of isolation of *Toxoplasma* from sheep, swine, and cattle. Although storage at deep-freeze temperatures eventually will kill even the more resistant cyst forms (Dubey, 1974), freezing is not as dependable as is adequate heating to destroy the infectivity. The high resistance of oocysts that are shed in cat feces make control difficult. The only place known to be free of *Toxoplasma* infection in man or animals is a Pacific island free of cats (Wallace et al., 1972). Freshly passed oocysts are not infective; hence, there is no danger from feces disposed of before sporulation by flushing down the toilet or by prompt incineration. The usual practice of permitting cats to run outdoors ensures sporulation of the oocysts because of the animal's habit of covering the feces with soil. The ripe oocysts will remain infective for at least a year within a relatively wide range of climatic conditions.

Cats may become infected more than once, but in subsequent infections, oocysts may not be shed. However, it is possible that if a cat's resistance is significantly lowered, it may be reinfected with oocysts from its own feces mixed into the soil months previously. Also, kittens can become infected by ingesting oocysts shed by the parent animal or by kittens from a previous litter. As cats also acquire infection by eating infected rodents, birds, or raw meat, exposure can be reduced by feeding them canned or home-cooked meat or commercially prepared dried food. Anyone coming in contact with soil contaminated with oocysts from cats runs the risk of infection. Protection of children's sand boxes is advisable as cats generally have access to play areas. Special care should be taken when a cat shows signs of an enteric infection. Cats with antibodies in the serum are likely to be resistant to infection and, hence, are safer pets. In view of the various routes of transmission, however, the possibilities for control are limited. Attempts to treat cats to prevent the shedding of oocysts in the feces have been unsuccessful (Dubey and Yeary, 1977).

THE MICROSPORA

In a classification proposed by the Society of Protozoologists' Committee on Systematics and Evolution (Levine et al., 1980), the group formerly referred to as microsporidia are treated as a phylum, designated Microspora, and defined as intracellular protozoan parasites with spores of unicellular origin, in an undivided membrane containing a single ameboid, uni- or binuclear germ discharged via a tubular polar filament. Most of the species of this group are parasites of invertebrates. Among the few that infect mammals, species of only one genus, *Nosema,* have been reported as parasites of man.

Nosema Nageli, 1857 (Producing microsporidiosis or nosematosis)

There have been five reports of finding microsporidian parasites in man, only one of which, according to Sprague (1974), provided adequate evidence to establish the diagnosis, although an encephalitozoon-like infection in a 9-year-old Japanese boy described by Matsubayashi et al. (1959) has been accepted by some authorities. The case accepted as authentic by Sprague was that reported by Margileth et al. (1973) in an immunologically compromised 4-month-old infant in Washington, D.C., who had diarrhea, malabsorption syndrome, and thymic alymphoplasia. Autopsy revealed *Pneumocystis* infection in the lungs and *Nosema* widely disseminated throughout the body, distributed intracellularly in clumps in ganglia and nerve fibers, and in smooth cardiac and skeletal muscles, in the parenchymal cells of the kidneys and liver, and in the muscle cells of arteries. They were identified merely as a species of *Nosema*.

The organisms present in preserved autopsy material were restudied by Sprague (1974), described briefly, and named *Nosema connori*. The spores in preserved material were described as broadly elliptical, 4 to 4.5 by 2 to 2.5 μm, with no internal structure evident. In hematoxylin and eosin- and Giemsa-stained sections, sporoblasts and immature spores, each with two nuclei, were evident, and the spore wall was essentially unstained. In some of the mature spores, the wall was acid-fast, but in some it was not. A notable diagnostic feature was a PAS-positive granule just inside the wall at the polar end of the mature spore (Strano et al., 1976).

PNEUMOCYSTIS

The parasites known as *Pneumocystis* are generally regarded as protozoa belonging to the class Sporozoea, subclass Coccidia, or to an undefined group of uncertain status possibly related to the microsporidia. Takeuchi (1980) observed that, in electronmicrographs, the trophozoites resemble protozoa and the cysts resemble fungi. Based on features seen in electronmicrographs and by freeze-fracture microscopy, Yoneda et al. (1982) concluded that *Pneumocystis carinii* is a protozoan, presumably a Rhizopoda. Although the genus *Pneumocystis* and the species *P. carinii* are well recognized and clearly defined, the status of other species of the genus is unsettled. In describing the

Pneumocystis found in man as new, naming it *P. jiroveci,* Frenkel (1976) noted that, although the *Pneumocystis* from the pulmonary alveoli of humans is serologically nonreactive in the indirect fluorescent antibody test with hyperimmune serum of rats infected with *P. carinii,* the occurrence of *P. carinii* in animals is of concern as a possible source of human infection. Potential animal sources of infection include mice, rats, rabbits, hares, dogs, cats, horses, pigs, and others.

Pneumocystis carinii
Delanoë and Delanoë, 1912
(Producing interstitial plasma cell pneumonia)

Historical and Geographical Notes. *Pneumocystis carinii* was first described by Chagas (1909) in smears of the lungs of guinea pigs infected with *Trypanosoma cruzi.* Carini (1910) found the same bodies in lung tissue of rats infected with *T. lewisi,* and Delanoë and Delanoë (1912) found them in Paris rats and assigned a specific name. Thirty-three years after it was first reported by Chagas (1909) in the guinea pig, *Pneumocystis* was reported in man. In 1942, Van der Meer and Brug described three human cases, two in infants and one in an adult who had nonsyphilitic, atypical interstitial pneumonia, which had been described by Ammich in 1938. Occasional cases were reported thereafter from Central Europe and, beginning in 1947, from the United States and Canada. More recently, human cases have been reported from Australia, Chile, South Africa, Israel, and Melanesia. In Japan, the first case was reported in 1961, and by 1979, the number had expanded to 292 cases (Yoshida and Ikai, 1979). Since 1980, a high incidence of *Pneumocystis* pneumonia has been observed in homosexual males (Waldhorn et al., 1982).

Pneumocystis carinii typically causes an interstitial plasma-cell pneumonia in premature or debilitated babies in the first 3 months of life (Gajdusek, 1976). Weisse and Wedler (1954), however, reported bodies identical with *Pneumocystis* in 23 of 152 persons, aged 5 days to 70 years, who had diverse symptoms but did not have interstitial pneumonia.

As was indicated in the classification of the sporozoa (p. 149), the taxonomic relationship of *Pneumocystis* to other organisms is undetermined. By ultrastructural evidence, Vavra and Kučera (1970) were convinced that *Pneumocystis* is not a protozoan, but rather an atypical fungus with yeast-like characteristics. Therefore, *Pneumocystis* was omitted from the revised classification of the protozoa prepared by Levine et al. (1980).

Morphology, Biology, and Epidemiology. *Pneumocystis carinii* is seen in two forms, trophozoites and cysts, both of which are found in pulmonary alveoli suspended in a matrix of material that, in sections, shows a characteristic honeycomb pattern (Fig. 13–13). The trophozoites are somewhat ameboid, up to 5 μm in diameter, with a nucleus that is evident on Giemsa- or hematoxylin-stained smears or sections. Electron micrographs show filopodia projecting from the surface of the trophozoites and usually one or two mitochondria, free ribosomes, endoplasmic reticulum, and some vacuoles, granules, and globules in the cytoplasm (Frenkel, 1976). The cysts, which are about 5 μm in diameter and have a relatively thick wall when mature, usually contain eight trophozoites, generally referred to as intracystic bodies (sometimes termed sporozoites), each with a single nucleus (Fig. 13–13,*C,D*).

The life cycle usually involves asexual reproduction of free trophozoites. The trophozoites then form precysts and thin-walled cysts that mature with the increasing thickness of the cyst wall and the production of eight trophozoites or sporozoites, which is apparently achieved by asexual cell division. When the intracystic bodies are liberated, the empty cysts collapse into irregular, often crescentic, bodies, and the trophozoites begin a new cycle of reproduction. Although the *Pneumocystis* in human infections, redescribed and named *P. jiroveci* by Frenkel (1976), may be distinct from the *P. carinii* from rats, their basic life cycles probably do not differ. Infection is acquired through contact with an infected host, and it also is naturally acquired through an airborne stage of the parasite (Hughes, 1982).

Based on ultrastructural studies on the *Pneumocystis* stages found in the lungs of young rats with cortisone-induced pneumonia, new aspects of the life cycle have been described. In the life cycle as described by Vossen et al. (1977, 1978), there are both intracellular and extracellular stages of the parasite. In the extracellular development, spherical daughter cells, which are formed by endodyogeny in thin-walled pneumocysts, are liberated by excystation. In thick-walled pneumocysts, eight daughter cells are formed and liberated in the same manner. It is believed that these daughter cells initiate the intracellular phase of the cycle, forming thin-walled pneumocysts inside the alveolar epithelial cells. When mature, these are set free by lysis of the epithelial cell, and the extracellular phase of a new cycle is initiated in the alveolar space. Cell reproduction by endodyogeny, which is a type of internal budding in which two daughter cells are formed within the cytoplasm of the mother cell and then break out, destroying it, also occurs in *Toxoplasma, Sarcocystis,* and some related groups.

Pathogenesis and Pathology. The presence and multiplication of the parasites in the lungs stimulates the exudation of serous fluid and histiocytes, lymphocytes, and plasma cells. Consequently, ventilation is impaired and death occurs due to

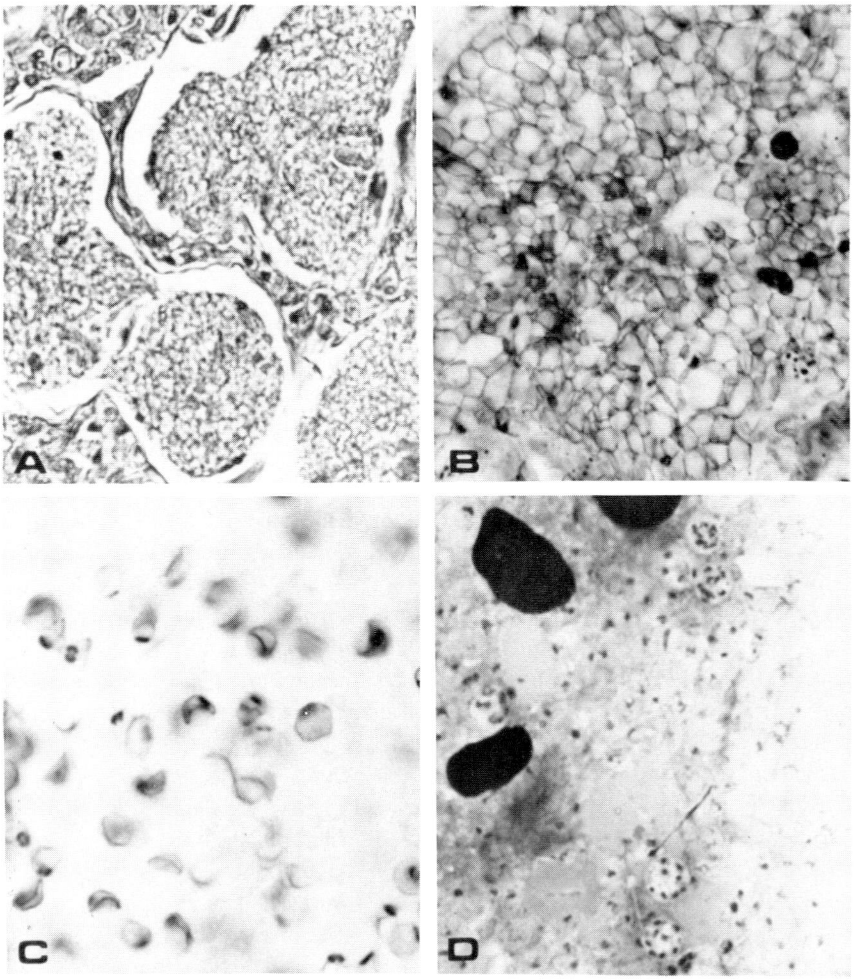

Fig. 13–13. *Pneumocystis carinii* in human lung. *A*, Section showing characteristic honeycomb pattern of the alveolar exudate. H & E stain. (× ca. 400.) *B*, Section prepared with Warthin-Starry stain to emphasize the honeycomb pattern of material in the alveoli. (× 1200.) *C*, Section prepared with Gomori's methenamine-silver stain to show the comma-shaped thickening in the cyst wall of the parasite. (× 1200.) *D*, Impression smear of lung showing several parasites with eight nuclei and numerous small uninucleate forms. Giemsa stain. (× 1200.) (From Faust, E.C., Beaver, P.C., and Jung, R.C. 1975. *Animal Agents and Vectors of Human Disease*, 4th ed. Philadelphia, Lea & Febiger. Courtesy of Dr. R.G. Yaeger.)

respiratory failure. At autopsy, the lungs are increased in weight and volume, and there is moderate thickening of the pleura. The cut surface is gray and shows consolidation and prominent interlobular septa. The interstitial tissues are thickened and heavily infiltrated with mononuclear and plasma cells. The alveolar walls are thickened, and the alveoli are filled with eosinophilic amorphous exudate containing histiocytes, lymphocytes, plasma cells, and the parasitic organisms.

It should be noted that, when pneumocystis pneumonia occurs as a consequence of agammaglobulinemia or immunosuppression, the characteristic exudate may be absent due to lack of B lymphocytes.

Symptomatology. The onset is insidious and the course gradual. The incubation period is about 1 to 2 months. The illness begins with a cough, which is typically nonproductive, and gradually progresses to serious loss of ventilatory capacity, respiratory failure, and cyanosis. The degree of respiratory distress far exceeds the degree of abnormality in clinical findings. The physical signs in the chest are moderately deranged, temperature is normal or only slightly elevated, and the white blood cell count is normal or slightly increased. An eosinophilia may be present.

Chest roentgenographs show symmetrical cloudiness with a ground-glass appearance that spreads bilaterally from the hila. Alternating areas of lobular collapse and emphysema may cause a honeycomb effect. Mediastinal emphysema and pneumothorax sometimes occur.

Diagnosis. In all instances, diagnosis must be based on the demonstration of the organisms in tissues or material from the lungs. Such material may be obtained as an ordinary cough specimen, by nasotracheal suction, by transtracheal needle suction, or by bronchoscopic aspiration (Lau et al., 1976). Aspiration of sputum is more successful if it is preceded by inhalation of acetylcysteine (Frenkel, 1976). Percutaneous pulmonary needle aspiration or biopsy, sometimes performed under fluoroscopic guidance, can provide diagnostic material (Castellino, 1976). However, because this carries a degree of risk of causing pneumothorax or intrapulmonary hemorrhage, open biopsy may be the safest and most sensitive method for obtaining a diagnostic specimen (Rifkind, 1976). *Pneumocystis* can be grown in cell cultures, but such procedures are impractical for diagnosis. Frequently, laboratory rodents are naturally infected; therefore, they cannot be used for diagnosis.

Cysts are best demonstrated by Gomori's methenamine silver stain in tissue sections or in smears fixed with acetone-alcohol. In Giemsa-stained sections or smears, cysts can be recognized by the relatively clear circular areas containing up to eight nuclei; with hematoxylin-eosin staining, three to eight cells stain purple or lavender, and the wall, which is unstained, contrasts with stained elements inside and outside the cyst. Cysts, but not trophozoites, can be recognized in Papanicolaou-stained impression smears, having much the same appearance as when stained with hematoxylin but less well differentiated. For quick screening, toluidine blue O can be used in place of the methenamine silver technique; only the cysts can be identified with these stains (Pintozzi et al., 1979). Fluorescent antisera have been used to detect cysts or trophozoites when scarce (Frenkel, 1976), and a concentration technique described by Yoshida and Ikai (1979) can be used to detect latent or persistent light infection. Among the serologic tests, an immunofluorescence test was found to be the most sensitive, yet it detected only a third of the infections (Kagan and Norman, 1976). However, serial propagation of the parasites in cell line cultures offers promise of application to studies on life cycles and diagnostic methods (Latorre et al., 1977). With antigen prepared from organisms grown in embryonic chick epithelial lung cultures that were used in counterimmunoelectrophoresis and indirect immunofluorescent antibody techniques, Pifer et al. (1978) obtained evidence of *P. carinii* infection in two-thirds of a sample of normal 4-year-old children in Tennessee.

Treatment. Antibiotics are useless. The treatment of choice is a combination of trimethoprim, 20 mg/kg body weight per day, and sulfamethoxazole, 100 mg/kg body weight per day, which is administered orally in four divided doses for 14 days. An alternative drug is pentamidine, 4 mg/kg body weight per day, which is administered intramuscularly for 12 to 14 days. Pentamidine may be obtained from the Parasitic Diseases Division, Centers for Disease Control, Atlanta, Georgia 30333.

Pneumocystosis was reported in a child with chronic granulomatous disease in whom cellular and humoral immunity were found to be normal (Pedersen et al., 1979). The administration of sulfamethoxazole and trimethoprim, along with a series of granulocyte transfusions, resulted in apparent cure.

Supportive measures are usually necessary and include administration of oxygen by intubation and continuous negative pressure-assisted ventilation.

Prognosis. The prognosis is very poor. Most cases are diagnosed at autopsy. When the disease is associated with an immune deficient state, the mortality tends to approach 100%. Mortality may be as low as 10% if the etiology is suspected early, proper diagnostic procedures are carried out, and appropriate treatment is given.

Prevention. There is no preventive measure available. The best that can be hoped for is for the disease to be anticipated in patients undergoing immunosuppressive therapy so that prompt treatment may be given.

REFERENCES

Aryeetey, M.E., and Piekarski, G. 1976. Serologische *Sarcocystis*-Studien an Menschen und Ratten. Z. Parasitenk., *50*:109–124.

Babb, R.R., Differding, J.T., and Trollope, M.L. 1982. Cryptosporidia enteritis in a health professional athlete. Am. J. Gastroenterol., *77*:833–834.

Barker, I.K., and Carbonell, P.L. 1974. *Cryptosporidium agni* sp. n. from lambs, and *Cryptosporidium bovis* sp. n. from a calf, with observations on the oocyst. Z. Parasitenk., *44*:289–298.

Beaver, P.C., Gadgil, R.K., and Morera, P. 1979. *Sarcocystis* in man: A review and report of five cases. Am. J. Trop. Med. Hyg., *28*:819–844.

Benenson, M.W., Takafuji, E.T., Lemon, S.M., Greenup, R.L., and Sulzer, A.J. 1982. Oocyst-transmitted toxoplasmosis associated with ingestion of contaminated water. N. Engl. J. Med., *307*:666–669.

Beverley, J.K.A. 1976. Toxoplasmosis in animals. Vet. Rec., *99*:123–127.

Bird, R.G. 1981. Protozoa and viruses. Human cryptosporidiosis and concomitant viral enteritis, pp. 39–47. In *Parasitological Topics. A Presentation Volume to P.C.C. Garnham, FRS on the Occasion of His 80th Birthday, 1981*. Edited by E.U. Canning. Society of Protozoologists Special Publication, no. 1.

———, Smith, M.D., and Bryceson, A.M. 1980. Cryptosporidial and viral enteritis in immunodeficient animals and man. Trans. R. Soc. Trop. Med. Hyg., *74*:110.

Boch, J., Laupheimer, K.E., and Erber, M. 1978. Drei Sarkosporidienarten bei Schlachtrindern in Süddeutschland. Berl. Munch. Tierarztl. Wschr., *91*:426–431.

Bonciu, C., Petrovici, M., and Panaitescu, D. 1981. Observations on a case of chronic myositis due to *Sarcosporidia*. Arch Roum. Pathol. Exp. Microbiol., *40*:361–363.

Brandborg, L.L., Goldberg, S.B., and Breidenbach, W.C. 1970. Human coccidiosis—a possible cause of malabsorption. N. Engl. J. Med., *283*:1306–1313.

Bunyaratvej, S., Bunyawongwiroj, P., and Nitiyanant, P. 1982. Human intestinal sarcosporidiosis: Report of six cases. Am. J. Trop. Med. Hyg., *31*:36–41.

Castellino, R.A. 1976. Percutaneous pulmonary needle diagnosis of *Pneumocystis carinii* pneumonitis, pp. 137–140. In *Symposium on Pneumocystis Carinii Infection*. National Cancer Inst. Monogr., *43*. Washington, D.C., G.P.O.

Chinchilla, M. 1980. Epidemiologia de la toxoplasmosis en Costa Rica: Importancia de los roedores domesticos. Rev. Biol. Trop., *26*:113–124.

Connal, A. 1922. Observations on the pathogenicity of *Isospora hominis*, Rivolta, emend. Dobell, based on a second case of human coccidiosis in Nigeria; with remarks on the significance of Charcot-Leyden crystals in the faeces. Trans. R. Soc. Trop. Med. Hyg., *16*:223–245.

Cook, E.B.M. 1961. Safety in the public health laboratory. Public Health Rep., *76*:51–56.

Dorfman, R.F., and Remington, J.S. 1973. Value of lymph-node biopsy in the diagnosis of acute acquired toxoplasmosis. N. Engl. J. Med., *289*:878–881.

Dubey, J.P. 1974. Effect of freezing on the infectivity of *Toxoplasma* cysts to cats. J. Am. Vet. Med. Assoc., *165*:534–536.

———. 1977. Persistence of *Toxoplasma gondii* in the tissues of chronically infected cats. J. Parasitol., *63*:156–157.

———, and Yeary, R.A. 1977. Anticoccidial activity of 2-sulfamoyl-4,4-diaminodiphenylsulfone, sulfadiazine, pyrimethamine and clindamycin in cats infected with *Toxoplasma gondii*. Can. Vet. J., *18*:51–57.

Dubin, I.N., and Wilcox, A. 1947. *Sarcocystis* in *Macaca mulatta*. J. Parasitol., *33*:151–153.

Eaton, R.D.P. 1966. Human coccidiosis in Saskatchewan. Can. Med. Assoc. J., *95*:1384–1385.

Faust, E.C., Giraldo, L.E., Caicedo, G., and Bonfante, R. 1961. Human isosporosis in the Western Hemisphere. Am. J. Trop. Med. Hyg., *10*:343-349.

Fayer, R. 1980. Epidemiology of protozoan infections: The coccidia. Vet. Parasitol., *6*:75–103.

———, Heydorn, A.O., Johnson, A.J., and Leek, R.G. 1979. Transmission of *Sarcocystis suihominis* from humans to swine to nonhuman primates *(Pan troglodytes, Macaca mulatta, Macaca irus)*. Z. Parasitenk., *59*:15–20.

Fletcher, A., Sims, T.A., and Talbot, I.C. 1982. Cryptosporidial enteritis without general or selective immune deficiency. Br. J. Med., *285*:22–23.

Frenkel, J.K. 1973. *Toxoplasma* in and around us. Bioscience, *23*:343–352.

———. 1976. Pneumocystosis, pp. 303–307. In *Pathology of Tropical and Extraordinary Diseases*. Edited by C.H. Binford and D.H. Connor. *An Atlas*, vol. 1. Washington, D.C., Armed Forces Institute of Pathology.

———, and Dubey, J.P. 1975. *Hammondia hammondi* gen. nov. sp. nov. from domestic cats, a new coccidian related to *Toxoplasma* and *Sarcocystis*. Z. Parasitenk., *46*:3–12.

———, Dubey, J.P., and Miller, N.L. 1970. *Toxoplasma gondii*: The oocyst, sporozoite, and infection of cultured cells. Science, *167*:892–896.

———, Heydorn, A.O., Mehlhorn, H., and Rommel, M. 1979. Sarcocystinae: *Nomina dubia* and available names. Z. Parasitenk., *58*:115–139.

———, and Ruiz, A. 1980. Human toxoplasmosis and cat contact in Costa Rica. Am. J. Trop. Med. Hyg., *29*:1167–1180.

Gajdusek, D.C. 1976. *Pneumocystis carinii* as the cause of human disease: Historical perspective and magnitude of the problem, pp. 1–10. In *Symposium on Pneumocystis Carinii Infection*. National Cancer Inst. Monogr., *43*. Edited by J.B. Robbins, V.T. DeVita, Jr., and W. Dutz. Washington, D.C., G.P.O.

Heydorn, A.O. 1977. Beiträge zum Lebenszyklus der Sarkosporidien. IX. Entwicklungszyklus von *Sarcocystis suihominis* n. spec. Berl. Munch. Tierarztl. Wschr., *99*: 218–224.

———, Gestrich, R., and Janitschke, K. 1976. Beiträge zum Lebenszyklus der Sarkosporidien. VIII. Sporozysten von *Sarcocystis bovihominis* in den Fäzes von Rhesusaffen *(Macaca rhesus)* und Pavianen *(Papio cynocephalus)*. Berl. Munch. Tierarztl. Wschr., *89*:116–120.

———, and Mehlhorn, H. 1978. Light and electron microscopic studies on *Sarcocystis suihominis*. 2. The schizogony preceding cyst formation. Zbl. Bakt. Hyg., I. Abt. Orig. A, *240*:123–134.

Hughes, W.T. 1982. Natural mode of acquisition for de novo infection with *Pneumocystis carinii*. J. Infect. Dis., *145*:842–848.

Jones, T.C., Kean, B.H., and Kimball, A.C. 1969. Acquired toxoplasmosis. N.Y. State J. Med., *69*:2237–2242.

Kagan, I.G., and Norman, L.G. 1976. Serology of pneumocystosis, pp. 121–125. In *Symposium on Pneumocystis Carinii Infection*. National Cancer Inst. Monogr., *43*. Edited by J.B. Robbins, V.T. De Vita, Jr., and W. Dutz. Washington, D.C., G.P.O.

Kan, S.P., Prathap, K., and Dissanaike, A.S. 1979. Light and electron microstructure of a *Sarcocystis* sp. from the Malaysian long-tailed monkey, *Macaca fascicularis*. Am. J. Trop. Med. Hyg., *28*:634–642.

Krick, J.A., and Remington, J.S. 1978. Toxoplasmosis in the adult—an overview. N. Engl. J. Med., *298*:550–553.

Laarman, J.J. 1968. Experiences with human *Isospora*. Acta Leidensia, *36*:145–148.

Latorre, C.R., Sulzer, A.J., and Norman, L.G. 1977. Serial propagation of *Pneumocystis carinii* in cell line cultures. Appl. Environ. Microbiol., *33*:1204–1206.

Lau, W.K., Young, L.S., and Remington, J.S. 1976. *Pneumocystis carinii* pneumonia. Diagnosis by examination of pulmonary secretions. J.A.M.A., *236*:2399–2402.

Levine, N.D. 1977a. Nomenclature of *Sarcocystis* in the ox and sheep and of fecal coccidia of the dog and cat. J. Parasitol., *63*:36–51.

———. 1977b. Taxonomy of *Toxoplasma*. J. Protozool., *24*:36–41.

———, Corliss, J.O., Cox, F.E.G., Deroux, G., Grain, J., Honigberg, B.M., Leedale, G.F., Loeblich, A.R., Lom, J., Lynn, D., Marinfeld, E.G., Page, C., Poliansky, G.,

Sprague, V., Vavra, J., and Wallace, F.G., 1980. A newly revised classification of the protozoa. J. Protozool., *27*:17–58.

———, and Tadros, W. 1980. Named species and hosts of *Sarcocystis* (Protozoa: Apicomplexa: Sarcocystidae). System. Parasitol., *2*:41–59.

Margileth, A.M., Strano, A.J., Chandra, R., Neafie, R., Blum, M., and McCully, R.M. 1973. Disseminated nosematosis in an immunologically compromised infant. Arch. Pathol., *95*:145–150.

Matsubayashi, H., Koike, T., Mikata, I., Takei, H., and Hagiwara, S. 1959. A case of *Encephalitozoon*-like body infection in man. Arch. Pathol., *67*:181–187.

———, and Nozawa, T. 1948. Experimental infection of *Isospora hominis* in man. Am. J. Trop. Med., *28*:633–637.

McCracken, A.W. 1972. Natural and laboratory-acquired infection by *Isospora belli*. South. Med. J., *65*:800–818.

Mehlhorn, H., and Heydorn, A.O. 1977. Light and electron microscopic studies of *Sarcocystis suihominis*. 1. The development of cysts in experimentally infected pigs. Zbl. Bakt. Hyg., I. Abt. Orig. A, *239*:124–139.

———. 1978. The Sarcosporidia (Protozoa, Sporozoa): Life cycle and fine structure. Adv. Parasitol., *16*:43–91.

———. 1979. Electron microscopical study on gamogony of *Sarcocystis suihominis* in human tissue cultures. Z. Parasitenk., *58*:97–113.

———, and Gestrich, R. 1975. Licht- und elektronenmikroskopische Untersuchungen an Cysten von *Sarcocystis fusiformis* in der Muskulatur von Kälbern nach experimenteller Infektion mit Oocysten und Sporocysten von *Isospora hominis* Railliet et Lucet, 1891. 1. Zur Entstehung der Cyste und der Cystenwand. Zbl. Bakt. Hyg., I. Abt. Orig. A, *231*:301–322.

Meisel, J.L., Perera, D.R., Meligro, C., and Rubin, C.E. 1976. Overwhelming watery diarrhea associated with a *Cryptosporidium* in an immunosuppressed patient. Gastroenterology, *70*:1156–1170.

Miller, F.H., Pizzuto, A.V., and McCauley, H. 1971. Human isosporosis: Two cases. Am. J. Trop. Med. Hyg., *20*:23–25.

Nime, F.A., Burek, J.D., Page, D.L., Holscher, M.A., and Yardley, J.H. 1976. Acute enterocolitis in a human being infected with the protozoan *Cryptosporidium*. Gastroenterology, *70*:592–596.

Pathmanathan, P., and Kan, S.P. 1981. Human *Sarcocystis* infection in Malaysia. Southeast Asian J. Trop. Med. Public Health, *12*:247–250.

Pedersen, F.K., Johansen, K.S., Rosenkvist, J., Tygstrup, I., and Valerius, N.H. 1979. Refractory *Pneumocystis carinii* infection in chronic granulomatous disease. Successful treatment with granulocytes. Pediatrics, *64*:935–938.

Piekarski, G., Heydorn, A.O., Aryeetey, M.E., Hartlapp, J.H., and Kimmig, P. 1978. Klinische, parasitologische und serologische Untersuchungen zur Sarkosporidiose *(Sarcocystis suihominis)* des Menschen. Immun. Infekt., *6*:153–159.

Pifer, L.L., Hughes, W.T., Stagno, S., and Woods, D. 1978. *Pneumocystis carinii* infection. Evidence for high prevalence in normal and immunosuppressed children. Pediatrics, *61*:35–41.

Pintozzi, R.L., Blecka, L.J., and Nanos, S. 1979. The morphologic identification of *Pneumocystis carinii*. Acta Cytol., *23*:35–39.

Rao, N.A., and Font, R.L. 1977. Toxoplasmic retinochoroiditis. Electronmicroscopic and immunofluorescence studies of formalin-fixed tissue. Arch. Ophthalmol., *95*:273–277.

Ravenel, J.M., Suggs, J.L., and Legerton, C.W. 1976. Human coccidiosis. Recurrent diarrhea of 26 years duration due to *Isospora belli:* A case report. J. S. Carolina Med. Assoc., *72*:217–219.

Reese, N.C., Current, W.L., Ernst, J.V., and Bailey, W.S. 1982. Human and calf cryptosporidiosis: A case report and results of comparative infections in mice and rats. Am. J. Trop. Med. Hyg., *31*:226–229.

Rifkind, D. 1976. *Pneumocystis carinii* pneumonia in renal transplant recipients, pp. 49–53. In *Symposium on Pneumocystis Carinii Infection*. National Cancer Inst. Monogr., *43*. Edited by J.B. Robbins, V.T. De Vita, Jr., and W. Dutz. Washington, D.C., G.P.O.

Robbins, J.B., De Vita, V.T., Jr., and Dutz, W., editors. 1976. *Symposium on Pneumocystis Carinii Infection*. National Cancer Inst. Monogr., *43*. Washington, D.C., G.P.O.

Rommel, M., and Heydorn, A.-O. 1972. Beiträge zum Lebendszyklus der Sarkosporidien. III. *Isospora hominis* (Railliet und Lucet, 1891) Wenyon, 1928, eine Dauerform der Sarkosporidien des Rindes und des Schweins. Berl. Munch. Tierarztl. Wschr., *85*:143–145.

———, Fischle, B., and Gestrich, R. 1974. Beiträge zum Lebendszyklus der Sarkosporidien. V. Weitere Endwirte der Sarkosporidien von Rind, Schaf und Schwein und die Bedeutung des Zwischenwirtes für die Verbreitung dieser Parasitose. Berl. Munch. Tierarztl. Wschr., *87*:392–396.

Routh, C.F., McCroan, J.E., Jr., and Hames, C.G. 1955. Three cases of human infection with *Isospora* in Georgia. Am. J. Trop. Med. Hyg., *4*:1–8.

Ruiz, A., and Frenkel, J.K. 1980. *Toxoplasma gondii* in Costa Rican cats. Am. J. Trop. Med. Hyg., *29*:1150–1160.

Sheffield, H.G., and Melton, M.L. 1970. *Toxoplasma gondii:* The oocyst, sporozoite, and infection of cultured cells. Science, *167*:892–893.

Sloper, K.S., Dourmashkin, R.R., Bird, R.B., Slavin, G., and Webster, A.D.B. 1982. Chronic malabsorption due to cryptosporidiosis in a child with immunoglobulin deficiency. Gut, *23*:80–82.

Spencer, H., et al., editors. 1973. *Tropical Pathology*. New York, Springer-Verlag.

Sprague, V. 1974. *Nosema connori* n. sp., a microsporidian parasite of man. Trans. Am. Microsc. Soc., *93*:400–402.

Stagno, S., Dykes, A.C., Amos, C.S., Head, R.A., Juranek, D.D., and Walls, K. 1980. An outbreak of toxoplasmosis linked to cats. Pediatrics, *65*:706–712.

Strano, A.J., Cali, A., and Neafie, R.C. 1976. Microsporidiosis, pp. 336–339. In *Pathology of Tropical and Extraordinary Diseases. An Atlas*, vol. 1. Edited by C.H. Binford and D.H. Connor. Washington, D.C., Armed Forces Institute of Pathology.

Syrkis, I., Fried, M., Elian, I., Pietrushka, D., and Lengy, J. 1975. A case of severe human coccidiosis in Israel. Israel J. Med. Sci., *11*:373–377.

Tabbara, K.F., and O'Connor, G.R. 1980. Treatment of ocular toxoplasmosis with clindamycin and sulfadiazine. Ophthalmology, *87*:129–134.

Tadros, W., and Laarman, J.J. 1976. *Sarcocystis* and related coccidian parasites. A brief review, together with a discussion on some biological aspects of their life cycles and a new proposal for their classification. Acta Leiden., *44*:1–137.

Takeuchi, S. 1980. Electronmicroscopic observations on *Pneumocystis carinii*. Jap. J. Parasitol., *29*:427–453.

Trier, J.S., Moxey, P.C., Schimmel, E.M., and Robles, E. 1974. Chronic intestinal coccidiosis in man: Intestinal morphology and response to treatment. Gastroenterology, *66*:923–935.

Tzipori, S., McCartney, E., Lawson, G.H.K., Rowland, A.C., and Campbell, I. 1981. Experimental infection of piglets with *Cryptosporidium*. Res. Vet. Sci., *31*:358–368.

Vavra, J., and Kučera, K. 1970. *Pneumocystis carinii* Delanoë, its ultrastructure and ultrastructural affinities. J. Protozool., *17*:463–483.

Vetterling, J.M., Jervis, H.R., Merrill, T.C., and Sprinz, H.

1971a. *Cryptosporidium wrairi* sp. n. from the guinea pig *Cavia porcellus*, with an emendation of the genus. J. Protozool., *18*:243–247.

———, Takeuchi, A., and Madden, P.A. 1971b. Ultrastructure of *Cryptosporidium wrairi* from the guinea pig. J. Protozool., *18*:248–260.

Vossen, M.E.M.H., Beckers, P.J.A., Meuwissen, J.H.E.T., and Stadhouders, A.M. 1978. Developmental biology of *Pneumocystis carinii*, an alternative view on the life cycle of the parasite. Z. Parasitenk., *55*:101–118.

———, Beckers, P.J.A., Stadhouders, A.M., Bergers, A.M.G., and Meuwissen, J.H.E.T. 1977. New aspects of the life cycle of *Pneumocystis carinii*. Z. Parasitenk., *51*:213–217.

Waldhorn, R.E., Tson, E., and Kerwin, D.M. 1982. *Pneumocystis carinii* pneumonia in a previously healthy adult. J.A.M.A., *247*:1860–1861.

Wallace, G.D., Marshall, L., and Marshall, M. 1972. Cats, rats, and toxoplasmosis on a small Pacific island. Am. J. Epidemiol., 95:475–482.

Walzer, P.D., Schnelle, V., Armstrong, D., and Rosen, P.P. 1977. Nude mouse: A new experimental model for *Pneumocystis carinii* infection. Science, *197*:177–179.

Weisse, K., and Wedler, E. 1954. Über das Vorkommen der Sogenannten "*Pneumocystis carinii*." Klin. Wochenschr., *32*:270–271.

Wenyon, C.M. 1923. Coccidiosis of cats and dogs and the status of the *Isospora* of man. Ann. Trop. Med. Parasitol., *17*:231–239.

Westerman, E.L., and Christensen, R.P. 1979. Chronic *Isospora belli* infection treated with co-trimoxazole. Ann. Intern. Med., *91*:413–414.

Work, K. 1971. Toxoplasmosis with special reference to transmission and life cycle of *Toxoplasma gondii*. Acta Path. Microbiol. Scandinavica., Sec. B, Suppl. 221:1–51.

Yoneda, K., Walzer, P.D., Richey, C.S., and Birk, M.G. 1982. *Pneumocystis carinii:* Freeze-fracture study of stages of the organism. Exp. Parasitol., *53*:68–76.

Yoshida, Y., and Ikai, T. 1979. *Pneumocystis carinii* pneumonia: Epidemiology in Japan, and cyst concentration method. Zbk. Bakt. Hyg., I. Abt. Orig. A, *244*:405–410.

Chapter 14

Malaria Parasites and Piroplasms

The malaria parasites of man are species of the genus *Plasmodium,* family Plasmodiidae, which, along with other intraerythrocytic sporozoa *(Hepatocystis, Haemoproteus, Leucocytozoon),* constitute the suborder Haemosporina. The piroplasms are species of the genera *Babesia* and *Entopolypoides,* family Babesiidae of the order Piroplasmida. Both groups—the malaria parasites and piroplasms—are classified as class Sporozoea, phylum Apicomplexa (see. p. 40).

THE HAEMOSPORINA

Family Plasmodiidae Mesnil, 1908

In this group, the macrogamete and microgamont develop independently, and the microgamont produces about eight flagellated microgametes. The zygote becomes a motile ookinete, and the sporozoites are naked. Development is heteroxenous, with merogony in a vertebrate and sporogony in an invertebrate host, and infection is transmitted by blood-sucking insects.

Genus *Plasmodium*

Marchiafava and Celli, 1885

In species of the genus *Plasmodium,* gamonts develop in erythrocytes, and schizogony (merogony) is also in erythrocytes and in various other tissues depending on the species. Pigment that is visible with the light microscope is formed. For species developing in mammals, the vectors are mosquitoes.

The genus *Plasmodium* may be classified into nine subgenera (Table 14–1). Three species of the subgenus *P. (Plasmodium)* and one of the subgenus *P. (Laverania)* cause malaria in man. The species of another subgenus, *P. (Vinckeia),* infect mammals other than primates.

In the subgenus *Plasmodium,* the gametocytes are round and tend to be produced 4 to 6 days after the peak of parasitemia. In *P. vivax* and *P. ovale,* exoerythrocytic forms persist in the liver, and perhaps in other tissues, and the merozoites produced by the exoerythrocytic schizonts are responsible for relapse. In the subgenus *Laverania,* exoerythrocytic schizogony is limited to one generation and exoerythrocytic forms do not persist. There is, therefore, no true relapse in *Laverania (P. falciparum)* infection. According to Garnham (1977), *P. malariae* also has only one generation of exoerythrocytic schizogony and, therefore, does not relapse. The subgenus *Laverania* is further distinguished from the subgenus *Plasmodium* by having crescentic gametocytes that persist in the circulating blood after the other erythrocytic forms have disappeared. The host mosquitos for both subgenera are species of *Anopheles.*

The plasmodia normally infecting man are *P. (Plasmodium) vivax, P. (Plasmodium) ovale, P. (Plasmodium) malariae,* and *P. (Laverania) falciparum.* Other plasmodia of the subgenus *Plasmodium* occurring naturally or experimentally in man are *P. schwetzi* of African chimpanzees and gorillas (Rodhain and Dellaert, 1955), *P. eylesi* of Malayan gibbons (Warren et al., 1965), *P. cynomolgi* of Oriental monkeys (Coatney et al., 1961; Schmidt et al., 1961), *P. cynomolgi bastianelli* of Oriental monkeys (Eyles et al., 1960), *P. simium* of black howler monkeys of Brazil (Deane et al., 1966), *P. inui* of Oriental monkeys (Das Gupta, 1938), *P. brasilianum* of New World monkeys (Contacos and Coatney, 1963), *P. shortti* of Indian and Ceylonese monkeys (Coatney et al., 1966), and *P. knowlesi* of Oriental monkeys (Ciuca et al., 1955; Chin et al., 1965).

The four species normally parasitic in man have been inoculated—usually with some difficulty and often only after splenectomy—into laboratory animals. For example, chimpanzees have been in-

Table 14–1. Genus *Plasmodium*.*

Subgenera	*Approx. no. species*	*Type species*	*Vertebrate host*	*Erythrocytic schizonts*	*Gametocytes*
Plasmodium	26	*malariae*	Primates	Large	Round
Vinckeia	15	*bubalis*	Antelopes, rodents, other mammals, not primates	Small	Round
Laverania	2	*falciparum*	Primates	Large	Elongate, crescentic
Haemamoeba	10	*relictum*	Birds	Large	Round
Giovannolaia	9	*circumflexum*	Birds	Large	Elongate
Novyella	5	*vaughani*	Birds	Small	Elongate or oval
Huffia	2	*elongatum*	Birds	Large	Elongate
Sauramoeba	15	*agamae*	Saurians	Large	—
Carinia	8	*minasense*	Saurians	Small	—

*After Garnham, P.C.C. 1966. *Malaria, Parasites and Other Haemosporidia*. Oxford, Blackwell Scientific Publications.

fected with *P. vivax,* although rhesus monkeys seem totally immune (Bray, 1957a; Garnham, 1966). Chimpanzees also have been infected with *P. ovale* (Bray, 1957b), which may be identical with *P. schwetzi* (Coatney, 1968). *Plasmodium malariae,* which seems to be identical to *P. rodhaini,* occurs naturally in chimpanzees (Garnham, 1966). *Plasmodium falciparum* has been transmitted to gibbons and to New World monkeys (Garnham, 1966; Geiman and Meagher, 1967).

There have been exchanges of plasmodia between man and animals, but they have usually been experimentally induced. Remarkably few instances of natural infections of man by animal plasmodia or of accidental laboratory infections have been reported. Whether malaria should be classified as a true zoonosis remains a question. Certainly the subject deserves continued investigation, but as yet, it does not seem to suggest a significant barrier to eradication of human malaria.

The importance of malaria as a cause of human morbidity and death exceeds that of any other parasitic disease. Notable treatises on malaria are those of Russell et al. (1963), Garnham (1966), Pampana (1969), Coatney et al. (1971), Kreier (1980), and Bruce-Chwatt (1980).

LIFE CYCLES AND METHODS OF REPRODUCTION OF MALARIA PARASITES OF MAN

Malaria parasites of man have two stages—an *extrinsic* phase in *Anopheles* and an *intrinsic* vertebrate phase in man. The extrinsic phase, in which sexual reproduction occurs, is referred to as ''definitive,'' and the intrinsic phase, in which asexual reproduction occurs, is referred to as ''intermediate'' (Fig. 14–1).

Invertebrate Phase. When mature *microgametocytes* (male) and *macrogametocytes* (female) are taken into the gut of a susceptible mosquito, they quickly mature into *gametes*. Microgametocytes having undergone multiple nuclear division mature by *exflagellation,* a process in which, in 10 to 12 minutes, the *microgametes* break away from the red cell and become motile (Fig. 14–1). The macrogametocyte develops into a *macrogamete;* its nucleus shifts to the surface, where a projection is formed. *Fertilization* occurs when a microgamete penetrates this projection, and the fertilized macrogamete is called a *zygote*. After about 20 minutes, the zygote puts out a pseudopodium into which its protoplasm flows to form a fusiform body, the *ookinete,* a motile vermicule stage. As the ookinete grows, its male and female nuclei fuse and it penetrates the brush border and body of a midgut epithelial cell to that cell's opposite side. There the ookinete secretes a thin wall and grows into a spherical *oocyst,* 50 μm or more in size, which extends into the insect's hemocoel. Depending on temperature, during a period of 4 to 15 days after the mosquito ingests gametocytes, the oocyst matures, with multiplication of nuclei and transformation of the cytoplasm into as many as a thousand separate bodies called ''sporozoites.'' The oocyst bursts, and sporozoites enter the mosquito's hemocoel, from which they disperse throughout the insect's body. Those that come into contact with the salivary glands bore into and through the cells and

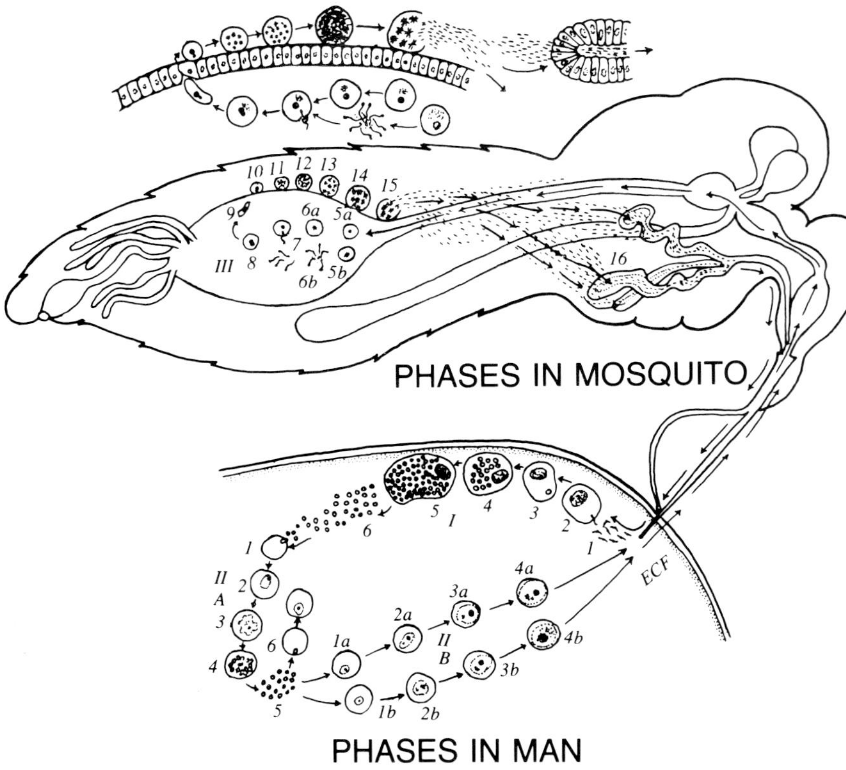

Fig. 14–1. Life cycle of the malaria parasite. I, Pre-erythrocytic phase. Sporozoites (1) introduced into man by the infected mosquito enter cells of the liver and, by asexual multiplication, or schizogony (2–5), produce merozoites (6). II, Erythrocytic phases. Merozoites from pre-erythrocytic schizonts invade red blood cells, producing trophozoites (1–5) leading to replicative schizogony (6), and formation of female (1a-4a) and male (lb–4b) gametocytes. III, Mosquito phases. Gametocytes, taken up with peripheral blood of man, reach the mosquito's midgut (5a, 5b), where they mature into female and male gametes (6a, 6b). One male gamete enters a female gamete (7), producing a zygote (8), which becomes the motile ookinete (9) and migrates to the outer stomach wall, where it forms the oocyst. The oocyst grows (10–14), producing a large number of sporozoites. When mature, the oocyst bursts (15), releasing the sporozoites, which reach the salivary glands (16), pass down the salivary ducts to the hypopharynx to be introduced into man when the mosquito takes its next blood meal. (Adapted from original of E.C. Faust.)

lodge in acinal ducts. Thereafter, when the mosquito injects saliva while obtaining a blood meal, it also injects sporozoites that enter their human host. Only female *Anopheles,* not male, take blood meals.

Sporozoites have a complex structure and a relatively thick cell membrane. Their shape varies from narrow and slightly curved *(vivax),* to thick *(malariae),* to sickle-shaped *(falciparum),* and they measure 10 to 14 μm in length. They have peripheral fibrils that may have locomotory function. Mitochondria are present, but they do not have pigment.

Vertebrate Phase. Sporozoites rapidly leave the bloodstream, which is generally entirely free of them within 1 hour. In rodent malaria the sporozoites have been demonstrated in hepatocytes 1 hour after inoculation (Meis et al., 1982). The classic demonstrations of Shortt and Garnham (1948) and colleagues (Shortt et al., 1949, 1951), and subsequent observations by others, have made it clear that, in primates, the sporozoites penetrate parenchyma cells of the liver. Here the next stage of development takes place. It is known as *primary exoerythrocytic schizogony* (EE schizogony), sometimes called pre-erythrocytic schizogony (Fig. 14–2).

The sporozoite becomes a round or oval *exoerythrocytic schizont,* and its nucleus divides repeatedly. No pigment is present. The schizonts of species infecting man vary in size from about 24 to 60 μm. EE schizogony results in liberation of large numbers of *exoerythrocytic merozoites* from the matured EE schizonts. Primary EE schizogony terminates when EE merozoites have invaded red blood cells. The first two or three generations of EE merozoites may reinvade parenchyma cells, but this has not been shown and there is some evidence

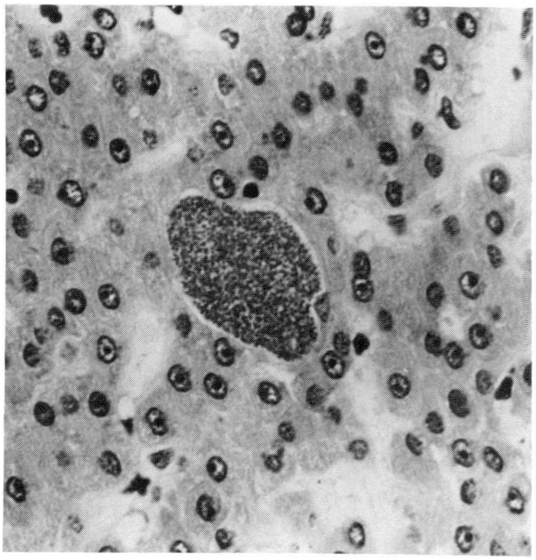

Fig. 14–2. *Plasmodium cynomolgi,* mature pre-erythrocytic schizont at 8 days of infection. Giemsa stain. (× 500.) (Courtesy of Dr. W.A. Krotoski.)

against it (Garnham, 1977). There is also no proof that late EE forms are not in sites other than the liver.

It has been reported that, in *P. vivax* and *P. ovale,* a dormant or *secondary exoerythrocytic schizozoite* (hypnozoite) occurs, with one or two or more generations of EE merozoites appearing after red cells have been invaded. Pre-erythrocytic forms of *P. cynomolgi* have been demonstrated in the liver at 48 hours but not at 24 hours. Hypnozoites and schizonts remain present up to 105 days (Krotoski et al., 1982). Such delayed EE schizogony does not occur in *P. falciparum,* and this may also be true for *P. malariae*. EE stages arising directly from sporozoites have been called "cryptozoites," but this term is now limited to avian parasites, as is the term "metacryptozoite," which is applied to the EE stage arising from cryptozoites and occurring prior to invasion of erythrocytes.

Blood Invasion. EE merozoites enter erythrocytes and reticulocytes, where they grow at the expense of the host cells. The merozoite has a selectively adhesive surface for attachment to the erythrocyte and an apical complex of rhoptries, micronemes, and polar rings for invasion; it also has the usual metabolic organelles, such as ribosomes and nuclear complex, and other structures (Fig. 14–3). Invasion of the erythrocyte by the merozoite has been described by Bannister (1977) as involving adhesion to the erythrocyte surface, attachment of the apical complex to the red cell membrane, formation of a parasitophorous vacuole, and engulfment of the merozoite by the red cell, during which the outer coat of the merozoite, at least in some species, disappears. In the red cell, the merozoite is vacuolated, ring-shaped, more or less ameboid, and uninucleate. It is now called *trophozoite,* a name it retains until its nucleus begins to divide. It feeds on hemoglobin, which it does not completely metabolize, so residues of globin and an iron porphyrin hematin remain. The latter is *malaria pigment,* a compound of hematin (ferrihemic acid) and protein. (The terms "melanin" and "hemozoin" are not recommended [WHO, 1963].)

The trophozoite grows until its nucleus divides by mitosis, its vacuole is filled in, its ameboid motility is curbed, and it becomes a mature *schizont*. This mature schizont then undergoes erythrocytic schizogony, dividing into *erythrocytic merozoites*. The red cell ruptures, and merozoites flow into the bloodstream. Many are destroyed by the host's forces of immunity, but others invade red cells in which a new cycle of erythrocytic schizogony takes place.

After two or three erythrocytic generations, the phenomenon of *gametocytogenesis* begins. Some of the intracellular merozoites do not become schizonts but develop into male *microgametocytes* or female *macrogametocytes*. Whether they develop from a special type of schizont, sometimes referred to as a "gamont," or are produced in response to a specific stimulus is unknown (Fig. 14–1).

BIOCHEMISTRY AND METABOLISM OF MALARIA PARASITES

Plasmodia are complex obligate intracellular parasites, and it has been difficult to determine how they ingest and metabolize their food. The studies of Rudzinska et al. (1960, 1965) and others have suggested that the parasite feeds on its host's cytoplasm by *pinocytosis*—i.e., minute incuppings or invaginations of its plasma membrane, apparently at random, engulf and close around some cytoplasm, forming a vacuole within which digestion takes place.

Other studies (Ladda et al., 1966) have suggested that the parasite feeds by way of a single food-ingesting organelle resembling the *micropyle* of a sporozoite. Aikawa et al. (1966) suggested that the parasite has a specific *cytostome* that ingests host cell cytoplasm. Whether it is due to

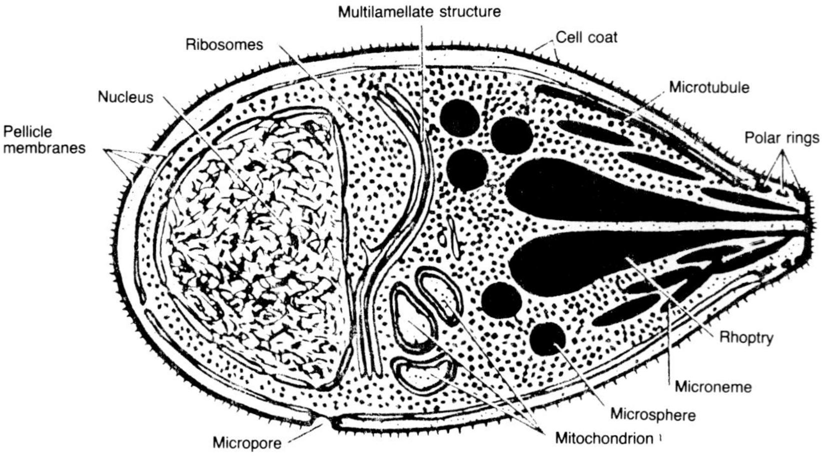

Fig. 14–3. *Plasmodium.* Diagram of a merozoite showing the principal features seen in electron micrographs of longitudinal section. (From Bannister, L.H., Taylor, A.E.R., and Muller, R. (editors). 1977. Invasion of red cells by *Plasmodium,* pp. 27–55. In Symposia of the British Society for Parasitology, vol. 15. Oxford, Blackwell Scientific Publications.)

pinocytosis or cytostomal activity, inclusions of host cell material in food vacuoles are a striking feature of electron microscope pictures of erythrocytic stages of plasmodia (Trager, 1966).

Precise nutrient needs for growth and multiplication of the parasites have not been determined. In general, the required energy of the parasite depends on phosphorylation of glucose, oxidative processes are maintained by oxyhemoglobin of the erythrocyte, globin of the hemoglobin is split by enzymes into amino acids and peptides that form the parasite's protein, and lipids are synthesized in relatively large quantities. Metabolites of the parasite derive from hemoglobin of the erythrocyte and from blood plasma. Characteristic morphologic alterations in the infected red cell are caused by the growing parasites. The size and shape of the cell are affected, and distinct intracellular dots, clefts, or bodies become evident in Romanowsky-stained thin smears. *P. vivax* causes the cell to enlarge and *P. ovale* causes subspherical shapes; both species (and *P. cynomolgi)* cause a coarse brick-red stippling of the cytoplasm referred to as *"Schüffner's dots"* (Plate II,5). These consist of a caveola with an alveolar complex of endocytotic vesicles that probably are pinocytotic in origin (Aikawa et al., 1975). Similar, but smaller, stippling caused by *P. malariae* are *Ziemann's dots,* and coarser dots caused by *P. falciparum* are called *"Maurer's dots"* or *"clefts."*

Glucose-6-Phosphate Dehydrogenase Deficiency Trait. A genetic deficiency known as the glucose-6-phosphate dehydrogenase (G6PD) deficiency trait is believed to confer some protection against *P. falciparum* infection, but the effect is controversial. Some 11% of male American blacks (Beutler, 1959), and numerous individuals in some highly malarious areas such as East Africa, inherit this deficiency. Low activity of this enzyme results in subnormal concentration of reduced glutathione in red cells and also in limitation of the hexose monophosphate shunt metabolic pathway (Geiman, 1951). In their metabolism, plasmodia use the hexose monophosphate shunt pathway (e.g., in production of essential ribose phosphate). Also, they require reduced glutathione for growth *in vitro* (McGhee and Trager, 1950). These findings might explain a basis for protection by G6PD deficiency against *P. falciparum* infection. Luzzatto et al. (1969) demonstrated that the parasite rate was higher in normal than in G6PD-deficient erythrocytes, and Bienzle et al. (1972, 1979) reported greater resistance against malaria of females heterozygous for G6PD deficiency. Miller and Carter (1976) have questioned the validity of this conclusion. No other explanation for the high frequency of the gene for G6PD deficiency in the malarious regions of Africa has been brought forward.

Duffy Factor. The presence of the Duffy blood factor, Fy^a, is thought to increase the individual's susceptibility to malaria rather than to protect him against it. It has been theorized that the molecule serving as the antigen in the Duffy-positive individual serves the malaria parasite as a site of attachment or penetration on the surface of the erythrocyte. Mathews and Armstrong (1981) have shown that ethnic groups in Africa with a high prevalence of Duffy factor also tended to have a high incidence of vivax malaria, while those in which the Duffy factor was uncommon had a much lower incidence.

Sickle-Cell Anemia Trait. Another inherited trait responsible for some resistance to malaria is sickle-cell anemia (Allison, 1954). The sickle-cell gene, which is common in some areas of south Europe, Asia, and Africa, results in an abnormal hemoglobin, the molecules of which, when deoxygenated, tend to collect in rigid, rod-like masses that distort the red cell into a sickle-like shape. *P. falciparum* schizonts appear to have difficulty in utilizing this abnormal hemoglobin so that their growth and schizogony are stunted. The practical results have been summarized by Allison (1963) as follows: (1) in children under 1 year of age, who are always highly susceptible to *P. falciparum,* sickle-cell heterozygotes are as frequently infected as normal infants, but parasites—and, consequently, mortality—remain low; (2) in the age group 1 to 4 years, incidence of infection and parasite counts are lower in sickle-cell trait carriers, and duration of attacks is shorter; (3) in adults, who are generally highly immune, the effect of the sickle-cell trait seems negligible, because it is overweighed by the effect of acquired immunity.

Other hemoglobins that may provide some protection against malaria are thalassemia hemoglobin and hemoglobin E, which protect against *P. vivax,* and human fetal hemoglobin, which protects against all plasmodia of man (Allison, 1963).

Adenosine triphosphate deficiency in nonimmune black males seems to lower morbidity and mortality levels in *P. falciparum* infections (Brewer and Powell, 1965).

In the mosquito, also, malaria parasite growth is influenced by the nutrition of the host. Many observers have noted that oocysts develop better when the host mosquito has a blood, rather than a glucose, diet. It is also possible that host specificity is determined in part by presence or absence of certain amino acids in the mosquito hemolymph (Garnham, 1966).

Culture of Plasmodia. In early attempts to grow malaria parasites *in vitro,* Bass and Johns (1912) succeeded in getting maturation of *P. vivax* and *P. falciparum* trophozoites to schizonts in defibrinated blood with dextrose added, but in whole blood cultures, they observed no reinvasion of erythrocytes by merozoites. In cultures free of leukocytes, however, the parasites developed and segmented, and most of the merozoites entered new red blood cells. Two to three successive generations were observed by Bass and Johns, but continuous cultivation was not attained until 1976, when Trager and Jensen devised a method that, with improvements, is now being used successfully in numerous laboratories throughout the world (Trager, 1980). In the first continuous cultures, *P. falciparum* was grown by a method that provided a slow flow of medium over a thin layer of human erythrocytes. With improvements in this continuous flow method, parasites in sufficient quantities for biologic studies and experimental testing of antimalarial drugs and immunizations have become available (Trager and Jensen, 1978; Trager, 1979). A method of *in vitro* maintenance of malaria parasites for the testing of *P. falciparum* for drug resistance was devised by Rieckmann et al. in 1968.

Plasmodium (Plasmodium) vivax
(Grassi and Feletti, 1890)
(Producing vivax or tertian malaria)

Synonyms. *Haemamoeba vivax* Grassi and Feletti, 1890; *Haemamoeba laverani* var. *tertiana* Labbé, 1894; *Plasmodium malariae tertianae* Celli and Sanfelice, 1891; *Plasmodium malariae tertianae* Kruse, 1892; *Haemosporidium tertianae* Lewkowicz, 1897; *Plasmodium malariae* var. *tertiana* Laveran, 1901; *Plasmodium tertianae* Billet, 1904.

History and Nomenclature. In 1880, Laveran first observed *Plasmodium vivax,* although not as a distinct species. Some of the intracellular pigmented bodies he described were undoubtedly *vivax* schizonts. In 1886, Golgi described it as a distinct species of plasmodium, but it was not given a specific name until 1890, when Grassi and Feletti designated it *Haemamoeba vivax.*

Distribution. Until successful malaria eradication in the United States, Europe, and other areas, which will be described later, *P. vivax* had nearly worldwide distribution between summer isotherms of 16° to 20° north and 20° south. Vivax infections, which are more common in temperate than in tropical regions, were known northwards to England, Siberia, and Manchuria, southwards to Argentina and South Africa, and in Australia. In some African areas, the local inhabitants appear to have always had strong immunity to vivax malaria. This plasmodium is generally not found at all in most areas between the Congo and Mauritania, which are still highly malarious (Escudié and Hamon, 1961).

Gametes. The microgametocyte of *P. vivax* usually exflagellates six microgametes, each 20 to 25 μm in length. The macrogametocyte becomes a macrogamete when it is free of its red cell and has developed a small projection. A microgamete penetrates this bulge, and the ookinete is formed. The process takes from 10 minutes to an hour in the insect's gut.

Ookinete. Vivax ookinetes are 15 to 22 μm in length and 3 μm in width. They are motile and penetrate the epithelium of the insect's midgut 24 to 48 hours after the blood meal has been ingested.

Oocysts. The minimum maturing period of vivax oocysts is 7 to 8 days at 30°C, and the period may be as long as 30 days at 17.5°C. Pigment granules are distributed irregularly throughout the oocysts.

Sporozoites. Vivax sporozoites are narrow and slightly curved and measure about 14 μm in length. Their complex morphology has been described by Sinden and Garnham (1973).

Exoerythrocytic Stages. Primary vivax EE schizogony is completed in from 7 (Bray, 1957a) to 8 (Garnham, 1966) days. After 8 days, the schizont is ovoid, approximately 40 μm in diameter, with blue-staining, somewhat clumped cytoplasm, irregular nuclei, and relatively large vacuoles. At maturity, the nucleus undergoes multiple division, vacuoles and clumps disappear, and cytoplasm condenses around nuclear particles, giving rise to as many as 10,000 merozoites. These EE merozoites are spherical, measure about 1.2 μm in diameter, and are released to enter the host's circulation.

Secondary EE schizonts have been observed 14 days after infection (Bray, 1957a) and as late as 9 months after infection (Rodhain, 1956). These secondary schizonts do not differ in morphology from primary schizonts (Fig. 14–2). Bray (1957a) estimated that tissue forms of *P. vivax* decline from 20,000 primary to 500 first-generation secondary forms per mm^3 of liver. The secondary EE cycle in the liver may persist, with periods of latency, for several years.

Erythrocytic Stages. Sporozoites leave the bloodstream within an hour after infection, and a *prepatent period,* which is usually about 8 to 17 days, ensues with a bloodstream free of plasmodia. In an Italian strain of *P. vivax,* the prepatent period was reported to be only 5 or 6 days (Monaldi and Raffaele, 1935). In so-called temperate-zone strains (e.g., the St. Elizabeth) and strains from Korea, northern Russia, and the Netherlands, the prepatent period may be much more prolonged—approximately 9 to 12 months. In such strains, the length of the prepatent period is dependent on the number of sporozoites inoculated. If over 1000 sporozoites are inoculated, the prepatent period is short.

The asexual cycle in *P. vivax* usually takes about 42 to 44 hours, but cycles of *P. vivax* are not precise. Some acceleration or retardation may occur. If two broods are present, one a day behind the other, the result will be daily release of erythrocytic merozoites. Asexual multiplication builds up to a peak in about 2 weeks following first schizogony. Thereafter, numbers rapidly decrease and, typically, a month's latency occurs, which is followed by a week of patency and then, again, latency. A long period of latency may follow two or three or more such episodes.

Gametocytes. Vivax gametocytes usually appear on the fifth day of patency but have been reported as early as the third day. Gametocytes mature in 4 days, circulating for several days. The peak of gametocytes follows that of asexual forms by some 4 to 6 days.

Morphology. ***Unstained Preparations.*** The *P. vivax* trophozoite appears within the red blood corpuscle as a small hyaline ring or disk, which becomes ameboid with characteristic rapid movement that justifies its name, "vivax." In a few hours, it begins to show delicate grains of pigment. As growth continues, the plasmodium enlarges and acquires more pigment. By the end of 24 to 36 hours, it fills about two-thirds of the infected erythrocyte. The organism at this time is still ameboid and irregular in shape. At the end of 36 hours, it fills most of the infected cell, the pigment tends to collect near the center, ameboid motion is lost, and radial striations appear, dividing the organism (now a maturing *schizont)* into 12 to 24 daughter cells or *merozoites.* At the end of approximately 48 hours, the organism fills the greatly enlarged infected corpuscle, and division into merozoites is complete.

The *gametocytes* are hyaline in appearance and, in unstained preparations, can be differentiated only by the fact that they fill the enlarged infected erythrocyte but show no evidence of segmentation.

Stained Preparations. In blood smears stained by Giemsa's, Wright's, or other Romanowsky technique, plasmodia cytoplasm stains robin's-egg blue and nuclear chromatin stains crimson or violet, while cytoplasm of the infected erythrocyte stains yellowish or salmon-pink (Plate II).

The earliest erythrocytic *trophozoite* is a delicate blue-stained ring of cytoplasm with a red chromatin dot. The erythrocyte is stained salmon-pink (see Plate II, *1, 2).* Frequently, two dots may be observed, either close together or separated, and this appearance has led some observers to suggest, incorrectly, binary division of these "ring forms."

As the plasmodium enlarges, the ring form becomes irregular and larger and contains dots or threads of red-stained chromatin and granules of dark pigment. The host corpuscle is somewhat enlarged. Orange-pink or pink spots, known as Schüffner's dots, may appear in the cytoplasm. This stippling is demonstrable when smears are stained with a Romanowsky preparation having a pH of 7.2 to 7.4. As the plasmodium enlarges, it appears as an oval or round, blue-stained body, the *schizont,* the cytoplasm of which contains first a few then several to many large granules or masses of red-stained, irregularly distributed chromatin and much dark-brown pigment in granules distributed throughout the cytoplasm. At this stage of development, the infected erythrocyte is pale and much enlarged, and the cytoplasm may be filled with Schüffner's dots.

Just before segmentation, the plasmodium fills its erythrocyte, chromatin is collected in irregular masses throughout the cytoplasm, and pigment granules are aggregated into two or three lumps (Plate II, *14*). The infected erythrocyte is greatly enlarged and is seen as a narrow rim of cytoplasm, often containing Schüffner's dots, surrounding the presegmenting plasmodium.

At segmentation, *P. vivax* occupies the entire enlarged, abnormally pale erythrocyte, and appears as a collection of oval or round blue bodies, each having a bright red or violet chromatin dot, situated at or near the periphery. These bodies are merozoites, typically 16 in number. Erratic nuclear division may produce 8 to 24 merozoites, but most often 12 to 18. After segmentation, merozoites are liberated in the blood plasma. Many enter red cells and appear as oval, blue-stained bodies with a red dot of chromatin near the center.

In all stages of *P. vivax* development before nuclear division, the stained plasmodia show a veil-like, unstained area or vacuole surrounding the chromatin dot or mass associated with it. This unstained area is most clearly seen in fully developed ring forms and the youngest schizonts, but, as nuclear division progresses, it is lost.

The earliest *gametocytes* of *P. vivax* are round or oval with blue-stained cytoplasm. Each has a spherical mass of chromatin that stains deep red or violet. As gametocytes grow, they preserve the round or oval shape with no appearance of vacuoles or of ameboid activity.

The *microgametocyte* cytoplasm stains pale blue while that of the *macrogametocyte* stains deep blue (Plate II, *23,24*). When fully developed, the microgametocyte fills the erythrocyte and has a large nucleus in which granules or fine fibrils of chromatin are arranged in the form of a spindle or loose skein lying in an unstained area within the pale cytoplasm. Coarse granules of pigment are more abundant in gametocytes than in asexual forms and are more irregularly distributed in the microgametocyte than in the macrogametocyte. In those microgametocytes that are about to exflagellate, chromatin may be divided into several masses collected near the periphery of the parasite; such forms may be mistaken for immature schizonts.

Microgametocytes undergoing exflagellation are irregular, pale blue bodies, from which several long thread-like, red or violet flagellated objects (the *microgametes)* project. Pigment is in an irregular mass, and chromatin, stained a deep red, is abundant.

Microgametes appear to consist almost entirely of red- or violet-stained chromatin. They have pointed extremities and are slightly thicker at the middle than toward the ends. Exflagellating microgametocytes and their microgametes are not seen in peripheral blood smears unless the blood is collected on a moist slide and examined 5 to 10 minutes later.

The macrogametocyte of *P. vivax* has a smaller nucleus than the microgametocyte, and chromatin is arranged in a compact mass near the periphery. The cytoplasm stains bright blue and the chromatin, deep red or violet. Pigment in fine granules is arranged in small masses or in a wreath-like manner near the periphery of the organism (Plate II, *24)*.

Double or triple infection of a red corpuscle by *P. vivax* sometimes occurs, and one may see a red cell containing a schizont and a gametocyte. Such multiple infections led Schaudinn to describe, mistakenly, parthenogenesis of the plasmodium.

Thick Blood Films. In well-stained, thick blood films (see Technical Appendix, pp. 751–752) that have been properly filmed and adequately dehemoglobinized, malaria parasites appear compact and not flattened as in thin films. Furthermore, there are usually no outlines of parasitized corpuscles to guide the search for plasmodia; thus, special training is necessary for thick film diagnosis of malaria, and especially for differential diagnosis of stages and of species (Plate III).

Young *trophozoites* (ring forms) of *P. vivax* may appear as unbroken rings, although they are considerably smaller than they appear in thin films. They usually appear as interrupted partial rings, with a distinct chromatin dot but with cytoplasm broken into several fragments. The cytoplasm sometimes may be a solid mass, so that it, with its chromatin mass, looks like a broad exclamation mark or a question mark. Occasionally, the chromatin mass may be in the center and the cytoplasm a single mass on either side, so that the entire plasmodium may suggest a bird in flight. More mature trophozoites have a larger chromatin body, a more substantial mass of cytoplasm, and characteristic pigment. *Schizonts* in early segmentation also have substantial cytoplasm and pigment and two to several red chromatin masses. More mature schizonts have numerous chromatin bodies arranged around clumped pigment. In the ripe schizont, each chromatin mass has a small bit of cytoplasm surrounding it.

Young *gametocytes* are more difficult to recognize, but they are usually as large as late trophozoites, have a more continuous cytoplasm and more pigment. Older gametocytes may be irregular in outline, with dispersed cytoplasm and abundant pigment.

Since hemoglobin of the red cells has been dissolved, there will be no Schüffner's granules to confirm the diagnosis and no opportunity to compare the size or color index of parasitized and uninfected cells. (For aid in microscopic diagnosis of malaria parasites see also Bruce-Chwatt, 1980; Wilcox, 1960; Russell et al., 1963; Garnham, 1966).

Plasmodium (Plasmodium) ovale Stephens, 1922 (Producing ovale malaria)

Synonyms. *Plasmodium vivax minutum* Craig, 1926, *nec* Emin, 1914.

Distribution. According to Garnham (1966), the first acceptable description of *P. ovale* was by MacFie and Ingram (1917) in the Gold Coast (Ghana), an area where it is now common. Stephens, in 1918, found it in a soldier from East Africa and named it *Plasmodium ovale,* now a generally accepted *vivax*-like but separate species of *Plasmodium*. *P. ovale,* which is the least common of the species infecting man, occurs mostly in tropical Africa, principally on the West Coast. It is endemic in Ethiopia.

Gametes. *P. ovale* gametes produce up to eight microgametes that are somewhat smaller in size than those of *P. vivax*.

Ookinetes. The ookinetes appear some 18 to 24 hours after the mosquito ingests gametocytes. These ookinetes are smaller in size than those of *P. vivax*.

Oocysts. The oocysts develop slowly, often requiring about 2 weeks to mature at 27°C, probably never less than 12 days at any temperature. Pigment granules are arranged in two rows that cross approximately at right angles, a unique characteristic.

Sporozoites. *P. ovale* sporozoites are plump, elongate, somewhat pointed at one end, and measure 11 to 12 μm in length.

Exoerythrocytic Stages. Primary *P. ovale* EE schizogony is completed 9 days after the sporozoites enter liver parenchymal cells. The schizont is unusually large and distinctive, confirming the validity of *P. ovale* as a distinct species. Relatively enormous nuclei are characteristic. The mature schizont is lobulated. Secondary EE schizogony occurs (Garnham, 1966).

Erythrocytic Stages. *P. ovale* EE merozoites begin to invade reticulocytes on the ninth day after infection. In this species, exoerythrocytic schizogony requires approximately 50 hours, a little longer than in *P. vivax*. Low density but unusual persistence of parasitemia characterize *P. ovale* infection. There may also be prolonged latency—up to 4 or more years.

Gametocytes. *P. ovale* gametocytes appear on the fifth day of parasite patency. In about 3 weeks, they become sufficiently numerous to infect mosquitoes.

Morphology. It is difficult to differentiate *P. ovale* from *P. vivax* in thin blood films, and it is even more difficult in thick films (see Fig. 14–4 and Table 14–2). In quick-dried, very thin, stained blood films, the ovale-infected red cell is moderately enlarged and presents a characteristic artifact—an oval shape with fimbriation of one or both ends. The artifact is not seen in unstained blood and is often not seen in the presence of uncontrolled tropical humidity.

Stippling with Schüffner's dots appears earlier and is more prominent than in *P. vivax*. Some observers, e.g., Moshkovski (1941), refer to ovale stippling as "James's dots," noting that the dots are fewer but more prominent and more violet than those in vivax-infected cells.

Plasmodium ovale parasites are fairly compact, generally smaller than corresponding stages of *P. vivax,* and display considerably less ameboid activity. In thick films, there is less cytoplasmic scatter and fragmentation than is seen in *P. vivax*. Mature schizonts display fewer and larger merozoites, averaging eight, each with a prominent nucleus. The stained gametocytes closely resemble those of *P. vivax*. The pigment seems darker than that of *P. vivax*.

Garnham (1966) summarizes diagnostic features

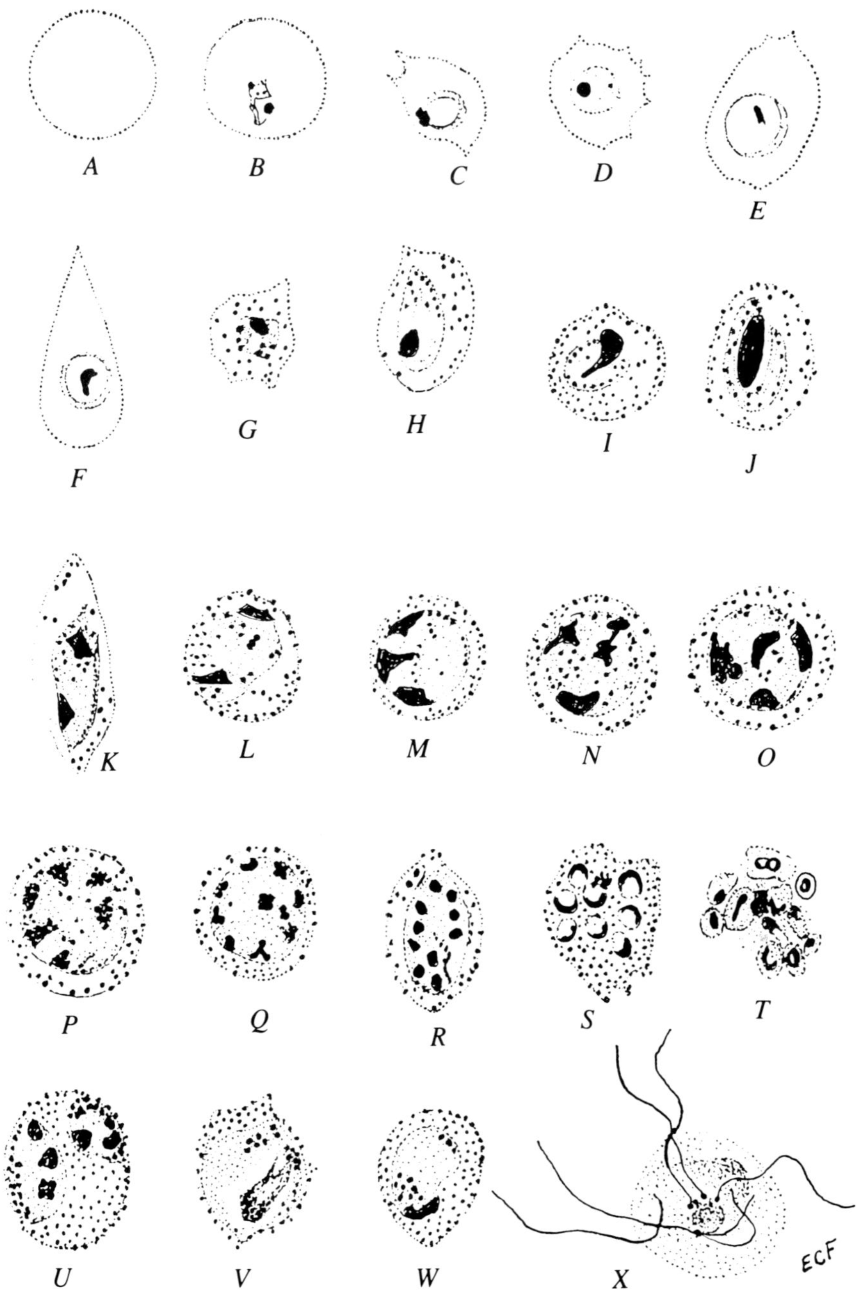

Fig. 14–4. *Plasmodium ovale* in thin-blood films. *A*, Uninfected red blood cell for comparison with *ovale*-infected cells. *B* to *J*, Stages in development of the trophozoite, showing oval irregular contours of many parasitized cells. *K* to *S*, Developing schizont. *T*, Merozoites that have escaped from exhausted red blood cell. *U*, Two developing schizonts in one red cell. *V*, Mature male gametocyte (microgametocyte). *W*, Mature female gametocyte (macrogametocyte). *X*, Exflagellation (maturation) of the microgametocyte, with production of about six microgametes. *G* to *W*, Schüffner's granules in the unoccupied stroma of the parasitized red cell. (*B–R* and *T*, adapted by E.C. Faust from Stephens, J.W.W., and Owen, D.U. 1927. *Plasmodium ovale*. Ann. Trop. Med. Parasitol., *21*. *S* and *U–X*, adapted by E.C. Faust from James, S.P., Nicol, W.D., and Shute, P.G. 1933. *Plasmodium ovale* Stephens, 1922. Parasitology, *25*.)

of *P. ovale* as follows: (1) large nuclei with clear-cut borders; (2) absence of much ameboidicity; (3) low average number of merozoites in primary attacks; (4) darker color of pigment; (5) smaller size.

Plasmodium (Plasmodium) malariae (Grassi and Feletti, 1892) (Producing malariae or quartan malaria)

Synonyms. *Haemamoeba laverani* var. *quartana* Labbé, 1894; *Haemamoeba malariae* Feletti and Grassi, 1899; *Plasmodium malariae quartana* Kruse, 1892; *Haemosporidium tertianae* Lewkowicz, 1897; *Plasmodium quartanae* Billet, 1904; *Plasmodium rodhaini* Brumpt, 1939.

Distribution. The quartan parasite is common in tropical Africa, Burma, Sri Lanka, and parts of India. It is prominent in certain areas of Malaysia and Indonesia. In the past, this species has been indigenous but not common throughout Europe. Its incidence in North America was spotty, and it has been quite uncommon in South America.

Gametes. *P. malariae* gametes are smaller than those of *P. vivax* and *P. falciparum*. Typically, the microgametocytes form eight microgametes but often there are fewer. The microgametes measure about 16 μm in length.

Ookinetes. Little is known about the *P. malariae* ookinetes, although they seem to be more dense than those of other species.

Oocysts. *Plasmodium malariae* does not develop easily in mosquitoes, and no satisfactory experimental vector has been found. Oocysts, usually numerous in *P. vivax,* are scanty in *P. malariae;* hundreds of cysts in the former sharply contrast with two or three in the latter. *P. malariae* oocysts mature in 13 days at 25°C but may not rupture until the sixteenth day (Garnham, 1966). In general, mature *P. malariae* oocysts rupture in 14 days at 30°C, 15 to 20 days at 25°C, and 30 days at 20°C.

Sporozoites. *P. malariae* sporozoites are thicker and coarser than those of other species that infect man. They average 13 to 14 μm in length.

Exoerythrocytic Stages. Primary *P. malariae* EE schizogony is probably completed in from 13 to 16 days. Secondary EE schizogony is now thought by some authorities not to occur. What used to be interpreted as late relapses may instead be recrudescences of prolonged, chronic, low-grade, asymptomatic infection (Garnham, 1977). Such recurrences of symptoms have been observed as late as 40 years after the initial infection, notably after splenectomy (Tsuchida et al., 1982).

Erythrocytic Stages. Malaria parasites in red cells are often scanty and sometimes not found in blood smears until after several typical quartan paroxysms have occurred. The parasite prefers to invade mature red cells, rarely reticulocytes. Schizogony occurs in 72 hours and usually 8, but occasionally up to 12, merozoites are liberated.

Gametocytes. As a rule, *P. malariae* gametocytes are slow to appear, sometimes only after 3 weeks. They have been observed, however, as early as the fifth day of parasite patency and appear, on the average, between the tenth and fourteenth days. Peak gametocyte density usually occurs about 6 days after that of the asexual parasites.

Vertebrate Hosts. *Plasmodium malariae* alone of the four species infecting man is a natural parasite of certain animals (e.g., chimpanzees). Rodhain (1948) demonstrated that chimpanzees were hosts of *P. malariae* in areas of West Africa from Sierra Leone to the Congo and that the *P. rodhaini* of chimpanzees was identical to *P. malariae*. There is complete cross-immunity between the two strains. The vector of the chimpanzee strain has not been demonstrated, so the danger to man is not clear. Although close association between man and chimpanzee is uncommon, *P. malariae-rodhaini* malaria is a potential zoonosis.

Morphology. ***Unstained Preparations.*** *P. malariae* appears in the blood corpuscle as a small, hyaline, ovoid or ring-like trophozoite. Ameboid activity is less than that of *P. vivax,* and at every stage, this species is smaller than the latter. Pigment is less, darker, and in larger grains than in *P. vivax*. In 68 to 72 hours, the *schizont* almost entirely fills the corpuscle and contains a moderate amount of dark pigment in large granules or small masses. The infected corpuscle is not enlarged. At segmentation, pigment usually collects at the center of the parasite, and faint striations in the cytoplasm divide the plasmodium into 6 to 12 segments. The segments, or *merozoites,* and the pigment are liberated in the plasma and appear as minute hyaline bodies.

Gametocytes in unstained preparations are morphologically similar to those of *P. vivax* but are smaller when fully developed, and the corpuscle containing them is not enlarged. Pigment is less in amount and is arranged in larger masses or, in the microgametocyte, in a wreath-like formation within the cytoplasm.

Stained Preparations. Young trophozoites of *P. malariae* resemble those of other plasmodia, but the cytoplasm stains a deeper blue than does that of *P. vivax,* has a thicker "bib" of cytoplasm, which shows little evidence of pseudopodial extension, and a relatively small vacuole (Plate IV).

Older trophozoites and schizonts, both about half-grown, are often ribbon-like in shape (the so-called "band forms,") and stretch directly across the infected corpuscle. Presegmenting and segmenting stages of *P. malariae* resemble those of *P. vivax* except that merozoites number 6 to 12, usually 8, in contrast to the 12 to 18, typically 16, vivax merozoites. *P. malariae* merozoites are notably symmetrical, often referred to as rosettes, "daisy-heads," or "marguerite" formations. The red corpuscle infected with *P. malariae* is not enlarged, but when schizonts are segmenting, they almost fill their host cells. Pigment in the segmenter is a compact mass at the center, surrounded by oval, blue-stained merozoites. The merozoite consists of a deep blue, oval or round mass of cytoplasm enclosing a bright red dot of chromatin usually surrounded by a small unstained area, the vesicular portion of the nucleus.

Gametocytes of *P. malariae* resemble those of *P. vivax* but they are smaller, their host cells are not enlarged, and they are less richly pigmented. Staining reactions of *microgametocytes* and *macrogametocytes* and nuclear morphology of the two species are much alike, so that one description answers well for gametocytes of both species (see pp. 180–181).

Red cells infected with *P. malariae* stain salmon-pink, and the eosinophilic granulations known as Schüffner's dots are not present. With prolonged staining or by the fluorescein-tagged antibody technique (Voller and Bray, 1962), however, a characteristic stippling of *P. malariae*-infected cells may be seen. This is Ziemann's stippling, which has smaller, more irregular, less prominent dots than the Schüffner's stippling in *P. vivax*-infected cells. In addition to structures evident by light microscopy, the trophozoites and schizonts of *P. malariae* have been observed to cause a microcrenation of the surface of the infected erythrocyte evident on electron micrographs (Smith and Theakston, 1970).

Thick Blood Films. Thick films of *P. malariae* are more readily diagnosed than are those of *P. vivax* because the plasmodia have more continuous cytoplasm and appear less shrunken (see Plate V). Early trophozoites may resemble question marks or thick exclamation marks, with chromatin centered between two fragments of cytoplasm. Usually, the cytoplasm is monomorphic, with its chromatin mass to one side. More mature trophozoites are compact and rounded, with a few distinct dots of pigment and an eccentric chromatin mass. Young *schizonts* are difficult to identify because their chromatin masses are poorly defined and their cytoplasm is rather irregular in contour although not dispersed, as it is in *P. vivax*. Ripe schizonts or segmenters generally consist of distinct *merozoites* around a pigment center, with apparently no residual cytoplasm to bind the organism together. *Gametocytes* are characterized by undivided chromatin masses in the female, granular dispersed chromatin in the male, and a wealth of characteristic pigment. Since, in thick films, the hemoglobin of parasitized cells has been dissolved, there are no distinct outlines to guide the diagnostician.

Plasmodium (Laverania) falciparum
(Welch, 1897)
(Producing malignant tertian, estivo-autumnal, or falciparum malaria)

Synonyms. *Oscillaria malariae* Laveran, 1881; *Plasmodium malariae* Marchiafava and Celli, 1885; *Laverania malariae* Feletti and Grassi, 1889; *Ematozoon falciforme* Antolisei and Angelini, 1890; *Haemamoeba laverani* Labbé, 1894; *Haematozoon falciforme* Thayer and Hewetson, 1895; *Haematozoon falciparum* Welch, 1897; *Haemamoeba praecox* Grassi and Feletti, 1890; *Haemamoeba immaculata* Grassi, 1891; *Plasmodium malariae quotidianae* Celli and Sanfelice, 1891; *Plasmodium malariae irregularis* Kruse, 1892; *Haemosporidium sedecimanae* Lewkowicz, 1897; *Haemosporidium undecimanae* Lewkowicz, 1897; *Haemosporidium vigesimo-tertianae* Lewkowicz, 1897; *Haemamoeba febris irregularis* Sakharov, 1892; *Plasmodium tropica* Koch, 1899.

The specific name *"falciparum"* was derived by Welch from *"falx"* (sickle or crescent) and *"parere"* (to bring forth), and it was not intended to mean "crescent-like" as has been suggested.

Distribution. *P. falciparum* is most prevalent in the tropics and subtropics, and "it still remains almost unchallenged as the greatest killer of the human race over most parts of Africa and elsewhere in the tropics" (Garnham, 1966). This plasmodium is also common in tropical and subtropical Asia. It used to be fairly common in the southern United States, along the Mediterranean shores, and elsewhere in the temperate zone, where it is now uncommon following the malaria eradication program.

Gametes. Microgametocytes of *P. falciparum* produce four to six and, at the maximum, eight microgametes by exflagellation, which begins in the host insect's gut about 10 minutes after ingestion at 33°C and after perhaps 2 hours at 14°C. Microgametes are slender and measure about 16 to 25 μm in length. Macrogametocytes become spherical and shed the host cell. Their nuclei approach the edge of the parasite, forming a protuberance into which a microgamete will penetrate.

Ookinetes. *P. falciparum* ookinetes form in 12 to 18 hours after the insect takes its blood meal. The ookinetes are more slender than those of *P.*

vivax; they measure about 2.5 μm in breadth and 11 to 13 μm in length.

Oocysts. *P. falciparum* oocysts are generally smaller than those of other malaria parasites of man and are "peculiarly light-reflecting and look like little globules of glass" (Garnham, 1966). They reach maturity in 9 days at 30°C and 23 days at 20°C. Pigment in the oocyst is black and coarse and often in double rows or chains.

Sporozoites. *P. falciparum* sporozoites are sickle-shaped, with equally pointed ends, and they average 10.5 to 12 μm in length. They invade the salivary glands of the host mosquito 9 days or more after the insect's blood meal and they remain infective for up to about 54 days. (For finer structure, see Sinden and Garnham, 1973.)

Exoerythrocytic Stages. In *P. falciparum,* EE schizogony appears to be limited to a single generation; there is no secondary EE schizogony. The EE schizont grows rapidly, usually coming to maturity between the fifth and sixth days. A fully grown EE schizont may measure some 60 μm or more in diameter and may liberate some 30,000 or more EE merozoites. (For further details, see Shortt et al., 1949, 1951; Jeffery et al., 1952; Bray, 1958, 1960; Garnham, 1966.)

Erythrocytic Stages. *P. falciparum* EE merozoites invade young and old red cells. Many infected reticulocytes are commonly seen. The young erythrocytic trophozoites are polymorphic, appearing not only as hair-like rings but as solid bodies, with odd tenuiform variations. Nuclei are prominent and may be in two parts that are opposite each other or side by side. Some *P. falciparum* rings characteristically lie along the edges of red cells and are called "appliqué" or "accolé" forms, although they are intracellular. Also characteristically, red cells may be invaded by several merozoites, even up to eight. This is not the result of binary fission but of multiple invasion; it may also be the result of two or more merozoites entering a red cell before they have completely separated.

Pigment appears as single grains or specks some 24 hours after the cell is infected, when the ring has become thicker and its vacuole is disappearing. As development of the parasite proceeds, host red cells usually shift from the peripheral circulation to that of the internal organs so that, in later stages of the typically 48-hour cycle, few parasites may be seen in the ordinary blood smear.

Schizogony results in the formation of an average of 16 merozoites, as few as 8 and perhaps as many as 32, although the latter number is exceptional and may be due to the presence of two segmenters in a single cell. The strain of the parasite and the location of the host cell during schizogony may affect the number of merozoites.

As the parasite develops, the host cell becomes darker, with a deep-red edging, and characteristically displays clefts that can be seen after Romanowsky staining, if special care is taken to prepare fresh thin films and a stain buffered to pH 7.2 to 7.4 is applied for a somewhat longer time than usual. This stippling is called "Maurer's clefts."

Falciparum infections may be overwhelming, with invasion of even 50% or more of the red cells. Death of the host is usual if 25% or more of the erythrocytes are infected. Peak density of parasites in falciparum malaria generally comes after about 10 days or more of parasite patency. Often, a rapid decline ensues, but there may be recrudescences at fortnightly or monthly intervals for some 9 months or sometimes longer, after which the infection ends. Since there is no secondary EE schizogony, there are no true or long-term relapses (i.e., recurrences). Cases of active falciparum infection after a long latent period of 2 years or longer in immunologically compromised persons remain unexplained (Duperval et al.,1979).

Gametocytes. *P. falciparum* gametocytes begin their development while their host red cells are in the spleen, bone marrow, and other viscera. At first, the gametocytes are round or ovoid, but they become spindle- , cigar- , or diamond-shaped. The gametocytes display scattered pigment; it is not clumped as it is in the schizonts.

P. falciparum gametocytes appear in the peripheral circulation after 8 to 11 days of parasite patency, and by then, they have assumed their typical crescentic shapes. They soon reach peak density and then decline in numbers, disappearing in about 3 months as a rule. The microgametocyte is more banana-shaped than crescentic, and is thicker and blunter than the female macrogametocyte, which stains a darker blue and has a more compact nucleus. The host cell of falciparum gametocytes may display unique corpuscular inclusions that were first reported by Garnham (1933) and described by Field and Shute as *Garnham's bodies.* These are thick filaments, bars, or loops that stain deep red.

Morphology. Schizogony of *P. falciparum* is completed in from 36 to 48 hours, usually 48 hours. The rhythm of schizogony in *P. falciparum* infec-

PLATE II

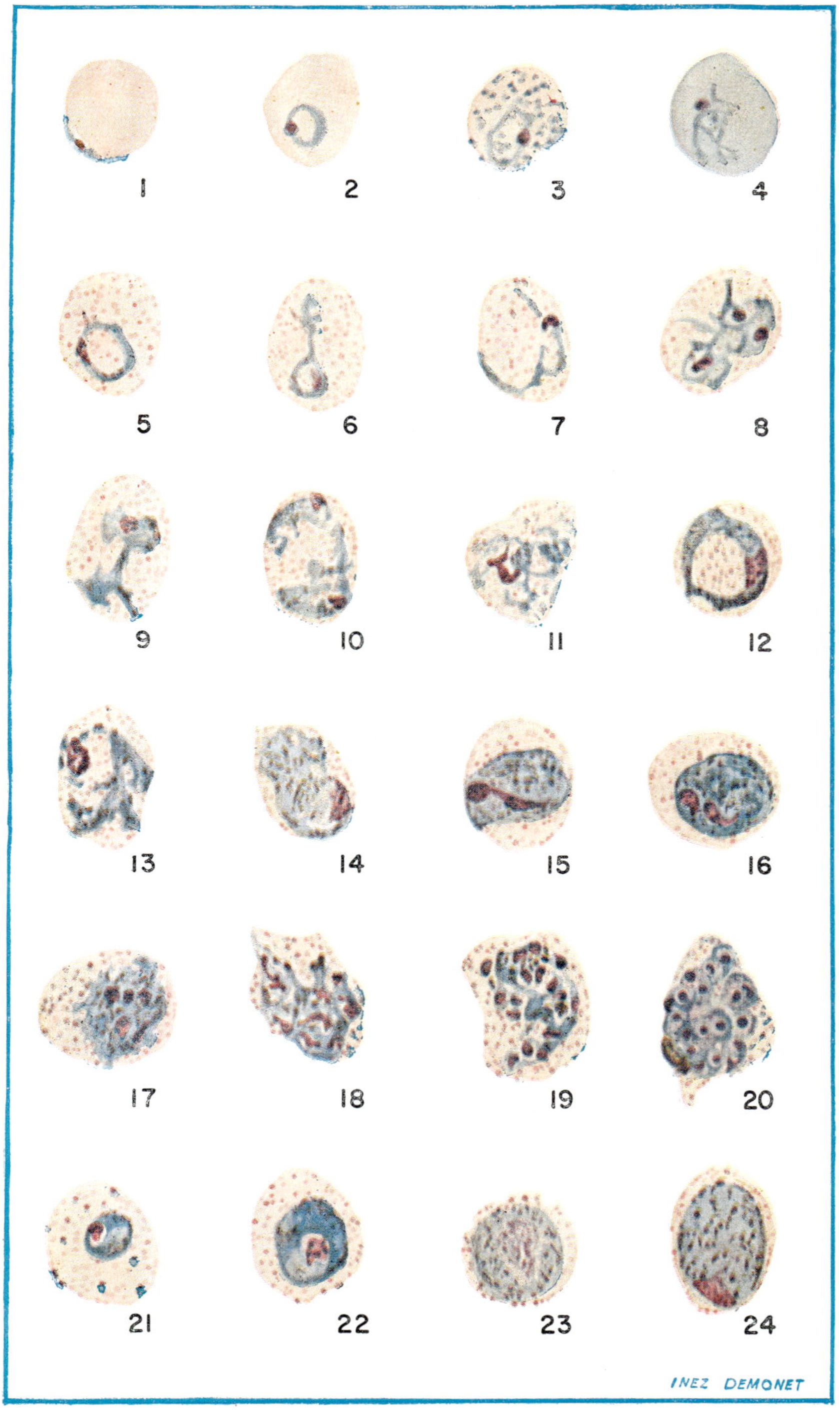

Stages of *Plasmodium vivax* in human erythrocytes from thin-film preparation. 1, Normal-sized red cell with marginal ring-form trophozoite. 2, Young signet-ring trophozoite in a red cell. 3, Slightly older ring form showing basophilic stippling. 4, Polychromatophilic red cell containing young parasite with pseudopodia. 5, Ring-form trophozoite showing pigment in cytoplasm in an enlarged cell containing Schüffner's stippling.* 6,7, Tenuous medium trophozoite forms. 8, Three ameboid trophozoites with fused cytoplasm. 9, 11–13, Older ameboid trophozoites in process of development. 10, Two ameboid trophozoites in one cell. 14, Mature trophozoite. 15, Mature trophozoite with chromatin apparently in process of division. 16–19, Schizonts showing progressive steps in division (''presegmenting schizonts''). 20, Mature schizont. 21, 22, Developing gametocytes. 23, Mature microgametocyte. 24, Mature macrogametocyte. (From Wilcox, A. 1960. Manual for the microscopical diagosis of malaria in man. National Institutes of Health Bulletin, *796.)*

**Schüffner's stippling does not appear in every infected cell, as would be indicated by these drawings, but it can be found with any stage from the fairly young ring form onward.*

PLATE III

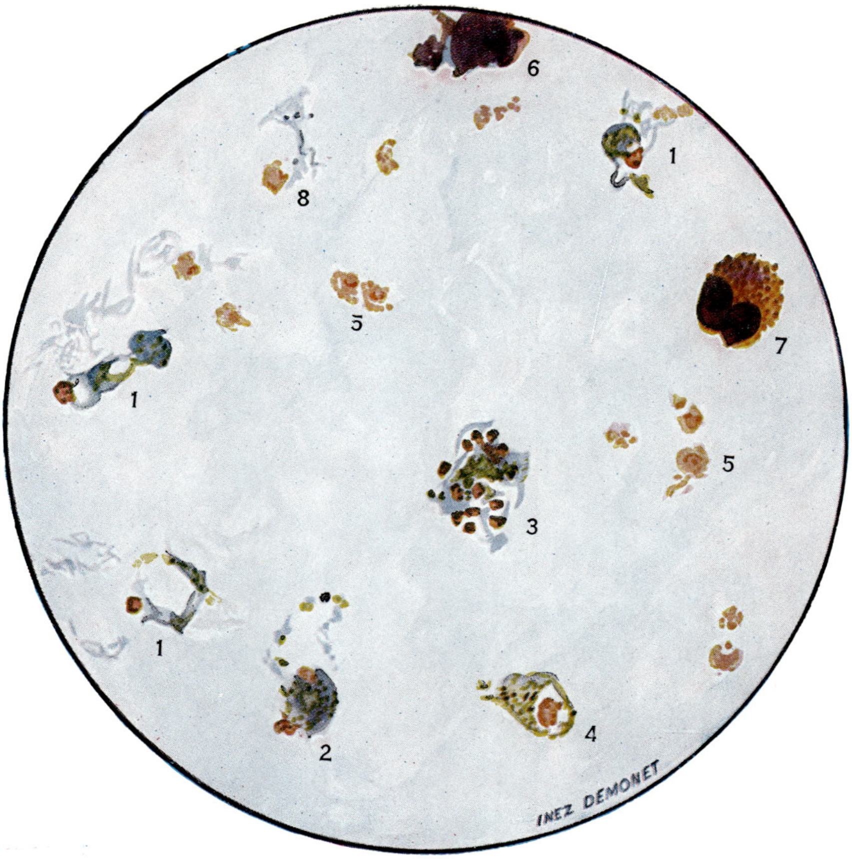

Plasmodium vivax, thick-film preparation. The film contains three developing trophozoites, an immature and a mature schizont, a microgametocyte, clumps of blood platelets, an eosinophil, and a neutrophil. 1, Ameboid trophozoites. 2, Schizont—two divisions of chromatin. 3, Mature schizont. 4, Microgametocyte. 5, Blood platelets. 6, Nucleus of neutrophil. 7, Eosinophil. 8, Blood platelet associated with cellular remains of young erythrocytes. (From Wilcox, A. 1960. Manual for the microscopical diagnosis of malaria in man. National Institutes of Health Bulletin, *796.)*

PLATE IV

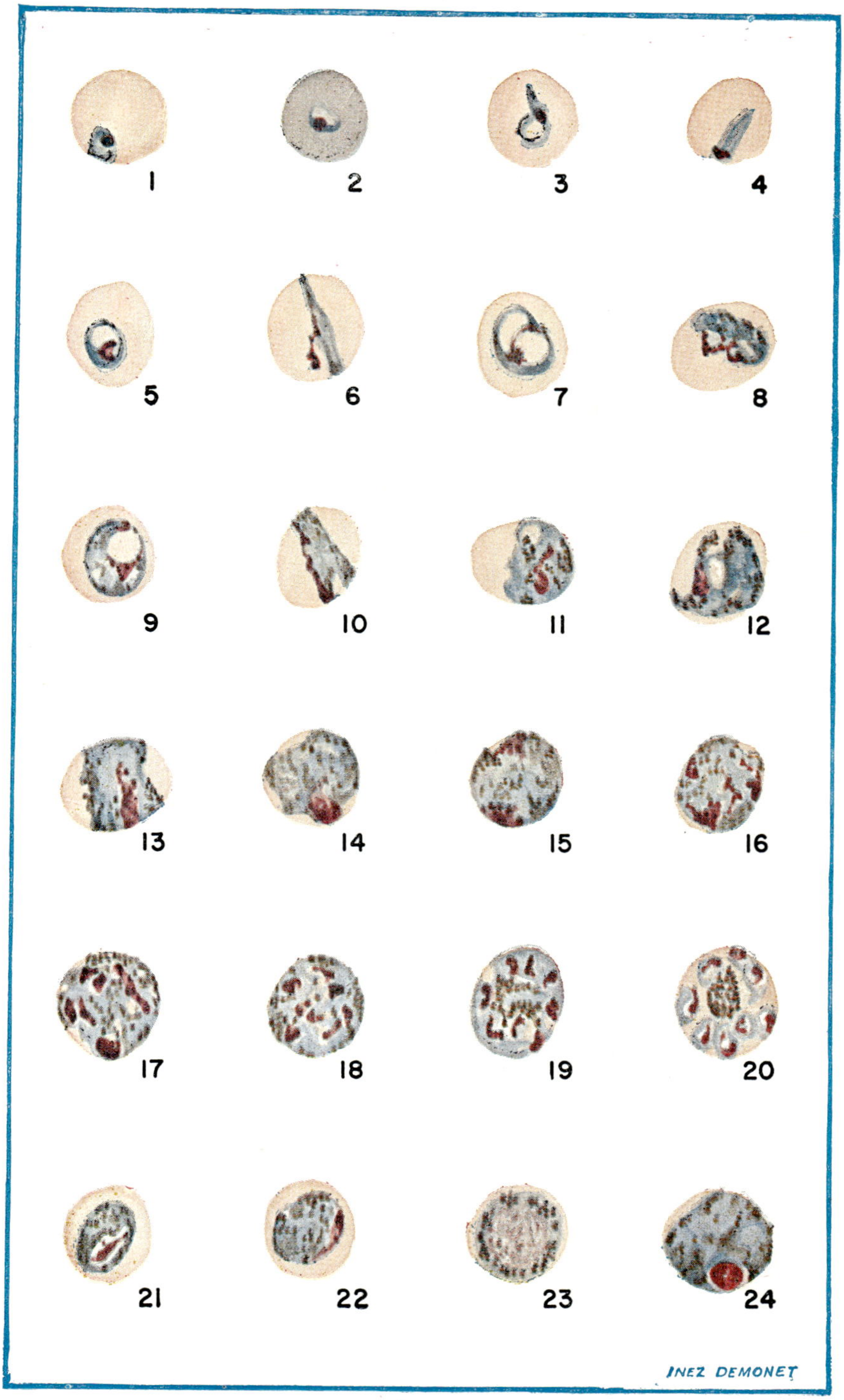

Stages of *Plasmodium malariae* in human erythrocytes from thin-film preparation. 1, Young ring-form trophozoite. 2–4, Young trophozoite forms showing gradual increase of chromatin and cytoplasm. 5, Developing ring form showing pigment granule. 6, Early band form—elongated chromatin, some pigment apparent. 7–12, Some forms that the developing trophozoite may take. 13, 14, Mature trophozoites—one a band form. 15–19, Phases in the development of the schizont (''presegmenting schizonts''). 20, Mature schizont. 21, Immature microgametocyte. 22, Immature macrogametocyte. 23, Mature microgametocyte. 24, Mature macrogametocyte. (From Wilcox, A. 1960. Manual for the microscopical diagnosis of malaria in man. National Institutes of Health Bulletin, *796.)*

PLATE V

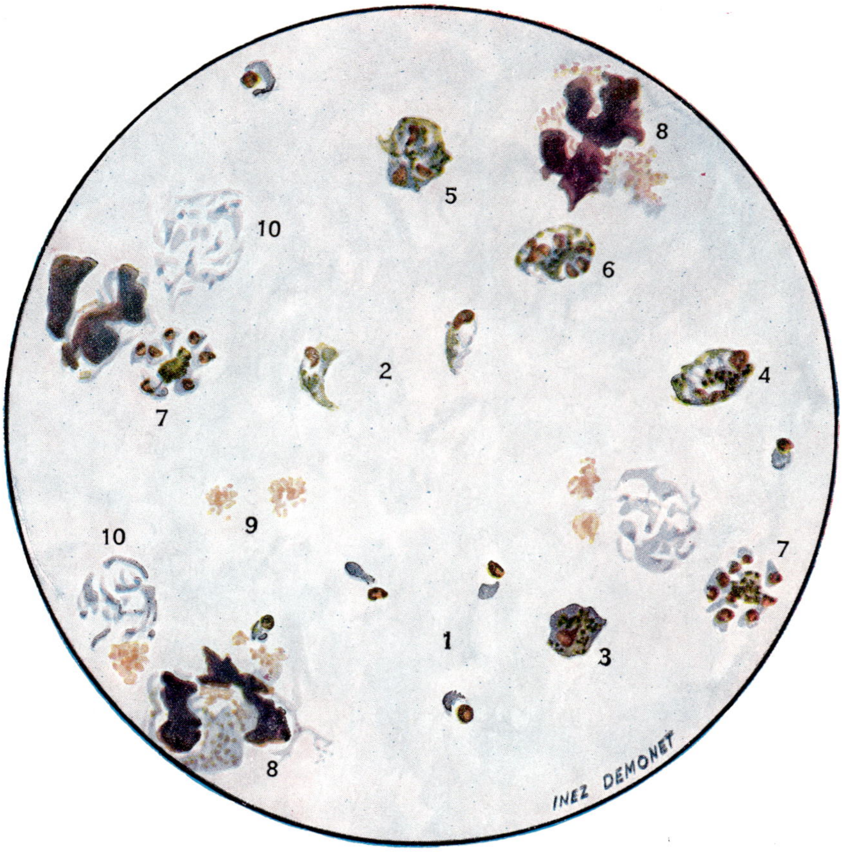

Plasmodium malariae, thick-film preparation. The film contains young and developing trophozoites, mature trophozoites, young and mature schizonts, neutrophilic leukocytes, blood platelets, and remains of red cell stroma. 1, Small trophozoites. 2, Growing trophozoites. 3, Mature trophozoites. 4–6, Schizonts (presegmenting) with varying number of divisions of the chromatin. 7, Mature schizonts. 8, Nucleus of leukocyte. 9, Blood platelets. 10, Cellular remains of young erythrocytes. (From Wilcox, A. 1960. Manual of microscopical diagnosis of malaria in man. National Institutes of Health Bulletin, *796.)*

PLATE VI

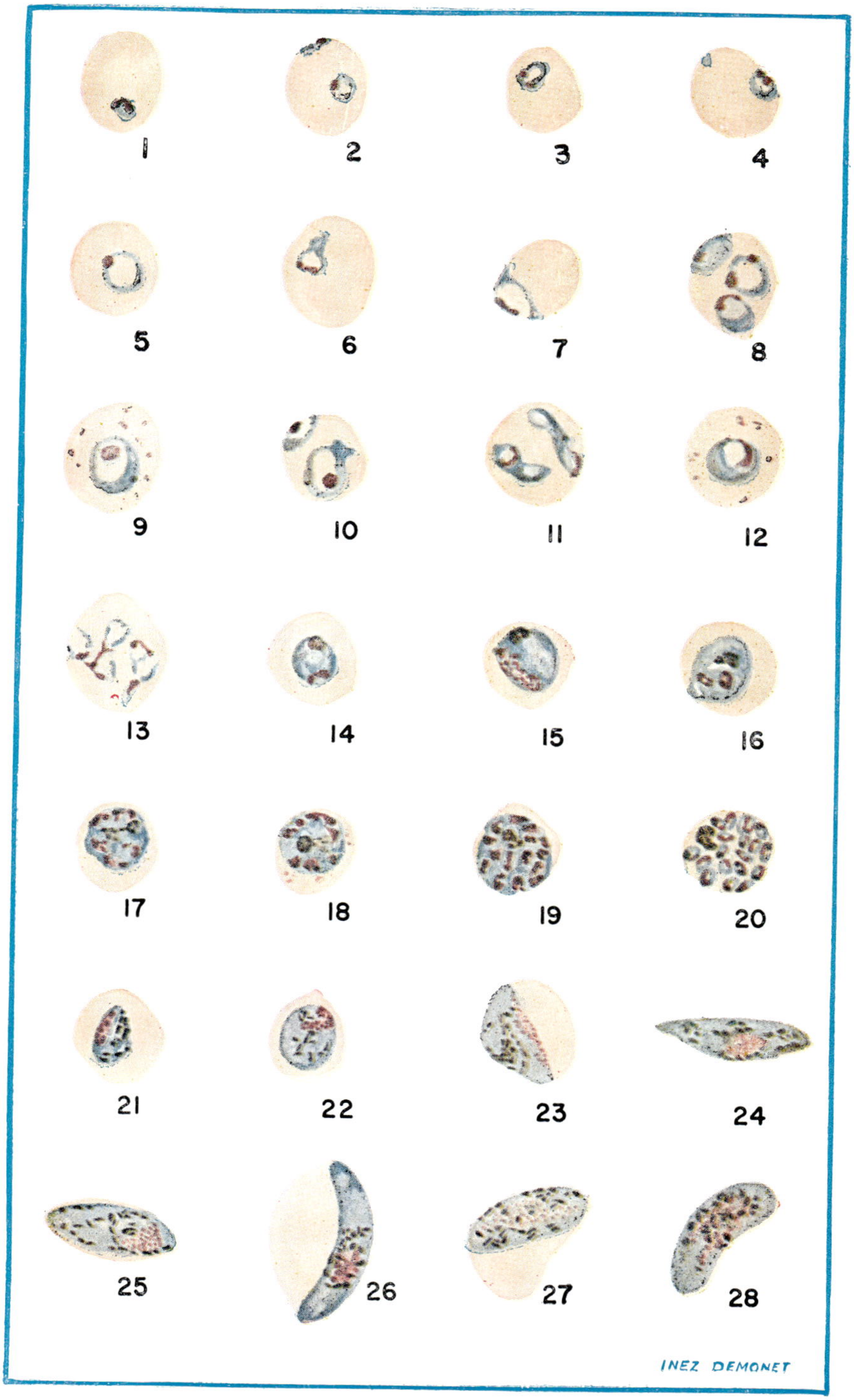

Stages of *Plasmodium falciparum* in human erythrocytes from thin-film preparation. 1, Very young ring-form trophozoite. 2, Double infection of single cell with young trophozoites, one a "marginal form," the other, "signet-ring" form. 3, 4, Young trophozoites showing double chromatin dots. 5–7, Developing trophozoite forms. 8, Three medium trophozoites in one cell. 9, Trophozoite showing pigment, in a cell containing *Maurer's spots*. 10, 11, Two trophozoites in each of two cells showing variation of forms that parasites may assume. 12, Almost mature trophozoite showing haze of pigment throughout cytoplasm. Maurer's spots in the cell. 13, Multiple rings. 14, Mature trophozoite, showing clumped pigment. 15, Parasite in the process of initial chromatin division. 16–19, Various phases of the development of the schizont ("presegmenting schizonts"). 20, Mature schizont. 21–24, Successive forms in the development of the gametocyte, usually not found in the peripheral circulation. 25, Immature macrogametocyte. 26, Mature macrogametocyte. 27, Immature microgametocyte. 28, Mature microgametocyte. (From Wilcox, A. 1960. Manual for the microscopical diagnosis of malaria in man. National Institutes of Health Bulletin, *796.)*

PLATE VII

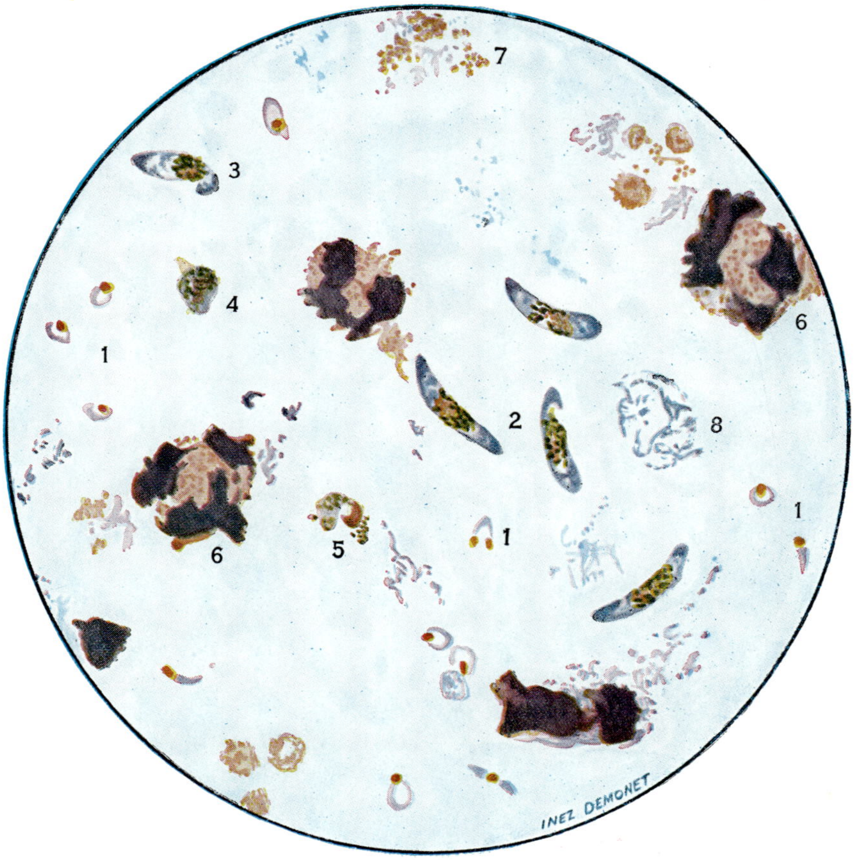

Plasmodium falciparum thick-film preparation. The film contains numerous ring-stage trophozoites, five normal gametocytes, one rounded-up and one degenerate gametocyte, leukocytes, blood platelets, and remains of red cell stroma. 1, Small trophozoites. 2, Gametocyte—normal. 3, Slightly distorted gametocyte. 4, "Rounded-up" gametocyte. 5, Disintegrated gametocyte. 6, Nucleus of leukocyte. 7, Blood platelets. 8, Cellular remains of young erythrocyte. (From Wilcox, A. 1960. Manual for microscopical diagnosis of malaria in man. National Institutes of Health Bulletin, *796*.)

tion is frequently less regular than in the other plasmodia infecting man.

Usually only young trophozoites, or "ring forms," and gametocytes of *P. falciparum* are seen in peripheral blood smears, because *P. falciparum*-infected cells tend to retreat to and linger in capillaries and blood sinuses of internal organs and bone marrow. In severe pernicious infections, however, pigmented and segmented schizonts may be found in films of peripheral blood.

Unstained Preparations. *Trophozoites* of *P. falciparum* appear first as hyaline, usually "ring-shaped" bodies, which are well defined. Multiple infection with two to six or even more plasmodia in a red cell is common. Ring forms increase slowly in size and, at the end of 24 hours, occupy about one-sixth of the infected corpuscle. They are ameboid, sending out and withdrawing minute pointed pseudopodia with no change in their position within the corpuscle. As the ring forms enlarge, an increase in hyaline cytoplasm develops, usually at one side of the ring, and in this portion, minute grains of pigment may sometimes be observed after 24 hours. The ring form is often delicate in appearance, consisting of a thread-like hyaline ring.

The *gametocytes* of *P. falciparum* are easily recognized in unstained preparations by their peculiar shape, the microgametocytes being shaped like a kidney or lima bean, while the macrogametocytes are typically crescentic in shape. They are hyaline in appearance and often present a double outline.

Stained Preparations. Staining reactions of *P. falciparum* are similar to those of the other plasmodia of man: cytoplasm stains blue; chromatin of the nucleus stains red or violet; and the vesicular portion of the nucleus, or vacuole, remains unstained (see Plate VI).

Trophozoites, or "ring forms," appear as typical blue-stained rings of cytoplasm lying within the red corpuscle and having, at one portion of the periphery, a red or violet dot of chromatin. The rings often present double dots of chromatin, which are close together or separated and sometimes at opposite sides of the ring. Multiple infection of erythrocytes is common. The larger ring forms present an expanded portion of blue cytoplasm usually opposite the chromatin dot, and in this portion, pigment is observed as fine black grains. The larger pigmented plasmodia may be seen in stained preparations in blood from pernicious infections or from placental blood; they have blue-stained cytoplasm with a small amount of chromatin, while in the early stages of segmentation, the chromatin is collected in small, irregular, red-stained clumps within the blue-stained cytoplasm. The maturing *schizonts* present a collection of minute, blue-stained bodies, each containing a red dot of chromatin. These daughter bodies surround a mass of pigment. The number of segments, or *merozoites,* is stated by different authorities to be from 8 to 32, the average being 16. The size and number of merozoites formed in a mature schizont apparently depend on the strain of parasite and on the organ in which the host cell lingers (Garnham, 1966).

In stained preparations, trophozoites of *P. falciparum* sometimes have irregular or bizarre shapes that resemble those of *P. vivax*. These forms seem to be produced when making the blood smears. Such forms (e.g., the so-called *P. tenue)* have been described incorrectly as new species.

The host cell may show reddish clefts or blotches called *"Maurer's clefts,"* which are probably due to cytoplasmic damage caused by the parasite. Infected cells often take a bluish color instead of the pink seen with other malaria parasites, and the periphery of the corpuscle frequently is crenated.

P. falciparum "appliqué forms" are common; the intracellular plasmodium appears as a blue-stained thread or blister of cytoplasm containing a red dot of chromatin and applied to the periphery of the corpuscle. Similar *P. vivax* and *P. malariae* forms are much less common.

Smears of peripheral blood in falciparum infections characteristically reveal few, if any, half-grown or maturing trophozoites, although Young (1963) reported that such forms were fairly common in the early stages of attacks that he observed. Sometimes the microscopist rules out falciparum malaria if the films show plasmodia larger than the signet-ring forms; this may be a mistake, because developing forms of *P. falciparum* in the peripheral circulation may be a danger signal of a severe infection calling for urgent treatment. Shute (1968) noted that never more than one or two lumps of pigment are present in the cytoplasm of a falciparum trophozoite, whereas in the other three species, all forms larger than rings contain 20 to 50 grains of pigment.

In stained preparations, *gametocytes* of *P. falciparum* appear at first as ovoidal or rounded, blue-stained bodies that are about 1.5 μm in diameter and have a red dot of chromatin at the center. As development continues, these bodies enlarge and gradually become crescentic until, when fully developed, they are definitely crescent- or kidney-like in shape. Corpuscle remnants appear as red-stained borders or pink-stained thin masses of cytoplasm connecting the poles of the crescent, forming a "bib." The average length of the crescents is about one and a half times the diameter of a red corpuscle, and the breadth is about one-half this diameter. The *microgametocyte* stains a pinkish-blue or deeper violet-red color, with red-stained chromatin collected in fine grains over a considerable portion of the cytoplasm, while the pigment is scattered granules and minute rods. The *macrogametocyte* stains deep blue or bluish-lavender, the red chromatin is collected in a compact mass near or at the center, and the pigment occurs in larger granules, sometimes surrounding the nucleus in a wreath-like manner. Microgametocytes are shaped like kidney or lima beans with bluntly rounded ends, while the macrogametocytes are typically more crescentic in shape with sharply rounded or pointed ends. The early stages in the development of the gametocytes are seldom observed in the peripheral blood, and only a small proportion of the crescentic bodies is found there; most of the crescentic bodies are present in the capillaries of the internal organs. (For exflagellation of a microgametocyte, see Fig. 14–4,X.)

Thick Blood Films. Although the asexual stages of *P. falciparum* are the smallest of the plasmodia infecting man, they are readily recognized and diagnosed in thick films (Plate VII). The young trophozoite is a complete or partial azure-stained ring, and the chromatin dot takes an intensely deep red stain. The more mature trophozoites and the schizonts are usually not found in peripheral blood. In placental blood, a mature trophozoite tends to become an irregular solid mass with obliteration of the vacuole and with a moderately large deep-red dot of chromatin that is eccentric in position. Schizonts consist of dense masses of purplish-red chromatin that are surrounded by cytoplasmic envelopes. The pigment is readily recognized as a few grains or one small mass of dark material. Mature gametocytes ("crescents") are frequently as characteristic as in thin films, but they may be falciform, ovoid, or irregular in contour. All have a wealth of pigment, which is either dispersed (male gametocytes) or clumped together (female gametocytes).

The hemoglobin of the parasitized cell has been dissolved, so that the outline of the erythrocyte is indistinct.

DIFFERENTIAL DIAGNOSIS OF THE MALARIA PARASITES OF MAN

Epidemiologically and clinically, it is important to identify the species of malaria parasites present in an infection. The principal diagnostic features of the four plasmodia of man are given in Table 14–2.

Differentiation of Gametocytes. Differentiation of gametocytes from schizonts is important from the standpoint of epidemiology and prophylaxis. *Gametocytes* of *P. vivax, P. malariae,* and *P. ovale* are easily identified by attention to the following rule: any rounded or ovoid plasmodium of these species that almost entirely fills the infected corpuscle and that, in properly stained specimens, does not show any division of the chromatin into *distinct masses* distributed throughout the cytoplasm is a gametocyte. The gametocytes of *P. falciparum* are identified by their crescentic or kidney-bean shape.

Microgametocytes of the four species are identified by the pale blue or dark violet-red staining of the cytoplasm, while the chromatin of the nucleus in those of *P. vivax, P. malariae,* and *P. ovale* is arranged in a loose mass of fibrils or granules, spindle-like in shape, lying in an unstained oval area, often stretching across the body of the gametocyte. The microgametocyte of *P. falciparum* is not completely crescentic but has broadly rounded ends and is shaped like a kidney or lima bean.

The *macrogametocytes* of the four species are identified by the deep-blue staining of the cytoplasm, while the chromatin of the nucleus in those of *P. vivax, P. malariae,* and *P. ovale* is collected in a dense mass lying in an unstained area either at the center or to one side of the body of the gametocyte. The macrogametocyte of *P. falciparum* is more slender than the microgametocyte and more typically crescentic with sharply rounded or pointed ends.

Geographical Distribution of Malaria. Since 1920, malaria has been reported as far north as the Dvina River, near Archangel, in the USSR (64°), and as far south as Cordova, in Argentina (32°). It has occurred as high as 2770 meters (9086 ft.) in the Cochabamba region of Bolivia, where *Anopheles pseudopunctipennis* is the vector, and at 2591 meters (8500 ft.) near Londiani in Kenya, Africa, where imported *An. gambiae* s.l. and *An. funestus* shelter in native huts. In Tadzhik, USSR

(Tadzhikistan), in south Central Asia, it is said that *An. superpictus* is a malaria vector at 2850 meters (9348 ft.). In contrast, the disease has also been known in the Dead Sea Basin, 400 meters (1312 ft.) below sea level. Of course, within these limits of altitude and latitude, many areas have never been infected; for example, the Hawaiian Islands, the Galapagos, Marquesas, Fiji, New Caledonia, New Zealand, Marshalls, and Carolinas are entirely without *Anopheles* and, thus, are not malarious. On Guam, *An. subpictus* appeared after the beginning of World War II, but up to the present time, malaria transmission has not taken place.

Very likely, malaria reached its all-time maximum world dissemination between 1855, when it was found in southern Canada, and 1922 to 1923, when it touched the Arctic Circle in Russia. In the period that followed, there was a downward trend in both distribution and prevalence, which was considerably more marked in some countries than in others, until around 1969, when efforts toward malaria eradication began to wane. In the 10-year period to 1979, the control programs in several countries were allowed to deteriorate, resulting in a marked resurgence of the disease. Tabulation of reports by the World Health Organization for the 6-year period from 1972 to 1977 showed an overall increase in the number of cases reported for the Americas (from 284,000 to 398,000), Southeast Asia (from 1,920,000 to 5,552,000), Europe (21,000 to 116,000), and the Western Pacific (from 171,000 to 214,000, excluding areas not reported). In the same period, the number of cases reported for the eastern Mediterranean declined from 855,000 in 1972 to 224,000 in 1977. The increase in the number of cases recorded in the European region was due largely to malaria having become epidemic in eastern Turkey in 1976. Because the malaria cases in most countries of Africa were not reported systematically, that region was omitted from the summary. However, it is well known that most of Africa south of the Sahara, excluding South Africa, is highly malarious. Also of significance in interpreting the above records is the observation that, with the change from malaria eradication to malaria control in many countries, the search for cases had been curtailed significantly (W.H.O., 1979). On the other hand, it must be stressed that, despite the notable resurgence of malaria in a large part of the world, satisfactory progress toward effective control has continued in Mexico and in parts of India and some other countries, in addition to areas in the eastern Mediterranean region. Among 143 countries or areas where malaria had been endemic and for which information was available in 1979, there were 37 countries or areas that had been freed from malaria and 16 others in which the malaria risk had become minimal.

Epidemiology. Epidemiologists deal with malaria as it appears, spreads, recedes, or disappears in a community. The disease is *endemic* where there is a constant measurable incidence of cases and natural transmission over a succession of years. It is *epidemic* where the incidence of cases in an area rises rapidly and notably above the usual level, or where the disease suddenly occurs in an area previously nonmalarious. (Unusual seasonal increased incidence is not termed an epidemic.) An epidemic becomes *pandemic* when it spreads far beyond usual limits.

Cases of malaria contracted locally are called *autochthonous;* when the infections are acquired outside the area and brought in, they are called *imported*. Cases proved to be locally derived from imported malaria are called *introduced*. Malaria cases are *indigenous* if they are natural to an area. In areas in which autochthonous cases are few and scattered, malaria is said to be *sporadic*. Infections acquired by parenteral inoculation (e.g., by blood transfusion) are *induced*. The prevailing frequency and intensity of endemic malaria are referred to as *malaria endemicity* (W.H.O., 1963).

Malaria is *holoendemic* when the spleen rate (frequency of splenomegaly) in children (2 to 9 years old) is constantly over 75% and adult populations have high tolerance; *hyperendemic* when the children's spleen rate is constantly over 50% and the rate is also high in adults; *mesoendemic* when the children's spleen rate is 11 to 50%; and *hypoendemic* when the spleen rate is 10% or less.

When prevalence is relatively steady from year to year and season to season, malaria is said to be *stable*, but when there are wide differences in incidence from year to year or season to season, malaria is *unstable*.

Malaria ceases to be endemic in any country or region in which no autochthonous case has occurred for 3 years—assuming, of course, that special efforts have been made to determine this fact. If the disease is introduced into a nonmalarious area and is naturally transmitted in 2 successive years, then, in the third and succeeding years of natural transmission, it may properly be defined as endemic.

Table 14–2. Comparative characters of plasmodia of man (stained thin smears).

Stage or Period	*Plasmodium vivax*	*Plasmodium ovale*	*Plasmodium malariae*	*Plasmodium falciparum*
Early trophozoite or ring	Relatively large; usually one prominent chromatin dot, sometimes two; often two rings in one cell	Compact; one chromatin dot; two rings in one cell uncommon	Compact; one chromatin dot; two rings in one cell rare	Small, delicate; two chromatin dots and two rings in red cell common; appliqué forms frequent
Late trophozoite	Large; markedly ameboid; abundant chromatin; prominent vacuole; pigment in fine rods	Small; compact; not ameboid; vacuole inconspicuous; pigment coarse	Smaller than vivax; compact; often band-shaped; not ameboid; vacuole inconspicuous; pigment coarse	Medium size; usually compact, rarely ameboid; vacuole inconspicuous; rare in peripheral blood after half grown; pigment granular
Young schizont or pre-segmenter	Large; somewhat ameboid; dividing chromatin masses numerous; pigment in fine rods	Medium size; compact; chromatin masses few; pigment coarse	Small; compact; chromatin masses few; pigment coarse	Small; compact; chromatin masses numerous; single pigment mass; rare in peripheral blood
Mature schizont or segmenter	Schizonts and merozoites large; pigment coalescent	Merozoites larger than in *P. malariae*; irregular rosette	Schizonts small, but merozoites large	Small merozoites; single pigment mass
Number of merozoites	12–24, usually 12–18	6–12, usually 8	6–12, usually 8	8–26, usually 8–18
Microgametocytes (usually smaller and less numerous than macrogametocytes)	Spherical; compact; no vacuole; undivided chromatin; diffuse coarse pigment; cytoplasm stains light blue	Similar to *P. vivax* but somewhat smaller; never abundant	Similar to *P. vivax* but smaller and less numerous	Crescents usually sausage-shaped; chromatin diffuse; pigment scattered large grains; nucleus rather large; cytoplasm stains pale blue

Macrogametocytes	Spherical, compact, larger than microgametocyte; smaller nucleus; pigment same; cytoplasm stains darker blue	Similar to *P. vivax* but somewhat smaller; never abundant	Similar to *P. vivax* but smaller and less numerous	Crescents often longer and more slender; chromatin central; pigment more compact; nucleus compact; cytoplasm stains darker blue
Pigment except in mature schizonts	Short, rather delicate rods, irregular scattered; not much tendency to coalesce	Similar to but somewhat coarser than *P. vivax;* sometimes clumped or in lateral bands	Seen in young rings; granules rather than rods; tendency toward peripheral scatter	Pigment granular; early tendency to coalesce; typical single solid mass in mature trophozoite; coarse "rice grains" in crescents
Alterations in the infected red cell	Enlarged and decolorized; Schüffner's dots usually seen	Enlarged; decolorized; prominent Schüffner's dots appear early; infected cells may be oval-shaped with fimbriated ends	Cell may seem smaller; fine stippling (Ziemann's dots) occasionally seen	Normal size; Maurer's dots (or "clefts") common; Garnham's bodies occasionally seen
Length of asexual phase	48 hours	about 48 hours	72 hours	36–48 hours, usually 48
Prepatent period, minimal	8 days	9 days	14 days	5 days, average 8–12
Usual incubation period	8–31 days, average 14	11–16 days	28–37 days, average 30	7–27 days, average 12
Interval between parasite patency and gametocyte appearance	3–5 days	5 to 6 days; appearance irregular and numbers few	10–14 days; appearance irregular and numbers few	8 to 11 days
Developmental period in mosquito	30 days or more at 17.5°C; 16–17 days at 20°C; 10 days at 25–30°C	16 days at 25°C; 14 days at 27°C	30–35 days at 20°C; 25–28 days at 22–24°C	22–23 days at 20°C; 10–12 days at 27°C (80°F)

Endemic and epidemic malaria result from a complex interplay of numerous factors, the most important being, of course, man as gametocyte carrier, the *Anopheles* mosquito as vector, and man as recipient of an infection. The following factors modify this chain of transmission from man to mosquito to man: the species and strain of plasmodium; immunity in man; bionomics or habits of man and mosquito; environmental conditions, such as temperature, relative humidity, rainfall, topography, soil, flora, and fauna; control measures applied to mosquito; and treatment of man by therapeutic and prophylactic drugs.

Man the Donor. From a practical standpoint, man is the only source for infecting mosquitoes with human malaria, although in areas of central Africa and Malaya, the possibility of human infections derived from chimpanzees or monkeys in nature has been demonstrated (pp. 174–175).

Man may infect man with malaria in the absence of the mosquito vector when blood containing erythrocytic trophozoites is transferred by injection or transfusion. There are also congenital infections, but they are relatively rare, even in highly endemic areas.

To be infective to mosquitoes, a man must be accessible to vector mosquitoes and must have malaria gametocytes in his peripheral blood. Gametocytes are not invariably present in malaria infections in sufficient numbers or in suitable condition to infect mosquitoes. Gametocytes appear in the blood some days after the first trophozoites are seen in red cells, the time lag being generally from 3 to 5 days in *P. vivax,* 5 to 6 days in *P. ovale,* 10 to 14 days in *P. malariae,* and 8 to 11 days in *P. falciparum*. Occasionally, gametocytes appear before the onset of symptoms. Numbers, age, sex distribution, and quality of gametocytes influence the infectivity of blood for mosquitoes. Green (1929) found that, to infect *A. maculatus,* peripheral blood must contain at least 1 gametocyte of *P. falciparum* to 200 leukocytes, 1 of *P. vivax* to 1000 leukocytes, and 1 of *P. malariae* to 330 leukocytes. He noted also that, when there were many more macrogametocytes than microgametocytes, fewer mosquitoes were infected. Young gametocytes and those that are overripe do not develop in mosquitoes. Finally, the quality of gametocytes appears to vary, so that some gametocyte-carrying individuals are "poor infectors" and others "good infectors." The basic reason for this is not apparent. Gametocytes appear and disappear in the peripheral blood without apparent reason and sometimes suddenly.

The chief seed-bed, or "gametocyte pool," of an area is often in the poorest section, where treatment of malaria is least satisfactory and the principal carriers are children aged 1 to 4 years who may not have symptoms of acute illness, although they generally have enlarged spleens.

The Mosquito Vector. The natural history of anophelines is an important part of the malaria formula, since the only natural vectors are certain species of the genus *Anopheles* (see Table 41–2, pp. 640–642). The developmental cycle of plasmodia in mosquitoes varies with external temperatures and does not proceed below 15.6°C (60°F) or much above 37.8°C (100°F). The cycle averages 10 days at 25°C (77°F) and 16 days at 20°C (68°F) for *P. vivax;* 22 days at 20°C for *P. falciparum;* and 30 to 35 days at 20°C for *P. malariae.*

Malaria infection in the mosquito is never transovarial. Sporozoites have remained infective in a mosquito for as long as 90 days. Plasmodia may remain dormant in a hibernating mosquito during the winter and renew their development in the insect in the spring.

The main reasons why a particular species of mosquito is a vector of human malaria are the following: (1) it is susceptible to infection by the parasite; (2) it is present in effective numbers near human habitations; (3) it repeatedly takes human rather than animal blood; and (4) on the average, it lives long enough for gametocytes to develop into sporozoites. Any factors of temperature, relative humidity, rainfall, topography, soil, flora, and fauna that influence the density of a susceptible species, the species' habits *vis-à-vis* man (e.g., indoor and outdoor feeding), its length of life, or the development of the plasmodium in its body will modify the malaria formula, increasing or decreasing the amount of malaria in an area.

The mosquito vector and gametocyte carrier are two important factors in the rise and fall of malaria incidence. A *critical density* level exists for each species of malaria vector; the species that have closest contact with man have the lowest critical density levels. When the population of the vector falls below this critical number in a locality, malaria transmission tapers off or ceases. Gametocytes, too, have a critical density in a community. If there are too few gametocyte carriers in an area, there will be no significant malaria transmission. The critical density of neither the vector nor the gametocyte is fixed or independent. Each depends largely on the other, in inverse ratio, but there will be a fluctuating minimum level below which malaria transmission is unlikely. When the two factors concurrently fall, the malaria transmission rate drops notably, and when the two factors rise concurrently, an epidemic may be expected (Macdonald, 1957).

Sporozoite indices in various species of vector anophelines vary greatly, depending on such factors as the degree of contact with man and the height of malaria endemicity at the time the index is taken. For example, *An. culicifacies,* the only vector in some endemic areas of South India, often had the low index of about 1 infection per 1000 dissections; yet, in times of epidemics in the Punjab, it has shown indices above 25%. In other endemic areas of south India, *An. fluviatilis* usually exhibited sporozoite indices 10 to 20 times higher than *An. culicifacies* in the same areas. The latter species is much more easily deviated to cattle for its blood meals.

Shute (1945) found that the number of sporozoites in the salivary glands may vary from 950 to more than 200,000 from the fourth to the eleventh day after infection and that about 3000 sporozoites are injected into a person bitten at each feeding.

Man the Victim. The extent to which man becomes a victim of malaria depends, in the first place, on how his own habits affect his accessibility to vector feeding. For example, location and type of housing in reference to vector habits, night travels in malarious areas, outdoor sleeping, migrations, and cooperation in control and eradication operations are all factors that may determine the presence or absence of infection. In the second place, the influence of immunity is a phenomenon that still requires elucidation.

Malaria *immunity* is the combination of host mechanisms that tends to prevent plasmodial infection, reinfection, or superinfection, that destroys *Plasmodium* or limits its multiplication, and that modifies its effect on the body and aids specifically in repair of tissue. This immunity may be *natural* or *acquired*. Natural immunity is genetic or innate, independent of previous infection. Acquired immunity is the result of antigenic stimulation by the parasite or its products; it may be *passive* (conferred by maternal transfer or by injection) or *active* (due to malaria infection). Active immunity may be *concomitant* (present while there is a parasitemia) or *residual* (persisting after eradication of parasites). *Premunition* is a specific immune response to malarial infection that results in clinical recovery and resistance to specific challenge but that permits the persistence of a low-grade parasitemia. *Tolerance* is a term applied to an infection immunity that lessens the effects on the host of a given quantum of parasites, often allowing infection without acute illness. Variations in virulence occur in different strains of the same *Plasmodium* as well as in different species.

Malaria immunity is specific for homologous strains of parasites but is ineffective against heterologous strains and species. A major protective mechanism for the malaria parasite against the immunity of the host is the continual variation in the parasite's antigenic structure. Since the host's defenses must be not only species and strain specific but also variant specific in order to be fully effective, the parasites are able to escape destruction by undergoing antigenic variation.

Since immunity to malaria is not transferred with passive transfer of serum, and since the development and transfer of protective immunity are dependent on viable thymus-derived (T) cells, Brown (1976) suggested that strain-specific immunity to malaria is mediated through sensitization of T cells. Increased phagocytic activity of the reticuloendothelial system, especially in the spleen and liver, has some protective effect but does not play a primary role in resistance. It seems likely that T cells function as "helper" cells in the synthesis of specific protective antibody by B cells.

It has been suggested that host immune mechanisms against EE stages are responsible for delayed relapse and, conversely, that failure of these mechanisms precipitates relapse. Because of the difficulties inherent in research on the EE stages, these events are largely speculation (Brown, 1976).

A malaria infection provokes an intense immunologic response in the host. The reactions are both nonspecific and specific. As soon as a significant parasitemia has appeared, the phagocytic activity of the reticuloendothelial system increases, especially in the spleen, liver, and bone marrow. Studies on carbon clearance (Lucia and Nussenzweig, 1969) show that this effect is nonspecific and is at its peak when parasitemia is high and declines to normal with decrease in parasitemia. With recrudescence, the activity increases once again. The response is probably due to macrophage activation mediated by T cells that are specifically sensitized to parasite antigen (North and Mackaness, 1973).

Malaria stimulates the rapid multiplication of splenic lymphocytes that are involved not only in activation of macrophages but also in synthesis of immunoglobulin.

Although not yet well defined, it seems that malaria antibodies closely associated with the 7S fraction of serum gamma globulin are developed during an infection. These seem to include lysins, agglutinins, precipitins, opsonins, ablastins, and complement-fixing and cytoplasm-modifying antibodies. The 7S gamma globulin fraction of serum has been observed at increased levels in the inhabitants of areas of highly endemic malaria. Experiments indicate that this hypergammaglobulinemia is, to a large extent, the result of malaria infections. This immune gamma globulin has been shown to have therapeutic and preventive action in malaria of children.

Malaria immunity is illustrated by the fact that in areas of high endemicity where it is estimated that all inhabitants are infected with sporozoites at least once a month and where all four species of plasmodia, including several strains of *P. falciparum,* infecting man are present, the parasite rates are high in the age group 4 months to 3 years but become low in the age group 10 to 15 years and remain low in adults, who rarely have acute attacks. Infants up to 3 months of age in these areas are relatively resistant to malaria, probably because of the passive transfer of protective antibodies from their mothers (Bruce-Chwatt, 1963; McGregor et al., 1965).

Acquired immunity protects against superinfection rather than against reinfection, and it does not seem to persist much beyond the time when the plasmodia are eradicated from the body. Such eradication may be difficult to demonstrate.

The resistance against malaria conferred by the sickle-cell hemoglobin trait, as well as the question of malaria and the glucose-6-phosphate dehydrogenase deficiency trait and the Duffy factor, has already been discussed (p. 178).

Pathogenesis. As far as is known, all of the pathogenicity of malaria is related to the erythrocytic infection. The inoculated sporozoites promptly leave the bloodstream without having caused apparent harm, and they lodge in the hepatic parenchymal cells, where EE schizogony takes place. Destruction of invaded liver cells occurs, but without noticeable host reaction or clinically evident ill effect. The EE merozoites invade red blood cells and pathogenic effects then become significant.

The disease caused by a malarial infection is a dynamic interplay between the mechanisms of multiplication and survival of the parasites and the defense mechanisms of the host. Because these mechanisms vary from species to species and even from strain to strain of *Plasmodium,* and with race, age, and immune status of the host, the nature of the disease, malaria, is quite protean.

The plasmodia that invade the red blood cells grow and segment at the expense of these host cells, which rupture when schizogony is complete. The debris of the ruptured cells, the released merozoites, and their metabolic byproducts, stimulate chemoreceptors of the temperature-regulating mechanism of the host to conserve heat. At first the amount of pyrogen released at one time is not enough to produce marked reaction, although it may cause prodromal symptoms (vide infra). As the number of invaded red cells increases and the asexual cycle of the parasite becomes more synchronized, the quantity of the pyrogen released at one time becomes sufficient to produce the characteristic chills and fever of a malaria attack.

The species of plasmodia differ greatly in their ability to multiply in the blood. *P. vivax* prefers to invade the youngest erythrocytes, whereas *P. malariae* prefers the older red cell. As a result, these species seldom parasitize more than 2% of the host's red blood cells. *P. falciparum,* on the other hand, invades erythrocytes of all ages and, thus, is capable of parasitizing a very high percentage of host cells. Furthermore, due to the varying number of merozoites produced in schizogony by the species of *Plasmodium, P. falciparum* multiplies more rapidly than *P. vivax,* which, in turn, multiplies more rapidly than *P. malariae* and *P. ovale*. The parasitemia of *P. falciparum* also tends to be higher, because more than one parasite frequently develops in a single erythrocyte.

The phenomenon of *dominance* appears to parallel the attributes of invasiveness and rapid multiplication. The term "dominance" refers to the tendency of *P. falciparum* to repress the parasitemia of *P. vivax* or *P. malariae* in mixed infections, while *P. vivax* dominates over *P. malariae*.

With each schizogony, the parasitized cells are destroyed, but there also is considerable destruction of unparasitized cells due to lysis and phagocytosis; phagocytosis especially occurs in the spleen and liver. In falciparum malaria with a very high parasitemia, hemolytic jaundice may be evident and the anemia may be severe.

However, the malignant character of falciparum malaria is not so much related to its rapid multiplication and invasiveness as to the manner in which it causes lesions in the human host. It has long been recognized that the erythrocytes parasitized by *P. falciparum* become "sticky," tending to adhere to one another and to the lining of the blood vessels as the parasites (excepting gametocytes) grow older. It is for this reason that only young rings and gametocytes are typically found in the peripheral blood film of patients with falciparum malaria. It is thought that the characteristic lesions are due to blockade of small vessels by "sticky" parasitized erythrocytes. The blockade causes stasis, then local anoxia, then increased vascular permeability, which allows plasma and unparasitized cells to leak into the perivascular space. This incidentally results in additional loss of erythrocytes. The selective attachment of falciparum malaria-infected erythrocytes to endothelial cells can be demonstrated in cultured human endothelial cells (Udeinya et al., 1981).

These factors combine to cause a decrease in circulating erythrocytes in the circulating blood volume, local tissue anoxia and edema. *P. falciparum* multiplies so rapidly that, in two or three asexual cycles, the number of infected red cells may reach a dangerous threshold, often without the production of a typical chill followed by fever.

Pathology. The *pathologic and microscopic anatomy* of malaria is characteristic. Owing to destruction of red corpuscles by plasmodia, the blood always shows anemia which, in long-continued infections, may resemble primary pernicious anemia. The leukocyte count is usually reduced except during febrile paroxysms, while large mononuclear leukocytes increase in number. Hemoglobin is usually reduced proportionally with the red corpuscles. Poikilocytosis and a pale appearance of the center of uninfected corpuscles are often present, and free pigment and pigmented leukocytes, as well as phagocytes containing plasmodia (Fig. 14–5), may occur in the peripheral blood and are always present when obtained from the spleen by puncture.

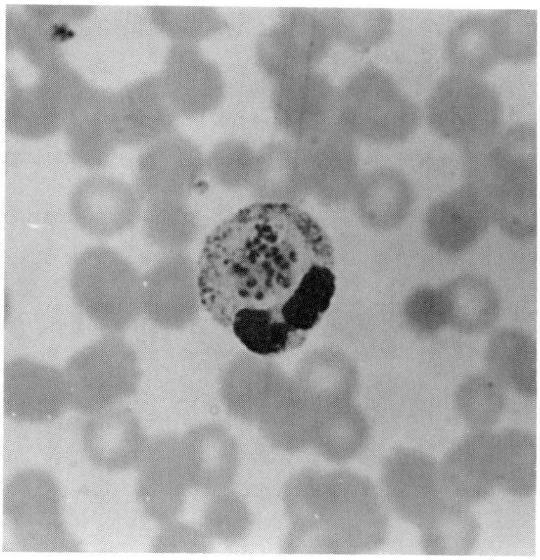

Fig. 14–5. *Plasmodium vivax*. Schizont in a neutrophil granulocyte. Giemsa stain. (× 1000.) (By P.C. Beaver. Preparation courtesy of Dr. R.G. Yaeger.)

Urine is usually normal in *P. vivax* or *P. malariae* patients, but in *P. falciparum* infection, especially in pernicious cases, the urine may sometimes show albumin and hyaline and granular casts, also occasionally seen in severe tertian or quartan infections.

The characteristic pathologic features of the viscera and brain in malaria are the presence of pigmentation; occurrence within the capillaries of free pigment, free plasmodia, and infected erythrocytes; phagocytic cells containing plasmodia, infected erythrocytes, and pigment; and sometimes thrombi produced by accumulation of these elements, particularly in *P. falciparum* infections.

The *brain* is congested, capillaries more manifest, and smaller arterioles tortuous, with thickenings and endarteritis. Their endothelial lining cells may be detached from the intima and there may be small hemorrhages in the parenchyma caused by clumping of parasites, which block and subsequently produce rupture of the capillaries (Fig. 14–6). The cortex may be grayish or slightly brownish.

The spleen, liver, and bone marrow are conspicuously affected. There, the plasmodia accumulate in large numbers even when they are not readily detectable in peripheral blood. The *spleen* is enlarged, soft in initial infections, hard in chronic infections, and usually grayish, dark brown, or even black in long-continued infections. Capillaries are filled with infected erythrocytes, free plasmodia, pigment, and macrophages containing infected erythrocytes and segmenting plasmodia; hemorrhages frequently occur. The *liver* is often pigmented, congested, and enlarged, and, microscopically, the same cell elements are observed in the capillaries as in the spleen, although macrophages are much less numerous. Likewise, *bone marrow*

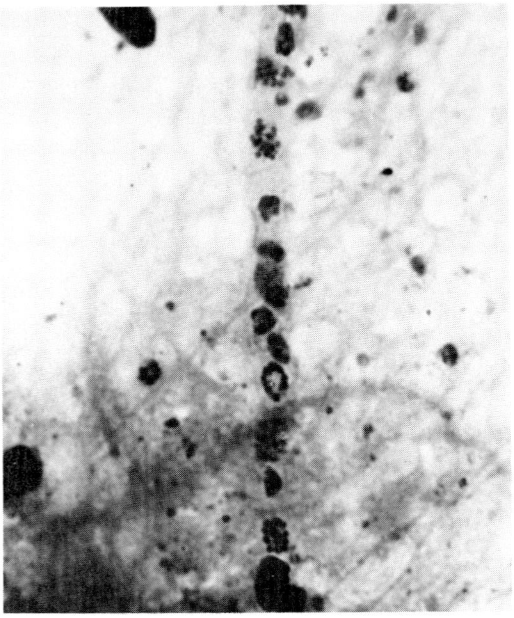

Fig. 14–6. *Plasmodium falciparum*. Brain smear showing numerous schizonts in blood capillary. Giemsa stain. (× 1000.) (By P.C. Beaver. Preparation courtesy of Dr. R.G. Yaeger.)

is affected by plugging of its capillaries, and the reticuloendothelial cells lining the capillaries are actively phagocytic. In acute falciparum malaria, the marrow becomes red and hyperplastic; in chronic cases, it is pale and shows evidence of lost compensatory power.

The *kidneys* are congested and may reveal acute or subacute glomerular nephritis, with hemorrhagic areas. Quartan malaria sometimes causes a focal, proliferative and membranous glomerulonephritis.

The *adrenals* may be markedly affected, with consequent hypotension and weakness. These organs seldom show pigmentation macroscopically, but the tubular epithelium is frequently pigmented.

The *lungs* may be pigmented, and the capillaries may carry free pigment, pigmented leukocytes, phagocytes, and infected erythrocytes.

The *stomach* and *intestines* sometimes show marked congestion of mucous membrane with slight pigmentation; ulceration, due to blocking of capillaries by pigment, infected erythrocytes, and macrophages, may be observed.

The *heart* may show fatty degeneration. Moreover, in falciparum infection, coronary occlusion, plugging, or thrombosis may result from the tendency of the infected red cells and free parasites to adhere to capillary walls. Anoxia of the myocardium due to coronary occlusion may be an important factor in fatal cases.

The pathology of chronic malaria is characterized by anemia, marked splenic enlargement and hardening ("ague cake"), enlargement of the liver, and pigmentation of the brain cortex, spleen, liver, and kidneys. The spleen and liver may be slightly enlarged in latent infections. Capillaries of these organs may contain a few infected erythrocytes, pigmented leukocytes, and free pigment.

Symptomatology. Symptoms of malaria typically consist of definite paroxysms of chills, fever, and sweating, which occur at regular intervals depending upon the time of segmentation of the infecting plasmodium. However, the chills and fever are not as significant as the periodicity. *P. vivax* and *P. ovale* complete their erythrocytic schizogony cycle in approximately 48 hours and produce a paroxysm on alternate days. *P. malariae* segments every 72 hours and produces a paroxysm at 3-day intervals, while *P. falciparum* segments more irregularly and produces a paroxysm every 36 to 48 hours. Some falciparum infections cause a febrile paroxysm every 24 hours.

The *incubation period* varies, depending on the host's resistance and the strain of *Plasmodium*. *P. vivax* symptoms usually appear in 10 to 15 days; *P. ovale* symptoms in 11 to 16 days; *P. malariae* symptoms in 28 to 37 days; and *P. falciparum* symptoms in 8 to 15 days. In all four species, the period of incubation may be prolonged to several weeks or even months. Following chemoprophylaxis, onset may be delayed for more than 2 years (Simmons et al., 1982). Long-term delay of over 10 months is characteristic of a strain of *P. vivax* in parts of China (Jiang et al., 1982).

Onset is usually sudden in vivax, ovale, and malariae malaria, beginning with a shaking chill that is succeeded by a fever reaching 104° or 105°F (40° or 40.6°C) and accompanied by symptoms of an acute febrile infection, such as headache, muscular pains, malaise, nausea, and increased pulse and respiration. After persisting for several hours, fever ends by crisis accompanied by severe sweating, especially in vivax malaria. Pernicious symptoms, such as coma, convulsions, or cardiac failure, rarely occur in vivax, ovale, or quartan malaria.

In falciparum infections, the chill is less pronounced but the febrile period is longer and intervals between paroxysms are shorter. In uncomplicated infections, a characteristic temperature curve is observed, with initial rise, stage of oscillation, pseudocrisis in which temperature may fall almost to normal, pre-critical rise, and the true crisis or fall. The entire paroxysm lasts about 36 hours and is repeated in 48 hours. Bradycardia is usual in falciparum fever periods. Pernicious symptoms frequently occur in falciparum malaria. Most deaths due to malaria, and the most marked histopathologic changes, are caused by this plasmodium. In some cases of pernicious malaria, little or no fever may be present, especially in the so-called ''algid'' malaria, which apparently results from medical shock.

Malaria may simulate many other diseases. Although chills, fever, and sweating are the main features in at least 80% of newly acquired infections, one or more of the following symptoms and signs may occur: headache, backache, pain in the back of the neck or generalized aching; muscular pain and stiffness; gastrointestinal symptoms such as gastric distress with vomiting, diarrhea, or constipation, or dysentery without marked tenesmus but usually with vomiting, occasionally a syndrome suggesting acute appendicitis; respiratory involvement with subacute bronchitis; jaundice and an icteric tinge of the conjunctivae; convulsions; coma.

Pernicious malaria has been classified according to symptomatology as comatose, algid, bilious, cardialgic, choleraic, delirious, dysenteric, eclamptic, hemorrhagic, hemiplegic, and pneumonic. The patient, for example, may be cold, pulseless, and perhaps unconscious (algid type), comatose or restless and, at times, even violent (cerebral type); bleeding from the skin or mucous membranes (hemorrhagic type); or vomiting, hiccoughing, suffering from choleraic symptoms, profuse diarrhea, or acute dysentery (gastrointestinal type). Because of its protean manifestations, malaria may at times mimic almost any disease syndrome.

Relapses. Clinicians have long been aware of malaria relapses. By W.H.O. (1963) definition, there are two types of renewed parasite and clinical activity (i.e., relapses). The first is called a *recrudescence,* and is due to an increase in a surviving population of erythrocytic plasmodia. The second is called a *recurrence,* or a ''true relapse,'' and is the reappearance of parasitemia with clinical symptoms after a significant period during which parasites have been absent from the peripheral blood. The mechanism by which true relapses occur is unknown. After the discovery of the EE schizogony in the liver by Shortt and Garnham, it was commonly believed that relapses in vivax and ovale malaria were due to a series of subsequent generations of EE schizonts in the hepatic parenchymal cells. Recent evidence suggests that such relapses are due to delayed schizogony of dormant sporozoites (hypnozoites) (Krotoski et al., 1980). If true, then the phenomenon that is called a ''true relapse'' may be more properly termed a ''delayed primary attack'' (Corradetti, 1981). In *P. vivax* and *P. ovale,* primary EE schizogony produces a first clinical attack that, after one or more paroxysms, becomes latent when the host has developed an immune barrier. This results in a parasite-free bloodstream but frequently does not terminate the existence of malaria parasites in the body, specifically in the liver parenchymal cells. At some later time, either due to a weakening in the host's immune mechanisms, or possibly due to an intrinsic timing of the dormant hypnozoites, EE schizogony again occurs. EE merozoites from secondary EE schizogony again invade red cells and erythrocytic schizogony is resumed with a recurrence of clinical symptoms (i.e., a ''true relapse''). These EE merozoites may stem from a continuous process of secondary EE schizogony or, as now seems more probable (Garnham, 1977), from dormancy of first invaders.

Plasmodium falciparum and possibly *P. malariae* have no secondary EE schizogony. The tissue phase does not persist after the first blood phase ends. Consequently, there are no recurrences or ''true relapses,'' although there are often recrudescences when, after a period of low density, the erythrocytic parasites become numerous enough to cause renewed clinical signs and symptoms. These are commonly called ''relapses,'' sometimes ''short-term'' relapses, although the term ''recrudescence'' is preferred.

Since EE schizogony appears to stem only from sporozoites, there will be no recurrences or ''true relapses'' in blood-induced malaria infections in which no sporozoites are involved.

Blackwater Fever. Blackwater fever appears to be a manifestation of repeated infections with falciparum malaria, which were inadequately treated with quinine. Sometimes resumption of quinine therapy or quinine treatment of a new attack is followed by massive destruction of uninfected red cells. The hemolysis probably is caused by erythrocytic auto-antibodies derived from previous infections and reacting with auto-antigens initiated by fresh erythrocytic infection with the same falciparum strain (Maegraith, 1963). The quininized and parasitized corpuscle acts as an antigen against which hemolysins are formed (Garnham, 1966). As other antimalarials have replaced quinine, blackwater fever has become uncommon.

Latent Malaria. As already mentioned, plasmodia may be found in the peripheral blood of a host who exhibits no acute symptoms of malaria. Such persons are said to have *latent malaria,* and they are of considerable importance in some areas because there are gametocytes in their peripheral blood; hence, they are "silent" carriers of malaria, infectious to mosquitoes.

Malaria in Children. In highly endemic areas, malaria is a major cause of death in children. Falciparum malaria is a serious childhood hazard. Often, typical symptoms do not occur. Drowsiness, vomiting, diarrhea, unproductive cough, and fever without periodicity may introduce an attack that may be characterized by (1) cerebral manifestations of convulsions or coma; (2) hyperpyrexia; or (3) medical shock. Renal involvement may cause oliguria followed by anuria. Splenic enlargement may be huge, with infarctions or even rupture. The liver also enlarges and there may be jaundice. Severe monocytic anemia is common after repeated falciparum attacks.

Vivax and ovale malaria are usually milder and do not have manifestations in children. Vivax malaria in the young often does not present classic periodic episodes of cold, hot, and sweating stages. Quartan malaria in children in endemic areas may induce a nephrotic syndrome. A focal, proliferative, and membranous glomerulonephritis leads to gross edema, massive proteinuria, and severe hypoproteinemia.

Congenital Malaria. Congenital malaria is comparatively rare, especially in highly endemic areas. Covell (1950) reviewed 125 published cases, 57 of which were vivax, 56 falciparum, and 12 malariae. He considered intrauterine transmission of malaria a well-established fact. For reasons unknown, congenital malaria is more common in nonimmune infected mothers than in highly immune mothers whose placentas are more frequently heavily loaded with *P. falciparum* schizonts (Fig. 14–7). Congenital malaria has been reported with each of the four species of *Plasmodium* infecting man. Three of four congenital infections reported by Quinn et al. (1980) were *P. vivax,* the fourth, *P. falciparum.* A double infection with *P. vivax* and *P. malariae* was reported in a 6-week-old infant by MacLeod et al. (1982).

Most observers believe that passage of the parasite to the fetus occurs only when the placental barrier has been injured. There is no evidence that infection ever takes place by passage of sporozoites or of EE forms to the fetus. Blood examination of mothers, of umbilical cords during delivery, and of babies within 5 hours after birth convinced Madecki and Kretschmar (1966) that so-called congenital malaria is usually connatal (i.e., malaria acquired during parturition).

Diagnosis. Although symptoms in vivax and quartan infections may be characteristic so that one is quite sure of the diagnosis, the diagnosis should always be confirmed by demonstration of plasmodia in peripheral blood. In falciparum and mixed infections, symptoms are notoriously unreliable; therefore, blood smear examination is essential.

The only certain diagnosis of malaria is by demonstration of the causative *Plasmodium.* History, clinical findings, and response to specific therapy may or may not be helpful. *Regardless of signs or symptoms, if there is a history of possible exposure to infection, even for a single night or for a few evening hours at an airport in a malarious area, the blood should be examined for malaria parasites.* Modern transportation and frequency of travel accentuate this advice. Deaths occur each year in nonmalarious countries from undiagnosed falciparum malaria.

While thin blood films are usually satisfactory, thick blood films are less time-consuming. For laboratory methods, refer to Technical Appendix, pages 751–752.

Since malaria may be transmitted by blood transfusion, in any case of fever of unknown origin in which there is a history of recent blood transfusion, the patient's blood should be examined for malaria parasites.

In infections with *P. vivax, P. ovale,* and *P. malariae,* all asexual stages, as well as the gametocytes, occur in peripheral blood; in *P. falciparum* infections, however, only the rings or young trophozoites and gametocytes may be present. Plasmodia are most numerous in the peripheral blood late in the febrile paroxysm; between the paroxysms, especially in *P. falciparum* infections, they are situated mostly within capillaries of internal organs, so that, at this time, repeated blood films may be required. A therapeutic test with an antimalarial such as chloroquine is sometimes useful,

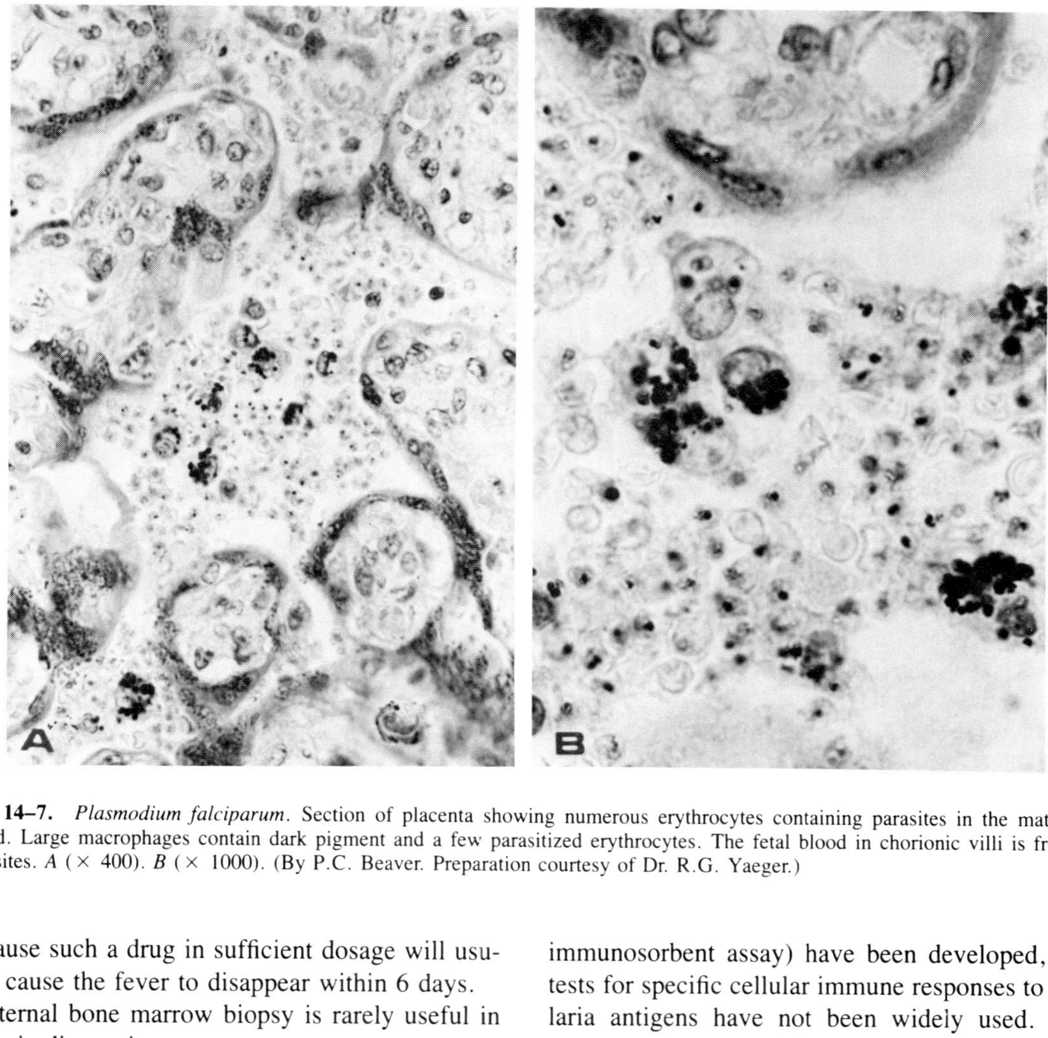

Fig. 14–7. *Plasmodium falciparum.* Section of placenta showing numerous erythrocytes containing parasites in the maternal blood. Large macrophages contain dark pigment and a few parasitized erythrocytes. The fetal blood in chorionic villi is free of parasites. *A* (× 400). *B* (× 1000). (By P.C. Beaver. Preparation courtesy of Dr. R.G. Yaeger.)

because such a drug in sufficient dosage will usually cause the fever to disappear within 6 days.

Sternal bone marrow biopsy is rarely useful in malaria diagnosis.

Conditions with which malaria may be confused clinically are tuberculosis, typhoid and paratyphoid fever, undulant fever, yellow fever, dengue, relapsing fevers, shigellosis or amebic dysentery, cholera, bronchopneumonia, influenza, septicemia, cerebral apoplexy, pyelonephritis, and uremia. By employing various laboratory methods one should be able to make a differential diagnosis.

Except in regard to asymptomatic blood donors with subpatent parasitemia, malaria diagnosis in clinical medicine is usually not difficult granted proper technique and training. Because of the numbers involved, however, diagnosis of malaria is a problem in malaria eradication programs and in surveys to measure incidence, to detect parasite carriers, to determine immunity levels, and to practice surveillance. Tests for the measurement of humoral immune responses (indirect fluorescent antibody, indirect hemagglutination, enzyme-linked immunosorbent assay) have been developed, but tests for specific cellular immune responses to malaria antigens have not been widely used. The greatest present need for such tests is for epidemiological assessment of the malaria status in endemic areas (W.H.O., 1979). Immunodiagnostic tests are being used for these purposes but only to a limited extent (W.H.O., 1979). Methods of concentrating parasitized cells for diagnosis have been described (Le Bras and Payet, 1978).

Treatment. Ideally, malaria treatment should, without toxicity to the patient, (1) promptly destroy all asexual forms of the parasite in order to cure the clinical attack; (2) destroy sporozoites and EE forms in order to prevent relapses; and (3) destroy gametocytes so that mosquito infections will be prevented. No single drug will answer all objectives with certainty, but a combination of drugs, as outlined below, will often meet the need. However, malaria therapy is complicated by the resistance of certain strains of *Plasmodium* to antimalarials.

Resistance of Malaria Parasites to Drugs. Drug

resistance may be defined as "ability of a parasite strain to multiply or to survive in the presence of drug concentrations that normally destroy parasites of the same species or prevent their multiplication. Such resistance may be relative (yielding to increased doses of the drug tolerated by the host) or complete (withstanding maximum doses tolerated by the host)" (W.H.O., 1963). Grade-1 to Grade-3 resistance (see below) has been reported for strains of *P. falciparum* acquired by visitors in the coastal areas of Kenya, Uganda, Tanzania, Malagasay, and the Comoros Islands (Weniger et al., 1982).

Most troublesome is the resistance of certain falciparum strains to the schizonticidal action of 4-aminoquinolines—chloroquine and amodiaquine. Such resistance has been observed since 1960 and is now known to exist in Southeast Asia (Cambodia, Malaysia, Laos, Thailand, Vietnam, southern China), Indonesia (Borneo and New Guinea islands), Philippines, Papua New Guinea, India (Assam State), Nepal, Bangladesh, Burma, Panama, French Guiana, Guyana, Surinam, Brazil, Bolivia, Colombia, Ecuador, and Venezuela (Lopez et al., 1979; Ross Institute, 1981). *P. vivax, P. ovale,* and *P. malariae,* as well as many strains of *P. falciparum* in areas of resistance, respond as usual to treatment by 4-aminoquinolines. Certain falciparum strains, however, are not only refractory to chloroquine and amodiaquine but also to other synthetic antimalarials and, in rare strains, apparently even to quinine.

The effectiveness of chloroquine and amodiaquine has been graded (W.H.O., 1967) as follows: (1) schizonts are cleared from the blood within 7 days of treatment and there is no subsequent recrudescence; (2) clearance of schizonts but followed by recrudescence (most common); (3) marked reduction in parasitemia but no clearance; (4) no marked reduction in parasitemia. Grade-2 resistance often may be overcome by increased dosage.

Falciparum strains resistant to chlorguanide, pyrimethamine, and quinacrine, and vivax strains resistant to chlorguanide and pyrimethamine, are known. Some falciparum strains are resistant to all of the drugs mentioned as well as to the 4-aminoquinolines (Peters, 1967, 1969; Powell and Tigertt, 1968; Pays and Hugues, 1981).

Treatment Regimens

I. *Curative Treatment*

A. Falciparum and quartan malaria, and transfusion or injection malaria of any species (if no sporozoites are involved): the best treatment in acute, uncomplicated cases is to administer chloroquine or amodiaquine as in 1, 2, or 3, which follow:

1. Chloroquine diphosphate or sulfate, initial dose 600 mg base (i.e., two 500-mg tablets of the salt), followed in 6 hours by 300 mg base, then 300 mg base daily for next 2 days.
2. Amodiaquine dihydrochloride, initial dose 600 mg base (i.e., three 261-mg tablets of the salt), then 400 mg base daily for 2 days.
3. If a patient comes from a highly endemic area and has some immunity because of previous attacks, the initial dose of 600 mg of chloroquine or of amodiaquine base may be sufficient to cure a mild falciparum attack.

These three simple treatment regimens will promptly alleviate clinical symptoms unless one is dealing with a chloroquine- or amodiaquine-resistant strain of *P. falciparum,* in which case treat as described in (6) or (17). Falciparum, quartan, and transfusion or injection malaria of any species usually will be completely cured.

B. Vivax and ovale malaria: in these types of malaria, since neither chloroquine nor amodiaquine will destroy EE parasites, it is recommended that to prevent relapses primaquine phosphate be added to the chloroquine or amodiaquine as follows:

4. Administer chloroquine or amodiaquine as described in (1), (2), or (3), but on the second day, also give 15 mg primaquine base (i.e., one 26.4-mg tablet of the salt) and repeat this dose of primaquine once daily for a total of 14 doses.
5. Administer chloroquine or amodiaquine as described in (1), (2), or (3); on the second day of treatment, also give 45 mg primaquine base in one dose, repeating this dose *once a week* for a total of eight doses of primaquine.

Caution. Primaquine alone does not promptly relieve clinical symptoms. This drug is rarely toxic in the preceding dosages, which should not be exceeded. Primaquine should not be used with quinacrine (Atabrine), which increases its toxicity. Primaquine is usually not recommended for children under 6 years of age. (See the following discussion of toxicity of primaquine.)

C. Emergency parenteral treatment: if oral administration of drugs is impossible (e.g., when there is coma or excessive vomiting), the following parenteral therapy is recommended:

6. Administer not more than 10 ml of a 5% solution of chloroquine salt (300 to 400 mg base) *intramuscularly*. A total of 900 mg base spread over 24 hours may be given in this way to an adult weighing about 150 pounds (68 kg). (Commercially prepared ampules of Aralen, Resochin, or Nivaquine B are available.) Dosages for children should be carefully adjusted to weight of patient and should not exceed 5 mg/kg body weight. Oral medication should be resumed as soon as possible and the total adult dose, parenteral plus oral, should be kept at 1500 mg of chloroquine base.

7. Chloroquine also may be given by slow infusion. Dilute with 150 ml glucose-saline, repeated in 12 hours if necessary. Dosage in children should not exceed 5 mg/kg body weight.

Where resistant strains of *P. falciparum* occur, it is wiser to use parenteral quinine, as described in (17).

II. *Suppressive Treatment*

D. Unless resistant strains of *P. falciparum* are involved, either chloroquine or amodiaquine in weekly doses will prevent acute attacks of malaria if given during exposure as follows:

8. Chloroquine diphosphate or sulfate, 300 mg base once weekly.

9. Amodiaquine dihydrochloride, 400 mg base once weekly.

Either regimen, (8) or (9), will prevent malaria of any type and will usually cure falciparum malaria. If vivax or ovale infections have occurred during the period of suppression, however, they may become manifest as delayed primary clinical attacks after the medication has ended. Therefore, it is recommended that, upon returning to a nonmalarious area and ending suppressive therapy, one should take primaquine, as follows, to achieve radical cure:

10. Suppressive chloroquine or amodiaquine during period of exposure as described in (8) or (9); then primaquine phosphate, 15 mg daily for 14 doses.

III. *Prophylaxis*

E. Pyrimethamine, taken as follows, may be a true prophylactic:

11. Pyrimethamine phosphate (Daraprim), 25 mg base once a week during period of exposure.

Pyrimethamine is not recommended for treating acute infections, but it rapidly renders gametocytes noninfectious, has been reported to act against sporozoites and EE forms, and is perhaps the most persistent of the antimalarials. Strains of *Plasmodium* resistant to pyrimethamine have, unfortunately, become rather common.

Another prophylactic drug that has wide use is chlorguanide (proguanil, Paludrine), taken as follows:

12. Chlorguanide, 100 mg salt once daily, or 200 to 300 mg salt once weekly.

IV. *Mass Treatment*

F. Sometimes mass malaria treatment is useful against falciparum malaria (e.g., in an eradication program as an adjuvant to insecticidal attack). The following regimens may be used:

13. Either chloroquine or amodiaquine in single oral doses of 600 mg base.

14. Chloroquine or amodiaquine as described in (13) and add either primaquine, 15 mg base per day for 14 days, or pyrimethamine, 100 mg base repeated once a month for 6 months. Primaquine is not recommended for children under 6 years of age, and pyrimethamine is not prescribed for infants under 1 year.

In a few special cases, it has been possible to use medicated "table" salt for mass treatment. Chloroquine diphosphate has been mixed with common salt at a concentration of 0.3 to 0.4%. Other antimalarials have also been tried in salt, but the method has proved practical in only a few cases.

V. *Alternative Treatment*

G. If chloroquine or amodiaquine is not available for treatment of acute malaria, one of the following regimens may be followed:

15. Quinacrine hydrochloride (Atabrine), 300 mg base three times the first day and twice on the second day, then 100 mg three times daily for 5 days.

16. Quinine sulfate or dihydrochloride, 650 mg every 8 hours for 10 days, plus py-

rimethamine 25 mg every 12 hours for six doses beginning on the first day.

For parenteral treatment, if chloroquine is not available as in (6) above, quinine may be used as follows:

17. Quinine dihydrochloride 650 mg (10 grains) in sterile normal saline solution, or other suitable diluent, injected *intravenously,* preferably by slow drip. The same dose should be repeated in 6 hours if necessary, but not more than three injections should be given within 24 hours and oral medication should be instituted as soon as possible. Quinine should be injected *very slowly,* at a rate not exceeding 65 mg (1 grain) per minute. Intramuscular injections of quinine are not recommended. In falciparum cerebral malaria, parenteral quinine may be preferable to chloroquine.

VI. *Treatment of Drug-Resistant Falciparum Malaria*

H. Resistant falciparum malaria usually responds to quinine treatment as in (16) or, if parenteral administration is needed, as in (17), but it is preferable to give other drugs concurrently, such as regimens (18) or (19).

18. Sulfadoxine 1.0 g plus pyrimethamine 50 mg (equivalent to two tablets of Fansidar) in a single dose.

19. Tetracycline 250 mg four times daily for 7 days. For chemoprophylaxis in areas where multiresistant *P. falciparum* occurs, follow (20) or (21).

20. Pyrimethamine 25 mg base, plus sulfadoxine 500 mg once weekly or double these doses every 2 weeks.

21. Daily oral dose of 25 mg diphenylsulfone *in addition* to weekly prophylaxis as described in (11).

Recently, qinghaosu (artemisinine), a sesquiterpene lactone that is the active antimalarial principle of a Chinese herb, *Artemisia annua* L., has undergone clinical trials against chloroquine-resistant falciparum malaria with promising results. Two active derivatives, artemether and artemate, have been synthesized. Preliminary reports indicate that artemate may be superior to quinine for cerebral malaria resistant to chloroquine, but complete knowledge regarding the toxicity and teratogenicity of this group of drugs is still lacking (Bruce-Chwatt, 1982).

Another synthetic antimalarial drug, mefloquine, has been developed. It is an aminoalcohol structurally similar to quinine, but it is effective against strains of falciparum resistant to all known drugs, including quinine (Trenholme et al., 1975). Other aminoalcohol drugs are undergoing tests (Canfield, 1980). Falciparum malaria that is resistant to mefloquine in vitro has been reported from the Philippines (Smrkovski, et al., 1982).

In nonimmunes with drug refractory falciparum infections, evidence indicates that doubling or even tripling chloroquine dosage will not significantly counter the ineffectiveness of the drug. In areas of refractory falciparum strains, however, increased chloroquine dosage may be useful in treating local residents who have considerable natural immunity and for whom a 10-day treatment regimen might not be practical.

To estimate dosages for children, the following formula is useful:

Dosage for children under 12

$$= \text{Adult dosage} \times \frac{\text{Age next birthday}}{\text{Age} + 12}$$

However, allowance should be made for variance from normal age-weight relationship.

Description of Antimalarial Drugs

1. *Chloroquine* U.S.P. XV (Aralen diphosphate, Avloclor diphosphate, Resochin diphosphate, Nivaquine sulfate), chemically 7-chloro-4-(4′ - dimethylamino - 1′ - methylbutylamino) quinoline. Chloroquine may occasionally cause slight gastric distress but is less toxic than quinacrine and does not discolor the skin. A powerful schizonticide, it rapidly clears the blood of nonresistant asexual parasites, promptly terminating clinical attacks. Hence, it is also an excellent suppressive. It does not directly affect EE forms or falciparum gametocytes.

   Each 500 mg of chloroquine diphosphate salt (Aralen, Avloclor, or Resochin) contains 300 mg of base. Each 200-mg tablet of chloroquine sulfate salt (Nivaquine B) contains 150 mg of base.
2. *Amodiaquine* U.S.P. XV (Camoquine dihydrochloride, Flavoquine dihydrochloride, Miaquine dihydrochloride, Basoquine), sim-

ilar to chloroquine in action on parasites and absence of toxicity. Chloroquine-resistant *P. falciparum* strains are resistant to amodiaquine. Each 261-mg tablet of amodiaquine dihydrochloride dihydrate salt (Camoquin) contains 200 mg base. Amopyroquin, an analogue of amodiaquine, is marketed as Propoquin dihydrochloride in sterile injectable form.

3. *Chlorguanide* (Proguanil monohydrochloride, Paludrine monohydrochloride). This drug is N′-(*p*-chlorophenyl)-N[5]-isopropyldiguanide, and it is relatively nontoxic to man. There is some evidence, though inconclusive, that it has some sporonticidal effect and has some direct action on the EE stages of *P. falciparum,* although it is not as effective as primaquine. Thus, it is a *causal prophylactic,* but it is not sufficiently rapid in its action on erythrocytic schizonts to warrant its use for treating overt malaria attacks. Moreover, in certain areas, strains of *P. falciparum* and *P. vivax* have developed strong tolerance to chlorguanide; this drug resistance is transmissible. Chlorquanide for suppression has the advantages of being less expensive and less toxic than most antimalarials.
4. *Pyrimethamine* (Daraprim, Malocide). Pyrimethamine chemically is 2,4-diamino-5-*p*-chlorophenyl-6-ethylpyramidine. This drug, like chlorguanide, has direct action on EE forms of *P. falciparum* and may also be sporonticidal. But it is slow in its effect on erythrocytic schizonts and, therefore, is not recommended for treatment of overt attacks. Like chlorguanide, in some areas, *P. vivax, P. malariae,* and *P. falciparum* have developed a lasting resistance that is transmissible. There is a cross-resistance between these two drugs. Marketed as Daraprim, each tablet contains 25 mg pyrimethamine base, but in the French Malocide, each tablet contains 50 mg of pyrimethamine base.
5. *Quinacrine hydrochloride* U.S.P. XV (Atabrine hydrochloride, Mepacrine hydrochloride, Mepacrine methane sulfonate). Quinacrine chemically is 2-methoxy-6-chlor-9-(4′-diethylamino-1′-methyl-butylamino) acridine. This drug, which was useful in World War II but is now of secondary importance, has rapid action against erythrocytic schizonts; thus, it not only effectively treats an overt attack but is also an effective suppressant of all forms of malaria, although it does not act against EE stages. It is more toxic than chloroquine and usually causes a temporary yellow discoloration of the skin. Minor intestinal disturbances are not unusual. When taken over long periods, it may be associated with skin lesions, such as a lichenoid dermatitis. Occasionally, transient psychosis may follow its use.

   Each 100-mg tablet of quinacrine hydrochloride contains 79 mg base. In the preceding sections, the dosage of quinacrine has been given in terms of the marketed tablet of 100-mg size (and not in terms of the base). Each 375-mg ampule of quinacrine methane sulfonate is equivalent to 300 mg quinacrine hydrochloride.
6. *Primaquine* (Neoquipenyl). Primaquine chemically is 6-methoxy-8-(4′-amino-1′-methyl-butylamino) quinoline. This drug acts on primary EE stages of *P. vivax* and *P. falciparum* and secondary EE stages of *P. vivax,* and, therefore, is prophylactic. It also destroys gametocytes of all species, but is not rapidly effective against erythrocytic trophozoites. It is a drug that must be used with reasonable caution. Although it is considerably less toxic than plasmochin, it has a relatively narrow margin of safety between effective and toxic doses, especially in blacks. The toxic symptoms may vary from mild gastric distress to severe hemolytic reactions that occur in the presence of glucose-6-phosphate dehydrogenase deficiency. Severe toxicity is rare in white or black patients on doses of not over 15 mg base daily for 2 weeks. Doses of 30 mg daily for 2 weeks are generally tolerated by white patients but not so uniformly by black patients. Quinacrine (Atabrine) tends to potentiate the toxicity of primaquine and should not be given concurrently, nor should a course of primaquine be started within 3 weeks of terminating a course of quinacrine.
7. *Proguanil* (chlorguanide, Paludrine). Proguanil hydrchloride is a slow-acting blood schizonticide but an effective causal prophylactic against susceptible strains of *P. falciparum.* Its wide use in several countries has been followed by rapid appearance of drug resistance.

8. *Chlorproguanil* (Lapudrine). Chlorproguanil resembles Proguanil in its action, wide use in some countries, and rapid development of drug resistance.
9. *Quinine*. Quinine is 6-methoxy-alpha-(5-vinyl-2-quinuclidyl-4-quinoline-methanol). This drug has a long history of usefulness against malaria. If a more effective and less toxic antimalarial such as chloroquine is not obtainable or if the parasite is resistant to chloroquine, quinine may still be used to subdue an attack. There is evidence, however, that a chloroquine-resistant strain of *P. falciparum* may also be resistant to quinine, although this occurs rarely. Quinine does not affect the EE forms or *P. falciparum* gametocytes.

   The chief toxic effects of quinine are grouped under the term *cinchonism,* a syndrome that may include one or more of the following: tinnitus, deafness, dizziness, amblyopia, photophobia, diplopia, headache, gastric distress, nausea, vomiting, diarrhea, and urticarial rash. In hypersensitive individuals, quinine may induce an angioneurotic edema, urticarial wheals, and sometimes hemorrhage from nose, uterus, or kidneys.

   Cinchonism in standard malaria treatment is often absent and, if present, mild. However, overdosage may cause permanent damage to retinal or auditory ganglion cells or to the optic nerve. The minimum lethal dose for an adult is about 8 g (120 grains).
10. *Newer Antimalarials*. Increasing problems with multiresistant strains of *P. falciparum* have led to experimentation with sulfones and sulphonamides, which are usually given in combination with a standard antimalarial. For example, sulfadoxene, an ultralong-acting drug with a half-life of 150 hours, is preferred in chloroquine-resistant falciparum malaria. Another promising drug is 5,6-dimethoxy-4-sulfanilamidopyrimidine (sulforthomidine, Fanzil, fanasil, sulformetoxine, sulphormethoxine) (Chin et al., 1966). Another such drug is 4,4′-diaminodiphenylsulfone (Diaphenylsulfone, dapsone, DDS). Others are sulfalene (Kelfizina), sulfadimethoxine (Levisul), sulfamethoxypyridiazine (Midicel), sulfadiazine (Diazine), and metachloridine (SN-11437). Each of these drugs has other names (W.H.O., 1967).

   Repository drugs (i.e., soluble, injectable drugs that form local depots in muscle or subcutaneous tissue from which the active principle is gradually released into the circulation), such as cycloguanil emboate (CI-501), cycloguanil pamoate (camolar), and sulfadiamine (Camilan, CI-556), have been tested, but none can be used successfully for control by mass drug administration (Peters, 1978).

Supportive Measures. Although many malaria patients do not require hospitalization if specific therapy is promptly given, most patients require rest and symptomatic treatment to relieve headache, provide sedation, or promote free bowel movement. High fever is best reduced by the specific malaria therapy but may be relieved by cold sponges and packs. Falciparum malaria frequently requires bed treatment at hospital or home.

In addition to malaria therapy, it may be advisable to use antivomiting drugs to replace lost body fluids and electrolytes or, in algid malaria, to promote rapid expansion of circulating blood volume (e.g., by plasma or glucose-saline). Rapid improvement in the patient with fulminant falciparum malaria may be achieved by exchange transfusion (Kramer et al., 1983). In cerebral malaria, blood transfusions are helpful if the hematocrit drops below 30%. Diphenylhydantoin, 400 mg daily, may be useful as an anticonvulsant. Dexamethasone sodium phosphate, 3 mg every 8 hours to reduce cerebral edema, and dextran-75, 1 unit every 12 hours for 3 days to reduce vascular sludging, as well as hydrocortisone, have been recommended (Hall, 1977).

Delirious patients must be carefully guarded to prevent self-injury.

In convalescence, a high protein, high vitamin diet, supplemented by a suitable iron formulation, is indicated.

Prognosis. So far as danger to life is concerned, the prognosis of uncomplicated infections with *P. vivax, P. ovale,* and *P. malariae,* even if untreated, is good, but in untreated or poorly treated infections with *P. falciparum,* the prognosis is often grave. Even in these infections, however, prompt treatment is followed by prompt recovery in most cases. The prognosis in individuals whose resistance has been markedly lowered is less favorable. In such patients, even usually benign tertian and quartan infections may be fatal. During World War

II, the case mortality of malaria among American troops was less than 0.1% in hospitalized cases.

Malaria in a quiescent stage frequently becomes activated by shock of major surgery, or in pregnancy at the time of still-birth or labor. Gilles et al. (1969) stated, "The single most important antenatal medical practice in a country where malaria is endemic is the administration of antimalarial drugs prophylactically throughout pregnancy."

Malaria Control and Prevention. Malaria control, for the average practitioner, involves (1) recognition, whatever the signs or symptoms, of the possibility of malaria infections in tourists or others with a history of exposure, however brief, in a malarious area; (2) proper treatment of the patient, with provision for avoiding or promptly dealing with relapses; (3) a report of all malaria cases by name and residence to local health department authorities; (4) practical advice to patient and family as to how to avoid malaria infections by use of bed nets, screening, insect repellents, boots, change of habits or of residence, use of prophylactic drugs and of household insecticidal sprays; and (5) active support of community malaria control measures. A practical method of immunization is not yet available. As stated by Cohen (1982), "The slow development of naturally acquired immunity in man and its incomplete nature have raised legitimate doubts about the feasibility of developing an effective vaccine against human malaria. . . Nevertheless, experimental vaccination against rodent, simian and human plasmodia has frequently induced far greater immunity than that engendered by the corresponding natural infection."

The appearance and growing frequency of strains of *P. falciparum* resistant to chloroquine in many tropical countries have presented the North American physician with a dilemma. The most effective and safest suppressive drug preparations are not available in the United States. Therefore, when preparing his patient for a visit to a country where resistant falciparum malaria may be encountered, it is necessary for the physician to: (1) prescribe chloroquine to use as a suppressive; (2) inform the traveler of the generic and common proprietary names of the preferred suppressive formulation to replace the chloroquine and, if possible, give advice as to where it may be obtained; and (3) admonish the traveler to seek medical aid for possible falciparum malaria if he should develop a febrile illness.

The spraying of residual insecticides has made possible a tremendous worldwide acceleration in malaria control. Nationwide eradication has become practicable in some countries, and in many areas, malaria has been eradicated. However, it remains endemic in essentially all tropical countries of the world (Fig. 14–8).

For information about malaria mosquito control, see pages 656 to 657.

ORDER PIROPLASMIDA
Wenyon, 1926

FAMILY BABESIIDAE
Poche, 1913

Babesia

Historical and Geographical Notes. The genus *Babesia* and family Babesiidae are named after Babes, who, in 1888, described *Babesia* parasites in the blood of cattle and sheep in Romania. The classic studies of Smith and Kilborne (1893) showed that *Babesia* was the cause of Texas fever, an acute, febrile, hemolytic disease of cattle in the southern United States, and that it was transmitted by ticks. These findings are of historic importance, not only because of the enormous impact of Texas fever, but more so because they introduced the concept of disease transmission by arthropods.

Babes named the organisms *Haematococcus*, and Smith and Kilborne named them *Pyrosoma*, but both of these names were invalid. Although a multitude of other names have been assigned, *Babesia* (Starcovici, 1893) has precedence. There has been an effort to assign the species to different genera for the large parasites *(Piroplasma)*, small parasites *(Microbabesia* and *Babesiella)*, and those that produce four daughter cells within erythrocytes *(Nuttalia)*, but the preferred practice today is to recognize one genus, *Babesia*, for intraerythrocytic organisms that have solid, pyriform shapes and tend to be arranged in pairs.

The taxonomic position of the *Babesia* has always been uncertain. Because of superficial similarities in form and their occurrence as intracellular blood parasites, they were usually placed in the Sporozoa, order Coccidia. This classification was brought into question because inclusion in the Sporozoa was based on sexual reproduction in the life cycle, which resulted in the formation of spores or oocysts. Since repeated attempts failed to demonstrate such a method of reproduction, the *Babesia* were transferred to the Superclass Sarcodina by the Committee on Taxonomy and Taxonomic Problems of the Society of Protozoologists in 1964. Dissatisfied with this exclusion of *Babesia* from the Sporozoa, Levine (1969, 1970) proposed a new subphylum, Apicomplexa, based on the presence, in at least one stage, of certain polar structures as revealed by electron microscopy. The *Babesia* do have the apical complex and, therefore, fall into the Apicomplexa. The class Sporozoa, subclass Piroplasmia (Levine, 1961), was proposed to include the Apicomplexa for which sexual reproduction leading to the production of spores or oocysts was unknown.

Human cases of babesiosis have been reported from Yugoslavia (Skrabalo and Deanovic, 1957; Harambasic et al., 1976), Ireland (Fitzpatrick et al., 1968), Scotland (Entrican et al., 1979), France (Bazin et al., 1976), and the United States (Scholtens et al., 1968; Parry et al., 1977; Ruebush et al., 1977; Stahl et al., 1978; Healy and Ruebush, 1980; Gombert et al., 1982). In two of the cases reported from Europe, diagnosis was made at autopsy, but in the case reported by Scholtens et al. (1968),

Fig. 14–8. Malaria. Approximate world distribution. Shading indicates areas where malaria transmission occurs or there is a significant risk of infection. Small circles indicate isolated sites of malaria transmission including oases (in Egypt), Cape Verde, Zanzibar and the Comoros, Maldive, Andaman, and Nicobar Islands. The status in mainland China was unknown. (Adapted by R.C. Jung from map in WHO Weekly Epidemiologic Record. January 1982. Atlanta, Centers for Disease Control, Malaria Surveillance Annual Summary for 1980.)

the patient was at first thought to be suffering from malaria and was treated with chloroquine with complete recovery. However, a stained blood film revealed the presence of erythrocytic parasites of unusual morphology, and *Babesia* antibodies were demonstrated in the patient's serum by both complement-fixation and tube latex-agglutination tests.

Until 1969, when the infection was observed in a nonsplenectomized patient on Nantucket Island, it was believed that human beings were not susceptible unless splenectomized. More than 20 cases of human infection have been reported to date. In most of these cases, the infections were acquired on islands that lie along the coast of southern New England, and the parasites were identified as *Babesia microti*. In the five cases reported from Europe, all patients were asplenic, all the illnesses were fatal or severe, and the parasites probably were *Babesia* species from cattle (Ruebush and Spielman, 1978).

Morphology and Life Cycle. The developmental changes in *Babesia* that take place in the tick after ingestion are not definitely known, but sexual union of the parasite with ookinete development (Riek, 1964) has been described. It is known that the parasites invade epithelial cells of the tick's gut. The organism enlarges, and, as it does so, its nuclear material becomes arranged in small dots throughout the cytoplasm. The cytoplasm becomes divided, a portion to each dot, forming small, spherical bodies. These open and unfold to form club-shaped bodies that are called "vermicules" (Riek, 1964), which escape from the ruptured host cell and, being motile, migrate throughout the body of the tick.

The vermicules, which are 11 to 15 μm in length depending on the species, stain deeply at the blunt end and have a pellicle consisting of an outer fine plasmalemma and a thick inner membrane. At one end, there is a polar ring from which a microtubule layer radiates. Clavate organelles are present. There is a nucleolus within the nucleus. Cytoplasmic structures include Golgi complex and endoplasmic reticulum, but there are no mitochondria. Electron micrographic studies have confirmed that the development of sporozoites in *Babesia* occurs inside the cells of the tick's salivary glands and that it is an agamogonic process (Schein et al., 1979).

The vermicules are detectable in the hemolymph of the tick in from 3 to 5 days after ingestion. They invade and multiply within hemocytes, ovaries, Malpighian tubules, muscle, and gut epithelium and invade oocytes of the female tick early and continue to do so during the period of oviposition. After oviposition, the vermicules multiply in the egg yolk and the merozoites invade the gut epithelium of the developing larva. Succeeding generations spread throughout the larva. After the larval tick commences to take vertebrate blood, the vermicules enter salivary gland cells, where they undergo schizogony, forming round pyriform bodies. These are like those found in vertebrate erythrocytes both in regard to external morphology and fine structure (vide infra).

Evidently, *Babesia* may be maintained in tick populations by transovarial transmission without passage through a vertebrate host. This has been demonstrated for 32 generations of *Rhipicephalus bursa* infected with *B. bovis* (Markov and Abramov, 1966), but not with other vectors and species of *Babesia*. *B. divergens* has been shown to persist into only a second generation of *Ixodes ricinus*. *B. bigemina* may not be transmitted transovarially in *Boophilus microplus* (Riek, 1964), and *B. bovis* does not persist in *Boophilus microplus* beyond the larval stage (Mahoney, 1977).

When the salivary forms of *Babesia* are inoculated into the vertebrate host, they enter the blood and invade erythrocytes, wherein they multiply asexually. These parasites, though typically pear-shaped, may be spherical, ovoid, spindle-shaped, or ameboid. They may be found singly, in pairs, or in multiples of two (Fig. 14–9). In the small species of *Babesia*, the pyriform bodies are 1.0 to 2.5 μm long; those of the large species are 2.5 to 5.0 μm long. At the blunt end of the organism is a polar ring and a number of clavate organelles forming the apical complex. After invasion of an erythrocyte, the pyriform body loses certain structures, including the apical complex. The young endoerythrocytic parasite is enveloped only by the thin outer membrane and ingests erythrocytic cytoplasm by pinocytosis. Vacuoles are formed in the cytoplasm of the parasite and, by electron micrography, may be seen to contain hemoglobin, which is digested without formation of residual pigment. There are structures covered with double membranes, which may serve functions similar to those of mitochondria, but typical mitochondria are absent.

After a period of feeding and growth, the full complement of organelles, including three-layered pellicle, reappears. The nucleus migrates toward the blunt end and develops two lobes that protrude from the cell surface, forming buds. As these enlarge, nuclear and cytoplasmic material from the parent cell are shared by the two daughter cells, which become pyriform, with blunt end distal.

The apical complex passes over to one of the daughter cells, but the other develops these organelles. The daughter cells then separate at the pointed ends. This process is a form of schizogony, and the resulting organisms are merozoites.

Pathogenesis and Symptomatology. The clinical picture produced by babesiosis in various mammalian hosts varies, depending on the virulence of the strain of *Babesia* and factors of resistance in the host, such as age, transplacental antibody, antibody from colostrum, and host specificity. Although man probably is not a suitable host for some species of *Babesia*, it is now well established that human beings may become infected with several species of *Babesia* without prior splenectomy. Furthermore, these infections may be fulminating or subclinical, just as in the usual mammalian hosts. Of 101 persons residing in a rural area in Mexico where equine, bovine, ovine, and canine babesiosis were known to be common, 38 had serum reactive to antigen from *B. canis*. *Babesia* was isolated by inoculation of blood from three of the asymptomatic persons into splenectomized hamsters (Osorno et al., 1976).

In the clinically apparent cases of human babesiosis, the incubation period apparently was 2 or 3 weeks, although published data on this point are inexact. There is, then, an acute onset of shaking chills, fever of 104°F (40°C), headache, and pain in muscles, abdomen, and back. Anemia may develop rapidly, and there may be detectable jaundice. Erythrocytes are destroyed not only by the emerging parasites but also by other mechanisms. In experimental animals, increase of liver cells and glomerulitis and a decrease in serum complement have been observed.

In human survivors, the febrile illness continued for several weeks and was followed by a prolonged carrier state. The course in the splenectomized patient is much more fulminating and typically ends fatally 5 to 8 days after onset of symptoms.

Epidemiology. The epizootiology of babesiosis involves a cycle of transmission by ixodid ticks from the infected mammalian reservoir to the uninfected. The exact nature of the cycle varies with species of vector, parasite, and reservoir host. Human beings become infected when bitten by an infected tick; thus,

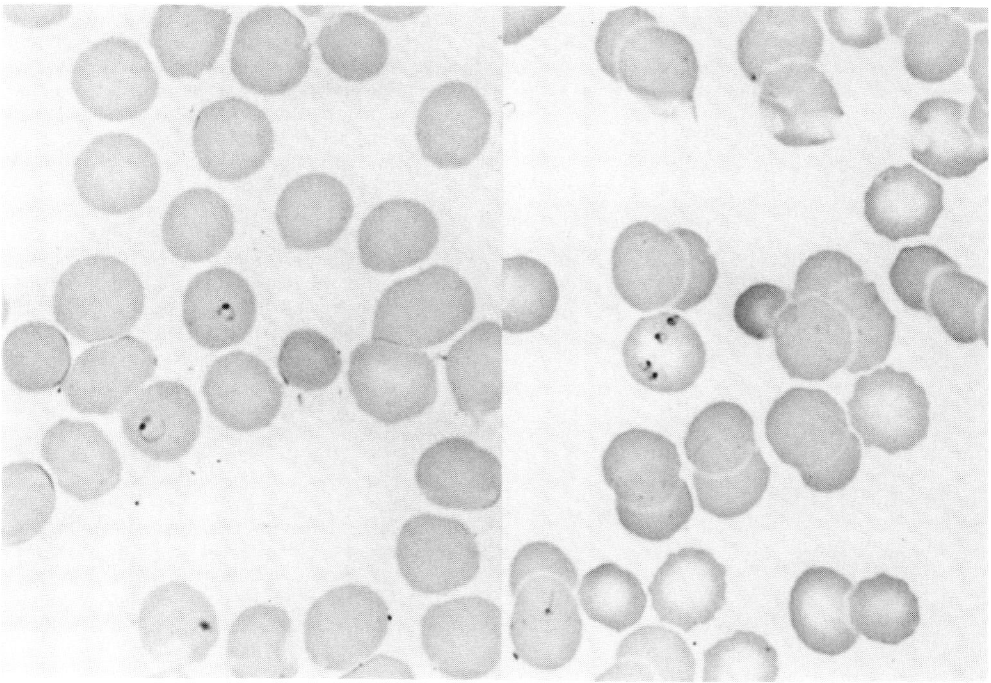

Fig. 14–9. *Babesia microti*. Ring stages in human blood smears. Two single parasites are near the center of the left field, and two pairs are in a single erythrocyte in the right field. Giemsa stain. (× ca. 1200.) (Photomicrographs by Centers for Disease Control, Atlanta, Ga. Courtesy of Dr. George Healy.)

the infections are likely to occur in persons who have been in rural areas where ticks are abundant. Inapparent human infections may be common in such areas (vide infra).

The specific nomenclature of *Babesia* has been confused by the renaming of the same parasites at different places and times. The species infecting human beings to date have been identified as *B. divergens*, *B. bovis* (bovine), *B. caballi* or *equi* (equine), *B. microti* (rodent), and a strain serologically related to but not identical with *B. canis* (canine).

Diagnosis. The diagnosis is suspected on the basis of the clinical picture of acute febrile onset with rapidly developing hemolytic anemia. The infected persons typically have a history of exposure to tick bites. Proof of the etiology is established by demonstration of the intraerythrocytic parasites in the blood film stained with Giemsa or Wright's stain. In many instances, erroneous interpretation of the blood film has led to a diagnosis of malaria, especially of infection with *Plasmodium falciparum*, because of the abundant small rings with several parasites in an erythrocyte. A distinguishing feature of *Babesia* is the absence of pigment.

Inoculation of blood into hamsters and examination of blood (in stained films) drawn from the tail at 1 month post-inoculation is an efficient method of detecting *Babesia microti* infection in wild rodents. The same procedure is recommended for human blood specimens when examination of Giemsa-stained thin blood films has given negative results (Etkind et al., 1980).

Indirect fluorescent antibody, indirect hemagglutination, and capillary agglutination tests are useful for distinguishing species of *Babesia* involved in individual cases and in epidemiologic studies but not in establishing the diagnosis of babesiosis.

Treatment. A number of patients with babesiosis have been treated with chloroquine because it was thought that they had malaria, and they have improved. Nevertheless, the value of this drug has not been definitely established, since parasites persist in the blood for a prolonged period in spite of treatment with chloroquine (Miller et al., 1978). Pentamidine isethionate has not been used to treat babesial infection in human beings but is known to be effective against babesiosis in animals. This is given in a dosage of 2 to 4 mg/kg of body weight intramuscularly daily for 15 days. It is available as an experimental drug from the Parasitic Diseases Division, Centers for Disease Control, Atlanta, Georgia. Diminazene, which is highly effective against babesiosis in animals, has been tried once in treatment of a human infection (Ruebush et al., 1979). Although the drug appeared to be effective, the patient developed Guillain-Barré syndrome after completion of the course of treatment. At present, therefore, pentamidine would be preferable. Exchange transfusions can be used in critically ill splenectomized patients; in most cases, *B. microti* infection can be effectively managed with symptomatic and supportive care (Gombert et al., 1982).

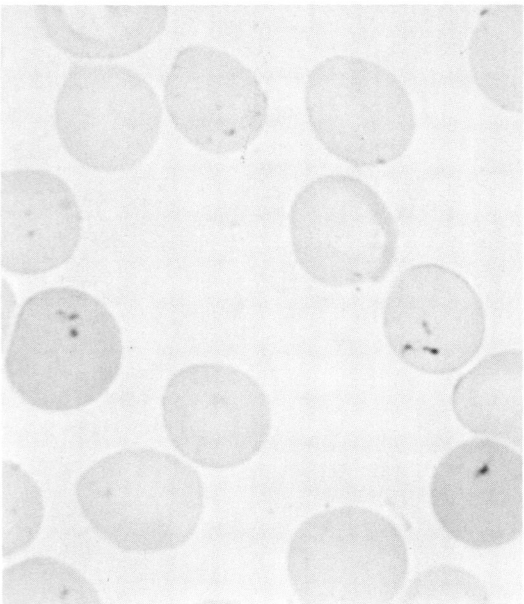

Fig. 14–10. *Entopolypoides.* Minute pleomorphic trophozoites in human erythrocytes; several of the cells contain two or three organisms. Giemsa stain. (× ca. 1800.) (Photomicrograph by Centers for Disease Control, Atlanta, Ga. Courtesy of Dr. George Healy.)

Prevention and Control. There is no practical way of controlling babesiosis in wild animals. Infection in domestic herds can be controlled by (1) eradication of foci of tick infestation by treatment of the environment and dipping of livestock; (2) immunization of livestock against babesiosis; and (3) restriction of movement of animals.

Babesiosis in human beings can be prevented only by avoiding exposure to tick bites in enzootic areas. Splenectomized persons should exercise special care, since they are particularly prone to this infection and have a markedly higher fatality rate.

Genus Entopolypoides

Along with *Babesia* and some other piroplasms, the minute ameboid intraerythrocytic organisms known as *Entopolypoides* are placed in the family Babesiidae. Hoare (1981) regards *Entopolypoides* as a synonym of *Babesia*. Until 1978, naturally occurring *Entopolypoides* infections had been reported only in nonhuman primates. In that year, two cases of infection with *Entopolypoides* were reported (Wolf et al., 1978). In one, the patient was a 54-year-old man who was born in Turkey. Soon after coming to the United States, he was found to have hepatosplenomegaly with hypersplenism, esophageal varices, and hypoprothrombinemia. His spleen was removed, and a liver biopsy showed portal and periportal fibrosis. The second patient was a 28-year-old merchant seaman who also had hepatic dysfunction. In both cases, the organisms were discovered in blood smears examined for malaria parasites. Serum samples were negative for antibodies to all the *Babesia* and *Plasmodium* species tested, except for *Babesia canis* at low titer in the second case. In the first case, chloroquine was given, 1.5 g the first day and 0.5 g daily for 3 days. Three months later, the parasitemia was unchanged and a serum sample was again negative for antibodies against *Babesia* and *Plasmodium* species. The second patient recovered fully without antimalarial treatment and the parasitemia decreased as his liver function improved.

The parasites in both cases were basophilic and distinctly smaller than the *Babesia* species reported in man. They were pleomorphic and without pigment, which is characteristic of the Babesiidae. They were further described and illustrated as small chromatin dots with minute circles or wisps of cytoplasm (Fig. 14–10). The chromatin was either single with straight or curved cytoplasmic extensions that appeared as commas or rings, or in two or more fragments on jointed stems of cytoplasm. Often, several of these small forms were seen in a single erythrocyte, although less numerous, larger forms with two or more chromatin dots or bars dispersed along the cytoplasm were also present. The larger forms appeared as rings, baskets, or irregular ameboid forms. The blood of both patients produced low-level parasitemias when injected intraperitoneally into rats and hamsters.

REFERENCES

Aikawa, M., Huff, C.G., and Sprinz, H. 1966. Comparative

feeding mechanisms of avian and primate malarial parasites. Milit. Med., *131*(suppl.): 969–983.

———, Miller, L.H., and Rabbage, J. 1975. Caveola-vesicle complexes in the plasmalemma of erythrocytes infected by *Plasmodium vivax* and *P. cynomolgi*. Unique structures related to Schüffner's dots. Am. J. Pathol., *79*:285–294.

Allison, A.C. 1954. Protection afforded by sickle-cell trait against subtertian malarial infection. Br. Med. J., *1*:290–294.

———. 1963. Inherited factors in blood conferring resistance to protozoa, pp. 109–122. In *Immunity to Protozoa. A Symposium*. Edited by P.C.C. Garnham, A.E. Pierce, and A. Roitt. Oxford, Blackwell Scientific Publications.

Bannister, L. 1977. Invasion of red cells by *Plasmodium*, pp. 27–55. In *Parasite Invasion. Symposia of the British Society for Parasitology*. Vol. 15. Edited by A.E.R. Taylor and R. Muller. Oxford, Blackwell Scientific Publications.

Bass, C.C., and Johns, F.M. 1912. The cultivation of malarial plasmodia *(Plasmodium vivax* and *P. falciparum) in vitro*. J. Exp. Med., *16*:567–579.

Bazin, C., Lamy, C., Piette, M., Gorenflot, A., Duhamel, C., and Valla, A. 1976. Un nouveau cas de babesiose humaine. Nouvelle Presse Med., *5*:799–800.

Beutler, E. 1959. The hemolytic effect of primaquine and related compounds: A review. Blood, *14*:103–139.

Bienzle, U., Ayeni, O., Lucas, A.O., and Luzzatto, L. 1972. Glucose-6-phosphate dehydrogenase and malaria. Greater resistance of females heterozygous for enzyme deficiency and of males with non-deficient variant. Lancet, *1*:107–110.

———, Guggenmoos-Holzmann, I., and Luzzatto, L. 1979. Malaria and erythrocyte glucose-6-phosphate dehydrogenase variants in West Africa. Am. J. Trop. Med. Hyg., *28*:619–621.

Bray, R.S. 1957a. Studies on malaria in chimpanzees. II. *Plasmodium vivax*. Am. J. Trop. Med. Hyg., *6*:638–649.

———. 1957b. Studies on malaria in chimpanzees. IV. *Plasmodium ovale*. Am. J. Trop. Med. Hyg., *6*:638–649.

———. 1958. Studies on malaria in chimpanzees. VI. *Laverania falciparum*. Am. J. Trop. Med. Hyg., *7*:20–24.

———. 1960. Observations on the cytology and morphology of the mammalian parasites. I. A process of apparent plasmotomy in the pre-erythrocytic phase of *Laverania falciparum*. Riv. Parassitol., *21*:267–376.

Brewer, G.J., and Powell, R.D. 1965. A study of the relationship between the content of adenosine triphosphate in human red cells and the course of falciparum malaria. A new system that may confer protection against malaria. Proc. Natl. Acad. Sci., *54*:741–745.

Brown, K.N. 1976. Resistance to malaria, pp. 268–295. In *Immunology of Parasitic Infections*. Edited by S. Cohen and E. Sadun. Oxford, Blackwell Scientific Publications.

———, and Brown, I.N. 1965. Immunity to malaria: Antigenic variation in chronic infections of *Plasmodium knowlesi*. Nature, *208*:1286–1288.

Bruce-Chwatt, L.J. 1963. Congenital transmission of immunity to malaria, pp. 89–108. In *Immunity to Protozoa. A Symposium*. Edited by P.C.C. Garnham, A.E. Pierce, and I. Roitt. Oxford, Blackwell Scientific Publications.

———. 1980. *Essential Malariology*. London, William Heinemann Medical Books, Ltd. New York, Springer Publishing Co.

———. 1982. Qinghaosu: A new antimalarial. Br. Med. J., *284*:767–768.

Canfield, C.J. 1980. Antimalarial aminoalcohol alternatives to mefloquine. Acta Trop., *37*:232–237.

Centers for Disease Control. 1978. Malaria in participants of a natural history safari to Kenya, Africa. Morbidity & Mortality Weekly Rep., *27*:214.

Chin, W., Contacos, P.G., Coatney, G.R., and Kimball, H.R. 1965. A naturally acquired quotidian-type malaria in man transferable to monkeys. Science, *149*:865.

———, Contacos, P.G., Coatney, G.R., and King, H.K. 1966. The evaluation of sulfonamides, alone or in combination with pyrimethamine, in the treatment of multiresistant falciparum malaria. Am. J. Trop. Med. Hyg., *15*:823–829.

Ciuca, M., Chelarescu, M., Sofletea, A., Constantinescu, P., Teriteanu, E., Cortez, P., Balanovschi, G., and Ilies, M. 1955. Contribution expérimentale a l'étude de l'immunité dans le paludisme. Bucharest, Editions de l'Academie de la Republique Roumaine, pp. 61–108.

Clark, H.C., and Tomlinson, W.J. 1949. The pathologic anatomy of malaria, pp. 874–903. In *Malariology*. Edited by W.F. Boyd. Philadelphia, W.B. Saunders Co.

Coatney, G.R. 1968. Simian malarias in man: Facts, implications, and predictions. Am. J. Trop. Med. Hyg., *17*:147–155.

———, Chin, W., Contacos, P.G., and King, H.K. 1966. *Plasmodium inui*, a quartan malaria parasite of Old World monkeys transmissible to man. J. Parasitol., *52*:660–663.

———, Collins, W.E., Warren, M., and Contacos, P. 1971. *The Primate Malarias*. Bethesda, Md. U.S. Department of Health, Education, and Welfare, National Institutes of Health.

———, Elder, H.A., Contacos, P.G., Getz, M.E., Greenland, R., Rossan, R.N., and Schmidt, L.H. 1961. Transmission of the M strain of *Plasmodium cynomolgi* to man. Am. J. Trop. Med. Hyg., *10*:673–678.

Cohen, S. 1982. Progress in malaria vaccine development. Br. Med. Bull., *38*:161–165.

Contacos, P.G., and Coatney, G.R. 1963. Experimental adaptation of simian malarias to abnormal hosts. J. Parasitol., *49*:912–918.

Corradetti, A. 1981. Relapses and delayed primary attacks in malaria. Trans. R. Soc. Trop. Med. Hyg., *76*:279–280.

Covell, G. 1950. Congenital malaria. Trop. Dis. Bull., *47*:1147–1167.

Das Gupta, B.M. 1938. Transmission of *Plasmodium inui* to man. Proc. Natl. Acad. Sci. India, *4*:241–244.

Deane, L.M., Deane, M.P., and Neto, J.F. 1966. Studies on transmission of simian malaria and report of a natural infection of man with *Plasmodium simium* in Brazil. Bull. W.H.O., *35*:805–808.

Duperval, R., Longpre, B., and Madarnas, P. 1979. Unexplained falciparum malaria in a patient with chronic lymphocytic leukemia. Can. Med. Assoc. J., *121*:1585–1588.

Entrican, J.H., Williams, H., Cook, I.A., Lancaster, W.M., Clark, J.C., Joyner, L.P., and Lewis, D. 1979. Babesiosis in man. A case from Scotland. Br. Med. J., *2*:474.

Escudié, A., and Hamon, J. 1961. Le paludisme en Afrique occidentale d'expression francaise. Med. Trop., *21*:661.

Etkind, P., Piesman, J., Ruebush, T.K., Spielman, A., and Juranek, D.D. 1980. Methods for detecting *Babesia microti* infection in wild rodents. J. Parasitol., *66*:107–110.

Eyles, D.E., Coatney, G.R., and Getz, M.E. 1960. Vivax-type malaria parasite of macaques transmissible to man. Science, *131*:1812–1813.

Fitzpatrick, J.E.P., Kennedy, C.C., McGeown, M.G., Oreopoulos, D.G., Robertson, J.H., and Soyannwo, M. 1968. Human case of piroplasmosis (babesiosis). Nature, *217*:861–862

Garnham, P.C.C. 1933. On some curious bodies found in erythrocytes containing immature crescents. Trans. R. Soc. Trop. Med. Hyg., *26*:401–403.

———. 1963. The ultrastructure of Haemosporidia and allied protozoa with special reference to the motile stages, pp. 427–433. In *Progress in Protozoology*. Prague, Proceedings of the 1st International Congress on Protozoology.

———. 1966. *Malaria Parasites and Other Haemosporidia.* Oxford, Blackwell Scientific Publications.
———. 1977. The continuing mystery of relapses in malaria. Protozool. Abstr., *1*:1–12.
Geiman, Q. M. 1951. The cultivation of malaria parasites, pp. 131–149. In *Parasitic Infections in Man.* Edited by H. Most. New York, Columbia University Press.
———, and Meagher, M.J. 1967. Susceptibility of a New World monkey to *Plasmodium falciparum* from man. Nature, *215*:437–439.
Gilles, H.M., Lawson, J.B., Sibellas, M., Voller, A., and Allan, N. 1969. Malaria and pregnancy. Trans. R. Soc. Trop. Med. Hyg., *63*:1.
Gombert, M.E., Goldstein, E., Benach, J.L., Tenenbaum, M., Grunwaldt, E., Kaplan, M., and Eveland, L. 1982. Human babesiosis. Clinical and therapeutic considerations. J.A.M.A., *248*:3005–3007.
Green, R. 1929. Observations on some factors influencing the infectivity of malarial gamete carriers in Malaya. Bull. Inst. Med. Res. Kuala Lumpur, no. 5 of 1925.
Hall, A.P. 1977. The treatment of malaria. Br. Med. J., *1*:323–328.
Harambasic, H., Krsnjavi, B., and Urbanke, A. 1976. The second case of human babesiosis (piroplasmosis) in Yugoslavia. Acta Parasitol. Iugoslav., *7*:37–41.
Healy, G.R., and Ruebush, T.K. 1980. Morphology of *Babesia microti* in human blood smears. Am. J. Clin. Pathol., *73*:107–109.
———, Walzer, P.D., and Sulzer, A.J. 1976. A case of asymptomatic babesiosis in Georgia. Am. J. Trop. Med. Hyg., *25*:376–378.
Hoare, C.A., 1981. On the status of *Entopolypoides* (Piroplasmida), pp. 112–116. In *Parasitological Topics. A Presentation Volume to P.C.C. Garnham, FRS on the Occasion of His 80th Birthday, 1981.* Edited by E.U. Canning. Society of Protozoologists, Special Publication.
Jeffery, G.M., Wolcott, G.B., Young, M.D., and Williams, D., Jr. 1952. Exoerythrocytic stages of *Plasmodium falciparum.* Am. J. Trop. Med. Hyg., *1*:917–926.
Jiang, J.B., Huang, J.C., Liang, D.S., Liu, J.X., Zhang, S.W., and Cheng, F.C. 1982. Long incubation of *Plasmodium vivax multinucleatum* as demonstrated in three experimental human cases. Trans. R. Soc. Trop. Med. Hyg., *76*:845–847.
Kramer, S.L., Campbell, C.C., and Moncrieff, R.E. 1983. Fulminant *Plasmodium falciparum* infection treated with exchange blood transfusion. J.A.M.A., *249*:244–245.
Krier, J.P., editor. 1980. Malaria, volumes, 1, 2 and 3. New York, Academic Press.
Krotoski, W.A., Krotoski, D.M., Garnham, P.C.C., Bray, R.S., Killick-Kendrick, R., Draper, C.C., Targett, G.A.T., and Guy, M.W. 1980. Relapses in primate malaria: Discovery of two populations of exoerythrocytic stages. Preliminary note. Br. Med. J., *1*:153–154.
———, Garnham, P.C.C., Bray, R.S., Krotoski, D.M., Killick-Kendrick, R., Draper, C.C., Targett, G.A.T., and Guy, M.W. 1982. Observations on early and late post-sporozoite tissue stages in primate malaria. I. Discovery of a new latent form of *Plasmodium cynomolgi* (the hypnozoite), and failure to detect hepatic forms within the first 24 hours after infection. Am. J. Trop. Med. Hyg., *31*:24–35.
Ladda, R., Arnold, J., and Martin, D. 1966. Electron microscopy of *Plasmodium falciparum.* I. The structure of trophozoites in erythrocytes of human volunteers. Trans. R. Soc. Trop. Med. Hyg., *60*:369–375.
Le Bras, J., and Payet, M. 1978. Concentration of parasitized erythrocytes in malaria diagnosis. Trans. R. Soc. Trop. Med. Hyg., *72*:552.
Levine, N.D. 1961. Problems in the systematics of the "Sporozoa." J. Protozool., *8*:442–451.
———. 1969. Taxonomy of the Sporozoa, pp. 365–366. In *Progress in Protozoology.* Prague, Proceedings of the 3rd International Congress on Protozoology.
———. 1970. Taxonomy of the Sporozoa. J. Parasitol., *56*, no. 4, Sect. II:208–209.
Lopez, C.E., Ruebush, T.K., II, and Schultz, M.G. 1979. Occurrence of malaria acquired during travel abroad among American civilians, 1970–1976. J. Infect. Dis., *139*:255–260.
Lucia, H.L., and Nussenzweig, R.S. 1969. *Plasmodium chabaudi* and *Plasmodium vinckei:* Phagocytic activity of mouse reticuloendothelial systems. Exp. Parasitol., *25*:319–323.
Luzzatto, L., Usanga, E.A., and Reddy, S. 1969. Glucose-6-phosphate dehydrogenase deficient red cells: Resistance to infection by malaria parasites. Science, *164*:839–841.
Macdonald, G. 1957. *The Epidemiology and Control of Malaria.* London, Oxford University Press.
MacLeod, C.L., West, R., Saloman, W.L., Heyneman, D., and Goldsmith, R.S. 1982. Double malaria infection in a six-week-old infant. Am. J. Trop. Med. Hyg., 31:893–896.
Madecki, O., and Kretschmar, W. 1966. Häufigkeit und Ursachen der angeborenen Malaria in West-Afrika. Zeitsch. Tropenmed. Parasitol., *17*:195–210.
Mahoney, D.F. 1977. *Babesia* of domestic animals, pp. 1–52. In *Parasitic Protozoa.* Vol. IV. Edited by J.P. Kreier. New York, Academic Press.
Markov, A.A., and Abramov, I.V. 1966. Reciprocal adaptation of certain blood parasites and their hosts. Proceedings of the 1st International Congress on Parasitology, *1*:268–269.
Mathews, H.M., and Armstrong, J.C. 1981. Duffy blood groups and vivax malaria in Ethiopia. Am. J. Trop. Med. Hyg., *30*:299–303.
McGhee, R.B., and Trager, W. 1950. The cultivation of *Plasmodium lophurae* in vitro in chicken erythrocyte suspensions and the effects of some constituents of the culture medium upon its growth and multiplication. J. Parasitol., *36*:123–127.
McGregor, I.A., Williams, K., Voller, A., and Billewicz, W.Z. 1965. Immunofluorescence and the measurement of immune response to hyperendemic malaria. Trans. R. Soc. Trop. Med. Hyg., *59*:395–414.
Meis, J.F., Verhave, J.P., Jap, P.H., and Meuwissen, J.H. 1982. The demonstration of *Plasmodium berghei* sporozoites in rat hepatocytes one hour after inoculation. Z. Parasitenkd., *67*:345–348.
Miller, L.H., and Carter, R. 1976. Innate resistance in malaria. Exp. Parasitol., *40*:132–146.
———, Neva, F.A., and Gill, F. 1978. Failure of chloroquine in human babesiosis *(Babesia microti).* Case report and chemotherapeutic trials in hamsters. Ann. Intern. Med., *88*:200–202.
Monaldi, T. de S., and Raffaele, G. 1935. Sulla fasa negativa del sangue e nell'infezione da *P. vivax.* Riv. Malariol., *14*:344–351.
Moshkovski, S.D. 1941. *Microbiological Methods in Investigations in Infectious Diseases.* Moscow, Medgiz.
North, R.G., and Mackaness, G.H. 1973. Immunological control of macrophage proliferation *in vitro.* Infect. Immun., *8*:67–73.
Osorno, B.M., Vega, C., Ristic, M., Robes, C., and Ibarra, S. 1976. Isolation of *Babesia* spp. from asymptomatic human beings. Vet. Parasitol., *2*:111–120.
Pampana, E.J. 1969. *Textbook of Malaria Eradication.* 2nd Ed. London, Oxford University Press.
Parry, M.F., Fox, M., Burks, S.A., and Richard, W.J. 1977. *Babesia microti* infection in man. J.A.M.A., *238*:1282–1283.

Pays, J.F., and Hugues, F.C. 1981. Polyrésistance des *P. falciparum* asiatiques à la chloroquine et à l'association pyriméthamine-sulfadoxine. À propos d'un nouveau cas. Bull. Soc. Pathol. Exot., *74*:406–413.

Peters, W. 1967. A review of recent studies on chemotherapy and drug resistance in malaria parasites of birds and animals. Trop. Dis. Bull., *64*:1145–1175.

———. 1969. Drug resistance in malaria. A perspective. Trans. R. Soc. Trop. Med. Hyg., *63*:25–40.

———. 1978. Current concepts in the treatment of malaria with particular reference to drug resistance. 18th Southeast Asian Ministers of Education Organization—Tropical Medicine Seminar, Kuala Lumpur, 2–5 August, 1977, pp. 55–57.

Powell, R.D., and Tigertt, W.D. 1968. Drug resistance of parasites causing human malaria. Annu. Rev. Med., *19*:81–102.

Quinn, T.C., Jacobs, R.F., Mertz, G.J., Hook, E.W., and Locklsey, R.M. 1982. Congenital malaria: A report of four cases and a review. J. Pediatr., *101*:229–232.

Rieckmann, K.H., Sax, L.J., Campbell, G.H., and Mrema, J.E. 1968. Drug sensitivity of *Plasmodium falciparum*. An in-vitro microtechnique. Lancet, *1*:22–23.

Riek, R.F. 1964. The life cycle of *Babesia bigemina* (Smith and Kilborne, 1903) in the tick vector *Boophilus microplus* (Canestrini). Aust. J. Agric. Res., *15*:802–821.

Rodhain, J. 1948. Contribution à l'étude des Plasmodiums des anthropoides africains. Transmission du *Plasmodium malariae* de l'homme au chimpanzé. Ann. Soc. Belg. Med. Trop., *28*:39–49.

———. 1956. Paradoxical behaviour of *Plasmodium vivax* in the chimpanzee. Trans. R. Soc. Trop. Med. Hyg., *50*:287–293.

———, and Dellaert, R. 1955. Contribution à l'étude de *Plasmodium schwetzi* E. Brumpt. L'infection à *Plasmodium schwetzi* chez l'homme. Ann. Soc. Belg. Med. Trop., *35*:757–776.

Ross Institute. 1981. Malaria prevention in travellers from the United Kingdom. Br. Med. J., *2*:214–216.

Rudzinska, M.A., Bray, R.S., and Trager, W. 1960. Intracellular phagotrophy in *Plasmodium falciparum* and *Plasmodium gonderi*. J. Protozool., *12*:563–576.

———, Trager, W., and Bray, R.S. 1965. Pinocytotic uptake and the digestion of hemoglobin in malaria parasites. J. Protozool., *12*:563–576.

Ruebush, T.K., Cassaday, P.B., Marsh, H.J., Lisker, S.A., Voorhees, D.B., Mahoney, E.B., and Healy, G.R. 1977. Human babesiosis on Nantucket Island. Clinical features. Ann. Intern. Med., *86*:6–9.

———, Rubin, R.H., Wolpow, E.R., Cassaday, P.B., and Schultz, M.G. 1979. Neurologic complications following the treatment of human *Babesia microti* infection with diminazene aceturate. Am. J. Trop. Med. Hyg., *28*:184–189.

———, and Spielman, A. 1978. Human babesiosis in the United States. Ann. Intern. Med., *88*:263.

Russell, P.F., West, L.S., Manwell, R.D., and Macdonald, G. 1963. *Practical Malariology.* 2nd Ed. London, Oxford University Press.

Schein, E., Mehlhorn, H., and Voigt, W.P. 1979. Electron microscopical studies on the development of *Babesia canis* (Sporozoa) in the salivary glands of the vector tick, *Dermacentor reticulatus*. Acta Trop., *26*:229–241.

Schmidt, L.H., Greenland, R., and Genther, C.S. 1961. The transmission of *Plasmodium cynomolgi* to man. Am. J. Trop. Med. Hyg., *10*:679–688.

Scholtens, R.G., Braff, E.H., Healy, G.R., and Gleason, N. 1968. A case of babesiosis in man in the United States. Am. J. Trop. Med. Hyg., *17*:810–813.

Shortt, H.E., Fairley, N.H., Covell, G., Shute, P.G., and Garnham, P.C.C. 1949. The pre-erythrocytic stage of *Plasmodium falciparum*. Br. Med. J., *2*:1006–1008.

———. 1951. The pre-erythrocytic stage of *Plasmodium falciparum*. Trans. R. Soc. Trop. Med. Hyg., *44*:405–419.

———, and Garnham, P.C.C. 1948. The pre-erythrocytic development of *Plasmodium cynomolgi* and *Plasmodium vivax*. Trans. R. Soc. Trop. Med. Hyg., *41*:785–795.

Shute, P.G. 1945. An investigation into the number of sporozoites found in the salivary glands of *Anopheles* mosquitoes. Trans. R. Soc. Trop. Med. Hyg., *38*:493–498.

———. 1968. Malaria in Britain. Br. Med. J., *1*:578.

Simmons, M., Dunn, A., Storer, A-M., and Bartholomew, A. 1982. Malaria. Br. Med. J., *284*:113.

Sinden, R.E., and Garnham, P.C.C. 1973. A comparative study on the ultrastructure of *Plasmodium* sporozoites within the oocyst and salivary glands, with particular reference to the incidence of the micropore. Trans. R. Soc. Trop. Med. Hyg., *67*:631–636.

Skrabalo, Z., and Deanovic, Z. 1957. Piroplasmosis in man. Report on a case. Doc. Med. Geogr. Trop., *9*:11–16.

Smith, D.H., and Theakston, R.D.G. 1970. Comments on the ultrastructure of human erythrocytes infected with *Plasmodium malariae*. Ann. Trop. Med. Parasitol., *64*:329.

Smrkovski, L.L., Buck, R.L., Alcantara, A.K., and Rodriquez, C.S. 1982. In vitro mefloquine resistant *Plasmodium falciparum* from the Philippines. Lancet, *2*:322.

Stahl, W., Jackson, R., Zaki, M.H., Turek, G., and Gaafar, H.A. 1978. Babesiosis. First case in New York State. N.Y. State J. Med., *78*:642–644.

Trager, W. 1966. Comments on ultrastructure. Milit. Med., *131*(suppl.):1009–1010.

———. 1979. *Plasmodium falciparum* in culture: Improved continuous flow method. J. Protozool., *26*:125–129.

———. 1980. Cultivating friends and parasites. Am. J. Trop. Med. Hyg., *29*:325–329.

———, and Jensen, J.B. 1976. Human malaria parasites in continuous culture. Science, *193*:673–675.

———. 1978. Cultivation of malarial parasites. Nature, *273*:621–622.

Trenholme, G.M., Williams, R.L., Desjardins, R.E., Frischer, H., Carson, P.E., Rieckmann, K.H., and Canfield, C.J. 1975. Mefloquine (WR 142,490) in the treatment of human malaria. Science, *190*:792–794.

Tsuchida, H., Yamaguchi, K., Yamamoto, S., and Ebisawa, I. 1982. Quartan malaria following splenectomy 36 years after infection. Am. J. Trop. Med. Hyg., *31*:163–165.

Udeinya, I.J., Schmidt, J.A., Aikawa, M., Miller, L.H., and Green, I. 1981. Falciparum malaria-infected erythrocytes specifically bind to cultured human endothelial cells. Science, *213*:555–556.

Warren, M., Bennett, G.F., Sandosham, A.A., and Coatney, G.R. 1965. *Plasmodium eylesi* sp. nov., a tertian malaria parasite from the white-handed gibbon, *Hylobates lar.* Ann. Trop. Med. Parasitol., *59*:500–508.

Weniger, B.G., Blumberg, R.S., Campbell, C.C., Jones, T.C., Mount, D.L., and Friedman, S.M. 1982. High-level chloroquine resistance of *Plasmodium falciparum* malaria acquired in Kenya. N. Engl. J. Med., *307*:1560–1562.

Wilcox, A. 1960. *Manual for the Microscopical Diagnosis of Malaria in Man.* U.S. Department of Public Health, Education and Welfare, PHS Publication No. 796. (Formerly NIH Bull. No. 180.) Washington, D.C., G.P.O.

Wolf, R.E., Gleason, N.N., Schoenbaum, S.C., Western, K.A., Klein, C.A., Jr., and Healy, G.R. 1978. Intraerythrocytic parasitosis in humans with *Entopolypoides* species (Family Babesiidae): Association with hepatic dysfunction and serum factors inhibiting lymphocyte response to phytohemagglutinin. Ann. Intern. Med., *88*:769–773.

World Health Organization, 1963. *Terminology of Malaria and Malaria Eradication*. Geneva.

———. 1967. Chemotherapy of malaria. Geneva, Tech. Rep. Ser., no. 375.

———. 1979. WHO Expert Committee on Malaria. Seventeenth report. Geneva, Tech. Rep. Ser., no. 640.

Young, M.D. 1963. Schizogonic forms of *Plasmodium falciparum* in the peripheral blood. *Progress in Protozoology*. Prague, Proceedings of the 1st International Congress on Protozoology, pp. 456–463.

Chapter 15

The Ciliate Protozoa

Balantidium Coli

Members of this group of Protozoa are classified in the phylum Ciliophora, which includes one-celled organisms that are provided with short thread-like extensions of their ectoplasmic membrane (cilia) during some stage of their life cycle. In most of the ciliates, cilia are present during all stages of development; in some, however, cilia are lacking in the definitive stage.

The cilia beat in a coordinated rhythm, one following the other, producing wave-like movements along the surface of the organism. In this way, the animal moves forward in a slightly spiral path, due to the somewhat oblique pattern of the cilia. By reversing the beat of the cilia, it may back up and change its direction.

A large macronucleus accompanied by a micronucleus is characteristically present in the ciliates; occasionally, several micronuclei are present. The former divides amitotically, the latter mitotically. In case of conjugation, the genetic characters are transmitted by the micronucleus, but there is some evidence that the macronucleus may contain the determinants for some of the phenotypic characters of the organism. These protozoa typically have a distinct *cytostome* (cell mouth) that leads into a *cytopharynx,* which is usually located near the anterior end; aborally, there is a less conspicuous *cytopyge* (anal pore). The cilia may cover the entire body and be arranged in longitudinal or obliquely spiralled parallel rows, or they may be confined to limited areas, such as around the cytostome (peristomal region). Many species of ciliates are free-living, while others live in a wide variety of habitats in the digestive tract of invertebrate and vertebrate hosts. The one species of medical interest is *Balantidium coli,* which belongs to the subclass Vestibuliferia, order Trichostomatida, family Balantidiidae.

Genus *Balantidium*

Claparède and Lachmann, 1858

Members of this genus are exclusively parasitic in the digestive tract of invertebrate or vertebrate hosts. They are more or less ovoidal in shape, have a conspicuous cytostome, a ciliated covering over the entire body, contractile vacuoles, a conspicuous, curved *macronucleus,* and a minute *micronucleus*. As in other ciliates, asexual multiplication is accomplished by transverse binary fission. Conjugation has been observed, and encystation is characteristic.

Balantidium coli

(Malmsten, 1857) Stein, 1862

(Producing balantidiasis)

Synonyms. *Paramoecium coli* Malmsten, 1857; *Leukophyra coli* Stein, 1860; *Holophyra coli* Leuckart, 1863.

Historical and Geographical Notes. *Balantidium coli* was first described by Malmsten (1857) from the dysenteric discharges of two patients and was soon observed by Leuckart (1861) and Stein (1862), who transferred the species to the genus *Balantidium,* which was created by Claparède and Lachmann (1858) for the ciliate discovered in the frog's intestine.

Balantidium coli has extensive distribution in hogs throughout temperate and warm climates and in various species of monkeys in the tropics. Human infection is moderately common in warm climates and typically sporadic in cooler areas, except in institutionalized communities with a low level of personal hygiene.

Morphology, Biology, and Life Cycle. *B. coli* is the largest protozoan parasitizing man; at times, it is visible to the unaided eye. It has a trophozoite and a cyst stage. In diarrheic or dysenteric stools, the trophozoite (Fig. 15–1) is comparatively large, ovoid, and covered with short cilia of relatively

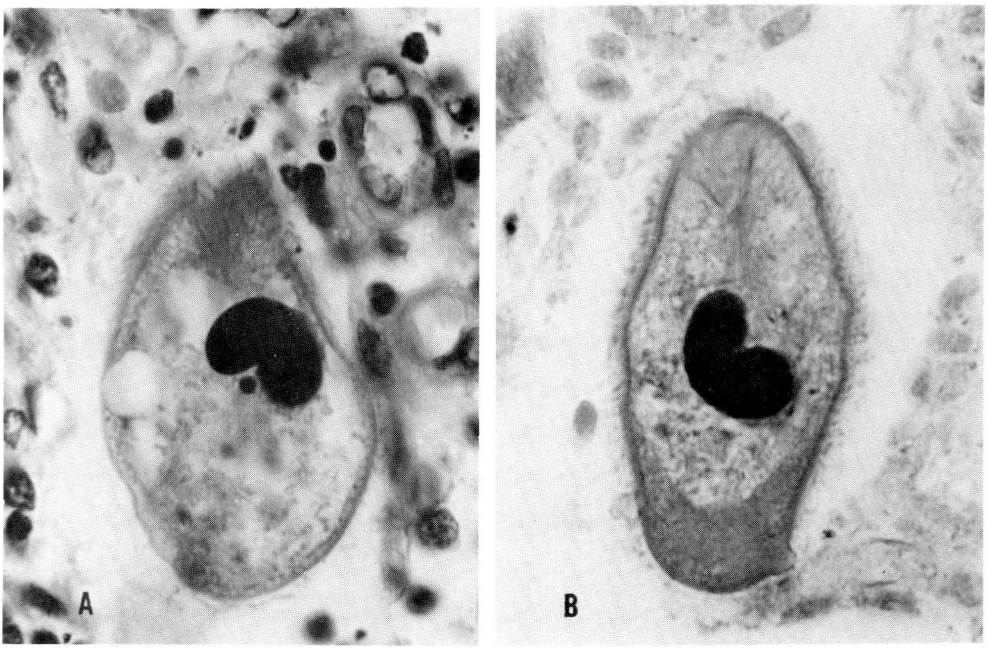

Fig. 15–1. *Balantidium coli.* Trophozoites in section of colon of man in Costa Rica. *A,* Showing cytostome, macronucleus, micronucleus and contractile vacuole. *B,* Showing cilia, macronucleus, margin of cytostome, and (protruding at bottom right) cytopyge. (× 1000.) (By P.C. Beaver.)

uniform length that, in the living organism, maintain a constant synchronized motion that vigorously propels the protozoan forward. The anterior end is somewhat pointed and, on one side of the longitudinal axis, has a deep, slightly curved, inverted conical depression, the *cytostome*. The posterior end is broadly rounded. The length of the trophozoites varies from 50 to 200 μm, and the breadth varies from 40 to 70 μm. When the ciliate is fixed in wet fecal films and stained with carmine or iron-hematoxylin, each of the cilia is found to originate from a basal granule that is immediately under the cell membrane. The peristomal cilia are somewhat longer. The cytoplasm contains numerous food vacuoles and one or two contractile vacuoles. At the posterior end, there is a small opening *(cytopyge)* in the cell membrane through which, in the living trophozoite, the residual contents of food vacuoles empty periodically (Fig. 15–1,B). The most distinctive objects within the cytoplasm are the two nuclei (Fig. 15–1,A). The large trophic nucleus (macronucleus) has the shape of a narrow bean and is densely packed with chromatin granules, which stain as a single mass. The smaller nucleus *(micronucleus)* lies in the center of the inner curvature of the macronucleus. It is a round mass that takes a very deep stain and is believed to function as a kinetic organelle.

The natural habitat of *B. coli* is the large intestine of man, monkeys, and pigs, where the trophozoites feed on cell debris of the intestinal wall, starch grains, bacteria, and mucus as lumen parasites (Fig. 15–2). Objects in food vacuoles may

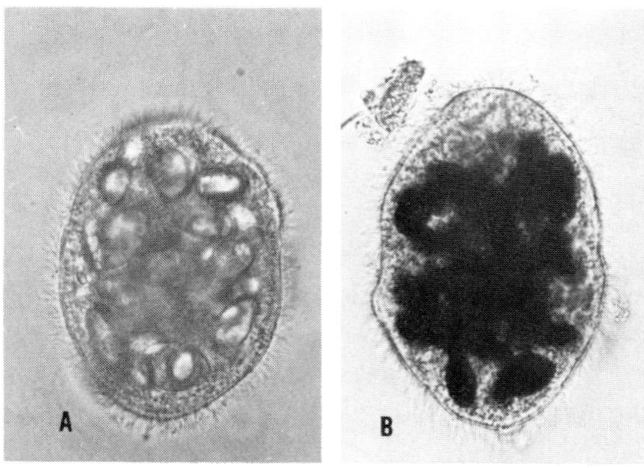

Fig. 15–2. *Balantidium coli.* Trophozoites in formalin-fixed feces from pig. The endoplasm is filled with food vacuoles containing starch grains that had been mistaken for the eggs of *Ascarops strongylina,* a nematode parasite found in the stomach of pigs. *A,* Unstained. *B,* Stained with dilute aqueous iodine. (× 300.) (By P.C. Beaver. Specimen courtesy of Dr. S. Varma and Dr. P.D. Malik.)

be mistaken for cell organelles or for cysts or eggs of other parasites (Varma et al., 1976). *Balantidium coli* reproduces by transverse binary fission, in which the micronucleus, then the macronucleus, divides and, finally, the cytoplasm separates into two daughter organisms. While conjugation (i.e., reciprocal exchange of nuclear material) has been observed in *B. coli,* this is uncommon (Svensson, 1955), and is probably not essential to continued life of the colony.

Encystation of the trophozoite occurs as it is being transported down the intestine, or it may take place after evacuation of a semiformed stool. In this process, the organism partially rounds up, then, without completely retracting the cilia, it secretes a tough cyst membrane (Fig. 15–3). Unlike encystation in the amebae, in *B. coli* this is not preceded by complete discharge of undigested foods; likewise, unlike most encysted amebae, *B. coli* does not have an increase in nuclei during this stage, so that only one organism emerges at the time of excystation. The organism within the cyst remains active for several hours in the fecal specimen. Cysts vary from around 45 to 75 μm in diameter.

Balantidium coli can be cultured in various nutrient media, with or without associated bacteria (see Technical Appendix, p. 759). Cysts in feces remain viable for only 1 to 2 days at room temperature and disappear entirely in about 1 week.

Epidemiology. *Balantidium coli,* or forms morphologically indistinguishable from *B. coli,* are common parasites of many simian hosts, including *Macaca mulatta,* the orangoutan, chimpanzee, New World monkeys, domestic and wild hogs, and wild rats in the USSR (Awakian, 1937). The high prevalence of this organism in domestic hogs, even in cool climates such as in Belfast, Ireland (74%), provides frequent opportunity for human exposure to the infection on the farm and in abattoirs through accidental ingestion of hog feces laden with viable *B. coli* cysts. In New Guinea, where pigs are the principal domestic animal, human infection among pig farmers and slaughterhouse workers is 28%; in contrast, in Moslem Egypt, where human exposure to hogs is uncommon, human balantidiasis is very rare. However, human balantidiasis is reported only as an incidental finding in most temperate areas of the world. Man is relatively refractory to infection with *B. coli* of porcine origin. Hence, in most surveys for human intestinal protozoa, including those in warm climates, *B. coli* has been found only incidentally. In contrast, once *B. coli* has become established in man, the infection may develop epidemically where poor environmental sanitation and low-grade personal hygiene prevail. Such situations have been reported in Brazil, Venezuela, and the Soviet Republic of Georgia, as well as in other areas. These situations have also occurred in mental hospitals in the United States.

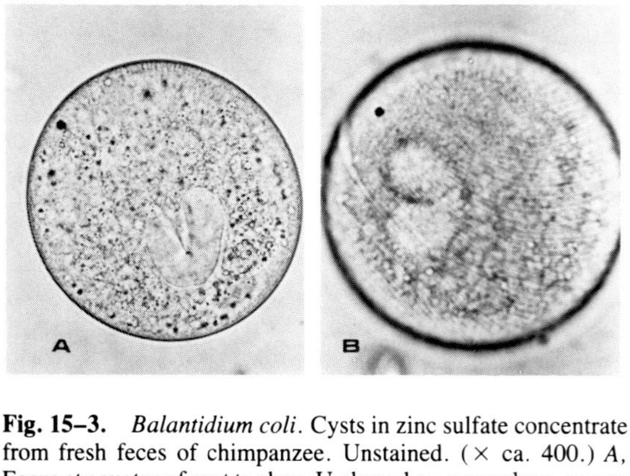

Fig. 15–3. *Balantidium coli*. Cysts in zinc sulfate concentrate from fresh feces of chimpanzee. Unstained. (× ca. 400.) *A*, Focus at equator of cyst to show U-shaped macronucleus, coarse granular endoplasm, and basal elements of cilia, evident under cyst membrane at 12 o'clock position. *B*, Focus at surface to show curved parallel lines of basal elements of cilia and large macronucleus. (By P.C. Beaver.)

Irrespective of their origin, the cysts of *B. coli* are relatively resistant to most unfavorable environmental conditions, but they rapidly succumb to desiccation and direct sunlight.

The cat and the monkey have been infected with *B. coli* from man, but attempts to transmit *B. coli* from the pig or monkey to man were unsuccessful. Nevertheless, it is generally assumed that the usual source of infection for man is the pig. This is especially true in Papua New Guinea, where it is reported that one of the most outstanding ecologic features of human populations in the Highlands is the close association with pigs (Radford, 1973), and on the Pacific Island of Truk, where an outbreak of 110 cases of balantidiasis in 1971 was reported (Walzer et al., 1973). In a fatal case reported from Ghana, there had been close association with pigs (Christian, 1974).

Pathogenesis, Pathology, and Symptomatology. Once established in man, *B. coli* can quickly become a tissue invader. After excystation, the freed trophozoites, on contact with the mucosal surface of the colon, may burrow into the mucosa and set up colonies. This action is greatly aided by the many cilia; it is assumed that lytic action supplements mechanical entry.

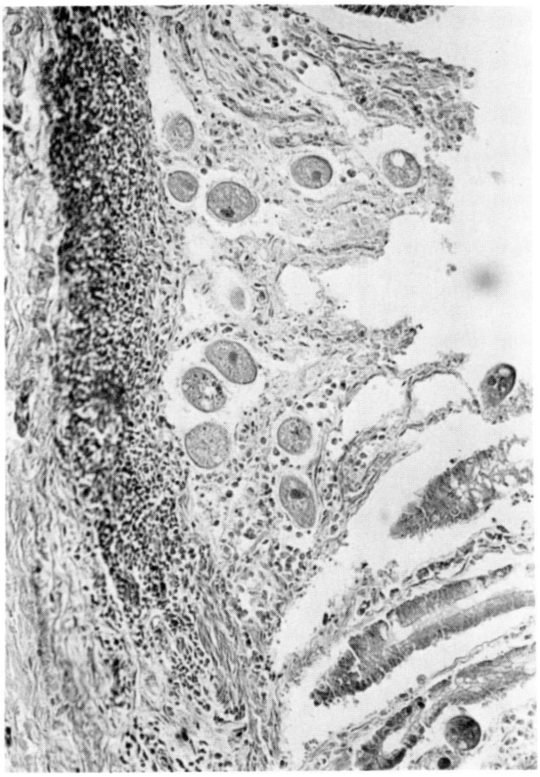

Fig. 15–4. *Balantidium coli*. Trophozoites in ulcer of colon from a 3-year-old child in Puerto Rico. (× 160.) (By P.C. Beaver. Preparation courtesy of Dr. G.C. Carrera.)

Although the characteristic early lesion in *Balantidium* infection generally resembles that of *Entamoeba histolytica,* the opening into the mucosa is of larger dimensions. Cellular infiltration in the vicinity of the ulcer is common and occurs early because of the ease with which bacteria enter the ulcer. Although *B. coli* is a pathogen, it may live in the lumen of the human bowel without producing symptoms. In most infected individuals, diarrhea is present, and in some individuals, a severe type of dysentery that is accompanied by ulceration of the colon develops. Within the tissues, the balantidia multiply and produce ulcers or subsurface abscesses in the mucous or submucous coats (Fig. 15–4) that sometimes extend to the muscular layer. The ulcers are round, ovoid, or irregular in shape with undermined edges; the ulcer floor is covered with pus and necrotic material. The abscesses are usually small and, when incised, are found to be filled with a mucoid material containing numerous balantidia. The mucous membrane between the ulcers may appear normal or it may be swollen and hemorrhagic. As in amebic infection, the ulcers may communicate by passages that are beneath the mucous membrane or upon the surface of the mucosa. Sections of infected intestine show round-cell infiltration, abscesses, hemorrhagic areas, coagulation necrosis in the walls of the ulcers, and numerous balantidia in nests within the tissues (Fig. 15–4).

In rare instances, *B. coli* invades extraintestinal sites. Correa Heanao (1947), in Colombia, reported two fatal cases of balantidial peritonitis following rupture of fulminating colonic ulcers. Maliwa and von Haus (1920), in Germany, recovered balantidia from the urinary tract of a female patient suffering from urethritis, cystitis, and pyelonephritis; Isaza Mejia (1955), in Colombia, incriminated *B. coli* in an inflammatory vaginitis in a 62-year-old woman. The extraintestinal infection was presumably secondary to colonic balantidiasis. In seven autopsy cases that were reported from Peru, all involved the mucosa and submucosa of the colon; four, the muscular layers; two, the subserosa with perforation; two, the appendix; and in one, the balantidia were adjacent to the mesenteric lymphatics without entering the lymphatic vessels or veins.

Symptomatology of infection with *B. coli* is similar to that of amebiasis. Many individuals present no symptoms, but in a majority, diarrhea or dysentery accompanied by abdominal colic, tenesmus, nausea, and vomiting is characteristic. Loss of appetite, headache, insomnia, muscular weakness, and loss in weight have all been observed in infections with this parasite. The dysentery may develop gradually or may appear precipitately following months of diarrhea. Physical examination may be negative, but usually there is some tenderness over the colon, and both skin and mucous membranes may appear anemic. In some cases, the symptoms are those of a severe dysentery of the amebic type, and the stools may contain much blood and mucus; in others, constipation may be present.

Diagnosis. Diagnosis depends on demonstration of *B. coli* in the feces of the patient. Motile trophozoites are present when the stools are diarrheic or dysenteric in character; cysts are found in semiformed and formed stools. The methods employed to demonstrate the parasite in the feces are similar to those used for the intestinal amebae (see p. 117).

Treatment. The drug of choice in the treatment of balantidiasis is tetracycline in a dosage of 500 mg four times daily for 10 days. As an alternative, diiodohydroxyquin may be used. Successful treatment with nitrimidazine (Botero, 1973) and with metronidazole (Garcia-Laverde and De Bonilla, 1975) has been reported. Following trial treatments with metronidazole, all of 20 patients were cured with total dosages ranging from 2.5 g during 5 days for children under 8 years of age to 12.5 g during 10 days for adults. *Balantidium* disappeared from the feces after 4 days of treatment in all cases. Ten patients were followed for 2 to 6 weeks, the others for 20 months. A dosage of 800 mg administered three times daily for 5 days appears to be adequate for adults (Nicholson, 1978).

Prognosis. In healthy individuals, infection with *B. coli* frequently disappears spontaneously or may become latent. Symptomless carriers of the parasite are usually freed of their infection by therapeutic measures. In debilitated individuals, infection with *B. coli* may be serious and even fatal (Alvarez and Garcia, 1967).

Prevention. Methods of prevention are similar to those employed for *Entamoeba histolytica* infections, but *B. coli*-infected pigs are probably the main danger, because infection of man in the first instance depends on contact with this animal or the contamination of food or drink with the feces of infected pigs.

Ciliate of Uncertain Status

Nyctotherus faba Schaudinn, 1899

Schaudinn (Jakoby and Schaudinn, 1899) described this ciliate from stools of a patient in Berlin. The organism is bean-shaped, slighlty flattened, and measures 26 to 28 μm in length, 16 to 18 μm in breadth, and 10 to 12 μm in thickness. The body is covered with short, delicate cilia, and cilia line the long stout peristome that extends along the concave side of the organism from the anterior end to the equator, where it terminates in a short esophagus that opens into the endoplasm. A single large contractile vacuole is situated posteriorly. Near the center of the body are a relatively large spherical macronucleus that is 6 to 7 μm in diameter and a minute spherical or ovoid micronucleus that is 1.0 to 1.5 μm in diameter.

Nothing is known about the life cycle of *N. faba* or its potential pathogenicity. Other species of *Nyctotherus* parasitize frogs, fishes, and insects but no pathogenic effect has been ascribed to them.

Coprozoic Ciliates

Many species of ciliates live in water and are apt to be found in fecal specimens that have become contaminated by water containing them. These free-living species usually belong to the genera *Chilodon* or *Cyclidium* and can be confused morphologically with *B. coli* or regarded as new species of *Balantidium* by one unaccustomed to the study of these organisms. When collecting specimens of human feces, one should take care that water exposed to the air is not mixed with the material, because from time to time, such mistakes will inevitably occur.

REFERENCES

Alvarez, V.R., and Garcia, T.R. 1967. Estudio de un caso mortal de balantidiosis humana. Rev. Invest. Salud. Publ. (Mex.), *27*:217–224.

Awakian, A.A. 1937. Studies on the intestinal protozoa of rats. II. Rats as carriers of *Balantidium*. Trans. R. Soc. Trop. Med. Hyg., *31*:93–98.

Botero, R.D. 1973. Effectiveness of nitrimidazine in treatment of *Balantidium coli* infections. Trans. R. Soc. Trop. Med. Hyg., *67*:145.

Christian, E.C. 1974. Fatal balantidiasis. Ghana Med. J., *10*:86–89.

Correa Henao. 1947. Bol. Clin. Univ. de Antioquia. (Cited by Isaza Mejia, 1955.)

Garcia-Laverde, A., and De Bonilla, L. 1975. Clinical trials with metronidazole in human balantidiasis. Am. J. Trop. Med. Hyg., *24*:781–783.

Isaza Mejia, G. 1955. Balantidiasis vaginal. Antioquia Med., *5*:488–491.

Jakoby, M., and Schaudinn, F. 1899. Ueber zwei neue Infusorien im Darm des Menschen. Cbl. Bakteriol., *25*:487–494.

Maliwa, E., and Von Haus, V. 1920. Ueber Balantidiuminfektion der Harnwege. Zeitschr. Urol., *14*:495–501.

Nicholson, N.W. 1978. Case report of *Balantidium coli* infection. E. Afr. Med. J., *55*:133.

Radford, A.J. 1973. Balantidiasis in Papua New Guinea. Med. J. Aust., *1*:238–241.

Svensson, R. 1955. On the resistance to heating and cooling of *Balantidium coli* in culture and some observations regarding conjugation. Exp. Parasitol., *4*:502–525.

Varma, S., Malik, P.D., and Lal, S.S. 1976. The feeding habits of *Balantidium coli*. Vet. Rec., *99*:256.

Walzer, P.D., Judson, F.M., Murphy, K.B., Healy, G.R., English, D.K., and Schultz, M.G. 1973. Balantidiasis outbreak in Truk. Am. J. Trop. Med. Hyg., *22*:33–41.

Section III

Helminths and Helminthic Infections

Chapter 16

Introduction to Helminths and Helminthic Infections

The word "helminth," from the Greek ελμινσ, means "worm" and originally referred to intestinal worms, but it is more usually interpreted to include both parasitic and free-living species of roundworms (phylum Nematoda), "hair snakes" or gordiid worms (phylum Nematomorpha), turbellarians, flukes and tapeworms (phylum Platyhelminthes), and thorny-headed worms (phylum Acanthocephala). By a still more liberal interpretation, it may also include the leeches, a class group of the phylum Annelida. All of these helminths belong to the Metazoa.

Free-living members of these groups are more nearly related to their ancestral prototypes than are parasitic species, which are frequently highly modified as an adaptation to parasitic existence. Thus, complete or partial loss of a digestive tract is presumably due to location of the parasite in a milieu where predigested food is abundant.

The *integument* of parasitic helminths may be hardened, tough and elastic, or relatively delicate, but in most cases, it is resistant to digestion while the organism is alive. Frequently, the integument is provided with spines, hooks, cutting plates, stylets, or other armature for attachment to, penetration of, or abrasion of host's tissues. Such modifications are particularly common in the regions of the mouth. There may also be muscular acetabula, or suckers (especially in trematodes and cestodes), by which the organism maintains its position at a particular site in the host. In some trematodes, the acetabula are also used for locomotion. At no time during the parasitic phase of its life cycle is the helminth's integument provided with cilia. Many—possibly a majority of—parasitic helminths are provided with secretory glands that usually open near the mouth, are lytic in function, serve to digest host's tissue for use as food, or enable the worm to migrate through the tissues to the site where it matures.

The greater part of the life cycle of many parasitic worms is lived under conditions that are essentially anaerobic. Storage of fats and glycogen is common and probably provides basic material from which oxygen is obtained for metabolic needs. In tapeworms, calcium salts are also stored, but their function is poorly understood. Eggs, embryos, or larvae are characteristically provided with concentrated foodstuffs that are utilized during the development of the young organism until it can obtain its own nourishment.

While most systems of parasitic worms are greatly reduced as a result of parasitism, the reproductive systems are greatly elaborated. In all but a few species, in which parthenogenesis prevails, both male and female organs are fully developed. The sexes are separate in a majority of roundworms, gordiid worms, and thorny-headed worms, and in a few trematodes; hermaphroditism is the rule in the remainder of trematodes and all cestodes. In the more highly developed cestodes, there is a strobila or chain of several to many proglottids (commonly, but inaccurately, called "segments"), which, when mature, have complete male and female organs; in a few species, the proglottids have two sets of male and female organs.

The *male worm*, or the *male system* of the hermaphrodite, has a testis (or multiple testes), a vas efferens for each testis, a vas deferens, and an ejaculatory duct or cirrus organ. Frequently, there may be a seminal vesicle and prostate glands, and sometimes, additional secondary copulatory structures. An ovary, oviduct, seminal receptacle, uterus, and vagina are usually present in the *female system*. In trematodes and cestodes, there is an ootype where the component parts of the egg are assembled and separate organs for the production of yolk material and shell. The end-product of all of these structures is the egg, which is produced in considerable numbers (at times, as many as 200,000 or more per female system each 24 hours). The need for providing the egg with yolk material and with a protective shell, as well as the production of enormous numbers of eggs, is due to vicissitudes encountered by eggs after they are shed from the uterus. With isolated exceptions, eggs, or larvae hatched from eggs, pass out of the definitive host's body into an external environment, where many of them perish before they become infective for the next appropriate host. Because of the variety of environmental factors that determine the future course of the organism and the hazards encountered, it is impossible to predict the chances of survival, but in all except the most favorable environments, it may be less than one in a million. Furthermore, many parasitic helminths require one or more intermediate hosts. In trematodes and a few cestodes that require intermediate hosts, additional multiplication takes place in these supplementary hosts. In the roundworms, however, no such propagation is known.

Among the unfavorable conditions that may be encountered by the egg or the hatched larva are the following: desiccation, fermentation, overgrowth by bacteria and fungi, and other unfavorable conditions in the environment that retard or prevent embryonation or hatching or kill the larva after it escapes from the eggshell; inability to find, or failure to be taken up by, an appropriate intermediate or definitive host; death of the intermediate host; and, finally, man-made barriers that kill the parasite during its intermediate stages or prevent its arrival in the appropriate definitive host. Under favorable circumstances, however, the helminth may encounter none of these vicissitudes and may develop in epidemic proportions, or it may be a serious chronic public-health problem for native populations.

The role of malnutrition in helminthic infections may be great, both with respect to the parasite and its host (Tripathy et al., 1971a, b; Freij et al., 1979; Stephenson, 1980). Alteration of diet or dietary deficiencies are important factors governing susceptibility to certain helminths, in spontaneous loss of the worms, or in partial or complete refractoriness to infection. Malnutrition interferes with some types of antibody production, may decrease inflammatory reaction and, in this way, lower resistance. Deficient diet combined with helminthiasis may cause anorexia, impaired liver function, fever, neutralization of digestive enzymes, failure to absorb important vitamins, and aggravation of hemorrhage.

The medical parasitologist, physician, and public-health worker are primarily concerned with the helminths that parasitize man, their modes of entry into the human body, sites in the body where they become established, their pathogenicity and symptomatology, their diagnosis, and approved methods for eradicating the organisms from the individual, the environment, and the community.

In some regions of the world, climatic and other environmental conditions, or human customs, favor the prevalence of a particular helminth—for example, the vesical blood-fluke *(Schistosoma haematobium),* the lung fluke *(Paragonimus westermani),* the beef tapeworm *(Taenia saginata),* or the pinworm *(Enterobius vermicularis).* In other areas, especially in warm, moist climates, infection with several intestinal or other helminths (i.e., *polyhelminthism*) is encountered in a large proportion of individuals in the population. A light infection with a single helminth species usually poses no serious problem. On the other hand, when they are present in great numbers, essentially all kinds of worms are harmful (Beaver, 1974). In communities with high prevalence of multiple infections, the clinical and public-health aspects are complex, requiring careful evaluation of each parasite in the causation of the clinical picture and multiple methods of attack to reduce the prevalence of the parasites in the individual and the community.

REFERENCES

Beaver, P.C. 1974. Biology of soil-transmitted helminths: The massive infection. Health Lab. Sci., *12*:116–126.

Freij, L., Meeuwissen, G.W., Berg, N.O., Wall, S., and Gebre-Medhin, M. 1979. Ascariasis and malnutrition. A study in urban Ethiopian children. Am. J. Clin. Nutr., *32*:1545–1553.

Stephenson, L.S. 1980. The contribution of *Ascaris lumbricoides* to malnutrition in children. Parasitology, *81*:221–233.

Tripathy, K., Garcia, F.T., and Lotero, H. 1971a. Effect of nutritional repletion on human hookworm infection. Am. J. Trop. Med. Hyg., *20*:219–223.

———, Gonzalez, F., Lotero, H., and Bolanos, O. 1971b. Effects of *Ascaris* infection on human nutrition. Am. J. Trop. Med. Hyg., *20*:212–218.

Chapter 17

Roundworms: Morphology, Biology, and Classification

The term "roundworm" includes not only true roundworms or nematodes (phylum Nematoda) but also "hair snakes" (phylum Nematomorpha). The Acanthocephala, which were placed with the Nematoda and Nematomorpha in the phylum Nemathelminthes for many decades, are a distinct phylum. The "hair snakes" or gordiid worms have relatively few species compared with the nematodes; they rarely, if ever, parasitize man.

NEMATODES OR TRUE ROUNDWORMS

Many species of nematodes are free-living forms, some of which are found in fresh or salt water, others in mud or soil. Other nematodes are parasites of roots, stems, and even the seeds of plants. Thousands of species of roundworms are parasites of invertebrate and vertebrate animals. Among the vertebrates, it is doubtful if any species exists that is not parasitized by one or more roundworms. It is not surprising, therefore, that the largest number of helminth parasites of man belongs to the group of the roundworms. Several of these, such as hookworms, *Ascaris lumbricoides, Trichinella spiralis,* and Bancroft's filarial worm *(Wuchereria bancrofti),* are important agents of human disease. In addition to those species that are primarily or exclusively human parasites, such as the pinworm *Enterobius vermicularis,* some important nematode parasites of man have a variety of mammalian hosts (e.g., *Trichinella spiralis*); and still others, such as *Micronema* species, are free-living, but under circumstances that are not yet understood, they may become parasitic. In addition, many species of nematodes are common parasites in domestic animals but relatively incidental human parasites. Thus, a wide range in species, in types of life cycles, and in biologic and host-parasite relationships is found among the roundworms that parasitize man.

Morphology. Nematode species are typically elongate-cylindroidal in shape, primitively bilaterally symmetrical but with a secondary triradiate symmetry at the anterior end. The adults vary in size from forms barely visible to the unaided eye to those as large in diameter as a lead pencil *(Dioctophyma renale)* or a meter in length *(Dracunculus medinensis)*. However, neither size alone nor even the number of the individual parasites infecting man necessarily provides a criterion of the damage that they may produce or the symptoms that they may provoke. A single pair of the minute Bancroft's filarial worm may at times cause more irreparable damage than scores of *Ascaris lumbricoides*.

The integument of nematodes (Fig. 17–1) consists of a non-nucleated cuticula secreted by the underlying hypodermis. It may be smooth, striated, bossed, or ornamented with spines. The nematode cuticle is a complex and varied structure with basically three main layers that may be subdivided into further layers. It is not an inert secretion but a metabolically active structure. Usually around the mouth, and at times in the cervical, anal, and caudal regions, there are sessile or pedunculated papillae, that are probably sensory in function. Under the cuticle is the hypodermis, which, in mature roundworms, appears as a syncytium of intermingled nuclei and fibers. Below it is the layer of somatic muscles. Arising from the hypodermis and projecting into the body cavity are four longitudinal cords, which are characteristic of the group.

Nematodes lack circular muscles. They move by

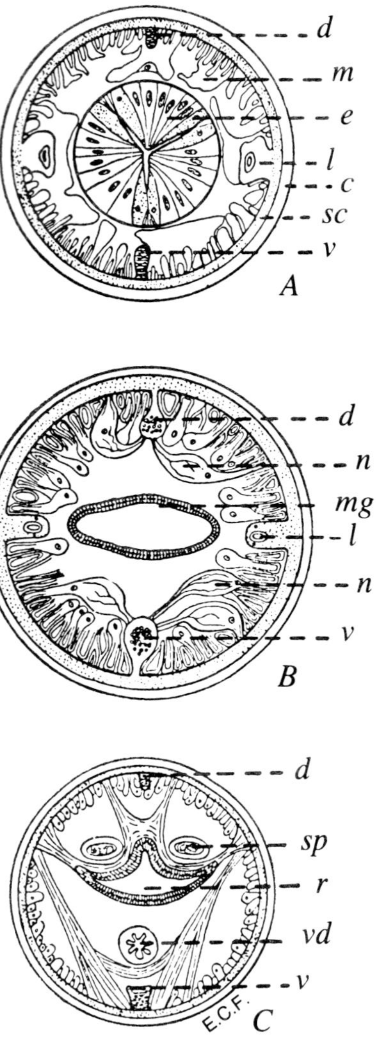

Fig. 17–1. *Ascaris lumbricoides*. Transverse sections through three regions of the body. *A*, Esophageal level. *B*, Mid-body region. *C*, Region of male rectum. *c*, Cuticle. *d*, Dorsal cord. *e*, Esophagus. *l*, Lateral cord and excretory canal. *m*, Muscle cell. *mg*, Mid-intestine. *n*, Nerve. *r*, Rectum. *Sc*, Subcuticula (hypodermis). *sp*, Spicule. *v*, Ventral cord. *vd*, Vas deferens. (*A* adapted from R.B. Goldschmidt. *B* adapted from G.P.H. Brandes. *C* adapted from K.G.F.R. Leuckart. In Faust, E.C. 1939. *Human Helminthology*. Philadelphia, Lea & Febiger.) (× ca. 10.)

sinuously flexing their bodies, slowly or rapidly, at times in figure-8 gyrations. Filariform larvae of *Strongyloides* are able to swim in water. The somatic muscle bands consist of a single layer of longitudinal nonstriated muscle cells with sarcoplastic processes that extend into the body cavity. These muscle bands are divided into four groups by the dorsal, ventral, and lateral cords and are designated as *meromyarian* or *polymyarian*. The meromyarian type (e.g., *Enterobius* and *Ancylostoma)* is the simpler and typically consists of very few (2 to 5) cells in a cross-section of each quadrant; in each quadrant of the polymyarian type (e.g., *Ascaris)*, there are numerous cells, the protoplasmic elements of which project into the body cavity. Muscle cells with contractile fibrils that are limited to the basal zone and in contact with the hypodermis are termed *"platymyarian"*; when the protoplasmic part of the muscle cell extends into the body cavity and the contractile fibrils are situated along the sides as well as at the base of the cell, the muscles are referred to as *"coelomyarian."*

The body cavity is a *pseudocele* (i.e., it is not lined with mesothelium). All of the viscera, including the digestive system, the excretory system (primitively), and the reproductive system, are suspended in the pseudocele.

At the anterior end of the nematode is the mouth, which is frequently, although not always, provided with spines, hooks, cutting plates, stylets, or other structures useful for attachment or for penetrating tissues. In passing from one larval stage to the next and from the last larval stage to the adult worm, the cuticula of the previous stage (i.e., the "sheath") is shed.

The *mouth* is frequently provided with a capsular thickening. The *oral* or *pharyngeal cavity* may be large and hollowed out, it may be short, or it may be long and capillary in type. Behind the oral opening (Fig. 17–2) is the *esophagus*, which characteristically varies in shape and structure in different groups, but, at its distal end, usually has an enlarged muscular bulb that is provided with strong valves capable of closing the opening into the midgut. Both the pharyngeal cavity and the esophagus are lined with cuticle. Typically, the nematode has a single dorsal and two subventral esophageal glands, each with a single nucleus, that probably aid in digestion by secreting lytic substances. The *midgut*, or intestine, is lined with a single layer of columnar cells. The *rectum* is situated just behind the intestine and connects with the posterior end of the latter. It is lined with cuticle and opens outwards through the anus or, in the male, through the *cloaca*. In the male worm, the spermatozoa are also discharged through the cloaca. At the caudal extremity of nonbursate species like *Ascaris*, there are three *rectal glands* in females, six in males. In bursate species, as the hookworms, they are possibly represented in the male by the *cement glands*, which aid in copulation.

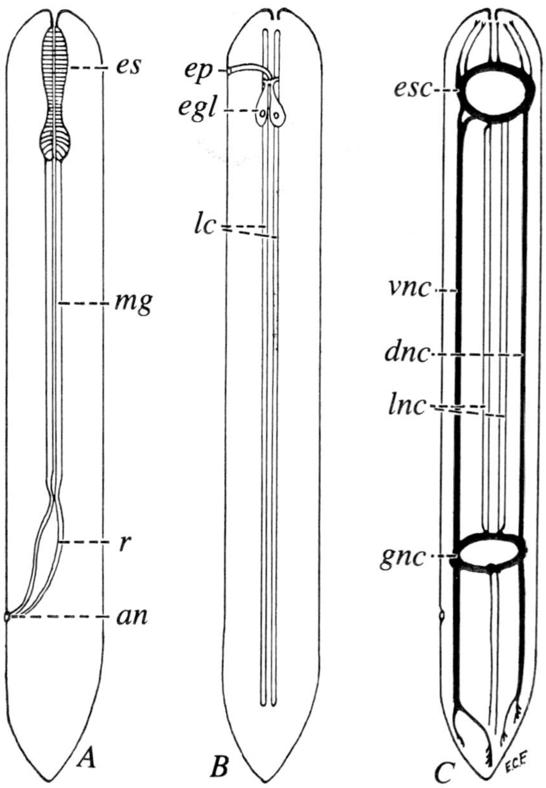

Fig. 17–2. Nematode organ systems viewed from left side. *A*, Alimentary tract. *B*, Excretory organs. *C*, Nerve cords and commissures. *an*, Anus. *dnc*, Dorsal nerve cord. *egl*, Excretory gland cell. *ep*, Excretory pore. *es*, Esophagus. *esc*, Circumesophageal nerve commissure. *gnc*, Genital nerve commissure. *mg*, Intestine. *lc*, Lateral excretory canals. *lnc*, Lateral nerve cords. *r*, Rectum. *vnc*, Ventral nerve cord. (Adapted by E.C. Faust.)

The *excretory system* (Fig. 17–2*B*) is highly modified from its primitive type, and the characteristic archetypic pair of lateral collecting tubules may be asymmetrical or may be represented by a single lateral or median gland cell. The excretory pore of present-day nematodes opens midventrally in the cephalic or cervical region. Weinstein and Haskins (1955) provided chemical evidence that supports the classical morphologic view that the physiologic function of this system is excretory.

Nematodes do not have a *circulatory system*.

The nervous system (Fig. 17–2*C*) consists of a dorsal, a ventral, and four lateral longitudinal trunks with transverse commissures. The most important commissure is the circumesophageal ring, which constitutes the nerve center. Nerve endings terminate in all of the important organs and in the integument, especially in sensory papillae.

There is probably a pair of laterally placed minute receptor organs called "amphids," or "lateral organs," in the cephalic or cervical region of all nematodes. In species without caudal glands, there is a pair of minute lateral postanal *phasmids*, or caudal "chemoreceptor organs."

Reproduction. All nematodes parasitic in man have separate sexes. The male is usually smaller than the female, and is commonly provided with special organs in the perianal or caudal regions that facilitate copulation. In the *Syngamus* species, the male and female may be permanently fused in copula. The male genital system (Fig. 17–3*A*) typi-

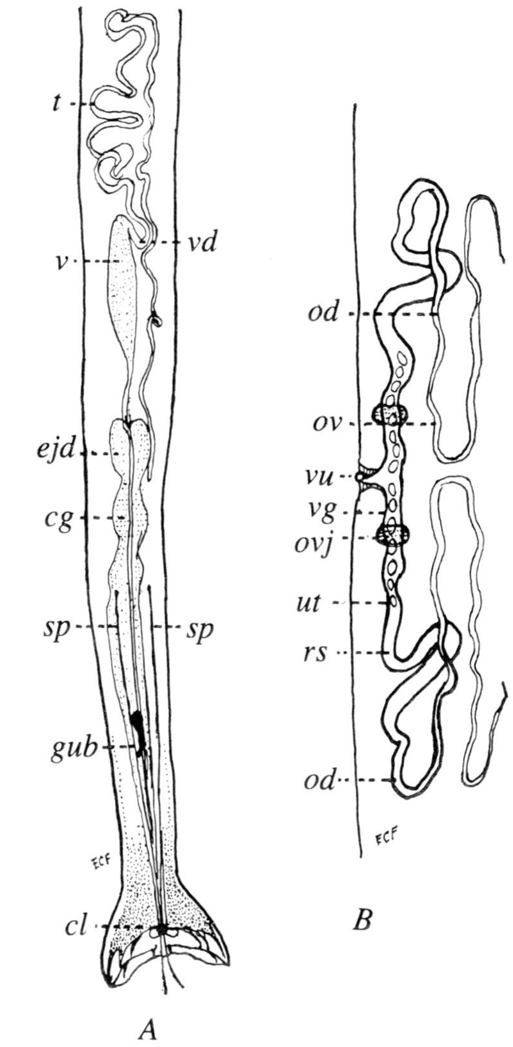

Fig. 17–3. *Ancylostoma duodenale*, male and female reproductive systems. *A*, Male organs viewed from ventral side. *B*, Female organs viewed from left side. *cg*, Cement gland. *cl*, Cloaca. *sp*, Copulatory spicule. *ejd*, Ejaculatory duct. *gub*, Gubernaculum. *od*, Oviduct. *ov*, Ovary. *ovj*, Ovejector. *rs*, Seminal receptacle. *v*, Seminal vesicle. *t*, Testis. *ut*, Uterus. *vd*, Vas deferens. *vg*, Vagina. *vu*, Vulva. (Adapted by E.C. Faust.)

cally consists of a single delicate tubule differentiated from its inner end outwards into *testis (f)*, *vas deferens (vd)*, *seminal vesicle (v)*, and *ejaculatory duct (ejd)*, successively. It is folded on itself one or more times in the body cavity and opens into the *cloaca (cl)*. The ejaculatory duct is lined with prostate or *cement glands (cg)*. One or two *copulatory spicules (sp)* and, frequently, a *gubernaculum (gub)* constitute the accessory copulatory apparatus. In bursate forms like the hookworms, the perianal and caudal cuticle is extended into an umbrella-like expanse that is reinforced by ribbed thickenings to serve as a copulatory disk (i.e, *bursa copulatrix)* around the vulvar opening of the female worm. The spermatozoa are usually ameboid rather than flagellate in type.

The *female genital* system (Fig. 17–3*B*) may be single *(Trichinella,* whipworm) or double (*Ascaris lumbricoides, Enterobius vermicularis,* hookworms). The *ovary* is a delicate thread-like tubule that gradually increases in diameter as it proceeds toward the oviduct. The tubule is differentiated into *oviduct (od)*, *seminal receptacle (rs)*, *uterus (ut)*, *ovejector (ovj)*, and *vagina (vg)*, respectively. In forms like the whipworm, the vulva is the outermost portion of the single tubule. When there is a twinned system, the two vaginas typically unite to form the *vagina vera,* which, in turn, leads to the *vulva (vu)*, which opens to the exterior. This vulvar opening is usually situated midventrally in the anterior half of the worm, but, in some species, it is near the anus. The generative tubules in *Strongyloides* and the whipworm are relatively short, and, at most, are reflected only twice; in *Ascaris,* however, these tubules have manifold coiling and, when dissected out, may have a total length measurement much greater than that of the worm.

During development in the ovary, the oocytes of *Ascaris lumbricoides* and other secernentean nematodes are attached to an axial protoplasmic rachis; their maturation and movement toward the oviduct are synchronous. Spermatozoa are stored in the oviducts and seminal receptacles and it is here that fertilization takes place. Fertilization immediately stimulates shell formation, in which three layers are formed by the egg itself and a fourth is added by the secretions of the uterus. First, the vitelline layer appears and beneath it is a chitinous layer, which is followed by a third that is formed under the second by a coalescence of refringent granules extruded from the cytoplasm (Fig. 17–4). Material secreted from the wall of the uterus adheres to the

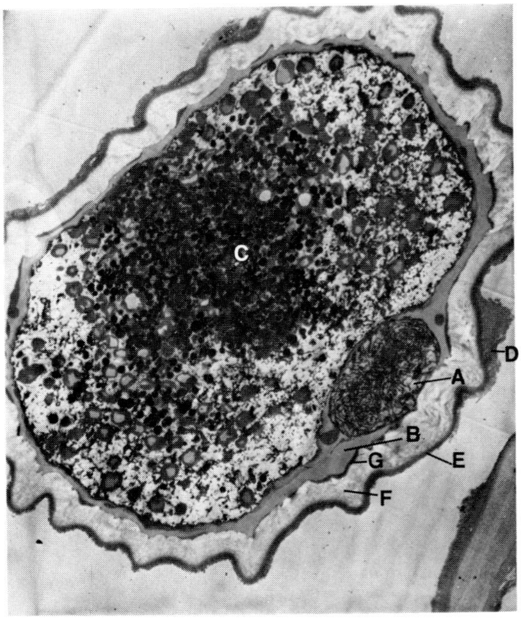

Fig. 17–4. *Ascaris lumbricoides,* eggshell formation. Fertilized egg with first polar body *(A)* confined within ascaroside layer *(B)*. Dense granules *(C)* have aggregated at center of the egg. *D,* External uterine layer. *E,* Vitelline layer. *F,* Chitinous layer. *G,* Dense material. (× 1500.) (From Foor, W.E. 1967. Ultrastructural aspects of oocyte development and shell formation in *Ascaris lumbricoides*. J. Parasitol., *53*:1245–1261.)

outer surface of the egg, forming the fourth layer, which, while the egg is in the uterus, remains albuminous and colorless; on contact with bile in the flow of intestinal contents, however, it hardens and becomes yellowish-brown. In *Trichuris* and some other nematodes, the outer eggshell is fully formed and deeply colored while still in the uterus. In strongyloid species, such as hookworms and related forms, the outermost layer is lacking (LeJambre and Georgi, 1970).

The daily output of eggs for each female varies widely among species but is more or less constant for individual species. *Strongyloides stercoralis,* for example, lays only a few eggs daily but places them within the epithelial lining of the intestine where essentially every egg produces a larva; at the other extreme, *Ascaris lumbricoides* may produce more than 200,000 eggs per female per day, with all being passed from the body unembryonated in the feces, and with relatively few developing to the infective stage in the outside environment. In *Trichinella* and *Dracunculus,* the eggs hatch in the uterus, and motile larvae are deposited by the female. In the filarial worms, a motile prelarval stage, the *microfilaria,* develops in the uterus and is discharged into the tissues, cavities, or vascular system of the host; in some species, the chitinous egg membrane is retained and elongated to serve as a sheath for the microfilaria.

The mature functional spermatozoan in the uterus of *Ascaris lumbricoides* consists of a clear ameboid anterior region and a conical posterior part containing numerous surface membrane specializations, dense mitochondria, a lipid-like refringent body, and a dense nucleus (Foor, 1968). On contact with the oocyte, the pseudopodia of the sperm and the cell membrane of the oocyte interdigitate and finally fuse (Fig. 17–5). In some nematodes, the spermatozoa in the seminal vesicle of the male differ markedly in form and appearance from those that are stored in the seminal receptacles of the female (Foor et al., 1971). The spermatozoa of *Dipetalonema viteae,* long and slender when deposited in the uterus, become rounded and ameboid as they are in *Ascaris* within 24 hours.

Mating generally is intermittent. In some species of parasitic nematodes, however, copulation occurs only once, as in the pinworm, *Enterobius;* in others, it is continual, as in *Syngamus,* with the male and female permanently joined or, as in *Trichosomoides crassicauda,* with the minute male residing within the uterus of the relatively large female. In a related group, *Anatrichosoma* sp., the male inserts the whole hind part of his body into the vagina and uterus of the female but does not maintain that position permanently (Little and Orihel, 1972). When ready to mate initially, and when periodic mating is needed to maintain a supply of sperm in the seminal receptacles, the female communicates to the male her readiness to mate (estrus) by secreting a sex pheromone, and she may, as in *Ascaris,* make temporary morphological changes in the region of the vulva to facilitate copulation (Beaver and Little, 1964). In some instances, mating behavior, as in heavy hookworm infection, may contribute significantly to the worms' pathogenicity (Beaver et al., 1964).

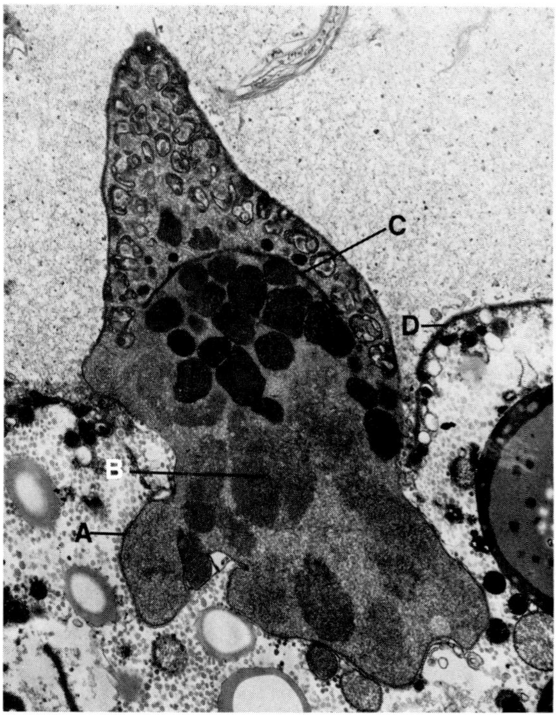

Fig. 17–5. *Ascaris lumbricoides,* fertilization. Micrograph showing interdigitations between an oocyte oolemma *(A)* and the plasma membrane of a penetrating sperm. Aggregations of dense material *(B)* are apparent in the anterior sperm cytoplasm. *C,* Refringent body. *D,* Extraneous coat. (×7500.) (From Foor, W.E. 1968. Zygote formation in *Ascaris lumbricoides* (Nematoda). J. Cell Biol., *39*:119–134.)

Development. Nematodes have five successive fundamental stages—namely, four larval stages and the adult, with growth and a molt of cuticle between the two stages. In the life cycle of most of the secernentean nematode parasites of man, the infective stage usually is the third-stage larva (Anderson, 1980).

The eggs (and larvae) of nematode species living in the digestive tract reach the outside world in the feces of the host. Eggs of nematodes of man that live in the lungs or kidneys may be ejected in sputum *(Syngamus)* or urine *(Dioctophyma renale).* The microfilariae of filarial worms circulate through the blood or migrate through the tissues *(Onchocerca volvulus)* and are removed from the cutaneous layers by blood-sucking insects. The female *Dracunculus* discharges living young through a ruptured cutaneous blister when this area of the host comes in contact with fresh water.

The fate of these eggs or larvae once they have left the human body depends on a variety of factors. In *Enterobius,* the eggs are almost fully embryonated when laid by the female worm. Such eggs are not commonly found in feces but usually are shed on the perianal and perineal skin. Scratching of the area disseminates the eggs on the skin or clothing, and they are then easily transferred to the mouth and swallowed, thus providing a mechanism for infection or reinfection. Eggs of *Ascaris,* the whipworm, and hookworms require a period for embryonation in soil. In the case of the first two species, hatching normally occurs only after the infective stage eggs have been swallowed and reach the small intestine. Hookworm eggs hatch in soil, the first-stage larvae feed, molt, and metamorphose into the filariform stage, which is infective for man by the cutaneous route. Similarly, the first-stage larvae of *Strongyloides stercoralis* feed in soil or in feces, where they may develop into adults and pass through one or more complete free-living cycles or may be transformed directly into infective filariform larvae. Microfilariae proceed to the first larval stage only after they have been taken up by the appropriate blood-sucking arthropod and have passed into its body tissues. Under favorable conditions, these larvae transform into an infective stage and then migrate to the mouth parts, ready to infect man when the vector takes its next blood meal.

Once introduced into the human body by the oral route, fully developed eggs hatch in the small intestine. The larvae may then proceed to the level of the bowel where the adults are usually found and develop directly into the adult worms *(Enterobius,* whipworm) or they may require a migration through the lungs *(Ascaris).* Hookworm and *Strongyloides* larvae enter the skin and travel to the lungs before arriving in the intestine. When introduced into the skin, infective larvae of the filarial worms migrate through the tissue for weeks or months; during this period of migration, they mature and eventually settle down in sites peculiar for each species, at which time they mate. After mating, the females produce and discharge microfilariae into the nearby tissues of the host.

In some nematodes, the larval stages become infective for the final host in an intermediate host or in the soil, and then may be passed through transport hosts in which no essential growth or development takes place, although infectivity is preserved. This kind of transmission through nonessential, but advantageous, intermediate hosts is called "paratenesis," and the transport hosts are referred to as "paratenic" hosts (p. 321). In a few species, the infective larvae, on entering the final mammalian host, accumulate in the tissues of the female for transmission to the offspring either prenatally through the placenta or, more commonly, by the transmammary route in the colostrum or milk, as in certain species of *Strongyloides, Ancylostoma,* and *Toxocara* (pp. 262, 274, 320).

CLASSIFICATION OF NEMATODES

The nematodes are unsegmented helminths with fundamental bilateral symmetry and secondary triradiate symmetry of the oral and esophageal structures. The body is elongate, cylindrical, and frequently filiform, with a digestive tract and functional mouth and anus, and a body cavity not lined with mesothelium (i.e., a pseudocele). Typically, the sexes are separate. In some classifications, the nematodes are given the rank of phylum or subphylum. Here, following in large part the classification outlined in the Commonwealth Institute of Helminthology (CIH) Keys to the Nematode Parasites of Vertebrates (Chabaud, 1974), they are ranked as class Nematoda, with two subclasses, the Adenophorea and the Secernentea, formerly called the Aphasmidia and Phasmidia, respectively. As noted by Olsen (1974), these terms were first proposed by von Linstow in 1905. Dougherty (1958) emended and proposed Secernentia to replace Phasmidia, and Chitwood (1958) proposed Adenophorea to replace Aphasmidia. Von Linstow (1909) defined his Secernentes to include *Ascaris, Oxyurus, Strongylus, Physaloptera,* and some others, and his Adenophori to include *Rhabditis* and related forms. *Trichocephalus, Trichosoma,* and *Hystrichis* which Chitwood placed in his *Adenophorea,* were in von Linstow's *Pleuromyarii.*

SUBCLASS ADENOPHOREA (APHASMIDIA)
Caudal papillae reduced or absent; excretory system without lateral canals; phasmids absent; esophagus cylindrical, or with esophageal glands free in pseudocele and forming stichosome or trophosome; eggs usually unsegmented with plug at either pole, or hatching in uterus; first-stage larva often with stylet, usually infective to final host.

Order **Enoplida** Chitwood, 1933

Superfamily **Trichuroidea** Railliet, 1916
Intestine and rectum well developed; stichosome present; sexes separate; monodelphic.
Trichinella spiralis, Trichuris trichiura, Capillaria hepatica, Capillaria philippinensis, Capillaria aerophila, Anatrichosoma cutaneum

Superfamily **Dioctophymatoidea** Railliet, 1916
Esophagus cylindrical, well developed; stichosome absent; tail of male modified to form terminal sucker; monodelphic.
Dioctophyma renale

Superfamily **Mermithoidea,** Wülker, 1924

Alimentary tract reduced, trophosome (extension of intestine overlapping esophagus) present; amphidelphic; male with one or two spicules; female reproductive system highly developed; larval mermithids parasites of arthropods, rarely present in human intestine by accidental ingestion of raw vegetables.

SUBCLASS SECERNENTEA (PHASMIDIA)
Caudal papillae usually numerous; excretory system with lateral canals; phasmids present; eggs without polar plugs, but may be operculate at one pole.

Order **Rhabditida** Chitwood, 1933
Cuticle smooth or annulated; amphids small; ovejector weakly developed; vagina transverse, without well-developed sphincter; esophagus with terminal bulb; caudal glands absent.

Superfamily **Rhabdiasoidea** Railliet, 1916
Cephalic papillae 6 in inner circle, 10, 6 or 4 in outer; excretory system H-shaped; oral stylet absent; parasitic female parthenogenetic (except in *Parastrongyloides*), with cylindrical and frequently long esophagus; parasitic generation alternating with free-living generation of males and females with rhabditoid esophagus.
Strongyloides stercoralis, S. fuelleborni

Superfamily **Rhabditoidea** Travassos, 1920
Stoma an open tube, amphids on sides of lateral lip region, rarely behind lips; excretory system with lateral canals; anterior part of esophagus cylindrical with procorpus (with or without corpus), isthmus and terminal bulb; valves in corpus or bulb, or in both; valve in bulb with plates.
Turbatrix aceti and species of *Micronema, Diploscapter, Rhabditis,* and *Pelodera*

Order **Tylenchida** Thorne, 1949

Superfamily **Tylenchoidea** Chitwood & Chitwood, 1937
Oral stylet present; dorsal esophageal glands empty into lumen near base of stylet; otherwise, closely resemble Rhabdiasoidea.
Species of *Ditylenchus, Meloidogyne,* and *Heterodera*

Order **Strongylida** Diesing, 1851
Male with caudal bursa; stoma without lips, with six lips, or with corona radiata; buccal capsule variable in size and form, surrounded by esophageal tissue; larvae with rhabditoid, adult with claviform, esophagus; excretory system H-shaped with two subventral glands; spicules similar; preinfective larvae free-living or parasitic in invertebrates.

Superfamily **Ancylostomatoidea** (Nicoll, 1927)
Buccal capsule large, strongly cuticularized; subglobular (not hexagonal in transverse section); lips and corona radiata absent; oral opening unarmed or with teeth and cutting plates.
Ancylostoma duodenale, A. ceylanicum, Necator americanus

Superfamily **Strongyloidea** Hall, 1916
Buccal capsule varied; oral opening sometimes hexagonal, sometimes surrounded by six lips or corona radiata; without teeth or cutting plates.
Ternidens deminutus, Oesophagostoma apiostomum, Oe. stephanostomum, Mammomonogamous laryngeus

Superfamily **Trichostrongyloidea** Cram, 1927
Buccal capsule absent or small and weakly cuticularized; bursa well developed, at least lateral lobes; cephalic vesicle often present; longitudinal cuticular ridges frequently present; first-stage larva often with valved esophageal bulb and conical tail ending in simple point.
Haemonchus contortus and species of *Trichostrongylus*

Superfamily **Metastrongyloidea** Cram, 1927
Mouth reduced, bursa often reduced; outer layer of cuticle frequently fluid-filled, forming tegmental sheath; heteroxenous, mainly in gastropods; first-stage larva usually with short undulating tail.
Angiostrongylus cantonensis, A. costaricensis, Metastrongylus elongatus

Order **Oxyurida** Weinland, 1858
Male without caudal bursa, reduced number of caudal papillae, generally only one spicule; excretory gland single, ventral; body short, relatively stout; esophagus with bulb; eggs often flattened on one side, embryonate in uterus; monoxenous; two molts in egg.

Superfamily **Oxyuroidea** Railliet, 1916
With characters of the order.
Enterobius vermicularis, Syphacia obvelata

Order **Ascaridida** Yamaguti, 1961
Anterior extremity triradiate; lips with lateral papillae; dorsolateral caudal papillae two or three pairs; esophagus not divided into muscular and glandular parts.

Superfamily **Ascaridoidea** Railliet and Henry, 1915
Generally large worms with three well-defined lips, sometimes separated by interlabia; esophagus simple and cylindrical or terminated by valves containing uninucleate gland cells; coelomyarian; eggs thick-shelled, not embryonated when deposited; first molt inside egg.
Ascaris lumbricoides, Toxocara canis, Toxacara cati, Lagochilascaris minor, species of *Anisakis* and *Phocanema*

Order **Spirurida** Chitwood, 1933
Anterior extremity bilaterally symmetrical; caudal papillae all ventral or ventrolateral; esophagus divided into anterior muscular and longer posterior glandular parts, division sometimes indistinct; larval stages develop in intermediate host.

Suborder **Camallanina** Chitwood, 1936
Larvae without cephalic hooks; tail generally long and pointed, usually with conspicuous phasmids with broad cavity and prominent pore; intermediate hosts copepods.

Superfamily **Dracunculoidea** Cameron, 1934
Buccal cavity weakly developed; internal labial papillae prominent; male small; female long, cord-like; esophagus and intestine rudimentary; vulva near middle of body, nonpatent in mature worm; uteri divergent; viviparous; larvae rhabditoid.
Dracunculus medinensis.

Suborder **Spirurina** Railliet and Henry, 1913
Larvae usually with cephalic hooks or spines; phasmids inconspicuous, pore-like; esophageal glands multinucleate; intermediate hosts invertebrates other than copepods (except Gnathostomatoidea).

Superfamily **Gnathostomatoidea** (Lane, 1923)
Pseudolabia trilobed; anterior extremity sometimes swollen into bulb.
Gnathostoma spinigerum, G. hispidum

Superfamily **Physalopteroidea** (Leiper, 1908)
Pseudolabia not lobed, with teeth on free border; cuticle behind pseudolabia often expanded to form collarette; intermediate hosts insects.
Physaloptera caucasica

Superfamily **Rictularioidea** (Railliet, 1916)

Pseudolabia absent; buccal cavity well cuticularized; oral opening hexagonal; two rows subventral cuticular spines along length of body.
Rictularia

Superfamily **Thelazoidea** (Railliet, 1916)
Pseudolabia absent; buccal cavity variable, sometimes long and cylindrical; caudal alae absent; vulva markedly anterior, or posterior; caudal papillae numerous, irregular.
Thelazia callipaeda, T. californiensis

Superfamily **Spiruroidea** Railliet and Henry, 1915
Pseudolabia absent; buccal cavity short, well cuticularized; caudal alae present; caudal papillae few, regular.
Gongylonema pulchrum, Spirocerca lupi

Superfamily **Acuarioidea** (Seurat, 1913)
Pseudolabia large; buccal cavity generally long, cylindrical; cephalic cuticle with ornamentation (cordons).
Cheilospirura

Superfamily **Filarioidea** Stiles, 1907
Body filiform; mouth simple, circular or slightly elongated dorsoventrally, surrounded by papillae; egg thin-shelled, with poorly differentiated larvae (microfilariae); spicules dissimilar; normal habitat tissues and body cavities of host; arthropod intermediate hosts.
Wuchereria bancrofti, Brugia malayi, B. timori, Onchocerca volvulus, Mansonella perstans, M. streptocerca, M. ozzardi, Loa loa, and zoonotic species of *Dirofilaria, Brugia, Dipetalonema, Onchocerca*

THE NEMATOMORPHA

Members of this group have a degenerate intestinal tract in adult forms and the body cavity is lined wholly or in part with mesothelium; there is a proboscis in the first larval stage. Species of human interest belong to the class Gordiacea.

CLASS GORDIACEA VON SIEBOLD, 1843 (SYN. GORDIDIDEA ORTLEPP, 1924)

These "hair snakes" or gordiid worms are commonly found as adults in fresh water, with larval stages in insects. Accidentally ingested by man, they may be passed in the feces or actively migrate from the anus or mouth; rarely, they have been reported (mistakenly?) as voided in the urine.

REFERENCES

Anderson, R.C. 1980. Fifty years of helminthology, pp. 459–468. In *Perspectives in World Agriculture*. Farnham Royal, Bucks, England, Commonwealth Agricultural Bureau.

Beaver, P.C., and Little, M.D. 1964. The genital girdle in relation to estrus and mating in *Ascaris lumbricoides*. J. Parasitol., *50*:128–130.

———, Yoshida, Y., and Ash, L.R. 1964. Mating of *Ancylostoma caninum* in relation to blood loss in the host. J. Parasitol., *50*:286–293.

Chabaud, A.G. 1974. Class nematoda. Keys to subclasses, orders and superfamilies. In *CIH Keys to the Nematode Parasites of Vertebrates*. Edited by R.C. Anderson, A.G. Chabaud, and S. Willmott. Farnham Royal, Bucks, England, Commonwealth Agricultural Bureau.

Chitwood, B.G., et al. 1937, 1938, 1940, 1941. *An Introduction to Nematology*, Sect. 1, Pts. 1–3 (rev. 1950); Sect. 2, Pts. 1–2. Baltimore and Marquette (Mich.), B.G. Chitwood.

Foor, W.E. 1967. Ultrastructural aspects of oocyte development and shell formation in *Ascaris lumbricoides*. J. Parasitol., *53*:1245–1261.

———. 1968. Zygote formation in *Ascaris lumbricoides* (Nematoda). J. Cell Biol., *39*:119–134.

———, Johnson, M.H., and Beaver, P.C. 1971. Morphological changes in the spermatozoa of *Dipetalonema viteae* in utero. J. Parasitol., *57*:1163–1169.

Goodey, J.B. 1963. *T. Goodey's Soil and Freshwater Nematodes*. New York, John Wiley & Sons.

Grassé, P.P. (editor). 1965. *Traité de Zoologie. Anatomie, Systématique, Biologie*. Vol. 4, Fascicule 3. Paris, Masson et Cie.

LeJambre, L.F., and Georgi, J.R. 1970. Influence of fertilization on ovogenesis in *Ancylostoma caninum*. J. Parasitol., *56*:131–137.

Little, M.D., and Orihel, T.C. 1972. The mating behavior of *Anatrichosoma* (Nematoda: Trichuroidea). J. Parasitol., *58*:1019–1020.

Olsen, O.W. 1974. *Animal Parasites—Their Life Cycles and Ecology*. Baltimore, University Park Press.

Weinstein, P.P., and Haskins, W.T. 1955. Chemical evidence of an excretory system of the filariform larva of *Nippostrongylus muris*. Exp. Parasitol., *4*:226–243.

Yamaguti, S. 1961. *Systema Helminthum*. Vol. 3. *The Nematodes of Vertebrates*. Parts 1 and 2. New York, Interscience Publishers Inc.

Chapter 18

Aphasmid Nematodes (Adenophorea)

SUPERFAMILY TRICHUROIDEA*

Trichinella spiralis
(Owen, 1835) Railliet, 1895
(Producing trichinosis or trichinelliasis)

Synonym. *Trichina spiralis* Owen, 1835

Historical and Geographical Notes. *Trichinella spiralis*, the etiologic agent of trichinosis, was first discovered in the encysted larval stage in the muscles of patients at autopsy in London by Peacock (1828), Hilton (1833), and Paget (1835). Richard Owen (1835) first described and named the encysted larvae (Campbell, 1979). Other human cases were soon reported in Europe and North America. Von Siebold (1844) and Dujardin (1945) suggested that the parasite was the larval stage of a roundworm. Joseph Leidy, of Philadelphia, first found the cysts in pig's flesh (1846) and suggested their similarity to those found in man. The investigations of Leuckart (1853–1860) and Virchow (1858–1860) proved that these cysts, when fed to certain experimental hosts, grew to adults in a few days and that the females in the duodenal wall produced living young that migrated to muscles and later encysted. In 1860, at autopsy of a 20-year-old German country woman who had died of "atypical typhoid," Friedrich von Zenker discovered the intestinal forms of the trichina worm. In his study, he found thousands of encysted larvae in the muscles of the cadaver and concluded that the infection had resulted from eating raw sausage. He emphasized the clinical importance of trichinosis. Several years elapsed before German investigators proved experimentally that raw or insufficiently cooked pork was responsible for trichinosis in man. In 1897, Brown demonstrated that hypereosinophilia was highly suggestive of trichinosis. During the latter part of the 19th century and the early part of the present century, the disease was recognized as an important public health problem. Within the past several years, it has again been given special consideration by health authorities, physicians, and medical parasitologists, particularly in the United States, where repeated outbreaks and a few deaths have occurred in people who had eaten trichinous pork.

Trichinosis has a cosmopolitan distribution but is much less important as an infection of man in the Tropics and the Orient than it is in Europe and the United States. According to his biographer, Virchow estimated the human incidence in Germany in his day to be about 90%. In 1898, Osler found only 0.6% of the diaphragms infected at routine autopsies in the Johns Hopkins Hospital. In 1916, Ransom estimated that 6% of American pigs were parasitized. Control measures have resulted in a striking decrease in prevalence of human infection in the United States during the past 35 years; yet, in recent years, approximately 100 or more cases annually have been reported in the United States. In 1978, there were 11 common-source outbreaks (Centers for Disease Control, 1979). Infection in Canada is not rare, but, in large part, is acquired from eating wild animals, not from eating pork (Schmitt et al., 1978; Eaton, 1979). In 1979 there were 135 cases in the United States, 28 of which were acquired from the meat of wild animals (Centers for Disease Control, 1980). The infection is present in Mexico.

The disease is widely distributed, although the incidence is relatively low in Holland, Germany, and Spain. The infection is so rare in Great Britain that a national survey commenced in 1971 was terminated at the end of 1979 (Charlton, 1980). There is high incidence in Lithuania, Poland, western U.S.S.R., Ukraine, Romania, Hungary, Bulgaria, Yugoslavia, and the maritime provinces of U.S.S.R. Recent outbreaks have been reported from Italy (Pampiglione et al., 1978), Laos (Sicard et al., 1976), Tanzania (Bura and Willett, 1977), and France (Boureé et al., 1979). Infections have been reported from Malaysia, Vietnam, and Thailand. Apparently, the only focus of any intensity east of Suez is Syria, where outbreaks are occasionally reported. In Central and South America, autochthonous infections have been recorded for Brazil, Ecuador, Uruguay, and Chile.

Morphology, Biology, and Life Cycle. When meat that contains viable infective-stage larvae of *T. spiralis* is eaten, the cysts are digested free from the tissues in the stomach and the larvae excyst there or in the duodenum and jejunum. They then invade the mucosal epithelium and rapidly develop through four larval stages, each of which is terminated by a molt (Ali Khan, 1966; Kozek, 1971a, b). They become mature worms early in the second day of infection. Mating may occur as early as 30 hours of infection, and most of the females will be inseminated during the second day (Kozek, 1976). Both males and females are minute and thread-like (Fig. 18–1) and adapted to living within the mucosal epithelium (Gardiner, 1976). The males measure approximately 1.2 mm in length by 60 µm in maximum diameter. The anterior end is relatively delicate, with cephalic papillae, and other structures are evident only in electron micrographs (Kim and Ledbetter, 1980). The posterior extrem-

*For superfamily characteristics, see page 228.

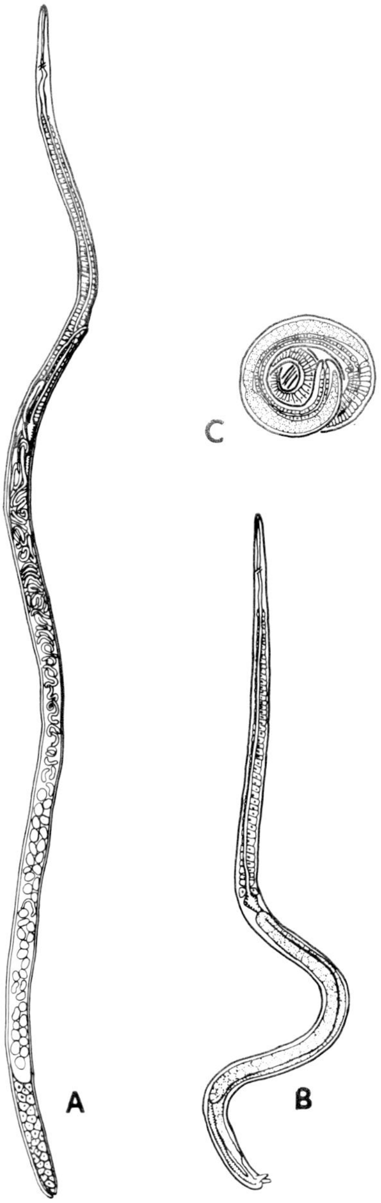

Fig. 18–1. *Trichinella spiralis*. Camera lucida drawings of stained specimens in permanent preparations. *A*, Adult female, (× 75.) *B*, Adult male. (× 75).) *C*, Coiled larva, (× 100.) (From Beaver, P.C., 1952. The detection and identification of some common nematode parasites of man. Am. J. Clin. Pathol., 22:481–494.)

ity bears a pair of conspicuous conical papillae, one on either side of the cloacal orifice. The cloaca is evertible during copulation. A single testis is situated a short distance from the posterior end of the body, extends forward to about mid-body, where it joins the genital tube, which, in turn, extends posteriorly to the cloaca. The female measures approximately 2.2 mm in length, nearly twice as long as the male and one and a half times as broad (Fig. 18–1). A single ovary is in the extreme posterior part of the body, with the oviduct, seminal receptacle, uterus, and vagina forming a nearly straight channel to the vulva, which is situated on the ventral side of the body about one-fourth the body length from the anterior end.

In both sexes, the anterior part of the body is largely filled with the stichosome, a compact cord of linearly arranged gland-like cells called "stichocytes." Anterior to the stichosome, the esophagus is muscular. Where it extends along and is somewhat embedded in the stichosome, the esophagus is extremely narrow; behind the stichosome, it joins the intestine, which is a relatively delicate, simple tube extending to the anus or cloaca. In both sexes, the mouth bears a protusible stylet. Thus, *T. spiralis*, by the size and shape of its body and by its mouth structures (Kim and Ledbetter, 1980), is well adapted to living in the mucosal epithelium of the crypts and villi (Fig. 18–2). This specialization was first described by Gardiner (1976) and confirmed by Despommier et al. (1978). Wright (1979) showed by electron micrography that, on moving from place to place, the worms pass through as well as between the cells of the mucosal epithelium.

Soon after insemination, the females begin to produce eggs that develop into minute larvae in the uterus. By the sixth day of infection, they begin depositing motile larvae, about 100 μm long by 6 μm wide, which migrate into the intestinal lymphatics or mesenteric venules and are distributed throughout the body. Larviposition continues for a period of about 4 weeks and may continue for up to 16 weeks. The natural life span of *T. spiralis* in the intestine, however, generally is about 1 month or less in the mouse, and it probably is about the same in man. During this time, up to about 500 larvae per female may be produced, the number depending somewhat on the number of worms present and the presence or absence of previous infection (the output per female is less in heavy infections and in hosts previously infected) (Chirasak, 1971; Kennedy, 1979). Males can inseminate several females, and females may mate two or more times during the course of infection (Kozek, 1976). Larvae deposited by worms in the intestine may become lodged in various locations, including the myocardium, brain, cerebrospinal fluid, and body cavities, from which they later re-enter the blood stream and finally reach the striated muscle where

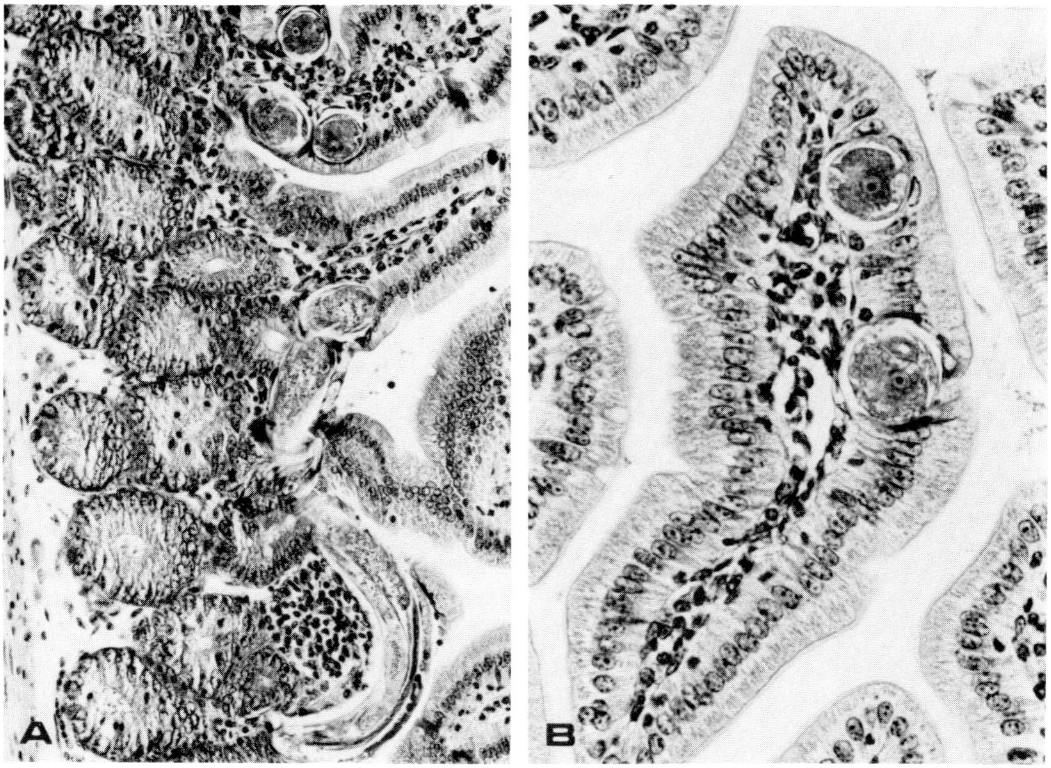

Fig. 18–2. *Trichinella spiralis*. Section of adult worm in mucosal epithelium of mouse small intestine at 30 hours of infection. *A*, One longitudinal and six transverse sections of a male worm at base of villi, (× 190.) *B*, Two transverse sections of a female in epithelium of a villus. (× 650). (From Gardiner, C.H. 1976. Habitat and reproductive behavior of *Trichinella spiralis*. J. Parasitol., *62*:865–870.)

they become encapsulated within the fiber. The greatest production of larvae is during the second week of infection, and the greatest invasion takes place in muscles relatively poor in glycogen. Encapsulated larvae in muscle develop essentially all of the morphologic features of the adult male and female, and the sexes are readily distinguished (Kozek, 1975).

Larvae reach full differentiation before encapsulation. At this stage, larvae in press preparations of infected muscle are inconspicuous, whereas later they are easily seen among the muscle fibers (Figs. 18–3 and 18–4).

The capsule is an adventitious ellipsoidal sheath resulting from round cell and eosinophilic infiltration around the tightly coiled larva (Fig. 18–1,*C*). The long axis of the cyst parallels that of the muscle fibers. The diaphragm, muscles of the larynx, tongue, abdomen, and intercostal spaces, biceps, psoas, pectoral, gastrocnemius, deltoid, and other muscles that are constantly active are most heavily infected. Growth of the larva to a length of 0.8 to 1 mm within its cyst capsule causes degeneration or swelling of the adjacent muscle fibers, a thickening and change in the sarcolemma, with hypertrophy and hyperplasia, fragmentation of fibers, and deposition of an inner capsular membrane followed by the formation from the endomysium of an outer fibrous capsule that is infiltrated with blood capillaries. The entire cyst wall results from host-tissue reaction and, therefore, is induced, rather than secreted, by the larva. The cyst wall is permeable to substrates and metabolic products. In man, encysted larvae may remain viable for many years, although calcification often occurs within 6 to 9 months (Fig. 18–5). Five larvae/g of body muscle can produce death, but as many as 1000/g have been recorded for persons who died of other causes (Gould, 1945). Schenone et al. (1967) reported a fatal heavy concurrent infection by *T. spiralis* and *Taenia solium* in a raw-pork eater; at autopsy, 1600 *T. spiralis* larvae/g of muscle were recovered by the digestion method.

Strain Variation. Although identical in morphology, the strains of *T. spiralis* that have been isolated from people or animals in different geo-

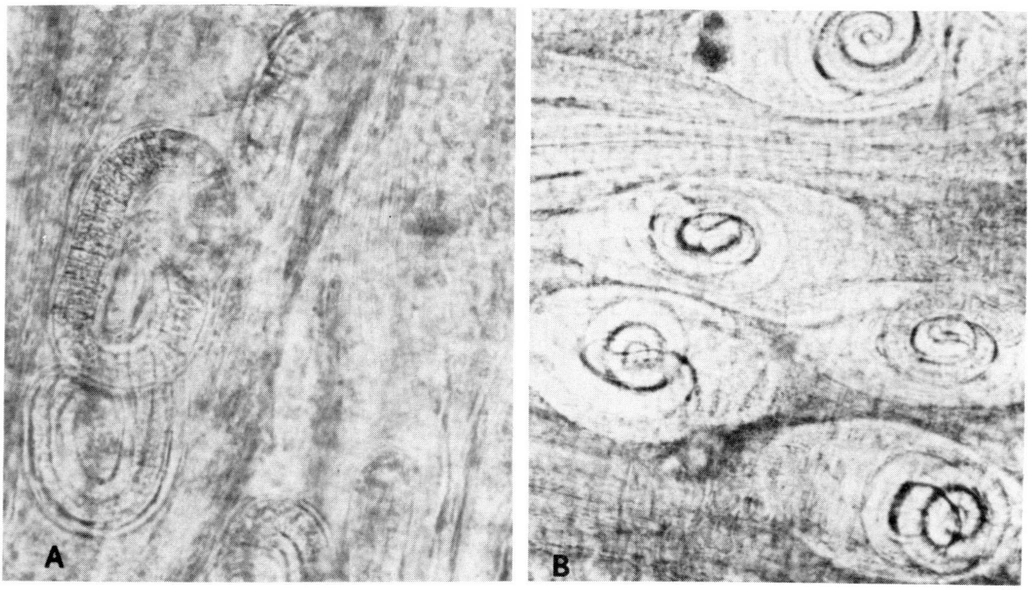

Fig. 18–3. *Trichinella spiralis*. Larvae in press preparations of muscle. *A*, Early infection (22 days) with larva fully differentiated but unencapsulated, *B*, Old infection (6 months) with larvae encapsulated. Unstained. (× 250 and × 100.) (By P.C. Beaver.)

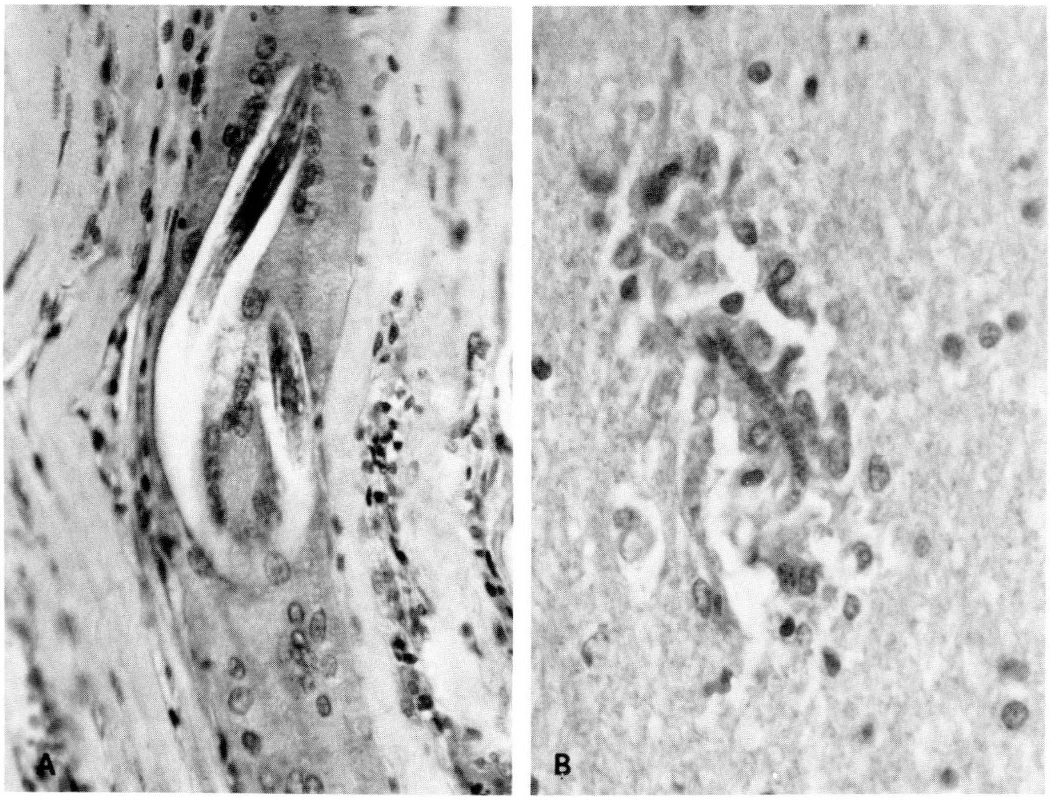

Fig. 18–4. *Trichinella spiralis*, larvae in human tissue. *A*, Muscle (× 345) and *B*, brain. (× 620) from autopsy 20 days after onset of illness. The larva in muscle is well differentiated and in early stage of encapsulation; in brain the larva, showing no growth or development, is surrounded by scattered glial cells in a small area of necrosis. (From Faust, E.C., Beaver, P.C., and Jung, R.C. 1975. *Animal Agents and Vectors of Human Disease*. Philadelphia, Lea & Febiger.)

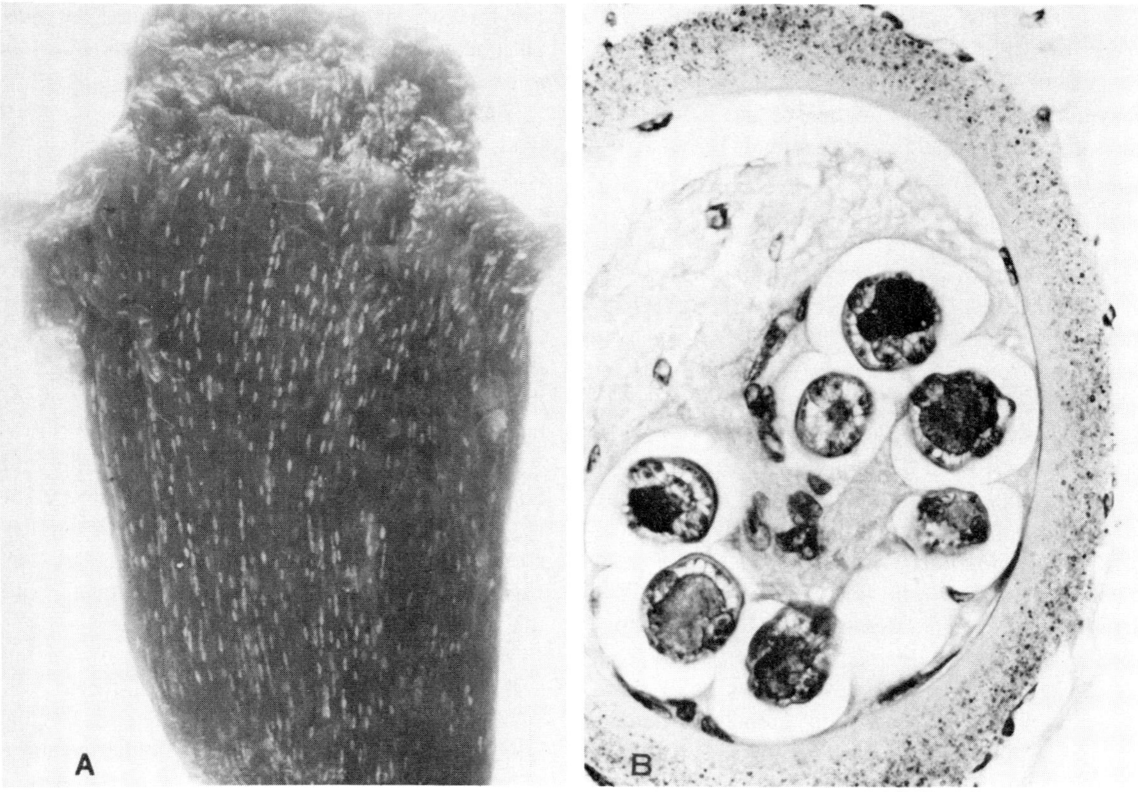

Fig. 18-5. *Trichinella spiralis*. Calcified cysts of unknown age in muscle of a woman who died with leukemia. *A*, Portion of muscle showing numerous fusiform cysts 1 to 2 mm in length. Unstained, slightly enlarged. (From Jacobson, E.S., and Jacobson, H.G. 1977. Trichinosis in an immunosuppressed human host. Am. J. Clin. Pathol., *68*:791–794.) *B*, Transverse section of a cyst from the same specimen showing minute stippling of calcareous deposits in outer portions of the wall, and seven transverse sections of a coiled larva normal in appearance lying in normal multinucleated matrix. (× 250.) (By P.C. Beaver. Specimen courtesy of Dr. E.S. Jacobson.)

graphic regions apparently differ in biology and pathogenicity. In Arctic strains, larvae remain infective in frozen meat (Eaton, 1979; Chadee and Dick, 1982), whereas those of temperate-zone strains are killed by freezing. An East African strain was reported to be relatively less infective to rats when it was first isolated, although not distinctly so when tested later (Kozar and Kozar, 1965). Some authorities have suggested raising four strains to the level of species *(T. spiralis, T. pseudospiralis, T. nativa, T. nelsoni),* the arguments for which were summarized and rejected by Madsen (1974, 1976). However, *T. pseudospiralis,* a species described by Garkavi in 1972 from the raccoon in North Caucasus, appears to be distinct in at least one important respect—it does not stimulate heavy encapsulation in any of the numerous hosts tested thus far. Therefore, it differs markedly from *T. spiralis* and may escape detection or recognition. According to Pawlowski and Ruitenberg (1978), the disease caused by *T. pseudospiralis* differs only slightly from that caused by *T. spiralis,* and in experimentally infected monkeys, *T. pseudospiralis* larvae caused the invaded muscle fibers to form a "pseudocapsule." A point stressed by Madsen (1974, 1976) is that man readily acquires infection by eating meat harboring any of the various strains of *Trichinella*. Presumably, this would hold true for *T. pseudospiralis*. A natural infection of *T. pseudospiralis* was reported in a hawk in California (Wheeldon et al., 1982).

Epidemiology. While both adults and encysted larvae of *Trichinella spiralis* develop within the same host, two hosts are required to complete the life cycle. There are three principal maintenance cycles: (1) *pig to pig*, in areas where the custom of feeding household garbage to pigs is widespread; (2) *rat to rat,* in the presence or absence of appreciable infection in pigs; and (3) *sylvatic*, perpetuated by carnivorous or omnivorous wild hosts. In nature, infection is thought to be commonly propagated by the black rat and the brown rat, which are cannibalistic. At times, 75% of examined rats have been found to be infected. Pigs, wild boars, bears, cats, dogs, and other mammals that eat flesh become infected. Witenberg (1964) cited heavy infections in cats from Poland, central U.S.S.R., Norway, Mexico, and U.S.A., and in dogs from Poland, Romania, Byelorussia, Manchuria, Norway, and U.S.A. Between February 1965 and February 1966, the diaphragms from 155 dogs and 95 cats from central New Jersey were examined for *T. spiralis;* infection rates were 2.6 and 3.2%, respectively (Hunt, 1967). Lukashenko and Brzesky (1962) reported that the following Arctic mammals were infected: walrus *(Odobenus rosmarus),* ring seal *(Pusa hispida),* polar bear, brown bear, polar fox *(Alopex lagopus),* ermine *(Mustela erminea),* sable *(Martes zibellina),* fox, wolf, wolverine *(Gulo gulo),* badger *(Meles meles),* silver fox, hedgehog, raccoon, dog *(Nyctereutes procyonides),* polecat *(Putorius putorius),* and several species of small rodents. Examination of wild animals in Canada (British Columbia) revealed *T. spiralis* in 15 species of mammals including three species of rodents. Most commonly infected were bears, mustelids, cats, and the coyote (Schmitt et al., 1978). In Alaska, Rausch (1953) demonstrated trichinosis in bears, dogs, wolves, foxes, wolverines, tree squirrels, white whale and harp seals, and concluded that polar bears are probably the usual source of human infection. Maynard and Pauls (1962), in reporting two epidemic outbreaks in Alaska, attributed the infection to eating the raw meat of black bear, while Roselle et al. (1965) reported six cases in New England that resulted from eating meat of a bear shot in Vermont. In Arizona, 4 of 51 black bears *(Ursus americanus)* examined in 1978 to 1979 were infected (Le Count, 1981). Kozar (1959, 1961), in Poland, attributed high incidence of trichinosis in the human population (38%) and epidemic outbreaks to eating sausage from pigs that had been fed the flesh and viscera of wild animals (foxes, badgers, martens, polecats, and wild boars). A survey of mammals in northeastern Iran detected *T. spiralis* in 105 of 125 jackals, 3 of 10 red foxes, and 9 of 100 dogs (Hamidi, 1979). Forrester et al. (1961) and Nelson et al. (1961) reported the first authentic outbreak of trichinosis in Africa south of the Sahara in native tribesmen in Kenya, which resulted from eating flesh of the wild boar *(Potamochoerus porcus).* In Tanzania, an outbreak causing two deaths and numerous infections was attributed to eating a warthog *(Phacochoerus aethiopicus)* (Bura and Willett, 1977).

Man usually acquires the infection from eating inadequately cooked or improperly processed infected lean pork or, incidentally, bear's tenderloin. Country sausage and certain types of European-style sausage prepared from a single heavily infected pig provide the most common source of clinical trichinosis. When meat from one infected pig is diluted with that of many uninfected ones, the dosage per pound of meat is proportionately reduced, resulting in a mild or asymptomatic type of trichinosis.

Outbreaks of trichinosis attributed to imported horsemeat were reported from France (Boureé et al., 1979) and Italy (Mantovani et al., 1980). A strain of *T. spiralis* isolated from the latter was used to demonstrate that horses are susceptible to the infection (Pampiglione et al., 1978).

Under conditions developed by man for raising and fattening pigs, garbage containing infected pork scraps or the trimmings from infected wild mammals probably constitutes the most common source of trichinous pork in the United States at the present time. Unprocessed or inadequately processed pork, especially in the form of "country sausage," constitutes the usual source of human infection (Schantz et al., 1977).

Another possible source of exposure for swine consists in their ingestion of larvae passed in the feces of infected carnivores, rats, and other hogs at the period when the excysted mature larvae are becoming established in the intestinal wall of these animals (Zimmerman et al., 1958; Olsen and Robinson, 1958). Noting that larvae remain infective in the decaying carcass, Madsen (1976) has stressed the probable importance of the carnivore carcass in maintaining the sylvatic cycle.

Annual summaries of outbreaks and isolated cases of trichinosis in the U.S.A. in the period 1976 to 1980 show the highest incidence to be in Massachusetts, California, Alaska, Pennsylvania, New Jersey, New York, Texas, and Connecticut. In 1980, there were 130 reported cases in the United States, 52 of which were in Louisiana and New Jersey (Centers for Disease Control, 1981).

In 1947, an extraordinary epidemic of trichinosis occurred in Greenland, with 300 cases and 33 deaths. Walrus meat was found to be the source of infection. Diagnosis was based on common symptoms, high eosinophilia, positive intradermal and serologic tests with trichina antigen, and postmortem demonstration of *Trichinella* larvae in muscle samples of one patient (Thorborg et al., 1948).

In contrast to the usual epidemiologic picture, in regions where customs, religion, or sanitary measures prevent people from eating pork, infection may exist in nature while decades may pass without a single human case. This has occurred in Israel, for example, where domestic and wild animals are infected yet no human case has been reported for more than 80 years (Witenberg, 1964).

Pathogenesis, Pathology, and Symptomatology. When encapsulated larvae escape from the cysts, they rapidly penetrate into the epithelium of the duodenal and jejunal mucosa (Fig. 18–2). The amount of damage to the intestinal lining is dependent on the number of invading organisms and the time and number of previous exposures. In 5 to 7 days, the females begin to discharge minute larvae into the tissues (Table 18–1).

The larvae may migrate through any of the tissues of the host's body, but they have a predilection for striated muscle, especially the muscles of the jaws, tongue, larynx, diaphragm, neck, and ribs. The sites of attachment of the muscles to the tendons and skeleton are more frequently and more heavily involved than are other parts. Also, the larvae tend to be found more in the superficial muscles or parts of muscles.

An inflammatory reaction appears in whatever tissues they pass through. With encapsulation, the local reaction subsides but is replaced by a systemic reaction. Encapsulation takes place only in striated muscle. Passage through the myocardium occurs to some degree with resultant inflammation, necrosis, and fibrosis of myocardial fibers, but there is no encapsulation of larvae in the heart.

The pathologic changes and the symptomatology are divided into three successive stages: (1) intestinal; (2) muscle invasion; and (3) convalescence. These correspond approximately to the stages of (1) incubation, or (intestinal) invasion; (2) acute, or larval migration; and (3) chronic, or encapsulation, respectively (Fig. 18–6).

During the *intestinal phase,* inflammation of the duodenal and jejunal mucosa caused by the pene-

Table 18–1. Synoptic tabulation of progressive biologic stages of development of *Trichinella spiralis,* and the corresponding clinical conditions.

Biological Stage	*Beginning or Onset*	*Clinical Conditions*
	Hours	
Ingested larvae excyst and enter intestinal epithelium	2–4	
	24	Gastrointestinal symptoms
Worms become mature and mate	30	
	Days	
Females deposit larvae, which invade skeletal muscles	6	
	7	Edema of face and fever
Maximum invasion of muscle fibers	10	Fever at maximum (40–41°C)
	11	Myositis and "rheumatic" pains
Decrease in larvipositing	14	Eosinophilia and circulating antibody
Larvae in muscles fully differentiated	17	
	20	Eosinophilia reaches maximum
Early encapsulation	21	Myocarditis or encephalitis appear
Intestine practically free of adults	23	
	26	Respiratory symptoms
	Months	
Encapsulation practically complete	1	
	2	Fever subsides
Maximum life of worms in intestine	3	Death from myocarditis or encephalitis most likely
Cyst calcification may begin	6	Slow convalescence
	8	Neurologic symptoms and myocarditis subside
	Years	
Cyst calcification may be complete	1	
Larvae possibly still viable within calcified capsules	6	

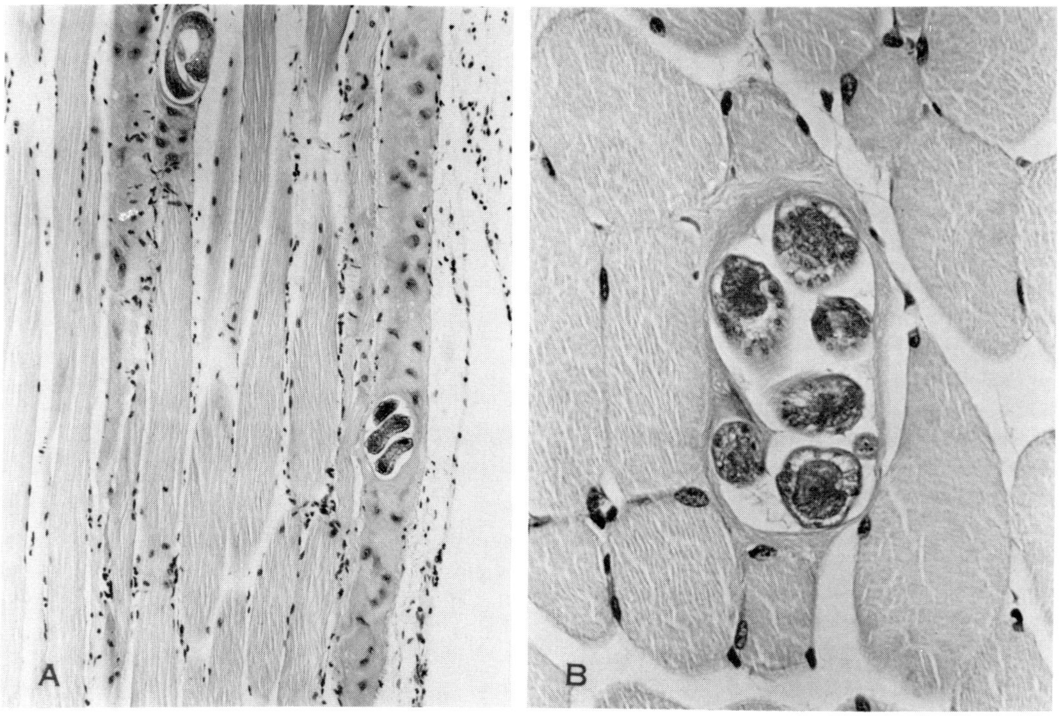

Fig. 18–6. *Trichinella pseudospiralis*. Larvae in muscle of mouse at 6 months of infection. In the invaded muscle fiber, there is degeneration of striated fibrils, proliferation of nuclei, and failure to develop a capsule around the larva. *A* (× 100). *B* (× 400). (By P.C. Beaver. Preparation courtesy of Dr. D.D. Despommier.)

tration and development of the adult worms may produce symptoms of malaise, nausea, diarrhea, and abdominal cramps. This is usually diagnosed as acute food-poisoning, especially since groups of persons are commonly simultaneously afflicted. In a minority of patients, vomiting may occur. Constipation, rather than diarrhea, may appear. Fever is usually absent. Death rarely occurs during the intestinal phase.

In the *stage of muscle invasion,* typical findings are fever, facial edema (particularly around the eyes, sides of the nose, and temples, and hands), pain, swelling, and weakness of involved muscles, and peripheral eosinophilia. Other less common symptoms include headache, flushing of the face, conjunctivitis, pruritus, diaphoresis, anorexia, and thirst. Lymph nodes frequently become enlarged and tender. Damage to muscles in this stage may cause difficulty in eye movement, breathing, chewing, swallowing, and speech, or in the use of the extremities.

Fever usually appears with the onset of muscle invasion, reaches its maximum in a few days to a week, then commonly persists for 2 to 7 days more at somewhat lower levels. The degree of fever is proportional to the severity of the infection; it exceeds 104°F (40°C) in fatal cases. Chilly sensations are common, but shaking chills are uncommonly seen. The fever is remittent and persists for a few days to 6 weeks, depending on the severity of the infection.

Eosinophilia of the blood typically appears during or at the end of the second week of infection, reaches a peak in the third or fourth week, then gradually declines over a period of months. Usually, the relative eosinophilia is at least 20% and commonly exceeds 50%, the maximum being around 90%.

Myocarditis, the most common grave manifestation of trichinosis, may appear as early as the second week but more often after the third week. Death from myocarditis usually occurs between the fourth and eighth weeks of infection. Encephalitis, meningitis, and neurological disturbances also may occur at this stage (Fig. 18–6).

Encapsulation begins at about the end of the third week, beginning the stage of *convalescence*. Fever generally subsides and muscular symptoms begin to decrease. If edema has been marked, diuresis may occur. Appetite returns to normal and malaise subsides.

Myocarditis may still be present at this stage, and physical exertion may precipitate congestive heart failure. Bronchopneumonia, vascular thrombosis, and encephalitis may also complicate convalescence.

Eventually, all symptoms subside, and the cyst wall and, finally, the larva itself calcify. The calcified cysts can be seen grossly in exposed muscle fibers (Fig. 18–5).

The relationship of the three successive biologic stages in the host to the typical pathology and symptomatology is illustrated in Figure 18–6.

While the aforementioned symptoms are characteristic of clinical trichinosis, the onset and progress of the disease may at times be sufficiently atypical as to lead to an inaccurate diagnosis, or, *in the majority of milder cases, no special train of symptoms may be present.* A relatively heavy infection may be an incidental finding at autopsy (Jacobson and Jacobson, 1977).

Diagnosis. ***Clinical Diagnosis.*** Patients with symptoms of trichinosis give a history of having eaten raw or inadequately cooked or improperly processed meat, usually pork. The disease frequently occurs epidemically in several members of the same group. In its early stage, it requires differentiation from acute food poisoning, cholera, and diarrhea or dysentery of other origin. Later, typhoid must be excluded. Many of the milder cases may suggest intestinal "flu" with rheumatic pains. A marked eosinophilia (30% or more), together with the other characteristic signs and symptoms, is suggestive of trichinosis. The ophthalmologist may often be the first person consulted because of an acute onset of badly swollen eyelids and severe conjunctivitis.

Specific Diagnosis. It is rarely possible to recover the adult worms and larvae in the feces during the diarrheic stage of the disease, or to recover the larvae in the blood, spinal fluid, or milk during the period of migration. These are almost never recovered in routine examination. Following arrival of the larvae in skeletal muscle, diagnosis can be specifically made by demonstration of free or encapsulated larvae in compressed small samples of deltoid, biceps, gastrocnemius or, most conveniently, pectoralis muscles obtained by biopsy. Muscle strips obtained by biopsy and digested in artificial gastric juice for several hours at 37°C provide a concentrate that is more diagnostic. Prost (1959) found that, in lightly infected pigs, trichinoscopy by examination of compressed muscle samples missed 25% of positives identified by the digestion method, and 1000 supposedly trichina-free pigs were found to be positive by muscle digestion. It has been observed repeatedly that digestion methods detect a high percentage of infections not detected by trichinoscopy (Bessonov et al., 1978).

Beck (1953) did a *xenodiagnosis* on each of 100 human diaphragms, first examining samples of the material microscopically, then feeding it to uninfected albino rats that were killed a month or more later. He recommended this technique as particularly useful in surveys to detect viable larvae.

The *Bachman intradermal test* (Bachman, 1928) with antigen in a 1:5000 to 1:10,000 dilution provides both a practical and highly specific test when carried out with good antigen by competent workers. In positive tests, irrespective of the intensity of the infection, a small white swelling immediately appears around the injected site, which is surrounded by an unraised, irregular wheal that is up to about 5 cm in diameter. The reading is made within 15 to 20 minutes, after which the reaction begins to fade. It is desirable to do a serologic test as a check on the intradermal test, since it may give better evidence that the infection has been contracted relatively recently. Widely used for routine laboratory diagnosis is the bentonite flocculation (BF) test. With this test, antibodies are detected after the third week of infection. In central reference laboratories, it is the standard test for the diagnosis of trichinosis (Kagan and Norman, 1976). Complement fixation and indirect hemagglutination tests are also considered useful.

Treatment. Although there are several compounds that have been shown to kill *Trichinella spiralis* or interfere with oviposition, there is no established specific treatment for trichinosis. The oldest of the active drugs is thiabendazole, which has been shown to kill *T. spiralis* when administered to mice or swine 24 hours after inoculation (Cuckler et al., 1963). There has been variable success with the drug in the treatment of human beings, and early trials were associated with a high incidence of severe allergic reactions (Kean and Hoskins, 1964; Rugiero et al., 1966). More recent reports indicate that thiabendazole in a dosage of 50 mg/kg body weight daily may prevent the appearance of symptoms if given from the second day after ingestion of infected meat, or greatly mitigate the illness if the drug is given between the fifth and ninth days after infection (Gerwel et al., 1974; Casali and Costa, 1977).

The administration of corticosteroid drugs or of ACTH would be appropriate treatment for the allergic reaction to the death of the parasites, but the immunosuppressive drug methotrexate has been shown to interfere with the trichinellicidal activity of thiabendazole when the two drugs were given concurrently to experimentally infected mice (Ozeretskovskaya et al., 1974).

Mebendazole has been shown to have a lethal effect on both the invasive and encapsulated phases of *Trichinella* in mice and rats (McCracken and Taylor, 1980), but its usefulness in human infection is yet to be established. In an outbreak of trichinellosis in eastern Siberia, mebendazole at a dosage of 300 mg daily for 5 to 7 days was reported to be effective (Morenets et al., 1981). The primary drawback in the treatment of human trichinosis has been that the diagnosis is usually made after invasion is well under way. Drugs effective only in the early stages may be ineffective against the older parasites, especially those that are encapsulated. If mebendazole given in a course of 14 days is effective against all stages in human infection, as reported by Thienpont et al. (1974) with regard to experimental infection in rats, that drug may become preferred in human trichinosis.

During the critical stage of muscular invasion the administration of ACTH or of adrenocorticosteroid drugs inhibits the severity of the disease and may protect the myocardium and central nervous system from significant damage. Otherwise, the treatment consists of supportive measures and symptomatic relief.

Prognosis. Prognosis is usually good in mild cases, grave in severe infections. Absence of eosinophilia or a low-grade eosinophilia is a poor prognostic sign.

Prevention. Smoking, salting, and drying of meat are not effective measures. Refrigeration at 5°F (−15°C) for not less than 20 days, at −10°F for 10 days, −20°F for 6 days, or deep freezing (−37°C) provides effective treatment of pork containing temperate-zone strains of the parasite (Kagan, 1959). In the United States, pork products customarily eaten raw are adequately processed only in the government-inspected abattoirs; the smaller plants are unsupervised. The most effective methods of control consist in not feeding raw garbage to hogs, extermination of rats around farms, and thorough cooking or deep freezing of all pork expected to be eaten by man. Freezing may not kill larvae in the meat of bears and other Arctic mammals (Chadee and Dick, 1982).

Striking reductions in prevalence, mortality, and severity of infection with *T. spiralis* in man and pigs in the United States have occurred in recent years, possibly in part due to legislation requiring heat treatment of garbage fed to swine as well as to extensive use of low-temperature storage of pork, increased use of prepared pork products, and education of the public in proper methods of cooking pork. The sporadic cases and outbreaks that still occur are the result of carelessness in feeding, processing, and cooking pigs raised for personal use.

Trichuris trichiura (Linnaeus, 1771) Stiles, 1901

Synonyms. *Trichocephalus trichiurus* (Linnaeus, 1771) Blanchard, 1895; *Trichocephalus dispar* Rudolphi, 1802.

Historical and Geographical Notes. This worm was first described by Linnaeus in 1771. The life cycle was first studied by Grassi (1887), and later by Fülleborn (1923) and Hasegawa (1934), while that of the related species, *T. vulpis* of the dog, was investigated by Miller (1941, 1947).

The human whipworm is cosmopolitan in distribution but is more common in the warm, moist regions of the world, where both the incidence and the intensity of the infection may be high. In certain hyperendemic areas, 90% of the population may be infected. In the southern United States, infection rates may reach 20 to 25%, but the worm burden is usually relatively low. In moist tropical countries, the observed number of worms per individual is frequently more than a hundred. While whipworm infection is more or less coextensive with ascariasis, the former is more prevalent in areas of high rainfall, high humidity, and dense shade.

Well-preserved eggs of this worm were obtained from feces in the rectum of a 9-year-old Inca girl, whose frozen body was recovered from a stone building at 17,568-foot altitude near Santiago, Chile. It was calculated that death had occurred 450 years earlier (Pizzi and Schenone, 1954).

Morphology, Biology, and Life Cycle. *Trichuris trichiura* lives attached to the wall of the cecum, and less commonly to the wall of the appendix, colon, or most posterior level of the ileum. Man is the only known natural host, although whipworms obtained from the pig and certain monkeys are morphologically similar. The flesh-colored worms (Fig. 18–7) are attenuated in the anterior three-fifths of the body and fleshy in the posterior portion. A broad, longitudinal *bacillary band* (Fig. 18–9,*A*, *B*) is on the ventral side of the anterior part of the body. This is an area of cuticular pores, each with an underlying tall, columnar hypodermal gland cell (Wright, 1968). The esophagus is a delicate capillary tubule, but it has been shown to have anterior muscles with a stylet and posterior muscles capable of dilating at its more distal level.

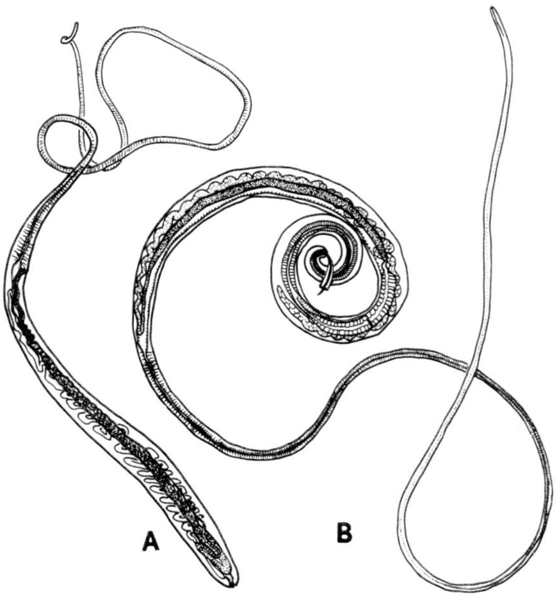

Fig. 18–7. *Trichuris trichiura*. *A*, Female, showing characteristic whip-like body, convoluted ovary and oviduct, straight uterus and vagina leading to vulva near junction of intestine and stichosome. *B*, Male, showing stout hind-body filled with reproductive organs, including convoluted testis, straight genital tube and extruded spicule, and stichosome and esophagus in the much longer, thin, anterior part of body (× ca. 7). (From Beaver, P.C. 1952. The detection and identification of some common nematode parasites of man. Am. J. Clin. Pathol., *22*:481–494.)

The male (Fig. 18–7*B*) measures 30 to 45 mm in length and has its caudal extremity coiled through 360° or more. The male genitalia consist of a long sacculate testis, vas deferens, and ejaculatory tubule, which empties into the cloaca. A single lanceolate spicule protrudes through a retractile sheath that has a bulbous termination covered with many small, recurved spines. The female measures 35 to 50 mm in length and is bluntly rounded at the posterior end. The female genitalia consist of a single sacculate ovary, oviduct, and uterus, which constricts as it joins the vagina, which in turn continues as a serpentine tubule to the vulva, which lies ventrally at the anterior extremity of the fleshy portion of the worm. The eggs (Fig. 18–8) are characteristically barrel-shaped and, in addition to a vitelline membrane, have a triple shell, the outermost layer of which is brown. They also have bipolar, unstained, intralaminar prominences that have the appearance of mucoid plugs. The eggs measure 50 to 54 μm by 22 to 23 μm. They are discharged and passed in the feces in the unsegmented stage and require at least 10 to 14 days for embryonation in the soil. The exact egg-laying ca-

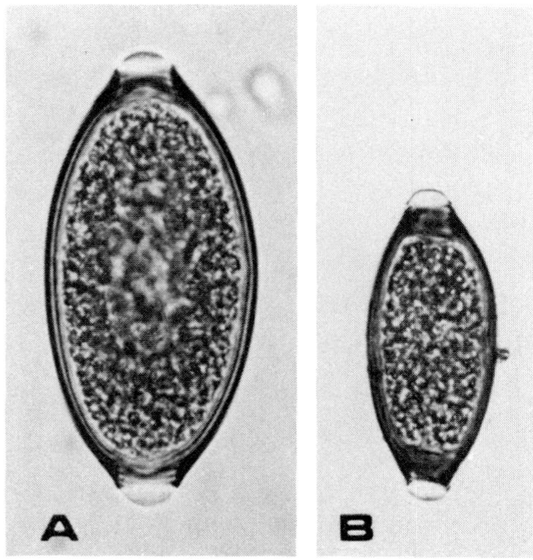

Fig. 18–8. *Trichuris* eggs with characteristic polar plugs and ovum. (× 600.) *A*, *T. vulpis*. *B*, *T. trichiura*. (Courtesy of Dr. M.D. Little.)

pacity of this worm has not been determined. Estimates range from 1000 upwards per female per day. Moisture is essential for their development. There is apparently no larval ecdysis within the eggshell. People become infected on swallowing the fully embryonated eggs. The larva hatches in the small intestine and enters the crypts of the lower intestine and colon, eventually entering the epithelium of the cecum and other parts of the large intestine (Fig. 18–9). Once it is firmly established in the epithelium at one site, the developing worm probably is unable to move to a different site (Panesar and Croll, 1980). It grows into an adult worm in about 3 months.

Trichuris eggs that have much greater than normal dimensions are occasionally seen in human feces. Along with an increase in size, abnormal shape and thickness of the shell have been reported in stools following treatment with mebendazole (Wagner and Peña Chavarria, 1974) and other drugs (Mazzotti et al, 1958). However, *Trichuris* eggs of large size occur when not associated with treatment. In some instances, the eggs were identified as those of *T. vulpis,* the dog whipworm, largely on the basis of size, although other differences have been noted (Kenney and Yermokov, 1980). *T. vulpis* eggs are 70 to 80 × 30 to 42 μm and have prominent, but relatively small, polar plugs. *T. trichiura* eggs as large as those of *T. vulpis* differ in the greater width of the polar plugs. In a study by Little (1968), eggs of large size (70–83 μm long) among others of normal size were recovered from a human stool and identified as *T. trichiura;*

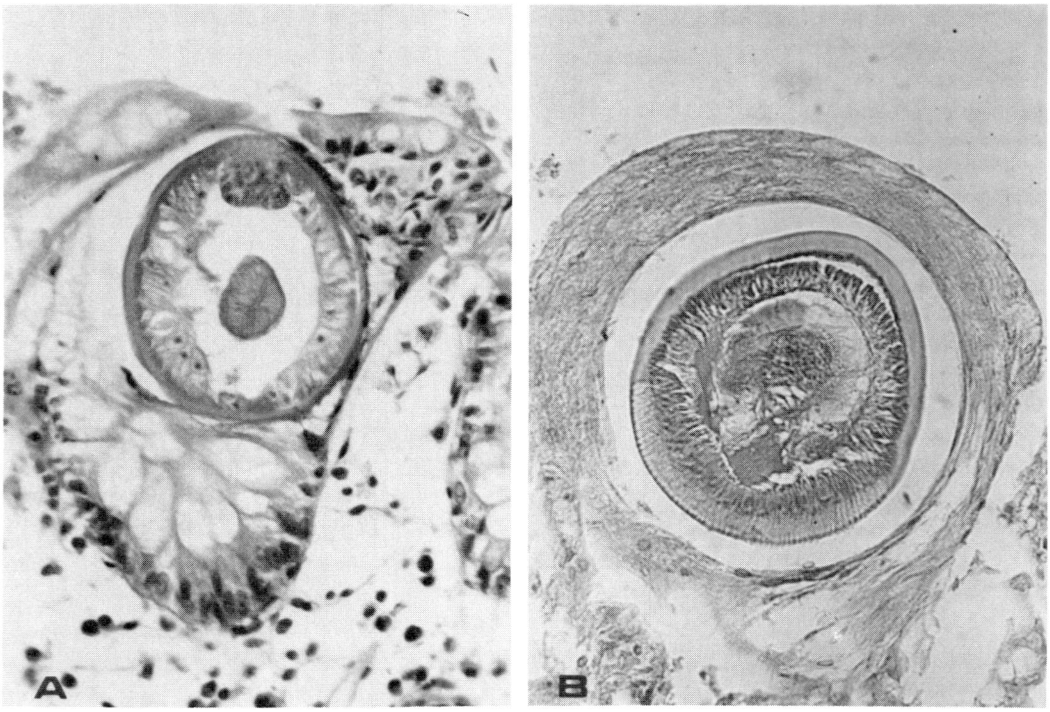

Fig. 18–9. *Trichuris trichiura.* Transverse sections of slender forebody in tunnel in mucosal epithelium of cecum. *A*, Section cut at level of muscular esophagus near mouth, anterior to stichosome, at which level there is little host reaction. *B*, Section at level of stichosome where the worm has induced host reaction to form fibrous tissue that supports it. Here the bacillary band extends over nearly half the worm's circumference, whereas in *A*, it is narrow. (× 320.) (By P.C. Beaver.)

when put in cultures, eggs of both sizes developed to the infective stage. When 22 of them were swallowed by a human volunteer, patent infection developed between 101 and 114 days later. Eggs from the volunteer measured 59 to 74 μm in length. In two of eight female *T. trichiura* expelled by treatment, a few large eggs (70 μm or more in length) as well as normal-sized eggs were present together in the same worm. This observation was confirmed by Corrêa et al. (1980).

Epidemiology. Conditions favorable for development of the eggs of *T. trichiura* consist of warm, shaded moist soil. Development to the infective stage requires about 3 weeks, after which an active first-stage larva is coiled inside the shell. Infection results from swallowing these eggs, which are obtained directly or indirectly from the soil. The eggs are much less resistant to desiccation and heat than are those of *Ascaris,* will not usually develop to the infective stage on hard clay, ashes, or cinders, and will not survive direct sun's rays, intense cold, a putrefying medium, or action of many chemical agents. Even in a moist atmosphere, a moderately dry film of feces will support survival for only a few days or weeks. Areas of high prevalence and heavy worm burden are usually those polluted by small children, who are often more commonly infected than are adults.

In areas of high endemicity, small children develop heavy infections (Jung and Beaver, 1951); yet, the greatest prevalence typically occurs in children of primary school age who contaminate dooryard soil with their feces and later pick up the infective eggs and in various ways transfer them to their mouths.

Authorities are uncertain as to whether *Trichuris* of the pig is the same or a different species from *T. trichiura* of man. Heavy natural infections with a *Trichuris* presumed to be *T. trichiura* produce diarrhea and dysentery in the chimpanzee. This animal is readily infected experimentally with *T. trichiura* from humans.

Pathogenesis, Pathology, and Symptomatology. The immature *Trichuris trichiura* threads its entire body in the epithelium of the colon. Electron micrography has shown that *T. muris* in the cecum of the house mouse creates, by a lytic action, a syncytium of the epithelial cells as its head advances and presumably feeds on the cellular elements in the syncytium. The body remains covered by the roof formed by the persisting luminal surfaces of the enterocytes. As the worm matures, its posterior portion is extruded or ruptures from the tunnel and hangs in the intestinal lumen, available for functions of copulation and oviposition.

Microscopically, relatively little tissue reaction to this process and only slight damage are found. The enterocytes converted into a syncytium are destroyed (Fig. 18–9). Their luminal surfaces show loss of height of the microvilli associated with loss of the filament bundles that support normal microvilli. There is no bleeding into or around the tunnel. Some infiltration of plasma cells, lymphocytes, and sometimes eosinophils is seen between the crypts.

In contrast, in the patient with whipworm dysentery, the mucosa can be grossly demonstrated to be edematous and friable. Anal sphincter tone is lost. The rectum tends to prolapse (Fig. 18–10). This often causes linear bleeding rents in the edematous mucosa.

The exact mechanism by which *Trichuris trichiura* affects the human host is not known, but two processes—mechanical and allergic—may be involved. Theoretically, it would seem that mechanical trauma would be an important cause of the disease produced by *Trichuris,* since the adult worm is threaded into the epithelium of the cecum. In fact, the mechanical damage to the intestine produced by this process is insignificant, although it is said that the appendiceal lumen may become blocked by a matted mass of worms. Hartz (1953) suggested that the presence of the worms in the mucous membrane irritates the nervous plexus of the mucosa, causing diarrhea and cramps.

The diffuse nature of the colonic inflammation associated with trichuriasis and the characteristic presence of eosinophils and Charcot-Leyden crystals in the colonic exudate suggest that the mechanism is allergic, but this has not been proved.

Bowie et al. (1978) have speculated on the existence of an unknown factor produced by *Trichuris* that is responsible for growth retardation and clubbing in the host *(vide supra).*

There has been considerable disagreement as to the pathogenicity of *Trichuris* and the characteristics of the syndrome that it produces. This is because the clinical picture produced by trichuriasis in the human host is dependent on the intensity of infection (worm burden), the duration of infection, and the age and nutrition of the host.

The great majority of infections with *Trichuris* are moderate or light. These infections are tolerated by the host with a few or no symptoms, and no abnormalities in laboratory findings except for the presence of eggs in the stools.

In heavy whipworm infections in children 2 to 6 years of age, a diffuse colitis that causes chronic diarrhea typically occurs. In the heaviest, or massive, infections, worms are attached to the mucosa of the rectum, and chronic dysentery with abdom-

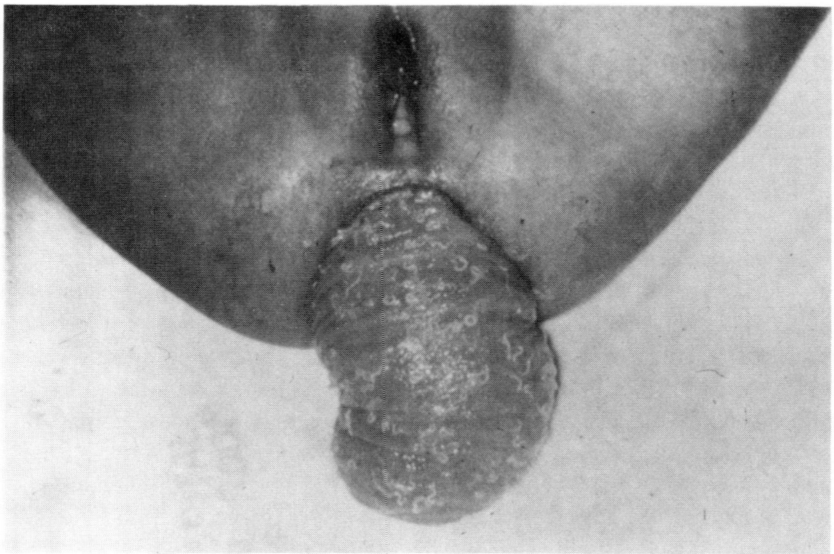

Fig. 18–10. *Trichuris trichiura.* Numerous adult worms attached to mucosa of prolapsed rectum of 2-year-old Louisiana child. Only the stout posterior ends of worms are visible. (From Jung, R.C., and Beaver, P.C. 1951. Clinical observations on *Trichocephalus trichiurus* (whipworm) infestation in children. Pediatrics, *8*:548–557.)

inal cramps and severe rectal tenesmus are produced (Jung and Beaver, 1951). If these patients are readily diagnosed and treated, no other signs or symptoms may appear. If, however, the condition is permitted to continue, there is interference with nutrition, the child loses weight or fails to gain weight, rectal prolapse occurs, and fingers may become clubbed (Gilman et al., 1976; Bowie et al., 1978).

Hypochromic anemia, which is seen in cases with prolonged massive infection, is due to the general malnutrition and blood loss from the friable colon and prolapsed rectum (Jung and Jelliffe, 1952), not directly to the extraction of blood from the host by the parasite, as suggested by Layrisse et al. (1967). Anemia does not occur in light infections, nor even in symptomatic infections, when there is adequate intake and assimilation of iron and protein (Hartz, 1953; Lotero et al., 1974; Greenberg and Cline, 1979). Although the presence of eosinophils is characteristic of the mucoid stools that are produced in dysentery caused by whipworms, eosinophilia of the peripheral blood is not always seen, nor is the degree of eosinophilia proportional to the intensity of infection. Peripheral eosinophilia is not unusual, although it seldom exceeds 15%.

Diagnosis. In massive infections that have been present for a long period of time, the clinical picture may resemble that of hookworm infection, but it is easily distinguished by the dysentery with marked tenesmus. The clinical pictures of amebic dysentery and whipworm dysentery may be identical, but whipworm dysentery tends to be more chronic, more associated with malnutrition and clubbed fingers, and more likely to cause rectal prolapse. In heavily infected children, there is a significant tendency to concurrent infection with *Entamoeba histolytica* and enteropathogenic bacteria (Jung and Beaver, 1951; Gilman et al., 1976).

Specific diagnosis consists in demonstration of the characteristic eggs in the patient's feces, but quantitation is essential to relate the infection to clinical signs and symptoms, since light infections are not expected to cause detectable disease. By means of the direct smear technique, infections are classified as "light" (less than 10 eggs per smear), "heavy" (over 50 per smear), or "massive" (too numerous to count and unequally distributed). Although concentration by centrifugal flotation or sedimentation may be necessary to detect very light infections, these infections are not of clinical importance except in the prepatent period. Eggs in dysenteric stools are bunched and often arranged in strands, indicating egg-laying by worms in the sigmoid and rectal areas. Charcot-Leyden crystals are abundant among the exudates and may become abundant in the diarrheal or dysenteric stools before the worms develop to the egg-laying stage (Fig. 18–11).

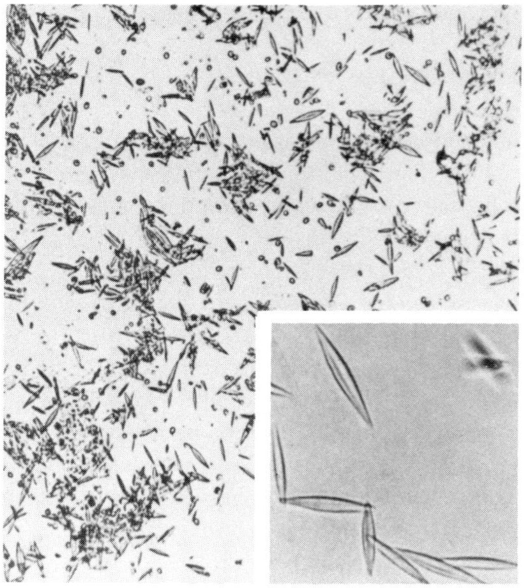

Fig. 18–11. *Trichuris trichiura.* Charcot-Leyden crystals concentrated from dysenteric stool of child with heavy worm burden. The crystals are dipyramidal and hexagonal with each of the six sides (see Fig. 21-10B) tapering from the base at an angle of 80° and meeting at the apex to form a 20° angle (× ca.100.) Inset (× 400). (From Hornung, M. 1962. Preservation, recrystalization and preliminary biochemical characterization of Charcot-Leyden crystals. Proc. Soc. Exp. Biol. Med., *110*:119–124.)

Of great value in the diagnosis of clinically significant infection is examination of the rectal mucosa by means of proctoscopy or during prolapse. In some situations, proctoscopy may be easier to accomplish than fecal examination and egg count. Invariably, worms will be seen when there is dysentery due to whipworm infection.

Large *Trichuris* eggs (i.e., up to 76 by 34 μm), which are sometimes found in human feces, are believed to be due to infection with the whipworm of dogs, *Trichuris vulpis* (Kenney and Eveland, 1978; Wagner, 1979). Equally large eggs occasionally may be produced by *T. trichiura* (vide supra).

Treatment. Treatment is by oral administration of mebendazole in a dosage of 100 mg twice daily for 3 days. Difetarsone and oxantel pamoate have been shown to be effective but are not available in the United States (Leary et al., 1974; Lee et al., 1976).

Prognosis. The prognosis is good in light infections and serious in heavy infections unless the worms are removed.

Prevention. Sanitary disposal of feces should reduce both the prevalence and intensity of the infection. Children can be trained not to defecate promiscuously on the ground. Attempts to destroy infective eggs in soil have been unsuccessful (Burden et al., 1976).

Capillaria hepatica (Bancroft, 1893) Travassos, 1919

Synonym. *Hepaticola hepatica* (Bancroft, 1893) Hall, 1916.

Biology and Life Cycle. *Capillaria hepatica* is a relatively common parasite of the rat (50–90% in some areas) and its relatives; it has also been reported from the muskrat, squirrel, beaver, North American prairie dog, European hare, dog, chimpanzee, capuchin monkey *(Cebus capucinus imitator)*, spider monkey *(Ateles geoffroyi)*, peccary *(Tayassu pecari spirudens)*, pig, cat, and many other species of mammals in all continents. It lives in the liver tissues, and the eggs are laid in the surrounding parenchyma (Fig. 18–12). The following are reports of authentic human cases of infection with *Capillaria hepatica:* (1) British soldier in India, at autopsy in 1923 (MacArthur, 1924); (2) a 17-month old Louisiana child in 1948 in New Orleans (McQuown, 1950); (3) a 7-year-old girl in Baltimore, studied clinically (liver biopsy) and postmortem (Otto et al., 1954); (4) a 60-year-old man in Turkey, at autopsy (Turhan et al., 1954); a 15-month old Hawaiian, liver biopsy and autopsy (Ewing and Tilden, 1956); (6) a 15-month-old South African studied clinically at Johannesburg, liver biopsy; patient was living 2 years later (Cochrane et al., 1957); (7) a 2-year-old Louisiana girl diagnosed at autopsy (Ward and Dent, 1959); (8) a 20-month-old North Carolina child with hepatomegaly and eosinophilia (liver biopsy), recovery (Calle, 1961); (9) a 5-year-old African child in Durban, at autopsy (Kallichurum and Elsdon-Dew, 1961); (10) first case reported from Mexico (Garcia et al., 1962); (11) first case reported from Brazil (Piazza et al., 1963). Twelve cases in Czechoslovakia have been reported, nine in adults and three in children (Slais, 1973); two cases in Italy (Cislagi and Radice, 1970; Pampiglione and Conconi, 1970), and one case in South Africa (Silverman et al., 1973). The nine adult cases in Czechoslovakia were diagnosed by finding the worm (in six) or characteristic lesions (in three). Failure of the worms to mature and lay eggs was unexplained. A similar case diagnosed by punch biopsy of the liver of an eccentric schoolteacher in England was reported by Sumner and Tinsley (1967) as visceral larva migrans. An infection in a Nigerian woman was diagnosed by punch biopsy of the liver (Attah et al., 1983). In addition, several instances of spurious human infection have been recorded from recovery of eggs of *C. hepatica* in the feces; in genuine human cases, eggs are not passed in the feces.

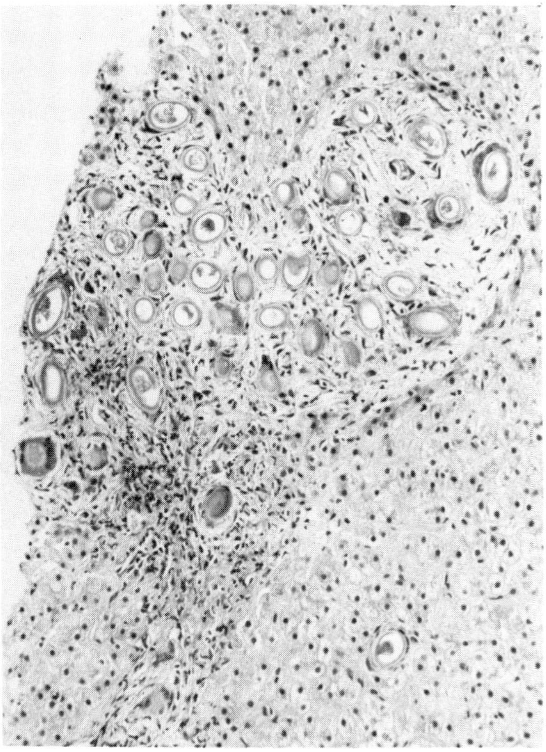

Fig. 18–12. *Capillaria hepatica.* Eggs cut at various angles in a punch biopsy of liver from an unreported case of a 19-month-old South African child with hypereosinophilia and enlargment of the liver. (× 130.) (From Faust, E.C., Beaver, P.C., and Jung, R.C. 1975. *Animal Agents and Vectors of Human Disease,* 4th ed. Philadelphia, Lea & Febiger.)

The adult worms, when dissected out of the tissues, are much more delicate than those of *Trichuris*. The female measures about 20 mm long by about 100 μm wide and the male is about half as long. The male has a slightly chitinized spicule within a protrusible membranous sheath, and the female has a protrusible vulvar opening in the esophageal region. The eggs (Fig. 18–13) resemble those of *Trichuris,* but the poles are less tapered and the outer shell is distinctly patterned. They measure 51 to 68 by 30 to 35 μm. If the infected liver of a natural host containing eggs is eaten by a predator or scavenger animal, the eggs, liberated by digestion of the tissues, are passed in the feces and develop to the infective stage in damp soil. Infection

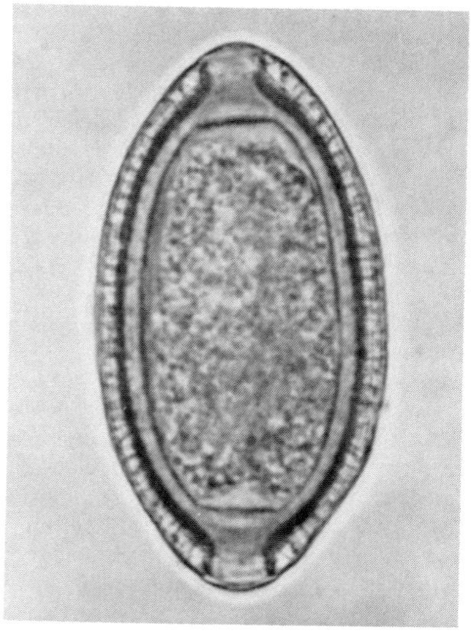

Fig. 18–13. *Capillaria hepatica.* Egg from feces of a cat that ate the liver of an infected rat. (× 1000.) (By P.C. Beaver.)

is acquired by ingesting these eggs directly in soil or in food or drink contaminated by such soil. The larva hatches from the egg in the intestine, migrates via the portal blood to the liver, and matures in about 1 month (Wright, 1961). Eggs are resistant to freezing temperatures and remain viable in soil for several months (Freeman and Wright, 1960). Among small rodents, cannibalism is a major factor in transmission.

Clinical Notes. The pathologic picture in genuine infection with this parasite is that of an acute or subacute hepatitis with hypereosinophilia. Symptoms and signs may suggest amebic involvement of the liver, Loeffler's syndrome, trichinosis, Hodgkin's disease, histoplasmosis, or visceral larva migrans. Diagnosis is made by liver biopsy or at necropsy. There is no known therapy.

Capillaria philippinensis Chitwood, Valesquez, and Salazar, 1968 (Producing intestinal capillariasis)

Historical and Geographical Notes. The first proved case of human infection with *Capillaria philippinensis* occurred in 1963, in a male patient from Ilocos Norte, Philippines, who died 3 days after admission to the Philippine General Hospital with a diagnosis of malabsorption syndrome. Eggs of *Capillaria* sp. were found in the stools, and autopsy showed parasitism of the large and small intestine, but the importance of this finding was not realized until 1967, when attention of the health authorities was drawn to the presence, in the area of Ilocos Sur, of a disease producing severe symptoms and death in adult males. During the subsequent investigations, *Capillaria* eggs were found in the feces of most patients (Cabrera et al., 1967). Chitwood et al. (1968) described the adults and developmental stages and assigned the name *Capillaria philippinensis* to the species. Later, it was discovered in Thailand, in an 18-month-old girl (Pradatsundarasar et al., 1973). The infection is now known to be widely distributed in the Northern Luzon area of the Philippines (Tidball et al., 1978) and in Thailand (Bhaibulaya et al., 1979; Prakitrittranon et al., 1980).

Morphology, Biology, and Life Cycle. In man, *Capillaria philippinensis* lives burrowed into the mucosa of the small bowel, predominantly the jejunum. The adults are small, males measuring 2.3 to 3.17 mm and females 2.5 to 4.3 mm in length. The male is characterized by ventrolateral caudal expansions and an extraordinarily long, smooth spicular sheath. The body of the female is divided into two almost equal parts: the anterior contains the esophagus and esophageal glands (stichosome), and the posterior contains the intestine and reproductive system with slightly prominent vulva. Females produce ova with thin shells and free larvae, as well as typical bioperculate thick-shelled eggs that pass from the host before development of the embryo. All stages of development are commonly observed in man and in experimentally infected monkeys and gerbils. Internal autoinfection appears to be a normal feature of the natural life cycle (Cross et al., 1972). Eggs passed in the feces require at least 10 to 14 days to embryonate (Bhaibulaya et al., 1979), and development to the infective stage in fish requires at least 3 weeks (Cross et al., 1972).

Epidemiology. No host other than man has been found to be naturally infected, and the exact manner in which man acquires the infection is unknown. Experimentally, infection is transmitted through small fish that serve as intermediate hosts; the infective larvae are reported to be located in the mucosa of the fish intestine (Cross et al., 1972; Bhaibulaya et al., 1979).

Pathogenesis, Pathology, and Symptomatology. *Capillaria philippinensis* invades the small bowel, predominantly the jejunum, and, depending on worm burden, causes a more or less severe enteropathy, which is manifested by derangement of intestinal function with malabsorption and loss of fluid, electrolytes, and plasma proteins into the intestinal tract.

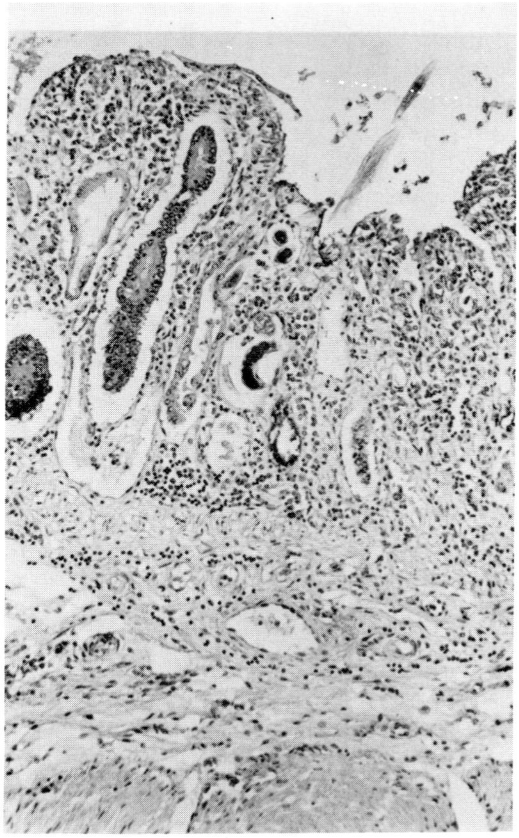

Fig. 18–14. *Capillaria philippinensis*. Adult female in mucosal glands of intestine obtained at delayed autopsy. (× 100.) (By P.C. Beaver. Tissue courtesy of Dr. John Cross.)

Most of the pathological processes are found in the small intestine. The wall of the small intestine is found to be thickened and indurated, and the surface vessels prominent. It is distended with copious watery fluid that, at autopsy, is found to contain abundant larval and adult *C. philippinensis*. Eggs, larvae, and adult worms are present in the glands and mucosa in the lower third of the duodenum and in the glands and adjacent tissues of the lamina propria of the jejunum and proximal third of the ileum (Fig. 18–14). Most of the worms are seen between the mucosa and the basement membrane of the glands. Some are found some distance from the glands in the lamina propria. Inflammatory cells in the tissue around the parasites are remarkably absent. A portion of the worm protrudes into the lumen of the gland or the bowel.

The secondary anatomic effects of electrolyte (especially potassium) depletion are diffuse and include vacuolization of myocardial sarcoplasm and of the cells lining the proximal convoluted tubules of the kidneys and pulmonary edema (Fresh et al., 1972).

Clinically, the disease is intestinal malabsorption with symptoms of chronic diarrhea and borborygmus. The latter symptom is the most constant feature, appearing before the diarrhea in case of relapse and persisting after diarrhea ceases in the case of cure. Abdominal pain also occurs, but it is usually not severe.

Voluminous, sprue-like stools are passed up to eight times daily and may amount to 2 L daily in some cases.

With the onset of diarrhea, the patient loses weight rapidly and becomes weak. There is wasting of muscles, abdominal distention, and edema. The patient eventually appears emaciated and cachectic. Serum proteins become depleted, as do the electrolytes.

Death may occur 2 weeks to 2 months after the onset of symptoms due to pneumonia, heart failure, hypokalemia, or cerebral edema (Watten et al., 1972).

Treatment. Thiabendazole given in high doses over a prolonged course is moderately effective (Whalen et al., 1971) but has been replaced as drug of choice by mebendazole, which is more effective and less toxic. The dosage is 400 mg daily, administered orally, for 20 to 30 days. The 30-day course is recommended in the case of relapses or treatment failures. The treatment is highly effective, with marked improvement occurring as early as the first day of treatment (Singson et al., 1975).

Prevention. Since infection is acquired by eating raw fish, in endemic areas fish should be thoroughly cooked before being eaten.

Capillaria aerophila (Creplin, 1839)

Synonym. *Thominx aerophila.*

Capillaria aerophila is a common parasite of cats, dogs, foxes, and some other carnivores in North America, Europe, and Asia. It has also been reported from South America (Buenos Aires) and Australia. It develops and remains in the epithelial lining of the trachea, bronchi, and bronchioles (Fig. 18–15). Males measure up to 18 mm by 70 μm, females up to 20 mm by 105 μm. Unsegmented eggs that measure about 65 by 35 μm, with characteristic polar plugs, are laid in tunnels in the mucosal epithelium, later sloughed into the air passages, coughed up, swallowed, and discharged in the feces. In moist soil at summer temperatures, the eggs become infective in about 6 weeks. Transmission is reported to be direct, but earthworms and possibly other animals may serve as paratenic or transport hosts (Skrjabin et al., 1957).

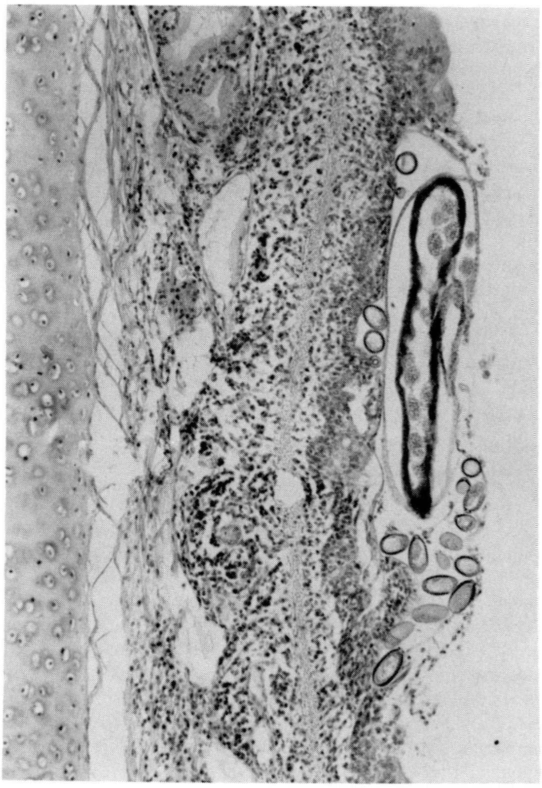

Fig. 18–15. *Capillaria aerophila*. Adult female and eggs in tunnel produced by the worm in epithelium of tracheal mucosa of a cat. (By P.C. Beaver. Preparation courtesy of Dr. L.H. Herman.)

Capillaria aerophila has been reported as a human parasite in seven cases in the U.S.S.R., one in Morocco, and one in Iran. The report of the Iran case by Aftandelians et al. (1977) included a review of the eight earlier cases. A tenth case was reported by Skipina (1976) in the Ukraine. Infection in man produces acute bronchitis and bronchiolitis, usually with asthma and productive cough. Eggs may be found in the sputum or feces, or both. In the Iran case, diagnosis was made by identifying the worm in a lung biopsy. Diethylcarbamazine, thiabendazole, and steroid were given, and relief of symptoms was rapid.

Anatrichosoma cutaneum (Swift, Boots, and Miller, 1922) Chitwood and Smith, 1956

Synonym. *Trichosoma cutaneum* Swift, Boots, and Miller, 1922.

This *Capillaria*-like worm is a parasite of the skin and nasal mucosa of monkeys *(Macaca mulatta)* in Asia and Africa (Allen, 1960; Orihel, 1970). Two human cases of creeping eruption caused by this worm have been reported from Japan (Morishita and Tani, 1960) and Vietnam (Le-Van-Hoa et al., 1963).

SUPERFAMILY DIOCTOPHYMATOIDEA*

Dioctophyma renale (Goeze, 1782) Stiles, 1901 (The giant kidney worm)

Synonym. *Eustrongylus gigas* (Rudolphi, 1802) Diesing, 1851.

Biology and Life Cycle. This giant worm, which was first obtained from a dog's kidney by Goeze (1782), has a fairly cosmopolitan distribution in mammals, including the dog, wolf, *Canis jubatus,* puma, glutton, raccoon, coati, mink, marten, skunk, weasel, otter, seal, cat, ox, and horse (Ehrenford and Snodgrass, 1955; Schacher and Faust, 1956). It has been widely reported from dogs and other reservoir hosts in the Americas and has been recovered from the dog in China. At least 13 cases of authentic human infection with the adult worm are known: in 5 cases, infection was discovered at autopsy; in 6 cases, worms were observed migrating from or were expelled through the urethra, and in one case a worm was discharged through the skin over an abscessed kidney; and in 1 case, eggs were found in the urine. In an additional case, an immature worm (Fig. 18–16) was identified in a subcutaneous nodule from the chest wall of a man in California (Beaver and Theis, 1979). The living adult worm (Fig. 18–17) has a blood-red color, is cylindrical in shape and slightly attenuated at both ends. Males measure 14 to 20 cm in length by 4 to 6 mm in diameter, and at the posterior extremity, have a bursa that is not supported by rays but a marginal and inner covering of papillae. There is a single copulatory bristle 5 to 6 mm long that usually projects through the bursa. Females are much larger (20–100 cm in length by 5–12 mm in diameter). The vulva is mid-ventral in position, near the anterior end of the worm. The fertile eggs (Fig. 18–18) are ellipsoidal, brownish-yellow in color, and have deeply sculptured depressions except at the poles; they measure 60 to 80 μm by 39 to 46 μm. The eggs

*For characteristics, see page 228.

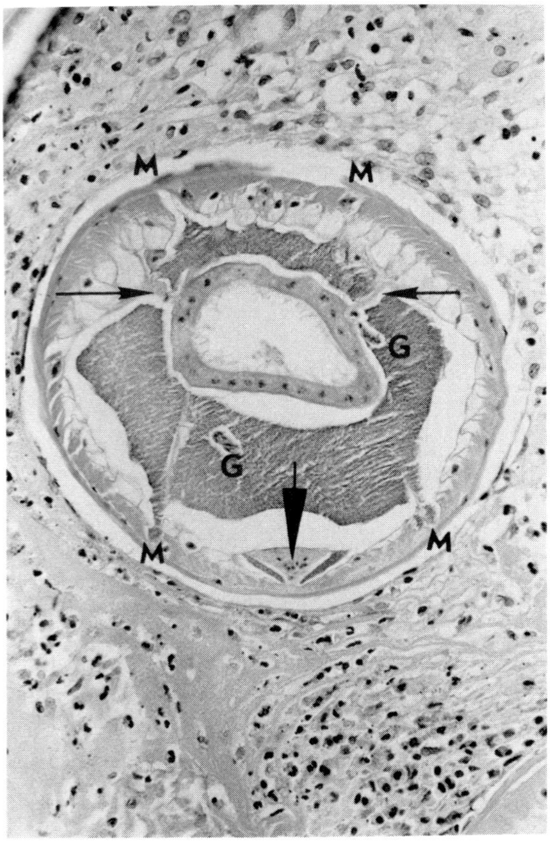

Fig. 18–16. *Dioctophyma renale*. Transverse section of larva in fibrous nodule deep in the subcutaneous tissues of the chest of a man in California, showing ventral cord (large arrow), dark precipitate between intestine and somatic muscles, mesenteries (small arrows), four somato-intestinal muscles *(M)*, and two portions of genital primordium *(G)*. (× 250.) (From Beaver, P.C., and Theis, J.H. 1979. Dioctophymatid larval nematode in a subcutaneous nodule from man in California. Am. J. Trop. Med. Hyg., *28*:206–212.)

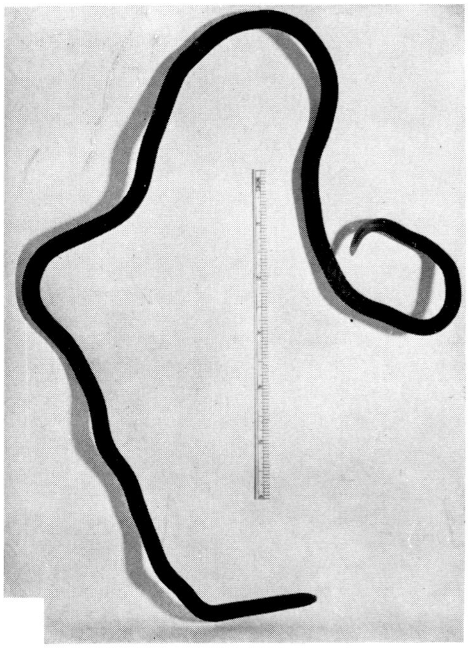

Fig. 18–17. *Dioctophyma renale*. Mature female from kidney of naturally infected dog. Scale is 60 mm. (Courtesy of Dr. T. B. Magath.)

from mated females usually contain two cells (Fig. 18–18). At 20 to 25°C, first-stage larval development is complete in about 1 month (Mace and Anderson, 1975). Eggs hatch in the intestine of aquatic oligochaetes, and after 2 months, the larvae are infective. Mink and other mammals can acquire infection directly by ingestion of an infected oligochaete, but more often the infective larvae are first taken up by tadpoles or frogs, or by fish *(Ictalurus)*, which serve as transport or paratenic hosts. In the mink, the prepatent period is about 5 months; in the dog, it is somewhat less (Mace and Anderson, 1975).

Clinical Notes. The worms live as adults in the pelvis of the kidney, usually the right kidney, or in body cavities. In the kidney, they gradually destroy the parenchyma, leaving only the capsule, which is greatly modified (Fig. 18–18). Their presence in the organ first causes renal colic with hematuria. All of the reported human cases, except one (Beaver and Theis, 1979), have involved the kidney. The presence of characteristic eggs in the urine provides specific diagnosis. Thorough cooking of freshwater fish and frogs should prevent man from acquiring the infection.

Eustrongylides Jaegerskiöld, 1909

Species of the genus *Eustrongylides* are parasites of the esophagus and proventriculus of fish-eating birds. The infective-stage larvae develop in minnows, small fish, and amphibia. They become large, reaching lengths of 10 cm or more. Although they are not sexually mature, their reproductive organs are fully differentiated. They are red or pink, and usually are found coiled in a fibrous capsule in the body cavity of the host. However, they also may be found unencapsulated and migratory in the tissues of mammals and other animals that have eaten an infected fish (Von Brand and Cullinan, 1943). This proved to be true in man. When three fishermen in Maryland (United States) swallowed live bait minnows, they all had severe abdominal pain, and two required abdominal surgery to remove worms that were migrating through the intestinal wall to the peritoneal cavity (Gunby, 1982). The detailed larval morphology of *Eustrongylides*, described by Panesar and Beaver (1979), closely resembles that of *Dioctophyma*.

SUPERFAMILY MERMITHOIDEA*

Mermithoid Worms

This group of true roundworms includes the so-called "cabbage snakes" (family Mermithidae), which are long, opaque nematodes with a pointed anterior end, a tapering body, and a smooth, finely striated cuticle. The esophagus is nonmuscular, and the intestine, when present, is modified into a nutrient storage organ. In the mature female, the anus is closed, but in males, the cloaca persists. The larval stages develop in the hemocoel of insects, particularly grasshoppers, while the adults are free-living in the soil. According to Christie (1936), eggs are deposited in the soil, hatch there, and the larvae migrate to the surface, climb onto vegetation, and enter the hemocoel of nymphal grasshoppers, where they require 1 to 3 months to mature before breaking out of the body and entering the soil.

Four cases of accidental human infection have been recorded:

*For characteristics, see pages 228 and 229.

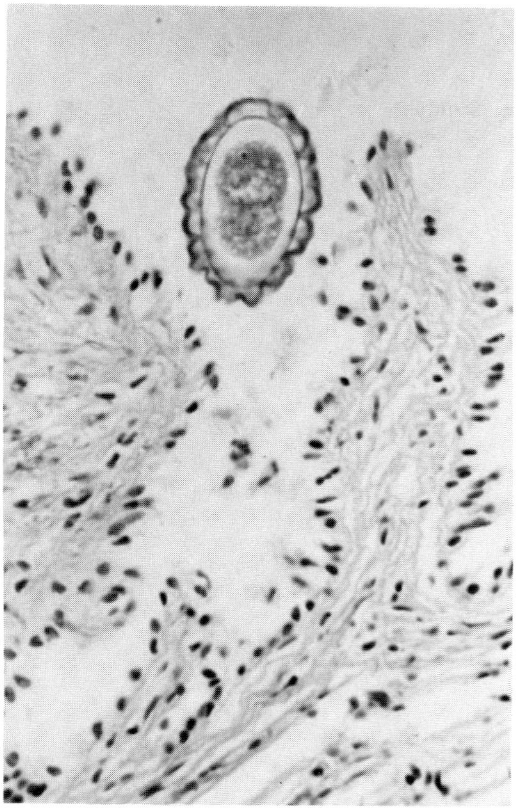

Fig. 18–18. *Dioctophyma renale*. Section of kidney capsule of mink from which two mature worms (male and female) were removed. Lining of capsule is deeply folded and covered with epithelium. Egg shows typical two-cell embryo and thick, deeply sculptured shell. (× 400.) (By P.C. Beaver.)

(1) an immature specimen, 14 cm long, obtained from a child's mouth *(Agamomermis hominis oris,* reported by Leidy in 1850); (2) a specimen 65 cm in length and 1.5 mm in diameter, discovered emerging from the urethra of an adult male *(Agamomermis restiformis,* reported by Leidy in 1880); (3) a specimen 56 cm in length and less than 1 mm in diameter said to have been passed by a woman suffering from uterine cancer (Baylis, 1927); and (4) an immature mermithid obtained from the feces of a 5-year-old Ecuadorian girl who complained of diarrhea and abdominal pain (Leon, 1946). The records are important because they demonstrate the possibility of confusing accidental or spurious infection with actual parasitism by a pathogenic nematode. Four additional equally questionable cases have been reported—two in which a worm was observed in the urethra, one in the urine, and one in the feces. These and the four others referred to above were reviewed by Poinar (1979), who concluded that none of the eight records of human infection by mermithid nematodes can be accepted as authentic.

REFERENCES

Aftandelians, R., Raafat, F., Taffazoli, M., and Beaver, P.C. 1977. Pulmonary capillariasis in a child in Iran. Am. J. Trop. Med. Hyg., *26*:64–71.

Ali Khan, Z. 1966. The postembryonic development of *Trichinella spiralis* with special reference to ecdysis. J. Parasitol., *52*:248–259.

Allen, A.M. 1960. Occurrence of the nematode, *Anatrichosoma cutaneum,* in the nasal mucosa of monkeys. Am. J. Vet. Res., *21*:389–392.

Attah, E., Nagarajan, S., Obineche, E., and Gera, S. 1983. Hepatic capillariasis. Am. J. Clin. Pathol., *79*:127–130.

Bachman, G.W. 1928. A precipitin test in experimental trichiniasis. J. Prev. Med., *2*:513–523.

Beaver, P.C. 1950. The standardization of fecal smears for estimating egg production and worm burden. J. Parasitol., *36*:451–456.

———. 1952. The detection and identification of some common nematode parasites of man. Am. J. Clin. Pathol., *22*:481–494.

———, and Theis, J.H. 1979. Dioctophymatid larval nematode in a subcutaneous nodule from man in California. Am. J. Trop. Med. Hyg., *28*:206–212.

Beck, J.W. 1953. Xenodiagnostic technic as an aid in the diagnosis of trichinosis. Am. J. Trop. Med. Hyg., *2*:97–101.

Bessonov, A.S., Uspenskii, A.V., and Shekhovtsov, N.V. 1978. Group diagnosis of trichinellosis in pigs under conditions of meat-packing plants, pp. 519–522. In *Trichinellosis*. Edited by C.W. Kim and Z.S. Pawlowski. Hanover, New Hampshire, University Press of New England.

Bhaibulaya, M., Indra-Ngarm, S., and Ananthapruti, M. 1979. Freshwater fishes of Thailand as experimental intermediate hosts for *Capillaria philippinensis*. Int. J. Parasitol., *9*:105–108.

Bourée, P., Bouvier, J.B., Passeron, J., Galanaud, P., and Dormont, J. 1979. Outbreak of trichinosis near Paris. Br. Med. J., *1*:1047–1049.

Bowie, M.D., Morison, A., Ireland, J.D., and Duys, P.J. 1978. Clubbing and whipworm infestation. Arch. Dis. Child., 53:411–413.

Bura, W.T., and Willett, W.C. 1977. An outbreak of trichinosis in Tanzania. East Afr. Med. J., *54*:185–193.

Burden, D.J., Whitehead, A., Green, E.A., McFadzean, J.A., and Beer, R.J.S. 1976. The treatment of soil infested with the human whipworm, *Trichuris trichiura*. J. Hyg., Camb., *77*:377–382.

Cabrera, B.D., Canlas, B., Jr., and Dauz, U. 1967. Human intestinal capillariasis. III. Parasitological features and management. Acta Med. Philip., *4*:92–103.

Calle, S. 1961. Parasitism by *Capillaria hepatica*. Pediatrics, *27*:648–655.

Campbell, W.C. 1979. History of trichinosis: Paget, Owen and the discovery of *Trichinella spiralis*. Bull. Hist. Med., *53*:520–552.

Casali, A.J., and Costa, E.A. 1977. Investigación clínica y anatomopatologica del tratamiento de la triquinosis aguda con tiabendazol. Bol. Chil. Parasitol., *32*:66–70.

Centers for Disease Control. 1981. *Trichinosis Surveillance. Annual Summary 1980*. Atlanta, U.S. Department of Health and Human Services.

Chadee, K., and Dick, T.A. 1982. Designation and freezing resistance of isolates of *Trichinella spiralis* from wild carnivores. J. Wildl. Dis., *18*:169–173.

Charlton, I.K. 1980. Termination of trichinosis survey. Meat Hygienist (1980), no. 25:4, 32.

Chirasak, K. 1971. Output of larvae and life span of *Trichinella spiralis* in relation to worm burden and superinfection in the mouse. J. Parasitol., *57*:289–297.

Chitwood, M.B., Velasquez, C., and Salazar, N.G. 1968. *Capillaria philippinensis* sp. n. (Nematoda: Trichinellida), from the intestine of man in the Philippines. J. Parasitol., *54*:368–371.

Christie, J.R. 1936. Life history of *Agamomermis decaudata,* a parasite of grasshoppers and other insects. J. Agric. Res., *52*:161–198.

Cislaghi, F., and Radice, C. 1970. Infection by *Capillaria hepatica*. First case report in Italy. Helv. Paediat. Acta, *25*:647–654.

Cochrane, J.C., Sagorin, L., and Wilcocks, M.G. 1957. *Capillaria hepatica* infection in man. S. Afr. Med. J., *31*:751–755.

Corrêa, L.L., Yamanaka, M.T., Corrêa, M., Silva, M., and Silva, R. 1980. Ocorrência de ovos grandes de *Trichuris trichiura* em fezes humanas. Rev. Inst. Adolfo Lutz, *40*:59–64. (Trop. Dis. Bull., *76*, Abstr. no. 1418.)

Cross, J.H., Banzon, T., Clarke, M.D., Basaca-Servilla, V., Watten, R.H., and Dizon, J.J. 1972. Studies on the experimental transmission of *Capillaria philippinensis* in monkeys. Trans. R. Soc. Trop. Med. Hyg., *66*:819–834.

———, Singson, C.N., Battad, S., and Basaca-Sevilla, V. 1980. Intestinal capillariasis: Epidemiology, parasitology and treatment, pp. 81–87. In *Health Policies in Developing Countries: Royal Society of Medicine International Congress and Symposium No. 24*. London, Academic Press and Royal Society of Medicine.

Cuckler, A.C., Campbell, W.C., and Egerton, J.R. 1963. The effect of thiabendazole on the migratory stages of certain tissue invading nematodes. Proceedings of the 7th International Congress on Tropical Medicine and Malaria, *2*:167–168.

Despommier, D.D., Sukhdeo, M., and Meerovitch, E. 1978. *Trichinella spiralis:* Site selection by the larva during the enteral phase of infection in mice. Exp. Parasitol., *44*:209–215.

Eaton, R.D.P. 1979. Trichinosis in the Arctic. Can. Med. Assoc. J., *120*:22.

Ehrenford, F.A., and Snodgrass, T.B. 1955. Incidence of canine dioctophymiasis (giant kidney worm infection) with a summary of cases in North America. J. Am. Vet. Med. Assoc., *126*:415–417.

Ewing, G.M., and Tilden, I.L. 1956. *Capillaria hepatica:* Report of a fourth case of true human infestation. J. Pediatr., *48*:341–348.

Forrester, A.T.T., Nelson, G.S., and Sander, G. 1961. The first record of an outbreak of trichinosis in Africa south of the Sahara. Trans. R. Soc. Trop. Med. Hyg., *55*:503–513.

Freeman, R.S., and Wright, K.A. 1960. Factors concerned with the epizootiology of *Capillaria hepatica* (Bancroft, 1893) (Nematoda) in a population of *Peromyscus maniculatus* in Algonquin Park, Canada. J. Parasitol., *46*:373–382.

Fresh, J.W., Cross, J.H., Reyes, V., Whalen, G.E., Uylangco, C.V., and Dizon, J.J. 1972. Necropsy findings in intestinal capillariasis. Am. J. Trop. Med. Hyg., *21*:169–173.

Garcia, F.R., Mendiola, G.J., and Biagi, F.F. 1962. Eosinofilia elevada con manifestaciones viscerales. IV. Primer caso de infeccion por *Capillaria hepatica* en Mexico. Bol. Hosp. Infant. Mex., *19*:473–479.

Gardiner, C.H. 1976. Habitat and reproductive behavior of *Trichinella spiralis*. J. Parasitol., *62*:865–870.

Garkavi, B.L. 1972. Species of *Trichinella* from wild carnivores. Vet. Mosk., *49*:90–91.

Gerwel, C., Pawlowski, Z., Kociecka, W., and Chodera, L. 1974. Thiabendazole in prophylaxis of clinical trichinelliasis. Acta Parasitol. Lithanica, *12*:159–162.

Gilman, R.H., Davis, C., and Fitzgerald, F. 1976. Heavy *Trichuris* infection and amoebic dysentery in Orang Asli children. A comparison of the two diseases. Trans. R. Soc. Trop. Med. Hyg., *70*:313–316.

Gould, S.E. 1945. *Trichinosis*. Springfield, Ill., Charles C Thomas.

Greenberg, E.R., and Cline, B.L. 1979. Is trichuriasis associated with iron deficiency anemia? Am. J. Trop. Med. Hyg., *28*:770–772.

Gunby, P. 1982. One worm in the minnow equals too many in the gut. J.A.M.A., *248*:163.

Hamidi, A.N. 1979. Trichiniasis among the animals in northeastern Iran, 1969, 1976, 1977. Bull. Soc. Pathol. Exot., *72*:254–257.

Hartz, Ph. D. 1953. Histopathology of the colon in massive trichocephaliasis of children. Doc. Med. Geogr. Trop., *5*:303–313.

Hornung, M. 1962. Preservation, recrystalization and preliminary biochemical characterization of Charcot-Leyden crystals. Proc. Soc. Exp. Biol. Med., *110*:119–124.

Hunt, G.R. 1967. *Trichinella spiralis* in dogs and cats. J. Parasitol., *53*:659.

Jacobson, E.S., and Jacobson, H.G. 1977. Trichinosis in an immunosuppressed human host. Am. J. Clin. Pathol., *68*:791–794.

Jung, R.C., and Beaver, P.C. 1951. Clinical observations on *Trichocephalus trichiurus* (whipworm) infestation in children. Pediatrics, *8*:548–557.

———, and Jelliffe, D.B. 1952. The clinical picture and treatment of whipworm infection. West Afr. Med. J., *1*:3–7.

Kagan, I.G. 1959. Trichinosis in the United States. Public Health Rep., *74*:159–162.

———, and Norman, L. 1976. Serodiagnosis of parasitic diseases, pp. 382–409. In *Manual of Clinical Immunology*. Edited by N.R. Rose and H. Friedman. Washington, D.C., American Society for Microbiology.

Kallichurum, S., and Elsdon-Dew, R. 1961. *Capillaria* in man. A case report. S. Afr. Med. J., *35*:860–861.

Kean, B.H., and Hoskins, D.W. 1964. Treatment of trichinosis with thiabendazole. A preliminary report. J.A.M.A., *190*:852–853.

Kennedy, M.W. 1979. Effects of the host immune response on the longevity, fecundity and position in the intestine of *Trichinella spiralis* in mice. Parasitology, *80*:49–60.

Kenney, M., and Eveland, L.K. 1978. Infection of man with *Trichuris vulpis*, the whipworm of dogs. Am. J. Clin. Pathol., *69*:199.

———, and Yermakov, V. 1980. Infection of man with *Trichuris vulpis*, the whipworm of dogs. Am. J. Trop. Med. Hyg., *29*:1205–1208.

Kim, C.W., and Ledbetter, M.C. 1980. Surface morphology of *Trichinella spiralis* by scanning electron microscopy. J. Parasitol., *66*:75–81.

Kozar, Z. 1959. Problems of trichinellosis in Poland and plan of control of the invasion. Wiadomosci Parazytol., *5*:280–291.

———. 1961. Some comments on the outbreak of trichinellosis in Mozina Province of Poznan. Wiadomosci Parazytol., *7*:589–594.

———, and Kozar, M. 1965. A comparison of the infectivity and pathogenicity of *Trichinella spiralis* strains from Poland and Kenya. J. Helminthol., *39*:19–34.

Kozek, W.J. 1971a. The molting pattern in *Trichinella spiralis*. I. A light microscope study. J. Parasitol., *57*:1015–1028.

———. 1971b. The molting pattern in Trichinella spiralis. II. An electron microscope study. J. Parasitol., *57*:1029–1038.

———. 1975. *Trichinella spiralis:* Morphological characteristics of male and female intestine-infecting larvae. Exp. Parasitol., *37*:380–387.

———. 1976. *Trichinella spiralis:* Size of F_1 generations produced in mice by transplanted females or from infections with known numbers of male and female larvae. Rev. Biol. Trop., *24*:175–183.

Layrisse, M., Aparcedo, L., Martinez-Torres, C., and Roche, M. 1967. Blood loss due to infection with *Trichuris trichiura*. Am. J. Trop. Med. Hyg., *16*:613–619.

Leary, P.M., Jones, C., Douglas, F., and Boyd, S.T. 1974. Difetarsone in outpatient treatment of *Trichuris trichiura* infestation. Arch. Dis. Child., *49*:486–488.

Le Count, A.L. 1981. A survey of trichinosis among black bears of Arizona. J. Wildlife Dis., *17*:349–351.

Lee, E.L., Iyngkaran, N., Grieve, A.W., Robinson, M.J., and Dissanaike, A.S. 1976. Therapeutic evaluation of oxantel pamoate (1,4,5,6-tetrahydro-1-methyl-2- trans-3-hydroxy-styryl pyrimidine pamoate) in severe *Trichuris trichiura* infection. Am. J. Trop. Med. Hyg., *25*:563–567.

Le-Van-Hoa, Duong-Hong-Mo, and Nguyen-Luu-Vien. 1963. Premier cas de capillariose cutanée humaine. Bull. Soc. Pathol. Exot., *56*:121–126.

Little, M.D. 1968. A strain of *Trichuris trichiura* having large eggs. Madison, Wis., Program of the 43rd Annual Meeting of the American Society of Parasitologists, June 16–21, 1968.

Lotero, H., Tripathy, K., and Bolaños, O. 1974. Gastrointestinal blood loss in *Trichuris* infection. Am. J. Trop. Med. Hyg., *23*:1203–1204.

Lukashenko, N.P., and Brzesky, W.W. 1962. Trichinellosis in wild animals in Siberia, Arctic and Far East USSR. Wiadomosci Parazytol., *8*:589–597.

MacArthur, W.P. 1924. A case of infestation in the human liver with *Hepaticola hepatica* (Bancroft, 1893) Hall, 1916. Proc. R. Soc. Med., *17*:83–84.

Mace, T.F., and Anderson, R.C. 1975. Development of the giant kidney worm, *Dioctophyma renale* (Goeze, 1782) (Nematoda: Dioctophymatoidea). Can. J. Zool., *53*:1552–1568.

Madsen, H. 1974. The principles of the epidemiology of trichinelliasis with a new view on the life cycle, pp. 615–638. In *Trichinellosis*. Proceedings of the 3rd International Conference on Trichinellosis. Edited by C.W. Kim. New York, Intext Educational Publishers.

———. 1976. The life cycle of *Trichinella spiralis* (Owen, 1835 (syns.: *T. nativa* Britov et Boev, 1972, *T. nelsoni* Britov et Boev, 1972, *T. pseudospiralis* Garkavi, 1972) with remarks on epidemiology and a new diagram. Acta Parasitol. Polon., *24*:143–158.

Mantovani, A., Filippini, I., and Bergomi, S. 1980. Indagini su un'epidemia di trichinellosi umana verificatasi in Italia. Parassitologia, *22*:107–134.

Maynard, J.E., and Pauls, F.P. 1962. Trichinosis in Alaska. A review and report of two outbreaks due to bear meat with observations on serodiagnosis and skin testing. Am. J. Hyg., *76*:252–261.

Mazzotti, L., Gonzalez, B.D., and Arredondo, A., 1958. Alteraciones morfológicas en los huevos de *Trichuris trichiura* de pacientes tratados con "Dithiazanine" (yoduro de 3-3′ dietiltiadicarbocianina). Rev. Med. (Mex)., *38*:332–335.

McCracken, R.O., and Taylor, D.D. 1980. Mebendazole therapy of parenteral trichinellosis. Science, *207*:1220–1222.

McQuown, A.L. 1950. *Capillaria hepatica;* report of genuine and spurious cases. Am. J. Trop. Med., *30*:761–767.

Miller, M.J. 1947. Studies on the life cycle of *Trichocephalus vulpis*, the whipworm of dogs. Can. J. Res., D., *25*:1–11.

Morenets, T.M., Bronshtein, A.M., and Tikhonova, L.V. 1981. Treatment with Vermox of trichinellosis caused by natural strains in Sakhalin. Med. Parazitol. Parazit. Bolezni, *50*:43–48. (Russian, with English summary.)

Morishita, K., and Tani, T. 1960. A case of *Capillaria* infection causing creeping eruption in man. J. Parasitol., *46*:79–83.

Nelson, G.S., Rickman, R., and Pester, F.R.N. 1961. Feral trichinosis in Africa. Trans. R. Soc. Trop. Med. Hyg., *55*:514–517.

Olsen, O.W., and Robinson, H.A. 1958. Role of rats and mice in transmitting *Trichinella spiralis* through their feces. J. Parasitol., *44*(sect. 2): 35.

Orihel. T.C. 1970. Anatrichosomiasis in African monkeys. J. Parasitol., *56*:982–985.

Otto, G.F., Berthrong, M., Appleby, R.E., Rawlins, J.C., and Wilbur, O. 1954. Eosinophilia and hepatomegaly due to *Capillaria hepatica*. Bull. Johns Hopkins Hosp., *94*:319–336.

Ozeretskovskaya, N.N., Pereverzeva, E.V., and Veriennikova, N.L. 1974. The effect of thiabendazole and methotrexate on experimental trichinellosis, pp. 499–513. In *Trichinellosis*. Proceedings of the 3rd International Conference on Trichinellosis. Edited by C.W. Kim. New York, Intext Educational Publishers.

Pampiglione, S., Baldelli, R., Corsini, C., and Mantovani, S. 1978. Infezione sperimentale del cavallo con larve di trichina. Parassitologia, *20*:183–193.

———, and Conconi, C. 1970. Primo caso di capillariosi epatica osservata nell'uomo in Italia. Parassitologia, *12*:125–134.

Panesar, T.S., and Beaver, P.C. 1979. Morphology of the advanced-stage larva of *Eustrongylides wenrichi* Canavan 1929, occurring encapsulated in the tissues of *Amphiuma* in Louisiana. J. Parasitol., *65*:96–104.

———, and Croll, N.A. 1980. The location of parasites within their hosts: Site selection by *Trichuris muris* in the laboratory mouse. Int. J. Parasitol., *10*:261–273.

Pawlowski, Z.S., and Ruitenberg, E.J. 1978. Is *Trichinella pseudospiralis* likely to be a human pathogen? Lancet, *1*:1357.

Piazza, R., Corrêa, M.O.A., and Fleury, R.N. 1963. Sobre um caso de infestacao humana por *Capillaria hepatica*. Rev. Inst. Med. Trop. S. Paulo, *5*:37–41.

Pizzi, T., and Schenone, H. 1954. Hallazgo de huevos de *Trichuris trichiura* en contenido intestinal de un cuerpo arqueológico Incaico. Bol. Chil. Parasitol., *9*:73–75.

Poinar, G.O. 1979. *Nematodes for Biological Control of Insects*. Boca Raton, Fla., CRC Press, pp. 225–231.

Pradatsundarasar, A., Pecharanond, K., Chintanawongs, C., and Ungthavorn, P. 1973. The first case of intestinal capillariasis in Thailand. S.E. Asian J. Trop. Med. Public Health, *4*:131–134.

Prakitrittranon, W., Eua-ananta, Y., Dhiensiri, T., and Bhaibulaya, M. 1980. Intestinal capillariasis from Maha Sarakham province, northeast Thailand: Report of a case. Southeast Asian J. Trop. Med. Public Health, *11*:496–497.

Prost, E. 1959. Effectiveness of trichinoscopical examination in abattoirs. Wiadomosci Parazytol., *5*:297.

Rausch, R. 1953. Animal-borne diseases. Public Health Rep., *68*:533.

Roselle, H.A., Schwartz, D.T., and Geer, F.G. 1965. Trichinosis from New England bear meat. Report of an epidemic. N. Engl. J. Med., *272*:304–305.

Rugiero, H.R., Ramo, H.R., Tchoulamjan, A., Guinzburg, J., and Golera, N. 1966. Triquinosis. Ensayo y tratamiento con una nueva droga. Rev. Asoc. Med. Argent., *80*:301–306.

Schacher, J.F., and Faust, E.C. 1956. *Dioctophyma renale* in Louisiana. J. Parasitol., *42*:533–535.

Schantz, P., Juranek, D.D., and Schultz, M.G. 1977. Trichinosis in the United States, 1975: Increase in cases attributed to numerous common-source outbreaks. J. Infect. Dis., *136*:712–715.

Schenone, H., Borel, H., Yutronic, T., Norambuena, L., and Carrasco, L. 1967. Triquinosis mortal y teniasis multiple. Bol. Chil. Parasitol., *22*:32–37.

Schmitt, N., Saville, J.M., Greenway, J.A., Stovell, P.L., and Hole, L. 1978. Sylvatic trichinosis in British Columbia. Public Health Rep., *93*:189–193.

Sicard, D., Fontan, R., Richard-Lenoble, D., and Gentilini, M. 1976. Human trichinelliasis. A recent epidemic at Vientiane (Laos). Study of 32 cases. Bull. Soc. Pathol. Exot., *69*:521–525.

Silverman, N.H., Katz, J.S., and Levin, S.E. 1973. *Capillaria hepatica* infestation in a child. S. Afr. Med. J., *47*:219–221.

Singson, C.N., Banzon, T.C., and Cross, J.H. 1975. Mebendazole in the treatment of intestinal capillariasis. Am. J. Trop. Med. Hyg., *24*:932–934.

Skipina, L.V. 1976. A case of human tominxosis. Med. Parasitol. Parasit. Dis. (Moscow), *45*:607.

Skrjabin, K.I., Shikhobalova, N.P., and Orlov, L.V. 1957. *Trichocephalidae and Capillariidae of Animals and Man and the Diseases Caused by Them*. Jerusalem (1970), Israel Program for Scientific Translations, pp. 513–543.

Slais, J. 1973. The finding and identification of solitary *Capillaria hepatica* (Bancroft, 1893) in man from Europe. Folia Parasitol. (Praha), *20*:149–161.

Sumner, D., and Tinsley, E.G.F. 1967. Encephalopathy due to visceral larva migrans. J. Neurol. Neurosurg. Psychiatry, *30*:580–584.

Thienpont, D., Vanparijs, O.F., and Vandesteene, R. 1974. Anthelmintic and histopathological effects of mebendazole on *Trichinella spiralis* in the rat, pp. 515–527. In *Trichinellosis*. Proceedings of the 3rd International Conference on Trichinellosis. Edited by C.W. Kim. New York, Intext Educational Publishers.

Thorborg, N.C., Tulinius, S., and Roth, H. 1948. Trichinosis in Greenland. Acta Pathol., *25*:778–794.

Tidball, J.S., Aguas, J.P., and Aldis, J.W. 1978. A new concentration of human intestinal capillariasis on western Luzon. Southeast Asian J. Trop. Med. Public Health, *9*:33–40.

Turhan, B., Unat, E.K., Yenermen, M., and Sumer, C. 1954. Insan Karacigerinde *Capillaria hepatica* (Bancroft, 1893) Travassos, 1915. Mikrobiol. Dergisi, *7*:149–159.

Virchow, R. 1865. Zur Trichinen-Lehre. Arch. Pathol. Anat., *32*:332–371.

Von Brand, T., and Cullinan, R. 1943. Physiological observations upon a larval *Eustrongylides*. V. The behavior in abnormal warmblooded hosts. Proc. Helm. Soc., *10*:29–33.

Wagner, E.D. 1979. Observations on "large" *Trichuris* eggs in man. Proc. Helm. Soc. Wash., *46*:135–137.

———, and Peña Chavarria, A. 1974. Morphologicaly altered eggs of *Trichuris trichiura* following treatment with mebendazole. Am. J. Trop. Med. Hyg., *23*:154–157.

Ward, R.L., and Dent, J.H. 1959. *Capillaria hepatica* infection in a child. Bull. Tulane Univ. Med. Fac., *19*:27–33.

Watten, R.H., Beckner, W.M., Cross, J.H., Gunning, J.-J., and Jarimillo, J. 1972. Clinical studies of capillariasis philippinensis. Trans. R. Soc. Trop. Med. Hyg., *66*:828–834.

Whalen, G.E., Rosenberg, E.B., Gutman, R.A., Cross, J., Fresh, J.W., Strickland, T., and Uylangco, C. 1971. Treatment of intestinal capillariasis with thiabendazole, bithionol, and bephenium. Am. J. Trop. Med. Hyg., *20*:95–100.

Wheeldon, E., Kock, M., Tucker, R., and Anderson, M. 1982. Trichinellosis in a Cooper's hawk. J. Am. Vet. Med. Assoc., *181*:1385–1386.

Witenberg, G.G. 1964. Helminthozoonoses. 1.Nematodiasis, pp. 529–601. In *Zoonoses*. Edited by J. Van der Hoeden. Amsterdam, Elsevier Publishing Co.

Wright, K.A. 1961. Observations on the life cycle of *Capillaria hepatica* (Bancroft, 1893) with a description of the adult. Can. J. Zool., *38*:167–182.

———. 1968. Structure of the bacillary band of *Trichuris myocastoris*. J. Parasitol., *54*:1106–1110.

———. 1979. *Trichinella spiralis:* An intracellular parasite in the intestinal phase. J. Parasitol., *65*:441–445.

Zimmerman, W.J., Hubbard, E.D., and Mathews, J. 1958. Studies on fecal transmission of *Trichinella spiralis*. J. Parasitol., *44*(sect. 2):34.

Chapter 19

The Rhabditida: *Strongyloides* and Related Forms

SUPERFAMILY RHABDIASOIDEA*

Strongyloides stercoralis
(Bavay, 1876) Stiles and Hassall, 1902
(The threadworm, producing strongyloidiasis or strongyloidosis)

Synonyms. *Anguillula stercoralis* Bavay, 1876; *A. intestinalis* Bavay, 1876; *Strongyloides intestinalis* (Bavay, 1876) Grassi, 1879.

Historical Notes. The etiologic agent of strongyloidiasis was first found by Normand (1876) in the feces of French colonial troops who had been suffering from uncontrollable diarrhea in Cochin-China. Five patients died, and, at autopsy, minute roundworms of a species apparently different from those seen in feces were obtained from the wall of the ileum and biliary and pancreatic ducts. The stercoral worms were designated by Bavay as *Anguillula stercoralis* and the intestinal worms as *A. intestinalis*. Grassi (1879), Perroncito (1880), and Leuckart (1882) demonstrated that these two "species" constituted parts of the same life cycle, which was heterogonic, with both a parasitic and a free-living generation. Later it was discovered that there were two types of life cycles—one with a direct (homogonic) development, the other with an indirect (heterogonic) development. Leichtenstern (1899) defended the view that the former was a temperate-zone strain and the latter a tropical strain.

In 1900, Askanazy demonstrated that the parasitic females live in the intestinal wall and discharge their progeny in that location rather than in the lumen of the bowel. Closely parallel studies on the life cycle of the hookworm by Looss (1899 and 1905), Durme (1902), Ransom (1907) and Fülleborn (1914) showed that the infective-stage larvae of *Strongyloides* usually entered the body by way of the skin, passed through the bloodstream to the lungs, emerged into the air sacs, ascended the respiratory tree to the epiglottis, were swallowed and, on arrival in the intestinal tract, developed into the parasitic generation. Kreis (1932) reported parasitic male worms, and Faust (1933) described development and differentiation of the two sexes of the parasitic generation. These observations on parasitic males have not been confirmed. Faust (1933, 1935) confirmed and amplified Fülleborn's finding (1914) that, at times, female worms entered, matured in, and produced progeny in the respiratory epithelium.

Fülleborn (1914) suggested that rhabditoid larvae in moist fecal material lodged on the perianal skin might readily transform into infective-stage larvae and invade the patient at this site, thus producing *autoinfection*. Nishigori (1928) and Faust (1932–1936) demonstrated that there was an internal type of autoinfection in which rhabditoid larvae, while still in the bowel, transform into filariform larvae and enter the wall of the colon, so that they are able to reach the lungs by an internal route. Meanwhile, Sandground (1926) contributed important data on the biology of *Strongyloides* and confirmed Fülleborn's work on host resistance to reinfection. Graham (1936), who developed strains of *S. ratti* after exposing experimental rats to a single infective-stage larva, concluded from his comparison of homogonic and heterogonic lines that the former produces a predominantly direct type and the latter a predominantly indirect type of progeny.

Strongyloides fuelleborni von Linstow, 1905, a relatively common intestinal parasite of the chimpanzee and African baboon, may produce a transitory infection in man. Buckley (1946) reported this species to be widespread among the inhabitants of villages in Zimbabwe, possibly more common than *S. stercoralis*. Wallace et al. (1948) reported a case of human infection with *S. fuelleborni* contracted in the Philippines. In both animals and man, eggs, instead of larvae, are eliminated with the feces. *S. fuelleborni* is morphologically similar to *S. stercoralis* but may be distinguished from it by a peculiar post-vulvar constriction in the free-living female (Witenberg, 1964; Little, 1966) (see p. 262).

Strongyloides myopotami Artigas and Pacheco, 1933 of nutria and *S. procyonis* Little, 1965 of raccoon produced dermatitis in the form of creeping eruption when third-stage larvae were applied to the skin of a human volunteer (Little, 1965).

Geographical Distribution and Prevalence. *Strongyloides stercoralis* is primarily adapted to warm climates, but it has been reported sporadically in temperate and cold regions. In general, its distribution parallels that of the human hookworms but, in some regions, its prevalence is considerably less than that of the hookworms and it is common in some areas where the hookworm does not occur. Few basic surveys have been made to establish its actual distribution in warm climates. Infection rates up to 20% or more of surveyed populations have been reported. It was contracted by a considerable number of European prisoners of war in Burma, Thailand, and Malaysia during World War II.

Morphology, Biology, and Life Cycle. Like earlier workers, Galliard (1950, 1951) was impressed with the extraordinary variability and adaptability of *Strongyloides stercoralis* under different environmental conditions, which resulted in what he termed "a multiplicity of biotypes."

The Free-Living Generation. Larvae are passed in the feces of infected persons. Development takes

*For superfamily characteristics, see page 229.

place in the feces or in soil provided with an excess of fresh water, which is often too water-logged for development of hookworm larvae. It is the basic life cycle of the parasite and one that is almost constantly present in warm climates where moisture and abundant fecal material in the soil favor free-living existence and multiplication. The rhabditoid male (Fig. 19–1,*C*) is broadly fusiform, measuring 0.7 to 1.0 mm in length by 40 to 50 μm in greatest transverse diameter. There are no caudal alae, but two spicules and a gubernaculum are developed; the tail portion is pointed and curved ventrad. The rhabditoid female (Fig. 19–1,*B*) is also stout, measures approximately 1.0 to 1.7 mm by 50 to 75 μm, and has a two-horned uterus and a short vulva that opens near the middle of the ventral side. The developing eggs fill the uterus and occupy a major portion of the body. The fertilized female discharges partially embryonated eggs, which complete their development in a few hours and hatch. The first-stage *rhabditoid larvae* (Fig. 19–1,*D*) have a characteristic shape (up to 380 μm long by 20 μm wide) and a muscular esophageal structure diagnostic for the genus—namely, a club-shaped anterior portion, post-median constriction, and posterior bulbus. There is a relatively conspicuous genital primordium on the ventral side halfway down the midgut. The buccal cavity is short and of small diameter. This larva feeds on organic debris in feces or soil, molts once, feeds again, and grows through three additional molts into a free-living adult. Under optimal conditions, this free-living phase possibly may be repeated. Under certain conditions not yet fully de-

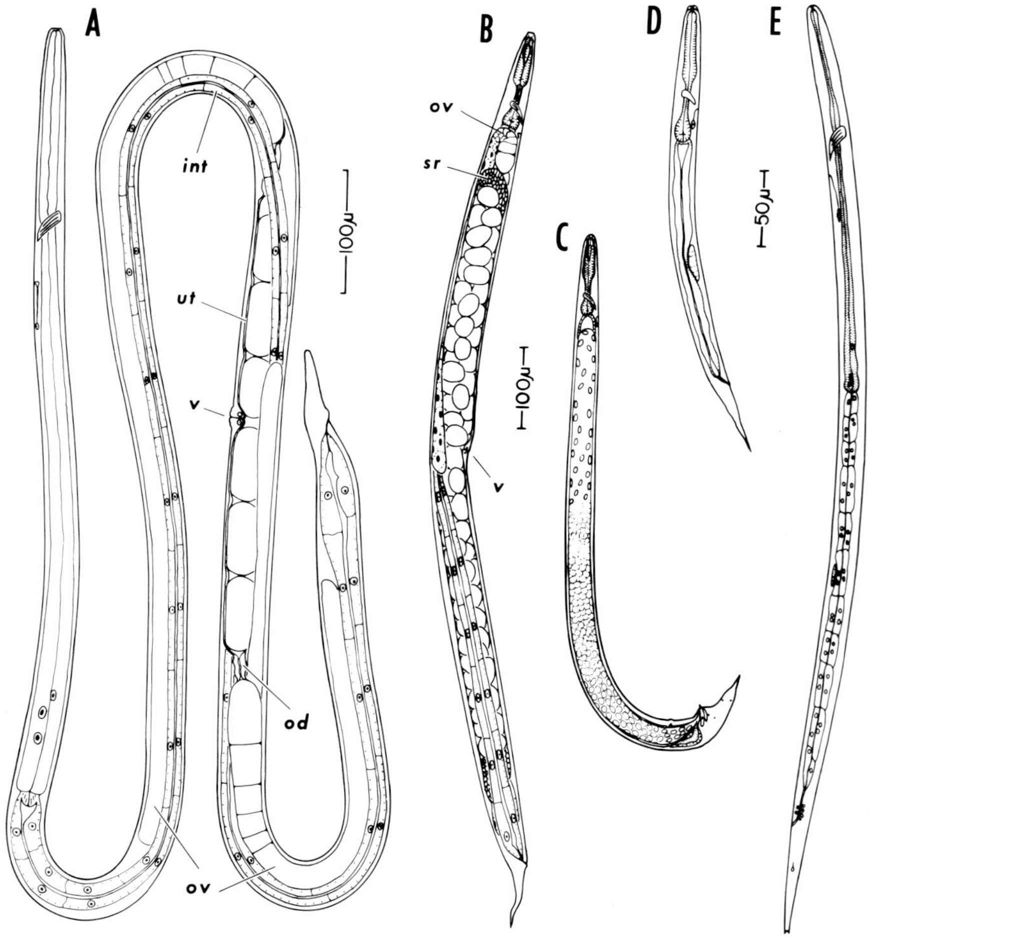

Fig. 19–1. *Strongyloides stercoralis*. *A*, Parasitic female. *B*, Free-living female. *C*, Free-living male. *D*, First-stage (rhabditoid) larva from fresh fecal specimen. *E*, Third-stage (filariform) larva from surface of fecal specimen after storage for 24 hours at room temperature. *int*, Intestine. *od*, Oviduct. *ov*, Ovary. *sr*, Seminal receptacle. *ut*, Uterus containing eggs. *v*, Vulva. (By Dr. M. D. Little. From Faust, E.C., Beaver, P.C., and Jung, R.C. 1975. *Animal Agents and Vectors of Human Disease*, 4th ed. Philadelphia, Lea & Febiger.)

fined, rhabditoid larvae metamorphose into the post-feeding infective-stage *filariform larvae* (Fig. 19–1,*E*), which are delicate and slender (up to 630 μm long by 16 μm wide), with a long esophagus and tail with a notched appearance. They are almost immediately infective and may remain alive in soil or water for several days. In water, they are able to swim, whereas hookworm larvae are not.

The Parasitic Generation. On contact with the skin of human beings, the filariform larvae penetrate to the small cutaneous blood vessels and are carried to the lungs. Here they break out of pulmonary capillaries into alveoli and migrate via the respiratory tree to the pharynx, then on to the intestine. Here they enter the mucosal epithelium, molt twice, and become mature females in about 2 weeks (Fig. 19–2). The usual location for development and egg-laying is the duodenum and upper jejunum. Rarely they invade the mucosa of the pylorus and stomach, and very rarely, except in immunosuppressed hosts, the larvae enter, mature, and produce eggs that hatch, all in the respiratory epithelium (Fig. 19–3). Little is known about the mechanism of entering and remaining within the mucosal or bronchial epithelium. It can be surmised that a burrowing motility serves that purpose in the adult worm, but it must be noted that the nonmobile egg, whose diameter is almost equal to that of the worm, also occupies this delicate, shallow habitat.

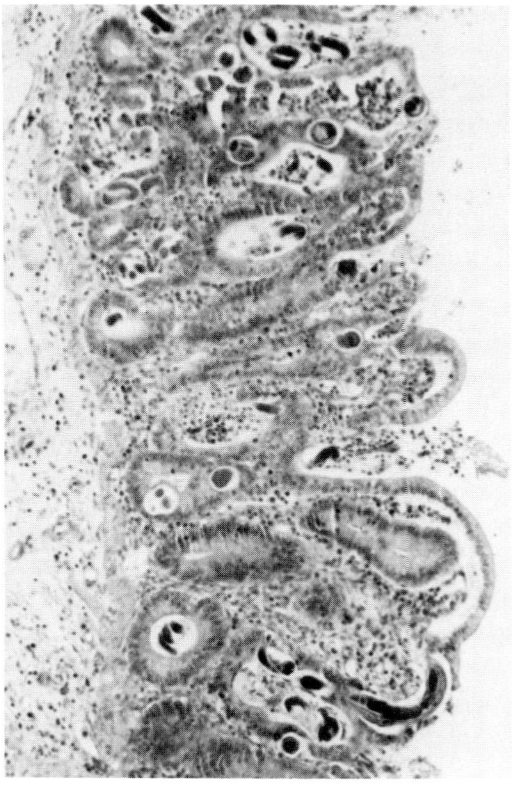

Fig. 19–2. *Strongyloides stercoralis*. Parasitic female and eggs in early and late stages of embryonation in intestinal epithelium, and rhabditoid larvae in lumen of glands. (× 100.) (By P.C. Beaver. Tissue courtesy of Dr. P. Hartz.)

Parasitic females (Fig. 19–1,*A*) are delicate, filiform worms, measuring up to 2.7 mm in length and 30 to 40 μm in width. The cylindrical esophagus extends through the anterior third or two-fifths of the body. The caudal extremity is pointed, and the anal opening is situated mid-ventrally, a short distance in front of the caudal tip. Paired uteri, oviducts, and ovarian tubules extend at right angles from the short vulva, one member of the pair going towards the anterior, the other toward the posterior end of the body. Upon maturing, females lay an undetermined number of partially embryonated eggs each day (Fig. 19–2). These thin-shelled, transparent, ovoid eggs, measuring 50 to 58 μm by 30 to 34 μm, usually hatch in the mucosal epithelium, and the escaping larvae migrate into the intestinal lumen and are discharged in the feces. The commonly parasitized levels of the bowel are duodenum and upper jejunum, but all levels have been found invaded from the pyloric wall of the stomach to the anus, including the appendix.

In cases of *autoinfection,* all or some of the rhabditoid larvae in the lumen of the intestine molt to the second rhabditoid stage, then metamorphose into infective filariform larvae *en transit* down the bowel (Fig. 19–4). They may produce reinfection by invading the mucosa of the lower portion of the ileum and the colon, travel via portal or accessory portal veins to the lungs, then return to the intestine and mature in the mucosa. At other times, filariform larvae are passed in freshly evacuated stools. Upon contact with the perianal and perineal skin, such larvae cause reinfection (Caplan, 1949; Napier, 1949) and, at times, have been associated with a radiating perianal "creeping eruption," which was described earlier by Fülleborn (1914). Less frequently, in patients with greatly lowered resistance or during immunosuppression, rhabditoid larvae in the bowel wall metamorphose into filariform larvae, may invade deeper layers, gain access to mesenteric venules, and initiate fatal reinfection by the internal route. Cases of human infection that have persisted for many years (during which time the patient has not had access to external sources of infection) are probably explained on the basis

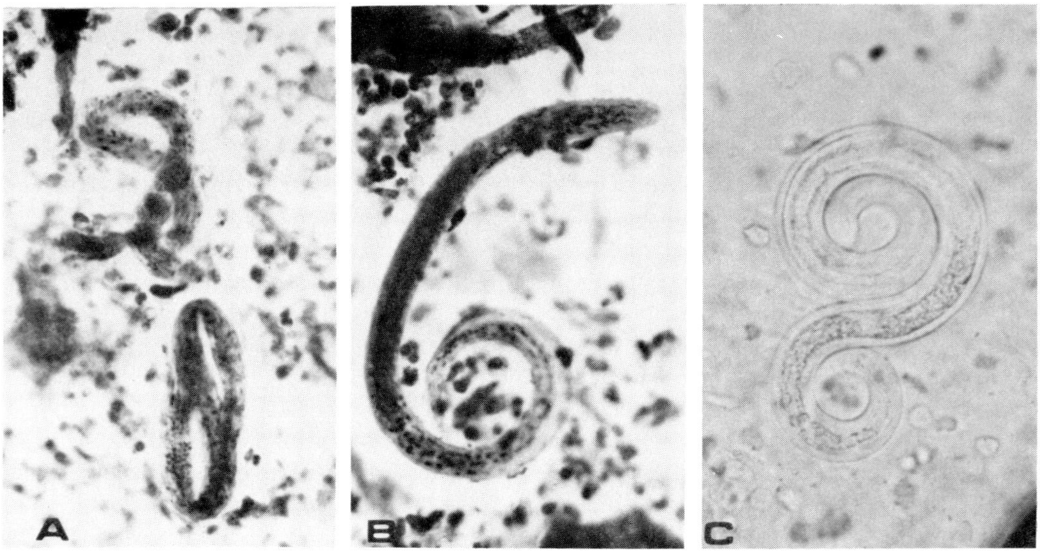

Fig. 19–3. *Strongyloides stercoralis.* Larvae in sputum of hyperinfected patients who were immunosuppressed. *A*. Rhabditoid larvae in egg. *B*, Advanced stage larva in Papanicolaou stained smear. (× 400.) *C*, Advanced rhabditoid larva in fresh smear. (X 400.) (By P.C. Beaver. Stained smear courtesy of Keith Humpherys. Fresh smear from Charity Hospital, New Orleans.)

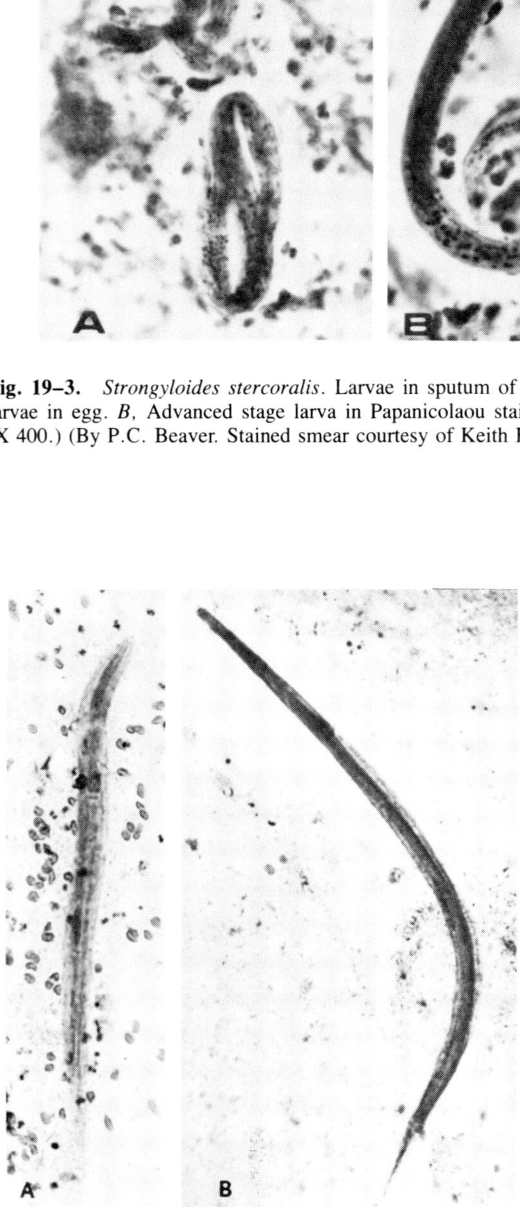

Fig. 19–4. *Strongyloides stercoralis.* Larvae in Giemsa-stained material aspirated from the colon in cases of dysentery, Durban, South Africa. *A*, Rhabditoid larva in the developmental stage usually seen in fresh fecal specimens. (× 185.) *B*, Filariform larva in early stage of development indicative of internal autoinfection. (× 185.) (By P.C. Beaver.)

of internal autoinfection, since it seems improbable that individual parasites survive for so long a period.

In summary, the complete life cycle of *Strongyloides stercoralis* consists of one or more of the following phases at the same or at different periods in development:

1. *Indirect development,* which is based essentially on a generation of free-living growth and transformation into the parasitic phase under certain environmental conditions that are primarily present in the moist tropics.
2. *Direct development* is primarily responsible for human infection, since it requires the parasitic phase for its continuation. It is the usual or predominant type in temperate climates.
3. *Autoinfection* provides filariform larvae for perianal infection or for internal reinfection without leaving the host.

Epidemiology. While man is the important host of *Strongyloides stercoralis,* dogs and cats have been found naturally infected with strains indistinguishable from those in man. The infective-stage larvae are developed in the soil or feces, from which man becomes infected typically through skin contact. Autoinfection may be responsible for strongyloidiasis in persons who have had no direct contact with infested soil for years.

In certain areas, strongyloidiasis tends to parallel hookworm infection both in geographic distribution and prevalence; in other areas, hookworm disease is highly prevalent and *Strongyloides* infection is relatively uncommon; in still other areas, as in tropical Brazil and Colombia, strongyloidiasis is the more prevalent and serious disease. Examination of about 1000 persons in a district of Cali, Colombia, detected *Strongyloides* infection in 14%; in males over 40 years of age, the rate was 30% (Faust and Giraldo, 1960). In Salvador-Bahia, Brazil, examination of 1000 patients from a middle-class general population revealed that 15% of the population were infected (Pedreira de Oliveira, 1978). In Surinam, a survey detected that 33% of approximately 400 inhabitants of an interior community were infected with *S. stercoralis* (this percentage includes all age groups); 6 of 36 persons less than 1 year old were infected (Terpstra, 1972). Strongyloidiasis tends to be particularly common in institutional groups, such as mental hospitals and prisons in warm, moist climates. It is more prevalent in adults than in children in the general population, but it has been reported as highly prevalent in homes for retarded children, even in temperate regions. Dogs are readily susceptible to infection with strains of *S. stercoralis* in humans and dogs, or humans can acquire infection through fecal contamination by infected dogs (Georgi and Sprinkle, 1974). The infection can be somewhat enzootic in urban localities where fecal contamination of soil by numerous dogs occurs daily (Jaskoski et al., 1982).

Zoonotic Strongyloidiasis. The various species of *Strongyloides* are morphologically similar in all of their developmental stages. Likewise, while the different species each have a characteristic host species, they are not strictly host specific. Therefore, species of *Strongyloides* whose natural hosts are various wild or domestic animals may at least invade the human skin, and some of them are known to produce patent infections in man. Since species diagnosis is difficult, some of the clinical and pathologic conditions attributed to *S. stercoralis* may have been caused by a different species.

Any of the numerous species that penetrate the skin may cause a dermatitis that is often linear and progressive, and thus referred to as "cutaneous larva migrans." Observations on a human volunteer showed that the larvae of *Strongyloides myopotami,* which lives in the nutria, and *Strongyloides procyonis,* which lives in the raccoon, readily penetrate the human skin (Little, 1965). In the former, inflammatory reaction with urticaria did not develop until after a number of re-exposures, and the lesions generally were more papular than linear; an intestinal infection did not develop (Fig. 19–5,*B*). The raccoon form produced linear lesions on first exposure, and a transient intestinal infection was detected by daily fecal cultures, which were positive on the thirteenth day (Fig. 19–5,*A*). Dermatitis acquired from wading in swampy coastal areas of Louisiana is common where the nutria and raccoon are numerous (Burks and Jung, 1960). Accidental exposures of laboratory workers resulting in papular and linear lesions caused by infective-stage larvae of *Strongyloides ransomi* of swine, of *Strongyloides papillosus* of sheep, goats, and cattle, and of *Strongyloides westeri* of horses have been reported (Roeckel and Lyons, 1977).

Pathogenesis, Pathology, and Symptomatology. In the great majority of cases of strongyloidiasis, the infection is benign, being discovered as an incidental finding. The only objectively discerned abnormality is usually peripheral eosinophilia. On the other hand, the disease may be severe.

The types of disease caused by strongyloidiasis may be classified as cutaneous, pulmonary, and intestinal.

Cutaneous. The initial penetration of the skin by filariform larvae of *Strongyloides* provokes little or no reaction, although, if the number of larvae is unusually large, there may be pruritus and erythema at the site of penetration. Sensitization of the host by a continued infection or repeated inoculations may be responsible for the development of an allergic response to the penetrating larvae. In a highly sensitized individual, the larvae may not be successful in promptly carrying out their migration to the lungs. Instead, they may migrate in the skin, causing a type of creeping eruption or *larva migrans*.

Larva Currens. In 1949, Caplan reported having seen, in a period of less than 3 years, several hundred cases of creeping eruption in ex-prisoners of war from the Far East. Without exception, larvae that were indistinguishable from those of *Strongyloides stercoralis* were found in the feces. In most cases, the creeping eruption lesion was first noticed near the anus, and it extended upward as far as the axillary folds and downward over the upper third of the thighs. An unusual feature of the linear urticarial lesion was its intermittent rapid progress of up to 10 cm per hour.

A similar type of creeping eruption had been described much earlier by Fülleborn (1926), who

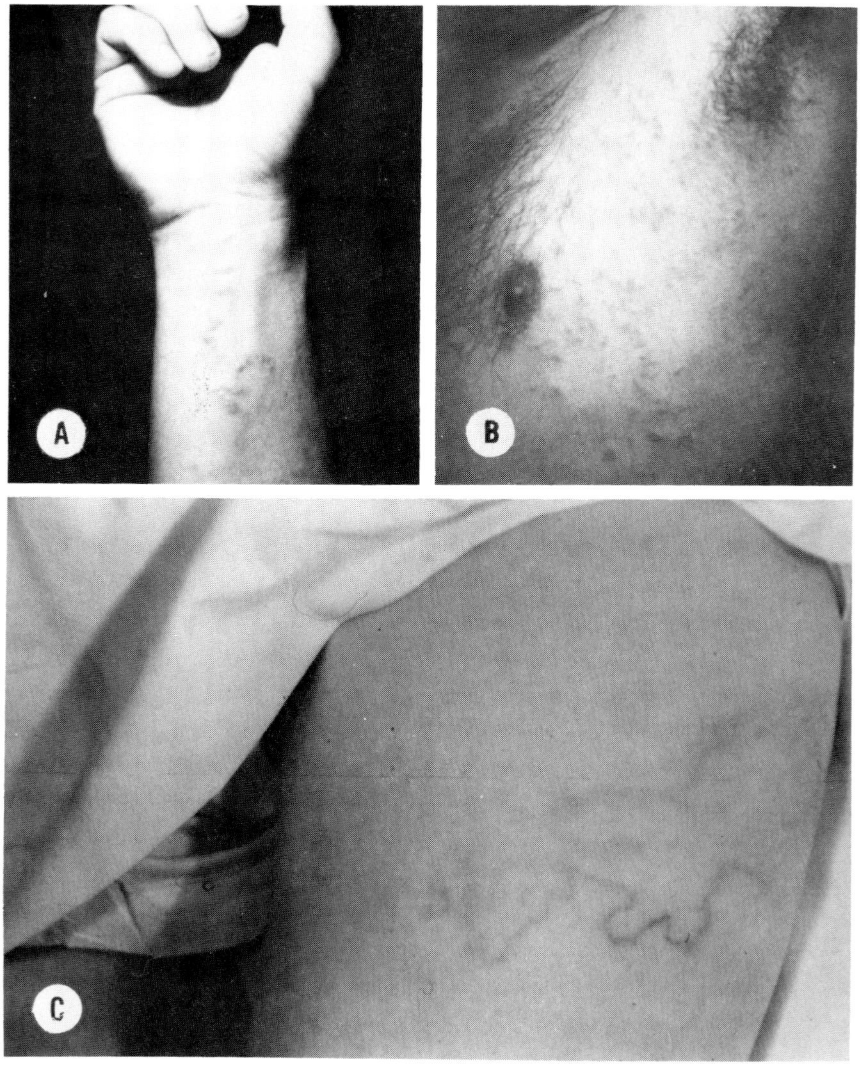

Fig. 19–5. *Strongyloides*, cutaneous lesions caused by *(A)* two migrating larvae of *S. procyonis* from raccoon, and *(B)* 30 larvae of *S. myopotami* from nutria in a volunteer. *C*, Reaction to rapidly moving larva referred to as "larva currens" on thigh of 44-year-old Texas oil-field worker whose stool and that of his wife contained larvae reported as *S. stercoralis*. (*A* and *B* from Little, M.D. 1965. Dermatitis in a human volunteer infected with *Strongyloides* of nutris and raccoon. Am. J. Trop. Med. Hyg., *14*:1007–1009. *C* from Stone, O.J., Newell, G.B., and Mullins, J.F. 1972. Cutaneous strongyloidiasis: Larva currens. Arch. Dermatol., *106*:734–736.)

demonstrated experimentally (on himself) that the causative organism was the infective-stage larva of *S. stercoralis*. Following Caplan's (1949) report, numerous other reports described large series of cases in which creeping eruption with rapidly progressing linear lesions was associated with *S. stercoralis* infection in the intestine.

The term *larva currens* (meaning "racing larva") was proposed by Arthur and Shelley (1958) who saw, in Philadelphia, a patient who had a single, rapidly progressing linear urticarial lesion that began on the thigh and was thought to be caused by a larval *Strongyloides*. The infection in that case, and in a similar one reported from Texas (Stone et al., 1972), may have been zoonotic. However, the term *larva currens* is now well established and is generally applied to all cases in which there is one or more rapidly progressing linear urticarial trails starting at or near the anus and known or presumed to be caused by a *Strongyloides* larva. Numerous such cases are still being seen in Great Britain (Gill and Bell, 1979, 1982), Australia (Grove, 1980), and France (Brumpt and Ho, 1973; Chestier et al., 1978). In essentially all such cases in Europe, the patients had been prisoners of war in 1942 to 1945 in the Far East. However, in a series of 295 cases of strongyloidiasis seen in Paris, only 5 had *larva currens,* and prisoner-of-war status was not characteristic of the group (Coulaud et al., 1980).

The lesion may take a very circuitous course or be relatively straight or interrupted. The onset is sudden, usually in the evening, and its progression is apparent within 1 hour. It is composed of two parts: the zone of progression, which is red, raised, tender, and 2 to 3 mm wide, and the older zone, which is less red, not as raised, and much wider (up to 12 mm) (Fig. 19–5,*C*). The lesions last 12 to 48 hours, then disappear without desquamation or pigmentation (Stone et al., 1972; Brumpt and Ho, 1973; Chestier et al., 1978).

It is possible that larva currens is caused by some species of *Strongyloides* other than *S. stercoralis,* such as *S. fuelleborni* in Southeast Asia or *S. procyonis* in the United States (Little, 1965; Pampiglione and Ricciardi, 1971; Hira and Patel, 1980).

Pulmonary Infection. Those migrating larvae that succeed in reaching the lungs break out of the pulmonary capillaries into the alveoli, causing small hemorrhages and cellular infiltration into the alveoli and bronchioles. Intensity of host-tissue reaction is proportional to the host's sensitization to the worms' products. A severe reaction may be manifested as pneumonia and may result in delay in the passage of the worms through the lungs and possibly in the death of some of them. During this phase of the infection, and also later, if there is endoautoinfection, larvae may be found in the patient's sputum.

Occasionally, larvae may lodge in the bronchial epithelium and develop to maturity there, producing a chronic bronchitis or asthmatic symptoms, with eggs and rhabditoid larvae in the sputum (Fig. 19–3) (Shiroma, 1964; Nwokolo and Imohiosen, 1973).

Intestinal Infection. In heavy infections, the mucosa may be honeycombed by the adult worms and larvae, and sloughing of extensive patches may occur. Usually there is remarkably little damage grossly evident in the segment of bowel mainly involved (duodenum and upper jejunum), even in heavy infections (Fig. 19–2).

The symptoms of intestinal infection usually resemble those of peptic ulcer. In a voluntary self-infection, Desportes (1944–1945) developed prodromal abdominal symptoms 20 days after skin inoculation that initially consisted of hunger pains, followed in a few days by profound lassitude and diffuse abdominal pain, which increased in frequency and intensity for up to 6 weeks, after which time it became less severe. Jones (1950) found that the most common complaint is abdominal pain that is rather sharply localized in the right upper abdominal quadrant. Milder et al. (1981) found that diarrhea and abdominal pain were equally common. Coulaud et al. (1980) reported that about 30% presented with abdominal pain and 13% with peptic ulcer symptoms. Radiographic findings may suggest Crohn's disease of the proximal small intestine (Brasitus et al., 1980).

Although chronic intestinal strongyloidiasis has been said to cause a malabsorption syndrome, Garcia et al. (1977) found that malabsorption in their group of patients was probably due to concomitant malnutrition rather than to the infection per se. In hyperinfection, intestinal malabsorption does occur (see below).

In chronic infection, the adult worms also may invade the stomach, causing gastritis and perhaps ulceration (Than, 1979). Granulomatous hepatitis due to the infection has also been reported (Poltera and Katsimbura, 1974).

The blood picture at the beginning of the intestinal phase is typically one of leukocytosis with a relative peripheral eosinophilia of 50 to 75% in some cases. In the chronic stage, lower degrees of eosinophilia are characteristic.

Strongyloidiasis tends to persist for many years. Of 125 former war prisoners of the Japanese during World War II, 11 (8.8%) still harbored *Strongyloides* in 1981 (Gill and Bell, 1981).

Hyperinfection. It is probable that autoinfection commonly occurs in human strongyloidiasis and that this phenomenon is responsible for the long duration of infections commonly seen (Faust and de Groat, 1940). The host's immune mechanisms and the parasite's reproductive mechanisms remain in balance so that neither is seriously affected. Under circumstances in which there is impaired cellular immunity, this balance is destroyed and the infection proliferates, along with the production of massive numbers of larvae that penetrate into all parts of the body. Such conditions include protein-calorie malnutrition, lepromatous leprosy, cancer, severe burns, irradiation sickness, and cirrhosis. Impairment of humoral immunity, such as with hypogammaglobulinemia, may also lead to hyperinfection and resistance to specific treatment (de Oliveira et al., 1981). In the past, such impairment of cellular immunity coexisting with strongyloidiasis was relatively uncommon and, correspondingly, hyperinfection was rare. Recently, due to the common use of adrenocorticosteroid drugs and other immunosuppressive agents, this condition is seen more commonly and is one of the important hazards of such treatment (Galliard, 1974; Purtilo et al., 1974; Humpherys and Hieger, 1979). Even local injection of corticosteroids carries this risk (West and Wilson, 1980). In such cases, peripheral eosinophilia tends to be absent, which is an unfavorable prognostic sign. Larvae develop into the filariform stage while still in the intestine, penetrate the intestinal wall, make the lung migration, and re-enter the intestine to develop into parasitic adult females. Ultimately, there are innumerable adult females occupying a much more extensive segment of bowel than they do in chronic infections.

Although there is impaired tissue reactivity, including diminished granuloma formation, the mechanical damage caused by the overwhelming number of worms leads to ulceration and necrosis of the small and large intestines, paralytic ileus, and even perforation. Migration through the lungs causes pneumonitis with dyspnea, cyanosis, and cough. Respiratory distress mimicking pulmonary edema is a common presentation (Weller et al., 1981). Nodular infiltration may be seen on chest x-ray (Venizelos et al., 1980) but, frequently, little evidence of infiltration is present on x-rays (Leelarasamee et al., 1978). Hemoptysis may occur. Involvement of the central nervous system leads to progressive lethargy, coma, and death (Meltzer et al., 1979). Meningitis, brain infarction, and brain abscess associated with disseminated strongyloidiasis have been reported (Masdeu et al., 1982).

If the patient does not die as a consequence of peritonitis due to a perforated bowel, or of brain damage or respiratory failure, the cause of death is likely to be overwhelming septicemia due to intestinal bacteria, which are probably introduced into the bloodstream by the migrating larvae (Leelarasamee et al., 1978). As in the case of overwhelming infections due to other agents, disseminated intravascular coagulation may occur.

The prognosis is extremely poor. Treatment with thiabendazole, which is highly effective in chronic intestinal strongyloidiasis, is relatively ineffective against massive hyperinfection (Dwork et al., 1975; Smith et al., 1977; Barr, 1978).

Diagnosis. Intestinal strongyloidiasis may be suspected on the basis of continued diarrhea with excess of mucus, pain in the upper abdomen, and high eosinophilia. Confirmation depends on demonstration of one or more stages of the parasite in specimens of feces, sputum, or duodenal aspirates.

Papanicolaou-stained smears of sputum or gastric aspirate (Yassin and Garret, 1980) may contain recognizable *S. stercoralis* eggs, rhabditoid larvae, filariform larvae, parasitic adult females (Humphreys and Hieger, 1979), eggs and larvae (Kenney and Webber, 1974; Wang et al., 1980), or larvae (Harris et al., 1980). The filariform larva may be identified in smears of ascitic fluid (Avagnina et al., 1980) and cerebrospinal fluid (Meltzer et al., 1979), as well as in feces and sputum. In cases of disseminated hyperinfection resulting from immunosuppression, larvae may be seen in sections of brain as well as in other tissues (Neefe et al., 1973) and must be distinguished from those of *Micronema, Toxocara,* and other types.

Recovery of active rhabditoid larvae (Fig. 19–1,*D* and 19–4,*A*) from the stool provides a ready diagnosis. Jones (1950) demonstrated larvae in only 27% of 952 stools of 100 *Strongyloides*-positive patients examined by direct film and zinc sulfate concentration techniques. In contrast, larvae were demonstrated in 71 patients subjected to duodenal aspiration. In their study of 67 patients in São Paulo, Brazil, Coutinho et al. (1954) diagnosed 57% by stool examination, an equal number by Baermann isolation, and only 36% by duodenal aspiration. They pointed out that duodenal drainage reveals only infection involving the duodenal region. In some cases, the presence and numbers of *Strongyloides* larvae in the feces vary remarkably from day to day.

If the stool has stood for some time at room temperature, *Strongyloides* larvae need to be differentiated from the first larval stage hatched from hookworm eggs. Less commonly, filariform larvae require specific identification (i.e., in cases of autoinfection); likewise, species of *Rhabditis* may contaminate the stool, or free-living adults may have developed in a stool that has stood for 24 hours. Fecal cultures made by methods described by Harada-Mori (Beaver et al., 1964), Dancescu (1968), and others occasionally may reveal infection when other methods fail, but they are most useful in distinguishing *S. stercoralis* infection from infection with other species (Technical Appendix, pp. 746 to 748).

Larvae may be found in unconcentrated fecal

films, but washing and concentrating the specimen by centrifugal sedimentation or centrifugal flotation with zinc sulfate solution of a specific gravity of 1.180 produces a richer yield. Some parasitologists let the fecal sample stand for a day or two in a stoppered bottle at room temperature, allowing viable larvae to migrate out of the stool and crawl toward the top of the container, where they may be seen with a hand lens. In pulmonary infections, the sputum will contain rhabditoid larvae, but rarely filariform larvae. Whitehall and Miller (1944) recovered larvae from the urine, while Redewill (1949) obtained eggs of the parasite from mucosal scrapings of eroded urethra in a female patient during cystoscopic examination. In cases with pleural effusion, the larvae may be found in the aspirate.

Although immunologic tests have not become available for routine clinical laboratory use, Kagan (1980) lists the indirect hemagglutination test as having been favorably evaluated for the diagnosis of strongyloidiasis; however, he notes that an antigen for the test is not available commercially in the U.S. or elsewhere. With antigen prepared from infective-stage larvae, Neva et al. (1981) found an enzyme-linked immunosorbent assay useful.

Radiologic studies in chronic intestinal strongyloidiasis show coarsening of the folds of the duodenum, which indicates edema of the mucosa or narrowing of the proximal small bowel with absence of the normal mucosal pattern, which is suggestive of Crohn's disease (Brasitus et al., 1980). In hyperinfection, one finds signs of thickening of the jejunal wall as well as signs of paralytic ileus and small intestinal and colonic ulceration. Occasionally, there is pulmonary infiltration.

Treatment. At present, the treatment of choice for chronic intestinal strongyloidiasis is with thiabendazole in a dosage of 25 mg per kilogram of body weight twice daily (maximum 3 g. per day) for 2 or 3 days (Pedreira, 1977, Schumaker et al., 1978). The same treatment is applicable to *larva currens* but is likely to require repetition. Side effects, including nausea, vomiting, and mental symptoms, are common and may require reduction of the daily dosage, with corresponding lengthening of the course. Estimated cure rates range from 55 to 100% (Grove, 1982).

Hyperinfection is preferably managed in an intensive care unit where changes in the patient's physiologic and chemical status can be constantly monitored and appropriate therapy administered. Specific treatment is with thiabendazole as described previously, but the drug should be given for at least 5 days. In patients with deficits in immune mechanisms, it may have to be given for much longer periods (de Oliveira et al., 1981). A three-day course of treatment with albendazole has been reported to provoke few side effects and to provide a cure rate of 77% (Coulaud et al., 1982).

Prognosis. Prognosis is generally good in chronic intestinal strongyloidiasis and *larva currens*. In the case of massive hyperinfection, the prognosis is extremely poor. In general, absence of eosinophilia is a poor prognostic sign.

Prevention. Care must be exercised to avoid contact with infective soil, long-voided feces, or contaminated surface water. All diagnosed cases should be treated specifically and, if necessary, repeatedly in order to minimize soil contamination and guard against internal or perianal autoinfection.

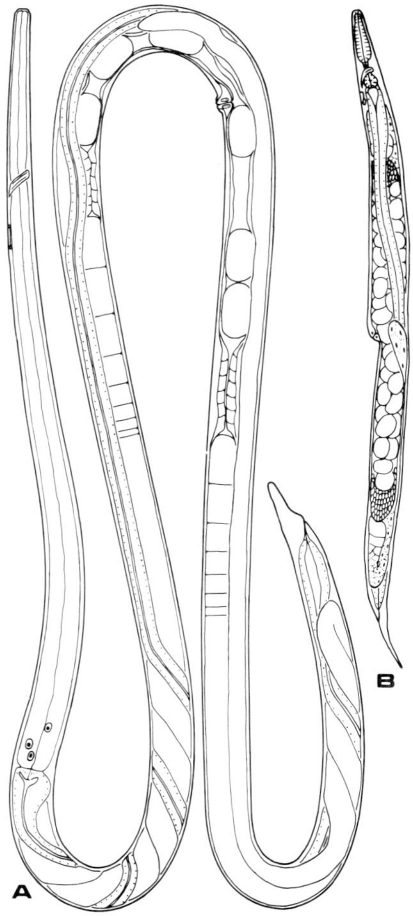

Fig. 19–6. *Strongyloides fuelleborni*. *A*, Parasitic female. *B*, Free-living female. (From Little, M.D. 1966. Comparative morphology of six species of *Strongyloides* of nutria and raccoon. Am. J. Trop. Med. Hyg., *14*:1007-1009.)

Before treatment with immunosuppressive drugs is undertaken, patients should be examined to rule out the presence of strongyloidiasis. This should be done even for patients who have never been out of the continental United States.

Strongyloides fuelleborni von Linstow, 1905

Strongyloides fuelleborni is a common parasite of Old World monkeys and apes. It occasionally is reported as a zoonotic infection in man, and the infection is endemic in some areas of Africa. In parts of Papua New Guinea, a *fuelleborni*-like species is endemic (Kelly et al., 1976; Ashford et al., 1981).

Strongyloides fuelleborni was first reported in man by Blackie (1932) in Southern Rhodesia (Zimbabwe). More recently, areas of endemic infection have been discovered in Zambia and northward in most of the central or middle African countries (Pampiglione and Ricciardi, 1972; Hira and Patel, 1980). *S. fuelleborni* and *S. stercoralis* occur together in some regions and occasionally are found together in the same person (Goldsmid, 1968). *S. fuelleborni* differs from *S. stercoralis* most markedly by shedding eggs rather than larvae in the feces. Eggs in fresh human feces usually are in early stages of cleavage and resemble those of hookworms, but they are distinctly shorter and somewhat smaller, measuring around 50 by 35 μm. The free-living female is markedly constricted posterior to the vulva, and the lips of the vulva are not prominent (Fig. 19–6,*B*); these features contrast with those of *S. stercoralis*. Also, the reflexed ovaries of the parasitic female are straight in *S. stercoralis*, spiraled around the intestine in *S. fuelleborni* (Little, 1966) (Figs. 19–6,*A* and 19–7).

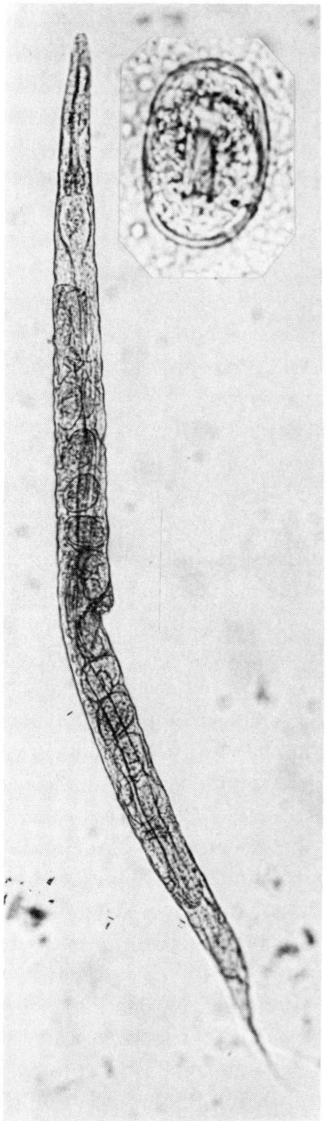

Fig. 19-7. *Strongyloides fuelleborni*. Free-living female with distinct post-vulvar constriction, from fecal culture. (× 140.) Inset, egg of parasitic female in fresh fecal smear. Unstained. (× 400.) (From Hira, P.R., and Patel, B.G. 1977. *Strongyloides fülleborni* infections in man in Zambia. Am. J. Trop. Med. Hyg., *26*:640–643.)

Transmammary transmission of *Strongyloides* larvae to human infants was first reported by Brown and Girardeau in 1977 in Zaire. They found *S. fuelleborni* infection in 34% of 76 infants under 200 days of age and in 44% of the general population. Examination of milk from nursing mothers revealed, in one case, three *Strongyloides* larvae in a 2-ml sample on the fourteenth postpartum day. This observation suggests that *S. stercoralis* may be similarly transmitted. The *fuelleborni*-like *Strongyloides* in Papua New Guinea causes an acute infection in infants. In one area, "swollen belly sickness" attributed to the infection is a common cause of death in infants (Vince et al., 1979). Since the infection is more commonly found in infants rather than in children or adults (massive infections occur as early as age 3 weeks), transmission through mothers' milk appears to be likely (Ashford et al., 1979). Transmammary infection has been reported for *S. ransomi* in swine (Moncol, 1975; Stewart et al., 1976), *S. papillosus* in sheep and cattle (Lyons et al., 1970; Nwaorgu, 1978), and *S. westeri* in horses (Lyons et al., 1973).

Free-Living Nematodes of the Superfamily Rhabditoidea

Nematodes of the order Rhabditida are small, free-living, have smooth or annulated cuticle, small pocket-like amphids, and an esophagus with a terminal bulb. In those species that are members of the suborder Rhabditina, superfamily Rhabditoidea, the stoma is an open tube, the excretory system has lateral canals, and the esophagus consists of a cylindrical anterior part containing no valve plates, a slender isthmus, and a terminal bulb that does contain valve plates. In the family Rhabditidae, which includes the genera *Rhabditis* and *Diploscapter*, the stoma is cylindrical, with 3 or 6 lips (in *Rhabditis*), or replaced by paired oral hooks (in *Diploscapter*); the ovaries are usually paired. In the Panagrolaimidae, which includes the genera *Micronema* and *Turbatrix*, the wall of the buccal cavity has three thickened bands, and a posterior portion is surrounded by the anterior end of the esophagus; the anterior part of the esophagus is cylindrical or has a swollen, valveless median bulb; the isthmus is narrow; there is a valvular apparatus in the bulbus; the ovary is prodelphic and reflexed, extending some distance towards the vulva; the postvulval sac is sometimes present; and in the male, if present, the testis usually is reflexed and the tail, without a bursa, bears 5 to 7 pairs of caudal papillae.

Rhabditis Dujardin, 1845

Species of *Rhabditis*, of which many have been described, resemble the free-living forms of *Strongyloides* sufficiently closely to require careful study in order to make a differential diagnosis. Their life cycles are typical of a free-living rhabditoid nematode. The males fertilize the females, which lay eggs that hatch and liberate rhabditoid larvae. These feed, grow, molt, and develop into free-living adults. There is no filariform larval stage. In the early rhabditoid stage, it is at times very difficult to distinguish these forms from those of *Strongyloides*. In the larger rhabditoid larvae and in adults, the longer, more thickly cuticularized buccal chamber, with a distinct annulus at the inner end and the bulbar swelling in the middle of the anterior club-shaped portion of the esophagus, are diagnostic of the genus (Fig. 19–8,*D*). In some species, there is a long filiform caudal appendage; in the males of certain species, caudal alae, at times supported by ribs, serve to distinguish these forms.

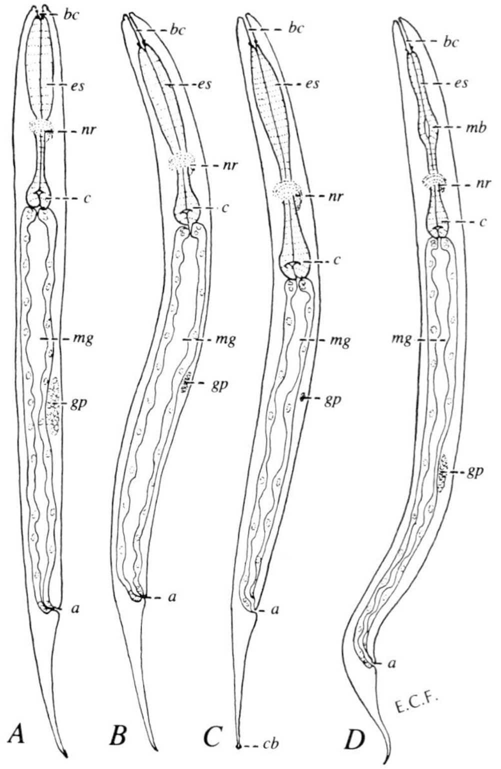

Fig. 19–8. Rhabditoid larvae of *A*, *Strongyloides;* *B*, hookworm; *C*, *Trichostrongylus;* and *D*, *Rhabditis*.(By E.C. Faust.)

Rhabditis species typically live in decaying organic debris but are able to accommodate themselves temporarily to ulcerated or damaged human tissues or to the lumen of various organs. Thus, *Rhabditis hominis*, *R. donbass*, *R. schachtiella*, and *R. gracilis* have been found on various occasions living and producing progeny in freshly passed human feces. There is no evidence, however, that these worms reside for any length of time in the human bowel; they are merely contaminators of food or drink. *R. pellio* has been found living in the vagina of women whose acid urine contained the larval stage, together with albumin, pus, and red blood cells. More nearly parasitic in its habits is *R. niellyi*, the larvae of which Nielly (1882) found on one occasion in itching cutaneous papules in a French youth. The lesions grossly resembled "craw-craw" of West Africa. Feng and Li (1950) diagnosed *Rhabditella axei* from the urine of two Chinese patients.

Pelodera Schneider, 1866

The genus *Pelodera*, earlier regarded as a synonym of *Rhabditis*, is now recognized as distinct (Anderson and Bain 1982). In *rhabditis*, the lips are closed, an esophageal collar is present, and the male spicules are separate. In species of *Pelodera*, the lips are distinct and open, the female tail is conical or dome-shaped with a short spine, and the male spicules are usually fused at the tip. In the dauer larval stage, a species identified as *Pelodera strongyloides* is commonly found in the orbit of burrowing wild rodents. The same species has been reported as colonizing in skin lesions of dogs, horses, cattle, and sheep (Levine, 1980). Human infection has been reported once, in an 11-year-old girl in Poland (Pasyk, 1978).

Diploscapter Cobb, 1913

Another rhabditoid nematode, *Diploscapter coronata*, which typically lives on decaying plants or other putrefying organic matter in the soil or in sewage beds, has been reported as a facultative parasite of man. *D. coronata* resembles *Rhabditis* in having a long capillary pre-esophageal chamber, a slight (if any) median esophageal swelling, and lips (the dorsal and ventral ones modified to form strong, outwardly directed, curved hooks with two points, and the lateral lips extend forward) (Goodey, 1963). Males are rarely found, and parthenogenetic propagation is probably the rule. S. Yokogawa (1936) found *D. coronata* in the urine of an elderly Japanese female, Witenberg (1951) reported three cases of urinary infection in Israel, and Chandler (1938) recovered living specimens of this species from the stomach contents of nine patients who were suffering from achlorhydria. In these instances, the worms must have been acquired from decaying vegetation.

Micronema Körner, 1954

Two fatal human infections with a saprophagous nematode identified as *Micronema deletrix* have been reported; the first

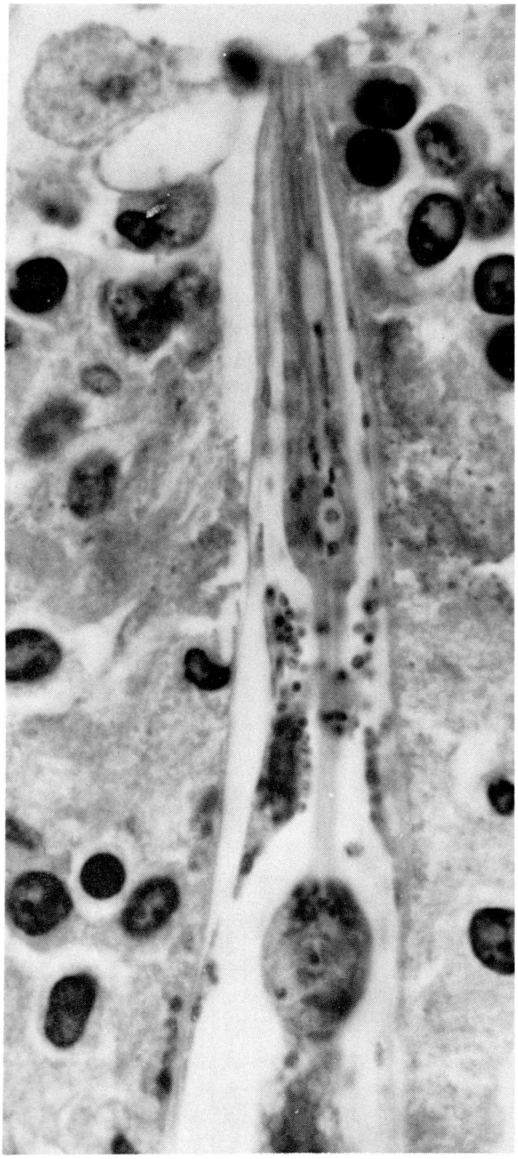

Fig. 19–9. *Micronema.* Anterior end of adult female identified as *M. deletrix* in section of brain of 5-year-old boy who died with meningoencephalitis 24 days after receiving multiple deep lacerations in a farm accident. Section shows typical stoma and esophagus. (× 1536.) (From Hoogstraten, J., and Young, W.G. 1975. Meningo-encephalomyelitis due to the saprophagous nematode, *Micronema deletrix*. Can. J. Neurol. Sci., 2:121–126.)

in a 5-year-old boy in south central Canada (Hoogstraten and Young, 1975), the second in a 47-year-old man in Texas (United States) (Shadduck et al., 1979). A third case, in which the invading nematode was identified only as *Micronema*, was reported from Washington, D.C. (Gardiner et al., 1981). In all cases, the infection produced a fatal meningoencephalitis, and the causative organisms were not discovered and identified until tissues were examined postmortem. In the first case, signs of central nervous system involvement began to develop 18 days after an unusual farm accident in which the child passed through the mechanism of a manure spreader, sustaining multiple deep lacerations of the face and other parts of the body, and penetrating wounds of the chest and abdomen. It was assumed that the worms invading the central nervous system had migrated from one or more of the lacerations. At autopsy, 24 days after the accident, innumerable worms in various stages of development were found in sections of tissues from all parts of the brain and spinal cord. Juvenile and larval worms seen in small numbers in the lungs were thought to have come from the brain or spinal cord. In the second case, the patient died 19 days after admission to hospital with a diagnosis of meningoencephalitis. The origin of the *Micronema* infection was not determined. Mature and larval worms were numerous in grossly visible lesions in all parts of the brain, particularly in the pons and upper medullary portions of the spinal cord. In the third case, the infection was thought to have spread to the brain and other organs after colonization in decubitus ulcers.

Inflammatory foci described as consisting of lymphocytes, swollen microglial cells or macrophages, and rare multinucleated giant cells were seen scattered throughout the leptomeninges, grey and white matter of the cerebral hemispheres, pons, medulla, midbrain, cerebellum, and spinal cord. In all cases, neutrophils and eosinophils were notably scarce. It was apparent that the parasites were reproducing in the tissues because mature and immature females, a few eggs, and countless numbers of larvae were present. Male worms were not seen in any case. The largest adult females were described as approximately 250 μm long in the first case, 300 to 380 μm long by 14 to 17 μm in the second, and up to 460 μm in the third. The buccal structures and esophagus of adult and larval forms were characteristically prominent, with a long, heavy-walled buccal cavity, a distinct procorpus that was somewhat swollen at the base to form a metacorpus, a relatively long slender isthmus, and a large pyriform bulbus containing well-formed valve plates (Fig. 19–9). Eggs in the uterus measured 10 to 15 μm by 30 to 41 μm and, in some instances, contained larvae, although the worms evidently were oviparous.

Although the worms in two cases were reported as *Micronema deletrix* Anderson and Bemrick, 1965, the identifications may have been somewhat presumptive. *M. deletrix* was described from a collection of many thousands of worms taken from bilateral tumors in the nares of a horse in Minnesota. A similar infection was reported in the upper jaw and maxillary sinuses of another horse in Minnesota (Johnson and Johnson, 1966). In two other *Micronema* infections in horses in Georgia (Stone et al., 1970) and Alabama (Powers and Benz, 1977), the worms invaded and reproduced in the central nervous system as in the human cases, and in a Florida horse, the brain and kidneys were the sites of infection (Rubin and Woodard, 1974). In all of these American reports and in one describing infection in the brain of a horse in Egypt (Ferris et al., 1972), the worms were identified as *Micronema deletrix*. Whether they differ from those reported as *Rhabditis gingivalis* Stephanski, 1953 in the jaw and gums of a horse in Poland (Stefanski, 1954) is uncertain. In another horse in Europe, unidentified rhabditoid nematodes were observed in granulomatous inflammatory lesions of the maxillae and nasal tissues and in the brain and other tissues (Linde-Sipman and Gruys, 1970).

Turbatrix aceti (Müller, 1783) Peters, 1927

The "vinegar eel" (*Turbatrix aceti*) has been recovered on several occasions from the urine or vaginal discharge of women who had introduced the living microscopic worms into the vagina when vinegar was used as a douche. Caballero (1937) recovered mature worms of both sexes as well as larvae of this worm from the urine of a woman in Mexico. Such findings have no known clinical significance.

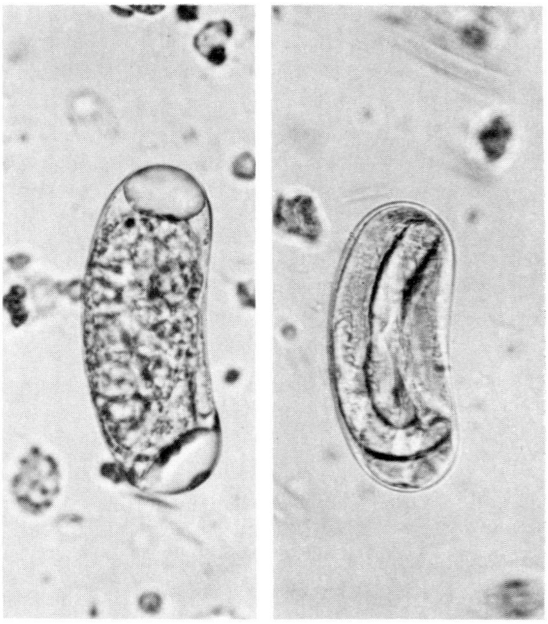

Fig. 19–10. *Heterodera- or Meloidogyne-* type eggs from human stool. Unstained. (× 400.) (By P.C. Beaver.)

Plant-Parasitic Nematodes of the Superfamily Tylenchoidea

Nematodes of the order Tylenchida have a protrusible spear in the stoma, a single lateral excretory canal, and an esophagus that consists of a procorpus, a corpus with crescentic valve plates, an isthmus, and a glandular bulbus. In species of the family Heteroderidae, the mature females are swollen to subspherical shape with two ovaries, and the vulva is usually subterminal; the males are vermiform. In species of the family Tylenchidae, the median bulb (corpus) of the esophagus is only vaguely evident; in the male, there is an adanal or subterminal bursa.

On one occasion, the stem or bulb eelworm, *Ditylenchus dipsaci,* a common parasite of onion bulbs, was reported from the vomitus of a patient who had eaten onions (Botkin, 1883).

Another plant parasite, *Meloidogyne javanica* (syn. *M. radicicola, Heterodera incognita, H. marioni, H. radicicola),* is commonly found in the roots and stems of many edible plants. When such infected plant tissues are eaten, the eggs and larvae are, at times, recovered from the feces. These eggs (Fig. 19–10) are thin-shelled, hyaline, elongate ovoid in shape with rounded ends, and may be slightly flattened on one side. They measure 82 to 120 μm by 24 to 43 μm and may be in the early stages of cleavage or may contain a fully developed, motile larva.

Eggs of *Heterodera radicicola* and related species of plant parasites have been mistaken for infertile *Ascaris* or hookworm eggs; such mistakes have resulted in unnecessary anthelmintic treatment (Keller, 1935). Eggs of this type (Fig. 19–9) were observed in a large number of soldiers and, since they were presumed to be those of an unknown species of *Oxyuris,* were given the name *Oxyuris incognita* (Kofoid and White, 1919). It was soon recognized that these were the eggs of *Heterodera* and related forms of plant nematode parasites (Sandground, 1923; Bacigalupo, 1936).

REFERENCES

Adam, M., Morgan, O., Persaud, C., and Gibbs, W. 1973. Hyperinfection syndrome with *Strongyloides stercoralis* in malignant lymphoma. Br. Med. J., *1*:264–266.

Anderson, R.C., and Bain, O. 1982. Keys to genera of the superfamilies Rhabditoidea, Dioctophymatoidea, Trichinelloidea and Muspiceoidea. In *CIH Keys to the Nematode Parasites of Vertebrates*. Edited by R.C. Anderson, A.G. Chabaud, and S. Willmott. Farnham Royal, England, Commonwealth Agricultural Bureaux.

Anderson, R.V., and Bemrick, W.J. 1965. *Micronema deletrix* n. sp., a saprophagous nematode inhabiting a nasal tumor of a horse. Proc. Helm. Soc. Wash., *32*:74–75.

Arthur, R.P., and Shelley, W. 1958. Larva currens. A distinctive variant of cutaneous larva migrans due to *Strongyloides stercoralis*. A.M.A. Arch. Dermatol., *78*:186–190.

Ashford, R.W., Hall, A.J., and Babona, D. 1981. Distribution and abundance of intestinal helminths in man in western Papua New Guinea with special reference to *Strongyloides*. Ann. Trop. Med. Parasitol., *75*:269–279.

———, Vince, J.D., Gratten, M.A., and Bana-Koiri, J. 1979. *Strongyloides* infection in a mid-mountain Papua New Guinea community. Papua New Guinea Med. J., *22*:128–135.

Avagnina, M.A., Elsner, B., Iotti, R.M., and Re, R. 1980. *Strongyloides stercoralis* in Papanicolaou-stained smears of ascitic fluid. Acta Cytol., *24*:36–39.

Bacigalupo, J. 1936. Presencia de huevos *Heterodera radicicola (Oxiurus incognita)* en las materias fecales humanas. Semana Med., no. 6.

Barr, J.R. 1978. *Strongyloides stercoralis*. Can. Med. Assoc. J., *118*:933–935.

Beaver, P.C., Malek, E.A., and Little, M.D. 1964. Development of *Spirometra* and *Paragonimus* eggs in Harada-Mori cultures. J. Parasitol., *50*:664–666.

Blackie, W.K. 1932. A helminthological survey of Southern Rhodesia. Med. Mem. Ser. No. 5, London School of Hygiene and Tropical Medicine.

Botkin, S.P. 1883. (Quoted in Braun-Siefert's *Die Tierischen Parasiten des Menschen,* Leipzig, Ier Teil., p. 329, 1925.)

Brasitus T.A., Gold, R.P., Kay, R.H., Magun, A.M., and Lee, W.M. 1980. Intestinal strongyloidiasis: A case report and review of the literature. Am. J. Gastroenterol., *73*:65–69.

Brown, R.C., and Girardeau, M.H.F. 1977. Transmammary passage of *Strongyloides* sp. larvae in the human host. Am. J. Trop. Med. Hyg., *26*:215–219.

Brumpt, L.C., and Ho, T.S. 1973. *Larva currens* seul signe pathognomonique de la strongyloidose. Ann. Parasitol., *48*:319–328.

Buckley, J.J.C. 1946. A helminthological survey in Northern Rhodesia. J. Helminthol., *21*:111–174.

Burks, J.W., and Jung, R.C. 1960. A new type of water dermatitis in Louisiana. South. Med. J., *53*:716–719.

Caballero, E. 1937. Un caso de parasitismo accidental por *Rhabditis pellio* en Mexico. An. Inst. Biol., Univ. Nac. Mex., *7*:373–384.

Caplan, J.P. 1949. Creeping eruption and intestinal strongyloidiasis. Br. Med. J., *1*:396.

Chandler, A.C. 1938. *Diploscapter coronata* as a facultative parasite of man, with a general review of vertebrate parasitism by rhabditoid worms. Parasitology, *30*:44–45.

Chestier, Y., Barbier, D., Leroy, P., Rozin, G., Houin, R., and Beaumont, J.-L. 1978. Syndrome de larva currens tardif chez un homme de 46 ans, 22 ans après le dernier séjour en zone endémique. Ann. Med. Interne, *129*:43–47.

Coulaud, J.P., Deluol, A.M., Cenac, J., and Rossignol, J.F. 1982. L'Albendazole dans le traitement de la strongyloidose. A propos de 66 observations. Bull. Soc. Path. Ex., *75*:530–533.

———, Le Mercer, Y., Tessier, S., and Mechali, D. 1980. Analyse épidémiologique, clinique et thérapeutique de 427 cas d'anguillulosea observés a Paris. Bull. Soc. Pathol. Exot., *73*:100–108.

Coutinho, J.O., Croce, J., Campos, R., Neto, V.A., and Fonseca, L.C. 1953–1954. Contribuição para o conhecimento da estrongiloidiase humana em São Paulo. Folia Clin. Biol., *20*:141–176; *21*:19–48, 94–120.

Dancescu, P. 1968. Investigations on the intensity of infection in a strongyloidiasis focus. The coal culture method. Trans. R. Soc. Trop. Med. Hyg., *62*:490–495.

de Oliveira, R.B. de, Voltarelli, J.C., and Meneghelli, U.G. 1981. Severe strongyloidiasis associated with hypergammaglobulinaemia. Parasite Immunol., *3*:165–169.

Desportes, C. 1944–1945. Sur *Strongyloides stercoralis* (Bavay, 1876) et sur les *Strongyloides* de primates. Ann. Parasitol., *20*:160–190.

Dwork, K., Jaffe, J.R., and Lieberman, H.D. 1975. Strongyloidiasis with massive hyperinfection. N.Y. State J. Med., *75*:1230–1233.

Faust, E.C. 1933. The development of *Strongyloides* in the experimental host. Am. J. Hyg., *18*:114–132.

———. 1935. The pathology of *Strongyloides* infection. Arch. Pathol., *19*:769–806.

———, and De Groat, A. 1940. Internal autoinfection in strongyloidiasis. Am. J. Trop. Med., *20*:350–375.

———, and Giraldo, L.E. 1960. Parasitological surveys in Cali, Departamento del Valle, Colombia. VI. Strongyloidiasis in Barrio Siloé, Cali, Colombia. Trans. R. Soc. Trop. Med. Hyg., *54*:556–564.

Feng, L.C., and Li, F. 1950. Two human cases of urinary infection with *Rhabditella axei* (Cobbold, 1884) Chitwood, 1933. Peking Nat. Hist. Bull., *18*:195–202.

Ferris, D.H., Levine, N.D., and Beamer, P.D. 1972. *Micronema deletrix* in equine brain. Am. J. Vet. Res., *33*:33–38.

Fülleborn, F. 1914. Untersuchungen über den Infektionsweg bei *Strongyloides* und *Ankylostomum* und die Biologie dieser Parasiten. Arch. Schiffs-u. Tropenhyg., *5*:26–80.

———. 1926. Experimentell erzeugte "creeping eruption." Dermat. Wochenschr., *83*:1474–1475.

Galliard, H. 1950–1951. Recherches sur l'infestation experimentale à *Strongyloides stercoralis* au Tonkin. Ann. Parasitol., *25*:441–473; *26*:67–84, 201–207.

———. 1974. Accion de los corticoides y los immunosupresores sobre las infecciones de *Strongyloides stercoralis*. Rev. Cub. Med. Trop., *26*:99–102.

Garcia, F.T., Sessions, J.T., Strum, W.B., Schweistris, E., Tripathy, K., Bolaños, O., Lotero, H., Duque, E., Ramelli, D., and Mayoral, L.G. 1977. Intestinal function and morphology in strongyloidiasis. Am. J. Trop. Med. Hyg., *26*:859–865.

Gardiner, C.H., Koh, D.S., and Cardella, T.A. 1981. *Micronema* in man: Third fatal infection. Am. J. Trop. Med. Hyg., *30*:586–589.

Georgi, J.R., and Sprinkle, C.L. 1974. A case of human strongyloidosis apparently contracted from asymptomatic colony dogs. Am. J. Trop. Med. Hyg., *23*:899–901.

Gill, G.V., and Bell, D.R. 1979. *Strongyloides stercoralis* infection in former Far East prisoners of war. Br. Med. J., *2*:572–574.

———, and Bell, D.R. 1982. Longstanding tropical infections amongst former war prisoners of the Japanese. Lancet, *1*:958–959.

Goldsmid, J.M. 1968. Studies on intestinal helminths in African patients at Harari Central Hospital, Rhodesia. Trans. R. Soc. Trop. Med. Hyg., *62*:619–628.

Goodey, T. 1963. *Soil and Freshwater Nematodes*, 2nd ed. London, Methuen & Co.

Graham, G.L. 1936. Studies on *Strongyloides*. I. *S. ratti* in parasitic series, each generation in the rat established with a single homogonic larva. Am. J. Hyg., *24*:71–87.

Grove, D.I. 1980. Strongyloidiasis in allied ex-prisoners of war in south-east Asia. Br. Med. J., *1*:598–600.

———. 1982. Treatment of strongyloidiasis with thiabendazole: an analysis of toxicity and effectiveness. Trans. R. Soc. Trop. Med. Hyg., *76*:114–118.

Harris, R.A., Musher, D.M., Fainstein, V., Young, E.J., and Clarridge, J. 1980. Disseminated strongyloidiasis: Diagnosis made by sputum examination. J.A.M.A., *244*:65–66.

Hira, P.R., and Patel, B.G. 1977. *Strongyloides fülleborni* infections in man in Zambia. Am. J. Trop. Med. Hyg., *26*:640–643.

———. 1980. Human strongyloidiasis due to the primate species *Strongyloides fülleborni*. Trop. Geogr. Med., *32*:23–29.

Hoogstraten, J., and Young, W.G. 1975. Meningo-encephalomyelitis due to the saprophagous nematode, *Micronema deletrix*. Can. J. Neurol. Sci., *2*:121–126.

Humpherys, K., and Hieger, L.R. 1979. *Strongyloides stercoralis* in routine Papanicolaou-stained sputum smears. Acta Cytol., *23*:471–476.

Jaskoski, B.J., Barr, V., and Borges, M. 1982. Intestinal parasites of well cared for dogs: An area revisited. Am. J. Trop. Med. Hyg., *31*:1107–1110.

Johnson, K.H., and Johnson, D.W. 1966. Granulomas associated with *Micronema deletrix* in the maxillae of a horse. J. Am. Vet. Med. Assoc., *149*:155–159.

Jones, C.A. 1950. Clinical studies in human strongyloidiasis. I. Semeiology. Gastroenterology, *16*:743–756.

Kagan, I.G. 1980. Serodiagnosis of parasitic diseases, Pages 724–750. In *Manual of Clinical Microbiology*. E.H. Lennette, A. Balows, W.J. Hausler, and J.P. Truant, eds. Washington, D.C., American Society for Microbiology.

Keller, A.E. 1935. The occurrence of eggs of *Heterodera radicicola* in human feces. J. Lab. Clin. Med., *20*:390–392.

Kelly, A., Little, M.D., and Voge, M. 1976. *Strongyloides fulleborni*-like infections in man in Papua New Guinea. Am. J. Trop. Med. Hyg., *25*:694–699.

Kenney, M., and Webber, C.A. 1974. Diagnosis of strongyloidiasis in Papanicolaou-stained sputum smears. Acta Cytol., *18*:270–273.

Kofoid, C.A., and White, A.W. 1919. A new nematode infection of man. J.A.M.A., *72*:567–569.

Kreis, H.A. 1932. Studies on the genus *Strongyloides* (Nematodes). Am. J. Hyg., *16*:450–491.

Leelarasamee, A., Nimmannit, S., Nakorn, S.N., Aswapokee, N., Aswapokee, P., and Benjasuratwong, Y. 1978. Disseminated strongyloidiasis: Report of seven cases. Southeast Asian J. Trop. Med. Public Health, *9*:539–542.

Leichtenstern, O. 1899. Zur Lebensgeschichte der *Anguillula intestinalis*. Cbl. Bakt., *2*:226–231.

Leuckart, R. 1882. Ueber die Lebensgeschichte der sogennanten *A. stercoralis* und deren Beziehungen zu der sogennanten *A. intestinalis*. Bericht. Verhandl. Sachs. Gesellsch. Wissensch., Math.-phys. Klasse, Leipzig, *34*:85–107.

Levine, N.D. 1980. *Nematode Parasites of Domestic Animals and of Man*. Minneapolis, Burgess Publishing Company.

Linde-Sipman, J.S. van de, and Gruys, E. 1970. Dor Rhabditoida veroorzaakte granulomen bij een paard. Tijdschr. Diergeneesk., *95*:242–247.

Little, M.D. 1965. Dermatitis in a human volunteer infected with *Strongyloides* of nutria and raccoon. Am. J. Trop. Med. Hyg., *14*:1007–1009.

———. 1966. Comparative morphology of six species of *Stron-*

gyloides (Nematoda) and redefinition of the genus. J. Parasitol., *52*:69–84.

Looss, A. 1905. Die Wanderung der *Ancylostomum*- und *Strongyloides*-Larven von der Haut nach dem Darm. Berne, Comptes Rendus 6me Congres International Zoologie, 1904, pp. 225–233.

Lyons, E.T., Drudge, J.H., and Tolliver, S.C. 1970. *Strongyloides* larvae in milk of sheep and cattle. Mod. Vet. Pract., *51*:68.

———, Drudge, J.H., and Tolliver, S.C. 1973. On the life cycle of *Strongyloides westeri* in the equine. J. Parasitol., *59*:780–787.

Masdeu, J.C., Tantulavanich., S., Gorelick, P.P, Maliwan, N., Heredia, S., Martinez-Lage, J.M., Rubino, F.A., Ross, E., and Mamdani, N. 1982. Brain abscess caused by *Strongyloides stercoralis*. Arch. Neurol., *39*:62–63.

Meltzer, R.S., Singer, C., Armstrong, D., Mayer, K., and Knapper, W.H. 1979. Antemortem diagnosis of central nervous system strongyloidiasis. Am. J. Med. Sci., *227*:91–98.

Milder, J.E., Walzer, P.D., Kilgore, G., Rutherford, I., and Klein, M. 1981. Clinical features of *Strongyloides stercoralis* infection in an endemic area of the United States. Gastroenterology, *80*:1481–1488.

Moncol, D.J. 1975. Supplement to the life history of *Strongyloides ransomi* Schwartz and Alicata, 1930 (Nematoda: Strongyloididae) of pigs. Proc. Helm. Soc. Wash., *42*:86–92.

Most, H., Yoeli, M., Campbell, W.C., and Cuckler, A.C. 1965. The treatment of *Strongyloides* and *Enterobius* infections with thiabendazole. Am. J. Trop. Med. Hyg., *14*:379–382.

Napier, L.E. 1949. *Strongyloides stercoralis* infection. Part II. J. Trop. Med. Hyg., *52*:46–48.

Neefe, L.I., Pinilla, O., Garagusi, V., and Bauer, H. 1973. Disseminated strongyloidiasis with cerebral involvement. A complication of corticosteroid therapy. Am. J. Med., *55*:832–838.

Neva, F.A., Gam, A.A., and Burke, J. 1981. Comparison of larval antigens in an enzyme-linked immunosorbent assay for strongyloidiasis in humans. J. Infect. Dis., *144*:427–432.

Nielly, M. 1882. Un cas de dermatose parasitaire non encore observee en France *(Anguillula leptodera)*. Bull. Acad. Med. Paris, *46*:395–402.

Nishigori, M. 1928. The factors which influence the external development of *Strongyloides stercoralis* and on autoinfection with this parasite. J. Formosan Med. Assoc., *No. 276*, 1–56.

Nwaorgu, O.C. 1978. Hypobiosis in *Strongyloides papillosus*. Warsaw, Proceedings of the 4th International Congress on Parasitology, p. 179.

Nwokolo, C., and Imohiosen, E.A.E. 1973. Strongyloidiasis of respiratory tract presenting as "asthma." Br. Med. J., *2*:153–154.

Pampiglione, S., and Ricciardi, M.L. 1972. Geographic distribution of *Strongyloides fülleborni* in humans in tropical Africa. Parassitologia, *14*:329–338.

Pasyk, K. 1978. Dermatitis rhabditidosa in an 11-year-old girl. A new cutaneous parasitic disease of man. Brit. J. Dermatol., *98*:107–112.

Pedreira de Oliveira, A. 1978. Studies of human strongyloidiasis. Rev. Patol. Trop., *7*:29–36.

Poltera, A.A., and Katsimbura, N. 1974. Granulomatous hepatitis due to *Strongyloides stercoralis*. J. Pathol., *113*:241–246.

Powers, R.D., and Benz, G.W. 1977. *Micronema deletrix* in the central nervous system of a horse. J. Am. Vet. Med. Assoc., *170*:175–177.

Purtilo, D.T., Meyers, W.M., and Connor, D.H. 1974. Fatal strongyloidiasis in immunosuppressed patients. Am. J. Med., *56*:468–493.

Redewill, F.H. 1949. *Strongyloides stercoralis* involving the genito-urinary tract. Urol. Cutan. Rev., *53*:609–614.

Roeckel, I.E., and Lyons, E.T. 1977. Cutaneous larva migrans, an occupational disease. Ann. Clin. Lab. Sci., *7*:405–410.

Rubin, H.L., and Woodard, J.C. 1974. Equine infection with *Micronema deletrix*. J. Am. Vet. Med. Assoc., *165*:256–258.

Sandground, J.H. 1923. *"Oxyuris incognita"* or *Heterodera radicicola*. J. Parasitol., *10*:92–94.

———. 1926. Biological studies on the life cycle of the genus *Strongyloides* Grassi, 1879. Am. J. Hyg., *6*:337–338.

Schumaker, J.D., Band, J.D., Lensmeyer, G.L., and Craig, W.A. 1978. Thiabendazole treatment of severe strongyloidiasis in a hemodialyzed patient. Ann. Intern. Med., *89*:644–645.

Shadduck, J.A., Ubelaker, J., and Telford, V.O. 1979. *Micronema deletrix* meningoencephalitis in an adult man. Am. J. Clin. Pathol., *72*:640–643.

Shiroma, Y. 1964. Studies on human strongyloidiasis on Okinawa, Ryukyu. Kagoshima Med. J., *34*:243–246.

Smith, S.B., Schwartzman, M., Mencia, F., Blum, E.B., Krogstad, D., Nitzkin, J., and Healy, G.R. 1977. Fatal disseminated strongyloidiasis presenting as acute abdominal distress in an urban child. J. Pediatr., *91*:607–609.

Stefanski, W. 1954. *Rhabditis gingivalis* sp. n., parasite trouvé dans un granulome de la gencive chez un cheval. Acta Parasitol. Polon., *1*:330–334.

Stewart, R.B., Stone, W.M., and Marti, O.G. 1976. *Strongyloides ransomi:* Prenatal and transmammary infection of pigs of sequential litters from dams experimentally exposed as weanlings. Am. J. Vet. Res., *37*:541–544.

Stone, O.J., Newell, G.B., and Mullins, J.F. 1972. Cutaneous strongyloidiasis: Larva currens. Arch. Dermatol., *106*:734–736.

Stone, W.M., Stewart, T.B., and Peckham, J.C. 1970. *Micronema deletrix* Anderson and Bemrick, 1965 in the central nervous system of a pony. J. Parasitol., *56*:986–987.

Terpstra, W.J. 1972. Intestinal parasites in Amerindians of the interior of Surinam. Doctoral thesis, Leiden, Rijksuniversiteit.

Than, K. 1979. *Strongyloides* gastritis—report of a case. J. Trop. Med. Hyg., *82*:21–22.

Venizelos, P.C., Lopata, M., Bardawil, W.A., and Sharp, J.T. 1980. Respiratory failure due to *Strongyloides stercoralis* in a patient with a renal transplant. Chest, *78*:104–107.

Vince, J.D., Ashford, R.W., Gratten, M.J., and Bana-Koiri, J. 1979. *Strongyloides* species infestation in young infants of Papua New Guinea: association with generalized oedema. Papua New Guinea Med. J., *22*:120–127.

Wallace, F.G., Mooney, R.D., and Sanders, A. 1948. *Strongyloides fülleborni* infection in man. Am. J. Trop. Med., *28*:299–302.

Wang, T., Reyes, C.V., Kathuria, S., and Strinden, C. 1980. Diagnosis of *Strongyloides stercoralis* in sputum cytology. Acta Cytol., *24*:40–43.

Weller, I.V.D., Copland, P., and Gabriel, R. 1981. *Strongyloides stercoralis* infection in renal transplant recipients. Br. Med. J., *282*:524.

West, B.C., and Wilson, J.P. 1980. Subconjunctival corticosteroid therapy complicated by hyperinfective strongyloidiasis. Am. J. Ophthalmol., *89*:854–857.

Whitehall, R., and Miller, M.H. 1944. Infestation of the genitourinary tract by *Strongyloides stercoralis:* A case report. Bull. Johns Hopkins Hosp., *75*:169–174.

Witenberg, G. 1951. Some unusual observations on helminthiasis in Israel. Harefuah, *41*:178–180.

———. 1964. Zooparasitic diseases. A. Helminthozoonoses. c. Strongyloidiasis, pp. 554–601. In *Zoonoses*. Edited by J. Van der Hoeven. Amsterdam, Elsevier Publishing Co.

Yassin, S.M.A., and Garret, M. 1980. Parasites in cytodiagnosis. A case report of *Strongyloides stercoralis* in Papanicolaou smears of gastric aspirate, with a review of the literature. Acta Cytol., *24*:539–544.

Yokogawa, S. 1936. A human case of accidental parasitism of *Diploscapter coronata*. (Cobb, 1893) Cobb, 1913. Zool. Mag., Tokyo, *48*:507–512.

Chapter 20

The Strongylida: Hookworms and Other Bursate Nematodes

THE HOOKWORMS

The order Strongylida includes the superfamily Ancylostomatoidea and three other groups of bursate nematodes: the Strongyloidea, Trichostrongyloidea, and Metastrongyloidea. Features common to these four superfamilies are (1) presence of a complex ovejector with well-developed sphincters in the female, (2) a caudal bursa and a pair of equal spicules in the male, and (3) a claviform esophagus and H-shaped excretory system in both sexes.

Superfamily Ancylostomatoidea

The Ancylostomatoidea differ from the most closely related superfamilies by having a well-developed subglobular buccal capsule bearing cutting plates or teeth and no corona radiata. The caudal bursa is symmetrical, with seven pairs of rays: two dorsal, three lateral, and two ventral. The two most common hookworms of man, *Ancylostoma duodenale* and *Necator americanus,* represent two distinct families that differ markedly in morphology and life cycle. In the *Ancylostoma* family (Ancylostomidae), the buccal capsule is armed with large ventral cutting teeth; in the *Necator* group (Bunostomidae), the buccal capsule bears ventral cutting plates. In *Ancylostoma,* at least some of the species are adapted to transmission through paratenic hosts, and development in the lungs of the final host is not essential, whereas in *Necator,* percutaneous entry with a phase of larval development in the lungs is essential.

Hookworm infection has been one of the major diseases of mankind in warm moist climates since prehistoric times, only exceeded by malnutrition and malaria in the production of human misery and economic loss. Even though the epidemiology of these three major diseases is quite dissimilar, they have frequently combined in the same populations to make life precarious.

Once man contracted hookworms, the development of family, clan, and village customs without the sanitary disposal of human excreta favored the habits of the hookworm during its extrinsic existence and provided equally well for human exposure to infection by this parasite. Elucidation of the etiology of hookworm infection, together with its epidemiology, has made possible development of public health programs focused on effective control.

Ancylostoma duodenale (Dubini, 1843) Creplin, 1845 (The "Old World" hookworm, producing ancylostomiasis)

Synonyms. *Agchylostoma duodenale* Dubini, 1843; *Dochmius duodenalis* (Dubini, 1843) Leuckart, 1867; *Ankylostomum duodenale* (Dubini, 1843) Bugnion, 1880.

Historical and Geographical Notes. Although this worm was probably referred to in the Ebers papyrus of ancient Egypt (1600 B.C.) and was mentioned by the Persian physician Avicenna (980–1037 A.D.), the first accurate description was provided by Dubini, in 1843, from material obtained at autopsy of a Milanese woman in 1838. In 1878, Grassi and Parona demonstrated that the infection could be readily diagnosed by recovery of the characteristic eggs in the stools of patients, particularly those suffering from "miner's anemia." Perroncito (1880) reported development of the free-living first-stage larvae hatched from eggs and their metamorphosis into filariform larvae. Leichtenstern (1886–1887) found that mature filariform larvae, when introduced into the intestinal tract, developed into adult worms in the small bowel. Arthur Looss (1896–1897), working in the Public Health Laboratory in Alexandria, Egypt, first accidentally infected himself by exposing his skin to filariform larvae. Later, by experimental studies on the dog (using a canine species of hookworm, *A. caninum*), he elucidated the complete route of migration and consecutive stages of development from the site of cutaneous invasion through the lungs, up the respiratory tree, over the epiglottis, and down into the small intestine. There the worms developed a mouth capsule, became attached to the intestinal wall, became sexually mature, and mated, and the females began to oviposit. All subsequent investigations on hookworms have been based on these fundamental studies of the free-living and parasitic stages of the worms.

Ancylostoma duodenale is essentially a parasite of southern Europe, the north coast of Africa, northern India, North China, and Japan (Fig. 20–1). It has been found in native Paraguayan Indians, in the hill tribes of Fukien, China, and in the aborigines of western Australia. Recently, the migration of infected persons from endemic areas has introduced this parasite into many new foci where *Necator americanus* is, or has been, the predominant human hookworm, namely, southern India, Burma, Malaysia, Philippines, Indonesia, Polynesia, Micronesia, and Portuguese West Africa, but it is almost invariably less common than *Necator americanus* in these localities.

Ancylostoma duodenale is the predominant hookworm of man in coastal Peru and Chile.

Morphology. The *adult worms* (Fig. 20–2) are relatively stout, somewhat cylindroidal, slightly constricted anteriorly, and have a cervical curvature so that the anterior end of the worm is directed dorsoanteriad. The living worms are pinkish or creamy gray in color, are covered with a tough cuticula, and are provided with a pair of lateral cervical papillae just behind the circumesophageal nerve ring. The males (Fig. 20–2,*A*) are 8 to 11 mm in length, and have a maximum diameter of 0.4 to 0.5 mm, while the females (Fig. 20–2,*B*) average 10 to 13 mm in length by 0.6 mm in diameter. The large buccal capsule (Fig. 20–3,*A*) is heavily reinforced by "chitinoid" material, is oval in contour, with a greater transverse diameter, and is composed of an outer articulated portion and a smooth, fused, inner portion. Situated on the ventral (i.e., upper) side of the capsule are bilaterally symmetrical cutting processes each consisting of two fused teeth, of which the outer tooth is the larger and the inner tooth is provided with a small, inconspicuous accessory median process. On the dorsal (i.e., lower) aspect of the capsule there is a dental plate with a median cleft. The depth of the capsule contains a pair of inconspicuous teeth. A dorsal esophageal gland opens through the dorsal dental plate. On the outer ventrolateral aspect of the stoma there are openings of a single pair of amphidial (cephalic) glands that extend posteriad as far as the middle of the worm. Likewise, extending from the excretory pore posteriorly is a pair of slender cervical (excretory) glands (Fig. 20–2).

Proceeding inward from the buccal cavity is the esophagus (Fig. 20–2, *e*), which is lined with cuticula, has a triradiate lumen, and extends posteriad about one-sixth of the body length. It is somewhat enlarged posteriorly and is provided at its posterior extremity with a tripartite valve. The wall of the esophagus contains a single dorsal and a pair of subventral glands. The midintestine (Fig. 20–2, *m*), which is noncuticularized, extends from the posterior limit of the esophagus to the short rectum, which is lined with cuticle. The food of the worm consists essentially of the mucous membrane of the host's intestine, free cells, and plasma. Owing to the pumping action of the esophagus, blood escaping from ruptured intestinal capillaries is passed undigested through the worm's intestine.

The excretory system is a complicated apparatus consisting of a "carrying cell," a "suspensory cell," and a pair of tubules that extend from the level of the buccal capsule to the subcaudal region. The excretory pore (Fig. 20–2,*B*, *ex p*) is midventral in position; it is situated just behind the nerve ring and just in front of the cervical papillae. Intimately associated with this system is the pair of cervical or excretory "gland" cells referred to previously (Fig. 20–2, *ex gl*) which are ventrally disposed, extend posteriad a considerable distance behind the esophagus, and drain into the excretory canals near the point of junction of the paired canals.

The *male worms* (Fig. 20–2,*A*) have the posterior portion of their body extended into a campanulate bursa, supported by fleshy rays with a pattern characteristic of the species (Fig. 20–3,*B*). Within the bursa is the cloaca into which both the rectum and genital canal open. The single testis (Fig. 20–2,*A*) originates as a fine tubule near the anterior end of the cement gland, and proceeds anteriad, winding its way back and forth on the side of the midintestine to the level of the nucleus of the anterior cervical gland. It then returns posteriad as a convoluted tubule, and in the midplane of the body expands into the broadly lenticular seminal vesicle (Figs. 17–3,*A*, *v*, 20–2, *sv*). The genital canal is then continued as a muscular ejaculatory duct, which receives the cementum from the pair of ventrally situated cement glands and opens posteriad into the cloaca. In its posterior portion the ejaculatory duct is provided with a heavy cuticular covering. The pair of long bristle-like copulatory spicules *(c sp)* (each 0.9 to 1 mm long) lies in a tubular pocket lateral to the duct. They are regulated by retractor and extensor muscles, as well as by the dorsally situated gubernaculum.

The *female worms* (Fig. 20–2,*B*) have a conoidal posterior end, with a subterminal, ventrally dis-

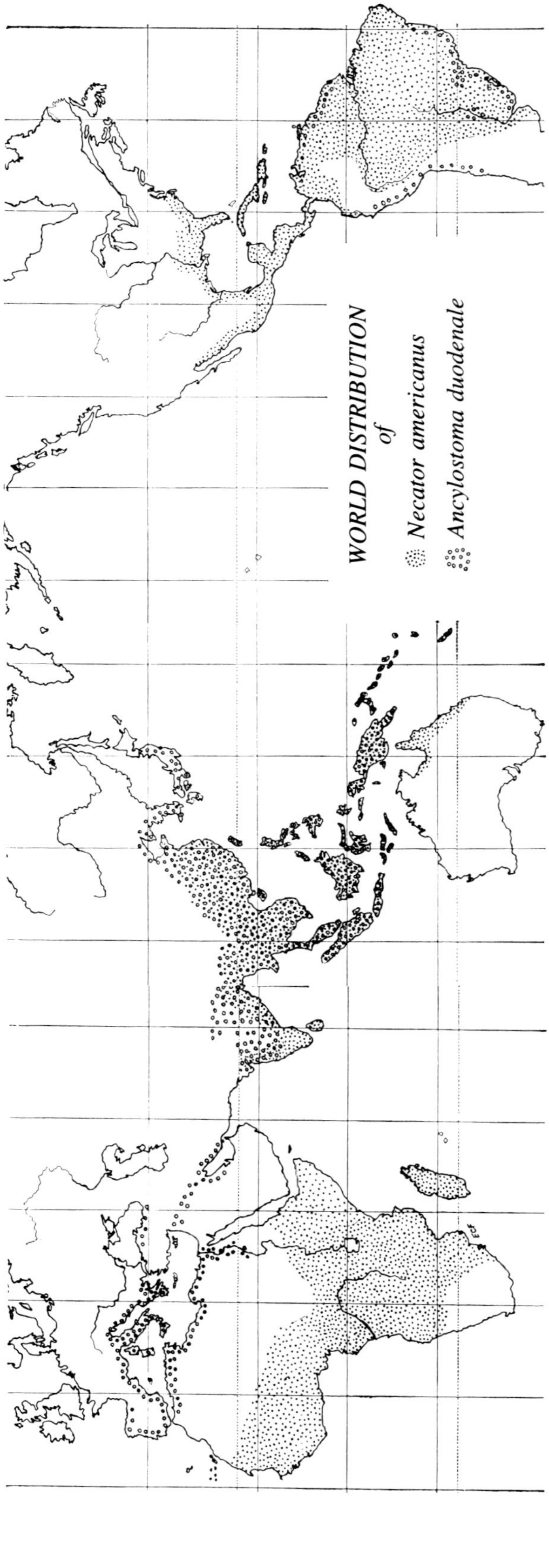

Fig. 20–1. Approximate world distribution of human hookworm infection. Not shown is an area endemic for both *A. duodenale* and *N. americanus* in the Caspian Sea region of Iran, from which *A. duodenale* extends southward to the Persian Gulf and westward to the Black Sea. (By E.C. Faust.)

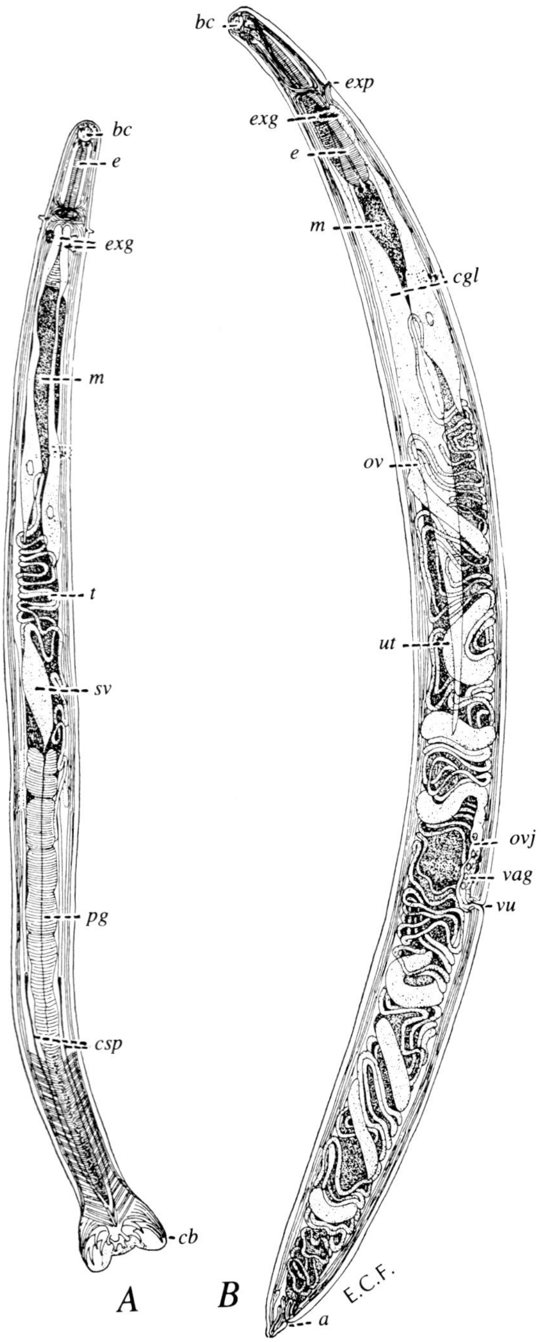

Fig. 20–2. *Ancylostoma duodenale*, adults. *A*, Male, ventral view. *B*, Female, lateral view. *a*, Anus. *bc*, Buccal capsule. *cb*, Copulatory bursa. *cgl*, Cephalic (amphidial) gland. *csp*, Copulatory spicules. *e*, Esophagus. *exg*, Excretory (cervical) gland. *exp*, Excretory pore. *m*, Midgut. *ov*, Ovary. *ovj*, Ovejector. *pg*, Prostate gland. *sv*, Seminal vesicle. *t*, Testis. *vag*, Vagina. *vu*, Vulva. (× 20.) (Adapted from original of A. Looss. In Faust, E.C. 1929. *Human Heminthology*, 1st ed. Philadelphia, Lea & Febiger.)

posed, anal opening. The vulva *(vu)* is situated midventrally at the beginning of the distal third of the body. The two ovarian tubules *(ov)* are intricately coiled back and forth over the intestinal tract, the anterior one in the middle third of the body, the posterior one in the distal third. The length of each member of this pair is approximately two-and-a-half to four times the total length of the female worm. These ovaries (*ov*) at their distal ends are continuous with the short attenuated oviducts, each of which in turn opens into its seminal receptacle and is succeeded by a relatively short uterus *(ut)* and, finally, by the muscular ovejector *(ovj)*, before the paired tubules join to form the short transverse vagina *(vag)*, which opens through the vulva *(vu)*. Insemination occurs when the male applies its bursa copulatrix around the female at the level of the vulva, inserts the pair of copulatory spicules, and secretes cementum to provide uninterrupted discharge of the spermatozoa. These travel, or are transported, up the female canals to the seminal receptacles. The spermatozoa of *Ancylostoma* are exceedingly small. In the seminal vesicle they have the appearance of small-rod bacteria about 2 μm long, and they have the same appearance in the genital organs of the female. Spermatozoa of *Necator* are only slightly larger.

The *eggs*, which are formed from a multinucleated mass of protoplasm in the ovarian tubules, are fertilized as they pass through the seminal receptacles. The vitelline membrane and shell are then added as a secretion from the uterine wall. The eggs are shuttled out through the ovejectors, passed through the vagina, and discharged into the lumen of the intestine of the host. The daily output of *A. duodenale* may average between 25,000 and 30,000 eggs. Although man is the optimal host of *Ancylostoma duodenale*, a morphologically similar worm has been reported from several simian hosts, from the pig, and from canines and felines.

The Life Cycle. ***Free-Living Phase.*** When discharged, fertilized eggs are usually in early cleavage stages. Their shell is thin, hyaline, and ovoidal, with bluntly rounded ends and have an average measurement of 60 by 40 μm (Fig. 20–4). A few hours after passage in the feces, eggs may reach the morula or tadpole stage; less commonly, the egg may contain a motile first-stage larva. In undiluted feces development is slow; when eggs are mixed with urine, death of the embryos occurs relatively soon. The condition most favorable to the embryos usually occurs in primitive commu-

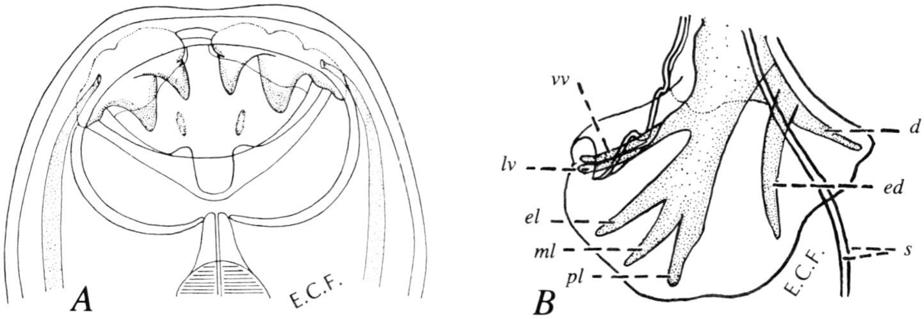

Fig. 20–3. *Ancylostoma duodenale. A,* Anterior end, showing buccal capsule with two pairs of teeth. (× 240.) *B,* Posterior end of male, lateral view, showing bursal rays and spicules; genital cone not shown. (× 80.) Bursal rays clockwise from dorsal *(d)* and externodorsal *(ed)* are posterolateral *(pl),* mediolateral *(ml),* externolateral *(el),* lateroventral *(lv),* and ventroventral (vv). Paired spicules *(s)* are plain and free at the tips. (By E.C. Faust.)

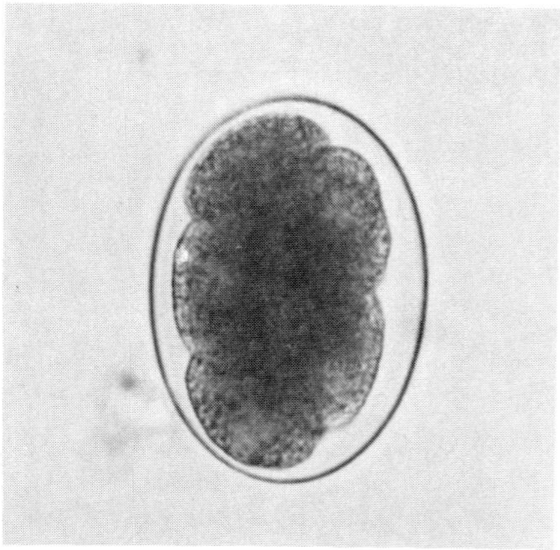

Fig. 20–4. Hookworm egg from human feces. (× 480.) Eggs of *Necator americanus* and *Ancylostoma duodenale* are indistinguishable. See Fig. 46–1, p. 736 (By P.C. Beaver.)

nities in which stools are frequently deposited in a moist, shady, warm place, where decaying vegetation covers a sandy loam. Here the eggs hatch in 1 or 2 days, and the emerging larvae (0.25 to 0.30 mm in length by 17 μm maximum diameter) begin to feed on bacteria and organic debris. Characteristic first-stage hookworm larvae (Fig. 20–5,*A*) have a long, narrow buccal chamber, flask-shaped muscular esophagus occupying the anterior third of the digestive tract, midgut, short rectum, and an inconspicuous genital primordium. They may be readily distinguished from *Strongyloides* larvae (Figs. 19–1,*D*, 19–4,*A* and 19–8,*A*) and those of *Rhabditis* (Fig. 19–8,*D*) and other free-living nematodes. After about 3 days of feeding and growth, the larva sheds its cuticula, then, as a second rhabditoid stage, grows to a maximum of 0.5 to 0.6 mm in length with corresponding increase in diameter. Usually between the fifth and eighth days the mouth becomes closed, the esophagus elongates, and the larva becomes transformed into the nonfeeding filariform type (third larval stage, Fig. 20–5,*B*). The old cuticula may be shed or may remain for some time as a protective covering around the young worm. This stage, which is infective for man, may remain viable in the soil for several weeks. It may migrate several inches below the top humus.

Parasitic Phase. On contact with the skin, filariform larvae secure lodgment in the hair follicles, or more commonly under scaling fragments of epidermis. They penetrate down to the deeper portion of the corium or to subcutaneous tissue, where they enter the superficial venules and are carried to and through the right heart to the lungs, where they break out of the capillaries into the air sacs. Without essential development in the lungs, they migrate and are carried up the respiratory tree to the epiglottis. They pass over into the digestive tract and down to the small intestine. After shedding the filariform cuticula (third molt), they develop a temporary buccal capsule with which they are able to attach themselves to the mucosa. They grow in size and become sexually differentiated. They then develop the definitive mouth capsule within the old one, which is shed together with the fourth cuticula, and the worms grow into adults. A period of about 5 weeks or more elapses from the time of exposure until the females first begin to oviposit. Mature filariform larvae of *A. duodenale,* when swallowed, develop into mature worms without the lung passage (Okamoto, 1961; Higo, 1962).

The ability of *A. duodenale* to infect by mouth as well as by the cutaneous route was confirmed in dogs (Nagahana and Yoshida, 1965). In the dog,

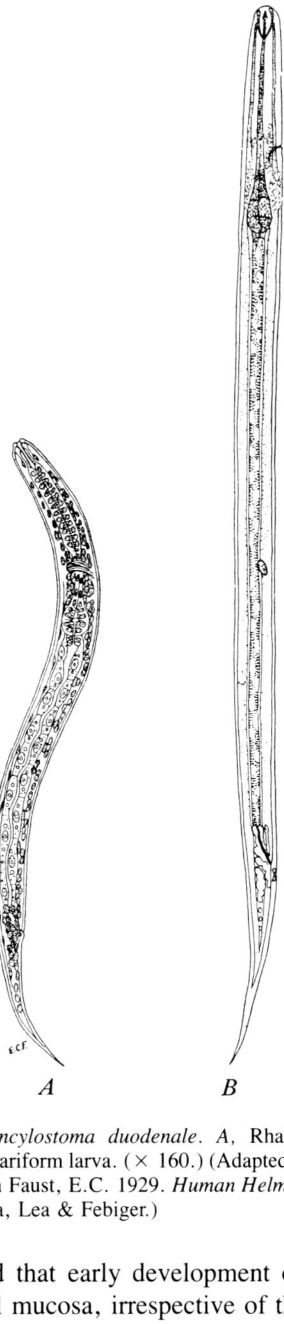

A B

Fig. 20–5. *Ancylostoma duodenale. A,* Rhabditoid larva. (× 310.) *B,* Filariform larva. (× 160.) (Adapted from original by A. Looss. In Faust, E.C. 1929. *Human Helminthology,* 1st ed. Philadelphia, Lea & Febiger.)

it was found that early development occurred in the intestinal mucosa, irrespective of the mode of entry, and that eggs appeared in the feces during the fourth or fifth week. In *A. duodenale,* however, and probably in other species of *Ancylostoma,* the prepatent period is somewhat unpredictable and may be as long as 40 weeks (Nawalinski and Schad, 1974). In experimental animals the larvae of *A. duodenale* are sequestered in the muscles (Soh, 1958), as are also the larvae of *A. caninum.* In the latter at least, and probably in *A. duodenale,* the larvae are located inside the muscle fiber and do not become encapsulated (Lee et al., 1975). Observations on experimentally infected dogs suggest that larvae sequestered in the muscles migrate to the intestine via the lungs and develop to maturity (Little, 1978.) Larvae also migrate from the muscles to the mammary glands and establish infection in newborn pups via colostrum and milk (Stone and Smith, 1973; Stoye, 1976). Galactogenic hookworm infection in humans has not been demonstrated, but it probably occurs in areas where *A. duodenale* or *A. ceylanicum* is endemic. The high incidence of neonatal infection with *A. duodenale* suggests the possibility of transmammary infection despite the failure to demonstrate the presence of larvae in the milk of nursing mothers (Nwosu, 1981). Failure to find hookworm larvae in human milk in regions of Africa where *Necator americanus* is the species present (Brown and Girardeau, 1977; Nwosu, 1978) could have been expected. A larva in human milk has been reported as *Necator* (Setasuban et al., 1980), but the identification is questionable. In *N. americanus* there is essential development in the lungs, and larvae do not persist in the tissues beyond the period of essential development (Nichols, 1956; Soh, 1958).

Many of the worms may be eliminated within a year, although some may persist for as long as 9 years. Continuous reinfection commonly occurs in endemic foci.

Ancylostoma ceylanicum Looss, 1911

Ancylostoma ceylanicum is one of the smaller species of the genus. The male averages 8 mm and the female 10 mm in length. In his original description Looss (1911) stated: "At the anterior edge of the mouth capsule there is, on each side, one large tooth and below, or behind this, on the side turned towards the middle line, a very small tooth of which the tip only looks out from below the large one; the arrangement of these teeth thus appears to resemble that in" [*A. malayanum* Alessandrini, 1905]. A restudy of the claims for validity of the species *ceylanicum,* as distinct from *braziliense* de Faria, 1910, Biocca (1951), and Biocca and LeRoux (1957), confirmed Looss' description of the dentition for *ceylanicum* and indicated further that the oral aperture is transversely oval, whereas that of *braziliense* tends to be somewhat smaller, and longer in the longitudinal plane (Fig. 20–6,*B,* 1 and *C,*1). Moreover, the bursa of the male *ceylanicum* (Fig. 20–6,*B,* 2) is broader, the lateral rays are more curved, the mediolateral and

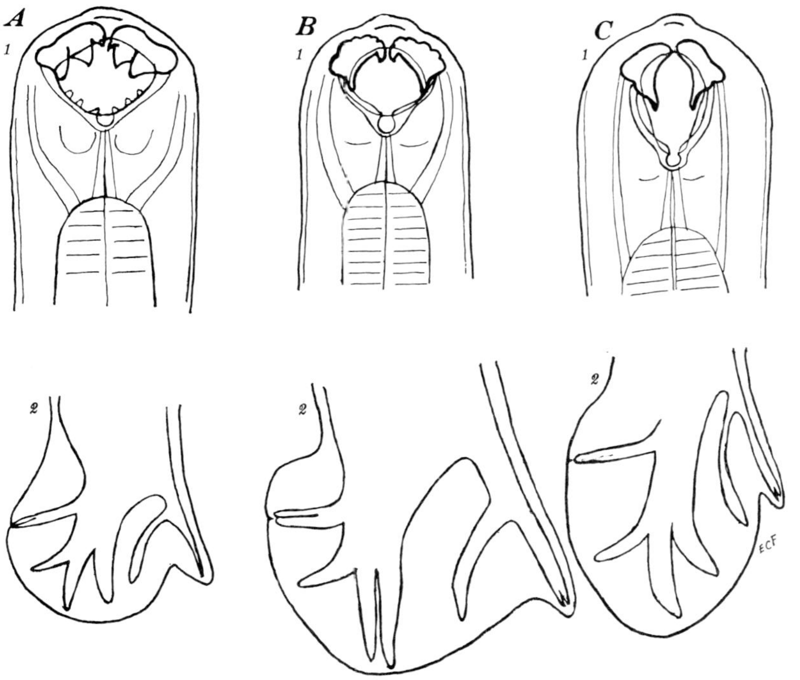

Fig. 20–6. Head (1) and bursa (2) of *A*, *Ancylostoma caninum*, *B*, *Ancylostoma ceylanicum*, and *C*, *Ancylostoma braziliense*. All drawn approximately to scale. (Adapted by E.C. Faust from Biocca, A, and LeRoux, P.L. 1957. Suddivisione del genere *Ancylostoma* (Dubini, 1843) in quattro sottogeneri. Accad. Naz. dei Lincei, Rend. Sci. Fis. Mat. Nat., *23*(ser. 8):470–477.)

externolateral rays are more divergent, the externodorsal rays originate nearer to the stem of the dorsal ray, and the ray supports are plumper than in *braziliense* (Fig. 20–6,*C*, 2). All these features are constant.

Ancylostoma ceylanicum was described by Looss (1911) from the intestine of a civet cat *(Viverricula malaccensis)* from Ceylon. This hookworm occurs typically in cats and their wild relatives in Southeast Asia. It probably also parasitizes these animals, and to a lesser extent dogs and man, throughout the Far East and elsewhere in tropical and semitropical climates of the world, invariably as an intestinal infection. Human infection has been reported in the Philippines (Velasquez and Cabrera, 1968), Taiwan (Yoshida et al., 1971), and India (Chowdhury and Schad, 1972). Experimental infection in man was reported by Yoshida et al. (1971). *A. ceylanicum,* erroneously identified as *A. braziliense,* was reported in military personnel returning to the Netherlands from New Guinea (Rep, 1966).

Much confusion has existed concerning the validity of the species *ceylanicum,* in part because Lane (1922), Yorke and Maplestone (1926), Brumpt (1949), and other authorities considered it to be the same as *braziliense*. Insofar as is known, *A. ceylanicum,* unlike *A. braziliense,* does not produce cutaneous larva migrans in man. Hence, in addition to its specific morphology, *A. ceylanicum* as an intestinal parasite needs to be distinguished from *A. duodenale* and *Necator americanus,* both of which are more common hookworms of man in most areas where *A. ceylanicum* is found. Yoshida and associates (1974) reported detailed comparative studies on *A. ceylanicum* and *A. braziliense*.

Ancylostoma braziliense de Faria, 1910

Synonym. Larva—*Agamonematodum migrans* Kirby-Smith, Dove and White, 1926.

First reported from cats and dogs in southern Brazil in 1910, this worm is present in wild and domesticated feline and canine hosts in numerous localities throughout the tropics and subtropics. Adult *A. braziliense* has been incorrectly reported from man in the Philippines, Indonesia, New Guinea, Ceylon, India, Thailand, southeast Africa, and Brazil. Presumably the parasite in all of these cases was *A. ceylanicum*.

Ancylostoma braziliense, like *A. ceylanicum,* is

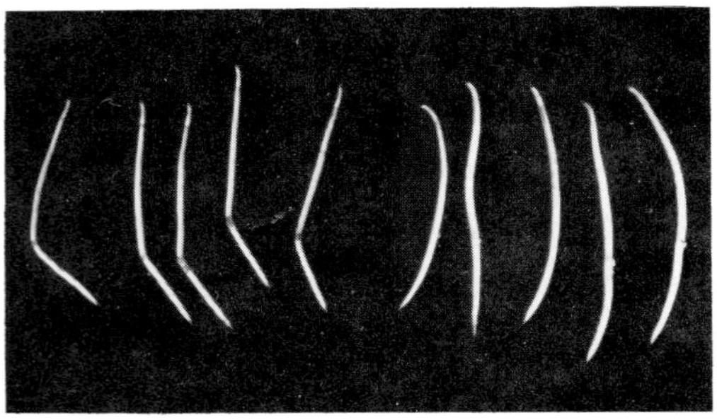

Fig. 20–7. *Ancylostoma ceylanicum* (right) and *Ancylostoma braziliense* (left). Adult females after fixation, showing angular flexion in the body of *A. braziliense* at the level of the vulva. (From Yoshida, Y. 1971. Comparative studies on *Ancylostoma braziliense* and *Ancylostoma ceylanicum*. I. The adult stage. J. Parasitol., *57*:983–989.)

one of the smaller species of the genus; the males measure approximately 8 mm in length by 0.27 mm in diameter, and the females 10 mm by 0.3 mm. In fixed specimens of *A. braziliense,* the females have a distinctive angular bend at the level of the vulva (Fig. 20–7). The buccal capsule is specifically diagnostic, having a pair of small, inconspicuous median teeth and a pair of larger outer teeth (Fig. 20–6*B*,8*A*). The bursa of the male is also distinct; it is small, almost as broad as long, and is supported by relatively short lateral rays (Fig. 20–8*B*). The eggs are not readily distinguishable from those of *A. duodenale.*

For pathological and clinical aspects of this infection, see page 281.

Ancylostoma caninum
(Ercolani, 1858) Hall, 1913
(The dog hookworm)

Ancylostoma caninum is a common parasite of dogs, particularly in the northern hemisphere. It has been reported once as a parasite of man in the Philippine Islands (Manalang, 1925) and on four other occasions (Deane, 1950; Witenberg, 1951).

The larval stage has been reported as causing creeping eruption (see p. 281), and recently it was found in human muscle (Little et al., 1983). A pest-control worker who had acquired extensive creeping eruption while working under a house in Louisiana developed a mild but persistent swelling and tenderness of the thigh muscles under the most severely affected area of the skin. A biopsy of the muscle contained a third-stage hookworm larva that was located in a muscle fiber. By its location and morphology, the larva was identified as *Ancylostoma caninum,* a species known to invade and persist for months in the skeletal muscle fibers of various mammals that may serve in nature as paratenic hosts. Larvae of other *Ancylostoma* species (*A. braziliense, A. tubaeforme*) also persist in the tissues of mammalian paratenic hosts, but they are found in the tissues of the nasopharynx, not in the muscles. The detailed morphology of the infective-stage larva as seen free and in tissues was described by Nichols (1956).

The adult male worm has an average length of 10 mm and a greatest diameter of 0.4 mm; the female measures 14 mm in length and 0.6 mm in greatest diameter. A wide buccal capsule bearing three pairs of ventral teeth is a diagnostic character of the species. The bursa is supported by long, moderately slender rays (Fig. 20–6*A*). The eggs of this species are similar to those of *A. duodenale* but are slightly larger (64 by 40 μm). This species has been used to investigate the host-parasite relationships of hookworms, much of which has been impracticable with the species found in man. Of interest in this connection is the work of Thorson (1956a), who found that extracts of the esophagus of *A. caninum* contain a proteolytic enzyme (casein substrate) that is inhibited by homologous immune dog serum. Thorson (1956b) also found that saline extracts of the esophagus, excretory glands, or amphidial glands of this worm delay the coagulation of the blood. Eiff (1966) found that the anticoagulant of *A. caninum* is secreted by the amphidial (cephalic) glands.

Unusual Records of Related Species

Although man is the only normal definitive host of *A. duodenale,* a female worm identified as *A. duodenale* was collected from a tiger in Calcutta (Baylis and Daubney, 1923). Worms so identified have been reported in the pig, the dog, and several species of wild mammals in captivity. Biocca (1951) remarked that closely related species found in animals may be reported erroneously as *A. duodenale.* Biocca and LeRoux (1957) divided the genus *Ancylostoma* into four subgenera: *A. Ancylostoma* to include *duodenale, caninum, paraduodenale,* and others; *A. Ceylancylostoma* for *ceylanicum* and *malayanum; A. afrancylostoma* for *braziliense;* and *A. amerancylostoma* for *pluridentatum.*

Ancylostoma malayanum (Alessandrini, 1905), a parasite of bears in India and Malaysia, is the largest species of the genus, the males measuring 12 to 15 mm by 0.6 mm, the females 15 to 19 mm by 0.6 mm. Its buccal capsule is much like that of *A. duodenale;* there are two pairs of ventral teeth, although the outer pair is longer and more slender. The copulatory spicules are long (3 mm). The eggs are indistinguishable from those of *A. duodenale. A. malayanum* was reported once from man (Yorke and Maplestone, 1926).

Cyclodontostomum purvisi Adams, 1933, a common hookworm parasite found in the large intestine of several species of wild rats in Malaysia, was reported once from man (Bhaibulaya and Indrangarm, 1975). In that case, two worms, a male and

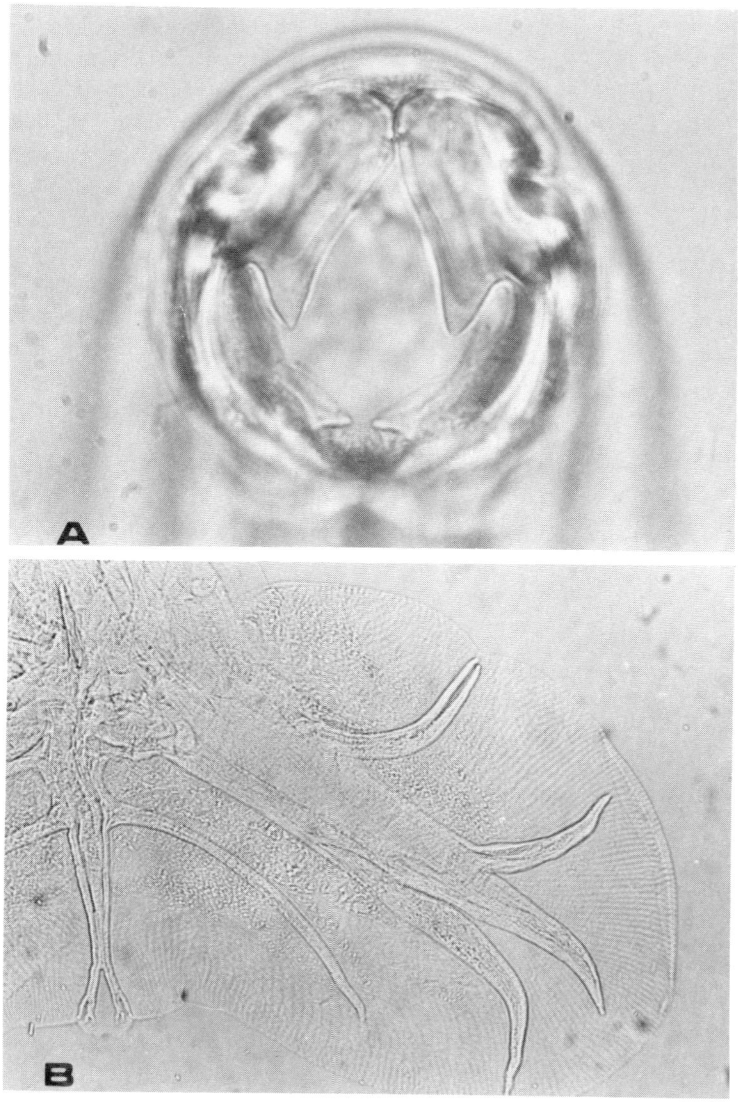

Fig. 20–8. *Ancylostoma braziliense,* North American strain. *A,* En face view of buccal capsule. (× 350.) *B,* Bursa, spread to show shape and relationship of rays and bursal lobes. (× 150.) (From Norris, D.E. 1971. Morphology of a North American strain of *Ancylostoma braziliense* Gomes de Faria, 1910. J. Parasitol., *57*:993–997.)

female, were recovered from the feces of a 47-year-old man who was being examined for suspected intestinal capillariasis.

Necator americanus (Stiles, 1902) Stiles, 1906 (The "American hookworm," literally the "American murderer," producing necatoriasis [uncinariasis of many clinical texts])

Synonyms. *Uncinaria americana* Stiles, 1902; *Ankylostomum americanum* (Stiles, 1902) v. Linstow, 1903; *Necator africanus* Harris, 1910; *N. argentinus* Parodi, 1920.

Historical and Geographical Notes. This species of hookworm was described by Stiles in 1902 as a new species after he had examined specimens sent to him by Allen J. Smith from Galveston, Texas. Soon afterward this hookworm was found to be widely distributed throughout the southern United States, the islands of the Caribbean, Central America, and northern South America. Then it was found to be the predominant hookworm of Central and South Africa, southern Asia, Melanesia, and Polynesia (Fig. 20–1). It was apparently not indigenous in the Amerind population but was introduced from Africa. This worm has been an important agent in the production of "laziness" in the poor white population of the southern United States and "tropical anemia" of Puerto Rico and other tropical areas in the western hemisphere.

Morphology. The *adult worms* (Fig. 20–9) are appreciably attenuated at the anterior extremity and have a small buccal capsule (Fig. 20–10,*A*) provided with a ventral and a dorsal pair of semilunar cutting plates *(cp, dp)* in place of teeth, which

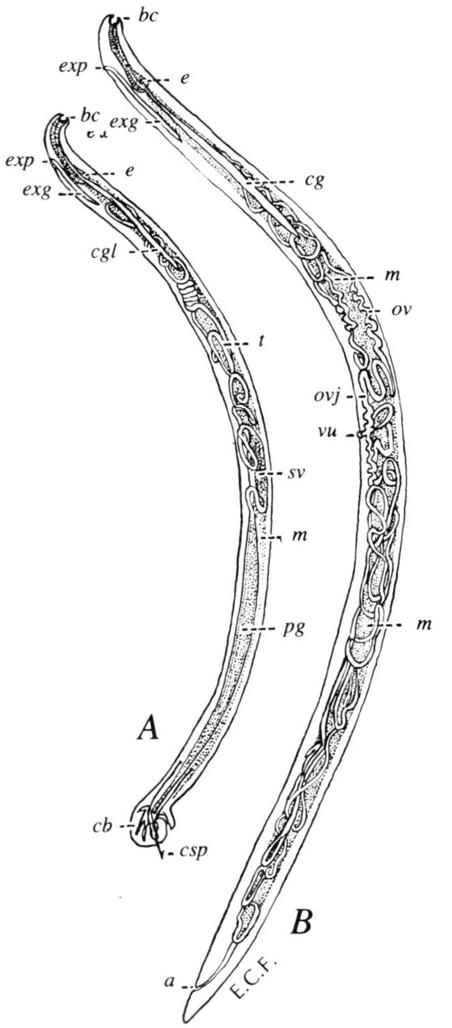

Fig. 20–9. *Necator americanus* adults, lateral view. *A*, Male. *B*, Female. *a*, Anal pore. *bc*, Buccal capsule. *cb*, Copulatory bursa. *cg*, and *cgl*, Paired cephalic gland. *csp*, Copulatory spicules. *e*, Esophagus. *exg*, Excretory gland. *exp*, Excretory pore. *m*, Midgut. *ov*, Ovary. *ovj*, Ovejector. *pg*, Prostate gland. *sv*, Seminal vesicle. *t*, Testis. *vu*, Vulva. (× 18.) (Adapted by E.C. Faust from Lane, C. 1932. *Hookworm Infections*. London, Oxford University Press.)

distinguish members of the genus *Ancylostoma*. They also have a pair of triangular subventral lancets *(ll)* and a subdorsal pair *(dt)* in the depth of the buccal cavity. The worms are strongly reflexed dorsalward at the anterior end. Opening to the outside at or through the amphids, which are located near the lateral margin of the buccal capsule, is a long cephalic or amphidial gland attached to the hypodermal lateral cord for about one-fourth the worm's length on either side of the body. Two more slender excretory glands, also referred to as cervical glands, ventral in position and connected to the tubular excretory system, extend from just anterior to the nerve ring posteriorly one-third to one-half the length of the body. A set of secretory structures, located within the muscular wall of the esophagus, consists of a large dorsal and two smaller subventral glands. Observations by McLaren (1974) clarified the anatomic relationship of amphidial glands, earlier described as emptying on the inside of the buccal capsule. The amphidial, excretory, and esophageal glands all serve an exodigestive function (McLaren et al., 1974). A potent anticoagulant is also secreted by the amphidial glands, serving to promote the flow of any blood or tissue fluids that are drawn into the buccal capsule through the worm's intestine (Eiff, 1966). Contraction of the muscular esophagus serves to draw tissues and fluids into the buccal capsule and thus anchors the worm to the intestinal wall.

The living worms are grayish yellow, at times with a reddish undertone. The males measure 7 to 9 mm in length by 0.3 mm in greatest diameter; the females, 9 to 11 mm by 0.4 mm. The bursa of the male is long and wide and is diagnostic for the species (Fig. 20–10,*B*). The copulatory spicules are long and slender, and fused at the tip, forming a delicate barb (Figs. 20–9,*A*, 10,*B*). In the female the vulva (Fig. 20–9,*B*) is situated either in the middle of the body or somewhat anterior to this position (Fig. 20–9). The eggs resemble those of *A. duodenale* but tend to be longer and somewhat more elongated (64 to 76 by 36 to 40 μm) (Fig. 20–4). Eggs produced by unmated females are characteristic in size but more variable in shape. The shell is relatively thin, and the contents unorganized (Palmer, 1941; Beaver, 1952, 1955; Komiya and Yasuraoka, 1966; Rep, 1972). Infertile eggs fail to appear in the concentrate by brine or zinc sulfate flotation and therefore are seldom seen. According to Palmer (1941), Brug (1910) was probably the first to report the occurrence and appearance of infertile hookworm eggs.

The Life Cycle. The *adult worms* live in the jejunum and upper levels of the ileum. They are apparently attached to the mucosa most of the time (Fig. 20–11), although there is detachment for movement from place to place during feeding (Kalkofen, 1970) and mating (Beaver et al., 1964). The life span of *N. americanus* exceeds that of *A. duodenale*, having been recorded as extending to 15 years (Palmer, 1955). The average duration of individual infections, however, is much less. Records of longevity vary from 4 to 20 years for *Ne-*

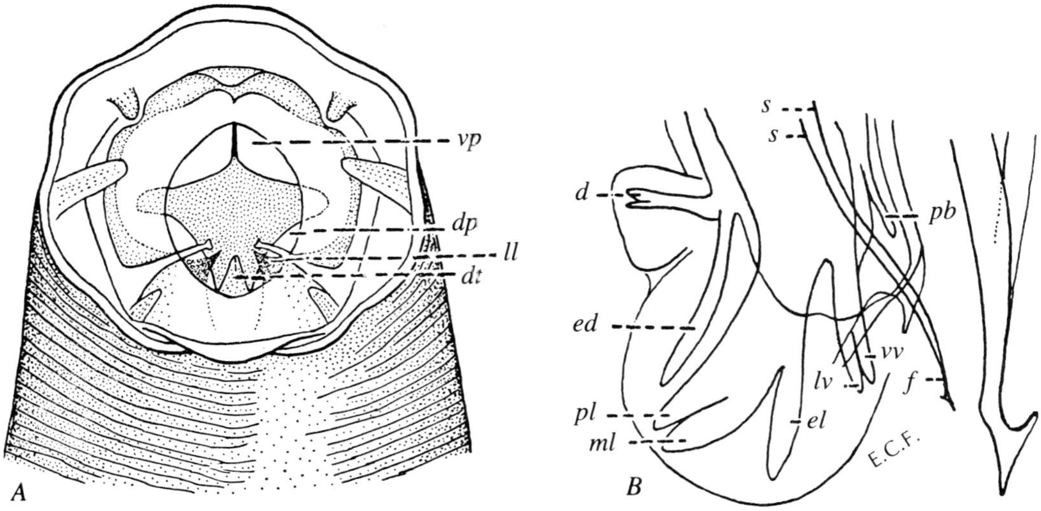

Fig. 20–10. *Necator americanus. A*, Anterior end, showing buccal capsule with cutting plates. *dp*, Dorsal cutting plate. *dt*, Dorsal tooth. *ll*, Lateral lancet. *vp*, Ventral cutting plate. (× 400.) *B*, Posterior end of male, showing bursal rays and spicules. At right, greatly enlarged lateral view of fused termination of the two spicules, ending in a barb. *d*, Dorsal ray. *ed*, Externodorsal ray. *f*, Fused terminus of spicules. *lv*, Lateroventral ray. *ml*, Mediolateral ray. *pb*, Prebursal ray. *pl*, Posterolateral ray. *vv*, Ventroventral ray. (× 80.) (Adapted from original of E.C. Faust.)

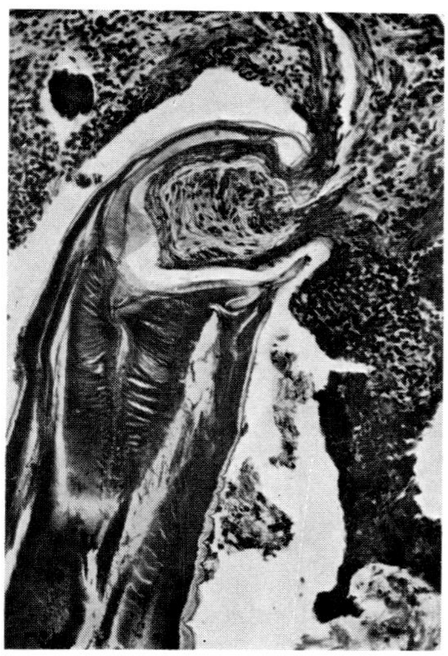

Fig. 20–11. *Necator americanus.* Adult worm attached to human jejunum, with mucosal tissues drawn into the buccal cavity. (Army Medical Museum Collection.)

cator as compared with 5 to 7 years for *A. duodenale* (Watson, 1960).

The life cycle is similar to that described for *A. duodenale (vide supra)* except that *N. americanus* requires a pulmonary migration with essential development in the lungs before it can develop to the adult stage in the intestine (Yoshida and Okamoto, 1960; Tanabe, 1962). Each fertile adult female produces thousands of eggs per day, such that in light infections the output per female is 100 to 500 eggs per gram of feces (Beaver, 1955). In moderate to heavy infections, the egg output per female is much less (Krupp, 1961).

Yoshida and Fukutome (1967) demonstrated the full cycle of *N. americanus* in rabbits, and Sen and Seth (1967) reported complete development in golden hamsters. Infective-stage larvae placed on the skin of rabbits migrated after 3 days to the lungs, where they grew and developed to the advanced third stage. The third ecdysis occurred on arrival in the intestine, and the fourth ecdysis on or after the twenty-fourth day of infection. Fertilized eggs were found in the uterus of the worms 60 to 63 days after exposure. In the hamster, eggs appeared in the feces as early as the sixth week. In human volunteers, eggs first appeared in the feces during the seventh week or early in the eighth week (Palmer, 1941; Turton, 1977). The shortest recorded prepatent period in chimpanzees was 42 days, the longest, 54 days, and in the patas monkey the corresponding periods were 42 and 49 days (Orihel, 1971).

Unusual Records of *Necator*. *Necator americanus* is a parasite of the upper levels of the small intestine of man, its natural host. A hookworm morphologically indistinguishable from *N. americanus* of man has been rarely recovered from several species of simian hosts *(Pan satyrus, Gorilla gorilla, Erythrocebus patas)* and from the rhinoceros, the pangolin *(Manis javanicus* in Indonesia and *M. pentadactyla pentadactyla* in Taiwan), and a rodent *(Coendu villosus)* in Africa. Under specially controlled conditions, it may develop to the adult stage in dogs and cats. *N. suillus* Ackert and Payne, 1922, a parasite of pigs in Central America and Trinidad, is a closely related species.

A small collection of hookworms from man in Iran was described and recognized as a *Necator* species distinct from *N. americanus,* but a name was not proposed for it (Ghadirian and Biocca, 1972).

Hookworm Infection and Disease

Epidemiology. The endemicity of hookworm infection depends on (1) the continuous presence of infection in the human population; (2) defecation habits ensuring that eggs of the hookworms are deposited in favorable locations for extrinsic development of the parasite; (3) appropriate conditions of environment (shade and a sandy humus); and (4) opportunity for the infective (filariform) stage to come in contact with the skin of man so that the parasitic phase may be reestablished. In many tropical and subtropical countries conditions are satisfactory for maintenance of the cycle throughout the greater part of the year, whereas in cooler climates certain conditions may be operating to favor extrinsic development and human exposure during the warmer months only. Prolonged dry or cold seasons are unsatisfactory for extrinsic development, but in many endemic areas promiscuous defecation or the use of night soil for agricultural or garden crops provides unusually good opportunities for inoculation of the soil and for exposure of the people.

Temperature is a major factor in determining the geographic distribution and prevalence of the two common species of hookworm. The eggs and developmental stages of *Necator* are more tolerant of high temperature than are those of *Ancylostoma*. The eggs and developing larvae of *Ancylostoma* are able to tolerate low temperatures that retard or terminate development of *Necator* (Svensson, 1925; Matsusaki, 1963).

The pattern of rainfall is also important. During periods of dampness produced by rain or by moisture condensation (dew), the infective larvae migrate to the soil surface. They then migrate upward on grass or other surfaces as high as moisture permits, and readily penetrate the skin on contact with any part of the body. During periods of drying, the larvae retreat with the receding water film and escape below the surface. However, a high percentage of the larvae fail to escape desiccation and perish. Thus, alternate wetting and drying hasten the destruction of larvae from desiccation and, in the absence of added contamination, the soil soon becomes noninfective during periods of frequent rainfall (Beaver, 1953). Flushing rains also remove larvae from some locations and concentrate them in lower areas (Beaver, 1975). During periods of drought or low temperatures, the infective larvae may remain viable in the soil for several months. The migration of larvae against the capillary flow of water is illustrated by their behavior in the Harada-Mori fecal culture (Beaver, 1975).

An important natural factor in the generation of infective hookworm larvae in the soil is the burying action of dung beetles (Miller, 1954). In regions of high hookworm endemicity during the season(s) of transmission, human stools that are deposited on the soil at night, if not dissipated by rain or eaten by dogs, pigs, or other coprophagous animals before daybreak, are buried by dung beetles. Stools deposited during the day quickly disappear underground, buried by beetles (Miller, 1954). It is probable that the well-documented correlation between high hookworm prevalence and open sandy soils is largely a consequence of such soils being especially favorable for the feeding and propagation of large dung-burying beetles (Beaver, 1975). Eggs of hookworms and other helminths ingested with the feces by dung beetles are destroyed by the chewing mouth parts (Miller, 1961; Miller et al., 1961), but in burying a stool, the beetles produce a soil-feces mixture that favors development of larvae (Fig. 20–12). Other insects may also contribute to the dissemination of hookworm eggs in feces (Oyerinde, 1976).

The methods that are employed to determine the amount of hookworm infection in a community and how it is maintained are relatively well standardized. A preliminary survey of a representative cross-section of the population is done to determine *prevalence* and *intensity* of the infection. This survey is accomplished by one or another of the egg-count technics from which the relative worm burden is estimated. As egg counts are merely estimates of egg output, they cannot be used to determine the number of worms actually present in the individual cases. This phase of the survey provides information on the amount of low-grade, moderate, and severe infection in the entire community. Closely correlated is a study of the prevalence and intensity of soil contamination in the area. For recovering hookworm larvae from the soil, the Baermann apparatus or the damp gauze-pad method of Beaver (1953) have been employed (Schad et al., 1973).

Pathogenesis and Symptomatology: Cuta-

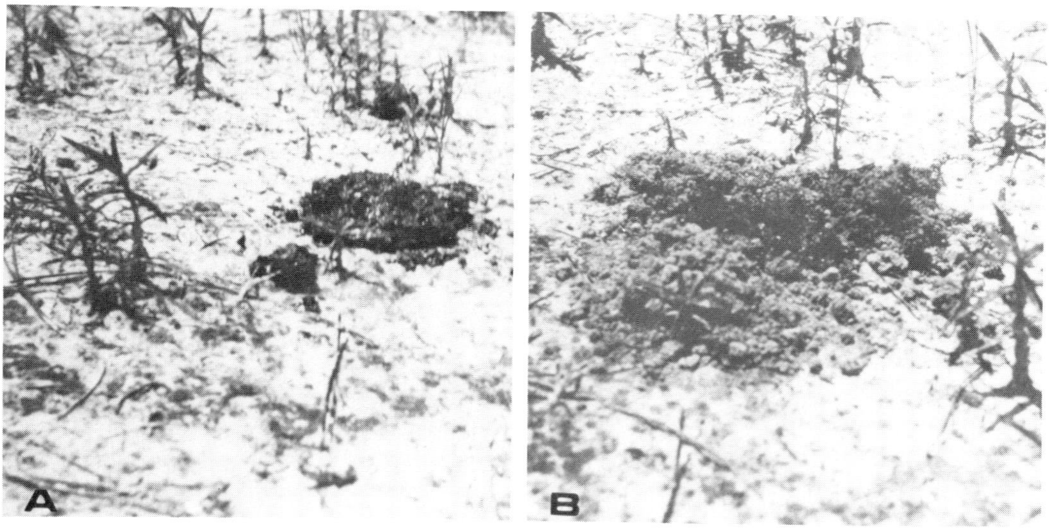

Fig. 20–12. Dung beetle activity. *A*, Human stool 15 minutes after deposit on sandy soil in early morning. A fecal ball has been formed and is being rolled away to be buried by tumble-bugs (*Canthon larvis*). *B*, The same stool 15 minutes later (30 minutes after deposit) has been buried and largely mixed with soil by large dung-burying species, *Phanaeus* and *Pinotus;* these beetles bury the feces by excavating the sand from beneath the stool, allowing it to fall below the surface. (From Miller, A. 1954. Dung beetles (Coleoptera, Scarobaeidae) and other insects in relation to human feces in a hookworm area of southern Georgia. Am. J. Trop. Med. Hyg., *3*:372–387.)

neous Lesions. The infective-stage larvae of hookworms actively enter skin wherever they make contact. At the site of entry mechanical damage to the skin is minimal, but the penetration of the larvae causes an itching sensation. In cases in which the number of larvae penetrating is great, this phase of the infection may be conspicuous because of the severe unrelieved pruritus. Within several hours allergic reaction to the worms or their products causes a pruritic, erythematous papular rash that may become vesicular. This reaction is known as "ground itch." Secondary bacterial infection is prone to occur if the lesions are abraded by scratching.

Occasionally *Necator* larvae may migrate in the skin, producing a "creeping eruption" similar to that produced by *A. braziliense* but of shorter duration and smaller dimensions (Beaver, 1964; Ball and Bartlett, 1969).

Creeping Eruption (Cutaneous Larva Migrans). Creeping eruption was recognized as a clinical entity long before a cause for it was known. Although the identity of the original discoverer is lost in antiquity, the earliest published report was that of Lee, an English physician, in 1874. Additional cases were reported, parasitic etiology being suspected in several, but not until 1896 in the fifth published case (Hammelstjerna, 1896) was an agent found, a *Gastrophilus* larva. Subsequent cases were reported from Europe and, beginning in 1902 (von Harlingin), from the United States. Following the studies on cutaneous larva migrans by Kirby-Smith and his collaborators in 1926, it became clear that a common causative agent of the condition in the southern United States is *Ancylostoma braziliense,* a hookworm of dogs and cats.

Cutaneous infection with *A. braziliense* in man has a wide distribution throughout the coastal areas of the United States from southern New Jersey to the Florida Keys on the Atlantic Coast, and along the Gulf of Mexico to Texas, extending inland in Texas as far as Dallas and San Antonio. Cases are most frequently seen in Florida, particularly in the area near Jacksonville. Creeping eruption known or presumed to be caused by *A. braziliense* has been reported from many subtropical and tropical coastal regions, including Brazil, Uruguay and Argentina, Spain, southern France, South Africa, India, the Philippines, and Australia.

Species of hookworms other than *A. braziliense* are also capable of producing cutaneous larva migrans in man. Known examples are the common cosmopolitan hookworm of dogs, *Ancylostoma caninum,* the European hookworm of dogs, *Uncinaria stenocephala,* the hookworm of cattle, *Bunostomum phlebotomum,* and others. *A. braziliense* is apparently the most common cause of the infection. It is noteworthy that transient creeping

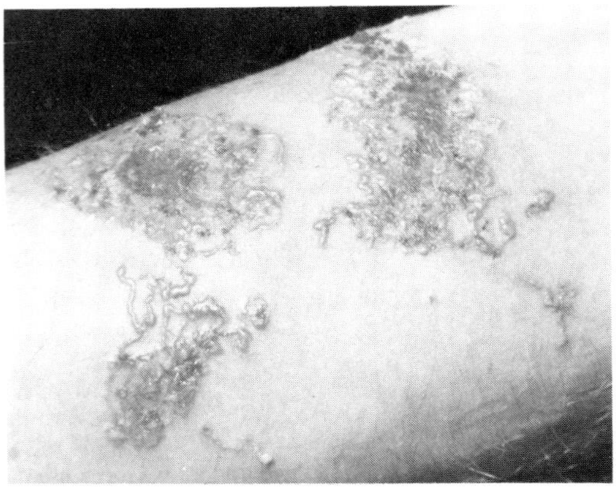

Fig. 20–13. *Necator americanus.* Larva migrans-like cutaneous lesion on the third day after exposure of a volunteer to about 200 larvae. A similar exposure 2 months earlier produced only slight erythema with no evident linear reaction. (From Beaver, P.C. 1964. Cutaneous larva migrans. Industrial Med. Surg., *33*:319–321.)

eruption can be caused by larvae of the human hookworm *Necator americanus* (Fig. 20–13) (Beaver, 1964; Ball and Bartlett, 1969). Since larvae in the skin are rarely recovered, and they are not readily or easily identified when found, the species involved in individual cases is usually unknown.

When the larvae of *A. braziliense* invade the skin, they produce itching papules, which in 2 or 3 days become serpiginous tunnels in the epidermis between the stratum germinativum and the corneum, accompanied by inflammation of the adjacent corium. The linear lesion, at first erythematous, soon becomes elevated and vesicular (Fig. 20–14). Movement of the larva through the skin may extend the lesion several millimeters per day, the older part of the tunnel becoming dry and crusty while the most advanced part is not yet evident (Fig. 20–15). Pruritus caused by the larva's movement and the reaction to it generally leads to scratching, which exposes the lesion to pyogenic organisms. Migration of larvae may continue for several weeks or months, resulting in extensive skin involvement. At times it becomes a serious illness.

Larvae of *A. braziliense, A. caninum,* and probably most other species of *Ancylostoma* are able to infect by mouth as well as by skin, and infection can be acquired through the ingestion of larvae in paratenic hosts. It is expected that, as in experimental animals, *Ancylostoma* larvae in man will invade the deep tissues as well as the skin, and those entering the skin and causing creeping eruption will later migrate to the deeper tissues. Creeping eruption, when extensive, is followed by pneumonitis, and the migrating larvae along with massive numbers of eosinophils and Charcot-Leyden crystals are found in the sputum (Muhleisen, 1953). *Ancylostoma* larvae also invade the cornea (Nadbath and Lawlor, 1965).

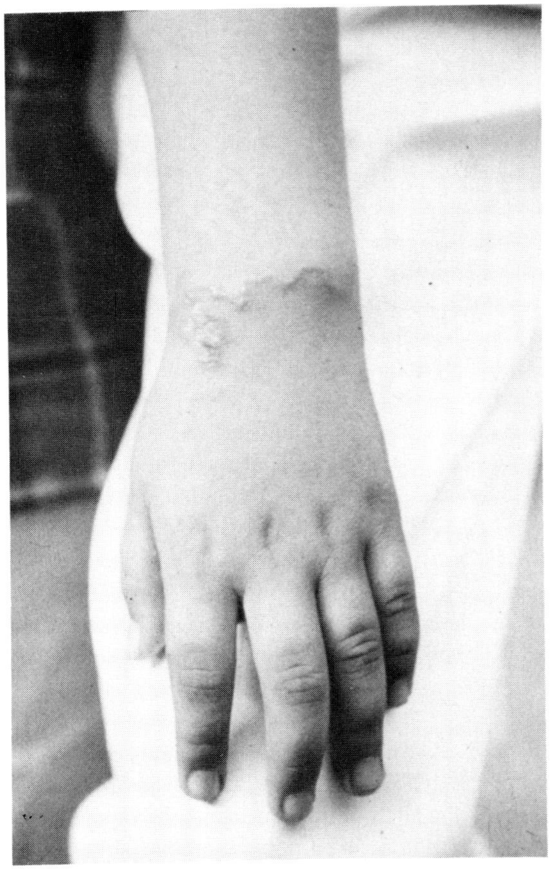

Fig. 20–14. Cutaneous larva migrans. Typical lesion produced by unidentified hookworm larva, probably *A. braziliense,* on arm of child in Merida, Mexico. (Courtesy of Dr. A. Muhlfordt and Dr. A. D'Alessandro.)

Diagnosis usually can be made on inspection of the lesions along with a characteristic history of exposure. Creeping eruption caused by hookworm larvae must be differentiated from that caused by fly larvae, such as *Hypoderma, Gasterophilus,* and others (see p. 690), and in Asia from that caused by *Gnathostoma* larvae (see p. 340).

Treatment for creeping eruption with thiabendazole was first reported to be effective by Stone and Mullins in 1965. It is now the treatment of choice, rarely failing to cure when administered in recommended dosage either by mouth or topically. When only one or a few larvae are involved, a

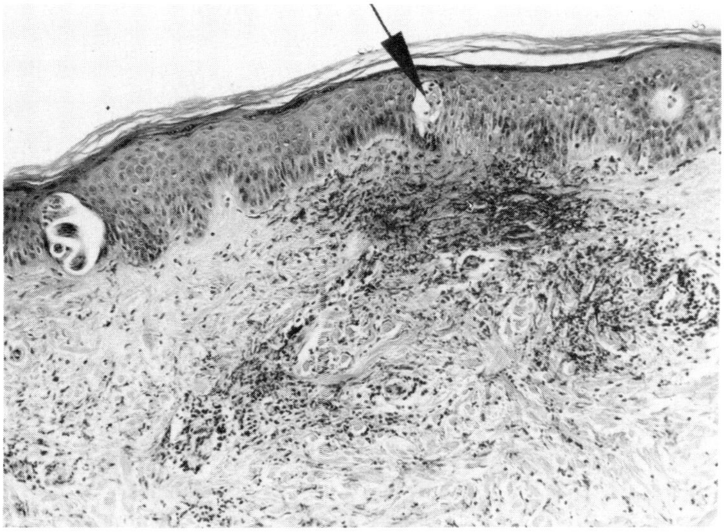

Fig. 20–15. *Ancylostoma braziliense*. Larva presumed to be *A. braziliense* in section of skin biopsy from a man with creeping eruption in Miami, Florida. (× 100.) As the area around the larva in the epidermis at left has no inflammatory infiltrate, whereas the trail behind the migrating larva (arrow) is marked by acute inflammatory reaction, attempts to obtain a biopsy specimen containing the larva are rarely successful. (By P.C. Beaver. Slide courtesy of Dr. Harvey Blank. From Faust, E.C., Beaver, P.C., and Jung, R.C. 1975. *Animal Agents and Vectors of Human Disease*, 4th ed. Philadelphia, Lea & Febiger.)

successful procedure is to remove them individually by producing a bleb by freezing the skin over the area around the advancing part of the lesion with ethyl chloride or solid carbon dioxide. As the larva is located in the epidermis (Fig. 20–15), it will be lost in the epidermal slough. Symptoms may be alleviated by antihistamines, antipruritics, sedatives, and anesthetic or protective coatings.

Infections producing creeping eruption are acquired from contact with moist or wet sandy or sandy-loam soil containing filariform larvae of hookworms generated from the feces of dogs and cats, usually in unprotected sandboxes, on bathing beaches, and under houses where workers lie prone to repair leaking water pipes.

Effective *preventive measures* are covering sandboxes when not in use, allowing no dogs or cats on bathing beaches, and spreading a waterproof sheet over damp work areas under houses or elsewhere if they may have been contaminated with dog or cat feces. As a further protective measure, and for other and better reasons, vagrant dogs and cats should be removed from areas of human habitation.

Pulmonary Lesions. The signs and symptoms observed following ingestion of *A. duodenale* larvae suggest that not only do larvae pass directly down to the small intestine where they develop, but others penetrate the mucous membranes of the mouth and pharynx carrying out a lung migration, as in the case of larvae penetrating the skin. The syndrome caused by these migrating larvae has been called Wakana disease in Japan. The initial symptoms, noted soon after ingestion, are nausea, vomiting, salivation, itching of the pharynx, and hoarseness. There follows an illness of several days' duration consisting of cough, dyspnea, hypereosinophilia, nausea, and vomiting. Infiltrations may be visible on chest x-ray. Harada (1962) has reported that these symptoms may be due to an allergic reaction to the larvae, which may occur even if the worms have been killed by boiling.

The migrating larvae of *N. americanus* and *A. duodenale* usually reach the pulmonary capillaries almost immediately after entry into the cutaneous venules, though they tend to remain in the skin for several hours after penetration. As they break out of the lung capillaries into the alveoli, they cause minute focal hemorrhages, but only in the case of massive migration of larvae simultaneously is pneumonitis of clinical grade usually produced. Hookworm larvae during pulmonary transit do not typically cause the high degree of sensitization characteristic of *Ascaris* or *Strongyloides*. Nevertheless, as in the case of Wakana disease of Japan, symptoms may be produced *(vide supra)*. Likewise, severe cough has been noted following heavy percutaneous infection with *A. duodenale* (Rogers and Dammin, 1946), and with massive creeping eruption presumably due to *A. braziliense* (Muhleisen, 1953).

Intestinal Infection and Hookworm Disease. On arrival in the small intestine, the worms attach themselves to the intestinal mucosa, first by a temporary head capsule, later by a permanent one, and begin to digest the portions of villi that are sucked into their mouths. Blood is sucked out of the tissues, but greater amounts of blood are lost by bleeding at sites of attachment. This may be facilitated by an anticoagulant substance, as has been observed to be produced by *A. caninum* (Eiff, 1966; Thorson, 1956; Spellman and Nossel, 1971). Occasionally, a worm may become embedded in the wall of the intestine (Piza and Biagi, 1963).

Previous sites of attachment are often abandoned and may remain for a time as open ulcers, but bleeding tends to cease shortly after the worm leaves a given site.

In *acute* heavy hookworm infections of the intestine, fatigue, nausea, and vomiting occur, followed by burning and cramping abdominal pain. These symptoms appear to be more common in *A. duodenale* infections than in infections with *N. americanus*. Diarrhea appears, with stools that vary from black to red depending on the amount and rate of blood loss. Weakness and pallor rapidly develop. Extremely heavy infections, especially in infants, may present as massive intestinal hemorrhage (De la Riva et al., 1981; McDowell, 1981), which may prompt unnecessary surgery or may cause death due to shock (Rogers and Dammin, 1946; Naik et al., 1976). During this period an eosinophilic leukocytosis and Charcot-Leyden crystals in the feces may persist for several weeks.

In *chronic* hookworm infection the signs and symptoms are essentially those of a profound iron-deficiency anemia. In varying degrees there are pallor, facial and pedal edema, dull expression, and listlessness or apathy. Hemoglobin levels may be reduced to 5 g/dl or lower, and the heart is enlarged. Functional (hemic) murmurs appear. In children, mental and physical development may be greatly retarded.

This clinical entity is called "hookworm disease," a term that has come to refer specifically to the microcytic, hypochromic anemia caused by intestinal hookworm infection and not to other manifestations (e.g., cutaneous or pulmonary).

The essential damage produced in intestinal hookworm infection is hemorrhage from the intestinal wall. In general this is proportional to the number of attached worms, but blood loss is disproportionately great when the worms are crowded, especially in the early period of infection when mating competition tends to be synchronous (Beaver et al., 1964). In such situations, especially in infants, the disease may present as massive intestinal hemorrhage caused by laceration of the mucosa as described previously. The typical case, however, involves chronic blood loss with inadequate replacement of iron and sometimes of protein. The result is an iron deficiency, microcytic hypochromic anemia.

The clinical severity of the infection is closely related to the worm burden. Thus, counts of less than 5 eggs per milligram of feces are seldom of clinical grade, whereas counts of more than 20 eggs per milligram are often associated with significant anemia. In massive infections, egg-counts exceed 50 per milligram of feces.

Good nutrition does not protect the individual from hookworm infection (Roche and Layrisse, 1966), nor does nutritional repletion appreciably affect an existing hookworm population (Tripathy et al., 1971). Hookworm *disease* can, however, be prevented or overcome by an adequate, well-balanced diet containing iron, other minerals, and animal protein if the hookworm infection is light or only moderately heavy. In heavy infections, adequate absorption of necessary iron and other nutrients is not possible even with highly supplemented diets. It was observed recently (Onwuliri et al., 1981), as earlier (Beaver, 1951; Krupp, 1961), that in the etiology of hookworm disease there is a threshold component, not merely a continuous function of worm burden. This is explained as due not to the worms' blood sucking, but to the bleeding from multiple points of laceration, which occurs when crowding causes constant movement and reattachment of worms in the overpopulated intestine. Blood losses from the worms' feeding activities may be relatively insignificant (Kalkofen, 1970). The conclusion that hookworms feed on mucosal tissues and generally do not ingest blood is supported by observations on the buccal capsule and associated glands of *Bunostomum trigonocephalum,* the hookworm of sheep (Wilfred and Lee, 1981).

It has been noted that the chronic anemia of hookworm disease causes physiologic compensations in the host (Porter, 1937), namely, increased pulmonary vital capacity, increased tolerance of tissue cells to anoxia, decreased systolic pressure, and reduced peripheral blood flow. In addition to dilatation, the heart may show increased collateral circulation of coronary arteries with consequent decreased risk of myocardial infarction.

The x-ray picture of chronic hookworm disease is one of intestinal hypermotility, dilatation of the proximal jejunum, and coarsening of the mucosal folds.

Diagnosis. Combined with a subnormal level of hemoglobin, a high number of hookworm eggs in the feces is a sign of hookworm disease. The threshold level of significant egg-count for well-nourished individuals is about 20 eggs per mg (20,000/g); for those with iron or protein deficiency, egg-counts of 5 per mg or higher have been classified as significant (Andrews, 1942a, 1942b). By physical examination alone, hookworm disease is not distinguishable from the anemia of malnutrition. Estimates of the number of eggs in the feces based on egg-counts made by any standard method are adequate for clinical purposes. The presence of Charcot-Leyden crystals in the feces in the absence of eggs may indicate heavy hookworm infection in the prepatent period (Rogers and Dammin, 1946). In fecal specimens stored for more than 24 hours at room temperature, hookworm larvae may develop and hatch. The free larvae must be differentiated from those of *Strongyloides* and free-living nematodes.

Treatment. In cases of hookworm infection without clinical disease or with only moderate anemia, treatment consists of administration of a specific anthelmintic. The current drugs of choice are mebendazole, 100 mg twice daily for 3 days, or pyrantel pamoate in a single dose of 11 mg per kilogram of body weight (maximum 1 g) (Migasena et al., 1978). Thiabendazole in a dosage of 25 mg per kilogram of body weight twice daily for 2 days is somewhat less effective.

Tetrachloroethylene, though largely supplanted in this country by the aforementioned drugs, which are less toxic, is still a useful drug and is the standard with which the newer anthelmintics are compared. Unfortunately it is not readily available in the United States for administration to humans. It is sold commercially in soft gelatin capsules as a veterinary remedy. It is given in a single dose of 0.12 ml per kilogram of body weight (maximum 5 ml) after fasting and without purgation (Carr et al., 1954; Migasena et al., 1978). Side effects are dizziness, nausea, and symptoms resembling alcoholic intoxication, such as slurred speech and staggering.

Bephenium hydroxynaphthoate has been employed with success in endemic areas of *Ancylostoma duodenale* infection (Goodwin et al., 1958), but is relatively ineffective in the elimination of *Necator.* Best results are obtained when it is given orally in the morning on an empty stomach in a single dose of 5 g for adults, 2 to 4 g for children depending on age, with no post-treatment purgation and with food permitted 3 hours after dosage. Principal side effects are nausea and vomiting in a small percentage of recipients in most areas, although in the southern United States, where *A. duodenale* is not found, these side effects were frequent (Jung and McCroan, 1960).

The symptoms of Wakana disease may be relieved by administration of adrenocorticosteroid drugs (Yoshida and Nakanishi, 1958).

Prevention and Control. Hookworm infection of clinical grade is most prevalent among poor people in rural communities where sanitary disposal of human feces is inadequate or lacking. When the hookworm life cycle and mode of infection became known, it was immediately apparent that the basic approach to control would be through the sanitary disposal of feces. In addition, when it was shown that certain drugs were effective in expelling hookworms from the intestine, a second approach to control was introduced. Diagnosis by microscopic examination of feces for the presence of eggs had been established much earlier. Beginning in 1910, these events led to extensive programs of control in the hookworm areas of southeastern United States and elsewhere, with three objectives: (1) fecal examination of all members of the community and treatment of all who were passing eggs, (2) education of the people regarding the transmission and the nature of hookworm disease, and (3) assistance in the construction of latrines and encouragement of their use and proper maintenance. Interruption of transmission remains today the basic objective of control programs, but in some instances greater emphasis has been placed on disease control than on infection control. In addition, the most effective community or country-wide programs have been integrated into the general health services as in Japan (Komiya and Kobayashi, 1965; Morishita, 1966) and the United States (Andrews, 1942; Eberson, 1980).

Periodic Treatment. In livestock production it has been shown that for general health maintenance the elimination of all worm parasites from all animals is not essential, and that with the scheduled use of anthelmintics worm burdens can be held to subclinical levels (Campbell, 1977).

Where transmission is seasonal, as it is in regions where the low temperatures of winter or the lack of moisture during dry season(s) eliminate the infective-stage larvae from the soil surface, annual or more frequent mass fecal examinations followed by treatment of all infected individuals will effectively reduce the number and intensity of hookworm infections. When combined with programs of education of the public and the proper disposal of human feces, such programs will eventually eliminate hookworm disease as a public health problem. In Japan, hookworm prevalence rates were in this manner reduced from 23% in 1926 to less than 10% in 1941, and less than 3% in 1961. In 1964 hookworm eggs were found in only 2.6% of the 6,031,346 stools examined (Morishita, 1966). Similar but more restricted and less sustained efforts with less success have been undertaken in Korea, Taiwan, and elsewhere. In general, periodic treatment programs have been less successful against hookworm infection than they have against ascariasis, and in control programs emphasizing sanitation it was observed that poorly maintained latrines increased rather than decreased the risk of hookworm infection (Stürchler et al., 1980).

Hookworm Disease Control. A program designed to be carried out by local health departments was set up in Georgia (United States) in 1940 (Andrews, 1942a,b). Annual selective surveys were conducted in the schools in all communities where hookworm anemia was a disease of major concern. The children with both hookworm infection and anemia were identified, and members of their families were similarly investigated. All the infected persons found to be anemic and others carrying a hookworm burden sufficient to cause anemia (5 or more eggs per mg feces) were treated, repeatedly if necessary, to eradicate or satisfactorily reduce the worm burden. In 1970 to 1971, 30 years after the program was started, examination of a random sample from the entire former endemic area showed that in those communities where the soil and customs of the people were most favorable for transmission, in those counties where in 1950 over 50% of the school children had been infected with hookworms and over 40% of the infected group had had moderate to heavy worm burdens (Beaver, 1951), only 125 of 1,039 children (12%) had hookworm infections and only one-fourth of the infections were moderate to heavy (Martin, 1972a). Among the strictly rural families, 16% of the white and 8% of the black members had hookworm infections (Martin, 1972b). The season of hookworm transmission in southeastern United States extends from March through November, but most of the infections are acquired during the period of summer rains, from mid-June to mid-September. Examination of children with apparent anemia in late autumn and early winter detects the most significant hookworm infections early. This kind of disease control in communities where other forms of transmission control are not feasible apparently has not been implemented elsewhere.

Disinfection of soil for the control of hookworm infection in dogs in kennels may be effective in some circumstances, but this approach has not been applied successfully to the control of human hookworm disease. Aqueous solutions of iodine in concentrations of 50 to 60 parts per million at temperatures of 15° to 30°C kill hookworm larvae in about 15 minutes or less (Thitasut, 1961). Natural food-preservative substances, such as salt, sugar, acetic acid, onion, garlic, mustard, pepper, and spices, are lethal to hookworm eggs and larvae (Soh, 1960). Garlic ingested in amounts of 15 to 21 grams daily was observed to have no significant effect on the egg production of either *Necator* or *Ancylostoma,* but a high proportion of the eggs in feces passed after ingestion of garlic failed to develop in Harada-Mori fecal cultures (Bastidas, 1969). In countries where the use of night soil is a necessity, composting of garbage mixed with night soil or proper storage of feces and urine before use can destroy the eggs of hookworms and some other parasites.

Although encouraging people to wear shoes as a protective measure against hookworm infection is recommended, shoes that would provide protection are either too expensive or, if made of rubber or plastic, unsuitable for the tropics. Hookworm infection is acquired only in wet situations, and expensive shoes generally would be removed to protect them from water damage and to avoid the discomfort of wet feet in wet shoes. Impermeable shoes are cheap and they are protective against hookworm larvae, but in hot, wet weather they are extremely uncomfortable and tend to cause troublesome dermatophytoses. The barefoot habit is based on such practical considerations, and is so strongly preferred in situations that favor hookworm transmission, that to encourage the purchase and habitual use of shoes as a protective measure is unrealistic.

Measurement of Prevalence and Intensity. Darling (1922) and Cort (1921–1925) emphasized the need for accurate information on the prevalence of infection and the relative frequency of heavy worm burdens in a given population, as well as the numbers and location of infective larvae resulting from fecal contamination of the soil. In early studies, the Baermann apparatus for isolating hookworm larvae from soil provided a means of determining the degree of soil infestation. Later this was replaced by the more efficient damp gauze-pad method (Beaver, 1953). Egg-counting techniques supplanted the cumbersome method of recovering and counting the worms expelled by anthelmintic treatment. First to be developed was the dilution technique (Stoll and Hausheer, 1926), followed by the standard direct smear (Beaver, 1950), and more recently by the Kato thick-smear technique (Martin and Beaver, 1968). Any of these procedures can be used satisfactorily to estimate the prevalence of heavy, moderate, and light infections.

Superfamily Strongyloidea

Ternidens deminutus Railliet and Henry, 1909

Ternidens deminutus (Fig. 20–13) is a relatively common parasite of the large intestine of several species of simian hosts in Africa, India, and Indonesia, and has been found on several occasions in man from Southeast and South Africa. Sandground (1931) found it to be common in natives of Zimbabwe (50 to 60% infection). Goldsmid (1971) found it in up to 87% of Africans in some communities and pointed out that it is not found in Europeans. Superficially *Ternidens* may be mistaken for a hookworm, but it can be readily distinguished by the anteriorly directed buccal capsule, which is guarded by a double crown of 22 stout bristles (Fig. 20–16,*A*). The generic name of the worm *(Ternidens)* refers to the three complex teeth in the hollow of the subglobose buccal cavity, each bearing three lamellae and arising from a lobe of the esophagus. On the anterior surface of the buccal capsule immediately around the mouth there are four knobbed submedian papillae. The transparent ovoid eggs closely resemble those of human hookworms but are appreciably larger (84 by 51 μm). The male worms in baboons have a mean length of 8 mm, in humans 10 mm. Females are slightly larger. The bursa is well developed, with relatively stout rays forming a characteristic pattern. The copulatory spicules are long, plain, and strong, and the gubernaculum is narrow and thick at its proximal end, broad and thin distally. The females have a protuberant vulva a short distance anterior to the anal opening. Third-stage larvae develop from eggs in soil or Harada-Mori cultures in 8 to 10 days. A pair of sphincter cells at the junction of the esophagus and intestine distinguishes third-stage (filariform) larvae of *T. deminutus* from other nematode larvae of this stage with which they might be confused (Goldsmid, 1967). The life cycle is incompletely known. Attempts to infect human volunteers and lower primates by mouth or by skin penetration were consistently unsuccessful (Goldsmid, 1971). These worms inhabit the wall of the large bowel, where they may produce ulceration and cystic nodules. Thiabendazole gives cure rates of over 90%. Pyrantel pamoate is equally effective and is relatively free of side effects (Goldsmid, 1972; Goldsmid and Saunders, 1972).

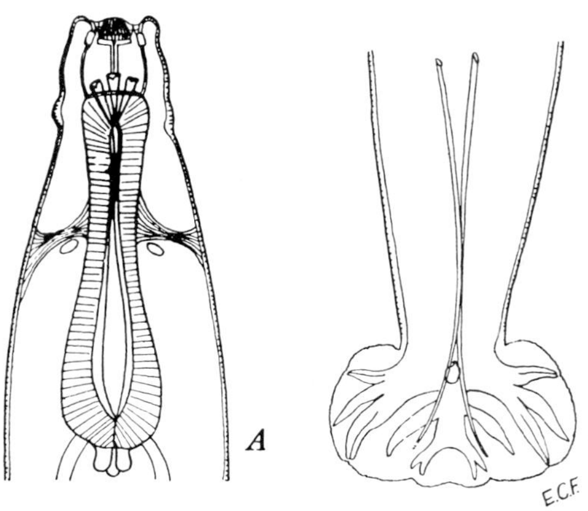

Fig. 20–16. *Ternidens deminutus. A,* Anterior end of body, showing buccal structures and esophagus. (× 35.) *B,* Posterior end of male, showing spicules and bursa.(× 35.) (Adapted by E.C. Faust from originals by A. Railliet and A.C. Henry. In Brumpt, E. 1927. *Précis de Parasitologie.* Paris, Masson et Cie.)

Oesophagostomum Species (Causing oesophagostomiasis or helminthoma)

Species of the genus *Oesophagostomum* are common parasites of monkeys and apes in Africa, Asia, and South America. Certain species are also common in swine and sheep, where they are referred to as "nodular worms." They produce large nodules in the wall of the intestine, mainly the cecum and colon. They resemble hookworms in size but differ markedly in the character of the buccal capsule which, as in *Ternidens* and other strongylids, is cylindrical and opens forward with a crown of bristles or setae (corona radiata) around the mouth.

On the ventral side at the level of the excretory pore there is a transverse groove and the cuticle between it and the mouth is perceptibly inflated (Fig. 20–17,*A*). The bursa is symmetrical and moderately developed (Fig. 20–17,*B, C),* the spicules long, slender, and equal. The vulva and short vagina are situated near the anus, and the two branches of the uterus extend forward from a heavy walled ovejector (Fig. 20–17,*D*). The eggs resemble those of the human hookworms (about 60 by 30 to 40 μm).

In the normal host the adult worms are attached to the mucosa of the cecum and colon. Eggs are in early cleavage when discharged and, as in hookworms, there are three free-living larval stages. Infection is acquired when the third-stage larvae are swallowed. The site of early development is the wall of the cecum, in large, firm encapsulating nodules. When development is complete, the worms emerge, attach to the mucosa, mate, and eventually produce eggs that are discharged in the feces.

In human infections the larvae produce tumor-like exudative nodules and abscesses, usually 1 to 2 cm in diameter, occasionally much larger, in the wall of the intestine, in various other organs of the abdomen, and in the abdominal wall (Gordon et al., 1969; Kaminsky and Ndinya-Achola, 1977). According to Anthony and McAdam (1972) who reviewed 34 cases seen

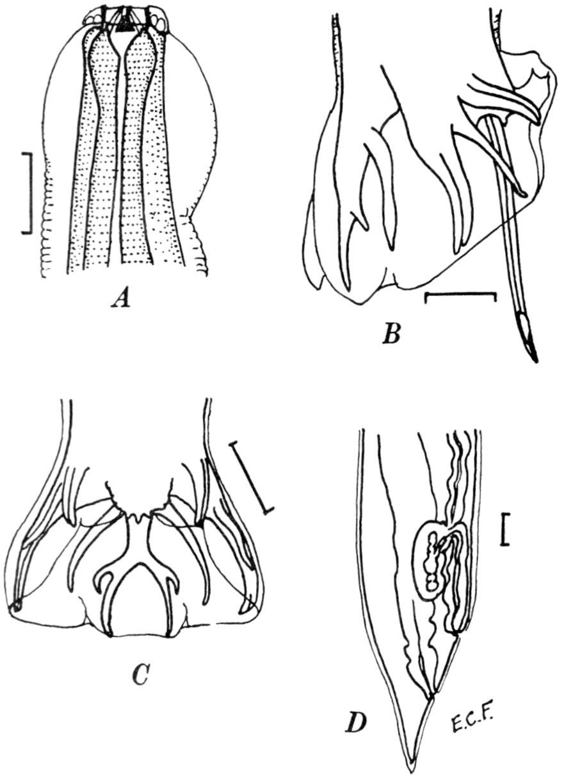

Fig. 20–17. *Oesophagostomum apiostomum. A,* Head-end of worm, showing external corona radiata, head papillae, and cuticular pattern. *B,* Posterior end of male, lateral view, showing copulatory bursa with fleshy rays and copulatory spicules. *C,* Posterior end of male, ventral view. *D,* Posterior end of female, lateral view, showing rectum and terminal genitalia. The line adjacent to each figure equals 0.1 mm. (Adapted from Lie Kian Joe, 1949. Helminthiasis of the intestinal wall caused by *Oesophagostomum apiostomun* [Willach, 1891] Railliet and Henry, 1905. Doc. Neerl, et Indon. de Morbis Tropicis, *1*.)

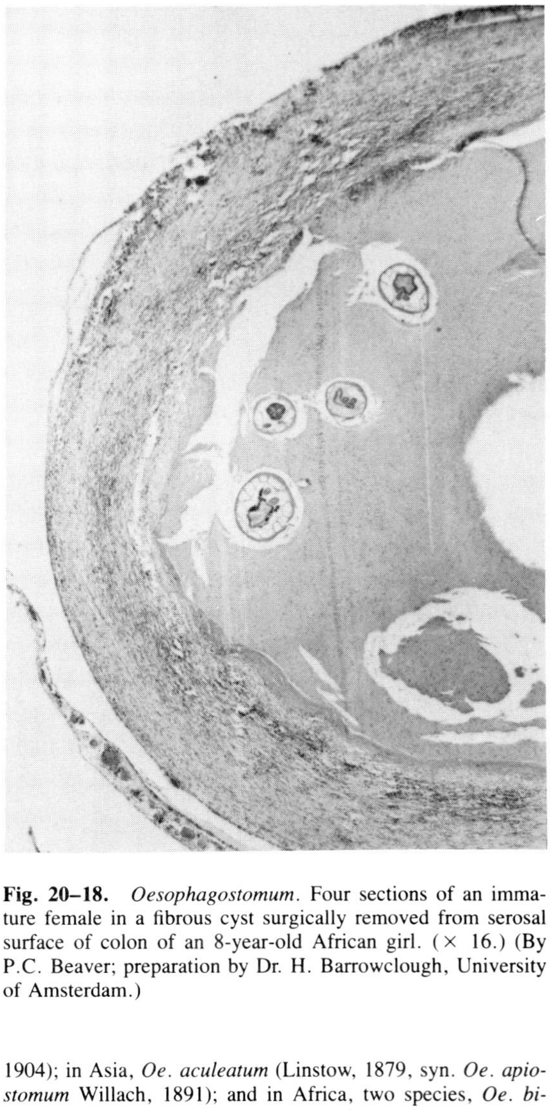

Fig. 20–18. *Oesophagostomum.* Four sections of an immature female in a fibrous cyst surgically removed from serosal surface of colon of an 8-year-old African girl. (× 16.) (By P.C. Beaver; preparation by Dr. H. Barrowclough, University of Amsterdam.)

in Uganda, the most common condition observed is abdominal pain caused by a solitary tumor-like mass or abscess (helminthoma) in the ileocecal region (Fig. 20–18). Other levels of the large and small intestines are affected less commonly and the nodules may be multiple (Barrowclough and Crome, 1979). Clinical diagnosis is difficult, even at laparotomy, and as carcinoma, appendicitis, ileocecal tuberculosis, and ameboma are frequently simulated, unnecessary radical surgery may be the result (Anthony and McAdam, 1972).

Most of the recorded cases of oesophagostomiasis were reported from East and West Africa, in both Africans and Europeans; individual cases have been reported from Indonesia and Brazil.

The worms are most frequently observed in sections of nodules, where by the character of the body wall and inner organs they can be readily distinguished from all other nematodes except hookworms (Fig. 20–18). Even with more favorable material at hand, the species involved in human infections have been only presumptively identified, usually as *Oe. apiostomum* or *Oe. stephanostomum*. A worm with the appearance of a schistosome in a section of nodular colon in one instance was presumed to be *Oe. brumpti* (Leoutsakos et al., 1977). Chabaud and Larivière (1958) concluded that the species found in man in South America probably is *Oe. stephanostomum* (Stossich, 1904); in Asia, *Oe. aculeatum* (Linstow, 1879, syn. *Oe. apiostomum* Willach, 1891); and in Africa, two species, *Oe. bifurcum* (Creplin, 1894) and *Oe. stephanostomum*. In cases reported by Haaf and Van Soest (1964) from Ghana, live worms were recovered from nodules in the large intestine and identified as *Oe. bifurcum*.

Oesophagostomiasis is generally self-limiting, occasionally requiring surgical intervention to relieve mechanical problems produced by a nodule or abscess (Barrowclough and Crome, 1979). No anthelmintic is known to be effective against the worms in tissues.

Mammomonogamus laryngeus (Railliet, 1899) Ryjikov, 1948

Synonyms. *Syngamus laryngeus* Railliet, 1899; *S. kingi* Leiper, 1913; *Cyathostoma* of St. John, Simmons and Gardner, 1929.

Morphology and Biology. Members of the genus *Mammomonogamus* are commonly found in the upper respiratory tract of certain mammals, including cattle and felines. The thick-walled buccal capsule is directed anteriad and is armed in its inner base with eight subequal teeth (Fig. 20–19). It is provided with a thick muscular wall down to its junction with the esoph-

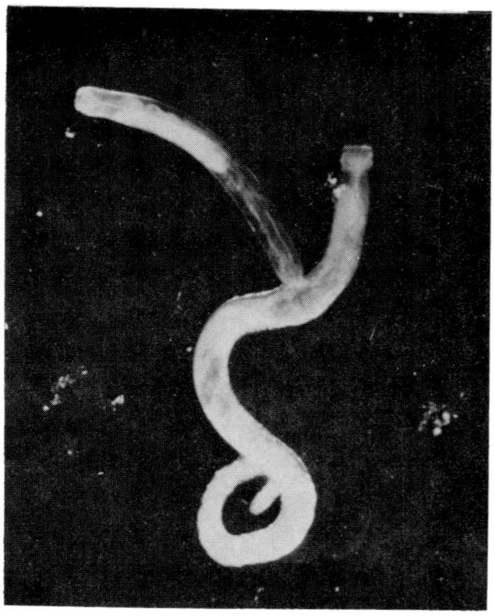

Fig. 20–19. *Mammomonogamus laryngeus*. Female and male, permanently in copula, removed at bronchoscopy from mucosa of bronchus of Australian woman who had been in Guyana. (× 8.) (From Birrell, D.J., Moorhouse, D.E., Gardner, M.A.H., and May, C.S. 1978. Chronic cough and haemoptysis due to a nematode, "*Syngamus laryngeus*." Aust. N.Z. J. Med., *8*:168–170.)

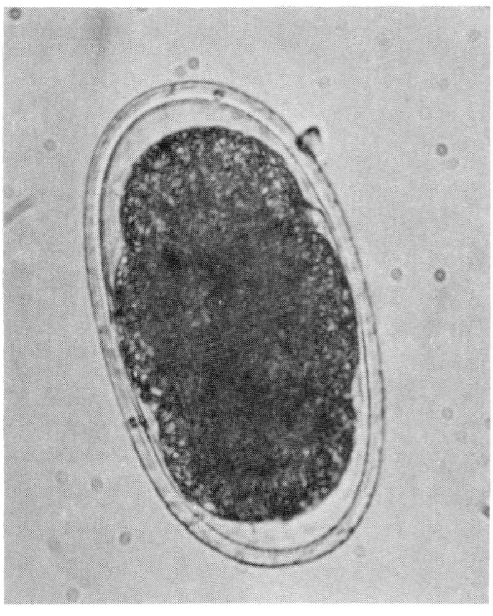

Fig. 20–20. *Mammomonogamus laryngeus*. Egg from feces of patient. (× 640.) (Courtesy of Dr. D.E. Moorhouse.)

agus. Immediately around the oral opening there is a thick cuticular ring, surrounding which there are a pair of broad dorsal and ventral petal-like lips and a pair each of dorsolateral and ventrolateral lips. In the notch between each two lips there is a minute papilla. The male worm is considerably smaller than the female, and the two are permanently joined *in copula* (Fig. 20–19). The copulatory bursa and its supporting rays are both generically and specifically characteristic. The copulatory spicules are typically short and thick. The vulva is situated a short distance in front of the middle plane of the body of the female, which has an acuminate posterior end. In the avian species of *Syngamus* the ovoid eggs, with characteristic sculpturing, are provided with polar caps, but in the mammalian species these caps are absent (Fig. 20–20). Although earthworms serve as intermediate host for avian species of *Syngamus*, the life cycle of *M. laryngeus* has not been elucidated.

Epidemiology. According to Mornex et al. (1980), 78 cases of human infection have been reported, mostly from tropical America. Fifty-one were reported from Martinique or from France in people who had been in Martinique. Others were reported from Puerto Rico, Dominica, St. Lucia, Trinidad, Guyana, Brazil, and Philippines. Cases in which the infection was acquired in Guyana (Birrell et al., 1978) and Dominica (Grell et al., 1978) but diagnosed elsewhere are typical. All these were apparently infections with the bovine species, *M. laryngeus*, which parasitizes the upper respiratory tract of cattle, water buffaloes, and goats. Buckley (1933) was inclined, however, to refer one specimen to *M. nasicola* of which goats are the usual hosts. Little is known of the life cycle, although the third larval stage is probably infective for the vertebrate host. Larval stages of *S. trachea*, the fowl gapeworm, have been found in earthworms and aquatic snails in addition to several types of arthropods.

Clinical Notes. The worms in the trachea produce irritative symptoms, usually accompanied by hemoptysis and at times asthma. Discharge of the worms in sputum after violent attacks of coughing and recovery of the characteristic eggs in sputum or feces provide means of diagnosis. Specific therapy in man has not been studied.

Superfamily Trichostrongyloidea

Trichostrongylus species

Members of this group are commonly parasitic in the digestive tract of herbivorous animals throughout the world. Most species occur only as incidental infections of man, but a few are of considerable clinical and public health importance.

Geographical Distribution. The following species of this genus have been reported from man: (1) *Trichostrongylus colubriformis*, Egypt, India, and Assam (extensive distribution), Armenia, the Atherton tableland of Australia, once in New Orleans, Louisiana, Japan, and Java; (2) *T. probolurus*, North Africa, Armenia, and Siberia, U.S.S.R.; (3) *T. vitrinus*, Egypt, Armenia, and Siberia; (4) *T. instabilis*, Armenia and Siberia; (5) *T. axei*, Armenia, Mauritius, Siberia, Japan, and Java; (6) *T. orientalis*, Japan, Korea, central and south China, Taiwan, Iran, and Armenia; (7) *T. skrjabini*, Armenia; and (8) *T. brevis*, Japan. Undiagnosed species of *Trichostrongylus* have been reported from man in Tunisia, from Europeans in the former Belgian Congo, from natives of Zimbabwe, Turkey, Iraq, Basra, eastern Saudi Arabia, from a Greek resident of St. Louis, Missouri, from people in Hawaii, Fiji, Chile, and Peru, and from Iraqi immigrants to Israel.

In Armenia, where six species of trichostrongylids have been identified as human parasites, Kalantarian (1927) reported a 15% prevalence; Watson (1953) found up to 25% infection with *T. colubriformis* in some localities of Iraq; Sahba et al. (1967) reported that this was the most common infection in a rural area of southwest Iran where a 69% prevalence was found; Lawless et al. (1956) found that 70% of the adult population in an Egyptian village were infected.

Otsuru (1962) found a 40% prevalence of trichostrongyles in some villages of Niigata Prefecture on the island of Hokkaido, Japan; the majority were *T. orientalis*, with a few each of *T. colubriformis* and *T. axei* and of a new species, *T. brevis*, so named because of the short copulatory spicules.

In extensive parasitologic surveys in Japan, Hunter et al. (1950–1953) reported *Trichostrongylus* sp. (*T. orientalis?*) as follows: Fukui Prefecture, 3.5%; Yamanashi Prefecture, 44.4%; and Shikoku Island, 3.8%, indicating that special environmental conditions favored human infection with this nematode in Yamanashi Prefecture.

In a summary of records in Iran, Ghadirian and Arfaa (1975) reported that seven species of *Trichostrongylus* are common in man and animals in most parts of the country, all of which are found in the region of Isfahan in central Iran where the overall prevalence was 67%, with *T. orientalis* and *T. colubriformis* the predominant species. Other species present were *T. vitrinus*, *T. axei*, *T. capricola*, *T. probolurus*, and *T. skrjabini*. All the species found in man (except an unidentified, possibly new species) were also found in sheep, goats, cattle, and camels. In Iran, as in other regions, *T. orientalis* is considered to be predominantly a parasite of man.

Morphology, Biology, and Life Cycle. These worms, which live embedded in the mucosa of the small intestine, are small and slender, with a minute, unarmed head, lack a distinct buccal capsule, and have a conspicuous ventral notch in the cervical region where the excretory pore opens. The males have a copulatory bursa with long lateral and poorly developed dorsal rays (Fig. 20–21). The spicules (Fig. 20–22) are brownish, stout, and sculptured differently in each species. They constitute the most useful characters for differential species diagnosis. The paired genital tubules in the female open through a common vulva, which is situated a short distance behind the mid-body level. The eggs are generically characteristic; they are oval-elongated, hyaline-shelled, and resemble those of hookworms (Fig. 20–4), except for the longer, narrower shape and more pointed ends (Fig. 20–23). *Trichostrongylus* eggs are usually discharged in the morula stage of embryonation and, under favorable conditions of moisture and temperature, may hatch within 24 hours. The hatched larvae are pseudorhabditoid in type (Fig. 19–8,*C*, p. 263) with a long capillary, pre-esophageal cavity, an esophagus with a long, ovoid, anterior portion, a constricted portion behind the esophageal nerve ring, and a typical posterior bulbous swelling. At the anal region they have a noticeable dorsal bend. The postanal portion is attenuated and terminates in a minute knob. There are three free-living larval stages with two ecdyses. They may develop to the infective stage in a minimum of 60 hours. These semifilariform third-stage larvae are resistant to desiccation.

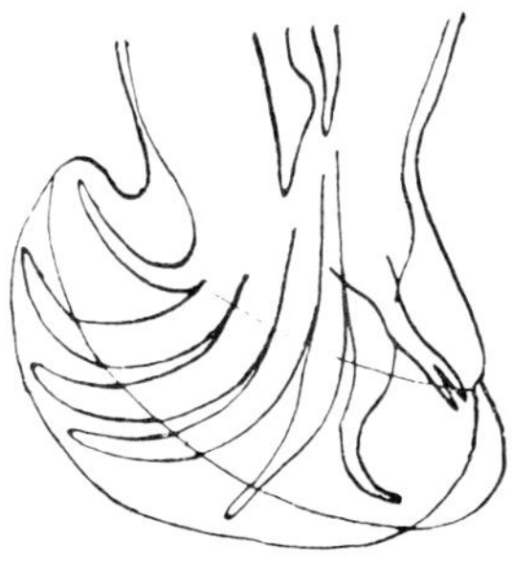

Fig. 20–21. *Trichostrongylus orientalis*. Male bursa, lateral view. (× 25.) (By E.C. Faust.)

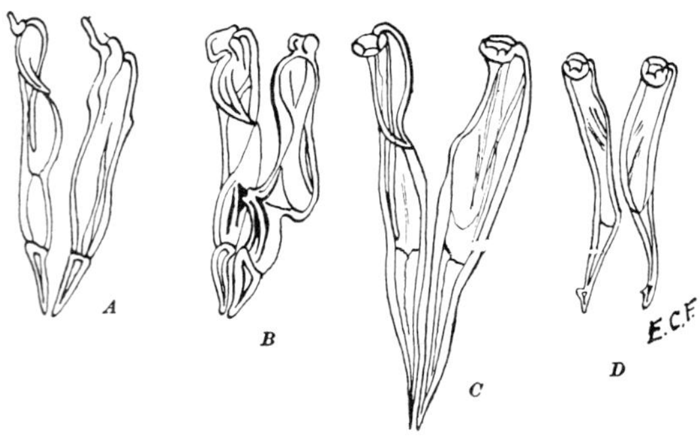

Fig. 20–22. *Trichostrongylus*. Spicules of species commonly found in man. *A*, *T. colubriformis*. *B*, *T. probolurus*. *C*, *T. vitrinus*. *D*, *T. orientalis*. (× 250.) (By E.C. Faust.)

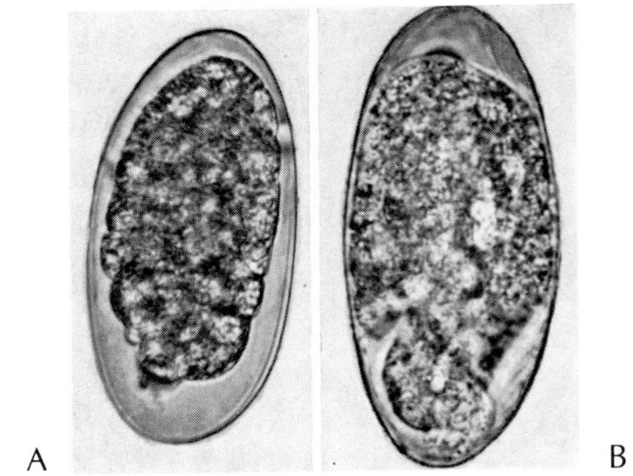

Fig. 20–23. *Trichostrongylus orientalis*. Eggs. *A*, As seen in fresh feces. *B*, With nearly developed rhabditoid larva. (× 500.) (Courtesy of Dr. T.B. Magath.)

While infective-stage larvae may at times be induced to penetrate the skin, infection is normally contracted by their ingestion as a contamination of vegetation. On reaching the small intestine after being swallowed, they migrate into the mucosa and develop to sexual maturity in about 3 to 4 weeks without requiring migration through the lungs. This parasitic development is divided into three stages by two additional ecdyses.

Pathogenesis, Pathology, and Symptomatology. In infections with *Trichostrongylus*, the primary lesion is produced by maturing and adult worms in sufficient numbers to cause traumatic damage to the mucosa of the small intestine. Desquamation with hemorrhage may occur at the sites of intestinal infection, but usually several hundred worms are necessary to provoke clinical manifestations.

Diagnosis. A diagnosis may be made by recovering the characteristic eggs (Fig. 20–23) in the feces, or by culturing feces and recovering the typical larvae. These require differentiation from similar stages of hookworms, and the larvae from those of *Strongyloides*. The eggs of *Trichostrongylus* are readily concentrated by brine and zinc sulfate techniques (see p. 742).

Treatment. Bephenium hydroxynaphthoate and 1-bromo-*B*.-naphthol (Otsuru, 1962) and thiabendazole (Markell, 1968) have proved effective. Pyrantel pamoate has been used extensively and with good success (Ghadirian and Arfaa, 1975).

Prevention. Infection can be prevented by cooking contaminated vegetable foods, particularly during or immediately following rainy periods. Since herbivorous animals are common reservoir hosts, grazing lands are constantly being reinfested. The resistance of both the eggs and larvae to desiccation complicates the problem.

Ostertagia species

Trichostrongyles of the genus *Ostertagia* have a delicate head, small buccal cavity, and cervical papillae. In the male the caudal bursa has two large lateral lobes joined by a small dorsal lobe; the ventral rays are close together; the external dorsals develop separately; the dorsal ray is bifurcated at its distal portion, each fork consisting of one or two short rami. The copulatory spicules are equal, short, and terminate in one, two, or three points; a gubernaculum may be present or absent. Prebursal papillae are present. The vulva of the female opens in the posterior fifth of the body. Members of this genus are oviparous and parasitize herbivorous mammals. The life cycle is essentially the same as that of *Trichostrongylus*.

Kasimov (1943) in Azerbaidjan, U.S.S.R., recovered a single male *Ostertagia ostertagi* (Stiles, 1892) Ransom, 1907 from a human necropsy and *O. circumcincta* (Stadelman, 1894) Ransom, 1907 from another case. He suggested that the infections were most likely accidental, possibly due to eating inadequately cooked abomasum of cattle, sheep, or goats containing the nodular stage.

Haemonchus contortus
(Rudolphi, 1803) Cobb, 1898
(The sheep wireworm)

The most common parasite of domestic sheep throughout the world is *Haemonchus contortus*. It is also a parasite of other herbivores, including domestic cattle. Human infections have been reported once by de Magalhães in Brazil and in three aborigines in Australia by Sweet (1924). In ruminants the worms live attached to the fourth stomach and occasionally to the duodenal mucosa. The adult worms are distinguished by a single curved lancet (i.e., "blood hooklet") in the depth of the small buccal cavity and by a pair of spine-like cervical papillae. The males (10 to 20 mm by 0.4 mm) have a homogeneous reddish color, while the females (18 to 30 mm by 0.5 mm) have ivory-white ovaries twisted around red intestines. The worms' heads are quite small, with a diameter of only 30 μm. In the male, the bursa has elongated lateral lobes, supported by long slender rays and an asymmetrical dorsal lobe with a bifid ray. The spicules (0.3 to 0.5 mm long) and gubernaculum (200 μm long) are characteristic of the species. The vulva in the posterior fifth of the female is covered by a conspicuous, posteriorly projecting, tongue-shaped process on its anterior aspect. The thin-shelled hyaline eggs measure 75 to 90 by 40 to 50 μm. The life cycle closely resembles that of *Trichostrongylus*, and infection is acquired through ingestion of vegetation on which the third-stage larvae have crawled. The prepatent period in sheep is about 3 weeks.

In sheep this infection produces marked anemia, emaciation, edema, and profound digestive disturbances due to the blood loss caused by the worms and their injection of anticoagulatory and hemolytic substances into the traumatized mucosa. Since eggs passed in the feces may be confused with those of other strongylate worms, specific identification can only be obtained by recovery of larvae from fecal cultures, or the adult worms after anthelmintic treatment, or at autopsy. Prophylaxis in reservoir hosts consists in periodic anthelmintic treatment of all infected animals to reduce the infestation in the soil. The free-living stages withstand cold but succumb to long or severe drought.

Superfamily Metastrongyloidea
(For the superfamily characteristics see p. 229)

Metastrongylus elongatus
(Dujardin, 1845) Railliet and Henry, 1911
(The swine lungworm)

Synonym. *Metastrongylus apri* (Gmelin, 1790) Railliet and Henry, 1907.

These worms are filiform in shape, have a rudimentary buccal cavity, and a mouth surrounded by a pair of lateral trilobed lips. The males measure 12 to 25 mm by 160 to 225 μm, and the females 20 to 58 mm by 400 to 450 μm. They commonly parasitize the respiratory tract of hogs, less commonly of sheep and cattle. Three human infections are on record: (1) a 6-year-old boy (respiratory tract, Diesing, 1845); (2) adult (respiratory tract, Rainey, 1855); and (3) a pork vender (digestive tract, Chatin, 1888). In the male the bursa is relatively small, bilobed, with a long anterolateral ray, swollen at the tip. The copulatory spicules are long and delicate (4 to 4.2 mm) and terminate in a distinct minute hook. The entire posterior end of the female is curved ventrad. The vulva is situated just anterior to the anus.

Thick-shelled, hyaline eggs, measuring 51 to 54 by 33 to 36 μm, are fully embryonated when deposited in the respiratory passages. They may hatch in the bronchi or in the digestive tract after being coughed up and swallowed, but according to Alicata (1934, 1935) they are usually evacuated as eggs in the definitive host's feces. Although the larvae may survive for 3 months in soil, they require a suitable species of earthworm as an intermediate host, in the calciferous glands, wall of the esophagus, proventriculus, or blood vessels of which they complete necessary development, with two ecdyses in about 12 days. On ingestion of the infective third-stage larvae by the definitive host, either in earthworms or after accidental escape from these annelids, the larvae burrow through the mesenteric lymph nodes, where they pass a third ecdysis, and migrate to the lungs where, after a final molt, they mature. The incubation period is 5 to 8 weeks.

In pigs the infection produces a pneumonitis and bronchitis, which may prove fatal in young animals. Dead worms may produce whitish nodular patches in the lungs. Diagnosis is based on recovery of characteristic eggs in the feces. No specific treatment is known. Personal cleanliness on the part of individuals in contact with infected reservoir hosts should prevent human infection.

Angiostrongylus cantonensis (Chen, 1935) Dougherty, 1946 (The rodent lungworm)

Synonym. *Pulmonema cantonensis* Chen, 1935.

Historical and Geographical Notes. *Angiostrongylus cantonensis* was originally described by Chen in 1935 from the bronchial tree, and from one specimen in the heart, of *Rattus norvegicus* and *R. rattus rattus,* caught in Canton, China. During the 1950s and early 1960s it was reported as a relatively common parasite in the lungs of domestic rats in Australia, Malaysia, Taiwan, New Caledonia, Fiji, Guam, Ponape, Truk, Tahiti, and Hawaii. Later it was reported from rats in Japan (Ohbayashi and Orihara, 1968), in North Sumatra and Central Java in Indonesia (Margono, 1970), and in Madras and Bombay, India (Sharma et al., 1981). Human infection with *A. cantonensis* was first reported by Nomura and Lin (1945) from Taiwan, in a 15-year-old boy with suspected meningitis. Four female and two male worms were recovered from the spinal fluid (Beaver and Rosen, 1964). As a human parasite, it has been identified also in Tahiti, New Caledonia, Vanuatu, Loyalty Islands, Thailand, Marshall Islands, Hawaii, the Caroline, Cook, and Society Islands, the Philippines, Sumatra (Beck et al., 1980), and in India (Sharma et al., 1981). It has been found in rats in Cuba (Aguiar et al., 1981), and several confirmed and presumptive infections in children in the same locality of Havana have been reported (Pascual et al., 1981a, b). In man, this infection has been associated with outbreaks of eosinophilic meningitis, and in three cases in Thailand has involved the eye. Perhaps the most noteworthy aspect of human angiostrongylosis is that a quarter of a century elapsed from the time *A. cantonensis* was discovered until it was recognized as the cause of an important, occasionally fatal, human disease.

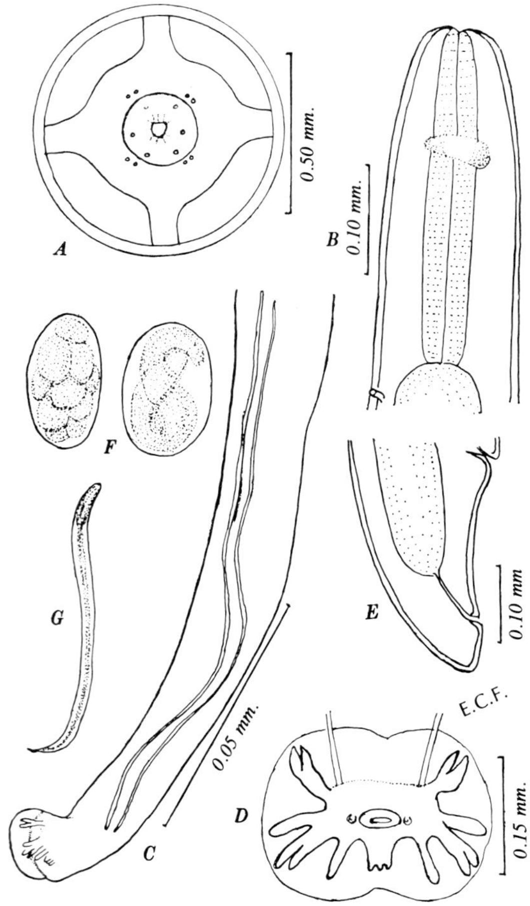

Fig. 20–24. *Angiostrongylus cantonensis. A,* Head-on view of buccal extremity. *B,* Lateral view of anterior end. *C,* Posterior end of male, lateral view showing bursa and the two copulatory spicules. *D,* Posterior view of bursa. *E,* Posterior extremity of female, lateral view. *F,* Embryonating eggs. *G,* Larva hatched from egg. *(A-E* adapted from Chen, H.T. 1935. Un noveau nematode pulmonaire *Pulmonema cantonensis* N.G., N. Sp. des rats de Canton. Annales de Parasitologie, *13.* [*F* and *G* from unpublished sketches by Dr. L.R. Ash.])

Morphology, Biology, and Life Cycle. *Angiostrongylus cantonensis* is a delicate filiform worm almost imperceptibly narrowed at both extremities. According to Chen (1935), it has a length of 17 to 25 mm and a maximum diameter of 0.26 to 0.36 mm. In the living state it has a pallid hue. The cuticula is smooth and is slightly thickened at the two extremities. The cephalic tip is simple, with three minute lips, of which the dorsal one has two submedian papillae and each of the two laterals one submedian and one lateral papilla. In addition, there are four pairs of papillae disposed symmetrically at quadrant distances from one another on the external border of the head (Fig. 20–24,*A*). Lacking a buccal capsule, the mouth opens directly into a muscular esophagus (Fig. 20–24,*B*). The *males* measure 16 to 19 mm in length by 0.26 mm in diameter. The small bursa copulatrix at the posterior extremity is directed ventrad (Fig. 20–24,*C*). Externally it has two broad lobes that coalesce medially. The fleshy bursal rays, as pictured by Chen (1935), are illustrated in Figure 20–24,*D*. The *females* measure 21 to 25 mm in length by 0.30 to 0.36 mm in diameter. The posterior end (Fig. 20–24,*E*) is shaped somewhat bluntly, with a subterminal anal pore on the ventral aspect and a vulvar opening approximately 0.20 mm anterior to the anus. The elongated ovoid eggs have a delicate hyaline shell, measure 46 to 48 by 68 to 74 μm, and are unembryonated at the time of oviposition. A single female may lay up to 15,000 eggs daily.

The life cycle of *A. cantonensis* was first described by Mackerras and Sandars in 1955 in Australia. The eggs hatch in the lungs of the rodent host and the larvae migrate up to the trachea. After being swallowed, the first-stage larvae are expelled in the feces (Fig. 20–24,*G*). They next infect a molluscan intermediate host in which they reach the third (infective) larval stage in about 2 weeks. When ingested by the rat (or man), infective larvae migrate to the brain and reach maturity in 4 weeks. Early development is in the brain (gray matter). After the final molt the young adults, about 2 mm long, migrate to the surface, at about 2 weeks of infection. They then migrate to the pulmonary arteries and after 2 more weeks begin laying eggs.

Numerous species of slugs in widely separated regions bordering on the Pacific Ocean and China Sea, as well as several land snails and a land planarian *(Geoplana septemlineata)* in Hawaii, are intermediate hosts of *A. cantonensis*. Natural infections have been found in land crabs *(Cardisoma hirtipes),* coconut crabs *(Birgus latro)* and freshwater prawns in Micronesia (Alicata, 1965), and in frogs *(Hyla aurea)* from market gardens in Noumea, New Caledonia (Ash, 1968). Experimentally, larvae from all natural invertebrate hosts have developed to adult worms when fed to rats. In nature, the following species of domestic rats have been found infected: *Rattus rattus rattus, R. rattus diarii, R. norvegicus, R. alexandrinus, R. exulans,* and *R. exulans concolor.*

Studies have been conducted to determine the course of the

infection in experimentally infected rats (laboratory and wild strains) and in rhesus monkeys. Rats fed not more than 10 larvae develop a chronic, lasting infection, but when 50 or more larvae are introduced, the rats die. The prevalence rate in rats in enzootic areas may be as high as 40 to 88%. When experimentally inoculated rats and monkeys were killed at successive intervals during the worms' development, migrating forms were found in the pulmonary arteries and veins and in the blood vessels of the central nervous system. Maturity of the worms was slower in the monkeys than in the rats, but in both types of experimental hosts there were granulomatous lesions in the central nervous system, with or without trapped worms.

Epidemiology. Although the life cycle pattern of *A. cantonensis* in the natural rodent and intermediate hosts has been amply described, it is not always possible to identify precisely the source of infection in individual cases in man. Hundreds of cases have now been reported; through June 1978 on Taiwan alone there were 259 known cases (Chen, 1979). There, and in Thailand, the infections were usually known or presumed to have been acquired by eating raw or inadequately cooked snails, usually the giant African land snail, *Achatina fulica*. Numerous other species of snails and slugs are known to be natural hosts and are possible sources of human infection.

From reviews of studies in New Caledonia (Ash, 1976), Papua New Guinea (Scrimgeour and Burke, 1979), Indonesia (Carney and Stafford, 1979), Malaysia (Lim and Ramachandran, 1979), Vietnam and Australia (Bhaibulaya, 1979), the Philippines (Garcia, 1979), Taiwan (Chen, 1979), Japan (Kojima et al., 1979; Otsuru, 1979), American Samoa (Beck et al., 1980), and India (Sharma et al., 1981), it is evident that *A. cantonensis* can develop to the adult stage in numerous mammals, and to the infective-stage larva in numerous land and aquatic mollusks. It can also survive in numerous species of shrimp and crabs which serve as paratenic hosts. This lack of host specificity, together with the well-known geographic mobility of rats and that of the large African land snail, *Achatina fulica,* has allowed *A. cantonensis* to become established throughout much of the tropical and subtropical regions of the world. Recently discovered new locations include Egypt (Yousif and Ibrahim, 1978), Cuba (Aguiar et al., 1981), and several Pacific islands, as already noted.

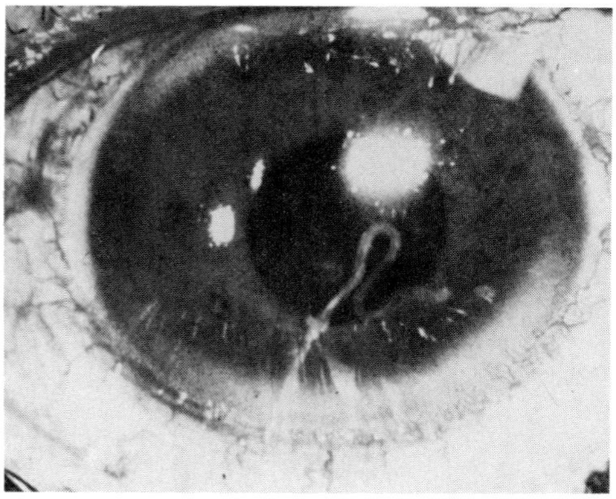

Fig. 20–26. *Angiostrongylus cantonensis*. Young adult male in the eye of a 34-year-old man in Bangkok. The worm was removed, described in detail, and identified specifically. (Courtesy of Dr. Kobchai Prommindaroj.)

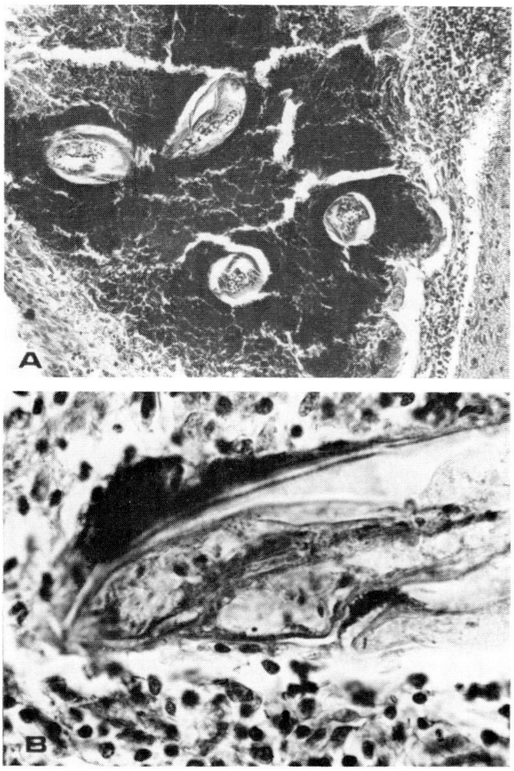

Fig. 20–25. *Angiostrongylus cantonensis. A,* Four sections of dead worm near surface of brain in necrotic fibrinoid material. (× 85.) *B,* Median sagittal section of female worm lying deep in brain tissue, showing anus near tip of tail and vulva just anterior to anus. (× 420.) (From Tangchai, P., Nye, S.W., and Beaver, P.C. 1967. Eosinophilic meningoencephalitis caused by angiostrongyliasis in Thailand. Am. J. Trop. Med. Hyg., *16*:454–461.)

Pathogenesis, Pathology, and Symptomatology. The first well-documented fatal case of human angiostrongylosis was reported by Rosen et al. in 1962. A 50-year-old male Filipino was admitted to a mental hospital in Hawaii in January 1958 and died there 2 years later. During the long hospitalization he had increasing confusion, incoherence, disorientation and impairment of memory and, terminally, coma. The infection was apparently acquired long after admission as the meningoencephalitis was first noted only 6 weeks before death. Postmortem examination revealed leptomeningitis and encephalomalacia involving the frontal and temporal lobes bilaterally. Sections revealed the presence of a nematode. On dissection of the brain tissues, intact and fragmented worms of both sexes were found and identified as immature adult *A. cantonensis*. Some were in the vascular and perivascular tissues of the cerebrum and cerebellum; others were in blood vessels of the meninges.

The second fatal case was reported from Thailand by Tangchai et al., in 1967. A 28-year-old woman with a history of low-grade fever and headache for 2 weeks and coma for 12 hours was hospitalized and died the next day. At autopsy a slight thickening of the leptomeninges over the basal portions of the brain was noted, but the external surfaces of the brain and spinal cord were otherwise not remarkable. However, routine sections of the brain revealed the presence of five worms, three of which were somewhat necrotic and surrounded by granulomatous reaction, whereas the others, apparently alive when fixed, were free in the tissues (Fig. 20–25). Large numbers of Charcot-Leyden crystals were observed in granulomatous meningeal lesions. Findings in this case, in 18 others reported from Thailand, 10 from Taiwan, and 1 each from Malaysia, Indonesia, and Vietnam were summarized by Punyagupta et al. (1975)

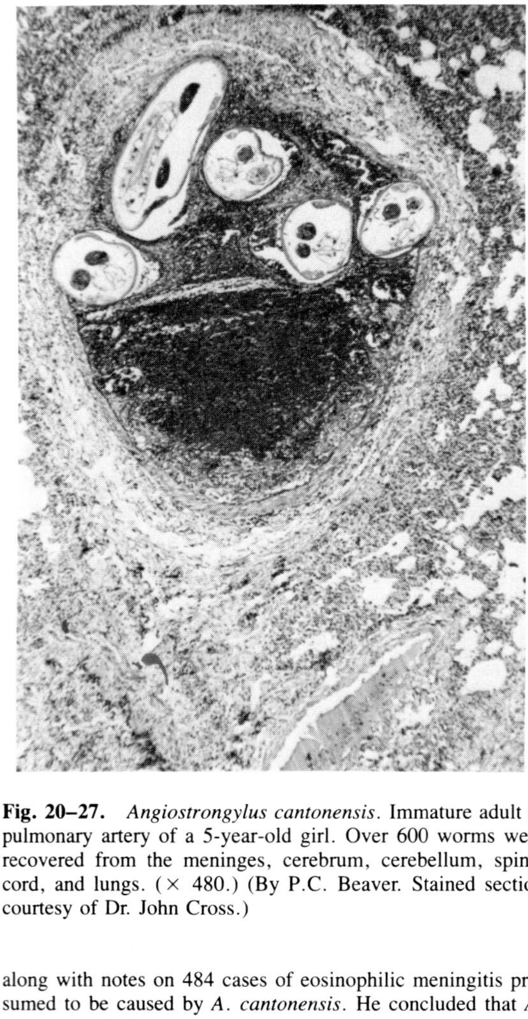

Fig. 20–27. *Angiostrongylus cantonensis*. Immature adult in pulmonary artery of a 5-year-old girl. Over 600 worms were recovered from the meninges, cerebrum, cerebellum, spinal cord, and lungs. (× 480.) (By P.C. Beaver. Stained section courtesy of Dr. John Cross.)

along with notes on 484 cases of eosinophilic meningitis presumed to be caused by *A. cantonensis*. He concluded that *A. cantonenis* is the most important cause of eosinophilic meningitis in man in Asia, there being hundreds of cases recognized each year. Although the disease is self-limiting and most of the cases are relatively mild, fatalities do occur. Ocular involvement with failing vision was reported in 47 of 484 cases (16%). In 7 cases an actively moving worm was visualized in the eye and removed (Prommindaroj et al., 1962), from the anterior chamber in 4 cases, and from the vitreous chamber in the others (Fig. 20–26). Associated complications were intraocular hemorrhage and retinal detachment.

In most cases the incubation period, when known, was less than 20 days; the longest was 36 days. The chief complaint in 96% of the cases was severe headache; in others it was convulsions, weakness of the arms or legs, paresthesia, vomiting, facial paralysis, stiffness of the neck, and fever. Pulmonary symptoms were essentially absent, although in two instances immature adult worms were observed in sections of lung tissue obtained at autopsy (Fig. 20–27) (Yii et al., 1968; Sonakul, 1978). Thus far, patent human infection evidenced by larvae in sputum or feces has not been reported.

Diagnosis. In endemic areas eosinophilic meningitis caused by *A. cantonensis* should be considered in patients with acute severe headache and signs and symptoms of meningitis or meningoencephalitis, particularly with low-grade fever, cranial nerve or ocular involvement, or psychosis. Eosinophilia of the blood has been observed in most cases, reaching a peak at around 30 days after exposure. Larvae, or more often young adult worms, may be found in cerebrospinal fluid obtained by using ordinary 20-gauge lumbar puncture needles. In 16 of 32 proven cases, the diagnosis was based on finding young worms in the cerebrospinal fluid. Eosinophilia of the cerebrospinal fluid ranging up to 90% has been a common finding. Additional cases in which worms were recovered in the cerebrospinal fluid were reported in a child in Hawaii (Kuberski et al., 1979) and a 13-year-old boy in Cuba (Pascual et al., 1981b).

Other helminthiases causing high eosinophilia, sometimes with central nervous system involvement, namely, cerebral cysticercosis, trichinosis, toxocariasis, and strongyloidiasis, must be ruled out. Punyagupta et al. (1975) have emphasized the need in Thailand to distinguish eosinophilic myeloencephalitis caused by *Gnathostoma spinigerum* from the eosinophilic meningitis of *Angiostrongylus cantonensis*. In the former there are nerve root pain, a paralysis of the extremities, or a sudden change in the sensations, and blood or xanthochromic spinal fluid is observed. Serologic tests have not generally been helpful (Punyagupta, 1979), although an enzyme-linked immunosorbent assay with antigen prepared from larvae recovered from the brain of rats appears to be promising (Tharavanij, 1979).

Treatment. Severe headache, which is the most frequent symptom of eosinophilic meningitis caused by *Angiostrongylus cantonensis,* is usually relieved dramatically by spinal puncture and removal of about 10 ml of cerebrospinal fluid. In some cases there is no further complaint, but usually two to four additional withdrawals of cerebrospinal fluid are necessary because of recurrence of headache. In about 70% of cases headache persists for longer than 1 month.

Since the disease is usually self-limiting, lasting only 4 to 6 weeks, administration of anthelmintic drugs is not necessary. Furthermore, killing worms in the central nervous system by such a treatment could theoretically greatly increase the inflammatory reaction and thus unnecessarily endanger the patient.

One might expect that corticosteroid drugs, because of their anti-inflammatory and immunosuppressive effects, would be helpful in mitigating the disease process. In fact, administration of oral prednisolone at a dosage of 30 to 60 mg daily in combination with analgesics has provided results similar to those in controls. Use of corticosteroids might still be recommended in severe cases with cranial nerve involvement.

Surgical removal of the parasite is required when it is found in the anterior or vitreous chamber of the eye.

Prognosis. The prognosis is generally good. Of 484 cases studied by Punyagupta et al. (1975), 41% were mild and did not require hospitalization. Of those patients admitted, 41% were able to be discharged within the first 5 days with improvement, and only 6% stayed in the hospital longer than 1 month.

Symptoms gradually disappear with recovery, the meningeal symptoms being the first to subside, then visual abnormalities, then parasthesia. Cranial nerve involvement is the last to recover, and permanent defects do occur.

Recurrence of the disease, or reinfection, has been observed in two cases.

The mortality rate is low: 1 of 484 cases in Thailand, 4 of 125 in Taiwan, and none of several hundred in the South Pacific (Rosen et al., 1962; Punyagupta et al., 1975). Most deaths have occurred between the second and fourth weeks of infection.

Control. No program of overall prevention has been proposed.

Angiostrongylus costaricensis Morera and Cespedes, 1971 (Causing abdominal angiostrongylosis)

Synonym. *Morerastrongylus costaricensis* (Morera and Cespedes, 1971) Chabaud, 1972.

Historical and Geographical Notes. *Angiostrongylus cos-*

taricensis was described by Morera and Cespedes in 1971. In the following year, Chabaud (1972) reviewed the relationships of eight genera in the subfamily Angiostrongylinae, including a new genus, *Morerastrongylus,* which he proposed for *costaricensis* and one other species *(A. andersoni* Petter, 1972) from the lungs of an African rodent. However, Anderson (1978) reduced *Morerastrongylus* Chabaud, 1972 to synonymy with *Angiostrongylus* Kamensky, 1905. He proposed a subgenus *Angiostrongylus (Parastrongylus)* (= *Pulmonema* Chen, 1935; = *Rattostrongylus* Schulz, 1951; = *Morerastrongylus* Chabaud, 1972) for *A. costaricensis, A. cantonenis,* and other species of *Angiostrongylus* having the externolateral ray joined to a common stalk with the other lateral rays, and *A. (Angiostrongylus)* for *A. vasorum* and others having the externolateral ray separate from the other lateral rays of the bursa. This anatomic feature (i.e., separation of the externolateral ray) is not readily recognized in some species.

When *Angiostrongylus costaricensis* was described as a new species, its natural host was unknown. It had been seen only as specimens dissected from fixed tissues, or in sections of resected portions of intestine from the ileocecal region from children in Costa Rica. While the report describing the new species was being published, the natural hosts of the parasite in Costa Rica were discovered. The adult worms were found in the arteries and arterioles of the cotton rat *(Sigmodon hispidus)* and the black rat *(Rattus rattus)* by Morera (1971), and the infective-stage larvae were found in the tissues of a common slug, *Vaginulus plebeius,* by Morera and Ash (1971). The albino rat, *Rattus norvegicus,* was found to be a relatively poor host.

Naturally infected mammals have been reported as follows: Panama—*Sigmodon hispidus, Rattus rattus, Zygodontomys microtinus, Liomys adspersus,* and *Oryzomys fulvescens;* El Salvador—*Sigmodon hispidus;* Costa Rica—*Sigmodon hispidus, Rattus rattus, Zygodontomys microtinus, Liomys adspersus, Oryzomys fulvescens,* and *Nasua narica bullata* (the coati); Texas—*Sigmodon hispidus;* Colombia—*Oryzomys caliginosus.* In man, *A. costaricensis* has been reported in scores of cases in Costa Rica where in the period 1966 to 1976 in one hospital 116 cases were recorded (Loria-Cortes and Lobo-Sanahuja, 1980). Elsewhere, human infection has been reported in one case each in Honduras (Sierra and Morera, 1972) and Colombia (Malek, 1981), two each in Venezuela (Zambrano et al., 1975) and Mexico (Zavala et al., 1974), three in El Salvador (Sauerbrey, 1977), and six in Brazil (Magalhaes et al., 1982; Rojas Ayala et al., 1982).

In addition to the common slug, *Vaginulus plebeius,* in Costa Rica, Honduras, and Panama, a related species, *Veronicella occidentalis,* in Colombia is known to be a suitable intermediate host (Malek, 1981). Other species of slugs and other types of terrestrial molluscs have not been ruled out as intermediate hosts.

Morphology. The adult worm is filiform, with a large dark intestine and white reproductive organs that are evident through a relatively thin, transparent body wall. The male measures up to 22 mm by 140 μm and the female 42 mm by 350 μm. In the male the esophagus has a maximum length of 225 μm, the anterior end of the testis is just posterior to the esophagus, and the spicules are striated and relatively short, 330 μm or less. The bursa is symmetrical, with the ventral rays fused except at the tips, and the laterals are widely separated from the ventrals. The medio- and posterolaterals are somewhat smaller than the externolateral and fused at the base; the externodorsal ray arises near the posterolateral, which it matches in length; the dorsal ray is short and stout, bearing two conical branches with a ventral papilla at the bifurcation. Three large papillae aligned in a forward-facing position are equally spaced along the posterior border of the cloacal orifice. In the female the tail is conical and sharply flexed ventrad with a minute projection at the tip; the anus is 60 to 65 μm, and the vulva, only slightly protruding, is less than 300 μm from the tip of the tail.

Life Cycle. When passed in the feces of the rat, first-stage larvae measure about 270 μm by 15 μm and are actively motile; the tail bears the notch characteristic of the metastrongyle group. When ingested by the slug intermediate host, the larvae migrate to the mantle and foot, grow, molt twice, at 4 days and at about 2 weeks, respectively, and become infective at 16 to 19 days, at which time a length of about 470 μm is reached. When the slug is eaten by a rat and the larvae are digested free, they migrate through the intestinal tissues to the lymphatics. Here they molt twice, on the fourth and fifth days, respectively. On about the tenth day or earlier, after attaining a length of about 4 mm, they migrate to the vessels of the ileocecal level of the intestine. By active penetration of the vessels' walls they enter the small arterioles and arteries, causing small hemorrhages at the sites of entry. By the eighteenth day egg-laying begins, and on the twenty-second to twenty-fourth day larvae appear in the rat's feces. The eggs when deposited are ovoid, about 90 μm long, and unembryonated with a thin transparent shell. Eggs discharged into the arterioles are carried into the wall of the intestine; here they develop to first-stage larvae which, on hatching, actively migrate to the intestinal lumen and enter the fecal stream to be deposited in the rat's feces. Larvae may also develop and hatch in the regional mesenteric lymph nodes (Morera, 1973).

Epidemiology. Human infection with *A. costaricensis* is acquired by ingestion of infected slugs. As raw slugs are not regarded as edible by the people in regions where human infections have been recorded, it is presumed that the ingestion in most instances is accidental. It has been suggested that infection may be produced by free larvae that leave the slug in the mucus and are present on soil or on other surfaces to be ingested with raw fruits or vegetables. In experimentally infected aquatic snails, infective larvae are shed for at least 30 days and survive in tap water for about 3 weeks (Ubelaker, 1980). The infective larvae are unable to penetrate skin.

Infections reported from Costa Rica were seen mostly in children, more often in boys than in girls, and with the highest incidence in October and November. Infection does occur, however, in adults as well, in people of various socioeconomic levels, and in both rural and urban communities.

Pathogenesis, Pathology, and Symptomatology. Eggs deposited in the intestinal wall via the mesenteric arterioles cause a granulomatous inflammatory reaction, with edema and marked thickening of the intestinal wall and a narrowing of the lumen. There is also a dense eosinophilic infiltration of the vessels occupied by the worms, with marked thickening of the walls. Frequently the vessel is partially occluded. Sections of the involved area, usually the cecum, appendix, and adjacent parts of the colon and ileum, show eggs in various stages of development and larvae both in the egg and free, in granulomas or areas of dense eosinophilic infiltration (Fig. 20–28). Sections may contain adult worms in functional or occluded vessels. Rarely, the worms may be located in the liver, causing conditions suggestive of visceral larva migrans (Morera et al., 1982).

The usual signs and symptoms of abdominal angiostrongyliasis are pain and tenderness and a palpable tumor-like mass in the right lower quadrant, fever, and in some cases diarrhea and vomiting. In most cases there is leukocytosis with eosinophilia ranging up to 60%. The clinical findings in general resemble those of appendicitis. Consequently, an operation is performed in a high proportion of cases.

Diagnosis. Larvae apparently are not discharged in the feces of man as they are in the rat, although larvae in eggs and free may be seen in sections of human tissue. A precipitin test described and recommended by Sauerbrey (1977) has not yet been widely used. Thus far, diagnosis has generally been based

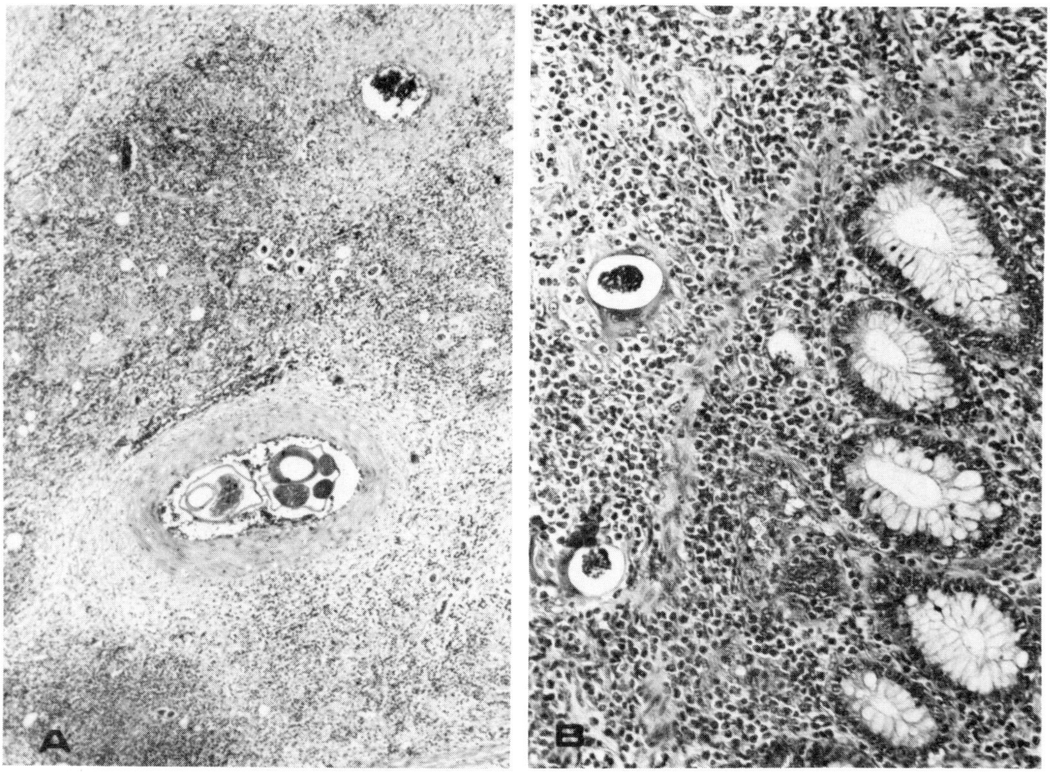

Fig. 20–28. *Angiostrongylus costaricensis.* Worm and eggs in section of appendix removed from a child in Costa Rica. *A,* Adult male in small artery and eggs in surrounding eosinophilic granulomatous tissue. (× 48.) *B,* Eggs with embryos in cleavage stages in submucosa, which is heavily infiltrated with eosinophils. (× 163.) (By P.C. Beaver. Stained sections courtesy of Dr. Pedro Morera.)

on the examination of surgical specimens. When tissues have not been available, the diagnosis has been based on clinical findings.

Treatment. Diethylcarbamazine has been tried, but it is not thought to be beneficial. Thiabendazole in a dosage of 75 mg/kg body weight daily for 3 days is currently recommended (Loria-Cortes and Lobo-Sanahuja, 1980). This dosage is likely to produce side effects and then must be reduced.

Prognosis. Few deaths have been attributable to *A. costaricensis* infection, although operation was necessary in three-fourths of the cases diagnosed (Loria-Cortes and Lobo-Sanahuja, 1980).

Control. An effective community program of rodent control in enzootic areas should help prevent infection in people living in those areas.

Angiostrongylus species

An unidentified species of *Angiostrongylus (Parastrongylus)* was recovered in the cerebrospinal fluid of a 10-year-old child in Ivory Coast, West Africa (Nozais et al., 1980). The child was admitted to the hospital with eosinophilic meningoencephalitis and was known to have eaten a portion of raw land snail *(Achatina* sp.). Only one worm, a 10-mm male, was recovered; it differed from all species of *Angiostrongylus* that have been described from Africa.

A rat lungworm found in the same host species and in the same areas of Australia as *Angiostrongylus cantonensis* is *Angiostrongylus mackerrasae* Bhaibulaya, 1968. The two species differ morphologically in the greater length of the spicules in *A. cantonensis* and in some other respects, and development in the rat is slightly more rapid in *A. cantonensis,* but their migratory patterns are the same. Another species found in the pulmonary arteries of rats in Southeast Asia is *Angiostrongylus malaysiensis* Bhaibulaya and Cross, 1971. A Southeast Asian species which, like *A. costaricensis,* lives in branches of the mesenteric artery of several species of rats is *Angiostrongylus siamensis* Ohbayashi, Kamiya, and Bhaibulaya, 1979. Of the three species of *Angiostrongylus* reported from Southeast Asian rats, only *A. cantonensis* has been identified as a cause of disease in man. Whether *A. malaysiensis* or *A. siamensis* can develop in man is unknown. In most instances these closely related species would be indistinguishable in tissue sections. *A. siamensis,* like *A. costaricensis,* is not highly host-specific (Kamiya et al., 1980).

REFERENCES

Aguiar, P.H., Morera, P., and Pascual, J. 1981. First record of *Angiostrongylus cantonensis* in Cuba. Am. J. Trop. Med. Hyg., *30*:963–965.

Alicata, J.E. 1934. Life history of *Metastrongylus salmi* and remarks on the eggs of the swine lungworms. Proc. Helm. Soc. Wash., *1*:12.

———. 1935. Early developmental stages of nematodes occurring in swine. U.S. Dept. of Agriculture Tech. Bull. No. 498.

———. 1965. Notes and observations on murine angiostrongylosis and eosinophilic meningoencephalitis in Micronesia. Can. J. Zool., *43*:667–672.

Anderson, R.C. 1978. Keys to genera of the superfamily Metastrongyloidea. CIH Keys to the Nematode Parasites of Vertebrates, No. 5. Commonwealth Agricultural Bureaux, Farnham Royal, Slough, U.K.

Andrews, J. 1942a. New methods of hookworm disease investigation and control. Am. J. Public Health, *32*:282–288.

———. 1942b. Modern views on the treatment and prevention of hookworm disease. Ann. Intern. Med., *17*:891–901.

Anthony, P.P., and McAdam, I.W.J. 1972. Helminthic pseudotumours of the bowel: Thirty-four cases of helminthoma. Gut, *13*:8–16.

Ash, L.R. 1968. The occurrence of *Angiostrongylus cantonensis* in frogs of New Caledonia with observations on paratenic hosts of metastrongyles. J. Parasitol., *54*:432–436.

———. 1976. Observations on the role of mollusks and planarians in the transmission of *Angiostrongylus cantonensis* infection to man in New Caledonia. Rev. Biol. Trop., *24*:163–174.

Ball, P.A.J., and Bartlett, A. 1969. Serological reactions to infection with *Necator americanus*. Trans. R. Soc. Trop. Med. Hyg., *63*:362–369.

Barrowclough, H., and Crome, L. 1979. Oesophagostomiasis in man. Trop. Geogr. Med., *31*:133–138.

Bastidas, G.J. 1969. Effect of ingested garlic on *Necator americanus* and *Ancylostoma caninum*. Am. J. Trop. Med. Hyg., *18*:920–923.

Baylis, H.A., and Daubney, R. 1923. A further report on parasitic nematodes in the collection of the Zoological Survey of India. Rec. Indian Mus., *25*:551–578.

Beaver, P.C. 1950. The standardization of fecal smears for estimating egg production and worm burden. J. Parasitol., *36*:451–456.

———. 1951. Hemoglobin determination in hookworm disease case finding. Am. J. Trop. Med., *31*:90–97.

———. 1952. The detection and identification of some common nematode parasites of man. Am. J. Clin. Pathol., *22*:481–494.

———. 1953. Persistence of hookworm larvae in soil. Am. J. Trop. Med. Hyg., *2*:102–108.

———. 1955. Observations on *Necator* infections resulting from exposure to three larvae. Rev. Iber. Parasitol., Tomo Extraordinario, 713–721.

———. 1964. Cutaneous larva migrans. Industrial Med. Surg., *33*:319–321.

———. 1975. Biology of soil-transmitted helminths: The massive infection. Health Lab. Sci., *12*:116–125.

———, and Rosen, L. 1964. Memorandum on the first report of *Angiostrongylus* in man, by Nomura and Lin, 1945. Am. J. Trop. Med. Hyg., *13*:589–590.

———, Yoshida, Y., and Ash, L.R. 1964. Mating of *Ancylostoma caninum* in relation to blood loss in the host. J. Parasitol., *50*:286–293.

Beck, M.J., Cardina, T.M., and Alicata, J.E. 1980. Eosinophilic meningitis due to angiostrongyliasis cantonensis in American Samoa. Hawaii Med. J., *39*:254–257.

Bhaibulaya, M. 1968. A new species of *Angiostrongylus* in an Australian rat, *Rattus fuscipes*. Parasitology, *58*:789–799.

———. 1979. Geographical distribution of *Angiostrongylus* and angiostrongyliasis in Thailand, Indo-china, and Australia, pp. 49–52. In *Studies on Angiostrongyliasis in Eastern Asia and Australia*. Edited by J.H. Cross. Special Publication, U.S. Naval Medical Research Unit No. 2, Taipei, Taiwan.

———, and Cross, J.H. 1971. *Angiostrongylus malaysiensis* (Nematoda: Metastrongylidae), a new species of rat lungworm from Malaysia. Southeast Asian J. Trop. Med. Public Health, *2*:527–533.

———, and Indrangarm, S. 1975. Man, an accidental host of *Cyclodontostomum purvisi* (Adams, 1933), and the occurrence in rats in Thailand. Southeast Asian J. Trop. Med. Public Health, *6*:391–394.

Biocca, A. 1951. On *Ancylostoma braziliense* (de Faria, 1910) and its morphological differentiation from *A. ceylanicum* (Looss, 1911). J. Helminthol., *25*:1–10.

———, and LeRoux, P.L. 1957. Suddivisione del genere *Ancylostoma* (Dubini, 1843) in quattro sottogeneri. Accad. Naz. dei Lincei, Rend. Sci. Fis. Mat. Nat., *23*(Ser. 8), 470–477.

Birrell, D.J., Moorhouse, D.E., Gardner, M., and May, C.S. 1978. Chronic cough and haemoptysis due to a nematode, "*Syngamus laryngeus*." Aust. N.Z. J. Med., *8*:168–170.

Brown, R.C., and Girardeau, M.H.F. 1977. Transmammary passage of *Strongyloides* sp. larvae in the human host. Am. J. Trop. Med. Hyg., *26*:215–219.

Brug, S.L. 1910. Het onbevruchte Ancylostomum-ei. Geneesk. Tijdschr. Ned.–Indie, *50*:220.

Brumpt, E. 1949. *Précis de Parasitologie*. 2 vols, 6th Ed. Paris, Masson et Cie.

Buckley, J.J.C. 1933. Some observations on two West Indian parasites of man. Proc. R. Soc. Med., *27*:134–135.

Campbell, W.C. 1977. The control of parasites: The role of drugs. Proc. Helm. Soc. Wash., *44*:17–28.

Carney, W.P., and Stafford, E.E. 1979. Angiostrongyliasis in Indonesia: A review, pp. 14–25. In *Studies on Angiostrongyliasis in Eastern Asia and Australia*. Edited by J.H. Cross. Special Publication, U.S. Naval Medical Research Unit No. 2, Taipei, Taiwan.

Carr, H.P., Pichardo Sarda, M.E., and Nuñez, N.A. 1954. Anthelmintic treatment of uncinariasis. Am. J. Trop. Med. Hyg., *3*:495–503.

Chabaud, A.G. 1972. Description de *Stefanskostrongylus dubosti* n. sp. parasite du potamogale et essai de classification des nematodes Angiostrongylinae. Ann. Parasitol., *47*:735–744.

———, and Larivière, M. 1958. Sur les oesophagostomes parasites de l'homme. Bull. Soc. Pathol. Exot., *51*:384–393.

Chatin, J.C.M. 1888. Le strongle paradoxale chez l'homme. Bull. Acad. Med. Paris, *52*:482–491.

Chen, E.R. 1979. Angiostrongyliasis and eosinophilic meningitis on Taiwan: A review, pp. 57–73. In *Studies on Angiostrongyliasis in Eastern Asia and Australia*. Edited by J.H. Cross. Special Publication, U.S. Naval Medical Reseach Unit No. 2, Taipei, Taiwan.

Chen, H.T. 1935. Un nouveau nematode pulmonaire, *Pulmonema cantonensis* n.g., n. sp. des rats de Canton. Ann. Parasitol., *13*:312–317.

Chowdhury, A.B., and Schad, G.A. 1972. *Ancylostoma ceylanicum:* A parasite of man in Calcutta and environs. Am. J. Trop. Med. Hyg., *21*:300–301.

Cort, W.W., et al. 1921–1925. Investigations on the control of hookworm disease. Am. J. Hyg., Vols. *1* to *5*. (34 papers.)

Darling, S.T. 1922. The hookworm index and mass treatment. Am. J. Trop. Med., *2*:397–447.

Deane, M.P. 1950. Helmintos eliminados por um grupo de residentes da Amazônia apos um tratamento pelo hexilresorcinol. Rev. Serv. Esp. Saude Pub., *3*:443–464.

De la Riva, H., Gomes-Escamilla, D., and Frati, A.C. 1981. Acute massive intestinal bleeding caused by hookworm. J.A.M.A., *246*:68.

Dubini, A. 1843. Nuovo verme intestinale umano *(Agchylostoma duodenale),* constitutente un sesto genere dei nematoidei proprii dell'uomo. Ann. Univ. di Med., Milano, *106*:5–13.

Eberson, F. 1980. Eradication of hookworm disease in Florida. J. Fla. Med. Assoc., *67*:736–742.

Eiff, J.A. 1966. Nature of an anticoagulant from the cephalic glands of *Ancylostoma caninum*. J. Parasitol., *52*:833–843.

Garcia, E.G. 1979. *Angiostrongylus cantonensis* in the Philippines: A review, pp. 53–56. In *Studies on Angiostron-*

gyliasis in Eastern Asia and Australia. Edited by J.H. Cross. Special Publication, U.S. Naval Medical Research Unit No. 2, Taipei, Taiwan.

Ghadirian, E., and Arfaa, F. 1975. Present status of trichostrongyliasis in Iran. Am. J. Trop. Med. Hyg., *24*:935–941.

———, and Biocca, E. 1972. Su un particolare *Necator* parassita dell'uomo rinvenuto in Iran. Parassitologia, *14*:137–142.

Goldsmid, J.M. 1967. *Ternidens deminutus* Railliet and Henry (Nematoda). Cent. Afr. J. Med., *13*:54–58.

———. 1971. *Ternidens deminutus:* A parasitological enigma in Rhodesia. Research Lecture Series No. 4, Faculty of Medicine, University of Rhodesia.

———. 1972. Thiabendazole in the treatment of human infections with *Ternidens deminutus* (Nematoda). S. Afr. Med. J., *46*:1046–1047.

———, and Saunders, C.R. 1972. Preliminary trial using pyrantel pamoate for the treatment of human infections with *Ternidens deminutus*. Tr. R. Soc. Trop. Med. Hyg., *66*:375.

Goodwin, L.G., Jayewardene, L.G., and Standen, O.D. 1958. Clinical trials with bephenium hydroxynaphthoate against hookworm in Ceylon. Br. Med. J., *2*:1572–1576.

Gordon, J.A., Ross, C.M.D., and Affleck, H. 1969. Abdominal emergency due to an oesophogostome. Ann. Trop. Med. Parasitol., *63*:161–164.

Grell, G.A.C., Watty, E.I., and Muller, R.L. 1978. *Syngamus* in a West Indian. Br. Med. J., *2*:1464.

Haaf, E., and Van Soest, A.H. 1964. Oesophagostomiasis in man in North Ghana. Trop. Geogr. Med., *16*:49.

Harada, Y. 1962. Wakana disease and hookworm allergy. Yonago Acta Med., *6*:109–118.

Higo, A. 1962. An experimental study on the migration route and the development of *Ancylostoma duodenale* in pups after cutaneous infection. J. Kyoto Pref. Med. Univ., *70*:851–874.

Hunter, G.W., III, et al. 1950–1953. Parasitological studies in the Far East. Jap. Med. J., *3*:359–364 (1950); *4*:113–124 (1951); Jap. J. Med. Sci. Biol., *6*:45–55 (1953).

Jung, R.C., and McCroan, J.E. 1960. Efficacy of bephenium and tetrachloroethylene in mass treatment of hookworm infection. Am. J. Trop. Med. Hyg., *9*:492–495.

Kalantarian, H. 1927. Trichostrongylosen des Menschen in Armenien, pp. 84–96. In Skrjabin's *Sammlung Helminthologischer Arbeiten*. Moscow.

Kalkofen, U.P. 1970. Attachment and feeding behavior of *Ancylostoma caninum*. Z. Parasitenkd., *33*:339–354.

Kaminsky, R.G., and Ndinya-Achola, J.O. 1977. *Oesophagostomum* sp. from Kenya. Identification through tissue sections. East Afr. Med. J., *54*:296–297.

Kamiya, M., Oku, Y., Kamiya, H., Ohbayashi, M., and Bhaibulaya, M. 1980. Abdominal angiostrongyliasis in laboratory animals caused by *Angiostrongylus siamensis* and *A. costaricensis*. Abstr. 10th Int. Congr. Trop. Med. Malaria, p. 260.

Kasimov, B. 1943. First case of *Ostertagia ostertagi* in man in Azerbaidjan. Med. Parasitol. Parasit. Dis. (Moscow), *12*:81.

Kirby-Smith, J.L., Dove, W.E., and White, G.F. 1926. Creeping eruption. Arch. Dermatol. Syphilol., *13*:137–173.

Kojima, S., Hata, H., Kobayashi, M., Yokogawa, M., Takahashi, N., Takaso, T., and Kaneda, J. 1979. Eosinophilic meningitis: A suspected case of angiostrongylosis found in Shizuoka Prefecture, Honshu, Japan. Am. J. Trop. Med. Hyg., *28*:36–41.

Komiya, Y., and Kobayashi, A. 1965. Techniques applied in Japan for the control of *Ascaris* and hookworm infections—a review. Jpn. J. Med. Sci. Biol., *18*:1–17.

———, and Yasuraoka, K. 1966. The biology of hookworms, pp. 5–114. In *Progress of Medical Parasitology in Japan*, Volume 3. Edited by K. Morishita, Y. Komiya, and H. Matsubayashi. Tokyo, Meguro Parasitological Museum.

Krupp, I.M. 1961. Effects of crowding and of superinfection on habitat selection and egg production in *Ancylostoma caninum*. J. Parasitol., *47*:957–961.

Kuberski, T., Bart, R.D., Briley, J.M., and Rosen, L. 1979. Recovery of *Angiostrongylus cantonensis* from cerebrospinal fluid of a child with eosinophilic meningitis. J. Clin. Microbiol., *9*:629–630.

Lane, C. 1922. *Ancylostoma braziliense*. Ann. Trop. Med. Parasitol., *16*:347–352.

Lawless, D.K., Kuntz, R.E., and Strone, C.P.A. 1956. Intestinal parasites in an Egyptian village of the Nile Valley with emphasis on the protozoa. Am. J. Trop. Med. Hyg., *5*:1010–1014.

Lee, K.T., Little, M.D., and Beaver, P.C. 1975. Intracellular (muscle-fiber) habitat of *Ancylostoma caninum* in some mammalian hosts. J. Parasitol., *61*:589–598.

Leoutsakos, B., Agnadi, N., and Kolisiatis, S. 1977. Rectal bleeding due to *Oesophagostomum brumpti:* Report of a case. Dis. Colon Rectum, *20*:632–634.

Lim, B.L., and Ramachandran, C.P. 1979. Ecological studies on *Angiostrongylus malaysiensis* (Nematoda: Metastrongylidae) in Malaysia, pp. 26–48. In *Studies on Angiostrongyliasis in Eastern Asia and Australia*. Edited by J.H. Cross. Special Publication, U.S. Naval Medical Research Unit No. 2, Taipei, Taiwan.

Little, M.D. 1978. Dormant *Ancylostoma caninum* larvae in muscle as a source of subsequent patent infection in the dog. Abstract 63rd Annual Meeting Am. Soc. Parasitol., p. 58.

———, Halsey, N.A., and Cline, B.L. 1983. *Ancylostoma* larva in a muscle fiber of man following cutaneous larva migrans. Am. J. Trop. Med. Hyg., *32*. In press.

Looss, A. 1898. Zur Lebensgeschichte des *Ankylostoma duodenale*. Cbl. Bakt., *24*:441–449, 483–488.

———. 1905, 1911. The Anatomy and Life History of *Agchylostoma duodenale* Dub. A monograph. Rec. Egyptian Govt. Sch. Med., *3*, 166 pp.; *4*, 446 pp.

Loria-Cortes, R., and Lobo-Sanahuja, J.F. 1980. Clinical abdominal angiostrongylosis. A study of 116 children with intestinal eosinophilic granuloma caused by *Angiostrongylus costaricensis*. Am. J. Trop. Med. Hyg., *29*:538–544.

Mackerras, M.J., and Sandars, D.F. 1955. The life history of the rat lung-worm, *Angiostrongylus cantonensis* (Chen) (Nematoda: Metastrongylidae). Aust. J. Zool., *7*:1–21.

Magalhaes, A.V. de, Andrade, J.E. de, Koh, I.H.J., Soares, M. do C., Alves, E., Tubino, P., Santos, F., and Raick, A.N. 1982. Novo caso de angiostrongilose abdominal. Rev. Inst. Med. Trop. São Paulo, *24*:252–256.

Malek, E.A. 1981. Presence of *Angiostrongylus costaricensis* Morera and Céspedes 1971 in Colombia. Am. J. Trop. Med. Hyg., *30*:81–83.

Manalang, C. 1925. Studies on ankylostomiasis in the Philippines. Trans. 6th Congr. Far Eastern Assoc. Trop. Med., Tokyo, pp. 351–368.

Margono, S. 1970. *Angiostrongylus cantonensis* in Djakarta, Indonesia. Southeast Asian J. Trop. Med. Public Health, *1*:158–159.

Markell, E.K. 1968. Pseudohookworm infection—trichostrongyliasis. Treatment with thiabendazole. N. Engl. J. Med., *278*:831–832.

Martin, L.K. 1972a. Hookworm in Georgia. I. Survey of intestinal helminth infections and anemia in rural school children. Am. J. Trop. Med. Hyg., *21*:919–929.

———. 1972b. Hookworm in Georgia. II. Survey of intestinal helminth infections in members of rural households of southeastern Georgia. Am. J. Trop. Med. Hyg., *21*:930–943.

———, and Beaver, P.C. 1968. Evaluation of Kato thick-smear technique for quantitative diagnosis of helminth infections. Am. J. Trop. Med. Hyg., *17*:382–391.

Matsusaki, G. 1963. Influence of low temperature, sunshine, desiccation and human excreta upon the eggs of *Ancylostoma duodenale* and *Necator americanus*. Part I. Low temperature. Yokohama Med. Bull., *14*:73–79.

McDowell, D.E. 1981. Intestinal bleeding caused by hookworm. J.A.M.A., *246*:2806.

McLaren, D.J. 1974. The anterior glands of adult *Necator americanus* (Nematoda: Strongyloidea). I. Ultrastructural studies. Int. J. Parasitol., *4*:25–37.

———, Burt, J.S., and Ogilvie, B.M. 1974. The anterior glands of adult *Necator americanus* (Nematoda: Strongyloidea). II. Cytochemical and functional studies. Int. J. Parasitol., *4*:39–46.

Migasena, S., Suntharasamai, P., and Harinasuta, T. 1978. Mebendazole, tetrachlorethylene and pyrantel pamoate in the treatment of hookworm infection. Ann. Trop. Med. Parasitol., *72*:199–200.

Miller, A. 1954. Dung beetles (Coleoptera, Scarabaeidae) and other insects in relation to human feces in a hookworm area of southern Georgia. Am. J. Trop. Med. Hyg., *3*:372–387.

———. 1961. The mouth parts and digestive tract of adult dung beetles (Coleoptera: Scarabaeidae), with reference to the ingestion of helminth eggs. J. Parasitol., *47*:735–744.

———, Chi-Rodriguez, E., and Nichols, R.L. 1961. The fate of helminth eggs and protozoan cysts in human feces ingested by dung beetles (Coleoptera: Scarabaeidae). Am. J. Trop. Med. Hyg., *10*:748–754.

Morera, P. 1971. Investigación del huésped definitivo de *Angiostrongylus costaricensis* (Morera y Céspedes, 1971). Bol. Chil. Parasitol., *25*:133–134.

———. 1973. Life history and redescription of *Angiostrongylus costaricensis* Morera and Céspedes, 1971. Am. J. Trop. Med. Hyg., *22*:613–621.

———, and Ash, L.R. 1971. Investigación del huésped intermediario de *Angiostrongylus costaricensis* (Morera y Céspedes, 1971). Bol. Chil. Parasitol., *25*:135.

———, and Céspedes, R. 1971. *Angiostrongylus costaricensis* n. sp. (Nematoda: Metastrongyloidea), a new lungworm occurring in man in Costa Rica. Rev. Biol. Trop., *18*:175–185.

———, Perez, F., Mora, F., and Castro, L. 1982. Visceral larva migrans-like syndrome caused by *Angiostrongylus costaricensis*. Am. J. Trop. Med. Hyg., *31*:67–70.

Morishita, K. 1966. *The Japan Association of Parasite Control and Its Activities*. Tokyo.

Mornex, J.F., Magdeleine, J., and De Thore, J. 1980. La syngamose humaine *(Mammonogamus nasicola)* cause de toux en Martinique: 37 observations récentes. Nouvelle Presse Med., *9*:3628.

Muhleisen, J.P. 1953. Demonstration of pulmonary migration of the causative organism of creeping eruption. Ann. Intern. Med., *38*:595–600.

Nadbath, R.P., and Lawlor, P.P. 1965. Nematode *(Ancylostoma)* in the cornea. A case report. Am. J. Ophthalmol., *59*:486–490.

Nagahana, M., and Yoshida, Y. 1965. Complete development and migratory route of *Ancylostoma duodenale* in young dogs. J. Parasitol., *51* (No. 2, Section 2), 52.

Naik, S.R., Mitra, S.K., and Mehta, S. 1976. Massive intestinal haemorrhage due to infection with *Ancylostoma duodenale*. J. Trop. Med. Hyg., *79*:2–5.

Nawalinski, T.A., and Schad, G.A. 1974. Arrested development in *Ancylostoma duodenale*: Course of a self-induced infection in man. Am. J. Trop. Med. Hyg., *23*:895–898.

Nichols, R.L. 1956. The etiology of visceral larva migrans. II. Comparative larval morphology of *Ascaris lumbricoides, Necator americanus, Strongyloides stercoralis* and *Ancylostoma caninum*. J. Parasitol., *42*:363–399.

Nomura, S., and Lin, H. 1945. First clinical case of *Hemostrongylus ratti* Yokogawa in man. Taiwan No. Ikai, *3*:589–592. (In Japanese; for English translation see Beaver and Rosen, 1964.)

Nozais, J.P., Moreau, J., Morlier, G., Kouame, J., and Doucet, J. 1980. Premier cas de meningite a eosinophiles en Cote-d'Ivoire avec presence d'un *Parastrongylus* sp. dans le liquide cephalo-rachidien. Bull. Soc. Pathol. Exot., *73*:179–182.

Nwosu, A.B.C. 1981. Human neonatal infections with hookworms in an endemic area of southern Nigeria: A possible transmammary route. Trop. Geogr. Med., *33*:105–111.

Ohbayashi, M., Kamiya, M., and Bhaibulaya, M. 1979. Studies on the parasite fauna of Thailand. I. Two new metastrongylid nematodes, *Angiostrongylus siamensis* sp. n. and *Thaistrongylus harinasutai* gen. et sp. n. (Metastrongyloidea: Angiostrongylidae) from wild rats. Jpn. J. Vet. Res., *27*:5–10.

———, and Orihara, M. 1968. Discovery of *Angiostrongylus cantonensis* (Chen, 1935) Dougherty, 1946 from a rat in Sapporo, northern Japan. Jpn. J. Parasitol., *17*:1–4.

Okamoto, K. 1961. An experimental study of the migration route and development of *Ancylostoma duodenale* in pups after oral infection. J. Kyoto Pref. Med. Univ., *70*:135–152.

Onwuliri, C.O.E., Nwosu, A.B.C., and Anya, A.O. 1981. Experimental *Ancylostoma tubaeforme* infection in cats: Changes in blood values and worm burden in relation to single infections of varying size. Z. Parasitenkd., *64*:149–155.

Orihel, T.C. 1971. *Necator americanus* infection in primates. J. Parasitol., *57*:117–121.

Otsuru, M. 1962. *Trichostrongylus brevis* sp. nov. from man (Nematoda: Trichostrongylidae). Acta Med. Biol., *9*:273–278.

———. 1979. *Angiostrongylus cantonensis* and angiostrongyliasis in Japan, pp. 74–117. In *Studies on Angiostrongyliasis in Eastern Asia and Australia*. Edited by J.H. Cross. Special Publication, U.S. Naval Medical Research Unit No. 2, Taipei, Taiwan.

Oyerinde, J.P.O. 1976. The role of the house fly *(Musca domestica)* in the dissemination of hookworm. Ann. Trop. Med. Parasitol., *70*:455–462.

Palmer, E.D. 1941. The course of the daily egg output during an early infection with the hookworm *Necator americanus*. Am. J. Hyg., *34*:1–12.

———. 1955. Course of egg output over a 15-year period in a case of experimentally induced Necatoriasis americanus. Am. J. Trop. Med. Hyg., *4*:756–757.

Pascual, J.E., Bouli, R.P., and Aguiar, H. 1981a. Eosinophilic meningitis in Cuba, caused by *Angiostrongylus cantonensis*. Am. J. Trop. Med. Hyg., *30*:960–962.

———, Aguilar, P.H., and Galvez, M.D. 1981b. Hallazgo del *Agiostrongylus cantonensis* en el líquido cefalorraquideo de un niño con meningoencefalitis eosinofílica. Rev. Cub. Med. Trop., *33*:92–95.

Piza, J.E., and Biagi, F. 1963. Localization of *Ancylostoma duodenale* within the intestinal wall. Parassitologia, *5*:189–192.

Porter, W.B. 1937. Heart changes and physiologic adjustment in hookworm anemia. Am. Heart J., *13*:550–579.

Prommindaroj, K., Leelawongs, N., and Pradatsundarasar, A. 1962. Human angiostrongyliasis of the eye in Bangkok. Am. J. Trop. Med. Hyg., *11*:759–761.

Punyagupta, S. 1979. Angiostrongyliasis: Clinical features and human pathology, pp. 138–150. In *Studies on Angiostrongyliasis in Eastern Asia and Australia*. Edited by J.H.

Cross. Special Publication, U.S. Naval Medical Research Unit No. 2, Taipei, Taiwan.

———, Juttijudata, P., and Bunnag, T. 1975. Eosinophilic meningitis in Thailand. Clinical studies of 484 typical cases probably caused by *Angiostrongylus cantonensis*. Am. J. Trop. Med. Hyg., *24*:921–931.

Rainey, G. 1855. Entozoon found in the larynx. Trans. Pathol. Soc. Lond., *6*:370–372.

Rep, B.H. 1966. Pathogenicity of *Ancylostoma braziliense*. IV. Blood loss caused by the worms in the prepatent period. Trop. Geogr. Med., *18*:329–352.

———. 1972. Unfertilized hookworm eggs. Trop. Geogr. Med., *24*:363–369.

Roche, M., and Layrisse, M. 1966. The nature and causes of "hookworm anemia." Am. J. Trop. Med. Hyg., *15*:1031–1102.

Rogers, A.M., and Dammin, G.J. 1946. Hookworm infection in American troops in Assam and Burma. Am. J. Med. Sci., *211*:531–538.

Rojas Ayala, M.A., Guerra, I.F., Schir, R.A., and Motizuki, A. 1982. Angiostrongiloidose abdominal. Apresentacão de um caso. Mem. Inst. Oswaldo Cruz, Rio de Janeiro, *77*:189–193.

Rosen, L., Chappell, R., Laqueur, G.L., Wallace, G.D., and Weinstein, P.P. 1962. Eosinophilic meningoencephalitis caused by a metastrongylid lungworm of the rat. J.A.M.A., *179*:620–624.

Sahba, G.H., Arfaa, F., and Bijan, H. 1967. Intestinal helminthiasis in the rural area of Khuzestan, south-west Iran. Ann. Trop. Med. Parasitol., *61*:352–357.

Sandground, J.H. 1931. Studies on the life history of *Ternidens deminutus*, nematode parasite of man, with observations on its incidence in certain regions of South Africa. Ann. Trop. Med. Parasitol., *25*:147–184.

Sauerbrey, M. 1977. A precipitin test for the diagnosis of human abdominal angiostrongyliasis. Am. J. Trop. Med. Hyg., *26*:1156–1158.

Schad, G.A., Chowdhury, A.B., Dean, C.G., Kochar, V.K., Nawalinski, T.A., Thomas, J., and Tonascia, J.A. 1973. Arrested development in human hookworm infections: An adaptation to a seasonally unfavorable external environment. Science, *180*:502–504.

Scrimgeour, E.M., and Burke, W.J. 1979. Angiostrongyliasis cantonensis. Aust. N.Z. J. Med., *9*:75–77.

Sen, H.G., and Seth, D. 1967. Complete development of the human hookworm, *Necator americanus*, in golden hamsters, *Mesocricetus auratus*. Nature, *214*:609–610.

Setasuban, P., Punsri, W., and Meunnoo, C. 1980. Transmammary transmission of *Necator americanus* larva in the human host. Southeast Asian J. Trop. Med. Public Health, *11*:535–538.

Sharma, K.D., Renapurkar, D.M., Bhopale, M.K., Nathan, J., Boraskar, A., and Chotani, S. 1981. Study of a focus of *Angiostrongylus cantonensis* in greater Bombay. Bull. Haffkine Inst., *9*:38–46.

Sierra, E., and Morera, P. 1972. Angiostrongilosis abdominal. Primer caso humano encontrado en Honduras (Hospital Evangelico de Siguatepeque). Acta Med. Costaric., *15*:95–99.

Soh, C.T. 1958. The distribution and persistence of hookworm larvae in the tissues of mice in relation to species and to routes of inoculation. J. Parasitol., *44*:515–519.

———. 1960. The effects of natural food-preservative substances on the development and survival of intestinal helminth eggs and larvae. II. Action on *Ancylostoma duodenale* larvae. Am. J. Trop. Med. Hyg., *9*:8–10.

Sonakul, D. 1978. Pathological findings in four cases of human angiostrongyliasis. Southeast Asian J. Trop. Med. Public Health, *9*:220–227.

Spellman, G.G., and Nossel, H.L. 1971. Anticoagulant activity of dog hookworm. Am. J. Physiol., *220*:922–927.

Stiles, C.W. 1902. A new species of hookworm *(Uncinaria americana)* parasitic in man. Am. Med., *3*:777–778.

Stoll, N.R., and Hausheer, W.C. 1926. Concerning two options in dilution egg counting: Small drop and displacement. Am. J. Hyg., *6*(March suppl.), 134–145.

Stone, O.J., and Mullins, J.F. 1965. Thiabendazole effectiveness in creeping eruption. Arch. Dermatol., *91*:427–429.

Stone, W., and Smith, F.W. 1973. Infection of mammalian hosts by milk-borne nematode larvae: A review. Exp. Parasitol., *34*:306–312.

Stoye, M. 1976. Ubersichtsreferat: Pränatale und galaktogene Helmintheninfektionen bei Haustieren. Dtsche. Tierarztl. Wschr., *83*:569–576.

Stürchler, D., Stahel, E., Saladin, K., and Saladin, B. 1980. Intestinal parasitoses in eight Liberian settlements: Prevalences and community anthelmintic chemotherapy. Tropenmed. Parasitol., *31*:87–93.

Svensson, R. 1925. Observations on the development and longevity of hookworm larvae in different temperature conditions. China Med. J., *39*:1–11.

Sweet, W.C. 1924. The intestinal parasites of man in Australia and its dependencies as found by the Australian Hookworm Campaign. Med. J. Aust., *1*:405–407.

Tanabe, K. 1962. An experimental study on the migration route and the development of *Necator americanus* in pups after cutaneous infection. J. Kyoto Pref. Med. Univ., *71*:513–537.

Tangchai, P., Nye, S.W., and Beaver, P.C. 1967. Eosinophilic meningoencephalitis caused by angiostrongyliasis in Thailand. Am. J. Trop. Med. Hyg., *16*:454–461.

Tharavanij, S. 1979. Immunology of angiostrongyliasis, pp. 151–164. In *Studies on Angiostrongyliasis in Eastern Asia and Australia*. Edited by J.H. Cross. Special Publication, U.S. Naval Medical Research Unit No. 2, Taipei, Taiwan.

Thitasut, P. 1961. Action of aqueous solutions of iodine on fresh vegetables and on the infective stages of some common intestinal nematodes. Am. J. Trop. Med. Hyg., *10*:39–43.

Thorson, R.E. 1956a. Proteolytic activity in extracts of the esophagus of adults of *Ancylostoma caninum* and the effect of immune serum on this activity. J. Parasitol., *42*:21–25.

———. 1956b. The effects of extracts of the amphidial glands, excretory glands and the esophagus of adults of *Ancylostoma caninum* on the coagulation of dog's blood. J. Parasitol., *42*:26–30.

Tripathy, K., Garcia, F.T., and Lotero, H. 1971. Effect of nutritional repletion on human hookworm infection. Am. J. Trop. Med. Hyg., *20*:219–223.

Turton, J.A. 1977. Studies on primary, secondary, tertiary and quaternary self-infections with *Necator americanus*. Parasitology, *75*:xxxvi.

Ubelaker, J.E. 1980. Experimental observations on the transmission of *Angiostrongylus costaricensis*. Proceedings 10th International Congress Tropical Med. Malaria, p. 162.

———, Bullick, G.R., and Caruso, J. 1980. Emergence of third-stage larvae of *Angiostrongylus costaricensis* Morera and Céspedes 1971 from *Biomphalaria glabrata* (Say). J. Parasitol., *66*:856–857.

Velasquez, C.C., and Cabrera, B.C. 1968. *Ancylostoma ceylanicum* (Looss, 1911) in a Filipino woman. J. Parasitol., *54*:430–431.

Watson, J.M. 1953. Human trichostrongylosis and its relationship to ancylostomiasis in southern Iraq, with comments on world incidence. Parasitology, *43*:102–109.

———. 1960. *Medical Helminthology*. Baltimore, Williams & Wilkins.

Wilfred, M., and Lee, D.L. 1981. Observations on the buccal

capsule and associated glands of adult *Bunostomum trigonocephalum* (Nematoda). Int. J. Parasitol., *11*:485–492.

Witenberg, G. 1951. Some unusual observations on helminthiasis in Israel. Harefuah, *41*:178–180.

Yii, C.-Y., Chen. C.-Y., Fresh, J.W., Chen, T., and Cross, J.H. 1968. Human angiostrongyliasis involving the lungs. Chin. J. Microbiol., *1*:148–150.

Yorke, W., and Maplestone, R.A. 1926. *The Nematode Parasites of Vertebrates*. London, J. & A. Churchill.

Yoshida, Y. 1971. Comparative studies on *Ancylostoma braziliense* and *Ancylostoma ceylanicum*. I. The adult stage. J. Parasitol., *57*:983–989.

———, and Fukutome, S. 1967. Experimental infection of rabbits with the human hookworm, *Necator americanus*. J. Parasitol., *53*:1067–1073.

———, Kondo, K., Kurimoto, H., Fukutome, S., and Shirasaka, S. 1974. Comparative studies on *Ancylostoma braziliense* and *Ancylostoma ceylanicum*. III. Life history in the definitive host. J. Parasitol., *60*:636–641.

———, and Nakanishi, Y. 1958. Experimental and clinical studies on the cortisone treatment of "Wakana-disease." Jpn. J. Parasitol., *7*:442–448.

———, Okamoto, K., and Chiu, J.C. 1971. Experimental infection of man with *Ancylostoma ceylanicum* Looss, 1911. Chin. J. Microbiol., *4*:157–167.

———, Okamoto, K., Higo, A., and Imai, K. 1960. Studies on the development of *Necator americanus* in young dogs. Jpn. J. Parasitol., *9*:735–743.

Yousif, F., and Ibrahim, A. 1978. The first record of *Angiostrongylus cantonensis* from Egypt. Z. Parasitenkd., *56*:73–80.

Zambrano P.Z., Diaz P.I., and Salfelder, K. 1975. Ileocolitis seudotumoral eosinofilica de origen parasitario. GEN, *29*:87–96.

Zavala Velazquez, J., Ramirez Baquedano, W., Reyes Perez, A., and Bates Flores, M. 1974. Angiostrongilosis costaricensis. Primeros casos mexicanos. Rev. Invest. Clin. (Mex.), *26*:389–394.

Chapter 21

Oxyuroidea and Ascaridoidea

SUPERFAMILY OXYUROIDEA

The Oxyuroidea are relatively stout nematodes with a distinct esophageal bulb. They are monoxenous. The male has few caudal papillae and a single spicule. The female produces eggs that contain an embryo when laid and that are often flattened on one side.

The only oxyurid parasite of man is *Enterobius vermicularis*. *Syphacia obvelata,* a pinworm of the mouse, was reported once from man (Riley, 1920). *Syphacia muris,* a pinworm of the rat, and *Aspiculuris tetraptera* of the mouse are of interest for the information gained from them in experimental studies on the biology of pinworms.

Enterobius vermicularis (Linnaeus, 1758) Leach, 1853 (The human pinworm or seatworm, producing enterobiasis or oxyuriasis)

Synonym. *Oxyuris vermicularis* (Linnaeus, 1758) Lamarck, 1816.

Historical and Geographical Notes. *Enterobius vermicularis* has been known since ancient times and has been the subject of hundreds of biological, epidemiologic, and clinical studies.

The worm has a cosmopolitan distribution. It is more common in temperate and cold climates than in warm climates because of less frequent bathing and more general use and infrequent changing of underclothing. In the moist tropics, where infants and young children are essentially unclothed, oxyuriasis is less prevalent than in cooler climates where children are crowded together in sleeping quarters and in schoolrooms. Various local surveys show rates of infection in school children ranging from 1% in Guam (Stoll et al., 1947) to 100% in Amsterdam during the years of World War II (Schüffner, 1944).

Morphology, Biology, and Life Cycle. The adult worms inhabit the cecum, appendix, and adjacent portions of the ascending colon. Whether they are usually attached to the mucosa or move freely over its surface is unknown. The usual location of adult pinworms of mice is in the layer of mucus between the mucosa and the fecal gut contents. The oral end lacks a true buccal capsule, but is provided with three lips and dorsoventral bladder-like expansions of the cuticula. The *male* (Fig. 21–1) measures 2 to 5 mm in length by 0.1 to 0.2 mm in greatest diameter and has its posterior end strongly curved ventrad. It has a single conspicuous copulatory spicule (70 μm long) but lacks a gubernaculum. A greatly reduced bursa, usually designated as "caudal alae," is supported by a pair of anterior pedunculated papillae and a pair of posterior papillae. In addition, there are three pairs of more median, postanal sessile papillae, and a pair of sessile papillae lateral in position to the anterior pedunculated pair (Fig. 21–2).

The *female* (Fig. 21–1) measures 8 to 13 mm in length and 0.3 to 0.5 mm in greatest diameter.

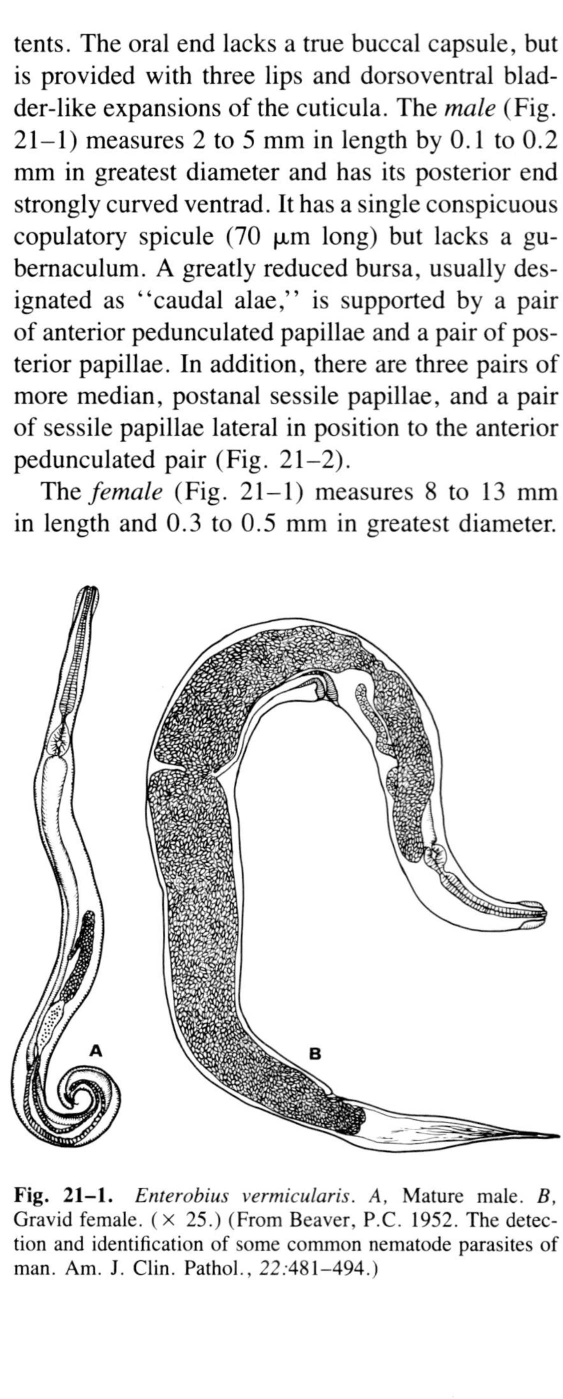

Fig. 21–1. *Enterobius vermicularis. A,* Mature male. *B,* Gravid female. (× 25.) (From Beaver, P.C. 1952. The detection and identification of some common nematode parasites of man. Am. J. Clin. Pathol., *22:*481–494.)

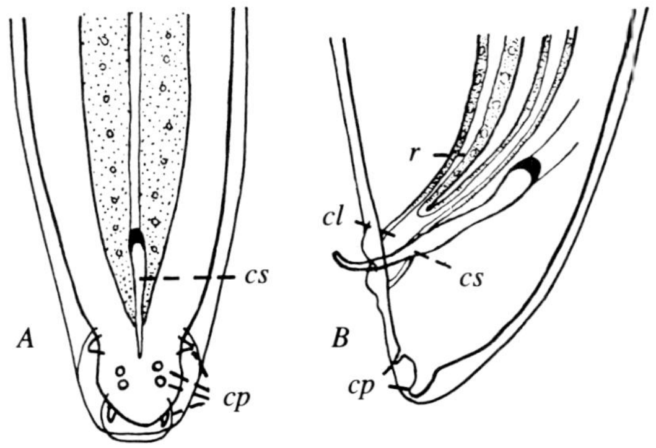

Fig. 21–2. *Enterobius vermicularis.* Posterior end of male. *A,* ventral view. (× ca. 200.) *B,* Lateral view. (× ca. 240.) *cl,* Cloaca. *cp,* Caudal papillae. *cs,* Copulatory spicule partially protruding from cloaca. (Adapted from Yorke, W., and Maplestone, P.A. 1926. *The Nematode Parasites of Vertebrates.* London, J. & A. Churchill, in Faust, E.C. 1929. *Human Helminthology,* 1st ed. Philadelphia, Lea & Febiger.)

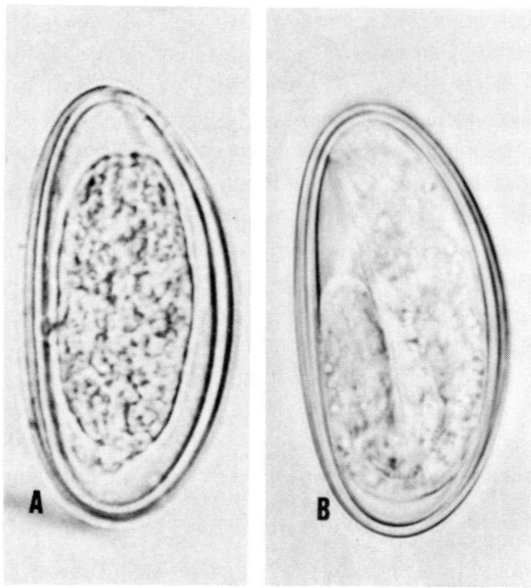

Fig. 21–3. *Enterobius vermicularis. A,* Egg from feces, containing an embryo in tadpole stage. *B,* Egg from anus, containing a larva that is well differentiated but not fully developed. (× 1000.) (By P.C. Beaver.) (See also Fig. 46–6, p. 751.)

The tail (i.e., postanal) portion is sharply pointed. The vulva opens midventrally just in front of the middle third of the body. The vagina is relatively long and extends posteriad some distance from the vulva before it joins the paired genital organs, consisting successively of uteri, oviducts, and ovarian tubules, which are coiled back and forth several times in the middle half of the body. In gravid females the uteri are tremendously distended, so that the entire body is packed with eggs. When this occurs, the female migrates down the colon and out the anus to crawl on the perianal and perineal skin and, in female patients, at times enters the vagina. The time required for full development, not known precisely, is reported to be 15 to 43 days after the ingestion of infective-stage eggs. The female oviposits as she crawls over the skin around the anus, and she usually deposits all her eggs at once and perishes. Although it is suspected that she may reenter the intestinal canal, this has not been proved. Worms in the anal canal or at the orifice, however, may retreat into the rectum. Males may also be seen emerging from the anus. The eggs deposited by a single female vary from 4,672 to 16,888 (Reardon, 1938), with a mean of 11,105. They are elongate-ovoidal, distinctly compressed laterally, flattened on one side, and measure 50 to 60 by 20 to 30 μm (Figs. 21–3, 46–6). The shell is relatively thick and colorless. Eggs are not commonly laid within the bowel, although at times they may be seen in the feces. No intermediate host is required for subsequent development, and usually eggs are infective within a few hours after they are deposited (6 hours at body temperature according to Cram, 1943). They are resistant to disinfectants, and under cool, moist conditions may remain viable up to 13 days.

When eggs containing infective larvae are swallowed and reach the intestine, they hatch. On reaching the cecal region, the larvae develop into adult worms. The time required for development to the egg-laying stage has been variously stated, ranging from 2 weeks to 2 months. It is probably about 1 month, as eggs are again deposited at the anus as early as 5 weeks after successful treatment (Jung and Beaver, 1953), and an effective second anthelmintic treatment, given 3 weeks after a first, will expel worms that are over 8 mm long but still immature (Cho et al., 1981). Because the prepatent period for *E. vermicularis* has not been precisely determined, observations on this common and unpredictably pathogenic parasite are often misinterpreted, especially with reference to the success or failure of treatment and prophylactic measures.

The early larval stages probably develop in the crypts of the mucosa as they do in the mouse pinworm, *Aspiculuris tetraptera,* and some other oxyurid species (Anya, 1966).

Epidemiology. Enterobiasis (oxyuriasis) is a group infection. It is more common in infected families or in asylums than in populations at large, and is more common in children than in adults. Families of two or more children, at any socioeconomic level, seldom escape having at least one bout with enterobiasis. It is several times more common in North American whites than in blacks (Cram, 1941). The methods of transmission are threefold. The most common means is direct anus-to-mouth by finger contamination. Soiled night clothes are another anus-to-mouth transmission hazard. These relatively direct sources of transfer of viable eggs to the mouth are responsible for continued heavy infection in an individual or group of persons who have similar habits. In addition, persons sleeping in the same bed or bedroom with carriers are exposed from viable eggs on soiled bed linen, table tops, door knobs, and other contaminated objects with which their hands make contact. In addition, airborne eggs that are dislodged from bed linens and clothes, get into the mouth or are inhaled, are swallowed, and provide a relatively small number of worms in a large group of individuals in contact with the contaminated air. This may explain the high percentage of persons found to be lightly infected. A fourth method, "retrofection," has been suggested (Schüffner and Swellengrebel, 1943), but has not been demonstrated. It is apparently trivial if it occurs at all. It is supposed that in a moist environment infective-stage eggs at times hatch on the anal mucosa, and the hatched larvae may migrate up into the bowel and develop into adult worms. There is no convincing evidence of this or of internal autoinfection.

The degree of infection varies greatly. As many as 5000 worms have been removed from a single patient, but the majority of cases average less than one migrating worm nightly.

Pathogenesis, Pathology, and Symptomatology. The most significant pathologic conditions are produced by gravid females that migrate from the anus onto the perianal and perineal skin to oviposit, causing a crawling sensation and pruritus. This condition results in scratching the area, at times causing scarification. In most infected persons there is no other clinical evidence of the infection.

Eosinophilia may or may not accompany this infection. It is presumed that at times the worms may produce appendicitis by invasion of the wall, but ordinarily they do not cause any appreciable damage to this organ (Fig. 21–4,*A*). However, *E. vermicularis* is often found in surgically removed appendices showing superficial and deep ulcerations and infiltrations with predominance of eosinophils (Fig. 21–4,*B*). Women are affected three times as frequently as men, and young persons (11 to 20 years of age) more frequently than older people. Bijlmer (1946) reported a necropsy case of a 40-year-old emaciated man in Holland in whom approximately 10,000 pinworms had penetrated the intestinal wall into the subserosa. In children, nervousness, insomnia, nightmares, and even convulsions have been attributed to enterobiasis. A frequent complaint among infected young children is sharp pain that they attribute to being bitten at the anus. The presence of many worms in the rectum may cause rectal discomfort. Occasionally, an adult worm may wander into the upper levels of the small intestine. Rarely, a dead pinworm has been found in a granuloma deep in the parenchyma of the liver (Little et al., 1973) and in the lung (Beaver et al., 1973).

In heavily infected girls, the infection is more frequently troublesome. Brady (1947) reported that 14 of 45 girls with pinworm infection had eggs of the worm on the vulva as determined by a single N-I-H swab examination, while 10 of 51 infected young girls had a mucoid vaginal discharge. Several cases are recorded in which gravid females migrated up the vagina and uterus into the fallopian tubes and either became encapsulated there or, having wandered into the peritoneal cavity, became encapsulated on the peritoneum. In the tubes they may produce symptoms of chronic salpingitis.

A large part of the recent literature on enterobiasis consists of case reports of urinary, genital, or abdominal disturbances in young girls (Simon, 1974; Broadbent, 1975; Nutting et al., 1980) or in older women (Kacker, 1973; Fitzgerald et al., 1974; Saffos and Rhatigan, 1977; Tsung and Loh, 1979). In males, tissue invasion has usually involved the colon (Shiraki et al., 1974).

The blood picture in pinworm infection is not remarkably altered except in cases of peritoneal invasion via the female genital tract when there may be a significant peripheral eosinophilia (Fatherree et al., 1951).

Diagnosis. Pinworm infection may be suspected by obtaining a history of any of the more striking manifestations described previously. *Specific diagnosis* is based (1) on recovery of characteristic eggs (Fig. 21–3) in feces, perianal scrapings, or swabs (Fig. 46–6), or from under fingernails, or (2) on capture of adult worms following enemas, or of females migrating from the anus, usually at night. Since one worm emerging from the anal

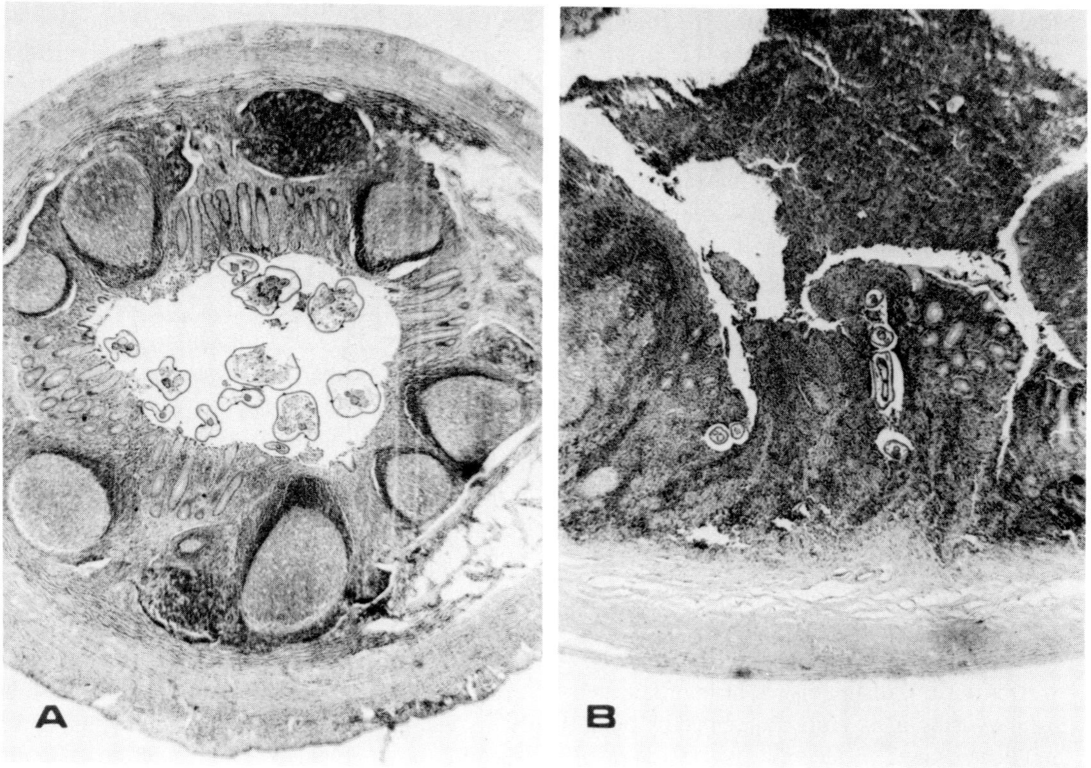

Fig. 21–4. *Enterobius vermicularis.* Sections of immature adult worms in the appendix. *A,* Numerous worms of both sexes, limited to the lumen, and causing no evident tissue changes. (× 20.) *B,* Two young males deeply embedded in an area of chronic inflammation in the submucosa. (× 25.) (By P.C. Beaver.)

canal may terminate an infection, several positive daily swabs are essential to prove that the infection persists. While examination of heavily infected children may reveal eggs at the anus daily for indefinite periods, in others eggs may be present intermittently for 1 to 3 or 4 days (Jung and Beaver, 1953; Cho et al., 1980).

Eggs are present in the feces in not more than 5% of infected individuals. At times a nest of them may be discovered in an ordinary fecal film; at other times only a few may be discovered after careful search of a specimen following concentration. Perianal and perineal swabbings, particularly in the perianal folds, yield more reliable findings.

In the period 1860 to 1937 many different kinds of tools were used to remove pinworm eggs from the anus and to transfer them to a glass slide for microscopic examination (Beaver, 1949). In 1937, Hall and colleagues utilized a nonadhesive cellophane swab, commonly referred to as the N-I-H swab (from National Institutes of Health), and demonstrated its superiority over previously used swabs for the recovery of *Enterobius* eggs. This swab in turn was supplanted by the transparent adhesive cellulose tape which, with various modifications, has remained in use to the present. Widely used for a few years was the wet pestle swab introduced by Schüffner and Swellengrebel in 1943. An interesting feature of the wet pestle method was its ability to demonstrate protozoan cysts as well as all common helminth eggs in addition to those of the pinworm. The adhesive Scotch cellulose tape was introduced by Graham in 1941 and improved by Jacobs the following year, but it did not become the method of choice until the use of toluene to clear the preparation was introduced (Beaver, 1949, 1952a).

The Scotch-tape swab consists of a 3- to 4-inch length of 3/4-inch width Scotch transparent cellulose adhesive tape held sticky side out over the rounded end of a wooden tongue depressor by means of the thumb and index finger. This technique has been simplified by mounting tape on a glass or plastic microscope slide for transport. When used, it is lifted and turned back over a tongue depressor. This method is particularly adapted to swabs taken from children at home. The mounted tape is firmly *pressed* first against one side and then the other side of the outer margin of the anal canal where it joins the perianal skin. The tape is then folded back on the slide sticky side down. For examination under the microscope, the tape is lifted as far back as the "exposed" area, a drop of toluene is added, and the tape is again pressed to the slide. If numerous eggs are trapped, they can be readily identified under low power of the microscope. When eggs are few, the toluene has the advantage of displacing air bubbles, clearing epithelial cells and debris, and dissolving the adhesive material of the tape so that any eggs in the preparation become highly refractive and conspicuous.

By any method, best results are obtained from perianal swabs in the morning, before a stool is passed and before the bath.

Mothers frequently discover worms evacuated after an enema or spontaneously crawling from the anus at night after the child has been put to bed. Such worms must be distinguished from proglottids of *Dipylidium caninum*. Worms obtained by mothers should be placed in alcohol (or vinegar) and brought to the clinic for confirmation.

Treatment. Several drugs are effective against infection with *E. vermicularis* and are reasonably free of side effects. Pyrantel pamoate is given in a single dose of 11 mg per kilogram of body weight (maximum, 1 gram); mebendazole in a single dose of 100 mg; piperazine citrate in a 7-day course of 65 mg per kilogram of body weight (maximum daily dose 2.5 grams); and pyrvinium pamoate in a single dose of 5 mg per kilogram of body weight (maximum 350 mg).

Although any one of these treatments can be expected to eliminate an infection with *E. vermicularis*, it is usually necessary to repeat the treatment after about 2 weeks to eliminate any infection acquired from ingestion of eggs that remained in the environment following the first treatment. Mebendazole and pyrvinium remove worms of all stages of development, but pyrantel or piperazine given in a single dose is ineffective against the early stages (Cho et al., 1981).

The decision to treat a child for pinworm infection should be based on evidence that the infection is continuing. This is determined by a series of cellulose adhesive tape tests (see preceding Diagnosis section). If only an occasional test is positive, there will be little benefit to the child from treatment. In group situations it is frequently found that only one or a few of a group of individuals with symptomatic pinworm infection have continual exiting of worms and egg deposition. Treatment of the entire group, followed by periodic retreatment of the more susceptible individuals, may provide prolonged freedom from infection.

Prognosis. Uncomplicated pinworm infection is never fatal. Periodic treatment as needed provides a favorable prognosis.

Prevention. Attempts to kill *Enterobius* eggs with hydrocyanic gas fumigation, paradichlorobenzene, and naphthalene have been ineffective, but sunlight and ultraviolet lamp radiation destroy eggs in the environment, and dry oven heat will sterilize nonwashable metal toys. The eggs are not killed by chlorination of water in swimming and wading pools.

Satisfactory control can usually be attained by personal and group hygiene, combined with individual or mass chemotherapy. Within the family, pinworm problems can be reduced or prevented by encouraging habits of cleanliness in the children, by regular use of closed sleeping garments, and by keeping the fingernails short and clean. Diagnosis and treatment of infected members of a household repeated as often as necessary will delay reinfection, but in usual circumstances total prevention is not a realistic goal.

Syphacia obvelata (Rudolphi, 1802) and *S. muris* (Yamaguti, 1935)

Syphacia obvelata is a common, cosmopolitan parasite of the mouse. For many years it was considered to be equally common in rats, but Yamaguti (1935) recognized the rat pinworm as a distinct species and originally described it as *Enterobius muris*. Later, after examining the male worm, he transferred it to the genus *Syphacia*. Two adult female *S. obvelata* and eggs were identified in the feces of an American child living in the Philippines (Riley, 1920).

Syphacia obvelata and *S. muris* live in the cecum and, as in *Enterobius*, the mature female migrates to the lower colon and emerges from the anus to deposit masses of eggs on the perianal skin. Eggs contain a larva when laid and develop to the infective stage in less than 24 hours. Eggs of *S. obvelata* measure approximately 130 by 36 μm; those of *S. muris* are much smaller (75 by 30 μm). The adult *S. muris* is also smaller, 4.0 vs. 5.2 mm in females and 1.2 vs 1.3 mm in males (Hussey, 1957).

A third species, *Aspiculuris tetraptera* (Nitzsch, 1821) Schulz, 1924 is common in mice. It lives in the cecum and upper colon and migrates downward when fully mature, but the female deposits eggs in the lower colon, not outside the body. Eggs are carried in the mucus capsule of the fecal pellets and require several days for development to infectivity. Larvae hatching from the eggs migrate into the crypts of the colon for early development (Mya, 1955; Anya, 1966). Other pinworms, including *Enterobius* and *Syphacia* species, may also migrate into the crypts for early development, as there is a characteristic biological tendency of oxyurids to have a transitory tissue phase.

In the adult stage *Syphacia*, and to a lesser extent *Aspiculuris*, ingest the intestinal contents of the host (Mya, 1955).

SUPERFAMILY ASCARIDOIDEA

Members of the order Ascaridida, superfamily Ascaridoidea, are generally relatively large, with three well-developed lips, cylindrical esophagus, and coelomyarian body muscles. The eggs are thick-shelled and not embryonated when deposited.

Genus Ascaris Linnaeus, 1758

Species of the genus *Ascaris* have a simple esophagus without a ventriculus; cervical alae are absent. The lips bear dentigerous ridges; interlabia are absent. In the male, the tail bears numerous preanal and few postanal papillae, caudal alae are absent, and the spicules are equal. The female is larger than the male, the vulva is anterior to the middle of the body, and the eggs, unsegmented when deposited, have a thick shell with an outer albuminous, mammillated layer. Numerous species have been described and assigned to the genus *Ascaris*, but only *Ascaris lumbricoides* Linnaeus, 1758 and *Ascaris suum* Goeze, 1782 are known to be important parasites of man.

Ascaris lumbricoides Linnaeus, 1758 (The large intestinal roundworm, causing human ascariasis)

Historical and Geographical Notes. *Ascaris lumbricoides* was observed and reported by many ancient peoples. The Romans referred to it as *Lumbricus teres*, indicating that it was confused with the common earthworm. It is the most cosmopolitan and most common of all human helminths, excepting perhaps *Enterobius vermicularis*. It flourishes in warm, moist climates or in moist, temperate regions where personal and community hygiene and environmental conditions combine to favor development and persistence of the infective-stage larvae in the egg in polluted soil.

Ascariasis is uncommon but present among urban populations in Belgium, France and Germany, Italy, Spain, and Portugal, but high prevalence, especially among children, is occasionally reported from rural communities. In four communities of northern France, overall average prevalence among school children in 1964 was 18%; in one group of 162 children it was 46% (Biguet et al., 1964).

In Japan and Taiwan, infection rates have been reduced to low percentages except in relatively remote rural communities. In the Philippines, Indonesia, and countries of Southeast Asia, the rates in rural and village populations often exceed 50% and occasionally approach 100%. In India and in Middle Eastern and African countries, infection rates are generally high among people of rural and village communities.

In most countries of Central and South America the average infection rates range up to around 45%. Foci of high prevalence among preschool and young children still persist in southeastern United States. A random-sample survey of approximately 200 white and 40 black households in southeastern Georgia revealed *Ascaris* infection in 1% of individuals of all age groups in the former, and in 12% of the latter (Martin, 1972).

Morphology, Biology, and Life Cycle. The body of *A. lumbricoides* is cylindrical, tapering gradually at the anterior end and somewhat less so at the posterior end (Fig. 21–5,*D, E*). A lateral

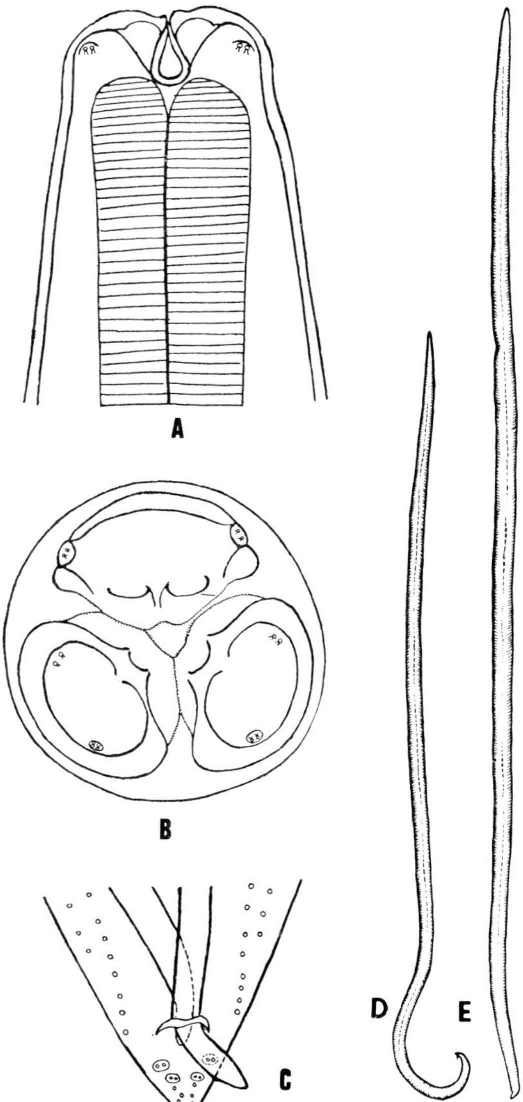

Fig. 21–5. *Ascaris lumbricoides*, morphology. *A*, Anterior extremity of mature worm, ventral view. (× 35.) *B*, Head-on view of worm, showing dorsal and two subventral lips and papillae. (× 45.) *C*, Posterior end of male, ventral view, showing copulatory spicules (one slightly extruded) and caudal papillae. (× 35.) *D*, Male with typical flexion of posterior end. *E*, Female. (*A–C* adapted by E.C. Faust from Yorke, W., and Maplestone, P.A. 1926. *The Nematode Parasites of Vertebrates*. London, J. & A. Churchill. *D* and *E* from Beaver, P.C. 1952. The detection and identification of some common nematode parasites of man. Am. J. Clin. Pathol., *22*:481–494.)

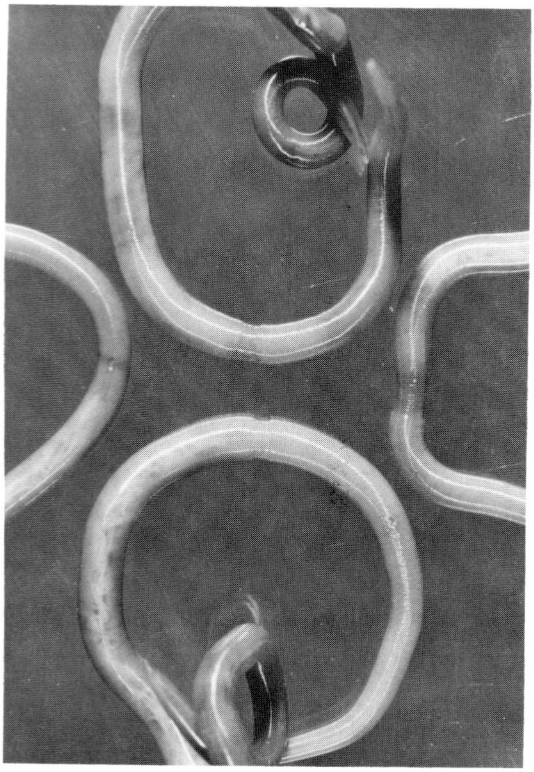

Fig. 21–6. *Ascaris lumbricoides*. Four females, each about 20-cm long, removed from the intestine at autopsy of a 16-year-old boy following heart surgery. No other worms were present and all of these had unfertilized eggs in the uterus and no sperm. The dark areas were pink or red before the worms were placed in formalin. The worms are arranged with the genital girdle towards the center of the photograph. About natural size. (From Beaver, P.C., and Little, M.D. 1964. The genital girdle in relation to estrus and mating in *Ascaris lumbricoides*. J. Parasitol., *50*:128–130.)

line can usually be seen as a whitish streak along the entire length of the flesh-colored body. The three lips, a broad median dorsal one and a pair of ventrolateral ones, are finely denticulated (Fig. 21–5,*B*). Each lip has minute paired papillae on its lateral margins. Centrally located among the lips is a small triangular buccal cavity. *Ascaris* is the largest of the common nematode parasites of man. The mature male measures 15 to 31 cm in length by 2 to 4 mm in diameter; the female, 20 to 35 cm, less commonly up to 49 cm, in length by 3 to 6 mm in diameter. The posterior end of the male is curved ventrad (Fig. 21–5,*D*). Its genitalia consist of a long single tubule made up of testis, vas deferens, and ejaculatory duct, which is tortuously coiled in the posterior half of the worm and opens into the cloaca, which is subterminal. A pair of equal or subequal, simple, cylindrical copulatory spicules measuring 2.0 to 3.5 mm in length, with pointed ends, is situated in a pocket dorsal to the genital tubule (Fig. 21–5,*C*). There is no gubernaculum. Numerous preanal and postanal groups of caudal papillae are symmetrically disposed in the pericloacal region of the male. The vulva of the female is situated midventrally near the junction of the anterior and middle thirds of the body. At times, a distinct groove referred to as the genital girdle encircles the body at the level of the vulva (Fig. 21–6). As this groove is present on female worms that have no spermatozoa in the seminal receptacles or have nearly exhausted the store of sperm, and are therefore laying unfertilized eggs, the genital girdle is apparently an indication of estrus, and is formed to facilitate mating (Beaver and Little, 1964). Extending from inside the vulva posteriad, a single vagina branches to form the pair of genital tubules, each consisting of uterus, seminal receptacle, oviduct, and ovary. These tubules are coiled through the middle and posterior thirds of the body. Their total length is several times that of the worm. They may contain at one time as many as 27,000,000 eggs. Daily egg-laying in light infections has been estimated to be approximately 200,000 for each female. In a study on children in Brazil, the average output per female worm was found to be 2925 eggs per gram (roughly 3/mg) of feces (Mello, 1974).

The fertilized egg (Fig. 21–7,*A*) is broadly ovoid, with a thick, transparent shell, consisting of a relatively nonpermeable innermost lipoidal vitelline membrane (not found in unfertilized eggs), a thick transparent middle layer, and an outermost coarsely mammillated albuminoid layer, usually bile-stained a golden brown. The vitelline membrane is inert. Because of its impermeability, it prevents toxic substances in the immediate environment from injuring the embryo. These eggs (Fig. 21–7) measure up to 75 μm in length and to 50 μm in lesser diameter; when spherical, they are approximately 60 μm in diameter. The ovum is unsegmented and contains a mass of coarse lecithin granules when the egg is discharged in the feces. Unfertilized eggs, measuring up to approximately 90 μm in length, are usually elongate with a relatively thin middle layer of the shell and often with little or no outer mammillated layer (Fig. 21–7,*B, C*). Produced by unmated females, they are frequently seen in the feces. The presence of one to several females in the absence of males is a common occurrence. As each unmated female produces eggs that are dis-

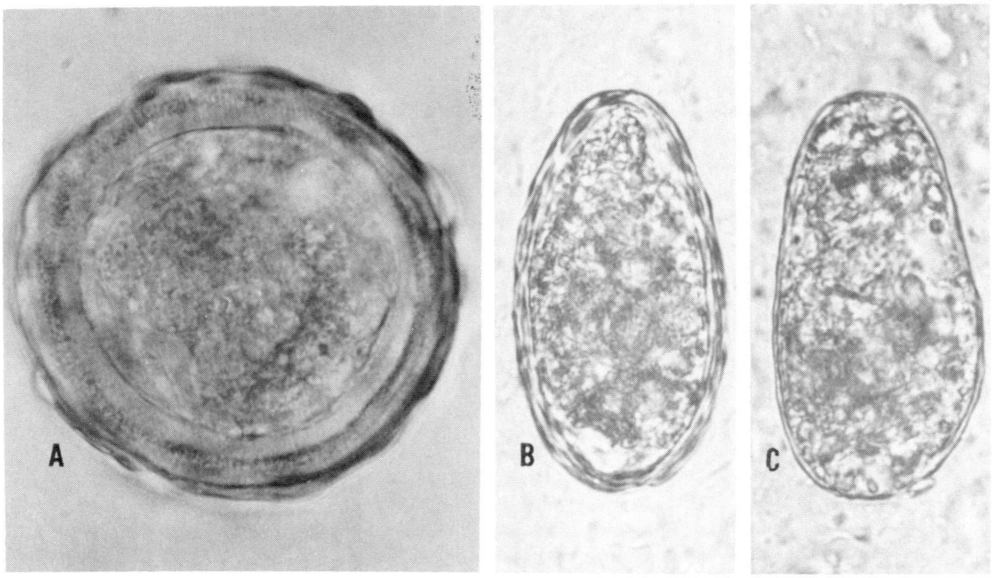

Fig. 21–7. *Ascaris lumbricoides* eggs. *A*, fertilized. (× 1000.) *B* and *C*, Unfertilized. (× 540.) (By P.C. Beaver.)

tinctive in size and appearance, when only one or a few such females are present the number of distinctly different types of eggs indicates the number of egg-laying worms in the intestine (Fig. 21–8). Internally, the unfertilized egg consists of a mass of disorganized, highly refractive granules of various sizes. Both fertile and infertile eggs at times lack the outer albuminoid coat.

Fertilized eggs require a period of incubation before they are infective. The percentage of fully embryonated eggs that develop and the rate of development depend on several environmental conditions. The eggs are resistant to desiccation, low temperature, and putrefaction of the medium and strong chemicals, remaining dormant under such conditions, but with a return of favorable factors they may be stimulated to accelerated embryonation. Eggs stored for approximately 4 months in the complete absence of oxygen remain in the one-cell stage, but when exposed to air at leat 75% of them develop into infective-stage larvae (Kmetec et al., 1963).

Moist shady locations with temperatures of 22 to 30°C favor rapid development of the embryo into a motile infective-stage larva. It is presumed that in the development of *Ascaris lumbricoides* and other Ascaridoidea, as in other nematodes, there are four larval stages and a fifth-stage adult, with a molt between the successive stages. It is well known that the first molt occurs in the egg, and for many years it was accepted that after molting but once the infective stage that emerges from the egg is a second-stage larva. However, several observers have reported two molts in the egg, which if confirmed as true molts would make the infective stage a third-stage larva. According to Schacher (1957), two molts within the *Ascaris* egg were first reported by Dubinin in 1943 in *Ascaris tarbagan*, and the significance of the observation was discussed by Mozgovoi in 1953. Schacher observed two molts within the egg of *Toxocara canis* and discussed whether in certain ascarids the infective form may be a third-stage rather than the generally accepted second-stage larva, but he questioned whether the two observed cuticles could be interpreted as evidence of a true molt rather than two layers of one swollen cuticle. Thust (1968) and Araujo (1972) have reported two molts in the eggs of *A. lumbricoides* and *A. suum*. More recently Maung (1978) reviewed the reports of others in the light of her own findings and concluded that in the developing larva in the egg there are without question two true molts and that the second, though commenced in the egg, may be completed during early migration in the intestinal wall or liver of the host. The minimum time required for complete development in the egg at 28°C is 18 days.

Eggs deposited in feces develop and survive better on heavy soil (clay) than on light soil (sandy), though in areas where the soil is sandy, transmission may nevertheless occur as a result of favorable conditioning of the soil by the domestic refuse that

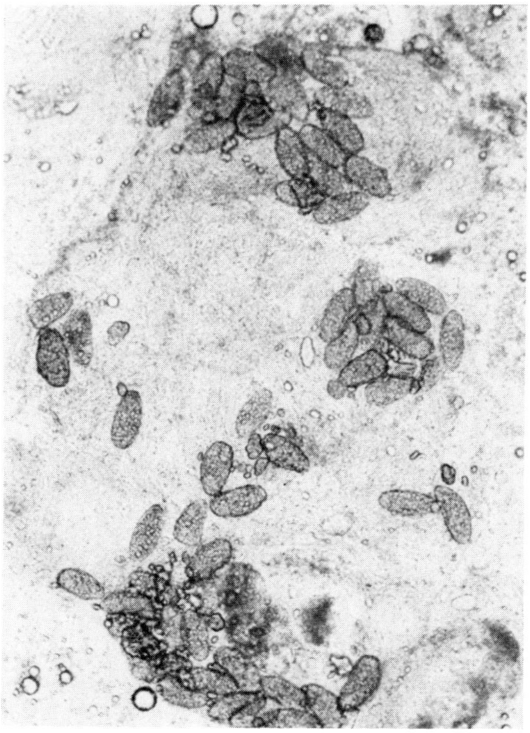

Fig. 21–8. *Ascaris lumbricoides*. Unfertilized eggs in a direct smear of dysenteric stool. By the clumped distribution and the uniform size and shape of the eggs, it is apparent that all were produced by one noninseminated female. Unstained. (× 80.) (By P.C. Beaver.)

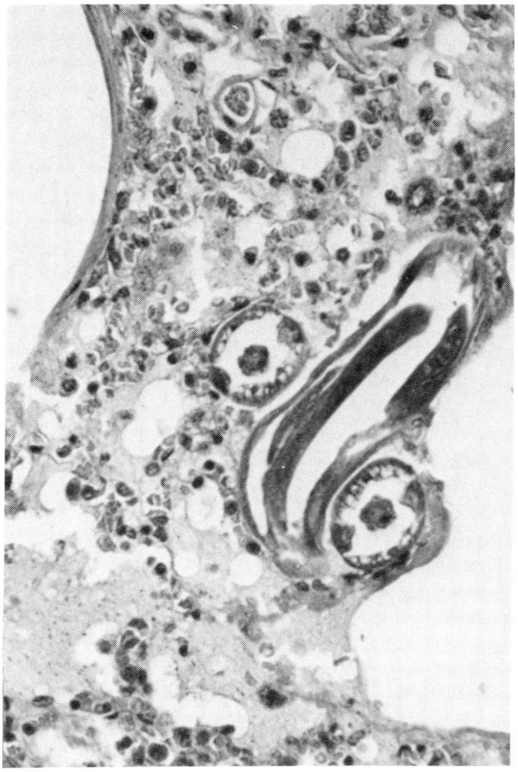

Fig. 21–9. *Ascaris lumbricoides*. Developing larva seen in oblique-longitudinal and two transverse sections in the interalveolar tissues of the lung of a 1-year-old child who died with bronchopneumonia on the third day of illness. Numerous other larvae were seen in other sections. (× 325.) (From Faust, E.C., Beaver, P.C., and Jung, R.C. 1975. *Animal Agents and Vectors of Human Disease*, 4th ed. Philadelphia, Lea & Febiger. Section courtesy of Armed Forces Institute of Pathology.)

people discard from their houses (Beaver, 1952b). Eggs do not hatch in the soil or enter the body via the cutaneous route. The only mode of natural infection is through eggs containing infective larvae taken into the mouth, possibly also the nasopharynx.

When infective eggs are swallowed, the larvae hatch in the stomach and upper small intestine, mostly the latter. Emergence from the egg is through a minute opening which the larva itself produces in the otherwise unaltered shell. The hatched larvae actively migrate into the intestinal wall, then into the portal circulation to the liver, the right heart, and on into the pulmonary vessels to the interalveolar tissues of the lungs (Fig. 21–9). Larvae emerging from the egg are about 260 μm in length by 14 μm in diameter. After growth and essential development in the lungs, which is completed on about the ninth day of infection or a few days later, the larvae migrate or are carried in the mucus flow through the respiratory passages to the throat and are swallowed. At that stage they measure 1.2 to 1.6 mm long by 36 to 39 μm wide. Larvae in bronchioles (Fig. 21–10,*A*) and in the stomach en route from the lungs to the intestine (Fig. 21–11) may reach a length of 1.7 mm and measure around 60 μm in width. Larvae from the tissues of experimental animals were described in detail by Nichols (1956b).

Including the 9 to 15 days in the lungs, development of *A. lumbricoides* from the hatched larva to the mature male and female in the intestine requires 8 to 12 weeks (Vogel and Minning, 1942). In experimental animals, during the first few hours or days after arrival from the lungs, the larvae lie in the tissues of the intestinal mucosa, where they molt and become fourth-stage larvae 2 mm or more in length. The worms then live free in the lumen and grow to a length of 1 cm or more when with the final molt at 3 to 4 weeks of infection they become adult males and females. At 8 or 9 weeks, when egg-laying begins, the females are generally 15 to 20 cm in length, though some may be smaller

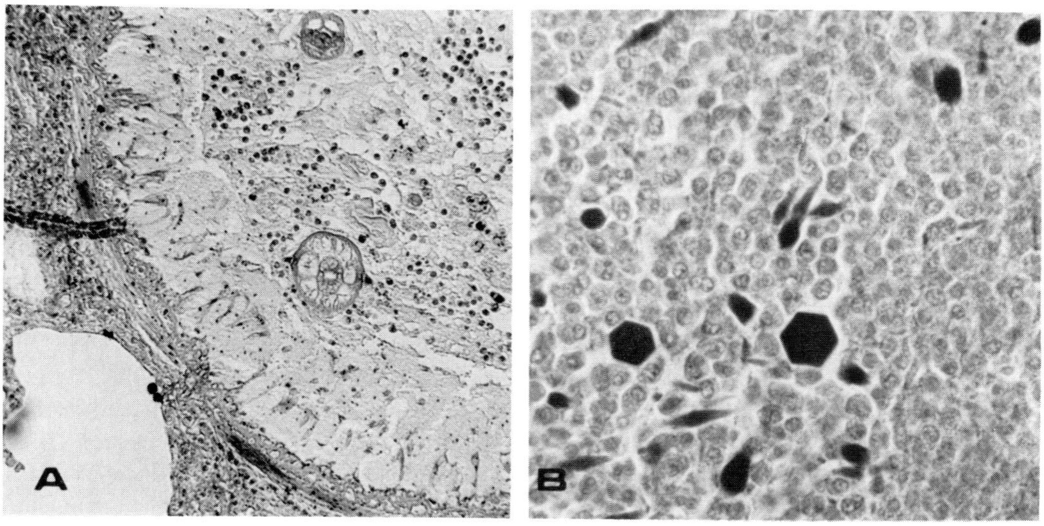

Fig. 21–10. *Ascaris lumbricoides. A,* Larva in stage of migration from lung to intestine, cut at levels of esophagus and mid-intestine, in human autopsy section of a 1-mm bronchiole filled with mucopurulent exudate. (× 160.) *B,* Charcot-Leyden crystals among eosinophilic leukocytes in a different bronchiole in the same case. (× 550.) (From Beaver, P.C., and Danaraj, T.J. 1958. Pulmonary ascariasis resembling eosinophilic lung. Autopsy report with description of larvae in the bronchioles. Am. J. Trop. Med. Hyg., *7*:100–111.)

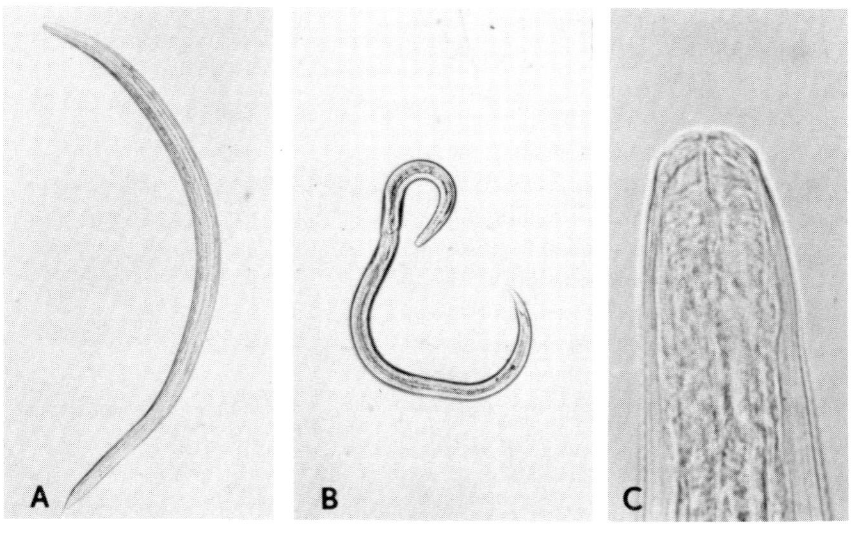

Fig. 21–11. *Ascaris lumbricoides.* Larvae recovered from gastric washings *(A, C)* and from sputum *(B)* of patients with severe seasonal pneumonitis and hypereosinophilia in Saudi Arabia. The larvae are about 1.2-mm long and have low, poorly developed lips, characteristic of the advanced third-stage larva. *A,* (× 50). *B,* (× 45). *C,* (× ca. 500). (Larvae received from Dr. S.P. Gelpi. From Faust, E.C., Beaver, P.C. and Jung, R.C. 1975. *Animal Agents and Vectors of Human Disease*. 4th ed. Philadelphia, Lea & Febiger.)

(Seo, 1981). The adult worms may live as long as 20 months (Hobo, 1956), but the usual life span is about 1 year. Growth continues in sexually mature worms. Females may reach a length of 45 cm and a diameter of 5 mm. Males rarely exceed 25 cm in length and 3 mm in diameter. Mating occurs periodically at undetermined intervals. Fever and probably other abnormal conditions cause worms to migrate from the intestine and from the body through both the mouth (or nose) and the anus. Male worms appear to be more responsive to these conditions than are females. Thus the sex ratio is frequently unbalanced in favor of females, and in children who have had a febrile illness it is not unusual for females only to be found in the intestine (Beaver and Little, 1964; Garcia, 1973; Seo et al., 1979).

Epidemiology. Human infection with *A. lum-*

bricoides is acquired by swallowing fully embryonated eggs from contaminated soil. As in ancylostomiasis, trichuriasis, and other soil-transmitted helminthiases of man and animals, the two principal factors in maintaining endemicity of ascariasis are favorable qualities of the soil and habitual or frequent fecal contamination of the soil. In some regions the use of human excreta to enhance fertility of soil provides the contamination necessary for transmission, and in others it is mainly the promiscuous defecation by small children. In some regions both of these sources contribute to the deposits of viable eggs that accumulate in the soil. In the tropics, people of all ages are commonly infected, whereas in subtropical countries infection rates are high in young children, low in adults.

Rainfall is also a basic factor in transmission, both in the amount of precipitation and in the pattern of its seasonal distribution. Rains not only provide essential moisture but they also favorably redistribute the eggs both horizontally and vertically. Horizontal transport spreads the eggs over wide areas, but it also tends to concentrate helminth eggs wherever puddles are formed. Among the particulate elements of soil, sand grains are larger and heavier than *Ascaris* eggs, and colloidal elements (clay) are lighter. Silt particles have about the same size and density as the eggs of *Ascaris*. Therefore, when sand, coarse silt, medium-sized silt, helminth eggs, fine silt, and clay are all suspended together in water as they are during and immediately after a heavy rain, these diverse particles sediment in strata in the order listed (Beaver, 1975). This sedimentation puts the *Ascaris* eggs under a blanket of clay and fine silt that protects them from the sun's rays and from desiccation, but near the surface where they rapidly develop to the infective stage and are readily picked up by dirt-eating children, or transported to favorable locations for being swallowed by others. Infective eggs are killed by drying but they are sufficiently resistant to desiccation to reach the table in viable form when splashed onto fruits and vegetables to be eaten raw. Splash of rain drops deposits eggs on surfaces 30 cm above ground level (Beaver, 1952b). In certain localities it is possible that *Ascaris* eggs may be airborne. When polluted soil containing infective eggs is dried, among the dust particles picked up and transported by air currents are the eggs of *Ascaris* and some other helminths. Because examination revealed the presence of *Ascaris* eggs in nasal mucus and washings from the nose, it has seemed probable that infection may be acquired by swallowing eggs that were inhaled to the pharynx in dust. Studies thus far have not shown that this is a significant mode of transmission.

Also important in maintaining endemicity of ascariasis is the coprophagy of various animals. Dung beetles bury human stools and eat human feces. Their mouth parts are adapted to crush all kinds of helminth eggs except those species for which the beetles serve as intermediate hosts, such as *Ascarops* and other spirurid nematodes whose eggs are both hard-shelled and minute (Miller, 1961). However, while burying and ingesting fecal matter, dung beetles also scatter and mix the feces with soil, which promotes transmission (Fig. 20–12). Other coprophagous animals serve to disseminate eggs. Young dogs avidly eat fresh human stools, as do certain vultures. The domestic pig likewise eats human stools, especially if garnished with maggots. As *Ascaris* eggs pass unchanged through the intestine of these animals, the result is that eggs in

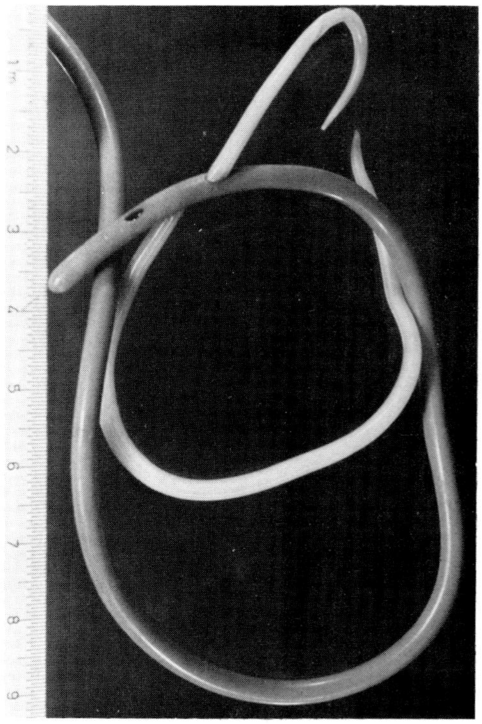

Fig. 21–12. *Ascaris lumbricoides*. Young, 15-cm sexually mature female threaded through the openings of a decompression tube withdrawn from a child under treatment for intestinal obstruction. By its probing tendency, an *Ascaris* occasionally finds its way into the biliary channels, causing abscesses in the liver, less frequently intrahepatic calculi. (Courtesy of Dr. Rupert Spillman. From Faust, E.C., Beaver, P.C., and Jung, R.C. 1975. *Animal Agents and Vectors of Human Disease*, 4th ed. Philadelphia, Lea & Febiger.)

stools placed in a socially acceptable location are transported and deposited at locations more advantageous to the parasite.

Eggs in clay or heavy loam soils, protected against desiccation and freezing, and from the heat and direct rays of the sun, remain infective for several months or possibly a few years. However, eggs that are deep enough in the soil to be preserved in viable and infective condition for years—10 years according to Brudastov et al. (1970), and even 15 years according to Krasnonos (1978)—would seldom have an opportunity to cause infection. Eggs in the upper layers of soil are eventually destroyed by fungi (Lysek, 1979) and probably by other kinds of soil-inhabiting organisms.

On rare occasions *A. lumbricoides* has been reported from dog, cat, sheep, muskrat *(Ondatra zibethica),* and squirrel *(Sciurus indicus, S. pygerythrus).* In some of these instances the observed parasite was probably a different species. Worms indistinguishable from *A. lumbricoides* have been reported in the chimpanzee, gorilla, orangutan, gibbon, and various macaques; heavy infections have killed chimpanzees (Orihel, 1971). However, early attempts to infect primates were unsuccessful. In 1971 and 1972, Orihel succeeded in getting patent experimental infections in 40% of an unstated number of chimpanzees inoculated with 50 to 750 infective *A. lumbricoides* eggs that were cultured from human feces. All the infected animals passed fertile eggs in the feces for several months. The prepatent period was about 9 weeks.

None of the primates or any other susceptible animal except the domestic pig is a significant reservoir of infection. Pigs are susceptible to infection with human *Ascaris,* and people can acquire infection with the pig *Ascaris, A. suum.* Sporadic *Ascaris* infections in people, usually children, occasionally are observed beyond the geographic limits of endemic ascariasis. In such localities where hogs are raised, it is presumed that the observed worms are *A. suum.* In Sri Lanka, although hog-raising is common, *A. suum* is reported to be absent while ascariasis in people is hyperendemic. As it has been shown that pigs, at least specific pathogen-free pigs, are susceptible to infection with *A. lumbricoides* (Galvin, 1968), it is remarkable that *Ascaris* infection is not seen in pigs in that country.

Pathogenesis, Pathology, and Symptomatology. The pathogenic effects of ascariasis are due to the following mechanisms: immune reactions of the host, mechanical effects of the adult worms, and the effects of adult worms on the host's nutrition.

In the stage of larval migration, the first passage of larvae through the liver and lungs provokes no remarkable pathologic changes or symptoms unless the number of larvae is immense. On reaching full development in the lungs, the larvae break out of the tissues into the alveoli. The occurrence of bronchial damage with sloughing of bronchial epithelium has been reported (Hoeppli, 1923; Vogel and Minning, 1942).

Subsequent migrations due to reinfection may cause intense tissue reaction in both liver and lungs even when the number of larvae involved is relatively small (Vogel and Minning, 1942). The pathologic changes noted in animals and presumed to be similar in man consist of eosinophilic infiltration and granuloma formation around and in the path of migrating larvae, and inflammation along the portal tract with infiltration of eosinophils in the acute phase. Later, fibrosis of the periportal and interlobular spaces occurs. Around the larvae in the tissues is found an intense local cellular reaction, with infiltration of eosinophils and aggregation of macrophages and epithelioid cells. These are manifestations of a cellular immune mechanism of the host, absent or not fully developed during the initial migration.

In the lungs this pathologic process is known as *Ascaris* pneumonitis. Heavy infections causing severe signs and symptoms in human volunteers have been described, first in the self-inoculation experiments of Koino (1922), and more recently by Brudastov et al. (1971). Severe pulmonary symptoms in humans after being inoculated with eggs of *Ascaris suum* were noted by Koino (1922) and Phills et al. (1972).

The signs and symptoms of *Ascaris* pneumonitis are the result of the aforementioned pathologic process coupled with an allergic reaction. They consist of dyspnea, often of the asthmatic type; a dry or productive cough; rales that are usually wheezing or coarse but rarely crepitant; moderate fever; transient eosinophilia; and x-ray findings of scattered, shifting mottled lung infiltrations suggestive of viral pneumonia. This picture of transient pulmonary infiltration, varying from day to day and spontaneously clearing in 3 to 14 days, and associated with marked peripheral eosinophilia, is called *Löffler's syndrome.* If sputum is produced, it characteristically contains abundant eosinophils and Charcot-Leyden crystals. The larvae may at times be seen in the sputum but are more easily recovered in gastric washings (Fig. 21–11) (Proffitt and Walton, 1962). Although Löffler's syndrome is also caused by other agents, both living and nonliving, in areas where ascariasis is endemic it and other nematode infections (hookworm, *Strongyloides*) are most frequently responsible. Pulmonary ascariasis is apt to be most evident and severe in endemic areas where transmission is sharply seasonal (Gelpi and Mustafa, 1967, 1968). Occasionally it is fatal, even in adults (Fig. 21–10) (Beaver and Danaraj, 1958; Piggott et al., 1970). In regions where transmission is more or less continuous throughout the year, clinically evident pulmonary ascariasis is rare (Spillmann, 1975).

The allergic manifestations of ascariasis are not limited to bronchospasm and pulmonary infiltrates but include also urticarial reactions. These may be pronounced during the pulmonary migration (Phills et al., 1972), or tend to appear as the pulmonary phase wanes (Vogel and Minning, 1942). Asthma and urticaria may continue during the intestinal phase of ascariasis and be relieved by expulsion of the worms (Chacko, 1970).

Intestinal ascariasis tends to be well tolerated unless the infection is heavy or the intake of nutrients by the host is inadequate. Nevertheless, as noted before, even a single worm has the potential of causing serious or even fatal disease. In the usual infection in children, there may be intermittent colicky cramps, loss of appetite, and fretfulness. The abdomen is often protuberant. The nutritional drain and space requirements of massive infections may be great, as in the case of a 2-year-old girl seen at autopsy in New Orleans. She weighed but 8 kg after having passed 80 worms in the hospital and an unrecorded number earlier, and still harbored 578 worms that weighed more than 800 grams with a volume of approximately 1 liter.

Intestinal ascariasis is associated with significant nutritional impairment, which is more or less proportional to the worm burden. The directly measurable effects are increased fecal nitrogen, increased fecal fat (steatorrhea), and impaired carbohydrate absorption as measured by the d-xylose test. These effects are attributable to abnormalities of the jejunal mucosa consisting of broadening and shortening of villi, elongation of crypts, decrease in villus-crypt ratio, and round cell infiltration of the lamina propria. Most of these abnormalities revert to normal soon after elimination of the worms (Tripathy et al., 1971, 1972).

The long-term effect of the malnutrition caused by ascariasis is impairment of growth. It has been noted that periodic deworming of children in highly endemic communities may result in improved growth and development (Stephenson, 1980; Stephenson et al., 1980). This result is dependent on other factors, especially average worm burden and food intake (Freij et al., 1979; Cerf et al., 1981).

From time to time there is spontaneous loss of worms associated with neither treatment nor present or previous illness. These events are of no special significance and are usually the consequence of the limited life span of the worm. If attended with an acute febrile illness of any kind, however, the active movement of worms from the intestine is highly significant, and is to be interpreted as a

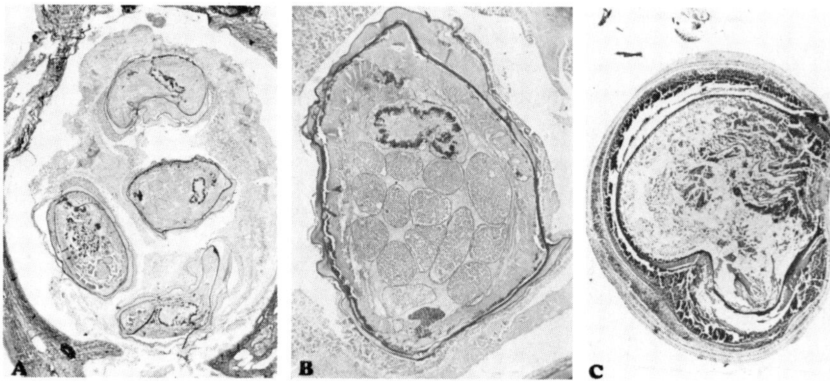

Fig. 21–13. *Ascaris lumbricoides*. Intrahepatic calculi formed in human bile ducts around nidi of calcified and fragmented adult worms. *A*, Transverse sections of four such segments of a worm. (× 6.) *B*, Higher magnification of one section showing by its microanatomy that the worm was a mature female. (× 20.) *C*, Calculus with worm tissues barely evident. (× 22.) (Sections courtesy of Dr. Alex Cobo. From Faust, E.C., Beaver, P.C., and Jung, R.C. 1975. *Animal Agents and Vectors of Human Disease*, 4th ed. Philadelphia, Lea & Febiger.)

direct consequence rather than the immediate cause of the febrile illness. When disturbed by fever (usually a temperature over 102°F, or 38.9°C) or other abnormal conditions, the worms migrate outward in both directions, or congregate in a closely packed mass that may tend to obstruct the bowel, causing mechanical ileus when it lodges at the ileocecal valve. In other movements the migrating adult worms have a proclivity for probing and forcing themselves into any aperture that they may encounter. This tendency is illustrated by the strangulation of a worm in a shoe eyelet that had been swallowed by a child (Beaver, 1964), and one threaded into a decompression tube (Fig. 21–12). Those that migrate upward may enter and block the biliary or pancreatic duct, resulting in acute biliary obstruction or acute pancreatitis (Louw, 1974; Schuster et al., 1977), or may form the nuclei of calculi (Cobo et al., 1964; Kalejaiye et al., 1977; Raney et al., 1970) (Fig. 21–13). Probably more frequently they force their way into the parenchyma of the liver where, usually in a day or so, they die (Fig. 21–14). Liver abscesses in children are probably caused more frequently by *A. lumbricoides* than by *E. histolytica*. Rupture of such an abscess through the diaphragm into the pleural cavity results in empyema containing recognizable eggs of *Ascaris*. Penetrating into hepatic or caval venous channels, worms may be carried as emboli to the brain, heart, and lungs (Fig. 21–15) (Sobrinho et al., 1972).

Migrating further up the alimentary tract the worms gain access to several other apertures. Most

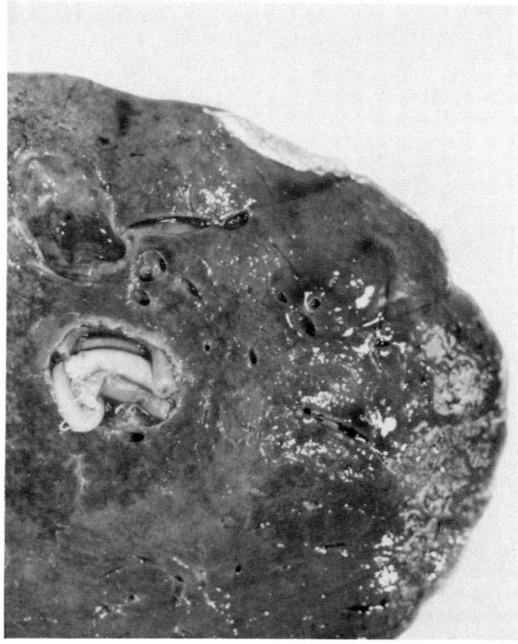

Fig. 21–15. *Ascaris lumbricoides*. Dead worm coiled and folded in parenchyma of liver. Worms migrating from the intestine via bile ducts to liver live only 1 or 2 days, although occasionally a worm may find its way into the hepatic vein and reach the heart and pulmonary artery. Slightly reduced. (Courtesy of Dr. D. E. Wykoff.)

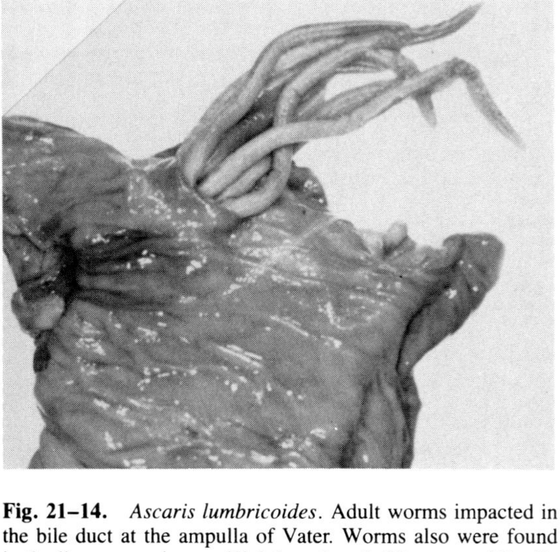

Fig. 21–14. *Ascaris lumbricoides*. Adult worms impacted in the bile duct at the ampulla of Vater. Worms also were found in the liver parenchyma. Slightly reduced. (Courtesy of Dr. D. E. Wykoff.)

commonly they are simply expelled or crawl from the mouth. They may enter the nasal meatus via the nasopharynx and exit through a naris. Rarely, a lacrimal duct may be entered by an immature worm (Roche, 1971), and the worm attempts to exit into the eye. Likewise a worm in the oropharynx may enter a eustachian tube and penetrate to the middle ear and through the tympanic membrane to the external auditory meatus (Shah and Desai, 1969). It is not uncommon for the trachea to be entered and the worm aspirated, causing respiratory obstruction (Mittal et al., 1976), lung abscess, or atelectasis.

Worms migrating downward become lodged in the appendix and may cause obstructive appendicitis (Smedresman, 1977). They may perforate the intestinal wall weakened by gangrene from prolonged ileus (Fig. 21–16), or perhaps even the intact wall (Ihekwaba, 1979). Eggs of a worm in the ulcerated appendix may become lodged in the submucosa and deeper layers of the organ (Fig. 21–17). In the peritoneal cavity the worms die and are absorbed within a few days. The eggs are more resistant and generally become widely scattered be-

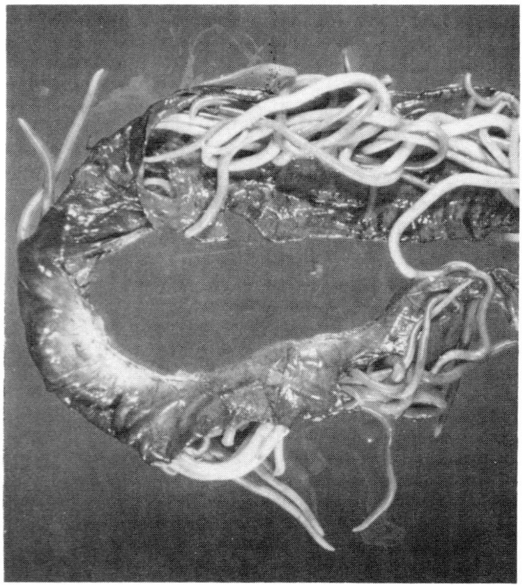

Fig. 21–16. *Ascaris lumbricoides*. Bolus of worms causing obstruction and necrosis of a segment of the ileum of a child in Louisiana. (Courtesy of Dr. W.H. Sternberg.)

fore becoming enclosed in granulomatous reactions (Reddy et al., 1975).

Worms may reach other sites such as the kidney via a fistula formed after intestinal perforation, or by squeezing between sutures of a surgical incision (Bustamante et al., 1977; Hakami et al., 1976).

The life span of the mature *A. lumbricoides* in the human intestine does not generally exceed a year. Yet in hyperendemic communities daily exposure is usual, so that new broods of larvae are migrating through the lungs and reaching the intestine to replace the previous mature ones as soon as they are lost. The interval between patent infections in such circumstances is 2 to 3 months (Yokogawa et al., 1956).

Ascariasis is fequently associated with whipworm infection and larval toxocariasis as well as diseases due to other causes. These complicate the clinical picture, preventing accurate appraisal of the role played by *Ascaris*.

Diagnosis. In the early stages of infection, if suspicion of ascariasis is strong based on clinical evidence of *Ascaris* pneumonitis (Löffler's syndrome), a diagnosis may be made by finding larvae in the sputum or, more readily, in gastric washings (Fig. 21–11) (Gelpi and Mustafa, 1967, 1968). Successful examination of gastric washings for migrating larvae of *Ascaris* was first reported by Proffitt and Walton (1962); in samples taken from a 2-year-old child 3 and 4 days after admission to hospital with pneumonia and hypereosinophilia, they were able to recover 7 and 10 larvae, respectively. However, as stated earlier, in hyperendemic areas where trasmission occurs throughout most of the year, Löffler's syndrome caused by *Ascaris* is rarely seen (Spillmann, 1975). It occurs most frequently in areas where transmission is highly seasonal, interrupted annually by a period of low temperatures or drought.

Once the infection is patent, the number of eggs produced by even one female, around 200,000 per day or roughly 1000 to 3000 per gram of feces (1 to 3/mg), is sufficient to make microscopic examination of feces highly effective. A direct fecal smear conscientiously made and examined should only fail to detect an infection when only males or immature worms are present. However, while false negative reports are uncommon, false positive ones are not. Human feces contain numerous artifacts that are frequently mistaken for *Ascaris* eggs. An outstanding example of extensive clinical studies based on false positive laboratory findings is one reported by Tullis (1970) in which 90% of 200 cases of bronchial asthma in a nonendemic area were erroneously attributed to ascariasis diagnosed by fecal examination. An interesting possibility for the detection and measurement of the quantity of infection with *Ascaris* was described by Soprunova et al. (1973), who reported that by gas chromatography certain volatile fatty acids produced by the worms can be detected in the urine. The application of such biochemical methods has not progressed beyond preliminary stages (Soprunov et al., 1981).

Ascariasis should be suspected as a cause when a child who is passing worms from the mouth or anus develops intestinal or biliary obstruction, or sudden empyema in the right pleural space. In intestinal obstruction, the bolus of worms can usually be felt as a sausage-shaped mass in the right lower abdominal quadrant. Frequently this mass can be visualized by x-ray, which also shows scattered worms in the characteristic pockets of gas above the fluid levels demonstrable in cases of ileus.

In the case of empyema due to transdiaphragmatic rupture of an *Ascaris* liver abscess, eggs are found in examination of the aspirate by thoracentesis.

Intestinal ascariasis is often diagnosed as a result of a radiographic study of the gastrointestinal tract with barium sulfate as a contrast medium (GI se-

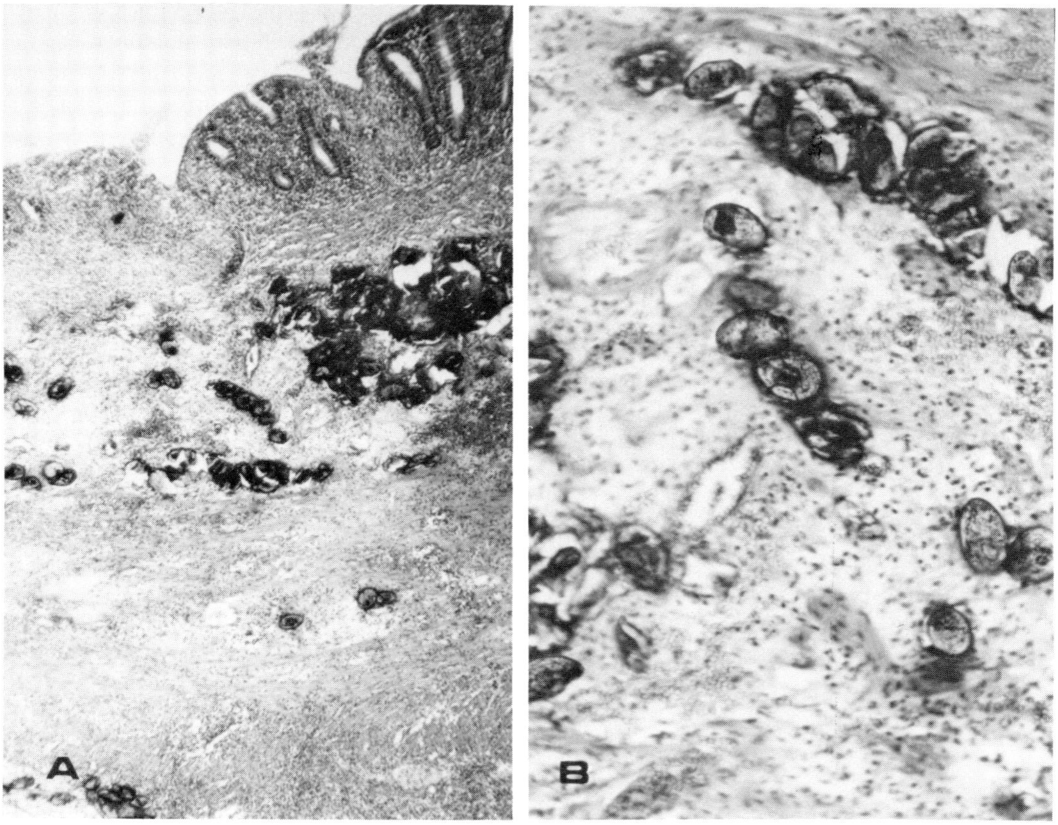

Fig. 21–17. *Ascaris lumbricoides*. Unfertilized eggs, in the wall of an appendix that had ruptured before surgical removal from a 59-year-old man. The patient's history of having been in the Orient 30 years ago led to an incorrect diagnosis of schistosomiasis. *A*, Section showing eggs in submucosa, muscularis, and subserosa. Most of the eggs are calcified. (× 48.) *B*, Eggs in submucosa. (× 150.) (By P.C. Beaver. Specimen courtesy of Dr. G.H. Houser.)

ries). The worms may be seen as curved, linear, negative shadows surrounded by the contrast medium (Fig. 21–18) or, if they have ingested the barium sulfate, they may be recognized from the finer linear shadows of their intestine after the contrast medium has passed from that segment of the host's intestine. A group of worms lying parallel to each other and filling the host's intestinal lumen produces an x-ray shadow of parallel lines that has been called the tramway sign because of the resemblance to the parallel tracks of tram (trolley car) lines.

Treatment. Several modern anthelmintics are more effective or less toxic than the old popular remedies, santonin, oil of chenopodium, and hexylresorcinol crystoids. The best of these are:

pyrantel pamoate, which is given in a single dose of 11 mg per kilogram body weight (maximum 1 gram),

mebendazole, which is given in a dosage of 100 mg twice daily for 3 days, and

piperazine citrate, which is given in a dosage of 75 mg per kilogram of body weight (maximum 3.5 grams) daily for 2 days. This dosage schedule may be altered by extending it to 7 days or by administering in a single dose.

Cyclobendazole is a new benzimidazole derivative that has promise as an agent againt *Ascaris* (Stürchler et al., 1980).

None of these treatments requires pre- or post-treatment purgation or fasting.

There is no conclusive evidence that any anthelmintic currently used against intestinal ascariasis is effective against the larval worms migrating through the liver and lungs.

The appearance of signs and symptoms of ileus in a child who is passing *Ascaris* by mouth or anus indicates obstruction due to a bolus of worms. This condition warrants preparation of the patient for possible surgical intervention, including: (1) before treatment, search should be made for the cause of the febrile illness which usually precipitates the

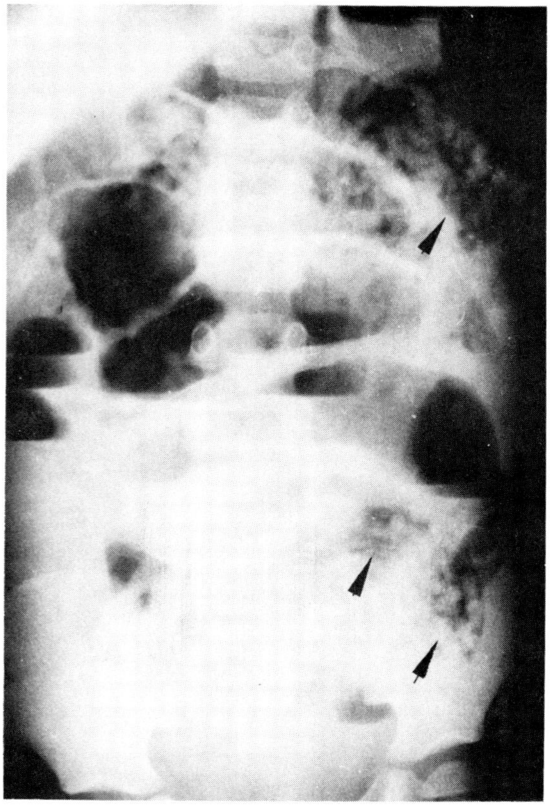

Fig. 21–18. *Ascaris lumbricoides.* Worms in the intestine, evident on roentgenogram of the abdomen. (Courtesy of Dr. A. D'Alessandro.)

complication, and appropriate treatment instituted; (2) fluid and electrolyte balance should be restored through intravenous infusions; (3) antipyretic measures should be carried out; (4) decompression should be accomplished by introduction of an intestinal tube (Miller-Abbott) with constant suction; and (5) after rehydration a single dose of piperazine citrate may be introduced into the intestinal tube and the tube clamped for 2 hours (Swartzwelder et al., 1957). In most instances the obstruction will be spontaneously resolved before surgical intervention is necessary. If the bolus and signs of obstruction persist in spite of these measures, laparotomy is indicated. The bowel is opened at the palpable bolus and the worms are removed with syringe forceps.

Obstructive jaundice in a child with intestinal ascariasis indicates the need for surgical intervention to remove the worms from the common bile duct. Administration of piperazine by tube as for intestinal obstruction *(vide supra)* may sometimes resolve the obstruction and obviate surgery.

Liver abscess due to *Ascaris* must be treated by drainage and administration of antibiotics.

Prognosis. In most infected persons the prognosis is excellent. In cases of acute ileus or other surgical types of complications, prognosis depends on the rapidity of the diagnosis and treatment, and the skill of the surgeon. It is less favorable in perforation of the bowel or invasion of the biliary system. At times the prognosis is grave when there is massive larval invasion of the lungs.

Prevention. As ascariasis endemicity is sustained by fecal contamination of the soil, the control and ultimate eradication of the infection in a community must depend largely on the installation and universal use of sanitary facilities in all parts of the community. That of course takes time, and can only be undertaken where costs can be met. Even in communities where flush toilets have become commonplace, ascariasis in regions where conditions of soil and climate favor transmission can remain prevalent among preschool children, maintained by their promiscuous defecation and natural inclination to eat dirt.

For children in communities that have sanitary facilities, and for all age groups where sanitation is substandard or essentially lacking, periodic treatment with an effective anthelmintic can markedly reduce the prevalence of infection. In theory, the parasite could be eradicated by periodic mass treatment if all worms were removed before any could reach full maturity. This would require a treatment schedule of every 40 to 50 days during the season or seasons of transmission—8 or 9 times annually in a region where new infections are acquired throughout the year. The logistics and costs of mass treatments make large-scale programs on such a schedule impractical. However, mass treatment of a population or of all children even once or twice annually has been found beneficial despite high rates of reinfection. In Japan, mass treatments once or twice a year were modified later to one or two annual treatments of only the infected individuals, and finally to one treatment annually of only the school children who were positive on fecal examination. In a period of 20 years these treatments reduced the prevalence of ascariasis from 60% in 1949 to 1.6% (Morishita, 1972). In a community in the Philippines where transmission was continuous throughout the year, a series of four mass treatments given every 3 months to all residents 1 year of age and older reduced the overall infection rate from 85% (512 of 606 people) to only 0.5%,

3 positive among 561 examined (Cabrera et al., 1975).

No practical means of killing *Ascaris* eggs in soil has been discovered, and their disappearance by natural processes is slow wherever the soil qualities favor survival. Therefore, to gain the maximum benefit of a periodic treatment program, the schedule must be continued for several years.

In communities where night soil (human feces and urine) is used as fertilizer of garden crops, raw vegetables (including roots, stems, leaves, and fruits that ripen in or near the ground) cannot be safely eaten raw or unprocessed. Aqueous iodine in sufficient strength (200 p.p.m.) to kill infective eggs and larvae of *Ascaris* and other helminths in 10 to 15 minutes does not affect the flavor or appearance of common types of rooty and leafy vegetables (Thitasut, 1961). Certain natural food preservative substances such as garlic and mustard oil are lethal to *Ascaris* eggs, but others such as brine, syrup (sugar), natural acids (including vinegar), oils, and spices have little or no effect (Soh, 1960).

Composting and various schedules of storage of human excreta may provide effective methods of ascariasis control by destroying the parasite eggs before the night soil is used. However, among the various parasite eggs and larvae to be considered in night soil treatment, none is more resistant to chemical and physical harm than are those of *Ascaris*.

Ascaris suum
Goeze, 1782
(The pig *Ascaris*)

Historical and Geographical Notes. The pig *Ascaris (A. suum)* and human *Ascaris (A. lumbricoides)* are morphologically so similar that until 1952, when Sprent described differences in the labial denticles, the two species were generally regarded as indistinguishable. At that time the validity of the two species rested almost entirely on the host difference. The chromosome numbers were thought to be different, but the findings of different workers were widely inconsistent, and even the findings of individual workers failed to give consistent results (Lysek, 1975). Likewise, while some workers confirmed and extended Sprent's observations (Abdulrachman and Joe, 1954; Ubelaker and Allison, 1972; Maung, 1973; Weise, 1973), others failed to find consistent differences in the labial denticles (Kurimoto, 1974).

Little (1968) observed that the denticles of both species changed markedly as the worms advanced in size and age, but when worms of similar size and age from man and pig were compared side-by-side, the denticles of worms from pigs were generally perceptibly larger. However, individual worms could not be identified as to source by denticle size alone since there was wide overlapping within each size range, and since shape as well as size varied with age. On the other hand, it was noted by Little (1968) that the copulatory spicules in worms of similar size from man and pig, though similar in size, had terminal parts that were different in shape. Those of *A. suum* were usually distinctly thinner and sharper at the tip than were those of *A. lumbricoides*. Prokopic (1979) confirmed that the denticles of young worms are small, sharply pointed, and conical, and in older worms they become larger and cylindrical as the tips gradually become ground off, especially those located in the middle of the lips. He concluded that the morphology of the denticles is too variable to be of taxonomic importance.

The biological and biochemical differences are apparently more distinct and more reliably demonstrated. Consistent differences were found by Kurimoto (1974) in the hemolymph electrophoresis bands and in the immunoelectrophoretic patterns. Galvin (1968) showed that in experimentally infected pigs the prepatent period was shorter (40 to 53 days) than that of *A. lumbricoides* (54 to 61 days), and that in both pigs and rabbits *A. suum* had a distinctly earlier period of tracheal migration from lungs to intestine. In addition, a higher percentage of *A. suum* developed to maturity in the pig, and the duration of patent infection was greater than was that of *A. lumbricoides*.

Morphology, Biology, and Life Cycle. As stated previously, *A. suum* and *A. lumbricoides* are so similar in morphology that an individual worm usually cannot be reliably identified as one or the other. The most distinctive feature that has been recognized thus far is the shape of the copulatory spicules in the male, those of *A. suum* being more uniformly narrow and sharply pointed at the tip (Little, 1968; Maung, 1973). In both species the shaft is cylindrical with tapering at the distal (posterior) end, i.e., transverse sections are round throughout, even at the tips. Differences are evident only in the shape of the tips, those in *A. suum* being conical with the apex forming an angle of 35 to 45 degrees, whereas in *A. lumbricoides* they are more broad and rounded at the tip.

The eggs, both fertilized (65 to 75 μm by 50 to 55 μm) and unfertilized (75 to 80 μm by 41 to 52 μm) (Roberts, 1934) are indistinguishable from those of *A. lumbricoides* (Fig. 21–7). Length-width measurements of the adult worms range up to about 28 cm by 3.4 mm in the male, and 35 cm by 5.0 mm in the female. At 30°C eggs are infective at 18 to 22 days of incubation, and under favorable conditions remain infective for several months. When swallowed, eggs hatch in the stomach and upper small intestine, mostly the latter, and the larvae migrate through the wall of the intestine to the liver and on to the lungs where, for a period of 8 days or more, essential development and growth occur. Larvae leaving the lungs to migrate or be carried in the mucus flow to the stomach via the trachea, throat, and esophagus are in the advanced third stage of development. After two molts in the intestine, one soon after arrival (to the fourth stage) and one during the fourth week of infection (to the immature adult), and further growth, development, and mating, egg laying begins. Eggs, always in the one-cell stage or unfertilized, usually appear in the feces during the eighth week of infection. The prepatent period may be as short as 40 days in the pig (Galvin, 1968) and 46 days in the rabbit (Berger, 1971). According to Roberts (1934), larvae emerging from the egg in the intestine are 230 to 270 μm long. On leaving the lungs and arriving in the intestine, the larvae have increased in length to about 1.5 mm. At 14 days of infection, the fourth-stage larvae measure 2.4 to 3.2 mm, at 21 days 4.5 to 6.5 mm. At 29 days, when the larvae are molting to the fifth stage, lengths of worms range from 16 mm to 25 mm, and at 62 days, when eggs first appear in the feces, lengths of worms range from 14.5 cm to 18.0 cm. Galvin (1968) observed eggs in the feces as early as 42 days, and the egg-laying females were 17 to 22 cm long. Berger (1971) recorded egg laying by a worm only 8 cm long.

Clinical Aspects. Patent experimental infections of *A. suum* in 7 of 17 adult human volunteers who swallowed 20, 21, or 25 eggs each were reported by Takata (1951). As the prepatent period in five of the volunteers was 25 to 29 days, which is far less than the expected time, Takata's findings need confirmation. An attempt by Tanigawa to infect three human volunteers with 100, 100, and 500 eggs, respectively, was unsuccessful (Morishita, 1972).

Human infection with adult *A. suum* has been reported occasionally, identified on the basis of size and shape of the denticles and a history of exposure to infection on farms where pigs were raised (Goldsmid, 1976) or where manure from a pig farm was used as a vegetable garden fertilizer (Crewe and Smith, 1971). A single mature male ascarid 20 cm long was passed by a laboratory worker who accidentally swallowed embryonated eggs of *A. suum* about 8 months previously (Jaskoski, 1961). Worms passed by children who were exposed to pig *Ascaris* in areas where human *Ascaris* is absent have been identified as *A. suum* on epidemiological evidences (Lord and Bullock, 1982).

Of greater importance are the human infections with *A. suum* in which the larvae invade the liver, lungs, and the intestine, but do not develop to maturity. It is well known that the larvae of *A. suum* as well as those of *A. lumbricoides* are able to make the pulmonary migration in mice, guinea pigs, and other laboratory animals. In the rabbit, development may proceed on to the mature adult stage (Berger, 1971). It is to be expected, therefore, that in exposure of man to *A. suum* infection, larvae would at least invade the lungs and make the migration from lungs to intestine. In view also of the observation that a human volunteer who swallowed only six infective eggs of *A. lumbricoides* developed apparent pneumonitis and other manifestations of reactions to the invading larvae (Vogel and Minning, 1942), it is to be expected that massive infections with *A. suum* larvae will produce severe disease. As stated, such was the result of malicious contamination of food with infective eggs of *A. suum* in four cases reported from Montreal, Canada (Phills et al., 1972). Four male students who unknowingly swallowed massive numbers of the infective eggs were hospitalized 10 to 14 days later with severe pneumonitis, blood eosinophilia, and asthma. In one of them generalized urticaria developed the day after admission and cleared in 48 hours. A few days after admission, active larvae were recovered in the sputum of one and in gastric washings of three of the victims. In two cases many immature worms were passed in the stools 4 weeks after ingestion of the eggs. This is the expected period of the fourth molt. Anthelmintic treatment 7 weeks after ingestion of eggs apparently expelled no worms. Massive infections of this magnitude in pigs result in the loss of larvae in the feces at 2 weeks and at 4 weeks, corresponding to the periods of the third and fourth molts (Roberts, 1934; Schwartz, 1959). In experimental infections in pigs, a high percentage of infective larvae develop to the mature adult stage when the number inoculated is 50 (Andersen et al., 1973; Jørgensen et al., 1975) or 100 (Galvin, 1968), but when 500 or more are in the inoculum, few worms or none develop to maturity.

No firm estimate can be given of the significance of *A. suum* as a cause of disease in man. It is possible that among people who frequently visit or live on farms where pigs are raised, *A. suum* is a significant occasional cause of illness. Naturally occurring outbreaks of pneumonia caused by *A. suum* in calves and yearling cattle have been reported (McCraw, 1973; McLennan et al., 1974).

Genus Toxocara Stiles, 1905

(Syn. *Belascaris* Leiper, 1907)

Species of the genus *Toxocara* have distinct cervical alae, the lips are prominent and usually bear fine dentigerous ridges; interlabia are absent; and the esophagus has a distinct posterior ventriculus, but there is neither a ventricular diverticulum nor an intestinal cecum. The male has a terminal digitiform caudal appendage and caudal alae are present in some species. Five pairs of caudal papillae are posterior to the cloaca, many anterior, and a subventral double papilla is on either side posterior to the cloaca. The female vulva is in the anterior part of the body; the eggs have a pitted surface.

The only species of *Toxocara* that are known to cause disease in man are *T. canis* and *T. cati*. According to Warren (1971), however, there are nine other species, all parasites of terrestrial mammals. As several of these hosts are closely associated with people, *Toxocara* species other than *T. canis* and *T. cati* may eventually be recognized as causing disease in man.

Toxocara canis

(Werner, 1782) Johnston, 1916

(The dog ascarid)

Biology and Life Cycle. The dog ascarid, *Toxocara canis*, is cosmopolitan in distribution although in some regions, particularly in the far north, it is partially replaced by a worm of similar size and appearance, *Toxascaris leonina* (Linstow, 1902). The adult worm has been reported from species of wild dog, coyote, wolf, jackal, dingo, hyena, and fox (Sprent, 1958). Records of adult worms from man (Bisseru et al., 1966) are questionable. In one such case, the worm was later identified as *Ascaris*, small and immature (Woodruff, 1981). Mature males generally measure 4 to 6 cm, females 6 to 10 cm, in length; maximum lengths of 13 cm and 20 cm, respectively, for males and females have been recorded (Warren, 1971). The cervical alae are much longer than broad (2 to 4 mm by about 0.2 mm) (Fig. 21–19). The esophagus is generally about 5 mm long including the ventriculus, which is about 0.5 mm in both length and width. The vulva is one-fifth to one-sixth the body length from the anterior end. The eggs are subglobose, about 85 by 75 μm, with pitted shell, and unembryonated when passed (Fig. 21–20).

In enzootic areas pups are usually infected prenatally, or through the mother's milk soon after birth (Stoye, 1976). They remain susceptible to reinfection until sexually mature, after

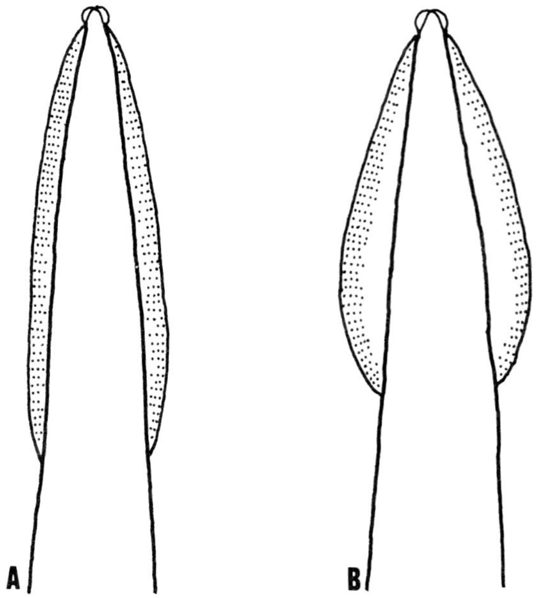

Fig. 21–19. *Toxocara*. Anterior end of adult worms showing cervical alae. *A*, *T. canis*. (× 15.) *B*, *T. cati*. (× 15.) (From Beaver, P.C. 1952. The detection and identification of some common nematode parasites of man. Am. J. Clin. Pathol., *22*:481-494.)

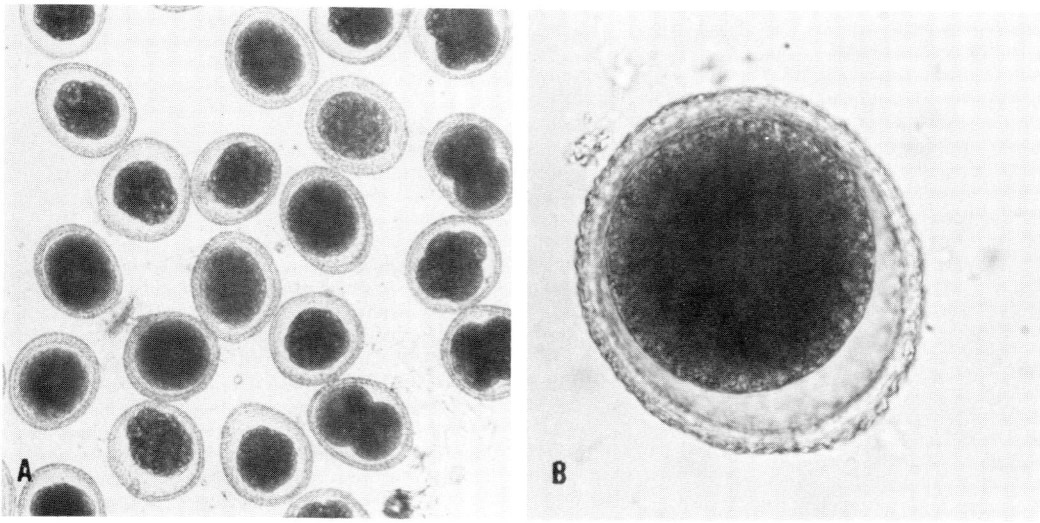

Fig. 21–20. *Toxocara canis*. Eggs washed from feces of an infected pup, showing the typical pitting of the outer shell membrane. Those shown in *A* were in a water culture for several days and some are in early cleavage stages. The ovum typically does not fill the eggshell. *A* (× 150) and *B* (× 540). (By P.C. Beaver.)

which time they acquire some resistance to reinfection, especially the females. Under favorable conditions of moisture and temperature, eggs deposited on soil develop to the infective stage in 3 to 4 weeks, and protected in the egg the larva remains infective for several months. Infection is acquired when eggs containing fully developed larvae are swallowed.

When larval *T. canis* infections are acquired by mice during pregnancy, some of the larvae migrate through the placenta to the fetuses and are present in the tissues of the newborn (Lee et al., 1976). However, most of the larvae remain in the tissues of the mother and become encapsulated. Mice, rats, hamsters, and all other mammals thus far tested are susceptible to infection. In all mammals except the dog and other canids, *T. canis* larvae migrate into the tissues, become encapsulated, undergo no growth or development, and remain viable and infective indefinitely—for the life of the host in small mammals. Experimental infections in monkeys *(Macaca mulatta)* persist for at least 9 years (Beaver, 1966). Larvae in the tissues of animals can be transferred from host to host through cannibalism or predation with no growth or development until a susceptible canine host is reached. This kind of transfer of infective-stage larvae through *paratenic* or transfer hosts is a common mode of transmission among the helminth parasites of carnivorous vertebrates. *T. canis* larvae also become encapsulated and persist for indefinite periods in the tissues of dogs, particularly in older ones and in sexually mature females where they later can reach the offspring through the placenta or through the milk. Larvae remaining in the mother's tissues after one pregnancy are able to infect one or more subsequent litters of pups (Douglas and Baker, 1966). Infection acquired by pups before or immediately after birth involves migration and development in the lungs. Infection may become patent with massive numbers of eggs appearing in the feces of pups less than 4 weeks of age and continuing for several months if not terminated by treatment. During the first 7 to 10 days after birth, sometimes longer, the pups' feces are ingested by the mother. Feces of heavily infected pups contain viable larvae that have failed to become established in the pups but readily develop to maturity in the intestine of the mother (Sprent, 1961).

In summary, infection of the dog with *T. canis* may be acquired in four ways: (1) by ingesting infective-stage eggs from soil, (2) by ingesting larvae in the tissues of paratenic hosts that were infected from soil or by cannibalism or predation, (3) by receiving larvae from the mother via the placenta or the milk, or both, and (4) by ingesting larvae passed in the feces of suckling young pups. Human infections with larvae of *T. canis* are acquired by ingestion of infective eggs from soil or possibly larvae in uncooked tissues of a paratenic host.

The importance of larval *T. canis* infection in the human host is considered in the following section under *visceral larva migrans* (see p. 325).

Toxocara cati (Schrank, 1788) Brumpt, 1927 (The cat ascarid)

Synonyms. *Ascaris mystax* (Zeder, 1800) Rudolphi, 1802; *Belascaris mystax* (Zeder, 1800) Castellani and Chalmers, 1910; *Belascaris cati* (Schrank, 1788) Railliet and Henry, 1911.

Biology and Life Cycle. The cat ascarid, *T. cati,* is cosmopolitan in geographic distribution and in parts of the United States and Europe, infection rates exceed 50%. It is prevalent in both wild and domestic species and in both young and older cats. A similar worm, *Toxascaris leonina,* is found in cats as well as in dogs. The adult *T. cati* has been reported from man 20 times, most recently from an infant in Vermont (von Reyn et al., 1978). Human infection with the adult worm probably cannot be acquired in the usual manner by ingesting infective larvae in the egg or in a paratenic host. It is therefore probable that in some instances the infections reported in people were spurious or based on erroneous observations (Rodan and Buckley, 1969; Wiseman and Lovel, 1969).

The male worms range up to 6 cm in length, the females up to 12 cm. The lips are characteristic of ascarids, one dorsal and two subventral. The cervical alae are broad, tapering gradually toward the anterior end, rounded posteriorly, giving the anterior end of the body a pyriform outline (Fig. 21–19). The esophagus is about 2 mm long and the ventriculus longer than wide. The caudal papillae of the male are specifically diagnostic. The female vulva is about one-fourth the body length from the anterior end. The eggs are subglobose, relatively thin-shelled, finely pitted, and measure about 75 by 65 μm. They are unembryonated when passed in the feces.

Under favorable conditions full development of the larva requires 3 to 4 weeks, after which time the eggs are infective. If protected against desiccation, direct sunlight, and high or low temperatures, they may remain infective in soil for several months. Cats acquire infection by ingesting larvae in eggs or in paratenic hosts. Some of the larvae may make the lung migration and develop to maturity in the intestine, but in mature cats most of the larvae migrate to the muscles. Prenatal infection with *T. cati* apparently does not occur, but transmission through the mother's milk is the usual mode of infection in newborn kittens. In the cat and in paratenic hosts larvae persist for periods of several months encapsulated or unencapsulated in the tissue, primarily in the muscle fibers.

Infections in older cats are usually light. Even in kittens infections are generally well tolerated although both, especially kittens, frequently discharge immature and mature worms in the feces and vomitus. If handled by children, such worms may be mistakenly interpreted as having come from the child (Swartzwelder, 1941). An adult *T. cati* unquestionably expelled from the anus of an 8-month-old child was probably acquired as a partially or fully developed adult expelled from a cat and ingested by the child (von Reyn et al., 1978).

Larval *T. cati* infections in humans causing *visceral larva migrans* is considered in a following section (see p. 325).

Genus Lagochilascaris Leiper, 1909

In species of *Lagochilascaris* the three lips have a heavy cuticular tegument, and they are separated from the body by a deep groove. Extending forward from the posterior rim of the groove are three narrow triangular interlabia which lie between the lips. Each lip has a deep median depression giving it a form like a harelip, a feature from which the generic name is derived. Lateral alae are present throughout the length of the body. The esophagus is simple. In the male the tail is conical, without alae; there are usually five pairs of postcloacal (caudal) papillae, and 25 or more precloacal papillae; spicules are subequal and alate. In the female the vulva is near midbody in position; eggs are spherical, with a thick, coarsely pitted shell.

Five species of *Lagochilascaris* are known: *L. minor* Leiper, 1909, the type species, which has been reported in man only but doubtless is enzootic in one of more wild host species; *L. major* Leiper, 1910, reported in the lion *(Felis leo)* in Africa (Durette, 1962) and in a cat in Argentina (Sprent, 1971); *L. turgida* (Stossick) Travassos, 1924 reported in the stomach of opossums *(Didelphis)* in South America; *L. buckleyi* Sprent, 1971 reported in *Felis concolor,* the American mountain lion; and *L. sprenti* Bowman, Smith and Little, 1983 reported in the stomach of opossums in Louisiana and an abdominal tumor of a raccoon *(Procyon lotor)* in Texas (reported as *L. major* by Craig et al. in 1980). An unidentified species was discovered in the larynx of an ocelot *(Felis pardalis)* in Costa Rica (Brenes and Frenkel, 1972). Only *L. minor* has been reported in man.

Lagochilascaris minor Leiper, 1909

Lagochilascaris identified as *L. minor* has been reported in man 16 times, from tissues of the head or neck in nearly all cases. The first 11 cases were summarized by Sprent (1971): 4 from Trinidad, 4 from Surinam, and 1 each from Tobago, Costa Rica, and Brazil. Five more recent cases have been reported, one each by Oostburg (1971) from Surinam, by Mondragon et al. (1973) from Colombia, by de Leão et al. (1978) from Para, Brazil, by Correa et al. (1978) from Mato Grosso, Brazil, and by Carrada-Bravo (1980) from Mexico. Colonies of adult males and females together with eggs and larvae in various stages of development were found in abscesses, in the subcutaneous tissues on the right or left side of the neck in 9 instances, in or over the mastoid process in 3, in the nasal cavity in 2, and in the tonsils in 2 cases. In some instances the parasites were recovered from more than one site; in two instances the site was described merely as subcutaneous tissues. The involvement was unilateral in nearly all cases, on the right in 6, left in 5, unstated in 4, in the nose in 1 case. Seven of the infected persons were female, six were male, and in three the host's sex was not reported.

According to Sprent (1971), the adult male worms measure 5 to 17 mm in length by 0.19 to 0.60 mm in width; the females, 60 to 20 mm by 0.2 to 0.81 mm. Lateral alae form a low keel along the sides at the lateral line. The lips are wider than long and bear no denticles (Fig. 21–21,*B*). Interlabia extend along roughly half the length of the lips. The esophagus, up to 1.5 mm long, has no ventriculus; the intestine is lined with high columnar cells with a conspicuous brush border. The lateral cords are heart-shaped in cross section in the anterior region but flat in the posterior half, containing four or more rows of nuclei in each column. The vulva is near the middle of the body, usually slightly posterior to the middle. Eggs, present in females 8 mm or more in length, are spherical or slightly ovoid, about 50 by 65 μm. The pits of the sculptured surface of the eggs measure about 5 μm in diameter (Fig. 21–21,*A*).

The transmission cycle is now known for *L. sprenti* (Smith et al., 1983). The adult worms, about 2 cm long, live in cavities or tunnels in the submucosa of the stomach of the opossum *(Didelphis virginiana)*. Eggs passed in the feces embryonate in or on the soil and contain an infective-stage larva after about 1 month of development. When ingested by mice or other small mammals, the larvae hatch in the intestine, migrate to the skeletal muscles, and grow to a length of about 4 mm. They become infective for the final host in about 1 month and persist indefinitely encapsulated in the tissues. When ingested by the final host, the larvae mature and produce eggs in about 3 weeks. Occasionally in an intermediate host one or more larvae may develop to maturity and produce eggs in abscesses in the tissues. It is probable that human infection with *L. minor* is acquired either by swallowing infective eggs from the environment or by ingesting infective larvae in the tissues of an intermediate host.

In the typical human infection, a tumor-like mass develops slowly over a period of several months, becomes tender then painful, and eventually pus containing worms is discharged from a fistula (Fig. 21–21,*C*).). Without treatment such episodes of drainage may continue intermittently for several years. With diethylcarbamazine treatment, the infection was apparently terminated in one case (de Leão et al., 1978), though in others treatment was unsuccessful. In one instance, despite high and continuous dosage of diethylcarbamazine, living worms were occasionally discharged in pus 4 months after treatment began (Draper, 1963). In this patient the abscess was still present after more than 10 years, and whenever diethylcarbamazine was given, worms were usually seen in the pus (Sprent, 1971). Treatment with thiabendazole failed in one case (Oostburg and Verma, 1968), but in another when given in high dosage (50 mg per kg per day for 5 days) and repeated after 1 week the mass decreased in size and the fistula closed (Oostburg, 1971). Treatment with thiabendazole in a later case was unsuccessful (Mondragon et al., 1973).

Anisakis and Related Forms

Biology and Life Cycle. Large larval ascaridoid nematode species of the genera *Anisakis* Dujardin, 1845, *Phocanema* Myers, 1959, *Terranova* Leiper and Atkinson, 1914, and *Contracaecum* Railliet and Henry, 1912 occasionally have been found in or attached to the wall of the throat, stomach, or

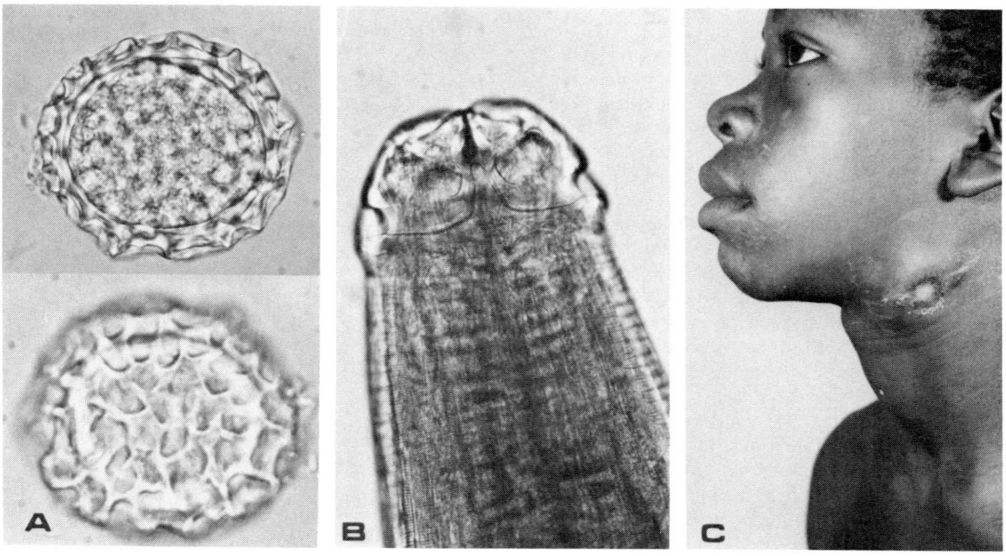

Fig. 21–21. *Lagochilascaris minor. A,* Eggs. (× 400.) *B,* Head of worm that —with many others—was removed from a fibrous nodule in an excised tonsil of a 26-year-old woman with a history of recurrent attacks of tonsillitis for 3 years. (× 105.) (From Mondragon, H., Cano, R.M., and Botero, D. 1973. Primer caso de infeccion humana por *Lagochilascaris minor* en Colombia. Antioquia Med., *23*:463–464.) *C,* Surinam boy with neck lesion from which pus, adult worms, larvae, and eggs had drained. (From Faust, E.C., Beaver, P.C., and Jung, R.C. 1975. *Animal Agents and Vectors of Human Disease,* 4th ed. Philadelphia, Lea & Febiger. Courtesy of Dr. B.F.J. Oostburg.)

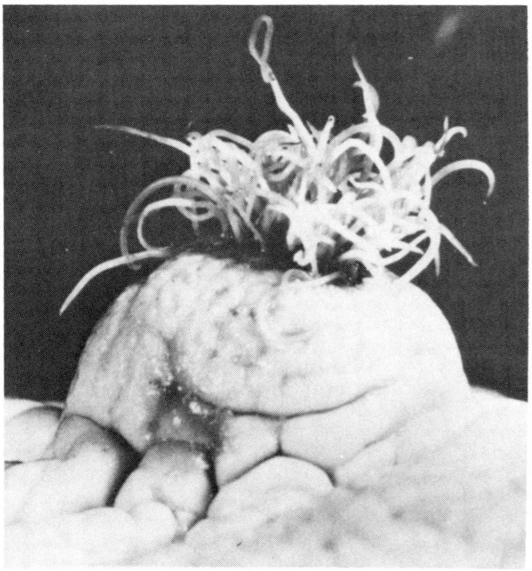

Fig. 21–22. *Anisakis.* Worms burrowed in center of swelling on mucosal surface of stomach wall of porpoise. (× ca. 2.) (Photograph by Dr. R. Vik. From Beaver, P.C. 1969. The nature of visceral larva migrans. J. Parasitol., *55*:3–12.)

intestine of man. The adult stages are found in the stomach of marine mammals. The eggs of these worms, passed in the host's feces, embryonate and hatch free in the water as microscopic second-stage larvae that are free-swimming and able to survive for 2 to 3 months (Hatsushika, 1979). When eaten by small, shrimp-like crustaceans (Euphausiidae), they develop into third-stage larvae that are infective to fish and squid. When eaten by the fish and squid hosts, the larvae migrate from the intestine to the tissues, mostly in the peritoneal cavity, and grow to macroscopic sizes up to 3 cm or more in length. After death of the fish host, larvae encapsulated in the body cavity tend to migrate into the muscles of the body wall (Smith and Wootten, 1975). By predation these larvae can be transferred from fish to fish. When the infected fish are eaten by whales, porpoises, dolphins, or other final hosts, the larvae penetrate the gastric mucosa where they live in clusters, becoming adult males and females with the anterior end of the body embedded in a crater-like tumor (Fig. 21–22).

When ingested by man in raw, freshly salted, or smoked marine fish, these large anisakid larvae penetrate the gastric or intestinal mucosa in the same manner as they would have done had they been eaten by a natural final host. The stomach is invaded about twice as often as the intestine. Occasionally the site of attachment and partial penetration is the throat. In the stomach and intestine the worm is usually found embedded in a dense eosinophilic granuloma deep in the wall below the mucosa. In the stomach a portion of the worm remaining in the lumen may be seen at operation or through a gastroscope.

The behavior of the worm, its location and persistence in the tissues, and the pathological and clinical conditions produced by it depend somewhat on the species involved. In Europe (the Netherlands), where anisakiasis was first recognized and 11 cases were reported (Van Thiel et al., 1960), the species, at first reported as *Eustoma,* was identified as *Anisakis marina* by Van Thiel (1966) and as *A. simplex* by Davey (1971). In most of those cases the worms were discovered at surgical intervention for relief of intestinal obstruction; there were two fatalities and one instance of peritonitis. Larvae of *Contracaecum osculatum* were identified in two eosinophilic granulomas of the intestine in a patient in Germany (Schaum and Müller, 1967). In Glasgow, U.K., a larval anisakine 23 mm long was recovered from an inflamed and edematous 8-cm segment of ileum resected from a woman who denied eating any seafood other than *Mytilus edulis,* a species of mussel (Watt et al., 1979).

Epidemiology. In Japan, since 1965 when the first cases of gastrointestinal anisakiasis were reported there (Asami et al., 1965), more than 500 cases have been recorded (Yoshimura et al., 1979). The first Japanese cases of *Terranova* infection in the stomach wall were reported in 1972, and 15 or more additional cases were reported soon thereafter (Suzuki et al., 1972; Koyama et al., 1973). In the vicinity of Kanazawa, a coastal city of midwestern Japan, 200 patients with a clinical diagnosis of gastrointestinal anisakiasis were seen in a period of less than 3 years. In 49 of them one or more worms were detected and identified, 26 by means of gastroscopic examination, 8 in the vomitus, 5 by surgical operation, and 8 in histopathological sections (Yoshimura et al., 1979); in each of two patients examined by gastroscopy, three *Anisakis* (type 1) larvae were recovered. Three of the larvae detected in vomitus and two recovered by gastroscopy were *Terranova*. In two instances an *Anisakis* larva was identified in sections of a cyst recovered from the omentum. Most patients in this series complained of gastric pain or colic, vomiting, and sometimes diarrhea within 24 hours after eating raw marine fish.

Human infection with anisakid larvae has been reported in the Western Hemisphere: twice from Massachusetts (Little and McPhail, 1972; Pincus et al., 1975), twice from California (Chitwood, 1975; Juels et al., 1975), and once each from Vermont (Kates et al., 1973), New York (Little and Most, 1973), Alaska (Lichtenfels and Broncato, 1976), Pennsylvania (Valdiserri, 1981), and Chile (Sapunar et al., 1976). Four cases were reported from California (Kliks, 1983). In the first case reported from Massachusetts, a *Phocanema-Terranova* type of larva 31 mm long was found in the abdominal cavity. In the second Massachusetts case an *Anisakis* larva was discovered in the wall of the cecum in an area of necrotizing eosinophilic granulomatous inflammation. In the Pennsylvania case a larva too necrotic to be identified to type was removed in a band of granulation tissue attached to the omentum and jejunum. In the other cases one or two worms were coughed up or removed from the throat. In the case reported from Chile and one from California, two larvae recovered from the throat were identified as *Anisakis*. In the other cases single larvae taken or coughed from the throat were identified as *Phocanema* or *Terranova*, or *Phocanema-Terranova* type.

Diagnosis. In the larval stage the various types of anisakid larvae are closely similar in most respects. Oshima (1972) recognized three types of larval *Anisakis* which he designated types I, II and III, one type of *Terranova* designated type A, one of *Contracaecum* designated type B, and a *Contracaecum* from squid; all are types that may be encountered in human tissues. He regarded *Phocanema* and *Terranova* as indistinguishable. The adults of *Phocanema* and *Terranova* differ in the character of the cephalic structures and male tail, but distinctive features of the larval stages have not been described (Little and McPhail, 1972).

Anisakid third-stage larvae in the fish host grow to a relatively large size, usually 2 to 3 cm long but often more than 3 cm long, and up to 1 mm wide. The somatic muscles are polymyarian and coelomyarian (Fig. 21–23). The mouth is surrounded by three lips (one dorsal, two subventral) and a ventral boring tooth; the lips and ventral boring tooth may be poorly differentiated. The esophagus is cylindrical with an anterior muscular corpus and a shorter glandular ventriculus that is joined to the intestine. The intestine is simple and relatively large; the anus is subterminal. A large renette cell extends along the left ventral side of the esophagus and anterior part of the intestine, and the excretory pore is ventral, situated at the level of the nerve ring, or between the lips just posterior to the boring tooth. Anisakid larvae recovered from humans may be third-stage, fourth-stage, or in an intermediate stage in the process of molting. The exsheathed fourth-stage larva differs from the third by

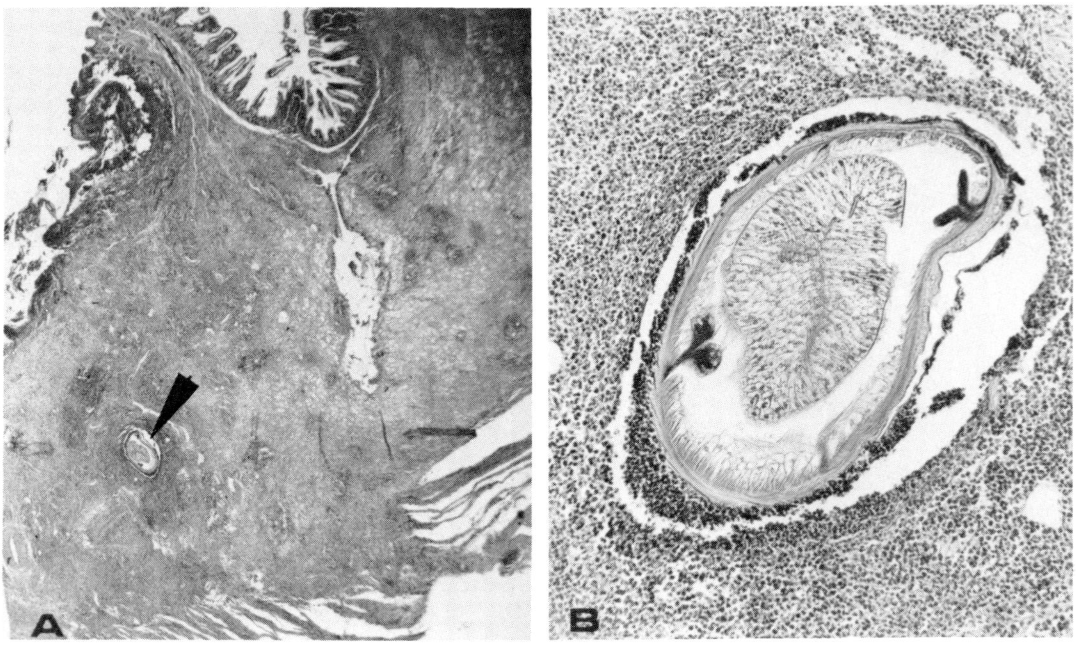

Fig. 21–23. *Anisakis. A*, Larva (arrow) in large eosinophilic granuloma in submucosa of lower ileum of a man in Kyushu, Japan. (× 10.) *B*, Same larva at higher magnification showing character of muscles, lateral cords, and intestine in transverse section. The section is slightly oblique and at a level posterior to the large renette cell. (× 100.) (Preparation courtesy of Dr. I. Miyazaki. From Faust, E.C., Beaver, P.C., and Jung, R.C. 1975. *Animal Agents and Vectors of Human Disease*, 4th ed. Philadelphia, Lea & Febiger.)

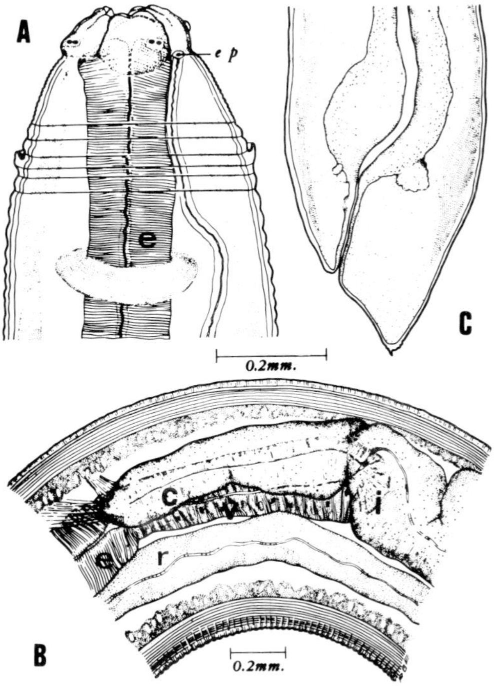

Fig. 21–24. *Phocanema.* Fourth-stage larva removed from throat of adult male Eskimo. *A,* Lateral view of head, showing bilobate lips with dentigerous ridges, excretory pore *(ep)* and position of nerve ring. *B,* Lateral view of posterior part of esophagus *(e),* ventriculus *(v),* anterior part of intestine *(i)* with intestinal cecum *(c),* and excretory gland or renette *(r)* ventral to ventriculus. C, Lateral view of tail. (From Lichtenfels, J.R., and Brancato, F.P. 1976. Anisakid larva from the throat of an Alaskan Eskimo. Am. J. Trop. Med. Hyg., *25*:691–693.)

having well-developed lips, no boring tooth, and evidences of developing sex organs.

In the common type (type I) *Anisakis* larvae (third-stage), the boring tooth is prominent, the ventriculus is simple, and the tail is short with a well-developed mucron. In the *Phocanema* and *Terranova* species there is a well-developed intestinal cecum extending anteriorly from the intestine along the ventriculus and posterior part of the esophagus, and the excretory pore empties at the base of the subventral lips (Fig. 21–24). *Contracaecum* larvae also have an intestinal cecum, as in *Phocanema,* but in addition there is a posteriorly directed appendage of the ventriculus, and the excretory pore empties at the level of the nerve ring (Oshima, 1972; Myers, 1975).

Distinctive features of the *Anisakis* larva in stained tissue sections are large body size, polymyarian-coelomyarian muscles, moderately thick cuticle, large thick-walled intestine, large renette cell, and large lateral cords that extend into the body cavity and are distinctly divided into dorsal and ventral columns each with several to numerous nuclei (Fig. 21–23) (Chitwood and Lichtenfels, 1972; Shiraki, 1974).

Visceral Larva Migrans

Definition. Visceral larva migrans is a syndrome caused by the migration of larval stages of *Toxocara canis* and certain other parasitic helminths of dogs, cats, and other carnivorous animals in the deeper parts of the body. It is characterized by long-persistent hypereosinophilia occurring together with hepatomegaly or pneumonitis, or both, hyperglobulinemia, and fever. Typically it is a disease of young children, usually mild, persisting for several months to a year or longer. A serious complication, more often seen in older children, is endophthalmitis caused by larvae that invade the eye producing an eosinophilic granulomatous reaction, usually in the retina. Usually the term visceral larva migrans includes involvement of the eye. Because the infection rarely results in concurrent ocular and systemic disease, the former may be referred to as ocular larva migrans (Glickman and Schantz, 1981).

Historical and Geographical Notes. The visceral larva migrans syndrome was described and named by Beaver and associates in New Orleans, Louisiana in 1952. Three children, 16, 26, and 30 months of age when admitted to hospital, were followed for periods of 2 to 3 years, the first beginning in August 1949. Liver biopsy revealed eosinophilic granulomatous trails and foci in all, and a larval nematode that was identified as *Toxocara,* probably *T. canis,* in one. Somewhat earlier Perlingiero and György (1947) in Philadelphia, Zuelzer and Apt (1949) in Detroit, Mercer and associates (1950) in Cleveland, and Behrer (1951) in St. Louis had examined liver biopsies containing typical lesions or larvae, or both, and had described the pathological and clinical aspects of the disease, erroneously attributed to *Ascaris lumbricoides.* Wilder (1950) had reported finding in the eyes of children, mostly from southeastern states, nematode larvae thought to be *Necator* or *Ancylostoma.* However, these were later identified as *Toxocara* by Nichols (1956a, b), whose classical studies had established morphological criteria for the recognition of common species of nematode larvae in tissue sections. A larva reported as possibly *A. lumbricoides* in the midbrain of a child in England by Beautyman and Woolf (1951) was later recognized as *Toxocara* by Beautyman et al. (1966). With the report of a larval *T. canis* in the eye of a child in California by Irvine and Irvine in 1959, and a succession of reports of *Toxocara* larvae in liver biopsies, within 10 years of the initial report more than 150 cases of visceral larva migrans had been recorded (Beaver, 1962), mostly from eastern and southern United States, but also from Europe, Philippines, Mexico, Australia, and South Africa. In 1981, it was stated that more than 1900 cases of toxocariasis had been reported from 48 countries throughout the world and from every region of the United States (Glickman and Schantz, 1981).

Etiology. Chief among the helminths known to cause visceral larva migrans is the dog ascarid, *Toxocara canis;* the cat ascarid, *Toxocara cati,* has been identified in a few cases. To some extent the terms toxocariasis and visceral larva migrans have become synonymous. However, the hookworms of dogs and cats, the *Ancylostoma* species that cause cutaneous larva migrans, also invade the deeper tissues and persist in much the same manner as the *Toxocara* species. Likewise, the *Gnathostoma* larvae that cause cutaneous larva migrans also cause a form of visceral larva migrans in humans, as do the sparganum of *Spirometra,* a tapeworm, and the mesocercaria of *Alaria*, a trematode. All of these are examples of helminth parasites of carnivorous mammals with life cycles involving, in part at least, transmission through paratenic hosts. Once the parasite in individual cases has been identified, however, the disease or infections caused by *Toxocara, Ancylostoma, Gnathostoma, Spirometra*, and *Alaria* may be referred to under their respective generic names or, as in the case of *Spirometra*, under the designation of the larval stage—sparganosis.

Toxocariasis

Biology and Life Cycle. The morphology, biology, and life cycles of *T. canis* and *T. cati* in dogs and cats, respectively, were described in an earlier section on *Toxocara* (see p. 320). When an infective-stage egg is ingested, the larva hatches in the small intestine, immediately penetrates the intestinal wall, and migrates to the liver. It may then move on to the lungs and to other parts of the body or remain in the liver; it may even migrate through the trachea and esophagus back to the intestine, but it remains an infective-stage larva with no perceptible change in size or morphology as long as it remains in the tissues of the human or any other host except the dog or cat. Intestinal infections with adult-stage *T. canis* and *T. cati* have been reported in humans, but such records are generally unreliable. In the few cases in which an adult worm was unquestionably passed from the anus or mouth of a child, circumstances have suggested that the worm had been ingested as a mature or, more likely, an immature adult, having been taken from the feces or vomitus of an infected dog or cat (von Reyn et al., 1978).

The larvae of *T. canis* are less than 0.5 mm long, with a maximum width of about 20 μm; those of *T. cati* are somewhat smaller. They are easily distinguished from other species of larvae that are encountered in human tissues, but they differ little from each other, *T. cati* being slightly smaller and its lateral alae somewhat taller (Nichols, 1956a).

Epidemiology. *Toxocara canis* and *T. cati* are among the most common helminth parasites of dogs and cats. Both species are cosmopolitan in distribution, although in some areas, particularly in the far north, a similar species, *Toxascaris leonina*, is more common in both dogs and cats. *T. leonina* is not known to cause disease in humans. As outlined on pages 320 to 322, infection with *Toxocara* is acquired in many different ways and nearly all pups and kittens are infected early in life. These animals, often living in the same households with toddler-age children, pass enormous numbers of eggs in the feces daily for many months. Although the feces of a small dog or cat will be dissipated in a short time, the eggs of their parasites accumulate in the environment. Where protected from direct sunlight and desiccation, the eggs develop to the infective stage in about 3 weeks at summer temperatures and persist in the soil for many months. Thus the soil in areas where infected dogs defecate soon becomes highly infective. Moreover, through the sorting action of rains, the eggs may be transported to rather distant locations and become highly concentrated in certain places. A few milligrams of soil taken from the surface at such locations may contain hundreds of infective eggs, making it possible for a dirt-eating child to ingest thousands of infective eggs, or for older children and adults to acquire light infections by ingesting imperceptible amounts of heavily contaminated soil. Examination of soil from parks and playgrounds in various cities of the United States, Canada, and Europe has demonstrated infective *Toxocara* eggs, which serve as sources of infection in children and contribute to the high rates of infection in dogs (Borg and Woodruff, 1973; Glickman and Schantz, 1981; Bozdech, 1981). The presence of dogs, especially pups, in the home is a high-risk factor for toxocariasis in children (Schantz et al., 1980). Ocular toxocariasis has been reported in siblings in families with dogs as household pets (Pollard, 1979).

Pathogenesis and Symptomatology. Many of the larvae reaching the liver and lungs, particularly the liver, remain there. After an undetermined period of migration, they become embedded in a matrix of epithelioid cells and encapsulated in dense fibrous tissue (Fig. 21–25). Others migrate or are transported in the blood to essentially all major organs of the body including the brain (Dent et al., 1956; Schoenfeld et al., 1964; Mikhall et al., 1974) and eye (Brown, 1970; Schantz et al., 1980), where they remain active for undetermined periods, moving from place to place. In the wake of their migrations they cause eosinophilic inflammation followed by granuloma formation (Figs. 21–26, 21–27).

As in other helminthic infections, the severity of the disease in visceral larva migrans depends primarily on the number of larvae invading the tissues, the particular tissue invaded, and the du-

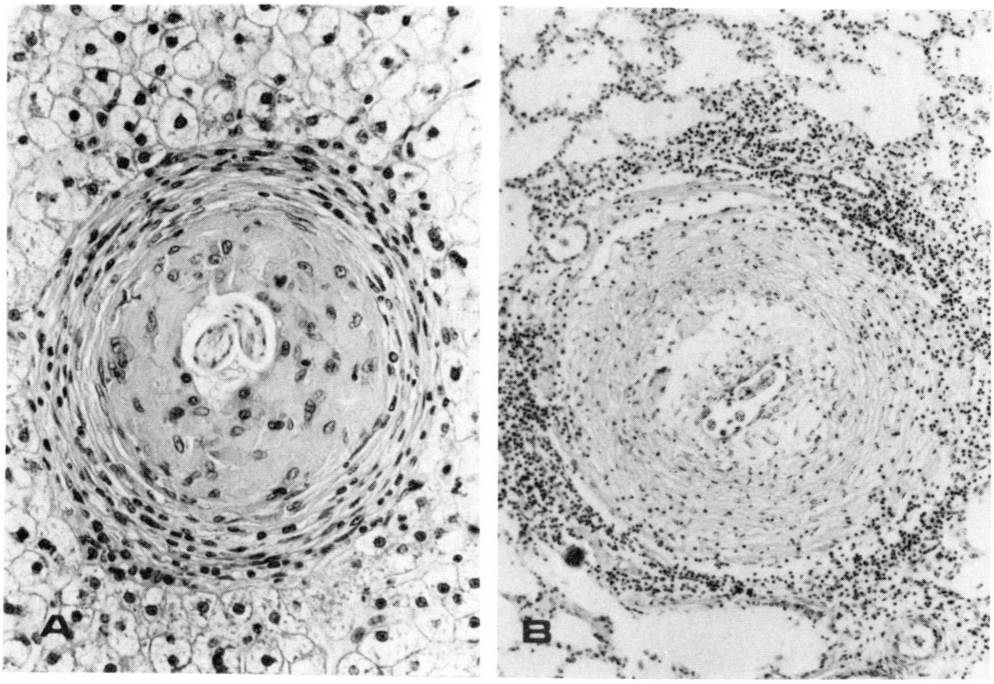

Fig. 21–25. *Toxocara canis,* encapsulated larvae. *A,* In liver of experimental *Macaca mulatta* at 9 months of infection. The larva rests in a matrix of epithelioid cells surrounded by a fibrous capsule with little inflammatory reaction. (× 250.) *B,* Larva in lung of 4-year-old child, an incidental finding at autopsy. The capsule is thick and dense, and inflammation is minimal. (× 100.) (By P.C. Beaver.)

ration of the infection. The most constant and striking clinical feature is a high, sustained eosinophilia of the blood. As few as 200 *T. canis* larvae in 2- to 3-year-old children may produce peripheral esoinophilia of 20 to 40% for more than a year, and with no detectable associated symptoms (Smith and Beaver, 1953). Cases with eosinophilia levels above 50% may be symptomatic. Levels in symptomatic cases may reach 90% of 100,000 leukocytes per mm^3, and absolute levels of eosinophilia above 25,000/mm^3 are not uncommon (Snyder, 1961). However, visceral larva migrans is only one of several known causes of hypereosinophilia (Chusid et al., 1975). Other typical findings are a history of pica, fever, hepatomegaly, and hyperglobulinemia. Less frequent manifestations are pulmonary infiltration, cough, neurological disturbances, and endophthalmitis.

Diagnosis. As toxocariasis, ascariasis, and trichuriasis are all soil-transmitted infections, they often occur together. The presence of *Ascaris* or *Trichuris* eggs in the feces suggests the probability of *Toxocara* larvae in the tissues. Elevated anti-A and anti-B (especially anti-A) isoagglutinin titers in the absence of any other helminthic infection suggests the presence of *Toxocara* larvae in the tissues (Huntley et al., 1969). Serodiagnostic tests have become increasingly reliable (Glickman et al., 1979; Schantz et al., 1980). A microprecipitation test on living *T. canis* larvae is considered by Huismans (1980) to be superior to other serological tests. Matthes and Buchwalder (1982) substituted frozen larvae for live larvae in the microprecipitation test and reported superior results. Skin tests are not generally useful. A reliable diagnosis of *Toxocara* infection can only be made when the larvae are identified in biopsy or autopsy tissue specimens, usually from the liver or in an enucleated eye. Biopsy is not generally recommended. The presence of larvae in autopsy tissues may be an incidental finding. Motile nematode larvae less than 500 μm in length observed in the eye may be *Toxocara,* but several other species are also known to invade the eye (Beaver et al., 1980). In such cases serological tests may be helpful. In patients with presumed ocular toxocariasis, greater amounts of antibody to larval *Toxocara* have been found in the aqueous humor than in the serum, suggesting localized antibody production (Felberg et al., 1981).

Nichols (1956a, b) described the diagnostic features of nematode larvae seen in stained microsec-

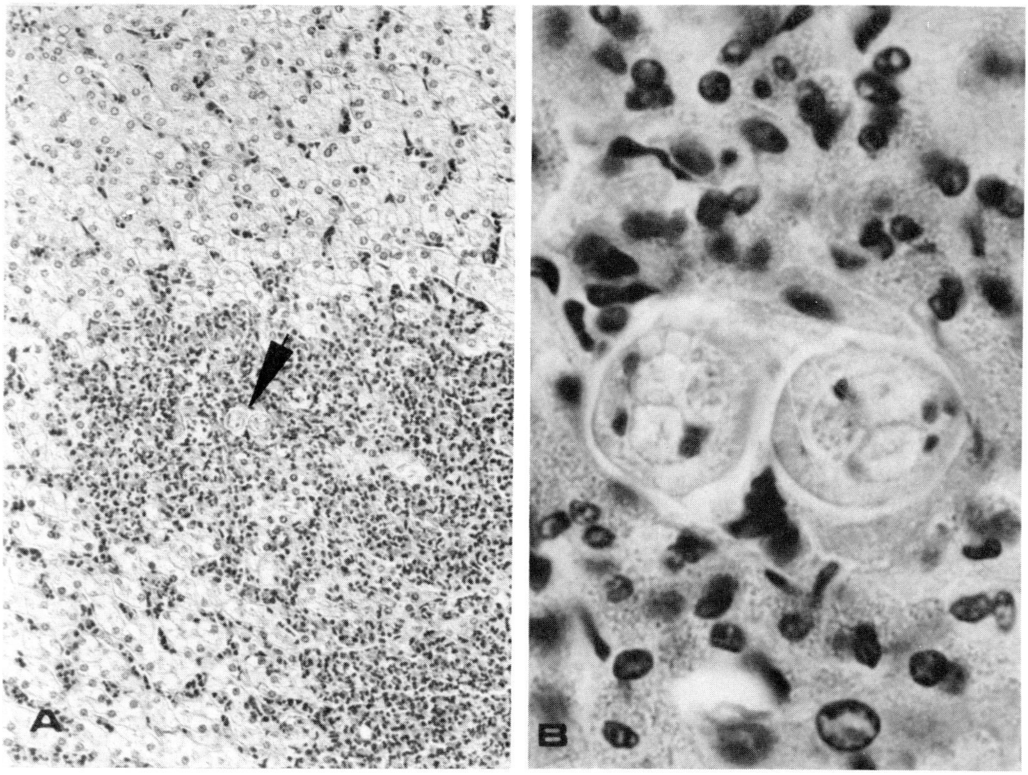

Fig. 21–26. *Toxocara* larva in liver biopsy from 2-year-old boy. *A,* In the migration phase of the infection, the area of inflammation, characteristically, is relatively great (measured in millimeters) as compared with the size of the larva (20 by 400 μ), which here is located within the inflammatory area (arrow), though it frequently is found in normal tissue a millimeter or more from the lesion. (× 150.) *B,* The structure of the larva, when seen in transverse sections (lateral alae, paired excretory columns, cord-like intestine) is distinctive. (× 1250.) (By P.C. Beaver. Section courtesy of Dr. John W. Walsh.)

tions of tissues. In addition to those of *T. canis* and *T. cati,* the larvae described by Nichols include *Ascaris lumbricoides, Necator americanus, Strongyloides stercoralis,* and *Ancylostoma caninum.*

Treatment. Though not known with certainty to be effective, thiabendazole in adult dosages ranging from 125 mg twice daily for 3 days (Gass et al., 1978) was apparently beneficial in some cases but not in others (Aur, 1971). In a dosage of 20 to 50 mg/kg body weight daily for 7 to 10 days, thiabendazole was reported to be effective in the treatment of three children, whereas in others there was no apparent effect (Pollard, 1979). Destruction of the larva by photocoagulation is recommended when it is visible in the eye and is favorably located (Gass et al., 1978; Huismans, 1977). Earlier, for patients with an active larva in the eye, thiabendazole 50 mg/kg daily for 2 or 3 days along with steroids was suggested (Brown, 1970).

Prognosis. In most cases, unless the eye is invaded, the eventual prognosis is favorable. Even when the eye is involved, the prospect of preserving the eye's usefulness by prompt diagnosis and effective treatment is good.

Prevention. Periodic administration of anthelmintics to household dogs and cats to maintain them free of mature worms is a basic safeguard for members of the household. Efforts can be made to prevent children from eating dirt, especially dooryard soil contaminated by household or neighborhood dogs. In some metropolitan, urban, and suburban communities, feral dogs can be eliminated and the promiscuous and away-from-home defecation by licensed dogs can be curbed.

Ancylostoma and Other Helminths Causing Visceral Larva Migrans

Ancylostoma. In cases of creeping eruption, larvae causing the cutaneous lesions may appear a few days later in the sputum (Muhleisen, 1953), demonstrating migration into the deeper organs. Following either oral or percutaneous inoculations in mice and other laboratory hosts, the larvae of *Ancylostoma caninum* may persist in the brain and skeletal muscles and remain infective for more than a year (Nichols, 1956b). Larvae of *Ancylostoma braziliense* and *Ancylostoma tubaeforme* persist

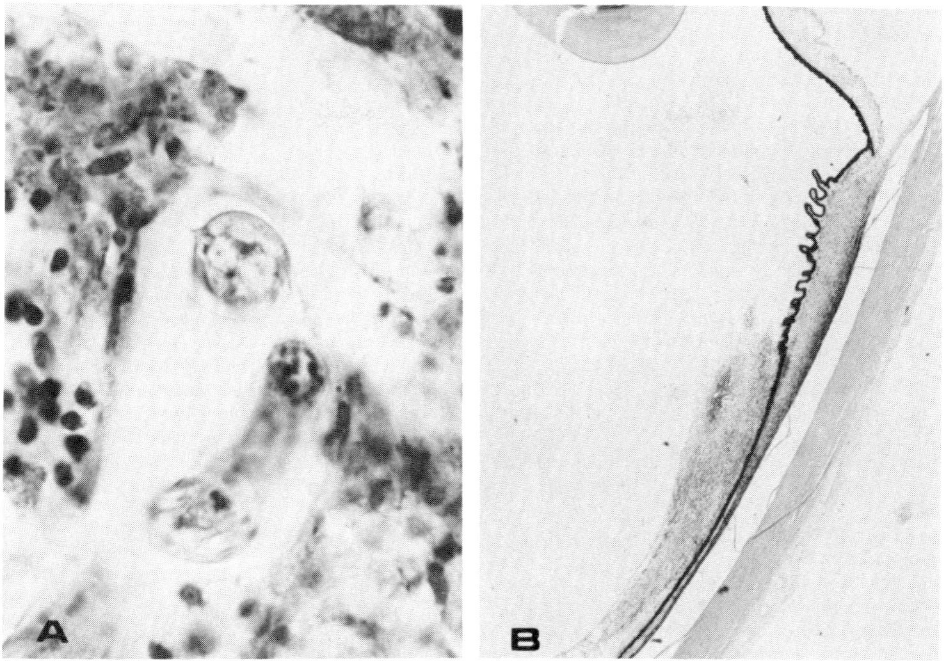

Fig. 21–27. *Toxocara canis,* larva in eye of child. *A,* Transverse and oblique sections of larva, showing the morphological features that made a specific identification possible. (× 650.) *B,* Granuloma surrounding larva in ora serrata. (× 17.) (From Faust, E.C., Beaver, P.C., and Jung, R.C. 1975. *Animal Agents and Vectors of Human Disease,* 4th ed. Philadelphia, Lea & Febiger. Preparation courtesy of Dr. A.R. Irvine.)

in the tissues of experimental animals for at least 18 months without apparent development (Norris, 1971). There are variations in migratory behavior among the different *Ancylostoma* species, but in *A. caninum* at least the larvae live free within the muscle fibers and do not become encapsulated. These observations call attention to the adaptation of *Ancylostoma* species for paratenic transfer in the life cycle, and suggest that finding the larvae of these common parasites of dogs and cats in the muscles (and central nervous system?) of people can be anticipated. Thus far, *Ancylostoma* larvae other than *A. duodenale* and those causing creeping eruption in the skin have been reported in man from the sputum and cornea (Beaver, 1966) and from the muscles (Little et al., 1983) (see p. 276).

Gnathostoma, Spirometra, and Alaria. The third-stage larva of *Gnathostoma,* which is a spirurid nematode, the sparganum (plerocercoid larva) of *Spirometra,* which is a pseudophyllidean cestode, and the mesocercaria (larva) of *Alaria,* a strigeid trematode, are parasites of carnivorous mammals. All are examples of helminths whose natural transmission involves paratenic hosts. In man, all of these larval helminths produce migratory, inflammatory, creeping-eruption-like lesions in the subcutaneous tissues, and they also invade and persist for long periods in the deeper organs. *Gnathostoma* larvae have been observed in nearly all organs of the body including the brain; spargana similarly involve various parts of the body where their presence is not apparent at the surface; and the mesocercaria of *Alaria* have been observed in the eye, lungs, and other deep organs as well as in the subcutaneous tissues. Thus, while these infections are referred to as gnathostomiasis, sparganosis, and *Alaria* infection, not as visceral larva migrans, the infections with these forms and with *Toxocara* are biologically similar in that their migratory behavior and long persistence in the tissues are the same in man as they are in the natural paratenic hosts.

Non-Paratenic Larval Invasion of Viscera

There is a basic biological difference between the kind of infection properly referred to as visceral larva migrans and the tissue-invasion phase of infections with natural endemic helminths that later develop to full maturity, such as *Ascaris, Necator,* and *Strongyloides.* Also to be distinguished from visceral larva migrans are the zoonotic infections in which the worm's relationship to the human host is the same as it is to the natural final host, and the worm may or may not develop to the adult stage. Notable examples of such forms are species of *Capillaria, Angiostrongylus, Anisakis,* and *Dirofilaria.* These, and others that undergo significant development and may even attain full sexual maturity in man, can be excluded from the list of helminths that cause visceral larva migrans (Beaver, 1969, 1978).

REFERENCES

Abdulrachman, S., and Joe, L.K. 1954. Morphological differences between *Ascaris* from man and pigs. Doc. Med. Geog. Trop., *6*:342–344.

Andersen, S., Jørgensen, R.J., Nansen, P., and Nielsen, K. 1973. Experimental *Ascaris suum* infection in piglets. Acta Pathol. Microbiol. Scand., Section B, *81*:650–656.

Anya, A.O. 1966. Studies on the biology of some oxyurid nematodes. II. The hatching of eggs and development of *Aspiculuris tetraptera* Schulz, within the host. J. Helminthol., *40*:261–268.

Araujo, P. 1972. Observações pertinentes as primeiras ecdises de larvas de *Ascaris lumbricoides, A. suum* e *Toxocara canis.* Rev. Inst. Med. Trop. S. Paulo, *14*:83–90.

Asami, K., Watanuki, T., Sakai, H., Imano, H., and Okamoto, R. 1965. Two cases of stomach granuloma caused by *An-*

isakis-like larval nematodes in Japan. Am. J. Trop. Med. Hyg., *14*:119–123.

Aur, R.J.A., Pratt, C.B., and Johnson, W.W. 1971. Thiabendazole in visceral larva migrans. Am. J. Dis. Child., *121*:226–229.

Beautyman, W., and Woolf, A.L. 1951. An *Ascaris* larva in the brain in association with acute anterior poliomyelitis. J. Pathol. Bacteriol., *63*:635–647.

———, Beaver, P.C., Buckley, J.J.C., and Woolf, A.L. 1966. Review of a case previously reported as showing an ascarid larva in the brain. J. Pathol. Bacteriol., *91*:271–273.

Beaver, P.C. 1949. Methods of pinworm diagnosis. Am. J. Trop. Med., *29*:577–587.

———. 1952a. The detection and identification of some common nematode parasites of man. Am. J. Clin. Pathol., *22*:481–494.

———. 1952b. Observations on the epidemiology of ascariasis in a region of high hookworm endemicity. J. Parasitol., *38*:445–453.

———. 1962. Toxocarosis (visceral larva migrans) in relation to tropical eosinophilia. Bull. Soc. Pathol. Exot., *55*:555–576.

———. 1964. *Ascaris* strangled in a shoe-eyelet. Am. J. Trop. Med. Hyg., *13*:295–296.

———. 1966. Zoonoses, with particular reference to parasites of veterinary importance, pp. 215–227. In *Biology of Parasites*. Edited by E.J.L. Soulsby. New York, Academic Press.

———. 1969. The nature of visceral larva migrans. J. Parasitol., *55*:3–12.

———. 1975. Biology of soil-transmitted helminths: The massive infection. Health Lab. Sci., *12*:116–125.

———. 1978. Visceral larva migrans after 25 years, pp. 40–44. In Proceedings of the Eighteenth SEAMEO-TROPMED Seminar: Current Concepts in the Diagnosis and Treatment of Parasitic and other Tropical Diseases in South East Asia. Kuala Lumpur, 2–5 August 1977. Edited by D.S.K. Tan. Published by Institute for Medical Research.

———, and Danaraj, T.J. 1958. Pulmonary ascariasis resembling eosinophilic lung. Autopsy report with description of larvae in the bronchioles. Am. J. Trop. Med. Hyg., *7*:100–111.

———, Kriz, J.J., and Lau, T.J. 1973. Pulmonary nodule caused by *Enterobius vermicularis*. Am. J. Trop. Med. Hyg., *22*:711–713.

———, and Little, M.D. 1964. The genital girdle in relation to estrus and mating in *Ascaris lumbricoides*. J. Parasitol., *50*:128–130.

———, Meyer, E.A., Jarroll, E.L., and Rosenquist, R.C. 1980. *Dipetalonema* from the eye of man in Oregon, U.S.A. Am. J. Trop. Med. Hyg., *29*:369–372.

———, Snyder, C.H., Carrera, G.M., Dent. J.H., and Lafferty, J.W. 1952. Chronic eosinophilia due to visceral larva migrans. Pediatrics, *9*:7–19.

Behrer, M.R. 1951. Hypereosinophilia with eosinophilic granuloma of liver associated with *Ascaris* infection. J. Pediatr., *38*:635–640.

Berger, H. 1971. Experimentally induced patent infections of *Ascaris suum* in rabbits. J. Parasitol., *57*:344-347.

Biguet, J., Rose, F., and Bonnel, R. 1964. Frequence de l'ascaridiose chez des enfants des écoles primaires dans des regions rurales du nord de la France. Bull. Soc. Path. Exot., *57*:558–564.

Bijlmer, E. 1946. Exceptional case of oxyuriasis of the intestinal wall. J. Parasitol., *32*:359–366.

Bisseru, B., Woodruff, A.W., and Hutchinson, R.I. 1966. Infection with adult *Toxocara canis*. Br. Med. J., *1*:1583–1584.

Borg, O.A., and Woodruff, A.W. 1973. Prevalence of infective ova of *Toxocara* species in public places. Br. Med. J., *4*:470–472.

Bowman, D., Smith, J., and Little, M.D. 1983. *Lagochilascaris sprenti* sp. n. (Nematoda: Ascarididae) from the opossum, *Didelphis virginiana (Marsupalia: Didelphidae)*. J. Parasitol. (In press.)

Bozdech, V. 1981. Zur Larven-Toxocarose des Menchen. I. Eifunde in Prager Parkanlangen. Angew. Parasitol., *22*:71–77.

Brady, F.J. 1947. Pinworm infection and trichinosis. Am. Pract., *1*:583–588.

Brenes M.R.R., and Frenkel, J. 1972. Discovery of *Lagochilascaris* sp. in the larynx of a Costa Rican ocelot *(Felis pardalis mearnsi)*. J. Parasitol., *58*:978.

Broadbent, V. 1975. Children's worms. Br. Med. J., *1*:89.

Brown, D.H. 1970. Ocular *Toxocara canis*. Part II. Clinical review. J. Pediatr. Ophthalmol., *7*:182–191.

Brudastov, A.N., Krasnonos, L.N., Lemelev, V.R., and Kholmukhamedov, S.K. 1970. Infectivity of *Ascaris lumbricoides* eggs to man and guinea-pigs after 10 years in the soil. Med. Parasitol. Parasit. Dis. (Moscow), *39*:447–451.

———, Lemelev, V.R., Kholmukhamedov, S.K., and Krasnonos, L.N. 1971. Clinical picture of the migration phase of ascariasis in self-infection. Med. Parasitol. Parasit. Dis. (Moscow), *40*:165–168.

Bustamante-Sarabia, J., Martuscelli-Q., A., and Tay, J. 1977. Ectopic ascariasis: Report of a case with adult worms in the kidney. Am. J. Trop. Med. Hyg., *26*:568–569.

Cabrera, B.D., Arambulo, P.V., III, and Portillo, G.P. 1975. Ascariasis control and/or eradication in a rural community in the Philippines. Southeast Asian J. Trop. Med. Public Health, *6*:510–518.

Carrado-Bravo, T. 1980. Avances recientes en el conocimiento de las enfermedades transmisibles en Mexico. Gac. Med. Mex., *116*:47–49.

Cerf, B.J., Rhode, J.E., and Soesanto, T. 1981. Ascaris and malnutrition in a Balinese village: a conditional relationship. Trop. Geogr. Med., *33*:367–373.

Chacko, D.D. 1970. Intestinal parasites and asthma. N. Engl. J. Med., *283*:101.

Chitwood, M. 1975. *Phocanema*-type larval nematode coughed up by a boy in California. Am. J. Trop. Med. Hyg., *24*:710–711.

———, and Lichtenfels, J.R. 1972. Identification of parasitic metazoa in tissue sections. Exp. Parasitol., *32*:407–519.

Cho, S.-Y., Hong, S.-T., Chai, J.-Y., and Seo, B.-S. 1980. Selection criteria of subjects for drug evaluation in enterobiasis. Korean J. Parasitol., *18*:45–53.

———, Hong, S-T., Kang, S-Y., and Song, C-Y. 1981. Morphological observation of *Enterobius vermicularis* expelled by various anthelmintics. Korean J. Parasitol., *19*:18–26.

Chusid, M.J., Dale, D.C., West, B.C., and Wolff, S.M. 1975. The hypereosinophilic syndrome. Analysis of fourteen cases with review of the literature. Medicine, *54*:1–27.

Cobo, A., Hall, R.C., Torres, E., and Cuello, C.J. 1964. Intrahepatic calculi. Arch. Surg., *89*:936–941.

Correa, M.O.A., Hyakutake, S., Brandi, A.J., and Monteiro, C.G. 1978. Novo caso de parasitismo humano por *Lagochilascaris minor* Leiper, 1909. Rev. Inst. Adolfo Lutz, *38*:59–65.

Craig, T.M., Robinson, R.M., McArthur, N.H., and Ward, R.D. 1980. *Lagochilascaris major* in a raccoon. J. Wildl. Dis., *16*:67–70.

Cram, E.B. 1941. Studies on oxyuriasis. IX. The familial nature of pinworm infestation. Med. Ann. Distr. Columbia, *10*:39–48.

———. 1943. Studies on oxyuriasis. XXVIII. Summaries and conclusions. Am. J. Dis. Child., *65*:46–59.

Crewe, W., and Smith, D.H. 1971. Human infection with pig *Ascaris (A. suum)*. Ann. Trop. Med. Parasitol., *65*:85.

Davey, J.T. 1971. A revision of the genus *Anisakis* Dujardin, 1845. (Nematoda: Ascaridata). J. Helminthol., *45*:51–72.

De Leão, R.N.Q., Leão Filho, J., Bragadias, L., and Calheiros, L.B. 1978. Infeccão humana pelo *Lagochilascaris minor* Leiper, 1909. Registro de um caso observado no Estado do Pará (Brasil). Rev. Inst. Med. Trop. S. Paulo, *20*:300–306.

Dent, J.H., Nichols, R.L., Beaver, P.C., and Carrera, G.M. 1956. Visceral larva migrans. With a case report. Am. J. Pathol., *32*:777–803.

Douglas, J.R., and Baker, N.F. 1966. Some host-parasite relationships of canine helminths, pp. 97–115. In *Host-Parasite Relationships*. Edited by J.E. McCauley. Corvallis, Oregon State University Press.

Draper, J.W. 1963. Infection with *Lagochilascaris minor*. Br. Med. J., *1*:931–932.

Durette, M.C. 1962. Nématodes parasites de mammifères et de reptiles. 4. Phasmidiens (2e note). Parc National de l'Upemba—Mission G.F. de Witte et al. (1946–1949), Fascicule 68(1).

Fatherree, J.P., Carrera, G.M., and Beaver, P.C. 1951. *Enterobius vermicularis* in the human uterus. Report of a case. Mississippi Doctor, *28*:159–161.

Felberg, N.T., Shields, J.A., and Federman, J.L. 1981. Antibody to *Toxocara canis* in the aqueous humor. Arch. Ophthalmol., *99*:1563–1564.

Fitzgerald, T.B., Mainwaring, A.R., and Ahmed, A. 1974. Pelvic peritoneal oxyuriasis simulating metastatic carcinoma. A case report. J. Obstet. Gynecol. Br. Commonw., *81*:248–250.

Freij, L., Meeuwisse, G.W., Berg, N.O., Wall, S., and Gebre-Medhin, M. 1979. Ascariasis and malnutrition. A study in urban Ethiopian children. Am. J. Clin. Nutr., *32*:1545–1553.

Galvin, T.J. 1968. Development of human and pig *Ascaris* in the pig and rabbit. J. Parasitol., *54*:1085–1091.

Garcia, E. G. 1973. Sex ratio of *Ascaris lumbricoides*. Acta Med. Philipp., *9*:152–154.

Gass, J.D., Gilbert, W.R., Jr., Guerry, R.K., and Scelfo, R. 1978. Diffuse unilateral subacute neuroretinitis. Ophthalmology, *85*:521–545.

Gelpi, A.P., and Mustafa, A. 1967. Seasonal pneumonitis with eosinophilia. A study of larval ascariasis in Saudi Arabs. Am. J. Trop. Med. Hyg., *16*:646–657.

———. 1968. *Ascaris* pneumonia. Am. J. Med., *44*:377–389.

Glickman, L., Cypess, R., Hiles, D., and Gessner, T. 1979. *Toxocara*-specific antibody in the serum and aqueous humor of a patient with presumed ocular and visceral toxocariasis. Am. J. Trop. Med. Hyg., *28*:29–35.

———, and Schantz, P.M. 1981. Epidemiology and pathogenesis of zoonotic toxocariasis. Epidemiol. Rev., *3*:230–250.

Goldsmid, J.M. 1976. Ascariasis in Rhodesia. Cent. Afr. J. Med., *22*:220–223.

Graham, C.F. 1941. A device for the diagnosis of *Enterobius* infection. Am. J. Trop. Med., *21*:159–161.

Hakami, M., Kharrad, M., and Mosavy, S.H. 1976. Escape of ascarides through herniorrhaphy wounds. Am. J. Proctol., *27*:47–48.

Hall, M.C. 1937. Studies on oxyuriasis. I. Types of anal swabs and scrapers, with a description of an improved type of swab. Am. J. Trop. Med., *17*:445–453.

Hatsushika, R. 1979. An experimental study on development and hatching of the eggs of *Anisakis physeteris* (Nematoda: Ascaridata). Kawasaki Med. J., *5*:1–7.

Hobo, B. 1956. Epidemiological studies on *Ascaris* infection among prisoners and the length of life of *Ascaris lumbricoides* in human host. Jpn. J. Nation's Health, *25*:1–14.

Hoeppli, R. 1923. Die durch Ascarislarven bei experimenteller Infektion im Tierkorper bewirkten anatomischen Veranderungen. Virchow's Arch. Pathol. Anat. Physio. Klin. Med., *244*:159–182.

Huismans, H. 1977. Über das Solitärgranulom bei okularer *Toxocara canis* Infektion (Hundespulwurm). Ophthalmologica, *174*:10–13.

———. 1980. *Toxocara canis* (Larva migrans visceralis) aus ophthalmologischer Sicht. Klin. Monatsbl. Augenheilk., *177*:584–600.

Huntley, C.C., Lyerly, A.D., and Patterson, M.V. 1969. Isohemagglutinins in parasitic infections. J.A.M.A., *208*:1145–1148.

Hussey, K.L. 1957. *Syphacia muris* vs. *S. obvelata* in laboratory rats and mice. J. Parasitol., *43*:555–559.

Ihekwaba, F.N. 1979. *Ascaris lumbricoides* and perforation of the ileum: A critical review. Br. J. Surg., *66*:132–134.

Irvine, W.C., and Irvine, A.R., Jr. 1959. Nematode endophthalmitis: *Toxocara canis*. Report of one case. Am. J. Ophthalmol., *47*:185–191.

Jacobs, A.H. 1942. Enterobiasis in children; incidence, symptomatology, and diagnosis, with a simplified Scotch cellulose tape technique. J. Pediatr., *21*:497–503.

Jaskoski, B.J. 1961. An apparent swine *Ascaris* infection of man. J. Am. Vet. Med. Assoc., *138*:504–505.

Jørgensen, R.J., Nansen, P., Nielsen, K., Eriksen, L., and Andersen, S. 1975. Experimental *Ascaris suum* infection in the pig population. Kinetics following low and high levels of primary infection in piglets. Vet. Parasitol., *1*:151–157.

Juels, C.W., Butler, W., Bier, J.W., and Jackson, G.J. 1975. Temporary human infection with a *Phocanema* sp. larva. Am. J. Trop. Med. Hyg., *24*:942–944.

Jung, R.C., and Beaver, P.C. 1953. Treatment of enterobiasis (pinworm infestation) with diphenan, egressin and gentian violet. Pediatrics, *11*:611–616.

Kacker, P.P. 1973. Vulvo-vaginitis in an adult with threadworms in the vagina. Br. J. Vener. Dis., *49*:314–315.

Kalejaiye, E.O., Solanke, T.F., Adekunle, O., and Ogunbiyi, O. 1977. Biliary lithiasis associated with ascariasis in a Nigerian woman. Arch. Surg., *112*:645–647.

Kates, S., Wright, K.A., and Wright, R. 1973. A case of human infection with the cod nematode *Phocanema* sp. Am. J. Trop. Med. Hyg., *22*:606–608.

Kliks, M.M. 1983. Anisakiasis in the western United States: Four new case reports from California. Am. J. Trop. Med. Hyg., *32*:526–532.

Kmetec, E., Beaver, P.C., and Bueding, E. 1963. Succinoxidase activities and cytochrome oxidase of extracts of *Ascaris* eggs. Comp. Biochem. Physiol., *9*:115–120.

Koino, S. 1922. Experimental infection with *Ascaris* in human body with special reference to the clinical symptoms of *Ascaris* pneumonia. Tokyo Iji Shinshi, October.

Koyama, T., Kumada, M., Suzuki, H., Ohnuma, H., Karasawa, Y., Ohbayashi, M., and Yokogawa, M. 1973. *Terranova* (Nematoda: Anisakidae) infection in man. II. Morphological features of *Terranova* sp. larva found in human stomach wall. Jpn. J. Parasitol., *21*:257–261.

Krasnonos, L.L. 1978. Many-year viability of ascarid eggs (*Ascaris lumbricoides* L. 1758) in soil of Samarkand. Med. Parasitol. Parasit. Dis. (Moscow), *47*:103–105.

Kurimoto, H. 1974. Morphological, biochemical and immunological studies on the differences between *Ascaris lumbricoides* Linnaeus, 1758 and *Ascaris suum* Goeze, 1782. Jpn. J. Parasitol., *23*:251–267.

Lee, K.T., Min. H.K., and Soh, C.T. 1976. Transplacental

migration of *Toxocara canis* larvae in experimentally infected mice. J. Parasitol., *62*:460–465.

Lichtenfels, J.R., and Brancato, F.P. 1976. Anisakid larva from the throat of an Alaskan Eskimo. Am. J. Trop. Med. Hyg., *25*:691–693.

Little, M.D. 1968. Comparison of human and pig *Ascaris*. Eighth International Congress Tropical Medicine Malaria, Abstracts & Reviews, pp. 181–182.

———, Cuello, C.J., and D'Alessandro, A. 1973. Granuloma of the liver due to *Enterobius vermicularis*. Report of a case. Am. J. Trop. Med. Hyg., *22*:567–569.

———, Halsey, N.A., and Cline, B.L. 1983. *Ancylostoma* larva in a muscle fiber of man following cutaneous larva migrans. Am. J. Trop. Med. Hyg., in press.

———, and McPhail, J.C. 1972. Large nematode larva from the abdominal cavity of a man in Massachusetts. Am. J. Trop. Med. Hyg., *21*:948–950.

———, and Most, H. 1973. Anisakid larva from the throat of a woman in New York. Am. J. Trop. Med. Hyg., *22*:609–612.

Lord, W., and Bullock, W. 1982. Swine *Ascaris* in humans. N. Engl. J. Med., *306*:1113.

Louw, J.H. 1974. Biliary ascariasis in childhood. S. Afr. J. Surg., *12*:219–225.

Lysek, H. 1975. On the problem of cytogenetic determination of *Ascaris* in man and in pig. Acta Univ. Palackianae Olomucensis, *74*:47–50.

———. 1979. To the problem of possible biological control of geohelminthoses. Helminthologia, *16*:107–113.

Martin, L.K. 1972. Hookworm in Georgia. II. Survey of intestinal helminth infections in members of rural households of southeastern Georgia. Am. J. Trop. Med. Hyg., *21*:930–943.

Matthes, H-F., and Buchwalder, R. 1982. Untersuchungen zum Larvenprazysitationstest mit *Ascaris suum*- und *Toxocara canis*-Larven. Angew. Parasitol., *23*:1–8.

Maung, M. 1973. *Ascaris lumbricoides* Linne, 1758 and *Ascaris suum* Goeze, 1782: Morphological differences between specimens obtained from man and pig. Southeast Asian J. Trop. Med. Public Health, *4*:41–45.

———. 1978. The occurrence of the second moult of *Ascaris lumbricoides* and *Ascaris suum*. Int. J. Parasitol., *8*:371–378.

McCraw, B.M. 1973. Reinfection of yearling calves with *Ascaris suum*. Can. J. Comp. Med., *37*:21–24.

McLennan, M.W., Humphris, R.B., and Rac, R. 1974. *Ascaris suum* pneumonia in cattle. Aust. Vet. J., *50*:266–268.

Mello, D.A. 1974. A note on egg production of *Ascaris lumbricoides*. J. Parasitol., *60*:380–381.

Mercer, R.D., Lund, H.Z., Bloomfield, R.A., and Caldwell, F.E. 1950. Larval ascariasis as cause of chronic eosinophilia with visceral manifestations. Am. J. Dis. Child., *80*:46–58.

Mikhael, N., Montpetet, V., Orizaga, M., Rowsell, H., and Richard, M. 1974. *Toxocara canis* infestation with encephalitis. Can. J. Neurol. Sci., *1*:114–120.

Miller, A. 1961. The mouth parts and digestive tract of adult dung beetles (Coleoptera: Scarabaeidae), with reference to the ingestion of helminth eggs. J. Parasitol., *47*:735–744.

Mittal, V.K., Dhaliwal, R., Yadav, R.V.S., and Sahariah, S. 1976. Fatal respiratory obstruction due to round worm. Med. J. Aust., *2*:210–212.

Mondragon, H., Cano, R.M., and Botero, D. 1973. Primer caso de infeccion humana por *Lagochilascaris minor* en Colombia. Antioquia Med., *23*:463–464.

Morishita, K. 1972. Studies on epidemiological aspects of ascariasis in Japan and basic knowledge concerning its control, pp. 3–153. In *Progress of Medical Parasitology in Japan*. Volume 4. Edited by K. Morishita, Y. Komiya, and H. Matsubayashi. Tokyo, Meguro Parasitological Museum.

Muhleisen, J.P. 1953. Demonstration of pulmonary migration of the causative agent of creeping eruption. Ann. Intern. Med., *38*:595–600.

Mya, M. 1955. Life cycle studies on the mouse pinworms *Syphacia obvelata* and *Aspiculuris tetraptera*. M.S. Thesis, Tulane University.

Myers, B.J. 1975. The nematodes that cause anisakiasis. J. Milk Food Technol., *38*:774–782.

Nichols, R.L. 1956a. The etiology of visceral larva migrans. I. Diagnostic morphology of infective second-stage *Toxocara* larvae. J. Parasitol., *42*:349–362.

———. 1956b. The etiology of visceral larva migrans. II. Comparative larval morphology of *Ascaris lumbricoides, Necator americanus, Strongyloides stercoralis,* and *Ancylostoma caninum*. J. Parasitol., *42*:363–399.

Norris, D. 1971. The migratory behavior of the infective-stage larvae of *Ancylostoma braziliense* and *Ancylostoma tubaeforme* in rodent paratenic hosts. J. Parasitol., *57*:998–1009.

Nutting, S.W., Murphy, F., and Inglis, F.G. 1980. Abdominal pain due to *Enterobius vermicularis*. Can. J. Surg., *23*:286–287.

Oostburg, B.F.J. 1971. Thiabendazole therapy of *Lagochilascaris minor* infection in Surinam. Report of a case. Am. J. Trop. Med. Hyg., *20*:580–583.

———, and Verma, A.A.O. 1968. *Lagochilascaris minor* infection in Surinam. Report of a case. Am. J. Trop. Med. Hyg., *17*:548–550.

Orihel, T.C. 1971. Primates as models for parasitological research, pp. 772–782. In *Medical Primatology*. Basel, S. Karger.

———, and Seibold, H.R. 1972. Nematodes of the bowel and tissues, pp. 76–103. In *Pathology of Simian Primates*. Edited by R.N.T.W. Fiennes. Basel, S. Karger.

Oshima, T. 1972. *Anisakis* and anisakiasis in Japan and adjacent area, pp. 301–393. In *Progress of Medical Parasitology in Japan,* Volume 4. Edited by K. Morishita, Y. Komiya, and H. Matsubayashi. Tokyo, Meguro Parasitological Museum.

Perlingiero, J.G., and György, P. 1947. Chronic eosinophilia: Report of a case with necrosis of liver, pulmonary infiltrations, anemia and *Ascaris* infection. Am. J. Dis. Child., *73*:34–43.

Phills, J.A., Harrold, A.J., Whiteman, G.V., and Perelmutter, L. 1972. Pulmonary infiltrates, asthma and eosinophilia due to *Ascaris suum* infestation in man. N. Engl. J. Med., *286*:965–970.

Piggott, J., Hansbarger, E.A., Jr., and Neafie, R.C. 1970. Human ascariasis. Am. J. Clin. Pathol., *53*:223–234.

Pincus, G.S., Coolidge, C., and Little, M.D. 1975. Intestinal anisakiasis. First case report from North America. Am. J. Med., *59*:114–120.

Pollard, Z.F. 1979. Ocular *Toxocara* in siblings of two families. Diagnosis confirmed by ELISA test. Arch Ophthalmol., *97*:2319–2320.

Proffitt, R.D., and Walton, B.C. 1962. *Ascaris* pneumonia in a two-year-old girl. Diagnosis by gastric aspirate. N. Engl. J. Med., *266*:931–934.

Prokopic, J. 1979. Scanning electron microscopic study of the morphology of *Ascaris suum* Goeze, 1782. Folia Parasitol., *26*:377–380.

Raney, R., Lilly, J., and McHardy, G. 1970. Biliary calculus of roundworm origin. Ann. Intern. Med., *72*:405–407.

Reardon, L. 1938. Studies on oxyuriasis. XVI. The number of eggs produced by the pinworm, *Enterobius vermicularis*, and its bearing on infection. Public Health Rep., *53*:978–984.

Reddy, C.R.R.M., Venkateswar Rao, D., Sarma, E.N.B., and

Swamy, G.M.N. 1975. Granulomatous peritonitis due to *Ascaris lumbricoides* and its ova. J. Trop. Med. Hyg., *78*:146–149.
Riley, W.A. 1920. A mouse oxyurid, *Syphacia obvelata,* as a parasite of man. J. Parasitol., *6*:89–92.
Roberts, F.H.S. 1934. The large roundworm of pigs, *Ascaris lumbricoides* L., 1758. Its life history in Queensland, economic importance and control. Bulletin No. 1, Animal Health Station, Yeerongpilly.
Roche, P.J.L. 1971. *Ascaris* in the lachrimal duct. Trans. R. Soc. Trop. Med. Hyg., *65*:540.
Rodan, K.S., and Buckley, J.J.C. 1969. Infection with adult *Toxocara cati*. Br. Med. J., *2*:188.
Saffos, R.O., and Rhatigan, R.M. 1977. Unilateral salpingitis due to *Enterobius vermicularis*. Am. J. Clin. Pathol., *67*:296–299.
Sapunar, J., Doerr, E., and Letonja, T. 1976. Anisakiasis humana en Chile. Bol. Chil. Parasitol., *31*:79–83.
Schacher, J.F. 1957. A contribution to the life history and larval morphology of *Toxocara canis*. J. Parasitol., *43*:599–612.
Schantz, P.M., Weis, P.E., Pollard, Z.F., and White, M.C. 1980. Risk factors for ocular larva migrans: A case-control study. Am. J. Public Health, *70*:1269–1272.
Schaum, E., and Müller, W. 1967. Die Heterocheilidiasis. Eine Infektion des Menschen mit Larven von Fisch-Ascariden. Dtsch. Med. Wochenschr., *92*:2230–2233.
Schoenfeld, A.E., Githnic, E., and Rosen, N. 1964. Granulomatous encephalitis due to *Toxocara* larvae (visceral larva migrans). Harefuah, *46*:337–340.
Schüffner, W. 1944. Die Bedeutung der Stabinfektion für die Oxyuriasis. Munch. Med. Wochenschr., Nos. 31–32, pp. 411–414.
———, and Swellengrebel, N.H. 1943. Retrofection in oxyuriasis. A newly discovered mode of infection with *Enterobius vermicularis*. J. Parasitol., *35*:138–146.
———. 1943. Eine zweizeitige method zum Nachveis von Oxyuris-Eiern. Ihre Leistung gegenüber dem americanischen NIH Wischer. Zbl. Bakt. I. Abt. Orig., *151*:71–80.
Schuster, D.I., Belin, R., Parker, J.C., Jr., Burke, J.A., and Jona, J.Z. 1977. Ascariasis—its complications, unusual presentations and surgical approaches. South. Med. J., *70*:176–178.
Schwartz, B. 1959. Experimental infection of pigs with *Ascaris suum*. Am. J. Vet. Res., *20*:7–13.
Seo, B-S. 1981. Ascariasis and its control problems in Korea. Seoul J. Med., *22*:323–341.
———, Cho, S.-Y., and Chai, J.-Y. 1979. Frequency distribution of *Ascaris lumbricoides* in rural Koreans with special reference on the effect of changing endemicity. Korean J. Parasitol., *17*:105–113.
Shah., K.N., and Desai, M.P. 1969. *Ascaris lumbricoides* from the right ear. Indian Pediatr., *6*:92–93.
Shiraki, T. 1974. Larval nematodes of family Anisakidae (Nematoda) in the Northern Sea of Japan—as a causative agent of eosinophilic phlegmone or granuloma in the human gastro-intestinal tract. Acta Med. Biol., *22*:57–98.
———, Otsuru, M., Kenmotsu, M., Kihara, T., Hisayasu, N., and Motoyama, N. 1974. Invasion of *Enterobius vermicularis* into the human tissues—a report of 4 cases of eosinophilic granulomas caused by adult *Enterobius* in the intestine and their adjacent tissue. Jpn. J. Parasitol., *23*:125–137.
Simon, R.S. 1974. Pinworm infestation and urinary tract infection in young girls. Am. J. Dis. Child., *128*:21–22.
Smedresman, P. 1977. *Ascaris lumbricoides* as an unusual cause of appendicitis in an 8-year-old girl. Clin. Pediatr., *16*:197.
Smith, J.L., Bowman, D., and Little, M.D. 1983. Life cycle and development of *Lagochilascaris sprenti* (Nematoda: Ascarididae) from opossums (Marsupalia: Didelphidae) in Louisiana. J. Parasitol. (In press.)
Smith, J.W., and Wootten, R. 1975. Experimental studies on the migration of *Anisakis* sp. larvae (Nematoda: Ascaridada) into the flesh of herring, *Clupea harengus* L. Int. J. Parasitol., *5*:133–136.
Smith, M.H.D., and Beaver, P.C. 1953. Persistence and distribution of *Toxocara* larvae in the tissues of children and mice. Pediatrics, *12*:491–497.
Snyder, C.H. 1961. Visceral larva migrans—ten years' experience. Pediatrics, *28*:85–91.
Sobrinho, V.M. da C., Ferreira, D., and Faria, M. 1972. Ascaridiase cardiaca. Rev. Goiana Med., *18*:215–226.
Soh, C.T. 1960. The effects of natural food-preservative substances on the development and survival of intestinal helminth eggs and larvae. I. Action on *Ascaris lumbricoides* eggs. Am. J. Trop. Med. Hyg., *9*:1–7.
Soprunov, F.F., Lurje, A.A., Laynis, Y.Y., Soprunova, N.Y., and Alieva, H.H. 1981. Metabolic end products of helminths: Their degradation and excretion by the host. Acta Tropica, *38*:449–460.
Soprunova, N.J., Soprunov, F.F., and Lurje, A.A. 1973. Nachweis von Helminthem-Metaboliten im Harn des Wirtes als ein neuer diagnostischer Test für Helminthiasen. Angew. Parasitol., *14*:11–17.
Spillmann, R. 1975. Pulmonary ascariasis in tropical communities. Am. J. Trop. Med. Hyg., *24*:791–800.
Sprent, J.F.A. 1952. The dentigerous ridges of the human and pig *Ascaris*. Trans. R. Soc. Trop. Med. Hyg., *46*:378.
———. 1958. Observations on the development of *Toxocara canis* (Werner, 1782) in the dog. Parasitology, *48*:185–198.
———. 1961. Post-parturient infection of the bitch with *Toxocara canis*. J. Parasitol., *47*:284.
———. 1971. Speciation and development in the genus *Lagochilascaris*. Parasitology, *62*:71–112.
———. 1971. A note on *Lagochilascaris* from the cat in Argentina. Parasitology, *63*:45–48.
Stephenson, L.S. 1980. The contribution of *Ascaris lumbricoides* to malnutrition in children. Parasitology, *81*:221–233.
———, Crompton, D.W.T., Latham, M.C., Schulpen, T.W.J., Nesheim, M.C., and Jansen, A.A.J. 1980. Relationships between *Ascaris* infection and growth of malnourished preschool children in Kenya. Am. J. Clin. Nutr., *33*:1165–1172.
Stoll, N.R., Chenoweth, B.M., Jr., and Peck, J.L., Jr. 1947. Low incidence of *Enterobius vermicularis* in natives of Guam, M.I. Puerto Rico J. Public Health Trop. Med., *22*:235–253.
Stoye, M. 1976. Ubersichtsreferat: Pränatale und galaktogene Helmintheninfektionen bei Hausteiren. Dtsch. Tierarztl. Wochenschr., *83*:569–576.
Stürchler, D., Stahel, E., Saladin, K., and Saladin, B. 1980. Intestinal parasitoses in eight Liberian settlements: Prevalences and community anthelminthic chemotherapy. Tropenmed. Parasitol., *31*:87–93.
Suzuki, H., Ohnuma, H., Karasawa, Y., Ohbayashi, M., Koyama, T., Kumada, M., and Yokogawa, M. 1972. *Terranova* (Nematoda: Anisakidae) infection in man. I. Clinical features of five cases of *Terranova* larva infection. Jpn. J. Parasitol., *21*:252–256.
Swartzwelder, J.C. 1941. *Toxocara cati* (cat ascarid) infection in man. J. Trop. Med. Hyg., *44*:61–62.
———, Miller, J.H., and Sappenfield, R.W. 1957. The effective use of piperazine for the treatment of human helminthiases. Gastroenterology, *33*:87–96.
Takata, I. 1951. Experimental infection of man with *Ascaris* of man and pig. Kitasato Arch. Exp. Med., *23*:49–59.
Thitasut, P. 1961. Action of aqueous solutions of iodine on fresh vegetables and on the infective stages of some com-

mon intestinal nematodes. Am. J. Trop. Med. Hyg., *10*:39–43.

Thust, R. 1968. Submikroskopische Untersuchungen über die Morphogenese des Integumentes von *Ascaris lumbricoides* L. 1758. Z. Wissehschaft. Zool., *178*:1–39.

Tripathy, K., Duque, E., Bolaños, O., Lotero, H., and Mayoral, L.G. 1972. Malabsorption syndrome in ascariasis. Am. J. Clin. Nutr., *25*:1276–1281.

———, Gonzalez, F., Lotero, H., and Bolaños, O. 1971. Effects of *Ascaris* infection on human nutrition. Am. J. Trop. Med. Hyg., *20*:212–218.

Tsung, S.H., and Loh, W.P. 1979. Invasion of the fallopian tube by *Enterobius vermicularis*. Ann. Clin. Lab. Sci., *9*:393–395.

Tullis, D.C.H. 1970. Bronchial asthma associated with intestinal parasites. N. Engl. J. Med., *282*:370–372.

Ubelaker, J.E., and Allison, V.F. 1972. Scanning electron microscopy of the denticles and eggs of *Ascaris lumbricoides* and *Ascaris suum*. Thirtieth Annual Proceedings Electron Microscopy Society America, 1972.

Valdiserri, R.O. 1981. Intestinal anisakiasis. Report of a case and recovery of larvae from market fish. Am. J. Clin. Pathol., *76*:329–333.

Van Thiel, P.H. 1966. The final hosts of the herringworm *Anisakis marina*. Trop. Geogr. Med., *18*:310–328.

———, Kuipers, F.C., and Roskam, T.H. 1960. A nematode parasitic to herring, causing acute abdominal syndromes in man. Trop. Geogr. Med., *12*:97–113.

Vogel, H., and Minning, W. 1942. Beiträge zur Klinik der Lungen-Ascariasis und zur Frage der flüchtigen eosinophilen Lungeninfiltrate. Beitr. Klin. Tuberkulose, *98*:624–654.

von Reyn, C.F., Roberts, T.M., Owen, R., and Beaver, P.C. 1978. Infection of an infant with an adult *Toxocara cati* (Nematoda). J. Pediatr., *93*:247–249.

Warren, G. 1971. Studies on the morphology and taxonomy of the genera *Toxocara* Stiles, 1905 and *Neoascaris* Travassos, 1927. Zool. Anzeiger, *185*:393–442.

Watt, I.A., Girdwood, R.W.A., McLean, N.R., Kissen, L.H., and Fyfe, A.H.B. 1979. Eosinophilic gastroenteritis associated with a larval anisakine nematode. Lancet, *2*:893–894.

Weise, R.W. 1973. *Ascaris suum:* A scanning electron microscope study. J. Parasitol., *59*:141–146.

Wilder, H. 1950. Nematode endophthalmitis. Trans. Am. Acad. Ophthalmol., *55*:99–109.

Wiseman, R.A., and Lovel, T.W.I. 1969. Human infection with adult *Toxocara cati*. Br. Med. J., *3*:454–455.

Woodruff, A.W. 1981. Not an infection with adult *Toxocara canis*. Br. Med. J., *283*:1124.

Yamaguti, S. 1935. Studies on the helminth fauna of Japan. Part 13. Mammalian nematodes. Jpn. J. Zool., *6*:433–457.

Yokogawa, M., Oshima, T., Sano, M., Kihata, M., Sato, S., and Komiya, Y. 1956. Experiments in the mass treatment of hookworm and ascariasis. Expulsion of young *Ascaris* worms. Bull. Inst. Public Health (Tokyo), *5*:2–6.

Yoshimura, H., Akao, N., Kondo, K., and Ohnishi, Y. 1979. Clinicopathological studies on larval anisakiasis, with special reference to the report of extra-gastrointestinal anisakiasis. Jpn. J. Parasitol., *28*:347–354.

Zuelzer, W., and Apt, L. 1949. Disseminated visceral lesions associated with extreme eosinophilia. Am. J. Dis. Child., *78*:153–181.

Chapter 22

The Spirurida: *Dracunculus* and Others

In the spirurid nematodes the cephalic structures are bilaterally symmetrical, and the esophagus is divided into anterior muscular and posterior glandular parts. The normal location of spirurids in the host is generally the esophagus or stomach, or in the tissues or tissue spaces. In Chabaud's (1975) classification of the order Spirurida there are two suborders: 1) Camallanina with two superfamilies, the Camallanoidea and Dracunculoidea, and 2) Spirurina with ten superfamilies, including the Filarioidea and six others in which there are representatives among the parasites of man. Those six and the Dracunculoidea *(Dracunculus)* are considered here.

SUBORDER CAMALLANINA

Superfamily Dracunculoidea

The Dracunculoidea include the family Dracunculidae and several families of fish parasites (Philometridae and others). Outstanding features of the Dracunculidae are extreme sexual dimorphism and complete atrophy of the anus in the adult female. The family Dracunculidae includes two genera: *Avioserpens* Wehr and Chitwood, 1934, which are parasites of birds, and *Dracunculus* (Reichard, 1759), which are parasites of birds and mammals. One species, *D. medinensis,* is found in man.

Dracunculus medinensis
(Linnaeus, 1758) Gallandant, 1773
(The guinea worm, medina worm, serpent worm, or dragon worm, producing dracunculiasis or dracontiasis)

Synonyms. *Gordius medinensis* Linnaeus, 1758; *Vena medinensis* (Linnaeus, 1758) Gallandant, 1773; *Filaria medinensis* (Linnaeus, 1758) Gmelin, 1790; *Fuellebornius medinensis* (Linnaeus, 1758) Leiper, 1926.

Historical and Geographical Notes. This worm and its lesion have been known since antiquity. Egyptian priests regarded the infection as an important clinical entity, and physicians during the Greek and Roman periods were familiar with it. Fedtschenko (1870) (reprinted in English translation, 1971) first studied this nematode carefully.

The distribution of *Dracunculus medinensis* includes the Nile Valley, central equatorial Africa (from Lake Chad to the Sudan and Uganda), and the northwest and west coasts of Africa (Mauritania, Senegal, and Upper Volta, Ghana, Togoland, Dahomey, Nigeria [common in western part] and the Cameroons), Arabia (especially on the Red Sea), Iran, Afghanistan, Pakistan, southeastern regions of the U.S.S.R., and India (extensive areas in Bombay, Madras, Rajasthan, and Hyderabad states). Cases are also known for Indonesia, probably imported from India. One autochthonous human case was reported from Korea in 1928. In North America a species reported as *Dracunculus insignis* (Leidy, 1858) Chandler, 1942 is found commonly in the raccoon *(Procyon lotor)* and the mink *(Mustela vison)* from Louisiana to Canada, and less commonly in other fur-bearing animals and dogs; and *Dracunculus lutrae* Crichton and Beverley-Burton, 1973 has been reported as common in the otter *(Lutra canadensis)* in Ontario (Crichton and Beverley-Burton, 1973, 1974, 1975). In China *D. medinensis* has been reported in dogs.

Morphology, Biology, and Life Cycle. *Dracunculus medinensis* females are elongated, cylindroidal, bluntly rounded at the anterior end, and recurved at the caudal extremity. The cuticula is smooth. Surmounting the anterior end (Fig. 22–1,*A, B)* is an oval shield, in the midst of which is the minute triangular mouth within an oval or quadrate prominence. It is surrounded by (1) an inner circle of six well-developed papillae, of which the two laterals are single but the two ventrals, and at times the two dorsals, may be twinned, and (2) an outer circle of four double papillae. A lateral pair of cervical papillae is situated just behind the nerve ring. Males of *D. medinensis* are known from a single specimen obtained from a natural infection of man in India (40 mm long), from specimens obtained by Moorthy (1937) and others (Muller, 1968) from experimentally infected dogs and monkeys (12 to 29 mm long by a maximum diameter of 0.4 mm). Females measure from 70 to 120 cm long (average less than 100 cm) by 0.7 to 1.7 mm in diameter. The posterior end of the male

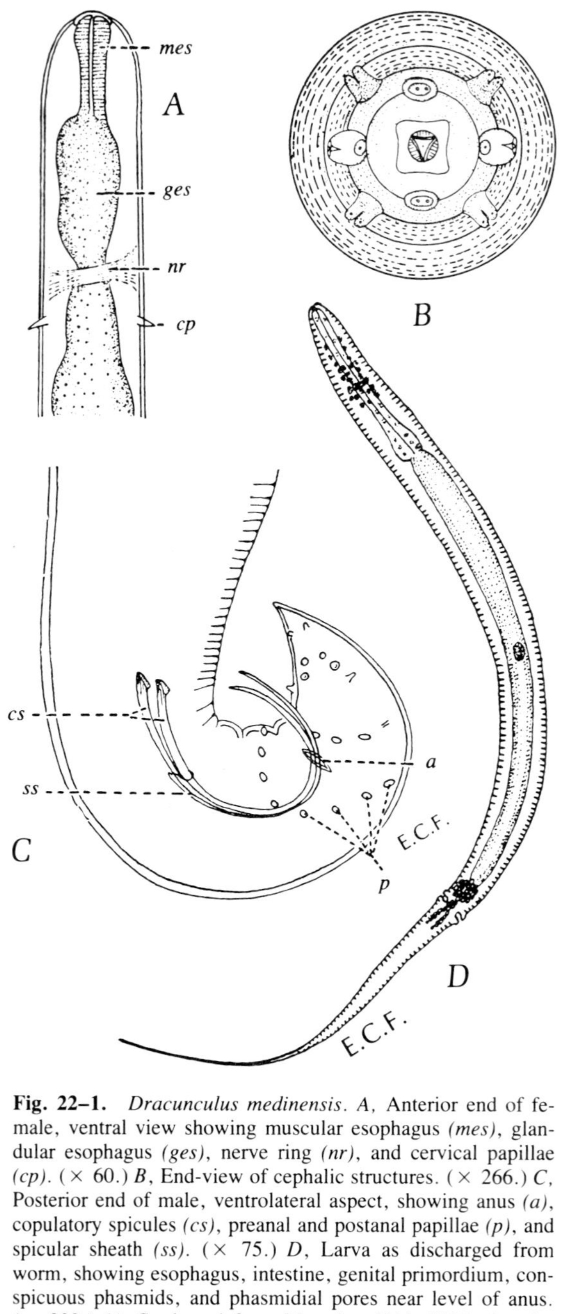

Fig. 22–1. *Dracunculus medinensis. A,* Anterior end of female, ventral view showing muscular esophagus *(mes)*, glandular esophagus *(ges)*, nerve ring *(nr)*, and cervical papillae *(cp)*. (× 60.) *B*, End-view of cephalic structures. (× 266.) *C*, Posterior end of male, ventrolateral aspect, showing anus *(a)*, copulatory spicules *(cs)*, preanal and postanal papillae *(p)*, and spicular sheath *(ss)*. (× 75.) *D*, Larva as discharged from worm, showing esophagus, intestine, genital primordium, conspicuous phasmids, and phasmidial pores near level of anus. (× 200.) *(A–C* adapted from Moorthy, V.N. 1937. A redescription of *Dracunculus medinensis*. J. Parasitol., *23*:220–224. *D*, from Faust, E.C. 1939. *Human Helminthology*, 2nd ed. Philadelphia, Lea & Febiger.)

(Fig. 22–1,*C)* is coiled on itself one or more times. There are ten pairs of caudal papillae *(p)*, four preanal and six postanal. The copulatory spicules are subequal, 490 to 730 μm long. The gubernaculum is 200 μm long. In the female the ovarian tubules, oviducts, and uteri are paired, and the vagina is unpaired. No vulva has been found in gravid female *D. medinensis;* in *D. insignis* and *D. lutrae* it is situated slightly postequatorially (Crichton and Beverley-Burton, 1973).

In experimentally infected dogs and monkeys, the larvae migrate through the tissues to the axillary and inguinal regions where fertilization takes place. Extensive migration of the female may occur late in the prepatent period (Muller, 1968). In the gravid female the ovaries are postfunctional and the uteri are highly coiled, distended masses filled with rhabditoid larvae and occupying the greater part of the body.

At the end of the prepatent period, about 12 months after the infected *Cyclops* is swallowed, the cephalic end of the worm approaches the skin and a papule is formed in the dermis; this changes into a blister within 24 hours, or may enlarge for several days before the blister develops. These lesions usually appear on the distal portion of the limbs. Eventually the blister ruptures. On contact with fresh water, a loop of the worm's uterus, prolapsed through a ruptured anterior part of its body wall or through its mouth, bursts open and discharges a swarm of motile first-stage rhabditoid larvae. This process is repeated until the entire brood has been discharged.

The freed larvae (Fig. 22–1,*D*) are wiry and vermiform, 500 to 750 μm long by 15 to 25 μm in the greatest diameter (anterior third of body). The tail is long and filiform, about one-third of the entire body length. The cuticula is conspicuously striated. The larva moves about with a stiff motion, briskly coiling and uncoiling the body. It has no active means of gaining entrance into the intermediate host; however, if an appropriate species of *Cyclops* is present in the water, some of the larvae are ingested and, on reaching the copepod's midintestine, break through the soft wall and come to lie in the celomic cavity (Fig. 36–2). More than five or six larvae usually cause death of the *Cyclops*. In suitable species, the larvae proceed to undergo metamorphosis with loss of the striated cuticula in about 8 days and the development of a delicate enveloping sheath 2 days later. Subsequently they become quiescent and show no inclination to leave the copepod.

Ingestion of an infected copepod in unfiltered water provides opportunity for human infection. On arrival in the duodenum, the larvae migrate through the wall of the digestive tract and reach the loose connective tissue, usually retroperitoneal in position, where they develop into adults. In experimentally infected dogs, cats, and monkeys the larvae migrate through the intestinal wall to the mesentery, then to the abdominal and chest wall, reach the subcutaneous tissue in about 3 weeks, and develop to full maturity and emerge through the skin in 329 to 360 days (Muller, 1968, 1972). Migration and development of *D. insignis* in the raccoon follow a similar pattern; sexual maturity is reached and the worms mate between 60 and 77 days after inoculation (Crichton and Beverley-Burton, 1975).

Dracunculus insignis is morphologically indistinguishable from *D. medinensis*, and whether they are distinct species remains unknown (Beverley-Burton and Crichton, 1976).

Epidemiology. Human infection results from swallowing infected cyclops, the necessary intermediate hosts. The cyclops become infected when the gravid female worms discharge their larval progeny into the water where the cyclops breed. Infection in reservoir hosts is not epidemiologically important.

In a 2-year survey of a Deccan village in India, Lindberg (1946a,b) found much higher incidence among those drinking from step wells (Fig. 22–2,*A*) than from curb wells (38.0 vs. 14.5%). The rate was significantly higher in males than in females of both categories. The prevalence increased with age to 85.6% in the 30- to 35-year age group, then decreased sharply. The number of worms varied from 1 to 50, although few patients had more than 6. One individual had 15 worms in a single year. Infection rates have declined in western India where water sources have improved, but in some villages infection may be found in 6% of the population (Johnson and Joshi, 1982). In southern Ghana, guinea worm infection is acquired almost exclusively from pond water during the dry season (Belcher et al., 1975). This is true also in western Nigeria, where over 13% of 8200 people in rural villages were infected. Those with infection were disabled for an average of 100 days, mostly at the peak of agricultural activity (Kale, 1977). In the arid region of western India, dracontiasis remains a major cause of disability (Sharma et al., 1979). Infection in Iran is acquired exclusively from large, enclosed cisterns (Fig. 22–2,*B*) locally called *berkeh* (Sahba et al., 1973).

Pathogenesis, Pathology, and Symptomatology. No pathologic process has been observed to occur with the penetration of the mature larvae ingested in infected *Cyclops* into the viscera and deep somatic tissues, and their development into mature worms in these locations. When the gravid female worms begin to migrate to the skin, host reaction to the products of the worms develops, which may be systemic or local. When the worm has reached the skin, has produced a cutaneous blister at the outer end of the tunnel (Fig. 22–3,*A*), and the blister has burst in contact with fresh water, symptoms may decrease, owing to the continued discharge of the worm's metabolites together with motile larvae. The incubation period is about 1 year. A few hours preceding the development of the local cutaneous lesion there may be pronounced systemic symptoms including erythema and urticarial rash, with intense pruritus, nausea, vomiting, diarrhea, dyspnea, giddiness, and syncope. In a few hours the local lesion develops. It is a reddish papule with a vesicular center and an indurated margin measuring 2 to 7 cm in diameter. These lesions are most common between the metatarsal bones on the soles of the feet or on the ankles (Figs. 22–3,4), but may be found on the hands or arms, trunk, buttocks, scrotum, knee joint, calf, thigh, shoulder, or even the angle of the jaw. The fluid in the blister is typically sterile yellow serum, containing many mononuclear, eosinophilic, and polymorphonuclear leukocytes. Beneath the blister is the anterior extremity of the female worm lying in a subcutaneous or deeper tunnel. On rupture of the blister the anterior portion of the worm is protruded up to about 5 cm and discharges larvae. Although pain and other symptoms of host reaction to the worm tend to lessen with the rupture of the blister, attempts to remove the worm by traction may reactivate the symptoms. This is especially true if the worm should break under traction. This event is typically followed by intense inflammatory reaction, with pain, swelling, and redness of cellulitis along the track of the worm.

Secondary infection of the lesion occurs commonly and seriously aggravates the condition (Fig. 22–4). Tetanus is not an uncommon complication.

Infection may be single or multiple. Sequelae include arthritis, synovitis, ankylosis, and contractures of the involved limbs. The average work loss among agricultural workers in southern Ghana is 5 weeks (Belcher et al., 1975), and in western Nigeria 100 days (Kale, 1977).

A gravid female worm was found in an abscess cavity in the pericardium of a 35-year-old Hindu man with constrictive pericarditis (Kinare et al., 1962). Donaldson and Angelo (1961) reported a

Fig. 22–2. *Dracunculus medinensis,* major sources of infection. *A,* A step well, in Rajasthan, India, illustrates a form of public water supply used in semiarid regions of India and Pakistan. The water level, very low in the illustrated well, fluctuates with the water-table. The plankton flora and fauna, including *Cyclops* species, are rich during the warmer months, the season of guinea-worm transmission. *B,* A *berkeh* in the Lar region of Iran, is a type of cistern adapted to near-desert conditions. Where rains are relatively light and limited to one period of only a few days annually, surface water is channeled into the berkeh, where, though densely turbid with soil and feces of man and animals at first, it soon becomes clear and potable through sedimentation and natural biological conditioning. The heavy roof and walls provide insulation against heat, evaporation, and sunlight, yet enough light is admitted to sustain a stable, moderately dense population of algae and zooplankton, including *Cyclops.* Berkehs range in size up to 20 meters in diameter, with the depth and diameter generally approximately equal. As they must provide water for all needs of man and animals, a dozen or more berkehs may be required by a community of 200 people. (From Faust, E.C., Beaver, P.C., and Jung, R.C. 1975. *Animal Agents and Vectors of Human Disease,* 4th ed. Philadelphia, Lea & Febiger.)

case of quadriplegia due to guinea-worm abscess in the spinal canal with recovery after laminectomy and removal of the worm. A guinea worm was removed from the subconjunctival space of the right eye of a 12-year-old girl in India (Verma, 1966). Two cases of extradural abscess caused by guinea worms, one of them fatal, were reported by Reddy and Valli (1967).

Diagnosis. A diagnosis is possible only with the onset of systemic symptoms, followed in a few hours by development of the cutaneous lesion. In most cases, patients appear at the clinic after the worm is patent. Direct evidence of infection can usually be depended on for specific diagnosis. At times the infection may be confused with carbuncles, deep cellulitis, gumma, onchocerciasis or loaiasis, acute focal myositis, focal periosteitis, or rheumatism. Dead calcified worms may be located by roentgen-ray examination.

Treatment. The systemic symptoms due to allergic reaction may be relieved by administration of epinephrine. It is reasonable to expect that systemic administration of corticosteroid drugs would be even more effective but there are no published records of such treatment. Muller (1971, 1979) has reported on the efficacy of topical hydrocortisone in experimentally infected rhesus monkeys and cats.

Three drugs have been demonstrated to be effective in relieving the signs and symptoms of dracontiasis: niridazole in a dosage of 25 mg/kg body weight daily (maximum 1.5 g per day) for 10 days (Kale, 1977); thiabendazole, 50 mg/kg body weight daily for 3 days (Raffier, 1969; Sastry et al., 1978); and metronidazole 400 to 500 mg daily for 10 to

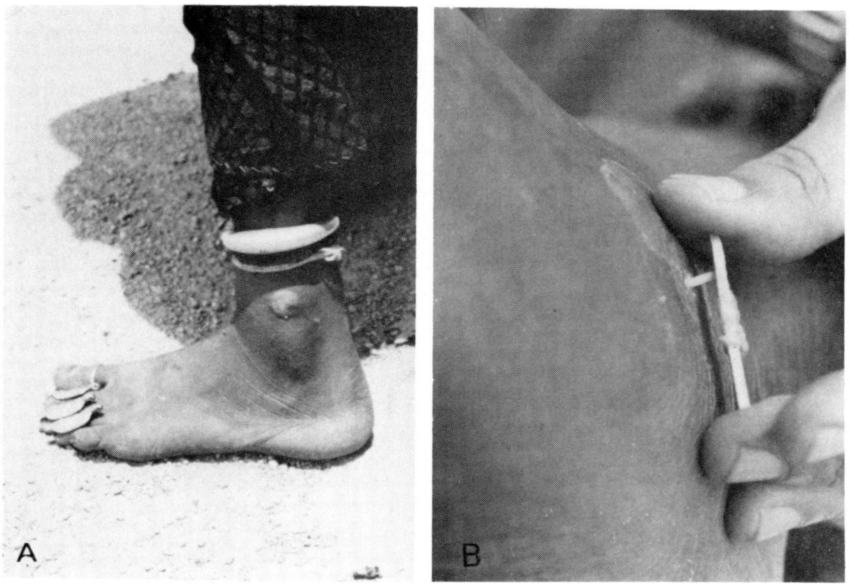

Fig. 22–3. *Dracunculus medinensis. A,* Bleb produced by gravid worm on ankle of woman in Rajasthan, India. On contact with water, the worm will emerge to discharge larvae. *B,* Extruded worm being drawn by gradual winding on a match stick from a tunnel opening at the ankle of a man in Singapore. (*A* and *B* from Faust, E.C., Beaver, P.C., and Jung, R.C. 1975. *Animal Agents and Vectors of Human Disease,* 4th ed. Philadelphia, Lea & Febiger. A by P.C. Beaver, *B* courtesy of Dr. John F. Schacher.)

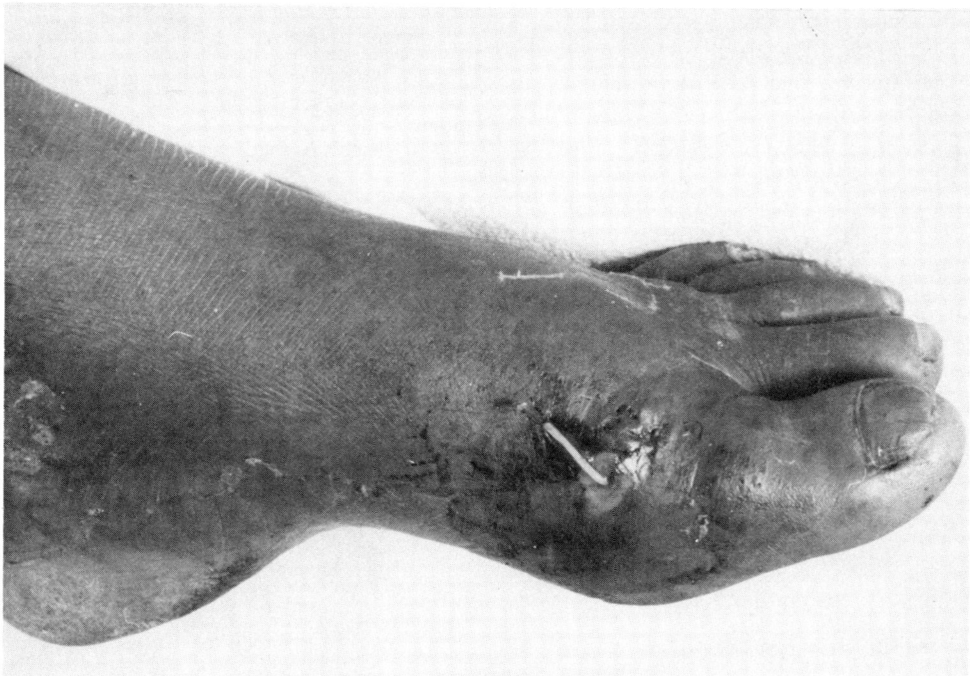

Fig. 22–4. *Dracunculus medinensis.* Female extruding from ulcer on foot about 10 days after initial emergence through a ruptured bleb. Much of the anterior part of the body has been broken off and discarded, yet much of the worm remains, extending far into the tissue of ankle and shin. Secondary infection has retarded the worm's outward progress. (Courtesy of Dr. T. J. Danaraj. From Faust, E.C., Beaver, P.C., and Jung, R.C. 1975. *Animal Agents and Vectors of Human Disease,* 4th ed. Philadelphia, Lea & Febiger.)

20 days (Pardanani et al., 1977; Sharma et al., 1979). Although administration of these drugs relieves pain and pruritus and facilitates removal of the worm by traction, there is no evidence that the drugs have any adverse effect on the worms. Preemergent worms exposed to these drugs subsequently emerge in the usual way, though with less severe lesions than in untreated patients. It appears that the mode of action of the drugs is by suppression of inflammation (Muller, 1971, 1979).

Administration of any one of the aforementioned drugs should be accompanied by removal of the worms by exerting gentle traction and rolling the protruded portion on a stick.

Mebendazole has been found to be effective in one trial (Shafei, 1976). The action of this drug differs from that of the others mentioned since it causes the worm's tissues to break down. The worm becomes fragmented and is discharged in bits and pieces.

In endemic areas of dracontiasis, local practice includes the surgical removal of the worms, which is performed by both physicians and nonprofessional healers. Fairley and Liston (1925) suggested removal of the worm by an operative technique that would reduce the likelihood of sepsis. Mirza (1932) described a local Indian barber's technique, which may be utilized to advantage if the head of the worm has not yet emerged from the blister

Prognosis. Usually the prognosis is good, unless secondary infection results in septicemia or severe arthritis producing physical deformity.

Prevention. In order to develop a practical program for the reduction or elimination of *Dracunculus medinensis* in a community, two steps must be taken: (1) infected individuals must not be allowed to bathe or wade in water used for drinking purposes, and (2) persons must be persuaded not to drink suspected water. Chemical control of the copepods transmitting the infection is possible (Muller, 1970). However, thus far the only effective method of control is to provide piped water or wells that are curbed or covered to prevent transfer of larvae from infected persons to the water. All other methods have been ineffective.

SUBORDER SPIRURINA

Including the Filarioidea, which are discussed in Chapter 23, there are seven superfamilies of the Spirurina with species that are found in man. However, except in the Filarioidea, all of these are natural parasites of animals and are only occasionally, or rarely, found in man. The classification and characteristics of the six superfamilies to which these zoonotic Spirurina belong are given in Chapter 17 (see p. 229).

Gnathostoma spinigerum Owen, 1836 (Producing gnathostomiasis)

Synonym. *Gnathostoma siamense* (Levinsen, 1890) Railliet, 1893.

Historical and Geographical Notes. In his review of the genus *Gnathostoma*, Miyazaki (1960) listed 19 named species of which he considered the majority to be synonyms of previously described ones, and possibly only the following to be valid species: *G. spinigerum* Owen, 1936; *G. hispidum* Fedtschenko, 1872; *G. turgidum* Stossich, 1902; *G. doloresi* Tubangui, 1925; *G. nipponicum* Yamaguti, 1941; *G. americanum* Travassos, 1925; and *G. procyonis* Chandler, 1942. Since then three new species have been added: *G. miyazakii*, Anderson, 1964 from the otter *(Lutra c. canadensis)* in Ontario, Canada; *G. malaysiae* Miyazaki and Dunn, 1965 from rats in Malaysia; and *G. vietnamicum* described by Le-Van-Hoa (1965) from the otter *(Lutra elioti)* in Vietnam. Daengsvang (1980) believes two other species to be valid, namely, *G. didelphis* Chandler 1932 from opossums in the United States and *G. brasiliense* Ruiz, 1952 from *Lutreolina crassicaudata* in Brazil. In enzootic areas *G. spinigerum* is the species most often causing human infection, but *G. hispidum* has been discovered as a human parasite on a few occasions. *G. spinigerum* was first described by Owen from gastric tumors of a tiger. It has since been reported from similar sites in the domestic cat, wild cat, lion (Zambia), leopard, mink (Pennsylvania, United States), and the dog in India, Thailand, Malaysia, China, Japan, and Australia (?). Yoshida discovered *G. nipponicum* (erroneously diagnosed as *G. spinigerum)* in Japanese mink (Miyazaki, 1954). Human cases have been reported from Thailand, Malaysia, China, Japan, India, Java, Israel, Vietnam, and the Philippines. Two cases of human gnathostomiasis acquired in Mexico have been reported (Peláez and Perez-Reyes, 1970). Except for the two cases of possibly spurious parasitism reported by Chandler (1927) from India, almost all human infections have been with the immature worms in the subcutaneous tissues. Thailand is the country of heaviest infection, both in human and in reservoir hosts.

Morphology, Biology, and Life Cycle. The adult worms in the type host *(Felis tigris)* vary in length from 11 to 25 mm for males and 25 to 54 mm for females. In smaller hosts they are considerably reduced in size. They are stout, reddish, slightly transparent nematodes with a subglobose cephalic swelling separated from the remainder of the worm by a cervical constriction. The worms are usually curved ventrad at both ends. One or more individuals inhabit a tumor which they produce in the wall of the stomach. The anterior half of the nematode is covered with leaf-like spines that are broader and tridented just behind the cervix and narrower and singly pointed more equatorially. These spines are species-characteristic for structure and number. The posterior half of the cuticula is aspinose except for a few small terminal spines. The cephalic portion of the body (Fig. 22–5,*B*) is covered with four to eight transverse rows of sharp, recurved hooks. The mouth is guarded by a pair of conspicuous fleshy lips. Four conspicuous cervical glands (Fig. 22–5,*A*) are arranged symmetrically around the esophagus. They fuse in pairs and open through two ducts that perforate the lips. The male has a pseudobursa, which is provided with four pairs of perianal nipple-shaped papillae (Fig. 22–5,*C*). The copulatory spicules are chitinoid rods measuring 1.1 mm and 0.4 mm, respectively. The vulva of the female is slightly postequatorial in position. The vagina is long and is anteriorly directed; the other genital tubules are paired. The eggs (65 to 70 by 38 to 40 μm) are ovoid, transparent, superficially pitted, have a mucoid plug at one end, and are unembryonated when laid (Fig. 22–5,*F*). According to Prommas and Daengsvang (1933) and Miyazaki (1954), embryonation is completed in 1 week at 27

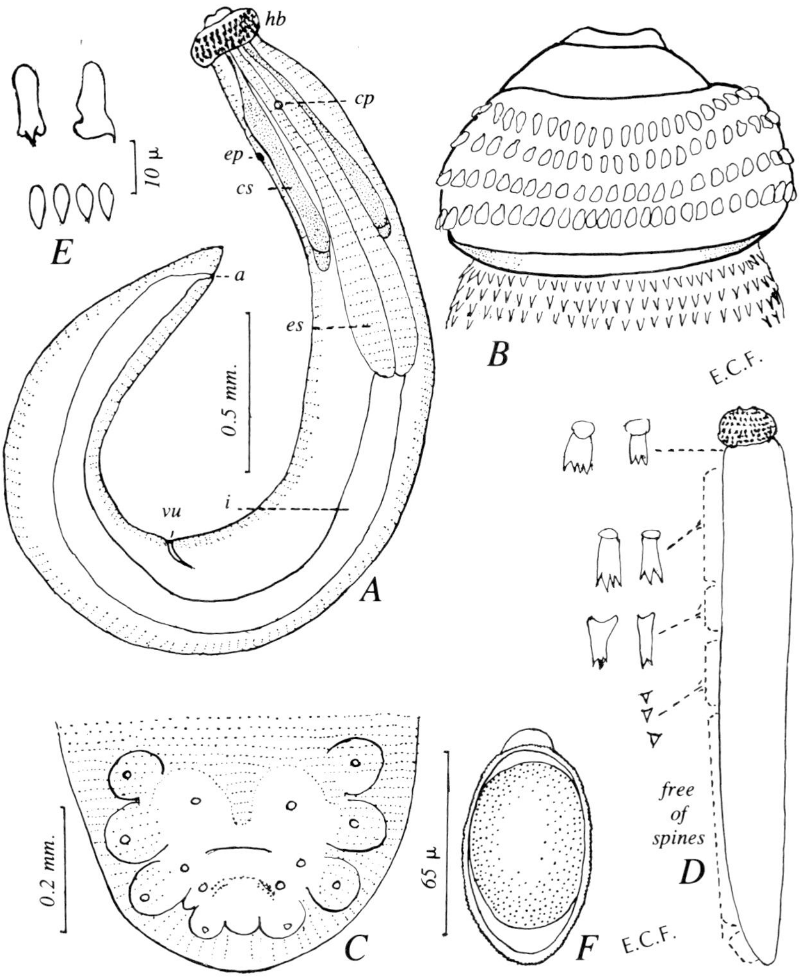

Fig. 22–5. *Gnathostoma spinigerum*. *A*, Third-stage larva, lateral view. *B*, Third-stage larva head bulb with four rows of hooklets. *C*, Posterior end of male, ventral view, with minute cuticular spines omitted. *D*, Diagram, showing types of spines at different levels of the body. *E*, Detail of spines on head bulb. *F*, Fertilized egg. *a*, Anus. *cp*, Cervical papilla, *cs*, Cervical sac (gland). *ep*, Excretory pore. *es*, Esophagus. *hb*, Head bulb. *i*, Intestine. *vu*, Vulva. (Adapted from Miyazaki, I. 1954. Studies on *Gnathostoma* occurring in Japan (Nematoda: Gnathostomidae). II. Life history of *Gnathostoma* and morphological comparison of its larval forms. Kyushu Mem. Med. Sci., *5*:123–140.)

to 31°C, whereupon the motile first-stage larva, measuring 223 to 275 by 13 to 17 µm and having a rounded anterior end provided with spines, emerges from the eggshell. When eaten by a species of *Cyclops (sensu lato)*, it forces its way into the hemocele of this crustacean and metamorphoses in 10 to 14 days into a second-stage larva (350 to 450 by 60 to 65 µm), which is provided with a head-bulb armed with four rings of spines and two pairs of cervical glands like the adult. Prommas and Daengsvang (1936) demonstrated that a second intermediate host is required. Chandler (1925) found immature worms with four rows of cephalic hooks encapsulated on the mesentery of the rock python and cobras in India. Prommas and Daengsvang (1936) and Daengsvang and Tansurat (1938) in Thailand, Africa, and elsewhere, and Yogore and Juliano (1951) in the Philippines, found freshwater fishes *(Clarias batrachus, Monopterus albus, Ophiocephalus striatus, Glossogobius giurus,* and *Therapon argenteus)* to harbor the encapsulated stage after feeding upon infected *Cyclops*. Daengsvang and Tansurat (1938) incriminated a frog *(Rana rugulosa)*. Completion of the experimental cycle by transfer to the definitive host has been demonstrated by Prommas and Daengsvang in 1937. In Japan, Miyazaki (1954) found the third-stage larva in crayfish, crabs, amphibia, reptiles, and many mammals that had eaten infected fish flesh, but only maturing to the adult stage in the stomach of dogs and cats following migration to the liver and other extraintestinal tissues. Both Miyazaki (1954), and Morita (1955) studied the species-distinguishing characters of larval stages of *Gnathostoma* occurring in Japan.

The life cycle in nature (Daengsvang, 1980a) is as follows: (1) Adult *G. spinigerum* lives in tumors in the stomach wall of cats, their wild relatives, and dogs. (2) Eggs extruded from the lesions and evacuated in feces into water embryonate and hatch. (3) The larvae are ingested by species of *Cyclops*, in which they transform into the second stage, with a cephalic bulb having four rows of single pointed spines. (4) When the infected *Cyclops* are eaten by fishes, frogs, or snakes, the larva develops into the third stage in the flesh of these animals. This is called "primary infection." (5) Other hosts that are not suited to be a definitive host, including man, may ingest the vertebrate host containing third-stage larvae and become infected. This is called "secondary infection." Since the larva undergoes no further development, such a host is paratenic. Third-stage larvae may thus be passed from one paratenic host to another as the predator of a smaller host becomes in turn the prey of a larger. Since advanced third-stage larvae in the copepod host are infective to monkeys, it is probable that human infection can be acquired by swallowing an infected copepod in drinking water (Daengsvang, 1971). (6) When the suitable definitive host eats either the infected second larval or the paratenic host, the infection develops to maturity in the stomach wall in about 6 months, completing the cycle.

Epidemiology. The epidemiology has been inadequately studied insofar as human exposure is concerned. However, eating inadequately processed fish containing the encapsulated third-stage larva is usually responsible for human gnathostomiasis. In Thailand, the domestic duck and the domestic chicken probably constitute the more important sources of infection in man (Daengsvang et al., 1966). Although information is too scant for significance, the infection tends to be more common in females than in males. It is possible that this may be due to transmission via skin penetration taking place during the handling of animal flesh in the preparation of food. Skin penetration has been demonstrated experimentally in animals (Daengsvang et al., 1970). Infection occasionally may be acquired directly from copepods swallowed in drinking water taken from ponds or streams (Daengsvang, 1971).

Pathogenesis, Pathology, and Symptomatology. The infection in reservoir definitive hosts, in which the adult worms are coiled inside tumors of the digestive tract, is referred to as *gnathostomiasis interna*. This condition has never been described in man, who is probably always a paratenic host. In human infections, the third-stage larvae, with either four or eight cephalic rings of spines, migrate in the tissues, their presence usually becoming evident when they reach the superficial layers. This latter condition is known as *gnathostomiasis externa*, and includes both (1) the development of abscess pockets or indurated nodules with abscessed centers, and (2) the formation of deep cutaneous or subcutaneous tunnels in which the worms migrate, causing *larva migrans* or creeping eruption. The lesions of the cutaneous larva migrans of gnathostomiasis differ from those of hookworm in that they are usually deeper, i.e., subcutaneous, or even more frequently in the muscles. They therefore appear as intermittent migratory swellings that are circumscribed and up to several centimeters in diameter. If the swelling is subcutaneous, there are usually signs of inflammation, redness, and pain which, except pain, are absent in the more common, deeper lesions. The swelling is hard and non-pitting, and lasts for 1 to 4 weeks. Then it disappears, appearing anew after a free interval of 1 week to several months in a new location not far from the previous one. The swellings appear most frequently on the upper extremity, shoulder, neck, thorax, face, scalp, abdominal wall, thigh, and dorsum of foot. The lesions may be relatively free of discomfort or be pruritic or painful. Pain may be mild or so severe as to disturb sleep (Daengsvang, 1980a).

The swelling is believed to be due to allergic reaction to the worm and its products, and to involve little mechanical damage (Miyazaki, 1960). The histologic picture is one of intense cellular infiltration with neutrophils, small mononuclear and plasma cells, and massive numbers of eosinophils.

The direction and rate of creeping are variable, having been observed from 1.8 mm per hour (Tamura, 1921) to 30 cm per hour (Daengsvang et al., 1970).

Any part of the body may be invaded, such as the mucous membrane of the mouth, pharynx, intestine, anus, or cervix, producing visceral larva migrans (Fig. 22–6). Worms have been removed from the urinary bladder and expelled from the respiratory tract by coughing (Fontan et al., 1975). Of greater importance is invasion of the central nervous system, which causes an encephalomyelitis. Entry to the central nervous system appears to be by migration along a peripheral nerve into the spinal cord, thence to the brain. Pain, paralysis, seizures, or coma may be produced, depending upon the location of the worm. Since eosinophils are found in the cerebrospinal fluid, the condition may be confused with eosinophilic meningitis caused by *Angiostrongylus cantonensis*. However, it differs from the latter in having a more severe, frequently fatal course with paralysis or coma and bloody or xanthochromic cerebrospinal fluid (Bunnag et al., 1970).

Invasion of the eye also occurs, probably by migration via the optic nerves, although penetration through the sclera may be possible (Tudor and Blair, 1971).

Diagnosis. The nonmigrating type of this infection must be differentiated from furunculosis or other localized bacterial processes; the migrating type must be distinguished from migrating larvae of hookworms (*vide supra*, p. 281) and cutaneous myiasis *(vide infra*, p. 690). In both types specific diagnosis should be confirmed after the worms have been removed and their characteristic features observed. Miyazaki (1960) recommended an intradermal test employing 0.5 ml of a 1:50,000 saline solution of an antigen prepared from larval or adult *Gnathostoma*, read in 15 minutes. The skin test probably has greatest value in supporting a clinical diagnosis when it is positive (Daengsvang, 1980a). Precipitin reaction with patient's serum is also specific. Clinically, the blood picture in patients with edema and leukocytosis in excess of 10,000 with high percentage of eosinophils is highly suggestive in enzootic areas.

Treatment. The only available treatment is incision of the lesion and removal of the larva. Surgical removal from the eye may save the patient's vision and prevent more serious consequences of migration to the brain (Bathrick et al., 1981). Effective chemotherapy for humans has not been found. For experimentally infected cats, ancylol (2, 6,-diodo-4-nitrophenol) given by subcutaneous injection has shown efficacy against the migrating larvae (Daengsvang, 1980b). Mebendazole in relatively high dosage has been recommended (Medical Letter, 1982).

Prognosis. Usually the prognosis is good except in cases involving the central nervous system.

Prevention. Man and reservoir hosts acquire the infection from ingestion of infected tissues of second intermediate hosts. In human infection the larvae must then migrate to the superficial tissues from the digestive tract. Percutaneous infection is possible. Larvae do not survive in well-cooked fish, and they can be killed by immersion in strong vinegar for 5 hours or longer. Immersion in lime juice or chilling at 4°C for 1 month is ineffective. In areas of potential exposure all uncooked freshwater fish and chicken should be proscribed.

Gnathostoma hispidum Fedtschenko, 1872

G. hispidum (syn. *Cheiracanthus hispidus)* is a relatively common parasite in the stomach wall of wild and domestic pigs in Europe and Asia and is reported for Australia (Heydon, 1929). It is stouter than *G. spinigerum*, has 12 rings of cephalic spines, has multidigitate cuticular spines extending to the caudal extremity, and has only one pair of small ventral alar papillae in the male. This worm has been found once in a progressive linear swelling of the left thenar eminence of a young female Japanese patient (Morishita, 1924), possibly twice as a human infection in India (Maplestone, 1929; Mukerji and Bhaduri, 1945), and once in a male Cantonese suffering from ocular gnathostomiasis (Chen, 1949).

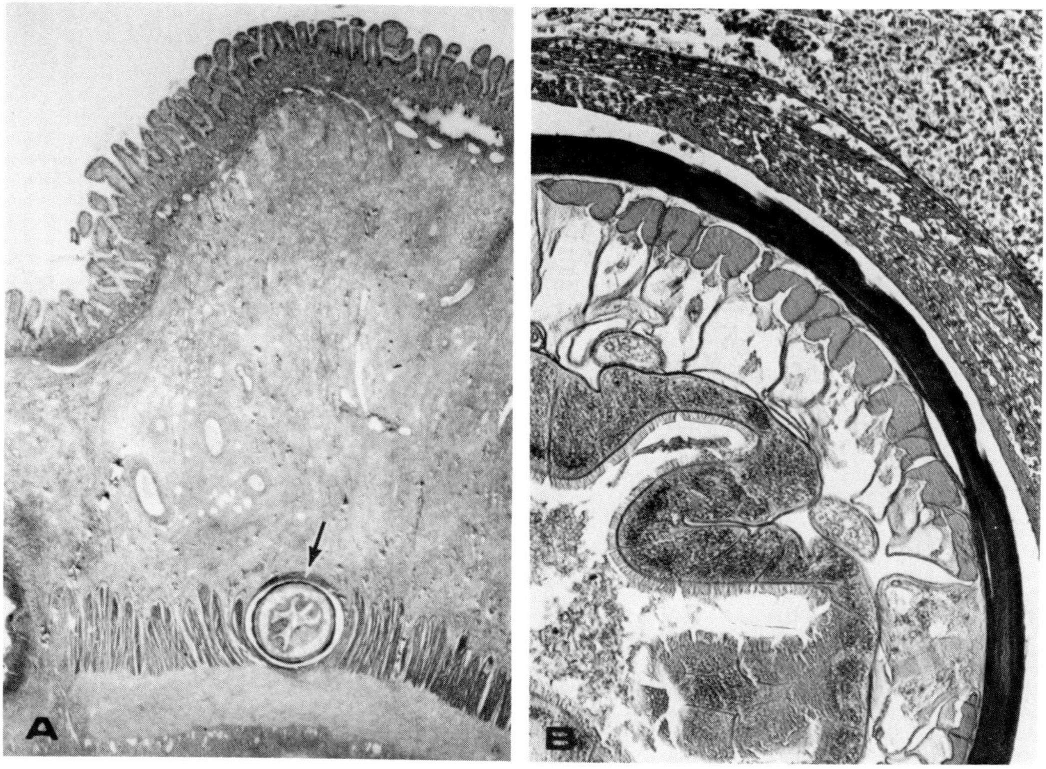

Fig. 22–6. *Gnathostoma spinigerum*. Larva in section of terminal ileum surgically removed from 46-year-old woman in Laos. *A*, Worm is between muscularis externa and edematous submucosa. Transverse section. (× 15.) *B*, Higher magnification showing characteristic thick cuticle, thin hypodermis, wide lateral cord, coelomyarian somatic muscles, and large, coarsely-granular intestinal cells with distinct brush border. Two sections of cord-like genital primordium are between intestine and muscles. (× 160.) (By P.C. Beaver.)

Physaloptera caucasica von Linstow, 1902 (Producing physalopteriasis

Synonym. *Physaloptera mordens* Leiper, 1907.

Historical and Geographical Notes. This worm was first described from the ileum of a patient in the Caucasus and later was found to be relatively common in natives of tropical Africa. Blackie (1932) reported one human case from Zimbabwe. In less than one year five cases of human physalopteriasis due to *P. caucasica* were diagnosed at Kinshasa, Zaire (Vandepitte et al., 1964). Faust and Martinez (1935) found eggs of a species of *Physaloptera* in the feces of a native Panamanian, but concluded that it was a case of spurious parasitism. Two cases of parasitism with *Physaloptera* sp. were reported from Colombia by Soriano-Lleras and Pan (1955), and one from India by Singh and Rao (1954). Witenberg (1951) reported eggs of this species in a patient's feces in Israel, and Hira (1976) reported the same in an adult in Zambia. The worm is also a normal parasite of African monkeys, which Leiper believed to be the reservoir hosts. The parasites live with their heads embedded in the wall of the digestive tract of the host, from the level of the esophagus to the ileum, but the liver is also at times parasitized. These nematodes are fairly large (males, 14 to 50 mm in length by 0.7 to 1 mm in diameter; females, 24 to 100 mm by 1.14 to 2.8 mm). Superficially they may be mistaken for immature ascarids, from which they may be distinguished by the cuticular collarette around the head, the pair (rather than triad) of fleshy lips surrounding the mouth, provided with a characteristic pattern of dental processes and papillas (Fig. 22–7,*A*,*B*,*C*), the asymmetrical bursa copulatrix of the male with its four pairs of long pedunculate papillae and six pairs (1 pair preanal, 5 pairs postanal) of sessile papillae (Fig. 22–7,*D*), the transversely bossed perianal cuticle of the male, and the unequal copulatory spicules with hooked distal ends. The vulva of the female is considerably pre-equatorial in position; the vagina is directed posteriad and branches at its outer end to form four uterine tubules, each with its oviduct and ovarian tubule. The smooth thick-shelled ovoid eggs (measuring 44 to 65 by 32 to 45 μm) are fully embryonated when laid (Fig. 22–7,*E*). The life cycle is unknown but Mönnig (1934) and Petter (1960) suggested that coprophagous beetles and cockroaches serve as intermediate hosts for the species of this genus. Second intermediate hosts, presumably vertebrates, may be either required or paratenic, and human infection may result from ingestion of infected first or other intermediate hosts. Of the many species of *Physaloptera* parasitizing vertebrates, *P. causasica* is the only form recorded from man.

Nicolaides et al. (1977) have reported gangrene of the distal small bowel due to infection with worms that were probably *Physaloptera*. The resected bowel showed ischemic necrosis, eosinophilic granulomata, and larval worms in the lumen and wall of blood vessels. In the cases studied by Vandepitte et al. (1964), symptoms included repeated vomiting, epigastric pain, and eosinophilia of the blood. Diagnosis was facilitated by using the formol-ether technique for concentration of feces and by

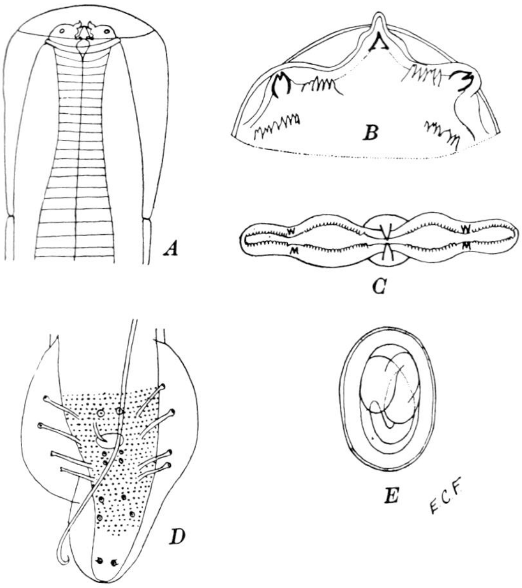

Fig. 22–7. *Physaloptera caucasica. A,* Anterior end of worm with integumentary collarette. (× 30.) *B,* Lateral view of inner lips, showing dental process. (× 210.) *C,* Head-on view of inner lip. (× 210.) *D,* Caudal end of male, showing asymmetrical alae, pedunculated and sessile papillae and copulatory spicules. (× 25.) *E,* Embryonated egg. (× 315.) (*A* adapted by E.C. Faust from Leiper, R.T. 1913. Observations on certain helminths of man. Trans. Soc. Trop. Med. Hyg., *6*. *B* adapted by E.C. Faust from Ortlepp, R.J. 1926. On the identity of *Physaloptera caucasica* v. Linstow, 1902, and *Physaloptera* Mordens Leiper, 1908. J. Helminthol., *4*. *C* and *E* adapted by E.C. Faust from Schulz, R.E. 1926. Sur la morphologie de *Physaloptera caucasica* v. Linstow, 1902, de l'homme. Ann. Parasitol., *4*. *D* adapted by E.C. Faust from von Linstow, O. 1902. Zwei neue Parasiten des Menschen. Cbl. Bakt. Parasitol., I. Abt. Orig., *31*.)

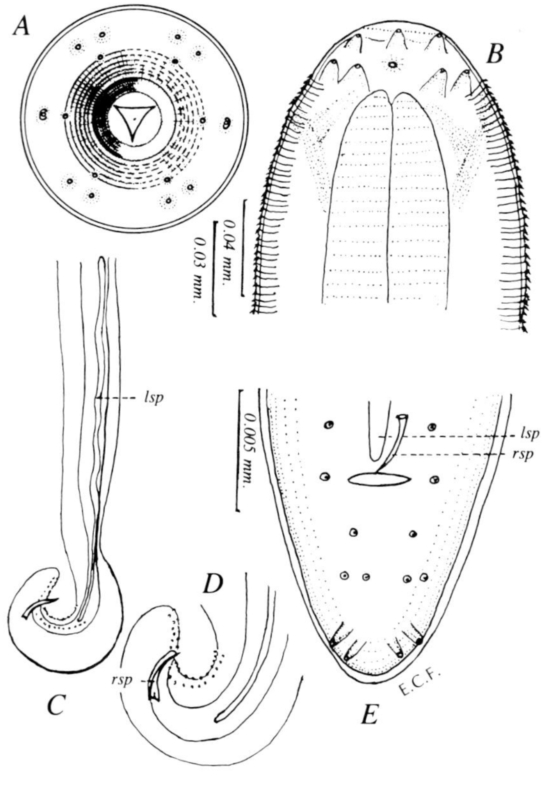

Fig. 22–8. *Thelazia callipaeda. A,* Head-on view of anterior end. *B,* Dorsal view of anterior end. *C,* Lateral view of posterior end of male. *D,* Extreme posterior tip of male, lateral view, showing papillae and right copulatory spicule. *E,* Posterior end of male, ventral view. *lsp,* left copulatory spicule; *rsp,* right copulatory spicule. (Adapted from Hsü, H.F. 1933. On *Thelazia callipaeda* Railliet and Henry, 1910. Infection in man and dog. Arch. f. Schiffs- u. Tropenhyg., *37*:363–369.)

examining the sedimented contents of the stomach and duodenum. Carbamazine derivatives seem to be of more use in treatment than other vermifuges.

Thelazia callipaeda Railliet and Henry, 1910 (The Oriental "eye worm," producing thelaziasis)

Morphology, Biology and Life Cycle. *Thelazia callipaeda* is a parasite of the conjunctiva of the dog, the rabbit, and man. It has been recovered from the dog in India, Burma, and in several localities in China; from the rabbit once in Peking, China, and from man at least seven times in China; twice from Korea; once from a 15-year-old girl in India; and once from Russia. Bhaibulaya et al. (1970) reported infections in two humans (a 4-month-old boy and a 47-year-old woman) and five dogs from northeast Thailand, and listed the cases reported earlier from Thailand and elsewhere. They also described the morphologically distinct features of *T. callipaeda* and *T. californiensis*. An infection in a 15-year-old boy not listed by Bhaibulaya et al. was reported from Thailand by Chaiyaporn and Phanich (1969). In Japan, Arizono et al. (1976) reported *Thelezia* infection in a 35-year-old man in Kyoto and in 11 of 292 dogs examined in the same area. According to Kagei et al. (1981) 30 human cases of thelaziasis have been reported in Japan, mostly in Kyushu. The species was originally described in the dog by Railliet and Henry in 1910.

The adult worms are thread-like, creamy-white, and range in size from 4.5 to 13 mm in length by 0.25 to 0.75 mm in diameter for males, and 6.2 to 17 mm in length by 0.3 to 0.85 mm in diameter for females. The oral end (Fig. 22–8,*A*) is provided with a round chitinoid capsule, an inner circlet of six sessile papillae, and an outer ring of four twinned, sessile papillae in addition to a pair of amphids. The male has a distinctly recurved posterior end (Fig. 22–8,*C, D*). The perianal papillae are variable in number; they consist of 6 to 10 pairs of sessile postanal papillae (Fig. 22–8,*D, E*). The vulva of the female worm opens ventrad a considerable distance pre-equatorially. The vagina is directed posteriad, as is the outer single stem of the uterus, which later bifurcates and is continued as twinned tubules into the oviducts and ovaries. The genital tubules are intricately coiled in the posterior half of the worm. The eggs are hyaline, thin-shelled and ovoid in contour (54 to 60 by 34 to 37 μm), fully embryonated when laid. Soon, however, the shell becomes greatly distended and spherical, with digital evagination on one side into which the larva crawls. Nothing is known of the larval development of *T. callipaeda*, but Krastin (1950) reported that *T. gulosa* and *T. skrjabini* of cattle use species of *Musca* as the intermediate host, and Burnett et al. (1957) found species of *Fannia* to be naturally and experimentally susceptible larval hosts. In reports of *Thelazia lacrymalis* in horse and *T. gulosa* and *T. skrjabini* in cattle in Kentucky, Lyons and Drudge (1975a, b) noted that the larvae of *Thelazia* species in the United States have been found in face flies.

Thelazia californiensis. In addition to *T. callipaeda,* nine species of the genus, all having a predilection for the conjunctiva, have been described from mammals and ten from birds. Kofoid and Williams (1935) recovered *T. californiensis* from one human subject in California, and Kofoid et al. (1937) reported a second human case from California. Five additional human infections (Hosford et al., 1942; Freidman, 1951; Lee and Parmelee, 1958) have been recorded from California. This species, also reported from dogs, cat, sheep, black bear, jack rabbit *(Lepus californicus),* coyote *(Canis latrans),* domestic fox, Columbian black-tailed deer *(Odocoileus memionis columbianus),* and mule deer, all from California, has six to seven asymmetrically disposed pairs of preanal papillae and normally three pairs of postanal papillae, as well as a pair of subterminal phasmids, and a gubernaculum associated with the pair of copulatory spicules.

Epidemiology. Factors in transmission have been inadequately studied.

Pathogenesis, Pathology, and Symptomatology. *Thelazia* produces damage to the tissues associated with the eye. Its presence in the conjunctival sac provokes excess flow of lacrimal fluid. Its repeated migration across the corneal conjunctiva irritates this layer and eventually results in scarification and fibrous opacity of the region.

The presence of adult worms in the conjunctival sac produces lacrimation. Excursions across the front of the cornea produce minute scratching of the surface, resulting in the development of opacities. Paralysis of the muscles of the lower eyelid, with ectropion, has been attributed to the worms. The presence and movement of the parasites in the conjunctival sac cause surprisingly mild symptoms consisting of excessive lacrimation, itching, or pain with the sensation of a foreign body in the eye. At times the pain may be severe.

Diagnosis. The presence of creamy-white threadworm masses coiled in the conjunctival sac or migrating over the cornea is suggestive. Removal with forceps and identification under the microscope are necessary for specific diagnosis.

Treatment. After application of topical anesthetic to the eye, worms may be easily removed with forceps. If all the worms are removed, symptoms promptly subside. If symptoms persist, the patient should be reexamined for the presence of worms overlooked in the first treatment.

Prognosis. On removal of the adult worms from the conjunctiva, irritating symptoms disappear, but superficial opacity of the cornea resulting from scarification produced by the worms may be permanent.

Prevention. This condition cannot be prevented until the life cycle has been elucidated. Since direct transmission from the reservoir host to man is unlikely, however, care not to ingest suspected arthropods or raw drinking water probably constitutes essential prophylaxis for the human host.

Gongylonema pulchrum Molin, 1857 (The scutate threadworm, producing gongylonemiasis

Synonyms. *Filaria labialis* Pane, 1864; *Gongylonema scutatum* (Leuckart, 1873) Railliet, 1892; *G. hominis* Stiles, 1921.

Historical and Geographical Notes. This thread-like nematode is a cosmopolitan parasite of ruminants, and has been diagnosed from pigs, bears, hedgehogs, monkeys, and occa-

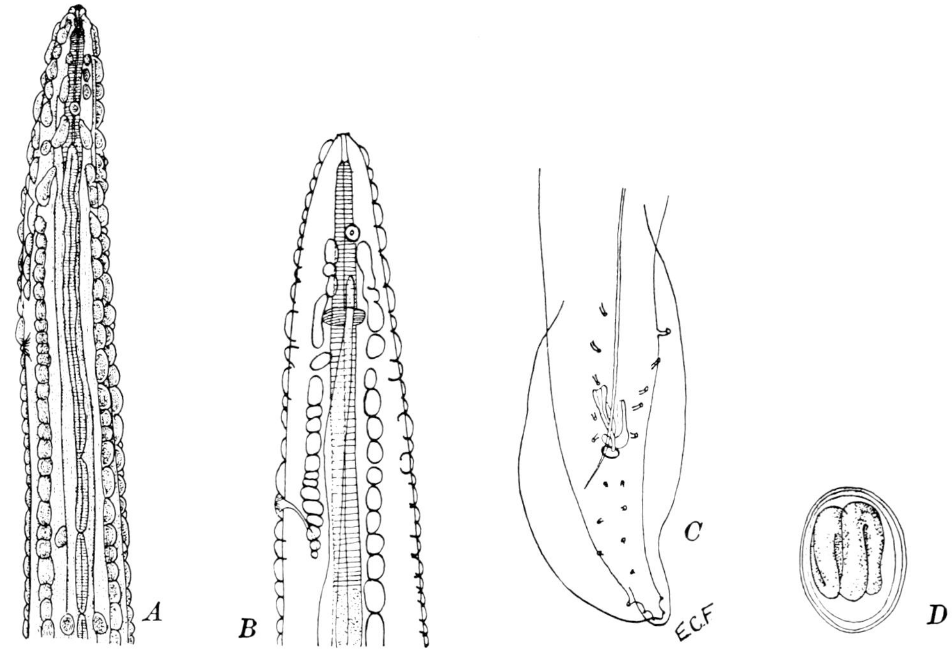

Fig. 22–9. *Gongylonema pulchrum. A*, Anterior end of worm from human host, lateral view, (× 76.) *B*, Anterior end of worm, lateral view. (× 80.) *C*, Posterior end of male, ventral view, showing alae, perianal and caudal papillae, spicules and gubernaculum. (× 80.) *D*, Egg with coiled larva. (× 375). (*A* adapted by E.C. Faust from Ward, H.B. 1916. *Gongylonema* in the role of a human parasite. J. Parasitol., *2*. *B* and *C* adapted by E.C. Faust from Baylis, H.A. 1925. On the species of *Gongylonema* (Nematoda) parasitic in ruminants. J. Comp. Pathol. Therap., *38*. *D* adapted by E.C. Faust from Fibiger, J. 1923. Die tierischen Parasiten der Haus- und Nutztiere sowie des Menschen. Vienna and Leipzig.)

sionally from man. It was first recovered as a human parasite in Italy by Pane in 1864. Since that time at least 28 additional infections in man have been reported, chiefly from Europe, Morocco, U.S.S.R., China, New Zealand, Sri Lanka, and the United States.

Morphology, Biology, and Life Cycle. In ruminants, probably the optimum hosts, males of *G. pulchrum* reach a length of 62 mm by 0.15 to 0.3 mm in diameter, while females are much larger (145 mm in maximum length by 0.2 to 0.5 mm in diameter). In man and in the pig the worms are considerably smaller. The anterior end (Fig. 22–9,*A*) is covered by numerous bosses or scutes, which are typically arranged in eight longitudinal series, two each in the four submedian fields. In addition, there is a pair of small lateral cervical papillae (Fig. 22–9,*B*), one on each side, about 0.1 to 0.2 mm from the anterior extremity. Just posterior to these papillae is a pair of cervical alae, which extend distal almost as far as the scutes. The excretory pore opens into a small crater-like prominence near the junction of the anterior (muscular) and the posterior (glandular) portion of the esophagus. The caudal end of the male is asymmetrically alate, there is a typical perianal pattern of papillae, the copulatory spicules are distinctly dissimilar, and the gubernaculum resembles a midiron with a short handle (Fig. 22–9,*C*). The caudal end of the female is asymmetrically bluntly conical. The vulva is thick-walled, slightly protuberant, and is situated some distance in front of the subterminal anus. The vagina is long, extending cephalad to the equatorial plane. The transparent, thick-shelled, broadly ovoid eggs (50 to 70 by 25 to 37 μm) are embryonated when laid (Fig. 22–9,*C*). They require ingestion by various species of dung beetles or cockroaches in the intestine of which they hatch. They then burrow into the hemocoel of these insect hosts and are encapsulated. The definitive host becomes infected by swallowing an infected insect. The larvae then become freed of their capsules and probably migrate within the wall of the stomach or duodenum up to the esophagus or buccal cavity.

Epidemiology. Human infection is acquired by ingestion of infected insects that serve as intermediate hosts.

Pathogenesis, Pathology, and Symptomatology. In man the damage produced by *Gongylonema* is due to maturing and adult worms migrating through the mucosa and submucosa of the buccal cavity. In animals the worms are found in long, raised, sinuate channels in the mucosa and submucosa of the esophagus, the buccal cavity, and the tongue. In the reported human cases the lips, gums, hard and soft palate, tonsil, and angle of the jaw have been the sites from which the worms were recovered. In most instances they were seen actively migrating in the submucosa. The patients were conscious of their presence and in at least one instance the patient himself removed the worm by applying slight traction with his fingers. In one patient an accompanying pharyngitis and stomatitis may have been produced by the parasites; in the others, only local irritation resulted. In two cases studied by Feng et al. (1955) there was repeated spitting of blood. In one of these patients esophagoscopy revealed erosive patches on the esophageal mucosa, from which blood was oozing. Biopsy sections revealed eggs of *G. pulchrum* in tunnels of the mucosa, similar to those recovered from expectorated blood. This clinical study suggests that gongylonemiasis in man typically involves the esophageal mucosa. There is no evidence that this species produces neoplasms such as are attributed to *Gongylonema neoplasticum* and *G. orientale*.

Diagnosis. Discovery of migrating thread-like worms in the buccal mucosa suggests this species. Specific identification requires examination of the worms under the microscope.

Treatment. The worms can be removed either with the fingers or by a needle inserted under a portion of the worm in its tunnel.

Prevention. All the human infections were probably acquired from accidentally swallowing cockroaches or other infected insects.

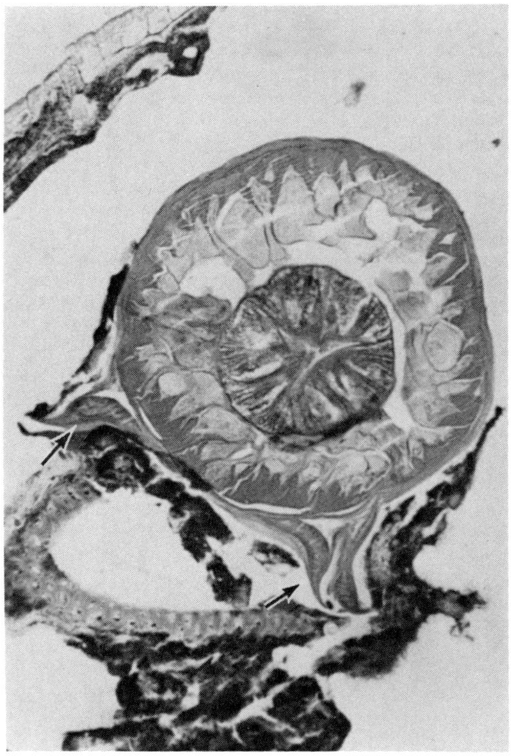

Fig. 22–10. *Rictularia*. Transverse section at level of esophagus of mature female of unidentified species in section of appendix of 90-year-old man in New York City. Subventral cuticular combs are evident (arrows). (× 40.) (From Kenny, M., Eveland, L.K., Yermakov, V., and Kassouny, D.Y. 1975. A case of *Rictularia* infection of man in New York. Am. J. Trop. Med. Hyg., *24*:596–598.)

Genus Rictularia
Froelich, 1802

Species of the genus *Rictularia* are generally parasites of rodents and bats. A distinctive morphological feature of the adult worm is the presence of two subventral rows of spines or cuticular combs. The eggs are thick-shelled and fully embryonated when deposited. Infection in man has been reported once (Kenney et al., 1975). In a routine postmortem examination of a 90-year-old man in New York City, a mature female was found in histopathologic sections of the appendix (Fig. 22–10).

Genus Cheilospirura
(Diesing, 1861)

A species of spiruroid worm (*Cheilospirura* sp.) belonging to the family Acuariidae was reported by Africa and Garcia (1936) from an ovoid tumor. Situated on the lower palpebral conjunctiva, the tumor was about 1 cm from the external canthus of the right eye of a Philippine farmer suffering from a chronic catarrhal conjunctivitis and keratitis.

Genus Spirocerca
(Railliet and Henry, 1911)

Male and female specimens of the species *Spirocerca lupi* (Rudolphi, 1809) were recovered post mortem from a somewhat premature infant (8 months) in Naples. The mother was symptomless until the sixth month of pregnancy. On the second postnatal day the infant was diagnosed clinically as having an intestinal obstruction, located by x-ray film in the terminal ileum, with complication of peritonitis. Exploratory laparotomy confirmed these findings. The baby died on the twelfth day. Autopsy revealed five nematodes that were 26 to 34 mm long by 2.5 mm in diameter and reddish, embedded in pockets of the terminal ileum. Examination demonstrated that the worms were mature and that the females contained many typical eggs characteristic of *Spirocerca lupi* (Biocca, 1959).

It was believed that the infection was acquired from the mother some time before the sixth month of pregnancy. The mother must have ingested infected coprophagous beetles, which are the intermediate hosts of this spiruroid worm, and the larvae must have migrated transplacentally to the fetus, in the digestive tissues of which they developed to maturity.

Spirocerca lupi is prevalent in dogs, the natural definitive hosts not only in the Mediterranean area but in the countries to the east as far as China and elsewhere throughout the world. In view of this prevalence, it is surprising that previous human infection has not been observed, since human exposure from ingestion of infected beetles must have occurred repeatedly throughout the enzootic area.

REFERENCES

Africa, C.M., and Garcia, E.Y. 1936. A new nematode parasite *(Cheilospirura* sp.) of the eye of man in the Philippines. J. Philip. Ids. Med. Assoc., *16*:603–607.

Arizono, N., Yoshida, Y., Kondo, K., Kurimoto, H., Oda, K., Shiota, T., Shimada, Y., Ogino, K., and Furuta, M. 1976. *Thelazia callipaeda* from man and dogs in Kyoto and its scanning electron microscopy. Jpn. J. Parasitol., *25*:402–408.

Bathrick, M.E., Mango, C.A., and Mueller, J.F. 1981. Intraocular gnathostomiasis. Ophthalmology, *88*:1293–1295.

Belcher, D.W., Wurapa, F.K., Ward, W.B., and Lourie, I.M. 1975. Guinea-worm in southern Ghana: Its epidemiology and impact on agricultural productivity. Am. J. Trop. Med. Hyg., *24*:243–249.

Beverley-Burton, M., and Crichton, V.F.J. 1976. Attempted experimental cross infections with mammalian guinea-worms, *Dracunculus* spp. (Nematoda: Dracunculoidea). Am. J. Trop. Med. Hyg., *25*:704–708.

Bhaibulaya, M., Prasertsilpa, S., and Vajrasthira, S. 1970. *Thelazia callipaeda* Raillief and Henry, 1910, in man and dog in Thailand. Am. J. Trop. Med. Hyg., *19*:476–479.

Biocca, E. 1959. Infestazione umana prenatale da *Spirocerca lupi* (Rud. 1809). Parassitologia, *1*:137–142.

Blackie, W.K. 1932. A helminthological survey of southern Rhodesia. Memoir Ser. No. 5, London School of Hygiene and Tropical Medicine.

Bunnag, T., Comer, D.S., and Punyagupta, S. 1970. Eosinophilic myeloencephalitis caused by *Gnathostoma spinigerum*. Neuropathology of nine cases. J. Neurol. Sci., *10*:419–434.

Burnett, H.S., Parmelee, W.E., Lee, R.D., and Wagner, E.D. 1957. Observations on the life cycle of *Thelazia californiensis* Price, 1930. J. Parasitol., *43*:433.

Chabaud, A.G. 1975. Keys to genera of the Order Spirurida. Part 2. Spiruroidea, Habronematoidea and Acuarioidea. CIH keys to the nematode parasites of vertebrates. No. 3. Commonwealth Institute of Helminthology, St. Albans, U.K.

Chaiyaporn, V., and Phanich, V. 1969. Thelaziasis in man in Thailand. J. Parasitol., *55*:941.

Chandler, A.C. 1925. A contribution to the life-history of a gnathostome. Parasitology, *17*:237-244.

———. 1927. The prevalence and epidemiology of hookworm and other helminthic infections in India. Part VI. Burma. Indian J. Med. Res., *14*:733–744.

Chen, H.T. 1949. A human ocular infection by *Gnathostoma* in China. J. Parasitol., *35*:431–433.

Crichton, V.F.J., and Beverley-Burton, M. 1973. *Dracunculus lutrae* n. sp. (Nematoda: Dracunculoidea) from the otter, *Lutra canadensis,* in Ontario, Canada. Can. J. Zool., *51*:521–529.

———. 1974. Distribution and prevalence of *Dracunculus* spp. (Nematoda: Dracunculoidea) in mammals in Ontario. Can. J. Zool., *52*:163–167.

———. 1975. Migration, growth, and morphogenesis of *Dracunculus insignis* (Nematoda: Dracunculoidea). Can J. Zool., *53*:105–113.

Daengsvang, S. 1971. Infectivity of *Gnathostoma spinigerum* larvae in primates. J. Parasitol., *57*:476–578.

———. 1980a. *A Monograph on the Genus* Gnathostoma *& Gnathostomiasis in Thailand.* Tokyo, Southeast Asian Medical Information Center.

———. 1980b. Chemotherapy of feline *Gnathostoma spinigerum* migration stage with multiple subcutaneous doses of ancylol. Southeast Asian J. Trop. Med. Public Health, *11*:359–362.

———, Sermswatsri, B., Youngyi, P., and Guname, D. 1970. Development of adult *Gnathostoma spinigerum* in the definitive host (cat and dog) by skin penetration of the advanced third-stage larvae. Southeast Asian J. Trop. Med. Public Health, *1*:187–192.

———, and Tansurat, P. 1938. A contribution to the knowledge of the second intermediate hosts of *Gnathostoma spinigerum* Owen, 1836. Ann. Trop. Med. Parasitol., *32*:137–140.

———, Thienprasitthi, P., and Chomcherngpat, P. 1966. Further investigations on natural and experimental hosts of larvae of *Gnathostoma spinigerum* in Thailand. Am. J. Trop. Med. Hyg., *15*:727–729.

Donaldson, J.R., and Angelo, T.A. 1961. Quadriplegia due to guinea-worm abscess. J. Bone Joint Surg., *43–A*:197–198.

Fairley, N.H., and Liston, W.G. 1925. *Studies on Guinea Worm Disease*. Collected papers from Indian J. Med. Res. and Indian Med. Gaz., 1924. Calcutta, Thacker, Spink & Co.

Faust, E.C., and Martinez, W.H. 1935. Notes on helminths from Panama. II. Rare human nematode eggs in the feces of individuals from the Chagres River, Panama. J. Parasitol., *21*:331–336.

Fedchenko (Fedtchenko), A.P. 1971. Concerning the structure and reproduction of the guinea-worm *(Filaria medinensis* L.). Am. J. Trop. Med. Hyg., *20*:511–523.

Feng, L.C., Tung, M.S., and Su, S.C. 1955. Two Chinese cases of *Gongylonema* infection. A morphological study of the parasite and clinical study of the cases. Chinese Med. J., *73*:149–162.

Fontan, R., Beauchamp, F., and Beaver, P.C. 1975. Sur quelques helminthiasis humaines nouvelles au Laos. I. Nematodes. II. Plathelminthes. Bull. Soc. Pathol. Exot., *68*:557–573.

Friedman, M. 1951. Thelaziasis der Conjunctiva, ein Nachtrat. Ophthalmologica, *122*:252–254.

Heydon, G.M. 1929. Creeping eruption or larva migrans in North Queensland and a note on the worm *Gnathostoma spinigerum*. Med. J. Aust., *1*:583–590.

Hira, P.R. 1976. Observations on helminthozoonoses in Zambia. East Afr. Med. J., *53*:278–286.

Hosford, G.N., Stewart, M.A., and Sugarman, E.I. 1942. Eye

worm *(Thelazia californiensis)* infection in man. Arch. Ophthalmol., *27*:1165–1170.

Johnson, S., and Joshi, V. 1982. Dracontiasis in Rajasthan. VI. Epidemiology of Dracontiasis in Barmer District, Western Rajasthan, India. Int. J. Epidemiol., *11*:26–30.

Kagei, N., Hayashi, S., Ishida, T., Yamaguti, T., Asami, K., and Takeuchi, T. 1981. A case of thelaziasis callipaeda in Tokyo, Japan. Jap. J. Parasitol., *30*:337–344 (In Japanese, English abstract.)

Kale, O.O. 1977. The clinico-epidemiological profile of guinea-worm in the Ibadan district of Nigeria. Am. J. Trop. Med. Hyg., *26*:208–214.

Kenney, M., Eveland, L.K., Yermakov, V., and Kassouny, D.Y. 1975. A case of *Rictularia* infection of man in New York. Am. J. Trop. Med. Hyg., *24*:596–598.

Kinare, S.G., Parulkar, G.B., and Sem, P.K. 1962. Constrictive pericarditis resulting from dracunculosis. Br. Med. J., *1*:845.

Kofoid, C.A., and Williams, O.L. 1935. The nematode *Thelazia californiensis* as a parasite of the eye of man in California. Arch. Ophthalmol., *13*:176–180.

———, Williams, O.L., and Veale, N.C. 1937. *Thelazia californiensis,* a nematode eye worm of dog and man, with a review of the thelazias of domestic animals. Univ. Calif. Publ. Zool., *41*:225–234.

Krastin, N.I. 1950. A study of the developmental cycle of the nematode *Thelazia gulosa* (Railliet & Henry, 1910), a parasite of the eyes of cattle. Dokl. Akad. Nauk. S.S.S.R., n.s., *70*:549–551.(Russian text; abstract in Helm. Abstr., *19*:90–91, 1951.)

Lee, R.D., and Parmelee, W.E. 1958. Thelaziasis in man. Am. J. Trop. Med. Hyg., *7*:427–428.

Le-Van-Hoa. 1965. Etude d'un nouveau gnathostome *G. vietnamicum* n. sp., chez *Lutra elioti* du Viet-Nam. Bull. Soc. Pathol. Exot., *58*:228–235.

Lindberg, K. 1946a. Enquete epidemiologique sur la dracunculose dans un village de Deccan (Inde). Bull. Soc. Pathol. Exot., *39*:303–318.

———. 1946b. Dracunculose dans l'état de Djodhpour (Radjpoutana), Inde. Bull. Soc. Pathol. Exot., *39*:318–328.

Lyons, E.T., and Drudge, J.H. 1975a. Occurrence of the eyeworm, *Thelazia lacrymalis,* in horses in Kentucky. J. Parasitol., *61*:1122–1124.

———. 1975b. Two eyeworms, *Thelazia gulosa* and *Thelazia skrjabini,* in cattle in Kentucky. J. Parasitol., *61*:1119–1122.

Maplestone, P.A. 1929. A case of human infection with a gnathostome in India. Indian Med. Gaz., *64*:273–285.

Medical Letter. 1982. *Handbook of Antimicrobial Therapy.* New Rochelle, New York, The Medical Letter.

Mirza, M.B. 1932. Dracontiasis (naru) in Shorapur. Proc. Muslim Assoc. Adv. Sci., November 1932, pp. 43–47.

Miyazaki, I. 1954. Studies on *Gnathostoma* occurring in Japan (Nematoda: Gnathostomidae). II. Life history of *Gnathostoma* and morphological comparison of its larval forms. Kyushu Mem. Med. Sci., *5*:123–140.

———. 1960. On the genus *Gnathostoma* and human gnathostomiasis, with special reference to Japan. Exp. Parasitol., *9*:338–370.

Mönnig, H.O. 1934. *Veterinary Helminthology and Entomology,* p. 235. Baltimore, Wm.Wood & Co.

Moorthy, V.N. 1937. A redescription of *Dracunculus medinensis.* J. Parasitol., *23*:220–224.

Morishita, K. 1924. A pig nematode, *Gnathostoma hispidum* Fedtschenko, as a human parasite. Ann. Trop. Med. Parasitol., *18*:23–26.

Morita, T. 1955. A study of the third-stage larva of *Gnathostoma spinigerum.* I. Morphological features. Igaku Kenkyu, *25*:411–431. (Japanese with English abstract.)

Mukerji, A.K., and Bhaduri, N.V. 1945. Gnathostome infection of the eye. Indian Med. Gaz., *80*:126–128.

Muller, R. 1968. Studies on *Dracunculus medinensis* (Linnaeus). I. The early migration route in experimentally infected dogs. J. Helminthol., *42*:331–338.

———. 1970. Laboratory experiments on the control of *Cyclops* transmitting guinea worm. Bul. W.H.O., *42*:563–567.

———. 1971. The possible mode of action of some chemotherapeutic agents in guinea worm disease. Trans. R. Soc. Trop. Med. Hyg., *65*:843–844.

———. 1972. Maintenance of *Dracunculus medinensis* (L.) in the laboratory and observations on experimental infections. Parasitology, *64*:107–116.

———. 1979. Guinea worm disease: Epidemiology, control and treatment. Bull. W.H.O., *57*:683–689.

Nicolaides, N.J., Musgrave, J., McGuckin, D., and Moorhouse, D.E. 1977. Nematode larvae (Spirurida: Physalopteridae) causing infarction of the bowel in an infant. Pathology, *9*:129–135.

Pardanani, D.S., Trivedi, V.D., Joshi, L.G., Daulatram, J., and Nandi, J.S. 1977. Metronidazole ('Flagyl') in dracunculiasis: A double blind study. Ann. Trop. Med. Parasitol., *71*:45–52.

Peláez, D., and Pérez-Reyes, R. 1970. Gnatostomiasis humana en América. Rev. Lat.-Am. Microbiol., *12*:83–91.

Petter, A.J. 1960. La blatte germanique *(Blattella germanica* L.), hôte intermédiaire probable d'*Abbreviata caucasica* (Linstow, 1902), nématode parasite des primates et de l'homme. C.R. Soc. Biol., *114*:87.

Prommas, C., and Daengsvang, S. 1933. Preliminary report of a study of the life cycle of *Gnathostoma spinigerum.* J. Parasitol., *19*:287–292.

———. 1936. Further report of a study on the life cycle of *Gnathostoma spinigerum.* J. Parasitol., *22*:180–186.

———. 1937. Feeding experiments of cats with *Gnathostoma spinigerum* larvae obtained from the second intermediate host. Indian Med. Gaz., *23*:115–116.

Raffier, G. 1969. Efficacy of thiabendazole in the treatment of dracunculiasis. Texas Rep. Biol. Med., *27*(Suppl.): 601–609.

Railliet, A., and Henry, A. 1910. Nouvelles observations sur les thélazies, nématodes parasites de l'oeil. C.R. Soc. Biol., *48*:783.

Reddy, C.R.R.M., and Valli, V.V. 1967. Extradural guinea-worm abscess. Am. J. Trop. Med. Hyg., *16*:23–25.

Sahba, G.H., Arfaa, F., Fardin, A., and Ardalan, A. 1973. Studies on dracontiasis in Iran. Am. J. Trop. Med. Hyg., *22*:343–347.

Sastry, S.C., Kumar, K.J., and Lakshminarayana, V. 1978. The treatment of dracontiasis with thiabendazole. J. Trop. Med. Hyg., *81*:32–35.

Shafei, A.Z. 1976. Preliminary report on the therapeutic effect of mebendazole in guinea worm infection. J. Trop. Med. Hyg., *79*:197–200.

Sharma, V.P., Rathore, H.S., and Sharma, M.M. 1979. Efficacy of metronidazole in dracunculiasis. A clinical trial. Am. J. Trop. Med. Hyg., *28*:658–660.

Singh, S.N., and Rao, V.R. 1954. On a species of *Physaloptera* causing a subcutaneous abscess in the neck of an Indian. J. Helminthol., *28*:155–158.

Soriano-Lleras, A., and Pan, C. 1955. Two cases of *Physaloptera* infection in man from Colombia. J. Parasitol., *41*:635.

Tamura, H., 1921. On creeping disease. Br. J. Dermatol. Syphil., *33*:81–84.

Tudor, R.C., and Blair, E. 1971. *Gnathostoma spinigerum*: An unusual cause of ocular nematodiasis in the Western Hemisphere. Am. J. Ophthalmol., *72*:185–190.

Vandepitte, J., Michaux, J.L., Fain, A., and Gatti, F. 1964.

Premières observations congolaises de physaloptérose humaine. Ann. Soc. Belg. Med. Trop., *44*:1067–1076.
Verma, A.K. 1966. Ocular dracunculosis. J. Indian Med. Assoc., *47*:188.
Witenberg, G. 1951. Some unusual observations on helminthiasis in Israel. Harefuah, *41*:178–180.
Yogore, M.G., and Juliano, S.J. 1951. Report of a case of human gnathostomiasis. Acta Med. Philip., *7*:239–247.

Chapter 23

The Filariae

Adult filariae (superfamily Filarioidea) live in the tissues or body cavities of a vertebrate host. Knowledge of their transcuticular uptake of nutrients and their metabolism is incomplete (Barrett, 1983). The females produce microfilariae that are unique because they are less differentiated than are the first-stage larvae of other nematodes. The microfilaria is essentially a highly motile, thread-like prelarva that in some species retains the egg membrane as a sheath, i.e., it becomes a "sheathed" form, whereas in others the microfilaria ruptures the egg membrane to become a naked "unsheathed" form. The egg membrane is shed, or not, usually while the microfilariae are still in the uterus (Fig. 23–1). On being extruded by the female, the microfilariae enter the blood or lymphatic vessels, and while circulating in peripheral blood or moving about in the cutaneous tissues they are ingested by blood-sucking arthropods. In an appropriate arthropod the microfilaria migrates through the wall of the digestive tract into the hemocoel, then into a specific suitable location (usually the thoracic muscles, malpighian tubules or fat bodies) for development through two distinct larval stages (first- and second-stage larvae) before reaching the infective stage, which is the third-stage larva. The infective-stage larva then migrates to the mouthparts and escapes into or onto the vertebrate host's skin when the arthropod host takes a blood meal.

Of the numerous species of the superfamily Filarioidea, approximately 100 have been studied with respect to their arthropod intermediate hosts, which include mosquitoes, various gnats, midges, flies, fleas and lice (Schacher, 1973). In species of some of the more primitive groups, eggs containing embryos are taken up by non-blood-sucking intermediate hosts, but none of these species is among the parasites of man. A critical period in the life cycle of the filaria is that concerned with its development in the arthropod host. Possibly most filariae produce some damage to the tissues of the vector, during either development, migration, or emergence from the mouthparts.

The species of filariae that are relatively common as parasites of man are the following:

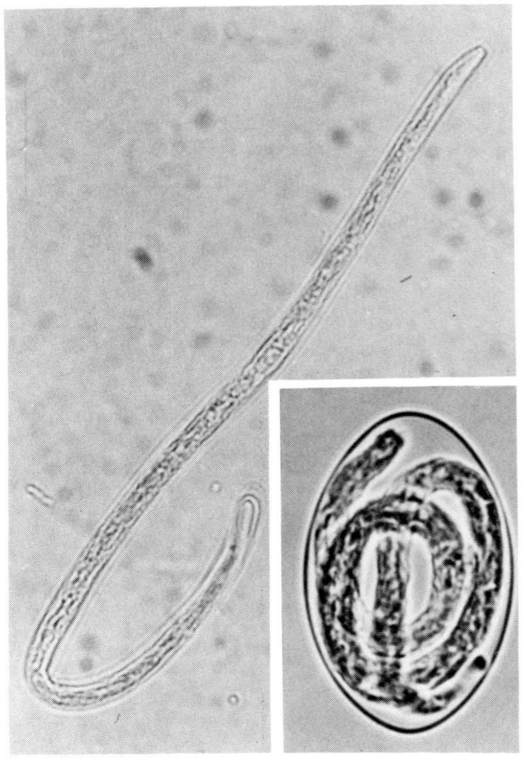

Fig. 23–1. Microfilariae from the ruptured uterus of a *Loa loa* that was taken from the conjunctiva of a resident of Nigeria. The coiled form in the egg membrane was from the middle part of the uterus. (× 700.) In the uncoiled microfilaria, from the anterior part of the uterus, the egg membrane has transformed into a sheath, most clearly evident at the flexed tip of the tail. (× 470.) (By P.C. Beaver.)

1. *Wuchereria bancrofti,* which lives in the lymphatics, produces "sheathed" microfilariae that exhibit nocturnal periodicity (except in the South Pacific), and is widely distributed throughout the tropics and subtropics.
2. *Brugia malayi,* which also lives in the lymphatics, produces "sheathed" microfilariae that in some areas exhibit nocturnal periodicity. It is an important parasite in eastern Asia, southwestern Pacific Islands, parts of India, and elsewhere. A closely similar species, *Brugia timori,* is found in people on the island of Timor and on neighboring islands.
3. *Onchocerca volvulus,* which lives in subcutaneous nodules, produces "unsheathed" microfilariae that distribute themselves in the cutaneous lymphatics (rarely are they found in the blood). It is widely distributed in tropical Africa, North Yemen, parts of Mexico, Guatemala, Venezuela, Brazil, Colombia, and Ecuador.
4. *Loa loa,* which lives in the cutaneous and subcutaneous tissues, produces "sheathed" microfilariae that exhibit diurnal periodicity, and is limited in distribution to tropical Africa.
5. *Mansonella* (syn. *Dipetalonema) perstans,* which lives in the body cavities, produces microfilariae that are "unsheathed" and nonperiodic, and occurs in tropical Africa and tropical South America.
6. *Mansonella* (syn. *Dipetalonema) streptocerca,* which lives in the skin and subcutaneous tissues, produces "unsheathed," nonperiodic microfilariae, and is found only in tropical Africa.
7. *Mansonella ozzardi,* which lives in subcutaneous tissues and possibly the body cavities, produces "unsheathed," nonperiodic microfilariae, and is found only in tropical America.

In addition, numerous filariae of animals are occasionally found in man in various parts of the world.

Wuchereria bancrofti
(Cobbold, 1877) Seurat, 1921
(Bancroft's filarial worm, producing Bancroft's filariasis, or filariasis bancrofti)

Synonyms. *Filaria sanguinis hominis* of Bush, 1872; *Filaria bancrofti* Cobbold, 1877; *Filaria nocturna* Manson, 1891; *Wuchereria pacifica* Manson-Bahr, 1941.

Historical and Geographical Notes. Ancient Hindu (Susruta, 600 B.C.) and Persian physicians (Rhazes, Avicenna) recognized the dramatic symptoms accompanying Bancroft's filariasis, which was referred to as elephantiasis arabicum. In 1863 Demarquay first demonstrated the microfilariae in hydrocele fluid and Wucherer (1866), in Brazil, first discovered their presence in chylous urine. In 1872 Lewis, in India, found the microfilariae in peripheral blood. Adult females were first recovered by the elder Bancroft in Brisbane (1876–1877), and adult males by Sibthorpe (Bourne, 1888). Patrick Manson first demonstrated that *Culex quinquefasciatus* (syn. *C. pipiens fatigans, C. p. quinquefasciatus, C. fatigans)* was an intermediate host (1878), and first described the nocturnal swarming of the microfilariae in peripheral circulation and diurnal concentration in the lungs (1879). Sir Philip Manson-Bahr (1912) and O'Connor (1923) demonstrated that in natives of Polynesia, strict nocturnal periodicity did not occur.

Bancroft's filariasis is indigenous throughout practically all the warm regions of the world, from about 41°N to about 28°S latitude in the Eastern Hemisphere and from about 30°N to 30°S in the Western Hemisphere (Fig. 23–2). In Europe, it was formerly confined to Barcelona, Hungary, Yugoslavia, and Turkey; it remains only in Turkey. In the Far East it extends coastwise from Shandong (Shantung) Province, China and southern Japan in the north, through the Ryukyu Islands, Vietnam, Thailand, Malaysia, Philippines, Borneo, Celebes, New Guinea, Sumatra, Flores, Soemba, Timor, Western New Guinea (Irian), the lesser islands of Indonesia, and throughout southern Micronesia. Highly endemic areas are on the Samoan, Ellice, Tokelau, and Fiji Islands in the South Pacific region. The disease has extensive distribution in Burma and India, is found around the coast of Saudi Arabia, Malagasy, Mauritius, Réunion. Comores (common), has limited foci in Egypt, is rare in Morocco, is prevalent in humid coastal regions of West Africa and tropical East Africa, is prevalent in equatorial Africa and at the mouth of the Congo, and is rare in South Africa. The early records in Korea and probably elsewhere in eastern Asia refer to *Brugia malayi,* which was first specifically distinguished in 1927. In the Western World, *W. bancrofti* is common along the coast in northern South America (northeast Brazil, the Guianas, Venezuela), but apparently does not extend far inland in the Guianas or Panama. It is common in the Greater and Lesser Antilles, and in parts of Central America along the coastal areas. It is possible, though doubtful, that autochthonous infections persist in Venezuela and Colombia. The periodic form has disappeared from the mainland of Australia, from North America, and from some of the islands in the Caribbean, but may be invading new areas in the growing towns of Asia, Africa, and South America.

Morphology, Biology, and Life Cycle. The adult worms are minute and thread-like in form, with a smooth cuticula. Although tapering towards both ends, their terminations are bluntly rounded. The head is slightly swollen and is surmounted by two rings of small sessile papillae (Fig. 23–3,*A*). The mouth is unarmed; a buccal vestibule is lacking. The male measures about 40 mm in length by 0.1 mm in diameter. Its caudal extremity (Fig. 23–3,*B*) is curved sharply ventrad. There are up to 12 pairs of perianal pedunculated papillae (8 preanal and 4 postanal) supporting narrow inconspicuous alae, and further caudad two pairs and one pair, respectively, of larger sessile papillae. The copulatory spicules (Fig. 23–3,*C*) are distinctly unequal and dissimilar; the gubernaculum is crescent-shaped. The female measures 80 to 100

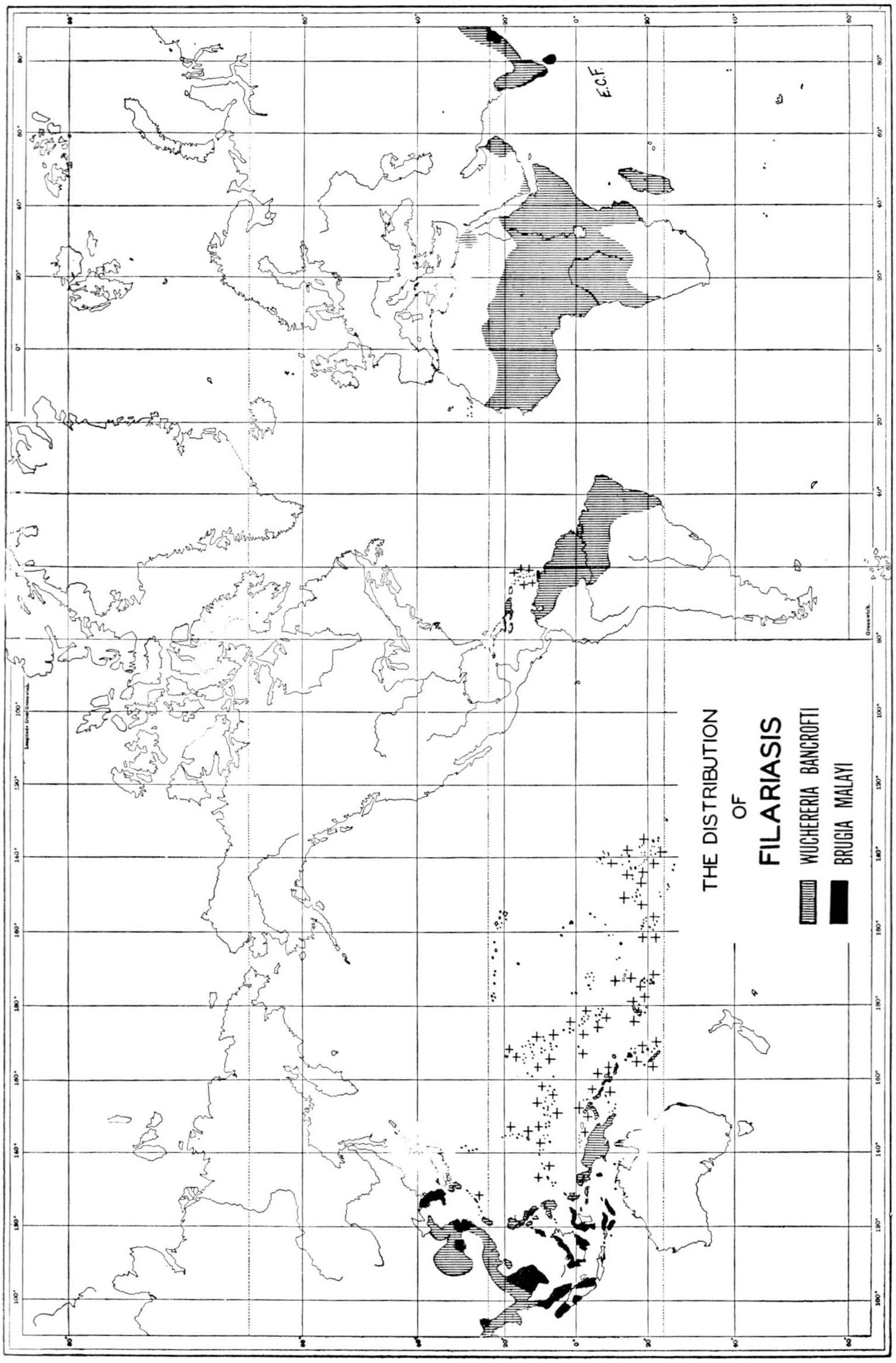

Fig. 23–2. World distribution of *Wuchereria bancrofti* and *Brugia malayi* (including periodic and subperiodic strains of both species). "+" indicates presence of one or the other species in Polynesia and the West Indies. (Adapted by P.C. Beaver from original of E.C. Faust.)

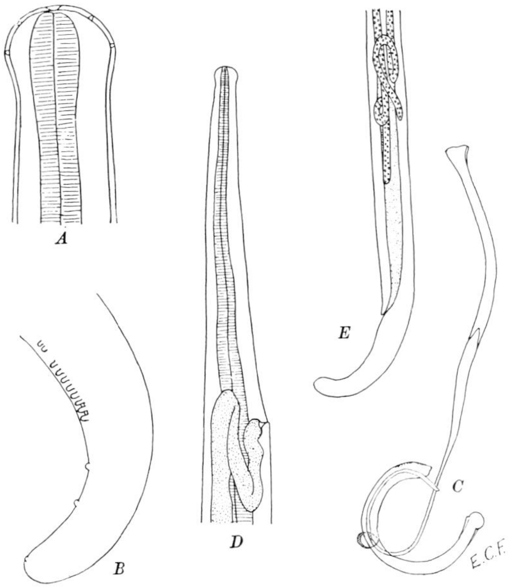

Fig. 23–3. *Wuchereria bancrofti. A,* Anterior end, ventral view showing distribution of papillae. (× 300.) *B,* Posterior end of male, lateral view, showing caudal papillae. *C,* Spicules and gubernaculum, highly magnified. *D,* Anterior end of female, lateral view, showing position of vulva. *E,* Posterior end of female, lateral view, showing loops of genital tube above and intestine extending to anus. (× 70.) (*A-E* adapted by E.C. Faust from Yorke, W., and Maplestone, P.A. 1926. *The Nematode Parasites of Vertebrates.* London, J. & A. Churchill.)

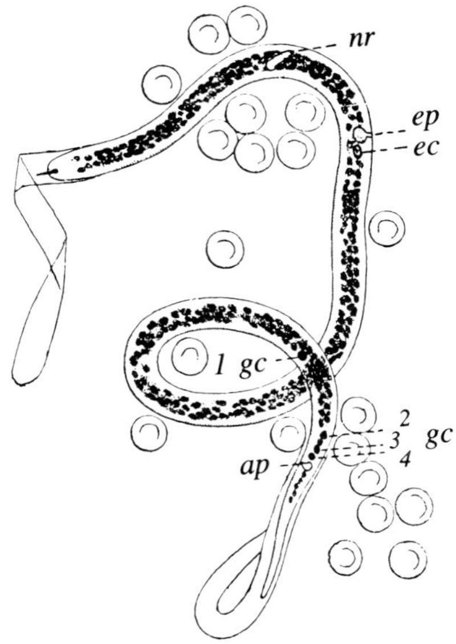

Fig. 23–4. *Wuchereria bancrofti.* Ensheathed microfilaria. *nr,* Nerve ring. *ep,* Excretory pore. *ec,* Excretory cell. 1, 2, 3, 4 *gc,* So-called "genital" or rectal cells. *ap,* Anal pore. (× 666.) (By E.C. Faust.)

mm by 0.24 to 0.30 mm. The vulva (Fig. 23–3,*D*) is cervical in position. Extending from the short vagina, the outer segment of the uterus is single, but the remainder of the female genitalia is paired. The young embryos in the inner portion of the uterus are confined within thin, hyaline, ovoid shells, 38 by 25 μm in size. As they progress toward the outer portion of the uterus, the shells become elongated to accommodate the uncoiling embryos and become known as the *"sheaths"* of the *microfilariae.* In this stage they are discharged from the worm.

Some investigators believe they have found sufficient morphological differences between the usual type of *W. bancrofti,* whose microfilariae exhibit nocturnal periodicity, and the South Pacific type in which periodicity is less evident. To differentiate them, the latter type is sometimes referred to as *W. pacifica.* It is more usual to refer to the former as *periodic* and the latter as *subperiodic.*

The *microfilariae* (Fig. 23–4) that migrate into lymph and blood channels average 244 to 296 μm in length by 7.5 to 10 μ in diameter. They are bluntly rounded anteriorly and pointed caudally. The cuticula is smooth. The microfilariae move about actively through the lymph or blood. Their internal structure can be studied only by the use of fixed stained preparations or by vital dyes, which show a central column of nuclei, among which are important anatomical "landmarks" used in differentiating this species from other microfilariae in blood and tissue films (Fig. 23–4). These "landmarks" consist of the nerve ring *(nr),* excretory pore *(ep),* excretory cell *(ec),* "G cells" (1 *gc,* 2–4 *gc*), and anal pore *(ap).* The percentage distances of these "landmarks" from the anterior end of the microfilaria provide a formula that is quite constant for the species (Table 23–1).

The morphological characters of head and tail of "sheathed" microfilariae of *W. bancrofti, B. malayi* and *Loa loa,* and of the "unsheathed" microfilariae of *Onchocerca volvulus, Mansonella perstans, M. streptocerca,* and *M. ozzardi* are distinctive.

Microfilarial Periodicity. In most regions of the world there is a marked periodicity in the circulation of *W. bancrofti* microfilariae in the blood. The numbers are high during a 4-hour period at midnight and scanty or absent during daylight hours. This type of circadian fluctuation is referred to as *nocturnal periodic.* In parts of the Pacific region, *W. bancrofti* microfilarial fluctuation is relatively slight, with an increase in numbers during the afternoon and evening, but with microfilariae

Table 23–1. Differences Between the Microfilariae of Periodic Strains of *Wuchereria bancrofti, Brugia malayi,* and *Loa loa,* All of Which Are Sheathed (Adapted from Feng, 1933)

Feature	*Mf. bancrofti*	*Mf. malayi*	*Mf. loa*
Periodicity	Nocturnal	Nocturnal	Diurnal
Length in thick films	244 to 296 μm	177 to 230 μm	250 to 300 μm
Excretory cell	Small (31%) near excretory pore (29%)	Large (37%) far behind excretory pore (31%)	Similar (37%) to *Mf. malayi*; pore (32%)
G cells	Small, similar size; G_2–G_4 far behind G_1; G_1, 70%	Larger, G_1 relatively near and larger than G_2–G_4; G_1, 68%	Similar to *Mf. malayi*; G_1; 69%
Anal pore	82%	82%	82%
Tail	Tapering to delicate point, no terminal nuclei	Often constricted between 2 terminal nuclei	Tapering gradually, caudal nuclei continuous with those of the trunk
Appearance	Graceful, sweeping curves	Stiff, with secondary kinks	Similar to *Mf. malayi*
Pathology	Regional lymphangitis and lymphadenitis; later, lymph varix, elephantiasis of lymphatics of scrotum as well as extremities	Similar to *W. bancrofti,* confined mostly to lymphatics of extremities	Fugitive swelling of subcutaneous tissue
Intermediate hosts (Vectors)	Optimum in many areas, *Culex quinquefasciatus* (syn. *C. fatigans*) others include *Anopheles* spp., *Aedes* spp.	*Mansonia* spp. *Anopheles* spp.	*Chrysops* spp.

present in the blood continuously throughout the day and night. This type, therefore, is referred to as *diurnal subperiodic*. In some other species the microfilariae are subperiodic with a significant increase during the night, but the term *subperiodic* when applied to *W. bancrofti* is understood to be diurnal. In general, west of 140°E longitude, only nocturnal periodicity occurs, between this meridian and 180°E longitude either type may occur, and east of 180°E longitude only the subperiodic type is present. Manson originally advanced the idea of adaptation of the microfilariae to the feeding habits of the mosquito vector, since the *Culex* mosquito, which he found to be the vector in China, was a "night-feeder." Manson-Bahr found later that the subperiodic type in Fiji was transmitted by a "daytime-feeder," *Aedes*.

In other filariae, and presumably in *W. bancrofti,* only a small part of the total microfilaria population is in the circulating blood, even at peak levels of microfilaremia (Pacheco and Orihel, 1968; Greenough and Buckner, 1969). A large part of the noncirculating population is sequestered in the lungs. Whether they remain in the blood and lymph capillaries and larger vessels or lie free in the tissues is unknown. As they are known to pass from maternal to fetal circulation, and they readily migrate from midgut to hemocoel through tissues of the vector, it is possible that they move freely in and out of the circulation in the vertebrate host (Beaver, 1970). Transfer of microfilariae from the mother's blood to that of the fetus may be a general occurrence. It is known to occur in *Dirofilaria immitis* and *Dirofilaria repens* in dogs (Mantovani and Jackson, 1966), *Dipetalonema viteae* in birds, *Meriones ibycus* (Geigy et al., 1967), *Mansonella perstans* in man (Zanetti and Lambrecht, 1948) (see Beaver, 1970) and *Wuchereria bancrofti* in man (Bloomfield et al., 1978).

It has been shown by electron microscopy that microfilariae move freely into and through cells of the vertebrate host's tissues, and may often be located intracellularly, especially in the liver and muscle cells (Mehlhorn et al., 1981).

The stimuli that send the microfilariae from the lungs and other tissues into the circulating blood and later back into the tissues are essentially unknown (Hawking et al., 1981). Body temperature and oxygen tension have been proposed as factors that determine the circadian rhythm (Hawking, 1967). The periodicity of *W. bancrofti* microfilariae can be reversed by reversing the periods of sleep and activity of the host (Hunter and Warren, 1950). The microfilariae of *Brugia malayi,* which in Pahang, Malaysia are subperiodic in man and cats, show marked nocturnal periodicity when transmitted to leaf monkeys *(Presbytis obscurus)* (Laing et al., 1961).

Distribution and Principal Vectors. Information on the geographic distribution and vectors of human filariasis was compiled by Hawking and Denham (1976) for the Pacific region, and by Hawking (1976, 1977, 1979) for the remainder of the world. These summaries were based largely on information received from members of the International Filariasis Association. No endemic focus of Bancroftian filariasis is known to exist in either North America or Europe, although such foci did exist earlier in South Carolina (southeastern United States), Portugal, Spain, Italy, and Greece. There is, however, a small focus in a Mediterranean coastal community in Turkey (Alanya), where the vector is *Culex pipiens (biotype molestus).*

Western Hemisphere. Bancroftian filariasis in the Western Hemisphere is nocturnal periodic, and is endemic in limited coastal areas of Costa Rica, Venezuela, Guyana, French Guiana, Surinam, and in northeastern Brazil. It also occurs in Haiti, the Dominican Republic, and other islands of the West Indies. *Culex quinquefasciatus* is generally a principal vector. Listed for Brazil are *Aedes scapularis, Anopheles aquasalis, A. bellator, A. darlingi, Coquillettidia juxtamansonia,* and *M. pseudotitillans.*

Africa and adjacent islands. W. bancrofti of the nocturnal periodic form occurs in much of the humid sub-Saharan areas of Africa as far south as Mozambique, and on the islands of Malagasy, the Comoros, Mascarenes, Seychelles, and the Chagos group. The distribution on the continent is markedly patchy and mainly rural. The chief vectors are *Anopheles gambiae* and *A. funestus. A. gambiae* is the vector in Tanzania (Bushrod, 1981). In addition to these, in some coastal communities of East Africa *Culex quinquefasciatus* is the main vector (Wijers and Kiilu, 1977). Listed also for Africa are *Anopheles melas, A. merus,* and *A. wellcomei.*

Asia. In Asia, *W. bancrofti* is distributed over much of the southeastern regions and extends northward to Japan and westward through India; a small focus remains in Turkey. It occurs in three forms. In most areas it is the nocturnal periodic form, transmitted by *C. quinquefasciatus.* However, in one province of Thailand (Kanchanaburi) it is subperiodic nocturnal, and among the aboriginal and rural populations of Peninsular Malaysia and Indonesia there is a "rural" form that is transmitted by anopheline and other mosquitoes. A similar rural form reported by Rozeboom and Cabrera (1965) in the jungles of Palawan in the Philippines is transmitted by *Anopheles flavirostris.*

Pacific Islands, New Guinea, and Australia. Bancroftian filariasis has apparently disappeared from the mainland of Australia. On the basis of common vectors and strains of the filaria, the Pacific area can be divided into four zones.

(1) *Micronesian* (Mariana, Caroline, Marshall, Gilbert Islands, Nauru, and Ocean Island), where the strain is periodic nocturnal and is transmitted by *Culex quinquefasciatus.*
(2) *Papuan* (New Guinea, the Bismarck Archipelago, the Solomon Islands, and Vanuatu), where the filaria is periodic and is transmitted by one or more of the following: *Anopheles farauti, An. koliensis, An. bancroftii, Mansonia uniformis, Aedes kochi, Culex annulirostris,* and *C. bitaeniorhynchus.*
(3) *Polynesian* (Fiji, Rotuma, the Ellice Islands, and other island groups extending eastward to the Tuamotu Archipelago and the Marquesas Islands), where the filaria is subperiodic and the vectors include one or more of the following species: *Aedes polynesiensis, Ae. pseudoscutellaris, Ae. rotumae,* and *Ae. cooki.*
(4) *New Caledonian* (New Caledonia and the Loyalty Islands), where the filaria is subperiodic and the vector is *Aedes vigilax.*

The Mosquito Phase of the Life Cycle. On arrival in a blood meal in the stomach of an appropriate host, the microfilariae lose the sheath in the first 2 to 6 hours. Some of them work their way, by use of a hook at the cephalic end of the body, through the wall of the proventriculus and cardiac portion of the midgut, and in the course of 4 to 17 hours have reached the thoracic muscles. During the next 2 days they metamorphose into a short, sausage-shaped organism measuring 124 to 250 μm in length by 10 to 17 μm in diameter (Bain, 1972). By the fifth or sixth day the tail has atrophied to a mere stump, the intestinal tract has become well differentiated, and a body cavity has developed, although the genital primordia are still inconspicuous. Meanwhile, the first true sheath has been shed. The larvae now measure 225 to 300 by 15 to 30 μm, and three subterminal caudal papillae appear. During the early part of the second week a second ecdysis occurs and the worms rapidly elongate into the filiform (third-stage) mature larvae, which measure 1.4 to 2 mm by 18 to 23 μm. Under optimal conditions this stage is reached in 10 or 11 days, when the larvae migrate through the hemocoel within the labium (i.e., proboscis sheath) down to the labella of the mosquito, (through the tip of which they escape, most frequently during flexure of the proboscis at the time the blood meal is being drawn up), and are ready to enter the skin by active migration through the puncture wound (Ewert, 1967; Ewert and Ho, 1967).

Of the species of mosquitoes that serve as *natural* biological transmitters of *W. bancrofti,* the following, as outlined previously, are particularly important: *Culex quinquefasciatus; C. pipiens molestus* and *C. pipiens pallens* (Curtis and Feachem, 1981); *Aedes aegypti; Ae. polynesiensis; Anopheles funestus, An. arabiensis; An. gambiae; An. punctulatus;* and *An. farauti.* Moreover, many other species of *Culex, Aedes, Mansonia,* and *Anopheles* have been shown to be satisfactory intermediate hosts by experimental tests, although these demonstrations do not necessarily indicate that all the species in which the larvae complete development are good transmitting hosts in nature (Table 41–4).

Most good mosquito vectors are predominantly anthropophilic (i.e., they prefer human blood). Their breeding sites are often near human habitations, and the adult females frequent places where blood meals from man are readily available.

Development in the Human Host. It is commonly accepted that the infective-stage larvae, after entering the skin, pass through peripheral lymphatics, in which they migrate and grow, then settle down in certain lymphatic vessels retrograde to lymph nodes, grow to maturity, and mate. This process is followed by production of microfilariae by the gravid females. The microfilariae discharged by mature female worms may be expected to appear in peripheral blood in 8 to 12 months. Filariasis without microfilaremia is not uncommon. The adult worms in infected individuals are coiled up in nodular dilatations of the lymphatic vessels, most frequently in the cortex of lymph nodes and in the testicular tissues. Man is the only known natural definitive host of *Wuchereria bancrofti.*

Development in Experimental Animals. Prior to 1979, numerous attempts had been made to infect experimental animals with *W. bancrofti,* all with essentially no success. However, partial development of larvae had been obtained in cats, monkeys, gibbons, and hamsters, and development to subadults (mid-fourth stage larvae) had been reported in the bird, *Meriones unguiculatus* (Ash and Schacher, 1971). In 1979, Cross and associates reported complete development of adult worms and microfilaremia in experimentally infected Taiwan monkeys *(Macaca cyclopis).* Infective larvae were taken from *Culex fatigans* and *Aedes togoi* that had fed on volunteers in Indonesia and on Kinman Island, China. The infections in the donor volunteers were periodic nocturnal. Larvae were introduced by subcutaneous inoculation, by foot puncture (method of Ewert, 1967), and by having the infected mosquitoes feed directly on the monkeys. Larvae or adult worms were recovered from 63% of 43 male and 29% of 35 female monkeys. Microfilariae were detected in the blood of 9 monkeys, all of which were male and some of which had been infected with both of the two strains of the parasite. Microfilaremia levels were low (maximum, 14/ml blood), but persisted for 9 to 16 months in 6 animals and for more than 21 months in 1 animal. One of the animals had been splenectomized and three were given antilymphocyte serum, but the remainder were given no treatment. A male *Macaca mulatta* and a male *M. irus* were also experimentally infected. Most of the worms recovered from the males (62 of 76) were located in the testes; those from females were recovered from the pelt and carcass, or from lymph nodes. The shortest prepatent period was 8 months (in an untreated *M. cyclopis),* the longest 18 months (also in an untreated *M. cyclopis).*

A year later Dissanaike and Mak (1980) reported having experimentally infected a splenectomized and immunosuppressed long-tailed macaque, *Macaca fascicularis.* Fully mature adult worms of a rural strain of *W. bancrofti* were obtained and described. At 231 days microfilariae could not be found in the blood, but the female worms contained microfilariae that appeared to be fully developed. Six complete males, two complete females, and parts of others of both sexes were recovered from the monkey, all from the lymphatic channels of the testicular region, the site where the larvae had been inoculated. Palmieri et al. (1983) reported that 4 of 5 silvered leaf monkeys *(Presbytis cristatus)* developed a microfilaremia at 206 to 285 days after inoculation with 50 to 202 *W. bancrofti* larvae. Similar findings were reported by Harinasuta et al. (1981).

Epidemiology. Man is inoculated by the infected

mosquito when mature larvae escape through the rupture of the distal end of the mosquito's proboscis into the puncture wound in the skin at the time of a blood meal. The mosquitoes have previously become infected from ingesting the microfilariae in peripheral blood of infected human subjects, in most endemic areas only during the night. Highly endemic areas have numerous people with microfilariae circulating in their blood, abundant breeding of appropriate mosquitoes near habitations or places of work of infected human beings ("carriers"), and susceptible human subjects repeatedly exposed to the bites of the vectors.

In such areas exposure begins early in childhood and continues throughout life. Infection gradually builds up with each decade, although it may not become clinically apparent. In coastal communities of East Africa, microfilaria rates increase with age, reaching more than 50% in the 60- to 69-year group (Wegesa et al., 1979). In the absence of re-exposure, infection with either subperiodic or periodic *W. bancrofti* generally disappears in less than 8 years (Mahoney and Aiu, 1970; Webber, 1977). The presence of microfilariae in the blood of a woman who was continuously outside an endemic area for 40 years was reported by Carme et al. (1977). Unfortunately she was treated without the finding being confirmed. In Tahiti 33% of persons over 5 years of age had filariasis bancrofti (Beye et al., 1952). Newcomers to such an endemic region, living under comparable conditions of exposure, may develop early symptoms, but usually do not develop a microfilaremia. Filariasis without microfilaremia was experienced by more than 12,000 men in the military services in the Pacific area during World War II (Beaver, 1970).

Important factors in transmission are the density of the vector mosquito population and the numbers of microfilariae circulating in the blood. Although the threshold levels cannot be stated precisely, it is often observed that in highly endemic areas the people receive thousands of infective bites per person per year. The fate of the larvae that do not develop to mature worms is unknown. The frequent presence of mature worms in people without microfilaremia is also unexplained. It is of interest that the number of infective larvae developing in the mosquito may be only a fraction of the microfilariae ingested. At low levels the percentage of microfilariae reaching the hemocoel is proportional to the number ingested, but when the rate of ingestion increases, the percentage reaching the hemocoel is limited (Bain, 1971). This has a sparing effect on the vector. It seems possible that significant transmission may still be accomplished when levels of microfilaremia have been reduced below levels detectable by usual methods (Desowitz and Hitchcock, 1974). On the other hand, the level of mosquito population necessary for sustained endemicity may be relatively high. In areas where malaria and filariasis are transmitted by the same vector, malaria control by antimosquito efforts may have a greater effect on the filariasis than it has on the malaria (Webber, 1977).

Pathogenesis. The manner in which *Wuchereria bancrofti* and *Brugia malayi* produce their pathogenic effects has been a subject of controversy since a filarial worm was first discovered to be the etiologic agent of tropical elephantiasis. It was not until an experimental model simulating the human infection became available that the pathogenic mechanisms of human filariasis could be elucidated. This model is the cat infected with *Brugia pahangi, B. malayi,* or *B. patei.* Although cats, and in fact all mammals other than man, do not develop elephantiasis due to filarial infection of the

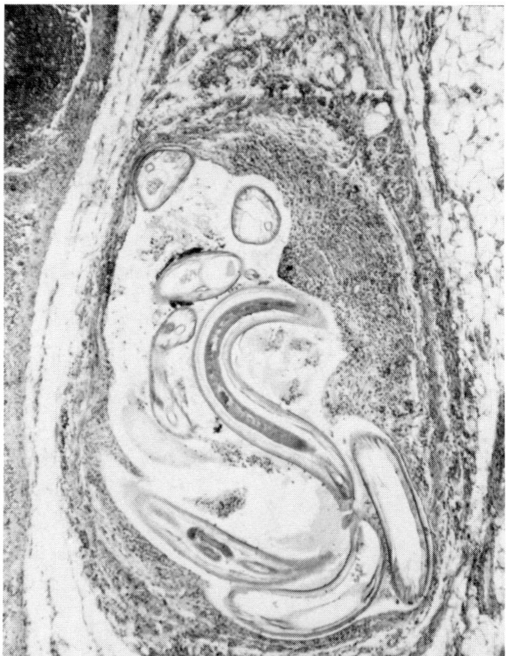

Fig. 23–5. *Wuchereria bancrofti.* Mature, unmated female in afferent lymphatic vessel partially occluded by inflammatory cells in enlarged, tender, axillary lymph node from 11-year-old Haitian girl in New York City. Examination of blood showed normal hemogram and no microfilariae. (× 48.) (By P.C. Beaver. Stained section courtesy of Dr. Harry Most.)

lymphatic system, this difference from the course of the human infection can be explained by the hydrodynamic forces brought about by the bipedal posture and man's greater size.

Based on research with the experimental model and on observations of human infection with *W. bancrofti* and *B. malayi,* the following is believed to be the pathogenesis of human filariasis (Ewert et al., 1972; Gooneratne, 1973).

After penetrating human skin, the infective larvae migrate directly to the lymphatic vessels or nodes, completing their migration within 2 or 3 days. Migration is usually to lymphatic spaces relatively close to the point of initial entry. The larvae undergo two molts in the lymph channels. The presence of the worms, their molting fluid and other products, and their cast-off sheaths provoke a reaction within the lymphatic vessels. At this early stage the mechanism may be through an immunological response of the host or may be primarily an irritant effect of the offending substances. The effect on the host, demonstrable by lymphography, is dilatation and slowing of the lymph fluid in vessels occupied by the worms.

As the infection ages, the worms become larger and immunological reactions definitely occur. Infiltration of plasma cells, eosinophils, and macrophages in and around the affected vessel becomes evident in histological sections (Figs. 23–5, 23–6). Usually there is lymphangitis with swelling, redness, and pain. Because of the chronic inflammation, the lymph valves proximal to the worms become damaged and incompetent. This damage contributes to the hydrostatic pressure already increased by the dilatation and consequent slowing of lymph flow (Fig. 23–7,*A*).

In addition, permeability of the walls of the vessels increases, which permits the leakage of fluid with high concentration of protein into the surrounding tissues. This results in the characteristic hard or brawny edema of filarial elephantiasis.

Regional lymph nodes proximal to the worms become initially enlarged, then shrink as fibrosis takes place. Proximal lymphatic vessels likewise become stenotic and obstructed. There is a secondary development of collateral channels and anastomoses which provide an eventual route for the lymph to enter the thoracic duct.

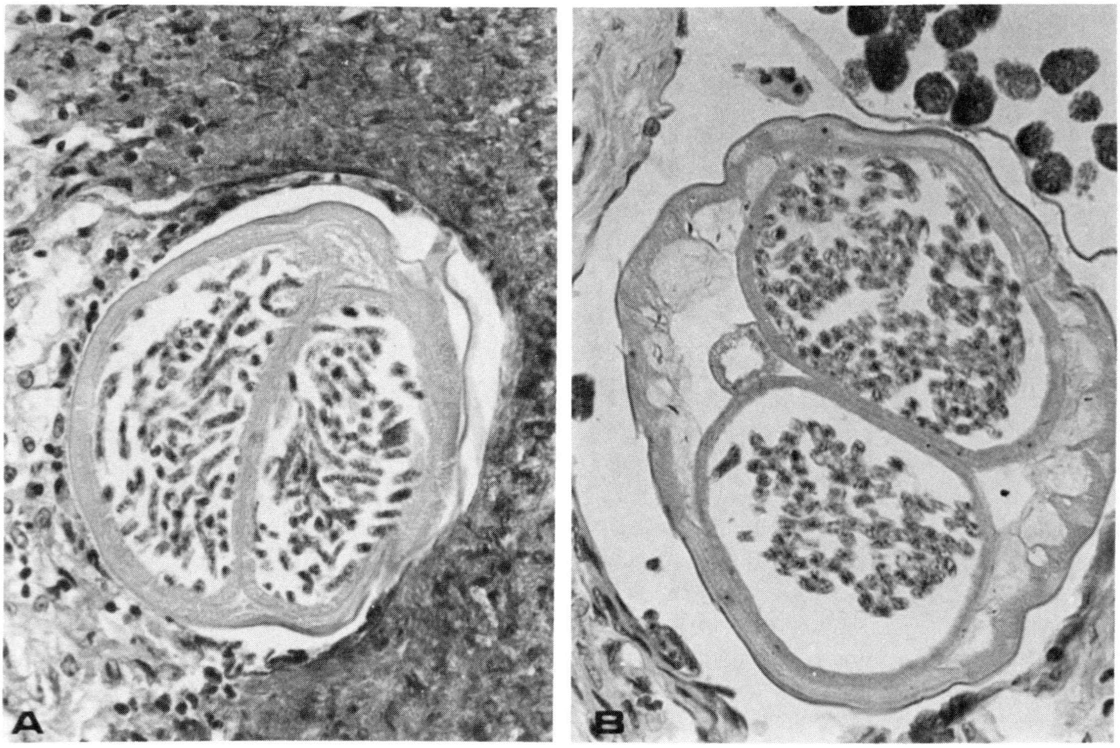

Fig. 23–6. *Wuchereria bancrofti*. Gravid females in lymphatic vessels of spermatic cord. *A*, Dead worm containing living microfilariae, partially encased in area of necrosis consisting mainly of eosinophilic fibrinoid cellular debris. (× 360.) *B*, Living worm in functional vessel, showing typical microstructure of cuticle with increased thickness over lateral cords, three or four muscle fibers per quadrant, small, thin-walled intestine, and large paired uteri filled with microfilariae. (× 360.) (By P.C. Beaver. Stained sections from collection of Dr. F.W. O'Connor, courtesy of Dr. H.F. Smetana.)

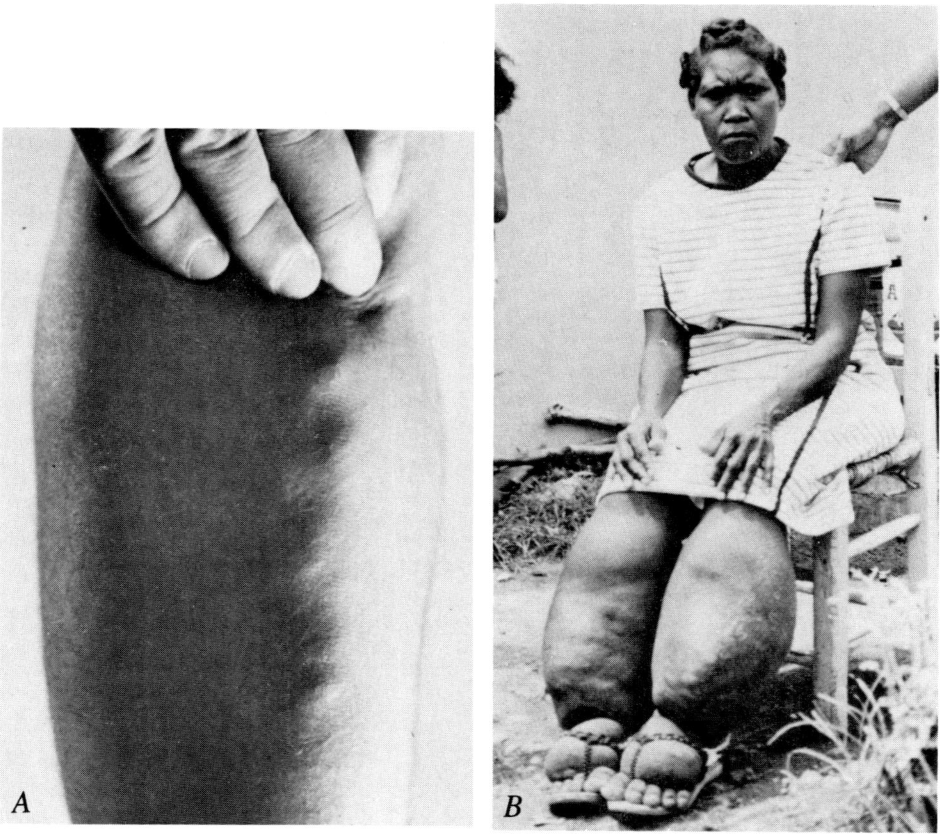

Fig. 23–7. *Wuchereria bancrofti. A*, Blockage caused by worm(s) in lymph vessel in forearm of 22-year-old male in Singapore General Hospital. (Courtesy of Dr. J.F. Schacher.) *B*, Woman in Dominican Republic with filarial elephantiasis. (From Vincent, A.L., et al. 1981. Filariasis in the Dominican Republic. Am. J. Trop. Med. Hyg., *30*:739–741.)

In the final stages there are dilated, tortuous lymphatic channels, proximally obstructed, fibrosed lymph nodes with obliterated lymph spaces, and lymphedema with high protein content. Fibroblasts migrate into the edematous spaces and generate a thickening of the subcutaneous connective tissue. The result is the classic picture of filarial elephantiasis. At times the involved part may become tremendously enlarged. The enlargement consists of lymph and fat in a matrix of fibrotic tissue covered by tightly stretched, thickened skin (Fig. 23–7,*B*).

In endemic areas repeated infections typically occur, which results in overlapping of the developmental sequence. It is presumed that such repetitions of the noxious effects of the infection and the presence of dead worms must result in even more damage to the lymphatic system (Fig. 23–8).

It is noteworthy that this disease process is not wholly dependent upon the death of the worms. Rather, it typically develops as a consequence of the presence of living worms distal to the sites of lymphangitis. Lymphangitis and subsequent events are not wholly dependent on the development of secondary infection with streptococci or other bacterial organisms, but the patient with lymph stasis is especially prone to develop such infection. Streptococcal secondary infection intensifies and hastens the lymphangitis (Bosworth et al., 1973).

Not all persons infected with *Wuchereria bancrofti* develop elephantiasis, whether or not they exhibit clinical signs of early infection such as lymphadenitis, lymphangitis, and eosinophilia. It is also possible for the late effects of lymph stasis to develop in an individual who has never had demonstrable microfilaremia. Microfilariae may be kept out of the peripheral blood by an immmunologic response in many persons, especially those from nonendemic areas (Beaver, 1970). Microfilariae may be able to escape the immune response of the typical host by the action of antigen-specific suppressor cells and suppressor factors (Piessens et al., 1980).

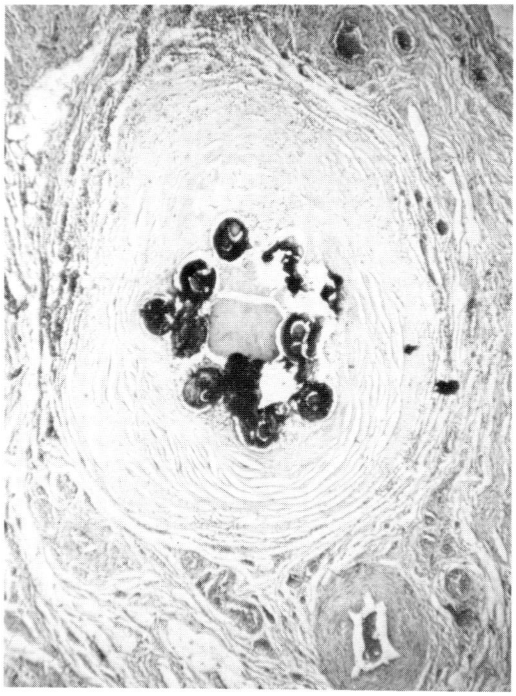

Fig. 23–8. *Wuchereria bancrofti*. Section of calcified female in fibrotic lymphatic of spermatic cord. (× 46.) (By P.C. Beaver. Stained section from collection of Dr. F.W. O'Connor, courtesy of Dr. H.F. Smetana.)

Pathology and Symptomatology. The course of the disease process in human filariasis is variable, depending upon interactions of the host and parasite. The spectrum of disease pictures varies from one of severe clinical manifestations without microfilaremia, which is disadvantageous to both parasite and host, to one of asymptomatic microfilaremia, which is advantageous to the parasite and well tolerated by the host.

The *biological incubation* or *prepatent period* in areas of endemic filariasis lasts 1 year or more (i.e., from the entry of the third-stage infective larvae into the skin until microfilariae first appear in peripheral blood). In native populations of highly endemic areas, this commonly occurs during childhood and is likely to be accompanied by no significant evidence of the infection. The immature worms typically provoke relatively little local tissue reaction.

Conversely, persons coming into an endemic area from one that is free from filariasis are apt to develop early manifestations of lymphangitis many months before the inoculated larvae become mature. The infection never becomes patent, i.e., microfilaremia never appears in such patients (Beaver, 1970).

When the worms settle down and develop into adults, the parturient females and microfilariae produce tissue reactions that constitute the *acute stage*. Host reaction to the worms consists of an accumulation of histiocytes, epithelioid cells, lymphocytes, plasma cells, giant cells, and eosinophils in the lumen of the vessel around the worms. Hyperplasia of the endothelium and perilymphatic cellular infiltration occur, not only around the worms but also proximally along the vessel to the regional lymph nodes in an ascending lymphangitis and lymphadenitis. Stenosis of the lymph vessel results.

Whenever such an acute inflammatory reaction takes place, there is a reddened, swollen, raised lymph tract immediately around and extending distally from the site of blockage (Fig. 23–7,*A*). It is exquisitely tender, painful, and usually associated with fever. It recurs periodically, in some individuals as frequently as monthly, in others less frequently. Exacerbations may be associated with physical exertion. In women they may accompany the menses. Not uncommonly, an attack is preceded by a prodrome of malaise or urticaria. Gradually attacks become less severe and involved lymph channels less painful. Huntington et al. (1944) referred to these attacks as "acute allergic filarial lymphangitis."

King (1944) stated that the earliest symptoms in young adult males develop 3 to 16 months after exposure; that the cardinal manifestations are lymphangitis, usually with associated lymphadenitis, and that in only 20% of cases is there mild fever of short duration. The lymphangitis may originate in an upper extremity, mostly epitrochlear, but eventually there is a concentration of lesions in the scrotum consisting of inflammation of the spermatic cord, epididymis, and testes.

In lightly exposed immigrants in Samoa, Webster (1946) found that lymphangitis with fever recurred annually. In natives of highly endemic areas, this stage of the infection may be asymptomatic. Since first exposure is likely to take place early in childhood, severe early involvement of the genitalia is usually lacking. Living worms may be present for years in dilated lymphatic channels without remarkable symptoms. These persons characteristically have microfilariae circulating in the blood and constitute the usual source of infection for the vector.

American military forces in the Samoan-Ellice-Wallis Islands hyperendemic area (1942–1944) were subjected to repeated exposure to heavily infected mosquitoes within a few weeks after their arrival and rapidly developed one or more of the following syndromes: (1) lymphangitis of an extremity; (2) lymphadenitis proximal to involved lymphatic vessels; or (3) acute inflammation of the scrotum and its contents. The first manifestation consisted of intense, at times exquisite pain, swelling and redness along a lymphatic vessel in an extremity, or funiculitis, with or without epididymitis, orchitis, or hydrocele. Occasionally two separate sites were involved simultaneously. Many of the patients showed no elevation of temperature and no severe constitutional symptoms, no leukocytosis, and no significant eosinophilia, but there was indication of generalized hyperplasia of the involved site. Occasionally there were manifestations of tropical pulmonary eosinophilia *(vide infra)*. The symptoms subsided following complete rest but tended to recur when physical activity was resumed. Microfilaremia was extremely rare in this group (Huntington et al., 1944; Wartman and King, 1944), although in some instances biopsies of glands revealed these embryos in lymphatic vessels in the immediate vicinity of the parent worms. In a few cases later on, scant numbers of microfilariae appeared in peripheral blood vessels. Ten to sixteen years later some of the infected individuals still experienced lymphedema and lymphadenitis two or three times a year, aggravated by exercise, especially during warm, damp days (Trent, 1963).

The *chronic stage* eventuates as the acute inflammation subsides and fibrosis advances. The worms die and are absorbed or become calcified, sometimes many years after initial inoculation. Typically the sites involved are the groin, with the development of nodular inguinal gland varicosities; the lower extremities, one or both of which become elephantoid; and the external genitalia, male or female (Fig. 23–7,*B*). The swelling is nonpitting. Microfilaremia is not likely to be present. The elephantoid mass consists of fat in a matrix of fibrous tissue containing distended lymph spaces.

Not all infected persons develop serious elephantoid lesions. The extreme form of filarial elephantiasis is seen in only a small percentage of infected persons in highly endemic regions. It is possible that secondary streptococcal or other bacterial infection (Bosworth et al., 1973) or some host factor is responsible for its appearance.

Organs and Tissues Involved in Lymphangitis, Adenitis, and Elephantiasis. Blockage and inflammation in the thoracic duct or median abdominal lymph vessels may result in lymphangitis of the scrotum, penis, or prepuce in the male and of the external genitalia in the female. If the groin is involved, "varicose groin glands" develop. Involvement and hypertension of lymphatics of the bladder or kidney may result in rupture of the lymph vessels concerned, with consequent lymphuria, more frequently chyluria. With similar tension of the tunica vaginalis, lymphocele or chylocele may develop; in the case of the peritoneum, lymphous or chylous ascites may result. Fibrosis and hypertrophy of affected tissues produce elephantiasis. About 95% of all cases involve one or both of the lower extremities or the scrotum (Fig. 23–7,*B*). In women the mammary gland is not infrequently involved. Three cases of ophthalmofilariasis, reportedly due to *W. bancrofti,* are on record. Anderson (1953) reported the finding of a well-preserved gravid female in the lymphatics of the parietal pericardium.

Tropical Eosinophilia. This syndrome of chronic pulmonary infiltrations, hypereosinophilia of the peripheral blood, chest pain, cough, and asthmatic attacks especially at night has been recognized as a distinct clinical entity since the early 1940s (Frimodt-Müller and Barton, 1940; Weingarten, 1943). A distinctive feature of the syndrome is the prompt relief of symptoms afforded by treatment with arsenical drugs or diethylcarbamazine. The syndrome has also been referred to as Weingarten's syndrome, eosinophilic lung, and tropical pulmonary eosinophilia. A series of studies and observations culminating in those of Danaraj et al. (1966) more precisely defined the syndrome to include the presence of microfilariae in the tissues of the lungs and their absence from peripheral blood. *Wuchereria bancrofti* and *Brugia malayi* are probably the etiologic agents in most cases of tropical eosinophilia, and most likely the adult worms are located in the pulmonary arteries or pelvic lymphatic vessels. Presumably the syndrome is due to a failure of suppression of host-immune mechanisms against the microfilariae, which consequently are screened out in the lungs and destroyed to the accompaniment of manifestations of an allergic reaction.

In recapitulation, the pathogenesis and clinical course of human filariasis vary, there being several distinct pictures:

1. Filariasis with microfilaremia.

(a) Relatively asymptomatic infection, with marked and prolonged microfilaremia.

(b) Recurrent lymphangitis and eventual chronic manifestations of lymph stasis.

2. Filariasis without microfilaremia.

(a) Early lymphangitis and lymphadenitis in persons moving from an area without filariasis into an endemic zone.

(b) Tropical pulmonary eosinophilia.

(c) Extreme filarial elephantiasis following a chronic course of recurrent lymphangitis, often complicated by cellulitis due to streptococcal or other bacterial infection.

Diagnosis. The diagnosis of Bancroft's filariasis may be based on demonstration of microfilariae in the blood, or on clinical evidence. Most commonly used for the detection of microfilaremia is the thick film containing 20 to 60 mm^3 blood, air-dried, laked, fixed, and stained with hematoxylin or Giemsa. Large amounts can be spread on slides in strips for easy examination. Active microfilariae may be found in lymph or chylous urine by microscopic examination of sediment after centrifugation. A method of concentrating microfilariae from 2-ml blood samples in 2% formalin was developed early by Knott (1939). More recently filtration of venous blood through microfilters has come into common use (Desowitz and Hitchcock, 1974). However, the necessity for venipuncture reduces the number of people who are willing to be examined (Webber, 1977). In East Africa, surveys requiring venipuncture are unacceptable (Wijers, 1977). In areas where microfilaremia is nocturnal or is present throughout the day but at low levels, mirofilariae may be induced to appear in the blood during the day by giving diethylcarbamazine in a single dose of about 2 mg per kg body weight. Usually peak numbers are reported to occur 15 to 20 minutes after administration of the drug, though drawing blood samples as late as 45 to 50 minutes after giving one 100-mg tablet to adults apparently has been satisfactory (McMahon et al., 1979; Wijeyaratne et al., 1982). For surveys, McMahon (1982) found examination of blood 50 minutes after a 2 mg/kg dose most satisfactory. However, the night positives detected by this method may be as low as 40% (Rajapakse, 1974). Microfilariae of *W. bancrofti* require differentiation from those of *B. malayi, Loa loa,* and other species (Fig. 23–4).

Skin tests with saline antigen extracts of microfilariae, adult worms, and third-stage larvae of *Brugia malayi* may provide useful epidemiological information (Grove et al., 1977). Immunodiagnostic methods now available are not sufficiently sensitive or specific to give reliable information on the prevalence or intensity of infection in either individual cases or in population surveys.

Where microfilariae cannot be demonstrated in the blood, diagnosis must depend on the history of lymphangitis following exposure in an endemic area or, at times, on biopsy of an enlarged lymph node immediately proximal to the involved lymphatic tract.

Treatment. *Diethylcarbamazine* (DEC) (1-diethyl-carbamyl-4-methylpiperazine dihydrogen citrate), first tried clinically in 1947, has become the drug of choice for treating active cases of Bancroft's filariasis (Hawking, 1955). It has the special advantage of oral administration and acts rapidly in causing disappearance of circulating microfilariae, which become congregated in visceral blood vessels where they are destroyed by phagocytic cells. Although the action of this drug on adult worms is much less dramatic, Hawking (1955) believes that the adult females are eventually sterilized. Extensive trials made by workers in Puerto Rico, St. Croix, Guyana, Egypt, Tahiti (Galliard and Mille, 1949; Kessel et al., 1953; March et al., 1960), and Tanzania and Uganda (Jordan, 1959) showed that when this drug is administered in the amount of 0.5 to 2.0 mg per kg body weight three times daily after meals for 2 to 3 weeks, it rapidly eliminates circulating microfilariae and tends to act more slowly on the adult worms. DEC is relatively well tolerated in Bancroft's filariasis. Among infected Polynesians the administration of DEC at times produces bullous eruptions, which are considered allergic reactions to rapid death of circulating microfilariae. Prescription of corticosteroids 1 to 14 days after appearance of these eruptions produces full recovery in 1 to 3 days (March and Laigret, 1958).

Other drugs have shown efficacy against microfilariae and sometimes also against adult worms. Killing microfilariae by arsenical drugs is the basis for the efficacy of this type of drug in tropical eosinophilia *(vide supra)*. Neither arsenicals nor antimonials are effective against adult worms in a dosage schedule that is safe for the host.

Suramin is moderately effective but is not acceptable because of the intravenous route of administration and greater toxicity than that of DEC. Of several new organophosphorus compounds, metrifonate shows the greatest promise. Administration of levamisole prior to DEC, earlier recommended, proved not to be useful (McMahon, 1981).

Surgical removal of elephantoid tissue by a modified Kondoleon operation temporarily reduces the member, although recurrence may be anticipated if collateral lymphatic circulation has not been established. During the acute attacks of lymphangitis, local compresses of magnesium sulfate solution are helpful and general palliative measures are indicated. The use of sulfonamides or antibiotics is helpful in cases in which septic complications are present.

The most useful nonspecific procedure in elephantiasis of the lower limbs is pressure bandaging (Knott, 1938), using 6-inch strips of bath toweling, painting with dextrin syrup, and covering with cotton elastic crepe bandage and an outer muslin bandage to keep out dirt. Exercise is required to prevent cyanosis and to hasten reduction of the lymphedema. The member gradually becomes more normal and in milder cases may assume normal size, contour, and skin texture. While the cutaneous lesions may be reduced in more advanced cases, the underlying fibrosis is not affected.

Prognosis. Once exposure has occurred and the infective-stage larvae have entered the lymphatics, the prognosis depends primarily on the following factors: (1) number of immature or mature worms that gained entrance to the patient's body; (2) their potentialities for mating; (3) opportunities for reexposure and therefore superinfection; (4) activity of the reticuloendothelial system and degree of sensitization of the patient; and (5) number of times acute cellular reaction occurs at the same anatomical location.

Following removal of the patient from the endemic area, the disease often runs a self-limiting course, provided medical, psychological, educational, and vocational activities are carefully planned and supervised; yet sequelae may persist for years (Trent, 1963). No impairment of sexual function has been demonstrated.

In chronic cases with clinical involvement, even with surgical intervention, eventual prognosis is relatively poor. In subclinical cases the outlook is fair, but continued exposure or gradual development of existing lesions to a chronic state may be anticipated at any time.

Prevention. For the individual who wishes to prevent infection in endemic areas of Bancroft's filariasis, the use of an insect repellent may be effective. One of the available repellents not only is protective against the insect bites but is lethal to infective *W. bancrofti* larvae (Hocking, 1967). Immunization has been widely and intensively studied but thus far has shown little promise even for animals.

Control measures applied to populations in endemic area may be directed against the parasite or the vector. Diethylcarbamazine, which came into general use as a filaricide more than 30 years ago, is effective in destroying microfilariae. In a high proportion of persons given adequate doses, the microfilaremia is removed permanently, indicating that the adult worms may also be killed. However, in some individuals microfilaremia persists, at least at low levels, and the microfilariae are infective to the vector mosquitoes. Nevertheless, the effectiveness of mass treatments repeated annually or at shorter intervals has been demonstrated in a number of places, notably in Japan (Sasa 1970), Tahiti (Laigret et al., 1980), Solomon Islands (Webber, 1979), and Samoa (Hawking and Denham, 1976). As diethylcarbamazine is relatively nontoxic, it can be safely added to salt and common food substances (Hawking and Marques, 1967; Fan et al., 1975; Rao et al., 1981; Zhong et al., 1981).

A first step in the prevention of filariasis by vector control is to determine which of the local species is chiefly responsible for transmission, and to obtain full and reliable information on its flight, feeding, and breeding habits. Spraying residual insecticides onto common resting sites has been a basic form of control for filariasis vectors, as for malaria vectors. A difficulty in this approach has been that many vector species have become resistant to the available insecticides. In some instances insecticides can be effectively used against the larval stages, or better, when feasible, the breeding sites may be removed or modified to prevent breeding. Major vectors in a large part of the endemic regions are mosquitoes of the *Culex pipiens* complex whose breeding sites are polluted ditches, ponds, and pools created in newly urbanized communities where sanitation generally is poorly developed. This results in increases in the number of breeding sites along with greater human population density. The construction of septic tanks and pit latrines, poorly planned, improperly installed, or badly maintained, often offers ideal habitats for the breeding of vector mosquitoes. In such situations general improvements in sanitation provide the most effective means of control (Curtis and Feachem, 1981).

Wuchereria kalimantani Palmieri, Purnomo, Dennis and Marwoto, 1980

Wuchereria kalimantani was described from 12 female and 9 male worms recovered from the inguinal lymph nodes and testes of 22 (5 male, 17 female) of 64 silvered leaf monkeys *(Presbytis cristatus)* taken from a rubber estate on South Kalimantan (Borneo), Indonesia (Palmieri et al., 1980). Like that of *W. bancrofti,* the microfilaria is sheathed, has no nuclei in the tip of the tail, and the cephalic space is short; the microfilaria is slightly smaller and its sheath stains reddish with Giemsa stain, instead of purple as in *W. bancrofti*. The adult female is also somewhat smaller, and its ovejector is longer than that of *W. bancrofti*. In the male, the left-to-right spicule ratio is greater and the postcloacal papillae are more variable than in *W. bancrofti*. The microfilaremia is sharply periodic and nocturnal. The natural vector of *W. kalimantani* is unknown. Experimentally it develops to the infective stage in *Anopheles subpictus* and *Aedes togoi*.

Wuchereria Species of Doubtful Status

Galliard and Brygoo (1955) described from people in the coastal area of Malagasy a microfilaria whose morphology was intermediate between *W. bancrofti* and *B. malayi*. Later this form was redescribed as a distinct strain by Galliard et al. (1955) under the designation *W. bancrofti* var. *vauceli* Galliard and Brygoo, 1955. The distinctive features of the new variety were shorter cephalic space, less compact innerkorper, and the size and position of the excretory and genital cells in relation to the excretory pore and anal pore, respectively. Galliard (1959) later raised this form to full species status.

On the basis of nuclear number in various parts of the microfilaria, Schacher et al. (1967) recognized affinities and differences between *W. bancrofti* from various geographic areas. Additional studies supportive of the concept (Schacher and Geddawi, 1969) led finally to the decription of a new species, *Wuchereria lewisi,* from man in Brazil, based entirely on distinctive morphological features of the microfilaria (Schacher, 1969). At the same time, *W. pacifica* Manson-Bahr, 1941 (= *W. bancrofti* var. *pacifica* Manson-Bahr and Muggleton, 1952) and *W. vauceli* (Galliard and Brygoo, 1955) Galliard, 1959 were considered to be synonyms of *W. bancrofti*. Schacher's published measurements were examined by Colless (1971), who made a principal components analysis, and by Campbell (1976), who found that the analysis showed complete separation of *W. lewisi* and *W. bancrofti,* and that the morphological differences resulted primarily from differences in the size of the individual microfilariae. *W. philippinensis* Low, 1941, said by Palmieri et al. (1980) to have been considered by Schacher (1969) as synonymous with varieties of *W. bancrofti*, seems not to have been discussed by Schacher.

Genus Brugia Buckley, 1958 (In honor of S.L. Brug)

In species of the genus *Brugia,* the body is distinctly dilated at the anterior end, the vulva is anterior to the base of the esophagus, the center section (lamina) of the left spicule is complex (in *Wuchereria* it is simple), and the microfilaria is sheathed and has two nuclei near the tip of the tail, one of which is at the extreme tip. All known species of the genus are parasites of the lymphatics (Buckley, 1960).

Brugia malayi (Brug, 1927) Buckley, 1958 (Malayan filarial worm, producing Malayan filariasis)

Synonyms. *Filaria malayi* Brug, 1927; *Microfilaria malayi* (Brug, 1927) Faust, 1929; *Filaria bancrofti* Cobbold, 1877, auct., *pro parte, Wuchereria malayi* (Brug, 1927) Rao and Maplestone, 1940.

Historical and Geographical Notes. This filaria was first observed by Lichtenstein in its microfilaria stage in blood films of natives of northern Sumatra and was described by Brug (1927) as a new species. Until 1940 the adult worms had never been found, but the specific characters of the microfilaria indicated beyond a doubt that it was a distinct species, probably related to *Wuchereria bancrofti*. Adult males and females were studied and described by Rao and Maplestone (1940), by Bonne et al. (1941), and by Buckley (1960).

This filaria is now known to parasitize man in considerable areas of India (including Kerala, Hyderabad, Orissa and Madhya Pradesh states, and southern Annam), Sri Lanka, Sumatra, Java, Ceram, Flores, Borneo, Celebes, Malaysia, Philippines, southern Thailand, North Vietnam, China (coastally and inland as far north as Shandong [Shantung] Province), South Korea, and Japan (Hachijo-Komisha Island). In China, *B. malayi* is found mainly in the southern hilly and mountainous regions, whereas *W. bancrofti* is distributed chiefly in the plains regions (Zhong et al., 1981).

Morphology, Biology, and Life Cycle. The adult filaria, species *malayi,* was first described by Rao and Maplestone (1940) from human material from India, and the following year by Bonne et al. from Indonesia. A more complete description was provided by Buckley and Edeson (1956) from natural infection of the Kra monkey, *Macaca irus*. On the basis of a study of this material, Buckley (1958) removed the species *malayi* from the genus *Wuchereria* and created for it and two related species *(B. pahangi* from wild and domestic cats, dogs, and monkeys in Malaysia, and *B. patei* in wild and domestic dogs and cats from East Africa) the genus *Brugia*. Six additional species of *Brugia* have been described, namely, *B. buckleyi* Dissanaike and Paramanthan, 1961 from heart and blood vessels of the hare in Sri Lanka; *B. ceylonensis* Jayewardene, 1962 from lymphatics of the dog in Sri Lanka; *B. guyanensis* Orihel, 1964 from lymphatics of the coatimundi in Guyana; *B. beaveri* Ash and Little, 1964 from lymphatics of the raccoon and bobcat in the United States; *B. tupaiae* Orihel, 1966 from lymphatics of the tree shrew in Malaysia; and *B. timori* Partono et al., 1977 from man in Timor, one of the islands of southeast Indonesia.

The adults of *B. malayi* bear a general resem-

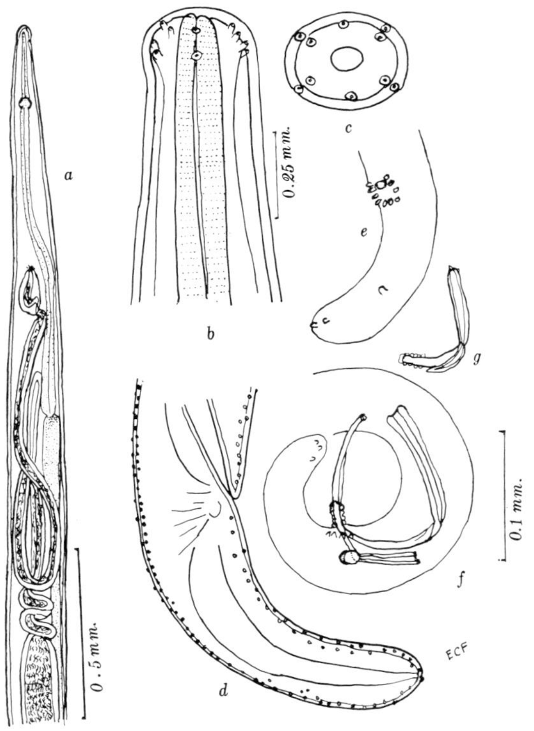

Fig. 23–9. *Brugia malayi. a,* Anterior part of female showing esophagus, vulva, vagina vera, and vagina uterina. *b,* Anterior extremity of female, lateral view, showing slightly bulged cephalic portion. *c,* Head-on view of anterior end showing pattern of perioral papillae. *d,* Posterior end of female showing cuticular bossing. *e,* Posterior tip of male, ventrolateral view showing perianal and postanal papillae. *f,* Posterior tip of male, lateral view, showing spicules and gubernaculum. *g,* Enlarged right spicule. (Adapted from Buckley, J.J.C., and Edeson, J.F.B. 1956. On the adult morphology of *Wuchereria* sp. *(malayi?)* from a monkey (*Macaca irus*) and from cats in Malaya, and on *Wuchereria pahangi* n. sp. from a dog and a cat. J. Helminthol., *30*:1–20.)

blance to those of *W. bancrofti;* they are delicate, whitish, thread-like filariae that live in dilated lymphatics. The tapering anterior end (Fig. 23–9,*a*) is provided with a nonlabiate mouth surrounded by two rows of minute papillae (Fig. 23–9,*b*), an inner row of six and an outer one of four, just as in *W. bancrofti,* except that in *B. malayi* they are slightly larger and more prominent. Mature females vary in length from 43 to 55 mm, and in breadth from 130 to 170 μm (compared with 80 to 100 mm and 0.24 to 0.3 mm, respectively, for the previously described material of *W. bancrofti*). Similarly, the mature males vary in length from 13 to 23 mm, and in breadth from 70 to 80 μm (compared with roughly 40 mm and 100 μm, respectively).

In the male the cuticula is smooth except that at the caudal end there are numerous ventral pseudostriations resulting from the coiling of this portion of the body in a left-handed curve. The adanal papillae (Fig. 23–9,*e*) vary from three to four, are relatively prominent, and are of unequal size. Immediately behind the anus there are a pair of papillae, three to four lateral adanal pairs, and immediately preanally an unpaired papilla. At the tip of the tail there are four to six small papillae, and between these and the adanal ones are two, one, or no papillae. The long left spicule (Fig. 23–9,*f*) consists of three parts: a proximal tubular portion, having a "pitted appearance," a short median nontubular portion, and a distal solid curved rod that may extend through the anogenital opening. The much shorter right spicule is curved ventrally and consists of a tubular proximal half and a narrower cylindrical distal corrugated portion that embraces the left spicule. The gubernaculum is more or less crescentic like that of *W. bancrofti*.

In the female the vulva (Fig. 23–9,*a*) consists of a transverse slit, within which the ovejector is folded back and forth before its junction with the vagina, a long, double-walled tube with a lumen of small diameter, which, in turn, merges with the uterus. Microfilariae are crowded together in coiled masses within the uterus and even the vagina in the mature paired female. These are discharged from the adult females and move into the visceral and then the peripheral circulation.

Essential differences between *Mf. malayi, Mf. bancrofti,* and *Mf. loa* are provided in Table 23–1. The most distinctive character of the microfilaria of *B. malayi* is the presence of two discrete nuclei in the tip of the tail (Fig. 23–10). In peripheral blood of patients, this embryo exhibits nocturnal periodicity. Feng (1936) studied developmental stages of this species in culicid mosquitoes. In the periodic strain in China they parallel closely those of *W. bancrofti,* with three true larval stages and two ecdyses. He demonstrated that the G cells of the microfilaria are not genital primordia but are the precursors of the rectum and associated parts.

Complete larval development in the susceptible mosquito requires a minimum of 6 days. The prepatent period in experimental animals was found to be 67 to 98 days (Orihel and Pacheco, 1966).

Epidemiology. Periodic *Brugia malayi* occurs in most of the filarial areas of Asia and in Indonesia on all the major islands as far east as the Island of Ceram (130°E longitude). The typical vector is *Mansonia,* but in Japan (one small island), Korea, and coastal China it is *Aedes togoi*. A subperiodic

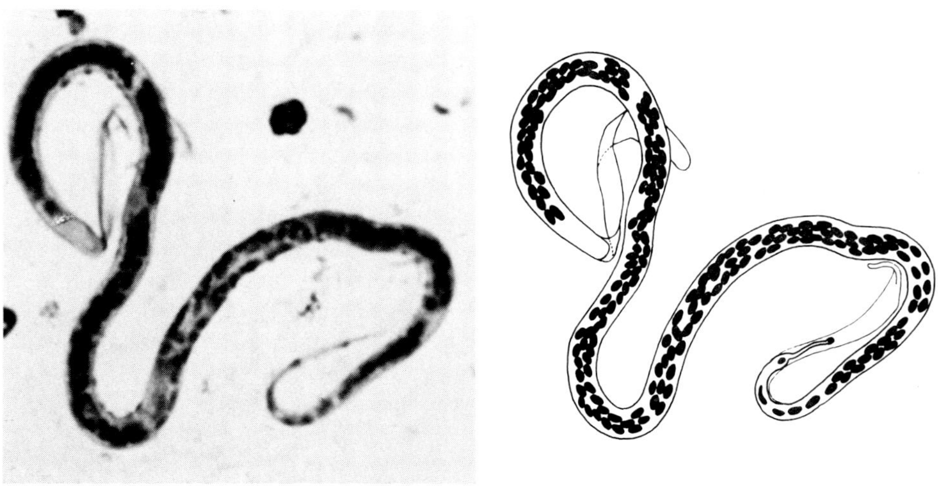

Fig. 23–10. *Brugia malayi.* Microfilaria in thick blood film stained with hematoxylin. Photomicrograph and camera lucida drawing. (× 1000.) (Courtesy of Dr. T.C. Orihel. From Faust, E.C., Beaver, P.C., and Jung, R.C. 1975. *Animal Agents and Vectors of Human Disease,* 4th ed. Philadelphia, Lea & Febiger.)

type that is slightly nocturnal occurs in parts of Malaysia, Thailand, South Vietnam, Sabak, Sumatra, Java, and the Philippines. In Korea *B. malayi* is the only filaria of man present. Some older records, prior to 1951, incorrectly indicated the presence of *W. bancrofti* in Korea. Natural vectors of the periodic type are *Aedes togoi, Mansonia uniformis, M. annulifera, M. indiana, M. dives, Anopheles campestris,* and *An. donaldi.* Those for the subperiodic type are *M. dives, M. bonneae, M. uniformis,* and *M. annulata.*

The role of biological vectors of *B. malayi* varies geographically and environmentally; the important transmitter in one area may not necessarily be the chief vector in another, even within relatively small areas such as adjacent river valleys. Human infection results from introduction of the mature larva into the puncture wound when an infected mosquito prepares to take a blood meal. Moreover, the periodic strains, which are responsible for most of Malayan filariasis in man, are not found as natural parasites of other animals, although experimentally they can be transmitted to cats and rhesus monkeys. In contrast, the subperiodic strains occur in a swamp-forest type of environment, the leaf monkey *(Presbytis)* is a natural host (75% infection), and species of *Mansonia* mosquitoes are the vectors in Malaysia, possibly also in swampy forests near human habitations in Borneo, Thailand, and other endemic areas.

In a filariasis survey of the Province of Palawan, Philippines, *Brugia malayi* was found to be restricted to areas of freshwater swamps, whereas *W. bancrofti* had a much wider distribution (Cabrera and Tamondong, 1966). Prevalence rates were approximately equal. The microfilarial rate of each species was higher in males than in females but the density rate was higher in *W. bancrofti* infection.

Experimentally in cats the microfilariae of the Malayan swamp environment are subperiodic as they are in man, but when the same strain of *B. malayi* is transmitted to monkeys (through susceptible mosquitoes) it becomes periodic. When transmitted back to cats following development in monkeys, this strain reverts to the subperiodic type.

Pathogenesis, Pathology, and Symptomatology. The pathogenic mechanisms of human filariasis due to *Brugia malayi* are essentially the same as those of *Wuchereria bancrofti.* Early lesions and their clinical manifestations in Malayan filariasis have been described by several French investigators, who studied the disease in North African and French soldiers who had been repeatedly exposed in nonurban areas in Vietnam and later repatriated. The syndrome discovered on the patients' return home was allergic in nature, with cardinal symptoms of high eosinophilia, bronchial asthma, and adenopathy. Microfilariae were never found in the blood of the French military personnel but were demonstrated from aspiration and biopsy of enlarged lymph nodes, at times associated with parent worms. Fugitive lymphangitis appeared in 16% of the patients.

Patients with more advanced Malayan filariasis may be asymptomatic with microfilariae in the peripheral blood at night, or they may have elephantoid enlargement of one or more limbs, occasionally the groin, rarely the scrotum. Rao and Maplestone (1940) obtained specimens of adult worms from a lymphatic cyst of the right forearm. The Indonesian workers (Bonne et al., 1941) described worms removed from inguinal and popliteal lymph nodes. Rao (1945) reported on 40 Indian cases, all with elephantiasis of the limbs, none with groin or genital lesions. Seo (1978) described the symptom complex of Malayan filariasis in which lymphadenopathy, lymphadenitis, and eventually elephantiasis, primarily affecting the lower limbs, are the prominent features.

Brugia malayi infection of the eye has been reported in two patients (Dissanaike et al., 1974; Joon-Wah et al., 1974). Living adult *B. malayi* may also be found in branches of the pulmonary artery, causing infarction (Beaver et al., 1971; Scully and McNeely, 1974).

Diagnosis. As in Bancroft's filariasis, specific diagnosis is made on recovery of active microfilariae of the particular type (Fig. 23–10) from the patient's blood, most frequently at night. Microfilariae of the periodic and subperiodic types of *B. malayi* in Giemsa-stained blood films differ in appearance in that those of the former tend to shed the sheath during the process of drying.

During the earlier, prepatent period the only evidence is clinical, primarily of an allergic nature, at times supported by biopsy of lymph nodes. In chronic infections the presence of elephantiasis, predominantly in an extremity, is suggestive.

Treatment. Specific treatment consists of administration of *diethylcarbamazine* (DEC) by mouth in a course as follows: 50 mg on day 1; 50 mg three times on day 2; 100 mg three times on day 3; 2 mg per kg three times daily on days 4 through 21. A second course of treatment may be required to kill the parent worms.

Prevention. In Kerala State, India, where *Mansonia annulifera* is the chief vector of *Brugia malayi,* removal of the water plant *Pistia stratioides* from the breeding grounds greatly reduced new infections. More recently in Sri Lanka, use of herbicides has been employed successfully to kill aquatic plants *(Pistia, Eichornia, Salvinia)* from the underwater stems of which the *Mansonia* vectors obtain oxygen, and in this way transmission has been effectively reduced. In urban areas, where *Anopheles* vectors are involved, intensive larvicidal and residual spraying with insecticides reduces exposure. In Malaysia, periodic mass treatment with diethylcarbamazine can be effective in the control of periodic, but not subperiodic, *B. malayi,* owing to zoonotic transmission of the latter (Mak et al., 1982).

Brugia timori
Partono, Purnomo, Dennis, Atmosoedjono, Oemijati and Cross, 1977
(The Timor filaria)

Synonym. *Timor microfilaria* of David and Edeson, 1964.

Historical and Geographical Notes. The Timor filaria was first recognized as distinct from *Brugia malayi* by David and Edeson (1965) during an investigation into filariasis on the island of Timor at the eastern end of the Indonesian archipelago. David had earlier observed that two kinds of microfilariae were present in the blood of military recruits from Portuguese Timor. One was *W. bancrofti;* the other was a species resembling *B. malayi*. Differences noted were in overall length, in cephalic-space ratio, and in failure of the sheath to stain with Giemsa. Detailed observations on the microfilaria were reported by David and Edeson (1965) and others (Purnomo et al., 1977), and eventually the adult worms were obtained from experimentally infected Mongolian jirds *(Meriones unguiculatus)* and described by Partono and associates in 1977. The infective larvae had been grown in *Aedes togoi* that were allowed to feed on people in a small village (Ae Bubu) in the eastern part of Flores Island.

Morphology, Biology, and Life Cycle. The microfilariae of *B. timori* are distinctly nocturnal periodic. The adults are normally found in the lymphatics, but those from experimentally infected jirds were recovered from the lungs, heart, and associated large vessels as well as the lymphatics of the testes. At 142 days of age the largest males were about 2 cm long by 70 μm wide; females were 3 cm by 100 μm. In both sexes the anterior end is expanded into a bulb-like head. The tail is long, tapered, and smoothly rounded. Four circumoral papillae are on either side, arranged in inner and outer circles. The esophagus is about 1 mm long with the border between muscular and glandular portions indistinct. In the male the coiled tail bears four or five subventral adanal papillae, a single large preanal and one pair of smaller postanal papillae, one pair of subventral intermediate papillae, and one pair of terminal caudal papillae. In addition, at the level of the anus there are one to five lateral papillae. The spicules are unequal and markedly dissimilar, as in *B. malayi,* the left spicule up to about 400 μm and the right one 142 μm in length, with a crescent-shaped gubernaculum about 30 by 4 μm. In the female, the vulva is anterior to the base of the esophagus, the ovejector somewhat pear-shaped about 160 by 58 μm, and the vagina inside the ovejector is looped. The tail, up to 196 μm long, bears numerous minute cuticular bosses. David and Edeson (1965) obtained fragments, including a head and a tail of two adult females in inguinal lymph nodes removed from patients. Bosses were observed on the tail.

The microfilariae of *B. timori* were compared with those of Indonesian strains of periodic and subperiodic *B. malayi* by Purnomo et al. (1977). As noted by others, they found marked distinctions in the failure of the sheath to stain with Giemsa, the long cephalic space with a length:width ratio of about 3:1, and a greater length of the body. In Giemsa-stained thick blood films, about 60% of the *B. timori* microfilariae were exsheathed, as is generally found in preparations of periodic *B. malayi* from Sulawesi (Celebes), but this percentage was markedly greater than the 1 to 2% found in subperiodic *B. malayi* from Kalimantan (Borneo). In Giemsa-stained films, *B. timori* microfilariae averaged 310 μm in length as compared with 264 μm and 247 μm in periodic and subperiodic *B. malayi,* respectively. In formalin-fixed specimens the relative average lengths were 341, 300, and 287 μm, respectively. Other differences noted were greater numbers of contiguous nuclei in the tail (5 to 8 vs. 4 to 5), and smaller nuclei in the distal portion of the tail, producing little if any bulge in the cuticle.

Epidemiology. *B. timori* is apparently limited in its distribution to the Lesser Sunda Islands in the Indonesian Archipelago extending from Wallace's line on the west (between Bali and Lombok) to Weber's line beyond Timor on the east. (Fig. 23–2). No other species of *Brugia* is known to occur in the same region. *Wuchereria bancrofti* does occur in the same region and is occasionally found in the same person. The natural vector of *B. timori* is *Anopheles barbirostris,* which breeds in rice fields and feeds at night both indoors and outside (Atmosoedjono et al., 1977). No animal reservoir has been discovered. Patent infections have not been produced in any experimental host. Infection rates are slightly lower in females and in children under 10 years of age, but in a survey of about 200 people on Flores, 8 of 37 children under 5 years were infected; in the same group microfilaremia rates were 31 and 20% for males and females, respectively (Dennis et al., 1976). *B. timori* was recently found in over 8% of about 1000 inhabitants of Alor, an island north of Timor. In the same group, *W. bancrofti* was found in 3% (Joesoef and Dennis, 1980).

Pathogenesis, Pathology, and Symptomatology. David and Edeson (1965) reported that on the island of Timor they found the lesions caused by *T. timori* few and mild. Elephantiasis was confined to edema of the legs below the knee. On the island of Flores, in a village of about 200 persons living in 32 households, Dennis et al. (1976) found evidences of filarial disease in 62%—lymphadenitis and lymphangitis in 56, scarring in 39, and elephantiasis in 14 of 99 males, and lymphadenitis and lymphangitis in 46, scarring in 27, and elephantiasis in 16 of 105 females. All households had one or more members with a history, or existing evidences, of lymphadenitis and retrograde lymphangitis of the legs, and 29 of 32 households had persons with characteristic scarring from draining abscesses in the chain of nodes along the great saphenous vein and its main tributaries (Fig. 23–11).

Diagnosis, Treatment, and Prevention. Except that morphological features of the microfilariae are different, the diagnosis of *B. timori* infection or disease is the same as for periodic *B. malayi*. Treatment and prevention are likewise similar for

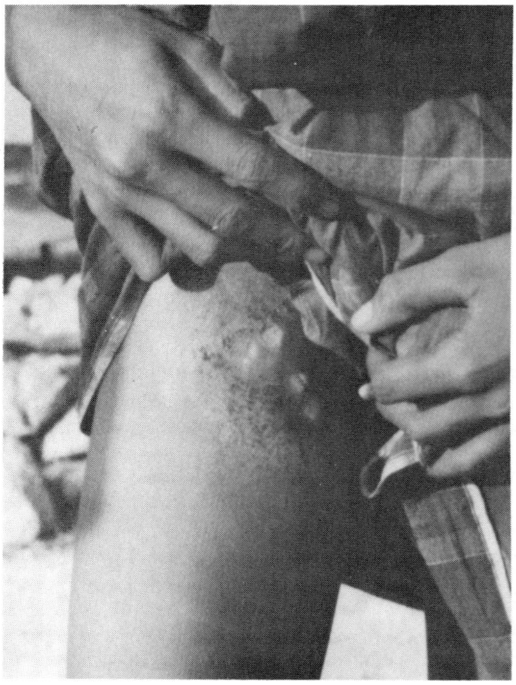

Fig. 23–11. *Brugia timori*. Draining abscesses caused by worms in lymph vessels along saphenous vein, referred to as "Bowo" by people in endemic areas on the Island of Flores, Indonesia. (Courtesy of Dr. F. Partono.)

the two species. Periodic mass treatment followed by two annual selective re-treatments and prompt treatment of all persons with acute clinical manifestations has been successful, the follow-up treatments being administered by a responsible leader of each community (Partono et al., 1981).

Zoonotic *Brugia* Infections

Nine species of *Brugia* have been described: *B. malayi* and eight others (see p. 364). *B. malayi* and *B. pahangi* in Malaysia and an unidentified species, possibly *B. beaveri*, in the United States are transmitted through vector mosquitoes from animals to man.

Brugia malayi. The subperiodic form of *B. malayi* in Malaysia occurs both in man and in a number of wild and domestic animals. Its presence in man and animals of the same localities makes control programs difficult.

Brugia pahangi. *B. pahangi* is a common parasite of dogs and cats in parts of Malaysia. It was described as *Wuchereria pahangi* by Buckley and Edeson in 1956. When the genus *Brugia* was described by Buckley in 1960, the species *W. pahangi* along with *W. malayi* and *W. patei* were transferred to the new genus. One of two volunteers inoculated with larvae from *Armigeres obturbans* that had fed on a *B. pahangi*-infected cat developed a microfilaremia that persisted for 56 days from the eighty-fourth day of infection (Edeson et al., 1960). Both volunteers experienced episodes of lymphangitis and lymphadenitis in the inoculated arm and hand. In some parts of Malaysia naturally acquired zoonotic *B. pahangi* infection could not readily be distinguished from endemic infection.

Brugia in the United States. Filarial worms reported as *Brugia* have been identified in sections of lymph nodes removed from six persons who could not have acquired the infections outside the north or northeastern region of the United States. The first to be identified was removed from the groin of a woman who had acquired the infection in the vicinity of New York City (Rosenblatt et al., 1962). The second was found in a node removed from behind the ear of a man who acquired the infection in New Jersey or Maryland (Schlesinger et al. 1977). Three others were reported by Coolidge et al. (1979). One was removed in a node from the chest wall of a woman who lived in Massachusetts but had visited New Hampshire and Maine, another was from the neck of a woman who had acquired the infection in Rhode Island, and one was from the chest wall of a 15-year-old boy whose exposure was limited to Massachusetts. In the sixth case a mature infertile female was found in a lymph node removed from the left groin of a 53-year-old man in northern Ohio (Gutierrez and Petras, 1982). In all cases the worms were sexually mature—single males in two, single females in three, and both sexes in one. Identification to species was not possible. The diameter of the worms was 48 to 55 μm in four cases, and 35 μm and 75 μm, respectively, in the other two. *B. beaveri* is the only described species known to occur in North America. It is a common parasite of the raccoon in Louisiana and its microfilariae have been identified in stained blood films from bobcats in Louisiana and Florida. It has not been reported in animals from the northeastern states. However, the range of its natural hosts extends to and beyond the states where infections in man were acquired. The diameter and other features of the worms evident in sections of tissue from man were indistinguishable from those of *B. beaveri* described by Ash and Little (1964).

Onchocerca volvulus
(Leuckart, 1893) Railliet and Henry, 1910
(The convoluted filaria, producing onchocerciasis, "river blindness.")

Synonym. *Onchocerca caecutiens* Brumpt, 1919.

Historical and Geographical Notes. This filaria was described by Leuckart from specimens obtained from a native of Ghana, West Africa. On the Pacific slope of Guatemala, Robles (1915) found an *Onchocerca*, which Brumpt (1919) described as *O. caecutiens*, but which is *O. volvulus*. The most important step in describing the biology of the disease was the demonstration of the black fly, *Simulium*, as the intermediate host and transmitter (Blacklock, 1926). The most significant clinical finding was the causative relationship of this filaria to blindness (Robles, unpublished report, 1915).

Ruiz Reyes (1952) presented evidence consistent with the view that onchocerciasis was brought to endemic areas in tropical America when blacks were imported to cultivate sugar cane and coffee.

In Africa, according to a WHO (1976) report, the northern boundary of endemic onchocerciasis extends from Senegal to about latitude 15°N in Mali, Niger, and Chad, and on eastward to Sudan, where it reaches its northernmost point in the Abu Hamed area at 19°15'N, and from there extends southward through western Ethiopia. Farther to the south, foci of the disease exist in Gambia, Guinea-Bissau, Guinea, Sierra Leone, Liberia, Ivory Coast, Upper Volta, Ghana, Togo, Benin, Nigeria, Cameroon, Equatorial Guinea, Central Africa Republic, Congo, Zaire, Rwanda, Burundi, and Uganda. The southernmost endemic areas are in the Cholo district of Malawi at latitude 17°S, in Tanzania at latitude 12°S, and in the Macha district of Zambia at latitude 16° 26'S (Beaver et al., 1983). In Kenya there is a small focus along the Ugandan border. In the Western Hemisphere *O. volvulus* has been found on certain coffee plantations in the western slope of Guatemala; at 600 to 2000 meters elevation in the southern states of Mexico, including Chiapas and Oaxaca; in northern Venezuela at an altitude of 250 to 1,500 meters (states of Monagas, Carabobo, and Aragua); and in southern Venezuela (Amazonas) along the Brazilian border (Rassi et al., 1977). It is prevalent among Indians of the Amazonas and Roraima regions of northwest Brazil (Moraes et al., 1978, 1979). It is also present in isolated communities in the mountains along the Pacific Coast in southwestern Colombia (Ewert et al., 1979) and northeastern Ecuador (Arzube, 1982; Guderian et al., 1982). There is a small focus of infection in Yemen, South Arabia (Fawdry, 1957; Omar et al., 1979; Büttner et al., 1982; Connor et al., 1983).

Morphology, Biology, and Life Cycle. Adult worms are commonly located in nodules in subcutaneous connective tissue (Fig. 23–12,*A*), although at times they are so deeply situated that they cannot be easily palpated. Nodules may appear on any part of the body but are most common in the region of the pelvic arch, at the junction of the long bones, and in the temporal or occipital regions of the scalp (tropical America). The living worms are white, opalescent, and transparent, with distinct transverse striations of the cuticula (Fig. 23–12,*B*). Nodules may contain one to several worms of one sex only, but usually both sexes are present and they may be numerous. Nodules containing up to 20 females and 13 males have been

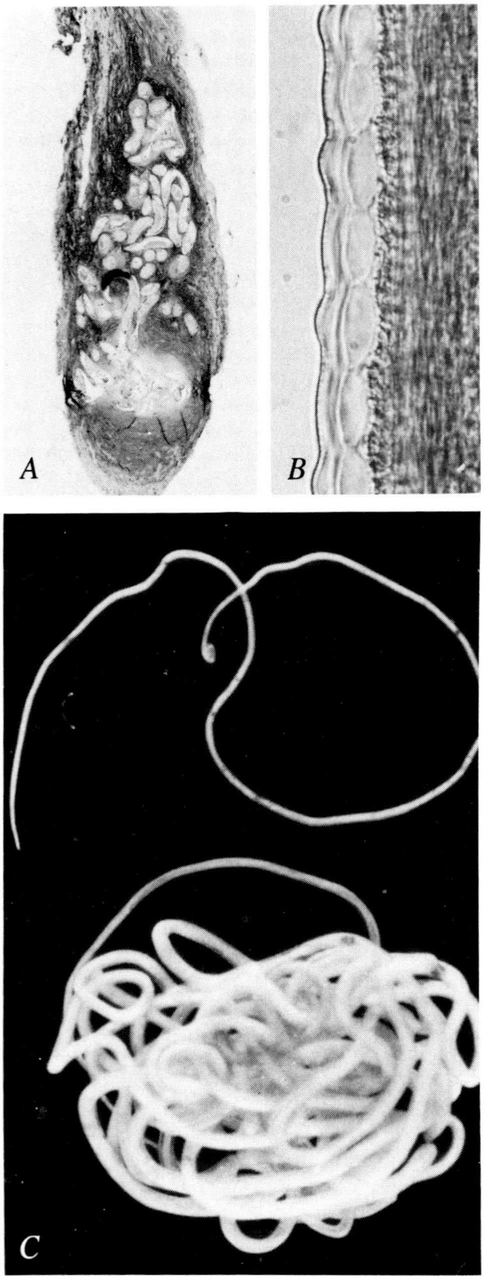

Fig. 23–12. *Onchocerca volvulus. A,* Section through a recently formed nodule removed from subcutaneous tissue over the iliac crest of an American woman missionary who had lived in a remote area of northern Brazil. (× 5.4.) *B,* Longitudinal section of cuticle showing pattern of external ridges and internal striae, on a worm in a nodule from a resident of Huixtla, Mexico. (× 600.) *C,* Mature male (above) and female digested out of a nodule. (× ca. 5.) (By P.C. Beaver.)

observed (Schulz-Key and Albiez, 1977). The worms are filiform and blunt at both ends. At the anterior extremity are eight small, submedian sessile papillae arranged in two rings, and a pair of large, oval, lateral papillae (Fig. 23–13,*A*). The worms are characteristically tightly coiled within the nodule (Fig. 23–12,*C*).

The males measure 19 to 42 mm in length by 130 to 210 μm in diameter; the females 33.5 to 50 cm by 270 to 400 μm. The male is tightly recurved ventrad at the posterior end and has several perianal and caudal papillae which vary considerably in number, size, and symmetry (Fig. 23–13,*B*). In the female the vulva opens slightly behind the posterior extremity of the esophagus. *In utero* the embryos are at first ovoid but later become elongated. On leaving the mother worm, the microfilariae escape from their "sheath." Early workers described the microfilariae of *O. volvulus* as falling into two size-classes, which they concluded were male and female (Blacklock, 1926; Wanson, 1950). Other observers, also noting differing size-classes, one with length measurements around 165 μm, the other 250 μm, attributed these differences to uneven

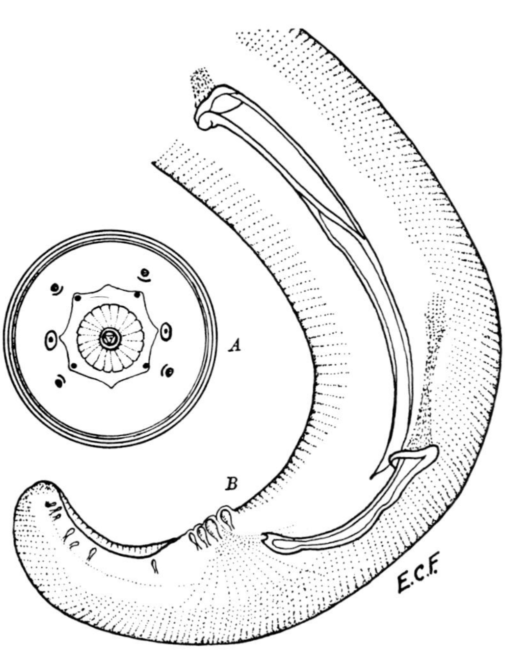

Fig. 23–13. *Onchocerca volvulus. A,* Head-on view of adult worm, showing papillae. (× ca. 660.) *B,* Posterior end, lateral view, showing copulatory spicules and caudal papillae. (× ca. 690.) (Adapted from Sandground, J.H. On the validity of the various species of the genus *Onchocerca* Diesing. In Strong, R.P., Sandground, J.H., Bequaert, J.C., and Ochoa, M.M. *Onchocerciasis, with Special Reference to the Central American Form of the Disease,* pp. 135–172. Cambridge, Harvard University Press.)

shrinkage in different areas of the smear (Gratama, 1966). A wide range in dimensions (220 to 360 by 5 to 9 μm) is frequently reported (Gibson et al., 1980). When microfilariae from skin are immobilized and stained without fixation, or fixed in 2% formalin for Knott's concentration, the length-width measurements are 315 to 360 μm by 6 μm (Bain, 1975). The anterior and posterior ends of the microfilariae are nuclei-free, as is the region of the excretory pore. These microfilariae occasionally are found in peripheral blood and in the urine and sputum, but they are typically found in the lymphatics of the connective tissues and cutaneous layers in the vicinity of the parent worms, as well as in the stratum germinativum and the corneal conjunctiva. Adult *O. volvulus* may live for 15 years and are capable of producing microfilariae for at least 9 to 10 years (Roberts et al., 1967). The life span of the microfilariae may be as long as 2 years (Duke, 1981).

The studies of Blacklock (1926) demonstrated that the black fly *(Simulium)* serves as intermediate host of this worm. This finding was confirmed by Bequaert (1928) in Africa, Strong (1931) in Guatemala, and Hoffmann (1930) and others in Mexico. Black flies are "pool feeders" and ingest blood and tissue fluids. Following uptake, microfilariae characteristically migrate from the anterior portion of the midgut through the hemocoel to the thoracic muscles. In the thoracic muscles development with two ecdyses occurs, requiring a minimum of 6 days after which the third true larval stage (filiform in type) migrates to the mouthparts and is ready to infect man when the fly takes the next blood meal. The prepatent period in man is 3 to 15 months.

The species of *Simulium* thus far incriminated are as follows: in Africa, *S. damnosum* complex and *S. neavei* group, both of which have an extensive distribution in tropical Africa, and possibly *S. albivirgulatum* on a limited basis; in Guatemala and southern Mexico, *S. ochraceum* and possibly *S. metallicum* and *S. callidum*; in Venezuela *S. metallicum* is the principal vector; in Colombia the chief vector is *S. exiguum,* and *S. metallicum* is a poor experimental host (Tidwell et al., 1980); and in Brazil the vector is *S. sanguineum*. These species are anthropophilic. In Guatemala, *S. metallicum* and *S. callidum* are zoophilic, feeding on cattle and horses as well as on humans.

Natural infections have been found in the spider monkey, *Ateles geoffroyi vellerosus,* in Mexico and in the gorilla in the Congo, but there is no conclusive evidence of any extra-human source of *O. volvulus* infection.

Epidemiology. The species of *Simulium* that serve as intermediate hosts and transmit the filaria to man at the time they take a blood meal obtain their infection from infected human beings in whose skin the microfilariae are migrating. The flies oviposit on stones, trailing vegetation or protruding objects in fast-flowing mountain streams, where the larval and pupal stages develop. Therefore, the disease is most common along rivers and smaller streams, which are ideal for *Simulium* breeding. The important breeding grounds of *S. damnosum* and *S. neavei* are not in the main river course but in the smaller, fast-flowing rivulets. The larvae and pupae of *S. neavei* attach themselves to crabs *(Potamonautes)*. This black fly is an important vector in Uganda. A related species, *S. woodi,* whose larvae and pupae also attach to crabs serves as a vector in Ethiopia, Malawi, and Tanzania. Species of the *S. neavei* group are largely confined to eastern Africa (Raybould and White, 1979).

In Mexico and Guatemala the disease is associated with the cultivation of coffee, yet onchocerciasis does not occur in all areas where the appropriate species of *Simulium* breed. In Oaxaca State, Mexico the endemic zone is inland from the Pacific Coast in relatively inaccessible mountain valleys.

In West Africa, in the arid northern savannah, human onchocerciasis is associated with severe eye lesions and blindness, whereas in the humid southern forest, although much onchocerciasis is present, eye lesions and blindness are rare. Duke and associates in 1966 carried out experimental studies on *Onchocerca volvulus* from forest and savannah zones and *Simulium damnosum* from various bioclimatic zones of West Africa (Duke, 1972). The results suggested that in this region two strains of *O. volvulus* and also two principal physiological strains of *Simulium damnosum* are present. These associations form two largely distinct parasite-vector complexes, one extending over the forest and Guinea-savannah zones and the other confined to the Sudan-savannah zones.

Studies on the bio-ecology of female *S. damnosum* in West Africa have shown that the fly's life span increases, but the area of dispersal around the breeding places decreases from the tropical forest to the savannah. In the savannah zone there is considerable risk of human infection near the breeding places where the longer-lived females are concentrated and a relatively low risk farther away. In the forest areas the risk of human infection is less but extends over a wide area. These variations are sufficient to explain the epidemiological differences in human infection in forest and savannah regions.

Experimentally, the microfilariae of the forest strain develop well in *S. damnosum* from the forest, but develop poorly in those from the savannah, and vice versa. Microfilariae of the savannah strain develop poorly in flies from the forest (Duke, 1972). There are corresponding differences in pathogenicity of the two strains of the parasite (Prost et al., 1980).

Pathogenesis, Pathology, and Symptomatology. The basic mechanism of disease in onchocerciasis, as in infections with *Wuchereria* and *Brugia,* is the provocation of the host's allergic reaction to the adult worms and to the microfilariae. The differences between the clinical pictures of onchocerciasis and Bancroft's or Malayan filariasis are due to the differing habitats of the parasites. *Onchocerca volvulus* adults reside typically in subcutis rather than in lymphatic vessels, and their microfilariae migrate in the skin rather than in the blood and lymphatic circulation. In those cases in which lymphatics are significantly involved, clinical features may appear in onchocerciasis resembling those of Bancroft's or Malayan filariasis.

The adult worms live singly, in pairs, or in coiled entangled masses in the deep fascia and in the subcutaneous tissue. Some may lie free in the tissue, undetectable except by search of the tissues at autopsy. Others provoke a fibroblastic reaction in the host which causes the worms to become enveloped in a fibrous scar forming the onchocercal nodule or *onchocercoma.* These nodules tend to occur over anatomic sites where the bone is superficial, such as the scalp, scapulae, ribs, elbows, trochanters, iliac crests, sacrum, and knees. Other nodules may be so deeply situated as to be unpalpable. Insufficient autopsy studies have been performed to estimate what proportion of adult worms are thus deeply situated or in internal viscera. A nongravid female has been found in the wall of the aorta in an African patient (Buck, 1974). Among African subjects with microfilariae of *O. volvulus* in skin snips, of those over 20 years of age 60% or more have palpable nodules, of those 10 to 19 years old 30 to 60% have nodules, and of those 5 to 10 years old, fewer than 30% have nodules. It is therefore evident that a high percentage of adult *Onchocerca* exist either without fibrous encapsulation or in situations too deep to be palpable.

The typical onchocercoma is a firm, round, or ovoid, nontender, well-circumscribed subcutaneous tumor (Fig. 23–14,*A*). They vary in diameter from 0.5 cm to 10 cm or more. Usually they are fully movable, but they may be attached to the periosteum, deep fascia, or skin. They may occur as isolated nodules or as large conglomerations. Small satellite nodules may surround such a mass.

In hyperendemic African villages the average number of palpable nodules in males over 40 years of age is 10, but some individuals have over 100. In Africa the nodules are commonly found about the pelvis, especially the iliac crests, trochanters, and lateral chest wall. Less commonly they occur about the knee, leg, and ankle or foot. In contrast, in Central America over 50% of the nodules are found on the head, 15% on the chest wall, and only 25% around or below the pelvis.

The nodule is composed of an outer wall of hyalinized and vascularized fibrous tissue that encloses the adult worms (Fig. 23–12,*A*). In some nodules the scar tissue is present throughout the nodule, restricting the worms to narrow tunnels. In others the worms reside in a soft center composed of granulation tissue, fibrin, precipitated immune complexes, polymorphonuclear neutrophils and eosinophils, macrophages, giant cells, lymphocytes, plasma cells, and Russell bodies. Each nodule contains one or more worms, nongravid or gravid, living or necrotic. Male worms may or may not be present. A thin layer of eosinophilic material, probably an immune complex, tends to form around degenerating worms. Calcification of dead worms may occur (Kershaw et al., 1955). Numerous microfilariae are present in the centers and walls of the nodules containing living adult worms (Gibson et al., 1980).

Onchocercal nodules per se usually cause no symptoms excepting their effect on physical appearance. At times a deep-seated bundle of adult worms or an onchocercoma may be situated in relation to a joint capsule or tendon sheath between muscles or against the periosteum. In such instances they may cause deep pain.

The other clinical manifestations of onchocerciasis are due to the effects of microfilariae on the host. These mostly affect the skin and eyes, but lymphatic and systemic lesions are also produced.

Onchocercal Dermatitis. The damage caused to the skin by microfilariae is probably by immune

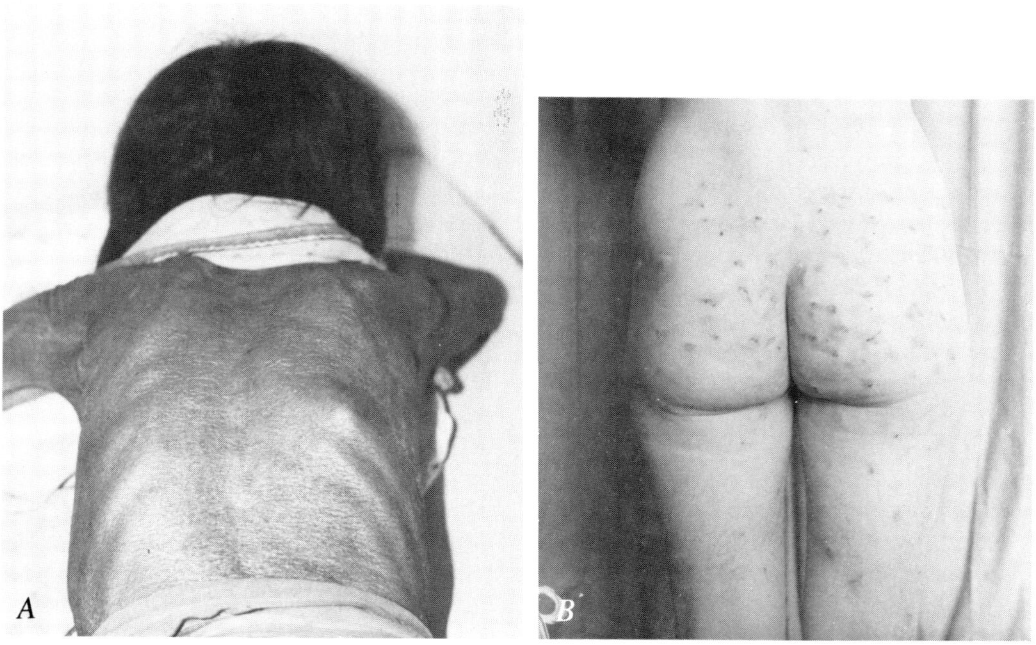

Fig. 23–14. *Onchocerca volvulus. A,* Subcutaneous nodule on chest wall of a child in Ecuador. (From Guderian, R.H., et al. 1982. Report on a focus of onchocerciasis in Esmeraldas Province of Ecuador. Am. J. Trop. Med. Hyg., *31*:270–274.) *B,* Diffuse pruritic rash of a type frequently associated with onchocercosis. (Courtesy of Dr. P.E.C. Manson-Bahr.)

and possibly directly irritative mechanisms. Certain microscopic features are consistently found in onchocercal dermatitis. These include: (1) *Fibrosis.* This is probably the earliest recognizable microscopic abnormality and continues throughout the course of infection. It is recognized by an increase of fibroblasts among collagen fibers and of reticulum fibers throughout the dermis, and the appearance of scar tissue in the dermal papillae. The reticulum fibers tend to encircle the blood vessels. (2) *Acid Mucosaccharide.* This is recognized as diffuse and focal deposits of alcian-blue-staining material between collagen strands and in proliferating stellate mesenchymal cells. (3) *Phagocytosed Melanin Pigment* in the upper dermis (Buck, 1974).

The earliest symptom is usually pruritus, which may be intense but sometimes does not occur. The first objective sign is altered pigmentation. Areas of hyperpigmentation occur either in a macular pattern or in irregular blotches. In the Western Hemisphere manifestations of the disease in the skin are mild or absent. Both pruritus and altered pigmentation tend to occur in the lower parts of the body in Africa. In West Africa the forest variety of onchocerciasis causes more severe cutaneous and lymphatic lesions than does the Sudanese savanna variety but less ocular damage *(vide infra)* (Duke and Moore, 1974). Rarely there may be hypopigmented macules resembling those of streptocerciasis, granuloma multiforme, leprosy, the treponamatoses, and some dermal mycoses.

Early in onchocercal dermatitis the histopathology consists of light infiltration with inflammatory cells around the dermal vessels and appendages. These cells include lymphocytes, histiocytes, plasma cells, and eosinophils but not neutrophils. There is edema of the collagen of the skin and increased numbers of fibroblasts between the collagen fibers. Microfilariae are present but may be scarce.

Acanthosis, hyperkeratosis, and focal parakeratoses are present, and macrophages containing ingested melanin are found in the upper dermis by the time there is detectable change in pigmentation.

Most of the numerous varieties of skin changes may be grouped into three main types:

(1) Acute pruritic rashes composed of small papules, sometimes with urticarial or vesicular lesions (Fig. 23–14,*B*). Excoriation due to pruritus leads to secondary infection.

(2) Hypertrophic, hyperpigmented thickened skin. *Erisipela de la costa* of the Western Hemisphere and *sowda* of Yemen are extreme examples.

(3) Lichenified, atrophic, depigmented lesions in chronic cases.

A more useful classification for an understanding of the immune mechanisms involved in onchocerciasis is: *generalized* or *localized*. The generalized disease is the form most commonly seen in endemic areas. Microfilariae tend to be abundant in the skin with no significant differences in number with respect to distance or direction from a nodule (Kale, 1982). Skin reaction may be inapparent or relatively severe but, when present, tends to be symmetrical. The most striking example of localized onchocerciasis is sowda *(vide infra)*. Although the skin reaction is marked in this condition, it is localized. Microfilariae are extremely scanty in skin snips. Sowda occurs predominantly in young adult males of North Yemen. In older males the condition tends to be generalized with "leopard skin" configuration. Some of the older males have earlier had sowda treated in the traditional manner by removal of the enlarged inguinal nodes. It is suggested by Connor et al. (1983) that generalized onchocerciasis occurs in persons whose immune reactions to the microfilariae are less marked, possibly because of prenatal experience with microfilarial antigen; in localized disease such as sowda, an intense host reaction to the microfilariae rapidly destroys them, markedly reducing their numbers and keeping the skin reaction localized.

The lesions have been described as follows: *craw craw* or *gale filaremia* is a rash of variously sized papules containing pus or clear fluid. Pruritus is usually severe and the resultant scratching leads to secondary infection. Microscopically the lesions are intraepithelial abscesses containing serous fluid, polymorphonuclear neutrophils, and fibrin. One or more microfilariae can usually be discovered within the abscess. A parakeratotic scale or crust forms over the primary lesion.

Edema may be localized and red superficially, causing the "peau d'orange" appearance or the diffuse, poorly pitting swelling of lymphedema. Swelling of a single extremity is characteristic. Microscopically the dermis is edematous and contains many microfilariae, some of which form clusters within dilated lymphatic vessels.

Lichenoid change consists of dryness and scaling in a pattern that has been called "lizard skin." Microscopically the keratin layers and keratin are separated from epidermis. Atrophy of the epidermis is present, and there is scarring and loss of elastic fibers in the dermis.

"Leopard skin" occurs in long-standing onchocerciasis and consists of localized spotty depigmentation. The white depigmented areas contain dark spots, the foci of persistent pigmentation around the pores and hair follicles. Histopathologically there is an absence of melanin in the affected areas of epidermis.

Atrophy is another late stage of onchocercal dermatitis. The skin is glossy and thin appearing, dry, and unevenly pigmented. Owing to loss of elasticity, it tends to be wrinkled and to hang in folds. Microscopic examination shows a thin keratin layer that separates easily from the epidermis. The epidermis is thin and the rete ridges are absent. The most striking feature in the dermis is the replacement of normal dermal structures with scar tissue. Reticular fibers are abundant, and elastic fibers are reduced, especially in the upper dermis. In the lower dermis the latter are thickened and knotted or tangled (Buck, 1974).

Sowda, from the Arabic word, "aswad," meaning "black," is the name of an important manifestation of localized onchocerciasis in Yemen. It occurs predominantly in young adult males, probably because of their habit of dress, which exposes the legs to the bites of the vector. Usually the condition is localized to one lower extremity, although in females and children, whose style of dress is different, it tends to occur on the arms or back. The affected area of skin is markedly pruritic, dark, and thickened, and shows a papular or pustular eruption. The most striking feature of the histopathology is marked, extensive inflammatory cell infiltration accompanied by edema and fibrosis. The inflammatory cells, consisting of abundant plasma cells, some lymphocytes and eosinophils, and a smaller number of histiocytes, are concentrated around the skin appendages and vessels, but also involve and thicken the dermis and extend into the subcutaneous fat. Neutrophils are few. Mast cells are numerous around small vessels and throughout the extracellular interfibrillar spaces. After treatment with DEC, degranulating mast cells may be found. Edema is diffusely present, and lymphatic channels of the upper dermis are dilated. There may be fibrinoid change of collagen and focal granulomas of epithelioid cells with giant cells (Büttner et al., 1982; Connor et al., 1983).

Lymphadenitis. The lymphadenitis that occurs in African onchocerciasis differs markedly from that in Yemen (sowda). In the African disease, any of the superficial lymph nodes may become palpably involved, but those draining the areas of onchocercal dermatitis are not affected. The reaction in the lymph nodes may be due to deposition of immune complexes carried to them via lymphatic channels (Connor et al., 1983). The nodes tend not to become matted or fixed, and are neither painful nor tender but are firm or hard. The histopathology consists of atrophy of the follicles with fibrosis of the capsule, trabeculae, and medulla of the node and, in the early and intermediate stages, with dilatation of sinusoids of the subcapsular level and of the medullary lymphatics. Histiocytes are found in the sinusoids, and plasma cells, eosinophils, and mast cells throughout the node. Macrophages containing ingested melanin pigment are present. Microfilariae are found in the nodes, some apparently viable and others degenerating and surrounded by a granulomatous reaction consisting of epithelioid cells, giant cells, eosinophils, and fibrous tissue. The medulla contains large numbers of eosinophils, plasma cells, mast cells, histiocytes, and a few giant cells and Russell bodies. In the late stage the capsule is thick and fibrotic, and the lymphoid tissue is largely replaced by scar tissue. The subcapsular sinusoids that are dilated in the earlier stages are frequently obliterated.

In *sowda (vide supra),* the onchocercal disease of Yemen, the lymph nodes present a distinct picture: large (up to 6 cm in diameter) and soft. The cut surfaces are homogeneous and pale yellow. Microscopically they show marked hyperplasia of the follicles. Germinal centers are large and numerous. Plasma cells and histiocytes, though plentiful, are fewer than in nodes of African patients. Microfilariae are not found in these nodes (Gibson and Connor, 1978).

A peculiar manifestation of onchocercal lymphadenitis in Africa is the hanging groin. Elephantiasis of an extremity and of the genitalia has been reported. These conditions do not occur in sowda.

Ocular Manifestations. The appearance and course of ocular lesions of onchocerciasis depend on the duration and severity of the infection, and the anatomic location of the adult worms. Males, who tend to be more heavily infected than females, therefore are more likely to develop eye lesions. Males also develop them at an earlier age than do females. While the same types of ocular lesions are found in all areas of endemic onchocerciasis in both Eastern and Western Hemispheres, in African savanna villages sclerosing keratitis and anterior uveitis are more common and more severe than in villages in the rain forest with comparable infection rates.

The specific lesions of the eyes are as follows:

Punctate keratitis is a whitish corneal opacity of about 0.5 mm diameter, with an indistinct edge when it occurs in the more superficial third of the stroma. The epithelium is intact. In deeper lesions the color is gray. Punctate keratitis is due to the reaction forming around a dead microfilaria. It is relatively symptomless, interferes little with vision, and disappears without residue.

Sclerosing keratitis is caused by the reaction to large numbers of microfilariae in the cornea. It consists of corneal haziness appearing as a white segment at the 2 to 4 or 8 to 10 o'clock positions. The lesion becomes vascularized with fine superficial vessels, and pigment migrates into it. The sclerosed lesion progresses from the sides and below. Eventually the pupil may be covered and blindness result, but frequently the central and upper areas remain relatively clear. The clearer areas are packed with innumerable microfilariae. When severe and advanced, sclerosing keratitis is frequently accompanied by anterior uveitis. Calcification of the cornea may occur in elderly patients.

Iridocyclitis occurs in the more heavily infected patients. It is variable in degree from mild uveitis without conjunctival injection and with scant precipitate and no synechiae in the anterior chamber to severe plastic inflammation with heavy infection, and precipitates on the corneal endothelium, lens capsule, and the base of the anterior chamber. Fibrosis retracts the iris, causing distortion of the pupil, and posterior and anterior synechiae are formed. Glaucoma may occur secondary to severe iridocyclitis, and eventually the globe may be destroyed. Atrophy of the iris is believed sometimes to be a result of onchocerciasis. This causes a loss of pigment of the iris frill and is sometimes accompanied by a fibrin deposit on the pupil margin.

Choroidoretinal lesions and optic atrophy also occur, usually in association with anterior segment lesions in heavy infections.

Symptomatology of ocular onchocerciasis begins with photophobia and progresses through gradual blurring of vision to blindness. Both sexes tend to be affected but usually not to the same degree (Buck, 1974; Gibson et al., 1980).

Diagnosis. Specific diagnosis may be made from puncture of a suspected tumor and discovery of microfilariae in the aspirate. At times the nodules may be inhabited only by females or males or by dead worms, in which case no microfilariae are present. Biopsy of a small piece of epidermis, placed on a slide in water or physiologic salt solution, is a satisfactory technique for demonstrating the presence of microfilariae. A sharp razor blade is usually employed, and the sample of skin may be so superficial that little or no bleeding occurs. More uniform specimens can be taken with a standard corneal-scleral punch. Any location on the body may be satisfactory for the biopsy, but the scapular region is usually preferred. Specimens taken around mid-day usually yield the highest number of microfilariae (Thomas et al., 1973; Duke and Moore, 1974), though differences between yields at different times of the day may be slight (Tada and Figueroa, 1974). Biopsy specimens are allowed to stand for 30 minutes or longer in water or saline and are examined for microfilariae that migrate from the tissue. For quantitative diagnosis the biopsies are incubated in tissue culture medium or buffered saline plus antibiotics for 24 hours (Collins et al., 1980; Awadzi et al., 1980). Microfilariae in specimens of skin or adult worms in nodules can be recovered alive and counted after digestion in collagenase (Schulz-Key et al., 1977; Schulz-Key, 1978).

In patients with ocular manifestations, small biopsied specimens of the corneal conjunctiva may yield the microfilariae. However, it is impossible to make an accurate diagnosis of the actual damage produced by the infection without a thorough ophthalmoscopic examination. Sometimes microfilariae can be demonstrated in the skin of persons with no palpable nodules. Only exceptionally are microfilariae accidentally found in peripheral blood. They must then be differentiated from microfilariae of other species normally circulating in the blood.

Immunological tests have not come into general use for the diagnosis of onchocerciasis. Usually conventional methods are adequate, with immunological methods used only to meet special requirements.

Treatment. *Nodulectomy* is a popular form of treatment because it is simple, easy to perform, and produces immediate results visible to and therefore appreciated by the patient. Unfortunately, as has been pointed out previously, since a large percentage of the microfilariae are coming from adult worms not dwelling in palpable nodules, the procedure may have little effect on the patient's total burden of adult worms or microfilariae. Nevertheless, removal of nodules from the head, especially in children, effectively reduces the risk and severity of ocular disease due to microfilarial invasion. In Yemen, the removal of enlarged regional lymph nodes is the traditional treatment for sowda (Connor et al., 1983).

Discovery of *diethylcarbamazine* (DEC) (Hetrazan) and its efficacy in Bancroft's filariasis (Mazzotti and Hewitt, 1948) and the clinical trials with *suramin* (Bayer 205, Germanin, Naphuride) suggested the potential value of these two drugs in the treatment of onchocerciasis. Both of these agents, even when administered in amounts considerably less than those well tolerated in Bancroft's filariasis, produce side effects of considerable intensity in *Onchocerca* patients, namely, severe pruritus, urticaria, and ocular changes (including temporary blindness). These reactions are interpreted as of foreign protein causation, due to the rapidly lethal effect of drugs on the microfilariae and initial damage to the parent worms. Antihistaminics administered concurrently frequently relieve the severe allergic symptoms, but corticosteroids are more effective. If the *Onchocerca* tumors are removed before chemotherapy is started, the ocular manifestations are minimized.

Treatment of onchocerciasis with suramin consists of the weekly administration of 1 g for 5 weeks as a 10% solution adminstered by vein (WHO Expert Committee on Onchocerciasis, 1954). DEC is given orally in one dose of 0.1 mg per kilo on day 1, then 0.1 mg per kilo thrice on day 2, then in doubled dosage every 3 days up to 4 to 5 mg per kilo thrice daily (maximum 400 mg per day) continued for 2 weeks. Ashburn et al. (1949) found that with the former treatment nodules removed 121 days later contained dead or dying worms. The lethal effects of DEC are much slower on the adult worms, although this drug rapidly destroys the circulating microfilariae. Hawking (1958) suggested preliminary treatment with DEC to kill circulating microfilariae followed by suramin to kill the adult worms. DEC has no effect on the nodule. The formation of a nodule during DEC therapy has been reported (Boye and Connor, 1982).

Administration of levamisole, either alone or in combination with DEC or mebendazole, does not appear to be useful. Clinical trials with mebendazole and flubendazole have shown that both are as effective as DEC in reducing numbers of microfilariae, with fewer systemic side effects, although the effect of DEC is more rapid. Reduction of microfilariae in skin snips and in the eye was more lasting with flubendazole. Mebendazole appears to interrupt embryogenesis of the adult female (Rivas-Alcalá et al., 1981; Dominguez-Vasquez, 1983). Thus, mebendazole and flubendazole may prove to be useful alternatives to DEC in the treatment of onchocerciasis.

Trichlorophone (Metrifonate), an organophosphorus compound, has been reported in Mexican trials to be effective against the microfilariae of *O. volvulus* and to kill some of the adult worms (Salazar et al., 1971). In African tests microfilaricidal activity was confirmed but no action on adult worms was demonstrated (Duke, 1972).

Prognosis. Prognosis is good when lesions do not endanger the eye. Ocular damage is the most dangerous aspect of onchocerciasis.

Prevention. The prevention of transmission can be accomplished most readily by application of insecticides to the streams where the vector *Simulium* is breeding. In a geographically isolated region, if the vector is eradicated no other measure is needed. This was demonstrated by Garnham and McMahon in Kenya (Garnham, 1954), where it was shown that for 15 years after eradication of the vector, *S. neavei,* there was no transmission and a gradual disappearance of the infection (Duke, 1972). In less isolated areas where *S. damnosum* is the vector, and reinvasion cannot be prevented, eradication has proved to be more difficult or impossible. Building dams for other purposes has eliminated breeding sites in some instances, cited by Nelson (1970), and in a few instances control by bush clearing and modification of habitats in other ways has been successful. In West African savanna areas a rapidly biodegradable organophosphorus compound, temephos (Abate), is being used in a long-term control program covering nearly 1 million km^2 in Benin, Ghana, Ivory Coast, Mali, Niger, Togo, and Upper Volta. Aerial application at a dosage of 0.05 ppm/10 min. of river discharge has proved successful, although there is evidence of incipient insecticide resistance. The program aims at providing protection for over 10 million people for a period of 15 to 20 years (Duke, 1981).

In areas such as Guatemala and Mexico where the vectors breed in small streams that are numerous and difficult to reach for applying insecticides, the chief effort has been toward disease control by surgical removal of nodules rather than by control of the vectors. Periodic denodulization campaigns have been carried out specifically for the prevention of blindness in Mexico, Guatemala, and Africa. Mass chemotherapy as a control measure is not practical because the available drugs, though effective in killing both the microfilariae (DEC) and adult worms (suramin), cause severe reactions.

Zoonotic *Onchocerca* Infections

A female *Onchocerca* in a nodular swelling, removed from the wrist of a woman in Illinois (United States), was reported by Beaver et al. (1974). The same report reviewed three other *Onchocerca* infections that had been acquired by people who had not lived in or visited an area where *Onchocerca volvulus* is endemic. The worm in each case was a mature female with no microfilariae in the uterus, located in a fibrous nodule. One was removed from the knee joint of a 25-year-old man in eastern Switzerland, another was from the conjunctiva and sclera of a 15-year-old girl in the Crimea (USSR), and the third was from the wrist of a middle-aged woman in Ontario, Canada whose only travel had been in England and Italy a few weeks before the wrist swelling was noticed (Ali-Khan, 1977). On the basis of its cuticular features, the worm in each case was readily identified as a species of *Onchocerca*. Five species of *Onchocerca* are known to occur in animals in some regions of the United States, Canada, and Europe: two in horses, three in cattle. In the horse, *O. cervicalis* is found in the ligaments of the neck, and *O. reticulata* lives in the ligaments of the joints and flexor tendons of the legs. The species in cattle and their normal location are: *O. gutturosa* in the sheath of the ligamentum nuchae, *O. lienalis* in the gastrosplenic ligament, and *O. stilesi* in the loose connective tissues overlying the dense elastic tissues of the stifle-joint ligaments.

In all species of *Onchocerca* the cuticle is thick, transversely striated, and reinforced externally by spiral or transverse ridges, often interrupted in the lateral fields. In the male, there is usually one such ridge for each stria of the cuticle, and in the female the ridges are farther apart with two, three, or four striae for each ridge, in a pattern that is constant for each species. In the female *O. reticulata, O. volvulus,* and *O. lienalis,* the external ridges are low and often inconspicuous, or even vague, and are located over alternate striae, i.e., there is one stria between adjacent ridges. In *O. stilesi* females the ridges are large and conspicuous, and there usually are two striae, occasionally only one, between adjacent ridges (Eberhard, 1979). In *O. gutturosa* and *O. cervicalis* the ridges are usually distinct and located over about every fourth stria, with three or occasionally two striae between adjacent ridges. As the arrangement of cuticular ridges and striae observed in the four zoonotic worms was indistinguishable from that in *O. cervicalis* and *O. gutturosa,* the specific identification of the zoonotic *Onchocerca* remains undetermined. *O. cervicalis* is transmitted by biting midges *(Culicoides), O. gutturosa* by black flies *(Simulium).* Both types of vectors are widely distributed.

Loa loa

(Cobbold, 1864) Castellani and Chalmers, 1913
(The loa worm or "eye worm," producing loiasis or "calabar swellings")

Synonyms. *Filaria oculi humani* Dujardin, 1845; *F. lacrymalis* Dubini, 1850; *F. oculi* Gervais and ven Beneden, 1859; *F. subconjunctivalis* Guyon, 1864; *Microfilaria diurna* Manson, 1891.

Historical and Geographical Notes. From 1770, when Mongin extracted this worm from the eye of a Negress in St. Domingo, West Indies, until 1844, there was a series of cases of "eye worm" infections among West African slaves recently imported into the Neotropical region. In Africa the first record was made by Guyon (ca. 1777) from Angola. The present-day areas of endemicity are confined to the forest belt of West and Central Africa, extending from Sierra Leone to Uganda and southward to Angola. They do not extend into the grasslands or the mangrove swamps. Infection rates are high in many communities of West and Central Africa, especially in Nigeria, Cameroon, and Zaire, where 50 to 90% prevalences have been observed (Hawking, 1977).

Morphology, Biology, and Life Cycle. Looss (1904) first gave careful study to this filaria. The body is whitish, and tapers somewhat toward the cephalic end, which is provided with one pair of lateral papillae and two pairs of submedian papillae, all sessile and of small size. The cuticula is ornamented with small bosses, which are lacking at the two extremities of the male. Males measure 30 to 34 mm in length by 0.35 to 0.43 mm in diameter; females, 40 to 70 mm by 0.5 mm.

The caudal end of the male curves ventrad and is provided with narrow alae. There are about eight pairs of perianal papillae, five pairs of large preanals, and three pairs of small postanals. These vary in size and symmetry. The copulatory spicules are unequal in length and somewhat dissimilar in shape. The vulva in the female opens in the cervical region. The microfilaria (Fig. 23–15) is sheathed, has a diurnal periodicity, and measures 250 to 300 μm long and 6 to 8 μm wide. Because the sheath stains poorly in ordinary blood films, it may be overlooked. (For its comparative anatomical "landmarks" see Table 23–1.) The optimum site for the microfilariae during their noncirculation phase is the lungs; they appear in peripheral blood only during the day. The microfilariae have been recovered from urine and sputum, and even spinal fluid (Janssens et al., 1958). The intermediate hosts are certain day-biting flies of the genus *Chrysops* (*C. dimidiata* and *C. silacea,* and other species of this genus may be involved), in which three larval stages are found during development in the fat body. In addition to man, the mandrill and several species of *Cercopithecus* have been found infected with a *Loa* worm morphologically similar to *Loa loa* in man. However, the monkey strain has larger adults, microfilariae with nocturnal periodicity, and utilizes species of *Chrysops* that are canopy-dwelling crepuscular forms. This may indicate a "radiative evolution" in separate host-vector complexes between the human and simian forms (Duke, 1972). The prepatent period in man is relatively long, probably about 6 months, and the adult worm may live for 17 years or more (Ziemann, 1926).

Experimental infections were established by Orihel and Moore (1975) in two patas monkeys (*Erythrocebus patas*) from Nigeria and a baboon (*Papio anubis*) that was born in the laboratory. The infective larvae had been grown in *Chrysops silacea* that had fed on a naturally infected man in Kumba, Cameroon and was transported by air to Louisiana. Larval development was observed in the

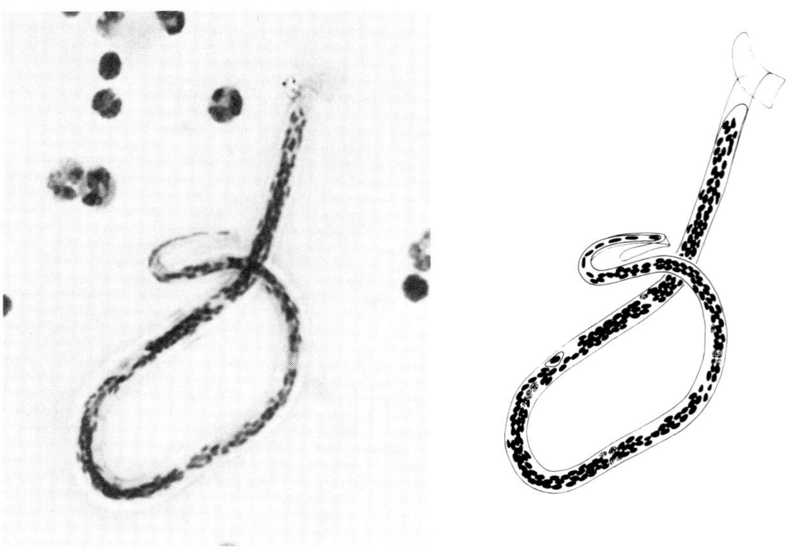

Fig. 23–15. *Loa loa*. Microfilaria in thick blood film stained with hematoxylin. Photomicrograph (× 530) and camera lucida drawing. (From Faust, E.C., Beaver, P.C., and Jung, R.C. 1975. *Animal Agents and Vectors of Human Disease*. 4th ed. Philadelphia, Lea & Febiger. Courtesy of Dr. T.C. Orihel.)

fat body of the flies and infectivity was reached in 9 days at 80°F. The prepatent period in the baboon was 146 days; in the patas monkeys, 135 and 148 days. Microfilaremia persisted beyond 40 weeks in the monkeys at levels of about 10,000 microfilariae per ml of blood. The baboon's microfilaremia began to decline at 13 weeks, but following splenectomy it returned to high levels and persisted at around 10,000 per ml beyond 40 weeks. American deer-flies *(Chrysops atlanticus)* produced large numbers of infective-stage larvae with full development in 9 or 10 days when fed on experimentally infected patas monkeys, and larvae from these flies produced patent infections in approximately 5 months in two patas monkeys (Orihel and Lowrie, 1975).

Epidemiology. Humans in endemic foci acquire infection when bitten by infected *Chrysops silacea* and *C. dimidiata,* which are the vectors of the human strain. Infection is most prevalent in small settlements situated in or at the margin of the forests, and though it may disappear with urbanization, it is favored by the presence of rubber plantations (Hawking, 1977).

Pathogenesis, Pathology, and Symptomatology. The adult worms normally live in the subcutaneous tissues, through which they migrate back and forth, provoking a temporary inflammation (i.e., "fugitive swelling" or "Calabar swelling") in the course of their wanderings (Fig. 23–16). However, at times they may migrate through all the viscera of the body. Swellings develop rapidly, may reach the size of a hen's egg, are at times painful, and last for 2 or 3 days. They are regarded as local reactions to sudden liberation of the worm's metabolites. Adults are particularly troublesome, although not necessarily painful, when passing in front of the eyeball or across the bridge of the nose, and in nervous patients may be responsible for a state of anxiety. Diagnosable adult worms have been surgically recovered from the back, axilla, groin, breast, penis, scalp, eyelids, and bulbar conjunctiva. There is frequently an associated eosinophilia, which may amount to 50 to 70% of the leukocytes.

At times loiasis produces remarkable periods of allergic reaction, particularly in Caucasian hosts, with reddened giant urticarial swellings of skin and mucous membranes, an associated fever, and high eosinophilia. The microfilariae of *Loa loa* frequently are not demonstrable in these cases, and it is possible that they are snared by the intense host reaction (see tropical eosinophilia, p. 361).

Ocular manifestations other than the traversing of the cornea beneath the conjunctiva have been ascribed to loiasis. In Uganda these have been of three types: granulomata in the bulbar conjunctiva, edema of the lids, and proptosis. The *conjunctival granulomata* present as solitary or multiple small yellow nodules about 2 mm in diameter. They are in the deep layers of the conjunctiva close to the episcleral tissue and consist of granulation tissue forming granulomata with central eosinophilic coagulation necrosis. The necrotic core presumably represents a worm track, and occasionally an adult worm is found in this necrotic material (Poltera, 1973). No pain is caused by these lesions, but occasionally they cause itching. *Painless edema of the eyelids* of sudden onset is called "bung-eye" in Uganda. It is frequently accompanied by itching but not by fever or any other constitutional disturbance. Edema of the orbital cellular tissue causes *proptosis,* which in Uganda is also called "bug-eye" or "bulge-eye." The onset is rapid. The condition is painless but usually associated with itching. Rapid resolution without residual effects is the rule (Owen and Hennessey, 1932; Nnochiri, 1972). Confirmation of the identification of the worms causing these eye lesions in Uganda is needed (see p. 386).

Diagnosis. Loiasis may be suggested by the presence of "fugitive swellings" in association with high eosinophilia in persons who have visited or lived in endemic areas. It can be specifically diagnosed from identification of the microfilaria in peripheral blood during the day, or by removal of the adult worms from the skin or conjunctiva.

The presence of mature unfertilized females is not uncommon, and such worms producing infertile, degenerate eggs and no microfilariae are known to cause typical Calabar swellings (Fig. 23–16,*B*) (Granz and Hartwig, 1963; Eveland et al., 1975). The conditions that cause the worms to induce acute reactions at unpredictable intervals are unknown.

Treatment. Complete recovery from the annoyance of these worms can be effected by their surgical removal. The most favorable time is when the worms are migrating through the corneal conjunctiva or across the bridge of the nose.

Suramin and particularly *diethylcarbamazine* (DEC) as administered in Bancroft's filariasis are specific chemotherapeutics (Hawking, 1955; Schneider, 1961), although caution must be observed in administering the drugs because of the tendency for allergic reactions to occur upon the

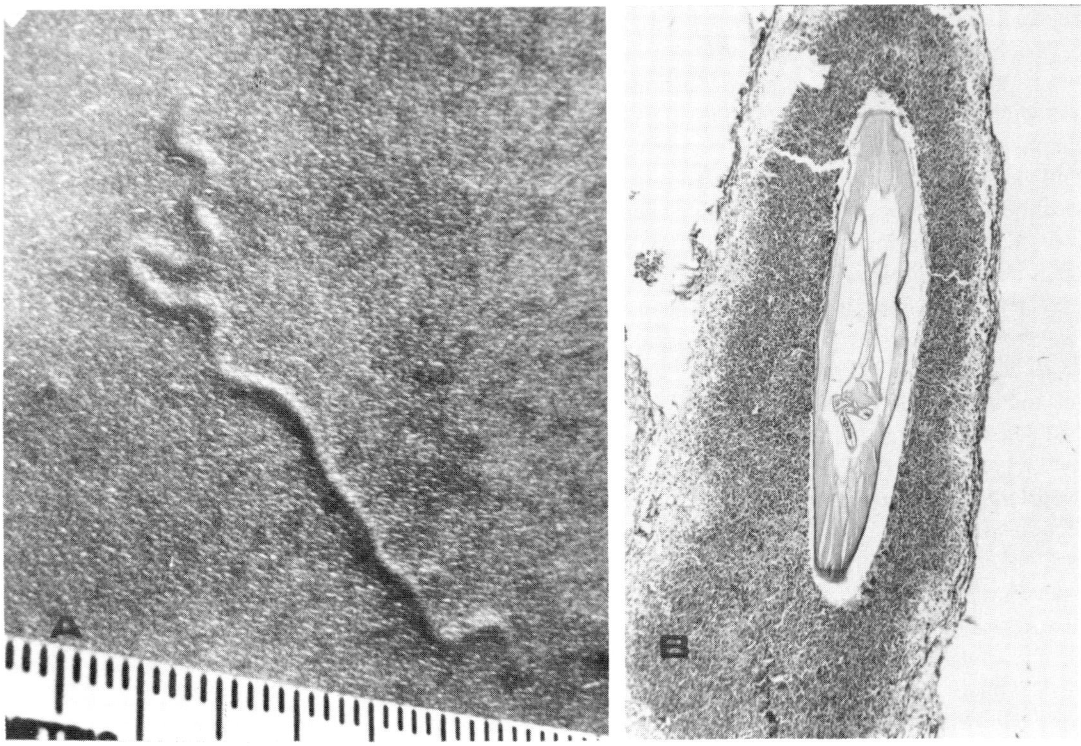

Fig. 23–16. *Loa loa. A,* Adult worm in outer layer of dermis on flank of American woman missionary on leave from station in Nigeria. One week previously, three worms had been removed from the eye; treatment with diethylcarbamazine had been started, then discontinued on second day when severe headache, fever, and rash developed. Midday microfilaria counts were more than $200/mm^3$. (Courtesy of Dr. P.E.C. Manson-Bahr.) *B,* Section of infertile female surrounded by intense tissue reaction in conjunctiva of Nigerian student in Germany with hypereosinophilia (67% of 12,000 WBC/mm^3) and history of frequent calabar swellings for 4 months and no mirofilaremia. (× 48.) (By P.C. Beaver. Stained section courtesy of Dr. W. Granz.)

death of numerous microfilariae. Such reactions are minimized by initiating treatment with minimal doses accompanied by administration of a corticosteroid drug. DEC has been tested prophylactically against *Loa loa* (Duke, 1963). In the amount of 200 mg on 3 consecutive days each month (adult dosage), complete protection is probably provided. Intolerance of an allergic nature is remedied by small initial doses, building up to a maximum, accompanied by antihistaminics or corticosteroids (Schneider, 1961).

Prognosis. The prognosis is good when the migrating worms can be removed or treatment can be tolerated.

Prevention. Preventive measures consist of (1) protection from *Chrysops* bites in endemic areas by application of repellents to exposed skin, (2) antilarval campaigns by clearing and draining forests to alter larval habitats, and (3) detection and DEC treatment of infected members of the population to destroy circulating microfilariae. The last of the these would seem to be the most practicable. However, this would depend on the presence or absence of reservoir hosts. Vector control on a wide scale is not possible.

Loa inquirenda. Maplestone (1938) designated as *Loa inquirenda* two unfertilized female worms, one 12 cm long removed in two parts from under the conjunctiva and one 14 cm long from the neck of a European woman in India. They differed from *Loa loa* in being two to three times as long, and in having a straight caudal end and a subterminal anal pore. A comparison with *Dirofilaria repens* was not made.

Genus Mansonella

(Faust, 1929) Orihel and Eberhard, 1982
(In honor of Sir Patrick Manson)

The description of *Mansonella* was based on published observations made on specimens that were limited in number and of poor quality. For that reason, when the *Dipetalonema* group was reclassified by Chabaud and Bain (1976), *M. ozzardi* was omitted as insufficiently known for taxonomic classification. When adequate material was acquired from experimentally infected monkeys *(Erythrocebus patas),* Orihel and Eberhard (1982) redescribed the species and redefined the genus. Meanwhile the genus *Tetrapetalonema* Faust, 1935 was defined, and as additional species were described it became evident that some of those assigned to the genus *Dipetalonema* could be regrouped into several subgenera, and that others should be placed in the genus *Tetrapetalonema*. These groupings were described by Chabaud and Bain (1976) who reclassified two familiar species as *Tetrapetalonema (Esslingeria) perstans* and *Tetrapetalonema (Esslingeria) streptocerca*. Orihel and Eberhard (1982) found that *Mansonella* and *Tetrapetalonema* were congeneric and reduced the latter to synonymy. This placed *T. perstans* and *T. streptocerca* in the genus *Mansonella*.

In species of the genus *Mansonella* (syn. *Tetrapetalonema)* the esophagus is narrow and poorly differentiated, the caudal extremity of the female generally bears four terminal lobes, and the preanal papillae of the male are closely associated with the perianal papillae. The main axis of the cephalic plate is dorsoventral, the spicules are moderately long with left to right spicule ratio 2.6 to 3.0, and the tail of the microfilaria is attenuated, terminating in an open hook or sickle-shaped curve, devoid of nuclei. In the *Dipetalonema* species the esophagus is well developed with muscular and glandular parts, the caudal extremity usually bears two well-developed petaloid appendages, and there are usually four pairs of regularly arranged preanal papillae.

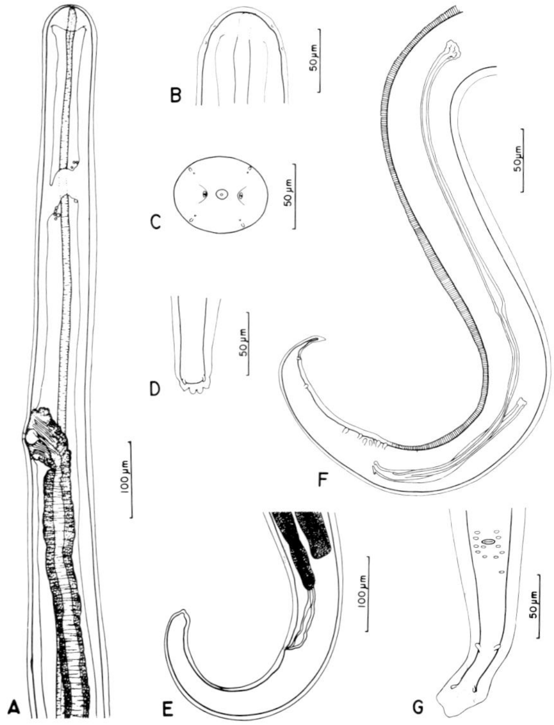

Fig. 23–17. *Mansonella ozzardi*. *A*, Anterior part of female, lateral view. *B*, Cephalic structures of female, lateral view. *C*, Cephalic structures, en face view. *D*, Tail of female, ventral view. *E*, Posterior end of female, lateral view. *F*, Posterior end of male, lateral view. (From Orihel, T.C., and Eberhard, M.L. 1982. *Mansonella ozzardi*. A redescription with comments on taxonomic relationships. Am. J. Trop. Med. Hyg., *31*:1142–1147.)

Mansonella ozzardi
(Manson, 1897) Faust, 1929
(Ozzard's filaria, producing mansonelliasis ozzardi)

Synonyms. *Filaria ozzardi* Manson, 1897; *Filaria demarquayi* Manson 1897; *Filaria tucumana* Bigheri and Araoz, 1917.

Historical and Geographical Notes. The microfilaria of this species was obtained by Ozzard from Carib Indians from the interior of Guyana, and was described as a new species by Manson. In 1929, when the genus *Mansonella* was described, *M. ozzardi* was designated the type and only species.

Mansonella ozzardi is a New World filaria, distributed widely in much of the region extending from northern Argentina through Bolivia, Brazil, Peru, the Guianas, Venezuela, Trinidad, Colombia (and probably Ecuador) to Panama and to Yucatan, Mexico. It may be absent in Central America north of Panama. It is present in the West Indies, notably in Haiti, Dominican Republic, Anguilla, Antigua, Guadeloupe, Dominica, Martinique, St. Lucia, and St. Vincent. There are no authentic records of autochthonous cases in the Old World.

Morphology, Biology, and Life Cycle. The worms in man were found in body cavities; in monkeys they were located in subcutaneous and connective tissues of the carcass. From man, only a posterior segment of the male, 38 mm long and 0.2 mm in maximum diameter, was collected. From the monkey, males are 24 to 28 mm long and 70 to 80 μm in diameter. The esophagus is about 1 mm long with the testis extending to that level. The tail, flexed into 1.5 to 2 coils, is tapered to a rounded tip and flattened dorsoventrally (Fig. 23–17,*G*). The left spicule, tubular and simple in structure, is 0.38 to 0.42 mm long; the right spicule, 0.14 to 0.16 mm long, is twisted at the midsection, broadened distally, with the end barbed (Fig. 23–17,*F*). The caudal alae are well developed with pericloacal papillae, usually five on the left side, six on the right. Caudal papillae are small and a rugose area extends about 0.5 mm anteriorly from the cloaca (Fig. 23–17,*F*). The females from man measure 65 to 81 mm in length, with a diameter of 0.21 to 0.25 mm. Those from monkeys are 32 to 62 mm long and 0.130 to 0.160 mm in diameter. Other features (Fig. 23–17,*A–E*) are: vulva anterior to base of esophagus, about 1 mm from anterior end;

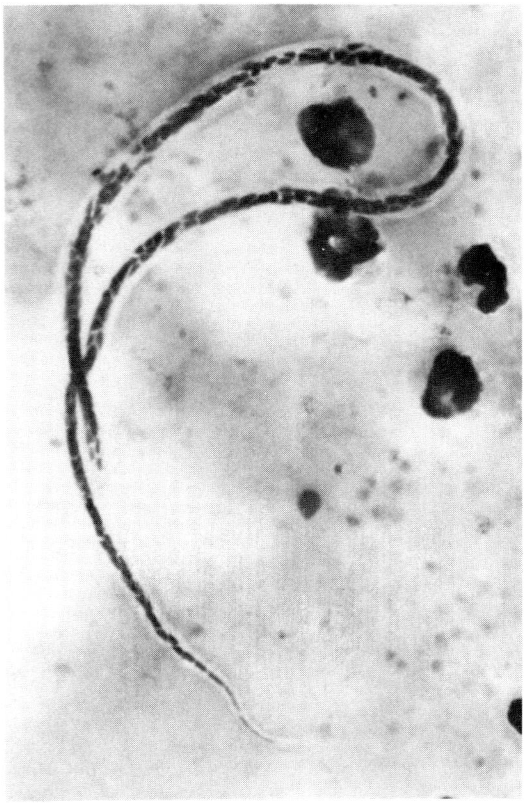

Fig. 23–18. *Mansonella ozzardi.* Microfilaria in thick blood film from Amerindian in Venezuela near Brazil border. Giemsa stain. (× 1000.) (By P.C. Beaver. Preparation courtesy of Dr. G.A. Godoy.)

ovejector large, ovoid; vagina directed posteriad, often coiled or looped; two uteri that fill body cavity and extend to near the posterior end; ovaries that extend to near the anus; tail tapered, 0.17 to 0.25 mm long, flexed ventrad, bearing four distinct terminal papillae. The nonperiodic unsheathed microfilaria (Fig. 23–17,*H*) somewhat resembles that of *M. perstans* in size but differs in the appearance of the tail, which tapers to a thin filament, contains a column of four to six ovoid or bar-like nuclei, and the tail itself beyond the nuclei is somewhat basophilic; also it usually forms a slight but distinct curve (Fig. 23–18). Experiments by Buckley (1934) incriminated *Culicoides furens* as the vector in St. Vincent. *C. furens* and probably *C. phlebotomus* are vectors in Haiti (Raccurt et al., 1980), and the latter also in Trinidad (Nelson and Davies, 1976). Species of *Culicoides* are apparently vectors in Surinam and Argentina, but in the drainage area of the Amazon, in Brazil (Shelley and Shelley, 1976), and in Colombia (Tidwell et al., 1980) *Simulium amazonicum* and *Simulium sanguineum*, respectively, are the principal vectors.

Epidemiology. *Mansonella ozzardi* infection is most commonly found in isolated populations of Amerindians. Where it is present at all, it tends to be highly prevalent, with infection rates increasing with age, approaching or exceeding 90% in adults (Moraes et al., 1978; Lawrence et al., 1979; Godoy et al., 1980b; Lightner et al., 1980).

Pathogenesis, Pathology, and Symptomatology. There are no manifest symptoms in infected individuals. The adult worms apparently produce little tissue reaction, although there are some reports of an associated hydrocele and enlarged lymph glands.

Diagnosis. Recovery of the nonperiodic unsheathed microfilaria from peripheral blood or biopsy specimens of skin and its specific identification constitute diagnosis. Although examination of skin samples may reveal infection in over 30% of cases, it is rarely as satisfactory as examination of blood, by either the thick film or Knott technique (Lightner et al., 1980).

Treatment. Diethylcarbamazine has no lethal or sterilizing effect on this filaria (Hawking, 1955).

Mansonella perstans (Manson, 1891) Orihel and Eberhard, 1982

Synonyms. *Acanthocheilonema perstans* (Manson, 1891) Railliet, Henry and Langeron, 1912; *Dipetalonema perstans* (Manson, 1891) York and Maplestone, 1926; *Tetrapetalonema (Esslingeria) perstans* (Manson, 1891) Chabaud and Bain, 1976.

Historical and Geographical Notes. This worm was first found by Daniels (1898) in aborigines of former British Guiana. Manson described Daniels' material and also found the microfilariae in the blood of natives in the Congo. The species was redescribed by Chabaud in 1952, and Chabaud and Bain in 1976 assigned it along with *M. streptocerca* to a new subgenus *(Esslingeria)*. Orihel and Eberhard in 1982 assigned it to the genus *Mansonella*.

Mansonella perstans is distributed extensively throughout tropical regions of Africa (Sierra Leone, Ghana, Nigeria, the Congo Basin, Cameroon Republic, Uganda, Tanzania, and Zambia), Algiers and Tunis, coastal South America from Venezuela to the Argentine, Panama (coastally), Trinidad, and nearby Surinam. In Africa it is frequently associated in patients with Bancroft's filaria, *Onchocerca volvulus* and *Loa loa,* and occasionally *Mansonella streptocerca;* in northern South America its distribution is at times contiguous with and may overlap that of *Mansonella ozzardi* (Orihel, 1967). Man is the important definitive host of this species, although *Pan satyrus* and other primates in Africa have been listed as reservoir hosts. Several related species of the genus are found in New World monkeys.

Morphology, Biology, and Life Cycle. The adult worms of this genus live in body cavities, most commonly in the peritoneal cavity, next in the pleural cavity, and rarely in the pericardium. They are delicate filariae with a smooth cuticula,

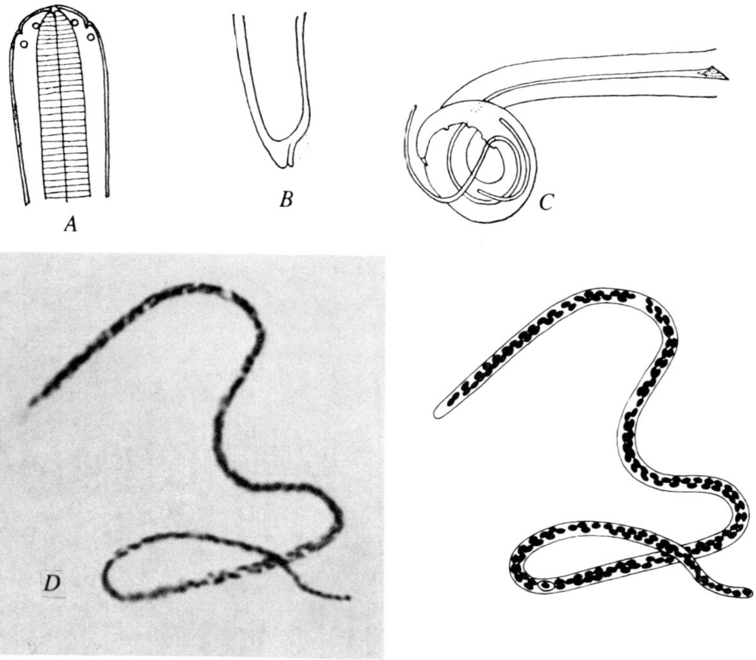

Fig. 23–19. *Mansonella perstans*. *A*, Anterior end of worm, showing distribution of papillae. *B*, Caudal end of female worm, lateral view. *C*, Caudal end of male, showing caudal papillae and copulatory spicules. *D*, Microfilaria in thick blood film. Hematoxylin stain. Photomicrograph and camera lucida drawing. (× 980.) (*A-C* adapted by E.C. Faust from Leiper, R.T. 1913. Observations on certain helminths of man. Trans. R. Soc. Trop. Med. Hyg., *6*. *D* from Faust, E.C., Beaver, P.C., and Jung, R.C. 1975. *Animal Agents and Vectors of Human Disease*. 4th ed. Philadelphia, Lea & Febiger. Courtesy of Dr. T.C. Orihel.)

a bluntly rounded anterior end surmounted by a shield bearing a pair of large lateral and two pairs of submedian papillae, and a caudal extremity curved ventrad in both sexes and bifurcated in the female to form two pairs of triangular nonmuscular flaps (Chabaud, 1952). The male measures about 45 mm by 60 μm; the female, 70 to 80 mm by 120 μm. The posterior end of the male is distinguished by having four preanal pairs of papillae and one postanal pair, and unequal rod-like copulatory spicules (Fig. 23–19,*C*). The vulva of the female lies in the cervical region. The microfilariae are not strictly periodic, but their numbers in the peripheral circulation are apparently fewer in the daytime than at night, i.e., they are subperiodic. They measure about 200 by 4.5 μm and are unsheathed (Fig. 23–19,*D*). The intermediate hosts in Africa are *Culicoides austeni* and *C. grahami* (Gordon and Duke, 1955).

Epidemiology. Human infection is produced by bites of infected *Culicoides* in endemic foci. The detailed epidemiology has never been studied. In hyperendemic regions in tropical Africa 90% of the individuals in some villages may harbor the parasite.

Pathogenesis, Pathology, and Symptomatology. Although there is no general agreement as to the pathogenicity of *M. perstans*, several authors have considered this parasite to be the cause of transient abdominal pain, subcutaneous swelling, rashes, pain in the limbs, and serous cavities and fatigue, especially in persons not indigenous to endemic areas (Stott, 1962; Adolph et al., 1962). Morenas (1929) studied a patient who had toxic edema of the left eyelid, dyspnea, precordial pain, and 50% eosinophilia attributed to the infection. Dubois and Van den Berghe (1948) stated that although the allergic dermatitis attributed to *M. perstans* has not been definitely proved, this filaria "has been incriminated as being largely responsible for the eosinophilia so widely found among natives" in endemic regions in Africa. Strohschneider (1956) attributed severe allergic phenomena in seven patients to the presence of this worm.

Diagnosis. Recovery of the specific microfilariae from peripheral blood day or night provides a diagnosis. They must be differentiated from those of *M. streptocerca* and *M. ozzardi,* which are also unsheathed. Adolph et al. (1962) employed hemagglutination and bentonite flocculation when microfilariae could not be demonstrated in blood films.

Treatment. Hawking (1955) reported that diethylcarbamazine (DEC) has no apparent action on the microfilariae, but may have a slight lethal or sterilizing effect on the adults. Adolph et al. (1962) administered DEC to microfilaria-negative patients, at times evoking temporary symptoms and positive blood films, followed by resolution of symptoms and apparent cure. Two Swedish patients who had acquired *M. perstans* infection in Zaire were successfully treated with mebendazole, 100 mg twice daily for 30 days (Wahlgren, 1982).

Prevention. Preventive measures have not been studied. The intermediate hosts breed in jungles and swamps. Their eradication at present is not practicable.

Mansonella streptocerca (Macfie and Corson, 1922) Orihel and Eberhard, 1982

Synonyms. *Agamofilaria streptocerca* Macfie and Corson,1922; *Microfilaria streptocerca* (Macfie and Corson, 1922) Stiles and Hassall 1926; *Acanthocheilonema streptocerca* (Macfie and Corson, 1922) Faust, 1949; *Dipetalonema streptocerca* (Macfie and Corson, 1922) Peel and Chardome, 1946; *Tetrapetalonema (Esslingeria) streptocerca* (Macfie and Corson, 1922) Chabaud and Bain, 1976.

Biological Notes. *Mansonella streptocerca* occurs only in West Africa, in the same areas as *M. perstans*. In some areas it is found together with *Onchocerca* and *Loa loa*. The microfilaria of this worm was first described by Macfie and Corson from biopsy of skin of natives of Ghana. It was present in 44% of a surveyed group, all of whom were in apparent good health. In 1938 one native of the Congo was found to harbor this filaria, and in 1939 three additional human infections were discovered. In 1946 Peel and Chardome for the first time discovered adults (two females and a fragment of another) in the cutaneous tissue of *Pan paniscus* and *Pan satyrus*.

Adult *M. streptocerca* were first described in human tissue by Meyers et al. in 1972. At the same time they also reported clinical observations on 40 patients with streptocerciasis in a rural community near the equator in western Zaire. Detailed observations on the adult female in tissue sections were reported by Neafie et al. in 1975, and the adult male was likewise described by Meyers et al. in 1977. These studies were followed by observations on the adult worms and microfilariae in the tissues after diethylcarbamazine therapy (Meyers et al., 1978).

The microfilariae measure 180 to 240 μm by 3 to 5 μm and are sheathless. They are readily rec-

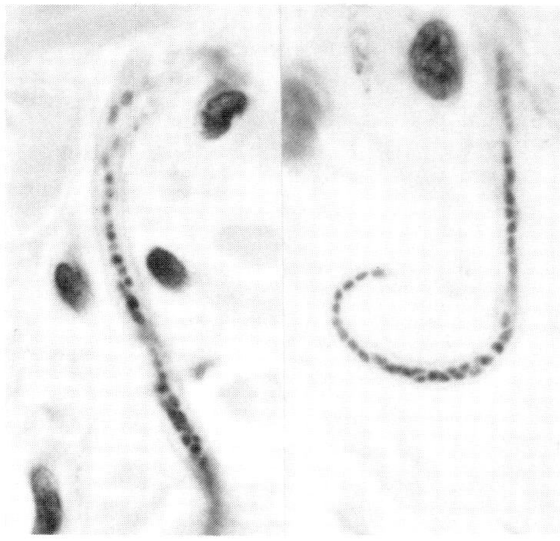

Fig. 23–20. *Mansonella streptocerca*. Microfilariae in tissue sections, showing characteristic pattern of nuclei at anterior (left) and posterior ends, and curved, "shepherd's-crook" configuration of tail. Giemsa and H & E stains, respectively. (× 1080.) (From Neafie, R.C., Connor, D.H., and Meyers, W.M. 1975. *Dipetalonema streptocerca* (Macfie and Corson, 1922): Description of the adult female. Am. J. Trop. Med. Hyg., *24*:264–267.)

ognized in tissue sections, as in stained blood films, by the patterns of nuclei in the head and tail (Fig. 23–20). The cephalic space is slightly longer than wide, and the first four or five nuclei are elongate and tend to be staggered with little or no overlap. The tail nuclei tend to be quadrate, arranged in a single column of 9 to 12 which extend to the tip, leaving a short clear caudal space. When fixed, the microfilariae are usually relatively straight except at the posterior end which is strongly bent in a shepherd's crook curve (Fig. 23–20). The anatomical landmarks, expressed as percentage distance from anterior end, are as follows: nerve ring 27, excretory pore 34, first G cell 70, and anal pore 86%.

The adult worms live in the dermis, usually less than 1 mm below the skin surface (Fig. 23–21,*A*). Features seen in transverse sections at midbody levels are (1) paired uteri that fill the body cavity, (2) intestine that is relatively minute, (3) dense somatic muscles, about ten per quadrant, (4) low lateral cords, and (5) cuticle 1 to 2 μm thick, slightly thicker over the lateral cords (Fig. 23–21,*B*). Diameters at the level of the vulva anteriorly, and at the ovaries near the posterior end, are approximately 50 μm; diameter at midbody is up to 85 μm. Maximum diameter of the male is 50 μm. The length of the female, according to Peel and Chardome (1946), is 27 mm. *Culicoides grahami* is probably the usual vector (Duke, 1958).

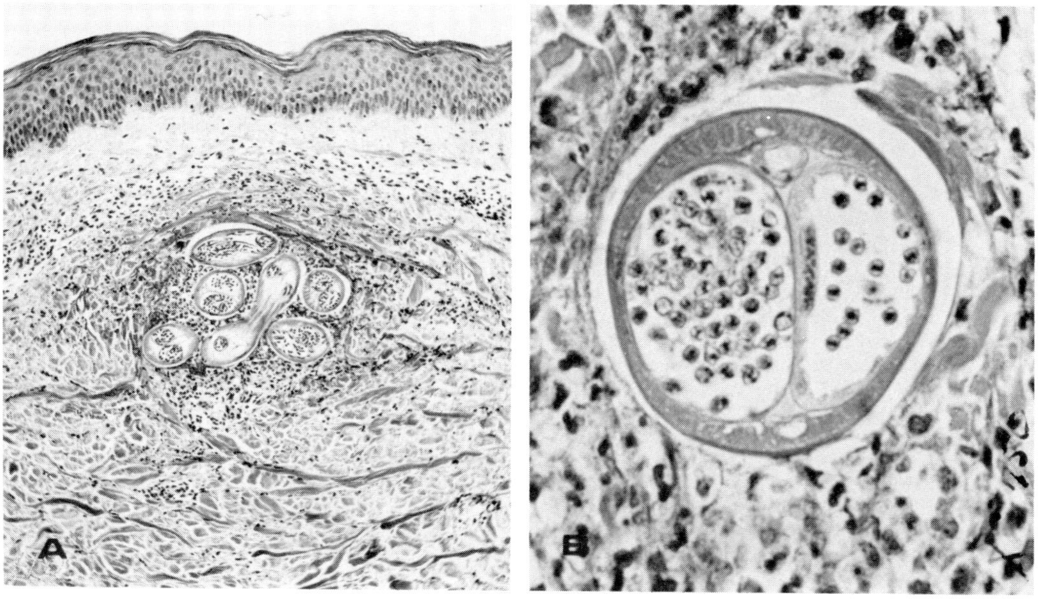

Fig. 23–21. *Mansonella streptocerca. A,* Sections of adult female in upper dermis. (× 100.) *B,* Transverse section of female cut at midbody level, showing large, paired uteri containing microfilariae, minute intestine, and low, dense muscles, about ten per quadrant. (× 680.) (From Neafie, R.C., Connor, D.H., and Meyers, W.M. 1975. *Dipetalonema streptocerca* (Macfie and Corson, 1922): Description of the adult female. Am. J. Trop. Med. Hyg., *24*:264–267.)

Diagnosis. The presence of streptocerciasis should be suspected in an individual who has been residing in an endemic area and who presents with the complaints of hypopigmented macules and itching. The primary diagnostic procedure is the examination microscopically of skin snips in saline solution as for onchocerciasis.

The condition must be distinguished from leprosy and granuloma multiforme, both of which cause hypopigmented macules resembling those of streptocerciasis, and from onchocerciasis, which causes a pruritic dermatitis accentuated by treatment with diethylcarbamazine.

The macules of streptocerciasis are distinguished from those of leprosy in that they are not anesthetic. This distinction is not always conclusive, however, since some leprotic macules, especially those on the face, may not be anesthetic. Differential diagnosis ultimately depends on biopsy and histopathologic examination.

Granuloma multiforme, a rare condition, must be differentiated from streptocerciasis by skin biopsy and histopathologic examination.

Hypopigmented macules do not tend to occur in onchocerciasis, whereas iridocyclitis and subcutaneous nodules seen in onchocerciasis are not features of streptocerciasis. The spotted depigmentation or "leopard skin" commonly seen in onchocerciasis does not occur in streptocerciasis. Final differentiation between the two filarial infections must be made by identification of the respective microfilariae or by biopsy of the skin and identification of the filariae in tissue sections. Concurrent infection with both filarial species may occur.

Treatment. Diethylcarbamazine administered as in treatment of Bancroft's filariasis is effective in streptocerciasis. With the treatment not only are the microfilariae destroyed but degenerative changes also occur in adult worms. The death and deterioration of the parasites with treatment result in about 24 hours in an exacerbation or appearance of pruritus and development of cutaneous papules which are at the sites of lightly coiled, degenerating adult worms. Other symptoms of systemic intoxication, such as chills, fever, nausea, vomiting, headache, and muscle or joint pains, may less commonly occur following treatment.

Dipetalonema, Zoonotic Species

American Records. An immature filaria, 21 mm in length, removed from the eye of a man in western Oregon, was identified as a female *Dipetalonema,* the third such case to be reported from the same locality (Beaver et al., 1980). In this and one of the earlier cases the worms were identified as probably *Dipetalonema arbuta* Highby, 1943, which lives in the body cavity of the porcupine *(Erethizon dorsatum),* or possibly *Dipetalonema sprenti* Anderson, 1953, which lives in the body cavity of the beaver *(Castor canadensis)*. Immature nematodes inside the eye have frequently been reported as having the body proportions of a filarid but for various reasons were not identified (Raymond et al., 1978).

A small adult filaria, 0.12 mm in diameter, in a nodule from the abdominal wall of a woman in Mississippi, and a similar worm, reported earlier (Jung and Harris, 1960) in a nodule from the arm of a man in Louisiana, were identified as a species of *Dipetalonema* or a related genus (Beaver and Orihel, 1965). The nodule in both cases was subcutaneous.

A microfilaria similar in size and morphology to that of *Dipetalonema reconditum* was found in the blood of a 12-year-old girl from a rural community of Alabama (Greene et al., 1978). She was being treated with prednisone for systemic lupus erythematosus. Only one microfilaria was ever seen, and it provided the only evidence that adult worms were present. In addition to *D. reconditum*, which is a common subcutaneous filaria of the dog, *D. interstitium*, which is a subcutaneous filaria of the gray squirrel, was noted to have a similar microfilaria.

Ugandan Eye Worm. Among patients attending eye clinics in parts of Uganda, East Africa, a common condition first reported by Owen and Hennessey (1932) is the presence of one to several small yellow nodules in the bulbar conjunctiva. Conditions often associated with the nodules are marked edema of the lids and proptosis, referred to as "bungeye" and "bulgeeye," respectively. In some cases all three conditions occur together, and usually associated with any one, or a combination of them, is a high eosinophilia. Examination of microsections of excised nodules often reveals a small adult filarial worm. When a worm is not seen, its previous presence is suggested by massive eosinophilic infiltration and necrosis. The worms are usually dead, and frequently too necrotic to be identified.

In a study by Poltera (1973), portions of a worm were seen in 18 of 40 excised nodules. The maximum diameter of eight worms that could be measured ranged from 53 to 105 μm. All were necrotic but two were gravid females with microfilariae that appeared to be sheathed. In addition, structures on the cuticle of some of the worms were described as cuticular bosses, leading to a questionable diagnosis of a *Loa* species (see p. 379). The size and evident morphology of the worms described and illustrated in this and the earlier study by Owen and Hennessey (1932) suggest a species with features seen in *Dipetalonema* or *Mansonella*, namely, small size (often 100 μm or less in width), relatively thick and smooth cuticle, and low, numerous, and dense somatic muscles. Such small worms in the conjunctiva would not be readily visible to the unaided eye.

Meningonema peruzzii Orihel and Esslinger, 1973

The description of *Meningonema peruzzii* by Orihel and Esslinger (1973) was based on a collection of specimens from three talapoin monkeys *(Cercopithecus talapoin)* from equatorial Guinea. The same species had been reported by Peruzzi from a *Cercopithecus* in Uganda in 1928. Special interest in this filaria was generated by the observation of Orihel (1973) that serious lesions affecting the brain of patients in Zimbabwe (Rhodesia) and attributed to *Dipetalonema (=Mansonella) perstans* were apparently caused by *M. peruzzii*.

The adult worms were found coiled in the subarachnoid spaces along the dorsum of the brain stem at the level of the medulla oblongata. The geographic range of *M. peruzzii* probably extends over all of Central Africa and its host range may include species of *Cercopithecus* other than *C. talapoin*. Among the outstanding morphological features of the adult worms are (1) an exceptionally thin body wall, (2) body width greatest near the anterior extremity, (3) esophagus short and not distinctly divided into muscular and glandular parts, and (4) an expanded intestine whose wall and contents are darkly pigmented. The female is 86 to 122 mm long and 0.5 mm in maximum width, at the level of the vulva near the base of the esophagus. It is quadridelphic annd opisthodelphic, with one or more of the four ovaries extending posteriorly beyond the anus into the tail. The male, 33 to 50 mm long, has short, simple spicules that are similar and slightly unequal, with a maximum length of less than 0.2 mm.

The microfilaria resembles that of *Mansonella perstans*, but differs in having a sheath. However, the sheath fits closely to the body and in living and formalin-fixed specimens (Fig. 23–22,*A*,*B*) it extends only slightly beyond the ends of the body, and therefore is not readily seen. It also may be inapparent in Giemsa-stained thick blood films (Fig. 23–22,*C*), though in hematoxylin-stained films it is relatively conspicuous.

In the description of *M. peruzzii* it was noted that the microfilariae were remarkably similar to those described from the cerebrospinal fluid in two cases reported as cerebral filariasis from Rhodesia (Dukes et al., 1968). Based on examination of photographic material from those cases, Orihel (1973) concluded that the microfilariae were indistinguishable from those of *M. peruzzii*, and that the species responsible for the central nervous system disease was in fact *M. peruzzii*.

The vector species and frequency of infection in man are unknown. In addition to those in the talapoin monkey in West Africa, natural infections occur in green monkeys from Uganda and Tanzania.

No evidence of pathogenicity of either the adult worms or the microfilariae was seen in monkeys, either at autopsy or in the living infected animals. In one of the two patients with microfilariae in the cerebrospinal fluid, the disease was described as acute encephalomyelitis; in the other, the illness was temporary and mild, characterized by headache, drowsiness, and fatigue.

In the monkey, microfilariae of *M. peruzzii* are present in the circulating blood during the day. Periodicity studies have not been reported. In humans microfilariae have been identified in the cerebrospinal fluid only.

Microfilaria semiclarum Fain, 1974

A microfilaria with the striking feature of a long clear region in the posterior half of the body was described by Fain (1974), who found it in 52 inhabitants of three villages in northwestern Zaire. In one village it was present in blood films from half the adult population. It was described as unsheathed and subperiodic with a slight increase during the day. It resembles somewhat the microfilaria of *M. perstans* in size (198 by 5.2 μm, average, in stained thick films) and, as noted, has a long clear region in the posterior half of the body, the feature that suggested its name. Its cephalic space is much longer than that of *M. perstans*. The adult stage is unknown. Presumably the natural host is a primate or carnivore. The microfilaria differs from that of all species known in animals of the African region. Its resemblance to *M. perstans* suggested placing it in the genus *Dipetalonema (=Tetrapetalonema=Mansonella)*.

Microfilaria bolivarensis Godoy, Orihel and Volcan, 1980

Microfilaria bolivarensis was disovered in the blood of American Indians in a remote jungle area along the upper Caura River, a tributary of the Orinoco River, near the Brazilian border in Venezuela (Godoy et al., 1980a). It is unsheathed and measures 256 by 7 to 8 μm in thick blood smears, and 300 by 8 μm in Knott samples. The cephalic end is bluntly rounded and the tail is relatively short and pointed. The terminal part is devoid of nuclei. The cephalic space is short and nearly as wide

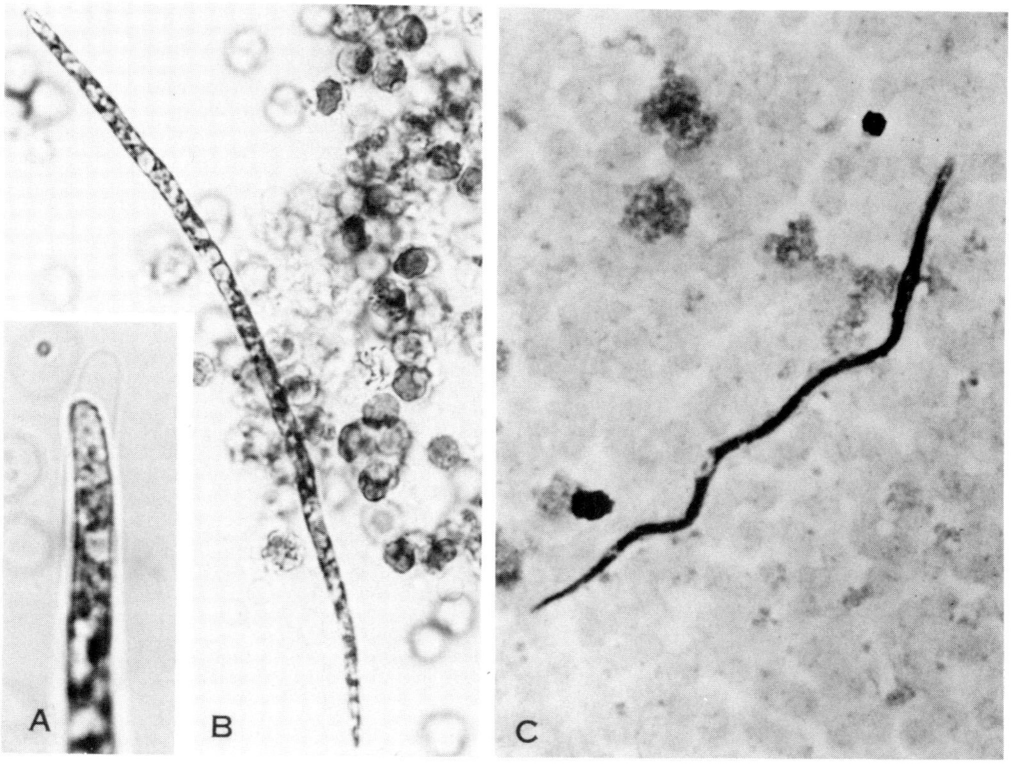

Fig. 23–22. *Meningonema peruzzii.* Microfilaria from blood of monkey (*Cercopithecus talapoin*). *A* and *B*, Fixed in 2 percent formalin. Hematoxylin stain. (× 1440 and × 645.) *C*, In thick blood film. Giemsa stain. (× 550.) In *A*, sheath is shown extending beyond anterior end of body. (From Orihel, T.C. 1973. Cerebral filariasis in Rhodesia—a zoonotic infection? Am. J. Trop. Med. Hyg., *22*:596–599.)

as long (Fig. 23–23,*A*). Microfilariae in Knott preparations have a pronounced bend in the body and the tail is sharply flexed (Fig. 23–22,*B*). Microfilariae of *Mansonella ozzardi* were found in people of the same communities and in the same individuals (Godoy et al., 1980b). *M. bolivarensis* resembles the microfilaria of *Onchocerca volvulus* but is slightly smaller and stouter and has a shorter cephalic space and less a compact nuclear column. *M. bolivarensis* is much larger than the microfilaria of *M. ozzardi*. It does not closely resemble any microfilaria reported in animals of the neotropical region.

Microfilaria rodhaini
Peel and Chardome, 1946

Microfilariae indistinguishable from *Mf. rodhaini* were found in skin biopsy specimens taken from 14 of 440 people examined for *volvulus* and *streptocerca* in Gabon (Richard-Lenoble et al., 1982). The microfilariae were 300 to 332 μm long by 2 to 2.5 μm wide in Giemsa-stained smears. *Mf. rodhaini* was described originally from *Pan paniscus* and *Pan satyrus* in Zaire.

Genus Dirofilaria
Railliet and Henry, 1911

The subfamily Dirofilariinae Sandground, 1921 is characterized by highly developed caudal alae and large papillae in the male. Among the natural filarial parasites of man, *Loa loa* is the only member of this subfamily. However, several species of *Dirofilaria* are transmitted from wild and domestic mammals to man.

In the genus *Dirofilaria* the mouth is without lips, cephalic papillae are inconspicuous, spicules are markedly dissimilar, the caudal papillae are large and numerous, and the vulva is posterior to the esophagus. The tail is short and rounded in both sexes. Microfilariae are unsheathed and circulate in the blood. Faust (1937) designated two subgenera—*Dirofilaria,* with *D. Dirofilaria immitis* Leidy, 1856 as type, and *Nochtiella,* with *D. Nochtiella repens* Railliet and Henry, 1911 as type. The *immitis* group was defined as large filariae in which the caudal papillae of the male show only slight asymmetry in number and distribution, and the larger of the copulatory spicules is not distinctly acuminate (sharply pointed). In the *repens* group, the species are relatively small, the caudal papillae of the male exhibit a distinct asymmetry, and the larger of the spicules is distinctly acuminate. Faust (1957) later added to the definition of the *Nochtiella* group the presence of longitudinal ridges on the outer surface of the cuticle, structures earlier de-

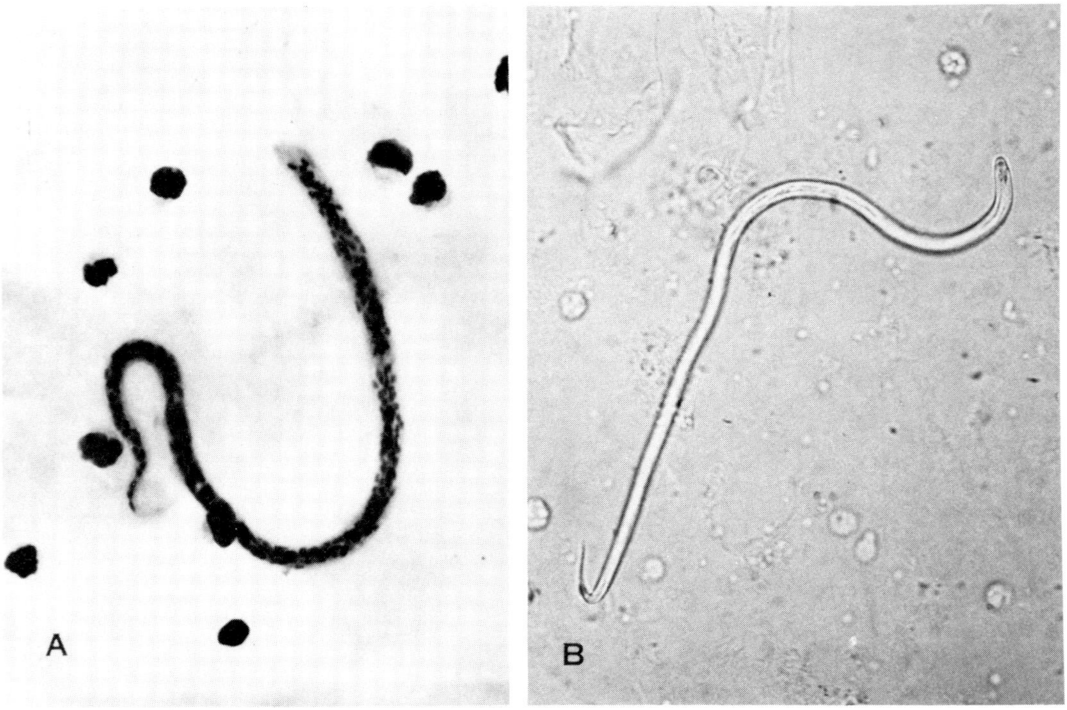

Fig. 23–23. *Microfilaria bolivarensis. A,* In thick blood smear. Giemsa stain. (× 500.) *B,* In 2 percent formalin, showing characteristic flexure of body and tail. Unstained. (× 400.) (From Godoy, G.A., Orihel, T.C., and Volcan, G.S. 1980. *Microfilaria bolivarensis:* a new species of filaria from man in Venezuela. Am. J. Trop. Med. Hyg., *29*:545–547.)

scribed by other workers. Anderson (1952) had suggested the possible use of these structures in determining taxonomic relationships of the *Dirofilaria* species. It should be noted that in *D. immitis* and probably in others of the subgenus *Dirofilaria,* as well as in the *Nochtiella* species, longitudinal ridges are present on the ventral side of the coiled part of the male hindbody.

Both *Nochtiella* and *Dirofilaria* types of infection have been reported in man. Although it is not possible in most cases to make a specific identification of the worm, usually a consideration of where the infection was acquired, where in the body the worm was located, along with the morphological features of the worm as seen in the cleared whole specimen or in histopathological sections make it possible to at least place the worm in one of the two subgenera. In most cases it is possible to give it a species type designation. Exceptions are those cases in which the worm is necrotic beyond recognition except as a *Dirofilaria,* or probably a species of that genus. In such cases, when the worm's location is the lungs it is presumed to be *Dirofilaria immitis,* the heartworm of dogs; if located in subcutaneous tissues, it has often been referred to as *Dirofilaria conjunctivae,* especially if its presence is associated with the eye. Specific types of *Dirofilaria* that have been reported in man are those indistinguishable from *D. immitis, D. repens, D. tenuis,* and *D. ursi.* The zoonotic aspects of filarial infections in man were reviewed by Dissanaike (1979). Features by which most of the different types can be recognized in tissue sections were described earlier (Beaver and Orihel, 1965; Orihel and Beaver, 1965; Beaver and Samuel, 1977).

Dirofilaria immitis
Leidy, 1856
(The dog heartworm)

Synonyms. *Filaria sanguinis* Cobbold, 1869; *D. louisianensis* Faust, Thomas and Jones, 1941.

Geographical Notes. *Dirofilaria immitis* is a common parasite of dogs in much of the tropical, subtropical, and warm temperate regions of the world. In the United States, it is most prevalent in the southern and southeastern states, but it is enzootic as far north as Massachusetts, Michigan, and Minnesota. Though this parasite is generally uncommon or absent in the western states, transmission is known to occur in California. Natural infections in the United States have been recorded in the timber wolf, red fox, gray fox, coyote, domestic and wild cats, bear, beaver, sea lion, harbor seal, muskrat, otter, nutria, and man. *D. immitis* is apparently rare or does not occur in

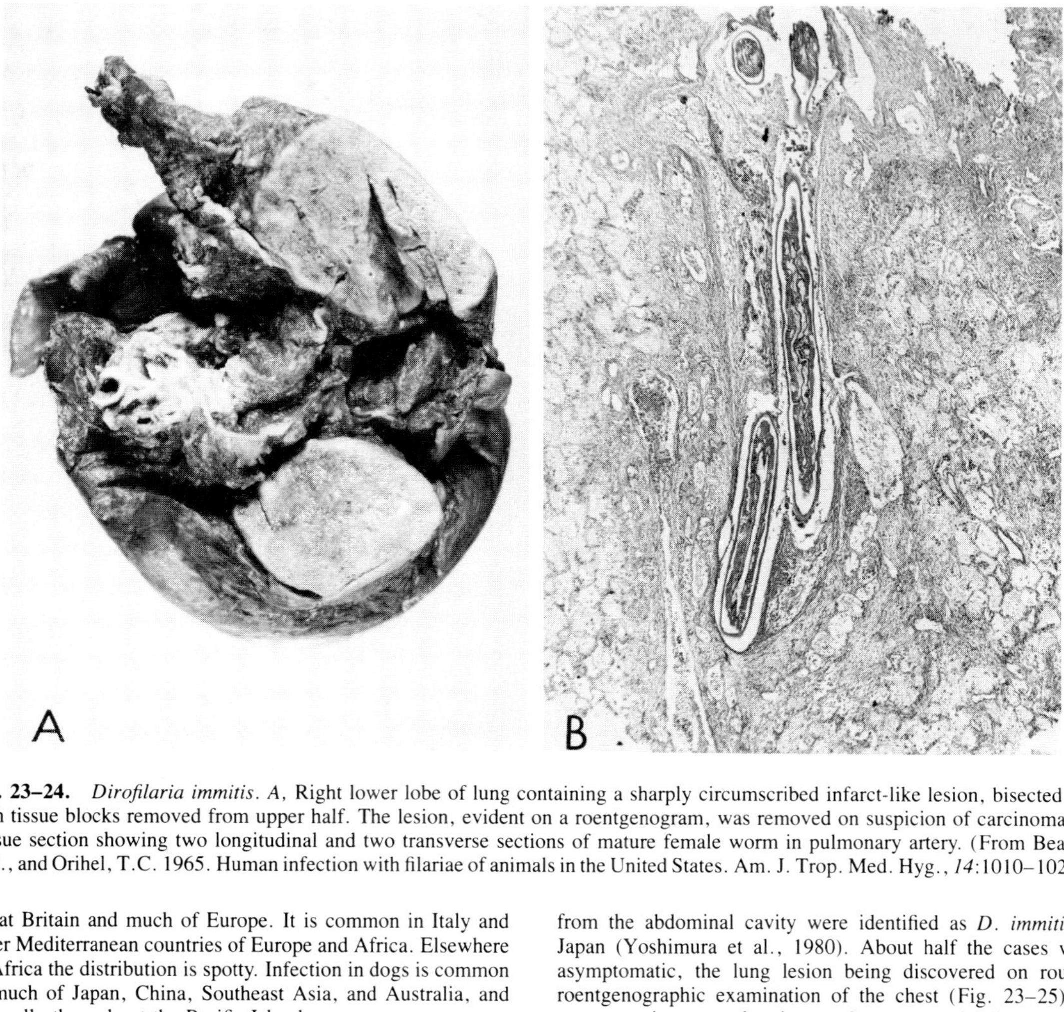

Fig. 23–24. *Dirofilaria immitis. A,* Right lower lobe of lung containing a sharply circumscribed infarct-like lesion, bisected and with tissue blocks removed from upper half. The lesion, evident on a roentgenogram, was removed on suspicion of carcinoma. *B,* Tissue section showing two longitudinal and two transverse sections of mature female worm in pulmonary artery. (From Beaver, P.C., and Orihel, T.C. 1965. Human infection with filariae of animals in the United States. Am. J. Trop. Med. Hyg., *14*:1010–1029.)

Great Britain and much of Europe. It is common in Italy and other Mediterranean countries of Europe and Africa. Elsewhere in Africa the distribution is spotty. Infection in dogs is common in much of Japan, China, Southeast Asia, and Australia, and generally throughout the Pacific Islands.

Morphology in the Dog. The mature adult *D. immitis* in the dog and most other hosts is usually found in the chambers and connecting large vessels of the right heart. The female measures 25 to 30 cm in length and approximately 1 mm in width; the male is smaller, 12 to 18 cm in length. The esophagus is divided into anterior muscular and posterior glandular portions. The vulva is about 2.7 mm from the anterior end, just posterior to the base of the esophagus. The microfilariae are sheathless, 300 to 325 μm long by about 7 μm wide. Microfilaremia is subperiodic nocturnal, and microfilaria levels are higher in summer than in winter.

Records in Man. *Dirofilaria immitis* has been recorded in man with increasing frequency since it was first reported in the vena cava of a woman in New Orleans (Faust et al., 1941). In most instances a single worm was found in an occluded branch of the pulmonary artery, surrounded by a large area of infarction (Fig. 23–24). Exclusive of some known or suspected duplications, there were at the end of 1980 at least 32 reports describing 60 cases of lung or heart infection with *D. immitis* that was presumed to have been acquired in the United States (Ciferri, 1982), 1 that was acquired in Colombia (reported as probably *W. bancrofti* by Faust et al., 1952), 1 in Brazil (Leonardi et al., 1977), 10 or more in Japan (Yamane et al., 1977; Kumada et al., 1980; Yoshimura et al., 1981), and 18 in Australia (Moorhouse et al., 1976). In two cases worms removed from the abdominal cavity were identified as *D. immitis* in Japan (Yoshimura et al., 1980). About half the cases were asymptomatic, the lung lesion being discovered on routine roentgenographic examination of the chest (Fig. 23–25). In symptomatic cases, the signs and symptoms leading to chest examination were, in order of frequency, persistent cough, chest pain, and hemoptysis, often with an associated eosinophilia. In the United States most of the cases have been reported from the coastal southeastern states from Texas to North Carolina, but about one-fourth of the reported infections were acquired in Pennsylvania, New York, Connecticut, or Massachusetts. One case was reported from California (Ciferri, 1982). Infection has been recorded more often in men than in women, by a ratio of 3:2, and the youngest patient was a woman 28 years of age. Only 15% were under 40 years of age; 24% were 60 or older.

Not all worms found in the pulmonary artery are *D. immitis.* Worms identified as *Wuchereria bancrofti*-like and *Brugia malayi*-like in a pulmonary artery surrounded by necrotic tissue have been reported (Beaver et al., 1971; Beaver and Cran, 1974). Other kinds of nematodes may also be seen in sections of lung tissue. Therefore, for an identification to be made the specific morphology of the parasite must be evident and correctly interpreted. Frequently the worm is too necrotic to be identified.

Dirofilaria immitis is a relatively large filaria. The male in tissue sections is usually 140 to 200 μm in diameter, while the females range up to about 300 μm. The cuticle is thick with two or three distinct layers, a smooth outer surface and a lateral inner thickening that is usually distinct, and lateral cords separating heavy somatic muscles into dorsal and ventral bands.

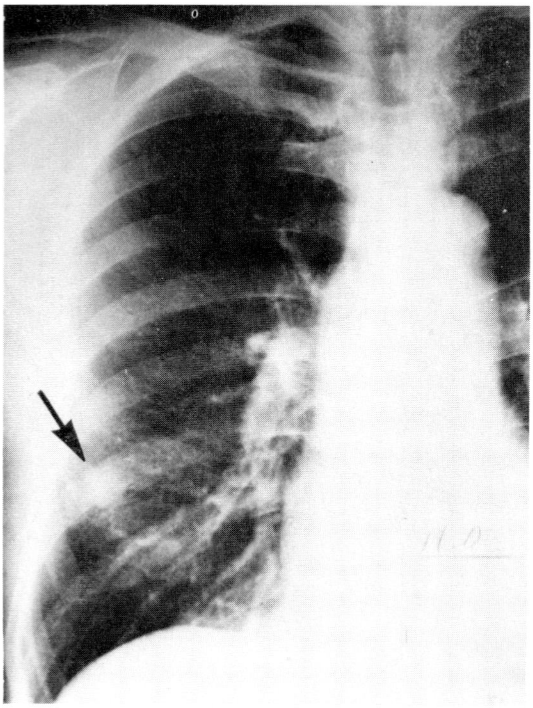

Fig. 23–25. *Dirofilaria immitis*. Coin lesion (arrow) seen on roentgenogram in periphery of right lung. (From Faust, E.C., Beaver, P.C., and Jung, R.C. 1975. *Animal Agents and Vectors of Human Disease*, 4th ed. Philadelphia, Lea & Febiger. Courtesy of Dr. D.M. Bradburn.)

The male genital tube is single at all levels. In the female, the anterior part of the genital tract is single, but the middle and hind parts are double (Beaver and Orihel, 1965; Chitwood and Lichtenfels, 1972; Ohmori et al., 1977).

The life cycle of *D. immitis* includes a period of development and growth in the subcutaneous and deeper tissues of the body; it is a young adult when it migrates to the heart (Orihel, 1961). In the dog, this early development requires 70 to 90 days, and development to full maturity after reaching the heart requires an additional period of about 3 months. Small *Dirofilaria* that possibly were immature *D. immitis* have been described in several cases: in the subcutaneous tissues on the brow of a 4-year-old child in Canada; the scrotum of a 44-year-old man in Chicago; and at the mastoid and temporal areas of the face of two women, 42 and 50 years of age, in Nebraska, all reviewed by Billups et al. (1980). In Japan, larger worms thought to be *D. immitis* were found in the subcutaneous tissues over the breast of a 52-year-old woman, the chest of a 34-year-old woman, and in others, reviewed by Kaneda et al. (1980). Worms with characteristics of *D. immitis* were found in the abdominal cavity in two women in Japan (Tada et al., 1979; Yoshimura et al., 1980).

Small *Dirofilaria* identified as *D. immitis* have also been reported from inside the eye of humans in Malaysia (Dissanaike et al., 1977) and Australia (Moorhouse, 1978), and in the eyes of dogs in seven cases; 2 in Australia, 2 in Virginia, and 1 each in Hawaii, Illinois, and Arkansas (Eberhard et al., 1977).

"Dirofilaria conjunctivae" (Addario, 1885) Desportes, 1939–1940

Synonyms. *Filaria conjunctivae* Addario, 1885; *F. labialis* of Pierantoni (nec *F. labialis* Pane, 1864 = *Gongylonema*, *fide* Sambon, 1924 and Brumpt, 1927); possibly *F. palpebralis* Pace, 1867; *F. peritonaei-hominis* Babes, 1879; *F. inermis* Grassi, 1887; *F. apapillocephala* Condorelli-Francaviglia, 1892; *Loa extraocularis* Skrjabin, 1917.

Filariae that Desportes (1939–1940) regarded as identical with the one described by Addario (1885), or in many ways resembling Addario's filaria, have been reported from man on numerous occasions from several countries in Europe, Asia, Africa, the United States, and Canada. The filariae have been removed from tumors or abscesses in various anatomical locations. The palpebral conjunctivae and other tissues of the eye are often affected, but such worms have also been found in the subcutaneous tissues of the arm, breast, neck, upper lip, and other sites. In 1965, Beaver and Orihel tabulated 24 cases of subcutaneous *Dirofilaria* infections in man in the United States in which the worms were assignable to *D. conjunctivae;* the majority of these were acquired in Florida. Numerous more recent cases in the United States have been recorded (Font et al., 1980). Similarly, numerous cases have been reported from Europe and elsewhere (Pinon et al., 1980; Bruijning, 1981).

Since patent infection did not develop in any of the observed cases, it was apparent that the worms identified as *D. conjunctivae* were natural parasites of animals, not of man. On the basis of morphological characteristics, Orihel and Beaver (1965) concluded that "*D. conjunctivae*" in the United States is frequently *D. tenuis,* a parasite of raccoons *(Procyon lotor),* whereas in Europe, the U.S.S.R., and Ceylon such infections are usually caused by *Dirofilaria repens* Railliet and Henry, 1911, a natural parasite of dogs in those areas. More recently it has become clear that several additional species of *Dirofilaria* can be recognized among the worms that formerly would have been identified as *D. conjunctivae*. In addition to *D. tenuis* and *D. repens,* the species which probably, or possibly, are among those found in man are *D. ursi,* a parasite of bears, and *D. striata* found in the subcutaneous tissues of wild cats *(Lynx rufus)* in the United States and Canada (Orihel and Beaver, 1974).

Dirofilaria (Nochtiella) repens Railliet and Henry, 1911

Synonyms. *Filaria conjunctivae* Addario, 1885; *F. acutiuscula* Molin, 1858).

Dirofilaria repens is a natural subcutaneous parasite of dogs in Europe, Africa, U.S.S.R., India, Sri Lanka, and Vietnam. It is also a parasite of cats in Vietnam, and it occurs in cats but not dogs in Malaysia; in Okinawa it was recorded as the species found in a man, though not yet recorded for either dogs or cats in that locality (MacLean et al., 1979). Females measure 13 to 18 cm and males 5 to 7 cm in length. The microfilariae are sheathless and measure 290 to 360 μm by 6 to 8 μm. Microfilaremia is subperiodic nocturnal. Mosquitoes of several species are natural intermediate hosts.

Human infection with *D. repens* was first reported by Skrjabin et al. (1930), who recovered a male worm from a subcutaneous nodule of the right lower eyelid of a woman in the U.S.S.R. Two additional cases in which the worm was identified as *D. repens,* and 26 in which it was identified as *D. conjunctivae* in regions where *D. repens* is enzootic, were tabulated by Faust in 1957. Since then numerous cases of subcutaneous dirofilariasis have been reported from Italy (Carneri et al., 1973; Pazzaglia et al., 1974; Pollavini et al., 1975), Greece (Tzamourans et al., 1976), France (Pinon and de Lalla, 1980), Yugoslavia (Jokovic and Nikolic, 1977), and elsewhere in Europe (Kiesselbach, 1967). In U.S.S.R. about 30 cases have been reported (Kovalev et al., 1971; Vodovozov et al., 1971). Other reports in which the worm was identified as *D. conjunctivae* or *D. repens* include 1 from Africa (O'Grady et

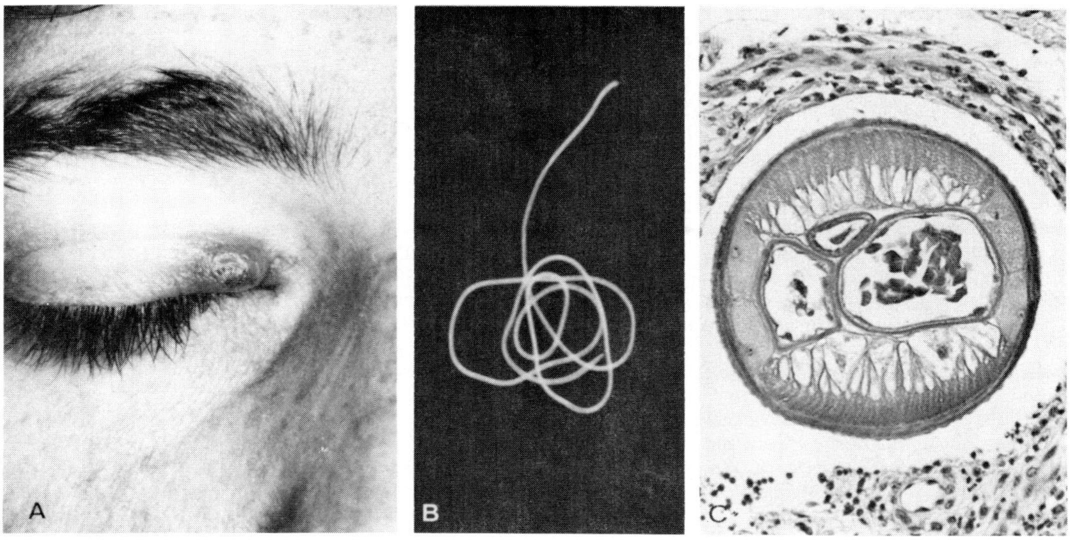

Fig. 23–26. *Dirofilaria "conjunctivae." A,* Nodule in upper lid of right eye of man in Texas. *B,* Sexually mature female worm released by incision of nodule shown in *A* and identified as *Dirofilaria tenuis. C,* Transverse section of worm with longitudinally ridged cuticle and other features characteristic of *D. tenuis* removed in a nodule from the thigh of a woman in Florida. (× 100.) (*A* and *B* courtesy of Dr. Whitman Rowland. *C* from Beaver, P.C., and Orihel, T.C. 1965. Human infection with filariae of animals in the United States. Am. J. Trop. Med. Hyg., *14*:1010–1029.)

al., 1962), 3 from Israel (Romano et al., 1976; Zweig et al., 1981), 12 from Sri Lanka (Dissanaike et al., 1972), and 1 from Okinawa (MacLean et al., 1979). A large female worm removed from the abdominal wall near the groin of a man in Wisconsin had been acquired in Greece at least 8 years earlier (Gardiner et al., 1978). The worm may be visible and active under the conjunctiva (Bruijning 1981), or coiled in a fibrous nodule, often surrounded by lymphoid tissue that may be mistaken for a lymph node (Zweig et al., 1981). When seen in motion under the conjunctiva, *D. repens* could be mistaken for *Loa loa,* which, however, is a worm of stouter proportions.

The distinctive features of *D. repens* in microsections are fine, regularly spaced, low, longitudinal ridges. The diameter varies from less than 200 μm in males to more than 400 μm in female worms. *D. repens, D. tenuis,* and other species of the *Nochtiella* subgenus are similar in cuticular morphology (Wong and Brummer, 1978).

Dirofilaria (Nochtiella) tenuis Chandler, 1942

Dirofilaria tenuis is a subcutaneous parasite of the raccoon *(Procyon lotor)* in the United States. The females measure 8 to 13 cm by 0.26 to 0.36 mm; the males, 4.0 to 4.8 cm by 0.19 to 0.26 mm. The vulva is just posterior to the base of the esophagus, and in some instances the vagina loops anteriorly over the posterior third of the esophagus. In microsections the cuticle, 5 to 8 μm thick, appears to have three layers, the outer of which bears longitudinal ridges that are smoothly rounded and spaced at intervals of about 10 μm. The microfilaria is sheathless; preserved in 2% formalin it measures 370 to 390 by 7 μm, and in thick films approximately 300 by 5 μm. Microfilaremia is periodic nocturnal.

In 1965, *D. tenuis* was redescribed and compared with the worms from 25 infections identified as *D. conjunctivae* in the United States (Orihel and Beaver, 1965). With the exception of one female from a woman in Washington near the Canadian border, all were indistinguishable from *D. tenuis.* The exceptional one was later classified with the *D. ursi* group. More than a dozen additional cases of subcutaneous dirofilariasis in which the worms were *D. tenuis*-like have been reported, 8 of which (reviewed by Font et al., 1980) involved the eye or eyelid (Fig. 23–26). Two were of special interest because the worms were mature females with microfilariae in the uterus; one was in a man from rural Mississippi (Jung and Espenan, 1967), the other in a man from Florida (Pacheco and Schofield, 1968).

Dirofilaria (Nochtiella) ursi Yamaguti, 1941

Dirofilaria ursi is a common parasite of the subcutaneous tissues of bears in Japan, Canada, and northern United States (Frechette and Rau, 1977; Manville, 1978). It was described in detail by Anderson (1952), who at the same time reviewed the genus *Dirofilaria.* Females are 117 to 224 mm long, 460 to 700 μm wide, and the vulva is postesophageal, 1.3 to 3.1 mm from the anterior end of the body; males are 51 to 86 mm by 330 to 460 μm. Microfilariae are unsheathed, 185 to 292 μm by 4.6 to 6.0 μm. The adult cuticle bears distinct longitudinal ridges that are generally straight with little branching, and delicate transverse striae that cut across the ridges.

The vectors of *D. ursi* are black flies (Simuliidae). The chief vector species in Ontario, Canada, is *Simulium venustum.* Larvae develop in the malpighian tubules and, at 23°C, reach infectivity in 9 days. The prepatent period in bears is 210 to 271 days. Larvae did not develop in any of the *Culicoides,* tabanids, or mosquitoes tested (Addison, 1980).

Worms resembling *D. ursi,* though not known definitely to be that species, have been reported from subcutaneous tissues of man in six cases: above the right breast of a woman in Washington State (Welty et al., 1963); from the scalp of a woman in Ontario (Anderson et al., 1968); from the breast of a woman in Quebec (Meerovitch et al., 1976); from the upper arm near the axilla of a woman in British Columbia (Beaver and Samuel, 1977); from the brow above the right eye of a woman near the Canadian border in Vermont (Christie, 1977);

and two worms, a male from the occipital region of the scalp and a female from the midsternal region of the chest, from a woman in upper Michigan, United States (Payan, 1978). The male worm in the last case measured 132 μm in diameter; the remaining worms (5 female, 1 male) were 175 to 200 μm in diameter. In all, the longitudinal ridges of the cuticle were indistinguishable from the type described and illustrated for *D. ursi*. However, the worms in human tissues, all nongravid, were smaller than the fully mature *D. ursi*. The cross-section microstructure of *D. ursi* agrees with that seen in the worms from human tissues (Uni et al., 1980).

Dirofilaria Species. A worm removed in a nodule from the lower eyelid of a farmer in North Carolina was unlike the *Dirofilaria* usually seen in subcutaneous nodules (Brumback et al., 1968). It was a mature infertile female about 250 μm in diameter, and the cuticle was smooth as in *D. immitis*. While it possibly was that species, there are other smooth-cuticle species that it might be, e.g., *D. lutrae,* a subcutaneous parasite of the American otter *(Lutra canadensis)* and *D. striata,* found in the subcutaneous tissues of American wild cats (Orihel and Beaver, 1974).

REFERENCES

Addison, E.H. 1980. Transmission of *Dirofilaria ursi* Yamaguti, 1914 (Nematoda: Onchocercidae) of black bears *(Ursus americanus)* by blackflies (Simuliidae). Can. J. Zool., *58*:1913–1922.

Adolph, P.E., Kagan, I.G., and McQuay, R.M. 1962. Diagnosis and treatment of *Acanthocheilonema perstans* filariasis. Am. J. Trop. Med. Hyg., *11*:76–88.

Ali-Khan, Z. 1977. Tissue pathology and comparative microanatomy of *Onchocerca* from a resident of Ontario and other enzootic *Onchocerca* species from Canada and the U.S.A. Ann. Trop. Med. Parasitol., *71*:469–482.

Anderson, R.C. 1952. Description and relationships of *Dirofilaria ursi* Yamaguti, 1941, and a review of the genus *Dirofilaria* Railliet and Henry, 1911. Trans. R. Canad. Inst., *29*:35–64.

———, and Bain, O., 1976. Keys to genera of the order Spirurida. Part 3. Diplotriaenoidea, Aproctoidea and Filarioidea. No. 3 of CIH Keys to the Nematode Parasites of Vertebrates. Edited by Roy C. Anderson, Alain G. Chabaud, and Sheila Willmott. Commonwealth Agricultural Bureau, Farnham Royal, Bucks, England.

———, Morse, P.G., and Scholten, T. 1968. *Dirofilaria (Nochtiella)* infestation in a human in Ontario. Can. Med. Assoc. J., *98*:788.

Anderson, W.A.D. 1953. *Pathology*, p. 395. 2nd ed. St. Louis, C.V. Mosby, Co.

Anteson, R.K., 1971. Occurrence of *Dipetalonema streptocerca* Macfie and Corson, 1922 in Ghana. Ghana Med. J., *10*:165.

Arzube, M.E., 1982. Oncocercosis en el Ecuador. Primer foco endemico descubierto en el pais, hallazgos clinicos, parasitologicos i entomologicos. Tropenmed. Parasitol., *33*:45–50.

Ash, L.R., and Little, M.D. 1964. *Brugia beaveri* sp. n. (Nematoda: Filarioidea) from the raccoon *(Procyon lotor)* in Louisiana. J. Parasitol., *50*:119–123.

———, and Schacher, J.F. 1971. Early life cycle and larval morphogenesis of *Wuchereria bancrofti* in the jird, *Meriones unguiculatus*. J. Parasitol., *57*:1043–1051.

Ashburn, L.L., Burch, T.A., and Brady, F.J. 1949. Pathogenic effects of suramin, Hetrazan and arsenamide on adult *Onchocerca volvulus*. Bol. Ofic. Sanit. Panam., *28*:1107–1117.

Atmosoedjono, S., Partono, F., Dennis, D.T., and Purnomo, A. 1977. *Anopheles barbirostris* (Diptera: Culicidae) as a vector of the Timor filaria on Flores Island. Preliminary observations. J. Med. Entomol., *13*:611–613.

Awadzi, K., Roulet, H., and Bell, D.R. 1980. The chemotherapy of onchocerciasis V. A standard method for the determination of microfilarial density in skin snips. Ann. Trop. Med. Parasitol., *74*:363–366.

Bain, O. 1971. Transmission des filarioses. Limitation des passages des microfilaires ingérées vers l'hémocèle du vecteur; interprétation. Ann. Parasitol., *46*:613–631.

———. 1972. Recherches sur la morphogénèse des filaires chez l'hôte intermédiaire. Ann. Parasitol., *47*:251–303.

———. 1975. Redescription de cinq especes d'onchocerques. Ann. Parasitol., *50*:763–788.

Barrett, J. 1983. Biochemistry of filarial worms. Helm. Abs., Ser. A. Animal and human helminthol., *52*:1–18.

Bartholomew, C.F., Nathan, M.B., and Tikasingh, E.S. 1978. The failure of diethylcarbamazine in the treatment of *Mansonella ozzardi* infections. Trans. R. Soc. Trop. Med. Hyg., *72*:423–424.

Beaver, P.C. 1970. Filariasis without microfilaremia. Am. J. Trop. Med. Hyg., *19*:181–189.

———, and Cran, I.R. 1974. *Wuchereria*-like filaria in an artery, associated with pulmonary infarction. Am. J. Trop. Med. Hyg., *23*:869–876.

———, Fallon, M., and Smith, G.H. 1971. Pulmonary nodule caused by a living *Brugia malayi*-like filaria in an artery. Am. J. Trop. Med. Hyg., *20*:661–666.

———, Hira, P.R., and Patel, B.G. 1983. Onchocerciasis in Zambia: report of *O. volvulus* in a child and its differentiation from *O. dukei* in cattle. Trans. R. Soc. Trop. Med. Hyg., *77*:162–167.

———, Horner, G.S., and Bilos, J.Z. 1974. Zoonotic onchocercosis in a resident of Illinois and observations on the identification of *Onchocerca* species. Am. J. Trop. Med. Hyg., *23*:595–607.

———, Meyer, E.A., Jarroll, E.L., and Rosenquist, R.C. 1980. *Dipetalonema* from the eye of a man in Oregon, U.S.A. A case report. Am. J. Trop. Med. Hyg., *29*:369–372.

———, Neel, J.V., and Orihel, T.C. 1976. *Dipetalonema perstans* and *Mansonella ozzardi* in Indians of southern Venezuela. Am. J. Trop. Med. Hyg., *25*:263–265.

———, and Orihel, T.C. 1965. Human infection with filariae of animals in the United States. Am. J. Trop. Med. Hyg., *14*:1010–1029.

———, and Samuel, W.M. 1977. Dirofilariasis in man in Canada. Am. J. Trop. Med. Hyg., *26*:329–330.

Bequaert, J. 1928. The insect carrier of *Onchocerca volvulus* in Liberia. Proceedings Fourth International Congress Entomology, *2*:605–607.

Beye, H.K., Edgar, S.A., Mille, R., Kessel, J.F., and Bambridge, B. 1952. Preliminary observations on the prevalence, clinical manifestations and control of filariasis in the Society Islands. Am. J. Trop. Med. Hyg., *1*:637–661.

Billups, J., Schenken, J.R., and Beaver, P.C. 1980. Subcutaneous dirofilariasis in Nebraska. Arch. Pathol. Lab. Med., *104*:11–13.

Blacklock, D.B. 1926. The development of *Onchocerca volvulus* in *Simulium damnosum*. Ann. Trop. Med. Parasitol., *20*:1–48.

Bloomfield, R.D., Suarez, J.R., and Malangit, A.C. 1978. Transplacental transfer of Bancroftian filariasis. J. Natl. Med. Assoc., *70*:597–598.

Bonne, C., Lie, K.J., Molenkamp, W.J.J., and Myeren, F.W. 1941. *Wuchereria malayi,* de Macrofilaria behoorende bij de *Microfilaria malayi*. Geneesk. Tijdschr. Ned.-Indie, *81*:1487–1501.

Bosworth, W., Ewert, A., and Bray, J. 1973. The interaction of *Brugia malayi* and streptococcus in an animal model. Am. J. Trop. Med. Hyg., *22*:714–719.

Boye, E.S., and Connor, D.H. 1982. An onchocercal nodule appearing during diethylcarbamazine therapy. Am. J. Trop. Med. Hyg., *31*:1123–1127.

Brug, S.L. 1927. Een nieuwe Filaria-soort *(Filaria malayi)*, parasiterende bij den Mensch. Geneesk. Tijdschr. Med.-Indie, *5*:750–754.

Bruijning, C.F.A. 1981. Human dirofilariasis: A report of the first case of ocular dirofilariasis in the Netherlands and a review of the literature. Trop. Geogr. Med., *33*:295–305.

Brumback, G.F., Morrison, H.M., and Weatherly, N.F. 1968. Orbital infection with *Dirofilaria*. South. Med. J., *61*:188, 192.

Brumpt, E. 1919. Une nouvelle filaire pathogene parasite de l'homme (*Onchocerca caecutiens* n. sp.). Bull. Soc. Pathol. Exot., *12*:464–473.

Buck, A.A., editor. 1974. *Onchocerciasis. Symptomatology, Pathology, Diagnosis*. Geneva, World Health Organization.

Buckley, J.J.C. 1934. On the development in *Culicoides furens* Poey of *Filaria (Mansonella) ozzardi* Manson, 1897. J. Helminthol., *12*:99–118.

———. 1958. A new genus, *Brugia*, for *Wuchereria* spp. of the *malayi* group. Abstr. Papers, Sixth International Congress Tropical Medicine & Malaria, p. 35.

———. 1960. On *Brugia* gen. nov. for *Wuchereria* spp. of the "*malayi*" group, i.e., *W. malayi* (Brug, 1927), *W. pahangi* Buckley and Edeson 1956, and *W. patei* Buckley, Nelson and Heisch, 1958. Ann. Trop. Med. Parasitol., *65*:245–251.

———, and Edeson, J.F.B. 1956. On the adult morphology of *Wuchereria* sp. *(malayi?)* from a monkey *(Macaca irus)* and from cats in Malaya, and on *Wuchereria pahangi* n. sp. from a dog and a cat. J. Helminthol., *30*:1–20.

Bushrod, F.M. 1981. The *Anopheles gambiae* Giles complex and Bancroftian filariasis transmission in a Tanzanian coastal village. Ann. Trop. Med. Parasitol., *75*:93–100.

Büttner, D.W., Laer, G.V., Mannweiler, E., and Büttner, M. 1982. Clinical, parasitological and serological studies on onchocerciasis in the Yemen Arab Republic. Tropenmed. Parasitol., *33*:201–212.

Cabrera, B.D., and Tamondong, C.T. 1966. Bancroftian and Malayan filiarasis in Palawan. Extent and distribution. Acta Med. Philip., *3*:20–36.

Campbell, N.A. 1976. A multivariate approach to variation in microfilariae: Examination of the species *Wuchereria lewisi* and demes of the species *W. bancrofti*. Aust. J. Zool., *24*:105–114.

Carme, B., Pichon, G., Merlin, M., and Laigret, J. 1977. Longevite des filaires lymphatiques: À propos d'une filaire *W. bancrofti* toujours fécode apres 40 ans d'existence. Med. Malad. Infect., *5*:254–257.

Carneri, I. de, Sacchi, S., and Pazzaglia, A. 1973. Subcutaneous dirofilariasis in man—not so rare. Trans. R. Soc. Trop. Med. Hyg., *67*:887–888.

Chabaud, A.G. 1952. Le genre *Dipetalonema* Diesing 1861; essai de classification. Ann. Parasitol., *27*:250–285.

———, and Bain, O. 1976. La lignee *Dipetalonema*. Nouvel essai de classification. Ann. Parasitol., *51*:365–397.

Chitwood, M.B., and Lichtenfels, J.R. 1972. Identification of parasitic metazoa in tissue sections. Exp. Parasitol., *32*:407–519.

Choyce, D.P. 1958. Some observations on the ocular complications and their relationship to blindness. Trans. R. Soc. Trop. Med. Hyg., *52*:112–121.

Christie, R.W. 1977. *Dirofilaria tenuis* in Vermont. N. Engl. J. Med., *297*:706–707.

Ciferri, F. 1982. Human pulmonary dirofilariasis in the United States: A critical review. Am. J. Trop. Med. Hyg., *31*:302–308.

Colless, D.H. 1971. An application of numerical methods to the systematics of microfilariae. Ann. Trop. Med. Parasitol., *65*:245–251.

Collins, R.C., Brandling-Bennett, A.D., Holliman, R.B., Campbell, C.C., and Darsie, R.F. 1980. Parasitological diagnosis of onchocerciasis: Comparisons of incubation media and incubation times for skin snips. Am. J. Trop. Med. Hyg., *29*:35–41.

Connor, D.H., Gibson, D.W., Neafie, R.C., Merighi, B., and Buck, A.A. 1983. Sowda—onchocerciasis in North Yemen: A clinicopathologic study of 18 patients. Am. J. Trop. Med. Hyg., *32*:123–137.

Coolidge, C., Weller, P.F., Ramsey, P.G., Ottesen, E.A., Beaver, P.C., and von Lichtenberg, F.C. 1979. Zoonotic *Brugia* filariasis in New England. Ann. Intern. Med., *90*:341–343.

Crewe, W. 1974. The persistence of *Acanthocheilonema perstans*. Proceedings Third International Congress Parasitology, *2*:621.

Cross, J.H., Partono, F., Hsu, M.K., Ash, L.R., and Oemijati, S. 1979. Experimental transmission of *Wuchereria bancrofti* to monkeys. Am. J. Trop. Med. Hyg., *28*:56–66.

Curtis, C.F., and Feachem, R.G. 1981. Sanitation and *Culex pipiens* mosquitoes: A brief review. J. Trop. Med. Hyg., *84*:17–25

Danaraj, T.J., Pacheco, G., Shanmugaratnam, K., and Beaver, P.C. 1966. The etiology and pathology of eosinophilic lung (tropical eosinophilia). Am. J. Trop. Med. Hyg., *15*:183–189.

David, H.L., and Edeson, J.F.B. 1965. Filariasis in Portuguese Timor, with observations on a new microfilaria found in man. Ann. Trop. Med. Parasitol., *59*:193–204.

Dennis, D.T., Partono, F., Purnomo, A., Atmosoedjono, S., and Saroso, J.S. 1976. Timor filariasis: Epidemiologic and clinical features in a defined community. Am. J. Trop. Med. Hyg., *25*:797–802.

Desowitz, R.S., and Hitchcock, J.C. 1974. Hyperendemic Bancroftian filariasis in the Kingdom of Tonga: The application of the membrane filter concentration technique to an age-stratified blood survey. Am. J. Trop. Med. Hyg., *23*:877–879.

Desportes, C. 1939–1940. *Filaria conjunctivae* Addario, 1885, parasite accidental de l'homme, est un *Dirofilaria*. Ann. Parasitol., *17*:380–404; 515–532.

Dissanaike, A.S. 1979. Zoonotic aspects of filarial infections in man. Bull. W.H.O., *57*:349–357.

———. Hock, Q.C., and Min, T.S. 1974. Mature female filaria, probably *Brugia* sp., from the conjunctiva of man in west Malaysia. Am. J. Trop. Med. Hyg., *23*:1023–1026.

———, Lykov, V.P., Sivayoham, I.S.S.R., Wijesekera, S.V.J., and Perera, M.C.S. 1972. Four more cases of human infection with *Dirofilaria (Nochtiella)*. Ceylon Med. J., *17*:105–112.

———, and Mak, J.W. 1980. A description of adult *Wuchereria bancrofti* (rural strain) from an experimental infection in the long-tailed macaque, *Macaca fascicularis* (syn. *M. irus*). *J. Helminthol.*, *54*:117–122.

———, and Paramanthan, D.C. 1961. On *Brugia (Brugiella subgen. nov.) buckleyi* n. sp., from the heart and blood vessels of the Ceylon hare. J. Helminthol., *35*:209–220.

———, Ramalingam, S., Fong, A., Pathmayokan, S., Thomas, V., and Kan, S.P. 1977. Filaria in the vitreous of the eye of man in Peninsular Malaysia. Am. J. Trop. Med. Hyg., *26*:1143–1147.

Dominguez-Vasquez, A., Taylor, H., Ruvalcaba-Macias, A., Murphy, R., Greene, B., Rivas-Alcala, A., and Beltran-Hernandez, F. 1983. Comparison of flubendazole and diethylcarbamazine in treatment of onchocerciasis. Lancet, *1*, No. 8317: 139–143.

Dubois, A., and Van den Berghe, L. 1948. *Diseases of the Warm Climates. Their Clinical Features. Diagnosis and Treatment.* New York, Grune and Stratton.

Duke, B.O.L. 1957. The reappearance, rate of increase and distribution of the filariae of *Onchocerca volvulus* following treatment with diethylcarbamazine. Trans. R. Soc. Trop. Med. Hyg., *51*:37–44.

———. 1958. The intake of the microfilariae of *Acanthocheilonema streptocerca* by *Culicoides milnei,* with some observations on the potentialities of the fly as a vector. Ann. Trop. Med. Parasitol., *52*:123–128.

———. 1963. Studies on the chemoprophylaxis of loiasis. II. Observations on diethylcarbamazine citrate (Banocide) as a prophylactic in man. Ann. Trop. Med. Parasitol., *57*:82–96.

———. 1977. Onchocerciasis. Br. Med. Bull., *28*:66–71.

———. 1981. The six diseases of WHO: Onchocerciasis. Br. Med. J., *283*:961–962.

———, and Moore, P.J. 1974. The concentration of microfilariae of a Guatemalan strain of *Onchocerca volvulus* in skin snips taken from chimpanzees over 24 hours. Tropenmed. Parasitol., *25*:153–159.

Dukes, D.C., Gelfand, M., Gadd, K.G., Clarke, V. de V., and Goldsmid, J.M. 1968. Cerebral filariasis caused by *Acanthocheiloma perstans.* Central Afr. J. Med., *14*:21–27.

Eberhard, M.L. 1979. Studies on the *Onchocerca* (Nematoda: Filarioidea) found in cattle in the United States. I. Systematics of *O. gutturosa* and *O. lienalis* with a description of *O. stilesi* sp. n. J. Parasitol., *65*:379–388.

———. Daly, J.J., Weinstein, S., and Farris, H.E. 1977. *Dirofilaria immitis* from the eye of a dog in Arkansas. J. Parasitol., *63*:978.

Edeson, J.F.B., Wilson, T., Wharton, R.H., and Laing, A.B.G. 1960. Experimental transmission of *Brugia malayi* and *B. pahangi* to man. Trans. R. Soc. Trop. Med. Hyg., *54*:229–234.

Eveland, L.K., Yermakov, V., and Kenney, M. 1975. *Loa loa* infection without microfilaremia. Trans. R. Soc. Trop. Med. Hyg., *69*:354–355.

Ewert, A. 1967. Studies on the transfer of infective *Brugia pahangi* larvae from vector mosquitoes to the mammalian host. Trans. R. Soc. Trop. Med. Hyg., *61*:110–113.

———, Balderach, R., and Elbihari, S. 1972. Lymphographic changes in regional lymphatics of cats infected with *Brugia malayi.* Am. J. Trop. Med. Hyg., *21*:407–414.

———, Corredor, A., Lightner, L., and D'Alessandro, A. 1979. Onchocerciasis focus in Colombia: Follow-up study after 12 years. Am. J. Trop. Med. Hyg., *2*:486–490.

———, and Ho, B.C. 1967. The fate of *Brugia pahangi* larvae immediately after feeding by infective vector mosquitoes. Trans. R. Soc. Trop. Med. Hyg., *61*:659–662.

Fain, A. 1974. Une nouvelle microfilaire parasitant le sang de deux muridae sauvages en Republique du Zaire (Nematoda: Filarioidea). Acta Zool. Pathol. Antverpiensia, No. 58, 103–108.

Fan, P.C., Wang, Y.C., Liu, J.C., Lo, H.S., and King, H.L. 1975. Control of Bancroftian filariasis by common salt medicated with diethylcarbamazine in Liehyu District (Little Kinmen), Kinmen (Quemoy) Islands, Republic of China. Ann. Trop. Med. Parasitol., *69*:515–516.

Faust, E.C. 1937. Mammalian heart worms of the genus *Dirofilaria,* pp. 131–139. In Festschrift Bernhard Nocht zum 80. Geburtstag. Hamburg.

———. 1957. Human infection with species of *Dirofilaria.* Zeitschr. Tropenmed. Parasitol., *8*:59–68.

———, Agosin, M., Garcia-Laverde, A., Sayad, W.Y., Johnson, V.M., and Murray, N.A. 1952. Unusual findings of filarial infections in man. Am. J. Trop. Med. Hyg., *1*:239–249.

———, Thomas, E.P., and Jones, J. 1941. Discovery of human heartworm infection in New Orleans. Parasitology, *27*:115–122.

Fawdry, A.L. 1957. Onchocerciasis in South Arabia. Trans. R. Soc. Trop. Med. Hyg., *51*:253–256.

Feng, L.C. 1936. The development of *Microfilaria malayi* in *A. hyrcanus* var. *sinensis* Wied. Chinese Med. J., Suppl. I, 345–367.

Font, R.L., Neafie, R.C., and Perry, H.D. 1980. Subcutaneous dirofilariasis of the eyelid and ocular adnexa. Report of six cases. Arch. Ophthalmol., *98*:1079–1082.

Frechette, J.L., and Rau, M.E. 1977. Helminths of the black bear in Quebec. J. Wildl. Dis., *13*:432–434.

Frimodt-Müller, C., and Barton, R.M. 1940. Pseudo-tuberculous condition associated with eosinophilia. Indian Med. Gaz., *75*:607–610.

Galliard, H. 1959. Filaires nouvelles du type *Bancrofti-malayi* chez l'homme et l'animal dans l'aire Afrique Orientale-Ocean Indien. Bull. Soc. Pathol. Exot., *52*:578–582.

———, and Brygoo, E.R. 1955. *Microfilaria bancrofti* var. *vauceli,* variété nouvelle de la cote sud-est de Madagascar. Bull. Soc. Pathol. Exot., *48*:473–475.

———, Brygoo, E.R., and Golvan, Y. 1955. Description de la microfilaire de *Wuchereria bancrofti* var. *vauceli* Galliard et Brygoo 1955. Ann. Parasitol., *30*:481–487.

———, and Mille, R. 1949. Un nouveau médicament antifilarien: le l-diéthyl-carbamyl-4-methyl-piperazine, experimenté à Tahiti. Bull. Soc. Pathol. Exot., *42*:304–313.

Gardiner, C.H., Oberdorfer, C.E., Reyes, J.E., and Pinkus, W.H. 1978. Infection of man by *Dirofilaria repens.* Am. J. Trop. Med. Hyg., *27*:1279–1281.

Garnham, P.C.C. 1954. The control of onchocerciasis. Ann. Soc. Belge Med. Trop., *34*:763–768.

Geigy, R., Aeschlimann, A., and Weiss, N. 1967. Transplacentäre Übertragung von Mikrofilarien der Art *Dipetalonema witeae* (Chabaud, 1952) bei *Meriones libycus* (Lichtenstein, 1823). Acta Trop., *24*:266–267.

Gibson, D.W., and Connor, D.H. 1978. Onchocercal lymphadenitis: clinicopathologic study of 34 patients. Trans. R. Soc. Trop. Med. Hyg., *72*:137–154.

———, Heggie, C., and Connor, D.H. 1980. Clinical and pathologic aspects of onchocerciasis, pp. 195–240. In *Pathology Annual,* Part 2, Vol. 15. Edited by S.C. Sommers and P.P. Rosen. New York, Appleton-Century-Crofts.

Godoy, G.A., Orihel, T.C., and Volcan, G.S. 1980a. *Microfilaria bolivarensis:* a new species of filaria from man in Venezuela. Am. J. Trop. Med. Hyg., *29*:545–547.

———, Volcan, G., Medrano, C., Teixeira, A., and Matheus, L. 1980b. *Mansonella ozzardi* infections among Indians of the southwestern part of the state of Bolivar, Venezuela. Am. J. Trop. Med. Hyg., *29*:373–376.

Goldsmid, J.M., and Rogers, S. 1979. A preliminary study on the treatment of filariasis due to *Dipetalonema perstans.* Central Afr. J. Med., *25*:51–52.

Gooneratne, B.W.M. 1973. A chronological lymphographic study of cats experimentally infected with *Brugia* filariasis from 5 days to 5 years. Lymphology, *6*:127–149.

Gordon, R.M., and Duke, B.O.L. 1955. Section cut through the skin of an African infected with *Acanthocheilonema perstans* and *A. streptocerca* at a time when a midge *Culicoides grahamii* was obtaining a blood meal. Trans. R. Soc. Trop. Med. Hyg., *49*:299.

Gratama, S. 1966. Onchocerciasis in the south-eastern territories of Liberia. With studies on the role of *Onchocerca volvulus* and *Wuchereria bancrofti* in the pathogenesis of hydrocele and elephantiasis. Acta Leidensia, *35*:1–135.

Greene, B.M., Otto, G.F., and Greenough, W.B. 1978. Circulating non-human microfilaria in a patient with systemic lupus erythematosus. Am. J. Trop. Med. Hyg., *27*:905–909.

Greenough, W.B., and Buckner, D. 1969. Removal of microfilariae from unanesthetized dogs by continuous flow centrifugation. Trans. R. Soc. Trop. Med. Hyg., *63*:259–262.

Grove, D.I., Cabrera, B.D., Valeza, F.S., Guinto, R.S., Ash, L.R., and Warren, K.S. 1977. Sensitivity and specificity of skin reactivity to *Brugia malayi* and *Dirofilaria immitis* antigens in Bancroftian and Malayan filariasis in the Philippines. Am. J. Trop. Med. Hyg., *26*:220–229.

Guderian, R.H., Leon, L.A., Leon, R., Fabian, C., Vasconez, C., and Johnston, T.S. 1982. Report on a focus of onchocerciasis in Esmeraldas Province of Ecuador. Am. J. Trop. Med. Hyg., *31*:270–274.

Gutierrez, Y., and Petras, R. 1982. *Brugia* infection in northern Ohio. Am. J. Trop. Med. Hyg., *31*:1128–1130.

Harinasuta, C., Sucharit, S., and Choochote, W. 1981. The susceptibility of leaf monkeys to Bancroftian filariasis in Thailand. Southeast Asian J. Trop. Med. Public Health, *12*:581–589.

Hawking, F. 1955. The chemotherapy of filarial infections. Pharmacol. Rev., *7*:279–299.

———. 1958. Chemotherapy of onchocerciasis. Trans. R. Soc. Trop. Med. Hyg., *52*:109–111.

———. 1967. The 24-hour periodicity of microfilariae: Biological mechanisms responsible for its production and control. Proc. R. Soc. B, *169*:59–76.

———. 1976. The distribution of human filariasis throughout the world. Part II. Asia. Trop. Dis. Bull., *73*:968–1016.

———. 1977. The distribution of human filariasis throughout the world. Part III. Africa. Trop. Dis. Bull., *74*:650–679.

———. 1979. The distribution of human filariasis throughout the world. Part IV. America. Trop. Dis. Bull., *76*:694–710.

———, and Denham, D.A. 1976. The distribution of human filariasis throughout the world. Part I. The Pacific region, including New Guinea. Trop. Dis. Bull., *73*:348–373.

———, Jennings, T., Louis, F.J., and Tuira, E. 1981. The mechanisms which affect the periodic cycle of Pacific *Wuchereria bancrofti* microfilariae. J. Helminthol., *55*:95–100.

———, and Marques, R.J. 1967. Control of Bancroftian filariasis by cooking salt medicated with diethylcarbamazine. Bull. W.H.O., *37*:405–414.

Hocking, B. 1967. Use of repellents and attractants in filariasis control. Bull. W.H.O., *37*:323–327.

Hoffmann, C.C. 1930. Über *Onchocerca* in Suden von Mexiko und die Weiterentwicklung ihrer Mikrofilarien in *Eusimulium mooseri*. Arch. f. Schiffs-u. Tropenhyg., *34*:461–472.

Hunter, G.W., III, and Warren, V.G. 1950. Studies on filariasis. VI. Observations of the reversal of microfilarial periodicity in a case of filariasis bancrofti. J. Parasitol., *36*:164–168.

Huntington, R.W., Jr., Fogel, R.H., Eichold, S., and Dickson, J.G. 1944. Filariasis among American troops in a South Pacific island group. Yale J. Biol. Med., *16*:529–537.

Janssens, P.G., Van Bogaert, L., Tverdy, G., and Wanson, M. 1958. Reflexions sur le sort des microfilaires de *Loa loa* l'organisme humain parasite. Manifestations viscerales provoquees por leur infiltration dans les tissus. Bull. Soc. Pathol. Exot., *51*:632–645.

Jayewardene, L.G. 1962. On two filarial parasites from dogs in Ceylon, *Brugia ceylonensis* n. sp. and *Dipetalonema* sp. inq. J. Helminthol., *36*:269–280.

Joesoef, A., and Dennis, D.T. 1980. Intestinal and blood parasites of man on Alor Island, Southeast Indonesia. Southeast Asian J. Trop. Med. Public Health, *11*:43–47.

Jokovic, B., and Nikolic, B. 1977. A new case of dirofilariasis in Yugoslavia. Acta Parasitol. Jugoslav., *8*:27–30.

Joon-Wah, M., Singh, D., Sukoharyong, J., and Sivanandam, S. 1974. *Brugia malayi* infection of the human eye: A case report. Southeast Asian J. Trop. Med. Public Health, *5*:226–229.

Jordan, P. 1959. A pilot scheme to eradicate Bancroftian filariasis with diethylcarbamazine. Trans. R. Soc. Trop. Med. Hyg., *53*:54–60.

Jung, R.C., and Espenan, P.H. 1967. A case of infection in man with *Dirofilaria*. Am. J. Trop. Med. Hyg., *16*:172–174.

———, and Harris, F.H. 1960. Human filarial infection in Louisiana. AMA Arch. Pathol., *69*:371–373.

Kale, O.O. 1982. Studies on the concentration of *Onchocerca volvulus* microfilariae in the skin in relation to the nodule. Tropenmed. Parasitol., *33*:43–44.

Kaneda, Y., Asami, K., Kawai, T., and Sakuma, M. 1980. A case of human infection with *Dirofilaria* in the subcutaneous tissue. Jpn. J. Parasitol., *29*:245–249.

Kershaw, W.E. 1955. The epidemiology of infections with *Loa loa*. Trans. R. Soc. Trop. Med. Hyg., *49*:143–150.

———, Beesley, W.N., and Crewe, W. 1955. Studies on the intake of microfilariae by their insect vectors, their survival, and their effect on the survival of their vectors. VI. Further observations on the intake of the microfilariae of *Loa loa* and *Acanthocheilonema perstans* by *Chrysops silacea* in laboratory conditions: The pattern of the intake of a group of flies. Ann. Trop. Med. Parasitol., *49*:114–120.

———, Duke, B.O.L., and Budden, F.H. 1954. The distribution of the microfilariae of *Acanthocheilonema streptocerca* in the skin of man. Trans. R. Soc. Trop. Med. Hyg., *48*:287.

Kessel, J.F., Thooris, G.C., and Bambridge, A. 1953. The use of diethylcarbamazine (Hetrazan or Notezine) in Tahiti as an aid in the control of filariasis. Am. J. Trop. Med. Hyg., *2*:1050–1061.

Kiesselbach, K. 1967. Ein Fall von Dirofilariosis in Deutschland. Dtsch. Med. Wochenshr., *92*:2353–2356.

King, B.G. 1944. Early filariasis diagnosis and clinical findings. A report of 268 cases in American troops. Am. J. Trop. Med., *4*:285–298.

Knott, J. 1938. The treatment of filarial elephantiasis of the leg by bandaging. Trans. R. Soc. Trop. Med. Hyg., *32*:243–252.

———. 1939. A method for making microfilarial surveys on day blood. Trans. R. Soc. Trop. Med. Hyg., *33*:191–196.

Kovalev, N.E., Zueva, V.K., and Mareuch, O.I. 1971. (On human dirofilariasis.) Medskaya Parasit., *40*:741–742.

Kumada, N., Kawamoto, F., Mizuno, S., Fujioka, H., Imaizumi, M., Suyama, M., Yamauchi, S., Yoshii, S., Takeura, S., Morise, M., and Sakai, S. 1980. Human case of pulmonary infarction due to *Dirofilaria immitis*. Jap. J. Parasitol., *29*:101–108.

Laigret, J., Fagneaux, G., and Tuira, E. 1980. Chimiothérapie de masse par la diéthylcarbamazine en doses espacées: Effets obtenus à Tahiti sur la microfilarémie à *Wuchereria bancrofti*, var. *pacifica*. Bull. W.H.O., *58*:779–783.

Laing, A.B.G., Edeson, J.F.B., and Wharton, R.H. 1961. Studies on filariasis in Malaya: Further experiments on the transmission of *Brugia malayi* and *Wuchereria bancrofti*. Ann. Trop. Med. Parasitol., *55*:86–92.

Lawrence, D.N., Erdtmann, B., Peet, J.W., Mello, J.A.N. de, Healy, G.R., Neel, J.V., and Salzano, F.M. 1979. Epidemiologic studies among Amerindian populations of Amazonia. II. Prevalence of *Mansonella ozzardi*. Am. J. Trop. Med. Hyg., *28*:991–996.

Leonardi, H.K., Lapey, J.D., and Ellis, F.H. 1977. Pulmonary dirofilariasis: Report of a human case. Thorax, *32*:612–615.

Lightner, L.K., Ewert, A., Corredor, A., and Sabogal, E. 1980. A parasitologic survey for *Mansonella ozzardi* in the Comisaria del Vaupes, Colombia. Am. J. Trop. Med. Hyg., *29*:42–45.

Looss, A. 1904. Zur Kenntniss des Baues der *Filaria loa* Guyot. Zool. Jahrb., Abt. Syst., *20*:549–574.

Macfie, J.W.S., and Corson, J.F. 1922. A new species of filarial larva found in the skin of natives in the Gold Coast. Ann. Trop. Med. Parasitol., *16*:465–471.

MacLean, J.D., Beaver, P.C., and Michalek, H. 1979. Subcutaneous dirofilariasis in Okinawa, Japan. Am. J. Trop. Med. Hyg., *28*:45–48.

Mahoney, L.E., Jr., and Aiu, P. 1970. Filariasis in Samoan immigrants to the United States. Am. J. Trop. Med. Hyg., *19*:629–631

Mak, J.W., Cheong, W.H., Yen, P., Lim, P., and Chan, W.C. 1982. Studies on the epidemiology of subperiodic *Brugia malayi* in Malaysia: problems in its control. Acta Trop., *39*:237–245.

Manson-Bahr, P.H. 1912. *Filariasis and Elephantiasis in Fiji*. Report to the London School of Tropical Medicine, London.

Mantovani, A., and Jackson, R.F. 1966. Transplacental transmission of microfilariae of *Dirofilaria immitis* in the dog. J. Parasitol., *52*:116.

Manville, A.M. 1978. Ecto- and endoparasites of the black bear in northern Wisconsin. J. Wildl. Dis., *14*:97–101.

Maplestone, P.A. 1938. A new filarial worm from a human being. Indian Med. Gaz., *73*:8–10.

March, H.N., and Laigret, J. 1958. The effect of cortisone and prednisone on bullous reactions following treatment of filariasis with diethylcarbamazine. Am. J. Trop. Med. Hyg., *7*:185–186.

———, Kessel, J.F., and Bambridge, B. 1960. Reduction in the prevalence of clinical filariasis in Tahiti following adoption of a control program. Am. J. Trop. Med. Hyg., *9*:180–184.

Marinkelle, C.J. 1973. First finding of *Dipetalonema perstans* in Colombia. Filariasis in Colombia III. Trop. Geogr. Med., *25*:51–52.

———, and German, E. 1970. Mansonelliasis in the Comisaria del Vaupes of Colombia. Trop. Geogr. Med., *22*:101–111.

Markell, E.K. 1964. Filarial infections in a California clinic. Ann. Intern. Med., *61*:1065–1070.

Mazzotti, L. 1948. Resultados negativos de la administracion del "Hetrazan" en dos pacientes infectados con *Mansonella ozzardi*. Med. Rev. Mex., *28*:317–318.

———, and Hewitt, R. 1948. Tratamiento de la oncocercosis por el cloruro de 1-dietilcarbamil-4-metil-piperazine (Hetrazan). Nota preliminar. Rev. Med. Mex., *28*:39–42.

McMahon, J.E. 1981. Chemotherapy with diethylcarbamazine and levamisole in Bancroftian filariasis. Tropenmed. Parasitol., *32*:250–252.

———. 1982. The examination-time/dose interval in the provocation of nocturnally periodic microfilariae of *Wuchereria bancrofti* with diethylcarbamazine and the practical uses of the test. Tropenmed. Parasitol., *33*:28–30.

———, Marshall, T.F. de C., Vaughan, J.P., and Kilstrup, N. 1979. Tanzania filariasis project: A provocative day test with diethylcarbamazine for the detection of nocturnally periodic *Wuchereria bancrofti* in the blood. Bull. W.H.O., *57*:759–765.

Meerovitch, E., Faubert, G., and Groulx, G. 1976. Zoonotic subcutaneous dirofilariasis in Quebec. Can. J. Public Health, *67*:333–335.

Mehlhorn, H., Steinhorst, U., Lindenblatt, B., Grüntzig, J., and Thomas, H. 1981. Electron microscopic studies on microfilariae of *Dipetalonema viteae* and *Litomosoides carinii:* The occurrence of intracellular parasites after treatment with metrifonate or diethylcarbamazine. Zbl. Bakt. Hyg., I. Abt. Orig., A. *251*:263–272.

Merrill, J.R., Otis, J., Logan, W.D., Jr., and Davis, M.B. 1980. The dog heartworm *(Dirofilaria immitis)* in man. An epidemic pending or in progress? J.A.M.A., *243*:1068–1070.

Meyers, W.M., Connor, D.H., Harman, L.E., Fleshman, K., Moris, R., and Neafie, R.C. 1972. Human streptocerciasis. A clinico-pathologic study of 40 Africans (Zairians) including identification of the adult filaria. Am. J. Trop. Med. Hyg., *21*: 528–545.

———, Moris, R., Neafie, R.C., Connor, D.H., and Bourland, J. 1978. Streptocerciasis: Degeneration of adult *Dipetalonema streptocerca* in man following diethylcarbamazine therapy. Am. J. Trop. Med. Hyg., *27*:1137–1147.

———, Neafie, R.C., and Connor, D.H. 1977. Onchocerciasis: invasion of deep organs by *Onchocerca volvulus*. Autopsy findings. Am. J. Trop. Med. Hyg., *26*:650–657.

———, Neafie, R.C., Moris, R., and Bourland, J. 1977. Streptocerciasis: Observation of adult male *Dipetalonema streptocerca* in man. Am. J. Trop. Med. Hyg., *26*:1153–1155.

Montestruc, E., Blache, R., and Laborde, R. 1950. Action du l-diéthylcarbamyl-4-methylpiperazine sur *Filaria ozzardi*. Bull. Soc. Pathol. Exot., *43*:275–278.

Moorhouse, D.E. 1978. *Dirofilaria immitis:* A cause of human intraocular infection. Infection, *6*:192–193.

———, Abrahams, E.W., and Stephens, B.J. 1976. Human pulmonary dirofilariasis associated with pleural effusion. Med. J. Aust., *2*:902–903.

Moraes, M.A.P., Almeida, M.M.R., Lovelace, J.K., and Chaves, G.M. 1978. *Mansonella ozzardi* entre indios ticunas do estado do Amazonas, Brasil. Bol. Ofic. Sanit. Panam., *85*:16–25.

———, Calheiros, L.B., Porto, M.A.S., Neves, R.N.A., and Shelley, A.J. 1978. Novas observações sobre o foco de oncocercose da área do Rio Toototobi, estado do Amazonas, Brasil. Bol. Ofic. Sanit. Panam., *84*:510–519.

———, Porto, M.A.S., Calheiros, L.B., and Shelley, A.J. 1979. Novas observações sobre o foco de oncocercose da aréa do Rio Auaris, territorio de Roraima, Brasil. Bol. Ofic. Sanit. Panam., *86*:509–516.

Morenas, L. 1929. Un cas de filariose dû à *Acanthocheilonema perstans* avec manifestations cliniques et grosse éosinophile. Bull. Soc. Pathol. Exot., *22*:325–330.

Nathan, M.B., Bartholomew, C.F., and Tikasingh, E.S. 1978. The detection of *Mansonella ozzardi* microfilariae in the skin and blood with a note on the absence of periodicity. Trans. R. Soc. Trop. Med. Hyg., *72*:420–422.

Neafie, R.C., Connor, D.H., and Meyers, W.M. 1975. *Dipetalonema streptocerca* (Macfie and Corson, 1922): Description of the adult female. Am. J. Trop. Med. Hyg., *24*:264–267.

Nelson, G.S. 1970. Onchocerciasis. Adv. Parasitol., *8*:173–224.

———, and Davies, J.B. 1976. Observations on *Mansonella ozzardi* in Trinidad. Trans. R. Soc. Trop. Med. Hyg., *70*:16–17.

Nichols, W.L., and Kershaw, W.E. 1954. Studies on the intake of microfilariae by their insect vectors, their survival, and their effect on the survival of their vectors. III. The intake of the microfilariae of *Acanthocheilonema perstans* by *Culicoides austeni* and *C. grahamii*. Ann. Trop. Med. Parasitol., *48*:201–206.

Nnochiri, E. 1972. The causal agent of the ocular syndrome of the "Kampala eye worm." East Afr. Med. J., *l0*:198.

O'Grady, F., Fawcett, A., and Buckley, J.J.C. 1962. A case of human infection with *Dirofilaria (Nochtiella)* sp. probably of African origin. J. Helminthol., *36*:309–312.

Ohmori, Y., Yamaguchi, T., and Ohishi, I. 1977. Morphological observations on the immature worm of *Dirofilaria immitis* in cross section. Jpn. J. Parasitol., *26*:323–335.

Omar, M.S., Franz, M., and Büttner, D.W. 1979. Some observations on onchocerciasis including sowda in the Yemen Arab Republic. Tropenmed. Parasitol., *30*:113–119.

Orihel, T.C. 1961. Morphology of the larval stages of *Dirofilaria immitis* in the dog. J. Parasitol., *47*:251–262.

———. 1964. *Brugia guyanensis* sp. n. (Nematoda: Filarioidea) from the coatimundi *(Nasua nasua vittata)* in British Guiana. J. Parasitol., *52*:162–165.

———. 1966. *Brugia tupaiae* sp. n. (Nematoda: Filarioidea) in tree shrews *(Tupaia glis)* from Malaysia. J. Parasitol., *52*:162–165.

———. 1967. Infections with *Dipetalonema perstans* and *Mansonella ozzardi* in the aboriginal Indians of Guyana. Am. J. Trop. Med. Hyg., *16*:628–635.

———. 1973. Cerebral filariasis in Rhodesia—a zoonotic infection? Am. J. Trop. Med. Hyg., *22*:596–599.

———, and Beaver, P.C. 1965. Morphology and relationship of *Dirofilaria tenuis* and *Dirofilaria conjunctivae*. Am. J. Trop. Med. Hyg., *14*:1030–1043.

———, and Beaver, P.C. 1974. Zoonotic filariasis in the United States. Proc. 3rd Int. Congr. Parasitol., *2*, B13.

———, and Eberhard, M.L. 1982. *Mansonella ozzardi*: A redescription with comments on taxonomic relationships. Am. J. Trop. Med. Hyg., *31*:1142–1147.

———, and Esslinger, J.H. 1973. *Meningonema peruzzii* gen. et sp. n. (Nematoda: Filarioidea) from the central nervous system of African monkeys. J. Parasitol., *59*:437–441.

———, and Lowrie, R.C., Jr. 1975. *Loa loa:* Development to the infective stage in an American deerfly, *Chrysops atlanticus*. Am. J. Trop. Med. Hyg., *24*:610–615.

———, and Moore, P.J. 1975. *Loa loa:* Experimental infection in two species of African primates. Am. J. Trop. Med. Hyg., *24*:606–609.

———, and Pacheco, G. 1966. *Brugia malayi* in the Philippine macaque. J. Parasitol., *52*:394.

Owen, H.B., and Hennessey, R.S.F. 1932. A note on some ocular manifestations of helminthic origin occurring in natives of Uganda. Trans. R. Soc. Trop. Med. Hyg., *25*:267–274.

Pacheco, G., and Orihel, T.C. 1968. Relationship between detectable microfilaremia and total population of microfilariae in monkeys infected with *Dirofilaria corynodes*. J. Parasitol., *54*:1234–1235.

———, and Schofield, H.L., Jr. 1968. *Dirofilaria tenuis* containing microfilariae in man. Am. J. Trop. Med. Hyg., *14*:180–182.

Palmieri, J.R., Purnomo, A., Dennis, D.T., and Marwoto, H.A. 1980. Filarid parasites of south Kalimantan (Borneo), Indonesia. *Wuchereria kalimantani* sp. n. (Nematoda: Filarioidea) from the silvered leaf monkey, *Presbytis cristatus* Eschscholtz 1921. J. Parasitol., *66*:645–651.

———, Connor, D.H., Purnomo, A., Dennis, D.T., and Marwoto, H. 1982. Experimental infection of *Wuchereria brancrofti* in the silvered leaf monkey *Presbytis cristatus* Eschscholtz, 1821. J. Helminthol., *56*:243–245.

Partono, F., Purnomo, A., Dennis, D.T., Atmosoedjono, S., Oemijati, S., and Cross, J.H. 1977. *Brugia timori* sp. n. (Nematoda: Filarioidea) from Flores Island, Indonesia. J. Parasitol., *63*:540–546.

———, Purnomo, A., Oemijati, S., and Soewarta, A. 1981. The long term effects of repeated diethylcarbamazine administration with special reference to microfilaraemia and elephantiasis. Acta Trop., *38*:217–225.

Payan, H.M. 1978. Human infection with *Dirofilaria*. Arch. Dermatol., *114*:593–594.

Pazzaglia, A., Sacchi, S., and Carneri, I. de. 1974. Un'affezione cutanea non rara in Italia: la dirofilariosi. Due nuovi casi pavesi. Minerva Dermatol., *109*:217–224.

Peel, E., and Chardome, M. 1946. Sur des filarides de chimpanzes *"Pan paniscus"* et *"Pan satyrus"* au Congo belge. Ann. Soc. Belge Med. Trop., *26*:117–156.

Peruzzi, M. 1928. Cerebral filariasis in five monkeys infected with trypanosomal meningitis. League of Nations Health Org., Final Report of the Int. Committee on Human Trypanosomiasis, p. 309–313.

Piessens, W.F., Ratiwayanto, S., Tuti, S., Palmieri, J.H., Piessens, P.W., Koiman, I., and Dennis, D.T. 1980. Antigen-specific suppressor cells and suppressor factors in human filariasis with *Brugia malayi*. N. Engl. J. Med., *302*:833–837.

Pinon, J.M., Dousset, H., Ologoudou, L., Sulahian, A., and Adnet, J.-J. 1980. Dirofilariasis of the breast in France. Am. J. Trop. Med. Hyg., *29*:1018–1019.

Pollavini, G., and de Lalla, F. 1975. Un caso autoctono di dirofilariasi sottocutanea nell'uomo. Ann. Sclavo, *17*:678–683.

Poltera, A.A. 1973. The histopathology of ocular loiasis in Uganda. Trans. R. Soc. Trop. Med. Hyg., *67*:819–829.

Prost, A. 1980. Le polymorphisme des onchocercoses humaines ouest-africaines. Ann. Parasitol., *55*:239–245.

———, Rougemont, A., and Omar, M.S. 1980. Caractères épidemiologiques, cliniques et biologiques des onchocercoses de savane et de forét en Afrique occidentale. Revue critique et éléments nouveaux. Ann. Parasitol., *55*:347–355.

Purnomo, A., Dennis, D.T., and Partono, F. 1977. The microfilaria of *Brugia timori* (Partono et al. 1977 = Timor microfilaria, David and Edeson, 1964): Morphologic description with comparison to *Brugia malayi* of Indonesia. J. Parasitol., *63*:1001–1006.

Raccurt, C., Lowrie, R.C., Jr. and McNeeley, D.F. 1980. *Mansonella ozzardi* in Haiti. I. Epidemiological survey. Am. J. Trop. Med. Hyg., *29*:803–808.

Rajapakse, Y.S. 1974. Diethylcarbamazine provocation of the microfilariae of *Wuchereria bancrofti*. J. Trop. Med. Hyg., *77*:182–184.

Rao, S.S. 1945. Filarial infection in Dhamda (Drug District, C.P.) due to *Wuchereria malayi*. Indian J. Med. Res., *33*:175–176.

———, and Maplestone, P.A. 1940. The adult of *Microfilaria malayi* Brug, 1927. Indian Med. Gaz., *75*:159–160.

Rao, C.K., Rao, P.P., Russel, S., Hamzakoya, K.K., Chandrasekharan, A., Roychowdhury, S.P., and Rao, Ch. K. 1981. Control of bancroftian filariasis with common salt medicated with diethylcarbamazine in Lakshadweep. Indian J. Med. Res., *73*:865–873.

Rassi, E., Monzon, H., Castillo, M., Hernandez, I., Perez, J.R., and Convit. J. 1977. Discovery of a new onchocerciasis focus in Venezuela. Bull. Pan. Am. Health Org., *11*:41–64.

Raybould, J.N., and White, G.B. 1979. The distribution, bionomics and control of onchocerciasis vectors (Diptera: Simuliidae) in Eastern Africa and the Yemen. Tropenmed. Parasitol., *30*:505–547.

Raymond, L.A., Gutierrez, Y., Strong, L.E., Wander, A.H., Buten, R., and Cordan, D. 1978. Living retinal nematode (filarial-like) destroyed with photocoagulation. Ophthalmology, *85*:944–949.

Rivas-Alcala, A., Taylor, H., Ruvalcaba-Macias, A., Mackenzie, C., Greene, B., Dominguez-Vasquez, A., Lugo-Pfeiffer, C., and Beltran, F. 1981. Chemotherapy of onchocerciasis: A controlled comparison of mebendazole, levamisole, and diethylcarbamazine. Lancet, *2*, No. 8245:485–490.

Roberts, J.M.D., Neumann, E., Göckel, C.W., and Highton, R.B. 1967. Onchocerciasis in Kenya 9, 11 and 18 years after elimination of the vector. Bull. Wld. Health Org., *37*:195–212.

Robinson, N.B., Chavez, C.M., and Conn, J.H. 1977. Pulmonary dirofilariasis in man. J. Thorac. Cardiovasc. Surg., *74*:403–408.

Rodger, F.C. 1959. *Blindness in West Africa*. London, H.K. Lewis & Co., Ltd.

Romano, A., Sachs, R., and Lengy, J. 1976. Human ocular dirofilariasis in Israel. Isr. J. Med. Sci., *12*:208–214.

Rosenblatt, P., Beaver, P.C., and Orihel, T.C. 1962. A filarial infection apparently acquired in New York City. Am. J. Trop. Med. Hyg., *11*:641–645.

Rozeboom, L., and Cabrera, B. 1965. Filariasis caused by *Wuchereria bancrofti* in Palawan, Republic of the Philippines. Am. J. Epidemiol., *81*:216–221.

Rubin, R.H., Austen, K.F., and Goetzl, E.J. 1975. Studies of immediate hypersensitivity in a patient with *Acanthocheilonema perstans* filarial infection. J. Infect. Dis., *131*(Suppl.):S98–S103.

Salazar, M.M., Gonzales, B.D., and Jurado, M.J. 1971. Trichlorophone treatment of onchocerciasis. Ann. Trop. Med. Parasitol., *65*:393–398.

Sasa, M. 1970. *Filariasis and Schistosomiasis. Recent Advances in Japan*. Baltimore, University Park Press.

———. 1976. *Human Filariasis: A Global Survey of Epidemiology and Control*. Baltimore, University Park Press.

Schacher, J.F. 1969. Intraspecific variation in microfilariae, with description of *Wuchereria lewisi* sp. nov. (Nematoda, Filarioidea) from man in Brazil. Ann. Trop. Med. Parasitol., *63*:341–351.

———. 1973. Laboratory models in filariasis: A review of filarial life cycle patterns. Southeast Asian J. Trop. Med. Public Health, *4*:336–349.

———, and Geddawi, M.K. 1969. An analysis of speciation and evolution in *Wuchereria bancrofti* by the study of nuclear constancy (eutely) in microfilariae. Ann. Trop. Med. Parasitol., *63*:67–82.

———, Geddawi, M.K., and Churchill, C.W. 1967. Nuclear number of microfilariae as a test for intraspecific groupings and evolution in *Wuchereria bancrofti*. J. Parasitol., *53*:892–893.

Schlesinger, J.J., Dubois, J.G., and Beaver, P.C. 1977. *Brugia*-like filarial infection acquired in the United States. Am. J. Trop. Med. Hyg., *26*:204–207.

Schneider, J. 1961. Notes sur la therapeutique des filarioses. Ann. Soc. Belge Med. Trop., *41*:346–366.

Schulz-Key, H. 1978. A simple technique to assess the total number of *Onchocerca volvulus* microfilariae in skin snips. Tropenmed. Parasitol., *29*:51–54.

———, and Albiez, E.J. 1977. Worm burden of *Onchocerca volvulus* in a hyperendemic village of the rain-forest in West Africa. Tropenmed. Parasitol., *28*:431–438.

———, Albiez, E.J., and Büttner, D.W. 1977. Isolation of living adult *Onchocerca volvulus* from nodules. Tropenmed. Parasitol., *28*:428–430.

Scully, R.E., and McNeely, B.U. (Eds.). Case records of the Massachusetts General Hospital. Case 26–1974. N. Engl. J. Med., *291*:35–42.

Seo, B.S. 1978. Malayan filariasis in Korea. Korean J. Parasitol., *16*(Suppl):118.

Sharp, N.A.D. 1927. A note on *Agamofilaria streptocerca* Macfie and Corson, 1922. Ann. Trop. Med. Parasitol., *21*:415–417.

———. 1928. *Filaria perstans;* its development in *Culicoides austeni*. Trans. R. Soc. Trop. Med. Hyg., *21*:371–396.

Shelley, A.J. 1975. A preliminary survey of the prevalence of *Mansonella ozzardi* in some rural communities on the river Purus, state of Amazonas, Brazil. Ann. Trop. Med. Parasitiol., *69*:407–412.

———, and Shelley, A. 1976. Further evidence for the transmission of *Mansonella ozzardi* by *Simulium amazonicum* in Brazil. Ann. Trop. Med. Parasitol., *70*:213–217.

Skrjabin, K.I., Algansen, A.J., and Schoulmann, E.S. 1930. Premier cas de *Dirofilaria repens* chez l'homme. Trop. Med. Vet., Moscow, *2*:9 (Russian text).

Sondergaard, J. 1972. Filariasis caused by *Acanthocheilonema perstans*. Arch. Dermatol., *106*:547–548.

Stott, G. 1962. Pathogenicity of *Acanthocheilonema perstans*. J. Trop. Med. Hyg., *65*:230–232.

Strohschneider, H. 1956. Oral and intraperitoneal treatment of *Acanthocheilonema (Dipetalonema) perstans* with diethylcarbamazine. Am. J. Trop. Med. Hyg., *5*:158–162.

Strong, R.P. 1931. *Onchocerca* investigations in Guatemala. N. Engl. J. Med., *204*:916–920.

Tada, I., and Figueroa M., H. 1974. The density of *Onchocerca volvulus* microfilariae in the skin at different times of the day in Guatemala. Jpn. J. Parasitol., *23*:220–225.

———, Sakaguchi, Y., and Eto, K. 1979. *Dirofilaria* in the abdominal cavity of a man in Japan. Am. J. Trop. Med. Hyg., *28*:988–990.

Thomas, D.B., Anderson, R.I., and MacRae, A.A. 1973. Daytime variation in the density of *Onchocerca volvulus* microfilariae in human skin. Bull. W.H.O., *49*:493–498.

Tidwell, M.A., Tidwell, M., and Muñoz de Hoyos, P. 1980. Development of *Mansonella ozzardi* in a black fly species of the *Simulium sanguineum* group from eastern Vaupes, Colombia. Am. J. Trop. Med. Hyg., *29*:1209–1214.

———, and Corredor, A. 1980. *Simulium exiguum*, the vector of *Onchocerca volvulus* on the Rio Micay, Colombia. Am. J. Trop. Med. Hyg., *29*:377–381.

Trent, S.C. 1963. Reevaluation of World War II veterans with filariasis acquired in the South Pacific. Am. J. Trop. Med. Hyg., *12*:877–887.

Tzamourans, N., Litsios, B., Papaevangelou, G., and Tsitouris, G. 1976. Deux nouveaux cas de dirofilariose humaine souscutanee en Grece. Arch. Inst. Pasteur Hellenique, *22*:61–73.

Uni, S., Kimata, I., and Takada, S. 1980. Cross-section morphology of *Dirofilaria ursi* in comparison with *D. immitis*. Jpn. J. Parasitol., *29*:489–497.

Vincent, A.L., de Gomez, M.V., Gonzalvo, A., Nayar, J., and Sodeman, W.A. 1981. Filariasis in the Dominican Republic. Am. J. Trop. Med. Hyg., *30*:739–741.

Vodovozov, A.M., Kovalev, R.P., and others. 1971. *Dirofilaria repens*. Med. Parasitol. Parasit. Dis. (Moscow), *40*:739–740.

Wahlgren, M. 1982. The successful treatment of *Dipetalonema perstans* filariasis with mebendazole. Ann. Trop. Med. Parasitol., *76*:557–559.

Wanson, M. 1950. Contribution a l'étude de l'onchocercose africaine humaine. Ann. Soc. Belge Med. Trop., *30*:667–683.

Wartman, W.B., and King, B.G. 1944. Early filariasis in American soldiers. Bull. U.S. Army Med. Dept. No. 76, 45–49.

Webber, R.H. 1977. The natural decline of *Wuchereria bancrofti* infection in a vector control situation in the Solomon Islands. Trans. R. Soc. Trop. Med. Hyg., *71*:396–400.

———. 1979. Eradication of *Wuchereria bancrofti* infection through vector control. Trans. R. Soc. Trop. Med. Hyg., *73*:722–724.

Webster, E.H. 1946. Filariasis among white immigrants in Samoa. U.S. Naval Med. Bull., *46*:186–192.

Wegesa, U., McMahon, J.E., Abaru, D.E., Hamilton, P.J.S., Marshall, T.F. de C., and Vaughan, J.P. 1979. Tanzania filariasis project. Survey methodology and clinical manifestations of Bancroftian filariasis. Acta Trop., *36*:369–377.

Weingarten, R.J. 1943. Tropical eosinophilia. Lancet, *1*:103–105.

Weller, P.F., Simon, H.B., Parkhurst, B.H., and Medrek, T.F. 1978. Tourism-acquired *Mansonella ozzardi* microfilaremia in a regular blood donor. J.A.M.A., *240*:858–859.

Wellman, F.C. 1907. Preliminary note on some bodies found

in ticks—*Ornithodoros moubata* (Murray)—fed on blood containing embryos of *Filaria perstans* (Manson). Br. Med. J., *2*:142–143.

Welty, R.F., Ludden, T.E., and Beaver, P.C. 1963. Dirofilariasis in man: Report of a case from the state of Washington. Am. J. Trop. Med. Hyg., *12*:888–891.

Wijers, D.J.B. 1977. Bancroftian filariasis in Kenya. I. Prevalence survey among adult males in the Coast Province. Ann. Trop. Med. Parasitol., *71*:313–331.

———, and Kiilu, G. 1977. Bancroftian filariasis in Kenya III. Entomological investigation in Mambrui, a small coastal town, and Jaribuni, a rural area more inland (Coast Province). Ann. Trop. Med. Parasitol., *71*:347–359.

Wijeyaratne, P.M., Singha, P., Verma, O.P., and Motha, B. 1982. Evaluation of the diethylcarbamazine provocative test in the diagnosis of *Wuchereria bancrofti* infections in the Nigerian savanna and the effects on *Dipetalonema perstans*. Trans. R. Soc. Trop. Med. Hyg., *76*:387–391.

Wong, M.M., and Brummer, M.E.G. 1978. Cuticular morphology of five species of *Dirofilaria:* A scanning electron microscope study. J. Parasitol., *64*:108–114.

World Health Organization 1954. Expert Committee on Onchocerciasis. Report. W.H.O. Tech. Rep. Ser. No. 87.

———. 1966. Expert Committee on Onchocerciasis. Second report. W.H.O. Tech. Rep. Ser. No. 335.

———. 1976. Epidemiology of onchocerciasis. Report of a W.H.O. expert committee. W.H.O. Tech. Rep. Ser. No. 597.

Yamane, Y., Yazaki, S., Fukumoto, S., Matsui, K., and Tanaka, A. 1977. A human case of pulmonary dirofilariasis. Yonago Acta Med., *21*:111–118.

Yoshimura, H., Akao, N., and Kamimura, K. 1981. Case reports of infections with *Dirofilaria immitis* in lung and *Wuchereria bancrofti* in epididymis, diagnosed by cross section morphology of worms in pathological specimens. Jpn. J. Parasitol., *30*:381–386.

———, Akao, N., Kondo, K., Ohnishi, Y., Kitagawa, M., and Kamimura, K. 1980. Human dirofilariasis in Japan. Case report and review of literature. Int. J. Zoonoses, *7*:107–114.

Zhong, H., He, L., and Cao, W. 1981. Present situation of filariasis in China. Chin. Med. J., *94*:567–584.

Ziemann, H. 1926. Ein Fall von *Filaria-loa*-infektion mit mindestens 17 jähriger Dauer. Arch. f. Schiffs- u. Tropen-Hyg., *30*:626–628.

Zweig, A., Karasik, A., and Hiss, J. 1981. *Dirofilaria* in a cervical lymph node in Israel. Hum. Pathol., *12*:939–940.

Chapter 24

Nematomorpha. Gordiacea or Hairworms

The Gordiacea belong to the phylum Nematomorpha. The adults are wiry worms, measuring 10 to 50 cm in length with a bluntly rounded anterior end. The caudal end of the male is either bifurcate or bears a dorsoventral groove; that of the female is either entire or trilobate. In the mature state they have an atrophied digestive tract, a true body cavity, gonads discontinuous with their ducts, and they lack the lateral line and excretory structures characteristic of nematodes. The females have a cloaca. Adults are free-living in water; larvae are parasites of insects. Human infection is accidental.

The sexually mature worms mate in water, where eggs are laid in strands. The larvae, which actively emerge from the eggs following embryonation, have a beaked proboscis provided with retractile stylets and three rows of large reversed spines. On penetration of the body wall of various species of Orthoptera and other insects, they reach the hemocoel where they molt and metamorphose into gordius-like worms. As they approach maturity they escape actively from the insect and become free-living.

Members of the class Gordiacea (so-called because they are at times coiled into a Gordian knot) belong to two recognized families: Gordiidae and Chordodidae. The *former* have a smooth cuticle. Body bristles, when present, are derived from cuticular fibers. The posterior end of the male has two postanal projections and a postanal crescent; that of the female is entire. All species of this family belong to the genus *Gordius*. In the *second family* the cuticle is rough with true areoles. Tubercles and bristles arise from a nonfibrous cuticle. The posterior end of the male is bifurcate or grooved, but without a postanal crescent; that of the female is either entire or trifurcate (Fig. 24–1). The recognized genera include: *Chordodes, Paragordius, Parachordodes, Neochordodes,* and *Pseudogordius*.

Gordius Worms as Human Parasites

Although earlier medical parasitologists and physicians attributed grave consequences to "hair snakes" in the body, their introduction into the digestive tract of man undoubtedly resulted from ingestion of free-living adults or adolescent worms still within their insect hosts in raw drinking water or food.

Gordiid worms identified as *Parachordodes raphaelis* were

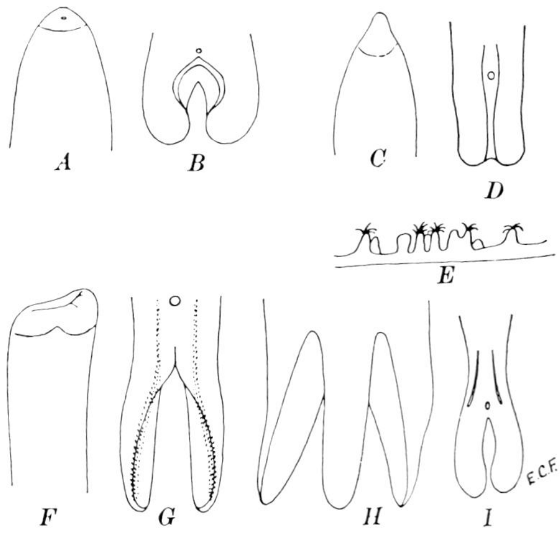

Fig. 24–1. Gordiacea, characteristic features. *A*, Anterior extremity. *B*, Posterior extremity of male *Gordius villoti*, enlarged. *C*, Anterior extremity. *D*, Posterior extremity of male *Chordodes*, enlarged. *E*, Section through papillae and cuticular hairs of *Chordodes*, greatly enlarged. *F*, Anterior extremity. *G*, Posterior extremity of male. *H*, Posterior extremity of female *Paragordius varius*, enlarged. *I*, Posterior extremity of male *Parachordodes*, enlarged. (*A-D* and *I* adapted from Camerano, L. 1897. Monografia dei Gordii. Mem. R. Accad. Sc. Torino, *47*:339–419. *E* from Romer, F. 1896. Beitrag zur Systematik der Gordiiden. Seuchenbergischen naturforschenden Gesellschaft, *23*, no. 2. *F* and *H* adapted from Stiles, C.W. 1907. Three new American cases of infection of man with horse-hair worms (species *Paragordius varius*), with summary of all cases reported to date. U.S. Hyg. Lab. Bull., no. *34*:53–58, and from May, H.G. 1920. Contributions to the life history of *Gordius robustus* Leidy and *Paragordius varius* (Leidy). Illinois Biol. Monogr., *5*, no. 2.)

reportedly passed *per urethram* by a girl in South Africa. Carvalho (1942) recorded the passage *per urethram* of a female *Paragordius esavianus* by a girl in the state of Espirito Santo, Brazil. Yeh and Jordan (1957) listed a third case in which a species designated as "*Pseudogordius tanganyikae*" was believed to have been passed *per urethram* by a 5-year-old European girl in East Africa. Many immature gordiid worms passed in the urine of a 45-year-old black man in Charleston, South Carolina, had presumably caused dysuria.

The species reported as having been recovered from man include: *Gordius aquaticus,* Europe and El Salvador; *G. setiger* (syn. *G. villoti),* Europe; *G. chilensis,* Chile; *Gordius* sp. (probably *robustus*), Florida; one adult female *Gordius* (probably *G. robustus*), Kansas City, Missouri; *Chordodes capensis,* East Africa; *Paragordius tricuspidatus,* France; *P. varius,* North America; *P. cinctus,* Transvaal, South Africa; *P. areolatus,* southeast Africa; *Parachordodes tolosanus,* Italy; *P. violaceus,* France; *P. pustulosus,* Italy; *P. alpestris,* France; *Pseudogordius tanganyikae,* East Africa, and *Neochordodes colombianus,* Colombia, South America (Faust and Botero, 1960). Reports from Japan (Ichihara et al., 1967), Brazil (Lucena and Campelo, 1975), Canada (Ali-Khan and Ali-Khan, 1977), and the United States (Burger, 1972) describe cases in which worms were recovered in vomitus, urine, feces, and urine, respectively, and review earlier similar cases. In some instances the stated origin of the worms from the human body was poorly documented. In no instance has pathogenicity of the worm been demonstrated.

Symptoms attributed to these pseudoparasites either were due to other causes or were psychological. In one instance (Sayad et al., 1936), a worm recovered from the eyelid of a man in Florida was reported as a *Gordius* (probably *G. robustus*), but on re-examination was found to be a *Dirofilaria* species (Beaver and Orihel, 1965).

REFERENCES

Ali-Khan, F.E.A., and Ali-Khan, Z. 1977. *Paragordius varius* (Leidy) (Nematomorpha) infection in man: A case report from Quebec (Canada). J. Parasitol., *63*:174–175.

Beaver, P.C., and Orihel, T.C. 1965. Human infection with filariae of animals in the United States. Am. J. Trop. Med. Hyg., *14*:1010–1029.

Burger, R. 1972. *Paragordius esavianus* passed per urethram. J. Urol., *108*:469.

Carvalho, J.C.M. 1942. Studies on some Gordiacea of North and South America. J. Parasitol., *28*:213–222.

Faust, E.C., and Botero, R.D. 1960. Extraordinario hallazgo de una nueva especie de *Neochordodes* (Gordiacea) en Colombia. Homenaje al Dr. Eduardo Caballero y Caballero, Mexico, D.F., p. 523–527.

Ichihara, A., Kamegai, S., Nonobe, H., and Machida, M. 1967. The second case of hair worm, *Gordius* sp. (Nematomorpha) vomited from human body in Japan. Res. Bull. Meguro Parasit. Mus., *1*:23–25.

Lucena, D.T. de, and Campelo, M. 1975. Accidental parasitism of man with two gordiids. Rev. Inst. Med. Trop. S. Paulo, *17*:390–393.

Sayad, W.Y., Johnson, V.M., and Faust, E.C. 1936. Human parasitization with *Gordius robustus*. J.A.M.A., *106*:461–462.

Yeh, L.S., and Jordan, P. 1957. A new gordiid worm parasitic in man. Ann. Trop. Med. Parasitol., *51*:313–316.

Chapter 25

Flatworms (Platyhelminthes). Introduction and Classification

The flatworms (phylum Platyhelminthes) include not only the exclusively parasitic flukes (Trematoda) and tapeworms (Cestoidea), but also the almost exclusively free-living Turbellaria, which are closely related to the parasitic flatworms. As a group, the flatworms are bilaterally symmetrical, usually compressed dorsoventrally, have an anteroposterior axis, and in their definitive stage characteristically have the space between the visceral and somatic organs and tissues, referred to as the "body cavity" in most Metazoa, filled with spongy mesenchymatous cells.

The tegument of digenetic trematodes and cestodes is a syncytial cytoplasmic covering continuous with nucleated portions of the cytoplasm situated in the parenchyma beneath the cortical muscle layers. The attachment end of the worm is usually referred to as the anterior or cephalic end. Except for rare instances, the digestive system in the trematodes ends blindly without a rectum or anus. In the cestodes the digestive system is lacking. All classes of flatworms have a bilaterally symmetrical excretory system, with collecting tubules and capillaries that terminate in "flame cells" *(solenocytes),* so named because they have a group of vibratile cilia that beat in unison and give the appearance of a flickering candle flame. All classes of flatworm lack a circulatory system. The nervous system consists of an esophageal commissure, together with anteriorly and posteriorly directed pairs of nerve trunks in the dorsal, ventral, and lateral planes. Many of these terminate in sensory processes, particularly in the anterior end. Some species have photosensitive organs on the dorsum near the anterior end, usually referred to as "eyespots." The sexual organs of flatworms are highly elaborated and complicated. Most species are hermaphroditic (i.e., they are *monecious*); in some, however, the sexes are separate (i.e., they are *diecious*). In most tapeworms the organism is a chain of numerous sexually independent units, termed proglottids, and in a few species a double set of reproductive organs is present.

Development in some of the Turbellaria and ectoparasitic trematodes is direct with only a single generation in the life cycle. In some forms metamorphosis is incomplete; in others there is a distinct larval stage. In a few larval tapeworms asexual budding occurs *(Taenia multiceps, Echinococcus).* Development may involve a single obligatory host for the completion of the life cycle, or may require two or more successive hosts. In the digenetic (endoparasitic) trematodes, however, the life cycle is more complicated and consists of at least three generations, one of which is developed in the definitive host and the other two in some species of mollusc.

CLASSIFICATION

Members of the phylum Platyhelminthes are many-celled animals, usually leaf-like or tape-like, rarely cylindrical, bilaterally symmetrical, with three body layers, lacking a body cavity and a circulatory system. When a digestive system is present, it is ordinarily without an anal opening and with a bilaterally symmetrical excretory system terminating in "flame cells" (solenocytes). The phylum consists of three classes: Turbellaria, Trematoda, and Cestoidea.

CLASS TURBELLARIA

Almost exclusively free-living organisms; rarely commensals or parasites; body of adult worm covered with a ciliated epithelium; with or without suckers; development usually direct, without metamorphosis; hermaphroditic.

CLASS TREMATODA

Exclusively parasitic organisms; definitive stage covered with a nonciliated integument; ciliated epithelium confined to larva hatched from egg; suckers usually present; digestive canal present except in sporocyst generation of the Digenea.

Subclass Monogenea

Ectoparasitic or in the bladder or respiratory organs of the host; suckers usually present, those at posterior end consisting of powerful adhesive disks usually provided with chitinoid hooks or anchors; excretory pores double, opening anteriad; development direct, with simple metamorphosis and single host. Parasites of cold-blooded aquatic vertebrates (fishes, amphibians, reptiles). No representative in man.

Subclass Aspidogastrea

Parasitic on or in the viscera of molluscs (or in the intestinal canal of cold-blooded aquatic vertebrates); development probably always direct, with ciliated or unciliated larva hatching from egg; definitive stage with or without alternation of hosts; oral sucker poorly developed, but venter provided with a powerful sucking disk, frequently divided into a series of sucking cups; intestine a simple blind sac. Parasites typically of molluscs. No representative in man.

Subclass Digenea

Almost all species endoparasitic; organs of attachment consisting of one or more suckers, of which one is circumoral; excretory pores open posteriad (single in definitive stage, double in larval stages); development complicated, with an alternation of three or more generations and an alternation of hosts, of which that harboring the intermediate stages is a mollusc; larva hatched from egg has ciliated epithelium.

Order Prosostomata

Mouth at or near anterior tip of body, surrounded by a sucker. *All trematodes parasitizing man belong to this order.*

Suborder Strigeata

Definitive stage monecious or diecious, in intestine or blood of vertebrates; anterior sucker almost always present; one or more ventral acetabula usually present; cercaria with a bifid tail; two pairs of flame cells in the miracidium.

Superfamily Strigeoidea Railliet, 1919

Definitive stage hermaphroditic; body usually divided into a flattened, curved, or cupped anterior portion, which serves for attachment, and a cylindroidal, ovoidal, or conical posterior portion, containing most of the genital organs; genital pore posterior; eggs operculate or with polar filament; cercariae with a true oral sucker and muscular pharynx; metacercariae encysted in molluscs, leeches, or lower vertebrates; definitive stage in vertebrates that feed on the second intermediate host. Representatives in man: *Alaria americana* and related species.

Superfamily Clinostomatoidea Dollfus, 1931

Definitive stage always hermaphroditic, flattened, apharyngeal; having an excretory system consisting of a primary collecting bladder, tubules, and flame cells, and a secondary network of ramified lacunae; eggs operculate; furcocercous cercariae developing in rediae in gastropod host; metacercariae encysted in fishes or frogs; adults in the mouth, esophagus, or respiratory tree of swimming and wading birds and of reptiles. No representatives in man except a few instances of accidental infection with *Clinostomum sp.*

Superfamily Schistosomatoidea Stiles and Hassall, 1926

Definitive stage monecious or diecious, living in portal blood of vertebrates; without muscular pharynx; with or without a ventral acetabulum; eggs nonoperculate; cercariae apharyngeal, with anterior sucker preoral in position; no encysted metacercarial stage; cercariae enter definitive host through the skin. Human representatives (all belonging to the type family Schistosomatidae): *Schistosoma haematobium, S. mansoni, S. japonicum,* and other species of this and related genera.

Suborder Paramphistomata

Definitive stage hermaphroditic; ventral acetabulum highly developed, posteroterminal or subterminal in position, behind the reproductive organs.

Superfamily Paramphistomatoidea Stiles and Goldberger, 1910

Acetabulum caudoterminal or subterminal; oral sucker and esophagus present; genital pore pre-equatorial; testes, one or two, usually preovarial; vitellaria lateral; eggs operculate; one pair of flame cells in the miracidium. Representatives in man: *Watsonius watsoni* (family Paramphistomatidae), *Gastrodiscoides hominis* (family Gastrodiscidae).

The Distomata

(The term Distomata is used here in a descriptive sense, not as a taxon.)

Definitive stage hermaphroditic; oral and ventral suckers present; reproductive organs mostly or completely posterior to ventral sucker; one pair of flame cells in the miracidium. This group contains thousands of species, typically parasitic in vertebrate hosts. Representatives in man are found in the three-superfamilies that follow.

Superfamily Echinostomatoidea Faust, 1929

Elongated medium-sized trematodes, living in the intestine and less commonly in the biliary passages of vertebrates; with a well-developed ventral sucker situated a short distance behind the oral sucker; producing large operculate eggs that are immature when laid; typically (in the family Echinostomatidae) with a collar of cervical spines of characteristic number and pattern; miracidia, with median double eyespot, developing directly into rediae, cercariae developing in second-generation rediae, with simple or fluted (i.e., keeled) tails; encysting in the tissues of molluscs, other invertebrates, vertebrates or on vegetation, which constitute a source of infection for the final host. Representatives in man (type family Echinostomatidae): *Echinostoma ilocanum, E. lindoense;* family Fasciolidae: *Fasciola hepatica, F. gigantica, Fasciolopsis buski.*

Superfamily Plagiorchioidea (Dollfus, 1930) emend. McMullen, 1937, emend. Faust, 1949

Small- to medium-sized trematodes, flattened or cylindroidal, living in biliary passages, pancreatic passages, intestine, or lungs of vertebrates; producing small- to medium-sized eggs, with thick operculum and fully developed miracidium when laid; miracidia metamorphosing into sporocysts; cercariae (styletted or nonstyletted) produced in second generation sporocysts or rediae, encysting in crustacean, insect, molluscan or other intermediate host, or on vegetation, which constitute the source of infection for the final host. Representatives in man: *Dicrocoelium dentriticum* (family Dicrocoeliidae); *Plagiorchis javensis, P. philippinensis, P. muris* (family Plagiorchiidae); *Paragonimus westermani* (family Troglotrematidae); *Spelotrema brevicaeca* (family Microphallidae).

Superfamily Opisthorchioidea Vogel, 1934, emend. Faust, 1949

Small- to medium-sized flukes, frequently spinose, with poorly developed musculature, with or without "eyespots" in adult stage. Cirrus pouch lacking; testes behind ovary; seminal receptacle present; metraterm and ejaculatory duct unite to form common genital duct. Eggs small, thick-shelled, operculate. Miracidia fully developed when oviposition occurs but hatch only following ingestion by appropriate mollusc. Cercariae developing in simple rediae without ambulatory appendages; pleurolophocercous or parapleurolophocercous, with eyespots, rudimentary acetabula, without stylet but having 2 or 3 rows of short, hook-like spines above the mouth. Cercariae encysting in fishes; adults in intestinal or biliary tract of mammals, birds, reptiles, or fishes. The described species from man include: *Opisthorchis felineus, O. viverrini, Clonorchis sinensis* (family Opisthorchiidae); *Heterophyes heterophyes, Metagonimus yokogawai* (family Heterophyidae).

CLASS CESTOIDEA

Exclusively parasitic organisms; adults hermaphroditic, covered with a nonciliated integument; ciliated epithelium, when present, confined to embryos (oncospheres) hatched from eggs; scolex provided with suckers and frequently hooks for attachment to host tissue; no digestive tract; body in most species divided transversely into separate, sexually complete units, called proglottids.

Subclass Cestodaria

Body not divided into proglottids; only a single set of reproductive organs; oncosphere hatching from egg has 10 hooklets. No species found in man.

Subclass Cestoda

Body typically with scolex and series of proglottids, each containing one set (rarely two sets) of male and female organs; oncosphere typically has 6 hooklets.

Order Pseudophylloidea

Scolex typically unarmed, with a dorsal and a ventral sucking organ (bothrium), never with 4 suckers or 4 proboscides. All representatives in man belong to the superfamily Bothriocephaloidea.

Superfamily Bothriocephaloidea Braun, 1903

Scolex with two elongated suckers or one apical sucker with proboscis or rostellum; uterine pore present; genital organs not degenerate at time of oviposition; egg operculate with single shell layer; mature embryo (oncosphere) ciliated; larvae require one or more intermediate hosts; adults in intestine of vertebrates.

Family Diphyllobothriidae Lühe, 1910. Scolex unarmed, of a variety of patterns, usually serving as an adhesive organ; opening of cirrus and vagina on the same surface, anterior to uterine pore. Representatives in man: *Diphyllobothrium latum, D. cordatum, Spirometra houghtoni, Diplogonoporus grandis, Digramma brauni, Ligula intestinalis; Sparganum mansoni, S. proliferum, S. baxteri, S. mansonoides.*

Order Trypanorhyncha

Scolex with 2 or 4 sucking grooves and also, at apex, 4 protrusile proboscides armed with many hooks; genitalia with lateral sex pores and abundantly developed vitellaria; adults in spiral valve of selachian fishes, rarely in ganoids. No representative in man.

Order Tetraphyllidea

Scolex with four flexible sucking cups; genitalia mostly similar to Trypanorhyncha; adults in intestine of fishes, amphibians and reptiles. No representative in man.

Order Diphyllidea

Scolex consisting of head and shaft; two bothria each dorsal and ventral on the head, appearing fused medially; rostellum having dorsal and ventral hooks; neck short; proglottids frequently become separated from strobila before maturity; sex pores open ventrally; adults in intestine of selachian fishes. No representative in man.

Order Cyclophyllidea

Scolex with four suckers, and usually a centrally placed, apical rostellum, frequently armed with hooks; sex pores, when patent, open laterally; vitellaria a single mass, usually posterior to the ovary. All species in this order found in man belong to the superfamily Taenioidea.

Superfamily Taenioidea Zwicke, 1841

Body flattened; suckers four, simple; eggs, nonoperculate, with one or more shell layers; oncosphere nonciliated; adults in intestine of vertebrates.

Family Anoplocephalidae Cholodkowsky, 1902. Scolex unarmed, without rostellum; suckers large, unarmed; neck region lacking. Representatives found in man: *Bertiella studeri, B. mucronata.*

Family Davaineidae Fuhrmann, 1907. Rostellum cushion-shaped, armed with numerous hammer-shaped hooks in two rows; suckers armed; uterus partitioned into egg capsules. Representative found in man: *Raillietina demerariensis.*

Family Linstowiidae Fuhrmann, 1907. Members of this family resemble those of the family Davaineidae in most respects but like the Anophocephalidae have an unarmed scolex. Representative found in man: *Inermicapsifer madagascariensis.*

Family Mesocestoididae Perrier, 1897. Scolex unarmed, without a rostellum; genital opening midventral; no uterine pore; eggs develop in a common stout-walled uterine capsule; adults in birds and mammals. Representative found in man: *Mesocestoides variabilis.*

Family Dilepididae Fuhrmann, 1907. Rostellum armed when present; suckers unarmed or rarely armed; uterus partitioned into egg capsules; genital organs single or occasionally double. Representative found in man: *Dipylidium caninum.*

Family Hymenolepididae Raillet and Henry, 1909. Proglottids usually broader than long; testes 1 to 4; genital pores unilateral; uterus persistent, sac-like. Representatives found in man: *Hymenolepsis nana, H. diminuta.*

Family Taeniidae Ludwig, 1886. Scolex usually armed; uterus with median longitudinal stem and lateral branches; genital pores lateral, irregularly alternating. Representatives found in man: *Taenia solium, T. saginata, T. taeniaeformis, T. multiceps, T. serialis, T. brauni, Echinococcus granulosus, E. multilocularis, E. oligarthrus, E. vogeli.*

REFERENCES

Faust, E.C. 1949. *Human Helminthology,* 3rd ed., pp. 70–71, 84–94, 255–257. Philadelphia, Lea & Febiger.

Skrjabin, K.I., et al. 1964. *Keys to the Trematodes of Animals and Man.* Translation by R.W. Dooley, Edited by H.P. Arai. Urbana, University of Illinois Press.

Wardle, R.A., and McLeod, J.A. 1952. *The Zoology of Tapeworms.* Minneapolis, University of Minnesota Press.

———, and Radinovsky, S. 1974. *Advances in the Zoology of Tapeworms,* 1950–1970. Minneapolis, University of Minnesota Press.

Yamaguti, S. 1958. *Systema Helminthum.* I. *The Digenetic Trematodes of Vertebrates.* New York, Interscience Publishers.

———. 1959. *Systema Helminthum. II. The Cestodes of Vertebrates.* New York, Interscience Publishers.

———. 1971. *Synopsis of Digenetic Trematodes of Vertebrates.* Tokyo, Keigaku.

Chapter 26

Trematodes. General Considerations and Molluscs of Medical Importance

GENERAL CONSIDERATIONS

The Worm. The sexually mature digenetic trematode, in the definitive host, is usually dorsoventrally flattened, although some species may be elongate-cylindroidal in shape and others may be ovoid in contour, or elongated at the anterior end and ovoid or hemispherical at the posterior end. These worms vary in size from species just visible to the naked eye, like *Heterophyes* and *Metagonimus,* to large fleshy species, like *Fasciola* and *Fasciolopsis*. The most characteristic external structures are the acetabula (i.e., suckers), which caused the early workers in this group to call them Trematoda, i.e., "body with holes." A more or less conspicuous sucker usually surrounds the mouth and, except in the group of the "monostomes" and a few schistosomatoid species, a second imperforate sucker is present along the median ventral line. Some species also have a characteristic genital sucker or genital cone.

In the sexually mature trematode, the body covering consists of *integument* laid down by the underlying mesenchymatous tissue, the true cuticula and epidermis having been shed or failed to develop. The surface is frequently provided with spines, which are usually more conspicuous in the anterior portion of the body.

Underlying the integument is *mesenchymatous tissue,* which fills up all space reserved for the body cavity in groups like the Annelida, Arthropoda, and vertebrates. Most cells comprising this tissue are undifferentiated, but some are glandular. In the midst of the mesenchyme are the excretory and genital systems, as well as the intestinal tract and the nervous system.

The *musculature* of digenetic trematodes consists of circular, oblique, and longitudinal fibers in the region just beneath the integument, dorsoventral fibers passing through the mesenchyme and, in addition, those developed in the acetabula, in the region of specialized hooks and spines and in the digestive tract, the excretory bladder, and the genital organs. By use of these muscles, the worm changes its form, attaches itself to the tissues of its host, moves about with a "measuring worm" movement, and performs many of its numerous processes.

Except in the gasterostomes, the mouth of the mature digenetic trematode is at, or near, the anterior end of the body. The typical *alimentary system* has the figure of an inverted Y, although in the gasterostomes it is unforked. Characteristically, there is an *oral cavity* surrounded by the *oral sucker.* Proceeding successively inwards there is the *prepharyngeal tube,* the muscular, more or less spherical *pharynx,* the *esophagus,* and the two *ceca* that end blindly. The bifurcation of the intestine occurs just in front of the ventral sucker in distomate species, although it also occurs in somewhat the same position in monostomes, which lack this acetabulum but may have had one primitively. In some species the ceca are straight tubes; in others they are highly branched. Food consists of the liquid and semiliquid material found in the region where the worm is located. Most of this food is already digested, but unusable material must be rejected through the mouth. In a few instances the digenetic trematode has a complete digestive tract with a functional anus. In general, nutrients provided by the host are absorbed through the tegument. Some species, and certain developmental stages of all species, do not have an intestinal tract and, therefore, evidently absorb all nutrients through the body surface.

Trematodes have no special respiratory organs. Oxygen, if needed, must be taken through the tegument. Although most species live under more or less aerobic conditions, some are able to live without oxygen. Schistosomes, for example, live in the blood and are known to use oxygen, but if their oxygen metabolism is interrupted, they revert to an anaerobic metabolism. All aspects of the physiology of trematodes have been reviewed by von Brand (1979).

The *excretory system* (Fig. 26–1) is bilaterally symmetrical and opens at the posterior end of the body, usually on the dorsal aspect. Phylogenetically and embryologically there is a pair of bladders, one for each side of the body, but in a mature worm these are fused into a median bladder, which is provided with a sphincter. Arising from the inner aspects of this bladder, usually from the anterolateral horns, is the pair of main collecting tubules. These may soon branch, each into anterior and posterior secondary tubules, or the primary tubule may proceed anteriad to the cephalic region and then return posteriad as the secondary canal. There may also be tertiary tubules. Eventually, however, there are terminal capillary twigs, each with a flame cell (i.e., *solenocyte)* at its inner extremity, provided with vibratile cilia that move in unison in the terminus of the capillary, presumably drawing fluid wastes into the system from the adjacent tissues. These flame cells may possibly also serve as a water-regulating mechanism. While this system is characteristic of all flatworms, in the digenetic trematodes the fundamental pattern for the different subordinal groups, with the elaboration of the pattern in families, is so constant that it constitutes an important basis, or at least a valuable clue, in determining a natural system of classification. In most species this pattern can be described in the living worm. In some groups, however, this is impracticable because the tissues are too opaque to see all of the solenocytes and their connecting capillaries. The first-stage larva *(miracidium)* of most digenetic flukes has a single pair of solenocytes and collecting tubules; that of the strigeoids and blood flukes, two pairs; that of the Aspidogastrea, three pairs. In the *cercaria* the pattern is based on this fundamental character, while that of the adult is typically an elaboration of the cercarial excretory pattern. In his classification of digenetic trematodes, La Rue (1957) emphasized the primary taxonomic importance of the excretory bladder, into which the collecting tubules empty.

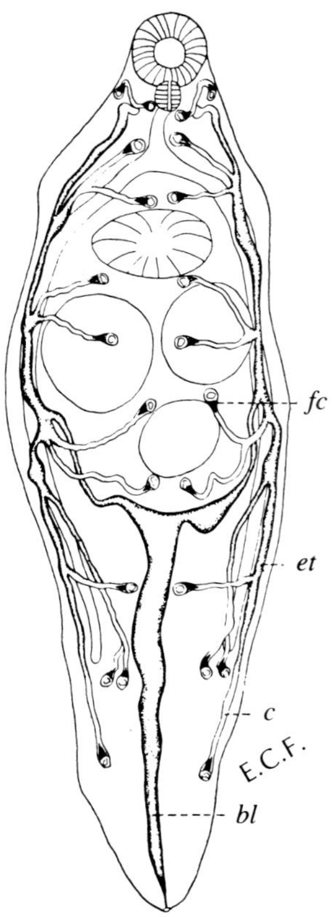

Fig. 26–1. Digenetic trematode *(Dicrocoelium)*. Excretory system. *bl,* Bladder. *c,* Capillary. *et,* Excretory tubule. *fc,* Flame cell. (From Faust, E.C. 1929. *Human Helminthology,* 1st ed. Philadelphia, Lea & Febiger.)

The *nervous system* is composed of a group of paired ganglion cells disposed like a saddle on the dorsum of the pharynx or esophagus, and a main pair each of posteriorly and anteriorly directed nerve trunks in the dorsal, ventral, and lateral fields. Special ganglia are found in the region of the important genital organs, and special sensory nerve endings, particularly those sensitive to light, are in the anterior part of the body.

Except in a few species like the blood flukes, the *reproductive organs* (Figs. 26–2 and 26–3) are hermaphroditic, possibly as a survival adaptation in case only a single worm has become established in the host. On the male side of the system the primary organ (testis) is usually paired (occasionally single *[Monorchotrema]* or multiple *[Schistosoma]*). From each testis a *vas efferens (ve)* leads to the single *vas deferens (vd),* which enlarges as it enters a *cirrus sac (cs)* to form the convoluted *seminal vesicle (sv)*. The duct (now known as the

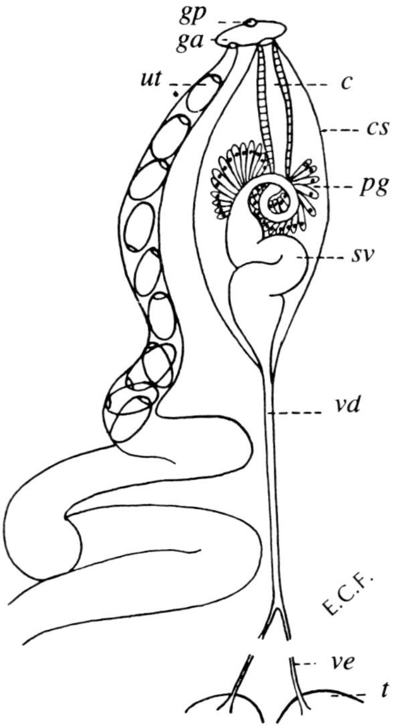

Fig. 26–2. Reproductive organs in region of genital pore of a digenetic trematode. *c*, Cirrus. *cs*, Cirrus sac. *ga*, Genital atrium. *gp*, Genital pore. *pg*, Prostate glands. *sv*, Seminal vesicle. *t*, Testis. *ut*, Uterus, with eggs. *vd*, Vas deferens. *ve*, Vas efferens. (From Faust, E.C. 1929. *Human Helminthology*, 1st ed. Philadelphia, Lea & Febiger.)

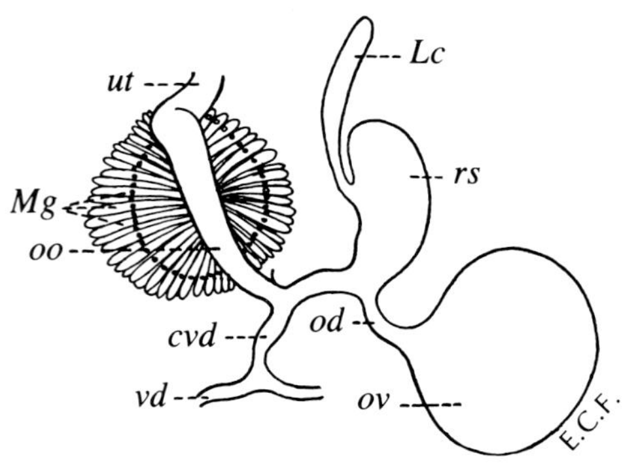

Fig. 26–3. Female reproductive organs of a digenetic trematode. *cvd*, Common vitelline duct. *Lc*, Laurer's canal. *Mg*, Mehlis' glands. *od*, Oviduct. *oo*, Ootype. *ov*, Ovary. *rs*, Seminal receptacle. *ut*, Uterus. *vd*, Vitelline duct. (From Faust, E.C. 1929. *Human Helminthology*, 1st ed. Philadelphia, Lea & Febiger.)

ejaculatory duct) then usually passes through a region of *prostate glands (pgl)* and terminates in a *cirrus organ (c)*, which is frequently muscular and protrusible through the common *genital atrium* and the *genital pore*. In distomate species this pore is typically situated in the vicinity of the ventral sucker. In the female system there is a single *ovary (ov)*, with its *oviduct (od)* leading into the *ootype (oo)*, where the components of the egg are put together. Paired multiple *vitelline glands*, commonly bisymmetrically disposed in the lateral fields, are connected by collecting ducts that lead into a pair of *transverse vitelline ducts* and then a *common vitelline duct (vd)*, which opens into the ootype. On the dorsal aspect of the ootype there may be a flask-shaped *seminal receptacle (rs)*, and an associated duct *(Laurer's canal, Lc)*, leading to, and at times opening through, the dorsal surface. The former is for storage of spermatozoa in the female system; the latter is believed to be homologous to the vagina of tapeworms. Both open into the oviduct before it enters the ootype. Surrounding the ootype is a cluster of acinus glands known as *Mehlis' glands (Mgl)*, believed by some workers to be *shell glands*. Other investigators regard the so-called vitelline glands as shell glands and Mehlis' glands as vitellaria or as lubricating glands. On the side of the ootype, opposite the common vitelline duct and the oviduct, is the inner end of the *uterus (ut)*, which coils in a simple or more tortuous pattern through the body of the *metraterm;* the latter opens through the genital atrium and the common genital pore.

In the process of egg production, the spermatozoa elaborated in the male system reach the seminal receptacle by passing successively through the genital atrium, the uterus, ootype, and inner end of the oviduct. (In species with a patent Laurer's canal, the spermatozoa may gain access directly into the seminal receptacle.) Most species are self-fertilized, although in many forms cross-fertilization is possible. The naked ovum is passed into the ootype, or in some species *(Fasciola hepatica)* into the proximal end of the uterus, is fertilized there, the vitelline and shell materials are added, and the egg is shuttled into the uterus. In some species the eggs become fully embryonated *in utero;* in others practically no development takes place until the egg is laid by the worm. In some species *(Clonorchis, Metagonimus)* the uterus contains a supply of eggs sufficient for several days of oviposition; in others *(Schistosoma, Fasciola)*, only a portion of the daily output is present in the uterus at any one time.

Life Cycle. When deposited by the adult worms, the *egg* directly or indirectly reaches the outside world *via* the feces *(Fasciola, Fasciolopsis, Clonorchis,* heterophyid forms, *Schistosoma japonicum, S. mansoni)*, urine *(S. haematobium)*, or

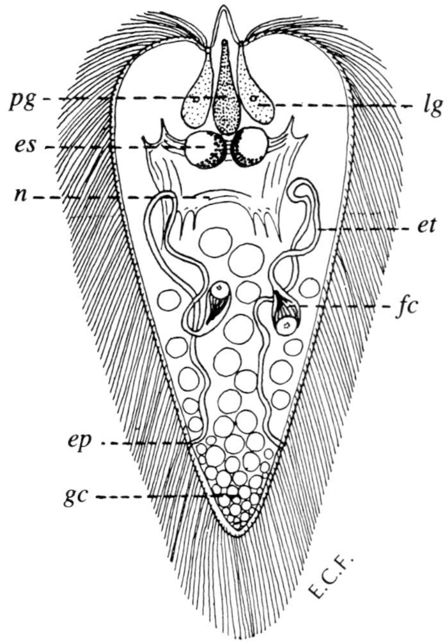

Fig. 26–4. Miracidium of *Fasciolopsis buski,* showing ciliated epithelium, lytic glands *(lg),* primitive gut *(pg),* nerve trunks *(n),* flame cells *(fc),* excretory tubules *(et),* excretory pore *(ep),* and germ cells *(gc).* (From Faust, E.C. 1929. *Human Helminthology,* 1st ed. Philadelphia, Lea & Febiger.)

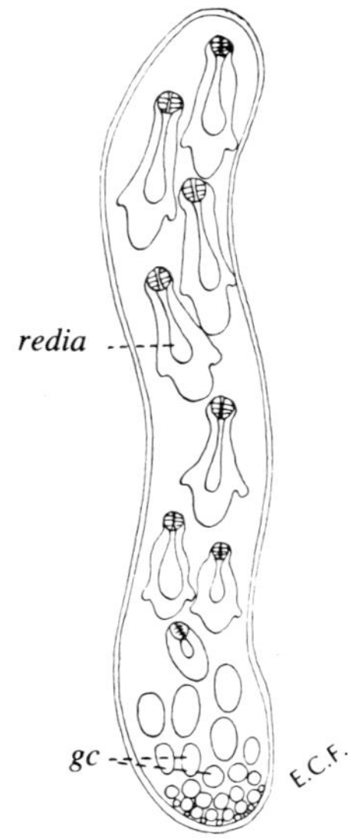

Fig. 26–5. First generation sporocyst, with second generation progeny (rediae) developing from germ cells *(gc)* in the brood cavity. (From Faust, E.C. 1929. *Human Helminthology,* 1st ed. Philadelphia, Lea & Febiger.)

sputum *(Paragonimus).* In some species *(Schistosoma* spp., *Clonorchis, Opisthorchis,* heterophyid forms) the eggs are fully embryonated or nearly so at the time they are discharged from the body. Schistosome eggs hatch soon after the excreta are diluted with water; opisthorchoid and plagiorchoid forms, only after ingestion by the molluscan host; *Fasciola, Fasciolopsis,* and *Paragonimus* eggs require a period for embryonation after deposition in water. The fully embryonated egg contains a *miracidium,* which is obovoid in contour and is covered by a ciliated epithelium. It has a characteristic basic excretory system *(vide supra),* a nervous system, penetration glands (frequently of two types), and a primitive nonfunctional gut (Fig. 26–4).

Miracidium. On hatching in water (schistosomes, *Fasciola, Fasciolopsis, Paragonimus),* the miracidium assumes a brief free-living, swimming existence, but is unable to take nourishment. Within 24 hours or less it must find the appropriate molluscan host to which it is adapted, or perish, since the mollusc is a necessary first intermediate host. (Opisthorchoid and plagiorchoid species hatch only in the appropriate snail after they have been ingested in the egg stage.)

Sporocysts and Rediae. In case the appropriate mollusc is present in the immediate vicinity of the free-swimming miracidium, the larva is attracted by mucous or other fluid discharges of the mollusc, attacks and quickly penetrates the soft tissues of the body (tentacles, head, foot, mantle, or pulmonary tissue), and in the course of a few hours it gains access to lymph channels. It loses its ciliated epithelium shortly after entering the host, and during the next few days becomes greatly simplified in its organization. In related but nonspecific molluscan hosts, penetration into the soft tissues may also occur, but the larva is soon destroyed by an enveloping infiltration of host cells. In all but a limited group (echinostomatoid species), the organism becomes a simple, elongated sac, without mouth or other distinct external or internal structures, but possibly always with a birth pore. It is now known as a *first-generation sporocyst* (Fig. 26–5), which has a large body cavity into which germ balls are proliferated. These embryos develop either into *second-generation sporocysts* (schistosomes) or into somewhat more complex daughter organisms, the *rediae* (named in honor of Francesco Redi). The rediae have a mouth, a muscular pharynx, a blind unforked cecum, a birth pore, and in some species a pair of locomotor appendages. During this development and multiplication the organisms have migrated through the lymph channels into the spaces surrounding the digestive gland and the hermaphroditic organ that lie in the spire of the gastropod molluscs (i.e., snails) in which almost all human trematodes develop during this intermediate period.

Cercaria. Within the second-generation sporocysts or rediae other germ balls are proliferated, which develop into tailed larvae, the *cercariae* (Fig. 26–6). On maturity, the cercariae escape from the parent organism, and then leave the snail host and become free-living in the water.

Although the various trematode groups differ in development, the number of second-generation organisms produced in the first-generation forms is relatively small, and their development and escape may be simultaneous from the mother organism,

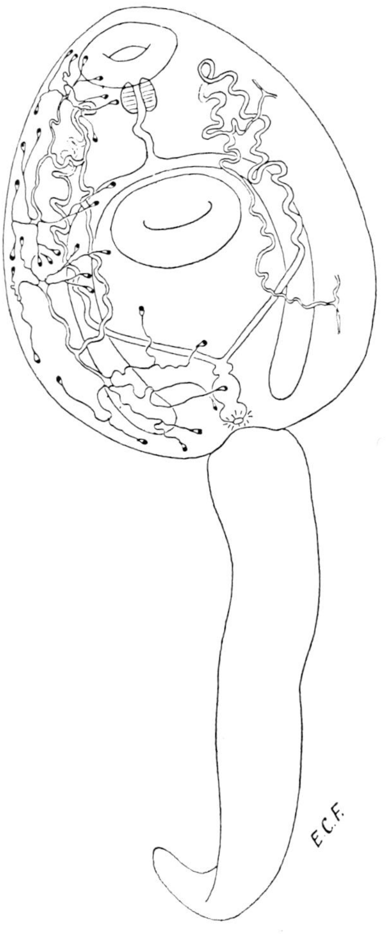

Fig. 26–6. Cercaria of a digenetic trematode *(Fasciola* sp.), showing excretory and digestive systems. The excretory system is composed of a median posterior bladder with pore, a pair of main collecting tubules that ascend to the anterior region and bend back to the postequatorial plane, paired secondary and tertiary tubules, capillaries and terminal flame cells. (From Faust, E.C. 1929. *Human Helminthology,* 1st ed. Philadelphia, Lea & Febiger.)

which at times succumbs with the birth of the offspring. However, in the next generation, in which cercariae are usually produced, the larvae are developed in numbers over a period of several days to many weeks, provided the snail survives. For *Schistosoma mansoni,* Faust and Hoffman (1934) found that a single miracidium may be responsible for as many as 100,000 to 250,000 cercarial progeny, and that most of this multiplication is due to the fertility of the second-generation sporocyst over a period of months. In *Fasciola* and *Fasciolopsis* the redia (second-generation organism) usually produces other rediae as well as cercariae so that one or more additional generations of rediae may be interpolated into the intramolluscan phase of the life cycle. In echinostome species sporocysts are not developed, and two or more generations of rediae are produced. These important differences in development may be summarized as follows:

1. Egg embryonated when laid, hatches on contact with water; miracidium actively enters appropriate mollusc; develops into first-generation sporocyst; produces second-generation sporocysts, which, in turn produce cercariae. Example: schistosomes.
2. Egg not embryonated when laid but matures in water and then hatches; miracidium, after entering mollusc, develops into first-generation sporocyst; produces rediae, which then produce cercariae. Example: *Paragonimus* (the lung fluke). (In *Fasciola* and *Fasciolopsis* the sequence is the same, except that one or more additional generations of rediae usually develop after the first rediae are produced.)
3. Egg not embryonated when laid but matures in water and hatches; miracidium, after actively entering mollusc, develops into redia; produces second-generation rediae, which then produce cercariae. Example: *Echinostoma* spp.
4. Egg embryonated when laid but hatches only when ingested by appropriate snail; miracidium develops into first-generation sporocyst; produces rediae, which then produce cercariae. Example: *Clonorchis, Metagonimus.*

The stage in the life cycle that emerges from the mollusc is the *cercaria.* Except for the relatively small group of species in which this larval stage is known as a *cercariaeum,* all cercariae have a tail, which may be vestigial *(Paragonimus),* may have a simple trunk *(Fasciola, Fasciolopsis),* may be provided with a fluted keel *(Clonorchis, Metagonimus),* or may be bifurcated in its distal portion (schistosomes and other forms). There may be a stylet or other larval piercing organ in the region of the mouth. The integument is nonciliated but is frequently spinose. Penetration glands and cystogenous glands are frequently present. Certain structures, such as the suckers, may resemble those of the adult, and the excretory system is similar but usually much simpler than that of the adult worm. The cercaria is a free-living aquatic form, which in species of blood flukes typically emerges from the mollusc in swarms during a certain period of the day. Cercariae cannot feed, and must find their next host or perish.

The Metacercaria. On arrival in or on the appropriate next host, the cercaria discards its tail, becoming a *metacercaria*.

The least complicated type of further development is taken by the blood flukes, which, by the aid of their penetration gland secretions, penetrate into the body of their definitive host, usually by the cutaneous route. After a migration through the bloodstream of the host, they come to reside in the portal or caval system where they mature.

The majority of digenetic trematode species, however, utilize a second intermediate host on which or in which they encyst. *Fasciola* and *Fasciolopsis* encyst on vegetation in aquatic habitats; *Echinostoma ilocanum,* in a second molluscan host; *Clonorchis, Opisthorchis,* and heterophyid species, in the tissues of freshwater (in some heterophyid species, saltwater) fishes; *Paragonimus,* in the soft tissues of crabs and crayfishes. By eating a second intermediate or transport host, raw or inadequately cooked, the definitive host acquires infection.

All encysted metacercariae excyst in the upper levels of the small intestine and migrate to the tissues or organs where they develop into mature worms. Heterophyid and echinostome forms, as well as *Fasciolopsis,* become attached to the intestinal wall and mature. *Clonorchis* and *Opisthorchis* migrate through the ampulla of Vater to the distal bile passages to complete development; *Fasciola* penetrates the intestinal wall and reaches the bile tracts after migrating through Glisson's capsule and the liver parenchyma; *Paragonimus* penetrates the intestinal wall and traverses the peritoneal cavity, the diaphragm, and the pleural cavity before it reaches its typical habitat in the lungs.

Pathogenesis, Pathology, and Symptomatology. The clinical picture of trematode infections depends primarily on three conditions: (1) size and number of worms present in the host, (2) organs or tissues parasitized, and (3) the degree to which parasites, including eggs, exist in the tissues rather than in the intestinal lumen or in cavities of the host. Many individuals of small species like *Metagonimus* and *Heterophyes,* when attached superficially to the mucosa of the small bowel, produce only mild changes; a similar number of *Fasciolopsis* in the small intestine may be responsible for extensive local damage and profound systemic toxemia. *Clonorchis* or *Opisthorchis* in moderate numbers in the distal biliary passages may produce considerable local tissue reaction, but the effect on the liver as a whole is not usually serious. Because of its greater size, the trauma produced by *Fasciola* in its migration through the liver substance and its location in the larger, proximal biliary tracts causes more profound damage.

The changes produced in the host may be local or systemic, usually both. The former consist of ulceration, sloughing of tissue, abscess formation, and frequently development of fibrous repair processes. These changes produce variable signs and symptoms depending on the amount and type of damage produced, the organ or tissues affected, and whether the damage is temporary or irreversible. Systemic manifestations are usually due to absorption of antigenic by-products of the worms, which frequently provoke a generalized leukocytosis and hypereosinophilia and, not uncommonly, allergic manifestations. Most damage is caused by the worms themselves in the immediate vicinity of their contact with host tissues. Although this type of damage is appreciable in the case of the schistosomes, the greater cumulative damage from their presence in the human body is due to perivascular infiltration of their eggs in the tissues. Responses by host tissue cells to the presence of the eggs are responsible for the important chronic clinical pictures produced in schistosomiasis. In all trematode infections it is possible to recognize (1) a prepatent or incubation period, (2) an acute stage, and (3) a chronic stage.

Epidemiology. Human trematode infections are largely confined to the Tropics and the Orient. An important exception is opisthorchiasis, which has an extensive distribution in East Germany, the U.S.S.R., and the Balkan States. The essential factor governing this distribution is the necessity of the appropriate molluscan intermediate host, without which no amount of individuals infected with the trematode can produce an endemic focus. Where susceptible snails are native to a region, the disease may be readily introduced and become endemic.

Second intermediate hosts also play a part in the transmission of certain trematode infections, but their choice by the larval trematode is much less specialized than is that for the mollusc.

The habits of the human populations in endemic foci not only are responsible for their own infections, but also are frequently the essential factors in propagating the disease. Bathing, wading, and washing in streams and pools where schistosome cercariae are present provide exposure to the parasites. Eating raw vegetation *(Fasciola, Fasciolopsis),* raw crabs or crayfish *(Paragonimus),* raw fish *(Clonorchis, Opisthorchis,* heterophyid species), or raw snails *(Echinostoma ilocanum),* constitutes the source for a large portion of human fluke infections. Insanitary disposal of human feces, urine, or sputum provides the source of pollution of water, from which the intermediate hosts become infected.

Finally, reservoir hosts constitute an additional source of danger. In most trematode infections present in large numbers of the human population, man is principally responsible for the propagation of the organism. In *Fasciola, Dicrocoelium,* the amphistome species, and several of the heterophyid forms, where man is only occasionally infected and other hosts are commonly involved, these reservoirs are the definitive hosts that maintain the life cycle.

Control and Prevention. Diseases of trematode etiology may be combated by the following methods:

1. Treatment of infected persons discovered in the population.
2. Disinfection or sanitary disposal of excreta.
3. Antimolluscan campaigns.
4. Education of the population to refrain from bathing in "infected" waters and from eating infected foods.

The public health aspect of the propagation of trematode infections has been particularly recognized in the case of the schistosome infections, which are increasing in their distribution and incidence, due to immigration of infected persons into new areas where susceptible snails are present. This topic is considered in detail in Chapter 27.

MOLLUSCS OF MEDICAL IMPORTANCE

The molluscs are divided into six classes, one of which, the Gastropoda, contains species of medical importance. The following outline of the gastropod groups that are directly injurious to man or are essential intermediate hosts of helminth parasites is based on the notes on families and genera outlined in the previous edition and on the more recent classification of Malek (1980).

Class Gastropoda
(Snails and Slugs)

The gastropods of medical importance are of two principal types (subclasses): the Prosobranchiata or operculate snails, which have a well-formed shell and usually have gills, and the Pulmonata or air-breathing snails and slugs. In the *operculate* groups the head-foot is often snout-like and bears one pair of nonretractile tentacles and one pair of eyes, the sexes are separate, and the eggs are usually laid in capsules, although some are ovoviviparous. In the *pulmonate* groups the shell may be reduced or absent as in the slugs, the head-foot bears two pairs of tentacles, all members are monoecious, and most are oviparous.

Two orders of the prosobranchs (Neogastropoda and Mesogastropoda) are of medical importance. In the order Neogastropoda, family Conidae, species of the genus *Conus* (poison cone shells) have two lanceolate teeth in each row of the radula; they are marine forms with beautiful shells, poison glands, and a stinging apparatus capable of discharging a highly toxic substance.

Aquatic and Amphibious Snails

In the order Mesogastropoda, there are six families with members that serve as intermediate hosts of medically important helminths:

Thiaridae (syn. Melaniidae). Fresh and brackish water species with broad snout, pedunculated eyes at outer base of tentacles, and a dextrally spiralled shell. *Thiara* spp. *(Paragonimus westermani, Metagonimus yokogawai,* and other heterophyid flukes in the Orient).

Pleuroceridae. Snout somewhat elongate, tentacles long and tapering, with eyes on swellings at outer base; shell usually turreted, dextral, smooth, striated, or tuberculated, aperture sometimes extended into a short canal. *Semisulcospira* spp. (*Paragonimus westermani* in the Orient); *Goniobasis plicifera silicula (Troglotrema salmincola,* in the Pacific Northwest, United States).

Potamidae (syn. Cerithiidae). Marine and brackish water species with broad, short, contractile rostrum, widely separated tentacles, eyes on short peduncles; shell many-whorled, turreted, often tuberculate or spinose; horny spiralled operculum. *Pironella conica,* along the Mediterranean coast of Egypt, and *Cerithidea cingulata* in Japan *(Heterophyes heterophyes;* also host of cercariae causing schistosome dermatitis); *Pyrazus ebeninus* in Australia (host of dermatitis-producing schistosome cercariae).

Pilidae (syn. Ampulariidae). Freshwater species with bifid snout, two long tentacles with pedunculate eyes at outer base; large turbinate umbilicate shell with large aperture and fitted operculum. *Pila* spp. (intermediate host of *Angiostrongylus cantonensis,* and second intermediate host of *Echinostoma ilocanum* in the Orient).

Synceridae (syn. Assimineidae). Small amphibious fresh and brackish water forms, with outer lip of aperture not thickened, body with a single pair of short eye stalks instead of tentacles. *Syncera luteola (Paragonimus iloktsuenensis,* in rodents, in China).

Hydrobiidae. Small forms with conical, dextral shell, a long snout and long thin tentacles. Two subfamilies: (1) Hydrobiinae which includes *Oncomelania* spp. *(Schistosoma japonicum,* in the Orient), and *Pomatiopsis lapidaria (Paragonimus kellicotti* and experimental *Schistosoma japonicum,* United States); and (2) Buliminae (syn. Bithyniinae), which includes *Parafossarulus* and *Bulimus* (syn. *Bithynia* and *Bithinia)* spp. *(Opisthorchis felineus,* in eastern and southeastern Europe and U.S.S.R.; *Clonorchis sinensis, Metagonimus yokogawai, Echinochasmus perfoliatus,* in the Orient).

The subclass *Pulmonata* contains three orders that are of medical interest: the Basommatophora, which are aquatic forms that have a single pair of contractile tentacles each with an eye at its base, and two orders of land snails and slugs, the Stylommatophora and Systellommatophora, both of which have two pairs of tentacles; the posterior pair has eyes at the tip.

In the pulmonate order Basommatophora there are four principal families with members that serve as essential intermediate hosts of helminth parasites of man:

Lymnaeidae. With oval or elongate, dextrally spiralled shell. Many species of *Lymnaea, Fossaria, Pseudosuccinea, Radix, Stagnicola* and others *(Fasciola hepatica, Fasciola gigantica,* several species of *Echinostoma, Plagiorchis;* freshwater dermatitis-producing schistosome cercariae).

Planorbidae. Body sinistral with pulmonary and genital openings on left side; long filiform tentacles, eyes at inner base of tentacles. Four subfamilies:

Planorbinae. With discoidal, orb-like shell. *Biomphalaria* spp. *(Schistosoma mansoni* in Africa and Near East); *Biomphalaria glabrata, B. straminea (S. mansoni* in tropical America); *Gyraulus* spp. *(Echinostoma ilocanum* in the Orient).

Helisomatinae. Usually large, discoidal shell with high whorls. *Helisoma* spp. *(Echinostoma revolutum* in United States, Mexico), *Planorbarius metidjensis* (syn. *dufourii), Schistosoma haematobium* in Portugal and Morocco.

Segmentininae. Small discoidal, shiny shell, with or without internal partitions. *Segmentina* spp., *Hippeutis* spp. *(Fasciolopsis buski, Echinostoma ilocanum,* in the Orient).

Bulininae. Shell resembling *Physa,* ovoid, globose or elongated, with spire terminally obtuse. *Bulinus (Bulinus)* spp. and *B. (Physopsis)* spp. *(Schistosoma haematobium* in Africa, Near East, Middle East); *Indoplanorbis exustus (Schistosoma spindale, S. nasale* in India, Malaysia, Sumatra).

Ancylidae (Limpets). Shell usually cap-like or patelliform; body with oval foot, and eyes at inner base of short, blunt tentacles. *Ferrissia tenuis (Schistosoma haematobium* in endemic area of India).

Physidae. Shell sinistrally spiralled. *Physa* spp. (*Echinostoma revolutum* in Orient; schistosome cercariae producing dermatitis from freshwater and marine shoreline snails).

Land Snails and Slugs

In the pulmonate order Systellommatophora there is one important family, the Veronicellidae, all members of which are slugs, mostly tropical. Several species in the South Pacific islands, China Sea area, Australia, and Cuba are natural hosts of *Angiostrongylus cantonensis,* and species in American tropical areas are hosts of *Angiostrongylus costaricensis.*

In the Stylommatophora, there are several common and important families of snails: the Helicidae, Helicellidae, Cionellidae, and Achatinidae. *Achatina fulica,* a species of the last-named family, is known as the giant African land snail; it has spread to the tropical areas of Asia and throughout the Pacific Islands where it is an important intermediate host of *Angiostrongylus cantonensis*. Species of the other families are molluscan hosts of *Dicrocoelium dendriticum*. Members of the family Limacidae are common slugs; species of *Limax* and *Deroceras* are intermediate hosts of lungworms of domestic mammals, and experimentally of *A. cantonensis*.

REFERENCES

Faust, E.C., and Hoffman, W.A. 1934. Studies on schistosomiasis mansoni in Puerto Rico. III. Biological studies. 1. The extra mammalian phases of the life cycle. Puerto Rico J. Public Health Trop. Med., *10*:1–47.

La Rue, G.R. 1957. The classification of the digenetic trematodes: A review and a new system. Exp. Parasitol., *6*:306–349.

Malek, E. 1980. *Snail-transmitted Parasitic Diseases,* Volume 1, pp. 98–101. Boca Raton, Florida, CRC Press.

von Brand, T. 1979. *Biochemistry and Physiology of Endoparasites*. New York, Elsevier/North Holland.

Chapter 27

Schistosomes or Blood Flukes

GENERAL CONSIDERATIONS

Genera and species of the Schistosomatoidea lack a muscular pharynx and produce nonoperculate eggs. The fork-tailed cercariae have a preoral sucker provided with organs of penetration that function both mechanically and lytically. The usual portal of entry into the definitive host is the skin. Members of this group parasitize fishes, turtles, birds, and mammals. Four species *(Schistosoma haematobium, S. mansoni, S. japonicum,* and *S. mekongi)* are important agents of disease in man, whereas others are incidental or potential human parasites. Species parasitizing man are diecious and inhabit the mesenteric-portal and caval venous blood vessels.

The schistosomes of man are similar in their basic life cycles and in their pathologic effects on mammalian host tissues, yet they differ in external and internal morphology of the adults, the shape of their eggs and larvae hatched from the eggs, in the groups of snails that they utilize as intermediate hosts, and in mammalian host susceptibility. Moreover, within each of the principal species parasitizing man there are physiologic strain differences, due perhaps to geographic isolation and molluscan host adaptations. Such differences have been demonstrated between (1) *S. japonicum* in Formosa (enzootic only) versus other areas of infection in the Far East (endemic and enzootic); (2) *S. mansoni* in Egypt, West Africa, Puerto Rico, Venezuela, and Brazil; and (3) *S. haematobium* in Egypt, West Africa, Morocco, and Portugal. It is convenient to refer to the *S. haematobium* complex, which includes not only *S. haematobium* but also *S. bovis, S. mattheei, S. intercalatum,* possibly *S. spindale* and *S. incognitum,* and the *S. mansoni* complex. The last complex includes *S. mansoni* and *S. rodhaini,* and the *S. japonicum* group, which includes *S. mekongi* and possibly *S. margrebowiei*. In the *S. haematobium* group, hybridization of species is readily demonstrated in the laboratory and probably occurs occasionally in nature, especially between some species (Taylor, 1970; Wright and Ross, 1980). In addition to the species that parasitize man and mammalian hosts as adult worms, there are others, for the most part adapted to avian hosts, which in the cercarial stage produce dermatitis in man. The interrelationship of the species that infect man and domestic and wild mammals offers opportunity for continued intensive investigation.

There are at least three functional aspects to host immunity in chronic schistosomiasis. One concerns the mechanisms involved in the development of the immunopathology associated with the disease; another relates to the establishment of protective immunity. The third deals with the persistence of the parasite within the host, i.e., the mechanism by which the parasite avoids detection or damage by the host's immune system. In the early stages of schistosomiasis, the host reaction is acute and intense; in the chronic phase of the disease, the host's inflammatory reactions are suppressed and the lesions are predominantly fibrotic (Nash, 1982).

"The pattern of distribution of bilharziasis is changing in two opposite directions: the increase of irrigation favours its spread, and control schemes to block transmission are now becoming feasible. Drugs are being improved, and their use increases. For these reasons, the pattern of schistosome infection is changing from equilibrium, or at least stability, to a variety of unstable situations, and the natural history of the disease is being altered to an unnatural history whose public health importance is even less understood" (World Health Organization, 1967).

Schistosoma japonicum Katsurada, 1904 (The Oriental blood fluke, producing schistosomiasis japonica or Oriental schistosomiasis)

Synonym. *Sinobilharzia japonica* Le Roux, 1958.

Historical and Geographical Notes. Schistosomiasis japonica was first mentioned by Fujii, a Japanese physician, in 1847. Baelz (1883) made the first epidemiologic survey in Okayama Prefecture, Japan, but attributed the disease to *Clonorchis sinensis*. Yamagiwa (1890), Kurimoto (1893), and especially Fujinami (1904) found eggs of the parasite in various organs of patients and ascribed the etiologic role in the infection to the eggs. Kasai (1903) first found the eggs in feces. Fujinami (1904) first found the female worm in the portal vein and Katsurada (1904) first described the adult worms, obtained from infected dogs and cats, and assigned to the parasite a new specific name, i.e., *japonicum*. Logan (1905) diagnosed the first case in China. The experimental work of Fujinami (1909), Miyagawa (1912–1913), and Miyairi and Suzuki (1913–1914) furnished proof for the life cycle of *S. japonicum* and provided a basis for Leiper's work on the blood flukes in Egypt (1915–1918). Details of the biology and pathogenesis of *S. japonicum* were described by Faust and Meleney in 1924.

This infection is confined to areas in the Far East (Fig. 27–1). In Japan five relatively small foci are involved. In China practically the whole of the Chang Jiang (Yangtze) Basin is studded with endemic foci from western Sichuan to the sea. In addition, the coastal area from northern Jiangsu Province to Quangdong, including river valleys extending far inland, the large island of Hainan off the coast of southeastern China, and the Mekong Valley in southwestern Yunnan Province are also affected. However, it remains to be determined whether the schistosome in people of the Mekong Valley in Yunnan Province is *S. japonicum* or *S. mekongi*. In Taiwan there is an enzootic region in the midwestern part of the island but no evidence of locally acquired human infection has been reported. In the Philippines infection is endemic on Leyte, Mindanao, Mindoro, Samar, and southwestern Luzon. A focus of infection was found by Brug and Tesch (1937) and confirmed by Faust and Bonne (1948) in the Paloe District of Celebes (Fig. 27–1). In 1958 an apparently newly infested area was discovered in Thailand, some distance south of Bangkok; since then infections have been reported from areas to the north and northeast of Bangkok (Lee et al., 1966). In addition, Barbier (1966) reported a new endemic focus in the Mekong Valley in western Laos and human infection has been reported from the Mekong River area in Cambodia (Audebaud et al., 1968). More recently the Mekong schistosome was described as a distinct species, *Schistosoma mekongi* Voge, Bruckner and Bruce, 1978.

Morphology, Biology, and Life Cycle. The adult worms (Fig. 27–2) superficially resemble those of *S. haematobium* and *S. mansoni,* but lack integumentary tuberculations. Males measure 12 to 20 mm by 0.5 to 0.55 mm in greatest diameter. Their integument is covered with minute acuminate spines, most prominent in the region of the suckers and the gynecophoral canal. There are 7 testes. Females average 26 mm in length by about 0.3 mm in greatest diameter. Their integument is also covered with minute spines. The ovary lies somewhat behind the midplane of the body. The vitellaria are confined to the lateral fields in the posterior fourth of the body. The ootype (Fig. 27–2,*B)* is situated just anterior to the equator. The uterus is a long straight tube, containing as many as 50 eggs at one time. The eggs in the proximal end of the uterus are hyaline, subspherical, or oval in lateral view. Near the genital pore they are pale yellow, more definitely ovoid. Near one end there is a depression from which there extends a small spinose process which, in the mature egg, looks like a recurved hook and can be seen only when the egg is properly oriented. The more mature uterine eggs measure about 67 by 50 μm; those discharged in the feces (Fig. 27–3) are 70 to 100 μm by 50 to 65 μm. Oviposition characteristically takes place by deposition of clumps of eggs into the small venules of the submucosa. The typical habitat of *S. japonicum* for ovipositing is the radicles of the superior mesenteric vein draining the small intestine. The oviposited eggs break through the submucosa and mucosa and are discharged into the lumen of the bowel. Gradually the thickening of the wall and formation of scar tissue impede their course through the tissues, so that the eggs are increasingly filtered out in the bowel wall. Viable eggs that are passed in the feces are usually mature when evacuated and require only sufficient dilution of the feces with water to hatch, allowing the miracidium to escape, shed its embryonic membrane, and swim about in the water. This larva (Fig. 27–4) closely resembles that of *S. haematobium* and *S. mansoni* and differs only in its smaller size and in certain minute internal structures. On coming in contact with an appropriate amphibious snail, all of which are small (Fig. 27–5) (namely, *Oncomelania nosophora* in Japan, and along the southern coast of China as well as in the southwestern province of Yunnan; *Oncomelania formosana* in Taiwan; *Oncomelania hupensis* in the Chang Jiang (Yangtze) Basin, as well as in south and southeastern China, and *Oncomelania quadrasi* in the Philippines), the miracidium enters the soft tissues of the mollusc and, in the course of 5 to 7 weeks, produces first- and second-generation sporocysts, then fork-tailed cercariae (Figs. 27–6 and 27–7). An experimentally acceptable molluscan host is *Pomatiopsis lapidaria* (Abbott, 1948), which has wide distribution in the United States. In Taiwan, Chiu (1965) produced a 70% infection rate with the zoophilic strain in the snail *Tricula chiui*. Extensive ecological studies of the snail hosts have been carried out in the Philippines, Japan, China, and Taiwan. DeWitt (1954) found that *S. japonicum* (Philippine strain) devel-

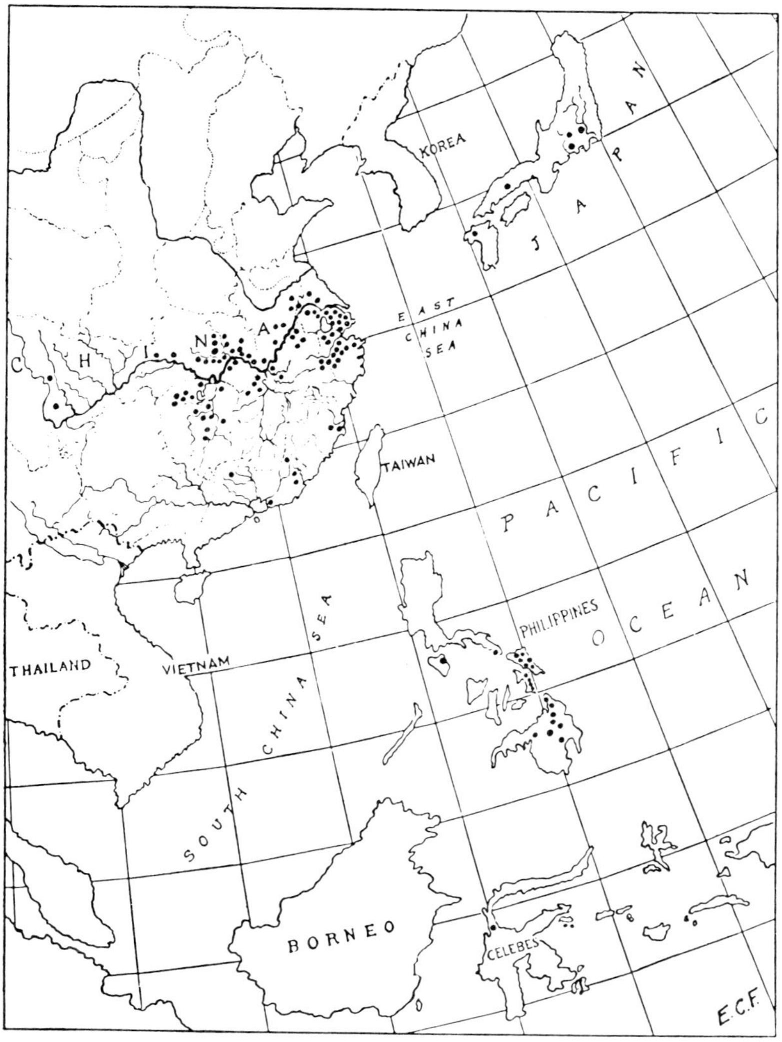

Fig. 27–1. *Schistosoma japonicum*. Map showing endemic foci of infection in Japan, China, Taiwan, the Philippines, and Celebes. (By E. C. Faust.)

oped only in *O. quadrasi,* the Chinese strain only in *O. hupensis,* the Japanese strain in *O. nosophora* and *O. hupensis,* and the Formosan strain in all except *O. hupensis*. The snails *O. nosophora* and *O. hupensis,* on the basis of cross-fertilization tests, appear to be biologically indistinguishable (Komiya and Kojima, 1959), indicating specific adaptation to the local snail species in different geographic areas. Moose and Williams (1963) demonstrated the presence of definite geographical or physiological strains of *Oncomelania formosana* in Taiwan, differing not only in their susceptibility to infection with the Formosan strain of *S. japonicum,* but also in morphology and habitats. However, snail susceptibility is not fully understood, and it has been shown that previously resistant snails may become infected if re-exposed (McQuay, 1953; Ruff and Yamaguchi, 1973).

The cercariae are discharged only when the snails are at or below the water level. Their measurements are as follows: body, 100 to 160 μm in length by 40 to 66 μm in transverse diameter; tail trunk, 140 to 160 μm by 20 to 35 μm; furci, 50 to 75 μm long. Internally their structure (Fig. 27–7) is almost identical with that of the cercarial stage of *S. haematobium* and *S. mansoni*. The cercaria of *S. japonicum* has 8 flame cells, that of *S. haematobium* and *S. mansoni* a similar number. On contact with the skin of mammals, the cercariae cast off their tails and, in the course of 24 hours, penetrate down to the cutaneous capillary beds as a result of the action of their penetration glands that secrete

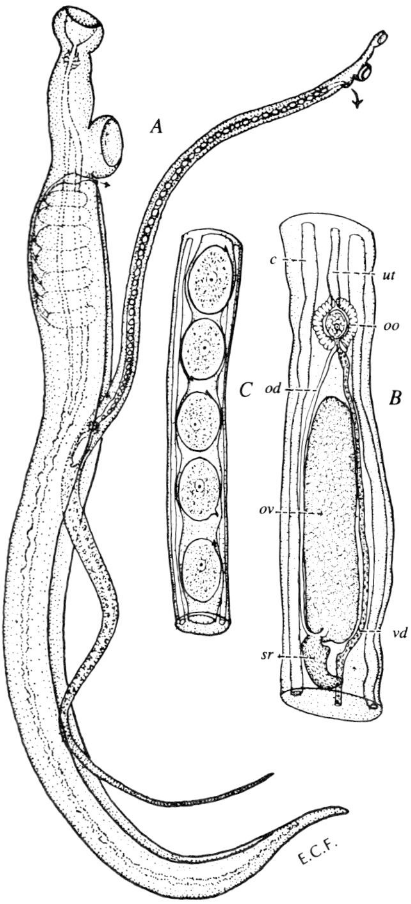

Fig. 27–2. *Schistosoma japonicum,* adult male and female. *A,* Worms in copula, with male on left, female extruded from gynecophoral canal of male. Sex organs of male shown just behind ventral sucker. *B,* Detail of female at level of ootype and ovary. *C,* Detail of female in inner portion of uterus. *c,* Cecum. *od,* Oviduct. *oo,* Ootype. *sr,* Seminal receptacle. *ut,* Uterus. *vd,* Vitelline duct. (From Faust, E.C. 1949. *Human Helminthology,* 3rd ed. Philadelphia, Lea & Febiger.)

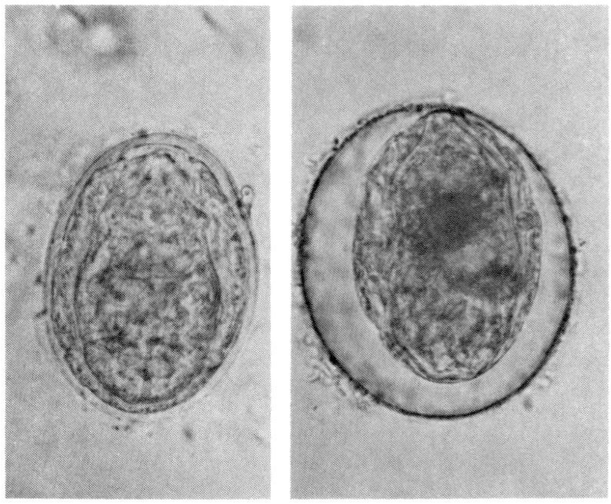

Fig. 27–3. *Schistosoma japonicum,* two forms of formalin-preserved eggs with fully embryonated miracidium. Note spine on upper right border of one and tissue debris characteristically adherent to shell of the other. (× 400.) (By P.C. Beaver.)

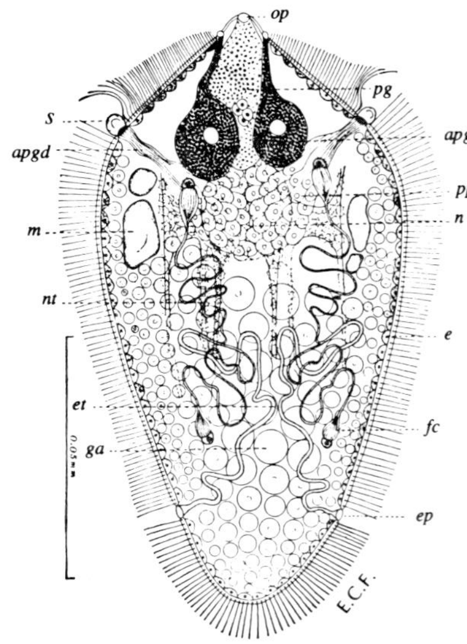

Fig. 27–4. *Schistosoma japonicum,* miracidium. *apg,* Anterior penetration gland. *apgd,* Anterior penetration gland duct. *e,* Epithelium. *ep,* Excretory pore. *et,* Excretory tubule. *fc,* Flame cell. *ga,* Genital anlage. *m,* Mass (unidentified). *n,* Nerve center. *nt,* Nerve trunk. *op,* Oral pore. *pg,* Primitive gut. *ppg,* Posterior penetration gland. *s,* Secretion. (× 500.) (From Faust, E.C., and Meleney, H.E. 1924. Studies on schistosomiasis japonica. Am. J. Hyg. Monogr. Series No. 3.)

proteolytic enzymes. Then they enter the venous circulation and are carried through the right heart to the lungs. Thence, according to one view after growth and essential development, they squeeze through the pulmonary capillaries and are carried through the left heart into the systemic circulation (Miller and Wilson, 1980). Individuals that by chance reach the intrahepatic portal circulation via the mesenteric artery and capillaries proceed to feed, grow, and migrate into the superior mesenteric venules. There is not full agreement on the route taken by the schistosomula in their migration from lungs to liver. It is possible that, as in *S. mattheei,* the schistosomula crawl against the bloodstream along the walls of the pulmonary arteries, right heart, and caval veins to enter the liver through the hepatic veins (Kruger et al., 1969). Schistosomula are able to reach the liver and develop to full maturity in mice after total ligation of the portal vein (Van Marck and Gigase, 1978). Within 5 to 6 weeks, depending on the strain of *S. japonicum,* the worms have mated and migrated to the mesenteric veins, and the females are laying eggs.

Epidemiology. Human feces and excreta of mammalian reservoir hosts commonly pollute water

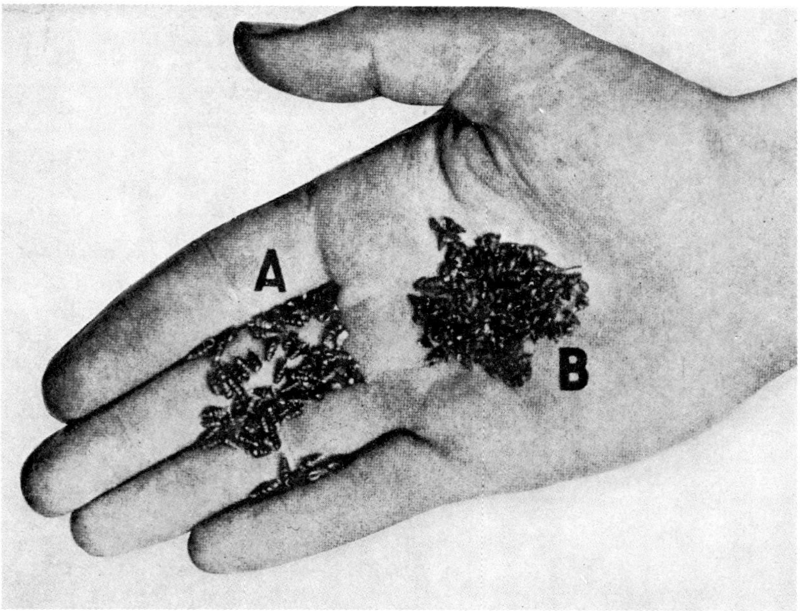

Fig. 27–5. *Oncomelania. A, O. nosophora* from Japan. *B, O. quadrasi* from Philippines. (From Abbott, R.T. 1948. Handbook of medically important mollusks of the Orient and the Western Pacific. Bull. Mus. Comp. Zool., Harvard College, *100*:246–328.)

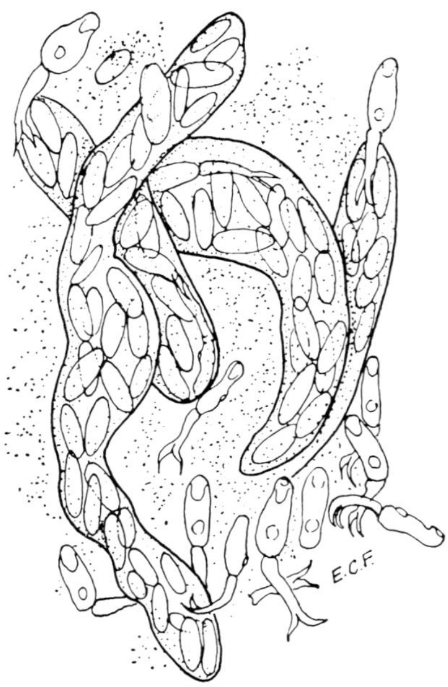

Fig. 27–6. *Schistosoma japonicum*. Second-generation sporocysts, with escaping cercariae dissected out of infected *Oncomelania quadrasi*, molluscan intermediate host in the Philippines. (× ca. 100.) (From Faust, E.C. 1949. *Human Helminthology*, 3rd ed. Philadelphia, Lea & Febiger.)

in which the *Oncomelania* snails breed. Dogs, cats, rats, mice, field mice, cattle, water buffaloes, pigs, horses, sheep, and goats (also *Felis bengalensis chinensis, Mustela davidina,* and *Meles meles*) are naturally infected with *Schistosoma japonicum* in many endemic foci. Extensive use of human feces for fertilizing garden and field crops greatly increases the chances that the snails will become infected. The small snails (Fig. 27–5) are operculate, amphibious in their habits, and are able to survive several months in a relatively dry state, meanwhile conserving the infection between periods of immersion in water. Mature viable eggs can survive for 80 days under winter conditions such as prevail in endemic areas of central China and Japan, and may then hatch and infect snails when they become activated the following spring. Infection persists indefinitely in man and animals, 47 years in one recorded instance (Hall and Kehoe, 1970).

Hsü et al. (1962) were able to obtain fertile first-generation worms after interstrain crosses of schistosomes from Japan, China, Taiwan, and the Philippines, and use of hybrid snail hosts from the four areas. They concluded that the parasites should be regarded as geographical strains. Two additional strains are now recognized: Thailand-Malaysia and

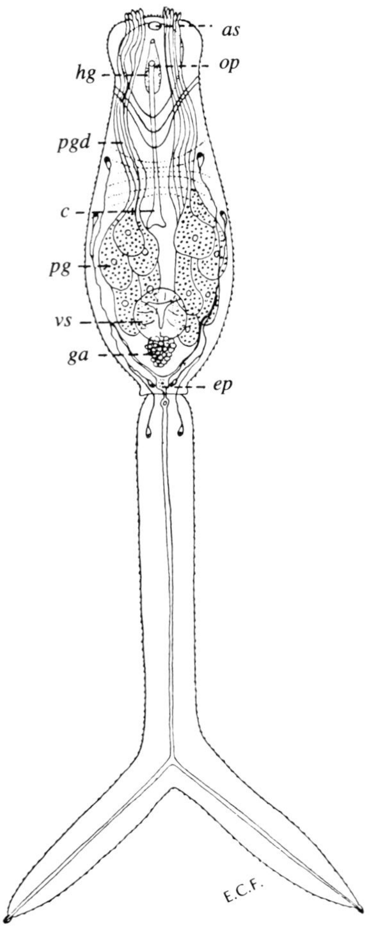

Fig. 27–7. *Schistosoma japonicum*, cercaria. *as*, Orifice of anterior sucker. *c*, Cecum. *pgd*, Penetration gland ducts. *pg*, Penetration glands. *hg*, Head gland. *ep*, Excretory pore. *ga*, Genital anlage. *op*, Oral pore. *vs*, Ventral sucker. (× 340.) (From Faust, E.C. 1929. *Human Helminthology*, 1st ed. Philadelphia, Lea & Febiger.)

Celebes. The apparent differences are mainly in snail host susceptibility (Cross and Lo, 1980).

In Japan the prevalence rates since 1960 have been steadily and markedly declining. In the four endemic areas of the main island of Honshu (Tone River Basin, Kofu Basin, Numazu, and Katayama Basin) a survey of approximately 19,000 inhabitants in 1973 detected eggs in the feces of only 124 persons. In the Chikugo River Basin on the island of Kyushu, where in 1948 eggs were detected in the feces of 156 (73%) of 214 people, in 1971 only 6 of 467 were positive for *S. japonicum* eggs. The youngest of the 6 infected people was 36 years of age (Yokogawa, 1976a,b). In China, schistosomiasis remains one of the main health problems among rural people in extensive endemic areas of the coastal provinces and far inland along the major rivers (Mao and Shao, 1982). In the Philippines the disease varies in intensity, from relatively low grade to 45% or higher in numerous communities on Samar, Leyte, and Mindanao (Lewert et al., 1979). In Celebes the disease is highly endemic in the Lake Lindu area where in 1971 fecal examination showed *S. japonicum* infection in 53% of 126 people ranging in age from 7 to 70 years (Pinardi et al., 1972), and in the Napu Valley where the reported infection rates were 8 and 12% in two villages and 7% in *Rattus exulans*, a wild rat (Carney et al., 1978).

Pathogenesis, Pathology, and Symptomatology. The lesions produced by infection with *Schistosoma japonicum*, as well as the sequence of their development, are practically the same as those in *S. mansoni* and *S. haematobium* infection. The organs and tissues most seriously involved are the intestinal tract and the liver. In all three infections the severity of the disease is in general dependent on the worm burden. Light infections may present few gross abnormalities, but the reactions of the host also vary.

There are three progressive stages in the pathologic processes: (1) incubation, (2) egg deposition and extrusion, and (3) tissue proliferation and repair. During the stage of migration and maturation *(incubation period)* the lesions consist of (1) dermatitis associated with penetration of the cercaria, conspicuous from the twenty-fourth to the thirty-sixth hour after invasion but not accompanied by remarkable cellular infiltration of the area in susceptible hosts; (2) traumatic and infiltrative changes in the lungs, at times with hemorrhages, and local accumulation of eosinophils, epithelioid cells, and giant cells around pulmonary blood vessels in the wake of migrating larvae; (3) acute hepatitis following arrival of the larvae and during their growth within the intrahepatic portal blood vessels; (4) hyperemia in the wall of the small intestine following arrival and during maturation of the worms in superior mesenteric venules; (5) trauma with hemorrhage as the eggs laid by female worms escape from the venules through the intestinal submucosa and mucosa into the intestinal canal; and (6) usually marked increase in circulating eosinophils as the sensitizing-toxic pathologic process develops from systemic absorption of the worms' metabolites.

Once oviposition has begun, the eggs accumulate in minute, short, sausage-shaped chains or clumps in radiating branches of the smaller mesenteric venules within the submucosa. Occlusion of blood arriving from the mesenteric venules, hypermotility of the parasitized segments of the intestine, and irritating mucoid secretions of maturing larvae within the egg shells cause the eggs to escape from the vessels into the perivascular tissues with extravasation of blood, to be discharged into the intestinal canal, and to be evacuated in the feces. Meanwhile an increasing number of eggs become sequestered in the submucosa and mucosa by infiltration of host cells, conspicuously eosinophils, with production of macroscopic granulomata called pseudotubercles. The diameter of a pseudotubercle is many times that of the egg or clump of eggs at the center that is responsible for the lesion. As egg-laying continues, the parent worms tend to abandon old sites and migrate to new mesenteric venules at different levels of the bowel, while eggs that become free in larger venules are carried into the intrahepatic portal vessel where they filter out perivascularly and provoke miliary pseudotubercle formation. Thus, the development of increasing granulomata in the intestinal wall and then in the liver adds substantially to the earlier lesions primarily of sensitizing-toxic nature and to traumatic damage due to extrusion of eggs into the intestinal canal. Moreover, worms occasionally migrate out of the portal into the caval venous system, with deposition of nests of eggs in pulmonary arterioles, or into the vertebral venous system, with oviposition in many ectopic sites, including the brain and spinal cord (Faust, 1948; Chang et al., 1957).

The cellular immune response of the host continues throughout the long life of worms, up to 47 years (Hall and Kehoe, 1970) if the patient survives, although it becomes less evident (i.e., reduced eosinophil count) during the chronic stage. On the other hand, continued development of pseudotubercles around eggs produces fibrosis, papillomatous growth, and stenosis of extensive segments of the intestinal tract and fibrosis of the liver, so that normal function of these organs is materially reduced. Studies in mice have shown that the presence of granulomata in the liver is associated with destruction, obstruction, and distortion of the intrahepatic portal venous system. Although the intrahepatic arteriovenous shunts that are seen in cirrhosis in the intestine have not been reported to date, numerous microvascular arteriovenous shunts are produced in mice experimentally infected with *S. mansoni* (Nagy et al., 1981). Such shunts play an important role in increasing the pressure in the portal venous system. Vascular changes such as those described are to some extent reversible after successful treatment of the infection (Andrade and Brito, 1981).

Ling et al. (1949) emphasized the protean clinical features of this disease, dependent chiefly on the toxic by-products of the worms, filtration of the eggs in tissues, and the subsequent tissue and systemic reaction. The manifestations include dysentery, hepatic fibrosis, splenomegaly, appendicitis, intestinal obstruction, pneumonitis, vascular occlusion, cerebral syndrome, myelitis, constitutional intoxication, and asyndromic forms.

Incubation Period. Urticarial rash accompanied by fever may develop as early as the fifth day after exposure. The prodromes occur about the end of the fourth week, at which time an intense urticaria, with edema of the subcutaneous areas and characteristic wheals of varying diameters in the skin and mucous membranes, may be occasionally observed. Diarrhea usually develops toward the end of this period.

Egg Deposition and Extrusion. This stage is ushered in by deposition of eggs into the tissues of the small intestine and their discharge into the lumen of the bowel. There may be profuse dysentery with eggs of the parasite in the feces. These eggs are frequently surrounded by a layer of cellular debris (Fig. 27–3). There is recurrent daily fever, epigastric pain with tenderness over the area, enlarged tender liver and spleen, and loss of appetite and weight. This acute attack terminates in 3 to 10 weeks, the fever subsides, and the patient feels nearly normal, although heavy exercise causes relapse. The blood picture changes from one of leukocytosis and hypereosinophilia to that of a secondary anemia, leukopenia, and reduction in the eosinophil count. The pathologic basis for these signs and symptoms is the infiltration of the eggs in the tissues of the intestinal wall, in the periportal tissues of the liver, and in other organs, where pseudotubercles develop around the eggs (Fig. 27–8). In the bowel the granulomata formed around eggs frequently break through into the intestinal lumen, with discharge of their contents. Repair occurs with scar tissue formation. Giant cells frequently develop within the pseudotubercles, which become transformed to fibrous nodules (Fig. 27–9). In the liver, spleen, and lymph nodes hematin pigment is commonly deposited in endothelial cells of the blood vessels, free macrophages, and similar

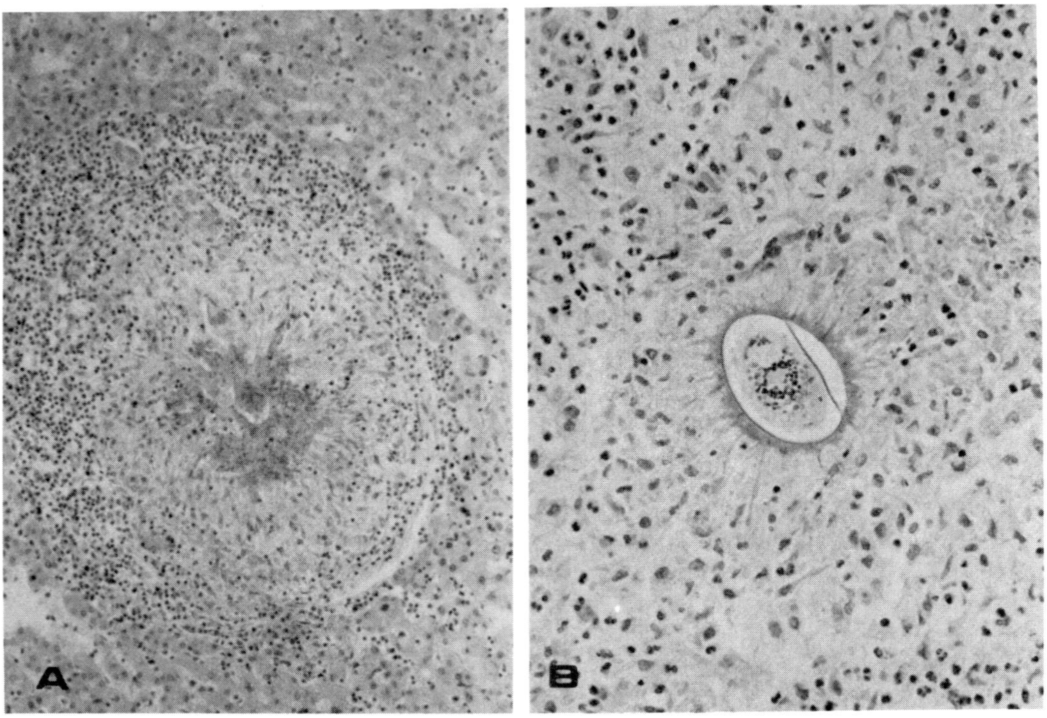

Fig. 27–8. *Schistosoma japonicum*. Eggs in liver of man in the Philippines. *A*, In a granuloma with chronic inflammatory cells at the periphery. (× 100.) *B*, Egg at center of granuloma with eosinophilic material radiating from the eggshell, known as the Hoeppli reaction. (× 250.) (By P.C. Beaver.)

phagocytic cells in the organizing pseudotubercles. The liver increases in size as a result of miliary lesions around infiltrated eggs. The spleen becomes congested, its fibrous reticulum increases, and consequently it becomes greatly enlarged. The mesenteric lymph nodes become modified by loss of active lymphoid tissue. All these changes are more pronounced in schistosomiasis japonica than they are in Manson's schistosomiasis, possibly because (1) the daily output of eggs in this infection is several times greater, and (2) the greater proximity of the worms to the liver in the superior mesenteric venules contributes to greater involvement of this organ.

Although the early intestinal lesions in schistosomiasis japonica are confined primarily to the small bowel, this is by no means exclusively the case in human infections. Both proctoscopic and postmortem evidence acquired during 1945 in a clinical study of American and Australian military forces who had acquired this infection on Leyte (Philippine Islands) late in 1944 or early in 1945 indicated that relatively early during the active stage of the disease some of the worms were in the inferior mesenteric radicles and at times in rectal venules, discharging eggs. At times they were producing only congestion of the venules; at other times they were causing development of pseudotubercles.

Tissue Proliferation and Repair. This stage gradually supersedes the second; frequently the two are not distinguishable because of repeated reinfection. The liver gradually decreases in size as its parenchyma becomes replaced by scar tissue; the spleen continues to enlarge. The mesentery and omentum frequently become thickened, binding down the colon and producing an upper and a lower abdominal distention (Fig. 27–10). The thoracic viscera are pressed upward by the increase in the abdominal contents. Weakness, skin pallor, and dyspnea on slight exertion are marked. Digestion is poor, emaciation is extreme, and dilatation of the superficial veins of the abdomen and thorax is common.

Ascites may or may not be conspicuous, depending on the degree of portal thrombosis and hepatic fibrosis. Infected children are retarded in both physical and mental development. Anemia is more conspicuous than in the previous stage. The intestinal wall becomes increasingly thickened and fibrosed, with involvement of the large bowel and

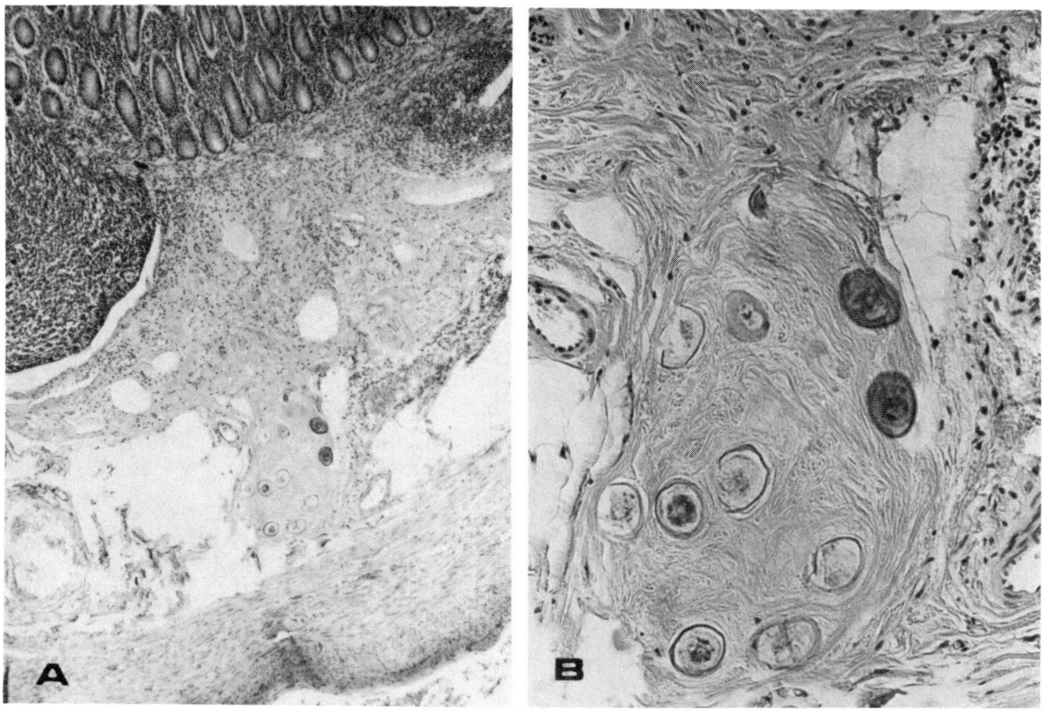

Fig. 27–9. *Schistosoma japonicum*. Eggs in wall of appendix removed incidental to abdominal surgery for an unrelated condition. *A*, Characteristic cluster in submucosa. (× 50.) *B*, Higher magnification to show fibrous nodule and absence of inflammation. (× 170.) (By P.C. Beaver.)

development of papillomata on the mucosa. The mesenteric and portal vessels become conspicuously thrombosed. Most significant is the development of portal fibrosis due to deposition of eggs about the portal triads. The severe fibrosis in heavy infections originally described as cirrhosis (Symmers, 1903) is frequently referred to as Symmers' fibrosis.

The granulomatous reaction about the eggs ultimately leads to continuous fibrosis of the periportal tissues. This fibrous tissue is typically white and hard and therefore also called *pipe-stem fibrosis* (von Lichtenberg et al., 1971). Eggs may reach the lungs, bladder, or even the capillaries of the brain. This third stage may last for many years or may be terminated by pneumonia or other infectious diseases.

Nearly 200 cases of cerebral involvement in schistosomiasis japonica have been reported in the medical literature from 1889 to the present. These were considered by Ariizumi (1963), who included 51 patients under his own supervision. At the time of clinical study they were classed symptomatically as (1) acute, (2) chronic with granulomas or diffuse lesions, (3) principally psychotic in nature, and (4) portal system encephalopathies. Ariizumi (1963) recovered *S. japonicum* eggs from 45 patients, and had positive intradermal tests for the remainder. One patient was operated on following headache, Jacksonian episodes, paresthesias, and poor vision. A walnut-sized tumor and several smaller growths surrounding the tumor were removed from the left parietal lobe of the brain. Treatment was undertaken with sodium antimony tartrate, a total of 500 ml 0.3% solution administered intravenously every other day over 50 days, with partial remission of symptoms. This physician recommended sodium antimony tartrate in neurological complications of schistosomiasis japonica and concluded that surgical intervention is not usually indicated. Chang et al. (1957) reported that 43% of all cases of schistosomiasis japonica exhibit acute or chronic cerebral involvement, including epileptic seizures.

Diagnosis. The acute febrile illness of early schistosomiasis must be distinguished from serum sickness, which it closely resembles (Hiatt et al., 1979). Typhoid fever and other acute bacterial and viral infections must also be ruled out. The accompanying peripheral eosinophilia may lead to suspicion of trichinosis. Later, the dysentery must be

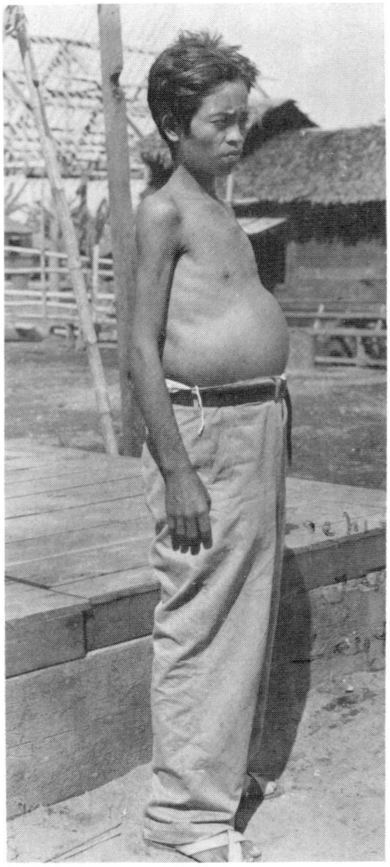

Fig. 27–10. *Schistosoma japonicum*. Third stage of infection, showing enlarged upper portion of abdomen with constricted middle region at belt level. (From Faust, E.C., Beaver, P.C., and Jung, R.C. 1975. *Animal Agents and Vectors of Human Disease*, 4th ed. Philadelphia, Lea and Febiger.)

distinguished from amebic or bacillary dysentery, hookworm disease, intestinal tuberculosis, carcinoma, and typhoid fever. Involvement of the appendix must be differentiated from appendicitis of bacterial or amebic origin; hepatomegaly, with tenderness in the right upper quadrant of the abdomen, must be differentiated from infectious hepatitis, amebic hepatitis, amebic liver abscess, or hydatid cyst of the liver; hepatic fibrosis must not be confused with cirrhosis or tuberculous peritonitis with ascites.

Specific diagnosis rests on recovery of characteristic eggs in the stools (Figs. 27–3, 27–8). Studies on chronic and on lightly infected cases indicate the advantage of concentration techniques for recovery of the eggs from feces. Proctoscopic aspiration and biopsy demonstrate eggs in some cases where stools are consistently negative and also provide information regarding abundance of eggs and relative numbers of living and dead eggs. Eosinophilia, positive serodiagnostic tests, together with the clinical history and evidence that the patient has lived in an endemic area, strongly suggest schistosomiasis even during the incubation period before eggs are recoverable in the feces (Lewert et al., 1980). In the late chronic case, where eggs cannot be found in stool concentrates or by other techniques, one of the recently refined serological tests should be given careful consideration (Yogore et al., 1978, 1979). A diagnosis of schistosomiasis should not be excluded unless negative serological tests, two or three negative stool examinations with concentration techniques, and a negative rectal biopsy have been obtained.

Treatment. *Schistosoma japonicum* is the most pathogenic of the human schistosomes and also the most resistant to treatment. Only tartar emetic in a lengthy course of intravenous injections has been dependably effective in curing the infection. For these reasons intravenous administration of tartar emetic has long been the treatment of choice in spite of the inconvenience and expense involved and the relatively high degree of toxicity of the drug.

Praziquantel, a new antischistosomal pyrazinoquinoline compound, administered orally in a single day, is effective and relatively well tolerated. It should therefore completely replace tartar emetic in the treatment of schistosomiasis unless some heretofore unobserved ill effects are noted when the drug is more extensively used. Oral administration of praziquantel in 3 doses each of 20 mg/kg at 4-hour intervals resulted in about 80% cures (Santos et al., 1979). Side effects occur in 50 to 60% of the patients receiving this dosage, but they are usually mild and transient (Ishizaki et al., 1979). They include abdominal discomfort, drowsiness, headache, backache, fever, sweating, and giddiness. In patients receiving a single dose of 50 mg/kg body weight, side effects are of somewhat greater frequency (70%) and efficacy somewhat less.

Oxamniquine and metrifonate, which are highly effective against schistosomiasis mansoni and schistosomiasis haematobia, respectively, are ineffective against schistosomiasis japonica. Treatment with niridazole achieves a reduction in egg count but does not eliminate the infection.

Prognosis. Prognosis is good only when patients are given specific treatment before or during the early part of the second period of the disease and are not reexposed. Prognosis is grave after the intestine and particularly the liver have become fibrosed.

Prevention and Control. Control of schistosomiasis japonica is complicated by the common use of human night soil for fertilizer in endemic areas. Promiscuous defecation on the part of the human population is considerable. Farmers and canal boatmen are principally affected, but persons of all ages, particularly children, are occasionally exposed to infection while bathing or wading in these areas, as are persons using infested water for laundry purposes. Urban populations are occasionally endangered when floodwaters wash infected snails into the canals that pass through towns or cities, as occurred in Shanghai following the 1931 Yangtze River floods. In Japan, domestic animals and field mice are important reservoirs; in the Philippines, dogs, cats, pigs, water buffaloes, and wild rats are commonly infected. These reservoir hosts constitute an important factor in the control of schistosomiasis japonica. Among 45 field rats in an area of Leyte, Philippines, 39 (86%) were found to be infected (Oshima et al., 1978). In another area of Leyte, 163 of 199 rats (82%) were infected (Kamiya et al., 1980).

Another factor is the snail host, which is operculate, amphibious, and able to withstand desiccation for a month or more. Infected snails can survive several weeks of drying, and on renewed contact with water again become active and shed infective cercariae. Of the currently available molluscicides, sodium pentachlorophenate (NaPCP) has been most widely used in Japan. This chemical kills both snails and their eggs, and excellent extension downstream has been achieved; however, it is irritating and potentially dangerous to the handler, and the required dose is relatively high compared with newer molluscicides. In addition, its activity may be reduced by bright sunlight. Copper sulfate has not provided satisfactory control. In Japan, cement lining of irrigation ditches inhabited by *Oncomelania* snails has been successful in reducing the snail population, and since 1972, Yurimin, a chemical developed in Japan, has completely replaced NaPCP (Yokogawa, 1976b). In China, along with treatment of people and domestic animals, major emphasis is placed on drainage and other physical methods of snail control (Mao and Shao, 1982). In an area of Japan where 72% of the people were infected in 1950, no child under 15 years was infected in 1978 (Hunter et al., 1982).

In the control of schistosomiasis japonica, attention to reservoir hosts is essential, especially to cows and buffaloes used in agriculture, stray dogs, and field rats. It is probable that rats play an important role in the maintenance of the disease in both Japan and the Philippines.

In areas where night soil is still used, and antimolluscan campaigns are not practicable, the most effective available measure is to allow the night soil to stand longer in containers, or to disinfect it with fertilizer salts like ammonium nitrate in order to kill the eggs before the material is spread on the fields.

Schistosoma mekongi
Voge, Bruckner, and Bruce, 1978
(The Mekong schistosome)

Historical and Geographical Notes. The early history of schistosomiasis in southeast Asia was reviewed by Schneider (1976) and by Harinasuta (1980). A *Schistosoma japonicum*-like infection was diagnosed in a patient from southern Thailand in 1950, and soon thereafter surveys demonstrated infection in about 50 people, 21 to 40 years of age, in two districts of the same area. In 1966, two cases from the northeastern region of Thailand were reported in which *S. japonicum*-like eggs were identified in sections of tissue (Lee et al., 1966). Further reports indicated that endemic schistosomiasis existed in northern and northeastern Thailand. In Laos, the first evidence of endemic schistosomiasis was detected in Paris, France in a student who 9 years earlier had lived on Khong Island in the Mekong River in southern Laos and northern Cambodia (Barbier, 1966). It was soon shown that Khong Island is a true endemic area of schistosomiasis with a significant number of people showing the usual signs and symptoms of severe disease. The snail host was identified as *Lithoglyphopsis aperta* by a research team from the Smithsonian Institution, United States, and the Faculty of Tropical Medicine, Mahidol University, Bangkok. The snail was observed to be aquatic, not amphibious, and infection was found to be common among people living in floating villages of raft houses on the Mekong River south of Khong Island in southern Laos and Cambodia. In 1973, infection with *S. japonicum*-like eggs in the tissues of an aborigine was discovered in northeastern Malaysia. A retrospective study of autopsy materials revealed *S. japonicum*-like eggs in tissues from nine individuals who had lived in the same area (Pahang and Perak provinces). It was suggested that the Malaysian worms were probably of the same strain as those in Thailand, different from the Mekong schistosome, which in 1978 was described as a new species. In the same year, the first living person with *S. japonicum*-like infection in Malaysia was reported by Murugasu et al. (1978). Reservoir hosts and naturally infected snails and wild rats were discovered in Malaysia by Greer et al. in 1980. Further studies may demonstrate that the Thailand-Malaysia and the Mekong schistosomes, along with those in Japan, China, Taiwan, Philippines, and Celebes (Sulawesi), all represent different geographic strains of *S. japonicum*. In the meantime the Mekong schistosome can be referred to as *S. mekongi*. The relationship of the schistosome of the Mekong River valley of Yunnan Province, China to that of the lower Mekong has not been studied.

Morphology, Biology, and Life Cycle. The original description of *S. mekongi* is based on specimens recovered from a naturally infected dog. The

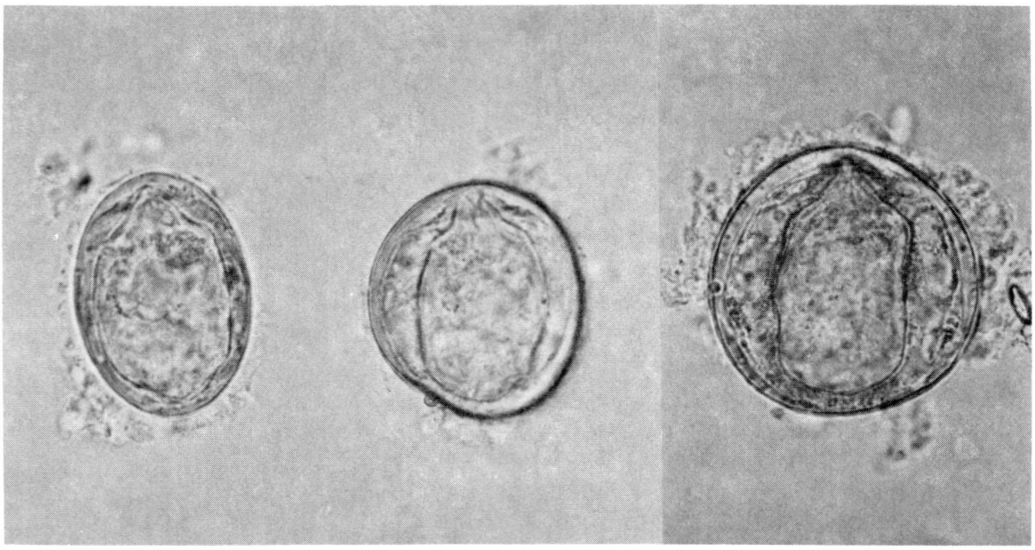

Fig. 27–11. *Schistosoma mekongi*. Showing range in size and shape of eggs digested from liver of experimentally infected mouse. A spine is evident on lower left border of egg in middle position. (× 400.) (By P.C. Beaver. Egg collection courtesy of Dr. John Bruce.)

mature female is 12 mm long, 0.23 mm wide at the widest level, in the posterior half of the body. The oral and ventral suckers measure 30 μm and 35 μm in diameter, respectively. The intestinal ceca extend from the esophagus at 0.22 mm from the anterior end to near the posterior end where they fuse. The ovary and ootype complex are in the middle fifth of the body, and the uterus, bearing numerous eggs, extends straight or slightly winding to the genital pore just anterior to the esophageal bifurcation. Vitellaria fill the posterior three-eighths of the body. Eggs are subspherical, 40 by 45 μm in diameter, with a small knob-like spine near one end (Fig. 27–11). The male is about 15 mm long by 0.4 mm maximum width, with the gynecophoral canal extending the full length of the body behind the acetabulum. The oral and ventral suckers measure 150 by 120 and 310 by 300 μm, respectively. There are seven testes, slightly lobed or smooth, 100 to 110 μm long by 80 to 110 μm wide. In both sexes the tegument over most parts of the body is spinose but without tuberculations. The worms are normally found in the mesenteric veins. In experimental infections egg production begins at 33 to 35 days of infection. In the original species description (Voge et al., 1978), *S. mekongi* was compared with *S. japonicum* and the following differences were noted: Embryonated eggs of *S. mekongi* are smaller and less elongate; the snail host is distinctly different in its biology, being aquatic instead of amphibious; and the prepatent period is longer, eggs being present in the liver of mice at 20 to 26 days in *S. japonicum,* at 35 days but not at 30 days or less in *S. mekongi*. Sornmani et al. (1973) reported that *S. mekongi* cercariae emerge from snails in daytime, mostly in early morning, 45 to 53 days after penetration of the snail by a miracidium, and that the prepatent period in dogs, mice, and hamsters is 43 to 49 days. Infection did not become patent in rabbits. This finding was confirmed by Byram and von Lichtenberg (1980) who also noted that *S. mekongi* is relatively more pathogenic than *S. japonicum* in the mouse and hamster. Kitikoon (1980) confirmed that the egg of *S.*

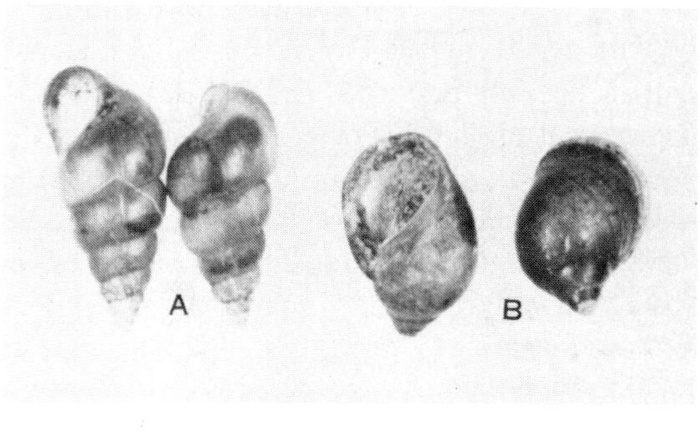

Fig. 27–12. Schistosomiasis vector snails. *A, Oncomelania nosophora,* intermediate host of *Schistosoma japonicum*. (× 4.) *B, Tricula aperta* (Syn. *Lithoglyphopsis aperta*), intermediate host of *Schistosoma mekongi*. Three races of *T. aperta* are recognized. Shown here is the alpha race, the largest. (× 4.) (By P.C. Beaver. B-specimens courtesy of Dr. John Bruce.)

mekongi is smaller and more spherical than that of *S. japonicum,* and he observed consistent differences in the shapes and pattern of epidermal plates and papillae on the miracidia. After extensive and detailed studies on the snail host of *S. mekongi,* Davis (1980) concluded that three distinct races (alpha, beta, and gamma) exist, and the correct designation for the species is *Tricula aperta* (Temcharoen, 1971). This snail, like *Oncomelania,* is small, the largest race of *T. aperta* being 3 to 4 mm long. Generally the adults are about 3 × 2 mm (Fig. 27–12).

Epidemiology. Thus far, natural infections detected in people and dogs, the only known reservoir hosts, were apparently acquired from exposure at the margins of the main channels of the Mekong River. The snail host appears to be limited to the main channels; searches for it elsewhere in Thailand and Cambodia have been unsuccessful (Sornmani et al., 1980). The highest rates of infection have been found in children 7 to 15 years of age. In an active transmission area on Khong Island, a survey by fecal examination in 1971 detected infection in 45 of 72 people, 35 of whom were 5 to 14 years of age.

Pathogenesis, Pathology, and Symptomatology. Only limited clinical studies have been made on people in the Mekong endemic area. Among 45 infected inhabitants of a Khong Island community, 25 were found to have hepatomegaly or splenomegaly; four of these had dilated superficial abdominal veins, and ascites was detected in two cases, one of which was severe. Apparently *S. mekongi* is no more, or less, pathogenic than is *S. japonicum*. Most of the children with *S. mekongi* infection were also infected with *Opisthorchis viverrini* which, along with malnutrition, could have contributed to the prevalence of hepatosplenomegaly.

Diagnosis, Treatment, and Prognosis. In its clinical aspects *S. mekongi* infection has not been shown to differ from *S. japonicum* infection. Skin test screening of over 20,000 Cambodian refugees in Thailand identified 1712 with positive reactions, 74 of whom were confirmed by formalin-ether fecal examination (Keittivuti et al. 1982). Nash et al. (1982) have reported efficacy of a one-day oral treatment with praziquantel in a dosage of 60 mg/kg body weight in three divided doses.

Prevention. Because the snail breeds in large flowing-water habitats, and dogs are important reservoir hosts, prevention of infection with *S. mekongi* presents a difficult and somewhat unique problem. Control measures have not been undertaken, although it has been shown that the snail host is sensitive to conventional molluscicides (Upatham et al., 1980), and the parasite in mice and monkeys is readily eliminated by treatment with amoscanate (Crawford et al., 1980).

Schistosoma mansoni Sambon, 1907 (Manson's blood fluke, producing schistosomiasis mansoni, or Manson's intestinal schistosomiasis, often called bilharziasis)

Synonyms. *Distoma haematobium* Bilharz, 1852 *pro parte; Schistosomum americanum* da Silva, 1909; *Afrobilharzia mansoni* (Sambon, 1907) Le Roux, 1958.

Historical and Geographical Notes. Lateral-spined eggs were first observed by Bilharz in female schistosomes recovered at autopsy from a patient in Cairo. Sonsino and Manson considered them specifically distinct from worms containing terminal-spined eggs, but Looss opposed these views. Castellani in Uganda (1902) and Manson (1902), González Martinez (1904), and Letulle (1904) in West Indies showed a different geographical distribution of the two types of worms, while the clinical picture of the patients with lateral-spined eggs indicated that the involvement was intestinal rather than vesical in type. On the basis of these differences, Sambon (1907) created the species *mansoni* for the newly differentiated type. The morphologic study of da Silva (1909) and autopsy studies of Flu (1911) and Risquez (1918) substantiated Sambon's idea, while Leiper's experimental work (1915-1918) proved the distinctness of *S. haematobium* and *S. mansoni*.

Schistosoma mansoni is widely distributed in Africa, South America, and islands of the Caribbean Sea (Fig. 27–13). Although uncommon from Cairo up to Khartoum, it is widespread in the Upper Sudan and occurs along the East African coast from Eritrea through Kenya, Tanzania, and Malawi to the Zambesi River and Zimbabwe and inland through Zambia and Tanzania to the Congo. In South Africa autochthonous cases have been reported from Natal and the Transvaal. It is present in Malagasy. In West Africa its distribution extends through Senegal, the Cameroons, Dahomey, and Guinea inland to Lake Chad, throughout the Congo Basin, Sierra Leone, and Liberia. It is reported in Libya. Many autochthonous cases have been found in the Yemen. In the Western Hemisphere, where it is the only human blood fluke, it is common in northeastern Brazil, where it has extensive distribution throughout most of the coastal and much of the inland areas. It is present in endemic foci near the coast in Venezuela, and is highly endemic along the coast in Dutch Guiana. In the West Indies it is endemic in Antigua, Guadeloupe, Martinique, St. Lucia, Nevis, Montserrat, Vieques, and in numerous foci in Puerto Rico. The former infection in St. Kitts apparently no longer exists. It occurs on the south coast of Santo Domingo. Extension of the distribution of *S. mansoni* has been reported from several areas in recent years, particularly Brazil (southern states and state of Maranhão); however, in Israel the *S. mansoni* focus in the Yarkon River area has completely disappeared, and the application of control measures has resulted in a reduced prevalence of the disease in both Puerto Rico and Venezuela. Large numbers of Puerto Ricans infected with *Schistosoma mansoni* reside in cities of the continental United States, particularly New York (Welton and Imperato, 1975).

Morphology, Biology, and Life Cycle. Super-

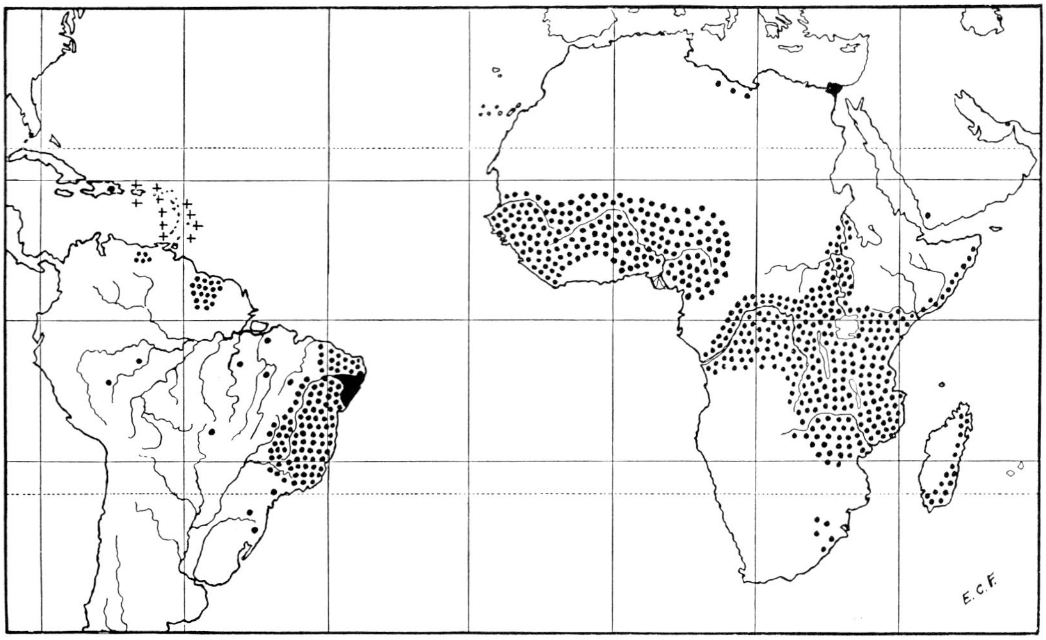

Fig. 27–13. *Schistosoma mansoni*. Map showing endemic foci of infection in Africa, South America, and the West Indies. The solid black areas in the delta region of the Nile River and in northeastern Brazil indicate areas of hyperendemicity. Foci in the Lesser Antilles are shown by a "+." (From Faust, E.C. 1949. *Human Helmintholgy*, 3rd ed. Philadelphia, Lea & Febiger.)

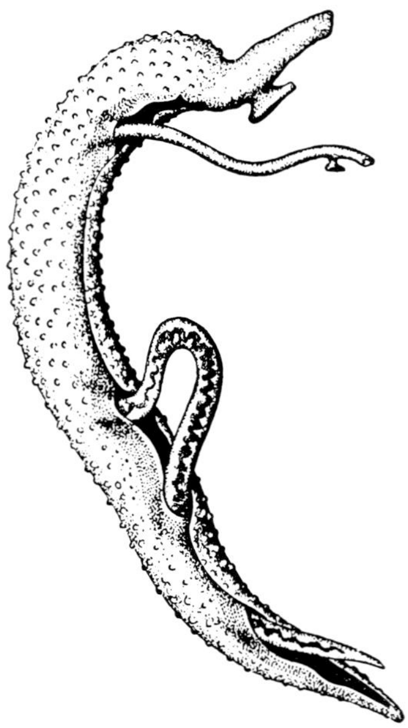

Fig. 27–14. *Schistosoma mansoni*, paired male and female. The lateral margins of the male interlock on the ventral side to hold the female in the gynecophoral canal. (× 10.) (From Gönnert, R. 1948. Die Struktur des Korperoberfläche von *Bilharzia mansoni* (Sambon, 1907). Z. Tropenmed. Parasitol., *1*:105–112.)

ficially the *adult worms* (Fig. 27–14) resemble those of *S. haematobium*, although both the male (6.4 to 12 mm long) and female (7.2 to 17 mm long) *S. mansoni* are smaller. The integumentary tuberculations of the male *S. mansoni* are also more conspicuous and the testes number 6 to 9. Gönnert (1948) observed that the lateral margins of the male interlock and are held in position by acuminate spines that are longer than those in other locations. Minute sensory papillae are distributed over the surface, and the tips of the tuberosities are covered with minute spines. In the female the ovary is situated in the anterior half of the body. At its posterior end there is a retort-shaped seminal receptacle. The vitellaria occupy lateral fields in the posterior half of the body. The uterus is conspicuously short and contains at most only a few lateral-spined eggs. Sexually mature worms are typically found in the mesenteric venules draining the large bowel and the most posterior segment of the ileum, although some may be found in the upper branches of the superior mesenteric vein, vesical plexus, and within the intrahepatic portion of the portal vessel. At the time of oviposition the females are characteristically held by the males in the small venules at or in the intestinal wall, where eggs are deposited into vessels, frequently in the submucosal layer of the wall.

In *S. mansoni* infection the eggs commonly break out of the venules in the submucosal and mucosal layers of the intestine and, escaping into the lumen of the bowel, are passed in the feces. They have a yellowish-brown transparent shell and characteristic lateral spine (Fig. 27–15), and measure 114 to 175 μm in length by 45 to 68 μm in transverse diameter. The shell is porous and densely covered with microvilli-like projections that give it a velvety appearance on scanning electron micrographs (Sakamoto and Ishii, 1976). After eggs are deposited in the tissues, the miracidium grows and develops, and the egg shell enlarges (Fig. 27–15,B), utilizing in the process exogenous metabolites and secreting enzymes that facilitate passage through tissues into the intestinal lumen (Stjernholm and Warren, 1974). Eggs are usually mature when discharged in the feces, absorbing water and hatching soon after being washed from the feces in water. The escaping miracidia are larger than those of *S. japonicum* or *S. haematobium*. The fine structure of the miracidium of *S. mansoni* has been described in detail (Pan, 1980). Newly emerged from the shell, they swim at a velocity of 690 to 750 cm/hr. They have a well-developed scanning ability and can locate and infect *B. glabrata* 86 cm distant horizontally. The free-swimming period does not ordinarily exceed 16 hours.

In lower Egypt, upper Egypt, Saudi Arabia, and Yemen, the commonly infected snail is *Biomphalaria alexandrina* (syn. *B. boissyi*). In Cameroons, Congo, Uganda, Kenya, Tanzania, and Malawi, *B. sudanica* is a common host. In the Sudan, Eritrea, Ethiopia, Kenya, Uganda, and Tanzania, *B. rüppellii* is a vector. On the west coast of Africa (Gambia, Guinea, Sierra Leone, Liberia, Mauritania, Senegal, Ghana, and Nigeria), *B. pfeifferi gaudi* is involved. In Tanzania, Zimbabwe, Zambia, South Africa, Mozambique, and Malagasy, *B. pfeifferi* is the common mollusc involved. In Uganda the species *B. choanomphala, B. smithi,* and *B. stanleyi* have also been found infected.

In the Western Hemisphere, *Biomphalaria glabrata* is the molluscan host in Puerto Rico, Vieques, Martinique, St. Lucia, Guadeloupe, Venezuela, and Dutch Guiana. It is also the sole or predominant host in parts of Brazil, although *Tropicorbis centrimetralis* has been found naturally infected in the states of Minas Gerais, Sergipe, and elsewhere in the North. Experimentally *Drepanotrema cultratus* in Venezuela and *Tropicorbis havanensis* collected in Louisiana have proved susceptible to infection. Still other species of *Tropicorbis* from Chile, Ec-

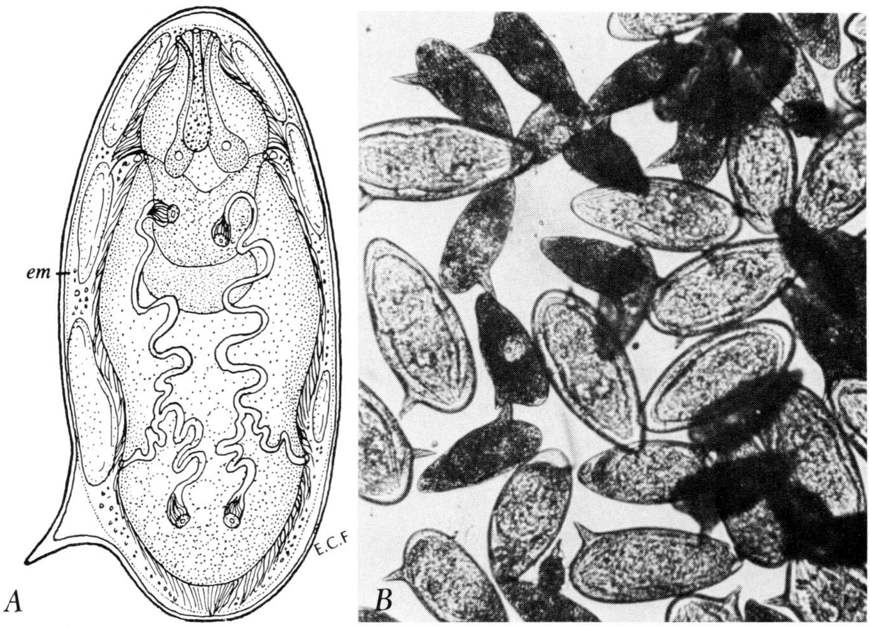

Fig. 27–15. *Schistosoma mansoni. A,* Mature egg with miracidium, dorsal view. *em,* Embryonic membrane. (× 500). *B,* Eggs obtained from worms that were cultivated in vitro, after 12 days of incubation. Eggs failing to develop remained as small as when laid, whereas the dimensions of eggs that developed a mature miracidium increased markedly in all parts except the spine. (× 180.) (*A,* From Faust, E.C. 1949. *Human Helminthology,* 3rd ed. Philadelphia, Lea & Febiger. *B,* Courtesy of Dr. T.H. Weller.)

uador, and Puerto Rico have been susceptible to experimental infection.

The life cycle within the snail parallels that of *S. japonicum*. Under favorable conditions a minimum period of 4 weeks is required. Pan (1965) studied the relationship between *S. mansoni* and *Biomphalaria glabrata* in aquarium-reared and bacteriologically sterile snails, and has provided detailed observations on the sites of development, routes of migration, and normal and abnormal histology of the parasite, and on the pathologic changes produced in the snail host. Successive generations of sporocysts may develop, at least in some snails (Theron, 1981). File (1975) observed that snails infected with a single miracidium could be superinfected with a miracidium of the opposite sex; thus, naturally infected snails tend to shed cercariae of both sexes. The emerging cercariae closely resemble those of *S. japonicum* (Fig. 27–7) (body, 185 to 230 μm by 75 to 110 μm; tail trunk, 185 to 300 μm by 60 to 75 μm; furci, 90 to 130 μm long) and have several pairs of penetration glands. Two pairs of these glands are preacetabular, while there are three or four postacetabular pairs. With the aid of substances from the penetration glands, the cercariae actively invade the skin, particularly at folds or minute wrinkles, or, in the case of experimental hosts, along hair shafts or under corneal scales, shedding their tails at the entry site. No morphological differences between the sexes of cercariae have been noted (Short and Cartrett, 1973).

The method of attack, invasion of the skin, and migration through the bloodstream to the intrahepatic portal system is similar to that of *S. japonicum*, with several exceptions. In experimental mice, Standen (1953) found larvae in the subcutaneous lymphatics 30 minutes after skin penetration; these schistosomula accumulated in regional lymphatics up to 5 days. Other observers report that after one to several days in the skin essentially all migrating schistosomula penetrate directly into veins and are carried more or less passively to the heart and lungs (Wilson et al., 1978; Wheater and Wilson, 1979; Wilson and Lawson, 1980). The same observers reported that schistosomula on reaching the lungs transform into slender forms with cuticular spines only at the ends of the body. They migrate actively within the pulmonary capillaries to reach the systemic circulation for passive transport to the liver, depending on chance to deliver them eventually to the mesenteric-portal vessels (Crabtree and Wilson, 1980; Miller and Wilson, 1980). Kruger et al. (1969) described a route of migration that does not depend on passive transport or chance delivery. They found that the schistosomula of *S. mattheei* in sheep apparently migrate by actively crawling along the walls of the pulmonary arteries, right heart, and caval and hepatic veins to the liver. In mice, 4- and 10-day-old schistosomula are able to find their way to the liver when injected into the subcutaneous tissues or the peritoneal cavity (Pereira et al., 1972).

S. mansoni may be readily kept in the laboratory for study. The Puerto Rican strain of *Biomphalaria glabrata* is easily maintained in culture, providing a reliable source of cercariae for infection of mice and hamsters (suitable hosts for experimental purposes). The laboratory rat is naturally refractory.

Epidemiology. Human infection is almost exclusively derived from human sources, although monkeys and baboons in endemic areas have occasionally been found infected. In Tanzania a tribe of wild baboons is sustaining the life cycle without human intervention (Fenwick, 1969). Natural infection of a nonprimate host was first reported in a gerbil in Egypt (Kuntz, 1952). *S. mansoni* in wild rodents in Brazil and in *Rattus rattus* in the Congo are not morphologically distinguishable from the form found in man, but in wild rodents in the Congo (Schwetz, 1954) they belong to a different variety, namely, *rodentorum* Schwetz, 1953. Rodents *(Gerbillus gerbillus* and *Mastomys coucha)* at times become infected from ingestion of infected snails (Luttermoser, 1963). This disease is perpetuated by infected persons who defecate in or near water courses containing the appropriate intermediate host. Sewage from homes in endemic foci, emptying into streams, adds to propagation of the infection. Infected individuals migrating to previously uninfected areas having susceptible snails initiate new centers of the disease. Extension of irrigation systems in endemic areas and the relocation of infected persons constitute grave hazards, making control programs more difficult. Within recent decades extension of the disease has been demonstrated in Africa and Brazil as a result of one or more of these activities. The disease is acquired by direct contact with water containing the cercariae. In lower Egypt, as in Puerto Rico, farmers are exposed in irrigation ditches. In Santos, Brazil, the sole focus of infection was found to be a series of ditches for growing salad greens (watercress, *Nasturtium officinale)*. In other parts of Brazil, as in

Venezuela, infection results from bathing and washing clothes in infested water.

Neither sand filtration nor aluminum sulfate clarification protects city water supplies from these cercariae, although chloramine, sodium hypochlorite, and gaseous chlorine kill them. Experimentally, diatomaceous silica filters remove them.

Pathogenesis, Pathology, and Symptomatology. The lesions produced in infection with *Schistosoma mansoni,* as well as the sequence of their development, are practically the same as those of *S. japonicum* infection. The smaller number of eggs produced by *S. mansoni* delays the production of granulomas around eggs that become lodged in the tissues, and the granulomata surrounding the eggs of *S. japonicum* are larger, more exudative and destructive, and lead to more residual scarring (von Lichtenberg et al., 1971). The organs and tissues most seriously involved are the colon and rectum, but eggs carried in the mesenteric current into the portal vessel filter out in the periportal tissues within the liver and set up pathologic processes leading to hepatic fibrosis. The vascular lesions, in particular those of the portal venous system, are also similar. As demonstrated in experimentally infected mice, these are destruction, obstruction, and distortion of the intrahepatic portal system and microvascular arteriovenous shunts in the intestine (Joky et al., 1978; Nagy et al., 1981). Successful treatment to remove the infection may be followed by regression of vascular lesions at least in part (Andrade and Brito, 1981).

As in schistosomiasis japonica, the symptoms in schistosomiasis mansoni are referable to three successive stages: (1) incubation, (2) egg deposition and extrusion, and (3) tissue proliferation and repair. The ultimate severity of the infection is dependent on the worm burden and host reactivity.

Incubation Period. The initial papular rash with pruritus, and at times vesiculation over the exposed skin, toward the end of the period usually is followed by diarrhea. The signs and symptoms of acute schistosomiasis mansoni resemble those of serum sickness (Hiatt et al., 1979). The most severe cases have sudden onset of high evening fever with diaphoresis, peripheral eosinophilia, hepatomegaly and splenomegaly, cough, abdominal pain, dysentery, weight loss, myalgia, arthralgia, and malaise.

Egg Deposition and Extrusion. This stage involves primarily the intestinal wall, the liver, and spleen. The initial egg extrusion, 5 to 7 weeks after exposure to infection, is accompanied by the classic schistosomiasis dysentery, with blood and mucus in the feces. The patient has abdominal pain and frequent stools, with lateral-spined eggs mixed with the feces or deposited in nonfecal stools consisting of blood, mucus, and exudates. As more and more tissue reaction takes place around eggs infiltrated in the intestinal wall, dysentery becomes less conspicuous, although it may readily be reactivated by exercise. The intestinal wall becomes inflamed, thickened, and fibrosed, with abscesses opening through the mucosa and with hyperemia of the peritoneal coat. As the condition progresses, it proceeds down the bowel, with development of papillomata and frequently with prolapse of the rectum. Meanwhile the mesenteric lymph nodes have become conspicuously enlarged by cellular reaction to infiltrated eggs. The liver and spleen are greatly enlarged and tender, the former due primarily to infiltration with eggs, the latter mostly on account of passive congestion. Eggs may at times escape into the tissues of the lungs, pancreas, spleen, kidneys, adrenals, myocardium, or even the spinal cord, and initiate pathologic processes in these organs with corresponding symptoms.

Tissue Proliferation and Repair. Fibrosis and thickening of the intestinal wall continue, with de-

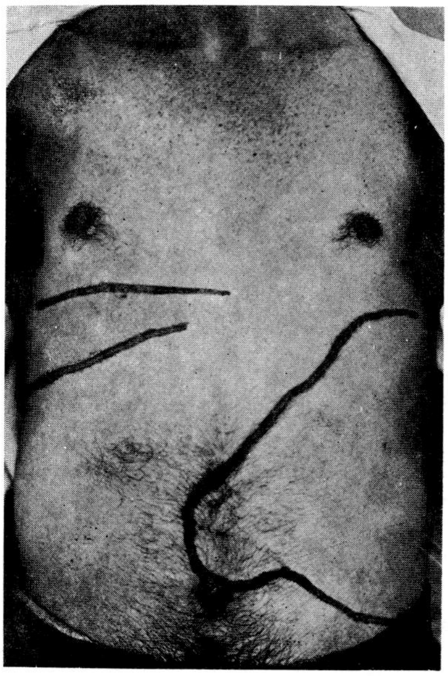

Fig. 27–16. *Schistosoma mansoni*. Early third (chronic) stage of infection. The liver has become somewhat cirrhosed, while the spleen is notably enlarged. Most of the pathologic disturbances are located in the liver and colon. A mild type of intestinal involvement is more frequent in schistosomiasis mansoni than in schistosomiasis japonica. (Courtesy of Dr. J.A. Pons.)

velopment of papillomata along the entire bowel characteristic of cases in hyperendemic areas in Africa, but seen rather infrequently at proctoscopic examination in Puerto Rico and Venezuela. Rapidly the tone of the wall becomes diminished, and fibrous constrictions encroach on the lumen. Although dysentery is not conspicuous during this stage, diarrhea assumes a more important role. Fissure in ano and piles due to *S. mansoni* infection have been observed (Warner, 1942). Fistulas frequently develop into the ischiorectal fossa, the perineum, the buttocks, or the urinary bladder. Ulceration and neoplasia of these areas are not rare. In perhaps 0.1% of infected individuals, lateral-spined eggs may reach the lumen of the bladder and be discharged in the urine. Meanwhile the liver has gradually become fibrosed and the spleen has become tremendously enlarged (Fig. 27–16). In the advanced stage, due to thickening of the large veins of the liver, to development of arteriovenous shunts in the liver and to microvascular arteriovenous shunts in the intestine (Nagy et al., 1981), and to pipe-stem periportal fibrosis, the development of ascites and esophageal varices, which may bleed profusely, frequently complicate the clinical picture.

Meira (1942) described pulmonary complications in schistosomiasis mansoni including a bronchopulmonary type simulating late tuberculosis, with endarteritis of pulmonary vessels, and cardiopulmonary form terminating in congestive heart failure. Schistosomal cor pulmonale is an obstructive vascular disease that involves other lung components to a minor degree. There is no diffuse fibrosis or vascular shunting. Typically it presents without cyanosis (Sadigursky and Andrade, 1982). Pons (1937), in Puerto Rico, recognized an intestinal type without essential involvement of the liver or spleen, but he did not find an exclusively hepatosplenic type. Despite high prevalence and intensity of infection, the degree of morbidity may be low in endemic areas (Pope et al., 1980).

Schistosomal myelopathy due to deposition of eggs in the spinal cord has been rarely demonstrated (Rosenbaum et al., 1972; Wood et al., 1976; Cohen et al., 1977; Queiroz et al., 1979), but it is probable that some cases are overlooked. Two spinal cord infections, discovered at autopsy of patients from Puerto Rico, were reported by Budzilovich et al. (1964). Eggs were randomly distributed in the spinal cord, indicating blood-borne transport from the original deposition sites, but in one focus in one cord some 25 eggs were found in linear arrangement along a vessel, suggesting the presence of an ovipositing worm. There were no inflammatory reactions to the eggs, hence no neurological signs, but elsewhere in the tissues typical reactions were observed. On the basis of these findings it seems likely that central nervous system deposition of eggs in schistosomiasis mansoni (without lesions) is more common than the few reports would suggest. *S. mansoni* may cause a papular dermatitis of the lower torso, genitalia, or perineum which may or may not be pruritic (MacDonald and Morrison, 1976; Wood et al., 1976; Walther, 1979). Occasionally, *S. mansoni* may cause lesions in the upper female genital tract (Arean, 1956), but gynecological schistosomiasis is more frequently caused by *S. haematobium*.

Newton et al. (1968) recovered a living male *Schistosoma mansoni* from the anterior chamber of the right eye of a Puerto Rican patient who had lived in New York City for more than 8 years.

Cheever (1968) described a technique for the quantitative recovery of *S. mansoni* from human cadavers; analysis of the data from 197 autopsies indicated that nearly all worm pairs were recovered. Evidence of heavy infection was most common in cases of Symmers' clay pipe-stem fibrosis of the liver. When present, schistosomal pulmonary arteritis and cor pulmonale were associated with Symmers' fibrosis. In chronic hepatosplenic schistosomiasis, immune complexes may be deposited in the glomerular capillaries and basement membrane leading to nephrosclerosis and renal failure (Falcão and Gould, 1975).

Diagnosis. During the prodromal phase toward the end of the prepatent period, specific diagnosis is not practicable except possibly with serologic tests. Acute fevers of other etiologies must be ruled out, including serum sickness, viral hepatitis, and typhoid fever. As soon as egg extrusion has begun, diagnosis is readily accomplished by recovery of the characteristic lateral-spined eggs (Fig. 27–15) in the stool. Clinically this infection requires differentiation in its early stages from food poisoning, trichinosis, amebiasis, and other involvements of the bowel, liver, and spleen. Fissure in ano and piles of *S. mansoni* origin must be considered. In its advanced stage Banti's syndrome must be ruled out. In patients with pulmonary complications resulting from schistosomiasis lesions, diagnosis is based on the clinical picture, visualization of the lesions by roentgen ray, and at times the recovery of eggs of *Schistosoma mansoni* in the sputum (Meira, 1942).

In dysenteric cases concentration methods of fecal examination are usually unnecessary. In chronic cases the eggs become increasingly less frequent and concentration techniques must be used since eggs may not be found in direct fecal films or in Kato thick fecal films. This process may be carried out by sedimentation of the screened stool in 0.5% glycerinated water, or ether concentration techniques. In patients with chronic schistosomiasis mansoni, rectal scrapings, aspirates, or biopsy via the proctoscope may be more rewarding than fecal examination (Fig. 27–17). When eggs cannot be isolated for examination, seroimmunodiagnostic tests can be useful (Hillyer et al., 1980). The use of monoclonal antibodies facilitates the detection of schistosomal antigen in the blood (Zodda et al., 1983). Demonstration of antigenemia may be useful not only in diagnosis, but also in evaluating the efficacy of treatment (Weltman, 1982). For community surveys the Kato technique is satisfactory (Jordan et al., 1981). Occasionally eggs may be found in urine samples (Pugh et al., 1980). Smithers (1960) reported a tissue-digestion technique for recovery of *S. mansoni* egg in a viable condition from liver and intestinal wall. Serologic and immunologic tests for schistosomiasis mansoni are discussed in Chapter 48.

Fig. 27–17. *Schistosoma mansoni.* Eggs encapsulated in granulomas in biopsy specimen from rectum viewed under pressure of coverglass. Unstained. (× ca. 50.) (By P.C. Beaver.)

Treatment. Chemotherapy should be carried out only if there is a positive diagnosis based on demonstration of eggs in feces or rectal snips. As in schistosomiasis japonica, tartar emetic is an effective treatment, but it is difficult to administer and relatively poorly tolerated; it is not, therefore, the drug of choice. Stibophen (Fuadin) is given intramuscularly in a course of 40 to 75 ml of a 6.3% solution in 10 to 16 injections. Niridazole (CIBA 32,644Ba or Ambilhar) is effective against schistosomiasis mansoni. The standard course of treatment calls for a dosage of 25 mg/kg per day for 5 to 10 days. Although side effects are less common than with other schistosomicidal drugs, niridazole causes central nervous system effects ranging from abnormalities in the electroencephalogram to epileptiform convulsions and psychosis, expecially in patients with decompensated liver disease. A newer drug of great interest is a nitroquinoline, oxamniquine, which is given orally. The optimum dosage has not been established. A recommended dosage is 15 mg/kg body weight in a single dose (Abramowicz, 1982). Since niridazole is more effective in the treatment of childhood schistosomiasis mansoni, it is preferred over oxamniquine for children. On the other hand, oxamniquine is preferred for adults. Side effects of niridazole are more significant (Kilpatrick et al., 1982). Abnormal electroencephalograms caused by oxamniquine have been reported (Krajden et al., 1982). Treatment with praziquantel is safe and effective in a single dose of 40 mg/kg. This dose is less than the amount required in the treatment of schistosomiasis japonica *(vide supra)*. Side effects, though frequent, are mild and transient (Saif and Abdel-Meguid, 1981). Oltipraz is a newer anthelmintic reported to be effective against schistosomiasis and well tolerated (Bella et al., 1982).

Prognosis. Prognosis is good to fair in early cases or in lightly infected patients who submit to specific therapy and are not re-exposed to infection. Prognosis is poor in chronic cases, particularly where fibrosis of the liver has become advanced.

Prevention and Control. The problem of control is essentially the same as for *S. japonicum*. However, it is considerably less likely that the disease will be spread to man from reservoir hosts.

An overall practical solution remains to be found. Schistosomiasis control projects in Egypt, Brazil, Ghana, Philippines, Zimbabwe, Tanzania, Venezuela, Saint Lucia, and elsewhere have demonstrated that the application of molluscicides under certain conditions can effectively reduce or completely interrupt transmission. However, snail control alone is seldom effective if not continued for long periods. It cannot be recommended on a wide scale because such a program not only is expensive but becomes objectionable from the standpoint of environmental modification through toxicity to other organisms (McCullough et al., 1980).

Biological control of snail hosts is attractive in concept but has had few successes in actual field application. The present status and future prospects of biological control of snail intermediate hosts of human schistosomes were reviewed by McCullough (1981). He noted that three general approaches to biological control of snails have been made or considered: the introduction and management of (1) predators, parasites, or pathogens; (2) interlarval trematode antagonists or predators; and (3) intramolluscan competition. Certain species of fish and insects eat snails, certain parasites tend to sterilize snails by destroying the reproductive organs, and certain micropathogens either kill the snail or attack the developmental stages of the schistosome. None of the aforementioned has offered promise of successful application in field trials. The possibility of mass cultivation of trematode parasites to be introduced continuously into the environment, to infect the same snail as the schistosome which it then displaces, has been advocated (Lie, 1971; Rysavy et al., 1974). That approach cannot be seriously considered, however, because of the relative vastness of the areas involved, as compared with those of laboratory demonstrations. A significant effect on schistosomiasis transmission is unattainable by such means even in a limited geographic area (Combes, 1982). The introduction of a large and strongly competitive snail to displace a vector snail has been used with apparent success in certain types of habitats in Puerto Rico. When introduced into ponds, *Marisa cornuarietis* displaced or effectively reduced the populations of *Biomphalaria glabrata,* and thus far there has been no evident adverse environmental effects. Similarly, in Brazil, it appears possible to drastically reduce if not eliminate *B. glabrata* by introducing other species of the same genus *(B. straminea* and *B. tenagophilic)* which serve poorly, if at all, as vectors of the schistosome. In a small man-made lake about 100 meters in diameter in Tanzania, three species of pulmonate snail *(Biomphalaria pfeifferi, Bulinus tropicus,* and *Lymnaea natalensis)* were eliminated within about 2 years after introduction of a *Marisa cornuarietis* colony (Nguma et al., 1982).

Treatment of patients and public health programs providing public baths, laundry tanks, and sanitary drainage canals greatly reduce exposure and provide a moderate amount of control. An intensive and long-sustained schistosomiasis control program in Saint Lucia has shown that whereas snail control by chemicals and improved drainage is expensive and difficult, an intensive case-finding campaign, followed by treatment of all infected persons, and the provision of individual household water supplies can be effective in halting transmission (Barnish, 1982; Jordan et al., 1982 a,b). Recent advances in chemotherapy make it clear that drug treatment will play an increasingly important role in control of both disease and transmission (World Health Organization, 1980; Webbe, 1981). Treated patients of all ages, children especially, are susceptible to reinfection (DeLima et al., 1980).

Schistosoma haematobium (Bilharz, 1852) Weinland, 1858

(The vesical blood fluke, causing vesical schistosomiasis or schistosomiasis haematobia, vesical or urinary bilharziasis, schistosomal hematuria)

Synonyms. *Distoma haematobium* Bilharz, 1852; *Bilharzia haematobia* (Bilharz, 1852) Cobbold, 1859; *Bilharzia capensis* Harley, 1864.

Historical and Geographical Notes. *Schistosoma haematobium* infection has been common in the Nile Valley for several millennia. Many of the soldiers of Napoleon suffered from its effects during the Egyptian campaign. The worms were first recovered in 1851 by Bilharz from the mesenteric veins of a native of Cairo. Shortly thereafter he demonstrated them to be the cause of hematuria in native laborers who were discharging terminal-spined eggs of this fluke in the urine. In 1864, Harley reported the organism from South Africa under the name *Bilharzia capensis.* Although Harley and Cobbold both believed some mollusc to be a necessary intermediate host, and although several workers concluded that the skin was the portal of entry for the infective stage of this organism, Looss (1894–1914) failed to elucidate the life cycle and concluded that there was no intramolluscan phase. Meanwhile discovery of lateral-spined eggs, usually in the feces, suggested to Manson (1893) that there were two species of human schistosomes in Egypt, and Sambon (1907) designated the second species with lateral-spined eggs as *Schistosoma mansoni.* Definite proof of the existence of two separate species was provided by Leiper (1915–1918), who showed (1) that one was typically intestinal, with lateral-spined eggs, and one typically vesical, with terminal-spined eggs; (2) that the adults were morphologically distinguishable;

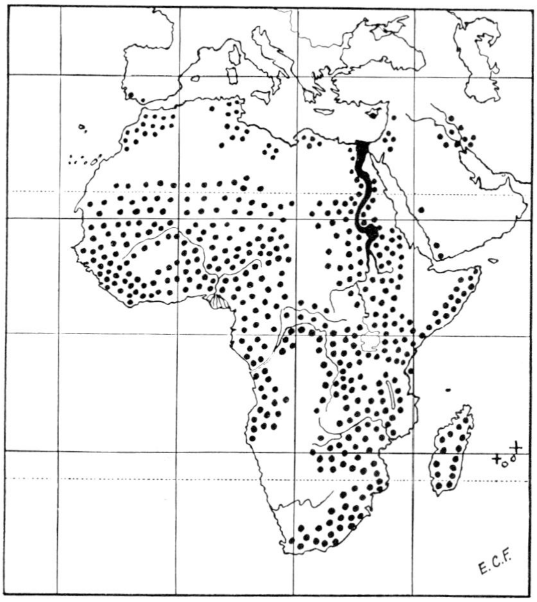

Fig. 27–18. *Schistosoma haematobium.* Map showing endemic foci. The solid area in the Nile Valley indicates extensive hyperendemicity. A focus in southwestern India is not shown. (From Faust, E.C. 1949. *Human Helminthology,* 3rd ed. Philadelphia, Lea & Febiger.)

(3) that they required a different molluscan intermediate host; and (4) that infection was acquired by the cutaneous route. The use of tartar emetic as a specific therapeutic in vesical schistosomiasis was first advocated by McDonagh in 1918 and demonstrated on a large scale by Christopherson in the same year.

While the original focus of vesical schistosomiasis was probably the headwaters of the Nile Valley, today there are extensive endemic areas, including the greater part of Africa (the Sudan, Ethiopia, east coast from Somaliland to the Cape; vast areas in Central Africa; West Africa from Lake Chad and Nigeria as far south as Angola; North Africa from Egypt to Morocco); the southern tips of Europe (southern Portugal, and the isle of Cyprus); western Asia (Lebanon, northern Syria, Arabia, Iraq, and Iran); and the islands on the east coast of Africa (Malagasy, Mauritius, and Réunion) (Fig. 27–18). An endemic focus in India, about 200 kilometers south of Bombay, was discovered by Gadgil and Shah in 1952 (Gadgil, 1963). Among a population of 1200 people, 250 were passing eggs in the urine. Efforts were made to eradicate the focus but in 1979 examination of 607 urine samples revealed eggs in 4, including 1 from an 8-year-old child (Gaitonde et al., 1981; Sathe et al., 1981).

The prevalence of schistosomiasis haematobia varies widely in different areas, even in different localities within highly endemic territory (namely, 11 to 95% in lower Egypt, 83% in villagers of all ages in middle Somalia, 33 to 94% in different districts of Tanzania, 85% in male Bantu school children, 0 to 60% in Zambia, and 60% in children and 21% in adults in Angola). With extensive irrigation projects under way in most endemic areas, the infection will undoubtedly spread and the prevalence increase in these areas in spite of control activities.

Morphology, Biology, and Life Cycle. The adult *S. haematobium* (Fig. 27–19) lives primarily in the vesical and pelvic plexuses of the venous circulation, less commonly in the portal blood-

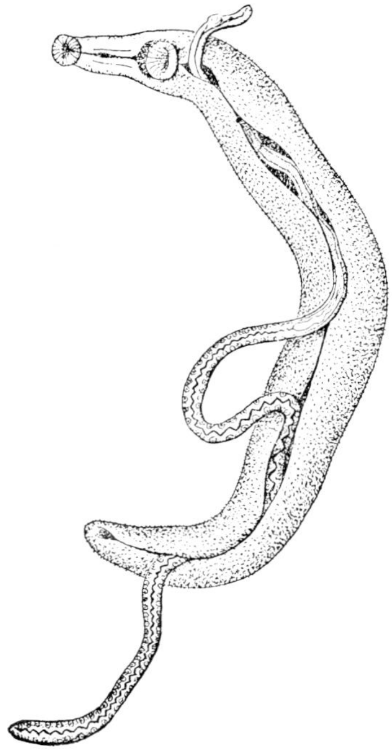

Fig. 27–19. *Schistosoma haematobium.* Female in the gynecophoral canal of the male. (× 12.) (Adapted by E.C. Faust from Looss, A. 1896. Recherches sur la faune parasitaire de l'Egypte. Premiere partie. Mem. Inst. Egypt, *3*:1–252.)

stream, and rarely in other venules. The male is the shorter, stouter organism, measuring 10 to 15 mm in length by 0.8 to 1 mm in breadth. It is covered with minute integumentary tuberculations and has two muscular suckers, of which the ventral one is the larger. Behind the ventral sucker the body of the male is infolded ventrad all the way to the caudal extremity to form the *gynecophoral canal,* in which the female is held during copulation and oviposition. In place of a muscular pharynx there is a conspicuous bulbous cluster of glands around the esophagus. The paired intestinal ceca extend to the equatorial plane and then fuse to form a single serpentine trunk that ends blindly near the caudal extremity. Situated in a group just behind the ventral sucker are 4 (or 5) testes, each with its vas efferens leading into the common vas deferens. This structure dilates to form a seminal vesicle which, in turn, opens ventrad through the genital pore just behind the ventral sucker. There is no prostate gland or penial organ. The female is long and slender, measuring about 20 by 0.25 mm. The suckers are small, subequal, and not conspicuously muscular. The integumentary tuberculations are usually confined to the extremities of the worm. The digestive tract is similar to that of the male.

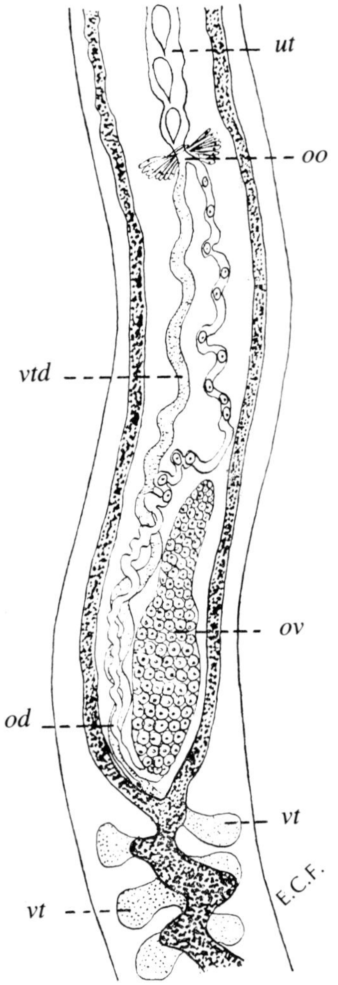

Fig. 27–20. *Schistosoma haematobium*. Reproductive organs of the female. *od*, Oviduct. *oo*, Ootype. *ov*, Ovary. *ut*, Inner end of uterus with eggs. *vt*, Vitellaria. *vtd*, Common vitelline duct. (From Faust, E.C. 1939. *Human Helminthology*, 2nd ed. Philadelphia, Lea & Febiger.)

The female reproductive organs are illustrated in Figure 27–20, except for the outer end of the uterus (and for a seminal receptacle, which has not been described for this species but is probably present at the posterior border of the ovary, near the origin of the oviduct). Naked ova leave the ovary *(ov)* at its posterior end and are passed through the oviduct *(od)* to the ootype *(oo)*, where they are fertilized, receive the vitelline material and the shell, and are then passed into the uterus *(ut)*. From 20 to 100 eggs develop *in utero* at one time.

Oviposition usually occurs in the small venules of the vesical and pelvic plexuses as well as in rectal venules. It occurs less commonly in pulmonary arterioles, the mesenteric-portal system, and in ectopic sites such as the prostate gland and subcutaneous tissues of the groin and scrotum (Barlow and Meleney, 1949), periumbilical cutaneous tissue (Hull and Hay, 1979), conjunctiva (Sparrow, 1966), and lacrimal gland (Jakobiec et al., 1977). These eggs are partly mature when laid, and each usually contains a motile miracidium by the time it is evacuated in the urine or feces. Due to pressure within the venule, piercing action of the spine, and probably also lytic substances that ooze out through the porous eggshell, the egg breaks through the wall of the vessel and, passing through the mucosa, escapes into the lumen, together with extravasated blood. The eggs are then discharged in the urine, especially at the end of micturition, or in dysenteric stools. They have a light yellowish-brown transparent shell with a distinct terminal spine and measure 112 to 170 μm by 40 to 70 μm (Fig. 27–21). Scanning electron micrographs show that the shell is porous and covered with microspines that are smaller and more rounded at the tips than are those of *S. mansoni* eggs (Sakamoto and Ishii, 1976).

On dilution of the urine or stool with 10 or more parts of water, the viable eggs soon hatch and the miracidia become free-swimming. On contact with an appropriate species of snail, they infect the snail and in the course of 4 to 8 weeks pass through first- and second-generation sporocysts and finally a generation of fork-tailed cercariae, which break out of the snail daily in swarms over a period of several weeks or months.

The snail hosts of *S. haematobium,* according to Malek (1980), are *Bulinus (Bulinus) truncatus* in Morocco, Algeria, Tunisia, Libya, Egypt, Sudan, Turkey, Syria, Saudia Arabia, Yemen, Iraq, and Iran; a subspecies, *B. (B.) truncatus rohlfsi,* in Cameroons, Ghana, Gambia, and Mauritania; *B. (B.) guernei* in Senegal; *B. (B.) senegalensis* in parts of Senegal and Gambia; *B. (Physopsis) globosus* in parts of Gambia and throughout most of the remaining regions of Africa (Portuguese Guinea, Sierra Leone, Liberia, Cameroons, Ghana, Nigeria, Sudan, Angola, South Africa, Zimbabwe, Tanzania, Uganda and Kenya); and *B. (Ph.) africanus* in Mozambique. The molluscan host in India is a limpet, *Ferrisia tenuis,* and in Portugal a snail, *Planorbarius metidjensis*.

The cercaria has an elongated ovoidal body (140 to 240 μm by 57 to 100 μm) and a tail consisting of a trunk (175 to 250 μm by 35 to 50 μm) and a pair of caudal furci (60 to 100 μm long). There is a large, blind, urn-shaped anterior sucker (56 to

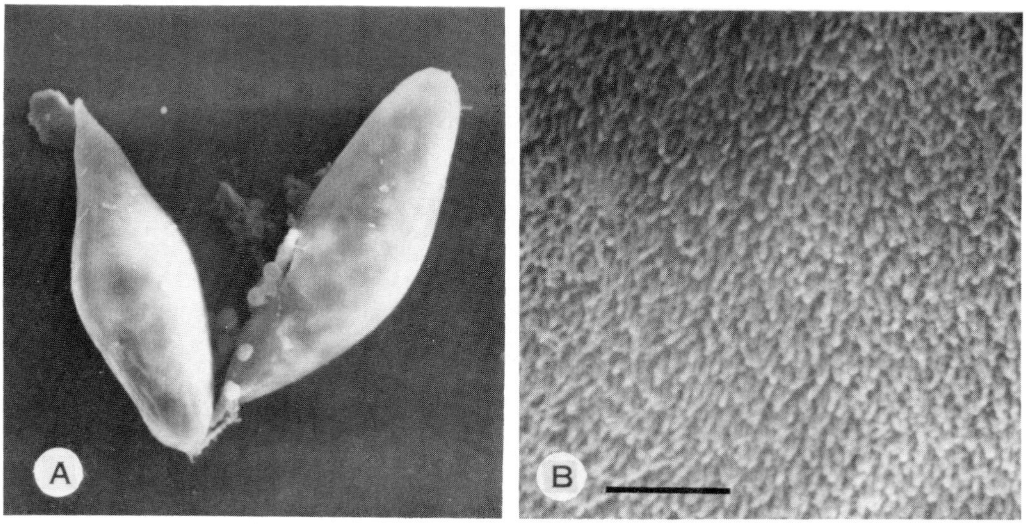

Fig. 27–21. *Schistosoma haematobium*. Scanning electron micrographs of eggs. *A*, Showing shape and surface texture. (× 400.) *B*, Surface of egg, covered with microvilli. Scale is 1 μm. (From Sakamoto, K., and Ishii, Y. 1976. Fine structure of schistosome eggs as seen through the scanning electron microscope. Am. J. Trop. Med. Hyg., *25*:841–844.)

60 μm by 39 to 64 μm) and a smaller ventral sucker, which is completely covered with minute spines. The great mass of the body is filled with five pairs of lytic or penetration glands, made up of two anterior pairs with granular oxyphilic material and three posterior pairs with more homogenous basophilic material. Individual ducts from these glands open in two bundles through the anterodorsal margin of the oral sucker.

On leaving the snail, the cercaria swims vigorously, at times sinking to the bottom, then swimming back to the surface film, where it may become attached temporarily. When people bathe or wade in cercaria-infested water, the cercariae come in contact with the skin and, as the water evaporates, enter the skin, casting off their tails. Their free-living existence is limited at most to 3 days, usually 24 hours or less, during which time they are unable to feed.

Skin penetration is effected by forceful introduction of the anterior end, accompanied by digestive action of penetration gland secretions. Entry below the epidermal surface is accomplished in less than 30 minutes. Usually within 1 or 2 days the larvae have entered peripheral venules, are transported to the right heart, and pass into pulmonary vessels. Some are temporarily sidetracked in peripheral lymph nodes but usually work their way through these into blood vessels. Several days are required for essential development of the larvae in preparation for migration to the liver. Whether migration is through the systemic circulation or through the right heart and caval veins to the hepatic veins and liver is unknown.

In the intrahepatic portal vessels the larvae begin to feed and grow, become sexually differentiated, and upon becoming adolescent about 20 days after skin penetration, migrate against incoming blood. They pass through inferior mesenteric veins, at times lodging and maturing in rectal veins, but commonly migrating through the hemorrhoidal and pudendal veins into the vesical and pelvic plexuses, where they arrive within 3 months after initial exposure of the skin. Some time before, or upon, arrival in vesical or pelvic venules, the worms become sexually mature, mate, and the fertile females begin laying eggs. Thus, the prepatent period is usually less than 3 months (average in man, 10 to 12 weeks). While the skin is the usual portal of entry into the body, the buccal mucosa is also suitable for reaching the venous circulation. However, the cercaria cannot survive passage through the normal stomach. In addition to man, the usual host, a monkey *(Cercocebus torquatus atys)*, baboons *(Papio doguera* and *P. rhodesiae)*, and the chimpanzee *(Pan satyrus)* have been found naturally infected. Experimentally, a number of feral rodents have been shown to be susceptible. The baboon, *Papio anubis*, is a suitable laboratory host (Jordan et al., 1967), and successful experimental infections have been produced in the monkey, *Cerco-*

cebus aethiops, and the chimpanzee, *Pan satyrus* (Vogel, 1967).

Epidemiology. Man is probably the only significant maintenance host of *S. haematobium*. Infection results from contact with water into which the fork-tailed cercariae of the schistosome have been discharged from infected snails. In endemic foci, as in parts of Egypt, practically the entire population is infected. The disease is increasing in distribution as snails are carried from infected foci into new irrigation projects or as appropriate snails in previously uninfected areas become infected from eggs discharged into the streams, pool, or canals by infected individuals. Farmers, women washing clothes in the streams, and children bathing or wading in the water are all subject to exposure. In most endemic foci children are more frequently and more heavily exposed and infected than are adults. In endemic foci of certain countries religious practices tend to increase pollution of the water and encourage exposure to infections.

Pathogenesis, Pathology, and Symptomatology. On contact with human skin the cercaria of *S. haematobium* burrows down to cutaneous blood vessels. Essentially no local damage to host tissue results. No local cellular reaction occurs on arrival from the skin via the right heart in the pulmonary vessels. Some break out with accompanying hemorrhage, while a leukocytic infiltration (predominantly eosinophilic) develops around the young worms. About the fifth day after skin penetration the young worms begin reaching the portal vein and liver. Approximately 3 weeks after entry into the skin the maturing worms begin migrating out of the portal vessel via the inferior mesenteric vein, hemorrhoidals, and pudendals into the vesical plexus. At 10 to 12 weeks of infection females begin to lay terminal-spined eggs into the venules. Some may take up residence in the rectal veins, and others may be carried in caval blood into pulmonary arterioles, where they mature. Groups of worms tend to occupy adjacent vessels resulting in an uneven deposition of eggs. In heavy infections they produce dense, localized deposits of eggs.

With discharge of eggs into the venules another type of lesion develops, namely, escape of the eggs through perivascular tissues into the lumen of the organ, accompanied by blood, exudates, and necrotic tissue cells. In vesical schistosomiasis this damage occurs primarily in the wall of the urinary bladder, secondarily in the distal portion of the ureter and in adjacent urinary and genital organs or the rectum, and later in the lungs and more remote organs and tissues.

In increasing numbers eggs of the parasite that have infiltrated in the tissues are retained in the tissues and become centers for pseudoabscess formation. Abscesses near the lumen of the bladder (or other organs) may break through and discharge their eggs, but others in increasing numbers proceed to fibrosis, with resultant miliary pseudotubercle formation. This eventually causes fibrosis of the entire organ, with mineral deposits in the mucosa and on its surface.

Involvement of the lower ureteral segments and trigone causes obstructive uropathy with consequent bacterial infection proximal to the obstruction, chronic pyelonephritis, and hydronephrosis (Smith et al., 1974).

Thus, the pathogenic effects of *S. haematobium* consist of (1) generalized and localized reaction to metabolites of growing and mature worms, (2) trauma, with hemorrhage, as eggs escape from the venules, (3) pseudoabscess and pseudotubercle formation around eggs lodged in perivascular tissues, and (4) obstructive uropathy.

The clinical aspects of this infection are divided into three periods: (1) *incubation,* (2) *egg deposition* and *extrusion,* and (3) *tissue proliferation and repair.*

Incubation Period. The prepatent period covers the time from cutaneous exposure to cercariae until eggs are first laid by the females, a period of about 10 to 12 weeks. The earliest manifestations are associated with the entry of cercariae into the skin and consist of a nettling irritation that varies in severity depending on intensity of exposure and on sensitiveness of the individual. Within a few hours, minute hemorrhages may be seen at the sites of invasion of blood vessels, but these disappear in a day or two, usually without local inflammation. Commonly no further indications of the infection develop until near the end of the prepatent period, when there is either a gradual or a sudden onset of toxic symptoms, consisting of anorexia, headache, malaise, generalized pain in the back and the limbs, and late afternoon fever, followed by rigors and night sweating. There is frequently a pronounced urticaria, at first on the limbs, later more generalized in distribution. The blood shows a leukocytosis, frequently with remarkable eosinophilia. The abdomen may be enlarged and tender, the liver and spleen hypertrophic. At times there is precordial pain and breathing becomes labored. In endemic populations where reinfection is common, this uncomplicated first stage is not usually seen.

Egg Deposition and Extrusion. The first appearance of eggs in the urine usually occurs in painless passage of a small amount of blood at the end of micturition. This semilatent condition may continue for months or years without subjective symptoms, but eventually there is an increasingly frequent burning sensation at the time of urination and between periods, and increased desire to urinate. Some patients may experience prodromes consisting of headache, backache, lassitude, late afternoon fever, and frequent urge to urinate, accompanied or followed by dull pains in the loins and suprapubic region, and bladder colic. Cystoscopic examination characteristically reveals hyperplasia and inflammation of the mucous membranes of the lower segment of the bladder, and initial papillomatous growths, with concretions of uric acid and oxalate crystals around the eggs as nuclei in the mucosa of the bladder. Likewise, inflammation and induration of the prostatic region may have developed as a result of infiltration of eggs. Meanwhile, the perivascular tissues of these areas have also become infiltrated with eggs surrounded by leukocytes, eosinophils, and giant-cells, leading to microabscess and pseudotubercle formation, and the trigonum vesicae may be inflamed, thickened, and superficially full of gritty particles (i.e., eggs). Most of the eggs are massed throughout the submucosa and muscularis of the organs (Fig. 27–22), and some remain in the venules, producing a proliferation of their endothelium, at times causing thrombosis. Although most of the eggs that reach the lumen of the bladder are viable, some undergo calcification. The bladder wall may be sloughing, or plaques of phosphatic deposits may be scaling off into the lumen.

Tissue Proliferation and Repair. Egg deposition and extrusion are followed by inflammation and pseudotubercle formation around eggs accompanied by (1) a generalized hyperplasia and fibrosis of the bladder wall and lower ureters with development of "sandy patches" in and on the bladder

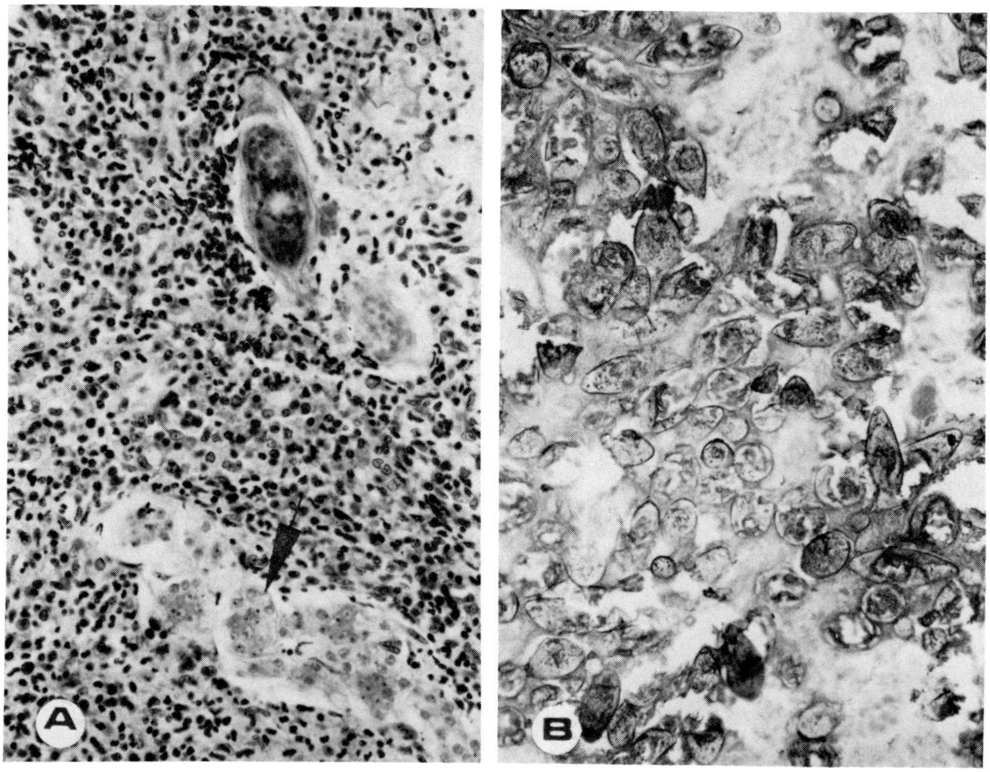

Fig. 27–22. *Schistosoma haematobium*, eggs in histological section of the bladder wall. *A*, Biopsy from a man with early infection showing one egg with a well-developed shell and miracidium, and several eggs that were freshly deposited with the shell only vaguely evident and the miracidium immature (arrow). (× 250.) *B*, From a man with late, heavy infection showing numerous calcified eggs in a matrix of fibrous tissue. (× 100.) (By P.C. Beaver. *A*, Preparation courtesy of Dr. G.C. Carrera. *B*, Preparation courtesy of Dr. J.F.B. Edeson and Dr. T.J. Danaraj.)

wall, as well as (2) secondary infection. The symptoms are those of chronic cystitis. Cystoscopic examination becomes increasingly difficult. Calculi in the bladder, originally formed of oxalates and uric acid crystals around eggs, clots of blood, or sloughed portions of papillomata, now increase in size as a result of phosphatic deposits. The urethra frequently becomes occluded as a result of the thickening of its wall by generalized hyperplasia or nodular swellings or from plugs of purulent debris that are being discharged from the bladder. The lower portion of the ureters are frequently involved, and the process may even ascend to the pelvis of the kidneys. Obstructive uropathy causes retrograde infection, hydronephrosis, and eventual renal failure. The lesion in the male may extend into the penis, with fibrosis of the sheath and elephantiasis of the organ caused by blockage of the scrotal lymphatics. Pyogenic organisms may invade the periurethral or perivesical tissues, causing abscesses that open into the bladder or produce fistulas into the rectum or scrotum. Purulent discharges may resemble those of gonorrhea. In the female the lesions are usually less pronounced, although the cervix, vagina, and vulva may be affected. Salpingitis is less common and uterine involvment is rare (Wright et al., 1982). The symptoms of this stage are those of extreme weakness, emaciation, and increased pain on micturition, which becomes more frequent but less in amount, finally with only small uncontrolled passages of blood and pus. With such profound involvement of the urogenital system the patient soon succumbs.

Ferguson in 1911 demonstrated that many of these patients have malignant growths of the bladder, usually the posterior wall, but at times the entire organ. The role of *S. haematobium* in the etiology of bladder cancer remains an open question (Lucas, 1982). In addition to involving the urogenital organs, eggs may be carried into the inferior mesenteric veins and produce schistosomal appendicitis or involvement of the cecum, colon, and particularly the rectum. In such cases the eggs may be passed in the feces. Small numbers of eggs are constantly swept back through the portal vessels into the liver, where abscess and pseudotubercle formation with consequent fibrosis occur in the periportal tissues. Likewise, eggs, and even worms, are carried through the hypogastric and common iliac veins, inferior vena cava, and the right heart to the lungs, where they filter out.

Cutaneous papular lesions about the lower torso and pelvis occur more frequently than with schistosomiasis mansoni.

Diagnosis. Specific diagnosis can be made only (1) after eggs are discharged into the lumen of the bladder and appear in the urine, or (2) after eggs are discharged into the lumen of the bowel and are present in the feces, or (3) from aspirated or biopsy material obtained through the cystoscope or proctoscope and examined microscopically for eggs after compression or sectioning. Hematuria in endemic areas is a suggestive sign. Digital examination through the rectum and cystoscopic examination are aids in clinical diagnosis. Specific diagnosis consists of recovery of the typical terminal-spined eggs (Fig. 27–23) in the urine, usually in the blood and pus discharged at the end of micturition or evacuated in the feces. Eggs in massive numbers may be discharged in the seminal fluid (Barlow and Meleney, 1949; Lembeli and Venkataramaiah, 1981). Immunodiagnostic tests are generally only group specific (Hillyer et al., 1980), and they are most useful in cases with symptoms during the late prepatent period. The disease must be differentiated from renal calculi, acute nephritis, benign and malignant growths of the bladder, hemoglobinuria, oxaluria, filarial elephantiasis of the penis, gonorrhea, tuberculosis of the urinary tract, acute and chronic bacterial cystitis, or other disease involving the colon and rectum.

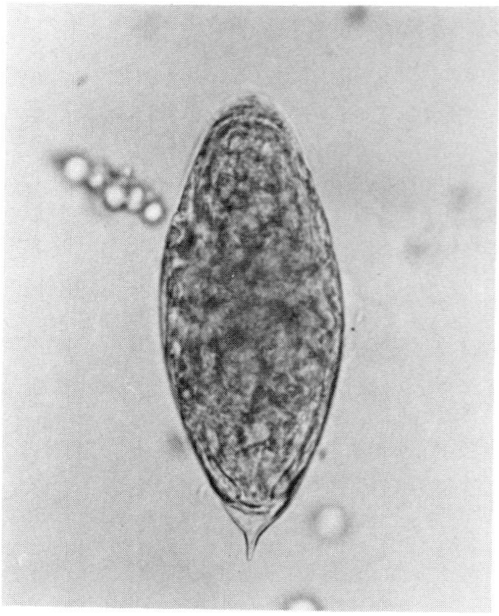

Fig. 27–23. *Schistosoma haematobium*. Egg from urine, containing fully developed miracidium with its anterior end at the egg's spine. Miracidia may face either pole of the egg. (× 400.) (By P.C. Beaver.)

Studies on the gibbon have shown that passage of scanty eggs in the urine is not necessarily an indication of a benign process. Abundant eggs may be trapped in the tissues causing profound damage when few eggs are excreted (Cheever and Duvall, 1981). Such variations in host reaction to the parasite may be due to an immune mechanism and may explain the differences between the course of the disease in visitors to an endemic area as compared with the native population.

Treatment. The treatment of schistosomiasis haematobia has changed greatly during the past decade with the advent of new drugs that are more consistently effective, less toxic, and administered orally rather than parenterally.

Metrifonate (Bilarcil) is an organophosphorus cholinesterase inhibitor. While not effective against schistosomiasis mansoni nor japonica, it is the leading drug for the treatment of vesical schistosomiasis because of its low cost, efficacy, and ready acceptance by the patients. The dosage is 5 to 15 mg per kilogram of body weight given at 2-week intervals for 3 doses (Jewsbury et al., 1977). Unfortunately the prolonged duration of the course of treatment requiring 4 weeks renders the regimen unsatisfactory for mass treatment. Teesdale et al. (1980) have reported that a single combined dose of metrifonate 12.5 mg per kilogram plus niridazole 25 mg per kilogram of body weight is as effective as the 3-dose course of metrifonate alone. Prophylaxis has been carried out in a highly endemic area with a dosage of 7.5 mg per kilogram of body weight given once every 4 weeks.*

Praziquantel is effective against vesical schistosomiasis as well as against schistosomiasis due to the other species of *Schistosoma*. For dosage see under *Schistosoma mansoni* (p. 433) (McMahon and Kolstrup, 1979; Davis et al., 1979).

Oxamniquine is not effective against vesical schistosomiasis, probably because *S. haematobium*, when under the influence of the drug, does not migrate to the liver.

In patients with secondary bacterial infection, antibiotics may be advantageously used locally or systemically, while corticosteroids may be useful in controlling inflammatory reaction to toxic byproducts of the eggs impacted in the tissues. In neoplasms or fistulas, surgical intervention is indicated. Advanced cases are not likely to benefit from specific therapy owing to the profound modification of the involved tissues.

*This drug is available in the United States from the Parasitic Disease Division, National Center for Disease Control, Atlanta, Georgia, 30333.

Prognosis. The prognosis is excellent to fair in uncomplicated cases with adequate treatment, but is frequently grave in untreated cases.

Prevention and Control. Attempts to control this infection in Egypt and elsewhere by extensive chemotherapeutic programs have not been successful, due to constant re-exposure. However, newer drugs that are effective in a single dose given by mouth and are acceptable as having low toxicity may prove to be useful in community-wide control programs (Davis et al., 1979; McMahon and Kolstrup, 1979).

Although the number of snails is greatly reduced by periodic drying of the irrigation canals in Egypt, some that find refuge in the crevices of mud or under vegetation survive. However, they have usually lost their infection during the period and require reinfection from human sources before they are again a serious menace; they soon produce susceptible progeny.

A three-fold program is needed here as in the other human schistosomiases: (1) eradication of the molluscan hosts, for at least one transmission cycle, by water management and molluscicidal campaigns adapted to each particular endemic area and the mores of its inhabitants; (2) sanitary improvement of the environment to reduce the pollution of snail habitats by schistosome eggs discharged in human urine and feces that provide the source of infection for the snails; and (3) effective treatment of infected persons, especially carriers, to lessen the likelihood of water pollution.

The only permanent advance in control of the disease may result from a long-time educational effort to obtain proper disposal of human excreta, since man, and not reservoir hosts, is primarily responsible for the perpetuation of the life cycle of this parasite.

OCCASIONAL AND POTENTIAL SCHISTOSOME PARASITES OF MAN

Several species of *Schistosoma*, other than *S. japonicum*, *S. mansoni*, *S. haematobium*, and *S. mekongi*, have been reported as human parasites. Related species that parasitize other mammalian hosts may produce infection in man. Though unimportant epidemiologically, zoonotic infections in individual cases are occasionally clinically important and are generally of great interest.

Schistosoma intercalatum Fisher, 1934

The name *S. intercalatum* was created by Fisher (1934) for

a schistosome found in several hundred cases of human infection in Zaire, in which terminal-spined eggs (140 to 240 μm by 50 to 85 μm) were recovered exclusively in the stools. Additional cases have been reported from Gabon (Pautrizel et al., 1966; Becquet, 1967) and from the Cameroon and Central Africa Republic (Deschiens and Delas, 1968). Also reported are cases of Europeans and Americans who acquired the infection in Africa (Wolfe, 1974). Symptoms and postmortem studies indicated that the worms, their eggs, and tissue reactions were confined to the mesenteric-portal system and the vesical venous circulation was not involved. Rectal biopsy may demonstrate eggs of *S. intercalatum* when fecal examinations are negative. *Bulinus (Physopsis) globosus, B. (P.) africanus,* and in the Cameroon *B. (P.) forskalii* are the known intermediate hosts of this schistosome. *S. intercalatum* morphologically resembles *S. mattheei* and may be merely a strain of this species.

Schistosoma bovis
(Sonsino, 1876) Blanchard, 1895

Schistosoma bovis was first described in cattle in Egypt. This member of the *S. haematobium* complex is a common parasite of the mesenteric-portal system of cattle, sheep, goats, and equines in southern Europe, Israel, Africa, and Iraq, and has been reported from baboons in Rhodesia (Blackie, 1932). The eggs (230 to 380 μm by 70 to 90 μm) are longer and much narrower than those of *S. haematobium* and have an equatorial bulge; also, the parent worms are larger and stouter. *Bulinus truncatus* is the natural molluscan host in most areas; *B. africanus* is a vector in Tanzania (Kinoti, 1964). Isolated reports of recovery of *S. bovis* eggs from human urine and stools in South Africa, Zimbabwe, and Zaire lack satisfactory proof. In Italy this species has been reported as causing cercarial dermatitis in man (Biocca, 1960).

Schistosoma mattheei
Veglia and Le Roux, 1929

Schistosoma mattheei is a natural parasite of the mesenteric-portal blood of sheep, goats, cattle, impala, zebra, monkeys and, rarely, man in South Africa (Transvaal), Zimbabwe, and Zambia. The terminal-spined eggs measure 210 to 240 μm by 40 to 70 μm. The snail hosts are *Bulinus (Physopsis) africanus* and *B. (P.) globosus*. Blackie (1932) reported eggs of this worm passed in human urine and adults associated with those of *S. haematobium* in two human autopsies. *S. mattheei* occurs in over 1% of persons infected with schistosomiasis in Zimbabwe and Kiva Zulu (Schutte et al., 1980). Eggs are passed in both feces and urine (Hira, 1975). Hybridization with *S. haematobium* occurs readily, and experimentally produced hybrids exhibit heterosis in infectivity to both snail and mammal hosts, growth, maturation, and egg output (Wright and Ross, 1980).

Schistosoma curassoni
Brumpt, 1931

The African schistosome for which the species name *S. curassoni* was created is probably identical with *S. mattheei*.

Schistosoma spindale
Montgomery, 1906

Schistosoma spindale is a common schistosome in the caval blood of cattle, water buffaloes, sheep, and goats in India, Malaysia, Zambia, and Sumatra. It occurs less frequently in South Africa. The eggs (160 to 400 μm by 70 to 90 μm) are long, fusiform, have a conspicuous triangular equatorial bulge, and have a short spinose termination. The molluscan host in India is *Indoplanorbis exustus*. There is no substantial evidence that *S. spindale* develops in man, but in Thailand it has been found to cause cercarial dermatitis (Harinasuta and Kruatrachue, 1964).

Schistosoma nasalis
Rao, 1932

Schistosoma nasalis, a relatively common parasite of cattle in India, develops in veins of the nasal mucosa, producing miliary pseudotubercles around eggs and causing severe chronic rhinitis with mucous discharge. The adults and eggs closely resemble those of *S. spindale*. Transmission occurs with *Indoplanorbis exustus, Lymnaea luteola,* and *L. acuminata* as molluscan hosts. There has been no record of human infection.

Schistosoma faradjei
Walkiers, 1928

The species name *Schistosoma faradjei* was proposed for a schistosome, the eggs of which were recovered from stools of five natives in Zaire. The eggs were relatively small and had no lateral spine; they were possibly eggs of *S. margrebowiei*.

Schistosoma rodhaini
Brumpt, 1931

Schistosoma rodhaini, a parasite of African wild rodents (Congo, Uganda, Kenya), produces eggs with a subterminal spine and a slight curvature of the shell in the reverse direction at the opposite pole. Natural infection of dogs and experimental infection of dogs, cats, and other laboratory mammals have been reported. *S. rodhaini* differs from *S. mansoni* in its wider range of susceptible hosts and greater virulence. Cercariae of *Schistosoma* shed by *Biomphalaria* cannot be assumed to be *S. mansoni* unless experimentally ruled out as possibly those of *S. rodhaini*. Le Roux (1954) and Taylor (1970) succeeded in hybridizing *S. rodhaini* and *S. mansoni,* with offspring producing eggs that were a blend of the two species. Human infection with *S. rodhaini* has been reported in Zaire.

Schistosoma incognitum
Chandler, 1926

Schistosoma incognitum, a natural parasite of the mesenteric-portal systems of Indian pigs (Bhalerao, 1934) and dogs, produces eggs 97 to 148 μm by 45 to 81 μm (Sinha and Srivastava, 1956) having a short, pointed terminal spine somewhat resembling those of *S. indicum* Montgomery, 1906. The eggs were twice recovered from human feces by Chandler (1926), who proposed the species name *incognitum* before the adult worms had been discovered in their natural hosts. The species has been reported in small rodents in Thailand, Sulawsi, and Java (Carney et al., 1977).

Schistosoma margrebowiei
Le Roux, 1933

Schistosoma margrebowiei is a parasite of horses, cattle, sheep, and various antelopes in South Africa and Zambia. As its egg resembles that of *S. japonicum,* it is probably this species that is occasionally reported as *S. japonicum* in South Africa (Malek, 1980). However, according to Southgate and Knowles (1977), the egg of this species has a small terminal spine.

SCHISTOSOME CERCARIAL DERMATITIS
(Swimmer's itch, clam digger's itch)

Historical, Geographical, and Biological Notes. In 1928,

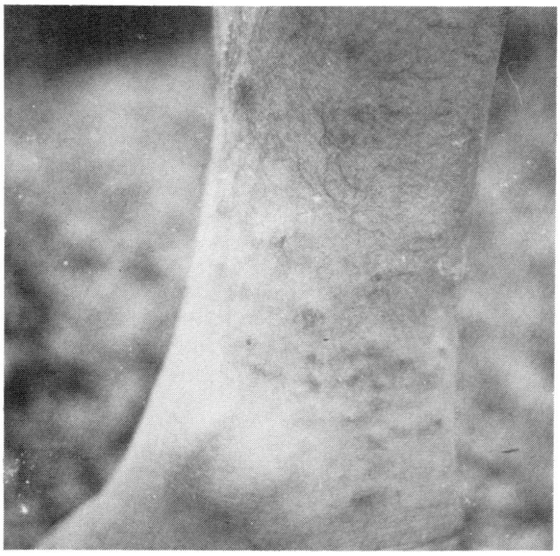

Fig. 27–24. Schistosome dermatitis of 7 days' duration on the ankle of a rice-field worker in the Caspian region of Iran, caused by *Orientobilharzia turkistanicum*, a cattle schistosome. (Courtesy of Dr. E.A. Malek.)

Cort showed that the cercariae of *Trichobilharzia ocellata* (La Valette) McMullen and Beaver, 1945 (syn. *Cercaria elvae* Miller, 1923), a nonhuman schistosome developing in the snail *Lymnaea stagnalis* var. *appressa*, and what was believed to be the same species in *L. emarginata-angulata* and *Physa parkeri*, were responsible for the production of papular eruptions of the skin of persons swimming in Douglas Lake, Michigan. As the water evaporated from the skin, a prickling sensation was experienced, followed by urticarial wheals. After one-half hour the initial dermatitis subsided leaving only a few macules. Several hours later, however, intense itching of the area developed, with edema and transformation of the papules into pustules (Fig. 27–24). The reaction was most intense between the second and third day following exposure. Olivier (1949) demonstrated that the cercariae of *T. stagnicolae* do not need an evaporating film of water on the skin to stimulate penetration; moreover, they are able to enter the skin from relatively cool water (17° to 23.5° C). Both Olivier (1949) and Macfarlane (1949) regarded the dermatitis as a sensitization phenomenon that may build up to a severe reaction following repeated exposure. Christenson and Greene (1928) found that *T. ocellata* was responsible for "swimmer's itch" in several Minnesota lakes. Cercaria-infested fresh water has also been found to be responsible for prickling dermatitis in Oregon, Washington, southern California, Montana, Colorado, Iowa, Canada, Germany, Switzerland, France, Wales, El Salvador, Cuba, Argentina, Burma, India, Japan, Australia, New Zealand, South Africa, and Alaska. The cercariae causing the dermatitis belong for the most part to avian species of blood flukes, particularly *Trichobilharzia* (McMullen and Beaver, 1945), *Gigantobilharzia* (Cort, 1950; Hunter et al., 1951), and *Ornithobilharzia* (Szidat, 1951). The molluscan hosts include species of *Lymnaea, Stagnicola, Physa, Planorbis, Polyplis,* and *Chilina*.

A second type of avian schistosome dermatitis is acquired along saltwater beaches, with marine molluscs serving as the intermediate hosts, and birds as definitive hosts. Cercaria-infested water has been the subject of investigation on the coasts of Connecticut, Rhode Island, New Jersey, Florida, California, and Hawaii. Stunkard and Hinchcliffe (1952) identified the adult worm of the cercaria that they studied as a species of *Microbilharzia* and Leigh (1955) a species *(huttoni)* of *Gigantobilharzia*.

A third type of cercarial dermatitis results from penetration into human skin by cercariae of mammalian species of blood flukes for which man is not a compatible definitive host. Buckley (1938) first reported an itching dermatitis in paddy workers in Malaysia due to skin penetration of the cercariae of *Schistosoma spindale* (see p. 442). Cases of dermatitis due to this species have also been reported from Thailand. In Italy *S. bovis* (see p. 442) is an etiologic agent. Alves (1953) reported a similar type of dermatitis from experimental exposure to cercariae of *S. mattheei*, and Cort (1950) listed *Schistosomatium douthitti* as a potential causal agent. *Heterobilharzia americana* Price, 1929 may be one of the etiologic agents causing dermatitis among oil field workers in southern Louisiana (Lee, 1962). A species causing dermatitis in Thailand and Iran is *Orientobilharzia turkestanicum*, a parasite of bovine hosts (Sahba and Malek, 1979).

Clinical Aspects of Schistosome Dermatitis. As water containing the dermatitis-producing schistosome cercariae drains off the exposed skin, the victim experiences a prickling or nettling sensation, at times accompanied by erythema of the invaded area and, in highly sensitive individuals, by local or generalized urticaria. This reaction is intensified by repeated exposure. The initial irritation characteristically soon subsides, leaving a minute macule at each site of penetration, but is followed shortly by intense itching of the region, with transformation of the macules into papules. The height of the reaction is usually reached on the second or third day, then gradually subsides, but symptoms are reactivated by rubbing of the involved areas.

Treatment consists of trimeprazine orally and local application of palliatives to reduce the itching, and sedation when indicated.

Control. Exposure to the freshwater type of the disease can be greatly reduced during the bathing season by use of copper sulfate or other suitable molluscicide dissolved along the shores of freshwater lakes or ponds in effective concentrations. This solution is only temporary since some of the snails survive and are reinfected the following season, frequently from eggs deposited in the feces of migratory birds.

No practical solution, even temporary, has been proposed to limit exposure on saltwater beaches.

REFERENCES

Abbott, R.T. 1948. Handbook of medically important mollusks of the orient and the western Pacific. Bull. Mus. Comp. Zool. Harvard College, *100*:246–328.

Abramowicz, M., (editor) 1982. *Handbook of Antimicrobial Therapy.* New Rochelle, New York, The Medical Letter.

Alves, W. 1953. Experimental cercarial dermatitis in man due to cercariae of *Schistosoma mattheei*. Trans. R. Soc. Trop. Med. Hyg., *47*:272.

Andrade, Z.A., and Brito, P.A. 1981. Evolution of schistosomal hepatic vascular lesions after specific chemotherapy. Am. J. Trop. Med. Hyg., *30*:1223–1227.

Arean, V.M. 1956. Manson's schistosomiasis of the female genital tract. Am. J. Obstet. Gynecol., *72*:1038–1053.

Ariizumi, M. 1963. Cerebral schistosomiasis japonica. Report of one operated case and fifty clinical cases. Am. J. Trop. Med. Hyg., *12*:40–55.

Audebaud, G., Tournier-Lasserve, C., Brumpt, V., Jolly, M., Mazaud, R., Imbert, X., and Bazillio, R. 1968. Premier cas de bilharziose humaine observe au Cambodge. Bull. Soc. Pathol. Exot., *71*:778–784.

Barbier, M. 1966. Determination d'un foyer de bilharziose arterior-veineuse au sud-Laos (Province de Sithadone). Bull. Soc. Pathol. Exot., *59*:974–983.

Barlow, C.H., and Meleney, H.E. 1949. A voluntary infection with *Schistosoma haematobium*. Am. J. Trop. Med., *29*:79–87.

Barnish, G. 1982. Evaluation of chemotherapy in the control of *Schistosoma mansoni* in Marquis Valley, Saint Lucia II. Biological Results. Am. J. Trop. Med. Hyg., *31*:111–115.

Becquet, R. 1967. Contribution à l'étude de la bilharziose a *Schistosoma intercalatum*. Ann. Soc. Belge Med. Trop., *47*:35–60.

Bella, H., Rahim, A., Mustafa, M., Ahmed, M., Wasfi, S., and Bennett, J. 1982. Oltipraz-antischistosomal efficacy in Sudanese infected with *Schistosoma mansoni*. Am. J. Trop. Med. Hyg., *31*:775–778.

Bhalerao, G. 1934. On the occurrence of *Schistosoma japonicum* Katsurada in India. Indian J. Vet. Sci. Anim. Husbandry, *4*:148–151.

Biocca, E. 1960. Osservazioni sulla morfologia e biologia del ceppo sardo di *Schistisome bovis* e sulla dermatitie umana da esso provocata. Parassitologia, *2*:47–54.

Blackie, W.K. 1932. A helminthological survey of Southern Rhodesia. Mem. Ser. London Sch. Hyg. Trop. Med., *5*:1–91.

Brug, S.L., and Tesch, J.W. 1937. Parasitaire wormen aan het Lindoe Meer (Oa Peloe, Celebes). Geneesk. Tijdschr. Ned.-Indie, *36*:2151–2157.

Buckley, J.J.C. 1938. On a dermatitis in Malaya caused by the cercariae of *Schistosoma spindale* Montgomery, 1906. J. Helminthol., *16*:117–120.

Budzilovich, G.N., Most, H., and Feigin, I. 1964. Pathogenesis and latency of spinal cord schistosomiasis. Arch. Pathol., *77*:383–388.

Byram, J.E., and von Lichtenberg, F. 1980. Experimental infection with *Schistosoma mekongi* in laboratory animals: Parasitological and pathological findings, pp. 125–159. In *The Mekong Schistosome*. Edited by J.I. Bruce and S. Sornmani. Whitmore Lake, Michigan, Malacol. Rev. Suppl. 2.

Carney, W.P., Purnomo, Van Peenen, P.F.D., Brown, R.J., and Sudomo, M. 1977. *Schistosoma incognitum* from mammals of Central Sulawesi, Indonesia. Proc. Helm. Soc. Wash., *44*:150–155.

———, Van Peenen, P.F.D., and Sudomo, M. 1978. A mammalian reservoir of *Schistosoma japonicum* in the Napu Valley, Central Sulawesi, Indonesia. J. Parasitol., *64*:1138–1139.

Chandler, A.C. 1926. A new schistosome infection in man, with notes on other human fluke infestations in India. Indian J. Med. Res., *14*:179–183.

Chang, Y.-C., Chu, C.-C., and Fan, W.K. 1957. Cerebral schistosomiasis—an observation of forty-five cases. Chinese Med. J., *75*:892–907.

Cheever, A.W. 1968. A quantitative post-mortem study of schistosomiasis mansoni in man. Am. J. Trop. Med. Hyg., *17*:38–64.

———, and Duvall, R.H. 1981. Bladder calcification and obstructive uropathy in a gibbon infected with *Schistosoma haematobium*. Am. J. Trop. Med. Hyg., *30*:604–608.

Chiu, J.K. 1965. *Tricula chiui:* A new snail host for Formosan strain of *Schistosoma japonicum*. J. Parasitol., *51*:206.

Christenson, R.O., and Greene, W.P. 1928. Studies on biological and medical apects of "swimmer's itch"; schistosome dermatitis in Minnesota. Minnesota Med., *11*:573–575.

Cohen, J., Capildo, R., Rose, F.C., and Pallis, C. 1977. Schistosomal myelopathy. Br. Med. J., *1*:1258.

Combes, C. 1982. Trematodes: antagonism between species and sterilizing effects on snails in biological control. Parasitology, *84*:151–175.

Cort, W.W. 1950. Studies on schistosome dermatitis. Am. J. Hyg., *52*:251–307.

Crabtree, J.E., and Wilson, R.A. 1980. *Schistosoma mansoni:* A scanning electron microscope study of the developing schistosomulum. Parasitology, *81*:553–564.

Crawford, K.A., Bruce, J.I., and Bueding, E. 1980. Effects of the schistosomicide amoscanate in mice and primates infected with *Schistosoma mekongi*, pp. 181–193. In *The Mekong Schistosome*. Edited by J.I. Bruce and S. Sornmani. Whitmore Lake, Michigan, Malacol. Rev. Suppl. 2.

Cross, J.H., and Lo, C.T. 1980. Susceptibility of new Taiwan foci of *Oncomelania hupensis* to geographic strains of *Schistosoma japonicum*. Southeast Asian J. Trop. Med. Public Health, *11*:374–377.

da Silva, L.C., Sette, H., Jr., Christo, C.H., Saez-Alquezar, A., Carneiro, C.R.W., Lacet, C.M., Ohtsuki, N., and Raia, S. 1981. Praziquantel in the treatment of the hepatosplenic form of schistosomiasis mansoni. Arzneimittel Forschung—Drug Research, *31*:601-603.

Davis, A., Biles, J.E., and Ulrich, A.-M. 1979. Initial experiences with praziquantel in the treatment of human infections due to *Schistosoma haematobium*. Bull. W.H.O., *57*:773–779.

Davis, G.M. 1980. Snail hosts of Asian *Schistosoma* infecting man: Evolution and coevolution, pp. 195–238. In *The Mekong Schistosome*. Edited by J.I. Bruce and S. Sornmani. Whitmore Lake, Michigan, Malacol. Rev. Suppl. 2.

De Lima e Costa, M.F., Katz, N., and Pinto Dias, J.C. 1980. Re-infection of patients in schistosomiasis mansoni endemic areas after specific treatment. IV. Observations taken in Belo Horizonte. Rev. Inst. Med. Trop. S. Paulo, *22*:97–107.

Deschiens, R., and Delas, A. 1968. Sur l'extension geographique en Afrique tropicale de la bilharziose humaine a *Schistosoma intercalatum*. C.R. Acad. Sci. Paris, *267*:1605–1606.

DeWitt, W.B. 1954. Susceptibility of snail vectors to geographic strains of *Schistosoma japonicum*. J. Parasitol., *40*:453–456.

El-Alamy, M.A., Habib, M.A., McNeeley, D.F., and Cline, B.L. 1981. Preliminary results of chemotherapy using praziquantel on a large scale in Qalyub Bilharziasis Project where simultaneous infection with *S. mansoni* and *S. haematobium* exists. Arzneimittel Forschung—Drug Research, *31*:612–615.

Falcão, H.A., and Gould, D.B. 1975. Immune complex nephropathy in schistosomiasis. Ann. Intern. Med., *83*:148–154.

Faust, E.C. 1948. An inquiry into ectopic lesions in schistosomiasis. Am. J. Trop. Med., *28*:175–199.

———, and Bonne, C. 1948. Mammalian blood flukes of Celebes. J. Parasitol., *34*:124–131.

———, and Meleney, H.E. 1924. Studies on schistosomiasis japonica. Am. J. Hyg. Monogr. Ser. No. 3.

Fenwick, A. 1969. Baboons as reservoir hosts of *Schistosoma mansoni*. Trans. R. Soc. Trop. Med. Hyg., *63*:557–567.

Ferguson, A.R. 1911. Associated bilharziosis and primary malignant disease of the urinary bladder, with observations on a series of forty cases. J. Pathol. Bacteriol., *16*:76–94.

File, S. 1975. Superinfection of *Biomphalaria glabrata* with *Schistosoma mansoni*. J. Parasitol., *61*:75–78.

Fisher, A.C. 1934. A study of the schistosomiasis of the Stanleyville district of the Belgian Congo. Trans. R. Soc. Trop. Med. Hyg., *28*:277–306.

Forsyth, D.M., and Bradley, D.J. 1966. The consequences of bilharziasis. Medical and public health importance in northwest Tanzania. Bull. W.H.O., *34*:715–735.

Gadgil, R.K. 1963. Human schistosomiasis in India. Indian J. Med. Res., *51*:244–251.

Gaitonde, B.B., Sathe, B.D., Mukerji, S., Sutar, N.K., Athalye, R.P., Kotwal, V.P., and Renapurkar, D.M. 1981. Studies on schistosomiasis in village Gimvi of Maharashtra. Indian J. Med. Res., *74*:352–357.

Gönnert, R. 1948. Die Struktur des Körperoberfläche von *Bilharzia mansoni* (Sambon, 1907). Z. Tropenmed. Parasitol., *1*:105–112.

Greer, G.J., Ow-Yang, C.K., Inder Sing, K., and Lim, H.K. 1980. Discovery of a site of transmission and hosts of a *Schistosoma japonicum*-like schistosome in Peninsular Malaysia. Trans. R. Soc. Trop. Med. Hyg., *74*:425.

Hall, S.C., and Kehoe, E.L. 1970. Prolonged survival of *Schistosoma japonicum*. Calif. Med., *113*(August):75–77.

Harinasuta, C. 1980. Introduction: Schistosomiasis in Asia, pp. 1–8. In *The Mekong Schistosome*. Edited by J.I. Bruce and S. Sornmani. Whitmore Lake, Michigan, Malacol. Rev. Suppl.

———, and Kruatrachue, M. 1964. Schistosomiasis in Thailand. Trans. R. Soc. Trop. Med. Hyg., *58*:195.

Hiatt, R.A., Sotomayor, Z.R., Sanchez, G., Zambrana, M., and Knight, W.B. 1979. Factors in the pathogenesis of acute schistosomiasis mansoni. J. Infect. Dis., *139*:659.

Hillyer, G.V., Ramzy, R.M.R., El Alamy, M.A., and Cline, B.L. 1980. Immunodiagnosis of infection with *Schistosoma haematobium* and *S. mansoni* in man. Am. J. Trop. Med. Hyg., *29*:1254–1257.

Hira, P.R. 1975. Observations on *Schistosoma mattheei* Veglia and Le Roux, 1929 infections in man in Zambia. Ann. Soc. Belge Med. Trop., *55*:633–642.

Hsü, S.Y. Li, Hsü, H.F., and Chu, K.Y. 1962. Interbreeding of geographic strains of *Schistosoma japonicum*. Trans. R. Soc. Trop. Med. Hyg., *56*:483–485.

Hull, P.R., and Hay, I.T. 1979. Peri-umbilical cutaneous schistosomiasis. A case report. S. Afr. Med. J., *56*:654.

Hunter, G.W. III, Ritchie, L.S., and Tanabe, H. 1951. The epidemiology of schistosome dermatitis ("koganbyo") in Japan. Trans. R. Soc. Trop. Med. Hyg., *45*:103–112.

———, Yokogawa, M., Akusawa, M., Sano, M., Araki, K., and Kobayashi, M. 1982. Control of schistosomiasis japonica in the Negatoishi area of Kurume, Japan. Am. J. Trop. Med. Hyg., *31*:760–770.

Ishizaki, T., Kamo, E., and Boehme, K. 1979. Double-blind studies of tolerance to praziquantel in Japanese patients with *Schistosoma japonicum* infections. Bull. W.H.O., *57*:787–791.

Jakobiec, F.A., Gess, L., and Zimmerman, L.E. 1977. Granulomatous dacryoadenitis caused by *Schistosoma haematobium*. Arch. Ophthalmol., *95*:278–280.

Jewsbury, J.M., Cooke, M.J., and Weber, M.C. 1977. Field trial of metrifonate in the treatment and prevention of schistosomiasis infection in man. Ann. Trop. Med. Parasitol., *71*:67–83.

Joky, A., Cornu, M., Lousi, D., and Grimaud, J.A. 1978. Polyvinylchloride (PVC) particles implantation in mouse liver. A technique for experimental study of schistosome egg-induced liver pathology. Experientia, *34*:547–549.

Jordan, P., Bartholomew, R.K., and Peters, P.A.S. 1981. A community study of *Schistosoma mansoni* egg excretion assessed by the Bell and a modified Kato technique. Ann. Trop. Med. Parasitol., *75*:35–40.

———, Bartholomew, R., Grist, E., and Auguste, E. 1982a. Evaluation of chemotherapy in the control of *Schistosoma mansoni* in Marquis Valley, Saint Lucia. I. Results in humans. Am. J. Trop. Med. Hyg., *31*:103–110.

———, Unrau, G., Bartholomew, R., Cook, J., and Grist, E. 1982b. Value of individual household water supplies in the maintenance phase of a schistosomiasis control programme in Saint Lucia, after chemotherapy. Bull. W.H.O., *60*:583–588.

Jordan, P., von Lichtenberg, F., and Goatly, K.D. 1967. Experimental schistosomiasis in primates in Tanzania. Bull. W.H.O., *37*:393–403.

Kamiya, H., Tada, Y., Matsuda, H., Tanaka, H., Blas, B.L., Nosenas, J.S., and Santos, A.T., Jr. 1980. Annual fluctuation of *Schistosoma japonicum* infection in field rats, *Rattus rattus mindanensis*, in Dagami, Leyte, Philippines. Jpn. J. Exp. Med., *50*:375–382.

Keittivuti, B., Keittivuti, A., and D'Agnes, T. 1982. Schistosomiasis in Cambodian refugees at Ban-Kaeng holding center, Prachinburi Province, Thailand. Southeast Asian J. Trop. Med. Public Health, *13*:216–219.

Kilpatrick, M.E., el Masry, N., Bassily, S., and Farid, Z. 1982. Oxamniquine versus niridazole for treatment of uncomplicated *Schistosoma mansoni* infection. Am. J. Trop. Med. Hyg., *31*:1164–1167.

Kitikoon, V. 1980. Comparison of eggs and miracidia of *Schistosoma mekongi* and *S. japonicum*, pp. 93–103. In *The Mekong Schistosome*. Edited by J.I. Bruce and S. Sornmani. Whitmore Lake, Michigan, Malacol. Rev. Suppl. 2.

Komiya, Y., and Kojima, K. 1959. The production of F_2 offspring from F_1 (hybrids) of *Oncomelania nosophora* and *O. hupensis*. J. Parasitol., *45*(sec. 2):23–24.

Krajden, S., Keystone, J.S., and Glenn, C. 1983. Safety and toxicity of oxamniquine in the treatment of *Schistosoma mansoni* infections, with particular reference to electroencephalographic abnormalities. Am. J. Trop. Med. Hyg., *32*: in press.

Kruger, S.P., Heitman, L.P., Van Wyi, J.A., and McCully, R.M. 1969. The route of migration of *Schistosoma mattheei* from the lungs to the liver in sheep. J. S. Afr. Vet. Med. Assoc., *40*:39–43.

Kuntz, R.E. 1952. Natural infection of an Egyptian gerbil with *Schistosoma mansoni*. Proc. Helm. Soc. Wash., *19*:123–124.

Lee, H.F. 1962. Susceptibility of mammalian hosts to experimental infection with *Heterobilharzia americana*. J. Parasitol., *48*:740–745.

———, Wykoff, D.E., and Beaver, P.C. 1966. Two cases of human schistosomiasis in new localities in Thailand. Am. J. Trop. Med. Hyg., *15*:303–306.

Leigh, W.H. 1955. The morphology of *Gigantobilharzia huttoni* (Leigh, 1953), an avian schistosome with marine dermatitis-producing larvae. J. Parasitol., *41*:262–269.

Leiper, R.T. 1915–1918. *Report on the Results of the Bilharzia Mission in Egypt*. London.

Lembeli, C.M., and Venkataramaiah, N.R. 1981. *Schistosoma haematobium* ova in semen. J. Urol., *125*:603.

Le Roux, P.L. 1954. Hybridization of *Schistosoma mansoni* and *S. rodhaini*. Trans. R. Soc. Trop. Med. Hyg., *48*:3.

Lewert, R.M., Yogore, M.G., Jr., and Blas, B.L. 1979. Schistosomiasis japonica in Barrio San Antonio, Basey, Samar, the Philippines. I. Epidemiology and morbidity. Am. J. Trop. Med. Hyg., *28*:1010–1025.

———, Yogore, M.G., Jr., Martin, W.R., and Blas, B.L. 1980. Schistosomiasis japonica in Barrio San Antonio,

Basey, Samar in the Philippines. IV. "Atypical" precipitates in the circumoval precipitin test—a reaction of recent infection. Am. J. Trop. Med. Hyg., *29*:431–434.

Lie Kian Joe. 1971. A possible biological control of schistosomiasis and other trematodes in snails. Proc. SE Asian Seminar on Parasitology & Tropical Medicine, Schistosomiasis & Other Snail-transmitted Helminthiasis, pp. 131–138.

Ling, C.C., Cheng, W.J., and Chung, H.L. 1949. Clinical and diagnostic features of schistosomiasis japonica. A review of 200 cases. Chinese Med. J., *67*:347–366.

Lucas, S.B. 1982. Bladder tumours in Malawi. Br. J. Urol., *54*:275–279.

Luttermoser, G.W. 1963. Infection of rodents with *Schistosoma mansoni* by ingestion of infected snails. J. Parasitol., *49*:150.

MacDonald, D.M., and Morrison, J.G.L. 1976. Cutaneous ectopic schistosomiasis. Br. Med. J., *2*:619–620.

Macfarlane, W.V. 1949. Schistosome dermatitis in New Zealand. Am. J. Hyg., *50*:143–167.

Malek, E.A. 1980. *Snail-transmitted Parasitic Diseases,* Volume 1. Boca Raton, Florida, CRC Press, pp. 179–307.

Mao, S.-P., and Shao, B.-R. 1982. Schistosomiasis control in the People's Republic of China. Am. J. Trop. Med. Hyg., *31*:92–99.

McCullough, F.S. 1981. Biological control of the snail intermediate hosts of human *Schistosoma* spp.: A review of its present status and future prospects. Acta Trop., *38*:5–13.

———, Gayral, P., Duncan, J., and Christie, J.D. 1980. Molluscicides in schistosomiasis control. Bull. W.H.O., *58*:681–689.

McMahon, J.E., and Kolstrup, N. 1979. Praziquantel: A new schistosomicide against *Schistosoma haematobium*. Br. Med. J., *2*:1396–1399.

McMullen, D.B., and Beaver, P.C. 1945. Studies on schistosome dermatitis. IX. The life cycles of three dermatitis-producing schistosomes from birds and a discussion of the subfamily Bilharziellinae (Trematoda, Schistosomatidae). Am. J. Hyg., *42*:128–154.

McQuay, R.M. 1953. Studies on variability in the susceptibility of a North American snail, *Tropicorbis havanensis,* to infection with the Puerto Rican strain of *Schistosoma mansoni*. Trans. R. Soc. Trop. Med. Hyg., *47*:56–61.

Meira, J.A. 1942. Estudo clínico das formas pulmonares da esquistosomiase mansônica. Arq. Cirurg. Clin. Exp., *6*:3–136.

Miller, P., and Wilson, R.A. 1980. Migration of the schistosomula of *Schistosoma mansoni* from the lungs to the hepatic portal system. Parasitology, *80*:267–288.

Miyagawa, Y. 1912. Route of migration of *Schistosoma japonicum* from the skin to the portal vein, and morphology of the young worm at the time of penetrating the skin. Tokyo Igakai Zasshi, 26, No. 5. (Japanese text).

Miyairi, K., and Suzuki, M. 1914. Der Zwischenwirt des *Schistosomum japonicum* Katsurada. Mitteil. Med. Fak. Kaiserl. Univ. Kyushu, *1*:187–197.

Moose, J.W., and Williams, J.E. 1963. Susceptibility of *Oncomelania formosana* from three different areas of Taiwan to infection with Formosan strain of *Schistosoma japonicum*. J. Parasitol., *49*:702–703.

Murugasu, R., Wang, F., and Dissanaike, A.S. 1978. *Schistosoma japonicum*-type infection in Malaysia—report of the first living case. Trans. R. Soc. Trop. Med. Hyg., *72*:389–391.

Nagy, B.A., File, S.K., and Smith, J.H. 1981. Changes in the enteric vasculature of mice infected with *Schistosoma mansoni*. Am. J. Trop. Med. Hyg., *30*:999–1009.

Nash, T.E., moderator. 1982. Schistosome infections in humans: perspectives and recent findings. Ann. Intern. Med., *97*:740–754.

———, Hofstetter, M., Cheever, A., and Ottesen, E. 1982. Treatment of *Schistosoma mekongi* with praziquantel: a double-blind study. Am. J. Trop. Med. Hyg., *31*:977–982.

Newton, J.C., Kanchanaranya, C., and Previte, L.R., Jr. 1968. Intraocular *Schistosoma mansoni*. Am. J. Ophthalmol., *65*:774–778.

Nguma, J.F.M., McCullough, F.S., and Masha, E. 1982. Elimination of *Biomphalaria pfeifferi, Bulinus tropicus* and *Lymnaea natalensis* by the ampullarid snail, *Marisa cornuarietis,* in a man-made dam in northern Tanzania. Acta Tropica, *39*:85–90.

Olivier, L. 1949. Schistosome dermatitis, a sensitization reaction. Am. J. Hyg., *49*:209–301.

Oshima, S., Yasuraoka, K., Irie, Y., Blas, B.L., Noseñas, J.S., and Santos, A.T., Jr. 1978. Final localization of *Schistosoma japonicum* in the lungs of field rats, *Rattus mindanensis,* in Leyte, Philippines. Jpn. J. Exp. Med., *48*:503–509.

Pan, C.-T. 1965. Studies on the host-parasite relationship between *Schistosoma mansoni* and the snail *Australorbis glabratus*. Am. J. Trop. Med. Hyg., *14*:931–976.

———. 1980. The fine structure of the miracidium of *Schistosoma mansoni*. J. Invertebrate Pathol., *36*:307–372.

Pautrizel, R., Tribouley, J., Gaubert, J., Verdaguer, S., and Louisia, G. 1966. A propos d'un cas de bilharziose a *Schistosoma intercalatum*. Bull. Soc. Pathol. Exot., *59*:325–332.

Pereira, L.H., Coelho, P.M.Z., Fonseca, J.J.A., Bredt, A., and Pellegrino, J. 1972. Migration of *Schistosoma mansoni* larvae in the albino mouse. Rev. Inst. Med. Trop. S. Paulo, *14*:306–309.

Pinardi, H., Carney, W.P., Clarke, M.D., Cross, J.H., Jusuf, A., Sulianti Saroso, J., and Oemijati, S. 1972. *Schistosoma japonicum* and intestinal parasites of the inhabitants of Lake Lindu, Sulawesi (Celebes); a preliminary report. Southeast Asian J. Trop. Med. Public Health, *3*:594–599.

Pons, J.A. 1937. Studies on schistosomiasis mansoni in Puerto Rico. V. Clinical aspects. P.R. J. Public Health Trop. Med., *13*:171–254.

Pope, R.T., Cline, B.L., and El Alamy, M.A. 1980. Evaluation of schistosomal morbidity in subjects with high intensity infections in Qalyub, Egypt. Am. J. Trop. Med. Hyg., *29*:416–425.

Pugh, R.N.H., Burrows, J.W., and Tayo, M.A. 1980. Malumfashi Endemic Diseases Research Project, XIV. Increasing schistosomiasis transmission. Ann. Trop. Med. Parasitol., *74*:569–570.

Queiroz, L.D.S., Nucci, A., Facure, N.O., and Facure, J.J. 1979. Massive spinal cord necrosis in schistosomiasis. Arch. Neurol., *36*:517–519.

Rosenbaum, R.M., Ishii, N., Tanowitz, H., and Wittner, M. 1972. Schistosomiasis mansoni of the spinal cord. Report of a case. Am. J. Trop. Med. Hyg., *21*:182–184.

Ruff, M.D., and Yamaguchi, S. 1973. Susceptibility of previously resistant *Oncomelania hupensis formosana* to reexposure with *Schistosoma japonicum*. Int. J. Parasitol., *3*:441–442.

Rysavy, B., Barus, V., Moravec, F., and Yousif, F. 1974. On some problems of the biological control of human schistosomes in Egypt. Folia Parasitol., *21*:161–168.

Sadigursky, M., and Andrade, Z. 1982. Pulmonary changes in schistosomal cor pulmonale. Am. J. Trop. Med. Hyg., *31*:779–784.

Sahba, G.H., and Malek, E.A. 1979. Dermatitis caused by cercariae of *Orientobilharzia turkestanicum* in the Caspian Sea area of Iran. Am. J. Trop. Med. Hyg., *28*:912–913.

Saif, M., and Abdel-Meguid, M. 1981. Observations on tolerance and efficacy of praziquantel in Egyptian patients infected simultaneously with *S. haematobium* and *S. mansoni*. Arzneimittel Forschung—Drug Research, *31*:604–608.

Sakamoto, K., and Ishii, Y. 1976. Fine structure of schistosome eggs as seen through the scanning electron microscope. Am. J. Trop. Med. Hyg., *25*:841–844.

Santos, A.T., Blas, B.L., Nosenas, J.S., Portillo, G.P., Ortega, O.M., Hayashi, M., and Boehme, K. 1979. Preliminary clinical trials with praziquantel in *Schistosoma japonicum* infections in the Philippines. Bull. W.H.O., *57*:793–799.

Sathe, B.D., Mukerji, S., Gaitonde, B.B., and Renapurkar, D.M. 1981. Reinvestigation of an old focus of schistosomiasis in Gimvi Village, District Ratnagiri, in Maharashtra State, India. Bull. Haffkine Institute, *9*:34–37.

Schneider, C.R. 1976. Schistosomiasis in Cambodia: A review. Southeast Asian J. Trop. Med. Public Health, *7*:155–165.

Schutte, C.H.J., Van Deventer, J.M.G., and Lamprecht, T. 1980. *Schistosoma mansoni* and *S. mattheei* infestation in northern Kwa Zulu. S. Afr. Med. J., *58*:66–70.

Schwetz, J. 1954. On two schistosomes of wild rodents of the Belgian Congo, *Schistosoma rodhaini* Brumpt, 1931 and *Schistosoma mansoni* var. *rodentorum* Schwetz, 1953 and their relationship to *S. mansoni* in man. Trans. R. Soc. Trop. Med. Hyg., *48*:89–100.

Short, R.B., and Cartrett, M.L. 1973. Argentophilic "papillae" of *Schistosoma mansoni* cercariae. J. Parasitol., *59*:1041–1059.

Sinha, P.K., and Srivastava, H.D. 1956. Studies on *Schistosoma incognitum* Chandler, 1926. I. On the synonymy and morphology of the blood fluke. Parasitology, *46*:91–100.

Smith, J.H., Kamel, I.A., Elwi, A., and von Lichtenberg, F. 1974. A quantitative post mortem analysis of urinary schistosomiasis in Egypt. I. Pathology and pathogenesis. Am. J. Trop. Med. Hyg., *23*:1054–1071.

Smithers, S.R. 1960. The isolation of viable schistosome eggs by a digestion technique. Trans. R. Soc. Trop. Med. Hyg., *54*:68–70.

Sornmani, S., Kitikoon, V., Schneider, C.R., Harinasuta, C., and Pathammavong, O. 1973. Mekong schistosomiasis. I. Life cycle of *Schistosoma japonicum*, Mekong strain in the laboratory. Southeast Asian J. Trop. Med. Public Health, *4*:218–225.

———, Kitikoon, V., Thirachantra, S., Harinasuta, C. 1980. Epidemiology of Mekong schistosomiasis, pp. 9–18. In *The Mekong Schistosome*. Edited by J.I. Bruce and S. Sornmani. Whitmore Lake, Michigan, Malacol. Rev. Suppl. 2.

Southgate, V.R., and Knowles, R.J. 1977. On *Schistosoma margrebowiei* Le Roux, 1933: The morphology of the egg, miracidium and cercaria, the compatibility with species of *Bulinus*, and development in *Mesocricetus auratus*. Z. Parasitenkd., *54*:233–250.

Sparrow, C.H. 1966. Bilharziasis in the human eye. Report of a case. Cent. Afr. J. Med., *12*:55–56.

Standen, O.D. 1953. The penetration of the cercariae of *Schistosoma mansoni* into the skin and lymphatics of the mouse. Trans. R. Soc. Trop. Med. Hyg., *47*:292–298.

Stjernholm, R.L., and Warren, K.S. 1974. *Schistosoma mansoni*: Utilization of exogenous metabolites by eggs *in vitro*. Exp. Parasitol., *36*:222–232.

Stunkard, H.W., and Hinchcliffe, M.C. 1952. The morphology and life history of *Microbilharzia variglandis* (Miller and Northrup, 1926) Stunkard and Hinchcliffe, 1951, avian blood-flukes whose larvae cause "swimmer's itch" of ocean beaches. J. Parasitol., *38*:248–265.

Symmers, W. St. C. 1903. Note on a new form of liver cirrhosis due to the presence of the ova of *Bilharzia haematobia*. J. Pathol. Bacteriol., *9*:237–239.

Szidat, L. 1951. Cercarias schistosómicas y dermatitis schistosómica humana en la Republica Argentina. Commun. Inst. Nac. Invest. Ci. Nat., *2*:129–150.

Taylor, M.G. 1970. Hybridization experiments on five species of African schistosomes. J. Helminthol., *44*:253–314.

Teesdale, C.H., Chitsulo, L., and Mkandawire, A.C. 1980. Concurrent dosage with niridazole and metrifonate for the treatment of bilharziasis in school children in Malawi. Cent. Afr. J. Med., *26*:202–205.

Theron, A. 1981. Dynamics of larval populations of *Schistosoma mansoni* in *Biomphalaria glabrata* II. Chronobiology of the intramolluscan larval development during the shedding period. Ann. Trop. Med. Parasitol., *75*:547–554.

Upatham, E.S., Thirachantra, S., and Sornmani, S. 1980. Sensitivity of the alpha and gamma races of *Lithoglyphopsis aperta* to conventional and controlled release molluscicides, pp. 263–276. In *The Mekong Schistosome*. Edited by J.I. Bruce and S. Sornmani. Whitmore Lake, Michigan, Malacol. Rev. Suppl. 2.

Van Marck, E.A.E., and Gigase, P.L. 1978. Effect of total portal vein ligation on the maturation of *Schistosoma mansoni* in mice. Ann. Soc. Belge Med. Trop., *58*:157–158.

Voge, M., Bruckner, D., and Bruce, J.I. 1978. *Schistosoma mekongi* sp. n. from man and animals, compared with four geographic strains of *Schistosoma japonicum*. J. Parasitol., *64*:577–584.

Vogel, H. 1967. Experimentelle Infektionen mit *Schistosoma haematobium* em Mangaben und Schimpansen. Ann. Soc. Belge Med. Trop., *47*:107–115.

von Lichtenberg, F., Sadun, E.H., Cheever, A.W., Erickson, D.G., Johnson, A.J., and Boyce, H.W. 1971. Experimental infection with *Schistosoma japonicum* in chimpanzees. Parasitologic, clinical, serologic, and pathological observations. Am. J. Trop. Med. Hyg., *20*:850–893.

Walther, R.R. 1979. Chronic papular dermatitis of the scrotum due to *Schistosoma mansoni*. Arch. Dermatol., *115*:869–870.

Warner, B.W. 1942. Anorectal and colonic manifestations of *Schistosoma mansoni* infestation (intestinal bilharzia). Case report. Am. J. Surg., *57*:168–172.

Webbe, G. 1981. The six diseases of WHO. Schistosomiasis: Some advances. Br. Med. J., *2*:1104–1106.

Weltman, J.K. 1982. Effects of praziquantel on antigenemia in murine schistosomiasis. Am. J. Trop. Med. Hyg., *31*:1294–1296.

Welton, J., and Imperato, P.J. 1975. Schistosomiasis in New York City. Bull. N.Y. Acad. Med., *51*:263–271.

Wheater, P.R., and Wilson, R.A. 1979. *Schistosoma mansoni*: A histological study of migration in the laboratory mouse. Parasitology, *79*:49–62.

Wilson, R.A., Draskau, T., Miller, P., and Lawson, J.R. 1978. *Schistosoma mansoni:* The activity and development of the schistosomulum during migration from the skin to the hepatic portal system. Parasitology, *77*:57–73.

———, and Lawson, J.R. 1980. An examination of the skin phase of schistosome migration using a hamster cheek pouch preparation. Parasitology, *80*:257–266.

Wolfe, M.S. 1974. *Schistosoma intercalatum* infection in an American family. Am. J. Trop. Med. Hyg., *23*:45–50.

Wood, M.G., Srolovitz, H., and Schetman, D. 1976. Schistosomiasis. Paraplegia and ectopic skin lesions as admission symptoms. Arch. Dermatol., *112*:690–695.

World Health Organization. 1967. Measurement of the public health importance of bilharziasis. Report of a WHO Scientific Group. Geneva, Tech. Rep. Ser. No. 372.

———. 1980. Epidemiology and control of schistosomiasis. Report of a WHO Expert Committee. Geneva, Tech. Rep. Ser. No. 643.

Wright, C.A., and Ross, G.C. 1980. Hybrids between *Schis-*

tosoma haematobium and *S. mattheei* and their identification by isoelectric focusing of enzymes. Trans. R. Soc. Trop. Med. Hyg., *74*:326–332.

Wright, E.D., Chiphangwi, J., and Hutt, M. 1982. Schistosomiasis of the female genital tract. A histopathological study of 176 cases from Malawi. Trans. R. Soc. Trop. Med. Hyg., *76*:822–829.

Yogore, M.G., Jr., Lewert, R.M., and Blas, B.L. 1978. Schistosomiasis japonica in Barrio San Antonio, Basey, Samar in the Philippines: II. Quantitative fecal examination and circumoval precipitin tests. Southeast Asian J. Trop. Med. Public Health, *9*:344–355.

———. 1979. Schistosomiasis japonica in Barrio San Antonio, Basey, Samar in the Philippines: III. The plasma circumoval precipitin test—procedure and use in epidemiologic studies. Southeast Asian J. Trop. Med. Public Health, *10*:23–31.

Yokogawa, M. 1976a. Review of prevalence and distribution of schistosomiasis in Japan. Southeast Asian J. Trop. Med. Public Health, *7*:137–143.

———. 1976b. Programme of schistosomiasis control in Japan. Southeast Asian J. Trop. Med. Public Health, *7*:322–329.

Zodda, D.M., Abdel-Hafez, S., and Phillips, M. 1983. Characterization of monoclonal antibodies against *Schistosoma mansoni*. Am. J. Trop. Med. Hyg., *32*:69–77.

Chapter 28

Amphistomate and Distomate Flukes

The superfamilies Strigeoidea, Clinostomatoidea, and Schistosomatoidea constitute the suborder Strigeata. In this group the adult worm usually has an anterior sucker and one or more ventral suckers, the cercaria has a forked tail, and the miracidium has two pairs of flame cells. The schistosomes were discussed in the previous chapter. The other two groups are inserted here for convenience. In each there is one family with one or more members of medical interest.

Family Diplostomatidae
Poirier, 1886

In the Diplostomatidae (superfamily Strigeoidea) the adult stage is found in the intestine of birds and mammals, not including man. In humans, it is a larval stage that causes disease. The adult stage is a small- to medium-sized worm with the body divided into distinct fore and hind parts. The forebody is broad and somewhat flat and bears the mouth and organs of attachment, whereas the hindbody is cylindrical and contains the main reproductive organs. The eggs are large, operculate, and unembryonated when laid. Eggs discharged in the feces of a final host embryonate and hatch when deposited in lakes and streams. On emerging from the egg, the free-swimming miracidium invades the tissues of an appropriate snail, producing two generations of sporocysts, the second of which in turn produces forked-tail cercariae. On emerging from the snail the cercariae penetrate the skin of a fish or larval frog (tadpole) and reach the final host when the fish or frog is eaten. On entering a tadpole, the cercaria of certain species does not become encysted as a metacercaria, but remains free in the tissues. This stage, referred to as a *mesocercaria,* is able to move from tail to legs during the metamorphosis from tadpole to adult frog. In addition, mesocercariae can be transferred through a succession of paratenic hosts when infected tadpoles or frogs are eaten by larger frogs or snakes, or by mammals other than the natural final host. In at least one species of this group, *Pharyngostomoides procyonis,* whose natural final host is the raccoon *(Procyon lotor),* mesocercariae in the tissues of the mother can be transferred to her newborn young through the milk, i.e., by the transmammary route of infection (Miller, 1981).

Human Infection With Mesocercariae

The first case of infection with a mesocercaria in man was reported by Shea et al. in 1973. An active worm about 0.3 mm long was seen in the retina of a 29-year-old woman, a resident of Ontario, Canada. It had caused blurred vision for 4 months and may have been responsible for intermittent visual disturbances of peripheral vision for 6 years. As the worm was destroyed in situ by argon laser treatment, it could be identified only as a mesocercaria. It may have entered the eye by direct penetration of the cornea, having been accidentally transported to the patient's conjunctiva during one of the many times over the years when she was preparing frog legs to be cooked and eaten. That hypothesis is supported by experimental evidence (Walters et al., 1975). According to Ashton et al. (1969), there are two questionable earlier reports of larval trematodes having been found in the lens of the human eye.

In 1976, a massive fatal infection with mesocercariae involving the lungs and other organs was reported from Ontario, Canada (Freeman et al., 1976; Fernandes et al., 1976). A 24-year-old man who presumably had eaten inadequately cooked frog legs was hospitalized with severe respiratory failure. Two days earlier he had had vague abdominal pain with malaise and, on the day of admission, cough, shortness of breath, and hemoptysis. A chest radiograph showed diffuse nodular pulmonary densities. A biopsy specimen of lung taken on the fourth day of illness revealed a mesocercaria and diffuse interstitial hemorrhage. By the sixth hospital day the hemorrhagic process had obliterated all respiratory air space. Autopsy revealed numerous mesocercariae in ascitic fluid and in sections of lung, liver, heart, kidney, pancreas, stomach, spleen, lymph nodes, fat, brain, and spinal cord. The mesocercariae were identified as *Alaria americana,* which in the adult stage is an intestinal parasite of foxes and other canines (Fig. 28–1).

A third case of mesocercaria infection in man was reported from Louisiana (Beaver et al., 1977). A 43-year-old man who had eaten baked raccoon and other small game noticed a small 1-cm intradermal swelling on the upper left thigh near the groin. When excised and sectioned it was found to contain a mesocercaria (Fig. 28–2). Three weeks later a second such swelling appeared at the iliac crest on the same side. It was removed, sectioned completely in series, and the worm contained in it was reconstructed from the sections. It was identified as an *Alaria* or *Alaria*-like species, which Shoop and Corkum (1981) later concluded was *Alaria marcianae* (La Rue, 1917) Walton, 1949. They found natural infections of the mesocercaria in the tissues of five amphibian, seven reptilian, and two mammalian species (raccoons and opossums).

Family Clinostomatidae
Lühe, 1901

Many species of the family Clinostomatidae, type genus *Clinostomum* Leidy, 1856, parasitize the throat and esophagus of birds that eat freshwater fish. Three cases of human laryngitis due to lodgement of *Clinostomum* in the pharynx have been reported: a Japanese woman (Yamashita, 1938), an Israeli patient (Witenberg, 1944), and a Japanese male office worker who admitted eating raw freshwater fish (Kamo et al., 1962).

Family Paramphistomatidae
Fischoeder, 1901

Paramphistome flukes are characterized by having a terminal oral sucker that may have a pair of lateroposterior pouches, and a terminal or subterminal acetabulum at the posterior end of the body. The testes are tandem or oblique, in the middle third of the body, anterior to the ovary.

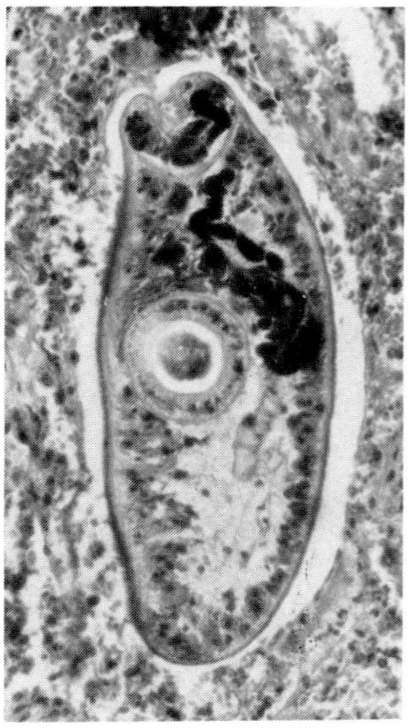

Fig. 28–1. *Alaria americana*. Longitudinal section of mesocercaria in lung of a man in Canada. (× 170.) (From Freeman, R.S., et al. 1976. Fatal human infection with mesocercariae of the trematode *Alaria americana*. Am. J. Trop. Med. Hyg., *25*:803–807.)

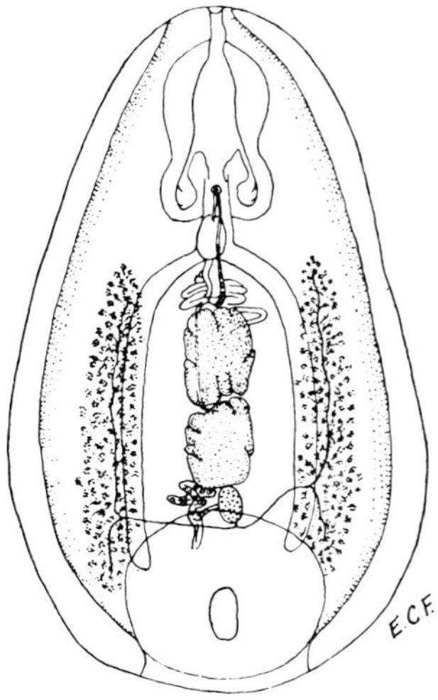

Fig. 28–3. *Watsonius watsoni*, adult worm, ventral view. (× 6.) (Adapted by E.C. Faust from Stiles, C.W., and Goldberger, J. 1910. A study of the anatomy of *Watsonius* (n.g.) *watsoni* of man. Hyg. Lab. Bull., no. 60.)

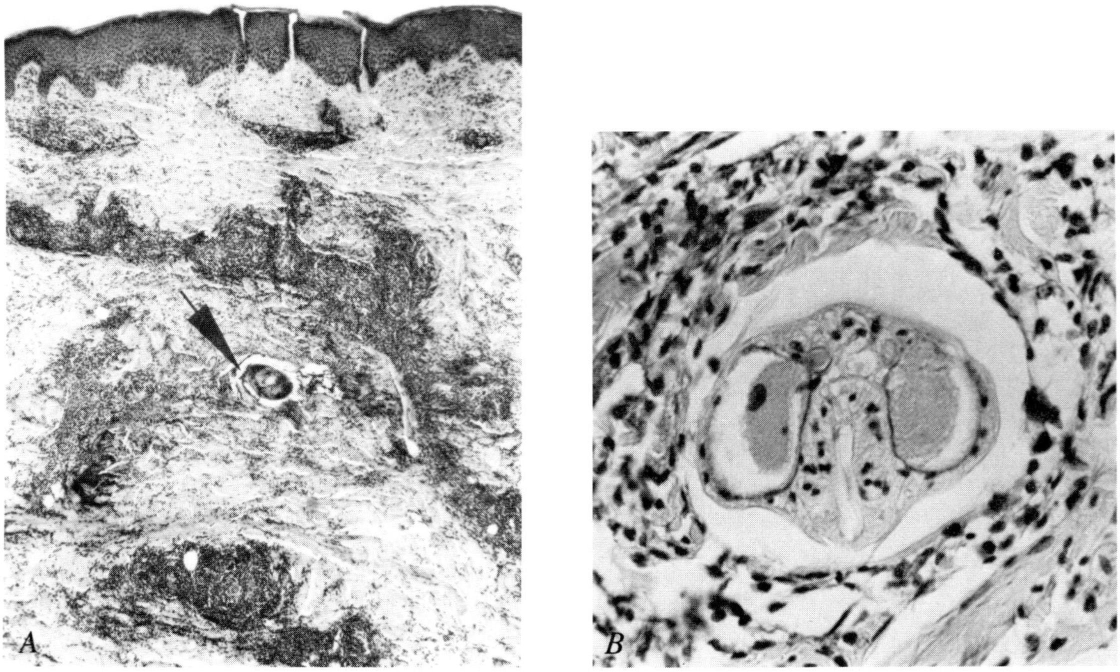

Fig. 28–2. *Alaria*-like mesocercaria, probably *Alaria marcianae*, in section of skin of a man in Louisiana. *A*, Although it appears to be deep in the skin (arrow), the worm is only 1 mm from the surface. (× 42.) (From Beaver, P.C., Little, M.D., Tucker, C.F., and Reed, R.J. 1977. Mesocercaria in the skin of man in Louisiana. Am. J. Trop. Med. Hyg., *26*:422–426.) *B*, Transverse section at level of acetabulum. On either side of the acetabulum is a large gland cell, and above each of these is an intestinal cecum (× 400).

Watsonius watsoni
(Conyngham, 1904) Stiles and Goldberger, 1910

This trematode was found only once in a human subject, at the autopsy of an emaciated West African Negro who died of a severe diarrhea. At autopsy many worms were recovered from the intestine, some attached to the duodenal and jejunal wall, others free in the lumen of the colon. Various species of primates in eastern Asia and Africa are natural hosts of this parasite. The worms (Fig. 28–3) have a pear-shaped body, are reddish-yellow when alive, and have a translucent glistening appearance. They are somewhat flattened dorsoventrally, are indented shallowly at the posterior end, measure 8 to 10 mm in length by 4 to 5 mm in breadth, and have a thickness of about 4 mm. Ventrally they are slightly concave, especially in the vicinity of the posterior sucker, which has a diameter of 1 mm. The subterminal oral sucker, which is sunken into the body, has a transverse diameter of 1.2 mm and a length of 1.6 to 2 mm, and is provided with a pair of posterolateral pouches.

The genital organs lie in the posterior two-thirds of the body. The two testes, squarish with notched margins, lie one in front of the other just anterior to the small ovoid ovary in the middle third of the body. The vas deferens is an intricately coiled tubule; the pars prostatica is bulbous, while the ejaculatory duct is a capillary tubule leading to the genital pore, situated ventral to the posterior margin of the oral sucker. The ootype lies immediately to the right of the ovary. The uterus coils back and forth many times between the ootype and the genital pore. The eggs, ovoid, operculate, light yellow, and measuring 122 to 130 μm by 75 to 80 μm, are immature when laid.

The life cycle is unknown, but infection is probably acquired by ingesting vegetation on which the metacercariae have encysted. Severe diarrhea apparently results from development of large numbers of the worms. Treatment and prevention have not been studied.

Gastrodiscoides hominis
(Lewis and McConnell, 1876) Leiper, 1913

Synonym. *Gastrodiscus hominis* (Lewis and McConnell, 1876) Fischoeder, 1902.

This trematode, discovered and described by Lewis and McConnell from the cecum of an Indian patient was redescribed by Stephens and Leiper. It is a common human parasite in Assam, and human infections have also been reported from Bengal, Bihar and Orissa in India, in Vietnam, the Philippines, Kasakstan in U.S.S.R., and from Indian immigrants in Guyana. The pig is a common reservoir host, while *Tragulus napu* in Malaysia, *Rattus brevicaudatus* in Java and Japan, and the rhesus monkey in India have also been found infected. The living worm (Fig. 28–4) has a bright pinkish color, is piriform in outline, and has a conical anterior and a discoidal posterior portion. In size it varies from 5 by 5 mm to 8 by 14 mm, the anterior portion having a length of about 2 mm. Slightly behind the middle of this portion in the median line is the prominent genital cone, which characterizes the genus. The conspicuous acetabulum, which occupies the ventral rim of the posterior part of the worm and has a deep cleft at its posterior end, measures 2.5 to 4.3 mm in diameter, depending on the degree of contraction. It is aspinose but has papillae on the ventral surface, especially around the genital atrium. Except for the outer portion of the uterus and of the male duct leading to the genital pore, all the reproductive organs lie in the region of the posterior disk. Near the anterior rim are the two obliquely situated, large, lobate testes. Cirrus pouch appears to be lacking. Unicellular prostate glands surround the seminal vesicle. The small spheroidal ovary lies in the center of the disk region. The vitellaria consist of finely granular fan-shaped bodies lying over the rim of the disk. The ootype (shell gland complex) lies to the right of the ovary. The uterus coils dorsal to the testes up to the genital pore where it opens posterior to the male duct. The ovoid, operculate eggs are immature when laid; they measure about 150 μm by 60 to 70 μm.

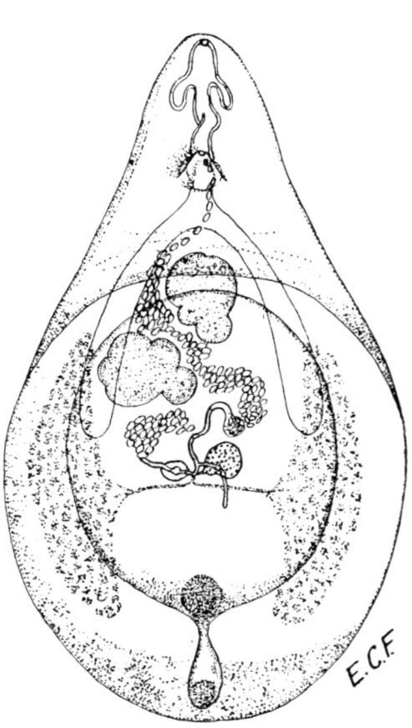

Fig. 28–4. *Gastrodiscoides hominis*. Adult worm, ventral view. (× 10.) (From Faust, E.C. 1929. *Human Helminthology*, 1st ed. Philadelphia, Lea & Febiger.)

The complete life cycle is unknown, but in India the planorbid snail *Helicorbis coenosus* has been found experimentally to serve as an intermediate host (Dutt and Srivastava, 1966). The presence of *G. hominis* in man, attached to the cecum and ascending colon, may produce a mucous diarrhea. Tetrachloroethylene is probably an efficient anthelmintic. Praziquantel should be tried. Prevention has not been studied.

Family Fasciolidae
Railliet, 1895

Members of the family Fasciolidae are relatively large distomes with the ventral sucker (acetabulum) close to the oral sucker, and the tegument often spinose, testes usually branched, vitellaria filling much of the lateral fields and confluent posterior to the testes, uterus short, and eggs large and undeveloped when laid; cercariae are produced in rediae. Species in this group that are parasites of man are *Fasciola hepatica* and *Fasciolopsis buski*. A less common species is *Fasciola gigantica*. A related species that is found in cattle and other herbivores, but not in man, is *Fascioloides magna*.

Fasciola hepatica
Linnaeus, 1758
(The sheep liver fluke, producing sheep liver rot or fascioliasis hepatica)

Synonyms. *Distoma hepaticum* L., 1758; *Fasciola californica* Sinitsin, 1933; *Fasciola halli* Sinitsin, 1933.

Historical and Geographical Notes. This was the first trematode to be described (de Brie, 1379). It was also the first for

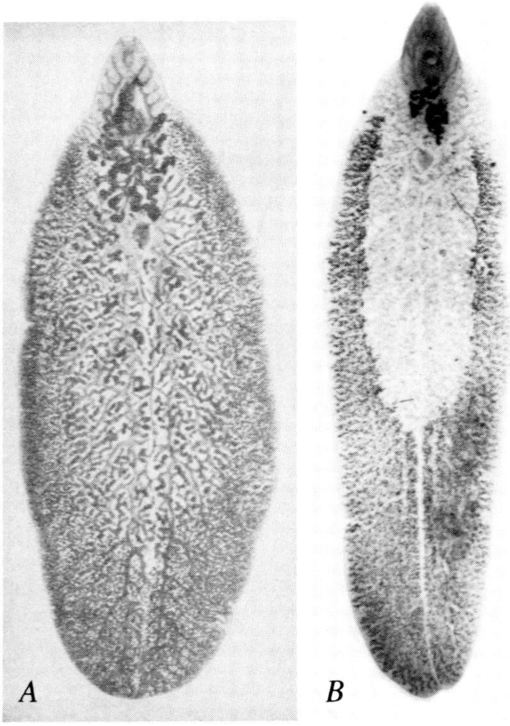

Fig. 28–5. *Fasciola. A, F. hepatica* from liver of sheep, dorsal view. (× 4.) *B, F. gigantica* from liver of water buffalo, ventral view. (× 3.5.) (*A*, Photograph by H.J. Van Cleave. *B*, Courtesy of Dr. M.D. Little.)

which the complete life cycle was elucidated (Leuckart, 1882; Thomas, 1883). Reinhard (1957) has given an historical account of the early investigations of this parasite. *F. hepatica* has a cosmopolitan distribution, and is prevalent in most sheep-raising countries. In the United States it is enzootic in extensive areas in the south, southeast, southwest, west and north central states, having been reported in cattle, sheep, goats, hogs, bison, rabbits, and deer. Human infections have been found in Venezuela, Uruguay, Argentina, Chile, Colombia, Mexico, Puerto Rico, Cuba, Costa Rica, Syria, Turkey, Japan, China, Soviet Russia (including Turkestan and Tashkent), Iran, Poland, Madeira, England, France, Italy, Corsica, Spain, Hungary, Romania, Salonika, the Dardanelles, Egypt, Algeria, Somaliland, South Africa, and Hawaii. A single autochthonous human infection has been reported for the continental United States (Norton and Monroe, 1961). It is an increasingly important parasite of man in Brazil and other Latin American countries (the vicuña is a reservoir in Peru), as well as in France, Algeria, and other Mediterranean countries. Approximately 50 cases have been recorded in Puerto Rico (Hillyer, 1981).

Morphology, Biology, and Life Cycle. *Fasciola hepatica* (Fig. 28–5) is a moderately fleshy fluke up to 30 mm in length by 13 mm in breadth; but it is relatively flat and leaf-like along the margins. It has integumentary scales that vary in size, pattern, and distribution; the posterior end is at times aspinose. At the anterior end there is a distinct conical projection. The posterior end is broadly pointed. The oral sucker measures about 1 mm in diameter and the nearby ventral sucker, 1.6 mm. In addition to a well-developed pharynx and a short esophagus, the digestive tract is characterized by having branched ceca that extend to the posterior end of the worm.

The two testes, which lie one behind the other in the second and third fourths of the body, are conspicuously dendritic. The cirrus pouch, situated over the ventral sucker, contains a posterior seminal vesicle, a median prostate region, and an anterior muscular cirrus organ, opening through a delicate precirral canal into the genital atrium, which lies just in front of the ventral sucker. The vitellaria, extending through the lateral fields, are highly branched, as is the dendritic ovary, which is situated on the right side in front of the anterior testis. There is a short Laurer's canal but no seminal receptacle. The ootype lies mesially behind the ovary; it is surrounded by a spherical mass of Mehlis' gland. The uterus is relatively short but is coiled rather intricately between the ootype and the genital pore, where it opens to the side of the precirral canal.

The eggs of *F. hepatica* (Fig. 28–6) are large, ovoid, operculate, light yellowish brown, and measure 130 to 150 μm by 63 to 90 μm. They are laid in the immature state in the biliary passages of the host and are evacuated in the feces. After development in water, for 9 to 15 days at an optimal temperature of 22° to 25°C, the miracidium escapes from the egg. If successful within 8 hours after emergence, the miracidium invades one of the var-

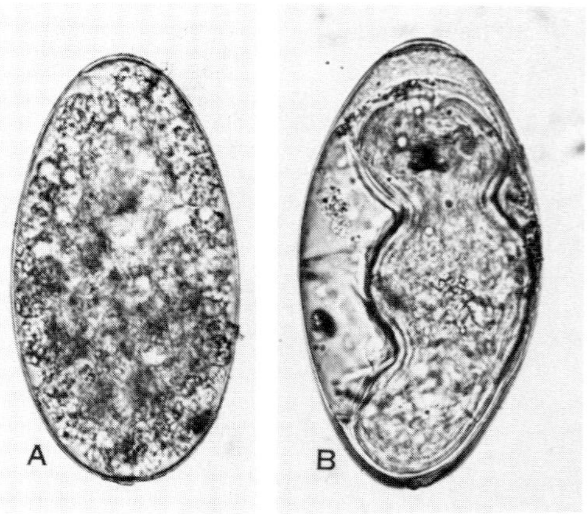

Fig. 28–6. *Fasciola hepatica*. Eggs from patient in Japan. *A*, Freshly recovered from bile. *B*, Containing miracidium after 10 days' incubation in water at 28°C. (× ca. 330.) (From Yoshida, Y., et al. 1974. A case of human infection with *Fasciola* sp. and its treatment with bithionol. Jpn. J. Parasitol., *23*:116–124.)

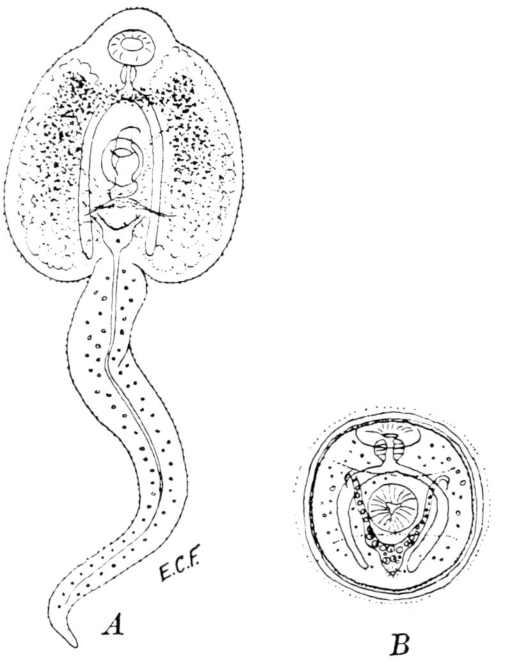

Fig. 28–7. *Fasciola hepatica. A,* Cercaria. *B,* Metacercaria in cyst. (× 100.) (From Faust, E.C. 1929. *Human Helminthology,* 1st ed. Philadelphia, Lea & Febiger.)

ious snail hosts including many species of *Lymnaea* and other genera of the family *Lymnaeidae* in different geographic areas. As it penetrates the snail's tissues, the miracidium sheds its ciliated cover. Within 3 weeks after penetration the sporocysts are producing first-generation rediae. In another week second-generation rediae and cercariae have developed. In some instances a third generation of rediae develop, the number depending on the size of the host snail (Rondelaud and Barthe, 1978). In the presence of water the mature cercariae (Fig. 28–7,*A*) swarm out of the snail, usually at night, swim about in the water for periods up to 8 hours and then, detaching their tails, encyst as minute white spherules (Fig. 28–7,*B)* on aquatic vegetation, bark, or other smooth surfaces. These cysts are viable for a long period in a moist atmosphere, but succumb quickly when dried.

Mammals eating such vegetation, particularly while moist, acquire the infection. The metacercariae excyst in the duodenum and migrate through the intestinal wall into the body cavity, thence through Glisson's capsule, traversing liver parenchyma to the biliary passages, where they settle down and grow to maturity. The prepatent period in calves is 8 weeks (De León et al., 1981); in man it may be 3 or 4 months. The duration of patency after a single exposure in cattle is usually less than 1 year (De León et al., 1981); in sheep egg production may continue for several years. One experimentally infected sheep passed eggs for 11 years (Chandler, 1955).

Epidemiology. Infection results from the ingestion of the encysted metacercariae attached to green vegetation or free at the bottom of shallow stands of fresh water. In Latin American countries, France, Algeria, and probably elsewhere, people frequently acquire the disease as a result of eating raw watercress *(Nasturtium officinale),* on which the metacercariae have encysted. Recent epidemics of human fascioliasis have been reported from France (Rondelaud, 1980), Egypt (Farag et al., 1979), and Brazil (Amaral and Busetti, 1979). Sporadic outbreaks have occurred in Japan (Yoshida et al., 1974), central Europe, U.S.S.R., and England (Facey and Marsden, 1960). Moderate epidemics have occurred among Europeans living in Algeria who eat watercress; native Algerians, who do not eat watercress, are not exposed (Coumbaras, 1966). Reservoir hosts (primarily sheep) play an important role in the propagation of this infection in Nature.

Pathogenesis, Pathology, and Symptomatology. The metacercarial larvae of *F. hepatica* that escape from cysts in the duodenum or jejunum of their definitive host normally produce no significant damage as they migrate through the intestinal wall into the peritoneal cavity and thence through Glisson's capsule, but traumatic and necrotic lesions are produced during migration through the liver parenchyma. Linear necrotic lesions 1 cm or more in diameter and heavily infiltrated with eosinophils are formed in the liver parenchyma by the young migrating flukes. Hyperplasia of the bile ducts commences while the immature flukes are still in the liver parenchyma possibly induced by the action of proline synthesized and released by the parasites (Isseroff et al., 1977). After arrival in the bile ducts, the larvae develop into adults, crowding the vessels and provoking inflammatory and adenomatous changes of the biliary epithelium with enveloping fibrosis of the ducts. If these processes continue and the number of worms is considerable, pressure atrophy of the liver parenchyma and extensive periductal fibrosis may ensue. Moreover, the gallbladder frequently undergoes the same pathologic changes as the bile ducts and may be invaded (Balci, 1975). In heavy infections the epithelium is eroded and the young worms may wander back into the liver parenchyma to produce abscess pockets and to "seed" the vital liver tissue with their eggs. Dawes (1961) demonstrated that the young flukes feed primarily on hepatic cells; although blood is also ingested, *F. hepatica* is not a true blood feeder. The adult flukes may feed on the hyperplastic ductal epithelium.

Larvae migrating through the peritoneal cavity may become lodged in ectopic foci, where abscesses or fibrotic lesions may develop. Human infections have been recorded in the blood vessels, lungs, subcutaneous tissue (abscess pockets), ventricles of the brain, and the orbit (Neghme and Ossandon, 1943). Newly born lambs have been found infected.

Symptoms recorded for human cases included colic and obstructive jaundice, with coughing and vomiting; generalized abdominal rigidity; acute epigastric pain and abdominal tenderness; urticaria; early leukocytosis and eosinophilia up to 60%; irregular fever; more or less persistent diarrhea; later, marked anemia; and rarely hemoglobinuria (Khorsandi, 1977; Farid et al., 1977; Jones et al., 1977). Cholelithiasis is a relatively common complication (Faiguenbaum et al., 1962).

In Lebanon and Armenia, ingestion of infected raw sheep and goat livers by the local population has caused pharyngeal fascioliasis, referred to as *halzoun* (i.e., suffocation). Temporary lodgement of adult worms on the pharyngeal mucosa causes edematous congestion of the soft palate, pharynx, larynx, nasal fossae, and eustachian tube, accompanied by dyspnea, dysphagia, deafness, and occasionally asphyxiation. Some of these cases may be due to a linguatulid (see p. 571).

Diagnosis. Whenever a patient has biliary symptoms, moderate to high eosinophilia, and has eaten watercress as a green salad, fascioliasis can be suspected. Specific diagnosis is based on recovery of the eggs (Fig. 28–6) in the patient's stool or from duodenal or biliary tract drainage. Early diagnosis is important so that treatment may be instituted before irreparable damage to the liver has occurred. The eggs are unembryonated and practically indistinguishable from those of *Fasciolopsis buski*. Although serodiagnostic tests are helpful, these have not yet been adapted for routine diagnosis. During invasion of the liver, fascioliasis must be distinguished from acute hepatitis of other etiology. Later it must be differentiated from cholecystitis, cholelithiasis, and cirrhosis resulting from other causes.

False fascioliasis, due to ingestion of infected livers, with passage of eggs in the feces, is at times mistaken for actual infection. This diagnosis may be ruled out by keeping the patient on a liver-free diet for 3 days or more. If eggs continue to be passed in *Fasciolopsis*-free areas, the infection is probably genuine.

Treatment. Treatment of choice is with *bithionol* orally in a dosage of 30 to 50 mg per kg body weight on alternate days for 10 to 15 doses. *Emetine hydrochloride* given intramuscularly in a dosage of 30 mg daily for 17 or 18 days, or 40 mg daily to a total of 5 mg per kg body weight, has been used with success. Favorable results with *hexachloroparaxylene* have been reported from Russia. Praziquantel is probably effective, as it is for other liver fluke infections, e.g., *Clonorchis* and *Opisthorchis* (Ambroise-Thomas et al., 1981; Bunnag and Harinasuta, 1980, 1981).

With adequate treatment of infections involving the liver, cholecystectomy is not usually necessary. Worms in ectopic foci should be visualized and, where feasible, should be excised. Pharyngeal fascioliasis is prevented by cooking of potentially infected livers.

Prognosis. The prognosis is fair in light infections, with persistent chemotherapy; it is grave in heavy infections or advanced involvement of the liver.

Prevention. Eventual extinction of the dangers of human infection with *F. hepatica* may be effected by eradication of the worms in the reservoir hosts. Adequate chemotherapy of infected animals, killing the molluscan hosts by use of molluscicides, and drainage of pastures will help. Deschiens et al. (1961) recommended for endemic-enzootic areas: sanitary protection of watercress beds; use of chemical fertilizer only for these beds; administrative control of the culture and sale of watercress; education of the local population to the danger of eating raw watercress; and destruction of the worms in reservoir hosts and the vector snails.

Related Species. *Fasciola gigantica* (Fig. 28–5,B) is larger and more lanceolate than *F. hepatica,* has a shorter cephalic cone, a larger ventral sucker, a more anterior position of the testes, and larger eggs (160 to 190 μm by 70 to 90 μm). It is a relatively common parasite of herbivorous mammals, particularly cattle, in Africa, Asia, and Hawaii, and has been recorded from man in Zimbabwe and Uganda in Africa, Tashkent in U.S.S.R., Vietnam, Iraq, and Hawaii. The life cycle, pathology, and symptomatology parallel those of *F. hepatica,* but *F. gigantica* employs different snails as intermediate hosts, development is slower, and metacercariae are more susceptible to desiccation.

Both *F. hepatica* and *F. gigantica* occasionally have been observed in ectopic locations. Doby and Beaucournu (1970) recorded from the literature 11 such cases of *F. hepatica*, 5 of *F. gigantica*, and 7 in which a specific identification was not made. In most instances a worm was discovered in a furunculoid or cystic mass excised from the subcutaneous or intermuscular tissues of the abdominal or chest wall. Perry et al. (1972) described a liver abscess due to *F. gigantica* in a native woman in Zimbabwe, the eggs being found in aspirated pus. Ongom (1980) reported finding a 16-mm *F. gigantica* in a subcutaneous swelling on the sternum of a 17-year-old male in Uganda, and he reviewed similar cases reported by others.

Fascioloides magna, the giant liver fluke of cattle and other hervivores in North America and northern Europe, responsible for extensive destruction of the liver parenchyma, has not been reported in man.

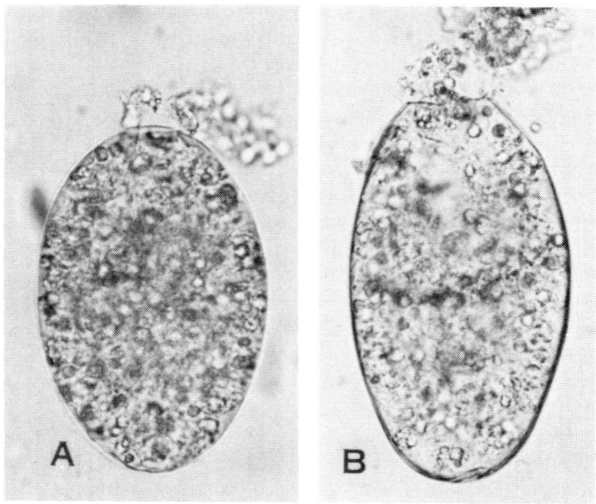

Fig. 28–9. *Fasciolopsis buski*. Eggs from feces of man in Taiwan. *A*, With operculum intact. *B*, With operculum detached by pressure on coverglass. (× 300.) (By P.C. Beaver.)

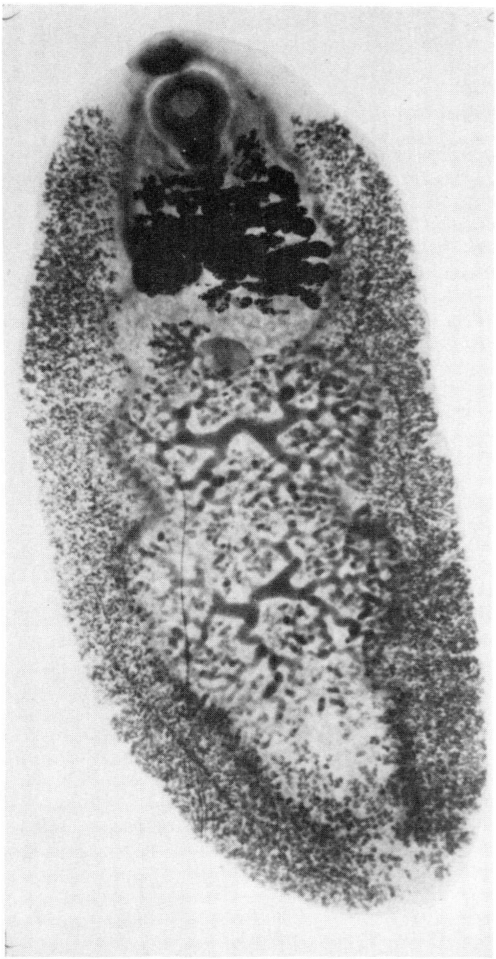

Fig. 28–8. *Fasciolopsis buski*. Adult worm from intestine of man in Shaohsing, China. (× 4.) Evident from anterior (top) to posterior are the small oral sucker, large ventral sucker, much coiled uterus, branched ovary and ovoid Mehlis' gland, and two branched testes filling much of the posterior half of the body. The lateral fields are filled with vitellaria. (Courtesy of Dr. M.D. Little.)

Fasciolopsis buski (Lankester, 1857) Odhner, 1902 (The giant intestinal fluke, producing fasciolopsiasis)

Synonyms. *Distomum crassum* Busk, 1859; *D. rathouisi* Porier, 1887; *Fasciolopsis rathouisi* (Porier, 1887) Ward, 1903; *F. fülleborni* Rodenwaldt, 1909; *F. goddardi* Ward, 1910.

Historical and Geographical Notes. First discovered by Busk, in 1843, in the duodenum of an East Indian sailor who died in London, this fluke has been featured under a variety of names, due to its size, different shapes, and differences in integumentary spine pattern.

F. buski is a common parasite of man and pig in Central and South China, Taiwan, Vietnam, Thailand, Borneo, Sumatra, Assam, Bengal, and other parts of the Orient. In the Canton, China area, dogs are occasionally infected, although they appear to be partially resistant to infection. Other domestic animals, with the possible exception of rabbits, are probably refractory.

Morphology, Biology, and Life Cycle. The fluke (Fig. 28–8), which is attached to the duodenal and jejunal wall, is a large fleshy worm at times broadly ovate, more usually elongate-ovoid, varying in size from 20 to 75 mm in length, 8 to 20 mm in breadth, and 0.5 to 3 mm in thickness. Typically the integument is spinose. There is no cephalic cone. The oral sucker measures about 0.5 mm in diameter; the nearby acetabulum, 2 to 3 mm. The intestine includes prepharynx, pharynx, esophagus, and unbranched ceca. The highly dendritic testes lie one behind the other in the posterior half of the worm. The elongate cirrus sac, arising from the vas deferens about halfway between the ootype and the ventral sucker, contains a posterior convoluted seminal vesicle, prostate glands surrounding the ejaculatory duct, a muscular cirrus canal lined with spines, and a precirral canal opening into the genital atrium, which lies immediately in front of the acetabulum. The ootype is situated in the midline near the equatorial plane. It is surrounded by a spherical mass of Mehlis' gland. The branched ovary lies to the right of the midline. Finely granular vitelline follicles extend in the lateral fields from the acetabulum to the caudal extremity. The uterus meanders between the ootype and metraterm, which lies just inside the genital pore. The large eggs, almost identical with those of *F. hepatica,* measure 130 to 140 μm by 80 to 85 μm, are operculate, and are unembryonated when laid (Fig. 28–9). They are deposited continuously in the intestinal lumen.

The life cycle of *F. buski,* worked out by Nakagawa (1921) on the pig and by Barlow (1925) on man, closely parallels that of *F. hepatica.* When the immature egg is discharged in feces and reaches fresh water, it requires 3 to 7 weeks at summer temperatures to complete its incubation, whereupon it hatches and the emerging miracidium swims about in the water. On contact with an appropriate snail, it penetrates the soft tissues, and within a period of a few weeks sporocysts, first- and second-generation rediae, and cercariae are produced in succession. The molluscan hosts in China, North Vietnam, and Taiwan include *Segmentina hemisphaerula* (syn. *S. coenosus, S. nitidella, S. calathus, S. largeillierti), Hippeutis cantori* (syn. *H. schmackeri),* and for Assam (India), *S. trochoideus.* The cercariae resemble those of *F. hepatica* except for minor details. In endemic foci they encyst on the seed pods of the water caltrop *(Trapa natans* in China, *T. bicornis* in Thailand and Bengal), the bulb of the "water chestnut" *(Eliocharis tuberosa),* and the roots of the lotus, water bamboo, and other aquatic vegetation, including *Valisneria sp., Salvinia natans,* and *Lemna polyrhiza.* People become infected from peeling off the hull or skin of these infested plants between their teeth and lips before swallowing the raw nut. Excysting in the duodenum, the metacercariae become attached to the intestinal wall and develop into adult

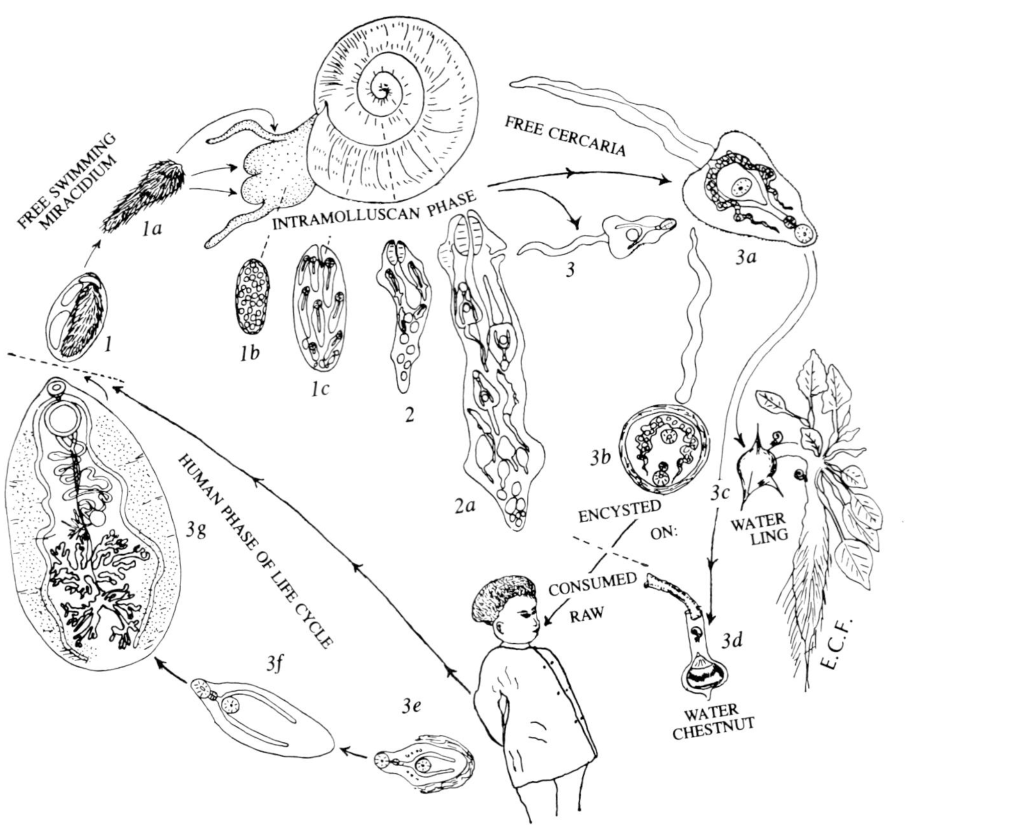

Fig. 28–10. *Fasciolopsis buski.* Diagram of life cycle. *1, 1a-1c,* First generation (i.e., egg to miracidium to sporocyst, which produces rediae). *2, 2a,* Second- and third-generation rediae, which produce cercariae. *3, 3a–3g,* Definitive generation (i.e., cercaria, to metacercaria, to adult worm, which produces eggs). (Adapted from Faust, E.C. 1929. *Human Helminthology,* 1st ed. Philadelphia, Lea & Febiger.)

worms in about 3 months. The life cycle is diagrammatically illustrated in Figure 28–10.

Epidemiology. Infection in man is associated with aquatic food plants such as roots, stems, or bulbs that are cultivated in endemic foci where the appropriate molluscan hosts abound. Deposition of human or pig feces containing the immature viable eggs of the worm directly or indirectly into shallow ponds or lakes in which the plant vectors are grown provides opportunity for the eggs to mature, hatch, and infect snails of the family Planorbidae, which are the only demonstrated molluscan hosts. Escape of the cercariae from the snail and their encystment on the plants follow in due sequence. Sadun and Maiphoom (1953) found that the rates of infection in Thailand are directly related to the amount of fecal contamination of the water in which the water caltrop *(T. bicornis)* is grown.

Pathogenesis, Pathology, and Symptomatology. The larvae of *Fasciolopsis buski* that escape from their cysts in the duodenum of man attach themselves to the duodenal or jejunal mucosa and in about 90 days develop into relatively large, fleshy worms. The damage that they produce is traumatic, obstructive, and toxic. At the sites of attachment the worms cause inflammation and ulceration of the mucosa. Large numbers of worms provoke increased secretion of mucus and may cause partial obstruction of the bowel.

In massive infections these changes may be less important than the profound intoxication and sensitization that result from absorption of the worm's metabolites into the system.

Although the worms are usually attached to the duodenal or jejunal wall, in heavy infections they adhere to the pyloric mucosa and may even be found attached to the wall of the large bowel. At each point of attachment there is a local center of inflammation and later of ulceration. The capillaries of the wall may be eroded, and hemorrhage may result. Abscesses may also develop in the wall. Large numbers of the parasites may produce acute ileus.

The unsanitary conditions that prevail where fasciolopsiasis is endemic lead to high prevalence of other parasitic infections, especially intestinal nematodes. In such a host population that suffers from the effects of multiple parasitoses, it is impossible to ascribe specific symptoms to the light or moderate infections with *F. buski,* which are the rule. This is not to say that such infections are harmless.

In heavy infections the course is more characteristic and grave, sometimes fatal (Sadun and Maiphoom, 1953; Sadun, 1957; Rahman et al., 1981).

The first *symptoms,* those of toxic diarrhea with hunger pains, manifest themselves toward the end of the incubation period. In light infections these may continue as the only conspicuous features of the disease. Heavier infections may simulate gastric ulcer, and the patient may become more asthenic. Meanwhile, generalized toxic and allergic symptoms appear, usually in the form of edema, particularly of the face, abdominal wall, and lower limbs. Ascites is common and there is generalized abdominal pain. The diarrhea, at first interrupted by periods of constipation, now becomes persistent, the stool is greenish-yellow, and it contains much undigested food. The appetite usually is normal, but anorexia, nausea, and vomiting occur. Eventually the skin becomes dry and harsh and prostration is extreme. Death is due to profound intoxication. Young (1935) found that 45% of his patients showed a leukocytosis due to an absolute eosinophilia, with a neutrophilic leukopenia. At times there is a lymphocytosis. There is no significant alteration in the erythrocytes.

Diagnosis. The diagnosis is based on the history of the case, particularly in endemic areas, and specifically on the recovery of the eggs (Fig. 28–9). Fascioliasis and the echinostomatiases *(vide infra)* must be excluded.

Treatment. Hexylresorcinol crystoids in a single dose of 0.4 g (1 to 7 years of age) to 1 g (13 years or over) was once considered the treatment of choice, but now the preparation is generally not available. Tetrachloroethylene administered as for hookworm infection is effective (Rahman et al., 1981).

Recently dichlorophen has been shown to eliminate the worms in about two-thirds of cases with reduction in egg count exceeding 80% (Idris et al., 1980). The dosage of dichlorophen administered orally was 1.5 g, 3.0 g, 4.0 g, for 9 to 15 kilogram, 15 to 22 kilogram, and 22 to 44 kilogram body weight, respectively. Niclosamide, sometimes used to treat this infection, was found to be ineffective. Praziquantel probably has become the drug of choice, given as for other trematodes (Rim et al., 1981a,b).

Prognosis. The prognosis is good to fair in early or lightly infected patients who are specifically treated, usually grave late in heavily infected ones.

Prevention. Immersion of vegetables in boiling water for a few seconds or even peeling and then washing them in clear water will provide a relatively satisfactory safeguard against individual infection. Since most endemic foci are perpetuated by use of night soil for fertilizer, prohibition of its use or sterilization of the feces is the fundamental solution to the problem.

Family Echinostomatidae
Poche, 1926

The distinctive feature of echinostomes is a head collar bearing one or two rows of spines. The tegument usually bears spines or scales, especially over the anterior part of the body; the suckers are well developed and usually near each other; the testes are tandem or somewhat diagonal, in the hindbody; the genital pore is preacetabular; the ovary is pretesticular and the uterus is coiled between the intestinal ceca on either side and anterior to the anterior testis; the eggs are large, operculate, and unembryonated when laid; and the cercariae develop in second- or third-generation rediae. Echinostomes are parasites of birds and mammals, usually in the intestine, and some species commonly develop in both groups (Beaver, 1937).

Among the numerous subgroups of echinostomes there are few with species that develop in man: In one group, the Echinostomatinae *(Echinostoma, Hypoderaeum, Paryphostomum, Artyfechinostomum,* and *Echinoparyphium),* the collar spines are arranged in two rows, interrupted ventrally but not dorsally; in the Himasthlinae *(Himasthla)* the collar spines are arranged in a single row, interrupted ventrally; and in a third group, the Echinochasminae *(Echinochasmus),* the collar spines are arranged in a single row, interrupted both dorsally and ventrally.

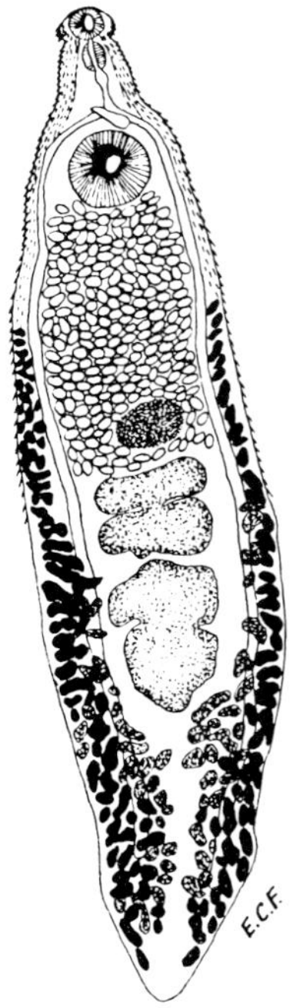

Fig. 28–11. *Echinostoma ilocanum.* Adult worm, ventral view. (× 20.) (Adapted by E.C. Faust from Odhner, T. 1911. *Echinostoma ilocanum* (Garr) ein neuer Menschen Parasit aus Ostasien. Zool. Anzeiger, *38*:65–68.)

Echinostoma ilocanum
(Garrison, 1908) Odhner, 1911

Synonyms. *Fascioletta ilocanum* Garrison, 1908; *Euparyphium ilocanum* (Garrison, 1908) Tubangui and Pasco, 1933.

Historical and Geographical Notes. Garrison first found the eggs of this fluke in native prisoners in Manila in 1907, and later recovered 21 adult worms after administration of oleoresin of aspidium. Tubangui in 1931 found that the Norway rat was a reservoir of the infection. Chen found 13.5% infection in native dogs in Canton.

Brug and Tesch reported this species of echinostome from the Paloë district of Celebes. Sandground and Bonne (1940) described the Celebes echinostome under a new name, *Echinostoma lindoensis.* Bonne et al. (1948) reported *E. ilocanum* from Java, especially in mental patients. Radomyos et al. (1982) recovered 1 to 109 *E. ilocanum* from 10 of 253 residents of northeastern Thailand who were treated with praziquantel for opisthorchiasis. They noted that *E. malayanum, E. revolutum,* and *Hypoderaeum conoideum* had been reported from man in the same region earlier.

Morphology, Biology, and Life Cycle. The worm (Fig. 28–11), which lives attached to the wall of the small intestine, is reddish-gray when alive, measures 2.5 to 6.5 mm in length by 1 to 1.35 mm in breadth, and 0.5 to 0.6 mm in thickness, and is attenuated posteriorly. At the anterior end it is provided with a circumoral disk (0.22 to 0.34 mm broad), somewhat separated from the body proper and surmounted with a crown of 49 to 51 spines, arranged as follows: 5 or 6 at each inner ventral angle, lateral to which there are 2 single, then 11 to 12 closely set ones, which are united across the dorsum by an irregularly attenuate row of 13 to 15 spines. The tegument is covered with plaque-like scales. The oral sucker (0.1 to 0.16 mm in diameter) lies in the center of the oral disk; the ventral sucker (0.4 to 0.46 mm in diameter) in the anterior fifth of the body proper. The deeply lobed testes lie one behind the other in the third fourth of the body; the ovary is just in front of the anterior testis. The vitellaria are medium-sized follicles in the lateral fields through the posterior three-fourths of the body. The uterine coils fill the intercecal field between the anterior testis and the ventral sucker.

The straw-colored operculate eggs (83 to 116 μm by 58 to 69 μm) are immature when passed in the feces (Fig. 28–12,*A*) but mature in water in 6 to 15 days (Fig. 28–12,*B*) and hatch. The miracidia penetrate the planorbid snails *Gyraulus convexiusculus* (Philippines, Java), *Hippeutis umbilicalis* (Philippines), and *G. prashadi* (India), entering through the mantle folds. They metamorphose into mother rediae and produce, in turn, daughter rediae and cercariae in 42 to 50 days.

The cercaria (Fig. 28–12,*C*) has a simple tail and a body resembling in miniature that of the adult worm, and has the same number of crown spines, but immature sex organs. After escape from the snail and a brief free-swimming existence, the cercaria may encyst (Fig. 28–12,*D*) in any freshwater mollusc, including the large species, *Pila conica* (Philippines) and *Viviparus javanicus* (Java), which are eaten raw by native populations and produce infection in these people.

Epidemiology. Human infection is acquired from eating raw snails containing encysted metacercariae.

Pathogenesis, Pathology, and Symptomatology. Inflammatory lesions may develop at the sites of attachment to the intestinal wall, together with a generalized toxic process. The clinical aspects have been minimally studied. Intestinal colic and diarrhea develop, owing to inflammatory reaction set up at the sites where the worms are attached to the intestinal wall.

Diagnosis. Based on recovery of the unembryonated egg (Fig. 28–12,*A*) in the host's stool, diagnosis requires differentiation from eggs of other echinostomes *(vide infra)* and from those of *Fasciola hepatica* and *Fasciolopsis buski.*

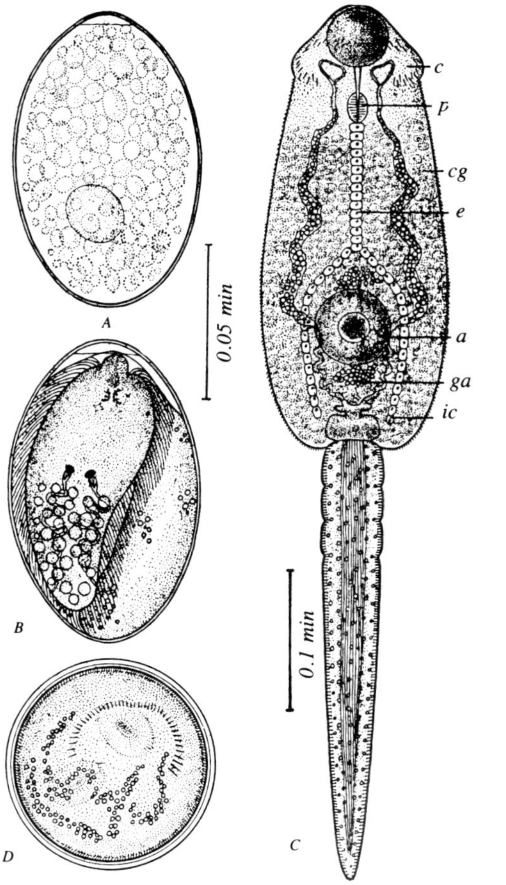

Fig. 28–12. *Echinostoma ilocanum*. Stages in the life cycle. *A*, Egg from feces. *B*, Embryonated egg with miracidium. *C*, Cercaria. *D*, Metacercaria. *a*, Acetabulum. *c*, Collarette. *cg*, Cystogenous glands. *e*, Esophagus. *ga*, Genital anlage. *ic*, Intestinal cecum. *p*, Pharynx. (Adapted from Tubangui, M., and Pasco, A. 1933. The life history of the human intestinal fluke, *Euparyphium ilocanum* (Garrison, 1908). Philip. J. Sci., *51*:581–603.)

Treatment. Praziquantel in a single dose of 40 mg/kg body weight was reported to be curative in all of nine cases; a patient given 25 mg/kg in the morning and in the evening of the same day also was cured (Radomyos et al., 1982).

Prevention. Individuals can avoid infection by not eating raw snails in endemic areas.

Other Echinostome Species Reported From Man

Echinostoma lindoense Sandground and Bonne, 1940 (syn. *Echinostoma ilocanum* of Brug and Tesch, 1937) closely resembles *E. revolutum* morphologically, especially in having 37 collar spines. In the Lake Lindoe district of Central Celebes the natives in the period 1937 through 1956 were heavily infected with this worm; as many as 249 specimens were passed following a single administration of tetrachloroethylene. The prevalence in some villages ranged from 24 to 96%. In recent years, due to changes in eating habits, infection has not been seen in people, though it is still prevalent in animals (Carney et al., 1980). *E. lindoense* has been found in Brazil where it utilizes *Biomphalaria glabrata* as its molluscan host (Lie Kian Joe, 1968). Rats and mice are experimentally good definitive hosts.

Echinostoma malayanum Leiper, 1911, was reported from two Tamil coolies at Singapore and at Kuala Lumpur, Malaysia, possibly Chiangmai, Thailand, and in the population around Batang, on the Sino-Tibetan border. This species was recovered once in North Sumatra from a 12-year-old girl who harbored 750 worms. The worm has only 43 crown spines, has a long, rather conspicuous cirrus pouch, and has relatively few, brownish-colored, operculate eggs, measuring 120 to 130 μm by 80 to 90 μm. *Lymnaea leuteola* is the first intermediate host in India. The cercariae encyst either in the same mollusc, in *Indoplanorbis exustus*, in *Gyraulus convexiusculus*, or in other snails. Dogs become infected when fed the encysted metacercariae. Man is an incidental host. The whole life cycle was demonstrated from eggs and adults obtained from a 5-year-old Indian girl in Kuala Lumpur by Lie and Virik (1963), with *Indoplanorbis* and *Lymnaea* used consecutively as molluscan hosts, and laboratory rats, white mice, and hamsters as the experimental definitive host. The child apparently became infected from eating naturally infected *Pila scutata*. The infection in the patient was light and caused no special gastrointestinal symptoms.

Echinostoma revolutum (Froelich, 1802) Looss 1899, normally a parasite of the small intestine of muskrats, of other mammals, and of ducks, geese, and fowl, was recovered from a Formosan woman following administration of *oleoresin* of *aspidium* (Anazawa, 1929). The circumoral spines consist of 37, of which 5 are found in a cluster on each side at the ventral angle. The operculate eggs measure 90 to 126 μm by 59 to 71 μm. In addition to the first intermediate molluscan host (species of *Lymnaea, Physa, Paludina, Segmentina,* and *Helisoma*), a second mollusc or tadpole is required for encystment of the metacercariae. Human infection results from eating salted or inadequately cooked snails or clams *(Corbicula producta)*. In Taiwan the prevalence in man has been estimated at 2.8 to 6.5%. Three additional human cases were reported from Indonesia (Bonne et al., 1948). Birds, dogs, and other mammals including mice are susceptible to laboratory infection, and the worm's size and morphology other than crown-spine pattern differ in the different hosts (Beaver, 1937).

Echinostoma cinetorchis Ando and Ozaki, 1923 and *E. macrorchis* A. and O., 1923, the metacercariae of which encyst in the tissues of tadpoles and frogs, are occasional human parasites in Japan and Taiwan, and were recovered incidentally in several necropsies from individuals from a rural area near Jakarta, Java (Bonne et al., 1948). Seo et al. (1980) redescribed *E. cinetorchis* from specimens collected after treatment of a 60-year-old Korean woman with bithionol (35 mg/kg body weight).

Psilorchis hominis Kifune and Takao, 1973, described as a psilostome, was found among about 30 mature *Echinostoma macrorchis* expelled from a Japanese after anthelmintic treatment. Twenty specimens were collected, all but two of which were more or less disintegrated. Whether the two on which the description was based were possibly echinostomes that had lost the crown spines apparently was not considered.

Euparyphium melis (syn. *E. jassyense*) (Leon and Ciurea, 1922), which was obtained by Leon, in 1916, from diarrheic stools of a Romanian patient, and again at autopsy of a Chinese patient (Hsü, 1940), has a head crown of 27 spines, of which 4 large ones are situated on each side at the ventral angle and the remaining 19 are disposed in an uninterrupted double row. The eggs are operculate and measure 132 to 154 μm by 79 to 85 μm. In the region of Douglas Lake, Michigan, Beaver (1939) found *Stagnicola emarginata angulata* to be the first intermediate host and tadpoles to serve as the host for the metacercarial stage. Yamaguti (1958) considered the American form to be a separate species and named it *Euparyphium beaveri*. This infection has not been studied clinically.

Hypoderaeum conoideum (Bloch, 1872) Dietz, 1909, a parasite of birds including ducks, geese, and fowl, is a common human parasite in northeast Thailand, where 55% of 254 persons examined were found to be infected (Yokogawa et al., 1965). Intermediate snail hosts include *Planorbis corneus, Lymnaea stagnalis, L. limosa, L. ovata,* and *L. palustris*. Human infection is probably acquired from eating raw snails, a common habit in Thailand.

Paryphostomum sufrartyfex (Lane, 1915) Bhalerao, 1931 (syn. *Artyfechinostomum sufrartyfex* Lane, 1915; *Euparyphium malayanum* Leiper, 1911; *Echinostoma sufrartyfex* Lane, 1915) was first obtained from an 8-year-old Assamese girl. One worm was vomited, 5 were passed following santonin, and 57 after administration of extract of aspidium. Preserved specimens averaged 9 mm long and 2.5 mm broad. Bhalerao (1931) redescribed this species from an Indian pig. The whole of the ventral surface and part of the dorsum are spinose. There is usually a constriction in the plane of the ventral sucker. The collar spines number 39 to 42, all of similar size except for one larger pair at the outer ventral angle of the disk. The oral sucker measures 0.13 to 0.2 mm, the ventral sucker, 1 mm in diameter. The deeply lobed testes lie in the third fourth of the body. The cirrus pouch is enormously enlarged, extending from a point 0.5 mm behind the posterior margin of the ventral sucker up to the genital pore. The eggs measure 90 to 125 μm by 60 to 75 μm. Lie Kian Joe (1963) believes this species to be synonymous with *Echinostoma malayanum*. The life cycle is unknown.

Artyfechinostomum mehrai, a probable synonym of *Paryphostomum sufrartyfex,* has been recorded as a human parasite on two occasions in India. The first case was a 14-year-old Hindu girl with a history of diarrhea, vomiting, anorexia, and loss of weight; worms were recovered from the vomit and from gastric contents (Raghunathan and Srinivasan, 1962). A fatal case in which death was due to malnutrition and anemia caused by the infection was reported by Reddy et al. (1964); several hundred worms were found in the intestine. This species utilizes the snail *Indoplanorbis exustus* as intermediate host. Human infection is probably acquired by eating raw snails.

Echinoparyphium recurvatum (von Linstow, 1873) (syn. *Echinoparyphium koidzumii* Tsuchimochi, 1924) is a cosmopolitan intestinal parasite of birds and mammals. According to Fain and Galal (1977), it has been recorded in man in Taiwan (identified as *E. koidzumi* by Morishita in 1929), Indonesia (in several autopsies by Bonne et al. in 1948) where it is commonly found in wild rats, again in Indonesia (in two autopsies by Lie Kian Joe in 1951), and twice in Egypt, recovered in the feces after treatment in one case and at autopsy in another. This species is relatively small, up to 3.7 mm by 0.6 mm. The body anterior to the large ventral sucker (400 μm in diameter) is strongly recurved ventrally and the tegument is spinose. The crown spines usually number 45, of which 4 or 5 are at each of the two ventral angles. It was also found in wild rat *(Arvicanthis niloticus)* in Egypt (Fain and Galal, 1977). The metacercariae encyst in tadpoles and frogs.

Echinoparyphium paraulum (Dietz, 1909), a natural parasite in the small intestine of ducks, geese, swans, and doves, has been recovered once in a human infection in U.S.S.R. (Skrjabin, 1938). It is probably a synonym of *Echinostoma revolutum*.

Himasthla muehlensi Vogel, 1933 specimens (5) were obtained by Mühlens, following use of oil of chenopodium from a German patient. He had lived for 6 years in Colombia but believed he had acquired the infection from eating several raw clams *(Venus mercenaria)* in New York City. These elongated worms measured 11 to 18 mm in length by 0.41 to 0.67 mm in breadth. Surmounting the reniform circumoral disk were 32 spines, of which 2 were situated on each side of the ventral angle and the remaining 28 were arranged in an uninterrupted horseshoe pattern around the sides and dorsum of the disk. The oral sucker measured 118 to 145 μm by 94 to 123 μm, the ventral sucker, 358 to 410 μm by 357 to 425 μm. The two testes were situated in the posterior part of the body and the transversely compressed ovary just in front of the anterior testis. In its proximal portion the uterus was broadly coiled and contained numerous irregularly ovoid eggs measuring 114 to 149 μm by 62 to 85 μm (Vogel, 1933).

The life cycle of several species involves marine operculate snails such as *Littorina* species as first intermediate host and bivalve molluscs such as *Mytilus* and *Mya* as hosts for the metacercariae (Stunkard, 1960, 1966). Gulls and other birds are natural final hosts. Clinical aspects have not been studied.

Echinochasmus perfoliatus (v. Ratz, 1908) Dietz, 1910 is a common parasite of the small intestine of dogs and cats in Hungary, Italy, Romania, the U.S.S.R., and the Far East. It has also been recovered from the pig and fox. In 1922, H. Tanabe reported a human infection in Japan that had been acquired from eating certain uncooked freshwater fishes. This echinostome varies greatly in size (0.5 to 12 mm by 0.1 to 2 mm). The entire tegument is spinose. The circumoral disk bears 24 spines in a single row which is discontinuous mid-dorsally as well as midventrally. The ventral sucker, which lies at the posterior end of the anterior third of the body, is appreciably larger than the oral sucker. The testes are large, subglobose, and situated in tandem in the third fourth of the body. Somewhat in front of the anterior testis, to the right of the midline, is the small globose ovary. The vitellaria extend from the level of the genital pore to the posterior end of the body. The uterus is relatively short, slightly coiled, and contains only a few eggs. The eggs measure 90 to 135 μm by 55 to 95 μm. The molluscan hosts are species of *Parafossarulus* (family Amnicolidae). Many fresh water fishes harbor the metacercariae, which are encysted only on the gills. The clinical aspects of human infection have not been studied.

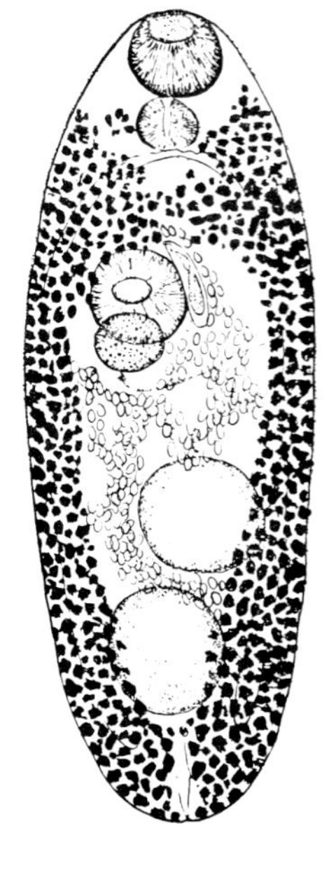

Fig. 28–13. *Plagiorchis javanensis*. Adult from intestine of man in Java. (× 40.) (From Sandground, J.H. 1940. *Plagiorchis javanensis* n. sp., a new trematode parasite of man. Rev. Med. Trop. Parasitol., 6:207–211.)

Family Plagiorchiidae
Lühe, 1901

Species of *Plagiorchis* occur in a wide range of vertebrate hosts, mostly in the intestine. The body is elongate, piriform, or ovoid. The testes are in the posterior third of the body, oblique, or opposite, and the ovary is anterior to the testes, usually near the acetabulum. The eggs are small and embryonated when laid. The cercariae, which develop in sporocysts, have a stylet in the dorsal wall of the oral sucker.

Plagiorchis philippinensis Sandground, 1940 (syn. *Plagiorchis* sp. of Africa and Garcia, 1937) was recovered by Africa and Garcia (1935a) at an autopsy in Manila, together with specimens of *Echinostoma ilocanum* and *Spelotrema brevicaeca,* from the small intestine of a native male Ilocanoan, where the inhabitants eat the grubs of certain insects believed to be the second intermediate hosts of this fluke.

Plagiorchis javensis Sandground, 1940 was obtained as a single specimen at postmortem of a native Javanese who had harbored a heavy infection of *Echinostoma ilocanum*. Figure 28–13 illustrates the characteristic features of the species and the genus.

Plagiorchis muris Tanabe, 1922, a natural parasite of several groups of birds, employs the snail *Lymnaea (Stagnicola) emerginata angulata* as both the first and second intermediate host; various aquatic insects also serve as second intermediate hosts. McMullen (1937) obtained experimental infection with this species in mice, rats, pigeons, and himself following feedings with the encysted metacercariae from the snail. *P. muris* has been reported once from man in Japan in a patient under treatment for heavy *Metagonimus yokogawai* infection (Asada et al., 1962).

Family Microphallidae
Travassos, 1920

The Microphallidae are minute intestinal parasites of a wide range of vertebrates. They are usually piriform and spinose, with small oral and ventral suckers, and short intestinal ceca. The testes are opposite or diagonal, the ovary usually pretesticular, and the eggs are small. In morphology this group resembles the Heterophyidae; in its life cycle it resembles the Plagiorchiidae.

Spelotrema brevicaeca (Africa and Garcia, 1935) Tubangui and Africa, 1939 (syn. *Heterophyes brevicaeca* Africa and Garcia, 1935) has been reported from man in the Philippines on several occasions. Its eggs were found in lesions in the heart, brain, and spinal cord of persons dying of acute cardiac dilatation (Africa and Garcia, 1935b, Africa et al., 1935, 1937a,b). The complete life cycle is unknown. Encysted metacercarial stages of *Spelotrema* sp. have been found in the crab, *Carcinus maenas* (Hockley, 1964).

S. brevicaeca is piriform to triangular in shape, thinner anteriorly than posteriorly, 0.5 to 0.7 mm long by 0.3 to 0.4 mm wide with a spinose cuticula. The acetabulum is slightly larger than the oral sucker; the latter is subterminal, 0.06 to 0.1 mm in diameter. The esophagus has a relatively thick wall and narrow lumen. The intestinal ceca are wide but short, not reaching posteriorly beyond the middle level of the acetabulum. The genital pore lies near the left margin of the acetabulum. The testes are globular to oval in shape, situated one on each side of the body behind the level of the acetabulum. The ovary is usually larger than the testes. The uterus is moderately long and loosely coiled, opening into the genital sinus. Vitellaria are prominent, mostly post-testicular. Eggs are small, yellowish, operculated, 15 to 16 by 9.4 to 10 μm.

Family Philophthalmidae

The Philophthalmidae are parasites that live in the conjunctival sac or cloaca, occasionally in the intestine, of birds. *Philophthalmus* sp. was reported once in man in Yugoslavia, causing symptoms suggestive of trachoma (Markovic, 1939).

Family Dicrocoeliidae
Odhner, 1910

The Dicrocoeliidae are small- to medium-sized flukes that live in the bile or pancreatic ducts of birds and mammals. They are elongate, with well-developed suckers. A notable feature is the position of the testes anterior to the ovary. The eggs are small, with a thick brown shell, and embryonated when laid.

Dicrocoelium dendriticum
(Rudolphi, 1818) Looss 1899
(The lancet fluke, producing dicroceliasis)

Synonyms. *Fasciola lanceolata* Rud., 1803 (homonym); *Fasciola dendritica* Rud., 1819.

Geographical Distribution and Prevalence. This fluke is a common parasite in the biliary passages of sheep, deer, and other herbivorous and omnivorous mammals in Europe (80% infections in sheep, South Poland), Turkey, northern Africa, northern Asia, Turkestan, parts of the Far East, and less commonly in North and South America. Numerous cases of human infection have been reported from the U.S.S.R. and elsewhere. Most of these are cases of false parasitosis resulting from people having eaten infected liver of sheep, with the evacuation of eggs in the feces, but genuine human cases have been diagnosed in Europe, Asia, and Africa.

Morphology, Biology, and Life Cycle. The worm (Fig. 28–14) is lancet-shaped, flat, and transparent. It measures 5 to 15 mm long by 1.5 to 2.5 mm broad. Its integument is aspinose. The acetabulum (0.5 to 0.6 mm in diameter) lies at the beginning of the second fifth of the body. The testes are slightly lobed, somewhat obliquely disposed, and situated anterior to the ovary, just behind the ventral sucker. The subglobose ovary lies to the right of the midline and somewhat anterior to the equator of

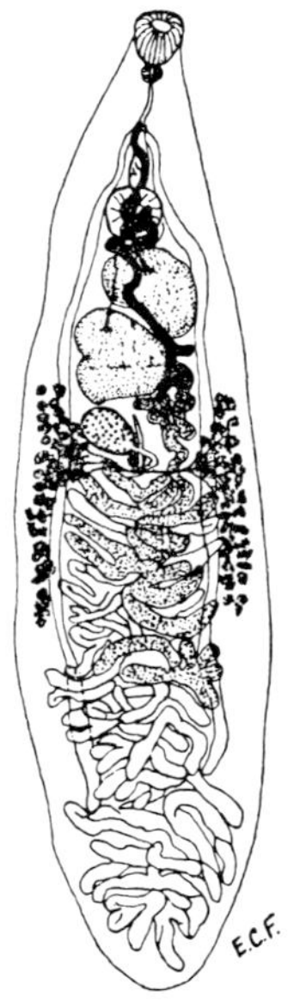

Fig. 28–14. *Dicrocoelium dendriticum*. Adult, ventral view. (× 10.) (Adapted from original by M. Braun. In Faust, E.C. 1929. *Human Helminthology*, 1st ed. Philadelphia, Lea & Febiger.)

the worm. The discrete vitelline follicles occupy lateral fields in the middle of the body. The ootype is situated somewhat anterior to the middle of the body. The uterus proceeds in intricate coils to the posterior end of the worm, coils back to the middle of the body, then ascends more loosely up to the genital pore. The eggs (Fig. 28–15) are thick-shelled, distinctly operculate, deep golden brown, and measure 38 to 45 μm by 22 to 30 μm. They are fully embryonated when passed in the feces, are resistant to desiccation, but do not hatch in water. However, on ingestion by appropriate land snails *(Zebrina detrita, Helicella candidula, H. ericetorum, Euomphalia strigella,* and *Abida frumentum* in Germany, *Z. detrita* in Switzerland, *Z. detrita* and *H. ericetorum* in Yugoslavia, *H. unifasciata* in the Moscow area and *Fruticocampylaea narzanensis* in central Caucasus, U.S.S.R., *H. ericetorum* and *Cochlicella acuta* in Scotland, and *Cionella lubrica* in New York State), the intermediate stages (two generations of sporocysts) develop with the eventual production of cercariae *(Cercaria vitrina* v. Linstow, 1887).

The cercaria (Fig. 28–16) is aspinose, measuring 700 by 70 μm when elongated and 400 by 200 μm when contracted. The oral sucker has a minute stylet set in its dorsal wall. There are 12 pairs of posterior and 3 pairs of anterior penetration glands. The long tail tapers distally to a small diameter.

Epidemiology. The cercariae emerge from the molluscan host when rains follow a long dry period. They migrate out of the daughter sporocysts to the snail's respiratory chamber, where up to several hundred of them agglomerate into slime balls. As the snail host crawls along, the slime balls are left behind on grass. In a series of careful studies, Krull and Mapes (1952, 1953) conclusively demonstrated that an ant, *Formica fusca,* is a required intermediate host for development of the metacercarial cyst in the United States, and that infection of the mammalian definitive host results from ingestion of infected ants. Vogel and Falcão (1954) confirmed this study for Germany, while in central Caucasus, U.S.S.R., Popov and Kalitina (1962) found *F. cinerea* and *F. picea* to be the intermediate hosts.

Pathogenesis, Pathology, and Symptomatology. Mechanical and toxic effects are less marked than in fascioliasis. The symptoms in human cases include chronic constipation and flatulent dyspepsia, with enlarged liver and systemic toxemia, diarrhea, and vomiting.

Diagnosis. The diagnosis is based on recovery of the characteristic eggs (Fig. 28–15) in the stool or by duodenal drainage, but care must be exercised to exclude spurious infection.

Treatment. Although success has been reported with the use of thymol and with stibophen (Fuadin) (Scheid et al., 1950), no drug for the treatment of human dicroceliasis has been proved dependable. Praziquantel and dichlorophen, which are effective against other trematodes, deserve a trial.

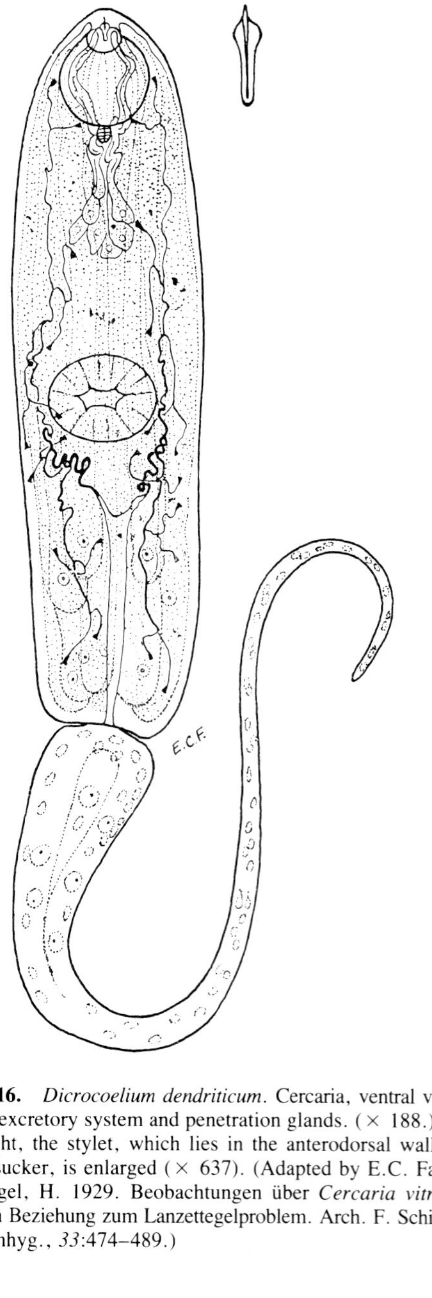

Fig. 28–16. *Dicrocoelium dendriticum.* Cercaria, ventral view showing excretory system and penetration glands. (× 188.) At upper right, the stylet, which lies in the anterodorsal wall of the oral sucker, is enlarged (× 637). (Adapted by E.C. Faust from Vogel, H. 1929. Beobachtungen über *Cercaria vitrina* und deren Beziehung zum Lanzettegelproblem. Arch. F. Schiffs- u. Tropenhyg., *33*:474–489.)

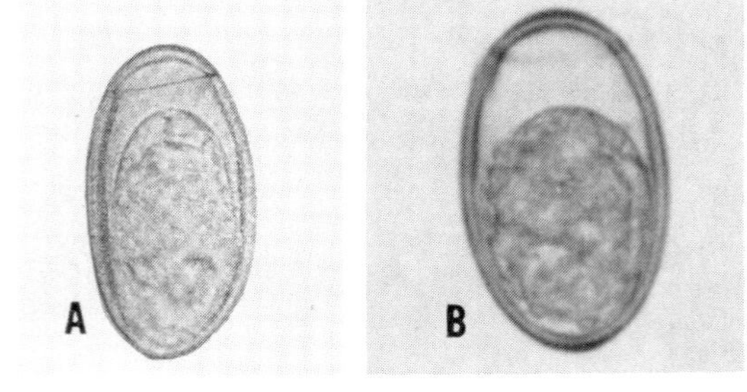

Fig. 28–15. *Dicrocoelium,* eggs. *A, D. dendriticum.* (× ca. 800.) *B, D. hospes,* from feces of man in Ghana. (× 860.) (*A* from Craig, C.F., and Faust, E.C. 1937. *Clinical Parasitology,* 1st ed. Philadelphia, Lea & Febiger. *B* by P.C. Beaver.)

Prevention. Fresh herbs collected from grazing areas for use as food for humans should be washed to remove ants.

Related Species. *Dicrocoelium hospes,* a common parasite of cattle in Africa, has often been reported as a human parasite. Most of these were cases of spurious parasitism due to ingestion of infected livers, but two genuine human infections in cattle boys have been reported from Ghana (Odei, 1966). This fluke is differentiated from *D. dentriticum* by its more uniform width, greater length, position of the testes, and shape of the uterus, which is arranged in an ascending and descending loop. The eggs are thick-shelled, distinctly operculate, deep golden brown, and measure 38 to 45 by 22 to 30 μm.

Eurytrema pancreaticum (Janson, 1899) Looss, 1907, a related fluke, is common in the pancreatic duct of hogs in south China and in the biliary and pancreatic passages of sheep, goats, cattle, and water buffaloes in the Orient and Brazil, less common in *Camelus bactrianus* (North China) and the monkey, *Macaca syrichta fascicularis*. The first intermediate hosts are land snails (species of *Bradybaena* in Malaysia, *Ganesella* species in China) and the second intermediate hosts are grasshoppers (Basch, 1965; Tang et al., 1979). According to Tang and Tang (1977), *E. pancreaticum* has been reported in man once each in Hong Kong and in Jiang-su Province, China and in at least six cases in Japan. Diagnosis was confirmed by recovering adult worms from the pancreatic ducts of man in two cases (Ishii et al., 1983). *E. pancreaticum* is much stouter and broader than *D. dentriticum* and it has slightly ruffled margins. The oral sucker is much larger than the ventral sucker. The two testes are deeply notched and are usually transversely disposed in the acetabular plane. The small, notched ovary lies slightly to one side of the midline some distance posteriad. The vitellaria are small, finger-like follicles lying in the lateral fields. The uterus is intricately coiled medial to the ceca from the subcaudal margin to the anterior face of the ventral sucker. The eggs are indistinguishable from those of *D. dendriticum*. The pathological and clinical aspects of *E. pancreaticum* infection in man were studied by Ishii et al. (1983).

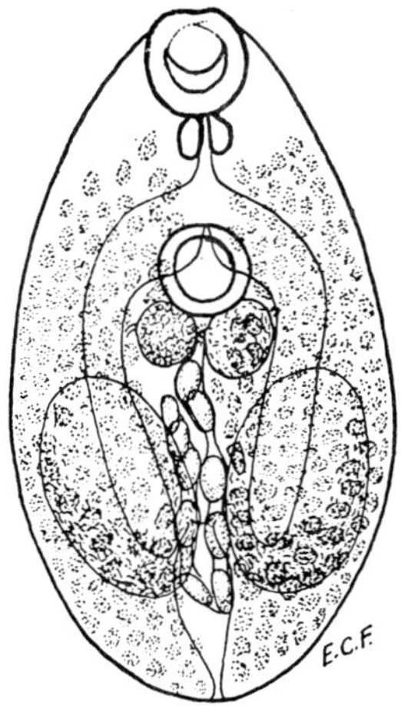

Fig. 28–17. *Troglotrema salmincola*. Adult, ventral view. (× 160.) (Adapted from Witenberg, G. 1932. On the anatomy and systematic position of the causative agent of so-called salmon poisoning. J. Parasitol., *18*:258–263.)

Family Troglotrematidae

Members of the family Troglotrematidae are usually thick-bodied and rounded, often with the ventral side relatively flat or concave. The acetabulum is near the middle of the body immediately anterior to the genital pore, and the integument bears spines or scales. Troglotrematids are parasites of birds and carnivorous mammals; their normal location is in the lungs, head sinuses, kidneys, or intestine. Those that are frequently or occasionally found in humans are species of *Troglotrema* and *Paragonimus*. A closely related family is Achillurbainiidae.

Troglotrema salmincola (Chapin, 1926) Witenberg, 1932 (Producing "salmon poisoning")

Synonyms. *Nanophyes salmincola* Chapin, 1926; *Nanophyetus salmincola* Chapin in Hall, 1927; *N. schikhobalowi* Skrjabin and Podjapolskaja, 1931.

Geographical Notes. *Troglotrema salmincola* has been associated with "salmon poisoning" in the dog on the Pacific coast of North America and Canada, and has also been found in the coyote, fox, raccoon, mink, and lynx in North America and reported from aborigines of eastern Siberia. Human infection is incidental to that in reservoir hosts.

Morphology, Biology, and Life Cycle. The minute, piriform worms (Fig. 28–17), which live in the intestinal wall of their host, measure 0.8 to 1.1 mm in length by 0.3 to 0.5 mm in breadth. The oral sucker (0.15 to 0.18 mm in diameter) is slightly larger than the ventral sucker, which lies midventrally. Two large, ovoid testes are situated side-by-side in the posterior third of the body. The vitellaria consist of numerous discrete follicles in the lateral fields extending from the plane of the esophagus to the posterior end of the worm. The uterus coils between the excretory bladder and the genital sinus, which lies beside the ovary just behind the ventral sucker. The eggs, developed a few at a time, are broadly ovoid, operculate, relatively thick-shelled, yellowish, and measure 60 to 80 μm by 34 to 50 μm. Bennington and Pratt (1960) found the molluscan host to be *Oxytrema silicula*. The cercaria, which develops in a redia, measures 270 μm in length by 80 μm in breadth, has a simple oral stylet, six penetration glands, and a microcercous tail. The free-swimming cercaria becomes attached to and later encysts primarily in the kidneys of salmon and trout, although many other species of fish are susceptible to the infection, and the Pacific giant salamander, *Dicamptodon ensatus*, has been found naturally infected (Gebhardt et al., 1966). When eaten raw, the metacercariae in these organs produce infection in the definitive host in whose small and large intestine the worms become deeply embedded. In a survey of an enzootic area in western Oregon, 79 of 149 wild animals, including 10 mammalian and 2 avian species, were found to harbor *T. salmincola*. Most heavily infected were the raccoon, *Procyon lotor*, and the spotted skunk, which are probably principal natural definitive hosts (Schlegel et al., 1968).

Clinical Aspects. The fluke *per se* is apparently not clinically important, but an associated disease-producing entity, a rickettsia, *Neorickettsia helmintheca*, produces severe, frequently fatal symptoms in dogs, coyotes, and foxes, and is possibly a human pathogen. In this infection the helminth replaces an arthropod as vector of the pathogenic agent (Philip, 1958). In dogs, after an incubation period of 6 to 10 days, there is a sudden onset, with complete loss of appetite and marked elevation of temperature, and the sensory apparatus is depressed. The temperature then drops to normal during the next week or 10 days. Purulent discharges and edema of the eyes occur during the first few days. Beginning on the fourth day vomiting is practically uncontrollable, especially after drinking water. Dysentery soon develops. Following the disappearance of these symptoms, the temperature frequently drops and death ensues within 24 to 48 hours. Recovery from one attack confers immunity. This severe syndrome has not been described for man.

Treatment. The old treatment was with oleoresin of aspidium. It is probable that a new drug, such as praziquantel, effective against other trematodes would be preferable. Abstinence from eating raw fish is indicated.

Paragonimus westermani (Kerbert, 1878) Braun, 1899 (The Oriental lung fluke, producing paragonimiasis)

Synonyms. *Distoma westermani* Kerbert, 1878; *D. ringeri* (Cobbold, 1880; *D. pulmonum* Baelz, 1880).

Historical and Geographical Notes. This fluke was first found by Kerbert in the lungs of two Bengal tigers that died in the Hamburg and Amsterdam Zoological Gardens. In 1879, a Portuguese resident of Formosa died of a ruptured aneurysm and on autopsy by Ringer was found to have a pulmonary parasite, which Patrick Manson recognized as a distomate fluke. A year later Manson found eggs in rusty-brown sputum of a Chinese patient who had lived in Formosa (Taiwan). These eggs resembled those expressed from Ringer's flukes. On examination of this material, Cobbold (1880) pronounced it a new human fluke which he named after Ringer. Meanwhile Baelz (1880) found the eggs in cases of hemoptysis in Japan and in 1883 discovered the flukes. Later Yamagiwa (1892), in Japan, and Musgrave (1907), in the Philippines, found the flukes in atypical foci in the body, including the brain and abdominal viscera. Japanese investigators, particularly Kobayashi and S. Yokogawa, contributed to elucidation of the complete life cycle.

Heavily infected areas are confined primarily to the Far East, including Japan, Korea, and Taiwan. Isolated foci exist in China and Manchuria, Vietnam, the Philippines, Thailand, Malaysia, Bengal, Malabar, Madras, and Assam. The infection is also recorded as endemic in New Guinea, the Solomon Islands, Java, Sumatra, American and British Samoa, Cameroon Republic, Zaire, Nigeria, Peru, Ecuador, Colombia, Costa Rica, Mexico, and Venezuela, but it is now known that the infection in some of these foci is caused by species other than *P. westermani*. Lung fluke infection reported for Matto Grosso State, Brazil was based on an error that originated with Diesing (1850). Human infection has not been observed in Venezuela within recent years. Reviews of the subject of human paragonimiasis were published by S. Yokogawa et al. (1960) and M. Yokogawa (1975, 1982).

Morphology, Biology, and Life Cycle. The adult worm (Fig. 28–18), which typically lives encapsulated in pockets of the lungs, is a plump, ovoid fluke, abruptly rounded anteriorly and somewhat more tapering posteriorly. It measures 7.5 to 12 mm in length, 4 to 6 mm in breadth, and 3.5 to 5 mm in thickness. Living worms have a reddish-brown hue; preserved specimens are gray or grayish-brown. The integument is provided with scale-like spines, entire or toothed, on all surfaces of the worm. The oral sucker and ventral sucker are subequal (0.75 to 0.8 mm in diameter). The latter is situated in the median plane somewhat anterior to the equator. The excretory bladder is long, slightly convoluted, reaching from the posterior extremity to the level of the pharynx. The irregularly, deeply lobed testes are situated nearly side-by-side and midway between the ventral sucker and the posterior end. There is no cirrus pouch and no true cirrus organ. The male system joins the female duct, the two opening jointly into the genital atrium just behind the acetabulum, slightly to the right of the midline. The ovary is a large, lobed organ on the right or left side at about the same level as the acetabulum. The vitellaria consist of extensively branched follicles situated in the lateral fields and overlying the ceca from the region of the pharynx to the caudal extremity. The main mass of the uterus consists of a tightly coiled rosette on the side opposite the ovary and slightly anterior to the plane of the ovary. It opens externally through the metraterm into the common genital sinus.

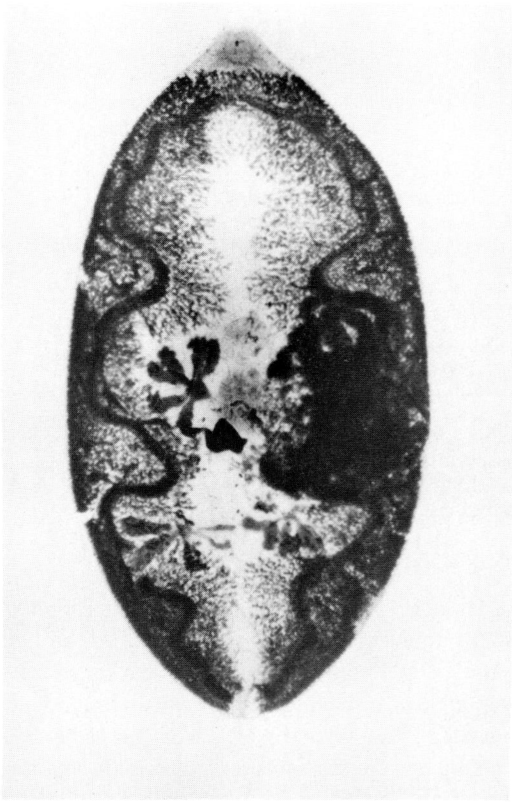

Fig. 28–18. *Paragonimus westermani*. Adult, ventral view. (× 8.5.) (Courtesy of Dr. M.D. Little.)

The eggs of *P. westermani* (Fig. 28–19,*A*) are broadly ovoid, with a distinct, relatively flattened operculum, are somewhat thickened at the abopercular end, are golden-brown, measure 80 to 118 μm by 48 to 60 μm, and are unembryonated when deposited. From the pulmonary pockets they escape through the bronchioles, are coughed up, voided in the sputum, or are swallowed and passed in the feces. In other anatomical sites they reach the outside world only when the capsules become abscessed and rupture.

On discharge from the body the eggs require at least 2 weeks for complete embryonation, where-

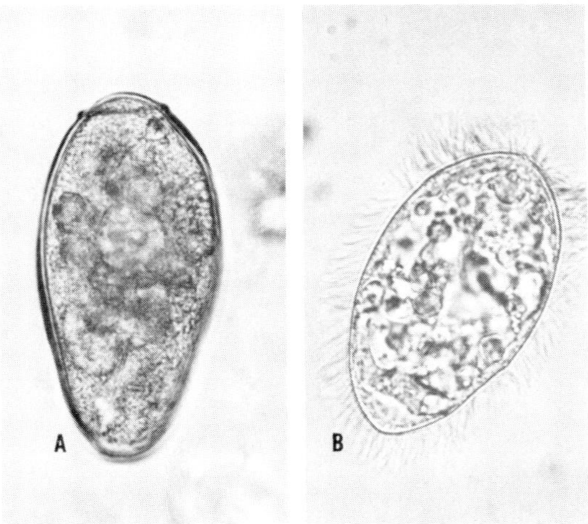

Fig. 28–19. *Paragonimus. A,* Egg of *P. westermani* from human feces. (× 400.) Eggs in feces or sputum contain a clear ovum or early embryo surrounded by granular vitelline cells. *B,* Miracidium hatched from an egg after 3 weeks of incubation in a Harada-Mori culture of feces from an infected animal. (× ca. 300.) (From Faust, E.C., Beaver, P.C., and Jung, R.C. 1975. *Animal Agents and Vectors of Human Disease,* 4th ed. Philadelphia, Lea & Febiger. Courtesy of Dr. M.D. Little.)

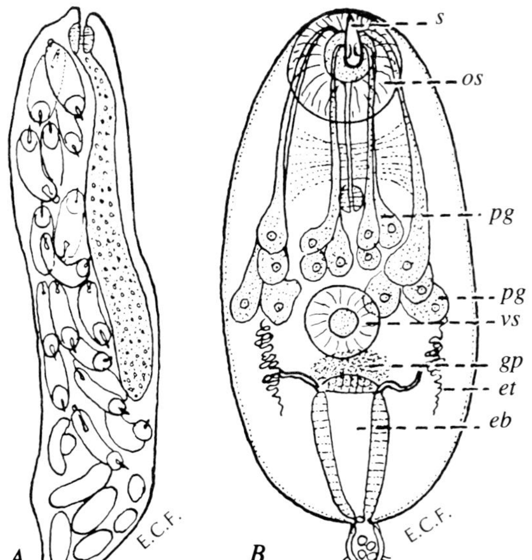

Fig. 28–20. *Paragonimus westermani. A,* Second-generation redia. (× 67.) *B,* Cercaria from snail host in Fukien Province, China. *eb,* Excretory bladder. *gp,* Genital primordium. *os,* Oral sucker. *pg,* Penetration glands. *s,* Stylet. *vs,* Ventral sucker. (× 260.) (Adapted by E.C. Faust from Tang, C.C. 1940. A comparative study of two types of *Paragonimus* occurring in Fukien, South China. Chinese Med. J., Suppl. 3:267–291.)

upon they hatch (Fig. 28–19,*B*) and the miracidia escape to become free-swimming.

The typical molluscan hosts are certain species of the families Pleuroceridae and Thiaridae: *Semisulcospira libertina* in Japan, Korea, Taiwan, and China; *Brotia asperata* in the Philippines; and *Brotia costula episcopalis* in Malaysia (Miyazaki, 1974). Within the snail during a period of several weeks a sporocyst and a redia generation are produced, followed by the development of cercariae, about 20 of which may be seen at one time within the redia (Fig. 28–20), in all stages of development. These cercariae are microcercous, with an ellipsoidal body measuring 200 to 220 μm in length by 50 to 80 μm in breadth, a minute oral stylet, and a knob-shaped tail that has several conspicuous posteriorly directed spines (Fig. 28–20,*B*). The oral sucker is larger than the ventral sucker. There are two types of penetration glands opening through the ducts at the side of the stylet: 4 pairs of large, deeply staining lateral glands, and 3 pairs of small, lightly staining median glands.

After leaving the snail, the cercariae (suspended in the fast-flowing stream which is the habitat of the intermediate hosts) may be carried long distances before reaching the crustacean host. When drawn into the gill chamber of a crayfish or crab, the cercariae invade the tissues and encyst to become metacercariae. Crabs and crayfish can also become infected by eating the infected snail. Encystment involves the formation of an inner cyst membrane produced by the cercaria and an outer fibrous membrane produced by the host. Several genera and numerous species of freshwater crabs serve as second intermediate hosts: species of *Eriocheir* and *Potamon* in Japan, Korea, Taiwan, and China; *Potamon* in Thailand; *Potamiscus, Parathelphusa,* and *Siamthelphusa* in Malaysia; and *Parathelphusa* in Taiwan, Philippines, and Ceylon. Hosts of the metacercaria also include species of crayfish, *Cambaroides* in Korea, far eastern USSR, and China, and *Procamburus* in Japan, and a species of shrimp, *Palaemon,* in Korea. The most significant vectors of human paragonimiasis are *Eriocheir japonicus* in Japan and Taiwan, *Eriocheir sinensis* in Korea and China, *Potamon denticulatus* in China, *Potamon smithianus* in Thailand, and *Cambaroides similis* in Korea.

Encysted metacercariae are located mainly in the vessels of the gills and in the muscles. The cyst wall is about 20 μm thick and the cyst diameter about 0.4 mm. The oral sucker, smaller than the ventral sucker, retains the stylet of the cercaria. A long period of development is necessary before the fully developed infective metacercaria is formed.

The size increases gradually, the esophagus and ceca of the digestive system become fully developed, and the excretory bladder elongates and becomes filled with large granules, all of which requires 6 to 8 weeks in summer. When infective, the metacercaria, freed from the cyst and fixed, measures 0.4 to 0.7 mm by 0.15 to 0.27 mm; measurements before fixation range from 0.8 to 1.1 mm in length and 0.27 to 0.38 mm in width. The entire integument is beset with minute spines. The stylet persists. Red granules scattered over all parts of the body near the surface give the infective-stage metacercaria a pinkish color.

When the definitive host (human or reservoir) ingests parasitized crab or crayfish meat (frankly raw, soaked in rice wine, or salted), the metacercariae excyst in the duodenum and migrate through the intestinal wall in 30 to 60 minutes, reaching the abdominal cavity in 3 to 6 hours. They enter and remain in the abdominal wall for several days before migrating through or around the diaphragm to the pleural cavity and lungs, finally arriving in the vicinity of the bronchioles, where they develop to adult worms usually in pairs in tissue capsules laid down by the host (Yokogawa et al., 1962). Lumsden and Sogandares-Bernal (1970) noted that the capsule surrounding the worms is largely lined by hypertrophied bronchiolar-like epithelium. Although pairing is usual, cross-fertilization is not essential (Fan and Chiang, 1970; Fan et al., 1974). Moreover, in mixed infections of two or more species, individuals of different species can develop normally together in the same cyst (Hatsushika, 1981). The circuitous route of migration allows the worms to lodge and mature in ectopic locations such as the mesentery, groin, pleura, or brain. The time required for completion of their development in the definitive host is 65 to 90 days. Infections in man may persist for 20 years.

Epidemiology. The natural definitive hosts other than man include the tiger, cat, wildcat, leopard, panther, fox, wolf, dog, pig, beaver, wolverine, "pencilled cat" *(Nyctereutes procyonides),* civet cat *(Viverra zibetha ashtoni),* Chinese lesser civet cat *(Viverricula indica pallida),* crab-eating mongoose *(Herpestes urva),* and Indian mongoose *(H. javanicus rubrifrons).* Infection is acquired from eating infected crabs or crayfish in endemic regions, which include India, Sri Lanka, Thailand, Indonesia, Malaysia, Philippines, China, Taiwan, Japan, Korea, and USSR. Records of its occurrence elsewhere are for the most part based on erroneous identification of different species. Although human exposure may result from eating raw crabs or crayfish harboring the encysted metacercariae, usually it is occasioned by eating the soft parts, including leg muscles, which have been previously placed in brine, vinegar, or wine. Such liquids coagulate the crustacean tissues but do not sterilize the cysts. Exposure may occur from contamination of fingers, kitchen utensils, and food with the metacercarial cysts in preparation of foods containing infected crabs. The juice of crabs in Japan, and of crayfish in Korea, is used as a remedy for fever and for diarrhea, and thus serves as a minor source of *Paragonimus* infection (Yokogawa, 1974). A noteworthy observation is that immature worms in the muscle tissue of the wild boar are infective, and human infection is probably acquired by eating raw meat from these animals (Miyazaki and Hirosi, 1976; Miyazaki, 1978). The eggs of the parasite, usually coughed up in the sputum and expectorated, or swallowed and passed in the feces, are the source of infection for the intermediate hosts. Human pollution of fresh water is probably less important than that produced by reservoir hosts.

Pathogenesis and Pathology. The introduction of Southeast Asian refugees into parts of the United States where edible freshwater crustacea are abundant creates a potential for autochthonous infection. Crayfish *(Cambarus clarkii)* in Louisiana serve as intermediate host for *P. kellicotti.* From the time the cysts of *Paragonimus westermani* are ingested in soft tissues of crabs or crayfishes until the larvae reach the sites where they settle down and develop into adult worms, there is no indication of any significant pathological process. The worms in tissues adjacent to, and usually opening into, the lungs typically provoke a granulomatous reaction that gradually proceeds to development of fibrotic encapsulation, except for the opening into the respiratory tract. In the abdominal cavity an abscess usually develops around the worm. In muscular tissue the worms themselves may provoke the formation of a cyst, while eggs that filter into adjacent tissues may become the centers for pseudotubercles. The worms reach and develop in ectopic foci because of the devious migration route from the intestine to the lungs.

In the lungs these flukes provoke leukocytic infiltration, followed by development of a broad layer of fibrous tissue on the outside, thus producing a thick cystic encapsulation of the parasite (Fig.

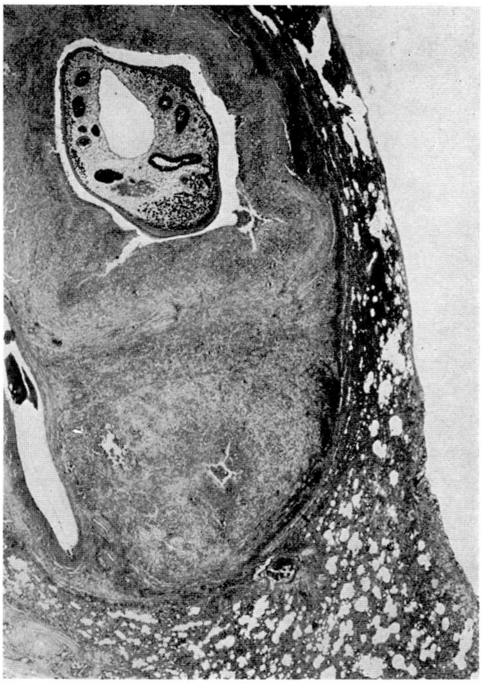

Fig. 28–21. *Paragonimus westermani*. Section of lung showing a worm (arrow) surrounded by leukocytic infiltration and fibrous encapsulation. (× ca. 5.) (From Faust, E.C. 1929. *Human Helminthology*, 1st ed. Philadelphia, Lea & Febiger.)

28–21). These cysts are most common in the deeper layers of the lungs and are usually about 1 cm in diameter. Typically there are small blood vessels within the matrix of the capsule and bronchial or bronchiolar exits. Immediately around the worm is an accumulation of blood-tinged, thick, purulent fluid, with flecks of rusty material, made up of clusters of golden-brown eggs. In addition to the lungs, these cysts may be found in the liver, intestinal wall, mesenteric lymph nodes, muscles, testes, brain, peritoneum, or pleura, where they are distinguished by their peculiar slate-blue color. About 30 cases of paragonimiasis of the greater omentum have been reported from Japan alone (Hatsushika and Endo, 1978). In these cases the worms or eggs are either in lymph nodes or in nodules consisting of fibrous tissue. Twenty-seven cases of spinal paragonimiasis have been reported, in the majority of which an extradural lesion in the lower thoracic area caused a spastic paraplegia (Oh, 1968). Musgrave (1907) recorded four fatal cases with marked cardiac involvement, and he grouped the lesions into four categories: (1) nonsuppurative, with eggs infiltrated in host tissue, leading to round cell and connective tissue reaction and usually to abscess formation; (2) tubercle-like, in which the abscess may contain caseous material; (3) suppurative; and (4) ulcerative, in which healing is only partially successful.

Symptoms. Paragonimiasis is usually insidious in its onset and mildly chronic in its course, but there may be an initial episode of chills and fever. Fulminating cases with fatal termination have been recorded. The symptoms are characteristically referable to the chest, abdomen, the lymph glands, or the brain.

In the *lungs* there is generalized or localized diffuse fibrosis, cystic dilatation of the bronchi, persistent pneumonia, and tubercle-like abscesses, with an accompanying persistent cough and hemoptysis, frequently attended by distressing chest pain (Fischer et al., 1980). The sputum is viscous, flecked with dark golden-brown particles (i.e., the eggs of the parasite), and is at times streaked with blood. Occasionally there is profuse hemoptysis following paroxysmal coughing. The manifestations are more severe after physical exertion. There are no characteristic auscultatory findings, but hyper-resonance is present. Lü et al. (1957) studied 195 cases in children in the Municipal Hospital at Ch'ang Ch'un, Manchuria, and reported that the most common symptoms were productive cough (at times with blood), chest pain, and night sweats. X-ray films of 100 patients demonstrated 51% with patchy, cloudy infiltration of the lungs, 35% with nodular shadows, 6% with calcified spots, 24% with pleural thickening, 10% with interlobular pleuritis, and 5% with pleural effusion. The physical signs may suggest bronchopneumonia, or bronchiectasis with pleural effusion.

In the *abdominal type* the worms produce less definite symptoms. Usually there is dull abdominal pain. Moderate rigidity and tenderness on deep palpation may be present. Lodged in the intestinal wall they provoke diarrhea, with eggs in the stools. *Glandular involvement* may be accompanied by fever; in the dermis or subcutaneous tissue abscesses are apt to develop. In the *brain* the worms are responsible for epilepsy of the Jacksonian type, such as occurs in cerebral cysticercosis (Mitsuno et al., 1952). In endemic foci in Japan and Korea, brain symptoms in children under 15 years of age have frequently been attributed to infantile paralysis, cerebral hemorrhage, encephalitis, or meningitis, whereas the actual causative agent was this fluke. The intelligence quotient was subnormal in

90% of patients with cerebral involvement tested in Korea (Oh and Jordan, 1967).

There is usually a local eosinophilic infiltration around the paragonimiasis abscesses, and peripheral eosinophilia may at times result (81% of cases seen by Lü et al. [1957]).

The disease may persist in chronic form for years after the patient has left an endemic area.

Diagnosis. Specific diagnosis is based on the finding of characteristic eggs (Fig. 28–19,*A*) in sputum, aspirated pleural effusion, feces, or from cutaneous lesions. The sputum is frequently blood-tinged and is peppered with rusty-brown flecks, consisting of clumps of eggs. Clinically the pulmonary type must be distinguished from bronchopneumonia, bronchiectasis, tuberculosis, bronchial spirochetosis, and pleural effusion. The cerebral type requires differentiation from nonparasitic brain tumors, hydatid cyst, cysticercosis, and encephalitis and meningitis of other etiology. The abdominal type is difficult to diagnose preoperatively, but in most cases complement-fixation tests are positive. Yokogawa et al. (1962) observed that in the complement-fixation test antibody titer is positive only during the life of the parasite, whereas the intradermal test with the same type of antigen (V.B.S.) may remain positive for many years following recovery from the infection. Hence the intradermal test, if positive, is not necessarily an index of active paragonimiasis (see Technical Appendix).

Generalized infection must be differentiated from filariasis bancrofti, histoplasmosis, toxoplasmosis, miliary tuberculosis, and myositis.

Treatment. Emetine hydrochloride is believed by some clinicians to be curative, but authorities are not in agreement.

Since 1962 the drug of choice in the treatment of paragonimiasis has been bithionol (2.2′ -thiobis [4.6 -dichlorophenol]) (Yokogawa et al., 1962). The drug is given orally in a dosage of 30 to 50 mg per kg body weight on alternate days for 10 to 15 doses (Cabrera and Fevidal, 1974). Temporary side effects of the drug include digestive disturbances and skin eruptions. Two recently studied drugs are praziquantel and niclofolan. Praziquantel in a dosage of 25 mg per kg body weight three times daily after meals for 1, 2, or 3 days was shown to cure 71.4, 89.5, and 100% of infections with no demonstrative side effects (Rim et al., 1981b). This drug promises to displace bithionol as the drug of choice. The patient may cough up and expectorate dead and living worms 8 to 20 days after treatment with praziquantel (Vanijanonta et al., 1981).

Niclofolan in a single dose of 2 mg per kg body weight is also highly effective. Side effects include transient rise in serum transaminase and blood urea levels, pain of muscles of the ischiosacral area and lower extremities, sweating, and lassitude. Niclofolan treatment of patients in Cameroon gave 100% cure rate (Ripert et al., 1981).

Clinical symptoms are usually less severe after 5 or 6 years, although the worms are known to persist in the tissues as long as 20 years or more. Infection is believed to confer partial immunity.

Prognosis. The prognosis is usually fair to good, except when worms become lodged in critical foci or when the disease develops to a generalized fulminating condition.

Prevention. Infection may be prevented by abstaining from eating raw, freshly salted, or inadequately cooked crabs or crayfish, or by contaminating fingers with the metacercarial cysts while preparing infected crustaceans for cooking. Since immersion of the infected crustacean host in rice wine or strong brine will not kill the cysts, it is imperative that these animals be boiled or fried in deep fat to insure against infection. Eliminating human paragonimiasis would probably have little effect on the prevalence of infection in reservoir animal hosts.

Other Species of Paragonimus

Yokogawa (1975) listed 31 species of *Paragonimus*, including *P. westermani*, but noted that some may not be valid, namely, *P. ringeri* Cobbold, 1880; *P. pulmonalis* Baelz, 1883; *P. edwardsi* Gulati, 1962; and *P. macacae* Sandosham, 1953. In addition to *P. westermani*, he listed six species that are known to develop and cause disease in man: *P. skrjabini* Chen, 1959; *P. miyazakii* Komo et al., 1961; *P. heterotremus* Chen and Hsia, 1964; *P. africanus* Voelker and Vogel, 1965; *P. mexicanus* Miyazaki and Ishii, 1968; and *P. peruvianus* Miyazaki, Ibanez, and Miranda, 1969. *P. peruvianus* is now regarded by Miyazaki et al. (1980) as identical with *P. mexicanus*. However, the following additional species have been described as parasites of man and animals: *P. ecuadoriensis* Voelker and Arzube, 1979 in Ecuador; *P. philippinensis* Ito et al., 1978 in the Philippines; and *P. hueitungensis* Chung et al., 1977 in China, bringing the total to eight. For some of them, however, the question of synonymy remains uncertain.

Species in Asia. *Paragonimus skrjabini* Chen, 1959 (syn. *P. szechuanensis* Chung and Tsao, 1962) is a natural lung fluke of cats and some other mammals. According to Zhong et al. (1981), *P. skrjabini* is a *nomen nudum* and the species is referred to correctly as *P. szechuanensis*. However, *P. skrjabini* was fully described by Chen in 1960. In man it has been reported in ectopic locations, especially in subcutaneous nodules, most commonly in people of Sichuan (Szechuan) and Yunnan provinces (Yokogawa, 1974). The snail hosts are *Assiminea lutea* and *Tricula gregoriana*, and the crab hosts are species of *Potamon*. The adult worm is relatively elongate, the oral sucker is smaller than the ventral sucker, the ovary is delicately branched, and both the ovary and testes are relatively small.

Paragonimus miyazakii Kamo, Nishida, Hatsushika, and Tomimura, 1961 is found in weasels, martens, dogs, and wild boars in southern Honshu, Shikoku, and Kyushu regions of Japan. The snail hosts are species of *Bythinella*, and the natural crab host is *Potamon dehaani*. The adult fluke is characterized by relatively large testes, moderately branched ovary, oral sucker only slightly smaller than the ventral one, and single body spines. The metacercaria is the largest of the Japanese species. Notable features of *P. miyazakii* infections in humans are the high frequency of pleural effusion, marked eosinophilia of the blood and pleural exudate and, in one case, recurrent spontaneous pneumothorax. Immunoserological tests can distinguish this infection from *P. westermani* infection (Kobayashi et al., 1975; Yokogawa et al., 1974; Yokogawa, 1975).

Paragonimus heterotremus Chen and Hsia, 1964 (syn. *P. tuanshanensis* Chung, Ho, Cheng, and Tsao, 1964) was first described in rats in China. Subsequently it was reported in dogs and cats in Thailand, in subcutaneous tissues of a 13-year-old boy in Thailand, and in the lungs of a 39-year-old man who died in Laos (Miyazaki and Fontan, 1970). According to Miyazaki (1974), this species is probably medically more important than *P. westermani* in some parts of Thailand, Laos, and China. The snail host in China is *Tricula gregoriana*. The crab hosts are *Potamon* species in China and Thailand. The adult fluke measures about 10 mm long by 5 mm wide, the cuticular spines are single, the diameter of the oral sucker is twice that of the ventral sucker, and the ovary is delicately branched. The testes are also branched, and much longer than the ovary. Zhong et al. (1981) consider *P. heterotremus* a synonym of *P. tuanshanensis*.

Paragonimus hueitungensis Chung, Hsu, Ho et al., 1977 was described from experimentally infected rats, cats, and dogs and immature forms were identified in biopsy specimens of migratory subcutaneous nodules from two children 4 and 6 years of age in Hunan Province in east-central China (Chung et al., 1977). The adult worm measures around 9 mm by 3 mm, the body spines are single, with occasional longitudinal splitting lateral to the ventral sucker, the ventral sucker is slightly larger than the oral sucker, the ovary has a relatively large central mass with 3 to 6 short, stout primary branches, each with 2 or 3 relatively stout secondary branches, and the testes are larger than the ovary and appear as an elongate mass with short and lumpy branches. The eggs average about 75 by 46 μm. The snail host is *Tricula cristella* and the crab hosts are species of *Sinopotamon* and *Isolapotamon*.

Paragonimus pulmonalis (Baelz, 1880) was listed as a synonym of *P. westermani* by Yokogawa (1975), but was regarded as distinct from *P. westermani* by Miyazaki (1978, 1979, 1981a). Miyazaki reported that *P. westermani* has diploid chromosomes and bisexual reproduction while *P. pulmonalis* has triploid chromosomes and reproduces by parthenogenesis. He also found that in *P. westermani* the eggs, larvae, and adults were smaller and were found mainly in tropical Asia, whereas *P. pulmonalis* is distributed mainly in Japan, Korea, and Taiwan. However, after studying the forms in Taiwan, Miyazaki and Chiu (1980) concluded that in both Taiwan and Japan *P. westermani* is distributed in mountainous areas whereas *P. pulmonalis* is in lower areas. Probably, until Miyazaki's interpretations have been confirmed, *P. pulmonalis* can be regarded as merely a variant of *P. westermani*. Yokogawa (1982) has suggested that until Miyazaki's theory is clarified the name *Paragonimus westermani* can be used, as formerly, but with the host and geographic area added, as for example, Japanese *P. westermani* from *Potamon dehaani*, or *P. westermani* from *Eriocheir* in Taiwan.

Paragonimus philippinensis Ito, Yokogawa, Araki, and Kobayashi, 1978 so closely resembles *P. westermani* that until 1978 it was not a recognized species. The first series of cases of human paragonimiasis in the Philippines was reported by Musgrave in 1907; he reported 17 cases, 8 of which were diagnosed at autopsy. This was followed by numerous reports of infection in people living on the Islands of Mindoro, Samar, Leyte, Cebu, Negros, Mindanao, and Luzon, all diagnosed as caused by *P. westermani*. In a study of the cercariae developing in freshwater snails in Leyte, Ito (1977) observed a species, which he called *Cercaria leyteensis*, and he noted that it was distinctly smaller than the cercaria of *P. westermani* from Japanese snails. Further studies showed distinct morphological, biological, and immunological characteristics. It was therefore described as new and named *P. philippinensis*. Its metacercariae readily infect laboratory rats and develop to full maturity in the lungs; *P. westermani* in Japan seldom forms lung cysts in rats and never develops to full sexual maturity. The morphology of the larvae and adult stages was described in detail by Ito et al. (1979) and Yokogawa et al. (1979). Miyazaki and Habe (1979) considered this form to be a subspecies and questionably named it *Paragonimus westermani filipinus*. Another species, *Paragonimus siamensis* Miyazaki and Wykoff 1965, originally found in northeast Thailand, was also reported from the Philippines (Cabrera and Vajrasthira, 1973), but this record was later discovered to be erroneous (Miyazaki, 1981*b*).

Species in the Western Hemisphere. First to be described in the Western Hemisphere was *Paragonimus rudis* (Diesing, 1850), from an otter that was captured in Brazil. This species doubtless has been seen, reported, and possibly even redescribed and renamed more than once, but the original description was inadequate and, therefore, *P. rudis* is generally regarded as a *species inquirenda* (Miyazaki et al., 1980). However, based on the descriptions of Braun in 1901 and Caballero in 1946, Yamaguti (1971) considered *P. rudis* a valid species. Next to be described was *Paragonimus kellicotti* Ward, 1908, commonly found in the lungs of mink in the United States and Canada. Sixty years later, two additional species were described: *Paragonimus caliensis* Little, 1968 and *Paragonimus mexicanus* Miyazaki and Ishii, 1968 from crab-eating mammals in Colombia and Mexico, respectively. More recently added to the list of American species are *P. peruvianus*, described by Miyazaki et al. (1969, 1971) in Peru from man and experimental cats; *P. amazonicus*, described by Miyazaki et al. (1973, 1974) from opossums in Peru; *P. inca*, described by Miyazaki et al. (1975) from felids and mustelids in Peru; and *P. ecuadoriensis*, described by Voelker and Arzube (1979), from a raccoon-like mammal *(Nasua nasua)* and experimentally infected cats in Ecuador. However, more detailed studies on the Central and South American species by Miyazaki et al. (1980) demonstrated that *P. peruvianus* is identical with *P. mexicanus*. The species that are known or presumed to be occasional parasites of man are *P. mexicanus*, *P. ecuadoriensis*, and possibly *P. kellicotti*. American paragonimiasis was reviewed in 1972 by Miyazaki, and in 1973 by Sogandares and Seed.

Paragonimus kellicotti Ward, 1908 is most commonly found in the lungs of the mink, *Mustela vison*, but it also has been recorded in several other species of mammal. Detailed morphology of the adult was described by Ward and Hirsch (1915), the developmental stages by Ameel (1932) and Ameel et al. (1951). The known distribution in the United States includes all midwestern states and South Carolina, Mississippi, and Louisiana. It is relatively common in parts of Canada. The first intermediate host is a small, amphibious operculate snail, *Pomatiopsis lapidaria*. Species of crayfish of the genus *Cambarus* serve as second intermediate hosts. The single reported human case was a 44-year-old German laborer who, for 20 years, had worked at railroad construction in the United States, had eaten crayfish prepared by a Chinese cook, and in 1896 returned to

Germany when he began coughing up blood (Abend, 1910). Eggs were found abundantly in the sputum and the description of them was consistent with that of *Paragonimus*. Hemoptysis continued until 1909. A report of autochthonous human paragonimiasis in Canada (Béland et al., 1969) apparently was based on erroneous interpretations.

Paragonimus mexicanus Miyazaki and Ishii, 1968 (syn. *P. peruvianus* Miyazaki, Ibanez, and Miranda, 1969) was described from common opossums in Colima, a Pacific coastal province of middle Mexico. Its known geographic range extends through Central America to Peru. It is the species found in man in Mexico, Costa Rica, possibly in Ecuador, and Peru. Its natural mammalian hosts are species of *Didelphis, Mephitis, Erocyon, Philander, Procyon, Nasua, Felis,* and *Canis*. The crab hosts are species of *Pseudothelphusa, Potamocarcinus, Ptychophallus,* and *Hypolobocera*. The snail host in Costa Rica is *Aroapyrgus costaricensis*. Eggs found in lung tissue from a 35-year-old man in Mexico reported by Martínez Báez and Jiménez Galán (1961) were identified as *P. mexicanus* by Miyazaki and Ishii (1968). Reported as *P. peruvianus*, two specimens were found in the lungs of a 36-year-old man in Peru who had eaten raw crabs in the jungle. Morera (1968) and Brenes et al. (1982) each reported two cases of paragonimiasis apparently caused by *P. mexicana* in Costa Rica. Adult *P. mexicanus* measures about 15 mm by 6 mm, the body spines are mostly single, the oral sucker is larger than the ventral sucker, the ovary is delicately branched, and the testes are larger than the ovary and less delicately branched. The eggs have a relatively thin shell and measure 60 to 70 μm long by 42 to 56 μm wide. *P. kellicotti* eggs are much larger and the shell is much thicker. It is the egg morphology that most clearly distinguishes these two species. The life cycle of *P. mexicanus* was described by Brenes et al. in 1980.

Paragonimus ecuadoriensis Voelker and Arzube, 1979 was described from naturally infected coati *(Nasua nasua)* and from experimentally infected cats. The crab host is *Hypolobocera aequatorialis*. The snail host is unknown. In the adult worm, which measures up to 10 by 5 mm, the oral sucker is larger than the acetabulum or nearly the same size, the body spines are single, the ovary is smaller than the testes, and both gonads have narrow lobes with little secondary branching. The species differs from others of the area by the shape of the gonads and the size and form of the eggs. The egg measurements average 51 by 91 μm in *P. ecuadoriensis*, 50 by 84 μm in *P. peruvianus* (= *P. mexicanus)*. The metacercariae of both species are found in the hepatopancreas of the crab, but those of *P. ecuadoriensis* are more frequently found in the muscles. Eggs in the sputum of two patients in western Ecuador (Jipijapa) were identified by Voelker and Arzube (1979) as *P. ecuadoriensis*.

Paragonimus caliensis Little, 1968 was described from the lungs of opossums *(Didelphis marsupialis* and *Philander opossum)* in Colombia, South America. Its distribution extends from Peru through Panama and Costa Rica to Mexico (Miyazaki, 1974). Adults are elongate, tapering toward the ends, and are characterized by a compact ovary with short, unbranched, or slightly branched lobes, small testes with unbranched lobes, and cuticular spines that are singly arranged. The eggs, measuring 70 to 92 μm by 38 to 54 μm, have a relatively thin, irregular shell. The metacercariae were found in the liver of crabs: *Strengeria* in Colombia and possibly *Pseudothelphusa chilensis* in Peru. The snail host in Colombia is *Aroapyrgus colombiensis*, a minute species with a remarkably heavy shell and body, giving it a density or weight comparable with that of coarse sand or fine gravel (Malek and Little, 1971). *P. caliensis* infection in man has not been demonstrated.

Species in Africa. Two species of *Paragonimus* have been described from man and wild animals in west Africa. A related trematode, *Poikilorchis congolensis*, is also an occasional parasite of man in parts of west and central Africa. Pulmonary paragonimiasis was reported in Nigeria earlier, but the parasite itself was unknown in Africa until 1965 when Voelker and Vogel described two new species—*P. africanus* in the Cameroons and *F. uterobilateralis* in the Cameroons, Liberia, and Nigeria.

Paragonimus africanus Voelker and Vogel, 1965 was described from the mongoose *(Crossarchus obscurus)* and dogs in natural infections, and from dogs, cats, *Mandrillus leucophaeus*, and *Cercopithecus aethiops* in experimental infections. The snail host is probably *Potadoma freethii* and the second intermediate hosts, freshwater crabs, are species of the genus *Sudanonautes*. In the adult fluke the oral sucker is larger than the ventral sucker, branches of the ovary are short and relatively delicate, the testes are much larger than the ovary, and the eggs measure about 90 by 50 μm. The metacercariae develop in spherical double-walled cysts about 450 μm in diameter, mainly in the muscles of the crabs. During and following the civil war in Nigeria (1967–1970), the incidence of pulmonary paragonimiasis was greatly increased in the southeastern part of the country (Nwokolo, 1972). In the lower Bakossi area of the Cameroon Republic, examination of sputum from 265 people, most of whom had a cough, revealed eggs of *Paragonimus* in 30 (Vogel and Crewe, 1965). All with the infection had a chronic cough with bloody sputum. The eggs could be recognized as those of *P. africanus*. The species of *Paragonimus* was not determined in one study of 100 patients by Nwokolo (1972), but in those whose infections were acquired near the Cameroon border the species was probably *P. africanus*.

Paragonimus uterobilateralis Voelker and Vogel, 1965 is found in the swamp mongoose *(Atilax paludinosus)* in Liberia, the dog in Cameroon, and in man in Nigeria, Liberia, and Guinea (Sachs and Voelker, 1982). It is an important human parasite in Nigeria (Voelker and Nwokolo, 1973) and its known distribution extends from Liberia to Cameroon (Nozais et al., 1980). The adult worm, roughly 8 by 3 mm, is much smaller than that of *P. africanus*, and its eggs are also smaller (about 70 by 40 μm). The suckers are nearly equal in size, the ovary is much smaller than the testes, and the uterus extends into both halves of the body. The crab hosts are species of *Sudanonautes (Potamonautidae)*. The metacercariae are not generally encysted, remaining free in the crab's muscles.

Paragonimus infection in Durban, South Africa was reported by Allanson (1975) who identified the eggs in material aspirated from the chest wall of a 9-year-old boy and in the feces of an 8-year-old girl.

Additional Species. Other species of *Paragonimus* described from reservoir or intermediate hosts include *P. bangkokensis* Miyazaki and Vajrasthira 1967 from Thailand; *P. compactus* (Cobbold, 1859) from India and Ceylon; *P. fukienensis* Tang and Tang, 1962 from China; *P. iloktsuenensis* Chen, 1940 from China, Japan, and Taiwan; *P. macrorchis* Chen, 1962 from China and Thailand; *P. ohirai* Miyazaki, 1939 from Japan; *P. proliferus* Hsia and Chen, 1964 from China; *P. siamensis* Miyazaki and Wykoff, 1965 from Thailand; and *P. yunnanensis* Ho, 1957 from China. An unidentified species of *Paragonimus* was found in wildcats *(Felis weidii* and *F. pardalis)* in Colombia (Little, 1968).

Family Achillurbainiidae

Species of *Achillurbainia* Dollfus, 1939 and *Poikilorchis*, Fain and Vandepitte, 1957

Achillurbainia. The genus *Achillurbainia* was described for a *Paragonimus*-like species, *A. nouveli* Dollfus, 1939, which was discovered in the frontal sinuses of a Malay panther *(Panthera pardus)* in a zoo in France. A second such observation was made in 1954 (Dollfus and Nouvel, 1957). Other species of the genus include *A. recondita* Travassos, 1942 which was found in the maxillary sinuses of a Brazilian opossum *(Didelphis* *marsupialis),* and *A. ratti* Miyazaki and Kwo, 1969 which lives in the trachea and bronchi of a giant Malay rat *(Rattus muelleri).* An unidentified species of *Achillurbainia* was reported from the parotid gland of a tree shrew *(Tupaia glis)* in Selangor, Malaysia (Ow-Yang and Wah, 1975).

Human infection with *A. nouveli* was reported by Chen (1965) who recovered a mature worm measuring 11 mm by 5 mm from a painless retroauricular nodule removed from a 10-year-old girl in south Kwangtung, China. Eggs identified as probably those of *A. recondita* were observed in granulomas scattered over the omentum and other peritoneal surfaces of a 19-year-old man in Honduras (Beaver et al., 1977).

Poikilorchis. Identified as *Poikilorchis congolensis,* worms or eggs in retroauricular cysts or abscesses were reported in four cases from Central Africa, eggs only being recovered from a young woman and two girls, and both a worm and eggs from a boy (Fain and Vandepitte, 1957 a,b; Vandepitte et al., 1957). In southern Nigeria, eggs of the same type were recovered from near the ear of a man (Yarwood and Elmes, 1943), and similarly from a man and two boys (Oyediran et al., 1975) (Fig. 28–22). In Asia, similar findings were reported in two boys in Sarawak (Lie et al., 1962; and Wong and Lie, 1965). One other species, *Poikilorchis nagahanai* Shimizu, 1979, was described from the paranasal sinuses of weasels *(Mustela siberica itatsi)* in Okayama, Japan.

In species of the genus *Achillurbainia* the outstanding feature is the presence of numerous small testes distributed over most of the body except the anterior-most part and that occupied by the female reproductive organs; vitellaria consisting of minute follicles extend over the same general area. The ovary is near the middle of the body left of the midline, and the uterus forms convolutions between the intestinal ceca anterior to the ovary. The genital pore is about halfway between the oral and ventral suckers. The esophagus is short, and the intestinal ceca, narrow and convoluted, extend to the posterior extremity. The acetabulum and oral sucker are about equally developed and of moderate size; the tegument is smooth. The features of *Poikilorchis* described as differing from *Achillurbainia* are irregularly shaped testes not extending anterior to the ventral sucker, and numerous folds of the ceca—features that seem insufficient for a generic distinction (Beaver et al., 1977). However, both genera are widely recognized (Yamaguti, 1971) and can be retained until better known. The eggs of species described thus far for both genera are roughly 65 to 70 μm by 34 to 40 μm, with a relatively thick amber wall and large operculum. Snail hosts have not been identified; second intermediate hosts are freshwater crabs (Miyazaki and Kannangara, 1970).

Family Opisthorchiidae
Braun, 1901

Opisthorchiidae, parasites of the bile ducts and gallbladder, are small to medium-sized, usually elongate, and relatively delicate, with weakly developed suckers. The testes, located posteriorly, may be branched or lobed; the ovary is near midbody, and the coiled uterus fills the area between the ovary and ventral sucker. The eggs are small and are fully embryonated when laid.

Opisthorchis felineus
(Rivolta, 1884) Blanchard, 1895
(The cat liver fluke, producing opisthorchiasis)

Synonyms. *Distoma felineum* Rivolta, 1884; *D. sibiricum*

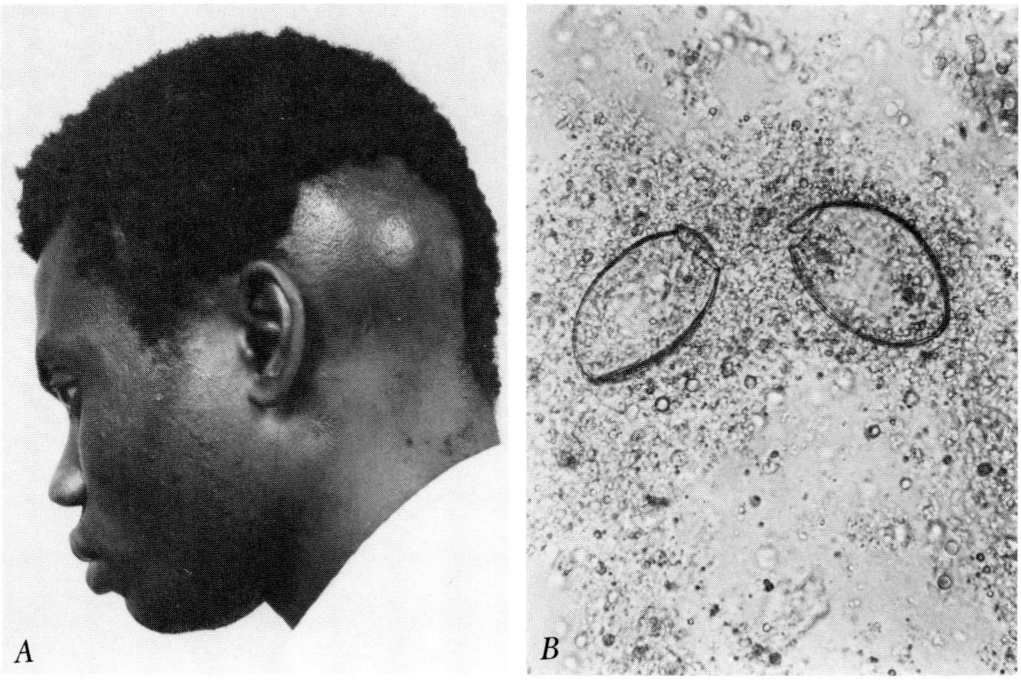

Fig. 28–22. *Paragonimus* or *Poikilorchis* (=*Achillurbainia)* species. *A,* Mass in left postauricular region, below which are two scars, in a Nigerian with a 3-year history of recurrent swelling and discharge from the area. *B,* Eggs in pus from abscess from mass after surgical removal. (From Oyediran, A.B.O.O., et al. 1975. Infection of the mastoid bone with a *Paragonimus*-like trematode. Am. J. Trop. Med. Hyg., *24:*268–273.)

Winogradoff, 1892; *D. winogradoffi* Jaksch, 1897; *O. tenuicollis* (Rudolphi, 1819) of Ejsmont, 1937.

Historical and Geographical Notes. Originally described from the cat (1884), this liver fluke was first reported in a human infection by Winogradoff from Tomsk, Siberia in 1892.

Opisthorchis felineus is a common fluke in southern, central, and eastern Europe (Spain, Italy, Albania, Greece, Macedonia; France, Switzerland, Holland, Germany; Poland, northern and lower Danube drainage, European U.S.S.R.), and in Asia (Turkey, Caucasus, Dnieper and Volga drainages, Siberia west of Ob River but including Novosibirisk and Tomsk), yet human infection is not known to occur east of Poland and the Danube delta.

Morphology, Biology, and Life Cycle. The adult worm, which lives attached by its two suckers in the biliary and pancreatic passages of its host, is lancet-shaped, is rounded posteriorly and attenuated anteriorly, is thin and transparent, and measures 7 to 12 mm in length by 2 to 3 mm in breadth (Fig. 28–23). The living worm in the bile tract is reddish-bile colored. The integument is smooth, although in immature worms there is a trace of the spines that are present on the integument of the cercaria. The two suckers are nearly equal in diameter (0.25 mm); the oral sucker is subterminal, while the ventral sucker is situated about one-fifth the body distance from the anterior end. The excretory bladder is a long sac-like tube extending as far forward as the ovary. The two testes are lobed and are situated obliquely to each other in the posterior fourth of the worm. The long, slightly coiled seminal vesicle terminates in a weakly muscular ejaculatory duct, which opens through the genital pore immediately in front of the ventral sucker. Cirrus sac, prostate gland, and cirrus organ are lacking. The small, oval, or slightly lobed ovary is median in position, just in front of the anterior end of the bladder. Immediately to the right is the ootype and nearby are the seminal receptacle and Laurer's canal. The vitellaria consist of numerous transversely compressed follicles disposed in the lateral fields in the middle third of the body. The uterus proceeds anteriad from the ootype as an intricately coiled tubule and terminates in the metraterm, beside the male genital opening.

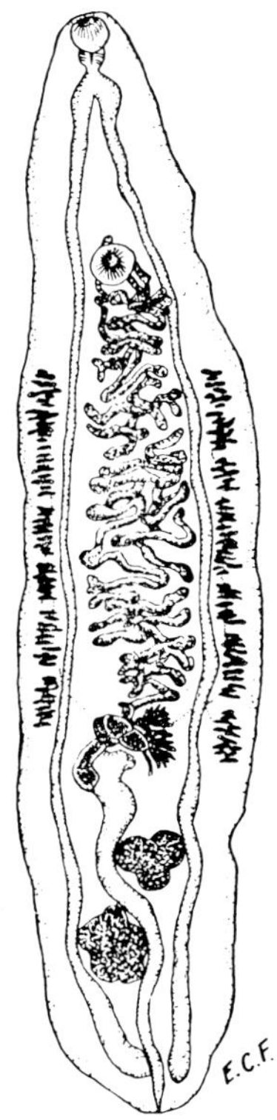

Fig. 28–23. *Opisthorchis felineus.* Adult, dorsal view. (× 10.) (Adapted by E.C. Faust from Stiles, C.W., and Hassall, A. 1908. Trematoda and trematode diseases. Bull. 37, Hyg. Lab. U.S. Public Health and Marine Hospital.)

The eggs (Fig. 28–24) are elongate ovoid in shape, with an operculum that fits into a thickened rim of the shell proper; they are light yellowish-brown and are about three times as long as broad (30 by 11 μm). At the posterior end of the shell there is usually a minute tubercular thickening. The miracidium, which has an asymmetrical internal organization, is fully developed when the egg is laid. Hatching does not occur in the water, but occurs only after ingestion by certain snails. The only known molluscan host is the amnicolid snail *Bithynia (Bulimus) leachi* in which successively the first- and second-generation sporocysts and rediae develop.

According to Vogel's (1934) observations, the cercariae leave the rediae before they are mature and about 2 months after infection the mature cercariae swarm out of the snail. They are lophocercous, with a pair of pigmented eye-spots, and have 10 pairs of penetration glands, each with a duct

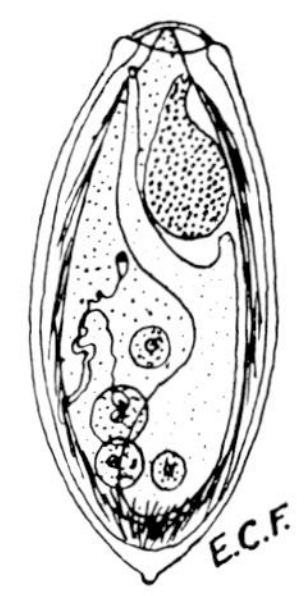

Fig. 28–24. *Opisthorchis felineus.* Egg with mature miracidium. (× 1200.) (Adapted by E.C. Faust from Faust, E.C., and Khaw, O.-K. 1927. Studies on *Clonorchis sinensis* (Cobbold). Am. J. Hyg. Monogr. Series No. 8.)

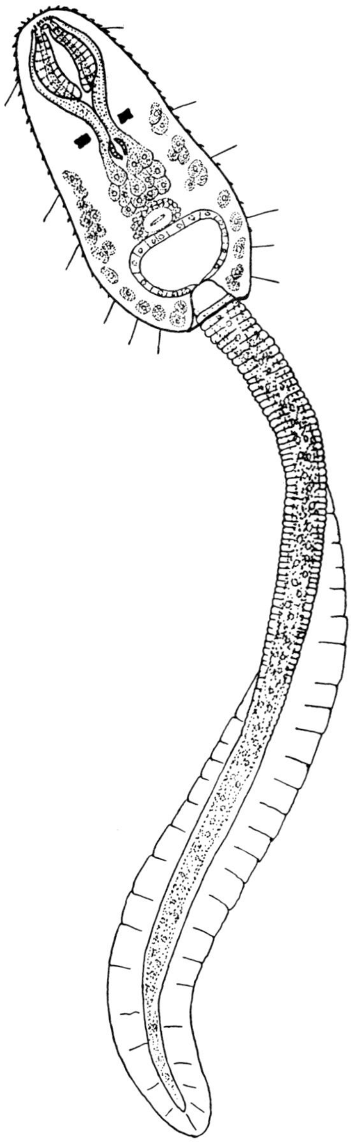

Fig. 28–25. *Opisthorchis felineus*. Cercaria, ventral view. (× 330.) (Adapted from Vogel, H. 1934. Der Entwicklungszyklus von *Opisthorchis felineus* (Riv.) nebst Bemerkungen über Systematik und Epidemiologie. Zoologica, *33*:1-103.)

opening dorsal to the mouth (Fig. 28–25). The proximal region of the tail is enveloped by an integumentary sheath, which is continued into a dorsal and a ventral rudder. Measurements are: body, 132 to 172 μm by 41 to 48 μm; tail, 440 to 500 μm long. Vogel (1934) believed that the cercariae attack the fish host only after they are disturbed by the fish, whereupon they become attached to the scales, drop their tails, and penetrate into the tissues. The following fishes, all species of Cyprinidae, have been incriminated: *Idus melanotus, Tinca tinca, Cyprinus carpio, Barbus barbus, Abramis brama, A. sapa, Alburnus lucidus, Aspius aspius, Blicca bjorkna, Leuciscus rutilis,* and *Scardinius erythophthalmus*. The first two are most commonly infected and usually serve as the sources of human infection. Upon ingesting infected fish, man and reservoir hosts become infected. The metacercariae excyst in the duodenum, migrate through the ampulla of Vater to the distal bile passages, become attached, and in 3 to 4 weeks begin to lay eggs. The entire life cycle requires a minimum of 4 months.

Epidemiology. In central, eastern, and southeastern Europe and adjacent U.S.S.R. the dog and cat, the red, silver, and polar foxes, domestic and wild swine, Norway rat, water rat *(Arvicola terestris),* wolverine, marten, beaver, rabbit *(Oryctolagus cuniculus),* polecat *(Putorius putorius),* gray seal *(Halichoerus grypus),* Caspian seal *(Phoca caspia),* bearded seal *(Erignathus barbatus),* and lion *(Felis leo)* in captivity are infected (Erhardt et al., 1962). The definitive host acquires infection from eating infected raw fish and, in turn, pollutes the water in which the intermediate hosts develop with feces containing the eggs of the parasite. In Europe the endemic foci are in the deltas of the larger streams and in Poland in dammed reservoirs in the Volga Basin.

Pathogenesis, Pathology, and Symptomatology. In the course of their development in the distal bile ducts, the maturing worms initiate inflammatory and proliferative changes of the biliary epithelium. These changes are accompanied by fibrosis of the distal biliary vessels. If the infection is heavy, the pathologic process may be extended to the proximal bile ducts and gallbladder and may result in periportal fibrosis.

The degree of clinical involvement depends largely on the mass and duration of the infection. Although considerable local damage may develop in the distal bile capillaries where the worms are usually lodged, in most human cases this does not appreciably affect liver function. Fifty to sixty worms cause no serious damage. Several hundred to a thousand worms cause moderate, painful enlargement of the liver and passive congestion of the spleen, with icterus and local eosinophilia in the wall of the bile ducts. In heavy infections the worms invade the pancreas, with digestive disturbances. Bile stones may form around eggs as nuclei and cause cholecystitis with colic. The appetite fails as the heavily infected patient becomes more toxic; the scar tissue formed around the bile tracts encroaches more and more on the liver cells and eventually on the portal vessels, with resultant collateral venous circulation, edema of the face and limbs and, at times, ascites. Such advanced liver disease due to opisthorchiasis occurs in only a small proportion of infections.

Diagnosis. Diagnosis is based on recovery of the typical eggs (Fig. 28–24) in the stool, or by duodenal intubation.

Treatment. For opisthorchiasis praziquantel has been reported as safe and effective (Bunnag and Harinasuta, 1980, 1981; Ambroise-Thomas et al., 1981). The optimum dose appears to be 40 mg per kilogram of body weight in a single dose after a meal (Bunnag and Harinasuta, 1981). Side effects include abdominal pain, vomiting, diarrhea, lassitude, myalgia, headache, and rashes. These are mild though they occur in about 80% of patients. With saline purgation at 4 hours after drug administration, flukes were passed in the feces within 1 to 2 hours.

Prognosis. Prognosis is usually good to fair; in heavily infected cases, it is poor.

Prevention. Only abstinence from eating raw or inadequately cooked fish prevents human infection in endemic areas. Since reservoir hosts, as well as man, pollute the waters containing the intermediate hosts, sanitary disposal of human feces does not solve the problem.

Other Species of Opisthorchis. *Opisthorchis viverrini* (Poirier, 1886) Stiles and Hassall, 1896 is distinguished from *O. felineus* by the greater proximity of its ovary and testes, the aggregation of its vitellaria into a few clusters of glandular material, and the different size and shape of its eggs (av. 27 by 15 μm). It is an important human parasite in northeastern Thailand, where 25% of the population is infected (Sadun, 1955), and 46% in the northernmost province of Udorn in this area is infected (Harinasuta and Vajrasthira, 1960). The life cycle has been studied in Thailand by Wykoff et al. (1965), who found the snail intermediate hosts to be *Bithynia goniomphalus, B. funiculata,* and *B. laevis*. The most important fish intermediate hosts are *Punteus orphoides, Hampala dispar,* and *Cyclocheilichthys siaja,* although other species were found to harbor the metacercariae. These workers were unable to distinguish *O. viverrini* from *O. felineus* on the basis of morphological characteristics of egg or adult, but found that the flame cell pattern of the cercaria as it leaves the snail host is distinctive in each species. Vajrasthira et al. (1961) studied the excretory system in the metacercaria, excysted after removal from the flesh of freshwater fishes *(Cyclocheilichthys* spp.) obtained in the hyperendemic-enzootic area in northeastern Thailand. Clinical studies by Harinasuta and Vajrasthira (1960) indicated high incidence of diarrhea, flatulence after meals, pain over the liver (usually enlarged), jaundice in some patients, and a moderate elevation of temperature (37.5° to 38.5°C). Based on extensive, detailed studies in various regions of Thailand, Sadun (1955) concluded that symptoms increase progressively as infections are gradually acquired and become older. At autopsy, a man who died with carcinoma of the bile ducts was found to harbor 5896 adult *O. viverrini*. However, Wykoff et al. (1966) studied clinical and biochemical findings on 921 persons with *O. viverrini* infections as compared with similar data from uninfected persons in northeastern Thailand and reported that in this area *O. viverrini* does not cause any specific symptoms. However, in a village of 309 people in the same area, the *O. viverrini* infection rate was 94% overall (100% for most age groups above 10 years) and 30% had heavy (10 to 50 eggs/mg) or very heavy (> 50 eggs/mg feces) infections (Upatham et al., 1982). Right upper quadrant abdominal pain was correlated with the intensity of infection. Other fish-eating mammals, including the civet cat and domestic dogs and cats, have been found infected in Thailand, even in the central part of the country where the disease is not endemic (Sadun, 1955). It is possible that the correct name of this worm is *O. tenuicollis* (Rudolphi, 1819).

Opisthorchis noverca Braun, 1902, a common parasite of the dog, wolverine, and pig in India, has been recorded twice as a human infection (McConnell, 1876). Its eggs are much larger (34 by 21 μm) and readily distinguishable from those of other species.

Opisthorchis guayaquilensis, described by Rodriguez et al. (1949), has been found in man in an isolated focus in Ecuador, where 4% of 214 persons in the village of P.O. Gomez, 32% of 31 persons in nearby rural communities, and 3% of the dogs were found infected. This species has been found to be identical with *Amphimerus pseudofelineus* (Ward, 1901) from the domestic cat in Nebraska, United States (Artigas and Perez, 1960–1962) and has been recovered from reservoir mammalian hosts in the United States, Panama, Ecuador, and southern Brazil.

Clonorchis sinensis
(Cobbold, 1875) Looss, 1907
(The Chinese liver fluke, producing chlonorchiasis)

Synonyms. *Distoma sinense* Cobbold, 1875; *D. spathulatum*

Leuckart, 1876; *D. endemicum* Ijima, 1886. Morgan (1927), Price (1940), and Dawes (1946) regard *Clonorchis* as having characteristics sufficiently like *Opisthorchis* to classify the species *sinensis* under the latter genus.

Historical and Geographical Notes. This fluke was first seen in 1874 and described in 1875 by McConnell in the bile passages of a Chinese carpenter in Calcutta. Although found in Japan in 1875, it was not described from that locality until 1883, when Baelz attempted to differentiate a pathogenic and a nonpathogenic variety. Heanley (1908) published the first record for the endemic region in south China.

The endemic areas include Japan, Korea, essentially all of China except the Northwest, Taiwan, and Vietnam. The heavy human infections are confined primarily to Okayama and Niigata Prefectures in Japan, Guangdong (Kwangtung) Province in South China, and North Vietnam. Although clonorchiasis is frequently diagnosed in Orientals in the Western Hemisphere, there is no evidence that the infection has become established in regions outside the China Sea area. Frozen, dried, or pickled fish shipped from endemic areas probably account for infection in persons who have never left the Hawaiian Islands.

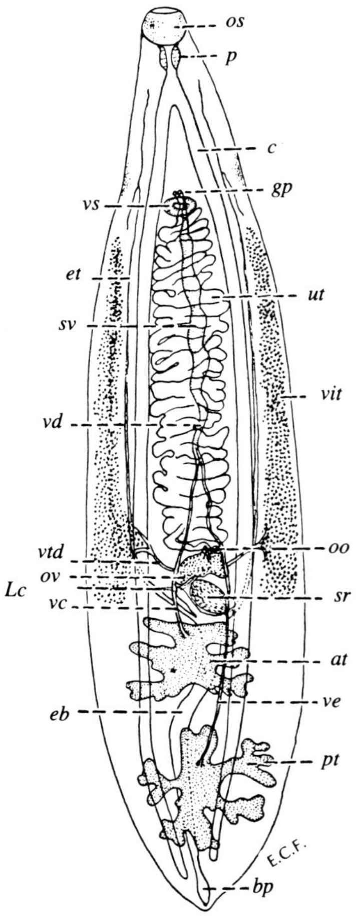

Fig. 28–26. *Clonorchis sinensis*. Adult, dorsal view. (× 8.) *at*, Anterior testes. *eb*, Excretory bladder. *c*, Intestinal cecum. *et*, Excretory tubule. *gp*, Genital pore. *Lc*, Laurer's canal. *oo*, Ootype. *ov*, Ovary. *p*, Pharynx. *sr*, Seminal receptacle. *sv*, Seminal vesicle. *pt*, Posterior testes. *ut*, Uterus. *vs*, Ventral sucker or acetabulum. *vd*, Vas deferens. *ve*, Vasa efferentia. *vit*, Vitellaria. *vtd*, Vitelline duct. (From Faust, E.C. 1939. *Human Helminthology*, 2nd ed. Philadelphia, Lea & Febiger.)

Morphology, Biology, and Life Cycle. The adult worm, which lives attached in the bile passages, occasionally in the pancreatic duct, is flat, transparent, and spatulate, attenuated anteriorly and somewhat rounded posteriorly (Fig. 28–26). It measures 10 to 25 mm in length by 3 to 5 mm in breadth. The integument of the adult worm is aspinose. The oral sucker is slightly larger than the ventral acetabulum, which is situated at the posterior end of the anterior third of the body. The excretory bladder is long, pouch-like, and slightly serpentine; the primary tubules arise some distance posterior to its blind anterior end. The most characteristic features of the worm are the large, deeply lobulated or branched testes, which are situated one behind the other in the posterior third of the body, with branches extending into the lateral fields. The common vas deferens enlarges in the equatorial region into the seminal vesicle, which ascends without marked curvature or torsion and terminates in a weakly muscular ejaculatory duct opening through the genital pore just anterior to the ventral sucker. Cirrus pouch, prostate glands, and cirrus organs are lacking. The small, slightly lobed ovary (Fig. 28–26,*A*) is situated medially at the junction of the posterior and middle thirds of the body. Nearby are the ootype, surrounded by Mehlis' gland, the retort-shaped seminal receptacle, and the delicate Laurer's canal. The vitellaria are delicate, granular aggregations in the lateral fields, extending through the equatorial third of the worm. The uterus arises from the ootype and proceeds forward in compact intricate coils to the genital atrium, where it opens beside the male duct.

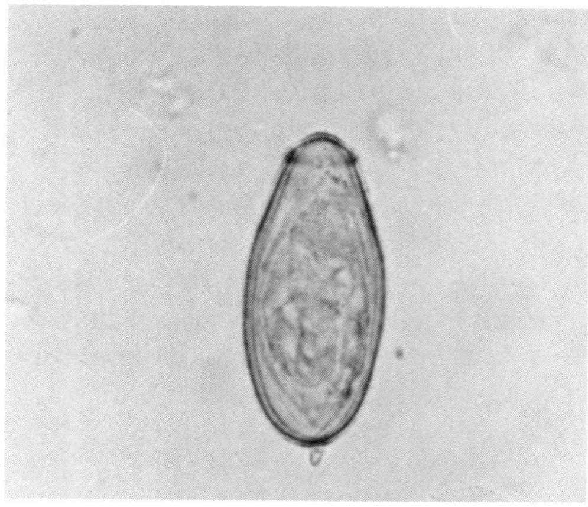

Fig. 28–27. *Clonorchis sinensis*. Egg with mature miracidium. (× 1000.) (By P.C. Beaver.)

The eggs (Fig. 28–27) are broadly ovoid, have a moderately thick, light yellowish-brown shell, and are provided with a distinct convex operculum, which fits into a rimmed extension of the shell. At the abopercular end there is frequently a small protuberance. The eggs measure 28 to 35 μm by 12 to 19 μm (average 29 by 16 μm) and are fully embryonated when discharged into the bile ducts. Like that of *Opisthorchis felineus,* the miracidium hatches only after ingestion by suitable species of operculate snails (*Parafossarulus, Bulimus, Semisulcospira, Alocinma,* and *Melanoides tuberculatus*). Within the mollusc there develop first-generation sporocysts and second-generation rediae. These latter, in turn, produce lophocercous cercariae with pigmented eyespots, penetration glands consisting of four inner and three outer pairs, and a dorsoventrally keeled tail (Fig. 28–28). After escape from the snail and a brief free-swimming existence, on coming in contact with certain freshwater fishes, the cercariae become attached, discard their tails, penetrate under the scales, and encyst there, in the skin or in the flesh. The period of essential development in the fish host is 23 days at summer temperatures. When fully developed, the eyespots disappear (Yoshimura, 1965). In the endemic-enzootic areas of the Sino-Japanese region, at least 80 species of fishes, mostly belonging to the family Cyprinidae, have been found naturally infected. In south China, where raw fish is considered a great delicacy by epicures, the ide, *Ctenopharyngodon idellus,* is the most common source of infection for the human population. When an infected fish is eaten, the metacercariae excyst in the duodenum, and usually enter the common bile duct through the ampulla of Vater, proceeding to the distal bile capillaries, where they mature in about 1 month. A period of about 3 months is required for the whole cycle.

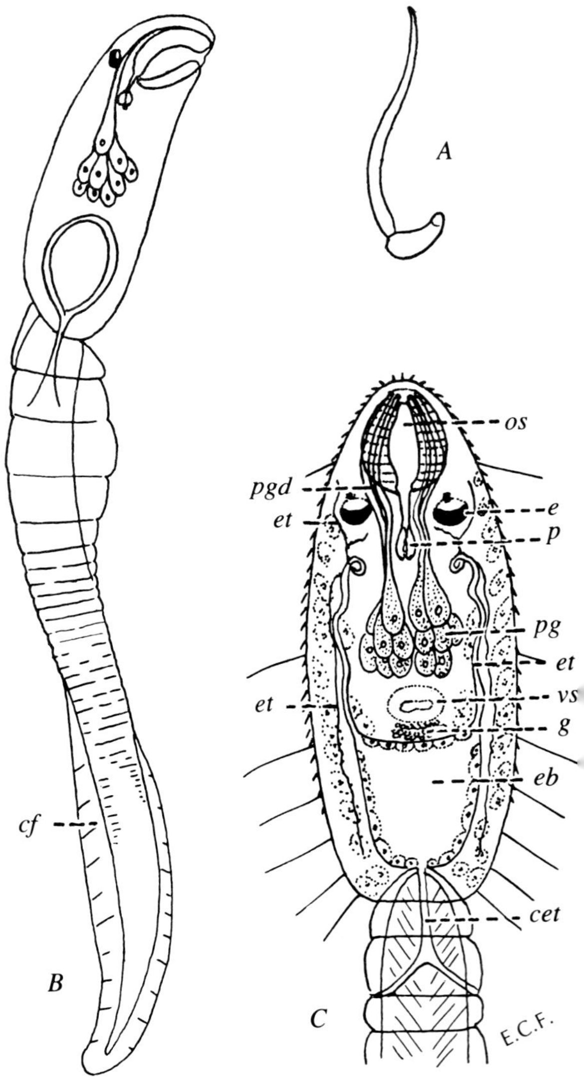

Fig. 28–28. *Clonorchis sinensis. A,* Cercaria as seen at rest in water. (× ca. 60.) *B,* Entire cercaria. (× 250.) *C,* Details of cercaria. (× 520.) *cet,* Caudal excretory tubule. *cf,* Caudal fin. *e,* Eyespot. *eb,* Excretory bladder. *et,* Excretory tubule. *g,* Genital primordium. *os,* Oral sucker. *pgd,* Penetration gland ducts. *pg,* Penetration glands. *p,* Pharynx. *vs,* Ventral sucker. (Adapted by E.C. Faust from Yamaguti, S. 1935. Ueber der Cercarie von *Clonorchis sinensis* (Cobbold). Z. Parasitenk., 8:183–187.)

Epidemiology. In addition to man, many reservoir hosts, including principally the dog and cat, are naturally infected in the Sino-Japanese areas. In North China, where autochthonous human infection is rare, these reservoir hosts are solely responsible for the perpetuation of the infection. The definitive host acquires the infection from eating raw, inadequately cooked, dried, salted, or pickled freshwater fish that harbor the encysted metacercariae. Natives of Hawaii, who have not been outside the Islands, are parasitized, presumably, from eating frozen, dried, or pickled fish imported from Japan or China. Pollution of water with night soil and feces of reservoir hosts provides for infection of the molluscan and subsequently of the fish hosts. Reservoir hosts contribute significantly to the infection rates, especially in Korea (Kim, 1974).

Pathogenesis, Pathology, and Symptomatology. When larvae of *C. sinensis* reach the distal bile ducts and mature, they initiate proliferative and inflammatory reactions in the biliary epithelium with which they make contact, and soon an encapsulating fibrosis of the ducts occurs. While a single exposure usually produces lesions of little clinical significance, repeated exposure provokes general involvement of the organ, with the lesions extending down to the larger bile ducts and gallbladder. Of 500 autopsy cases with *Clonorchis* infections, 9% had cirrhosis (Hou, 1955), and approximately the same percentage of the series of 102 post mortems of Germer et al. (1955) also had cirrhosis. Eight percent developed biliary obstruction with blockage of the common bile duct by adult worms or stones, or both.

Presence of the worms in distal bile passages provokes considerable local tissue reaction, i.e., proliferation of the biliary epithelium, and later its desquamation, as a result of both local trauma and local and remote toxic irritation; crypt formation in the bile ducts and formation of new bile capillaries; periepithelial fibrosis; periportal connective tissue hyperplasia; and fibrous tissue development around "graves of eggs" infiltrated in the liver parenchyma (Faust and Khaw, 1927) (Fig. 28–29). These changes are progressive; their total effect on liver function depends on the number of worms present from a single infection or from continuous reinfection over a period of years, and length of the infection, which in man may persist for 15 to 20 years or more after the patient has left an endemic focus. The average intensity of infection is probably less than 200 worms. On the other hand, heavy infections are not infrequent and account for rapidly developing hepatic damage with acute symptoms. Sambuc and Baujean (1913) recovered 21,000 worms from one patient in Indochina.

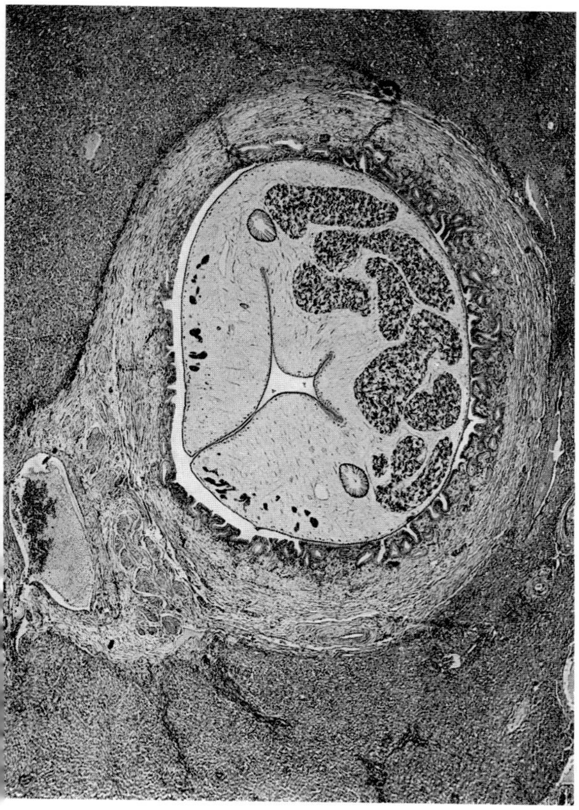

Fig. 28–29. *Clonorchis sinensis*. Mature worm in bile duct, where it has induced proliferation of the lining epithelium and fibrous elements of the wall. (× ca. 25.) (Adapted from original of E.C. Faust.)

Clonorchis sinensis and *Opisthorchis viverrini* have been linked epidemiologically with neoplasms of the bile ducts. Since marked hyperplasia of the epithelium is characteristic of ducts occupied by the worms, it seems plausible that while the factors causing carcinoma of the bile ducts are multiple, the initial step in the malignant transformation may be the high turnover of cells in the hyperplastic biliary epithelium induced by the worms (Flavell, 1981).

The *symptoms* evoked fall into three stages: (1) mild, essentially symptomless; (2) progressive, with irregular appetite, fullness in the abdomen, diarrhea, edema, and hepatomegaly; and (3) severe, with a syndrome associated with portal cirrhosis. Icterus is not constantly present. Otto (1935) stressed the importance of catarrhal cholangitis due to occlusion of the larger bile passages by sticky masses of eggs and by tissue proliferation. He also attributed symptoms of systemic toxemia, including palpitation of the heart and tachycardia, vertigo, tremors, tetanic cramps, and mental depression, to impairment of the detoxifying properties of the liver caused by the worms and their byproducts. Koenigstein (1949) studied the disease in representatives of about 20,000 Jewish refugees in Shanghai, some 20 to 30% of whom became infected in February to March, 1946, after eating pickled fish. One group was observed during the prodromal and acute period less than a month after exposure and 3 to 4 weeks before eggs were detected in the feces. The onset was at times gradual, at times sudden, with chills and fever up to 40°C. There was an enlarged, tender liver, and slight icteric tinge to the sclerae; splenomegaly in some cases; and eosinophilia in 10 to 40%. A few weeks later the chronic stage, with the classical symptoms of cholecystitis and hepatitis, was observed. Ling (1949) reported a fatal case with symptoms suggesting gallbladder disease and secondary hepatitis, with paroxysmal pain over the epigastrium radiating to both axillae, obstructive jaundice, enlarged tender liver, and polymorphonuclear leukocytosis. Although the peripheral blood picture is not significantly modified except for a frequent eosinophilia, experimental studies indicate a hyperplasia of the bone marrow with respect to both eosinophils and the reticuloendothelial cells.

In contrast to these studies on clonorchiasis during the earlier symptomatic period, the evidence of Strauss (1962), based on 57 infected Orientals and 48 Caucasians examined in San Francisco, together with noninfected controls of both racial groups, suggests that many of the symptoms associated with the infection may be due to other causes.

Diagnosis. The diagnosis is based on the recovery of characteristic eggs (Fig. 28–27) either in the feces or by duodenal aspiration. Advanced cases must be differentiated from adenocarcinoma of the liver, hydatid cyst, amebic liver abscess, beriberi, and other types of hepatic disease.

Treatment. Chloroquine diphosphate has been accepted as the drug of choice in the treatment of clonorchiasis, given in a dosage of 250 mg three times daily for 6 weeks. However, the treatment frequently fails to cure, and the side effects, including optic neuropathy, prompted the search for a more satisfactory drug. Praziquantel promises to be a more effective and safer drug (Horstmann et al., 1981; Rim et al., 1981a,b). At 75 mg per kg body weight in 3 doses 4 to 6 hours apart there were no failures among 46 patients and the side effects were mild (Ambroise-Thomas et al., 1981). Likewise, 34 patients given praziquantel in a dosage of 14 mg per kg three times daily for 5 days were cured, and passage of flukes in the feces began as early as 10 hours after the first dose (Liu Yuehan et al., 1982). A single dose of 40 mg per kg given after a meal may be equally effective (Bunnag and Harinasuta, 1981).

In the case of heavy infections complicated by the development of obstructive jaundice requiring surgical intervention, cholecystectomy with choledocholithotomy, exploration of the common duct, and a drainage procedure such as sphincteroplasty or choledochoduodenostomy should be performed (Sullivan and Koep, 1980).

Prognosis. Good in light infections, the prognosis is less satisfactory in heavy or chronic infections.

Prevention. Thorough cooking of all freshwater fish will assure protection to the human population. In south China and Indochina, where fish is consumed raw after seasoning with condiments, educational efforts should be effective in breaking this habit, and thus should reduce infection in this population group. Temporary storage of night soil is recommended. Eggs of *C. sinensis* in night soil die within 2 days at 26°C or 5 days at 4 to 8°C, or within 1 week at normal temperatures (Komiya, 1966). In Japan and elsewhere, prevalence rates have declined as a result of water pollution by industrial wastes and agricultural insecticides that kill the snails, land developments that modify habitats, and public health education that modifies the people's habits (Komiya and Suzuki, 1964).

Family Heterophyidae

Odhner, 1914

The heterophyids are small or minute intestinal parasites of birds and mammals. They are generally less than 1 mm long, piriform or ovoid, with spiny integument. Associated with the ventral sucker is a genital pore surrounded by a sucker-like gonotyl. The testes, variously arranged, along with the ovary are in the posterior part of the body. The uterus, usually much coiled, is widely and variously distributed. The eggs are small and embryonated when laid.

Heterophyes heterophyes

(v. Siebold, 1852) Stiles and Hassall, 1900

(Producing heterophyiasis)

Synonyms. *Distoma heterophyes* v. Siebold, 1852; *Heterophyes aegyptiaca* Cobbold, 1866; *Heterophyes nocens* Onji and Nishio, 1915.

Historical and Geographical Notes. This minute fluke was first found by Bilharz in 1851 at the autopsy of a native of Cairo.

The worm is common in the Nile Delta, is present in Turkey, and has been found in several foci in the Far East (Japan, southern Korea, central and south China, Taiwan, and the Philippines). It has been reported from the dog in India (Sen, 1965). Because of the difficulty of differentiating this species in the egg stage from the many other species of heterophyid flukes that man acquires from eating raw fish, there are no accurate data on the incidence of this infection.

Morphology, Biology, and Life Cycle. *Heterophyes heterophyes* lives deep in the wall of the small intestine. It is an elongate, piriform worm, with a broadly rounded posterior end and a more attenuate anterior end (Fig. 28–30). It measures 1 to 1.7 mm in length by 0.3 to 0.4 mm in breadth. The integumentary scales are narrow and closely set; they are most numerous in the anterior portion of the body. The oral sucker is minute (ca. 90 μm in diameter); the ventral sucker (230 μm in di-

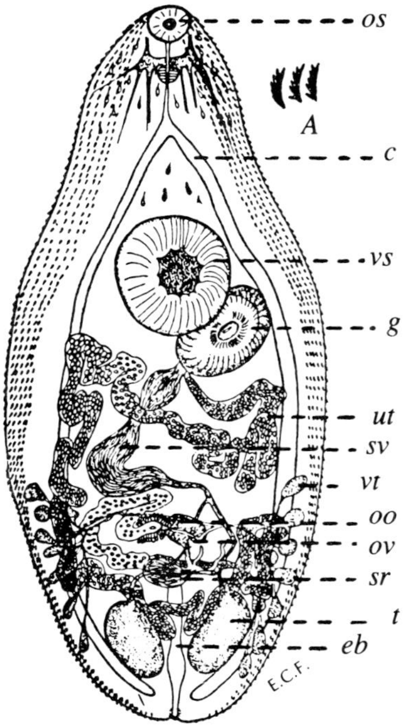

Fig. 28–30. *Heterophyes heterophyes,* ventral view. (× 50.) *c,* Cecum. *eb,* Excretory bladder. *g,* Gonotyl. *oo,* Ootype. *os,* Oral sucker. *ov,* Ovary. *sr,* Seminal receptacle. *sv,* Seminal vesicle. *t,* Testis. *ut,* Uterus. *vt,* Vitellaria. *vs,* Ventral sucker. *A,* Detail of spines of genital sucker. (Adapted by E.C. Faust from Looss, A. 1894. *Ueber den Bau von* Distomum heterophyes *v. Sieb. und* Distomum fraternum *n. sp.* Kassell, Th. G. Fisher & Co.)

ameter) is thick-walled, muscular, and lies in the anterior part of the middle third of the body. A genital sucker situated on the left posterior border has an average diameter of 150 μm and is provided with 60 to 90 multidigitate spines. The two oval testes lie in a slightly oblique plane in the subcaudal region. The retort-shaped seminal vesicle originates in the posterior part of the equatorial third of the worm and proceeds obliquely anteriorly, being continued as the muscular ejaculatory duct as the tubule approaches the genital sucker. An enveloping cirrus sac and a cirrus organ are lacking. Prostate glands surround the outer end of the tube. The small, subglobose ovary and the adjacent ootype are situated medially in the anterior portion of the posterior third of the body. About 14 large, polygonal vitelline follicles are also present in each of the lateral fields of this portion of the worm. Seminal receptacle and Laurer's canal are present. The uterus coils intricately through the intercecal field to the metraterm, which lies next to the male genital opening in the genital atrium at the base of the genital sucker.

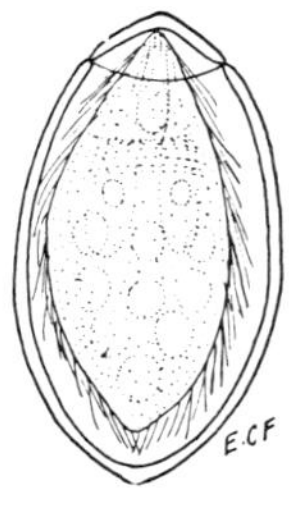

Fig. 28–31. *Heterophyes heterophyes,* egg with mature miracidium. (× ca. 1000.) (Adapted from A. Looss. In Faust, E.C. 1929. *Human Helminthology,* 1st ed. Philadelphia, Lea & Febiger.)

The eggs (Fig. 28–31) are minute, operculate, ovoid, light brown, and measure 28 to 30 μm by 15 to 17 μm. When laid, the egg contains a fully developed miracidium with a bilateral symmetry of its internal organs. It hatches, however, only after ingestion by the appropriate freshwater snail *(Pironella conica* in Egypt; *Cerithidea cingulata* in Japan), in which a sporocyst and one or two redial generations are developed and from which cercariae eventually emerge. The cercariae closely resemble those of *Metagonimus yokogawai.* They encyst in the mullet *(Mugil cephalus), Tilapia nilotica,* and *Aphanius fasciatus* in Egypt, and in the mullet and *Acanthogobius* sp. in Japan, where a brackish-water snail *(Cerithidea cingulata,* syn. *Tymphonotonus microptera)* is the molluscan host. On ingestion of these fishes uncooked, the definitive host becomes infected.

In the vicinity of Kobe, Japan, *Heterophyes katsuradai* Ozaki and Asada, 1925, a closely related species, broader and more rounded in contour, with a relatively enormous ventral sucker and with smaller eggs (25 to 26 μm by 14 to 15 μm), has been recovered after anthelmintic treatment from patients suffering with diarrhea. Infection is acquired from eating raw mullet.

Epidemiology. In addition to its presence in man, *H. heterophyes* has been found as a natural infection in the cat, dog, fox, and other fish-eating mammals in endemic areas. Infected fishermen, in particular, pollute the water with their feces, thus providing a human source for the extrahuman phases of the life cycle. Infection of the definitive host results from eating raw or undercooked fish.

Pathogenesis, Pathology, and Symptomatology. In the average infection with *Heterophyes heterophyes,* there is mild inflammatory reaction at sites where the minute worms become attached to the intestinal mucosa or burrow into the mucosa.

The worms produce mild irritation, accompanied by colicky pains and mucous diarrhea, with production of excess mucus and a superficial necrosis of the mucous coat. Africa and his associates (1935b, 1937a,b) reported that eggs of this worm in man in the Philippines may filter through the intestinal wall, be picked up by the mesenteric lymphatics, and be filtered out in cardiac valves and myocardium. There they provoke myocarditis leading to cardiac failure, or they may be carried to the brain. Up to 15% of fatal heart cases in the Philippines may result from heterophyid myocarditis (Kean and Breslau, 1964). Deschiens et al. (1958) reported two cases with neurological manifestations, in which surgical intervention demonstrated adult *H. heterophyes* encapsulated in the brain tissue. There frequently is significant diarrhea in *Heterophyes* infection.

Diagnosis. Eggs in the feces are differentiated with great difficulty from other heterophyid eggs, and they somewhat resemble those of *Clonorchis* and *Opisthorchis*.

Treatment. Tetrachloroethylene, as prescribed for hookworm infection, has been recommended. In view of its efficacy in metagonimiasis *(vide infra)*, niclosamide may be preferable.

Prognosis. The prognosis is usually good.

Prevention. Abstinence from eating raw or salted mullet or other infected freshwater fish is indicated.

Metagonimus yokogawai
Katsurada, 1913
(Yokogawa's fluke, producing metagonimiasis)

Synonyms. *Loxotrema ovatum* Kobayashi, 1912; *Metagonimus ovatus* Yokogawa, 1913; *Loossia romanica* Ciurea, 1915; *Loossia parva* Ciurea, 1915; *Loossia dobrogiensis* Ciurea, 1915.

Historical and Geographical Notes. This worm was first described by Katsurada as *Heterophyes yokogawai*, but later by the same investigator as *Metagonimus yokogawai* from material obtained by Yokogawa from man and experimental dogs in Formosa (1911). In highly endemic areas a considerable percentage of the native population is infected.

This heterophyid is probably the most common in the Far East (Sino-Japanese areas and Maritime Provinces of the U.S.S.R.), the northern provinces of Siberia, and the Balkan States. It has also been reported from man in Spain.

Morphology, Biology, and Life Cycle. The fluke (Fig. 28–32), in its habitat, size (1 to 2.5 mm by 0.4 to 0.75 mm), and shape, more or less resembles *H. heterophyes* (Fig. 28–30). The most distinctive feature concerns the ventral sucker (66 to 165 μm long, 55 to 114 μm wide), which is deflected to the right of the midline with its long axis in a diagonal plane. Closely associated with this acetabulum is the genital opening, the muscular outer rim of which is fused with it. The eggs (26 to 28 μm by 15 to 17 μm; Fig. 28–33) can be distinguished from those of *H. heterophyes* only with great difficulty. The first intermediate hosts *(Semisulcospira libertina,* and possibly related species) ingest the fully embryonated, unhatched eggs. A first sporocyst generation and two redial generation's are then developed. Finally, there erupt from the snail characteristic lophocercous (i.e., keeled) cercariae, with pigmented eyespots, abundant penetration glands, and a distinctive anterior end, having an anteroventral mouth and a pair of clustered openings for the penetration gland ducts. Intermediate between these ducts and the mouth

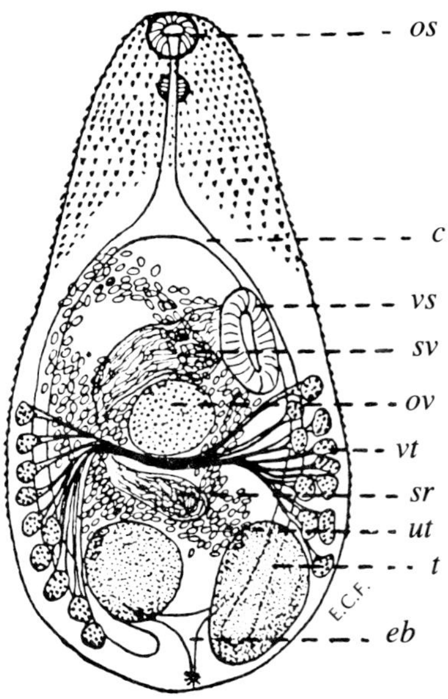

Fig. 28–32. *Metagonimus yokogawai*, ventral view. (× 36.) *c*, Cecum. *eb*, Excretory bladder. *os*, Oral sucker. *ov*, Ovary. *sr*, Seminal receptacle. *sv*, Seminal vesicle. *t*, Testis. *ut*, Uterus. *vt*, Vitellaria. *vs*, Ventral sucker. (Adapted from original of E.C. Faust.)

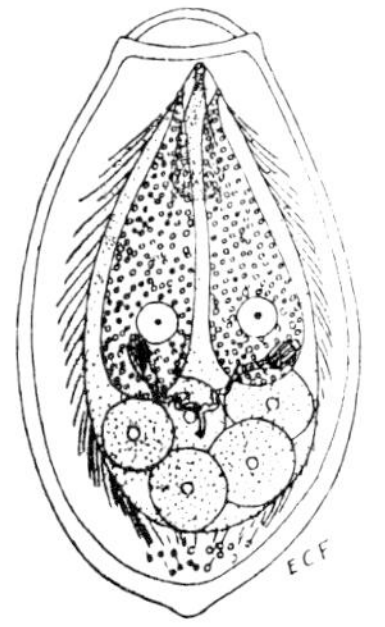

Fig. 28–33. *Metagonimus yokogawai*. Egg with mature miracidium. (× 1300.) (Adapted from original of E.C. Faust.)

are specialized spines. Swimming about in the water, the cercariae attack and become encysted under the scales, in the skin, or in the flesh of freshwater fishes, of which the following have been incriminated: *Plectoglossus altivelis, Odontobutis obscurus, Salmo perryi,* and *Tribolodon hakonensis,* all of which are valuable food fishes. Ingestion of the raw flesh of these fishes provides the means of human infection.

Epidemiology. Man becomes infected primarily from eating the uncooked Oriental freshwater trout *(Plectoglossus altivelis)* in the Far East, and other freshwater fishes, especially *Odontobutis obscurus* and *Salmo perryi,* in this and other endemic foci. Fish-eating mammals and even the pelican are reservoir hosts of the infection. Pollution of water in which the intermediate hosts live provides the original sources of infection.

Pathogenesis, Pathology, and Symptomatology. A moderate degree of inflammation develops around the sites in the intestine where the minute worms are attached. This inflammation is frequently accompanied by an appreciable amount of microscopic sloughing or necrosis of the surrounding mucosal cells.

The sucker attachments of the flukes irritate the mucosa, producing an excess of mucus and superficial erosion of the cells. At times they cause eosinophilic and neutrophilic infiltration of the wall, particularly around eggs that are deposited in the tissues. Or, they infiltrate in the intestinal capillaries and lymphatics and are carried to the myocardium, brain, spinal cord, and other tissues where granulomatous changes may be provoked by their presence. Usually the *symptoms* consist of mild diarrhea, but the extent of the symptoms depends on the number of worms present, the depth of their penetration into the wall, and the individual reaction of the patient. Africa et al. (1937a,b) stated that once the worms invade the mucosa they remain there until they die.

Diagnosis. The diagnosis is based on recovery of the characteristic heterophyid eggs in the feces. Specific diagnosis can be made only after recovery of evacuated flukes following specific anthelmintic treatment.

Treatment. Of several drugs tried for efficacy in the treatment of metagonimiasis, tetrachloroethylene and niclosamide were the most effective. Since niclosamide was better tolerated, it is preferable. The dosage in children was 100 to 125 mg per kg body weight (Rim, 1975). In view of its efficacy against other flukes and its relative lack of side effects, praziquantel given in a dosage as for fascioliasis is probably the drug of choice.

Prognosis. The prognosis is good.

Prevention. Abstinence from eating raw or inadequately cooked freshwater fishes is indicated.

Other Heterophyid Parasites of Man

Potentially all species of the family Heterophyidae may become human intestinal parasites. Incidentally, in Nature and by experimental tests, the following species have been reported from man: *Centrocestus armatus* (Tanabe, 1922); *C. formosana* (Nishigori, 1924); *Haplorchis pumilio* (Looss, 1896 [syn. *Monorchotrema taiokui* Nishigori, 1924]); *H. microrchia* (Katsuta, 1932); *H. yokogawai* (Katsuta, 1932); *H. taichui* (Nishigori, 1924); *Diorchitrema formosanum* (Katsuta, 1932); *D. amplicaecale* (Katsuta, 1931); *D. pseudocirratum* Witenberg, 1929; and *Metagonimus minutus* Katsuta, 1932. *Haplorchis taichui* was reported in man in Thailand (Kliks and Tantachamrun, 1974). Chen (1944) found metacercariae of *Haplorchis* and *Centrocestus* in *Rana limnocharis* and *Bufo melanostictus* in the vicinity of Hong Kong, while Vazquez-Colet (1943) claimed to have found young shrimp of the genus *Penacus* infected with encysted metacercariae of *Haplorchis yokogawai. Stellantchasmus falcatus* is reported to be established in Hawaii, where it is contracted from eating raw mullet, and two human infections with this species have been reported from Kochi Prefecture in Japan (Kagei et al., 1964). Likewise, Alicata (personal communication) found *D. pseudocirratum* in native Hawaiians who had eaten raw mullet caught locally. All these worms are contracted from eating raw, infected fish that mostly but not exclusively inhabit fresh water.

Minute heterophyids identified as *Pygidiopsis summa* Onji and Nishio, 1916 were recovered in large numbers (up to 4,000) from 8 adults in Korea after they were treated with bithionol and purged with magnesium sulfate (Seo et al., 1981). According to Yamaguti (1975), this species occurs in fish-eating birds and, experimentally, in the cat, dog, and rat. The adult worm is piriform, less than 1 mm long (up to 0.76 by 0.44 mm), and the eggs measure 21 to 23 by 11 to 14 μm. Cercariae develop to rediae in *Tympanotomus microptera,* and the metacercariae are found encysted in the wall of the esophagus or intestine of small fishes.

The intestinal lesions produced by other small heterophyids are similar to those described for *Heterophyes heterophyes* and *Metagonimus yokogawai*. Africa and Garcia (1935b) and Africa et al. (1935, 1937b) found eggs of *Haplorchis yokogawai, H. pumilio, H. taichui,* and *Diorchitrema pseudocirratum* in cardiac lesions of persons dying of cardiac failure and, in one case, found an adult heterophyid in the epicardial layer of the heart. Likewise, eggs tentatively diagnosed as those of *Haplorchis pumilio* were found by Africa et al. (1937a) in sections of the spinal cord, with loss of motor and sensory function at the levels where the lesions were located. Methods of collecting and identifying heterophyids, and the taxonomy of the *Haplorchis* group, were reported by Pearson and Ow-Yang (1982).

REFERENCES

Abend, L. 1910. Ueber Haemoptysis parasitaria. Dtsch. Arch. Klin. Med., *100*:501–511.

Africa, C.M. 1937a. Ova in the spinal cord of man. Philip. J. Sci., *62*:393–399.

———. 1937b. Evidence of intramucosal invasion in the life cycle of *Haplorchis yokogawai* (Katsuta, 1932) Chen, 1936 (Heterophyidae). J. Philip. Ids. Med. Assoc., *17*:737–743.

———, and Garcia, E.Y. 1935a. *Plagiorchis* sp., a new parasite of the human intestine. Festschr. Skrjabin, pp. 9–10.

———, and Garcia, E.Y. 1935b. Heterophyid trematodes of man and dog in the Philippines with descriptions of three new species. Philip. J. Sci., *57*:253–267.

———, Garcia, E.Y., and de Leon, W. 1935. Intestinal heterophyidiasis with cardiac involvement. Philip. J. Public Health, *2*:1–22.

———, de Leon, W., and Garcia, E.Y. 1937a. Heterophyidiasis. V. Ova in the spinal cord of man. Philip. J. Sci., *62*:395–399.

———. de Leon, W., and Garcia, E.Y., 1937b. Heterophyidiasis. VI. Two more cases of heart failure associated with the presence of eggs in sclerosed veins. J. Philip. Ids. Med. Assoc., *17*:605–609.

Allanson, G.R. 1975. Human *Paragonimus* infestation—second South African report. S. Afr. J. Med. Lab. Technol., *21*(2):4–5.

Amaral, A.D., and Busetti, E.T. 1979. Fasciolose hepatica humana no Brasil. Rev. Inst. Med. Trop. S. Paulo, *21*:141–145.

Ambroise-Thomas, P., Coullier, A., and Wegner, D.G.H. 1981. Le praziquantel dans le traitement des distomatoses hepatiques extreme-orientales a *Clonorchis sinensis* ou *Opistorchis viverinni*. Bull. Soc. Pathol. Exot., *74*:426–433.

Ameel, D.J. 1932. Life history of the North American lung fluke of mammals. J. Parasitol., *18*:264–268.

———, Cort, W.W., and Van der Woude, A. 1951. Development of the mother sporocyst and rediae of *Paragonimus kellicotti* Ward, 1908. J. Parasitol., *37*:395–404.

Anazawa, K. 1929. First instance of *Echinostomum revolutum* found in man and its course of infection. J. Med. Assoc. Formosa, No. 288, 10–13.

Artigas, P. de T., and Perez, M.D. 1960–1962. Considerações sôbre *Opisthorchis pricei*, Foster, 1939, *O. guayaquilensis* Rodriguez, Gomez et Montalvan, 1949, e *O. pseudofelineus* Ward, 1901. Descrição de *Amphimerus pseudofelineus minutus* n. subsp. Mem. Inst. Butantan, *30*:157–166.

Asada, J.L., Otagaki, H., Morita, D., Takeuchi, T., Sakai, Y., Honishi, T., and Okahashi, K. 1962. A case report on the human infection with *Plagiorchis muris* Tanabe, 1922 in Japan. Jpn. J. Parasitol., *11*:512–516.

Ashton, N., Brown, N., and Easty, D. 1969. Trematode cataract in fresh water fish. J. Small Anim. Pract., *10*:471–478.

Balci, S. 1975. Human fascioliasis. Gall bladder invasion by flukes in a five-year-old boy. Clin. Pediatr., *14*:1068–1069.

Barlow, C.H. 1925. The life cycle of the human intestinal fluke *Fasciolopsis buski* (Lankester). Am. J. Hyg. Monogr. Ser. No. 4.

Basch, P.F. 1965. Completion of the life cycle of *Eurytrema pancreaticum* (Trematoda: Dicrocoeliidae). J. Parasitol., *51*:350–355.

Beaver, P.C. 1937. Experimental studies on *Echinostoma revolutum* (Froelich) a fluke from birds and mammals. Contr. No. 491, Zool. Lab. Univ. Illinois.

———. 1939. The life cycle of *Euparyphium melis* (Trematoda: Echinostomidae). J. Parasitol., *25*(suppl.):19.

———, Duron, R.A., and Little, M.D. 1977. Trematode eggs in the peritoneal cavity of a man in Honduras. Am. J. Trop. Med. Hyg., *26*:684–687.

———, Little, M.D., Tucker, C.F., and Reed, R.J. 1977. Mesocercaria in the skin of man in Louisiana. Am. J. Trop. Med. Hyg., *26*:422–426.

Béland, J.E., Boone, J., Donevan, R.E., and Mankiewicz, E. 1969. Paragonimiasis (the lung fluke). Report of four cases. Am. Rev. Respir. Dis., *99*:261–271.

Bennington, E., and Pratt, I. 1960. The life history of the salmon-poisoning fluke, *Nanophyetus salmincola* (Chapin). J. Parasitol., *46*:91–100.

Bhalerao, G.D. 1931. Trematode parasites of pigs in Bengal. Rec. Indian Mus., *33*:475–482.

Bonne, C., Bras, G., and Lie Kian Joe. 1948. Five human echinostomes in the Malayan Archipelago. Med. Maandbl., *23*:456–465.

Brenes, R., Rodriguez-O, B., Solano, G.V., Obando, E.M.O., and Ruiz S., P. J. 1982. Cerebral hemorrhagic lesions produced by *Paragonimus mexicanus*. Report of three cases in Costa Rica. Am. J. Trop. Med. Hyg., *31*:522–526.

———, Zeledón, R., and Rojas, G. 1980. Biological cycle and taxonomic position of a Costa Rican *Paragonimus* and the present status of *Paragonimus* from the New World. Brenesia, *18*:353–366.

Bunnag, D., and Harinasuta, T. 1980. Studies on the chemotherapy of human opisthorchiasis in Thailand. I. Clinical trial of praziquantel. Southeast Asian J. Trop. Med. Public Health, *11*:528–531.

———, and Harinasuta, T. 1981. Studies on the chemotherapy of human opisthorchiasis. III. Minimum effective dose of praziquantel. Southeast Asian J. Trop. Med. Public Health, *12*:413–417.

Cabrera, B.D., and Fevidal, P.M. 1974. Studies on *Paragonimus* and paragonimiasis in the Philippines. III. Prevalence and treatment of human paragonimiasis with bithionol in Jaro, Leyte, Philippines. Southeast Asian J. Trop. Med. Public Health, *5*:39–45.

———, and Vajrasthira, S. 1973. Endemicity of *Paragonimus siamensis* Miyazaki and Wykoff 1965, the second species of !ung fluke found in the Republic of the Philippines. Southeast Asian J. Trop. Med. Public Health, *4*:509–518.

Carney, W.P., Sudomo, M., and Purnomo. 1980. Echinostomiasis: A disease that disappeared. Trop. Geogr. Med., *32*:101–105.

Chandler, A.C. 1955. *Introduction to Parasitology*. 9th Ed. New York, John Wiley & Sons, Inc.

Chen, H.T. 1944. Some parasitic diseases in Hong Kong with emphasis on those caused by helminths. Biol. Bull. Fukien Christian Univ., *4*:1–32.

———, 1960. Taxonomic consideration of *Paragonimus*, including morphological notes on *P. skrjabini* Chen. Acta Zool. Sinica, *12*:27–36.

———. 1965. *Paragonimus, Pagumogonimus* and a *Paragonimus*-like trematode in man. Chinese Med. J., *84*:429–530.

Chung, H.-L., Hsu, C.-P., Ho, L.-Y., Kao, P.-C., Lan, S., and Chiu, F.-H. 1977. Studies on a new pathogenic lung fluke—*Paragonimus hueit'ungensis* sp. nov. Chinese Med. J., *3*:379–394.

Coumbaras, A. 1966. La distomatose hépatique en Algérie. Ann. Parasitol., *41*:71–77.

Dawes, B. 1946. *The Trematoda; With Special Reference to British and Other European Forms*. Cambridge University Press.

———. 1961. On the early stages of *Fasciola hepatica* penetrating into the liver of an experimental host, the mouse: A histological picture. J. Helminthol., R. T. Leiper Supplement, 1961, pp. 41–52.

De Leon, D., Quiñones, R., and Hillyer, G.V. 1981. The prepatent and patent periods of *Fasciola hepatica* in cattle in Puerto Rico. J. Parasitol., *67*:734–735.

Deschiens, R., Collomb, H., and Demarchi, J. 1958. Distomatose cerebrale a *Heterophyes heterophyes*. Abstr. 6th Int. Congr. Trop. Med. Malaria, p. 265.

———, LeCorroller, Y., and Mandoul, R. 1961. Enquête sur les foyers de distomatose hépatique de la Vallée du Lot. Ann. Inst. Pasteur, *10*:5–12.

Doby, J.-M., and Beaucournu, J.-C. 1970. A propos des formes erratiques et abortives de deux parasitoses non exceptionnelles chez l'homme (distomatose par *Fasciola* et hypodermose). Difficultés du diagnostic clinique différentiel. Bull. Soc. Pathol. Exot., *63*:227–241.

Dollfus, R. 1939. Distome d'un abcés palpébro-orbitaire chez une panthère. Possibilité d'affinités lointaines entre ce distome et les Paragonimidae. Ann. Parasitol., *17*:209–235.

———, and Nouvel, J. 1957. Une enigme pour la spécificité parasitaire et la biogéographie: Le cas d'*Achillurbainia* (Trematoda Digenea). Union Int. Sci. Biol. Ser. B, *32*:228–229.

Dutt, S.C., and Srivastava, H.D. 1966. The intermediate host and the cercaria of *Gastrodiscoides hominis* (Trematoda: Gastrodiscidae). J. Helminthol., *40*:45–52.

Erhardt, A., Germer, W.D., and Hörning, B. 1962. Die Opisthorchiasis, hervorgerufen durch dem Katzenleberegel *Opisthorchis felineus* (Riv.). Parasitol. Schriftreihe, *15*: 1–170.

Facey, R.V., and Marsden, P.D. 1960. Fascioliasis in man: An outbreak in Hampshire. Br. Med. J., *2*:619–625.

Faiguenbaum, J., Feres, A., Donckaster, R., Atias, A., Jarpa, A., Niedmann, G., Donoso, F., Rubio, M., and Neruane, J. 1962. Fascioliasis (distomiasis) hépatica humana. Bol. Chil. Parasitol., *17*:7–12.

Fain, A., and Galal, A. 1977. Presence du trematode *Echinoparyphium recurvatum* (von Linstow, 1873) chez l'homme et chez un rongeur en Egypte. Ann. Soc. Belge Med. Trop., *57*:583–587.

———, and Vandepitte, J. 1957a. A new trematode, *Poikilorchis congolensis*, n. g., n. sp., living in subcutaneous retroauricular cysts in man from the Belgian Congo. Nature, *179*:740.

———. and Vandepitte, J. 1957b. Description du nouveau distome vivant dans des kystes ou abcès rétroauriculaires chez l'homme au Congo Belge. Ann. Soc. Belge Med. Trop., *37*:251–261.

Fan, P.C., and Chiang, C.H. 1970. Exposure of kittens and puppies to single metacercariae of *Paragonimus westermani* from Taiwan. J. Parasitol., *56*:48–54.

———, Lin, L.H., Ku, C.C., and Lin, H.C. 1974. Locality and maturity of single worm of *Paragonimus westermani* (Korea strain) and its egg producing capacity and distribution in domestic cats. Yonsai Rep. Trop. Med., *6*:81.

Farag, H.F., Barakat, R.M.R., Ragab, M., and Omar, E. 1979. A focus of human fascioliasis in the Nile Delta, Egypt. J. Trop. Med. Hyg., *82*:188–190.

Farid, H., Jalayer, T., Beheshti, G., and Sadeghi, A. 1977. Report of a human case with *Fasciola hepatica* in common bile duct. Iran. J. Public Health, *6*:98–101.

Faust, E.C., and Khaw, O.-K. 1927. Studies on *Clonorchis sinensis* (Cobbold). Am. J. Hyg. Monogr. Ser. No. 8.

Fernandes, B.J., Cooper, J.D., Cullen, J.B., Freeman, R.S., Ritchie, A.C., Scott, A.A., and Stuart, P.F. 1976. Systemic infection with *Alaria americana* (Trematoda). Can. Med. Assoc. J., *115*:1111–1114.

Fischer, G.W., McGrew, G.L., and Bass, J.W. 1980. Pulmonary paragonimiasis in childhood. A cause of persistent pneumonia and hemoptysis. J.A.M.A., *243*:1360–1362.

Flavell, D.J. 1981. Liver-fluke infection as an aetiological factor in bile-duct carcinoma of man. Trans. R. Soc. Trop. Med. Hyg., *75*:814–824.

Freeman, R.S., Stuart, P.F., Cullen, J.B., Ritchie, A.C., Mildon, A., Fernandes, B.J., and Bonin, R. 1976. Fatal human infection with mesocercariae of the trematode *Alaria americana*. Am. J. Trop. Med. Hyg., *25*:803–807.

Gebhardt, G.A., Millemann, R.E., Knapp, S.E., and Nyberg, P.A. 1966. "Salmon poisoning" disease. II. Second intermediate host susceptibility studies. J. Parasitol., *52*:54–59.

Germer, W.D., Yong, M.H., Schulze, W., Jeltsch, R., and Orrahood, M.D. 1955. Die Klinik der Clonorchiasis. Zeitschr. Hyg. Infektionskr., *141*:132–145.

Harinasuta, C., and Vajrasthira, S. 1960. Opisthorchiasis in Thailand. Ann. Trop. Med. Parasitol., *54*:100–105.

Hatsushika, R. 1981. Experimental studies on mixed infection with *Paragonimus miyazaki* Kamo et al., 1961 and other species of *Paragonimus* in an animal. Kawasaki Med. J., *7*:113–125.

———, and Endo, A. 1978. A case of paragonimiasis in greater omentum. Kawasaki Med. J., *4*:165–172.

Hillyer, G.V. 1981. Fascioliasis in Puerto Rico: A review. Bol. Assoc. Med. P. R., *73*:94–101.

Hockley, A.R. 1964. Observations on the metacercaria of *Spelotrema*. Proc. 1st Int. Congr. Parasitol., *1*:525.

Horstmann, R.D., Feldheim, W., Feldmeier, H., and Dietrich, M. 1981. High efficacy of praziquantel in the treatment of 22 patients with *Clonorchis/Opisthorchis* infections. Tropenmed. Parasitol., *32*:157–160.

Hou, P.C. 1955. The pathology of *Clonorchis sinensis* infection of the liver. J. Pathol. Bacteriol., *70*:53–64.

Hsü, H.F. 1940. *Euparyphium hassyense* Leon & Ciurea (= *E. melis* Schrank) found at autopsy of a Chinese. Chinese Med. J., *58*:552–558.

Idris, M., Rahman, K.M., Muttalib, M.A., and Azad Khan, A.K. 1980. The treatment of fasciolopsiasis with niclosamide and dichlorophen. J. Trop. Med. Hyg., *83*:71–74.

Ishii, Y., Koga, M., Fujino, T., Higo, H., Ishibashi, J., Oka, K., and Saito, S. 1983. Human infection with the pancreas fluke, *Eurytrema pancreaticum*. Am. J. Trop. Med. Hyg., *32* (in press).

Isseroff, H., Sawma, J.T., and Reino, D. 1977. Fascioliasis: Role of proline in bile duct hyperplasia. Science, *198*:1157–1159.

Ito, J. 1977. Studies on the fresh water cercariae in Leyte Island, Philippines. 3. Cercariae from Thiaridae. Jpn. J. Exp. Med., *47*:223–248.

———, Yokogawa, M., Araki, K., and Kobayashi, M. 1978. Studies on the morphology of larval and adult lung fluke in the Philippines, with a proposition of new name, *Paragonimus philippinensis* n. sp. Jpn. J. Parasitol., *27*:97–112.

———, Yokogawa, M., Araki, K., and Kobayashi, M. 1979. Further observations on the morphology of adult lung fluke, *Paragonimus philippinensis* Ito, Yokogawa, Araki and Kobayashi, 1978. Jpn. J. Parasitol., *28*:253–259.

Jones, E.A., Kay, J.M., Milligan, H.P., and Owens, D. 1977. Massive infection with *Fasciola hepatica* in man. Am. J. Med., *63*:836–842.

Kagei, N., Oshima, T., Ishikawa, K., and Kihata, M. 1964. Two cases of human infection with *Stellantchasmus falcatus* Onji et Nishio, 1915 (Heterophyidae) in Kochi Prefecture. Jpn. J. Parasitol., *13*:472–478.

Kamo, H., Ogino, K., and Hatsushika, R. 1962. A unique infection of man with *Clinostomum* sp., a small trematode causing acute laryngitis. Yonago Acta Med., *6*:37–40.

Kean, B.H., and Breslau, R.C. 1964. Cardiac heterophyidiasis. *Parasites of the Human Heart*. New York, Grune & Stratton, pp. 95–103.

Khorsandi, H.O. 1977. Obstructive jaundice due to *Fasciola hepatica*. Report of two cases. Bull. Soc. Pathol. Exot., *70*:626–628.

Kifune, T., and Takao, Y. 1973. Description of *Psilorchis hominis* sp. nov. from man (Trematoda: Echinostomatoidea: Psilostomidae). Jpn. J. Parasitol., *22*:111–115.

Kim, D.C. 1974. Ecological studies of *Clonorchis sinensis*—endemicity and propagation of clonorchiasis in high and low endemic areas in Korea. Yonsei Rep. Trop. Med., *5*:3–44.

Kliks, M., and Tantachamrun, T. 1974. Heterophyid (Trematoda) parasites of cats in North Thailand, with notes on a human case found at necropsy. Southeast Asian J. Trop. Med. Public Health, *5*:547–555.

Kobayashi, A., Suzuki, S., Horiuchi, K., Yokogawa, M., and Araki, K. 1975. Four human cases of paragonimiasis miyazakii. Jikeikai Med. J., *22*:127–135.

Koenigstein, R.P. 1949. Observations on the epidemiology of infections with *Clonorchis sinensis*. Trans. R. Soc. Trop. Med. Hyg., *42*:503–506.

Komiya, Y. 1966. *Clonorchis* and clonorchiasis. Adv. Parasitol., *4*:53–106.

———, and Suzuki, N. 1964. Biology of *Clonorchis sinensis*, pp. 551–645. In *Progress of Medical Parasitology in Japan*. Vol. 1. Edited by K. Morishita, Y. Komiya, and H. Matsubayashi. Tokyo, Meguro Parasitological Museum.

Krull, W.H., and Mapes, C.R. 1952. Studies on the biology of *Dicrocoelium dendriticum* (Rudolphi, 1819) Looss, 1899 (Trematoda: Dicrocoeliidae), including its relation to the intermediate host, *Cionella lubrica* (Müller). V. Notes on infections of *Dicrocoelium dendriticum* in *Cionella lubrica*. Cornell Vet., *42*:339–351.

———, and Mapes, C.R. 1953. Studies on the biology of *Dicrocoelium dendriticum* (Rudolphi, 1819) Looss, 1899 (Trematoda: Dicrocoeliidae), including its relation to the intermediate host, *Cionella lubrica* (Muller). IX. Notes on the cyst, metacercaria and infection in the ant, *Formica fusca*. Cornell Vet., *43*:389–409.

Leuckart, R. 1882. Zur Entwicklungsgeschichte des Leberegels. Arch. Naturgesch., *1*:80–119.

Lie Kian Joe. 1963. Studies on Echinostomatida in Malaya. III. The adult *Echinostoma malayanum* Leiper, 1911 (Trematoda) and the probable synonymy of *Artyfechinostomum sufrartyfex* Lane, 1915. Z. Parasitenkd., *23*:124–135.

———. 1968. Further studies on the life history of *Echinostoma lindoense* Sandground and Bonne, 1940 (Trematoda: Echinostomatidae) with a report of its occurrence in Brazil. Proc. Helm. Soc. Wash., *35*:74–77.

———, and Virik, H.K. 1963. Human infection with *Echinostoma malayanum* Leiper, 1911 (Trematoda: Echinostomatidae). J. Trop. Med. Hyg., *66*:77–82.

———, Williams, H.I., Miyazaki, I., and Kai, W.S. 1962. A subcutaneous retro-auricular abscess in a Dyak boy in Sarawak, probably caused by a trematode of the genus *Poikilorchis*, Fain and Vandepitte, 1957. Med. J. Malaya, *17*:37–40.

Ling, C.C. 1949. A fatal case of *Clonorchis sinensis* infestation. Chinese Med. J., *67*:414–420.

Little, M.D. 1968. *Paragonimus caliensis* sp. n. and paragonimiasis in Colombia. J. Parasitol., *54*:738–746.

Liu Yue-han, Qui Zhong-da, Wang Xiao-gen, Wang Qi-nan, Qu Zhen-gi, Chen Rong-xin, Liu Ji-bo, Zhang Cheng-da, and Qin Shou-an. 1982. Praziquantel in *Clonorchis sinensis:* A further evaluation of 100 cases. Chinese Med. J., *95*:89–94.

Lü, C.-H., et al. 1957. Clinical observations of 195 cases of *Paragonimus* in children. Chinese J. Pediatr., *8*:143–146.

Lumsden, R.D., and Sogandares-Bernal, F. 1970. Ultrastructural manifestations of pulmonary paragonimiasis. J. Parasitol., *56*:1095–1109.

Malek, E.A., and Little, M.D. 1971. *Aroapyrgus colombiensis* n. sp. (Gastropoda: Hydrobiidae), snail intermediate host of *Paragonimus caliensis* in Colombia. Nautilus, *85*:20–26.

Markovic, A. 1939. Der erste Fall von Philophthalmose beim Menschen. Arch. Ophthalmol. (Berlin), *140*:515–526.

Martínez Báez, M., and Jiménez Galán, A. 1961. Un caso de trematodiasis pulmonar registrado en Mexico. Rev. Inst. Salubr. Enferm. Trop., *21*:101–114.

McConnell, J.F.P. 1876. On the *Distoma conjunctum* as a human entozoon. Lancet, *1*:343–344.

McMullen, D.B. 1937. An experimental infection of *Plagiorchis muris* in man. J. Parasitol., *23*:113–115.

Miller, G.C. 1981. Helminths and the transmammary route of infection. Parasitology, *82*:335–342.

Mitsuno, T., Takeya-Siko, Inanaga, K., and Zimmerman, L.E. 1952. Cerebral paragonimiasis. A neurosurgical problem in the Far East. J. Nerv. Mental Dis., *116*:685–714.

Miyazaki, I. 1972. *Lung Flukes in the Western Hemisphere*. Tokyo, Overseas Cooperation Agency.

———. 1974. Lung flukes in the world. Morphology and life history, pp. 101–135. In *A Symposium on Epidemiology of Parasitic Diseases*. Tokyo, International Medical Foundation of Japan.

———. 1978. Two types of the lung fluke which has been called *Paragonimus westermani* (Kerbert, 1878). Med. Bull. Fukuoka Univ., *5*:251–263.

———. 1979. On the newly proposed lung fluke, *Paragonimus pulmonalis* (Baelz, 1880). Med. Bull. Fukuoka Univ., *6*:267–276.

———. 1981a. *Paragonimus pulmonalis* (Baelz, 1880) found from the human brain in 1906. Med. Bull. Fukuoka Univ., *8*:17–20.

———. 1981b. Taxonomical studies on the lung fluke occurring in the Philippines. Med. Bull. Fukuoka Univ., *8*:153–158.

———, and Chiu, J.-K. 1980. Examination of the so-called "*Paragonimus westermani*" in Taiwan. Med. Bull. Fukuoka Univ., *7*:277–279.

———, and Fontan, R. 1970. Mature *Paragonimus heterotremus* found from a man in Laos. Jpn. J. Parasitol., *19*:109–113.

———, Grados, O., and Uyema, N. 1973. A new lung fluke found in Peru, *Paragonimus amazonicus* sp. n. (Trematoda: Troglotrematidae). Jpn. J. Parasitol., *22*:48–54.

———, and Habe, S. 1979. *Paragonimus westermani filipinus* Miyazaki, 1978. stat. n. occurring at Jaro, Leyte, the Philippines. Med. Bull. Fukuoka Univ., *6*:447–462.

———, and Hirosi, H. 1976. Immature lung flukes first found in the muscle of the wild boar in Japan. J. Parasitol., *62*:836–837.

———, Ibanez, N., and Miranda, H. 1969. On a new lung fluke found in Peru. *Paragonimus peruvianus* sp. n. (Trematoda: Troglotrematidae). Jpn. J. Parasitol., *18*:123–130.

———, Ibanez, N., and Miranda, H. 1971. Studies on the metacercaria of *Paragonimus peruvianus* (Trematoda: Troglotrematidae). Jpn. J. Parasitol., *20*:425–430.

———, and Ishii, Y. 1968. Studies on the Mexican lung flukes, with special reference to a description of *Paragonimus mexicanus* sp. nov. (Trematoda: Troglotrematidae). Jpn. J. Parasitol., *17*:445–453.

———, and Kannangara, G. 1970. Metacercariae of *Achillurbainia* sp. from mountain crabs, *Parathelphusa rugosa* in Ceylon (Trematoda: Achillurbainiidae). Jpn. J. Parasitol., *19*:537–540.

———, Kifune, T., and Lamothe-Argumedo, R. 1980. *Taxonomical and Biological Studies on the Lung Flukes of Central America*. Department of Parasitology, School of Medicine, Fukuoka University.

———, Mazabel, C., and Grados, O. 1974. Mature *Paragonimus amazonicus* first found from a water opossum in Peru. J. Parasitol., *60*:547–548.

———, Mazabel, C., Grados, O., and Uyema, N. 1975. Studies on the lung fluke in Tingo Maria, Peru, with special reference to the description of *Paragonimus inca* sp. n. (Trematoda: Troglotrematidae). Med. Bull. Fukuoka Univ., *2*:303–311.

Morera, P. 1968. Trematodiasis pulmonar. Estudio de dos casos encontrados en Costa Rica. Acta Med. Costaric., *11*:225–237.

Musgrave, W.E. 1907. Paragonimiasis in the Philippine Islands. Philip. J. Sci., B, *2*:15–63.

Nakagawa, A. 1921. On the life cycle of *Fasciolopsis buski* Lankester. Kitasato Arch. Exp. Med., *4*:159–167.

Neghme, A., and Ossandon, M. 1943. Ectopic and hepatic fascioliasis. Am. J. Trop. Med., *23*:545–550.

Norton, R.A., and Monroe, L. 1961. Infection by *Fasciola hepatica* acquired in California. Gastroenterology, *41*:46–48.

Nozais, J.P., Doucet, J., Dunan, J., and N'Dri, G.A. 1980. Les paragonimoses en Afrique Noire. A propos d'un foyer récent de Côte-d'Ivoire. Bull. Soc. Pathol. Exot., *73*:155–163.

Nwokolo, C. 1972. Endemic paragonimiasis in Eastern Nigeria. Clinical features and epidemiology of the recent outbreak following the Nigerian civil war. Trop. Geogr. Med., *24*:138–147.

Odei, M.A. 1966. A note on dicrocoeliasis and *Fasciola gigantica* infection in livestock in northern Ghana, with a record of spurious and of genuine *Dicrocoelium hospes* infections in man. Ann. Trop. Med. Parasitol., *61*:215–218.

Oh, S.J. 1968. Spinal paragonimiasis. J. Neurol. Sci., *6*:125–140.

———, and Jordan, E.J. 1967. Findings of intelligence quotient in cerebral paragonimiasis. Jpn. J. Parasitol., *16*:436–440.

Ongom, V.L. 1980. Episternal abscess due to fascioliasis in an Etesot in Uganda. Trans. R. Soc. Trop. Med. Hyg., *74*:417.

Otto, J.H. 1935. Clinical, pathophysiological and therapeutic aspects of human clonorchiasis. Trans. 9th Congr. Far Eastern Assoc. Trop. Med., *1*:543–560.

Ow-Yang, C.K., and Wah, M.J. 1975. A remarkable trematode from the parotid gland of *Tupaia glis*. Southeast Asian J. Trop. Med. Public Health, *8*:449.

Oyediran, A.B.O.O., Fajemisin, A.A., Abioye, A.A., Lagundoye, S.B., and Olugbile, A.O.B. 1975. Infection of the mastoid bone with a *Paragonimus*-like trematode. Am. J. Trop. Med. Hyg., *24*:268–273.

Pearson, J.C., and Ow-Hang, C.K. 1982. New species of *Haplorchis* from southeast Asia, together with keys to the *Haplorchis* group of heterophyid trematodes of the region. Southeast Asian J. Trop. Med. Public Health, *13*:35–60.

Perry, W., Goldsmid, J.M., and Gelfand, M. 1972. Human fascioliasis in Rhodesia. Report of a case with a liver abscess. J. Trop. Med. Hyg., *75*:221–223.

Philip, C.B. 1958. A helminth replaces the usual arthropod as vector of a rickettsia-like disease. Proc. 10th Int. Congr. Entomol., *3*:651–653.

Popov, K.K., and Kalitina, Z.I. 1962. The development of *Dicrocoelium dendriticum* on high altitude grazing grounds of central Caucasus. Zool. Zhurnal, *41*:1793–1797.

Radomyos, P., Bunnag, D., and Harinasuta, T. 1982. *Echinostoma ilocanum* (Garrison, 1908) Odhner, 1911, infection in man in Thailand. Southeast Asian J. Trop. Med. Public Health, *13*:265–269.

Raghunathan, V.S., and Srinivasan, T. 1962. *Artyfechinostomum mehrai* infestation. J. Indian Med. Assoc., *38*:485–487.

Rahman, K.M., Idris, M., and Azad Khan, A.K. 1981. A study of fasciolopsiasis in Bangladesh. J. Trop. Med. Hyg., *84*:81–86.

Reddy, D.B., Ranganaykamma, I., and Vankataratnam, D. 1964. *Artyfechinostomum mehrai* infestation in man. J. Trop. Med. Hyg., *67*:58–59.

Reinhard, E.G. 1957. Landmarks of parasitology. I. The discovery of the life cycle of the liver fluke. Exp. Parasitol., *6*:208–232.

Rim, H.-J. 1975. Studies on the chemotherapy of trematode infections in Korea. Korea Med. J., *12*:425–457.

———, Lyu, K.-S., Lee, J.-S., and Joo, K.-H. 1981a. Clinical evaluation of the therapeutic efficacy of praziquantel (Embay 8440) against *Clonorchis sinensis* infection in man. Ann. Trop. Med. Parasitol., *75*:27–33.

———, Chang, Y.-S., Lee, J.-S., Joo, K.-H., Suh, W.-H., and Tsuji, M. 1981b. Clinical evaluation of praziquantel (Embay 8440; Biltricide®) in the treatment of *Paragonimus westermani*. Korean J. Parasitol., *19*:27–37.

Ripert, C., Carrie, J., Ambroise-Thomas, P., Baecher, R., Kum, N.P., and Same-Ekobo, A. 1981. Étude épidemiologique et clinique de la paragonimose au Cameroun. Résultats du traitement par le niclofolan. Bull. Soc. Pathol. Exot., *74*:319–331.

Rodriguez, J.D., Gomez, L.F., and Montalvan, C.J.A. 1949. El *Opisthorchis guayaquilensis* (una nueva especie de *Opisthorchis* encontrada en el Ecuador). Rev. Ecuatoriana Hig. Trop., *6*:11–24.

Rondelaud, D. 1980. Epidemiological studies on the human distomatosis (*Fasciola hepatica* L.) in the region of Limousin, France. The species of eaten plants and the host snails. Ann. Parasitol., *55*:393–405.

———, and Barthe, D. 1978. Arguments et propositions pour une nouvelle interprétation de l'évolution de *Fasciola hepatica* L. dans *Lymnaea (Galba) truncatula* Müller. Ann. Parasitol., *53*:201–213.

Sachs, R., and Voelker, J. 1982. Human paragonimiasis caused by *Paragonimus uterobilateralis* in Liberia and Guinea, West Africa. Tropenmed. Parasitol., *33*:15–16.

Sadun, E.H. 1955. Studies on *Opisthorchis viverrini* in Thailand. Am. J. Hyg., *62*:81–115.

———. 1957. Fasciolopsiasis and opisthorchiasis as helminthic components of tropical public health. Am. J. Trop. Med. Hyg., *6*:416–422.

———, and Maiphoom, C. 1953. Studies on the epidemiology of the human intestinal fluke, *Fasciolopsis buski* (Lankester), in Central Thailand. Am. J. Trop. Med. Hyg., *2*: 1070–1084.

Sambuc, E., and Baujean, R. 1913. Un cas de cachexie aqueuse chez l'homme (distomatose hépato-pancréatique) avec syndrome pseudo-béribérique. Bull. Soc. Med.-Chir. Indo-Chine, *4*:425–429.

Sandground, J.H. 1940. *Plagiorchis javensis* n. sp., a new trematode parasite of man. Rev. Med. Trop. Parasitol. Havana, *6*:207–211.

———, and Bonne, C. 1940. *Echinostoma lindoensis* n. sp., a new parasite of man in the Celebes, with an account of its life history and epidemiology. Am. J. Trop. Med., *20*:511–535.

Scheid, G., Mendheim, H., and Amenda, R. 1950. Die Lanzettegelinfektion (Dierocoeliasis) beim Menschen nebst Mitteilung eines neuen Falles. Z. Tropenmed. Parasitol., *2*:142–150.

Schlegel, M.W., Knapp, S.E., and Millemann, R.E. 1968. "Salmon poisoning" disease. V. Definitive hosts of the trematode vector, *Nanophyetus salmincola*. J. Parasitol., *54*:770–774.

Sen, H.G. 1965. *Heterophyes heterophyes* from a dog in the western part of India. Trans. R. Soc. Trop. Med. Hyg., *59*:610.

Seo, B.-S., Cho, S.-Y., and Chai, J.-Y. 1980. Studies on intestinal trematodes in Korea. I. A human case of *Echinostoma cinetorchis* infection with an epidemiological investigation. Seoul J. Med., *21*:21–29.

———, Hong, S.-T., and Chai, J.-Y. 1981. Studies on intestinal trematodes in Korea. III. Natural human infections of *Pygidiopsis summa* and *Heterophyes heterophyes nocens*.

Seoul J. Med., *22*:228–235. (Abstr. in Helm, Abstr., *51*:101, 1982.)

Shea, M., Maberley, A.L., Walters, J., Freeman, R.S., and Fallis, A.M. 1973. Intraretinal larval trematode. Trans. Acad. Ophthalmol. Otolaryngol., *77*:OP784–791.

Shimizu, M. 1979. A new trematode, *Poikilorchis nagahanai* sp. n. found from the weasel in Japan (Trematoda: Achillurbainiidae). Jpn. J. Parasitol., *28*:385–391.

Shoop, W.L., and Corkum, K.C. 1981. Epidemiology of *Alaria marcianae* mesocercariae in Louisiana. J. Parasitol., *67*:928–931.

Skrjabin, K.I. 1938. *Echinostoma paraulum*—nouveau parasite de l'homme. Med. Parasitol. Parasit. Dis. (Moscow), *7*:129–138.

Sogandares-Bernal, F., and Seed, J.R. 1973. American paragonimiasis. Curr. Top. Comp. Pathol., *2*:1–56.

Strauss, W.G. 1962. Clinical manifestations of clonorchiasis—a controlled study of 105 cases. Am. J. Trop. Med. Hyg., *11*:625–630.

Stunkard, H.W. 1960. Further studies on the trematode genus *Himasthla* with descriptions of *H. mcintoshi* n. sp., *H. piscicola* n. sp., and stages in the life-history of *H. compacta* n. sp. Biol. Bull., *119*:529–549.

———. 1966. The morphology and life history of the digenetic trematode, *Himasthla littorinae* sp. n. (Echinostomatidae). J. Parasitol., *52*:367–372.

Sullivan, W.G., and Koep, L.J. 1980. Common bile duct obstruction and cholangiohepatitis in clonorchiasis. J.A.M.A., *243*:2060–2061.

Tanabe, H. 1922. *Echinochasmus perfoliatus* (Ratz) found in Japan. J. Okayama Med. Assoc., No. 287, 1–20.

Tang, C., Cui, G., Dono, Y., Wang, Y., Nulimajabu, Lu, H., Zhang, C., Chen, M., Sun, G., and Qian, Y. 1979. Studies on the biology and epidemiology of *Eurytrema pancreaticum* (Janson, 1889) in Heilungkiang Province. Acta Zool. Sinica, *25*:234–242.

Tang, Z., and Tang, C. 1977. The biology and epidemiology of *Eurytrema coelomaticum* (Giard et Billet 1892) and *Eurytrema pancreaticum* (Janson, 1889) in cattle and sheep in China. Acta Zool. Sinica, *23*:267–282.

Thomas, A.P.W. 1883. The life history of the liver fluke *(Fasciola hepatica)*. Q. J. Microsc. Sci., *23*:99–133.

Upatham, E.S., Viyanant, V., et al. 1982. Morbidity in relation to intensity of infection in opisthorchiasis viverrini: study of a community in Khon Kaen, Thailand. Am. J. Trop. Med. Hyg., *31*:1156–1163.

Vajrasthira, S., Harinasuta, C., and Komiya, Y. 1961. The morphology of the metacercaria of *opisthorchis viverrini* with special reference to the excretory system. Ann. Trop. Med. Parasitol., *55*:413–418.

Vandepitte, J., Job, A., Delaisse, J., and Tabary, M.J. 1957. Quatre cas d'abces retro-auriculaires chez des Congolais, produits par un nouveau distome. Ann. Soc. Belge Med. Trop., *37*:309–316.

Vanijanonta, S., Radamyos, P., Bunnag, D., and Harinasuta, T. 1981. Pulmonary paragonimiasis with expectoration of worms: A case report. Southeast Asian J. Trop. Med. Public Health, *12*:104–106.

Vazquez-Colet, A. 1943. *Stictodora palmifera* nov. sp. with notes on *Stictodora manilensis* Africa and Garcia. Abstr. Sci. Papers Conf. Med. Sci. Commem. Establ. R. Philippines.

Voelker, J., and Arzube, R.M. 1979. Ein neuer Lungenegel aus der Küstenkordillere von Ecuador: *Paragonimus ecuadoriensis* n. sp. (Paragonimidae: Trematoda). Tropenmed. Parasitol., *30*:249–263.

———, and Nwokolo, C. 1973. Human paragonimiasis in Eastern Nigeria caused by *Paragonimus uterobilateralis*. Z. Tropenmed. Parasitol., *24*:323–328.

———, and Vogel, H. 1965. Zwei neue *Paragonimus*-Arten aus West-Afrika: *Paragonimus africanus* und *Paragonimus uterobilateralis* (Troglotrematidae: Trematoda). Z. Tropenmed. Parasitol., *16*:125–148.

Vogel, H. 1933. *Himasthla meuhlensi* n sp., ein neuer menschlicher Trematode der Familie Echinostomidae. Zbl. Bakt. Parasitol. I. Abt. Orig., *127*:385–391.

———. 1934. Der Entwicklungszyklus von *Opisthorchis felineus* (Riv.) nebst Bemerkungen über Systematik und Epidemiologie. Zoologica, *33*:1–103.

———, and Crewe, W. 1965. Beobachtungen uber die Lungenegel-Infektion in Kamerun (Westafrika). Z. Tropenmed. Parasitol., *16*:125–148.

———, and Falcão, J. 1954. Ueber den Lebenzyclus des Lanzettegels, *Dicrocoelium dendriticum* in Deutschland. Z. Tropenmed. Parasitol., *5*:275–296.

Walters, J.C., Freeman, R.S., Shea, M., and Fallis, A.M. 1975. Penetration and survival of mesocercariae (*Alaria* spp.) in the mammalian eye. Can. J. Ophthalmol., *10*:101–106.

Ward, H.B., and Hirsch, E.F. 1915. The species of *Paragonimus* and their differentiation. Ann. Trop. Med. Parasitol., *9*:109–163.

Witenberg, G. 1944. What is the cause of the parasitic laryngopharyngitis in the Near East (halzoun)? Acta Med. Orientalia, *3*:191–192.

Wong, S.K., and Lie, K.J. 1965. Another periauricular abscess from Sarawak, probably caused by a trematode of the genus *Poikilorchis,* Fain and Vandepitte. Med. J. Malaya, *19*:229–230.

Wykoff, D.E., Chittayasothorn, K., and Winn, M.M. 1966. Clinical manifestations of *Opisthorchis viverrini* infections in Thailand. Am. J. Trop. Med. Hyg., *15*:914–918.

———, Harinasuta, C., Juttijudata, P., and Winn, M.M. 1965. *Opisthorchis viverrini* in Thailand—the life cycle and comparison with *O. felineus*. J. Parasitol., *51*:207–214.

Yamagiwa, K. 1892. Ueber die Lungendistomenkrankheit in Japan. Virchow's Arch. Pathol. Anat., *127*:446–456.

Yamaguti, S. 1958. *Systema Helminthum* Volume 1. *The Digenetic Trematodes of Vertebrates*. New York, Interscience.

———. 1971. *Synopsis of Digenetic Trematodes of Vertebrates,* Vol. 1. Tokyo, Keigaku Publ. Co.

———. 1975. *A Synoptical Review of Life Histories of Digenetic Trematodes of Vertebrates*. Tokyo, Keigaku.

Yamashita, J. 1938. *Clinostomum complanatum,* a trematode parasite new to man. Annot. Zool. Japon., *17*:563–566.

Yarwood, G.R., and Elmes, B.G.T. 1943. *Paragonimus* cyst in a West African native. Trans. R. Soc. Trop. Med. Hyg., *36*:347–351.

Yokogawa, M. 1974. Epidemiology and control of paragonimiasis, pp. 137–149. In *Parasitic Diseases*. Edited by M. Sasa. Tokyo, International Medical Foundation of Japan.

———. 1975. *Paragonimus* and paragonimiasis. Iran. J. Public Health, *4*:42–51.

———. 1982. Newly introduced questions concerning *Paragonimus westermani*. J. Formosan Med. Assoc., *81*:774–780.

———, Araki, K., Saito, K., Momose, T., Kimura, M., Suzuki, S., Chiba, N., Kutsumi, H., and Minai, M. 1974. *Paragonimus miyazakii* infections in man first found in Kanto District, Japan—especially, on the methods of immuno-serodiagnosis for paragonimiasis. Jpn. J. Parasitol., *23*:167–179.

———, Cabrera, B.D., Kobayashi, M., Hata, H., Tazaki, T., and Araki, K. 1979. Comparative studies on Philippine *Paragonimus westermani* and Japanese *Paragonimus westermani*. Acta Med. Philip., *15*:119–131.

———, Harinasuta, C., and Charoenlarp, P. 1965. *Hypoder-*

aeum conoideum (Bloch, 1782) Dietz, 1909, a common intestinal fluke of man in the north-east Thailand. Jpn. J. Parasitol., *14*:148–153.

———, Okura, T., Tsuji, M., Iwasaki, M., and Shigeyasu, M. 1962. Chemotherapy of paragonimiasis with bithionol. III. The follow-up studies for one year after treatment with bithionol. Jpn. J. Parasitol., *11*:103–116.

———, Tsuji, M., and Okura, T. 1962. Studies on the complement fixation test for paragonimiasis as the method of criterion of cure. Jpn. J. Parasitol., *11*:117–122.

———, Yoshimura, H., Sano, M., Okura, T., and Tsuji, M. 1962. The route of migration of the larva of *Paragonimus westermani* in the final host. J. Parasitol., *48*:525–531.

Yokogawa, S., Cort, W.W., and Yokogawa, M. 1960. *Paragonimus* and paragonimiasis. Exp. Parasitol., *10*:81–137, 139–205.

Yoshida, Y., Matsuno, K., Kondo, K., Arizono, N., Akashi, Y., Uematsu, T., Yoshikawa, K., and Mori, K. 1974. A case of human infection with *Fasciola* sp. and its treatment with bithionol. Jpn. J. Parasitol., *23*:116–124.

Yoshimura, H. 1965. The life cycle of *Clonorchis sinensis:* A comment on the presentation in the seventh edition of Craig and Faust's *Clinical Parasitology.* J. Parasitol., *51*:961–966.

Young, S.T. 1935. The blood picture in fasciolopsiasis *(F. buski).* Trans. 9th Congr. Far Eastern Assoc. Trop. Med., *1*:563–566.

Zhong, H., He, L., Xu, Z., and Cao, W. 1981. Recent progress in studies of *Paragonimus* and paragonimiasis control in China. Chinese Med. J., *94*:483–494.

Chapter 29

Cestodes or Tapeworms

GENERAL CONSIDERATIONS

The Adult Worm. The adult tapeworms of man (Fig. 29–1) consist of a chain of a few to many egg-producing units, technically known as *proglottids*. The proglottids develop from the distal end of a *scolex,* which is attached to the intestinal wall of its host. The entire chain of proglottids is the *strobila*. Typically the following regions of the organism may be recognized: (1) the *scolex,* or attachment organ; (2) the *"neck,"* or region of growth, immediately behind the scolex; (3) *immature proglottids,* developing from the distal end of the "neck"; (4) *mature proglottids;* and (5) *gravid proglottids*.

The scolex is a more or less distended muscular organ, which in the case of the Pseudophyllidea (e.g., *Diphyllobothrium*) is narrowly spindle-shaped with a midventral and a middorsal elongated sucking groove, without hooks or other accessory attachment structure. In the Cyclophyllidea (including most species parasitic in man), the scolex is more or less transversely quadrate, is provided with a cup-like sucker at each of the four angles, and in many species has an anterior *rostellum* armed with spines or hooks, the type, number, and arrangement of which are specifically characteristic (Fig. 29–2). The "neck" is constricted and in a few species is essentially lacking. It has no special anatomical features but is the portion of the worm from which proglottids are primarily proliferated. The scolex and "neck" are important structures, since infection persists as long as these portions of the worm remain attached to the host's intestinal wall, even though the greater portion of the strobila may have become free and been evacuated. From the distal portion of the "neck" arise transverse divisions that indicate the earliest separation into proglottids. Proceeding distalward these proglottids become more and more distinctly differentiated both externally and internally. The region of *immature proglottids* is one where these units are just being differentiated and where the sexual organs are still immature. Farther removed from the scolex are the *mature proglottids*. These are larger units, each of which contains one (in some species two) full set of male and female genital organs (i.e., each is hermaphroditic). The *gravid proglottids* are those farthest away from the scolex; in them the primary genitalia have been greatly reduced or have atrophied, and each unit is given over to development of the uterus with its eggs.

The strobila is not fundamentally metameric or segmented in the manner seen in annelids, although there is repetition of genital organs as well as transverse commissures and excretory tubules. In its simplest form a complete worm may consist of scolex, neck, and a single immature, single mature, and single gravid proglottid *(Echinococcus granulosus)*. In its more complicated form there may be hundreds of units of each stage of proglottid development *(Taenia saginata)*.

The adult tapeworm is typically located in the middle portion of the small intestine of its host. The scolex is strictly an organ for attachment and orientation of the strobila. It is not for support against peristalsis, and it does not serve for procurement of food, which is absorbed from the medium in which the organism lives. There is no specialized alimentary or digestive system. Nutrients are absorbed through the worm's integument, which serves a protective function and also provides a membrane for molecular interchange with the environment (von Brand, 1979). Permeability and exchange functions of the integument are augmented by topographical features that maximize surface area (Lumsden, 1966, 1975a). At the surface is a layer of microvilli (microtriches) over

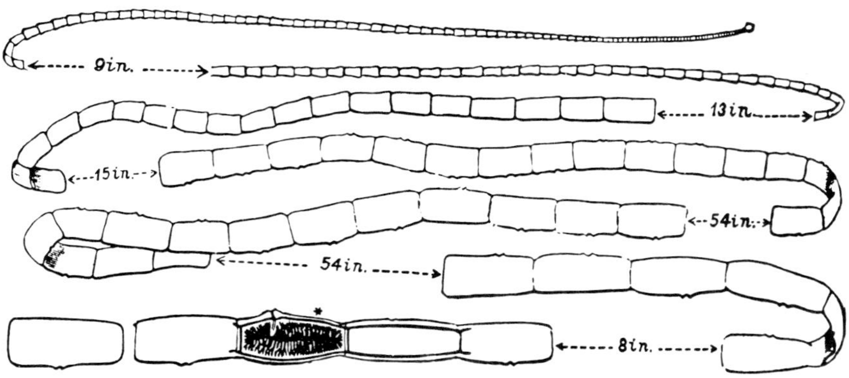

Fig. 29–1. *Taenia saginata.* complete worm with scolex, "neck," immature, mature and gravid (*) proglottids. About natural size. (Adapted from R. Leuckart, in *Parasiten des Menschen.)*

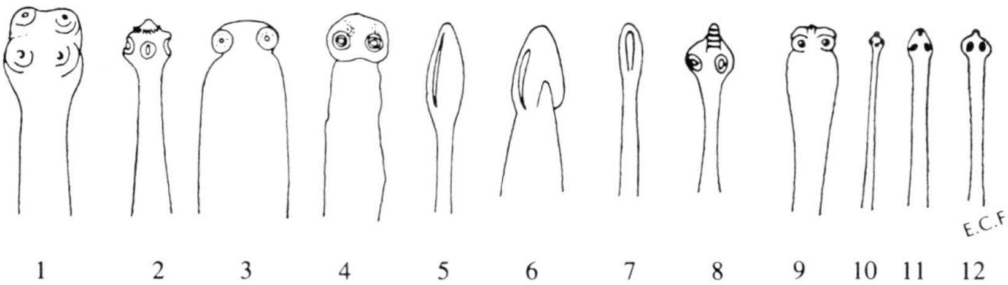

Fig. 29–2. Scolex and adjacent part of various tapeworms occurring in man. *1, Taenia saginata. 2, Taenia solium. 3, Taenia africana (=T. saginata). 4, Taenia confusa (=T. saginata). 5, Diphyllobothrium latum. 6, Diphyllobothrium cordatum. 7, Spirometra mansoni. 8, Dipylidium caninum. 9, Inermicapsifer madagascariensis. 10, Hymenolepis nana. 11, Hymenolepis diminuta. 12, Bertiella studeri. (× ca. 6.) (Adapted from original of E.C. Faust.)*

the entire outer surface, which is comparable with the brush border of the host's intestinal epithelium. Other histological analogies with the intestinal epithelium have also been shown by electron microscopy. The active transport of enzymes and presumably the secretion of substances that inactivate host enzymes are functions of the integument. Dead or moribund tapeworms, but not the living ones, are digested by host enzymes (Lumsden, 1975b).

The *excretory system* is primitively like that of the trematodes, with flame cells, capillaries, and collecting tubules. Typically, each worm has a pair of main dorsal and ventral longitudinal collecting tubules, situated near the lateral margins of the worms, each emptying through a bladder in the terminal proglottid, but in the mature worm only one pair of lateral longitudinal trunks is seen. These trunks arise from a complicated anastomosing network in the scolex, have a transverse anastomosis at the distal margin of each proglottid, and empty through the open end of the tubes at the distal end of the worm. The *nervous system* is confined primarily to the scolex, where there is a complex set of ganglia with connecting commissures, as well as apical nerves that are both sensory and motor in function. Typically, five pairs of nerves arise from the central nervous system and extend distalward through the proglottids: one pair of main lateral and two pairs of accessory lateral nerves, as well as two pairs of submedian nerve fibers. These provide coordination of the entire strobila. The transverse commissures are at the proximal rather than the distal end of each proglottid.

The *genital organs* are elaborately developed in each proglottid (Fig. 29–3). The *testes* are usually multiple and distributed throughout the mature proglottid. *Vasa efferentia* lead into a convoluted *vas deferens* which, as it approaches the common atrium *(m)* on the median ventral surface (Pseudophyllidea) or on the lateral margin (Cyclophyllidea), is differentiated into *pars prostatica* and a *cirrus organ,* the two being surrounded by a *cirrus sac*. In the female system a *vagina* proceeds inward from the region of the genital atrium to the *ootype*. At its inner end there is usually a *seminal receptacle* and a *spermatic duct*. The *ovary* is commonly a bilobed organ posterior to the equatorial plane of each proglottid. The *oviduct,* which receives the

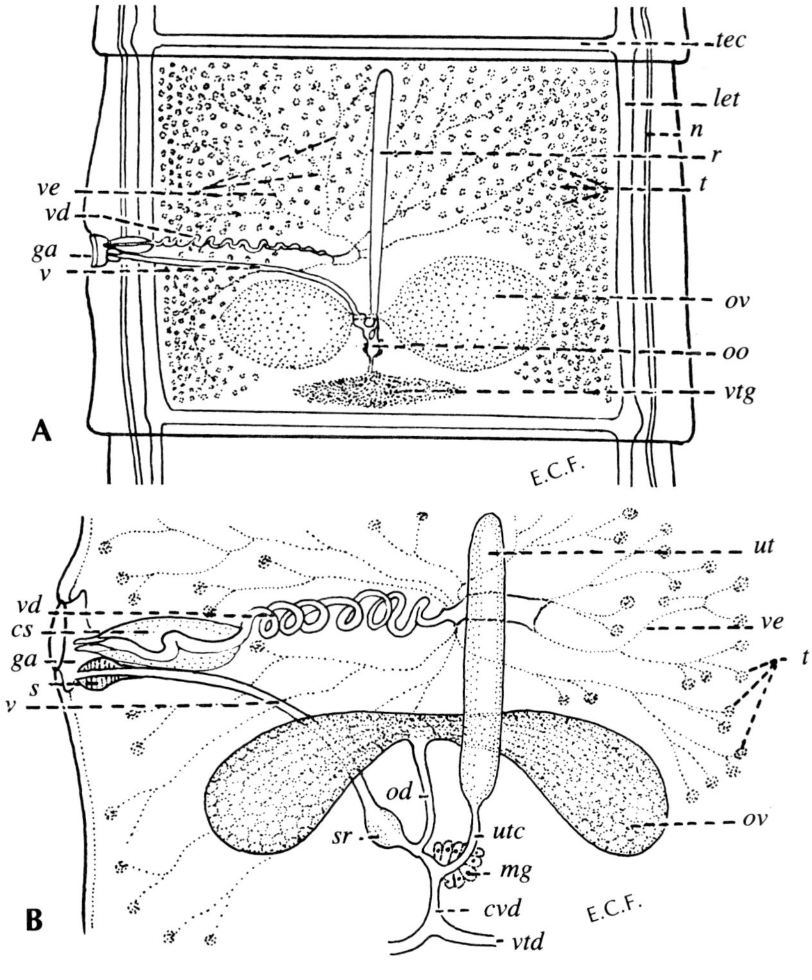

Fig. 29–3. *Taenia saginata. A,* Proglottid with major organs shown. *B,* Genital organs, enlarged for details. *cs,* Cirrus sac. *cvd,* Common vitelline duct. *ga,* Male and female genital atrium. *let,* Lateral excretory trunk. *mg,* Mehlis' gland. *n,* Lateral nerve trunk. *od,* Oviduct. *oo,* Ootype. *ov,* Ovary. *s,* Sphincter of vagina. *sr.* Seminal receptacle. *t,* Testes. *tec.,* Transverse excretory canal. *ut,* Uterus. *utc,* Uterine canal. *v,* Vagina. *vd,* Vas deferens. *ve,* Vas efferens, *vtg,* Vitelline gland, *vtd,* Vitelline duct. (From Faust, E.C. 1939. *Human Helminthology,* 2nd ed. Philadelphia, Lea & Febiger.)

spermatic duct, opens into the ootype. The *vitellaria,* which may consist of a bilobed mass (*Taenia* spp.), a single mass *(Dipylidium),* or multiple follicles *(Diphyllobothrium),* are connected with *vitelline ducts* which, together with the vagina and oviduct, empty through a *common vitelline duct* into the ootype (Fig. 29–3,*B*). Surrounding the ootype is a cluster of cells called *Mehlis' gland.* The *uterus,* arising from the anterior face of the ootype, may open through a pore on the median ventral aspect of the proglottid *(Diphyllobothrium)* or may end blindly *(Taenia, Dipylidium).* In the former type it becomes piled or coiled; in *Taenia* it develops branching lateral arms (Fig. 29–4). The shape or pattern of the gravid uterus is usually species-diagnostic.

Eggs are assembled in the ootype and are stored in the uterus. They consist of fertilized ovum and yolk material, enveloped in an embryonic membrane, outside of which is the characteristic shell (the *embryophore*). Eggs of pseudophyllidean species (e.g., *Diphyllobothrium)* are ovoid, operculate, and immature when discharged in the feces; those of cyclophyllidean species (e.g., *Taenia, Hymenolepis, Dipylidium*) are essentially spherical, nonoperculate, and almost fully embryonated when they are discharged from the proglottid.

By coordination of the entire body, the tapeworm is able to maintain its position in the intestine. Under conditions of starvation, intoxication, or increased peristalsis, equilibrium may be lost

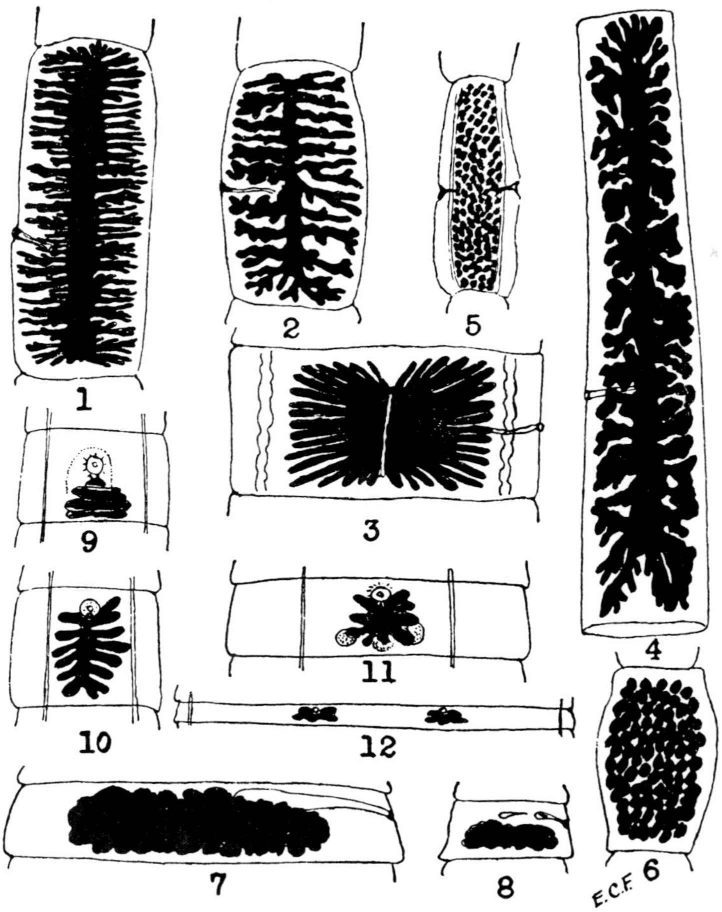

Fig. 29–4. Gravid proglottids of human tapeworms, characteristic uterine patterns. *1, Taenia saginata. 2, Taenia solium, 3, Taenia africana* (= *T. saginata*). *4, Taenia confusa* (= *T. saginata*). *5, Dipylidium caninum. 6, Inermicapsifer madagascariensis. 7. Hymenolepis diminuta. 8, Hymenolepis nana. 9. Spirometra mansoni. 10, Diphyllobothrium cordatum. 11, Diphyllobothrium latum. 12, Diplogonoporus grandis.* (From Faust, E.C. 1939. *Human Helminthology,* 2nd ed. Philadelphia, Lea & Febiger.)

and the greater portion of the worm separated from the scolex and evacuated spontaneously.

The number of strobilae in a single host varies with the size of the worm (usually solitary in the case of *Taenia saginata* and *T. solium*), age of the host, and host-parasite compatibility. Immune mechanisms of the host apparently do not cause destruction of the parasites once they are established in the tissues, although eventually they lose vitality, die, and frequently become calcified. The presence of adult or larval tapeworms produces high resistance to additional infection. Therefore, as was noted by Williams (1979), cestode infections offer unique advantages for experimental studies on concomitant immunity.

In microsections of a proglottid the outermost layer, the tegument, appears to be homogeneous and noncellular, though electron-microscopy has shown it to be part of a syncytium composed of a surface layer of microtriches and an inner subtegumental layer of nuclei and cytoplasm extending into the cortical portion of the proglottid. Interspersed between the tegument and subtegumental nuclei are two thin layers of muscle fibers: an outer layer of circular fibers and an inner layer directed longitudinally. Internally the proglottid is filled with a mesenchyme, or parenchyma, which is a loose meshwork of branched cells and fluid-filled spaces. Forming a distinct layer between the cortical and medullary portions of the parenchyma, in cyclophyllidean cestodes, is a strongly developed main muscle system of longitudinal and transverse fibers, and within the medulla are dorsoventral fibers (Fig. 29–5). In the pseudophyllidean species the longitudinal fibers are in bands scattered throughout the parenchyma forming no evident boundary between cortical and medullary parenchyma (Fig. 30–9). Usually present in the parenchyma of both

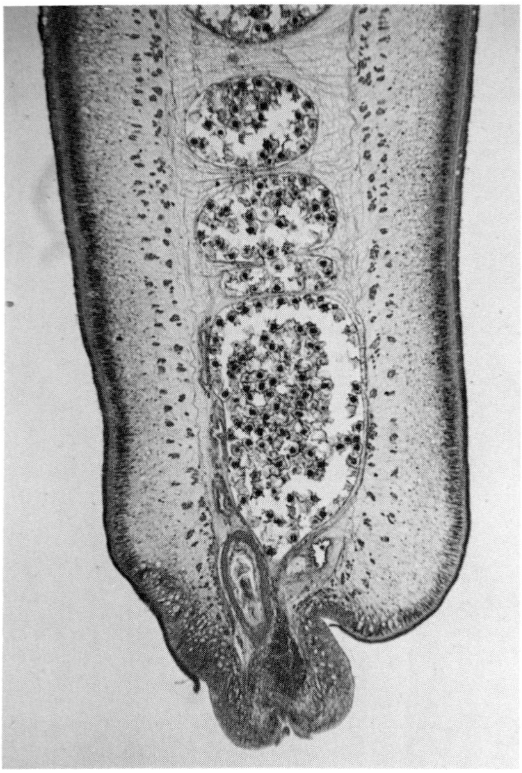

Fig. 29–5. Proglottid of cyclophyllidean cestode *(Bertiella studeri)*, transverse section near level of genital pore showing microstructure of the tegument and underlying subtegument, cortical layer of dense parenchyma and more open fibrous medullary layer containing the uterus (filled with eggs) and mostly dorso-ventral muscle fibers. Between the cortical and medullary layers are the main longitudinal muscle fibers, arranged in two layers, and the underlying transverse fibers, seen as wavy lines between the uterus and longitudinal muscles. Structures at bottom are a piriform cirrus sac and darkly stained glandular vagina. (× 50.) (By P.C. Beaver.)

larval and adult cestodes are spheroidal or discoidal bodies called *calcareous corpuscles*. The function of these characteristic structures is unknown. Observed differences in the shape and composition of calcareous corpuscles are more related to the species of worm than to the species of host or location within the host (von Brand et al, 1967). The main longitudinal excretory channels and lateral nerve trunks are located at the lateral margins of the medullary parenchyma or at a comparable location when the two zones are indistinct.

Life Cycle. The fully embryonated egg (almost mature when discharged from the uterus of Cyclophyllidea, mature only following incubation in water in Pseudophyllidea) contains an oncosphere (i.e., "hooked ball"). The oncospheres of human tapeworms typically have 3 pairs of hooklets (i.e., they are *hexacanth*), rarely 5 pairs or multiples of 3 pairs. These hooklets may be morphologically similar or each pair may be structurally different. There is also a pair of unicellular penetration glands, each with a pore opening at the anterior end. Secretions of these glands possibly assist the hooklets in penetrating the intestinal wall of the intermediate host. In *Diphyllobothrium* and its relatives, the mature oncosphere is covered with a ciliated epithelium, and is called a coracidium; in the cyclophyllidean forms cilia are lacking. In the former, hatching of the ripe egg occurs in water; in the latter species, it occurs only after ingestion by the appropriate intermediate host. In the former, which is more primitive, two intermediate hosts are required; in the latter, one of these hosts has been lost and, in *Hymenolepis nana* in man, both intermediate hosts have been dispensed with. In one sense, though, the definitive host serves additionally as intermediate host, since larval development occurs in the intestinal mucosa of this host.

Directly or eventually hexacanth embryos generally develop into one of the following types: (1) *plerocercus*, solid, relatively globular, with the scolex invaginated into the body of the larva; (2) *plerocercoid*, solid, elongated, with the head free or invaginated only into the neck; (3) *cysticercoid*, with a rather small body anteriorly, into which the head is invaginated, and frequently with a solid, elongated caudal portion, and (4) *cysticercus* (true "bladder worm"), with the head invaginated into the proximal portion of the bladder. The cysticercus type may contain only a single scolex, as in the taenias. There may be multiple scolices developed from the inner germinal wall (*coenurus* type), or the germinal layer may produce many scolices and develop many daughter bladders (or brood capsules), each of which, in turn, produces many scolices internally (echinococcus or "hydatid" type). The *tetrathyridium* is a plerocercoid-like form with a well-developed scolex bearing four suckers (Baer, 1951).

The definitive host becomes infected with the adult tapeworm from ingestion of the mature larva in its intermediate host tissue. Man is the definitive host of most tapeworms to which he is susceptible. Exceptions are *Echinococcus granulosus* and *Multiceps* spp., of which the dog is the usual definitive host and man an intermediate host, and also *Taenia solium*, of which man may be an intermediate as well as the only known definitive host.

Pathology and Symptomatology. Pathogenicity

of adult tapeworm infection may be attributed to passive obstruction, migration to unusual sites, and competition with the host for nutrients, which in the case of *Diphyllobothrium latum* may cause a macrocytic anemia.

The clinical aspects of tapeworm infections in man fall into two categories: (1) those due to adult worms and (2) those due to larval worms. In those species such as *Taenia saginata* in which the gravid proglottids actively migrate from the anus of the host, the discomfort and annoyance of these crawling worms constitute the usual complaints of the patient. Otherwise the adult worms are usually well tolerated by the human host, the exceptions being described under the respective parasites.

Larval infections are more likely to be serious because the larvae may lodge in critical foci in the brain, eye, or elsewhere or may grow into large masses producing pressure effects.

Distribution of Human Tapeworm Infections. Tapeworm infections in man are less limited in their distribution than are human fluke infections. They are frequently not restricted to any specialized group of intermediate hosts, as species of molluscs that may be local in their distribution or confined by certain meteorologic conditions. They seldom have a true free-living existence (except for the ciliated oncosphere of *Diphyllobothrium latum*). Some of them are dependent primarily or exclusively on man as definitive host for the continuation of their life cycles *(T. saginata, T. solium, Hymenolepis nana, Diphyllobothrium latum);* in other cases man is incidental to continued propagation of the parasite *(Echinococcus granulosus, Hymenolepis diminuta, Dipylidium caninum).*

Prevention. Ingestion of the infected flesh of the intermediate host, raw or inadequately cooked, constitutes the source of human infection with *Diphyllobothrium latum, T. saginata,* and *T. solium.* Thorough cooking of this flesh prevents infection of individuals. Sanitary disposal of human feces containing the viable eggs of these tapeworms protects communities. Improper personal hygiene with respect to contact with feces of man *(Taenia solium* and *Hymenolepis nana),* feces of dogs *(Echinococcus granulosus),* or the insect vermin of domestic animals *(Dipylidium caninum, H. diminuta)* accounts for all of the remaining tapeworms except a few that are rarely reported for man. In all known tapeworm infections, except certain varieties of sparganosis, in which exposure is topical, the portal of entry is the mouth. Hence, strict care not to swallow raw or inadequately cooked beef, pork, or fish, or food or water contaminated with feces or vermin, will insure protection of the individual. Sanitary disposal of human excrement and that of domestic animals constitutes the fundamental public health problem.

REFERENCES

Baer, J.G. 1951. *Ecology of Animal Parasites.* Urbana, University of Illinois Press.

Lumsden, R.D. 1966. Cytological studies on the absorptive surfaces of cestodes. I. The fine structure of the strobilar integument. Z. Parasitenkd., *27*:355–382.

———. 1975a. The tapeworm tegument: A model system for studies on membrane and function in host-parasite relationships. Trans. Am. Microsc. Soc., *94*:501–507.

———. 1975b. Surface ultrastructure and cytochemistry of parasitic helminths. Exp. Parasitol., *37*:267–339.

von Brand, T. 1979. *Biochemistry and Physiology of Endoparasites.* New York, Elsevier North-Holland.

———, Nylen, M., Martin, G., and Churchwell, F. 1967. Composition and crystallization patterns of calcareous corpuscles of cestodes grown in different classes of hosts. J. Parasitol., *53*:683–687.

Williams, J.F. 1979. Recent advances in the immunology of cestode infections. J. Parasitol., *65*:337–349.

Chapter 30

Pseudophyllidean Tapeworms

The pseudophyllidean tapeworms have a scolex bearing two bothria (sucking organs), one each on the dorsal and ventral surfaces. Representatives in man belong to the superfamily Bothriocephaloidea, family Diphyllobothriidae. In this group the cirrus and vagina are medial and open on the same surface anterior to the uterine pore.

Diphyllobothrium latum (Linnaeus, 1758) Lühe, 1910 (The broad or fish tapeworm, producing diphyllobothriasis)

Synonyms. *Taenia lata* Linnaeus, 1758; *Bothriocephalus latus* (L., 1758) Bremser, 1858; *Dibothriocephalus latus* (L., 1758) Lühe, 1899; *Bothriocephalus taenioides* León, 1916; *Dibothriocephalus minor* Cholodkowsky, 1916.

Historical and Geographical Notes. *Diphyllobothrium latum* is a common parasite of man in northern Italy (around Lakes Como, Maggiore, and Varese), Switzerland, parts of Germany, and in the Baltic countries. The highest incidence of human infection occurs in Finland, Sweden, and Lithuania; next, northern U.S.S.R., central Siberia, lower Volga, and North America. It is also a relatively common parasite of man in the vicinity of Lake Tiberias in Palestine, and in northern Manchuria and Japan, and has become indigenous to Uganda. It has been reported once in man in Czechoslovakia (Catár et al., 1967), once from the Philippines, and among the aborigines in Taiwan. It is common in the northern parts of Japan. Several foci are known for North America, including particularly the lake regions of northern Minnesota, southeastern Manitoba, and the Lake Nippigon district of Ontario. *D. latum* is probably the most common tapeworm of man in eastern Canada. However, in northern Michigan where formerly it was endemic, it is no longer present in man or in animals (Peters et al., 1978). In Finland the prevalence since 1956 has declined from 20% to around 2% (Bylund, 1976). The infection has become established in *Salmo lacustris* and *S. irideus* in the Andean Lake district about 500 miles south of Santiago, Chile.

Morphology, Biology, and Life Cycle. When freshly expelled from the human intestine, the worm (Fig. 30–1) is ivory colored. It measures up to 10 meters or more in length with as many as 3,000 or more proglottids (Führmann, 1930). Irrespective of length, the worm is generally folded, forming several loops that lie in contact with the intestinal mucosa. The scolex (Fig. 30–2) is small, spatulate, with a dorsal and a ventral median longitudinal sucking groove. It measures about 1 mm in cross-section diameter by 2.5 mm in length. It is extremely mobile and may be seen in a wide range of shapes both in the living state and when preserved (Hatsushika et al., 1979). Immediately behind the scolex is an unsegmented neck region, several times the length of the head. More distal are the newly formed proglottids, which become mature and finally gravid. About four-fifths of the entire worm consists of maturing and mature proglottids. The typical mature proglottid (Fig. 30–3) is somewhat broader than long and is practically filled with reproductive organs. The testes *(t)* are minute multiple follicles situated in both lateral fields in the dorsal plane of the body. The vasa efferentia unite mesially just in front of the ootype to form the vas deferens *(vd)*, which proceeds as a convoluted tubule to the upper border of the common genital atrium. Near its outer terminus is a seminal vesicle and a muscular cirrus organ *(c)*. In the posterior third of the proglottid, ventral in position, is the symmetrically bilobed ovary *(ov)*. Between its two lobes is the ootype *(o)*, surrounded by Mehlis' gland. A vagina *(vg)* proceeds directly up from the ootype and opens externally just below the male genital pore. At its inner end it is joined by the oviduct and by the common vitelline gland duct, which receives the yolk material from the minute multiple follicles situated ventral to the testes in the lateral fields. The uterus arises from the ootype and proceeds in rosetted convolutions up to the uterine pore *(up)*, which lies ventrally some distance to the side and behind the vaginal pore. Usually an inner (i.e., more distal) and an outer (i.e., more proximal) portion of the uterus may be clearly distinguished. Spermatozoa fertilize the naked ovum in the ootype; yolk material and shell

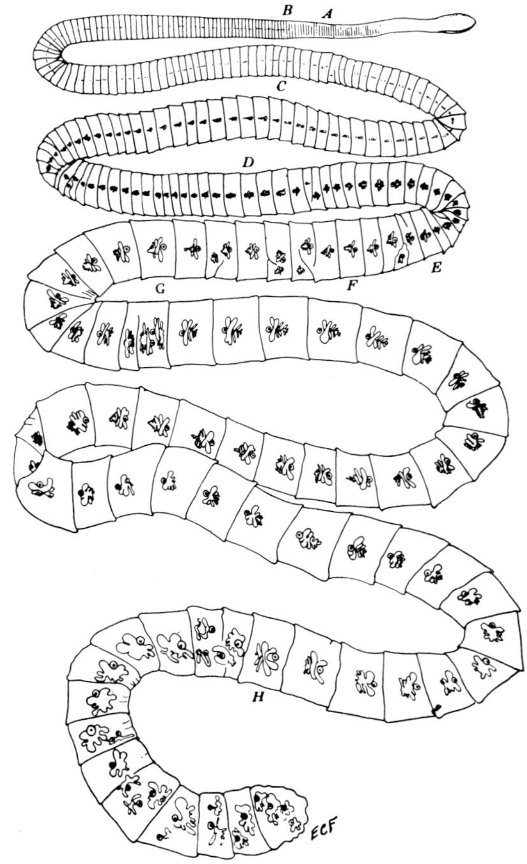

Fig. 30–1. *Diphyllobothrium latum.* Complete strobila with terminal proglottid still attached, obtained from a dog 24 days after experimental infection by feeding sparganum larvae in a fish *(Salmo irideus). A to F,* Progression in differentiation and maturation of reproductive organs. *G,* Proglottids with organs fully developed but not producing eggs. *H,* Proglottids producing eggs. Accessory proglottids are formed at several levels of the worm. (× ca. 2.) (Adapted from Faust, E.C. 1952. Some morphologic characters of *Diphyllobothrium latum.* Ann. Inst. Med. Trop. Lisboa, *9:*1277–1300.)

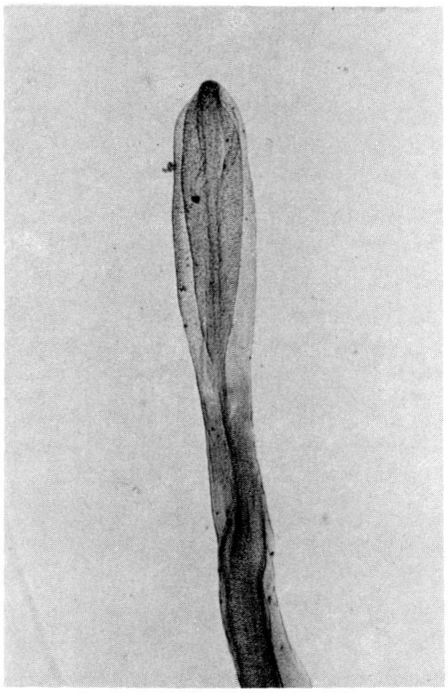

Fig. 30–2. Scolex of adult *Spirometra* in stained preparation, lateral view slightly rotated. (× 25.) The scolex of *Diphyllobothrium latum* is similar. (By P.C. Beaver. Specimen from the Faust Helminthological Collection.)

are then added, and the egg is passed into the uterus. In this group of tapeworms there is no clear distinction between mature and gravid proglottids; egg production and egg laying continue over a considerable period of time. Nor do the distalmost proglottids separate from the strobila. As they cease to function, they gradually disintegrate and are shed in strands of variable length.

Eggs are discharged periodically through the uterine pore of each functional proglottid. A single worm may discharge as many as 1,000,000 eggs daily.

The eggs of *D. latum* (Fig. 30–4) are broadly ovoid, operculate, moderately thick-shelled, light golden-yellow, and each contain an immature em-

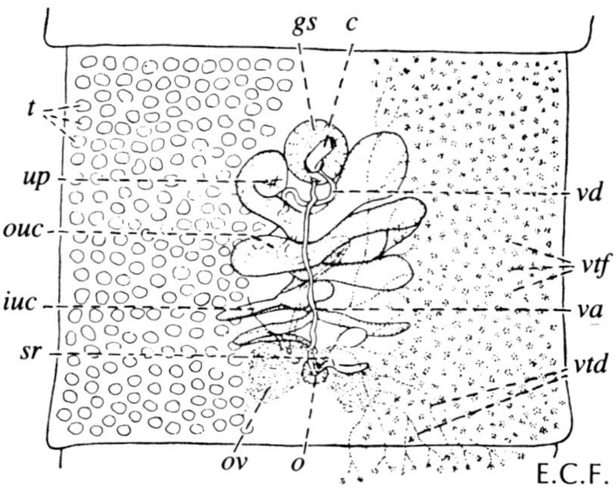

Fig. 30–3. *Diphyllobothrium latum.* Mature proglottid, ventral view. *c,* Cirrus. *gs,* Male genital sucker, *iuc,* Inner uterine coils. *o,* Ootype. *ouc,* Outer uterine coils. *ov,* Ovary. *sr,* Seminal receptacle. *t,* Testes, shown only in left field but symmetrically present also in right field. *up,* Uterine pore. *vd,* Vas deferens. *va,* Vagina. *vtf,* Vitelline follicles, shown only in right field but symmetrically present in left field. *vtd,* Vitelline ducts. (× 15.) (Adapted from Faust, E.C. 1952. Some morphologic characters of *Diphyllobothrium latum.* Ann. Inst. Med. Trop. Lisboa, *9:*1277–1300.)

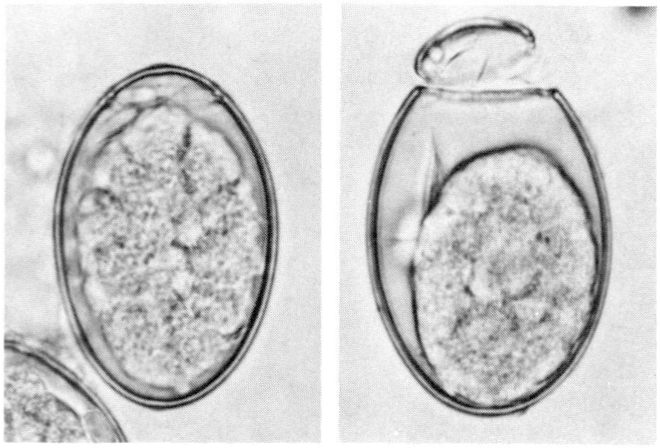

Fig. 30–4. *Diphyllobothrium latum.* Eggs with operculum intact and forced open by coverglass pressure. (× 500.) (From Faust, E.C., Beaver, P.C. and Jung, R.C. 1975. *Animal Agents and Vectors of Human Disease.* 4th ed. Philadelphia, Lea & Febiger.)

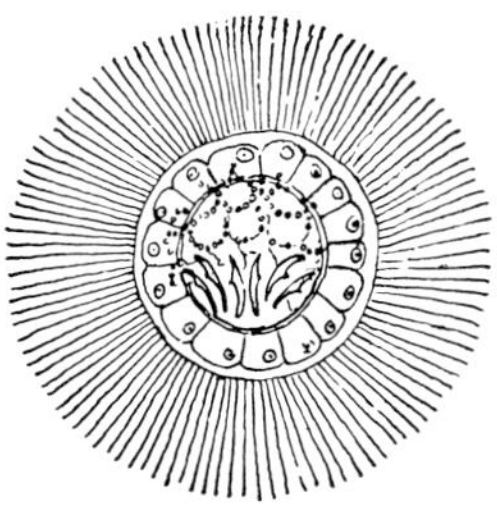

Fig. 30–5. *Diphyllobothrium latum.* Free swimming coracidium. (× 500.) (Adapted from Janicki, C. and Rosen, F. 1917. Le cycle evolutif du *Dibothriocephalus latus* L. Bull. Soc. Neuchateloise des Sci. Nat., *42*:19–53.)

bryo when discharged in the feces. They measure 58 to 76 μm by 40 to 51 μm (average 66 by 44 μm), are quite resistant to chemicals, but rapidly become nonviable when subjected to desiccation or putrefaction. In a water medium at 15° to 25°C, they require 11 to 15 days for development. Eggs also develop and survive well in environments such as those provided by damp soil and ooze at the water margin of lakes, ponds, and streams (Beaver et al., 1964). Upon maturing, the hexacanth embryo *(oncosphere)* escapes through the opercular opening of the shell, casts off its embryonal envelope, and as a ciliated embryo *(coracidium)* swims about in the water (Fig. 30–5). Within about 12 hours it must be ingested by an appropriate copepod in order to continue its development. The demonstrated copepod hosts include: *Diaptomus vulgaris* (commonly), *D. gracilis, D. graciloides, D. denticornis* (Norway), *Cyclops strenuus* (rarely), *C. furcifer* and *C. vicinus* in Europe, and *D. oregonensis, D. sicilis,* and *D. siciloides* in North America. In the midgut of these copepods the embryo casts off its ciliated coat and by use of its three pairs of hooklets and probably also an anterior pair of penetration glands, penetrates into the hemocoel where, in the course of 2 or 3 weeks, it becomes transformed into a *procercoid* larva measuring up to 550 μm in length and still retaining three pairs of hooklets in the *cercomer,* the caudal vestige of the invader (Fig. 30–6).

If the infected copepod is now ingested by a freshwater fish, the procercoid larva works its way through the tissues of the fish and comes to lie free between the muscle fibers, where it grows into a *plerocercoid* larva or *sparganum* measuring up to 20 mm or more in length. This larva (Fig. 30–7) is glistening, opaque white, and unsegmented. Characteristically it rests in the muscles with the head bent back on the body. Larger edible fishes

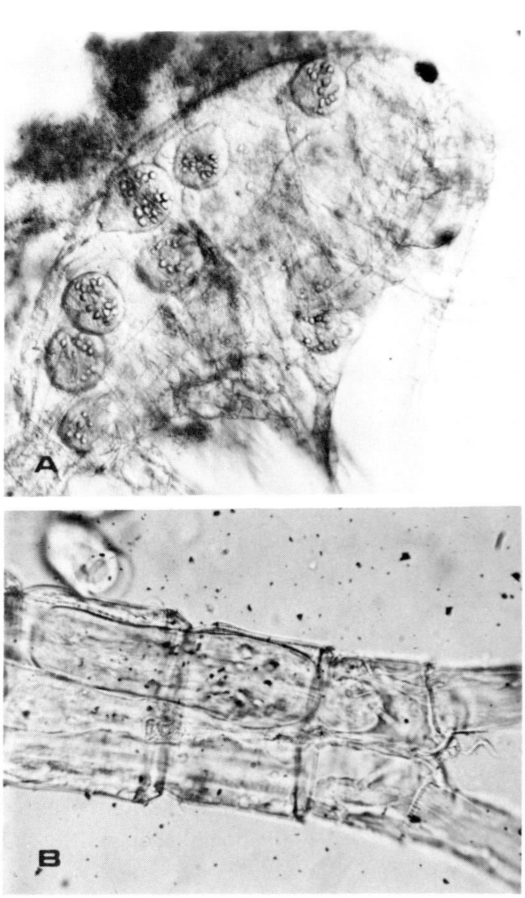

Fig. 30–6. Procercoids of *Spirometra mansonoides,* indistinguishable from those of *Diphyllobothrium latum,* in copepod hemocoel. *A,* Eight larvae in early development, each bearing several calcareous corpuscles. *B,* Fully developed infective procercoid, ready for development to plerocercoid in the fish host. The anterior end (left) bears minute spines; at the posterior end, a rounded cercomere bears three pairs of larval hooks. (Courtesy of Dr. J.F. Mueller.)

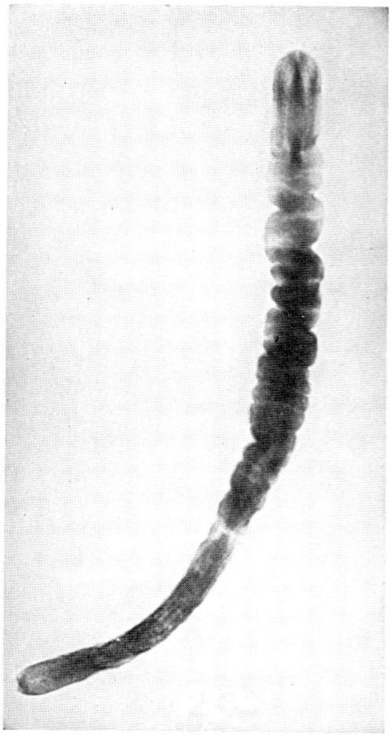

Fig. 30–7. *Diphyllobothrium latum.* Plerocercoid larva (sparganum) from flesh of fish (wall-eyed pike, *Stizostedion vitreum*). (× 12.) (From Magath, T.B. 1929. Experimental studies on *Diphyllobothrium latum*. Am. J. Trop. Med., 9:17–48.)

usually acquire the infection by devouring infected smaller species or their own young, and the spargana migrate to the flesh of these larger hosts. Important food fishes that have been found naturally infected are the following: European pike *(Esox lucius lucius)*, European perch *(Perca fluviatilis)*, lawyer *(Lota lota)*, salmon *(Salvelinus umbla)*, Swiss lake trout *(Salmo trutta)*, lake trout *(S. lacustris)*, white fish (*Coregonus* spp.), grayling *(Thymallus thymallus)*, ruff *(Acerina cernua)*, eel *(Anguilla anguilla)*, all from Europe; cherry salmon *(Oncorhynchus masou)*, pink salmon *(O. gorbuscha)* from Japan; European barbel *(Barbus vulgaris)* from Lake N'gami in Africa; and from North America, barred pike *(Esox lucius estor)*, wall-eyed or blue pike *(Stizostedion vitreum)*, sand pike or sauger *(S. canadense griseum)*, and American burbot *(Lota maculosa)*. In addition, the rainbow trout *(S. irideus)* and lake trout *(S. lacustris)* are infected in the Lake District of Chile and Argentina. These and other freshwater fishes also harbor spargana other than those of *D. latum,* which cannot be differentiated in the sparganum stage. Strictly marine fishes have never been incriminated, and freshwater fishes of Chile, Argentina, Africa, and Australia are apparently unacceptable hosts.

On eating the raw or insufficiently cooked flesh of infected fishes, man, the optimal definitive host, becomes infected. During the fourth week or later, usually in the course of 5 or 6 weeks, eggs begin to appear in the feces. The dog is easily infected but the worms are smaller in this host, the life of the parasite is much shorter than in man, and fewer of the eggs are viable. Cameron (1945) suggested that *D. latum* of North America may have been indigenous as a parasite of the brown bear.

Epidemiology. Human infection depends on (1) the presence of human or animal definitive hosts, (2) the presence of suitable intermediate hosts, (3) the food habits of the population, and (4) the amount of pollution of natural fresh waters by definitive hosts.

Many mammals other than man have been found infected with *D. latum*. These include the domestic dog, wild canines such as *Dusicyon gymnocercus gymnocercus, Urocyon cinereoargentatus, Canis lupus occidentalis;* domestic cat, and wild felines such as *Felis concolor, F. onca hernandesii, F. wiedii,* leopard *(F. pardus), Mellivora capensis;* mongoose, walrus, seals and sea lions, foxes, mink, bears, and domestic pig. Nevertheless, man is primarily responsible for the propagation of the infection in most endemic areas. He seeds the water with eggs that mature, hatch, and produce infection in the first intermediate hosts (copepods) which, in turn, on ingestion by susceptible fishes, produce infection in the fish. Finally man ingests the infected fish flesh, or possibly fish roe, raw or inadequately cooked, and acquires the infection. Food preparations containing raw fish, including salmon from cold fresh water, can be a source of infection. Infected refrigerated fish shipped to distant communities, as New York City, are the source for human infection hundreds of miles from endemic areas. In areas where human infection has become rare or is no longer present, infection in fish intermediate hosts has apparently disappeared also, indicating that wild mammalian hosts do not long sustain the natural transmission cycle (Peters et al., 1978).

Pathogenesis, Pathology, and Symptomatology. The pathogenic effects of *Diphyllobothrium latum* depend in part on the mass of the worm, the type and amount of its by-products absorbed by the host, the host's susceptibility to these foreign substances, and the kind and amount of essential metabolic intermediates absorbed by the worm thus depriving the host (von Bonsdorff, 1956). Some infected people show no evidence of its pathogenicity. In others there may be mechanical obstruction of the bowel by a large number of the worms that become tangled together, or they may at times produce diarrhea, abdominal pain, or anemia.

In man the worm may be present as a single strobila or in multiple. Records of hundreds of feet of worms evacuated after anthelmintic treatment almost invariably indicate multiple infection. The worms may remain active for many years or may be discharged spontaneously after a short sojourn in the intestine. Mistaking it for a part of himself, a patient may become disturbed by the passing of a large strand of tapeworm.

Large numbers of *D. latum* may be harbored for decades without causing ill effects, but a small number of carriers suffer from tapeworm pernicious anemia (''bothriocephalus anemia''). More than 70% of cases of tapeworm pernicious anemia are reported from Finland. The overall incidence among carriers is low (0.01 to 0.02%), although in some localities it may reach 2%. Finns belong to the Finno-Ungrian race; the condition is almost unknown in people of the Germanic and Latin races and in the inhabitants of Romania and Japan (Rees, 1967). *D. latum* is most frequently attached to the wall of the ileum, less commonly to that of the jejunum, seldom at other levels, and has been found in the gallbladder. Infection at the jejunal level causes impairment of the interaction between the extrinsic and intrinsic factors of Castle; hence, the pernicious type of anemia. Few of the reported cases originating in North America have had an accompanying history of ''bothriocephalus anemia.'' In the average case a slight leukocytosis with eosinophilia is the only significant blood finding.

Clinical and experimental observations of von Bonsdorff (1956), by means of microbiologic assays and radioactive tracer techniques have shown that *Diphyllobothrium latum* competes with its host for the available supply of vitamin B_{12}, causing pernicious anemia when the available supply of this essential vitamin are denied the host's metabolic demands. Nyberg (1963) illustrated the uptake of vitamin B_{12} by both host mucosa and parasite in the presence or absence of tapeworm pernicious anemia.

Diagnosis. A diagnosis is usually based on the recovery and identification of the characteristic eggs or proglottids in the feces (Fig. 30–4). In most cases in Japan the patients request treatment after having recognized the tapeworm strobila expelled in the feces (Tomita et al., 1979; Yokogawa et al., 1979). Occasionally, portions of the worm up to several meters long may be vomited (von Bonsdorff, 1947, 1978). Eggs are usually numerous and readily demonstrated in a direct fecal smear. The Kato technique is also effective in demonstrating the eggs (Khodakova and Lagunova, 1979). In rare instances the eggs need to be distinguished from those of *Diplogonoporus grandis,* a species that thus far has been found in man only in Japan, and from those of *Diphyllobothrium cordatum* and other species (Rausch and Hilliard, 1970).

Treatment. Treatment of choice is with *niclosamide* given orally in a single dose of 2 g chewed thoroughly. If this drug is not available, *quinacrine hydrochloride* may be used. The drug is administered on an empty stomach in the amount of 0.5 g (adult dose) with an equal amount of sodium bicarbonate. If the worm has not been evacuated in 2 hours, saline purgation should be prescribed.

Recent studies (Bylund et al., 1977; Espejo, 1977; Groll, 1980; Lumbreras et al., 1982) have shown praziquantel to be as effective as niclosamide against *Diphyllobothrium* infections in man in a single dose of 5 to 10 mg per kilogram of body weight. In Japan, bithionol (30 mg/kg in one or two doses) is commonly prescribed.

Prognosis. In patients having no predisposition to pernicious anemia the prognosis is good; it is fair in patients with pernicious anemia, provided the worms are removed by anthelmintic treatment and the anemia is cared for by modern methods.

Prevention. Thorough cooking of all suspected freshwater fish is indicated. Freezing for 24 to 48 hours at a temperature of −18°C in a deep-freeze kills any plerocercoids that may be present (von Bonsdorff, 1978). Barriers could be erected to prevent infected fish from being shipped out of endemic areas, unless previously subjected to freezing temperatures for at least 24 hours. In the Jewish populations ''gefüllte fish'' and minced fish spread on bread should be used only after thorough heating or freezing. Jewish housewives frequently become infected from tasting these delicacies while preparing them for the table. Dogs that appear to be reservoirs of the infection should be ''dewormed'' periodically.

Other Species of *Diphyllobothrium*

Diphyllobothrium pacificum (Nybelin, 1931) Margolis, 1956, a parasite of fur seals, has been reported from Japan (Kamo et al., 1982) and in seven human cases from northern Peru (Baer et al., 1967). All seven persons ate only sea fish, and had never left the coastal area of Peru. A group of 31 patients with *D.*

pacificum infection in Lima, Peru were successfully treated with praziquantel (Lumbreras et al., 1982). This species is distinguished by the position of the genital pore, almost in the middle of the proglottid. Eggs are thick-shelled, operculate, and considerably smaller than those of *D. latum,* measuring 50 to 60 μm long by 36 to 40 μm in diameter. In addition to those reported from Peru, cases have been reported from Chile (Sagua et al., 1976).

Diphyllobothrium cordatum (R. Leuckart, 1863) Faust, 1929 is a common parasite of the seal, walrus, and dog in Greenland and Iceland, of the dog in Japan, and of the bear in Yellowstone National Park, Wyoming. One human infection (adult female, Greenland) is recorded. This worm has a distinctly heart-shaped head, almost complete absence of a neck, and greatly compressed proglottids. The eggs (75 by 50 μm) are practically indistinguishable from those of *D. latum.*

Diphyllobothrium ursi Rausch 1954 was described from the brown bear, *Ursus arctos,* on Kodiak Island, Alaska. It has been reported from the black bear, *Ursus americanus,* in Alaska (two cases) and Canada (one case) (Margolis et al., 1973). The plerocercoid stage of *D. ursi* develops on the serosal surface of the stomach of sockeye salmon, not in the muscles as does *D. latum.* This largely explains the relatively rare occurrence of the infection in man. Other *Diphyllobothrium* species found in man in Alaska include *D. latum, D. dendriticum, D. lanceolatum,* and *D. dalliae,* all reported by Rausch and Hilliard (1970).

Diphyllobothrium yonagoensis Yamane, Kamo et al., 1981, whose description was based on a single worm, 465 cm in length by 13 mm maximum width, was expelled from a 41-year-old man in Japan after treatment with bithionol. Its distinctive features include a knob-like scolex (1.2 × 1.1 × .07 mm) with deep bothria, a short neck (3 mm), and the presence of two sets of genital organs in about one-fourth of the prolottids. It resembles *D. stemmacephalum* and *D. fuhrmanni,* which occur in marine mammals.

Other described species such as *D. parvum* (Stephens, 1908) Faust, 1929, recovered once by Elkington from a Syrian immigrant to Tasmania; *D. strictum* Talysin, 1932, obtained from a patient on an island in Lake Baikal, Siberia; and *D. tungussicum* Podjapolskaja and Gnedina, 1932, obtained from the tribes of northern Siberia, are all probably variants of *D. latum.*

Genus Spirometra
Mueller, 1937

Tapeworms of the genus *Spirometra* are characterized by a piling of the uterine coils as distinguished from a rosetted arrangement. Species of this genus have been found as adults in a variety of mammals, especially cats and dogs.

The species *S. houghtoni* has been described twice as a human intestinal parasite, once from Shanghai and once from Kiunkiang, China (Faust et al., 1929). The strobila is much smaller and more delicate than that of *D. latum* (85 to 100 cm long). The sulci of the scolex are poorly developed. The distalmost proglottids are rectangular, slightly broader than long (3 to 3.5 mm by 2.7 to 3.2 mm). The multiple testes are much larger than the numerous vitellaria, and the uterus has 4 to 7 compactly piled outer coils, while the inner coils are rosetted. The eggs somewhat resemble those of *D. latum* but are ellipsoidal, have a rounded, conical operculum, and measure 57 to 66 μm by 33 to 37 μm. Although the life cycle is unknown, a *Cyclops* is probably the first intermediate host and some vertebrate other than a fish is presumably the second intermediate host. The sparganum stage is capable of parasitizing man. Clinical aspects of this infection have not been studied.

Related species of the genus *Spirometra,* including *S. mansoni, S. decipiens, S. ranarum,* and *S. erinacei,* are frequent intestinal parasites of canine and feline hosts in Asia and have occasionally been reported from other parts of the world. *Spirometra mansonoides* is the common, and perhaps the only, counterpart in the Western Hemisphere of the various species of *Spirometra* in the Eastern Hemisphere. Species of *Cyclops* serve as first larval hosts; several groups of vertebrates other than fish are hosts for the sparganum stage. Gan (1949) made a thorough experimental study of *S. ranarum* in Indonesia. Spargana of these several species have been frequently reported in human infections (see the section on *Sparganum* and sparganosis). The adult stage of *Spirometra erinacei* has been recorded as a parasite of man in five cases (Suzuki et al., 1982). The spargana of *Spirometra mansonoides* and *Spirometra erinacei* produce a growth hormone-like substance that markedly stimulates growth in infected mice and other small mammals (Mueller, 1974; Shiwaku and Hirai, 1982). Experiments with *S. mansonoides* in hamsters suggest that the sparganum growth factor may function in a manner similar to endogenous growth hormone (Phares, 1982).

Diplogonoporus grandis
(R. Blanchard, 1894) Lühe, 1899

This tapeworm, normally a parasite of whales but reported from 55 Japanese patients (Kamo et al., 1971; Morishita, 1972; Kagei et al., 1981), measures 1.4 to 5.9 meters in length, has gravid proglottids measuring 15 to 25 mm in breadth by 0.54 mm in length and two sets of genital organs in each proglottid. The twin genital pores and uterine pores are found in ventral grooves on each side of the midline. The scolex is small (about 1 mm maximum width) with prominent dorsal and ventral grooves. The eggs measure 63 to 68 μm by 50 μm. Fishes are the second intermediate hosts. The worms may produce colicky pain, alternating diarrhea and constipation, progressive secondary anemia, accelerated pulse rate, and lassitude.

Digramma brauni
(Leon, 1907) Joyeux and Baer, 1929

Two Romanian patients were found parasitized by this worm, which measured about 12 cm long, had incomplete transverse "segmentation," slit-like dorsal and ventral sulci, an inconspicuous cervical region, and double genital organs for each proglottid. Joyeux and Baer (1929) concluded that this parasite was probably an immature avian species, accidentally acquired from eating raw fish. Infection was believed to cause anemia.

Genus Ligula
Bloch, 1782

Ligula intestinalis (Goeze, 1782) Gmelin, 1790 and *Braunia jasseyensis* Leon,1908 have been reported respectively from Romanian and French patients. The former species is a natural parasite of fish-eating birds, which Joyeux and du Noyer (1931) described from the vomitus of a patient. Joyeux and Baer (1929), however, suggested that the latter is really an immature specimen of *Ligula,* acquired by ingesting parasitized fish that had become infected from ingesting infected copepods *(Diaptomus gracilis* and *Cyclops strenuus).*

Sparganum and Sparganosis

Historical and Biological Data. In 1882, Manson recovered a sparganum from tissues of a native at autopsy in Amoy, China. Similar larvae have been recovered on numerous occasions since Manson's day from patients in China, Japan, Taiwan, Korea, Vietnam, Thailand, less commonly in India (Datta et al., 1982), Africa (*S. baxteri* Sambon, 1907, and *Sparganum* sp. of de Meillon and Leech, 1943), Queensland, Australia (Sandars, 1954), Java and Sumatra, Holland, Guyana, Puerto Rico, Belize

(British Honduras), Uruguay, Colombia, Texas, Arkansas, Louisiana, Oklahoma, Missouri, Tennessee, Virginia, Mississippi, Alabama, Georgia, Florida, Illinois (possibly acquired in Arkansas), and New York. Approximately 65 cases have been reported from various parts of the United States, mostly in the southeastern region (Norman and Kreutner, 1980). Numerous cases have been diagnosed and not reported (Gardiner et al., 1979). Autochthonous cases have been reported from Malagasy (Brygoo and Catala, 1961), Venezuela (Garcia et al., 1961), and Italy (Bianchi, 1964). Morphologically indistinguishable forms are common in the subcutaneous connective tissue and between the muscles of frogs, lizards, snakes, birds, and several species of mammals. These spargana, when removed from reservoir hosts and fed to dogs and cats, develop into several distinct but closely related species of the genus *Spirometra*. It is, therefore, impossible to assign a species name to the sparganum without experimental completion of its life cycle in a dog or cat and a study of the adult strobila which matures in the small intestine of these experimental hosts. *Sparganum mansoni* is the name usually assigned to this larval stage in Asia.

Mueller (1937, 1938) demonstrated that *Spirometra mansonoides* (which has a procercoid stage in *Cyclops leuckarti, C. viridis, C. bicuspidatus,* and possibly other species of this genus, a sparganum stage in mice and rhesus monkeys, and the definitive stage in the cat, the bobcat, and less commonly in the dog) is the common species of *Spirometra* in the United States. It is likely, therefore, that some or possibly all of the cases of sparganosis of the nonproliferating type reported from this area belong to *S. mansonoides* (Mueller et al., 1963).

Morphology and Biology. Human sparganosis may result from (1) ingestion of infected *Cyclops* containing the procercoid stage, the penetration of these larvae through the intestinal wall, and their migration to the subcutaneous tissues and muscle fascia, where they develop into typical spargana; (2) ingestion of a plerocercoid in raw flesh of an amphibian, reptile, bird, or mammal; or (3) local application of the flesh of an infected vertebrate to the skin, conjunctiva, or vagina, resulting in the migration of the sparganum out of the flesh into the human tissues. Since spargana develop in pigs, human infection occasionally may be acquired by eating raw pork (Corkum, 1966). These spargana, are ivory-white and ribbon-like, and measure up to 30 cm in length by about 3 mm in breadth, with an anteroposterior polarity, a pseudosucker but no clearly evident scolex at the anterior end, and no true transverse segmentation. Located in the scolex position in plerocercoids of *D. dendriticum* are specialized structures including a terminal frontal pit and two types of glands referred to as frontal glands (Gustafsson and Vaihela, 1981). One type, the "green" glands, empty into the frontal pit, are located in the cortical and medullary parenchyma, and extend into the neck. The other type, the "golden" glands, are limited to the scolex region and are associated with the nervous tissues. The green glands, active only during the early phase of infection or cultivation, disappear entirely within 3 days; the golden glands remain and are still present in the mature worm. In histopathological sections the features of a pseudophyllidean cestode are evident: large bilateral longitudinal excretory channels and numerous bundles of muscle fibers scattered throughout the parenchyma (Fig. 30–8).

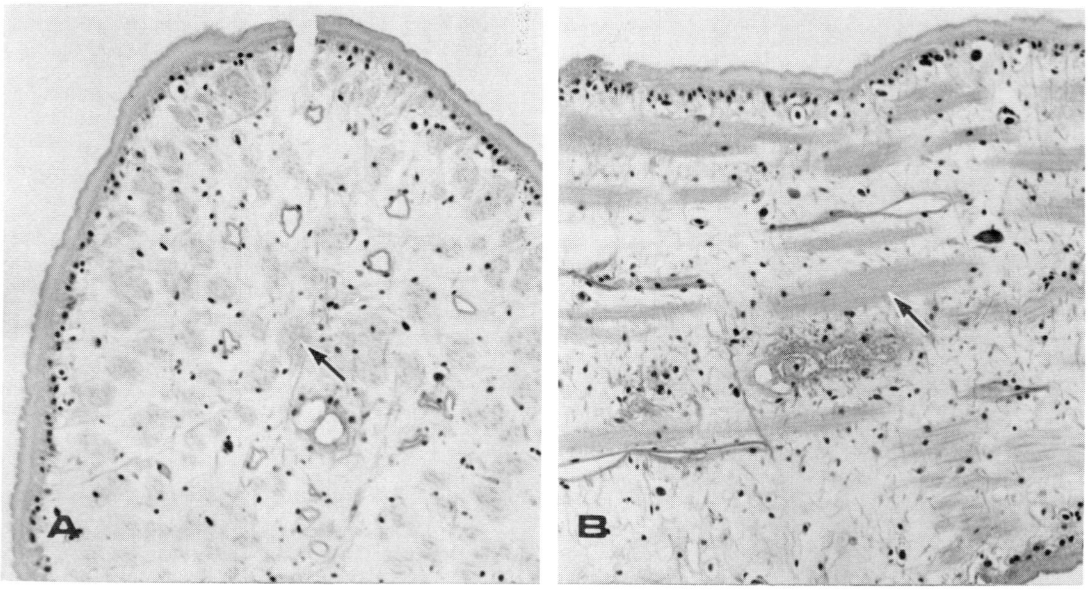

Fig. 30–8. *Spirometra mansonoides*. Sparganum from subcutaneous tissue of labium majus of a woman in South Carolina. *A*, Transverse section showing characteristic pattern of noncellular tegument, cellular subtegument, and parenchyma bearing numerous bundles of muscle fibers (arrow) and excretory canals. *B*, Longitudinal section showing bundles of muscle fibers (arrow) distributed throughout the parenchyma. (× 170.) (By P.C. Beaver.)

The living larvae are highly mobile and their behavior in certain hosts is unpredictable. When ingested by cats, procercoids in copepods become plerocercoids in the muscles, not adults in the intestine; plerocercoids, when ingested by either dogs or cats, may migrate to the muscles, or may become adult worms in the intestine; and plerocercoids injected into the peritoneal cavity of cats usually migrate to the muscles and subcutaneous tissues, but occasionally one or more may migrate from such sites to the intestine to become egg-laying adults (Corkum, 1973). Bonne (1932) reported recovery of two mature spargana from an infarcted pulmonary artery. Bonne and Lie Kian Joe (1940) found two spargana in the wall of the jejunum. The same parasite has been found in native Javanese monkeys and pigs (Bonne, 1942). All of the human material, when grown to the adult stage in kittens, was found to be *Spirometra ranarum*.

Sparganum proliferum (Ijima, 1905) Stiles, 1908, is a proliferating larva of a pseudophyllidean tapeworm, the adult form of which is unknown. It is elongate in shape, with an anteroposterior polarity, an apical end that may be capable of inversion and eversion, and lateral, sometimes branched proliferating processes that may become detached from the parent worm and develop into separate organisms. Five infections have been reported from Japan and one (in a fisherman) from Manatee, Florida. In these patients the spargana were recovered by the thousands in the subcutaneous tissues, intermuscular tissues, and viscera, including the brain in three cases (Fig. 30–9,*A*). In three other patients acephalic proliferating cestode larvae of a different type were found only in the deep organs: in the abdominal cavity in one, the chest cavity in one, and in both chest and abdominal cavities in the third case (Fig. 30–9,*B*). These were reported from Pennsylvania (United States), Taiwan, and Paraguay, respectively. These nine cases were reviewed by Beaver and Rolon (1981). A tenth case, resembling the one reported from Florida by Stiles in 1908, was reported from Venezuela by Moulinier et al. in 1982. Proliferating spargana and other forms of acephalic proliferating cestode larvae have been reported in lower primates, dogs, cats, and other mammals (Kuntz et al., 1970; Beaver and Rolon, 1981).

Pathogenesis, Pathology, and Symptomatology. In Asia where the species commonly found in man is *Spirometra mansoni,* the sparganum is usually in the subcutaneous tissues of various parts of the body, but other sites, such as the orbit, brain, and abdominal viscera, have been recorded (Mineura and Mori, 1980). In one instance a living sparganum was found free in the peritoneal cavity

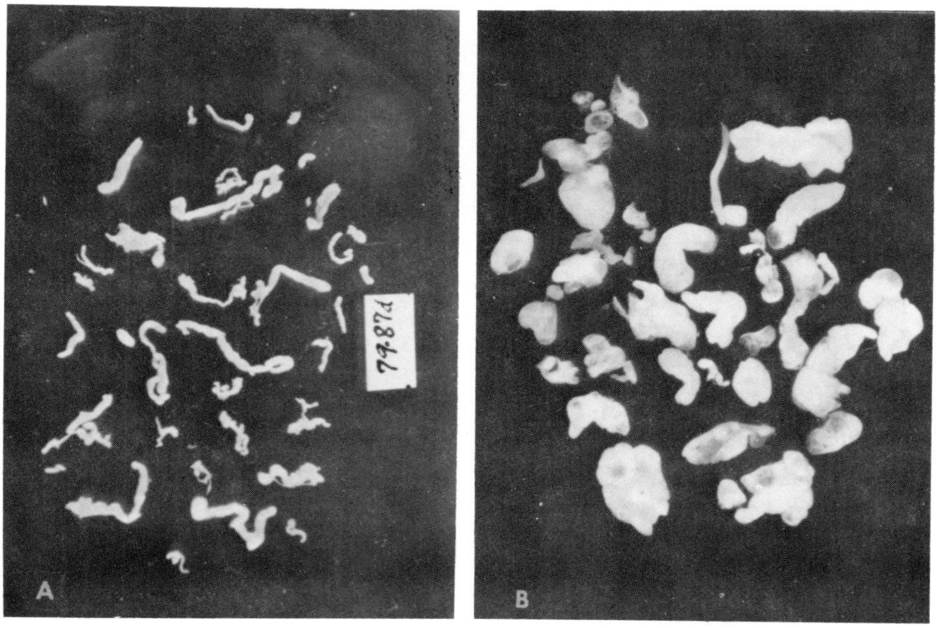

Fig. 30–9. *Sparganum proliferum. A,* Parasites from subcutaneous tissues of a man in Florida; this case was reported by Stiles in 1908 and by Gates in 1909. Natural size. (Specimens from U.S. National Museum.) *B,* Parasites dislodged from cut surfaces of a large mass from the abdominal cavity of a man in Paraguay. (× 3.) (From Beaver, P.C. and Rolon, F.A. 1981. Proliferating larval cestode in a man in Paraguay. A case report and review. Am. J. Trop. Med. Hyg., *30*:625–637.)

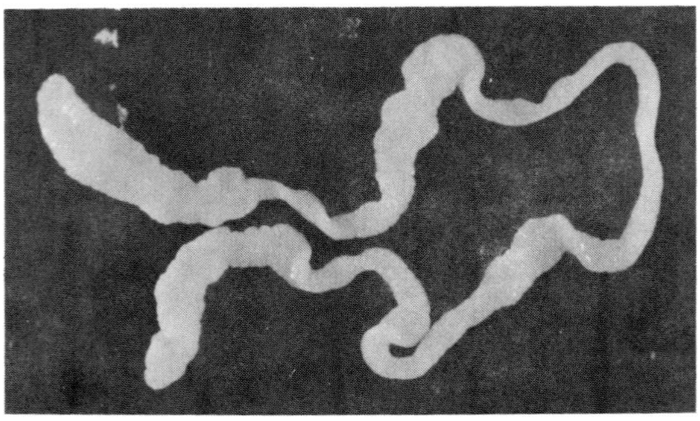

Fig. 30–10. *Sparganum,* probably *S. mansoni.* Live specimen in saline after removal from the abdominal cavity of a man in Thailand. (× 2.) (From Khamboonruang, C., Premasathian, D., and Little, M.D. 1974. A case of intra-abdominal sparganosis in Chiang Mai, Thailand. Am. J. Trop. Med. Hyg., *23*:538–539.)

(Khamboonruang et al., 1974) (Fig. 30–10). In Korea, where more than 60 cases have been reported, 70% of the infections were in males, and in 80% of the cases raw snake meat had been eaten as a remedy for certain diseases or physical deficiencies (Cho et al., 1975). In the usual case only one or a few spargana have been detected; in a case reported from Taiwan 17 of the worms were removed from the eye, nose, and neck of a man over a period of 2 years (Wang and Cross, 1974).

The presence of a sparganum in the immediate vicinity of the eye (ocular sparganosis), a condition not uncommon in Vietnam, is characterized by intense pain, irritation, and edematous swelling of the eyelids, with excessive lacrimation. Subconjunctival infection causes nodule formation around the worm. Retrobulbar invasion leads to lagophthalmia and ulceration of the cornea. In the proliferating type the invaded tissues become honeycombed and nodular. Itching pustules develop over and around the larvae in subcutaneous tissues.

In cases of sparganosis presumed to be caused by *Spirometra mansonoides,* the most common presentation is that of a subcutaneous mass or nodule, 1 to 2 cm in diameter, which may or may not be tender (Norman and Kreutner, 1980). The mass, tender or not, may be migratory and occasionally may disappear and reappear at a new location. Peripheral eosinophilia is usually unremarkable or absent but may be striking (Campbell and Beals, 1977). The tender mass or fibrous cyst caused by a sparganum is an eosinophilic granulomatous lesion of the type commonly caused by helminths in tissue (Taylor, 1976).

Diagnosis. Even in regions where the infection is seen most frequently, diagnosis is not usually made until after surgical removal and identification of the worms free or in histopathological sections (Fig. 30–9).

Treatment. Treatment has been by surgical removal of the worm, usually without preoperative diagnosis. In Vietnam, for ocular sparganosis, Cornet (1933) recommended injection of 2 to 4 ml of 40% ethyl alcohol with procaine (free of epinephrine) into the lesion to kill the worms *in situ,* after which they may be allowed to be absorbed. Keller (1937), however, reported marked success from intravenous use of novarsenobenzol. Tarsorrhaphy was also employed to preserve the cornea until the worms had been absorbed or were discharged on the dressing. In the more common type, in which the worms are single or at most few, and are located in subcutaneous tissue, surgical removal under local anesthesia leads to rapid recovery. In the proliferating type, on account of profound involvement of the tissues, except for palliative measures, treatment has been impracticable. In the most recent case, treatment with mebendazole was ineffective, and the patient's tolerance of praziquantel was poor (Torres et al., 1981).

Prevention. To avoid infection in endemic areas, drinking water (possibly containing infected *Cyclops*) should be boiled or filtered. Patients with ulcerated or inflamed areas of the skin, conjunctiva or vagina should avoid the local poulticing of these lesions with the flesh of frogs or other vertebrates likely to be infected with the sparganum. Ingestion of raw tadpoles, frog, snake, fowl or mammalian flesh, should be avoided. The unbranched sparganum may live more than 9 years in the human host.

REFERENCES

Baer, J.G., Miranda, C.H., Fernandez, R.W., and Medina, T.J. 1967. Human diphyllobothriasis in Peru. Z. Parasitenkd., *28*:277–289.

Beaver, P.C., Malek, E.A., and Little, M.D. 1964. Development of *Spirometra* and *Paragonimus* eggs in Harada-Mori cultures. J. Parasitol., *50*:664–666.

———, and Rolon, F.A. 1981. Proliferating larval cestode in a man in Paraguay. A case report and review. Am. J. Trop. Med. Hyg., *30*:625–637.

Bianchi, L. 1964. Sparganosi umana. Osservazioni spora un caso autoctono in alta Italia. Proc. 1st Int. Congr. Parasitol., *2*:820–822.

Bonne, C. 1932. A few remarks on two rare parasitic diseases in the Malayan Archipelago. A. Chromoblastomycosis. B. Sparganosis. Trans. 8th Congr. Far Eastern Assoc. Trop. Med., *2*:184–188.

———. 1942. Researches on sparganosis in the Netherlands East Indies. Am. J. Trop. Med., *22*:643–645.

———, and Lie Kian Joe. 1940. Darmwandhelminthiasis teweegebracht door Spargana. Geneesk. Tijdschr. Ned.-Indie, *80*:2788–2792.

Brygoo, E.R., and Catala, P. 1961. Premier cas de sparganose humaine à Madagascar. Arch. Inst. Pasteur Madagascar, *29*:75–77.

Bylund, B.G. 1976. The epidemiology and control of the broad fish tapeworm. Duodecim, *92*:646–648.

———, Bang, B., and Wikgren, K. 1977. Tests with a new compound (praziquantel) against *Diphyllobothrium latum*. J. Helminthol., *51*:115–119.

Cameron, T.W.M. 1945. Fish-carried parasites in Canada. (1) Parasites carried by freshwater fish. Can. J. Comp. Med., *9*:245–254, 283–286, 302–311.

Campbell, E.W., and Beals, C. 1977. Striking eosinophilia in sparganosis. Case report. Postgrad. Med., *62*:138–140.

Catár, G., Sobota, K., Kvasz, L., and Hruzík, J. 1967. The first non-imported case of diphyllobothriosis in Czechoslovakia. Bratisl. Lek. Listy, *47*:241–244.

Cho, S.Y., Bae, J., Seo, B.S., and Lee, S.H. 1975. Some aspects of human sparganosis in Korea. Korean J. Parasitol., *13*:60–77.

Corkum, K.C. 1966. Sparganosis in some vertebrates of Louisiana and observations on a human infection. J. Parasitol., *52*:444–448.

———. 1973. The duality of host function in spirometrid tapeworm infections. Proc. La. Acad. Sci., *36*:64–70.

Cornet, E. 1933. Essai de traitement de la sparganose retrobulbaire. Bull. Soc. Med.-Chir. Indo-Chine, *11*:452–455.

Datta, K., Datta S., Sharma, R., and Goyal, S. 1982. Sparganosis. J. Indian Med. Assoc., *79*:78–80.

de Meillon, B., and Leech, R.B. 1943. A sparganum from an East African native. S. Afr. Med. J., *17*:289–290.

Espejo, H. 1977. Tratamiento de infecciones por *Hymenolepis nana, Taenia saginata, Taenia solium* y *Diphyllobothrium pacificum* con praziquantel (Embay 8440). Bol Chil. Parasitol., *32*:39–40.

Faust, E.C., Campbell, H.E., and Kellogg, C.R. 1929. Morphological and biological studies on the species of *Diphyllobothrium* in China. Am. J. Hyg., *9*:560–583.

Führmann, O. 1930. Dritte Klasse des Cladus Plathelminthes: Cestoidea. Handb. Zool., *2*:141–256.

Gan, K.H. 1949. Research on the life cycle of *Diphyllobothrium ranarum*. Doc. Neerl. Indones. Morbis Trop., *1*:90–92.

García Ocampo, N., Padrón Lizardo, O., and d'Almaral, D. 1961. Primer caso de esparganosis humana en Venezuela. Rev. Venezol. Sanit. Asist. Social, *26*:69–72.

Gardiner, C.H., Rutland, E.D., Jr., and Watkins, R.W. 1979. Case for diagnosis. Milit. Med., *144*:600, 602.

Groll, E. 1980. Praziquantel for cestode infections in man. Acta Trop., *37*:293–296.

Gustafsson, M.K.S., and Vaihela, B. 1981. Two types of frontal glands in *Diphyllobothrium dendriticum* (Cestoda, Pseudophyllidae) and their fate during the maturation of the worm. Z. Parasitenkd., *66*:145–154.

Hatsushika, R., Kihara, T., Shimizu, M., and Oyama, F. 1979. Some observations on moving patterns of the vital scolex of *Diphyllobothrium latum* (Cestoda: Diphyllobothriidae). Kawasaki Med. J., *5*:151–156.

Joyeux, C.E., and Baer, J.G. 1929. Les cestodes rares de l'homme. Bull. Soc. Pathol. Exot., *22*:114–136.

———, and du Noyer, R. 1931. Presence d'une larve de ligule dans des vomissements. Marseille Med., *68*:235–236.

Kagei, N., Hayashi, S., Kato, K., and Masu, H. 1981. On the second case of diplogonoporiasis grandis discovered in Kanagawa Prefecture and anthelminthic effect of paromomycine sulphate. Jpn. J. Parasitol., *30*:457–465. (Japanese, English abstr.)

Kamo, H., Hatsushima, R., and Yamane, Y. 1971. Diplogonoporiasis and diplogonadic cestodes in Japan. Yonago Acta Med., *15*:234–246.

———, Maejima, J., Yazaki, S., Otsuru, M., Hasegawa, H., Kuniyoshi, S., and Asato, R. 1982. Occurrence of human infection with *Diphyllobothrium pacificum* (Nybelin, 1931) Margolis, 1956 in Japan. Jpn. J. Parasitol., *31*:165–170.

Keller, M. 1937. Sur une nouvelle methode de traitement de la sparganose oculaire. Ann. l'Ecole Super. Med. Pharm. Indo-Chine, *1*:77–89.

Khamboonruang, C., Premasathian, D., and Little, M.D. 1974. A case of intra-abdominal sparganosis in Chiang Mai, Thailand. Am. J. Trop. Med. Hyg., *23*:538–539.

Khodakova, V.I., and Lagunova, A.I. 1979. Comparative efficacy of some diagnostic laboratory methods in diphyllobothriasis. Materialy Nauchnoi Konferentsii Vsesoyuznogo Obshchestva Gl'mintologov (Tsestody i Tsestodozy), *31*:146–150.

Kuntz, R.E., Myers, B.J., and Katzberg, A.A. 1970. Sparganosis and "proliferative" spargana in vervets *(Cercopithecus aethiops)* and baboons (*Papio* sp.) from East Africa. J. Parasitol., *56*:196–197.

Lumbreras, H., Terashima, A., Alvarez, H., Tello, R., and Guerra, H. 1982. Single dose treatment with praziquantel (Cesol R, EmBay 8440) of human cestodiasis caused by *Diphyllobothrium pacificum*. Tropenmed. Parasitol., *33*:5–7.

Magath, T.B. 1929. Experimental studies on *Diphyllobothrium latum*. Am. J. Trop. Med., *9*:17–48.

Margolis, L., Rausch, R.L., and Robertson, E. 1973. *Diphyllobothrium ursi* from man in British Columbia—first report of this tapeworm in Canada. Can. J. Public Health, *64*:588–589.

Mineura, K., and Mori, T. 1980. Sparganosis of the brain. J. Neurosurg., *52*:588–590.

Morishita, K. 1972. *Diplogonoporus grandis* (R. Blanchard, 1894), pp. 489–512. In *Progress of Medical Parasitology in Japan,* Vol. 4. Edited by K. Morishita, Y. Komiya, and H. Matsubayashi. Tokyo, Meguro Parasitological Museum.

Moulinier, R., Martinez, E., Torres, J., Noya, O., de Noya, B.A., and Reyes, O. 1982. Human proliferative sparganosis in Venezuela: Report of a case. Am. J. Trop. Med. Hyg., *31*:358–363.

Mueller, J.F. 1937. A repartition of the genus *Diphyllobothrium*. J. Parasitol., *49*:294–296.

———. 1938. Studies on *Sparganum mansonoides* and *Sparganum proliferum*. Am. J. Trop. Med., *18*:303–328.

———. 1974. The biology of *Spirometra*. J. Parasitol., *60*:3–14.

———, Hart, E.P., and Walsh, W.P. 1963. Human sparganosis in the United States. J. Parasitol., *49*:294–296.

Norman, S.H., and Kreutner, A., Jr. 1980. Sparganosis: Clinical and pathologic observations in ten cases. South. Med. J., *73*:297–300.

Nyberg, W. 1963. The effect of changes in nutrition on the host-parasite relationship. *Diphyllobothrium latum* and human nutrition, with particular reference to vitamin B_{12} deficiency. Proc. Nutr. Soc., *22*:8–14.

Peters, L., Cavis, D., and Robertson, J. 1978. Is *Diphyllobothrium latum* currently present in northern Michigan? J. Parasitol., *64*: 947–949.

Phares, C.K. 1982. The lipogenic effect of the growth factor produced by plerocercoids of the tapeworm, *Spirometra mansonoides,* is not the result of hypothyroidism. J. Parasitol., *68*:999–1003.

Rausch, R.L., and Hilliard, D.K. 1970. Studies on the helminth fauna of Alaska. XLIX. The occurrence of *Diphyllobothrium latum* (Linnaeus, 1758) (Cestoda: Diphyllobothriidae) in Alaska, with notes on other species. Can. J. Zool., *48*:1201–1219.

Rees, G. 1967. Pathogenesis of adult cestodes. Helm. Abstr., *36*:1–23.

Sagua, H., Miranda, E., Fuentes, A., and Vladillo, V. 1976. *Diphyllobothrium pacificum* (Nybelin, 1931) Margolis, 1956. Primeros dos casos de infección humana en el norte de Chile. Bol. Chil. Parasitol., *31*:33.

Sandars, D.F. 1954. A sparganum from a Queensland woman. Med. J. Aust., *2*:817.

Shiwaku, K. and Hirai, K. 1982 Growth-promoting effect of *Spirometra erinacei* (Dudolphi, 1819) plerocercoids in young mice. Jpn. J. Parasitol., *31*:185–195.

Suzuki, N., Kumazawa, H., Hosogi, H., and Nakagawa, O. 1982. A case of human infection with the adult of *Spirometra erinacei* (Rudolphi, 1819) Faust, Campbell and Kellogg, 1929. Jpn. J. Parasitol., *31*:23–26.

Taylor, R.L. 1976. Sparganosis in the United States. Report of a case. Am. J. Clin. Pathol., *66*:560–564.

Tomita, S., Tongu, Y., Sakumoto, D., Suguri, S., Itano, K., Inatomi, S., and Kawaguchi, K. 1979. Epidemiological survey for *Diphyllobothrium latum* in Okayama Prefecture. Jpn. J. Parasitol., *28*:317–321.

Torres, J.R., Noya, O.O., Noya, B.A., Mouliniere, R., and Martinez, E. 1981. Treatment of proliferative sparganosis with mebendazole and praziquantel. Trans. R. Soc. Trop. Med. Hyg., *75*:846–847.

von Bonsdorff, B. 1947. In which part of the intestinal canal is the fish tapeworm found? Acta Med. Scand., *129*:142–155.

———. 1956. *Diphyllobothrium latum* as a cause of pernicious anemia. Exp. Parasitol., *5*:207–230.

———. 1978. The broad tapeworm story. Acta Med. Scand., *204*:241–247.

Wang, L.T., and Cross, J.H. 1974. Human sparganosis on Taiwan. J. Formosan Med. Assoc., *73*:173–177.

Yamane, Y., Kamo, H., Yazaki, S., Fukumoto, S., and Maejima, J. 1981. On a new marine species of the genus *Diphyllobothrium* (Cestoda: Pseudophyllidea) found from a man in Japan. Jpn. J. Parasitol., *30*:101–111.

Yokogawa, M., Niimura, M., Kobayashi, M., Hata, H., Tazaki, T., Tokita, K., and Kojima, S. 1979. Epidemiological survey for diphyllobothriasis latum in China Prefecture and treatment with bithionol. Jpn. J. Parasitol., *28*:133–138.

Chapter 31

Cyclophyllidean Tapeworms

Tapeworms of the order Cyclophyllidea have a scolex with four suckers and usually a rostellum that is located at the apex and frequently armed with hooks. Members of the group that are parasites of man belong to the superfamily Taenioidea in which the body is distinctly flattened, the suckers simple, and the eggs nonoperculate.

FAMILY ANOPLOCEPHALIDAE

Bertiella studeri (Blanchard, 1891) Stiles and Hassall, 1902

Synonyms. *Bertia studeri* Blanchard, 1891; *B. satyri* Blanchard, 1891; *Bertiella satyri* (Blanchard, 1891) Stiles and Hassall, 1902.

Historical and Geographical Notes. This tapeworm was first obtained from an orangutan *(Pongo pygmaeus pygmaeus)* in Borneo and has since been found in other primates, including the bonnet monkey *(Macaca radiata), M. syrichta syrichta, M. syrichta fascicularis, Cercopithecus aethiops pygerythrus, C. nictitans schmidti,* and *Hylobates hoolock*. In the Philippines the dog has been found infected. At least 29 human infections with this species have been reported, 4 from Mauritius, 8 from India, 7 from Sumatra, 2 from Java, and 1 each from Borneo, Singapore, East Africa, St. Kitts, B.W.I., the Philippines, Malaysia, Sri Lanka, and Yemen.

Morphology, Biology, and Life Cycle. The worm has a total length measurement of 275 to 300 mm and a maximum breadth of 10 mm. The subglobose scolex (measuring 475 μm in diameter) is distinctly set off from the cervical region (Fig. 31–1,*A*). Apically there is a rudimentary unarmed rostellum. There are four oval suckers (220 by 150 μm): two on the dorsal and two on the ventral surface of the scolex. Mature proglottids measure about 6 mm broad by 0.75 mm long. The arrangement of the genital organs is illustrated in Figures 29–5 and 31–1,*B*. Although the male organs persist for some time after the proglottids become gravid, the uterus finally comes to fill practically the entire proglottid. These segments are shed in groups of about 20 at a time. The eggs (Fig. 31–1,*C*) have an irregular, ovoid contour (45 to 46 μm by 49 to 50 μm); there is a delicate middle envelope and an inner shell with a bicornuate protrusion on one side.

The life cycle was described by Stunkard (1940), who fed eggs from gravid proglottids to 24 species of mites and recovered fully developed cysticercoids from *Scheloribates laevigatus* and *Galumna* spp. These larvae are spherical, ovoid, or piriform, measure 0.1 to 0.15 mm in diameter, and have a small cercomer.

Epidemiology. Infection is apparently acquired from accidental ingestion of certain mites containing the cysticercoid larva of the worm.

Clinical Aspects. The worms are evacuated after administration of niclosamide or other anthelmintics (Shoura and Morsy, 1974; Dissanaike et al., 1977), or passed spontaneously (Edirisinghe and Cumararajan, 1976). Their presence in the intestine may provoke no manifest symptoms or digestive disturbances. The worms are diagnosed by the peculiarity of the eggs (Fig. 31–1,*C*) evacuated in the feces, or on recovery of a chain of characteristic proglottids.

The majority of these infections have been in children. An infection with *Bertiella* in a 6-year-old boy was reported by Thompson et al. (1967). The infection probably had been acquired in Mombasa, East Africa almost 2 years previously. Proglottids were passed intermittently in the feces for 6 months; the boy was otherwise asymptomatic.

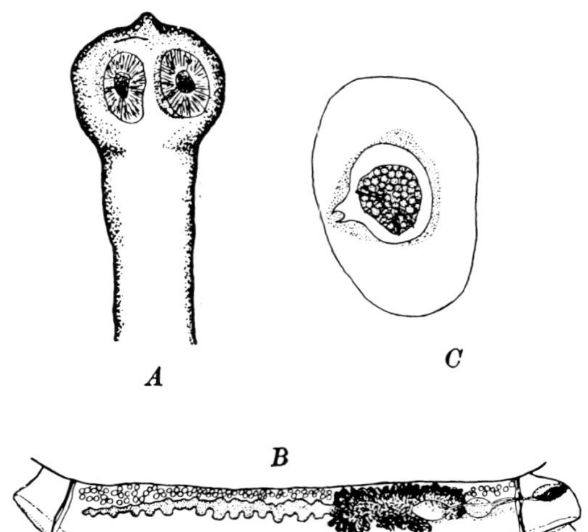

Fig. 31–1. *Bertiella studeri*. *A*, Scolex. (× 38.) *B*, Mature proglottid. (× 15.) *C*, Egg. (× 440.) (*A* and *B* adapted from original of A.C. Chandler. *C* adapted from original of R. Blanchard, In Faust, E.C. 1929. *Human Helminthology*, 1st ed. Philadelphia, Lea & Febiger.)

Bertiella mucronata (Meyer, 1895) Stiles and Hassall, 1902

Synonym. *Bertiella studeri* (Blanchard, 1891) of Stunkard, Koivastik, and Healy, 1964.

Distribution and Biology. This species of *Bertiella* is indigenous to warm climates of Latin America. First described from a howler monkey *(Alouatta nigra)* in Paraguay, it has since been reported from *Alouatta caraya* in northern Argentina and from three captive chimpanzees *(Callithrix nigrifrons)* in Cuba. Human infections have been recorded twice in Brazil (Costa et al., 1967) and once each in Cuba, Argentina, and Paraguay, and from a child in northcentral United States (Fargo, North Dakota), who may have been exposed by contact with an imported monkey (Koivastik, 1963). As described by D'Alessandro et al. (1963), all proglottids are several times as wide as they are long. A single genital complex develops in each proglottid, with irregularly alternating marginal openings of the seminal vesicle and vagina. In general morphology, *B. mucronata* is similar to *B. studeri*. The most distinctive feature is the character of the vagina, which in *B. studeri* is glandular in its outer portion only (Stunkard, 1940), whereas in *B. mucronata* it is thick-walled and glandular along its full length from the outer margin of the ovary to the genital atrium (Fig. 31–2). Other differentiating characteristics include more numerous testes and smaller eggs. Stunkard et al. (1964), in describing the worm from the North American child whose infection was reported by Koivastik, failed to recognize these differences. The worm, 45 cm long, was evacuated following quinacrine therapy.

Intermittent abdominal pain and vomiting were the symptoms.

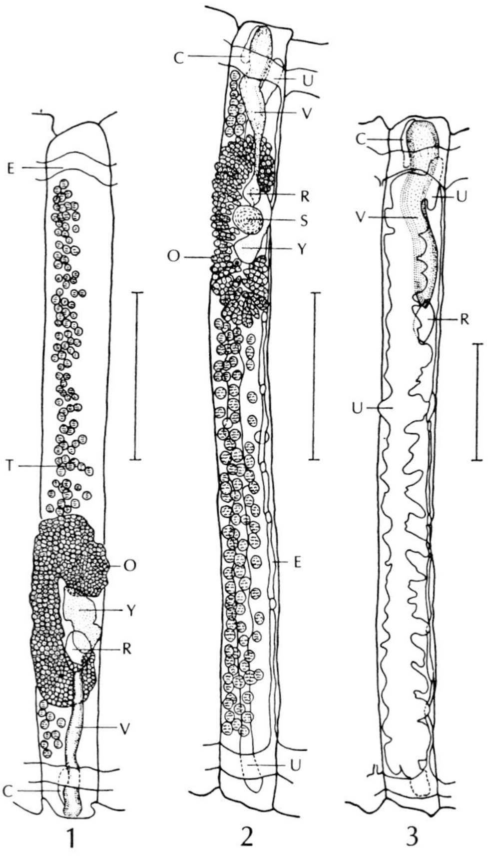

Fig. 31–2. *Bertiella mucronata,* camera lucida drawings of proglottids as seen in stained whole mounts. 1, Juvenile. 2, Mature. 3, Gravid. *C,* Cirrus sac. *E,* Excretory canal. *O,* Ovary. *R,* Seminal receptacle. *S,* Shell gland. *T,* Testes. *U,* Uterus. *V,* Vagina. *Y,* Vitellaria. Magnification reference lines. 1 mm. (From D'Alessandro, A., Beaver, P.C., and Masi Pallares, R. 1963. *Bertiella* infection in man in Paraguay. Am. J. Trop. Med. Hyg., *12*:193–198.)

FAMILY LINSTOWIIDAE

Inermicapsifer madagascariensis (Davaine, 1870) Baer, 1956

Synonyms. *Taenia madagascariensis* Davaine, 1870; *Davainea madagascariensis* (Davaine, 1870) Blanchard, 1891; *Anoplocephala (?) arvicanthidis* Kofend, 1917; *Raillietina madagascariensis* (Davaine, 1870) Joyeux and Baer, 1929; *R. cubensis* Kourí, 1938, *R. davainei* Dollfus, 1940; *R. loeche-salavezi* Dollfus, 1940; *R. kouridovali* Dollfus 1940; *Inermicapsifer arvicanthidis* (Kofend, 1917) Baer 1925; *I. cubensis* (Kourí, 1939) Baer, 1940; *I. cubanensis* Lopez-Neyra, 1943.

Historical and Geographical Notes. First reported by Grenet (1870) from an 18-month-old child in Mayotte (Comoro Islands northwest of Malagasy), this parasite has since been recognized from man in Malagasy (Nose Be Island), Congo, Mauritius, Réunion, Kenya, and Venezuela (Bailenger and Carcenac, 1970). In Havana, Cuba, it has often been found in man since its first recognition in 1935. In Africa this species is common in rodents south of the Sahara (Goldsmid and Muir, 1972). Baer (1956) indicated that *I. madagascariensis* outside of Africa is wholly a parasite of man.

Morphology, Biology, and Life Cycle. This cestode is long and narrow (Fig. 31–3) ranging up to 42 cm in length and 2.6 mm in width. A single scolex up to 610 μm wide bears four round unarmed suckers each about 185 μm in diameter (Fig. 31–3). There is no rostellum or other armature. Proglottids bearing the beginnings of the uterus are broader than long. Mature ones are approximately square, whereas gravid segments in which the uterus has been replaced by egg capsules are longer than broad (3 mm × 2.6 mm). The genital pores are unilateral, and open at or near the middle of the proglottid. The cirrus pouch contains a muscular cirrus and measures up to 150 μm in length by 57 μm in width varying with the maturity of the segment. The uterus develops as a branched sacculate structure but breaks down, the eggs becoming enclosed in capsules. There are upward of 150 to 175 capsules each containing a half dozen or more individual eggs 35 to 50 μm in diameter.

The life cycle of *I. madagascariensis* is unknown, though by analogy with related organisms of the genera *Raillietina* and *Choanotaenia* an arthropod intermediate host is certainly involved in transmission from rodent to rodent and man, and from man to man.

Epidemiology. Although the method and source of infection with this tapeworm are unknown, it seems certain that some arthropod vector is involved in transmission. Children, mostly 1 to 3 years of age, and young adults have been found infected. Rodents in Africa harbor this parasite.

Clinical Aspects. The disturbance and lesions produced by this tapeworm have not been adequately assessed. Symptoms are usually slight or absent, but some patients complain of loss of appetite, loss of weight, abdominal pain, or irritability (Goldsmid and Muir, 1972).

Diagnosis rests upon recovery of rice grain-like proglottids or strobilae from the stool, but these together with the eggs or egg capsules are not readily distinguishable from those of related

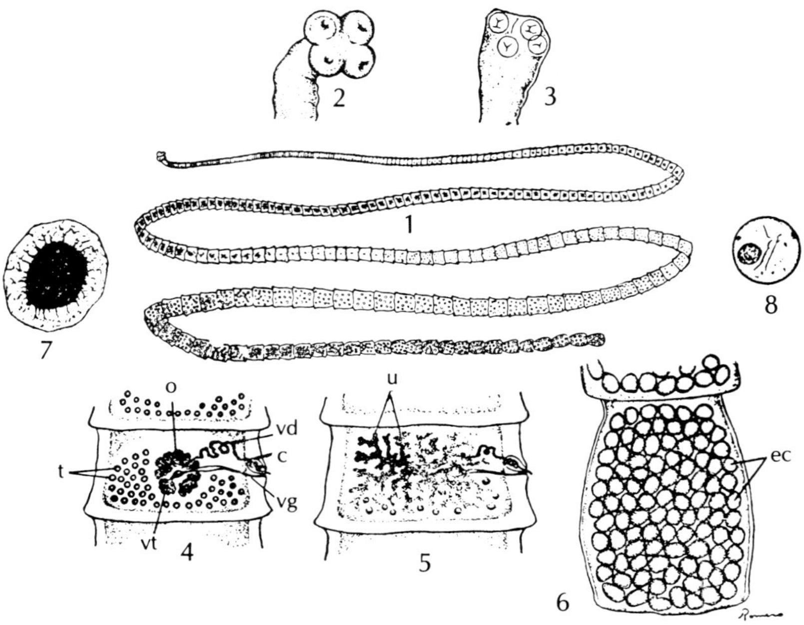

Fig. 31–3. *Inermicapsifer madagascariensis. 1,* Strobila. *2* and *3,* Scolex. *4,* Mature proglottid. *5,* Young gravid proglottid. *6,* Gravid proglottid filled with egg capsules. *7,* Egg capsule. *8,* Egg with small hexacanth larva inside. *c,* Cirrus sac. *ec,* Egg capsule. *o,* Ovary. *t,* Testes. *u,* Uterus. *vd,* Vas deferens. *vg,* Vagina. *vt,* Vitellarium. (Adapted from Kouri, P. 1940. A new human helminthic infection in Cuba. J. Parasitol., *26*:179–181.)

genera of cestodes (namely, *Raillietina*). In *Inermicapsifer*, the genital pore is in the middle of the lateral margin of the proglottid, whereas in *Raillietina* it is in the anterior third.

Niclosamide has been used satisfactorily for treatment (Horstmann et al., 1978; Hira, 1975).

FAMILY MESOCESTOIDIDAE

Mesocestoides Species

Specimens of *Mesocestoides*, probably *M. variabilis*, were passed by a 13-month-old child at Nacogdoches, Texas (Chandler, 1942). This constituted the first record of human infection by a member of this family of tapeworms, although related species have been reported from monkeys, baboons, the opossum, raccoon, fox, other carnivores, mice, and birds of prey. Chandler (1949) stated that a second human case was found in Denmark in a native of Greenland. A third case, in a European child living in Ruanda Burundi, was recorded by Fain and Herin (1954), and four additional human infections have been reported

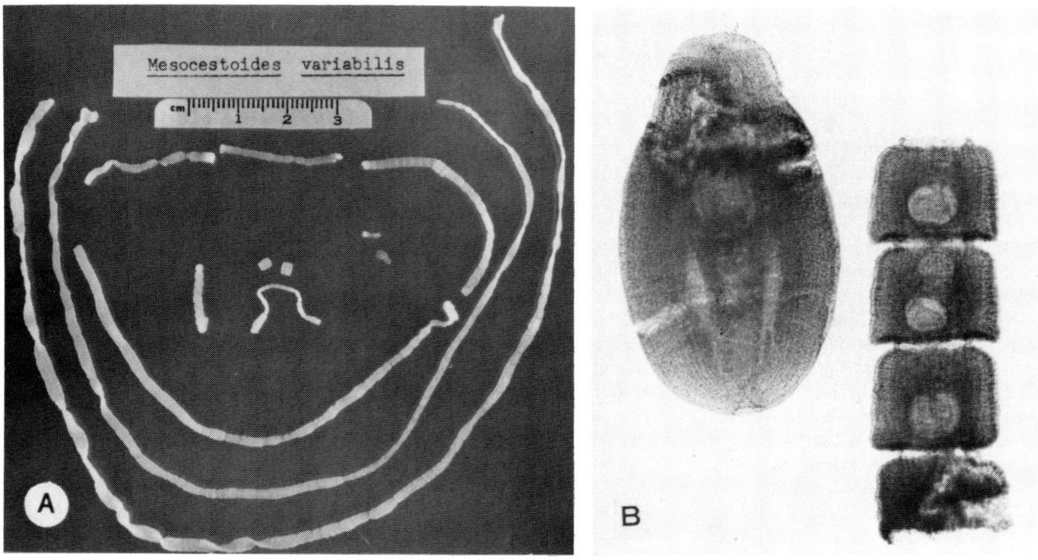

Fig. 31–4. *Mesocestoides variabilis. A,* Strobilae and proglottids expelled from a child in New Jersey after treatment with atabrine and later with niclosamide. *B,* Individual proglottids in which the parauterine organ is evident. (From Gleason, N.N., Kornblum, R., and Walzer, P. 1973. *Mesocestoides* (Cestoda) in a child in New Jersey treated with niclosamide (Yomesan[R]). Am. J. Trop. Med. Hyg., *22*:757–760.)

from North America, in Missouri, New Jersey, Ohio, and Mississippi (Hutchinson and Martin, 1980) (Fig. 31–4). Ten or more human infections with *Mesocestoides* (probably *M. lineatus*) have been reported from Japan (Kagei et al., 1974), and a single human infection has been found in Korea (Choi et al., 1967). The larval stages of species of this genus are found in the body cavity and tissues of reptiles, birds, and small mammals.

In Chandler's (1942) case the individual strobila had an estimated average length of 40 cm and a breadth of 2 mm. The child had harbored the worms for several weeks, was fretful, had a poor appetite, had abdominal colic, and was anemic. Infection was probaby acquired from eating inadequately cooked meat of some animal that serves as intermediate host for the second larval stage, a plerocercus type known as the *tetrathyridium*. Niclosamide, quinacrine, paromomycin, and other drugs have been used successfully to expel the worm (Gleason et al., 1973; Hutchison and Martin, 1980). In one case treatment with quinacrine failed repeatedly and the worm was passed spontaneously 6 weeks after the last treatment (Gutierrez et al., 1978).

FAMILY DILEPIDIDAE

Dipylidium caninum (Linnaeus, 1758) Railliet, 1892 (The double-pored dog tapeworm, producing dipylidiasis or dog tapeworm infection)

Synonyms. *Taenia canina* L., 1758; *T. cucumerina* Bloch, 1782; *Dipylidium cucumerinum* (Bloch, 1782) Leuckart, 1863; *D. sexcoronatum* v. Ratz, 1900. Venard (1938) recognized only two other species of the genus *Dipylidium*, namely *D. buencaminoi* Tubangui, 1925, from the dog, Manila, and *D. otocyonis* Joyeux, Baer and Martin, 1936, from *Otocyon megalotis*, North Somaliland.

Geographical Notes. *Dipylidium caninum* is a common tapeworm of the dog and cat throughout the world, and has been reported from the wild cat *(Felis silvestris)*, jungle cat *(Felis catus constantina)*, *Felis catus ocreata*, Indian palm cat *(Paguma leucomystax grayi)*, civet cat, hyena, jackal, dingo, fox and, from time to time, from man (almost invariably in children). Human infections have been reported from most countries of Europe, from the Philippines, China, Japan, Rhodesia, Argentina and the United States.

Morphology, Biology, and Life Cycle. The strobila of *D. caninum* consists of a chain of elliptical proglottids and measures 10 to 70 cm in length. The scolex (Fig. 31–5,*A*) is small, rhomboidal, with a transverse diameter of 250 to 500 μm, has four deeply cupped, oval suckers, and a median apical, club-shaped rostellum, capable of protrusion to a length of 185 μm or of complete invagination into the scolex. This rostellum is armed with one to seven circlets of spines, depending possibly on the age of the worm. Each spine has a short, curved arm and a large rounded base. The anterior spines are the largest and the posterior ones the smallest. The neck of the worm is short and slender. Immature proglottids are first broader than long and later squarish in outline. Mature and gravid proglottids are typically pumpkin-seed-shaped. Each is provided with a double set of reproductive organs with a genital atrium on each lateral margin and has a maximum width of 3.2 mm (Fig. 31–5,*B*). Gravid proglottids are filled with polygonal uterine blocks through the median field. Each block contains 8 to 15 eggs enclosed in an embryonic membrane.

The eggs (Fig. 31–5,*C,D*) are spherical, thin-shelled, and hyaline except for a brick-red tinge, measure 25 to 40 μm in diameter, and have delicate hooklets measuring 12 to 15 μm in length. Gravid proglottids separate singly or in groups from the strobila and frequently wander down the bowel and out of the anus. Disintegration of these proglottids does not commonly occur within the bowel, but at times groups of eggs within the embryonic membrane are passed in the feces.

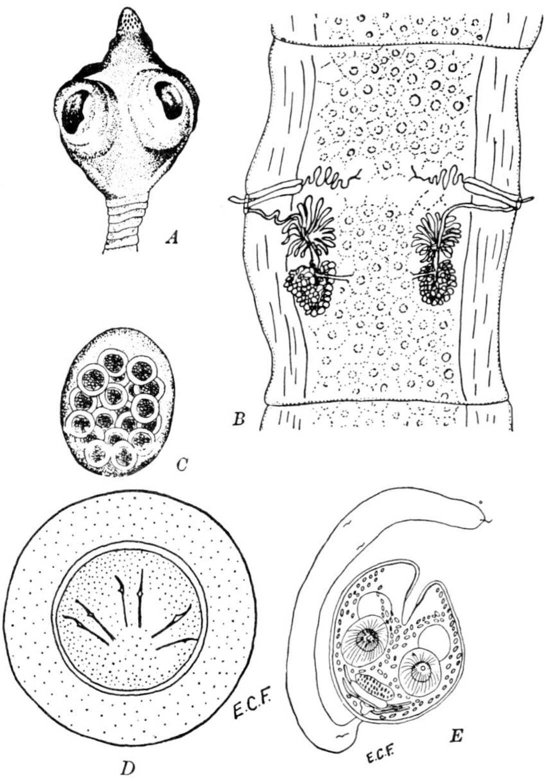

Fig. 31–5. *Dipylidium caninum*. *A*, Scolex, greatly enlarged. *B*, Mature proglottid, enlarged. *C*, Cluster of eggs in egg membrane. *D*, Egg. (× 900.) *E*, Cysticercoid larva, greatly enlarged. (*A* and *C* adapted by E.C. Faust from Stiles, C.W. 1925. Illustrated key to the cestode parasites of man. Hyg. Lab. Bull., no. 25. *B* and *D* by E.C. Faust. In Faust, E.C. 1939. *Human Helminthology*, 2nd ed. Philadelphia, Lea & Febiger. *E* adapted by E.C. Faust from G.B. Grassi and G. Rovelli. In Braun, M.G.C.C., and Seifert, O. 1925. *Die tierischen Parasitien des Menschen*. Wurzburg.)

When a group of the membrane-enclosed eggs is freed from the uterus, it assumes an ovoid to spherical contour (Fig. 31–5,*C*).

Eggs in capsules or proglottids are ingested by the larval stage of the dog flea *(Ctenocephalides canis)*, the cat flea *(C. felis)*, or the human flea *(Pulex irritans)*, all of which are common ectoparasites of the dog and cat. The dog louse, *Trichodectes canis*, has also been incriminated as a suitable host for *Dipylidium*, although Zimmermann (1937) considered that this species is not *D. caninum* but *D. sexcoronatum*. On the other hand, Venard (1938) stated that *D. sexcoronatum* is synonymous with *D. caninum*, in which case *Trichodectes* is a proven intermediate host of *D. caninum*. Grassi and Rovelli (1888) first discovered that the eggs hatch in the intestine of these insects and migrate into their hemocoel, where they develop into procercoid, later cysticercoid larvae (Fig. 31–5,*E*). Chen (1934) found that many fleas die as a result of this invasion but that amebocytes and other phagocytic cells in the hemocoel of the insect also destroy many of the larvae during the procercoid stage. Some survive metamorphosis of the insect to the adult stage, which, when ingested by the mammalian host, produces infection that matures in 3 to 4 weeks.

Epidemiology. Practically all human cases have been in children, suggesting either greater exposure or age tolerance. While infection in man is not common, especially considering the available sources of infection in dogs and cats, it is by no means rare.

Pathogenesis, Pathology, and Symptomatology. The degree of pathogenicity is associated primarily with the number of worms present. The symptoms evoked in children, who are seldom infected with more than a single worm, are mild and consist of slight intestinal disturbances, including indigestion and loss of appetite. Proglottids migrating from the anus of a child frequently cause anxiety in the parents.

Diagnosis. Diagnosis is based on recovery of the characteristic eggs (Fig. 31–5,*D*) in capsules (Fig. 31–5,*C*) or, more usually, the proglottids passed in the feces or wandering out of the anus.

Treatment. Treatment consists of administration of niclosamide (Jones 1979) or quinacrine, as recommended for other tapeworms. In a 5-year period (1973–1977), 43 patients infected with *D. caninum* in the United States were treated successfully with niclosamide (Yomesan) on a dosage schedule of 2 g in a single dose for adults, 1.5 g for children weighing more than 34 kg, and 1 g for children weighing 11 to 34 kg (Jones, 1979).

Prevention. Human beings become infected from accidental ingestion of infected insect hosts. Periodic administration of taeniafuges to dogs and cats reduces exposure. Insecticide dusting of dogs to kill fleas may be highly effective.

FAMILY DAVAINEIDAE

Raillietina celebensis (Janicki, 1902) Fuhrmann, 1920

Synonyms. *Davainea formosana* Akashi, 1916; *Davainea celebensis* Janicki, 1902.

Distribution and Biology. *Raillietina celebensis* is a common parasite of rats in the Orient, Indonesia, and Australia. Human infections, mostly in young children, have been reported from Japan, Taiwan, China, and Australia, and probably from the Philippines, Thailand, and elsewhere. Tang and Tang (1964) agreed with Baer and Sandars (1956) in concluding that *R. garrisoni* Tubangui, 1931, *R. sinensis* Hsu, 1935, *R. murium* Joyeux and Baer, 1936, *R. formosana* (Akashi, 1916), and *R. siriraji* Chandler and Pradatsundarasar, 1957 are synonyms of *R. celebensis*. The adult worm reaches lengths of about 40 cm and has aspinous suckers. The uterus contains up to 400 egg capsules each with 1 to 4 large eggs about 90 by 46 μm. The intermediate hosts, long suspected of being cockroaches, were shown by Tang and Tang (1964) to be species of ants. A species of *Cardiocondyle*, tentatively identified as *C. nuda* Mayer, which is widely distributed in China and the Indo-Malayan region, was experimentally infected, and the developing and mature cysticercoids were described. The cysticercoids reach maturity in the body cavity of the ant at about 38 days of infection. Gravid proglottids were carried into the nest by adult ants and fed to the larval ants; the young adults developing from the larval ants were found to harbor 1 to 6 cysticercoids measuring about 0.29 by 0.22 mm and bearing rostellar hooks 19 to 22 μm in length. Infection in children (or rats) is acquired by ingesting cysticercoids in infected ants. When mature, small white proglottids 3 mm or less in length and sluggishly active are seen on the surface of fresh stools. Usually the infection is asymptomatic or associated with a mild disorder such as diarrhea, and the symptoms, if any, disappear rapidly after treatment with niclosamide (Rougier et al., 1981).

Raillietina demerariensis (Daniels, 1895) Dollfus, 1939–1940

Synonyms. *Taenia demerariensis* Daniels, 1895; *Davainea madagascariensis* of Davila, 1922; *Raillietina quitensis* L.A. Leon, 1935; *R. luisaleoni* Dollfus, 1939; *R. equatoriensis* Dollfus, 1939; *R. leoni* Dollfus, 1939.

Distribution and Biology. This species was first recovered and described by Daniels (1895) from inland Guyana (Demerara), and has been reported in 5% of patients from the rural areas near Quito, Ecuador. Three cases have also been reported from Cuba. The adult worms measure up to 60 cm in length by 3 mm in width and comprise about 5000 proglottids. The scolex has ovoid suckers less than 0.5 mm in size, surrounded by a row of persistent hooklets, and a retractile rostellum, armed with two circlets of hooks. The free proglottids have the shape of a rice grain and contain 200 to 250 egg capsules, each about 200 μm in diameter. The ovoid eggs have conspicuously large hooklets.

Other Species of *Raillietina*

Raillietina asiatica (v. Linstow, 1901) Stiles and Hassall, 1926 is a worm of somewhat doubtful identity and relationship. A single strobila, having about 750 proglottids but without a scolex, was passed by a patient in Ashabad, northern Iran. Sixty to seventy eggs of unstated size were described for each egg capsule.

Raillietina garrisoni Tubangui, 1931, a parasite originally described from the rat in the Philippines, was reported from a 2-year-old patient on the Philippine Islands by Africa and Garcia (1935). In the period from 1959 to 1975, at least 18 cases of *Raillietina* infection were diagnosed in the Philippines (Jueco, 1975). All were in children, usually 3 years of age or younger, and none was symptomatic.

Raillietina siriraji Chandler and Pradatsundarasar, 1957 is a cestode infecting rats in Bangkok, Thailand. Cockroaches are the usual intermediate hosts. Pradatsundarasar (1960) reported 9 human infections, all in children, 2 of whom had ingested cockroaches. The first record of this worm in Thailand was that of Leuckart (1891) who considered it to be *Taenia madagascariensis* Davaine, 1870.

Until more careful, detailed study has been made of *R. asiatica* and *R. garrisoni*, and comparisons made with *R. siriraji* and other species, it will not be possible to know if they represent one or more valid species.

FAMILY HYMENOLEPIDIDAE

Hymenolepis nana (v. Siebold, 1852) Blanchard, 1891 (The dwarf tapeworm, producing hymenolepiasis nana)

Synonyms. *Taenia murina* Dújardin, 1845; *T. nana* v. Siebold, 1852; *Hymenolepis fraterna* Stiles, 1906 (murine type). Some authors (Yamaguti, 1959) follow Spassky in placing *H. nana* in the genus *Vampirolepis* Spassky, 1954.

Historical and Geographical Notes. This species was discovered by Bilharz in 1851 in the small intestine of a native boy in Cairo. Grassi and Rovelli (1887, 1892) first worked on the life cycle and demonstrated that no intermediate host was required.

Hymenolepis nana is practically cosmopolitan in its distribution, but is more common in warm than in cold climates. It is particularly common in southern Europe, U.S.S.R., and India and is the most frequently encountered human tapeworm in the southern United States and Latin America. It is the common tapeworm of man in Hawaii and in certain other Pacific island groups, but is uncommon on Guam.

Morphology, Biology, and Life Cycle. The entire worm (Fig. 31–6,*A*) is small, measuring up to 40 mm in length (often shorter, rarely longer) by a maximum of 1 mm in diameter. The size of the strobila is usually inversely proportional to the number of worms present in the host. The minute scolex (0.32 mm in diameter) is rhomboidal, has four suckers (80 μm in cross-section), and a short rostellum, armed with 20 to 30 spines in one ring, capable of invagination into the apex of the organ (Fig. 31–6,*B*). There is a long slender cervical region. Beginning with short, narrow, immature proglottids, those more distal in position become increasingly broader. At the distal end, however, the general contour of the proglottids is rounded. The maximum size of the proglottids (subdistal in position) is 0.15 to 0.3 mm in length and 0.8 to 1 mm in breadth. The eggs (Fig. 31–6,*C*) are set free by gradual disintegration of the distalmost proglottids. They are spherical or subspherical, hyaline, measure 30 to 47 μm in diameter, and contain an oncosphere that is enclosed in an inner envelope with two polar thickenings, from each of which arise 4 to 8 polar filaments. Within the oncosphere are three pairs of lancet-shaped hooklets.

Fully embryonated eggs, recently passed in the feces, on ingestion by mice hatch in the stomach or small intestine. The free oncospheres penetrate into the villi of the anterior part of the small intestine and metamorphose into cysticercoids (Fig. 31–7,*A*). These larvae then migrate out into the lumen of the small intestine, become attached by their scolices to the mucosa, and in the course of

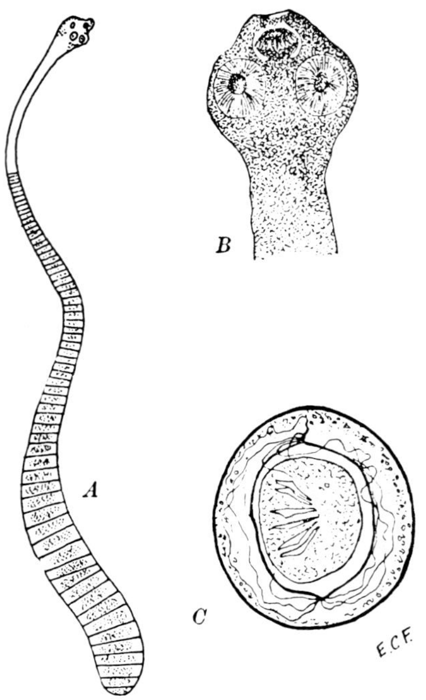

Fig. 31–6. *Hymenolepis nana. A,* Complete strobila. (× 10.) *B,* Scolex, greatly enlarged. *C,* Egg. (× 466.) (A by E.C. Faust. *B* adapted from R. Blanchard, and *C* adapted from C.E. Joyeux. In Brumpt, E. 1927. *Precis de Parasitologie.* Paris, Masson et Cie.)

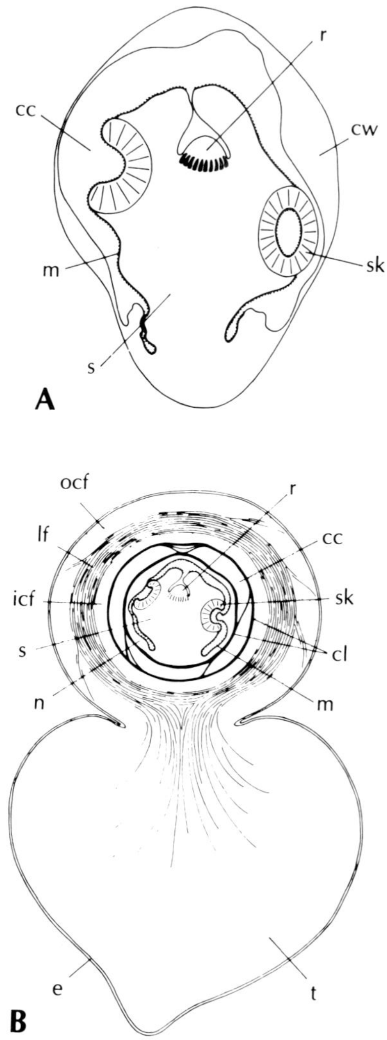

Fig. 31–7. *Hymenolepis nana,* cysticercoids. *A,* From intestinal villus of a mouse. *B,* From insect host. *cl,* Cyst lining. *cc,* Cyst cavity. *cw,* Cyst wall. *e,* Epidermal syncytium. *lf,* Longitudinal fibers. *icf,* Inner circular fibers. *m,* Microtriches. *n,* Neck. *ocf,* Outer circular fibers. *r,* Rostellum. *s,* Scolex. *sk,* Sucker. *t,* Tail. (From Caley, J. 1975. A comparative study of the two alternative larval forms of *Hymenolepis nana,* the dwarf tapeworm, with special reference to the process of encystment. Z. Parasitenkd., *47:*217–235.)

2 weeks or more develop into mature worms. Moreover, in experimental mice the oncospheres may penetrate into the intestinal lymph nodes, in which they may remain for a month before emerging into the intestinal lumen to become attached and to develop into mature worms. Experimental infection of various fleas and meal worms was first accomplished by Bacigalupo (1931), who demonstrated that *Ctenocephalides canis, Xenopsylla cheopis, Pulex irritans, Tenebrio molitor,* and *T. obscurus* in Argentina are capable of transmitting the murine strain. The cysticercoid developing in insect intermediate hosts differs morphologically from that developing from the egg in the final host's intestine (Fig. 31–7). Whereas the former has a three-layered protective outer membrane and tail-like appendage, the cysticercoid in the villus lacks these structures (Caley, 1975). Grassi first suggested that the human and murine strains were identical. Saecki (1920), Uchimura (1922), and Kiribayashi (1933) infected children with eggs from murine sources while Woodland (1924) infected mice with eggs from human sources. Shorb (1933) found that mouse and rat strains were physiologically as different from one another as they were from the human strain. It may be concluded that young children are at times susceptible to infection from murine sources, but are less endangered from these sources than from human infection; that adults are less susceptible than children; and that there is gradual development of resistance to infection and a definite resistance to superimposed infections (Bailey, 1951). Continued heavy infections in man are explained on the basis of internal autoinfection. Heyneman (1962) found that autoinfection can develop in 50 to 70% of nonimmune mice fed cysticercoids of *H. nana*. An exposure to 10 to 25 larvae may produce 1500 or more worms, but previous infection confers resistance to superinfection.

Epidemiology. This is the only human tapeworm typically utilizing no intermediate host; infection is direct from person to person. Infection is more common in children than in adults and is more often seen in family and institutional groups than in the population as a whole. It is diagnosed in about 1% of fecal examinations made in rural southern United States. In Argentina, Bacigalupo (1931) found 9% incidence in small children and 0.7 to 2.3% in adults. Payne et al. (1956) found high incidence in children and adults in high Andean villages of Peru. The variety in murine hosts *(H. nana fraterna)* is infective for man, but man is probably the usual source of human infection. Infection rates tend to be high in regions of high temperature and low rainfall (Biswas et al., 1978).

Pathogenesis, Pathology, and Symptomatology. *Hymenolepis nana,* even in large numbers, is commonly well tolerated. The mechanism by which symptoms are usually produced is an allergic reaction. Donckaster and Habibe (1958) studied the general symptom complex in 43 patients without other parasitoses: 86% complained of headache, dizziness, anorexia, inanition, pruritus of nose and anus, periodic diarrhea, and abdominal distress; 65% were restless and irritable; 2 (5%) had epileptiform convulsions; and one-third had eosinophilia in excess of 5%. A tapeworm identified as *H. nana* was found in a tumor removed from the chest wall of a 75-year-old Japanese woman (Mory et al., 1967) However, the identification of this worm is questionable (Bamrungphol et al., 1972).

Overwhelming internal autoinfection has been observed in immunosuppressed mice, with hyperinfection of the intestine by the adult worms and invasion of the mesenteric lymph nodes and liver by cysticercoids. Apparently internal autoinfection occurs in humans and may be a complication in patients given immunosuppressive treatments (Lucas et al., 1980).

Diagnosis. Diagnosis is based on identification of the characteristic eggs (Fig. 31–6,*C*).

Treatment. Treatment of choice until recently has been with *niclosamide* in a dosage of 2.0 g chewed thoroughly once daily for 5 to 7 days. Praziquantel in a single oral dose of 15 mg per kg body weight after breakfast is well tolerated and superior in efficacy to niclosamide (Espejo, 1977; Groll, 1980; Baranski et al., 1980; Schenone 1980; Rim et al., 1978). The direct effect of praziquantel on *H. nana,* consisting of vacuolization of the tegument of the neck region, has been reported by Becker et al. (1980).

Prevention. Since human infection is caused by ingestion of eggs in contaminated food or water, personal hygiene, particularly in large families and in institutional homes, is of primary importance. The optimum stage of viability of the eggs is immediately after their discharge from the bowel. The possibility of infection from rats and mice must not be overlooked (Biswas et al., 1978).

Hymenolepis diminuta
(Rudolphi, 1819) Blanchard, 1891
(The rat tapeworm, producing hymenolepiasis diminuta)

Synonym. *Taenia diminuta* Rudolphi, 1819.

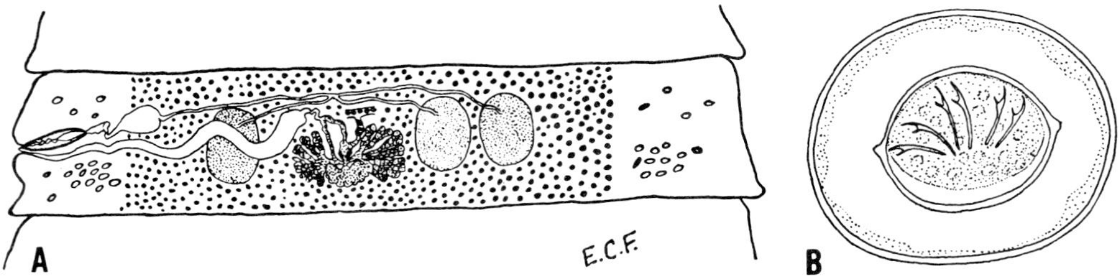

Fig. 31–8. *Hymenolepis diminuta. A,* Mature proglottid, greatly enlarged. *B,* Egg. (× 500.) (*A* adapted by E.C. Faust from F. Zschokke. In Stiles, C.W. 1925. Illustrated key to the cestode parasites of man. Hyg. Lab. Bull., no. 25. *B* by E.C. Faust. In Faust, E.C. 1939. *Human Helminthology,* 2nd ed. Philadelphia, Lea & Febiger.)

Geographical Distribution. *Hymenolepis diminuta* is a common parasite of the rat, mouse, *Praomys tullbergi jacksoni, Grammomys surdaster,* and *Apodemus sylvaticus.* Asada (1923) recovered it once from a dog. It has been reported rather infrequently from man. Its distribution is cosmopolitan in the reservoir hosts, and human cases have been reported from Belgium, Italy, U.S.S.R., Japan, China, Philippines, India, Africa, Argentina, Brazil, Venezuela, Colombia, Ecuador, Granada, Martinique, Nicaragua, Cuba, and the United States. Of the more than 200 authentic human cases, most have occurred in India, U.S.S.R., Japan, Italy, and certain areas of the southern United States (Tennessee, Georgia, and Texas), although one or more cases have been reported from each of the following additonal states: Alabama, Arkansas, California, Florida, Indiana, Louisiana, Minnesota, Nebraska, North Carolina, Oklahoma, and Virginia, as well as the District of Columbia.

Morphology, Biology, and Life Cycle. The mature worm measures 20 to 60 cm in length and increases gradually in width from 0.5 mm in the cervical region to about 3.5 mm near the distal end. The scolex is small, rounded, and has four deeply cupped suckers and an apical invagination cavity, into which the unarmed rostellum can be retracted. The terminal proglottids measure 0.75 mm in length by 3.5 mm in width. Each has three ovoid testes and the genital organs are arranged as shown in Figure 31–8. Gravid proglottids become detached from the strobila, disintegrate, and discharge their eggs which are passed in the feces. These eggs (Fig. 31–8) are subspherical, have a slightly yellowish, transparent outer membrane, measure 60 to 79 μm by 72 to 86 μm, and have an inner membrane around the oncosphere which has two polar thickenings but no polar filaments. Between the two membranes there is a colorless, gelatinous matrix. The six lanceolate hooklets of the oncosphere when at rest are arranged in a fan-shaped pattern. The eggs are somewhat resistant to desiccation, putrefaction, and chemicals but are sensitive to temperatures of 60°C or more.

Various arthropods serve as obligatory intermediate hosts; these are all coprozoic or scavenger in their habits during their larval or adult stage. They include the lepidopterans *Tinea granella, T. pellionella, Aglossa dimidiata* (both larva and imago), *Aphornia gularis,* and *Asopia farinalis* (both larva and adult); the earwig *Anisolabis annulipes;* the myriapods *Fontaria virginiensis* and *Julus* sp.; the larval stage of fleas *Nosopsyllus fasciatus, Orchopeas wickhami, Xenopsylla cheopis, Ctenopsyllus musculi, Ctenocephalides canis* and *Pulex irritans;* and the beetles *Akis spinosa, Scaurus striatus, Tenebrio molitor* (larva), *Dermestes peruvianus, Geotrupes stercosus, Tribolium castaneum, Ulosonia parvacornis, Aphodius distinctus,* and *Stegobium paniceum.* In the intestine of these arthropods the egg hatches and the oncosphere penetrates into the hemocoel, where it metamorphoses into a cysticercoid larva resembling that of *H. nana* (Fig. 31–7,*B*).

Among the tapeworms, *Hymenolepis diminuta,* more than any other species, has served as a living test subject in metabolic studies. The ratio of adult worms to the number of cysticercoid larvae ingested depends on the number of scolices that evaginate, become attached to the intestinal mucosa, and mature. Read (1951) obtained suggestive evidence that there is a factor, present in host tissue, which limits the size of the worms as their number increases.

Epidemiology. Accidental ingestion of the parasitized intermediate hosts causes infection of the definitive host. All age groups appear to be susceptible.

Pathogenesis, Pathology, and Symptomatology. As in the case of other adult tapeworms, *H. diminuta* is ordinarily well tolerated by the human host.

Diagnosis. Diagnosis is based on recovery of the characteristic eggs (Fig. 31–9). The worms are at times evacuated spontaneously.

Treatment. Treatment of choice is with *niclosamide* given as for *Diphyllobothrium latum* infection (see p. 498). In a 5-year-period (1973–1977), 43 patients infected with *H. diminuta* in the United States were treated with niclosamide (Yomesan). Of 19 patients examined at 3 months after treatment, 17 were cured and the 2 others were cured by a second treatment in the same dosage of 2 g in a single dose daily for 5 to 7 days for adults, less for children (Jones, 1979). In view of the efficacy against other cestodes, praziquantel is probably effective against *H. diminuta* as well. (*Vide supra* under treatment of *Diphyllobothrium* infection.)

Prevention. Man may become infected from accidentally swallowing the ectoparasites of the murine host or from ingesting insects infecting precooked cereals. Care to avoid such exposure is indicated. Anti-rat campaigns may have decreased the likelihood of human exposure.

FAMILY TAENIIDAE

Genus Taenia

Linnaeus, 1758

The genus *Taenia* according to Verster (1969) is defined as follows: Taeniidae of large size; rostellum usually present, armed with one or two crowns of hooks; testes numerous, confluent anterior to ovary; adults in the intestine of carnivorous mammals and man, rarely in birds; larval stage a monocephalic cysticercus, strobilocercus, or tetrathyridium, or a polycephalic coenurus (or other proliferative form). In most of the *Taenia* species the genital ducts pass from medulla to cortex *be-*

tween the longitudinal excretory vessels, but in *T. taeniaeformis* and certain species found in mustelids and viverids they pass from medulla to cortex ventral to the longitudinal excretory vessels. Besides the type species, *Taenia solium,* there are about 30 species of the genus.

Taenia solium
Linnaeus, 1758
(The pork tapeworm, producing taeniasis solium or pork tapeworm infection)

Synonyms. *Taenia cucurbitina* Pallas, 1766; *T. pellucida* Goeze, 1782; *T. vulgaris* Werner, 1782; *Taenia armata umana* Brera, 1808; *Taenia africana* von Linstow, 1900 *pro parte*.

Historical and Geographical Notes. Knowledge of this "armed" tapeworm of man dates back to the time of Hippocrates and possibly to Moses, although it was not specifically differentiated from the beef tapeworm, *T. saginata,* until the time of Goeze (1782). The Greeks (Aristophanes and Aristotle) mentioned the larval stage *(Cysticercus cellulosae)* in the tongue of swine and described it as resembling a hailstone. Gessnet (1558) and Rumler (1588) reported the larval stage in man. Küchenmeister (1855) and Leuckart (1856) first worked out the life cycle and demonstrated that the bladder worm in the tissues of the pig was the larval stage infective for man.

T. solium has a cosmopolitan distribution and is an important human parasite wherever raw or insufficiently cooked pork is eaten. It is most commonly found in the Slavic peoples (e.g., Czechs, Croats, and Serbs), and is frequently found in Mexico and other continental Latin American countries, as well as in North China, Manchuria, Pakistan, and India. In recent decades *T. solium* infection in the United States has become uncommon. Although human infection has never been reported from Iran, cysticerci were found in 0.02 to 0.03% of domesticated and 4% of wild swine slaughtered (Afshar, 1967).

Larval infections with *T. solium* in the human host, though rare in the United States (Jung et al., 1981; Keane, 1980), are relatively common in populations harboring the adult worm. During the first half of the nineteenth century, 2% of human autopsies in Berlin showed these cysticerci. With reduction of the adult infection in man and of the larvae in pigs, the incidence of human cysticercosis in Europe became less frequent, but in Africa, India, and China cysticercosis is common. In Mexico in 1944, Mazzotti reported that 2% of over 4000 stools examined contained *Taenia* eggs (for the most part those of *T. solium);* that 4.34% of 128,025 hogs slaughtered in a little over 2 years were "measled"; that 2.9% of 450 autopsies in Guadalajara revealed cysticercosis; and that 25% of 100 cerebral tumors that came to operation proved to be due to *Cysticercus cellulosae*. A three-year study of cysticercosis in a slaughterhouse in each of six Latin American countries (Costa Rica, El Salvador, Guatemala, Honduras, Nicaragua, and Panama) showed that 2.13% of all hogs slaughtered were infected, representing a heavy economic loss (Garrick, 1967). Examination of 420,000 pigs from 9 states in Brazil revealed *C. cellulosae* in 3.5% (Francis et al., 1979). In Nigeria, *Taenia* infection is prevalent (0.6 to 0.9%), but cysticercosis in man is rare and in pigs uncommon (0.4% or less) except in two areas where roughly 2% and 18% of the slaughtered pigs were found to be infected (Dada, 1980a,b.c).

Morphology, Biology, and Life Cycle. The adult *Taenia solium* lives in the small intestine with its scolex attached to the mucosa and its body folded back and forth in the lumen. Usually only one worm is present but numbers up to 25 have been recorded (Pineda et al., 1972). It attains a length of 2 to 7 meters. The scolex (Fig. 31–9,*A*) is roughly quadrate, with a diameter of about 1 mm, has four large, deeply cupped suckers (0.5 mm diameter) and a conspicuous, rounded rostellum, armed with a double row of large and small hooks, numbering 22 to 36 and measuring 140 to 200 μm and 100 to 150 μm, respectively, in length. The cervical region is short and only about one-half as thick as the scolex. Immature proglottids are broader than long, mature ones are nearly square, and gravid ones are longer than broad. The total number of distinct proglottids is under 1000. The mature proglottid (Fig. 31–9,*B*) differs from that of *T. saginata* in only a few essential details (Verster, 1967). The testes consist of 150 to 200 follicles, distributed throughout the dorsal plane. Numerous vasa efferentia join directly to form a common vas deferens, which extends as a convoluted tubule to the genital atrium on the lateral margin, becoming enlarged near its outer terminus to form the pars prostatica and the cirrus organ. A genital pore with a muscular sphincter guards the atrium. These openings alternate irregularly from side to side in adjacent proglottids. Directed mesad and then posteriad from the genital pore, the vaginal tube is on the lower side of the vas deferens, ending internally in the ootype. The ovary, situated in the posterior third of the proglottid, consists of two large symmetrical lobes and an accesory lobe on the side of the genital pore. A narrow, elliptical group of follicles, median in position behind the ovary, constitutes the vitellaria. The oviduct is joined by a common vitelline duct and the vagina before entering the ootype, which lies midway between the two large lobes of the ovary and the vitellaria, and which is surrounded by Mehlis' gland. The uterus arises from the anterior face of the ootype and at first proceeds toward the anterior border of the proglottid as a blind club-shaped organ. As it becomes filled with eggs, lateral extensions, which branch once or twice, are formed. The characteristic number of these main lateral branches is 7 to 13 (usually 9) for each side of the main uterine trunk (Fig. 29–4,*2*.) The terminal gravid proglottids become separated from the strobila and are passed in the feces or migrate from the bowel to the outside.

The eggs escape from the uterus through the ruptured wall at the anterior end after the ripe proglottid becomes free. These eggs are spherical or

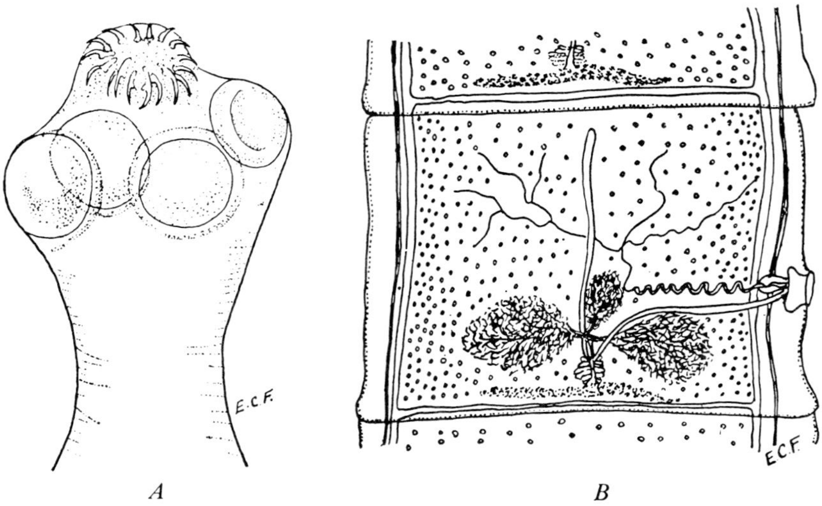

Fig. 31–9. *Taenia solium*. *A*, Scolex. (× 40.) *B*, Mature proglottid. (*A* adapted from photograph by L. Szidat, 1934. *B* adapted from R. Leuckart, *Parasiten des Menschen*.)

subspherical in shape, measure 31 to 43 μm in diameter, are pale buff to walnut-brown in color, and cannot be distinguished from those of *T. saginata* (Fig. 31–16). The shell is thick-walled, consisting of many truncated prisms cemented together, and is originally provided with a thin, hyaline embryonic membrane. Within the shell there is a fully developed oncosphere that has 3 pairs of hooklets.

On escape from the ruptured uterus of the gravid proglottid and after deposition on the soil, the eggs may remain viable for many weeks. On ingestion by hogs or man, rarely other mammals, and passage into the duodenum or jejunum, the shells disintegrate. Hatching of *T. solium* eggs occurs only when the eggs are first affected by gastric juice and then come in contact with the intestinal juice (Gönnert et al., 1967). The emergent oncospheres, by means of their hooklets, and possibly aided by lytic secretions, penetrate through the intestinal wall into mesenteric venules and are carried throughout the body. They typically filter out between the muscles, where, in the course of 60 to 70 days, they metamorphose into cysticerci, which have a subspherical to ovoid, milky-white bladder with a head invaginated into the bladder (Fig. 31–10). Early in its development it induces the host to form a fibrous capsule around it (Fig. 31–17). This bladder worm *(Cysticercus cellulosae)* usually measures about 5 mm long by 8 to 10 mm in breadth, except in the cavities of the brain and the subarachnoid and subdural spaces where it may grow to a volume of 60 ml or more (Berman et al., 1981). Man readily becomes infected with the adult tapeworm when cysticercus larvae are ingested in raw or inadequately cooked pork. In the stomach the larva is digested out of the pork flesh; in the upper levels of the small intestine the head evaginates (Fig. 31–11), becomes attached to the wall of the small intestine, and in 5 to 12 weeks develops into a mature worm. More than one worm may be harbored, although only one is usually present. The worms may live 25 years or more. Malformed proglottids or even malformed heads are at times developed. *T. solium* and *T. saginata* may occur in the same patient.

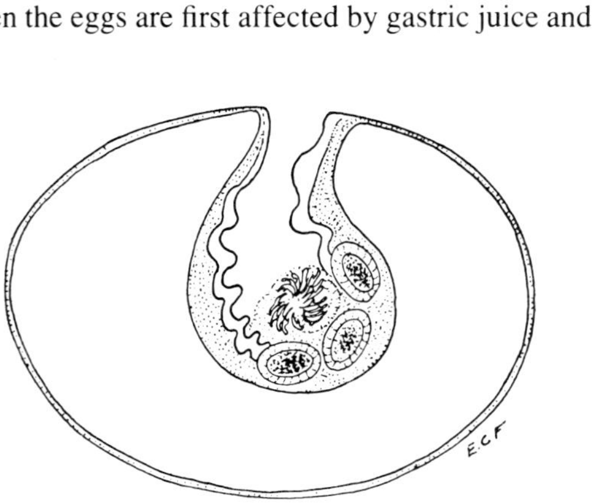

Fig. 31–10. *Taenia solium*. Diagrammatic representation of a cysticercus *(Cysticercus cellulosae)* with scolex at distal end of scolex canal (neck) within the bladder. Greatly magnified. (Adapted from original of E.C. Faust.)

Epidemiology. Man is the usual definitive host, but natural infection with the cysticercus stage has

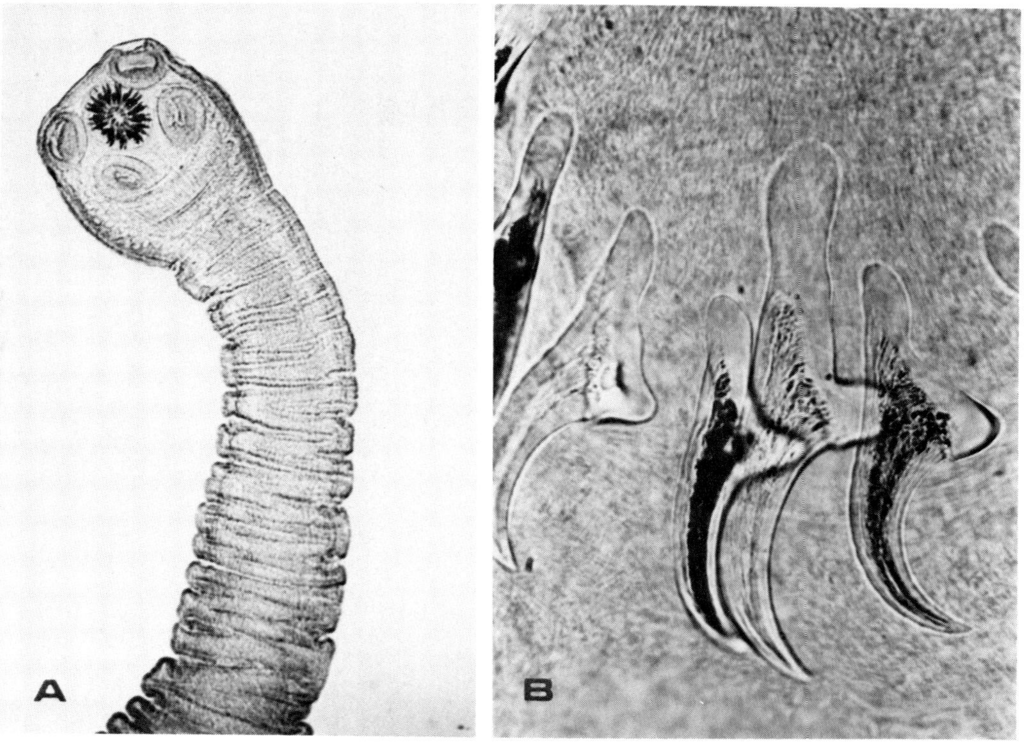

Fig. 31–11. *Taenia solium*, evaginated cysticercus from pork. *A*, Scolex and neck; the latter showing pseudosegmentation. (× ca. 25.) *B*, Rostellar hooks. (× ca. 400.) (By P.C. Beaver. Specimen courtesy of Dr. Hugo Schenone.)

been found in the lar gibbon *(Hylobates lar lar)* in Thailand, and experimental infection with the adult stage was produced in the same host (Cadigan et al., 1967). In the golden hamster after whole-body irradiation or chemical immunosuppression, the cysticerci develop to sexually mature worms but do not lay eggs (Verster, 1974). Inadequately heated or frozen pork is the exclusive source for human infection with the adult worm, although man and primates (less frequently sheep, dogs, and cats) harbor the cysticercus stage. It has also been reported in a Cape fur seal and one other marine mammal (De Graff et al., 1980). Human infections with *C. cellulosae* number into the tens of thousands, due to (1) ingestion of the eggs in contaminated food or water *(heteroinfection)*, (2) self-contamination by people who have the adult worm in their intestine, or *possibly* (3) internal autoinfection, whereby the eggs are carried by reversed peristalsis back to the duodenum or stomach, hatch, and after migration into the somatic or visceral tissues produce cysticercosis. In New Guinea in 1972 and 1973 an outbreak of severe burns among adults and older children called attention to a high prevalence of *Taenia solium* infection in humans and pigs. The burns occurred when people with cerebral cysticercosis rolled into open fires during epileptic seizures (Gajdusek, 1978). In some instances the seizures resulted from heavy primary invasion of the brain, not from inflammation caused by dead cysticerci in old infections. It is believed that *T. solium* was only recently introduced into the area. It is probably endemic in Irian Jaya too (Tjahjadi et al., 1978).

Pathogenesis, Pathology, and Symptomatology. The adult *Taenia solium* may cause slight irritation at the site of its attachment to the mucosa, or may infrequently produce intestinal obstruction. The cysticercus stage, as a somatic infection in man, induces fibrous encapsulation. As the larva begins to die, a pronounced cellular reaction occurs. When this is located in vital organs or tissues, acute and sometimes fatal sequelae develop.

The Adult Worm. Ordinarily the parasite produces no serious damage or grave manifestations, but at times it causes vague abdominal discomfort, hunger pains, chronic indigestion, persistent diarrhea, or alternating diarrhea and constipation. In some patients, its presence in the intestine may be responsible for loss of appetite. A moderate eosinophilia (up to 13%) may be present. It is possible that there may be symptomatic migration of proglottids from the anus *(vide infra* under Diagnosis).

The Cysticercus (Producing Cysticercosis Cellulosae). The larval stage of *T. solium* has been found in practically every organ and tissue of the body. The symptoms produced vary according to the number of cysticerci present and their location. The most frequent location is in the subcutaneous and intermuscular tissues (Fig. 31–12); next, in the eye; then in the brain, where they may develop in ventricles, the brain tissue (Fig. 31–13), or in superficial cyst capsules. They also occur in the heart, liver, lungs, and abdominal cavity. They rarely develop in the spinal cord. Except in the eye, the ventricles of the brain, and the subarachnoid space of the cranium and rachis, the cysticercus is surrounded by a fibrous capsule that is usually excised by the surgeon along with the larva, though, unlike the hydatid, the cysticercus is easily shelled out of its fibrous capsule of host tissue. The presence of the growing larva provokes a typical

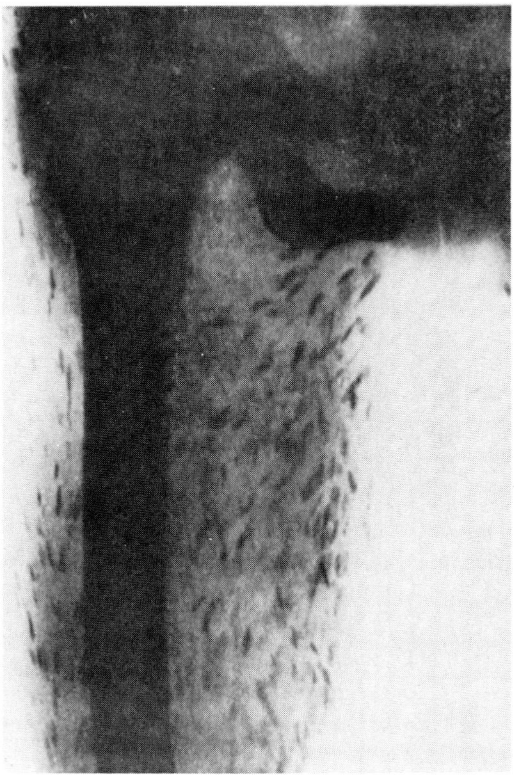

Fig. 31–12. *Taenia solium.* Numerous calcified cysticerci in the thigh tissues of man. (From Faust, E.C., Beaver, P.C., and Jung, R.C. 1975. *Animal Agents and Vectors of Human Disease,* 4th ed. Philadelphia, Lea & Febiger. Courtesy of Dr. R.R. Evans.)

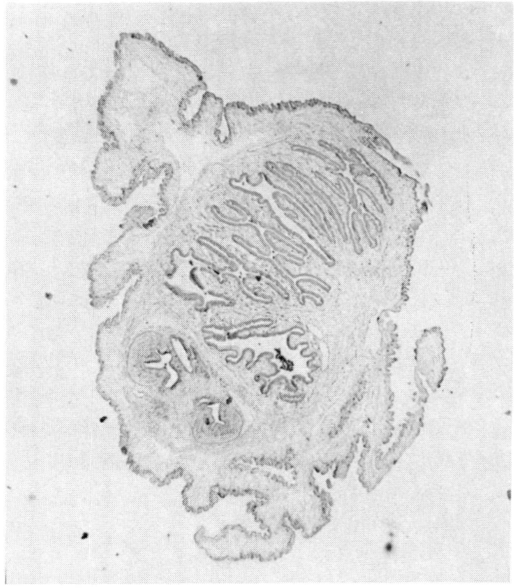

Fig. 31–13. *Taenia solium,* cysticercus. *A,* Section of larva removed from cortex of brain. (× 25.) (By P.C. Beaver.)

sequence of local cellular reactions, including infiltration of neutrophils and eosinophils, lymphocytes, plasma cells, and at times giant cells, followed by fibrosis and necrosis of the capsule, with eventual caseation or calcification of the larva.

Following lodgment of the precysticercus larva in the brain, little disturbance may be noted during its life, but as soon as the organism begins to lose vitality and a part of it dies, tissue reactions take place around it and a great variety of brain symptoms may develop. This may produce a rapidly fatal outcome. In other cases it appears that central nervous signs and symptoms result from the invasion of the brain by the living larvae (Gajdusek, 1978). Dixon and Smithers (1934) stated that "in every case of epilepsy occurring in a patient with no family history of fits and no previous history of fits in childhood the probability of cysticercosis should be entertained."

Obrador (1948) distinguished cysticercus-produced cerebral lesions anatomically as follows: cortical, single or multiple vesicular, basilar (particularly in the cisterna magna), cerebellopontine, anterior and middle fossae, mixed general, and diffuse. The studies of Ramirez and Verástegui in 1953, and others earlier, indicated the fairly common occurrence of epileptiform seizures in patients with cysticercosis. Nevertheless, cerebral cysticercosis is not necessarily accompanied by epilepsy, but may be responsible for disordered behavior, transient paresis, intermittent obstructive hydrocephalus, dysequilibrium, meningoencephalitis, and failing vision.

Wilber et al. (1980) have described the cerebrospinal fluid findings in five patients with cerebral cysticercosis. Only two of the patients showed CSF eosinophilia. There is usually pleocytosis, the most common cell being the lymphocyte. The authors considered the most significant cells to be atypical mononuclear lymphoid cells and the most distinctive features of the CSF to be atypical, reactive lymphoid cells in conjunction with a *mixed* cellular population.

Ocular cysticercosis has been reported occasionally from India, eastern Europe, and Latin America (Guillory and Zinn, 1980). Localization of the cysticercus in the eye may be subchoroidal, subretinal, subhyaloidal, intravitreous, or in the anterior chamber. Intraocular infection may be bilateral and multifocal (Topilow et al., 1981). Cysticerci also lodge in the eyelid and under the conjunctiva (de Buen 1970; Jampol et al., 1973; Perry and Font 1978). In the vitreous and anterior chamber, the living cysticerci are unencapsulated and may actively change shape, with periodic eversion and introversion of the scolex. Although symptoms may consist solely of discomfort from shadows cast by the larva in front of the retina, unless it is removed it will eventually produce uveitis, iritis, retinitis, choroidal atrophy, palpebral conjunctivitis, or cyst formation, depending upon its location. Upon the death of the parasite, an intense inflammatory reaction develops that may gravely endanger the eye.

Cysticercus racemosus. Occurring predominantly at the base of the brain, and at times extending to the spinal cord, is a proliferating form of cysticercus referred to as *Cysticercus racemosus* or *racemose cysticercus*. According to Slais (1967, 1970), Virchow in 1860 first drew attention to this peculiar form of cerebral cysticercus, although it was Zenker who in 1882 gave it the designation *Cysticercus racemosus*. It consists of several bladders of various sizes forming a grape-like mass with channels or stalks connecting the different parts. It may invade the brain tissue from other locations, but for the most part is located in the fourth ventricle and subarachnoid space (Fig. 31–14). A report from Peru described 100 cases of cysticercosis of the subarachnoid space of the spine, 92 of which had cervical involvement, the remainder involving the lower segments (Rocca and Neira, 1980). The 92 cases of cervical location were among 350 cases of intracranial infection. Cysticerci in the subarachnoid space of the rachis, as in that of the cranium, tend to proliferate into the racemose type. In one patient 62 free cysts were removed from the lumbosacral segment by surgical flushing of the subarachnoid space. Apparently in some instances the cysticercus may begin its development in the brain tissue, and the scolex later degenerates while its bladder wall continues to grow and to proliferate into the subarachnoid space to form a racemose pattern (Berman et al., 1981). Occasionally the bladder wall of an ordinary viable cysticercus when preserved in a brain specimen may become wrinkled and folded in ways to give the impression of proliferations with arborization when seen in sections. However, in such cases a well-developed typical scolex and scolex canal can be found, dem-

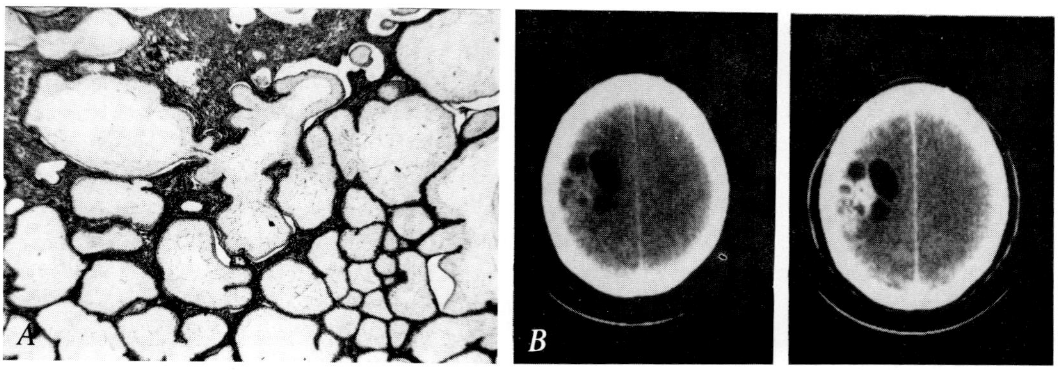

Fig. 31–14. *Cysticercus racemosus. A,* Section of proliferating cysticercus in brain of a Polish refugee who died in Prague; a case reported earlier (Faust, 1949) as *Sparganum proliferum.* (× 20.) (By P.C. Beaver. Section courtesy of Dr.Herman Šikl.) *B,* Tomograms (CAT) of proliferating cysticercus in human brain, showing the lesion prior to (left) and following contrast injection, demonstrating both solid and cystic components. (From Jung, R.C., et al. 1981. Racemose cysticercus in human brain. A case report. Am. J. Trop. Med. Hyg., *30*:620–624.)

onstrating that it is not a proliferative type of cysticercus (Fig. 31–13).

Although *C. racemosus* is generally presumed to be an aberrant cysticercus of *T. solium,* there is a possibility that what may be called a racemose cysticercus, in some instances at least, is a coenurus. The racemose pattern has been observed in larval cestodes which, by the presence of characteristic scolices, were known to be the coenurus type of larvae. In the racemose cysticercus the scolex, if found at all, is dead and recognized only by the presence of rostellar hooks (Berman et al., 1981). It is possible that eventually histological distinctions between the cysticercus and coenurus may be found (Slais 1967, 1970). Meanwhile a racemose larval cestode in which no scolex can be demonstrated can be referred to as *Cysticercus racemosus,* though it may be a cysticercus other than that of *T. solium* or a sterile coenurus (Jung et al., 1981).

Diagnosis. ***The Adult Worm.*** The presence of characteristic *Taenia* eggs in the stool does not permit differential diagnosis from *T. saginata*. Recovering the gravid proglottids and counting the main lateral arms of the uterus (7 to 13 on each side, as distinguished from 15 to 20 for *T. saginata*) constitute the only known specific pretreatment diagnosis. On evacuation of the scolex following medication, the armed rostellum differentiates this worm from *T. saginata*. When stained and cleared, the mature proglottid can be distinguished from that of *T. saginata* by the presence of an accessory lobe of the ovary and the absence of a vaginal sphincter (Fig. 31–9). In the gravid proglottid little of the ovary remains and the vaginal sphincter may not be evident.

According to Verster (1967), the gravid proglottids of *T. saginata* may leave the host spontaneously; for that reason infected people become aware of the presence of the parasite, whereas *T. solium* does not make its presence known in this way and is detected on examination of the stools only. However, the proglottids of *T. solium* are usually actively motile in the freshly passed stool and would appear to be capable of migrating from the colon to the outside of the body (Yoshino, 1934). Therefore, the question should be more thoroughly investigated, and it cannot be presumed that a proglottid forcing the anal sphincter is in all instances that of *T. saginata,* as has been stated (Sterba and Dykova, 1979). In unpublished notes, M.D. Little has recorded that during a 20-day period from diagnosis to treatment a house painter in Colombia passed 102 *T. solium* proglottids in or on the feces. The number ranged from 0 to 14 per day. On 3 days no stool was passed. On the second and fourteenth days motile proglottids (2 and 1, respectively) were collected by the patient from his thighs, and he believed that others had likewise been on the buttocks and thighs. Perianal Scotch tape preparations were negative for *Taenia* eggs on 3 days, positive once. The original generic diagnosis had been made by demonstrating eggs in the feces.

The Cysticercus. Except in endemic foci, diagnosis usually awaits excision of the larva and its microscopic examination. Its invaginated scolex, with its four suckers and anterior circle of hooks, is exactly the same as that of the adult worm (Fig. 31–11). The larvae may be present singly, but more frequently occur in multiples, up to several hundred (Fig. 31–12). Cysticercosis in the subcutaneous tissues or somatic musculature strongly suggests brain involvement. After calcification of the larvae, roentgenograms may be helpful in locating the lesions, especially those in the muscles and central nervous system (Figs. 31–12, 31–14). CT (computerized tomographic) scans of the brain accurately delineate the lesions and may eventually permit preoperative diagnosis. At times the CT findings in cerebral cysticercosis are diagnostic, the criteria being numerous small cystic structures, all about the same size and shape. The presence of typical calcification is supportive of the diagnosis (Priestly and Wiles, 1979). In racemose cysticercosis cerebri, the CT scan cannot distinguish the lesion caused by the parasite from other partially solid partially cystic lesions due to tumors. Calcification is common only in the scolex region of the larva, but at times the cyst wall may appear like an enveloping egg shell. Cysticercosis of the brain must be differentiated from hydatid cyst *(vide infra)* or other types of cyst, tumors of the brain, and hereditary epilepsy. A history of intestinal taeniasis solium is helpful in arriving at a diagnosis. Serologic tests may be useful in some cases, but cross-reactions between cysticercosis and hydatidosis are frequent (Schantz et al., 1980).

The diagnosis of ocular cysticercosis is suspected when a whitish or grayish mass is seen in the anterior or vitreous chamber with impaired vision. The characteristic movements and scolex of the worm are pathognomonic (de Buen 1970).

Subretinal cysticerci are also readily recognizable by the presence of a single scolex in a translucent cyst with undulating motility (Bartholomew, 1975). In those instances in which inflammatory reaction produces cloudiness of the ocular humors, interfering with visualization of the parasite, ultrasonography is of great value in identifying the offending parasite (Guillory and Zinn, 1980).

Treatment. ***The Adult Worm.*** Treatment is with praziquantel (Espejo 1977; Rim et al., 1979, Groll, 1980), or niclosamide, given as for infection with *Diphyllobothrium latum* (see p. 498). Mebendazole has been reported as effective in a dosage of 300 mg twice daily for 3 days (Peña-Chavarria et al., 1977). Since cysticercosis possibly may result from internal autoinfection, it is imperative that patients harboring the adult worm be *specifically diagnosed early* and the worm removed as soon as possible.

The Cysticercus. Excision is indicated wherever possible and as soon as possible. In the event of ocular cysticercosis, operation for removal of the cyst rather than enucleation of the eye is indicated. To forestall loss of the organ, the cysticercus should be removed while it is living (Lech, 1949).

Present therapeutic trials have shown that praziquantel in a dosage of 25 mg/kg body weight daily for 3 or 4 days is effective in destroying cysticerci, including those in the brain. The course must be repeated in some cases (Rim et al., 1980; Botero and Castaño, 1981). Side effects due directly to the drug, usually mild and reversible, include headache, dizziness, abdominal discomfort, nausea, and occasional skin rashes. In cerebral cysticercosis, side effects may be more severe and may be minimized by administration of a corticosteroid during the course of treatment with praziquantel.

Dramatic improvement has been reported in a case of cutaneous cysticercosis after treatment with metrifonate in a dosage of 635 mg per day orally for 6 days (Tschen et al., 1981). Metrifonate has been used successfully in the treatment of schistosomiasis.

Satisfactory results have also been reported in four patients with the use of specific antibodies labeled with radioactive iodine administered intravenously (Skromme-Kadlubik and Gonzalez, 1977).

Prognosis. The prognosis is good in infections with the adult worm, but grave in cases where racemose cysticercosis has developed. However, surgical removal of one or more encapsulated cysticerci from the brain has resulted in complete and permanent recovery from symptoms in some patients (Jung et al., 1981; Berman et al., 1981).

Prevention. Prevention involves (1) personal hygiene and (2) general sanitary measures. People must understand that eating raw pork may expose them to pork tapeworm (as well as trichinosis, p. 231), and that fecal contaminations make it possible to acquire cysticercosis either from themselves or from others infected with the adult worms. They should avoid eating raw vegetables grown on soil irrigated by sewage water. Persons known to be infected should be given specific treatment. Rigid sanitary laws regarding the disposal of human excreta in endemic foci and rigid inspection of pork in all slaughterhouses with condemnation of "measly pork" would eventually eliminate this infection. In São Paulo, Brazil, examination of the brains of all hogs slaughtered increased by 16.3% the efficiency of the routine inspection of cysticercosis (Mucciolo et al., 1968). Pickled or salted pork is not necessarily safe. Thorough cooking at 65°C is lethal for the cysticerci. Biagi et al. (1965) recommended refrigeration at −20°C for at least 12 hours as an effective method of destroying the larvae. Verster et al. (1976) found that carcasses infected with cysticerci can possibly be made fit for marketing by exposure to gamma radiation at levels that do not affect the quality of the meat.

Taenia saginata
Goeze, 1782
(The beef tapeworm, producing taeniasis saginata or beef tapeworm infection)

Synonyms. *Taenia solium* Linnaeus, 1767, *pro parte; Taenia thermis* Brera, 1802; *Taenia mediocanellata* Küchenmeister, 1852; *Taeniarhynchus mediocanellata* Weinland, 1858; *Taenia confusa* Ward, 1896; *Taenia africana* von Linstow, 1900, *pro parte; Taenia hominis* von Linstow, 1904; *Taenia tonkinensis* Railliet and Henry, 1905; *Taenia philippina* Garrison, 1907; *Taenia bremneri* Stephens, 1908; *Taenia cylindrica* Leon, 1972.

Historical and Geographical Notes. The common "unarmed" tapeworm of man was undoubtedly observed in ancient times but was not specifically differentiated from *T. solium* until the time of Goeze (1782). The larval stage *(Cysticercus bovis)* was apparently first observed in cattle by Wepfer, in 1675. In 1861, Leuckart first demonstrated experimentally that cattle serve as the intermediate host of this worm (Leuckart, 1863).

T. saginata has a cosmopolitan distribution and in most countries its prevalence is considerably greater than that of *T. solium*. It is particularly prevalent in Mohammedan countries and in Ethiopia, where *T. solium* is practically unknown, and is not infrequent in the United States. It occurs commonly in Mexico and is relatively prevalent in South America and East and West Africa. *T. saginata* is endemic throughout Germany, Austria, Belgium, Luxembourg, Holland, Great Britain, Switzerland, Hungary, Yugoslavia (unusually high prevalence), and Italy. In England and Wales there has been a marked reduction in cys-

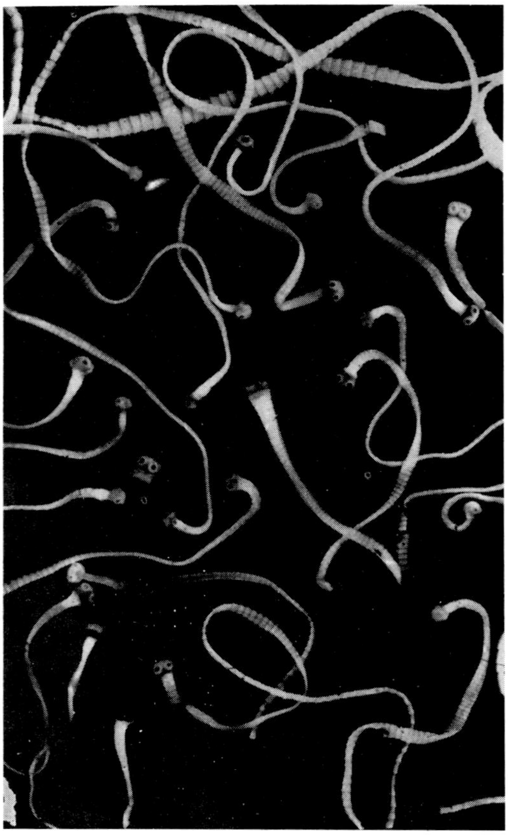

Fig. 31–15. *Taenia saginata.* Collection of 28 strobilae showing characteristic scolex, neck, and immature proglottids, recovered from a patient in Beirut, Lebanon. About natural size. (From Faust, E.C., Beaver, P.C., and Jung, R.C. 1975. *Animal Agents and Vectors of Human Disease,* 4th ed. Philadelphia, Lea & Febiger. Courtesy of Dr. D.A. Berberian.)

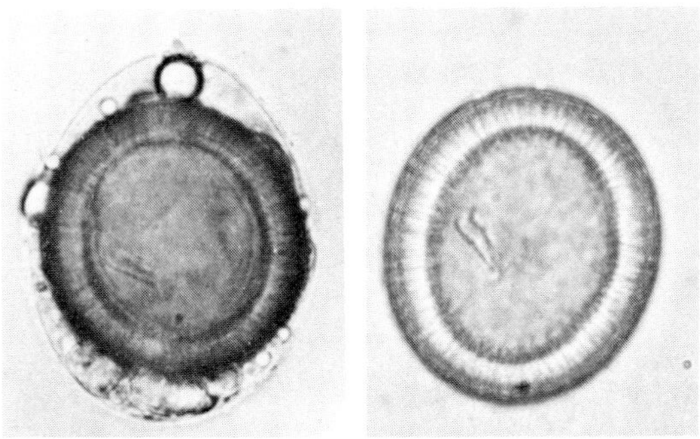

Fig. 31–16. *Taenia saginata. A,* Egg freshly shed from motile proglottid, with outer egg membrane intact. *B,* Egg after having shed its outer membrane. One of the three pairs of booklets of the hexacanth larva is evident. (× ca. 1000.) (By P.C. Beaver.)

ticercosis in slaughtered cattle, from 2 per 1000 in 1969 to 5 per 10,000 for the period since 1974 (Blamire et al., 1980).

Morphology, Biology, and Life Cycle. *Taenia saginata* is the principal representative of the subgenus *Taeniarhynchus,* which is characterized by having a scolex without a rostellum bearing hooks. The adult worms (Fig. 29–1), living with their heads embedded in the mucosa of the small bowel, are considerably longer than *T. solium* because of the greater number of proglottids developed and the greater length of the gravid proglottids. Under favorable conditions they may attain a length that can be expanded to 25 meters or more, but usually they measure 5 meters or less and have from 1000 to 2000 proglottids at one time. The head (Figs. 29–2, 31–15) is quadrate in cross section, with a maximum diameter of 1.5 to 2 mm and is provided with four hemispherical suckers of 0.7 to 0.8 mm in diameter, situated at the four angles of the head and serving as the sole organs of attachment. The apical region is characteristically somewhat concave and superficially pigmented. The neck when relaxed or extended is not more than one-half as broad as the head and several times its length; when contracted it may be as wide as the scolex. The neck is followed by the usual series of immature, mature, and gravid proglottids. Anomalous forms with six suckers on the scolex, triradiate symmetry (Thatcher, 1972), or with bilateral genital pores as in *Dipylidium caninum* have been reported in Turkey (Merdivenci, 1964) and in Korea (Cho et al., 1967). The mature proglottids are somewhat broader than long (maximum breadth, 12 mm), and the gravid ones considerably narrower and longer (5 to 7 mm broad, 20 mm long, but capable of extension to 4 cm). The genital organs in the mature proglottids are in general like those of *T. solium* but differ conspicuously in having about twice as many testes (300 to 400) and a well-developed vaginal sphincter, and in lacking the accessory ovarian lobe. In the gravid proglottids (Fig. 29–4,*1*) found in the distal fifth of the worm, the testes and ovary become spent and the uterus has 15 to 20 main lateral branches (av. 18), characteristics that are specifically diagnostic. The eggs (31 to 43 μm in diameter) cannot be distinguished from those of *T. solium,* other species of *Taenia,* or from those of *Multiceps* and *Echinococcus*. They are developed singly in thin, hyaline capsules which are shed soon after leaving the proglottid (Fig. 31–16).

When mature eggs are ingested by cattle and reach the level of the duodenum, the oncospheres hatch. They penetrate the intestinal wall, reach the mesenteric venules or lymphatics, and are carried through the circulation. They are filtered out in

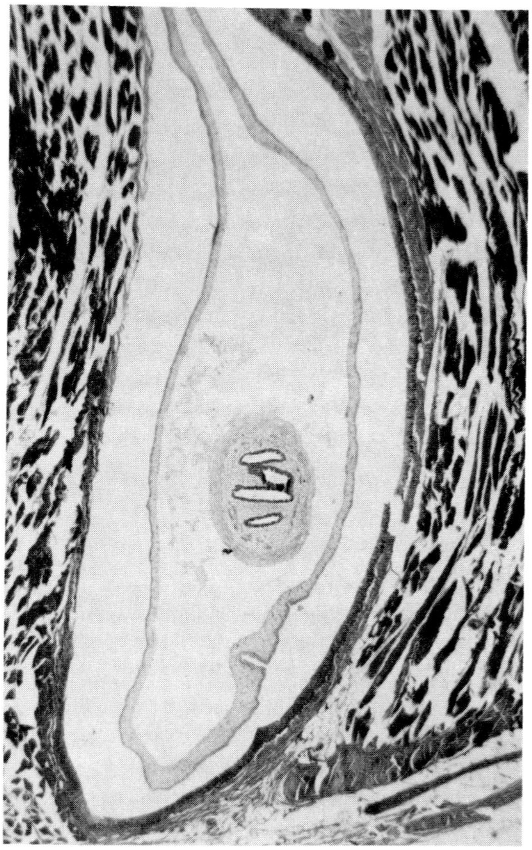

Fig. 31–17. *Taenia saginata*, cysticercus *(Cysticercus bovis)* in connective tissue capsule in muscle of a cow. The bladder of the cysticercus is free from the capsule and the inverted scolex canal ("neck") is seen in transverse section in the center of the bladder. (× 25.) (By P.C. Beaver.)

striated muscle, particularly the pterygoid muscles, those of the tenderloin region, and the myocardium, where, in 60 to 75 days, they develop into the bladder worm *(Cysticercus bovis)*. This worm, like the adult worm, is characterized by having an unarmed scolex. The mature cysticerci (Fig. 31–17) are ovoid in shape, milky-white, opalescent, and measure 7.5 to 10 mm in breadth by 4 to 6 mm in length. In addition to cattle, the buffalo, giraffe, and llama are recorded as natural hosts, and kids and sheep as experimental hosts of the cysticercus stage. Since *T. saginata* occurs in man in the reindeer-breeding regions of northern USSR where cattle are absent, reindeer may be intermediate hosts of the parasite (Krotov, 1961). Cases of human infection with *Cysticercus bovis* are probably all inaccurately diagnosed, except possibly the case reported by Fontan (1919), in which the larvae were taken from the mammary gland of a patient who harbored the adult worm, the case reported by De Rivas (1937), and the one from a lymph node of the mesoappendix (Niño, 1950). On ingesting infected meat, man, the sole definitive host, becomes infected, with a prepatent period of 10 to 12 weeks.

While infection with *T. saginata* usually consists of a single worm, multiple worms, occasionally as many as a dozen or more (Fig. 31–15), may develop simultaneously.

Epidemiology. Human infection is acquired from eating raw beef containing the viable cysticercus larvae. Cattle become infected from grazing on ground polluted by sewage or human feces containing the eggs of the parasite. Under suitable conditions the eggs may remain viable on exposed pastures for several weeks. Of 40 cows exposed to sewage sludge on a pasture in Virginia, 18 were found at slaughter to have cysticerci in the tissues, more in cardiac muscle than elsewhere (Hammerberg et al., 1978). Coprophagous insects and other animals disperse the eggs widely where they are picked up by cattle (Goddeeris, 1980). Cysticercosis in federal-inspected cattle in the United States increased from 0.14% in 1912 to 0.37% in 1930, since which time it has not appreciably decreased.

Pathogenesis, Pathology, and Symptomatology. In spite of its large size, *Taenia saginata* is ordinarily well tolerated. At times it is responsible for intestinal disturbances. In folklore it is imputed to cause hunger and weight loss in its host by utilizing host nutrients, but this is not generally so. At times the worm may produce acute intestinal obstruction; at other times individual proglottids may become lodged in the appendiceal lumen and initiate acute appendicitis. In one case a *Taenia* proglottid, probably *T. saginata,* was found in the uterus, causing postmenopausal bleeding (Schacher and Hajj, 1970). Toward the end of the prepatent period, diarrhea and hunger pains frequently develop and loss of weight may occur. The most common symptoms are the discomfort and embarrassment occasioned by the crawling of proglottids from the anus. Proglottids of *T. saginata* have a strong tendency to crawl from the anus during the day when its host is active, contrary to the habits of *Dipylidium caninum* and *Enterobius vermicularis* which emerge at night or at periods of inactivity during the day.

Diagnosis. Identification of the eggs can be used to indicate that the patient is infected with a species of *Taenia,* but it is impossible to diagnose the species on this basis. Furthermore, some patients have no eggs in the stool, but all patients are from time to time, frequently daily, passing gravid proglottids. Examining these proglottids (Fig. 29–4,*1* and *2*), pressed between two slides, and using a hand lens to determine the number of main lateral arms of the uterus constitutes the only known pretreatment diagnosis that is specific (15 to 20, usually 18 main lateral branches on each side of the main uterine stem in *T. saginata;* 13 or less, usually 9, in *T. solium).* At times the proglottids passed in the feces have shed most of their eggs, and the distinctive character of the uterus has been lost. Transparent adhesive (Scotch) tape pressed against perianal folds or the seat of underpants and examined as for eggs of *Enterobius* reveals eggs of *Taenia* deposited in or around the anal canal by migrating proglottids.

Treatment. *Niclosamide* or *quinacrine hydrochloride* should be given as for infection with *Diphyllobothrium latum.* These drugs may be supplanted by praziquantel, which is well tolerated and highly effective in a single dose of 10 mg per kg of body weight after breakfast (Espejo, 1977; Groll, 1980). Japanese workers (Yokogawa et al., 1962) have employed *bithionol,* administered per os in the amount of 40 to 60 mg per kg body weight in successfully treating *T. saginata* infection. Mebendazole has also been reported to be effective against taeniasis in a dosage of 300 mg twice daily for 3 days (Peña Chavarría et al., 1977).

Prognosis. With specific therapy the prognosis is good. This infection is less dangerous than pork tapeworm infection, since cysticercosis need not be anticipated.

Prevention. All beef to be eaten by man should be inspected for cysticerci. Inspection procedures for bovine cysticercosis do not always detect infection (Dewhirst et al., 1967). Thorough cooking of beef insures complete protection. The critical thermal point for the cysticerci is 56°C. The practice of prescribing raw or rare beef or beef juice for persons suffering from anemia or tuberculosis, or for pregnant women, has been responsible for *T. saginata* infection in some patients. Cattle should not be allowed to feed or graze on ground polluted by human feces or sewage (Hammerberg et al., 1978). However, cattle grazed on pastures irrigated with effluent from a lagooning process of sewage treatment apparently do not acquire infection (Arundel and Adolph, 1980). Jepsen and Roth (1950) reported that eggs remained viable 16 days in untreated sewage, 71 days in liquid manure, 33 days in water, and 159 days on grass in open air. Froyd (1962) demonstrated viability of *T. saginata* eggs 65 to 168 days after placing them in the refrigerator at 4 to 5°C in a 1:10,000 solution of thimerosal (Merthiolate). Workers at cattle feedlots should be periodically examined for *Taenia* infection. In an outbreak of cysticercosis acquired by cattle in a California feedlot, over 4% of the animals in a 9-month period were found to be infected when slaughtered. An infected worker at the feedlot was the source of infection (Slonka et al., 1978).

Taenia taeniaeformis (Batsch, 1786) Wolffügel, 1911

Synonyms. *Taenia infantis* Bacigalupo, 1922; *Hydatigera taeniaeformis* (Batsch, 1786); *Taenia crassicollis* (Rud., 1810); *Cysticercus fasciolaris.*

The adult stage of *Taenia taeniaeformis* is commonly found in the cat, which becomes infected from eating rats or other rodents that harbor the strobilocercus stage (a larval strobila with a cysticercus bladder remaining attached to the posterior end). Two human infections with a strobilocercus embedded in the tissues of the liver have been recorded: one reported as *Taenia infantis* by Bacigalupo in 1922, in a 5-year-old child in Argentina, and one in a 77-year-old man in Czechoslovakia (Sterba and Barus, 1976).

Taenia longihamatus (Morishita and Sawada, 1966)

Originally reported as a species of *Multiceps, Taenia longihamatus* was described on the basis of three strobilae, complete with scolex, expelled by a 3-year-old girl in Mie Prefecture in Japan. Three similar worms were passed by a 55-year-old woman in Mie Prefecture following treatment for *Ascaris* and hookworm infection. The most distinctive feature recorded for this species was the rostellum, armed with exceptionally large hooks, the long and short hooks measuring 0.418 and 0.209 mm, respectively. Other features include the following: strobila muscular, 60 to 100 mm long by 3 to 4 mm wide; rostellar hooks number 20 in each circle; testes in mature proglottids number 150 to 180; gravid uterus has 20 to 23 branches on either side of the median stem; eggs 25 to 28 μm in diameter. Two additional infections were reported, but without description, from Mie Prefecture. Morishita and Sawada (1966) also reported an unidentified species of *Multiceps* obtained from a 4-year-old girl living in Okinawa. The worm in that case was 160 mm long, 4 mm wide, and the length of its long rostellar hooks was 0.38 to 0.39 mm. In the length of the rostellar hooks, these worms resemble the species described as *Multiceps macracantha* by Clapham in 1942. In the five cases reported by Morishita and Sawada, the worms were collected by others, not by the authors (Morishita, 1972).

Taenia crassiceps (Zeder, 1800) Rudolphi, 1810

Taenia crassiceps is a common parasite of the red fox (*Vulpes vulpes* L.) in Europe, and its larval stage, *Cysticercus longicollis,* is commonly found under the skin and in the body cavities of various species of small rodents and the mole. In North America, the adult worm is found in the arctic fox in Alaska where the life cycle involves lemmings and voles as hosts of the larval stages, and in the red fox and various rodents in

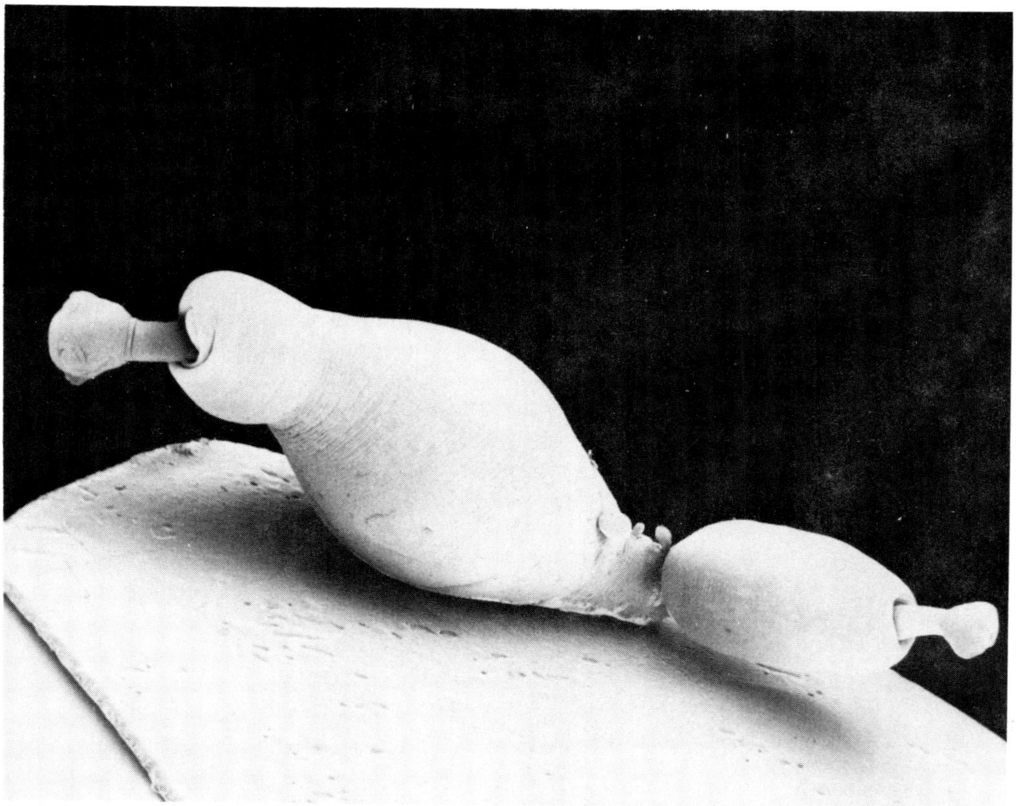

Fig. 31–18. *Taenia crassiceps*. Cysticercus from peritoneal cavity of mouse, showing at left, partially everted scolex, at right, one mature bud with scolex and several small buds. Scanning electron micrograph. (× ca. 20.) (Courtesy of Dr. Ming Ming Wong.)

Canada and northern United States. Experimentally, infection has been reported in the dog, reaching maturity in 5 to 6 weeks (Freeman, 1962). The infective larval forms (metacestodes) develop from eggs ingested by rodents, and these forms reproduce asexually by exogenous and occasionally endogenous budding at the pole opposite the scolex. These forms in turn differentiate a scolex at one end and proliferate buds at the other (Fig. 31–18).

Proliferating larval *T. crassiceps* has been reported once in a monkey in a zoological garden in Switzerland (Baer and Scheidegger, 1946), and in a 17-year-old girl in Ontario, Canada (Freeman et al., 1973; Shea et al., 1973). The infection in the monkey was massive, involving the pectoral and psoas muscles, retroperitoneal muscles, and spinal canal. In the young woman in Canada, the infection was apparently limited to the eye. One large cysticercus 5.8 mm long with the scolex evaginated was removed from the retina along with 29 free or attached buds and acephalic bladders (Fig. 31–19,*A*). The scolex was armed with characteristic spines, providing the basis for specific identification (Fig. 31–19,*B*). Some of the unidentified acephalic proliferating larval cestodes that have been reported in the pleural and thoracic cavities of dogs possibly were this species (Beaver and Rolon, 1981).

Taenia multiceps
Leske, 1780
(The "gid worm," producing cerebral and ocular coenurosis)

Synonyms. *Multiceps multiceps* (Leske, 1780) Hall, 1919; *Multiceps gaigeri* Hall, 1916; *Multiceps skrjabini* Popov, 1937: *Coenurus cerebralis* (Batsch, 1786) Rudolphi, 1808.

Morphology and Biology. The adult *T. multiceps,* 40 to 60 cm in length, has a piriform scolex about 0.8 mm in diameter with four strong suckers and a double circle of 22 to 30 rostellar hooks. The hooks are of two sizes, 150 to 170 μm and 90 to 130 μm in length, respectively. Gravid proglottids are 8 to 10 mm long by 3 to 4 mm wide. The gravid uterus has 18 to 26 slightly branched lateral arms extending from the main stem. The eggs range from 31 to 36 μm in diameter. Definitive hosts are the dog, wolf, and fox. The larval stage, a coenurus, is found in herbivorous mammals (sheep, goats, chamois, cattle, horse, gazelles, antelopes), as well as in monkeys *(Macaca mulatta, M. silenus),* and results from ingestion of eggs shed by proglottids expelled from dogs. Oncospheres hatch in the intestine of intermediate hosts, penetrate the intestinal wall, and eventually lodge in various tissues of the body, but in this species the larva usually develops in the brain or spinal cord. Here it develops into a *coenurus,* which is a bladder worm with multiple scolices, each in an inverted scolex-canal extending from the wall of the bladder to the interior, but capable of being everted to the exterior (Figs. 31–20, 31–21, 31–22, 31–23). Each scolex is a replica of that of the adult worm, and is capable of developing into an adult worm if eaten by a dog or other suitable host.

Historical Notes. The first authentic human infection with a coenurus was reported by Brumpt in 1913. The patient in that case, a Paris locksmith, gave a history of aphasia and fits, and at autopsy was found to have in the cerebrum a degenerate coenurus and a viable one bearing 75 scolices. Additional human

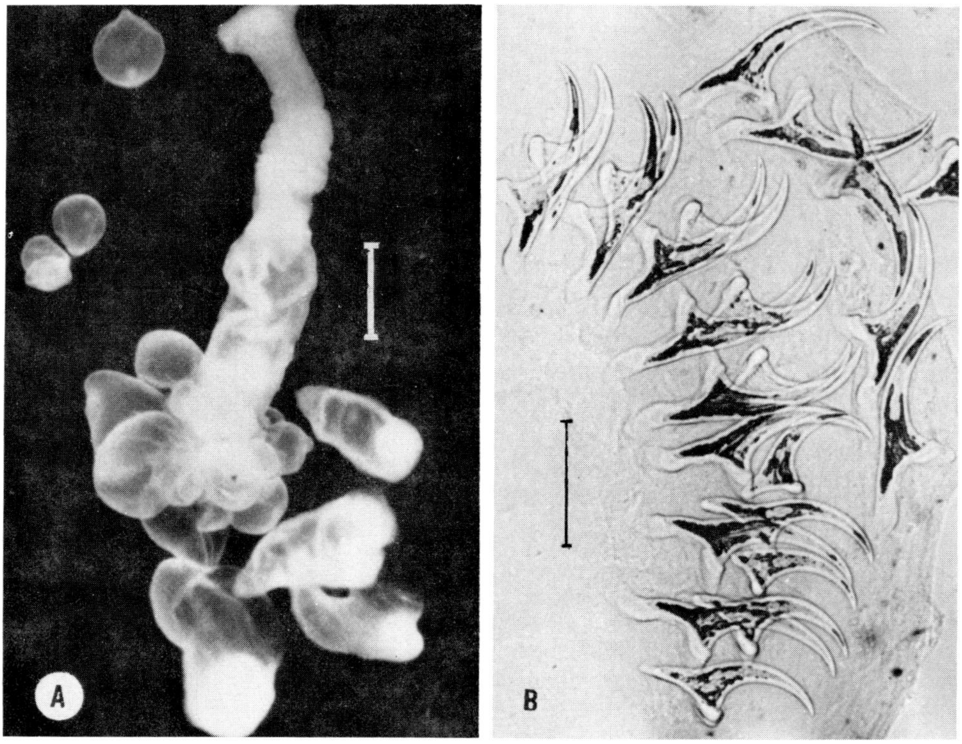

Fig. 31–19. *Taenia crassiceps. A,* Budding cysticercus removed from the retina of a woman in Canada. *B,* Rostellar hooks, showing characteristic features of the species. Scale is 1 mm in *A,* O.1 mm in *B.* (From Freeman, R.S., et al. 1973. Intraocular *Taenia crassiceps* (Cestoda). Part II. The parasite. Am. J. Trop. Med. Hyg., *22*:493–495.)

infections have been reported from France, Africa, England, Brazil, and the United States. Usually the infection is located in the brain, in subcutaneous or intermuscular connective tissues, or in the eye. Determination of the particular species to which a coenurus in human infection belongs is usually difficult, confusing, and at times practically impossible. A cerebral coenurus that is unilocular and has its protoscolices distributed in clusters (Fig. 31–20) rather than in a linear pattern is probably *T. multiceps*.

In sheep-raising regions of Europe and Asia, the economic losses from coenurus infection are considerable. Treatment of contact dogs about every 2 months is a common practice as a means of control.

Taenia serialis (Gervais, 1847) Baillet, 1863

Synonyms. *Multiceps serialis* (Gervais, 1847) Stiles and Stevenson, 1905; *Multiceps packii* Christenson, 1929; *Taenia antarctica* Fuhrmann, 1922; *Taenia laruei* Hamilton, 1940.

The adult *T. serialis* lives in the intestine of dogs and other canines. The larval stage, a coenurus, develops in the subcutaneous and intramuscular connective tissues of hares and rabbits, occasionally in rodents. *T. serialis* has been regarded as a synonym of *Taenia multiceps*, but Verster (1969) and others (Becklund, 1970) consider it to be a valid species. Although in features of the scolex they are indistinguishable, these authors note that the two species differ in intermediate host preferences, and that *T. serialis* has a well-developed vaginal sphincter, not present or poorly developed in *T. multiceps*. In the Western Hemisphere four cases of coenurus infection in man have been reported, two in which the parasite was in the brain. The first involving the brain was in a 2-year-old boy in California (Johnstone and Jones, 1950) (Fig. 31–21), the other in a 2-year-old boy in South Dakota (Barlow and Church, 1969; Hermos et al., 1970). In the first of the two other cases, the coenurus was removed from the muscles of the neck of a 62-year-old woman in Texas (Orihel et al., 1970); the second was from the breast of a 38-year-old woman in Ontario, Canada (Benger et al., 1981). In one of the two from the brain, the parasite was identified as *T. multiceps* (syn. *T. serialis*), and in the other it was reported as *T. multiceps* in the belief that only that species could develop a coenurus *(C. cerebralis)* in the central nervous system. However, it is improbable that *T. multiceps* has been present in the United States or adjacent parts of Canada for many years (Becklund, 1970). Larval *T. multiceps*, originally designated *Coenurus cerebralis*, and found only in the brain of sheep and some other ruminants, was never common or widely distributed in the United States, and there has been no record of it in sheep in this country for over 50 years. In experimentally infected mice, the coenuri of *T. serialis* develop in the brain as well as in subcutaneous and intermuscular sites, and the form of the cysts is that of *T. serialis* in both locations (Esch, 1964). The question of whether the coenurus infections in people living in Africa are *T. serialis* or some other species rests on the taxonomic status given to the species found in that part of the world (see *T. brauni*).

Taenia brauni (Setti, 1897) Fain *et al., 1956*

The adult stage of *Taenia brauni* was described from material taken from the intestine of a dog in Eritrea, northeast Africa, and the coenurus, from the intermediate host, a gerbil *(Gerbillus*

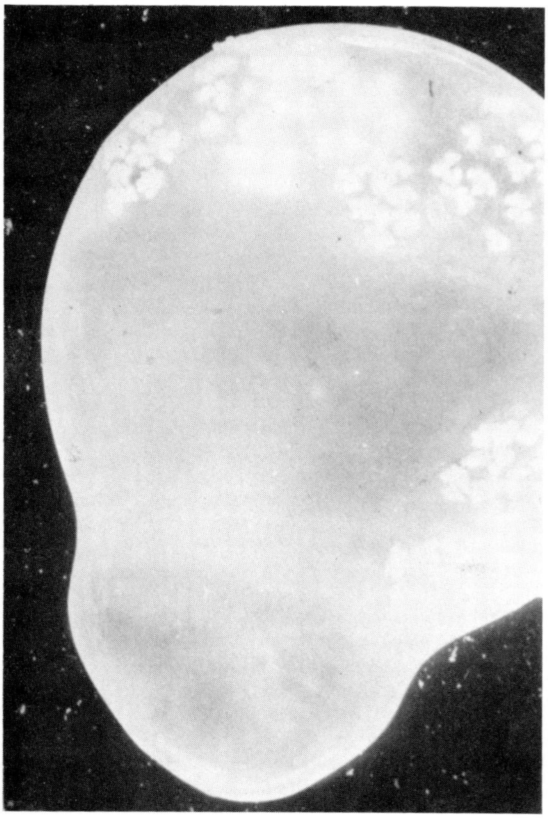

Fig. 31–20. *Taenia multiceps,* coenurus from brain of sheep in Lebanon. Protoscolices are arranged in clusters, and the clusters are clustered, not linear. (× ca. 3.) (Courtesy of Dr. Gordon Sweatman.)

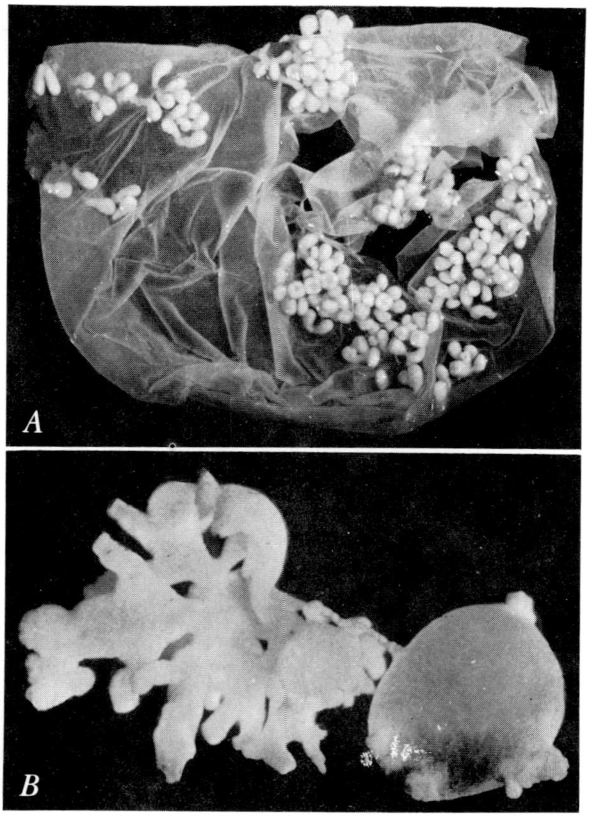

Fig. 31–21. *Coenurus,* reported as probably *Multiceps multiceps (= Taenia multiceps)* from base of brain and brain stem of a 2-year-old boy in California. *A,* Bladder form with protoscolices more clustered than linear in distribution. (× 3.5.) *B,* Racemose form (at left) and a daughter vesicle bearing several clusters of protoscolices. (× 6.) (Courtesy of Dr. Herbert Johnstone.)

pyramidum). Verster (1969) was unable to find valid morphological differences between this species and *T. serialis,* but because they showed slight differences in their intermediate host preferences, it was deemed advisable to retain the two as new subspecies *(T. serialis serialis* and *T. serialis brauni).* However, until its taxonomic status is more clearly established, it will be convenient to consider *T. brauni* a valid species. The neck of the mature worm is short and broad, and the rather weak rostellum bears approximately 30 hooks in two rows, of which the larger are 95 to 140 μm long, the smaller, 70 to 90 μm long. Ten infections with the coenurus have been recorded in native children in Rwanda (Fain et al., 1956; Vanderick et al., 1964). Approximately 30 cases have been reported from Uganda (Templeton 1968, 1970; Williams and Templeton, 1971), and several others have been reported from the same general area in Kenya and Zaire. All of the cases seen in this area were considered by Williams and Templeton (1971) to conform to *T. brauni,* the coenurus of which is found in animals of that region. In 11 cases reported from East Africa, and in 2 cases each in South Africa and Ghana, the coenurus was in or on the eye (Manschot, 1976). In no case reported from tropical Africa has the central nervous system been involved, a circumstance that favors the view that *T. brauni* is a valid species distinct from *T. serialis* and *T. multiceps.*

Taenia glomerata (Railliet and Henry, 1915) Brumpt, 1932

Synonyms. *Coenurus glomeratus* (Railliet and Henry, 1915) Turner and Leiper, 1919; *Multiceps glomeratus* Railliet and Henry, 1915.

The adult stage of this species and the definitive host are unknown. The coenurus was first described from the gerbil. A human infection with the coenurus was reported by Turner and Leiper (1919) from a cyst taken from the intercostal muscle of a native of northern Nigeria. The cyst measured 1 by 2 cm, had a delicate transparent wall, and had about 35 scolices, each invaginated and bearing 32 long and short rostellar hooks measuring 90 to 100 μm and 65 to 70 μm in length, respectively. A coenurus that possibly was this species was reported by Taramelli and Dubois (1931) from man in eastern Zaire. A third human infection, possibly of this species, was reported in the triceps muscle of a man from northern Nigeria. The coenuri in these cases are conceivably assignable to *T. serialis* or *T. brauni.*

Coenurosis

Definition. As described by Wardle and McLeod (1952), the coenurus is a larval cestode bladder up to the size of a hen's egg bearing multiple protoscolices, and peculiar in being able to bud off daughter bladders either internally, floating in the bladder fluid, or externally, attached by stalks. There is wide disagreement regarding the taxonomy of the *Taenia* species

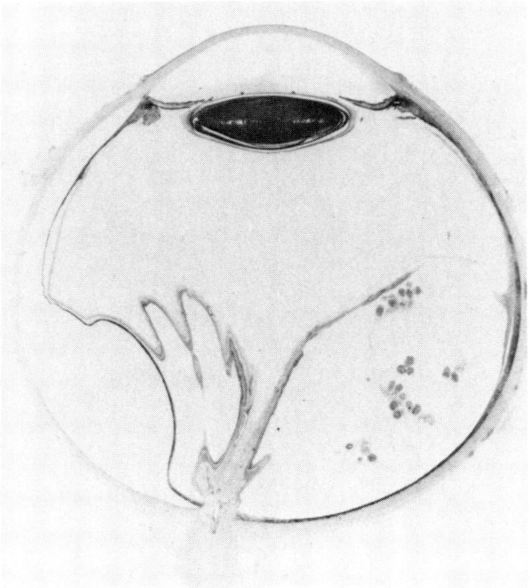

Fig. 31–22. *Coenurus,* in section of eye removed from a woman in Ghana. The parasite, lying between choroid and displaced retina, bears numerous protoscolices that are somewhat linear in disposition. The delicate cyst membrane is barely evident. (By P.C. Beaver. Section courtesy of Dr. W.A. Manschot.)

whose infective stages are coenuri. Moreover, the intermediate hosts and larval stages are still unknown for numerous species of *Taenia*. It is well known that among the cestode parasites of carnivorous mammals the larval stages are less host-specific than are the adult worms. Thus, man can be counted among the susceptible hosts of the larval stages of several species of *Taenia*. In most of the cases thus far reported in man, the coenuri appear to be either *T. multiceps* or *T. serialis,* but other species, described or not, also may be represented.

Geographical Notes and Biology. In 1968, Templeton referred to 36 cases of human *Coenurus* infection reported from England, France, the United States, and Africa (Zaire, Kenya, Uganda, Rwanda, Nigeria, and South Africa) and added 14 cases from Uganda. Two years later he reported an additional 14 cases of coenurus infection seen in Uganda, bringing the total to 64, mostly from East Africa and South Africa (Templeton, 1970). The following year 13 cases involving the eye, 5 of which were new cases, were reported from Uganda (Williams and Templeton, 1971). Four additional subcutaneous infections were reported from Kenya (Kaminsky et al., 1978) and four from Ghana. In two of these the coenurus was removed from subcutaneous tissue (Addae, 1972), and in two the coenurus was removed from the eye (Manschot, 1976). Whereas in most cases reported from tropical African countries, the coenurus was located in subcutaneous or muscle tissue, or more often in the eye (Fig. 31–22) or orbit (Fig. 31–23), especially in children, in other regions the coenuri have usually been found in the central nervous system of adult patients. In some of the tropical African cases the scolices were described as haphazard in disposition but typically they were arranged in rows, usually with a few in each row, but up to 30 in exceptional cases (Figs. 31–22 and 31–23). The coenuri in the tropical African cases were probably *T. brauni,* especially those in East Africa (Williams and Templeton, 1971).

Coenurus cysts in the eye and subcutaneous tissues are unilocular. In the central nervous system they are frequently multilocular, described in one case as racemose cysts, some of which had coenurus-type scolices (Fig. 31–23,*B*) (Barlow and Church, 1969), and in another report of the same case, as a grape-like mass of translucent, thin-walled cysts of a coenurus (Hermos et al., 1970). A similar parasite had been found earlier in the brain of a California child (Fig. 31–21) (Johnstone and Jones, 1950). In both instances racemose-like cysts bearing no protoscolices together with one or more cysts bearing protoscolices were distributed in clusters, some of which were linear in disposition (Fig. 31–21). These findings were discussed in two case reports of racemose cysticercus infections, and it was

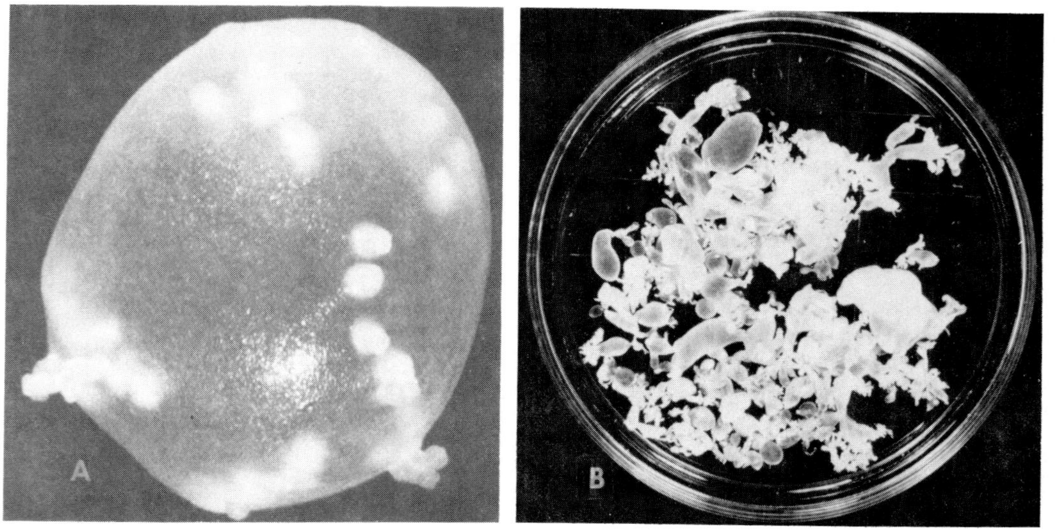

Fig. 31–23. *A, Coenurus,* removed from subconjunctival tissue of a 1-year-old boy in Uganda. Protoscolices, some of which project from the outer surface, are linear in disposition. (× 6.) (By P.C. Beaver. Specimen courtesy of Dr. A.C. Templeton.) *B,* Proliferating forms of a coenurus removed from the posterior fossa of a 2-year-old boy in South Dakota. Based on the presence of mature scolices with characteristic hooks, the parasite was identified as *Taenia multiceps*. Natural size. (From Barlow, J.F., and Church, B.G. 1979. Coenurosis in the brain of a child from South Dakota. S.D. J. Med., *2:37–44.*)

suggested that the so-called *Cysticercus racemosus* in some instances may be a sterile coenurus (Jung et al., 1981).

Pathogenesis and Symptomatology. Although the principal mechanisms producing disease in coenurosis are the space-occupying and pressure effects, other symptoms have been noted that are presumably due to allergic or toxic reactions. For example, a Canadian woman with coenurosis of a breast suffered severe and incapacitating symptoms including recurrent rash, pyrexia, night sweats, and lymphadenopathy leading to a diagnosis of, and treatment for, lymphoma (Benger et al., 1981).

Subcutaneous or intramuscular localizations produce swellings that may or may not be painful. Localization in the eye impairs or destroys vision (Williams and Templeton, 1971; Manschot, 1976).

Racemose cysticercosis, which possibly is cerebral coenurosis, may involve the ventricular system (Patharangkura et al., 1980), the subarachnoid space, choroid plexus, or cerebrum. Most commonly it is located in a large space such as the fourth ventricle or in the foramen magnum invading the fourth ventricle and extending down into the cord. In such a situation it appears to the surgeon as a bunch of grapes, consisting of numerous cysts attached to a central mass by stalks. The signs and symptoms are due to the increased intracranial pressure (internal hydrocephalus), although if the high pressure continues pressure necrosis occurs. Typically these symptoms begin with headache and vomiting. Visual defects appear, depending on pressure on the optic chiasma. Papilledema is found on funduscopic examination.

Localization in a cerebral hemisphere produces symptoms consistent with the area involved, such as localized paresis or paralysis or focal seizures (Jung et al., 1981).

Prospective diagnosis is assisted by CT scan and ultrasonography, but to date these modalities do not distinguish between the coenurus and a partly cystic tumor.

Treatment to date has consisted of surgical removal of the parasite. In peripheral locations and in the eye this procedure is usually successful. Surgical extirpation of the parasite from the calvarium is difficult. Temporary improvement tends to occur when flow of cerebrospinal fluid through a blocked channel is reestablished, but remnants of the parasite left behind regrow and symptoms recur. The prognosis is thus poor.

Chemotherapy of coenurosis has not yet been successful. Trials of mebendazole in treatment of infected sheep gave unpromising results (Verster et al., 1978). An adequate trial of praziquantel has not been reported.

Genus Echinococcus
Rudolphi, 1801

Species of the genus *Echinococcus* are adapted to hosts having an essentially obligatory predator-prey relationship. The final host is a carnivore. Under natural conditions the larval form occurs in one or more species of herbivorous mammals upon which the final host is dependent for food. In Nature, *E. granulosus* and *E. multilocularis* are relatively host specific. Although where the two species are sympatric the final and intermediate hosts of one species may be exposed to infection by the appropriate stage of the other, they do not share an intermediate host. Under natural conditions the adult stage of *E. granulosus* is known to develop normally in only a single species, the wolf. Under conditions created by man the domestic dog has replaced the wolf as the final host of *E. granulosus*, and domestic ungulates replace the wild cervids as intermediate hosts (Rausch, 1967).

Four species, *E. granulosus, E. multilocularis, E. vogeli*, and *E. oligarthrus* have been reported in the larval stage in man. The report of *E. oligarthrus* was apparently erroneous. *E. granulosus* and *E. multilocularis* are widely distributed, whereas *E. vogeli* thus far has been found only in Central America and northwestern South America (Fig. 31–24).

Echinococcus granulosus
(Batsch, 1786) Rudolphi, 1805
(The hydatid worm, producing unilocular echinococcosis or hydatid disease)

Synonyms. *Taenia visceralis socialis granulosus* Goeze, 1782; *Hydatigena granulosa* Batsch, 1786.

In 1963 Rausch and Nelson reviewed the species of *Echinococcus* and concluded that the status of *E. felidis* Ortlepp, 1937 and *E. patagonicus* Szidat, 1960 was uncertain, but that both were possibly conspecific with *E. granulosus*. Considered to be synonyms of *E. granulosus* were *E. cameroni* Ortlepp, 1934; *E. intermedius* Lopez-Neyra and Soler, 1943; *E. longimanubrius* Cameron, 1926; *E. lycaontis* Ortlepp, 1934; *E. minimus* Cameron, 1926; and *E. ortleppi* Lopez-Neyra and Soler, 1943.

Historical and Geographical Notes. Although Hippocrates, Aretaeus, and Galen were familiar with hydatid cysts, Redi (1684), P.J. Hartmann (1685), and Tyson (1691) first suspected their animal nature. Pallas (1766) first mentioned the similarity of hydatids in man and other mammals. Goeze (1782) first studied the scolices of the larva and recognized their relationship to those of *Taenia*. P.J. Hartmann (1695) and later Rudolphi (1808) first studied the adult worms in the dog's intestine. Von Siebold (1852), followed by Haubner, Leuckart, Küchenmeister, and Nettleship, fed scolices of cysts from domestic animals to dogs and observed development of the adult worms in this host's intestine. Later Naunyn in Germany (1863), Krabbe in Iceland (1863), and Thomas in Australia (1885) obtained adult worms in dogs from scolices of human origin. According to Costa (1960), hydatid disease was introduced into South America with dogs of Nordic whalers who landed in Uruguay in the eighteenth century. Infected sheep could have been a contributing source also.

The most extensive and intensive enzootic endemic regions are those of the sheep- and cattle-raising countries, including South Australia, Tasmania, New Zealand, the northern and southern parts of Africa, Tanzania, and the southern half of South America, particularly Argentina, Paraguay, Chile, Uruguay, and southern Brazil. In addition, human infection occurs frequently throughout northern, southern, and eastern Europe, Siberia, Turkestan, Mongolia, North China, southern Japan, North Vietnam, the Philippines, Syria, Lebanon, Iraq, Iran, Israel, and Arabia.

The New Zealand National Hydatid Council reported in 1979 that of 166,753 fecal samples collected from dogs throughout the country, only 112 (0.07%) were positive for *E. granulosus*, and roughly 10% of ewe livers contained hydatid cysts. New cases of hydatid infection in humans dropped from 377 in the 1950 to 1953 period to only 72 in the 1974 to 1976 period. In Sardinia, where the infection had been especially prevalent, a survey in 1978 detected *E. granulosus* in approximately 2% of stray dogs and 24% of sheep dogs (Arru and Nieddu, 1978).

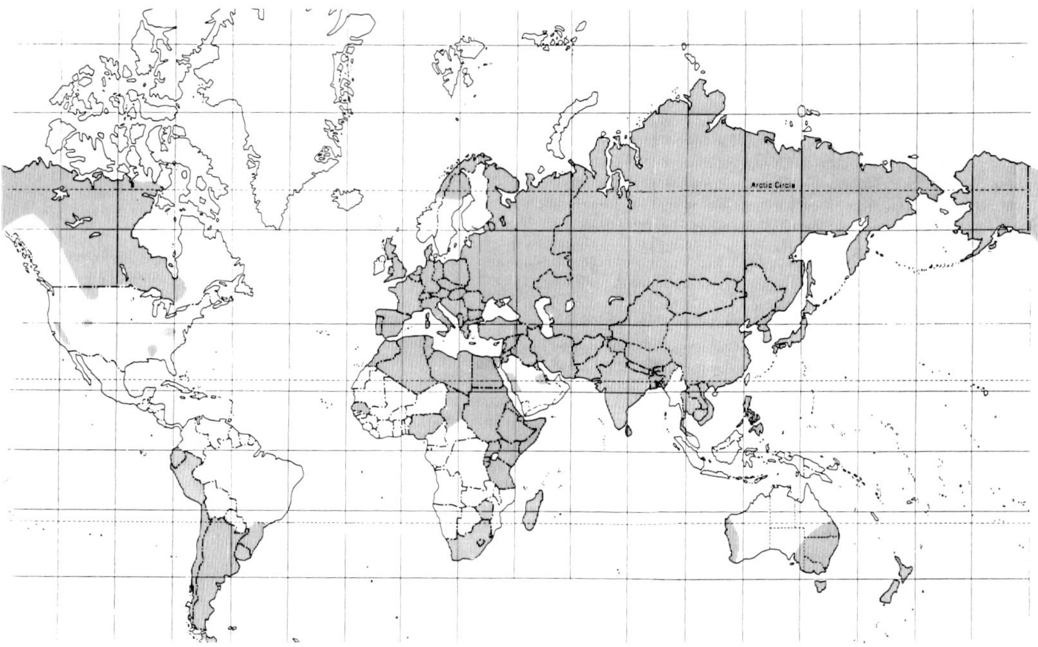

Fig. 31–24. Map showing the approximate distribution of *Echinococcus granulosus* in man and reservoir hosts. (Adapted by R.C. Jung from original of Matossian, R.M., Rickard, M.D., and Smyth, J.D. 1977. Hydatidosis: A global problem of increasing importance. Bull. W.H.O. *55*:499–507.)

In Iran the infection is prevalent in most parts of the country; in one year in the hospitals of Isfahan 59 people were admitted for surgical removal of hydatid cysts. A survey in western Iran in 1973 detected infection in 16 of 34 stray dogs, 28 of 550 sheep, and 12 of 116 cattle (Sabbaghian et al., 1975). In Nigeria, the infection detected in different parts of the country ranged from 50 to 94% of slaughtered camels, up to 60% of sheep, 81% of goats, and 15% of cattle (Dada, 1980a); in stray dogs the prevalence rates ranged from 1 to 6% (Dada, 1980b).

In England and Wales in the period from 1966, when reporting first began, to 1972 there were 77 reported infections in people. Hydatid cysts are seen frequently in cattle and horses as well as in sheep in Great Britain, particularly in Wales (Blamire et al., 1980). Areas where high rates of infection have resulted in organized control programs include Iceland, Cyprus, New Zealand, Tasmania, Falkland Islands, Bulgaria, Argentina, Uruguay, Peru, and parts of U.S.S.R. Such programs have been, or are expected to be, introduced in Spain, Greece, Yugoslavia, and Sardinia (Thornton, 1980).

In 1900 about 33% of autopsies in Iceland provided evidence of hydatid cyst. By 1913 the incidence had decreased to 16.6%. Between 1930 and 1944 only about 5% of 1231 postmortems at Reykjavik showed infection, and this was mostly in the higher age groups. In 1953 examination of 145 dogs and 20,000 sheep in Iceland provided no evidence of infection.

The first human infection with hydatid cyst in the United States was diagnosed in 1808. Through 1940 there were a total of 519 reported cases, 95% in immigrants, and a few cases are diagnosed each year. A focus of transmission was identified among Basque sheep farmers in California (Schwabe et al., 1972). The disease is endemic also in Utah where up to 1974 at least 33 people had contracted the infection and one had died; in five locations surveyed in 1970 to 1972, 2 to 33% of the dogs were infected (Spruance et al., 1974).

Morphology, Biology, and Life Cycle. The adult *Echinococcus granulosus* (Fig. 31–25), which lives attached to the mucosa of the small bowel of the dog and other canine hosts, is a minute worm, measuring 3 to 6 mm in length. It has a piriform scolex (300 μm in diameter), provided with 4 suckers and armed with 28 to 50 (usually 30 to 36) hooks, a short neck, and usually only one immature proglottid, one mature proglottid, and one gravid proglottid. The terminal proglottid is the broadest and longest; the mature one usually the narrowest. In the gravid unit (Fig. 31–25) the uterus with its lateral evaginations comes to resemble a loosely twisted coil. The uterus bursts either before or after evacuation of the gravid proglottids from the bowel, discharging the relatively few eggs. Adults of *E. granulosus* live 5 to 29 months in dogs. Natural infections in dogs tend to be heavy, with portions of the small intestine densely populated (Fig. 31–26).

These eggs cannot be distinguished from those of *Taenia* species in the dog. When swallowed by a suitable intermediate host such as sheep, goat, camel, hog, ox, or man, the eggs hatch in the duodenum and the oncospheres migrate through the intestinal wall, enter the mesenteric venules, and become lodged in various organs and tissues. The first and most important site is the liver, where, on

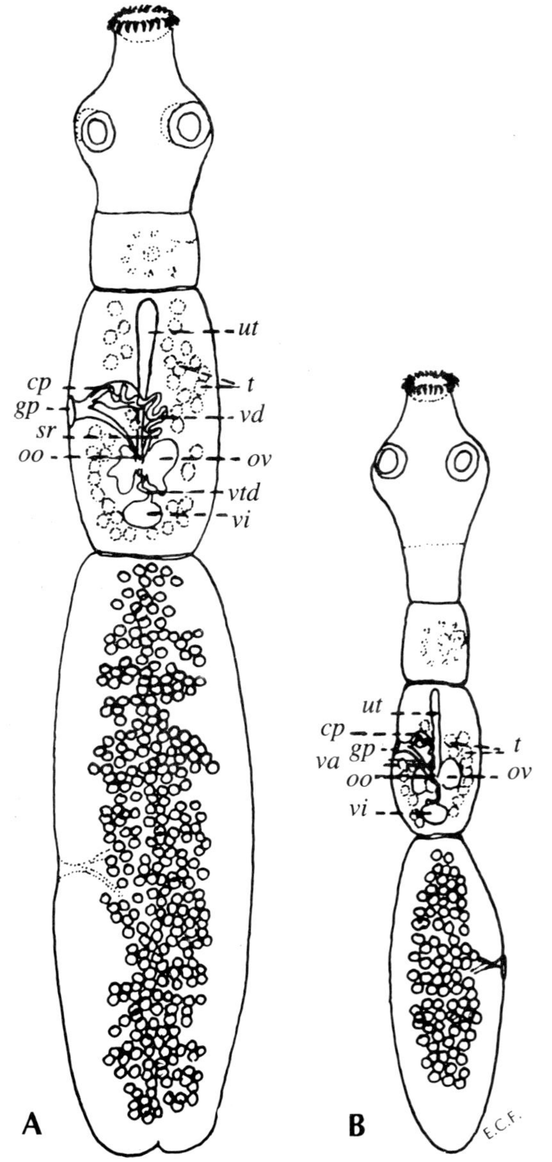

Fig. 31–25. *Echinococcus*, adult worms, the larvae of which produce two of the three known types of hydatid cysts in man. *A, E. granulosa* from small intestine of dog, Yugoslavia. *B, E. multilocularis*, from fox, South Germany. *cp*, Cirrus pouch. *gp*, Genital pore. *oo*, Ootype. *ov*, Ovary. *sr*, Seminal receptacle. *t*, Testes. *ut*, Uterus. *va*, Vagina. *vd*, Vas deferens. *vi*, Vitellaria. *vtd*, Vitelline duct. (× 25.) (Adapted from original of E.C. Faust.)

the average, about 70% of all infections in man are located. Next in importance are the lungs, and then various other foci. By the fourth day the larvae reach a diameter of 40 μm, when they begin to develop a cystic cavity. By the end of 3 weeks the young cyst has attained a diameter of 250 μm and has stimulated considerable host tissue reaction around it.

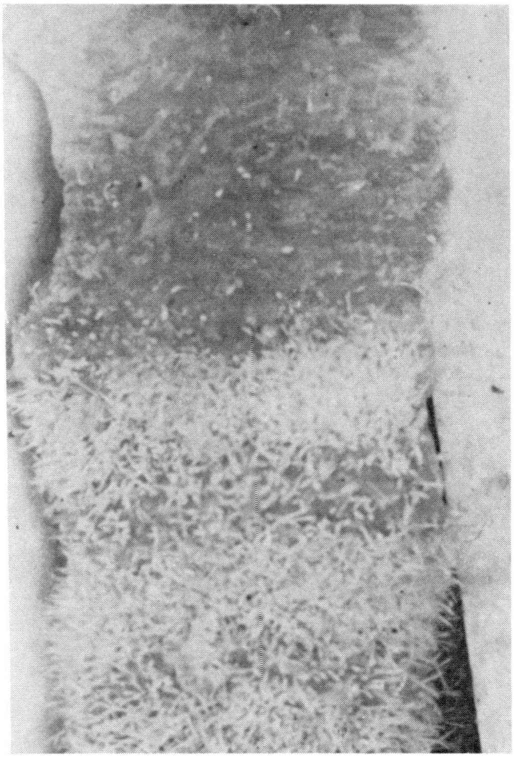

Fig. 31–26. *Echinococcus granulosus*. Adult worms in the intestine of a naturally infected dog in Lebanon. (× ca. 1.5.) (Courtesy of Dr. Gordon Sweatman.)

Immediately around the larva are host cells arranged radially with infiltrated foreign-body giant cells and eosinophils. An outer layer of fibroblasts, with many eosinophils and newly formed blood vessels, constitutes a second zone. Enveloping these is an insulating layer of fibrous tissue grading into normal tissue. By the fifth month the hydatid has reached a diameter of 1 cm and its intrinsic wall has become differentiated into (1) an outer friable, milky-opaque, non-nucleated *laminated layer* and (2) an inner nucleated *germinal layer* (Fig. 31–27). That the fibrous cyst wall is formed by the host while the laminated membrane is generated by the parasite is shown by the formation of the latter on cysts grown in vitro (Heath and Osborn, 1976). From this inner layer masses of cells are budded out into the cystic cavity. These cells become vacuolated and later stalked. These are the brood capsules, which may remain attached or may be set free into the fluid of the cystic cavity. From the inner wall of these capsules the protoscolices develop and invaginate as they become fully mature and infective (Fig. 31–27,*B*). The free brood cap-

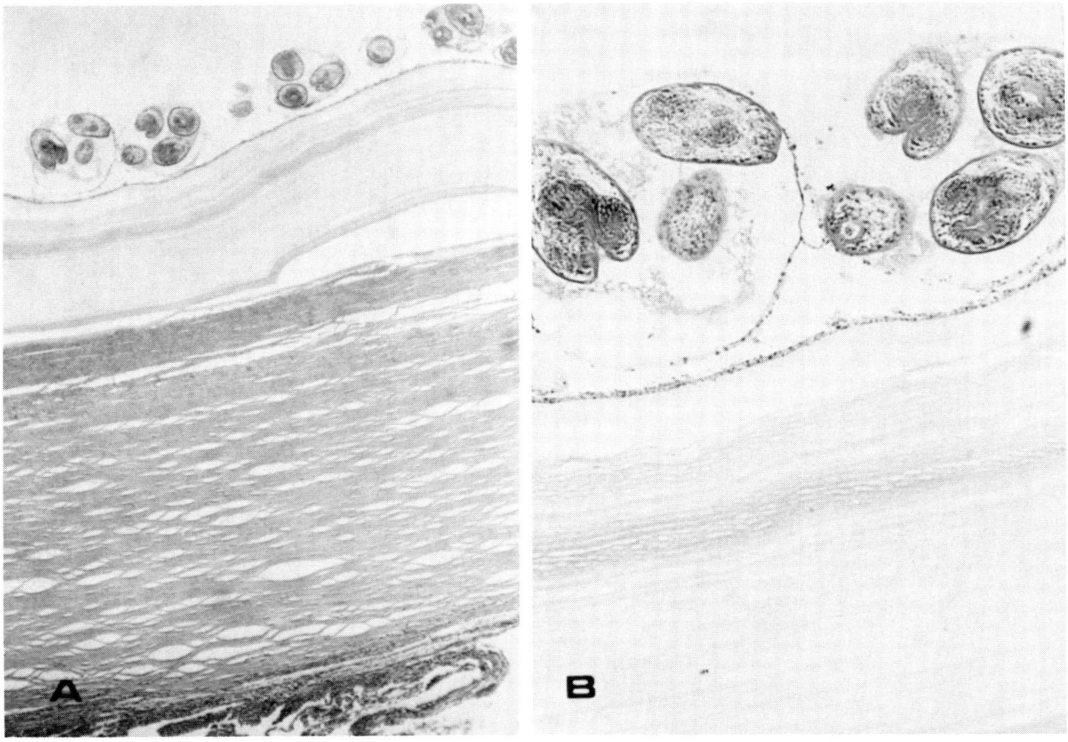

Fig. 31–27. *Echinococcus granulosus*. Section of a unilocular hydatid cyst from lung, showing brood capsules with protoscolices, a thin cellular germinal membrane, thick, noncellular laminated membrane, and much thicker, fibrous host-tissue capsule. *A* (× 40). *B* (× 140). (From Faust, E.C., Beaver, P.C., and Jung, R.C. 1975. *Animal Agents and Vectors of Human Disease,* 4th ed. Philadelphia, Lea & Febiger.)

sules and the free scolices are collectively referred to as "hydatid sand." Some cysts are sterile and never produce brood capsules; some become sterile by bacterial invasion or calcification. In other cysts the brood capsules never produce scolices; hence they are called *acephalocysts*. Usually, daughter cysts are produced within the mother cavity (Fig. 31–28). Unilocular cysts usually are less than 5 cm in diameter, but may reach a diameter of 20 cm and contain nearly 2 liters of fluid (Lenda et al., 1982).

This type of cyst is *unilocular* and constitutes by far the most prevalent human type of hydatid cyst. In certain localities, however, including particularly southern Germany, Switzerland, the Tyrol, eastern U.S.S.R., and west-central Siberia (Fig. 31–24), and in isolated instances elsewhere, another structural type of hydatid cyst develops in man. This was described by Virchow, in 1855, and is usually referred to as *alveolar hydatid* (see p. 533 under *Echinococcus multilocularis)*. However, cysts of *E. granulosus,* occasionally in man, and not uncommonly in suboptimal animal hosts, develop into morphological types resembling *E.*

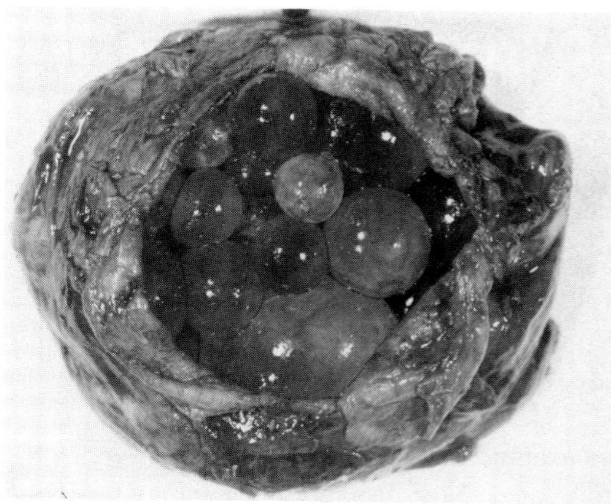

Fig. 31–28. *Echinococcus granulosus*. Unilocular hydatid cyst, 12 cm in diameter and containing numerous daughter cysts ranging up to 7 cm in diameter, removed from the liver of a Greek seaman from whom two similar cysts had been removed 6 years earlier. (From Faust, E.C., Beaver, P.C., and Jung, R.C. 1975. *Animal Agents and Vectors of Human Disease,* 4th ed. Philadelphia, Lea & Febiger.)

multilocularis cysts and are usually sterile (Toro et al., 1980).

A variety of unilocular cyst is *osseous hydatid,* which is found in bone. Because of confinement by the dense osseous tissue, the laminated layer is poorly developed and the organism migrates along the bony canal as a naked protoplast, frequently eroding the limiting osseous tissue.

Epidemiology. Man can be infected with only the larval or hydatid stage of *E. granulosus.* The dog is the usual *definitive host,* although the wolf, jackal, dingo, hyena, puma, jaguar, Arctic fox, and domestic cat have been found infected in Nature. Records of human hydatid infection number tens of thousands. Nevertheless, human infection is much less common than that in reservoir hosts. In addition to sheep, cattle, and pigs, which are the *common reservoirs* of the hydatid, naturally infected mammals include: horses, camels, goats, monkeys *(Macaca syrichta fascicularis, M. mulatta mulatta, M. sylvana, Papio comatus comatus),* the drill baboon *(Mandrillus leucophaeus),* lemurs, the African elephant, argali *(Ovis ammon ammon),* antelope *(Tetracercus quadricornis),* moose *(Alces alces alces* and *A. alces americanus),* reindeer *(Rangifer tarandus),* caribou *(Rangifer arcticus),* elk *(Cervus canadensis),* coast deer *(Odocoileus columbianus),* white-tailed deer *(O. virginianus),* bison *(Bison bison),* vicuña *(Lama vicugna)* of the Andes in South America, giraffe, gazelle, tapir, zebra, kangaroo, mongoose, and others. In northwest Canada the natural cycle involves the wolf and moose. Dogs that eat the infected viscera of moose provide in their feces the source of human exposure. In the comparable sylvatic cycle in Australia, the larval host is usually the wallaby, to a lesser extent the kangaroo and rabbit, with the dingo as the definitive host.

The definitive host feeds on the viscera of infected larval hosts and thereby becomes infected, frequently acquiring countless numbers of the adult worms (Fig. 31–26). The intermediate host swallows eggs passed in the feces of the definitive host and infection with the hydatid results.

In southern Australia, where a high percentage of the dogs harbor the adult worm, up to 2% of the human population in certain districts are infected with the hydatid. In New Zealand, hydatid disease, although much reduced, is still a major health problem, with about 70 new clinical cases a year. The most heavily infected districts in Europe appear to be Upper Pomerania, Greece, Italy, and Spain. Throughout southwestern Asia and northern Africa where sheep, goats, and camels, and to a lesser extent cattle and pigs, are slaughtered, hydatid infection in the people remains a health problem. In England the horse is a significant reservoir, hydatid cysts being found in approximately half of the farm horses and a third of the thoroughbreds in some areas (Wood, 1977). However, among approximately 1200 horses examined at an abattoir in South Yorkshire, only 5% were found to have cysts (Connor and Hizzard, 1979). In other parts of the world the infection in horses is rare or does not occur even in localities where it is widespread in other farm animals (Kaloyanov and Dobrev, 1979). Williams and Sweatman (1963) proposed that the horse strain in England be named *E. granulosus equinus,* a subspecies distinct from *E. granulosus granulosus.*

Distinct subspecies or strains appear to have developed in other hosts and in other geographic areas. Recognized by some authorities are ovine and porcine strains as well as the equine strain, each being able to develop in other hosts, but less readily. In addition, *E. granulosus canadensis* may differ from *E. granulosus granulosus* in certain other respects (Krotov, 1979). Rausch (1967) early recognized two major strains of *E. granulosus*: a northern strain, with Holarctic distribution, and a European strain, nearly cosmopolitan in distribution as a result of shipment of domestic animals from Europe and elsewhere to essentially all parts of the world. The natural cycle of the northern strain involves the wolf and large deer *(Rangifer* and *Alces),* and that of the other strain involves mainly the dog and domestic ungulates. In a district of northern Kenya where the prevalence of hydatid disease in the people is remarkably high, observations suggest that infection in dogs may be acquired from eating infected human corpses (French et al., 1982).

Most hydatid cysts in man are probably acquired in childhood. This may be due in part to greater susceptibility, but it is undoubtedly associated with close contact with infected dogs. Frequently the unilocular cyst may grow for 5 to 20 years before it is diagnosed, i.e., it may be almost as old as its host. Hydatid cysts tend to be more common in members of the same family than in the general population.

Pathogenesis, Pathology, and Symptomatology. The damage produced by the hydatid cyst of *Echinococcus granulosus* in the human body is mainly mechanical. The young cysts that develop from embryos lodged in vital centers may soon interfere with functions of the organ, with damaging, even fatal results. Yet in favorable, relatively unconfined locations, the benign unilocular cyst may grow to tremendous size without ever producing symptoms or it may become a physical burden to the patient and at times rupture, precipitating an anaphylactic reaction. Osseous cysts (1% of the total) that develop in syncytial growths erode bony structure, which frequently gives way without warning and results in permanent injury to the patient .

Hydatid disease due to *E. granulosus* is always serious, although the gravity of the infection depends on the type of tumor, whether unilocular or osseous, as well as the organ or tissue in which it first becomes implanted. An embryo that comes to rest in a favorable site and develops normally to a unilocular cyst, with the usual inflammatory and fibrous tissue reaction around it, may produce symptoms 5 to 20 years later, only after it has reached sufficient size to press upon the adjacent organs. Under these conditions the fluid within the cyst is localized, and only a minimum seeps out into the systemic circulation. However, incomplete insulation or rupture of the cyst may produce serious allergic or anaphylactic reactions, occasionally bronchial or biliary abscess, arterial embolism due to lodgment of capsular fragments, or their dissemination through the circulation as viable metastases to other foci in the body. In other cases, the young cyst may develop in a brain or orbital capillary or on a heart valve and may soon produce dangerous effects potentially fatal to the patient. Osseous hydatid may produce rapid erosion of the bone, with multiple fracture or crumbling of the bony structure. Hydatid of the lung is invariably intracapsular. A bronchogram shows the bronchi and bronchioles surrounding the cyst. It may rupture spontaneously, and the contents may be coughed up, with rapid recovery or asphyxiation. At times the cyst becomes calcified. Extrapulmonary cysts within the chest tend to produce bulging of the thoracic cage but no erosion of the ribs.

Earlier statistical analyses indicated that at least half the human cases of hydatid disease involved the liver. More recent reports from Greece, Italy, Peru, Chile, Great Britain, and Canada indicated significant prevalence of pulmonary hydatid.

Diagnosis. Large unilocular cysts frequently are not diagnosed until middle life or later. On the other hand, unilocular cysts, lodged in foci where serious damage or dysfunction results, are more likely to be diagnosed in early life, when most of the exposure occurs. Cysts may become secondarily infected and may suppurate, producing an added risk. Hydatid cyst should be especially considered in patients with (1) abdominal masses, particularly involving the liver and without evidence of other origin; (2) obscure thoracic enlargements, particularly at the base of the lungs, where tests for tuberculosis are negative; and (3) lesions of bones suggesting tuberculosis or other inflammatory processes. The diagnostic aids include:

1. The *Roentgenogram*. A roentgenogram is frequently helpful in hydatid cysts of the lungs or other thoracic involvement, in extrahepatic abdominal cysts, and in cases where the long bones are involved. In Uruguay mass surveys using miniature chest x-ray films proved to be of considerable epidemiologic value (W.H.O., 1967).

Bronchography is frequently useful in determining the nature of a lung shadow. Angiography of the liver shows a "rim-sign," contrast dye blockage, curved displacement of vessels, and stretching of hepatic arteries. Although angiography does not always provide accurate differential diagnosis between malignant liver disease and alveolar hydatid, early venous filling and arteriovenous shunts are not characteristic of alveolar hydatid (Langer et al., 1979). CT (computerized tomographic) scan and ultrasonographic techniques are highly effective in locating and delineating cysts but do not establish the diagnosis.

2. *Hydatid Thrill*. Although it constitutes a specific diagnostic sign in unilocular hydatid cysts of the abdominal viscera, hydatid thrill is frequently difficult to elicit.

3. *Exploratory Cyst Puncture*. Cyst puncture is an accurate method but is dangerous, owing to the likelihood of anaphylactic shock resulting from escape of hydatid fluid. This test should not be attempted unless the surgeon is immediately prepared to perform a radical operation for removal of the cyst. A technique of freezing the cyst wall for safe aspiration has been described by Saidi and Nazarian (1972).

4. *Eosinophilia*. Generalized eosinophilia is present in 20 to 25% of cases. This sign is suggestive along with other clinical evidence but it is not pathognomonic.

Immunologic methods of diagnosis of hydatid disease can be helpful (Iacona et al., 1980), but the interpretation of such tests requires experience and caution (Schantz et al., 1980).

Treatment. Nonsurgical procedures are usually unsuccessful. Surgical technique is generally helpful only in patients with unilocular cysts in operable sites. The operative procedure commonly employed for abdominal cysts is as follows: (1) Expose the adventitia surrounding the cyst by incision over the most prominent or most dependent part of the tumor. (2) Thoroughly wall off the exposed surface of the wound. (3) Aspirate the contents through a large-caliber needle or trocar connected with a closed-suction apparatus. (4) Inject 10 to 50 ml of 10% formalin solution and withdraw the fluid in 5 minutes. (5) Incise through the adventitia down to the actual cyst. (6) Separate the cyst from the adventitia and remove the cyst and its contents. (7) Swab the adventitia with 10% formalin, allowing a few cubic centimeters to remain in the site. (8) Obliterate the cavity (capitonnage) by intracapsular sutures wherever possible. (9) Close the adventitia by a double row of catgut sutures. (10) Close the cavity without open drainage, anchoring the adventitia to the tissue beneath the line of incision. This technique is preferred to marsupialization, which requires longer hospitalization.

In recent years treatment of multiple and inoperable hydatid disease by biological methods (i.e., desensitization of the host and death of the cyst, with gradual absorption) has gained many followers in South America but is not always a successful method of treatment.

During the past 5 years studies have been made of drugs of the benzimidazole group, including mebendazole, flubendazole, albendazole, and fenbendazole as to their efficacy against hydatidosis (Schantz et al., 1982). Although they offer some promise, it has not yet been conclusively shown that they will eradicate an infection. Mebendazole, for example, checks proliferation and spread of hydatid cysts but does not terminate infection in experimentally infected *Meriones unguiculatus*. Furthermore, host reaction to the dead or dying parasite may be severe (Murray-Lyon and Reynolds, 1979). Failure of mebendazole therapy was reported in a series of 16 cases of hepatic hydatid (Okello and Chemtai, 1981). Albendazole may produce regression of cysts, but it is teratogenic and embryotoxic in experimental animals (Morris et al., 1983).

Prognosis. The prognosis is fair in the unilocular type, where the cyst is in an operable site and can be removed without spillage of the cyst fluid or hydatid sand into the operative cavity. However, recurrence of the disease, due primarily to the development of secondary cysts from scolices spilled from the primary cyst at the time of operation, may be anticipated in 50% of the cases cared for by experienced surgeons. In the case of patients with unilocular cysts lodged in critical centers, the prognosis is usually grave.

Prevention. Since human infection results from contact with infected dogs or from ingesting eggs of the parasite in dirt, on vegetables, on utensils, or from contaminated clothing, strict personal hygiene is recommended. The eggs of *E. granulosus* are resistant to some disinfecting agents (Meymarian and Schwabe, 1962). Dogs should be prevented from eating the carcasses of slaughtered animals in endemic areas. Whenever large numbers of stray dogs are killed in order to control outbreaks of rabies, there is a decline for several seasons in the development of *E. granulosus* hydatid infection in man and other hosts.

In Cyprus, at the start of a recent control program, the annual incidence of surgical cases of echinococcosis was 13 per 100,000 and approximately half of a large population of dogs carried the infection. After 9 years of intensive control measures, the prevalence in dogs was less than 0.2%, infection in sheep was present in the oldest animals only, and the incidence of hydatid infection in the people had declined likewise (Polydoreau, 1980). These favorable results were accomplished largely by elimination of stray dogs, periodic treatment of pet and working dogs, and the control of abattoirs. Because treated dogs are readily reinfected (Wikerhauser et al., 1980), periodic mass retreatment, or routine examination and treatment of infected dogs, is an essential part of the echinococcosis control program. In general, anthelmintics that are effective against other cestodes can be used for *E. granulosus* in dogs. Arecoline hydrobromide (1 to 2 mg/kg) was commonly used in the past. Praziquantel (5 mg/kg) is effective against both immature and adult stages of cestodes, and it probably is now the drug of choice for the treatment of tapeworm infection in dogs and cats (Wikerhauser et al., 1976; Soulsby, 1982). Education of the public is also essential.

Echinococcus multilocularis
(Leuckart, 1863) Vogel, 1955
(The hydatid worm producing alveolar hydatid disease)

Synonyms. *Taenia echinococcus multilocularis* Leuckart,

1863; *Echinococcus alveolaris* Klemm, 1883; *Echinococcus sibiricensis* Rausch and Schiller, 1954.

Historical and Geographical Notes. For more than a century pathologists, including Virchow (1855), recognized a type of hydatid cyst in human necropsy material morphologically different from the typical unilocular cyst. Leuckart (1863) designated this variant as *T. echinococcus multilocularis* and Klemm (1883) as *E. alveolaris*. Other distinguished workers, including Dew (1953), supported the unitarian concept of Cameron (1927), who described how an alveolar type might arise from a young unilocular cyst. Cameron (1926) and Ortlepp (1934) recognized several different species of *Echinococcus* in the adult stage in wild carnivores in South Africa and suggested that these species might produce hydatid cyst in man, but these investigators provided no experimental evidence supporting this view. The investigations of Rausch and Schiller (1951), Thomas et al. (1954), and Fay (1973) on St. Lawrence Island, Alaska, incriminating tundra mice as intermediate hosts and dogs, foxes, and wolves as definitive hosts of an alveolar type of hydatid, focused attention anew on the subject. The research of Vogel (1955) in South Germany indicated that this type of alternating host relationship prevails in the endemic alveolar hydatid area of central and eastern Europe.

Alveolar hydatid in man is most frequently found in southern Germany, Switzerland, the Tyrol, northwestern China and U.S.S.R. to the Bering Sea, but cases have been reported from Iceland, northern Germany, Italy, France, Poland, the Balkans, Uruguay, Argentina, Australia, New Zealand, Rebun Island off the coast of Hokkaido, midwestern Japan, England, in reservoirs from Banks Island in the Canadian western Arctic, St. Lawrence Id., Alaska (Fay, 1973), mainland Alaska (Rausch, 1956), Bering Island, Middle Kuriles (Rausch, 1958), North Dakota (Rausch and Richards, 1971), Minnesota (Gamble et al., 1979), and Montana, United States (Seesee and Worley, 1976; Eastman and Worley, 1979), and from Turkey (Fig. 31–29).

Morphology, Biology, and Life Cycle. The adult *Echinococcus multilocularis,* which lives attached to the mucosa of the small bowel of the canine definitive host, is appreciably smaller than *E. granulosus:* 1.2 to 3.7 mm vs 3.0 to 6.0 mm long (Figs. 31–25, 31–30). In the sexually mature proglottid the genital pore is anterior to the equatorial plane. The number of testes is 16 to 26 rather than 45 to 65, and they are situated posterior to the cirrus pouch. The uterus lacks lateral outpockets often present in *E. granulosus* (Vogel, 1955; Rausch, 1956). The eggs are typically *Taenia*-like and are highly resistant to cold (Schiller, 1955). The prepatent period in dogs is 30 to 35 days (Sakamoto et al., 1971), in foxes, 32 days (Fay, 1973).

Posselt (1904) infected a dog by feeding scolices obtained from a human alveolar cyst. Study of this material by Vogel (1955) indicated that they cor-

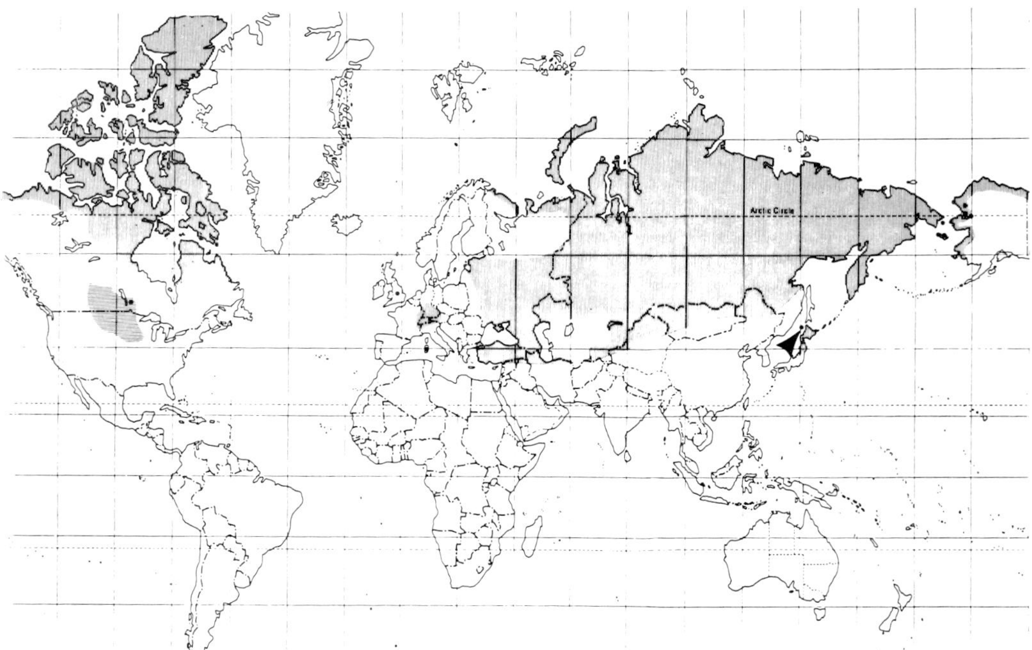

Fig. 31–29. *Echinococcus multilocularis*. Approximate world distribution. The vertical shading in the higher latitudes indicates the distribution of the arctic fox and, elsewhere, areas of enzootic *E. multilocularis* infection in red foxes. These species are the usual natural definitive hosts. The horizontal shading indicates the area in Canada and northern United States where the infection has been recorded in foxes and rodents since 1960. The black dots show the locations of sporadic cases of human infection in Alaska, Canada, England, and Sardinia. Numerous other cases have been reported earlier from Central and Eastern Europe and Asian U.S.S.R. The arrow points to Rebun Island in the Sea of Japan, where the rate of human infection is 1 percent. (Adapted by R.C. Jung from Rausch, R. 1967. On the ecology and distribution of *Echinococcus* sp. (Cestoda: Taeniidae), and characteristics of their development in the intermediate host. Ann. Parasitol., *42*:19–63.)

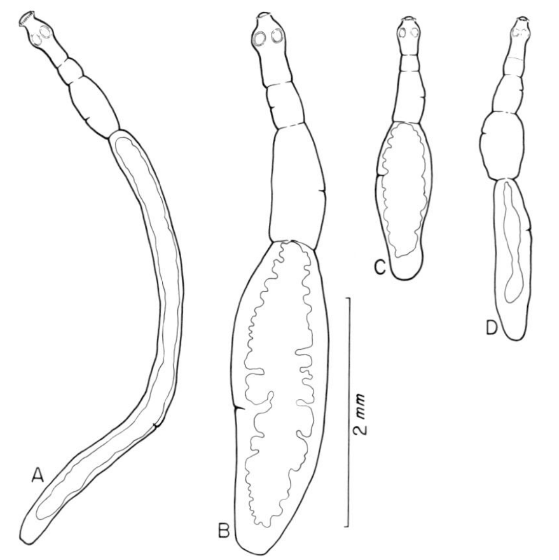

Fig. 31–30. *Echinococcus vogeli* and three other species. Drawings to the same scale to show relative size and general features. *A, E. vogeli* from naturally infected bush dog. *B, E. granulosus,* from a dog, 67 days after experimental infection. *C, E. oligarthrus,* from jaguarundi, natural infection. *D, E. multilocularis,* from arctic fox, 37 days after experimental infection. (From Rausch, R., and Bernstein, J.J. 1972. *Echinococcus vogeli* sp. n. (Cestoda: Taeniidae) from the bush dog, *Speothos venaticus* (Lund). Z. Tropenmed. Parasitol., *23*:25–34.)

respond closely to those he had obtained from a naturally infected red fox *(Vulpes vulpes)* and experimentally infected dog. In south Germany, Vogel found the field mouse *(Microtus arvalis)* to be the natural host of the alveolar hydatid stage. On Rebun Island, Japan, the red-backed vole and red fox or feral dogs are the larval and adult reservoirs (Rausch and Yamashita, 1957). In Alaska, Rausch and Schiller (1951) found the larval worms in the tundra vole *(Microtus oeconomus)* and the adults in dogs, foxes, and wolves. On St. Lawrence Island, Alaska, Thomas et al. (1954) reported the cystic stage in field mice *(Clethrionomys rutilis),* vole *(Microtus oeconomus),* ground squirrel *(Citellus undulatus),* and shrew *(Sorex jacksoni),* mostly in the liver, and adult worms in the Arctic fox *(Alopex lagopus).* Leiby and Nickel (1968) were able to produce infection in the deer mouse *(Peromycus maniculatus)* by feeding this host ground beetles previously exposed to fresh feces from a fox *(Vulpes vulpes)* naturally infected with *E. multilocularis.* Larval growth occurs through evagination with hollow protuberances of the cyst wall, and by exogenous budding of solid and bare extensions of the germinal membrane with progressive invasion of the organ (90% in the liver) (Vogel, 1978). The rate of proliferation and patterns of development and pathology differ in the various hosts (Ohbayashi et al., 1971).

Epidemiology. According to Vogel (1955), man becomes infected sporadically in Europe from eggs of *E. multilocularis* in the feces of foxes, by direct contaminative contact with polluted ground, or when picking strawberries, huckleberries, or cranberries. However, the sources of infection are generally unknown. Natural infections in the dog and cat have been reported from the endemic area of Baden-Württemberg in southern Germany (Eckert et al., 1974). In Fukui, a midwestern province of Japan where infection in animals is unknown, a fatal infection was reported in man. About 170 proven cases have been reported from Japan, all but 5 from Hokkaido and the northernmost part of the main island of Honshu (Yoshimura et al., 1979). Reports from China describe numerous cases in the northwestern provinces, especially Qinghai (Han et al., 1981) and Gansu (Jiang, 1981). In U.S.S.R. the endemic foci are in eastern and western Siberia, in the middle-east republics (Khirghiz, Uzbek), Kazakkstan, parts of Upper and Middle Volga, and the Ural Caucasian republics (Azerbaijan, Armenian, Georgian S.S.R.), and Moldavian S.S.R. (Lukashenko, 1971). The main sources of infection in those regions are working dogs, drinking water from melted ice, and the skins of fur animals. Infections are probably acquired most frequently in childhood and become symptomatic after many years.

Pathogenesis, Pathology, and Symptomatology. The alveolar hydatid in man consists of invasive parasite tissue, grossly appearing amorphous, and composed of numerous minute irregular spaces containing crumpled hyaline membrane and little or no fluid. The entire irregular mass is enclosed in a relatively avascular fibrous adventitia. Scolices, abundant in the vole, are rarely present in man (Fig. 31–31). There is a lack of encapsulation in some areas. Central necrosis and cavitation commonly occur. The alveolar hydatid is characterized by deficiency in the production of a hyaline membrane and by host tissue reaction consisting of infiltration of eosinophils and chronic inflammatory cells and fibrosis.

The hepatic lesion containing the alveolar hydatid appears grossly like an invasive neoplasm. It has been mistaken for carcinoma even after micro-

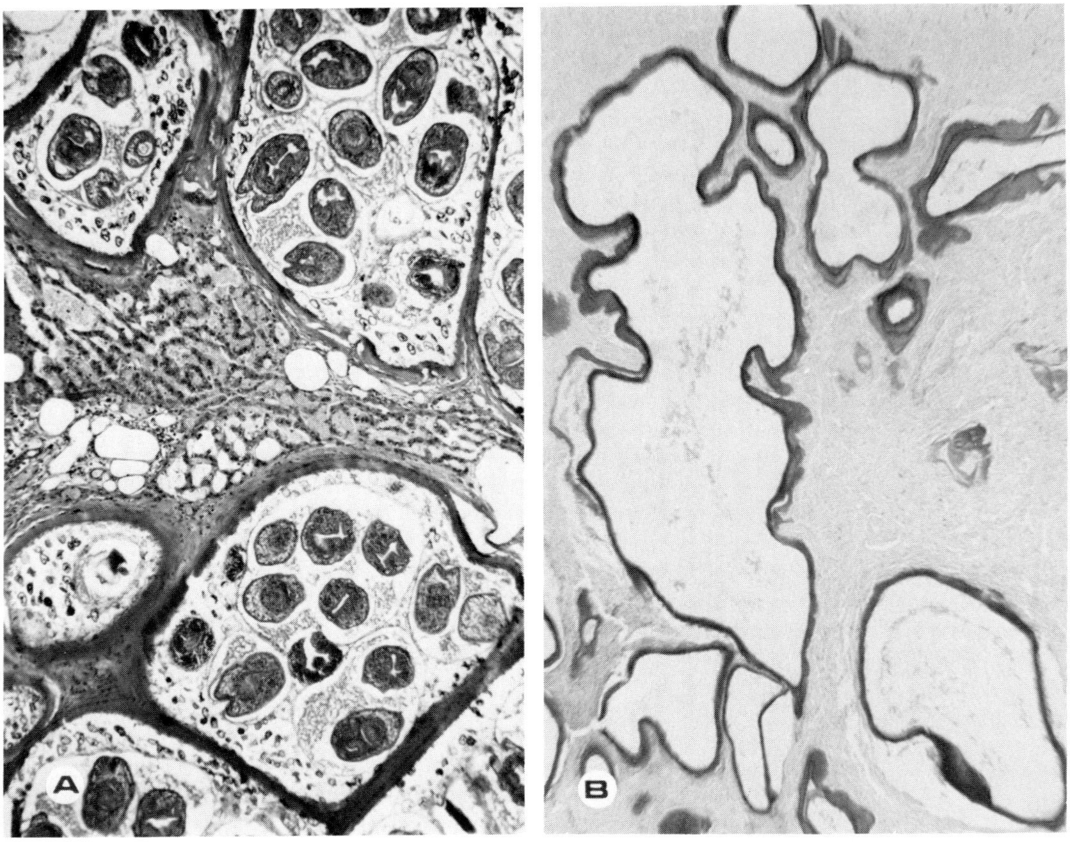

Fig. 31–31. *Echinococcus multilocularis. A,* Alveolar hydatid in liver of a natural host, the vole, showing normally developed protoscolices in brood capsules with thin laminated membrane and thick germinal membrane bearing abundant calcareous corpuscles. (× 160.) *B,* Histological appearance of alveolar hydatid in a section of human liver, showing absence of protoscolices and calcareous corpuscles; the laminated membrane is darkly stained with periodic acid Schiff stain. (× 100.) (From Wilson, J.F., and Rausch, R. 1980. Alveolar hydatid disease. A review of clinical features of 33 indigenous cases of *Echinococcus multilocularis* infection in Alaskan Eskimos. Am. J. Trop. Med. Hyg., *29*:1340–1355.)

scopic examination of sections by pathologists unfamiliar with the disease.

The course of the disease resembles that of a slowly growing carcinoma. It is among the most lethal of all the helminthic infections of man. Among untreated patients, 70% progress to death (Wilson and Rausch, 1980). In the course of the infection the liver becomes enlarged, sometimes to a marked degree. Destruction of liver parenchyma may lead to hepatic failure. Obstructive effects from the growing mass and contraction of cicatricial tissue produce biliary and portal obstruction. The resulting picture of malnutrition, ascites, and evidence of portal hypertension resembles that associated with hepatic cirrhosis.

Alveolar hydatidosis resembles carcinoma also in its ability to metastasize. Common sites of metastasis include lung and brain. The growth may extend through the diaphragm, chest, and abdominal wall.

Diagnosis. The diagnosis is likely to be missed not only by the examining physician but even by the surgeon at laparotomy and the pathologist if they are not familiar with the condition. Diagnosis in such cases is difficult because in man *E. multilocularis* rarely produces protoscolices and identification must depend on recognition of the characteristic germinal and laminated membranes (Fig. 31–31). These cysts are said to be "sterile" but, when transferred to a normal host, even small fragments of germinal membrane obtained in a biopsy specimen are capable of generating normal cysts with infective protoscolices and characteristic calcareous corpuscles (Rausch and Wilson, 1973).

Immunologic tests described for *E. granulosus* hydatids are equally applicable to those caused by *E. multilocularis*. However, the false-positive results for hydatidosis in cases of neoplasms may cause greater confusion in areas where *E. multilocularis* is endemic, owing to the carcinoma-like character of the lesion and disease. A simplified counterimmunoelectrophoresis method that may eliminate false positives was reported by Mansueto et al. (1980).

A positive serologic test is a useful screening procedure that should prompt further investigation to detect the infection. Early infections are more susceptible to cure either by surgical extirpation or by medical treatment *(vide infra)*.

Blood counts and blood chemistries are not very helpful. Eosinophilia is not the rule. The principal chemical abnormalities are those related to the damaged liver.

Abdominal roentgenograms show hepatomegaly and diffuse, small (2 to 4 mm in diameter) radiolucencies outlined by calcific densities. Liver scan with technetium sulfur-colloid demonstrates a filling defect. Angiography is helpful in delineating the lesion but does not differentiate it from carcinoma. However, early venous filling and arteriovenous shunts seen in carcinoma are not characteristic of alveolar hydatid. CT (computerized tomography) scan is also useful in ascertaining the size and shape of the lesion, and ultrasound depicts cavities. These tests are not useful in establishing etiologic diagnosis nor are they reliable in evaluating resectability (Wilson and Rausch, 1980).

Treatment. Surgical removal is the only reliably satisfactory method of treatment provided that the disease is diagnosed early enough that the entire hepatic lesion can be resected. Of seven patients having such a resection, all survived (Wilson and Rausch, 1980).

Recent studies have shown that mebendazole or a related drug may control or eliminate the infection. The drug is given in a dosage of 40 mg per kg body weight per day continuously for years. Average minimal survival was 8.2 years compared with 5 years in the untreated patient. Average minimal survival in the patients with hepatic resection was 14.7 years.

Prognosis. The prognosis is almost invariably grave.

Prevention. Alveolar hydatid is usually diagnosed postmortem in persons of middle age or older. Because of the slow growth of the parasite, it seems probable that exposure occurred in childhood. Hence, emphasis must be placed on the control of infection in dogs and cats having contact with children. One must take into account the infectivity of eggs in freshly passed feces as well as in contaminated soil. As eggs appear in the feces 30 to 35 days after infection, dogs should be treated about every 25 to 30 days (Sakamoto et al., 1971).

Echinococcus vogeli
Rausch and Bernstein, 1972

Historical and Geographical Notes. The original description of *E. vogeli* in 1972 was based on a collection of adult worms obtained by anthelmintic treatment of a mature bush dog *(Speothos venaticus* [Lund]) following the discovery of a proglottid during routine fecal examination at a zoo in Los Angeles. The animal had been recently captured in northwestern Ecuador. The geographic range of the bush dog (family Canidae, subfamily Simocyoninae) extends over Central America and western South America from Panama to Paraguay. The larval stage of *E. vogeli* was first identified by D'Alessandro et al. (1979) in surgical specimens from two patients in Colombia, the identification being based on examination of adult worms obtained by feeding parts of the cysts to dogs. Studies by Rausch et al. (1981) and D'Alessandro et al. (1981) demonstrated that the natural intermediate host of *E. vogeli* is the paca, *Cuniculus paca*, as had been predicted by Rausch and Bernstein (1972) on the basis of known prey-predator relationships of the paca and bush dog.

Morphology, Biology, and Life Cycle. The strobila of *E. vogeli* is relatively long and slender, measuring up to about 5.6 mm long, with the scolex and three segments (immature, mature, and gravid) of similar width, about 0.3 mm (Fig. 31–30). The rostellum is notably large, up to 0.188 mm wide, with 28 to 36 hooks arranged in two rows and measuring 49 to 57 μm and 30 to 47 μm (average, 53 and 47 μm) long. The gravid segment is slender, with almost parallel lateral margins and nearly round cross section, and measures roughly 3 to 4 mm long by 0.3 mm wide. In the mature segment, the genital pore is lateral and slightly posterior in position. The cirrus sac extends to or beyond the midline, and persists in gravid segments. The testes, about 56 in number, are distributed in essentially all space between the longitudinal excretory canals not occupied by other organs. The ovary, slightly lobed, occupies most of the central field in the posterior third of the segment. The gravid uterus is sac-like, extending through most of the segment's length. The eggs measure about 35 μm in diameter.

Details of morphology and development of the larval stage were described by Rausch et al. (1981). In the paca and nutria the larva is a polycystic hydatid formed by endogenous proliferation of the germinal laminated membranes forming folds and pockets within the primary vesicle. Essentially all larvae are in the liver, mostly somewhat protruding beneath Glisson's capsule. The cyst is a delicate germinal membrane covered by a relatively thick laminated membrane, which in turn is held in a connective tissue capsule 70 to 250 μm thick. However, the wall of living cysts is translucent, with brood capsules clearly evident beneath it. The laminated membrane is variable in thickness (20 to 400 μm). The germinal layer in both vesicle and brood capsules is extremely thin and contains essentially no calcareous corpuscles though they are numerous in protoscolices. Brood capsules measure 0.4 to 1.5 mm in diameter and contain numerous protoscolices (up to nearly 500) with average measurements of 175 by 133 μm. Development of the larva to the stage containing infective protoscolices requires several months. At 90 days in an experimental nutria, the primary vesicles were 5 mm or less in diameter and contained no protoscolices. Adult worms mature and produce gravid proglottids in less than 129 days in the dog (Rausch et al., 1981). The life span of this stage is unknown. The larval stage may show signs of degeneration in the paca, but this animal has a relatively long natural life, estimated to be about 12 years.

Epidemiology. The natural cycle of infection is maintained in the paca and bush dog. Infection in man could be acquired from contact with environments contaminated by feces from bush dogs, but such occurrences would be rare. While it has little if any importance in maintaining infection in the paca, the domestic dog is probably the chief source of infection in man. Fifteen cases of polycystic hydatid disease have been reported from Costa Rica, Panama, Colombia, Ecuador, and Venezuela. In ten cases the parasite was examined in detail; all were *E. vogeli*.

Pathogenesis, Pathology, and Symptomatology. In the natural host where exogenous proliferation does not occur, the larvae of *E. vogeli* are essentially nonpathogenic except when the infection is massive. In man, however, the parasite is invasive and is spread by exogenous budding, producing a pathologic process similar to that of hepatic carcinoma (D'Alessandro et al., 1979). Among 13 cases of proven polycystic hydatid disease, 8 had had an undiagnosed tumor in or near the liver, 1 a subcutaneous mass in the chest wall, and 2 had cysts in the chest cavity. The cysts produced by *E. vogeli* in man have been relatively large, filled with fluid, and often with numerous brood capsules filled with protoscolices.

Diagnosis. Usually the diagnosis has been based on examination of surgical specimens or tissues obtained at autopsy. In one case, an immunoelectrophoresis test was positive after polycystic hydatidosis with hepatic, pericardial, and pulmonary involvement had been diagnosed by examination of tissue (Varela-Diaz et al., 1978). Serodiagnostic tests soon after operation in two other cases and two years after surgery in one, showed that *E. vogeli* shares antigens with other *Echinococcus* species (D'Alessandro et al., 1979).

Treatment. Surgical removal of the invaded tissues has been successful in some cases, and in others it was probably successful, although follow-up was too brief for adequate evaluation. A favorable experience with continuous mebendazole therapy in nonresectable cases of infection with *E. multilocularis* suggests consideration of medical therapy for *E. vogeli* infection (Wilson and Rausch, 1980).

Prognosis. Without surgical or medical treatment the prognosis would be unfavorable in most cases. In one case, however, it was reported that the parasite had apparently died within 5 years following a diagnostic biopsy (D'Alessandro et al., 1979).

Prevention. Other than educational efforts, no public health measure against *E. vogeli* would be feasible.

Echinococcus oligarthrus (Diesing, 1863) Luke, 1910

Synonym. *Echinococcus cruzi* Brumpt and Joyeux, 1924.

Sousa and Lombardo (1965) reported a case of polycystic hydatid infection identified as *Echinococcus oligarthrus* in a 56-year-old native of the Province of Darien, Panama. However, based on comparative studies on the *Echinococcus* infections in man (D'Alessandro et al., 1979) and animals in Colombia and elsewhere in Central and South America (Rausch et al., 1978, 1981), it was concluded that no infection of man by the larval *E. oligarthrus* had been confirmed (D'Alessandro et al., 1981). Alveolar hydatid infections in man reported as *E. oligarthrus* or unidentified, from Ecuador, Panama, and Venezuela were apparently *E. vogeli*.

Known hosts for the larvae of *E. oligarthrus* are the agouti (*Dasyprocta* spp.), the spiny rat (*Proechimys* spp.), and the paca (*Cuniculus paca*). Hosts recorded for the adult worm are 6 species of cats: jaguarundi cat (*Felis yagouaroundi*), puma (*F. concolor*), jaguar (*F. onca*), ocelot (*F. pardalis*), Geoffroy's cat (*F. geoffroyi*), and pampas cat (*F. colocolo*) The anatomical location of the larvae is mostly extrahepatic, in the intermuscular connective tissues, especially in the legs and loin, and in the viscera of the pleural and peritoneal cavities. The larvae are structurally similar to those of *E. vogeli*, but in *E. oligarthrus* the large and small rostellar hooks are shorter (30 to 33 and 25 to 27 μm vs 39 to 42 and 32 to 34 μm, respectively), the brood capsules are smaller and more delicate (average, 0.39 by 0.35 mm) and contain fewer protoscolices (6 to 30), and the protoscolices are smaller (up to 163 by 142 μm). The laminated membrane is relatively thin, and the germinal membrane is relatively thick, generally bearing numerous calcareous corpuscles. In the adult stage *E. oligarthrus* is the smallest of the four known species, measuring up to 2.9 mm long by 0.4 mm wide (Fig. 31–30). In both larval and adult stages a reliable character for the specific identification of *Echinococcus* species is the size of rostellar hooks, as there is no indication that these structures increase in length after protoscolices reach full development and after the adult reaches sexual maturity (Rausch et al., 1978).

REFERENCES

Addae, R.O. 1972. Human coenurosis—the first reported case in Ghana. Ghana Med. J., *11*:424.

Africa, C.M., and Garcia, E.Y. 1935. A rat tapeworm (*Raillietina garrisoni*) transmissible to man. Philippine J. Public Health, *1*:44.

Afshar, A. 1967. Cysticercosis in Iran. Ann. Trop. Med. Parasitol., *61*:101–103.

Arru, E., and Nieddu, A.M. 1978. Echinococcosis of the dog in Sardinia. Nuovo Progr. Vet., *33*:434–438.

Arundel, J.H., and Adolph, A.J. 1980. Preliminary observations on the removal of *Taenia saginata* eggs from sewage using various treatment processes. Aust. Vet. J., *56*:492–495.

Asada, J. 1923. Studies on the *Hymenolepis diminuta* Rudolphi, which finds the host in a dog. Okayama Igakki Zasshi, *407*:876–880. (English abstract in Jpn. Med. World, *4*:40.)

Bacigalupo, J. 1922. Sobre una nueva especie de Taenia, *Taenia infantis*. Semana Med., *26*:726.

———. 1931. Evolution de l'*Hymenolepis fraterna* Stiles, chez *Pulex irritans* L., *Xenopsylla cheopis* Rothschild et *Ctenocephalus canis* Curtis. Ann. Parasitol., *9*:339–343.

Baer, J.G. 1956. The taxonomic position of *Taenia madagascariensis* Davaine, 1870, a tapeworm parasite of man and rodents. Ann. Trop. Med. Parasitol., *50*:152–156.

———, and Sandars, D.F. 1956. The first record of *Raillietina (Raillietina) celebensis* (Janicki, 1902) Cestoda) in man from Australia with a critical survey of previous cases. J. Helminthol., *30*:173–182.

———, and Scheidegger, S. 1946. Un cas interessant de tetraplegie d'origine parasitaire. Schweiz. Z. Pathol. Bakteriol., *9*:61–66.

Bailenger, J., and Carcenac, F. 1970. Une nouvelle localisation géographique d'*Inermicapsifer madagascariensis* (Davaine, 1870) Baer 1956: La Réunion. Bull. Soc. Pathol. Exot., *53*:242–244.

Bailey, W.S. 1951. Host-tissue reactions to initial and superimposed infections with *Hymenolepis nana* var. *fraterna*. J. Parasitol., *37*:440–444.

Bamrungphol, V., Little, M.D., Virayavejakul, A., Sujtanond, M., and Chaiyaporn, V. 1972. Juvenile strobilate tapeworm from the human spinal cord. Report of a case. Am. J. Trop. Med. Hyg., *21*:435–439.

Baranski, M.C., Gomes, N.R., de Godoy, O.F., da Silva, A.F. Kotaka, P.I., Giovannoni, M., and Carneiro, F.M. 1980. Terapêutica da teníase e da hymenolepiase nana com dose oral unica de praziquantel. Estudio da eficácia, tolerância e segurança. Rev. Inst. Med. Trop. S. Paulo, *22*:82–88.

Barlow, J.F., and Church, B.G. 1969. Coenurosis in the brain of a child from South Dakota. South Dakota J. Med., *22*(November):37–44.

Bartholomew, R.S. 1975. Subretinal cysticercosis. Am. J. Ophthalmol., *79*:670–673.

Beaver, P.C., and Rolon, F.A. 1981. Proliferating larval cestode in a man in Paraguay. A case report and review. Am. J. Trop. Med. Hyg., *30*:625–637.

Becker, B., Mehlhorn, H., Andrews, P., and Thomas, H. 1980. Scanning and transmission electron microscope studies on the efficacy of praziquantel on *Hymenolepis nana* (Cestoda) in vitro. Z. Parasitenkd., *61*:121–133.

Becklund, W.W. 1970. Current knowledge of the gid bladder worm, *Coenurus cerebralis* (=*Taenia multiceps*) in North American domestic sheep, *Ovis aries*. Proc. Helm. Soc. Wash., *37*:200–204.

Benger, A., Rennie, R.P., Roberts, J.T., Thornley, J.H., and Scholten, T. 1981. A human coenurus infection in Canada. Am. J. Trop. Med. Hyg., *30*:638–644.

Berman, J.D., Beaver, P.C., Cheever, A.W., and Quindlen, E.A. 1981. Cysticercus of 60-milliliter volume in human brain. Am. J. Trop. Med. Hyg., *30*:616–619.

Biagi, F., Velez, G., and Gutierrez, M.L. 1965. Destrucción de los cisticercos en la carne de cerdo parasitada. Bol. Of. Sanit. Panam., *58*:303–307.

Biswas, H., Arora, R.R., and Sehgal, S. 1978. Epidemiology of *Hymenolepis nana* infection in a selected rural community. J. Commun. Dis., *10*:170–174.

Blamire, R.V., Goodhand, R.H., and Taylor, K.S. 1980. A review of some animal diseases encountered at meat inspections in England and Wales, 1969-1978. Vet. Rec., *106*:195–199.

Botero R., D., and Castaño, M. 1981. Cisticercosis: tratamiento con praziquantel. Trib. Med. Colombia, *63*:31–36.

Cadigan, F.C., Jr., Stanton, J.S., Tanticharoenyos, P., and Chaicumpa, V. 1967. The lar gibbon as definitive and intermediate host of *Taenia solium*. J. Parasitol., *53*:844.

Caley, J. 1975. A comparative study of the two alternative larval forms of *Hymenolepis nana*, the dwarf tapeworm, with special reference to the process of encystment. Z. Parasitenkd., *47*:217–235.

Cameron, T.W.M. 1926. Observations on the genus *Echinococcus* Rudolphi, 1801. J. Helminthol., *4*:13–22.

———. 1927. Some modern biological conceptions of hydatid. Proc. R. Soc. Med., *20*:302–311.

Chandler, A.C. 1942. First record of a case of human infection with tapeworms of the genus *Mesocestoides*. Am. J. Trop. Med., *22*:493–497.

———. 1949. *Introduction to Parasitology*, 8th ed., p. 353. New York, John Wiley & Sons.

Chen, H.T. 1934. Reactions of *Ctenocephalides felis* to *Dipylidium caninum*. Z. Parasitenkd., *6*:603–633.

Cho, K.-M., Hong, S.-O., Kim, C.-H., Soh, C.-T., Kim, S.-H., Kim, O.-N., Kim, S.-H., and Yun, H.-C. 1967. Studies on taeniasis in Cheju-Do. J. Korean Modern Med., *7*:455–461.

Choi, W.Y., Kim, B.C., and Choi, H.S. 1967. The first case of human infection with tapeworms of the genus *Mesocestoides* in Korea. Korean J. Parasitol., *5*:60–64.

Connor, R.J., and Hizzard, P. 1979. Equine hydatidosis. Vet. Rec., *104*:417.

Costa, H.M. de A., Correa, L., and Brener, Z. 1967. Novo caso humano de parasitismo por *Bertiella mucronata* (Meyner, 1895) Stiles and Hassall, 1902 (Cestoda–Anoplocephalidae). Rev. Inst. Med. Trop. S. Paulo, *9*:95–97.

Costa, L. 1960. Conceito e programa da luta contra a hidatidose no Brasil. Rev. Brasil. Malariol. Doencas Trop., *12*:119–128.

Dada, B.J.O. 1980a. Taeniasis, cysticercosis and echinococcosis/hydatidosis in Nigeria: I. Prevalence of human taeniasis, cysticercosis and hydatidosis based on a retrospective analysis of hospital records. J. Helminthol., *54*:281–286.

———. 1980b. Taeniasis, cysticercosis and echinococcosis/hydatidosis in Nigeria. II. Prevalence of bovine and porcine cysticercosis, and hydatid disease in slaughtered food animals based on retrospective analysis of abattoir records. J. Helminthol., *54*:287–291.

———. 1980c. Taeniasis, cysticercosis and echinococcosis/hydatidosis in Nigeria. III. Prevalence of bovine and porcine cysticercosis and hydatid cyst infection based on joint examination of slaughtered food animals. J. Helminthol., *54*:293–297.

D'Alessandro, A., Beaver, P.C., and Masi Pallares, R. 1963. *Bertiella* infection in man in Paraguay. Am. J. Trop. Med. Hyg., *12*:193–198.

———, Rausch, R.L., Cuello, C., and Aristizabal, N. 1979. *Echinococcus vogeli* in man, with a review on polycystic hydatid disease in Colombia and neighboring countries. Am. J. Trop. Med. Hyg., *28*:303–317.

———, Rausch, R.L., Morales, G.A., Collet, S., and Angel, D. 1981. *Echinococcus* infections in Colombian animals. Am. J. Trop. Med. Hyg., *30*:1263–1270.

Daniels, C.W. 1895. *Taenia demerariensis*. Br. Guiana Med. Ann., *7*:96–98.

de Buen, S. 1970. Cisticercosis ocular. Gac. Med. Mexico, *100*:137–144.

De Graff, A.S., Shaughnessy, P.D., McCully, R.M., and Verster, A. 1980. Occurrence of *Taenia solium* in a Cape fur seal *(Arctocephalus pusillus)*. Onderstepoort J. Vet. Res., *47*:119–120. (Helm. Abstr., *50*, No. 618).

de Rivas, D. 1937. Cysticercus bovis in man. *Papers on Helminthology Published in Commemoration of the 30-Year Jubileum of K.J. Skrjabin*, Moscow, pp. 569–570.

Dewhirst, L.W., Cramer, J.D., and Sheldon, J.J. 1967. An analysis of current inspection procedures for detecting bovine cysticercosis. J. Am. Vet. Med. Assoc., *150*:412–417.

Dissanaike, A.S., Thomas, V., and Nagappan, N. 1977. *Bertiella studeri* (Blanchard, 1891) Stiles and Hassall, 1902 infection in a child—first case from Malaysia. Southeast Asian J. Trop. Med. Public Health, *8*:421–422.

Dixon, H.B.F., and Smithers, D.W. 1934. Epilepsy in cysticercosis *(Taenia solium)*. A study of seventy-one cases. Q.J. Med., *3*:603–616.

Donckaster, R., and Habibe, O. 1958. Contribución al estudio de la infección por *Hymenolepis nana*. Bol. Chil. Parasitol., *13*:9–11.

Eastman, K.L., and Worley, D.E. 1979. The muskrat as an intermediate host of *Echinococcus multilocularis* in Montana. J. Parasitol., *65*:34.

Eckert, J., Müller, B., and Partridge, A.J. 1974. The domestic cat and dog as natural definitive hosts of *Echinococcus (Alveococcus) multilocularis* in Southern Federal Republic of Germany. Tropenmed. Parasitol., *25*:334–337.

Edirisinghe, J.S., and Cumararajan, S.M. 1976. The first record of *Bertiella studeri* infection in a child from Sri Lanka. Ceylon Med. J., *22*:137–140.

Esch, G.W. 1964. The effects of cortisone and pre-infection starvation in the establishment of larval *Multiceps serialis* in mice. J. Elisha Mitchell Sci. Soc., *80*:114-120.

Espejo, H. 1977. Tratamiento de infecciones por *Hymenolepis nana*, *Taenia saginata*, *Taenia solium* y *Diphyllobothrium pacificum* con praziquantel. Bol. Chil. Parasitol., *32*:39–40.

Fain, A., Denisoff, N., Homans, L., Questiaux, G., Van Laere, L., and Vincent, M. 1956. Cénurose chez l'homme et les animaux du à *Taenia brauni* Setti au Congo Belge et au

Ruanda Urundi. II. Rélation de huit cas humains. Ann. Soc. Belge Med. Trop., *36*:679–696.

———, and Herin, V. 1954. Notes à propos d'un cas d'infestation humaine par un *Mesocestoide* à Astrida (Ruanda-Urundi). Ann. Soc. Belge Med. Trop., *34*:893–900.

Fay, F.H. 1973. The ecology of *Echinococcus multilocularis* Leuckart, 1863 (Cestoda: Taeniidae) on St. Lawrence Island, Alaska. I. Background and rationale. Ann. Parasitol., *48*:523–542.

Fontan, C. 1919. *Cysticercus bovis* chez l'homme localise à la region mammaire. *Taenia* inerme de l'intestin. Parasitisme adulte et larvaire chez le même sujet. Gaz. Hospital, *92*:183.

Francis, M., Machado de Mendonca, J., Bezerra Brito, D., and Carvalho Rodriguez, P. 1979. Cisticercose bovina, suina e humana e suas relações com as teniases humanas. At. Soc. Biol. Rio de Janeiro, *20*:1–6.

Freeman, R.S. 1962. Studies on the biology of *Taenia crassiceps* (Zeder, 1800) Rudolphi, 1810 (Cestoda). Can. J. Zool., *40*:969–990.

———, Fallis, A.M., Shea, M., Maberley, A.L., and Walters, J. 1973. Intraocular *Taenia crassiceps* (Cestoda). Part II. The parasite. Am. J. Trop. Med. Hyg., *22*:493–495.

French, C.M., Nelson, G., and Wood, M. 1982. Hydatid disease in the Turkana District of Kenya. I. The background to the problem with hypotheses to account for the remarkably high prevalence of the disease in man. Ann. Trop. Med. Parasitol., *76*: 425–437.

Froyd, G. 1962. Longevity of *Taenia saginata* eggs. J. Parasitol., *48*:279.

Gajdusek, D.C. 1978. Introduction of *Taenia solium* into West New Guinea with a note on an epidemic of burns from cysticercus epilepsy in the Ekari people of the Wissel Lakes area. Papua New Guinea Med. J., *21*:329–342.

Gamble, W.A., Segal, M., Schantz, P.M., and Rausch, R.L. 1979. Alveolar hydatid disease in Minnesota. First human case acquired in the contiguous United States. J.A.M.A., *241*:904–907.

Garrick, D. 1967. Estado actual de la brucelosis, tuberculosis, rabia y cysticercosis en Centro-america y Panama. Bol. Ofic. San. Panam., *63*:143–150.

Gleason, N.N., Kornblum, R., and Walzer, P. 1973. *Mesocestoides* (Cestoda) in a child in New Jersey treated with niclosamide (Yomesan). Am. J. Trop. Med. Hyg., *22*:757–760.

Goddeeris, B. 1980. The role of insects in dispersing eggs of tapeworms, in particular *Taeniarhynchus saginatum*. I. Review of the literature. Ann. Soc. Belge Med. Trop., *60*:195–201.

Goldsmid, J.M., and Muir, M. 1972. *Inermicapsifer madagascariensis* (Davaine, 1870) Baer, 1956 (Platyhelminthes: Cestoda) as a parasite of man in Rhodesia. Cent. Afr. J. Med., *18*:205–207.

Gönnert, R., Meister, G., Strufe, R., and Webbe, G. 1967. Biologische Probleme bei *Taenia solium*. Zeitschr. Tropenmed. Parasitol., *18*:76–81.

Groll, E. 1980. Praziquantel for cestode infections in man. Acta Trop., *37*:293–296.

Guillory, S.L., and Zinn, K.M. 1980. Intravitreal *Cysticercus cellulosae:* Ultrasonographic and fluorescein angiographic features. Bull. N.Y. Acad. Med., *56*:655–661.

Gutierrez, Y., Buchino, J.J., and Schubert, W.K. 1978. *Mesocestoides* (Cestoda) infection in children in the United States. J. Pediatr., *93*:245–247.

Hammerberg, B., MacInnis, G.A., and Hyler, T. 1978. *Taenia saginata* cysticerci in grazing steers in Virginia. J. Am. Vet. Med. Assoc., *173*:1462–1464.

Han, F., Liu, R., Zhong, T., and Zhang, Z. 1981. Echinococcosis alveolaris in Qinghai Provence. Chin. Med. J., *94*:391–395.

Heath, D.D., and Osborn, P.J. 1976. Formation of *Echinococcus granulosus* laminated membrane in a defined medium. Int. J. Parasitol., *6*:467–471.

Hermos, J.A., Healy, G.R., Schultz, M.G., Barlow, J., and Church, W.G. 1970. Fatal human cerebral coenurosis. J.A.M.A., *213*:1461–1464.

Heyneman, D. 1962. Studies on helminth immunity. I. Comparison between lumenal and tissue phases of infection in the white mouse by *Hymenolepis nana*. Am. J. Trop. Med. Hyg., *11*:46–53.

Hira, P.R. 1975. Human and rodent infection with the cestode *Inermicapsifer madagascariensis* (Davaine, 1870) Baer, 1956 in Zambia. Ann. Soc. Belge Med. Trop., *55*:321–325.

Horstmann, R., Bienzle, U., Kern, P., and Voelker, J. 1978. Tapeworm infestation with *Inermicapsifer madagascariensis*. Tropenmed. Parasitol., *29*:406–408.

Hutchison, W.F., and Martin, J.B. 1980. *Mesocestoides* (Cestoda) in a child in Mississippi treated with paromomycin sulfate (Humatin). Am. J. Trop. Med. Hyg., *29*:478–479.

Iacona, A., Pini, C., and Vicari, G. 1980. Enzyme-linked immunosorbent assay (ELISA) in the serodiagnosis of hydatid disease. Am. J. Trop. Med. Hyg., *29*:95–102.

Jampol, L.M., Caldwell, J.B.H., and Albert, D.M. 1973. *Cysticercus cellulosae* in the eyelid. Arch. Ophthalmol., *89*:319–320.

Jepsen, A., and Roth, H. 1950. Parasitological and bacteriological problems of sewage disposal. Nord. Vet. Med., *2*:967–991.

Jiang, C. 1981. Liver alveolar echinococcosis in the Northwest. Report of 15 patients and a collective analysis of 90 cases. Chin. Med. J., *94*:771–778.

Johnstone, H.E., and Jones, O.W. 1950. Cerebral coenurosis in an infant. Am. J. Trop. Med., *30*:431–441.

Jones, W.E. 1979. Niclosamide as a treatment for *Hymenolepis diminuta* and *Dipylidium caninum* infection in man. Am. J. Trop. Med. Hyg., *28*:300-302.

Jueco, N.L. 1975. *Raillietina* (a rat tapeworm) infection in young children in the Philippines. Acta Med. Philip., *11*:49–50.

Jung, R.C., Rodriguez, M.A., Beaver, P.C., Schenthal, J.E., and Levy, R.W. 1981. Racemose cysticercus in human brain. A case report. Am. J. Trop. Med. Hyg., *30*:620–624.

Kagei, N., Kihata, M., Shimizu, S., Urabe, M., and Ishii, A. 1974. The 10th case of human infection with *Mesocestoides lineatus* (Cestoda: Cyclophyllidea) in Japan. Jpn. J. Parasitol., *23*:380–382.

Kaloyanov, Z., and Dobrev. B. 1979. Occurrence of hydatidosis in odd-toed ungulates in the Shumen region. Vet. Sbirka, *77*:28–29.

Kaminsky, R.G., Gatei, D.G., and Zimmermann, R.R. 1978. Human coenurosis from Kenya. East Afr. Med. J., *55*:355–359.

Keane, J.R. 1980. Cysticercosis acquired in the United States. Ann. Neurol., *8*:643.

Kiribayashi, S. 1933. Studies on the growth of *Hymenolepis nana*, with special reference to the possibility of differentiation of *H. nana* var. *fraterna*. Taiwan Igakkai Zasski, *32*:1175–1190.

Koivastik, T. 1963. Anoplocephalid tapeworm infection in a child. Journal-Lancet, *83*:63–65.

Krotov, A.I. 1961. The epidemiology of *Taenia saginata*. Med. Parasitol. Parasit. Dis. (Moscow), *30*:98–99.

———. 1979. On subspecies and strains of the genus *Echinococcus* Rudolphi, 1801. Med. Parasitol. Parasit. Dis. (Moscow), *48*:22–26.

Langer, M., Langer, R., and Rittmeyer, K. 1979. Angiographische Kriterien bei Leberechinokokkose. Fortschr. Geb. Röntgenstr. Nuklearmed., *131*:151–156.

Lech, Jr. 1949. Ocular cysticercosis. Am. J. Ophthalmol., *32*:523–547.

Leiby, P.D., and Nickel, M.P. 1968. Studies on sylvatic echinococcosis. I. Ground beetle transmission of *Echinococcus multilocularis* Leuckart, 1863, to deer mice, *Peromyscus maniculatus* (Wagner). J. Parasitol., *54*:536–537.

Lende, S., Enstad, B., and Henricksen, T. 1982. An unusually large unpunctured hydatid cyst. Ethiop. Med. J., *20*: 221–222.

Leuckart, R. 1863. *Die menschlichen Parasiten und die von ihnen herruhrenden Krankheiten*. Vol. I. Leipzig u. Heidelberg.

Lucas, S.B., Hassounah, O., Muller, R., and Doenhoff, M.J. 1980. Abnormal development of *Hymenolepis nana* larvae in immunosuppressed mice. J. Helminthol., *54*:75–82.

Lukashenko, N.P. 1971. Problems of epidemiology and prophylaxis of alveococcosis (multilocular echinococcosis): A general review—with particular reference to the U.S.S.R. Int. J. Parasitol., *1*:125–134.

Manschot, W.A. 1976. *Coenurus* infestation of eye and orbit. Arch Ophthalmol., *94*:961–964.

Mansueto, S., Migneco, G., La Cascia, C., and Miceli, M.D. 1980. Malignancies and false-positive serological results for hydatidosis. Br. Med. J., *2*:1284.

Mazzotti, L. 1944. Datos sobre la cisticercosis en México. Rev. Inst. Salub. y Enferm. Trop., *5*:283–292.

Merdivenci, A. 1964. An abnormal *Taenia saginata* with double genital pores. J. Parasitol., *50*:476–477.

Meymarian, E., and Schwabe, C. 1962. Host-parasite relationships in echinococcosis. VI. Resistance of the ova of *Echinococcus granulosus* to germicides. Am. J. Trop. Med. Hyg., *11*:360–364.

Morishita, K. 1972. Rare human tapeworms reported from Japan, pp. 465–488. In *Progress of Medical Parasitology in Japan*. Vol. 4. Edited by K. Morishita, Y. Komiya, and H. Matsubayashi. Tokyo, Meguro Parasitological Museum.

———, and Sawada, I. 1966. On tapeworms of the genus *Multiceps* hitherto unrecorded from man. Jpn. J. Parasitol., *15*:495–501.

Morris, D.L., Dykes, P.W., Dickson, B., Marriner, S., Bogan, J., and Burrows, F. 1983. Albendazole in hydatid disease. Br. Med. J., *286*: 103–104.

Mory, Y., Shirayama, T., Agui, Y., Fujiwara, A., and Nishiura, T. 1967. A case of chest wall tumor brought on by *Hymenolepis nana*. Bull. Osaka Med. Sch., *13*:52–54.

Mucciolo, P., Gissoni, R.H., and Spitz, W.B. 1968. Inspeção de cérebro no diagnóstico da cisticercose suina. Bol. Of. Sanit. Panam., *64*:219–224.

Murray-Lyon, I.M., and Reynolds, K.W. 1979. Complications of mebendazole treatment for hydatid disease. Br. Med. J., *2*:1111–1112.

New Zealand National Hydatids Council. 1979. *19th Annual Report for the year ended 31 March 1979*. Wellington, New Zealand, Ministry of Agriculture and Fisheries.

Niño, F.L. 1950. Tratamiento de la teniasis por *Taenia saginata* con sales de acridina. Prensa Med. Argent., *39*:547–549.

Obrador, S. 1948. Clinical aspects of cerebral cysticercosis. Arch. Neurol. Psychol., *59*:457–468.

Ohbayashi, M., Rausch, R.L., and Fay, F.H. 1971. On the ecology and distribution of *Echinococcus* spp. (Cestoda: Taeniidae), and characteristics of their development in the intermediate host. II. Comparative studies on the development of larval *E. multilocularis* Leuckart, 1863, in the intermediate host. Jpn. J. Vet. Res., *19*(suppl.):1–53.

Okello, G.B.A., and Chemtai, A.K. 1981. Treatment of hepatic hydatid disease with mebendazole: Report of 16 cases. East Afr. Med. J., *58*:608–610.

Orihel, T.C., Gonzalez, F., and Beaver, P.C. 1970. Coenurus from neck of Texas woman. Am. J. Trop. Med. Hyg., *19*:255–257.

Ortlepp, R.J. 1934. *Echinococcus* in dogs from Pretoria and vicinity. Onderstepoort J. Vet. Sci. Anim. Industry, *3*:97–108.

Patharangkura, P., Jariya, P., and Tandhavadhana, C. 1980. *Cysticercus racemosus* in the fourth ventricle: Report of two cases. Southeast Asian J. Trop. Med. Public Health, *11*:532–534.

Payne, E.H., Gonzalez-Muraburu, L., and Schleicher, E.M. 1956. An intestinal parasite survey in the High Cordilleras of Peru. Am. J. Trop. Med. Hyg., *5*:696–698.

Peña Chavarría, A., Villarejos, V.M., and Zeledón, R. 1977. Mebendazole in the treatment of taeniasis solium and taeniasis saginata. Am. J. Trop. Med. Hyg., *26*:118–120.

Perry, H.D., and Font, R.L. 1978. Cysticercosis of the eyelid. Arch. Ophthalmol., *96*:1255–1256.

Pineda, A., Botero, D., and Bravo, C. 1972. Taeniasis. Presentación de 74 casos, 2 de ellos con infección múltiple. Antioquia Med., *22*:417–422.

Polydoreau, K. 1980. The control of echinococcosis in Cyprus. World Anim. Rev., *33*:19–25.

Posselt, R. 1904. Futterungsversuche mit Alveolar-Echinococcus. Fortschr. Vet. Hyg., *1*:348.

Pradatsundarasar, A. 1960. Nine cases of *Raillietina* infection in Bangkok. J. Med. Assoc. Thailand, *43*:56–58.

Priestley, S.E., and Wiles, P.G. 1979. Cysticercosis-induced epilepsy diagnosed by computerized tomography. J. Soc. Occup. Med., *29*:149–150.

Ramírez, A., and Verástegui, P. 1953. Patología del campo visual en la cisticercosis cerebral. Mem. Cong. Cien. Mexicano, *8*(Cien. Med.):198–199.

Rausch, R. 1956. Studies on the helminth fauna of Alaska. XXX. The occurrence of *Echinococcus multilocularis* Leuckart, 1863, on the mainland of Alaska. Am. J. Trop. Med. Hyg., *5*:1086–1092.

———. 1958. Recent studies on hydatid disease in Alaska. Parassitologia, *2*:391–398.

———. 1967. On the ecology and distribution of *Echinococcus* spp. (Cestoda: Taeniidae), and characteristics of their development in the intermediate host. Ann. Parasitol., *42*:19–63.

———, and Bernstein, J.J. 1972. *Echinococcus vogeli* sp. n. (Cestoda: Taeniidae) from the bush dog, *Speothos venaticus* (Lund). Z. Tropenmed. Parasitol., *23*:25–34.

———, D'Alessandro, A., and Rausch, V.R. 1981. Characteristics of the larval *Echinococcus vogeli* Rausch and Bernstein, 1972 in the natural intermediate host, the paca, *Cuniculus paca* L. (Rodentia: Dasyproctidae). Am. J. Trop. Med. Hyg., *30*:1043–1052.

———, and Nelson, G.S. 1963. A review of the genus *Echinococcus* Rudolphi, 1801. Ann. Trop. Med. Parasitol., *57*:127–135.

———, Rausch, V.R., and D'Alessandro, A. 1978. Discrimination of the larval stages of *Echinococcus oligarthrus* (Diesing, 1863) and *E. vogeli* Rausch and Bernstein, 1972 (Cestoda: Taeniidae). Am. J. Trop. Med. Hyg., *27*:1195–1202.

———, and Richards, S.H. 1971. Observation on parasite-host relationships of *Echinococcus multilocularis* Leuckart, 1863. in North Dakota. Can. J. Zool., *49*:1317–1330.

———, and Schiller, E.L. 1951. Hydatid disease (Echinococcosis) in Alaska and the importance of rodent intermediate hosts. Science. *113*:57–58.

———, and Wilson, J.F. 1973. Rearing of the adult *Echinococcus multilocularis* Leuckart, 1863, from sterile larvae from man. Am. J. Trop. Med. Hyg., *22*:357–360.

———, and Yamashita, J. 1957. The occurrence of *Echino-*

coccus multilocularis Leuckart, 1863, in Japan. Proc. Helm. Soc. Wash., *24*:128–133.

Read, C.P. 1951. The crowding effect in tapeworm infections. J. Parasitol., *37*:174–178.

Rim, H.-J., Park, C.-Y., Lee, J.-S., Joo, K.-H., and Lyu, K.-S. 1978. Therapeutic effects of praziquantel (Embay 8440) against *Hymenolepis nana* infection. Korean J. Parasitol., *16*:82–87.

———, Park, S.-V., Lee, J.S., and Joo, K.-H. 1979. Therapeutic effects of praziquantel (Embay 8440) against *Taenia solium* infections. Korean J. Parasitol., *17*:67–72.

———, Song, K.-W., Joo, K.-H., Lee, J.-S., and Kim, J.-J. 1980. An epidemiological note on the taeniasis in Korea. Korean J. Parasitol., *18*:235–240.

———, Won, C.-R., and Chu, J.-W. 1980. Studies on the human cysticercosis and its therapeutic trial with praziquantel (Embay 8440). Korea Univ. Med. J., *17*:459–472.

Rocca, E.D., and Neira, B. 1980. Cysticercosis of the rachis. Int. Surg., *65*:31–35.

Rougier, Y., Legros, F., Durand, J.P., and Cordoliani, Y. 1981. Four cases of parasitic infection by *Raillietina (R.) celebensis* (Kanicki, 1902) in French Polynesia. Trans. R. Soc. Trop. Med. Hyg., *79*:121.

Sabbaghian, H., Hoghooghi, N., and Ghadirian, E. 1975. A survey on the prevalence of echinococcosis in Shahre-Kord, Iran. Bull. Soc. Pathol. Exot., *68*:574–578.

Saeki, Y. 1920. Experimental studies on the development of *Hymenolepis nana*. Jikwa Zasshi, No. 238, 203–244

Saidi, F., and Nazarian, I. 1972. Surgical treatment of hydatid cysts by freezing of cyst wall and instillation of 0.5 per cent silver nitrate solution. N. Engl. J. Med., *284*:1346–1350.

Sakamoto, T., Orihara, M., Sarashina, T., Ishimoto, Y., and Kamiya, H. 1971. Studies on pharmaco-therapy against larval and adult multilocular echinococcosis. I. Anthelmintic and ovicidal effects of drugs against adult *Echinococcus multilocularis*. Jpn. J. Parasitol., *20*:120–131.

Schacher, J.F., and Hajj, S.N. 1970. *Taenia* proglottid in the human uterus. Am. J. Trop. Med. Hyg., *19*:626–628.

Schantz, P.M., Shanks, D., and Wilson, M. 1980. Serologic cross-reactions with sera from patients with echinococcosis and cysticercosis. Am. J. Trop. Med. Hyg., *29*:609–612.

———, Van den Bossche, H., and Eckert, J. 1982. Chemotherapy for larval echinococcosis in animals and humans: Report of a workshop. Z. Parasitenkd., *67*:5–26.

Schenone, H. 1980. Praziquantel in the treatment of *Hymenolepis nana* infections in children. Am. J. Trop. Med. Hyg., *29*:320–321.

Schiller, E.L. 1955. Studies on the helminth fauna of Alaska. XXVI. Some observations on the cold-resistance of eggs of *Echinococcus sibricensis* Rausch and Schiller, 1954. J. Parasitol., *41*:578–582.

Schwabe, C.W., Ruppanner, R., Miller, C.W., Fontaine, R.E., and Kagan, I.G. 1972. Hydatid disease is endemic in California. Calif. Med., *117*(November):13–19.

Seesee, F.M., and Worley, D.E. 1976. The occurrence of *Echinococcus multilocularis* Leuckart, 1863 (Cestoda: Taeniidae) in the red fox, *Vulpes vulpes* L., in southwestern Montana. Proc. Montana Acad. Sci., *36*:145–149.

Shea, M., Maberley, A.L., Walters, J., Freeman, R.S., and Fallis, A.M. 1973. Intraocular *Taenia crassiceps* (Cestoda). Trans. Am. Acad. Ophthalmol. Otolaryngol., *77*:778–783.

Shorb, D.A. 1933. Host-parasite relations of *Hymenolepis fraterna* in the rat and the mouse. Am. J. Hyg., *18*:74–113.

Shoura, M.J., and Morsy, T.A. 1974. A case of *Bertiella* infection in an immigrant from Yemen. J. Kuwait Med. Assoc., *8*:55–56.

Skromme-Kadlubik, G., and Gonzalez, C. 1977. Treatment of cerebral cysticercosis in man with radionuclides. Preliminary report. Salud Publ. Mex., *19*:651–653.

Slais, J. 1967. The morphology of *Cysticercus racemosus* and the determination of the cysticercus species. Folia Parasitol., *14*:27–34.

———. 1970. *The Morphology and Pathogenicity of the Bladder Worms* Cysticercus cellulosae *and* Cysticercus bovis. Prague, Publishing House of the Czechoslovak Academy of Sciences.

Slonka, G.F., Matulich, W., Morphet, E., Miller, C.W., and Bayer, E.V. 1978. An outbreak of bovine cysticercosis in California. Am. J. Trop. Med. Hyg., *27*:101–105.

Soulsby, E.J.L. 1982. *Helminths, Arthropods and Protozoa of Domesticated Animals*. 7th ed. Philadelphia, Lea & Febiger.

Sousa, O.E., and Lombardo, J.D. 1965. Informe de un caso de hidatidosis en sujeto nativo panamēno. Primer cas autoctono. Arch. Med. Panamenos, *14*:79–86.

Spruance, S.L., Klock, L.E., Chang, F., Fukushima, T., Andersen, F.L., and Kagan, I.G. 1974. Endemic hydatid disease in Utah. Rocky Mountain Med. J., *71*:17–23.

Sterba, J., and Barus, V. 1976. First record of *Strobilocercus fasciolaris* (Taeniidae-larvae) in man. Folia Parasitol., *23*:221–226.

———, and Dykova, I. 1979. Symptomatology of taeniasis caused by *Taenia saginata*. Folia Parasitol., *26*:281–284.

Stunkard, H. 1940. The morphology and life history of the cestode *Bertiella studeri*. Am. J. Trop. Med., *20*:305–332.

———, Koivastik, T., and Healy, G.R. 1964. Infection of a child in Minnesota by *Bertiella studeri* (Cestoda: Anoplocephalidae). Am. J. Trop. Med. Hyg., *13*:402–409.

Tang, C.C., and Tang, C.T. 1964. *Raillietina (R.) celebensis* (Janicki, 1902), its development in intermediate host, epidemiology and taxonomy. Acta Parasitol. Sinica, *1*:1–13.

Taramelli, N., and Dubois, A. 1931. Un cas de coenurose chez l'homme. Ann. Soc. Belge Med. Trop., *11*:151–154.

Templeton, A.C. 1968. Human coenurus infection. A report of 14 cases from Uganda. Trans. R. Soc. Trop. Med. Hyg., *62*:251–255.

———. 1970. Anatomical and geographical location of human coenurus infection. Trop. Geogr. Med., *23*:105–108.

Thatcher, V.E. 1972. An anomalous *Taenia saginata* from a human case in Colombia. Proc. Helm. Soc. Wash., *39*:264–267.

Thomas, L.J., Babero, B.B., Gallicchio, V., and Lacey, R.J. 1954. Echinococcosis on St. Lawrence Island, Alaska. Science, *120*:1102–1103.

Thompson, C.D., Jelland, C.H., and Buckley, J.J.C. 1967. Human infection with a tapeworm, *Bertiella* sp., probably of African origin. Br. Med. J., *3*:659–660.

Thornton, H. 1980. A symposium on hydatidosis. Cent. Afr. J. Med., *26*:244–246.

Tjahjadi, G., Subianto, D.B., Endardio, S., and Margono, S.S. 1978. Cysticercosis cerebri in Irian Jaya, Indonesia. Southeast Asian J. Trop. Med. Public Health, *9*:247–251.

Topilow, H.W., Yimoyines, D.J., Freeman, H.M., Moo Young, G.A., and Addison, R. 1981. Bilateral multifocal intraocular cysticercosis. Ophthalmology, *88*:1166–1172.

Toro, M., Duran, M.L., Fanego, J., and Reyes, R.M. 1980. Hidatidosis pseudo-alveolar. Rev. Iber. Parasitol., *40*:301–313.

Tschen, E.H., Tschen, E.A., and Smith, E.B. 1981. Cutaneous cysticercosis treated with metrifonate. Arch. Dermatol., *117*:507–509.

Turner, M., and Leiper, R.T. 1919. On the occurrence of *Coenurus glomeratus* in man in West Africa. Trans. R. Soc. Trop. Med. Hyg., *13*:23–24.

Uchimura, R. 1922. On *Taenia nana* and *Taenia murina*. Jpn. Med. World, *2*:212.

Vanderick, F., Fain, A., Langi, S., and Van Balen, H. 1964.

Deux nouveaux cas de cenurose humaine à *Taenia brauni* au Ruanda, avec une localisation orbitaire du cenure. Ann. Soc. Belge Med. Trop., *44*:1077–1080.

Varela-Diaz, V.M., Coltorti, E.A., and D'Alessandro, A. 1978. Immunoelectrophoresis tests showing *Echinococcus granulosus* arc 5 in human cases of *Echinococcus vogeli* and cysticercosis-multiple myeloma. Am. J. Trop. Med. Hyg., *27*:554–557.

Venard, C.E. 1938. Morphology, bionomics, and taxonomy of the cestode *Dipylidium caninum*. Ann. N.Y. Acad. Sci., *37*:273–328.

Verster, A. 1967. Redescription of *Taenia solium* Linnaeus, 1758 and *Taenia saginata* Goeze, 1782. Z. Parasitenkd., *29*:313–328.

———. 1969. A taxonomic revision of the genus *Taenia* Linnaeus, 1758 *s. str.* Onderstepoort J. Vet. Res., *36*:3–58.

———. 1974. The golden hamster as a definitive host of *Taenia solium* and *Taenia saginata*. Onderstepoort J. Vet. Res., *41*:23–28.

———, Du Plessis, T.A., and Van den Heever, L.W. 1976. The effect of gamma radiation on the cysticerci of *Taenia solium*. Onderstepoort J. Vet. Res., *43*:23–26.

———, Tustin, R.C., and Reinecke, R.K. 1978. An attempt to treat the larval stage of *Taenia multiceps* and a resume of its neural and extraneural distribution in sheep. Onderstepoort J. Vet. Res., *45*:257–259.

Vogel, H. 1955. Ueber den Entwicklungszyklus und die Artzuhöorigkeit des europaischen Alveolarechinococcus. Dtsch. Med. Wochenschr., *80*:931–932.

———. 1978. Wie wächst der Alveolarechinokokkus? Tropenmed. Parasitol., *29*:1–11.

Wardle, R.A., and McLeod. J.A. 1952. *The Zoology of Tapeworms*. Minneapolis, University of Minnesota Press.

Wikerhauser, T., Brglez, J., Kuticic, V., and Kozelj, B. 1976. The efficacy of praziquantel against immature stages of *Echinococcus granulosus* in the dog. Acta Parasitol. Jugoslav., *7*:35–36.

———, Džakula, N., Brglez, J., Koželj, B., Maržan, B., and Rapić, D. 1980. Instražwanje stečenog imuniteta pasa protiv ehinokohoze. I. O. imunogenom djelovanju primarne invazije. Acta Parasitol., Iugosol., *11*: 69–72.

Wilber, R.R., King, E.B., and Howes, E.L., Jr. 1980. Cerebrospinal fluid cytology in five patients with cerebral cysticercosis. Acta Cytol., *24*:421–426.

Williams, P.H., and Templeton, A.C. 1971. Infection of the eye by tapeworm *Coenurus*. Br. J. Ophthalmol., *55*:766–769.

Williams, R.J., and Sweatman, G.K. 1963. On the transmission, biology and morphology of *Echinococcus granulosus equinus*, a new subspecies of hydatid tapeworm in horses in Great Britain. Parasitology, *53*:391–407.

Wilson, J.F., and Rausch, R.L. 1980. Alveolar hydatid disease. A review of clinical features of 33 indigenous cases of *Echinococcus multilocularis* infection in Alaskan Eskimos. Am. J. Trop. Med. Hyg., *29*:1340–1355.

Wood, E.D. 1977. The horse and the hydatid tapeworm. Meat Hygienist, No. 23, 10, 12, 14–15.

Woodland, W.N.F. 1924. On the development of the human *Hymenolepis nana* (Siebold 1852) in the white mouse; with remarks on "*H. fraterna*," "*H. longior*" and *H. diminuta*. Parasitology, *16*:424–435.

World Health Organization. 1967. Joint FAO-WHO Expert Committee on Zoonoses. Third Report. Tech. Rep. Ser. No. 378, Geneva, Switzerland.

Yamaguti, S. 1959. *Systema Helminthum*. Volume II. *The Cestodes of Vertebrates*. New York, Interscience Publishers, Inc.

Yokogawa, M., Yoshimura, H., and Okura, T. 1962. The treatment of *Taenia saginata* with bithionol. Jpn. J. Parasitol., *11*:39–44.

Yoshimura, H., Kondo, K., Akao, N., Ohnishi, Y., and Nakaya, S. 1979. Case report of *Echinococcus multilocularis* infection from the Midwestern Province of Japan. Int. J. Zoonoses, *6*:111–114.

Yoshino, K. 1934. On the evacuation of eggs from the detached gravid proglottids of *Taenia solium* and on the structure of its eggs. J. Med. Assoc. Formosa, *33*:47–58.

Zimmermann, H.R. 1937. Life history studies on cestodes of the genus *Dipylidium* from the dog. Z. Parasitenkd., *9*:717–729.

Chapter 32

Acanthocephala (Thorny-Headed Worms)

The Acanthocephala, which were formerly included in the Nemathelminthes, are now recognized as a separate phylum. This phylum is placed in immediate sequence to the Cestoidea, to indicate that the Acanthocephala are more closely related to the tapeworms than to any other group in the Animal Kingdom. However, it can be argued that the closest affinities of the acanthocephalan are with the aschelminths (rotifers, kinorhynchs, priapulids) with links to the nematodes (Morris and Crompton, 1982). The body wall of the Acanthocephala has a completely different type of structure from that of other helminths. The presence of canals in the tegument and the arrangement of deeper layers of the body wall are distinctive features. Nutrients are absorbed through the body wall.

These exclusively parasitic worms have a *proboscis* that is retractile into a proboscis sheath and armed with spines, and a *body proper;* they lack a digestive tract and have a body cavity without an epithelial lining (pseudocoel); their embryos are spinose; their larvae require an intermediate host; and the adults are always parasites in the intestine of vertebrates.

Internally the elongate, more or less cylindrical adults have one or more pairs of elongate *lemnisci,* which extend posteriad from the region of the anterior plane of the proboscis sheath into the loose parenchymatous matrix of the body. A *nerve mass* is situated in or on the proboscis sheath.

The sexes are separate. The genital pore is at the posterior extremity. In the male (Fig. 32–1,*A*) it is surrounded by a campanulate *bursa (b).* Two *testes (t), cement glands (cg), cement receptacle (cgr),* and a *suspensory ligament (sl)* comprise the male genital apparatus. In the female (Fig. 32–1,*B*) a *suspensory ligament (sl)* extends from the posterior end of the sheath to the uterine bell. The *ovary* first breaks up into egg balls or *floating ovaries* from which a large number of eggs develop. These *eggs,* which are provided with three (at times probably four) enveloping membranes, lie free in the body parenchyma *(fo),* the ones with partially matured embryos being removed from the body cavity and introduced into the uterus by means of a muscular *selective apparatus* or *bell (ub).*

With the exception of the egg stage, the Acanthocephala are parasitic during their entire life cycle, with no free-living phase. The eggs, which are passed in the feces of the vertebrate host, are infective for the intermediate host. In order to develop further, the fully matured eggs must be ingested by certain species of arthropods, in which they hatch and develop through a number of stages (Van Cleave, 1947) as follows (Fig. 32–2): (1) The *acanthor,* or first larval stage, which hatches from the egg in the intestine of the arthropod intermediate host, is provided with rostellar hooklets that are employed in boring through the gut wall into the hemocoel. (2) In the hemocoel the acanthor metamorphoses into a simplified second-stage larva, the *acanthella,* which gradually, without definitely recognizable instars, acquires a proboscis and rudiments of other structures of the mature worm. (3) The last immature stage is the *juvenile (cystacanth),* in which the rudiments of structures become recognizable as those of the adult worm. This is the last stage in the arthropod host. On ingestion of the infected arthropod, the appropriate vertebrate host acquires the infection, the worm develops to maturity and mates, and egg laying is begun.

Van Cleave (1953) recognized two classes: the Eocanthocephala and the Metacanthocephala. Two species of thorny-headed worms recorded from man, *Macracanthorhyncus hirudinaceus* and *Moniliformis moniliformis,* belong to the latter class, order Archiacanthocephala. These are mostly large worms

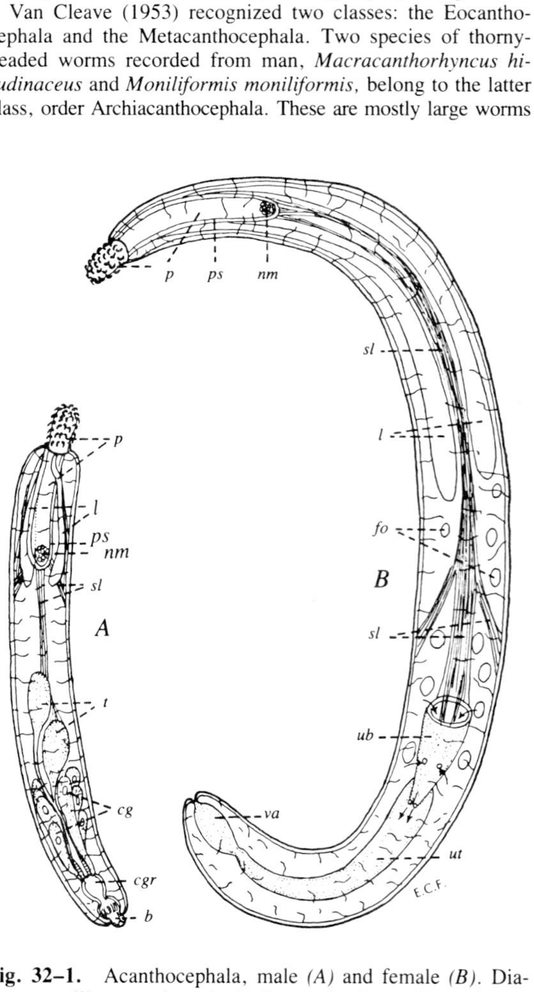

Fig. 32–1. Acanthocephala, male *(A)* and female *(B).* Diagrams to illustrate the principal reproductive and attachment organs. *ub,* Uterine bell. *b,* Bursa. *cg,* Cement glands. *cgr,* Cement gland receptacle. *l,* Lemnisci. *nm,* Nerve mass. *fo,* Floating ovaries. *p,* Proboscis. *ps,* Proboscis sheath. *sl,* Suspensory ligaments. *t,* Testes. *ut,* Uterus. *va,* Vagina. (Adapted from original of E.C. Faust.)

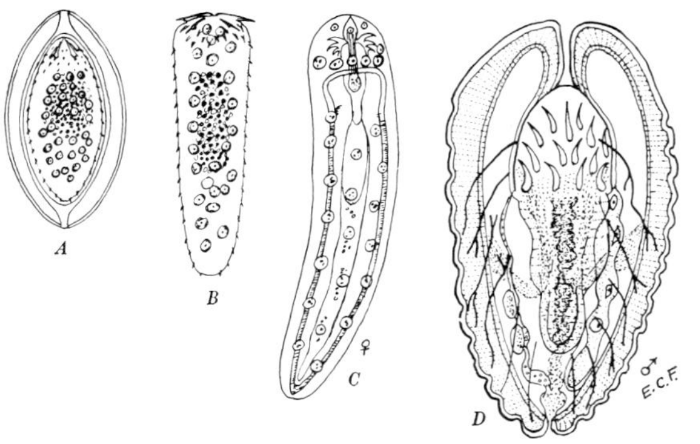

Fig. 32–2. *Macracanthorhynchus hirudinaceus.* Chief developmental stages, enlarged but not drawn to scale. *A,* Egg, fully embryonated. *B,* Acanthor, with rostellar hooks in relaxed position. *C,* Acanthella, female. *D,* Cystacanth, male. (Adapted by E.C. Faust from Van Cleave, H.J. 1947. A critical review of terminology for immature stages in acanthocephalan life histories. J. Parasitol., *33*:118–225.)

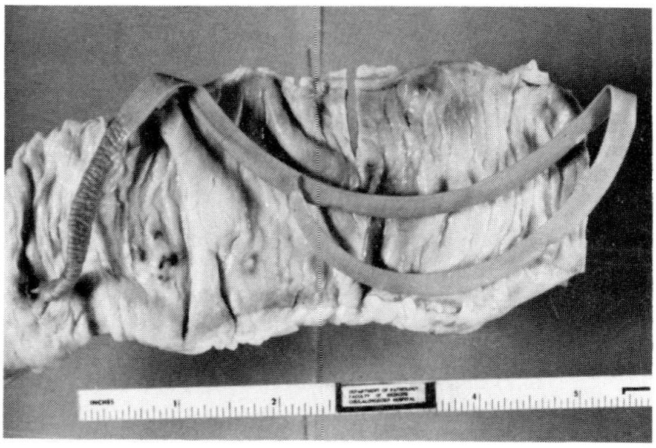

Fig. 32–3. *Macracanthorhynchus hirudinaceus,* female, attached to the jejunum; this was an incidental finding at autopsy of a woman in Thailand. The worm, round in transverse section and reddish in color when alive, collapsed when fixed, and the vagina (at posterior extremity) became everted. (From Faust, E.C., Beaver, P.C., and Jung, R.C. 1975. *Animal Agents and Vectors of Human Disease,* 4th ed., Philadelphia, Lea & Febiger.)

that live in terrestrial hosts, have spirally arranged spines on the proboscis, and no spines elsewhere on the body. A third form, an unidentified species of *Bolbosoma,* was recently recorded as a parasite of man (Tada et al., 1983; Beaver et al., 1983). It too is in the class Metacanthocephala but belongs to a different order, the Palaeacanthocephala, which are parasites of marine mammals and bear cuticular spines on the trunk as well as on the proboscis. Other palaeacanthocephalans found in man, each recorded once, are *Acanthocephalus rauschi* from the peritoneal cavity of an Alaskan Eskimo, *Acanthocephalus bufonis* recovered at autopsy from the intestine of a man in Indonesia, and *Corynosoma strumosum* recovered from the feces of an Alaskan Eskimo after treatment for other parasites. The latter two records possibly are spurious. The natural host of *A. rauschi* is unknown; *A. bufonis* is a common parasite of amphibians in Asia, and *C. strumosum* is common in seals but found also as immature worms in fish-eating mammals and birds (Schmidt, 1971).

Macracanthorhynchus hirudinaceus (Pallas, 1781) Travassos, 1917 (The giant leech-like acanthocephalan)

Synonyms. *Echinorhynchus gigas* Bloch, 1782; *Gigantorhynchus gigas* (Bloch, 1782) Hamann, 1892; *G. hirudinaceus* (Pallas, 1781) Railliet, 1893.

Morphology, Biology, and Life Cycle. This worm ranges from milky-white to red, is cylindroidal, slightly flattened dorsoventrally, and has a pseudosegmentation of the cuticula (Fig. 32–3). The males measure 5 to 10 cm by 3 to 5 mm; the females, 20 to 65 cm by 4 to 10 mm. The proboscis, which is embedded in the intestinal wall of the host, is provided with five or six series of recurved spines arranged more or less in symmetrically alternating rows. The posterior end of the male is extended into a campanulate bursa copulatrix; that of the female is bluntly rounded. The symmetrically ovoid eggs (Fig. 32–4), which measure 80 to 100 μm in length by about one-half that in diameter, are fully embryonated when laid and have three embryonic envelopes. Various species of beetles serve as appropriate intermediate hosts in which the embryos hatch on ingestion of the eggs, penetrate into the body cavity, and develop into mature larvae (cystacanths). On swallowing the infected beetle, the final host acquires the infection. In the pig this worm

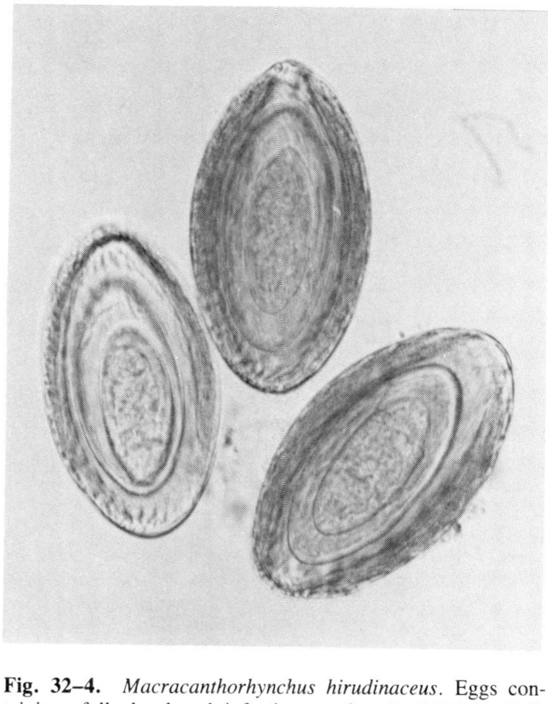

Fig. 32–4. *Macracanthorhynchus hirudinaceus.* Eggs containing a fully developed, infective acanthor. (× 340.) (By P.C. Beaver.)

has practically a cosmopolitan distribution. The wild boar and the peccary are also natural hosts.

Epidemiology. Few human infections have been reported in the literature. An immature female worm was recovered post mortem from a 9-year-old child who died of leukemia in Prague (Lambl, 1859). Brumpt (1949) considered that the worm may have been an intestinal parasite of a fish eaten by the child. Lindemann stated that in his day (1865) the infection was common in the Volga Valley, where Schneider (1871) found the beetle *Melolontha* was eaten raw. In recent years only one human case has been found in Russia, in a 5-year-old boy who had been observed to catch and squash beetles (Skrinnik et al., 1958). A third human infection was recorded in a 32-year-old woman who died of rheumatic heart disease in Thailand; at autopsy a female *M. hirudinaceus* was recovered from the intestine (Pradatsundarasar and Pechranond, 1965). A fourth case, the second for Thailand, was reported in a 26-year-old man; the worm, about 10 cm long, was removed in a resected segment of the terminal ileum (Kliks et al., 1974). An additional human infection in Thailand was reported by Tesana et al. (1982) in a 12-year-old boy admitted to hospital with acute abdominal pain caused by peritonitis and perforation of the lower ileum. Although eggs of *M. hirudinaceus* were recovered from 2 of 1236 human stools in Brazil, it is entirely possible that these were instances of false parasitism due to ingestion of the intestines of parasitized hogs.

Clinical Data. As a result of multiple infection by this worm, the hog suffers local trauma and inflammation at the site of attachment to the intestinal wall, and not infrequently perforation of the wall. At autopsy of the aforementioned Thai woman, who had died of other causes, it was found that a *M. hirudinaceus* in the upper jejunum had caused several ulcers, the most recent being 5 mm in diameter. The head of the worm had penetrated the bowel wall to the serosa. Symptoms may include abdominal pain and tenderness, anorexia, and nausea (Kliks et al., 1974).

Leng et al. (1983) reported that, in the last few years, many human cases have occurred in several provinces of China (Hebei, Quandong, Liaoning, and Shandong). The infections, mostly in children, are acquired through eating raw beetles, and the clinical manifestations are those of intestinal obstruction and perforation. Intestinal perforation occurred in over 200 cases. Eggs were present in the feces of none. In most cases the worms were discovered at operation or autopsy. Surgery was sometimes avoided by treatment with piperazine citrate to expel the worms. Several species of beetles (*Scarabaeidae, Cerambycidae*) harbor the infective larvae. In Lianoning Province, eggs were present in the feces of 16% to 60% of pigs.

Moniliformis moniliformis (Bremser, 1811) Travassos, 1915 (The moniliform acanthocephalan)

Synonym. *Echinorhynchus moniliformis* Bremser, 1811.

Morphology, Biology, and Life Cycle. The body of this worm is chalky or creamy-white, elongated, and attenuated at both ends. It is somewhat cylindrical but appears segmented because of the bead-like pseudosegmentation of the cuticle. The cylindrical proboscis is armed with 12 to 15 rows of recurved hooks, 7 to 8 hooks to each row. The males measure up to 10 cm in length and have a campanulate bursa; the females measure up to 27 cm in length. The eggs are ellipsoidal (85 to 118 μm by 40 to 52 μm) and are provided with three characteristic envelopes, the outer of which is smooth or covered with a thin wrinkled membrane, but otherwise similar to that of *Macracanthorhynchus* (Fig. 32–4). The embryos are striated and spinose. Beetles and cockroaches serve as intermediate hosts, in the body cavity of which the larvae mature in cystic capsules.

Epidemiology. On ingestion of infected insects, the final host acquires infection. In the rat the biological incubation period is 22 to 38 days (Sita, 1949). Murine species are the natural hosts in practically all parts of the world. Solitary human infections have been reported from Italy, the Sudan, Java, Israel, Iran, Iraq, Zambia, Colombia, and the United States. Infection was usually found in young children and several worms were passed in some cases (Sahba et al., 1970; Moayedi et al, 1971; Ayala et al., 1973; Goldsmid, 1974; Al-Rawas et al., 1977).

Clinical Data. Calandruccio experimentally infected himself with *M. moniliformis* (Grassi and Calandruccio, 1888). Although 5 weeks were required for the worms to become sexually mature, beginning on the nineteenth day serious gastrointestinal pain was experienced along with diarrhea, exhaustion, somnolence, and tinnitus. Administration of *extract of aspidium* removed all of the worms; the symptoms completely disappeared after 2 days. Signs and symptoms in other reported cases, in descending order of frequency, have been loss of appetite, diarrhea, abdominal pain, fever, protuberant abdomen, vomiting, jaundice, and cough. Retarded development and irritability have been noted in children.

Bolbosoma

Species of the genus *Bolbosoma* occur normally in marine animals only, cetaceans being the natural definitive hosts and fish and microcrustacea serving as second and first intermediate hosts, respectively. A distinctive morphological feature of *Bolbosoma,* as the name implies, is a bulb-like inflation or expanded region of the anterior extremity of the body bearing large spines that are usually arranged as two separate collars or spine fields (Van Cleave, 1953).

Two human infections with *Bolbosoma* have been reported, both from the province of Kyushu, Japan. In one, the worm was discovered in sections of a granulomatous mass excised from the wall of the colon of a 16-year-old boy in Kagoshima (Beaver et al., 1983). In the other case, the entire worm was recovered from the peritoneal cavity of a man in Kumamoto (Tada et al., 1983).

REFERENCES

Al-Rawas, A.Y., Mirza, M.Y., and Shafig, M.A. 1977. First finding of *Moniliformis moniliformis* (Bremser 1811) Travassos 1915 (Acanthocephala: Oligacanthorhynchidae) in Iraq from human child. J.Parasitol., *63*:196–197.

Ayala, S., Sánchez, C.E. de, Amézquita, M., and Gómez, C. de. 1973. Infecciones poco frecuentes vistas en los laboratorios de parasitologia. Acta Med. Valle, *4*:53–54.

Beaver, P.C., Otsuji, T., Otsuji, A., Yoshimura, H., Uchikawa, R., and Sato, A. 1983. An acanthocephalan, probably *Bolbosoma*, from the peritoneal cavity of man in Japan. Am. J. Trop. Med. Hyg., *32*: (in press).

Brumpt, E. 1949. *Précis de Parasitologie,* 6th ed. Paris, Masson et Cie.

Goldsmid, J.M. 1974. Human infection with *Moniliformis* sp. in Rhodesia. Ann. Trop. Med. Parasitol., *68*:363–364.

Grassi, G., and Calandruccio, S. 1888. Ueber einen Echinorhynchus welcher auch im Menschen parasitirt und dessen Zwischenwirth ein *Blaps* ist. Centralbl. Bakteriol., *3*:521–525.

Kliks, M., Tantachamrun, T., and Chaiyaporn, V. 1974. Human infection by an acanthocephalan *Macracanthorhynchus hirudinaceus* in Thailand: New light on a previous case. Southeast Asian J. Trop. Med. Public Health, *5*:303–309.

Leng, Y-J., Huang, W-D., and Liang, P-N. 1983. Human infection with *Macracanthorhynchus hirudinaceus* Travassos, 1916 in Guangdong Province, with notes on its prevalence in China, Ann. Trop. Med. Parasitol., *77*:107–109.

Moayedi, B., Izadi, M., Maleki, M., and Ghadirian, E. 1971. Human infection with *Moniliformis moniliformis* (Bremser, 1811) Travassos, 1915 (syn. *Moniliformis dubius).* Report of a case in Isfahan, Iran. Am. J. Trop. Med. Hyg., *20*:445–448.

Morris, S.C., and Crompton, D.W.T. 1982. The origins and evolution of the Acanthocephala. Biol. Rev., *57*:85–115.

Pradatsundarasar, A., and Pechranond, K. 1965. Human infection with the acanthocephalan *Macracanthorhynchus hirudinaceus* in Bangkok: Report of a case. Am. J. Trop. Med. Hyg., *14*:774–776.

Sahba, G.H., Arfaa, F., and Rastegar, M. 1970. Human infection with *Moniliformis dubius* (Acanthocephala) (Meyer, 1932) (syn. *M. moniliformis),* (Bremser, 1811) (Travassos, 1915) in Iran. Trans. R. Soc. Trop. Med. Hyg., *64*:284–286.

Schmidt, G.D. 1971. Acanthocephalan infections of man, with two new records. J. Parasitol., *57*:582–584.

Schneider, A.F. 1871. On the development of *Echinorhynchus gigas*. Ann. Mag. Nat. Hist., *4*:441–443.

Sita, E. 1949. The life cycle of *Moniliformis moniliformis* (Bremser, 1811), Acanthocephala. Current Sci. (India), *18*:216–218.

Skrinnik, M.R., Likhotinskaya, M.V., and Ocheret, A.M. 1958. A case of *Macracanthorhynchus* infection in man. Med. Parasitol. Parasit. Dis. (Moscow), *27*:450–451.

Tada, I., Otsuji, Y., Kamiya, H., Mimori, T., Sakaguchi, Y., and Makizumi, S. 1983. The first case of a human infected with an acanthocephalan parasite, *Bolbosoma* sp. J. Parasitol., *69*:205–208.

Tesana, S., Mitrchai, J., Chunsuttwat, S. 1982. Acute abdominal pain due to *Macracanthorhynchus hirudinaceus*: A case report. Southeast Asian J. Trop. Med. Public Health, *13*:262–264.

Van Cleave, H.J. 1947. A critical terminology for immature stages in acanthocephalan life histories. J. Parasitol., *33*:118–125.

———. 1953. *Acanthocephala of North American Mammals*. Illi. Biol. Monogr. 23, Nos. 1–2, Urbana, University of Illinois Press.

Chapter 33

Leeches (Hirudinea)

GENERAL CONSIDERATIONS

The leeches constitute a class of the phylum Annelida, which exhibit true segmentation, i.e., *metamerism*. They are akin to the oligochete annelids, but they have suckers, median genital orifices, and "jaws" and lack integumentary setae or bristles. Although some species eat earthworms, snails, insect larvae, and other organisms, leeches are primarily blood suckers and, like ticks, have a mechanism adapted for the engorgement of relatively large amounts of blood. They vary in size from small macroscopic to several inches in length, and in shape from elongated cylindrical or ovoid to broadly ovoid or piriform. Their dorsal side is convex and the ventral side is flattened or concave. Some leeches are aquatic, others terrestrial, and still others amphibious in their habits.

The body of the leech consists of 34 segments or *somites,* with a similar number of median nerve ganglia and rows of sensory papillae. Externally there are two to many annulations for each segment (Mann, 1962).

MORPHOLOGY, BIOLOGY, AND LIFE CYCLE OF LEECHES

Leeches have two muscular suckers, one circumoral in position, the other at the posterior end of the body. These serve for attachment and aid in locomotion.

The body of the leech is capable of extraordinary contraction and extension. It is covered with a thin, smooth *cuticula,* which is from time to time cast off in patches. Immediately beneath is the *epidermis,* consisting of wedge-shaped cells that are internally separated from one another by blood capillaries. From the epidermis are produced many unicellular glands, disposed in the underlying connective tissue but opening to the surface through long ducts. There are special glands, usually situated in the ninth, tenth, and eleventh segments *(clitellar somites)* which secrete cocoon-forming material.

In addition to the superficial annulations, the body surface is characterized by having numerous pairs of minute sensory *papillae* in the middle annulus of each segment. These papillae are usually more numerous on the dorsal than on the ventral aspect of the worm. Opening through the ventral surface of segments 7 to 23, on the annulus immediately in front of the median annulus of each segment, are 17 pairs of excretory apertures, the *nephridiopores*. On the dorsum of each of the first five segments in most leeches there is a pair of *eyespots* (Fig. 33–1,*h*), which are modified sensory papillae, but in some species the number of eyespots is reduced. Except for the genus *Acanthobdella,* leeches bear no bristles. A few species have external gills.

Leeches may be unpigmented or they may be provided with brown, black, or other pigments so distributed as to form longitudinal stripings or ornate, bilaterally symmetrical patterns. Moreover, the lateral pouches of the digestive tract, when distended with blood, may provide a beautifully patterned picture that is visible externally.

Leeches are quite muscular, owing to an outer layer of *circular muscles* next to the epidermis, an underlying, much thicker layer of *longitudinal muscles, dorsoventral fibers,* and *radial fibers*. Internal to the two main muscular layers is the so-called *botryoidal tissue,* which consists of branched cavities surrounding the digestive tract. The walls of these spaces consist of large cells filled with black pigment. The botryoidal tissue communicates with both the blood vascular system and the greatly reduced body cavity. All of the unoccupied interstices between the epidermis and the digestive tract are filled with *connective tissue,* consisting of cells and fibers in a gelatinous matrix.

Digestive System. In one order of leeches (the Rhynchobdellida), the oral end is provided with a protrusile proboscis. In another order (the Gnathobdellida) there is no proboscis but a group of three muscular *jaws* (Fig. 33–1), one mediodorsal and two ventrolateral in position within the oral sucker. Each jaw is like a hard cushion, is covered with chitin, and is frequently provided with numerous serrated denticles. By means of powerful muscles these three jaws operate to produce a characteristic triradiate wound in the victim's skin. The small mouth cavity leads into the muscular *pharynx (ph)* (somites 4 to 8), surrounding which are numerous unicellular *salivary glands.* These latter open near the buccal cavity and serve to secrete an anticoagulant *(hirudin),* which is injected into the bite wound and prevents clotting of ingested blood. The pharynx communicates distally with the extensive, thin-walled *crop,* which reaches from the levels of the eighth to the eighteenth somite and has 11 pairs of pouches or diverticula *(di)*. The crop is capable of tremendous distension when the leech takes a full blood meal. In it a portion of the blood meal may be stored for many months. Immediately behind the crop is the tubular stomach or *midgut (mg),* in which digestion takes place. Its anterior end is dilated and the wall of its distal portion is spirally infolded. When blood passes from the crop to the stomach, its color changes from reddish-brown to green. Behind the stomach is the short *intestine,* which, in turn, leads into a short *rectum* and opens externally through a small *anal pore* (a), anterior and dorsal to the posterior sucker (Fig. 33–1).

Excretory System. This system consists of 17 pairs of *nephridia (n)* situated in segments 7 to 23 (Fig. 33–1) which resembie those of the earthworm but are more complex. Each nephridium *(n)* consists of *(1)* a sinuously looped glandular tubule, with a ciliated inner funnel, which is at times occluded; (2) a central duct running through the cells of the tubule and many branched communicating ductules, (3) a dilated vesicle at the outer end of the primary duct, and (4) a terminal nephridiopore that opens on the ventral surface of the worm. Liquid wastes are conveyed from the body cavity to the exterior

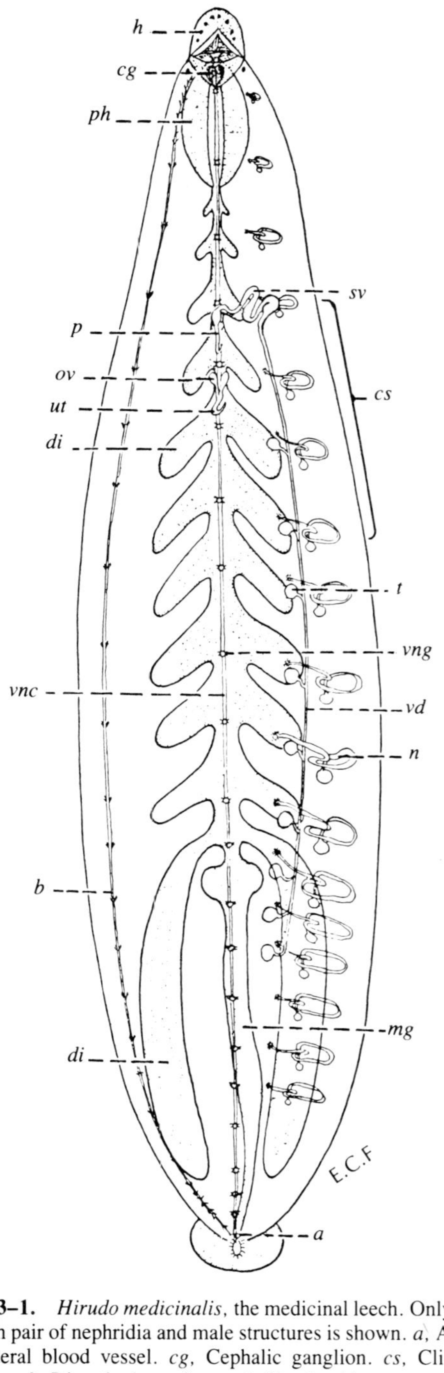

Fig. 33–1. *Hirudo medicinalis*, the medicinal leech. Only one of each pair of nephridia and male structures is shown. *a*, Anus. *b*, Lateral blood vessel. *cg*, Cephalic ganglion. *cs*, Clitellar somites. *di*, Diverticulum of crop. *h*, Head, with eyespots. *mg*, Mid-gut. *n*, Nephridium. *ov*, Ovary. *p*, Penis. *ph*, Pharynx. *sv*, Seminal vesicle. *t*, Testis. *ut*, Uterus. *vd*, Vas deferens. *vnc*, Ventral nerve cord. *vng*, Ventral nerve ganglion. (× ca. 2.) (From Faust, E.C. 1949. *Human Helminthology*, 3rd ed. Philadelphia, Lea & Febiger.)

by means of the cilia present at the inner end of each nephridium and by muscular contraction of its wall.

Blood Vascular System. The vascular system consists of (1) *blood vessels* with muscular walls and (2) *blood sinuses* without muscular walls. The former are composed of a pair of lateral trunks *(b)*, which unite at the anterior and posterior ends of the worm and also send off metameric dorsal and ventral branches, some of which anastomose with one another. The terminal vessels end in capillaries in the body wall, nephridia, and gonads. The blood sinuses, which represent a greatly reduced body cavity, consist of a dorsal and a ventral trunk. These also have anterior and posterior connections and metameric branches, ending in terminal capillaries that constitute their only communication with the blood vessels. The circulating blood contains plasma, at times colored with hemoglobin, and a small number of colorless corpuscles.

Nervous System. The central nervous system consists of a series of partially fused paired ganglia *(vng)* united by twinned nerve cords *(vnc)*, which lie within the ventral blood sinus (Fig. 33–1). At the anterior end of the system there is a conspicuously large cephalic or subesophageal ganglion *(cg)* (representing five fused pairs), which is united by circumesophageal commissures with the small dorsally situated brain. This latter lies above the anterior end of the pharynx. Nerves arising from the ganglia innervate the more important organs and tissues of the body, including the pairs of eyespots on the dorsal side of the anteriormost segments, the sensory papillae, suckers, and genitalia.

Reproductive Organs. Leeches are hermaphroditic. Each worm (Fig. 33–1) has 1 to 10 or more pairs of small hollow, spherical *testes (t)*. A small *vas efferens*, arising from each testis, enters one of the paired *vasa deferentia (vd)*, which continue as paired *seminal vesicles (sv)*, each usually provided with a *prostate gland*, an *ejaculatory duct*, and a muscular *penis* (*p*). The two ejaculatory ducts enter a common *bursa copulatrix* or *genital atrium*. The *ovaries (ov)* consist of a single pair of coiled, filamentous sacs that are continuous with their ducts. The two ducts unite to form a common convoluted *oviduct*, which is continued as a muscular *uterus (ut)* and opens through a short vaginal tube in the midventral line, one segment behind the male genital opening (usually stated to open on somite 9).

Reproduction and Life Cycle. In some leeches insemination is accomplished when one leech implants onto the cuticle of another a horny pocket or *spermatophore*, from which spermatozoa issue forth, migrate through the tissues of the recipient and reach its ovary. In the group to which the medicinal leech *(Hirudo medicinalis)* belongs, reciprocal copulation takes place by the introduction of the penis of each into the vagina of the other and the reciprocal deposition of a spermatophore. Thus, in either type, fertilization takes place before the eggs are laid. Some species of aquatic leeches deposit a few eggs at a time in small tough capsules that are attached to submerged objects. Other leeches surround their eggs with a cocoon, which is then deposited in or among submerged objects, on the bottoms of lakes and ponds, or in moist earth. The capsule or cocoon is secreted by glands of the clitellum and hardens on contact with water. Species of still another group carry both their eggs and their young around with them. As soon as the leechling is able to suck blood, it leaves the parent and takes up an independent existence.

CLASSIFICATION OF LEECHES

The leeches are divided into three orders: the Gnathobdellida Vaillant, 1890, the Rhynchobdellida Blanchard, 1897, and the Pharyngobdellida Johansson, 1913. Forms of special medical importance are found in the Gnathobdellida.

Gnathobdellida Vaillant, 1890.

This group contains species having a smooth cuticula, a mouth lacking a proboscis but usually armed with three jaws or pseudojaws, frequently armed with denticles, and a spoon-like sucker.

MEDICAL IMPORTANCE OF LEECHES

Throughout the years leeches have been regarded as having a twofold medical importance: (1) as a medical aid and (2) as a detrimental or dangerous predator of man.

Leeches as a Medical Aid. Use of leeches for bloodletting was common until the middle of the nineteenth century. Until recently the so-called "medicinal leech" *(Hirudo medicinalis* or a closely related species) was employed in Europe and America for the partial exsanguination of patients suffering from ailments ranging from common colds to cancer. This custom was so universally accepted as a part of medical art that by analogy the physician himself was referred to as a leech.

With gradual recognition that bloodletting was usually harmful to the patient, the use of the leech as a medical aid has been almost completely abandoned.

Leeches Injurious to Man. Travelers through tropical rain forests of Ceylon, Borneo, and East Indian islands, or explorers who pass through the humid valleys of the Himalayas, Assam, Burma, Indochina, or the Chilean Andes, provide colorful accounts of the scourge of bloodthirsty terrestrial leeches that lurk on every stone, leaf, and stem, spring onto the wayfarer, painlessly insert their denticled jaws in his skin, and produce trickling springs of blood from each puncture site. Moreover, thirsty travelers throughout northern Africa and Western Asia, as well as natives in parts of southern Europe, who unwarily drink water from a spring or brook, may acquire infestation of the upper digestive or respiratory tract with the aquatic leech, *Limnatis nilotica,* or one of its close relatives. This aspect of leech infestation is considered under the sections on *external* and *internal hirudiniasis.*

External Hirudiniasis

Etiology and Epidemiology. Although species of aquatic leeches, commonly ectoparasitic on aquatic vertebrates (e.g., fishes, frogs, turtles, molluscs), frequently attach themselves to the skin of humans with whom they come in contact and avidly suck blood, terrestrial leeches are most notorious in this respect. They commonly live in the tropical rain forests, temporarily attached to shrubs from which they drop upon unsuspecting mammals or people who brush against the herbage. More than a dozen species of terrestrial leeches that attack man have been described from Asia, Polynesia, Oceania, Australia, Malagasy, and South America. These species include *Haemadipsa zelanica* (Sri Lanka, India, Pakistan, Malaysia, and Borneo); *H. montana* (India, Pakistan, and Burma); *H. sylvestris* (Burma, Malaysia, Borneo, Java, Sumatra); *H. ornata* (Assam); *H. picta* (Malaysia, Borneo); *Phytobdella catenifera* (Malaysia); *H. talagalla* (Philippines); *H. javanica* (Java); *H. japonica* (Japan); *H. fallax, H. morsitans,* and *H. vagans* (Malagasy); and *Dinobdella ferox* (Sri Lanka, India, Burma, Thailand, and China). A specimen identified as *Hirudo sjoestedi* was reported as having embedded itself in the thigh of a woman in Zimbabwe (Goldsmid, 1977).

Chin (1949) published a note with reference to the leeches that he had previously reported (1941) from the human nose in Guizhou (Kweichow) and Yunnan Provinces, China. In identifying these leeches as immature specimens of *Dinobdella ferox,* Moore (1959) commented that *Dinobdella ferox* (Blanchard) is widely distributed in India, Burma, southeast China, and Taiwan (Formosa) and is a serious pest of domestic and some wild mammals. It is known to attack man by taking up its abode in the air passages.

It breeds at springs, wells, and small streams (and probably some larger bodies of water) frequented by cattle, horses, and other mammals. The egg capsules are deposited in the mud and the newly hatched young, 12 to 20 mm long, abound in the water on aquatic plants and other smooth surfaces. If one places a hand in the water, many of the little leeches instantly attach to it and rapidly move up the arm. The same happens to the muzzle of a cow or horse or the face of a man drinking leech-infested water. The young leeches quickly enter the nostrils or mouth and attach themselves to the wall of the respiratory passages, usually far back in the pharynx or larynx. They grow rapidly, reach a large size, and often do much damage.

Walton (1955), in reporting this species from Malaysia and Borneo, surmised that it depends on wild mammals for its food, since it is found in great numbers in streams remote from human habitations.

Pathogenesis, Pathology, and Symptomatology. Although the puncture is painless, the wounds from which the worms have been removed continue to bleed for hours and heal slowly, even when not secondarily infected.

Treatment and Prevention. When fully engorged, the land leeches drop off. The removal may be hastened by applying a few drops of brine, alcohol, or strong vinegar to the site, or a match flame skillfully applied to the worm. They may be pulled off by applying steady gentle traction. If the bleeding from the bites continues for some time, each may be stanched with a styptic pencil. The wounds should be bathed for several days with mild antiseptic lotions such as boracic acid or calamine lotion to prevent sepsis.

Persons traveling through areas infested with land leeches should wear knee-high waterproof leather boots and closely woven pants, or should liberally apply such repellents as dimethyl phthalate, Rutgers-612, or indalone to all parts of the skin and clothing liable to infestation. Ribbands (1946) reported that dimethyl phthalate, when applied to clothing at the rate of 4 ml per square foot, is completely repellent for land leeches for as long as 6 days. Audy and Harrison (1954) found that in field tests the United States Army all-purpose repellent M-1960 (equal parts of (1) n-butylacetanilid, (2) 2-butyl-2-ethyl-3-propanediol, and (3) benzyl benzoate, with 10% Tween 80 emulsifier), when impregnated into clothing adapted for field service, completely protected the wearers and killed leeches trapped in the clothing. In field tests in Sabah (north Borneo), in which subjects wore clothing impregnated with M-2065 and M-2066 (each containing approximately 29% each of undecylenic acid, N-propylacetanilide, and N-butyl-4-cyclohexane-1,2-carboximide, and lindane 3% and 1% respectively, in 10% Tween 80 as emulsifier), under conditions of constant tropical rains, these two repellents were found to be as protective against land leeches as M-1960 (Walton et al., 1956). In tests of available insecticides and molluscicides against *Hirudo nipponia,* chlordane, diazinon, dinex, and I.C.I. 24223 (an experimental molluscicide) were the most effective. Field use of these agents to control leeches is impractical because of their lethal effects on fish and other aquatic life (Kimura and Keegan 1966).

Internal Hirudiniasis

Etiology and Epidemiology. Internal hirudiniasis is due to aquatic leeches taken into the nose or mouth in drinking water, or gaining entrance to the genitourinary tract from wading in deep water. Species of several genera of aquatic leeches have been incriminated in internal hirudiniasis, but those that have caused the greatest variety of symptoms and have produced the most suffering are members of the genus *Limnatis.* One widely distributed species of this genus, *L. nilotica,* is deservedly notorious.

When engorged, *Limnatis nilotica* is a large worm, measuring 8 to 15 cm in length by 1 to 1.5 cm in greatest breadth. The body is broad posteriorly and more or less pointed anteriorly. The mouth is surrounded by a relatively weak sucker, the upper lip of which has a longitudinal furrow on its inner aspect. The three jaws are armed with a total of more than 100 denticles and are provided with sensory papillae. The powerful posterior sucker is at least twice as large as the oral one. The dorsum of the body is typically a dark olive green and the venter, dark gray. At times there are dark longitudinal stripes on the dorsum.

L. nilotica lives in quiet brooks, streams, freshwater ponds, and lakes in southern Europe (Portugal, Spain, France, Italy, Greece, Bulgaria), in northern Africa (Egypt, Ethiopia, Tunisia, Algeria, Morocco), the Azores and the Canary Islands, and western Asia (Turkey, Armenia, Syria, Israel, Iran, Baluchistan, Afghanistan, Chinese Turkestan, and even within the frontiers of India). The species reported from the environs of Singapore is *L. maculosa,* while *L. africana* has been identified from Senegal and the Congo basin, *L. mysomelas* from Senegal, and *L. granulosa* from India.

Internal hirudiniasis, produced by undiagnosed species of leeches, has been reported on several occasions from Java and Sumatra. Once an aquatic leech, *Haemopis cavillina,* was found fixed to the sclerocorneal limbus of an Italian subject.

The small, young *Limnatis nilotica* get into the mouth in raw drinking water and frequently become attached to the mucous membrane of the pharynx, nasopharynx, epiglottis, and esophagus. By deep inhalation they may be carried to the larynx or even to the trachea or bronchi. They readily puncture mucous surfaces and proceed to engorge with blood. Hemoptysis caused by *L. nilotica* infestation of the vocal cords has been reported from Iraq (Almallah, 1968). Most of the many persons infested each year are relieved by home remedies (gargling with strong salt solution, inhalation of pungent odors, and the successful use of a tough twig provided with a single reversed thorn), and only individuals with the most serious cases seek a physician. Since the bite is painless, many persons with nasal, nasopharyngeal, or pharyngeal infestation are unaware of the presence of the worm and present with the symptom of epistaxis, hemoptysis, or difficult breathing (Hii et al., 1978; Keegan et al., 1970). Although the majority of the patients probably have only pharyngeal or nasopharyngeal involvement, the literature reports more than half the patients suffering from infestation of the larynx and vocal cords; 16%, the trachea; 16%, the nasopharynx; 6%, the esophagus, and the remainder, infestation in multiple sites.

Pathogenesis, Pathology, and Symptomatology. At the site of attachment the leech secretes *hirudin* and proceeds to draw out blood far in excess of its maximum needs. While this bloodletting itself is a relatively painless procedure, physical obstruction caused by the worm frequently produces a feeling of pressure, pain, and uncomfortable sensation emanating from the parasitized focus, together with functional disturbance of the affected organ. Trauma and invasion of the wound by pathogenic microorganisms give rise to inflammation of the involved mucosa and at times result in submucous abscesses.

If the leech has entered the mouth and has anchored itself to the mucous membrane of the upper respiratory or digestive passage, epistaxis, hemoptysis, or hematemesis may result, depending on the organ infested.

Leeches in the nares may cause persistent headache. When they are lodged in the larynx or on the vocal cords, there may be continuous coughing, with a slimy, bloody discharge. There may be pain in the chest, dyspnea with or without cyanosis, hoarseness, and at times complete loss of speech. In the larynx and trachea leeches may produce suffocation, occasionally resulting in death. If they are lodged on the epiglottis or in the esophagus, difficulty in swallowing and nausea are experienced.

Persons wading or bathing in fresh water at times suffer from leech infestation of the genitourinary tract. Woolnough (1928) reported uncontrolled hemorrhage from the vagina of a 3-year-old girl in Australia, caused by leech bite. Goldsmid (1977) reported *Hirudo intermedia* attached to the wall and extruding from the vagina of a woman in Zimbabwe. Hamilton (1933) cited leech infestation of the labium majus which simulated uterine hemorrhage. Several physicians in India, Algeria, and Italy have reported leech infestation of the male urethra. One jute washer in India observed a leech entering the external meatus of the urethra and was unable to prevent its entry by traction (Ghosh, 1933).

In 1903 Kuwahara reported the recovery of young specimens of *Limnatis japonica* 2 to 3 cm long from the conjunctiva of a patient. The worms had caused hemorrhage, photophobia, and an excessive flow of tears. A similar case involving *L. nilotica* was recorded by Gilkes (1957).

Limnatis nilotica is more likely to be the causal agent of "halzoun" (suffocation) than is the sheep liver fluke *Fasciola hepatica (vide supra,* p. 454).

Treatment. Leeches attached to the skin are usually pulled off by the victim without medical consultation. Leeches that cling more obstinately can be induced to detach by application of a small flame, lighted cigarette, insect repellent, or table salt to the worm. Continued bleeding from the attachment site may be controlled with a styptic pencil. Removal of leeches from the nasopharynx or lower in the respiratory tract or the esophagus is best done under visualization of the worm with the appropriate endoscopic instrument. Forceps may be required and the procedure can be facilitated by application of topical anesthesia. For leech infestation of the genitourinary tract, Indian practitioners have found that strong saline solution irrigation of the involved organ has proved helpful in evacuating or killing the worm.

Prevention. Since the majority of patients suffering from internal hirudiniasis become infested from drinking unfiltered water, it is important that individuals or troops, passing through or quartered in regions where *Limnatis nilotica* and its relatives abound, be required to drink only water that has been filtered or at least strained through several layers of cheesecloth. An even sounder dictum is to boil all suspected water.

REFERENCES

Almallah, Z. 1968. Internal hirudiniasis as an unusual cause of haemoptysis. Br. J. Dis. Chest, *62*:215–218.

Audy, J.R., and Harrison, J.L. 1954. Field tests of repellent M-1960 against leeches. Med. J. Malaya, *8*:240-250.

Chin, T.H. 1949. Further note on leech infestation in man. J. Parasitol. *35*:215.

Ghosh, M.M. 1933. A leech in the male urethra. Indian Med. Gaz., *68*:574.

Gilkes, M. 1957. Leech-bite of the cornea. Br. J. Ophthalmol., *41*:124.

Goldsmid, J.M. 1977. Leech infestation of man in Rhodesia. Trans. R. Soc. Trop. Med. Hyg., *71*:86.

Hamilton, C.S.P. 1933. Leech bite of labium majus. Indian Med. Gaz., *68*:87–88.

Hii, J.L.K., Kan, S.K.P., and Yong, K.S.A. 1978. A record of *Limnatis maculosa* (Blanchard) (Hirudinea: Arynchobdellida) taken from the nasal cavity of man in Sabah, Malaysia. Med. J. Malaysia, *32*:247–248.

Keegan, H.L., Radke, M.G., and Murphy, D.A. 1970. Nasal leech infestation in man. Am. J. Trop. Med. Hyg., *19*:1029–1030.

Kimura, T., and Keegan, H.L. 1966. Toxicity of some insecticides and molluscicides for the Asian blood sucking leech, *Hirudo nipponia* Whitman. Am. J. Trop. Med. Hyg., *15*:113–115.

Kuwahara, Y. 1903. Ueber lebende Hirudineen im Bindehautsack des menschlichen Auge. Cbl. Prakt. Augenheilk., *27*:262–263.

Mann, K.H. 1962. *Leeches (Hirudinea). Their Structure, Physiology, Ecology and Embryology.* New York, Pergamon Press.

Moore, J.P. 1959. Hirudinea, pp. 542–557. In *Ward and Whipple's Fresh-Water Biology,* 2nd ed. Edited by W.T. Edmondson. New York, John Wiley & Sons.

Ribbands, C.R. 1946. Experiments with leech repellents. Ann. Trop. Med. Parasitol., *40*:314–319.

Walton, B.C. 1955.The "nasal-leech" *Dinobdella ferox* from Borneo and Malaya. J. Parasitol., *41*(sec. 2):32.

———, Traub, R., and Newson, H.D. 1956. Efficacy of the clothing impregnants M-2065 and M-2066 against terrestrial leeches in North Borneo. Am. J. Trop. Med. Hyg., *5*:190–196.

Woolnough, S.J. 1928. Intractable bleeding from a leech bite. Med. J. Aust., *1*:115.

Section IV

Arthropods and Human Disease

Chapter 34

Arthropods and Human Disease

GENERAL CONSIDERATIONS

Members of the phylum Arthropoda (insects and their allies) constitute the largest assemblage of species in the Animal Kingdom and one of the most important sources, other than man himself, of human pathogens. Medical importance of the arthropod group is due not so much to the number of species of pathogens that they transmit, although the list is formidable, as it is to the morbidity and mortality resulting from these diseases, and their extensive distribution over the face of the globe. From ancient times these infections have produced untold economic loss and death in the human population, and many of them have at times reached pandemic proportions.

ARTHROPODS AS HOSTS AND AS BIOLOGICAL VECTORS OF MICROORGANISMS

In considering arthropods as biological vectors of human pathogens, one should remember that, in many instances, these microorganisms are so well adapted to the arthropod host that they produce little or no tissue damage. Transmission to man and other higher animals is accidental and frequently incidental to the perpetuity of the microorganism. Moreover, many other species of microorganism (including several protozoan forms, a few helminths, many species of bacteria, rickettsiae, and viruses) are normal parasites of the arthropod and are not known to be transmissible to man or other vertebrates.

Since arthropods are medically important primarily because of their capacity to transmit pathogens, it is appropriate to provide synoptic information as to how significant and how extensive this field of microbiology is. From the viewpoint of the etiologic agents, the material is summarized in the following paragraphs. From the viewpoint of arthropods as hosts and vectors, information is presented later in this chapter in Table 34–1.

Protozoa

The leishmaniae, trypanosomes, malaria parasites, and *Babesia* spp. are considered in Section II.

Helminths

The more important arthropod-transmitted helminths are the filaria worms, lung flukes, and some of the tapeworms, presented in appropriate places in Section III.

Bacteria

Arthropods are responsible for transmission of many bacteria pathogenic to man. In a few instances, these microorganisms may undergo propagative development in a biological vector, i.e., multiplication and subsequent transmission to a vertebrate host. The relationship may also be a more casual one involving contamination of food with enteric bacteria by house flies or cockroaches, or placement of pyogenic bacteria in or around open wounds by *Hippelates* gnats.

Yersinia pestis and Francisella tularensis. The etiologic agent of plague *(Yersinia pestis)* is transmitted almost exclusively by rodent fleas following propagation in the blocked gut of the vector. The relationship between *Francisella tularensis,* the causal agent of tularemia, and its vectors, i.e., ticks, bloodsucking flies, and possibly fleas, is more uncertain. It is probable that this disease is more frequently contracted by direct contact with infected tissues of reservoir hosts or from a contaminated water source.

Rickettsiae and Related Forms. These pathogens belong to the order Rickettsiales or Chlamydiales and are small, rod-shaped, spherical, or pleomorphic, gram-negative microorganisms living and multiplying in living tissues or cell-free media.

Rickettsiaceae, or rickettsiae, are prokaryotic microorganisms intimately associated with arthropods and, with one exception, have not as yet been grown in cell-free media. They have a distinct cell wall, a definite cytoplasmic boundary, and a more condensed center. Many species have been reported and described as symbionts of arthropods, including ticks, mites, biting and sucking lice, assassin bugs, mosquitoes, flies, and fleas. None of these symbiotic species is known to be transmissible to man or laboratory animals.

The *pathogenic rickettsiae* responsible for disease in man are grouped into three genera: *Rickettsia* da Rocha Lima, 1916; *Coxiella* (Philip) Philip, 1948; and *Rochalimaea* (Macchiavello) Krieg, 1961. The genus *Rickettsia* includes the *spotted fever group, typhus group, scrub typhus,* and *rickettsialpox.* In the genus *Coxiella* there is only one described species, *C. burnetii,* causing Q fever. The genus *Rochalimaea* includes the causative agent of trench fever, *R. quintana.* This rickettsia grows in the lumen of the gut of its arthropod vector (body louse) and can be cultured in cell-free media.

Bartonellaceae are small, frequently pleomorphic, gram-negative microorganisms that stain well with Giemsa's stain. They parasitize red blood cells and in some species they also multiply in cells of the reticuloendothelial system of man and other vertebrates. Unlike the majority of the rickettsiae, these microorganisms can be cultivated in vitro on nonliving media. The one species that infects man is *Bartonella bacilliformis,* the causative agent of bartonellosis (Carrión's disease, Oroya fever, verruga peruana).

Chlamydiaceae (order Chlamydiales) are minute pleomorphic, frequently coccoid bodies, which are gram-negative intracytoplasmic parasites having a characteristic developmental cycle. Several species of this family are important as pathogens of man, responsible for trachoma, lymphogranuloma venereum, psittacosis, ornithosis, and meningopneumonitis.

Spirochaetaceae (order Spirochaetales). Members of this assemblage are slender and sinuous, have one or more complete spirals, are motile, measure up to 500 μm in length, may or may not have an axial filament or lateral crista, and may or may not manifest transverse segmentation. They stain well with Giemsa stain or by silver impregnation, and when living are readily seen with darkfield illumination. This family contains the genera *Borrelia* and *Treponema.* Species of *Borrelia* produce relapsing fever in man and other warm-blooded vertebrates. *Treponema pertenue,* the causative agent of yaws, can apparently be transmitted mechanically by chloropid flies. A treponema-like spirochete may also be the causative agent of Lyme disease as well as erythema chronicum migrans, a persistent skin lesion associated with tick bites.

Viruses

Viruses are obligate intracellular parasites of protozoa, bacteria, fungi, and higher plants and animals. Replication of infectious particles involves adsorption of the virus to the cell surface, penetration, and production of new virions by using the synthetic machinery of the parasitized cell. The site of synthesis depends upon the viral nucleic acid composition (DNA or RNA).

Fifty or more families and groups of viruses have been established on the basis of the kind and strandedness of nucleic acid making up the viral genome, and the presence or absence of a lipoprotein envelope surrounding the virion. Arthropod-borne viruses (arboviruses) causing clinical illness include representatives from four families: the Reoviridae, Togaviridae, Rhabdoviridae, and Bunyaviridae.

ARTHROPODS AS VECTORS OF PATHOGENS

In their role as transmitters or *vectors* of various pathogens to man, arthropods vary in the intimacy of their association with the disease-producing organism. In their simplest relationship they may be only *mechanical vectors* of the etiologic agent, but most important arthropod-transmitted diseases employ the arthropod as *biological vector,* requiring a period of incubation or development in this host. For example, most species of trypanosomes undergo a period of metacyclic multiplication in the arthropod before becoming infective for man. In some infections in which incubation, either with simple multiplication of the organism or with a required metacyclic phase, takes place in the body of the arthropod, the intestinal lumen of this host serves as the incubating site. In others, the gut must be breached. In many rickettsial and in all viral infections, the pathogens invade the hemocoel and become intracellular inclusions in most organs of the arthropod's body. In the *Plasmodium* life cycles associated with human malaria, the sexual phase of the parasite is completed in the gut lumen of the anopheline mosquito, with sporogony taking place after the motile zygote (ookinete) has penetrated the gut wall. In filarial infections, involving mosquitoes and other biting flies, the larval nematodes migrate through the wall of the digestive tract to the thoracic muscles or fat body, where they undergo a twofold metamorphosis before they migrate to the head and are introduced into the tissues of the vertebrate (definitive) host via the mouthparts.

Although with few exceptions the arthropods of medical interest ingest the pathogenic organisms, the methods by which they convey the organisms to man are various. Nonbloodsucking flies may deposit a vomit-drop containing the pathogens on human food or drink (enteric infections of man). Other species of arthropods may obtain the parasitic organisms in a blood meal from an infected person and deposit them in a vomit-drop in the puncture wound (plague), or in fecal pellets near the puncture wound (epidemic and endemic typhus; trench fever; Chagas' disease) made in the skin of an uninfected person. Contamination from infected hemolymph may occur when the arthropod is crushed on the skin (epidemic typhus; epidemic relapsing fever). During the act of feeding, soft ticks produce infective fluids from the coxal glands that contaminate the bite wound or skin. Still other arthropods discharge the pathogens (malaria sporozoites; virus particles) through the hypopharynx in minute droplets of salivary secretion at the time they procure a blood meal, or (filaria larvae) from the proboscis (labium and mouth area [labrum-epipharynx]) thus enabling the pathogens to migrate out and reach the vicinity of the puncture wound.

The more important diseases of man, in which arthropods serve as biological vectors, together with the stages of incubation or development in the arthropod host and the methods of transmission to man, are summarized in Table 34–1.

SOURCES OF PATHOGENS TRANSMITTED BY ARTHROPODS

In human malaria, although certain simian plasmodia have been transmitted to man experimentally by mosquitoes, man himself is the only certain source from which the anopheline mosquito obtains the infection. Similarly, the filarial worms that infect man are almost invariably derived from human sources, although *Brugia malayi* (subperiodic form) and possibly *Brugia pahangi* may at times be derived from sylvatic reservoirs. Domestic animal filariasis also poses a zoonotic problem since several species of *Dirofilaria* can produce lung lesions and subcutaneous nodules (Dissanaike, 1979). The causative agent of bubonic plague is typically obtained by fleas from rodents. *Rickettsia prowazekii* passes from man to man, with the body louse as the transmitting arthropod, but murine typhus, caused by *R. typhi,* is spread by rodents and their arthropod ectoparasites, and man is only incidentally involved. Yellow fever and many of the viral encephalitides are natural diseases of certain simian and avian hosts, respectively. In certain helminthic infections transmitted by arthropods, man is only incidentally involved. All the arthropod-transmitted spiruroid nematodes are incidental human parasites, as are the accidentally ingested gordius worms.

In many arthropod-associated diseases, the etiologic agents were undoubtedly parasites of their invertebrate hosts long before man came into the life cycle. In some cases the parasite has been so long and so well adjusted to the arthropod that it produces no obvious injury, namely, *Rickettsia rickettsii* and *Borrelia duttoni* in ticks, and *R. tsutsugamushi* in trombiculid mites. In the tick and in the mite the respective parasites are even transmitted congenitally (vertically) and require no vertebrate host for at least several generations. On the other hand, *R. prowazekii* in the body louse can cause extensive and often fatal damage to this ectoparasite, suggesting that the vector-pathogen relationship is an imperfect one.

Instances of transovarial and trans-stadial transmission of viruses, rickettsiae, spirochetes, and protozoa by ticks (or mites for rickettsiae) have been well documented. More recent accounts of the vertical transmission of viruses by mosquitoes (Culicidae) and sand flies (Psychodidae) indicate that these Diptera, as the Acari, may serve as both vector and reservoir to a certain extent. Transovarial transmission of important human pathogens has been demonstrated in mosquitoes, with several serotypes of the California group *(Bunyavirus)* as well as the viruses causing Japanese encephalitis, dengue, yellow fever, and St. Louis encephalitis (all in the genus *Flavivirus*). Vertical transmission of the etiologic agents of vesicular stomatitis *(Vesiculovirus)*, phlebotomus fever *(Phlebovirus)*,

Table 34–1. Arthropods that are Biological Vectors of Important Human Pathogens

Class Group	*Order*	*Arthropod Taxon*	*Disease*	*Etiologic Agent*	*Relation of Arthropod to Pathogen and to Human Infection*
CRUSTACEA	COPEPODA	*Cyclops* spp.	Dracontiasis (Medina worm infection)	*Dracunculus medinensis*	Intermediate host, with no multiplication of pathogen; man acquires infection by swallowing water containing infected *Cyclops*
		Diaptomus spp., *Cyclops* spp.	Diphyllobothriasis (fish tapeworm infection)	*Diphyllobothrium latum*	First intermediate host, with no multiplication of pathogen; source of infection for freshwater fishes that produce infection in man
		Cyclops spp.	Gnathostomiasis	*Gnathostoma spinigerum*	First intermediate host, with no multiplication of pathogen; source of infection for second intermediate host which is then ingested by man, a paratenic host
		Cyclops spp.	Sparganosis	*Spirometra erinacei* and *S. mansonoides*	First intermediate host with no multiplication; man acquires infection by swallowing an infected copepod
	DECAPODA (e.g., crayfishes, crabs)	Freshwater crabs and crayfishes	Paragonimiasis (pulmonary distomiasis, endemic hemoptysis)	*Paragonimus westermani*; other *Paragonimus* spp.	Second intermediate host, with no multiplication of pathogen; man acquires infection by eating infected crabs and crayfish
DIPLOPODA ("millipedes")		*Julus* sp., *Fontaria virginiensis*	Rat tapeworm infection	*Hymenolepis diminuta*	Pathogen develops from embryo to larva in hemocoel of arthropod; definitive host infected by swallowing infected adult arthropod

ARACHNIDA (ticks, mites)	PARASITIFORMES (ticks and gamasid mites)	*Amblyomma* spp., *Dermacentor* spp., *Haemaphysalis* spp., *Hyalomma* spp., *Ixodes* spp.	Spotted fevers (tick typhus)	*Rickettsia rickettsii* and related forms	Pathogen multiplies in epithelial cells of gut, hemolymph, or ovary; man inoculated through "bite" of tick
		Rhipicephalus sanguineus, *Dermacentor andersoni*, *Haemaphysalis* spp., others	Q fever	*Coxiella burnetii*	Pathogen multiplies in tick midgut cells; vertically transmitted in tick; man may be inoculated through "bite" of tick or through contaminated tick feces, but usually from contact with infected animals
		Amblyomma americanum, *Dermacentor* spp., and many other ticks	Tularemia	*Francisella tularensis*	Pathogen multiplies in gut and malpighian tubules of tick; vertically transmitted in some species; man inoculated through "bite" or crushing of tick
		Ornithodoros spp.	Endemic relapsing fever	*Borrelia duttoni*, others	Pathogen multiplies in tissues of tick outside gut; man inoculated through "bite" of tick or via contaminated coxal fluid
		Ixodes ricinus, *I. dammini*	Babesiosis	*Babesia bovis*, *B. divergens*, *B. microti*	Pathogen multiplies in gut cells, ovary, and other tissues; vertically transmitted; human infection occurs through "bites" of infected immature ticks
		Ixodes ricinus, *I. persulcatus*, *Dermacentor* spp., *Haemaphysalis* spp., *Hyalomma*, spp., others	Tick-borne encephalitis, Kyasanur Forest disease; Colorado tick fever; Crimean-Congo hemorrhagic fever	Viruses belonging to the genera *Flavivirus*, *Orbivirus*, and *Nairovirus*	Pathogen multiplies in gut cells and other tissues; transmitted trans-stadially and transovarially; man is infected from "bite" of tick

Table 34–1. *Continued*

Class Group	*Order*	*Arthropod Taxon*	*Disease*	*Etiologic Agent*	*Relation of Arthropod to Pathogen and to Human Infection*
		Liponyssoides sanguineus	Rickettsialpox	*Rickettsia akari*	Pathogen multiplies in gut of mite; vertical transmission occurs in the vector; human infection occurs through "bite" of nymphs or adults
	ACARIFORMES (trombiculid mites)	*Leptotrombidium* spp.	Chigger-borne rickettsiosis (scrub or mite typhus)	*Rickettsia tsutsugamushi*	Pathogen multiplies in gut cells, hemocytes, and other tissues of mite; trans-stadially and trans-ovarially passaged; human infection occurs from "bite" of larval mite
	ANOPLURA (sucking lice)	*Pediculus humanus humanus* (body louse)	Classical typhus fever	*Rickettsia prowazekii*	Pathogen multiplies in epithelium of louse's midgut; man inoculated through contaminated feces or crushing of louse on skin
		Idem	Trench fever	*Rochalimaea quintana*	Pathogen multiplies in mid-gut lumen of louse; man inoculated by feces or by crushing louse on skin
		Idem	Epidemic relapsing fever	*Borrelia recurrentis*	Pathogen multiplies in tissues of louse outside gut; man inoculated by crushing louse on skin
INSECTA (insects)	MALLOPHAGA (chewing lice)	*Trichodectes canis* (dog louse)	Dog tapeworm infection	*Dipylidium caninum*	Pathogen develops from embryo to larva, without multiplication, in hemocoel of louse; man inoculated by swallowing infected louse

DICTYOPTERA (cockroaches)	*Periplaneta americana, Blatta,* and *Blatella* spp.	*Gongylonema* infection	*Gongylonema pulchrum*	Pathogen matures from embryo to larva without multiplication, in hemocoel; man inoculated by swallowing infected insect
	Idem	Acanthocephaliasis	*Moniliformis moniliformis*	*Idem*
DERMAPTERA (earwigs)	*Anisolabis annulipes*	Rat tapeworm infection	*Hymenolepis diminuta*	*Idem*
HEMIPTERA (true bugs)	*Triatoma* spp., *Panstrongylus* spp., *Rhodnius* spp., others (cone-nosed bugs)	Chagas' disease	*Trypanosoma cruzi*	Pathogen transforms in midgut-hindgut of bug; man inoculated through feces of bug rubbed into conjunctiva or skin
DIPTERA	*Anopheles* spp. (mosquitoes)	Malaria	*Plasmodium vivax, P. malariae, P. falciparum, P. ovale*	Pathogen completes sexual cycle, then multiplies by sporogony in mosquito; man inoculated through "bite" of mosquito
	Anopheles spp., *Mansonia* spp., *Aedes* spp.	Malayan filariasis	*Brugia malayi*	Pathogen matures from embryo to infective larva, without multiplication in thoracic muscles of mosquito; man inoculated through "bite" of mosquito
	Anopheles barbirostris	Timor filariasis	*Brugia timori*	*Idem*
	Culex spp., *Aedes* spp., *Anopheles* spp., *Mansonia uniformis*	Bancroft's filariasis	*Wuchereria bancrofti*	*Idem*
	Aedes spp., *Culex* spp., *Culiseta* spp., *Mansonia* spp., *Psorophora* spp.	Equine encephalitides, epidemic polyarthritis, others	Viruses belonging to the genus *Alphavirus*	Pathogen multiplies in gut cells and other tissues of mosquito; man inoculated through "bite" of mosquito
	Aedes spp., *Haemagogus* spp., *Culex* spp.	Yellow fever, dengue, Japanese encephalitis, others	Viruses belonging to the genus *Flavivirus*	*Idem;* transovarial transmission occurs in some cases

Table 34–1. *Continued*

Class Group	*Order*	*Arthropod Taxon*	*Disease*	*Etiologic Agent*	*Relation of Arthropod to Pathogen and to Human Infection*
		Culex spp., *Aedes* spp.	Dengue-like illness, California encephalitis, others	Viruses belonging to the genus *Bunyavirus*	*Idem;* transovarial transmission occurs in the California group
		Culicoides spp. (gnats)	Mansonelliasis	*Mansonella perstans, M. streptocerca, M. ozzardi*	Pathogen matures from embryo to infective larva, without multiplication, in thoracic muscles of fly; man inoculated through "bite" of gnat
		Simulium spp. (black gnats, black flies, coffee flies)	Onchocerciasis Mansonelliasis	*Onchocerca volvulus* *Mansonella ozzardi*	*Idem* *Idem*
		Phlebotomus spp., *Lutzomyia* spp. (sand flies)	Visceral leishmaniasis; cutaneous leishmaniasis; mucocutaneous leishmaniasis	*Leishmania donovani, L. tropica, L. braziliensis*	Pathogen multiplies in midgut or hindgut of fly; man inoculated through "bite" of fly
		Idem	Bartonellosis (Carrion's disease)	*Bartonella bacilliformis*	
		Idem	Sand fly fever	Viruses of the genus *Phlebovirus*	Pathogen multiplies in tissues of fly; vertical transmission occurs; man inoculated through "bite" of fly
		Chrysops spp. (mango flies)	Loiasis (filariasis)	*Loa loa*	Pathogen matures from embryo to infective larva, without multiplication, in fat body of fly; man is inoculated through "bite" of fly
		Chrysops spp. (deer flies) and *Tabanus* spp. (horse flies)	Tularemia	*Francisella tularensis*	Pathogen occurs on proboscis and in gut of fly; man inoculated through "bite" of fly

	Glossina spp. (tsetse flies)	African trypanosomiasis	*Trypanosoma brucei gambiense, T. b. rhodesiense*	Pathogen multiplies in midgut and migrates to salivary glands of fly; man inoculated through "bite" of fly
SIPHONAPTERA (fleas)	*Xenopsylla cheopis*, other rodent fleas; possibly *Pulex irritans* (human flea)	Plague	*Yersinia pestis*	Pathogen multiplies in gut of flea; man inocuated primarily through "bite" of flea
	Xenopsylla cheopis and possibly other fleas	Murine typhus fever	*Rickettsia typhi* (syn. *R. mooseri*)	Pathogen multiplies in epithelium of midgut of flea; man inoculated through "bite" and feces of flea
	Xenopsylla cheopis and other rat fleas	Rat tapeworm infection	*Hymenolepis diminuta*	Pathogen develops from embryo to larva in hemocoel of flea; man inoculated by swallowing infected flea
	Ctenocephalides canis, C. felis, Pulex irritans	Dog tapeworm infection	*Dipylidium caninum*	*Idem*
COLEOPTERA (beetles)	Many species	Rat tapeworm infection	*Hymenolepis diminuta*	Pathogen develops from embryo to larva in hemocoel of beetle; man inoculated by swallowing infected beetle
	Idem	*Gongylonema* infection	*Gongylonema pulchrum*	*Idem*
	Idem	Acanthocephaliasis	*Macracanthorhynchus hirudinaceus, Moniliformis moniliformis*	*Idem*
LEPIDOPTERA (moths and butterflies)	A few species of grain moths	Rat tapeworm infection	*Hymenolepis diminuta*	Pathogen develops from embryo to larvae in hemocoel of larval moth; man inoculated by swallowing infected insect

and possibly the Changuinola group *(Orbivirus)* also occurs in species of *Lutzomyia* or *Phlebotomus* (Tesh, 1980).

ARTHROPODS AS ETIOLOGIC AGENTS OF DISEASE

In addition to their importance in the transmission of pathogens to man, arthropods themselves play a significant part as disease-producing organisms. Certain species, like the larval stage of the myiasis-producing flies, the chigoe flea, and the sarcoptic mite, invade the tissues of man and produce serious lesions. Others, by mechanical means, and by toxins introduced into the skin, cause troublesome and at times serious manifestations. Trauma results when a tick introduces its hypostome and chelicerae into the skin preparatory to obtaining a blood meal from man. Certain ticks produce fatal paralysis in some individuals, presumably due to toxins in their secretions. Venoms introduced into the skin by the bite of a "black widow" spider, or the sting of a scorpion or bee, may at times produce both local reactions and profound systemic shock. Even minute bloodsucking flies, in depositing droplets of saliva in the skin, may provoke serious allergic reactions. House dust mites may cause rhinitis or asthma by serving directly as allergens or by concentrating highly antigenic materials. Finally, the fear of arthropods (acaro- or entomophobia) can be extremely important as a severe neurosis.

GENERAL REFERENCES

Bettini, S. editor. 1978. *Arthropod Venoms*. (Handbook of Experimental Pharmacology, Volume 48). New York, Springer Verlag.

Buchanan, R.E., and Gibbons, N.E. editors. 1974. *Bergey's Manual of Determinative Bacteriology*. Baltimore, Williams & Wilkins.

Bulla, L.A., Jr., and Cheng, T.C. editors. 1975. *Pathobiology of Invertebrate Vectors of Disease*. Vol. 266 of Ann. N.Y. Acad. Sci. New York, New York Academy of Sciences.

Dissanaike, A.S. 1979. Zoonotic aspects of filarial infections in man. Bull. W.H.O., *57*:349–357.

Harwood, R.F., and James, M.T. 1979. *Entomology in Human and Animal Health*, 7th ed. New York, Macmillan Publishing Co.

Kreier, J.P. editors. 1980. *Malaria. Volume 1: Epidemiology, Chemotherapy, Morphology and Metabolism. Volume 2, Pathology, Vector Studies, and Culture*. New York, Academic Press.

Matthews, R.E.F. 1982. Classification and nomenclature of viruses. Intervirology, *17*:1–199.

Molyneux, D.H. 1977. Vector relationships in the Trypanosomatidae. Adv. Parasitol., *15*:1–82.

Orihel, T.C. 1975. The peritrophic membrane: Its role as a barrier to infection of the arthropod host, pp. 65–73. In *Invertebrate Immunity*. Edited by K. Maramorosch and R.E. Shope. New York, Academic Press.

Sasa, M. 1976. *Human Filariasis*. Baltimore, University Park Press.

Tesh, R.B. 1980. Vertical transmission of arthropod-borne viruses of vertebrates, pp. 122–137. In *Vectors of Disease Agents*. Edited by J.J. McKelvey, B.F. Eldridge, and K. Maramorosch. New York, Praeger Scientific Publishers.

Theiler, M., and Downs, W.G. 1973. *The Arthropod-Borne Viruses of Vertebrates*. New Haven, Yale University Press.

Chapter 35

Biology and Classification of Arthropods of Medical Importance

Arthropods are triploblastic, segmented, bilaterally symmetrical invertebrates that have paired, articulated appendages (''Arthropoda,'' from αρθρον = joint and πούς = foot). Typically, arthropods have an exoskeleton composed of secretions of chitin, a nitrogenous polysaccharide, and various proteins, some of which are cross-linked, enhancing hardness. In some forms, the exoskeleton is impregnated with calcium salts. This outer skeleton is divided into scleral plates that are separated by flexible membranes so that movement is possible. The thickness of parts of the exoskeleton also may vary for greater flexibility.

The important organs and structures of a generalized arthropod are shown in Figure 35–1. There is a complete *digestive tract,* with a mouth at one end and an anus at the other. This tract is divided into three main parts: (1) an anterior *foregut* lined with chitin, consisting of a *buccal cavity,* muscular *pharynx, esophagus,* and *proventriculus* used in the ingestion and trituration of food; (2) a nonchitinized *midgut* for the digestion and absorption of food; and (3) a posterior portion lined with chitin, usually consisting of *hindgut* and *rectum,* for the accumulation and discharge of feces. Metabolic wastes are collected from the body cavity by means of tubes (malpighian tubules) that empty into the junction of the mid- and hindgut, or excretion of ammonia may occur via the gills in some aquatic forms. The *blood system* consists of a ''closed'' and an ''open'' portion. The former, with a pumping organ (the *heart*), *aorta,* and *paired vessels,* is dorsal in position. The latter, consisting of the hemal cavity (i.e., *hemocoel),* communicates with the heart. The *central nervous system* consists of a dorsal *''brain''* with *circumesophageal commissures* in the head, and a *nerve cord* with *ganglia* for each segment, ventral in position beneath the digestive tract. Arthropods respire by means of gills, booklungs, tracheae, or directly through the cuticle. The *skeletal musculature* is generally striated. *Sexes* are almost always separate. Except in primitive groups, postembryonic development usually involves some degree of metamorphosis, that is, a change in form resulting in distinct stages. During periods of growth the entire exoskeleton is shed and renewed.

The molting hormones of arthropods are *ecdysones,* which are steroids secreted by certain endocrine glands. These hormones are responsible for the periodic shedding of the exoskeleton in a unique growth process by which parts of the old cuticle are digested and reabsorbed, followed by *ecdysis. Juvenile hormones,* typically found in insects, are terpenoid compounds that act to maintain juvenility during growth by the larval or nymphal stage and may also later exert a gonadotropic effect. In addition to the ecdysones and juvenile hormones, neurosecretory cells produce peptides and polypeptides that aid in a variety of physiological functions, including initiation of the molting cycle and cuticular tanning, diuresis, and regulation of sugar and lipid levels in the hemolymph.

Arthropods exhibit a high degree of cephalization and generally have well-developed sense organs. The arthropod eye varies greatly in complexity, ranging from a simple condition with only a few photoreceptors to a complex arrangement of simple eyes or ommatidia that form the compound eye. When present, the antennae can perceive acoustical, chemical, and electromagnetic stimuli. As a result of these structures as well as more

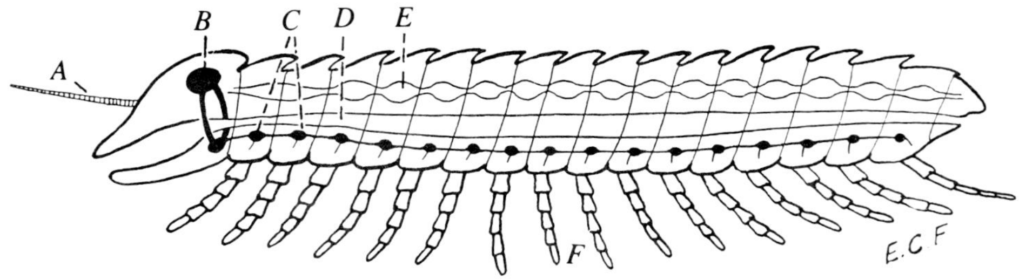

Fig. 35–1. Primitive arthropod, lateral view. *A*, Antenna. *B*, Dorsal nerve ganglion or "brain." *C*, Ventral nerve cord with ganglion for each segment. *D*, Digestive tract. *E*, Dorsal vessel. *F*, Segmented legs of left side. (Adapted by E.C. Faust from Schmeil, O., and Folsom, J.W. 1909. *Entomology, with Special Reference to Its Biological and Economic Aspects*. Philadelphia, P. Blakiston's Son & Co.)

specialized receptor organs, many arthropods exhibit complex and occasionally unique behavioral patterns, including those that bring them into close proximity with man.

Specialized Morphology and Phylogenetic Origin

In some primitive arthropods (now extinct) all of the body segments (somites, metameres) were similar to one another and the exoskeleton was relatively soft. The paired appendages, arising from ventrolateral aspects of the body covering, consisted of a basal portion (*protopodite,* Fig. 35–2, *pr*), an inner axial continuation, the *endopodite (en),* and an outer portion, the *exopodite (ex),* attached to the distal segment of the protopodite. As the exoskeleton became harder on the dorsum and venter, it assumed a more distinct form, with a dorsal chitinuous *tergum (te)* and a ventral *sternum (st)*. The *pleuron,* or lateral portion of the exoskeleton *(pl),* at first remained membranous, although it frequently gave rise to semihardened *pleurites*. Later it also became hardened, calcified, and fused with the tergum. From the basal segment of the protopodite (i.e., *coxopodite*) there developed *epipodites* (Fig. 35–2, *e*) or gills. Such paired appendages were primitively repeated on each metamere and served the purposes of walking, swimming, and respiration. In the process of specialization for the procurement of food and for defense, these appendages became greatly modified, particularly in the region of the head, where they became transformed into sense organs *(antennae)* and jaws *(mandibles),* and in the thorax and abdomen, where they developed into walking legs or biramous paddle-like swimming appendages. This diversification is illustrated in the lobster or crayfish. Meanwhile, the tergal portion of the head became fused into one plate, or the head and thoracic portion *(cephalothorax)* developed a common tergal plate. Aquatic chelicerate arthropods, differing in general body makeup and limb configuration from the early Crustacea, became terrestrial and gave rise to the arachnids. A third form, which also was terrestrial, developed uniramous appendages from lobopodia (hollow body outgrowths) and acquired mandibulate mouthparts in the process of cephalization. This ancestral form was the likely progenitor of the insects and their myriapod allies. With additional modifications in certain groups the external evidence of segmentation became obscured (spiders, ticks, mites) or the articulated appendages were lost during the larval stages (flies, fleas, bees, and other holometabolous insects).

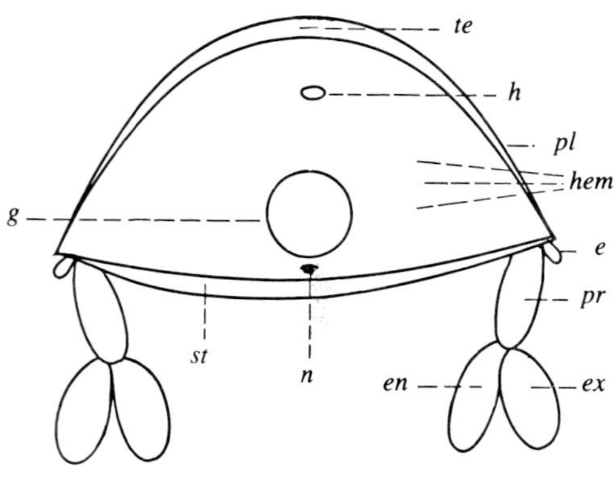

Fig. 35–2. Primitive arthropod. Diagram of a cross section of the body. *e*, Epipodite. *en*, Endopodite. *ex*, Exopodite. *g*. Gut. *h*, Heart. *hem*, Hemocoel. *n*, Ventral nerve cord. *pl*, Pleuron. *pr*, Protopodite. *st*, Sternum. *te*, Tergum. (By E.C. Faust.)

The great diversity of form within the Arthropoda has led some to conclude that it is a polyphyletic assemblage. Based on morphological and embryological evidence, Manton (1977) and Anderson (1979) suggest that the extant taxa be subdivided into three phyla: Crustacea, Uniramia (insects, millipedes, centipedes), and Chelicerata (spiders, scorpions, ticks, mites). However, since important fossil records are unavailable, a definite conclusion regarding the true phyletic origin of the Arthropoda must wait.

Because of diverse adaptations to various modes of life, the Phylum Arthropoda has become the largest in the Animal Kingdom, with the number of known species greater than that of all the other groups combined. Economically, arthropods constitute a great hindrance to man's progress, although many are useful. Medically, many species provide a constant menace, and the conquest of several of the most deadly human diseases has been or will be made possible only by control of the arthropod vectors.

Life Cycles

Typically the life cycle of an arthropod begins when the egg produced by the female is fertilized by a spermatozoon from the male, but in a few instances the egg proceeds with its development without fertilization (i.e., by *parthenogenesis*). The postembryonic developmental pattern may vary, showing gradual or abrupt changes at each ecdysis. Within the phylum, a general system of classification can be established according to the number of distinct forms that occur (Clarke, 1973). Arthropods having *monophasic* life histories usually exhibit a number of molts with changes in size occurring gradually. In some of the millipedes and centipedes, somites and limbs are added. *Diphasic* life histories are characterized by a considerable change in body form at ecdysis as well as the usual small changes through successive molts. Examples can be seen within the Crustacea in the development of certain copepods where, following a series of molts, the nauplius larva changes to the copepodid stages. In some Acari (for example, Ixodida), a *triphasic* life history can be seen where molts result in a larva-nymph-adult development sequence. Arthropods having *polyphasic* life histories exhibit at least four distinct forms, such as those seen in the complex series of changes evident in decapod Crustacea or some endopterygote insects that have active triungulin larvae.

Classification

For the clinical parasitologist it is both impracticable and unnecessary to present an exhaustive classification of the 10 or more recognized classes of arthropods. It is desirable, however, to outline the main groups of the phylum that are medically important, and to describe briefly the characters of these groups. The classification scheme presented in the following section is basically that used by Borror, DeLong, and Triplehorn (1981) for the major groups in the phylum Arthropoda and the class Insecta with the following exceptions: the Pentastomida is included here as a class and the Dictyoptera is recognized as an order of the Insecta. The general classification of the Arachnida and within that class, the Acari, follows that of Krantz (1978).

ARTHROPODS OF MEDICAL IMPORTANCE

Phylum Arthropoda v. Siebold and Stannius, 1845

Subphylum *Chelicerata* Heymons, 1901. Forms lacking antennae and usually possessing 6 pairs of appendages: a pair of fanglike chelicerae (the principal trophic appendages), a pair of pedipalps (leglike structures) behind the chelicerae, and 4 pairs of legs behind the pedipalps. The body is usually divided into two divisions: the prosoma (cephalothorax) and opisthosoma (abdomen). There are three classes of Chelicerata of which one (Arachnida) is of medical importance.

Class *Arachnida* Lamarck, 1815. Forms with body divided into prosoma and opisthosoma; without antennae; most adults with 4 pairs of legs; respiration by gills (primitive forms), booklungs, tracheae, or through the cuticle; gonads open near the base of the abdomen.

Subclass *Scorpiones* Latreille, 1810 (True Scorpions). Body elongate; prosoma unsegmented; anterior portion of the opisthosoma broad as the prosoma, composed of 7 segments; posterior portion of the opisthosoma narrowed, with 6 flexible segments of which the terminal one is differentiated into a piriform poison organ ending in a hollow stinger-spine. Representatives: *Buthus, Tityus, Androctonus.*

Subclass *Araneae* Lamarck, 1818 (Spiders). Prosoma distinctly constricted from the opisthosoma, which is usually unsegmented; first pair of appendages (chelicerae) 2-jointed, with opening of poison gland through hollow-fanged tip; with 3 pairs of spinnerets preanal in position, each connected with a silk gland. Representatives: widow spiders, brown recluse spiders, funnel-web spiders.

Subclass *Acari* Leach, 1817 (Ticks and Mites). Prosoma and opisthosoma united, not externally segmented; mouth parts contained in a discrete anterior gnathosoma; adults with 4 pairs of legs, larvae with 3 pairs of legs.

Order *Parasitiformes* Reuter, 1909 (*sensu* Zachvatkin, 1952). Forms with 1 to 4 pairs of stigmata (a pair of respiratory spiracles) posterior to the coxae of the second pair of legs; the stigmata may be associated with simple encircling plates or grooves (peritremes) of varying length and complexity.

Suborder *Ixodida* van der Hammen, 1968 (Ticks). Hypostome modified into a piercing organ with ventral recurved teeth; skin leathery, venter furrowed; adults with tracheae and with 1 pair of stigmatal plates near the bases of legs III or IV. Dorsum of the tarsus of leg I with a distinct sensory pit (Haller's organ). Representatives: *Argas, Ornithodoros, Ixodes, Dermacentor.*

Suborder *Gamasida* van der Hammen, 1968. Medically important mites possessing a small, toothless hypostome; eyes lacking; tracheae present, opening above or behind third coxa; opisthosoma without annulations or ventral furrows: body often with leathery shield. Representatives: *Ornithonyssus bacoti, Dermanyssus gallinae*.

Order *Acariformes* Zachvatkin, 1952. Mites without visible stigmata posterior to the coxae of the second pair of legs; some forms with simple sensilla sensory organs or optically active setae and sensory structures located on the anterior portion of the prosoma (propodosoma).

Suborder *Actinedida* van der Hammen, 1968. Forms having 3- to 5-segmented palps; propodosomal sensory organs often present. Some species with paired stigmata at or near the base of the chelicerae or on the propodosoma. Representatives: *Demodex folliculorum, Pyemotes tritici, Leptotrombidium* spp., *Cheyletiella* spp.

Suborder *Acaridida* van der Hammen, 1968. Forms having 2-segmented palps and no specialized sensory organs on the propodosoma. Respiration is through the integument so that there is no evidence of tracheae, stigmata, or peritremes. Representatives: *Sarcoptes scabiei, Tyrophagus* spp., *Acarus* spp., *Glycyphagus* spp., *Dermatophagoides* spp.

Subphylum *Mandibulata* Snodgrass, 1938. Forms having 1 or 2 pairs of antennae and a pair of mandibles, the principal trophic appendages homologous with the pedipalps of the Chelicerata, on the second postoral segment. Body form and the number of legs are highly variable.

Class *Crustacea* Pennant, 1777. Forms having typically 2 pairs of preoral antenniform appendages and at least 3 pairs of postoral jaw-like appendages; chiefly aquatic species, breathing entirely through gills.

Subclass *Copepoda* Latreille, 1831 (Copepods). Small forms, without carapace; having 1 pair of maxillae and 4 or 5 pairs of biramous legs.

Order *Eucopepoda* Claus, 1875 (Water "Fleas"). Lacking compound eyes; females carry egg sacs. Representatives: *Diaptomus, Cyclops*.

Subclass *Malacostraca* Latreille, 1802. Large forms, typically with 5 cephalic, 8 thoracic, and 7 abdominal segments, and appendages on all but 1 abdominal segment.

Order *Decapoda* Latreille, 1802. Carapace covering the thorax; first pair of chelipeds usually large, with distinct claws; thoracic legs usually lack exopodites. Representatives: Crayfishes, lobsters, crabs.

Class *Pentastomida* Heymons, 1926 (Tongue worms). Body elongate, tongue-like, cylindrical or moniliform, with many pseudosegments, but no separation into head, thorax, and abdomen; head with median subterminal mouth and 2 pairs of hollow retractile claws; eyes lacking; sexes separate, male smaller than female; developmental stages include embryo, nymph, and adult. Representatives: *Linguatula serrata, Armillifer armillatus, A. moniliformis*.

Class *Diplopoda* Latreille, 1802 (Millipedes). Forms with 2 pairs of legs and 2 pairs of spiracles for the greater number of body segments; gonads open behind the second pair of legs.

Class *Chilopoda* Latreille, 1802 (Centipedes). Forms usually with a single pair of legs and 1 pair of spiracles for each of the postcephalic segments of the body; first pair of legs modified to form poison claws; gonads open on penultimate abdominal segment. Representatives: *Lithobius, Scolopendra*.

Class *Insecta* Linnaeus, 1758 (Insects). Forms with distinct head, thorax, and abdomen; with 1 pair of antennae; mouthparts modified for chewing (mandibulate) or sucking (haustellate); typically with 2 pairs of wings on the thorax, which consists of 3 segments; abdomen composed typically of 10 evident segments, of which the terminal one is modified for sexual functions; with tracheate respiration; opening of the genital organs is near the caudal end of the body. Some groups develop without metamorphosis *(ametabolous);* some have simple or incomplete metamorphosis *(hemimetabolous);* still others have complete metamorphosis *(holometabolous)*.

Subclass *Apterygota* (Brauer, 1885) Lang, 1889. Wingless insects, believed to be descended from wingless ancestors. This group includes 4 orders (sometimes considered to be classes): Collembola, Protura, Diplura, and Thysanura. A few species of Collembola may rarely be annoying, causing a dermatitis.

Subclass *Pterygota* (Brauer, 1885) Lang, 1889. Winged and secondarily wingless insects descended from winged ancestors. This group includes 23 or more orders (depending on the authority consulted), of which the following are of medical importance.

Order *Anoplura* Leach, 1815 (Sucking lice). Wingless, with body compressed dorsoventrally; tarsi 1-clawed, apposed by a toothed projection of the tibia; with mouth parts adapted for piercing and sucking; with simple metamorphosis (i.e., egg—nymph—adult).

Order *Dictyoptera* Leach, 1818 (Cockroaches). Flattened dorsoventrally, with smooth tough integument; typically with 2 pairs of wings; head flexed downward and backward when at rest; antennae filiform, many-segmented; typically cursorial, but with ability to fly; mouth parts adapted to biting and chewing; primarily nocturnal in habits; often synanthropic; with simple metamorphosis (i.e., egg (in ootheca)—nymph—adult). Cockroaches are ancient insects and highly gregarious.

Order *Hemiptera* Linnaeus, 1758 (True bugs). Typically with 2 pairs of wings, with basal part of front wings thickened, more or less leathery, but wingless in some families (i.e., bedbugs); body somewhat compressed dorsoventrally; tarsi 2-clawed; mouth parts adapted for piercing and sucking; beak arising from front part of head; metamorphosis simple (i.e., egg—nymph—adult).

Order *Coleoptera* Linnaeus, 1758 (Beetles). Typically with 2 pairs of wings, of which the first pair is greatly thickened to form a protective cover or *elytra;* mouth parts adapted for chewing; metamorphosis complete (i.e., egg—larva—pupa—adult).

Order *Hymenoptera* Linnaeus, 1758 (Bees, wasps, ants). Typically with 2 pairs of membranous wings; with mouth parts adapted for chewing or for chewing and sucking; abdomen of female usually provided caudally with a "sting"; metamorphosis complete (i.e., egg—larva—pupa—adult).

Order *Lepidoptera* Linnaeus, 1758 (Moths and butterflies). Typically with 2 pairs of membranous wings covered with overlapping scales; mouth parts adapted for sucking; metamorphosis complete (i.e., egg—larva—pupa—adult).

Order *Diptera* Linnaeus, 1758 (True flies). With 1 pair of wings attached to the second thoracic segment; second pair of wings replaced by minute club-shaped "balancing organs" *(halteres);* adult mouth parts adapted for sucking, sponging, and rasping; metamorphosis typically complete (i.e., egg—larva—pupa—adult).

Order *Siphonaptera* Latreille, 1825 (Fleas). Wingless, with body strongly compressed laterally, with an especially long pair of hind legs adapted for jumping; with mouth parts adapted for piercing and sucking; metamorphosis typically complete (i.e., egg—larva—pupa—adult).

In addition to these insect orders, the Orthoptera Latreille,

1796, suborders Caelifera and Ensifera, comprising the grasshoppers and crickets, are occasional intermediate hosts of incidental human helminthic infections.

GENERAL REFERENCES

Anderson, D.T. 1979. Embryos, fate maps, and the phylogeny of arthropods, pp. 59–105. In *Arthropod Phylogeny.* Edited by A.P. Gupta. New York, Van Nostrand-Reinhold.

Borror, D.J., DeLong, D.M., and Triplehorn, C.A. 1981. *An Introduction to the Study of Insects,* 5th ed. Philadelphia, W.B. Saunders.

Clarke, K.U. 1973. *The Biology of the Arthropoda.* London, Edward Arnold Ltd.

Krantz, G.W. 1978. *A Manual of Acarology,* 2nd ed. Corvallis, Oregon State University Book Stores.

Manton, S.M. 1977. *The Arthropoda. Habits, Functional Morphology and Evolution.* Oxford, Clarendon Press.

Richards, O.W., and Davis, R.G., editors. 1977. *Imms' General Textbook of Entomology,* 10th ed, Vols. I and II. London, Chapman and Hall.

Service, M.W. 1980. *A Guide to Medical Entomology.* London, Macmillan Press.

Smith, K.G.V., ed. 1973. *Insects and Other Arthropods of Medical Importance.* London, British Museum (Natural History).

Sutherland, D.W.S. 1978. *Common Names of Insects and Related Organisms.* Spec. Publ. 78–1. College Park, Maryland, Entomological Society of America.

Chapter 36

Crustacea, Linguatulids, Millipedes, Centipedes, Scorpions, Spiders, Ticks, and Mites

CRUSTACEA OF MEDICAL IMPORTANCE

Copepods (Eucopepoda). Copepods (both free and parasitic forms) are almost microscopically small, simple, aquatic organisms, lacking a carapace, with one pair of maxillae and five pairs of biramous swimming legs. They lack compound eyes. The free-living species are elongated, more or less convex on the dorsum, piriform through the cephalic and thoracic regions, narrowed in the abdominal portion, and have two caudal rami (Fig. 36–1). The genera *Diaptomus* and *Cyclops* are medically important. In *Diaptomus* the right fifth leg of the male terminates in a more or less sickle-shaped hook; in *Cyclops* the fifth legs are rudimentary. In both genera there is a median (i.e., cyclopean) eye. In both genera the female carries her eggs in a pair of conspicuous brood sacs extending out from the abdomen. After hatching, the organism passes through three distinct larval stages before it reaches the adult form. This results in an increase in the number of appendages and in the number of thoracic and abdominal segments, as well as in the structural complexity of the appendages. Species of copepods inhabit both freshwater and saltwater bodies, including lakes, ponds, and even temporary pools, throughout the world, but they are more commonly found in still water than in running streams.

Species of *Diaptomus* are important first intermediate hosts of *Diphyllobothrium latum* (*vide supra,* p. 494), and one species is suspected of serving as the intermediate host of *Drepanidotaenia lanceolata*. Many species of *Cyclops* have been incriminated as intermediate hosts of *Dracunculus medinensis* (see p. 335), of *Diphyllobothrium latum,* and of other members of this genus of tapeworms (see p. 494–499).

In *Dracunculus* the first-stage larvae, which have been shed into pools by the gravid female worm from cutaneous lesions in man, are actively ingested by cyclops (both larvae and adults), and soon penetrate into the hemocoel (Fig. 36–2) where they develop further during a 2-week period (Muller, 1979). When the parasitized cyclops is accidentally ingested in contaminated water (Belcher et al., 1975), the larvae migrate through

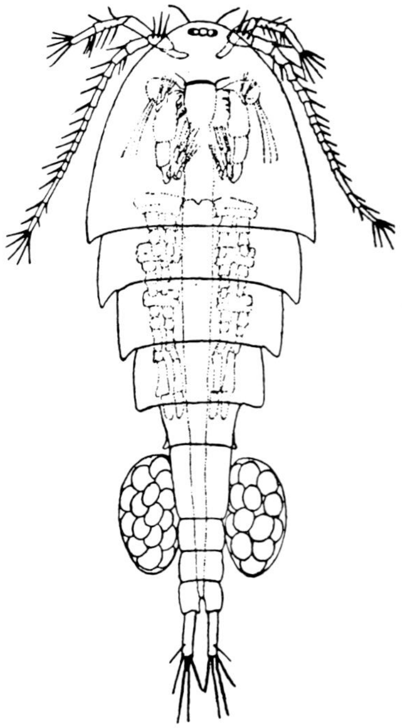

Fig. 36–1. *Cyclops (C. coronatus).* Female, dorsal view. (Adapted from original of E.C. Faust.)

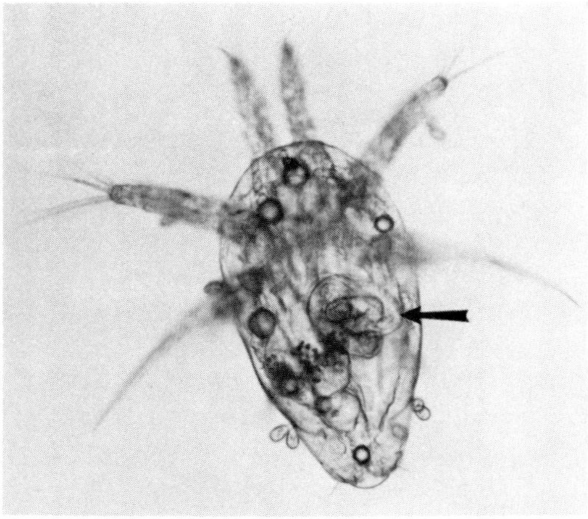

Fig. 36–2. *Cyclops vernalis,* living nauplius with larva of *Dracunculus medinensis* in its hemocoel. (Photograph by Ralph Muller. From Schmidt, G., and Roberts, L.S., 1981. *Foundations of Parasitology,* 2nd ed. St. Louis, C.V. Mosby Co.)

the intestinal wall and develop into mature worms in the human tissues (*vide supra,* p. 335).

In the case of *Diphyllobothrium latum* and related species, the copepods ingest the ciliated embryos hatched from the eggs. After shedding the ciliated epithelium, the embryos actively migrate through the intestinal wall into the copepod's hemocoel, where they develop into procercoids. The copepod hosts of *D. latum* constitute *important food* for fishes. In species of the related genus *Spirometra,* the parasitized copepods are *accidentally ingested* in drinking water by the second intermediate hosts (practically all species of vertebrates other than fishes).

One or more species of *Cyclops* in Asia have been incriminated as first intermediate hosts of *Gnathostoma spinigerum (vide supra,* p. 340).

Only freshwater copepods of the genera *Diaptomus* and *Cyclops* are known to serve as intermediate hosts of parasites infective for man.

Decapods. This group includes the prawns, shrimps, lobsters, crayfishes, and crabs, the largest and most familiar crustaceans. The cephalothorax is always covered by a carapace. The three anterior pairs of thoracic appendages are modified into biramous maxillipeds, but the more conspicuous five posterior pairs are developed into uniramous legs.

The crayfishes are important as second intermediate hosts of *Paragonimus (vide supra,* p. 464). In Japan and Korea species of *Cambaroides* and *Procambarus* are natural hosts of *Paragonimus westermani;* in North America species of *Cambarus* have been found naturally infected with cysts of *P. kellicotti.* These crayfishes are freshwater forms.

Crabs incriminated in the transmission of *P. westermani* are also freshwater species, but may migrate into brackish water. They include species of *Potamon, Parathelphusa, Pseudothelphusa, Eliocheir, Sesarma,* and possibly other genera. Rivercrabs of the genus *Potamonautes* may act as phoretic hosts for the larvae and pupae of the *Simulium neavei* complex, vectors of *Onchocerca volvulus* in East Africa.

Prawns in the South Pacific are believed to be paratenic hosts of the nematode *Angiostrongylus cantonensis,* which causes cerebral damage in exposed human hosts (*vide supra,* p. 292).

LINGUATULIDS OR PENTASTOMES ("TONGUE WORMS")

Biology. The members of this small group of arthropods are endoparasites of vertebrates. The adult "worms" are linguiform or moniliform, with an external pseudosegmentation, and have a mouth and two pairs of clawed rudimentary appendages at the anterior end of the body. They have no specialized respiratory or circulatory system. The adults live in the respiratory passages and body cavities of reptiles, birds, and mammals, where the females lay eggs that are discharged from the respiratory tract in sputum and mucus. When swallowed by the larval host, these eggs hatch and *4-* or *6-legged larvae* emerge in the intestine and migrate to the liver, spleen, lymph nodes, or lungs, and there transform into nymphs that become encapsulated.

Taxonomic Position. We consider the Pentastomida to be a class within the Arthropoda. However, the taxonomic status of this highly modified group has varied considerably over the past 50 years, reflecting different perceptions of their phylogenetic relationship with other invertebrates. According to Riley et al. (1978), the pentastomes have been considered highly modified annelids or descendants from an annelid-like ancestor; a distinct phylum, similar in rank to Onychophora and Tardigrada; or arthropods. These authors suggest that the Pentastomida should be placed within the Crustacea as a subclass near the Branchiura. Evidence includes the presence of spiral cleavage during embryogenesis (a crustacean characteristic), similarity in sperm structure to Branchiura, and a hypothetical evolutionary host pathway from fish to piscivorous reptiles and, secondarily, to birds and man. This scheme appears likely and should be given serious consideration.

Clinical Aspects. On several occasions "tongue worms" have been reported from man as larval host. *Armillifer armillatus* larvae and nymphs have been recovered from the liver, spleen, and lungs (visceral pentastomiasis), and from the eye (Gratama and von Thiel, 1958; Polderman and Manschot, 1979) of man in Africa (Egypt, Gambia, Senegal, Ghana, Zimbabwe, Cameroon Republic, Zaire). *A. moniliformis* has been recovered once each from humans in Manila, Indonesia, and Tibet (Fig. 36–3). Adult male and female *A. armillatus* are shown in Figure 36–4. Fain and Salvo (1966) recorded 8 cases of pentastomiasis produced by *A. grandis* in man in Zaire. A fatal case has recently been reported from Ivory Coast (Cagnard et al., 1979). A nymphal *Pentastoma najae* was once recovered from a human host in India. These infestations were probably acquired from ingesting raw infected snakes, the usual intermediate hosts.

Linguatula serrata, commonly found as an adult or nymph in the upper respiratory passages of dogs and other mammals, and as a larva or encapsulated nymph in herbivores, has been recorded from man in Africa, Switzerland, Germany, Greece, Turkey, Brazil, Chile, Colombia, Panama, Cuba, and United States (Virginia, Mississippi, Texas)(Fig. 36–5). The home slaughter of both domestic and wild herbivores may contribute to infection (FAO/WHO, 1979). Lodgment of the immature stages in the pharynx produces a syndrome known as *halzoun* or *marrara* (nasopharyngeal pentastomiasis), a well-known clinical entity in Lebanon and the Sudan (Schacher et al., 1969). Hunter and Higgins (1960) and Rendtorff et al. (1962) reported nymphs of *L. serrata* causing visual damage due to invasion of the eye in their North American patients.

In Malaysian aborigines, infection with porocephaloid species was found in 45.4% of adults in a series of 30 consecutive autopsies (Prathap et al., 1969). The liver and lungs were most frequently infected; in three cases more than one organ was

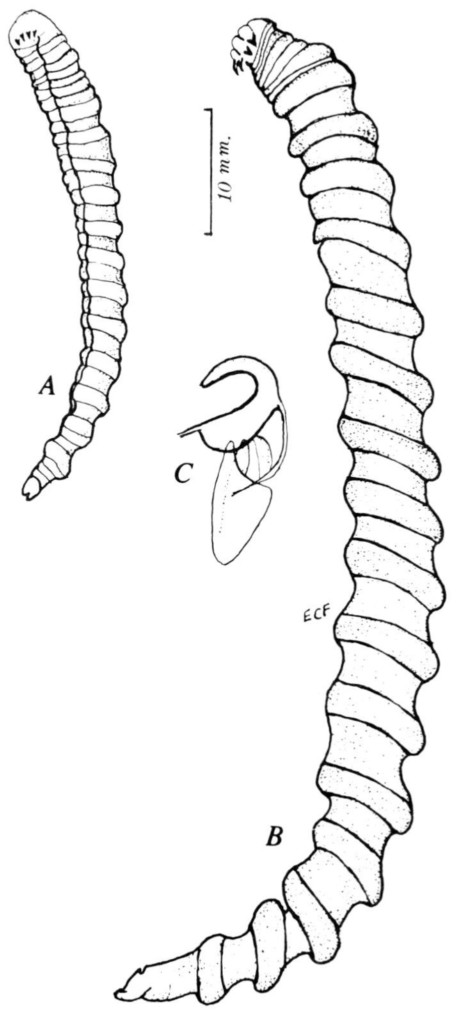

Fig. 36–4. *Armillifer armillatus,* adult. *A,* Male. *B,* Female. *C,* Detailed drawing of anterior hooklet. (Adapted from Fain, A. 1961. Les pentastomides de l'Afrique Centrale. Annales de la Musée Royal de l'Afrique Centrale, ser. 8, no. 92, pp. 67, 70.)

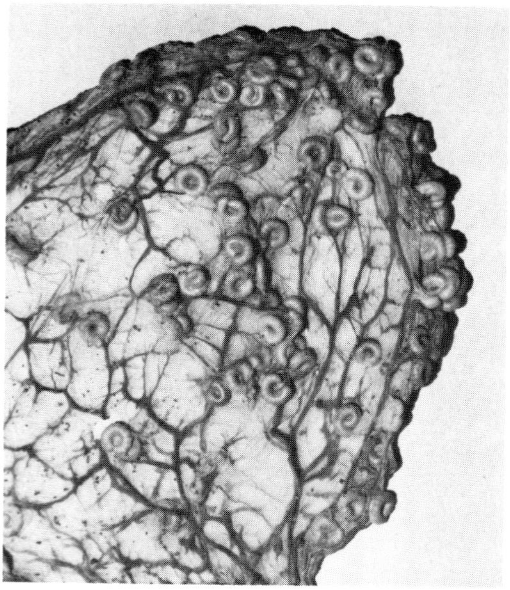

Fig. 36–3. *Armillifer* sp. Nymphs in omentum of East African vervet, *Cercopithecus aethiops.* (× ca. 1.5.) (Courtesy of Dr. R.E. Kuntz.)

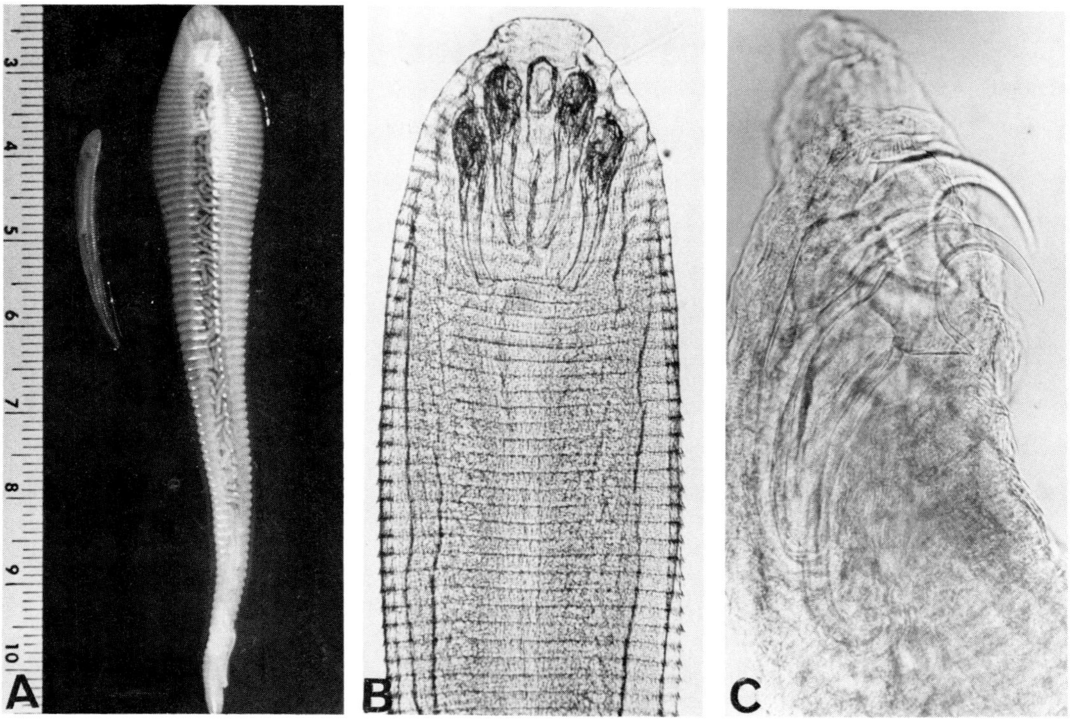

Fig. 36–5. *Linguatula serrata. A,* Adult male (left) and female, ventral view (centimeter scale). *B,* Nymph, anterior end, ventral view. (× 40.) *C,* Head, lateral view showing two hooks at right side of mouth. (× 110.) (From Faust, E.C., Beaver, P.C., and Jung, R.C. 1975. *Animal Agents and Vectors of Human Disease,* 4th ed. Philadelphia, Lea & Febiger. Courtesy of Dr. J.F. Schacher.)

parasitized. Three types of lesion were observed; the encysted pentastomid nymph, the necrotic pentastomid granuloma, and the "cuticle" granuloma.

Usually the encapsulated larvae or nymphs produce no symptoms and are discovered only at autopsy; occasionally these parasites in the lungs produce marked congestion. In rare instances, multiple infection has occasioned symptoms of severe abdominal pain and icterus when the larvae were migrating from the intestine to the other viscera. Cannon (1942) reported the death of a native Nigerian woman of 50 years after intestinal obstruction by a large number of nymphal *A. armillatus.* The adult "worm" has also been reported from the upper respiratory tract of man (nares). In three patients undergoing abdominal surgery in Nigeria, parasitic cysts found on the mesentery were later diagnosed as live nymphs of *Armillifer* (Dönges, 1966). Pentastomiasis in Africa was reviewed by Self et al. (1975). According to Fain (1975), although human infections with five species of three genera of pentastomids have been reported, the only species known to have developed in man are *L. serrata,* a cosmopolitan species, *A. armillatus* and *A. grandis* in Africa, and *A. moniliformis* in Asia. The species reported as *Porocephalus moniliformis* in Malaysian aborigines (Prathap et al., 1969) actually is *Armillifer moniliformis.*

MILLIPEDES (DIPLOPODA)

Millipedes are not venomous in the usual sense since they have no fangs, only a pair of biting mandibles and a fused pair of maxillae, the gnathochilarium. Radford (1975) reviewed the clinical features of the burning dermatitis and sloughing of the skin caused by contact with giant tropical millipedes and their secretions, which they squirt for a distance of several inches. The secretions are elaborated in special exocrine defense glands that are serially arranged as segmented pairs (Eisner et al. 1978). These secretions may contain hydrogen cyanide as well as an array of other aliphatic acids, aldehydes, esters, phenols, and quinones. Species of *Julus* and *Polydesmus* have been reported from the digestive and urinary tracts of man (Garzia, 1938), and species of *Julus* and *Fontaria virginiensis* are acceptable intermediate hosts of *Hymenolepis diminuta (vide supra,* p. 511).

CENTIPEDES (CHILOPODA)

Centipedes (Fig. 36–6) have a venom gland in

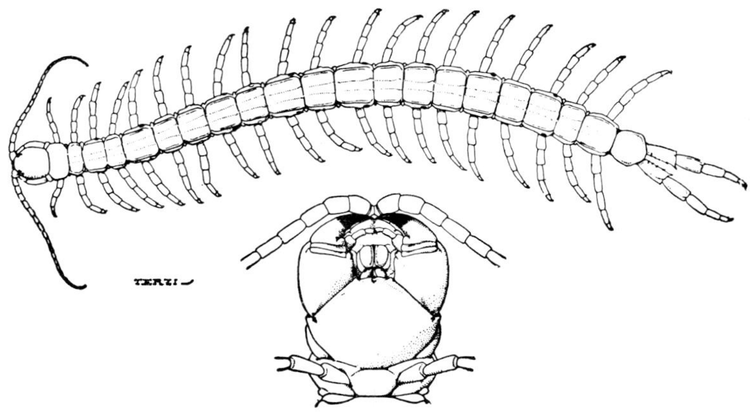

Fig. 36–6. *Scolopendra morsitans.* Dorsal view of entire centipede and anterior view of head. (From Castellani, A., and Chalmers, A.J. 1919. *Manual of Tropical Medicine.* London, Bailliere, Tindall and Cox.)

the subterminal joint of each of the first pair of trunk appendages (the maxillipeds). Ducts leading to the curved terminal claws of these appendages, which serve as fangs, open slightly dorsal to the tip. The venom is an acidic, opalescent fluid, slightly miscible in water. The venom proper constitutes only a small fraction of the total discharge from the venom glands, the function of which is primarily digestive.

Centipedes are not generally considered to be dangerous. Aside from a single report of the death of a 7-year-old girl bitten on the head (Minelli, 1978), their importance lies in the occasional painful bite they are able to inflict. Some persons, particularly small children, may be especially sensitive to the toxins of centipedes capable of puncturing the human skin, but there is no evidence that these arthropods are to be shunned. The larger forms, which may reach a length of up to 25 cm or more and include species of *Scolopendra, Lithobius,* and *Geophilus,* are troublesome because of the greater amount of venom they secrete. Should a centipede bite prove painful, palliatives such as compresses of sodium bicarbonate or saturated solution of Epsom salts (magnesium sulfate) should be applied locally. Control consists of removing such typical harborages as piles of rubbish or vegetable refuse (Busvine, 1980).

Species of *Geophilus* and *Lithobius* have been recovered on many occasions from the human nares and frontal sinuses, *Geophilus* once from the urethra, and species of *Geophilus, Himanterium, Stigmatogaster, Chaetechelyne, Lithobius,* and *Scutigera* on ten occasions from the digestive tract (Garzia, 1938). In each reported case, the organism was an accidental invader of the body.

SCORPIONS (SUBCLASS SCORPIONES)

Biology. True scorpions are characterized by the large pedipalps, which terminate in stout claws; a condensed and unsegmented cephalothorax or *prosoma;* and the division of the abdomen *(opisthosoma)* into a broad anterior 7-segmented portion *(mesosoma)* and a narrow posterior 5-segmented portion *(metasoma)* with a piriform telson (i.e., caudal extremity) ending in a hooked stinger (Fig. 36–7). Venom is discharged from a pair of poison glands through the tip of the stinger. These glands are situated in the enlarged portion of the telson. Scorpions sting to immobilize their prey, then crush them between the coxal (basal) segments of their pedipalps. The other four pairs of cephalothoracic appendages are developed as walking legs. Secondary sexual characteristics are largely species-limited, though males commonly have a longer posterior abdomen due to the lengthening of caudal segments IV and V. Adult scorpions have a pair of sensory combs *(pectines)* attached ventrally to the second abdominal segment. Most scorpions have a pair of eyes located medially on the prosoma, and a group of 2 to 5 ocelli on each anterolateral margin. Respiration is through four pairs of booklungs situated on the ventral side of mesosomal segments III to VI.

Scorpions are viviparous. After several months of gestation the young are born with an enveloping membrane, which facilitates parturition. Immediately after birth the mother removes the membrane with her pedipalps, and the young scorpions crawl

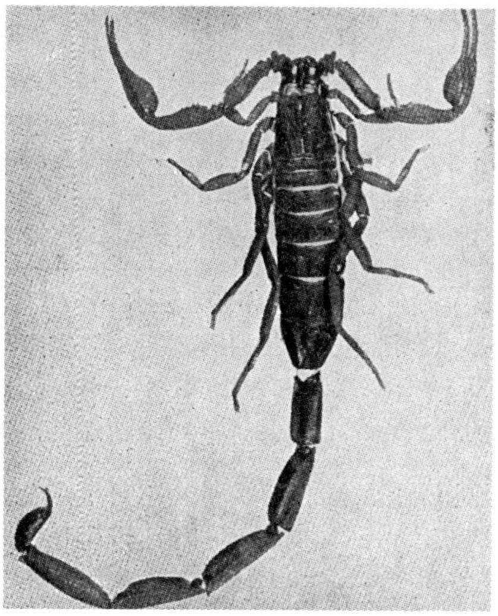

A

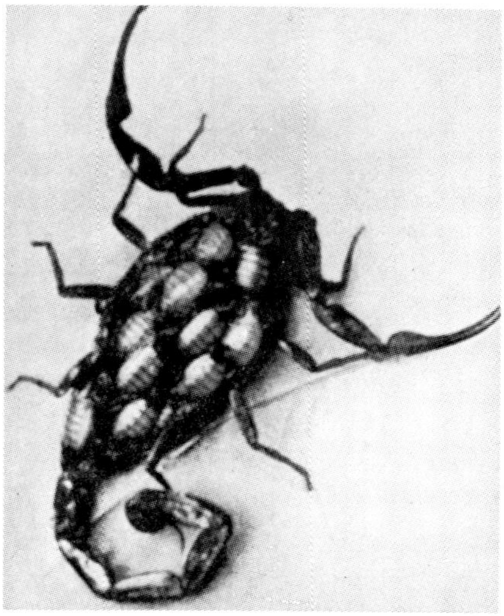

B

Fig. 36–7. Scorpion, *Centruroides* sp. *A*, Male. (× 1.) *B*, Female *C. sculpturatus* with newly born young. (× 1.) (*A* adapted by E.C. Faust from C.C. Hoffman. In Brennemann, J. 1936. *Practice of Pediatrics*. Hagerstown, W.F. Prior Company. *B* adapted from Stahnke, H.L. 1944. Scorpions of the United States. *Turtox News*. Chicago, General Biological Supply House.)

onto the mother's back, where they remain attached for some time (Fig. 36–7,*B*). These arthropods are terrestrial in their habitat and nocturnal or photophobic in their habits, hiding in dark places during daylight. When moving they carry the large pedipalps horizontally in front and the posterior abdomen arched upward or dragging behind. Prey usually consists of spiders and large insects.

Scorpion Envenomation

Epidemiology. Humans come in contact with scorpions by accidentally stepping on them with bare feet or by brushing an arm or hand over them, usually when they are concealed in dark places. This is particularly apt to occur on the dirt floors of adobe huts in dry tropical regions. In the moist tropics, scorpions invade human habitations during the rainy season and hide in boots, shoes, clothing, or towels, or lurk near water containers, in bathtubs, or in vegetation near human habitations.

Scorpions generally considered to be dangerous have a wide distribution (Keegan, 1980). In the New World, several species of *Centruroides* (southwestern United States, Mexico, Venezuela) and *Tityus* (Trinidad, Argentina, Brazil, Guyana, Venezuela) are important. Old World genera include species of *Androctonus* (Algeria, Egypt, Iraq, Israel, Morocco, Turkey), *Buthus* (Algeria, Jordan, Morocco, Sudan), and *Parabuthus* (South Africa) as well as *Hemiscorpion lepturus* (Iraq), *Leiurus quinquestriatus* (Egypt, Israel, Turkey), and *Mesobuthus gibbosus* (Turkey).

Children of preschool age are most commonly exposed and most seriously affected; the death rate due to scorpionism has been reported as high as 25% in this age group in Trinidad (Waterman, 1957). In adults the fatality rate is estimated to range from 0.25 to 1.80% (Bücherl, 1971*a*).

Pathology and Symptomatology. Some small species are unable to penetrate human skin with their stingers; others are notorious for their ability to pierce even the sole of the foot and inject venom. The large species are most likely to produce serious lesions with accompanying grave symptoms, particularly in children under 5 years of age, but at times also in adults. The most common effects by far are local pain and fright.

The more venomous scorpions produce an immediate, intense, aching pain, with burning sensations radiating from the site; lymphadenitis; systemic symptoms that frequently include generalized numbness, throbbing, and twitching, especially of the fingers, toes, ear lobes, and tip of the nose and chin; profuse sweating; excess salivation; glossopharyngeal involvement, so that speech and swallowing become difficult; vomiting; twitching and muscle spasm, especially of the abdomen; and convulsions. Mental disturbances and hallucinations, strabismus, temporary hemiplegia, monoplegia or blindness, rapid, weak pulse, hypotension, extreme thirst, dysuria or anuria, and elevation of temperature to 40°C or above, together with weak, shallow, but rapid respiration, and pulmonary edema are additional manifestations that may gradually subside after 2 to 8 hours. In some cases, however, the symptoms increase and the patient dies of respiratory paralysis, possibly within 1 to 2 hours after being stung. There may be toxic effects on the myocardium that cause electrocardiographic changes. Acute pancreatitis with abdominal pain, hyperglycemia, and glycosuria and elevated serum amylase occur with stings by *Tityus* (Reid, 1978).

Scorpion venoms vary in appearance and contain, in differing amounts depending on the species, neurotoxins, hemolysins, hemorrhagins, leukolysins, agglutinins, coagulants, enzymes, lecithin, cholesterin, a cardiac toxin, and a vascular toxin (Bettini, 1978). The hemorrhagin fraction (probably an enzyme) is responsible for hemorrhage and necrosis of tissue at the site of injection, and the neurotoxin causes the nervous and paralytic symptoms responsible for fatal cases of scorpion sting.

Diagnosis. Scorpion envenomation may be suspected when a patient presents himself with an aching, throbbing pain radiating from an inflamed, at times slightly indurated site, showing a single point of skin penetration. These symptoms are frequently accompanied by lymphadenitis medial to the site and by systemic symptoms of ascending motor paralysis. Scorpion envenomation must be distinguished from arachnidism, tick paralysis, bee sting, and other clinical entities that it may resemble.

Treatment. Local pain may be relieved by local anesthetic administered topically or injected. Suitable parenteral or oral analgesic may be used. In case of severe poisoning general supportive treatment is indicated with management of systemic manifestations as needed. The need for maintenance of airway, assisted respiration, and maintenance of blood pressure by intravenous infusion and hydrocortisone should be anticipated. For persons manifesting systemic intoxication, antivenin, when available, should be administered. (This is not produced commercially in the United States.) Several of the antivenins of the more dangerous scorpions are cross-protective (Whittemore et al., 1961; Potter and Northey, 1962). Standardized antivenin is administered intramuscularly. The possibility of sloughing, gangrene, or bacterial invasion of the puncture wound should be anticipated.

The following systemic therapy is recommended: 5 or 10% glucose in normal saline solution by intravenous infusion, followed by injection of insulin if there is hyperglycemia; treatment of shock with intravenous hydrocortisone; administration of atropine and trimethobenzamide, prochlorperazine, or benzquinamide to control vomiting; and a barbiturate or diazepam for convulsions and as a sedative.

Prognosis. In regions where the more venomous scorpions abound, prognosis is poor in young children. It is better in older children and in adults, except the aged and debilitated. It also depends on the size of the scorpion, the amount of venom present in the poison glands at the time of stinging, and whether the venom is introduced directly into the skin or through protective clothing. The venom is most potent during the warm breeding season, the period when most envenomation occurs. Scorpions that inject only hemorrhagic toxin are not usually dangerous; those having neurotoxic venom must be treated with respect. When an adequate amount of potent antiserum for neurotoxic envenomation is administered soon after the sting, the prognosis is usually excellent.

Prevention. Scorpions are susceptible to chlorinated hydrocarbons, carbamates, and organophosphorus compounds (Darrow, 1977). Area control of scorpions can be obtained by using outdoor sprays of 2% carbaryl, 10% granular diazinon, or 0.5% *gamma* HCH (lindane). If necessary, indoor hiding places can be treated by using malathion or lindane. Such mechanical methods as raising the floor level of the house and slightly separating the entrance steps from the foundation have also proved somewhat effective in barring access to domiciles (Ennik, 1972). This type of abatement should be supplemented with the removal of piles of debris, lumber, or other resting sites near houses (Keegan, 1980).

The whip scorpions (subclass Uropygi), in which all of the abdominal segments are broad and terminate in a caudal flagellum, and the pseudoscorpions or book scorpions (subclass Pseudoscorpiones), with a segmented broad abdomen, are harmless.

SPIDERS (SUBCLASS ARANEAE)

Biology. ***Anatomy.*** Both the cephalothorax *(prosoma)* and abdomen *(opisthosoma)* of spiders are

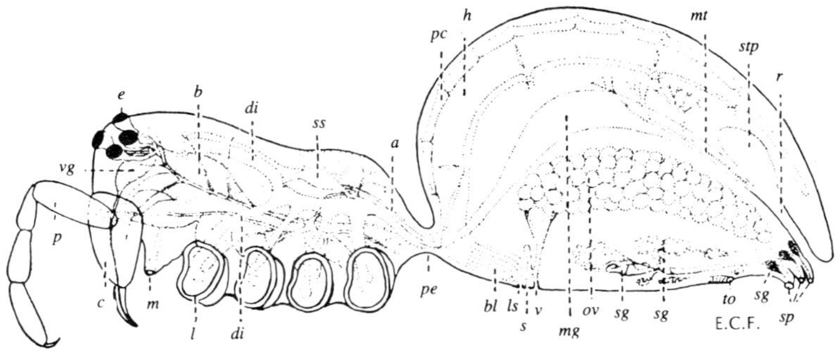

Fig. 36–8. Spider, diagram of lateral view showing external and internal organs. *a*, Aorta. *b*, Brain. *bl*, Book-lung. *c*, Chelicera. *di*, Diverticulum of midgut. *e*, Eyespot. *h*, Heart. *l*, Leg attachment. *ls*, Lung slit. *m*, Mouth. *mg*, Midgut. *mt*, Malpighian tubule. *ov*, Ovary. *p*, Pedipalp. *pc*, Pericardial cavity. *pe*, Pedicel. *r*, Rectum. *s*, Opening of spermatheca. *sg*, Silk gland. *sp*, Spinnerets. *ss*, Sucking stomachs. *stp*, Stercoral pocket. *to*, Tracheal opening (spiracle). *v*, Vagina. *vg*, Venom gland. (Adapted by E.C. Faust from Comstock, J.H. 1913. *The Spider Book*. Garden City, Doubleday, Page & Co.)

superficially unsegmented, although the head may be superficially separated from the thorax by a cervical furrow or groove (Fig. 36–8). The dorsal surface of the prosoma is covered by a leathery carapace. Hairs or bristles of various types may be distributed over the integument. The head bears several pairs of simple eyes near its front. The *mouthparts* consist of a pair of two-segmented *chelicerae* lying just above the preoral chamber; an *upper lip (rostrum*, i.e., ''labrum''); a median plate-like *epipharynx* on the ventral side of the labrum, with a *pharynx* that conveys fluids to the esophagus; a median *lower lip* (not homologous with the labium of insects); and the pedipalp coxal lobes that form the sidewalls to the entrance of the mouth. The chelicerae have a short, broad, basal segment and a clawed terminal segment, through the tip of which the poison gland opens. The pedipalps consist of six segments, of which the distal five are collectively termed the ''palp.'' In the males the palps are greatly modified for the transfer of sperm to the female. The *cephalothorax* or *prosoma* bears four pairs of walking legs, each consisting of seven segments. The *abdomen* is usually sacculate and often lacks traces of segmentation, except at the caudal extremity. Ventrally it has a transverse epigastric furrow and a subcaudal furrow. It is joined to the prosoma by a slender stalk or *pedicel*. Typically, there are one or two pairs of *spiracles*, the so-called lung-slits, on the venter of the abdomen, leading to primitive book-lungs. In addition, most true spiders have a paired *tracheal respiratory system* opening through a single (rarely paired) spiracle, just in front of the spinnerets. The external genital opening lies in the midline in the epigastric furrow. The silk glands are located in the abdomen and connect with the *spinnerets* (usually three pairs), which open ventrally at the subcaudal end of the body. They are complicated and morphologically varied, and occupy the ventral side of the abdomen posterior to the epigastric furrow (Fig. 36–8).

The poison or *venom glands* consist of a pair of elongate sacs lying at the base of the chelicerae or within the prosoma. The venom is discharged through a duct opening at the tip of each cheliceral fang.

Reproduction, Development, and Behavior. The fertilized eggs are laid in masses and range in number from a few to hundreds. The eggs are usually protected by a spun egg sac or cocoon. The eggs soon hatch, but the immature spiders remain within the cocoon for some time, even over the entire winter season. To permit growth, molting occurs recurrently up to 8 or 9 times with larger species molting a greater number of times. The first molt of the ''black-widow'' spider occurs before the spiderling escapes from the cocoon. Its development through these molting stages is comparable to the incomplete metamorphosis of insects and consists mostly of increase in size. During the last molt, however, the external genital organs of both sexes appear for the first time.

Spiders are terrestrial and occupy a wide variety of habitats. However, a few species are capable of diving beneath the surface of pools and streams to capture prey. All spiders are carnivorous and many are cannibalistic. They enmesh their prey in their webs, or ambush and pounce upon them. By introducing their venomous fangs into the bodies of their victims, they kill or render them helpless and then, at their leisure, suck out the nourishing juices.

Spider Envenomation (Arachnidism)

Although all true spiders produce venom, few have fangs powerful or sharp enough to pierce human skin, or venom potent enough to produce more than a transitory local irritation when introduced into the skin. On the basis of the type of tissue damage produced, venomous spiders may be classified as those that cause *systemic arachnidism* and those that primarily cause *necrotic arachnidism*.

Systemic Arachnidism

Species. Spiders causing systemic arachnidism include the large hairy, tropical tarantulas (family Theraphosidae), some forms of which can inflict a painful and occasionally dangerous bite (and they may also have urticating hairs), and the following family groups of somewhat smaller species capable of injecting potent venoms: (1) the comb-footed spiders (family Theridiidae), genus *Latrodectus,* containing the widow spiders, which can be classified either as a series of more or less distinct species (Levi, 1959) or as a group of closely related species including polytypic forms in the *L. mactans* subspecies complex (Bücherl, 1971b); *L. mactans mactans* ("black widow") from the United States to Chile (Fig. 36–9); *L. m. hasselti,* from New Zealand to India; *L. m. tredecimguttatus,* throughout the Mediterranean littoral; *L. m. cinctus (= L. indistinctus),* from the southern portions of Africa; *L. m. menavodi,* in Malagasy and Seychelles; *Latrodectus hystrix,* in Yemen; *L. dahli,* in Iran; *L. pallidus* from Asia Minor, and southern U.S.S.R.; *L. geometricus* (the "brown widow"), from southern Florida and widespread in the tropics (Levi, 1959), and *L. curacaviensis* (sometimes considered a species complex; see McCrone and Levi, 1964); (2) *Atrax robustus* and *A. formidabilis* ("funnel-web spiders"), from Australia, and *Trechona* spp. from South America (family Dipluridae) (Gray and Sutherland, 1978); (3) "wolf spiders" (genus *Lycosa,* family Lycosidae); (4) "wandering spiders," (genus *Phoneutria,* family Ctenidae), from Brazil, northern Argentina, Uruguay (Schenberg and Pereira Lima, 1978); and (5) *Harpactinella* (Barychelidae), from South Africa. None of these species is primarily domestic but will seek moisture under stone walls, around fence posts, under lumber piles, in cellars, and on the underside of privy seats, where their webs accidentally come in contact with unprotected human skin.

Other spiders worth noting include species of *Chiracanthium* (Clubionidae) in the United States, Japan, Europe, and Australia (Bettini and Brignoli, 1978) and *Steatoda* in eastern Europe (Maretic, 1978; Maretic and Lebez, 1979).

Widow Spiders. Members of the genus *Latrodectus* have received various taxonomic treatments. According to Gertsch (1979), the widow spiders found in the United States consist of a group of five closely related species that overlap geographically but can be separated by coloration and the presence of webs in relatively distinct habitats. *Latrodectus mactans* (Fig. 36–9), the southern black widow, has a shiny black abdomen with a red hourglass on the venter. It occurs northward to southern New England and typically lives under debris near buildings where the female spins a web made of coarse, tough fibers. *Latrodectus variolus,* the northern black widow, is widely distributed in the eastern United States, extending into western Florida and eastern Texas. The ventral hourglass of this species is divided into two transverse bands. In Florida, *L. variolus* often spins a web in the branches of a tree. The western black widow, *L. hesperus,* ranges from Oklahoma and Kansas into the Southwest. The webs of this species can be found in a variety of domestic and peridomestic habitats. Female coloration resembles that of *L. mactans*. The red widow, *L. bishopi,* and the brown widow, *L. geometricus,* occur in Florida. The former is found in sand-scrub habitats where it spins a web in vegetation 3 to 4 feet above ground. The abdomen of the adult female is usually spotted with brilliant red or yellow marks on the dorsum. The brown widow is mainly domestic, placing its web in or near buildings. It is usually drably colored gray or light brown, and the female has a ventral dull orange or reddish hourglass.

Because of the wide distribution of spiders belonging to the genus, latrodectism is a cosmopolitan disease. Where suitable habitats are available and human-spider contacts are frequent due to agricultural practices, envenomation by feral *Latrodectus* spp. can reach epidemic proportions (Maretic, 1978). Where domestic species of *Latrodectus* are more common, persons suffering from spider bite may come from a variety of backgrounds. Following recovery from one serious attack, the individual usually experiences a complete systemic immunity to subsequent bites, although there may be a more marked local reaction, with sloughing, on later occasions.

Fig. 36–9. *Latrodectus mactans*, black-widow spider. Female with egg-sac or cocoon. (× 3.8.) (Adapted from Needham, G.H. 1958. *Turtox News*. Chicago, General Biological Supply House.)

Pharmacological studies show black-widow spider venom to be a transparent, oily liquid composed of components that are destroyed by boiling for 5 minutes or heating to 75°C for 20 minutes, as well as by the action of proteolytic enzymes. Evidence therefore points to the venom being proteinaceous in nature. Studies indicate that *Latrodectus* venoms are neurotoxic, acting at both adrenergic and cholinergic junctions (Bettini and Maroli, 1978).

Pathology and Symptomatology. When a person accidentally brushes against a *Latrodectus, Atrax, Lycosa,* or *Ctenus* (almost invariably the female) or touches the web, the spiders defend themselves by attacking and injecting toxin into the skin. The effective fraction of the inoculum is a nonhemolytic neurotoxin. A sharp pain soon develops at the site of the bite, but there is little or no immediate swelling; later the area becomes red and swollen. The member burns and aches severely. Following lymphatic absorption and vascular dissemination, dizziness, weakness and tremor in the legs, and abdominal cramps characteristically develop. There is dull pain in the muscles of the chest, back, and abdomen, and frequently extreme prostration. The board-like rigidity and spastic contractions of the muscles, particularly those of the abdomen, suggest tetanus or acute appendicitis, although there is usually no local tenderness over the appendix. Other symptoms include hypertension, profuse cold perspiration, nausea, vomiting, severe headache, moderate rise in temperature, constipation, acute urinary retention, reduced heart beat and feeble pulse, cyanosis, anxiety, labored breathing and speech, insomnia, restlessness with light stupor, convulsions (in children), and at times complete prostration, shock, or delirium. Local necrosis at the site of the bite may suggest tropical phagedenic ulceration or dermal leishmaniasis. There is typically a moderate leukocytosis.

Diagnosis. At the site of the bite of *Latrodectus* there are two small red spots where the chelicerae have penetrated the skin. Diagnosis is dependent on a history of possible exposure and clinical signs and symptoms. Differentiation must especially be made from acute surgical abdomen.

Treatment. Treatment is essentially as for scorpion sting. Parenteral administration of diazepam (Valium) is indicated for allaying muscle spasm, pain, and fright. Later hot baths and intravenous injection of 10 ml of a 10% solution of calcium gluconate and of diazepam once daily for 2 or 3 days may be required to maintain control of symptoms. Neostigmine (Prostigmin) is reported to be effective in the relief of many of the symptoms (Schenone, 1959).

When there is objective clinical evidence of "black-widow" spider envenomation, 2.5 ml of the standardized antivenin should be administered *intramuscularly.** If symptoms are not relieved in an hour or two, the dose may be repeated. Before administration of the antivenin, the patient should first be tested for sensitivity and, if necessary, desensitized before injection. Kirby-Smith (1945) stated that this amount of antivenin equals in neu-

*Hyperimmune horse serum ("antivenin Latrodectus Mactans") is produced by Merck Sharp & Dohme.

tralizing power the venom of 750 black widows, and that following its administration the patient should experience relief in a few hours unless considerable time has elapsed between the bite and the treatment.

Prognosis. Depending on the amount and potency of the venom absorbed and the rapidity with which both symptomatic treatment and antivenin are administered, the prognosis is good to poor. Although numerous deaths have resulted from *Latrodectus* envenomation, uncomplicated recovery usually occurs; convalescence rarely may last as long as 2 months (Murray, 1946).

Prevention. Where spiders are a problem in homes, good housekeeping can often be the simplest and most effective control method. The destruction of webs by regular dusting, and the careful removal of trash, old clothing, and external debris around homes and in adjacent sheds and barns where spiders might live are recommended (Pratt, 1979).

Necrotic Arachnidism

Species. Spiders producing necrotic arachnidism belong to the genus *Loxosceles* (family Scytodidae). *L. laeta* and *L. rufipes* are widely distributed in southern South America, although most clinical reports of loxoscelism caused by these species have come from Uruguay, Chile, and Argentina. In the United States, human loxoscelism is most commonly associated with *L. reclusa,* although *L. unicolor (= L. deserta)* and *L. arizonica* have been incriminated in California and Arizona (Russell et al., 1969). These species are typically found inside human habitations, often in closets among clothes,

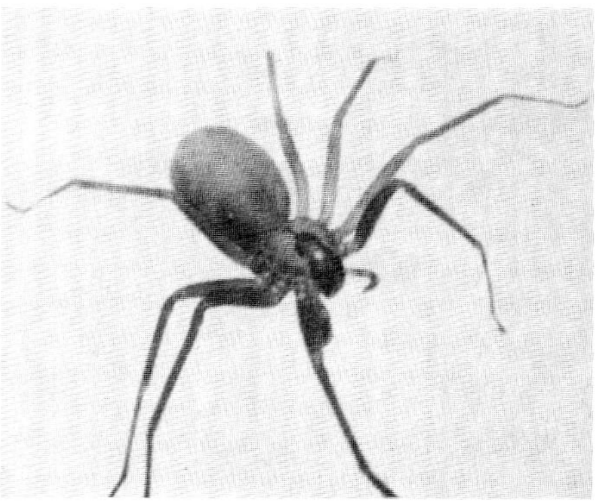

Fig. 36–10. *Loxosceles reclusa,* female. (× 2.) (From Atkins, J.A., et al. 1958. Necrotic arachnidism. Am. J. Trop. Med. Hyg. Courtesy of Professor Curtis W. Wingo, University of Missouri.)

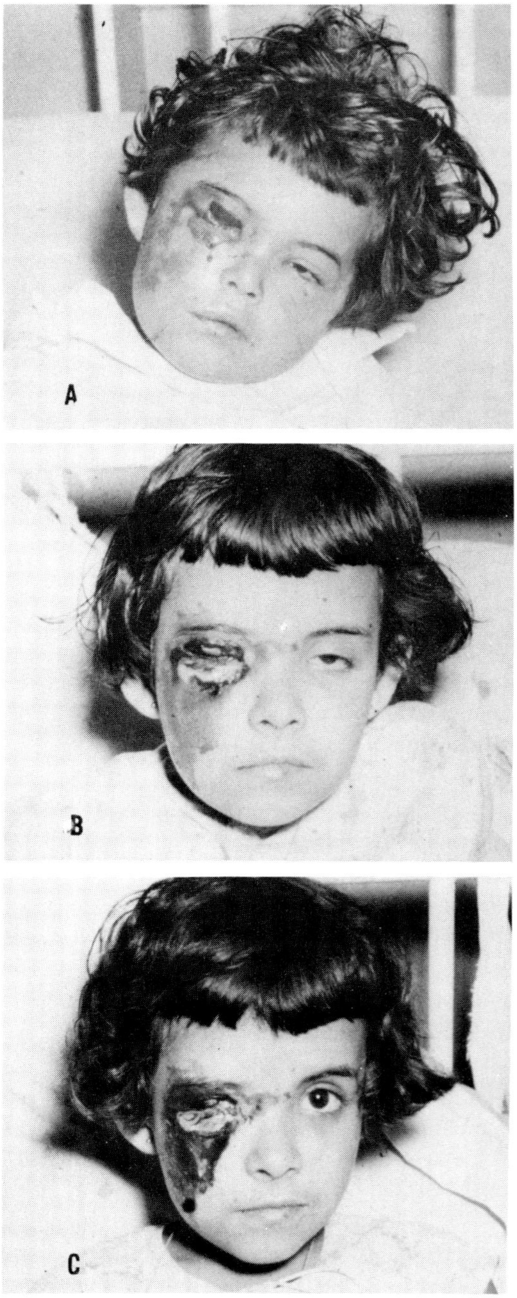

Fig. 36–11. *Loxosceles laeta.* Lesions produced on face of 7-year-old girl in Chile following envenomation. *A,* 20 hours after bite. *B,* 48 hours after bite. *C,* 21 days after bite. (Courtesy of Dr. Hugo Schenone.)

behind picture frames, in basements, under porches, in barns, or in stacks of debris, where they spin irregular webs. Spiders of this genus also occur in a wide variety of natural habitats. They are of medium size (0.9 to 1.5 cm long), yellowish-to-medium- or dark-brown, somewhat pubescent, and have a distinct constriction between cephalothorax and abdomen (Fig. 36–10).

Pathology and Symptomatology. When species of *Loxosceles* (family Scytodidae) accidentally come in contact with the unprotected skin, as when a person removes clothing from an infested closet, it bites, resulting in the immediate production of severe, sharp pain where the spider's chelicerae puncture the skin (Foil and Norment, 1979). The bitten area soon becomes swollen and ischemic. On the face, the injured area is frequently around one eye, and at first appears as a puffy, thick, reddened wheal with a necrotic center, so that the eye is swollen shut (Fig. 36–11,*A*). This first turns violaceous, then black, and becomes dry (Fig. 36–11,*B*). Later cutaneous necrosis develops (Fig. 36–11,*C*), and the area sloughs, leaving a deep granular center. Slow healing eventually takes place, with cicatricial tissue and sometimes deep disfigurement. If the bite is on a limb (Fig. 36–12), there may be separation of the superficial skin layers from the deeper tissue to form vesicles or blebs at each envenomated site. On the upper limb, extensive deep necrosis into the somatic musculature and sloughing may occur. If the venom introduced into the skin is small in amount and of low potency, there may be only mild systemic reaction with scarlatiniform rash; if it is considerable in amount and of high potency, a viscerocutaneous form of loxoscelism may occur. In addition to the aforementioned complications, systemic reactions consisting of hemolytic anemia, hemoglobinuria, hematuria, jaundice, fever, and sensorial involvement generally appear within the first 6 to 24 hours after the bite (Schenone and Suarez, 1978). This form of loxoscelism has a high mortality rate.

Diagnosis. The diagnosis is based on the history, revealing possible exposure to the bite of *Loxosceles*, and on the characteristic necrotic lesions.

Treatment. For necrotic arachnidism, no specific chemotherapeutics have been developed. If the patient sees the physician within a few hours after the envenomation, the administration of corticosteroids may ameliorate the early symptoms, control the toxic necrosis, and reduce the likelihood of disfiguring scars.

Prevention. Aerosol sprays containing pyrethroids or similar pesticides can be used indoors as contact toxicants. Residual sprays are often ineffective because many spiders tend to remain in or close to their webs and thereby avoid contact. However, the careful use of certain residual chemicals (diazinon, propoxur, and lindane) as water suspensions or as insecticidal dusts may be effective in controlling domestic infestations. (See also under systemic arachnidism, p. 578.)

TICKS (ORDER PARASITIFORMES: SUBORDER IXODIDA)

Definition. Ticks are a small group of acarine

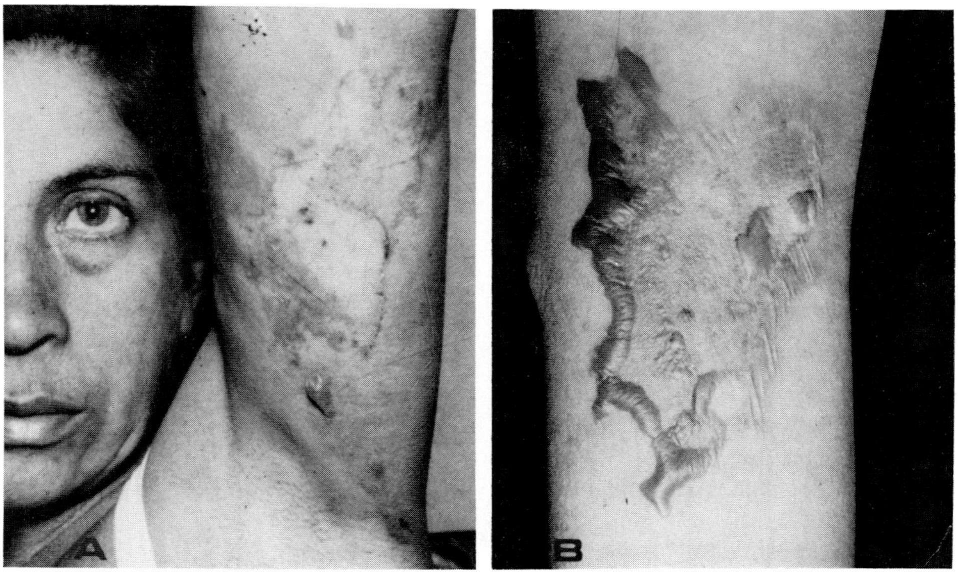

Fig. 36–12. *Loxosceles laeta*. Tissue necrosis resulting from envenomation. *A*, On upper arm of woman at 7 days. *B*, On flexor surface of arm of woman at 3 days. (Courtesy of Dr. Hugo Schenone.)

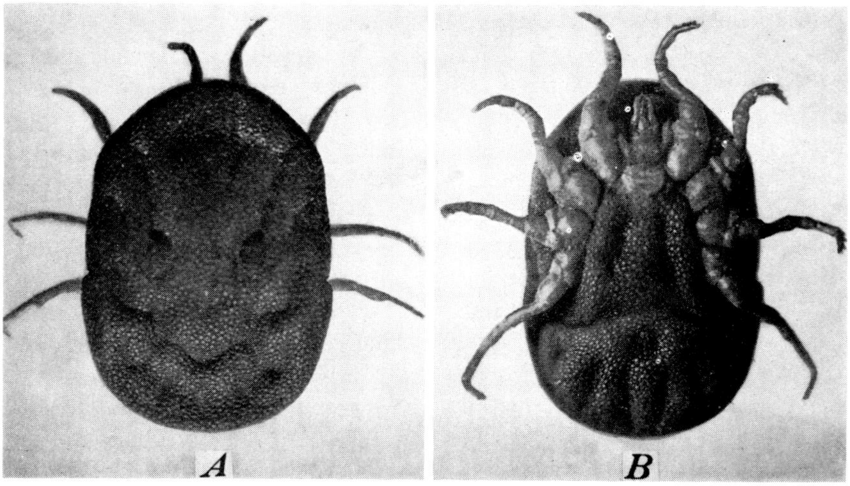

Fig. 36–13. *Ornithodoros moubata*, female. *A*, Dorsal view. *B*, Ventral view. (× 6 2/3.) (From Jirovec, O. 1948. *Parasitologie pro Lekare*. Prague, Melantrich.)

ectoparasites adapted exclusively to blood sucking for procurement of nourishment from mammals, birds, and reptiles. These Acari are developmentally heteromorphic and have a *larval* (three pairs of legs) and a *nymphal* (four pairs of legs) stage. As adults they usually measure a centimeter or more in length when engorged. They are differentiated from mites by having an *armed hypostome* and a sensory structure *(Haller's organ)* located on the dorsal aspect of tarsus I. A pair of *tracheal spiracular openings* are situated just behind the basal segment of the third or fourth pair of walking legs. There are two major families: the Argasidae (soft-bodied ticks) and the Ixodidae (hard-bodied ticks).

In the Argasidae (Fig. 36–13) there is no hard dorsal plate *(scutum),* the mouth parts *(capitulum* or *gnathosoma)* are situated ventral to the anterior extremity, and the spiracles are usually located directly behind the third pair of coxal segments. Like many other ectoparasites, soft ticks hide in cracks or crevices and feed when the host is in the lair, usually taking blood twice or more in the nymphal and adult stages and ovipositing after each adult feeding. Some species of *Otobius* and *Antricola* may also infest wandering hosts (Krantz, 1978).

In the Ixodidae (Fig. 36–14) the dorsal plate or shield covers the entire surface of the male, but only the anterior portion of the female, nymph, and larva. The capitulum extends forward beyond the dorsal shield, and the spiracles are situated behind and somewhat lateral to the fourth pair of coxal segments. In each stage of development (larva, nymph, adult) ixodid ticks take only a single blood meal from the host. The size of the blood meal taken by males is limited by the scutum, which prevents expansion of the body. The amount of blood ingested, therefore, is much smaller than that taken by the female. These limited quantities of blood are used for spermatogenesis and as an energy source for mating. Once attached and inseminated, females undergo a rapid phase of engorgement, using proteins from the blood meal for reproductive purposes. In most Ixodidae the stay on the host is quite prolonged. Attachment is accomplished by means of the hypostome, which is inserted into the wound created by the chelicerae. Many ixodids produce a cementlike substance from the salivary glands that also aids in anchoring the feeding tick.

External Anatomy. Ticks are peculiarly characterized by a gnathosoma or *capitulum* (Fig. 36–15), which is attached to the anterior end of the body of Ixodidae and is underneath the anterior dorsal surface in Argasidae. Its basal portion consists of a broad chitinous ring, the posterior constricted portion of which constitutes the connection with the body. This *basis capitulum (bc)* carries the mouth parts, consisting of the toothed *hypostome (h),* a more or less median, finger-like process, ventral to the mouth in position; a pair of *chelicerae (c, cd),* arising directly from the oral opening on the dorsolateral aspects of the hypostome and situated in sheaths that are the direct continuation of the capitular base; and a pair of *pedipalps,* or palps, that arise from the ventro-

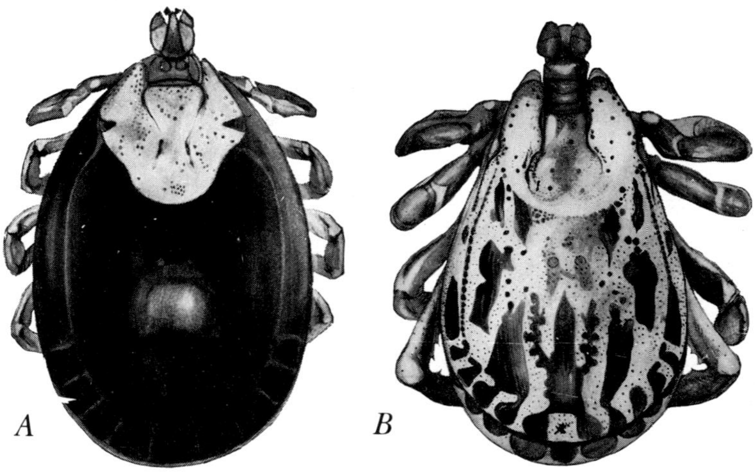

Fig. 36–14. *Dermacentor andersoni*, dorsal view. *A*, Female. *B*, Male. Enlarged. (From Stitt, E.D. 1929. *Diagnostics and Treatment* of *Tropical Diseases*. New York, The Blakiston Division, McGraw-Hill.)

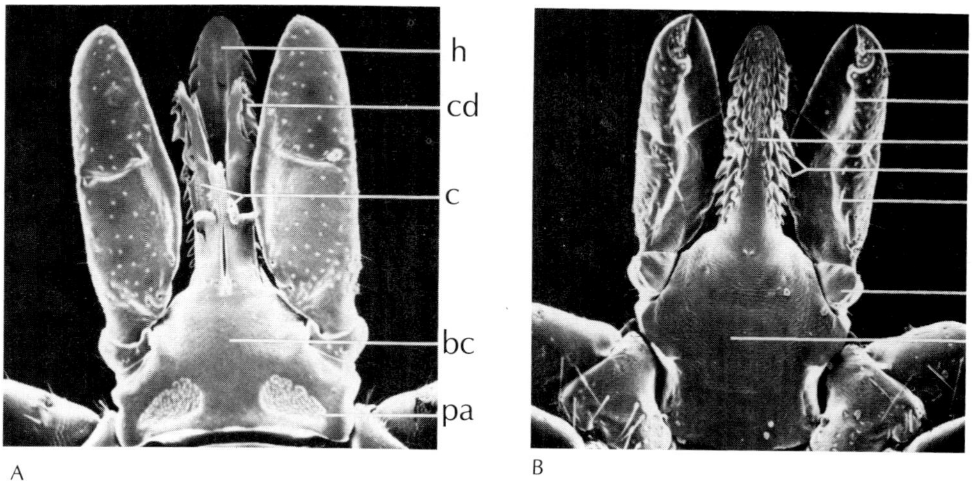

Fig. 36–15. *Ixodes affinis*, female. *A*, Dorsal. *B*, Ventral, views of the gnathosoma. (× ca. 90.) *bc*, Basis capitulum. *c*, Chelicerae. *cd*, Cheliceral digits. *d*, Denticles of the hypostome. *h*, Hypostome. *pa*, Porose area. *pl*, Segments 1 to 4 of palp. (From Keirans, J.E., and Clifford, C.M. 1978. The genus *Ixodes* in the United States: A scanning electron microscope study and key to the adults. J. Med. Entomol., Suppl. 2:41.)

lateral margins of the capitular base. The hypostome has a smooth convex base and is provided on its ventral surface (outer portion) with longitudinal rows of teeth or *denticles (d)* pointing backward with a median longitudinal fissure between the rows of teeth. These tooth patterns can be of diagnostic aid in the determination of species. The hypostome serves as an effective anchor in host tissue. Unless great care is used in dislodging the ticks, the hypostome is broken off in the skin and later produces considerable tissue reaction (Fig. 36–18). The chelicerae, each with its pair of sharply toothed terminal digits *(cd)* attached to a long cylindrical shaft, cut the host's tissue and provide the way for insertion of the entire capitulum (except for the palps and the basis capitulum) into the wound. In the Argasidae and Ixodidae, the paired palps each consist of four segments, of which the terminal one is small and set into a depression of the penultimate one (Fig. 36–15, *pl* 1–4). They serve as counter-anchors when the tick is feeding. The *porose area (pa)* on the basis capitulum of ixodid females is secretory in function and aids in coating newly deposited eggs.

The dorsal surface of the tick's body is characterized by having sculpturing, furrows, or color patterns that are frequently helpful in diagnosis. The ventral surface presents several important structures (Fig. 36–13,*B*). From the lateral aspects of the anterior half or three-fifths of the body the four walking legs arise, their basal (coxal) segments forming shield-like plates. The *genital orifice* lies in the midline in the anterior half, while the *anal pore* is situated some little distance in front of the posterior extremity. As previously mentioned, the *spiracular opening* on each side is found posterior to the third coxal segment in soft-bodied ticks and posterior to the fourth pair in hard-bodied ticks. The openings are located on chitinized *stigmal plates* or *stigmata*. Grooves divide the venter more or less into fields.

Internal Anatomy (Fig. 36–16). The *digestive tract* is usually described as starting anteriorly with a tubular mouth cavity formed by the apposition of the chelicerae with the hypostome and broadening out posteriorly where the pair of salivary ducts opens into it (Balashov, 1972). A tubular stylet lies within this cavity and is believed to serve as the sucking organ for the removal of liquid nourishment from the host and for its transfer to the muscular *pharynx,* which is provided with a chitinous lining. Bertram (1939) found this stylet to be a pouch, ''which together with the closure of the hypostomal gutter leading to the pharynx, blocks the pharyngeal orifice during contraction of the pharynx, thus preventing the regurgitation of blood into the wound.'' Insofar as this tongue-like process ''plays an essential part'' in the discharge of saliva, it is analogous to the hypopharynx of insects.

Behind the pharynx there are in direct continuation with one another a short, thin-walled *esophagus* (which passes through the ''brain''); a short, thin-walled *midgut* with numerous branched ceca or *lateral diverticula,* which occupy a large part of the body cavity and become greatly distended when the tick becomes engorged with blood, thus serving as a food-storage organ during nonfeeding periods; a *small intestine,* which is functional in ixodid species and most argasids (except *O. mou-*

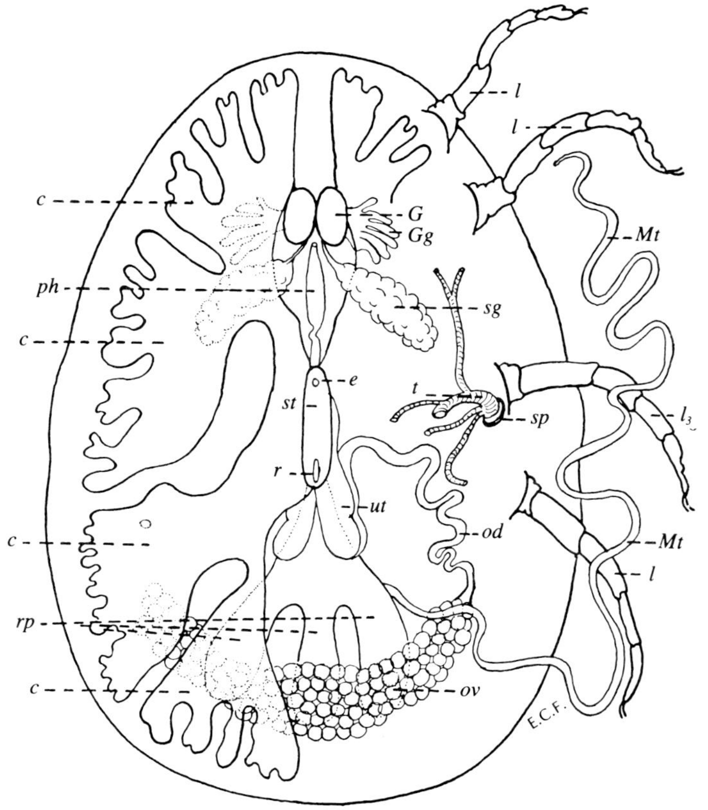

Fig. 36–16. Argasid tick, internal anatomy. Paired organs are shown only for one side. *c,* Cecal pouches. *e,* Esophagus. *G,* Gene's organ. *Gg,* Gland of Gene's organ. *l,* Legs of nymph or adult. *Mt,* Malpighian tubule, dissected out of position. *od,* Oviduct. *ov,* Ovary. *ph,* Pharynx. *r,* Rectum. *rp,* Rectal pouches. *sg,* Salivary gland. *sp,* Spiracle. *st,* Stomach. *t,* Main tubules of tracheal system. *ut,* Uterus. (Adapted by E.C. Faust.)

bata); and a *rectal sac-rectum* complex, into the ventral side of the anterior surface of which opens a pair of coiled capillary *malpighian tubules*, through which digestive wastes (mainly guanine) are accumulated and eliminated via the rectum. These tubules also serve, to a lesser degree, as osmoregulatory organs. In the Ixodidae the rectum and anus apparently do not normally function for the discharge of solid feces. The pair of aciniform *salivary glands* that lie in the anterior half of the body produce anticoagulins. These anticoagulins pass through a pair of ducts into the buccal cavity and, when injected into host tissue, enable the tick to obtain a continuous flow of blood.

In the *male system*, the testes may be completely separated, partially fused along the medial surface, or, in the Argasidae, fused posteriorly. The testes join with a pair of *vasa deferentia*, the members of which connect to form a common lobular *seminal receptacle*, in which spermatophores are formed. This latter organ opens outward directly through the genital orifice. In the *female system* (Fig. 36–16), there are a single obliquely disposed *ovary;* a pair of long, coiled *oviducts* arising from the ends of the ovary; a common bicornuate *uterus;* and a *vagina*, which leads to the genital orifice.

Reproduction and Development. At the time of mating the male seeks out a female attached to a host or, less frequently, off the host (Argasidae and many species of *Ixodes*), crawls between her and her support, and, apposing his ventral surface with hers, distends the vagina by means of his mouthparts, discharges a spermatophore, and with his mouthparts attaches it to the female's genital aperture (Oliver, 1974). Oviposition occurs off the host. As the globular eggs are extruded through the genital orifice of the female, a special organ *(Gené's organ)* situated anterodorsally covers them with a waxy secretion that prevents desiccation. Thus, the oviposited mass is always in front of the female's body. Some ixodid ticks may lay thousands of eggs in one batch.

Stages of embryonation have been described by Aeschlimann (1958, 1961). Upon hatching, an ixodid larva with three pairs of legs ("seed tick") emerges and actively crawls about, seeks out its host, obtains a blood meal, and then typically drops off (in a few species it remains on its host). In a few species, larval ecdysis occurs before hatching. The larva then molts to produce the nymph, which has four pairs of legs but lacks the external openings of the genital organs. After the nymph feeds, it drops off the host and transforms into the adult tick. In hard-bodied ticks there is only one nymphal stage; in soft-bodied species there may be as many as five molts within the nymphal stage, each with a separate blood meal. Engorgement may require only a few minutes or several hours, depending on the species.

Feeding and molting by ixodid ticks are sequential acts and may involve the alternation of several hosts as blood sources (Balashov, 1972). All three stages of "one-host" ticks engorge on the same animal, for example, *Boophilus annulatus* on bovine hosts. Other ticks may use two or even three hosts. The Rocky Mountain wood tick *(Dermacentor andersoni)*, a three-host species, normally completes its development in about 2 years, using small mammals for the larval and nymphal stages, followed by engorgement on larger animals, including man.

Table 36–1 provides a key to representative families and genera of ticks.

TICKS AND HUMAN DISEASE

In addition to the importance of ticks as the agents of diseases of great economic significance in domestic animals (i.e., Texas cattle fever, East Coast fever of South Africa, heartwater fever of South Africa, cattle anaplasmosis, canine piroplasmosis, and cattle and fowl spirochetosis), ticks are important as causative agents of human disease and as vectors of human pathogens. In this respect they play a threefold role: (1) Several species of ticks are required *hosts and transmitters to man* of important pathogenic organisms; (2) the *salivary secretions* of certain ticks, when introduced into the human skin when the tick takes a blood meal, produce *systemic toxemia* (i.e., tick paralysis), which may cause death; and (3) by the penetration of their mouthparts into the skin of humans, ticks produce *feeding lesions*.

Ticks are responsible transmitters (i.e., biological vectors) of a variety of human pathogens, including viruses, rickettsiae, bacteria, and protozoa. Certain key physiological and ecological facets of the tick life cycle confer a high vector potential to these acarines (Sonenshine, 1974). These include persistent and often lengthy bloodsucking by all stages and both sexes, a wide host range, transovarial and trans-stadial passage of some pathogens, longevity, and a relatively high reproductive potential.

Table 36–1. Key to Families and Genera of Ticks

1 (2)—	Lacking a dorsal shield (scutum) in all stages; capitulum ventral, subterminal, usually concealed from dorsal surface—family Argasidae (soft-bodied ticks . . . . . *a*
a (b)	Margin of body thin or flattened; usually exhibiting a suture separating dorsal from ventral side . . . . . *Argas*
b (a)	Margin of body lacking a sutural separation between dorsal and ventral side . . . . . *c*
c (d)	*Nymphs* with spinose integument, with well-developed hypostome; *adults* with granular integument and vestigial hypostome . . . . *Otobius*
d (c)	*Nymphs* with mammillated or tuberculous integument lacking spines; hypostome well-developed in both nymphs and adults . . . . . *e*
e (f)	Hypostome dorsally scoop-like, basally broad . . . . . *Antricola*
f (e)	Hypostome not dorsally scoop-like, not obviously broadened basally . . . . . *Ornithodoros*
2 (1)—	Having a dorsal shield (scutum) which in larvae, nymphs, and adult females extends only over a portion of anterior dorsal surface, but in adult males covers entire dorsal surface; capitulum always visible from dorsal aspect—family Ixodidae (hard-bodied ticks) . . . . . *a*
a (n)	With eyes . . . . . *b*
b (e)	Scutum usually ornate (i.e., having symmetrical color pattern dorsally in adult) . . . . . *c*
c (d)	Palps short; hypostomal denticles (teeth) in 6 or 8 longitudinal rows on a side, occupying almost entire length of hypostome . . . . . *Dermacentor*
d (c)	Palps long and thin; hypostomal denticles in 6 or 8 longitudinal rows on a side, occupying only distal part of hypostome . . . . . *Amblyomma*
e (b)	Scutum usually inornate . . . . . *f*
f (g)	Eyes not marginal . . . . . *Hyalomma*
g (f)	Eyes marginal . . . . . *h*
h (m)	Having festoons (i.e., posterior marginal rectangular areas on dorsal and ventral aspects) . . . . *i*
i (j)	Ventral shields or plates present in males, lacking in females . . . . . *Rhipicephalus*
j (i)	Ventral shields or plates lacking in both males and females . . . . . *k*
k (l)	Spurs on fourth-leg coxa very long . . . . . *Rhipicentor*
l (k)	Spurs on fourth-leg coxa short or lacking . . . . . *Anocentor (Otocentor)*
m (h)	Without festoons . . . . . *Boophilus*
n (a)	Without eyes . . . . . *o*
o (p)	Usually ornate . . . . . *Aponomma*
p (o)	Inornate . . . . . *q*
q (r)	With festoons; palps short, usually massive, conical when closed; anal groove posterior to anus . . . . . *Haemaphysalis*
r (q)	Lacking festoons; palps long or short but not conical when closed; anal groove surrounding anus anteriorly . . . . . *Ixodes*

Note: The family Nuttalliellidae with the single species *Nuttalliella namaqua* Bedford has been omitted, as well as several minor genera in the Ixodidae and Argasidae. For additional information, see USDA Agricultural Handbook No. 485 (1976) and Krantz (1978).

Tick-Borne Rickettsial Infections

Rickettsiae are microorganisms with a fine structure and chemical composition similar to that of gram-negative bacteria (Weiss, 1973) (Fig. 36–17). However, with the exception of *Rochalimaea quintana* (the causative agent of trench fever), rickettsiae grow intracellularly and cannot be cultured on artificial media. Multiplication by transverse binary fission occurs within definite areas of the cytoplasm, depending upon the taxon. Some species are also able to invade the nucleus. Rickettsiae transmitted to humans by ticks include those belonging to the spotted fever group and Q fever.

Rocky Mountain Spotted Fever

Rocky Mountain (New World) spotted fever (RMSF) (fiebre petequiél, Tobía fever, São Paulo fever, Choix fever), is caused by *Rickettsia rickettsii*. The rickettsial fevers of the spotted fever group are acute exanthematous diseases which are clinically similar and caused by closely related rickettsiae. The fundamental lesion is destruction of the endothelial cells of the smaller blood vessels that are invaded by the causative organisms. The infections are transmitted by various species of hard (Ixodid) ticks that are widely distributed throughout the world. The causative agents are intracellular, arthropod-transmitted, pleomorphic organisms measuring about 0.3 μm which invade nuclei as well as cytoplasm of host cells. Complement fixation tests using group-specific spotted fever antigens become positive in the second week of illness; the Weil-Felix reactions with Proteus OX-19 and Proteus OX-2 are less consistently positive. It is not feasible to distinguish the rickettsiae of this group in the clinical diagnostic laboratory, nor is it ordinarily necessary since the geographic ranges of the specific rickettsiae do not overlap.

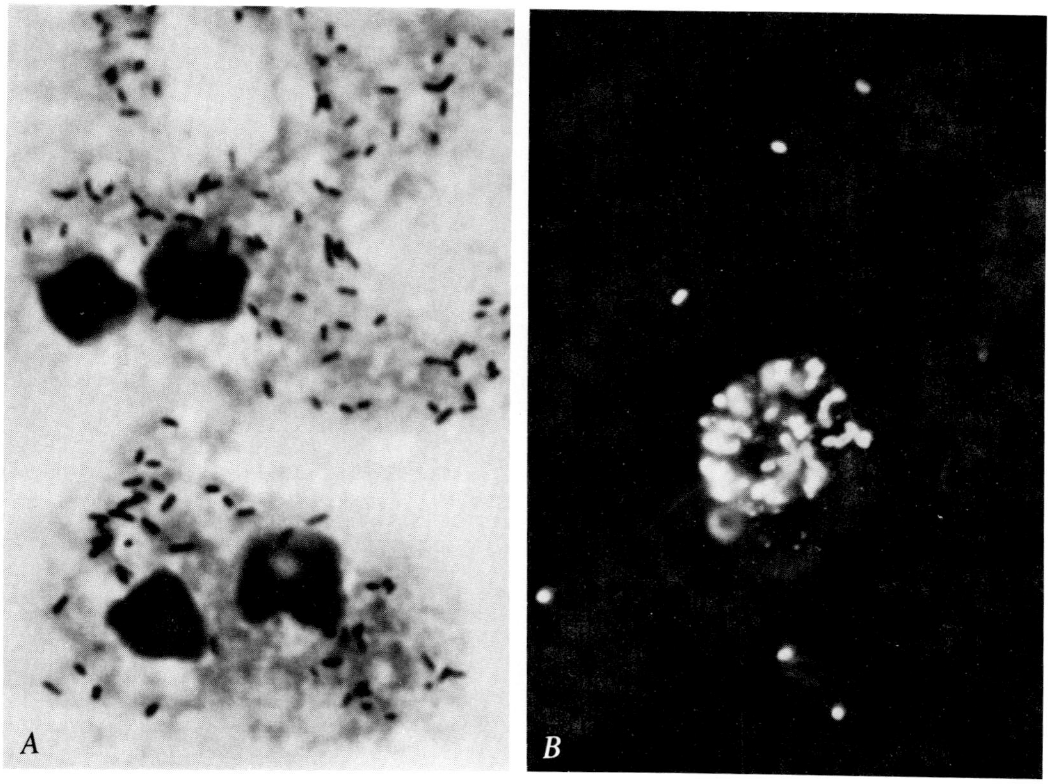

Fig. 36–17. *Dermacentor variabilis,* hemocytes with rickettsiae in cytoplasm. *A,* Gimenez stain. (× ca. 2800.) *B,* Fluorescent antibody staining with anti-*R. rickettsii* conjugate. (× ca. 1400.) (From Burgdorfer, W. 1977. Tick-borne diseases in the United States: Rocky Mountain spotted fever and Colorado tick fever. Acta Trop., *34*:103–126.)

Geographical Distribution. Rocky Mountain spotted fever occurs throughout the United States, more in the southeastern states than in the Rocky Mountain region. Its incidence is highest in wooded suburban areas of the Piedmont Plateau.

The disease referred to as *Choix fever* (also fiebre manchado) is present in rural areas of the Mexican states of Sinaloa and Sonora (in the northwest), Coahuila and Durango (north-central), and Vera Cruz (south Gulf coast). It is also indigenous in Panama. This disease is called *fiebre petequiél* in the Colombian region of Tobía, and *São Paulo fever* in the Brazilian states of São Paulo, Rio de Janeiro, and Minas Gerais, where it is endemic.

Epidemiology. Spotted fever occurs in rural areas where ixodid ticks and their rodent hosts are found, but dogs may bring infected ticks into suburban-urban areas. The principal ticks that convey spotted fever to man are *Dermacentor andersoni,* the Rocky Mountain wood tick of Western areas; *D. variabilis,* the American dog tick of Eastern states; and *Amblyomma americanum,* the Lone Star tick of Texas and Oklahoma. Other species considered to be potential vectors in the United States include *A. maculatum, D. occidentalis, Ixodes scapularis, I. pacificus,* and *I. cookei* (Burgdorfer, 1977). Neotropical vectors include *Rhipicephalus sanguineus* in Mexico and *A. cajennense* in Panama, Colombia, and Brazil. Ticks that propagate the disease among rodents, rabbits, and hares include *Haemaphysalis leporispalustris, D. andersoni,* and *D. parumapertus.* As a rule, only the adult tick feeds on man, although occasionally nymphs take blood from children. Human infection is generally incurred as the tick feeds, but there are instances of the disease following contamination of skin while crushing ticks removed from pets, or from fresh tick feces. A period of 4 to 6 hours of attachment of the tick is required for the rickettsiae to become activated and infection to occur. Once infective, a tick probably remains so throughout its life.

Philip (1958, 1961) stated that the following mechanisms operate in an enzootic-endemic area in maintaining spotted fever in nature: transovarial passage of the rickettsia through successive generations of ticks; starting of new lines through simultaneous feeding of infected and noninfected ticks; transmission of the enzootic lines by ticks that do not bite humans; and venereal passage of the rickettsia from infected male ticks to the eggs of uninfected females.

The morbidity rates of RMSF began to increase in 1960, especially in the eastern and southeastern United States, with a high incidence among children and women (Burgdorfer, 1977). This rise in reported cases (1,163 in 1980; MMWR—Centers for Disease Control) is related to infestation of household pets (particularly dogs) with the vector tick, the shifting of populations to suburbia, and the creation of new tick habitats by allowing cultivated lands to revert to forest for recreational purposes (Hoogstraal, 1981). The infection is not directly transmissible from man to man. One attack confers longlasting immunity.

Clinical Aspects. Rocky Mountain spotted fever is characterized by sudden onset, after an incubation period of 3 to 10 days, moderate to high fever, headache, chills, and conjunctivitis. The febrile illness lasts about 2 to 3 weeks. The rash is maculopapular, appears about the third day, and is centripetal, beginning on the palms and soles and spreading rapidly to most of the body. Petechial and ecchymotic hemorrhage is common. The fatality rate varies greatly with the strain of rickettsia involved but averages about 20% in the absence of specific chemotherapy. With appropriate, timely treatment employing a tetracycline antibiotic or chloramphenicol, the prognosis is good. The use of sulfonamides is contraindicated.

Prevention. This program is twofold: (1) prevention of the rickettsia-infected tick from introducing the organisms into the human skin, and (2) prophylactic vaccination.

The disease may be avoided by keeping out of tick-infested areas, such as low growths of bushes where rodents are likely to serve as common hosts for the ticks. Where dog ticks *(Dermacentor variabilis* or *Rhipicephalus sanguineus)* are the transmitters, this advice is difficult to follow because dogs bring gravid ticks to human habitations. However, the use of tick collars on dogs and cats may appreciably lower the number of ticks per pet. For persons obliged to work in enzootic areas, dressing properly by tucking shirt tails inside trousers and covering trouser legs with high socks is recommended. Impregnating clothing with tick repellents is also advised. When one is removing ticks from pets, a forceps or piece of strong paper should be used. Hands should be carefully washed after contact with ticks.

Tick eradication is difficult, but the virulence of spotted fever requires that strict preventive measures be employed. Area spraying with acaricides provides temporary control during the warm season of tick activity (see Chap. 45).

Prophylactic vaccination should be carried out against spotted fever in all enzootic-endemic areas, and constitutes the only assured safeguard against acquiring a fatal infection. Vaccination, first developed by Parker and his colleagues in 1930 (Parker, 1942), prevents infection in about 90% of exposed individuals, and even those who become infected have a mild form of the disease. The vaccine (a concentrate of dead rickettsiae) is administered in two subcutaneous injections of 1 ml each for children under 10 years, or two 2-ml injections for older children and adults, given 5 days apart. Revaccination should be done each year.

Related Spotted Fevers

Boutonneuse Fever (escarro nodulaire). This fever is caused by *Rickettsia conorii*.

Geographical Distribution. This disease is widely distributed throughout Africa, India, Europe, and the Middle East bordering the Mediterranean, Black, and Caspian Seas. It is transmitted by the tick *Rhipicephalus sanguineus* in Europe, the Mediterranean area, and India; in East Africa by other species *(Rhipicephalus simus, Haemaphysalis leachi, Amblyomma variegatum);* and in South Africa by *R. evertsi* and *A. hebraeum* (rural) and *H. leachi* (suburban). Rodents appear to be the chief animal reservoir of *R. conorii*, with dogs playing a significant role in bringing infected ticks into close contact with humans.

Clinical Aspects. This is a mild to moderately severe febrile illness characterized by a primary lesion (tache noire), which usually appears at the site of the tick bite. This lesion is an ulcer 2 to 5 mm in diameter with a black center and red halo. There is regional lymphadenopathy. After an incubation period of about 5 to 7 days there is a sudden onset of chills and fever which last from a few days to 2 weeks. On about the fourth or fifth day of illness a maculopapular, centripetal rash appears first on the palms and soles. It spreads to become generalized and lasts 6 to 7 days. Even without specific therapy the fatality rate is less than 3%. With administration of tetracycline antibiotic or chloramphenicol, the fever lasts no more than 2 days. The use of sulfonamides is contraindicated.

Prevention. Factors in prevention are essentially the same as for Rocky Mountain spotted fever in the eastern United States, since in both areas infected dog ticks are the important vectors. Measures directed toward disinfestation of house dogs, destruction of stray dogs, and removal of ticks from human beings soon after they become attached constitute sound prophylaxis. Vaccines prepared from

R. rickettsii do not provide protection against infection with *R. conorii*.

Queensland Tick Typhus. Caused by *Rickettsia australis*, Queensland tick typhus is a mild disease clinically similar to boutonneuse fever. It is found in Queensland, Australia and is probably transmitted by *Ixodes holocyclus*, which infests small marsupials and wild rodents in forest and scrub areas. An urban focus occurs in Sydney (Campbell et al. 1979).

Siberian Tick Typhus (North Asian Tick-Borne Rickettsiosis). Caused by *Rickettsia sibirica*, this illness is clinically similar to boutonneuse fever and is transmitted by ticks of the genera *Dermacentor* and *Haemaphysalis*, which infest wild rodents. It is found in Asiatic USSR and the Mongolian People's Republic. As in New World spotted fever, transmission to humans occurs via adult tick bites in grassy meadows and brush-covered hillsides. The natural cycle also includes rodents and immature *Haemaphysalis*, *Hyalomma*, and *Dermacentor* ticks as well as *Rhipicephalus sanguineus*. The causative agent, *Rickettsia sibirica*, is mildly pathogenic.

Q Fever
(Nine-mile fever of western Montana, Balkan grippe)

History and Geographical Distribution. *Q* or *"query" fever* was first reported by Derrick as an obscure illness among abattoir employees in Brisbane, Australia, in 1937. This disease, produced by the rickettsia *Coxiella burnetii* (syn. *Rickettsia diaporica*), has since been found to have a wide distribution throughout the world, and is chiefly associated with livestock industries (Weyer, 1978).

Epidemiology. Q fever has a small mammal-tick cycle with transmission from domestic animals to humans. Transmission from person to person is uncommon. In humans, Q fever is primarily a respiratory infection contracted either by inhalation of droplets or dust particles contaminated directly or indirectly by infected cattle, sheep, and goats, or by ingestion of contaminated milk (Marmion, 1954). Infected domestic animals generally show no signs of infection. Occasionally, humans may be infected directly by a biting tick or by inhaling the rickettsial organism in dried tick feces.

Q fever of humans is most common in persons exposed to airborne infection from diseased animals, i.e., stockmen. There is also a higher incidence in those whose work takes them to farms where they are exposed to airborne infection (e.g., rural postmen, agricultural machinery salesmen) and in those to whom agricultural workers go for service (e.g., village barbers and grocers). In the United States Q fever has been contracted by stock handlers and slaughterhouse workers, by herders of sheep and goats in northern California, by dairy workers and others living in the vicinity of dairy herds, and by consumers of raw milk. More cases of Q fever occur in males than in females. Direct transmission from person to person is rare. The infection confers a longlasting, probably permanent, immunity.

Clinical Aspects. Q fever is an acute, febrile illness, frequently with sudden onset, following an incubation period of 2 to 3 weeks depending on the size of the infecting inoculum. Symptoms typically include chilly sensations, headache (especially behind the eyes), debility, malaise, and sweats. Usually there is a pneumonitis with nonproductive cough, chest pain, and paucity of physical findings. Severity of the illness varies from inapparent to grave, but even in untreated patients the mortality rate is less than 1%. With appropriate treatment, which is as for spotted fever, the prognosis is excellent. Complications of endocarditis and hepatitis have been reported.

Diagnosis. Diagnosis is based on residence in endemic foci, on the history of contact with livestock, on the clinical course, on the development of the organisms in laboratory animals and chick-embryo tissue cultures, and differentially by the specific agglutination or complement-fixation reaction. Inoculated guinea pigs develop an enlarged spleen. The host's serum agglutinates suspensions of *C. burnetii* (in dilutions of 1 to 10) prepared from infected mouse spleen.

Prevention. Epidemiologic data suggest the need to guard against direct or indirect contact with infected animals, with individuals carrying infested dust particles, or with contaminated fomites.

Tick-Borne Bacterial Infections

Ixodid and argasid ticks serve as vectors of several pathogenic bacteria, including the causative agents of relapsing fever and tularemia.

Relapsing Fever (Recurrent Fever)

Definition. Relapsing fever is a systemic disease caused by various strains of the spirochete, *Borrelia duttoni*, and several related species. It is characterized by episodes of fever lasting 2 to 9 days separated by afebrile periods of 2 to 4 days. Each attack of fever tends to develop suddenly and defervescence is by crisis. The total number of relapses varies from two to over ten. Evanescent petechial rashes commonly occur during the initial attack. There are two clinical and epidemiological varieties. The louse-borne variety (*vide infra,* p. 616), which is epidemic, lasts 13 to 16 days and is the more severe, with a mortality rate sometimes exceeding 50% in untreated individuals. The tick-borne variety is endemic, the milder of the two, and tends to last longer. Its mortality rate is 2 to 10% when untreated.

Geographical Distribution. Relapsing fever is widely distributed throughout the world, and is transmitted by different species of *Ornithodoros* in different areas (Burgdorfer, 1976). Most authorities assign a different species name to each strain of the spirochete transmitted by a different species of tick; certainly each such strain is immunologically distinct (Johnson, 1977; Hoogstraal, 1979a). It was first described from southeastern Africa, where it was found to be transmitted by *O. moubata sensu lato,* which is distributed throughout Africa. It is present in Tunis, Morocco, Spain, and southeastern Portugal, where *O. erraticus* is involved; from Cyprus through Israel, Syria, Jordan, Iraq, and Iran to the Northwest frontier of India, where *O. tholozani* (syn. *O. papillipes*) is the transmitter; in the neighborhood of Tiflis, Iran, where *O. verrucosus* has been found naturally infected; in the high tablelands of central Asia, where *O. tholozani* is stated to be the transmitter; in Venezuela, Colombia, and Panama, where *O. rudis* (syn. *O. venezuelensis*) and *O. talaje* are the transmitters; in Guatemala, where *O. talaje* is the only described species; in Zacatecas State, Mexico, where *O. turicata* is naturally infected; in Kansas, New Mexico, Texas, and Oklahoma, where *O. turicata* and *O. talaje* are the vectors; in California, Oregon, Washington, Idaho, Nevada, eastern Utah, Colorado, Wyoming, and Montana, where *O. hermsi* or *O. parkeri* is the infected tick; in Arizona, where *O. talaje* has been found naturally infected; and in British Columbia, where *O. hermsi* may be the vector.

Epidemiology. The first demonstration of the transmission of tick-borne relapsing fever was made in Africa, in 1905, by Dutton and Todd. Within hours after ingestion of the spirochetes in a blood meal from a human or appropriate reservoir host, the organisms disappear from the tick's digestive tract and make their way to the hemocoel and organs of the body, including the coxal glands and coxal reservoirs (adult ticks), salivary glands (nymphal ticks), and ovaries, in which they multiply. They have been demonstrated occasionally in the malpighian tubules but not in the feces. In a few days or weeks, depending on temperature, new broods of spirochetes are abundant in these organs and body fluids. Infection of susceptible mammals may occur solely through the bites of infected ticks or via the coxal fluid, which is shed from the body through slit-like pores during blood feeding.

The spirochetes in this fluid may either enter the puncture wound or penetrate the unbroken skin. Once the tick has become infected, it remains so for years and transmits the spirochetes vertically by transovarial passage. Thus, this spirochete perpetuates itself symbiotically without need for human or other vertebrate hosts (Hoogstraal, 1979*a*). Tick-borne relapsing fevers are usually considered as not naturally infective for the louse.

Reservoir vertebrate hosts of the strains of tick-borne relapsing fever infective for humans probably include numerous species of monkeys, squirrels, chipmunks, rats, hedgehogs, and possibly bats and other cave-inhabiting mammals likely to be parasitized by species of *Ornithodoros* which are good hosts for the relapsing fever spirochete. In East Africa tick-borne relapsing fever is primarily a disease of humans, who serve as the mammalian reservoir. In North Africa, the eastern Mediterranean region, central Asia, tropical and subtropical America, and many parts of North America, relapsing fever is primarily a disease of rodents only incidentally transmitted to humans by argasid ticks (Burgdorfer, 1976).

In tick-borne relapsing fever, as in the louse-borne type, several species of the spirochetal agent have been described on the basis of slight differences in the clinical picture, differences in geographical distribution, and (in the tick-borne type) different species of argasid ticks incriminated as unique transmitters of particular strains of the spirochete.

Clinical Aspects. Diagnosis is by demonstration of the causative spirochete in darkfield preparations of fresh blood, stained thick films of blood, or intraperitoneal inoculation of laboratory rats or mice with subsequent examination of their blood.

Treatment is by administration of a tetracycline antibiotic.

For a consideration of louse-borne relapsing fever, see p. 616.

Prevention. Persons may usually avoid exposure by keeping out of caves or living quarters where *Ornithodoros* abounds. Likewise, in endemic areas the handling of squirrels and chipmunks or other potentially infected mammals should be avoided. Flame torches applied to stone walls in which ticks may hide may be indicated. In the infected foci in the California mountains, the task is more difficult, since the ticks lurk like bedbugs in the chinks between timbers of human habitations. Similarly, in the tropics, where homes are made of bamboo or adobe and thatch, a practical way to eliminate exposure has not been found. Because species of soft-bodied ticks, including *Ornithodoros,* are relatively sensitive to propoxur, lindane, malathion, and carbaryl, dusting of these acaricides into the hiding places of ticks may provide considerable protection.

Tularemia

Definition. Tularemia is an infectious bacterial disease caused by *Francisella tularensis*, involving primarily an inflammatory enlargement and ulceration of the skin and lymph nodes nearest the site of exposure. At times it manifests a generalized septic condition with fever, or is characterized by abdominal involvement.

Geographical Distribution and Epidemiology. The disease is frequent in fall and winter, particularly among individuals who skin and dress rabbits, the common reservoirs of the infection. It is also transmitted during spring and summer by several types of blood-sucking arthropods, including ticks, deer flies and, possibly, mosquitoes and fleas. It has widespread distribution in the United States and Mexico, and is known to occur in Canada, Alaska, several European countries, North Africa, Israel, Turkey, Iran, extensive areas in the U.S.S.R., and Japan.

Inoculation may occur through the bite, or by crushing an infected arthropod on the skin, or possibly via contaminated tick feces. The pathogen is transovarially transmitted in the tick, and the microorganism is present both in the gut and the hemocoel. Trans-stadial passage also occurs.

Species of *Haemaphysalis, Rhipicephalus, Ixodes, Amblyomma*, and possibly *Otobius* (Stark, 1979) transmit the organism from rabbit to rabbit or fowl to fowl, while *Dermacentor andersoni* and *D. variabilis*, probably *D. occidentalis* and *Amblyomma americanum*, transmit it to humans in the United States. *Dermacentor pictus, D. marginatus, I. ricinus*, and *R. rossica* are the principal tick vectors in the U.S.S.R.; *H. flavis, I. japonensis*, and *I. nipponensis* are important in Japan (Hopla, 1974).

Clinical Aspects. The characteristic early lesion of tularemia consists of an inflamed area 2 to 3 cm in diameter, proceeding to central necrosis and ulceration. This is followed by the development of enlarged, reddened, moderately firm, regional lymph nodes, which later become soft and fluctuant as a result of abscess formation. The total white count may be normal but is usually increased to as much as 16,500. There is at least one record of congenital transmission.

Diagnosis is made specifically by obtaining a positive agglutination test with the patient's blood. Treatment consists of supportive measures and specific therapy with tetracycline, chloramphenicol, or streptomycin or dihydrostreptomycin administered intramuscularly, with surgical drainage, and requires 6 to 8 weeks of hospitalization for recovery.

Prevention. The disease can be prevented by avoiding the bites of blood-sucking arthropods and not exposing the skin to infection when dressing rabbits.

Tick-Borne Viral Infections

Ticks are disseminators of several kinds of human viral pathogens that cause encephalitis, hemorrhagic fever, or severe myalgia with fever. Transmission occurs directly by injection of infective saliva or occasionally by contamination from crushing infected ticks between the fingers. These arboviruses are zoonotic agents and usually use mammals or birds as parts of the reservoir. The more common tick-borne viruses of public health importance include one or more representatives of the Togaviridae, Reoviridae, and Bunyaviridae (Matthews, 1982).

Encephalitis and Hemorrhagic Syndrome

Tick-borne flaviviruses include antigenically similar forms capable of causing encephalitis and/or hemorrhagic fevers. Human virulence varies by antigenic strain and geography, with mortality rates reaching 25% in some localities. These single-stranded RNA viruses (Togaviridae; *Flavivirus*) are transmitted by ixodid ticks (Hoogstraal, 1973; 1981).

TICK-BORNE ENCEPHALITIDES
(Russian spring-summer encephalitis [RSSE], central European encephalitis [CEE], diphasic milk fever or diphasic meningoencephalitis, louping ill, and Powassan encephalitis)

Geographical Distribution and Epidemiology. Two antigenic types of tick-borne encephalitis virus (TBE) occur. The Far Eastern strain (see RSSE virus *sensu stricto* in Berge, 1975) is found principally in the deep taiga forests of the Soviet Union where it is transmitted primarily by *Ixodes persulcatus*. The immature stages of this tick are opportunistic feeders on small, medium, or large mammals, with adults feeding on larger hosts. In addition to these animal reservoirs, this virus can also be passed transovarially and trans-stadially in *I. persulcatus*. The European strain of TBE virus occurs throughout eastern Europe, western U.S.S.R., Austria, and France, where it is transmitted principally by *Ixodes ricinus*. Small mammals and birds may serve as hosts for this tick.

Increased human exposure to the bites of *Ixodes persulcatus* and *Ixodes ricinus* has occurred as a result of habitat alteration through the clear-cutting of forests for agricultural and industrial purposes and the recreational use of forested pathways. Transmission via unpasteurized goat or sheep milk or milk products also occurs in foci having high *I. ricinus* densities.

Powassan virus occurs in Canada and the northern and western United States, where it occasionally causes cases of human encephalitis. Rodents and hares constitute the basic reservoir. Humans seldom become infected owing to the sylvatic cycle of the virus and the apparent zoophilic preference of the tick vectors. Virus isolations have been made from *Dermacentor andersoni, Ixodes spinipalpis, I. cookei*, and *I. marxi*. Louping ill occurs in Ireland, Scotland, and northern England.

Clinical Aspects. The group of diseases caused by these viruses resemble clinically the mosquito-borne encephalitides, with certain exceptions. Russian spring-summer disease often causes flaccid paralysis with residua, especially of the shoulder girdle. Central European encephalitis is longer in duration, averaging 3 weeks, and presents with an initial febrile stage lacking central nervous system manifestations followed by defervescence and, after

4 to 10 days, a second febrile phase with meningoencephalitis. Fatality and residua are less common than with the Russian spring-summer disease. Louping ill in man also presents a diphasic fever curve and is mild. Specific identification is not generally feasible; group identification is by serological tests of viral isolation in suckling mice or tissue culture.

Omsk Hemorrhagic Fever and Kyasanur Forest Disease

Viruses causing Omsk hemorrhagic fever (OHF) and Kyasanur Forest disease are also tick-borne flaviviruses, though they rarely cause disease of the central nervous system. Instead, both are etiologic agents of hemorrhagic fevers (Simpson, 1978). Several species of *Dermacentor* and *Ixodes* found in rodent burrows in southwestern Siberia are believed to be involved in the transmission of OHF virus to humans, although infected muskrat carcasses and pelts have been incriminated as more likely fomites. Water contaminated by feces and urine from these rodents may constitute another source of infection. The virus causing Kyasanur Forest disease is transmitted to humans by *Haemaphysalis* ticks, particularly *H. turturis* and *H. spinigera*. Tick-human contact occurs during wood gathering on the periphery of the forest. The virus persists in nymphs of *H. spinigera,* which are most important in transmitting this pathogen to humans and langur monkeys. The emergence of this viral disease as a public health problem has resulted from ecological changes due to deforestation and increased agricultural practices, notably an increase in cattle production that served to increase tick populations.

Crimean-Congo Hemorrhagic Fever

Definition. Crimean-Congo hemorrhagic fever (CCHF) is a serious and occasionally fatal disease characterized by sudden onset of fever and chills with severe pains in the limbs and loins (Simpson, 1978). A fine petechial rash often covers the entire body, with bleeding from the nose, gums, buccal mucosa, stomach, uterus, intestines, and lungs. Some central nervous system involvement may also occur. The course of the disease varies from mild to severe depending to a certain extent upon the route of infection, with mortality rates ranging from 15 to 30% in tick-borne infections to 40 to 80% in intrahospital outbreaks (Goldfarb et al., 1980). The etiologic agent of CCHF is a single-stranded RNA virus belonging to the genus *Nairovirus* (Bunyaviridae) (Casals and Tignor, 1980).

Geographical Distribution and Epidemiology. Strains of CCHF virus have been isolated from India, Pakistan, Afghanistan, Iran, Asian and European areas of the U.S.S.R., Bulgaria, Yugoslavia, Greece, Hungary, France, 8 mid-African countries (Senegal to Ethiopia), and Egypt. Enzootic foci are geobotanically diverse and are associated with ixodid ticks, of which 27 taxa have been incriminated as possible reservoirs or vectors (Hoogstraal, 1979b). Hazara virus, isolated from *Ixodes redikorzevi* in Pakistan, is serotypically aligned within the CCHF serogroup.

Maintenance of CCHF virus within enzootic foci may involve ticks having one, two, or three hosts, and includes species of *Boophilus, Rhipicephalus, Hyalomma, Ixodes, Haemaphysalis, Dermacentor,* and *Amblyomma*. Vertebrate hosts are birds and wild and domestic mammals. Keeping livestock (cattle and sheep) is particularly important in the production of large populations of certain two-host ticks (e.g., members of the *Hyalomma marginatum* complex), some species of which readily attack humans. Ground-feeding birds support tick populations as aviremic hosts and serve as porters in distributing ticks both intra- and intercontinentally. CCHF virus can also be passaged vertically (transovarially and trans-stadially), thus ensuring temporary maintenance in the absence of suitable vertebrate hosts.

Transmission to humans may take place by infective bite or through contamination from crushed ticks, either during removal from the body or while shearing sheep. Contagions such as bloody discharges from acutely infected patients in both domestic and hospital situations, and contaminated carcasses in abattoirs, have also been incriminated, with case fatality rates significantly higher.

Epidemics of CCHF have been recognized since 1945 in Europe and 1967 in Africa (Goldfarb et al., 1980). Factors giving rise to these outbreaks center around landscape alterations that promote the development and maintenance of large numbers of ticks and their hosts within enzootic foci, and the importation of susceptible humans, either as military personnel or as agriculturalists.

Colorado Tick Fever

Definition. The causative agent of Colorado Tick fever (CTF) is a double-stranded RNA virus belonging to the genus *Orbivirus* (family Reoviridae). CTF is characterized by a sudden onset of chills, headache, fever, and severe general myalgia (Burgdorfer, 1977). The initial febrile period is usually followed by a 2- to 3-day remission with a second episode of fever. CNS involvement may occur in a few cases, especially in children, though the disease is usually benign. The virus may persist in some patients for up to 120 days, thus facilitating diagnosis (Hughes et al., 1974; Burgdorfer, 1977).

Geographical Distribution and Epidemiology. CTF occurs in the western United States and in the Canadian provinces of Alberta and British Columbia. Cases are usually reported during the warmer months of March through August, reflecting an increase in human-tick contact through recreational and occupational activities. The Rocky Mountain wood tick, *Dermacentor andersoni,* is considered the principal vector, although CTF virus has been isolated from *D. albipictus, D. parumapertus, D. occidentalis, Otobius lagophilus,* and *Haemaphysalis leporispalustris* (Burgdorfer, 1977). Several rodent species serve as hosts for the immature stages of the tick and as virus amplifiers. Since transovarial passage of CTF virus apparently does not occur, the virus overwinters in quiescent nymphs and is transmitted to rodent hosts in the spring when the ticks become active. Newly hatched larval ticks also become infected from viremic hosts during this time and the virus is then passaged trans-stadially at the larval-nymphal molt where it can again overwinter. Trans-stadial transmission can also occur at the nymphal-adult molt, producing infective adults. Because of its persistence in human hosts, CTF virus may also be transmitted by transfusion.

Diagnosis. Diagnosis is by isolation of the virus in inoculated suckling mice or tissue culture with erythrocytes. Complement-fixing and neutralizing antibodies do not appear for 2 weeks or more.

A second virus having serological affinities (complement-fixation and neutralization tests) with CTF virus has been isolated from *Ixodes ricinus* in the Neckar Valley, Germany (Rehse-Kupper et al. 1979). This single isolate, termed Eyach virus, has been placed with the CTF virus serogroup, but little is known about its ecology.

Other Tick-Borne Viral Fevers

Other tick-borne viruses causing fevers are Dugbe and Nairobi sheep disease in Africa and Ganjam in India, members of the genus *Nairovirus* (Bunyaviridae); Kemerova, Tribec, and Lipovnik in Siberia and eastern Europe (Reoviridae: *Orbivirus),* and Quaranfil (serologically ungrouped) in Egypt, Iraq, Kuwait, Iran, and Yemen.

Tick-Borne Protozoal Infections

Babesiosis

Until recently, human babesiosis was considered a rare but threatening zoonotic disease associated principally with splenectomized individuals in Europe and Russia (Garnham, 1980). The causative agents, *Babesia bovis* and *B. divergens* (which Hoare [1980] considered synonymous with *B. bovis*) are tick-borne piroplasms of cattle that cause red-water fever. In splenectomized humans, however, these parasites cause a fatal disease.

Human babesiosis in North America is caused by *B. microti,* a parasite of rodents (Ruebush, 1980). This piroplasm appears to be widespread in North America, though 16 of the 21 cases reported as of 1980 were from patients on four neighboring islands along the New England coast. *Babesia microti* infections have not proved fatal (only one case has been reported from a splenectomized patient), though this babesiosis is severe in older individuals. Parasitemias may be prolonged, leading to the possibility of transmission by blood transfusion.

The northern deer tick, *Ixodes dammini,* is presumed to be the vector in the New England area (Spielman et al., 1979) with pathogen transmission to humans occurring through the bite of infected nymphs. *Ixodes dammini* ordinarily feeds on rodents and other small animals in its larval and nymphal stages, and on the white-tailed deer as an adult.

Tick Paralysis

Tick paralysis has been reported from every continent (Murnaghan and O'Rourke, 1978). Some 43 tick species in 10 genera of both Ixodidae and Argasidae are able to induce paralysis in humans, other mammals, or birds (Gothe et al., 1979). Paralysis may develop from a tick implanted on any part of the body, but the manifestations are most serious when the tick injects its toxin on the back of the neck or along the spinal column. The syndrome is that of a generalized toxemia, with elevation of temperature to 40°C (104°F), rapid ascending flaccid paralysis, difficulty in swallowing and in respiration, and death. This toxicosis may develop within 24 hours or may not appear for several days or more following the bite. The disease appears to be due to salivary secretions that prevent the nerve impulse from traversing terminal motor nerve fibers. The toxin is most highly concentrated in the adult female tick. There is little change in the leukocyte picture.

The species particularly involved in tick paralysis in North America is the Rocky Mountain wood tick, *Dermacentor andersoni,* which is widely distributed throughout the Rocky Mountain states and Canada. It is most common, however, in western Montana, Idaho, and Oregon, where epizootics have developed in livestock, and numerous human cases of tick paralysis have been recorded with several deaths. Cases of human paralysis in the northwestern United States occur most frequently in the spring and early summer, when the ticks are most avid for blood. Other cases of tick paralysis have been reported from Georgia, Tennessee, Louisiana, and the Atlantic seaboard, following human infestation with *Dermacentor variabilis, Amblyomma maculatum,* or *A. americanum.* Species of *Ixodes, Rhipicephalus,* and *Hyalomma* have produced similar symptoms in Australia, South Africa, and Europe (Crete, Germany, France, Yugoslavia). Argasid ticks, particularly species of *Ornithodoros,* have been shown to have the toxic principle in greater or lesser amounts.

Where ixodid ticks abound, care should be taken to prevent young children from wandering from cultivated ground into underbrush or bushes where they are liable to be attacked. Should such excursions occur, the child should be immediately stripped of clothing and carefully examined for ticks, which should be removed as soon as possible. The type of motor paralysis in tick paralysis requires differentiation from poliomyelitis and other paralytic diseases.

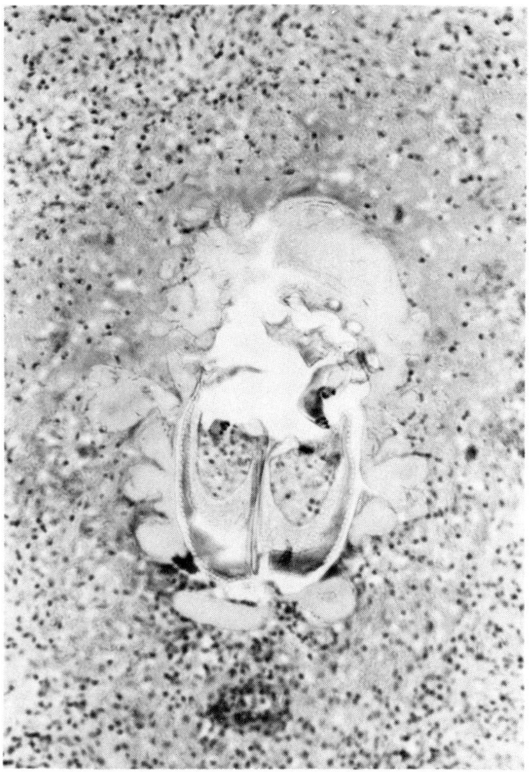

Fig. 36–18. Hard tick mouth parts in cyst-like swelling excised from the scrotum, where a partially engorged tick had been removed. (× 100.) (By P.C. Beaver. Specimen courtesy of Dr. John H. Arrington.)

Feeding Lesions

Any species of tick that attaches itself to human skin produces trauma by the insertion of its mouth parts into the skin layers. Its salivary secretions inhibit coagulation of the blood and cause histolysis at the feeding site. The presence of the mouth parts and extended feeding activity provoke an inflammatory infiltration of the perivascular tissues of the corium, with local hyperemia, edema, and hemorrhage, together with a thickening of the stratum corneum. Frequently, additional damage is caused by attempting to pull out the tick from its anchorage. A few drops of chloroform or ether placed on the head of the tick will usually cause it to relax its hold and will allow it to be easily removed by traction. The anterior portion of the tick should be grasped with forceps, close to the skin, and pulled *straight* out, with a steady pressure. Detached mouth parts remaining in the tissues when a tick is mechanically removed during engorgement cause foreign body reactions that may be severe (Fig. 36–18).

Other Tick-Associated Diseases

Erythema Chronicum Migrans

A tick-associated dermatologic condition known as *erythema chronicum migrans* (ECM) is prevalent in central Europe and Scandinavia, where it is associated with the bites of *Ixodes* ticks, particularly *I. ricinus*. The skin lesion begins as a red macule or papule that expands to form a large ring with central clearing. Lesions usually last about 3 weeks without treatment. The characteristic skin lesion has also been reported from several states in the northeastern United States (Mast and Burrows, 1976), Wisconsin (Dryer et al., 1979), California, and Oregon. It is associated with *Ixodes dammini* in the east and midwest and with *I. pacificus* in California and Oregon. An etiologic agent has not been identified (a spirochete is suspected); the lesion can be resolved quickly with antibiotic therapy (Steere et al., 1980b).

Lyme Disease

A form of inflammatory (oligoarticular) arthritis, termed Lyme disease after Old Lyme, Connecticut where it was first observed (Steere et al., 1977), occurs in the United States and is associated with bites of the aforementioned *Ixodes* species. Clinically, it is preceded by ECM and may occur intermittently for several years or may become chronic in the knees. Initially confused with juvenile rheumatoid arthritis, the illness is characterized by symmetric swelling and pain in a few large joints, some of them with an edematous papule that develops into an expanding red annular lesion. Malaise and myalgia are commonly present. The attacks are recurrent, with remission of up to 2 months. Aseptic meningitis, encephalitis, and cranial or spinal neuropathies, especially unilateral or bilateral Bell's palsy, occur in a few cases. Although the etiologic agent has not been proved conclusively, there is evidence that it is a treponema-like spirochete. Such spirochetes have been isolated from *Ixodes dammini,* and longlasting cutaneous lesions appeared in rabbits on which infected ticks had been fed. Sera of patients with Lyme disease contained antibodies to the spirochete (Burgdorfer et al., 1982).

Lyme carditis, an associated cardiac abnormality of Lyme disease, is clinically similar to acute rheumatic fever with the exceptions that complete heart block may be more common, myopericardial involvement tends to be milder, and valves seem not to be affected (Steere et al., 1980a). Arthritis and cardiac involvement as sequelae to tick bites are as yet unknown in Europe. Over 100 cases, mostly in juveniles, have been reported from New England coastal towns. A total of 226 cases of suspected Lyme disease with onset in 1980 were reported to CDC. The infections were probably acquired in 11 states.

Diagnosis. One bases diagnosis on the association of the typical ECM skin lesions with arthritic manifestations, in some cases neurologic or cardiac abnormalities, and epidemiological evidence.

Treatment. Treatment is with penicillin or other antibiotics (Steere et al., 1980b).

Tick Control. For effectiveness of toxicants in the control of ticks, see Chapter 45, p. 718.

MITES OF MEDICAL IMPORTANCE

Mites, as distinguished from ticks (suborder Ixodida), are usually only a millimeter in greater length or breadth and have a hypostome unarmed with tooth-like anchoring processes. Parasitic forms of medical importance may feed on blood, lymph, digested tissues, or sebaceous secretions on or near the surface of the skin. Others tunnel subcutaneously, causing an intense pruritus. Certain mites may serve as both reservoir and vector of an important rickettsial agent, *R. tsutsugamushi*. One group, the house dust mites, either produces or concentrates potent allergens commonly found within the home.

Because of the increasing numbers of species and the great ecological diversity of the Acari, as well as a lack of fossil records, the higher classification of this subclass continues to be an evolving process (Krantz, 1978). Current acarological systematics therefore reflect a variety of viewpoints, and it should be understood that there are several taxonomic schemes for ordinal-subordinal designations. The system employed here is that used by Krantz. It includes two orders (Parasitiformes, Acariformes) with discussion of medically important representatives found within three suborders: Acaridida, Actinedida, and Gamasida. Recent synonymies of these suborders are also given.

Order Parasitiformes: Suborder Gamasida (= suborder Mesostigmata of Krantz, 1970)

Several families of medical and economic significance are found in this suborder. The family Macryonyssidae includes *Ornithonyssus* spp., and

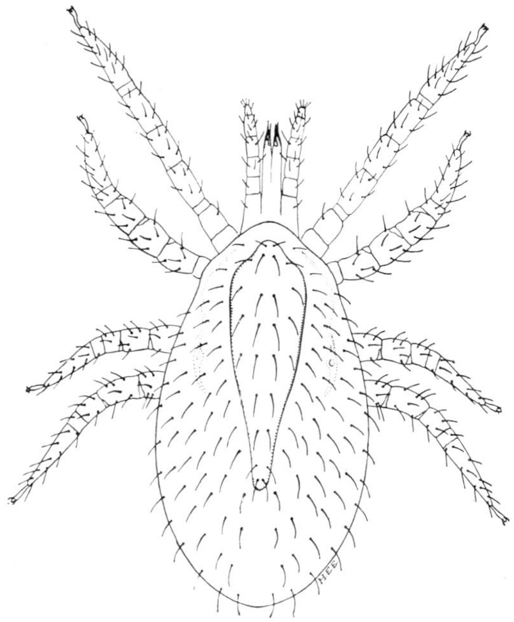

Fig. 36–19. *Ornithonyssus bacoti*. Greatly enlarged. (From Ewing, H.E. 1929. *External Parasites*. Springfield, Ill., Charles C Thomas.)

the family Dermanyssidae contains two genera of importance, *Dermanyssus* and *Liponyssoides* (= *Allodermanyssus*).

Ornithonyssus bacoti (Fig. 36–19) is the common rat mite of warm climates and is common in the southern United States. Persons employed in granaries, food supply houses, stockyards, and other places frequented by rats are greatly annoyed and discomforted by this mite. It feeds on blood or serum and drops off its host after each blood meal. Dove and Shelmire (1932) showed that each mite requires at least four blood meals from larva to adult. Apparently a temperature of 24° to 26°C with a relative humidity of 47% is optimum for development. The life cycle is completed in about 11 to 16 days; the adults survive 62 days and the females lay about 99 eggs each.

The attacks of *O. bacoti* and other species of this genus (*O. bursa*, *O. nagayoi*, and *O. sylviarum*) commonly produce a dermatitis that may be localized or generalized depending on the abundance and distribution of the mites on the whole body. *O. bursa*, a bird mite, is a common pest in the southern United States, migrating into homes from abandoned nests in the eaves of houses or from birdhouses. Dermatitis, with urticarial wheals, papules, vesicles, and even hemorrhagic necrosis, develops at the sites of the puncture wounds, and secondary infection from scratching the lesions frequently occurs, particularly in children. The mites tend to feed in sites where there is constricting clothing or a fold of skin, thereby producing clusters of lesions.

The local pruritus and pain due to the "bite" of these mites may be reduced by topical applications in mild cases of phenolated camphor in a petrolatum base or, in more severe reactions, by corticosteroid creams or ointments.

Dermanyssus gallinae, a common ectoparasite of fowls, temporarily infests human beings and produces troublesome skin lesions at the sites where it sucks blood or tissue juices. *Liponyssoides (= Allodermanyssus) sanguineus,* an ectoparasite of mice and rats, has been reported from several foci in the Eastern Hemisphere and from several locations in the United States (from Boston to Washington, D.C., Indianapolis, and Tucson). In studying the life cycle of this mite, Fuller (1954) described the following stages: egg, larva, protonymph, and deutonymph. The larva did not feed, the two nymphal stages each fed once, and the adults of both sexes fed several times. Unfed and fed adults survived a maximum of 51 and 63 days, respectively.

Rickettsialpox

Definition. Rickettsialpox is a mild exanthematic disease related immunologically to Rocky Mountain spotted fever and caused by *Rickettsia akari*. The natural vector is *Liponyssoides sanguineus,* a mite of domestic mice and rats.

Geographical Distribution. Rickettsialpox occurs enzootically in domestic mice and rats, and in humans in endemic (or at times epidemic) form in Manhattan and the Bronx (Huebner et al., 1946), Boston, Hartford (Connecticut), and Cleveland (Ohio). It has likewise been found in U.S.S.R., transmitted by this same mite species. In at least one locality in New York City an epidemic of rickettsialpox resulted from the removal of a garbage dump. Mice that had used the dump as a home and food supply moved into nearby apartments, where these rodents with their ectoparasites started the epidemic (Modell, 1963).

Clinical Aspects. At the site where the infected mite bites, an initial eschar develops, resembling a small vesicle or pustule, followed usually by mild symptoms including fever, sweating, regional adenopathy, and a maculate rash readily confused with ecchymoses of chickenpox. Diagnosis made presumptively on the dermatitis and fever must be confirmed by specific complement-fixation test and/or inoculation of clean laboratory mice with material from the lesions. Although specific therapy is probably not needed, chloramphenicol* or tetracycline drugs as administered in spotted fever (*vide supra,* p. 588) hasten convalescence. Prognosis is favorable.

Prevention. Use of acaricides and rodenticides is advised to reduce the reservoirs and their ectoparasites.

Order Acariformes: Suborder Actinedida (= suborder Prostigmata of Krantz, 1970)

Follicular or Demodicid Mites (Demodicidae)

Two species of *Demodex* (Fig. 36–20) commonly parasitize humans, each within a defined niche. *Demodex folliculorum* is found in the follicles of simple hairs above the level of the sebaceous gland, and *D. brevis* occurs in the sebaceous glands of the vellus hairs (Desch and Nutting, 1972). Both species are typically located in the epidermis of the nose, forehead, and adjacent regions of the face. Each is a low-grade pathogen by virtue of its feeding on the contents of cells in the follicular epithelium and sebaceous glands by puncturing the cell with two needle-like chelicerae and injecting salivary secretions (Nutting, 1975).

The gross signs of demodicidosis may include a mild rosaceous pruritus and a fibrous tissue response around the offending organism. *Demodex* mites have also been associated with a dermatitis of the scalp (Miskjian, 1951).

Lindane (gamma hexochlorocyclohexane gammexane), 1% in a vanishing cream, massaged into the lesions is probably the treatment of choice.

Predaceous Mites (Pyemotidae)

Some of the species of this group are predaceous on insects that attack grain crops. Of particular interest, both biologically and medically, is *Pyemotes tritici* (Moser, 1975), the straw itch mite, which feeds on the larvae of several insects infesting seeds, grain, and plant stems. The young female is elongate-piriform in shape. After fertilization the eggs are retained in her abdomen, hatch there, and develop into mature mites before they are discharged. During this period of "pregnancy," the abdomen of the mother becomes tremendously inflated to accommodate the progeny, which may number up to 200 or 300. Optimum development apparently takes place at about 28°C, with cyclical reproduction about every 4 months, the mite accommodating itself to new broods of graminivorous insects on which it feeds.

Persons in contact with grains, straw, or hay

*Note the risk of bone marrow depression.

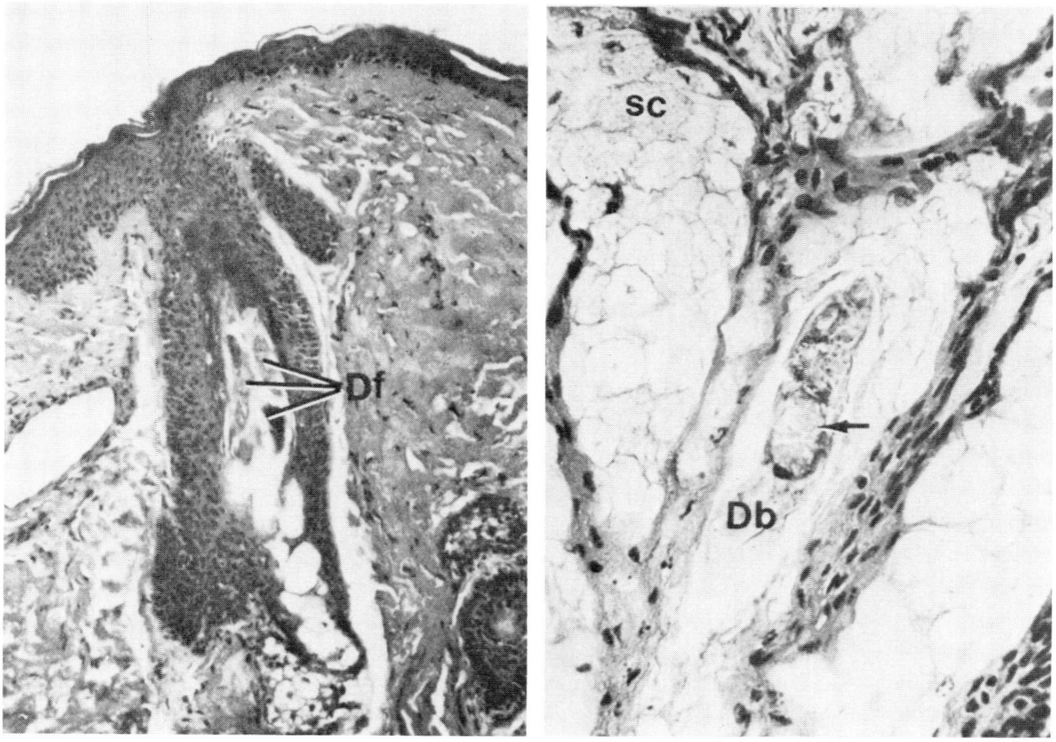

Fig. 36–20. *Demodex folliculorum* and *D. brevis*. *A*, *D. folliculorum* (Df) in hair follicle. *B*, *D. brevis* (Db) in sebaceous gland (*sc*, sebaceous gland cells). (× ca. 225.) (From Desch, C., and Nutting, W.B. 1972. *Demodex folliculorum* (Simon) and *D. brevis* Akbulatova of man: Redescription and reevaluation. J. Parasitol., *58*:169–177.)

harboring *P. tritici*, particularly threshing crews or persons sleeping on unsterilized straw mattresses, may be attacked by hordes of these organisms. Bites by young female mites produce pale pink to bright red spots on the skin, followed by petechial hemorrhage and erythema, later followed by a wheal, a central vesicle, and a pustule at each site of attack. In some cases these bites cover most of the body. The lesions produce an annoying burning pruritus, with fever and sweating. Vulvitis has been reported. The dermatitis is often confused with scabies, pediculosis, and allergy. Other complications may include headache, fever, nausea, and albuminuria (Treat, 1975).

Applications of warm water containing oatmeal flakes or Linit starch and mild antiseptics should be placed locally over the areas involved. Little or no protective immunity is developed to this mite; persons may become reinfested several times during the same season. The most helpful preventive measure is disinfestation of granaries infested with the insects on which the mites prey.

Cheyletiellid Mites Causing Dermatitis

Several species of *Cheyletiella* (family Cheyletiellidae) associated with pets occasionally cause a pruritic dermatitis in man (Cohen, 1980). These nonburrowing mites, which infest dogs *(C. yasguri)*, cats *(C. blakei)*, and rabbits *(C. parasitovorax)*, can be passed by direct contact with these animals (van Bronswijk and deKreek, 1976). The pattern of papular lesions is typically diffuse on the arms, chest, and abdomen, and resembles the transient dermatitis due to *Sarcoptes scabiei* (var. *canis*) and *Notoedres cati*.

Trombiculid Mites, Harvest Mites, Chiggers (Trombiculidae)

Biology. Mites of the family Trombiculidae have spiracular openings of the tracheate system at the side of the gnathosoma (capitulum). They are of particular medical importance because their larvae attack humans and other vertebrates, producing severe dermatitis, frequently followed by allergic reactions, and because certain species transmit the causative agent of "mite typhus" (tsutsugamushi disease, scrub typhus). These mites are commonly called harvest mites (chiggers, red mites, "red bugs," tsutsugamushi [meaning "dangerous bug"],

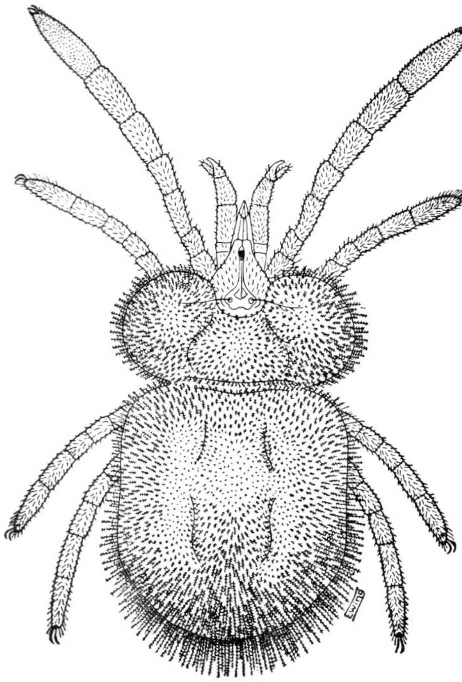

Fig. 36–21. *Eutrombicula alfreddugesi* (American "red bug"), adult. Greatly enlarged. (From Ewing, H.E. 1929. *External Parasites*. Springfield, Ill., Charles C Thomas.)

Fig. 36–22. *Leptotrombidium akamushi,* larva. Greatly enlarged. (Adapted from Nagayo, M., et al. 1921. Five species of tsutsugamushi (the carrier of Japanese river fever) and their relation to the tsutsugamushi diseases. Am. J. Hyg., *1*.)

or Kedani mite of Japan), various species of which are distributed extensively throughout the world. Most species are reddish or are spotted with red, orange, or black, but some are colorless. The nymphs and adults are either predaceous on other arthropods, the juices of which they extract, or are wholly vegetable feeders. Only the 6-legged larvae are ectoparasitic on mammals, birds, and reptiles. The nymphs and adults of this family may be recognized because of a well-marked constriction of the body just anterior to the third pair of legs (Fig. 36–21). They lack stalked eyes and lay their eggs singly. The adults of some species reach a length of 1.25 mm, but most of them are smaller. Adults are completely clothed with hairs; those on the body have multiple barbs in some species (Fig. 36–26). Many forms designated by species names are known only in their larval stage, and as a result their taxonomy is confused.

Eggs of the trombiculids are laid on the ground or on the stems or leaves of low grass and bushes. After an initial embryonic stage, the mite enters a quiescent phase within the egg for a short time (*prelarval* or *deutovum* stage), then hatches as a 6-legged larva (Fig. 36–22). Depending upon the species, it may attack birds, reptiles, mammals (Traub and Wisseman, 1974), or even scorpions or snails (Krantz, 1978). On smaller vertebrate hosts, such as mice, the ears are a favorite locality for these larvae. The hypostome and chelicerae are sunk into the skin of the host and a prolonged meal is taken. The larva then drops to the ground and transforms into a quiescent *protonymph (nymphochrysalis)* followed by an active, 8-legged form, the *deutonymph,* which preys on eggs of other arthropods or softbodied invertebrates. After passing through a second quiescent nymphal stage (the *tritonymph* or *imagochrysalis*), the adult mites are produced and the life cycle is repeated. Johnston and Wacker (1967) discuss postembryonic development of chigger mites.

Trombiculids and Disease

Species of trombiculids have a dual role in the production of disease: (1) a few species serve as both reservoir and vector of the etiological agent of scrub typhus, and (2) certain species of trombiculids produce an irritating dermatitis in humans.

Scrub Typhus
(Japanese river fever, Tsutsugamushi disease, Kedani fever, Sumatra fever, New Guinea mite typhus, Queensland coastal fever)

Definition. Scrub typhus, caused by *Rickettsia tsutsugamushi,* is an acute exanthematous rickettsial disease characterized by a primary skin lesion consisting of a punched-out ulcer covered by eschar at the site of the attachment of the vector mite. The fundamental lesion is invasion and destruction of the endothelial lining of the smaller blood vessels. There is an acute onset of fever from 6 to 21 days (usually 10 to 12 days) after inoculation, with headache, diaphoresis, conjunctival infection, and lymphadenopathy. Within the first week of illness a red maculopapular rash appears on the trunk, extends centrifugally to the extremities, and fades in a few days. It is not petechial. Pneumonitis is a common complication. Without specific therapy the duration of fever is about 2 weeks. The fatality rate varies from 1 to 40% depending on geographic site, rickettsial strain, and susceptibility of the patient. A benign Japanese variety of the disease is called Shishito fever.

Geographical Distribution. This disease occurs throughout most of Japan; the southern half of Korea; Taiwan and the Pescadores; the Philippines; southern China and Tibet; extensive areas in Indochina, northern Thailand, Burma, Pakistan, India; Sri Lanka, Maldive Islands, Andaman and Nicobar Islands; Malaysia; Indonesia, eastern and western parts of Borneo, northern Celebes; New Guinea, and numerous islands off the north coast; New Britain; Bougainville and New Georgia in the Solomon Islands; Espiritu Santo (northern Hebrides Islands); and Queensland, northeastern Australia.

Epidemiology. *Rickettsia tsutsugamushi* is commonly present in field mice, wild rats *(Rattus)*, shrews, and voles. In Japan the vole *Microtus montebelli* is particularly important, whereas wild rats in special environmental conditions, as in the suburbs of Calcutta or farmyards on Taiwan, may play a role in the epidemiology of this disease. The marsupial bandicoot is a reservoir in New Guinea. At times domestic or wild birds may harbor the etiologic agent or may be infested with infected trombiculids.

These vertebrates may become rickettsemic and serve as donors (Traub et al., 1975). During engorgement, the larval mite induces the formation of a *stylostome* or feeding tube (Fig. 36–23) and acquires *R. tsutsugamushi*. Three types of stylostome may be formed, depending upon the mite species and the contents of its saliva. Those stylostomes that cause severe inflammation, penetrate deeper into the host tissue, and leave residual granulomatous foci are more likely to yield rickettsiae (Hase et al., 1978). Once in the mite, the rickettsiae pass through the intestinal wall and enter the hemocoel and extraintestinal tissue, where they multiply and remain viable through the nymphal and adult stages of this generation. They are passed through the egg to the larva of the second generation, when again they have an opportunity to be inoculated into a susceptible mammal.

Relatively few of the many species of trombiculid mites described from the enzootic areas of scrub typhus have been found naturally infected. The epidemiologic patterns in enzootic-endemic areas, and even in relatively nearby regions, differ remarkably. Where there have been outbreaks of chigger-borne typhus, species of *Leptotrombidium* are believed to serve as the major vectors. Within this genus, the *L. (Leptotrombidium) deliense* group constitutes the major vector assemblage (Traub and Wisseman, 1974). *Leptotrombidium deliense* occurs over a wide geographic area (Assam, Burma, China, India, Malaya, New Guinea, Pakistan, Pescadores, Philippines, Thailand, and Queensland, Australia). It has been found to be naturally infected over this range on numerous occasions. *Leptotrombidium fletcheri* is considered to be a vector in Malaya, New Guinea, and the Philippines, whereas *L. arenicola* is restricted to Malaya. *Leptotrombidium akamushi (sensu stricto)*, the classical vector of *R.tsutsugamushi* in Japan, is found along major river systems, whereas *L. scutellare* occurs in the Mt. Fuji area. *Leptotrombidium pallidum* is also a vector in Japan and Korea, as well as a secondary vector in the Primorye region of the USSR, where *L. pavlovskyi* is considered the primary vector.

Year-round transmission occurs in tropical environments. Humans are exposed in enzootic areas when they enter high savanna grass, scrub growth, or the periphery of jungles in which the mammalian reservoirs and their trombiculid ectoparasites live or, probably much less frequently, when the infected mammalian reservoirs make excursions from their natural habitats to adjacent cultivated land.

In Pakistan, rickettsial infection in rodents has been found in a number of unusual habitats and geographical areas, including alpine terrain at 10,500 feet elevation in the Kaghan Valley of the Himalayas, areas where snow is on the ground for 9 to 10 months of the year, the mountain deserts of Gilgit Agency at 36° N latitude, in the semidesert of Multan, and in the plains of Lahore District.

Diagnosis. This disease requires differentiation from typhus fever, both louse-borne and murine varieties (see pp. 614 and 705), spotted fever (see p. 586), Q fever (see p. 589), relapsing fevers (see pp. 589 and 616), Weil's disease, and other febrile and exanthematous infections. Diagnosis is based on a history of having been in an endemic area at the season when larval mites were active, on symptoms, and in particular on specific agglutination of *Proteus O* of the XK (K for Kingsbury) strain by the patient's serum but failure of the serum to agglutinate the strain X19 or X2. A progressively rising titer is diagnostically valuable, but this does not typically occur until the fever begins to abate. Specific complement-fixation is also available. Different strains of *Rickettsia tsutsugamushi* behave differently to specific antigens prepared from various strains.

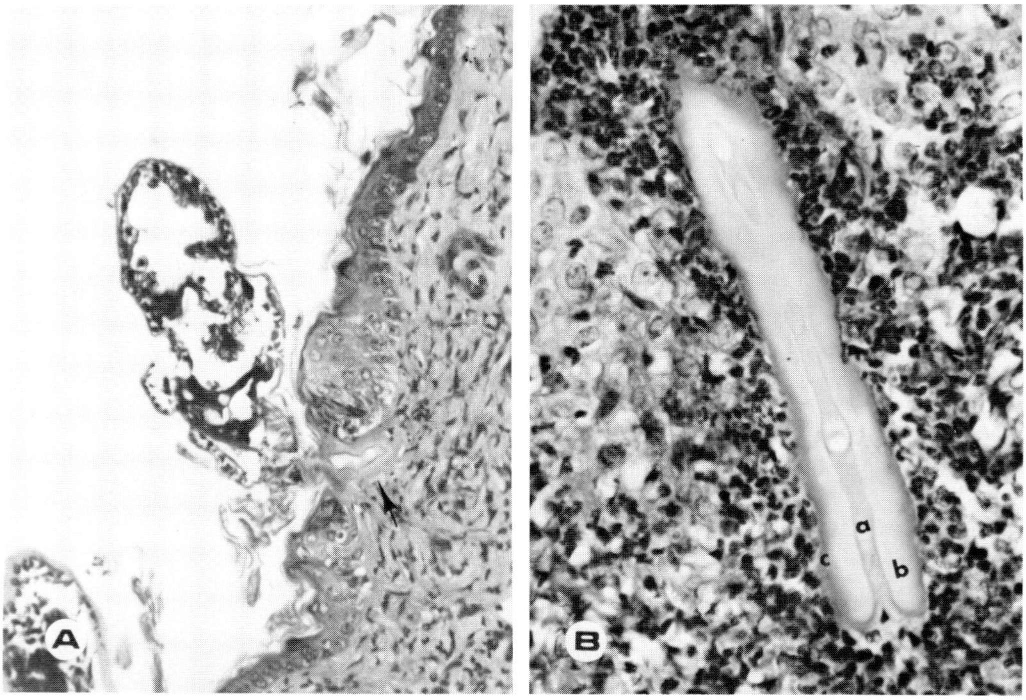

Fig. 36–23. *Leptotrombidium fletcheri*. *A*, Larva attached to mouse skin; an arrow points to the forming stylostome. (× ca. 210.) *B*, Mesenchymal stylostome opening into the dermis of host tissue 24 hours after attachment. *a*, Central canal. *b*, Chromophobic tube. *c*, Eosinophilic coat. (× ca. 375.) (From Hase, T., Roberts, L.W., Hildebrandt, P.K., and Cavanaugh, D.C. 1978. Stylostome formation by *Leptotrombidium* mites (Acari: Trombiculidae). J. Parasitol., *64*:712–718.)

Treatment. Treatment is with one of the tetracycline antibiotics or chloramphenicol continued until the patient is afebrile, which is usually within 2 days. Use of sulfonamides is contraindicated. A second course of the antibiotics should be given after an interval of 6 days to prevent recrudescence. Prognosis is excellent with specific treatment.

Prevention. Prevention varies, depending on the reservoir hosts of the disease. In Japan, the endemic foci are restricted primarily to the flood plains of certain rivers where the mite *T. akamushi* primarily parasitizes small rodents. In other enzootic-endemic areas the disease is more widely distributed. It is found in the tropical forests or jungles, and particularly in scrub or jungle grasses, such as thatch grass *(Imperata cylindrica),* or even in dry inland areas (Pescadores).

In sparsely populated regions this infection is almost solely enzootic in wild rodents and certain of their trombiculid mites (*T. deliense* and related species), and human exposure in such areas produces the disease in epidemic form. For nonimmunized natives in these areas scrub typhus constitutes a serious handicap to productive agriculture. When previously unexposed individuals enter an enzootic area, the number who contract the disease depends on the amount of infection in the reservoir hosts, the prevalence of the infected mite transmitters, the abundance of larvae to transmit the pathogen, and the number of exposures per individual.

In prevention the principal attack must be on the mite transmitter. This can be effected by spraying infested areas with suitable acaricides, after which jungle grass can be cut down with considerable safety. The mites are sensitive to drying and sunlight. Similarly, where the infected rodent population and their mites live in burrows, in caves, or under rocks, these toxicants may be hand-power sprayed into these habitats. For protection of persons whose work requires temporary exposure, a rapidly acting acaricide such as benzyl benzoate can be mixed with a repellent such as dimethyl (or dibutyl) phthalate (Busvine, 1980).

Trombiculid Dermatitis

Species of *Eutrombicula, Neotrombicula, Schoengastia, Euschoengastia, Acomatacarus, Siseca, Blankaartia,* and their close relatives attack humans to obtain a tissue-juice meal, attaching themselves to the skin or to the base of hairs, with the capitulum inserted into the hair follicle. Here they secrete salivary juices, and the combination of trauma and digestive secretions causes disintegration of the contents of cells. This disintegration produces a marked local reaction that usually appears 12 to 24 hours after attachment. The lesion first appears as a minute red elevation on the skin, accompanied by intense painful pruritus (trombiculosis). Soon a wheal develops around the site, frequently with extravasation of blood. Usually the skin becomes excoriated as a result of scratching, with serous exudation, and frequently a pustule develops at the site. Some persons are more sensitive to these bites than are others, and the lesions may remain painful for a week or more. The regions most commonly attacked are the ankles, legs, external genitalia, groin, and waistline, but the mites may reach the axilla or the breasts. Once attached to the skin, the mites may be individually removed by using a sharp needle, but the duration of the dermatitis is not measurably affected by this procedure.

Eutrombicula alfreddugesi has a widespread distribution from southern New York westward to South Dakota and extending southward, with increased prevalence, to the states on the Gulf of Mexico; it also occurs in Arizona and California. *E. splendens* has essentially the same distribution, but with lesser intensity except in Florida. *E. batatas* is widely distributed throughout tropical America, has been collected from Georgia, Florida, Alabama, Louisiana, and Texas, and has been reported as attacking man in California. In Europe, *Neotrombicula autumnalis* larvae are common pests in the fall during the harvest season. Many other dermatitis-producing trombiculids have been described from other areas (Vercammen-Grandjean, 1968; Vercammen-Grandjean and Langston, 1971, 1976; Brennan and Goff, 1977; Nadchatram and Dohany, 1974; Nadchatram et al., 1980).

Phenolated camphor solutions in mineral oil applied locally provide some relief, but topical preparations of corticosteroids are most effective in inhibiting the host reaction. Clothing impregnated with an emulsion of benzyl benzoate (45 parts), dibutyl phthalate or dimethyl phthalate (45 parts), and a detergent emulsifier in water (10 parts) is recommended as a prophylactic.

Suborder Acaridida (= Astigmata of Krantz, 1970)

Sarcoptid Itch Mites or Mange Mites (Sarcoptidae)

Of the several genera of acaridid mange mites infesting the skin of mammals (i.e., *Psoroptes* of sheep, cattle, rabbits; *Notoedres* of cats and rats; *Chorioptes* of horses, cattle, and other large domestic mammals; *Knemidocoptes* of poultry; *Otodectes* of dogs and cats; and *Sarcoptes* of various domestic mammals), only one species, *Sarcoptes scabiei* (Linnaeus, 1758) Latreille, 1802, commonly causes disease in humans, although mange mites of domestic animals occasionally cause a pruritic rash in man.

Sarcoptes scabiei, the etiologic agent of human sarcoptic mange, was first definitely proven to cause human mange by Raspail and his Corsican student, Renucci, in 1834, though its presence in the skin had been shown much earlier (Bruce-Chwatt, 1978).

Morphology. *S. scabiei* is a microscopic ar-

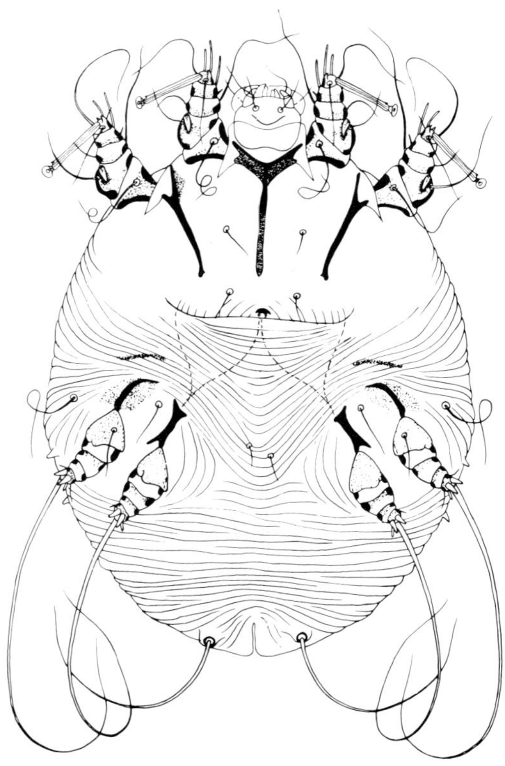

Fig. 36–24. *Sarcoptes scabiei*. Female, ventral view. (From Fain, A. 1968. Etude de la variabilité de *Sarcoptes scabiei* avec une révision des Sarcoptidae. Acta Zool. Pathol. Antverp., *47*:1-196.)

thropod of ovoid contour, with chelate chelicerae, no special respiratory apparatus, and its anterior two pairs of legs well separated from its posterior two pairs. The anterior legs terminate in delicate stalked discs; the posterior legs both end in long-hair bristles in the female, while in the male the third pair has bristles and the fourth pair discs. The males and females are also differentiated by their size and genitalia. The female measures 330 to 450 μm in length by 250 to 350 μm in breadth (Fig. 36–24), and the male, 200 to 240 μm by 150 to 200 μm. On the dorsal surface of the body there are numerous more or less parallel ridges, except for a small, smooth median area just behind the head, several mid-dorsal tooth-like spines, and in other areas longer, finger-like spines and long and short hairs. The number, type, and position of these several integumentary structures are of diagnostic importance. The ventral surface is smooth except for a few hairs and bristles. The legs are supported at their bases by chitinous rods *(epimera)*.

Development. These mites live in cutaneous burrows several millimeters to a few centimeters in length, where the fertilized female usually develops 1 or 2 eggs at a time within her body. The female is found characteristically at the blind end of the tunnel, with her oviposited eggs behind her. Egg laying lasts for up to 8 weeks (i.e., the lifetime of the female), during which time several large, thin-shelled, ovoid, transparent eggs are deposited at 2- to 3-day intervals. Within 3 to 5 days the eggs hatch and 6-legged larvae emerge. The larvae migrate to the skin surface and then pass through the nymphal stage, reaching adulthood approximately a week after having hatched as larvae. Mating ensues, and both males and females burrow back into the skin. The variety of *S. scabiei* infesting man is believed to be both morphologically and physiologically distinct from the varieties of this species found in other mammals (Fain, 1968). It has a cosmopolitan distribution.

Epidemiology. Although sarcoptic mange of humans is commonly acquired from contact with infested persons, people are temporarily susceptible to infestation with nonhuman strains of *S. scabiei* from infested horses, sheep, goats, cattle, and dogs.

Pathology and Symptomatology. The lesions in the skin are developed rapidly, owing to the short life cycle of the mite and the rapid production of successive generations of progeny. These mites may burrow into the skin on any part of the body, but the usual sites are the interdigital spaces, backs of the hands, elbows, axillae, groin, breasts, umbilicus, penis, shoulder blades, small of the back, and buttocks. The burrows are made in the epidermis, with an external opening through the horny layer. Although these tunnels run more or less parallel to the surface, their inner ends are sloped somewhat deeper than the portion near the orifice. As they feed on the tissues, the mites deposit minute fecal pellets believed to provoke the characteristic vesiculations of the skin and the excruciating pruritus, which is intensified by warmth and perspiration of the affected area. Hence the outstanding clinical diagnostic symptom is the intense itching. The first objective evidence is a small, usually reddish, slightly elevated track on the surface of the skin. The orifice can be identified through a hand lens. Frequently one can also see a minute vesicle at the far end of the tunnel, near which is the grayish glistening female at the end of the burrow. Scratching commonly causes weeping and bleeding of the lesion, provides an opportunity for pyogenic infections to enter, and spreads the infestation to other sites. The usual case seen by the physician has multiple lesions, with vesiculation and papule formation. Pustules, excoriation, and secondary infection of the lesions, sometimes with a weeping eczema, are common complications. Occasionally scabies presents as urticaria (Chapel et al., 1981).

A peculiar form of the infestation, called Norwegian scabies, occurs in certain probably anergic individuals (Youshock and Glazer, 1981). Keratotic excrescences occur on the body and extremities and contain abundant mites (Fig. 36–25). Pruritus is absent (Burks et al., 1956). Cases of Norwegian scabies are highly contagious and have often been the cause of hospital epidemics (Fain, 1978).

Diagnosis of Sarcoptic Mange. The diagnosis is based clinically on the characteristic lesions and their topographic distribution; etiologically, on recovery of the mite. To recover the mite, the physician must first find the slightly enlarged distal end of the burrow. When this is opened aseptically, the mite appears as an opalescent, white speck visible to the naked eye, and can be removed with a sharp, flat needle.

Treatment. *Sulfur treatment* for sarcoptic mange was found effective long before the mite was definitely established as the etiologic agent. In ancient Mesopotamia and later in Rome, baths in sulfurous hot springs were known to be curative for eczematous mange. The most commonly used parasi-

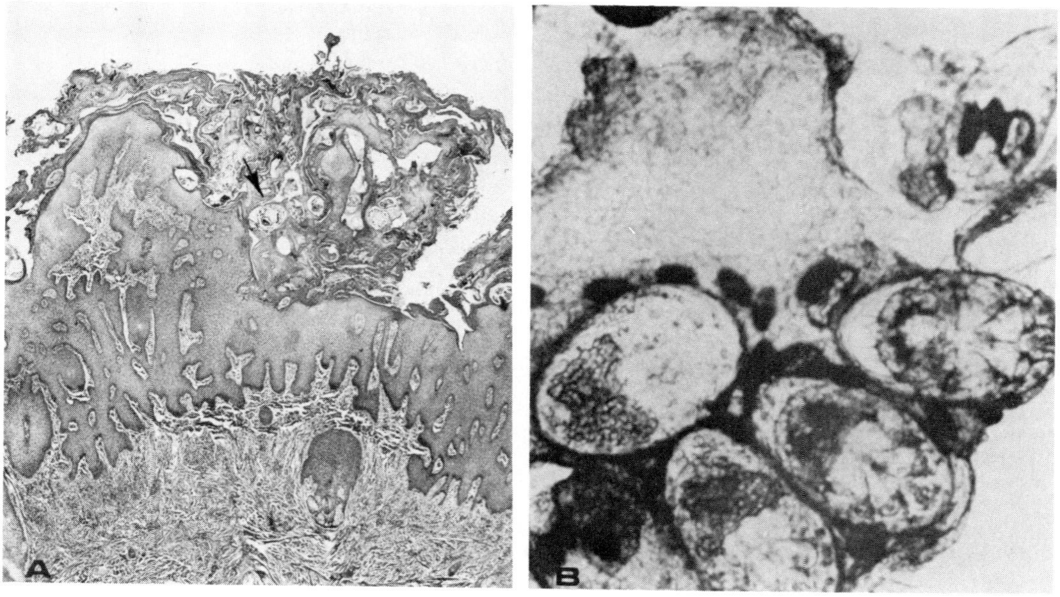

Fig. 36–25. *Sarcoptes scabiei. A,* Histological section of skin from patient suffering from Norwegian scabies, showing mites in the hyperkeratotic horny layer. (× ca. 15.) *B.* Skin scrapings showing developing larvae in eggs and characteristic black fecal pellets from burrows. (× ca. 200.) (*A* from Binford, C.H., and Connor, D.H. 1976. *Pathology of Tropical and Extraordinary Diseases,* vol. 1. Washington, D.C., Armed Forces Institute of Pathology [negative no. 73–5339]. *B* courtesy of Dr. Adam Ewert.)

ticide for this infestation is a 5% suspension of flowers of sulfur in lanolin. The modern effective scabicide consists of the application of lindane (gamma isomer of hexachlorocyclohexane) 1% in a lotion or ointment base. This treatment usually produces no dermatitis or sensitivity and provides 100% clinical cures in one to three applications.

Before treatment with any parasiticide, the infested part of the patient's body should be thoroughly scrubbed all over with soap, soaked in a bath of warm water to expose the lesions, and then dried. The sites where the lesions occur are then thoroughly rubbed with the lotion, which is left on overnight. In the morning, freshly sterilized clothing is put on and all soiled clothing and bed linen are boiled. In mild cases one application is curative, but in more severe involvement a second application of the lotion between the sixth and tenth days may be required to kill new broods hatched from eggs not reached by the first application. For small children with tender skin, it may be necessary to dilute the lotion one-half with some inert base.

Prevention. Sarcoptic itch or mange is most common in armies, jails, and other groups, such as isolated families, where personal hygiene is neglected. Although exposed individuals occasionally acquire sarcoptic infestation from contact with dogs, cats, cows, sheep, horses, or other infested domestic animals, practically all patients become infested from contact with other human beings. All individuals should be given prompt treatment and their clothing thoroughly sterilized by boiling.

Other Acaridid Mites Causing Dermatitis

The families Acaridae, Glycyphagidae, and Carpoglyphidae contain several genera of mites that frequently contaminate stored foods and secondarily cause a dermatitis in persons handling infested products. Species of *Acarus* (cheese mite), *Tyrophagus* (copra mite), *Glycyphagus* (grocery mite), and *Carpoglyphus lactis* may get on the hands, migrate under microscopic scales of the horny layer of the skin, or into cracks of the epidermis, and produce temporary pruritus. Washing the affected members thoroughly with kerosene-soapsuds, or local application of 1% lindane lotion or ointment should remove and kill these organisms, which are not known to suck blood. Occasionally these mites are swallowed in food and are discovered in the microscopic examination of the feces or urine, rarely in sputum. Although they may become lodged for a brief time in anatomical locations, they are not true tissue parasites.

There is no proof that any of these mites is able to produce ''creeping eruption'' in cutaneous burrows. On the other hand, should any of these mites or those of other superfamilies be inhaled and become temporarily resident in the respiratory passages, it is entirely possible that they may be responsible for asthmatic bronchitis (Ingram et al., 1979).

Allergenic Mites in House Dust

House dust often contains allergenic mites belonging to the family Pyroglyphidae. Several species can be found in dust and vacuum samples from floors, furniture, mattresses, and bedding (Arlian et al., 1979). House dust mites are nidicolous and feed on organic debris, including sloughed human skin, fungi, spilled food, and pollen. These mites are particularly prevalent in beds, though carpeted floors near beds or couches may also have large numbers. The most common species are the American house dust mite, *Dermatophagoides farinae,* and the European house dust mite, *D. pteronyssinus* (Fig. 36–26). Both of these Acari can concentrate dust allergens and may also produce allergic reactions in humans (Wharton, 1976). It is estimated that 4% of the population in the United States suffer from house dust allergy.

Mite levels in the home can be controlled by exclusion of dust from the bedrooms and furniture of sensitive individuals. This can best be accomplished by thorough housecleaning, including weekly vacuuming of beds and overstuffed furniture.

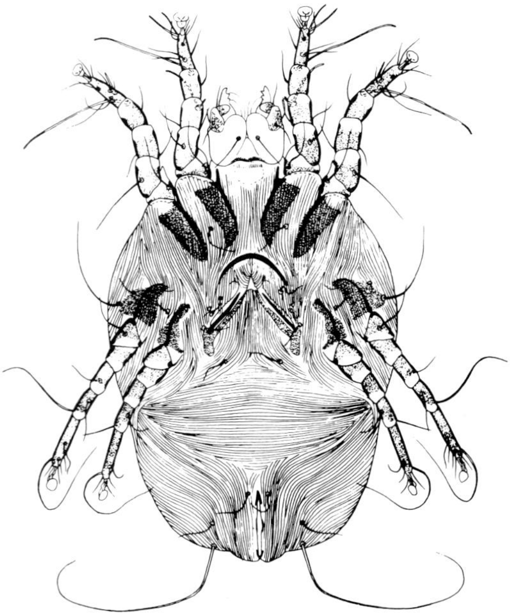

Fig. 36–26. *Dermatophagoides pteronyssinus.* Female, ventral view. Enlarged. (From Fain, A. 1966. Nouvelle description de *Dermatophagoides pteronyssinus* (Trouessart, 1897). Importance de cet acarien en pathologie humaine (Psoroptidae: Sarcoptiformes). Acarologia, *8*:302–327.)

REFERENCES

Aeschlimann, A. 1958. Développement embryonnaire d'*Ornithodorus moubata* (Murray) et transmission transovarienne de *Borrelia duttoni.* Acta Trop., *15*:15–64.

———. 1961. Complement a l'étude de l'embryologie d'*Ornithodorus moubata* (Murray). Acta Trop., *18*:58–60.

Arlian, L.G., Bernstein, I.L., Johnson, C.L., and Gallagher, J.S. 1979. Ecology of house dust mites and dust allergy, pp. 185–195. In *Recent Advances in Acarology,* Vol. 2. Edited by J.G. Rodriguez. New York, Academic Press.

Balashov, Y.S. 1972. Bloodsucking ticks (Ixodoidea)—vectors of diseases of man and animals. Misc. Publ. Entomol. Soc. Am., *8*:161–376.

Belcher, D.W., Wurapa, F.K., Ward, W.B., and Lourie, I.M. 1975. Guinea worm in southern Ghana: Its epidemiology and impact on agricultural productivity. Am. J. Trop. Med. Hyg., *24*:243–249.

Berge, T.O., editor. 1975. *International Catalogue of Arboviruses Including Certain Other Viruses of Vertebrates.* U.S. Dept. of Health, Education and Welfare, Publ. No. (CDC) 75-8301.

Bertram, D.S. 1939. The structure of the capitulum in *Ornithodoros.* Ann. Trop. Med. Parasitol., *33*:229–258.

Bettini, S., editor. 1978. *Arthropod Venoms.* (Handbook of Experimental Pharmacology, n.s., Vol. 48.) New York, Springer-Verlag.

———, and Brignoli, P.M. 1978. Review of the spider families, with notes on the lesser-known poisonous forms, pp. 103–120. In *Arthropod Venoms.* Edited by S. Bettini. New York, Academic Press.

———, and Maroli, M. 1978. Venoms of Theridiidae, genus *Latrodectus.* A. Systematics, distribution and biology of species; chemistry, pharmacology and mode of action of venom, pp. 149–184. In *Arthropod Venoms.* Edited by S. Bettini. New York, Academic Press.

Brennan, J.M., and Goff, M.L. 1977. Keys to the genera of chiggers of the Western Hemisphere (Acarina: Trombiculidae). J. Parasitol., *63*:554–566.

Bruce-Chwatt, L.J. 1978. Origin of medical entomology. Trans. R. Soc. Trop. Med. Hyg., *72*:211–212.

Bücherl, W. 1971a. Classification and biology of scorpions, pp. 339–342. In *Venomous Animals and Their Venoms. 3. Venomous Invertebrates.* Edited by W. Bücherl and E.E. Buckley. New York, Academic Press.

———. 1971b. Spiders, pp. 197–277. In *Venomous Animals and Their Venoms. 3. Venomous Invertebrates.* Edited by W. Bücherl and E.E. Buckley. New York, Academic Press.

Burgdorfer, W. 1976. The epidemiology of the relapsing fevers, pp. 191–200. In *The Biology of Parasitic Spirochetes.* Edited by R.C. Johnson. New York, Academic Press.

———. 1977. Tick-borne diseases in the United States: Rocky Mountain spotted fever and Colorado tick fever. Acta Trop., *34*:103–126.

———, Barbour, A.G., Hayes, S.F., Benach, J.L., Grunwaldt, E., and Davis, J.P. 1982. Lyme disease—a tick-borne spirochetosis? Science, *216*:1317–1319.

Burks, J.W., Jr., Jung, R., and George, W.M. 1956. Norwegian scabies. A.M.A. Arch. Dermatol., *74*:131–140.

Busvine, J.R. 1980. *Insects and Hygiene.* London, Chapman and Hall.

Cagnard, V., Nicolas-Randegger, J., Akribi, A.D., Rain, B., Nozais, J.P., Nomel, P.E., Ette, M., Doucet, J., and N'Dri, G.A. 1979. Pentastomose généralisée et mortelle a *Armillifera grandis* (Hett, 1915). Bull. Soc. Pathol. Exot., *72*:345–352.

Campbell, R.W., Abeywickrema, P., and Fenton, C. 1979. Queensland tick typhus in Sydney: A new endemic focus. Med. J. Aust., *1*:350.

Cannon, D.A. 1942. Linguatulid infestation of man. Ann. Trop. Med. Parasitol., *36*:160–166.

Casals, J., and Tignor, G.H. 1980. The *Nairovirus* genus: Serological relationships. Intervirology, *14*:144–147.

Chapel, T.A., Krugel, L., Chapel, J.,and Segal, A. 1981. Scabies presenting as urticaria. J.A.M.A., *246*:1440–1441.

Cohen, S.R. 1980. *Cheyletiella* dermatitis. Arch. Dermatol., *116*:435–437.

Darrow, D.I. 1977. Laboratory evaluation of twelve selected insecticides against the scorpion *Centruroides vittatus* (Arachnida: Scorpionida). J. Med. Entomol., *14*:393–394.

Desch, C., and Nutting, W.B. 1972. *Demodex folliculorum* (Simon) and *D. brevis* Akbulatova of man: Redescription and reevaluation. J. Parasitol., *58*:169–177.

Dönges, J. 1966. Parasitäre Abdominalcysten bei Nigerianern. Tropenmed. Parasitol., *16*:252–256.

Dove, W.E., and Shelmire, B. 1932. Some observations on tropical rat mites and endemic typhus. J. Parasitol., *18*:159–168.

Dryer, R.F., Goeliner, P.G., and Carney, A.S. 1979. Lyme arthritis in Wisconsin. J.A.M.A., *241*:498–499.

Eisner, T., Alsop, D., Hicks, K., and Meinwald, J. 1978. Defensive secretions of millipeds, pp. 41–72. In *Arthropod Venoms*. Edited by S. Bettini. New York, Academic Press.

Ennik, F. 1972. A short review of scorpion biology, management of stings and control. Calif. Vector Views, *19*:69–80.

Fain, A. 1966. Nouvelle description de *Dermatophagoides pteronyssinus* (Trouessart, 1897). Importance de cet acarien en pathologie humaine (Psoroptidae: Sarcoptiformes). Acarologia, *8*:302–327.

———. 1968. Etude de la variabilité de *Sarcoptes scabiei* avec une révision des Sarcoptidae. Acta Zool. Pathol. Antverp., *47*:1–196.

———. 1975. The Pentastomida parasitic in man. Ann. Soc. Belge Med. Trop., *55*:59–64.

———. 1978. Epidemiological problems of scabies. Dermatology, *17*:20–30.

———, and Salvo, G. 1966. Pentastomose humaine produite par des nymphes d'*Armillifer grandis* (Hett) en République Démocratique du Congo. Ann. Soc. Belge Med. Trop., *46*:675–682.

FAO/WHO Expert Committee on Parasitic Zoonoses. 1979. Geneva, WHO Tech. Rep. Ser. No. 637.

Foil, L.D., and Norment, B.R. 1979. Envenomation by *Loxosceles reclusa*. J. Med. Entomol., *16*:18–25.

Fuller, H.S. 1954. Studies of rickettsialpox. III. Life cycle of the mite vector, *Allodermanyssus sanguineus*. Am. J. Hyg., *59*:236–239.

Garnham, P.C.C. 1980. Human babesiosis: European aspects. Trans. R. Soc. Trop. Med. Hyg., *74*:153–155.

Garzia, G. 1938. Un caso di parassitismo accidentale da *Himanterium gabrielis* (L.) nell'intestino di un bambino. Riv. Parassitol., *2*:305–313.

Gertsch, W.J. 1979. *American Spiders*. New York, Van Nostrand Reinhold Co.

Goff, M. L., Loomis, R., Welbourn, W., and Wrenn, W. 1982. A glossary of chigger terminology (Acari: Trombiculidae). J. Med. Entomol., *19*:221–238.

Goldfarb, L.G., Chumakov, M.P., Myskin, A.A., Kondratenko, V.F., and Reznikova, O.Y. 1980. An epidemiological model of Crimean hemorrhagic fever. Am. J. Trop. Med. Hyg., *29*:260–264.

Gothe, R., Kunze, K., and Hoogstraal, H. 1979. The mechanisms of pathogenicity in the tick paralyses. J. Med. Entomol., 16:357–369.

Gratama, S., and von Thiel, P.H. 1958. Ocular *Armillifera armillatus* in man. Doc. Med. Geogr. Trop., *9*:374–376.

Gray, M.R., and Sutherland, S.K. 1978. Venoms of Dipluridae, pp. 121–148. In *Arthropod Venoms*. Edited by S. Bettini. New York, Academic Press.

Harwood, R.F., and James, M.T. 1979. *Entomology in Human and Animal Health*. New York, Macmillan Publishing Co.

Hase, T., Roberts, L.W., Hildebrandt, P.K., and Cavanaugh, D.C. 1978. Stylostome formation by *Leptotrombidium* mites (Acari: Trombiculidae). J. Parasitol., *64*:712–718.

Hoare, C.A. 1980. Comparative aspects of human babesiosis. Trans. R. Soc. Trop. Med. Hyg., *74*:143–148.

Hoogstraal, H. 1973. Acarina (ticks), pp. 89–103. In *Viruses and Invertebrates*. Edited by A.J. Gibbs. New York, American Elsevier Publishing Co.

———. 1979a. Ticks and spirochetes. Acta Trop., *36*:133–136.

———. 1979b. The epidemiology of tick-borne Crimean-Congo hemorrhagic fever in Asia, Europe, and Africa. J. Med. Entomol., *15*:307–417.

———. 1981. Changing patterns of tickborne diseases in modern society. Annu. Rev. Entomol., *26*:75–99.

Hopla, C.E. 1974. The ecology of tularemia. Adv. Vet. Sci. Comp. Med., *18*:25–53.

Huebner, R.J., Jellison, W.L., and Pomerantz, C. 1946. Rickettsialpox—a newly recognized rickettsial disease. IV. Isolation of a rickettsia apparently identical with the causative agent of rickettsialpox from *Allodermanyssus sanguineus*, a rodent mite. Public Health Rep., *61*:1677–1682.

Hughes, L.E., Casper, E.A., and Clifford, E.M. 1974. Persistence of Colorado tick fever virus in red blood cells. Am. J. Trop. Med. Hyg., *23*:530–532.

Hunter, W.S., and Higgins, R.P. 1960. An unusual case of human porocephaliasis. J. Parasitol., *46*:68.

Ingram, C.G., Jeffrey, I.G., Symington, I.S., and Cuthbert, O.D. 1979. Bronchial provocation studies in farmers allergic to storage mites. Lancet, *2*:1330–1332.

Johnson, R.C. 1977. The spirochetes. Annu. Rev. Microbiol., *31*:89–106.

Johnston, D.E., and Wacker, R.R. 1967. Observations on postembryonic development in *Eutrombicula splendens* (Acari-Acariformes). J. Med. Entomol., *4*:306–310.

Keegan, H.L. 1980. *Scorpions of Medical Importance*. Jackson, University Press of Mississippi.

Keirans, J.E., and Clifford, C.M. 1978. The genus *Ixodes* in the United States: A scanning electron microscope study and key to the adults. J. Med. Entomol., Suppl. 2.

Kirby-Smith, H.T. 1945. Specific treatment of black widow spider bite. South. Med. J., *38*:696–698.

Krantz, G.W. 1978. *A Manual of Acarology*, 2nd ed. Corvallis, Oregon State Book Stores.

Levi, H.W. 1959. The spider genus *Latrodectus* (Araneae, Theridiidae). Trans. Am. Microsc. Soc., *78*:7–43.

Lewis, J.G.E. 1981. *The Biology of Centipedes*. Cambridge, Cambridge University Press.

Maretic, Z. 1978. Venoms of Theridiidae, genus *Steatoda*, pp. 213–216. In *Arthropod Venoms*. Edited by S. Bettini. New York, Academic Press.

———, and Lebez, D. 1979. *Araneism*. Belgrade, Nolit Publishing House.

Marmion, B.P. 1954. Q fever: Natural history and epidemiology of Q fever in man. Trans. R. Soc. Trop. Med. Hyg., *48*:197–207.

Mast, W.E., and Burrows, W.M., Jr. 1976. Erythema chronicum migrans and "Lyme arthritis." J.A.M.A., *236*:2392.

Matthews, R.E.F. 1982. Classification and nomenclature of viruses. Intervirology, *17*:1–199.

McCrone, J.D., and Levi, H.W. 1964. North American widow spiders of the *Latrodectus curacaviensis* group (Araneae: Theridiidae). Psyche, *71*:12–27.

Minelli, A. 1978. Secretions of centipedes, pp. 73–86. In *Arthropod Venoms*. Edited by S. Bettini. New York, Academic Press.

Miskjian, H.G. 1951. Demodicidosis (*Demodex* infestation of the scalp). Arch. Derm. Syphilol., *63*:282–283.

Modell, W. 1963. Hazards of new drugs. Science, *139*:1180–1185.

Moser, J.C. 1975. Biosystematics of the straw itch mite with special reference to nomenclature and dermatology. Trans. R. Entomol. Soc. London, *127*:185–191.

Muller, R. 1979. Guinea worm disease: Epidemiology, control, and treatment. Bull. W.H.O., *57*:683–689.

Murnaghan, M.F., and O'Rourke, F.J. 1978. Tick paralysis, pp. 419–464. In *Arthropod Venoms*. Edited by S. Bettini. New York, Academic Press.

Murray, R.E. 1946. *Spiders, Ticks and Mites, Including Species Harmful to Man in Australia and New Guinea*. Australia, Serv. Publ. Sch. Public Health and Trop. Med., No. 6.

Nadchatram, M., and Dohany, A.L. 1974. A pictorial key to the subfamilies, genera and subgenera of Southeast Asian chiggers (Acari, Prostigmata, Trombiculidae). Bull. Inst. Med. Res., Malaysia, No. *16*:1–67.

———, Goff, M.L., and Thanalukshumi, P. 1980. The genus *Schoengastia* (Acari: Trombiculidae) in the Asiatic-Pacific region. J. Med. Entomol., *17*:268–281.

Nutting, W.B. 1975. Pathogenesis associated with hair follicle mites (Acari: Demodicidae). Acarologia, *17*:493–507.

Obenchain, F.D., and Galun, R., editors. 1982. *Physiology of Ticks*. New York, Pergamon Press.

Oliver, J.R., Jr. 1974. Symposium on reproduction of arthropods of medical and veterinary importance. IV. Reproduction in ticks (Ixodoidea). J. Med. Entomol., *11*:26–34.

Parker, R.R. 1942. Rocky Mountain spotted fever: Results of fifteen years vaccination. Proc. 6th Pacific Sci. Congr., *5*:589–596.

Philip, C.B. 1958. Some epidemiological considerations in Rocky Mountain spotted fever. Public Health Rep., *74*:595–600.

———. 1961. Arthropod vectors in relation to the reservoir mechanism of microbial agents of animal diseases. Acta Trop., *18*:257–262.

Polderman, A.M., and Manschot, W.A. 1979. *Armillifer armillatus* located within the lacrimal caruncle. Acta Leiden., *47*:71–77.

Potter, J.M., and Northey, W.T. 1962. Immunological evaluation of scorpion venoms. Am. J. Trop.Med. Hyg., *11*:712–716.

Prathap, K., Lau, K.S., and Bolton, J.M. 1969. Pentastomiasis: A common finding at autopsy among Malaysian aborigines. Am. J. Trop. Med. Hyg., *18*:20–27.

Pratt, H.D. 1979. Spider identification and control. Pest Control, *47*:24–32.

Radford, A.J. 1975. Millipede burns in man. Trop. Geogr. Med., *27*:279–287.

Rehse-Kupper, B., Casals, J., Danielova, V., and Ackermann, R. 1979. Eyach virus: The first relative of Colorado tick fever isolated in Germany, pp. 233–238. In *Recent Advances in Acarology*, Vol. 2. Edited by J.G. Rodriguez. New York, Academic Press.

Reid, H.A. 1978. Bites and stings, pp. 239–246. In *Price's Textbook of the Practice of Medicine*. Oxford, Oxford University Press.

Rendtorff, R.C., DeWeese, M.W., and Murrah, W. 1962. The occurrence of *Linguatula serrata*, a pentastomid, within the human eye. Am. J. Trop. Med. Hyg., *11*:762–764.

Riley, J., Banaja, A.A., and James, J.L. 1978. The phylogenetic relationships of the Pentastomida: The case for their inclusion within the Crustacea. Int. J. Parasitol., *8*:245–254.

Ruebush, T.K. 1980. Human babesiosis in North America. Trans. R. Soc. Trop. Med. Hyg., *74*:149–152.

Russell, F.E., Waldron, W.G., and Madon, M.B. 1969. Bites by the brown spiders *Loxosceles unicolor* and *Loxosceles arizonica* in California and Arizona. Toxicon, *7*:109–117.

Schacher, J.F., Saab, S., Germanos, R., and Boustany, N. 1969. The aetiology of halzoun in Lebanon: Recovery of *Linguatula serrata* nymphs from two patients. Trans. R. Soc. Trop. Med. Hyg., *63*:854–858.

Schenberg, S., and Pereira Lima, F.A. 1978. Venoms of Ctenidae, pp. 217–246. In *Arthropod Venoms*. Edited by S. Bettini. New York, Academic Press.

Schenone, H. 1959. Aspectos prácticos en la clínica del síndrome del latrodectismo y su tratamiento con Neostigmina (Prostigmina). Bol. Chil. Parasitol., *14*:80–82.

———, and Suarez, G. 1978. Venoms of Scytodidae. Genus *Loxosceles*, pp. 247–278. In *Arthropod Venoms*. Edited by S. Bettini. New York, Academic Press.

Self, J.T., Hopps, H.C., and Williams, A.O. 1975. Pentastomiasis in Africans. Trop. Geogr. Med., *27*:1–13.

Simpson, D.I.H. 1978. Viral hemorrhagic fevers of man. Bull. W.H.O., *56*:819–832.

Smith, K.G.V., editor. 1973. *Insects and Other Arthropods of Medical Importance*. London, British Museum (Natural History).

Sonenshine, D.E. 1974. Vector population dynamics in relation to tick-borne arboviruses: A review. Phytopathology, *64*:1060–1071.

Spielman, A., Clifford, C.M., Piesman, J., and Corwin, M.D. 1979. Human babesiosis on Nantucket Island, USA: Description of the vector, *Ixodes (Ixodes) dammini* n. sp. (Acarina: Ixodidae). J. Med. Entomol., *15*:218–234.

Stark, H.E. 1979. Review of tularemia in Utah and the Great Basin. Great Basin Nat., *39*:103–121.

Steere, A.C., Batsford, W.P., Weinberg, M., Alexander, J., Berger, H.J., Wolfson, S., and Malawista, S.E. 1980a. Lyme carditis: Cardiac abnormalities of Lyme disease. Ann. Intern. Med., *93*:8–16.

———, Malawista, S.E., Hardin, J.A., Ruddy, S., Askenase, P.W., and Andiman, W.A. 1977. Erythema chronicum migrans and Lyme arthritis: The enlarging clinical spectrum. Ann. Intern. Med., *86*:730–733.

———, Malawista, S.E., Newman, J.H., Spieler, P.N., and Bartenhagen, N.H. 1980b. Antibiotic therapy in Lyme disease. Ann. Intern. Med., *93*:1–7.

Traub, R., and Wisseman, C.L., Jr. 1974. The ecology of chigger-borne rickettsiosis (scrub typhus). J. Med. Entomol., *11*:237–303.

———, Jones, M.R., and O'Keefe, J.J. 1975. The acquisition of *Rickettsia tsutsugamushi* by chiggers (trombiculid mites) during the feeding process. Ann. N.Y. Acad. Sci., *266*:91–114.

Treat, A.E. 1975. Adult Prostigmata, pp. 239–270. In *Mites of Moths and Butterflies*. Ithaca, Cornell University Press.

United States Dept. of Agriculture, Handbook No. 485. 1976. *Ticks of Veterinary Importance*. Washington, D.C., Superintendent of Documents, U.S. Government Printing Office.

van Bronswijk, J.E.M.H., and deKreek, E.J. 1976. *Cheyletiella* (Acari: Cheyletiellidae) of dog, cat and domesticated rabbit, a review. J. Med. Entomol., *13*:315–327.

Vercammen-Grandjean, P.H. 1968. *The Chigger Mites of the Far East (Acarina: Trombiculidae and Leeuwenhoekiidae). An Illustrated Key and a Synopsis; Some New Tribes, Genera and Subgenera*. Washington, D.C., U.S. Army Med. Res. & Develop. Command.

———, and Langston, R.L. 1971. *The Chigger Mites of the World*. Volume 8. *Guntherana Complex. Section A, Genus* Guntherana. San Francisco, George Williams Hooper Foundation, University of California.

———. 1976. *The Chigger Mites of the World.* Vol. 3. *Leptotrombidium Complex. (Sections A, B, C.)* San Francisco, George Williams Hooper Foundation, University of California.

Waterman, J.A. 1957. Some notes on scorpions: Poisoning in Trinidad. Carib. Med. J., *19*:113–128.

Weiss, E. 1973. Growth and physiology of rickettsiae. Bacteriol. Rev., *37*:259–283.

Weyer, F. 1978. Progresses in ecology and epidemiology of rickettsioses. Acta Trop., *35*:5–21.

Wharton, G.W. 1976. House dust mites. J. Med. Entomol., *12*:577–621.

Whittemore, F.W., Jr., Keegan, H.L., and Borowitz, J.L. 1961. Studies of scorpion antivenin. 1. Paraspecificity. Bull. W.H.O., *25*:185–188.

Wikel, S.K. 1982. Immune responses to arthropods and their products. Annu. Rev. Entomol., *27*:21–48.

Youshock, E., and Glazer, S.D. 1981. Norwegian scabies in a renal transplant patient. J.A.M.A., *246*:2608–2609.

Chapter 37

Insects of Medical Importance: General Considerations

Insects are arthropods that, as adults, typically have three pairs of legs on the thorax, and for this reason have frequently been referred to as *hexapods*. They are the most numerous of all the arthropods and comprise about 70% of all the known species of animals. Insects may have arisen from trilobites, an extinct group of the Phylum Arthropoda; a more plausible theory associates primitive insects with millipedes and centipedes. These several classes of arthropods probably arose from a common parent stock of many-legged, wingless animals.

The Primitive Insect. Evidence from various fields (morphological, embryological, phylogenetic, and paleontological) indicates that the primitive or ancestral insect (Fig. 37–1) probably had the following characteristics: The head consisted of six fused segments, of which the second, fourth, fifth, and sixth bore appendages in the adult stage (i.e., antennae, mandibles, maxillae, and lower lip, or labium). In addition, there were two unpaired, nonappendicular structures: the labrum or upper lip, and a median tongue-like hypopharynx. The thorax consisted of three distinct segments, each with one pair of ambulatory appendages. The thoracic segments later developed wings from their dorsolateral angles. Those on the prothorax became vestigial and later disappeared. The abdomen consisted of 11 segments (11 regular segments and a terminal telson, which rarely persists as a discrete region), each with a pair of appendages. Upon the eleventh segment was a pair of pointed caudal organs: the cerci. The digestive system was made up of a foregut (with buccal cavity, pharynx, esophagus, and crop), a sacculate midgut or "stomach," and a hindgut emptying through an anus. Paired salivary glands, lying in the thorax, had ducts that opened into the buccal cavity. Malpighian tubules (1 to 3 pairs) emptied into the anterior end of the hindgut. Respiration was carried on by 10 pairs of metamerically repeated tracheal trees, each pair opening through spiracles on the lateral aspects of the ventral side of the second and third thoracic segments and the first eight abdominal segments. The organisms were sexually differentiated. The male had a pair of testes, each testis with its vas deferens and accessory glands, and a common ejaculatory duct opening through the genital pore; the female had a pair of ovaries, each with its oviduct and accessory glands, a single vagina, and single spermatheca. Metamorphosis from hatched egg to adult was lacking *(ametabolous)* or was simple.

Insect Specializations Related to Human Disease. From the more primitive general type, extensive and various modifications in form, mode of development, the character of the mouthparts, and nutritional requirements have evolved in the several orders and lesser groups of the class. With few exceptions, the forms of greatest medical importance belong to the more specialized, more highly developed groups. The primitive mouthparts were adapted to grasping and chewing. From this basic mandibulate stock evolved insects having *haustellate* or sucking mouthparts. This general pattern of development led to orders whose nymphs and adults have piercing-sucking mouthparts (bugs, lice) or whose larvae are mandibulate and adults are haustellate (many flies, fleas, butterflies, and moths). Within the Diptera (true flies), the variation in haustellate mouthparts is considerable, ranging from the feeding fascicle of mosquitoes, composed of six stylets, to the sponging muscoid type

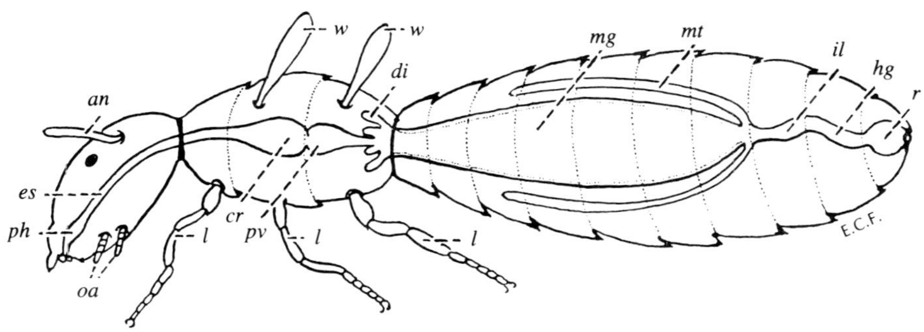

Fig. 37–1. Schematic representation of a primitive winged insect, viewed from the left side. *an,* Antenna. *cr,* Crop. *di,* Food diverticulum. *es,* Esophagus. *hg,* Hindgut. *il,* Ileum. *l,* Thoracic legs. *mg,* Midgut. *mt,* Malpighian tubule. *oa,* Oral appendages. *ph,* Pharynx. *pv,* Proventriculus. *r,* Rectum. *w,* Wings. (Adapted by E.C. Faust.)

typical of house flies. In medically important species of nematocerous and brachycerous Diptera, males have reduced mouthparts and confine their feeding to nectars and other floral secretions.

There are two distinct modes, based on the general type of haustellate mouthparts, in which blood is obtained. When blood is withdrawn directly from a capillary or small vein, the feeding mode is referred to as *solenophagous;* when blood is acquired from a pool formed by the slashing movements of the mouthparts within the capillary bed, the term *telmophagous* is used (Lavoipierre, 1965). These methods of blood acquisition must be considered when evaluating the vector competence of an insect, in terms of both acquiring and transmitting pathogens. The role of blood digestion is also of epidemiological consequence. As noted by Richards (1975), the gut of hematophagous insects shows ultrastructural diversity from one group to another, and may contain a series of membranes that act as barriers to the movement of a pathogen across it. In addition, bloodfeeding insects produce an array of digestive enzymes (Gooding, 1975) that also may inactivate pathogens.

The type of metamorphosis exhibited by medically important insects can be significant in assessing the pathogen transmission potential of a population. This is particularly true where the vector species is *hemimetabolous,* i.e., when it exhibits a simple pattern of metamorphosis having stages consisting of the egg, nymph (with several instars), and adult. A typical example can be seen in the Triatominae (''cone-nosed'' bugs), some species of which are intermediate hosts of *Trypanosoma cruzi,* the causative agent of Chagas' disease. All of the nymphs and both adult sexes are hematophagous and are largely lair ectoparasites. *Holometabolous* orders have a more complicated developmental pattern, consisting of the egg, larva (with several instars), pupa, and adult. The larval and adult stages are active but usually occur in completely different habitats. Some of the groups of medical importance, particularly the flies, have become adapted to an aquatic or moist environment during their larval stage, and a knowledge of these developmental sites constitutes an essential background for any measure designed to control these organisms.

The physiological-biochemical processes of insects associated with maintenance, reproduction, and growth require the usual carbohydrates, lipids, and proteins (Rockstein, 1978). Convergence on blood feeding and the utilization of various components in the blood meal are worthy of mention. As reviewed by Friend and Smith (1977), feeding by hematophagous insects is the culmination of a number of interrelated physical and chemical factors. Following host-seeking behavior, which is cued by various stimuli received by specialized chemical and mechanical receptors, the following discrete events can be detected: probing, skin penetration, location of blood, engorgement, and cessation of feeding. Many hematophagous Diptera convert blood proteins to vitellogenins (yolk proteins) for uptake by the ovary. The site of protein synthesis is the abdominal fat body, an organ equivalent in metabolic function to the vertebrate liver. Proline, a heterocyclic amino acid, is removed from the blood meal and used as an energy source by species of *Glossina* (tsetse flies). Insects deaminize catabolic products of protein metabolism with uric acid as the end product, which collects in the malpighian tubules, is concentrated into a pasty substance, and is discharged into the hind gut for evacuation with the digestive wastes. In addition, insects have special and unique problems during growth, associated with the periodic shedding and synthesis of a new exoskeleton, an activity mediated largely by hormonal interplay. The insect cuticle, which forms the exoskeleton, is laminated and constructed by the sequential secretions of epidermal cells. The cuticle also lines the foregut, hindgut, and tracheal (respiratory) system.

Insect sense organs exhibit a great diversity for reception of external and internal stimuli. As described by Richards and Davies (1977), sensory perception can occur by the stimulation of mechanoreceptors, auditory organs (specialized mechanoreceptors), chemoreceptors, visual organs, and temperature and humidity receptors. Each sense organ or *sensillum* is composed of a cuticular or epidermal structure that transduces the stimulus so that one (or more) sensory neuron (primary sense cell) is activated.

The sensory hairs of mechanoreceptors are widely distributed over the insect body and are especially prominent on the antennae, tarsi, and cerci. The more typical auditory receptors include the tympanal organ of the abdomen, foretibia, or antennae (Johnston's organ). Chemoreceptor sensilla are abundant on various parts of the body, including the labella and tarsi, ovipositor, antennae, and maxillary palps. These sensilla vary in the thickness of the sensillar cuticle as well as in general shape, including organs that are hair-like (sensilla trichoidea), peg-like (sensilla basiconica; sensilla coeloconica), cone-like (sensilla styloconica), or plate-like (sensilla placodea). Temperature and humidity receptors occur on the antennae and maxillary palps of some hematophagous Diptera (for instance, *Aedes aegypti)* where they aid in orientation to a host. Visual organs in the Insecta consist chiefly of compound eyes and more simple dorsal and lateral ocelli (stemmata).

By virtue of these varied and specialized organs, insects can exhibit many patterns of behavior. Responses may be directed toward or away from stimuli in a simple linear fashion, as in the orientation to light (Borror et al., 1981). However, activities such as mating, feeding, oviposition, and selection of habitat involve a series of different acts that are rhythmic and often motivated by physiological state. Such rhythms are usually directed by natural light regimens so that greatest activity occurs during diurnal, nocturnal, or crepuscular periods. This circadian rhythm is important epidemiologically when vector populations are assessed, particularly in filariasis and malaria field work.

Control of Insects. Within recent years much thought and effort have been given to the control of insects of economic and medical importance. In the field of insect toxicology, beginning with DDT and other chlorinated hydrocarbons, remarkable progress in insect control has been achieved. Prolonged use of insecticides has resulted, however, in resistance to these toxicants—either true physiological resistance due to survival of strains capable of detoxifying the chemical agent, or behavioristic avoidance of the agent by the insect. This has been particularly notable in campaigns to control filth flies, some of the important mosquitoes, body lice, and fleas. This aspect of the problem of arthropod-borne agents of human disease is considered in Chapter 45, p. 718.

Classification of Insects. The classification of insects is outlined in Chapter 35, p. 565.

GENERAL REFERENCES

Borror, D.J., DeLong, D.M., and Triplehorn, C.A. 1981. *An Introduction to the Study of Insects,* 5th ed. Philadelphia, W.B. Saunders.

Busvine, J.R. 1980. *Insects and Hygiene,* 3rd ed. London, Chapman and Hall.

Chapman, R.F. 1976. *The Insects. Structure and Function,* 2nd ed. New York, Elsevier-North Holland Publishing Co.

Friend, W.G., and Smith, J.J.B. 1977. Factors affecting feeding by bloodsucking insects. Annu. Rev. Entomol., *22*:309–332.

Gooding, R.H. 1975. Digestive enzymes and their control in haematophagous arthropods. Acta Trop., *32*:96–111.

Harwood, R.F., and James, M.T. 1979. *Entomology in Human and Animal Health,* 7th ed. New York, Macmillan Publishing Co.

Lavoipierre, M.M.J. 1965. Feeding mechanism of bloodsucking arthropods. Nature, *208*:302–303.

Richards, A.G. 1975. The ultrastructure of the midgut of hematophagous insects. Acta Trop., *32*:83–95.

Richards, O.W., and Davies, R.G. 1977. *Imm's General Textbook of Entomology,* 10th ed. *Vol. 1. Structure, Physiology and Development.* London, Chapman and Hall.

Rockstein, M., editor. 1978. *Biochemistry of Insects.* New York, Academic Press.

Wilkinson, C.F., editor. 1976. *Insecticide Biochemistry and Physiology.* New York, Plenum Press.

Chapter 38

Biting Lice, Sucking Lice, and Cockroaches

Lice are relatively small, wingless insects. They are distinctly flattened dorsoventrally, have easily recognized 3- to 5-jointed antennae, and have three pairs of large conspicuous legs, each ending in a tarsal claw.

The order Mallophaga (literally, insects that are "wool-eaters") are biting lice with strongly chitinized, toothed, mandibulate jaws, a head broader than the thorax and, primitively, three distinct thoracic segments. Their thoracic spiracular openings are *ventrally* situated, while their first pair of abdominal spiracles opens on the second visible abdominal segment. They are ectoparasites of birds and mammals, feeding on feathers, hairs, and epidermal scales. No species lives on man.

The order Anoplura (literally, insects that have an "unarmed tail") are sucking lice (sometimes referred to as the Siphunculata) that have mouthparts adapted to sucking their host's blood and cutaneous juices, a head narrower than the thorax, and indistinct external thoracic segmentation. Their thoracic spiracular openings are *dorsally* situated, while their first pair of abdominal spiracles opens on the third visible abdominal segment. They are exclusively ectoparasites of mammals.

Both groups of lice exhibit incomplete metamorphosis. They lay operculate eggs that are cemented to hairs, feathers, or clothing. They normally spend their entire lives on their hosts. Some workers regard these two groups as subdivisions of the same order, but their differences appear to be sufficient to warrant separation into distinct orders.

THE BITING LICE (MALLOPHAGA)

The Mallophaga are of considerable economic importance, sometimes seriously infesting domestic mammals and birds. No species is of primary importance in human medicine, although the biting louse of the dog, *Trichodectes canis,* is an intermediate host of *Dipylidium caninum* (see Section II, p. 508), and could conceivably transmit the infection to humans.

THE SUCKING LICE (ANOPLURA)

Morphology. Sucking lice (Fig. 38–1) have legs provided with claws adapted for clinging to hairs or fibers, and spiracular openings in the dorsolateral position. Except in one family group, the head is rectangular-ovoid from the dorsal aspect. In species found on humans and monkeys, there is a pair of distinct eyespots, anterior to which is a pair of 3- to 5-jointed antennae. Between the eyes and the antennae is the *frons,* anterior to which is the *cly-*

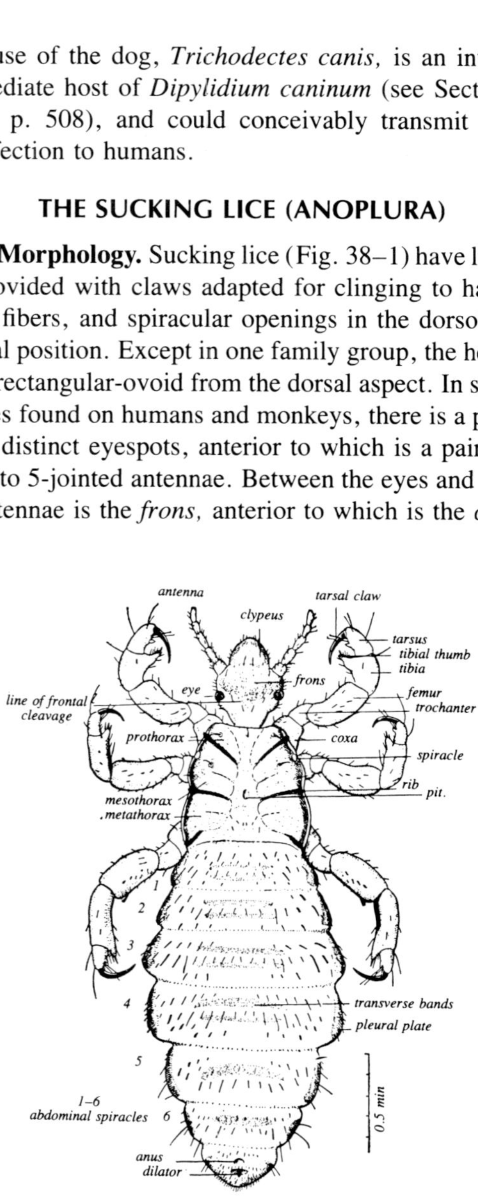

Fig. 38–1. *Pediculus humanus.* Male, dorsal view. (Adapted from G.H.F. Nuttall. *In* Matheson, R. 1932. *Medical Entomology.* Springfield, Ill., Charles C Thomas.)

peus. The *genae* (cheeks) are lateral to the clypeus. The distal prolongation of the clypeus constitutes the *labrum*, or upper lip, which is armed internally with six pairs of minute denticles, the *prestomal teeth*, which are everted when the louse feeds and serve to anchor its mouth to the skin of the host. The lightly chitinized food channel *(buccal funnel)* leads inward to the pharynx, and to the ventrally situated *stylet sac*, which contains the piercing or stabbing apparatus. This apparatus consists of a dorsal pair of closely opposed stylets *(maxillae)*, a median slender tubular stylet (the *hypopharynx*) through which the salivary duct passes, and a ventral, fused, gutter-like structure, the *labium*, which supports the other piercing parts. When taking a blood meal, the louse first inserts its prestomal teeth into the epidermis, forces the stylets of the stabbing apparatus into the skin, deposits droplets of saliva in the wound, and with its pharyngeal pump draws blood or tissue juice up through the buccal funnel into the pharynx. Lavoipierre (1965) demonstrated that the louse is a vessel feeder *(solenophage)* and that this method is similar to the feeding mechanism of a triatomid bug.

The three *thoracic segments* are externally indistinct; their dorsal and ventral surfaces are poorly chitinized, but the lateral (pleural) plates are well developed. Each of the six legs consists of the five segments characteristic of insects. The terminal tarsus is unjointed and is provided with a conspicuous, articulated, hooked claw that is apposed to a thumb-like extension of the tibia (fourth segment), enabling the louse to grasp hairs or fibers readily.

The *abdomen* is 9-segmented, but in some species the anterior segments have become fused. The pleural plates are more highly developed than the tergal and sternal ones. The abdomen of the male is narrower than that of the female and is rounded posteriorly. In the female, the terminal segment is inverted V-shaped, and on its ventral side it has a pair of blunt processes *(gonopods)* that grasp a hair or fiber onto which an egg is later deposited and glued. The *spiracles* consist of one prothoracic pair that opens dorsolaterally between the first and second pairs of legs, and six abdominal pairs that open on the pleural plates.

The *digestive tract* internal to the buccal funnel (Fig. 38–2) is made up successively of a two-chambered muscular *pharynx (ph)* with an intermediate and a posterior valve; a long capillary *esophagus (es)*; a *midgut (mg)* that is dilated anteriorly and constricted posteriorly; a *hindgut;* and

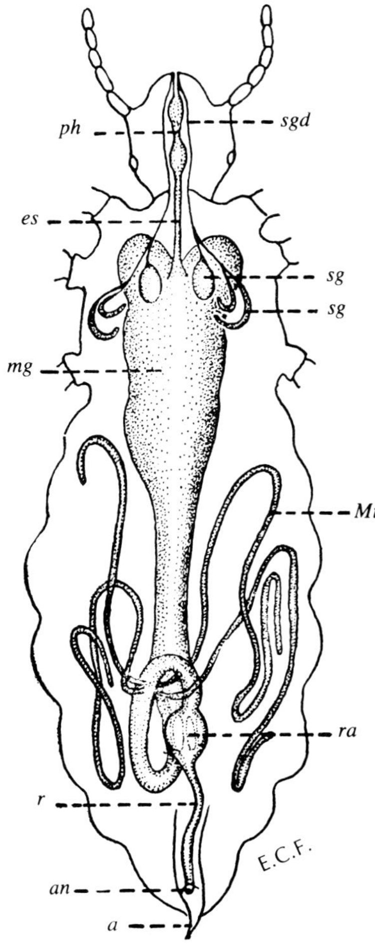

Fig. 38–2. *Pediculus humanus*, male. Internal anatomy. *a*, Aedeagus. *an*, Anus. *es*, Esophagus. *mg*, Midgut. *Mt*, Malpighian tubule. *ph*, Pharynx. *r*, Rectum. *ra*, Rectal ampulla. *sg*, Salivary glands. *sgd*, Salivary gland duct. (Adapted from G.H.F. Nuttall. *In* Matheson, R. 1932. *Medical Entomology*. Springfield, Ill., Charles C Thomas.)

a *rectum (r)* with a median ampullar distention *(ra)*. The rectum loops forward under the posterior end of the midgut and then proceeds backward to open through the *anus (an)*, in the median dorsal region of the eighth segment. There are four *malpighian tubules (Mt)*, which enter the intestine some little distance posterior to the junction of the mid- and hindgut. The four *salivary glands* consist of one pair of more medially disposed reniform bodies and one pair of bifurcated elongate tubules *(sg)*. They secrete an anticoagulin.

The *male genitalia* include a pair of twin-lobed *testes*, each with a delicate *vas deferens, accessory gland*, and dilated tubular *seminal vesicle;* an unpaired *ejaculatory duct;* and the terminal chitinized *aedeagus* with its *intromittent* or *penial organ.*

The *female genitalia* consists of a pair of digitate *ovaries* each with a short *oviduct,* and a short *common oviduct* that passes back to the genital opening situated between the gonopods. There is no seminal receptacle.

Development. Eggs are laid within a day or two after the female louse becomes sexually mature and is fertilized. Head lice deposit their eggs (Fig. 38–3,*A*) on the hairs of the head and neck, near the base of the hair shaft. Pubic lice (Fig. 38–3,*B*) commonly oviposit on the hairs of the pubic region, less commonly on those of the chest, axilla, eyebrows, or eyelashes. Body lice usually attach their eggs to the fibers of clothing, especially in the seams of woolen and quilted cotton garments, and occasionally on the hairs of the chest and axilla. These minute white eggs ("nits") are ovoidal, operculate, and are cemented to the hair or fiber, a few each day. At body temperature they hatch in 4 to 14 days. The first nymphal instar closely resembles the adult. They feed and molt in 2 to 4 days. The second and third nymphal instars follow, and the adults emerge from the last nymphal skins 24 to 27 days after the eggs have been laid. The adults live about 30 days, and each female may lay 50 to 300 eggs, depending on climate and species concerned.

Although lice are dependent on their hosts for food and warmth, they are active travelers and transfer from one host to another of the same species by the slightest bodily contact. Human body lice rapidly leave patients with high temperatures, while at low temperatures they become inactive or immobile. Head lice may survive for several days off their victims.

Both the human head louse and body louse feed frequently, discharging comparatively large pellets of dark red excreta during the process.

Types of Human Lice. There are three types of human lice: (1) head louse (*Pediculus humanus capitis* de Geer), (2) body louse (*P. humanus humanus* Linn.), and (3) pubic or "crab" louse (*Phthirus pubis* Linn.) (Fig. 38–4). The head louse and body louse are often considered unstable environmental subspecies of *P. humanus* Linn. (Clay, 1973), but evidence from double infestations suggests that gene exchange by populations of the two forms is limited. Each "species" is allopatric in distribution as well as ecologically separated by virtue of the different microenvironments it occupies (Schaeffer, 1978). *Phthirus pubis* differs not only in its characteristic location on the body, but is so distinctive in appearance that it can never be confused with species of *Pediculus* (compare Figs. 38–1 and 38–4). Pubic lice remain attached by their mouthparts to the same site for many days, sucking from time to time but apparently without becoming engorged. *Phthirus pubis* commonly occurs on the perianal, axillary, chest and body hair, eyelashes, and eyebrows. Rarely it is found on the scalp (Mueller, 1973).

Other sucking lice of potential medical importance are those parasitizing rodent hosts, including

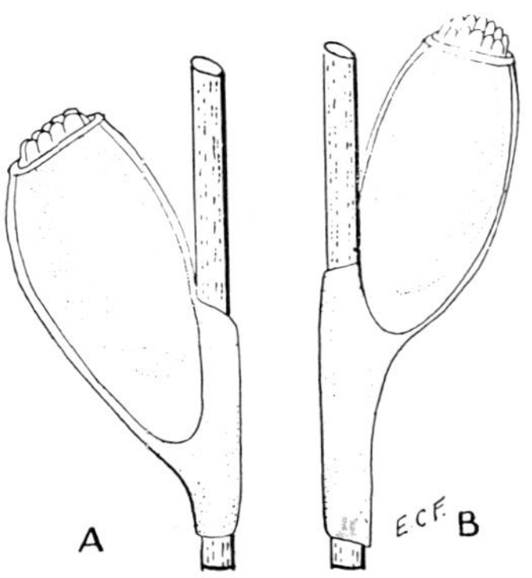

Fig. 38–3. Eggs of human lice. *A, Pediculus humanus* var, *capitis.* B, *Phthirus pubis.* Greatly enlarged. (*A* adapted by E.C. Faust from W.S. Patton and A.M. Evans. 1929. *Insects, Ticks, Mites and Venomous Animals.* London, H. Grubb Ltd. *B* adapted by E.C. Faust from R. Matheson. 1932. *Medical Entomology.* Springfield, Ill., Charles C Thomas.)

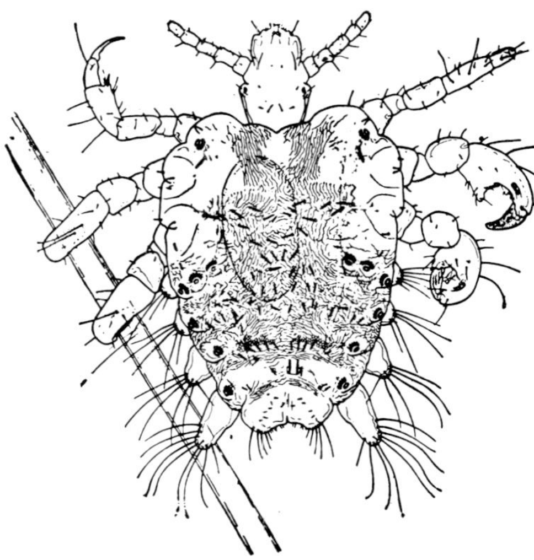

Fig. 38–4. *Phthirus pubis.* Adult attached to hair, dorsal view. (Adapted by E.C. Faust from Martini, E. 1923. *Textbook of Medical Entomology.* Jena, Gustav Fischer.)

Polyplax spinulosa, which transmits *Rickettsia typhi,* the causative agent of murine typhus, from rat to rat.

LICE AND HUMAN DISEASE

Lice play a twofold role in the production of disease in humans: (1) Body lice, and possibly head and pubic lice, transmit the causative agent of epidemic typhus fever; louse-borne relapsing fever is also associated with *P. humanus.* (2) Lice produce an irritating dermatitis *(pediculosis).*

Louse-Borne Typhus Fever

(Exanthematic typhus, classical European typhus, epidemic typhus, tabardillo of Mexico *pro parte)*

Definition and History. Louse-borne typhus is an acute, highly infectious, febrile, exanthematous disease that manifests itself in epidemic form. This disease took a tremendous toll of the native Amerind peoples of Mexico, Central America, and northern South America after their conquest by Columbus and his successors. It contributed to the devastating results of Napoleon's retreat from Moscow in 1812, and caused the death of millions of soldiers and civilians in Serbia, Poland, and the Ukraine during the 1914–1918 conflict. Before the Spanish Revolution it was an important cause of death in Morocco, and was widely distributed in Spain. During World War II it not only developed to epidemic proportions in Poland, the Ukraine, and the Balkans, but became widely disseminated throughout southern, central, and western Europe. In past decades, epidemic typhus was the most formidable of all rickettsial diseases. In famine-stricken populations and among soldiers and refugees it caused devastating epidemics due to communal infestation with the body louse, the efficient vector. Since 1945, epidemic typhus has become a less serious public health problem, although reservoirs remain in Asia (India, Pakistan), in some of the upland areas of Africa (Ethiopia, Rwanda, and Burundi), and in Latin America.

Brill's disease, first observed by Brill in New York in 1898, was found by Zinsser (1937) to be a mild recrudescent form of louse-borne typhus.

Etiology. The agent of louse-borne typhus is *Rickettsia prowazekii,* first isolated and described by da Rocha Lima in 1916. It is gram-negative, coccus-like or short rod-like, measures 300 to 500 mμ, and is an obligatory parasite of the cytoplasm of the host's cells, especially the endothelial lining of smaller blood vessels. Because of immunologic and epidemiologic differences between louse-transmitted and flea-transmitted typhus fevers, the agent of the former is referred to as *R. prowazekii* and of the latter as *R. typhi.*

Geographical Distribution. Louse-borne typhus fever is *cosmopolitan* in its distribution wherever groups of human beings are crowded together under conditions of poor sanitation and malnutrition. Historically, it has been especially *prevalent in winter,* when change of clothing and bathing were difficult to carry out in congested quarters. In the Western Hemisphere, louse-borne typhus occurs in parts of Mexico, Guatemala, and other countries of Central America, in the Pacific Coast countries of South America (Chile, Peru, Ecuador, Colombia), and is prevalent in Bolivia. In the Eastern Hemisphere, it has an extensive distribution in Ethiopia, Rwanda, Burundi, and Uganda. Endemic foci are also distributed irregularly in western Asia and eastern-southeastern Europe.

Epidemiology. Nicolle, in 1909, experimentally produced "classical" or "exanthematic" typhus in the monkey and demonstrated that it was transmitted by the human body louse. In 1911 da Rocha Lima and von Prowazek discovered minute pleomorphic bacteria-like bodies in the intestinal epithelium of infected lice. Louse-borne typhus fever is *epidemic* in its expression, and is transmitted from person to person by human lice, principally the body louse *Pediculus humanus humanus.* The head louse can become infected with *Rickettsia prowazekii* and disseminate this pathogen in its feces (Murray and Torrey, 1975). However, the roles of *Pediculus humanus capitis* and *Phthirus pubis* in the epidemiology of typhus appear to be limited (Weyer, 1952). When a human body louse feeds on a typhus-infected person toward the end of the incubation period or during the first few days of patent disease, the louse ingests the rickettsiae. In the stomach of the louse the rickettsiae enter the epithelial lining cells, where they multiply. About a week later, after tremendous swelling and then disintegration of these cells, the rickettsiae are discharged into the intestinal lumen of the louse, or, by rupture of the gut wall, may enter the hemocoelic cavity. Thereafter, they may be passed in the louse's feces, or may be rubbed into the conjunctivae or broken skin, or may contaminate the host's skin when the louse is crushed. Since human lice typically feed only on human beings, a closed cycle of human-louse-human is established. The louse's feces are infective from the second day after the louse has ingested an infected blood meal. Once rickettsiae are ingested, the louse remains infected for life. *Rickettsia prowazekii* is pathogenic for *Pediculus humanus humanus* as a result of the extensive destruction of the cells of the midgut (Shkolnik and Zatulovsky, 1971; Silverman et al., 1974). The organisms remain viable for months in dried louse feces, but congenital transmission of this rickettsia has not been demonstrated.

Human infection may be acquired from an infected louse by *depositing louse feces* on broken or, possibly, uninjured skin; *transferring infective dejecta* on soiled fingers to conjunctivae; *crushing the louse* on excoriated or, possibly, unbroken skin; contaminating mucous membranes with *clothing soiled with dried louse feces;* or *inhaling* air-borne rickettsiae.

Louse-borne typhus is epidemiologically and immunologically distinct from flea-borne typhus, and the two varieties of the disease are not reciprocally transmissible.

Clinical recovery from typhus does not necessarily mean that the rickettsiae have been eradicated. After years of quiescence, they may again become activated, producing a milder type of the disease (Brill's disease). Thus, humans may serve as carriers of *R. prowazekii* between epidemic outbreaks, and louse-infested inapparent cases in various parts of the world may be sources from which epidemics arise.

Epidemics of typhus are favored by cold weather, crowding, extreme droughts that restrict bathing and laundering, and by famine and malnutrition that reduce resistance to infection.

Recently there have been increasing reports of antibodies against *R. prowazekii* in domestic animals in various endemic countries (Weyer, 1978). The agent has been isolated from ixodid ticks and from the blood of goats and sheep in Ethiopia and donkeys in Egypt, suggesting that there is an extrahuman cycle of epidemic typhus in such animals. The isolation of six strains of a rickettsial agent identical to *R. prowazekii* was reported from eastern flying squirrels *(Glaucomys volans)* collected in Florida and Virginia, United States (Bozeman et al., 1975). Sonenshine et al. (1978) proposed a sylvan cycle of *R. prowazekii* during the winter months that involves flying squirrels and their ectoparasites, principally *Neohaematopinus sciuroptes* and *Orchopeas howardii,* a squirrel flea, as possible vectors. An epidemiological investigation following the detection of antibodies to *R. prowazekii* in the sera of patients from the eastern United States led McDade et al. (1980) to suggest that the histories of these patients were consistent with a rickettsial disease spread by a vector from an extrahuman source, probably gray or flying squirrels.

Clinical Aspects. Louse-borne or epidemic typhus is an acute exanthematous rickettsial disease caused by *Rickettsia prowazekii.* After an incubation period of 1 to 2 weeks, averaging 12 days, there is an onset, usually sudden but sometimes gradual, of chills, fever, headache, prostration, and generalized aches in joints and muscles. A macular rash appears on the fifth or sixth day of illness, at which time toxic manifestations are marked. After about 2 weeks of fever, the untreated disease terminates by rapid lysis. The fatality rate varies from 10 to 40% in untreated cases and increases with age. The illness may be mild, sometimes without a rash, especially in previously vaccinated persons. Brill-Zinsser disease, first observed by Brill in New York in 1898, was found by Zinsser (1937) to be a mild recrudescent form of louse-borne typhus. The diagnosis is by complement-fixation reaction with group-specific typhus antigen, which becomes positive in the second week of illness. The Weil-Felix reaction with Proteus OX-19 also becomes positive but is less specific. Treatment is with a tetracycline antibiotic or chloramphenicol. The use of sulfonamide drugs is contraindicated. With appropriate treatment prognosis is good.

Severity of the disease depends on the virulence of the particular strain of the rickettsia, as well as constitution, state of nutrition, and age of the patient. In epidemics the mortality has ranged from 10 to 80%. For individuals over 50 years of age the prognosis is usually grave; in children the disease is frequently mild and may be abortive. Once a person has recovered from typhus, he has considerable resistance to subsequent infection. If patients are diagnosed early in the disease and have the benefit of modern chemotherapy and good nursing care, the prognosis is good in 90 to 95% of the cases.

Prevention. Control measures consist of (1) isolation and delousing of infested persons and clothing, and (2) immunization.

Delousing. Clothing may be deloused by placing it in steam chambers or in heated ovens, both efficient techniques that require bulky apparatus and damage clothing; or clothing may be fumigated with ethyl formate (Busvine, 1980), an effective and practical chemical that, although slightly flammable, need not endanger human life if properly employed.

While an infested individual's clothing is being deloused, hot baths and treatment of louse bites with antiseptic lotions are indicated. Chemicals for louse disinfection are effective in killing lice (and usually their nits) but do not kill the typhus rickettsiae that may have been deposited on clothing in minute fecal dejecta of the louse. Heat will guarantee such sterilization of fomites. In the past, control of widespread infestation by *Pediculus humanus humanus* by applying chlorinated insecticidal dusts directly to the clothing and underwear has met with phenomenal success. However, owing to the lipid solubility of these compounds and their subsequent accumulation within the body, as well as the occurrence of resistance, other insecticides should be considered (*vide infra,* under "Prevention of Pediculosis," p. 617).

Immunization. Prophylactic protection of exposed persons by vaccination is highly satisfactory. Several methods for obtaining and preparing living or killed vaccine have been used. The living rickettsiae have potential danger because the recipient may develop a severe form of typhus fever. Fox et al. (1955) and Montoya et al. (1958) tested the immunizing properties of living avirulent *Rickettsia prowazekii* (strain E) on North American volunteers and in endemic populations in Peru. With rare exceptions, the inoculum was well tolerated and produced satisfactory immunity when challenged by virulent strains of the organism.

A practical technique for obtaining large quantities of killed rickettsiae is the chick-embryo tissue culture of Cox (1942). It provides a highly efficient vaccine when administered subcutaneously in 3 doses of 1 ml each at intervals of 7 to 10 days. In epidemic areas a single 1-ml dose is given every 6 months thereafter to reactivate the immune bodies.

Trench Fever

(Wolhynia fever, five-day fever)

Definition. Trench fever is an acute, infectious, febrile disease believed to be caused by the rickettsia *Rochalimaea quintana* (syn. *Rickettsia pediculi, R. wolhynica*).

Geographical Distribution. Trench fever occurred in epidemic form on all European battlefronts during 1914 to 1918. Except for possible sporadic cases reported from the western Ukraine, the disease then disappeared from medical observation and record until 1942 to 1943, when it again appeared among soldiers on the European military fronts. It now occurs in limited foci in Europe, Asia, Africa, and the highlands of Latin America.

Epidemiology. The disease occurs in louse-infested persons living in confined quarters. Lice become infected from feeding on infected people (from the first day of clinical manifestations for about 1 year). In 5 to 9 days the louse becomes infective; the fecal dejecta provides the inoculum. Unlike *R. prowazekii,* the rickettsiae are not pathogenic to the louse; instead, they live and multiply in the lumen of the midgut, and not intracellularly, as in *R. prowazekii* infection. Congenital transmission from louse to louse has not been demonstrated.

Clinical Aspects. Following an incubation period of 10 to 30 days, trench fever has a sudden onset, with severe headache, prostration, hyperesthesia of the skin, and pain in the back and lower extremities (hence the name "shin-bone fever"). Patients experience intermittent fever every 4 or 5 days (hence the designation "febris quintana") and, in the majority of cases, a roseate macular exanthema lasting from a few hours to a few days. The clinical picture is quite variable: the fever span and peak are not constant, no lasting immunity develops, and the rickettsemia may last for months or even years beyond the period of clinical manifestations. The rickettsiae in circulating blood may become so scarce at times, though, that lice feeding on the patient may not become infected, yet may develop infection at a subsequent feeding.

Diagnosis is made on the characteristic symptoms, the history of louse infestation in crowded quarters, and the epidemic appearance of the disease. Laboratory confirmation of trench fever may be made by feeding clean lice on patients, then transferring the infection to clean laboratory mice or monkeys and demonstrating rickettsemia without symptoms. The administration of chloramphenicol or, preferably, tetracycline drugs should promptly relieve the pain and cause reduction of the temperature to normal, as in typhus fever *(vide supra).* The disease is seldom fatal, but is incapacitating.

Prevention. *Delousing* (see pp. 615 and 617) is effective as a control measure.

Louse-Borne Relapsing Fever

Definition. Louse-borne relapsing fever is an acute, febrile, infectious disease produced by a spirochete, *Borrelia recurrentis.*

Geographical Distribution. Louse-borne relapsing fever exists sporadically in parts of South America, Europe, and Asia (Burgdorfer, 1976). It is endemic in Ethiopia and the Sudan (Sholdt et al., 1979).

Epidemiology. Louse-borne relapsing fever is transmitted from person to person by the body louse, *Pediculus humanus humanus,* and is *epidemic* in character (i.e., it has a human-louse-human cycle, as distinguished from tick-borne *enzootic-endemic* relapsing fever; see p. 589). Lice that feed on patients with relapsing fever of the louse-transmitted type, particularly during the febrile periods, take up the spirochetes from the peripheral blood. The organisms reaching the louse's intestine pass within 2 hours into the hemocoel and are distributed throughout the body of the louse where they multiply as spirochetes; after 6 days or less they are numerous in the hemolymph. Apparently they never return to the digestive tract, cannot be passed via the louse's salivary fluids or fecal dejecta, and infect humans only when the louse is crushed on the skin. Although the louse remains infective for life, the spirochetes are not vertically transmitted. Epidemics occur during winter months when lice populations burgeon due to poor personal hygiene. Overcrowding during wars or natural catastrophes also leads to lousiness and relapsing fever epidemics (Johnson, 1977).

Clinical Aspects. Louse-borne relapsing fever is clinically similar to the tick-borne disease.

Other Louse-Associated Pathogens

Lice have been experimentally incriminated in the mechanical transmission of a wide variety of bacteria (through their feces) and fungi that produce epidermophytosis. However, with the possible exception of *Salmonella* spp., the natural transmission of these agents appears extremely unlikely.

Pediculosis

Definition. Infestation with lice is *pediculosis*. It occurs most commonly among prisoners in jails, groups of soldiers under wartime conditions, and beggars. However, it may occur in other population groups in whom bathing and change of clothing are infrequent, or where persons of clean habits come in contact with either pediculous individuals or their clothing, headgear, or bedding. Eggs or motile stages of the lice may even be present in or on public conveyances. Human body lice most frequently occur in cool to cold climates where heavy clothing is worn and bathing is infrequent, but this is not always the case, as for example among Bedouins and others in North Africa.

Body lice produce distinctive cutaneous lesions on any part of the body where the nymphs or adults feed. *Head lice* are often located at the back of the neck, although they are frequently scattered over the scalp, more commonly in children than adults. Saliva produced in the salivary glands of the louse, when introduced into the puncture wound at each feeding site, produces a roseate elevated papule, accompanied by intense pruritus, invariably causing a desire to scratch the affected site. Frequently, the skin becomes scarred with parallel stripes, indurated, and bronzed by pigment (i.e., *vagabond's disease)*. The saliva and fecal dejecta of the louse induce a dermal hypersensitivity consisting of a purpuric element due to the act of feeding, and the development of an inflammatory reaction following sensitization. If the lesions become secondarily infected by bacteria and assume an eczematous condition, they may be confused with sarcoptic mange. The irritation may be somewhat reduced by application of a soothing lotion, but no more than temporary relief can be afforded unless the eggs and motile stages are killed.

Prevention. The prevention of pediculosis and of louse-borne diseases resolves itself into two lines of attack: (1) *treatment of individuals* or small groups and (2) *mass control*. Hardly a village or schoolhouse exists in any part of the world where someone has not been infested with head lice. Body lice are more closely associated with poverty and filth, particularly in cold climates. Pubic lice are commonly acquired in jails, brothels, and other similar institutions, but may also be transferred from person to person in dormitories, locker rooms of clubs, and from mats on gymnasium floors. Most pubic lice infestations result from sexual intercourse.

For *head lice (Pediculus humanus capitis),* the infested scalp is treated by use of a shampoo or ointment (1% lindane or *gamma* HCH) which is thoroughly rubbed into the infected areas.

For *body lice (Pediculus humanus humanus),* decontamination by thorough dusting with an insecticidal powder kills the active stages, including young lice as soon as they emerge from the egg cases. Since body lice have developed resistance to DDT due to prolonged and repeated application, safe, effective alternatives for disinfestation currently include inert dusts with 1% malathion, *gamma* HCH (lindane) or permethrin (Busvine, 1980). Bedding can be fumigated with ethyl formate.

For *pubic lice (Phthiris pubis), gamma* HCH (lindane) treatment, as recommended for head lice, is effective.

When large groups, as in jails, refugee camps, or army barracks, are infested with lice, mass delousing with insecticide dusting powder is indicated, and should be repeated until the danger of reinfestation is over. All bedding should be fumidated, using necessary precautions.

COCKROACHES (DICTYOPTERA)

This relatively small group of insects, with simple metamorphosis from egg to adult, has survived without remarkable morphologic change from ancestors that lived in swamps during the Carboniferous times, approximately 250 million years ago. They are recognized by their smooth, shiny appearance, often light- to chestnut-brown color, dorsoventral flattening, multiple-segmented long filiform antennae, and biting-chewing mouthparts. Like grasshoppers, their outer pair of wings is leathery, the inner pair membranous. They are easily distinguished from other orthopteran insects because they are typically runners, rather than hoppers, although on occasion they are competent fliers. The name "cockroach" is presumably derived from the Spanish name "cucaracha."

Cockroaches are omnivorous but prefer starches and sugars. Synanthropic species like moist environments, are gregarious in their habits, and feed primarily at night, discharging their feces and vomitus as they eat. They produce eggs in tough cases (oothecae), which they may carry for a time but eventually deposit in dark places. Development from egg through nymph to adult requires 32 to 90 days, depending on the species and the temperature.

Domestic cockroaches (particularly *Blattella germanica, Blatta orientalis,* and *Periplaneta americana*) have been incriminated as intermediate hosts of *Hymenolepis diminuta, Raillietina* spp., *Gongylonema pulchrum,* and *Moniliformis moniliformis,* typically in localities infested with domestic rats, the common definitive hosts of these helminths.

As transit hosts, cockroaches harbor a wide variety of microbial pathogens within the intestinal tract or on the body surface. Roth and Willis (1957) noted that the more common domiciliary species naturally acquire, maintain, and excrete enteric viruses (polio) and several kinds of bacteria. In total, about 40 species of pathogenic bacteria have been isolated from naturally contaminated cockroaches, suggesting at least a peripheral role for these insects in the movement of such pathogens.

Since cockroaches feed on filth and feces, it has long been thought that they play an important role in mechanical transmission of infection with fecal organisms. Live traps baited with fresh human feces are effective in attracting cockroaches. However, studies with *Periplaneta americana* showed that *Salmonella* spp. do not readily become established in the intestinal tract of that insect (Klowden and Greenberg, 1976). In fact, the ingestion of a massive number of bacilli, far exceeding that conceivably encountered under natural conditions, is required for passage of the organism in the feces of cockroaches even without multiplication (Jung and Shaffer, 1952). Ash and Greenberg, on the other hand, found the German cockroach *Blattella germanica* to be a more effective mechanical transmitter, but even with this species of cockroach the minimum dose of bacilli required to result in 50% of the insects' feces containing *Salmonella* during the first 10 days following ingestion was 1.6×10^7. This dose probably exceeds the maximum occurring under natural conditions.

Aspergillus fumigatus and *A. niger* have also been isolated from cockroaches, as have cysts of several enteric protozoans (*Giardia* and *Entamoeba* spp.). An interesting report proposes the transport of *Toxoplasma gondii* from feces of the domestic cat by cockroaches (Chinchilla and Ruiz, 1976).

In addition to their role as pathogen disseminators, cockroaches may serve as a potentially important source of contactant, inhalant, injectant, and ingestant allergens (Roth and Alsop, 1978). Typical symptoms associated with these antigens include itching, dermatitis, localized necrosis, asthma, and hay fever. A thermostabile allergen has been found in food partially consumed by cockroaches, suggesting that this antigen should be incorporated with routine diagnostic allergens, particularly where patients live in cockroach-infested environments (Bernton and Brown, 1967).

Cockroach infestation in a home is facilitated by introduction of these pests with grocery supplies, incomplete household hygiene, poor insulation around water pipes, or flight entry at night through opened doors or screens. Kerosene solutions or emulsions of insecticides sprayed on cockroach hiding places, or sprinkled as a dust, provide good control, but care must be taken to protect food, dishes, and cooking utensils from contamination. Suitable insecticides include fenitrothion, propoxur, or dioxacarb, used at 1%; bendiocarb at 0.3%; and chlorpyrifos at 0.5% (Busvine, 1980). Propoxur, applied as a 1% dust to undisturbed places such as ducts and wall spaces, is also satisfactory.

REFERENCES

Ash, N., and Greenberg, B. 1980. Vector potential of the German cockroach (Dictyoptera: Blattellidae) in dissemination of *Salmonella enteritidis* serotype *typhimurium*. J. Med. Entomol., *17*:417–423.

Bernton, H.S., and Brown, H. 1967. Cockroach allergy. II. The relation of infestation to sensitization. South. Med. J., *60*:852–855.

Bozeman, F.M., Masiello, S.A., Williams, M.S., and Elisberg, B.L. 1975. Epidemic typhus rickettsiae isolated from flying squirrels. Nature, *255*:545–547.

Burgdorfer, W. 1976. The epidemiology of the relapsing fevers, pp. 191–200. In *The Biology of Parasitic Spirochetes*. Edited by R.C. Johnson. New York, Academic Press.

Busvine, J.R. 1980. *Insects and Hygiene*, 3rd ed. London, Chapman and Hall.

Chinchilla, M., and Ruiz, A. 1976. Cockroaches as possible transport hosts of *Toxoplasma gondii* in Costa Rica. J. Parasitol., *62*:140–142.

Clay, T. 1973. Phthiraptera (lice), pp. 395–397. In *Insects and*

Other Arthropods of Medical Importance. Edited by K.G.V. Smith. London, British Museum (Natural History).

Cox, H.R. 1942. Cultivation of rickettsiae in the embryonic tissues of developing chicks. Proc. 6th Pacific Sci. Congr., *5*:605–610.

Fox, J.P., Jordan, M.E., Conwell, D.P., and Robinson, T.A. 1955. Immunization of man against epidemic typhus by infection with avirulent *Rickettsia prowazeki* (strain E). Am. J. Hyg., *61*:174–182.

Johnson, R.C. 1977. The spirochetes. Annu. Rev. Microbiol., *31*:89–106.

Jung, R.C., and Shaffer, M.F. 1952. Survival of ingested *Salmonella* in the cockroach *Periplaneta americana*. Am. J. Trop. Med. Hyg., *1*:990–998.

Klowden, M.J., and Greenberg, B. 1976. *Salmonella* in the American cockroach: Evaluation of vector potential through dosed feeding experiments. J. Hyg., *77*:105–111.

Kluge, H.A. 1980. *Management of Head Lice (Anoplura: Pediculidae): An Integrated Approach*. Pest Management Papers No. 20. Burnaby, B.C., Canada, Simon Fraser University.

Lavoipierre, M.M.J. 1965. Feeding mechanism of bloodsucking arthropods. Nature, *208*:302–303.

McDade, J.E., Shepard, C.C., Redus, M.A., Newhouse, V.F., and Smith, D.J. 1980. Evidence of *Rickettsia prowazekii* infections in the United States. Am. J. Trop. Med. Hyg., *29*:277–284.

Montoya, J.A., Jordan, M.E., Fox, J.P., and Gelfand, H.M. 1958. Immunization against epidemic typhus with avirulent *R. prowazeki* (strain E). Recent observations with reference to a field trial in Peru. Abstr. 6th Int. Congr. Trop. Med. Malaria, 201.

Mueller, J.F. 1973. Pubic lice from the scalp hair; a report of two cases. J. Parasitol., *59*:943–944.

Murray, E.S., and Torrey, S.B. 1975. Virulence of *Rickettsia prowazeki* for head lice, pp. 25–34. In *Pathobiology of Invertebrate Vectors of Disease*. Edited by L.A. Bulla, Jr., and T.C. Cheng. Ann. N.Y. Acad. Sci., Vol. 266.

Roth, L.M., and Alsop, D.W. 1978. Toxins of Blattaria, pp. 465–487. In *Arthropod Venoms*. Edited by S. Bettini. New York, Springer-Verlag.

———, and Willis, E.R. 1957. *The Medical and Veterinary Importance of Cockroaches*. Smith. Misc. Coll. 134, No. 10.

Schaeffer, C.W. 1978. Ecological separation of the human head lice and body lice (Anoplura: Pediculidae). Trans. R. Soc. Trop. Med. Hyg., *72*:669–670.

Shkolnik, L.Y., and Zatulovsky, B.G. 1971. Electron microscopy of vaccine and Breinl strains of *Rickettsia prowazeki* in louse gut cells. Acta Virol., *15*:102–106.

Sholdt, L.L., Holloway, M.L., and Fronk, W.D. 1979. *The Epidemiology of Human Pediculosis in Ethiopia*. Special Publication. Jacksonville, Florida, Navy Disease Vector Ecology and Control Center.

Silverman, D.J., Boese, J.L., and Wisseman, C.L., Jr. 1974. Ultrastructural studies of *Rickettsia prowazeki* from louse midgut cells to feces: Search for "dormant" forms. Infect. Immun., *10*:257–263.

Sonenshine, D.E., Bozeman, F.M., Williams, M.S., Masiello, S.A., Chadwick, D.P., Stocks, N.I., Lauer, D.M., and Elisberg, B.L. 1978. Epizootiology of epidemic typhus *(Rickettsia prowazekii)* in flying squirrels. Am. J. Trop. Med. Hyg., *27*:339–349.

Weyer, F. 1952. Die experimentelle Infektion der Filzlaus *Phthirus pubis* L. mit *Rickettsia prowazeki* und *R. quintana*. Z. Tropenmed. Parasitol., *3*:302–309.

———. 1978. Progresses in ecology and epidemiology of rickettsioses. Acta Trop., *35*:5–21.

Zinsser, H. 1937. The rickettsial diseases: Varieties, epidemiology, and geographical distribution. Am. J. Hyg., *25*:430–463.

Chapter 39

True Bugs (Hemiptera)

True bugs, the Hemiptera, have a hinged proboscis (3- or 4-jointed labium), and four stylets that are adapted for piercing and sucking. Most true bugs have two pairs of wings, of which the anterior pair *(hemelytra)* is thickened basally and membranous distally, and the posterior pair is membranous throughout. The first thoracic segment (prothorax) comprises the greatest part of the thorax. The Hemiptera usually have two or three tarsal segments, with the terminal segment bearing two small claws. The abdomen is large and more or less dorsoventrally flattened. The species of medical importance are bloodsucking forms (suborder Gymnocerata), having conspicuous, freely movable antennae. Compound eyes are always present; ocelli occur in cone-nosed bugs (family Reduviidae; subfamily Triatominae) but are lacking in bedbugs (family Cimicidae).

Biology. The mouthparts arise from the anterior end of the head. They consist of an inconspicuous *labrum,* a proboscis *(labium),* a pair of grooved *maxillae,* which with a pair of blade-like *mandibles* armed with minute teeth at the distal extremity form the *sucking tube* or *fascicle.* A short cone-shaped *hypopharynx* directs the flow of saliva. When the bug takes a blood meal, it (1) moves the proboscis from a horizontal to a vertical position, (2) inserts the mandibular lancets, bending back the distal end of the labium to secure deeper penetration of the mandibles and maxillae into the skin, and (3) proceeds to suck up blood. When the bug is replete with blood, the sucking apparatus is withdrawn from the skin, the labium is straightened out and fitted around the fascicle, and the entire proboscis is returned to its resting position beneath the body. Geigy and Kraus (1952) provided detailed structural and functional information on the sucking apparatus of the bug *Rhodnius prolixus.* Friend and Smith (1971) examined salivation and the pumping mechanism used to move blood along the food canal in this insect.

The *digestive system* internal to the buccal cavity consists of a small muscular *pharynx;* a relatively short, thin-walled muscular *esophagus* of small diameter, lacking diverticula; a *midgut,* arising directly from the posterior end of the esophagus (in bedbugs without an intervening proventriculus) and consisting of a sacculate anterior portion and a longer, tubular, posterior portion; and a short sacculate *rectum* (the *hindgut*), which is devoid of papillae. Four capillary *Malpighian tubules* open into the anterior end of the rectum.

There are two pairs of *salivary glands.* One pair of minute bodies (the ovoid salivary glands) has long ducts that empty into the anterior ends of a second, larger, more median pair (the cardiac salivary glands). From each of these latter glands a long duct proceeds toward the base of the hypopharynx, where it joins its mate to empty into a minute *salivary pump.* Saliva is then passed into the proximal end of the hypopharynx. It flows down a minute channel in the maxillary tube, through which it is discharged into the skin of the victim. The secretion of the ovoid salivary glands in cone-nosed bugs is believed to contain a potent anticoagulin. In addition to paired thoracic and abdominal *spiracular openings,* there is a *respiratory stigma* on the distal joint of each tarsus.

The *male reproductive system* consists of a pair of lobulated *testes,* each with a short *vas deferens;* a pair of *branched accessory glands,* each with a reservoir opening into the distal end of the vas deferens; a common *collecting chamber;* a muscular intromittent or *penial organ;* and a curved chitinous *aedeagus.* In the *female reproductive system* each of the two *ovaries* is composed of seven digitate ovarioles. A short *oviduct* and a *seminal receptacle* on each side, together with the *com-*

mon oviduct, complete the system. In addition, a peculiar slit on the right side of the fifth sternal plate in female bedbugs leads into a spherical accessory "spermatheca," known as the *organ of Berlese,* into which the male introduces relatively long spermatozoa during copulation. In the bedbug, the aedeagus is apparently never introduced into the common oviduct. Spermatozoa work their way in clusters out of the organ of Berlese, traverse the hemocoel, and enter the seminal receptacles, from which they pass up to the ovarioles through the wall of the oviduct.

Development. In the bedbug, embryonation is well advanced before the eggs leave the ovaries. The eggs are coated with a thin gelatinous secretion that firmly fixes them in cracks and crevices where they are deposited by the female bug. In warm climates they hatch in 4 to 8 days and pass through five nymphal instars before reaching maturity. Each stadium (period between molts) requires approximately 1 week. Both sexes and all nymphal instars take blood. Nymphs feed within 24 hours after molting, taking proportionately larger blood meals as they age. Mating occurs 1 to 2 days after adult emergence. Oviposition begins within several days following blood feeding. Batches of 2 to 8 eggs are laid approximately every day over a period of several months if a blood source is available. The eggs are placed in crevices of furniture, floors, and cracks in the walls, where females hide in the daytime. To insure fertility, supplemental inseminations are also required.

In triatomine bugs, the habits and sequence of events are generally similar except that the time required for each stage is much longer. Only one complete blood meal between molts is required. The adults begin to feed within several days after emergence, and eggs are laid within 1 to 2 weeks.

As lair ectoparasites, members of the Cimicidae feed on various mammals and birds, including humans, domestic animals, bats, domestic fowl, and a variety of wild avian hosts. Most cone-nosed bugs frequent human habitations only when they are deprived of a blood supply from other sources or are attracted by light, but some species are primarily domestic. It should also be noted that both the Triatominae and Cimicidae have become adapted to surviving long periods of starvation.

CLASSIFICATION OF BUGS OF MEDICAL IMPORTANCE

Bedbugs (Cimicidae). Species of this family

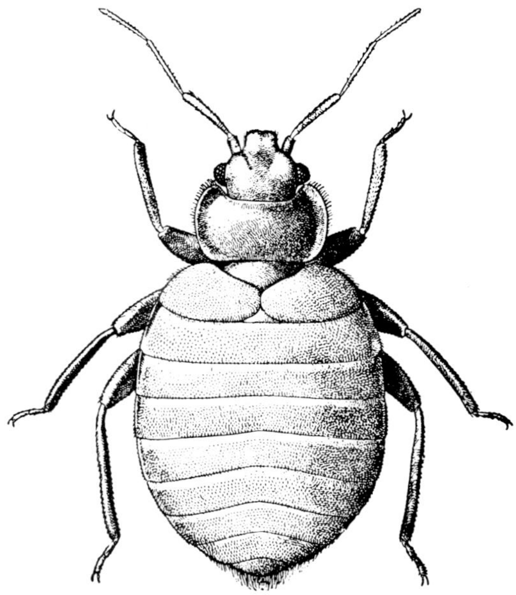

Fig. 39–1. *Cimex lectularius.* Adult female, dorsal view. (× 10.) (Adapted from M.E. MacGregor. In Byam, W., and Archibald, R.G. 1921–1923. *Practice of Medicine in the Tropics.* New York, Oxford University Press.)

(Fig. 39-1) have at most only vestigial wings. The last joints of the tarsi are entire and have the claws inserted at the tips. The antennae are four-jointed and the proboscis is three jointed. Ocelli are lacking. The body is ovate and distinctly dorsoventrally flattened. Only two species of the type genus *Cimex (C. lectularius* Linnaeus and *C. hemipterus* Fabricius) are commonly found in human habitations. The former is the common bedbug of Temperate Zones; the latter, of the Tropics. According to Lewis (1958), *C. lectularius* is found in cities and along well-traveled areas in the Sudan, whereas *C. hemipterus* is more rural. The two species are seldom found in the same house. Both species are mahogany-brown, about 5 mm long by 3 mm broad, and covered with many short hair-like spines.The two species are easily differentiated by the presence in *C. lectularius* of distinct, wing-like, anterolateral expansions of the pronotum, a feature that is lacking in *C. hemipterus.* A third bug, *Leptocimex boueti* (Brumpt), may occur as a human pest in West Africa, but more commonly feeds on bats.

Cone-Nosed Bugs *(Reduviidae: Triatominae).* These species differ from the Cimicidae in having a larger size, well-developed wings, ocelli always present posterior to the eyes, a more or less cone-shaped head, and an elongate, ovoidal abdomen that is not conspicuously flattened. The eggs are

Fig. 39–2. *Triatoma* spp. Stages in the life cycle of cone-nosed bugs. *A*, Egg of *T. infestans*. *B*. Fifth instar nymph of *T. dimidiata*. *C*, Adult female of *T. dimidiata* (pinned specimen). Enlarged. (From Lent, H., and Wygodzinsky, P. 1979. Revision of the Triatominae (Hemiptera, Reduviidae), and their significance as vectors of Chagas' disease. Bull. Am. Mus. Nat. Hist., *163*:127–250.)

normally not provided with glutinous material to fasten them into cracks and crevices. There are five nymphal instars (Fig. 39–2).

Members of the subfamily Triatominae are hematophagous Reduviidae that can be divided into 5 tribes containing 14 genera and 111 species (Lent and Wygodzinsky, 1979). Several are important biological vectors of *Trypanosoma cruzi*, the etiological agent of Chagas' disease. This subfamily is mainly Neotropical and occasionally Nearctic in distribution, though a few indigenous representatives occur in the Orient and extreme northeastern Australia. *Triatoma rubrofasciata* is also well dispersed by man, occurring in port cities throughout Africa and the Orient.

Morphological characters such as the location of ocelli, shape of the head, relative length of segments of the proboscis (sometimes referred to as the rostrum), and the relative body size are useful in making taxonomic designations at the tribe and generic levels. First- and fifth-instar nymphs can also be separated taxonomically to the generic level on the basis of relative length of setae, color, and the appearance of the dorsal side of the abdominal segments.

All species of Triatominae are hematophagous, with the majority feeding on mammalian or avian blood (Schofield, 1979). Orientation to a host is directed by an interaction of several stimuli, including heat, carbon dioxide, and possibly odor. Engorgement is regulated by the relative amounts of nucleotides, such as adenosine triphosphate (ATP), in the blood. Time required for a full blood meal and number of interrupted feeds during the process of engorgement in relation to defecation behavior are of epidemiological significance, because *Trypanosoma cruzi* is a cavitary parasite voided in the feces of the bug (Brener, 1973; Brener and Alvarenga, 1976). Zeledon et al. (1977) compared the feeding and defecation patterns of *Rhodnius prolixus, Triatoma infestans,* and *T. dimidiata*. They observed that *R. prolixus* exhibited a tendency to defecate sooner and more frequently after a blood meal than the other two species, with the nymphs surpassing the adults in number of defecations. Females of all three species defecate faster than do males. The investigation suggested that this physiological-behavioral trait and the vector densities within houses are the principal factors that influence both the incidence and prevalence of Chagas' disease.

Other Biting Hemiptera. In addition to species of Cimicidae and Triatominae (Reduviidae), 13 other families of Hemiptera and 4 from the order Homoptera have been incriminated as having bitten or annoyed humans (Ryckman, 1979; Ryckman and Bentley, 1979). In some cases, where trauma is inflicted by predaceous bugs that secrete proteolytic enzymes, the bite can be painful, with tissue destruction resulting in necrosis and ulceration. The annoying bites of phytophagous bugs (Homoptera) are much less severe, causing transitory reactions.

BUGS AND HUMAN DISEASE

Bedbugs as Vectors of Pathogens. Bedbugs may, mechanically or otherwise, transmit pathogenic microorganisms to man, but there is no convincing evidence that they are natural biological vectors for the transmission of disease-producing agents. Experimentally they are good incubators of a variety of microbial pathogens, but have not been found to transfer the organisms to humans or laboratory animals efficiently. Thus, with the exception of hepatitis B virus, there is little to suggest that *Cimex* species are centrally involved in the epidemiology of any microbial agent. Recent findings in West and South Africa suggest that *Cimex hemipterus* (Wills et al., 1977) and *C. lectularius* (Jupp and McElligott, 1979) may be naturally infected with hepatitis B virus and may transmit it mechanically or via contaminated feces (Ogston and London, 1981).

Dermatitis and Host Reaction to Bedbugs. Both *C. lectularius* and *C. hemipterus* commonly crawl out of their hiding places to obtain blood meals at night. Nymphs require 6 to 9 minutes, adults 10 to 15 minutes, for engorgement, after which they withdraw from their victim and retreat quickly to their hiding places. While some people are undisturbed by their bites, others are seriously affected, not only by development of edema, inflammation, and an indurated ring at the site of the feeding, but also by asthma and anxious concern.

Triatominae as Vectors. Autochthonous cases of Chagas' disease have been reported from practically every country of the Western Hemisphere south of Canada (Zeledon and Rabinovich, 1981). The list of bugs found to be naturally infected with *Trypanosoma cruzi* is impressive, containing about 66 species belonging to the genera *Triatoma, Panstrongylus, Rhodnius, Eratyrus, Psammolestes, Cavernicola, Parabelminus,* and *Dipetalogaster.* Of these taxa, those considered important vectors include species that have the ability to disperse from sylvatic habitats and become adapted to living in human dwellings. *Triatoma infestans* is the principal vector of *T. cruzi* in portions of southern Peru, Bolivia, Chile, Argentina, Uruguay, Paraguay, and Brazil. *Panstrongylus megistus* is important in Brazil and to a lesser extent in Paraguay and Argentina. *Rhodnius prolixus* is a primary vector in northern South America, Central America (excluding Panama and Costa Rica), and parts of Mexico. Other vector species include *R. pallescens* in Panama and *T. dimidiata* in Central America and Mexico. With the exception of *Triatoma sanguisuga,* whose nymphs are occasionally found in human dwellings, there are no domestic species of Triatominae in the United States. Because *T. protracta, T. rubida,* and *T. gerstaeckeri* enter houses, however, their potential as zoonotic vectors should at least be considered where *T. cruzi* is well established in surrounding rodent and marsupial populations.

Skin Reactions to Reduviid Bites. Species of the Triatominae and many other species of the Reduviidae can inflict painful wounds, with local swelling and induration, oozing of blood from the puncture site, and sometimes generalized urticaria. The reduviid bug, *Reduvius personatus,* has a potent salivary toxin, which causes severe, temporary, localized pain when introduced into the puncture wound by this bug. *Arilus cristatus,* the wheel bug, is also important in this regard. Others, including *Panstrongylus megistus,* have evolved a technique for puncturing the victim's skin and withdrawing blood with little or no pain. The lesions are commonly produced on exposed parts of the skin, particularly the face, outer canthus of the eye, and not uncommonly the lips (hence the occasional name of "kissing-bug"). Lavoipierre et al. (1959) described in detail the successive steps in the act of skin penetration and uptake of blood. Even when the puncture is not attended by pain or allergic manifestations, pyogenic organisms may contaminate the wound.

PREVENTION OF BUG INFESTATION

Bedbugs. When *bedbugs* have invaded a domicile, only thorough extermination measures will be effective. For heavy infestations in a large residence or hospital, fumigation with pressurized aerosols of pyrethroids or dichlorvos may be used if proper precautions are taken for the occupants. However, the application of residual insecticides, usually organochlorines, is invariably required (Busvine, 1980). As an alternative to these chemicals, or where resistance to this class of insecticides is a factor, malathion (1 to 2%) or propoxur (1%) may be substituted. Infested beds and mattresses require special attention. The bed springs and slats should be removed and covered with a wet spray. Application to the mattress should be light; it should never be soaked. Sheets and blankets should be thoroughly laundered or dry cleaned. Old-fashioned plaster, covered with wallpaper, constitutes a favorite site for these bugs. It should be torn off, replaced by modern cement plaster, sized, and painted. Baseboards and cracks in the floor should be examined and treated if necessary.

Triatomine Bugs. Because of the long-range flying propensities of *cone-nosed bugs* and their general distribution in rural areas rather than in cities, effective control has previously been impracticable. Proper screening of houses or the use of bed nets would help prevent the bugs from obtaining blood meals from humans, but in adobe and thatched huts, where these bugs most commonly hide, such measures are not ordinarily feasible. The removal of litter and the use of plaster inside to reduce crevices are recommended where possible. Control by interior spraying with HCH (the *gamma* isomer of benzene hexachloride), DDT, and malathion has been variable. Resistance to these chemicals is growing, which suggests that the persistent pyrethroids be substituted whenever possible (Busvine, 1980).

REFERENCES

Brener, Z. 1973. Biology of *Trypanosoma cruzi.* Annu. Rev. Microbiol., *27*:347–382.

———, and Alvarenga, N.J. 1976. Life cycle of *T. cruzi* in the vector, pp. 83–88. In PAHO Sci. Publ. No. 318, *New Approaches in American Trypanosomiasis Research.* Washington, D.C., Pan American Health Organization.

Busvine, J.R. 1980. *Insects and Hygiene,* 3rd ed. New York, Chapman and Hall.

Friend, W.G., and Smith, J.B.B. 1971. Feeding in *Rhodnius prolixus:* Mouthpart activity and salivation, and their correlation with change of electrical resistance. J. Insect Physiol., *17*:233–243.

Geigy, R., and Kraus, C. 1952. Russel und Stechakt von *Rhodnius prolixus.* Acta Trop., *9*:272–276.

Jupp, P.G., and McElligott, S.E. 1979. Transmission experiments with hepatitis B surface antigen and the common bedbug *(Cimex lectularius* L.). S. Afr. Med. J., *56*:54–57.

Lavoipierre, M.M.J., Dickerson, G., and Gordon, R.M. 1959. The manner in which triatomid bugs obtain a blood meal. Ann. Trop. Med. Parasitol., *53*:235–250.

Lent, H., and Wygodzinsky, P. 1979. Revision of the Triatominae (Hemiptera, Reduviidae), and their significance as vectors of Chagas' disease. Bull. Am. Mus. Nat. Hist., *163*:127–520.

Lewis, D.J. 1958. Hemiptera of medical interest in the Sudan Republic. Proc. R. Entomol. Soc. Lond. (A), *33*:43–47.

Ogston, W.C.,and London, W.T. 1981. Excretion of hepatitis B surface antigen by the bedbug *Cimex hemipterus* Fabr. Trans. R. Soc. Trop. Med. Hyg., *74*:823–825.

Ryckman, R.E. 1979. Host reactions to bug bites (Hemiptera, Homoptera): A literature review and annotated bibliography. Part I. Calif. Vector Views, *26*:1–24.

———, and Bentley, D.G. 1979. Host reactions to bug bites (Hemiptera, Homoptera): A literature review and annotated bibliography. Part II. Calif. Vector Views, *25*:25–49.

Schofield, C.J. 1979. The behaviour of Triatominae (Hemiptera: Reduviidae): A review. Bull. Entomol. Res., *69*:363–379.

Wills, W., London, W.T., Werner, B.G., Pourtaghva, M., Larouze, B., Millman, I., Ogston, W., Diallo, S., and Blumberg, B.S. 1977. Hepatitis-B virus in bedbugs *(Cimex hemipterus)* from Senegal. Lancet, *2*:217–219.

Zeledon, R., Alvardo, R., and Jiron, L.F. 1977. Observations on the feeding and defecation patterns of three triatomine species (Hemiptera: Reduviidae). Acta Trop., *34*:65–77.

———, and Rabinovich, J. E. 1981. Chagas' disease: An ecological appraisal with special emphasis on its insect vectors. Annu. Rev. Entomol., *26*:101–133.

Chapter 40

Diptera: Anatomy and Classification

Flies (order Diptera, i.e., insects with two wings) are typically equipped in the adult stage with a single pair of membranous wings attached to the dorsolateral angle of the second thoracic segment. The second pair of wings is modified into a pair of minute club-shaped *halteres* ("balancing organs"). All flies have a complete metamorphosis, i.e., egg, larval stage with several instars, pupa, and adult.

External Anatomy of the Adult Fly

The Head. The head (Fig. 40–1) is a separately movable, relatively large part of the body, connected with the thorax by a cervical constriction. It is covered with numerous immovable chitinous plates *(sclerites)* separated by sutures or grooves. These plates are relatively constant in size, shape, and position in closely related species, but vary considerably in different suborders and families. The most conspicuous feature on the head (Fig. 40–1,*A*) is the pair of compound eyes *(CE)*, which may be contiguous in the midfrontal line *(holoptic)* or separated by the unpaired frontal plate *(dichoptic)*. On the upper median aspect (the *vertex, VV)* are the *ocelli* or simple eyes *(Oc)*, usually three in number, disposed at the angles of an inverted triangle. Below the *frons (Fr)*, or frontal plate, is the unpaired *clypeus (Cl)*, and to the sides of the clypeus are the cheeks or *genae (Ge)*, which may bear the cheek bristles. Extending from behind the cheeks and eyes are the two *occipital plates*, which are contiguous with the median *occipital-central plate* on the dorsum behind the *vertex*. On the dorsoposterior aspect of the head (Fig. 40–1,*B*) is a pair of lateral plates and below the *foramen (F)*, through which the internal organs extend into the thorax, is a median *postgenal plate (Pgn)*. In many groups a median *epicranial suture* extends from the dorsal aspect of the frons upward between the vertex plates and backward to the *occiput (O)*. In those forms that emerge from the pupal case by the extrusion of a bladder-like organ (the *ptilinum, Pt)* through a U-shaped opening (suborder Cyclorrhapha), a corresponding suture is located above the antennae and separates the frons below from the vertex above.

On the anterior aspect of the head, contiguous with and below the clypeus, is the median unpaired *labrum-epipharynx* (upper lip, *LE*). For adult mosquitoes the term "labrum" and not "labrum-epipharynx" is currently used (Harbach and Knight, 1980). The *antennae (An)*, situated in a pair of foramina between or mesoventral to the compound eyes, are movable multisegmented appendages that vary greatly in different families (Fig. 40–2). Their structure constitutes an important basis for classification.

The mouthparts are adapted for lapping or sponging, or for piercing and sucking. In the former type, as illustrated by the house fly *(Musca domestica)*, they consist of three parts: *rostrum* (or beak), *haustellum*, and *labella*. The rostrum, which belongs to the head proper, is shaped like an inverted pyramid anchored in front to the clypeus above the labrum. Through its posterior side runs the pharyngeal tube, and attached to its anterolateral aspects is the pair of *maxillary palps* (Fig. 40–1,*MP*). The haustellum, which is contiguous with and distal to the rostrum, consists of a *labrum-epipharynx* (upper lip), *hypopharynx*, and *labium*. Mandibles and maxillae are lacking. The anterior labrum-epipharynx (labrum external, epipharynx internal) and posterior hypopharynx *(H)* form the food channel, which joins the outer end of the pharynx. The salivary duct passes down through the hypopharynx. The labium *(La)*, or lower lip, is a cylindrical structure on the posterior aspect of the haustellum, partially surrounding the hypo-

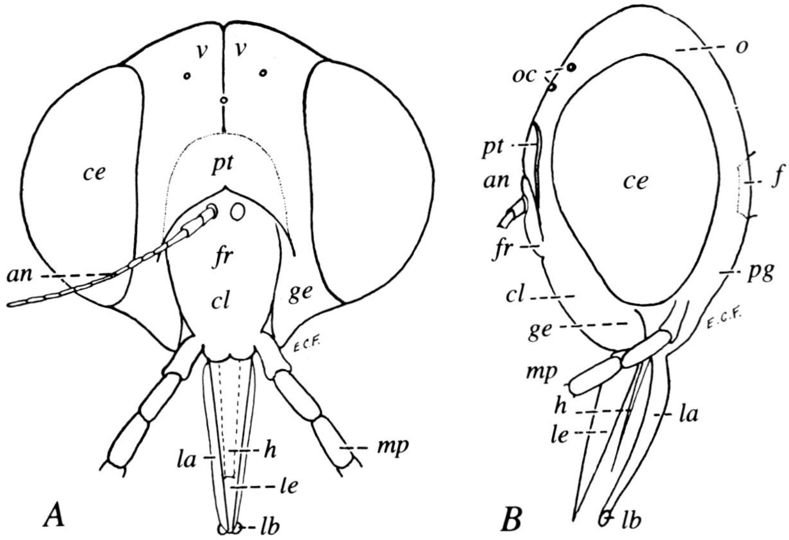

Fig. 40–1. Head of a hypothetical fly. *A*, Frontal view. *B*, Lateral view. *an*, Antenna. *ce*, compound eye. *cl*, Clypeus. *f*, Foramen. *fr*, frons. *ge*, Gena or cheek. *h*, Hypopharynx. *la*, Labium. *lb*, Labella. *le*, Labrum-epipharynx. *mp*, Maxillary palp. *o*, Occiput. *oc*, Ocelli. *pg*, Postgena. *pt*, Ptilinum. *v*, Vertex. (Adapted by E.C. Faust.)

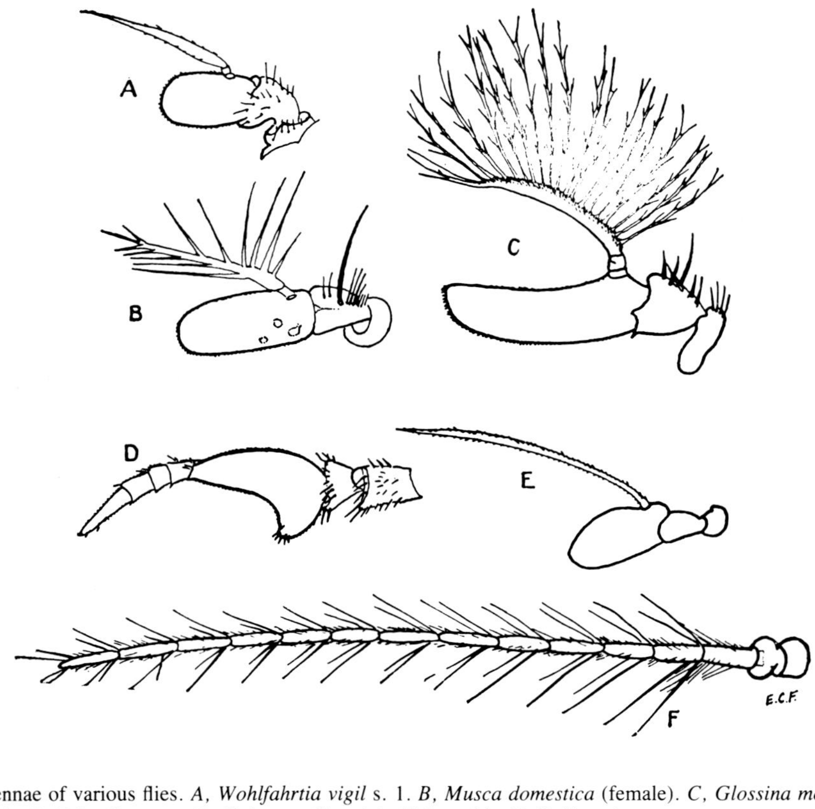

Fig. 40–2. Antennae of various flies. *A*, *Wohlfahrtia vigil* s. 1. *B*, *Musca domestica* (female). *C*, *Glossina morsitans* (male). *D*, *Tabanus* sp. (male). *E*, *Eristalis tenax*. *F*, *Culex* sp. (female). (Adapted by E.C. Faust.)

pharynx and labrum-epipharynx. At its external (distal) extremity the labium ends in a pair of oval lobes, the *labella (Lb)*, each provided with numerous pseudotracheae, which are actually chitinous food troughs. The pseudotracheae pass from the rim toward the inner margin and discharge food at the mouth opening of the food canal. When at rest, the labellar lobes approximate each other; when feeding, they are expandable and can be molded to the feeding surface. In some muscoid flies *(Glossina, Stomoxys)* several rows of prestomal teeth occur between every two rows of pseudotracheae.

In the female mosquito (Fig. 41–3), which has mouthparts adapted for piercing the skin and sucking blood, the labrum (an inverted trough that forms the food meatus), the lanceolate hypopharynx *(H)* with its enclosed salivary duct, and a pair of styletted mandibles *(Md)* and maxillae *(Mx)* all enter the puncture wound. The labium, or proboscis sheath with the labella *(La)* placed on the host's skin just behind the puncture site, buckles back when the other mouthparts are inserted into the skin. In the mosquito the capillary tube formed by the labrum and hypopharynx is not completely enclosed, but with the powerful suction of the pharynx serves as a relatively efficient tube to remove blood from the host's skin to the fly's stomach.

Thorax. The first (pro-) and third (meta-) thoracic segments are usually reduced, while the second or mesothoracic segment, which bears the wings, is conspicuously developed. The chitinous plates or sclerites of the thorax are quite complicated, and the homologies of the terga, pleura, and sterna with other insects are occasionally controversial. For a recent interpretation, see McAlpine et al. (1981).

The wings are thin, membranous, double extensions of the tergal integument, occasionally naked, but in some species covered or patterned with fine hairs, spines, or scales. In flies, as in other insects with complete metamorphosis, the wings arise from buds or *imaginal discs* and are usually not apparent externally until emergence of the adult from the pupal case. The wing veins, the patterns of which are important for classification, consist primitively of small tracheae each enclosed in a chitinous tubule. Embryologically they carry blood through the wing. In many species there is an axillary lobe, near the base, on the posterior margin of the wing, separated from the wing proper by a deep cleft. This lobe may be greatly expanded and folded on itself. The third thoracic segment has a pair of *halteres* in place of a second pair of wings. The 6 thoracic legs are many-jointed *(coxa, trochanter, femur,* and *tibia* with 1 joint each, the distal *tarsus,* consisting of 5 segments, with a pair of terminal articulated *claws, pulvilli* with hairs, and in some species a median bristled or feathered *empodium).*

Abdomen. The basal segments of the abdomen are frequently broader than the terminal portion. The anal segment *(proctiger)* consists of a pair of *cerci* and the anus. In most flies the seventh to tenth segments of the female form the *ovipositor;* the ninth and tenth segments of the male form the *hypopygium* or external genitalia. The abdomen bears no metameric appendages in the adult fly.

Contact chemoreceptors in such flies as *Phormia regina, Calliphora vicina,* and *Stomoxys calcitrans* consist of sensory hairs on the legs and aboral surface of the labellum, and interpseudotracheal papillae on the oral surface of the labellum (Adams et al., 1965; Stürkow et al., 1973; Felt and Berg, 1976). They are sensitive to a variety of chemicals, with 1 or 2 pronged sensilla responding to stimuli. In *Simulium venustum* females, the ventral surface of the tarsomeres of each mesothoracic leg has bifurcate sensilla, which appear to have secondarily acquired an olfactory function (McIver et al., 1980). Presumed olfactory receptors of carbon dioxide and other substances have also been found in sensory pits of maxillary palpal segment 3 in this hematophagous black fly (Mercer and McIver, 1973).

Internal Anatomy

Digestive Tract. In most flies the pharyngeal tube is capable of considerable dilatation. The chief or only pumping organ is the muscular pharynx. In mosquitoes the pharynx is at the inner end of the passage and is separated from the tube by a valve that prevents back flow of blood. With relatively few exceptions, the esophagus bifurcates soon after passing into the thorax, one branch leading to the large *crop* or food reservoir, usually situated ventrally in the anterior portion of the abdomen, and the other leading to the *"proventriculus."* This, in turn, empties into the midgut, with an anterior "chyle stomach" or *ventriculus* and a posterior tubular *proximal intestine.* The *malpighian tubules,* which serve as excretory organs, filter the hemolymph and empty into the *ileal* portion of the hindgut. The *colonic* portion of the hindgut continues posteriorly into the *rectum,* which is provided with 2, 4, or 6 papillae. At the anterior end of the digestive tract is a pair of *salivary glands.*

In mosquitoes these glands are trilobed and lie in the thorax. Their paired ducts join to form a common median tubule passing through the hypopharynx.

Nervous System. The nervous system conforms in general to that of the typical insect, consisting of the connected sub- and supraesophageal ganglia, and a ventral nerve cord.

Circulatory and Respiratory Systems. In the more highly developed flies the heart consists of 4 large chambers and an anterior smaller one. Each chamber has a pair of dorsolateral openings guarded by valves, through which the blood enters the organ. Anteriorly an aortic tubular prolongation of the heart leads to the thorax and head. Its distal end or ends are open, so that the blood passes into the hemocoel. In addition to the paired, arborescent tracheal system, provided with mural annulations, and the delicate, terminal, nonannulated tracheoles, which carry oxygen to all the cells of the insect's body, many of the flies have highly developed expansible air sacs, especially in the abdomen.

Reproductive Systems. The reproductive systems do not differ essentially from the systems described for insects as a group (see p. 608).

Development of Flies

Most flies are oviparous, but certain forms, particularly the flesh flies, the tsetse flies, and the Pupipara (Nycteribiidae, Streblidae, Hippoboscidae), are viviparous. In the flesh flies larvae are deposited as first instar forms; in the other two groups the larvae are developed one at a time *in utero,* and are mature and ready to pupate when deposited. The larvae lack legs but are provided with flexible protruberances, with or without spines. The number of recognizable body segments commonly present in dipterous larvae varies from 12 in the Nematocera and 11 in the Brachycera (Teskey, 1981) to as many as 13 in some muscoid forms (Zumpt, 1965). The larval head also varies in shape and complexity. Larvae of mosquitoes and their allies have mandibulate mouthparts and a conspicuous head, well separated from the thorax; in house fly larvae the head is greatly reduced and the mouth hooks, correspondingly simplified, work in a vertical plane. This atrophy of the head structure and mouthparts is related to the saprobiotic feeding habits of these species.

In certain dipterous larvae the tracheal system is primitive, with a pair of spiracular openings for all *(holopneustic)* or most of the body segments *(peripneustic).* In most groups it is *amphipneustic* (i.e., only the first and last pairs of spiracles are open), begining with the second instar. In mosquito and cattle bot fly *(Hypoderma)* larvae, only the last pair of abdominal spiracles is open *(metapneustic).* The abdominal spiracles of some of the metapneustic and amphipneustic types are of great value in classification because of their genera- and species-specific patterns. The larvae of the Simuliidae and Ceratopogonidae have no spiracular openings *(apneustic).* Aquatic larvae frequently have accessory organs of respiration, consisting of "blood gills" (midges), caudal tracheal gills (mosquitoes), or rectal tracheal gills *(Simulium* spp., *Eristalis* spp.).

Flies usually have three or four larval instars. Between each stage or instar an ecdysis or molting of the cuticle occurs. After the last instar the skin is shed and the pupa develops (mosquitoes and their allies), or the pupal stage is passed inside the last larval skin *(puparium).* In mosquitoes and other aquatic Nematocera, the pupa respires through a pair of spiracles opening on the first thoracic segment *(propneustic).* In the house fly and other Cyclorrhapha, pupae respire through a pair of special spiracles arranged at a particular level of the body, depending on the species.

Emergence of the Adult Fly

Following the metamorphosis of the fly within the pupal skin or within the puparium, the adult *(imago)* emerges from its pupal covering, usually through a middorsal T-shaped split on the mesothoracic segment, or by pushing out a circular plate at the anterior end of the puparium. Upon complete emergence the wings dry, after which the adult assumes an aerial existence.

CLASSIFICATION OF DIPTERA

The order Diptera is usually divided into three suborders: Nematocera, Brachycera, and Cyclorrhapha. Formerly the first two suborders were grouped together under the Orthorrhapha, because in both groups the adult escapes from the pupal covering through a T-shaped opening. This opening is formed by a longitudinal split on the mesothorax and a crosswise split at the anterior end of the longitudinal split (rarely with a dehiscence more caudal, between the seventh and eighth abdominal segments). In a recent classification of the Diptera by McAlpine et al. (1981), two suborders were recognized (Nematocera, Brachycera) with the infraorder Muscomorpha included in the Brachycera and serving in the place of the suborder Cyclorrhapha.

Suborder *Nematocera*

Flies in which the antennae are typically multisegmented, elongated, and slender (i.e., "thread-like horns"); maxillary palps having from 1 to 5 segments, usually 4; larva with a well-developed head and horizontally biting mandibles; pupa obtect (covered with a hard chitinous case). In some families the antennae are long and bear whorls of hairs, and both the legs and abdomen are long and slender; in other families the antennae are shorter than the head and thorax and lack whorls of hair at the nodes, the legs are relatively short and the abdomen moderately stout. The families of medical importance are the following:

Family Culicidae (the mosquitoes). Small forms, with long, slender abdomen; with long, narrow wings; with antennae plumose in the males, pilose in the females; usually with a long, slender proboscis; with a distinctive fringe of scales on the margins of the wings and usually patterned scales on the wing veins; larvae and pupae aquatic. All species of medical importance belong to the subfamilies Anophelinae and Culicinae *(vide infra,* Chapter 41).

Family Psychodidae (the "moth" or "sand" flies). Minute forms with legs, body, and wings clothed with long hairs, at times accompanied with scales; wings short and broad, usually pointed at the apex, without apparent cross veins and sloping at rest. Certain species in the subfamily Phlebotominae are blood suckers and are transmitters of several types of human pathogens.

Family Ceratopogonidae (= Heleidae) (Punkies, "no-see-ums"). Small, delicate forms, occasionally mistaken for mosquitoes, but usually smaller; antennae plumose in male, pilose in the female; head small, usually concealed by the thorax; mouthparts adapted for sucking blood or hemolymph from other insects; veins tending to be ovate, at times hairy but lacking scales, with anterior veins near costal margin stouter, more distinct than the most posterior ones. Members of several genera *(Culicoides, Leptoconops, Forcipomyia)* are vicious blood suckers and species of one *(Culicoides)* serve as intermediate hosts of certain filarial worms of man.

Family Simuliidae (the black flies, coffee flies, buffalo gnats). Small, stoutly built, hump-backed flies, with short legs and long mandibles; with antennae subdivided and appearing 11-jointed, hardly longer than head; males holoptic; with broad, hairless, scaleless wings having thickened anterior veins; other veins less conspicuous or vestigial. Females of several genera *(Simulium, Prosimulium, Cnephia)* are vicious blood suckers, especially in bright sunlight, and some species of *Simulium* are transmitters of the human filarial worm, *Onchocerca volvulus.*

Suborder *Brachycera*

Flies in which the antennae are short (i.e., "short-horned" flies), usually 3-jointed, with the terminal joint elongated; with arista of antenna terminal when present; with short, 1- or 2-jointed, extended maxillary palps. Only one family of this group is medically important.

Family Tabanidae (the horse flies, mango flies, deer flies). Medium to large, thick-set, bulky flies, with a head as wide as or wider than the thorax, convex in front, with large, conspicuous, brilliant-colored eyes, holoptic in male, dichoptic in female; antennae 3-jointed, having 4 to 8 annulations, without an arista. The females suck blood of mammals as well as nectar, and produce appreciable trauma at the site where they puncture the skin. Several species are transmitters of mammalian trypanosomes. A few species of the genus *Chrysops* transmit the filarial worm, *Loa loa,* to humans, and other species are vectors of *Francisella tularensis,* the causative agent of tularemia.

Family Rhagionidae (snipe flies). Medium to large flies having a somewhat rounded head, long tapering abdomen, and relatively long legs. Species of several genera *(Symphoromyia, Spaniopsis)* are hematophagous and can, at times, be extremely pestiferous. These flies have mouthparts similar to those of the Tabanidae and are vicious biters.

Suborder Cyclorrhapha

Flies in which the adult escapes through a circular opening in the anterior end of the pupal case; a frontal suture through which the ptilinum (organ of emergence) is pushed out may or may not be present; the antennae are three-segmented with an arista or style on the last segment.

Series I. *Aschiza.* Forms *without* a frontal suture; ptilinum always lacking. Of the 4 recognized families, species of 2 (Syrphidae, Phoridae) are medically important in the accidental production of myiasis.

Series II. *Schizophora.* Forms having a frontal suture; ptilinum always present. This group includes the greatest number of "circular-seamed" flies, many of which are of medical importance. Some of the families of small-sized forms (acalyptrate species) have only a rudimentary or vestigial axillary (i.e., basal) lobe to the wings, so that they do not fold over and conceal the halteres. The Chloropidae *(Hippelates* spp., *Siphunculina* spp.) or eye gnats belong to this acalyptrate group. Of more general medical importance are certain calyptrate species in which the axillary lobe of the wings is well developed. Differentiation of these families can be difficult, although certain characters provide a superficial distinction. Included in this section are the following families and familiar representatives:

Family Muscidae. *Musca domestica* (common house fly); *Musca sorbens* (eye fly); *Fannia canicularis* (lesser house fly); *Fannia scalaris* (latrine fly); *Muscina stabulans* (false stable fly); *Glossina* spp. (tsetse flies); and *Stomoxys calcitrans* (stable fly).

Family Calliphoridae. Many species, namely: *Calliphora vomitoria, C. vicina* (blue bottle flies); *Lucilia caesar, Phaenicia sericata* (green bottle flies); *Phormia regina* (black blow fly); *Cordylobia anthropophaga* and *C. rodhaini* ("tumbu" flies); *Auchmeromyia senegalensis* ("Congo floor maggot"); *Cochliomyia macellaria, C. hominivorax,* and *Chrysomya bezziana* (screwworm flies); and *Pollenia rudis* (cluster fly).

Family Sarcophagidae. *Wohlfahrtia vigil vigil, W. v. opaca, W. magnifica,* and *Sarcophaga haemorrhoidalis* (flesh flies).

Family Oestridae. *Oestrus ovis* (sheep nasal bot); *Rhinoestrus purpureus.*

Family Hypodermatidae. *Hypoderma bovis, H. lineatum* (warble or bot flies of cattle).

Family Gasterophilidae. *Gasterophilus intestinalis, G. nasalis, G. haemorrhoidalis* (horse bots).

Family Cuterebridae. *Dermatobia hominis* (tórsalo; human bot fly).

Series III. *Pupipara.* Forms that suck blood, adapted to an ectoparasitic existence, mostly on mammals and birds; body covering tough, leathery, with some dorsoventral flattening; with or without wings; with short, nonretractile mouthparts; with highly developed, toothed, or spined claws on the broadly separated legs; larviparous, the larvae pupating soon after deposition. This group includes the species of *Hippobosca,* which infest dogs, cattle, horses, and other large mammals, and *Melophagus ovinus* ("ked"), a bothersome ectoparasite of sheep. None of the species of this group is primarily a human parasite, though some can inflict a painful bite.

The Pupipara are occasionally classified in the superfamily Muscoidea or Hippoboscoidea as part of the Schizophora.

REFERENCES

Adams, J.R., Holbert, P.E., and Forgash, A.J. 1965. Electron microscopy of the contact chemoreceptors of the stable fly, *Stomoxys calcitrans* (Diptera: Muscidae). Ann. Entomol. Soc. Am., *58*:909–917.

Felt, B.T., and Berg, J.S.V. 1976. Ultrastructure of the blowfly chemoreceptor sensillum *(Phormia regina)*. J. Morphol., *150*:763–783.

Greenberg, B. 1971. *Flies and Disease*. Vol. I. *Ecology, Classification and Biotic Associations*. Princeton, New Jersey, Princeton Univ. Press.

———. 1973. *Flies and Disease*. Vol. II. *Biology and Disease Transmission*. Princeton, New Jersey, Princeton Univ. Press.

Harbach, R.E., and Knight, K.L. 1980. *Taxonomists' Glossary of Mosquito Anatomy*. Marlton, New Jersey, Plexus Publishing.

———, and Knight, K.L. 1981. Corrections and additions to *Taxonomists' Glossary of Mosquito Anatomy*. Mosquito Syst., *13*: 201–217.

James, M.T. 1947. *The Flies That Cause Myiasis in Man*. Washington, D.C., U.S. Dept. Agric., Misc. Publ. No. 631.

McAlpine, J.F., Peterson, B.V., Shewell, G.E., Teskey, H.J., Vockeroth, J.R., and Wood, D.M. (coordinators). 1981. *Manual of Nearctic Diptera*. Vol. 1. Monograph 27 (Agriculture Canada).

McIver, S., Siemacki, R., and Sutcliffe, J. 1980. Bifurcate sensilla on the tarsi of female black flies, *Simulium venustum* (Diptera: Simuliidae): Contact chemosensilla adapted for olfaction? J. Morphol., *165*:1–11.

Mercer, K.L., and McIver, S.B. 1973. Studies on the antennal sensilla of selected blackflies (Diptera: Simuliidae). Can. J. Zool., *51*:729–734.

Stürkow, B., Holbert, P.E., Adams, J.R., and Anstead, R.J. 1973. Fine structure of the tip of the labellar taste hair of the blow flies, *Phormia regina* (Mg.) and *Calliphora vicina* R.-D. (Diptera, Calliphoridae). Z. Morphol. Tiere, *75*:87–109.

Teskey, H.J. 1981. Morphology and terminology—larvae, pp. 65–88. In *Manual of Nearctic Diptera*, Vol. 1. J.F. McAlpine, B.V. Peterson, G.E. Shewell, H.J. Teskey, J.R. Vockeroth, and D.M. Wood, coordinators. Monograph 27 (Agriculture Canada).

Zumpt, F. 1965. *Myiasis in Man and Animals in the Old World. A Textbook for Physicians, Veterinarians and Zoologists*. London, Butterworth & Co.

———. 1973. Diptera parasitic on vertebrates in Africa south of the Sahara and in South America, and their medical significance, pp. 197–205. In *Tropical Forest Ecosystems in Africa and South America: A Comparative Review*. Edited by B.J. Meggers, E.S. Ayensu, and W.D. Duckworth. Washington, D.C., Smithsonian Institution Press.

Chapter 41

Bloodsucking Diptera: Mosquitoes, Gnats, Midges, and Flies

The term "bloodsucking flies" or hematophagous Diptera is one of convenience; it is not based on taxonomic grounds. It cleaves down through the suborders, families, and subfamilies of Diptera (Gooding, 1972), and in many instances separates females from males of the same species, denoting the individuals that have mouthparts adapted for piercing the skin and for sucking blood or tissue juices. It has, therefore, both a morphologic and physiologic basis, and even at times has taxonomic value. From a medical standpoint, the use of the term is amply justified since many of the species or individuals with these adaptations not only are pests, but also are responsible vectors of human and zoonotic pathogens. Included here are the mosquitoes (Culicidae), sand flies (Psychodidae), gnats and biting midges (Ceratopogonidae), black flies (Simuliidae), horse flies and deer flies (Tabanidae), stable flies (Muscidae), and tsetse flies (Glossininae).

Among the nematocerous flies in which only the female ingests blood, a large percentage of the proteinaceous component is utilized for reproduction. In *Aedes aegypti,* the fat body produces vitellogenin, the major egg yolk protein, in response to hormonal stimulation (Hagedorn and Kunkel, 1979). Nectar sugars from flowering plants may also serve as energy sources for reproduction, maintenance, or flight in mosquitoes (Magnarelli, 1978), black flies (Cupp and Collins, 1979), and moth flies (Young et al., 1980). In *Glossina* spp., both sexes exhibit strict hematophagy and utilize proline from the blood meal as a substrate for energy needs (Bursell et al., 1974). A significant proportion of the blood meal is also used by "pregnant" females to produce a "milk" that is rich in proteins, amino acids, and fats.

MOSQUITOES (FAMILY CULICIDAE)

Definition. Mosquitoes (Fig. 41–1) are slender, delicate forms that are distributed worldwide and number at least 3071 species (Knight and Stone, 1977; Knight, 1978). The larval and pupal stages are always aquatic. In the related families Chaoboridae and Chironomidae, the proboscis is relatively short and not adapted for piercing; the wing veins are not completely covered with scales; and the larvae are detritus feeders or are predaceous. When abundant, these nonbiting midges can be a nuisance or allergenic. On the other hand, the mouthparts of the members of the family Culicidae are greatly elongated and in the females (with few exceptions) are adapted for piercing and sucking blood. The wings have scales over their veins. The three-segmented antennae are long with the third segment (flagellum) comprised of 13 to 14 flagellomeres covered with whorls of hairs so that the antennae of the male appear plumose, those of the female, pilose. The wing venation (Fig. 41–2) of a culicid is characterized by a series of long veins (excluding costa and subcosta) several of which are forked. The areas of wing membrane delimited by veins or by veins and wing margin are referred to as cells and are of taxonomic significance.

External Anatomy of Adult Mosquitoes

The Head. The head of a culicid mosquito is subglobose, connected with the thorax by a slender neck. The compound eyes are prominent and ocelli are indistinct but present between the antennal sockets and the postfrontal sutures (Harbach and Knight, 1980). In the male, except in the genus *Uranotaenia* and in the tribe Sabethini, the maxillary palps are nearly as long as the proboscis. In

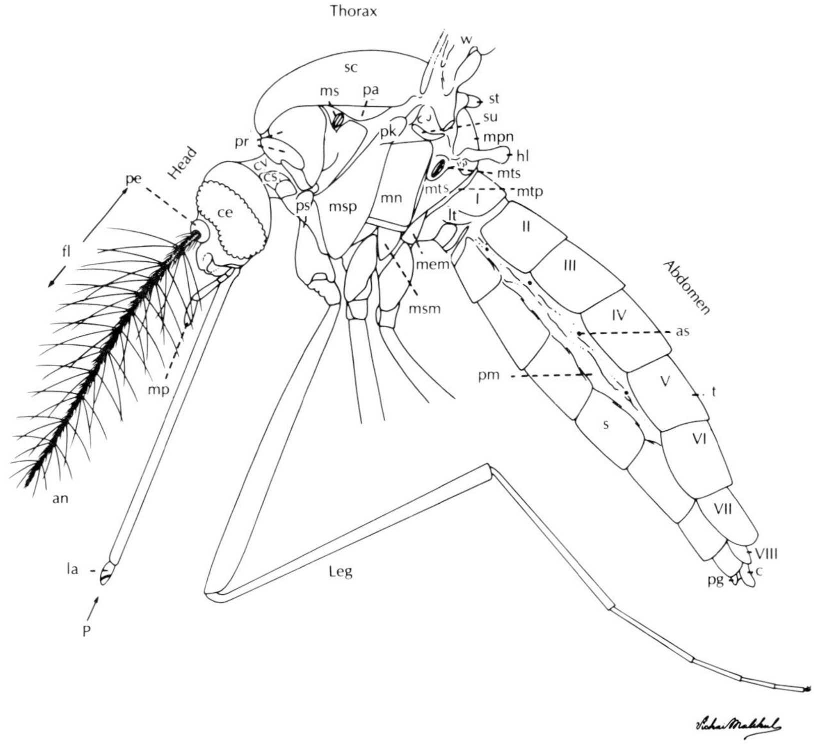

Fig. 41–1. Diagram of an adult female mosquito, lateral aspect. *as*, Abdominal spiracle. *an*, Antenna. *c*, Cercus. *ce*, Compound eye. *cs*, Cervical sclerite. *cv*, Cervix. *fl*, Flagellum. *hl*, Halter. *la*, Labellum. *lt*, Laterotergite. *mem*, Metameron. *mn*, Mesepimeron. *mp*, Maxillary palpus. *mpn*, Mesopostnotum. *msp*, Mesepisternum. *ms*, Mesothoracic spiracle. *msm*, Mesomeron. *mtp*, Metapostnotum. *mts*, Metathoracic spiracle. *p*, Proboscis. *pa*, Paratergite. *pe*, Pedicel. *pg*, Postgenital lobe. *pk*, Prealar knob. *pm*, Pleural membrane. *pr*, Pronotum. *ps*, Proepisternum. *s*, Sternum. *su*, subalare. *sc*, Scutum. *st*, Scutellum. *t*, Tergum. *w*, Wing. (By J.N. Belkin, 1962. From Harbach, R.E., and Knight, K.L. 1980. *Taxonomists' Glossary of Mosquito Anatomy.* Marlton, N.J., Plexus Publishing.)

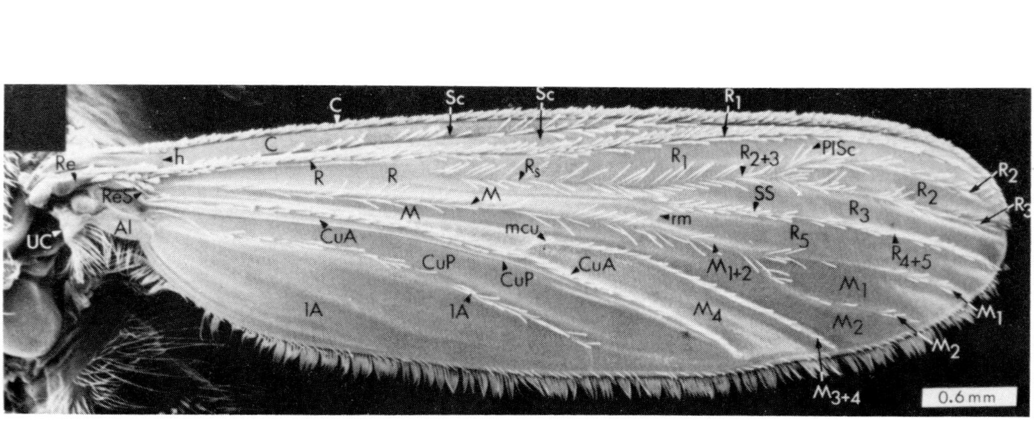

Fig. 41–2. Wing of *Culiseta inornata*, dorsal aspect. *Al*, Alula. *C*, Costa and cell. *CuA*, Cubitus anterior and cell. *CuP*, Cubitus posterior and cell. *h*, Humeral crossvein. *M*, Media and cell. M_1, Media-one and cell. M_{1+2}, Media-one-plus-two. M_2, Media 2 and cell. m_{3+4}, media-three-plus-four. M_4, cell. *mcu*, Mediocubital crossvein. *PlSc*, Plume scale. *R*, Radius and cell. R_1, Radius-one and cell. R_2, Radius-two and cell. R_{2+3}, Radius-two-plus-three. R_3, Radius-three and cell. R_{4+5}, Radius-four-plus-five. R_5, Cell. *Re*, Remigium. *ReS*, Remigial seta. *Rs*, Radial sector. *rm*, Radiomedial crossvein. *Sc*, Subcosta and cell. *SS*, Squame scale. *UC*, Upper calypter. *1A*, Anal vein and cell. (From Harbach, R.E., and Knight, K.L. 1980. *Taxonomists' Glossary of Mosquito Anatomy.* Marlton, N.J., Plexus Publishing.)

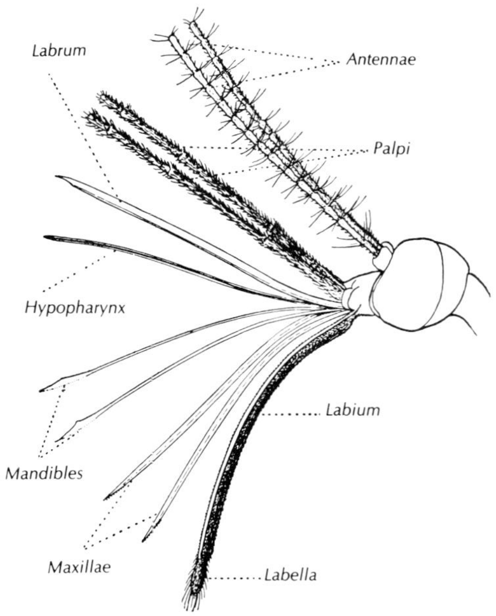

Fig. 41–3. Mouthparts of a female mosquito *(Anopheles quadrimaculatus)*, dissected out of the sheath. Note that the mandibles have finely serrated, blade-like tips for cutting, and that the maxillae have pointed chitinous tips with fewer and coarser serrations than the mandibles. (From Fox, C. 1928. *Insects and Diseases of Man*. Philadelphia, Blakiston Company.)

the female the palps are long in the subfamily Anophelinae, short in the subfamily Culicinae. When dissected out of the labium (i.e., proboscis sheath), the mouthparts of a female *Anopheles* mosquito, together with the antennae, appear as illustrated in Figure 41–3. The fascicle is composed of six stylets: paired maxillae and mandibles; a median hypopharynx, which is penetrated by the salivary canal; and the labrum, which is anterior to the mandibles and forms the food meatus. The mechanical forces responsible for insertion of the stylets result from a complex interaction between several skeletal structures of the head, a protractor muscle, and retractor muscles that aid indirectly in pushing the mandibles into the skin (Wenk, 1980). The cibarial pump, which lies underneath the clypeus and joins with the upper surface of the hypopharynx, and the pharyngeal pump are responsible for ingestion of blood. In the male the mandibles are reduced or lacking, the maxillae may be vestigial, the labium may be terminally blunt, and the hypopharynx may be fused, thus rendering the proboscis unsuitable for bloodfeeding.

Thorax. The thorax is elongate-ovoid or rectangular from the dorsal aspect, and is covered for the most part by the *mesonotum* (the conspicuous dorsal scleratization of the second thoracic segment), which is separated posteriorly from the *scutellum* (the dorsoposterior area of the second thoracic segment) by a transverse suture. In *Anopheles* this *scutellum* is arcuate; otherwise, it is trilobate. The dorsal thoracic sclerite of the prothorax is transversely subdivided into an *antepronotum* and a *postpronotum;* the third dorsal sclerite *(metanotum),* which lies behind the scutellum, is inconspicuous. The pattern of the lateral sclerites *(pleura* or *pleurites),* and especially the number and disposition of their hairs, provides a feature that specialists use in grouping species of Culicidae into their proper genera. The three pairs of legs are borne by the ventral sclerites, which form more or less the apex of the wedge-shaped thorax. These legs are long and slender, and the tarsi of the hind legs are particularly long. The terminal tarsal joint is provided with a pair of claws, which are modified and at times toothed in the male and sometimes in the female. The wings arise from the posterolateral angle of the mesonotum; when at rest they lie dorsally over the abdomen. At times their scales provide distinctive color patterns of diagnostic importance.

Abdomen. The abdomen is elongated, more or less cylindrical, and consists of ten segments, although only eight are clearly visible, owing to the sexual modification of the terminal two segments. In the female the ninth is greatly reduced, and at its junction with the eighth bears the genital opening. The tenth, which is also reduced and narrowed, bears the anus and terminal cerci. The ninth and tenth segments of the male are twisted 180 degrees and are modified into the external organs of reproduction, i.e., the *genitalia,* which have a relatively complicated structure (Fig. 41–4), and are relied on by specialists for obtaining an accurate identification of the species. The abdominal segments are covered by tergal and sternal plates, separated by an expansible pleural membrane on each side, allowing distention of the abdomen when a considerable amount of food has been ingested. In addition to the scales on the wings, the whorls of hairs on the antennae, and hairs on the thorax, these integumentary ornaments also appear in varying numbers and patterns on the head, legs, and abdomen.

Respiration in the adult mosquito is completely tracheate. The paired spiracular openings are present as follows: one pair each on the second and

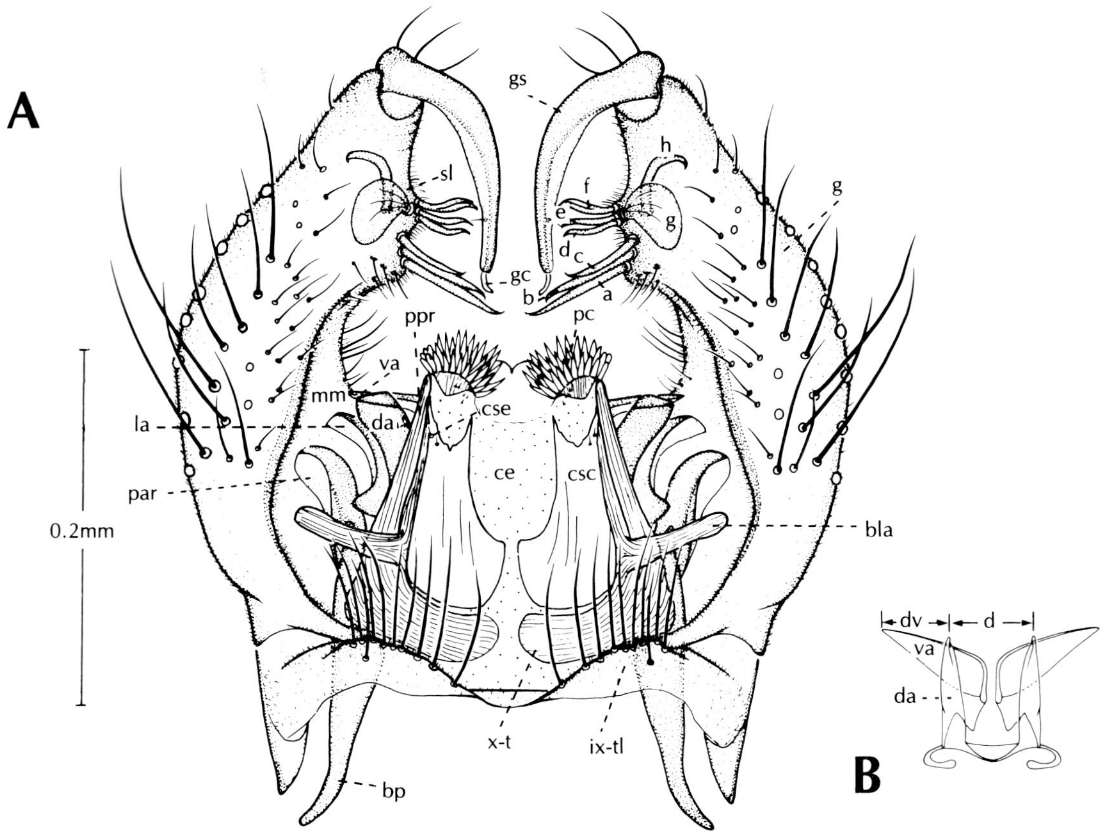

Fig. 41–4. *A,* Male genitalia of *Culex (Culex) pipiens* Linnaeus (dorsal aspect-prerotation sense). *a–h,* Setae a-h. *bla,* Basal lateral arm. *bp,* Basal piece. *ce,* Cercus. *csc,* Cercal sclerite. *cse,* Cercal seta. *da,* Dorsal arm. *g,* Gonocoxite. *gc,* Gonostylar claw. *gs,* Gonostylus. *la,* Lateral arm. *mm,* Mesal membrane. *par,* Paramere. *pc,* Paraproct crown. *ppr,* Paraproct. *sl,* Subapical lobe. *va,* Ventral arm. *ix-tl,* Tergum IX lobe. *X-t,* Tergum X. *B,* Dorsal *(da)* and ventral *(va)* arms of the lateral plates of male *Culex* terminalia illustrating measurements used to calculate the ratio dv/d for separation of members of the *Culex pipiens* complex. (From Harbach, R.E., and Knight, K.L. 1980. *Taxonomists' Glossary of Mosquito Anatomy.* Marlton, N.J., Plexus Publishing.)

third thoracic segments and one each on the first eight abdominal segments.

Internal Anatomy

Digestive Tract. The region just within the base of the proboscis is the buccal cavity, which opens inward through a valve into the muscular pharyngeal bulb (Fig. 41–5). This, in turn, is succeeded by a short esophagus with its pair of dorsal diverticula and its posterior muscular valve. A distensible, ventral diverticulum or "crop" lies in the abdomen and serves as a reservoir for water and nutrients. The nonchitinized midgut extends from the midplane of the thorax through most of the abdomen; the posterior portion is capable of considerable expansion and serves as a "stomach." The peritrophic membrane, a tube-like structure produced by the midgut epithelial cells in response to distension, serves to keep the liquid portion of the blood meal in the gut (Freyvogel and Stäubli, 1965). In the last few abdominal segments, the midgut joins the hindgut, a tubular structure with a bladder-like rectum that opens subterminally through the anus. At the junction of the midgut and hindgut, the long, coiled malpighian tubules empty into the digestive tract. These excretory organs filter the hemolymph and are important in the elimination of nitrogen-containing substances. They also function in osmoregulation shortly after the ingestion of a blood meal in the elimination of fluids (diuresis). The pair of salivary glands is situated in the first segment of the thorax. Each gland consists of three tubules (2 long and 1 short), lined with a single layer of secretory cells. The ducts, which originate within the tubules, unite to form a common lateral duct. The two lateral ducts soon

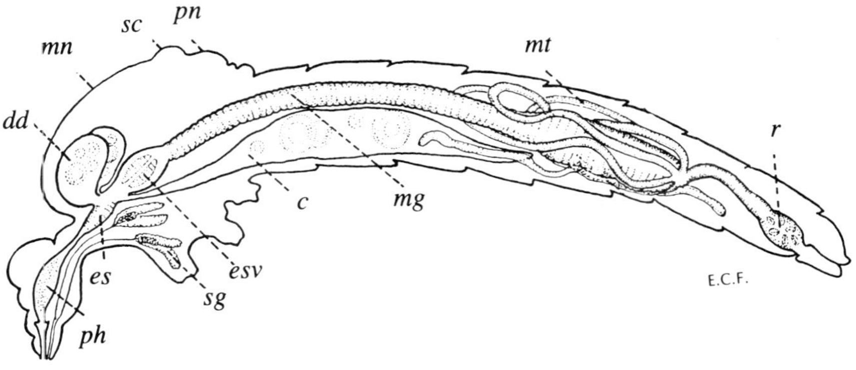

Fig. 41–5. Internal anatomy of a mosquito. *c*, Crop. *dd*, Dorsal diverticula. *es*, Esophagus. *esv*, Esophageal valve. *mg*, Mid-gut. *mn*, Mesonotum. *mt*, Malpighian tubule. *ph*, Pharynx. *pn*, Postnotum. *r*, Rectum, with papillae. *sc*, Scutellum. *sg*, Salivary gland. (Adapted by E.C. Faust from Fox, C. 1925. *Insects and Diseases of Man*. Philadelphia, Blakiston Company.)

join to form a common duct, which passes through a minute salivary pump and then passes down on the ventral side of the head into the hypopharynx. The saliva is expelled in minute droplets into the puncture wound before blood is withdrawn from the host.

Host Seeking and Reproduction. Mosquitoes may respond to environmental stimuli in a variety of ways depending largely upon their age and physiological state. Juvenile hormone induction of biting behavior has been shown in *Culex pipiens* and *Culex quinquefasciatus*. The absence of this hormone also prevents follicular development of the ovary in diapausing *C. pipiens* and restricts the mosquito to a diet of plant juices (Meola and Petralia, 1980). Questing females of other species respond to visual cues (Bidlingmayer and Hem, 1980) and have a keen sense of smell, being particularly sensitive to plumes of odors exuded from man and other mammals (McIver, 1982). Carbon dioxide also serves to guide the mosquito toward a host and probably intermingles with host odor; it may also be important at close range when combined with warm, moist convection currents (Gillies, 1980). For a review of factors affecting feeding by hematophagous insects, including mosquitoes, see Friend and Smith, 1977. The essential female organs of reproduction consist of: (1) a pair of ellipsoid ovaries with the anterior end connected to a delicate filament which, in turn, is attached to the alary muscles as far forward as the second abdominal segment; (2) a pair of short lateral oviducts; and (3) a common oviduct that opens mesad externally at the posterior border of the eighth abdominal segment. At its outer extremity this oviduct enlarges into an atrium, and receives the capillary spermathecal duct and the duct of the bursa copulatrix. The spermatheca is a small globose body filled with spermatozoa after mating has occurred (Ronquillo and Horsfall, 1969).

In the male the internal genital system consists of a pair of broadly spindle-shaped testes, a pair of long vasa efferentia, each with a vas deferens connecting to an outer seminal vesicle, and a common ejaculatory duct, into which opens a pair of accessory glands (Horsfall and Ronquillo, 1970). At its termination the ejaculatory duct is chitinized and joins an external intromittent organ *(aedeagus)*. In newly emerged males, abdominal segments 9 and 10 must rotate 180° prior to mating. Fertilization of females usually occurs within 1 to 2 days after they have emerged. Typically, in many species there is a prenuptial swarming of the males in considerable numbers several feet above or below a specific marker, occurring about sundown on a quiet evening, or in a protected spot. Females apparently enter the swarm and are grasped by the males, the caudal extremities of the mating couple unite, and copulation occurs. Many mosquito species are *eurygamous* and require a considerable space for swarming and mating; on the other hand, *Aedes aegypti* mates readily in confined quarters *(stenogamous)*. Usually only one insemination occurs during the life of the female, the male dying soon after copulation.

In Nature a female may survive for several weeks to many months, or may die from the hazards of oviposition or unfavorable biotic conditions. Some temperate zone species of *Culex* and *Anopheles* hibernate as adults, undergoing gonotrophic dissociation ("the cessation of egg production despite the continued taking of blood meals;" see Washino, 1977) in response to decreasing day length. Conversely, nulliparous *Anopheles arabiensis* appear capable of aestivation in the semiarid parts of the Sudan, bloodfeeding, and undergoing a gradual advance in ovarian development as the dry season progresses (Omer and Cloudsley-Thompson, 1970). Inseminated gravid or parous *Culiseta inornata* females also aestivate in the Colorado Desert of southeastern California (Barnard and Mulla, 1978).

Oogenesis and Oviposition

Ovarian development can be completed after ingestion of one or more blood meals *(anautogeny)* or in some cases without intake of proteinaceous materials by the adult *(autogeny)*. In mosquitoes, follicles bearing developing oocytes pass through a series of well-characterized stages that can be used for age-grouping of populations (Detinova, 1962). Sperm from the spermatheca fertilize eggs shortly after ovulation as they pass down the common oviduct. Unfertilized eggs may also be oviposited, but will not develop. Eggs are placed in water or on moist substrates where water is likely to accumulate later. The eggs are deposited one by one in the chosen site until a batch of several to many dozen has been discharged, depending on the species and the vitality of the mosquito. The number of gonotrophic cycles may vary; as many as nine have been recorded for *Anopheles farauti*, a vector of *Plasmodium vivax* in New Guinea (Spencer, 1979).

Anopheline eggs and those of many *Aedes* spp. are laid singly on the surface of the water or slightly above the water line (Fig. 41–6,*B-E*); eggs of *Culex* are accumulated in floating rafts (Fig. 41–6,*A*). After the chorion has hardened, anopheline eggs survive moderate desiccation for a few hours to 3 days or more, those of some species of *Aedes* and *Psorophora* for many months or even years. Overwintering most commonly occurs in the egg stage. Eggs remain latent during the winter months and at least some of them hatch with the return of warm conditions.

Development

Egg. Hatching occurs in water (fresh or brackish, depending on the adaptation of the species). The period of embryogeny varies with both the temperature and the particular species. Under favorable conditions *Culex pipiens* eggs hatch in 36 to 48 hours, whereas *A. quadrimaculatus* requires only 24 hours at 35°C but 30 hours at temperatures under 33.8°C. The mature first-stage larva within the eggshell emerges through a slit produced by the special chisel-shaped larval organ, the *hatching spine*, which is situated on the dorsum of the head and is lost with the first molt. For an account of the embryogeny of *Aedes aegypti*, see Raminani and Cupp, 1975 and 1978.

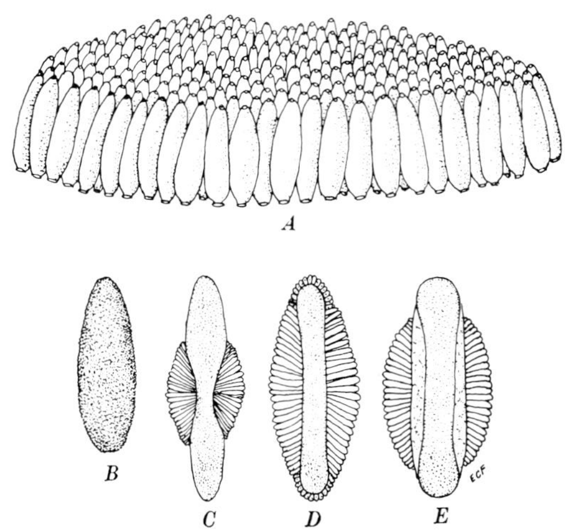

Fig. 41–6. Mosquito eggs. *A*, Raft of *Culex* eggs. *B*, Egg of *Aedes*. *C-E*, Eggs of three different species of *Anopheles*. (From Faust, E.C., Beaver, P.C., and Jung, R.C. 1975. *Animal Agents and Vectors of Human Diseases*, 4th ed. Philadelphia, Lea & Febiger.)

Larva. The larval body regions consist of head, thorax, and abdomen (Fig. 41–7,*A-C*). The head is more or less quadrato-ovoid in shape, and is somewhat dorsoventrally flattened. It has a pair of short, shaft-like antennae, frequently provided with spines, and a complex set of mouthparts, consisting of a median ventral, triangular labium, a pair each of maxillary plates and mandibular processes, a lateral palatal brush, and a pair of maxillary palps lateral to the maxillae. The mouth brushes set up a current of water, bringing small particulate material within reach of the mandibles, which crush and chew it, while the maxillae roll it into a ball to be swallowed. The anterior dorsal sclerite of the head is the frontoclypeus, which characteristically bears three pairs of *clypeal hairs* that are of diagnostic importance, particularly in anopheline species. The ocular, trans-sutural, antennal, subantennal, and frontal hairs are of lesser diagnostic value.

The *thorax* is quadrate, compressed dorsoventrally, and is broader than long. On its dorsal and

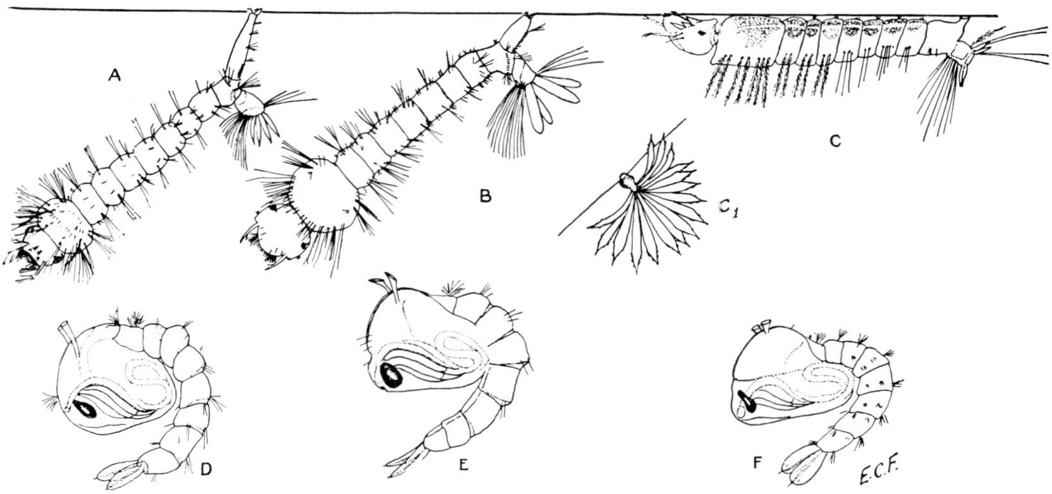

Fig. 41–7. Larvae and pupae of mosquitoes. *A-C*, Larvae, showing breathing position at surface film. *D-F*, Pupae. *A, D, Culex pipiens*. *B, E, Aedes aegypti*. *C, F, Anopheles quadrimaculatus*. C_1, Palmate hair tuft from dorsum of third abdominal segment of *A. quadrimaculatus*. (By E.C. Faust.)

lateral aspects it bears numerous hairs of some diagnostic importance.

The *abdomen* is more or less cylindrical but tapers somewhat caudally. It is composed of 10 segments (9 are apparent), the last two of which in culicines (terminal one only in anophelines) are bent ventrally at an obtuse angle and bear both dorsal and ventral brushes and four anal (or tracheal) gills. In the culicine mosquitoes there is a *siphon* arising from the eighth segment, with a terminal *spiracular apparatus* bearing the openings of the pair of tracheal tubes. It is bent dorsally at an angle, giving a bifurcated appearance to the caudal extremity of these forms. In anopheline species, however, this siphon is not present, and the pair of spiracular tracheal openings is found on a dorsal stigmal plate on the anterior border of the ninth segment. A few culicines, such as *Coquillettidia* and *Mansonia*, have a special cutting apparatus on the siphon which enables them to pierce the tissues of submerged plants, from which they obtain their oxygen. Most culicine larvae have a series of teeth on the siphon, collectively called the *pecten*. A patch of spines on each side of the eighth segment is referred to as the *comb*. The number of teeth, or spines, and their detailed structure and disposition in the pecten and comb, constitute valuable aids in specific diagnosis of these larvae. In anophelines the pecten is represented by a row of teeth ventrolateral to the stigmal plate.

Most anophelines have a pair of *palmate setae* (''float hairs''), made up of numerous leaflets, on the dorsum of certain of the first to seventh abdominal segments as well as in the posterior corners of the thorax (Fig. 41–7, *C*, C_1). These, together with certain other hairs, are used to identify species.

Larval mosquitoes typically obtain oxygen for respiration when their siphons (culicines) or stigmal plates (anophelines) are protruded through the surface film, but they may also utilize dissolved oxygen in well-aerated water by means of their anal gills and cuticle, especially in the earlier instars. The type of external opening of the tracheal tubes governs to a considerable extent the feeding habits of the larvae. Anopheline larvae (Fig. 41–7, *C*) lie suspended by the palmate setae parallel with and just below the surface film in order to obtain free oxygen from the air. When they are feeding in the surface film, the head is turned at an angle of 180 degrees with the body, so that the ventral side of the head is uppermost. On the other hand, culicine species (Fig. 41–7, *A, B*) hang at an angle from the surface and feed through a broad arc, with the body as the radius. The food consists of minute plankton forms, both living and dead, organic debris, and probably even nonparticulate nourishment in the colloidal or dissolved state. Ten days or more are usually required to complete growth. There are four larval instars, with a molt between each. Molting occurs by the larva emerging through a middorsal split of the skin along the thorax and abdomen. The fourth instar is most useful for species identification. For identification of species in North America, the reader is referred to Wood et al., 1979 and Darsie and Ward, 1981.

Pupa. When the fourth larval instar is ready to transform into a pupa, it bends its head downward, ceases to feed, and becomes sluggish. In the course of an hour or more a median dorsal slit occurs on the thorax, and the comma-shaped pupa emerges. It does not feed and, although it can swim and dive, it is mainly inactive. Except when it is disturbed, the pupa floats at the surface, taking in air through its conspicuous thoracic respiratory trumpets. As it grows older, the pupa becomes darker in color, paralleling the profound internal metamorphosis that is taking place, and becomes more buoyant. The greater part of the body (Fig. 41–7, *D-F)* consists of a pear-shaped cephalothorax, which possesses a pair of compound eyes and near its upper lateral margins the mesothoracic breathing trumpets. In culicines (Fig. 41–7, *D, E)* these trumpets are usually long and narrow *(Culex)* or obliquely truncated *(Aedes)*; in anophelines (Fig. 41–7, *F*), they are short and funnel-shaped. As the adult mosquito matures, the head, wings, and legs can be seen within the cuticular covering. The tail of the "comma" is composed of eight abdominal segments, with a terminal pair of paddles or fins. After 1 to 5 days the pupa comes to lie quietly at the surface of the water. The thoracic cuticle now splits and the thorax appears, followed by the head; the abdomen apparently produces the pressure against the pupal skin necessary for this emergence. The wings are next unfolded and soon dry; soon afterward the abdomen emerges, and finally the legs are freed and stretch out. In 5 to 10 minutes the animal is ready for flight.

Classification and Identification

The differential characters of the eggs, larvae, pupae, and adults of the culicine and anopheline subfamilies are indicated in Table 41–1.

Mosquitoes are divided into three subfamilies: Anophelinae, Toxorhynchitinae, and Culicinae. The subfamily Anophelinae contains the genera *Chagasia, Bironella,* and *Anopheles*. The genus *Anopheles* contains the subgenera *Stethomyia, Anopheles, Nyssorhynchus, Kerteszia, Lophopodomyia,* and *Cellia*. The subfamily Toxorhynchitinae consists of one genus, *Toxorhynchites,* which has three subgenera. Species of *Toxorhynchites* are huge, handsome forms that have a proboscis adapted only for sucking plant juices. Their larvae are predaceous and may prove useful as biological control agents of vector species (Trpis, 1973). The subfamily Culicinae consists of 10 tribes, with three *(Aedini, Culicini, Mansoniini)* containing the preponderance of vector genera.

The accurate identification of species of mosquitoes, as of many other insects, can be made only by persons who have had special training in this subject. It is not expected that the average physician, public health officer, or medical student

Table 41–1. Identification of Some Common New World Mosquito Genera in Egg, Larval, Pupal, and Adult Stages.

	Subfamily Culicinae	*Subfamily Anophelinae*
EGGS	Laid in rafts *(Culex)* or singly in water or where water is likely to appear *(Aedes, Psorophora)*, or in water held by plants (Fig. 41–6, *A,B*).	Naviculate, with air sacs (floats), laid singly in water (Fig. 41–6, *C,D,E*).
LARVAE	When breathing, usually lie obliquely or vertically to water surface; they always have a well-developed siphon through which the spiracles open (Fig. 41–7, *A,B*).	Lie parallel to surface film when breathing or feeding, with head revolved 180°, ventral side up; palmate float hairs present on certain of the first to seventh abdominal segments; certain of the anterior abdominal segments bear antepalmate tufts (Fig. 41–7, *C*).
PUPAE	Breathing trumpets long and tubular *(Culex)*, or short and truncated *(Aedes)*; paddles at tip of abdomen have a large terminal hair (Fig. 41–7, *D,E*).	Breathing trumpets short and broad at margin; paddles at tip of abdomen have an accessory hair in addition to a terminal hair (Fig. 41–7, *F*).
ADULTS	Rest with proboscis not in line with abdomen; body usually humped; wings not usually spotted or dappled; scutellum at posterior end of mesonotum usually trilobate; palps of females less than half as long as proboscis.	Most species rest with proboscis in a straight line with body, at an oblique angle from resting surface; wings usually spotted or dappled; scutellum arcuate; palps of female usually as long as proboscis.

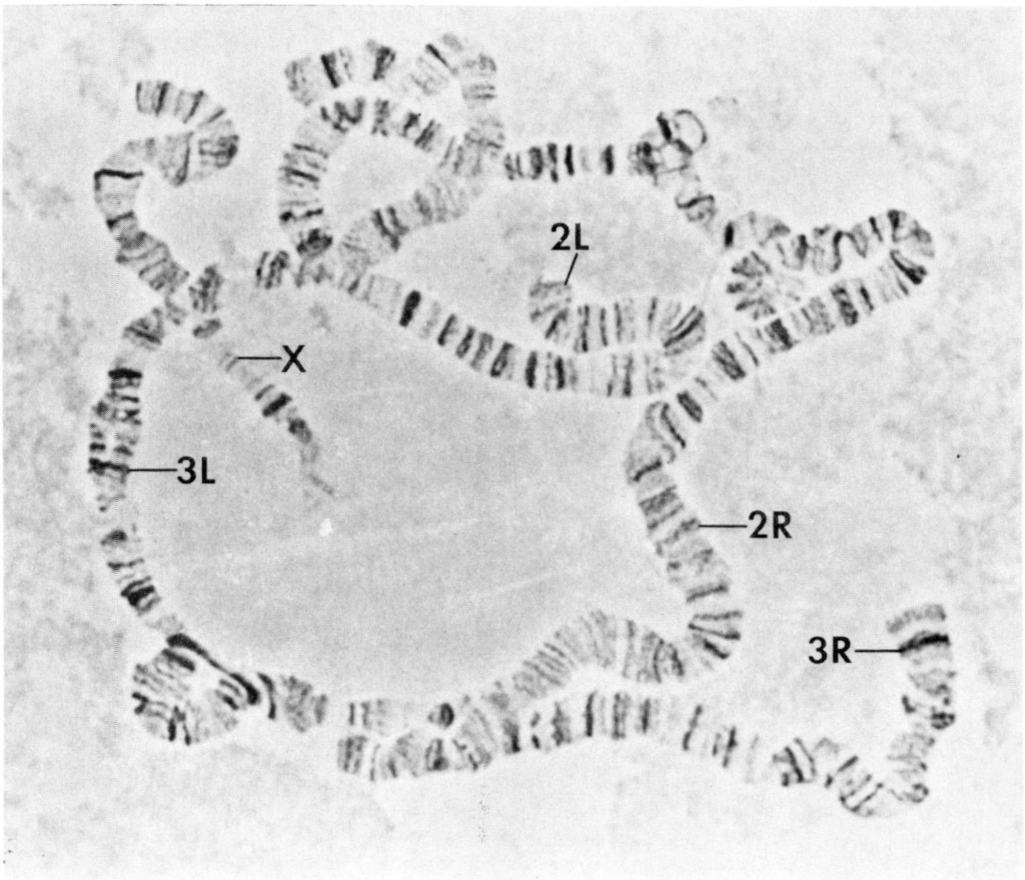

Fig. 41–8. Chromosome complement of *Anopheles (Nyssorhynchus) darlingi*. Araraquara, Brazil population. X = X chromosome. 2L, 2R = autosome 2, left and right arms. 3L, 3R = autosome 3, left and right arms. (× 450.) (Courtesy of Dr. R.D. Kreutzer, Youngstown State University.)

will need or acquire proficiency in specific identification of these forms. Students especially interested in this subject are referred to Darsie and Ward (1981). For broader information concerning anatomy, bionomics, and taxonomy, see Harbach and Knight, 1981; Mattingly, 1973; Harwood and James, 1979; and Russell et al., 1963. One should also remember that in many geographical locations a range of modern techniques is required for final taxonomic identification due to the presence of polytypic species and sibling species complexes (White, 1979). Additionally, genetic variability within species populations or demes (Fig. 41–8) can affect vector behavior (Coluzzi et al., 1977; Trpis and Hausermann, 1978) as well as susceptibility to infection by human pathogens (Hardy et al., 1983; Trpis et al., 1981). However, it is still desirable to learn the few relatively simple criteria by which *Anopheles*, *Culex*, and *Aedes* may be differentiated from one another and to understand the important roles that they play in the propagation and perpetuation of human diseases.

MOSQUITOES AND DISEASE

Mosquitoes have a twofold role in producing human disease: (1) They constitute the most important group of arthropods that transmit pathogenic microorganisms to man, including the causative agents of malaria, filariasis bancrofti, filariasis malayi, filariasis timori, and a large number of arboviral diseases (see page 643). (2) Mosquitoes are annoying pests which frequently produce an irritating dermatitis.

Malaria

Mosquitoes are the usual definitive hosts of mammalian and avian species of *Plasmodium*. Typically, the causative agents of mammalian malaria are only transmitted by *Anopheles* mosquitoes, but sporozoites of *Plasmodium cynomolgi bastianelli* and scanty oocysts of *P. falciparum* have developed experimentally in *Mansonia uniformis. Mansonia crassipes* may transmit *Plasmodium traguli* of Malaysian chevrotains (Garnham, 1966). Malaria parasites and infections in man are described in Section II (pp. 174 to 205).

Table 41–2. Principal Anopheles Vectors of Human Plasmodia.

This table includes 76 *Anopheles* species that merit consideration as vectors of human plasmodia (Mattingly et al., 1973; Harwood and James, 1979; Wernsdorfer, 1980). Specific names and geographical distribution conform to Knight and Stone (1977) and Knight (1978). Other data on systematics and bionomics are from various sources, including Horsfall (1955) and Russell et al. (1963). In certain cases, control has altered the status of some species to that of potential vectors. Bionomic characters are listed as follows: *Adults*—usually: **1,** feed on man indoors; **2,** feed on animals; **3,** feed on either man or animals; **4,** feed on man out-of-doors; **5,** rest indoors in daytime; **6,** rest out-of-doors in daytime; **7,** rest either indoors or outdoors. *Aquatic Stage*—usually found: **8,** in fresh water in household utensils; **9,** in puddles; **10,** in pools, ponds, lakes; **11,** in seepages, springs; **12,** in wells, cisterns; **13,** in ricefields; **14,** in running streams, canals; **15,** in brackish water in swamps, marshes; **16,** in shade; **17,** in sunshine; **18,** in either shade or sunshine; **19,** in epiphytic bromeliads; **20,** in tree holes.

I. Nearctic Region

A. crucians Wiedemann: Eastern U.S., Mexico, Central America; 4, 6, 10, 16; very low vector potential.

A. freeborni Aitken: Western U.S., southwest Canada, Mexico; 1, 5, 9, 10, 11, 13, 14, 16; formerly a vector in California and western U.S.

A. punctipennis (Say): Southern Canada, eastern U.S.; 3, 4, 6, 9, 10, 14, 16; very low vector potential.

A. quadrimaculatus Say: U.S., southern Canada, Mexico; 1, 5, 10, 17; formerly the main vector in U.S.

A. walkeri Theobald: Eastern half of U.S., southern Canada; 1, 4, 6, 10; very low vector potential.

II. Neotropical Region

A. albimanus Wiedemann: Texas and Florida to Uruguay, Antilles; 3, 6, 10, 11, 13, 15, 17; important wherever it occurs.

A. albitarsis Lynch Arribálzaga: Guatemala to Uruguay, Trinidad; 3, 6, 8, 10, 13, 17; vector in Brazil, Colombia.

A. aquasalis Curry: Nicaragua to Brazil; 3, 5, 13, 15; vector in lesser Antilles, Trinidad, Brazil.

A. aztecus Hoffman: Mexico; 1, 5, 10, 11, 14; vector at high elevations.

A. bellator Dyar and Knab: Brazil, Colombia, Surinam, Trinidad, Venezuela; 1, 4, 6, 19; vector in Trinidad and Brazil.

A. cruzii Dyar and Knab: Brazil and Venezuela; 3, 6, 19; vector in Brazil.

A. darlingi Root: Mexico to Argentina; 1, 5, 10, 16; important vector wherever found.

A. nuneztovari Gabaldón: Colombia, Bolivia, Guianas, Brazil; 1, 4, 7, 10; important incidental vector in Colombia and Venezuela.

A. pseudopunctipennis Theobald: Antilles, southern U.S. to Argentina; 3, 7, 10, 11, 16; important vector where found in abundance.

A. punctimacula Dyar and Knab: Mexico to Argentina; 3, 5, 10, 14, 16; incriminated in Panama and Colombia.

A. vestitipennis Dyar and Knab: Mexico, Central America, Colombia, W. Indies: 1, 5, 10, 11, 14, 16; comparatively rare but incriminated in Br. Honduras and Cuba.

III. Palearctic Region

A. algeriensis Theobald: Widely distributed throughout Europe and the Mediterranean region to U.S.S.R.; 3, 4, 10, 14; low vector potential.

A. atroparvus Van Thiel: Europe, littoral; 1, 5, 10, 13, 15, 17; formerly a vector in Netherlands, Germany, Spain, Portugal, Romania.

A. claviger (Meigen): Denmark, Germany, Great Britain, Spain, Corsica, Sardinia, Cyprus, southwestern U.S.S.R., Morocco, Israel, Syria, Lebanon, Iraq, Iran, Afghanistan; 3, 7, 10, 11, 12, 18; has supported endemic malaria in Cyprus, Israel, Lebanon, Syria, and around Baku.

A. dthali Patton: Middle East, Ethiopia, S. Arabia, Oman, Yemen, Aden, Somalia, Iraq, Iran, Pakistan, India, Tunisia, Algeria; 1, 5, 9, 11, 12, 14, 18; possible vector in Saudi Arabia, Sinai, Somalia.

A. hispaniola (Theobald): Mediterranean region; 3, 7, 10, 14, 17; chief vector in Canaries and incidental vector in Morocco and Algeria.

A. labranchiae Falleroni: Italy, Spain, Corsica, Sardinia, Sicily, Morocco, Algeria, Tunisia; 1, 5, 10, 11, 12, 13, 15, 17; usually important vector wherever found.

A. maculipennis Meigen: Continental Europe. S.W. Asia to Persian Gulf; 3, 5, 10, 13, 14, 17; vector in Caucasia, Trans-Caucasia, N. Iraq, and some areas of Greece, Yugoslavia, and Bulgaria.

A. messeae Falleroni: Northern Palearctic Region; 2, 7, 10, 17; formerly a vector in northern U.S.S.R., parts of Hungary, and eastern Romania.

A. multicolor Cambouliu: N. Africa, Cyprus, Middle East, Pakistan, Afghanistan; 1, 7, 10, 12, 15, 18; probable vector in oases of Egypt.

Table 41–2 *Continued*

A. sacharovi Favre: U.S.S.R., Austria, Italy, Cyprus, Greece, Middle East, Iraq, Iran; 1, 5, 10, 11, 15, 18; a vector wherever found.
A. sergentii (Theobald): N. Africa, Middle East, Pakistan; 1, 5, 10, 11, 13, 14, 17; vector in Israel, Jordan, Saudi Arabia, oases in N. Africa east of Algeria.
A. superpictus Grassi: Mediterranean region, Middle East, U.S.S.R., Iraq, Afghanistan, Pakistan; 1, 5, 10, 13, 14, 17; important vector wherever found.

IV. Oriental Region

A. aconitus Dönitz: Oriental region; 3, 5, 10, 13, 14; vector in Indonesia, Thailand, Cambodia.
A. annularis Van der Wulp: Oriental region; 3, 7, 10, 13, 14; considered an incidental vector in India (Assam).
A. balabacensis Baisas: Philippines, Indonesia (Java, Borneo), Malaysia, Indochina Peninsula, India (Assam); 1, 4, 7, 9, 10, 16; vector in Indonesia, Malaysia, Burma, and Assam.
A. barbirostris Van der Wulp: Oriental region; 3, 4, 10, 11, 13, 17; vector in Indonesia and Malay Peninsula.
A. campestris Reid: Malaysia, Thailand; 1, 5, 12, 13, 18; vector in Malaysia.
A. culicifacies Giles: China to Iran, Oman, Bahrain; 3, 5, 10, 11, 12, 13, 14; vector in India, Pakistan, Afghanistan, Sri Lanka.
A. dirus Peyton and Harrison: Thailand, Bangladesh; 1, 4, 9 (footprints), 10 transitory pools, 11, 16; an important sylvatic vector.
A. donaldi Reid: Thailand, Malaysia, Borneo; 1, 3, 6, 10, 16; minor vector in Borneo and Malaysia.
A. flavirostris (Ludlow): Philippines, Borneo, Sumatra; 3, 6, 11, 14, 18; important vector in the Philippines, Java, N. Celebes.
A. fluviatilis James: China to Iraq, Taiwan, U.S.S.R., Saudi Arabia, Bahrain, Oman; 1, 6, 10, 11, 14; important vector in India, Pakistan, and Nepal.
A. hyrcanus (Pallas): Central and northern Asia, northern Mediterranean; 1, 3, 10, 13, 16; generally zoophilic but an incidental vector when associated with more anthropophilic species.
A. jeyporiensis James: India, Burma, Indochina Peninsula; 1, 6, 10, 11, 13, 14; vector in S. China.
A. karwari (James): Oriental region, N. Guinea; 3, 7, 10, 11, 14, 17; vector in New Guinea.
A. kochi Dönitz: Oriental region; 3, 7, 9, 10, 11, 13; vector in Greater Sundas.
A. kweiyangensis Yao and Wu: China; 1, 13, 14; important vector in Szechuan Province.
A. lesteri Baisas and Hu: Philippines, Burma, Malaysia, China; 3, 7, 10, 13; important vector in lower Yangtze Valley.
A. letifer Sandosham: Malaysia, Sumatra, Borneo; 1, 6, 9, 10, 16; important vector in Malaysia.
A. leucosphyrus Dönitz: Sumatra, Borneo, Malaysia, Sri Lanka, Burma, Taiwan; 1, 5, 9, 12; important vector in Borneo; incidental vector in Indonesia.
A. maculatus Theobald: Oriental region; 3, 6, 11, 13, 14, 17; important vector in Malaysia, Sumatra, Java, Indochina Peninsula.
A. minimus Theobald: Oriental region, Ryukyu-Retto: 1, 5, 10, 11, 13, 14; important vector wherever found.
A. nigerrimus Giles: India, Pakistan, Sri Lanka, Burma, Thailand, Malaysia, China; 2, 6, 10, 13; vector in Celebes, Java, Sumatra, Thailand.
A. pattoni Christophers: China; 3, 14; vector in N. China.
A. philippinensis Ludlow: Philippines, Borneo, Malaysia, Indochina Peninsula, Indonesia, China, Burma, India; 3, 7, 10; important vector in Bengal and Bangladesh, Ganges Delta.
A. pulcherrimus Theobald: From India to Israel, U.S.S.R., Saudi Arabia; 3, 5, 10, 11, 13; vector in Central Asia and Pakistan.
A. sinensis Wiedemann: China to Afghanistan, Taiwan, Japan, Ryukyu-Retto; 2, 6, 10, 13; vector in Japan, Taiwan, Manchuria, central China, Indochina; incidental vector in Thailand.
A. stephensi Liston: China to Iraq, Bahrain, Oman, Saudi Arabia; 1, 5, 8, 12; important vector in India, Pakistan, Afghanistan, Iran, Iraq, Persian Gulf littoral.
A. subpictus Grassi: From China to Iran, Timor, Bismarck Archipelago, New Guinea; 3, 5, 9, 10, 11, 14, 15; incidental vector in Indonesia, important in Celebes.
A. sundaicus (Rodenwaldt): Indonesia, Taiwan, China, Burma, Indochina, Thailand, Malaysia, Singapore, Timor, Andaman Is., India; 3, 7, 15; important vector wherever found.
A. tessellatus Theobald: Oriental region, Moluccas, New Guinea; 3, 7, 9, 14; has been a vector in Taiwan.
A. umbrosus (Theobald): Malaysia, Indonesia, Indochina Peninsula, Philippines, India, Thailand; 1, 6, 10, 14, 16; possible vector in Malaysia.
A. varuna Iyengar: India, Sri Lanka, Burma, Nepal; 3, 5, 10, 14; vector in India.

V. Australian Region

A. annulipes Walker: Australia, New Guinea; 1, 5, 8, 9, 14; has had a part in transmission in Australia.
A. bancroftii Giles: Australia, New Guinea, Bismarcks; 3, 5, 10, 16; important vector in northern New Guinea.
A. farauti Laveran: New Hebrides, Australia, Solomons, Bismarcks, New Guinea, Moluccas; 1, 6, 9, 10, 11, 14, 18; important vector wherever it occurs.
A. koliensis Owen: Solomons, New Guinea; 1, 5, 9, 17; important vector in parts of its range, i.e., northeast Guadalcanal.
A. punctulatus Dönitz: New Guinea, Bismarcks, Solomons, Moluccas; 3, 6, 9, 10, 14, 18; important vector wherever it occurs.

Table 41–2 *Continued*

VI. Ethiopian Region

A. arabiensis Patton (formerly *A. gambiae*, species B): Across sub-Saharan Africa to Kenya and the Sudan, southern Arabia, Malagasy, Mauritius, Zanzibar; 1, 3, 4, 7, 10 (pools), 11. A principal vector in arid or montane areas.*

A. brunnipes (Theobald): widely distributed in tropical Africa; 3, 7, 9, 10, 14, 17; an incidental vector.

A. funestus Giles: widespread and often abundant over Ethiopian region, altitudes up to 6000 feet in Kenya and Uganda and 7200 feet in Ethiopia; found in Malagasy and Comoros; 1, 5, 10, 11, 14; an important vector.

A. gambiae Giles (Formerly *A. gambiae*, species A): widespread in the forest and humid savanna zones of tropical Africa, including Malagasy, Mauritius, Réunion, Comoros, Cape Verde Islands, Sao Thome, Principe, and Equatorial Guinea; 1, 3, 7, 9, 10, 11, 13, 17; an important vector.*

A. gambiae (species D): Semliki Forest area of Uganda-Zaire; 1, 3, 4, 7, 15 (mineral water swamps), 17; a localized vector.*

A. hancocki Edwards: chiefly in West Africa forest and savanna zones; 1, 5, 11, 14, 17.

A. hargreavesi Evans: mainly a West Africa forest zone species; 1, 3, 7, 14, 17.

A. melas Theobald: the West Coast saltwater member of the *A. gambiae* complex; 1, 3, 4, 7, 15, 17; an important vector.*

A. merus Dönitz: the East Coast saltwater member of the *A. gambiae* complex; found also in Pemba, Comoros, and Mauritius; 1, 3, 4, 7, 15, 17; an important vector.*

A. moucheti Evans: found in lowland forest belt from Cameroon and Gabon through Congo basin to Uganda; 1, 5, 10, 14, 18; an important vector where abundant.

A. nili (Theobald): found in most areas of the Ethiopian region; 3, 7, 14, 16; an important vector.

A. pharoensis Theobald: widespread except in forest canopy area; found in Malagasy, also in the Nile delta, and in western Arabia; 1, 3, 4, 7, 10, 13, 14, 17; a known vector in Egypt and probably an incidental vector elsewhere.

*The *Anopheles gambiae* complex has been shown to consist of at least four freshwater and two saltwater species (White, 1974; Mattingly, 1977). The latter now have specific status as *A. melas* and *A. merus*, both major vectors of *Plasmodium* spp. The freshwater species have been almost fully differentiated and are currently designated as *A. arabiensis*, *A. gambiae*, *A. quadriannulatus* (formerly species C), and *A. gambiae*, species D. *A. quadriannulatus* occurs in southern Africa and Zanzibar and is not a vector.

Knight and Stone (1977) and Knight (1978) have listed 372 species of *Anopheles*. Table 41–2 includes 76 species that merit consideration as vectors of human plasmodia. Many others have been found naturally infected but are not considered dangerous. Designation of a mosquito as a vector requires proof of infection in Nature. A species may be capable of transmitting the infection under controlled laboratory conditions and yet be influenced by certain biotic factors that prevent its becoming infected under natural conditions. For example, *Anopheles atropos*, a brackish-water species that is distributed along the Gulf Coast and South Atlantic Coast of the United States, has been found to be a good experimental host but is as yet not known to carry the plasmodia in the natural state. Furthermore, a species that is an important transmitter in one country or locality may be unimportant in another. In northern Argentina, *A. pseudopunctipennis* is extremely important, whereas in North America it is apparently harmless. In Malaysia, *A. maculatus* is an important carrier, but in Assam it is not. In Indonesia, *A. aconitus* of rice fields is a vector, but in India it has little importance. Furthermore, where the three common species of human malarial parasites *(Plasmodium vivax, P. malariae,* and *P. falciparum)* are all present in the same area, all of the plasmodia are usually transmitted by the same anopheline. Again, one species of *Anopheles* may be important in maintaining endemic malaria in a locality, while another species may be responsible for epidemic outbreaks as a result of rapid seasonal or sporadic increase in numbers of the second species (e.g., *A. sinensis* in Indonesia).

In order to be efficient vectors, *Anopheles* females must possess four characteristics (McGregor, 1976): They must feed frequently upon humans, be moderately susceptible to gametocytes, live long enough for the *Plasmodium* parasite to complete development, and be present in adequate numbers to maintain transmission. These biological characteristics must also be considered in conjunction with such important abiotic factors as temperature, rainfall, and humidity when evaluating the differing ecological states of malaria, i.e., *stable* or *unstable* (Russell et al., 1963).

Blood Preferences. Some of the differences in malarial infection and transmission rates may be due either to a willingness to engorge on human blood *(anthropophilism)* or animal blood *(zoophilism)* or to the relative abundance of hosts. Therefore, surveys to determine feeding preference of mosquitoes for human blood may be misleading unless one checks the availability of domestic animals for blood supply. Hence, the use of a *human* *blood index* can establish the status of a mosquito as a known, suspected, or potential vector (Garrett-Jones et al., 1980). This ratio is a relative measure of human host utilization based on serological identification of single blood meals. In addition, interrupted feeding that results in two or more partial meals may be of possible epidemiological importance (Boreham, 1975), particularly where utilization of different host species (for instance, humans and bovines) could result in lowered transmission rates and possibly *zooprophylaxis* (loss of sporozoites to an inhospitable host). Conversely, where multiple feeding is "unmixed," but from two hosts of the same species having differing states of receptivity to *Plasmodium* spp. (for instance, infants versus adults), the effects could be of profound significance in the amplification of the parasite population.

Indexes of Malaria in the Mosquito. Malaria infection in the anopheline mosquito is determined by a dissection of the suspected females from endemic foci and examination of the stomach wall for oocysts or "sporocysts" and of the salivary glands for sporozoites. The salivary gland index (i.e., *sporozoite rate*) is considered the more valid criterion, since it indicates the amount of complete development of the plasmodium in the mosquito. In highly endemic areas the sporozoite rate of naturally infected anophelines has varied from 0.1 to 30.0%, depending on a variety of factors.

Since unfed, *nulliparous* mosquitoes are not infected, only those individuals having imbibed blood should be considered in a sporozoite survey. However, some species of *Anopheles* may require multiple blood meals and, following a single meal, can be recognized as "pregravids" by their partially developed ovaries (Gillies, 1954). Ordinarily, *parous* flies (those individuals having completed a gonotrophic cycle) can be determined by noting several changes in the reproductive tract, including an increase in the size of the ampulla (the confluence of the lateral and medial oviducts), the presence of follicular relics (dilatations of the follicular tube containing cellular debris), and mature ova retained from previous gonotrophic cycles. From a practical standpoint, in making malaria surveys, it should be remembered that certain malaria parasites of animals may develop in *Anopheles* mosquitoes.

While the role played by the more widely distributed anophelines primarily responsible for malaria has been known for several years, nearly every year one or more additional species, at times of considerable importance in restricted areas, are added to the list of transmitters. In India several species of *Anopheles*, each in a special epidemiological relation, are known to be important hosts. In the United States only two species are of primary importance, in Sri Lanka only one. As more intensive taxonomic and genetic studies are carried out, anophelines previously believed to be innocuous may be found to be important, and vice versa.

Comprehensive keys and systematic notes for the identification of the more common anophelines of the world have been published (Russell et al., 1963; Gillies and De Meillon, 1968; Reid, 1968; Harrison and Scanlon, 1975; Harrison, 1980; others cited in Knight and Stone, 1977 and Knight, 1978). Miles (1979) has constructed a biochemical key to adults of the *Anopheles gambiae* complex using horizontal starch gel electrophoresis. Kitzmiller (1976) reviewed mosquito cytogenetics and included references to salivary gland chromosome maps of *Anopheles*. Methods of control are discussed on pages 656 to 657.

Arboviral Diseases

Casals (1967) defined the arboviruses (arthropod-borne viruses) as animal viruses naturally transmitted between vertebrates and hematophagous arthropods. These viruses multiply in the tissues of both the arthropod and vertebrate host. In mosquitoes, the viremic blood meal is passed to the midgut where the epithelial cells soon become infected. Following an initial "eclipse phase" of replication within the gut, progeny virions escape into the hemocoel and spread throughout the body (Murphy et al., 1975). Salivary gland infection then occurs. After a suitable period of incubation, infectious virions are available for discharge in the saliva (Fig. 41–9). In the tissues of the vertebrate host the virus may be essentially harmless causing asymptomatic infection or may produce disease, at times with fatal termination. Under certain circumstances the virus may be mechanically transmitted, but for all practical purposes biological amplification in the susceptible arthropod and the vertebrate host is essential for the epidemiological cycle.

Classification of Arboviruses. Mosquito-associated viruses that are biologically transmitted can be conveniently divided into two families—Togaviridae and Bunyaviridae—based on genome characteristics (number of molecules of single-stranded

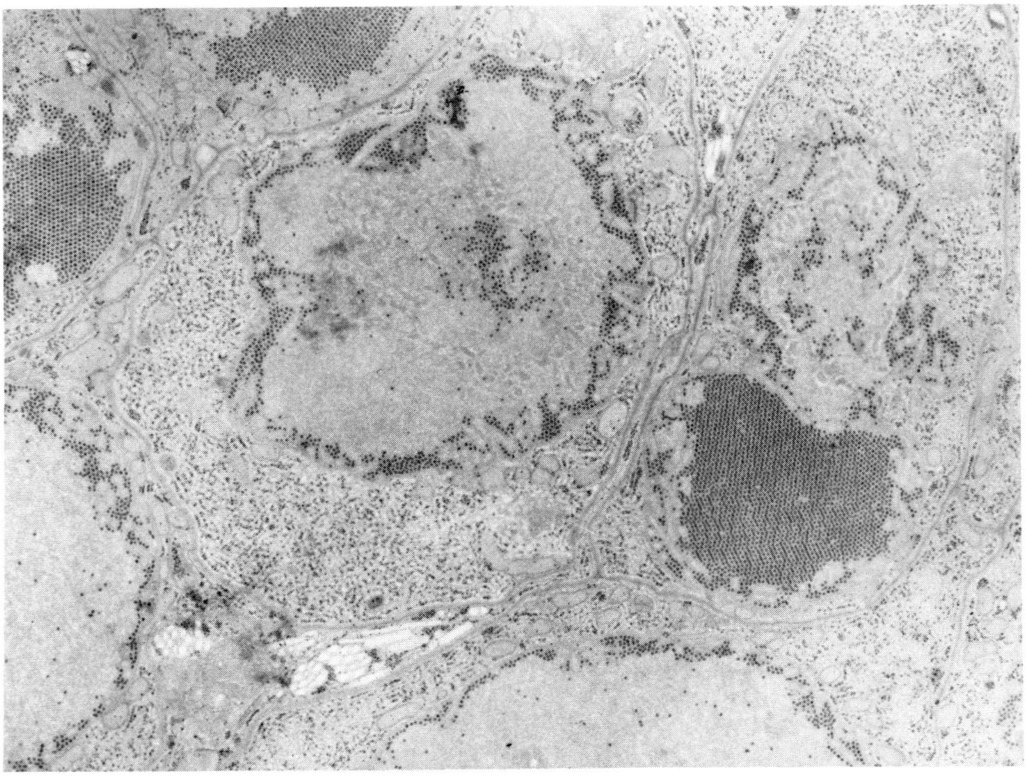

Fig. 41–9. Saint Louis encephalitis virions in *Culex pipiens* salivary gland cells. (× ca. 14,000.) (From Whitfield, S.G., Murphy, F.A., and Sudia, W.D. 1973. St. Louis encephalitis virus: An ultrastructural study of infection in a mosquito vector. Virology, *56*:70–87.)

RNA and method of translation), general morphology, and antigenic properties (Matthews, 1982). Further resolution to the genus and species levels can be made by using such serological techniques as hemagglutination inhibition, complement fixation, and neutralization. A number of these viruses occur commonly and are of clinical importance because they cause outbreaks of mild febrile illness, encephalitis, hemorrhagic fever, arthritis, and myalgia (Table 41–3). Most are zoonotic pathogens that use wild birds and mammals as reservoir hosts. The more important representatives of each family, their vectors, transmission cycles, and distribution are discussed in the following section.

Togaviridae: Genus *Alphavirus*

Eastern Equine Encephalomyelitis (EEE)

The viral agent of EEE causes a severe and frequently fatal encephalitis in humans and horses. It is distributed principally in the New World, ranging from southern Canada to northern South America (Fig. 41–10). In the eastern United States, EEE virus is maintained in a bird-mosquito cycle involving *Culiseta melanura* and a variety of avian species. This mosquito is highly ornithophilic and is the principal enzootic vector in freshwater swamps and wet forests. *C. melanura* may also disperse from these sylvatic foci and establish infections in domestic and peridomestic birds (Morris et al., 1980). The mechanisms associated with the epizootic-epidemic phase are still somewhat ill-defined, though it is believed that *Aedes taeniorhynchus, Ae. sollicitans, Ae. vexans, Ae. canadensis,* and *Coquillettidia perturbans* may serve as vectors owing to their more catholic feeding habits.

Western Equine Encephalomyelitis (WEE)

WEE virus also affects humans and equines. Human infection is more frequent than with EEE with a higher ratio of inapparent infections (Downs, 1976). The principal vector in the United States is *Culex tarsalis,* a mosquito whose aquatic stages are often associated with irrigated crops (Mitchell, 1977). Passerine birds normally serve as hosts with nestling House Sparrows *(Passer domesticus)* act-

Table 41–3. Some Mosquito-Transmitted Viruses of Clinical Importance.*

TOGAVIRIDAE

Genus *Alphavirus* (25 species)

Chikungunya (acute arthritis, often with rash). East Asia, India, Africa, Philippines.

Eastern equine encephalomyelitis (neurotropic, febrile). United States, Mexico, northern South America, eastern Europe, southeast Asia.

Mayaro (febrile). Northern South America, Trinidad.

O'Nyong-nyong (febrile). Sudan, Uganda, Kenya, Tanzania, Zaire.

Semliki Forest (neurotropic). North and south Africa, Philippines, India, Malaysia, Borneo, southeast Asia.

Venezuelan equine encephalomyelitis (neurotropic, myalgic). Argentina, Brazil, northern South America, Central America, Mexico, Florida (United States).

Western equine encephalomyelitis (neurotropic, febrile). North America, Mexico, eastern South America.

Ross River (epidemic polyarthritis). Northern and eastern Australia, New Guinea, Fiji, American Samoa.

Genus *Flavivirus* (26 mosquito-transmitted species)

Dengue types 1, 2, 3, and 4 (arthritic, myalgic, and with rash; serotypes 1, 3, and 4 occasionally hemorrhagic with shock, epidemically in southeast Asia with high morbidity and appreciable mortality, especially in children). Both New and Old World Tropics; active in southeast Asia and the Caribbean Region.

Ilhéus (neurotropic). Brazil, Venezuela, Colombia, Panama, Central America.

Japanese encephalitis (neurotropic, febrile, with occasional encephalitic complications). Japan, south to Philippines, and west to the Indian subcontinent.

Murray Valley encephalitis (neurotropic, febrile). Southeast and northeast Australia, New Guinea.

Rocio (neurotropic, febrile). Recently epidemic in São Paulo State, Brazil.

St. Louis (neurotropic, febrile). United States, Trinidad, Panama, Jamaica, Venezuela, Argentina.

Wesselsbron (severe influenza-like). South and central Africa, Thailand.

West Nile (dengue-like, febrile). Central, southern, and north Africa (Egypt), France, Israel, U.S.S.R. to India.

Yellow Fever (viscerotropic). East, central, and west Africa, tropical America.

BUNYAVIRIDAE

Genus *Bunyavirus*

Bunyamwera group (23 species; usually cause febrile illness).

Bunyamwera (febrile with rash). East, central, and west Africa.

Calovo. Czechoslovakia, Austria, Yugoslavia.

Germiston. South and central Africa.

Guaroa. Colombia, Brazil, Panama.

Ilesha. Central and western Africa.

Wyeomyia. Panama, northeastern South America.

C Group (14 species; febrile illness with occasional myalgia)

Apeu. Brazil.

Caraparu. Panama, Brazil (possibly Trinidad, Surinam).

Itaquí. Eastern Brazil.

Madrid. Panama.

Marituba. Eastern Brazil.

Murutucu. Eastern Brazil, French Guiana.

Nepuyo. Mexico, Guatemala, Honduras, Panama, Brazil, Trinidad.

Oriboca. Eastern Brazil, Surinam, French Guiana, Trinidad.

Ossa. Panama.

Restan. Trinidad, Surinam.

California Group. Originally isolated in California, this serological group now includes 14 species. Several may cause a febrile disease and, in some instances, an encephalitis of mild to moderate severity in children and adults.

California encephalitis. Western United States.

Inkoo. Finland.

Jamestown Canyon. Central and eastern United States.

La Crosse. Midwestern and eastern United States.

Tahyna. France, Italy, eastern Europe, U.S.S.R., eastern Africa.

Genus *Phlebovirus*

Rift Valley fever (febrile and hemorrhagic illness, with occasional ocular complications). Egypt, eastern and northern Africa.

*For a more complete classification of arboviruses, see Matthews (1982).

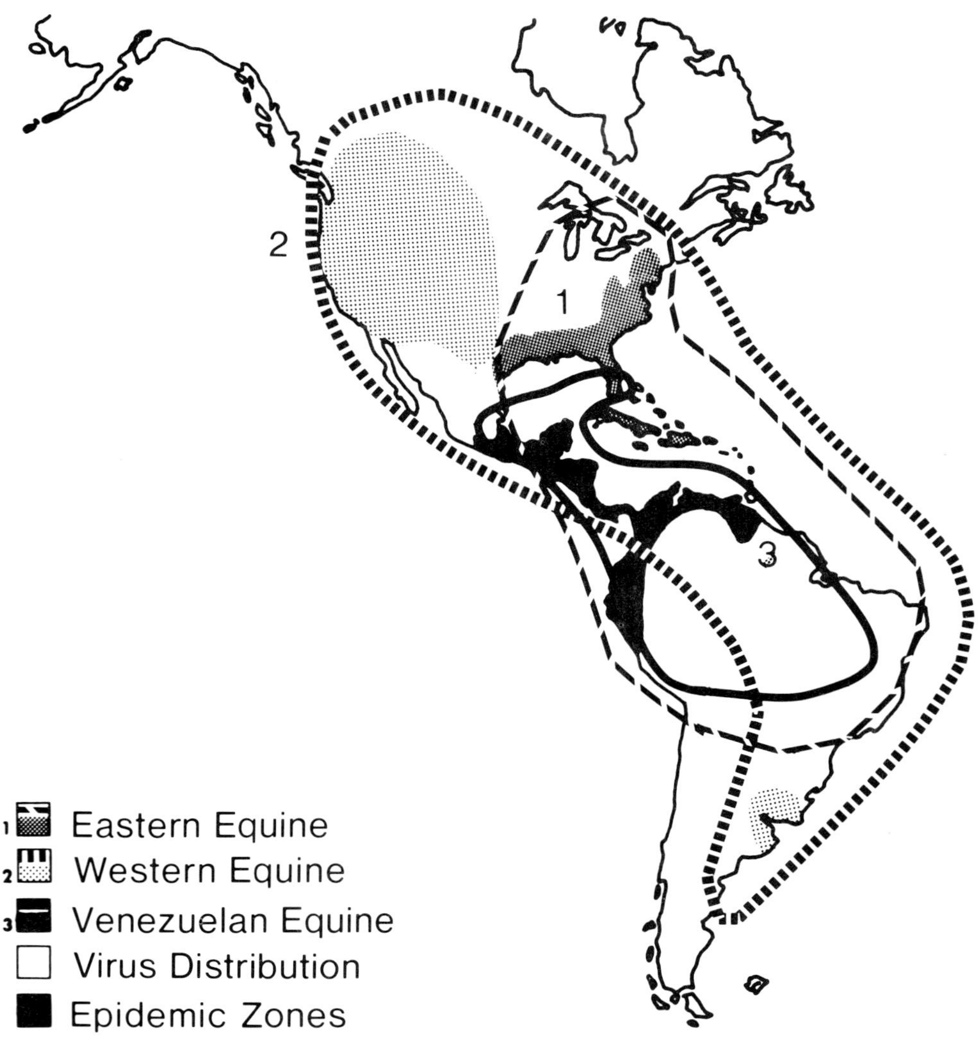

Fig. 41–10. Principal distribution of alphaviruses causing human encephalitis. (From Mitchell, C.J. 1977. Arthropod-borne encephalitis viruses and water resource developments. Cah. O.R.S.T.O.M., Ser. Entomol. Med. Parasitol., *15*:241–250.)

ing as excellent virus amplifiers. WEE virus is found principally in the western and north-central United States, southern Canada, Mexico, and eastern South America.

Venezuelan Equine Encephalomyelitis (VEE)

VEE virus is widely distributed throughout the Americas and exists as a complex of four antigenically distinct strains (I to IV), with five subtypes within strain I (IA to IE). Subtypes IA, IB, and IC are equine virulent and also cause a mild to moderately severe encephalitis in humans. These subtypes are explosively epizootic-epidemic, produce high viremias, and have a wide range of primary and incidental mosquito vectors, including species of *Aedes, Psorophora, Mansonia, Anopheles,* and possibly *Deinocerites pseudes* (Sudia and Newhouse, 1975). The remaining strains and subtypes are more highly focalized and primarily use species of *Culex (Melanoconion)* as vectors and small mammals as amplifiers (Galindo, 1972). *Culex portesi, Cu. aikenii sensu lato,* and *Cu. opisthopus* (= *taeniopus)* are vectors of enzootic VEE virus when humans enter the focus (Cupp et al., 1979).

Chikungunya Fever

The viral agent of chikungunya fever is transmitted in Africa by *Aedes africanus* and *Ae. aegypti,* which also is a vector in India and southeast Asia. Clinical signs include a sudden onset of fever and crippling joint pains that may completely incapacitate the patient (Simpson, 1978). It is suspected that monkeys may serve as hosts for this virus.

Ross River Fever

Ross River virus causes epidemic polyarthritis, an illness characterized by fever, polyarthralgia, and rash (Tesh et al., 1981). Its distribution includes Australia, New Guinea, American Samoa, and Fiji. Humans apparently serve as virus reservoirs in American Samoa with *Aedes polynesiensis* and *Ae. oceanicus* as probable vectors. In Australia, *Aedes vigilax* and *Culex annulirostris* are transmitters with wild rodents serving as probable reservoirs (Gard et al., 1973).

Togaviridae: Genus *Flavivirus*

Yellow Fever

History and Geographical Distribution. Yellow fever is an acute, febrile, infectious disease of variable severity and short duration caused by YF virus, the type species for the genus *Flavivirus*. This disease was not recognized as a distinct entity until the seventeenth century. During the eighteenth and nineteenth centuries the disease occurred epidemically in port cities of southern Continental Europe and southern England, as well as on the Atlantic Coast and Gulf Coast of the United States. It was repeatedly carried inland from southern ports of the United States, but invariably died out when cold weather developed. The last epidemic in the United States occurred in New Orleans in 1905. It was prevalent throughout the year in the West Indies, in many of the Gulf and Caribbean cities of Mexico, Central America, and northern South America, and in the Atlantic ports of Brazil. Known to exist on the West Coast of Africa, it developed to epidemic proportions when trading and slave ships touched at these ports.

The year 1900 marked a turning point in the history of yellow fever. Although Carlos J. Finlay of Havana had long advocated mosquito transmission of yellow fever, not until 1900 did critically controlled experiments in Cuba by the U.S. Army Yellow Fever Commission (Doctors Reed, Carroll, Lazear, and Agramonte) provide conclusive proof in support of Finlay's hypothesis that the domestic breeder *Aedes aegypti* (the "Stegomyia" mosquito) was the transmitting agent. By 1925 yellow fever was believed to be eradicated from the globe, but soon thereafter epidemics broke out in South America and in West Africa. Patient epidemiologic studies by several groups of workers led to the discovery that yellow fever existed in forest or jungle habitats. In these areas wild mosquitoes and reservoir hosts perpetuated the disease without need for man or the domestic *Aedes aegypti*. When man entered enzootic zones and was bitten by an infected mosquito, he was inoculated and frequently carried the disease back to urban areas, where it became epidemic.

Today, by new techniques and continuous efforts, yellow fever has been eradicated from most urban communities in the Western Hemisphere, but it persists in areas in South and Central America and Central Africa (Fig. 41–11). In South America, sylvatic (so-called "jungle") yellow fever exists throughout vast stretches of the Amazon River valley, from Bolivia northward, including this portion of Peru, Ecuador, Colombia, and Venezuela; likewise it occurs in the Caribbean drainage of Colombia, Venezuela, and the Guianas. In Central America it broke out progressively from Panama to Costa Rica in 1948 to 1949, was responsible for minor epidemics in natives living in villages contiguous to the enzootic area, and by 1955 had reached northern Honduras. In December 1955 and in January 1956 it was diagnosed in howler monkeys from the north coast of Honduras near Esparta and the Montagua River valley in Guatemala. In 1962 there were 52 cases reported from the Americas, 1 from Brazil, 30 from Colombia, 20 from Peru, and 1 from Venezuela. In Africa, sylvatic yellow fever extends from 10° S. latitude (except Angola) to 15° N. latitude, comprising a belt from the Atlantic to Ethiopia. Its potential distribution is much greater than its present-day extent, since yellow fever may develop in any territory where *Aedes aegypti* is present, provided the virus is introduced in a mosquito, monkey, or man.

Epidemiology. Yellow fever is primarily a sylvatic disease of warm moist climates. In such zones the virus exists in certain monkeys and other reservoir hosts. In South America day-biting wild species of *Aedes* and *Haemagogus* are primarily involved, and man becomes infected when he enters the jungle or possibly at times in his home a few yards from the edge of the forest (Pinheiro et al., 1981a). *Haemagogus* prefers high sylvatic foliage, but descends to ground level where there is sunlight, a road, or a small canyon (Fig. 41–12). In West Africa, species belonging to the *Aedes furcifer-taylori* group and possibly *Ae. luteocephalus, Ae. metallicus,* and *Ae. vittatus* are sylvatic vectors (Cornet et al., 1979; Port and Wilkes, 1979; Germain et al., 1980). In East Africa, however, the situation is slightly different. Man does not usually become infected in the jungle but from bites of *Aedes simpsoni* and possibly other semidomestic mosquitoes that are infected from marauding monkeys that leave the treetops for cultivated food, such as bananas. This is *enzootic yellow fever,* the "jungle" yellow fever, in which the wild mammalian hosts are not true reservoirs but in which the virus persists in the mosquitoes during their lifetime (Fig. 41–13).

Experimental laboratory evidence indicates that YF virus can be passaged transovarially in *Haemagogus equinus* (Dutary and LeDuc, 1981), a sylvatic vector in the New World, as well as *Aedes aegypti* (Beaty et al., 1980), an urban vector in both Africa and South America. Male mosquitoes infected with YF virus have also been collected from enzootic foci (Cornet et al., 1979). Thus, the mosquito vectors may, in a broad sense, be regarded as true reservoirs of infection. Human contact with the forest or, in the case of marauding monkeys in East Africa, infection of semidomestic mosquitoes, permits infected mosquitoes to inoculate man, who carries the infection to his home,

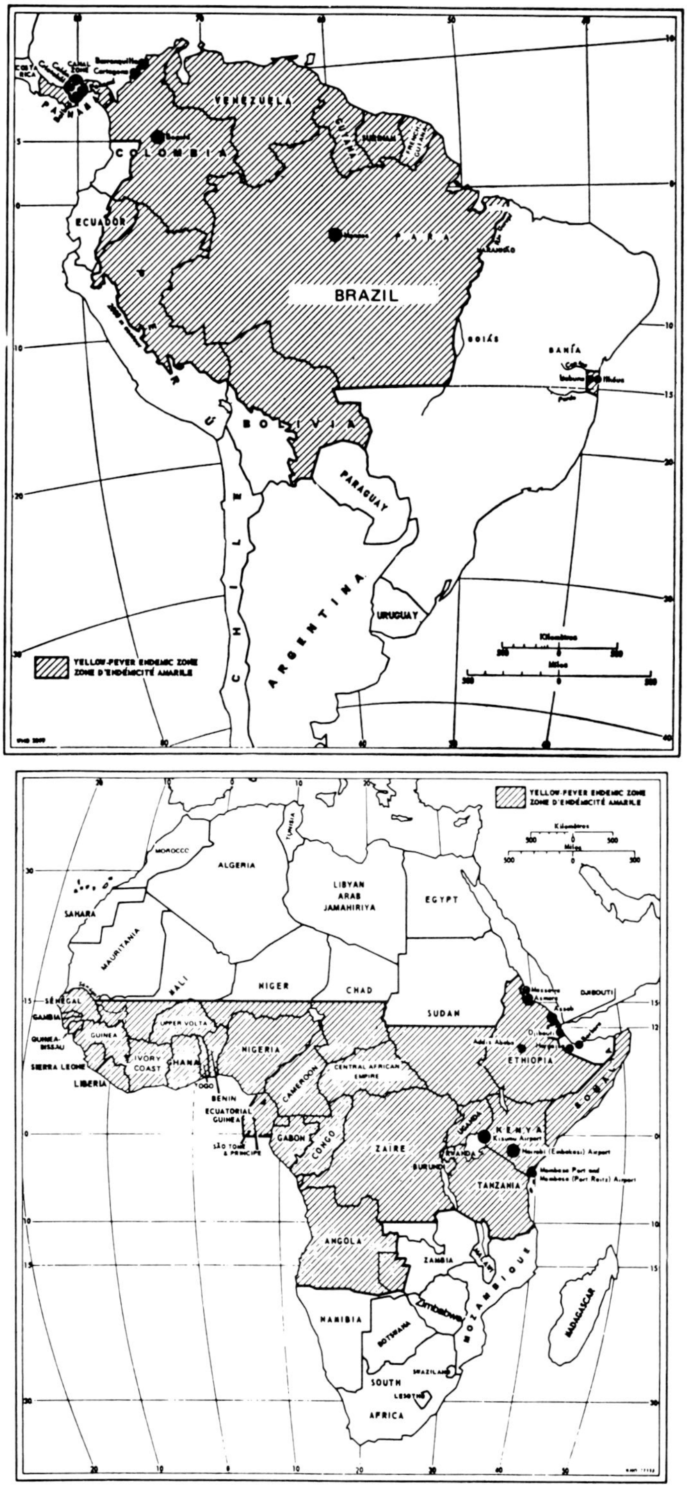

Fig. 41–11. Map showing yellow fever endemic zones in the Americas (*A*) and Africa (*B*). (From *Health Information for International Travel*. 1980. Centers for Disease Control. Atlanta, Ga.)

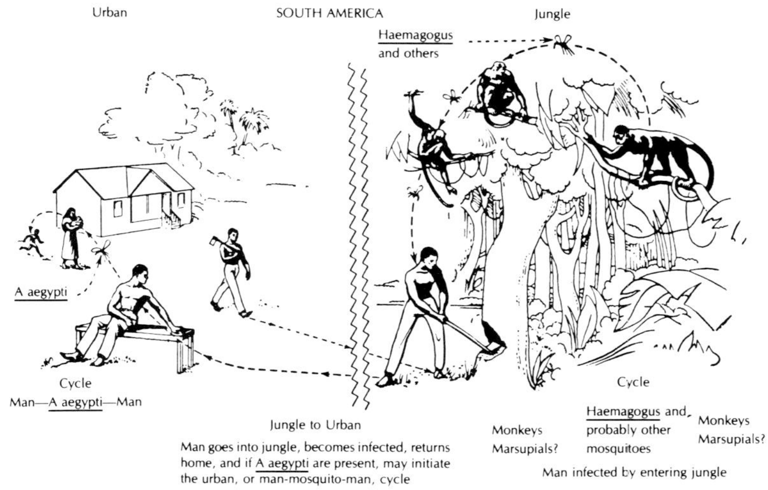

Fig. 41–12. Diagram illustrating the epidemiology of enzootic yellow fever in tropical America and the means of human exposure. (From Strode, G.K. [editor]. 1951. *Yellow Fever.* New York, McGraw-Hill.)

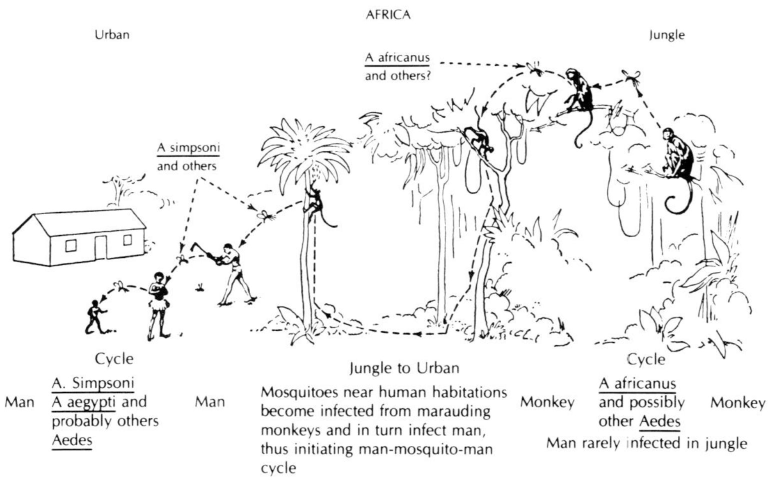

Fig. 41–13. Diagram illustrating the epidemiology of enzootic yellow fever in East Africa, and the means of human exposure. (From Strode, G.K. [editor]. 1951. *Yellow Fever.* New York, McGraw-Hill.)

whence it may be transported to urban centers where *Aedes aegypti* is found. Urban yellow fever is *epidemic yellow fever.* Classical urban yellow fever still exists along the west coast of Africa. The recent epidemic in The Gambia (Monath et al., 1980) showed the potential severity of this disease.

If susceptible mosquitoes take a blood meal from exposed wild mammals or human beings during the last 6 to 10 hours of the incubation period or during the first 3 or 4 days after symptoms develop, there is an incubation period of 8 to 14 days (usually 12 days) in the mosquito, after which the mosquito becomes infective and remains so for the rest of its life. Incubation in the mosquito is accelerated by warm weather and is prolonged during cooler periods. The multiplied virus is present in most tissues of the mosquito, which shows no detectable ill effects. The virus is usually inoculated into man in minute droplets of saliva introduced into human skin at the time the insect is feeding. Infection may also be transmitted through contact with the blood of an infected person during the first 3 or 4 days of the disease. Most of the laboratory infections in man have been contracted in this manner during handling of infected animals.

Pathogenesis. Yellow fever virus is primarily viscerotropic. The liver is the organ that first manifests marked destructive changes. In experimental animals degeneration of hepatic Kupffer cells appears within 24 hours. By the fourth day the virus produces toxic coagulation necrosis and nuclear degeneration. These hyaline necrotic cells lose all affinity for basic dyes and stain a bright cerise; they are the so-called "Councilman bodies." Progressing from the center of the lobule the necrosis becomes more intense in the midzone, but toward the periphery the destruction is frequently less active. Here some normal parenchyma may remain. The midzonal necrosis and fatty degeneration, together with intermediate normal parenchyma, constitute a pathologic picture characteristic of yellow fever.

In addition to the aforementioned changes, there are minute hemorrhages as the virus destroys the walls of the hepatic blood vessels. The Kupffer cells undergo granular degeneration, with bile elements and iron pigment in the cytoplasm, and the sinuses become infiltrated with leukocytes. Extensive hemorrhages may occur beneath Glisson's capsule. The condition referred to by pathologists as "boxwood" liver is due to extensive fatty changes.

In severe cases, rapid, extensive necrosis of the hepatic parenchyma almost immediately affects metabolic activities and leads to hypoglycemia, an increase in guanidine-like materials, and coagulation defects.

About the end of the second or third day toxic nephrosis develops. The tubules contain hyaline and granular casts, their lining develops necrotic and fatty degeneration, and the glomeruli become enlarged and congested. Albuminuria with albumin up to 25% of total volume occurs. Peptonuria is found in fatal cases.

Cloudy swelling and degeneration of the myocardium, including the conducting bundles, are characteristic. The stomach and intestines undergo extensive necrosis of the mucosa, with profuse hemorrhage, as do all the mucous membranes.

Symptomatology. The asymptomatic incubation period is 3 to 6 days, rarely longer. The subsequent illness may be inapparent, mild, moderately severe, or malignant. In mild yellow fever the only symptoms may be headache and fever of abrupt onset. Nausea, epistaxis, mild albuminuria, and Faget's sign may be noted. Mild yellow fever lasts only 1 to 3 days and resembles influenza except for the lack of coryza.

Moderately severe and malignant cases have three distinct clinical phases: invasion, remission, and intoxication. There is acute onset with severe malaise, intense headache, and muscle pains, accompanied by temperature of 39° to 40°C without bradycardia. The conjunctivae are congested; the skin is hot, dry, and congested. Thirst is intense. The pulse is elevated to about 120 but soon falls progressively as Faget's sign begins to appear by the second day of illness. The urine, normal at first, contains albumin at the end of the second day or later. Bleeding from the gums and nasal mucosa may occur. Usually there is nausea and vomiting, intense epigastric pain, extreme irritability, anxiety, and commonly prostration. There is a characteristic leukopenia.

On the second to fourth day the temperature declines, with lessening of subjective symptoms, although in the malignant form severe hemorrhages, anuria, and delirium may occur. However, the temperature soon returns to 39° to 40°C as a result of intense intoxication, with occurrence of toxic symptoms and increase in hemorrhage and in albumin (or peptone) in the urine. Jaundice appears about the third day, though it often is not conspicuous even in fatal cases. As the temperature increases, the pulse falls to as low as 40 per minute (Faget's sign). With extensive hematemesis, marked peptonuria, or anuria, the prognosis becomes grave. Delirium and agitation commonly appear shortly before death, which is prone to occur between the fourth and sixth days. Patients remain febrile in the third stage for 5 to 7 days, but the other signs of intoxication may last up to 2 weeks. In patients who have a favorable prognosis, the temperature gradually declines to normal and a protracted convalescence begins.

Recovery, with practically complete regeneration of necrotic tissue, almost always leads to a lifelong immunity to subsequent infection.

Diagnosis. Except during epidemics, when the patient manifests a characteristic picture of the disease, the diagnosis is based on three established laboratory procedures: 1) isolation of the virus from blood within the first 3 days of illness, 2) demonstration of increased titer of neutralizing antibody, and 3) demonstration of the typical hepatic histopathology on liver biopsy or postmortem viscerotomy. However, physicians attending a patient who has been in an endemic focus or has had a possible exposure should be mindful of the diagnostic significance of Faget's sign coupled with early albuminuria, jaundice, and hemorrhage, especially from the mucous membranes.

Four weeks after onset of the disease the patient's serum may be tested for neutralizing antibody in mice inoculated intracerebrally with the neurotropic (or other) strain of the virus. If the mouse survives, there is substantial proof that the patient has had the disease *(mouse protection test).* Moreover, this same test has been employed, as has the complement-fixation test, to determine the immunity titer following vaccination, while hemagglutination has been employed extensively in surveys to determine the presence of antibody developed to yellow fever and other arthropod-borne viruses.

Yellow fever requires differentiation from viral hepatitis, dengue fever, malaria, typhoid fever, Weil's disease, the relapsing fevers, and, at times, the typhus group of fevers.

Treatment. Treatment of yellow fever is entirely symptomatic and supportive. Complete rest in bed is the first essential, especially during the period of remission. Under primitive conditions a patient should not be moved. If not vomiting, he should drink a liberal amount of water supplemented by citrus fruit juices. A liquid diet high in carbohydrates is indicated. To counteract the hypoglycemia, there should be intravenous infusion of sterile glucose solution. Atropine may relieve gastrointestinal symptoms. Circulatory and electrolyte abnormalities should be detected and corrected. If disseminated intravascular coagulation is evident, administration of heparin should be considered.

Prognosis. With good nursing the prognosis is fair to good (5 to 10% fatality rate).

Prevention. Theoretically, it is possible to prevent yellow fever by destruction of the arthropod transmitters and vaccination of people. Destruction of the arthropod vectors is practical only in so far as the domestic mosquito, *Aedes aegypti,* is concerned. Persistent efforts in this respect have been successful in some areas. Protection has been afforded primarily through vaccination. Fresh concentrated vaccine (0.5 ml of a 1 to 10 dilution) is given subcutaneously. Revaccination is required every 6 years, although protection may last much longer, even for life.

A broad program of yellow fever control should include the following:

1. Continuous surveys of yellow fever distribution, including viscerotomy and immunity studies.
2. Anti-*Aedes aegypti* measures directed toward eradication wherever possible in all enzootic-endemic or potentially infectible areas.
3. Vaccination of travelers to and from enzootic-endemic areas, exposed rural populations, and urban communities where *A. aegypti* still breeds.
4. International regulations, including anti-*A. aegypti* control of ports and airports, disinsectization of aircraft, and quarantine of potentially infected travelers.

Dengue

History and Geographical Distribution. Dengue may develop in any warm, moist climate where transmitting mosquitoes exist. In recent years it has occurred epidemically in Hawaii, Philippines, Thailand, Malaysia, Vietnam, southwest Pacific region, Indonesia, and Africa. During this past decade it has raged in the Caribbean and along the coast of northern South America.

The infectious febrile disease caused by dengue virus is highly incapacitating. In epidemics of hemorrhagic types 3 and 4 it may cause high morbidity and even high mortality, especially in children. This flavivirus is contracted from bites of infected mosquitoes, usually *Aedes aegypti,* but occasionally other *Aedes* species. The disease was first described by Benjamin Rush in Philadelphia, Pennsylvania in 1780, and its viral etiology was established by Ashburn and Craig in 1907.

Epidemiology. Dengue virus is transmitted to man by the bites of certain mosquitoes, of which *Aedes aegypti* is the most common, although other species (including *A. albopictus, A. scutellaris,* and *A. polynesiensis*) have also been incriminated. The mosquito becomes infected only during the first 3 days of the patient's illness and is not infective for man until after an 11-day incubation period; however, the interval can be as short as 8 days. Once the mosquito reaches the infective stage, it is infective until it dies. The virus is found in all of its tissues, but human infection typically occurs when the mosquito injects saliva into the skin. Vertical transmission in the vector has been experimentally demonstrated (Tesh, 1980). Although monkeys may be involved in rural situations, man is apparently the principal host.

Pathogenesis, Pathology, and Symptomatology. The dengue virus probably penetrates into all tissues of the human body but has special predilection for parenchymatous organs and the endothelium of blood capillaries. Here by toxic action it produces degenerative changes of the cells and hemorrhage.

Dengue has an incubation period of 4 to 10 days. An abrupt onset of symptoms occurs, with a chill or chilly sensation, usually with excruciating pains in the joints, loins, epigastric region, and just behind the eyeballs; redness and swelling of the face; a rapid rise in temperature to 103° or rarely 104°F

(39.5° or 40°C) lasting 4 to 7 days; elevated pulse; and moderate to extreme prostration. Following this acute stage there is characteristically an abrupt drop in temperature, accompanied by profuse sweating, diarrhea, pyuria, and at times epistaxis. There is a distinct leukopenia, with relative lymphocytosis and monocytosis, and granulocytopenia during convalescence. Although the patient feels weak and dizzy, his pain becomes greatly ameliorated. Within a few hours, the fever returns in about 66% of cases (hence "saddle-back" designation for the temperature curve). The patient experiences extreme malaise, nervous depression, and the development of the relatively characteristic pathognomonic sign of a terminal macular, maculopapular, or scarlatiniform rash, which progresses rapidly from fingers and toes up the limbs to the trunk and at times covers the whole body. In most cases it lasts 2 or 3 days, then recedes and is followed by an itching or prickling desquamation. Frequently the cervical and other lymph nodes are enlarged. Occasionally the conjunctivae and skin appear icteric. The period of convalescence is slow, and for weeks or months afterward there may be nervousness, insomnia, relative bradycardia, and rheumatic pains. Many mild cases occur in which symptoms are atypical, or are so slight as to go unnoticed.

Dengue virus also causes a hemorrhagic fever (DHF), a severe illness that is endemic in most of tropical Asia and has occurred in tropical America. It is characterized by abnormal vascular permeability, hypovolemia, and impairment of blood clotting in addition to high fever and nonspecific constitutional symptoms. It is recognized mostly in children, although adults are at times affected. After a few days of fever, some patients experience a sudden deterioration of condition manifested by signs of circulatory failure and shock. This is the dengue shock syndrome (DSS). DHF, especially with DSS, has a significant mortality rate (10 to 50%). Serologic tests show a rise in titer of anti-dengue antibody, usually of the anamnestic type (IgG).

DHF may be caused by types 1, 2, 3, or 4 dengue virus (group B togavirus), but more likely by types 3 and 4. Most cases of DSS are seen during the second of two successive dengue infections due to two different types.

Diagnosis. Sudden onset, with severe pain in the joints and back of the eyeballs, the "saddle-back" fever curve, when it occurs, and the type of rash described previously are relatively pathognomonic, as is also the accompanying leukopenia, with monocytosis. Similar symptoms are produced by West Nile and chikungunya arboviruses. There is almost invariably a marked "shift to the left" in the Schilling count. Dengue must be differentiated from yellow fever, especially during the first 2 days of illness, from phlebotomus fever, malaria, measles, and scarlatina during the period of rash. In jaundiced patients, Weil's disease and relapsing fevers must be ruled out. Undulant fever, typhoid, influenza, and meningitis must be excluded.

Dengue virus can be propagated safely for diagnosis by intrathoracic inoculation of the test specimen into either sex of *Toxorhynchites amboinensis*, a large nonbloodsucking mosquito, or other species of *Toxorhynchites*. After 14 days of incubation at 32°C, virus may be easily detected by fluorescent antibody staining of head squashes using specific mouse antiviral serum (Rosen, 1981).

Treatment. Treatment is symptomatic.

Prognosis. Prognosis is excellent, except in the aged and debilitated, when DHF is not involved.

Prevention. Screening all early cases so that susceptible mosquitoes will not become infected and, most important, antimosquito campaigns directed against the breeding of the domestic *Ae. aegypti* or other *Ae. (Stegomyia)* spp., are effective preventive methods.

In hyperendemic areas in the Tropics most natives contract the disease and develop considerable tolerance. Epidemics may be permanently eliminated by reducing the mosquito vector below the threshold of transmission, i.e., they become so scarce that the chances of their becoming infected and then transmitting the infection to man are insignificant.

Saint Louis Encephalitis (SLE)

This virus causes illnesses in humans that vary from an influenza-like syndrome to acute encephalitis (Brinker and Monath, 1980). Increasing disease incidence and case-fatality rates are associated with age, with encephalitis significantly more frequent in persons above the age of 60.

In the United States, SLE virus is principally associated with species of *Culex (Culex)*. Members of the *Culex pipiens* complex *(C. pipiens, C. quinquefasciatus)* serve as vectors in the East and South, *C. nigripalpus* in Florida, and *C. tarsalis* in the

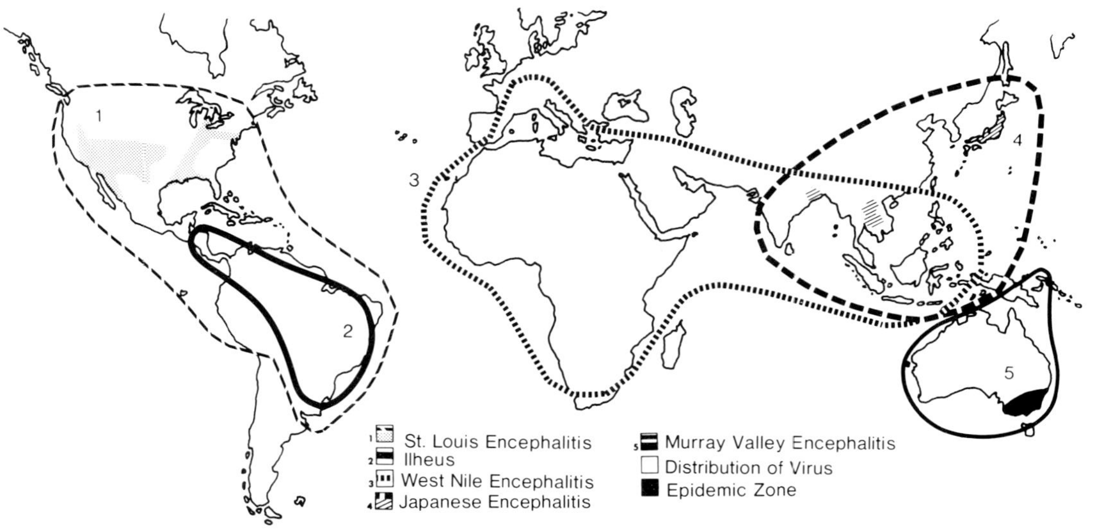

Fig. 41–14. Principal distribution of flaviviruses causing human encephalitis. (From Mitchell, C.J. 1977. Arthropod-borne encephalitis viruses and water resource developments. Cah. O.R.S.T.O.M., Ser. Entomol. Med. Parasitol., *15*:241–250.)

West. *Culex salinarius* and *C. restuans* may serve as "accessory" vectors in the East, whereas *C. peus* may play the same role in the western United States (Mitchell et al., 1980). Elsewhere over its range (Fig. 41–14), SLE virus has been isolated from other *Culex* species as well as from *Sabethes chloropterus, Trichoprosopon* spp., and *Coquillettidia venezuelensis*. However, the vector status of these species is unproven.

Birds serve as the amplifiers in the sylvatic or domestic cycle of SLE virus. In addition, in temperate zones, hibernating mosquitoes may play an important role in the overwintering of SLE virus (Bailey et al., 1978; Mitchell et al., 1979).

Rocio Meningoencephalitis

This flavivirus was first described in São Paulo State, Brazil in 1975 as the causative agent of epidemic meningoencephalitis. Although the transmission cycle remains essentially unknown, it is believed that *Culex (Culex)* spp. may serve as vectors with birds and possibly mammals acting as part of the reservoir (Mitchell et al., 1981). A single isolate of Rocio virus was also made from a pool of *Psorophora ferox*.

Japanese Encephalitis (JE)

JE virus has a wide distribution throughout much of Asia to the Indian subcontinent. Principal vectors include species of *Culex* whose aquatic stages commonly occur in rice fields. Agricultural practices coupled with the presence of swine as virus amplifiers typify the rural or peridomestic setting for Japanese encephalitis (Mitchell, 1977). *Culex tritaeniorhynchus* is an important vector in Japan. It blood-feeds on a variety of hosts, including ciconiiform birds that act as virus donors. Experimental evidence also supports the theory that JE virus can be passaged transovarially in *Cu. tritaeniorhynchus* and may therefore overwinter in hibernating adults (Rosen et al., 1980). Other important vectors include members of the *Culex vishnui* complex, *Cu. gelidus, Cu. fuscocephalus,* and *Cu. annulus*.

JE virus has been responsible for extensive outbreaks of summer encephalitis that affect all age groups, with mortality centered in the elderly or very young (Downs, 1982). However, serious disease can be avoided by use of an inactivated virus vaccine.

Murray Valley Encephalitis (MVE)

MVE virus is similar serologically to JE, Saint Louis, and West Nile viruses. The epidemiological pattern also involves a bird-mosquito cycle with *Culex annulirostris* serving as the principal epidemic vector in irrigated areas of the Murray Valley of Australia. MVE virus has also been isolated in New Guinea. Although infection is usually mild, acute encephalitis may occur with high fatality rates in the young.

West Nile Encephalitis

This virus is widely distributed throughout parts of southern Europe, northern Africa, and into southeast Asia. Encephalitis may occur in infected horses. The disease in humans ranges from mild febrile illness to occasional encephalitis in the elderly. West Nile virus ordinarily cycles between birds by means of *Culex* spp. vectors, including *Culex univittatus* in the Nile Delta and *Culex modestus* in the Rhone Delta of southern France (Mitchell, 1977).

Bunyaviridae: Genus *Bunyavirus*

C Group Viruses

Viruses of the C group cause a mild febrile illness in humans with occasional dengue-like symptoms. All have been isolated from mosquitoes, principally species of *Culex,* and particularly *Cu. (Melanoconion)* spp. Nonhuman hosts include rodents and possibly canopy-dwelling vertebrates (Woodall, 1979). This group of 14 viruses is generally distributed from Panama to Brazil.

California Group

This group of 14 serologically related viruses utilizes a similar maintenance cycle involving mammals as the principal vertebrate hosts and mosquitoes (frequently the genus *Aedes*) as vectors (LeDuc, 1979). Distribution occurs in both temperate and tropical zones as well as the arctic region of the world. Several viruses are readily transmitted transovarially in *Aedes* spp. (Watts et al., 1973), an important epidemiological factor. Three viruses of the group have been associated with human illness. La Crosse type is particularly important as a causative agent of summer encephalitis in the north-central United States (Downs, 1982).

Bunyaviridae: Genus *Phlebovirus*

Rift Valley Fever (RVF)

Until recently this virus was largely limited to eastern and southern sub-Saharan Africa where it caused illness in domestic animals and occasionally in humans. However, several epizootics have occurred in the Sudan (Eisa et al., 1980), Egypt (Meegan et al., 1979), and on a more limited basis in Nigeria (Tomori, 1980). In Egypt, Rift Valley fever usually presents as an acute febrile illness, but fatal encephalitic or hemorrhagic forms have also been seen (Laughlin et al., 1979). An ocular form causing retinitis may occur (Siam and Meegan, 1980).

Culex pipiens appears to be a vector of RVF virus in Egypt with a variety of other mosquitoes and biting flies *(Culicoides, Simulium)* as possible biological or mechanical vectors (Meegan et al., 1980).

Other Mosquito-Transmitted Diseases of Man

Filariasis

Three species of filaria *(Wuchereria bancrofti, Brugia malayi, B. timori)* are transmitted to man by mosquitoes, which constitute the obligatory intermediate hosts, and in which larval growth and metamorphosis occur. When the larvae reach maturity in the thoracic muscles of the appropriate mosquito they migrate to the head and pass down through the proboscis-sheath (labium), escaping to the skin surface of the patient when the mosquito takes the next blood meal. Infective larvae then enter the skin through the puncture wound, using the small drop of mosquito hemolymph deposited on the skin following rupture of the labium as an antidesiccant (Denham and McGreevy, 1977). For details of these infections see pages 350 to 369.

Species of mosquitoes that transmit human filariae can be grouped into five genera (Table 41–4). Their times of biting activity and distribution in relation to human habitations are important factors in considering the ecological setting for each filariasis as well as possible control measures. Many night-feeding species of *Anopheles,* certain species of *Mansonia* and *Aedes,* and members of the *Culex pipiens* complex are important vectors of periodic *W. bancrofti* and *B. malayi*. In their absence, subperiodic forms of both parasites are adapted to using diurnally active *Aedes* or zoophilous *Mansonia* (White, 1979; Denham and McGreevy, 1977). *Anopheles barbirostris,* an endophilic species, is the principal vector of *Brugia timori*.

Some filarial worms are zoonotic pathogens (Dissanaike, 1979). The subperiodic form of *Brugia malayi* (and possibly *B. pahangi*) is a sylvatic filaria that primarily parasitizes monkeys and carnivores, including dogs and cats. Its transmission to humans by *Mansonia* mosquitoes has been documented, and it has become an important pathogen in Indonesia, Malaysia, the Philippines, and Thailand. Human infection with *Dirofilaria immitis,* the dog heartworm, leads to focal pulmonary infarction with granuloma formation (Merrill et al.,

Table 41–4. Principal Mosquito Vectors of Human Filariae.*

	Filaria species	*Location*	*Vectors*
A.	*Wuchereria bancrofti* (nocturnally periodic)	Pacific and New Guinea	*Aedes kochi; Anopheles bancroftii; An. farauti, An. koliensis, An. punctulatus; Culex annulirostris; C. bitaeniorhynchus, C. quinquefasciatus; Mansonia uniformis.*
		India, Sri Lanka, Maldive Islands	*Anopheles nigerrimus, An. philippinensis; Culex bitaeniorhynchus, C. quinquefasciatus.*
		China, Japan	*Aedes togoi; Anopheles jeyporiensis, An. leucosphyrus, An. minimus, An. sinensis; Culex pipiens* (subspecies *pallens), C. quinquefasciatus.*
		Southeast Asia	*Aedes kochi, Ae. niveus, Ae. poicilius; Anopheles aconitus, An. barbirostris, An. donaldi, An. farauti, An. flavirostris, An. koliensis, An. letifer, An. leucosphyrus, An. maculatus, An. punctulatus, An. subpictus, An. whartoni; Culex annulirostris, C. bitaeniorhynchus, C. quinquefasciatus; Mansonia uniformis.*
		Egypt and eastern Mediterranean	*Culex pipiens* (biotype *molestus).*
		Sub-Saharan Africa and neighboring islands	*Anopheles arabiensis, An. funestus, An. gambiae, An. melas, An. merus, An. pauliani, An. wellcomei; Culex antennatus, C. quinquefasciatus.*
		South America and the Caribbean	*Aedes scapularis; Anopheles aquasalis, An. bellator, An. darlingi; Culex quinquefasciatus; Mansonia titillans.*
B.	*Wuchereria bancrofti* (nocturnally subperiodic)	Thailand	*Aedes niveus.*
C.	*Wuchereria bancrofti* (diurnally subperiodic)	Polynesia and eastern Pacific	*Aedes cooki, Ae. fijiensis, Ae. oceanicus, Ae. polynesiensis, Ae. pseudoscutellaris, Ae. rotumae, Ae. samoanus, Ae. tabu, Ae. tongae, Ae. upolensis, Ae. vigilax.*
D.	*Brugia malayi* (nocturnally periodic)	India	*Anopheles nigerrimus; Mansonia annulifera, M. indiana, M. uniformis.*
		Southeast Asia	*Anopheles barbirostris, An. campestris, An. donaldi, An. nigerrimus; Mansonia annulata, M. annulifera, M. bonneae, M. dives, M. indiana, M. uniformis.*
		China	*Aedes togoi; Anopheles lesteri, An. sinensis.*
		Korea	*Aedes togoi.*
E.	*Brugia malayi* (nocturnally subperiodic)	Southeast Asia	*Anopheles barbirostris, An. venhusi; Coquillettidia ochracea; Mansonia annulata, M. annulifera, M. bonneae, M. dives, M. indiana, M. uniformis.*
F.	*Brugia timori*	Flores (Lesser Sunda Islands of Indonesia)	*Anopheles barbirostris.*

*This information was taken from WHO Tech. Report Series No. 542 (1974), Sasa (1976), and White (1979).

1980). This filaria is transmitted by a relatively large number of mosquitoes in a domestic or peridomestic environment. *Dirofilaria tenuis,* a parasite of raccoons, has also been recovered from subcutaneous infections in humans (Beaver and Orihel, 1965).

Many of the mosquitoes that are associated with urban filariasis either are domestic (*Culex quinquefasciatus, C. pipiens, Aedes scutellaris* group) or are found in bodies of water not far from human habitations (*Mansonia* spp., *Anopheles* spp.). Measures previously outlined, particularly the prevention of breeding near homes and adequate screening of habitations, will be most useful in controlling the infections.

Myiasis

Mosquitoes of the species *Psorophora (Janthinosoma) lutzi, P. ferox,* and *P. cyanescens* serve as mechanical transmitters to the human skin of the eggs of the tropical botfly, *Dermatobia hominis,* the female of which, at the time of oviposition, cements a batch of 16 to 54 eggs on the abdomen of a female *Psorophora.* When the mosquito alights on man or some other mammal to take a blood meal, the body warmth causes the eggs of the fly to hatch, and the first-stage larvae then penetrate the skin of the mammal (*vide infra,* p. 685).

Mosquito Dermatitis

The minute droplets of saliva, which blood-sucking mosquitoes inject into a puncture wound before removing blood from the peripheral vessels, usually produce a local reaction lasting from a few hours to several days. In highly susceptible individuals they may produce generalized symptoms similar to those of serum sickness. This reaction is allergic in type, due to the introduction of foreign proteins into the body.

PREVENTION AND CONTROL OF MOSQUITO-BORNE DISEASES

Attempts to prevent mosquitoes from transmitting pathogens to man vary in complexity from the simple use by individuals of chemicals on the skin, or of bed-nets or screens, to the massive efforts of a nation or a group of nations to eradicate a vector such as *Aedes aegypti* from the Americas or to eradicate a disease such as malaria from the world.

Control Measures. The measures involved may be classified as follows (see Chap. 45, p. 718):

1. *Individual protection* by site selection of homes or towns, by nets and screens, by chemical repellents, and by avoidance of undue exposure, e.g., not sleeping out-of-doors.
2. *Chemotherapy,* e.g., in the case of malaria, reducing the reservoir of gametocytes in a community by mass therapy and drug prophylaxis so that fewer mosquitoes are infected.
3. *Source control,* chiefly by treatment or elimination of larval sites.
4. *Reduction in longevity of mosquitoes,* i.e., adult control by space and residual sprays, gases, or vapors so that fewer mosquitoes live long enough to become infective.
5. *Diversion of mosquitoes from man to animals,* e.g., cattle. This is *zooprophylaxis,* a method seldom practiced deliberately but often naturally effective.
6. *Vector eradication,* only practical in special circumstances, e.g., the successful eradication of *Anopheles gambiae* from Brazil from 1939 to 1940.
7. *Genetic control.* Successful application of the sterile-male technique for control of screwworms and fruit flies has stimulated research on the possible use of similar methods against mosquitoes. If adequate numbers of vigorous and competitive sterile males are introduced systematically into a natural population, it will soon cease to exist. Experiments to use cytoplasmic incompatibility in *Culex quinquefasciatus* have been sufficiently successful to warrant preliminary field tests. Cytoplasmic incompatibility is a condition in which crosses between various insect populations are sterile. A cytoplasmic factor transmitted from one generation to the next through the egg prevents fusion of the sperm pronucleus after its entry into the egg. Large-scale genetic control of mosquitoes in the field is still only theoretical, but it has definite promise for the future.

Factors. Obviously, an understanding of the epidemiology of a disease is an essential factor in practical control. Involved are (1) *circumstantial epidemiology,* i.e., the circumstances in which a disease occurs, such as static or epidemic, coastal or upland, and urban or rural; (2) *biological epidemiology,* i.e., the biology, morphology, and physiology of the vector, breeding, resting and feeding habits, and genetics; and (3) *mathematical epidemiology,* i.e., the pertinent mathematical relationships involved in malaria transmission, so that the circumstantial and biological epidemiological data may be integrated in order to provide a rational understanding on which to base control measures

(Macdonald et al., 1968; Wernsdorfer, 1980). In addition, malaria surveys should take account not only of the aforementioned considerations but of such human factors as public health organization, population, distribution and movements, sleeping and housing customs, and taboos.

Among the measures for individual protection are screens with mesh no larger than size 18 and bed-nets made of smooth fibers so that moving air is not impeded. Chemical repellents may also be used on bed-nets or on the skin (see Chap. 45).

Mass therapy and drug prophylaxis have been described in Chapter 14.

The reduction in numbers of mosquitoes by attacking the aquatic stages is still a useful measure in many circumstances. Larvicides such as Paris green, crude oils, or growth regulators may be used. The use of organic phosphorus compounds is also effective initially, but can result in development of resistance. Elimination or modification of larval sites may constitute the most practical measures in some situations.

Malaria Eradication by Mosquito Control. Most authorities agree today that in theory malaria can be eliminated from many parts of the world by mosquito control and that other methods may be uneconomical. Although *nationwide malaria eradication* remains a goal, a majority of malarious countries now strive only for *control*. The objective is not mosquito eradication, except in rare cases where this difficult task is demonstrably worth the extra expense.

An important measure for malaria control being employed throughout the world is the use of a residual insecticide because (1) it has a prolonged killing effect on adult mosquitoes when it is applied to surfaces upon which the insects later walk or rest and (2) its use is economically feasible even in tropical rural areas. This method of residual spraying greatly reduces the reproduction rate of malaria. The mosquitoes that rest on the treated surfaces do not live long enough to permit development of sporozoites.

The enormous success of DDT sprays has tended to obscure the fundamental value of draining and filling in malaria control programs. Any type of malaria control by antilarval measures is less likely to be economically feasible in rural areas, especially in the Tropics. For urban areas, for organized agricultural communities, as on tea gardens and rubber estates, and for the protection of industrial projects, the more permanent measures of draining and filling still have great importance. Any measure that at the same time prevents mosquito breeding and improves land for agricultural or urban use still deserves consideration in malaria control programs. Any but the simplest drainage requires the attention of competent sanitarians or engineers.

The emergence of mosquito resistance to insecticides such as DDT and related organochlorines as well as to organophosphorus compounds constitutes a serious problem in some areas. So much research has been reported on the biochemistry, physiology, and genetics of resistance that the phenomenon is well understood, but no simple solution has been found. The only practical alternative is the use of new insecticides. For this reason, the World Health Organization set up in 1960 a worldwide program for evaluation and testing of new insecticides. This program of basic and applied research has made progress and offers hope that the resistance barriers can be removed.

Harmful ecological side effects from the widespread use and abuse of DDT against agricultural and forest insects and pest mosquitoes have aroused conservationists, generated questions among scientists about long-term effects on the human body, and led some state and national governments to limit or ban its use.

Details of epidemiological and technical aspects of mosquito control and malaria eradication are found in Russell et al., 1963; WHO, 1972, 1973, 1974b, 1975; and Wernsdorfer, 1980. For a review of malaria control attempts from a global viewpoint, see Brown et al. (1976).

Viral Diseases and Mosquito Control. Urban epidemics of these infections result from transmission of the respective viruses by domestic mosquitoes into nonimmune groups in areas where the vectors are commonly found. Therefore, mosquito production should be minimized by an effective control program. In such areas, periodically inspecting and draining domestic water receptacles and gutters and screening cisterns can effect a rapid diminution of the breeding of these mosquitoes. In addition, routine control by the use of chemical larvicides can be implemented with space-spraying utilized for those mosquitoes that escape the larvicides and emerge as adults. When epidemics of Saint Louis encephalitis occur, ground or aerial ultra low volume (ULV) applications of insecticides have proved effective in killing a high percentage of adult mosquitoes and in preventing reinfestation from untreated areas (Breeland et al.,

1980). In the southern United States, epidemics of yellow fever and dengue have always subsided with reduction in breeding of *Ae. aegypti* as cold weather developed, but this mosquito has so extensive a distribution throughout this and surrounding territories that it still constitutes a potential menace throughout most of the year. Thus, preventive measures such as disease and mosquito surveillance coupled with organized control are required.

For *enzootic* and *epidemic yellow fever* in forested areas and nearby settlements the following measures have been advised: (1) viscerotomy of corpses of persons who die in these areas, to disclose the actual distribution of the disease; (2) anti-*Ae. aegypti* measures in cities and towns, both in endemic areas and in threatened regions; and (3) vaccination of populations in towns and isolated communities exposed to the virus, as well as of travelers in these areas (see p. 651).

SAND FLIES (FAMILY PSYCHODIDAE)

Morphology, Biology, and Life Cycle. The medically important members of this family belong to the genera *Lutzomyia* and *Phlebotomus* (subfamily Phlebotominae). *Lutzomyia* occurs in the New World tropics, whereas *Phlebotomus* is the dominant genus in the Palearctic Region, extending as far south as the Ethiopian Region (Lewis, 1973). The *adults* (Fig. 41–15) of these species are small, humpbacked, tawny in color, with conspicuous black eyes and hairy body, wings, and legs. The oval-lanceolate wings are carried erect, and the male terminalia are long and prominent. Only the females feed on blood, commonly on that of mammals, but at times on lower terrestrial vertebrates. Feeding usually occurs at night and the flies hide in dark, often damp places in the daytime. Blood meals are required for ovarian development, although a few species are autogenous. Sand flies are telmophagous and have mandibles that cut the host's skin, permitting insertion of the tip of the fascicle (Lewis, 1975a). Following insertion, the mandibles also slash the dermis, leading to blood pool formation. The hypopharynx, which is guided by the labrum and labium, is placed in the wound and injects saliva. The outer teeth of the maxillae (laciniae) serve to anchor the mouthparts. Sugars are also imbibed (Young et al., 1980) and may play an important role in the development of leishmanias in vector species (Killick-Kendrick, 1979). Many sand flies are capable of extended flight but take characteristic hops when they are attracted to light or are near their resting places.

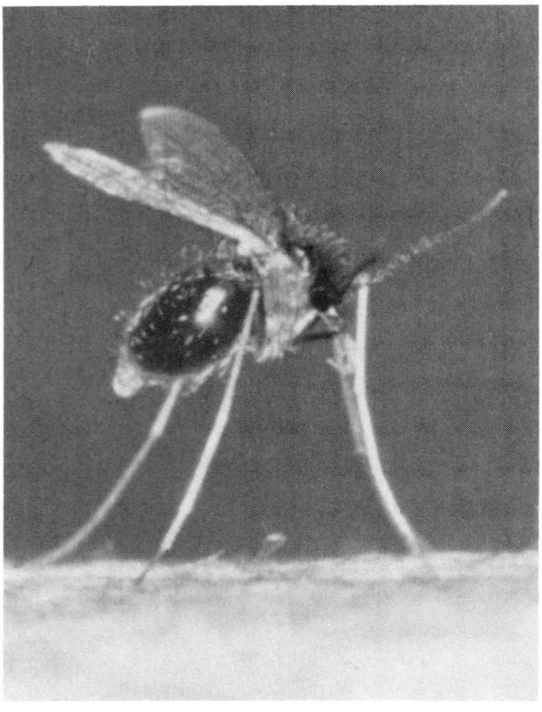

Fig. 41–15. Adult female *Phlebotomus papatasi* feeding on man. (Courtesy of Dr. Robert Tesh, Yale University.)

The long ovoid eggs (Fig. 41–16, *A*) are coated with a secretion from the accessory glands and are laid in batches of about 50 in cracks and crevices in dark, moist, soil microhabitats. After 9 to 12 days they hatch, and the young *larvae* (Fig. 41–16, *B*) begin to feed. They have small, short antennae, lack eyes, have mandibulate mouthparts, and have 3 distinct thoracic and 10 abdominal segments, of which the first 7 abdominal segments have ventral pseudolegs, the ninth segment bears a pair of long, wiry, caudal bristles, and the tenth has a pair of shorter anal bristles. Larvae feed on organic debris, and occasionally on dried leaves. After 4 larval instars and a pupal stage (Fig. 41–16, *C*), the adults emerge. Time for completion of the life cycle is greatly dependent upon ambient temperature. In northern climates, larvae may undergo diapause.

Sand Flies and Disease

Phlebotomus Fever
(Sandfly fever, pappataci fever)

Definition and Etiology. Phlebotomus fever is an acute infectious disease, characterized by fever, malaise, severe headache, orbital pain, and a leukopenia. It is caused by several

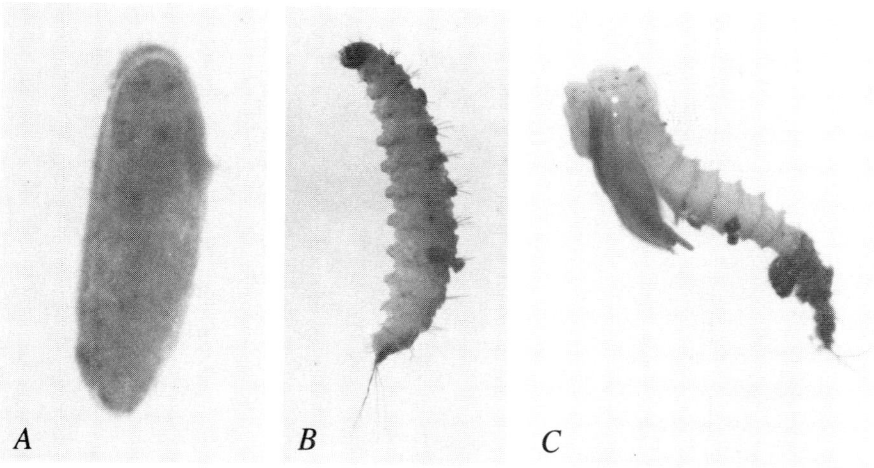

Fig. 41–16. Immature stages of *Phlebotomus papatasi*. *A*, Egg. (× ca. 300.) *B*, Larva. *C*, Pupa. (Photographs by Howard Lyon, Cornell University.)

different viral serotypes of the Phlebotomus fever group (Bunyaviridae: Genus *Phlebovirus*).

Historical and Geographical Notes. Doerr, Franz, and Taussig in 1908 demonstrated that the disease was caused by a filtrable virus and transmitted via sand flies. Phlebotomus fever is primarily but not exclusively prevalent in the moist lowlands of subtropical climates. It is especially prevalent in the Eastern Hemisphere between 20° and 45° north latitude, and is known to occur in Italy as far north as the Po valley, in Sicily, Yugoslavia on the Adriatic coast, Greece, Malta, Crete, Cyprus, Egypt (endemic in the Nile Valley), Israel, Syria, Iraq, Iran, the Crimea and Transcaucasia, and the northwest and central provinces of India. It is probably endemic in China along the coast from Tientsin to Hong Kong, in Burma, Sri Lanka, the Poona region of India, Aden and the Red Sea coast of Arabia, Sudan, the Mediterranean coast of Africa, southern coastal France, Corsica, Sardinia, Gibraltar, and the Atlantic coast of Portugal. Serotypes have also been isolated from the New World Tropics (Tesh et al., 1974).

Epidemiology. The disease develops epidemically during the summer and fall, following rains and increased density of the usual vectors, *Phlebotomus papatasi* and *P. sergenti*. These sand flies become infected from blood of infected humans, taken 1 or 2 days before the onset of symptoms and not later than 24 hours thereafter—the period when the virus is circulating in the blood. Following an incubation period of 7 to 10 days the fly remains infective for life. Only females transmit the infection.

With few exceptions, sand flies feed between dusk and daylight, usually at ground levels, seldom higher. Unfed females penetrate 18-mesh wire screening and ordinary mosquito netting. It is uncommon for them to bite a person in motion or in a direct breeze. They move in short jerky hops along walls and ceilings. Larvae of *P. papatasi* develop in dark, damp mossy soil between stones or in cracks of north wall exposures, so that their breeding places are difficult to discover. In warm weather the cycle from egg-laying to adult requires 4 to 6 weeks.

Sandfly fever virus is readily passed transovarially so that the vector also acts as a reservoir (Tesh and Chaniotis, 1975). This is an important epidemiological factor since blood virus titers are often low and of relatively short duration, hence limiting the dissemination of the virus into the vector population.

Pathogenesis, Pathology, and Symptomatology. The virus of phlebotomus fever is inoculated into the skin when infected sand flies take a blood meal. It becomes well distributed throughout visceral as well as somatic tissues, producing discomfort and pain in abdominal viscera, in skeletal musculature, and in joints, where there seems to be considerable inflammation. Moreover, it stimulates production of neutrophilic leukocytes but destroys the older neutrophils, so that the total number is about normal, but with an excess of immature forms.

After an incubation period of 2 to 6 days in symptomatic cases there is a sudden onset at times preceded for a day or more by constipation, abdominal distress, dizziness, and general malaise. A chilly sensation may be present, but there is no shaking chill. There is a sudden rise in temperature, usually reaching 102° to 104.5° F (38.8° to 40.3°C) and lasting about 2 to 4 days (occasionally only 1 day, rarely up to 9 days), with a peak on the first or second day. Defervescence is gradual. Rarely there is a second rise after the temperature has returned to normal, mimicking somewhat the "saddle-back" fever curve of dengue (see p. 651). The pulse is usually increased during the first day of the fever, then returns to normal more rapidly than the temperature.

There is typically severe frontal headache, the eyeballs are painful on touch or with movement, and there is postorbital pain, with a burning sensation in the eyes, injected conjunctivae, and photophobia. The patient experiences pain and stiffness in the neck, back, sacroiliac region, joints, and extremities. Anorexia and nausea are common, and vomiting may occur. There is characteristically no true rash. Profuse sweating occurs. The respiratory tract is normal on physical examination.

As the fever recedes there is great fatigue and weakness. Bradycardia may occur with subnormal blood pressure. Diarrhea is not uncommon at this time. Convalescence may require only a few days or it may be prolonged. The disease confers a relatively lasting immunity.

Diagnosis. The disease must be diagnosed almost exclusively on epidemiological and clinical grounds. It occurs epidemically during the breeding season of *Phlebotomus papatasi* and other common vectors. Sandfly fever requires particular differentiation from dengue. Sandfly fever must also be distinguished from acute rhinopharyngitis, gastroenteritis, the first chills and fever of malaria, and other acute febrile diseases.

The urine and spinal fluid in sandfly fever are negative for abnormalities and for the virus. The red blood cells are normal. However, there is a pronounced relative and absolute reduction in the number of mature neutrophils associated with a similar increase in the immature neutrophils. The monocytes are usually normal in number and appearance.

Treatment. Treatment is entirely symptomatic.

Prognosis. The prognosis is excellent.

Prevention. Dense vegetation should be cleaned out and debris and organic material removed from around dwellings to discourage the breeding of sand flies. In many situations, spraying indoors and around the exterior of houses with a residual insecticide can be effective. The use of fine-mesh bed-nets can also be worthwhile.

Vesicular Stomatitis

Several strains of vesicular stomatitis virus (VSV) (Rhabdoviridae: *Vesiculovirus)* cause an influenza-like illness in humans. In Panama, the Indiana strain of VSV infects a wide variety of animals and is transmitted by *Lutzomyia* spp. (Tesh et al., 1971). It can also be vertically transmitted to the F_1 progeny and is probably passed from generation to generation (Tesh, 1980).

Bartonellosis (Carrión's disease, Oroya fever, verruga peruana)

Definition and Etiology. Bartonellosis is an infection characterized by an insidious onset, frequently an acute hemolytic febrile stage which is at times fatal, with pain in the long bones and joints, and a severe anemia; later by hemangiomatous, warty eruptions on the skin, which may be nodular or miliary and confluent. It is produced by a minute, pleomorphic, bacillus-like organism, *Bartonella bacilliformis,* which invades primarily the red blood corpuscles and secondarily the reticuloendothelial cells.

Historical and Geographical Notes. It is probable that this disease existed in certain inter-Andean valleys in Peru and other northwestern countries of South America in pre-Columbian days. The earliest written account referring definitely to the disease was that of Gago de Vadilla, in 1630 (Rebagliati, 1942). It first received publicity in the 1870s during the building of the Central Railway from Lima to Oroya, Peru, when a severe epidemic broke out among the workmen resulting in 7000 deaths. In 1885 Carrión infected himself experimentally and died of the disease which bears his name. Townsend began a study of its epidemiology in 1913; Strong, Tyzzer, Sellards, Brues, and Gastiaburú studied it in 1913 (published in 1915); and Noguchi and Battistini announced the successful cultivation of the isolated etiological agent *Bartonella bacilliformis* in 1926. Hertig (1937–1942) made an intensive investigation of its insect transmission in the endemic areas in Peru, while Pinkerton and Weinman (1937) studied the pathogenesis and histopathology of the disease.

In addition to the endemic zones in the Rimac and Santa Eulalia valleys above Lima, in Peru, two other foci have been located, one in Colombia (Department of Nariño) near the Ecuadorean border (Patiño Camargo, 1939) and the other in the Province of Loja in Ecuador (Montalván, 1940). All of these regions are in elevated Andean valleys between 800 and 3000 meters above sea level.

Epidemiology. Bartonellosis is confined to certain mountain valleys in regions where the sand fly, *Lutzomyia verrucarum,* and possibly other species of sand flies are responsible for its transmission. The habits of *L. colombiana* are so nearly like those of *L. verrucarum* that the former species may be the transmitter of this pathogen in areas outside the range of the latter. A single night's exposure in an endemic zone may result in infection, and a stay of 1 to 3 months almost invariably produces the disease in newcomers. Most natives probably contract the infection early in childhood and, if they recover, have permanent immunity. Although monkeys are good experimental hosts, no natural reservoirs have been discovered.

Pathogenesis, Pathology, and Symptomatology. The etiologic agent, *Bartonella bacilliformis,* invades red blood cells in which it multiplies, causing destruction of the cells. In this way, just as in malaria, the toxic by-products of the pathogen, together with the increased destruction of red cells, produce a febrile reaction and are responsible for anemia, which is usually normocytic but at times macrocytic, and may reach as low a level as 1,000,000 r.b.c. per cu. mm. Moreover, the microorganisms invade the cells of the reticuloendothelial system, causing hypertrophy of these cells in the lymph nodes, Kupffer cells of the liver, and fixed histiocytes in the spleen, bone marrow, kidneys, adrenal glands, pancreas, thyroid, and testes. They also parasitize the endothelial lining cells of the blood and lymph vessels, which may be so distended by clusters of the parasites that the infected cells may be detected on low-power microscopic examination (Pinkerton and Weinman, 1937). The verrucous eruptions are sequelae to the lesions in the reticuloendothelium and the circulatory system.

Following an incubation period of 10 days to 3 weeks or more, there is usually an insidious onset, with manifestations of dizziness, headache, thirst, anorexia, and prostration. The fever is a low-grade septic type, ranging from 99° to 102°F (37.8° to 39°C) and lasting from 1 or 2 days to 2 months or more. This febrile condition is almost invariably associated with a severe anemia and with a definite reduction in the hemoglobin and in the number of circulating red blood cells, which, however, are normal or slightly increased in size. Unless there is a secondary infection there is no leukocytosis. During the period of acute anemia jaundice is common and there may be bleeding from the mucous membranes, with epistaxis, hematemesis, or melena. Once the fever has subsided, it rarely recurs.

If the fever is of short duration, there may be some little lapse of time before the appearance of the verrucous eruption, but if the febrile condition is prolonged the eruptions may gradually appear before defervescence. Typically they develop in the skin and subcutaneous tissues immediately over bony prominences and joints, but in some cases they are distributed generally over the body surface and may be miliary in character. They may reach 1 cm or more in diameter and may be so numerous as to be confluent, or they may be confined to the subcutaneous tissues. They are hemangiomatous and tend to break down, ulcerate, and become granulomatous or to become secondarily infected and remain as indolent weeping lesions. At times considerable hemorrhage from the eruptions occurs, aggravating the anemic state. Not infrequently the patient gives a history of development of the eruptions without a preceding acute febrile illness. The verrucous condition may last for weeks or months. The lesions may be completely absorbed, leaving brawny vestigia. Recurrence of the eruption after healing is not uncommon.

Many persons in endemic foci have a low-grade manifestation of Carrión's disease, and others are probably asymptomatic carriers of the parasites. Recovery from the disease in any of its forms confers lasting immunity.

Diagnosis. Because of the relatively limited geographical distribution of bartonellosis, living or traveling in endemic foci suggests the possibility of this disease, particularly if fever and anemia or verrucous eruptions are present. During the acute febrile stage, the pathogen can frequently be found in red blood corpuscles. In these patients, and frequently in those with milder symptoms, positive blood cultures may be obtained using the culture medium developed by Geiman (1941). In patients with verrucous eruption, biopsy of a lesion may reveal the organisms in microscopic section.

Treatment. Broad-spectrum antibiotics such as tetracyclines and chloramphenicol in the amount of 4 g daily for 4 days, then 2 g daily for 4 to 6 days (De Vault, 1952), provide rapid clinical improvement. Chloramphenicol is not recommended as first choice because of its occasional tendency to depress the bone marrow.

Prognosis. Prognosis is poor to good, depending on the degree of anemia, the significance of pyogenic secondary invaders, and other factors that determine the patient's ability or inability to resist the infection and develop permanent immunity. With modern chemotherapy the prognosis is much better.

Prevention. Sand flies can be controlled by the periodic application of residual insecticides.

Leishmaniasis

Leishmania donovani, L. tropica, and *L. braziliensis* have been considered in detail in Section II (pp. 70, 58 and 64). Phlebotomine species known or suspected of transmitting human leishmanias include 29 species of *Phlebotomus* and 27 species of *Lutzomyia* (Killick-Kendrick, 1978). The more common Old World vectors are *Phlebotomus ariasi, P. papatasi, P. perfiliewi, P. perniciosus,* and *P. sergenti* in the Mediterranean region; *P. alexandri* in Central Asia and Turkey; *P. caucasicus* group and *P. papatasi* in the U.S.S.R.; *P. argentipes* from India to Vietnam; the *P. chinensis* group in Iran, Pakistan, and China; and *P. langeroni orientalis, P. longipes,* and *P. martini* in the Sudan, Ethiopia, and Kenya. New World species are comprised of *Lutzomyia olmeca olmeca* in Belize and Central America, *Lu. flaviscutellata* in the Amazon Basin and Trinidad, *Lu. trapidoi* in Panama and neighboring countries, and *Lu. peruensis* in Peru.

For a discussion of the development of *Leishmania* in phlebotomine sand flies, see Killick-Kendrick, 1979.

Sand Fly Bite

The bite of a sand fly produces a local, frequently indurated, inflammatory lesion, with a wheal of 1 or 2 cm in diameter. At the time of the puncture there is a painful needling sensation, and later for several hours, days, or even weeks, an irritating local pruritus. Some individuals experience a marked systemic allergic reaction, with rise in temperature, nausea, and malaise.

GNATS AND BITING MIDGES (FAMILY CERATOPOGONIDAE)

Morphology, Biology, and Life Cycle. The important blood-sucking members of this group belong to three genera *(Culicoides, Leptoconops, Forcipomyia)* of which *Culicoides* is medically significant. The *adults* of this genus are small dark flies (Fig. 41–17), with long, delicate antennae (generally with 15 visible segments) which are pilose in both sexes. They have long maxillary palps; a slightly humped thorax; short, broad, delicately hairy wings with spotted pigmentation; and relatively long legs. Only the females have mouthparts adapted for sucking blood. The paired mandibles have an apical row of teeth for cutting the skin with the maxillae (laciniae) possessing retrorse denticles. The well-developed labrum and hypopharynx also enter the wound (Downes, 1971). Species in the adult stage are differentiated by wing characters (venation; pigmentation; number, type, and distribution of microtrichia and macrotrichia), number and shape of the spermathecae, and mouthpart configuration. *Eggs* are laid in a single row in decaying vegetation in water, in crab-holes, in tree-holes, in the scum of a wet sand bed, on aquatic plants, or even in leaf mold or manure piles. Some also breed in salt water, others in shallow lakes. The *larvae* (Fig. 41–18) are vermiform, smooth-bodied, have a head with mandibular mouthparts and 3 thoracic and 9 abdominal segments, are apneustic (i.e., lack spiracles), and usually possess anal gills and caudal hooks. When disturbed, they swim about with an active serpentine movement. The chrysalis-like *pupa* is elongate and has a pair of long, conspicuous, thoracic breathing trumpets, numerous integumentary tubercles and spurs, and fleshy terminal processes used for anchoring.

Fig. 41–17. *Culicoides variipennis,* adult female taking a blood meal. (× ca. 25.) (Courtesy of Lorry Dunning, University of California, Davis.)

Geographical Distribution. Species of *Culicoides* have a wide distribution throughout the Tropics and the Temperate Zones.

Medical Importance of *Culicoides*

Filariasis. In Africa, *C. austeni* and *C. grahami* have been incriminated in the transmission of the nonperiodic filarial worm, *Mansonella perstans,* and *M. streptocerca. Culicoides furens* has been found to transmit *Mansonella ozzardi* in Central and South America and Haiti (Lowrie and Raccurt, 1981). *C. phlebotomus* is the vector in Trinidad (Nathan, 1981).

Oropouche Virus Infection. *Culicoides paraensis* appears to be the main vector of Oropouche virus (Bunyaviridae: *Bunyavirus* [Simbu group]), the causative agent of a benign febrile illness in the Amazon basin of Brazil and in Trinidad (Pinheiro et al., 1981b, 1982).

The Bite. Swarms of these minute flies attack

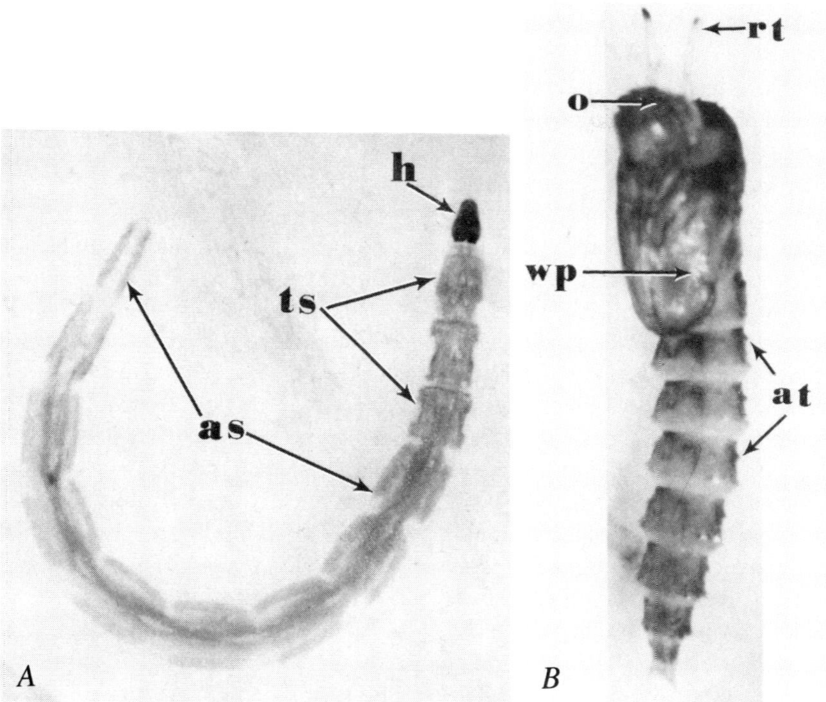

Fig. 41–18. *Culicoides variipennis*, immature stages. *A*, Dorsal view of larva. *as*, Abdominal segments I–IX. *h*, Head. *ts*, Thoracic segments. *B*, Ventro-lateral view of pupa. *at*, Abdominal tubercles. *o*, Operculum. *rt*, Respiratory trumpet. *wp*, Wing pad. (Courtesy of Brad Mullens, Cornell University.)

man (or mammals), usually at dusk when the wind has died down. They bite on the exposed extremities, around the head and neck, or even in between the head hairs, and produce irritating local lesions. In some persons these lesions disappear in a few hours, but in other individuals they persist for days, with appreciable systemic reaction.

Control. The unfed adults readily pass through 18-mesh screening and mosquito netting. Spraying the walls and screens of buildings with a residual insecticide or outdoor fogging greatly reduces annoyance by these gnats. However, these are only temporary measures and, where possible, control efforts should be directed toward elimination of larval habitats.

BLACK FLIES, BUFFALO GNATS, AND COFFEE FLIES (FAMILY SIMULIIDAE)

Morphology, Biology, and Life Cycle. The *adult* flies of this family (Fig. 41–19) are minute, stout-bodied species, with a humped back, short stout legs, relatively short antennae (9 to 12 segments), 5-jointed recurved maxillary palps, conspicuous eyes (holoptic in male, dichoptic in female), large unspotted wings, and inconspicuous

Fig. 41–19. *Simulium damnosum s. l.*, the African vector of *Onchocerca volvulus*, feeding on man. (Photograph by Dr. D.H. Connor [AFIP Negative 72–4519E]. From Gibson, S.W., Heggie, C., and Connor, D.H. 1980. Clinical and pathologic aspects of onchocerciasis. Pathol. Annu., Part 2, *15*:195–240.)

genitalia in the males. Only the females suck blood. The mandibles are blade-like and have apical teeth used for slashing the skin. The maxillae (laciniae) are also toothed and enter the feeding wound. The labrum serves as an anchoring device. The aquatic stages are found in running (i.e., oxygenated) water, such as mountain streams. Adults mate soon after emergence with the males usually forming swarms.

The females lay batches of many minute, triangular-shaped *eggs* at or just below the water level, attached by a gelatinous fluid to bare rocks, debris, or blades or stems of grass. As many as 500 may be oviposited in one mass. These hatch within a few days or in some temperate species serve as an overwintering stage. The subcylindrical *larvae* (Fig. 41–20), which are enlarged at both ends, have a pair of conspicuous, fan-shaped "mouth" brushes (cephalic fans), a pair of minute eyespots and slender antennae, a median ventral prothoracic pseudoleg armed with hooklets, anal gills, and a terminal hooked disc used for attachment. There usually are 7 larval instars, covering a period of 6 days or more, following which a pocket-shaped cocoon is spun by the pharate *pupa* by secretion of the salivary (i.e., silk) glands. The sclerotized pupa (Fig. 41–20) bears a pair of conspicuous, branched, external breathing tubes. The pupal stage is transient and usually lasts for several days. There are approximately 1270 described species of black flies, with medically important forms belonging to the genus *Simulium*.

Geographical Distribution. Black flies are practically worldwide in distribution and number hundreds of species. Four genera (*Simulium, Prosimulium, Austrosimulium,* and rarely *Cnephia*) contain man-biting species (Crosskey, 1973).

Medical Importance of Simuliidae

Filariasis. The obligatory role that certain species of *Simulium* play as transmitters of *Onchocerca volvulus* in Africa, southwestern Mexico, Guatemala, eastern Venezuela, northern Brazil, Colombia, and Ecuador has been previously described (see p. 369). However, it is worth noting the vector species in each location. The *Simulium damnosum* complex is widely distributed throughout sub-Saharan Africa and consists of a group of cytotypically defined sibling species (for a review, see Rothfels, 1979), several of which are important vectors. *Simulium damnosum sensu stricto* and *S. sirbanum* transmit *Onchocerca volvulus* in the Sudan and Guinea savannahs, whereas *S. soubrense, S. squamosum, S. yahense,* and possibly *S. sanctipauli* are vectors in the forests of West Africa (Townson and Meredith, 1979). Species of the *S. neavei* group, including *S. ethiopiense, S. neavei,* and *S. woodi,* serve as vectors in East Africa (Raybould and White, 1979). The larval and pupal stages of this group are phoretic on riverine crabs or freshwater prawns. *Simulium albivirgulatum* has been found naturally infected with *O. volvulus* and appears to be a vector in a focus in Zaire (Fain et al., 1980).

In the New World, *S. metallicum* is the principal

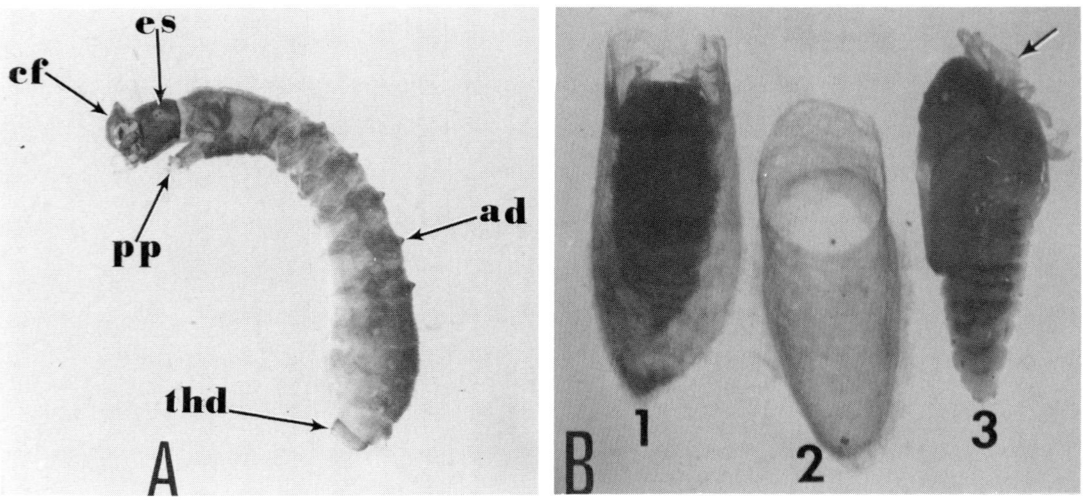

Fig. 41–20. *Simulium damnosum s. l.*, immature stages. *A*, Larva (probably *S. yahense*). *ad*, Abdominal tubercles. *cf*, Cephalic fan (closed). *es*, Eye spots. *pp*, Prothoracic pseudo-leg ("proleg"). *thd*, Terminal hooked disc. *B*, Pupa. *1*, Dorsal view of a pupa enclosed within the cocoon. *2, 3*, Lateral view of a pupa *(3)* removed from the pocket-shaped cocoon *(2)*. Note the pupal gills (arrow). (Courtesy of Howard Lyon, Cornell University.)

vector in Venezuela, but is of incidental importance in Mexico and Guatemala (Lewis and Ibañez de Aldecoa, 1962) where *Simulium ochraceum* is an inefficient but effective transmitter owing to its highly anthropophilic nature (Collins, 1979a,b). *Simulium exiguum* also exhibits limited host efficiency but is the apparent vector in endemic foci in Colombia (Tidwell et al., 1980b). In northern Brazil, *S. sanguineum* is apparently the vector (Townson and Meredith, 1979).

Mansonella ozzardi, a culicoid-associated filaria (see p. 381), can also develop in black flies. Tidwell et al. (1980a) described its growth to the third larval stage in *Simulium sanguineum sensu lato*. This species appears to be a vector in Colombia. *M. ozzardi* also developed to the infective stage in *Simulium amazonicum s.s.* and a related species *(S. argentiscutum)* in the Brazilian Amazon, where it is the probable vector (Shelley et al., 1980).

The Bite. Swarms of females attack man and domestic and wild mammals, usually during the daytime. Although their bite is not particularly painful, it is followed by an intensely itching, raised, ulcerative lesion at the site, due to irritation produced by their salivary secretions in the puncture wound. The saliva may contain several low-molecular-weight proteins that promote agglutinin activity or serve as an anticoagulant (Poehling et al., 1976; Yang and Davies, 1974). Susceptible persons may be annoyed for weeks with an almost unbearable pruritus, or practically incapacitated when the attacked members of the body become enormously swollen. "Black fly fever" is a common malady among children in the rural northeastern United States during the early portion of the black fly season (Jamnback, 1973). A hemorrhagic syndrome among immigrants living along the Transamazon Highway in Brazil was attributed to the bites of black flies (Pinheiro et al., 1974). Several deaths occurred.

Control. Because of the aquatic habitat in which *Simulium* larvae develop, control can frequently be difficult owing to physical factors as well as to concern for nontarget organisms living in the lotic environment. In addition, since adults at times fly long distances from these habitats and ordinarily do not bite indoors, attack on this stage may be only temporarily satisfactory. Even some repellents do not prevent certain anthropophilic species from biting bare legs or legs covered with thin-meshed stockings. Persons who are especially susceptible should wear clothing that is "puncture proof" and head masks of fine-meshed mosquito netting or bolting cloth when venturing into the areas where black flies are abundant.

A massive control effort is currently under way in an *O. volvulus* zone of West Africa that embraces 7 countries and covers an area of 700,000 sq km (Davies et al., 1978). The plan of attack is to lower the intensity of filaria transmission to a level that does not lead to impaired vision. If necessary, this effort will be continued for up to 20 years to break the transmission cycle. Vector control currently consists in larviciding at breeding sites with aerially applied Abate (temephos) at a dosage of 0.05 ppm/10 min of river discharge.

HORSE FLIES, GADFLIES, DEER FLIES, AND MANGO FLIES (FAMILY TABANIDAE)

Morphology, Biology, and Life Cycles. *Adult* tabanid flies are medium-sized to large, stoutly built flies, with large, prominent, iridescent eyes, but usually dull-colored bodies that are practically free of bristles. There is a large pair of powerful wings. The head (Fig. 41–21) is large and broad. The eyes are holoptic in males, dichoptic in females. The antennae are 3-segmented, but the terminal segment is composed of several joints. Only the females have mouthparts adapted to securing a blood meal. In some species of *Tabanus* a female

Fig. 41–21. *Chrysops dimidiata*, adult female. (From Grunberg, K. In Martini, E. 1923. *Textbook of Medical Entomology*. Jena, Gustav Fischer.)

feeds for 10 minutes and may take 0.2 ml of blood. They are particularly active on a warm, bright, moist day. The mandibles are blade-like, the maxillae toothed. The proboscis is short and powerful in the genus *Tabanus,* of medium length in *Chrysops,* and long in *Pangonia*. Oviposition usually takes place by extrusion of large batches (100 to 800) of elongate ovoid *eggs,* which are attached to plants or rocks overhanging standing water or a moist substrate. The *larvae* hatch after 5 to 7 days, drop into the water, and burrow into the mud or sand.

Larvae are cylindroidal, but taper at both ends. Their heads are dark, quite attenuated or even pointed, with mouthparts adapted to seizing and extracting the juices of aquatic insects and other animals on which the larvae prey. One year or longer may be required for development to the pupal stage. The larvae migrate to drier ground to pupate.

The *pupa* is chrysalis-like, without evident thoracic segmentation and with a pair of mesothoracic spiracles. The abdomen is segmented; each segment is divided into a dorsal, 2 lateral, and a ventral area by longitudinal grooves. Segments 2 to 7 each bear a lateral spiracle and have a posterior fringe of spines. After a period of 1 to 3 weeks the adults emerge and mate.

The family Tabanidae is relatively large, numbering about 3000 species. Classification of adults is based primarily on the structure of the antennae, secondarily on the wing pattern. Although the genus *Tabanus* contains the largest number of species, *Chrysops* is the only genus of distinct medical importance. *Chrysops* species (deerflies) are about the size of a house fly or larger. They have the third segment of the long antennae divided into 5 annulations, while the first and second antennal segments are both long. Their hind tibiae are terminally spurred, their wings are usually blackbanded, and they have ocelli or secondary eyespots.

Development of the ovaries of *C. silacea* produces a drain of the food reserves of the fly, so that blood meals and nectars from flowers must be taken in to guarantee normal survival for 3 to 4 weeks.

Medical Importance of Tabanids

Filariasis. *Chrysops dimidiata* and *C. silacea* (mango flies) in Africa are the intermediate hosts and transmitters of the filaria worm, *Loa loa* (see p. 377).

Tularemia. *Chrysops* spp. have been found naturally infected with *Francisella tularensis* and are presumed to be vectors in the western United States and the U.S.S.R. For a discussion of the role of tabanids in the epidemiology of this disease, see Krinsky (1976).

The Bite. Females of many species of Tabanidae will attack man and in taking a blood meal produce an ugly, painful wound that heals slowly. In addition, the saliva may be allergenic.

Control. No satisfactory larvicidal measures have been developed. Swamp drainage has reduced the numbers of *Tabanus* in some areas of northern Minnesota. Where *Loa loa* is a problem, clearing forests and draining the forest floor may effectively alter the larval habitat for control (WHO, 1974a).

BLOODSUCKING MUSCID FLIES (FAMILY MUSCIDAE)

Only a few genera of the muscid flies are bloodsuckers. Of the subfamily Stomoxyinae (=Stomoxydinae), only the biting house fly, dog fly, or stable fly, *Stomoxys calcitrans,* and the horn fly, *Haematobia irritans,* are cosmopolitan in their distribution. The former species frequently attacks man; the latter, which is a cattle pest, rarely bites man.

Biology and Life Cycle of Stomoxys Calcitrans. *Stomoxys calcitrans* somewhat resembles the common house fly *(Musca domestica)* but may be easily distinguished by its proboscis, which is designed for sucking blood (Fig. 41–22). Both males and females are vicious biters. The nonretractable haustellum is long and chitinized and projects conspicuously in front of the head. Inside the terminal portion of the proboscis are prestomal teeth, which

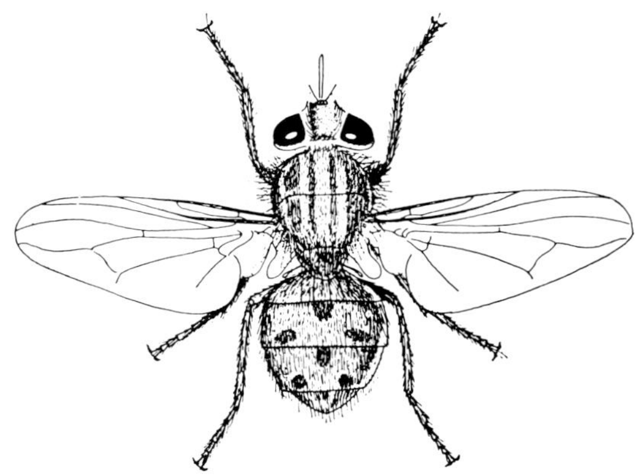

Fig. 41–22. *Stomoxys calcitrans.* Adult, dorsal view. (× ca. 6.) (Adapted from Edwards, F. W., et al., 1939. In Horsfall, W.R. 1962. *Medical Entomology.* New York, The Ronald Press.)

are exserted at the time of feeding and rip open a hole in the skin through which the entire biting apparatus, including the labium (i.e., lower lip), is introduced. This produces a sharp stabbing pain. (When so-called "house flies" "bite," they are usually *S. calcitrans.)* The antennae are short, club-shaped, with 6 segments, of which the fourth to sixth comprise the lateral arista. The terminal sixth is long and spinose and is provided with unbranched bristles only on its dorsal surface. The fourth vein of the wing is curved, not angled. At least three blood meals are required by the female before oviposition occurs. This species breeds in decaying hay or straw, grass (including marine grass, in Florida), rotting grain, or stable manure, laying its ovoid *eggs* (40 to 75) one at a time rather than in masses. After 2 to 5 days the *larvae* hatch and begin to feed on moist decaying vegetable matter. The larval period covers 2 to 3 weeks. The larvae, which reach a length of 20 mm, are more elongate than those of the house fly and have a pair of posterior spiracles, widely separated and somewhat triangular in outline (Fig. 42–2, *10*). The *pupae* are chestnut-brown, elongate-ovoid, measure 5 to 7 mm long, and have the same spiracular pattern as the older larvae. They are found near the edges of breeding places or in the ground below, where there is less moisture. After 6 to 20 or more days the adults emerge. Along the Florida coast this fly breeds in piles of decaying seaweed, so that bathing resorts at times have been forced to close during the fly season.

Medical Importance. The stable fly may possibly be a short-term mechanical vector of bacteria, viruses, or protozoans, but there is no real proof that it serves significantly in this capacity. *Stomoxys calcitrans* can be extremely pestiferous, often exhibiting an aggressive biting behavior.

Control. Control of the breeding of the fly is often difficult because the eggs and larvae are not found in groups and are not attacked appreciably by natural enemies. Piles of hay, straw, or other plant refuse should be protected from rain, and manure piles should be kept on a screened or slatted bottom about 50 cm above a concrete basin containing water about 10 cm in depth. Insecticides, such as the synthetic pyrethroids (see Chapter 45, p. 722), are effective against adults, particularly when resting sites are treated.

TSETSE FLIES (SUBFAMILY GLOSSININAE)

Species of the genus *Glossina* are treated here as a subfamily within the Muscidae, although Potts (1973) and Pont (in Crosskey, 1980) consider the tsetse flies to be a distinct family (Glossinidae). Evidence for this is based on comparisons of wing venation, the relative location of abdominal spiracles, and comparative aspects of sperm storage and transfer with other biting muscids such as *Stomoxys* (Pollock, 1971). Indeed, it has been proposed that *Glossina* spp. evolved from gasterophiloid flies which had functional mouthparts as adults (Pollock, 1973).

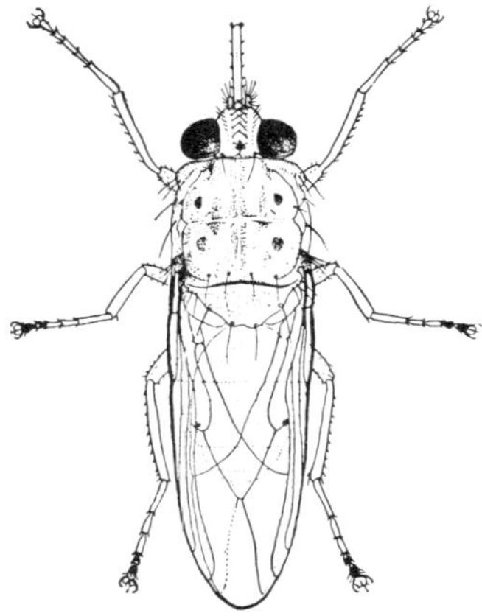

Fig. 41–23. *Glossina* sp., adult in resting position. (Adapted from Matheson, R. 1932. *Medical Entomology.* Springfield, Ill., Charles C Thomas.)

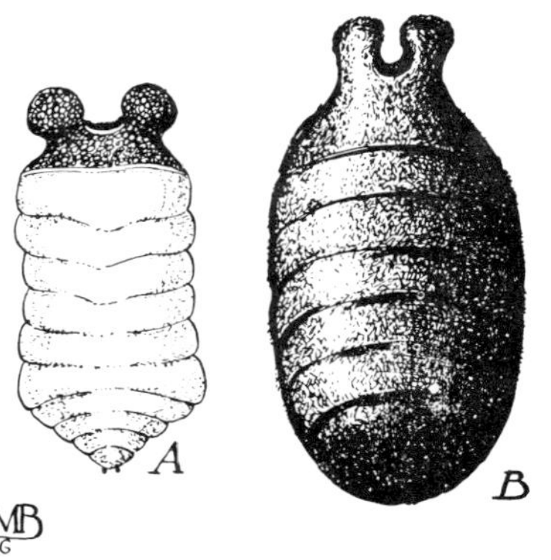

Fig. 41–24. *Glossina* sp. *A*, Larva. *B*, Puparium. (*A* adapted by E.C. Faust from Patton, W.S., and Evans, A.M. 1929. *Insects, Ticks, Mites and Venomous Animals.* London, H. Grubb Ltd. *B* courtesy of C.M. Buchanan.)

Morphology, Biology, and Life Cycle. Tsetse flies (*Glossina* spp.) are honey-brown dipterans about the size of a house fly. When at rest, the proboscis extends in front of the head, and the wings, which have a characteristic venation (the discal or "hatchet" cell is cleaver-shaped), fold over one another covering the abdomen (Fig. 41–23). The antennae are like those of *Stomoxys* except that they are somewhat larger and the bristles of the arista are branched. The maxillary palps are located at the proximal end of the proboscis. The proboscis consists of a short, pyramidal rostrum, a haustellum with a bulbous proximal region, a distal portion that ends in a stylet, and inconspicuous labella with inner rasps of teeth. As in *Stomoxys,* the labium enters the puncture wound made by the everted labellar teeth. Both males and females are voracious blood feeders and are usually telmophagous. *Eggs* produced in the ovary pass into the uterus one at a time. The *larvae* hatch and secure nourishment from the secretion of the milk gland. The posterior spiracles through which the larva respires are situated near the external *os* of the vulva. After 3 larval instars, usually requiring 9 to 12 days, the fully grown larva is expelled, crawls a short distance to a suitable location (usually a shaded site), and pupates. The *pupal stage* requires 20 to 30 days or more, depending on field temperature and species. The characteristic larva and pupa are illustrated in Figure 41–24, *A* and *B*.

Tsetse flies are peculiar to the continent of Africa south of the Sahara. The adults are relatively long-lived. They fly considerable distances, usually along a preferred flight line, and live in areas termed "fly belts," which are determined by their breeding habits and by other specific biotic adaptations. This general association with different kinds of vegetation in sylvatic, lacustrine, riverine, or savannah habitats is correlative with *Glossina* species (Potts, 1973).

The genus can be divided into three groups on the basis of differences in morphology of the male and female genitalia and karyotypes (Jordan, 1977). Within the *G. palpalis* group (subgenus *Nemorhina), G. palpalis* inhabits the tropical rainforest belt of Africa from Senegal to Angola on the west coast. *G. fuscipes fuscipes* occurs inland to the eastern border of Uganda. *G. tachinoides* has a more limited zone, from Guinea on the west coast to Lake Chad and along the northern border of Cameroon. All three are riverine. Members of the *G. morsitans* group (subgenus *Glossina)* are generally more limited to central and eastern Africa and are usually found in the savannah. However, *G. morsitans* extends from the southern Sudan, through parts of Uganda and Tanzania into Zambia and Angola, in a high savannah area, where there are abundant wild game animals. It is also found in West Africa, extending in a central belt from Senegal to the Sudan. *G. pallidipes* has a spotty distribution in eastern Zaire, Mozambique, Tanzania, Uganda, and Kenya. Populations of *G. swynnertoni* are concentrated in northeastern Tanzania.

Medical Importance. Tsetse flies are of medical and veterinary importance in vast areas of Africa. Ecological knowledge of these flies is essential for achieving successful control and eradication of the

Fig. 41–25. Typical habitat of *Glossina palpalis,* common transmitter of *Trypanosoma brucei gambiense.* (Adapted from Hoare, C.W.A. 1962. Reservoir hosts and natural foci of human protozoal infections. Acta Trop., *19.)*

Fig. 41–26. Typical habitat of *Glossina morsitans*, common transmitter of *Trypanosoma brucei rhodesiense*. (Adapted from R. Geigy and A. Herbig. In Hoare, C.W.A. 1962. Reservoir hosts and natural foci of human protozoal infection. Acta Trop., *19*.)

trypanosomiases with which they are associated. These infections have been considered in detail in Section II, pages 79 and 86. *G. palpalis* (Fig. 41–25), which is limited to areas with relatively permanent surface water, is the important transmitter of *T. gambiense*, and *G. morsitans* (Fig. 41–26), of *T. rhodesiense*, whereas *G. tachinoides* and *G. swynnertoni* are generally of secondary importance, but locally may be the essential vectors. In East Africa, *G. morsitans* is replaced in dry areas by *G. longipennis* of the *fusca* group (subgenus *Austenina)* and *G. pallidipes*, and in residual forest patches along rivers by *G. brevipalpis* and *G. austeni*. Although the usual transmission occurs after an incubation period in the fly, mechanical conveyance of the trypanosome can occur within a few hours during an interrupted blood meal.

Control. Control of these flies constitutes a most perplexing and important public health problem, on account of the breeding habits of the flies, their extensive distribution throughout regions of Africa where they can obtain abundant blood meals from animals, and the longevity of the adults. Suggestions have been made to kill off all the game animals near human habitations in areas infested by *G. morsitans*, in order to limit "fly food." This elimination of natural hosts would then cause rapid reduction in the tsetse population. Another plan, which at times has been necessary as a protective measure, has been to move entire villages outside the *Glossina morsitans* belt. However, the implementation of these strategies on a large scale appears to be ecologically and sociologically unacceptable.

Periodic clearing of the bush around villages, fords, and boat landings, as well as along the sides of roads, is advisable for *G. palpalis*, but not so practicable for *G. morsitans*. In the case of *G. pallidipes*, the following preventive steps have been pursued: (1) clearing the riparian bush, in which the flies reside, for a width of 1000 yards, (2) thinning thickets surrounding this protective barrier, (3) trapping the flies in concentrated areas, and (4) patrolling infested zones. These habitat alterations coupled with the selective use of insecticides for adult control can be effective, particularly if carried out during the dry season. The careful application of residual compounds to diurnal or nocturnal resting sites on vegetation or treatment of extensive areas with aerosol sprays from fixed-wing aircraft or helicopter have proved successful

(G. Matthews, 1979). Several of the synthetic pyrethroids appear promising as antitsetse materials (Spielberger et al., 1979), particularly when used in conjunction with conical traps (Laveissiere and Couret, 1980; Lancien, 1981). In addition, when pregnant females are treated with 0.5 μg of diflubenzuron, a growth regulator, the resulting offspring are unable to pupariate (Jordan et al., 1979).

REFERENCES*

Akey, D.H., and Potter, H.W. 1979. Pigmentation associated with oogenesis in the biting fly *Culicoides variipennis* (Diptera: Ceratopogonidae): Determination of parity. J. Med. Entomol., *16*:67–70.

Arzube, R.M.E. 1982. Oncocercosis en el Ecuador. Primer foco descubierto en el pais, hallazgos clinicos, parasitologicos i entomologicos. Tropenmed. Parasitol., *33*:45–50.

Ashburn, P.M., and Craig, C. 1907. Observations on the etiology of dengue fever. Med. Rec., *71*:949.

Auriault, M. 1979. Contribution a l'étude biologique et écologique de *Culicoides grahamii* (Austen, 1909) (Diptera: Ceratopogonidae). IV. Dynamique des populations. Cahiers, O.R.S.T.O.M., ser. Entomol. Med. Parasitol., *17*:31–36.

Axtell, R.C. 1979. Principles of integrated pest management (IPM) in relation to mosquito control. Mosquito News, *39*:709–718.

Bailey, C.L., Eldridge, B.F., Hayes, D.E., Watts, D.M., Tammariello, R.F., and Dalrymple, J.M. 1978. Isolation of St. Louis encephalitis virus from overwintering *Culex pipiens* mosquitoes. Science, *199*:1346–1349.

———, and Gould, D.J. 1975. Flight and dispersal of Japanese encephalitis vectors in northern Thailand. Mosquito News, *35*:172–178.

Bailey, D.L., Kaiser, P.E., and Lowe, R.E. 1981. Population densities of *Anopheles albimanus* adults and larvae inside and outside cotton-growing areas in El Salvador. Mosquito News, *41*:151–154.

Baker, J.R. 1977. Factors affecting transmission of vector-borne blood parasites. Acta Trop., *34*:7–19.

Baldry, D.A.T., and Chaudhury, M.F.B., editors. 1980. Epidemiology of African trypanosomiasis. Insect Sci. Appl., *1*:1–121.

Bang, Y.H., Bown, D.N., and Arata, A.A. 1980. Ecological studies on *Aedes africanus* (Diptera: Culicidae) and associated species in southeastern Nigeria. J. Med. Entomol., *17*:411–416.

Barnard, D.R., and Mulla, M.S. 1978. The ecology of *Culiseta inornata* in the Colorado Desert of California: Seasonal abundance, gonotrophic status, and oviparity of adult mosquitoes. Ann. Entomol. Soc. Am., *71*:397–400.

Barr, A.R. 1974. V. Reproduction in Diptera of medical importance with special reference to mosquitoes, pp. 35–40. In *Symposium on Reproduction of Arthropods of Medical and Veterinary Importance*. Edited by J.H. Oliver, Jr. J. Med. Entomol., *11*:1–45.

Barreto, P., Trapido, H., and Lee, V.H. 1970. Onchocerciasis in Colombia. Entomological findings in the first observed focus. Am. J. Trop. Med. Hyg., *19*:837–841.

Beaty, B.J., Tesh, R.B., and Aitken, T.H.G. 1980. Transovarial transmission of yellow fever virus in *Stegomyia* mosquitoes. Am. J. Trop. Med. Hyg., *29*:125–132.

*References listed here include those cited in the text and others of general interest concerning bloodsucking Diptera.

———, and Thompson, W.H. 1978. Tropisms of La Crosse virus in *Aedes triseriatus* (Diptera: Culicidae) following infective blood meals. J. Med. Entomol., *14*:499–503.

Beaver, P.C., and Orihel, T.C. 1965. Human infection with filariae of animals in the United States. Am. J. Trop. Med. Hyg., *14*:1010–1029.

Belkin, J.N. 1977. *Quinquefasciatus* or *fatigans* for the tropical (southern) house mosquito (Diptera: Culicidae)? Proc. Entomol. Soc. Wash., *79*:45–52.

Bellec, C., and Hébrard, G. 1980. Les lieux de repos des adultes du complexe *Simulium damnosum* (Diptera, Simuliidae). 2. Etude de la distribution spatio-temporelle. Cahiers O.R.S.T.O.M., ser. Entomol. Med. Parasitol., *18*:277–289.

Bidlingmayer, W.L. 1974. The influence of environmental factors and physiological stage on flight patterns of mosquitoes taken in the vehicle aspirator and truck, suction, bait and New Jersey light traps. J. Med. Entomol., *11*:119–146.

———, and Hem, D.G. 1980. The range of visual attraction and the effect of competitive visual attractants upon mosquito (Diptera: Culicidae) flight. Bull. Entomol. Res., *70*:321–342.

———, and Hem, D.G. 1981. Mosquito flight paths in relation to the environment. Effect of the forest edge upon trap catches in the field. Mosquito News, *41*:55–59.

Birley, M.H., and Rajagopalan, P.K. 1981. Estimation of the survival and biting rates of *Culex quinquefasciatus* (Diptera: Culicidae). J. Med. Entomol., *18*:181–186.

Boorman, J. 1974. The maintenance of laboratory colonies of *Culicoides variipennis* (Coq.), *C. nubeculosus* (Mg.) and *C. riethi* Kieff. (Diptera: Ceratopogonidae). Bull. Entomol. Res., *64*:371–377.

Boreham, P.F.L. 1975. Some applications of bloodmeal identifications in relation to the epidemiology of vector-borne tropical diseases. J. Trop. Med. Hyg., *78*:83–91.

———, and Lenahan, J.K. 1976. Methods for detecting multiple blood-meals in mosquitoes. Bull. Entomol. Res., *66*:671–679.

Bown, D.N., and Bang, Y.H. 1980. Ecological studies on *Aedes simpsoni* (Diptera: Culicidae) in southeastern Nigeria. J. Med. Entomol., *17*:367–374.

———, Bang, Y.H., Knudsen, A.B., Arata, A.A., and Fabiyi, A. 1980. Forest types and arbovirus vectors in the Mamu River Forest Reserve of southeastern Nigeria. Mosquito News, *40*:91–102.

Brady, J. 1974. Calculation of anopheline man-biting densities from concurrent indoor and outdoor resting samples. Ann. Trop. Med. Parasitol., *68*:359–361.

Braverman, Y., and Hulley, P.E. 1979. The relationship between the numbers and distribution of some antennal and palpal sense organs and host preference in some *Culicoides* (Diptera: Ceratopogonidae) from southern Africa. J. Med. Entomol., *15*:419–424.

Bray, R.S. 1974. *Leishmania*. Annu. Rev. Microbiol., *28*:189–217.

Breeland, S.G. 1972. Methods for measuring anopheline densities in El Salvador. Mosquito News, *32*:62–72.

———. 1980. A bibliography to the literature of *Anopheles albimanus* (Diptera: Culicidae). Mosquito Syst., *12*:50–150.

———, Taylor, R.T., and Mitchell, C.J. 1980. Vector control, pp. 605–622. In *St. Louis Encephalitis*. Edited by T.P. Monath. Washington, D.C., American Public Health Association.

Briegel, H., Lea, A.O., and Klowden, M.J. 1979. Hemoglobinometry as a method for measuring blood meal sizes of mosquitoes (Diptera: Culicidae). J. Med. Entomol., *15*:235–238.

Brinker, K.R., and Monath, T.P. 1980. The acute disease, pp.

503–534. In *St. Louis Encephalitis*. Edited by T.P. Monath. Washington, D.C., American Public Health Association.

Brown, A.W.A., Haworth, J., and Zahar, A.R. 1976. Malaria eradication and control from a global standpoint. J. Med. Entomol., *13*:1–25.

Bruce-Chwatt, L.J. 1980. *Essential Malariology*. London, W. Heinemann Medical Books, Ltd.

Bryan, J.H. 1979. Observations on the member species of the *Anopheles gambiae* complex in the Gambia, West Africa. Trans. R. Soc. Trop. Med. Hyg., *73*:463–466.

———, and Smalley, M.E. 1978. The use of ABO blood groups as markers for mosquito biting studies. Trans. R. Soc. Trop. Med. Hyg., *72*:357–360.

Bursell, E., Billing, K.C., Hargrove, J.W., McCabe, C.T., and Slack, E. 1974. Metabolism of the bloodmeal in tsetse flies. Acta Trop., *31*:297–320.

Bushrod, F.M. 1981. The *Anopheles gambiae* Giles complex and Bancroftian filariasis transmission in a Tanzanian coastal village. Ann. Trop. Med. Parasitol., *75*:93–100.

Cabrera, B.D., and Arambulo, P.V. III. 1977. Malaria in the Republic of the Philippines. A review. Acta Trop., *34*:265–279.

Campbell, C.C., Collins, R.C., Huong, A.Y., and Figueroa Marroquin, H. 1980. Quantitative aspects of the infection of *Simulium ochraceum* by *Onchocerca volvulus:* The relation of skin microfilarial density to vector infection. Tropenmed. Parasitol., *31*:475–478.

Carlson, D.A., and Service, M.W. 1980. Identification of mosquitoes of *Anopheles gambiae* species complex A and B by analysis of cuticular components. Science, *207*:1089–1091.

Carme, B., and Laigret, J. 1979. Longevity of *Wuchereria bancrofti* var. *pacifica* and mosquito infection acquired from a patient with low level parasitemia. Am. J. Trop. Med. Hyg., *28*:53–55.

Casals, J. 1967. Classification of arboviruses and delineation of clinically important types. Proceedings of the First International Conference on Vaccines Against Viral and Rickettsial Diseases of Man, pp. 243–251. Washington, D.C., Pan American Health Organization.

Chaniotis, B.N. 1975. A new method for rearing *Lutzomyia trapidoi* (Diptera: Psychodidae), with observations on its development and behavior in the laboratory. J. Med. Entomol., *12*:183–188.

———, Correa, M.A., Tesh, R.B., and Johnson, K.M. 1974. Horizontal and vertical movements of phlebotomine sandflies in a Panamanian rain forest. J. Med. Entomol., *11*:369–374.

Charlwood, J.D. 1980. Observations on the bionomics of *Anopheles darlingi* Root (Diptera: Culicidae) from Brazil. Bull. Entomol. Res., *70*:685–692.

Chow, C.Y. 1973. Filariasis vectors in the Western Pacific Region. Tropenmed. Parasitol., *24*:404–418.

Christensen, H.A., and Herrer, A. 1973. Attractiveness of sentinel animals to vectors of leishmaniases in Panama. Am. J. Trop. Med. Hyg., *22*:578–584.

———, and Herrer, A. 1980. Development of a Panamanian strain of *Leishmania mexicana* in co-indigenous *Lutzomyia sanguinaria* and *Lu. gomezi* (Diptera: Psychodidae). J. Med. Entomol., *17*:188–189.

———, Herrer, A., and Telford, S.R., Jr. 1972. Enzootic cutaneous leishmaniasis in eastern Panama. II. Entomological investigations. Ann. Trop. Med. Parasitol., *66*:55–66.

Collins, R.C. 1979a. Onchocerciasis transmission potentials of four species of Guatemalan Simuliidae. Am. J. Trop. Med. Hyg., *28*:72–75.

———. 1979b. Development of *Onchocerca volvulus* in *Simulium ochraceum* and *Simulium metallicum*. Am. J. Trop. Med. Hyg., *28*:491–495.

Coluzzi, M., Sabatini, A., Petrarca, V., and DiDeco, M.A. 1977. Behavioural divergences between mosquitoes with different inversion karyotypes in polymorphic populations of the *Anopheles gambiae* complex. Nature, *266*:832–833.

———, Sabatini, A., Petraca, V., and DiDeco, M.A. 1979. Chromosomal differentiation and adaptation to human environments in the *Anopheles gambiae* complex. Trans. R. Soc. Trop. Med. Hyg., *73*:483–497.

Cornet, M. 1974. Caractères morphologiques utilisés pour l'identification des *Culicoides* (Diptera: Ceratopogonidae). Cahiers O.R.S.T.O.M. Ser. Entomol. Med. Parasitol., *12*:221–229.

———, Robin, Y., Heme, G., Adam, C., Renaudet, J., Valade, M., and Eyraud, M. 1979. Une poussée épizootique de fièvre jaune selvatique au Senégal oriental. Isolement du virus de lots de moustiques adultes males et femelles. Med. Malad. Infect., *9*:63–66.

Crewe, W. 1961. The rate of development of larvae of *Loa loa* in *Chrysops silacea* at Kumba and the effect of temperature upon it. Ann. Trop. Med. Parasitol., *55*:211–216.

Crosskey, R.W. 1973. Diptera-Nematocera. Simuliidae (blackflies), pp. 109-153. In *Insects and Other Arthropods of Medical Importance*. Edited by K.G.V. Smith. London, British Museum (Natural History).

———, editor. 1980. *Catalogue of the Diptera of the Afrotropical* Region. London, British Museum (Natural History).

Cupp, E.W., and Collins, R.C. 1979. The gonotrophic cycle in *Simulium ochraceum*. Am. J. Trop. Med. Hyg., *28*:422–426.

———, Scherer, W.F., and Ordonez, J.V. 1979. Transmission of Venezuelan encephalitis virus by naturally infected *Culex (Melanoconion) opisthopus*. Am. J. Trop. Med. Hyg., *28*:1060–1063.

Curtis, C.F., and Feachem, R.G. 1981. Sanitation and *Culex pipiens* mosquitoes: A brief review. J. Trop. Med. Hyg., *84*:17–25.

Dang, P.T., and Peterson, B.V. 1980. Pictorial keys to the main species groups within the *Simulium damnosum* Theobald complex occurring in West Africa (Diptera: Simuliidae). Tropenmed. Parasitol., *31*:117–120.

Darsie, R.S., Jr., and Ward, R.A. 1981. *Identification and Geographical Distribution of the Mosquitoes of North America, North of Mexico*. Suppl. to Mosquito Systematics #1, American Mosquito Control Association, Fresno, California.

Davies, J.B., LeBerre, R., Walsh, J.F., and Cliff, B. 1978. Onchocerciasis and *Simulium* control in the Volta River Basin. Mosquito News, *38*:466–472.

Denham, D.A., and McGreevy, P.B. 1977. Brugian filariasis: Epidemiological and experimental studies. Adv. Parasitol., *15*:243–309.

Detinova, T.S. 1962. *Age-grouping Methods in Diptera of Medical Importance*. WHO Monogr. Ser. No. 47, Geneva.

De Vault, V.T. 1952. The treatment of bartonellosis with antibiotics. Ann. N.Y. Acad. Sci., *55*:1222–1227.

Dipeolu, O.O. 1976. Species distribution and abundance of *Culicoides* Latreille (Diptera: Ceratopogonidae) in Nigeria. Bull. Entomol. Res., *66*:685–693.

Disney, R.H.L. 1970. The timing of the first blood meal in *Simulium damnosum* Theobald. Ann. Trop. Med. Parasitol., *64*:123–128.

Dissanaike, A.S. 1979. Zoonotic aspects of filarial infections in man. Bull. W.H.O., *57*:349–357.

Downe, A.E.R., and Archer, J.A. 1975. The effects of different blood-meal sources on digestion and egg production in

Culex tarsalis Coq. (Diptera: Culicidae). J. Med. Entomol., *12*:431–437.

Downes, J.A. 1971. The ecology of blood-sucking Diptera: An evolutionary perspective, pp. 232–258. In *Ecology and Physiology of Parasites*. Edited by A.M. Fallis. Toronto, University of Toronto Press.

Downing, J.D., and Crans, W.J. 1977. The Ehrenberg pigeon trap as a sampler of *Culex* mosquitoes for St. Louis encephalitis surveillance. Mosquito News, *37*:48–53.

Downs, W.G. 1982. Arboviruses, pp. 95–126. In *Viral Infections of Humans–Epidemiology and Control*. 2nd edit. Edited by A.S. Evans. New York, Plenum Medical Book Co.

Duke, B.O.L. 1954. The transmission of loiasis in the forest-fringe area of the British Cameroons. Ann. Trop. Med. Parasitol., *48*:349–355.

———. 1973. Studies on factors influencing the transmission of onchocerciasis. VIII. The escape of infective *Onchocerca volvulus* larvae from feeding "forest" *Simulium damnosum*. Ann. Trop. Med. Parasitol., *67*:95–99.

———. 1975. The differential dispersal of nulliparous and parous *Simulium damnosum*. Tropenmed. Parasitol., *26*:88–97.

Dutary, B.E., and LeDuc, J.W. 1981. Transovarial transmission of yellow fever virus by a sylvatic vector, *Haemagogus equinus*. Trans. R. Soc. Trop. Med. Hyg., *75*:128.

Edman, J.D., Webber, L.A., and Schmid, A.A. 1974. Effect of host defenses on the feeding pattern of *Culex nigripalpus* when offered a choice of blood sources. J. Parasitol., *60*:874–883.

Eisa, M., El Sid, E.D.K., Shomein, A.M., and Meegan, J.M. 1980. An outbreak of Rift Valley fever in the Sudan—1976. Trans. R. Soc. Trop. Med. Hyg., *74*:417–419.

Eldridge, B.F., and Bailey, C.L. 1979. Experimental hibernation studies in *Culex pipiens* (Diptera: Culicidae): Reactivation of ovarian development and blood-feeding in prehibernating females. J. Med. Entomol., *15*:462–467.

Elliott, R. 1972. The influence of vector behavior on malaria transmission. Am. J. Trop. Med. Hyg., *21*:755–763.

Fain, A., Wery, M., and Tilkin, J. 1980. *Simulium albivirgulatum* Wanson et Henrard, 1944, vecteur d'*Onchocerca volvulus* dans la region de la Cuvette Centrale, au Zaire. Note preliminaire. Ann. Soc. Belge Med. Trop., *60*:285–289.

Feliciangeli, M.D. 1980. The phlebotomine sand flies (Diptera: Psychodidae) of Venezuela. J. Med. Entomol., *17*:245–264.

Foster, W.A. 1972. Studies on leishmaniasis in Ethiopia. III. Resting and breeding sites, flight behaviour, and seasonal abundance of *Phlebotomus longipes* (Diptera: Psychodidae). Ann. Trop. Med. Parasitol., *66*:313–328.

———, Boreham, P.F.L., and Tempelis, C.H. 1972. Studies on leishmaniasis in Ethiopia. IV. Feeding behaviour of *Phlebotomus longipes* (Diptera: Psychodidae). Ann. Trop. Med. Parasitol., *66*:313–328.

Francy, D.B., Rush, W.A., Montoya, M., Inglish, D.S., and Bolin, R.A. 1981. Transovarial transmission of St. Louis encephalitis virus by *Culex pipiens* complex mosquitoes. Am. J. Trop. Med. Hyg., *30*:699–705.

Freyvogel, T.A., and Stäubli, W. 1965. The formation of the peritrophic membrane in Culicidae. Acta Trop., *22*:118–147.

Friend, W.G., and Smith, J.J.B. 1977. Factors affecting feeding by bloodsucking insects. Annu. Rev. Entomol., *22*:309–331.

Fuller, G.K. 1978. Distribution of *Glossina* (Diptera: Glossinidae) in southwestern Ethiopia. Bull. Entomol. Res., *68*:299–305.

Galindo, P. 1972. Endemic vectors of Venezuelan encephalitis, pp. 249–253. In PAHO Sci. Publ. No. 243, *Venezuelan Encephalitis*. Washington, D.C., Pan American Sanitary Bureau.

Gard, G., Marshall, I.D., and Woodroofe, G.M. 1973. Annually recurrent epidemic polyarthritis and Ross River virus activity in the coastal area of New South Wales. II. Mosquitoes, viruses, and wildlife. Am. J. Trop. Med. Hyg., *22*:551–560.

Garms, R. 1973. Quantitative studies on the transmission of *Onchocerca volvulus* by *Simulium damnosum* in the Bong Range, Liberia. Tropenmed. Parasitol., *24*:358–372.

———. 1975. Observations on filarial infections and parous rates of anthropophilic blackflies in Guatemala, with reference to the transmission of *Onchocerca volvulus*. Tropenmed. Parasitol., *26*:375–380.

———, and Vajime, C.G. 1975. On the ecology and distribution of the species of the *Simulium damnosum* complex in different bioclimatic zones of Liberia and Guinea. Tropenmed. Parasitol., *26*:375–380.

———, Walsh, J.F., and Davies, J.B. 1979. Studies on the reinvasion of the Onchocerciasis Control Programme in the Volta River Basin by *Simulium damnosum* s. l. with emphasis on the south-western areas. Tropenmed. Parasitol., *30*:345–362.

Garnham, P.C.C. 1966. *Malaria Parasites and Other Haemosporidia*. Oxford, Blackwell Scientific Publications.

Garrett-Jones, C., Boreham, P.F.L., and Pant, C.P. 1980. Feeding habits of anophelines (Diptera: Culicidae) in 1971–78, with reference to the human blood index: A review. Bull. Entomol. Res., *70*:165–185.

Geiman, Q.M. 1941. New media for the growth of *Bartonella bacilliformis*. Proc. Soc. Exp. Biol. Med., *47*:329–332.

Gemetchu, T. 1974. The morphology and fine structure of the midgut and peritrophic membrane of the adult female *Phlebotomus longipes* Parrot and Martin (Diptera: Psychodidae). Ann. Trop. Med. Parasitol., *68*:111–124.

Germain, M., Francy, D.B., Monath, T.P., Ferrara, L. Bryan, J., Salaun, J.-J., Heme, G., Renaudet, J., Adam, C., and Digoutte, J.-P. 1980. Yellow fever in the Gambia, 1978-1979: Entomological aspects and epidemiological correlations. Am. J. Trop. Med. Hyg., *29*:929–940.

Gillett, J.D. 1972. *The Mosquito: Its Life, Activities, and Impact on Human Affairs*. Garden City, New York, Doubleday & Co.

———, and Connor, J. 1976. Host temperature and the transmission of arboviruses by mosquitoes. Mosquito News, *36*:472–477.

Gillies, M.T. 1954. The recognition of age-groups within populations of *Anopheles gambiae* by the pre-gravid rate and the sporozoite rate. Ann. Trop. Med. Parasitol., *48*:58–74.

———. 1974. Methods for assessing the density and survival of bloodsucking Diptera. Annu. Rev. Entomol., *19*:345–362.

———. 1980. The role of carbon dioxide in host-finding by mosquitoes (Diptera: Culicidae): A review. Bull. Entomol. Res., *70*:525–532.

———, and de Meillon, B. 1968. *The Anophelinae of Africa South of the Sahara (Ethiopian Zoogeographic Region)*, 2nd ed. South Afr. Inst. Med. Res. Publ. 54.

Gingrich, J.B., Ward, R.A., Macken, L.M., and Esser, K.M. 1981. Some phenomena associated with the development of *Trypanosoma brucei rhodesiense* infections in the tsetse fly, *Glossina morsitans*. Am. J. Trop. Med. Hyg., *30*:570–574.

Gooding, R.H. 1972. Digestive processes of haematophagous insects. I. A literature review. Quaest Entomol., *8*:5–60.

———. 1981. Genetic polymorphism in three species of tsetse flies (Diptera: Glossinidae) in Upper Volta. Acta Trop., *38*:149–161.

Gould, D.J., Edelman, R., Grossman, R.A., Nisalak, A., and Sullivan, M.F. 1974. Study of Japanese encephalitis virus in Chiangmai Valley, Thailand. IV. Vector studies. Am. J. Epidemiol., *100*:49–56.

Grady, G.F., Maxfield, H.K., Hildreth, S.W., Timperi, R.J., Jr., Gilfillan, R.F., Rosenau, B.J., Francy, D.B., Calisher, C.H., Marcus, L.C., and Madoff, M.A. 1978. Eastern equine encephalitis in Massachusetts, 1957–1976. A prospective study centered upon analyses of mosquitoes. Am. J. Epidemiol., *107*:170–178.

Grimstad, P.R., Ross, Q.E., and Craig, G.B., Jr. 1980. *Aedes triseriatus* (Diptera: Culicidae) and La Crosse virus. II. Modification of mosquito feeding behaviour by virus infection. J. Med. Entomol., *17*:1–7.

Grunewald, J. 1976. The hydro-chemical and physical conditions of the environment of the immature stages of some species of the *Simulium (Edwardsellum) damnosum* complex (Diptera). Tropenmed. Parasitol., *27*:438–454.

Gubler, D.J., and Bhattacharya, N.C. 1974. A quantitative approach to the study of Bancroftian filariasis. Am. J. Trop. Med. Hyg., *23*:1027–1036.

———, Nalim, S., Tan, R., Saipan, H., and Sulianti Saroso, Jr. 1979. Variation in susceptibility to oral infection with dengue viruses among geographic strains of *Aedes aegypti*. Am. J. Trop. Med. Hyg., *28*:1045–1052.

———, and Rosen, L. 1976. A simple technique for demonstrating transmission of dengue virus by mosquitoes without the use of vertebrate hosts. Am. J. Trop. Med. Hyg., *25*:146–150.

———, and Rosen, L. 1977. Quantitative aspects of replication of dengue viruses in *Aedes albopictus* (Diptera: Culicidae) after oral and parenteral infection. J. Med. Entomol., *13*:469–472.

Gutsevich, A.V., Monchadskii, A.S., and Shtakel'berg, A.A. 1974. *Fauna of the U.S.S.R. Diptera. Mosquitoes. Family Culicidae*, Vol. 3, No. 4. Jerusalem, Israel Program for Scientific Translations.

Hacker, C.S., Ling, W.-W., Bartholomew, P.H., and Crovello, T.J. 1977. An application of mathematical modelling to the study of reproductive adaptations in the yellow fever mosquito, *Aedes aegypti*. J. Med. Entomol., *13*:485–492.

Hagedorn, H.H., and Kunkel, J.G. 1979. Vitellogenin and vitellin in insects. Annu. Rev. Entomol., *24*:475–505.

Hamon, J. 1974. Onchocerciasis vectors in the Western Hemisphere. A. Vector biology and vector-parasite relationships, pp. 58–68. In PAHO Sci. Publ. No. 298, *Research and Control of Onchocerciasis in the Western Hemisphere*. Washington, D.C., Pan American Health Organization.

Harbach, R.E., and Knight, K.L. 1980. *Taxonomists' Glossary of Mosquito Anatomy*. Marlton, New Jersey, Plexus Publishing.

———, and Knight, K.L. 1981. Corrections and additions to Taxonomists' Glossary of Mosquito Anatomy. Mosquito Syst., *13*:201–217.

Hardy, J.L., Houk, E.J., Kramer, L.D., and Reeves, W.C. 1983. Intrinsic factors affecting vector competence of mosquitoes for arboviruses. Annu. Rev. Entomol., *28*:229–262.

Hargrove, J.W. 1977. Some advances in the trapping of tsetse (*Glossina* spp.) and other flies. Ecol. Entomol., *2*:123–137.

Harrison, B.A. 1980. Medical entomology studies—XIII. The *Myzomyia* series of *Anopheles (Cellia)* in Thailand, with emphasis on intra-interspecific variations (Diptera: Culicidae). Contr. Am. Entomol. Inst., *17*(No. 4).

———, and Scanlon, J.E. 1975. Medical entomology studies—II. The subgenus *Anopheles* in Thailand (Diptera: Culicidae). Contr. Am. Entomol. Inst., *12*(No. 1).

Harwood, R.F., and James, M.T. 1979. *Entomology in Human and Animal Health*. New York, Macmillan Publishing Co.

Hayes, C.G. 1979. Vector competence of colonized *Culiseta melanura* (Diptera: Culicidae) for western equine encephalomyelitis virus. J. Med. Entomol. *15*:253–258.

———, Basit, A., Bagar, S., and Akhter, R. 1980. Vector competence of *Culex tritaeniorhynchus* (Diptera: Culicidae) for West Nile virus. J. Med. Entomol., *17*:172–177.

Hobbs, J.H., Lowe, R.E., and Schreck, C.E. 1974. Studies of flight range and survival of *Anopheles albimanus* Wiedemann in El Salvador. I. Dispersal and survival during the dry season. Mosquito News, *34*:389–393.

Hoch, A.L., Peterson, N.E., LeDuc, J.W., and Pinheiro, F.P. 1981. An outbreak of Mayaro virus disease in Belterra, Brazil. III. Entomological and ecological studies. Am. J. Trop. Med. Hyg., *30*:689–698.

Hoogstraal, H., Meegan, J.M., Khalil, G.M., and Adham, F.K. 1979. The Rift Valley fever epizootic in Egypt 1977–78. 2. Ecological and entomological studies. Trans. R. Soc. Trop. Med. Hyg., *73*:624–629.

Horsfall, W.R. 1955. *Mosquitoes: Their Bionomics and Relation to Disease*. New York, Ronald Press.

———, and Ronquillo, M.C. 1970. Genesis of the reproductive system of mosquitoes. II. Male of *Aedes stimulans* (Walker). J. Morphol., *131*:329–358.

Houk, E.J., Obie, W., and Hardy, J.L. 1979. Peritrophic membrane formation and the midgut barrier to arboviral infection in the mosquito, *Culex tarsalis* Coquillett (Insecta, Diptera). Acta Trop., *36*:39–45.

Jaenson, T.G.T. 1980. Mating behaviour of females of *Glossina pallidipes* Austen (Diptera: Glossinidae). Bull. Entomol. Res., *70*:49–60.

Jamnback, H. 1973. Recent developments in control of blackflies. Annu. Rev. Entomol., *18*:281–304.

Javadian, E., Tesh, R., Saidi, S., and Nadim, A. 1977. Studies on the epidemiology of sandfly fever in Iran. III. Host-feeding patterns of *Phlebotomus papatasi* in an endemic area of the disease. Am. J. Trop. Med. Hyg., *26*:294–298.

Johnson, B.K., Gichogo, A., Gitau, G., Patel, N., Ademba, G., Kirui, R., Highton, R.B., and Smith, D.H. 1981. Recovery of O'nyong-nyong virus from *Anopheles funestus* in western Kenya. Trans. R. Soc. Trop. Med. Hyg., *75*:239–241.

Jordan, A.M. 1974. Recent developments in the ecology and methods of control of tsetse flies (*Glossina* spp.) (Dipt., Glossinidae)—a review. Bull. Entomol. Res., *63*:361–399.

———. 1977. Systematics, pp. 13–22. In *Tsetse: The Future for Biological Methods in Integrated Control*. Edited by M. Laird. Ottawa, Canada. International Development Research Centre.

———, Trewern, M.A., Borkovec, A.B., and DeMilo, A.B. 1979. Laboratory studies on the potential of three insect growth regulators for control of the tsetse fly, *Glossina morsitans morsitans* Westwood (Diptera: Glossinidae). Bull. Entomol. Res., *69*:55–64.

Jumali, Sunarto, Gubler, D.J., Nalim, S., Eram, S., and Sulianti Saroso, J. 1979. Epidemic dengue hemorrhagic fever in rural Indonesia. III. Entomological studies. Am. J. Trop. Med. Hyg., *28*:717–724.

Jupp, P.G., McIntosh, B.M., and Dos Santos, I. 1981. Laboratory vector studies on six mosquito and one tick species with chikungunya virus. Trans. R. Soc. Trop. Med. Hyg., *75*:15–19.

Kan, S.-P., and Beng-Chuan, H. 1973. Development of *Brugia pahangi* in the flight muscles of *Aedes togoi*. Ultrastructural changes in the infected muscle fibers and the infecting filarial larvae. Am. J. Trop. Med. Hyg., *22*:179–188.

Kanada, T., Joo, C.Y., and Choi, D.W. 1975. Epidemiological studies on Malayan filariasis in an inland area in Kyungpook, Korea. (2.) The periodicity of the microfilariae and the bionomics of the vector. Mosquito News, *35*:513–517.

Kay, B.H. 1979. Age structure of populations of *Culex an-*

nulirostris (Diptera: Culicidae) at Kowanyama and Charleville, Queensland. J. Med. Entomol., *16*:309–316.

———. 1980. Towards prediction and surveillance of Murray Valley encephalitis activity in Australia. Aust. J. Exp. Biol. Med. Sci., *58*:67–76.

———, Boreham, P.F.L., and Edman, J.D. 1979. Application of the "feeding index" concept to studies of mosquito host-feeding patterns. Mosquito News, *39*:68–72.

———, Carley, J.G., Fanning, I.D., and Filippich, C. 1979. Quantitative studies on the vector competence of *Aedes aegypti, Culex annulirostris* and other mosquitoes (Diptera: Culicidae) with Murray Valley encephalitis and other Queensland arboviruses. J. Med. Entomol., *16*:59–66.

Kettle, D.S. 1977. Biology and bionomics of bloodsucking ceratopogonids. Annu. Rev. Entomol., *22*:33–51.

Khamala, C.P.M., and Kettle, D.S. 1971. The *Culicoides* Latreille (Diptera: Ceratopognidae) of East Africa. Trans. R. Entomol. Soc. Lond., *123*:1–95.

Killick-Kendrick, R. 1978. Recent advances and outstanding problems in the biology of phlebotomine sand flies. Acta Trop., *35*:297–313.

———. 1979. Biology of *Leishmania* in phlebotomine sand flies, pp. 395–460. In *Biology of the Kinetoplastida,* Vol. 2. Edited by W.H.R. Lumsden and D.A. Evans. New York, Academic Press.

Kitzmiller, J.B. 1976. Genetics, cytogenetics, and evolution of mosquitoes. Adv. Genetics, *18*:315–433.

Knight, K.L. 1978. *Supplement to a Catalog of the Mosquitoes of the World.* College Park, Maryland, Entomological Society of America.

———, and Stone, A. 1977. *A Catalog of the Mosquitoes of the World,* 2nd ed. College Park, Maryland, Entomological Society of America.

Krafsur, E.S., and Garrett-Jones, C. 1977. The survival in nature of *Wuchereria*-infected *Anopheles funestus* Giles in north-eastern Tanzania. Trans. R. Soc. Trop. Med. Hyg., *71*:155–160.

Kreutzer, R.D., Kitzmiller, J.B., and Rabbani, M.G. 1975. The salivary gland chromosomes of *Anopheles argyritarsis* compared with those of certain other species in the subgenus *Nyssorhynchus*. Mosquito News, *35*:354–365.

Krinsky, W.L. 1976. Animal disease agents transmitted by horse flies and deer flies (Diptera: Tabanidae). J. Med. Entomol., *13*:225–275.

Kuhlow, F., and Zielke, E. 1978. Dynamics and intensity of *Wuchereria bancrofti* transmission in the savannah and forest regions of Liberia. Tropenmed. Parasitol., *29*:371–381.

Lacey, L.A., and Mulla, M.S. 1979. Factors affecting feeding rates of black fly larvae. Mosquito News, *39*:315–319.

Laird, M., editor. 1981. *Blackflies. The Future for Biological Methods in Integrated Control*. London, Academic Press.

Lancien, J. 1981. Description du piège monoconique utilisé pour l'élimination des glossins en Republique Populaire du Congo. Cah. O.R.S.T.O.M., Ser. Entomol. Med. Parasitol., *19*:235–238.

Langley, P.A. 1976. Initiation and regulation of ingestion by hematophagous arthropods. J. Med. Entomol., *13*:121–130.

———. 1977. Physiology of tsetse flies *(Glossina* spp.) (Diptera: Glossinidae): A review. Bull. Entomol. Res., *67*:523–574.

——— (reporter). 1981. Workshop No. 7. Digestion in haematophagous arthropods. Parasitology, *82*:95–116.

———, Huyton, P.M., Carlson, D.A., and Schwarz, M. 1981. Effects of *Glossina morsitans morsitans* Westwood (Diptera: Glossinidae) sex pheromone on behaviour of males in field and laboratory. Bull. Entomol. Res., *71*:57–63.

———, and Pimley, R.W. 1975. Quantitative aspects of reproduction and larval nutrition in *Glossina morsitans* Westw. (Diptera: Glossinidae) fed *in vitro*. Bull. Entomol. Res., *65*:129–142.

Laughlin, L.W., Meegan, J.M., Strausbaugh, L.J., Morens, D.M., and Watten, R.H. 1979. The Rift Valley fever epizootic in Egypt 1977–1978. 3. Observations on the spectrum of human illness occurring during the intercurrent epidemic. Trans. R. Soc. Trop. Med. Hyg., *73*:630–633.

Laveissiere, C., and Couret, D. 1980. Lutte contre les glossins riveraines a l'aide de pièges biconiques unprégnés d'insecticide en zone de savane humide. 2. Resultats quantitaifs obtenus lors des premiers essais. Cah. O.R.S.T.O.M., Ser. Entomol. Med. Parasitol., *18*:209–221.

LeDuc, J.W. 1979. The ecology of California group viruses. J. Med. Entomol., *16*:1–17.

Lee, R.M.K.W., and Davies, D.M. 1979. Feeding in the stable fly, *Stomoxys calcitrans* (Diptera: Muscidae). I. Destination of blood, sucrose solution and water in the alimentary canal, the effects of age on feeding, and blood digestion. J. Med. Entomol., *15*:541–554.

Lewis, D.J. 1971. Phlebotomid sandflies. Bull. W.H.O., *44*:535–551

———. 1973. Phlebotomidae and Psychodidae (sand-flies and moth-flies), pp. 155–179. In *Insects and Other Arthropods of Medical Importance*. Edited by K.G.V. Smith. London, British Museum (Natural History).

———. 1974. The biology of Phlebotomidae in relation to leishmaniasis. Annu. Rev. Entomol., *19*:363–384.

———. 1974b. The phlebotomid sandflies of Yemen Arab Republic. Tropenmed. Parasitol., *25*:187–197.

———. 1975a. Functional morphology of the mouth parts in New World phlebotomine sandflies (Diptera: Psychodidae). Trans. R. Entomol. Soc. Lond., *126*:497–532.

———. 1975b. The *Lutzomyia flaviscutellata* complex. J. Med. Entomol., *12*:363–368.

———. 1978. The phlebotomine sandflies (Diptera: Psychodidae) of the Oriental region. Bull. Br. Mus. Nat. Hist. (Entomol.), *37*:217–343.

———, and Ibañez de Aldecoa, R. 1962. Simuliidae and their relation to human onchocercosis in northern Venezuela. Bull. W.H.O., *27*:449–464.

———, Young, D.G., Fairchild, G.B., and Minter, D.M. 1977. Proposals for a stable classification of the phlebotomine sandflies (Diptera: Psychodidae). Syst. Entomol., *2*:319–332.

Lord, R.D. 1974. History and geographic distribution of Venezuelan equine encephalitis. Bull. Pan. Am. Health Org., *8*:100–110.

Lowrie, R.C., Jr., and Raccurt, C. 1981. *Mansonella ozzardi* in Haiti. II. Arthropod vector studies. Am. J. Trop. Med. Hyg., *30*:598–603.

Macdonald, G., Cuellar, C.B., and Foll, C.V. 1968. The dynamics of malaria. Bull. W.H.O., *38*:743–755.

Magnarelli, L.A. 1978. Nectar-feeding by female mosquitoes and its relation to follicular development and parity. J. Med. Entomol., *14*:527–530.

Mahmood, F., and Reisen, W.K. 1981. Duration of the gonotrophic cycles of *Anopheles culicifacies* Giles and *Anopheles stephensi* Liston, with observations on reproductive activity and survivorship during winter in Punjab Province, Pakistan. Mosquito News, *41*:41–50.

Marinkelle, C.J. 1980. The control of leishmaniases. Bull. W.H.O., *58*:807–818.

Matthews, G.A. 1979. *Pesticide Application Methods*. London, Longman.

Matthews, R.E.F. 1982. Classification and nomenclature of viruses. Intervirology, *17*:1–199.

Mattingly, P.F. 1973. Diptera-Nematocera. Culicidae (mosquitoes), pp. 37–107. In *Insects and Other Arthropods of*

Medical Importance. Edited by K.G.V. Smith. London, British Museum (Natural History).

———. 1977. Names for the *Anopheles gambiae* complex. Mosquito Systematics, *9*:323–328.

———, Crosskey, R.W., and Smith, K.G.V. 1973. Summary of arthropod vectors, pp. 497–532. In *Insects and Other Arthropods of Medical Importance*. Edited by K.G.V. Smith. London, British Museum (Natural History).

Mayrink, W., Williams, P., Coelho, M.V., Dias, M., Martens, A.V., Magalhaes, P.A., DaCosta, C.A., Falcao, A.R., Melo, M.N., and Falcao, A.L. 1979. Epidemiology of dermal leishmaniasis in the Rio Doce Valley, State of Minas Gerais, Brazil. Ann. Trop. Med. Parasitol., *73*:123–137.

McCrae, A.W.R. 1972. Age-composition of man-biting *Aedes (Stegomyia) simpsoni* (Theobald) (Diptera: Culicidae) in Bwamba County, Uganda. J. Med. Entomol., *9*:545–550.

McDonald, G. 1980. Population studies of *Culex annulirostris* and other mosquitoes (Diptera: Culicidae) at Mildura in the Murray Valley of Southern Australia. J. Aust. Entomol. Soc., *19*:37–40.

McGreevy, P.B., Bryan, J.H., Oothuman, P., and Kolstrup, N. 1978. The lethal effects of the cibarial and pharyngeal armatures of mosquitoes on microfilariae. Trans. R. Soc. Trop. Med. Hyg., *72*:361–368.

McGregor, I.A. 1976. Epidemiology and control of malaria, pp. 208–222. In *Epidemiology and Community Health in Warm Climate Countries*. Edited by R. Cruickshank, K.L. Standard, and H.B.L. Russell. London, Churchill Livingstone.

McIntosh, B.M. 1975. Mosquitoes as vectors of viruses in southern Africa. Entomol. Mem. Dept. Agric. Tech. Serv. Republic S. Afr., *43*:1–19.

———, and Jupp, P.G. 1979. Infections in sentinel pigeons by Sindbis and West Nile viruses in South Africa, with observations on *Culex (Culex) univittatus* (Diptera: Culicidae) attracted to these birds. J. Med. Entomol., *16*:234–239.

McIver, S.B. 1982. Sensilla of mosquitoes (Diptera: Culicidae). J. Med. Entomol., *19*:489–535.

———, Wilkes, T.J., and Gillies, M.T. 1980. Attraction to mammals of male *Mansonia (Mansonoides)* (Diptera: Culicidae). Bull. Entomol. Res., *70*:11–16.

McKinney, R.M., Spillane, J.T., and Holden, P. 1972. Mosquito blood meals: Identification by a fluorescent antibody method. Am. J. Trop. Med. Hyg., *21*:999–1003.

McLintock, J. 1978. Mosquito-virus relationships of American encephalitides. Annu. Rev. Entomol., *23*:17–37.

Meegan, J.M., Hoogstraal, H., and Moussa, M.L. 1979. An epizootic of Rift Valley fever in Egypt in 1977. Vet. Rec., *105*:124–125.

———, Khalil, G.M., Hoogstraal, H., and Adham, F.K. 1980. Experimental transmission and field isolation studies implicating *Culex pipiens* as a vector of Rift Valley fever virus in Egypt. Am. J. Trop. Med. Hyg., *29*:1405–1410.

Meola, R.W., and Petralia, R.S. 1980. Juvenile hormone induction of biting behavior in *Culex* mosquitoes. Science, *209*:1548–1550.

Merrill, J.R., Otis, J., Logan, W.D., Jr., and Davis, M.B. 1980. The dog heartworm *(Dirofilaria immitis)* in man. J.A.M.A., *243*:1066–1068.

Miles, S.J. 1978. Enzyme variation in the *Anopheles gambiae* Giles group of species (Diptera: Culicidae). Bull. Entomol. Res., *68*:85–96.

———. 1979. A biochemical key to adult members of the *Anopheles gambiae* group of species (Diptera: Culicidae). J. Med. Entomol., *15*:297–299.

Mitchell, C.J. 1977. Arthropod-borne encephalitis viruses and water resource developments. Cah. O.R.S.T.O.M., Ser. Entomol. Med. Parasitol., *15*:241–250.

———, Bowen, G.S., Monath, T.P., Cropp, C.B., and Kerschner, J.A. 1979. St. Louis encephalitis virus transmission following multiple feeding of *Culex pipiens pipiens* during a single gonotrophic cycle. J. Med. Entomol., *16*:254–258.

———, Francy, D.B., and Monath, T.P. 1980. Arthropod vectors, pp. 313–379. In *St. Louis Encephalitis*. Edited by T.P. Monath. Washington, D.C., American Public Health Association.

———, Monath, T.P., and Cropp, C.B. 1981. Experimental transmission of Rocio virus by mosquitoes. Am. J. Trop. Med. Hyg., *30*:465–472.

Molineaux, L., and Gramiccia, G. 1980. *The Garki Project. Research on the Epidemiology and Control of Malaria in the Sudan Savanna of West Africa*. Geneva, World Health Organization.

Monath, T.P. 1979. Arthropod-borne encephalitides in the Americas. Bull. W.H.O., *57*:513–533.

———, Craven, R.B., Adjukiewicz, A., Germain, M., Francy, D.B., Ferrara, L., Samba, E.M., N'Jie, H., Cham, K., Fitzgerald, S.A., Crippen, P.H., Simpson, D.I.H., Bowen, E.T.W., Fabiyi, A., and Salaun, J.-J. 1980. Yellow fever in the Gambia, 1978–1979: Epidemiologic aspects with observations on the occurrence of Orungo virus infections. Am. J. Trop. Med. Hyg., *29*:912–928.

Montalván C., J.A. 1940. Un foco de bartonelosis en el Ecuador. Informe preliminar. Bol. Of. Sanit. Panam., *19*:154.

Moore, C.G., Cline, B.L., Ruiz-Tiben, E., Lee, D., Romney-Joseph, H., and Rivera-Correa, E. 1978. *Aedes aegypti* in Puerto Rico: Environmental determinants of larval abundance and relation to dengue virus transmission. Am. J. Trop. Med. Hyg., *27*:1225–1231.

Morris, C.D., Zimmerman, R.H., and Edman, J.D. 1980. Epizootiology of eastern equine encephalomyelitis virus in upstate New York, USA. II. Population dynamics and vector potential of adult *Culiseta melanura* (Diptera: Culicidae) in relation to distance from breeding site. J. Med. Entomol., *17*:453–465.

Murphy, F.A., Whitfield, S.G., Sudia, W.D., and Chamberlain, R.W. 1975. Interactions of vector with vertebrate pathogenic viruses, pp. 25–48. In *Invertebrate Immunity*. Edited by K. Maramorosch and R.E. Shope. New York, Academic Press.

Nathan, M.B. 1981. Transmission of the human filarial parasite *Mansonella ozzardi* by *Culicoides phlebotomus* (Williston) (Diptera: Ceratopogonidae) in coastal north Trinidad. Bull. Entomol. Res., *71*:97–105.

Nelson, R.L., Tempelis, C.H., Reeves, W.C., and Milby, M.M. 1976. Relation of mosquito density to bird:mammal feeding ratios of *Culex tarsalis* in stable traps. Am. J. Trop. Med. Hyg., *25*:644–654.

Nicholas, W.L. 1953. The bionomics of *Culicoides austeni*, vector of *Acanthocheilonema perstans* in the rain-forest of the British Cameroons, together with notes on *C. grahamii* and other species which may be vectors in the same area. Ann. Trop. Med. Parasitol., *47*:187–206.

———, Kershaw, W.E., and Duke, B.O.L. 1955. Studies on the epidemiology of filariasis in West Africa with special reference to the British Cameroons and the Niger Delta. VII. Further records on the distribution of *Culicoides* spp., with a note on the taxonomic status of *C. austeni*. Ann. Trop. Med. Parasitol., *49*:455–460.

———, Kershaw, W.E., Keay, R.W.J., and Zahra, A. 1953. Studies on the epidemiology of filariasis in West Africa with special reference to the British Cameroons and the Niger Delta. III. The distribution of *Culicoides* spp. biting man in the rain-forest, the forest fringe and the mountain grasslands of the British Cameroons. Ann. Trop. Med. Parasitol., *47*:95–111.

Olson, J.G., Reeves, W.C., Emmons, R.W., and Milby, M.M.

1979. Correlation of *Culex tarsalis* population indices with the incidence of St. Louis encephalitis and western equine encephalomyelitis in California. Am. J. Trop. Med. Hyg., *28*:335–343.

Omer, S.M., and Cloudsley-Thompson, J.L. 1970. Survival of female *Anopheles gambiae* Giles through a 9-month dry season in Sudan. Bull. W.H.O., *42*:319–330.

Orihel, T.C., and Beaver, P.C. 1965. Morphology and relationship of *Dirofilaria tenuis* and *Dirofilaria conjunctivae*. Am. J. Trop. Med. Hyg., *14*:1030–1043.

Pan American Health Organization. 1972. *Venezuelan encephalitis*. Scientific Publication No. 243. Washington, D.C., Pan American Sanitary Bureau.

Panday, R.S. 1977. *Anopheles nuneztovari* and malaria transmission in Surinam. Mosquito News, *37*:728–737.

Parsonson, L.M., Della-Porta, A.J., and Snowdon, W.A. 1981. Developmental disorders of the fetus in some arthropod-borne virus infections. Am. J. Trop. Med. Hyg., *30*:660–673.

Partono, F., Purnomo, Pribadi, W., and Soewarta, A. 1978. Epidemiological and clinical features of *Brugia timori* in a newly established village, Karakuals, West Flores, Indonesia. Am. J. Trop. Med. Hyg., *27*:910–915.

Patino-Camargo, L. 1939. Bartonellosis en Colombia. Rev. Fac. Med., Bogota, *7*:467–501.

Pinheiro, F.P., Bensabath, G., Costa, D., Jr., Maroja, O.M., Lins, Z.C., and Andrade, A.H.P. 1974. Haemorrhagic syndrome of Altamira. Lancet, *1*:639–642.

———, Travassos da Rosa, A.P.A., and Moraes, M.A.P. 1981a. An epidemic of yellow fever in central Brazil, 1972–1973. II. Ecological studies. Am. J. Trop. Med. Hyg., *30*:204–211.

———, Travassos da Rosa, A.P.A., and Gomes, M.L.C. 1982. Transmission of Oropouche virus from man to hamster by the midge *Culicoides paraensis*. Science, *215*:1251–1252.

———, Travassos da Rosa, A.P.A., Travassos da Rosa, J.F.S., Ishak, R., Freitas, R.B., Gomes, M.L.C., LeDuc, J.W., and Oliva, O.F.P. 1981b. Oropouche virus. I. A review of clinical, epidemiological, and ecological findings. Am. J. Trop. Med. Hyg., *30*:149–160.

Pinkerton, H., and Weinman, D. 1937. Carrion's disease. I. Behavior of the etiological agent within cells growing or surviving in vitro. Proc. Soc. Exp. Biol. Med., *37*:587–590.

Poehling, H.M., Wolfrum, D.I., and Neuhoff, V. 1976. Mikroelektrophorese von Proteinen aus Speicheldrüsen und Hämolymphe verschiedener Simuliidenarten in Polyacrylamidgradienten-Gelen. Entomol. Exp. Appl., *19*:271–286.

Pollock, J.N. 1971. Origin of the tsetse flies: A new theory. J. Entomol. (B), *40*:101–109.

———. 1973. A comparison of the male genitalia and abdominal segmentation in *Gasterophilus* and *Glossina* (Diptera), with notes on the gasterophiloid origin of the tsetse flies. Trans. R. Entomol. Soc. Lond., *125*:107–124.

Port, G.R., and Boreham, P.F.L. 1980. The relationship of host size to feeding by mosquitoes of the *Anopheles gambiae* Giles complex (Diptera: Culicidae). Bull. Entomol. Res., *70*:133–144.

———, and Wilkes, T.J. 1979. *Aedes (Diceromyia) furcifer/taylori* and a yellow fever outbreak in the Gambia. Trans. R. Soc. Trop. Med. Hyg., *73*:341–344.

Potts, W.H. 1973. Diptera-Cyclorrhapha. Glossinidae (tsetse-flies), pp. 209–249. In *Insects and Other Arthropods of Medical Importance*. Edited by K.G.V. Smith. London, British Museum (Natural History).

Powell, J.R., Tabachnick, W.J., and Arnold, J. 1980. Genetics and the origin of a vector population: *Aedes aegypti*, a case study. Science, *208*:1385–1387.

Quillevere, D., and Sechan, Y. 1978. Morphological identification of females of the *Simulium damnosum* complex in West Africa: Differentiation of *S. squamosum* and *S. yahense*. Trans. R. Soc. Trop. Med. Hyg., *72*:99–100.

Ramalingam, S. 1975. Vectors of *Wuchereria bancrofti* and *Brugia malayi* in the South and Southeast Asian regions: Their distribution, biology and control. Ceylon J. Med. Sci., *24*:1–27.

Raminani, L.N., and Cupp, E.W. 1975. Early embryology of *Aedes aegypti* (L.) (Diptera: Culicidae). Int. J. Insect Morphol. Embryol., *4*:517–528.

———, and Cupp, E.W. 1978. Embryology of *Aedes aegypti* (L.) (Diptera: Culicidae): Organogenesis. Int. J. Insect Morphol. Embryol., *7*:273–296.

Rathburn, C.B., Jr. 1979. Insecticides for the control of mosquitoes and other Diptera. Mosquito News, *39*:58–63.

Raybould, J.N., and White, G.B. 1979. The distribution, bionomics and control of onchocerciasis vectors (Diptera: Simuliidae) in eastern Africa and the Yemen. Tropenmed. Parasitol., *30*:505–547.

Rebagliati, R. 1942. Verruga peruana. Proceedings of the American Scientific Congress (1940), *6*:215.

Reid, J.A. 1968. *Anopheline Mosquitoes of Malaya and Borneo*. Stud. Inst. Med. Res. Malaysia, *31*.

Reisen, W.K., Aslam, Y., Siddiqui, T.F., and Khan, A.Q. 1978. A mark-release-capture experiment with *Culex tritaeniorhynchus* Giles. Trans. R. Soc. Trop. Med. Hyg., *72*:167–177.

———, and Aslamkhan, M.A. 1979. A release-recapture experiment with the malaria vector, *Anopheles stephensi* Liston, with observations on dispersal, survivorship, population size, gonotrophic rhythm and mating behaviour. Ann. Trop. Med. Parasitol., *73*:251–269.

———, and Boreham, P.F.L. 1979. Host selection patterns of some Pakistan mosquitoes. Am. J. Trop. Med. Hyg., *28*:408–421.

———, and Mahmood, F. 1980. Horizontal life table characteristics of the malaria vectors *Anopheles culicifacies* and *Anopheles stephensi* (Diptera: Culicidae). J. Med. Entomol., *17*:211–217.

———, Mahmood, F., and Parveen, T. 1980. *Anopheles culicifacies* Giles: A release-recapture experiment with cohorts of known age with implications for malaria epidemiology and genetical control in Pakistan. Trans. R. Soc. Trop. Med. Hyg., *74*:307–317.

Roberts, R.H. 1976. The comparative efficiency of six trap types for the collection of Tabanidae. Mosquito News, *36*:530–535.

Romoser, W.S. 1974. Peritrophic membranes in the midgut of pupal and pre-blood meal adult mosquitoes (Diptera: Culicidae). J. Med. Entomol., *11*:397–402.

Ronquillo, M.C., and Horsfall, W.R. 1969. Genesis of the reproductive system of mosquitoes. I. Female of *Aedes stimulans* (Walker). J. Morphol., *129*:249–280.

Rosen, L. 1981. The use of *Toxorhynchites* mosquitoes to detect and propagate dengue and other arboviruses. Am. J. Trop. Med. Hyg., *30*:177–183.

———, Shroyer, D.A., and Lien, J.C. 1980. Transovarial transmission of Japanese encephalitis virus by *Culex tritaeniorhynchus* mosquitoes. Am. J. Trop. Med. Hyg., *29*:711–712.

Rothfels, K.H. 1979. Cytotaxonomy of black flies (Simuliidae). Annu. Rev. Entomol., *24*:507–539.

Russell, P.F., West, L.S., Manwell, R.D., and Macdonald, G. 1963. *Practical Malariology*, 2nd ed. London, Oxford University Press.

Samarawickrema, W.A., and Laurence, B.R. 1978. Loss of filarial larvae in a natural mosquito population. Ann. Trop. Med. Parasitol., *72*:561–565.

Sasa, M. 1976. *Human Filariasis*. Baltimore, University Park Press.

Saul, S.H., Grimstad, P.R., and Craig, G.B., Jr. 1977. Identification of *Culex* species by electrophoresis. Am. J. Trop. Med. Hyg., *26*:1009–1012.

Scherer, W.F., Cupp, E.W., Lok, J.B., Brenner, R.J., and Ordonez, J.V. 1981. Intestinal threshold of an enzootic strain of Venezuelan encephalitis virus in *Culex (Melanoconion) taeniopus* mosquitoes and its implications to vector competency and vertebrate amplifying hosts. Am. J. Trop. Med. Hyg., *30*:862–869.

Schmidt, J.R., Schmidt, M.L., and Said, M.L. 1971. Phlebotomus fever in Egypt. Isolation of phlebotomus fever viruses from *Phlebotomus papatasi*. Am. J. Trop. Med. Hyg., *20*:483–490.

Schowalter, T.D., and Klowden, M.J. 1979. Blood meal size of the stable fly, *Stomoxys calcitrans*, measured by the HiCN method. Mosquito News, *39*:110–112.

Service, M.W. 1976. *Mosquito Ecology. Field Sampling Methods*. London, Applied Science Publishers Ltd.

———. 1977a. A critical review of procedures for sampling populations of adult mosquitoes. Bull. Entomol. Res., *67*:343–382.

———. 1977b. Methods for sampling adult Simuliidae, with special reference to the *Simulium damnosum* complex. Tropical Pest Bull., No. 5.

Shelley, A.J., Luna Dias, A.P.A., and Moraes, M.A.P. 1980. *Simulium* species of the *amazonicum* group as vectors of *Mansonella ozzardi* in the Brazilian Amazon. Trans. R. Soc. Trop. Med. Hyg., *74*:784–788.

Siam, A.L., and Meegan, J.M. 1980. Ocular disease resulting from infection with Rift Valley fever virus. Trans. R. Soc. Trop. Med. Hyg., *74*:539–541.

Simpson, D.I.H. 1978. Viral haemorrhagic fevers of man. Bull. W.H.O., *56*:819–832.

———, Smith, C.E.G., Marshall, T.F. de C., Platt, G.S., Way, H.J., Bowen, E.T.W., Bright, W.F., Day, J., McMahon, D.A., Hill, M.N., Bendell, P.J.E., and Heathcote, O.H.U. 1976. Arbovirus infections in Sarawak: The role of the domestic pig. Trans. R. Soc. Trop. Med. Hyg., *70*:66–72.

Snow, W.F., and Wilkes, T.J. 1977. Age composition and vertical distribution of mosquito populations in the Gambia, West Africa. J. Med. Entomol., *13*:507–513.

Sommerman, K.M. 1977. Biting fly-arbovirus probe in interior Alaska (Culicidae) (Simuliidae)—(SSH: California complex) (Northway: Bunyamwera group). Mosquito News, *37*:90–103.

Spencer, M. 1979. Age grouping of female *Anopheles farauti* populations (Diptera: Culicidae) in Papua New Guinea. J. Med. Entomol., *15*:555–569.

Spielberger, U., Na'Isa, B.K., Koch, K., Manno, A., Skidmore, P.R., and Coutts, H.H. 1979. Field trials with the synthetic pyrethroids Permethrin, Cypermethrin and Decamethrin against *Glossina* (Diptera: Glossinidae) in Nigeria. Bull. Entomol. Res., *69*:667–689.

Sudia, W.D., Fernandez, Z., Newhouse, V.F., Sanz, B.R., and Calisher, C.H. 1975. Arbovirus vector ecology studies in Mexico during the 1972 Venezuelan equine encephalitis outbreak. Am. J. Epidemiol., *101*:51–58.

———, and Newhouse, V.F. 1975. Epidemic Venezuelan equine encephalitis in North America: A summary of virus-vector-host relationships. Am. J. Epidemiol., *101*:1–13.

———, Newhouse, V.F., Calisher, C.H., and Chamberlain, R.W. 1971. California group arboviruses: Isolations from mosquitoes in North America. Mosquito News, *31*:576–600.

Takahashi, M. 1976. The effects of environmental and physiological conditions of *Culex tritaeniorhynchus* on the pattern of transmission of Japanese encephalitis virus. J. Med. Entomol., *13*:275–284.

———, and Suzuki, K. 1979. Japanese encephalitis virus in mosquito salivary glands. Am. J. Trop. Med. Hyg., *28*:122–135.

Taylor, B. 1975. Observations on malaria vectors of the *Anopheles punctulatus* complex in the British Solomon Islands Protectorate. J. Med. Entomol., *11*:677–687.

Taylor, P. 1979. The construction of a life-table for *Glossina morsitans morsitans* Westwood (Diptera: Glossinidae) from seasonal age-measurements of a wild population. Bull. Entomol. Res., *69*:553–560.

Tempelis, C.H. 1975. Host-feeding patterns of mosquitoes, with a review of advances in analysis of blood meals by serology. J. Med. Entomol., *11*:635–653.

———, Reeves, W.C., and Nelson, R.L. 1976. Species identification of blood meals from *Culex tarsalis* that had fed on passeriform birds. Am. J. Trop. Med. Hyg., *25*:744–746.

Tesh, R.B. 1980. Vertical transmission of arthropod-borne viruses of vertebrates, pp. 122–137. In *Vectors of Disease Agents*. Edited by J.J. McKelvey, B.F. Eldridge, and K. Maramorosch. New York, Praeger Scientific Publishers.

———, and Chaniotis, B.N. 1975. Transovarial transmission of viruses by phlebotomine sandflies, pp. 125–134. In *Pathobiology of Invertebrate Vectors of Disease*. Edited by L.A. Bulla, Jr., and T.C. Cheng. Ann. N.Y. Acad. Sci., *266*.

———, Chaniotis, B.N., Carrera, B.R., and Johnson, K.M. 1972. Further studies on the natural host preferences of Panamanian phlebotomine sandflies. Am. J. Epidemiol., *95*:88–93.

———, Chaniotis, B.N., and Johnson, K.M. 1971. Vesicular stomatitis virus, Indiana serotype: Multiplication in and transmission by experimentally infected phlebotomine sandflies *(Lutzomyia trapidoi)*. Am. J. Epidemiol., *93*:491–495.

———, Chaniotis, B.N., Peralta, P.H., and Johnson, K.M. 1974. Ecology of viruses isolated from Panamanian phlebotomine sandflies. Am. J. Trop. Med. Hyg., *23*:258–269.

———, McLean, R.G., Shroyer, D.A., Calisher, C.H., and Rosen, L. 1981. Ross River virus (Togaviridae: *Alphavirus)* infection (epidemic polyarthritis) in American Samoa. Trans. R. Soc. Trop. Med. Hyg., *75*:426–431.

———, Saidi, S., Javadian, E., Loh, P., and Nadim, A. 1977. Isfahan virus, a new vesiculovirus infecting humans, gerbils, and sandflies in Iran. Am. J. Trop. Med. Hyg., *26*:299–306.

———, Saidi, S., Javadian, E., and Nadim, A. 1977. Studies on the epidemiology of sandfly fever in Iran. I. Virus isolates obtained from *Phlebotomus*. Am. J. Trop. Med. Hyg., *26*:282–287.

Thomas, A.W., and Gooding, R.H. 1976. Digestive processes of hematophagous insects. VIII. Estimation of meal size and demonstration of trypsin in horse flies and deer flies (Diptera: Tabanidae). J. Med. Entomol., *13*:131–136.

Thompson, B.H. 1976. The intervals between the blood-meals of man-biting *Simulium damnosum* (Diptera: Simuliidae). Ann. Trop. Med. Parasitol., *70*:329–341.

Thompson, W.H., and Beaty, B.J. 1978. Venereal transmission of LaCrosse virus from male to female *Aedes triseriatus*. Am. J. Trop. Med. Hyg., *27*:187–196.

Tibayrenc, M., Carriou, M.L., Corneau, B., and Pajot, F.-X. 1980. Etude allozymique chez *Lutzomyia umbratilis* (Diptera: Psychodidae), vecteur de la leishmaniose en Guyana francaise. Cahiers O.R.S.T.O.M., Ser. Entomol. Med. Parasitol., *18*:67–70.

Tidwell, M.A., Tidwell, M., and Munos de Hoyos, P. 1980a. Development of *Mansonella ozzardi* in a black fly species of the *Simulium sanguineum* group from Eastern Vaupes, Colombia. Am. J. Trop. Med. Hyg., *29*:1209–1214.

———, Tidwell, M., Munos de Hoyos, P., and Corredor, A.

1980b. *Simulium exiguum*, the vector of *Onchocerca volvulus* on the Rio Micay, Colombia. Am. J. Trop. Med. Hyg., *29*:377–381.

Tomori, O. 1980. Rift Valley fever virus infection in man in Nigeria. J. Med. Virol., *5*:343–350.

Townson, H., and Meredith, S.E.O. 1979. Identification of the Simuliidae in relation to onchocerciasis, pp. 145–174. In *Problems in the Identification of Parasites and Their Vectors*. Edited by A.E.R. Taylor and R. Muller. Symp. Br. Soc. Parasitol. Vol. 17. Oxford, Blackwell Scientific Publications.

Trpis, M. 1973. Interaction between the predator *Toxorhynchites brevipalpis* and its prey *Aedes aegypti*. Bull. W.H.O., *49*:359–365.

———. 1978. Genetics of hematophagy and autogeny in the *Aedes scutellaris* complex (Diptera: Culicidae). J. Med. Entomol., *15*:73–80.

———, Duhrkopf, R.E., and Parker, K.L. 1981. Non-Mendelian inheritance of mosquito susceptibility to infection with *Brugia malayi* and *Brugia pahangi*. Science, *211*:1435–1437.

———, and Hausermann, W. 1978. Genetics of house-entering behavior in East African populations of *Aedes aegypti* (L.) (Diptera: Culicidae) and its relevance to speciation. Bull. Entomol. Res., *68*:521–532.

Vajime, C.G., and Dunbar, R.W. 1975. Chromosomal identification of eight species of the subgenus *Edwardsellum* near and including *Simulium (Edwardsellum) damnosum* Theobald (Diptera: Simuliidae). Tropenmed. Parasitol., *26*:111–138.

Vale, G.A. 1979. Field responses of tsetse flies (Diptera: Glossinidae) to odours of men, lactic acid and carbon dioxide. Bull. Entomol. Res., *69*:459–467.

———. 1980. Flight as a factor in the host-finding behaviour of tsetse flies (Diptera: Glossinidae). Bull. Entomol. Res., *70*:299–307.

Vesenjak-Hirjan, J., Porterfield, J.S., and Arslanagic, E., editors. 1980. *Arboviruses in the Mediterranean Countries*. 6th FEMS Symposium. New York, Gustav Fischer Verlag.

Walker, A.R., and Boreham, P.F.L. 1976. Saline as a collecting medium for *Culicoides* (Diptera: Ceratopogonidae) in blood feeding and other studies. Mosquito News, *36*:18–20.

Wallace, H.G., Lim, T.W., Rudnick, A., Knudsen, A.B., Cheong, W.H., and Chew, V. 1980. Dengue hemorrhagic fever in Malaysia: The 1973 epidemic. Southeast Asian J. Trop. Med. Public Health, *11*:1–13.

Walsh, J.F., Davis, J.B., and LeBerre, R. 1979. Entomological aspects of the first five years of the Onchocerciasis Control Programme in the Volta River Basin. Tropenmed. Parasitol., *30*:328–344.

Ward, R.D., Lainson, R., and Shaw, J.J. 1973. Further evidence of the role of *Lutzomyia flaviscutellata* (Mangabeira) as the vector of *Leishmania mexicana amazonensis* in Brazil. Trans. R. Soc. Trop. Med. Hyg., *67*:608–609.

———, Lainson, R., and Shaw, J.J. 1978. Some methods for membrane feeding of laboratory reared neotropical sandflies (Diptera: Psychodidae). Ann. Trop. Med. Parasitol., *72*:269–276.

———, Pasteur, N., and Rioux, J.A. 1981. Electrophoretic studies on genetic polymorphism and differentiation of phlebotomine sandflies (Diptera: Psychodidae) from France and Tunisia. Ann. Trop. Med. Parasitol., *75*:235–245.

Washino, R.K. 1977. The physiological ecology of gonotrophic dissociation and related phenomena in mosquitoes. J. Med. Entomol., *13*:381–388.

Watts, D.M., and Eldridge, B.F. 1975. Transovarial transmission of arboviruses by mosquitoes: A review. Med. Biol., *53*:271–278.

———, Pantuwatana, S., DeFoliart, G.R., Yuill, T.M., and Thompson, W.H. 1973. Transovarial transmission of LaCrosse virus (California encephalitis group) in the mosquito, *Aedes triseriatus*. Science, *182*:1140–1141.

Wenk, P. 1980. How bloodsucking insects perforate the skin of their hosts, pp. 329–335. In *Fleas—Proceedings of the International Conference on Fleas (1977)*. Edited by R. Traub and H. Starcke. Rotterdam, A.A. Balkema.

Wernsdorfer, W.H. 1980. The importance of malaria in the world, pp. 1–93. In *Malaria. Volume 1. Epidemiology, Chemotherapy, Morphology, and Metabolism*. Edited by J. Kreier. New York, Academic Press.

Wetzel, H. 1980. The use of freeze-dried blood in the membrane feeding of tsetse flies *(Glossina p. palpalis*, Diptera: Glossinidae). Tropenmed. Parasitol., *31*:259–274.

White, G.B. 1973. Comparative studies on sibling species of the *Anopheles gambiae* Giles complex (Dipt., Culicidae). III. The distribution, ecology, behaviour and vectorial importance of species D in Bwamba County, Uganda, with an analysis of biological, ecological, morphological and cytogenetical relationships of Ugandan species D. Bull. Entomol. Res., *63*:65–97.

———. 1974. *Anopheles gambiae* complex and disease transmission in Africa. Trans. R. Soc. Trop. Med. Hyg., *68*:278–298.

———. 1977. Man-biting species of *Chrysops* Meigen, *Culicoides* Latreille and *Simulium* Latreille in Ethiopia, with discussion of their vector potentialities. Trans. R. Soc. Trop. Med. Hyg., *71*:161–175.

———. 1979. The identification of mosquitoes as vectors of malaria and filariasis, pp. 103–143. In *Problems in the Identification of Parasites and Their Vectors*. Edited by A.E.R. Taylor and R. Muller. Symp. Br. Soc. Parasitol., Vol. 17. Oxford, Blackwell Scientific Publications.

Whitfield, S.G., Murphy, F.A., and Sudia, W.D. 1973. St. Louis encephalitis virus: An ultrastructural study of infection in a mosquito vector. Virology, *56*:70–87.

Wijers, D.J.B., and Kiilu, G. 1977. Bancroftian filariasis in Kenya. III. Entomological investigations in Mambrui, a small coastal town, and in Jaribuni, a rural area more inland (Coastal Province). Ann. Trop. Med. Parasitol., *71*:347–359.

Wilkinson, R.N., Gould, D.J., Boonyakanist, P., and Segal, H.E. 1978. Observations on *Anopheles balabacensis* (Diptera: Culicidae) in Thailand. J. Med. Entomol., *14*:666–671.

Wilton, D.P. 1981. Light-trap response and the DV/D ratio in the *Culex pipiens* complex (Diptera: Culicidae). J. Med. Entomol., *18*:284–288.

Wood, D.M., Dang, P.T., and Ellis, R.A. 1979. *The Insects and Arachnids of Canada. Part 6. The Mosquitoes of Canada. Diptera: Culicidae*. Hull, Quebec, Canadian Government Publishing Centre.

Woodall, J.P. 1979. Transmission of group C arboviruses (Bunyaviridae), pp. 123–138. In *Arctic and Tropical Arboviruses*. Edited by E. Kurstak. New York, Academic Press.

World Health Organization. 1972. *Planning Manual for Malaria Eradication and Malaria Control Programmes*. Geneva, WHO.

———. 1973. *Manual on Larval Control Operations in Malaria Programmes*. Geneva, WHO.

———. 1974a. WHO Expert Committee on Filariasis. Report. Tech. Rep. Ser. No. 542, Geneva.

———. 1974b. Equipment for Vector Control. Geneva, WHO.

———. 1975. *Manual on Practical Entomology in Malaria. Part I. Vector Bionomics and Organization of Anti-Malaria Activities. Part II. Methods and Techniques*. Geneva, WHO.

———. 1976. Epidemiology of onchocerciasis. Tech. Rep. Ser. No. 597.

———. 1979. The African trypanosomiases. Tech. Rep. Ser. No. 635.

Yang, Y.J., and Davies, D.M. 1974. The saliva of adult female blackflies (Simuliidae: Diptera). Can. J. Zool., *52*:749–753.

Young, C.J., Turner, D.P., Killick-Kendrick, R., Rioux, J.A., and Leaney, A.J. 1980. Fructose in wild-caught *Phlebotomus ariasi* and the possible relevance of sugars taken by sandflies to the transmission of leishmaniasis. Trans. R. Soc. Trop. Med. Hyg., *74*:363–365.

Young, N.A., and Johnson, K.M. 1969. Antigenic variants of Venezuelan equine encephalitis virus: Their geographic distribution and epidemiologic significance. Am. J. Epidemiol., *89*:286–307.

Zumpt, F. 1973. *The Stomoxyine Biting Flies of the World. Diptera: Muscidae. Taxonomy. Biology, Economic Importance and Control Measures*. Stuttgart, Gustav Fischer.

Chapter 42

Filth Flies and Myiasis-Producing Flies

With few exceptions, filth flies have mouthparts that are not adapted for piercing and sucking blood. Instead, the adults have sponging-sucking mouthparts so that they depend on lapping up dissolved substances or finely particulate organic debris for their food. Many of these flies imbibe serous exudates from open wounds or at sites where blood-sucking flies have been feeding or they feed on lachrymal and sebaceous secretions. The filth-feeding habits of the adults, and particularly of the larvae, bring them in contact with many types of bacteria and other pathogenic microorganisms, while some species in their larval stage invade human tissue, producing myiasis. A few of the more common and more important types are considered, and others are referred to in diagrams or figures. Characteristics of mature larvae (i.e., "maggots") are illustrated in Figure 42–1, and patterns of the posterior spiracles, valuable in identifying the larva, are shown in Figure 42–2. A number of the adults are illustrated in Figure 42–3.

THE HOUSE FLY AND RELATED SPECIES (FAMILY MUSCIDAE)

Musca domestica Linnaeus, 1758 (The House Fly)

This species has an almost cosmopolitan distribution.

Anatomy and Behavior of the Adult. *M. domestica* is a medium-sized fly (males 5.8 to 6.5 mm, females 6.5 to 7.5 mm in length), of a general dusky-gray color (Fig. 42–3, *10*). The head is broad, the frons straw-colored and mixed with dark brown, the antennae brown, and the maxillary palps black. The squarish-ovoid thorax is patterned dorsally with four equally broad, dark, longitudinal stripes. The piriform abdomen is generally yellowish, with a dorsal longitudinal, dark-brown stripe, while the terminal segment is dark brown. The legs are dark brown. The wings are transparent and nearly straw-colored at the base. The fourth longitudinal vein of the wing bends abruptly upward, nearly meeting the third longitudinal vein. The squamae (i.e., axillary basal lobes) are large and opaquely yellowish. The wing spread is 13 to 15 mm. The antennae (Fig. 40–2,*B*), like those of *Stomoxys calcitrans* and *Glossina* spp., are more or less club-shaped (segments *i-iii*), with a lateral arista consisting of the 3 terminal segments. Segment *iii* is oval-elongated, bears 3 sensory pits, and externally near its base supports the arista. The long, spinose, terminal sixth segment has spinose branches on both its anterior and posterior margins. The mouthparts have been previously considered (see Chapter 40, pp. 625 to 627). The salivary glands consist of a pair of long, glistening white tubules, with short ducts joining to form the common salivary duct, which dilates into a small chamber before emptying into the base of the proboscis.

The muscular esophageal diverticulum ("crop" or food reservoir) is filled with recently imbibed liquid food. After a full feeding, the fly can wing its way to a more comfortable site and regurgitate the meal into the proventriculus and midgut at its leisure. Frequently some of this ingested food is allowed to flow back to the outer end of the proboscis, where it is known as a "vomit drop." This habit is important in the mechanical transfer of pathogenic microorganisms from a dirty feeding site to human food, since *M. domestica* and several other species of the genus are omnivorous.

The "foot" or terminal segment at the end of each leg of the house fly is provided with a pair of horny claws, a pair of ventral cushions or pulvilli, each with many glandular hairs (which enable the fly to walk on smooth surfaces and upside down

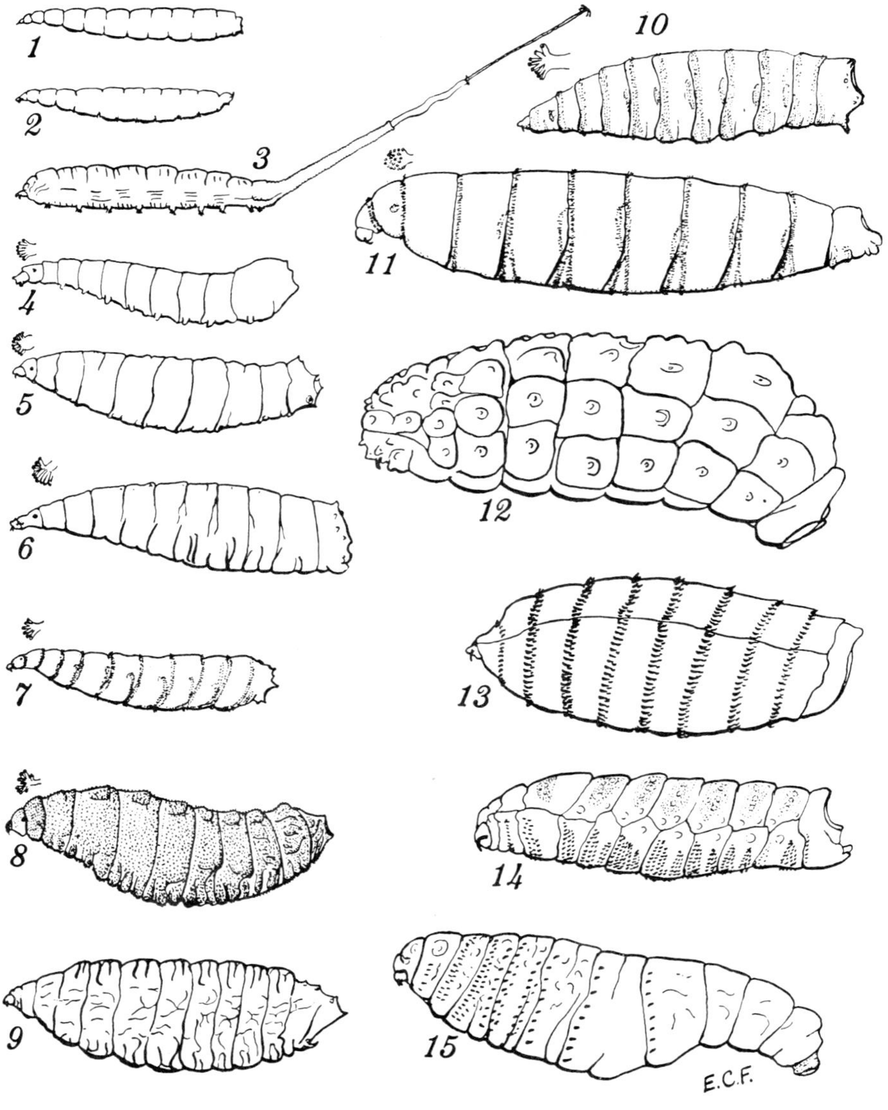

Fig. 42–1. Fly larvae, mature stage. Lateral view of some of the more common and important species. *1, Sepsis* sp. *2, Piophila* sp. *3, Eristalis tenax. 4, Musca domestica. 5, Phaenicia sericata. 6, Calliphora vicina. 7, Chrysomya bezziana. 8, Cordylobia anthropophaga. 9, Auchmeromyia senegalensis. 10, Cochliomyia hominivorax. 11, Wohlfahrtia magnifica. 12, Hypoderma bovis. 13. Gasterophilus intestinalis. 14, Oestrus ovis. 15, Dermatobia hominis.* (× 3⅓.) Anterior spiracle at upper left of *4–8, 10,* and *11.* (Adapted by E.C. Faust.)

on ceilings), and a single, median ventral, feathered spine, the empodium.

The posterior five abdominal segments of the female are telescoped into the end of the abdomen and are modified into a tubular ovipositor. At the time of oviposition this structure, which is as long as the abdomen when it is extended, is exserted. The elongate pearly-white *eggs* (1 mm in length) are then deposited in batches of 100 to 150. Several such batches are laid by each female during her lifetime. Favorable sites include dung of domestic animals, human feces, garbage and other decaying animal and vegetable matter, and at times clean food prepared for human consumption.

Development. The eggs hatch in approximately 24 hours at usual summer temperatures and the *larvae* (maggots) feed ravenously on organic matter immediately around them. There are three larval instars. Larval development requires 4 to 8 or more days. The last larval instar (Fig. 42–1,*4)* is about 12 mm long, milky-white, of greatest diameter posteriorly, and tapering to the head end. There are 12 apparent body segments, including the "head" *(i.e., gnatho-* or *pseudocephalon),* which is usu-

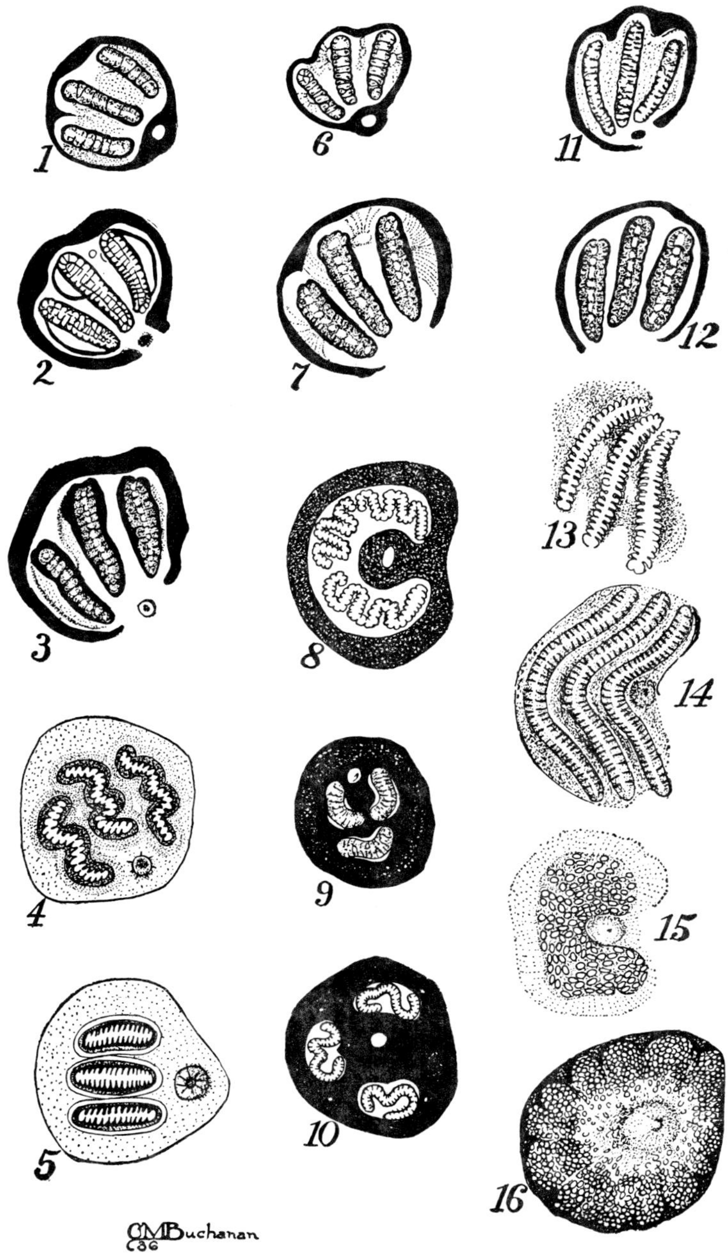

Fig. 42–2. Posterior spiracular pattern of representative fly larvae (left spiracular plate only). *1, Calliphora vomitoria, 2, Chrysomya bezziana. 3, Cochliomyia macellaria, 4, Cordylobia anthropophaga. 5, Auchmeromyia senegalensis. 6, Phaenicia sericata. 7, Phormia regina. 8, Musca domestica. 9, Muscina stabulans. 10, Stomoxys calcitrans. 11, Sarcophaga fuscicauda. 12, Wohlfahrtia vigil* sensu lato. *13, Dermatobia hominis. 14. Gasterophilus intestinalis. 15, Hypoderma bovis. 16, Oestrus ovis.* (Adapted by E.C. Faust.)

ally retracted within the anterior end of the body, and a truncate anal segment (James, 1947). There are no legs. The "head" bears a pair of dorsal lobes, each with a sensory papilla; a pair of clawed mouth hooks, used for boring into food, clinging to objects, and for progressing; and a food channel, through which the food (liquid organic matter) is introduced into the pharynx within. The first thoracic segment bears a pair of small lateral papillae that contain the anterior spiracles. At the posterior end of the abdomen is a pair of D-shaped posterior spiracles with long sinuous slits and a central button (see Fig. 42–2,*8*).

After reaching its full larval development, the

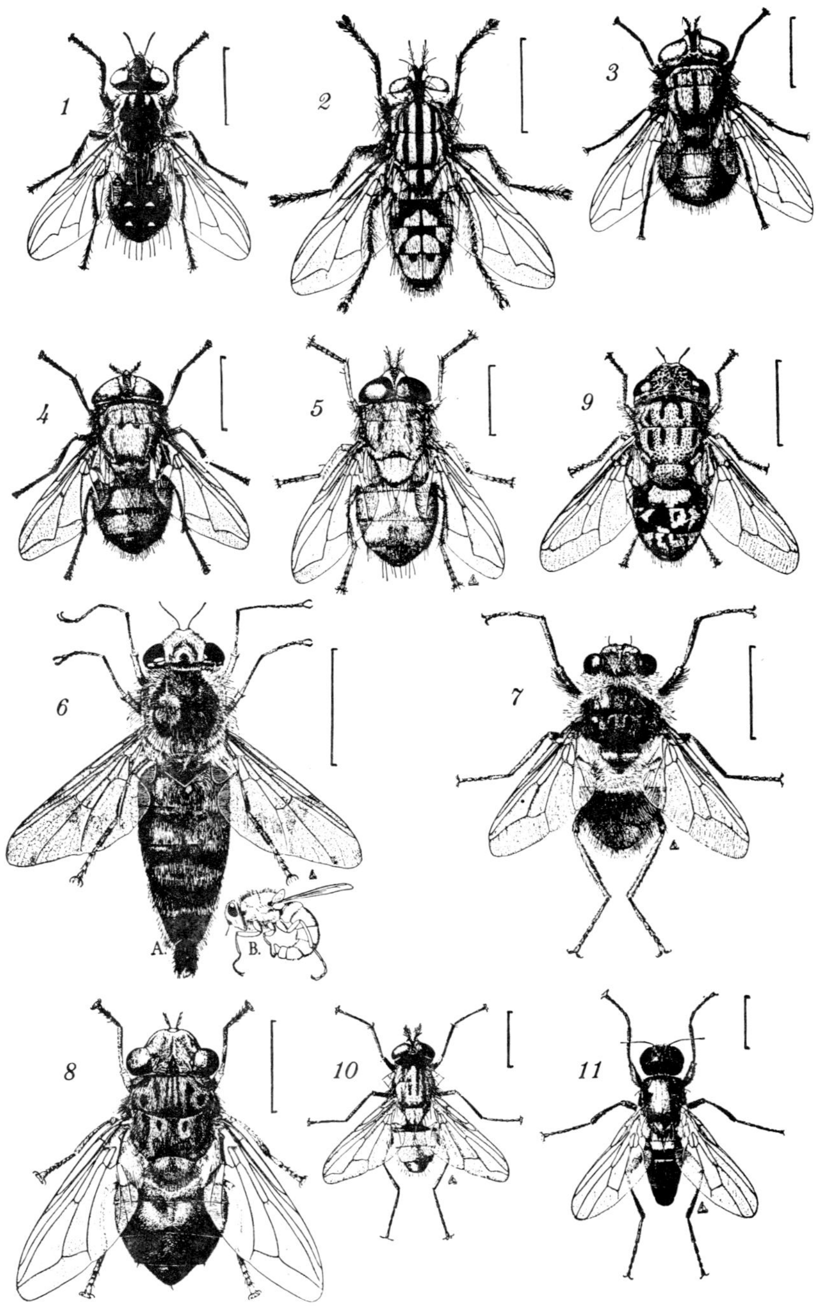

Fig. 42–3. Myiasis-producing flies. Adults of important or relatively common species. *1, 3–9,* Involved in specific myiasis (*1, Wohlfahrtia vigil* sensu lato, female; *3, Cochliomyia hominivorax,* female; *4, Chrysomya bezziana,* female; *5, Cordylobia anthropophaga,* male; *6, Gasterophilus intestinalis,* female; 7, *Hypoderma lineatum,* female; *8, Dermatobia hominis,* female; *9, Oestrus ovis,* female). *2,* Involved in semispecific myiasis (*Sarcophaga haemorrhoidalis,* male). *10, 11,* Involved in accidental myiasis (*10, Musca domestica,* male; *11, Piophila casei,* female). The line adjacent to each drawing indicates the natural length of the fly. (From James, M.T. 1947. *The Flies That Cause Myiasis in Man.* Misc. pub. no. 631. Washington, D.C., Entomology and Plant Quarantine, U.S. Dept. Agriculture.)

maggot crawls into a drier location, draws in its anterior end, and from the third larval cuticle forms an elongate-ovoidal pupal case around itself, thus becoming the *pupa* or "resting stage." Originally light yellow, the pupal case soon turns dark brown. In warm regions, complete pupation requires only 4 or 5 days, after which the adult emerges through a circular seam on the head end of the puparium. Mating takes place within a few days after emergence (2 days in the male, 3 days in the female), and 2 or 3 days later the first batch of eggs is laid. Sperm from the first mating are stored in the seminal receptacle of the female, and are sufficient for the rest of its life; subsequent matings are ineffectual. Under optimal conditions, a complete life cycle requires about 2 weeks.

Fannia canicularis Linnaeus, 1761 (The Lesser House Fly)

Fannia canicularis is a common, grayish, nonmetallic domestic fly that measures 5 to 6 mm in length, has 3 dark longitudinal stripes on the dorsum of the thorax, has a bare antennal arista, and has a distinct venation of the wing. (In contrast to *M. domestica,* the fourth longitudinal vein does not approach vein 3 at an abrupt angle.) It feeds and breeds in decaying vegetable and animal refuse. The *eggs* hatch in about 1 day. Under favorable conditions the larval period requires about 1 week and the pupal stage the same period of time. The *larvae* are characterized by a dorsoventral flattening out and by branched fleshy processes extending out from the sides of the body. (In the related species *F. scalaris,* the latrine fly, the fleshy processes of the larva are more "feathered" in appearance.)

Calliphorid Flies (Family Calliphoridae)

Cochliomyia (= Callitroga) macellaria (Fabricius, 1775) (Secondary Screwworm)

Cochliomyia macellaria is distributed throughout the Western Hemisphere. It requires specific differentiation from *C. hominivorax* (southern United States) and *Chrysomya bezziana* (India, Africa, Philippines), both of which are more important medically, as well as from other species of this and related genera: *Cordylobia anthropophaga, Auchmeromyia senegalensis,* species of *Calliphora, Cynomya, Phormia,* and *Phaenicia,* and those of the true flesh flies (family Sarcophagidae).

Anatomy and Behavior of the Adult. *C. macellaria* is 10 to 13 mm long, has a metallic blue-green color with a silvery sheen, and has three conspicuous longitudinal dark stripes on the dorsum of the thorax like the pattern of the primary screwworm, *Cochliomyia hominivorax* (Fig. 42–3,*3*). The large eyes of the male are close together (i.e., nearly holoptic), those of the female far apart. The antennae differ only in minor structural details from those of *M. domestica.* Their third segment is dark on the frontal side, while the two basal segments are orange. The maxillary palps are small, slender, and orange-colored. The squamae of the wings are dark and bear a few white hairs on the upper surface near the attachment end. The body is covered with numerous stiff black hairs. Laake et al. (1936) found that this species is difficult to distinguish from *C. hominivorax;* however, in *C. macellaria,* they state that "the underside of the abdomen along the midline is covered by a dense whitish pruinosity which, on the last visible segment of the abdomen, appears as two lateral whitish spots when the specimen is viewed from above." This character is absent in *C. hominivorax.* Adults feed on semiliquid or liquid matter derived from wounds or from moribund or decaying flesh. Females lay their eggs (up to 1000 or more) in batches of a few hundred, at 2- or 3-day intervals, in the flesh of dead animals.

Development. Depending upon temperature, the *eggs* hatch within 24 hours and the *larvae* feed ravenously on liquid exudates of wounds or on decaying flesh, which they penetrate by means of their powerful oral hooks (hence the name "screwworm"). There are three larval instars, requiring only 3 to 6 days for complete development. The third instar is 12 to 15 mm long. The general character of the *Cochliomyia* larva (Fig. 42–1,*10, C hominivorax)* is similar to that of *M. domestica,* but it can be distinguished readily by bands of small spines, each with 1 to 3 points, usually 2, on the anterior border of the body segments. These spines are dorsally interrupted on segment 10, and on segments 11 and 12 are confined to the ventral and ventrolateral surfaces. The specific pattern of the posterior spiracles is shown in Figure 42–2,*3*. When mature, the larvae crawl out of the carcass, bury themselves in loose topsoil, and pupate in 2 or 3 days. The *pupa* is elongate-ovoid, dark brown, and retains the bands of spines of the third larval instar. The pupal period requires 3 days or longer (about one-half that of *C. hominivorax).* A day or two after emerging, the adults mate and oviposition soon takes place. Under optimal conditions a complete life cycle requires 12 days (that of *C. hominivorax*, 24 days).

Species of *Chrysomyia* Robineau-Desvoidy, 1830

Medically important members of this genus include *C. bezziana*, the Old World screwworm, and several saprobiotic species that Zumpt (1965) considered to be secondary screwworms. Although these species have traditionally been thought to be Old World in distribution, Laurence (1981) and Gagné (1981) have noted that several African and Oriental species have become established in the Americas, particularly *C. chloropyga*, *C. albiceps*, and *C. megacephala* in Brazil. *C. megacephala* has also expanded its range from the Orient to include parts of West and South Africa.

The Flesh Flies (Family Sarcophagidae)

Sarcophaga haemorrhoidalis (Fallen, 1816–1817)

This large, grayish fly has a nearly worldwide distribution.

Anatomy and Behavior of the Adult. *S. haemorrhoidalis* has a length of 10 to 14 mm, five black longitudinal stripes on the dorsum of the thorax, and a checkered light and dark upper abdominal surface (Fig. 42–3,*2*). The antennal arista is spinose on both anterior and posterior margins of its basal half, bare on its distal half. The fourth vein of the wing bends sharply forward before proceeding to the margin of the wing, where it ends near the outer termination of the third vein. The hypopygia of the male are large, conspicuous, chitinized structures, somewhat reddish, and of diagnostic importance. These flies feed on feces or decaying meat or fish (cooked or uncooked). The females, which are *larviparous*, as are the other species of this genus and of *Wohlfahrtia* of this same family, deposit in these media 40 to 80 first-stage *larvae* that have hatched *in utero*.

Development. The larvae penetrate decaying or decayed material (or possibly invade clean or injured mucosa), and develop to maturity, with three successive instars. These larvae (Fig. 42–1,*11*, *Wohlfahrtia magnifica*) resemble those of *Cochliomyia* and *Calliphora*, with an attenuated anterior end and wide abdominal segments, but may be distinguished by a more distinctly narrowed posterior end; a pit-like depression within which the posterior spiracles are located; the pattern of these spiracles (see Fig. 42–2,*11*); relatively large, fan-shaped anterior spiracles in the third instar, with 15 to 18 short, closely massed, finger-like processes arranged in two irregular rows, bearing the spiracular openings; and delicate, finely pointed, dark spines, situated in bands at the anterior margin of each body segment. The *pupa* is dark brown, elongate-ovoid, and has a distinctive posterior pit at the base of which the spiracles are found.

Other Flesh Flies of Medical Importance

Flesh flies of medical importance include *Sarcophaga carnaria* (Linnaeus, 1758), *Wohlfahrtia magnifica* (Scheiner, 1862), *W. vigil opaca* (Coquillett), and *W. v. vigil* (Walker).

Differentiation of species of *Sarcophaga* is difficult. Adults of the related genus *Wohlfahrtia* may be readily distinguished from species of *Sarcophaga* by the spotted rather than checkered pattern of the dorsal surface of the abdomen (Fig. 42–3,*1* and *2*).

The Botflies (Families Cuterebridae, Gasterophilidae, Hypodermatidae, and Oestridae)

Dermatobia hominis (Modeer, 1786) Say, 1882 (Human Botfly)

This fly, which has a Neotropical distribution from Central Mexico on the Gulf coast to Argentina, is a member of the warble or botfly group, which includes the sheep nasal bot *(Oestrus ovis)*, the cattle bots *(Hypoderma* spp.), the horse bots *(Gasterophilus* spp.), the Russian gadfly *(Rhinoestrus purpureus)*, and species of other genera less commonly parasitizing humans.

Anatomy and Behavior of the Adult. The adult *D. hominis* is a large, stout fly, measuring 15 to 18 mm in length, with a yellowish face, black frons having a grayish pruinosity, a dark bluish-black thorax with a grayish pruinosity, and a more or less diamond-shaped abdomen of metallic blue with a violet tinge (Fig. 42–3,*8*). Both males and females are dichoptic, with relatively small eyes. The third antennal segment is long and bright orange. The arista is spinose only on its anterior side. The legs are orange. The wings are dark brown and have a venation somewhat resembling that of *Sarcophaga*, except that the angle of the fourth longitudinal vein is not conspicuous or abrupt. The mouthparts are vestigial, since the adults never feed.

The female derives nourishment for her eggs from the food stored up during the larval stage. When ready to oviposit, the female captures a mosquito or other bloodsucking arthropod, and glues to its abdomen a batch of 14 to 25 small, elongate-ovoid eggs.

In summarizing the characteristics of the adult female *D. hominis,* Bates (1942) stated that it is diurnal in its flying and seeks out insects of moderate size and moderately active habits on which to deposit its eggs. "Both *Dermatobia* and its vectors seem to be forest insects, though *Dermatobia* will occasionally leave the forest accompanying a host. The *Dermatobia* abundance may differ from forest to forest with no obvious relation to the abundance of cattle (usually the chief host) in the vicinity."

Development. When the phoretic vector of the *Dermatobia* eggs alights on a warm-blooded host, the young *larvae* (ready to hatch) emerge from the egg membranes through a small opercular opening, and in less than an hour invade the skin, frequently through the puncture wound made by the vector. Each larva produces a separate lesion. Common vectors are several species of *Psorophora* (subgenus *Janthinosoma),* and *Culex* among the mosquitoes, *Stomoxys calcitrans* of the bloodsucking muscoids, *Fannia* spp. of the nonbloodsucking muscoids, and the tick, *Amblyomma cajennense.*

Within the host's skin layers (Fig. 42–4) the larva feeds, grows, and molts twice. The recently hatched first-stage larva is elongate-cylindroidal, tapering somewhat posteriorly. The mature third-stage larva (Fig. 42–1,*15)* measures 18 to 24 mm in length and is characteristically inverted flask-shaped. It has well-developed oral hooks and a dark brown anterior cuticle. The dark brown anterior spiracles each consist of two rows of six small, dark brown processes, located on the mesothorax. Near the anterior border of the second and third thoracic segments, and more conspicuously on the first 4 to 6 abdominal segments (and on some abdominal segments also near the middle), are short, stout, posteriorly directed spines or hooklets that anchor the larva in the skin. The terminal two abdominal segments are covered with minute, inconspicuous, anteriorly directed spines. The posterior spiracles lie in a small, deep cleft and usually are concealed in the shrunken specimen. Each spiracular plate has three somewhat curved slits (Fig. 42–2,*13),* directed ventrad and somewhat mesad. Larval development requires from 46 days (Dunn, 1930) to as long as 3 months, after which the larva works its way out of the skin, falls to the ground, and pupates in the soil. The *pupa* requires 14 to 24 days for metamorphosis, following which the adults emerge and copulate. The females soon oviposit, apparently only once, and then die.

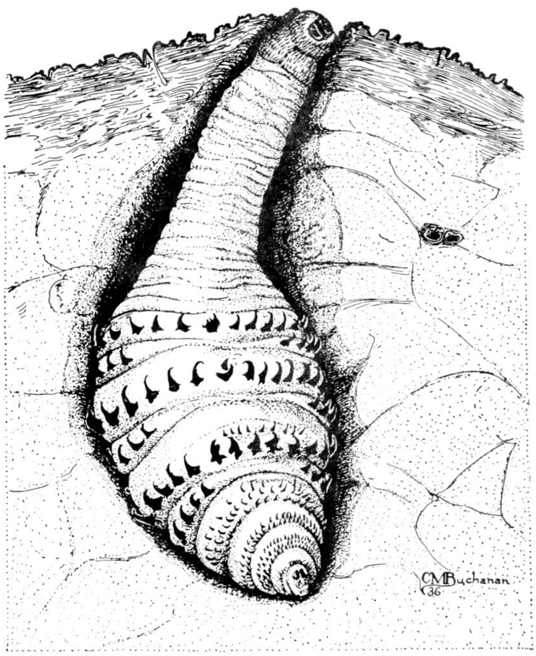

Fig. 42–4. *Dermatobia hominis,* maturing larva in human skin. The posterior end of the larva is at the outer end of the tunnel, so that breathing may occur through the posterior spiracles. (Drawing by C.M. Buchanan for E.C. Faust, 1936.)

The flies are found in the coastal areas as well as on the inland plains and in the jungles of tropical America, frequently near forests (Zumpt, 1973). Domestic and wild mammals are most commonly parasitized by the larvae. Humans are usually infested when associated with domestic animals.

Gasterophilus Species

These horse botflies, *Gasterophilus nasalis* (Linnaeus, 1758), *G. intestinalis* (de Geer, 1776), *G. haemorrhoidalis* (Linnaeus, 1758), and other less common species, are large, brown, hairy forms, with a naked antennal arista and attenuated posterior abdominal segments (Fig. 42–3,*6). G. nasalis* females usually attach their eggs near the base of the hairs of the host. In other species of the genus eggs are cemented more distally on the hair shaft. In *G. nasalis* and *G. haemorrhoidalis* the anatomical areas chosen are on the lower lip or between the mandibular branches of the lower jaw;

in *G. intestinalis,* the areas are on the inner side of the forelegs and shoulders (Georgi, 1980). The eggs hatch and the spinose larvae are licked off by the host, and pass from the mouth through the tissues to the stomach, or may invade the perioral skin and migrate down through the tissues to the stomach or anterior levels of the small bowel, where they attach to the wall while undergoing growth (Figs. 42–1,*13,* and 42–2,*14*). Eventually they are passed *per anum* to pupate in the ground. Accidentally they get on exposed human skin and produce creeping myiasis.

Hypoderma Species

Cattle botflies, *Hypoderma bovis* (Linnaeus, 1758) and *H. lineatum* (Villers, 1789), are large, stout, hairy, bee-like, usually tawny-colored forms, with functionless or rudimentary mouthparts lacking maxillary palps (Fig. 42–3,*7*). The females usually alight on cattle, to whose hairs they attach one or more of the eggs (up to 100 or more on one host). The larvae (Fig. 42–1,*12*), which bear numerous flat tubercles and spines, hatch; invade the skin, usually through a hair follicle, and migrate through somatic and visceral tissues. The left posterior spiracle of the second-stage larva is illustrated in Figure 42–2,*15*. After several molts requiring 6 months or more, the mature larvae (25 to 28 mm long) congregate in subcutaneous tissues on the backs of cattle, where the "warbles" can be recognized by the presence of raised indurated lumps, due to their adventitious encapsulation. In a month or more they work their way out of the capsule and through the skin layers, drop to the ground, and pupate. The perforations produced in the skins of cattle when the larvae leave the host render the hides worthless commercially. These species infest other herbivorous mammals and occasionally parasitize humans.

Oestrus and *Rhinoestrus* Species

The sheep botfly, *Oestrus ovis* (Linnaeus, 1758), and the Russian gadfly, *Rhinoestrus purpureus* (Brauer, 1858), are large dark-gray flies with dark spots on the dorsum of the thorax and abdomen, and are covered with a moderate amount of light-brown hairs (Fig. 42–3,*9*). The females dash at their victims and deposit hatched larvae in the nares, on the conjunctiva, and occasionally on the lips and in the mouth of sheep and other mammals, including humans. These larvae crawl up into the nasal passages to the frontal sinuses, where they

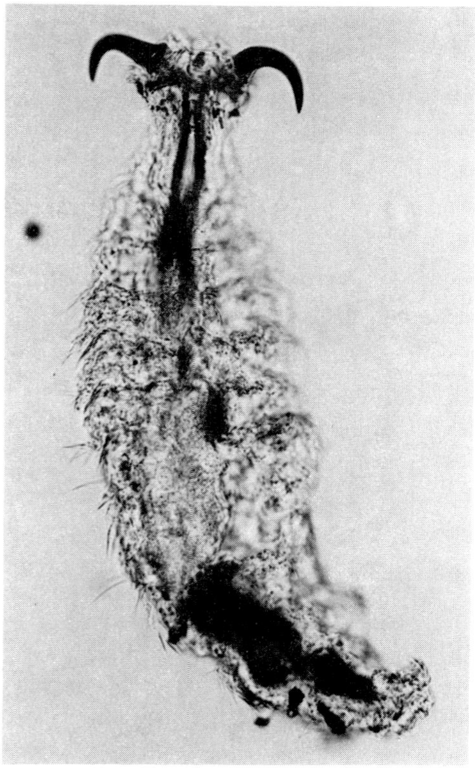

Fig. 42–5. *Oestrus ovis,* first-stage larva removed from the conjunctiva at the inner canthus of the eye of a child. (× 156.) (By P.C. Beaver.)

molt and grow rapidly to a length of 2 to 3 cm (Georgi, 1980). They are characterized by a pair of sharp, powerful dark brown oral hooks and by tufts of numerous brown hooks with spinose tips on the anterior margin of each body segment (Fig. 42–5). The posterior spiracles (Fig. 42–2,*16*) are found in the eighth segment directed ventroposteriad. When these larvae have reached complete development, they pass out of the tissues, drop to the ground, and pupate. Larval development requires several weeks, or first-stage larvae may remain dormant in the host during the cold months.

Eye Gnats (Chloropidae)

Members of this family are minute, acalyptrate forms (i.e., without axillary basal lobes to the wings), with short antennae having an almost spheroid third segment, an arista that is completely or essentially bare, and short, dark-colored, bare wings, a glistening dark thorax, and pale legs. In species of medical interest the mouthparts are considerably modified, so that the spinose tips of the six pseudotracheal rings (of each of the two labella) act as cutting or scarifying instruments. The pseudotracheae empty directly into the prestomium and are widely opened, allowing the entry of a range of particles (Greenberg, 1973).

These flies have a relatively huge esophageal diverticulum and commonly regurgitate droplets of blood or serous exudates, which appear as numerous vomit drops, deposited as "fly specks" within the first 2 or 3 hours following a meal. The flies develop in, but the adults do not feed upon, decaying vegetable matter, feces, and urine. The larval stage may take as little as 6 to 14 days, depending upon the species and climatic conditions. Pupation occurs near the surface of the larval site. For a further discussion of the life cycles of this family, see Greenberg (1973).

FLIES AND HUMAN DISEASE

As described in Chapter 41, bloodsucking flies are important vectors of infections that they inoculate into the human host by their mouthparts. Certain nonbloodsucking flies function as mechanical vectors of infectious disease and, as larvae, they may also be direct agents of disease.

Nonbloodsucking Flies As Mechanical Vectors

The nonbloodsucking flies, particularly the filth flies, carry pathogenic organisms on the external surfaces of their feet and mouthparts or within their alimentary tract. These organisms may then be transferred to man's food or drink by contact or deposited in the fly's vomitus or feces. The fly's role as vector is thus nonbiological or mechanical in contrast to the biological role of the bloodsucking flies in whose tissues or body fluids there is essential development of the etiologic agent.

Conditions Under Which Flies Serve as Vectors

Flies that mechanically transmit pathogens to humans are species that, as adults, feed indiscriminately on human feces and human food, or on lacrimal and cutaneous secretions and discharges from open wounds or ulcers. Although microorganisms ingested by fly larvae about to pupate may be retained through the pupal stage to the emerging adults, this occurs commonly only in the spore stage of the bacteria of tetanus or anthrax. Most microorganisms are carried either on soiled external hairs, bristles, foot pads, and external mouthparts, or in the digestive canal of the fly, later to be deposited in vomit drops or fecal drops. Flies that conform to these conditions include *Musca domestica* and related species of this genus *(M. sorbens), Fannia canicularis, F. scalaris, Muscina stabulans, Lucilia caesar, Phaenicia sericata, Chrysomya megacephala, Phormia regina, Calliphora vicina, Sarcophaga haemorrhoidalis, Drosophila* spp., *Eristalis tenax, Pollenia rudis,* and the eye gnats, *Hippelates* and *Siphunculina*.

Mechanically Transmitted Pathogens

Enteric Infections. Most of the pathogens that flies transmit mechanically belong to the enteric group. Experimental and epidemiologic evidence has incriminated *Musca domestica,* the most serious offender, in the following important enteric infections: *Salmonella typhi, S. enteritidis, Shigella dysenteriae* complex, *Vibrio comma,* and *Entamoeba histolytica*. Control of house flies and other domestic flies not only has brought typhoid epidemics under control, but in other instances has prevented their outbreak. Bacillary dysentery and salmonellosis have been similarly shown to be spread by flies. The causative organisms survive in the fly's gut for at least 4 or 5 days. Cholera vibrios remain viable on the legs of flies for 30 hours. They survive in the intestine of the fly for 24 to 120 hours, and may contaminate food or water by vomit drops up to 24 hours. The incrimination of flies in amebiasis is considered in Section II (p. 107). Larvae of *Musca domestica* and *Phaenicia cuprina* ingest *Ascaris* eggs, which survive metamorphosis through the pupal stage and are passed in fecal droppings of adult flies.

Filth flies probably play a role in the transmission of poliovirus, Coxsackievirus, and echovirus in rural and other fly-infested areas. However, it is likely that the fly-borne aspect of the total transmission cycle of these enteric viruses is minor (Greenberg, 1973).

Nonenteric Infections. In nonenteric infections, domestic flies have been associated with the spread of *Mycobacterium tuberculosis, Yersinia pestis, Bacillus anthracis,* and *Brucella abortus*. The incidence of trachoma and related eye diseases falls sharply when *Musca sorbens* is under control and rises again when such control is lost.

The genera *Hippelates* (New World) and *Siphunculina* (Old World) are nuisance synanthropes that are also believed to mechanically transmit the causative agents of tropical ulcers (*Streptococcus pyogenes, Staphylococcus* spp.), yaws *(Treponema pertenue),* and acute bacterial conjunctivitis. Species of medical importance include *Siphunculina funicola,* which feeds on lachrimal secretions, serous discharges from sores, and perspiration (India, Sri Lanka, Indonesia); *Hippelates collusor* (California; southwestern United States) and *Hippelates pusio* (southeastern United States), which feed on open wounds, sores, and mucous membranes, including the conjunctiva; and *H. flavipes, H. peruanus,* and *H. flavidus* (Mexico, Central and South America, West Indies). In addition to humans, these flies feed on chronic ulcers, mucopurulent secretions, sweat, and hair oils of domestic mammals, especially dogs. They are strongly phototactic.

Davis and Pittman (1950) isolated two types of *Hemophilus* that are the causal agents of acute conjunctivitis in the United States and are believed to be transmitted mechanically by *Hippelates pusio* in the southern United States. The infection may be treated effectively with streptomycin. Kumm (1935) and Kumm and Turner (1936) associated *H. flavipes* with the spread of yaws in Jamaica. This fly and related forms are also capable of transmitting several other types of ulcerative bacteria (Bassett, 1970; Taplin et al., 1967).

Control Of Nonbiting Flies

To control nonbiting flies, two types of efforts are used: (1) those designed to reduce the population of filth flies (i.e., house flies and their allies), and (2) those directed against the breeding of myiasis-producing flies.

Filth flies breed in animal manure, human feces, garbage and decaying organic matter around the house, in city dump heaps, on the seashore and on the farm, and in other exposed decaying or fermenting materials on which both the adult flies and their larvae feed. On farms or in small villages, where there are many domestic animals, where manure is stored in piles on the ground, and where unprotected outdoor privies are still in use, filth flies constitute both a nuisance and a threatening danger during the summer months. Under these latter circumstances, it is practically impossible to keep flies out of the house. Small children are particularly liable to be victims. In small or isolated communities, the danger of breeding flies needs to be impressed on all members of the family.

Human habitations should be well screened against flies, and children, in particular, should not be allowed to sleep in unscreened quarters, especially during the daytime. Flypaper and flytraps may be used to catch adults. Garbage should be kept in closed fly-proof containers and completely incinerated at frequent intervals. Dairy cattle and their stalls should be kept free of flies by efficient insecticidal sprays, although care must be taken that cumulative concentration of the chemical in milk, cream, and beef fats does not reach the level of toxicity for human beings who consume these products. Even more important are measures designed to prevent fly breeding in manure piles in the barnyard and in outdoor privies (which should have a tightly closed back and screened air vents, and should be treated periodically with disinfectants and fly repellents).

Although insecticides cannot take the place of sanitation in filth fly control, certain modern insecticides are useful and practical supplements. Several of the chlorinated hydrocarbon insecticides have been used as 1% water emulsions or as wettable powders sprayed twice a week over breeding media for fly larva control, and also as residual sprays on surfaces where adult flies rest. Unfortunately, fly populations may develop resistance to insecticides, particularly the chlorinated hydrocarbons, but also, in an increasing degree, the organic phosphates.

Manure should be disposed of by spreading it on the fields at least twice a week, when this is practical. It may be treated chemically, but this procedure is expensive. The most efficient and practical breeding preventive is to place the manure on a slatted platform over a concrete basin containing a shallow covering of water. The larvae drop from the manure into the basin and are drained out once each week through an outlet pipe that leads to a cesspool. Dead animals should be buried deeply as soon as they die to prevent the development of flesh flies in the carcasses.

Flies and Human Myiasis

Definition and Classification. Myiasis is "the infestation of live human and vertebrate animals with *dipterous* larvae which, at least for a certain period, feed on the host's dead or living tissue, liquid body substances, or ingested food" (Zumpt, 1965). Myiasis-producing flies have been classified in three categories: specific, semispecific, and accidental. The first group includes species of flies having larvae that are obligatory tissue parasites, namely, a majority of the botflies and several calliphorids. The second group contains species that usually deposit their eggs or larvae in decaying flesh or vegetable matter, but also at times on morbid tissues. The third group consists of those that usually deposit their eggs in excrement or decaying organic matter, but at times lay them in foodstuffs. The various types of myiasis are classified anatomically by the locations of affected tissues, e.g., intestinal, cutaneous (including subcutaneous), nasal, ophthalmic, or aural.

Pseudomyiasis refers to accidentally ingested dead or living fly larvae without necessarily associated pathology or symptomatology, while *sanguinivorous myiasis* denotes superficial attachment during blood sucking (Zumpt, 1965).

The more important myiasis-producing flies are briefly considered in relation to the types of lesions that they produce.

Cutaneous, Ophthalmic, Nasal, and Aural Myiasis

***Dermatobia hominis* (Tórsalo; "Human Botfly," producing Cutaneous Myiasis).** The manner in which the eggs of this fly are carried to humans or other warm-blooded vertebrates, hatch, and invade the skin has been described. (see p. 686). Lesions produced by the larvae are commonly on unprotected skin (hands and wrists, ankles, neck, and face). In Venezuela, Dao (1954) reported 104 cases of human infestation, most frequent on scalp, legs, forearms, and face, including eyelids, of children and adults. However, Dondero et al. (1979) noted that over half of the furuncular lesions in six cases of cutaneous myiasis caused by *D. hominis* occurred on skin surfaces normally protected by clothing. In describing scalp infestation of a Mayan in Guatemala, Dalmat (1955) referred to published accounts of palpebral involvement in Brazil and Panama, and cerebral infestation in Panama, in which larvae had migrated from the scalp through the anterior fontanelle and entered the left hemisphere of a 1½-year-old child. A similar case of cerebral invasion by *D. hominis* has been reported in Brazil (Rossi and Zucoloto, 1973). This dermatozoonosis originated in a 5-month-old child as cutaneous myiasis; however, a single larva penetrated through the scalp and fibrous portion of the bregmatic fontanelle, entering the right cerebral hemisphere (Fig. 42–6).

The tunnel produced is more or less perpendicular to the surface and is occupied by a single larva (Fig. 42–4). At the time the skin has just been invaded, one may observe a small, slightly elevated, somewhat itchy pimple that can rarely be distinguished from a mosquito bite. Soon, however, the growth and movements of the larva, destruction of local tissue, and accumulation of toxic by-products produce throbbing pain and an almost insufferable pruritus, which invariably provokes scratching. Occasionally, the posterior end of the larva protrudes a short distance through the mouth of the tunnel, at which time it can be definitely diagnosed. Usually the mouth of the lesion is patent and exudes seropurulent fluid. The lesion frequently resembles a local streptococcal infection and may cause an associated lymphangitis or lymphadenitis, particularly if it is situated on the ankle. The larva remains in the skin throughout its development of 6 weeks or more, after which it crawls out and drops on the ground to pupate.

The only known therapeutic procedure, other than applying local palliatives, is surgical removal.

For a review of the literature on the bionomics, distribution, taxonomy, and medical importance of *Dermatobia hominis*, see Guimarães (1967) and Catts (1982).

***Gasterophilus* Species (Horse Bots or Warble Flies).** The three most common species of *Gasterophilus* are *G. intestinalis, G. nasalis, and G. haemorrhoidalis*. Human infestation with these larvae (Fig. 42–1,*13*) is relatively uncommon. Rarely, the larvae in man reach the stomach and develop in this location, causing gastric irritation, with nausea and vomiting. Usually the lesion in humans is cutaneous, originating as a swelling at the site where the newly hatched larva enters the skin of the face, trunk, or limbs. It takes the form of a serpiginous tunnel, in which the larva is situated in the lower layers of the epidermis, with the stratum germinativum or corium at its base. In this tunnel the larva may advance several millimeters in 24 hours, over a period of several months. Grossly, the lesion bears a striking resemblance to that produced by *Ancylostoma braziliense* (Section III, p. 281), and similarly produces an aggravating pruritus, but it has no discharge unless secondarily infected. Differential diagnosis may be made by massaging a small amount of a clear mineral oil over the infested skin, so that the larva may be located and identified by the black transverse bands of spines on its body segments. It may then be removed by using a sharp needle, without a local anesthetic. Specific identification can be made by studying the detailed structure under a dissecting microscope. In the Western Hemisphere "creeping eruption" due to species of *Gasterophilus* is practically coextensive with that of *A. braziliense*. Al-

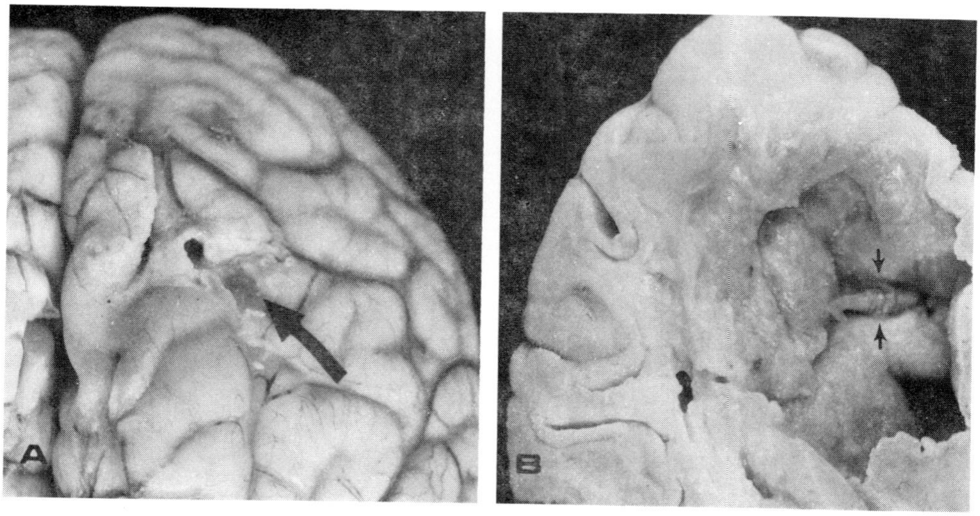

Fig. 42–6. *Dermatobia hominis,* larva in brain of infant in Brazil. *A,* Frontal lobe of cerebrum with opening where the larva entered. *B,* Horizontal section of cerebrum showing the larva in the anterior horn of the right lateral ventricle. (From Rossi, A., and Zucoloto, S. 1973. Fatal cerebral myiasis caused by the tropical warble fly, *Dermotobia hominis.* Am. J. Trop. Med. Hyg., *22*:267–269.)

though cases are usually isolated, they appear at times in epidemic proportions. Ophthalmomyiasis interna due to *G. intestinalis* has been reported once. An unusual case in which an infarcted pulmonary granuloma was attributed to a horse bot (probably a second-stage larva of *G. intestinalis)* has also been recorded (Ahmed and Miller, 1969).

***Hypoderma bovis* and *H. lineatum* (Cattle Bots or Warble Flies).** In these infestations of humans the lesion is less serpiginous, since the larvae (Fig. 42–1,*12)* bore more deeply into the subcutaneous tissues and produce an inflamed swelling more resembling a furuncle. More pain and less pruritus attend the lesion. Nevertheless, the larva migrates through the tissues and in the course of a month may be found many centimeters from a previous site. Thus, while the original site of invasion is usually an exposed part, the lesion noted later may be on any part of the body. The indicated therapeutic procedure is a cruciform incision through the lesion, removal of the larva, and cleaning and dressing the wound. It is important to extract the larva before it dies and causes a suppurative process of greater magnitude.

Ophthalmomyiasis interna has been reported on several occasions as caused by larvae of *H. bovis* and *H. lineatum.* The two species may be differentiated, after surgical removal from the lesion, by clearing the larvae in a 10% caustic potash solution for several hours and studying the oral hooks. In *H. bovis* these hooks are bifurcated at the tip, and in *H. lineatum* they have a single terminal point and a toothed process somewhat back from the tip. *H. bovis* is the common species in the North Temperate Zone and *H. lineatum* somewhat farther south, but the distribution of the two species is partly coextensive. Persons most usually infested reside near cattle where flies are prevalent. *H. diana,* a warble of deer, may also cause human myiasis.

***Oestrus ovis* (Sheep Bot or Nasal Fly).** Gravid females of this species dart onto the conjunctiva (usually the inner canthus), onto the outer nares, onto the lips, or into the buccal cavity, and deposit their first-stage larvae which, by means of large, claw-like oral hooks and spines, rapidly bore into the mucous membrane. Near the eye, the larvae may burrow into the eyelid, conjunctival sac, or lachrimal duct (external ophthalmomyiasis), but rarely into the eyeball (internal ophthalmomyiasis). The former causes great irritation; the latter may lead to optic atrophy and may require enucleation of the organ. From the nares the larvae may reach the nasopharynx. Although human ophthalmomyiasis is common in U.S.S.R., northern Africa, and Israel, it is rare in western Europe (Le Fichoux et al., 1981) and the United States, and is only occasionally seen in South Africa (Stulting and Meyer, 1981). An incidence of 10 per 100,000 population has been reported in Libya. Since sheep are kept within residential areas, cases have not been confined to shepherds (Dar et al., 1980). Infestation of the nasal sinuses with *Oestrus ovis* larvae (Fig. 42–1,*14*) may result in congestion of the passages, with continuous secretion, accompanied by pruritus of the parasitized region and distressing frontal headache. In the pharynx, they cause considerable inflammation, difficulty in swallowing and, at times, vomiting. Larvae in the external membranes of the eye may be removed aseptically with a dissecting needle or fine forceps; those in the nares will probably respond to chlorobutanol (Chloretone) or ephedrine drops. Gargles with strong salt solutions will probably cause pharyngeal forms to be discharged. Except in cases of ophthalmomyiasis interna, larvae spontaneously leave the tissues after a few days. Nasopharyngeal myiasis requires differentiation from internal hirudiniasis.

***Rhinoestrus purpureus* (Russian Gadfly).** This fly, which is distributed extensively throughout southern and eastern Europe, northern Africa, and Asia Minor, commonly deposits its larvae in nares of horses and cattle. Cases of human ophthalmomyiasis caused by this fly have been reported from U.S.S.R., together with similar infestation with *Oestrus ovis*. The larvae of these two species are not readily differentiated.

***Cordylobia anthropophaga* and *C. rodhaini* (Tumbu Flies).** These calliphorine species are found in tropical Africa (Zumpt, 1973). *C. anthropophaga* lays its eggs on dry sand polluted with the excrement of animals, or clothing soiled with feces or urine. In about 3 days the larvae hatch and invade unbroken skin (usually the feet) of humans and other animals that come in contact with them, producing a furuncular swelling at the site of invasion. The larvae (Fig. 42–1,*8*) mature rapidly and usually leave the human host in 8 or 9 days unless previously extracted. They may be identified by their pair of toothed, spade-like, oral hooklets and by several pairs of posteriorly directed fleshy processes arising from the eighth abdominal segment, as well as by the character of the posterior spiracles, which lack a distinct chitinous rim (Fig. 42–2,*4*). Introduction of a small amount of mineral oil into the lesion hastens spontaneous exit of the larvae. Dogs are commonly infected as are a wide range of feral hosts, including rats.

***Cordylobia* (=*Stasisia*) *rodhaini*.** This species closely resembles *C. anthropophaga* both as an adult and as a larva; however, human infestation with *C. rodhaini* is relatively rare since it is usually confined to antelopes and the giant rat, *Crocetomys* sp. (Smith, 1973).

***Chrysomya bezziana* (Old World Screwworm).** This calliphorine fly of the Oriental and Ethiopian regions is bluish-green, with dark thoracic stripes and dark transverse abdominal bands. The female lays her eggs in one batch in a cutaneous ulcer or cut on any exposed part of the body, or on the gums, in the nares, nasal sinuses, conjunctiva, ear, or vagina. The larvae hatch in 8 to 10 hours, burrow into the tissues, molt in 12 to 18 hours, burrow more deeply, reach the third instar (Fig. 42–1,*7*) on the second or third day, and usually leave the wound on the sixth or seventh day, occasionally as late as the tenth to fourteenth day. In the sinuses they may erode the bone and produce a disfiguring lesion. These lesions are foul-smelling, frequently secondarily infected, and have a mucopurulent discharge. Human cases have been encountered frequently in India and in the Congo (Fain et al., 1959), though the incidence is far higher in the Oriental region.

***Cochliomyia* (=*Callitroga*) *hominivorax* (Primary Screwworm).** This greenish-blue calliphorine species of tropical and subtropical regions of the Western Hemisphere is an obligatory parasite of warm-blooded animals. The eggs are deposited in small batches on injured or unbroken skin. The larvae (Fig. 42–1,*10*) usually hatch within 24 hours and invade the skin, producing festering, frequently deep, disfiguring wounds. Several persons in the southern United States have become infested, and some have died as a result of the lesions. During a one-year period (1972–1973) in Trinidad and Tobago, 22 human cases of wound myiasis produced by *Cochliomyia hominivorax* were reported; 2 were fatal due to brain damage (Murray and Thompson, 1976).

***Cochliomyia macellaria* (Secondary Screwworm).** This species, which requires specific differentiation from *C. hominivorax*, commonly deposits its eggs in decaying or moribund flesh of cattle, horses, and sheep, rarely in exposed wounds of humans (i.e., it is saprozoic rather than parasitic). It is probable that most human infestations referred to this species were due to *C. hominivorax*.

***Auchmeromyia senegalensis* (=*luteola*) (Congo Floor Maggot).** This calliphorine species of tropical Africa tolerates a wide range of temperature, humidity, and altitude. Zumpt (1973) noted that this maggot is the only ectoparasitic fly adapted to feeding on humans and the only bloodsucking fly larva that parasitizes humans. Related species are found in the burrows of ant bears (aardvarks), warthogs, and carnivores. It lays small batches of eggs on the floor of native huts or on dry sand. The larvae hatch within 2 days and await an opportunity to infest the skin of persons, particularly those asleep on the bare earth, from whom they suck blood. They are distinguished by having five pairs of fleshy posterior processes and characteristic posterior spiracles (Figs. 42–1,*9* and 42–2,*5*). They feed on blood that exudes from the wounds produced by their oral hooklets and body spines, so that the fed larvae have a reddish-black gut. The sanguivorous habits of this fly are chiefly nocturnal.

***Calliphora*, *Phaenicia*, and *Phormia* species.** Species of *Calliphora (C. vomitoria, C. vicina)*, or *Phaenicia (P. sericata, P. cuprina)*, *Phormia regina*, and rarely other less-common calliphorine species, have been found in human wounds, feeding primarily on moribund flesh and bones, but at times attacking adjacent healthy tissues, and even penetrating unbroken skin. *Phaenicia sericata* has also been obtained in large numbers incubating in human nares (Ryckman and Halstead, 1952).

***Wohlfahrtia magnifica* (Old World Flesh Fly).** This flesh fly is a large grayish species, with large black spots on the dorsal surface of the abdomen. It occurs in the Mediterranean region, the Near East, Iran, Arabia, Iraq, and throughout southern U.S.S.R. Gravid females deposit first-instar larvae in lesions of the skin, in the nasal sinuses, on the papillae of the tongue, in the external auditory meatus, in sore eyes, and in the vagina. As many as 70 larvae have been found in the human eye. The larvae are large and produce serious disfigurement unless promptly removed. Related species are recorded from the same territory in which *W. magnifica* occurs.

***Wohlfahrtia vigil vigil* (Nearctic Flesh Fly).** This fly is also larviparous, depositing its larvae on uninjured skin. A boil-like, traumatic myiasis results. In children in the northern United States and Canada, clinicians have reported larvae of this species in superficial swellings of the head and neck (Walker, 1922). Permanent disfigurement was prevented by surgical removal of the larvae. The characteristic posterior spiracles of the mature larvae are illustrated in Figure 42–2,*12*. Females are particularly attracted to young children and young animals for larviposition, although they do not enter homes.

***Wohlfahrtia vigil opaca*.** A closely related form with a distribution in the western United States, *Wohlfahrtia vigil opaca* has been observed as the agent of furuncular myiasis in children in Utah, Colorado, and California (Dong, 1977).

***Sarcophaga haemorrhoidalis*, *S. fuscicauda*, and *S. carnaria*.** These flesh flies larviposit on decaying flesh and human excrement, but also utilize ulcerated human tissues *(S. fuscicauda)*, including the orbit, face, and neck. They may larviposit on food later eaten by people *(S. haemorrhoidalis)* and may become temporary parasites in the human bowel.

Eggs of other flies may be deposited accidentally in exposed lesions of the skin and hatch. Larvae develop in these situations.

Intestinal Myiasis

Not uncommonly, gravid female flies (e.g., families Muscidae, Calliphoridae, Anthomyidae, Syrphidae, Psychodidae, Phoridae, Sepsidae, Trypetidae, and Stratiomyidae) deposit eggs on food to be eaten by humans, or larviposit (family Sarcophagidae) on stale food. Although the eggs are frequently digested or fail to hatch, or the larvae are killed en transit down the intestinal tract, they may survive, become temporarily lodged in the crypts and folds, and develop through part or all of their larval period before being evacuated.

Urogenital Myiasis

Although most reports of fly larvae in the urine are probably referable to a contaminated container for the specimen, several cases, in which physicians vouch for the passage of these larvae from the urethra, apparently constitute true accidental myiasis. In female subjects, in whom either the bladder or vagina has become infested with fly larvae, the opportunity for the organisms to enter these organs from the external meatus is appreciable (Disney and Kurahashi, 1978; Biery et al., 1979).

Lung Myiasis

Thirty third-instar larvae of *Megaselia spiracularis* (Phoridae) were recovered from the suction fluid of an open lung biopsy performed on a 19-year-old Japanese man (Komori et al., 1978). The patient had a history of cough and purulent sputum. X-rays showed multiple ill-defined shadows about 2 to 4 cm in diameter in the middle and lower lobes. Following recovery of the larvae on the third postoperative day, there was a gradual clearing of the lungs and a reduction of the white blood cell count. This unusual form of myiasis probably resulted from the accidental inhalation of either a gravid female fly or airborne eggs.

Control of Myiasis-Producing Flies

The warble flies and botflies constitute a special problem. In areas where species of *Gasterophilus* are prevalent, horses should be frequently groomed. A more effective measure, however, is the application of an emulsion spray of malathion (0.12% in water) to parts of the animals most commonly harboring the eggs, in order to kill developing embryos (Georgi, 1980). *Hypoderma* infection of cattle can be controlled by spraying exposed animals with a 4% formulation of Coumaphos immediately after adult fly activity ends. Application during this "safe period" avoids destruction of larvae in the esophageal region *(H. lineatum)* or spinal canal *(H. bovis)*. *Dermatobia* in cattle and horses may be treated similarly. Dilute concentrations of dichlorvos or fenthion may be sprayed directly into the nostrils for control of sheep nasal bots *(Oestrus ovis)*.

Calliphorine larvae that breed in lesions of domestic animals should be removed by skilled hands and the lesions cleaned and dressed with antiseptic powder or ointment. A bait-toxicant system effective against *Cochliomyia hominivorax* and *C. macellaria* has been described by Coppedge et al. (1980). Adult flies of both sexes are attracted by swormlure-2, a chemically defined bait, and they are intoxicated by 2% dichlorvos. A genetic control mechanism using mass-reared, sterile males has also proven effective for *C. hominivorax* over most of its range in the continental United States and neighboring islands (Richardson et al., 1982).

REFERENCES

Ahmed, M.J., and Miller, A. 1969. Pulmonary coin lesion containing a horse bot, *Gasterophilus*. Am. J. Clin. Pathol., *52*:414–419.

Bassett, D.C.J. 1970. *Hippelates* flies and streptococcal skin infection in Trinidad. Trans. R. Soc. Trop. Med. Hyg., *64*:138–147.

Bates, M. 1942. Mosquitoes as vectors of *Dermatobia* in eastern Colombia. Ann. Entomol. Soc. Am., *36*:21–24.

Biery, T.L., Clegern, R.W., and Hart, W.M. 1979. Two cases of phorid (Diptera: Phoridae) myiasis in Texas. J. Med. Entomol., *15*:122–123.

Catts, E.P. 1982. Biology of New World bot flies: Cuterebridae. Annu. Rev. Entomol., *27*:313–338.

Coppedge, J.R., Brown, H.E. Goodenough, J.L., Tannahill, F.H., Snow, J.W., Petersen, H.D., and Hofmann, H.C. 1980. Field performance of a new formulation of the screwworm adult suppression system. J. Econ. Entomol., *73*:411–414.

Dalmat, H.T. 1955. Cutaneous myiasis of the scalp due to *Dermatobia hominis* (Linnaeus) (Diptera: Cuterebridae). Am. J. Trop. Med. Hyg., *4*:334–335.

Dao, L. 1954. Miasis cutánea a *Dermatobia hominis* en Venezuela. Acta Med. Venezolana, *2*:4–12.

Dar, M.S., Amer, M.B., Dar, F.K., and Papazotos, V. 1980. Ophthalmomyiasis caused by the sheep nasal bot, *Oestrus ovis* (Oestridae) larvae, in the Benghazi area of eastern Libya. Trans. R. Soc. Trop. Med. Hyg., *74*:303–306.

Davis, D.J., and Pittman, M. 1950. Acute conjunctivitis in the south. Commun. Dis. Center, U.S. Public Health Ser. Bull., April 1950, pp. 18–19.

Disney, R.H.L., and Kurahashi, H. 1978. A case of urogenital myiasis caused by a species of *Megaselia* (Diptera: Phoridae). J. Med. Entomol., *14*:717.

Dondero, T.J., Jr., Scaffner, W., Athanasiou, R., and Maguire, W. 1979. Cutaneous myiasis in visitors to Central America. South. Med. J., *72*:1508–1511.

Dong, L. 1977. First report of human myiasis due to *Wohlfahrtia vigil opaca* in California and a review of the distribution of the subspecies in the state. Calif. Vector Views, *24*:13–17.

Dunn,L.H. 1930. Rearing of the larvae of *Dermatobia hominis* Linn. in man. Psyche, *37*:327–342.

Fain, A., Magis, P., Verdin, G., Donkers, J., and Gobbels, P. 1959. Sur deux cas de myiases humaines produites par *Chrysomya bezziana* Villeneuve, au Congo Belge. Ann. Soc. Belge Med. Trop., *39*:763–766.

Gagné, R.J. 1981. *Chrysomya* spp., Old world blow flies (Diptera: Calliphoridae), recently established in the Americas. Bull. Entomol. Soc. Am., *27*:21–22.

Georgi, J.R. 1980. *Parasitology for Veterinarians*, 3rd ed. Philadelphia, W.B. Saunders Co.

Greenberg, B. 1971. *Flies and Disease*. Vol. I. *Ecology, Classification and Biotic Associations*. Princeton, New Jersey, Princeton Univ. Press.

———. 1973. *Flies and Disease*. Vol. II. *Biology and Disease Transmission*. Princeton, New Jersey, Princeton Univ. Press.

Guimarães, J.H. 1967. *A Catalogue of the Diptera of the Americas South of the United States. 105: Family Cuterebridae*. Sao Paulo, Dept. Zool., Secr. Agricultura.

James, M.T. 1947. *The Flies That Cause Myiasis in Man*. Washington, D.C. U.S. Dept. Agric. Misc. Publ. No. 631.

Komori, K., Hara, K., Smith, K.G.V., Oda, T., and Karamine, D. 1978. A case of lung myiasis caused by larvae of *Megaselia spiracularis* Schmitz (Diptera: Phoridae). Trans. R. Soc. Trop. Med. Hyg., *72*:467–470.

Kumm, H.H.W. 1935. The digestive mechanism of one of the West Indian "eye gnats," *Hippelates pallipes* Lowe. Ann. Trop. Med. Parasitol., *29*:283–298.

———, and Turner, T.B. 1936. The transmission of yaws from man to rabbits by an insect vector, *Hippelates pallipes* Loew. Am. J. Trop. Med., *16*:245–272.

Laake, E.W., Cushing, E.C., and Parish, H.E. 1936. The biology of the primary screw worm fly, *Cochliomyia americana*, and a comparison of its stages with those of *C. macellaria*. Washington, D.C., U.S. Dept. Agric. Tech. Bull. No. 500.

Laurence, B.R. 1981. Geographical expansion of the range of *Chrysomyia* blowflies. Trans. R. Soc. Trop. Med. Hyg., *75*:130–131.

Le Fichoux, Y., Marty, P., Denis, G., Couturier, P., and Dellamonica, P. 1981. Un cas d'ophthalmomyiase externe à *Oestrus ovis* Linne, 1758 contractée sur la plage de Nice. Acta Trop., *38*:461–468.

Murray, V.I.E., and Thompson, K. 1976. Myiasis in man and other animals in Trinidad and Tobago (1972–73). Trop. Agric., *53*:263–266.

Richardson, R.H., Ellison, J.R., and Averhoff, W.W. 1982. Autocidal control of screwworms in North America. Science, *215*:361–370.

Rossi, M.A., and Zucoloto, S. 1973. Fatal cerebral myiasis caused by the tropical warble fly, *Dermatobia hominis*. Am. J. Trop. Med. Hyg., *22*:267–269.

Ryckman, R.E., and Halstead, B.W. 1952. Report of a case of nasal myiasis by the green-bottle fly, *Phaenicia sericata*, in San Bernardino County, California. Am. J. Trop. Med. Hyg., *1*:711–712.

Smith, K.G.V. 1973. Calliphoridae and Sarcophagidae (blowflies and flesh-flies), pp. 271–277. In *Insects and Other Arthropods of Medical Importance*. Edited by K.G.V. Smith. London, British Museum (Natural History).

Stulting, A.A., and Meyer, H. 1981. External ophthalmomyiasis caused by *Oestrus ovis*. S. Afr. Med. J., *60*:709–710.

Taplin, D., Zaias, N., and Rebell, G. 1967. Skin infections in a military population. Develop. Indust. Microbiol., *8*:3–12.

Walker, E.M. 1922. Some cases of cutaneous myiasis, with notes on the larvae of *Wohlfahrtia vigil* (Walker). J. Parasitol., 9:1–5.

Zumpt, F. 1965. *Myiasis in Man and Animals in the Old World. A Textbook for Physicians, Veterinarians and Zoologists.* London, Butterworth & Co.

———. 1973. Diptera parasitic on vertebrates in Africa south of the Sahara and in South America, and their medical significance, pp. 197–205. In *Tropical Forest Ecosystems in Africa and South America: A Comparative Review.* Edited by B.J. Meggers, E.S. Ayensu, and W.D. Duckworth. Washington, D.C., Smithsonian Institute Press.

Chapter 43

Fleas (Siphonaptera)

Fleas (order Siphonaptera, or Aphaniptera) are apterous insects with a prominent proboscis. They are partially host-specific ectoparasites that attach temporarily to the skin of mammals and birds to suck blood, and are particularly abundant on nest-building rodents. Some species, like the human flea *(Pulex irritans),* the Oriental rat flea *(Xenopsylla cheopis),* and the northern rat flea *(Nosopsyllus fasciatus),* have an extensive distribution; others are more or less limited to restricted geographical areas or to regions where particular hosts are found (Traub, 1980).

Morphology. Adults (Fig. 43–1) are small, wingless, medium-to-dark brown, shiny insects that are compressed laterally, have mouthparts adapted to piercing the skin and sucking blood, and have conspicuous legs, of which the hind pair is especially long and is adapted for jumping. Fleas are descended from winged, probably Mecoptera-like ancestors (Rothschild, 1975).

The *head* is compressed from side to side, is usually longer than high, and its chitinous covering (i.e., sclerites) lacks evidence of its original component parts. For convenience, the head is divided into two portions, obliquely separated by grooves in which antennae are situated. In front, the dorsal portion is called the *frons;* the ventral portion is termed the *gena* (i.e., cheek). *Simple eyes,* when present, lie just in front of the antennae. The region behind the antennal groove is called the *occiput.* In some species the ventral margin of the cheeks is beset with dark brown teeth, the *genal comb* or *ctenidium* (Fig. 43–6,*D,1*). A few species have, in addition, a *pronotal comb* at the posterior margin of the dorsal covering of the first thoracic segment, while other species have the pronotal comb without the genal comb (Fig. 43–6,*C,1*). At times there is a seta or bristle in front of the eyes and another seta on the gena near the genal comb, as well as several occipital ones. These combs and bristles serve as a means of protection, to facilitate passage through hair, and to maintain position on the host (Traub, 1980a). Each of the pair of *antennae* consists of a club-shaped organ, composed of three joints, lying in the *antennal groove* or fossa. The piercing-sucking mouthparts consist of a pair of anteriorly disposed, segmented *maxillary palps,* with more or less rounded distal ends; a pair of posteriorly disposed *labial palps* of essentially similar structure; a pair of triangular, blade-like *maxillae* (in part, the *stipes)* on the lateral aspects; a pair of rudimentary *mandibles;* and an unpaired, stylet-shaped *epipharynx* which, with the needle-like parts of the maxillae (the *laciniae)* forms the food channel (Fig. 43–2).

The characteristic three *thoracic segments* are readily distinguished on both the dorsal and ventral aspects of the flea. Each of the powerful legs, which are attached to pleural plates, consists of a large flat coxa; a small trochanter; a more or less ovoid, flattened femur; a tibia, which is narrow proximally and widened distally; and five tarsal segments, the terminal one of which bears a pair of curved claws (Fig. 43–1).

The *abdomen* is conspicuously compressed laterally, and consists of ten segments, of which the terminal three are modified in relation to the anus and external genitalia. The ninth segment on its dorsal side is small, cordate, and provided with a unique saddle-shaped structure called the *pygidium* (i.e., the "rump"), which may be sensory in function. The anus opens on the tenth segment. On the posterior margin of the dorsum of the seventh segment is a pair of more or less conspicuous bristles, the *antepygidial bristles.* Two pairs of spiracles are on the thorax, and a pair is on each of the first through the eighth abdominal segments. In the male there are two pairs of small claspers on the ninth

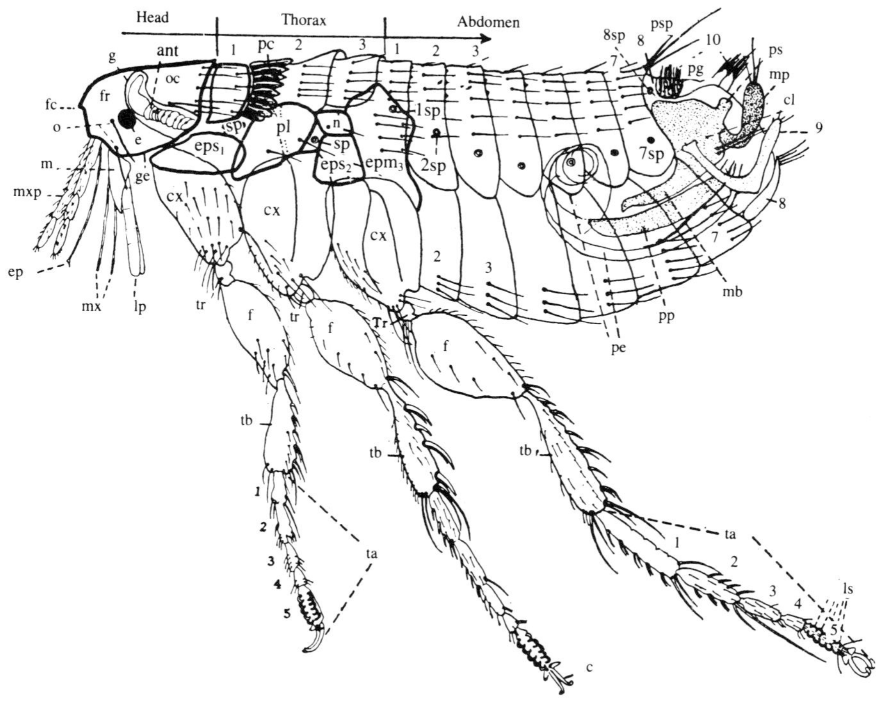

Fig. 43–1. Male flea. Lateral view showing the structures used in identification. *ant,* Antenna in antennal groove *(g), c,* Claw. *cl,* Clasper with its movable finger *(mp)* and the immovable process *(ps). cx,* Coxae. *e,* Eye. *ep,* Epipharynx. *epm3,* Epimeron of the third thoracic segment. *eps1* and *eps2,* Episterna of the first and third thoracic segments. *f,* Femur. *fc,* Frontal notch. *fr,* Frons. *g,* Groove for antenna. *ge,* Gena on which the gena comb is located when present. *lp,* Labial palpi. *ls,* Lateral spines on the last tarsal joint. *m,* Maxillary plate (stipes). *mb,* Manubrium of the clasper. *mp,* Movable process of the clasper. *mx,* Maxillae (lacinea of the maxillae). *mxp,* Maxillary palpi. *n,* A portion of the tergum of the third thoracic segment. *o,* Upper bristle is the ocular, lower the oral. *oc,* Occiput. *pc,* Prothoracic comb or ctenidium. *pp,* Penis plate. *pe,* Penis with the coil of springs. *pg,* Pygidium. *pl,* Pleuron of the second thoracic segment. *ps,* Immovable process of the clasper. *psp,* Antepygidial bristles or spines. *sp,* Spiracles of the meso- and metathoracic segments. *1sp–8sp,* Spiracles of the abdomen. *ta,* Tarsus, each with five segments (1–5). *tb,* Tibia. *tr,* Trochanter. (The numerals indicate the terga, 1–3, 1–10, and the sterna, 2–9, of the thorax and abdomen.) (From Matheson, R. 1950. *Medical Entomology,* 2nd ed. Ithaca, Comstock Publishing Company.)

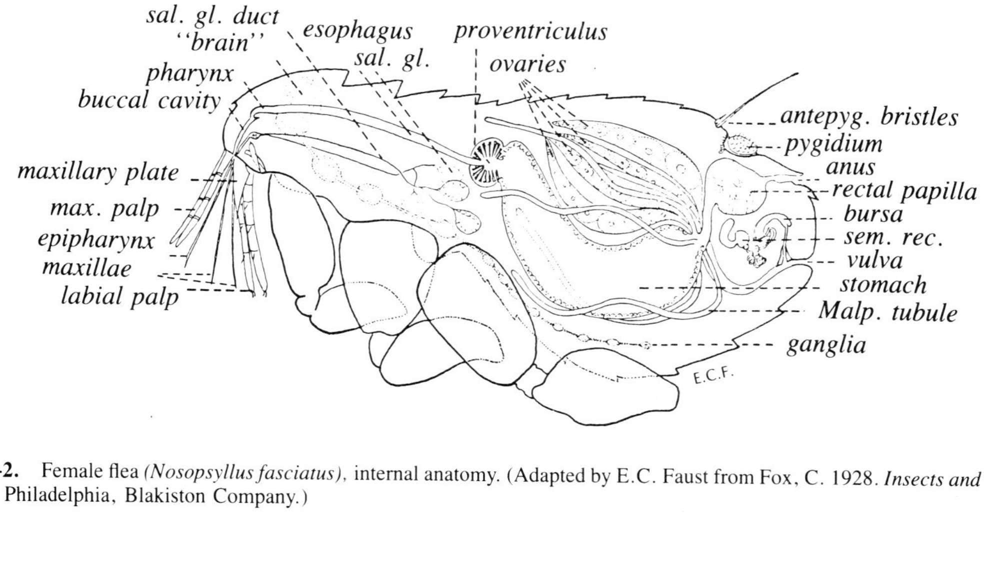

Fig. 43–2. Female flea *(Nosopsyllus fasciatus),* internal anatomy. (Adapted by E.C. Faust from Fox, C. 1928. *Insects and Diseases of Man.* Philadelphia, Blakiston Company.)

segment behind the pygidium. Extending inward from the region between the ninth and tenth segments is the long, coiled, complicated, chitinous *aedeagus* or cirrus organ, the presence of which makes sex determination easy. In the female the chitinous *spermatheca* or seminal receptacle, situated approximately in the midplane of the seventh segment, is readily seen in cleared mounts. This is a valuable organ for species differentiation (Fig. 43–6).

The *digestive tract* of the flea (Fig. 43–2) consists of a moderately long, muscular *pharynx*, which opens externally into the *buccal cavity* at the base of the hypopharynx; a long capillary *esophagus;* a short muscular *proventriculus,* provided with long chitinous teeth on its median portion; a large pouch-like "stomach" or *midgut;* a capillary *hindgut;* and a small pouched *rectum*. There is no esophageal diverticulum for storage of undigested food such as is found in the flies. During a blood meal the proventricular spines break up the ingested red blood cells, and then the homogenized food is passed into the midgut by synchronized peristaltic movement of proventriculus and midgut (Munshi, 1960). There are two pairs of small, globoidal *salivary glands,* each with a short duct that joins its mate and proceeds outward as one of a pair of long capillary ducts. These ducts unite just before they enter the base of the hypopharynx, where the salivary secretion is controlled by a small pump. Four delicate *malpighian tubules* empty into the intestine at the junction of the midgut and the hindgut. The *male organs* consist of a pair of pouch-like *testes,* each with a delicate *vas deferens;* four *accessory glands;* a *seminal vesicle;* an *ejaculatory duct;* and the complicated *aedeagus*. The *female organs* consist of a pair of multidigitate *ovaries;* a common *oviduct;* a retort-shaped *spermatheca,* which opens through a small duct into the vagina; and the *vagina,* which continues as the *vulva* that opens posteriorly between the eighth and ninth abdominal sclerites. Insemination can occur before ingestion of the initial blood meal or after a lengthy period of blood feeding on both or one sex of a host (Rothschild, 1975). Spermatozoa are introduced into the female reproductive tract by means of extrudable penis rods.

Development. The minute, ovoid, glistening, white eggs are deposited a few at a time by the female after insemination and a blood meal, and usually drop to the ground or floor where, after 3 to 10 days, depending on humidity and tempera-

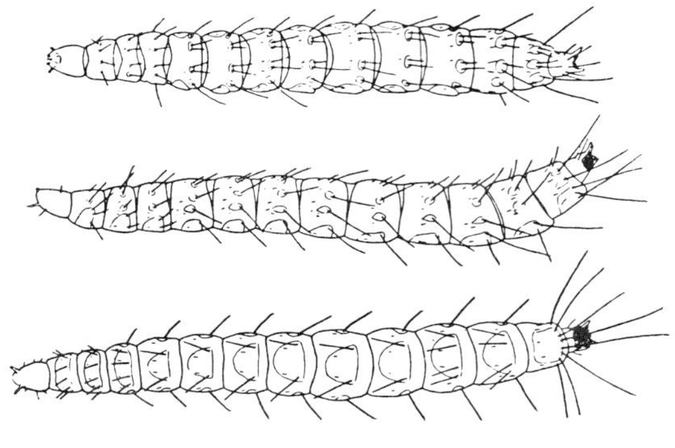

Fig. 43–3. *Xenopsylla cheopis,* larvae. Enlarged. (From Bacot, A.W., and Ridewood, W.G. 1914. Observations on the larvae of fleas. Parasitology, 7:157–175.)

ture, they hatch to elongate, legless larvae (Fig. 43–3). The larvae begin to feed, by use of their mandibulate mouthparts, on available organic debris and dejecta of the adult flea, which contains an abundance of dried blood. There are three larval instars, requiring as short a time as 7 days *(Xenopsylla cheopis* in India) to more than 100 days *(Ctenocephalides canis* in Europe) to develop into a pupa, depending on species, food conditions, temperature, and humidity. The mature third-instar larva spins a cocoon around itself and pupates. The pupal stage may be as short as one week or may last nearly a year. When this period has been completed, the adult emerges through a longitudinal slit on the middorsum of the thorax and breaks its way through the loose meshes of the cocoon. Rapid emergence in response to slight mechanical stimuli has been noted for some species. The adult can survive for a year or more, and for a considerable period of time even when unfed.

CLASSIFICATION OF FLEAS

The order Siphonaptera, formerly divided into two suborders, Fracticipita and Integrecipita, is now classified into three superfamilies (Pulicoidea, Malacopsylloidea, and Ceratophylloidea), consisting of 15 or 16 families (Holland, 1964; Smit, 1973; Lewis, 1972, 1973, 1974, 1974a, 1974b, 1975). Criteria for classification include the number of rows of bristles on the dorsum of the abdomen; presence or absence of eyes; presence or absence of a pronotal or genal comb, and the number of teeth in each; number of tubercles or bristles on various parts of the body; presence or absence of antepygidial bristles; relative size and shape of the tarsal joints; shape of the antennae; number of segments of the labial palps; shape of the spermatheca in the female; structure of the male terminalia; and the structure of the mesopleuron (Hopkins and Rothschild, 1953, 1956, 1962, 1966, 1971).

MEDICALLY IMPORTANT FLEAS

Medically important fleas are found in the following families: Tungidae *(Tunga penetrans);* Pulicidae *(Pulex irritans, Ctenocephalides canis, C. felis, Xenopsylla cheopis, X. brasiliensis, X. astia, Synosternus pallidus, Hoplopsyllus anomalus);* Ceratophyllidae *(Ceratophyllus niger, Nosopsyllus fasciatus, Diamanus montanus);* and Leptopsyllidae *(Leptopsylla segnis).*

***Tunga penetrans* (Linnaeus, 1758) Jarocki, 1838 (Chigoe, Jigger, Sand flea, or "Nigua" of Latin America).** The frons of this species is sharply angled and proportionately larger than the head of most other fleas (Fig. 43–4). The maxillae are long, stiff, and distinctly serrated. The three thoracic segments are markedly foreshortened. There is no suture from the base of the antennal groove to the vertex. The inseminated adult female (Fig. 43–5) burrows into the skin of man and other mammals ("tungosis"), but the tip of the abdomen remains in contact with the outer surface so that respiration can take place. Eggs are ejected into the environment where the larvae develop in dry, sandy soil. The presence of the engorged female causes a nodular swelling with subsequent ulceration; occasionally larvae may also develop in the epidermis. *T. penetrans* is distributed along the southern portion of the Gulf of Mexico and throughout Tropical America, its original home, as far south as Paraguay. It has been reported from the East and West African subregions of the Ethiopian region (Lewis, 1972), Pakistan, and India (Connor, 1976).

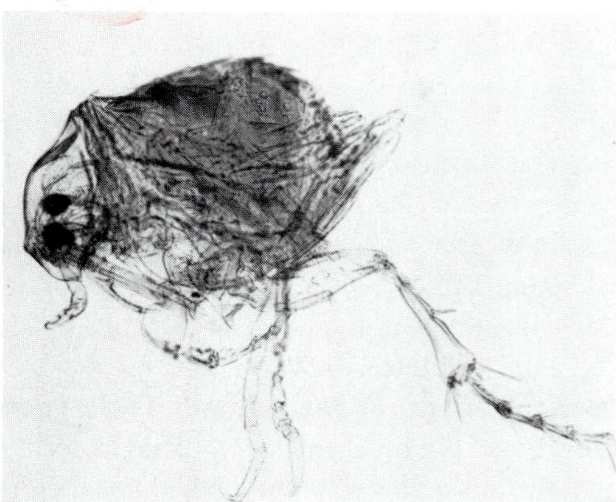

Fig. 43–4. *Tunga penetrans,* whole mount of a nongravid specimen. (× ca. 90.) (From Binford, C.H., and Connor, D.H. 1976. *Pathology of Tropical and Extraordinary Diseases,* vol. 1. Washington, D.C., Armed Forces Institute of Pathology [negative no. 54–16364–14].)

***Pulex irritans* Linnaeus, 1758 (Human Flea).** The head of *P. irritans* (Fig. 43–6,*A,1)* is normally rounded; the labial palps are strongly chitinized; the maxillary laciniae extend about one half the length of the forecoxae, differentiating it from *P. simulans,* an ectoparasite of wild mammals; the thoracic segments are not foreshortened; genal and pronotal combs are lacking; a pair of conspicuous eyespots is present; the mesopleuron is narrow, without a median vertical suture; the antepygidial bristles are short (Fig. 43–6,*A,2);* the outer plate of the male clasper is large and distinct; the seminal receptacle (Fig. 43–6,*A,3)* of the female is not strongly chitinized; and males as well as females feed exclusively on the skin. This species is cosmopolitan in distribution, and is found primarily on humans and dogs, but also on rats, cats, hogs, and cattle.

***Xenopsylla cheopis* (Rothschild, 1903) Rothschild, 1909 (Oriental Rat Flea).** This species (Fig. 43–6,*B,1–3)* resembles *Pulex irritans,* with the following important exceptions: (1) the mesopleuron is wide and provided with a median vertical suture; (2) the antepygidial bristles are long and stout; and (3) the seminal receptacle of the female is larger and more conspicuously chitinized, with a sharp caudal angle. *Xenopsylla cheopis* is common on rats in the Tropics. It requires differentiation from (1) *X. brasiliensis* (Baker, 1904), a widespread species in which the antepygidial bristles in the male are situated at the end of a long pedicle, and the caudal end of the seminal receptacle is long and narrow throughout its length, and from (2) *X. astia* Rothschild, 1911, distributed throughout the Indian subcontinent, in which the ninth sternal plate of the male is ribbon-shaped, and the caudal end of the seminal receptacle is bulbous at its base, narrowed at its distal end.

X. cheopis is cosmopolitan, owing its distribution to synanthropic murine rodents (Lewis, 1972).

***Nosopsyllus fasciatus* (Bosc, 1800) Jordan, 1933 (Northern Rat Flea).** In members of this genus (Fig. 43–6,*C,1–3),* there is a conspicuous pronotal comb but no genal comb. The eyespots are well developed. In *N. fasciatus,* the comb has approximately 18 teeth. Below the eye are three equidistant genal bristles. The terminal segment of the antenna is distinctly jointed. The common hosts

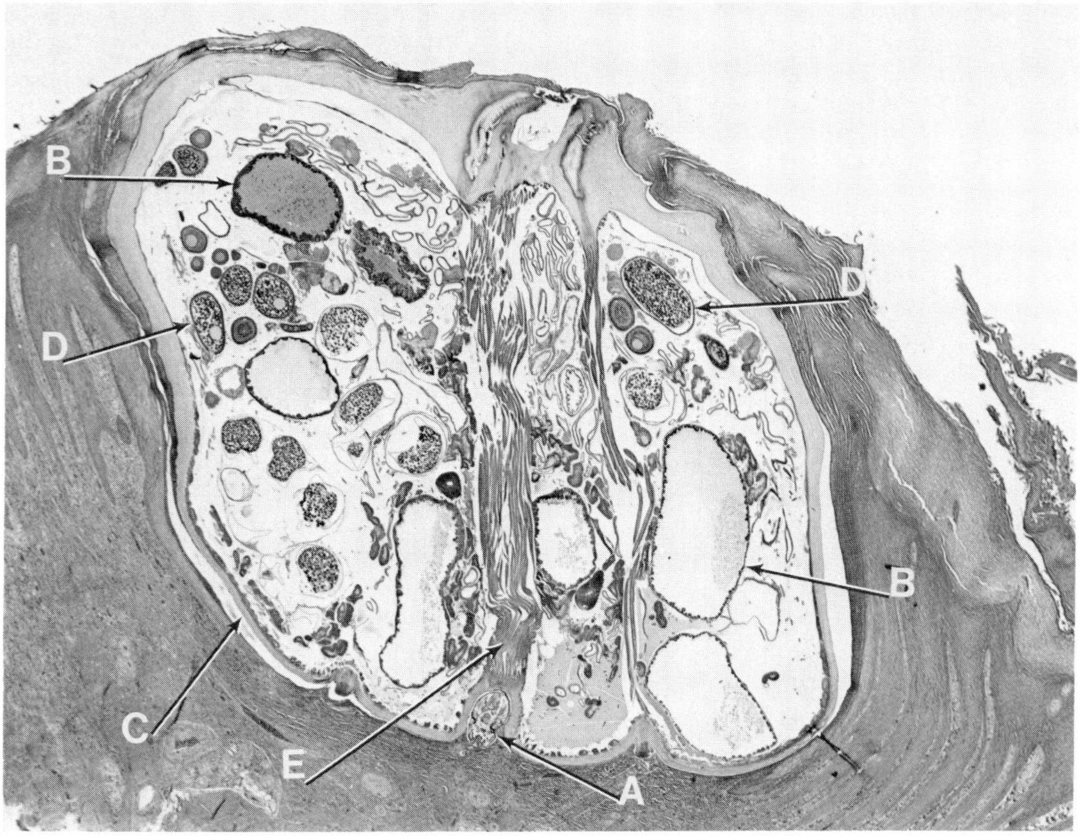

Fig. 43–5. *Tunga penetrans,* vertical section through the long axis of an embedded female showing the hole in the keratin through which eggs, excrement, and air pass. The head of the flea *(A)* is situated within the dermis. *B,* Distended digestive tube. *C,* The thickened cuticle. *D,* Developing ovary. *E,* Hypertrophied internal muscle band. (× ca. 24.) (From Binford, C.H., and Connor, D.H. 1976. *Pathology of Tropical and Extraordinary Diseases,* vol. 1. Washington, D.C. Armed Forces Institute of Pathology [negative no. 70-7115].)

of this flea are small rodents, and it is distributed throughout the Temperate Zones of the world. It requires differentiation from (1) *Diamanus montanus* (Baker, 1895), a parasite of the California ground squirrel, which has less conspicuous eyes, about 16 teeth on the pronotal comb, and three genal spines that are not equidistant; from (2) *Oropsylla silantiewi* (Wagner, 1898), which parasitizes the tarbagan *(Marmota marmota sibirica)* of Mongolia; and from (3) the *Citellophilus tesquorum* (Wagner, 1898) species complex on ground squirrels in Asiatic U.S.S.R.

Ctenocephalides canis **(Curtis, 1826) Stiles and Collins, 1930 (Dog Flea).** Members of this genus (Fig. 43–6,*D,1-3)* have both genal and pronotal combs. In *C. canis* the head is short with a more or less rounded frons. The front tooth of the genal comb is much shorter than the second. This flea is almost cosmopolitan in its distribution on humans and feral and domestic carnivores. It requires differentiation from *C. felis* (Bouche, 1835), the cat flea, which has a longer, narrower head, and in which the front tooth of the genal comb is not conspicuously shorter than the second. *Ctenocephalides felis* and *C. canis* are indiscriminate in host selection, attacking both dogs and cats, with *C. felis* occurring more frequently on dogs in North America than *C. canis.* Both species can also be household pests of man. The use of flea collars containing dichlorvos, an organophosphorus material with fumigant qualities, is effective in lowering adult flea infestations on pets.

Leptopsylla segnis **(Schönherr, 1811) (European Mouse Flea).** The head is subangulate at the frons; there is a pair of blunt-toothed genal combs but no pronotal comb; eyespots are lacking; a pair of short blunt spines is located just lateral to the angle of the frons. This flea is common on mice and rats, but is known from all continents and a myriad of hosts (Lewis, 1974).

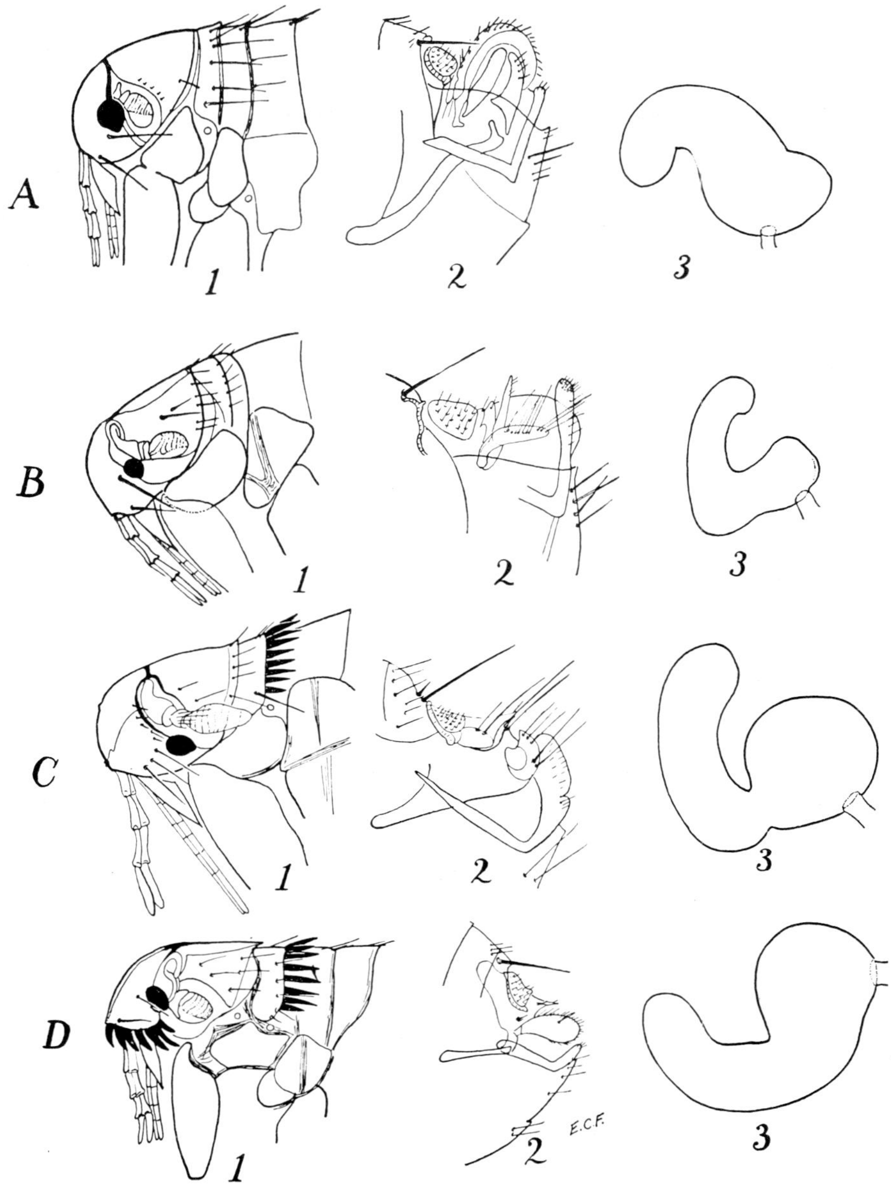

Fig. 43–6. Head *(1)*, caudal extremity of the male *(2)* and seminal receptacle *(3)* of fleas. *A*, *Pulex irritans*. *B*, *Xenopsylla cheopis*. *C*, *Nosopsyllus fasciatus*. *D*, *Ctenocephalides canis*. (Adapted by E.C. Faust.)

***Hoplopsyllus anomalus* (Baker, 1904).** This species, found on ground squirrels and hares in the southwestern United States, has a pronotal comb but no genal comb. The distal end of the antenna is somewhat enlarged and has segmental sutures extending less than halfway around it.

***Synosternus pallidus* (Taschenberg, 1880).** This species is rather closely related to the species of *Xenopsylla*. It is a predominant "house and body flea" in North Africa, the Near East, and the Middle East, extending to Afghanistan, and parts of Central Asia, south into the Indian peninsula. *S. pallidus* is usually parasitic on carnivores and larger insectivores.

FLEAS AND DISEASE

Fleas are harmful to humans in two ways: (1) They are hosts and transmitters of several disease agents, and (2) their bites produce an irritating dermatitis.

Fleas As Vectors

Plague ("Black death," peste)

Definition. Plague is an acute infectious disease that is sometimes contagious and, when untreated, often causes high mortality. On cutaneous inoculation the etiologic agent primarily produces inflammation of the lymphatics and localizes in the lymph glands *(bubonic type)* where it causes caseation necrosis, but secondarily involves the bloodstream *(septicemic type)* and the lungs *(pneumonic type)* with petechial and profuse hemorrhage. The extensive dark red mottling of the skin and mucous membranes in fatal cases is the basis of the popular designation of the disease, "black death."

Etiology. Plague is caused by *Yersinia pestis* (= *Y. pseudotuberculosis* spp. *pestis)*, a gram-negative, pleomorphic, bipolar staining bacillus, which is nonmotile, does not form spores, but at times produces a capsule. It was discovered by Yersin, in 1894, at the beginning of the extensive pandemic of the disease that developed in Hong Kong in that year.

History. Plague is a disease of antiquity, which in past centuries has swept through the human population in devastating epidemics or pandemics, such as the "black death" that visited Europe during the fourteenth and seventeenth centuries. The last pandemic, which supposedly originated in Yünnan Province, China, reached Hong Kong in 1894, Calcutta and Bombay in 1896. From India it was carried to almost every country throughout the world, principally on ships that harbored plague-infected rats. It reached the Western Hemisphere in 1899 (La Plata river ports in South America), by 1900 had arrived in San Francisco, by 1914 in New Orleans, and later other port cities of the United States. Knowledge of the etiology and transmission of plague enabled quarantine officials to set up effective barriers against its importation by ship, but enzootic plague in wild rodents has thus far resisted attempts at control and appears to be spreading.

Geographical Distribution. Plague is widely but unequally distributed throughout the world, typically occurring in desert and steppe zones. Heavy foci of human plague exist in Manchuria, parts of southeastern China, Thailand, Indonesia, Burma, India, Iran, East Africa, and Malagasy, while lesser foci have been reported from southern Africa. In Vietnam, where plague has been known to exist since 1898, it continues to be a major public health problem. Sporadic human cases have occurred in the Hawaiian Islands of Maui and Hawaii. Active foci of human plague are also present in northeastern Brazil, northern Argentina, Peru, and Ecuador, and a small area is reported for Venezuela. Enzootic foci in wild rodents exist on the northern border of China, in western Iran, on the eastern border of Zaire, in South Africa, eastern Brazil, northern Argentina, Bolivia, Peru, Ecuador, and Venezuela. One of the largest known areas of wild rodent plague is in the western United States and extends from the Pacific coast eastward into the Great Plains and into the southern part of the Canadian province of Alberta. During the 1970s, 105 cases occurred in eight western states, the largest number in the continental United States in any decade since 1900 to 1909 (Kaufman et al., 1980).

Epidemiology. In nature, plague is enzootic among wild rodents and is transmitted from infected to healthy rodents by fleas (Fig. 43–7). Probably the usual infection in semidomestic rats is in the brown rat *(Rattus norvegicus)*, less frequently the Alexandrine rat *(R. alexandrinus)* and other semidomestic species. When conditions become favorable for infection of the black rat *(R. rattus)*, which is more closely associated with human habitations, the disease can break out epidemically in the human population. The multimammate rat, *Mastomys natalensis*, also colonizes human dwellings and is an important reservoir in arid regions of East Africa (Akiev, 1982).

In the foci of wild rodent plague (rural or campestral plague), there are primary and secondary reservoirs (Harwood and James, 1979). The former are important in maintaining and spreading *Y. pestis*, sometimes to the human population that encroaches on the enzootic or epizootic areas of the disease. Rabbits also may act as passive carriers, as may apparently healthy dogs that have contact with infected rodents. Camels have been shown to be the source of some human infections.

There is no age immunity, although children are less commonly attacked by the disease because they are less often exposed.

***Flea Transmission of* Yersinia pestis.** The work of Simond in Bombay (1898), Bautier and Raybaud in Marseilles (1902), Liston in Bombay (1905), the Plague Investigation Commission in India (1907–1916), and many other groups of investigators produced cumulative evidence indicating the direct role of fleas, particularly on semidomestic and domestic rats, in the transmission of plague bacilli.

The flea usually feeds several times a day, becoming fully engorged either at one time or by interrupted feeding. After taking a blood meal from a plague-infected host, the flea may transfer the bacilli to an uninfected host in any of the following ways: (1) Plague bacilli in the intestine of the flea may multiply and be deposited in fecal dejecta in or near the puncture wound of the new host; (2) infected fleas may be crushed on the body, allowing the bacteria to gain entrance; or (3) multiplication of the bacilli in the midgut of the flea may produce a plug-like mass of blood and bacilli that "blocks" the chitinous teeth of the proventriculus. This renders the flea unable to fill its midgut with new blood, so that it wanders from one host to another, attempting to feed and contaminating the new host with bacilli. Fleas may attempt to feed on an infected host and then on an uninfected one, mechanically transferring the bacilli directly from one to the other. "Blocked" fleas are most commonly responsible for the transmission of plague bacilli from rodent to rodent, from rodent to human, and from human to human, while the remaining routes of transmission are least likely to produce infection (Bibikova, 1977).

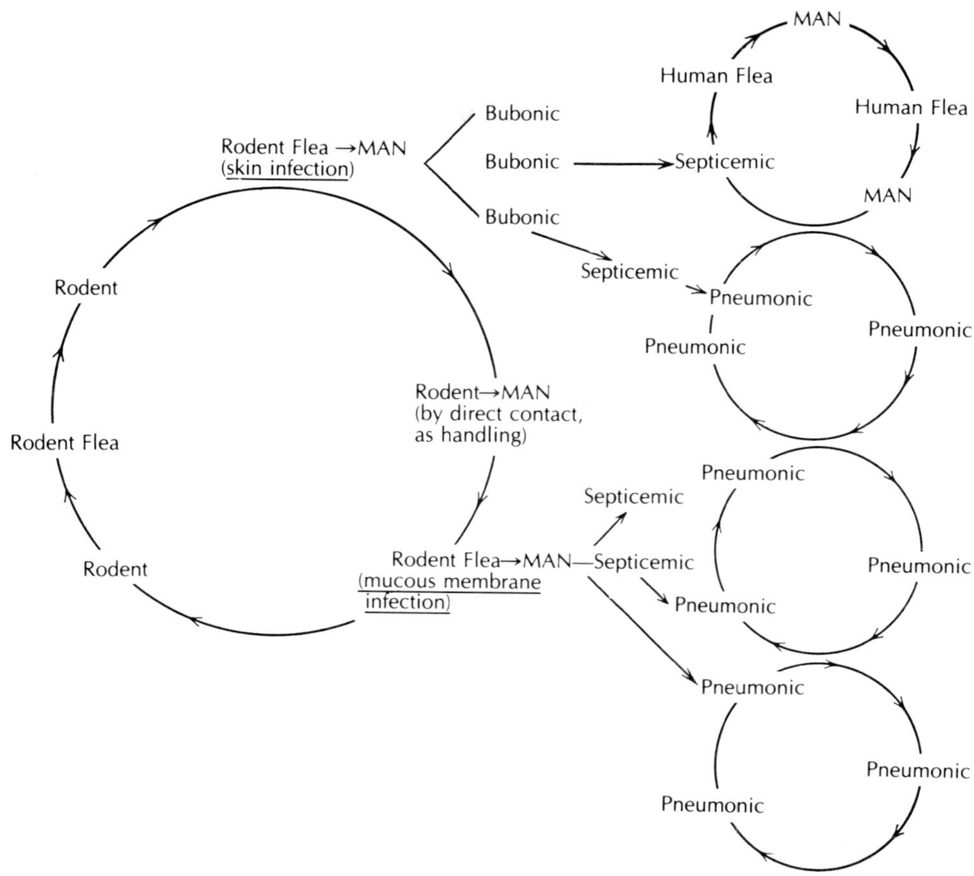

Fig. 43–7. Plague, transmission cycles. On left, *enzootic (campestral)* cycle; on right, potential *epidemic* cycles in man. (By E.C. Faust.)

The flea that "blocks" most readily in endemic areas is *Xenopsylla cheopis*, and this species is known to be the most common transmitter of *Y. pestis*. In this flea the bacilli multiply rapidly at warm temperatures up to 27.5°C, thus hastening the "blocking" process. Blocking in *X. cheopis* occurs as a result of the formation of fibrin from ingested blood induced by a *Y. pestis* coagulase and a trypsin-like enzyme produced by the flea (Cavanaugh, 1971). Fibrin accumulates in the proventriculus and serves as the lodgment site for the bacilli. However, at temperatures above 27.5°C, these bacteria produce a fibrinolytic factor that, with the trypsin-like flea enzyme, digests the fibrin plug so that *Y. pestis* is rapidly eliminated. This physiological process has been suggested as the means governing the decline of *X. cheopis*-associated plague epidemics that occur with the advent of hot weather. *Xenopsylla astia* and *X. brasiliensis* are relatively poor transmitters both naturally and experimentally, although in some areas they appear to be more commonly responsible than *X. cheopis*. *Nosopsyllus fasciatus*, a relatively good experimental transmitter, plays a minor role in epizootics of plague, probably owing to the relative freedom from plague of regions where this flea predominates, and the fact that its period of maximum seasonal prevalence in Europe falls outside the plague season. Likewise, *Pulex irritans*, *Ctenocephalides canis*, *C. felis*, and *Leptopsylla segnis*, with relatively cosmopolitan distributions, have not proven themselves to be important natural transmitters of plague. Other vectors worth noting include *Diamanus montanus* and *Hoplopsyllus anomalus*, which are commonly found on the California ground squirrel, and *Synosternus pallidus*, a West African species (Smit, 1973).

Enzootic Plague. Enzootic plague exists in domestic and semidomestic murine hosts (i.e., murine plague) and in wild rodents, especially in extensive nonurban areas (i.e., rural or campestral plague) (Fig. 43–7). In the United States, available evidence suggests that wild rodent enzootic plague became established in the ground squirrel and other nonurban rodents in California as a direct sequel to the San Francisco earthquake and fire of 1906. Infected rats were driven out of their harborages in the city, and soon thereafter ground squirrels in the nearby suburban and country districts were found suffering from the disease. The enzootic area spread north and south and then east into the Sierras. By 1935 it had reached the Rocky Mountains; by 1939 it had extended north and south along the Continental Divide all the way from Montana to southern New Mexico, and by 1943 to 1944 was extending eastward into the Great Plains areas of North Dakota, Montana, Wyoming, eastern Colorado, New Mexico, and Oklahoma.

Some of the rodents and other small mammals found infected in nature serve as relatively quiescent reservoirs of plague, whereas others disseminate the infection and are responsible for epizootic outbreaks. Ground squirrels have been primarily responsible for the rapid spread of the disease, because they are widely distributed, the ranges of different species overlap, and they tend to congregate in large numbers when conditions are favorable.

Diagnosis. A diagnosis is easily made in an epidemic, although the first human cases may be overlooked or misdiagnosed. The plague bubo is a low, rounded mound, as distinguished from the more elevated venereal bubo of lymphopathia venereum. Identification by culture of the etiologic agent obtained from the contents of the bubo, from blood, or from sputum is conclusive evidence. In culture, *Yersinia pestis* produces delicate dewdrop transparent colonies on agar within 24 to 48 hours. Penicillin, which is not specific for the plague organism, controls contaminants when it is added to the culture medium. Inoculation of a guinea pig with material containing *Y. pestis* provides valuable confirmation. Infected rodents or pooled specimens of fleas yield evidence of the existence of enzootic plague. Goldenberg (1968) has provided a review of the methods used in diagnosis of plague in the laboratory of the federal organization responsible for surveillance and control of plague in the United States.

In Brazil, survey results indicated that retrospective diagnoses of human bubonic plague could be obtained through the hemagglutination test, in a microtechnique method (Thiago de Mello, 1967). Positive results were obtained even as late as a year after manifestation of the disease.

Treatment. Streptomycin is the drug of choice, given intramuscularly 0.5 g every 3 or 4 hours until temperature is normal, thereafter 1.0 g daily in divided doses up to a total of 15.0 g. Early cases of bubonic plague can be effectively treated with sulfadiazine or sulfamerazine, in a total dosage of 10 g the first week. An initial intravenous dose of 2 g in glucose solution is recommended when streptomycin is not available. The tetracycline drugs are also effective in reducing the symptoms.

Good nursing and symptomatic treatment are important. Myocardial insufficiency may persist for several weeks after the acute illness.

Pollitzer (1949) indicated that the sulfa drugs, especially sulfamerazine, taken prophylactically by contacts in a dosage of 3 g daily over a period of 6 days provide considerable promise that pneumonic plague will not be contracted.

Prognosis. Today, streptomycin and the tetracycline drugs have lowered the fatality rate to 5% or less in bubonic and 10% in septicemic plague.

Prevention. ***Quarantine.*** Following the pandemic of 1894 to 1905, based on the knowledge that rats with their rat fleas were the common source of bubonic plague in the human population, international quarantine regulations were established that require ships from a plague-infected port to be kept in quarantine at ports of entry until danger of introduction of the disease is past. The ship is thoroughly fumigated, usually with cyanide gas (KCN, ½ ounce per 100 cubic feet, for 4 hours) to kill all rats on board. Likewise, when a ship docks, circular rat guards of galvanized iron are set up immediately on all ropes that tie the ship to the wharf, to prevent rats from passing down or up the ropes. These measures have been largely responsible for the insignificant number of cases imported from one country to another by ships.

Rat Control. See Chapter 45, p. 727.

Insecticides. Relatively simple control measures that have proved to be eminently satisfactory and safe in limiting the spread of plague have been the application of organochlorine (1% *gamma* HCH [lindane]) or 1% malathion to rat nests and burrows to kill their fleas. A substitute for these that does not present environmental hazards is carbaryl (Sevin). This insecticide is used as a dust or spray (see Chapter 45, p. 721). The insecticide campaign *must* precede the rodenticide to prevent the fleas that leave the dying rats from attacking human victims.

Wild Rodent Plague. Measures that guard against

introduction of plague by rats are not usually applicable to plague in wild rodents. Wherever people come in contact with the latter or become infested with plague-infected ectoparasites of infected wild rodents, they may acquire the disease. Thus, there is an apparent increase in plague in areas in the western United States where plague-infected rodents have been discovered. This extensive enzootic focus, and similar foci in South America, Africa, and Asia, constitute constant danger zones, from which the disease may reach domestic rodents and flare up epidemically in the human population. With relatively good international quarantine routinely established, the infection in wild rodents constitutes the major public health concern.

Immunization. Two types of vaccine are employed: (1) killed, highly virulent cultures of *Yersinia pestis,* originally prepared by Haffkine, and (2) relatively avirulent live cultures, originally employed by Otten in Indonesia, then by workers in Malagasy and South Africa. There are no reliable statistical data indicating the value of vaccine prophylaxis, or the relative efficiencies of the two types. There is, however, considerable evidence (1) that both are useful, provided immunization is begun at least 5 days before exposure and is repeated at least every 6 months, and (2) that vaccination under these conditions may not prevent the individual from acquiring plague on exposure but will greatly reduce his chances of developing fatal infection.

Murine Typhus Fever
(Endemic typhus, sporadic typhus, tabardillo of Mexico *pro parte)*

Definition and Etiology. Endemic or murine typhus is an acute, infectious, febrile disease of relatively mild form and low mortality. It is caused by a rickettsia, *Rickettsia typhi* (formerly *R. mooseri),* that is morphologically indistinguishable from *R. prowazekii,* of louse-borne typhus fever. it is also known as Toulon typhus, shop or urban typhus, and sporadic typhus.

Geographical Distribution. During earlier decades in the United States, endemic typhus fever extended along the Atlantic Coast from New York to Florida, with increased incidence from north to south, and a preponderance of cases south of a line extending westward from the southern border of Virginia. It was prevalent along the Gulf Coast and inland to Louisville, Memphis, and St. Louis, and had been diagnosed in Massachusetts, Connecticut, Ohio, Michigan, Illinois, Wisconsin, Kansas, Oklahoma, Montana, Arizona, and California (mainly in Los Angeles and San Diego Counties). States with especially high incidence were Georgia, Texas, and Alabama.

The disease is well established in Mexico and sometimes intergrades imperceptibly into epidemic louse-borne typhus fever. The etiologic agent in rats and humans has been identified in Peru, Colombia, Venezuela, Chile, and northern Argentina; southern and western Europe, the U.S.S.R.; North, West, and South Africa; Israel; Sumatra, Indochina, Thailand, central and north China, Manchuria, and Hawaii; and probably in other geographical regions (Weyer, 1978). In many areas endemic typhus fever has not been differentiated from epidemic typhus.

Epidemiology. The epidemiology of this zoonosis was first elucidated by Maxcy (1926, 1929) and amplified by Dyer et al. (1931) and their associates in the U.S. Public Health Service. Commensal rats *(Rattus* spp.) serve as the common reservoir of the infection, which is transmitted from rat to rat by rat fleas, commonly the Oriental rat flea, *Xenopsylla cheopis,* and probably less frequently by *X. astia, Leptopsylla segnis,* and *Nosopsyllus fasciatus* (Traub et al., 1978). The rat louse, *Polyplax spinulosa,* has been suggested as a natural transmitter from rat to rat, but it does not bite humans. Experimentally, the rat mite, *Ornithonyssus bacoti,* was found to be a transmitter, but natural infections of this arthropod have not been found. Although the rickettsia of murine typhus is not common in species of wild mammals and birds, it has been reported in the opossum, cotton mouse, eastern cottontail rabbit, southern fox squirrel, house mouse, old field mouse, swamp rice rat, Florida skunk, southern weasel, dog, and blue jay *(Cyanocitta cristata).*

When the rat flea takes a blood meal from an infected rat during the stage of rickettsemia, on arrival in the flea's midgut the microorganisms enter some of the epithelial cells of the wall and multiply enormously (Fig. 43–8). In 10 to 12 days or less the rickettsiae escape from the intestinal wall into the lumen of the flea's gut. The rat flea shows no evidence of disease due to *R. typhi* infection (Ito et al., 1975). The pathogens are discharged in an infective stage in the flea's feces. This is an important epidemiological consideration since, in suitable circumstances, dried feces may contain *R. typhi* that remain infective for 5 years or more (Traub et al., 1978). Routes of infection may be respiratory or contaminative by way of the mucous membranes. Rubbing feces or crushed bodies of infected fleas into the puncture wound made by the flea constitutes another route of infection, since inoculation does not occur by the proboscis route. There is also some likelihood that accidental ingestion of infected fleas, fluids from their crushed bodies, or their feces serves as a means of infection. Clothing soiled with feces of infected fleas may be another source for dispersal of the rickettsiae.

Murine typhus is typically sporadic in the human population and occurs mostly during summer and

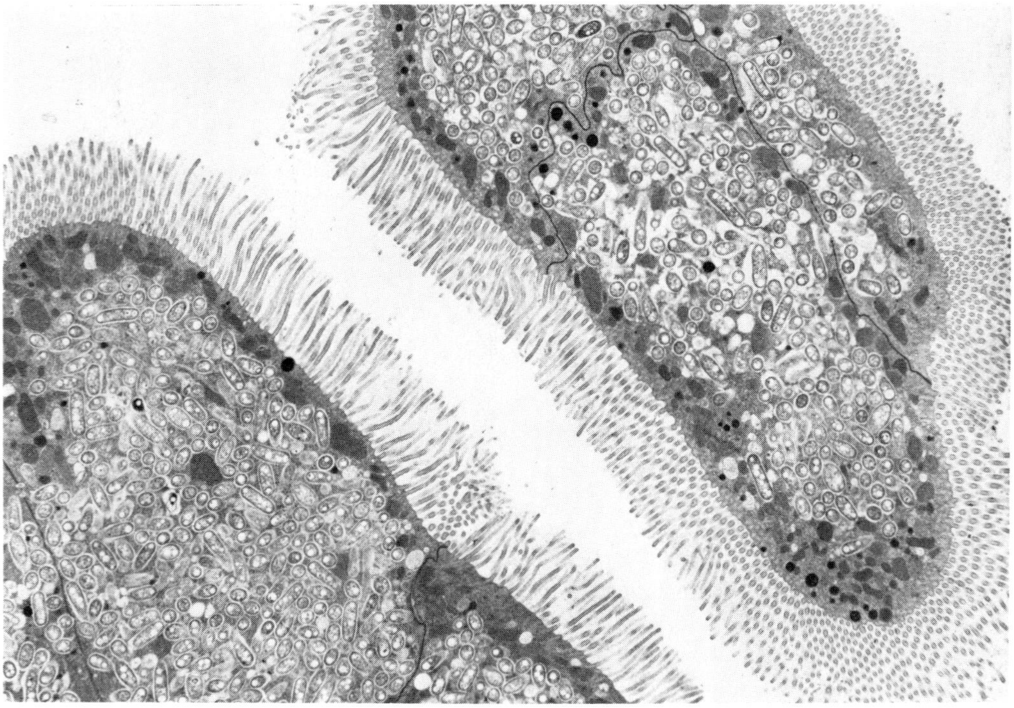

Fig. 43–8. *Rickettsia typhi,* massive numbers in section of midgut epithelial cells of *Xenopsylla cheopis.* (× ca. 4600.) (From Ito, S., Vinson, J.W., and McGuire, T.J., Jr. 1975. Murine typhus rickettsiae in the Oriental rat flea. Ann. N.Y. Acad. Sci., *266*:35–60.)

early fall. It is common among workers on docks, in grain elevators, or in provision houses, or wherever people work in rat-infested areas. Most cases occur in male subjects. In contrast, epidemic louse-borne typhus occurs in the louse-infected poverty-stricken, or other groups crowded together without opportunity for personal or group hygiene.

Clinical Aspects. The sequence of events in the development of murine typhus in man is similar to that in epidemic typhus fever. The lesions are similar, but the process is not so intense and the damage to the smaller blood vessels is usually less profound (see Clinical aspects of louse-borne typhus fever, p. 615). The syndrome is usually milder than in epidemic typhus, the fever may rise gradually rather than suddenly, and the exanthem may be slight or essentially inapparent. Mental symptoms are less severe and delirium may be wholly lacking. Specific diagnosis of typhus fever may be made after the sixth day of the disease by the Weil-Felix reaction. Typhus fever is differentiated from epidemic typhus by the complement-fixation test. Treatment is similar to that of louse-borne typhus. In mild cases only 3 or 4 weeks' rest in bed with usual nursing care may be needed. With appropriate treatment, the prognosis is usually good. The mortality rate in the United States has been less than 1%, and deaths occur mostly in persons over 50 years of age.

Prevention. Energetic campaigns to ratproof public and private buildings in endemic foci constitute an important attack on the problem of control. Associated with this must be the poisoning and trapping of domestic rats. In these respects measures to prevent endemic typhus are also guarantees against plague. The use of residual insecticides to kill rat fleas and other arthropod ectoparasites in enzootic areas is also important as a complementary means.

Fleas as Vectors of Other Pathogens

Ctenocephalides canis, C. felis, and *P. irritans* are the intermediate hosts of the dog tapeworm, *Dipylidium caninum* (see p. 508); *C. canis, P. irritans, N. fasciatus, X. cheopis, Leptopsylla segnis,* and other murine fleas are intermediate hosts of *Hymenolepis nana* and *H. diminuta.* A squirrel flea, *Diamanus montanus,* and rabbit flea have been experimentally incriminated as transmitters of tularemia, but there is no evidence that humans contract the disease through these fleas. *Orchopeas howardii,* a squirrel flea, has been found to be naturally infected with a rickettsial agent identical to *Rickettsia prowazekii* (Sonenshine et al., 1978). Since it occasionally bites humans, this flea may be a peripheral vector in what could be a hidden, sylvatic cycle of epidemic typhus. Eskey et al. in 1949 demonstrated transmission of *Salmonella enteritidis* by the fleas *X. cheopis* and *N. fasciatus,* via fecal contamination, and in 1951 they found both of these fleas doubly infected in nature with *Salmonella typhimurium* and *Yersinia pestis.*

Flea Dermatitis

The "bites" of fleas are practically innocuous to many persons, but cause considerable discomfort to others, on whose skin a roseate, raised lesion, frequently edematous, and at times indurated or even pustular, develops. Flea bites are typically most abundant on the lower extremities. Local application of carbolated petrolatum is usually palliative. Flea dermatitis is a worrisome problem for the family physician and the dermatologist, and the following repellents have been recommended; benzyl benzoate (impregnated in clothing), dimethyl phthalate, and undecenoic acid (2% with 1% pyrethrum).

In the case of the chigoe, *Tunga penetrans,* which invades the epidermis of humans, pigs, dogs, and other mammals, both males and females suck blood, but only the gravid female burrows into the skin and engorges until she frequently becomes as large as a pea (Fig. 43–5). The sites usually attacked are the soles of the feet and the interdigital spaces, but no area of the body is immune. In the Tropics of the Western Hemisphere and Central Africa, barefoot natives may be infested with dozens of these cutaneous parasites, each of which produces a discrete, festering sore that usually becomes secondarily infected and may lead to extensive ulceration, abscesses, or even gangrene. Human infestation is usually acquired from contact with soil or sand contaminated by pigs or dogs, from whose feet the eggs drop to the ground, where the larvae, pupae, and unfed adults develop. Whenever a person becomes infested with chigoes, the flea should be removed aseptically with a needle, and the wound bathed, treated with a topical antiseptic, and dressed. Wearing shoes prevents most infestations.

CONTROL OF FLEAS AND FLEA-BORNE DISEASES

Flea control, as far as human diseases are concerned, may be divided into two categories: (1) *fleas of dogs, cats, and humans* around the home, and (2) *fleas of rats and other rodents.*

Dog Fleas. This problem requires the disinfestation of dogs and cats, by dusting them periodically with rotenone (1%), malathion (4%), or pyrethrins (1%). Floors, rugs, carpets, pillows, and other domestic articles with which the dogs are associated and in which fleas may breed and lurk should be sprayed with kerosene solutions or emulsions of 3% malathion or 1% diazinon.

Rodent Fleas. Plague and murine typhus are primarily associated with rodent fleas. Hence control of these diseases requires anti-rodent flea measures. Many countries exercise strict quarantine against ships arriving from foreign ports, with fumigation of the entire ship to kill rats. In California and other western states, local authorities, in conjunction with the U.S. Public Health Service, quarantine foothill areas where ground squirrels and other nondomestic rodents are infected with plague. In Mongolia, human plague epidemics in years past have been initiated by fur trappers who approach or handle sick tarbagans (marmots), and from these contacts the disease has been carried to nearby villages in Manchuria or North China. Establishing quarantine barriers for fur hunters has usually enabled epidemics of plague to be forestalled. The appearance of murine typhus sporadically, rather than in epidemic form, provides a more difficult control problem for public health authorities. It involves the fumigation of warehouses, ratproofing of granaries, starvation of rats wherever possible by keeping garbage as well as clean food in covered or ratproof receptacles, and continuous efforts to trap rats in stores or markets where food is conserved or sold.

Public health officers in the southern United States, where murine typhus was formerly a serious clinical and public health problem, demonstrated the value of several kinds of organochlorines in killing rodent fleas by dusting or blowing the powder or an emulsion into rat holes and runs. A complete program includes bleeding trapped rats and testing the serum for antibodies to typhus. Detection of such antibodies is an indication for intensification of the dusting or spraying for fleas. Organochlorine insecticide applications are equally valuable in protecting people during epizootics of plague. However, when possible, residual insecticides other than organochlorines should be substituted (Gratz, 1980).

REFERENCES

Akiev, A.K. 1982. Epidemiology and incidence of plague in the world, 1958-1979. Bull. W.H.O., *60*:165–169.

Bibikova, V.A. 1977. Contemporary views on the interrelationships between fleas and the pathogens of human and animal diseases. Annu. Rev. Entomol., *22*:23–32.

Cavanaugh, D.C. 1971. Specific effect of temperature upon transmission of the plague bacillus by the Oriental rat flea, *Xenopsylla cheopis*. Am. J. Trop. Med. Hyg., *20*:264–273.

Connor, D.H. 1976. Tungiasis, pp. 610–614. In *Pathology of Tropical and Extraordinary Diseases*. Vol. 2. Edited by C.H. Binford, and D.H. Connor. Washington, D.C, Armed Forces Institute of Pathology.

Dyer, R.E., Rumreich, A., and Badger, L.F. 1931. Typhus fever: A virus of the typhus type derived from fleas collected from wild rats. Public Health Rep., *46*:334–338.

Eskey, C.R. 1938. Recent developments in our knowledge of plague transmission. Public Health Rep., *53*:49–57.

———, Prince, F.M., and Fuller, F.B. 1949. Transmission of *Salmonella enteritidis* by the rat fleas *Xenopsylla cheopis* and *Nosopsyllus fasciatus*. Public Health Rep., *64*:933–941.

———, Prince, F.M., and Fuller, F.B. 1951. Double infection of the rat fleas *X. cheopis* and *N. fasciatus* with *Pasteurella* and *Salmonella*. Public Health Rep., *66*:1318–1326.

Goldenberg, M.I. 1968. Laboratory diagnosis of plague infection. Health Lab. Sci., *5*:38–45.

Gratz, N.B. 1980. Problems and developments in the control of flea vectors of disease, pp. 217–240. In *Fleas—Proceedings of the International Conference on Fleas* (1977). Edited by R. Traub and H. Starcke. Rotterdam, A.A. Balkema.

Harwood, R.F., and James, M.T. 1979. *Entomology in Human and Animal Health,* 7th ed. New York, Macmillan Publishing Co.

Holland, G.P. 1964. Evolution, classification and host relationships of Siphonaptera. Annu. Rev. Entomol., *9*:123–146.

Hopkins, G.H.E., and Rothschild, M. 1953, 1956, 1962, 1966, 1971. *An Illustrated Catalogue of the Rothschild Collection of Fleas (Siphonaptera) in the British Museum (Natural History)*. Vols. I–V. London, British Museum (Natural History).

Ito, S., Vinson, J.W., and McGuire, T.J. Jr. 1975. Murine typhus rickettsiae in the Oriental rat flea. Ann. N.Y. Acad. Sci., *266*:35–60.

Kaufman, A.F., Boyce, J.M., and Martone, W.J. 1980. Trends in human plague in the United States. J. Infect. Dis., *141*:522–524.

Lewis, R.E. 1972. Notes on the geographical distribution and host preferences in the Order Siphonaptera. Part 1. Pulicidae. J. Med. Entomol., *9*:511–520.

———. 1973. Notes on the geographical distribution and host preferences in the Order Siphonaptera. Part 2. Rhopalopsyllidae, Malacopsyllidae, and Vermipsyllidae. J. Med. Entomol., *10*:255–260.

———. 1974. Notes on the geographical distribution and host preferences in the Order Siphonaptera. Part 3. Hystrichopsyllidae. J. Med. Entomol., *11*:147–167.

———. 1974a. Notes on the geographical distribution and host preferences in the Order Siphonaptera. Part 4. Coptopsyllidae, Pygiopsyllidae, Stephanocircidae and Xiphiopsyllidae. J. Med. Entomol., *11*:403–413.

———. 1974b. Notes on the geographical distribution and host preferences in the Order Siphonaptera. Part 5. Ancistropsyllidae, Chimaeropsyllidae, Ischnopsyllidae, Leptopsyllidae and Macropsyllidae. J. Med. Entomol., *11*:525–540.

———. 1975. Notes on the geographical distribution and host preferences in the Order Siphonaptera. Part 6, Ceratophyllidae. J. Med. Entomol., *11*:658–676.

Marshall, J.D., Jr., Quy, D.V., and Gibson, F.L. 1967. Symptomatic pharyngeal plague infection in Vietnam. Am. J. Trop. Med. Hyg., *16*:175–177.

Maxcy, K.F. 1926. Clinical observations on endemic typhus (Brill's disease) in southern United States. Public Health Rep., *41*:1213–1220.

———. 1929. Typhus fever in United States. Public Health Rep., *44*:1735–1742.

Munshi, D.M. 1960. Micro-anatomy of the proventriculus of the common rat flea *Xenopsylla cheopis* (Rothschild). J. Parasitol., *46*:362–372.

Pham Trong, Tran Quy Nhu, and Marshall, J.D. Jr. 1967. A mixed pneumonic bubonic plague outbreak in Vietnam. Milit. Med., *132*:93–97.

Pollitzer, R. 1949. Recent trends in the treatment and control of plague. Acta Trop., *6*:30–40.

Rothschild, M. 1975. Recent advances in our knowledge of the Order Siphonaptera. Annu. Rev. Entomol., *20*:241–259.

Smit, R.G.A.M. 1973. Siphonaptera (Fleas), pp. 325–373. In *Insects and Other Arthropods of Medical Importance*. Edited by K.G.V. Smith. London, British Museum (Natural History).

Sonenshine, D.E., Bozeman, F.M., Williams, M.S., Masiello, S.A., Chadwick, D.P., Stocks, N.I., Lauer, D.M., and Elisberg, B.L. 1978. Epizootiology of epidemic typhus *(Rickettsia prowazekii)* in flying squirrels. Am. J. Trop. Med. Hyg., *27*:339–349.

Thiago de Mello, M. 1967. La prueba de hemagglutinación para el diagnóstico de la peste. Bol. Of. Sanit. Panam., *62*:151–156.

Traub, R. 1980. The zoogeography and evolution of some fleas, lice and mammals, pp. 93–172. In *Fleas—Proceedings of the International Conference on Fleas. (1977)*. Edited by R. Traub, and H. Starcke. Rotterdam, A.A. Balkema.

———. 1980a. Some adaptive modifications in fleas, pp. 33–67. In *Fleas—Proceedings of the International Conference on Fleas (1977)*. Edited by R. Traub, and H. Starcke. Rotterdam, A.A. Balkema.

———, Wisseman, C.L., Jr., and Farhang-Azad, A. 1978. The ecology of murine typhus—a critical review. Trop. Dis. Bull., *75*:237–317.

Weyer, F. 1978. Progresses in ecology and epidemiology of rickettsioses. Acta Trop., *35*:5–21.

Chapter 44

Insects Causing Vesication, Urtication, and Envenomation

In addition to allergic and toxic manifestations produced by bites of bloodsucking flies, fleas, lice, and bugs, certain species of beetles, moths, bees, wasps, and ants cause a variety of local and systemic reactions in the human body when they introduce their secretions or body fluids onto or into the tissues (Levine and Lockey, 1981).

BEETLES (ORDER COLEOPTERA)

Definition. Beetles (Coleoptera), literally insects with "sheathed wings," typically have two pairs of wings, of which the members of the first pair are thickened to form sheathed covers, or *elytra,* that characteristically come together in a median dorsal line over all or part of the abdomen. The mouthparts are adapted for chewing. Metamorphosis is complete.

Beetles as Hosts of Human Parasites. Several species of beetles serve as intermediate hosts of helminths that occasionally infect humans (i.e., *Gongylonema pulchrum,* p. 345; *Macracanthorhynchus hirudinaceus,* p. 545; *Moniliformis moniliformis,* p. 546; and *Hymenolepis nana* and *H. diminuta,* p. 509).

Beetle Infestation (Canthariasis). Several cases of *canthariasis (scarabiasis),* or temporary infestation of the digestive and urinary tracts of humans with larval or adult beetles, have been reported from widely scattered localities. Other examples include nasal canthariasis due to oviposition of eggs within the human nares. Larval beetles have also been recovered from the human eye (Smith, 1973). Intestinal infestation with the dung beetle, *Onthophagus unifasciatus,* caused enteritis in a 3-year-old child in Calcutta (Bhattacharya, 1966). More often, certain species of adult beetles produce blisters on the human skin or cause severe irritation of the intestinal tract when accidentally swallowed. These vesications are due to body fluids that contain toxic substances.

Vesicating Beetles. Blister beetles (family Meloidae) (Fig. 44–1,*A*) contain cantharidin, a volatile terpene found in the hemolymph and at times amounting to about 2.5% of the dry weight of the beetle. It has a pungent odor and a weakly acid taste, and is unjustifiably used as a powerful rubefacient, less commonly in small amounts as a diuretic. Commercially, it is obtained from the "Spanish fly," *Lytta* (syn. *Cantharis) vesicatoria,* and as a powder has frequently been used as an aphrodisiac.

The lesions accidentally produced by crushing blister beetles on the skin and discharging their body fluids, or by external therapeutic use of cantharidin, consist of slowly forming blisters that involve only the outer layers of the skin (Frazier and Brown, 1980). Toxic doses of cantharidin taken internally give rise to severe gastrointestinal irritation, with nausea, vomiting, diarrhea, severe cramps, and occasionally collapse. In Southeast Africa, species of paussid beetles (suborder Adephaga) enter human habitations, attracted by night light, and on contact with the skin secrete a highly acid liquid that produces a painful blister. Additional reports on the annoying blisters produced by these beetles have come from Europe, the U.S.S.R., Gambia, Kenya, Zaire, Malaysia, Indonesia, Guatemala, Ecuador, and particularly Brazil.

Members of the genus *Paederus* (Fig. 44–1,*B),* of the worldwide family of rove beetles (Staphylinidae), have an irritating alkaloid *(pederin)* in

Fig. 44–1. Vesicating beetles. *A, Lytta vesicatoria* ("Spanish fly"). *B, Paederus* sp. (rove beetle). (Photographs by Howard Lyon, Cornell University.)

their body fluids. It causes a necrotizing lesion of human skin, with chronic desquamation, which heals slowly. This substance is present in all parts of the beetle's body. Rove beetles are most abundant during spring and early summer and cause most annoyance when the air is humid, as after heavy rains. Their lesion develops primarily on the most exposed parts of the skin, such as the face, neck, and hands. If the fluid gets into the eyes, there is commonly intense pain and swelling ("Nairobi eye"). Discharge of the fluid by the beetle is entirely involuntary and occurs when the insect is crushed. Major outbreaks of vesicular dermatitis caused by contact with rove beetles have been reported from Israel, southern Africa, and Uganda (Harwood and James, 1979).

The toxic hemolymph of the larva of the chrysomelid beetle, *Diamphidia simplex,* consists of a powerful hemolytic agent. South African Bushmen use this substance to poison the tips of their arrowheads. Other vesicating Coleoptera include several species of *Sessinia* (coconut beetles) and *Oxacis,* both of which cause blistering or dermatitis (Harwood and James, 1979).

Calamine lotion, applied locally to the cutaneous lesions, is palliative. Persons with sensitive skin should be warned not to handle or crush beetles. If the beetles get on the skin they may be safely blown off. Well-screened quarters prevent the beetles from getting on the body at night and prevent accidental crushing.

Allergy to Beetles. Dermestid beetle larvae may at times produce asthma in persons closely associated with them.

Beetles as Vectors of Pathogens. Many species of beetles that frequent human fecal deposits on the soil rapidly break up these deposits, scattering the eggs of *Ascaris,* whipworms, and hookworms over a considerable radius.

MOTHS AND BUTTERFLIES (ORDER LEPIDOPTERA)

Definition. Moths and butterflies (Lepidoptera), literally insects with "scaly wings," typically have two pairs of membranous wings covered with overlapping scales, and mouthparts adapted for sucking by virtue of a sharp modification of the maxillae into a coiled, extensible tube. The Lepidoptera exhibit complete metamorphosis. The larvae ("caterpillars") of most species have chewing mouthparts and are phytophagous; other larval forms feed on stored grain or meal or on woolen or silk cloth. Taxonomic subdivisions have been made based on characteristics of the wing (suborders Jugatae and Frenatae), antennae (suborders Rhopalocera—the butterflies and skippers, and Heterocera—the moths), or a combination of adult and larval characters (suborders Zeugloptera, Dacnonypha, Monotrysia, Ditrysia) (Common, 1975).

Lepidopteran Dermatitis

Etiology and Epidemiology. There are records of adult moths belonging to the genus *Hylesia* Hübner (Saturniidae) causing an urticating dermatitis ("lepidopterism"). The mucous membranes of the eye, the upper respiratory system, and the oropharynx can also be affected (Delgado Quiroz, 1978). Nettling is due to poison-containing larval hairs that cling to the adults as they emerge from the cocoon. The moths seldom come in direct contact with humans but are attracted to lights, flying about indoors and shedding the urticating setae in great clouds. This lepidopteran dermatitis has been recorded only in South America, paralleling the geographic distribution of *Hylesia*. Another instance of lepidopterism includes urtication associated with the Douglas fir tussock moth, *Orgyia pseudotsugata* (Perlman et al., 1976). Forest workers are at risk when they come in contact with egg masses (and larvae) that have been covered with body hair from the female.

Two families of butterflies (Nymphalidae; Morphoidae) and at least 11 families of moths have larvae equipped with specialized poison hairs or spines used in protection from enemies (Delgado Quiroz, 1978; Harwood and James, 1979) (Table 44–1). When these hairs enter the skin or the spines penetrate the epidermis, they produce urtication or vesication, depending on the species of caterpillar, the susceptibility of the victim, and the area of skin affected. Dermatitis and the general symptomatology associated with contact with larval Lepidoptera has been designated *erucism* (from the Latin *eruca* or caterpillar) (Pesce and Delgado Quiroz, 1966).

There are two types of poison hairs of caterpillars: a *primitive* type and a distinctly *modified* type. In the former (Fig. 44–2,*A*), a hair (or a group of several to many distinct hairs) is supplied with venom produced by a single glandular hypodermal cell at the base. In the second type (Fig. 44–2,*B*), there is a true spine, frequently heavily chitinized, either lined or filled with several hypodermal cells that elaborate the toxic substance. These true spines are tipped with a blunt or pointed process and may be provided with branched poison spines. In some caterpillars the poison hairs or spines may be isolated into tufts or may be borne on tubercles or plates.

Poison hairs used in building cocoons for the pupal stage, those scattered in nests of the cocoons, and those blown by the air and reaching the skin or mucous membranes of humans may have vesicating properties for several months. Hairs that have apparently been blown from nests in trees into drinking water have caused serious inflammation of the buccal mucosa. In other instances poison hairs have been inhaled into the upper respiratory tract.

Little is known of the chemical composition of erucic toxins. Histamine and several general kinds of enzymes have been identified, including proteases and hyaluronidases (Schmidt, 1982).

Pathogenesis and Symptomatology. In addition to the burning, stinging sensation experienced by the victim immediately after introduction of the poison substance into the skin, the affected site usually becomes erythematous (Fig. 44–3,*A*). Later it may become elevated and whitish, with a reddened area up to 5 cm in diameter and a peripheral zone of reddish macules 2 cm or beyond. The entire member of the body may become swollen; urticarial wheals may appear at various sites on the body; and serious systemic manifestations of toxemia may develop, including severe muscle cramps, intense headache, nervousness, and tachycardia. When the poison hairs come in contact with the conjunctivae, cornea, or iris, excruciatingly painful local lesions may be produced, with the development of pseudotubercles around the hairs. Bishop and Morton (1968) in Australia, described 132 cases of keratoconjunctivitis caused by hairs from caterpillars of the genus *Anthela*. The hairs became embedded in the conjunctiva, causing mechanical irritation of the cornea and a conjunctival reaction

Table 44–1. Some of the More Important Urticating Caterpillars, Their Poison Hairs, and Geographical Distribution*

Family of Lepidoptera and Type of Poison "Hairs"	*Common Name*	*Scientific Name*	*Geographical Distribution*
NYMPHALIDAE (4-footed butterflies)			
Long, heavily chitinized spines, having short, stubby terminal and branched spines, with tips set off by a chitinous ring	Mourning cloak	*Nymphalis antiopa*	North America, Europe
	Peacock's mirror	*Nymphalis io*	Europe
		Probably all other species with stiff spines	
MORPHOIDAE	Morpho butterfly	*Morpho hercules*	South America
MEGALOPYGIDAE (flannel moths)		*Aidos amanda*	Brazil
		Carama spp.	Brazil
Spine type, having spine short, lined with hypodermal cells; spines arranged radially on elevated ridges	White flannel moth	*Lagoa crispata*	Eastern and Southern United States
	Flannel moth	*Lagoa pyxidifera*	Southern United States
	Puss caterpillar	*Megalopyge opercularis*	Southern United States
	Tata-rana ("fire-body")	*Megalopyge* spp.	Upper Amazon (Brazil)
		M. krugi	Puerto Rico
		M. urens	Uruguay, Argentina
		M. lanata	Tropical America
		Norape crenata	United States
LIMACODIDAE (Syn. *EUCLEIDAE*) (slug-caterpillar moths)			
Spine type, having shaft of spines lined with hypodermal cells	Saddle-back caterpillar	*Sibine stimulea*	United States
		Parasa chloris	Eastern United States
		P. indetermina	Eastern United States
		P. hilarata	China, Korea
	Hag moth	*Phobetron pithecium*	Eastern United States
		Adoneta spinuloides	Eastern United States
		Sisyrosea textula	
NOTODONTIDAE (Syn. *THAUMETOPOEIDAE*) (processionary caterpillars)			
Modified barbed hairs	Oak processionary	*Anaphe panda*	Europe
		Thaumetopoea processionea	Europe
	Pine processionary	*T. pinivora*	Europe, Africa
		T. pityocampa	
LYMANTRIIDAE (Syn. *LIPARIDAE*) (tussock moths)			
Modified, single hair type in *Euproctis*, with setose hairs in groups, arising from cuticular cups; primitive hair type in *Hemerocampa* and *Psilura*, with hairs borne in dense groups on tussocks or tubercles	Gypsy moth	*Lymantria dispar*	Europe, United States
	Brown-tail moth	*Euproctis chrysorrhoea*	Europe, United States
		Calliteara pudibunda	Europe
	Golden-tail moth	*Euproctis similis*	United States
		E. similis	Canada, Europe
	Chinese tussock moth	*E. flava*	China, Japan
	White-marked tussock moth	*Orygia leucostigma*	North America
		Leucoma salicis	
	Satin moth	*Lymantria monacha*	North America, Europe
	Nun moth		Europe

Table 44–1. *Continued*

ARCTIIDAE (tiger moths)			
Primitive hair type, with hairs borne on dorsal tufts	Hickory tiger moth	*Halysidota caryae*	North America
	Harlequin milkweed caterpillar	*Euchaetes egle*	North America
		Arctia caja	Europe
		Lithosia caniola	Europe
		Lithosia griseola	Europe
NOCTUIDAE (owlet moths)			
Primitive hair type in *Apatela;* hairs scattered among dorsal hair tufts; slender, sharp spine in *Catocala,* breaking off in skin	American dagger	*Acronicta lepusculina*	United States
		Acronicta oblinita	United States
		Catocala sp	
SATURNIIDAE (Syn. *HEMILEUCIDAE)* (giant silkworms)			
Seta-bearing and pointed spines lined with hypodermal cells; spines distributed on dorsal and lateral tubercles; pointed spines function as a hypodermic syringe		*Hylesia irritans*	Fr. Guiana
		H. nigricans	Argentina, Patagonia
		H. volex	Peru
	Io moth	*Automeris io*	United States
	Maia moth	*Hemileuca maia*	Eastern and Central United States
	Range caterpillar	*H. oliviae*	Southwestern United States
		H. nevadensis	Western and Southwestern United States
		H. lucina	Southwestern United States
		Pseudohazis hera	Western United States
		Pseudohazis eglanterina	Western United States
		Coloradia pandora	Western United States

*Based on Herms and James, 1961. In addition, Harwood and James (1979) list Bombycidae, Lasiocampidae, Lithosiidae, and Sphingidae as families containing some urticating species.

probably accentuated by toxic factors. Randel and Doan (1956) reported a series of cases in the Panama Canal Zone resulting from contact with the poison hairs of *Megalopyge lanata*. Usually there are no distinct systemic reactions to caterpillar urtication, but in particularly sensitive individuals the following symptoms have been observed; nervousness, anxiety, tremors, convulsion, anorexia, nausea, vomiting, dyspnea, and bradycardia.

Treatment. There is no approved specific therapeutic agent to counteract the effects of vesication produced by the poison hairs of caterpillars. Local application of cool water sometimes partially relieves the burning, nettling pain. Application of calamine lotion to the injured skin is a moderately satisfactory palliative. Topical corticosteroid ointment is more effective and is indicated in more severe cases. At times of systemic reaction, supportive treatment is recommended to allay the pain, and oral or parenteral corticosteroid or antihistaminic drugs to counteract allergic complications. The lesion heals slowly, like a chemical burn. Repeated contacts with poison hairs of caterpillars may be followed by increased susceptibility.

Prevention of Urtication. The only effective preventive measures consist in refraining from handling caterpillars, avoiding vegetation on which they feed, and destroying plants in certain areas where they breed and plants on which they depend for food.

Other Effects of Lepidoptera

Asthma. In addition to the urticating properties of certain caterpillars, contact with eggs or larvae of the range caterpillar *(Hemileuca olivae)* may cause asthmatic attacks with coryza, coughing, and severe wheezing (Randolph, 1934). Contact with *Euproctis chrysorrhoea* L., the brown-tail moth, has produced hay fever and asthma in sensitized patients in England (Blair, 1979).

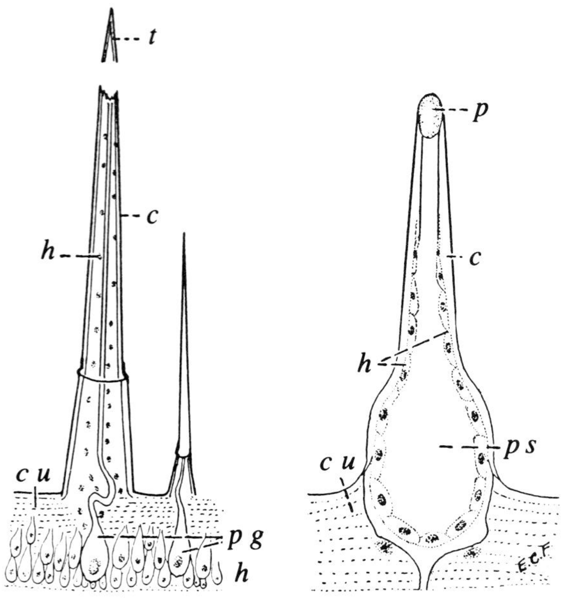

Fig. 44–2. Poison hairs of caterpillars. *A*, Primitive type. *B*, Modified type. *c*, Chitinous covering of spine. *cu*, Cuticula. *h*, Hypodermis. *p*, Canal plug. *pg*, Poison gland cell. *ps*, Poison sac. (Adapted by E.C. Faust from Gilmer, P.M. 1925. A comparative study of the poison apparatus of certain lepidopterous larvae. Ann. Entom. Soc. Am., *18*:203–239.)

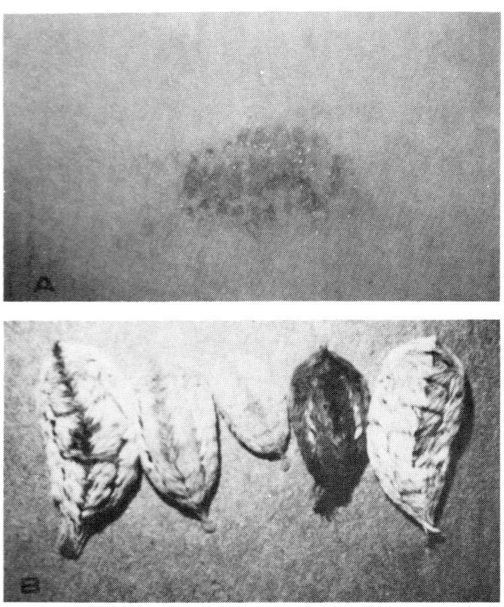

Fig. 44–3. *Megalopyge* sp (puss caterpillars). *A*, Lesion on thigh of child 48 hours after contact with a puss caterpillar. (× 1.) *B*, Specimens collected in southern Texas, showing variation in size and pigmentation. (× 1.) (Courtesy of Dr. Adam Ewert.)

Eye Feeding by Lepidoptera. Some species of moths in Africa and Asia have been reported feeding on secretions from the eyes of wild and domestic animals as well as sleeping humans (Büttiker, 1966). In *Calyptra eustrigata*, a noctuid moth, the mouthparts consist of two linked stylets armed with a series of hooks, barbs, and spines (Bänziger, 1980). Insertion of the stylets beneath the skin is possible due to a peculiar oscillatory torsion that forces the mouthparts downward. Blood is engorged following placement of the stylets, suggesting that pathogens may be mechanically transmitted by this moth.

Lepidoptera as Intermediate Hosts. Some of the Microlepidoptera are acceptable intermediate hosts of *Hymenolepis diminuta* (see p. 511).

BEES, WASPS, HORNETS, AND ANTS (ORDER HYMENOPTERA)

Definition. Members of the order Hymenoptera, literally insects with "membranous wings," typically have two pairs of membranous wings, held together on each side by a row of minute hooklets situated on the anterior border of the hind wings, which are considerably smaller than the fore wings. The mouthparts are adapted for chewing or for chewing and sucking or lapping food. In the aculeate ("stinging") Hymenoptera, the caudal extremity is usually provided with a sting apparatus, which is a modified ovipositor. Metamorphosis is complete.

The Sting Apparatus. The *venom apparatus* in *Apis mellifera* (family Apidae), the common honeybee, consists of: (1) a tubular *acid gland* that discharges its product through long coiled ducts that unite before entering the inner end of the *poison sac;* (2) an unpaired tubular *Dufour's gland* (sometimes called an *alkaline* gland) that empties into the outer tubular end of the poison sac; (3) a strong *muscular bulbus,* immediately distal to the neck of the poison sac; and (4) *the sting,* which is made up of *(a)* a dorsal *sheath,* barbed at its distal extremity, *(b)* a pair of ventrolateral *lancets,* having sharp, recurved teeth and fitted by grooves to the lower side of the sheath on which they slide up and down, and *(c)* a pair of lateral finger-like *sting palps* (Snodgrass, 1956). The sheath and lancets make the puncture wound, and the mixed secretions from the two types of glands pass down the semiclosed poison canal, formed by the sheath and the lancets, into the wound. In addition to bees, there are about 15,000 species of aculeate wasps in the world, of which 95% are solitary and non-aggressive (Akre and Davis, 1978). Of the remainder, the social Vespidae are by far the most important medically, particularly genera in the subfamily Vespinae (*Vespula* and *Dolichovespula*—yellowjackets; *Vespa*—hornets)(Akre et al., 1980).

Envenomation

When provoked to sting, the females of the honeybee (and some wasps) leave the posterior tip of their abdomen, together with the entire sting apparatus, in the skin, where the muscles attached to the apparatus continue to contract for some time, forcing the sheath and lancets more deeply into the wound and discharging more venom. The stinger is retained by the barbed tip, which acts as an anchor (Fig. 44–4). Unlike most bees, the bumblebee usually does not leave the sting in the wound and may use it several times.

The venoms of bees and other stinging Hymenoptera are complex mixtures of enzymes, polypeptides, and various smaller molecules, some of which may be pharmacologically active (histamine; 5-hydroxytryptamin) or neurotoxic (O'Connor and Peck, 1978; Edery et al., 1978). Hyaluronidase, found in *Apis mellifera* and many vespid wasps, aids in spreading venom through tissues by degradation of the intercellular cement. Phospholipases also occur. They synergize catabolism of cell membrane phospholipids as well as cause hemolysis. The mellitins, a family of peptides composing about 50% of *Apis* venom, act in concert with phospholipase A. Fractions of this group of compounds may also function as noncompetitive inhibitors of acetylcholinesterase. Apamin, a basic peptide component of bee venom, is neurotoxic and may be synergistic with phospholipase A.

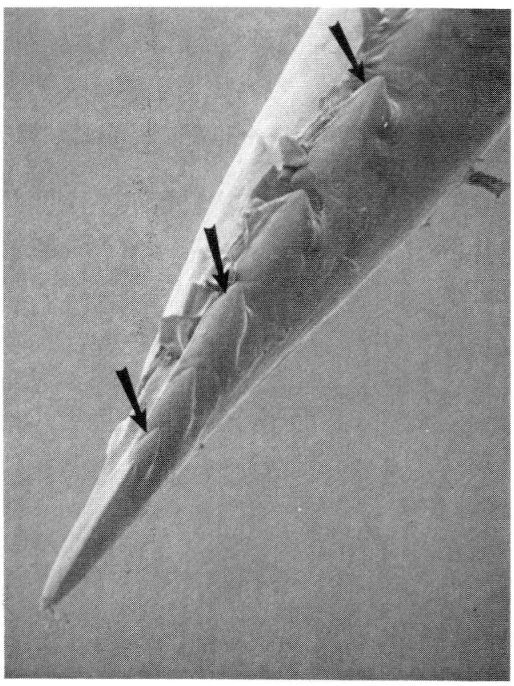

Fig. 44–4. Sting shaft of a worker honeybee. Note barbs (arrows) along the shaft's tip. Scanning electron micrograph, lateral view. (Courtesy of E.W. Cupp.) (× 600.)

Pathology and Symptomatology. The lesion produced by stings of the honeybee, bumblebee, hornet, and wasp may be a slight local inflammation that disappears within a few hours. At other times the injured member of the body becomes greatly inflamed and edematous, and the local reaction is accompanied by appreciable or profound systemic symptoms. The introduction of the venom into the conjunctiva is likely to be particularly troublesome.

Some individuals are hypersensitive to bee or wasp venoms and suffer anaphylactic reactions when stung. In these cases, death can result within less than 1 hour, often following a single sting. There are more deaths from anaphylaxis following wasp or bee stings in the United States and Great Britain than from all other venomous bites or stings. The findings of a number of postmortem examinations, as summarized by Edery et al. (1978), include edema of the lungs and larynx; hyperemia of the viscera and brain; petechial bleeding in pleural, epicardial, and endocardial membranes, and in the mucosae of the respiratory and gastrointestinal tracts; and cyanosis of the skin, serous membranes, and meninges. Rarely, cerebral infarction, including necrosis of the basal ganglia with extrapyramidal syndromes, may occur after insect-sting anaphylaxis (Gale, 1982).

Studies have shown that bees and wasps differ in the types of antigens found in their venoms. Phospholipase A is the major allergen of honeybees (Sobotka et al., 1976), whereas vespid venoms contain three major proteins that serve as allergens: a hyaluronidase, a phospholipase and a protein of unknown function, designated antigen 5, which appears to be the major allergen (Lichtenstein et al., 1979). The coupling of these antigens with cell-bound IgE triggers the release of pharmacologically active compounds (e.g., histamine and serotonin), leading to anaphylactic shock.

Although persons hypersensitized to hymenopteran venoms may be temporarily saved by the administration of epinephrine, they should be desensitized before additional exposure to sting is anticipated. In the past, this was attempted by injections of filtered, whole-body extracts of various Hymenoptera. An immunotherapeutic regimen using pure venoms instead of whole-body extracts now has been shown to protect allergic persons against anaphylactic reactions, presumably by elevation of antigen-specific IgG "blocking" antibody (Lichtenstein et al., 1979; Golden et al., 1980). The time course for desensitization with pure venoms can be relatively short when a "rush" regimen that causes a rapid rise in antivenom IgG and no difference in IgE is used.

Ants (family Formicidae) form a huge and important component of the Hymenoptera. Within this family, some species in the subfamilies Ponerinae, Dorylinae, and Myrmicinae may sting viciously (Harwood and James, 1979). Lofgren et al. (1975) discussed the "fire ant" problem, noting that *Solenopsis invicta* and *S. richteri* were introduced into the United States at Mobile, Alabama, around 1945 and 1918, respectively. Since that time, *S. invicta* Buren (the red imported fire ant) has spread throughout the southeastern United States, extending from the Carolinas to Texas. Mounds of this ant in an unconfined area have a diameter of 24 inches and a height of 18 to 24 inches. Mature colonies may include up to 230,000 workers.

Ants are also capable of secreting the products of two other exocrine glands in addition to the venom gland. These are the mandibular glands and Dufour's gland, both of which may contain a variety of hydrocarbons (Blum and Hermann, 1978). Several species of "fire ants" *(Solenosis invicta, S. richteri)* cause a fiery sensation immediately after stinging, commonly followed by an extremely pruritic vesicle at the site of injury. The cutaneous lesion where the hemolytic toxin of the ant has been introduced consists of a painful wheal that transforms in 24 hours into a vesicle, then usually a pustule. In a period of 3 to 8 days the vesicle or pustule breaks down, discharging a clear or purulent fluid, or it may be absorbed with development of a covering eschar. The pus is due to the pyogenic nature of the toxin, not to secondary infection. The venom of *S. invicta* is unlike most bee and wasp venoms, containing five alkaloidal compounds that, in addition to pustule formation, may cause systemic, allergic reactions. In patients with allergic manifestation, treatment consists of parenteral administration of 1:1000 epinephrine for children (0.1 to 1.0 ml depending on age and weight), and diphenhydramine HCl (Benadryl) for adults (1.5 ml parenterally).

In the past, control of fire ant populations consisted of residual soil surface treatments or bait toxicants. However, owing to the possibility of harmful environmental side effects, toxicants such as diazinon should be introduced directly into the mounds where the colonies are concentrated.

In the Tropics, certain species of stinging ants are aggressive, inflict wounds that produce agonizing pain, and may even endanger the life of persons attacked. The stings of large ants of central and northern South America, particularly the tucandeira *(Paraponera clavata),* cause excruciatingly painful lesions. The sting of these primitive giant-sized black ants, which have a matriarchal mode of life, is about 6 mm long. Their food consists of small beetles, caterpillars, spiders, and similar animal life. Males are reddish-brown, about half as large as the females. The venom is composed of histamine, 5-hydroxytryptamine, and acetylcholine. It produces an agonizing, throbbing, painful swelling about 18 cm in diameter, which, however, abates in a few hours unless the victim has an allergic tendency (Zahl, 1963). The lesions produced by tropical foraging ants (Genus *Monomorium)* in India, Africa, and the islands of the South Pacific are of an entirely different nature, being produced by the bites of their mandibular jaws. In the bull ant, *Myrmecia gulosa,* the venenating apparatus contains histamine, hyaluronidase, a component with kininlike activity, and a hemolytic factor.

Treatment and Control. No satisfactory treatment has been discovered for stings of the various species of Hymenoptera. When the sting apparatus has been left in the wound, it should be removed by scraping with a knife blade or needle, to prevent discharge of additional venom from the poison reservoir. It should not be grasped with forceps or fingers. Injection of epinephrine is sometimes necessary to counteract anaphylaxis. In cases of systemic shock, advanced life support measures may be needed. For relief of local discomfort, calamine lotion is helpful. Topical antihistaminics should not be used since they cause sensitization and photosensitization. Oral antihistaminics are helpful. If the local lesion is severe, a topical corticosteroid cream or ointment is indicated, provided there is no infection. Whenever possible, appropriate clothing should be worn to keep these insects from contacting the skin. Several insecticides are available for domestic wasp control, including malathion, dichlorvos, propoxur, and pyrethrins.

REFERENCES

Akre, R.D., and Davis, H.G. 1978. Biology and pest status of venomous wasps. Annu. Rev. Entomol., *23*:215–238.

———, Greene, A., MacDonald, J.F., Landolt, P.J., and Davis, H.G. 1980. *Yellowjackets of America North of Mexico.* U.S. Dept. of Agriculture Handbook No. 552. Washington, D.C., Superintendent of Documents, U.S. Government Printing Office.

Bänziger, H. 1980. Skin-piercing blood-sucking moths. III. Feeding act and piercing mechanism of *Calyptra eustrigata* (Hmps.) (Lep., Noctuidae). Bull. Entomol. Soc. Suisse, *53*:127–142.

Bhattacharya, N.C. 1966. Scarabiasis. Bull. Calcutta Sch. Trop. Med., *14*:86–87.

Bishop, J.W., and Morton, M.R. 1968. Caterpillar-hair keratoconjunctivits. Med. J. Aust., *2*:995–997.

Blair, C.P. 1979. The browntail moth, its caterpillar and their rash. Clin. Exp. Dermatol., *4*:215–222.

Blum, M.S., and Hermann, H.R. 1978. Venoms and venom apparatuses of the Formicidae: Myrmeciinae, Ponerinae, Dorylinae, Pseudomyrmecinae, Myrmicinae, and Formicinae, pp. 801–869. In *Arthropod Venoms.* Edited by S. Bettini. New York, Springer-Verlag.

Büttiker, W. 1966. Biological notes on eye-frequenting moths from N. Thailand. Bull. Soc. Entomol. Suisse, *39*:151–179.

Common, I.F.B. 1975. Evolution and classification of the Lepidoptera. Annu. Rev. Entomol., *20*:183–203.

Delgado Quiroz, A. 1978. Venoms of Lepidoptera, pp. 555–612. In *Arthropod Venoms.* Edited by S. Bettini. New York, Springer-Verlag.

Edery, H., Ishay, J., Gitter, S., and Joshua, H. 1978. Venoms of Vespidae, pp. 691–772. In *Arthropod Venoms.* Edited by S. Bettini.New York, Springer-Verlag.

Frazier, C.A., and Brown, F.K. 1980. *Insects and Allergy and What to do about Them.* Norman, University of Oklahoma Press.

Gale, A.N. 1982. Insect-sting encephalopathy. Br. Med. J., *284*:20–21.

Golden, D.B.K., Valentine, M.D., Kagey-Sobotka, A., and Lichtenstein, L.M. 1980. Regimens of Hymenoptera venom immunotherapy. Ann. Intern. Med., *92*:620–624.

Harwood, R.F., and James, M.T. 1979. *Entomology in Human and Animal Health,* 7th ed. New York, Macmillan Publishing Co.

Herms, W.B., and James, M.T. 1961. *Medical Entomology.* New York, Macmillan Co.

Levine, M.I., and Lockey, R.F. editors. 1981. *Monograph on Insect Allergy.* Hartland, Wisconsin, Parker Printing.

Lichtenstein, L.M., Valentine, M.D., and Sobotka, A.K. 1979. Insect allergy: The state of the art. J. Allergy Clin. Immunol., *64*:5–12.

Lofgren, C.S., Banks, W.A., and Glancey, B.M. 1975. Biology and control of imported fire ants. Annu. Rev. Entomol., *20*:1–30.

O'Connor, R., and Peck, M.L. 1978. Venoms of Apidae, pp. 613–660. In *Arthropod Venoms.* Edited by S. Bettini. New York, Springer-Verlag.

Perlman, F., Press, E., Googins, G.A., Malley, A., and Poarea, H. 1976. Tussockosis: Reactions to Douglas fir tussock moth. Ann. Allergy, *36*:302–307.

Pesce, H., and Delgado Quiroz, A. 1966. Lepidopterismo y erucismo. Epidemiologia y aspectos clinicos en el Peru. Mem. Inst. Butantan, *33*:829–834.

Randel, H.W., and Doan, G.B. 1956. Caterpillar urticaria in the Panama Canal Zone. Report of 5 cases, pp. 111–116. In *Venoms.* Washington, D.C., American Association for the Advancement of Science Publ. No. 44.

Randolph, H. 1934. Allergic response to dust of insect origin. J.A.M.A., *103*:460–462.

Schmidt, J.O. 1982. Biochemistry of insect venoms. Annu. Rev. Entomol., *27*:339–368.

Smith, K.G.V. 1973. Coleoptera (beetles) and other insects, pp. 413–416. In *Insects and Other Arthropods of Medical Importance.* Edited by K.G.V. Smith. London, British Museum (Natural History).

Snodgrass, R.E. 1956. *Anatomy of the Honey Bee.* Ithaca, New York, Comstock Publ. Assoc.

Sobotka, A.K., Franklin, R., Adkinson, N.F., Jr., Valentine, M.D., and Lichtenstein, L.M. 1976. Allergy to insect stings. II. Phospholipase A: The major allergen in honeybee venom. J. Allergy Clin. Immunol., *57*:29–40.

Zahl, P.A. 1963. Deadlier than the male. Am. Scholar, *32*:282–290.

Chapter 45

Control of Arthropods, Molluscs, and Rodents as Vectors and Reservoirs

A program for control of any living organism requires its species identification and knowledge of its geographical distribution, the local and more general conditions under which it exists and propagates among competing fauna and flora, its physiological processes, and its adaptive potentials under natural and abnormal conditions. Thus, programs undertaken to control vectors and reservoirs of disease must take into account their relations to associated organisms, so as not to upset the balance in nature. Toxicants employed to kill an offending species may provide opportunity for competitors to replace the offender, and present an equally serious problem. Toxicants may also endanger human food supplies. Finally, they may be directly harmful to the human population.

In any management system for control, it is essential to know how the arthropod, mollusc, or rodent is dangerous or damaging to human health and economy. Is the offender merely an annoyance? Does it produce severe irritation or allergic reactions? Does it cause envenomation? Is it an important mechanical or biological vector of a microorganism pathogenic to humans and domestic animals? To put the answers to these questions in their proper perspective, a program to be focused toward control of medically important vectors or reservoirs should determine what role the organism plays in human (and animal) welfare, the significant ecological interactions among the vector, pathogen, and susceptible vertebrate population; what additional information needs to be elucidated before control can be effectively (and safely) undertaken; and what results may be reasonably expected from the program (Metcalf, 1975).

In addition to epidemiologic information concerning the specific organism(s) to be controlled, reliable meteorologic data (i.e., maximum, minimum, and average seasonal and annual temperature and rainfall) for the area under consideration are needed. These environmental factors may differ appreciably from those of other areas previously studied, and may require complete reassessment of the methods to be employed. Moreover, since control is to be programmed for the better health and economy of a particular human population, knowledge of their socioanthropology, educational level, and standards of living must be considered. How much will this group of people understand, and how will they cooperate? Is the project planned as a short, intensive campaign, or will it require prolonged efforts to achieve satisfactory results? In any case, the cost should be estimated and sufficient financing obtained in advance, so that the program will not need to be discontinued for lack of funds.

In most projects involving regional or extensive area control, advice and assistance of the World Health Organization are essential for coordinating efforts and for securing experienced supervisory personnel.

Control factors are of two types: *natural* and *artificial*.

Natural Control

There are factors or conditions that in nature regulate the populations of arthropods, molluscs, and rodents, as well as other animals and plants with which they are associated. If the pathogens to be controlled are involved directly or indirectly in human disease, their cycles of transmission are often completed in nondomestic reservoir hosts, such as birds, monkeys, wild ungulates, rodents, hogs, canines, or felines. Human contact with the *parazoonotic cycle* exposes man to these pathogenic agents, so that they may then become *euzoonotic*. Humans are highly susceptible to many of these diseases; to other diseases humans have come to be exposed by domestication of wild reservoirs and through adaptation of anthropophilous vectors to the parazoonotic cycle. Need for control has resulted primarily from human encroachment on these diseases in their natural environment

Artificial Control

Methods for artificial control may be considered under the following categories, to be employed separately or conjointly, as opportunity or practicability may direct: (1) quarantine, (2) biological control, (3) planned engineering projects, (4) mechanical control, (5) physical control, (6) chemical control, (7) control by sterilization, (8) genetic control, and (9) chemotherapy.

Quarantine. This method has been employed since the end of the Middle Ages in an attempt to prevent the introduction of pathogens into previously uninfected human populations. Today international regulations are usually effective in preventing the introduction of jungle yellow fever virus into *Aedes aegypti*-infested areas, in delousing immigrants crossing political borders from endemic typhus and relapsing fever areas into uninfected zones, and in controlling bubonic plague through destruction of infected rodents and rodent fleas. Quarantine is today likewise employed in the form of insecticide disinfestation of airplanes before their arrival at foreign airports and in disinfestation of ships to prevent introduction of rats harboring plague-infected fleas. In addition, people and their clothing are dusted with insecticides to kill body lice. Immunization against yellow fever still constitutes an important protective measure, obviating the need for quarantining potentially exposed people.

With regard to trematode infections (in which molluscs are essential intermediate hosts), the most imminent hazard that should be eliminated by quarantine regulations is the introduction of human cases of schistosomiasis into nonendemic areas where susceptible molluscan hosts are already established. This danger is particularly serious in countries where infected populations are migrating from poverty-stricken, overcrowded zones into new agricultural regions, and where irrigation is derived from water courses in which susceptible snails breed.

Biological Control. Except for the use of larvivorous fish in mosquito control, the elimination of vectors and intermediate hosts of disease-producing organisms by applied biological methods has met with only limited success (Bay et al., 1976; National Academy of Sciences, 1976; Laird, 1981). However, a vast array of potentially useful biological control agents for arthropod vectors has been identified (Roberts and Strand, 1977). In addition, a new strain (var. *israelensis)* of the crystalliferous, spore-forming bacterium *Bacillus thuringiensis* has recently been shown to produce a powerful insecticidal toxin that is effective against dipterous larvae, particularly mosquitoes (Goldberg and Margalit, 1977) and black flies (Undeen and Colbo, 1980). Mosquitoes of the genus *Toxorhynchites* are found in tree holes and artificial containers of water as larvae. This stage is predaceous, and the larvae of medically important mosquitoes such as species of *Aedes (Stegomyia) aegypti* and *Aedes triseriatus* often serve as prey. The larvae of *Toxorhynchites* devour the larvae of *Ae. aegypti*. Studies are under way to determine whether propagation and release of adult *Toxorhynchites* mosquitoes, which do not feed on human blood, are effective in controlling *Ae. aegypti* (Gerberg and Visser, 1978). In molluscicidal programs, the introduction of the snail *Marisa cornuarietis* (L.) to control breeding of *Biomphalaria glabrata* was successful in pilot tests in Puerto Rico, but with the potential liability that this competing snail, a voracious feeder on vegetation, may get out of control and destroy agricultural crops (Berg, 1973; Bay et al., 1976). The Japanese weasel, *Mustela sibirica itatsi,* has been somewhat successful in controlling rodents when used in conjunction with rodenticides.

Planned Engineering Projects. These projects refer to programs for human betterment, including safe water supplies, sanitary sewage and garbage disposal, planning of urban renewal and resettlement to eliminate or reduce arthropod and rodent infestation of homes, and modern practices in agriculture and animal husbandry to increase and improve human food supplies. All these programs, if properly supervised and intelligently used by the people to be benefited, should reduce the sources of harmful arthropods, molluscs, and rodents and thus will control a large number of transmissible pathogens.

Mechanical Control. Hand-catching tsetse flies

in trypanosome-endemic areas of West Africa has been temporarily helpful in reducing transmission of these flagellates. Trapping mosquitoes has had some value in the past in preventing transmission of malaria. Clearing jungle grass in enzootic-endemic scrub typhus areas to eliminate trombiculid mites has been effective in preventing development of this infection in military forces, and destruction of thickets harboring tsetse flies around African villages has appreciably reduced human exposure to trypanosomiasis. In schistosomiasis areas, the mechanical screening of snails to prevent their transportation from primary to lateral irrigation channels reduces the likelihood of dispersal of these molluscs, particularly when molluscicidal agents are employed concurrently at control barriers. A more costly but more effective mechanical deterrent to snail breeding is lining the sides of main irrigation canals with concrete slabs to prevent growth of aquatic vegetation essential for growth of the snails (McJunkin, 1970).

Physical Control. Sunlight, ventilation, and excess heat or cold are all potentially useful in limiting development of some arthropods and molluscs. During midday heat, venomous scorpions in dry tropical regions lurk in dark corners of moist dirt floors of native huts or under piles of lumber or rubbish. The brown spider, *Loxosceles laeta,* of South America hides among garments in dark closets. Molluscs thrive among aquatic growth that serves as their food and shelter along the banks of streams and ponds. Clearing protective vegetation in standing water renders these sites unfavorable for development of some *Anopheles* species. Likewise, the elimination of larval habitats by draining, filling, or disposing of such water containers as barrels, cans, and tires is important for the control of certain *Aedes* and *Culex* species. Clothing infested with body lice may be sterilized by dry or steam heat.

Chemical Control. During the last few decades increasing reliance has been placed on chemical agents (both natural and synthetic) for the control of injurious and vector arthropods, rodents and, more recently, molluscs. In most cases these toxicants serve as contact and stomach poisons; hence, considerable investigation has been directed to the physiologic effect of these chemicals on the species to which they are applied. Some insecticides and acaricides that were highly efficacious when introduced have come to be less dependable, owing to the natural selection of resistant strains. Populations of such medically important arthropods as muscoid flies, mosquitoes, fleas, body lice, ticks, and mites are genetically heterogeneous. While the majority may succumb to chemical treatment, a few survivors with resistant genes are left to establish populations capable of rapidly detoxifying pesticides (Plapp et al., 1979). The selection of physiological resistance may be a result of campaigns against a particular species population (Georghiou and Taylor, 1977) or it may be caused by operational aspects, such as improper dose levels and treatment times (WHO, 1975; Georghiou and Taylor, 1977a). Intensive use of pesticides in agroecosystems may also induce resistance in vector populations. This phenomenon has been noted particularly in Central America where various kinds of insecticides are used in cotton growing (National Academy of Sciences, 1976). Likewise, some species of *Anopheles* mosquitoes have developed a behavioral avoidance reaction to insecticides sprayed on inside walls of human habitations (WHO, 1976). In some parts of the world rats have become resistant to warfarin and other anticoagulants. Similar difficulties may be anticipated when new synthetic molluscicides come into more general use.

Certain chemical agents serve as temporary repellents when applied to exposed skin or dusted or impregnated in garments, thus preventing the bites of mosquitoes, black flies, fleas, trombiculid mites (chiggers), and ticks, as well as leeches in tropical jungles.

Control by Sterilization. The development of an irradiation technique for sterilizing males of the screwworm fly, *Cochliomyia hominivorax,* thus preventing the development of progeny of this myiasis-producing species in the southern United States and other regions of the Americas, has opened a new vista for economic growth of the cattle industry, and is a prophylactic measure for man himself, since the larvae of this fly in past decades also caused appreciable and sometimes fatal human myiasis. Several chemicals (chemosterilants) show promise of producing sexual sterility in arthropods without some of the practical limitations of radiation (La Brecque and Smith, 1968). Chemosterilants fall into three general groups: (1) alkylating agents, including such materials as tepa, thiotepa, metepa, apholate, and tretamine; (2) antimetabolites, including aminopterin, methotrexate, 5-fluorouracil, and 5-fluoroorotic acid; and (3) miscellaneous compounds (e.g., thiourea, ethylene urea, calcium arsenate, and triphenyl tin hydroxide). Insects are usually treated as pupae or adults, although treatment of larvae has also been successful. In males, sterility may be caused by the induction of dominant lethal mutations in the sperm, aspermia, or sperm inactivation. Sterility in females may result from infecundity or the occurrence of dominant lethal mutations in the egg. Changes in mating behavior may also occur.

Genetic Control. Although practical methods for large-scale use of genetic control have not yet been worked out, this is potentially one of the most efficient tools for control of arthropod vectors and can result in elimination of a target species without affecting others in the area (Whitten and Pal, 1974). Factors leading to reproductive failure include hybrid sterility, cytoplasmic incompatibility, meiotic drive, distorted sex ratios, and chromosome translocations (Pal and La Chance, 1974). Several limited field investigations, particularly with mosquitoes, have demonstrated the feasibility of genetic methods for vector control. Other arthropods, such as tsetse flies, house flies, and ticks, may also prove amenable to genetic control.

Chemotherapy. The use of chemotherapeutic agents to cure disease and to alleviate human suffering is as ancient as other practical applications of human knowledge. Yet drug therapy as a preventive measure is of relatively recent origin. Possibly the earliest such development was in malaria, where administration of antimalarial drugs (principally cinchona derivatives), combined with screening the infected individuals from *Anopheles* mosquitoes, provided evidence of effective control of the disease. More recently, mass therapy has been employed in combination with anti-*Aedes* measures for the near eradication of Bancroft's filariasis, with special evidence of success on the island of Tahiti in the South Pacific. However, chemotherapy as a single public health tool for combating diseases involving arthropods or molluscs is never likely to provide a *therapia magna sterilans* for an entire population. In many endemic areas of disease, the epidemiology of the responsible agents and their vector hosts is relatively well known, but practical methods of application of this information have yet to be developed.

The following section of this chapter summarizes currently available data concerning chemicals employed in the control of arthropods as agents and vectors of human disease, molluscs injurious to man and those that are essential intermediate hosts of human helminths, and rodents as vectors and reservoirs of human disease. Consideration is focused on those toxicants that have been particularly helpful in controlling these animals.

CHEMICAL AGENTS EMPLOYED IN CONTROL OF ARTHROPODS

General Consideration

The agent or combination of agents to be selected for a particular control project depend on many factors that must be carefully weighed: the species of arthropod against which the agent is to be directed, its stage of development and the environment in which it occurs (whether air, water, vegetation, or soil), its size, structure, feeding and respiratory habits, and the probable ways in which various toxicants will affect its metabolism. Based on these data, the chemical is selected for outdoor or indoor use, and the vehicle is chosen for maximum efficacy as determined from previous tests in the laboratory and under natural conditions.

Toxicants used for control of medically important arthropods may serve primarily as contact poisons, fumigants, or repellents, or they may serve two or more of these purposes simultaneously. The delivery system for each toxicant must be carefully considered. It may best be incorporated as a dust in an inert powder to be applied to the human body or dispersed through a sprayer. As a wettable powder it may be diluted with water. Frequently it may need to be dissolved or suspended in a petroleum or other organic vehicle. It may be incorporated in an inert carrier for slow release into an aqueous environment or incorporated into a bait. Or, it may be in the form of an emulsion to be sprayed in minute droplets as an aerosol, as a technical-grade pesticide to be applied in ultralow amounts, or in granular form. Two or more chemicals may be combined, one for immediate "knock-down" effect, the other for residual effect. There may be a synergistic advantage of such a combination. If the arthropod has developed marked resistance to a toxicant previously employed, an effective substitute must be selected.

For satisfactory results, *fumigants* should be dispersed in tightly closed quarters, and the operator should guard against lethal fumes by gas mask and protective clothing. Persons who employ spraying techniques should be aware of the toxic potential of the agent, and should especially prevent its ingestion, inhalation, or skin contact. Recommended repellents are not usually harmful when applied to the exposed skin, but when impregnated in clothing for more lasting effect they should be confined to outer garments and socks.

Spraying of insecticides may be divided, for convenience, into *space spraying* and *residual spraying*. In both cases the toxicant may be dispensed within a home or over extensive areas of land or water. Space spraying is intended primarily to contact and knock down or kill flying insects within the confines of an enclosure or in an open area. An aerosol mist, containing minute drops of a uniform size, has usually settled down and served its purpose within an hour. Fogging with insecticides in the atmosphere serves the same purpose: killing flying insects that come in contact with the sprayed droplets. Ultralow-volume applications of concentrated insecticides (90% or more of active ingredients) have proved to be economical and efficient for large-scale space spraying (Lofgren, 1972; Mount, 1979). Residual spraying places the toxicant on the walls, ceiling, and floors of human and domestic animal quarters, so that the agent remains there for some time as a contact poison for arthropods that alight on or crawl over the treated surfaces. The degree of contact exposure and the length of useful time of the spray depend to a great extent on the porosity or smoothness of a wall or other sprayed surface and on its exposure to weather conditions. Porous walls absorb the toxicant and hold it for a considerably longer time than smooth surfaces. Incorporation of an insecticide into a surface paint has been attempted to overcome rapid loss of effectiveness. Residual spraying with hand sprayer, stirrup pump, pneumatic pressure applicator, or applicator gun is widely employed for dispersing the toxicants over grasslands, forested regions, and swamps and other aquatic habitats, to kill insect larvae (Matthews, 1979). Where the area to be covered is extensive or is practically inaccessible by jeep or motorboat, spraying by helicopter or fixed-wing aircraft is the only practical technique (WHO, 1977).

Detailed knowledge of the biochemical pathways used by insects for detoxification is necessary for understanding resistance mechanisms. Wilkinson (1976) summarized recent progress in this field, particularly regarding metabolism of insecticides, an important factor in the production of resistance. Aspects of mammalian toxicology were also discussed.

The World Health Organization has established a program for evaluating and testing insecticides to find new compounds that can overcome the problem of insecticide resistance without further contamination of the environment (Wright, 1971). Each new compound is subjected to a series of seven evaluation stages: three in the laboratory followed by four under field conditions. Results are published by WHO via a mimeograph series and in the Bulletin of the World Health Organization.

Inorganic Materials

Copper sulfate (blue vitriol; $CuSO_4{\cdot}5H_2O$). Blue crystalline solid, soluble in nonalkaline water; in dilute solutions harmless to humans and other vertebrates; limited value as mosquito larvicide; in dilutions of 0.6 ppm with 4.0 ppm $Ca(OCl_2)$ is lethal to *copepods* and molluscs.

Copper aceto-arsenite (Paris green; approx. $[3Cu(AsO_2)_2{\cdot}Cu(C_2H_3O_2)_2]$. Heavy green powder, slightly soluble in water (floats); poisonous when ingested but harmless to plants and vertebrates as used; widely employed in pre-DDT era as *mosquito* (especially *anopheline) larvicide,* as dust, 1 to 5% by volume, or 10 to 15% (hand dusters), or 22.5% by weight in slaked lime or road dust (airplane dusting); dispersed on water surface 1 lb/acre every 5 to 7 days as long as larvae are present.

Phosphorus (P_4). Waxy fuming solid, with garlic-like odor, slightly soluble in water, highly toxic if ingested; produces severe skin burns; employed as paste to kill *cockroaches,* 1 to 2% in moist starch base, smeared on surface of peeled raw Irish potato.

Sulfur (S_8). Yellow solid, insoluble in water, soluble in CS_2 and organic solvents; essentially harmless to vertebrates; classical therapeutic agent for *sarcoptic mange,* 1 to 15% in lanolin-petrolatum ointment; mostly superseded by lindane ointment (pure *gamma* HCH, see p. 723).

Naturally Occurring Organic Compounds

Pyrethrum (Persian and Dalmatian insect powder). Powder or resinous extract obtained from dried flower heads of *Chrysanthemum cinerariaefolium, C. roseum,* and *C. carneum,* containing 4 active principles, pyrethrins I and II and cinerins I and II, volatile non-nitrogenous organic esters comprising 1 to 2% of crude substance (Casida, 1973); soluble in organic solvents; nontoxic to vertebrates unless ingested; inhalation may produce sneezing and bronchial irritation. In an aerosol spray pyrethrum produces a quick "knock-down" of small arthropods by nerve action but has essentially no residual qualities. Combined with more lasting insect toxicants in powder, aerosol spray, or ointment, pyrethrum has been an important ingredient for combating *adult mosquitoes,* many other *biting flies, roaches, ants, bedbugs, fleas, body* and *pubic lice,* and *larvae* (maggots) of *myiasis-producing flies.* Applied to skin in mineral oil, pyrethrum has a moderately repellent effect against *biting flies, fleas,* and *chiggers.*

The development of synthetic pyrethroids has become increasingly important because they possess such desirable properties as higher insecticidal activity, low mammalian-toxicity, and a relative persistence beyond that of the naturally occurring pyrethrums. Synthesis of these newer compounds has involved varying the alcohol components of the pyrethrum molecules and the addition of halogens to the acid portions (Busvine, 1980). The first synthetic pyrethroid developed commercially was *allethrin,* (+)-3-allyl-2-methyl-4-oxocylcopent-2-enyl(+)-*cis,trans*-chrysanthemate. This was followed by *resmethrin,* 5-benzyl-3-furylmethyl (+)-*cis,trans*-chrysanthemate; *permethrin, cis* and *trans* isomers of 3-phenoxybenzyl-3-(2,2-dichlorovinyl)-2,2-dimethylcyclopropanecarboxylate; and *decamethrin*, alpha-cyano-3-phenoxybenzyl-3- (2,2 - dibromovinyl) -2,2-dimethylcyclopropanecarboxylate. The latter two compounds have residual insectidal activity comparable to that of most organophosphorus or carbamate insecticides. As a group, all are effective against the aforementioned arthropods, as well as ixodid ticks (Heller-Haupt et al., 1979). For a discussion of the chemistry of these compounds see Elliot and Janes, 1978.

Rotenone (principal insecticidal ingredient of derris root and cubé; $C_{23}H_{22}O_6$). Ingestion or inhalation by humans produces numbness of buccal mucosa, anorexia, muscle tremors, and tachypnea, and can cause death due to respiratory paralysis. As an insecticidal agent, rotenone is relatively harmless to mammals, causing at most conjunctivitis, but it is highly toxic to fishes. When sprayed on insects, it acts as a slow but effective contact and stomach poison. It is most effective when combined with pyrethrum for rapid "knock-down" and with chlorinated hydrocarbons for synergistic results.

Soaps and Detergents. Natural and synthetic salts of fatty acids, soluble in water; have some insecticidal effect, depending on their fatty acid components and surface tension; principal insecticidal value consists in their use as emulsifying agents for more potent insecticides.

Chlorinated Hydrocarbon Compounds

The use of chlorinated hydrocarbons in general has become increasingly regulated owing to the accumulation of these materials in the biosphere and the occurrence of resistance by major vector groups. However, because of their ease of synthesis, relatively low mammalian toxicity, and competitive cost (Brooks, 1974, 1974a), several still find limited use in public-health entomology.

Chlorophenothane [DDT, 2,2-*bis*-(p-chlorophenyl)-1,1,1-trichloroethane; $C_{14}H_9Cl_5$]. A white crystalline, stable, odorless solid with low volatility; insoluble in water, soluble in most organic solvents; is absorbed and stored by body fats; is secreted in milk, hence is contraindicated as a cattle spray and for insecticidal use in cattle barns; can be employed safely by humans if exposure is avoided. As insecticide has slow, nonrepellent, excitatory contact action, but high residual properties; was one of the most widely used insecticides against *adult* and *larval mosquitoes, moth flies, black flies, lice, fleas, ticks, chiggers, triatomine bugs, roaches, ants,* and *scorpions;* formulated as a dust, dissolved in kerosene or fuel oil for spraying on walls, water surfaces, piles of decaying refuse; incorporated in briquettes to be placed in swiftly flowing streams to kill *Simulium* larvae, in emulsion or suspension for application to porous surfaces; or incorporated in paint or varnish to cover smooth surfaces. Resistance or partial resistance to DDT and related chlorinated hydrocarbons has developed in house flies, several genera of mosquitoes, lice, fleas, and ticks.

DDT has now been restricted or banned by numerous state and national governments because of its widespread and prolonged presence in human and animal tissues and the uncertainty as to chronic toxic effects. Many conservationists oppose any further use of DDT because of its accumulation in delicate food chains. Public-health officials have been disappointed by the appearance of resistance to DDT found in strains of insects of several genera. The World Health Organization, however, emphasizes the excellent safety record of DDT in public-health use over the past 20 years, noting that the effects of dermal exposure have been minimal because the compound is poorly absorbed through the skin; furthermore, contaminated food represents the major source of DDT intake for the general population (WHO, 1979).

Methoxychlor (2,2-*bis*-(p-methoxyphenyl)-1,1,1-tri-chloroethane; $C_{16}H_{15}O_2Cl_3$). Solid white analogue of DDT, insoluble in water, soluble in many organic solvents; much less toxic to mammals than DDT, but poisonous for fishes. As residual spray, safe to use in dairy barns; toxic to *bedbugs* and *house flies;* valuable as *larvicide* for *mosquitoes* that have developed DDT resistance; also effective against *Simulium* larvae.

Lindane or *HCH*, formerly called BHC (*gamma* isomer of hexachlorocyclohexane, Gammexane; $C_6H_6Cl_6$). First of a group of six related compounds to gain repute as a valuable toxicant for medically important arthropods. Use of this material, like DDT, is highly regulated. Stable white or tan solid (depending on degree of purity), with musty odor; almost insoluble in water, soluble in kerosene and diesel oil; decomposes in alkalis; a complex of five isomers, of which the *gamma* is most insecticidal; pure *gamma* HCH is known as *lindane* (less toxic to most warm-blooded mammals and fishes than DDT); contact and stomach poison, with appreciable fumigant and residual effect, due principally to its vapor penetration into arthropods through cuticle or via tracheae, affecting nerve endings. As residual toxicant applied to walls, it kills by contact and penetration of its slowly vaporizing particles, which continue to volatilize for fairly long periods after spraying or other application to absorbent surfaces (e.g., wood, brick, or adobe), but residual action is shorter than that of DDT and analogues.

As a powder in an inert vehicle, lindane has been used against *body lice, head lice, fleas,* and *roaches.* Dusted on water surfaces it kills *mosquito larvae* (1 to 10 lb/acre), and on vegetation it kills *chiggers* (8 lb/acre) and *soft-bodied ticks* (but is less effective than DDT). Solutions and emulsions are employed for residual spraying against *adult flies, gnats,* and *mosquitoes* (100 mg *gamma* isomer/sq ft), *triatomine bugs* (500 mg/sq ft), and the spider *Loxosceles* (1% acetone solution as an aerosol space spray). Impure compound has offensive odor that contaminates poultry, meat, and milk following spray application.

In an ointment, 1% lindane cures human *sarcoptic* and *demodicid* mange in 1 to 3 applications without skin irritation.

Organophosphorus Compounds

Organophosphorus compounds exhibit a variety of biological activities. Some groups have insecticidal, acaricidal, rodenticidal, or anthelmintic properties. The anticholinesterase mode of intoxication for these compounds is fairly well understood as is their synthetic chemistry (Eto, 1974). These chemicals lack the advantage of long residual action, but for that reason do not accumulate significantly in the environment (Busvine, 1980). Certain compounds must be used with caution owing to their high acute toxicity for humans and other vertebrates.

Parathion [*O,O*-diethyl-*O*-(p-nitrophenyl)phosphorothioate]. Brown to yellow liquid; slightly volatile; garlic-like odor; sparingly soluble in water (2 ppm); soluble in ether, aromatic hydrocarbons, some vegetable oils, and alcohols; insoluble in kerosene; many times more poisonous for mammals than DDT. A powerful contact insecticide and acaricide, it has been employed as a *filth-fly larvicide*. More extensively it is used to impregnate sugar-baited cotton cords suspended in cattle barns to kill adult flies. Because of its highly dangerous toxic properties, it should not be used in malaria control.

Malathion [*O,O*,-dimethyl-*S*-1,2-di(carboethoxy)ethyl phosphorodithioate]. A yellow or clear brown liquid with unpleasant mercaptan odor; only slightly soluble in water. In doses of 2.0 g per m^2 every 3 or 4 months, has proved to be a successful spray against *Anopheles* when sprayed in huts. Used routinely by ULV (ultralow-volume) technique as an outdoor space spray for other kinds of mosquitoes and biting flies. Also has been used effectively in doses of 225 to 450 g per hectare or 0.2 to 0.4 lb per acre against *Culex* larvae resistant to chlorinated hydrocarbons. When impregnated on baited cotton cords hung in cattle sheds, it has controlled biting and nonbiting filth flies. Has been used for *body lice* as a 1% dust or for *head lice* as a 0.5% lotion.

As a substitute for HCH in triatomine control, undiluted malathion ULV concentrate applied by knapsack sprayer (2.5 cc/m^2) in adobe houses proved effective against adults, nymphs, and eggs (Martinez et al., 1975). Although less toxic than other organophosphorus insecticides, it must be used with proper safeguards.

Diazinon. Diazinon contains 90% [*O,O*-diethyl-*O*-(2-isopropyl-4-methyl-6-pyrimidyl) phosphorothioate]. Highly volatile dark brown liquid with unpleasant odor; almost insoluble in water. Has been used successfuly as residual spray against *Anopheles* and DDT-resistant filth flies. Less effective as a mosquito larvicide than malathion. When used in a granular form (14%; 2.2 kg AI/ha),* diazinon was 90% effective against the Lone Star tick, *Amblyomma americanum,* for up to 6 weeks (Mount et al., 1976). Should not be used for direct spraying in dairy barns, in food-processing plants, or in houses.

Fenthion. Fenthion contains 95 to 98% *O,O*-dimethyl-*O*-(4-methylthio)-*m*-tolyl) phosphorothioate. Brown liquid with garlic odor; practically insoluble in water. Other names are Baytex, Entex, and Tiguvon. Used in formulations of 50% emulsifiable concentrate, 20 and 40% (Baytex-40) water-dispersable powders, and as a ULV spray. For use against *mosquito larvae,* it is formulated in granules containing 2% of the toxicant. Not suitable for routine residual spraying. Care should be exercised when using near poultry or other birds. Shown to be effective against *filth flies, mosquitoes, roaches, ticks, bedbugs, fleas,* and *ants.*

Fenitrothion or Folithion. This compound is *O,O*-dimethyl-O-(3-methyl-4-nitrophenyl) phosphorothioate, another brown liquid organophosphorus insecticide. Can be used safely as a residual spray against *Anopheles,* but has a highly objectionable odor. Has also been used as a larvicide for both *black flies* (Jamnback, 1976) and *mosquitoes.*

Dichlorvos (DDVP) or *O,O*-dimethyl-*O*-2,2-dichlorovinyl phosphate. Highly volatile organophosphorus liquid that kills *mosquitoes* by fumigant action of its vapor, which may persist for a long time. Other names are Nuvan and Vapona. The insecticidal action is produced either by allowing the liquid to evaporate in a room or by using a solid formulation (e.g., Meatan wax-type or Shell I,P,625). As a pelletized formulation with a bait attractant for adult screwworms *(Cochliomyia hominivorax, C. macellaria),* 2% dichlorvos reduced populations of both these species by 84 to 90% (Coppedge et al., 1980). As a household insecticide, dichlorvos is effective against *ants, cockroaches, fleas,* and *wasps,* as well as *spiders* and *centipedes.*

Abate (temephos) or *O,O,O,O*-tetramethyl-*O,O*-thiodi-*p*-phenylene phosphorothioate. Another useful organophosphorus compound, this is an excellent *mosquito* larvicide regardless of genus or resistance to the chlorinated hydrocarbon insecticides. Abate has been formulated into sand granules impregnated with a 1% solution of the toxicant. The granules are applied once monthly in doses approximating 1 ppm of the active product, which, gradually released into the water, reaches a concentration of about 0.5 ppm. Abate is also an effective larvicide for *Simulium* spp. (Jamnback, 1976), and is currently being used as the primary insecticide by the Onchocerciasis Control Programme in West Africa. When formulated as an emulsifiable concentrate and applied at a rate of 10% with a saltwater carrier (0.78 kg/ha), it proved 98% effective against *Culicoides molestus* (Ferguson and Cawthorne, 1980). Abate is also effective against the larvae of the Phlebotominae and has been recommended as a control agent for *Cyclops* owing to its low mammalian toxicity (Muller, 1979).

*AI (active ingredient)/ha (hectare = 10,000 m^2 or 2.47 acres).

Naled (Dibrom) or *O,O*-dimethyl-*O*-1,2-dibromo-2,2-dichloroethyl phosphate. A yellow liquid with a slightly pungent odor, this compound is practically insoluble in water and is usually formulated as an emulsifiable concentrate or applied as technical material by ULV application. As a space spray, Naled is effective against adult *mosquitoes* and other *flies.* It is fast-acting with some fumigant qualities.

Chlorpyrifos [*O,O*-diethyl-*O*-3,5,6-trichloro-2-pyridynyl) phosphorothioate]. This compound is effective both as an insecticide and an acaricide. Mount et al. (1978) found this compound to be efficacious against chigger mites as a concentrated spray (1.6 kg AI/ha) and in granular form (1.1 and 1.3 kg AI/ha). Control of *ticks* had been demonstrated earlier (Mount et al., 1976) using lower concentrations for both the spray (0.31 kg AI/ha) and granular formulation (0.54 kg AI/ha). For *mosquito control,* chlorpyrifos can be used as a larvicide (0.0125 to 0.05 lb AI/a), or applied as a space spray for adults (0.025 to 0.05 lb AI/a). Chlorpyrifos methyl, a homologue, is also a promising compound for control of *black fly* larvae (Jamnback, 1976).

Chlorphoxim or (*O*-chlorophenyl) glyoxylonitrile oxime *O-O*-diethyl phosphorothioate. This insecticide is promising because of its relatively low mammalian toxicity and ability to induce toxic effects in otherwise organophosphorus-resistant insects. When fed as a bait (0.25% by weight) to field populations of *Dipodomys spectabilis,* the banner-tailed kangaroo rat, fleas were completely reduced within 3 days by the systemic action of this compound (Miller et al., 1977). Other sensitive insects include *mosquitoes* and *black flies.*

Carbamate Compounds

These synthetic pesticides are relatively recent additions to public health entomology. Though their molecular structure differs from that of the organic phosphorus compounds, the carbamates are also anticholinesterases (Kuhr and Dorough, 1976). Because they are degradable, biomagnification is also unlikely. As a group, the number of carbamates commercially available is significantly smaller than the organophosphorus insecticides.

Baygon (propoxur, aprocarb) or 2-isopropoxyphenyl *N*-methylcarbamate. This is an effective indoor residual insecticide for *cockroaches, ants, bugs, mosquitoes, fleas, spiders,* and *millipedes.* Baygon has been formulated into a water-dispersible powder and into an emulsion concentrate. Its action persists for many weeks. It can be applied as a space spray for mosquitoes by using a ULV concentrate (Baygon MOS). Baygon is also an effective acaricide for both *chigger mites* (Mount et al., 1975) and *ixodid ticks* (Mount et al, 1976).

Carbaryl (Sevin) or 1-naphthyl *N*-methylcarbamate. This low-hazard, moderately residual contact and stomach poison is available in a variety of formulations. It has been used as a space spray for *mosquitoes* as well as for *ticks, fleas,* and *mites* on dogs and cats. It is highly toxic to *honeybees.*

Bendiocarb (2,3-isopropylidenedioxyphenyl *N*-methylcarbamate). An effective household pesticide, Bendiocarb is particularly useful against *ants, bedbugs, cockroaches, fleas, spiders,* and *wasps.* Bendiocarb has rapid knock-down with good residual activity (up to 10 weeks of control).

Insecticide Synergists

Insecticide synergists, though usually nontoxic themselves, increase the potency of the insecticides with which they are combined through their ability to block enzymatic detoxification (Wilkinson, 1976). Synergists may act as modifiers by extending the activity spectrum of an insecticide or by increasing its selectivity. However, the principal advantage of the use of synergists lies in the control of insects that have gained some degree of resistance (Wilkinson, 1981). Synergists can be classified biochemically as those molecules whose structure closely resembles that of the insecticide they potentiate (analogue synergists) or as inhibitors of microsomal oxidation. Consequently, selection of synergists should be based on a knowledge of the mode of detoxification of the insecticide to be used. This is particularly true for both the organophosphates and carbamates, which undergo extensive oxidative detoxification.

As an example, *piperonyl butoxide* (3,4-methylenedioxy-6-propylbenzyl *n*-butyl diethyleneglycol ether); $C_{19}H_{30}O_5$) is a commonly used synergist. It is a pale yellow, oily, odorless liquid, soluble in organic solvents. Only slightly insecticidal per se, it has high synergistic properties when combined with pyrethrum, synthetic pyrethroids, rotenone, or several carbamates by increasing knock-down, and the lethal effects and residual capacity of these insecticides. In dusts, sprays (wettable powder, solution, emulsion), and gas aerosols (1% with 0.16% pyrethrums) or slurries, piperonyl butoxide has been employed to control *adult flies, gnats, mosquitoes, cockroaches, fleas, lice,* and *rodent mites*. In addition to derivatives of methylene dioxybenzene (the 1,3 benzodioxoles), the best known and most important class of insecticide synergists, other promising compounds include several series of aryl propynyl ethers, propynyl oxime ethers, propynyl phosphonate esters, and the benzylthiocyanates (Wilkinson, 1976).

Petroleum Products

Several petroleum oils and distillates can be used for effective control of insects, particularly mosquitoes. These include fuel oil, diesel oil, mineral oil, Flit MLO (a light oil with a high spreading pressure), kerosene, and xylene. These materials may be crude or fractioned saturated and unsaturated chains and ring hydrocarbons that are insoluble in water and useful per se as contact insecticides, ovicides, larvicides, and repellents. Their effectiveness varies with viscosity and volatility. However, their greatest insecticidal value consists of their incorporation with more potent toxicants.

Insect Growth Regulators

As an alternative to the more traditional pesticide groups already mentioned, many new chemicals having distinctly different modes of action have been formulated during the past decade. These compounds, termed insect growth regulators or insect development inhibitors (National Academy of Sciences, 1976), represent a "third generation" of insecticides, following the botanical derivatives and the more recent synthetic insecticides such as the organochlorines, carbamates, and organophosphorus materials. Generally, these materials cause reproductive dysfunction or inhibit the development of an insect or tick at some point before it becomes an adult. Thus, by their selectivity of action at a critical morphogenetic step, these materials are highly specific in action and relatively nonpersistent. This quality, while ecologically acceptable since biomagnification is avoided, requires special formulations, particularly in the use of slow-release capsules or other carriers.

The principal group of compounds being investigated are the juvenile hormone mimics which, by appearing at an inopportune time, may block insect metamorphosis and stagnate a potential disease vector in its egg, larval, or pupal stage. To date, extremely low concentrations have been effective against *flies* and *lice* (Chamberlain, 1975), *black flies* (Thompson and Adams, 1979), *mosquitoes* (McGovern et al., 1980), and *ants* (Edwards and Clarke, 1978). *Fleas* also appear to be susceptible. *Ticks,* while differing in the basic structure of their endocrine systems, are sensitive to juvenile hormone analogues (Obenchain, 1979). **Methoprene,** or isopropyl (2E,4E)-11-methoxy-3,7,11-trimethyl-2,4-dodecadienoate, is registered for use against *mosquitoes*. Numerous field tests have proved its efficacy, even against species resistant to traditional insecticides. When formulated in sand as a granular mixture and applied twice weekly, this JH mimic proved effective against asynchronous broods of *Culex nigripalpus*, a vector of Saint Louis encephalitis virus (Rathburn et al., 1980).

Diflubenzuron (Dimilin, TH-6040) or 1-(4-chlorophenyl)-3-(2,6-difluorobenzoyl) Urea acts as a growth regulator by inhibiting the activity of chitinase, thus preventing the normal deposition of cuticle. It is effective as a larvicide for *mosquitoes* and other *flies*. In dosages of 0.5 and 1.0 ppm, 80 to 100% control of *Culex pipiens fatigans* (= *Cu. quinquefasciatus)* larvae in polluted drains was achieved (Sharma et al., 1979). A large number of aliphatic amines also inhibit development and attack the egg, larval, and pupal stages of mosquitoes (Mulla et al., 1970; Cline and Hall, 1973).

Precocene II (6,7-dimethoxy-2,2-dimethyl-3-chromene). A natural product extracted from *Ageratum* spp., Precocene II has been shown to cause the corpora allata of insects to atrophy, thus preventing the production of juvenile hormone (Bowers et al., 1976). While still largely experimental, this and related compounds have been shown to exert a powerful morphogenetic effect on *triatomid bugs* (Tarrant et al., 1982), *mosquitoes* (Cupp et al., 1977), *ticks* (Leahy and Booth, 1980), and the human *body louse* (Feldlaufer and Eberle, 1980).

Fumigants

Methyl Bromide (CH_3Br). Colorless gas, odorless at low concentrations but poisonous to humans at inapparent concentrations; nonflammable in air, burns in O_2; irritating to skin and mucous membranes, vapor highly toxic on inhalation; slowly but highly insecticidal and rodenticidal. Standard formulation: 1 to 2.5 lb/1000 cu ft for 12 to 16 hours to destroy insects in homes, warehouses, ships, freight cars, and storage vaults; dispensed in fire extinguisher tanks for use in tightly closed quarters; also to disinfest clothing of *body lice* and *nits* (eggs), 30 minutes at 24 to 32°C, without danger of damaging cotton or woolen fabrics.

Chloropicrin (Trichloronitromethane; $CCl_3 \cdot NO_2$). Colorless or yellowish nonflammable liquid with strong odor, vaporizing at 112°C, having good penetrating properties; nonbleaching, nontarnishing, slightly corrosive. Highly irritating to human mucous membranes, especially the eyes ("tear gas"), lethal on strong inhalation. Highly insecticidal in closed containers; kills *grain* and *household insects,* including *lice* (formulated at 1 to 3 lb/1000 cu ft).

Ethylene Dichloride ($CH_2Cl \cdot CH_2Cl$) and Ethylene Oxide (C_2H_4O). The former is a colorless, oily, pleasant-smelling liquid, the latter a colorless gas vaporizing above 10.5°C with a faint sweetish odor; both are flammable, nonbleaching, not dangerously toxic, and slowly insecticidal. To reduce fire hazard, ethylene dichloride is combined with carbon tetrachloride, and ethylene dioxide with solid CO_2 in cylinders. It is employed as sprays to destroy *lice* in clothing, *bedbugs, grain mites,* and *beetles*.

Hydrocyanic Gas (Cyanogas; HCN). Colorless, highly volatile liquid with slight almond odor, changes to extremely poisonous gas at 26°C, requiring maximum operational precautions. Available in cylinders or in form of inert discs impregnated with calcium cyanide; a potent insecticide: 0.3 lb HCN or 75 lb 88% $Ca(CN_2)$ dust/1000 cu ft kills *household insects, bedbugs, rats;* in special vacuum chambers for 30 minutes is an effective pediculicide.

Sulfur Dioxide (SO_2). Colorless nonflammable gas having good penetrating ability; bleaches, tarnishes, and corrodes. A highly poisonous insecticide when dispensed from compressed liquid in cylinders (5 to 10% by volume in air) or from burning sulfur (4 to 8 lb/1000 cu ft); kills *household insects,* including *bedbugs*.

Repellents

Arthropod-borne infections can be prevented if individuals

are protected from the bites of vector species. This is frequently accomplished by the use of chemical barriers that act as repellents. These materials can be applied directly on the skin, impregnated in protective clothing, used as space repellents on wide mesh netting (Grothaus et al., 1974), or loosed into the immediate environment as a fog or aerosol (Means, 1973). A high degree of personal protection can also be afforded when clothing is treated with a synthetic pyrethroid that acts as a contact toxicant rather than a chemical repellent (Schreck et al., 1977, 1978). To date, permethrin has been shown to be effective against several kinds of mosquitoes, *Stomoxys calcitrans*, *Xenopsylla cheopis*, and *Amblyomma americanum*, the Lone Star tick (Schreck et al., 1978a, 1980).

N,N-diethyl-m-toluamide ($C_{12}H_{17}NO$). Organic liquid, soluble in water and several organic solvents; as mosquito repellent it can be applied to human skin without causing irritation. When impregnated in outer garments and socks (75 g/1.5 liters water containing 0.5% Tween 80 or Triton-x-100, saturating clothing, then drying), it repels biting flies and fleas. Commercially available as "Off" and "Deet."

Ethyl-hexanediol (Rutgers 612; $C_8H_{18}O_2$). Somewhat oily, slightly viscous clear liquid, sparingly soluble in water, soluble in alcohols and castor oil. Applied to skin does not produce irritation but on contact with eyes causes conjunctivitis. Standard effective repellent against *chiggers, mosquitoes, black flies*, and other *biting flies*, especially when combined with dimethyl phthalate or dimethyl carbamate and Indalone in proportion of 6/2/2; impregnated in clothing repels *ticks* and *fleas*.

Dimethyl Phthalate (DMP; $C_{10}H_{10}O_4$). Slightly aromatic, clear, oily liquid, with heavy vapor, somewhat irritating to mucous membranes and eyes, at times a CNS depressant. Common ingredient of insect repellents, but may be used alone to repel *mosquitoes* and *biting midges*. In laboratory tests found to be an effective repellent against tropical rat mite, *Ornithonyssus bacoti*.

Butylethyl-propandiol and Butylacetanilide. These materials are combined with benzyl benzoate in amounts of 30% each, with 10% detergent emulsifier (Tween 80 or Triton-x-100) for Formula M-1960, U.S. Army standard clothing impregnant as repellent against *ticks* and *chiggers*. Only trousers, blouse, and socks are treated (2 g/0.1 sq m cloth).

Benzyl Benzoate ($C_6H_5COOCH_2C_6H_5$). In crude form present in Peru and Tolu balsams; benzyl ester of benzoic acid occurs in leaflets or as oily liquid having faintly pleasant aromatic odor and sharp bitter taste; insoluble in water and glycerol, miscible with most organic solvents. Has been applied in 5% emulsion to skin as repellent for many arthropods, and topically as lotion or emulsion for *sarcoptic mange, demodicid* and *otodectic mange* of domestic animals, and *canine pediculosis*.

Insecticide Poisoning

Poisoning by chlorinated hydrocarbons is treated with barbiturates, especially phenobarbital, to which, in cases of BHC poisoning, atropine should be added. Poisoning by organophosphorus insecticides is treated with atropine, as is poisoning by the carbamates. In all cases prompt medical attention should be obtained. If the poisoning is due to contact, clothes should be removed and the body thoroughly washed with soap and water.

Insecticides should be kept out of reach of children, illiterate persons, and domestic animals. Spraying operatives should be well-trained in safety precautions, especially as regards clothing and bathing, and proper disposal of pesticides.

AGENTS EMPLOYED IN CONTROL OF MOLLUSCS

General Considerations

On the basis of their temporary or more lasting effectiveness, molluscicides can be divided into two groups: *nonresidual* and *residual toxicants*. Depending on the species of mollusc to be controlled and the field conditions under which it lives, the following factors should be considered in the selection of a suitable molluscicidal agent: its solubility in water and other solvents; the effect of alkalinity in the water and of sunlight in reducing its potency; its immediate or slowly developing toxicity for the mollusc and its eggs; effects of the agent on fishes, other animals, and plants in the same aquatic environment; and the hazards to people and to the economy of the community as a result of its use. Also to be considered are the amount and cost of the toxicant required to kill a high percentage of the molluscs and their eggs—in other words, the optimal agent and its dosage needed, on the basis of solubility, rapidity of action, and toxic effectiveness, to bring about control without seriously or permanently endangering the environmental balance or the human population. In some programs it may be desirable to combine nonresidual and residual molluscicides to achieve this end. Coupled with these requirements is the need for information concerning the most satisfactory method of application of the toxicant, whether by continuous drip feed, as in an irrigation canal, by hand or high-pressure sprayer, or by airplane dispersal. All these factors must be considered in planning any molluscicidal program (Malek, 1980; Malek and Cheng, 1974).

Molluscicides of Plant Origin

Endod. (Obtained by drying and grinding the ripe berries of *Phytolacca dodecandra*.) The active principle is a highly water-soluble triterpene glycoside; as a molluscicide, it has an LC_{90} of about 20 ppm (Lemma, 1970). Endod is also toxic to leeches, schistosome cercariae, and miracidia.

Saponin. (A glucoside from the fruit, kernel, bark, roots, and branches of the tree *Balanites aegyptiaca*, also *Sapindus saponaria*, *Swartzia madagascariensis*, and *Tephrosia vogeli*, indigenous to tropical Africa, Arabia, Egypt, and Israel.) When crushed in water, the fruit produces a foamy colloidal suspension, toxic to fishes and other cold-blooded animals, not to mammals. Field trials have demonstrated that saponin is lethal for *Bulinus*, *Physopsis*, *Biomphalaria*, and *Lymnaea*.

Rotenone. (Commonly obtained from the plant *Derris elliptica*.) This compound yields 8% toxic product in powder form.

Although commonly employed as an insecticide, it has molluscicidal properties as well.

Chemical Molluscicides

Inorganic Compounds

Copper (cupric) Sulfate ($CuSO_4 \cdot 5H_2O$). Blue transparent crystals, blue granules, or light blue powder, soluble in water (1 g/3 ml). Long employed as fungicide; first demonstrated by Chandler (1920) to be useful in control of *Lymnaea* and other snail hosts of *Fasciola* spp. and liver flukes of sheep and cattle. Rapidly becomes ineffectual in alkaline waters and those with high organic content; nonresidual; effectiveness requires clearing aquatic vegetation before use. Somewhat toxic to fish but essentially harmless to man in concentrations employed; highly corrosive to metal containers and applicators. Still widely employed in control of pulmonate snails *(Biomphalaria* in Venezuela, *Biomphalaria* and *Bulinus* spp. in Egypt, *B. (Physopsis)* spp. in South Africa); is only slightly ovicidal in concentrations used for adult snails; hence, requires at least two applications within a short period.

Calcium Arsenate ($Ca_3 \cdot [AsO_4]_2$) and Calcium Arsenite (of Variable Composition). White powder or granules, used in several countries as molluscicides; the arsenite has been sprayed at the rate of 5 to 10 g/sq m against *Oncomelania hupensis* in *Schistosoma japonicum* endemic areas of continental China, with 85% kill reported.

Zinc Oxide (ZnO). White or yellowish white amorphous powder, odorless, insoluble in water; contact and stomach toxicant for aquatic snails; essentially harmless to fishes, aquatic crustacea and vegetation, humans, and domestic animals. Laboratory tests indicate its lethal effect on *Bulinus contortus* and *Biomphalaria glabrata* in 15 days. using 30 g/100 l water; relatively expensive as residual molluscicide in the field.

Organic Compounds

Sodium Pentachlorophenate (NaPCP, Santobrite, Dowicide; from basic C_6Cl_5OH). Needle-like crystals; dust causes sneezing; moderately soluble in water and organic solvents. Originally used as herbicide; poisonous for fishes (immediately lethal) and other aquatic animals; toxic for humans when absorbed through skin or inhaled as dust, especially during preparation of spraying solutions. May be dispersed over snail habitats by manual or power sprayer. In deep bodies of water NaPCP tends to be trapped at (warmer) surface level, where it is exposed to sunlight and soon loses effective molluscicidal concentration; low concentration over longer periods would otherwise probably be preferable; low concentrations stimulate snail's respiration, high concentrations inhibit respiration; less toxic in alkaline than in slightly acid waters. Superior to $CuSO_4$ because vegetation does not need to be cleared beforehand from snail habitats; sprayed in dilution of 5 to 7 ppm twice yearly, NaPCP has been used effectively in Puerto Rico against *Biomphalaria glabrata*; in Africa against *Biomphalaria, Bulinus,* and *Lymnaea;* and in South America against *Biomphalaria*. NaPCP has proved a valuable molluscicide, in spite of its demonstrated drawbacks.

Copper Pentachlorophenate (Copper Salt of Pentachlorophenol). Insoluble in water, dispensed as a mixture half-and-half with NaPCP, combining the two products in the field just before application; a highly effective molluscicide, nonirritating to the snails, so that they do not crawl out of the treated area, and has long residual effect, which is not claimed for NaPCP.

Niclosamide (Bayluscide) (Ethanolamine salt of 5,2′-dichloro-4′nitro-salicylic anilide). Yellowish, odorless, almost tasteless powder, sparingly soluble in water, more soluble in organic solvents; molluscicidal properties reduced in highly alkaline water, not seriously affected by ultraviolet radiation under 12 hours. As 70% wettable powder in concentration of 0.3 to 1.0 ppm, produces 100% lethal effect on *Biomphalaria glabrata, B. pfeifferi, B. sudanicus, Bulinus contortus, B. coulboisi, B. forskalii, B. tropicus, B. (Physopsis) globosus, Oncomelania hupensis, O. quadrasi, Lymnaea natalensis, L. stagnalis, L. palustris, Radix ovata,* and *Bythinia tentaculata,* and 50% on *Physa* sp. These results are comparable to those requiring 10 ppm of $CuSO_4$ and NaPCP. Although frogs succumb at 1 ppm of niclosamide, most aquatic arthropods survive. Miracidia and cercariae are killed at minimal lethal concentrations for the snail hosts. Aquatic plants and warm-blooded animals are not damaged by exposure to 1 ppm or less for a considerable period. Niclosamide fails to irritate the snail or to cause excess mucus secretion, but stimulates it to retract into its shell. Like NaPCP, niclosamide inhibits O_2 uptake, and unlike NaPCP, the reaction produced by niclosamide is irreversible. Eggs of *Lymnaea* and *Biomphalaria* are killed at doses lethal for adult snails.

Ziram (Dimethyldithiocarbamate of Zinc). White powder, capable of forming flammable dust; insoluble in water, slightly soluble in organic solvents. Formerly employed as a fungicide, has definite molluscicidal properties in concentrations of 0.5 to 2.0 ppm, producing 100% kill of *Biomphalaria pfeifferi, B. gaudi, Bulinus sp.,* and *Lymnaea caillaudi* in 48 hours. Action is slower than that of copper sulfate, but remains effective in presence of high concentration of organic matter, diffusing through aquatic vegetation to reach snails; is stable in alkaline waters. At concentrations below 1 ppm, it is nontoxic to other aquatic fauna, domestic animals, and humans; also inhibits development of snail eggs, which die in egg masses.

Frescon, WL 8008 (*N*-tritylmorpholine). Can be formulated in a variety of ways, including emulsifiable concentrates, water-dispersible powders, and granules. Highly toxic to snail intermediate hosts of the Schistosomatidae; a concentration as low as 0.015 mg/liter was sufficient to produce 100% mortality in *Bulinus truncatus* in 7.5 days in the Sudan (Amin et al., 1976). Inactive against snail eggs. Applications for periods of 2 to 4 weeks at low concentration may be necessary to kill adult snails and destroy young snails as they hatch. Environmental factors have little effect on performance, but hydrolysis and loss of activity may occur in acid water. Has low toxicity for mammals.

CONTROL OF RODENTS AS VECTORS AND RESERVOIRS OF DISEASE

Rodents and Human Diseases. Rodents serve as effective reservoirs for a variety of human pathogens. Some are transmitted by ectoparasitic arthropods, others by handling infected rodents or their tissues, or contamination of food, water, or air by rodent feces or urine. The more important rodent-associated diseases include plague, murine typhus, leptospirosis, salmonellosis, and lymphocytic choriomeningitis. Norway rats *(Rattus norvegicus* Berk) are also aggressive and may inflict painful bites, thereby transmitting several bacterial agents capable of causing "rat-bite" fever. Rats and mice may also serve as hosts for such parasitic worms as *Trichinella spiralis, Hymenolepis nana* and *H. diminuta,* and *Angiostrongylus cantonensis*.

Satisfactory rodent control depends largely on

the removal of food and shelter. A control program for commensal rodents should include elimination of harborage, removal of food supply, rodent proofing of buildings, and destruction of rats and mice by trapping and poisoning (Brooks and Rowe, 1979). In ports, barriers should be used to prevent rats on ships from reaching the shore.

Rodenticides. Fumigants may be used to kill rodents and their ectoparasites in buildings, ships, or burrows in the soil. Fumigants most commonly used are *calcium cyanide* ($Ca(CN)_2$), *methyl bromide* (CH_3Br), *chloropicrin* (CCl_3NO_2), *hydrogen phosphide* (H_3P), *carbon dioxide* (CO_2), *carbon monoxide* (CO), and *sulfur dioxide* (SO_2)(Brooks and Rowe, 1979).

Most control efforts, however, involve the use of toxic substances incorporated into a bait or distributed as a powder. Based on their rapidity of action, these rodenticides are classified as acute or chronic. This latter group includes the anticoagulants which, if ingested in suitable amounts over several days, cause fatal internal hemorrhages. A number of these compounds are available and are usually considered the chemicals of choice for rodent control since they are relatively safe for humans and other nontarget animals. Some of the more common anticoagulants include *Warfarin,* (3-α-acetonylbenzyl)-4-hydroxycoumarin; *Pindone,* 2-pivalyl-1,3-indandione; *Fumarin,* 3-(α-acetonylfurfuryl)-4-hydroxycoumarin; *Diphacinone,* 2-diphenylacetyl-1,3-indandione; and *Difenacoum,* (3-*p*-diphenyl-1,2,3,4-tetrahydronaph-1-yl)-4-hydroxycoumarin, a recently discovered compound that is moderately effective against warfarin-resistant rats (Hadler and Shadbolt, 1975). *Brodifacoum,* 3-(3-[4′-bromobiphenyl-4-yl]-1,2,3,4-tetrahydronaph-l-yl)-4-hydroxy-coumarin, and *Bromadiolone,* (hydroxy-4′-coumarinyl-3′)-3-phenyl-(bromo-4-biphenyl-4′)-1-propanol-1), are also new, potent hydroxycoumarin derivatives that are effective at field concentrations of 0.005% against the Norway rat (Rennison and Dubock, 1978; Marsh, 1977).

Acute rodenticides are ordinarily used for short-term efforts. They should not be used more than twice a year in order to avoid selection of bait or toxic "shy" populations. Although these materials vary in toxicity to humans, they should be handled with care and by skilled workers. *Zinc phosphide, Red Squill* (a powder derived from the plant *Urginea maritima),* and *Norbormide,* 5(α-hydroxy-α-2-pyridylbenzyl)-7-(α-2-pyridylbenzylidene)-5-norbornene-2,3-dicarboximide, are moderately to minimally hazardous. *Sodium fluoroacetate* (1080), a highly toxic substance, can be used on a restricted basis, such as in sewers, ships, and buildings, where complete control of the rodenticide is possible.

Chemosterilants. A number of these steroidal and nonsteroidal compounds have been investigated, but methods for their use on a large scale have not yet been perfected (Marsh and Howard, 1973). It has been suggested that they should be used in conjunction with environmental management practices and poisons where the prebaiting technique is employed, so that rodents failing to eat a lethal dose of the poison may have reduced reproducing potential for some time afterwards (Brooks and Rowe, 1979).

Biological Methods. Efforts have been made to reduce the rat populations by utilizing their natural enemies in limited areas. The Japanese weasel, *Mustela sibirica itatsi,* has been introduced into the Ryukyu islands with some success on those islands where weasels and rodenticides were utilized together (World Health Organization, 1968).

REFERENCES

Amin, M.A., Fenwick, A., Osgerby, J.M., Warley, A.P., and Wright, A.N. 1976. A large-scale snail control trial with Trifenmorph in the Gezira Irrigation Scheme, Sudan. Bull. W.H.O., *54*:573–595.

Bay, E.C., Berg, C.O., Chapman, H.C., and Legner, E.F. 1976. Biological control of medical and veterinary pests, pp. 457–479. In *Theory and Practice of Biological Control.* Edited by C.B. Huffaker and P.S. Messenger. New York, Academic Press.

Berg, C.O. 1973. Biological control of snail-borne diseases: A review. Exp. Parasitol., *33*:318–330.

Bowers, W.S., Ohta, T., Cleere, J.S., and Marsella, P.A. 1976. Discovery of insect anti-juvenile hormones in plants. Science, *193*:542–547.

Brooks, G.T. 1974. *Chlorinated Insecticides. Vol. I. Technology and Application.* Cleveland, CRC Press.

———, 1974a. *Chlorinated Insecticides. Vol. II. Biological and Environmental Aspects.* Cleveland, CRC Press.

Brooks, J.E., and Rowe, F.P. 1979. X. Commensal rodent control. WHO Mimeograph Series in Vector Biology and Control (WHO/VBC/79.726).

Busvine, J.R. 1980. *Insects and Hygiene.* London, Chapman and Hall.

Casida, J.E., editor. 1973. *Pyrethrum. The Natural Insecticide.* New York, Academic Press.

Centers for Disease Control. 1981. *Insecticides for the Control of Insects of Public Health Importance.* Atlanta, U.S. Department of Health and Human Services, Public Health Service.

Chamberlain, W.F. 1975. Insect growth regulating agents for control of arthropods of medical and veterinary importance. J. Med. Entomol., *12*:395–400.

Chandler, A.C. 1920. Control of fluke diseases by destruction of the intermediate host. J. Agric. Res., *20*:193–208.

Cline, R.E., and Hall, M.H. 1973. Larvicidal activities of N,N′-dialkylakylenediamines and other fatty diamines in oil films and water dispersions. J. Econ. Entomol., *66*:697–702.

Coppedge, J.R., Brown, H.E., Goodenough, J.L., Tannahill, F.H., Snow, W.J., Petersen, D.H., and Hofmann, H.C. 1980. Field performance of a new formulation of the screwworm adult suppression system. J. Econ. Entomol., *73*:411–414.

Cupp, E.W., Lok, J.B., and Bowers, W.S. 1977. The developmental effects of 6,7-dimethoxy-2,2-dimethyl chromene on the preimaginal stages of *Aedes aegypti* (Diptera: Culicidae). Entomol. Exp. Appl., *22*:23–28.

Edwards, J.P., and Clarke, B. 1978. Eradication of Pharaoh's ants with baits containing the insect juvenile hormone analogue methoprene. Int. Pest Control, *20*:5–10.

Elliott, M., and Janes, N.F. 1978. Synthetic pyrethroids—a new class of insecticides. Chem. Soc. Rev., *7*:473–505.

Eto, M. 1974. *Organophosphorus Pesticides: Organic and Biological Chemistry.* Cleveland, CRC Press.

Feldlaufer, M.F., and Eberle, M.W. 1980. Insecticidal effect of precocene II on the body louse *Pediculus humanus.* Trans. R. Soc. Trop. Med. Hyg., *74*:398–399.

Ferguson, K.J., and Cawthorne, J.H. 1980. Boom spray application of insecticide for control of estuary breeding biting midges *(Culicoides,* Ceratopogonidae). Mosquito News, *40*:217–221.

Georghiou, G.P., and Taylor, C.E. 1977. Genetic and biological influences in the evolution of insecticide resistance. J. Econ. Entomol., *70*:319–323.

———, and ———. 1977a. Operational influences in the evolution of insecticide resistance. J. Econ. Entomol., *70*:653–658.

Gerberg, E.J., and Visser, W.M. 1978. Preliminary field trial for the biological control of *Aedes aegypti* by means of *Toxorhynchites brevipalpis,* a predatory mosquito larva. Mosquito News, *38*:197–200.

Goldberg, L.J., and Margalit, J. 1977. A bacterial spore demonstrating rapid larvicidal activity against *Anopheles sergentii, Uranotaenia ungiculata, Culex univittatus, Aedes aegypti* and *Culex pipiens*. Mosquito News, *37*:355–358.

Grothaus, R.H., Gouck, H.K., Weidhaas, D.E., and Jackson, S.C. 1974. Wide-mesh netting, an improved method of protection against blood-feeding Diptera. Am. J. Trop. Med. Hyg., *23*:533–537.

Hadler, M.R., and Shadbolt, R.S. 1975. Novel 4-hydroxycoumarin anticoagulants active against resistant rats. Nature, *253*:275–276.

Heller-Haupt, A., Varma, M.G.R., Crook, S., and Radalowicz, A. 1979. The effect of synthetic pyrethroids on some African Ixodidae, pp. 85–93. In *Recent Advances in Acarology. Vol. II*. Edited by J.G. Rodriguez, New York, Academic Press.

Jamnback, H. 1976. IV. Simuliidae (blackflies) and their control. WHO Mimeograph Series in Vector Biology and Control (WHO/VBC/76.653).

Kuhr, R.J., and Dorough, H.W. 1976. *Carbamate Insecticides: Chemistry, Biochemistry and Toxicology*. Cleveland, CRC Press.

LaBrecque, G.C., and Smith, C.N., editors. 1968. *Principles of Insect Chemosterilization*. New York, Appleton-Century-Crofts.

Laird, M., editor. 1981. *Biocontrol of Medical and Veterinary Pests*. New York, Praeger Publishers.

Leahy, M.G., and Booth, K.S. 1980. Precocene induction of tick sterility and ecdysis failure. J. Med. Entomol., *17*:18–21.

Lemma, A. 1970. Laboratory and field evaluation of the molluscicidal properties of *Phytolacca dodecandra*. Bull. W.H.O., *42*:597–612.

Lofgren, C.S. 1972. Ultralow volume (ULV) application of insecticides. Am. J. Trop. Med. Hyg., *21*:819–824.

Malek, E.A. 1980. *Snail-Transmitted Parasitic Diseases, Volumes I and II*. Cleveland, CRC Press.

———, and Cheng, T.C. 1974. *Medical and Economic Malacology*. New York, Academic Press.

Marsh, R.E. 1977. Bromadiolone, a new anticoagulant rodenticide. EPPO Bull., *7*:495–502.

———, and Howard, W.E. 1973. Prospects of chemosterilant and genetic control of rodents. Bull. W.H.O., *48*:309–316.

Martinez, A., Cichero, J.A., Alania. I.R., and Gonzalez, F.F. 1975. Control of *Triatoma infestans* (Klug) with malathion concentrate. J. Med. Entomol., *11*:653–657.

Matthews, G.A. 1979. *Pesticide Application Methods*. London, Longman.

McGovern, T.P., Dame, D.A., Focks, D.A., and Cameron, A.L. 1980. Laboratory and field evaluation and structure-activity relationships of some aryl terpenoid ethers as insect growth regulators against *Anopheles quadrimatulatus, Anopheles albimanus*, and *Aedes taeniorhynchus*. J. Econ. Entomol., *73*:443–448.

McJunkin, F.E. 1970. *Engineering Measures for Control of Schistosomiasis*. Washington, D.C., Office of Health, Bureau of Technical Assistance, Agency for International Development.

Means, R.G. 1973. Preliminary evaluation of the effectiveness of Mosquito Beater[R], a granular repellent, against mosquitoes and blackflies. Mosquito News, *33*:542–544.

Metcalf, R.L. 1975. Pest-management strategies for the control of insects affecting man and domestic animals, pp. 529–564. In *Introduction to Insect Pest Management*. Edited by R.L. Metcalf and W.H. Luckman. New York, John Wiley and Sons.

Miller, B.E., Bennett, W.C., Graves, G.N., and Wheeler, J.R. 1977. Field studies of systemic insecticides. II. Evaluation of chlorphoxim for control of fleas on five rodent species. J. Med. Entomol., *14*:161–166.

Mount, G.A. 1979. IX. Ultra-low-volume application of insecticides. A guide for vector control programmes. WHO Mimeograph Series in Vector Biology and Control (WHO/VBC/79.734).

———, Grothaus, R.H., Baldwin, K.F., and Haskins, J.R. 1975. ULV sprays of Propoxur for control of *Trombicula alfreddugesi*. J. Econ. Entomol., *68*:761–762.

———, ———, Reed, J.T., and Baldwin, K.F. 1976. *Amblyomma americanum:* Area control with granules or concentrated sprays of Diazinon, Propoxur and Chlorpyrifos. J. Econ. Entomol., *69*:257–259.

———, ———, ———, and ———. 1978. Area control of chigger mites with granules and concentrated sprays of Chlorpyrifos. J. Econ. Entomol., *71*:27–28.

Mulla, M.S., Darwazch, H.A., and Gillies, P.A. 1970. Evaluation of aliphatic amines against larvae and pupae of mosquitoes. J. Econ. Entomol., *63*:1472–1475.

Muller, R. 1979. Guinea worm disease: Epidemiology, control, and treatment. Bull. W.H.O., *57*:683–689.

National Academy of Sciences. 1976. *Pest Control: An Assessment of Present and Alternative Technologies. Vol. V. Pest Control and Public Health*. Washington, D.C. National Academy of Sciences.

Obenchain, F.D. 1979. Non-acaricidal chemicals for the management of acari of medical and veterinary importance, pp. 35–43. In *Recent Advances in Acarology, Vol. II*. Edited by J.G. Rodriguez. New York, Academic Press.

Pal, R., and La Chance, L.E. 1974. The operational feasibility of genetic methods for control of insects of medical and veterinary importance. Annu. Rev. Entomol., *19*:269–291.

Plapp, F.W., Jr., Browning, C.R., and Sharpe, P.J. 1979. Analysis of rate of development of insecticide resistance based on simulation of a genetic model. Environ. Entomol., *8*:494–500.

Rathburn, C.B., Boike, A.H., Hallmon, D.F., and Cotterman, S.G. 1980. Small plot field tests of Methoprene for control of asynchronous broods of *Culex nigripalpus* Theob. in Florida. Mosquito News, *40*:19–23.

Rennison, B.D., and Dubock, A.C. 1978. Field trials of WBA 8119 (PP 581, Brodifacoum) against warfarin resistant infestations of *Rattus norvegicus*. J. Hyg. Camb., *80*:77–82.

Roberts, D.W., and Strand, M.A., editors. 1977. Pathogens of medically important arthropods. Bull. W.H.O., *55*(suppl. 1).

Schreck, C.E., Posey, K., and Smith, D. 1978. Durability of Permethrin as a potential clothing treatment to protect against blood-feeding arthropods. J. Econ. Entomol., *71*:397–400.

———, Smith, N., Weidhaas, D., Posey, K., and Smith, D. 1978a. Repellents vs. toxicants as clothing treatments for protection from mosquitoes and other biting flies. J. Econ. Entomol., *71*:919–922.

———, Snoddy, E.L., and Mount, G.A. 1980. Permethrin and repellents as clothing impregnants for protection from the lone star tick. J. Econ. Entomol., *73*:436–439.

———, Weidhaas, D.E., Smith, N., and Posey, K.H. 1977. Chemical treatment of wide-mesh net clothing for personal protection against blood-feeding arthropods. Mosquito News, *37*:455–462.

Sharma, V.P., Batra, C.P., and Brooks, G.D. 1979. Laboratory and field evaluations of a growth-regulating compound (TH-6040) against *Culex pipiens fatigans*. J. Med. Entomol., *15*:506–509.

Tarrant, C., Cupp, E.W., and Bowers, W.S. 1982. The effects of precocene II on reproduction and development of triatomine bugs (Reduviidae: Triatominae). Am. J. Trop. Med. Hyg., *31*:416–420.

Thompson, B.H., and Adams, B.G. 1979. Laboratory and field trials using Altosid[R] insect growth regulator against black flies (Diptera: Simuliidae) of Newfoundland, Canada. J. Med. Entomol., *16*:536–546.

Undeen, A.H., and Colbo, M.H. 1980. The efficacy of *Bacillus thuringiensis* var. *israelensis* against blackfly larvae (Diptera: Simuliidae) in their natural habitat. Mosquito News, *40*:181–184.

Whitten, M.J., and Pal, R. 1974. *The Use of Genetics in Insect Control*. New York, Elsevier.

Wilkinson, C.F., editor. 1976. *Insecticide Biochemistry and Physiology*. New York, Plenum Press.

———. 1983. Role of mixed-function oxidases in insecticide resistance. pp. 175–205. In *Pest Resistance to Pesticides*. Edited by G.P. Georghiou and T. Saito. New York, Plenum Press.

World Health Organization. 1968. The use of weasels in rat control. WHO Chronicle, *22*:254–256.

———. 1975. Ecology and control of vectors in public health. Tech. Rep. Ser. No. 585.

———. 1976. Resistance of vectors and reservoirs of disease to pesticides. Tech. Rep. Ser. No. 561.

———. 1977. Engineering aspects of vector control operations. Tech. Rep. Ser. No. 603.

———. 1979. *Environmental Health Criteria. 9. DDT and Its Derivatives*. Geneva, WHO.

Wright, J.W. 1971. The WHO Programme for the Evaluation and Testing of New Insecticides. Bull. W.H.O., *44*:11–22.

Section V

Technical Appendix

Chapter 46

Examination of Specimens for Parasites

GENERAL CONSIDERATIONS

Laboratory diagnosis is a basic step in the evaluation of disease processes, at times confirming a presumptive clinical diagnosis, at other times providing evidence of an unsuspected agent of disease. Accurate laboratory diagnosis requires special training, skill, and sound judgment in recognizing true parasite objects and differentiating them from spurious objects and artifacts.

Examination of specimens for the diagnosis of parasitic infection involves two steps: first, the *detection* of a parasite object in the specimen, then its *identification*. Detection requires skills in carrying out procedures known to be effective in bringing the parasite objects into view grossly or under magnification. Identification requires skill in recognizing that the object observed is in fact a parasite, or a stage in the life cycle of a parasite (trophozoite, cyst, egg, larva, or nymph). The ability to distinguish parasite objects from nonparasite artifacts requires experience. Even well-trained technicians with limited experience tend to register more false-positive than false-negative reports because numerous objects resembling parasites may be seen in feces, urine, blood, sputum, and tissues. Examination of soil samples or fecal specimens contaminated with soil or water is especially difficult because many of the common free-living organisms and their cysts, eggs, and larvae closely resemble comparable stages of true parasites.

For a reliable parasitology diagnostic service, the facilities, equipment, and supplies must be appropriate and adequate. The most basic item is a microscope that is in good condition and is equipped with a suitable light source and clear, clean lenses. It is more important to have a microscope in good working condition than to have one of the latest design, equipped with modern accessories. A desirable accessory is an ocular micrometer and a means of calibrating it. Other essential items of equipment include a centrifuge, incubator, drying oven, and a means of sterilizing glassware and instruments.

In previous chapters, diagnosis was discussed in relation to the stages to be looked for in specimens, and the diagnostic features by which they are recognized and differentiated from other species of parasites, or from nonparasite objects. In this chapter and those to follow are descriptions of standard techniques of demonstrating the presence of parasitic infections, and certain guides to the selection of techniques that are appropriate for particular purposes, or references to relevant literature.

Calibration of the Microscope. Size is an important characteristic of all living things. Because the microscope magnifies the size of objects, and with the several lens combinations (oculars and objectives) objects are seen at several different magnifications, even highly experienced microscopists cannot reliably estimate the size of objects thus viewed. Therefore, for the recognition and description of microscopic parasites and their various stages, the microscope must be provided with a calibrated micrometer scale.

A micrometer disc that fits into the 10× ocular (eyepiece) is generally available. On it is marked an arbitrary scale with the smallest units measuring approximately 1, 2, and 10 μm *(micrometers,* formerly called *microns)* under oil-immersion (usually about 90×), high-dry (about 45×), and low-power (about 10×) objectives, respectively. Generally available also is a stage micrometer slide bearing a scale that is 1 mm long, divided into 0.1- and 0.01-mm intervals. This can be used to "measure" the units of the ocular micrometer scale for each of the objectives. When a stage micrometer is not available, a standard blood-cell counting chamber can be used to calibrate the ocular micrometer scale. The shortest dimension marked on standard blood-cell counting chambers in 50 μm or 0.05 mm. The number of units on the ocular scale that is equal to the 50 μm on the counting chamber can be used to compute the value of the whole of the ocular scale and each of its subunits under oil-immersion, high-dry, and low-power magnifications, respectively.

When an ocular micrometer scale is not available, a rough calibration can be done by installing a short eyelash pointer in the ocular and using the blood-cell counting chamber to measure its length under each magnification. To install a pointer, one can remove the upper lens of the ocular, place a small spot of glue on the "shelf" inside the ocular, and with a fine pointed forceps place on the glue a hair cut to the proper length to reach from the shelf about one-third of the distance to the center of the opening. This can serve as both a pointer and a micrometer scale having a known length at each magnification.

EXAMINATION OF FECES

In addition to microscopic examination for parasites, a fecal examination involves recognizing and recording gross elements and distinctive pathological odors or colors; unusual quantities of normal elements or abnormal elements, such as blood, cellular exudates, tissue fragments, Charcot-Leyden crystals, and undigested food; and the residues or products of special medications such as mineral oil, bismuth, iron, magnesium, or barium. Description of the fecal sample examined as well as the presence or absence of parasites can be helpful in assessing the reliability and significance of the report, as well as the condition of the patient. Stool specimens usually are described with reference to three qualities: consistency, color, and elements. A fourth characteristic, odor, would assist in interpreting the other findings if it could be accurately described, but the only term generally used in describing the odor of a stool specimen is "foul-smelling," which is too subjective to be useful. A stool classification and code that originated at the Tulane Tropical Medicine Clinic around 1948 makes it possible to record the chief characteristics of individual specimens with a three-digit code number (Table 46–1). With consistency graded from hard to watery (1 to 7), color from black to clay, or "other" (1 to 8), and elements graded from pulp and fiber nearly pure, to mucus with scanty feces or "other" (1 to 8), normal specimens that are formed, soft, or mushy in consistency, brown in color, and mostly colloidal would be coded as a combination of 2, 3, and 4, such as 333, 234, or 323.

Collection of the Specimen

Specimens should be collected in a clean container without contamination with urine, water, or other extrinsic substances. The amount of feces requested may range from the whole stool, or series of stools over a specified period, to milligram amounts scraped from the gloved finger after rectal examination. For routine microscopic examination, a 2- to 5-gram specimen protected from drying is adequate. The specimen should be free of oil and other nonfecal substances such as barium or bismuth. Specimens obtained by purgation or enema are generally less satisfactory than those taken from the normally passed stool; the chief advantage of such means of collection is ready availability on schedule.

Table 46–1. *Stool Classification and Code*

Consistency	*Color*	*Elements*
1. Hard, resists puncture	1. Black	1. Pulp and fiber nearly pure
2. Formed, can be punctured	2. Dark brown	2. Conspicuously fibrous
3. Soft, can be cut with applicator	3. Brown	3. Fiber moderate to scanty
4. Mushy, can be reshaped	4. Pale brown	4. Pure colloidal feces
5. Loose, shapes to container	5. Yellow	5. Feces with scanty mucus
6. Diarrheic, flows	6. Green	6. Feces with much mucus
7. Watery, pours	7. Clay	7. Mucus with scanty feces†
	8. Other*	8. Other (blood, barium)

*Color, such as blue, red, or purple, can indicate special medication.

†A specimen containing gross blood and mucus, with or without pus, would be recorded as "dysenteric" and would not be coded for consistency or color.

A formed stool specimen can be kept overnight at moderate or low room temperature without loss of diagnostic value, but specimens of diarrheic and dysenteric stools should be examined promptly or preserved. Others that cannot be examined within 3 or 4 hours of collection should be stored under refrigeration or preserved chemically in formalin, polyvinyl alcohol (PVA), merthiolate-iodine-formalin (MIF), or Schaudinn's fixative for examination by a special technique.

Choice of Method

For the microscopic examination of feces for parasitic infection, there is no all-purpose technique. Nor is there any "best" technique for any one purpose. For reliable diagnosis of intestinal parasites, a combination of several techniques is required. Reliability of the results of an examination depends less on the technique selected than on how well the procedures of the selected technique are carried out. Investigator experience and technician training are factors of great importance (Kaminsky, 1978). Some form of quality control should be a part of the routine of every diagnostic laboratory (Ayala and Sánchez, 1974; Petithory, 1979).

Direct Examination

Direct Saline Smear. The direct fecal smear, or saline suspension, is almost universally relied upon to a large extent for the diagnosis of intestinal parasitic infections. Frequently, a concentration technique is also used for further examination of the same specimen. For the detection of clinically significant infections, however, a direct smear, if properly prepared and skillfully examined, is adequate in most cases. In the hospital or clinic, the direct smear examination should be a routine procedure regardless of what other kind(s) of examination of the specimen may be made.

To prepare the direct fecal smear, place 1 drop of saline (0.85% NaCl) on the center of a clean slide. With an applicator select about 2 mg of feces (1 mg is approximately 1 cubic mm) and, without spreading the drop of saline, stir and daub until nearly the entire 2-mg sample is suspended. Remove any gross fibers or particles and cover with a 22 × 22 mm glass or plastic coverslip. Avoid air bubbles by drawing one edge of the coverslip slightly into the suspension and lowering it almost to the slide before letting it fall. If the preparation is unsatisfactory, it should be discarded. No time should be wasted on a preparation that can only yield doubtful results if negative. The ideal smear is one that contains a maximum of observable fecal elements without any object being obscured if it is within the size range of the smallest protozoa (about 5 μm).

To examine the smear, begin at one corner of the preparation and systematically examine successive adjacent swaths with the low power of the microscope (about 100×). All parasite objects to be found in feces, protozoa as well as eggs and larvae, are readily seen at 100× magnification if the light and focus are proper. With the light properly adjusted and the field in sharp focus, individual bacteria and brownian movement can be clearly seen. When a parasite-like object comes into view, it can be more closely examined and identified under the high power. The oil immersion cannot be used effectively on a direct-smear preparation.

Trophozoites and cysts of protozoa are lens-like in shape and clear. Therefore, they tend to shine and to be more conspicuous in the microscope field than are other objects of similar size. To identify protozoa, it is usually necessary to stain them. The stain most widely used is aqueous iodine solution (1% aqueous solution of potassium iodide saturated with iodine crystals). The stain can be placed at the edge of the cover glass and allowed to flow into the film, or it can be added to a new preparation before it is covered. When the stain is applied to the edge of the cover glass, its flow tends to sweep bacteria and other small particles away from the protozoa leaving them in a clear field and more readily detected. Helminth eggs and larvae are readily identified without stain, but the iodine stain can be used advantageously in the examination of larvae as it immobilizes and kills them and stains some parts differentially. However, for larvae the stain must be used sparingly since it tends to overstain.

Because the oil immersion lens cannot be used effectively on wet-mount preparations, the small species of intestinal protozoa can be only presumptively diagnosed in direct smear or other temporary wet preparations. This is especially true when the organisms are both minute and present in small numbers. Therefore, unless the small amebae have been identified on permanently stained fecal films, findings should be accepted with reservation. An exception, of course, is the finding of hematophagous *Entamoeba histolytica* in the motile trophozoite stage.

Kato Thick Smear. A technique of fecal examination generally referred to as the "Kato thick smear" was introduced by Kato and Miura in 1954. It was evaluated, standardized, and adapted to large-scale routine annual fecal examination of school children in Japan by Komiya and Kobayashi (1966). It soon was recognized as highly efficient for the detection and identification of the eggs of *Necator, Ancylostoma, Trichuris, Ascaris* and some other helminths, including *Schistosoma mansoni*. It was improved and adapted for quantitative as well as qualitative diagnosis by Martin and Beaver (1968).

Materials. The special materials required are:

1. Coverslips made of wettable cellophane* of medium thickness, cut into 22 × 30 mm rectangles.
2. Glycerine-malachite green solution: 10 ml pure glycerine and 1 ml 3% aqueous malachite green in 100 ml water. Allow the cellophane coverslips to soak in the glycerine mixture for at least 24 hours. (Although the malachite green reduces eyestrain at the microscope, it is not essential.)
3. Stainless steel bolting cloth, 105-mesh,† cut into 4-cm squares. A bolting cloth made of a sturdy synthetic fiber and of similar mesh can be used in place of the stainless steel.

Procedure.

1. With applicator, transfer about 50 mg of feces to a microscope slide. (A 4-mm cube of feces weighs about 65 mg.) For fibrous specimens, put about 2 g of feces on a disposable sheet of paper, place over it a square of the bolting cloth, then remove a sample of screened feces with an applicator stick by pressing the cloth into the feces and scraping the applicator across the surface of the cloth. Hard, formed fecal specimens require addition of a few drops of water and mixing to produce a mushy or pasty consistency.
2. Cover the fecal sample with a cellophane coverslip, turn the preparation upside down on a flat absorbent surface, such as a sheet of thick soft paper on a table top, and press it until the fecal film covers an area 20 to 25 mm in diameter. When the amount of feces placed on the slide is excessive, the excess usually flows from under the coverslip and is discarded with the soiled sheet of paper.
3. Allow the preparation to stand face up for about 1 hour at room temperature and moderate relative humidity, or for 20 to 30 minutes in a dry incubator or under an electric lamp. This clears the feces but leaves helminth eggs uncleared (Fig. 46–1). Since over-clearing causes the collapse and disappearance of delicate thin-walled eggs of hookworms and some other species, the preparation must be examined immediately upon clearing or, to avoid over-clearing, the drying process can be temporarily stopped by turning the slide facedown on a flat, smooth surface.
4. Examine the entire film under low magnification. With experience, microscopists can detect and identify eggs of all common species of helminths under scanning-lens magnification (35-40×).

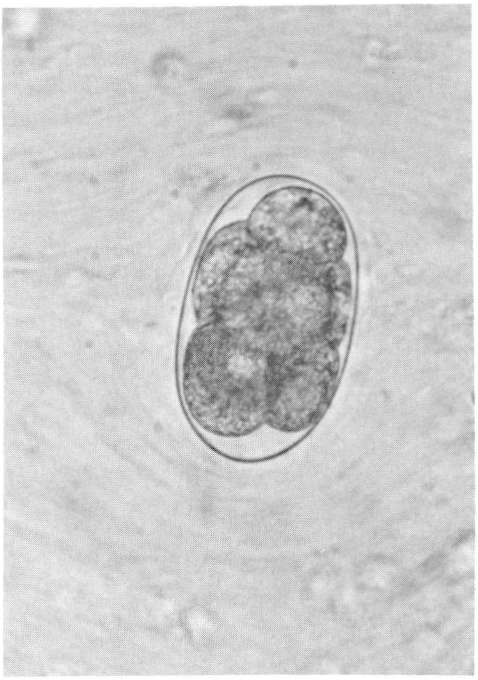

Fig. 46–1. Egg of *Necator* after clearing of feces in a Kato thick smear. (× 540.) (From Faust, E.C., Beaver, P.C., and Jung, R.C. 1975. *Animal Agents and Vectors of Human Disease,* 4th ed. Philadelphia, Lea & Febiger.)

The Kato thick-smear technique has been widely used and evaluated. Numerous surveys and special studies based on results from this simple procedure have been reported. It is the method of choice for the diagnosis of intestinal helminths. After it was shown to be satisfactory for quantitative diagnosis (Martin and Beaver, 1968), it was used extensively in studies on schistosomiasis mansoni (Katz et al., 1972; Peters et al., 1980; Bartholomew et al., 1981). (See the section on egg-count techniques.) It is unsuitable for the diagnosis of protozoa or

*No. 124 PD, E.I. DuPont de Nemours, Inc., Film Department, Wilmington, Delaware

†W.S. Tyler Co., Cleveland, Ohio

helminth larvae. It is also unsuitable for examining hard or liquid stools, and micro-bubbles of flatus are troublesome when numerous, as in frothy stools.

Permanent, Stained, Fresh Fecal Smear

The permanent, stained fecal smear frequently is needed for a reliable identification of protozoa, especially *Entamoeba histolytica* and other small amebae. Such preparations offer two major advantages: 1) The diagnostic features are more sharply differentiated than they are in other types of preparations, and they can be viewed at higher magnification. Only in a permanent stained smear can the organisms be examined effectively under high magnification of the oil-immersion lens. 2) The second advantage of the permanent stained smear is that two or more microscopists, in the same laboratory or in distant ones, can examine the same preparation, even the same individual organism, for confirmation or correction of a diagnosis. This is an advantage not generally appreciated and too seldom used. A third advantage of the permanent preparation is that, like an EKG or x-ray film, it can be kept for future examination as part of the patient's record. Two different stains are in general use: iron hematoxylin and trichrome. The iron hematoxylin has been a standard stain used by protozoologists for more than a century. The original descriptions of all intestinal protozoa of man were based primarily on morphological features evident in iron-hematoxylin stained fecal smears. The trichrome stain was developed by Gomori in 1950 for tissue differentiation and was adapted by Wheatley in 1951 for intestinal protozoa. It is preferred by some workers for routine use in the diagnostic laboratory because the procedure requires less time, and presumably demands less care and skill. For fine morphological detail, iron hematoxylin is preferred. Either stain can be used for fresh specimens or for specimens preserved in PVA or Schaudinn's fixative solution.

Iron-Hematoxylin-Stained Smear. The smear is made on 3- × 1-inch slides that are clean and free of grease. Provision must be made for marking the slide with a glass-etching pencil, unless slides with a frosted end are used. A watery specimen cannot be used, and hard, formed feces must be made soft enough to be spread by adding water or saline as needed. For good results it is important to have the fecal specimen fresh. The smear must be of proper thickness, and well fixed and stained.

Materials and Procedure. With an applicator or small (No. 3) brush, make a thin smear of the feces on a clean slide clearly marked for identification of the specimen. The ideal smear is made of one-brush-stroke thickness of feces of a pasty consistency. When an applicator is used, the feces should be spread with the side, not the end, of the shaft.

In the steps outlined here, the film must not be allowed to dry at any time, and the time schedule must be adhered to except where the smears are in the fixative or in 70% alcohol. Usually several slides with smears are processed simultaneously in special staining jars, the slides being transferred from jar to jar through the steps outlined.

1. Fix the smears in Schaudinn's solution 15 minutes or longer. (Schaudinn's solution is 200 ml saturated aqueous solution of mercuric chloride in distilled water, plus 100 ml 95% ethyl alcohol, to which 15 ml glacial acetic acid is added just before the fixative is to be used.)
2. Immerse smears successively in 70% alcohol, 70% alcohol to which enough iodine has been added to give it the color of urine, 70% alcohol, and 50% alcohol, 2 to 5 minutes in each.
3. Wash in running water 2 to 10 minutes.
4. Immerse in 2% aqueous ferric ammonium sulfate ($FeNH_2(SO_4)_2 . 12H_2O$) solution 5 to 15 minutes. (The solution should be made of clear purple crystals and discarded when a precipitate becomes evident.)
5. Wash in running water 3 to 5 minutes.
6. Stain in 0.5% aqueous hematoxylin 5 to 15 minutes. (The stain is made by adding to 100 ml distilled water, 5 ml of stock solution, which is 10 g of hematoxylin powder in 100 ml of 95% ethyl alcohol allowed to ripen several weeks in a loosely stoppered clear glass flask or bottle.)
7. Wash in running water (or several changes of water) for 2 to 5 minutes.
8. Differentiate in saturated aqueous solution of picric acid 10 to 15 minutes.
9. Wash in running water 10 to 15 minutes.
10. Dehydrate by immersion for 2 to 5 minutes each in 50%, 70%, 80%, and 95% alcohol and 5 minutes each in two changes (or steps) of absolute alcohol.
11. Clear the stained smears in two changes or steps of toluene or xylol, 3 to 5 minutes each.
12. Cover with mounting medium (neutral Canada balsam or a substitute) and thin (No. 1) cover glass.

Fecal films properly fixed, stained, cleared, and mounted will remain in good condition indefinitely.

Trichrome-Stained Smear. ***Materials.*** The trichrome stain is made of 0.6 g of Chromotrope 2R, 0.3 g of light-green SF, and 0.7 g of phosphotungstic acid. These dry components are mixed with 1.0 ml of glacial acetic acid and allowed to stand for 30 minutes, then diluted with 100 ml of distilled water.

Procedure. Steps in preparing stained smears of fresh feces are as follows:

1. Prepare a fecal smear as described for hematoxylin-stained smears.
2. Fix in Schaudinn's solution 15 minutes or longer.
3. Wash in 70% alcohol 1 to 5 minutes.
4. Wash in 70% alcohol, containing enough iodine to give it a yellow color, 1 to 5 minutes.
5. Wash in two changes of 70% alcohol 2 to 5 minutes each.
6. Stain in trichrome solution 10 minutes.
7. Differentiate in acid alcohol (99 parts 90% alcohol: 1 part glacial acetic acid) for 2 to 3 seconds.
8. Rinse in absolute alcohol (dip several times to remove acid).
9. Dehydrate in absolute alcohol, 2 changes, 2 to 5 minutes each.
10. Clear and mount (as for iron hematoxylin).

Preserved Fecal Specimens

Choice of Method. When large numbers of fecal specimens are being collected in the field, at large institutions, or even at hospitals and clinics where it is impractical to establish a routine of prompt examination, fecal specimens can be preserved and examined later with little sacrifice of reliability in most instances. Four different preservatives are in general use: formalin solution, Schaudinn's solution, polyvinyl alcohol (PVA) containing Schaudinn's solution, and merthiolate-iodine-formaldehyde (MIF). Each offers certain advantages and presents certain limitations. A precaution that applies to all of them is that the amount of the preservative for a given amount of feces must be adequate, and the preservative and specimen must be adequately mixed with no large particles remaining.

Specimen Collection. A satisfactory preserved specimen is 1 ml of feces mixed with 14 ml of preservative in a 20-ml screw-cap vial. As a guide for measuring the amount of feces, the vial can bear attached to its side an adhesive label of suitable width that is placed at the appropriate height to indicate at its bottom margin the 14-ml level and at its top the 15-ml level. A vial of any other size can be prepared in the same manner. The minimum recommended specimen is a 1-ml sample of feces in 7 ml of preservative, in a 10-ml vial. For quantitative diagnosis (egg counts), it is necessary to standardize the quantity of feces put into the preservative. The collector can be instructed to add just enough feces to bring the level of the preservative from the bottom margin of the label to the top margin. Enough space should remain above the fluid in the vial for adequate mixing by vigorous shaking. The label should indicate by code clearly and permanently the source of the specimen.

Formalin-preserved specimens are widely used for surveys of helminth infections, and MIF is excellent for preserving protozoa. However, neither of them is suitable for making permanent stained smears to be examined for protozoa. PVA and Schaudinn's solution are used especially for the diagnosis of protozoa. Specimens preserved in these solutions can be used also for the diagnosis of helminth infections by concentration methods, although they are less satisfactory for that purpose than are fresh or formalin preserved specimens.

PVA (Polyvinyl Alcohol) Fecal Smear. The PVA method of preserving and fixing fecal smears on slides was introduced by Brooke and Goldman in 1949. Owing to difficulties in preparing the solution and to occasional failure of the film to adhere to the slide, PVA is less widely used now than formerly. However, it still is highly recommended for surveys and as a routine procedure in the diagnostic laboratory (Markell and Quinn, 1977; Garcia and Ash, 1979; Markell and Kuritsubo, 1981). It is especially effective in detecting trophozoites in diarrheic and mushy stools that would escape detection by other techniques. Even formed stools may contain trophozoites of amebae and no cysts (Goldman and Brooke, 1953).

Materials. The PVA fixative-preservative solution, available commercially, is prepared as follows:

1. In a 500-ml beaker mix 62.5 ml of 95% ethyl alcohol, 125 ml of saturated aqueous solution of mercuric chloride, 10 ml of glacial acetic acid, and 3 ml of glycerine.
2. Add 10 g of PVA powder, without stirring.
3. Cover the beaker and allow to soak overnight.
4. Heat slowly to 75°C. When this temperature is reached, remove the mixture from the heat and

stir until a homogeneous, slightly milky solution is obtained (usually about 30 seconds).

PVA-preserved fecal material should be allowed to fix at room temperature for at least 30 minutes. The recommended proportions of 3 parts or more of PVA to 1 part of feces will give highly variable proportions of sediment and supernatant fluid. When on standing the sediment or the fixative is greatly in excess of 1 to 1, a balance can be obtained by adding or withdrawing a suitable quantity of the PVA preservative.

Procedure.

1. To prepare the film for staining, the specimen is mixed thoroughly. With the shaft of an applicator stick, two or three drops of the specimen are spread over a 22-mm length of the slide at the middle area, extending the smear to both edges of the slide. A variation recommended by Garcia and Ash (1979) is to pour some of the suspension onto a paper towel and allow it to stand for 3 minutes to absorb excess PVA. Then, with an applicator stick or brush, make a film of the material from the towel on a slide.
2. The prepared film is thoroughly dried in an incubator for several hours or overnight at room temperature.
3. The slide with the dry film is then transferred to 70% alcohol with iodine added to remove the fixative and then stained as described for the fresh fecal smear.

Note: When placed in water and aqueous hematoxylin stain, the PVA film may be lost from the slide. With trichrome staining, this problem does not occur.

Schaudinn's Preserved Specimen. When preservation of fecal samples in PVA fixative is impractical or fails to provide satisfactory permanent stained smears, an alternative procedure, recommended by Scholten (1972) and Scholten and Yang (1974), is to fix and preserve the specimen in Schaudinn's solution.

Materials. The Schaudinn's fixative slightly modified (Scholten and Yang, 1974) is, for a 1-liter stock, 45 g of mercuric chloride, 310 ml of 95% ethyl alcohol, 50 ml of glacial acetic acid, and 15 ml of glycerol, added to 625 ml of distilled water. In an unpublished evaluation study in Colombia, the following procedures gave highly satisfactory results on specimens preserved up to 1 year.

Collection of Specimen. With an applicator, collect from the stool specimen a 1-ml sample, measured by displacement, in a 20-ml screw-cap vial containing 14 ml of Schaudinn's solution. Break up the sample into small particles and stir until no large masses remain. Close the vial and shake vigorously for 20 to 30 seconds, longer if necessary to reduce fecal particles to less than rice-grain size.

Procedure. Prepare smears for permanent staining as follows:

1. Shake specimen to mix the sediment and let stand for 5 to 10 seconds to allow settling of sand and other heavy particles.
2. Pour 1 ml into a 15-ml conical centrifuge tube and add water to fill the tube.
3. Stir and strain through surgical gauze into a small (3 oz, about 80 ml) paper cup.
4. Rinse centrifuge tube to remove adhering coarse material.
5. Discard gauze, pour filtrate from cup back into centrifuge tube, and spin for 2 minutes at full speed in a clinical centrifuge.
6. Pour off supernatant, "flick" tube with thumb to mix sediment, adding 2 to 3 drops of water if necessary to suspend sediment, fill with water, and centrifuge 2 minutes again to remove the chemical preservative.
7. Pour off supernatant completely (i.e., invert the tube and allow it to drain for 10 seconds to remove all water).
8. Note the volume of sediment and add an equal amount of Mayer's albumen (equal parts of egg white and glycerin thoroughly mixed and filtered through gauze).
9. With disposable pipette, mix sediment thoroughly and transfer 2 drops to the center of a slide already marked to identify the specimen.
10. Spread the suspension over an area about 2 cm square.
11. Allow smears to dry at room temperature in the air, protected from dust and insects, for 1 to 2 hours. A glossy sheen appears when the smear is dry enough to be stained. If the smear is too wet, it will wash off in the stain; if it is too dry, the amebae will shrivel.
12. Stain with iron-hematoxylin or trichrome; when properly done, the former gives superior results.

MIF (Merthiolate-Iodine-Formaldehyde) Preserved Specimen. The MIF solution was introduced by Sapero et al. in 1951 and described more fully by Sapero and Lawless in 1953. It also is called TIF (for thimerosal-iodine-formaldehyde). It is prepared in two separate stock solutions to be mixed immediately before use.

Solution I, made up in 480-ml amounts, consists of 250 ml of distilled water, 200 ml of thimerosal (Merthiolate) (No. 99 Lilly, 1: 1000), 25 ml of formaldehyde U.S.P., and 5 ml of glycerol. Solution II is Lugol's solution (5% iodine in 10% potassium iodide solution in distilled water). Both solutions are stored in tightly stoppered brown glass bottles. For immediate use the two solutions are combined, 94 parts of solution I with 6 parts of solution II. Into a vial containing 9 ml of MIF, 1 ml (1 g) of the fecal specimen, measured by displacement, is added and mixed by stirring and shaking until completely suspended.

This preservative is useful for field collections. All microscopic parasite cysts, eggs, and larvae are well preserved and adequately stained for diagnosis as accurately as is possible in wet smear preparations. Organisms are well preserved for a year or more. When specimens are preserved in measured amounts, the material is suitable for quantitative diagnosis of helminth infections at high and moderate levels of worm burden (Dunn 1968, 1972). Although a technique of concentrating cysts and eggs of parasites in MIF-preserved specimens has been described (referred to as MIFC), experience has shown that it is unreliable for specimens that have been stored more than a few days (Dunn, 1968).

Concentration Methods For Fecal Specimens

The concentration and separation of cysts of protozoa and eggs of helminths from other elements of the fecal specimen can be a great advantage in diagnosis. This can be accomplished by sedimentation, flotation, or a combination of the two. Sedimentation is accomplished by suspending the fecal specimen in water or an aqueous solution and allowing natural settling to occur, or by accelerating the process by centrifugation. Flotation involves suspending the specimen in a medium of greater density than that of the cysts and eggs, which by their buoyancy become concentrated at the surface.

Sedimentation in Water. By sedimentation in water, either by gravity or centrifugation, helminth eggs in a fecal specimen are concentrated because the fecal elements that are small and of low density are decanted in the supernatant fluid. Resuspending and straining the sediment through gauze removes fiber and other particles larger than eggs, thus reducing the sediment to eggs and particles of about the same size and density. The proportion of eggs harvested in the sediment and the amounts of nonparasite elements remaining with the eggs in the sediment depend on the accuracy of timing of sedimentation, skill in pouring off the supernatant suspension, and the efficacy of the straining procedures. Except for the collection of reference specimens and teaching material, sedimentation in water is not generally used. It is relatively ineffective as a concentration technique.

Ether Centrifugal Sedimentation Methods

To increase the efficacy of centrifugal sedimentation for the recovery and concentration of helminth eggs from feces, Telemann in 1908 reported the successful use of hydrochloric acid to dissolve certain components, and the use of ether to dissolve and separate certain other elements from the fecal suspension, leaving in the sediment only the eggs and other particles of similar size and density. Apparently unaware of Telemann's experience, Carles and Barthelemy in 1917 described a citric acid-formalin-ether technique that they found to be especially effective in concentrating cysts of protozoa. Likewise, Cropper and Row in 1917 described a saline-ether technique that, with modification, was used with success by Boeck in 1917. In subsequent years it was further modified by substituting formalin for the saline, and it was extensively used by Boeck and Stiles (1923) to concentrate both protozoan cysts and helminth eggs. Reintroduced by Ritchie (1948), and modified and refined by others (Ridley and Hawgood, 1956; Knight et al., 1976), the formalin-ether technique is now widely used. It is satisfactory for concentrating the eggs of helminths and moderately satisfactory for concentrating the cysts of protozoa. It is most useful for detecting the eggs of schistosomes in feces.

The unique feature of this technique is the use of formalin for fixation and preservation of the fecal specimen. Other solutions used in combination with ether, not yet mentioned, are acetic acid (de Rivas, 1928), merthiolate-iodine-formaldehyde (Blagg et al., 1955), and phenol-alcohol-formalin (Burrows, 1967). The formalin and MIF solutions are advantageous when preservation of the specimen for later examination is important. However, concentration of cysts and eggs from MIF-preserved material is unreliable and therefore not to be recommended (Dunn, 1968).

The function of ether in all of these concentration techniques is sometimes mistakenly said to be the removal of fats and oils. The function of ether was correctly stated by Cropper and Row (1917) who noted that the fecal debris absorbs ether and consequently becomes lighter than water. When separation is complete, the fecal debris lies in a mass at the top of the saline, immediately below the column of ether. The cysts are not affected by the ether, and consequently remain in the saline beneath. Critical observers have noted that after a few days or weeks the eggs and cysts in preserved specimens, especially those in MIF, do absorb or adsorb ether, and are lost in the layer of debris that forms a plug under the ether. Because ether is flammable and explosive, considerable risks are involved in its storage, use, and disposal. A substitute for it is ethyl acetate which, according to Young et al. (1979) and Truant et al. (1981), is satisfactory in all respects.

Formalin-Ether Sedimentation. Formalin-ether centrifugal sedimentation can be used for fresh stools or for specimens collected earlier and preserved in formalin.

Procedure for Fresh Specimens.

1. In a paper cup, beaker, or any suitable container, comminute about 1 ml (1 g) of feces in just enough physiological saline (or water) to form a semifluid paste. Fill a 15-ml centrifuge tube with saline or water, add it to the comminuted specimen, and stir into a suspension. Pour the suspension into the centrifuge tube.
2. Rinse the paper cup and, through gauze held over its brim, strain the suspension back into it. Rinse the centrifuge tube with about 1 ml of water and pour through the gauze.
3. Transfer the sediment back to the centrifuge tube, fill to the level required for balance, centrifuge 1 minute at about 2000 r.p.m., and decant supernatant fluid.
4. Add 10 ml of 10% formalin, mix, and allow suspension to stand 10 minutes or longer for fixation.
5. Add 3 ml of ether, stopper the tube, and shake vigorously for 30 seconds.
6. Centrifuge 2 minutes at about 1500 r.p.m. Four layers will have formed: ether at the top, a plug of debris, then formalin, and sediment.
7. Ring the sides of the tube with an applicator to loosen the plug, then pour off all except the sediment.
8. If the fluid running down the sides of the tube is insufficient to suspend the sediment, add one or two drops, or whatever amount is needed, and transfer a drop of the suspension to a slide, add a drop of iodine stain if protozoa are to be identified, mix, cover with coverglass, and examine the entire preparation.

Note 1. Because the elements present in feces vary widely, the amount and character of the sediment yielded from a 1-g sample will also vary. When the amount is only enough for one or two preparations, all of it should be examined; when the amount is large, the bulk may be largely due to sand, coarse silt, or bar-shaped residues from the outer coat of beans or other legumes. In such cases where it is impractical to examine the whole of the sediment, a suitable amount of it may be drawn into the column of a pipette and held vertically for 10 to 15 seconds to let heavy elements settle and fall from the tip before the remainder is transferred to a slide for examination. This not only enriches the yield, but improves the visibility of the preparation to be examined.

Note 2. Instead of pouring off the upper layers as described in step 7, some technicians prefer to insert a pipette along the wall of the tube, cautiously breaking through the plug at the ether-formalin interface. They withdraw the sediment, then, holding the pipette in a horizontal position, draw it through a fold of gauze or paper towel to clean the outer surface before expelling the suspension of sediment onto a slide.

Procedure for Preserved Specimens. For specimens preserved in 10% formalin, take a portion estimated to contain about 0.5 ml of sediment, dilute to 15 ml, strain through gauze, and begin with step 3 described previously.

PVA-preserved specimens can be concentrated by the FE technique as follows: Mix the specimen, transfer a portion estimated to contain 0.5 ml of sediment to a centrifuge tube, dilute with an equal amount (or more) of saline solution (or water, or 10% formalin), strain through gauze, centrifuge for 2 minutes at 1500 to 2000 rpm, decant, resuspend and centrifuge, and then begin with step 3 described previously.

Acid-Ether Sedimentation. The advantage attributed to acids used as a suspension medium in acid-ether concentration techniques is their ability to dissolve mucus and various other substances, resulting in a cleaner sediment. Acid-ether techniques are unsatisfactory for the diagnosis of protozoa. They are especially useful in the detection of schistosome and other trematode eggs. The acid most commonly used is hydrochloric, in a concentration of 15% (40 ml of concentrated HCl in 60 ml of distilled water). The steps in the concentration procedure are essentially the same as for formalin-ether applied to the fresh specimen, i.e., suspend in acid, strain, centrifuge, resuspend in acid, add equal quantity of ether, stopper, and shake about 1 minute. Remove stopper and centrifuge, pour off all except the sediment, transfer sediment to a slide, and examine under low power for eggs and larvae of helminths.

Flotation Methods

In contrast to sedimentation, in which cysts and eggs of parasites settle and are concentrated at the bottom because they have greater density than the suspending medium, flotation concentrates the eggs and cysts at the top because their density (specific gravity) is less than that of the suspending medium. In addition to its density, a basic requirement of the suspending medium is that it neither dehydrate the cysts and eggs, nor be absorbed by them. To buoy the cysts and eggs, the medium must not only have the required specific gravity, but it must not affect the specific gravity of the cysts and eggs by causing them to expand or shrink.

Brine Flotation. The floating medium generally employed for floating the eggs of nematode parasites, such as hookworms, trichostrongyles, *Ascaris,* and *Trichuris,* is brine, a saturated solution of sodium chloride that has a specific gravity of approximately 1.20. Brine is unsuitable for trematode eggs, most types of cestode eggs, and for the cysts of nearly all intestinal protozoa.

The procedure in brine flotation is simple. A sample of feces is thoroughly mixed and suspended in 10 times its volume of brine, strained through gauze to remove coarse elements, transferred to a test tube that is filled with the suspension almost to the brim, and allowed to stand free of air currents and vibration for 20 to 30 minutes. Eggs are removed by means of a wire loop (about 5 mm diameter) to a glass slide for examination under a microscope. Usually two or three loops are taken and spread on a slide at about the width of the low-power field of the microscope and examined without a cover glass. Alternatively, instead of yielding the eggs with a loop, the tube may be filled completely, so that a clean slide placed over the mouth of the tube closes it without trapping air beneath it. After standing with no vibration for 20 to 30 minutes, the slide is lifted without tilting, reversed in a gentle but rapid movement to bring the wet side on top, and examined under the microscope. A cover glass is not needed if the preparation is examined promptly. With delay, crystallization of salt will interfere with the examination.

Zinc Sulfate Centrifugal Flotation. Zinc sulfate centrifugal flotation was developed by Faust et al. (1938) for the concentration of protozoan cysts as well as helminth eggs in a readily diagnosable condition. It effectively concentrates cysts of protozoa, eggs of the nematodes, and small tapeworms, but it is unreliable for nematode larvae and is unsuitable for infertile eggs of *Ascaris* and the eggs of most trematodes and large tapeworms. However, the eggs of *Clonorchis, Opisthorchis,* and some other trematode eggs are satisfactorily concentrated by flotation on zinc sulfate solution (Phillipson, 1962). Relatively poor results are obtained with specimens containing excessive amounts of fats and fatty acids. Although the zinc sulfate technique is highly efficient when properly done, the recommended procedures as described originally offered numerous opportunities for error. Procedures now generally recommended (Healy et al., 1970; Garcia and Ash, 1979) were designed and tested to minimize such errors (Beaver, 1952a).

Procedure.

1. With a wooden applicator transfer about 0.3 ml (slightly more if the feces contain much fiber) to a small test tube (12 mm × 100 mm) containing about 2 ml of water.

2. Using the same applicator as a stirring shaft, and an adapted small high-speed drill-motor as the propelling force, if available, comminute the sample thoroughly and add water to fill the tube. Then remove the applicator. When the applicator is removed, the water will be lowered to just the proper level for centrifuging. (Any of the small inexpensive high-speed drills such as are manufactured for wood carving or glass marking can be easily adapted to this purpose by removing the hood from a small vaccine-bottle type of rubber stopper and inserting it into the drill chuck after securing the instrument solidly in a vertical position where it can be used conveniently [Beaver, 1956]. This and all other materials and equipment ideally should be located over a sink. Water and zinc sulfate solution should be delivered through medicine-dropper-sized spouts attached to supply bottles that are elevated so that the force of the stream can be applied to stir the suspension as the fluids are added to the tubes. With these facilities, contaminating the hands and spilling fluids on the outside of the tubes can be

avoided; thus, the centrifuge cups can be kept clean, dry, and well balanced.) If a mechanical stirrer is not available, mixing can be done manually.

3. Balance tubes in a centrifuge and spin at full speed (2200 to 2500 rpm) for 1 minute.

4. Pour off all the supernatant fluid, add about 1 ml zinc sulfate solution (specific gravity 1.18), and mix well by striking the thumb against the bottom of the tightly held tube. Add more zinc sulfate solution to fill the tube, allowing the contents to be stirred by the stream's force as the solution is added. Repeated washing of the sample until the supernatant remains clear has been recommended but is unnecessary if the original sample is not too large.

5. Hold a strip of gauze over the rim of a small paper cup, crimp it slightly to "bag" the gauze, and strain the suspension from the tube into the cup. Discard the gauze and return the suspension to the tube through a narrow lip formed by crimping the brim. Straining is not essential for certain types of stool, but it requires so little time and equipment that it is definitely worthwhile as a routine procedure.

6. Add zinc sulfate solution to raise the level of the suspension just 3 mm below the rim of the tube, and spin as before for 1 minute. Do not remove the tube from the centrifuge until step 7 has been completed.

7. Place about one-third drop of iodine stain (1% potassium iodide saturated with crystalline iodine) on a clean slide. With a fine (28-gauge) wire loop (5 to 6 mm) bent 90° or less from the shaft, applied directly to the surface film in a horizontal position, remove a sample from the center and rinse it into the iodine without spreading the material more than necessary for good visibility. If the centrifuge is needed immediately, the tube may be removed (with as little disturbance as possible to the surface film) to a stand where additional surface-film samplings may be taken later if desired.

8. Without a cover glass, examine the entire preparation under low power of the microscope, focusing on those objects that come to rest on the slide. Floating debris can be shifted by gentle blowing. The high power of most microscopes can be used for identification of objects without a cover glass. When prolonged study is required, add iodine and additional looped samples if material is scanty, and use a cover glass to prevent evaporation and to preserve the stain.

Note: Poor results will be obtained (1) if the stool contains substances such as fatty acid crystals that form a layer above the eggs and cysts in the surface film; (2) if after washing and straining, the sample exceeds 0.3 ml or (3) the zinc sulfate solution is diluted with water that was not completely poured off in step 4; (4) if the tube is too full, (5) if the tube is not well balanced in the centrifuge, or (6) the centrifuge is not allowed to come to rest without disturbing the surface film; or (7) if the sample is not "looped off" and examined promptly after centrifugation. Cysts of protozoa will collapse after long exposure to the zinc sulfate solution. If the preparation is centrifuged only 1 minute and looped off promptly and transferred to iodine stain, almost all kinds of cysts and eggs, even those of some trematodes, will be yielded without collapse.

Time will be wasted (1) if one mixes the feces in water manually; (2) if the sample is too large, requiring it to be washed more than once; (3) if the suspension is centrifuged longer than 1 minute; (4) if the preparation to be examined microscopically is spread over a large area; or (5) if one searches for parasites with the high, rather than the low, power of the microscope.

Zinc sulfate flotation is satisfactory for formalin-preserved specimens containing protozoan cysts and helminth eggs other than those of schistosomes or other trematodes (Bartlett et al., 1978).

Schistosome Egg-Hatching

When mature viable schistosome eggs come into contact with water, the miracidia may hatch within a few minutes, or in 1 or 2 hours, and swim to the surface. Some species (not *S. japonicum*) are also attracted to light. Therefore, to detect the presence of schistosome eggs in quantities of feces larger than can be examined by other methods, or to concentrate miracidia hatched from eggs known to be present in feces, urine, macerated tissues, or postmortem intestinal scrapings, a technique of concentrating hatched miracidia is useful. McMullen and Beaver (1945) recommended using a sidearm flask, but for routine purposes an Erlenmeyer flask will serve.

Sidearm Flask. The specimen to be examined, 5 to 10 g of colloidal feces or up to 50 g of fibrous feces, is suspended in an equal amount of water (aerated or boiled and cooled tap water) and placed in a 1-liter distillation flask with a vertical sidearm (Fig. 46–2). Water is added until it reaches a level just below that of the sidearm (A). The suspension is allowed to sediment until the topmost 1-cm layer

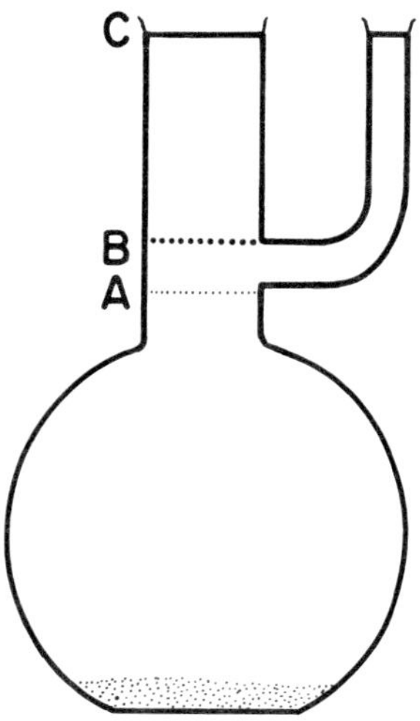

Fig. 46–2. Side-arm flask used for the hatching and concentration of schistosome miracidia. *A, B,* and *C,* critical water levels (see text). (From Faust, E.C., Beaver, P.C., and Jung, R.C. 1975. *Animal Agents and Vectors of Human Disease,* 4th ed. Philadelphia, Lea & Febiger.)

is clear. Then the flask is filled with water in two steps. First, the water is poured slowly into the sidearm until the level is above the sidearm (B). Second, the flask is filled to the brim by rapidly pouring water into the sidearm (C). This leaves the water in the sidearm clear, because all particles that rise from the bottom due to low density or gas bubbles go to the water surface in the neck of the flask. A dark cloth or paper bag with a hole of suitable size and position is then placed over the flask so that the arm protrudes while the rest is covered. The covered flask is placed so that the arm is facing a window or lamp. Miracidia that swim upward and are attracted to light will be concentrated at the surface in the sidearm. The results can be improved by prior washing, straining, and concentrating the specimen by sedimentation in physiological saline. The saline is used instead of water to avoid hatching of miracidia during the process.

Test Tube With Filter. In China, a different hatching technique has been developed (Mao and Shao, 1982). A sample of washed fecal sediment in a test tube is covered with a loose cotton ball, over which water is added and allowed to stand. The cotton filter prevents the water from becoming turbid, but allows the hatched miracidia to find their way to the upper surface of the water. Fecal sediment containing the eggs can be concentrated by screening in saline solution through 60- to 80-mesh metallic gauze, then washed into and through a nylon conical bag made of 260-mesh nylon cloth. This can be examined microscopically in a direct smear, after concentration, or by the hatching technique.

Quantitative Diagnosis

The only techniques of fecal examination that can be used for both qualitative and quantitative diagnosis are the direct saline smear and the cellophane-covered thick smear. All concentration techniques are unreliable for egg counts because the yields are variable. This is due to differences in the character of the specimens and to other factors, including the proficiency of technicians. The earliest attempts to determine relative worm burden by estimating egg output were based on direct-smear egg counts (Smillie, 1921). Techniques of counting the eggs in a sample of the diluted specimen seemed theoretically more accurate because the sample could be measured precisely by displacement. In addition, it was believed that eggs in the fecal specimen were not randomly distributed, and therefore counts made on a small sample from a large diluted specimen were thought to be more accurate than those made on a sample of equal size taken directly from the fecal specimen (Stoll, 1923a,b; Stoll and Hausheer, 1926a,b). However, it was shown later that egg counts made on direct smear samples were as reliable as those made by dilution methods (Beaver, 1949a). It was further shown that direct smear preparations could be made to contain specific amounts of feces and thus could be compared with counts made by dilution methods (Beaver 1950). Moreover, it was shown that counts expressed in terms of output of eggs in the feces per day were far more variable than those expressed as eggs per gram, or per milligram, or per direct smear (Martin, 1965). This was explained by showing that due to reflux of the contents of the cecum and upper colon, eggs contributed to the fecal stream by worms in the small intestine and upper large intestine are eliminated in the feces over a period of 1 to 7 or 8 days. Thus, eggs in a stool passed on a given day were laid by worms 1 day to 1 week earlier. Helminth eggs remain in the intestine for an average of about 3 days, and by the mixing action of the colon are randomly distributed in the feces. This has been demonstrated for *Schistosoma mansoni* eggs (Woodstock et al., 1971) as well as for hookworm eggs (Martin, 1965). Therefore, the variation in day-to-day egg counts in terms of eggs per unit of feces is largely due to statistical variation such as occurs predictably in estimates based on small samples taken from a large universe. For these reasons, it is important to emphasize that although levels of egg output are referred to as *egg counts*, they actually are *estimates* that have statistical probabilities of error and variation. Even the average of several counts made on the same specimen will have a predictable range of variation. Thus, a single egg count made by the most precise method and by the most reliable technician can only indicate roughly the level of worm burden.

The fact that egg counts actually are only rough estimates of egg output does not suggest that such estimates are not useful or even essential for certain interpretations. When one is dealing with either individual patients or large populations of people, it is helpful to know whether a given helminth infection is light, moderate, or heavy. Any attempt to classify infections in more than five levels (very light, light, moderate, heavy, and very heavy) will be misleading in most instances. A common error is to classify the infections in individuals with the highest of a series of counts as heavy when in fact all are light or range only from light to moderate. The levels of egg count that indicate clinical significance differ for the various species of helminths and for people of different age groups and nutritional status. For hookworm and whipworm infections, egg counts of less than 5/mg are light and those of more than 20 are heavy; for *Ascaris* infection, the corresponding counts are four or five times greater.

Direct-Smear Egg Counts. For the usual clinical evaluation of individual patients, the direct smear made in the manner described for qualitative diagnosis is suitable for egg counts. The counts can be reported as eggs per direct smear. For rough comparison with counts made by other techniques, the direct smear can be estimated as containing the equivalent of 2 mg of formed feces. Smears prepared by experienced technicians seldom contain less than 1 mg or more than 2 mg of formed feces.

Materials. The calibration of direct smears, or the preparation of direct smears containing known amounts of feces, involves the use of a photoelectric light meter that is adapted to measuring the turbidity of the smear. A wooden block 18 mm in thickness and of any convenient diameter is fitted to the window of the light meter and a 16-mm hole is drilled into the center of the block. This block serves as a platform for the microscope slide on which the smear is made, and it provides a mask that reduces the window to a convenient size for preparing and spreading the smear. An electric lamp is suspended directly over the reduced window and made adjustable so that arbitrary whole-number readings on the meter can be obtained.

To *calibrate* the meter, first prepare a solution of sodium sulfate (2N Na_2SO_4) and one of barium chloride (N/1 $BaCl_2$), and to each solution add pure glycerine, 2 parts solution to 1 part glycerine. For a 1-mg standard suspension, combine 1 part of the barium chloride with 6 parts of sodium sulfate. For a 2-mg standard, combine 1 part barium chloride with 3 parts sodium sulfate. Each proportion gives a white suspension of barium sulfate ($BaSO_4$). Next, with a clean slide on the platform and centered over the window, adjust the light to give a whole-number footcandle reading. Place 1 drop (1/20 ml) of the 1-mg standard suspension on the slide above the window and spread just to cover the window. Note and record the meter reading. Repeat until a reliable average is obtained, starting each time with the same arbitrarily selected whole-number reading through the clear slide. The average amount of light reduction produced by the standard suspension is the same as that produced by an even suspension of 1 mg of formed feces in 1 drop of water spread evenly over the 16-mm window. Next, repeat this procedure with the 2-mg standard suspension. In the original calibration of the standard direct smear, the light was adjusted to give a reading of 20 footcandles through the clear glass, and the 1-mg and 2-mg standard suspensions gave readings of 16 and 13 footcandles, respectively.

Procedure. To prepare a standard fecal smear for egg counting, place a clean slide over the window on the meter platform and adjust the light to give the predetermined "zero" reading. Next, put a drop of water on the slide over the window and stir (do not smear) feces into the water until the fecal suspension, when spread just to cover the window, reduces the light to equal that of the 2-mg standard barium sulfate suspension. Next, cover with a 22 × 22 mm cover glass and systematically count the eggs in the entire preparation. Record the count as eggs/mg. When a count is expected to be high, the 1-mg smear is preferred.

Dilution Egg Counts. Materials required for doing dilution egg counts by the procedures described by Stoll and Hausheer (1926b) and redescribed by Stoll (1963) are (1) N/10 (0.4%) sodium hydroxide (NaOH); (2) special Stoll flask* marked at 60 ml and 56 levels and fitted with a rubber stopper; (3) special pipette* calibrated to deliver 0.075 ml; (4) coarse glass beads or BB shot; (5) slide and 22 × 30 mm cover glass; and (6) wood applicator and soft paper towel or cellulose wipes.

Procedure.

1. Fill flask to the 56-ml mark with sodium hydroxide solution.
2. With applicator, add feces to bring contents to the 60-ml mark.
3. Add beads or shot to flask leaving room for stopper and small air space. Stopper and shake vigorously.
4. Allow to stand 12 to 24 hours with occasional shaking, always dashing beads against the rubber stopper.
5. Shake vertically for 20 to 30 seconds and with pipette take exactly 0.075 ml.
6. Wipe the outside of the pipette to avoid adding excess to the sample, and empty the pipette onto a slide.
7. Cover with a 22 × 30 mm cover glass and systematically count all eggs present.
8. Multiply the count by 200 and record as eggs/g, uncorrected.
9. Correct for consistency of fecal specimen by multiplying the uncorrected count as follows: hard-formed (difficult to puncture), ×1; mushy-formed (can be cut with applicator), ×2; mushy-diarrheic (takes shape of container), ×3; diarrheic (liquid but not watery), ×4. (Watery and dysenteric stools are unsuitable for egg counts. See Table 46–1 on page 734.)
10. Report the corrected count as eggs/g or divide by 1,000 and record as eggs/mg f.s.b. (formed stool basis). Counts are more accurately interpreted when recorded as eggs/mg rounded off to the nearest whole number (Beaver, 1950; Stoll, 1963).

Thick-Smear Egg Counts. The Kato cellophane-covered thick smear as modified by Martin and Beaver (1968) can be used for making egg counts (see p. 735). For simultaneous qualitative and quantitative diagnosis of the common intestinal nematode infections and routine quantitative diagnosis of *Schistosoma mansoni* and *Schistosoma japonicum* infections, the Kato thick smear generally is the preferred technique. Though not essential for reliable estimates of egg output, a modification recommended by Katz et al. (1972) is to use a calibrated template constructed of disposable cardboard (or plastic) of such thickness as to contain about 50 mg of feces in a 6-mm punched-out circular hole. After the fecal sample is pressed into the hole while held at the center of a slide, the template is lifted and discarded, the sample is spread under a cellophane coverslip, and the eggs are counted in the usual way. A further modification referred to as the "quick Kato smear" was introduced and tested by Peters et al. (1980). They used stainless steel templates that were constructed to deliver on the slide a 20-mg sample. By using the smaller sample that dried and cleared within about 15 minutes, they were able to make duplicate slides and examine them almost immediately. For large surveys or studies requiring the examination of large numbers of specimens, the 50-mg smear made in the least complicated manner has proved satisfactory (Martin, 1972a, b; Bartholomew et al., 1981). A difficulty in using a template is that air bubbles may be trapped in the measured sample and much of the sample may adhere to the template. A kit of disposable materials for the Kato smear as modified by Katz et al. (1972) is commercially available.

Dilution-Filtration Egg Counts. In 1963 Bell described a dilution technique in which schistosome eggs can be concentrated by filtering a suspension of feces through filter paper on which the eggs are stained and counted. The procedures are complicated and time-consuming, and special equipment is required. When first described, the technique offered the special advantage of yielding the eggs from a relatively large sample (100 or 200 mg). Another advantage was that the filter paper bearing the eggs could be stored for later examination. With the more recent development of Kato's cellophane thick-smear technique, the special features of Bell's dilution-filtration technique are no longer advantageous. Two, or even four, Kato smears, each containing 50 mg of feces, can be processed and counted in a relatively short period of time, and the smears containing eggs of schistosomes, whipworms, and some other helminths (not including the hookworms) can be stored for later examination when necessary.

Fecal Culture for Diagnosis of Nematode Larvae

Hookworm-like eggs observed in fecal specimens may be difficult or impossible to diagnose,

*Flask and pipette are available from A.H. Thomas Co., Philadelphia, PA.

whereas the larvae developing from them are readily identified. A common practical problem is the differentiation of *Necator* infections from those of *Ancylostoma*. For the culturing of larvae from eggs, a clean and simple test-tube method requiring only readily available materials was described by Harada and Mori (1955) and variously modified by others (Sasa et al., 1958; Hsieh, 1963). The method can be used also for the incubation of *Spirometra* and *Paragonimus* eggs (Beaver et al., 1964). For producing large quantities of larvae, cultures in charcoal are preferred. Refrigerated specimens are unsuitable for culturing hookworm larvae (Komiya et al., 1960).

Test-Tube Cultures. *Harada-Mori* cultures (Fig. 46–3) are set up as follows:

1. Spread a thin (1 to 2 mm) film of feces on one side of the middle third of a 13- by 120-mm strip of filter paper.
2. Place it in a 15-ml conical-tip centrifuge tube containing about 3 ml distilled water, forcing the paper to about the 1-cm level, and press the clean side against the side of the tube.

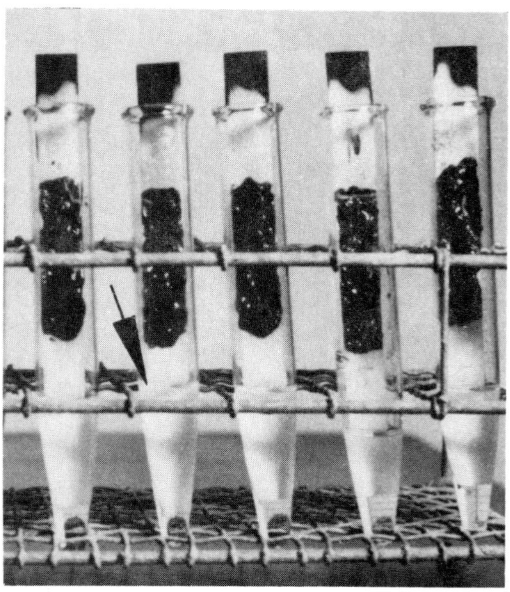

Fig. 46–3. Harada-Mori test-tube fecal culture for the diagnosis of hookworm and other intestinal nematode larvae. The cultures shown here are 3 days old, and by the upward flow and evaporation of water, the fecal film has been leached and the soluble elements have been deposited in the dark zone at the top of the filter-paper strip. The water level (shown by an arrow) is maintained by periodic replacement. After about 24 hours of leaching, the odor of feces disappears and is replaced by the aroma of damp soil. (From Beaver, P.C., Malek, E.A., and Little, M.D. 1964. Development of *Spirometra* and *Paragonimus* eggs in Harada-Mori cultures. J. Parasitol., *50*:664–666.)

3. Store the culture at room temperature (24 to 28°C) in the dark for 7 to 10 days, adding water daily as needed to keep the water level well above the bottom end of the filter paper.

The upward capillary flow of water through the paper and fecal film keeps the feces moist and carries soluble elements of the feces to the top of the paper where they either evaporate or accumulate in a dark deposit. On reaching the infective stage, the larvae migrate downward against the upward flow of water. When they reach the water pool, they sink to the bottom where they can be seen with a hand lens and taken up with a pipette for microscopic examination. The filter-paper strip can be transferred to a clean tube filled with water to harvest any larvae that have remained in the fecal film. Some species of larvae have a tendency not to migrate downward. Before removal, the larvae can be immobilized by immersing the tube in hot water (50 to 60°C) or by adding enough 20% acetic acid to the tube to kill the larvae. Iodine can be used to kill larvae, but it tends to overstain.

For identification, the larvae are transferred to a slide and mounted under a cover glass. They should be examined first under the scanning lens. If more than one species is present, the different kinds may be immediately evident when seen together at low magnification.

Charcoal Cultures. A charcoal culture is made by mixing 1 part of soft or softened (with water)

Fig. 46–4. Damp pad isolation of hookworm larvae from soil or from charcoal cultures. The dampened pad after contact with infested soil or culture is forced into a sedimentation flask containing tepid water, placing the contaminated side inwards so that the pad becomes a fluted cone enclosing the contaminated surface. The active larvae work through the pad, fall through the water, and are concentrated at the bottom of the flask. (From Beaver, P.C. 1953. Persistence of hookworm larvae in soil. Am. J. Trop. Med. Hyg., 2:102–108.)

feces with 5 to 10 parts of fine, granular (not powdered) animal charcoal, which is made damp throughout and packed into a covered but not airtight container. Excess water must be avoided, or removed. After 7 to 10 days, on reaching the infective stage of development, the larvae migrate to the surface where they can be removed by a method described for recovering hookworm larvae from soil (Beaver, 1953). A thick, 12- to 15-ply pad of soft cotton cloth (cheesecloth or absorbent gauze) of appropriate size to fit the culture container is dampened with warm water, pressed onto the culture surface, and the cover replaced. Most of the larvae migrate into the pad within about 30 minutes. The pad is then picked up at the center with forceps and transferred upside down to a sedimentation flask filled to near the brim with warm water (Fig. 46–4). The surface of the pad that was touching the charcoal is now uppermost. The larvae promptly fall to the bottom of the flask while the adherent charcoal (or soil) is held on the upper surface of the pad. Most of the larvae can be removed with a pipette after 20 to 30 minutes.

Filter Paper Slant Culture. A technique developed by Little (1966) for culturing the free-living stages of *Strongyloides* may also be used for culturing other kinds of intestinal nematodes.

Procedure.

1. On strips of filter paper having the same dimensions as a slide, cover one inch at one end with feces to a depth of 1 to 2 mm, and at the other end record with pencil the specimen identification.
2. Place the strip on a slide in a 10-cm × 15-mm petri dish with the smeared end resting on a glass tube, 6 to 7 cm long and 8 to 10 mm in outer diameter.
3. Add water to a depth reaching the filter paper, cover, and store at room temperature.
4. At 48 to 72 hours, and with a dissecting microscope and strong incident light, examine the surface of the fecal smear for free-living adult *Strongyloides*. At 24 hours with direct development, or at 3 to 4 days with indirect development, infective larvae of *Strongyloides* may be seen in the clear water. Likewise, at 7 to 10 days, if eggs are present, the infective larvae of hookworm and trichostrongyles will be found in the water.

Identification of Nematode Larvae

For the identification of infective-stage (filariform) nematode larvae from test-tube, filter-paper slant, or charcoal cultures of human feces, the key in Table 46–2 and illustrations (Fig. 46–5) prepared by Little (1981) will be helpful.

Worms Migrating from the Anus

Female pinworms migrate at night onto the perianal folds to lay eggs. Mature and immature *Ascaris* are frequently passed spontaneously or migrate from the anus. Such worms should be preserved in alcohol (70%) or formalin (5 to 10%) and identified. Tapeworm proglottids *(Taenia* or *Dipylidium),* singly or in strands, are discharged periodically in the stools or actively migrate from the anus. Tapeworms can most readily be identified from the moist proglottids. Patients should be instructed not to wrap them in toilet paper or allow them to dry. However, dry proglottids usually can be rehydrated in water and identified.

Anal Scrapings and Swabs

Amebiasis cutis of the perianal area may be diagnosed by demonstrating motile trophozoites of *E. histolytica* in material scraped from ulcers and examined in saline suspension on a slide under a cover glass.

Enterobius infection is usually diagnosed by demonstrating the presence of eggs at the anus. Vix in 1860 (reviewed by Beaver, 1949) apparently first recommended the use of an anal scraper or swab to obtain material for microscopical examination for pinworm eggs. Graham (1941) and Jacobs (1942) independently introduced a *Scotch cellulose adhesive tape technique* for obtaining eggs of *Enterobius* from the perianal area. A 3-inch length of the tape is held adhesive-side-out on the end of a wooden tongue blade by the thumb and index finger. After the tape is pressed firmly against the perianal folds between the dry and moist zones, it is placed adhesive-side-down on the slide for examination. When eggs are not immediately evident, the tape is marked with a wax pencil to indicate the area bearing perianal material, one end is lifted and turned back to expose that area, a drop of toluene is placed on the slide, and the tape is dropped back in place (Fig. 46–6). The toluene clears essentially everything except eggs and hair (Beaver, 1949). Eggs of other helminths frequently are seen in these preparations.

Recovery of Worms from Stools

Either of two methods may be used for the examination of feces for adult worms: (1) *straining*

Table 46–2. *Key for the Identification of Filariform Larvae from Human Coprocultures*

1a.	Oesophagus about one-half length of body; body slender (14–17 μm), lacking cuticular sheath; tip of tail not pointed, appearing notched	*Strongyloides (S. stercoralis* or *S. fuelleborni)**
1b.	Oesophagus about one-fourth length of body; cuticular sheath present; body thicker than 20 μm	2
2a.	Intestinal lumen straight	3
2b.	Intestinal lumen zigzagged, not straight	4
3a.	Body (not including sheath) about 500 to 600 μm long; tail (anus to tip) less than 73 μm long (50 to 72 μm); intestine, at esophagointestinal junction, as wide as esophageal bulb, buccal "spears" conspicuous, parallel throughout length, about 15 μm long; conspicuous transverse striations present on sheath in tail region	*Necator americanus*
3b.	Body (not including sheath) about 600 to 700 μm long; tail more than 73 μm long (75 to 94 μm); intestine, at esophagointestinal junction, narrower than esophageal bulb; buccal "spears" inconspicuous, about 10 μm long; transverse striations on sheath in tail region inconspicuous	*Ancylostoma duodenale*
4a.	Sheath relatively thin (thinner than cuticle of larva); pair of elongate sphincter cells present between esophagus and first pair of intestinal cells; tip of larva's tail pointed; posterior end of sheath elongate, tapering to thread-like tip; body 630 to 730 μm long by 29 to 35 μm wide	*Ternidens deminutus*
4b.	Sheath relatively thick (thicker than cuticle of larva), no sphincter cells between esophagus and intestine, tip of tail rounded or blunt	5
5a.	Posterior end of sheath relatively short, not tapering to fine point (distance from tip of larva's tail to tip of sheath less than distance from anus to tip of larva's tail)	*Trichostrongylus* sp.
5b.	Posterior end of sheath relatively long, tapering to a fine point (distance from tip of larva's tail to tip of sheath greater than distance from anus to tip of larva's tail)	*Oesophagostomum* sp.

*Filariform larvae of *Strongyloides stercoralis* and *S. fuelleborni* cannot easily be differentiated (Little, 1966). Diagnosis must be made on the basis of stages occurring in fresh feces (rhabditoid larvae of *S. stercoralis;* eggs of *S. fuelleborni* containing a developing embryo) or on morphology of free-living female worms from culture (prominent constriction of body behind vulva in *S. fuelleborni;* body not markedly constricted in *S. stercoralis*) (Little, 1966; Pampiglione & Ricciardi, 1971).

Note: Adult and larval stages of free-living nematodes (e.g., *Rhabditis* spp., *Pelodera* spp., and *Rhabditoides* spp.) may occur in cultures of feces if contaminated with soil, and must be differentiated from larvae of the human parasites.

It is possible that the first-stage larvae of *Angiostrongylus costaricensis* may occur in the feces of some infected individuals and may appear in coprocultures. These larvae are 260 to 290 μm long and 14 to 15 μm wide, much smaller than any filariform larva mentioned in this key. See Morera (1973) for morphological features of this larva.

an aqueous suspension of the specimen through wire sieves, or (2) *sedimentation-decantation*.

For satisfactory straining of specimens, at least two sieves are required: one coarse enough to pass small- to medium-sized worms but fine enough to retain larger elements (10 to 20 mesh), and one fine enough to retain small worms and the scolex of tapeworms (40 to 50 mesh). The specimen is stirred into a watery suspension, diluted, and poured through the two sieves, the coarse sieve nested above the fine one. Then it is washed by holding it under a tap or by pouring water through the sieves. The worms, if any, can be collected directly or after being transferred to a flat-bottomed glass bowl or pan by reverse flushing of the fine screen. Large worms retained on the coarse sieve will be evident, but the material should be suspended in water and examined since medium sized worms may be held in a felt-like mass or looped over a fiber.

Materials. The sedimentation-decantation method used by Looss for the collection of hookworms, slightly modified, generally is satisfactory (Beaver, 1952a). It requires only such items of equipment as are usually available: (1) two large vessels of equal capacity (2 to 5 liters), (2) a flat-bottomed glass dish or a pan (of uniform dark color), (3) a medicine dropper pipette, (4) a straight dissecting needle with 1 to 2 mm of the tip bent about 90°, (5) 10% formalin, and (6) a vial or jar of suitable size for collection and storage of the worms.

Procedure. Steps in the procedure are as follows:

1. Fill one vessel with water and stir occasionally to accelerate warming to room temperature and loss of dissolved air.

2. In another vessel of equal size, stir the fecal specimen into just enough water to form a uniform suspension, add water to fill the vessel, and allow it to stand with frequent gentle stirring of *the surface film* for 30 minutes.

3. Decant and discard supernatant fluid.

4. Holding the first vessel high, to produce a brisk stirring effect, pour the contents onto the sediment in the second vessel.

5. Repeat steps 1 and 2 at approximately 10-minute intervals until the supernatant fluid remains clear.

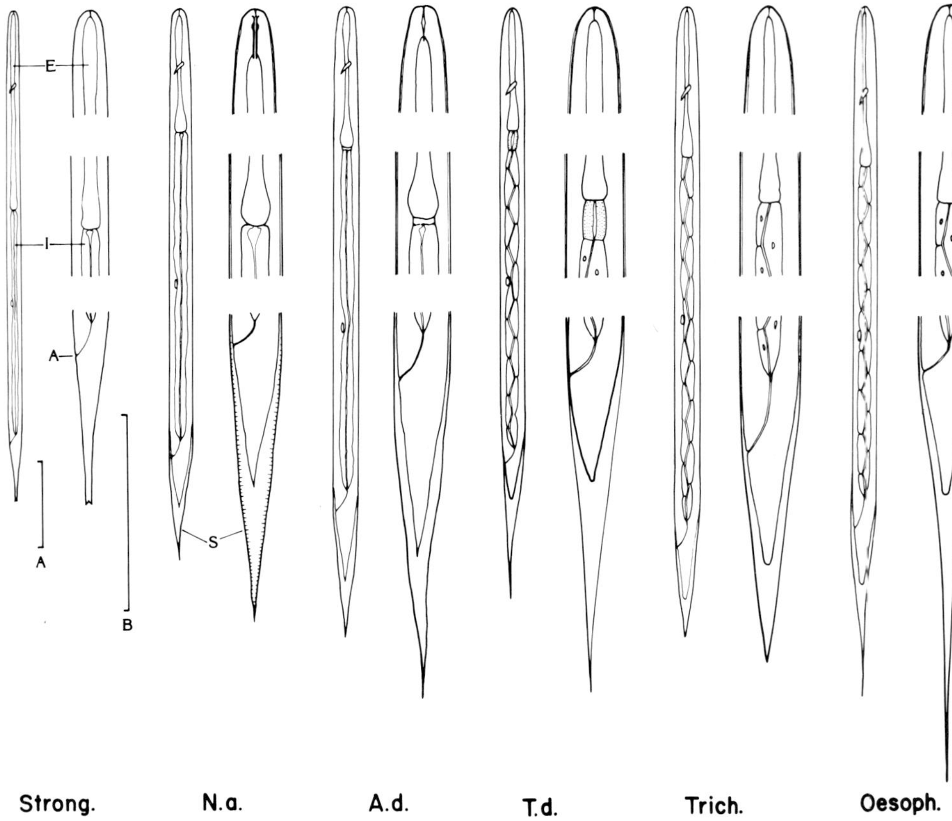

Fig. 46–5. Diagnostic features of filariform nematode larvae found in cultures of human feces. *Strong.*, *Strongyloides* sp. *N.a.*, *Necator americanus*. *A.d.*, *Ancylostoma duodenale*. *T.d.*, *Ternidens derminutus*. *Trich.*, *Trichostrongylus* sp. *Oesoph.*, *Oesophagostomum* sp. *A*, Anus. *E*, Esophagus. *I*, Intestine. *S*, Sheath. Scales are 100 μm, *A* for whole, *B* for parts of the larvae. (From Little, M.D. 1981. Differentiation of nematode larvae in coprocultures: Guidelines for routine practice in medical laboratories. W.H.O. Tech. Rep. Ser. No. 666, Intestinal Protozoan and Helminth Infections: 144–150.)

6. Transfer portions of the sediment in suitable amounts to a shallow pan or glass dish. Under bright light and over a dark surface collect the worms with a pipette or hook-pointed needle. Magnification may be helpful for worms as small as male *Enterobius*, *Hymenolepis nana*, and *Strongyloides*.

7. Preserve the worms in 10% formalin.

EXAMINATION OF URINE

Trichomonas vaginalis is frequently recovered in urine sediment from both male and female patients who are infected.

Urine regularly is examined for the eggs of *Schistosoma haematobium*. The specimen is collected in a sedimentation glass, and the eggs are allowed to sink to the bottom along with erythrocytes and pus cells. The sediment is then pipetted onto a slide and examined microscopically. A rapid and accurate technique of quantifying *S. haematobium* eggs after collection on Nuclepore filters has been described (Peters et al., 1976).

In Bancroft's filariasis with chyluria and occasionally in other filarial infections, the microfilariae may be discharged in the urine and recovered from centrifuged specimens. On rare occasions, *Strongyloides* larvae have been reported in the urine.

SPUTUM AND GASTRIC WASHINGS

In pleuropulmonary amebiasis, the rupture of the amebic abscess into a bronchus characteristically results in coughing up the contents containing blood, necrotic tissue, and trophozoites of *Entamoeba histolytica*. Similarly, rupture of a pulmonary hydatid cyst is followed by discharge of its contents, in which there are fragments of the laminated membrane and usually free scolices of *Echinococcus granulosus*. Sputum is commonly examined for the eggs of the lung fluke, *Paragon-*

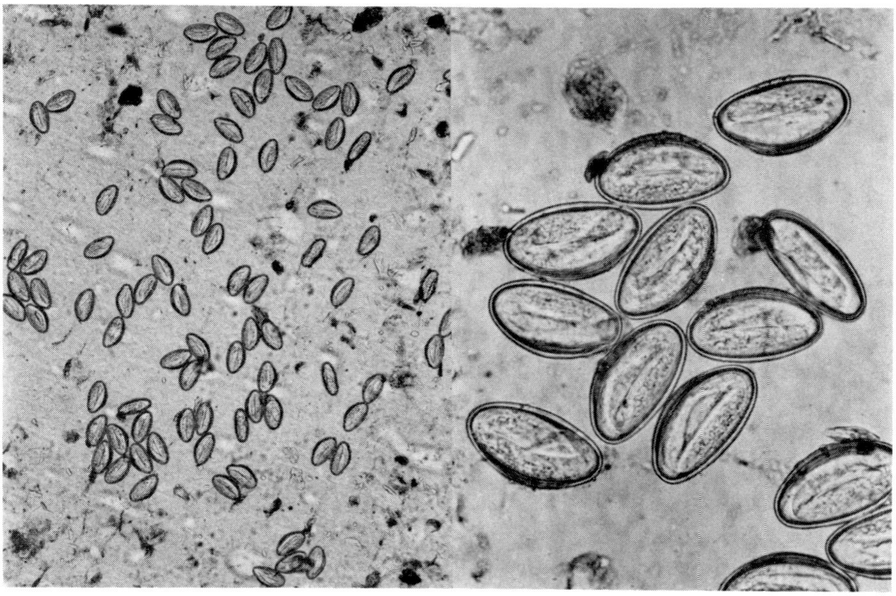

Fig. 46–6. Eggs of *Enterobius* on perianal adhesive tape preparation after clearing with toluene. The eggs contain infective-stage larvae and are seen to have sides of greater and lesser curvature and equatorial diameters of greater and lesser length; i.e., they have the general shape of a lima bean. Unstained. (× 100 and × 400.) (By P.C. Beaver.)

imus westermani. Occasionally, eggs of blood flukes or larvae of *Strongyloides* are coughed up.

In the pulmonary phase of infection with *Ascaris, Necator, Ancylostoma,* and *Strongyloides,* the migrating larvae along with eosinophils and Charcot-Leyden crystals may be demonstrated in the sputum, or more frequently, in the washed sediment from gastric washings (Proffitt and Walton, 1962). For examination, the sputum specimen is spread in a petri dish where flecks of material suspected of bearing eggs (brown streaks) or pus and Charcot-Leyden crystals (opaque white or tan bronchial casts) can be selected for examination under a cover glass on a slide. *Paragonimus* eggs can be concentrated by first digesting the sputum in 5% sodium hydroxide for 2 hours or longer, then washing by sedimentation in water. Usually *Paragonimus* eggs are detected in the feces.

EXAMINATION OF BLOOD

Next to feces, the blood provides the most common medium for recovery of various stages of animal parasites. From this source a diagnosis is routinely made of malaria, babesiosis, African trypanosomiasis, and most types of filariasis; less frequently of Chagas' disease; and rarely of kala-azar and toxoplasmosis. In any particular infection, the parasites may not be consistently present.

Protozoa in Blood

The standard procedure for diagnosis of protozoa in the circulating blood is to make microscope slide preparations. There are, in general, two types of slide preparations: the thin film and the thick film. The two are for different purposes, each superior for its appropriate use.

The *thin film* ideally is one cell thick, with erythrocytes lying flat on the glass surface and all cellular elements somewhat flattened so that their diameters are greater than in life. The hemoglobin as well as other proteins in the blood should be fixed before staining so that after they are stained all elements, including the erythrocytes, will remain intact and recognizable. The thin blood film is useful for studying details in the blood cells and blood parasites. Its primary disadvantage is that the amount of blood in the sample is small.

The *thick film,* many cells thick, contains 6 to 20 times as much blood per unit area as a thin film. It is spread over a roughly rectangular area 15 by 12 mm. It is necessary that the thick film *not* be fixed before staining since the erythrocytes must be dehemoglobinized during the procedure. The disruption of the erythrocytes and the loss of their hemoglobin from the slide permit the remaining structures, including blood parasites, to be seen microscopically even when lying deep in the film. The thick film is suited for rapid diagnosis of parasitemias too low to be detected in the thin film. It is not suited for fine anatomical studies.

It is important that the slides used for blood films be thoroughly cleaned since any adhering substance, such as grease, will cause the films not to adhere and the stains to vary in color. Fresh blood from a finger prick is preferred, although blood from a hypodermic syringe is acceptable if promptly used. Blood without anticoagulant is preferred.

The selected site for puncture (usually the *side* of the terminal phalanx of a finger, or in an infant an earlobe or heel) is wiped thoroughly with 70% alcohol, then *dried* with a clean gauze sponge. The puncture is made with a suitable sharp lancet, preferably a disposable one for single use to avoid risk of hepatitis B.

For the thin film, a slide is brought into contact with a small drop of blood at the puncture site in the midline of the slide about 2 cm from one end. An end of a second slide is then used as a spreader immediately to draw the blood out into a smoothly tapering film most of which is one cell thick. The film is allowed to dry thoroughly, in a drying oven if necessary, as in humid conditions, at about 37°C overnight. Before it is stained the film must be *fixed* by covering it with absolute methyl (or ethyl) alcohol for 2 to 3 minutes to prevent dehemoglobinization. Staining is then carried out with a suitable stain, such as Giemsa's. Since Wright's stain is a solution in absolute alcohol, fixation is carried out when the stain is poured on the slide. Staining occurs after the solution is diluted with neutralized (about pH 7) water. If Wright's stain is allowed to absorb water from the air, e.g., when the bottle is left open, the fixation phase will be unsatisfactory.

Thick films should be stained (or at least dehemoglobinized) as promptly as possible to assure total dehemoglobinization. The slides should be placed with blood film down on a rectangular shallowly curved plate. The arc of curvature is such that the center of the slide is about 3 mm from the surface of the plate. Giemsa's stain poured onto the plate is drawn under the slides by capillary attraction. As dehemoglobinization takes place, the debris falls away from the slides onto the plate and the cells absorb the stain. After drying, the slide is ready for examination with oil-immersion objective, but if it is to be preserved, both thick and thin smears should be covered with a thin layer of mounting medium under a No. 1 or No. 0 cover glass.

Microfilariae in Blood

In most instances microfilariae in the peripheral blood (or in cutaneous tissue fluid, namely, those of *Onchocerca volvulus* or *Mansonella [*syn. *Dipetalonema] streptocerca)* are satisfactorily stained by the Giemsa technique. The thick film is recommended. For best results the use of an anticoagulant should be avoided (Partono and Idris, 1977). After the material has been spread over a circular area about 20 mm in diameter on an absolutely clean microscope slide and has been thoroughly dried, the slide is placed vertically in the diluted Giemsa stain. In 30 to 60 minutes the staining is complete. The slide is removed, washed gently in tap water, and dried. Overstaining is not likely to occur.

Hematoxylin-Stained Blood Film. Some workers prefer hematoxylin stain. The dried thick film is dehemoglobinized in 0.5% HCl, washed in 1% lithium carbonate solution, and stained with hematoxylin (Bullard's or Delafield's).

Procedure.

1. Without drying, fix in 70% ethyl alcohol preheated to 55 to 60°C, or dry thoroughly and fix in a 1:1 mixture of ether and absolute methyl alcohol for 5 to 10 minutes.
2. Stain in undiluted hematoxylin for 10 minutes.
3. Wash in running tap water for 5 minutes (or in a slightly basic solution of lithium carbonate).
4. Dry thoroughly and add mounting medium and a cover glass.

Concentration of Microfilariae (Knott's Method). Defibrinated and dehemoglobinized blood, lymph, or chylous urine may be concentrated by centrifuging for about 1 minute at moderate speed, the supernatant fluid decanted, and the sediment examined for microfilariae. These substances may be vitally stained or the film air-dried, fixed, and permanently stained.

Knott (1939) modified this technique as follows: 2 ml of blood is thoroughly mixed with 10 ml of a 2% solution of formalin (2 ml commercial formaldehyde solution and 98 ml water) and allowed to stand for 10 minutes or longer, then centrifuged at moderate to high speed for 2 minutes. The supernatant fluid is decanted, the sediment is stained in bulk with a 1: 10,000 aqueous solution of either methylene blue or azure II, and the sediment is examined microscopically for microfilariae.

Membrane Filtration. Membrane filtration

techniques are effective in detecting microfilariae when present in small numbers in the blood. However, microfilariae collected in this manner are unsuitable for demonstrating the morphology for species identification. The technique recommended by Desowitz and Southgate (1973) is widely used.

Materials and Procedures.

1. Draw a 1-ml specimen of venous blood into a 10-ml syringe containing 0.1 ml of a 5% solution of sodium citrate.
2. In the same syringe draw 9 ml of a 10% solution of Teepol in physiological saline and shake gently for 1 minute.
3. Remove the needle and attach a Swinney filter holder containing a 25-mm membrane filter* of 5 μm porosity placed over a supporting filter paper pad of the same size, moistened with saline.
4. With gentle, steady pressure force the blood through the filter.
5. Wash the filter three times by passing 10 ml of physiological saline through it.
6. Remove the filter and stain for 5 minutes in hot, but not boiling, Harris' hematoxylin, then "blue" it briefly in running tap water.
7. Dry the membrane thoroughly, then cover it with mounting medium and cover glass on a microscope slide for examination. The preparation can be cleared and examined under a film of immersion oil, but this is not recommended if a permanent mounting may be needed later. Most solvents used for removing immersion oil cause warping of the filter disc.

EXAMINATION OF ASPIRATES

Recovery of protozoa or helminths from aspirated material often constitutes an important aid to diagnosis.

Proctoscopic aspirates and scrapings are useful in confirming suspicion of amebic ulcers in the lower sigmoid colon and rectum, but extreme caution must be exercised, since host tissue cells are often mistaken for nonmotile trophozoites or cysts of *Entamoeba histolytica*.

Duodenal aspiration may be useful in demonstrating infections with *Giardia lamblia, Strongyloides stercoralis, Fasciolopsis buski,* and parasites in the gallbladder or biliary tract. However, when biliary infection is suspected, it is preferable to obtain samples of pure bile. A device consisting of nylon yarn coiled inside a weighted gelatin capsule to be swallowed by the patient, and withdrawn when it has uncoiled and reached the duodenum, has been described (Beal et al., 1970). Microscopic examination of the adherent mucus may reveal infections not detected by stool examinations, especially giardiasis and strongyloidiasis (Bezjak, 1972; Goldsmid and Davies, 1978). In some cases the failure to find cysts or larvae in the feces may have been due to lack of skill or effort.

Aspirates from liver and lung lesions are particularly helpful in obtaining material for diagnosis of amebic liver abscess as well as hydatid cyst. In the liver abscess, amebic trophozoites are more likely to be found in material aspirated from the wall of the abscess than from the necrotic center. Enzymatic liquefaction of viscous aspirates greatly facilitates the search for amebae. Streptodornase in sterile physiologic saline solution, 50 units/ml, is added to the fresh aspirate, 1 part enzyme solution to 5 parts aspirate. The mixture is incubated for 30 minutes at 37°C, with stirring two or three times. The liquefied material is then transferred to tubes and centrifuged. Amebae in the sediment are motile and can be grown in preconditioned culture medium.

Aspirates from lymph nodes, spleen, liver, bone marrow, and *spinal fluid* frequently provide diagnostic evidence of African trypanosomiasis, visceral leishmaniasis (kala-azar), Chagas' disease, and toxoplasmosis. The material obtained should be used in part for direct unstained cover glass preparations to demonstrate motile organisms (trypanosomes), in part for impression smears that are stained (with Giemsa), and in part for culturing of organisms that may be present.

BIOPSY MATERIAL

Biopsy is frequently useful in the diagnosis of parasitic infections, and may be the most convenient and at times the only method of confirming clinical suspicion of an infection. If a mass removed surgically contains a parasite, it is likely to be at the center. When the presence of such a parasite is suspected, the mass should be cut at its middle not quite fully through, with care that no parts are lost as the two halves of the mass are opened on the remaining hinge of tissue. The specimen should then be embedded for sectioning with the two cut surfaces in the same plane and in the side of the block where sectioning will begin. Sectioning should be done serially with the entire rib-

*Millipore Corporation, Bedford, Massachusetts

bon saved. In skip-sectioning, the object of interest is often missed.

Skin. Biopsies of skin can be used to demonstrate amebiasis cutis, cutaneous leishmaniasis, and microfilariae of *Onchocerca volvulus, Mansonella* (syn. *Dipetalonema) streptocerca,* and *Mansonella ozzardi*. In the first two infections, a representative portion of the entire lesion should be removed; this should be fixed and processed for sections to be stained. For microfilariae only small, superficial biopsies are required. Thin samples of skin are placed in a drop of tepid physiologic saline solution or water, mounted with a cover glass, and examined for microfilariae emerging from the tissue. For identification when there is doubt, the cover glass can be removed to allow the preparation to dry. It is then fixed in absolute methyl alcohol and stained with Giemsa or hematoxylin in the usual manner.

Lymph Node. Biopsy of an enlarged superficial lymph node may demonstrate the organisms of African trypanosomiasis, kala-azar, Chagas' disease, or toxoplasmosis. Impression smears should be made for Giemsa staining, but a representative portion should be fixed, sectioned, and stained with hematoxylin and eosin.

Muscle. A muscle biopsy may demonstrate *Trichinella* larvae and, at times, the cysticercus of *Taenia solium*. Some of the tissue may be compressed between glass slides and examined directly for *Trichinella* cysts, but a portion should be digested in artificial gastric juice at 37°C and the larvae concentrated by centrifugation. If the specimen is adequate, a portion can be fed to a laboratory rat for evidence of viability of larvae that may be present. Detection and identification of larvae of *Ancylostoma* and *Toxocara* in muscle biopsy are best accomplished by complete serial sectioning and staining of the entire sample. If a cysticercus is discovered in the specimen, it can be dissected out of its capsule, ruptured, and compressed between a slide and thick cover glass to demonstrate the characteristic scolex with four suckers and circle of rostellar hooks.

Colon and Rectum. Amebic colitis may be diagnosed by biopsy of the colon or rectum, but usually this is done by proctoscopic aspiration. On the other hand, proctoscopic biopsy is a valuable procedure for demonstrating eggs of *Schistosoma mansoni, S. japonicum,* and *S. haematobium* that have been deposited in the lower mesenteric and rectal venules and have filtered into the tissues of the sigmoid colon and rectum. It is particularly valuable in chronic cases of intestinal schistosomiasis and as a follow-up of treatment, when stools are frequently negative for eggs of the parasite. Similarly, biopsy of the bladder may be useful in diagnosing *S. haematobium* infection.

AUTOPSY MATERIAL

For parasites in tissues obtained at autopsy, the methods of fixation and staining that are routinely used in the pathology laboratory generally are satisfactory. Too frequently in the sectioning of tissues the presence of a parasite is not suspected, and the practice of skip-sectioning results in the loss of much or all of diagnostically useful material (see the preceding discussion of biopsy material).

For routine identification of worms that are collected free of the tissues, fixation and preservation in 10% formalin are generally satisfactory. Small nematodes can be more easily examined and more accurately described if fixed in glacial acetic acid and stored in 70% ethyl alcohol containing 5% glycerine. A general fixative that is satisfactory for most kinds of worms is a mixture of equal parts of 10% formalin and 95% ethyl alcohol containing 5% acetic acid (5 parts glacial acetic acid and 95 parts formalin-alcohol). The fixative should be heated to about 63°C and the worms added with swirling of the fluid.

At the U.S. National Museum, all kinds of worms and tissues are stored in a preservative made up of 92 parts 70% ethyl alcohol, 5 parts glycerine, and 3 parts formaldehyde.

ENVIRONMENTAL MATERIAL

Parasite cysts and eggs can be recovered from clothing, vegetables, and other fomites and from contaminated water and soil. However, in general, the universe from which such samples are taken is immense in comparison with the usual sample of feces, urine, sputum, blood, or tissue. In addition, whereas eggs in feces are randomly distributed, their distribution in soil is highly unequal and nonrandom. Therefore, specimens taken for examination from the environment are usually collected in a manner that is more selective than random, and the more skillfully the samples are selected, the more likely they are to give positive results.

Procedures for the examination of specimens from fomites, water, or soil follow the same principles as those used in the examination of feces: suspension in water, sedimentation-decantation repeated until the supernatant is clear, and further concentration by flotation, by sedimentation with ether, or by a combination of these techniques.

Recovery of Parasite Objects from Fomites. To demonstrate the presence of eggs or cysts on fresh vegetables or other fomites, the first step is to remove them from adhesive surfaces. This is done first by soaking in water to which a detergent has been added, then flushing, brushing, and stirring to put the eggs into suspension. The next step is to strain out coarse particles and fiber and allow the suspension to settle, then decant, or collect the sediment by centrifugation. Finally, the eggs and cysts can be concentrated and examined by the zinc sulfate flotation or formalin ether technique.

Recovery of Parasite Objects from Water. It is usually impractical to attempt recovery of eggs and cysts from large bodies of water or open streams. If such recovery is undertaken at all, filtration probably is the method of choice (Jakubowski and Ericksen, 1979).

Recovery of Parasite Objects from Soil. No practical method has been developed for the recovery of protozoan cysts from soil, although it has been shown that cysts of *Entamoeba histolytica* mixed into soil experimentally can be recovered by zinc sulfate flotation (Beaver and Deschamps, 1949).

Sedimentation and Flotation. Helminth eggs have been isolated from soil in numerous studies (Spindler, 1929; Headlee, 1936; Beaver, 1952b). Based on published reports, the following steps have been recommended (WHO, 1967):

1. To each volume of soil add 2 volumes of dilute sodium hypochlorite (3 parts bleaching fluid to 10 parts water) and allow to stand for 30 minutes, with frequent stirring to break the adhesiveness of the eggs and soil particles.
2. Dilute in 10 to 20 parts of water, stir or shake, and allow to stand just as long as macroscopic particles can be seen falling to the bottom (10 to 15 seconds), then decant to a clean container and discard the sediment.
3. Strain out the coarse particles by pouring through a coarse mesh cloth (gauze) or metal strainer.
4. Concentrate and wash by gravity or centrifugal sedimentation.
5. Separate by brine (or zinc sulfate) flotation as for feces, but instead of using a loop to collect the eggs at the meniscus, pour off the surface film to a clean centrifuge tube, dilute, centrifuge (or allow to settle), and examine the sediment.

Variations of these procedures have been reported by Caveness and Jensen (1955), Dada (1979), Quinn et al. (1980), and Bozděch (1981). Woodruff and Shah (1976) used zinc sulfate flotation to separate the eggs and then concentrated them further on a filter pad.

Baermann Funnel Method. Larvae of hookworms can be recovered by the Baermann sieve-in-funnel technique, or by the damp pad method of Beaver (1953). In the former method, a large glass funnel about 15 cm or more in diameter is placed vertically in a support frame, and a rubber tube of suitable size and length with pinch clamp is fitted to the funnel spout. A sieve of appropriate size to rest in the funnel well below its brim is placed in the funnel. The soil sample is placed in a closed cloth bag which in turn is placed in the sieve. The pinch clamp is closed and warm water (30°C) is poured into the funnel over its brim (not onto the bag of soil) to bring the water level just above the soil sample. Since hookworm larvae cannot swim, they fall to the bottom when suspended in water. After 2 or 3 hours, depending on the size of the soil sample, the larvae can be released into a centrifuge tube or sedimentation flask by opening the pinch clamp. A major problem with this technique is that the hookworm larvae are hidden among a multitude of larval and adult free-living nematodes, some of which may be mistaken for hookworm larvae.

Damp Pad Method. The damp pad technique of soil sampling is preferred because it largely eliminates the problem of free-living nematodes and requires no special apparatus. Pads of 12 to 15 layers of gauze about 12 cm in diameter are dampened and pressed onto damp or dampened soil, covered with a plastic sheet to prevent evaporation, and allowed to remain for 20 to 30 minutes (20 for soil already damp, 30 or more for dry soil needing to be wetted). With long forceps or tongs the pad is rolled into the plastic sheet for transport. In the laboratory the pads are unrolled and quickly placed upside down (soil-bearing side up) in a sedimentation flask filled to near the brim with warm water (Fig. 46–4). The larvae quickly fall from the pad to the bottom of the flask while adherent soil particles remain on the pad. Larvae can be removed with a long pipette for microscopic examination on a slide, with or without a cover glass. For close examination, the larvae can be immobilized with heat, dilute aqueous iodine, or acetic acid. This technique of soil sampling for infective-stage hookworm larvae has proved satisfactory for extensive epidemiological studies (Beaver, 1953; Schad et al., 1973; Nawalinski et al., 1978).

REFERENCES

Ayala, S.C., and de Sánchez, C.E. 1974. Evaluación de los diagnósticos coproparasitológicos realizádos en los laboratórios clinícos de la ciudad de Cali. Acta Med. Valle, *5*:114–121.

Bartholomew, R.K., Peters, P.A.S., and Jordan, P. 1981. *Schistosoma mansoni* in St. Lucian and Kenyan communities—a comparative study using the Kato stool examination technique. Ann. Trop. Med. Parasitol., *75*:401–405.

Bartlett, M.S., Harper, K., Smith, N., Verbanac, P., and Smith, J.W. 1978. Comparative evaluation of a modified zinc sulfate flotation technique. J. Clin. Microbiol., *7*:524–528.

Beal, C.B., Viens, P., Grant, R.G.L., and Hughes, J.M. 1970. A new technique for sampling duodenal contents. Demonstration of upper small-bowel pathogens. Am. J. Trop. Med. Hyg., *19*:349–352.

Beaver, P.C. 1949a. Quantitative hookworm diagnosis by direct smear. J. Parasitol., *35*:125–135.

———. 1949b. Methods of pinworm diagnosis. Am. J. Trop. Med., *29*:577–587.

———. 1950. The standardization of fecal smears for estimating egg production and worm burden. J. Parasitol., *36*:451–456.

———. 1952a. The detection and identification of some common nematode parasites of man. Am. J. Clin. Pathol., *22*:481–494.

———. 1952b. Observations on the epidemiology of ascariasis in a region of high hookworm endemicity. J. Parasitol., *38*:445–453.

———. 1953. Persistence of hookworm larvae in soil. Am. J. Trop. Med. Hyg., *2*:102–108.

———. 1956. Small electric mixer with expendable stirring rod. Am. J. Clin. Pathol., *26*:1490–1491.

———, and Deschamps, G. 1949. The viability of *E. histolytica* cysts in soil. Am. J. Trop. Med., *29*:189–191.

———, Malek, E.A., and Little, M.D. 1964. Development of *Spirometra* and *Paragonimus* eggs in Harada-Mori cultures. J. Parasitol., *50*:664–666.

Bell, D.R. 1963. A new method for counting *Schistosoma mansoni* eggs in faeces. Bull. W.H.O., *29*:525–530.

Bezjak, B. 1972. Evaluation of a new technic for sampling duodenal contents in parasitologic diagnosis. Am. J. Dig. Dis., *17*:848–850.

Blagg, W., Schloegel, E.L., Mansour, N.S., and Khalaf, G.I. 1955. A new concentration technic for the demonstration of protozoa and helminth eggs in feces. Am. J. Trop. Med. Hyg., *4*:23–28.

Boeck, W.C., 1917. A rapid method for the detection of protozoan cysts in mammalian faeces. Univ. Calif. Publications Zool., *18*:145–149.

———, and Stiles, C.W. 1923. Studies on various intestinal parasites (especially amoebae) of man. Hygienic Lab. Bull., *123*:1–202.

Bozděch, V. 1981. Zur Untersuchung des Stuhls auf parasitäre Stadien. Angew. Parasitol., *22*:6–12.

Brooke, M.M., and Goldman, M. 1949. Polyvinyl alcohol-fixative as a preservative and adhesive for protozoa in dysenteric stools and other liquid materials. J. Lab. Clin. Med., *34*:1554–1560.

Burrows, R.B. 1967. A new fixative and technics for the diagnosis of intestinal parasites. Am. J. Clin. Pathol., *48*:342–346.

Carles, J., and Barthelemy, E. 1917. Procede special d'homogenisation et de tamisage pour collecter les kystes dysenteriques contenus dans les selles. Compt. Rend. Soc. Biol., *80*:402–403.

Caveness, F.E., and Jensen, H.J. 1955. Modification of the centrifugal-flotation technique for the isolation and concentration of nematodes and their eggs from soil and plant tissue. Proc. Helm. Soc. Wash., *22*:87–89.

Cropper, J.W., and Row, R.W.H. 1917. A method of concentrating *Entamoeba* cysts in stools. Lancet, *2*:179–182.

Dada, B.J.O. 1979. A new technique for the recovery of *Toxocara* eggs from soil. J. Helminthol., *53*:141–144.

De Rivas, D. 1928. An efficient rapid method of concentration for the detection of ova and cysts of intestinal parasites. Am. J. Trop. Med., *8*:63–72.

Desowitz, R.S., and Southgate, B.A. 1973. Studies on filariasis in the Pacific: 2. The persistence of microfilaraemia in diethylcarbamazine treated populations of Fiji and Western Samoa: Diagnostic application of the membrane-filtration technique. Southeast Asian J. Trop. Med. Public Health, *4*:179–183.

Dunn, F.L. 1968. The TIF direct smear as an epidemiological tool. Bull. W.H.O., *39*:439–449.

———. 1972. Intestinal parasitism in Malayan aborigines (Orang Asli). Bull. W.H.O., *46*:99–113.

Faust, E.C., D'Antoni, J.S., Odom, V., Miller, M.J., Peres, C., Sawitz, W., Thomen, L.F., Tobie, J., and Walker, J.H. 1938. A critical study of clinical laboratory technics for the diagnosis of protozoan cysts and helminth eggs in feces. Am. J. Trop. Med., *18*:169–183.

Garcia, L.S., and Ash, L.R. 1979. *Diagnostic Parasitology. Clinical Laboratory Manual*, 2nd ed. St. Louis, C.V. Mosby Co.

Goldman, M., and Brooke, M.M. 1953. Protozoans in stools unpreserved and preserved in PVA-fixative. Public Health Rep., *68*:703–706.

Goldsmid, J.M., and Davies, N. 1978. Diagnosis of parasitic infections of the small intestine by Enterotest duodenal capsule. Med. J. Aust., *1*:519–520.

Gomori, G. 1950. A rapid one-step trichrome stain. Am. J. Clin. Pathol., *20*:661–663.

Graham, C. 1941. A device for the diagnosis of *Enterobius* infection. Am. J. Trop. Med., *21*:159–161.

Harada, Y., and Mori, O. 1955. A new method for culturing hookworm. Yonago Acta Med., *1*:177–179.

Headlee, W.H. 1936. The epidemiology of human ascariasis in the metropolitan area of New Orleans, Louisiana. Am. J. Hyg., *24*:479–521.

Healy, G.R., Kagan, I.G., and Melvin, D.M. 1970. Parasitic infections, pp. 724–790. In *Diagnostic Procedures for Bacterial, Mycotic and Parasitic Infections*, 5th ed. Edited by H.L. Bodily, E.L. Updyke, and J.O. Rosen. New York, American Public Health Association.

Hsieh, H.C. 1963. A test-tube filter-paper method for the diagnosis of *Ancylostoma duodenale, Necator americanus* and *Strongyloides stercoralis*. WHO Tech. Rep. Ser. No. 255, pp. 27–30.

Jacobs, A. 1942. Enterobiasis in children. Incidence, symptomatology, and diagnosis with a simplified Scotch cellulose tape technic. J. Pediatr., *21*:497–503.

Jakubowski, W., and Ericksen, T.H. 1979. Methods for detection of *Giardia* cysts in water supplies. In *Waterborne Transmission of Giardiasis*. Edited by W. Jakubowski and J.C. Hoff. Proceedings of a Symposium, September 18–20, 1978. Cincinnati, U.S. Environmental Protection Agency, Environmental Research Center.

Kaminsky, R. 1978. Analysis of selected techniques for surveys of intestinal parasites. East Afr. Med. J., *55*:45–53.

Kato, K., and Miura, M. 1954. Comparative examinations. Jpn. J. Parasitol., *3*:35.

Katz, N., Chaves, A., and Pellegrino, J. 1972. A simple device for quantitative stool thick-smear technique in schistosomiasis mansoni. Rev. Inst. Med. Trop. S. Paulo, *14*:397–400.

Knight, W.B., Hiatt, R.A., Cline, B.L., and Ritchie, L.S. 1976. A modification of the formol-ether concentration technique for increased sensitivity in detecting *Schistosoma mansoni* eggs. Am. J. Trop. Med. Hyg., *25*:818–823.

Knott, J.I. 1939. A method for making microfilarial surveys on day blood. Trans. R. Soc. Trop. Med. Hyg., *33*:191–196.

Komiya, Y., and Kobayashi, A. 1966. Evaluation of Kato's thick smear technic with a cellophane cover for helminth eggs in feces. Jpn. J. Med. Sci. Biol., *19*:59–64.

———, ———, Sugiyama, T., and Kutsumi, H. 1960. Comparative studies on the technics for stool examination for hookworm ova during the cold climate, particularly on the recovering rate of hookworm ova with brine flotation technic and on culturing one with filter paper. Jpn. J. Parasitol., *9*:480–485.

Little, M.D. 1966. Comparative morphology of six species of *Strongyloides* (Nematoda) and redefinition of the genus. J. Parasitol., *52*:69–84.

———. 1981. Differentiation of nematode larvae in coprocultures: Guidelines for routine practice in medical laboratories. WHO Tech. Rep. Ser. No. 666, *Intestinal, Protozoan and Helminth Infections*, pp. 144–149.

Mao, S.P., and Shao, B.R. 1982. Schistosomiasis control in the People's Republic of China. Am. J. Trop. Med. Hyg., *31*:92–99.

Markell, E.K., and Kuritsubo, R.A. 1981. Intestinal protozoan infections: Prevalence in the San Francisco Bay area. West. J. Med., *135*:188–190.

———, and Quinn, P.M. 1977. Comparison of immediate polyvinyl-alcohol (PVA) fixation with delayed Schaudinn's fixation for the demonstration of protozoa in stool specimens. Am. J. Trop. Med. Hyg., *26*:1139–1142.

Martin, L.K. 1965. Randomness of particle distribution in human feces and the resulting influence on helminth egg counting. Am. J. Trop. Med. Hyg., *14*:747–759.

———. 1972a. Hookworm in Georgia. I. Survey of intestinal helminth infections and anemia in rural school children. Am. J. Trop. Med. Hyg., *21*:919–929.

———. 1972b. Hookworm in Georgia. II. Survey of intestinal helminth infections in members of rural households of southeastern Georgia. Am. J. Trop. Med. Hyg., *21*:930–943.

———, and Beaver, P.C. 1968. Evaluation of Kato thick-smear technique for quantitative diagnosis of helminth infections. Am. J. Trop. Med. Hyg., *17*:382–391.

McMullen, D.B., and Beaver, P.C. 1945. Studies on schistosome dermatitis. IX. The life cycles of three dermatitis-producing schistosomes from birds and a discussion of the subfamily Bilharziellinae (Trematoda: Schistosomatidae). Am. J. Hyg., *42*:128–154.

Morera, P. 1973. Life history and redescription of *Angiostrongylus costaricensis* Morera and Céspedes. Am. J. Trop. Med. Hyg., *22*:613–621.

Nawalinski, T., Schad, G.A., and Chowdhury, A.B. 1978. Population biology of hookworms in children in rural West Bengal. II. Acquisition and loss of hookworms. Am. J. Trop. Med. Hyg., *27*:1162–1173.

Pampiglione, S., and Ricciardi, M.L. 1971. The presence of *Strongyloides fülleborni* von Linstow, 1905, in man in Central and East Africa. Parassitologia, *13*:257–269.

Partono, F., and Idris, K.N. 1977. Some factors influencing the loss of microfilariae from stained thick blood films. Southeast Asian J. Trop. Med. Public Health, *8*:158–164.

Peters, P.A., El Alamy, M., Warren, K.S., and Mahmoud, A.A.F. 1980. Quick Kato smear for field quantification of *Schistosoma mansoni* eggs. Am. J. Trop. Med. Hyg., *29*:217–219.

———, Mahmoud, A.A.F., Warren, K.S., Ouma, J.H., and Siongok, T.K.A. 1976. Field studies of a rapid accurate means of quantifying *Schistosoma haematobium* eggs in urine samples. Bull. W.H.O., *54*:159–162.

Petithory, J. 1979. Le controle de qualité en parasitologie. Bull. Soc. Pathol. Exot., *72*:386–395.

Phillipson, R.F. 1962. Flotation technique for *Opisthorchis* and *Clonorchis* eggs. Trans. R. Soc. Trop. Med. Hyg., *56*:174.

Proffitt, R.D., and Walton, B.C. 1962. *Ascaris* pneumonia in a two-year-old girl. Diagnosis by gastric aspirate. N. Engl. J. Med., *266*:931–934.

Quinn, R., Smith, H.V., Bruce, R.G., and Girdwood, R.W.A. 1980. Studies on the incidence of *Toxocara* and *Toxascaris* spp. ova in the environment. 1. A comparison of flotation procedures for recovering *Toxocara* spp. ova from soil. J. Hyg. (Camb.), *84*:83–89.

Ridley, D.S., and Hawgood, B.C. 1956. The value of formol-ether concentration of faecal cysts and ova. J. Clin. Pathol., *9*:74–76.

Ritchie, L.S. 1948. An ether sedimentation technique for routine stool examinations. Bull. U.S. Army Med. Dept., *8*:326.

Sapero, J.J., and Lawless, D. 1953. The "MIF" stain-preservation technic for the identification of intestinal protozoa. Am. J. Trop. Med. Hyg., *2*:613–619.

———, ———, and Strome, C.P.A. 1951. An improved iodine-staining technique for routine laboratory diagnosis of intestinal protozoa. Science, *114*:550–551.

Sasa, M., Hayashi, S., Tanaka, H., and Shirasaka, R. 1958. Application of test-tube cultivation method on the survey of hookworm and related human nematodes infection. Jpn. J. Exp. Med., *28*:129–137.

Schad, G.A., Chowdhury, A.B., Dean, C.G., Kochar, V.K., Nawalinski, T.A., Thomas, J., and Tonascia, J.A. 1973. Arrested development in human hookworm infections: An adaptation to a seasonally unfavorable external environment. Science, *180*:502–504.

Scholten, Th. 1972. An improved technique for the recovery of intestinal protozoa. J. Parasitol., *58*:633–634.

———, and Yang, J. 1974. Evaluation of unpreserved and preserved stools for the detection and identification of intestinal parasites. Am. J. Clin. Pathol., *62*:563–567.

Smillie, W.G. 1921. A comparison of the number of hookworm ova in the stool with the actual number of hookworms harbored by the individual. Am. J. Trop. Med., *1*:389–395.

Spindler, L.A. 1929. On the use of a method for the isolation of *Ascaris* eggs from the soil. Am. J. Hyg., *10*:157–164.

Stoll, N.R. 1923a. Investigations on the control of hookworm disease. XV. An effective method of counting hookworm eggs in feces. Am. J. Hyg., *3*:59–70.

———. 1923b. Investigations on the control of hookworm disease. XVIII. On the relation between the number of eggs found in human feces and the number of hookworms in the host. Am. J. Hyg., *3*:156–179.

———. 1963. Dilution egg-counting technique for hookworm, *Ascaris, Trichuris* and other parasites. WHO Tech. Rep. Ser. No. 255, pp. 21–26.

———, and Hausheer, W.C. 1926a. Accuracy in the dilution egg counting method. Am. J. Hyg., *6*:80–133.

———, and ———. 1926b. Concerning two options in dilution egg counting: Small drop and displacement. Am. J. Hyg., *6*:134–145.

Telemann, W. 1908. Eine Methode zur Erleichterung der Auffindung von Parasiteneiern in den Faeces. Dtsch. Med. Wochenschr., *34*:1510–1511.

Truant, A.L., Elliott, S.H., Kelly, M.T., and Smith, J.H. 1981. Comparison of formalin-ethyl ether sedimentation, formalin-ethyl acetate sedimentation, and zinc sulfate flotation techniques for detection of intestinal parasites. J. Clin. Microbiol., *13*:882–884.

Wheatley, W.B. 1951. A rapid staining procedure for intestinal amoebae and flagellates. Am. J. Clin. Pathol., *21*:990–992.

Woodruff, A.W., and Shah, A.I. 1976. Improved method of recovering *Toxocara* species ova from soil. Br. Med. J., *2*:621–622.

Woodstock, L., Cook, J.A., Peters, P.A., and Warren, K.S. 1971. Random distribution of schistosome eggs in the feces of patients with schistosomiasis mansoni. J. Infect. Dis., *124*:613–614.

World Health Organization. 1967. Control of ascariasis. WHO Tech. Rep. Ser. No. 379.

Yang, J., and Scholten, T. 1977. A fixative for intestinal parasites permitting the use of concentration and permanent staining procedures. Am. J. Clin. Pathol., *67*:300–304.

Young, K.H., Bullock, S.L., Melvin, D.M., and Spruill, C.L. 1979. Ethyl acetate as a substitute for diethyl ether in the formalin-ether sedimentation technique. J. Clin. Microbiol., *10*:852–853.

Chapter 47
Culture Methods

GENERAL CONSIDERATIONS

Culture methods constitute a substantial segment of present-day study of animal parasites, especially species of protozoa. The purposes of culture are manifold: (1) for accurate diagnosis of the organism, as a supplement to other methods or to provide positive diagnosis when routine methods have failed; (2) for obtaining a rich yield of the organism to use as antigen in immunologic diagnosis; (3) for teaching purposes, particularly when clinical material is not available; (4) as a source for inoculating susceptible experimental animals; (5) for *in vitro* screening of drugs, and (6) for investigating the physiology of the organism.

Ideal *in vitro* cultivation of parasites for investigational or practical purposes is in an axenic environment, i.e., free of living associates such as bacteria. In the culture of the blood and tissue flagellates (leishmanias and trypanosomes), bacterially sterile media are required. For the relatively few helminths that have thus far been studied *in vitro,* this is likewise most important. Some of the more useful methods of culture for diagnosis are described or referred to in this chapter. Detailed descriptions of techniques and media used for *in vitro* cultivation of parasites in research studies have been compiled by Taylor and Baker (1978). Methods for the *in vitro* cultivation of malaria parasites, *Theileria,* trypanosomes, *Leishmania,* schistosomes, and filariae were described and discussed at a workshop held in Nairobi, Kenya, in 1979 (Workshop, 1980).

CULTIVATION OF PROTOZOA

Amebae and Other Intestinal Protozoa

To initiate a culture of *Entamoeba histolytica* or other intestinal amebae, fresh stools uncontaminated with urine are required. Formed stools containing cysts or liquid specimens with living trophozoites may be employed. In the latter case, elements of mucus and tissue detritus are more likely to contain the organism than the diluted feces. To prevent rapid overgrowth by enteric bacteria, it is helpful to add penicillin when the stool sample is inoculated into the culture medium, and it is necessary to introduce considerably more inoculum than in bacteriologic cultures. Usually the inoculum should consist of representative portions of the several components of the stool, saline-purged specimen, enema specimen, proctoscopic aspirate, liver aspirate, or other material obtained from the patient. About 50 mg of formed or mushy feces, or up to 0.5 ml of liquid feces or washed sediment, are inoculated into the medium, and the culture is incubated at 37°C. Subcultures are usually made at 48-hour intervals. At times no evidence of the suspected organism is found in the original culture, but subcultures may be positive. In addition, examination of sediment from cultures after fixation in PVA and staining of permanent smears, as for feces, may reveal the presence of amebae not found otherwise (Norman and Brooke, 1955).

The following media are selected from the many satisfactory ones that have been developed:

Balamuth's Monophasic Medium. Balamuth's (1946) all-liquid medium is commonly used for the isolation and maintenance of *Entamoeba histolytica* and some other intestinal protozoa. It is satisfactory for *Balantidium coli,* although the quality and amount of starch may need to be adjusted.

Preparation.

1. Weigh 288 g dehydrated egg yolk.
2. Add 288 ml distilled water and 1000 ml 0.85% saline. With an electric mixer (or similar instrument), emulsify until suspension is smooth.
3. Heat over an open flame in the top part of a double boiler, stirring constantly until coagulation begins (5 to 10 minutes).

4. Heat the mixture over boiling water, stirring occasionally until coagulation is complete (about 20 minutes). Add 160 ml distilled water to offset loss from evaporation.

5. Filter the mixture through a muslin bag. When the bag cools, it may be squeezed gently to obtain the maximum volume of filtrate.

6. Measure filtrate and add 0.85% saline to bring volume to 1000 ml.

7. Dispense the filtrate into two 1-liter Erlenmeyer flasks. Autoclave for 20 minutes at 15 lbs (121°C).

8. Place the flasks in the refrigerator overnight or until the solution is thoroughly chilled.

9. Filter while cold through a Buchner funnel, preferably using negative pressure to facilitate filtration. Use two pieces of Whatman qualitative filter paper of suitable size and pour the mixture in small amounts through the funnel, replacing the papers frequently.

10. Measure the filtrate. Add an equal volume of Balamuth's buffer solution.

11. Add 0.5% crude liver extract (Lilly, No. 408) (5 ml per liter of medium).

12. Dispense in 5- to 7-ml amounts in tubes; autoclave for 20 minutes at 15 lbs (121°C). Add 5–10 mg of sterile rice powder and incubate for 24 hours at 37°C before using. The medium may be left in flasks in large amounts, autoclaved, and stored in the refrigerator. It can be kept for a month or more without deterioration, although any accumulated sediment should be removed by filtration. Reautoclaving produces no harmful effects, and may even be beneficial.

Solutions.

1. Balamuth's buffer solution

(a) 1M K_2HPO_4	174.180 g
Distilled water to make	1000.0 ml
(b) 1M KH_2PO_4	136.092 g
Distilled water to make	1000.0 ml

2. Mix these solutions in a ratio of 4.3 parts 1M K_2HPO_4 to 0.7 parts 1M KH_2PO_4. To one part of this mixture add 14 parts of distilled water to prepare the M/15 buffer. This final solution is the one used in step 10 above in the directions for preparation of the medium.

Boeck and Drbohlav's Diphasic Medium. The egg-slant diphasic medium of Boeck and Drbohlav (1925), variously modified, is often used for routine primary isolation of *E. histolytica*. The concentrate from saline centrifugal sedimentation of feces rather than the feces is recommended, although inocula of 50 mg of fresh untreated feces have given satisfactory results (Robinson, 1968).

Preparation.

1. Emulsify 4 whole hens' eggs with 50 ml of Ringer's solution.

2. Filter the emulsion through several thicknesses of cheesecloth or surgical gauze.

3. Dispense 2- to 3-ml amounts into 16 × 150 mm test tubes.

4. Place several layers of paper in the bottom of the autoclave and slant the tubes at an angle of 30°. (Place the tubes in baskets so that the rows are not more than 2 or 3 deep.)

5. Inspissate for 10 min at 15 lb in the autoclave. (Set valves according to directions that accompany the autoclave.) Cool.

6. Overlay each slant with approximately 5 ml of buffered saline (pH 7.0), or normal saline containing 0.5% crude liver extract (Lilly, No. 408).

7. Sterilize slants in the autoclave for 20 min at 15 lb (121°C). Wrap the baskets in paper to minimize breaking of the slants during sterilization.

8. Add one loopful (about 10 mg) of sterile rice powder to each sterile overlaid slant.

9. Incubate medium for 24 hrs at 37°C just prior to use, to demonstrate sterility.

Solutions.

1. Stock Ringer's solution

NaCl	70.0 g
CaCl	3.0 g
KCl	2.5 g
Distilled water	1000 ml

Dilute the stock solution 1:10 with neutral distilled water for use.

2. Buffered saline (pH 7.0)

0.85% saline	450.0 ml
M/15 KH_2PO_4	19.0 ml
M/15 Na_2HPO_4	31.0 ml
M/15 KH_2PO_4, prepared as follows:	
KH_2PO_4	9.07 g
Distilled water to make	1000 ml
M/15 Na_2HPO_4, prepared as follows:	
Na_2HPO_4	9.46 g
Distilled water to make	1000 ml

Axenic Cultivation of *Entamoeba histolytica*. In 1961 Diamond first reported the successful cultivation of *Entamoeba histolytica* in the absence of bacteria. In 1965, a monophasic clear liquid medium was described, and this medium was further improved to give rapid growth of dense populations of amebae (Diamond and Bartgis, 1965; Diamond et al., 1978). Axenic cultivation has been an essential component of numerous immunological and biochemical studies on *E. histolytica*.

Axenic Cultivation of Giardia. *Giardia lamblia* was first successfully maintained in culture in association with a yeast *(Candida)* by Karapetyan in 1960. Isolates of *Giardia* from the rabbit and chinchilla were established in cultures by Meyer and Pope in 1965, again in association with growing colonies of a yeast, *Saccharomyces cerevisiae*. In 1970, Meyer reported the successful axenic cultivation of *Giardia* isolated from the rabbit, chinchilla, and cat. In 1976 he described the methods by which he was able to establish cultures of *G. lamblia* in the presence of intestinal fungi, separate the *Giardia* from the fungi, and grow the *Giardia* axenically. The medium used for establishing the colonies was human serum in a peptone broth containing glucose and L-cysteine hydrochloride (HSP-1). For separation and axenic growth, this medium was combined with culture medium NCTC-135 (Gibco) and glutathione-cysteine. More recently, Visvesvara (1980) and Gillin and Diamond (1981) have established luxuriant axenic colonies of *G. lamblia* in a medium developed by Diamond for the axenic cultivation of *Entamoeba histolytica*.

In Vitro Excystation of *Giardia*. A two-step procedure for *in vitro* excystation of *Giardia lamblia* as a test of viability was described by Rice and Schaeffer (1981). To a 0.5-ml suspension of washed cysts in a 15-ml screw-cap centrifuge tube, add in order 5 ml of warm (37°C) hydrochloric acid (pH 2.0), 2.5 ml of warm Hanks' balanced salt solution supplemented with 29 mM L-cysteine hydrochloride and 67 mM glutathione, and 2.5 ml of warm 0.1 M sodium bicarbonate. Immediately cap the tube, mix the solutions, and incubate the mixture for 30 minutes in a water bath at 37°C. In the second step, add 10 ml of 0.5% trypsin dissolved in Tyrode solution adjusted to pH 8, centrifuge the mixture, resuspend the sediment in 0.5 ml of fresh excystment medium, and transfer the cyst suspension to depression slides which are then covered, sealed, and incubated in the inverted position for 1 hour at 37°C. This procedure gave excystation rates ranging from 40 to 95% with an average of 87% for cysts from symptomatic patients.

Genitourinary Protozoa: *Trichomonas vaginalis*

Trussell and Johnson's Medium. Several media are available commercially for the isolation of *Trichomonas vaginalis*. A simple medium that gives good growth is that of Trussell and Johnson (1941):

1. Proteose peptone 2.0%
   NaCl 0.5%
   Make up in distilled water, adjust to pH 6, and autoclave.
2. Normal human serum 5.0%
   Sodium thioglycolate 0.1%

Simplified Trypticase Serum (STS) Medium. The STS medium developed by Kupferberg et al. (1948) is suitable for the isolation of *T. vaginalis* from urethral and vaginal exudates. It also is used to maintain bacteria-free cultures of the flagellate for teaching and research. Bacteria can be eliminated by serial transfer in medium containing 500 units of penicillin and 500 mg of dihydrostreptomycin sulfate per ml.

Materials.

Trypticase . . . . 20.0 g
Cysteine hydrochloride . . . . 1.5 g
Maltose . . . . 1.0 g
Agar (Difco) . . . . 1.0 g
Distilled water, q. s. . . . . 950.0 ml

Procedure. These materials when mixed are heated to boiling to dissolve the agar, and filtered while hot through filter paper. As an indicator of oxygen tension, 0.6 ml of 0.5% methylene blue is added. After being cooled to 46°C, the mixture is adjusted to pH 6.0 (with HCl or NaOH), if necessary, then brought back to 950 ml with distilled water, tubed in 9.5-ml volumes, autoclaved at 15 lb pressure for 15 minutes, and allowed to cool. The medium is made complete by adding 0.5 ml of sterile, undiluted human serum to each tube, giving a final volume of 10 ml.

Cerebrospinal Protozoa: *Naegleria and Acanthamoeba*

Culbertson's Medium. A satisfactory medium for the primary isolation of *Naegleria* and *Acanthamoeba* from cerebrospinal fluid or brain tissue is that of Culbertson et al. (1968).

Preparation.

1. Prepare 1.5% Bacto agar in distilled water and pour 8-ml amounts into plates.

2. Add 0.1 ml saline suspension of live *Escherichia coli* at 50°C and mix into agar, or place that amount of the suspension in the center of a hardened agar plate and spread over the agar surface.

3. Invert the plate to allow evaporation of excess moisture.

Inoculation.

1. Transfer 0.05 to 0.1 ml of the inoculum to the center of the plate. To avoid irregular spreading and excess moisture, which may interfere with examination, place a sterile antibiotic disc in the center of the bacteria-laden agar plate. If the inoculum is expected to be bacteria-free, or nearly so, moisten the disc with 0.05 ml distilled water and place 0.05 ml of suspension of tissue emulsion, tissue particles, or exudate on the disc. If the inoculum is expected or known to be contaminated with bacteria, place 0.05 ml of a solution containing 25 μg neomycin and 2.5 μg nystatin on the disc, then add the inoculum.
2. Seal the plate and incubate at 37°C.
3. If available material is sufficient, make similar preparations and incubate at 23°C.

Examination. The unopened, inverted plate can be examined with the scanning lens (2.5×) or low power (10×) objective of the microscope, scanning the area around the disc for small, more or less uniform, round or ovoid amebae extending radially from the disc. If amebae are seen, transfer some to fresh plates to maintain the strain. Transfer some of the amebae to a microscope slide in water under a thin cover glass and examine under higher magnification for morphological features and for motile flagellate forms.

Willaert's Medium. The simplest method of culturing the species of *Naegleria* and *Acanthamoeba* is on non-nutrient or low-nutrient agar plates, or test-tube slants, seeded with heat-killed or live bacteria, usually *Escherichia coli*, as described previously. For isolation from cerebrospinal fluid, the CGVS axenic medium of Willaert (1971) (casitone-glucose-vitamins containing 10% fetal calf serum) may be preferred (Stevens et al., 1981).

Blood and Tissue Protozoa

The inoculum consists of blood, aspirates, or small biopsies or postmortem samples from glands, spleen, liver, or bone marrow obtained aseptically from patients suspected of having infection with *Leishmania donovani, L. braziliensis, L. tropica, Trypanosoma gambiense, T. rhodesiense, T. cruzi*, or other hemoflagellates. To insure bacterial sterility, it is desirable to introduce 500 units of penicillin into each tube at the time of inoculation.

NNN Medium. This classic method of Novy and MacNeal (1904) was developed for cultivation of the leishmanias, but is equally satisfactory for *Trypanosoma cruzi*.

Preparation. The formula is as follows:

Agar	14 g
Sodium chloride	6 g
Distilled water	900 ml

Mix and bring to the boiling point, then distribute in tubes and sterilize in the autoclave.

The tubed medium is melted and then cooled to 48°C. To each tube of medium one-third of its volume of sterile defibrinated rabbit blood is added. This is mixed thoroughly with the medium by rotating the tube, after which the tubes are slanted and allowed to cool. This is best done on ice, as more water of condensation is obtained. It is in this supernate at the bottom of the tubes that the organisms develop most rapidly and in greatest numbers. Before using, the tubes should be tested for sterility by placing them in the incubator at 37°C for 24 hours.

Hockmeyer's Medium. Schneider's commercially prepared insect cell culture medium, supplemented by the addition of 30% heat-inactivated fetal calf serum and with 100 IU penicillin and 100 μg streptomycin per ml added, was reported by Hockmeyer et al. (1981) as being simple to use and satisfactory for the diagnosis and management of visceral leishmaniasis. Culture tubes (16 × 10 mm) containing 3 ml of freshly prepared medium (or vaccine vials containing freeze-dried medium to be reconstituted with 3 ml distilled water) are inoculated with splenic aspirates, incubated at 26°C, and examined microscopically daily for the presence of promastigotes. If positive, organisms usually are detected after 2 or 3 days of incubation.

Weinman's Medium. The medium described by Weinman (1944) was developed for culture of *Trypanosoma gambiense*. It is a modification of the Noguchi-Battistini medium for culturing *Bartonella bacilliformis*.

Preparation. The medium consists of (1) NaCl 8 g and distilled water to make up 900 ml; (2) citrated human plasma 100 ml and human hemoglobin (made by mixing 1 part of blood with 3 parts of distilled water) 20 ml. The (1) is autoclaved, (2) is added, and the medium is adjusted to pH 7.4 to 7.5. It is dispensed in rubber-stoppered test tubes and, when inoculated, incubated at 26° to 28°C. The cultures become positive in 7 to 10 days.

Tobie's Diphasic Medium. The medium described by Tobie et al. (1950) was developed for cultivation of African trypanosomes, but it is also satisfactory for cultures of *Trypanosoma cruzi, T. rangeli, Leishmania tropica, L. braziliensis, L. donovani,* and many species of hemoflagellates which are not known to parasitize man.

Preparation.

1. *Solid Phase.* 1.5 g Bacto-beef (Difco), 2.5 g Bacto-peptone (Difco), 7.5 g Bacto-agar (Difco), and 4 g NaCl are dissolved in 500 ml distilled water. Adjust to pH 7.2 to 7.4 with NaOH and autoclave at 15 lb pressure for 20 minutes. Cool to about 45°C, then add to the medium whole inactivated rabbit's blood in the proportion of 25 ml blood to 75 ml base. (To prevent coagulation of the blood, 0.5% sterile sodium citrate may be employed.)

2. *Liquid Overlay.* This sterile Locke's solution is composed of 8 g NaCl, 0.2 g KCl, 0.2 g $CaCl_2$, 0.3 g KH_2PO_4, and 2.5 g dextrose in 1000 ml distilled water.

Procedure. Dispense base in 5-ml amounts into test tubes or 25-ml amounts into flasks, with the former slanted until the medium has solidified. Then add the liquid overlay in the amounts of 2 ml and 10 to 15 ml, respectively. Stopper the containers with sterile cotton plugs. The medium is intended for inoculation with the flagellates before the overlay has evaporated. This medium, incubated at 24° to 25°C for 10 to 14 days, reportedly provides a yield of organisms many times that obtained by Weinman in his medium *(vide supra).*

Cultivation of Malaria Parasites. Approximately 70 years after Bass and Johns (1912) reported the *in vitro* cultivation of malaria parasites *(Plasmodium vivax* and *Plasmodium falciparum),* the maintenance of *P. falciparum* in continuous culture in human erythrocytes *in vitro* was achieved by Trager and Jensen (1976, 1978). They used medium RPMI 1640 in a continuous-flow system in which human erythrocytes were in a shallow stationary layer covered by a shallow layer of medium. The medium was made to flow slowly and continuously over the layer of settled red cells, under an atmosphere with 7% carbon dioxide and low oxygen (1 to 5%). This culture system is now widely used for the production of antigen and for other purposes. Long-term cultivation has been achieved with other *Plasmodium* species that have 24-, 48-, and 72-hour erythrocytic cycles of reproduction (Nguyen-Dinh et al., 1980).

Xenodiagnosis. Triatomid bugs have been employed in the diagnosis of *Trypanosoma cruzi* infections where it has not been possible to demonstrate the organisms in blood films. In addition, xenodiagnosis has proved useful in identifying animal reservoirs where standard serologic techniques are questionable. Laboratory-reared bugs are either fed directly upon a patient or are fed the patient's blood through a membrane. At 10 to 30 days after the blood meal, the intestinal contents of the bugs are examined for flagellates. These may then be inoculated into mice for further confirmation. First-instar nymphs of *Dipetalogaster maximus* have been used effectively to detect circulating *T. cruzi* in patients having chronic infections (Marsden et al., 1979). In Brazil, this bug proved to be more efficient than the local vector *(T. infestans)* and, because of the use of the first instar, more cost-effective since rearing efforts were diminished.

IN-VITRO DEVELOPMENT OF HELMINTHS

Attempts have been made to grow parasitic helminths in nutrient media, in some instances with considerable success (Weinstein, 1966). However, *in-vitro* cultivation of helminths is not applicable to problems of diagnosis. Fecal cultures for the diagnosis of hookworm and other nematode larvae are described in Chapter 46 (pp. 746–748).

DEVELOPMENT OF ARTHROPODS IN THE LABORATORY

For diagnostic purposes, it is at times desirable to culture certain arthropods from immature to adult stages. Although there are keys for the species identification of the larval stages of many Diptera, beetle larvae, and certain other arthropods of medical importance, such identifications usually are made by experts in taxonomic entomology. Many of these arthropods can develop and be maintained in the laboratory if they are needed for teaching or research. The common species of mosquitoes, particularly *Aedes aegypti* and *Culex quinquefasciatus,* and some species of *Anopheles,* are good examples. Details of insectary management and methods for rearing mosquitoes are provided by Gerberg (1970) and Foster (1980). For more information on arthropod culture, see page 776.

REFERENCES

Balamuth, W. 1946. Improved egg yolk infusion for cultivation of *Entamoeba histolytica* and other intestinal protozoa. Am. J. Clin. Pathol., *16*:380–384.

Bass, C.C., and Johns, F.M. 1912. The cultivation of malarial plasmodia (*Plasmodium vivax* and *Plasmodium falciparum*) *in vitro*. J. Exp. Med., *16*:567–579.

Boeck, W.C., and Drbohlav, J. 1925. The cultivation of *Entamoeba histolytica*. Am. J. Hyg., *5*:371–407.

———, and Stiles, C.W. 1923. Concentration methods. Hyg. Lab. Bull. No. 133, pp. 69–73.

Culbertson, C.G., Ensminger, P.W., and Overton, W.M. 1968. Pathogenic *Naegleria* sp.—study of a strain isolated from human cerebrospinal fluid. J. Protozool., *15*:353–363.

Diamond, L.S., and Bartgis, I.L. 1965. Axenic cultivation of *Entamoeba histolytica* in a clear liquid medium. Proceedings Second International Congress on Protozoology. Int. Congr. Ser. No. 91. Excerpta Medica Foundation, p. 102.

———, Harlow, D.R., and Cunnick, C.C. 1978. A new medium for the axenic cultivation of *Entamoeba histolytica* and other *Entamoeba*. Trans. R. Soc. Trop. Med. Hyg., *72*:431–432.

Foster, W.A. 1980. Colonization and maintenance of mosquitoes in the laboratory, pp. 103–151. In *Malaria. Volume 2. Pathology, Vector Studies and Culture*. Edited by J.P. Kreier. New York, Academic Press.

Gerberg, E.J. 1970. *Manual for Mosquito Rearing and Experimental Techniques*. Fresno, California. American Mosquito Control Association, Bull. No. 5.

Gillin, F.D., and Diamond, L.S. 1981. *Entamoeba histolytica* and *Giardia lamblia:* Growth responses to reducing agents. Exp. Parasitol., *51*:382–391.

Hockmeyer, W.T., Kager, P.A., Rees, P.H., and Hendricks, L.D. 1981. The culture of *Leishmania donovani* in Schneider's Insect Medium: Its value in the diagnosis and management of patients with visceral leishmaniasis. Trans. R. Soc. Trop. Med. Hyg., *75*:861–863.

Karapetyan, A.E. 1960. Methods of *Lamblia* cultivation. Tsitologiia, *2*:379–384.

Kupferberg, A.B., Johnson, G., and Sprince, H. 1948. Nutritional requirements of *Trichomonas vaginalis*. Proc. Soc. Exp. Biol. Med., *67*:304–308.

Marsden, P.D., Barreto, A.C., Cuba, C.C., Gama, M.B., and Ackers, J. 1979. Improvements in routine xenodiagnosis with first instar *Dipetalonema maximus* (Uhler 1894) (Triatominae). Am. J. Trop. Med. Hyg., *28*:649–652.

Meyer, E.A. 1976. *Giardia lamblia*: Isolation and axenic cultivation. Exp. Parasitol., *39*:101–105.

———, and Pope, B.L. 1965. Culture *in vitro* of *Giardia* trophozoites from the rabbit and chinchilla. Nature, *207*:1417–1418.

Nguyen-Dinh, P., Campbell, C.C., and Collins, W.E. 1980. Cultivation *in vitro* of the quartan malaria parasite *Plasmodium inui*. Science, *209*:1249–1251.

Norman, L., and Brooke, M.M. 1955. The effectiveness of the PVA-fixative technique in revealing intestinal amebae in diagnostic cultures. Am. J. Trop. Med. Hyg., *4*:479–482.

Novy, F.G., and MacNeal, W.J. 1904. On the cultivation of *Trypanosoma brucei*. J. Infect. Dis., *1*:1–30.

Rice, E.W., and Schaeffer, F.W. 1981. Improved in vitro excystation procedure for *Giardia lamblia* cysts. J. Clin. Microbiol., *14*:709–710.

Robinson, G.L. 1968. The laboratory diagnosis by culture of human parasitic amoebae. Trans. R. Soc. Trop. Med. Hyg., *62*:285–294.

Stevens, A.R., Shulman, S.T., Lansen, T.A., Chichon, M.J., and Willaert, E., 1981. Primary amoebic meningoencephalitis: A report of two cases and antibiotic and immunologic studies. J. Infect. Dis., *143*:193–199.

Taylor, A.E.R., and Baker, J.R., editors. 1978. *Methods of Cultivating Parasites in Vitro*. London, Academic Press.

Tobie, E.J., von Brand, T., and Mehlman, B. 1950. Cultural and physiological observations on *Trypanosoma gambiense*. J. Parasitol., *36*:48–54.

Trager, W., and Jensen, J.B. 1976. Human malaria parasites in continuous culture. Science, *193*:673–675.

———, and ———. 1978. Cultivation of malarial parasites. Nature, *273*:621–622.

Trussell, R.E., and Johnson, G. 1941. Physiology of pure culture of *Trichomonas vaginalis*. III. Fermentation of carbohydrates and related compounds. Proc. Soc. Exp. Biol. Med., *47*:176–178.

Visvesvara, G.S. 1980. Axenic growth of *Giardia lamblia* in Diamond's TPS-1 medium. Trans. R. Soc. Trop. Med. Hyg., *74*:213–215.

Weinman, D. 1944. Cultivation of *Trypanosoma gambiense in vitro* in cell-free medium. Proc. Soc. Exp. Biol. Med., *55*:82–83.

Weinstein, P.P. 1966. The *in vitro* cultivation of helminths with reference to morphogenesis, pp. 143–154. In *Biology of Parasites*. Edited by E.J.L. Soulsby. New York, Academic Press.

Willaert, E. 1971. Isolement et culture in vitro des amibes du genre *Naegleria*. Ann. Soc. Belg. Med. Trop., *51*: 701–708.

Workshop on *The* in vitro *Cultivation of Pathogens of Tropical Diseases*. 1980. *Tropical Diseases Research Series* No. 3. Published on behalf of United Nations Development Program/World Bank/World Health Organization, Special Programme for Research and Training in Tropical Diseases. Basel, Schwabe and Co. AG.

Chapter 48

Immunologic Diagnosis

GENERAL CONSIDERATIONS

The most reliable way to diagnose a parasitic infection is by detection and identification of the infecting organism. Failure to recover a parasite, however, does not exclude the possibility of infection. Furthermore, microscopic examination of excreta or host tissues is expensive and time-consuming, and it requires more training and experience than does serological testing. The securing and proper handling of specimens for microscopic examination may be interfered with because of aesthetic attitudes or superstition or, in the case of excreta, because of the requirements of sanitation.

It is therefore often necessary to rely on immunodiagnostic tests for supporting diagnostic information in the analysis of individual cases of suspected parasitic infection. Even more important are the immunodiagnostic tests in the field of epidemiologic surveys.

The principal hurdle that must be passed in the development of useful immunodiagnostic procedures is the preparation of satisfactory antigens. Since animal parasites are complex organisms, the extracts of parasites contain a mixture of substances, some that are antigenic and some that are not. The antigenic substances may be specific or shared by other organisms. Phylogenetically related species, and occasionally even unrelated species, contain or elaborate substances that stimulate the production of antibodies in the host. These substances cross-react because of similarities in the antigens. Sometimes the method of antigen extraction so alters the molecular structure of the antigen that cross-reaction results. It is important, therefore, that harsh methods of extraction not be used.

When it is difficult to obtain an adequate supply of a certain parasite for the extraction of antigen, a more abundant, phylogenetically related organism may be used instead to obtain antigenic extracts that cross-react with the subject infection. For example, antigen prepared from the dog heartworm *Dirofilaria immitis* is commonly used in serodiagnosis or skin testing for human filariasis.

The antigenic makeup of the various stages of the same species of parasite varies. It is important for achieving optimum sensitivity and specificity in an immunologic test that the antigen used be that, or at least cross-react with that, which provokes antibody production in the infected host. For example, the serological tests for infection with *Ascaris lumbricoides* or *Toxocara* (visceral larva migrans) have relatively low sensitivity and specificity when the antigens are prepared from the *adult* worm, since it is the larvae that provide the principal antigenic stimulus to the host.

The principal types of immunodiagnostic tests are intradermal and serological. Although widely employed because of their simplicity of procedure, low cost, and usually rapid results, intradermal tests are subject to serious limitations. They usually lack specificity, are difficult to standardize, are difficult to apply in a uniform manner, and are subject to variations in host reactivity derived from mechanisms unrelated to the infection for which the test is applied. Intradermal reactions may be of the immediate or delayed hypersensitivity types. Tests producing immediate reactions are commonly employed in the diagnosis of helminthic infections. Delayed hypersensitivity reactions are used as diagnostic tests for protozoan infections such as leishmaniasis, trypanosomiasis, and toxoplasmosis.

There are many serological tests for the diagnosis of parasitic diseases. Kagan (1980) lists 19 tests for at least 24 infections. These include complement fixation (CF), bentonite flocculation (BF), indirect hemagglutination (IHA), latex agglutination (LA), indirect immunofluorescence (IF), gel double diffusion (DD), immunoelectrophoresis (IE), countercurrent electrophoresis (CEP), and enzyme-linked immunosorbent assay (ELISA).

A rapid semiautomated fluorometric test marketed under the name of FIAX has been developed, based on the soluble antigen fluorescent antibody (SAFA) technique of Toussaint and Anderson (1965). The method permits processing of numerous specimens with results being read 1 hour after introduction of the specimens into the system (Bergquist, 1981).

Immunodiagnostic tests that are acceptable in epidemiologic studies in which a statistical result is desired may not be acceptable to the clinician, who is dealing with individuals. The judgment as to reliability of a test by a clinician is based on the proportion of false positives (and false negatives) among the specimens submitted for diagnosis. If the proportion of true cases among specimens submitted is small, the test may appear to be highly nonspecific. For example, if of 100 patients tested only two have the disease in question, an adequately sensitive test will identify these two positives. If the test is also relatively specific, i.e., 98%, we may expect two false positives among the same group of 100 patients. Thus, of four positives, to the clinician 50% are false (Ridley, 1981).

IMMUNOLOGIC TESTS FOR PROTOZOA

Amebiasis. Craig in 1929 prepared an antigen for the complement-fixation test by extraction with alcohol of abundant trophozoites of *E. histolytica,* either from culture or from dysenteric material in the stools of experimentally infected dogs. Although he reported high levels of sensitivity and specificity, results obtained by other workers employing his technique have been contradictory or equivocal.

Powell (1968) developed a capillary-tube precipitin test with antigen derived from axenically cultured *E. histolytica* and reported it to be satisfactory for use in individual cases requiring rapid diagnosis in emergencies. Gore and Sadun (1968) described a soluble antigen fluorescent antibody test (SAFA) for amebiasis that produced a high degree of sensitivity and reproducibility of results; however, cross-reactions occurred with sera from patients with visceral leishmaniasis and other infectious diseases. Gel diffusion has been found useful in differentiating precipitin systems in amebiasis, but it is not highly sensitive (Maddison, 1965).

Highly specific antigens from axenically grown *E. histolytica* are currently available. When they are used in appropriate tests, highly sensitive and specific reactions are obtained with sera from patients with amebic liver abscess. Titers tend to be lower with sera from patients with amebic dysentery, in which tissue invasion is less extensive, and tests tend to be negative in the nondysenteric cyst passer.

Tests currently in use include the indirect hemagglutination (IHA) described by Kessel et al. (1965), indirect immunofluorescence (IF) (Ambroise-Thomas and Kien-Truong, 1972), countercurrent electrophoresis (CEP) (Krupp, 1977), immunoelectrophoresis (IE), and enzyme-linked immunosorbent assay (ELISA) (Bos et al., 1975).

Malaria. Although a correlation has been shown between human and simian parasitemia and positive reactions by complement-fixation, precipitin reaction, specific and nonspecific flocculation, and a specific tyrosin-staining test, none of these methods is an adequate substitute for examination of thick blood films in routine diagnosis of malaria. The indirect hemagglutination test has potential usefulness in delineating the extent of malaria transmission. This test has been carried out with the use of sheep erythrocytes treated with formalin, tannic acid, glutaraldehyde or double aldehyde, and then sensitized with malaria antigen (Stein and Desowitz, 1964; Farshy and Kagan, 1972).

The indirect fluorescence test is the most widely used test for malaria at present (Kagan, 1980). It has been used for serodiagnosis of malaria with a false-positive rate of 1% (high specificity) and sensitivity of about 95%. For optimum specificity and sensitivity, human homologous malaria antigen should be used (Kagan, 1980). The ELISA test also has been shown to be sensitive and specific (Voller et al., 1977).

Leishmaniasis. The intradermal reaction has been tested in all three types of leishmaniasis and has potential diagnostic usefulness. The antigen is obtained by physiologic saline extraction of cultured leishmaniae. Positive diagnosis may be obtained before the organisms themselves are demonstrable in the tissues. The reaction is of the delayed type, but usually appears within 24 hours. In 1926 Montenegro demonstrated the diagnostic value of this test, with later confirmation by Pessôa and Pestana and Battistini and Herrer for *L. bra-*

ziliensis, and Dostrovsky and Sagher for *L. tropica*. Its particular diagnostic value in *L. braziliensis* infection is after extension of the primary lesion to the mucocutaneous junctions where the organisms may be difficult to demonstrate.

Visceral leishmaniasis can be diagnosed with a fair degree of accuracy by means of the indirect hemagglutination, indirect immunofluorescence (IF), complement-fixation, and direct agglutination tests. Sensitivity is relatively low in tests of serum from patients with cutaneous leishmaniasis. The IF test with amastigote antigen has been reported to give good results (Walton et al., 1972; Edrissian and Darabian, 1978). Other techniques reported to be useful in diagnosis of cutaneous leishmaniasis are the direct agglutination test with trypsinized culture forms (Allain and Kagan, 1975), complement fixation (Baldelli et al., 1975; Mansueto et al., 1975), and countercurrent electrophoresis (in Sudanese mucocutaneous leishmaniasis) (Abdalla, 1977).

African Trypanosomiasis. In human trypanosomiasis caused by *Trypanosoma gambiense* and *T. rhodesiense*, the serum IgM immunoglobulin level is consistently raised. This finding has led to the development of techniques of estimating IgM levels in dried blood to be used as a screening test for African trypanosomiasis (Cunningham et al., 1967). Used in conjunction with the fluorescent antibody test devised by Sadun et al. (1963) and later modified by Bailey et al. (1967), this technique has provided a diagnostic method that makes it possible to survey large populations as well as to diagnose the individual case (Lucasse, 1970; Latif and Adam, 1973).

American Trypanosomiasis. Complement fixation, agglutination, intradermal reactions, fluorescent antibody, and enzyme-linked immunosorbent assay tests have all been used in the diagnosis of Chagas' disease.

Complement fixation constitutes a valuable method for the diagnosis of Chagas' disease, particularly in chronic infections. As originally developed by Guerreiro and Machado, antigen was obtained as an aqueous extract of the spleens of experimentally infected puppies, and the test was conducted by the usual complement-fixation technique. Others have refined the procedure by obtaining antigen from cultured organisms.

Mayer and Pifano (1941) and Mazza et al. (1942) developed an *intradermal test*, employing as antigen cultured *T. cruzi* concentrates killed in 0.5% phenol in physiologic salt solution.

A simple and rapid indirect *fluorescent antibody test* for the diagnosis of trypanosomiasis utilizing trypanosomes in blood smears as antigen was described by Sadun et al. (1963). Although there were cross-reactions with sera from patients with other species of trypanosomes, there were relatively few with sera from patients with nontrypanosomal diseases, indicating a high degree of specificity. Toussaint et al. (1965) developed a *soluble antigen fluorescent antibody technique* that is also highly specific and has some advantages over the whole-organism indirect fluorescent antibody tests.

The indirect hemagglutination test is also sensitive and specific and can be carried out with eluates from filter-paper blood samples. This makes it well suited to field surveys and epidemiological studies. The enzyme-linked immunosorbent assay (ELISA) procedure also offers this advantage.

A direct agglutination test introduced by Vattuone and Yanovsky (1971) is highly sensitive in reactions with sera from patients with acute Chagas' disease.

Toxoplasmosis. A highly specific *complement-fixation test* was developed by Sabin (1949), who employed concentrates of the organism grown on chorioallantoic membranes of chick embryos. This test reveals active infection.

The *Sabin-Feldman dye test* is more commonly performed. As modified by Jacobs and Cook (1954), the test is carried out as follows:

Peritoneal exudates of mice, infected 3 days earlier with *Toxoplasma gondii* by the intraperitoneal route, are removed by paracentesis. The organisms are placed as a thin wet film on a test plate in a small amount of normal human serum, and a similar test plate is prepared with serum of a suspected case of toxoplasmosis. A drop or two of alkaline methylene blue solution (pH 10.8) is next mixed with the film to be tested on each plate. In the control plate, the cytoplasm and nuclear material of all cells are rather deeply stained; in *Toxoplasma*-positive serum, neutralizing antibodies will prevent the free organisms from taking the stain.

The dye test becomes positive earlier than the complement-fixation reaction, with long-persisting high titers developing in the second week. The complement-fixation test, however, is not positive until after one month of infection and lasts only a short time.

Direct agglutination test for *Toxoplasma* has been

developed by Fulton and Turk (1960). A purified suspension of the organism, obtained from peritoneal fluid of infected cotton rats and freed from other cells, is suspended in buffered saline with 1% formalin. To 0.4 ml of serum to be tested, an equal part of the suspension is added. The tube is incubated at 37°C for 2 hours and left overnight at 2°C. In round-bottomed 12 × 75 mm tubes, the readings are as follows: thick bottom ring, neg.; thin ring and carpet, +; carpet only, + +; carpet withgranules and rolled edges, + + +.

The indirect hemagglutination (IHA) and indirect immunofluorescence (IF) tests have come into popular use in recent years because employing a killed antigen does not result in laboratory infections and is cheaper and easier to perform. The IHA tends not to detect acute infections.

The ELISA (Voller *et al.*, 1976) and the fluorimmunoassay (FIAX) (Walls and Barnhart, 1978) have been shown to be equivalent in sensitivity to the IF test.

The *intradermal test* in toxoplasmosis gives results similar to those for the dye test, i.e., it indicates neutralizing antibodies in the subject's serum. This diagnostic procedure has been employed widely in various population groups to determine the distribution of human toxoplasmosis. The high prevalence of positives in most of these surveys has caused some epidemiologists to question the interpretation of the test.

Pneumocystosis. The most useful serological tests for diagnosis of pneumocystosis are complement-fixation and indirect immunofluorescence. With these procedures, there are about 15% false-negative results. Antigens for both tests are prepared from infected human or rat lung (Kagan and Norman, 1977).

IMMUNOLOGIC TESTS FOR HELMINTHS

Hookworm Infection. The latex agglutination (LA), double gel diffusion (DD), indirect immunofluorescence (IF), complement-fixation (CF), and indirect hemagglutination (IHA) tests have been evaluated in the serodiagnosis of hookworm infection. Results of tests on serum of a single experimentally infected person indicated that the CF test may be positive early in infection but not late. IF tests were positive early and late. IHA and DD reactions were negative (Ball and Bartlett, 1969). The IHA, DD, and circumoval precipitin tests are probably not of value in the diagnosis of hookworm infection (Sood et al., 1972).

Strongyloidiasis. Brannon and Faust (1949) described an intradermal test and a precipitin test with an antigen made from larvae of *Strongyloides fuelleborni*. Since then, in spite of difficulties with coprological diagnosis and the obvious allergenic manifestations of strongyloidiasis, little has been reported on serodiagnosis of the condition until recently. Grove and Blair (1981) have reported an immunofluorescence test (IFA) with *S. ratti* or *S. stercoralis* larvae as antigen, which is highly sensitive. A titer of 1:4 or greater was found in 98% of known infections among ex-prisoners of war in Southeast Asia, and was relatively specific with only 2% positives among control subjects. Although higher titers were produced with *S. stercoralis* used as antigen than with *S. ratti*, the correlation between the two tests was excellent. Thus, when *S. ratti* larvae are more readily available, they may be used for diagnosis of the human infection.

Carroll et al. (1981) and Neva et al. (1981) used an enzyme-linked immunosorbent assay (ELISA) for diagnosis of human strongyloidiasis. The sensitivity was not as high as that reported by Grove and Blair (1981) for the IFA test (82 to 84% of known infections when *S. ratti* was used as antigen and 84% with *S. stercoralis* larvae), but was still better than a single microscopic examination of feces. The latter sensitivity has been reported as only 66% among the same subject population (Grove, 1980).

Trichinosis. *Trichinella* antigen is obtained from laboratory-infected animals (rats, rabbits, guinea pigs), from the muscles of which the larvae are obtained by peptic digest, then concentrated by centrifugation and lyophilized.

Oliver-González (1941) observed that there are two types of precipitin reaction in trichinosis: one that is antilarval and one that is antiadult. The latter forms a precipitate *in vitro* around the mouth, vulva, and anus of adult trichinae, is detectable 15 days after infection, reaches its maximum about the twenty-fifth to thirty-fifth day, and terminates on the fiftieth day. The antilarval type of antibody produces a precipitate that reaches a maximum between the forty-fifth and sixtieth day.

For intradermal tests, stock antigen prepared by the method of Bachman is diluted 1 to 50 to secure a 1 to 5,000 dilution (see p. 239).

As a serological test, bentonite flocculation (BF) is adequately sensitive and specific for diagnosis of most cases of trichinosis. Experiments with pigs suggest that it is not sensitive enough to diagnose light infections (Scholtens et al., 1966). The indirect immunofluorescence test, however, is both sensitive to that degree and specific (Kagan and Quist, 1970).

The CF, LA, IHA, DD, and countercurrent electrophoresis (CEP) tests are all useful to some degree. The BF test is the standard procedure of the National Center for Disease Control. The IHA and the ELISA are highly sensitive and specific.

Antibodies become detectable by the CF and BF tests after about 3 weeks of infection. Titers by BF reach a peak after several weeks, then decline slowly until after 3 years almost all sera are negative.

Serological methods of diagnosis do not replace diagnosis by demonstration of the larvae. For diagnosis of light infections, the digestion method is superior to serological procedures.

Filariasis. Except in loiasis and onchocerciasis, in which the adult worm is frequently isolated, diagnosis of filariasis may be difficult, based on either clinical findings or the identification of microfilariae. However, microfilaremia is frequently lacking. In bancroftian filariasis, the presence of microfilaremia appears to have an inverse relationship to intracutaneous and precipitin reactions.

The *intradermal test* for filariasis is a group reaction. Antigen can be prepared from any true filarial worm. Commonly used is the dog heartworm, *Dirofilaria immitis*. Antigen prepared from microfilariae of *Wuchereria bancrofti* in a dilution of 1:8,000 was reported to be highly specific for skin-testing *W. bancrofti*-infected patients, but it produced false positives in allergic individuals (Hunter, 1961).

More recent studies on intradermal tests for the diagnosis of bancroftian filariasis with antigen made from adult *Dirofilaria immitis* showed the procedure to be of limited value (Weller et al., 1980). Although the distribution of immediate reactions was similar to that found in other endemic populations, the immediate reactions lacked both the sensitivity and specificity to identify individuals with microfilaremia or other signs of filariasis. Delayed hypersensitivity reactions likewise showed no correlation with the presence or absence of the stigmata of filariasis. An intradermal test for *Onchocerca volvulus* with antigen prepared from the excretory-secretory products obtained from *in vitro* cultivation of the microfilariae was reported to have a high sensitivity with low levels of cross-reactivity (Schiller et al., 1980).

For serological tests for filariasis, the IHA and BF tests are carried out with antigen prepared from *Dirofilaria immitis* adults as standard procedures. Specificity is not great and sensitivity is variable, depending on the type of case and other factors. Titers may be high in loiasis and onchocerciasis, but tend to be low in infections with *Mansonella perstans*.

Cross-reactions are usual among all of the species of filariae with the IF test, but when frozen sections of *Onchocerca volvulus* are used as antigen, it is at least as sensitive for diagnosis of onchocerciasis as skin biopsy (Ten Eyck, 1973a, b). *Dipetalonema viteae,* a filaria of rodents, may be useful as antigen, but for the diagnosis of bancroftian filariasis, sections of *Wuchereria bancrofti* provide a more sensitive test.

The *rosette test* and the *migration inhibition test* are techniques for evaluating cellular rather than humoral immunity. Pinon and Gentilini (1972) have reported the detection of human filarial infections by these tests, which were missed by the IF and IHA tests. The tests of cellular immunity cannot be regarded as dependable until further refinement of antigen is achieved (Bloch-Michel and Waltzing, 1973).

Ascariasis. The practical applications of immunologic reactions to *Ascaris* antigen concern their use in the diagnosis of visceral larva migrans *(vide infra)*. Symptomatic sensitization to *Ascaris* may occur not only from infection with the parasite, but also from other contacts, such as by workers in abattoirs or in parasitological laboratories. A highly purified antigen, possibly the principal allergen producing sensitization to *Ascaris,* has been prepared from *A. suum* (Hussain et al., 1973).

The *intradermal test* consists in placing a drop of body fluid of *Ascaris lumbricoides* or *A. suum* on a scarified area of the skin. Sensitized individuals have an immediate local reaction, consisting of an erythematous wheal at the site, and frequently extensive lymphatic and systemic involvement. The more alarming symptoms disappear in 1 or 2 hours, but generalized edema may persist for some days. *Ascaris* sensitization does not necessarily mean infection with *Ascaris* at the time of the test, but may be the result of a previous infection or, in the case of workers in a laboratory, merely contact with fresh or preserved worms.

Toxocariasis. Visceral larva migrans caused by

Toxocara canis, the common ascarid of dogs, stimulates the production of antibodies that may be demonstrated by the indirect hemagglutination test with antigen prepared from triturated whole *T. canis* (Jung and Pacheco, 1960), or from the perienteric fluid of *Ascaris suum* (Oliver-González et al., 1969). These tests are not highly specific. An extract of embryonated *Toxocara* eggs provides a more sensitive and specific test when used as antigen. In the enzyme-linked immunosorbent assay (ELISA), superior results have been obtained with this larval antigen (Cypess et al., 1977).

Schistosomiasis. Immunodiagnostic techniques for schistosomiasis do not approach, with regard to reliability, the demonstration of eggs in excreta or biopsy. These techniques are used primarily for epidemiological studies, in which light or past infections may be more easily detected than by microscopy, or for treatment evaluation.

Because of the great efforts that have been expended to control the infection, there has been a proliferation of various immunodiagnostic tests for schistosomiasis (Kagan and Pellegrino, 1961). In addition to the intradermal test, these include many serological techniques: complement-fixation (CF), bentonite flocculation (BF), indirect hemagglutination (IHA), cholesterol-lecithin, indirect immunofluorescence (IF), double gel diffusion (DD), the enzyme-linked immunosorbent assay (ELISA), the Cercarien Hüllen Reaktion (CHR), and the circumoval precipitation (COP) reaction. Most of the tests are sensitive, but lack of specificity is to some degree a problem with all. For example, cross-reactivity with trichinosis is common.

The *intradermal test* is a schistosome-group reaction. The antigen may be obtained from molluscs infected with mammalian schistosomes, from pools of cercariae, from concentrates of eggs obtained from feces or digested out of tissues, or from adult schistosomes removed from an experimentally infected laboratory mammal. The dried antigen may be employed as a 0.5 or 1% saline extract, sterilized by passage through a Seitz filter, and stored in sterile ampules in a refrigerator.

Vogel and Minning (1949) demonstrated the development of pericercarial (Cercarien-hüllen, CHR) membranes around *Schistosoma* cercariae placed in sera of experimental subjects parasitized by the schistosome.

In early serological tests, Oliver-González et al. (1955), as a result of extensive serologic studies on *Schistosoma mansoni,* demonstrated in the sera of infected individuals cercarial precipitins, cercarial agglutinins, circumoval precipitins, and adult worm precipitins in much lower titer in a small percentage of sera. In recent infections the cercarial precipitins and agglutinins were present in high titer; in chronic cases of 10 years or over they were low or lacking. Circumoval precipitins were consistently present but more active in infections between days 40 and 50 following exposure.

The IF and CF tests are currently the most useful for routine use, but the ELISA is superior in some respects. When the IF test is used with frozen sections of adult worms for antigen, specificity is increased (Wilson et al., 1974). Sensitivity is adequate for screening purposes. The CF test, on the other hand, is more highly specific and less sensitive. It tends to be positive in cases of active clinical infection and therefore is useful in diagnosing the clinically significant infections. It tends not to give positive results with sera from children nor from patients with old infections.

The complement-fixation reaction is particularly valuable in suspected cases of schistosomiasis (1) during the latter part of the incubation period before the eggs are produced, (2) in chronic cases in which the walls of the intestine and bladder have become fibrosed and eggs do not pass from the mesenteric veins or vesical plexus into the lumen of these organs, and (3) in unisexual infections. Kagan (1968) concluded that, with the proper antigen, complement-fixation is probably the best serologic procedure for the diagnosis of clinical schistosomiasis.

Several authors have reported the ELISA to be equal or superior to IF, IHA, and CF (Schinski et al., 1976; Deelder et al., 1977; McLaren et al., 1978).

The *circumoval precipitin (COP) test,* first described by Oliver-González in 1954, has been used successfully for the diagnosis of cerebral schistosomiasis (Reyes and Yogore, 1963), and extensively used for epidemiological studies in the Philippines (Yogore et al., 1979) and Puerto Rico (Hillyer and Rivera-Marrero, 1980). Globular blebs or segmented linear precipitates form on the wall of the schistosome egg when incubated in the serum of infected individuals (Fig. 48–1). Segmentation of the precipitate is especially evident on scanning electron micrographs (Fig. 48–2). In recently acquired infections, a distinctive, nonsegmented type of precipitate is formed (Lewert et al., 1980).

Eggs for the test are harvested from the liver of

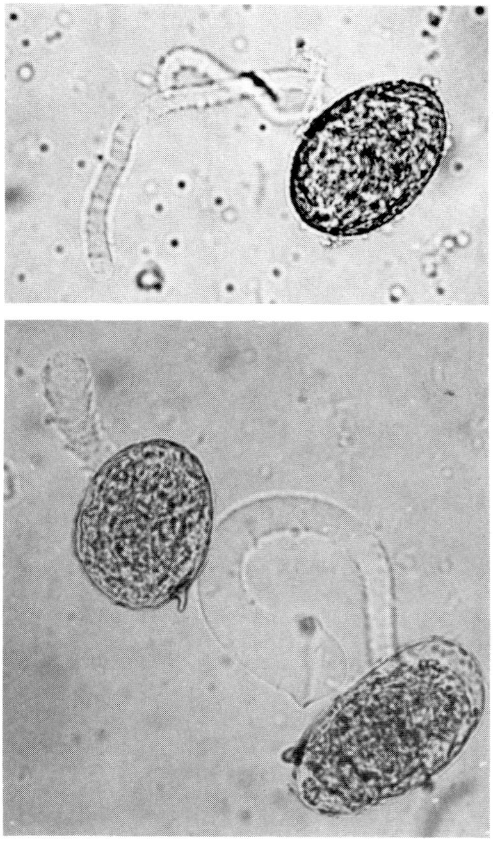

Fig. 48–1. Circumoval precipitin reaction. Segmented long precipitates on eggs of *Schistosoma japonicum* in plasma of man with chronic infection in the Philippines. Reactions in cases of recent infection produce unsegmented, usually short, stalks of precipitate with vacuolation at the distal end. (× 300.) (Courtesy of Dr. Robert Lewert.)

infected mice or rabbits. The tissue is homogenized in a Waring blender, digested in trypsin with added collagenase (1 mg/2 g of tissue), strained to remove fibrous tissue, and washed by centrifugation in saline. Eggs can be used as freshly removed from tissue or after being lyophilized, stored, and rehydrated. Blood samples collected on strips of filter paper can be used as the source of serum eluate (Yogore et al., 1978).

The test is done on an ordinary microscope slide with the preparations under a propped and sealed cover glass (Yogore et al., 1968) or on one with a sealed chamber (Garcia et al., 1981). About 100 eggs from a heavy suspension are placed on the slide, 2 drops of test plasma or serum that has been inactivated at 56°C for 30 minutes and merthiolated (1:10,000) are pipetted onto the eggs, and a cover glass is placed over the preparation, care being taken to avoid trapping air bubbles. The preparation is sealed by brushing the edges with a 1:3 mixture of hot petroleum jelly and paraffin, stored flat at room temperature, and examined at 24 and 48 hours. Since precipitates do not form on eggs containing immature or degenerate miracidia, a suitable preparation must contain numerous eggs, and in a positive test several eggs will be seen bearing clearly evident precipitates. Special modifications of the procedure suitable for collecting and testing sera from large numbers of people have been described (Yogore et al., 1979: Garcia et al., 1981).

Paragonimiasis. Intradermal tests with acid-soluble purified antigen were used by Watten et al. (1960) in a survey for paragonimiais in endemic villages in Taiwan. Reconstituted lyophilized antigen was used, and the test was read at 15 to 20 minutes. Significant reactions were produced in about 12% of the people in areas where the crab *Eliocheir japonicus* was found infected with *Paragonimus* metacercariae. The intradermal test for paragonimiasis remains positive 10 to 20 years after complete recovery, whereas results of complement-fixation and agglutination tests become negative 3 to 6 months after cure. Sawada et al. (1968) prepared a highly purified antigen by gel filtration and chromatographic methods that provided specific reactions in both intradermal and complement-fixation tests. Yokogawa (1966, 1969) has reviewed the literature on both intradermal and serological tests.

Antigen for complement-fixation tests, as originally developed by Ando (1922), was prepared by physiologic saline extraction of macerated adult *P. westermani* obtained from human cases at autopsy or from naturally or experimentally infected animals. A purified acid-insoluble, alkaline-soluble antigen in complement-fixation tests in Korea confirmed egg-positive infections and gave no false positives or cross-reactions to other trematode infections (Sadun, 1958). The indirect fluorescence, countercurrent electrophoresis, and indirect hemagglutination tests also are useful in diagnosis of paragonimiasis (Kagan, 1980).

Fascioliasis. A study by Stork et al. (1973) showed that of 137 children who were passing *Fasciola* eggs, testing provided positive results in 61% by the intradermal tests, 49% by the indirect fluorescence (IF), 48% by immunoelectrophoresis (IE), and 14% by complement-fixation (CF). With miracidia used as antigen, Azevedo and Rombert (1964) obtained superior sensitivity in the indirect

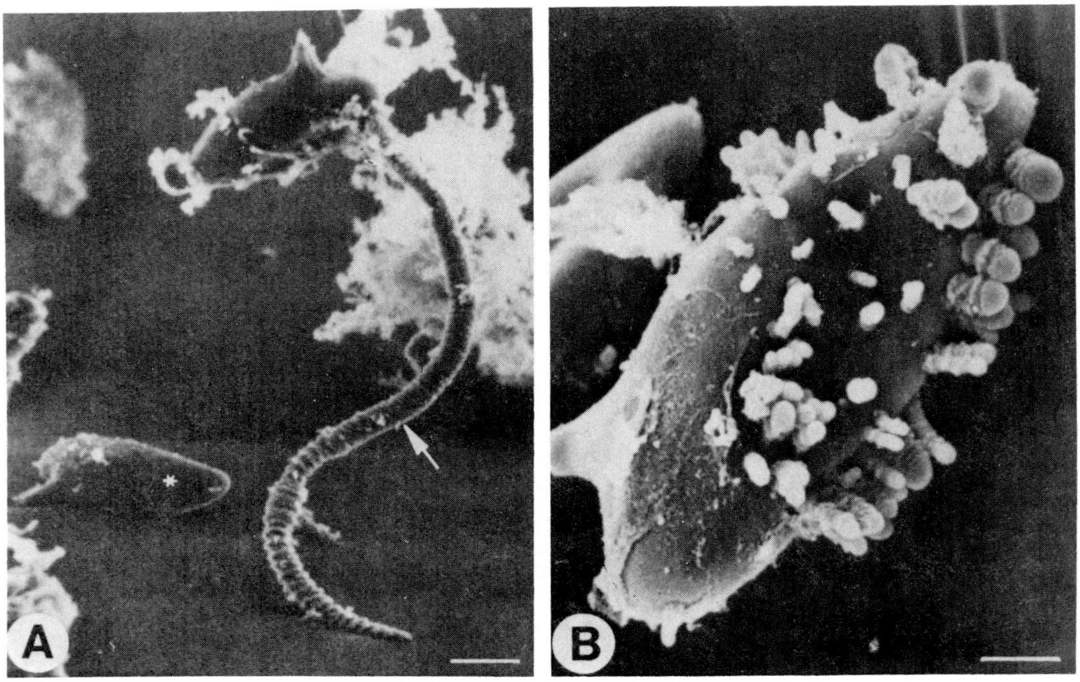

Fig. 48–2. Circumoval precipitin reaction (COP) on eggs of *Schistosoma mansoni*. *A*, Long COP formation showing what appears to have been pulsatile growth; the egg marked with an asterisk shows no reaction. *B*, Egg showing reaction at multiple sites on the surface. Scanning electron micrographs. Scale is 30 μm in *A*, 10 μm in *B*. (From Ford, J.W. Hillyer, G.V., and Connelly, T.G. 1980. The circumoval precipitin reaction on *Schistosoma mansoni* eggs as seen by electron microscopy. Am. J. Trop. Med. Hyg., *29*:1246–1248.)

fluorescence test against sera from infected humans and animals. As a test of cure in experimental animals, countercurrent electrophoresis has been reported as satisfactory (Hillyer and Diaz, 1976).

Clonorchiasis. Precipitin, complement-fixation (CF), double gel diffusion (DD), and indirect hemagglutination (IHA) tests have been evaluated in clonorchiasis. High levels of cross-reactivity have been noted, especially with paragonimiasis. Intradermal tests have also been used with a defatted adult extract fractionated by gel filtration and ion-exchange chromatography (Sawada et al., 1964). Pacheco et al (1960) found that the IHA test with a saline extract of ether-treated worms provided a sensitive reaction lacking cross-reactions with *Schistosoma* and *Paragonimus* infections.

Echinococcosis. Since 1911 when Casoni first described the specificity of the *intradermal test* for hydatid infection, it has been a useful diagnostic procedure. Antigen for the Casoni test is sterile hydatid fluid drawn from unilocular hydatid cysts from sheep, pigs, cattle, or man. The fluid is filtered, incubated to test its sterility, and stored in sealed ampules under refrigeration. For the test, 0.2 ml of the antigen is introduced intradermally on the upper arm, and is controlled by the injection of a like amount of sterile physiologic salt solution on the opposite arm. The control fades almost immediately while the tested site in positive cases develops a typical wheal within 30 minutes. A positive reaction indicates that infection is or has been present.

The precipitin test is the simplest and oldest of the serological tests. In sensitivity and specificity it is not equal to other more modern tests. These include the CF, IHA, IF, IE, BF, LA, DD, CEP, and ELISA.

In general, serological tests are sensitive but relatively nonspecific. That is, the tests tend to be positive in a high percentage of proved infections. High titers are good evidence of infection, but low titers are frequently due to cross-reaction with other conditions. The IHA, IF, and IE tests are the most commonly used and the aforementioned features apply to them. The IHA test, for example, provides low titers in cases of collagen disease, cirrhosis, schistosomiasis, and especially cysticercosis. The IHA test is more likely to be positive in cases of living hepatic hydatid than with dead hepatic cysts or living cysts in the lungs. With protoscolices from viable cysts used as antigen, the IF test is capable of detecting 93% of infections.

In the IE and DD tests a particular arc identified as "band 5" has been associated with sera from patients with *Echinococcus granulosus, E. vogeli* infections, and cysticercosis. More highly purified antigens applied to the CEP and ELISA tests have been reported to provide results both specific and highly sensitive.

Cysticercosis. For *intradermal tests* for cysticercosis in man, fluid obtained from cysticerci in domestic animals is used as antigen, and the test is done as in the Casoni test with a saline control.

As serological tests for cysticercosis, the CF, IHA, DD, CEP, IE, and IF have been used. Results with the CF have been inconsistent and with the DD insensitive. Good correlation between results with the CEP and the observed clinical condition was reported by Beltran and Gomez-Priego (1973).

IMMUNOLOGIC TESTS INVOLVING ARTHROPODS

Although Blacklock and Thompson (1923) demonstrated that invasion of the tissues of laboratory animals by the maggots of the tumbu fly *(Cordylobia anthropophaga)* induces an immune response to subsequent attack, this phenomenon generally has not been employed as a clinical laboratory test in this or other arthropod infestations. The most notable use that has been made of the sensitization reaction to arthropod metabolites concerns the venom of wasps and honeybees. Persons who develop hypersensitivity to hymenopteran sting may be desensitized and thus freed of the risk of fatal anaphylactic shock (see Chapter 44, p. 715).

Precipitin Tests for Blood Ingested by Arthropods. Malariologists and other workers frequently need to know whether a mosquito or other bloodsucking arthropod is selective in the animals on which it feeds, to determine whether a particular species is dangerous as a transmitter of infectious agents from person to person or from a zoonotic reservoir to humans. Precipitin tests and other standard methods of blood sampling provide the methods for obtaining this information. Essentials of this technique were published as early as 1923 by Bull and King, and later by Rice and Barber (1935).

Tempelis (1975) evaluated host-feeding patterns of mosquitoes and discussed advances in serological techniques for analysis of blood meals. He concluded that the precipitin test and its variations ("ring test"; capillary tube; agar gel diffusion; microplate) remain the basic serological tools for blood meal identification, with the passive hemagglutination inhibition test a promising adjunct for discrimination between closely related vertebrates, particularly birds (Tempelis et al., 1976). Use of the hemoglobin crystallization method (Washino and Tempelis, 1983) for blood meal identification is also a worthwhile nonserologic supplement where closely related hosts might otherwise be difficult to separate.

REFERENCES

Abdalla, R.E. 1977. Immunodiffusion, counterimmunoelectrophoresis and immunofluorescence in diagnosis of Sudan mucosal leishmaniasis. Am. J. Trop. Med. Hyg., *26*:1135–1138.

Allain, D.S., and Kagan, I. 1975. A direct agglutination test for leishmaniasis. Am. J. Trop. Med. Hyg., *24*:232–236.

Ambroise-Thomas, P., and Kien-Truong, T. 1972. Fluorescent antibody test in amebiasis. Am. J. Trop. Med. Hyg., *21*:907–912.

Ando, A. 1922. Ueber die Komplementbindungs-Reaction bei Distomiasis Pulmonum. II. Mittheilung. Jpn. J. Zool., *1*:18.

Azevedo, J.F., and Rombert, P.C. 1964. O diagnóstico pela imunofluorescéncia. Sua aplicacão em parasitología. Rev. Inst. Med. Trop. S. Paulo, *6*:35–45.

Bailey, N.M., Cunningham, M.P., and Kimber, C.D. 1967. The indirect fluorescent antibody technique applied to dried blood, for use as a screening test in the diagnosis of human trypanosomiasis in Africa. Trans. R. Soc. Trop. Med. Hyg., *61*:696–700.

Baldelli, R., Battelli, G., Caporale, V., and Semproni, G. 1975. A micro-CFT method for the diagnosis of visceral leishmaniasis. Nuovi Ann. Ig. Microbiol., *26*:179–190.

Ball, P.A.J., and Bartlett, A. 1969. Serological reactions to infection with *Necator americanus*. Trans. R. Soc. Trop. Med. Hyg., *63*:362–369.

Beltran, F., and Gomes-Priego, A. 1973. Evaluación de los contraimmunolectroforesis (CIEF) para la detección de anticuerpos en la cysticercosis experimentál y humana. Antioquia Med., *23*:472–473.

Bergquist, N.R. 1981. Rapid serology (FIAX) for screening human sera for antibodies to parasites. Parasitology, *82*(Part 4):52–54.

Blacklock, D.B., and Thompson, M.G. 1923. A study of the tumbu-fly, *Cordylobia anthropophaga* Grünberg, in Sierra Leone. Ann. Trop. Med. Parasitol., *17*:443–510.

Bloch-Michel, H., and Waltzing, P. 1973. Le phénomène des rosettes appliqué aux filarioses. Etude chez 56 patients. Nouv. Presse Med., *6*:382.

Bos, H.J., Van der Eyk, A.A., and Steerenberg, P.A. 1975. Application of ELISA in the serodiagnosis of amoebiasis. Trans. R. Soc. Trop. Med. Hyg., *69*:440.

Brannon, M.J.C., and Faust, E.C. 1949. Preparation and testing of a specific antigen for diagnosis of human strongyloidiasis. Am. J. Trop. Med., *29*:229–239.

Carroll, S.M., Karthigasu, K.T., and Grove, D.I. 1981. Serodiagnosis of human strongyloidiasis by an enzyme-linked immunosorbent assay. Trans. R. Soc. Trop. Med. Hyg., *75*:706–709.

Craig, C.F. 1929. The technique and results of a complement fixation test for the diagnosis of *Entamoeba histolytica*. Am. J. Trop. Med., *9*:277–296.

Cunningham, M.P., Bailey, N.M., and Kimber, C.D. 1967. The estimation of IgM immunoglobulin in dried blood, for use as a screening test in the diagnosis of human trypanosomiasis in Africa. Trans. R. Soc. Trop. Med. Hyg., *61*:688–695.

Cypess, R.H., Karol, M.H., Zidian, J.L., Glickman, L.T., and Gitlin, D. 1977. Larvae-specific antibodies in patients with visceral larva migrans. J. Infect. Dis., *135*:633–640.

Deelder, A.M., Ruitenberg, E.J., Kornelis, D., and Steerenberg, P.A. 1977. *Schistosoma mansoni:* Comparison of the immunoperoxidase techniques DASS and ELISA for human diagnosis. Exp. Parasitol., *41*:133–140.

Edrissian, G.H., and Darabian, P. 1978. A comparison of enzyme linked immunosorbent assay (ELISA) and indirect fluorescent antibody (IFA) in serodiagnosis of cutaneous and visceral leishmaniasis. 4th Int. Congr. Parasitol., Short Commun. E, 80.

Farshy, D.C., and Kagan, I. 1972. Use of stable sensitized cells in an improved indirect microhemagglutination test for malaria. Infect. Immun., *7*:680–682.

Ford, J.W., Hillyer, G.V., and Connelly, T.G. 1980. The circumoval precipitin reaction on *Schistosoma mansoni* eggs as seen by electron microscopy. Am. J. Trop. Med. Hyg., *29*:1246–1248.

Fulton, J.D., and Turk, J.L. 1960. Direct agglutination test for *Toxoplasma gondii*. Trans. R. Soc. Trop. Med. Hyg., *54*:7–8.

Garcia, E.G., Tapales, F.P., Valdez, C.A., Mitchell, G.F., and Tiu, W.U. 1981. Attempts to standardize the circumoval precipitin test (COPT) for schistosomiasis japonica. Southeast Asian J. Trop. Med. Public Health, *12*:384–395.

Gore, R.W., and Sadun, E.H. 1968. Soluble antigen fluorescent antibody test for amebiasis. Exp. Parasitol., *22*:316–320.

Grove, D.I. 1980. Strongyloidiasis in Allied ex-prisoners of war in South East Asia. Br. Med. J., *280*:598–601.

———, and Blair, A.J. 1981. Diagnosis of human strongyloidiasis by immunofluorescence, using *Strongyloides ratti* and *S. stercoralis* larvae. Am. J. Trop. Med. Hyg., *30*:344–349.

Hillyer, G.V., and de Diaz, A.L. 1976. Use of immunologic techniques to detect chemotherapeutic success in infections with *Fasciola hepatica*. Am. J. Trop. Med. Hyg., *25*:307–311.

———, and Rivera-Marrero, C. 1980. Circumoval precipitin antigens for the diagnosis of schistosomiasis. I. Development of an antiserum reactive with *Schistosoma mansoni* eggs by the circumoval precipitin test. Am. J. Trop. Med. Hyg., *29*:1249–1253.

Hunter, G.W., III. 1961. Skin testing for filariasis with an homologous antigen. Proc. 6th Int. Congr. Trop. Med. Malaria, *2*:479–483.

Hussain, R., Bradbury, S.M., and Stejan, G. 1973. Hypersensitivity to *Ascaris* antigens. VIII. Characterization of a highly purified allergen. J. Immunol., *111*:260–268.

Jacobs, L., and Cook, M.K. 1954. Variations in the dye test for toxoplasmosis. Am. J. Trop. Med. Hyg., *3*:860–867.

Jung, R.C., and Pacheco, G. 1960. Use of a hemagglutination test in visceral larva migrans. Am. J. Trop. Med. Hyg., 9:185–191.

Kagan, I.G. 1968. Serologic diagnosis of schistosomiasis. Bull. N.Y. Acad. Med., *44*:262–277.

———. 1980. Serodiagnosis of parasitic diseases, pp. 724–750. In *Manual of Clinical Microbiology*, 3rd ed. Edited by E.J. Lennette, A. Balows, J.P. Hausler, Jr., and J.P. Truant. Washington, D.C., American Society for Microbiology.

———, and Norman, L., 1977. The laboratory diagnosis of *Pneumocystis carinii* pneumonia. Health Lab. Sci., *14*:155–163.

———, and Pellegrino, J. 1961. A critical review of immunological methods for the diagnosis of bilharziasis. Bull. W.H.O., *25*:611–674.

———, and Quist, K.D. 1970. An evaluation of five serologic tests for the diagnosis of trichinosis in lightly infected swine, pp. 271–277. In *H.D. Srivastava Commemoration Volume*. Edited by S.K. Singh and B.K. Tandon. Iznatnagan, V.P., India, India Veterinary Research Institute.

Kessel, J.F., Lewis, W.P., Pasquel, C.M., and Turner, J.A. 1965. Indirect hemagglutination and complement-fixing tests in amebiasis. Am. J. Trop. Med. Hyg., *14*:540–550.

Krupp, I.M. 1977. Definition of the antigenic pattern of *Entamoeba histolytica,* and immunoelectrophoretic analysis of the variation of patient response to amebic disease. Am. J. Trop. Med. Hyg., *26*:387–392.

Latif, B.M.A., and Adam, K.M.G. 1973. Differentiation of *Trypanosoma brucei, T. rhodesiense* and *T. gambiense* by the indirect fluorescent antibody test. Bull. W.H.O., *48*:401–407.

Lewert, R.M., Yogore, M.G., Jr., Marin, W.R., and Blas, B.L. 1980. Schistosomiasis japonica in Barrio San Antonio, Basey, Samar in the Philippines. IV. Atypical precipitates in the circumoval precipitin test—a reaction of recent infection. Am. J. Trop. Med. Hyg., *29*:431–434.

Lucasse, C. 1970. Fluorescent antibody tests applied to serum of patients with Gambian sleeping sickness. Trop. Geogr. Med., *22*:227–236.

Maddison, S.E. 1965. Characterization of *Entamoeba histolytica* antigen-antibody reaction by gel diffusion. Exp. Parasitol., *16*:224–235.

Mansueto, S., Migneco, G., and La Cascia, C. 1975. The complement-fixation test with BCG in the diagnosis of leishmaniasis. Boll. Ist. Sieroter. Milan, *54*:140–144.

Mayer, M., and Pifano, F. 1941. O diagnóstico da molestia de Chagas por intradermoreacção com cultura de *Schizotrypanum cruzi*. Brasil. Med., *55*:317–319.

Mazza, S., Basso, G., and Basso, R. 1942. Enfermedad de Chagas en primer periodo diagnostica exclusivamente por biopsia de ganglio linfatico con hallazgo de parasitos leishmaniformes. Invest. Enferm. de Chagas, Mision de Estud. Patol. Reg. Argentina, No. 63, Buenos Aires.

McLaren, D., Draper, C.C., Roberts, J.M., Minter-Godebloed, E., Ligthart, G.S., Teesdale, C.H., Amin, M.A., Omer, A.H.S., Bartlett, A., and Voller, A. 1978. Studies on the enzyme-linked immunosorbent assay (ELISA) test for *Schistosoma mansoni* infections. Ann. Trop. Med. Parasitol., *72*:243–253.

Neva, F.A., Gam, A.A., and Burke, J. 1981. Comparison of larval antigens in an enzyme-linked immunosorbent assay for strongyloidiasis in humans. J. Infect. Dis., *144*:427–432.

Oliver-González, J. 1941. The dual antibody basis of acquired immunity in trichinosis. J. Infect. Dis., *69*:254–270.

———. 1954. Anti-egg precipitins in the serum of humans infected with *Schistosoma mansoni*. J. Infect. Dis., *95*:86–91.

———, Hurlbrink, P., Conde, E., and Kagan, I.G. 1969. Serologic activity of antigen isolated from the body fluid of *Ascaris suum*. J. Immunol., *103*:15–19.

———, Ramos, F.L., and Coker, C.M. 1955. Serological reactions against egg antigens as an aid in the evaluation of therapy in schistosomiasis. Am. J. Trop. Med. Hyg., *4*:908–912.

Pacheco, G., Wykoff, D.E., and Jung, R.C. 1960. Trial of an indirect hemagglutination test for the diagnosis of infections with *Clonorchis sinensis*. Am. J. Trop. Med. Hyg., 9:367–370.

Packchanian, A. 1935. Agglutination and precipitation tests for the diagnosis of *Trypanosoma cruzi* infection (Chagas' disease). J. Immunol., *29*:35.

Pinon, J.M., and Gentilini, M. 1972. Intéret de l'utilisation du teepol dans les reactions d'immunofluorescence indirecte en vue de l'elimination des reáctions croisees dans le diagnostic des filarioses. Bull. Soc. Pathol. Exot., *65*:306–308.

Powell, S.J. 1968. The capillary-tube precipitin test: A rapid serologic aid to clinical diagnosis in invasive amebiasis. Am. J. Trop. Med. Hyg., *17*:840–843.

Reyes, V.A., and Yogore, M.G. 1963. On cerebral schistosomiasis with a new diagnostic test. Philippine J. Surg., *18*:172–176.

Rice, J.B., and Barber, M.A. 1935. Malaria studies in Greece. A modification of the Uhlenhuth-Weidanz precipitin test for determining the source of blood meals in mosquitoes and other insects. J. Lab. Clin. Med., *20*:876–883.

Ridley, D.S. 1981. Serodiagnosis in management of clinical parasitic infections. Parasitology, *82*(part 4):50–51.

Sabin, A.B. 1949. Complement fixation test in toxoplasmosis and persistence of the antibody in human beings. Pediatrics. *4*:443–453.

Sadun, E.H. 1958. The public health significance of paragonimiasis in the Far East. Abstr. 6th Int. Congr. Trop. Med. Malaria, pp. 53–54.

———, Duxbury, R.E., Williams, J.S., and Anderson, R.I. 1963. Fluorescent antibody test for the serodiagnosis of African and American trypanosomiasis in man. J. Parasitol., *49*:385–388.

Sawada, T.Y., Nagata, Y., and Takei, K. 1964. Studies on the substance responsible for the skin tests on clonorchiasis. Jpn. J. Exp. Med., *36*:315–322.

———, Takei, K., Sato, S., and Matsuyama, G. 1968. Studies on the immunodiagnosis of paragonimiasis. III. Intradermal skin tests with fractionated antigens. J. Infect. Dis., *118*:235–239.

Schiller, E.L., D'Antonio, R., and Figueroa Marroquin, H. 1980. Intradermal reactivity of excretory and secretory products of onchocercal microfilariae. Am. J. Trop. Med. Hyg. *29*:1215–1219.

Schinski, V.D., Clutter, W.C., and Murrell, K.D. 1976. Enzyme and ^{125}I-labelled anti-immunoglobulin assays in the immunodiagnosis of schistosomiasis. Am. J. Trop. Med. Hyg., *25*:824–831.

Scholtens, R.G., Kagan, I.G., Quist, K.D., and Norman, L. 1966. An evaluation of tests for the diagnosis of trichinosis in swine and associated quantitative epidemiologic observations. Am. J. Epidemiol., *83*:489–500.

Sood, P., Prakash, O., and Bhujwala, R.A. 1972. A trial of hemagglutination, circumoval precipitin and gel diffusion tests in hookworm infection. Indian J. Med. Res., *60*:1132–1133.

Stein, B., and Desowitz, R. S. 1964. The measurement of antibody in human malaria by a formolized sheep cell hemagglutination test. Bull. W.H.O., *30*:45–49.

Stork, M.G., Venables, G.S., Jennings, S.M.F., Beesley, J.R., Bendez, P., and Capron, A. 1973. An investigation of endemic fascioliasis in Peruvian village children. J. Trop. Med. Hyg., *76*:231–235.

Tempelis, C.H. 1975. Host-feeding patterns of mosquitoes, with a review of advances in analysis of blood meals by serology. J. Med. Entomol., *11*:635–653.

———, Reeves, W.C., and Nelson, R.L. 1976. Species identification of blood meals from *Culex tarsalis* that had fed on passeriform birds. Am. J. Trop. Med. Hyg., *25*:744–746.

Ten Eyck, D.R. 1973a. Comparison of biopsy and fluorescent antibody staining techniques in the detection and study of onchocerciasis in an Ethiopian population. Am. J. Epidemiol., *98*:283–288.

———. 1973b. *Onchocerca volvulus* and *Wuchereria bancrofti:* Fluorescent antibody staining of frozen homologous sections for diagnosis. Exp. Parasitol., *34*:154–161.

Toussaint, A., and Anderson, R.I. 1965. Soluble antigen fluorescent-antibody technic. Appl. Microbiol., *13*:552–557.

———, Tarrant, C.J., and Anderson, R.I. 1965. An indirect antibody technique using soluble antigens for serodiagnosis of *Trypanosoma cruzi* infection. Proc. Soc. Exp. Biol. Med., *120*:783–785.

Vattuone, N.H., and Yanovsky, F. 1971. *Trypanosoma cruzi:* Agglutination activity of enzyme-treated epimastigotes. Exp. Parasitol. *30*:349–355.

Vogel, H., and Minning, W. 1949. Hüllenbildung bei Bilharzia-Cercaria im Serum bilharzia-infizierter Tiere und Menschen. Zbl. Bakt. Apt. I A, Orig., *153*:99–105.

Voller, A., Bidwell, D.E., Bartlett, A., and Edwards, T. 1977. A comparison of isotopic and enzyme-immunoassays for tropical parasitic diseases. Trans. R. Soc. Trop. Med. Hyg., *71*:431–437.

———, Bidwell, D.E., Bartlett, A., Fleck, D.G., Perkins, M., and Oladchin, B. 1976. A microplate enzyme immunoassay for toxoplasma antibody. J. Clin. Pathol., *29*:150–153.

Walls, K.W., and Barnhart, E.R. 1978. Titration of human serum antibodies to *Toxoplasma gondii* with a simple fluorometric assay. J. Clin. Microbiol., *7*:234–235.

Walton, B.C., Brooks, W.H., and Arjona, I. 1972. Serodiagnosis of American leishmaniasis by indirect fluorescent antibody test. Am. J. Trop. Med. Hyg., *21*:296–299.

Washino, R.K., and Tempelis, C.H. 1983. Mosquito host bloodmeal identification: Methodology and data analysis. Annu. Rev. Entomol., *28*:179–201.

Watten, R.H., Kuntz, R.E., and Liu, H.Y. 1960. Use of purified antigens for the detection of *Paragonimus westermani* and *Clonorchis sinensis* in peoples of Hsinchu Hsien, Taiwan (Formosa). J. Formosan Med. Assoc., *59*:364–370.

Weller, P.F., Ottesen, E.A., and Heck, L. 1980. Immediate and delayed hypersensitivity skin test responses to the *Dirofilaria immitis* filarial skin test (Sawada) antigen in *Wuchereria bancrofti* filariasis. Am. J. Trop. Med. Hyg., *29*:809–814.

Wilson, M., Sulzer, A.J., and Walls, K.W. 1974. Modified antigens in the indirect immunofluorescent test for schistosomiasis. Am. J. Trop. Med. Hyg., *23*:1072–1076.

Yogore, M.G., Jr., Lewert, R.M., and Blas, B.L. 1978. Schistosomiasis japonica in Barrio San Antonio, Basey, Samar in the Philippines. II. Quantitative fecal examination and circumoval precipitin tests. Southeast Asian J. Trop. Med. Public Health, *9*:344–355.

———, Lewert, R.M., and Blas, B.L. 1979. Schistosomiasis japonica in Barrio San Antonio, Basey, Samar in the Philippines. III. The plasma circumoval precipitin test—procedure and use in epidemiologic studies. Southeast Asian J. Trop. Med. Public Health, *10*:23–31.

———, Lewert, R.M., and Silan, R.B. 1968. The circumoval precipitin (COP) test in schistosomiasis japonica. Am. J. Trop. Med. Hyg., *17*:65–71.

Yokogawa, M. 1966. *Paragonimus* and paragonimiasis. Adv. Parasitol., *3*:99–158.

———. 1969. *Paragonimus* and paragonimiasis. Adv. Parasitol., *7*:375–387.

Chapter 49

Entomologic Methods

COLLECTION AND PRESERVATION OF ARTHROPODS

The technique of collecting arthropods varies, depending on the type or group concerned. Centipedes, scorpions, ticks, spiders, and assassin bugs are often brought into the laboratory by interested persons for identification. If placed in a dry container, they are usually not damaged when they arrive. Sarcoptoid mites and human lice are commonly obtained by general practitioners or dermatologists in charity clinics, and may be preserved in small vials containing alcohol. Bedbugs can be obtained by careful search in infested homes. Adult fleas can be removed from dogs and cats by touching them with a camel's hair brush moistened with xylol and removing them with a fine forceps to a vial or jar. Rats caught in traps may be placed in an airtight chloroform chamber for removal of rodent fleas, mites, and lice. Flea larvae are readily obtained by sieving the sweepings from the floors of dwellings occupied by infested persons, dogs, or cats.

Adult living female mosquitoes, sand flies, black flies, and ceratopognoids may be captured uninjured by carefully placing a wide-mouthed bottle over the exposed skin when these blood-sucking flies are beginning to feed. Then, when they have left the feeding site, a small piece of cardboard is slipped over the mouth of the bottle, the bottle is placed upright, and a cork or cap is substituted for the pasteboard. Questing females may also be readily collected with hand-held aspirators of various designs (Martin, 1977). In field situations, adults can be collected directly from diurnal resting sites by means of powerful suction devices or caught in a variety of baited or unbaited traps (Bram, 1978).

Many species of mosquitoes may be readily reared through several generations in the laboratory, so that students may have complete demonstrations of their life cycles (Belkin et al., 1965). Most species of anophelines require larger, better controlled breeding quarters, which are practical only in special laboratories under expert direction. However, eggs, larvae, and pupae may be taken from aquatic habitats and brought in for observation and study. For these and other insects, collecting nets, adapted for use in catching adults and larvae, are indispensable. Filth flies, which breed in garbage, human and animal manure, and decaying meat, may be obtained as adults from insanitary latrines, garbage piles, and dead animals lying on the ground, or the gravid females may be induced to oviposit (or larviposit) in the laboratory, on decaying food, or in manure as culture media. Larvae recovered from human lesions or excreta may be bred out on appropriate culture media, or may be fixed in hot 70% alcohol-glycerin, 10 to 1 parts by volume, and identified as larvae.

Large arthropods with hard exoskeletons may be killed in cyanide gas chambers or test tubes containing cotton plugs soaked in ethyl acetate. These specimens may then be preserved in 70% alcohol. Since many arthropods will shrink extensively if allowed to dry out, care should be taken in using containers with tight-fitting tops. Rohlf (1957) decribed a method and procedure for preservation of soft-bodied insects, such as tipulid, beetle and moth larvae, and spiders. The rationale is to replace body fluids and fats with ethyl alcohol and ether. Then, by heating the specimen moderately to prevent contraction by the pressure of the vaporizing ether inside the exoskeleton, the specimen is left dry and hard. Truman (1968) preserved larval mosquitoes, black flies, and sand flies as well as adult mosquitoes with a simple acetone treatment that prevented distortion or shriveling. No insects or other arthropods should be stored in formaldehyde solution, since this preservative causes disintegration of the exoskeleton.

For mounting delicate insects originally collected in alcohol, critical point drying has proved successful for nematocerous Diptera and other small acalyptrate flies (Gordh and Hall, 1979). The advantages of this technique include batch processing of specimens without collapse or breakage of appendages due to brittleness and retention of life-like pigmentation. Other medically important arthropods have been critical point dried with success, including spiders, mites, lepidopterous larvae, dipterous pupae, and cockroach oothecae.

The scales of delicate adult insects such as mosquitoes, are badly damaged by rough handling. These insects should be killed with ethyl acetate or cyanide, mounted with entomologic pins before they completely dry, and placed in suitable entomologic boxes, in case they are needed for demonstration. For transportation to specialists for identification, unmounted specimens of small biting Diptera (including mosquitoes) should preferably be placed in a well-buffered pill box between pieces of nonabrasive paper made of cotton fiber. When field collections are being made, the addition of chlorocresol to the bottom of the pill box before paper and specimens are added will prevent the deterioration of biological material and allow the collector to hold specimens for long periods (Tindale, 1961).

EXPERIMENTAL USE OF ARTHROPODS IN THE LABORATORY

Unless a laboratory worker has had considerable training and has become skillful in rearing and handling arthropods, it usually is not profitable or expedient for him to attempt experimental problems. For a study of the venomous arthropods, the investigator should have sufficient pharmacologic and clinical training to allow him to control and evaluate his observations. For experimental malaria studies, not only must he develop a technique for rearing the particular anopheline mosquito utilized and know the malarial parasites with which he is dealing, but he must be able to dissect mosquitoes dexterously in order to examine their "stomachs" and salivary glands for infection (Vanderberg and Gwadz, 1980). Similar skill is required for conducting studies on blood-sucking flies in the experimental transmission of filariae. When the diseases under consideration are dangerous to human life, as yellow fever, plague, and typhus, investigations should not be undertaken until the worker has had expert training in a laboratory where similar studies have been conducted (Subcommittee on Arbovirus Laboratory Safety, 1980).

Up-to-date information on the most useful techniques of colonization or maintenance of medically important arthropods in the laboratory can be found in recent journals of parasitology and medical entomology (Wade, 1976). Some helpful references for the more important arthropod groups include those dealing with mosquitoes (Gerberg, 1970; Klunker, 1979; Foster, 1980), sand flies (Killick-Kendrick et al., 1977; Ward, 1977), black flies (Brenner et al., 1980), *Culicoides* (Jones et al., 1969; Linley, 1969), tsetse flies (Itard and Jordan, 1977), triatomines (Gardiner and Maddrell, 1972; Langley and Pimley, 1978), human lice (Busvine, 1979; Blommers, 1979), fleas (Cavanaugh et al., 1972; Rutledge et al., 1979), mites (Traub et al., 1975; Wharton, 1976; Krantz, 1978), and ticks (Kaiser, 1966; Gladney et al., 1970; Drummond et al., 1971; U.S.D.A., 1976). For basic information on colonization of bedbugs, cockroaches, and muscoid flies, see the respective chapters in Smith (1966). Singh (1977) also provides comprehensive information on dietary requirements of several arthropod groups.

MOUNTING ARTHROPODS AS MICROSCOPIC PREPARATIONS

Some of the smaller arthropods and early larval stages of larger ones are sufficiently clear and transparent that *in toto* mounts may be made by passing them through dehydrating and clearing agents into Canada balsam or Caedax, a clearer mountant than balsam. Others, which have tissues that are too dense for satisfactory slide preparations without special treatment, should be placed in 10% cold aqueous potassium hydroxide solution for 24 hours. If rapid diagnosis is essential, the specimens may be punctured with finely pointed needles, placed in the caustic solution and heated for a few minutes, care being exercised not to cause their disintegration. After thorough washing, they are ready for dehydration and clearing. For sectioning of arthropods or arthropod members, the organism should be carefully punctured to allow rapid penetration of the fixing fluid. A number of fixatives are available, as are clearing, dehydrating, and embedding agents (Barbosa, 1974).

Chloral Hydrate Mounting Medium. Probably the simplest practical method for killing, fixing, clearing, and mounting small arthropods (such as *Cyclops* and *Diaptomus,* mites, unfed stages of ticks, fleas, lice, bedbugs, mosquito larvae, pupae and adults, other small bloodsucking or filth flies, as well as their dissected diagnostic parts) consists in using a modification of Berlese's gum chloral fluid. Hoyer's medium, a widely used variant, is prepared as follows (Krantz, 1978):

Distilled water	50 ml
Chloral hydrate	200 g
Gum arabic (amorphic)	30 g
Glycerin	20 ml

This mixture should be filtered through clean cheesecloth. Living specimens or those previously fixed and preserved in 70% alcohol may be utilized. The mounts should have little slivers of glass between the slide and cover glass to prevent too much compression of the mounted object. The preparation clears almost immediately and may be utilized as a temporary mount, or it may be heated in a drying oven at 37° to 45°C for permanency. Air bubbles in the preparation will be minimized if one drop of the medium is placed on the slide, the object to be mounted is centered therein, and a second drop of the medium is placed on the lower side of the cover glass to be superimposed on the object. The coverslip should then be thickly ringed with Glyptal, a waterproofing paint used for electrical wiring (Travis, 1968).

CARE OF DEMONSTRATION COLLECTIONS OF ARTHROPODS

Demonstrations and mounted specimens that are preserved as dry mounts must be kept in tight boxes away from dust, light, and vibrations. Precautions must be constantly taken to prevent psocids ("booklice") and larvae of dermestid beetles from eating the specimens. Crystals of paradichlorobenzene, placed in a little cardboard or paper cup in one corner of each box and renewed three or four times a year, will guard against such danger. Insecticides such as dichlorvos may also be used.

IDENTIFICATION OF ARTHROPODS

Accurate specific identification of arthropods usually requires highly specialized training and detailed knowledge of the taxonomy of the particular group in question, since the differentiation of species frequently depends upon minute technical details of structure. However, the nonspecialist can learn to recognize the major classes and orders of arthropods (Borror et al., 1981; Gertsch, 1979; Harwood and James, 1979) and, with this orientation as a starting point, seek the identity of specimens by the use of appropriate reference works (for example, see Darsie and Ward, 1981). An appreciation of the complexity inherent in the study of arthropods guards against many pitfalls and hasty, unwarranted conclusions. It is best to consider any identification approximate and tentative until it has been confirmed by a competent specialist. This caution is particularly indicated where the exact identity of the arthropod is of critical importance, as in experimental work or epidemiologic surveys.

For purposes of detailed identification, the material should be well preserved and properly mounted *(vide supra).* A hand lens of ×10- to ×14-power is useful for relatively superficial examination, but for detailed work both a binocular dissecting microscope (Greenough type with magnifications of about ×10 to ×100) and the usual compound microcope are practically indispensable. The dissecting microscope enables examination, with incident light, of opaque specimens that are living, mounted on pins, or immersed in fluid. The higher magnifications require a suitably intense source of light. Pinned specimens, e.g., adult Diptera, may be conveniently positioned on a cork, lump of Plasticene, or L-shaped holder of balsa wood or sheet cork. Alcoholic specimens, e.g., ticks, are examined in an open dish in water or alcohol that should completely cover the object to avoid confusing reflections. The compound microscope is used to study, by transmitted light, cleared specimens or parts mounted on slides (e.g., mites, lice, fleas, mosquito larvae, and genitalia). When available, a scanning electron microscope can be used for the identification of mites (Krantz, 1978) and ticks (Keirans and Clifford, 1978; Corwin et al., 1979).

Arthropods are systematically identified by the use of "keys" designed for the purpose, whereby matching the specimen with a series of increasingly specific alternatives places it within smaller and smaller categories of species groups, leading finally to species identification. This identification is then checked against more detailed morphological descriptions, with which the specimen should agree in every essential respect.

Biochemical methods of identification are beginning to be useful in some instances, e.g., differentiation of strains of *Aedes aegypti* and of species within the *Anopheles gambiae* complex (see

p. 642), as well as members of the *Simulium damnosum* group (see p. 663).

GENERAL REFERENCES

Barbosa, P. 1974. *Manual of Basic Techniques in Insect Histology.* Amherst, Massachusetts, Autumn Publishers.

Belkin, J.N., Hoge, C.L., Galindo, P., Aitken, T.H.G., Schick, R.X., and Powder, W.A. 1965. Mosquito studies (Diptera, Culicidae). II. Methods for the collection, rearing and preservation of mosquitoes. Contrib. Am. Entomol. Inst., *1*:19–78.

Blommers, L. 1979. Insecticidal tests on immature head lice, *Pediculus capitis* (Anoplura)—a new technique. J. Med. Entomol., *16*:82–83.

Borror, D.J., DeLong, D.M., and Triplehorn, C. A. 1981. *An Introduction to the Study of Insects,* 5th ed. Philadelphia, W.B. Saunders.

Bram, R.A. 1978. *Surveillance and Collection of Arthropods of Veterinary Importance.* U.S. Dept. of Argiculture Handbook No. 518. Washington, D.C., Superintendent of Documents, U.S. Government Printing Office.

Brenner, R.J., Cupp, E.W., and Bernardo, M.J. 1980. Laboratory colonization and life table statistics for geographic strains of *Simulium decorum* (Diptera: Simuliidae). Tropenmed. Parasitol., *31*:487–497.

Busvine, J.R. 1979. On rearing human lice. Antenna, *3*:89–91.

Cavanaugh, D.C., Starks, H.E., Marshall, J.D., Jr., and Rust, J.H., Jr. 1972. A simple method for rearing fleas for insecticide testing in the field. J. Med. Entomol., *9*:113–114.

Corwin, D., Clifford, C.M., and Keirans, J.E. 1979. An improved method for cleaning and preparing ticks for examination with the scanning electron microscope. J. Med. Entomol., *16*:352–353.

Darsie, R.S., Jr., and Ward, R.A. 1981. *Identification and Geographical Distribution of the Mosquitoes of North America, North of Mexico.* Suppl. No. 1, *Mosquito Systematics.* Fresno, California, American Mosquito Control Association.

Drummond, R.O., Whetstone, T.M., Ernst, S.E., and Gladney, W.J. 1971. Oviposition of the American dog tick (Acarina: Ixodidae). Ann. Entomol. Soc. Am., *64*:1305–1309.

Foster, W.A. 1980. Colonization and maintenance of mosquitoes in the laboratory, pp. 103–151. In *Malaria. Volume 2. Pathology, Vector Studies, and Culture.* Edited by J.P. Kreier. New York, Academic Press.

Gardiner, B.O.C., and Maddrell, S.H.P. 1972. Techniques for routine and large-scale rearing of *Rhodnius prolixus* Stal (Hem., Reduviidae). Bull. Entomol. Res., *61*:505–515.

Gerberg, E.J. 1970. *Manual for Mosquito Rearing and Experimental Techniques.* Bull. No. 5. Fresno, California, American Mosquito Control Association.

Gertsch, W.J. 1979. *American Spiders,* 2nd ed. New York, Van Nostrand Reinhold.

Gladney, W.J., Drummond, R.O., Whetstone, T.M., and Ernst, S.E. 1970. Effect of age on the attachment rate of the parasitic stages of the Lone Star tick, *Amblyomma americanum* (Linnaeus) (Acarina: Ixodidae) in the laboratory. J. Med. Entomol., *7*:92–95.

Gordh, G., and Hall, J.C. 1979. A critical point drier used as a method of mounting insects from alcohol. Entomol. News, *90*:57–59.

Harwood, R.F., and James, M.T. 1979. *Entomology in Human and Animal Health,* 7th ed. New York, Macmillan Co.

Itard, J., and Jordan, A.M. 1977. Mass rearing using animals for feeding, pp. 125–140. In *Tsetse. The Future for Biological Methods in Integrated Control.* Edited by M. Laird. Ottawa, Canada, International Development Research Centre.

Jones, R.H., Potter, H.W., and Baker, S.K. 1969. An improved larval medium for colonized *Culicoides variipennis.* J. Econ. Entomol., *62*:1483–1486.

Kaiser, M.N. 1966. The subgenus *(Persicargas* (Ixodoidea, Argasidae, *Argas).* 3. The life cycle of *A. (P.) arboreus,* and a standardized rearing method for argasid ticks. Ann. Entomol. Soc. Am., *59*:496–502.

Keirans, J.E., and Clifford, C.M. 1978. The genus *Ixodes* in the United States: A scanning electron microscope study and key to the adults. J. Med. Entomol., *Suppl. 2,* 1–149.

Killick-Kendrick, R., Leaney, A.J., and Ready, P.D. 1977. The establishment, maintenance and productivity of a laboratory colony of *Lutzomyia longipalpis* (Diptera: Psychodidae). J. Med. Entomol., *13*:429–440.

Klunker, R. 1979. Uberblick über die in-vitro-Fütterung blutsaugender Arthropoden. Angew. Parasitol., *20*:88–108.

Krantz, G.W. 1978. *A Manual of Acarology,* 2nd ed. Corvallis, Oregon State University Book Stores.

Langley, P.A., and Pimley, R.W. 1978. Rearing triatomine bugs in the absence of a live host and some effects of diet on reproduction in *Rhodnius prolixus* Stal (Hemiptera: Reduviidae). Bull. Entomol. Res., *68*:243–250.

Linley, J.R. 1969. Studies on larval development in *Culicoides furens* (Poey) (Diptera: Ceratopogonidae). I. Establishment of a standard rearing technique. Ann. Entomol. Soc. Am., *62*:702–711.

Martin, J.E.H., compiler. 1977. *The Insects and Arachnids of Canada. Part 1. Collecting, Preparing and Preserving Insects, Mites, and Spiders.* Hull, Quebec, Government Publishing Center.

Rohlf, F.J. 1957. A new technique in the preserving of soft-bodied insects and spiders. Turtox News, *35*:226–229.

Rutledge, L.C., Lawson, M.A., Young, L.L., and Moussa, M.A. 1979. Rearing of *Diamanus montanus* (Siphonaptera: Ceratophyllidae) and *Hoplopsyllus anomalus* (Siphonaptera: pulicidae). J. Med. Entomol., *15*:304.

Singh, P. 1977. *Artificial Diets for Insects, Mites and Spiders.* New York, IFI/Plenum.

Smith, C.N., editor. 1966. *Insect Colonization and Mass Production.* London, Academic Press.

Smith, K.G.V., editor. 1973. *Insects and Other Arthropods of Medical Importance.* London, British Museum (Natural History).

Subcommittee on Arbovirus Laboratory Safety of the American Committee on Arthropod-borne Viruses. 1980. Laboratory safety for arboviruses and certain other viruses of vertebrates. Am. J. Trop. Med. Hyg., *29*:1359–1381.

Tindale, N.S. 1961. The chlorocresol method for field collecting. J. Lepid. Soc., *15*:195–197.

Traub, R., Wisseman, C.L., Jr., Jones, M.R., and O'Keefe, J.J. 1975. The acquisition of *Rickettsia tsutsugamushi* by chiggers (trombiculid mites) during the feeding process. Ann. N.Y. Acad. Sci., *266*:91–114.

Travis, B.V. 1968. Glyptal—a useful slide ringing compound. J. Med. Entomol., *5*:24.

Truman, J.W. 1968. Acetone treatment for preservation of adult and larval mosquitoes. Ann. Entomol. Soc. Am., *61*:779–780.

United States Department of Agriculture. 1976. *Ticks of Veterinary Importance.* Handbook No. 485. Washington, D.C. Superintendent of Dcouments, U.S. Government Printing Office.

Vanderberg, J.P., and Gwadz, R.W. 1980. The transmission by mosquitoes of plasmodia in the laboratory, pp. 153–234. In *Malaria. Volume 2. Pathology, Vector Studies,. and Culture.* Edited by J.P. Kreier. New York, Academic Press.

Wade, J.O. 1976. A new design of membrane feeder incorporating an electrical blood stirring device. Ann. Trop. Med. Parasitol., *70*:113–120.

Ward, R.D. 1977. The colonization of *Lutzomyia flaviscutellata* (Diptera: Psychodidae), a vector of *Leishmania mexicana amazonensis* in Brazil. J. Med. Entomol., *14*:469–476.

Wharton, G.W. 1976. House dust mites. J. Med. Entomol., *12*:577–621.

Index

Page numbers in *italics* refer to figures; page numbers followed by "t" refer to tables.